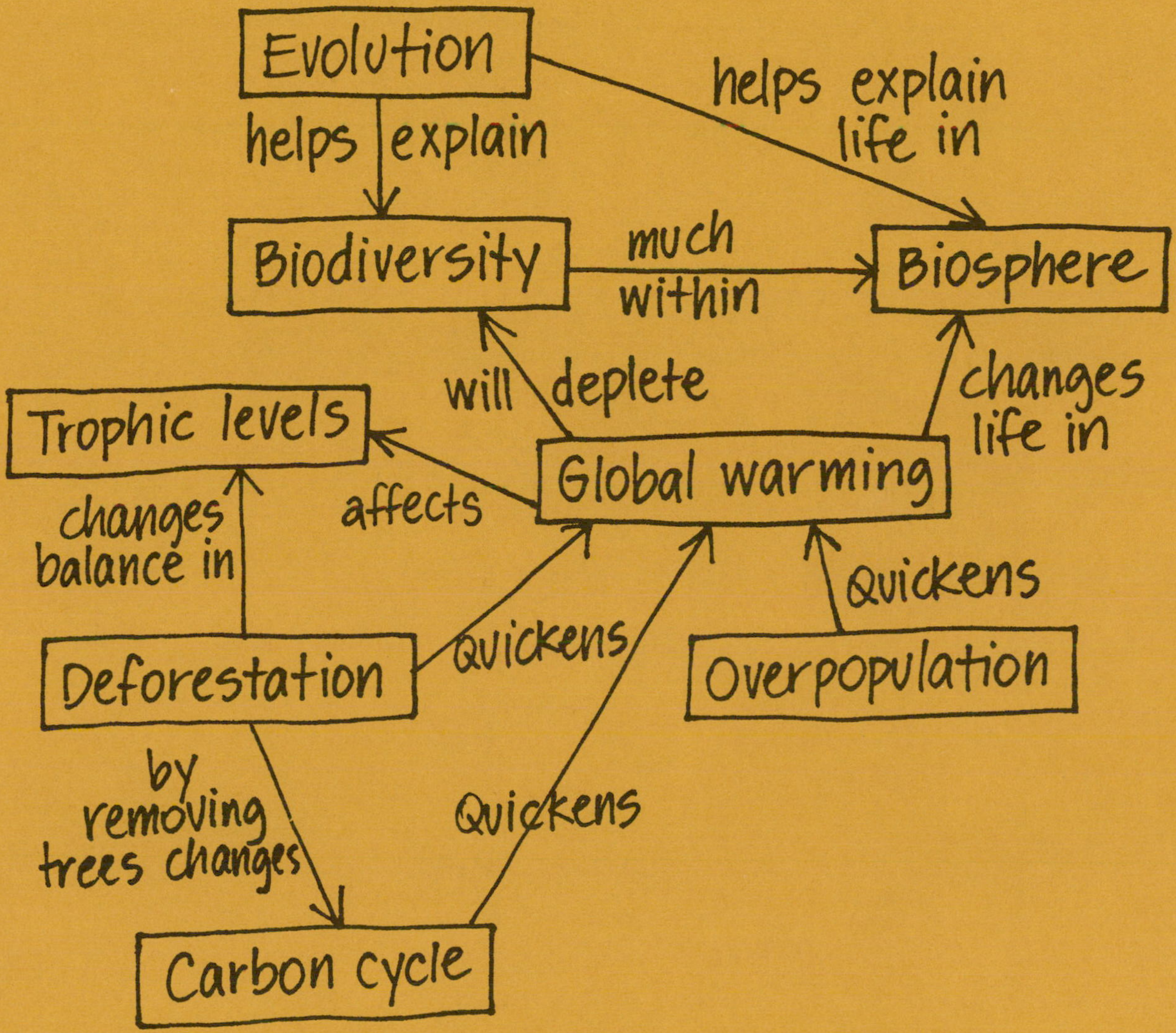

Evolution
helps explain
helps explain life in
Biodiversity
much within
Biosphere
will deplete
changes life in
Trophic levels
affects
Global warming
changes balance in
Quickens
Quickens
Deforestation
Overpopulation
by removing trees changes
Quickens
Carbon cycle

JOHN H. POSTLETHWAIT

University of Oregon

JANET L. HOPSON

University of California, Santa Cruz

McGRAW-HILL, INC.

New York St. Louis San Francisco Auckland
Bogotá Caracas Lisbon London Madrid
Mexico City Milan Montreal New Delhi
San Juan Singapore Sydney Tokyo Toronto

The Nature of Life

This book is printed on acid-free paper.

1 2 3 4 5 6 7 8 9 0 VNH VNH 9 0 9 8 7 6 5

ISBN 0-07-050750-3

This book was set in Melior by GTS Graphics.
The editors were Kathi M. Prancan, Deena Cloud, and Jack Maisel;
the designer was Gayle Jaeger;
the production supervisor was Janelle S. Travers.
The photo editor was Safra Nimrod;
the photo researchers were Deborah Hershkowitz and Elyse Rieder.
Von Hoffmann Press, Inc., was printer and binder.

Library of Congress Cataloging-in-Publication Data

Postlethwait, John H.
The nature of life / John H. Postlethwait. Janet L. Hopson.—3rd ed.
p. cm.
Includes bibliographical references and index.
ISBN 0-07-050750-3
1. Biology. 2. Life (Biology) I. Hopson, Janet L. II. Title.
QH308.2.P67 1995
574—dc20 94-37994

Cover photograph: Male panther chameleon (Chamaeleo pardalis). (Photo by James H. Carmichael, Jr./Image Bank.)

INTERNATIONAL EDITION

When ordering this title, use ISBN 0-07-113600-2.

This is dedicated to my enthusiastic teachers:

Sam Postlethwait
Jack Otten
James Hawker
Al Chiscon
Joe Vanable
Howard Schneiderman
Carroll Williams

JHP

For Jerry Dommer, my inspiration

JLH

ABOUT THE AUTHORS

JOHN H. POSTLETHWAIT is Professor of Biology at the University of Oregon where he teaches General Biology for nonmajors and Genetics and Evolution, the introductory course for biology majors. He received his Ph.D. degree in Developmental Biology from Case Western Reserve University, and was a post-doctoral fellow at Harvard University. His research on the genetic mechanisms of embryonic development is supported by the National Institutes of Health and the American Heart Association. For three one-year periods, Dr. Postlethwait conducted research supported by Fullbright grants at the Institut für Molekular Biologie in Salzburg, Austria, the Laboratoire de Génétique Moléculaire des Eucaryotes in Strasbourg, France, and the Imperial Cancer Research Fund in Oxford, England. A recipient of the Ersted Distinguished Teaching Award, Dr. Postlethwait encourages active participation of undergraduates in research and includes them as coauthors on publications—see, for example, *Science 264*, 699–703 (1994) "A Genetic Map for the Zebrafish,"—on which nine undergraduates are coauthors. His love of teaching dates from 1963 when he was first an undergraduate teaching assistant. Currently Dr. Postlethwait is participating in the federally funded University of Oregon Workshop Biology program for innovative teaching of biology.

JANET L. HOPSON is an active freelance science writer and for many years has taught in the Science Communication Program, University of California at Santa Cruz. She has also taught writing courses at the University of California at Berkeley and Mills College. She holds B.A. and M.S. degrees from Southern Illinois University and the University of Missouri. Coauthor of three other biology textbooks for McGraw-Hill, *Biology* and *Essentials of Biology* with Norman K. Wessells, and *Biology! Bringing Science to Life* with John H. Postlethwait and Ruth C. Veres, she has also written a trade book on the human sense of smell. She won the Russell L. Cecil Award for magazine writing from the Arthritis Foundation and has published dozens of articles in national magazines and newspapers, including *Smithsonian*, *Psychology Today*, *Science News*, *Science Digest*, *Outside*, and others. Her biography is included in *Who's Who in Science and Engineering*.

CONTENTS IN BRIEF

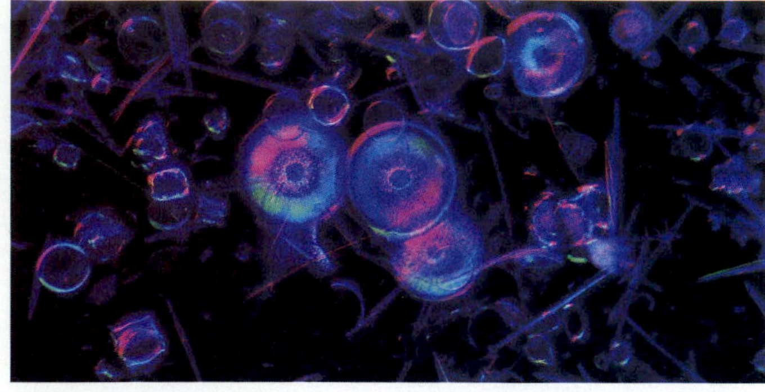

Part Four
How Animals Survive 561

Part Five How Plants Survive 755

Part Six
Ecology: Organisms and the Environment 831

CONTENTS

Chapter 14 The Human Life Cycle 332

Part Two: Apply and Decide

Part Three: Evolution and Life's Diversity 363

Chapter 15 Nature's Evidence for the Evolution of Living Things 364

Chapter 16 The Genetic Basis for Evolution 386

Chapter 28 Animal Nutrition and Digestion 636

Chapter 29 Excretion: Balancing Water and Salt 658

Chapter 30 Hormones and Other Molecular Messengers 676

PREFACE

Biology is an exciting but challenging subject, both for students to learn and for professors to convey. Successful students and teachers have both come to appreciate two key elements in learning biology: active participation in one's own learning, and the forging of links between new knowledge and a student's prior experience. The central focus of *The Nature of Life* is to provide the active student participation and the connections to prior knowledge necessary for meaningful mastery of biology. Since its first publication in 1989, *The Nature of Life* has focused on active learning and student involvement. This current edition, however, takes that charge to new levels and brings the reader to the science of life in innovative and, we predict, unusually successful ways.

- We began by retaining our most successful features from the first two editions—approaches that were product-tested by thousands of students and professors in the United States and several foreign countries, and polished through countless rounds of reviewing, editing, and revising.
- To motivate and excite nonmajors, we generated 42 stories—one per chapter, and most of them new to this edition. These stories are vehicles for capturing and holding attention, as well as for posing central questions about the natural world and organizing basic biological concepts. Educational research shows that students show more interest in basic subjects presented this way and retain more of what they've learned.
- We created an art program full of orienting icons, unique process diagrams, and colorful photos, then closely coordinated them to the text so readers could verbalize and visualize biological structures and activities simultaneously.
- We updated the science where appropriate to reflect the exciting developments in this fast-moving field.
- We incorporated current social issues and topics of student relevance throughout the chapters and in the numerous boxed essays.
- We retained several end-of-chapter study tools, such as Connections and Highlights in Review, to help students integrate what they have learned and test their own comprehension.
- We continued to focus on evolution and the environment as major bookwide themes, and to emphasize the process of science—not just individual facts and ideas.

This current edition has a large number of new features, based on input from dozens of teachers and other colleagues, and designed to expand and improve student participation and learning dramatically.

- We have redesigned the book's graphic elements from front to back, making *The Nature of Life* not just more inviting to read and study but helping users find the book's major parts quickly and distinguish the hierarchical importance of chapter sections more easily.
- We have added hundreds of new photographs and enlarged most of the rest; our purpose was to help students remember details more easily, as well as to better enjoy the beauty and fascination of biological images.
- We have reconfigured many of the book's diagrams to maximize the student's ability to identify structures, comprehend where they occur, sort the more relevant details from the less relevant ones, and follow the steps of sequential processes more effectively.
- We now display a list of Messages on each chapter opener page to serve as a conceptual roadmap, a set of chapter objectives, and a framework for review.
- We include Concept Challenges at the end of each major section within every chapter. These questions require students to stop, reflect upon, and apply the concepts they have just covered to interesting real-world problems.
- We have added Active Learning Boxes that walk a student through the scientific investigation of biological problems. This helps students to see exactly how facts and concepts are applied via the scientific method to real issues and involves them in that process first-hand.
- We have developed several new Concept Integrators, that is, large hand-painted illustrations that visually link various concepts and facts from a chapter discussion.
- We have created Apply and Decide essays that pose societal issues, remind students of the relevant biological principles from preceding chapters, show how students can apply those principles to the issue, and ask them to make a personal decision on the issue.
- We have replaced each chapter's study questions from the second edition with a vastly improved set that necessitates far more critical thinking and application of facts and principles and far less recitation. Answering these questions requires real understanding. The questions will not only help the student to better prepare for evaluation, but will enhance retention of the material.
- Finally, we explain early in the book how students can construct Concept Maps, then present several Concept Mapping exercises to help students understand relationships between topics, construct knowledge in personally meaningful ways, and pinpoint weak areas in their grasp of facts or concepts.

While the enthusiastic, thoughtful teacher and the engaging textbook are the central tools for biology education, multimedia software can integrate moving images, sound, and text to show how organisms look and act and how sequential events take place in the microscopic domain of a living cell. This edition of *The Nature of Life* comes supported by a range of technological study aids that we think will be both effective and entertaining, and will promote active student participation.

The Nature of Life helps students learn biology not only by the pedagogical tools just listed, but by the logical approach to biology subject matter taken by most biology teachers.

Our Approach to Biology

The Nature of Life takes a hierarchical approach to biology, beginning with the fundamental shared features of all life, moving to the specific features displayed by various groups of organisms, and then examining the interactions of groups of organisms with their environments. Chapter 1 introduces the main themes of the book, including evolution, the environment, energy, reproduction, and the process of science. Part I considers the fundamental principles that unify all life: Chapters 2 and 3 cover the molecular and cellular basis of life, and Chapters 4, 5, and 6 describe how organisms acquire and use energy.

Part II focuses on the many aspects of reproduction. Chapters 7 through 10 discuss how inherited traits are passed from parent to offspring. Chapters 11 and 12 explore the exciting world of biotechnology and the frontiers of human genetics. Chapters 13 and 14 probe the fascinating mechanisms by which genetic instructions and cellular mechanisms change an egg cell into a fish, fly, frog, or a human baby.

Once familiar with the cellular and genetic features that *unify life* from Parts I and II, the student is ready to investigate biology's grandest idea: the mechanisms of evolution, and how descent with modification can explain the *diversity of life*. Part III, new to the 3d edition, begins in Chapter 15 with an expanded investigation of the evidence for evolution in the natural world. This is followed in Chapter 16 by a grounding in the role genes play in evolution. Chapter 17 presents the fascinating theories for life's origins and history, and Chapter 18 places the world's current biodiversity crisis in an evolutionary context.

The book's evolutionary theme continues as we survey the evolution and lifestyles within the diverse kingdoms of organisms, from bacteria through protists, fungi, plants, and animals in Chapters 19 through 22. This section ends with a new chapter, Chapter 23, that describes the evolutionary descent of our own species. The four new chapters in Part III plus the five retained from the second edition build on the evolutionary themes that permeated our earlier editions, and significantly increase the breadth and depth of coverage of this topic which is so central to all levels of biological thought.

Parts IV and V investigate how the bodies of animals and plants function in ways that allow them to survive in their particular environments. Chapter 24 introduces general concepts that recur throughout the study of animal anatomy and physiology. The next nine chapters analyze individual physiological systems in animals and show how organisms adapt to sometimes bizarre environments and how the reader's own physiological systems work in familiar and extreme situations. Of particular interest to students may be Chapter 26 on the immune system; Chapters 31 and 32 on nervous systems, the brain, and behavior; and Chapter 33 on exercise and the body's musculoskeletal support.

Part V, expanded in this edition, demonstrates the intriguing anatomy and physiology of plants and the interplay with their environments. Chapter 34 describes basic plant architecture, then Chapters 35 and 36 explore how plants obtain nutrients, how they reproduce and develop, and how plant biotechnologists are harnessing plants to improve human health and nutrition. Chapter 37 on plant growth regulators ends the section by returning to the bookwide theme of how organisms interact with and respond to their physical and biological surroundings.

The backgrounding in Parts I through V prepare the student to thoroughly absorb what is arguably the most important part of the book, Part VI: the interactions of organisms with their environments. Our hierarchical approach is again evident in this section, beginning with a discussion of population biology in Chapter 38, enlarging the scope to biological communities and ecosystems in Chapters 39 and 40, then expanding the focus still further in Chapter 41 to the biosphere, all living organisms and their environments on Earth. Chapter 41 emphasizes the global changes now being triggered by human activities, and encourages students to act personally and perhaps professionally to help maintain a cleaner world with sustainable resources. This ecological theme continues into Chapter 42, the book's final chapter, as it investigates animal behavior in natural environments, including human behavior and how it will inevitably shape the world of the future.

A skilled, enthusiastic teacher will always be the most important element in a student's introduction to biology. A textbook such as *The Nature of Life*, however, can be a crucial tool in the student's exploration by engaging him or her, showing biology's relevance to daily campus life, and by conveying the excitement of discovery in life science. We believe this dramatically revised edition will enhance student participation and learning in demonstrable ways, and hope you will find that to be true.

Supplementary Materials

A comprehensive and completely integrated package of supplementary materials accompanies *The Nature of Life,* Third Edition.

THE SECRET OF LIFE

The Secret of Life is a multifaceted project that was developed in conjunction with WGBH, Boston to support and complement the third edition of *The Nature of Life*. *The Secret of Life* package includes the WGBH telecourse of the same name, and accompanying Faculty and Study Guides; an interactive videodisc with figures from the textbook and accompanying teacher's guide; and a videotape containing eight 15-minute modules, each dealing with a different concept in biology, and accompanying teacher's guide.

Telecourse The 13-unit college telecourse is intended as an introductory biology course for nonmajors with an emphasis on molecular biology. Each course unit builds understanding for those that follow, with early units addressing life at the cellular level and progressing to how animal systems function, how genes determine certain traits, and how life's species fit into the large picture of our planet. The telecourse materials include:

- ***The Nature of Life***, Third Edition
  John Postlethwait / Janet Hopson
- **Telecourse Programs***
  WGBH, Boston and BBC-TV.*

Eight one-hour television programs that reveal current trends in molecular biology, illustrate scientists at work, and convey challenges and opportunities in this growing field.

- **The Telecourse Study Guide and Faculty Guide**
  Joan Jolly / Michelle Barg, The Cadmus Group, Inc.

These guides provide the link for both professor and student between the telecourse and the textbook. Chapters in the two guides follow that of the 13-unit telecourse. Among its features, the Study Guide provides students with an overview before viewing each unit, as well as followup exercises for after viewing the program. The Faculty Guide features information on additional readings; text references to *The Nature of Life*, Third Edition; expanded treatments of difficult concepts, and a number of questions for discussion, writing exercises, and exams.

The Secret of Life Video Modules
WGBH, Boston and BBC-TV

WGBH has produced eight 15-minute video modules that illuminate the biological universe with unique stories and animation. Based on "The Secret of Life" television series, these modules span topics from archaebacteria and the basics of life to viruses, evolution, modern scientific technology and the biodiversity crisis. Each module concludes with a series of stimulating questions for class discussion.

- **Teacher's Guide to Accompany *The Secret of Life* Video Modules,**
  Gail Patt, Boston University

The Secret of Life Videodisc
WGBH, Boston

A two-sided videodisc will be available as a companion to *The Nature of Life*, Third Edition. Side A of the videodisc will provide 30 minutes of film clips, animation, and still images, covering cells, genetics, interactions and ecology, and research techniques. Side A will also contain 300 stills from the third edition of *The Nature of Life.* Side B of the videodisc will contain 4 of the 8 video modules from *The Secret of Life* series. Topic coverage will include biotechnology, human reproduction, portraits of modern science and research, and human genetics. The videodisc provides menu access and editor software for users as well as a user's guide featuring bar codes for level-one use.

- **User's Guide to Accompany *The Secret of Life* Videodisc**
  WGBH, Boston

FOR THE INSTRUCTOR:

Instructor's Manual/Test Bank/Visiquizzes
Dennis Todd, University of Oregon

For this edition, the instructor's manual has been combined with the text bank/visiquizzes to form a more unified teaching tool for the professor. The revised instructor's manual includes an updated multi-media reference list as well as the integration of eye-catching icons for quicker text and supplement cross-referencing.

The test bank has been revised and expanded to include 1000 *new* questions for a total of 3000. The visiquiz section will contain new black line illustrations from the third edition text for use in quizzes or exams.

* Check your local PBS station for telecourse air dates.
* Corporate funding for *The Secret of Life* is provided by the Upjohn Company.

Computerized Test Bank The printed test bank is available in a computerized format for IBM 5¼ inch, IBM 3½ inch, and MacIntosh computers.

Overhead Transparencies This package of 300 four-color acetate transparencies provide enlarged text illustrations and labels for use in lecture. These overhead transparencies will also be available electronically on *The Secret of Life* videodisc.

Instructor's Manual to Accompany Exploring The Living World: A Lab Manual for Biology

Anton E. Lawson, Arizona State University

The instructor's manual that accompanies this unique approach to lab investigations is designed to provide supportive material for the professor using methods of scientific reasoning in his or her lab course. Instructors are provided with background information about each of the 40 investigative labs as well as teaching tips which include advanced preparations, explorations, term introductions, and concept applications.

Primis Apply and Decide Case Studies

Janet L. Hopson

As a followup to this new text feature, the Apply and Decide case studies have been expanded to provide instructors with additional real-life studies that are tied into concepts presented in the text. By providing 25 original case studies on McGraw-Hill's electronic database system, Primis, professors will have the ability to incorporate additional case study topics over and above the six in the textbook. These can be used for class discussion, reading, or writing assignments.

FOR THE STUDENT:

Critical Thinking Workbook

Gail Patt, Boston University

In this revision the workbook has been updated to reflect the new organization of the main text. A new feature to the workbook is the addition of relevant popular press articles for each part opener to aid students in making the connection between biology and everyday life.

Exploring the Living World: A Laboratory Manual for Biology

Anton Lawson, Arizona State University

This new lab manual features 40 investigative labs for introductory biology students. The author has taken an alternative approach to traditional lab manuals. With an emphasis on scientific reasoning, students using this manual are asked to come up with causal questions and proposed explanations before actually carrying out the laboratory experiment. They are also periodically asked to develop their own experiments using the materials provided.

Exploring the Living World: Primis Edition

Anton Lawson, Arizona State University

This traditionally published lab manual will also be available on the Primis electronic database for individual instructor organization and selection.

SECOND EDITION SUPPLEMENTS:

For the Instructor:

- Laboratory Prep Guide to Accompany Hands-on Biology
- *The Nature of Life* Videotape

For the Student:

- Hands-on Biology
- Laboratory Manual to Accompany *The Nature of Life* revised edition
- Biopartner Tutorial Software
- Biology Write Now!

Acknowledgments

Insightful critiques improve any product, and a textbook is no exception. We appreciate the many teachers, subject area specialists, and students whose comments helped us revise this edition and improve its utility in teaching biology. A complete list of those participants follows. Here, we would like, in particular, to thank those reviewers who read the entire manuscript for this third edition: Steve Muzos, Sandra Steingraber, and Rob Tyser. Their comments and suggestions have been enormously valuable.

We would also like to express special gratitude to those subject area specialists who helped us update and streamline subjects that are sometimes challenging for students. Katherine Milton contributed substantial material for the new Chapter 23 on human evolution; Mark Schlessman focused on the plant biology in Chapters 34 to 37. Tom Sharkey commented on our discussion of photosynthesis. And George Barlow helped orient Chapter 42 on animal behavior.

Our hearty thanks go to Gail Patt, who composed the end-of-chapter study aids, which go far beyond the simple recall of facts and definitions to higher-level learning of concepts, applications, and problem-solving.

Finally, we would like to acknowledge members of the McGraw-Hill publishing team whose skills and dedication made such a difference in this third edition: Kathi Prancan, Suzanne Thibodeau, Denise Schanck, Jack Maisel, Gayle Jaeger, Art Ciccone, Safra Nimrod, and Deena Cloud. Their work, and that of all our contributors and reviewers, is going toward better biology education for today's students—a very worthy goal.

John H. Postlethwait
Janet L. Hopson

Reviewers: *W. Sylvester Allred,* Northern Arizona University; *Lisa M. Baird,* University of San Diego; *George Barlow,* University of California–Berkeley; *Stephen T. Bishoff,* University of South Carolina; *Thomas R. Campbell,* L. A. Pierce College; *Steven D. Carey,* University of Mobile; *Mary Colavito-Shepanski,* Santa Monica College; *Wade L. Collier,* Manatee Community College; *Jerold Davis,* Cornell University; *Roger M. Davis,* University of Maryland–Baltimore; *Paul H. Demchick,* Barton College; *Jean DeSaix,* University of North Carolina–Chapel Hill; *Cory R. Etchberger,* Pennsylvania State College–Berks; *Richard T. Fraga,* Lane Community College; *Sally Frost-Mason,* University of Kansas; *Chandler Fulton,* Brandeis University; *Elizabeth B. Gardner,* Pine Manor College; *Dana Geary,* University of Wisconsin; *Rebecca German,* University of Cincinnati; *Herbert H. Grossman,* Penn State University at Berks; *Laszlo Hanzely,* Northern Illinois University; *Lewis Held,* Texas Tech; *David Hickey,* Lansing Community College; *Martin D. Hollingsworth,* Tallahassee Community College; *Andrew Lapinski,* Reading Area Community College; *Siu-Lam Lee,* University of Massachusetts–Lowell; *Lionel O. Leon,* Okaloosa Walton Community College; *Anne Morris-Hooke,* Miami University; *Steven J. Muzos,* Austin Community College; *Murray Paton Pendarvis,* Southeast Louisiana University; *Gail Patt,* Boston University; *Peter Penderson,* Cuesta College; *Robert W. Phelps,* San Diego Mesa College; *David M. Polcyn,* California State–San Bernardino; *Gary A. Polis,* Vanderbilt University; *Alan F. Posey,* University of Arkansas at Pine Bluff; *Joanne Rosinski,* Sweet Briar College; *Walter Sakai,* Santa Monica City College; *Edwin Daniel Schreiber,* Mesa Community College; *Thomas D. Sharkey,* University of Wisconsin–Madison; *Sandra Steingraber,* Columbia College; *Barbara Y. Stewart,* Swarthmore College; *John F. Stolz,* Duquesne University; *Lloyd Tertinen,* University of Wisconsin–EauClaire; *Robin W. Tyser,* University of Wisconsin–La Crosse; *Thomas E. Weaks,* Marshall University; and *Rebecca Westbrooks,* Southeastern Community College. **Focus group participants:** *Wade L. Collier,* Manatee Community College; *Jerold Davis,* Cornell University; *Jean DeSaix,* University of North Carolina–Chapel Hill; *Dana Geary,* University of Wisconsin; *John Greening,* College of the Sequoias; *Kenneth D. Laser,* Salem State College; *David O. Norris,* University of Colorado; *Gary Peterson,* South Dakota State University; *Helen Pigage,* Elmhurst College; *Dorothy Puckett,* Kilgore College; *Rob Reinsvold,* University of Northern Colorado; *Julia Riggs,* Victoria College; *Gerald Summers,* University of Missouri; *Pamela Tabery,* Northampton Community College; and *Walter P. Trost,* Chicago State University.

Our thanks also go to the following individuals who served as reviewers or focus group participants on previous editions of this text. *Dr. Laura Adamkewicz,* George Mason University; *Olukemi Adewusi,* Ferris State University; *Dean A. Adkins,* Marshall University; *Kraig Adler,* Cornell University; *Dr. John U. Aliff,* Glendale Community College; *Joanna T. Ambron,* Queensborough Community College; *Steven Austad,* Harvard University; *Robert J. Baalman,* California State University, Hayward; *Aimée H. Bakken,* University of Washington; *Stuart S. Bamforth,* Tulane University; *Sarah F. Barlow,* Middle Tennessee State University; *R. J. Barnett,* California State University, Chico; *Joseph A. Beatty,* Southern Illinois University, Carbondale; *Nancy Benchimol,* Nassau Community College; *Dr. Rolf W. Benseler,* California State University, Hayward; *Gerald Bergtrom,* University of Wisconsin, Milwaukee; *Dr. Dorothy B. Berner,* Temple University; *Dr. A. K. Boateng,* Florida Community College, Jacksonville; *Jack Bostrack,* University of Wisconsin, River Falls; *William S. Bradshaw,* Brigham Young University; *Jonathan Brosin,* Sacramento City College; *Howard E. Buhse, Jr.,* University of Illinois, Chicago; *John Burger,* University of New Hampshire; *F. M. Butterworth,* Oakland University; *Guy Cameron,* University of Houston; *Ian M. Campbell,* University of Pittsburgh; *John L. Caruso,* University of Cincinnati; *Brenda Casper,* University of Pennsylvania; *Doug Cheeseman,* De Anza College; *Dr. Gregory Cheplick,* University of Wisconsin; *Dr. Joseph P. Chinnici,* Virginia Commonwealth University; *Carl F. Chuey,* Youngstown State University; *Dr. Simon Chung,* Northeastern Illinois University; *Charlotte Clark,* Fullerton College; *Norman S. Cohn,* Ohio University; *Paul Colinvaux,* Ohio State University; *Scott L. Collins,* University of Oklahoma; *Dr. August J. Colo,* Middlesex County College, *G. Dennis Cooke,* Kent State University; *Jack D. Cote,* College of Lake County; *Gerald T. Cowley,* University of South Carolina; *Louis Crescitelli,* Bergen Community College; *Orlando Cuellar,* University of Utah; *Thomas Daniel,* University of Washington; *David Darda,* Central Washington University; *J. Michael DeBow,* San Joaquin Delta College; *Loren Denny,* Southwest Missouri State University; *Ron DePry,* Fresno City College; *Dr. Kathryn A. Dickson,* California State University, Fullerton; *Thomas Dolan,* Butler University; *Patrick J. Doyle,* Middle Tennessee State University; *Leslie Drew,* Texas Tech University; *Helen Dunlap,* Millersville University of Pennsylvania; *Douglas J. Eder,* Southern Illinois University, Edwardsville; *Dr. David W. Eldridge,* Baylor University; *Lynne Elkin,* California State University, Hayward; *Paul R. Elliott,* Florida State University; *Eldon Enger,* Delta College; *Gauhari Farooka,* University of Nebraska, Omaha; *Marvin Fawley,* North Dakota State University; *Ronald R. Fenstermacher,* Community College of Philadelphia; *Edwin Franks,* Western Illinois University; *C. E. Freeman,* University of Texas, El Paso; *Lawrence D. Friedman,* University of Missouri, St. Louis; *Grace Gagliardi,* Bucks County Community College; *Dr. Ric A. Garcia,* Clemson University; *Wendell Gauger,* University of Nebraska, Lincoln; *Dr. S. M. Gittleson,* Fairleigh Dickinson University; *E. Goudsmit,* Oakland University; *John S. Graham,* Bowling Green State University; *Shirley Graham,* Kent State University; *Thomas Gregg,* Miami University; *Alan Groeger,* Southwest Texas State University; *Gregory Grove,* Pennsylvania State University; *Thaddeus A. Grudzien,* Oakland University; *James A. Guikema,* Kansas State University; *Richard Haas,* Cal State University, Fresno; *Madeline M. Hall,* Cleveland State University; *Robert W. Hamilton,* Loyola University of Chicago; *Earl L. Hanebrink,* Arkansas State University; *Dr. John P. Harley,* Eastern Kentucky University; *Richard C. Harrel,* Lamar University; *Marcia Harrison,* Marshall University; *T. P. Harrison,* Central State University; *Maurice E. Hartman,* Palm Beach Community College; *Dr. Karl H. Hasenstein,* University of Southwestern Louisiana; *Martin A. Hegyi,* Fordham University; *Dr. John J. Heise,* Georgia Institute of Technology; *H. T. Hendrickson,* University of North Carolina, Greensboro; *Frank Heppner,* University of Rhode Island; *T. R. Hoage,* Sam Houston State University; *Kurt G. Hofter,* Florida State University; *Dr. Rhodes B. Holliman,* Virginia Polytechnic Institute and State University; *Harry L. Holloway,* University of North Dakota; *E. Bruce Holmes,* Western Illinois University; *Jerry H. Hubschman,* Wright State University; *B. Hunnicutt,* Seminole Community College; *Robert N. Hurst,* Purdue University; *Hadar Isseroff,* State University of New York College, Buffalo; *Dr. Ira James,* California State University,

Long Beach; *Ursula Jando,* Washburn University of Topeka; *Dr. Wilmar B. Jansma,* University of Northern Iowa; *Dr. Margaret Jefferson,* California State University, Los Angeles; *Norma G. Johnson,* University of North Carolina, Chapel Hill; *Clyde Jones,* Texas Tech University; *Dr. Ira Jones,* California State University, Long Beach; *Dr. Patricia P. Jones,* Stanford University; *Dr. Craig T. Jordan,* University of Texas, San Antonio; *Maurice C. Kalb,* University of Wisconsin, Whitewater; *Bonnie Kalison,* Mesa College; *Judy Kandel,* California State University, Fullerton; *Arnold Karpoff,* University of Louisville; *Jerry Kaster,* University of Wisconsin, Milwaukee—Center for Great Lakes Studies; *L. G. Kavaljian,* California State University, Sacramento; *Thomas L. Keefe,* Eastern Kentucky University; *Robin C. Kennedy,* University of Missouri, Columbia; *Donald R. Kirk,* Shasta College; *R. Koide,* Pennsylvania State University; *Mark Konikoff,* University of Southwestern Louisiana; *Eliot Krause,* Seton Hall University; *Barbara S. Lake,* Central Piedmont Community College; *Jim des Lauvérs,* Chaffey College; *Elmo A. Law,* University of Missouri, Kansas City; *Tami Levitt-Gilmarr,* Pennsylvania State University; *Daniel Linzer,* Northwestern University; *J. R. Loewenberg,* University of Wisconsin, Milwaukee; *Dr. Robert Lonard,* University of Texas, Pan American; *Sharon R. Long,* Stanford University; *Carmita E. Love,* Community College of Philadelphia; *C. E. Ludwig,* California State University, Sacramento; *James Luken,* Northern Kentucky University; *Dr. Ann S. Lumsden,* Florida State University; *Dr. Bonnie Lustigman,* Montclair State College; *Edward B. Lyke,* California State University, Hayward; *Douglas Lyng,* Indiana University–Purdue University, Ft. Wayne; *George L. Marchin,* Kansas State University; *Philip M. Mathis,* Middle Tennessee State University; *Mrs. Margaret L. May,* Virginia Commonwealth University; *Edward McCrady,* University of North Carolina, Greensboro; *Bruce McCune,* Oregon State University; *Dr. John O. Mecom,* Richland College; *Tekié Mehary,* University of Washington; *Richard L. Miller,* Temple University; *Phyllis Moore,* University of Arkansas, Little Rock; *Carl Moos,* State University of New York, Stony Brook; *Doris Morgan,* Middlesex County College; *Donald B. Morzenti,* Milwaukee Area Technical College; *Steve Murray,* California State University, Fullerton; *Robert Neill,* University of Texas, Arlington; *Paul Nollen,* Western Illinois University; *Kenneth Nuss,* University of Northern Iowa; *William D. O'Dell,* University of Nebraska, Omaha; *Dr. Joyce K. Ono,* California State University, Fullerton; *James T. Oris,* Miami University; *Clark L. Ovrebo,* Central State University, Edmond; *Charles Page,* El Camino College; *Gail Patt,* Boston University; *Kay Pauling,* Foothill College; *Eugene C. Perri,* Bucks County Community College; *Dr. Chris E. Petersen,* College of DuPage; *Richard Petersen,* Portland State University; *Joel B. Piperberg,* Millersville University of Pennsylvania; *David Polcyn,* California State University, San Bernardino; *Michael Pollock,* Mount Royal College, Canada; *Jeffrey Pommerville,* Glendale Community College; *David I. Rasmussen,* Arizona State University; *Daniel Read,* Central Piedmont Community College; *Dr. Don Reinhardt,* Georgia State University; *Louis Renaud,* Prince George's Community College; *Jackie Reynolds,* Richland College; *Jennifer H. Richards,* Florida International University; *Thomas L. Richards,* California State Polytechnic University; *Tom Rike,* Glendale Community College, California; *C. L. Rockett,* Bowling Green State University; *William D. Rogers,* Ball State University; *Hugh Rooney,* J. S. Reynolds Community College; *Wayne C. Rosing,* Middle Tennessee State University; *Deborah D. Ross,* Indiana University–Purdue University, Ft. Wayne; *Frederick C. Ross,* Delta College; *A. H. Rothman,* California State University, Fullerton; *Mary Lou Rottman,* University of Colorado, Denver; *Dr. Donald J. Roufa,* Kansas State University; *Dr. Michael Rourke,* Bakersfield College; *Chester E. Rufh,* Youngstown State University; *Mariette Ruppert,* Clemson University; *Charles L. Rutherford,* Virginia Polytechnic Institute and State University; *Dr. Milton Saier,* University of California, San Diego; *Lisa Sardinia,* San Francisco State University; *A. G. Scarbrough,* Towson State University; *Dan Scheirer,* Northeastern University; *Randall Schietzelf,* Harper College; *Dr. Fred Schreiber,* California State University, Fresno; *Robert W. Schuhmacher,* Kean College of New Jersey; *Joel S. Schwartz,* College of Staten Island; *Erik P. Scully,* Towson State University; *Roger S. Sharpe,* University of Nebraska, Omaha; *Stanley Shostak,* University of Pittsburgh; *Dr. Jane R. Shoup,* Purdue University, Calumet; *J. Kenneth Shull, Jr.,* Appalachian State University; *C. Steven Sikes,* University of South Alabama; *Christopher C. Smith,* Kansas State University; *John O. Stanton,* Monroe Community College; *D. R. Starr,* Mt. Hood Community College; *Guy Steucek,* Millersville University of Pennsylvania; *Joseph D. Stogner,* Ferrum College; *Raymond Tampari,* Northern Arizona University; *Dr. Ruth B. Thomas,* Sam Houston State University; *Bert Tribbey,* California State University, Fresno; *Nancy C. Tuckman,* Loyola University of Chicago; *Dr. Spencer Jay Turkel,* New York Institute of Technology; *William A. Turner,* Wayne State University; *John Twente,* University of Missouri, Columbia; *C. L. Tydings,* Youngstown State University; *John Tyson,* Virginia Polytechnic Institute and State University; *Richard R. Vance,* University of California, Los Angeles; *C. David Vanicek,* California State University, Sacramento; *Harry van Keulen,* Cleveland State University; *Linda Van Thiel,* Wayne State University; *Roy M. Ventullo,* University of Dayton; *Judith A. Verbeke,* University of Illinois, Chicago; *Leonard S. Vincent,* Fullerton College; *Dr. Ronald B. Walter,* Southwest Texas State University; *Stephen Watts,* University of Alabama, Birmingham; *T. Weaver,* Montana State University; *Dr. Joel D. Weintraub,* California State University, Fullerton; *Marion R. Wells,* Middle Tennessee State University; *James White,* New York City Technical College; *Joe Whitesell,* University of Arkansas, Little Rock; *Fred Whittaker,* University of Louisville; *Roberta Williams,* University of Nevada, Las Vegas; *Kenneth A. Wilson,* California State University, Northridge; *Chuck Wimpee,* University of Wisconsin, Milwaukee; *Mala Wingerd,* San Diego State University; *Richard Wise,* Bakersfield College; *Gary Wisehart,* San Diego City College; *Dan Wivagg,* Baylor University; *Thomas Wolf,* Washburn University of Topeka; *Paul Wright,* Western Carolina University; *Richard P. Wurst,* Central Connecticut State University; *Edward K. Yeargers,* Georgia Institute of Technology; *Linda Yasui,* Northern Illinois University.

The Nature of Life

CHAPTER 1

The Nature of Life

THE EARTH SUMMIT: ISSUES IN MODERN BIOLOGY

In 1954, a plant researcher for Eli Lilly and Company, one of the world's largest pharmaceutical manufacturers, made a fateful discovery. In his laboratory in Switzerland, Dr. Gordon Svoboda extracted two compounds with distinct anticancer properties from an exotic tropical flower. He isolated both compounds from the delicate pink petals of *Catharanthus roseus*, the rosy periwinkle of Madagascar [FIGURE 1.1]. In the four decades since Svoboda's discovery, the drugs made from these compounds have saved the lives of many children and young adults who might have died from Hodgkin's lymphoma and leukemia, two forms of cancer. Despite the life-saving benefits, the drugs and their source have become controversial symbols of a complex international problem. These symbols garnered considerable attention at a historic event in the summer of 1992.

That year official delegations from 177 countries met in Rio de Janeiro, Brazil, for the largest biology class ever held. At the United Nations Conference on Environment and Development, better known as the Earth Summit, politicians and scientists carried out a massive discussion of how biology affects people's lives. The urgent issues on the agenda included pollution, overpopulation, resource use, economic development, global climate change, and the imminent loss of many of earth's species and natural habitats. World leaders were concerned about their people's current and future physical and economic well-being. They left the conference agreeing that the quality of life cannot be separated from the quality of the environment. They learned, too, that biology and ecology cannot be separated from politics, and that's where the rosy periwinkle comes in.

This small flower exemplifies the vast untapped biological resources in rain forests and other remote regions of the earth, including the oceans. But the representatives at the Earth Summit disagreed strongly on whether these remote regions, such as the rain forests, should be preserved or exploited. The rich countries of North America, Europe, and Asia have relatively small populations, but they use a disproportionate amount of the world's energy and other resources and contribute a disproportionately large amount to its pollution. At the Earth Summit, representatives of these countries argued for vigorous preservation of areas like tropical rain

FIGURE 1.1
Catharanthus roseus, Madagascar's rosy periwinkle.

forests, since such regions naturally release large amounts of oxygen into the atmosphere, they house a tremendous diversity of species, and they contain huge numbers of organisms that, like the rosy periwinkle, may be of benefit to humankind.

At the same time, the large and rapidly growing populations of the developing countries are pressing in on the rain forests. People want to harvest the trees for lumber and to clear the land for farms and homes. Some peoples indigenous to these otherwise uninhabited areas also attended the Earth Summit. They live in the earth's few remaining natural areas much as their ancestors did for generations, but their ways of life are disappearing along with the rain forests and other natural habitats. Besides trying to preserve their native lands, they wanted to receive compensation, through patents and royalties, for any commercial exploitation of indigenous plants, including the rosy periwinkle. The governments of countries such as Madagascar and Brazil, which contain large undeveloped areas, made similar financial claims. Pharmaceutical companies, in turn, pointed out that without sufficient return on investment, they would be unable to mount the huge research and development efforts required to bring a life-saving drug to market.

Delegates to the Earth Summit, therefore, faced a major question: How can humanity balance the need to retain natural environments and the resources they provide with the rapidly expanding human population and the rights of native peoples? The current human population of 5.5 billion people will swell to 11 billion within our lifetimes. Can 11 billion people share this small planet with flowers such as the rosy periwinkle growing in obscure tropical forests? What about plants, animals, and other organisms with no specific utility other than their small roles in nature?

The issues raised by the Earth Summit require an understanding of **biology**, the study of life. Our introduction to this chapter, and our detailed discussions of biology throughout this book, will often center on topics of concern at the Earth Summit. These topics, which represent three of the themes of this book, include the interactions of organisms and their environments, reproduction and development, and the role of energy in life.

The Nature of Life has two other basic themes that are related to *biology*: The first is the process of *evolution,* the change in life forms over long periods of time. The theory of evolution is central to modern biology, particularly in terms of changes involving the genetic (hereditary) mechanisms that operate in an observable and testable fashion. The final theme is the scientific way of knowing, a commonsense approach to learning about and surviving in the natural world.

In each chapter of this book, we end the introduction with a brief description of the chapter's contents and organization to help orient you to the material. We begin this chapter by discussing the basic characteristics that living things share and that differentiate them from nonliving things. We will see, in general, how living organisms use energy to maintain their complex organization. Next we will see why reproduction is so fundamental to each type of living organism, and how reproduction relates to evolution. Then we will learn how, through the mechanisms of evolution, changes appear among groups of organisms and how new kinds of organisms arise. We will go on to see how the scientific way of knowing, the scientific method, can be used to investigate the natural world as well as to make decisions about complex everyday problems. Finally, we will see how biological science is helping people solve some of the world's most urgent issues.

Let's begin, now, to explore life science. ❑

MESSAGES

1 Living organisms are highly ordered, and maintaining order requires energy and materials from the environment.

2 Individual organisms eventually die, but life continues via reproduction and heredity.

3 Over long periods of time, life forms change due to evolution. Natural selection—the ability of individuals with certain traits to reproduce more successfully than those lacking the traits—is a mechanism of evolution.

4 Science is a way of learning how nature functions. Scientists perform tests that might rule out possible explanations for a particular phenomenon. Explanations that have been repeatedly tested and not disproven are tentatively accepted.

5 The principles of biology can be applied to world problems such as overpopulation, disease, environmental degradation, famine, and drug addiction.

Characteristics of Living Organisms

What is a living organism? How can we tell if something is alive or dead? What, in fact, is the nature of life itself? These questions have never been more important because today's biologists and physicians have unprecedented abilities to sustain the human body and individual organs on life-support machines; to freeze human embryos for later use; and to change and merge hereditary traits of microbes, plants, and animals. Perhaps, one day, this list will extend to generating life in a test tube and to creating hybrids between computers and living things. In the broadest sense, biological science probes the question, What is life?, and it takes this entire book to provide an answer.

As this fascinating story unfolds, the five themes of the book—energy, reproduction, evolution, environment, and the scientific way of knowing—will occur again and again. In this chapter, the themes help organize a set of nine observable characteristics that define life. The characteristics of living things are:

1. **Order.** Each structure or activity is in a specific relationship to all other structures and activities.
2. **Metabolism.** Organized chemical steps break down and build up molecules, making energy available or building needed parts.
3. **Motility.** Using their own power, organisms move themselves or their body parts.
4. **Responsiveness.** Organisms perceive the environment and react to it.
5. **Reproduction.** Organisms give rise to others of the same type.
6. **Development.** Ordered sequences of progressive changes result in increased complexity.
7. **Heredity.** Organisms have units of inheritance called genes that are passed from parent to offspring and control physical, chemical, and behavioral traits.
8. **Evolution.** Through evolution, species acquire new ways to survive, to obtain and use energy, and to reproduce.
9. **Adaptations.** Specific structures and behaviors suit life forms to their own environment.

Many nonliving entities have one or more of the characteristics of life; if you consider the life traits of movement, energy use, and growth, you can observe that waves move, flames use energy, and crystals grow. Only living organisms, however, display *all* the characteristics at some point during their individual life cycle or species history. For example, the petals of a periwinkle flower move as the bud unfolds; the plant supporting the flower captures energy from the sun; and the plant originally emerged from a seed and grew in size.

Let's begin, now, to consider the defining characteristics of life, one by one, in groupings that match the list above, and that reflect our bookwide themes of energy, reproduction, the environment, and evolution.

Energy and Organization in Living Things

Many of the issues that delegates discussed at the Earth Summit centered around how people acquire and use energy. The developed countries of Europe, North America, and Asia use more than their share of global energy supplies—mainly in the form of oil. And many scientists have concluded that the burning of oil creates carbon dioxide and other waste products which contribute to the global climate change we call global warming. Human beings, in general, may use more total energy in more forms than other kinds of living organisms, but energy use is a shared trait, an essential feature of all life forms. Living systems use energy to maintain organization. Our discussions of life's characteristics, therefore, begin with those that require energy expenditure.

ORDER

Living organisms possess a degree of **order**—a structural and behavioral complexity and regularity—far greater than the finest Swiss clock, the fastest racing car, or anything else in the nonliving world. The eye of a fly, for example, and the spiral-packed seeds of a sunflower head consist of highly organized units repeated and arranged in precise geometric arrays [FIGURE 1.2]. In fact, some biologists suggest that the most highly organized structure yet discovered in the universe is the human brain, such as the one allowing you to read and understand this page.

Order is apparent in both the visible and microscopic organization of living things. This organization forms a hierarchy, beginning with the individual organism and moving toward both higher and lower levels of organization, [as shown in the Concept Integrator on page 10]. An **organism** is an independent individual that shows the characteristics of life. Each elephant and each tree in FIGURE 1.3 is a single organism. Each organism, in turn, is made up of **organ systems**, groups of body parts arranged so that together they carry out a particular function within the organism. The skeletal system, for example, supports an elephant's body.

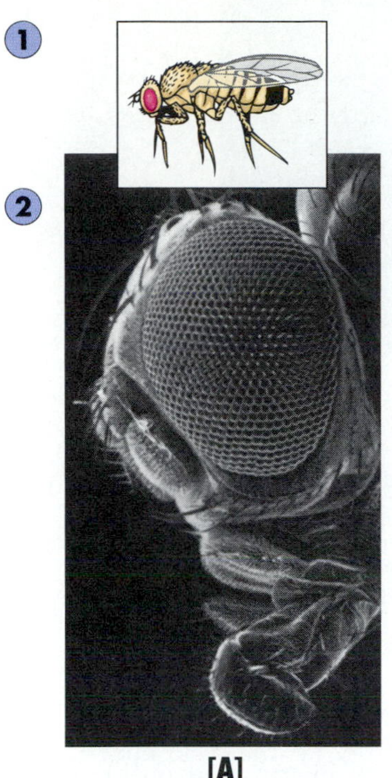

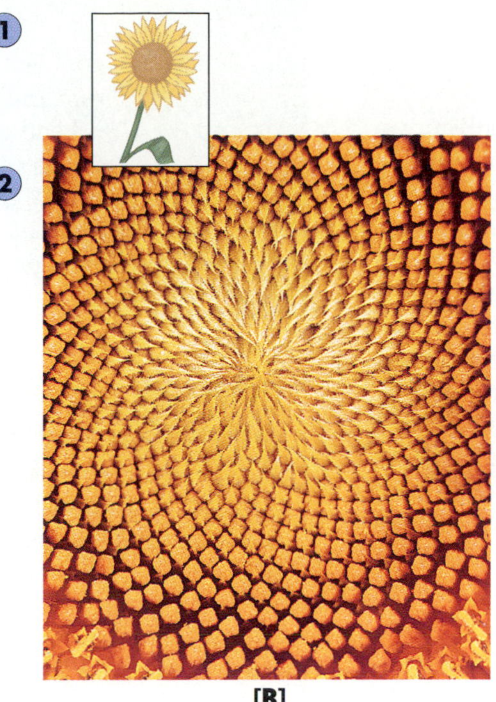

FIGURE 1.2

Order in Life and Nonlife.

[A] A fruit fly's eye and [B] the tightly packed seeds of a sunflower form geometric arrays that reveal the inherent order in living things.

Organ systems are made up of **organs**, structures that perform specialized functions for the organ system. Two examples are a single bone that supports part of an elephant's leg, and the leaves that produce sugars for a tree.

Each organ is made up **tissues**, groups of similar cells that carry out the function of the organ. For example, blood (a tissue) flows in tubes, or vessels, through the bone, providing food and oxygen for the bone cells.

Tissues are made up of **cells**, the simplest entities that have all the properties of life. Cells contain within them structures known as **organelles**, which perform the functions necessary for the life of the cell.

The organization of living things is important because it identifies the various levels at which we can examine and understand biology. For example, to understand the significance of the cancer-fighting drugs made from periwinkle flowers (organism level), a biologist wants to know how the plant makes the compounds without its own tissues becoming poisoned (cell and tissue levels). He or she also wonders how the drugs affect the flower's reproductive abilities or ability to survive.

FIGURE 1.3

Order Reigns at Every Level in the Living World.

A female elephant and her baby on the African savanna symbolize levels of organization. The elephants' bodies are made up of highly ordered cell parts, cells, tissues, organs, and organ systems that function together smoothly. This family is part of a population within a diverse community, including the egrets searching for insects stirred up by the elephants' heavy footsteps and the grasses they graze and trample. The community, in turn, is part of the savanna ecosystem, with its expansive plain, mountain range, and arid climate.

METABOLISM

To understand the life characteristic of metabolism, we can consider the strawberry frog, a brightly colored animal that inhabits the damp cloud forests of Central America [FIGURE 1-4]. This frog has two large, dark eyes, muscles and bones that allow its legs to hop and paddle, and a dazzling red skin that covers its exterior. As time passes, some of this organization can start to deteriorate. While climbing a tree one day in search of ants, the frog may slip and crash to the forest floor, puncturing its skin, bruising its legs, and becoming more disorganized.

This kind of wear and tear is not unique to living things. Rocks, stars, rivers, and mountain ranges all tend to become more disordered with passing time. You need only look as far as the nearest heap of papers on your desk and the accumulation of dirty clothes and dust balls in the corner to witness the tremendous tendency toward disorder. To combat a housekeeping disaster, you must expend energy to sort the papers and clean up the dust. Likewise, in an injured frog, broken skin is replaced and wounded muscles heal when the animal's body extracts energy and materials from the insects it catches and eats, then uses the energy and materials to generate new skin and healthy muscle tissue. In general, living things counteract the disorder that comes with time by taking energy and materials from their environment and using them for repair, growth, and other processes needed for survival.

The intricate order within an organism's body—whether it is a brilliantly colored tropical frog or a thorny desert tree such as the acacia—can only be maintained through an expenditure of energy. For that reason, organisms require a constant supply of energy and building materials. The maintenance of order, as well as growth, replacement of worn-out or damaged body parts, and the basic life processes are accomplished by a highly complex series of chemical reactions within the cells of the organism. The sum of all such reactions is the organism's **metabolism**.

FIGURE 1.4

Strawberry Frog in a Cloud Forest.

Dendrobates pumilio, a tiny resident of the Central American jungle, has flaming red skin. The frog carries a newly hatched tadpole on its back to a pool of water trapped in a leaf. This behavior ultimately helps ensure the survival of new generations of strawberry frogs.

FIGURE 1.5

Metabolism: Conversion of Energy and Materials.

The spoked leaf of a fan palm acts as a solar collector. Leaf cells take in the sun's energy and trap it in a chemical form that metabolic processes employ for growth, repair, and movement—activities that help maintain order within the plant.

Once an organism has taken in energy-containing food molecules or has manufactured its own food using sunlight and gases from the air, metabolism changes the food molecules. Food molecules taken from the environment tend to be large and complex, and these molecules are broken down by metabolism into smaller and simpler molecules that can be used by the cells. Metabolism also rearranges these simpler molecules and builds new ones, which are used for repair and growth. Some food molecules are broken down by a series of reactions that gradually releases their chemical energy. This energy is used by the cells to maintain themselves and the organism as a whole. Light striking a palm leaf [FIGURE 1.5] initiates a series of metabolic steps that causes the plant to make high-energy compounds, move organelles around inside cells, produce new leaves, and break down worn-out cells and tissues.

MOTILITY

Energy released by metabolic reactions allows materials to be moved to areas where they can be used to replace worn parts, increase size, and generally combat disorganization. Self-propelled movement, or *motility,* is characteristic of many kinds of organisms. Even organisms as simple as bacteria can move on their own. Organisms such as plants, which cannot move from place to place,

FIGURE 1.6
Animal Movement Can Be Swift and Elegant.
These impalas, residents of savanna and woodland, leap through the African sunshine.

nonetheless show various kinds of movement. For example, water carrying dissolved materials moves through the plant, and the material that fills the living cells is in constant motion. The flowers of some plants open in the morning and close at night. Animals, of course, have elevated movement to an art form in their pursuit of food, their escape from enemies, and their social relations [FIGURE 1.6].

RESPONSIVENESS

To survive, an organism must respond to changes in its environment. It must be able to detect food, water, or enemies, and to react appropriately. The response can be an instantaneous one, as when a moth hears the high-pitched whine of a swooping bat, or it can be gradual, as when trumpeter swans detect the shortening days of autumn and respond by flying to warmer regions. Plants, too, respond to their environments, conserving water during times of drought and capturing sunlight when it is available. Some plants even show instantaneous responses, reacting to external signals rapidly enough to trap an unsuspecting fly [FIGURE 1.7].

FIGURE 1.7
Responsiveness: The Life Characteristics of Sensing and Reacting.
This Venus flytrap has reacted to the featherlight step of a housefly by closing rapidly and imprisoning the hapless insect.

The life characteristics of metabolism, motility, and responsiveness allow an organism to obtain energy and materials from the environment and use them to maintain order and organization within their bodies. Nevertheless, as organisms age, they can no longer stop the disorganization that inevitably occurs, and eventually all individual organisms die. Life continues to exist, however, because organisms perpetuate themselves, and three characteristics of life are directly involved in the perpetuation of species: reproduction, development, and heredity.

Reproduction of Living Things

While all individual organisms die, many generate new individuals like themselves during their lifetimes. If they didn't, organisms of their type would disappear. The life process by which organisms give rise to others of the same kind is called **reproduction**. The two basic modes of reproduction are asexual and sexual.

MODES OF REPRODUCTION

Asexual Reproduction In **asexual reproduction**, there is a single parent, and the offspring are all identical to the parent and to each other. Asexual reproduction occurs most commonly in one-celled organisms, and the new cells are produced when the parent cell simply splits in half, producing two identical daughter cells.

Sexual Reproduction Most complex organisms reproduce sexually. In **sexual reproduction**, the offspring generally contain hereditary material from two different parents and are not identical either to the parent or to each other. The first cell of the new individual is formed when two specialized cells called the egg and sperm unite.

Many sexually reproducing organisms have special adaptations for reproduction. A male golden toad, for example, increases his chances of reproducing if he advertises his location and amorous intent by a distinctive song attractive to females of his species. In another example, tube-shaped scarlet gilia flowers produce stores of energy-rich nectar, and both the flaming red petals and the nectar attract hummingbirds [FIGURE 1.8]. The iridescent birds feed on the nectar and inadvertently assist in the plant's reproduction. The hummingbirds' feathers pick up sticky pollen grains, which contain the flower's sperm, and when the bird deposits the pollen on the next plant it visits, these sperm can reach the eggs of another flower and lead to the formation of a new plant generation.

FIGURE 1.8

Botanical Want Ad: The Scarlet Gilia's Brilliant Red Tube and Sweet Nectar Attract Hummingbirds and Ensure the Plant's Reproduction.

This *Gilia aggregata* is visited by a broad-tailed hummingbird. The animal laps up nectar with a long tongue, gains energy, and inadvertently carries pollen grains to the next flower it samples.

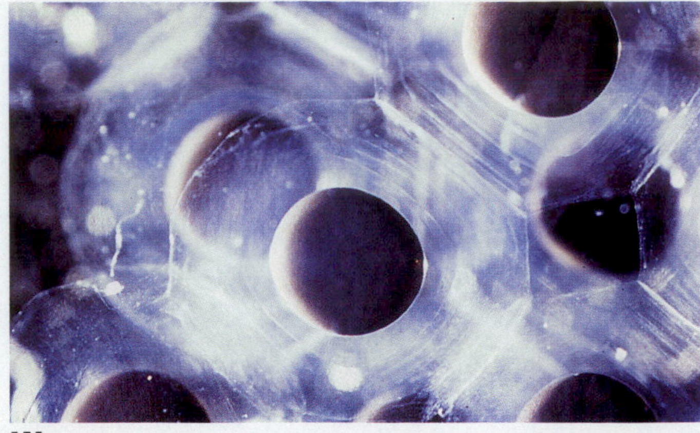

[A]

FIGURE 1.9

Development: A Sequence of Growth, a Consequence of Reproduction.

[A] A European common frog (*Rana temporaria*) begins as a fertilized egg cell. After a few hours, divisions in that original cell produce an embryo.
[B] Development continues during metamorphosis, producing a tadpole, and the physical transformation is completed by the time [C] the adult forms.

DEVELOPMENT

When an organism reproduces, the young start out smaller, and usually much simpler in form, than the parents. The offspring then grow in size and increase in complexity, a process known as **development** [FIGURE 1.9]. Eventually, they themselves can become parents.

One of the most intriguing questions in all biology is how a fertilized egg cell develops into the millions of cells of various types that function as a viable organism. The answer lies in the remarkable process of heredity.

GENES: THE UNITS OF HEREDITY

Identical twins make a convincing argument that some kind of information must direct the development of each individual in a fairly precise manner so that two very similar organisms result. When we contrast a set of identical twins with their different-looking brothers and sisters, however, we can see that there are also variations in hereditary information. The units of inheritance that control the development of specific physical, chemical, and behavioral traits in a living organism are the **genes**.

Genes are made of a remarkable molecule called DNA [FIGURE 1.10]. Each of your cells contains 46 of these twisting, threadlike molecules, and each DNA molecule in your cells contains about 4000 genes. The DNA in

[B]

[C]

eggs and sperm carries hereditary information from the parents that is passed on to their offspring. Specific genes, inherited from parents, determine whether a strawberry frog has big, bright red splashes of color or small, pale pink patches; whether a rosy periwinkle produces a lot of the cancer-stopping drug or only a little; and whether a person has blue, brown, or hazel eyes.

DNA has the unique ability to make identical copies of itself within a cell. This process, called *replication*, makes it possible for one cell to divide, yielding two identical cells. In this way, organisms can grow and also replace damaged or worn-out cells, the cells can expand and the organism can grow, and eventually, reproduction can take place and whole new organisms can form.

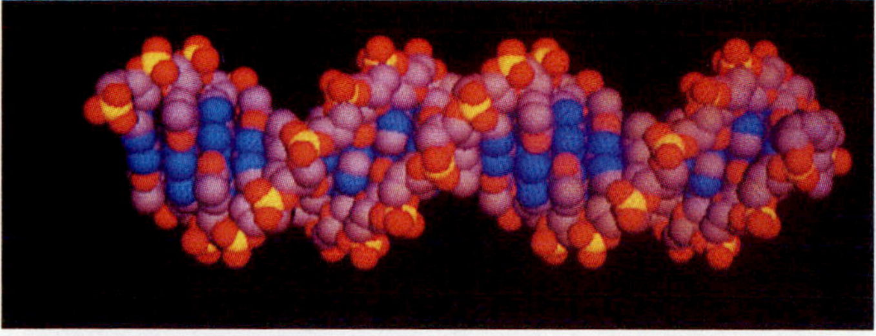

FIGURE 1.10

DNA and Genes.

Multicolor computer graphic of a small portion of a DNA molecule. DNA is the repository of genetic information for reproduction and cell function in all living cells.

Reproduction allows life to continue even though the bodies of individual organisms become disorganized over time and die, and as we saw, the processes of development and heredity—fundamental characteristics of living things—are completely integral to reproduction.

CONCEPT CHALLENGE

An automobile uses energy, is highly organized, is motile, and responds when the accelerator is pushed. Why wouldn't you consider an automobile to be alive?

The Hierarchy of Life

Picture, for a moment, some of your favorite organisms. You may envision your dog playfully returning a frisbee on a grassy lawn. Or a flock of geese flying in a vee-shaped formation. Or a gnarled pine tree growing atop a rocky summit. Regardless of whether an organism lives alone, lives as a member of a group, or lives attached to or inside of another organism, it must interact with its physical and biological environment. The branch of biology that studies the relationships between living organisms and their environment is **ecology**.

Ecologists study individual organisms, but also several levels of the hierarchy of life *above* the level of the organism. Recall from our discussion of the hierarchy of life *below* the level of the organism that living things are composed of organs, which are made up of cells, which are in turn made up of molecules. When an ecologist or a beginning student of biology considers life and the environment, he or she must focus on the higher orders of organization that include and impinge upon any given organism.

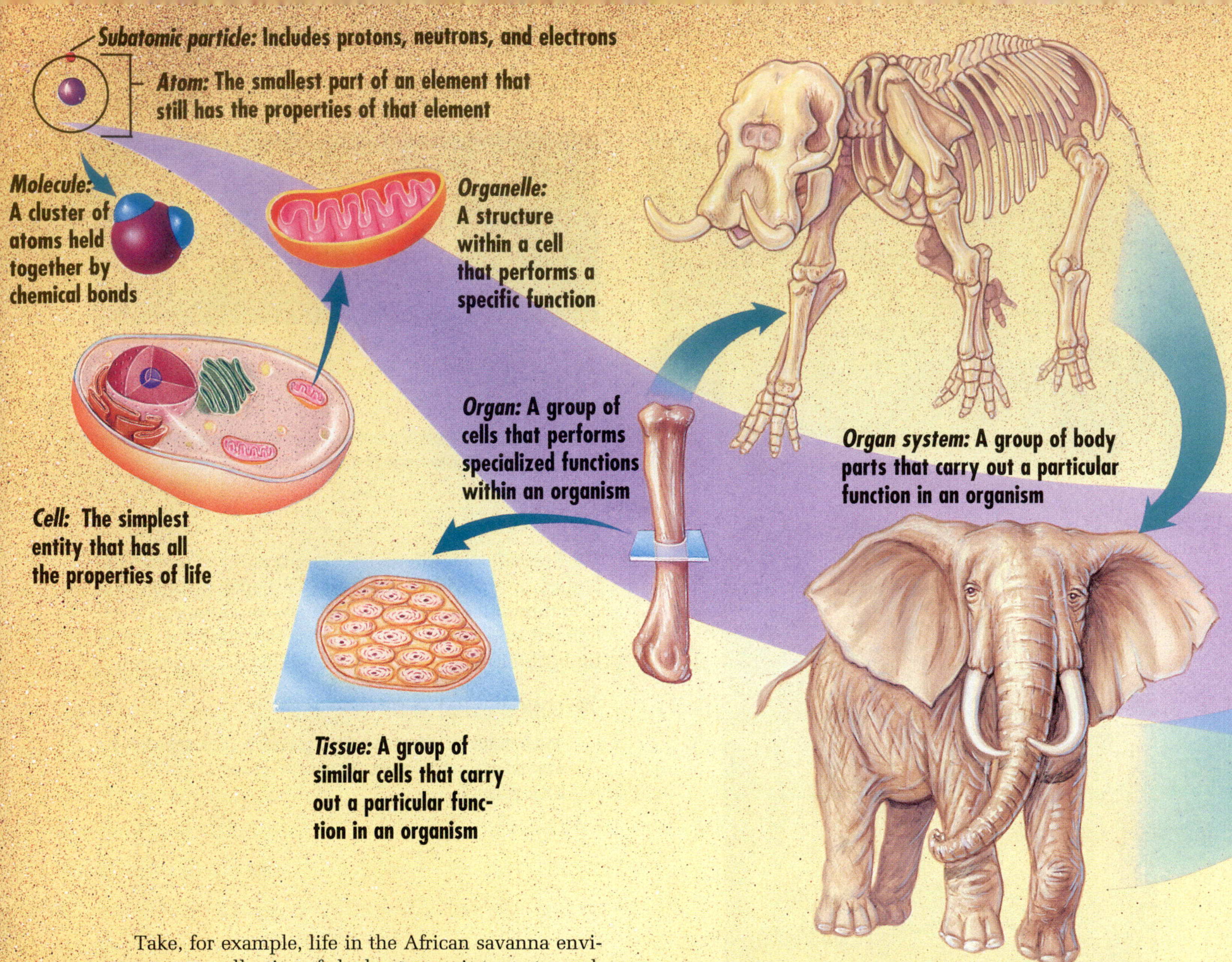

Take, for example, life in the African savanna environment—a collection of elephants, acacia trees, tussock grasses, and arid plains [FIGURE 1.3]. The hierarchy of life on the savanna proceeds from small to large in the following sequence [see Concept Integrator above.): **Organisms** are individual independent living things: an elephant is an organism and so is an acacia tree. Groups of a particular type of organism that live in the same area and actively interbreed with one another are called **populations**—for example, an elephant herd or a field of grass. All the populations that live in a particular area make up a **community**—for example, the plants, animals, and other organisms at a certain location on the savanna. The living community together with its nonliving physical surroundings is called an **ecosystem**. The savanna ecosystem includes elephants, the egrets that pick insects off their skin, and the coarse grass they chew and trample, as well as the water in clouds, the sandy soil underfoot, and the hot African sunshine. All the ecosystems of the earth make up the **biosphere**, that portion of the earth on which life exists. The biosphere is a relatively thin layer about 6½ to 13 kilometers (6 to 8 miles)* thick, and includes unimaginably remote places that nevertheless still teem with life, including the deepest parts of the ocean floor, deep-sea vents spewing superheated water, and the tops of earth's highest mountains where the air is thin and freezing cold. A huge variety of organisms exists in these and the more familiar and hospitable forests, meadows, lakes, marshes, and grasslands we know in our own environments. Biologists have tried to make sense of life's spectacular variety by classifying organisms into various groups.

*km (mi); see the abbreviations chart on the inside back cover.

Concept Integrator: The Hierarchy of Life

Classification of Living Things

Put on a mask and snorkel, and plunge into the warm, blue waters off Hawaii and count the varieties of fish you see in just a few minutes [FIGURE 1.11]—perhaps butterfly fish, bird wrasses, striped Moorish idols with streaming back fins, or intensely turquoise parrot fish. This same dazzling variety is evident in the colorful songbirds of eastern forests, the butterflies of midwestern weed fields, and the wildflowers of the Rocky Mountains.

To help make sense of this vast array of different organisms, biologists have created a system for categorizing organisms, separating living things into groups according to their similarities. Brightly colored tropical frogs, for example, are more similar to bullfrogs than to elephants, but frogs and elephants are more similar to each other than to mushrooms or grass. To make sense of this *diversity* of life, biologists divide organisms into the following categories:

Species (SPEE-seez; both singular and plural) are groups of structurally similar individuals that all descend from the same initial group and that have the potential to breed successfully with one another in nature.

Several related and similar species make up a **genus** (JEE-nuhss; plural, *genera*). Just as a person has a family name and a given name, biologists refer to each species by a two-part name: (1) The first part is the genus; for example, the rosy periwinkle is assigned to the genus *Catharanthus,* while the strawberry frog is assigned to the genus *Dendrobates,* along with the closely related play-actor frog and several other species. (2) The second part of the name distinguishes among several species within one genus. The rosy periwinkle is *Catharanthus roseus*, while the strawberry frog is *Dendrobates pumilio* and the related play-actor frog is *Dendrobates histrionicus.* After once mentioning the complete two-part name, biologists may subsequently abbreviate the genus, referring to *C. roseus* or *D. pumilio.*

Just as species are grouped into genera, genera are grouped into **families**, families into **orders**, orders into **classes**, and classes into **phyla** or **divisions**. In each successive grouping, similar organisms are arranged by broader and broader criteria.

Ultimately, biologists assign the dozens of divisions and phyla—and the millions of life forms they include—into **kingdoms**.

KINGDOMS OF LIVING THINGS

You can perform a simple exercise in classification by quickly writing down the names of about 10 organisms as they pop into your mind. Now take the organisms on your list and try grouping them into the broadest categories possible. What categories did you choose? How many categories were there? If you are like most people, you would have listed and grouped organisms in just two main categories—plants, such as roses and maple trees, and animals, such as dogs, cats, and flies.

In fact, the ancient Greeks divided all living things into those same two kingdoms, and this division into plants and animals was sufficient for most biologists and naturalists throughout most of recorded history. Not un-

FIGURE 1.11

Diversity Under the Waves: Rainbow of Fish on a Tropical Reef.

This coral reef is home to lavishly colored parrot fish, striped sergeant majors, and several other bright species. These represent only a few of earth's more than 30,000 fish species—a stunning variety that is itself only a small fraction of our planet's millions of species of living things.

til the mid-twentieth century did the American biologist R. H. Whitaker propose that the growing number of organisms discovered and studied by life scientists be divided into a five-kingdom scheme based on cell structure. In the late 1980s, studies of DNA supported a six-kingdom scheme: Animalia, Plantae, Fungi, Protista, Eubacteria, and Archaebacteria [FIGURE 1.12]. The five-kingdom scheme lumped Eubacteria and Archaebacteria together in the kingdom Monera, but the six-kingdom scheme better represents the family tree by which all living organisms appear to be related to each other. While your list of organisms may have included a couple of members of the kingdom Fungi (mushrooms and yeasts), it probably didn't include any members of the kingdoms Protista, Eubacteria, or Archaebacteria because the vast majority of species in these kingdoms are microscopic and are unfamiliar to most people. Biologists define kingdoms by the evolutionary relationships and cell structure of the species within them, and the ways in which these species obtain energy.

Three of the six kingdoms contain large, familiar organisms with complex cells. Members of the largest kingdom, **Animalia**—including insects, earthworms, snails, elephants, egrets, and human beings, to name but a few—consist of many cells, and get their energy and materials by ingesting other organisms, alive and dead. Members of the kingdom **Plantae**, such as ferns and maple trees, are usually green and many-celled, and they generate their own food from the energy in sunlight and the materials in air and water. The kingdom **Fungi** includes some single-celled forms like yeasts, and many-celled forms like mushrooms. Fungi obtain energy by absorbing materials from the environment after decomposing other living or dead organisms or their parts.

Most of the members of the other three kingdoms are invisible to the naked eye. The **Protista**, or *protists*, include both one-celled and many-celled species. Numerous kinds of algae are protists, as are amoebas and many other organisms that swim in drops of pond water. (Some biologists prefer to call this group *Protoctista*.)

The remaining two kingdoms consist of single-celled organisms whose cells are much simpler in structure than the cells of animals, plants, fungi, or protists. The kingdom **Eubacteria** (*eu-* means "true") includes small, single-celled microscopic organisms such as the organisms that turn milk into yogurt or those that cause strep throat. The kingdom **Archaebacteria** (*archae-* means "ancient") are single-celled organisms about the same size and shape as Eubacteria, but with some different cellular components (see CHAPTER 19). They live in very harsh environments, such as the inside of a cow's intestine, or in acidic, sulfurous hot springs. Many Archaebacteria obtain their energy from sulfur compounds, and some produce methane (swamp gas) as a waste product.

Although members of each of the six kingdoms of life play important roles in the environment, people generally think of the familiar plants and animals as the most significant groups. The delegates to the Earth Summit in Rio, for example, focused on saving species from the kingdoms Animalia and Plantae. Losing species from the other four kingdoms would change life on earth just as drastically, though. This is because members of the fungal, protist, and bacterial kingdoms are largely responsible for recycling materials in the environment. Have you ever wondered what happens to all the fingernails that people clip off and throw away? Silly as it may sound, fingernails would start to pile up in the environment if it weren't for the recycling activities of fungi and eubacteria. The important point here is that all life is interconnected. Environmentalists may focus on saving beautiful and inspiring animals such as the jaguar of the Amazon basin or the spotted owl of the Pacific Northwest. But for the natural world to go on functioning as we know it, we must preserve the widest possible spectrum of life's diversity—even organisms we have yet to discover and name.

THE UNITY AND DIVERSITY OF LIFE

While biologists distinguish the six kingdoms according to their cell structures and ways of obtaining energy and materials, the kingdoms share a great number of genetic similarities. These similarities suggest that all living things on earth arose from a single group of ancestral cells present at the dawn of life. The similarities stand as powerful evidence for the *unity* of life, a unity both of origin and of basic day-to-day functioning. In the chapters to come you will see dozens of examples of common life mechanisms at the molecular level, and you will also see how these similarities unite our planet's incredible diversity of life forms at a fundamental level. A question about these underlying connections has long intrigued biologists: What mechanisms can account for the unity of life at the cellular and genetic levels, and yet still foster the vast diversity of life which we group into six large kingdoms? That mechanism is the unifying theme for all life science: evolution.

CONCEPT CHALLENGE

By applying the rules of taxonomy, determine which two of the three following organisms are more closely related, even if their names are unfamiliar to you: *Felis domesticus, Musca domesticus, Felis concolor. Defend your choice.*

box 1.1
People in Biology

Darwin, Wallace, and Evolution by Natural Selection

History has not dealt equally with Charles Darwin and Alfred Russel Wallace. Yet these two men, living in the same country during the same era, proposed remarkably similar theories of evolution by *natural selection*—a central theme of this book and a unifying concept in all of biological science. To understand this historical oversight, as well as the stunning achievement evolutionary theory represents, one must consider both the social and scientific tenor of the mid-nineteenth century.

Darwin and Wallace both lived in Victorian England, an era of courtly manners, rigid social roles, and inflexible stratification by social class. In contrast to the social constraints of the day, there was excitement and progressive thinking in many areas of science, including a growing belief that the natural world is in a constant state of change.

For hundreds of years prior to the time of Darwin and Wallace, most scientists believed that all species originated about 6000 years ago and remained unchanged in form throughout history. With the careful study of fossils in the late eighteenth and early nineteenth centuries, however, biologists Georges Cuvier, Georges Louis Leclerc de Buffon, Jean Baptiste Lamarck, and others began to realize that extinct organisms were unlike those alive in the 1800s, and that the form of plants and animals seemed to change, or evolve, over the years. Simultaneously, geologists such as James Hutton and Charles Lyell were investigating rock layers and other evidence of mountain building, erosion, and volcanic eruptions and suggested that our planet must be millions—not thousands—of years old, and must be undergoing slow but continual change.

Darwin had also read the writings of Thomas Malthus (1766–1834), who pointed out that populations tend to grow geometrically (2,4,8,16, etc.) and that only wars, famine, and disease keep population growth in check. Within this social and scientific context, Darwin and Wallace separately made their own observations of unfamiliar organisms half a world away. Independently, they drew nearly identical conclusions about the mechanisms of evolutionary change within the organisms on our changing planet.

Darwin was a wealthy upper-class gentleman who had studied for the clergy at Cambridge University but was passionately interested in natural history [FIGURE 1]. Upon graduating, he agreed to serve as a social companion to Captain Fitzroy of the HMS *Beagle* on a five-year voyage (1831–1836) to map the coastline of South America. Throughout the voyage, Darwin had ample opportunity to explore the jungles, plains, and mountains of South America. A seminal stop on the trip was the Galápagos, a cluster of black volcanic islands 1000 kilometers (600 miles) west of Ecuador with a small but unique set of plants and animals.

FIGURE 1 Darwin (1809–1882).

Throughout his travels, Darwin was continually impressed by the variable shapes and colors that the members of a single species often showed, and he pondered the origin and significance of such variety. He returned to England with the idea of natural selection already half formed and notebooks full of observations to support the contention, but he continued studying variation in domesticated species for nearly 20 years.

Darwin had written a short summary of his theory in 1842, but it remained unpublished as late as 1858. That changed abruptly, however, when Alfred Russel Wallace's work in the arena came to light [FIGURE 2]. Younger than Darwin, Wallace had sailed to South America himself in 1848 to collect plants,

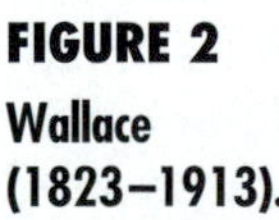

FIGURE 2 Wallace (1823–1913).

insects, and other natural specimens. Later Wallace visited Singapore and Borneo to make similar collections. Like Darwin, he had read the works of Malthus and Lyell, and had noticed on his travels the striking variations among populations of individual species. And like Darwin, Wallace had returned and devised a theory of evolution by natural selection to explain what he saw (even choosing the term *natural selection* independently!). Unlike Darwin, however, Wallace was a working-class man, and perhaps because he had far less to lose and few fears of public disapproval, he was more eager to promote his ideas.

By 1858, Wallace had published four papers containing portions of a theory of evolution that included the mechanism of natural selection, the notion of descent with modification, and the idea of survival of the fittest. Wallace sent Darwin his fourth paper in 1858, and Darwin—realizing that his life's work was being scooped—rushed to prepare an abstract. Friends read this abstract before the Linnean Society later that year along with Wallace's published work. Then the following year, Darwin released his monumental and well-documented book, *On the Origin of Species by Means of Natural Selection.*

As biologist William V. Mayer points out*, the simultaneous publication of nearly identical theories in today's highly competitive scientific world would probably lead to "recrimination and lawsuits." But products of the gracious Victorian era, Darwin and Wallace instead approached their situation with courtesy and generosity, each fully acknowledging each other's contribution at every opportunity. Darwin and Wallace, in fact, became personal friends, Darwin arranging for the impoverished Wallace to receive a government pension, and Wallace serving as a pallbearer at Darwin's funeral.

The intellectual stimulation each scientist gave the other provided generations of biologists with their broadest conceptual framework for understanding the living world. Now, over 130 years after the independent contributions of Darwin and Wallace and the thorough documentation of evolutionary principles, their insights still foster lively experimental inquiry into the mechanisms of life.

*William V. Mayer, "Wallace and Darwin," *The American Biology Teacher* 49 (1987): 406–410

Darwin, C. *On the Origin of Species: A Facsimile of the First Edition.* Cambridge, Mass: Harvard University Press, 1975

Evolution, Adaptation, and Natural Selection

Forms of living organisms change over time. We know this partly from the fossilized imprints of early organisms [FIGURE 1.13]. The older a fossil, the less similar it is to present-day forms, and this provides good evidence of the continued change in living species over millennia. Using fossils, analysis of DNA, and other kinds of evidence, biologists can trace the history of a group of organisms. All groups alive today are like the youngest branches of a family tree. For example, the strawberry frog and the common frog are amphibious cousins [FIGURE 1-14]. Reaching farther back in time, frogs and salamanders would share a common ancestor, and even farther back in the history of this lineage, frogs would share a common ancestor with snakes, fish, beetles, button mushrooms, rosy periwinkles, and bacteria. Tracing the evolutionary tree of life to more remote periods, all organisms would eventually share a common ancestor in the ancient past. This family-tree concept helps explain the origin of species: All living things are descended from a common ancestor, and arose as the result of genetic modifications in species that lived before them in a process of change called **evolution**.

The amazing diversity of plants, animals, and other living organisms that the Earth Summit delegates sought to protect is a product of evolution, as is each individual species.

ADAPTATION

As groups of organisms evolve, genetic modifications (mutations) arise that allow them to cope with their environments in more efficient ways. Different species, for example, have different ways of extracting energy and materials from their surroundings. Because these differences help the organism adapt to their own special ways of life, they are called **adaptations**. One adaptation in an adult frog, for example, is a long, sticky tongue folded over at the tip [FIGURE 1.15]. When an ant or fly ventures near, the frog quickly unfurls its tongue, mires the victim in the sticky coating, retracts its tongue, and devours the insect.

Not all adaptations relate to the intake of energy and materials. Some adaptations improve the organism's ability to grow; others, to reproduce, move, live in a group, or attract a mate. Some adaptations are truly wondrous and bizarre. Certain flowers, for instance, so closely resemble female wasps that male wasps try to "copulate" with

FIGURE 1.12

The Family Tree of Life.

[A] These newly divided *Streptococcus* cells (microscopic bacteria, magnification 20,750×) represent kingdom Eubacteria. **[B]** This *Methanobacterium ruminantium* cell (magnification 30,000×), which inhabits a cow's intestines, is from the kingdom Archaebacteria. **[C]** The kingdom Protista also contains microscopic single-celled organisms, but they are generally much larger and more complex than bacteria. This *Paramecium* cell, with its exterior covering of beating hairlike organelles, is a protist. **[D]** Fungi, such as this *Hygrophorus conicus* mushroom decomposing leaf litter on the forest floor, make up a fourth kingdom. **[E]** The plant kingdom includes a large group of many-celled organisms that generate their own food molecules and are often conspicuous and graceful, such as this pine. **[F]** The largest of the six kingdoms contains several million species of animals. Most of these lack backbones, but some, like this arctic loon, have an internal skeleton and complex behavior patterns.

[E] Plantae

[F] Animalia

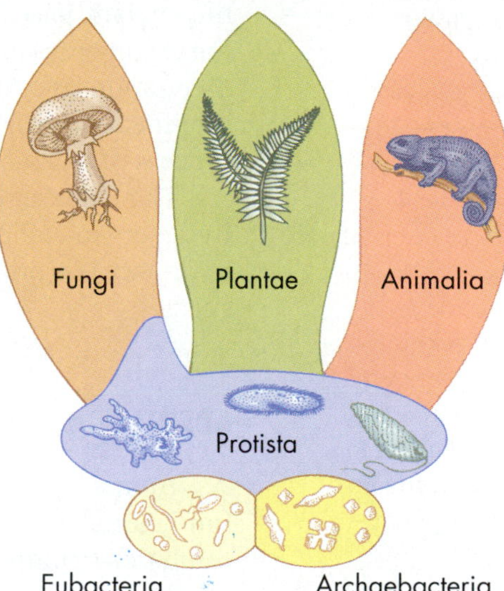

[D] Fungi

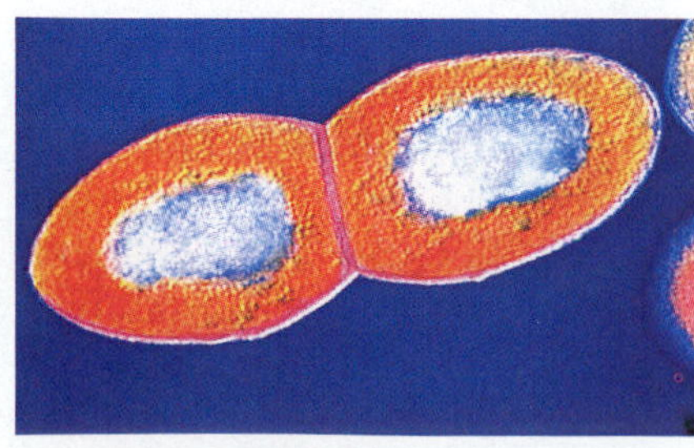

[A] Eubacteria

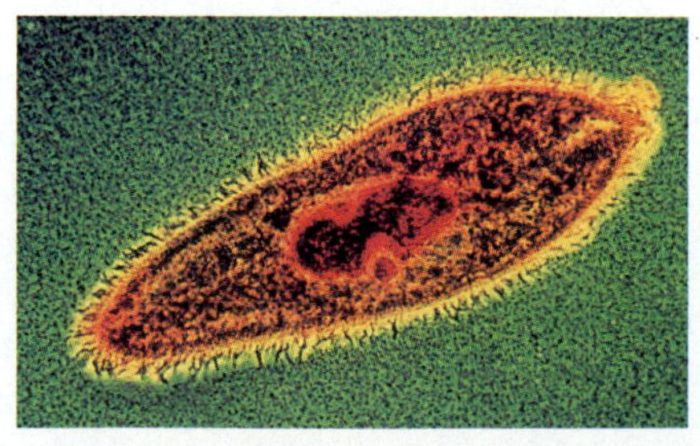

[C] Protista

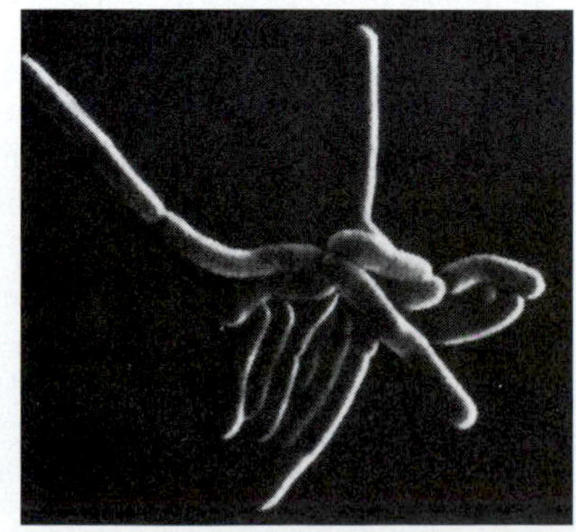

[B] Archaebacteria

them [FIGURE 1.16]. Pollen clings to the insects' bodies, and as they move on to "mate with" new "females," they carry the pollen from one flower to the next.

Adaptation and the broader theme of evolution so pervade the way biologists think about, understand, and explain life that an evolutionary perspective will appear in virtually every chapter in this book. The next section provides an overview of evolution by natural selection, which can serve as a useful framework until we explore evolution in greater detail in CHAPTERS 15 to 17.

[A]

[B]

FIGURE 1.13

Floral Ancestor in Stone: Fossil Gentian Compared with Modern Flower.

[A] This 50-million-year-old fossil of a seven-petaled flower bears a strong resemblance to [B] a modern gentian. Biologists use such comparisons in the study of evolution.

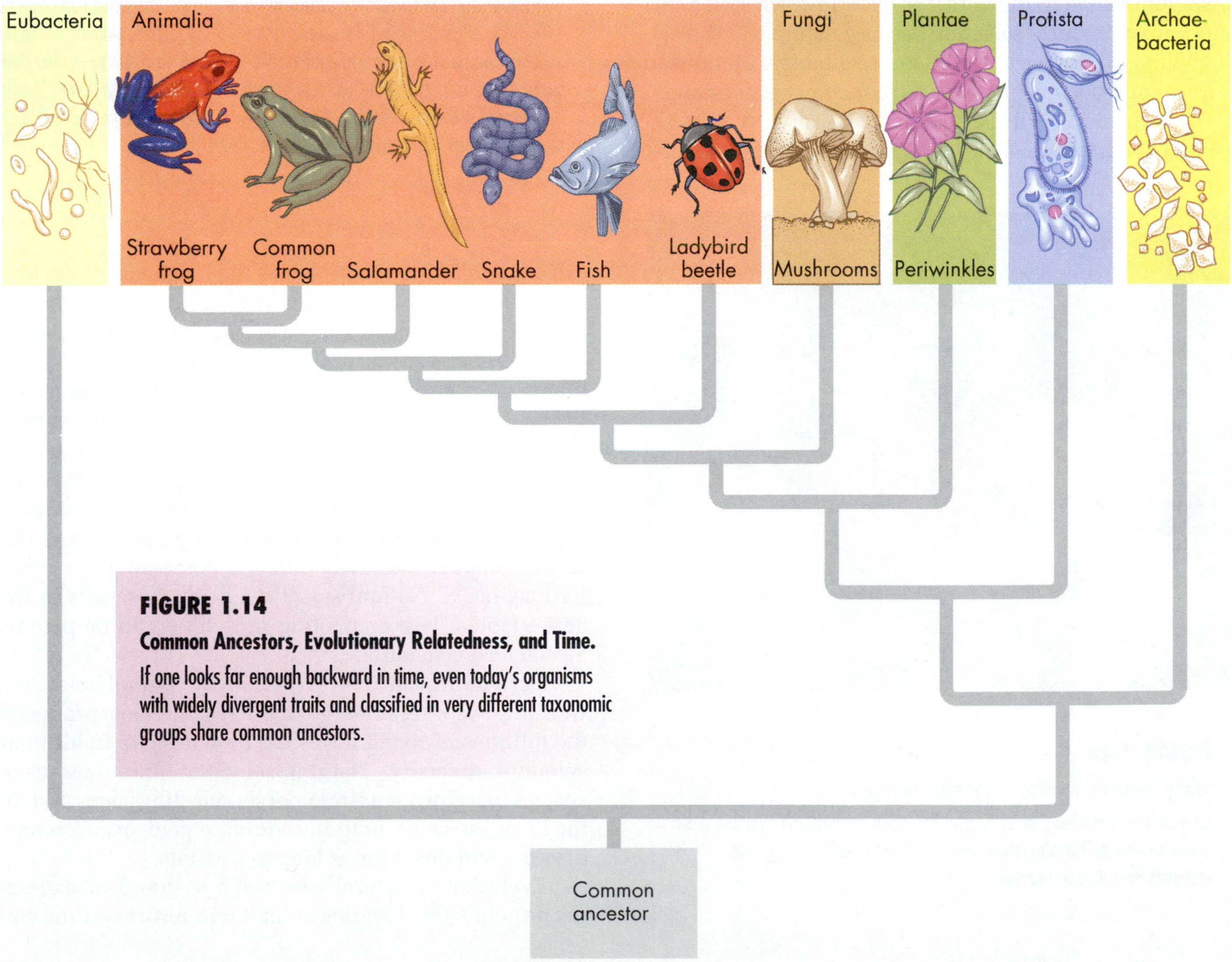

NATURAL SELECTION

On November 24, 1859, a drab little green book with a cumbersome title was published in London. Its first edition of 1250 copies was bought by booksellers in one day, and it has not been out of print since. The author was Charles Darwin [see BOX 1.1]. The title of the book was

Eubacteria
Animalia
Fungi
Plantae
Protista
Archaebacteria
Strawberry frog
Common frog
Salamander
Snake
Fish
Ladybird beetle
Mushrooms
Periwinkles
Common ancestor

FIGURE 1.14

Common Ancestors, Evolutionary Relatedness, and Time.

If one looks far enough backward in time, even today's organisms with widely divergent traits and classified in very different taxonomic groups share common ancestors.

On the Origin of Species by Means of Natural Selection, or The Preservation of Favoured Species in the Struggle for Life. Today we refer to his book as *The Origin of Species.*

The Origin of Species proposed these major ideas:

1. Living things are changing constantly, producing new species that did not exist when life on earth first originated.

2. All living organisms, no matter how different they seem, evolved from a single common ancestor.

3. Within any given population, variations in shape, size, vital processes, and behavior exist among individuals of the same species [FIGURE 1.17]. These variations can be inherited from one generation to the next.

From these ideas, Darwin concluded that individuals possessing variations that help them survive and reproduce in a competitive, changing environment have an advantage over members of their species that lack such traits. Darwin thought this last idea provided the mechanism for change in living things over time. He called this mechanism **natural selection** because nature was "selecting" the individuals with the most suitable variations to survive and become the parents of the next generation.

The principle of natural selection, now widely accepted as a main mechanism behind evolution in nature, can explain adaptations such as the long neck of giraffes. In a population of giraffes browsing on trees in the savanna many thousands of years ago [FIGURE 1.18], some of the giraffes probably would have had long necks and others short necks, each based on preexisting variations of the hereditary units called genes within the giraffe population. If there were too few low-hanging leaves around to feed all the giraffes in the population, then the long-necked giraffes could reach more food, harvest more energy and materials, and survive to produce more offspring—many, like their parents, possessing the genes for long necks. Natural selection, in short, results in the best adapted, best-competing individuals to be parents for the next generation.

Darwin argued that evolution by natural selection, working over long stretches of time, could have produced the millions of living forms we now observe in all their splendid diversity. Natural selection fine-tunes each species to its own environment by selecting genes that, in the broadest sense, help individual organisms overcome disorder and death for as long as possible.

Evolution by natural selection is so grand an organizing principle for all biology that it will resurface time and

FIGURE 1.16

Adaptations for Reproduction.

The orchid *Cryptostylus letochila* reproduces with the help of the wasp *Lissopimpla excelsia.* Natural selection has favored the accumulation of genes in the orchid that cause its flower to resemble a female wasp. The male wasp attempts to copulate with the orchid flower, and in the process picks up pollen from the orchid. When the wasp flies to another flower of the same species, he transfers the pollen to the female parts of the new flower. The flower's sperm in the pollen then fertilizes the orchid's eggs, and the resulting cell will become part of an orchid seed, and perpetuate the orchid species.

FIGURE 1.15

Sticky Tongues: Feeding Adaptation in Frogs.

An adult frog has a long, sticky tongue that it flips forward with lightning speed to entrap insects. This adaptation allows the frog to extract energy and materials from its environment.

FIGURE 1.17

Natural Variation in the Strawberry Frog.

As these photos show, not all strawberry frogs are a vivid scarlet. **[Left]** Some have red bodies and large dark spots; and **[Right]** some have green bodies and small spots. Natural selection acts on such variation, "selecting" those individuals best equipped to survive in the local environment as parents for the next generation.

again in this book, in the many magazine and newspaper articles you may read on life science, and in any future biology courses you may take.

CONCEPT CHALLENGE

Darwin presented three main ideas: that species could change over time, that evolution occurs by descent with modification, and the notion of natural selection. How do those three ideas relate to each other?

FIGURE 1.18

Natural Selection on the Savanna: Giraffes with Genetically Determined Long Necks Have Advantages for Survival and Reproduction.

Amid tall trees and long-necked neighbors, short-necked giraffes would have died and left few offspring, while long-necked giraffes would have survived, reproduced more often, and left more long-neck genes in the population.

Science As a Way of Knowing

An important result of the Earth Summit was a renewed focus on ecology: Participants realized that before we can decide how to preserve biological diversity, how to block climate change, and how to promote economic development without damaging the environment, we need to know more about how living things interact with each other and with the nonliving world around them. Delegates to the conference came away with a conviction that the world's nations must train more scientists to study ecology and the principles of nature in general.

Let's look now at the process of science and how scientists have coaxed, prodded, and pried the secrets of nature from living organisms. It is important to keep in mind that behind every fact and concept we present, there were people in laboratories or field stations engaged in the often tedious and frustrating, but sometimes joyful and exciting, pursuit of knowledge about what makes living things look and act the way they do. So let's look more closely at the fundamental principles behind the scientific process, as well as the general methods for problem solving that scientists employ, and that, in fact, we all use daily.

FUNDAMENTAL PRINCIPLES: CAUSALITY AND UNIFORMITY

A lightning bolt flashes in a cloud-darkened sky. A man eating in a restaurant becomes enraged and abusive when the waitress tells him they are out of apple pie. Modern scientists assume that events like these are due to natural causes, a principle they call **causality**. The ancient Greeks, on the other hand, believed that thunderbolts arose when the god Zeus hurled them at the earth, and that an emotionally disturbed person was possessed by evil spirits. Today's scientists may not yet fully understand the natural causes of a phenomenon like mental illness, but they firmly believe that it is based upon natural causes and that by applying the scientific process, they someday will be able to understand those causes.

A second fundamental principle of science is the **uniformity** of phenomena in time and space. Scientists contend that the laws of nature, for example, the speed of light, the laws of gravity, and so on, operate the same way today in Columbus, Ohio, as they did in Olduvai Gorge, in East Africa, 1 million years ago, or in the cloud of dust and gases that existed before our solar system formed billions of years ago. The principle of uniformity is impor-

FIGURE 1.19

Brightly Colored Animals and the Scientific Process.

Like strawberry frogs, coral snakes [A] are poisonous. One could generalize that all brightly colored animals are poisonous, and from this deduce that brightly colored black chinned red salamanders [B] are also poisonous—which, in fact, they are. But this fact, while consistent with the hypothesis, does not prove it. Tests can only *disprove* hypotheses. Since a test can show that cardinals [C] are bright red but not poisonous, the hypothesis is clearly incorrect. It can, however, be modified to this: Brightly colored reptiles and amphibians are often poisonous.

[A]

[B]

tant to biologists because events that led to life's origin and diversity occurred long before humans lived to observe them. Yet a biologist can feel confident that the same natural laws operating today functioned in the same way at the dawn of time, as life began, and all during the evolution of life on earth.

THE POWER OF SCIENTIFIC REASONING

Resting on the twin pillars of causality and uniformity is the power of scientific reasoning. We have seen that Darwin stated two facts from which he drew a grand conclusion: that natural selection is a mechanism of evolution. This kind of thinking is called **inductive reasoning**, or generalizing from specific cases to arrive at broad principles. Inductive reasoning is a particularly creative, intuitive, and exciting part of science. It's like creating a sonnet or sculpture or sonata. The instant when the mind leaps from previously isolated facts to a broad, unifying generalization is as thrilling, some scientists say, as starting down the steepest slope of a roller coaster.

The counterpart to inductive reasoning is **deductive reasoning**, or analyzing specific cases on the basis of preestablished general principles. Let's see how inductive reasoning and deductive reasoning differ in application. A person who knows, for example, that bright red strawberry frogs and red-and-yellow coral snakes [FIGURE 1.19A] are poisonous might conclude through inductive reasoning that all brightly colored organisms are dangerous. Such a tentative generalization is called a **hypothesis**. If that same person then encounters a flashy, black-chinned red salamander [FIGURE 1.19B], the person might, using deductive reasoning, assume that this brilliant salamander, too, is poisonous and avoid it.

While inductive and deductive reasoning help shape how scientists think, the final and equally crucial element in the scientific process is the method scientists use to try to test their generalizations.

TESTING GENERALIZATIONS

The person who generalized from strawberry frogs and coral snakes to form the hypothesis that all brightly colored organisms are poisonous might one day see a friend eating a bowl of red raspberries and start waving his or

[C]

her arms at the confused diner. If the friend survived the raspberries, however, that would *disprove* the hypothesis, or show it to be false, and a new hypothesis would be needed—something like: Bright red *animals* (not all organisms) are poisonous. But then, what about cardinals [FIGURE 1.19C]? A quick test could show that these scarlet birds are also nonpoisonous, requiring still further modification of the hypothesis.

A good scientist is always skeptical about hypotheses and is ready to discard or modify them when data from actual tests disprove them. Some hypotheses, however, have so far survived all attempts at disproof. A broad general hypothesis that is tested repeatedly but never disproved comes to be accepted as a **theory**, a general principle about the natural world, like the theory of gravity, the cell theory, or the theory of evolution. (Keep in mind that scientists use the term "theory" very differently than many lay people do. The popular phrase "It's just a theory" implies that something is an untested idea.)

To describe and understand nature accurately, scientists combine the kinds of creative thinking, reasoning, and testing we have been discussing into a series of steps called the **scientific method**:

1. They *ask a question* or identify a problem to be solved based on observations of the natural world.

2. They *propose a hypothesis,* a possible answer to the question or a potential solution to the problem.

3. They *make a prediction,* a statement of what they will observe in a specific situation if the hypothesis is correct.

4. They *test the prediction* by performing an experiment or making further observations.

While these steps may sound regimented, they are really little more than an organized commonsense approach—one you use regularly in your own life. Let's say that one evening you observe that your desk lamp isn't working. You would pose a question (Step 1): "What made my desk lamp go out?" And you would probably create a hypothesis (Step 2): "Maybe the light bulb burned out." Next you would make a prediction (Step 3): "If the bulb burned out, then when I replace it with a working bulb, the lamp should light." Finally, you would perform an experiment (Step 4): You would remove a bulb from a floor lamp that works, screw it into your desk lamp, and watch what happened.

When you performed the test, you automatically included what scientists call a **control**, a check of the experiment based on keeping all factors the same except for the one in question. Here, the control was the borrowed bulb that worked in the floor lamp. If the borrowed bulb failed to work in the desk lamp, you could conclude, based on your control, that a burned-out bulb was not the problem. You would next discard the faulty-bulb hypothesis and ask new questions: "Is the lamp itself broken? Is something wrong with the wiring to the wall socket?" You could then make new hypotheses, new predictions, and perform new tests until you had ruled out all but one hypothesis.

If you continued to generate tests to try ruling out this last hypothesis, but it continued to make correct predictions time after time, you could begin to believe in its validity, and discard it only after some future test showed the hypothesis to be false. Since we can often disprove hypotheses but can rarely prove one to be true, *scientific knowledge is nearly always tentative and conditional,* relying on the best answer yet available. This is unsatisfying to many, because people, in general, like clear-cut truths (or at least an illusion of them).

No tool is more powerful for understanding the natural world than the scientific method. It does not apply, however, to matters of religion, politics, culture, ethics, or art. These valuable ways of approaching the world and its problems proceed along different lines of inquiry and experience. Nevertheless, many of the world's complex problems have underlying biological bases, and their solutions demand knowledge of biological principles.

➤ CONCEPT CHALLENGE

Scientists often say that scientific knowledge is "provisional." What do they mean by that?

How Biological Science Can Help Solve World Problems

The Earth Summit in Rio focused on just a few of the world's problems, including climate change, biodiversity, sustainable development, and the world's forests. There are other pressing problems as well, including overpopulation, famine, war, crime, drug addiction, AIDS, cancer, heart disease, pollution, ozone depletion, acid rain, and changes in climate. While these problems tend to have multiple roots, their biological bases may yield to biological solutions, and these solutions, in turn, will help ease the associated pressures on society. For this reason, people in all fields must understand the biological bases of the world's problems.

Many of our most vexing problems stem from our species' enormous and burgeoning population. Five and one-half billion people are currently straining our planet's environmental resources, and by the year 2000, we will number about 6 billion. This immense population is accumulating because the human species, like all others, is finely honed to obtain energy and materials and to reproduce. The resulting human adaptations—the abilities to reason, communicate, and manipulate the physical world—have been so successful that our single species is busily exhausting the limited resources that support all

FIGURE 1.20

Deforestation in Madagascar: Landscapes Effaced, Healing Plants Lost.

Hills in Madagascar's central plateau lie denuded of their original forests, with precious topsoil washing away each time it rains. The jungles were cleared for firewood and for the creation of new farmland, but after a few years, the land's productivity plummeted, and farmers cleared new jungle. Thousands of species may be lost to this deforestation process.

life on our small planet. Many observers believe that our future security and quality of life are threatened less by war among nations than by the burden the sheer crush of humanity places on natural resources.

Take, for example, the plight of Madagascar, a large island off the southeast coast of Africa. In the past 35 years, half of Madagascar's forests have been leveled to provide fuel and to uncover farmland for the impoverished and rapidly growing population. Heavy rains, however, cause the cleared hills to erode, bleeding the rich red topsoil into the rivers and destroying the land's ability to grow crops [FIGURE 1.20]. Farmers plant clove tree seedlings on the cleared slopes and then wait seven years before the first full crop of the pungent clove buds are ready for harvest and export, simply hoping that the slope will not wash away before harvest. Whole villages depend on the developed world's appetite for vanilla ice cream, cola drinks, and other foods that contain extracts from clove buds. Yet the village children go hungry because cloves are not an adequate food source, and not much else is available to eat.

The exploitation of Madagascar's forests and wildlife began about 1500 A.D. when humans first came to the island. As the human population grew, many species became extinct, including the "elephant bird," an astonishing animal that stood 3 m (10 feet) tall, weighed 500 kg (half a ton), and laid 10-k (22-pound) eggs. The destruction has since accelerated, and with it has come the loss of hundreds of species, including plants that produce potentially life-saving drugs such as the drugs isolated from Madagascar's pink-petaled periwinkle that fight childhood leukemia. Observers fear that many such plants with potential uses in medicine and agriculture will perish along with Madagascar's tropical forests before they are ever studied and applied. And worse, they fear that Madagascar's own people are not even benefiting substantially from the wholesale clearing of forested lands.

Two decades ago, one might have predicted a similar fate for the Central American nation of Costa Rica. Although this small country's national debt is one of the highest in the world per citizen, the leaders are committed to sustainable development based on the application of modern biological principles and practices. Among other things, Costa Ricans are replanting deforested hillsides with fast-growing tropical hardwood trees. They are also growing nutritionally improved food crops, attempting to set aside up to 15 percent of their land as natural preserves for their tropical species, and instituting up-to-date health care and family-planning practices whenever possible. Costa Rica's success is a hopeful sign for the future: clearly these are biological solutions for social problems of biological origin.

In the chapters that follow, you will find many discussions of how biology is helping to solve world problems. We are, in fact, in the midst of a revolution in the biological sciences, with exciting new information surfacing weekly in the fight against cancer, heart disease, AIDS, infertility, and obesity. Researchers are making rapid advances in gene manipulation to create new drugs, crops, and farm animals; in exercise physiology to improve human performance; in the diagnosis of genetic diseases; and in the transplantation of organs, including brain tissue. Across all frontiers of biological science, at all levels of life's organization—from molecules to the biosphere—scientists are learning the most profound secrets of how living things use energy to overcome disorganization and reproduce to overcome death. In studying the living world, you are embarking on an adventure of discovery that will not only excite your imagination and enrich your appreciation of the natural world, but will also provide a basis on which you can contribute intelligently to the difficult choices society must make in the future.

Connections

All living things share a mutual interdependence, a common origin, and two fundamental problems: the tendencies toward disorder and death. Each organism must conquer the challenge of disorganization by extracting energy and materials from its environment and using them for growth, repair, and continued activity. While a frog captures ants and recycles the energy stored in the insects' muscles, a periwinkle plant captures energy directly from the sun and absorbs materials from the air and soil. As different as these solutions might be, animals, plants, and all other organisms share a basic set of characteristics that allow them to combat the challenges to survival, and these attest to the common origin and to the unity of all living things.

Because life's characteristics and challenges are common to all species, people, as living organisms, are powerless to escape them. To save ourselves and our world from current ecological problems, biologists and citizens in all fields have begun massive global efforts such as the Earth Summit in Rio, and they must continue to learn about life processes and the interactions of organisms with their environments. The underlying principles of biological science begin with the laws of chemistry and physics and how they govern the atoms and molecules in living things—the subject of our next chapter.

KEY TERMS

adaptation, 15
Animalia, 13
Archaebacteria 13
biology, 2
causality, 19
control, 21
deductive reasoning, 20
development, 8
Eubacteria, 13
evolution, 15
Fungi, 13
gene, 8
hypothesis, 20
inductive reasoning, 20
motility, 6
natural selection, 18
order, 12
Plantae, 13
prediction, 21
Protista, 13
reproduction, 7
responsiveness, 7
scientific method, 21
theory, 21
uniformity, 19

HIGHLIGHTS IN REVIEW

1 Living organisms are highly ordered; maintaining order requires energy and material from the environment.

a] The rosy periwinkle and the problems discussed at the Earth Summit illustrate five main themes in biology: the importance of energy in maintaining order; the consequences of reproduction; the evolutionary principles that created life's diversity; the interactions of diverse organisms with each other and the environment; and the utility of the scientific process in learning how nature functions.

b] Biologists have compiled a list of characteristics common to all living things. Order, metabolism, motility, and responsiveness relate to an organism's system for overcoming disorder; reproduction, development, and genes relate to perpetuation; and evolution and adaptation relate to interactions with the environment.

c] Living things exhibit a hierarchy of biological organization that includes the biosphere, ecosystems, communities, populations, individual organisms, organ systems, organs, tissues, cells, organelles, molecules, atoms, and finally, subatomic particles.

d] Living things exhibit metabolism, or the breaking down, rearrangement, and use of energy compounds, and the synthesis of other compounds.

e] Self-propelled movement is a diagnostic feature of life, even in plants and other organisms that appear to remain stationary.

f] All organisms are responsive; they detect information about their surroundings and react in appropriate ways.

2 Individual organisms eventually die, but life continues via reproduction and heredity.

a] Development, or change and growth during the life cycle is a key characteristic of life.

b] Living things have genes, or units of heritable information, that pass from one generation to the next and control specific physical, chemical, and behavioral activities within the organism.

3 Over long periods of time, life forms change due to evolution. Natural selection—the ability of individuals with certain traits to reproduce more successfully than those lacking the traits—is a major mechanism of evolution.

a] Adaptations, or particular innovations of structure and function, allow organisms to survive in their particular environments.

b] Life as a whole is characterized by unity, manifested by shared molecular-level characteristics, and diversity, the fantastic variety of living things all related by descent from common ancient progenitors.

c] Biologists categorize each living thing, placing each into a genus and a species, and into increasingly inclusive groups, including families, orders, classes, phyla or divisions, and six main kingdoms—Eubacteria, Archaebacteria, Protista, Fungi, Plantae, and Animalia.

4 Science is a way of learning how nature functions. Scientists perform tests that might rule out possible explanations for a particular phenomenon. Explanations that have been repeatedly tested and not disproven are tentatively accepted.

a] Biologists apply the process of science to their study of the natural world. This includes the principles of causality and uniformity, as well as inductive and deductive reasoning and the testing of hypotheses. The formal scientific method involves asking a question about some puzzling phenomenon, proposing a hypothesis, predicting an observable result, then testing the prediction through experimentation.

5 The principles of biology can be applied to world problems such as overpopulation, disease, environmental degradation, famine, and drug addiction.

a] Biological research and applications have the potential to help solve many of the world's current problems, a number of which stem from overpopulation and the strain that our sheer numbers place on the environment.

b] We live amidst a revolution in the biological sciences, and citizens in all fields must become informed so that they can contribute to societal choices.

UNDERSTANDING THE FACTS AND CONCEPTS

For Questions 1–5, match each of the phrases below with the one item in the following list that is most appropriate. Any item may be used once, more than once, or not at all.

a] order
b] metabolism
c] motility
d] responsiveness
e] reproduction
f] development
g] genetic mechanism
h] evolution
i] adaptation
j] all of these

1 The defining characteristics of life.

2 The chemical reactions characteristic of living things.

3 The way hereditary units work.

4 Self-induced movement.

5 The acquisition or modification of structures or of functions that work well in a particular environment.

As above, answer Questions 6–10, using the following list:

a] adaptation	d] natural selection
b] evolution	e] kingdom
c] mutation	f] none of these

6 A category of classification of living things.

7 A change in the structure of a gene.

8 A newly developed structure or behavior in an organism.

9 The gradual accumulation of mutations that leads to changes in the kinds of organisms living on earth.

10 The effect of environmental factors on adaptations.

As above, answer Questions 11–16, using the following list:

a] ecosystem	d] *a* and *b* are both correct
b] community	e] *b* and *c* are both correct
c] population	f] *a*, *b*, and *c* are correct

11 Includes only the members of a given species.

12 Includes only the members of several species.

13 Includes both living and nonliving elements in a given area or location.

14 Includes only living elements.

15 The most inclusive of the categories listed.

16 The most restrictive of the categories listed.

As above, answer Questions 17–21, using the following list:

a] protists	d] animals
b] fungi	e] all the preceding apply
c] plants	f] none of these

17 Multicellular decomposers.

18 Composed of complex cells.

19 Motile and multicellular.

20 A single kingdom that includes both unicellular and multicellular organisms.

INTEGRATE AND APPLY WHAT YOU HAVE LEARNED

1 It is conceivable that the drugs that are isolated from rare plants can be produced in the laboratory by new techniques of genetic engineering.
- a] What ecological problems might be avoided by doing this?
- b] What kinds of problems might be engendered?

2 Are the properties of each hierarchical level of life the sum of the properties of the lower levels, or do novel ones arise and become apparent? Explain.

3 A student wonders if his or her philodendron plants really need a weekly dose of nitrogen fertilizer, as the fertilizer package recommends. Explain how this student might find out, using the scientific method.

4 Poisonous species of butterflies often are brilliantly colored. Nonpoisonous species are known that resemble poisonous ones very closely. Explain how natural selection might account for this phenomenon of mimicry.

5 What is the most important mechanism within organisms that leads to evolutionary change, and what role, if any, does the environment play?

ANALYSIS

1 Considering the characteristics of motility and responsiveness, select the most appropriate statement made about them below.
- a] Motility and responsiveness are the same.
- b] Motility is self-induced motion, whereas responsiveness does not necessarily involve motion.
- c] Motility requires the use of energy derived from metabolism, whereas responsiveness has no energy dependency.
- d] Responsiveness as a characteristic is subject to natural selection, but motility as a trait is not.
- e] Responsiveness may or may not involve motility.

2 Suppose that a particular organism is unable to extract either energy or material substance from its environment. Which of the following would be a consequence?
- a] The organism would die sooner or later because its molecular order would decline below the level essential for metabolism.
- b] The organism could not grow, but could reproduce.
- c] It would have to recycle its wastes and use them repeatedly as food.
- d] It would shrink to microscopic size as it absorbs its own body cells for food.
- e] It would maintain itself in a state of suspended animation so long as it did not reproduce or grow.

3 If evolution proceeds by the selection of new adaptations that arise from genetic mutations, and if mutations arise as "mistakes" in the genes, does this mean that evolution proceeds from mistakes?
- a] No. New adaptations do not usually, if ever, come from new mutations or changed genes.
- b] No. Mistakes are not utilized, since evolution leads to more perfect organisms.
- c] No. Natural selection does not work on organisms, genes, or adaptations.
- d] Yes. However, what may be mistakes in one kind of environment may prove to be advantageous if the environment should change.
- e] Yes. But the mistakes that enable evolution to proceed are not ones that preadapt the organism.

4 Suppose you have formed a hypothesis and you have performed an experiment that supports it. Which of the following procedures would be most likely to produce further support?
- a] Repeat your previous experiment.
- b] Have someone else repeat your previous experiment.
- c] Repeat the experiment, but change the controls.
- d] Repeat the experiment under different conditions, such as place, temperature, light, humidity, etc.
- e] Make a new prediction based on your hypothesis and test it by a different experimental procedure.

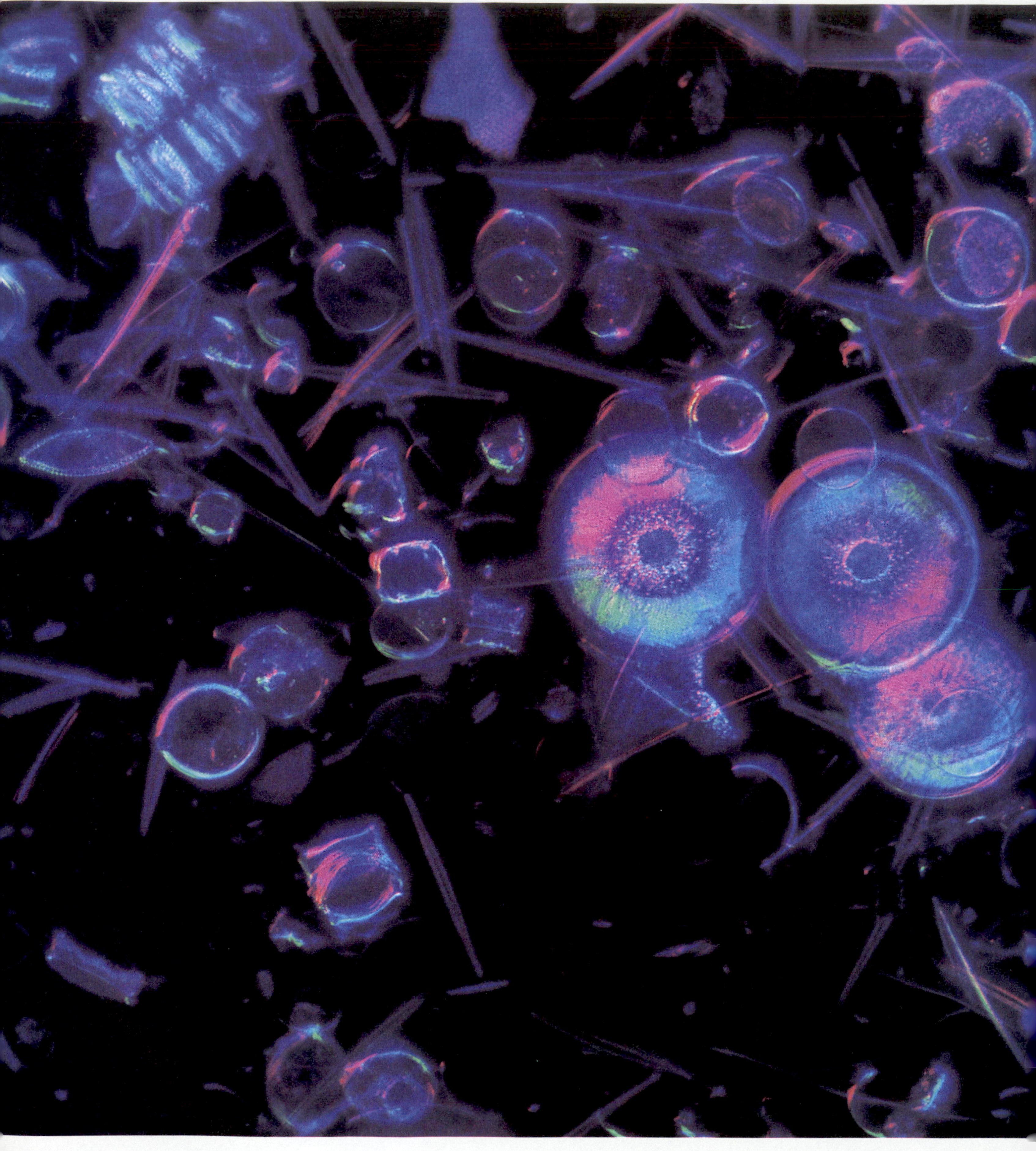

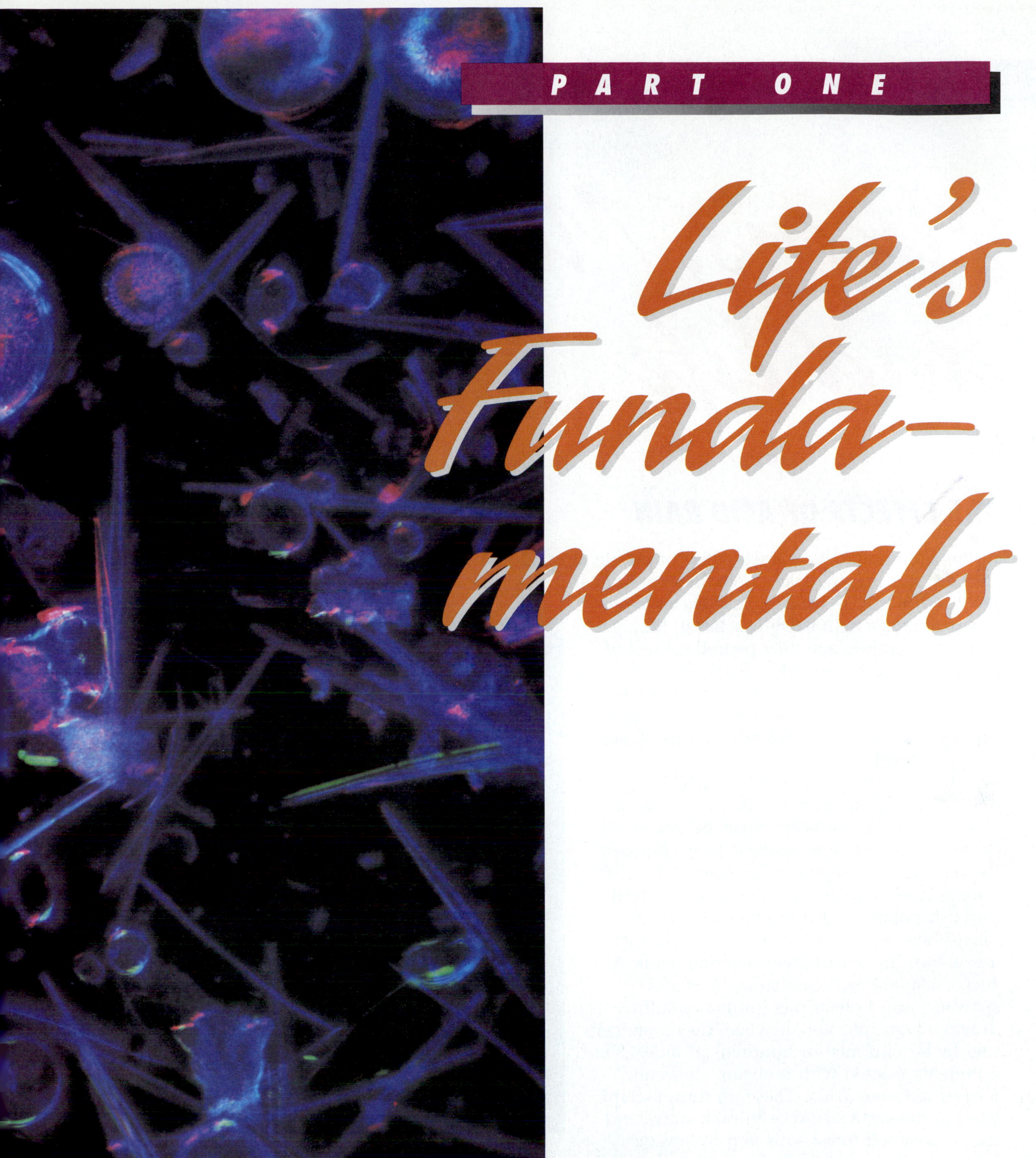

PART ONE

Life's Funda-mentals

CHAPTER 2

Atoms, Molecules, and Life

THE EFFECTS OF ACID RAIN

In the spring of 1976, a small group of people traveled from Winnipeg, Canada, to a pristine lake in Northwestern Ontario province, some 800 km (1300 mi) north of Minneapolis. There, amidst a vast tract of virgin evergreen forest, they performed a startling act: they poured gallons of sulfuric acid into the cold, clear waters of the lake [FIGURE 2.1]. This was not part of a terrorist plot. It was a deliberate deed by ecologists hoping to understand acid rain and its effects on aquatic wildlife.

FIGURE 2.1

Ecologist releases acid from the white box into Lake 223 in Ontario to test the effects of acid rain.

The research group, led by David W. Schindler, was far from the first to notice or study acid rain. For years, sport fishers in Scandinavia and New England had witnessed the decline of trout and other game fish in remote lakes and streams. Scientists realized that fish populations dwindled most often when their aquatic home was downwind from a coal-burning electricity-generating plant. At such installations, towering smokestacks spewed tons of chemicals (primarily sulfur dioxide) high into the air where they were carried by air currents for hundreds of miles. The pollutants reacted with moisture aloft and formed airborne acids. These, in turn, precipitated to the earth's surface in rain, snow, and fog, and turned forest soils and waterways acidic. While biologists had investigated the effects of acid rain on individual fish and frog species in the laboratory, no team before

Schindler's had studied the effects of acid rain on an entire ecosystem—a group of living organisms interacting with each other and the physical environment.

To do this, Schindler's group chose a clean, remote lake that was *not* downwind from a power plant, Lake 223. Then they spent two years monitoring the health and population sizes of various aquatic plants and animals, including algae, gnats, trout, and small shrimp-like animals. This gave them needed background data on the normal state—a so-called *base line.* Then, every spring for eight years, they dumped acid into Lake 223 to simulate acid rain, and carefully noted the changes in the lake's wildlife. (They also monitored a nearby lake but left it untouched, as a standard of comparison, or control, for this experiment.)

Some of the results were dramatic but unsurprising. In 1979, after three years of acid dumping, the lake trout still looked fairly plump and healthy. By 1982, however, after six years of acid treatment, the trout were starving, scrawny, and suffering from various degenerative diseases. It was the *reason* for this starvation that was so unexpected: the inconspicuous insects, tiny shelled animals, and other species trout feed upon had begun to die off at far lower acid levels than anyone predicted.

MESSAGES

1 The properties of life are based on atoms and molecules. Atoms are the fundamental units of matter.

2 The organization of an atom's protons, neutrons, and electrons determines its properties.

3 The special properties of water—such as why ice floats, why water dissolves many substances, and why water acts like a weak acid—depend on the water molecule's chemical and electrical properties.

4 Life is based on carbon atoms. Four classes of large carbon-containing molecules—carbohydrates, lipids, proteins, and nucleic acids—are especially important in understanding the structure and function of living things.

To perform their important studies, the Canadian ecologists had to know how to identify dozens of species of algae, gnats, and fish. But they had to know much more—they had to understand the chemical nature of acids and to comprehend how acids interact with other chemical constituents of the soil, rocks, lake water, and the plants and animals in and around the lake. In short, to understand acid rain and its effects on living things, they had to understand chemistry, our subject in this chapter.

Chemistry is the study of the structure and behavior of atoms and molecules, the particles that make up all living and nonliving matter. We will see in this chapter that, in general, the atoms and molecules in rocks, dirt, air, and other nonliving things differ in kind and proportion from those in living things. Organisms contain four main classes of biological molecules, including three you are probably familiar with—carbohydrates, proteins, and fats—and a fourth, the nucleic acids, including DNA. Living things also contain large amounts of water, a very special molecule. To understand biology, we must examine the basic chemical components of living systems and explore the close relationship between the structure of these chemicals and their functions. Only then can we appreciate that the materials making up living things, with all their marvelous diversity of shapes and behaviors, are subject to the same physical laws that govern all matter in the universe.

In this chapter, we will see how the structure of atoms and molecules gives them their particular properties and why the water molecule is so special for life. We will also consider the natures of the four major types of biological molecules. Along the way, you will learn why acid rain harms plants and animals, and how people are combatting this environmental problem. You will also learn why a "permanent" wave keeps hair curled longer than a curling iron; how one property of water contributes to a baby's first breath; why you should avoid certain kinds of fats in the diet; and how one natural protein may be related to alcoholism. ❑

Elements and Compounds

Ancient Greek philosophers realized that some materials, such as rocks, wood, and soil, are composed of more than one substance, while other materials, such as chunks of iron, gold, silver, aluminum, and sulfur, cannot be decomposed into different constituents by chemical processes. Scientists later studied the pure substances and called them *elementary substances*, or **elements**. They gave each known element a chemical symbol; for example, the symbols for the elements already mentioned are Iron (Fe), gold (Au), silver (Ag), aluminum (Al), and sulfur (S) (see TABLE 2.1). Substances made up of two or more elements that are chemically combined they called **compounds**.

In the 1820s, the English chemist Michael Faraday was able to show the essential differences between elements and compounds through a set of experiments. Faraday took common table salt, heated it to the boiling point, passed electricity through it, and produced the soft, silvery metal he called *sodium* (Na), plus the corrosive, greenish-yellow gas *chlorine* (Cl). Since neither substance could be decomposed further, Faraday knew they were both elements.

Chemists have discovered 92 naturally occurring elements and have created 13 more in the laboratory. The properties of the many elements vary widely. For example, helium is a colorless, odorless gas; lithium a soft, whitish metal; calcium a white powder; copper a reddish metal; sulfur a yellow solid; mercury a silver liquid; and iron a hard, dark, shiny metal.

Although the earth contains 92 elements, scientists have found that living things are made up, for the most part, of just a few [FIGURE 2.2A]. Geologists studying the elemental composition of the earth's crust, for example, have discovered that 98 percent of the planet's surface layer consists of the eight elements listed in FIGURE 2.2B, headed by oxygen (O), silicon (Si), and aluminum (Al). In fact, the characteristics of the earth and the life it supports are different not so much because they contain different elements but because they have radically different *proportions* of elements.

A typical organism consists mainly of the six elements carbon, hydrogen, nitrogen, oxygen, phosphorus, and sulfur. Trace amounts of nine other elements also occur in living organisms, and these are listed in FIGURE 2.2A. In our later discussion of water and carbon, we will see why living tissue is a unique and special form of matter and why it is so unlikely that life could be based mainly on the common crustal elements of the earth, silicon and aluminum.

Atoms and Molecules

After discovering that matter is composed of different elements, scientists began to investigate what makes each element distinct: What, for example, is the fundamental difference between the element oxygen and the element gold? In the 1800s, the English chemist John Dalton concluded that each element is composed of identical parti-

TABLE 2.1 Some Common Elements and Their Atomic Structures (In order of atomic number)

Element and Symbol	Atomic Number	Number of Protons	Number of Neutrons	Number of Electrons	Atomic Mass*
Hydrogen (H)	1	1	0	1	1
Carbon (C)	6	6	6	6	12
Nitrogen (N)	7	7	7	7	14
Oxygen (O)	8	8	8	8	16
Sodium (Na)	11	11	12	11	23
Phosphorus (P)	15	15	16	15	31
Sulfur (S)	16	16	16	16	32
Chlorine (Cl)	17	17	18	17	35
Potassium (K)	19	19	20	19	39
Calcium (Ca)	20	20	20	20	40

* These numbers represent the most common atomic mass of the element, rounded off to the nearest whole number.

FIGURE 2.2

Elements in Earth and Living Organisms.

If analyzed at the level of atoms, our planet has a very different composition from the living organisms that inhabit it. **[A]** The zebras, oxpeckers, and other organisms that live on that crust are mostly hydrogen, oxygen, and carbon atoms. **[B]** The crust at our planet's surface, containing rock, soil, sand, and other materials, is mostly oxygen, silicon, and aluminum atoms in various compounds.

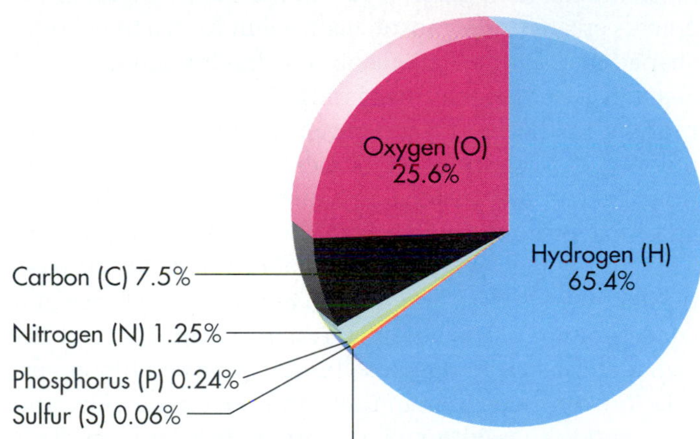

[A] Composition of living things

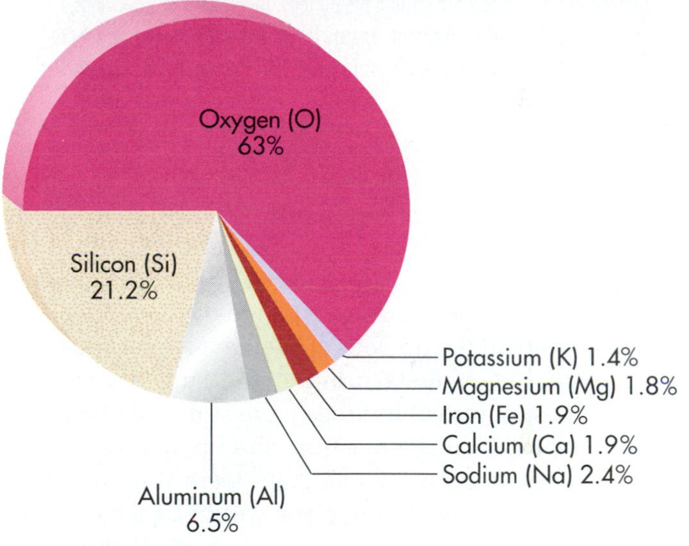

[B] Composition of the earth

cles. He called these particles *atoms* (from the Greek atomos, "indivisible"). **Atoms** are the smallest particles into which an element can be divided and still display the chemical properties of that element. Particles that consist of a chemical combination of two or more atoms are called molecules.

According to Dalton's atomic theory, a gold nugget consists of billions of gold atoms, and a test tube of pure oxygen gas consists of billions of oxygen atoms. Dalton, however, did not have the means to probe into the submicroscopic nature of atoms and elements. Later generations of chemists found that the properties of elements are based on the internal organization of their atoms.

STRUCTURE OF ATOMS

Chemists since Dalton have discovered that all atoms are composed of the same three types of subatomic particles: protons, neutrons, and electrons. A **proton** is a positively charged particle, and chemists have defined the *mass* of a proton (roughly speaking, how heavy it is) as about 1 atomic mass unit (amu). A **neutron** is a neutral particle with no electrical charge; neutrons also have a mass of about 1 atomic mass unit. **Electrons** are very small, negatively charged particles with a mass almost 2000 times *less* than that of a proton or neutron.

Using a simplified model of atoms, we can think of

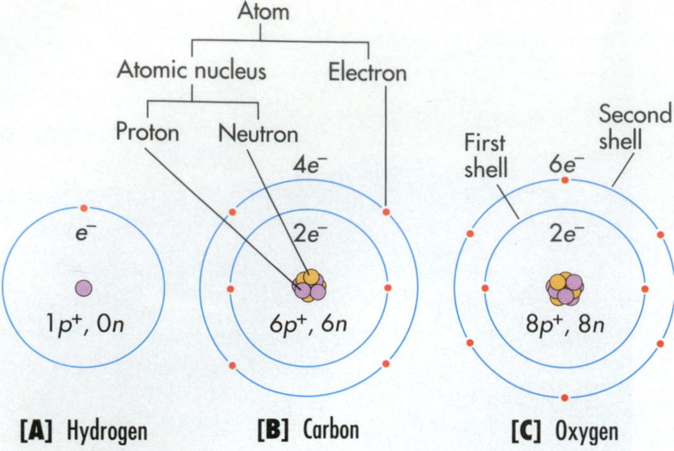

FIGURE 2.3

A Model of the Atom.

In each atom, negatively charged electrons (here, small gold dots) orbit the positively charged nucleus, which contains one proton (brown spheres) for each electron and a set of neutrons (tan spheres), usually close to the number of protons. A balance between opposing physical forces keeps the electrons moving at a given distance around the nucleus. This system of static rings, called the *Bohr model*, represents energy levels, but is not an accurate representation of the position of electrons in space. **[A]** Hydrogen. **[B]** Carbon. **[C]** Oxygen.

an atom as resembling a miniature solar system. At the center is the atomic **nucleus**, which contains protons and neutrons and accounts for most of the atom's mass. A specific number of electrons, equal to the number of protons, orbit about the nucleus at a relatively great distance, somewhat like planets orbiting the sun. FIGURE 2.3A shows the structure of hydrogen, the simplest atom, with its single proton, single electron, and no neutron.

If a carbon atom [FIGURE 2.3B] were the size of the Houston Astrodome, the nucleus would be a small marble on the 50-yard line. In reality, an atom is minute: About 1 million carbon atoms could sit side by side on the period ending this sentence.

ATOMIC NUMBER AND ATOMIC MASS

Since all atoms consist of the same three types of particles arranged in similar ways, what gives them different properties? Why is a chunk of the element carbon black and solid, while the element oxygen is a clear, colorless gas? The answer is that each type of atom contains a unique number of protons in its nucleus. All carbon atoms have six protons, for example, and all oxygen atoms have eight protons [FIGURE 2.3B and C]. The number of protons in an atom is its **atomic number**, and it identifies which element the atom is [TABLE 2.1].

Because each proton and each neutron contribute 1 atomic mass unit, the combined number of protons and neutrons in the nucleus is the weight of the atom. This number is the **atomic mass number**, or simply, **atomic mass**. (The mass of electrons is so small, it can be neglected in determining atomic mass.) Thus, the atomic number of oxygen is 8 and its atomic mass is 16 (8 protons + 8 neutrons; see TABLE 2.1).

An element often has an equal number of protons and neutrons, but not always. For example, the nucleus of the carbon atom usually contains 6 protons and 6 neutrons (for an atomic mass of 12), but phosphorus has 15 protons and 16 neutrons (atomic mass 31). The number of electrons in an atom equals the number of protons (and thus the atomic number): 1 for hydrogen, 6 for carbon, 8 for oxygen, and so on [review FIGURE 2.3].

Physicians use atomic structure in the diagnosis of major diseases. Chemists discovered many decades ago that in certain atoms, such as hydrogen and phosphorus, the protons whirl like tops. This spinning creates small magnetic fields in the nucleus. In the 1970s, medical engineers created a diagnostic instrument from a huge, ring-shaped magnet that can encompass the human body and generate a tremendous magnetic field. The magnets cause protons in the body's hydrogen atoms (and in certain other types of atoms) to align and spin in the same direction. Then, when a second magnetic field is applied, the spinning protons flip back to their normal alignment. This flipping gives off a signal that can be detected electronically, recorded, and used to reveal the presence and concentration of the atoms. Physicians can interpret the signal and use it to diagnose tumors, blocked blood vessels, broken bones, or injured muscles, ligaments, and other soft tissues without having to perform surgery. This technique is known as magnetic resonance imaging (MRI). FIGURE 2.4 shows an MRI device and a magnetic resonance image of a person's brain.

ELECTRONS AND ENERGY LEVELS

Because a nucleus contains both positively charged protons and uncharged neutrons, it has an overall positive charge. In contrast, the orbiting electrons have a negative charge. The attraction between these positive and negative charges pulls the electrons toward the nucleus, but the rapid movement of the orbiting electrons tends to throw them outward, away from the nucleus, like a rock tied to a twirling string.

Scientists picture the electrons in an atom as occurring in *shells*, at specific distances from the nucleus. The energy level of the electrons increases as their distance from the nucleus increases. Each shell can contain a certain maximum number of electrons, as follows. The shell

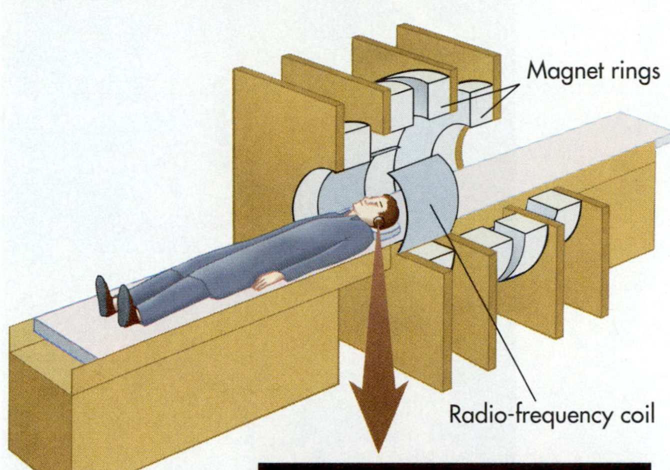

[A] Medical imaging by MRI

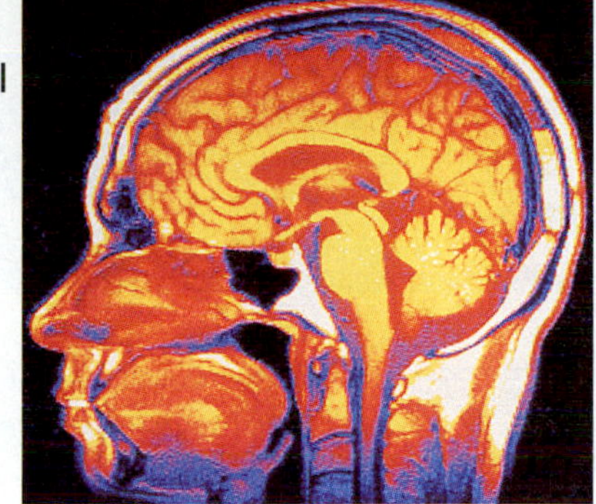

[B] MRI reveals vivid details of the brain's soft tissue

FIGURE 2.4

Atomic Structure and Human Medicine.

Physicians use magnetic resonance imaging (MRI) as a diagnostic tool. **[A]** MRI reveals vivid details of the human brain. **[B]** Detail of image produced by the MRI.

nearest the nucleus can hold either one electron, as in hydrogen [FIGURE 2.3A], or two, as in helium. The second shell can accommodate up to eight electrons, and will be filled before any electrons appear in the higher-energy third shell (which also becomes stable when it contains eight electrons). Subsequent shells also become filled with set numbers of electrons, and tend to fill in order. The shell-filling rules are important for understanding how atoms join together to form molecules.

VARIATIONS IN ATOMIC STRUCTURE: ISOTOPES AND IONS

Atomic bombs, acid rain, and the actions of your nerve cells all depend on slight exceptions to the standard structure of atoms. One set of variants, the isotopes, results from differences in the number of neutrons, while another set of variants, the ions, results from variations in the number of electrons.

Isotopes While the number of protons in the nucleus of a given kind of atom always remains the same, the number of neutrons can vary. Atoms with the same number of protons but different numbers of neutrons are different **isotopes** of the same element. A good example of an important isotope is uranium-235 (^{235}U), the explosive component of an atomic bomb. ^{235}U has three fewer neutrons in its nucleus than ^{238}U, the most common form of uranium. ^{235}U is **radioactive**—its nucleus is unstable and eventually breaks down, or decays. In radioactive decay, the atom emits energy and sometimes particles as it changes to a more stable form (of the same or a different element). The most common carbon isotope is ^{12}C with six neutrons. Other carbon isotopes are ^{13}C and ^{14}C, with seven and eight neutrons, respectively. Both ^{12}C and ^{13}C are stable, nonradioactive forms, but ^{14}C is radioactive.

Under certain circumstances the neutrons lost from a decaying ^{235}U atom can strike other uranium atoms and cause them to fly apart, releasing more energy and more neutrons, which can strike still more atoms. This can lead to a rapid-fire chain reaction that instantaneously and explosively releases an enormous amount of energy [FIGURE 2.5]. In contrast, other unstable isotopes, such as ^{14}C, with two extra neutrons, are used by scientists in

FIGURE 2.5

Bomb Blasts and Variation in Atom Structure.

The presence of unusual numbers of neutrons in an atom can make it unstable. Uranium atoms with three more neutrons than the common type of uranium atom tend to fall apart. When the fragments of a disintegrating atom strike other atoms, they too will break apart. If enough of these atoms are packed into a small space, this instability can result in an explosive chain reaction.

box 2.1 Biology Applied

Isotopes in Action: How Old Is the Ice Man?

In September, 1991, hikers high on a ridge separating Austria from Italy found a man's head and shoulders sticking out of a chunk of melting ice [FIGURE 1]. They thought this startling figure might have been the body of a less fortunate fellow hiker who succumbed to bad weather—and in a way, they were right. They quickly reported their discovery to the police, and the authorities arrived and excavated the body. To everyone's surprise, however, they found fur and leather clothing instead of Goretex, and a flint dagger instead of a Swiss army knife. *How old was this mummified hiker?*

University researchers at Oxford, England, and Zurich, Switzerland, used radioactive carbon-14 (^{14}C) to determine the age of the ice man.

Let's see how ^{14}C dating works. Radiation from the sun produces a small but constant amount of ^{14}C in the upper atmosphere. Plants take in ^{14}C along with the more abundant ^{12}C and use it to synthesize biological molecules during photosynthesis. When animals eat the plants, a small amount of the ^{14}C ends up in *their* bodies. As a result, the percentage of ^{14}C in all *living* plants and animals is the same as the percentage of ^{14}C in the atmosphere. When an organism dies, however, it stops incorporating ^{14}C; instead, ^{14}C begins to disappear, as it changes, or decays, into nitrogen-14 (^{14}C has six protons and eight neutrons, while ^{14}N has seven protons and seven neutrons). As the ^{14}C changes to ^{14}N, its concentration relative to ^{12}C decreases. Scientists have determined that half of the original amount of ^{14}C decays in 5730 years. (This is known as the *half-life* of ^{14}C.) So in 11,460 years, there will be one-fourth as much, and by 17,190 years, only one-eighth will remain. By calculating the concentration of ^{14}C in a small chunk of the ice man's flesh, the European scientists concluded that the ice man lived about 5300 years ago, and that he was the oldest, best-preserved, and most complete human body of that antiquity yet found.

FIGURE 1
The Ice Man.

Anthropologists got a richer understanding of prehistoric people by studying the artifacts found with the ice man. These included leather boots stuffed with grass for insulation, a wood-frame backpack, rope, antibiotic fungus probably carried as medicine, and an ax with a cast copper blade. Scientists now believe that on an early autumn evening some 5300 years ago, the ice man was hiking across the mountains. He ate a meal that included fresh autumn berries, set his belongings against nearby rocks, made a fire, then laid down to sleep. Apparently the weather changed abruptly, but he was so exhausted that he failed to wake up and feel the cold. His body remained stretched out, not curled up, and the most likely cause of death was freezing. Desiccated by dry autumn winds, and covered by heavy winter snows, the ice man did not decay for five millennia. Only in 1991, after record summer warmings, was the ice man "reintroduced" to a far different world.

experiments to trace the molecules that contain them. Biologists use ^{14}C to determine the age of dead organisms [see BOX 2.1].

Ions While isotopes result from variations in neutrons, ions result from variations in electrons. An **ion** is an atom with a specific positive or negative electrical charge because the number of its electrons does not equal the number of its protons. An understanding of ions can help us appreciate how acid rain harms living things and the environment. You will see the mechanisms of ionization in greater detail later in this chapter, but for now, it is important to know that an ion—specifically the hydrogen ion—is the chemical species that gives the acidity to acid rain.

FIGURE 2.6

Ions: Variation in Number of Electrons.

[A] A hydrogen atom has one negative electron and one positive proton, and hence the atom has no electrical charge. **[B]** A hydrogen ion has lost its electron, and so consists of a single proton, with a net charge of plus one. **[C]** The number of electrons equals the number of protons in a sodium atom. **[D]** A sodium ion has lost the single electron in its outer energy shell. Thus it has one fewer electron than protons. The result is a positively charged ion possessing an outer shell filled with eight electrons.

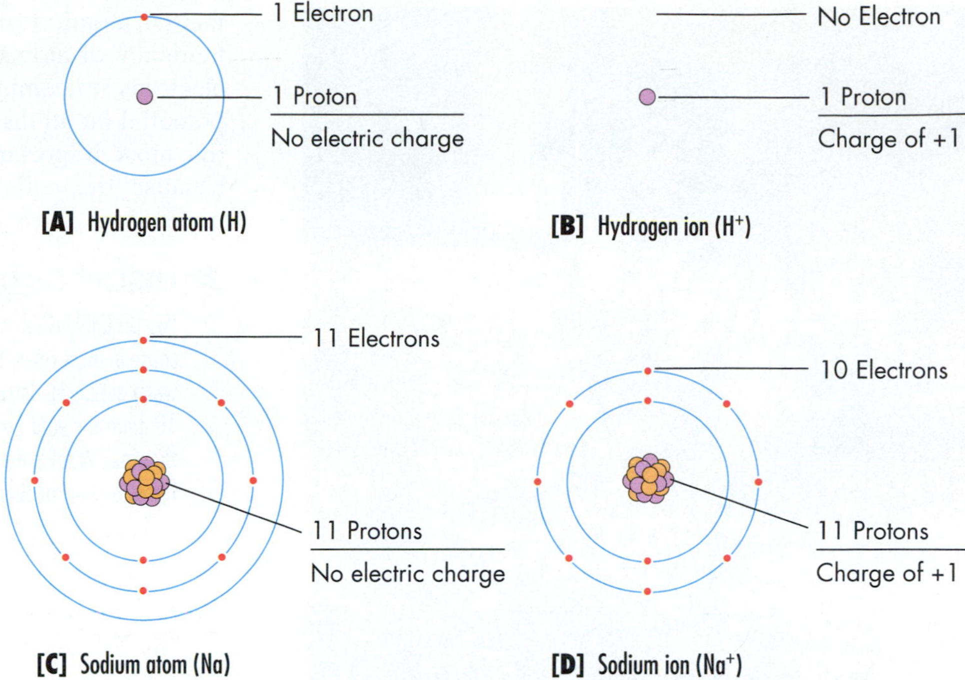

The most common form of hydrogen atom has one proton and one electron; the electrical charges of these two subatomic particles cancel each other so that the atom has no net charge [see FIGURE 2.6A]. A hydrogen ion, on the other hand, is missing its electron, and so it has simply one proton [FIGURE 2.6B]. Rain and lakes become acidic only after accumulating many billions of hydrogen ions.

One can think of a hydrogen ion as nature's simplest chemical—nothing but a proton, one of the three major subatomic particles. Yet when this "simplest chemical" accumulates in a lake such as Lake 223, its high concentrations can kill small organisms that serve as the food for other, larger living things, and this loss of a food source disturbs the entire ecosystem.

Another ion important to living cells is the sodium ion. If a neutral sodium atom (Na), having 11 protons and 11 electrons, loses an electron, it becomes a sodium *ion* (Na^+), with 10 electrons and 11 protons, and hence has an overall positive charge [FIGURE 2.6C and D]. A sodium ion has different properties from a sodium atom.

The rapid movement of sodium ions into and out of your nerve cells causes electrical signals that a researcher can detect and record with sensitive equipment (see the wave patterns in FIGURE 2.7). The movement of sodium ions into and out of brain cells is slightly different in a person meditating and one not meditating.

You have seen that the properties of the elements emerge from both the structure of the atomic parts and the way those parts are arranged (see TABLE 2.2 for a sum-

TABLE 2.2 Fundamental Concepts of Atomic Structure: A Summary

- **Elements** are substances that cannot be broken down to simpler substances by ordinary chemical means.
- **Atoms** are the fundamental units of elements; a visible quantity of an element contains billions of atoms of that element.
- Atoms contain an **atomic nucleus**, consisting of positively charged **protons**, and uncharged **neutrons**. Around the nucleus swirl negatively charged **electrons**.
- Each atom of an element contains the same number of protons and electrons, and hence is not electrically charged. The number of neutrons in an atom may be equal to or different than the number of protons.
- Atoms of different elements contain different numbers of protons.
- Two **isotopes** of an element contain the same number of protons, but different numbers of neutrons.
- An **ion** of an atom contains the same number of protons as other atoms of that element, but different numbers of electrons.
- A **molecule** is a group of two or more atoms held in a specific arrangement by energy links called chemical **bonds**.
- A **covalent bond** is a pair of electrons shared by two atoms. A **hydrogen bond** is the attraction of a hydrogen atom of one molecule to an atom of another molecule.

[A] Person meditating

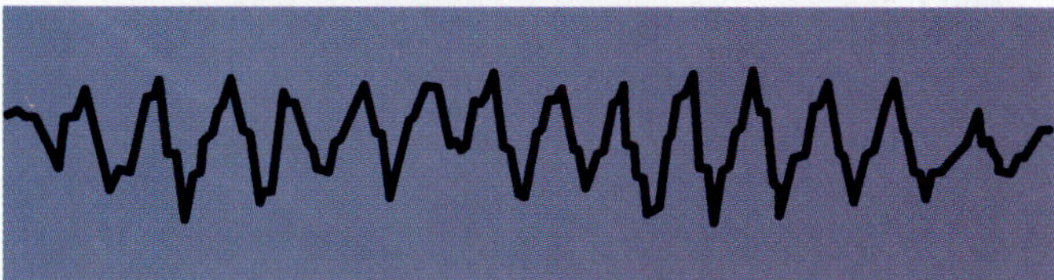

[B] Brain wave pattern of person meditating shows alpha waves

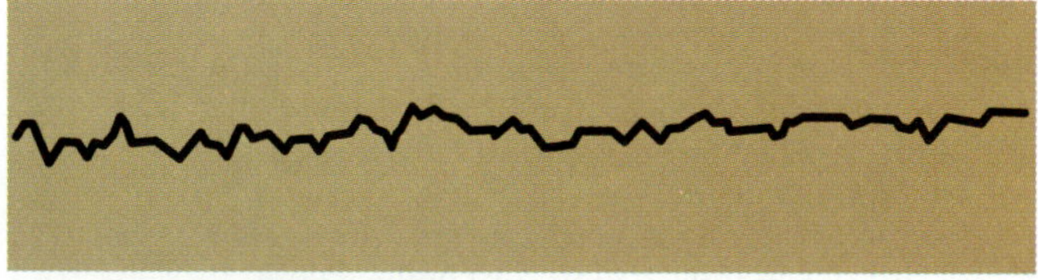

[C] Brain wave pattern of person resting

FIGURE 2.7

Ions, Nerve Impulses, and Meditation.

Sodium ions (Na^+) [review FIGURE 2.6D] are crucial to nerve cell functioning and electrical activity in the brain, which can be monitored and recorded. A person meditating with eyes half-closed **[A]** has a very distinctive brain wave pattern showing alpha waves **[B]**. The brain wave pattern of the same person while not meditating is quite different **[C]**. The brain waves shown here are actually from a Japanese Zen Buddhist priest, during and after meditating.

mary of atomic structure). From here we can examine the tendency of atoms, based on their outermost shells of electrons, to combine and react with other atoms. This potential for atoms combining into molecules is perhaps the most important aspect of atoms we will discuss because the cells of living things are composed of molecules.

CONCEPT CHALLENGE

You probably drink millions of ions of the element fluorine each day because in most cities, fluorine ions (fluoride) are added to the drinking water supply to decrease tooth decay. Ions of fluorine have 9 protons, 10 neutrons, and 10 electrons. What is the atomic number for this ion of fluorine? What is the atomic mass? What electrical charge would be found on ions of fluorine?

Chemical Bonds

Molecules are particles that consist of two or more atoms linked by an attractive force called a **chemical bond**. The atoms in a molecule may be identical or different. The bonds that link them are not actual physical objects, like the coupling between railroad cars, nor are they solid, like hardened glue. Instead, they are links of pure energy, usually based on shared or donated electrons. Bonds act like invisible springs; once a bond between two atoms is formed, it requires energy to pull the atoms apart or to push them closer together. The next three sections describe different types of bonds—covalent bonds, hydrogen bonds, and ionic bonds.

COVALENT BONDS

The most common type of chemical bond forms when two atoms share a pair of electrons. A shared pair of electrons constitutes a **covalent bond** [FIGURE 2.8], as in a water molecule, where two hydrogen atoms are joined with one oxygen atom. The *chemical formula* for water, H_2O, shows this relationship. As the two hydrogen atoms approach the oxygen atom, each positively charged nucleus begins to attract electrons orbiting the other two nuclei [FIGURE 2.8A]. Eventually, the electron orbits overlap and fuse. You can count the number of electrons shared in the outer shells in FIGURE 2.8B—eight for oxygen and two for each of the hydrogens—and see that all outer shells of each atom are filled. Atoms such as carbon can share more than one pair of electrons, producing double and triple bonds.

In some molecules, the electrons spend as much time orbiting one nucleus as the other, and the electrical

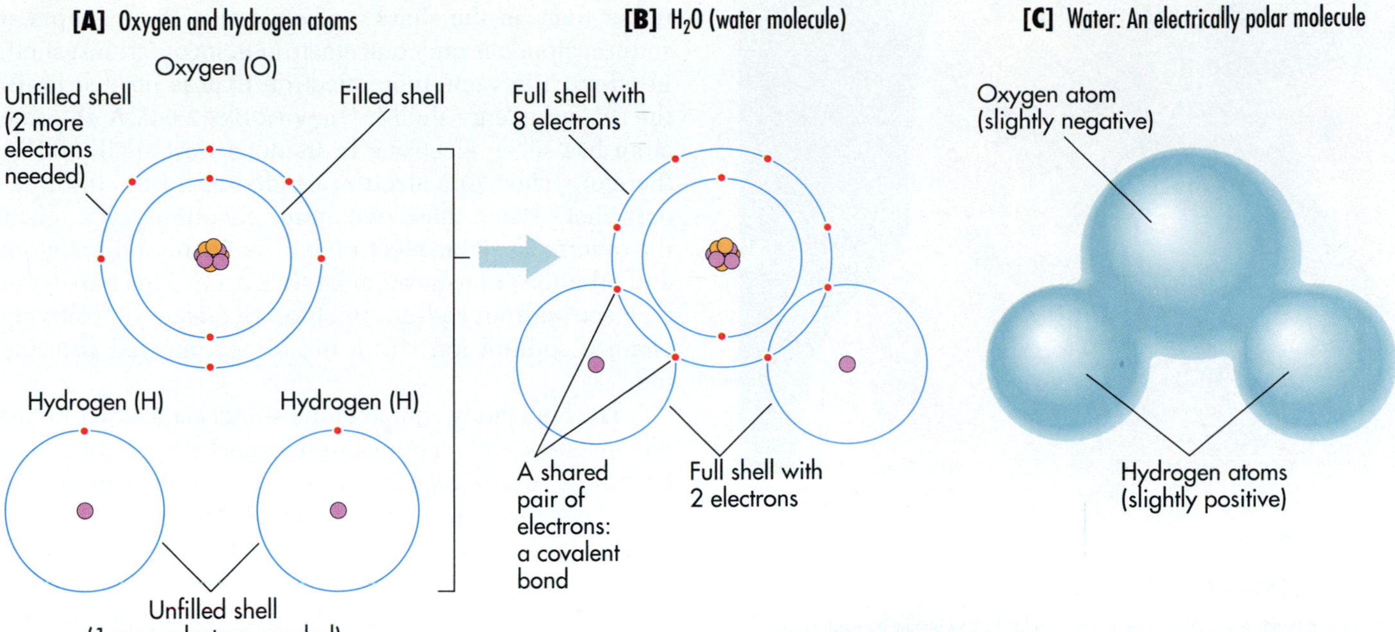

FIGURE 2.8

Covalent Bonds in a Water Molecule.

[A] The oxygen atom has only six electrons in its outer shell and needs two to be filled. Hydrogen atoms have one electron in their outer shell and need one to be filled. **[B]** In the H_2O molecule, the oxygen nucleus with its six protons attracts electrons from the two hydrogen molecules, the outer shells overlap, and the atoms share electron pairs in two covalent bonds. This sharing fills the outer shells of all three atoms. **[C]** The electrons in a water molecule spend more time orbiting the oxygen atom than the two hydrogens. This causes the oxygen atom to act as if it is slightly negatively charged and the two hydrogen atoms to act as if they are slightly positively charged.

charge is evenly distributed about both ends, or *poles*, of the molecule. Because of this equal sharing, the molecule is said to be **nonpolar**. In a molecule like water, however, the electrons spend more time orbiting the oxygen than the hydrogen [FIGURE 2.8C]. This leaves the oxygen pole of the molecule with a slightly negative charge, and the hydrogen pole of the molecule with a slightly positive charge. In such a case, the molecule is said to be **polar**. The three-dimensional shape of a water molecule reflects this polarity. Rather than the linear arrangement H—O—H, the two hydrogens are displaced toward one end, a bit like the corners of a triangle:

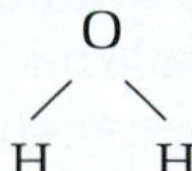

This shape and polarity can lead to the formation of another kind of chemical bond—the hydrogen bond.

HYDROGEN BONDS

A polar molecule can interact with another polar molecule because of the slight charges at their poles. In liquid water, for example, the molecules form the structure shown in FIGURE 2.9, with a hydrogen bond from one wa-

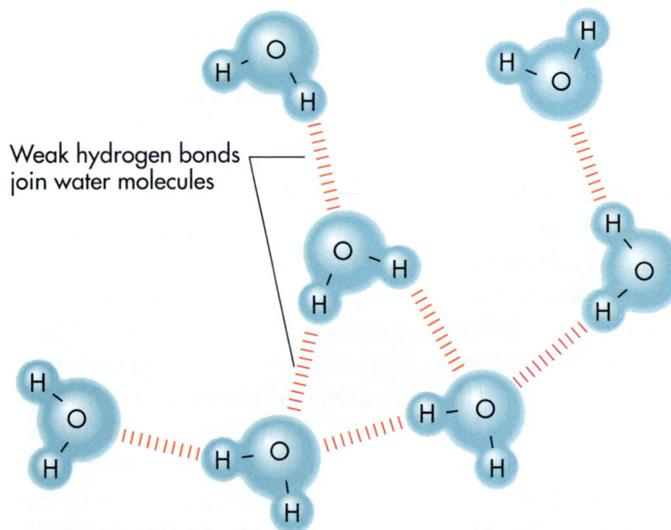

FIGURE 2.9

Hydrogen Bonds.

A hydrogen atom from one water molecule can be attracted to the oxygen atom in a different water molecule because the hydrogen has a partial positive charge, and the oxygen has a partial negative charge. This type of bond, called a hydrogen bond, is weak and easily broken. This weak bond helps explain properties of water that are crucial for living organisms, as well as interactions that maintain the structure of proteins and other biological molecules.

FIGURE 2.10

Hydrogen Bonds and Curling Hair.

Hydrogen bonds are relatively weak and can be broken readily by heat. Each protein molecule in a person's hair is linked to the neighboring molecules by hydrogen bonds. When hair is wrapped around a curling iron, the heat breaks the original hydrogen bonds. When the curled hair cools, new hydrogen bonds form, linking new molecules together and giving the hair a curled, but more stressed, form. Unfortunately, these new hydrogen bonds slowly break under the strain—especially in damp weather—and allow the hair to return to its original form.

ter molecule electrically attracting an oxygen from a different water molecule. This attraction of a hydrogen atom from one molecule to an atom (usually oxygen or nitrogen) of another molecule is called a **hydrogen bond**.

Hydrogen bonds are much more easily broken and re-formed than covalent bonds. Some of water's unusual properties (such as the tendency for ice to float) are based on hydrogen bonds, and some important biological molecules (such as the protein molecules that form your hair) have millions of these bonds, contributing to the shape and function of the molecule. Using an iron to straighten or curl your hair is a good demonstration of how easily hydrogen bonds in protein molecules can be broken and re-formed. The application of heat breaks the thousands of hydrogen bonds between the protein molecules that make up your hair—bonds that held it in its original shape [FIGURE 2.10]. As the hair cools in the shape of the iron, new hydrogen bonds form between the protein molecules and temporarily maintain a different shape with curls or waves.

IONIC BONDS

The third type of chemical bond depends on the complete transfer of electrons from one atom to another, rather than on the sharing of electrons. For example, a sodium atom has only one electron in its outermost shell. In effect, it has one more electron than is needed to fill the second energy shell [review FIGURE 2.6C]. A chlorine atom has seven electrons in its outermost shell, and is, therefore, short one electron needed to fill the third energy shell. When these two atoms encounter each other, the outermost single electron can leave the sodium atom and orbit the chlorine atom [FIGURE 2.11]. This transfer of an electron from sodium to chlorine creates a positively charged sodium ion and a negatively charged chlorine ion.

The oppositely charged ions attract each other, rather like magnets, and an **ionic bond** forms between them that holds the atoms together. These same attractions can cause the formation of a regular latticework of many sodium and chloride ions in a *crystal.* Ionic bonds are much stronger than hydrogen bonds, but not as strong as covalent bonds. When ionically bonded compounds are dissolved in water, they often dissociate (break down) into their component ions, as when table salt dissolves in a pot of soup.

Our discussion of bonding helps show how the properties of substances emerge from the internal organization and electrical activities of their constituent atoms. Salt is an excellent example of such emergent properties, with ionic crystals emerging from the structure of NaCl, and visible salt crystals emerging from the structure of submicroscopic ionic crystals. The water molecule is another example: This single remarkable chemical entity makes up most of the matter in most living organisms, and its structure and bonding properties make it uniquely suited to biological systems.

Life and the Chemistry of Water

Our planet and its life forms are inseparable from the simple compound H_2O, water. Fully three-quarters of the earth's surface is covered by water. Life began in an aquatic realm, and hundreds of thousands of species still live in water today. Since water is transparent, light can penetrate it for many meters, providing energy that allows plants and algae to synthesize sugars and thereby support much aquatic life. What's more, organisms themselves—whether in the sea or on land—are composed of 50 to 90 percent water.

What makes water so special chemically and so necessary to life? The answer is that hydrogen bonding be-

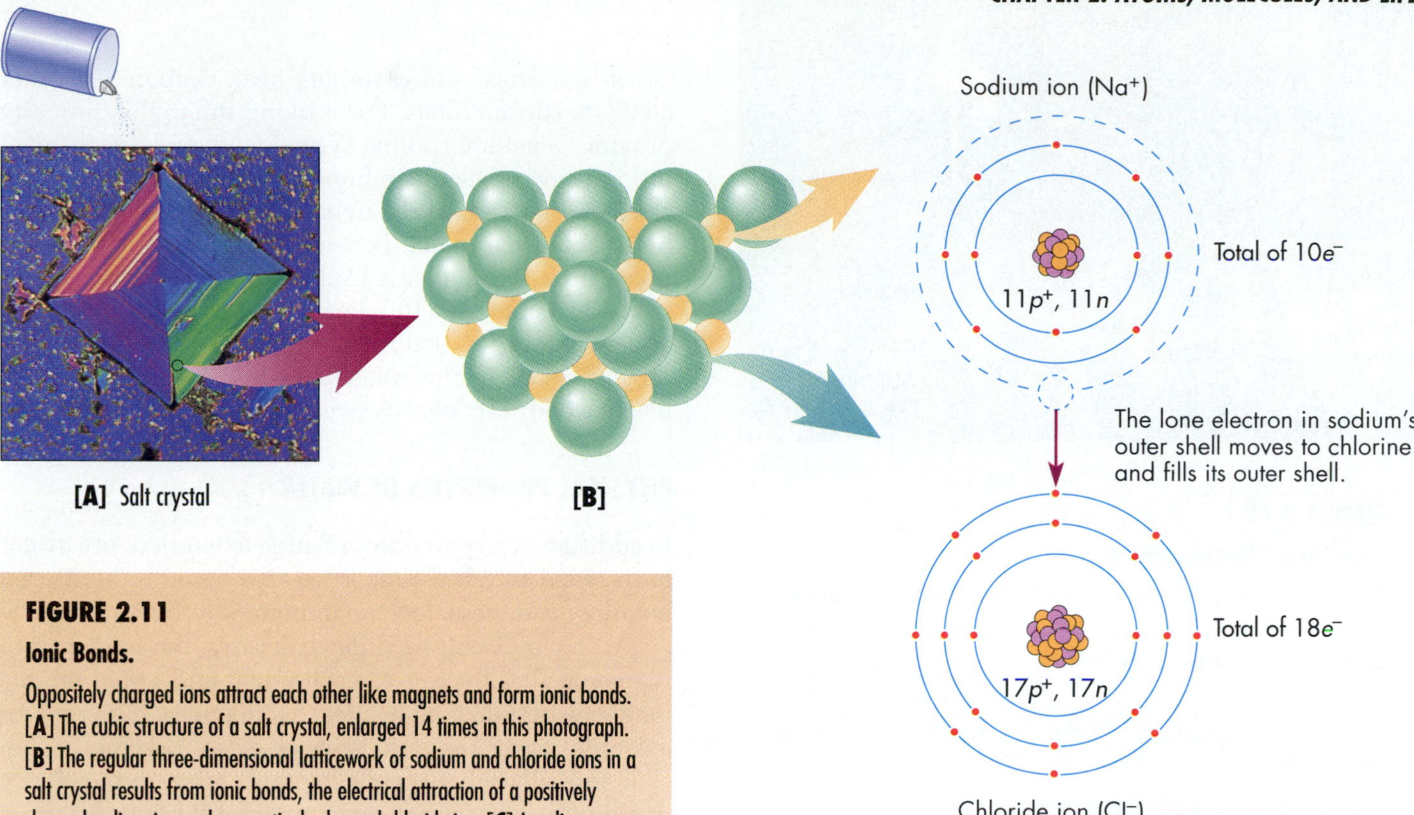

FIGURE 2.11

Ionic Bonds.

Oppositely charged ions attract each other like magnets and form ionic bonds. [A] The cubic structure of a salt crystal, enlarged 14 times in this photograph. [B] The regular three-dimensional latticework of sodium and chloride ions in a salt crystal results from ionic bonds, the electrical attraction of a positively charged sodium ion and a negatively charged chloride ion. [C] A sodium atom contains one lone electron in its third energy shell, while a chlorine atom lacks just one electron to complete its third shell. If the lone outer electron (here, colored red) leaves the sodium atom and joins the chlorine atom, then both will have filled outer shells. The result is a sodium ion with a positive charge (because it has one more proton than electrons) and a negatively charged chloride ion (having one more electron). These ions attract each other and form an ionic bond.

tween water molecules and other substances makes life possible. Recall that water is a polar molecule, with positive and negative ends that can attract the oppositely charged portions of other water molecules [review FIGURE 2.9]. This attraction has a profound effect on the properties of water.

WATER, TEMPERATURE, AND LIFE

Anyone who has had frostbite knows that water behaves differently at different temperatures and that these behaviors can have a significant impact on living cells. When a person is caught in a blizzard without gloves, the liquid water in his or her fingers turns into ice crystals, which can tear apart cells and cause pain, dysfunction, and even loss of the fingers. It is fortunate for living things that water usually remains liquid, within the normal temperature range found on earth, instead of assuming one of its other physical forms—gas (water vapor) or solid (ice).

Living things are also lucky that water is slow to heat; that is, the amount of heat needed to raise the temperature of a certain volume of water (its *specific heat*) is greater than that for most other liquids. This property is a direct result of water's hydrogen bonds. Much of the heat energy applied to water—say, the sunshine striking Lake 223—goes into stretching or breaking hydrogen bonds instead of raising the water temperature. This is important to living things because it helps ensure the relatively constant external and internal environments they need. Oceans, large lakes, and rivers are all slow to change temperature, and thus aquatic life does not have to cope with rapid temperature changes. On land, too, an organism's temperature must remain fairly constant. The fact that living things are composed mostly of water means that body temperature, especially of large organisms like dinosaurs, crocodiles, or large sea bass, also changes slowly.

A tennis player perspiring at the end of a vigorous match can appreciate another important temperature-related property of water: An unusually high amount of heat is required to turn liquid water into water vapor. Before a water molecule can evaporate, it must jostle about rapidly enough to fly off the surface of a water droplet. Before it can really jostle, the hydrogen bonds linking the molecule to its neighbors must be broken by the absorption of heat energy.

FIGURE 2.12

Sweating and Hydrogen Bonds.

On the skin of a sweaty horse, water molecules absorb heat from the animal's body, hydrogen bonds can break, and the water can evaporate. This bond-breaking and evaporation cool the horse.

FIGURE 2.13

Hydrogen Bonds Help Lift Water to the Tree Tops.

In a living tree, such as this huge curtain fig tree (*Ficus virens*) in North Queensland Australia, a continuous column of water molecules stretches from the tree's roots to its topmost branches. This column, up to 100 m (330 ft) tall, depends on the bonding of water molecules to each other by shared hydrogen atoms. When a water molecule at the surface of a leaf evaporates, it pulls on the next water molecule down the column. This molecule then pulls on the next, and so on, moving the entire column of water molecules from root to leaf.

In the process of absorbing heat, evaporating water cools its surroundings. For a living thing, this property provides a natural cooling system, and explains the evolution of sweat glands in horses, people, and some other mammals [FIGURE 2.12]. Envision a tennis player during a hotly contested match beneath the blazing sun. Sweat pours from sweat glands in the skin and covers parts of the body surface in a thin layer. As the sweat evaporates, a large amount of body heat goes into breaking the hydrogen bonds in the water molecules. As a result, the player's body is cooled as sweat evaporates from the skin.

PHYSICAL PROPERTIES OF WATER

In addition to its properties relating to temperature, water has several physical properties also based on hydrogen bonding, and these, too, are important to living things.

Water molecules exhibit *cohesion*, the tendency of like molecules to cling to each other, and *adhesion*, the tendency of unlike molecules to cling to each other—for example, water to paper, soil, or glass. Together, cohesion and adhesion account for **capillarity**, the tendency of a liquid substance to move upward through a narrow space against the pull of gravity, such as between the fibers of a paper towel being used to wipe up a spill, or up the inside of a narrow glass tube. Capillarity plays a part in the upward transport of water in many kinds of plants [FIGURE 2.13]. Without the cohesion of water molecules to each other and adhesion to the walls of narrow tubes, there would be no tall trees such as redwoods or eucalyptus.

One physical property of water must be overcome the moment a newborn baby draws its first breath. This property, called **surface tension**, is the tendency of molecules at the surface of liquid water to cohere to each other, but not to the air molecules above them. The water molecules in a newborn's lungs tend to bind to each other and to the tissue more strongly than to air; for this reason, they tend to pull the walls of the lungs together, which could cause those delicate organs to collapse [FIGURE 2.14]. The body, however, produces special molecules called *surfactant*, which function like detergent molecules to lower the surface tension of the water in the lungs, thus preventing the collapse of the lungs. Premature babies born before their lungs are capable of generating surfactant sometimes die because their lungs collapse.

Water has yet another physical property with biological implications: the tendency of ice (solid water) to float in liquid water [FIGURE 2.15A]. The hydrogen bonds in ice are fairly rigid, creating an open latticework that holds water molecules farther apart than the more easily broken bonds of the liquid form [FIGURE 2.15B]. Thus, ice is

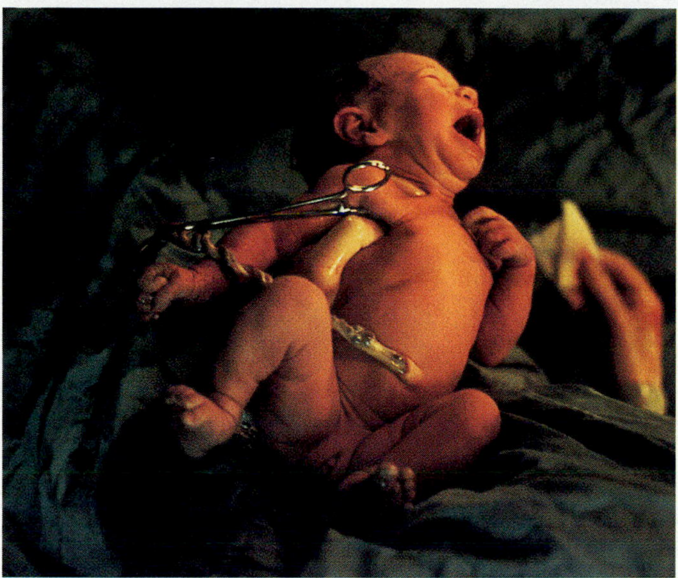

FIGURE 2.14

A Baby's First Breath: Fighting Surface Tension.

In the womb, a baby's lungs have no access to air. The fine air sacs are filled with fluid, and the hydrogen bonding of water molecules to each other and to the interior surfaces of the sacs tends to pull the walls of the sacs together, making it difficult for air to enter and expand the sacs. Most baby's lungs secrete a substance that acts like a detergent. This substance is capable of breaking hydrogen bonds of water molecules on the interior surfaces of the lung sacs, and permitting the lungs to expand on the first breath.

less dense than liquid water, and floats in it. This is a unique property; other substances become more dense when they freeze. If ice did not become less dense, many lakes would freeze solid in winter, and only the top few centimeters would thaw in summer. Instead, with a frozen top layer insulating the lower depths, plants and animals can survive winter in the chilly—but liquid—water beneath the ice.

CHEMICAL PROPERTIES OF WATER

Although washing dishes is no fun, it can—if one is in the right frame of mind—serve as a reminder of the chemical properties of water so important to living cells. Dishwater—in fact, all water—is a **solvent**, a substance capable of dissolving other molecules. Dissolved substances are called **solutes**. Water can dissolve polar compounds, such as table sugar, and most kinds of ionic compounds, such as table salt. When polar molecules of sugar—say, syrup on dirty plates, or glucose molecules inside cells—become surrounded by water molecules, hydrogen bonds form. With an ionic solute like salt, the component ions dissociate, and each becomes surrounded by an oriented cloud of water molecules [FIGURE 2.16A].

Compounds such as sugar and salt that dissolve readily in water are called **hydrophilic**, or "water-loving," compounds. In contrast, nonpolar compounds, such as cooking oils and animal fats on dirty dinner dishes, do not dissolve readily, and are called **hydrophobic**, or "water-fearing," compounds. Instead of dissolving in water, hydrophobic compounds such as cooking oil form a boundary, or *interface* with the water. The oil molecules cohere to other oil molecules in a film at the surface

FIGURE 2.15

Why Ice Floats.

[A] Many organisms would perish if it weren't for the simple fact that ice floats. If ice sank in water, the bottoms of lakes and seas would remain frozen most of the year. **[B]** Frozen water is less dense than liquid water because the hydrogen bonds in ice create a rigid, open latticework. Hence, this less dense material floats in the more dense water.

[A] Why does ice float on water?

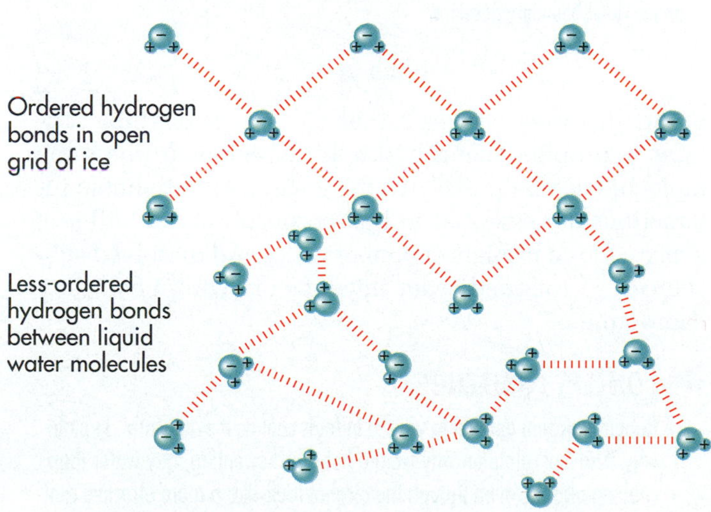

[B] Hydrogen bonds generate an open lattice in ice.

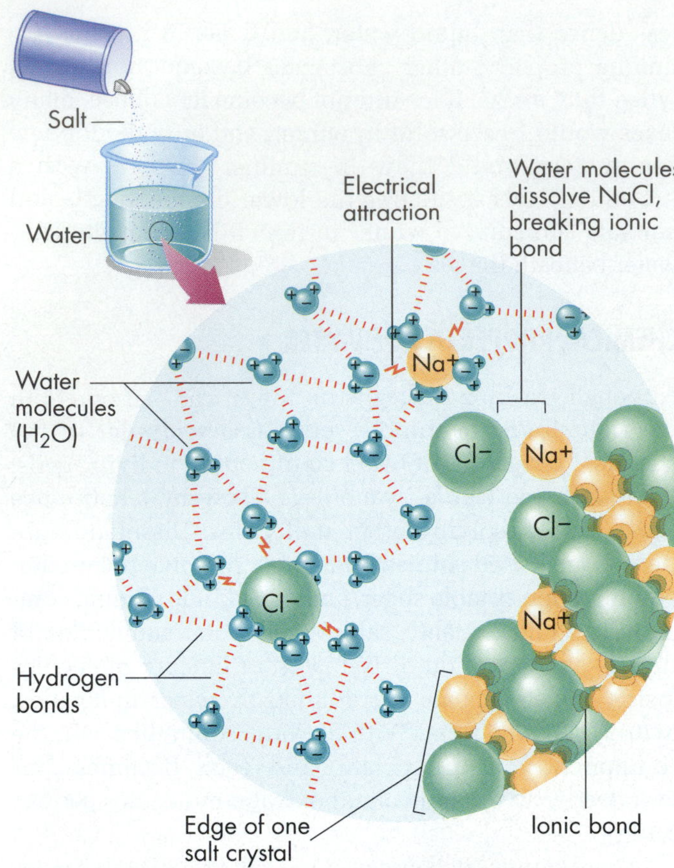

[A] Salt dissolves in water

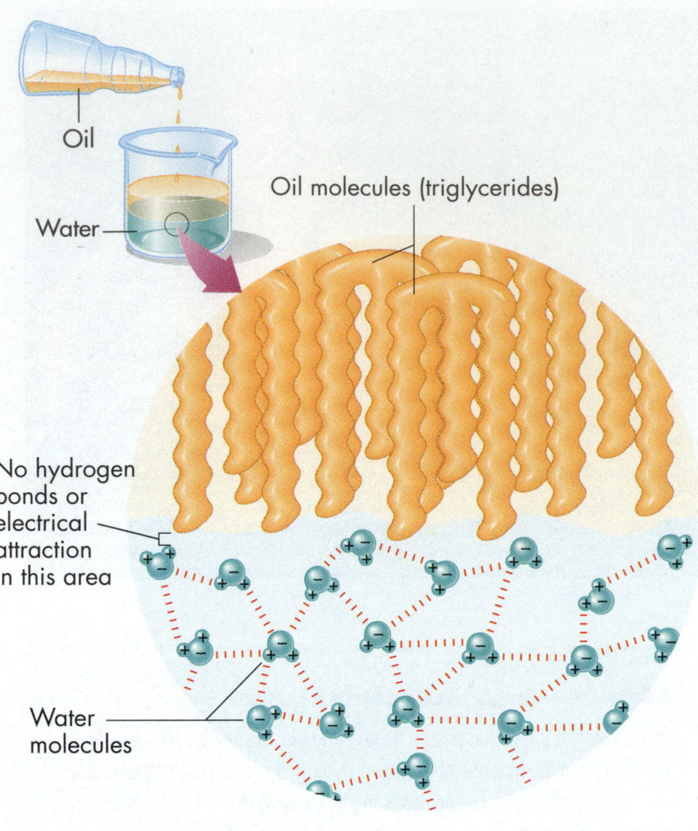

[B] Oil and water don't mix

FIGURE 2.16

Hydrophilic Compounds Dissolve Readily, Hydrophobic Ones Do Not.

[A] Salt is hydrophilic, or "water loving," because its charged ions form bonds with water molecules. **[B]** Oil is hydrophobic, or "water fearing," because its chemical structure is such that hydrogen bonds do not usually form, and hence an oil/water boundary develops.

of the dishwater [FIGURE 2.16B]. (A detergent which contains hydrophobic and hydrophilic regions in the same molecule can help dissolve the grease.) Hydrophobic interactions are essential to life, because, as you will see, every cell is a miniature pool of water and dissolved substances, surrounded by an envelope of hydrophobic fatty compounds.

CONCEPT CHALLENGE

Rubbing alcohol applied to your skin feels cool as it evaporates. Explain why. Also, speculate on why we are better off sweating salty water than rubbing alcohol, even though the alcohol feels like a more effective cooling agent for our bodies. As you speculate, consider the fact that it requires more energy to evaporate a milliliter of water than it does to evaporate a milliliter of rubbing alcohol.

Acids and Bases

Water has another property with significant implications for living things: Its molecules have a slight tendency to break down, or dissociate, into a positively charged hydrogen ion (H^+) and a negatively charged hydroxide ion (OH^-):

$$H_2O \rightarrow H^+ + OH^-$$

This breakdown, however, is a relatively rare event in pure water. By definition, an **acid** is any substance that gives off hydrogen ions when dissolved in water, thereby increasing the H^+ concentration of the solution. Lemon juice and vinegar are acidic solutions, and thus contain many hydrogen ions.

A **base** is any substance that accepts hydrogen ions when they are dissociated in water. An ammonia-containing window cleaner like Windex is a basic solution, with few hydrogen ions but many hydroxide ions. The concentration of hydrogen ions is important to living cells because many of the chemical reactions that drive life's processes—the digestion of foods, for example—depend on specific concentrations of these ions.

THE pH SCALE

Biologists measure hydrogen ion concentration on the **pH scale**,* which ranges from 1 to 14 [FIGURE 2.17]. On this scale, water has the neutral value of 7, in the middle of the scale. Acidic solutions have pH values between 0 and 7, and basic solutions have pH values between 7 and 14. The pH inside most cells stays fairly neutral, between about 6.5 and 7.5, and it is only within this narrow range that many vital cellular reactions take place at optimum speed. The pH scale is constructed so that the difference between one pH unit and the next *lower* unit represents a 10-fold *increase* in the concentration of hydrogen ions. For example, a cola drink at pH 3 has 10 times the concentration of hydrogen ions as tomato juice, at pH 4.

FIGURE 2.17 shows the pH of common solutions. Notice that one of the entries at pH 4 is acid rain. This pollutant forms when oxides of sulfur given off by automobile engines and industrial smokestacks dissolve in water droplets in the atmosphere, forming sulfuric acid. When this acid falls to earth in acid rain, snow, or fog, it has a pH as low as 3 or 4 in areas downwind of the industrialized and heavily populated regions. The acid rain burns the leaves and tender twigs of trees [FIGURE 2.18A], stunts the growth of fish [FIGURE 2.18B and C], and can essentially eliminate most life forms from small lakes and streams, as you saw earlier with experimental Lake 223. To understand how acid rain causes its harmful effects on living organisms, we first need to understand the structure of biological molecules, a topic we will consider a bit later in this chapter.

BUFFERS

Have you ever suffered from acid indigestion after eating too much spicy food too fast? This uncomfortable condition is caused when cells in your stomach secrete large quantities of hydrochloric acid (HCl). While the acid speeds up the digestion of food, it can also irritate your stomach lining. When hydrochloric acid dissolves in water (in your stomach, or elsewhere), it releases H^+ into the solution:

$$HCl \rightarrow H^+ + Cl^-$$

As an antidote to acid indigestion, you may have taken some bicarbonate (HCO_3^-)—the active ingredient in many over-the-counter heartburn remedies. Bicarbonate can act as a **buffer**, a substance that regulates pH by "soaking up" or "doling out" hydrogen ions as needed: When hydrogen ion concentrations are high, buffers bind to H^+, and when hydrogen ion concentrations are low,

*The *p* stands for "potential," and the *H* stands for "hydrogen."

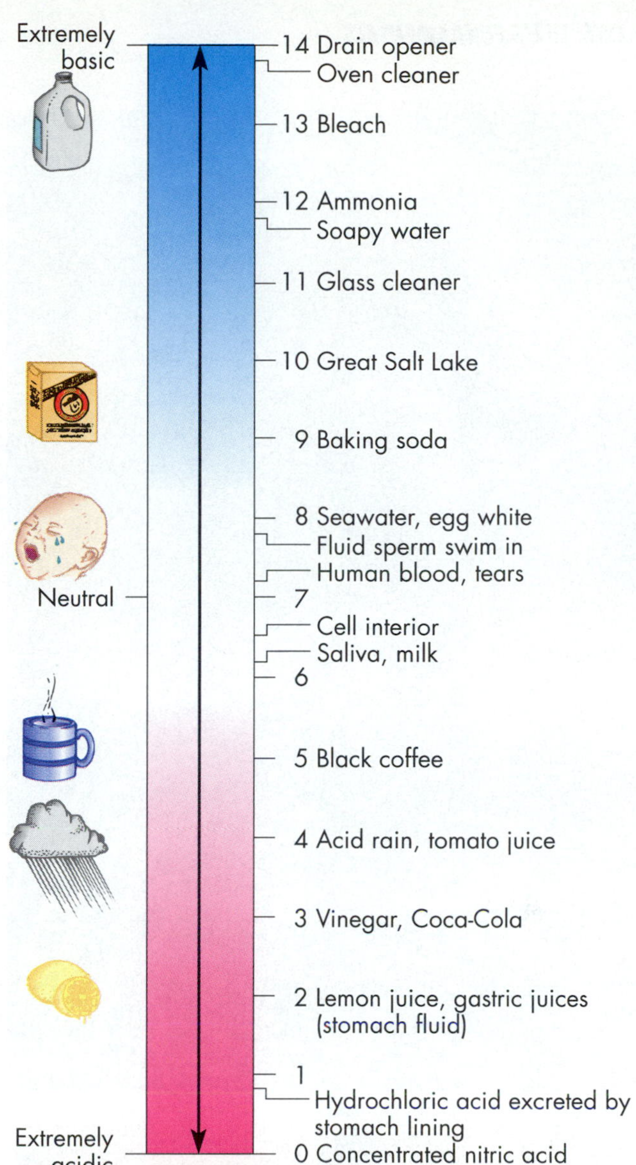

FIGURE 2.17

The pH Scale.

This scale is a representation of the hydrogen ion concentration in common biological and nonbiological substances. The more hydrogen ions, the more acidic a substance, the fewer hydrogen ions, the more basic. Why does the *most* acidic solution have the *lowest* number on the pH scale? And why does each pH unit represent a 10-fold difference in hydrogen ion concentration? Chemists measure the amount of substances in units called *moles*, and the concentration of substances in moles per liter (mol/L). The pH scale indicates the concentration of hydrogen ions in this concentration unit. A pH of 1 is 0.1 concentration units, a pH of 2 is 0.01 concentration units, and a pH of 10 is 0.000,000,000,1 concentration units. As you can see, the pH is equivalent to the number of decimal places to the right of the decimal point, accounting for the factor of ten difference between pH's. Since low pH values have few decimal places to the right of the decimal point, they have great concentrations of hydrogen ions, accounting for their great acidity. In a solution, the concentration of hydrogen ion (H^+) times the concentration of hydroxide ion (OH^-) is always the same. Whenever the concentration of hydrogen ions goes up, the concentration of hydroxide ions goes down, and vice versa.

[A]

[B]

[C]

FIGURE 2.18

Acid Rain and a Dead Forest.

[A] Fraser fir trees stand like skeletons at the summit of Mount Mitchell, North Carolina, stark reminders of the environmental consequences of air pollution. [B] In 1979, after only three years of acidification, the trout in Lake 223 were still plump and healthy-looking. [C] By 1982, the small animals on which the trout depended for food had been killed by the acid, and the gamefish were beginning to show serious signs of starvation, disease, and other physical distress.

buffers release H^+. Some lakes are naturally protected against the harmful effects of acid rain because substances in rocks on the lake bottoms act as buffers. These materials soak up the hydrogen ions entering the lake in acid precipitation before they harm living organisms.

When basic, or **alkaline**, substances dissociate in water, instead of giving off hydrogen ions, they combine with them, resulting in an excess of OH^-. For example, NaOH dissociates into sodium and hydroxide ions:

$$NaOH \rightarrow Na^+ + OH^-$$

The sodium ions dissolve in the water, and some of the freed hydroxide ions combine with free H^+ to form H—OH (or, more familiarly, H_2O), leaving fewer hydrogen ions than before NaOH was added to the water.

You have seen in this section that water—the substance of ocean waves, waterfalls, snow drifts, clouds, and icebergs—is also the major ingredient of living things. And you've also seen that the molecular properties of water, based on the atomic structures of oxygen and hydrogen, and on the versatile hydrogen bond, make it possible for people to stay cool even in the desert, for water to ascend to the treetops, and for life to occur in its myriad forms. But before we can truly understand the chemistry of life, we must consider another set of materials that makes up living things: compounds containing carbon.

CONCEPT CHALLENGE

Coke, Pepsi, and other carbonated beverages are made by forcing carbon dioxide gas at high pressure into flavored water. Some of the carbon dioxide molecules combine with water molecules, creating bicarbonate and hydrogen ions. Would you expect these drinks to have a pH greater than 7 or less than 7? Why?

Carbon Compounds

Despite the preponderance of oxygen and hydrogen in living organisms, fully 18 percent of a person's weight comes from carbon atoms, and for a large tree, the figure can approach 50 percent. Carbon atoms are so prevalent and so important to living things that the study of compounds containing carbon—called **organic** ("related to organisms") **chemistry**—is usually separated from the study of **inorganic** ("lifeless") **chemistry**, which deals with all substances not containing carbon.

Interestingly, of the two divisions of chemistry, organic chemistry includes a far greater number of compounds. Why? Because the structure of the carbon atom is such that carbon can form millions of different combi-

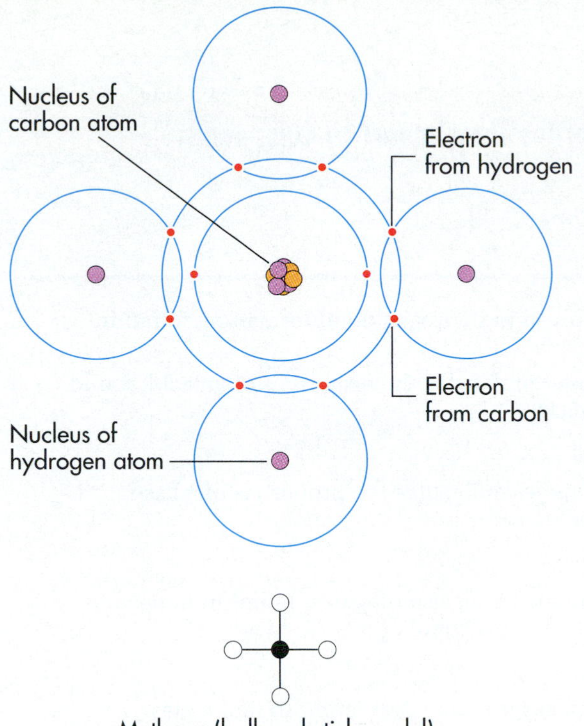

FIGURE 2.19

Carbon Can Bond with up to Four Other Atoms.

A carbon atom has four electrons in its outer shell, and thus requires an additional four electrons to fill its outer shell with eight electrons. Each of the four hydrogen atoms in this methane molecule shares a pair of electrons with the carbon atom. The sketch also shows a ball-and-stick model for methane.

nations with other atoms. Bonding versatility is carbon's key characteristic, and it explains why crustal elements such as silicon and aluminum, which form relatively few compounds, could never have formed the basis of life, with all its many manifestations.

Four classes of organic compounds—*carbohydrates, lipids, proteins,* and *nucleic acids*—are vital to the structure and function of living things. Together with a few other materials, these four compounds account for the diverse shapes, colors, textures, and other characteristics of organisms.

CARBON BACKBONES

A carbon atom has six electrons, and as you saw in FIGURE 2.3B, two of these electrons fill the first energy level and four occupy half of the eight available spots in the second energy level. Because a carbon atom is "looking for" four more electrons to fill its second energy level, carbon can form covalent bonds with up to four atoms at a time. Methane gas (CH_4), sometimes called *swamp gas,* is a simple example [FIGURE 2.19]. Many compounds in living cells, however, are larger and have a backbone of several carbon atoms bonded to each other in long, straight chains, branched chains, or rings [FIGURE 2.20]. The straight-chain molecule in FIGURE 2.20A, a component of some gasolines, causes your car engine to knock because it burns too rapidly in the car's cylinders. In contrast, the

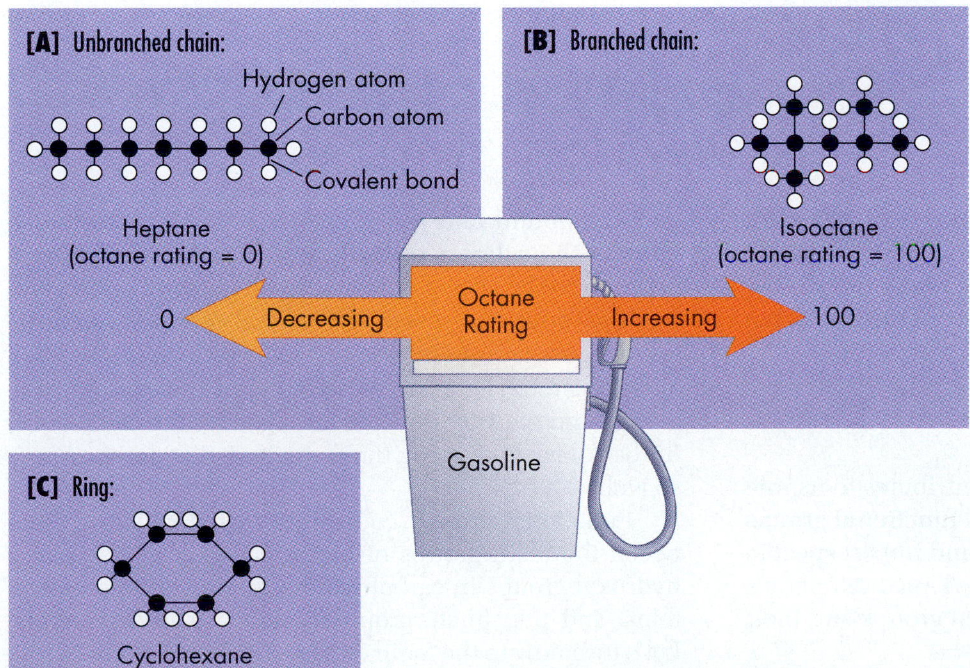

FIGURE 2.20

Carbon Chains and Car Engine Knocking.

Carbon atoms can join into long, straight chains [**A**], branched chains [**B**], or rings [**C**]. The eight-carbon branded chain in [**B**] is called *isooctane.* It burns slowly in an automobile engine due to its branches. This provides a smooth force that powers the engine's cylinder, which moves the car. In contrast, the seven-carbon straight chain in [**A**], called *heptane,* burns very rapidly in a car engine, and causes knocking or pinging of the engine, and hence reduces efficiency. On the gas pump you can read the gasoline's octane rating, a measure of its ability to prevent knocking. Isooctane has a rating of 100, but heptane a rating of zero. The next time you buy gas, check out the octane rating, and try to visualize the fraction of your gas that might be branched or unbranched carbons.

TABLE 2.3 Several Common Functional Groups and the Properties They Impart to Compounds

Name, Symbol, and Structure of Group		Characteristics
Hydroxyl group (OH)	R—O—H	Polar, water-soluble, involved in hydrogen bond formation; found in alcohols
Carboxyl group (COOH)	R—C(=O)—O—H	Weak acid (H^+ donor); gives molecules the properties of an acid; found in organic acids like vinegar
Amine group (NH_2)	R—N(—H)—H	Weak base (H^+ acceptor); gives molecules the properties of a base; found in proteins and urine
Aldehyde group (COH)	R—C(=O)—H	Polar, water-soluble, common in sugar molecules; found in molecules exhaled on breath after drinking alcohol
Keto group (CO)	R, R—C=O	Polar, soluble, common in sugar molecules; found in molecules expelled in urine when on high-fat diet
Methyl group (CH_3)	R—C(H)(H)—H	Hydrophobic (insoluble); gives molecules this property; found in fats and oils
Phosphate group (PO_4)	R—O—P(=O)(O—H)—O—H	Acidic; gives molecules properties of an acid; found in molecules used in energy flow within organisms; occurs in DNA molecules

R = rest of molecule

branched chain molecule called isooctane [FIGURE 2.20B], burns more slowly and hence more efficiently because of its branches. The octane rating on a gas pump reflects the fuel's carbon chain content and efficiency with respect to pure isooctane.

FUNCTIONAL GROUPS

While an organic molecule's shape contributes to its role in a cell, small clusters of atoms called **functional groups** often hang from the carbon backbone and impart specific chemical behaviors to the molecule. TABLE 2.3 shows seven biologically important functional groups and their properties.

FIGURE 2.21 shows the structure of an interesting carbon compound and the functional groups it contains. *Allin* is the substance inside a garlic bulb that is converted into the strong odor we associate with garlic. Notice that allin has several functional groups, including the amine group (—NH_2), the hydroxyl group (—OH), and the carboxyl group (—COOH). The presence of these groups and the way they are arranged on the carbon-sulfur backbone give garlic the characteristic odor we know so well.

Functional groups contribute to the difference between the main groups of biological molecules—many hydroxyl groups in carbohydrates, amine groups in proteins, and phosphate groups (—PO_4) in nucleic acids. Let's turn now to the main classes of the molecules of life: carbohydrates, lipids, proteins, and nucleic acids.

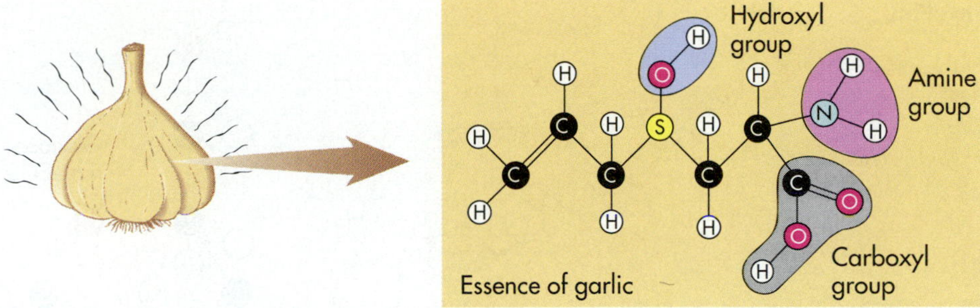

FIGURE 2.21

Functional Groups at Work in an Organic Compound.

One complex substance in garlic, called allin, has three functional groups that give the molecule (and the whole garlic bulb) the properties we recognize: a distinctive, strong smell with staying power on the breath.

Carbohydrates

Plants and fungi are made up mostly of water and carbohydrates. Considering the earth's great forested expanses, its grassy plains, and all the aquatic plants of lake and sea, by far the most abundant carbon compounds in living organisms are carbohydrates, and they serve both as structural components of cells and as energy reserves to fuel life's processes. The term *carbohydrate* means, literally, "hydrate (water) of carbon," and indeed, **carbohydrates** generally contain carbon, hydrogen, and oxygen in a ratio of 1:2:1 (CH_2O). For example, the carbohydrate that makes grapes taste sweet, the sugar glucose, has the formula $C_6H_{12}O_6$. Let's look now at both simple and more complex carbohydrates.

MONOSACCHARIDES: SIMPLE SUGARS

The simplest carbohydrates are called **monosaccharides**. Three common monosaccharides are glucose, fructose, and ribose. (The suffix *-ose* usually stands for sugar.) The simple sugars **glucose** and **fructose** have a sweet taste [FIGURE 2.22A and C]. Fructose, for example, is called fruit sugar because it is the compound that gives many kinds of fruit their sweet flavors. Honey contains mostly glucose and fructose. Glucose is the universal cellular fuel, which is broken down by virtually all living things, for the energy stored in its chemical bonds. Glucose and its conversion products will figure prominently in CHAPTER 5, when we discuss energy.

Glucose and fructose share the same molecular formula, $C_6H_{12}O_6$, but their different properties result from the different arrangements of the atoms in their molecules.

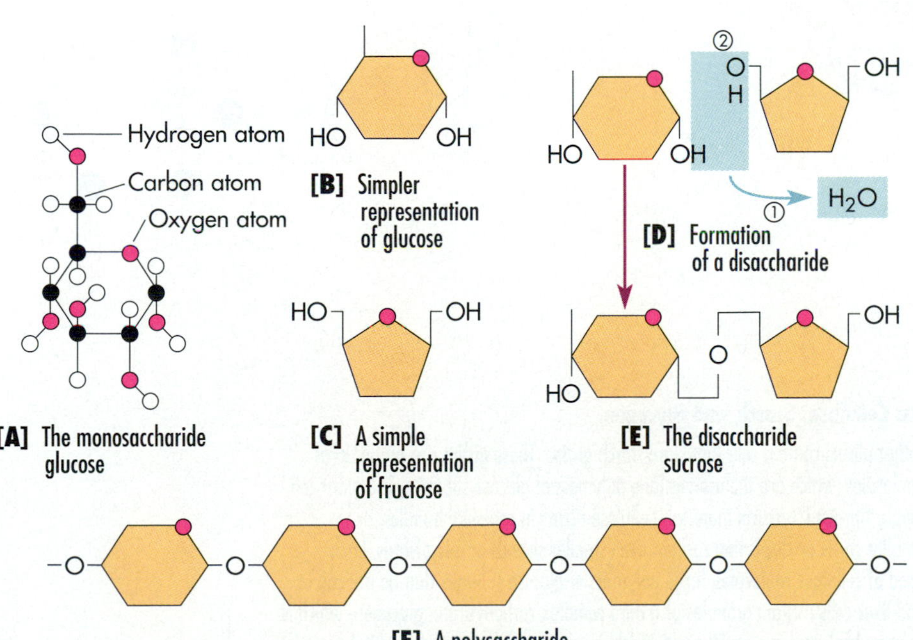

FIGURE 2.22

Sugars and Other Carbohydrates.

[A] Glucose, the monosaccharide (single sugar) that fuels most living cells, is made up of six carbons to which are bonded several hydroxyl groups (—OH). Glucose can form a ring made up of five carbons and an oxygen atom. [B] A simpler representation of the glucose molecule. [C] Fructose, fruit sugar, is a six-carbon sugar, but its ring structure contains just four of those carbons plus an oxygen. [D] A disaccharide, or double sugar, can form by means of a *condensation reaction*, which involves two steps: An enzyme removes oxygen and hydrogen from one monosaccharide and a hydrogen from the other monosaccharide. These atoms then combine and form water (1). Then the enzyme links the oxygen left over on one sugar to the other sugar (2), and a disaccharide is formed. [E] The disaccharide sucrose, common table sugar, consists of a glucose and a fructose joined together. [F] Many sugars joined together by reactions shown in [D] produce a polysaccharide such as starch.

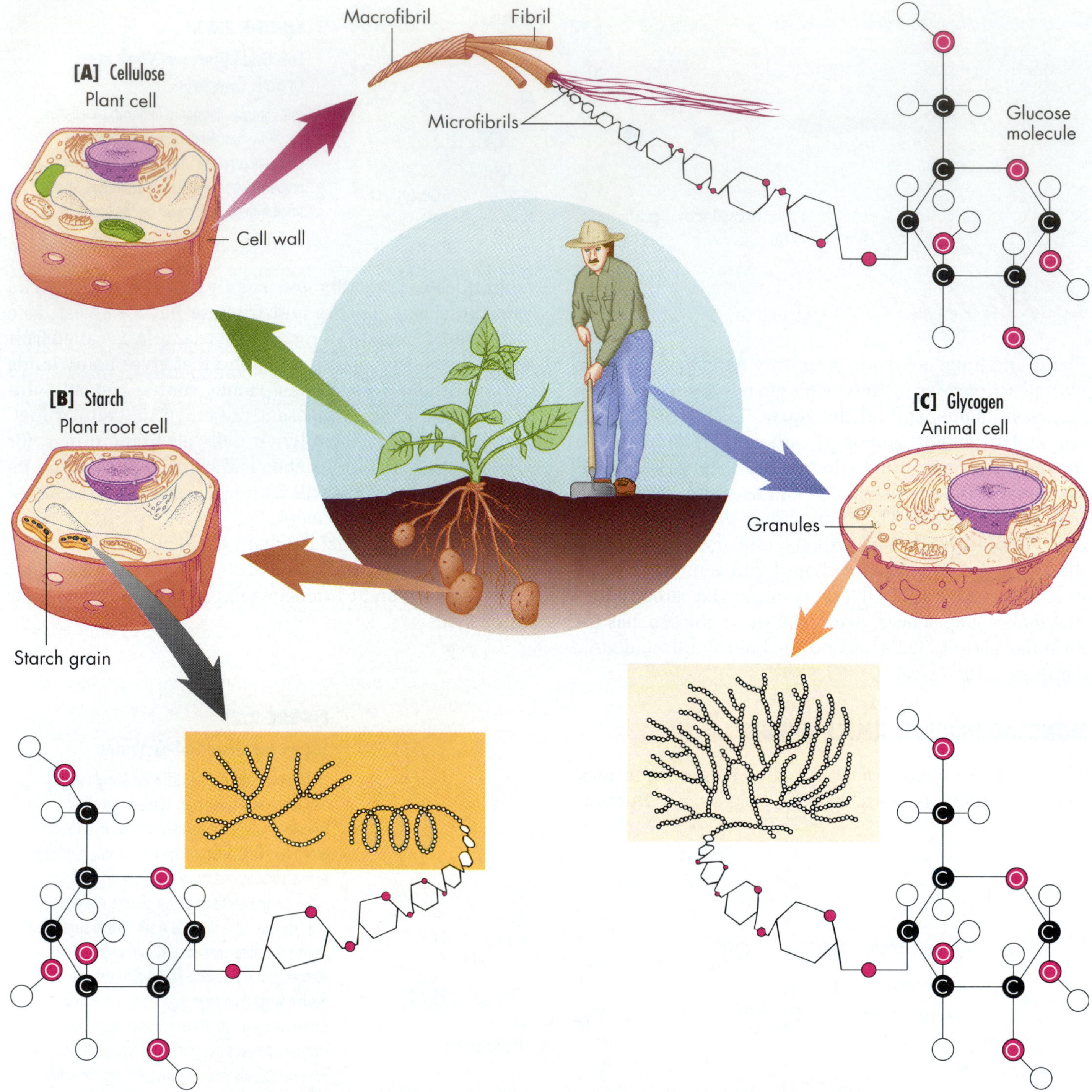

FIGURE 2.23

Complex Carbohydrates: Cellulose, Starch, and Glycogen.

The roots of potatoes and other plants contain cells that store starch grains. These grains are made up of branched or coiled starch molecules, which are themselves long polymers of glucose subunits. The plant cell walls contain cellulose, a tough, fibrous structural material. Cellulose occurs in cablelike bundles, or macrofibrils, made up of smaller cords or fibrils that contain even smaller strands or microfibrils. Each microfibril strand is composed of cellulose molecules, large polymers of glucose subunits that do not coil or twist. A person's muscles and liver cells contain granules of a third complex carbohydrate, glycogen, which is very highly branched and can be broken down rapidly and supply quick energy for muscular activity.

DISACCHARIDES: DOUBLE SUGARS

Disaccharides consist of two simple sugars joined together (*di* means "two") [FIGURE 2.22D and E]. The linking reaction joins two simple sugar subunits and releases a water molecule. For example, glucose bonded to fructose yields **sucrose**, or table sugar. Disaccharides are the common form in which sugars are transported inside plants. Sucrose is abundant in the saps of sugarcane, maple trees, and sugar beets—our major sources of sugar for refining. **Maltose**—two joined glucose subunits—gives barley seeds a sweet taste; beer brewers ferment the sugars in barley and other grains into alcohol.

POLYSACCHARIDES: MANY SUGARS

Polysaccharides are carbohydrates formed when many molecules of simple sugars are joined into long chains [FIGURE 2.22F]. A polysaccharide molecule is like a string of beads, where each bead is a simple sugar, often glucose. In general, a long molecular chain of similar subunits is called a **polymer**, and the individual subunits are called **monomers**. Polysaccharides, then, are polymers made up of many monomers, in this case, simple sugar units. As you will soon see, other types of major biological molecules are also polymers made up of different types of monomers.

The polysaccharides starch and glycogen are polymers of glucose that serve as the energy reserves of plants and animals, respectively. Cellulose and chitin (KYE-tin) are structural polysaccharides that give form and rigidity to plants, insects, and certain other organisms.

Starch is a polysaccharide composed of long chains of glucose subunits [FIGURE 2.23B]. The bonds between the glucose subunits allow the chains to coil and form granules that are stored in plant tissues and seeds. Most humans consume a diet that is mainly starch, obtained from the seeds of rice, wheat, corn, and other cereal plants grown agriculturally.

In humans and other animals, energy is stored in the muscles and liver in the form of highly branched **glycogen** molecules [FIGURE 2.23C]. Some chemists think that this branching allows for the faster liberation of glucose molecules, and hence, of energy. An animal suddenly confronted with a fierce and hungry predator needs a tremendous surge of energy to make a rapid escape.

Cellulose has a structure somewhat similar to that of starch: long chains of glucose units connected by bonds [FIGURE 2.23A]. But unlike starch, cellulose is a tough, fibrous structural material (the stuff of wood and paper). The orientation of the bonds between the glucose subunits in cellulose prevents twisting and adds rigidity to the chains. This stiffness makes cellulose an ideal component for the cell walls in crisp leaves and woody stems. Since people cannot digest cellulose, it provides us with no calories, although it does supply dietary fiber. Recently, food technologists have learned to process cellulose into a flour they call "fluffy cellulose." It has no calories, and can be mixed with starch and baked into dietic cakes and bread.

The structural carbohydrate **chitin** is a major component of the hard outer shells of insects, crabs, lobsters, and similar animals [see CHAPTER 21], as well as the cell walls of certain bacteria, and mushrooms and other fungi [see CHAPTERS 19 and 20].

In later chapters we will talk more about the importance of carbohydrates—as structural materials in plant cell walls, as nutrient storage compounds in most kinds of organisms, and as a medium for energy exchange in all forms of life.

Lipids

Lipids include **fats**, such as bacon fat, lard, and butter; **oils**, such as corn oil, coconut oil, and olive oil; **waxes**, such as beeswax and earwax; **phospholipids**, which are important components of cell membranes; and **steroids**, including certain vitamins, some hormones, and cholesterol (the clogger of blood vessels). Lipids can serve as waterproof coverings around cells and, like the carbohydrates, as energy-storage molecules. However, the chemical structures of the energy-storing lipids are quite different from those of carbohydrates, and these differences account for the unique properties of each type of lipid molecule. In general, lipids are not readily soluble in water, because they are composed mainly of carbon and hydrogen and tend not to form hydrogen bonds with water.

TRIGLYCERIDES: FATS AND OILS

Fats and **oils** have similar structures and serve as energy-storage molecules. A fat or oil molecule is composed of a three-carbon **glycerol** molecule bonded to three **fatty acids**, long chains of carbon and hydrogen atoms with an acid functional group (carboxyl group) on the end [FIGURE 2.24]. Because three fatty acids are joined to the three carbons of the glycerol, fats and oils are usually called **triglycerides**.

You have probably heard warnings of how saturated fats in the diet are linked to heart disease, and perhaps you have wondered about the difference between saturated and so-called polyunsaturated fats. The fatty acids of **saturated fats** have no double bonds and are saturated

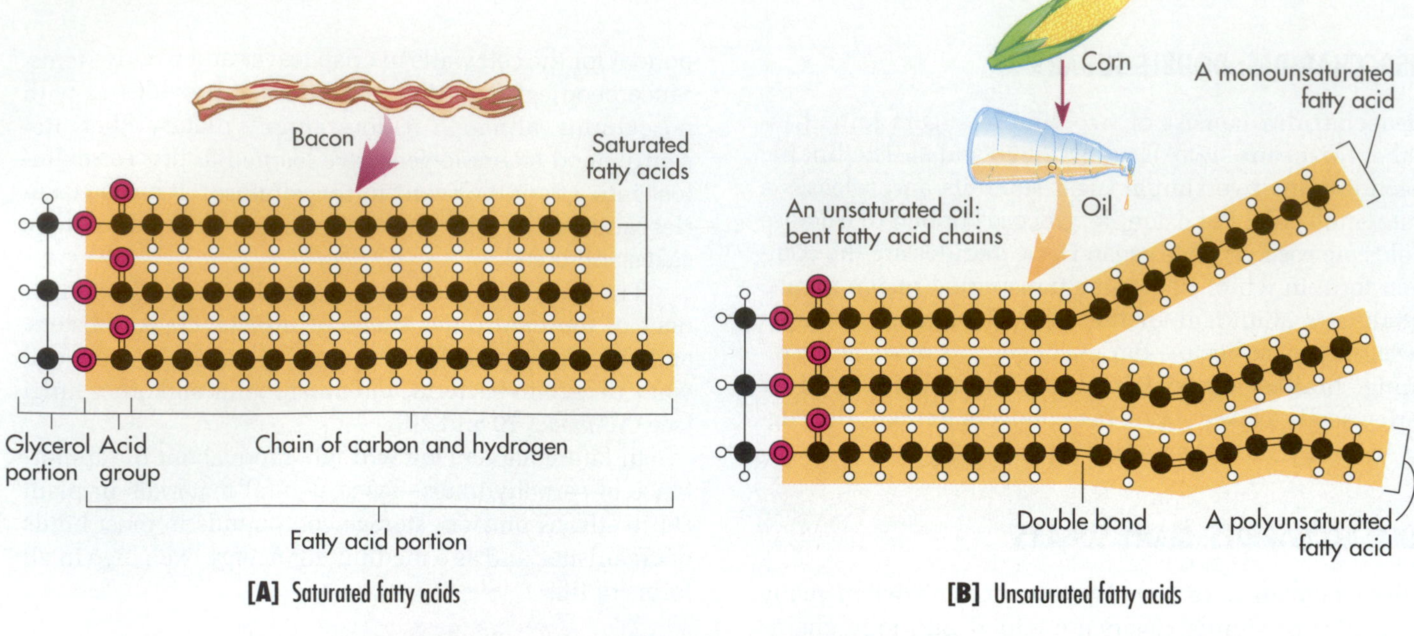

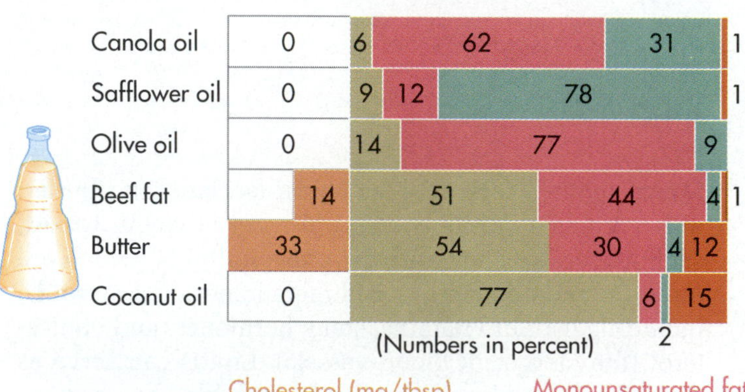

FIGURE 2.24

Triglycerides: Fat and Oils.

[A] A fat such as the solid white part of bacon contains mostly straight-chain fatty acids with no single bonds. These fatty acids are saturated with hydrogen. A diet high in saturated fats appears to contribute to diseases of the heart and blood vessels. **[B]** An oil such as this liquid corn oil generally contains unsaturated fatty acids. At some positions in these fatty acid molecules, adjacent carbon atoms are joined by two covalent bonds (a double bond). This decreases the number of hydrogen atoms that can bond to these two adjacent carbons, because each carbon can make a total of just four bonds. **[C]** Not all oils are equally healthful in the diet, since they contain different percentages of saturated and unsaturated fats and may or may not contain cholesterol.

with hydrogen; that is, all their carbon atoms are bonded to at least two hydrogen atoms and no more hydrogen atoms can be added to the molecules [FIGURE 2.24A]. Saturated fatty acids are straight-chain molecules, and they pack together tightly, as in the dense, white fat in a slab of bacon. Because the long, straight chains can pack together closely, a saturated fat is *solid* at room temperatures.

In contrast, **unsaturated fats** have one or more positions along the polymer chain where two adjacent carbons are linked by a *double bond*, two shared pairs of electrons. As a result of the double bonds, there are fewer hydrogens bonded to the carbons at this spot. The double bonds cause the fatty acid molecules to bend and kink. Thus, they cannot pack closely together, and the result is a slippery *liquid* at room temperatures [FIGURE 2.24B]. Nutritional research suggests that it is much healthier to cook with small amounts of vegetable oils such as canola, safflower, or olive oil (which are high in unsaturated fatty acids), than to consume saturated fats such as butter, solidified margarines, or lard [FIGURE 2.24C].

Monounsaturated fats have one double bond. They tend to lower the blood level of low-density lipoprotein (LDL), the "bad cholesterol" component. **Polyunsaturated fats** have several double bonds, and lower the total count of cholesterol in the blood [see FIGURE 2.24B].

WAXES

The soft but solid substance called **wax** is part of the lipid family, but its structure is different from that of triglycerides. Waxes have long-chain fatty acids combined with long-chain alcohols other than glycerol. This structure gives waxes a sticky, solid, waterproof character. Wax molecules are insoluble in water and can pack together to form solid masses and layers that serve as important waterproof coatings on leaves, bark, some fruits [FIGURE 2.25], and feathers; as the structural materials in the hon-

FIGURE 2.25

Waxes: Lipids That Form Solid Masses and Layers.

Waxes give plums their delicate whitish blush and help keep citrus fruit juicy.

eycombs of beehives; and as protective coatings for the ear canal.

Researchers have noticed that the needles of fir and pine trees are less likely to show the effects of exposure to acid rain than are the leaves of hardwood trees. They think the presence of heavy waxes probably provides the needles with some resistance to acid rain.

PHOSPHOLIPIDS

Another class of lipids, the **phospholipids**, has properties essential to a living cell. A phospholipid molecule has two fatty acid chains attached to a glycerol molecule, rather than three as in fats and oils, but it has a phosphate-containing group in place of the third chain [FIGURE 2.26A]. This group creates a water-soluble (hydrophilic) region on an otherwise insoluble molecule. The overall shape of a phospholipid molecule can be visualized as a soluble ball or "head" (the phosphate-containing group) on an insoluble pair of sticks or "tails" (the fatty-acid chains) [FIGURE 2.26B].

When phospholipid molecules are added to water, they form single layers, double layers, or spheres, with the polar "heads" oriented toward the water, and the nonpolar "tails" oriented away from it. This characteristic behavior is crucial to living things, for as you will see in CHAPTER 3, every cell is surrounded by a double layer of phospholipids [FIGURE 2.26C], and this vital barrier allows a cell to maintain its watery contents and integrity as a living unit, while still exchanging materials with the fluid environment surrounding it.

Acid rain affects fish directly by interfering with the usual exchange of materials across the phospholipid membranes of the cells lining the fish's gills. Excess hydrogen ions alter the ability of gill cells to transport other materials, such as sodium and chloride ions. Without a proper balance of these salt ions, fish become sick and die [review FIGURE 2.18B and C].

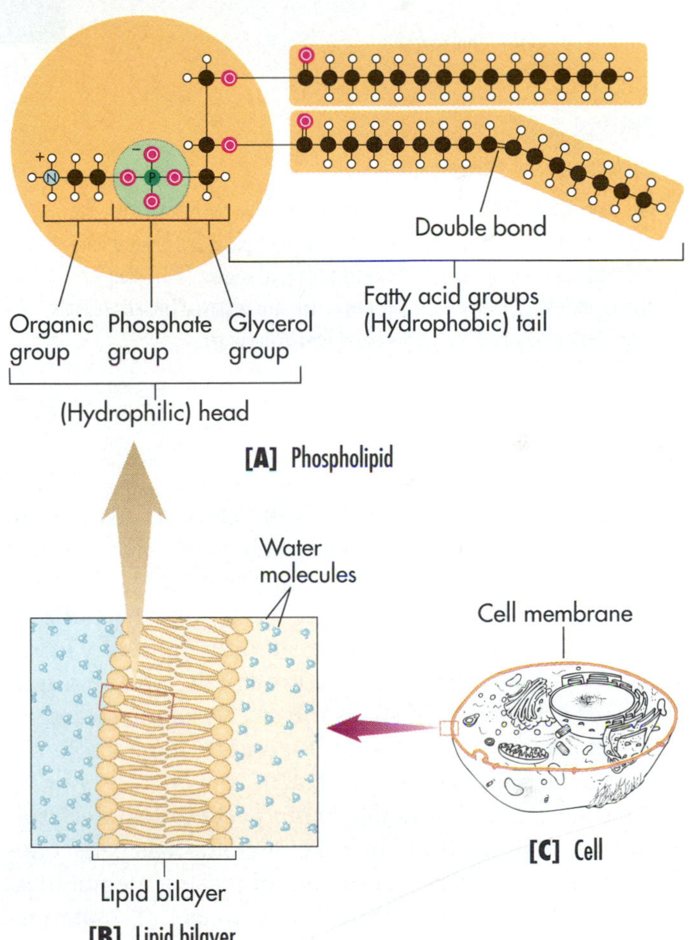

FIGURE 2.26

Phospholipids as Waterproof Barriers in Living Organisms.

[A] Each phospholipid has a head (containing a phosphate group) and a tail (made up of two fatty acids). [B] The head is water-soluble (hydrophilic), but the tail is not. When phospholipid molecules are surrounded by water, they can form double layers (bilayers), with their heads oriented toward the water molecules and their tails oriented away from the water, toward the inside of the "lipid sandwich." [C] The membranes surrounding living cells contain a double layer of phospholipids.

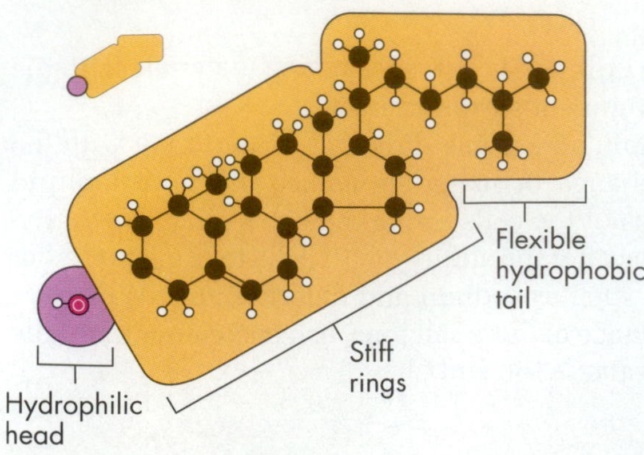

[A] Cholesterol

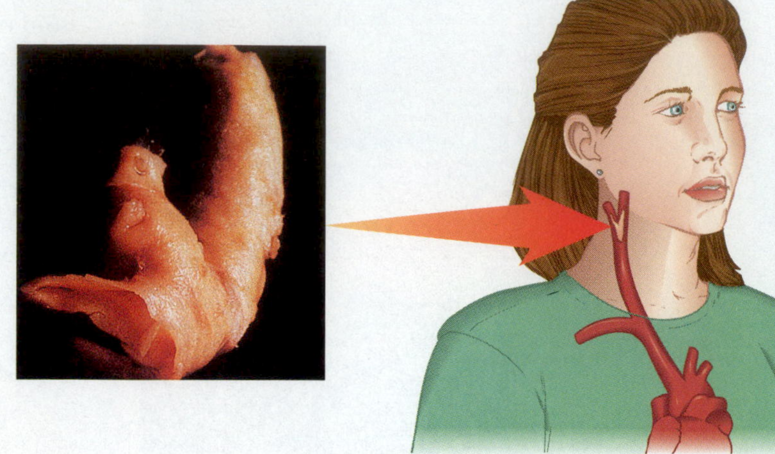

[B] Cholesterol plug in artery

FIGURE 2.27

Cholesterol and Plugged Arteries.

[A] Cholesterol is a common steroid that occurs within cell membranes. But cholesterol deposits can build up into solid clots and plugs that block arteries and lead to serious diseases of the heart and blood vessels. **[B]** A plug, or plaque, of solid cholesterol and other materials was removed from a patient's major neck artery to allow greater blood flow to the brain.

STEROIDS

The last major class of lipids is the **steroids**, which are composed of four interconnected rings of carbon atoms with functional groups attached [FIGURE 2.27A]. Steroids are insoluble in water, but can be dissolved in oils or lipid membranes. Examples of steroids include various regulatory molecules that pass through the phospholipid layers of cells easily, such as the sex hormones estrogen and testosterone [see CHAPTER 14] and the fat-soluble vitamins A, D, and E [see CHAPTER 22].

FIGURE 2.27A shows the common steroid **cholesterol**. Regularly manufactured by plant and animals cells, cholesterol is a necessary component of plasma membranes, affecting fluidity and thus the passage of materials through the membranes. In some people, however, a diet high in saturated fats from dairy products, meat, and eggs can lead to a buildup of cholesterol deposits in blood vessels, which in turn can lead to abnormally high blood pressure and the risk of strokes and heart attacks [FIGURE 2.27B; see CHAPTER 25].

CONCEPT CHALLENGE

The seeds of our domesticated corn and wheat contain large amounts of starch and other carbohydrates, while the seeds of almonds, walnuts, and other cultivated nut trees contain large amounts of lipid. These carbohydrates and lipids have certain characteristics in common that have proven especially useful to seeds and the tiny plant embryos they contain. What are these shared traits?

Proteins

The complexities of our bodies and the rich diversity of life in general—the millions of organisms of different shapes, colors, textures, and life-styles—could never be accounted for by carbohydrates and lipids alone. Carbohydrates and lipids are important for cell structure and for energy storage and use, but they simply do not contain enough different types of molecules to account for life's vast diversity. In contrast, **proteins** come in such a wide variety of forms (at least 10 to 100 million different kinds in the spectrum of the earth's organisms) that they can easily explain the myriad shapes and functions of living things.

There are several types of proteins:

1. *Structural proteins* form cell parts, such as the keratin fibers in hair and the contractile proteins in muscles.

2. *Regulatory proteins* control cell processes. The so-called *ras* proteins are regulators which, if altered slightly by mutation, can cause cancer.

3. Some proteins are hormones, or molecular messengers. Growth hormone, for example, helps govern how tall a person will grow.

4. *Transport proteins* carry other substances throughout the body. The hemoglobin protein in red blood cells, for example, transports oxygen.

5. *Antibodies* are proteins that protect some types of animals from disease.

6. *Enzymes* are important proteins that facilitate virtually all the life processes that go on in cells.

Moreover, the specialized shapes and functions of different cell types depend on protein. In other words,

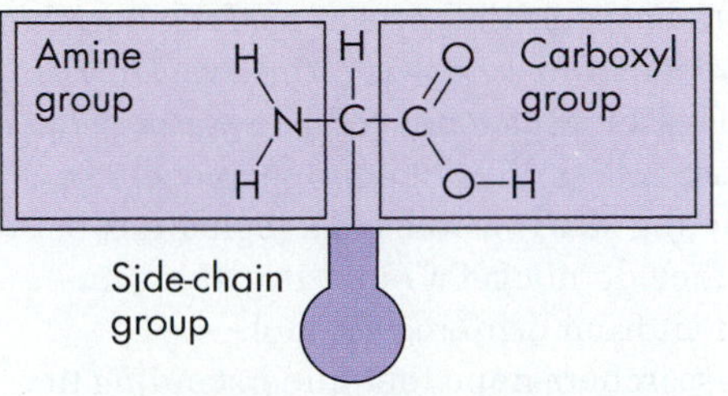

[A] Components of an amino acid

FIGURE 2.28

Amino Acids: Protein Building Blocks.

[A] Each amino acid has an amine group (NH_2) on one end and an acidic carboxyl group (COOH) on the other end. The third "group" is always a hydrogen atom, shown here at the top of the molecule. Only the fourth group, the side-chain group, varies from one amino acid to another. The side-chain group can be as small as the single hydrogen atom found in the amino acid glycine or as complex as a double-ring structure or chain. **[B]** Five of the 20 amino acids commonly found in proteins.

Glycine (Gly)	Phenylalanine (Phe)	Cysteine (Cys)	Glutamic Acid (Glu)	Arginine (Arg)
Side chain: H	Side chain: CH_2–benzene ring	Side chain: CH_2–SH	Side chain: CH_2–CH_2–C(=O)O^-	Side chain: CH_2–CH_2–CH_2–N–H–C=NH_2^+–NH_2
Glycine has the simplest side-chain group, a hydrogen atom. The animal protein collagen has many glycines.	Phenylalanine and six other amino acids have side-chain groups that are insoluble in water. Phenylalanine is metabolized to a poison in people with the genetic disease phenylketonuria.	Cysteine and eight other amino acids have side-chain groups that are soluble in water. Cysteine is important in holding protein chains together. The unpleasant smell of a home permanent breaks and makes these cysteine bonds in hair protein.	Glumatic acid has a negatively charged side-chain group. Glutamic acid with a sodium ion is monosodium glutamate, the flavor enhancer.	Arginine is a positively charged amino acid because of its side-chain group. Arginine helps rid the body of dangerous nitrogen wastes.

[B] Representative amino acids

because different types of cells have different sets of proteins, the cells in your eyes that detect light are different from the cells in your liver that detoxify poisons.

AMINO ACIDS: BUILDING BLOCKS OF PROTEINS

Proteins are long chains of subunits called **amino acids**, which have the general structure shown in FIGURE 2.28A. Each amino acid has three chemical groups bonded to one central carbon atom. The term *amino acid* comes from the two attached functional groups, the *amine* group (—NH_2) and the *acidic* carboxyl group (—COOH) which all amino acids share. The third group is different for each specific amino acid. FIGURE 2.28B shows five of the 20 amino acids common in protein. The amino acids of a protein are joined by covalent bonds in an order that is distinctive for each protein and determined by the organism's genes. Long chains of amino acid subunits are called **polypeptides**. (Notice the recurrence of the monomer-polymer theme in our discussions of the structures of biological molecules; just as polysaccharides are polymers and their monomers are simple sugars, proteins are polymers made up of amino acid monomers.)

box 2.2
Human Impact

ADH and Alcoholism

Alcoholism researchers suspect there may be a genetic component to this serious disease because children who have an alcoholic parent are predisposed to problem drinking, even if they are adopted and raised by nonalcoholic parents.

In some people, the liver protein alcohol dehydrogenase (ADH) appears to work more rapidly than in others to convert alcohol into a compound called *acetaldehyde*, which is even more toxic to the body than ethyl alcohol itself. In many Asians, the ADH enzyme converts alcohol into this toxin so rapidly that fairly soon after taking a drink, the drinker feels flushed and nauseated, and his or her heart pounds. As you might imagine, this uncomfortable sensation acts as a strong negative reinforcement to excessive drinking. The risk of alcoholism thus tends to be lower in people with fast-acting ADH than in people with slow-acting ADH, who have a higher risk because they do not get a rapid negative reinforcement from drinking alcohol.

Researchers hope that understanding the structure of ADH may someday lead to a better understanding of alcoholism, with its tragic consequences for the drinker, the family, and society in general, as well as leading to new methods of prevention and treatment.

The 20 types of amino acids function as subunits in a biological alphabet, forming complex proteins much as the 26 letters of our alphabet can be combined to form a nearly infinite array of words. Just as the arrangements of letters in a word determine its identity, and therefore its meaning, the identity of a protein—its shape, properties, and functions—depends on the exact order of its amino acid "letters."

Most proteins contain between 100 and a few thousand amino acid subunits. Even in a short polypeptide, the potential number of unique amino acid sequences that could be formed by the various combinations of 20 amino acids would equal the astonishingly huge number 20^{100}. (For comparison, 20^5 is 3.2 million.)

Actually, not every potential sequence of amino acids occurs in nature, just as many random combinations of letters fail to form real words. Nevertheless, there are tens of millions of unique sequences of amino acids in the living world, each spelling out a specific type of protein and together making possible life's remarkable diversity. The human body alone contains about 50,000 different proteins, each with a unique structure.

PROTEIN STRUCTURE

Many proteins consist of a single chain of amino acids, and the way that chain bends, folds, and forms loops determines the protein's function. To illustrate various aspects of protein structure, let's consider the liver protein called *alcohol dehydrogenase*, or *ADH* [FIGURE 2.29]. Besides being a useful teaching tool for understanding proteins, ADH also helps us see why alcohol has poisonous effects on the body.

The ADH protein helps break down alcohol in the liver, and its structure and activity may be relevant to the disease of alcoholism, which affects perhaps 1 out of 10 adults who drink alcohol [see BOX 2.2 above].

We'll use the ADH molecule to illustrate the various levels of protein structure. The primary level of protein structure is the linear sequence of amino acids [FIGURE 2.29A]. The ADH protein is a long chain of 375 amino acids, arranged in a unique and consistent order. The first five of these amino acids (methionine, serine, threonine, alanine, and glycine), for example, start at the blue tip in FIGURE 2.29A and follow along the yellow thread at the top.

As the long, single chain twists and loops, we can detect areas along the chain with regular folding patterns, representing a secondary level of protein structure. One pattern of the secondary structure is a coiled, "loop-the-loop" form (purple in FIGURE 2.29B), which is called the *alpha helix* (*α helix*). Another pattern consists of parallel rows of the polypeptide chain folded into an accordion-like sheet called a *beta sheet* (*β sheet;* shown in blue in FIGURE 2.29B). The coil and sheet are structural motifs found in many different proteins. Together, they cause the protein to assume a specific and rather rigid shape in three dimensions, which is its tertiary structure [FIGURE 2.29C]. *Disordered loops* (shown in yellow) [FIGURE 2.29B] tend to lie between areas of α helices and β sheets. As you can see from FIGURE 2.29C, the shapes of the disordered loop structures vary, depending on the loop's position in the protein.

Note that in the ADH molecule, the springlike α helices form rigid sides, while the parallel strands making up the β sheets look like a slightly twisted sheet of cardboard; the disordered loops run between helical and sheet regions and hold things together. The helices and sheets are held in their positions by weak hydrogen bonds, as well as other types of bonds.

In a lake that has been polluted by acid rain, the hydrogen ions may interfere with hydrogen bonding in proteins. Think of what happens when you put lemon juice, which is an acid, into nonfat cow's milk, which is a solu-

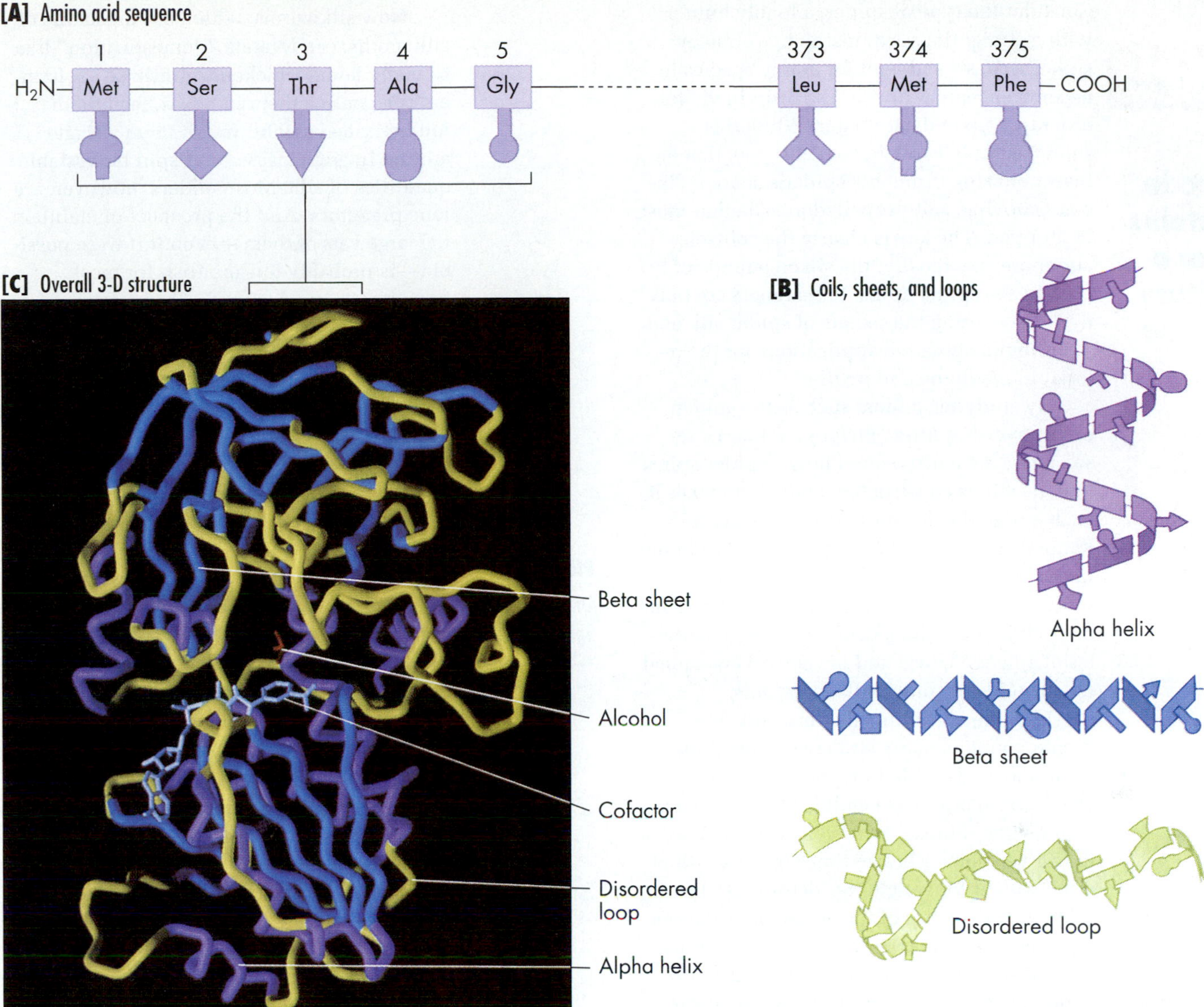

tion of proteins dissolved in water: the milk curdles. The curds in the curdled milk are clumps of protein molecules in which the helices, sheets, and loops have assumed new shapes. These shape changes occur when the extra hydrogen ions in the acid break weak bonds in the proteins. Since a protein's function depends on its shape, the functions of these clumped proteins have been destroyed by the acid. Similar shape changes take place among the proteins inside living cells of certain lake organisms. Particularly vulnerable are the proteins in cell membranes that regulate the passage of substances into and out of cells. In the Canadian lake that Schindler and his colleagues acidified on purpose, some of the plants and animals—particularly the ones that trout and other fish ate—may have been especially sensitive to acid-induced changes in protein structure. This is probably why these species tended to die off while others remained.

FIGURE 2.29

Levels of Protein Structure.

Alcohol dehydrogenase, or ADH, is an alcohol-degrading enzyme. **[A]** The linear sequence of amino acids in the polypeptide chain of ADH specifies the protein's shape and function. The figure shows only the first few and last few amino acids of this 375–amino acid protein. **[B]** Sections of the polypeptide chain fold up into regular patterns, such as the α helix (purple corkscrews) or the β sheet (blue corrugated regions). These patterns are maintained by hydrogen bonds and are connected by disordered loops (yellow portions), which have various shapes characteristic for each different protein. **[C]** The helices, sheets, and loops of the polypeptide chain are folded into a three-dimensional shape that is specific for that protein's function. In this case, the folding pattern creates a pocket deep in the molecule that interacts with ethyl alcohol (the red "V"), the intoxicating component of beer, wine, and liquors. The enzyme begins the metabolic pathway of alcohol in the body. The action of the enzyme is enhanced by a cofactor (called NAD, shown here in pale blue), which helps the enzyme metabolize alcohol.

box 2.3
Biology Applied

Miracle Proteins from a Solitary Spinner

Consider the typical spider, a lonely hunter with a straightforward, instinctive strategy: spin a web, sit and wait for a prey species to become ensnared, devour the drop-in visitor, or wrap it up and eat it later. While this sounds simple enough, consider, too, that an insect buzzing through a spider's snare is like a fast-moving, self-propelled missile that must be stopped. The key is clearly the cobweb, and more specifically, the silken strands of protein the spider spins. Researchers are only now discovering the secrets of spider silk and planning imaginative applications for the material in medicine and textiles.

By studying spiders such as the golden orb weaver *Nephilia clavipes* [FIGURE 1], researchers have discovered how a spider spins, how its silk is constructed, and what makes it such a remarkable substance. For example, Randolph Lewis at the University of Wyoming in Laramie has watched hundreds of orb weavers pump raw, liquid silk from glands in their abdomens and extrude it through batteries of spigots. Lewis and his coworkers named the proteins in spider silk *spideroins.*

Like the fibroin in silkworm silk, spideroin contains many stretches of the same 15 amino acids folded into beta pleated sheets. Spideroin has significantly more such regions than fibroin, however, and together, they give spider silk five times the strength of steel, weight for weight. Spideroin also has numerous stretches of the amino acid alanine repeated over and over and over. These fold up into springy alpha helices, and the coils appear to give spideroin its unusually great elasticity—up to 4 times the molecule's original length. This stretching allows the strands of a cobweb to slow down a hurtling insect by absorbing the energy of its forward motion. But another property is needed—and provided—by spider silk: a slow recoil, or return to original strand length. This slow recoil keeps the prey in the web, whereas a fast recoil would catapult the insect backward in its flight path and to safety, out of the hungry spider's grasp.

All in all, the various types of spider silk in a typical cobweb are strong, stretchy, waterproof, and chemically stable, and even if a strand blows into a person's nose or eyes, it won't trigger an allergic or other immune response. In contrast, silkworm silk is rather weak, especially when it's wet, and it can trigger allergies. For these reasons, researchers would love to produce commercial quantities of spider silk.

Now silkworms, which are the larvae of silk moths, can tolerate "domestication" like so many laying chickens or milk cows: farmers can confine them in boxes, feed them mulberry leaves, and watch them wriggle around in great masses and spin harvestable quantities of silk. Most spiders, however, are lone predators. And the prospect of maintaining large vats of them—even if it were possible—is probably too ghoulish for most farmers.

FIGURE 1
Orb Weaver, Spinning Web

Luckily, in this era of genetic engineering, researchers can accomplish the once inconceivable. Workers in several laboratories are unraveling the genetic information that directs a spider's silk production, and with a fascinating goal: If they succeed in isolating all the genes involved, they may be able to transfer them into a more tractable species, such as the silkworm, or even the cotton plant. Souped-up silkworms or super-charged cotton might then churn out tons of spider silk, and its harvest could lead to new products. These might include new kinds of medicinal sutures, or artificial ligaments and tendons that are strong and stretchy and don't trigger rejection by the body. Others have proposed silky wound dressings for burn victims or coatings for heart valves, vein grafts, or other implants, all of which will elude immune rejection. Some even envision high-tech spider-silk fabrics for bulletproof vests, designer dresses, and the cables on suspension bridges. Reaping crops of spider silk large enough for clothing—not to mention road projects—would be utterly impossible without modern biological techniques for probing protein structure and then genetically engineering "spider-silk factories" with other animals or plants. Yet who knows? Even with these modern tools, the difficult goal may prove as elusive as a spider and as gossamer as its web.

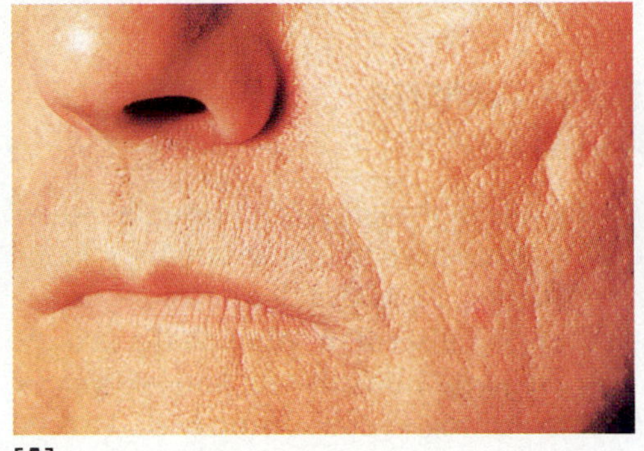
[A]

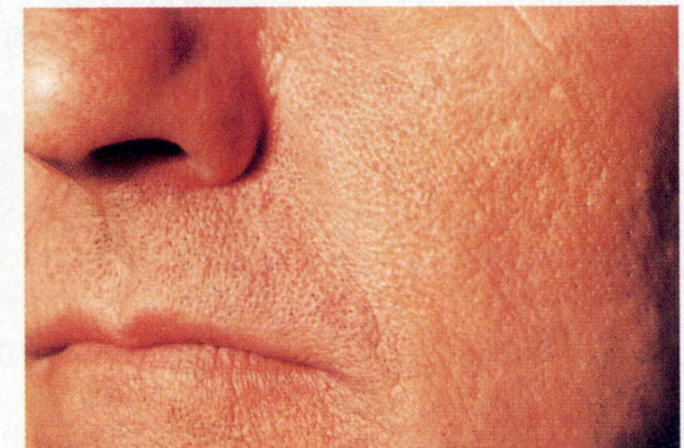
[B]

FIGURE 2.30
Collagen Injections: Before and After.
This man's facial scar [A] all but disappears [B] after cosmetic collagen injections.

SPECIFIC PROTEIN SHAPES

Different proteins have different sequences of amino acids, and hence different three-dimensional shapes. For example, *keratin*, the protein in hair, fingernails, horns, claws, and feathers, has many helical regions. The fact that weak hydrogen bonds stabilize the helix shape explains why hair can be stretched or curled when warm or wet [review FIGURE 2.10]. Heat and water disrupt the hydrogen bonds and allow the helix to unfold and the polypeptide chain to stretch out. When the keratin cools, helices re-form, and the hair takes on the new shape of the brush, roller, or curling iron it was coiled around. Permanent-wave lotions induce new strong covalent bonds to form—hence the longer-lasting effect.

Fibroin, the protein in silk, occurs largely as pleated sheets. The many bonds cross-linking the polypeptide chains in fibroin result in a very strong fiber. Stronger still is the silk protein in spider webs (so-called spideroin; see BOX 2.3, page 56), which researchers are studying as a new wonder fiber. The keratin in hair, fibroin in silk, and spideroin are *fibrous proteins*, as is *collagen* (KAHL-uh-jehn), the most abundant protein in the animal kingdom. Collagen gives strength, flexibility, and shape to skin, tendons, ligaments, cartilage, and bones. Some physicians now inject collagen to fill in scars, dents, and wrinkles left by acne, accidents, or aging [FIGURE 2.30]. It should be noted, however, that collagen injections are not always successful, and in fact can cause serious problems.

Other proteins, called *globular proteins*, tend to ball up into rounded shapes. The alcohol dehydrogenase protein is a globular protein, and so is hemoglobin, the iron-containing protein that carries oxygen in red blood cells. Hemoglobin also illustrates the fact that some proteins are made up of two or more folded polypeptide chains that fit together like the pieces of a three-dimensional puzzle. In proteins like hemoglobin, with two or more separate polypeptide chains, each folded chain touches the others at specific places, giving rise to a characteristic overall shape [FIGURE 2.31]. When assembled this way, hemoglobin forms an active protein.

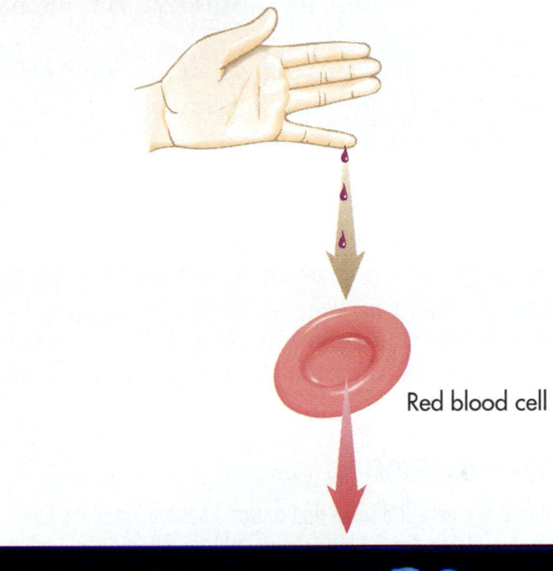

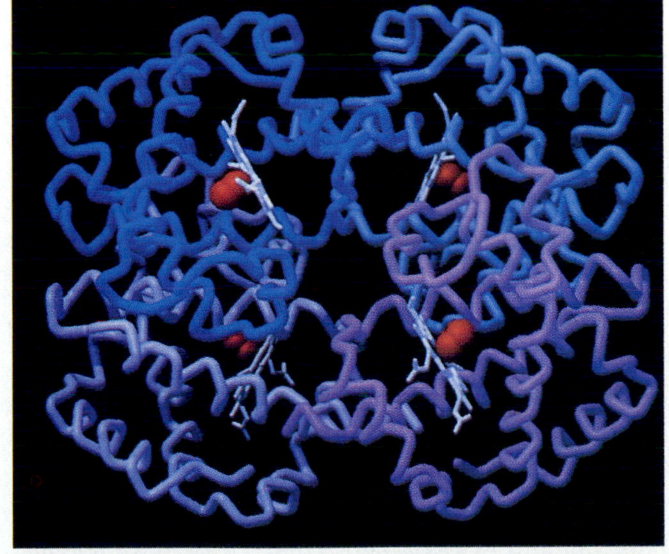

FIGURE 2.31
Hemoglobin.
Some proteins consist of more than one polypeptide chain. Hemoglobin, the protein in red blood cells that transports oxygen (O_2, the four pairs of red balls), has four chains, each about 150 amino acids long. Each of the four chains is shown in a slightly different color. Can you identify the helical regions in each of these four chains? A nonprotein part of the molecule is shown in pale blue. This part is called *heme*, contains iron, and helps the protein bind oxygen.

As a final note concerning protein structure, note that both ADH [FIGURE 2.29] and hemoglobin [FIGURE 2.31] have structures (shown in light blue) that are not made up of amino acids, and so are not links in the polypeptide chain. These added-on substances, called *cofactors*, help the proteins function. You will see later that the reason some vitamins are so essential in the diet is that they act as cofactors for certain enzymes.

There is a chain of logic linking our discussions of basic chemistry at this chapter's beginning, the details of protein structure just described, and the shapes and functions of specific parts within organisms. We can summarize that chain this way:

1. The functional groups that occur on amino acids [review FIGURES 2.21 and 2.28] determine the behavior of each of the 20 different amino acids found in proteins.

2. The order of amino acids within a protein molecule determines the protein's shape.

3. The shape of the protein molecule determines its function.

4. Collectively, the functions of a cell's or organism's proteins determine what the organism looks like and how it lives.

CONCEPT CHALLENGE

Cystic fibrosis is a genetic disease that causes deterioration of the lungs and premature death. Cystic fibrosis is caused by a change in a single protein in the body. In most people, this protein has 1480 amino acids, but in people with cystic fibrosis, it contains only 1479 amino acids. Given what you know about the importance of the amino acid sequence to the structure and function of the protein, speculate on what happens during this condition?

Nucleic Acids and Nucleotides

The final class of biological molecules contains the nucleic acids and their subunits, the nucleotides. **Nucleic acids**, including DNA and RNA, carry the chemical "code of life" and bear genetic information from one generation to the next. Nucleic acids are long polymers made up of monomers called **nucleotides**. Nucleotides also have their own role as energy transfer compounds in cells. Let's examine the nature of the nucleotide subunits, then see how they are combined into long chains in nucleic acids.

NUCLEOTIDES: ENERGY TRANSFER AND BUILDING BLOCKS

The *nucleotides* both transfer energy from one chemical compound to another and serve as the monomers (subunits) that make up nucleic acids. Nucleotides are three-part molecules—they consist of a phosphate group, which contains the elements phosphorus and oxygen; a five-carbon sugar; and a nitrogen-containing base. FIGURE 2.32 shows a common and important nucleotide called ATP (adenosine triphosphate), which actually has three phosphate groups. The ATP molecule is the main energy-transfer substance within cells. Cells harvest energy in the form of ATP when they break down sugars and other molecules, and they spend energy in the form of ATP when they perform activities, such as contracting muscle cells. Although appetite and taste are our conscious reasons for eating, the metabolic reason is to renew our supply of ATP. We will discuss ATP further in CHAPTERS 4, 5, and 6.

Another kind of nucleotide, called *cyclic AMP* (cyclic adenosine monophosphate), is a signal molecule that turns certain cellular activities on and off. Nucleotides called *coenzymes* are transport compounds necessary for energy harvest and the building of new cellular structures. The coenzyme in the ADH protein is a nucleotide, and is shown in pale blue in FIGURE 2.29. We will discuss coenzymes further in CHAPTERS 5 and 6.

Besides their important roles in cellular energy functions, nucleotides have an equally crucial role in the structure of hereditary information because they are the building blocks of nucleic acid molecules.

NUCLEIC ACIDS: INFORMATION STORAGE AND PROCESSING

You resemble your parents because of the nucleic acid molecules in each of your cells and in theirs. **Deoxyribonucleic acid (DNA)** stores hereditary information, and **ribonucleic acid (RNA)** helps in the use of that information. In addition, some RNA molecules can act to speed up certain chemical reactions much as enzymatic proteins do.

The structure of a single strand of DNA can be imagined as a chain with links of four different colors, each color representing one of four different nucleotides, depending on which kind of nitrogenous base is present. The names of these nucleotides are abbreviated A, T, C, and G [FIGURE 2.32]. DNA molecules are very long—an average DNA molecule in one of your cells contains 43 million nucleotide pairs.

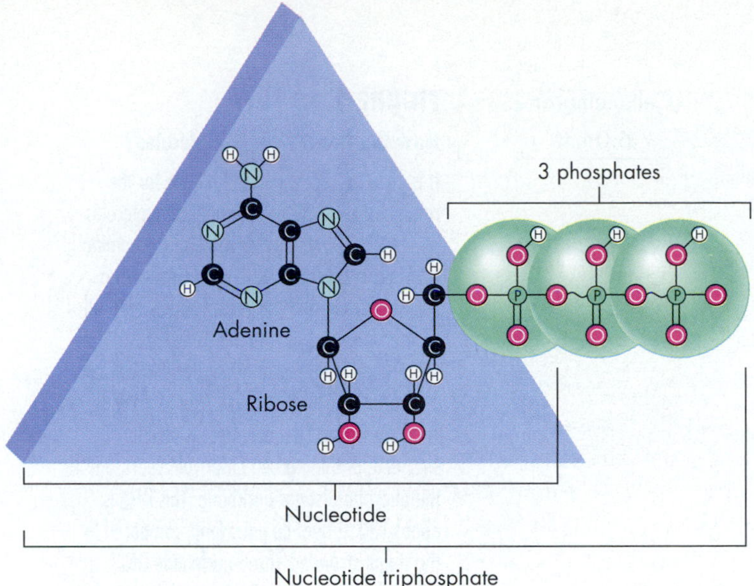

[A] ATP

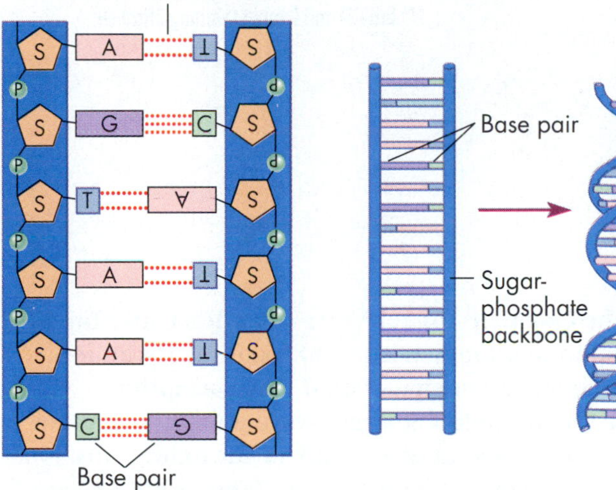

[B] Short length of DNA molecule

[C] DNA molecule twists to form a double helix

FIGURE 2.32

ATP and DNA.

[A] ATP is a nucleotide with a tail of three phosphates. In general, nucleotides contain a phosphate group, a ribose or deoxyribose sugar (here, we show ribose), and a nitrogen-containing base. There are five types of bases: adenine, guanine, cytosine, thymine, and uracil—each with slightly different chemical compositions. (Here, adenine is shown.) [B] Nucleotides join together, forming a single strand of DNA. A DNA molecule in a cell generally consists of two of the polymer chains twisted about each other in a double helix, with the outer coiled "rails" formed by deoxyribose sugars connected by phosphate "bridges," and the "ladder rungs" formed by pairs of bases (adenine and thymine, or guanine and cytosine) attached by hydrogen bonds. The precise order of the "rungs" carries information for the structure of enzymes and other proteins. In turn, enzymes act on raw materials to build all four kinds of biological molecules—carbohydrates, lipids, proteins, and nucleic acids—as well as the cell parts constructed of them.

In double-stranded DNA—the form in which genetic information is stored—there are two nucleotide chains bonded together like a ladder whose sides are held together by the rungs. The DNA "ladder" is coiled to form a spiral, or double helix [FIGURE 2.32].

The key to DNA and RNA structure lies in the order of the nitrogen bases in these long molecules of inheritance: like the order of letters in a sentence, the order of the bases carries crucial information for constructing and maintaining cells. The specific sequence of the different kinds of nucleotides in nucleic acids determines the exact order of amino acids in proteins, which, as you saw, determines the proteins' shapes and functions. The steps of this process are a fascinating subject, one which we investigate in CHAPTER 10.

What is the relationship of a DNA molecule, with its millions of base pairs, to a gene? Most geneticists define a gene as a portion of a DNA molecule that specifies how an RNA molecule will be constructed. Many RNA molecules, in turn, carry the information specifying the structure of a particular protein. Generally, a gene is several thousand base pairs long, so a single DNA molecule with its millions of base pairs contains thousands of genes.

In summary, nucleic acids are central to the life of a cell. By specifying the structure of proteins, they also dictate cell shape and cell function. This is because structural proteins and enzyme proteins produce the specific carbohydrates and lipids that make up an organism's cells and hence its body. Your DNA, by determining the shape of your proteins, has played a central role in shaping the way you look and the way your body functions.

THE SIZES OF MOLECULES

In this chapter we have discussed molecules as small as water, and as large as proteins and nucleic acids. But all, of course, are far below our ability to see except with the most sophisticated microscopes. To help you keep the size relationships among various molecules in perspective, FIGURE 2.33 shows the relative dimensions of several we have discussed. You can see that an amino acid is many times larger than a water molecule, and that an average protein like hemoglobin is much larger still. The typical protein has a greater circumference than a DNA molecule, but at this magnification, the average DNA molecule in one of your cells would be 60 km (37 mi) long, a truly colossal molecule. Individual lipid molecules in a membrane are of moderate size, but when many such molecules combine to form a membrane, they can enclose an entire cell. At this magnification, an average animal cell would be about 35 m (114 ft) in diameter.

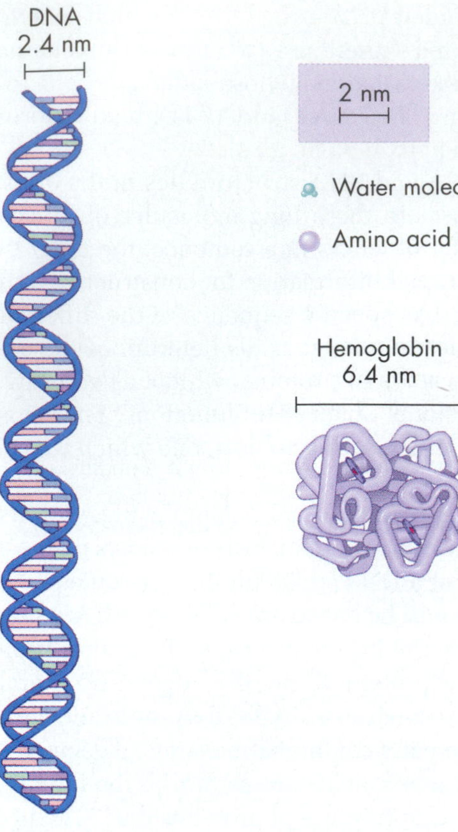

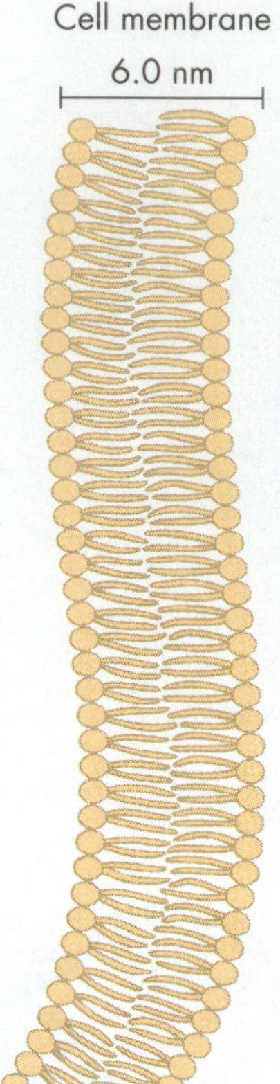

FIGURE 2.33

How Big Are Different Molecules?

It is good to have a general feeling for the relative sizes of different biological molecules. This sketch shows a water molecule, an amino acid, a protein (hemoglobin), a portion of a DNA molecule, and the phospholipids of a cell membrane, all drawn to the same scale. At this magnification, a single bacterial cell would be four m (13 ft) long and 1.5 m (5 ft) wide. The DNA molecule shown here is double-stranded, and the green outer rails represent the phosphate-sugar backbone. This DNA is about 130 nucleotide pairs long, compared to the single-stranded, three-nucleotide DNA shown in FIGURE 2.32B. DNA from just one of your 23 pairs of chromosomes is about 40 million nucleotide pairs long. This would be 50 km (31 mi) long at this magnification.

Connections

Ultimately, all aspects of life depend on the interactions of atoms and molecules. This is true even for the ecosystem, one of the highest levels of biological organization, as the phenomenon of acid rain demonstrates so well. By deliberately acidifying Lake 223, Canadian researchers showed that an accumulation of the simplest chemical—the hydrogen ion, a single proton—can severely disturb an entire ecosystem. An understanding of ecology, how organisms interact in the environment, therefore requires an understanding of how chemicals interact. Air pollutants react with moisture in the air, producing airborne acid droplets. Hydrogen ions in the acid enter lakes in acid precipitation, and interact with the proteins in living organisms. This interaction can alter protein structure, which in turn alters protein function. This can alter cell function, and the activity of the whole organism to the point of death. When a few key lake-dwelling species die in this way—even inconspicuous gnats, larvae, or shrimplike species a hiker or angler would seldom notice—the food chain, the hierarchy of who eats whom may become disrupted. The result can be starved game fish and a disturbed ecosystem.

The chemistry of acid rain also illustrates the intersection of biology with public policy. By understanding the chemistry of acid rain formation, scientists and policy makers were able to formulate laws that mandated decreased sulfur oxide emissions. Because of these cutbacks, many lakes were far less acidic by the early 1990s than ecologists working in the early 1980s had predicted they would be.

In the next chapter we will see how the molecules we discussed in this chapter interact and build functioning cells.

CONCEPT MAPPING

Draw a concept map relating the following terms. Link the concepts with arrows and a connecting word or short phrase

1 atom, atomic mass, atomic nucleus, chemical bond, electrical charge, electron, ion, molecule, neutron, proton

2 amino acid, cholesterol, energy storage, hereditary information, lipid, monomer, monosaccharide, nucleic acid, nucleotide, plasma membrane, polymer, polysaccharide, protein

KEY TERMS

acid, 42
adenosine triphosphate, 58
alkaline, 44
amino acid, 53
atom, 31
atomic mass, 32
atomic number, 32
base, 42
buffer, 43
capillarity, 40
carbohydrate, 47
cellulose, 49
chitin, 49
compound, 30
covalent bond, 36
disaccharide, 49
electron, 31
element, 30
fat, 49
fatty acid, 49
functional group, 46
glucose, 47
glycerol, 49
glycogen, 49
hydrogen bond, 38
hydrophilic, 41
hydrophobic, 41
inorganic chemistry, 44
ion, 34
ionic bond, 38
isotope, 33
lipid, 49
molecule, 36
monomer, 49
monosaccharide, 47
neutron, 31
nonpolar, 37
nucleic acid, 58
oil, 49
organic chemistry, 44
pH scale, 43
phospholipid, 51
polar, 37
polymer, 49
polypeptide, 53
polysaccharide, 49
protein, 52
proton, 31
radioactive, 33
solute, 41
solvent, 41
starch, 49
steroid, 52
sucrose, 49
surface tension, 40
wax, 50

HIGHLIGHTS IN REVIEW

1 The properties of life are based on atoms and molecules. Atoms are the fundamental units of matter.

a] Each element consists of atoms with a specific number of protons in the nucleus (hydrogen has one, carbon has six). Different isotopes of an element contain different numbers of neutrons (^{12}C has six neutrons, ^{14}C has eight neutrons). An ion of an element has different numbers of electrons and protons (a hydrogen ion has one proton and no electrons; a sodium ion has 11 protons but 10 electrons).

b] The distribution of electrons orbiting around atoms helps us understand the shapes and properties of those atoms and why certain elements like neon are inert while others like sodium are violently reactive.

c] Atoms can bond together, forming molecules. Two atoms sharing a pair of electrons are joined by a strong bond, called a covalent bond. A positively charged ion can be bonded to a negatively charged ion in an ionic bond. A hydrogen atom of one molecule can be attracted to an atom of another molecule in a hydrogen bond.

2 The organization of an atom's protons, neutrons, and electrons determine its properties.

a] Water is the universal solvent and can dissolve other substances by surrounding and separating the component materials with a layer of oriented water molecules. Compounds that dissolve easily in water are hydrophilic; water-insoluble compounds are hydrophobic.

3 The special properties of water—such as why ice floats, why water dissolves many substances, and why it acts like a weak acid—depend on the water molecule's chemical and electrical properties.

a] A substance that donates hydrogen ions is an acid; a substance that accepts hydrogen ions is a base. A buffer releases hydrogen ions when these ions are in short supply, and accepts hydrogen ions when they are in high concentration.

b] Scientists measure the concentration of hydrogen ions on the pH scale. A solution with pH of 0 to 7 is acidic and a solution with pH 7 to 14 is basic. Cells usually have a pH around neutral, between 6.5 to 7.5.

4 Life is based on carbon atoms. Four classes of large carbon-containing molecules—carbohydrates, lipids, proteins, and nucleic acids—are especially important in understanding the structure and function of living things.

a] Carbohydrates, such as sugars and starches, store energy and provide structural support. Lipids, such as fats and oils, do not dissolve well in water; they store energy, form waterproof coverings of cells, and act as chemical signals. Proteins are long chains of amino acids; they carry out many cellular functions by speeding up chemical reactions, transporting substances, and building cell parts. Nucleic acids are long chains of nucleotides; they store hereditary information and help translate that information into the structure of proteins.

b] Carbohydrates include simple sugars (monosaccharides) like glucose; double sugars (disaccharides) like sucrose; and long-chain polymers (polysaccharides) like starch, glycogen, cellulose, and chitin.

c] Lipids consist mainly of carbon and hydrogen and do not dissolve well in water. Lipids include energy-storing triglycerides (fats and oils); phospholipids (which form cell membranes); and sterols (which include cholesterol and sex hormones).

d] Proteins regulate processes (growth hormone); transport materials (hemoglobin); prevent diseases (antibodies); build structures (hair); and speed up chemical reactions (enzymes like alcohol dehydrogenase). The way cells look and function depends on the proteins they contain.

e] Proteins are long chains of amino acids (a hemoglobin chain has about 150). There are 20 different types of amino acids commonly found in proteins. Each different protein has a specific sequence of amino acids. The sequence of amino acids determines the protein's shape in three dimensions, and the protein's shape determines its function.

f] Nucleic acids (DNA and RNA) store and transmit the cell's hereditary information. DNA and RNA are long chains of nucleotides. Some nucleotides such as adenosine triphosphate (ATP) provide the cell's energy currency. Other nucleotides are coenzymes, the small units that help certain enzymes perform their functions.

UNDERSTANDING THE FACTS AND CONCEPTS For Questions 1–10 match each of the statements below with the one item in the following list that is most appropriate. Any item may be used once, more than once, or not at all.

a] molecular	e] polar
b] covalent	f] nonpolar
c] ionic	g] hydrophilic
d] hydrogen	h] hydrophobic

1 The basic kind of bond that joins elements together to form compounds.

2 A molecule in which the constituent electrons tend to be more concentrated at one part of the molecule than at another.

3 A weak chemical bond that makes possible life as we know it.

4 The kind of bond that is responsible for most of the special properties of water.

5 The kind of bond that is formed by the sharing of electrons between two atoms.

6 A structural character of the water molecule due to the distribution of its orbital electrons.

7 The kind of bond that is formed between two atoms when one of them transfers one or more of its electrons to the other.

8 The kind of bond that forms when the outer orbitals of two atoms overlap, resulting in electron sharing.

9 The bond that characterizes crystalline structure.

10 The term that applies to compounds, such as oils and fats, that resist dissolution in water.

As above, answer Questions 11–20 using the following list:

a] carbohydrate	e] $—NH_2$
b] lipid	f] —OH
c] protein	g] —COOH
d] nucleic acid	h] $—PO_4$

11 DNA

12 $C_6H_{12}O_6$

13 Carboxyl group

14 A polymer of amino acids

15 Triglycerides

16 Chitin

17 Steroids

18 Antibodies

19 Enzymes

20 One of its monomers can become modified to form ATP.

21 Identify each of the molecules *a* through *f* diagrammed below and describe in your own words what the figure portrays.

INTEGRATE AND APPLY WHAT YOU HAVE LEARNED

1 A water strider can do something that few other living creatures can do—walk on the surface of water. What property of water enables the insect to do this?

2 Why would life not be possible if cells were bounded by a layer of completely water-soluble molecules instead of by phospholipids?

3 List the functions of proteins, and explain why this category of biomolecules is so versatile.

4 How do carbohydrates, lipids, proteins, and nucleic acids differ with respect to constituent atoms and functional groups?

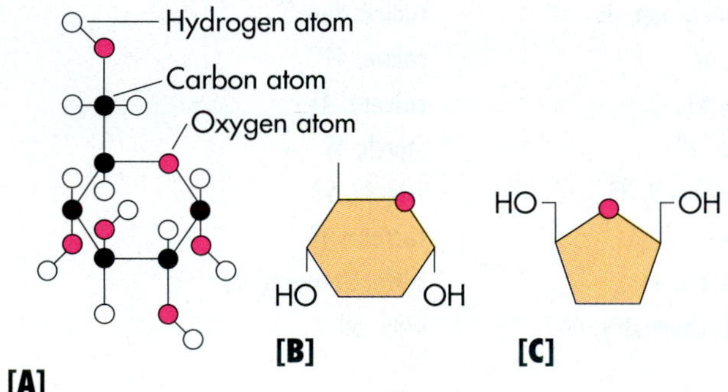

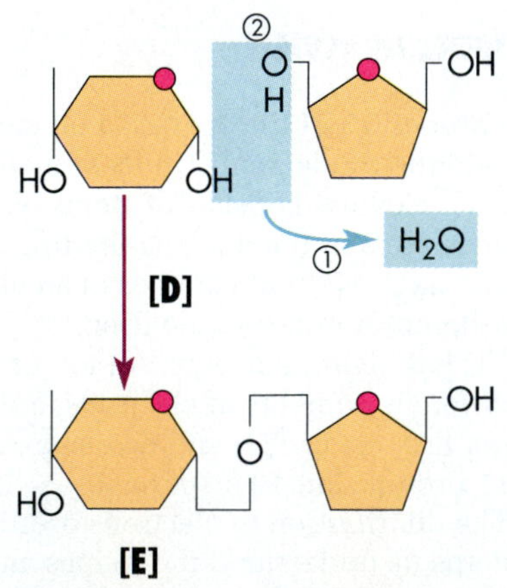

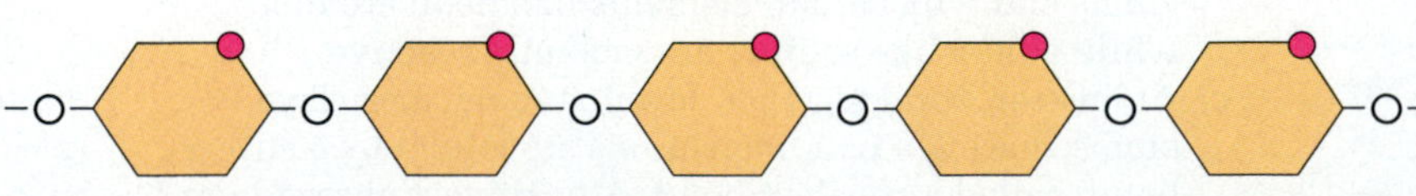

5 Cellulose, starch, and glycogen are all polymers of glucose. How can you account for their different properties?

ANALYSIS Choose the one item in the accompanying list that correctly answers each of the following questions.

1 Why can you conclude that an element's physical properties are determined by the number of protons in its atomic nucleus?

- **a]** The existence of isotopes.
- **b]** The existence of ions.
- **c]** Atoms gain or lose electrons without changing their physical properties.
- **d]** Each kind of element has atoms with a unique number of protons.
- **e]** All above are correct.

2 Golds of different carat weight have some different physical properties. For example, a 24-carat gold ring is shinier and much softer than an otherwise identical ring made from 14- or 10-carat gold. How can you explain this?

- **a]** There must be at least three fundamentally different kinds of gold atoms.
- **b]** There must be at least three different gold elements.
- **c]** At least three of these golds must not be pure elemental gold.
- **d]** None of these rings is made of pure gold.
- **e]** Each kind of ring is made of gold atoms that are joined by different kinds of chemical bonds.

3 When an egg is boiled, the egg white becomes rubbery. Given the fact that egg white consists of the protein albumin and water, which of the following is the most likely explanation for the change?

- **a]** Heat causes weak bonds in the protein to break and stronger ones to form in their place.
- **b]** Heat causes the albumin to change from a globular to a fibrous state.
- **c]** The albumin becomes less fluid because its water is driven off by evaporation.
- **d]** Heat changes the order of amino acids in the albumin, which alters the shape of the protein.
- **e]** Heat-induced changes in the functional groups of albumin lead to the change in consistency.

4 If you constructed a linear protein from 500 amino acid molecules, how many molecules of water would be produced?

- **a]** 500
- **b]** 501
- **c]** 499
- **d]** 498
- **e]** zero

5 Suppose the ice man shown in BOX 2.1 had, in fact, been an unfortunate hiker who met his demise only last year. What would you predict about the relative amounts of ^{12}C and ^{14}C in his remains?

- **a]** The amounts of ^{14}C and ^{12}C would be essentially equal.
- **b]** The ^{14}C concentration would be about the same as that of the air last year or today.
- **c]** There would be significantly more ^{14}C than ^{12}C in the body.
- **d]** The level of ^{14}C would have been significantly lower than that which was found in the actual ice man.
- **e]** There would be essentially no ^{14}C in the hiker.

CHAPTER 3

Cells: The Basic Units of Life

A CHILD, A DISEASE, A CELL DEFECT

About one hundred years ago, two anxious parents brought their little girl to the American neurologist Bernard Sachs. At birth, their baby appeared to be healthy, with normal, pretty features. At about 3 months of age, however, the little girl became listless and her eyes rolled about in a strange way. After another month or so, her muscles became so weak that she could no longer hold her head up or make any voluntary movements. She showed no preference for her mother, and no interest in toys or people's voices. During her first year, she would occasionally follow bright objects with her eyes, but then became totally blind. Her eye doctor could see that a cherry-red spot had developed on each of her retinas. The baby's hearing, however, remained acute, and any sudden sound would startle her. Her response was always the same: she would quickly stiffen, then extend her arms and legs in a mechanical way. Over time, the child's head and skull bones became larger than usual. Eventually, she grew weaker and stopped feeding. She caught pneumonia, and at the age of 2, she died.

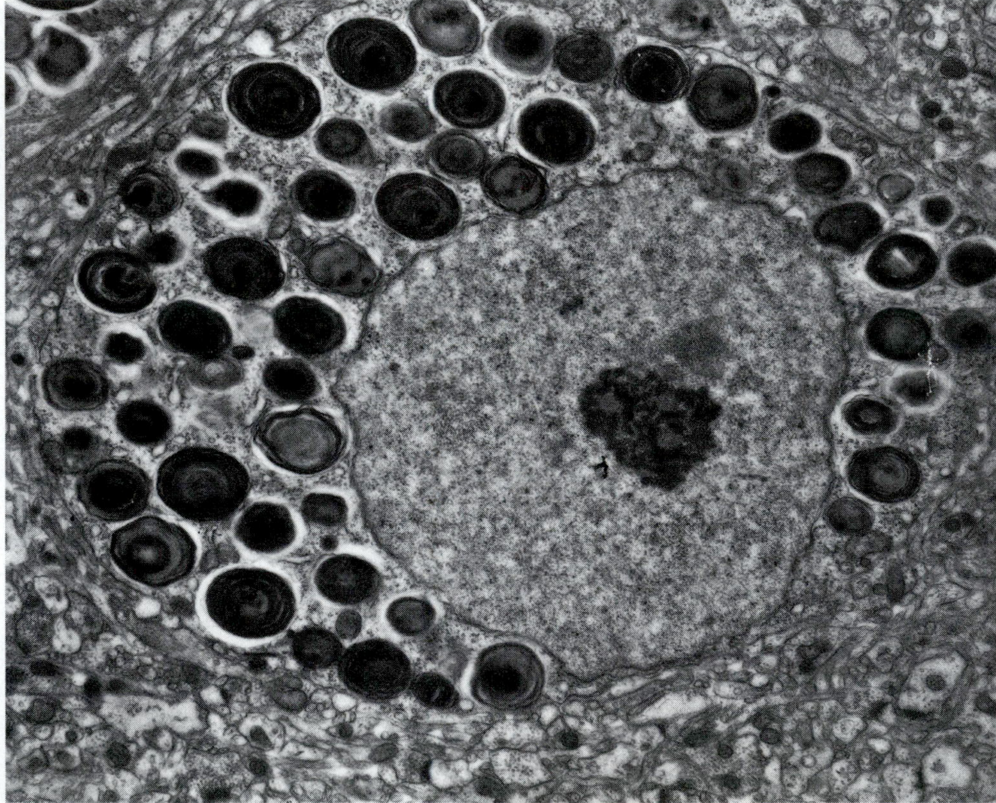

FIGURE 3.1

An Abnormal Brain Cell.

From a child with Tay-Sachs Disease the cell is filled with hundreds of little particles each consisting of whorls of membranes. What might cause such an altered cell?

The neurologist Dr. Sachs encountered several more children with the same physical signs during the next few years, including a sibling of that first little girl and four children in another family. When he consulted the medical literature, Sachs found that an English physician, Warren Tay, had already described this disease. From Tay's evidence and Sachs's observations, it appeared to be an inherited condition. Today, physicians call this Tay-Sachs disease. It is particularly prevalent in Jews from northeastern Poland and in some groups of French Canadians of Quebec. Among these people, up to one baby in 2000 can be born with the disease. Modern genetic studies confirmed that children can inherit Tay-Sachs disease from their parents if both mother and father carry a certain hereditary factor (gene), even though the parents show no symptoms of the disease. Many more studies were then needed to elucidate what kind of genetic defect could give rise to problems in the eyes, muscles, and brain.

Researchers in the 1980s made a key discovery that ultimately led to an understanding of Tay-Sachs disease. They found that the brain cells of Tay-Sachs patients contained abnormal structures and these, in turn, could explain the disease symptoms. It became apparent that an understanding of Tay-Sachs as well as of many other puzzling diseases requires a working knowledge of cells, the basic living units that make up every organism on earth and the main topic in this chapter. Because of their central role in all life processes, we study cells early in our consideration of biological science.

FIGURE 3-1 shows a cell from the brain of a child who died of Tay-Sachs disease. You can see that it is filled with dozens of whorled structures. Such structures are not present in normal cells, and as we investigate the structure and function of healthy cells, we will see why. In fact, you will encounter a string of clues in this chapter concerning the cause of Tay-Sachs disease. To solve the mystery, you can mark these or list them in a notebook. With each new clue, you can try to formulate a hypothesis for what the basic cellular defect may be and how it can cause the symptoms of the disease.

In our investigations of cell biology in this chapter, we will examine prokaryotic cells—simple cells such as bacteria that lack a distinct central nucleus—as well as eukaryotic cells—the basic cell type in plants, animals, fungi, and protists, which do have a true nucleus. Our overview will provide details about the all-important nucleus and the way it directs cellular activities, as well as information on organelles, the tiny structures within the cytoplasm that carry out specific activities.

This chapter will begin with a discussion of how early naturalists discovered cells. Then we will consider the differences between the prokaryotes and eukaryotes in detail. Next we will discuss the features common to many cells. And finally, we will focus on those features specific to specialized cell types.

Along the way, we will encounter the longest and largest cells in the living world; you'll see how penguins can stand on ice without getting frozen feet; and you'll learn about the "suicide bags" in some cells and the whorls in others. ❑

MESSAGES

1 Cells are the fundamental units of life, the smallest units that can independently acquire energy, metabolize, develop, and reproduce.

2 A lipid-containing plasma membrane surrounds all cells and separates the cell's interior from its surroundings. Within the cell is a region packed with DNA, the hereditary material. The space between the plasma membrane and the DNA is filled with cytoplasm, which contains structures that carry out life processes.

3 Eukaryotic cells contain a membrane-bound nucleus, as well as many other membrane-enclosed organelles. Prokaryotic cells lack a nucleus and other membrane-bound organelles.

The Discovery of Cells

We tend to take for granted the fact that modern cell biologists can look at brain cells from a Tay-Sachs patient, compare them with normal human brain cells, and point out structural differences. But studying cell structures, both normal and abnormal, is actually an amazing feat. Most cells are too small to be seen by the naked eye, and throughout most of human history, we had no idea they even existed. The discovery of cells awaited the invention of microscopes just 300 years ago. By 100 years ago, at the time that Dr. Sachs began treating the little girl in our chapter introduction, other scientists had just begun to theorize about what cells do in the human body. The discovery and basic theory of cells are monuments to human inventiveness.

[A] Hooke's microscope

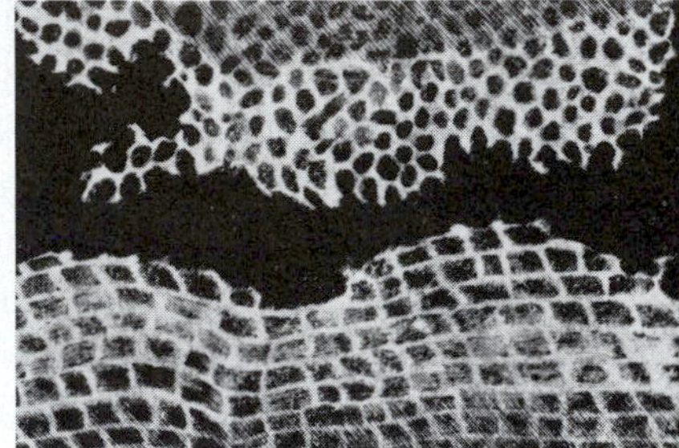

[B] Hooke's drawing of cork

FIGURE 3.2
Discovery of Cells.
[A] Robert Hooke, a seventeenth-century British scientist, used a simple microscope to explore common objects in a new way. **[B]** Hooke observed and drew a slice of cork, with its small chambers. These subunits of a formerly living thing he called "cells"; they were the first cells to be discovered.

THE PIONEERS: HOOKE AND LEEUWENHOEK

About 1663, Robert Hooke, one of England's greatest scientists, directed an instrument-maker to construct a microscope based on imported instruments Hooke had seen and used [FIGURE 3.2A]. Hooke began to focus his new instrument on everyday objects—the point of a pin, the edge of a razor, the surface of a nettle leaf, the body of a flea—and was astonished by the fine detail he could see.

When Hooke looked at a thin slice of cork through his microscope, he saw what he called "pores," or "cells," which reminded him of the small rooms inhabited by monks [FIGURE 3.2B]. We know today that Hooke was not seeing living cells, but only their remains—specifically, the tough cell walls that surround each plant cell.

Anton van Leeuwenhoek, a Dutch cloth merchant, was a contemporary of Hooke. Working in Delft, Holland, in the mid-1600s, he had far greater success at seeing living cells in action, despite his smaller, simpler, handheld microscopes. Considered the greatest early microscopist, Leeuwenhoek was unsurpassed in his day at grinding lenses and slicing and mounting specimens for viewing. Some of his instruments, which were about 5 centimeters (2 inches) long, had lenses no bigger than a grain of sand, but could magnify objects up to 500 times. When some of his original specimens were recently found at the Royal Society of London, scientists realized that the 300-year-old ultrathin slices he had made of cork, feathers, the pith of an elder bush, and the optic nerve of a cow were admirable specimens, even when viewed through a twentieth-century microscope. (Modern microscopes are described in BOX 3.1 on page 68).

WHAT IS A CELL?

Hooke was apparently the first person to see cells, but he could not fully define what he was observing. Modern biologists know that a cell is the smallest entity completely surrounded by a membrane and capable of reproducing itself. It is also the smallest unit displaying all the properties of life listed in CHAPTER 1, including the orderly chemical activities of metabolism, the capacity of self-propelled motion, the ability to reproduce and develop, and the potential to evolve over many generations. As Hooke and Leeuwenhoek discovered, cells often have the overall shape of a small box or ball, and closer examination shows that cells have three basic components:

1. A *plasma membrane*, a surface envelope of lipid and protein that controls the passage of materials into and out of the cell.

2. A central *nuclear region* that controls all the cell's functions and stores the DNA, repository of the cell's hereditary information.

3. The *cytoplasm*, a gel-like substance filling the cell between the plasma membrane and the DNA-storing nuclear region. Suspended in the cytoplasm of many cells are small uniquely structured bodies, or organelles, that carry out specialized functions, including energy harvest and protein synthesis.

An understanding of the Tay-Sachs disease we discussed in the introduction requires a knowledge of how all three cell parts function. We need to understand the nuclear region, because Tay-Sachs is an inherited disease; the cytoplasm, because the whorls visible in FIGURE 3.1 accumulate in that zone; and the plasma membrane, because it gives rise to the cytoplasmic whorls, as we will discover later in the chapter.

WHY ARE CELLS SO SMALL?

Why weren't cells discovered before the time of Hooke and Leeuwenhoek? The answer is that cells are so small people cannot see them with the unaided eye. Their discovery therefore awaited the invention of microscopes. The average cell in your body is just one-fifth the thickness of the paper in this book. Why are cells that tiny? The answer involves an important biological concept: the relationship between a cell's surface area and the volume of its active, interior cytoplasm. Biologists call this its **surface-to-volume ratio**.

A cell's active cytoplasm uses materials taken up from the cell's surroundings, and in turn, produces wastes. In general, the greater the volume of active cytoplasm, the more materials will be used and the more wastes will be produced. A cell exchanges materials and wastes with the outside environment across the plasma membrane—a boundary, gated wall, and raincoat all in one. The greater the surface area of this plasma membrane, the more rapidly substances can be exchanged with the environment. When a cell increases in size, its volume increases more rapidly than its surface area [FIGURE 3.3]. As a result, the cell's use of materials and production of wastes increases more rapidly than its ability to exchange these items with the surroundings. Thus, larger cells would be unable to take in needed materials and export wastes quickly enough to survive. An analogy can be seen in a pile of wet laundry, which, if left in a heap, takes a long time to dry because its surface area is small compared with its volume. But with the items of clothing separated and hung on a line, the exposed surface area is large, while the volume is unchanged; under these circumstances, water evaporates quickly, and the laundry dries in a short time.

In the same way, a cell that was very large and had no special modifications would exchange materials with its environment too slowly to survive, whereas many small cells can have the same total volume as one large one but still enough surface area for the rapid exchanges that sustain life. Thus a cell in an elephant's liver is the same size as a cell in a mouse's liver; the elephant's liver simply has many more cells than the mouse's.

One way around this surface-to-volume constraint on size is through altered cell shape or contents. A long, thin cell, such as a nerve cell that reaches from a giraffe's spine down to its hoof, can have the same volume as a round or cube-shaped cell, but a greatly expanded surface area. An egg yolk survives even though it is large and roundish because the active cytoplasm is flattened into a thin sheet just below the outer membrane. In contrast, the immense bulk of cytoplasm in the yolk interior is quite inactive metabolically; the yolk consists mainly of storage granules of lipid and protein. Likewise, each spindle-shaped sack of juice inside an orange or grapefruit section is a single cell with a very thin layer of cytoplasm [FIGURE 3-4].

The surface area of a cell can also be expanded with slender, fingerlike extensions of the cell's outer membrane. FIGURE 3.5 shows the kinds of cells that line the human small intestine. The numerous extensions, called *microvilli*, greatly expand the cell's surface area, allowing it to absorb nutrients with speed and efficiency.

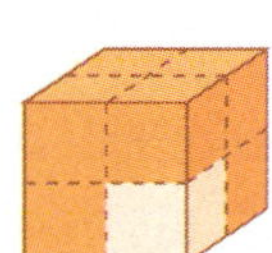
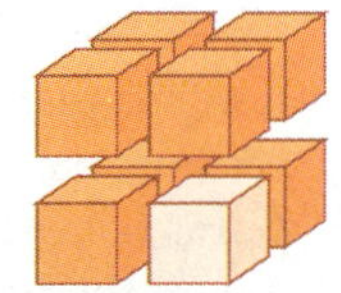
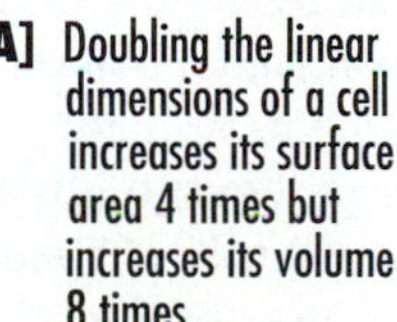

[A] Doubling the linear dimensions of a cell increases its surface area 4 times but increases its volume 8 times.

[B] A large cube made up of small cubes maintains the original surface-to-volume ratio.

FIGURE 3.3

Why Are Cells So Small? Surface-to-Volume Ratios.

The life of a cell depends on the exchange of materials across its surface. The greater the cell volume, the more surface area required. Most cells are microscopic, and their surface-to-volume ratios are favorable. **[A]** Doubling the linear dimensions of a cube increases its volume eight times while increasing its surface area only four times. **[B]** A large cube made up of many small cubes still has the same surface-to-volume ratio as each individual cube. That's why a large organism can survive. It has more cells than a small organism, but the cells are roughly the same size.

THE CELL THEORY

Because of the surface-to-volume constraint we just discussed, then, cells are small. One of the consequences of this smallness is the necessity for large organisms such as people and oak trees to be made up of trillions of cells. About 150 years ago, biologists were trying to understand how cells could be so small and numerous and yet function in such coordinated ways. Their goal was to combine the facts they had learned about cells up to that time into a comprehensive understanding of the cell's signifi-

box 3.1
Biology Applied

Microscopes: Tools for Studying Cells

How do we know that eukaryotic cells—which are usually far too small to see with the unaided eye—have a membrane-bound nucleus, chloroplasts, or any of the other organelles discussed in this chapter? Much of what we know about cells comes from biologists peering into microscopes of three basic types and witnessing the beauty inside and on the surfaces of the cells.

Light microscopes are instruments containing optical lenses that *refract*, or bend, light rays so that an object appears larger than it really is [FIGURE 1A]. Because it can illuminate and magnify, biologists use the light microscope extensively to locate cells in tissues, to observe the behavior of living cells [FIGURE 1B], and to detect cell organelles that can be stained bright colors, such as the nucleus, chloroplasts, and mitochondria [FIGURE 1C].

If a specimen is thin enough for light to pass through, a light microscope can magnify its physical details to more than 2000 times normal size. Colored stains can heighten the contrast between various structures in the specimen, making them easier to distinguish. Stains are usually used in combination with fixatives—agents that preserve the cells or tissues so that they remain unchanged. However, stains and fixatives kill cells, so to view living cells, biologists use differential interference, contrast, or *Nomarski* light microscopes that augment the differences in light refraction between unstained structures so that the contrast between them is bright even without stains [FIGURE 1D].

In the light microscope, magnification is limited by *resolving power*, the ability of the human eye to distinguish adjacent objects as distinct and separate. Two dots brought closer and closer together will eventually be perceived as merging at a distance of about 0.1 millimeter; that is the lower limit of the

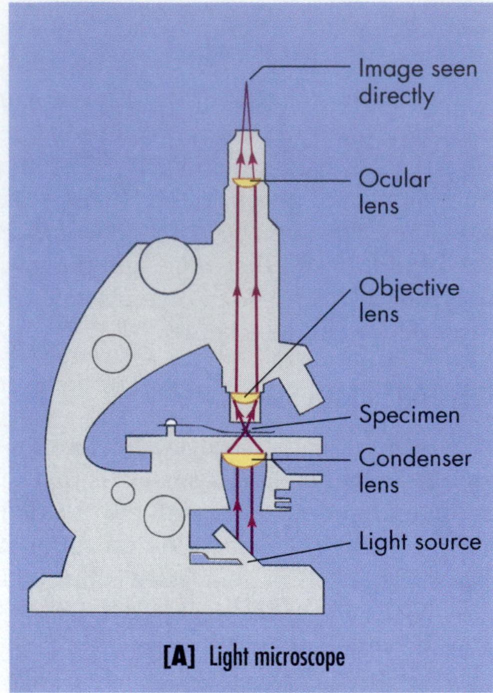

[A] Light microscope

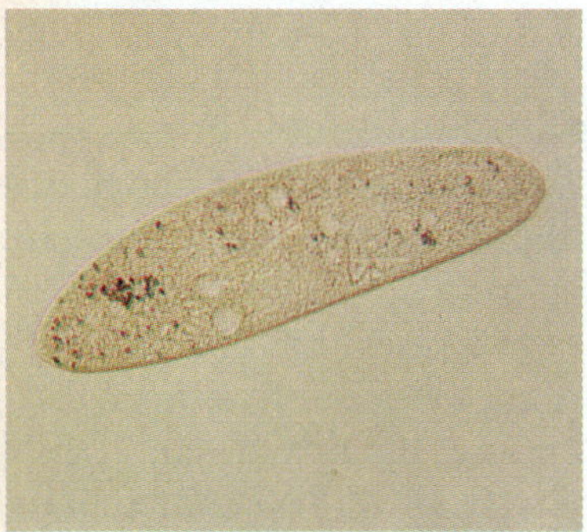

[B] Bright field micrograph of living *Paramecium* cell.

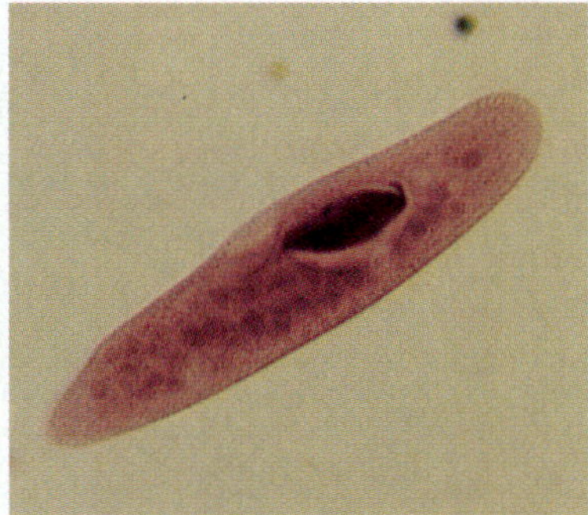

[C] Bright field micrograph of *Paramecium*, stained.

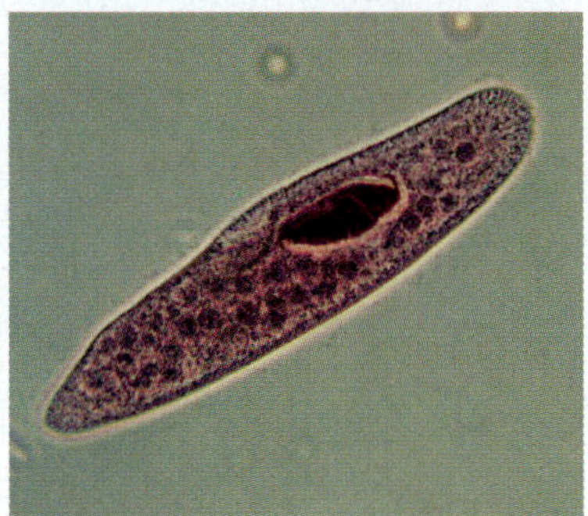

[D] Differential interference contrast micrograph of *Paramecium*.

---

cance to life. Their efforts resulted in the **cell theory**. According to modern cell theory:

1. All living things are made up of one or more cells.

2. Cells are the basic living units within organisms, and the chemical reactions of life take place within cells.

3. All cells arise from preexisting cells.

Let's explore briefly the importance of each tenet. First, there are no living organisms that do not consist of cells. Many organisms, such as bacteria, consist of a single cell, but most living things, from humans and trees to moths and mushrooms, are made up of several hundred to several trillion cells. Viruses, which cause diseases like flu and AIDS (acquired immune deficiency syndrome) are not cells, and most biologists would agree that eye's resolving power. (An analogy is the way the headlights of an approaching car appear as a single spot when the car is far away, but as two individual lights when the vehicle gets closer.) With a good light microscope, resolving power can be extended some 500 times to about 0.2 μm, or 200 nm. But this in itself represents a lower limit determined by the length of the electromagnetic waves in visible light.Red light has wavelengths of about 750 nm, and violet light has wavelengths of about 400 nm. Objects with a diameter of less than 200 nm, or half the shortest wavelength of visible light, are no longer visible with the light microscope.

It was the desire to see finer details in cells that led to the invention of the electron microscope. **Electron microscopes (EMs)** use beams of electrons with wavelengths 100,000 times shorter than visible light to view objects, thus magnifying them that much more. These microscopes also use magnets rather than ground glass lenses to focus the electron beams, and the beams must be generated in a vacuum. To use a **transmission electron microscope (TEM),** the microscopist must first cut an ultrathin slice of a specimen with a piece of broken glass or a diamond knife, then "stain" the specimen with heavy metal such as lead or uranium to increase contrast between the inner structures. Some electrons are scattered or absorbed by structures in the specimen while others pass through and strike a small fluorescent screen [FIGURE 1E]. The result is a greatly enlarged image of the object, which can be photographed and studied for its fine detail [FIGURE 1F].

A second type of electron microscope, the **scanning electron microscope (SEM),** has a greater depth of field and thus allows scientists to see a specimen's outer surface in three dimensions. The specimen is coated with a thin layer of gold or other metal, and electrons are then aimed to scan it rapidly [FIGURE 1G]. The surface atoms become excited and emit secondary electrons, which are detected by a screen to reveal the object's surface characteristics [FIGURE 1H].

Electron microscopes opened a new and marvelous realm to biologists and have provided much of our current knowledge of cell structure. TEMs have revealed a level of complexity and detail in the cell never suspected until the middle of the twentieth century. And SEMs have shown us a beautiful and sometimes bizarre vision of the intricate shapes of living cells and organisms.

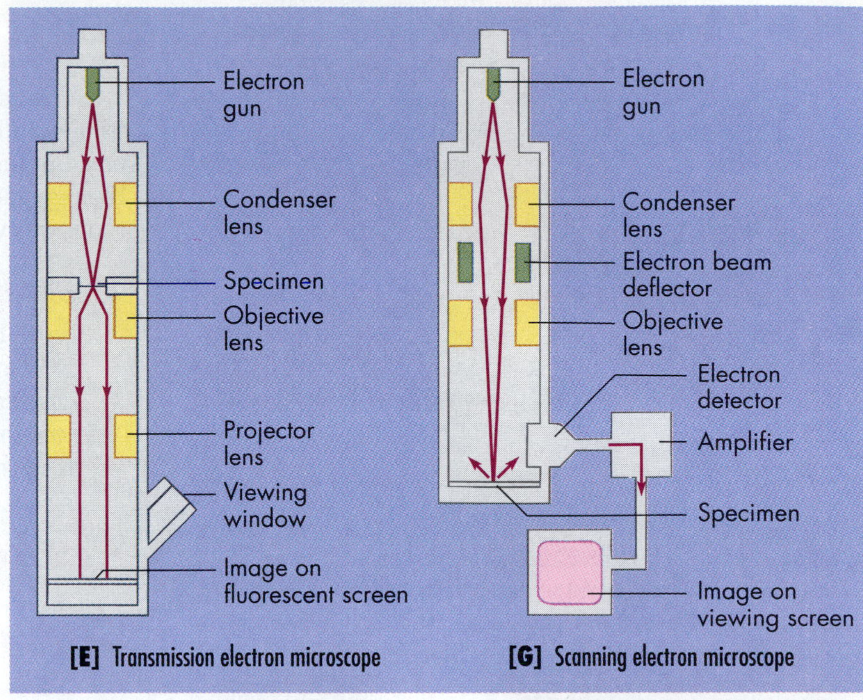

[E] Transmission electron microscope

[G] Scanning electron microscope

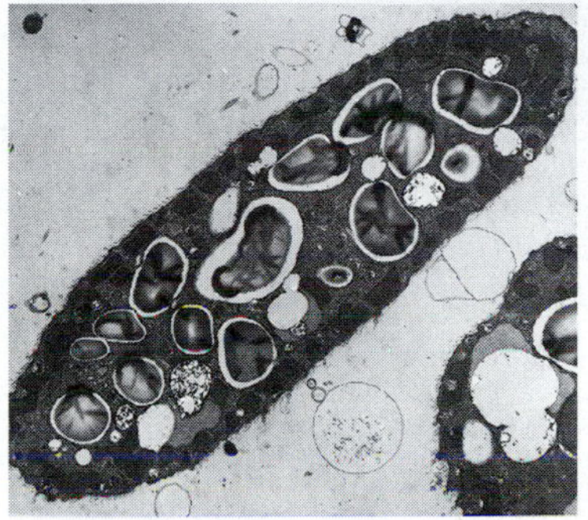
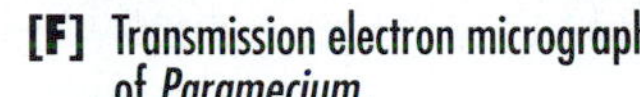

[F] Transmission electron micrograph of *Paramecium*.

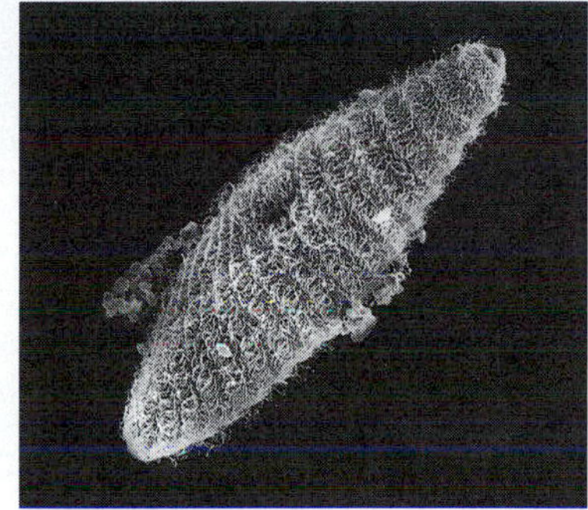

[H] Scanning electron micrograph of *Paramecium*.

they are not alive because they do not carry on metabolism on their own and because they can reproduce only within cells.

Second, biologists concluded that cells are the basic units of life because the individual components of cells themselves lack the properties of life. Once a cell biologist separates out a cell's nuclear region or various components of its cytoplasm, these structures can no longer carry out life functions or replicate.

Thirdly, new cells that arise today are always the products of the division of an existing cell. Each cell in your body can be traced back to a single fertilized egg cell generated when your mother's egg cell fused with your father's sperm cell. These sperm and egg, in turn, came from other cells in your parents' bodies; each of your parents arose from a single fertilized egg cell produced by your grandparents, and so on back in time.

FIGURE 3.4

A Familiar Cell Visible to the Human Eye.

While we can see very few cells without a microscope, the cells that store orange juice in an orange segment are exceptions. The cytoplasm of each such cell is a thin sheet lying next to the plasma membrane. More than 95 percent of the cell's contents is a single vacuole filled with juice.

The cell theory, like all good biological theories, has tremendous power to explain life's observable phenomena, but there is much more to learn about the cell's basic biology, including the characteristics of the two main cell types in living organisms.

CONCEPT CHALLENGE

The tiny organism called *Physarium* lives on rotting plant material. Each cell is surrounded by a single plasma membrane, but unlike most eukaryotic cells, it contains many nuclei within its cytoplasm. Does *Physarium* support the cell theory and help prove it, or does it contradict that theory and prove it incorrect?

Prokaryotic and Eukaryotic Cells

In the three centuries since Hooke and Leeuwenhoek first saw cells, biologists have examined cells from thousands of different kinds of organisms.

They have found that regardless of which living thing they study, the component cells are one of two basic types, prokaryotic or eukaryotic [see TABLE 3.1].

The feature that most readily distinguishes these two types of cells is the presence or absence of a cell nucleus. Sitting prominently within eukaryotic cells is a roughly spherical, membrane-enclosed body, the nucleus, which contains DNA, the cell's hereditary material. In contrast, the DNA within a prokaryotic cell is neither surrounded by a membrane nor separated from the rest of the cell's contents [FIGURE 3.6]. Eukaryotic cells with their prominent nuclei make up all the large organisms with which we are most familiar—the members of the animal, plant, and fungus kingdoms, as well as the protists [FIGURE 3.7]. Each member of the remaining two kingdoms, the eubacteria and archaebacteria, is made up of a single prokaryotic cell lacking a cell nucleus. Let's examine these two cell types more closely now, beginning with the simpler prokaryotic cells.

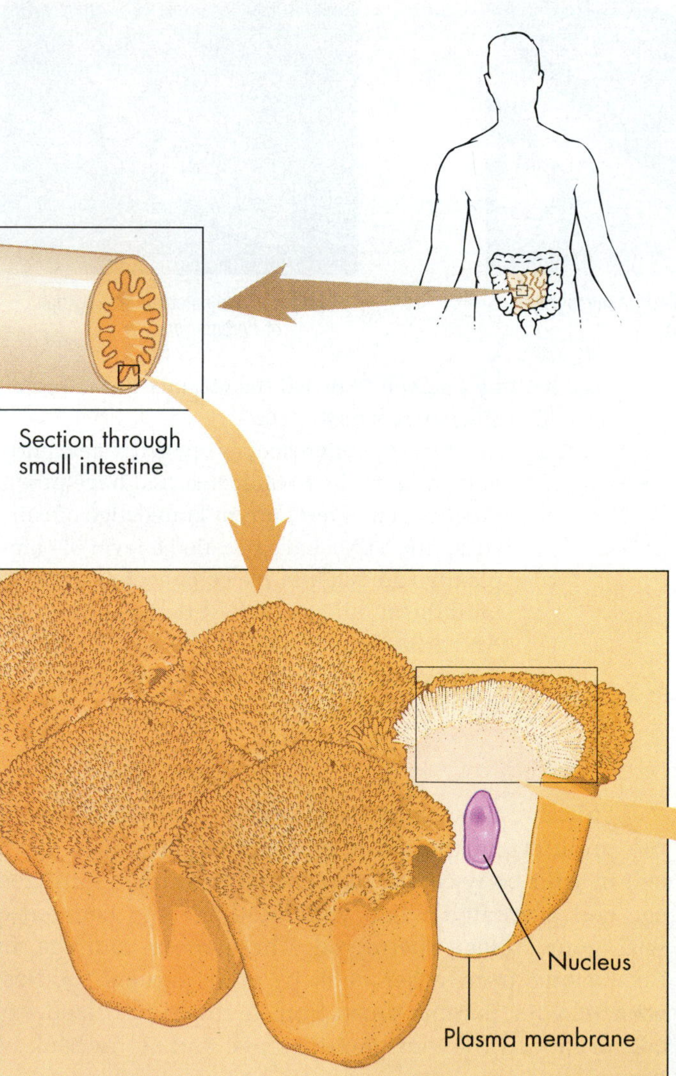

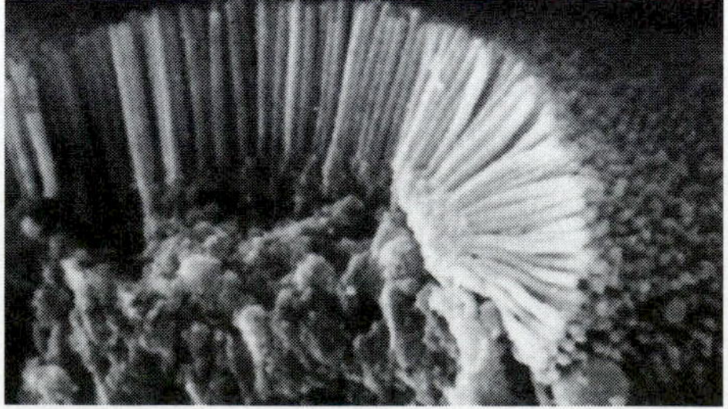

FIGURE 3.5

Microvilli: One Way to Expand Cell Surface.

High-powered magnification of the surface of intestinal cells reveals microscopic, fingerlike projections called microvilli. These expand the surface area, allowing the cells to absorb nutrients quickly.

TABLE 3.1 Components of Prokaryotic, Plant, and Animal Cells

Structure	Function	Present in Prokaryotic Cells	Present in Eukaryotic Cells	
			Plant Cells	Animal Cells
Cell surface				
Cell wall	Protects cell; maintains cell shape	✔	✔	
Extracellular matrix	Surrounds and protects cell			✔
Plasma membrane	Protection; communication; regulates passage of materials	✔	✔	✔
Flagella	Cell movement; only found in a few cells in each group	✔	✔	✔
Cilia	Cell movement; present only in certain cells			✔
Links between cells				
Pili	Mating	✔		
Junctions	Cell-to-cell communication; prevent fluid leakage; strengthen tissues		✔	✔
Genetic machinery				
DNA	Contains genetic information	✔	✔	✔
A single circular chromosome	Contains genes that govern cell structure and activities	✔		
Several linear chromosomes	Contain genes that govern cell structure and activity		✔	✔
Nuclear envelope	Surrounds genetic material		✔	✔
Cytoplasm	Gel-like interior of cell	✔	✔	✔
Ribosomes	Manufacture proteins	✔	✔	✔
Cytoskeleton	Aids in support and movement in cell		✔	✔
Endoplasmic reticulum	Intracellular transport		✔	✔
Golgi apparatus	Packages materials		✔	✔
Lysosomes	Contain enzymes; aid in cell digestion; play a role in programmed cell death		✔	✔
Microbodies	Storage		✔	✔
Mitochondria	Provide cellular energy		✔	✔
Chloroplasts	Capture sunlight; produce energy for cell		✔	
Central vacuole	Maintains cell shape; stores materials and water		✔	

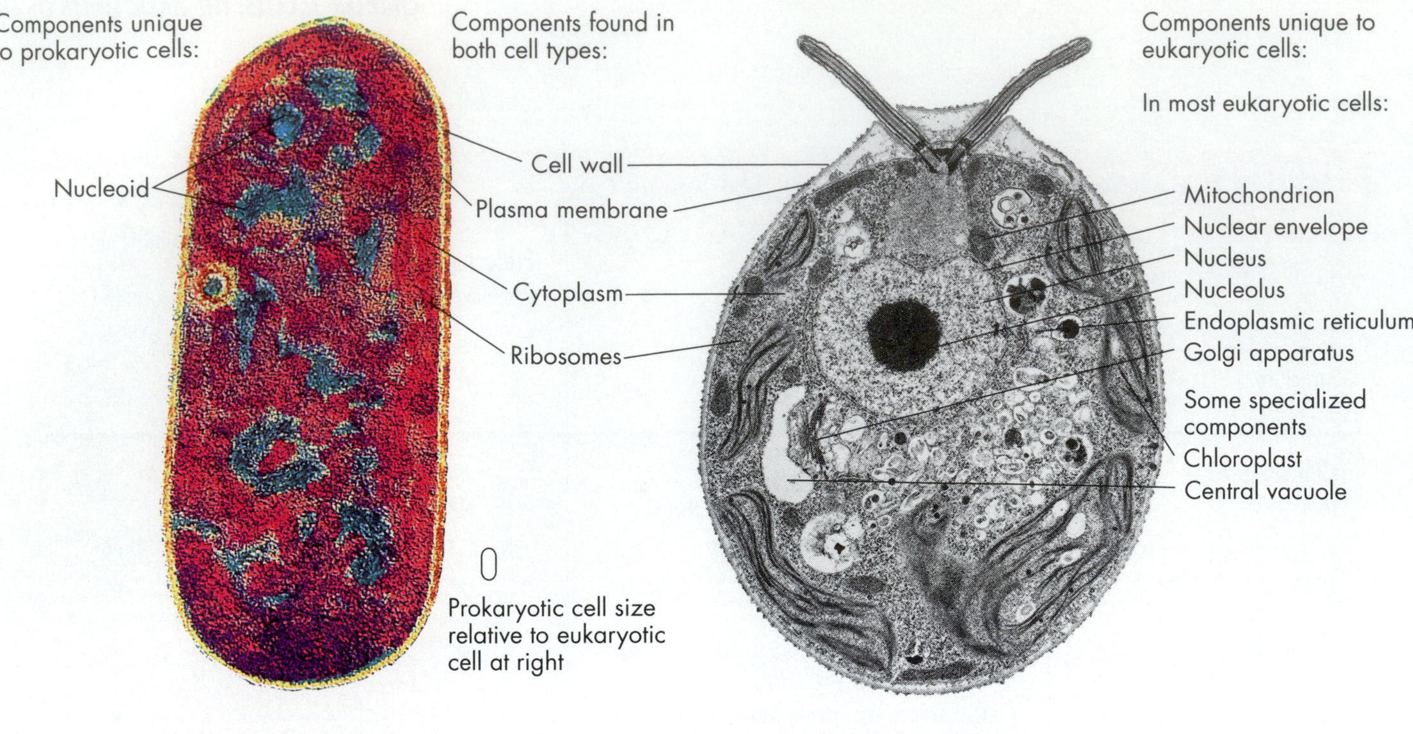

FIGURE 3.6

Absence or Presence of Nucleus Defines Cell Type.

[A] A prokaryotic cell, such as this bacterium (*Clostridium perfringens*, the agent that causes gas gangrene when a wound becomes infected, has a cell wall and a region of the cytoplasm filled with hereditary material. [B] A eukaryotic cell, such as this cell of the green alga *Chlamydomonas reinhardtii*, an inhabitant of ponds, is generally larger and more complex than a prokaryotic cell. The hereditary material of the eukaryotic cell is separated from the cytoplasm by an envelope, and the cell contains many other organelles that carry on specific functions. Eukaryotic cells have a nucleus; prokaryotic cells don't.

PROKARYOTIC CELLS

Each bacterium classified as a member of the kingdom Eubacteria or the kingdom Archaebacteria consists of a small, simple cell called a **prokaryotic cell** (pro-care-ee-OTT-ik; from *pro-*, "before," and the Greek *karyo,* "kernal," or nucleus) [FIGURE 3.6A]. An organism made up of a prokaryotic cell is called a **prokaryote**. As we have seen, the major distinguishing feature of a prokaryote is the lack of a nucleus. Although a prokaryote's DNA is not separated from the cell contents by a membrane, it is generally concentrated toward the center of the cell in a region called the **nucleoid**.

Prokaryotes are the smallest cells, and some bacteria are no more than 0.2 micrometer* in length. For comparison, this page is about 100 μm thick. Other prokaryotic cells, for example, a typical bacterium like *Escherichia coli* (normally a harmless inhabitant of the human intestinal tract), have a volume about 10 times greater than the smallest prokaryote and are about 2 μm long. Hun-

* μm (one-millionth of a meter).

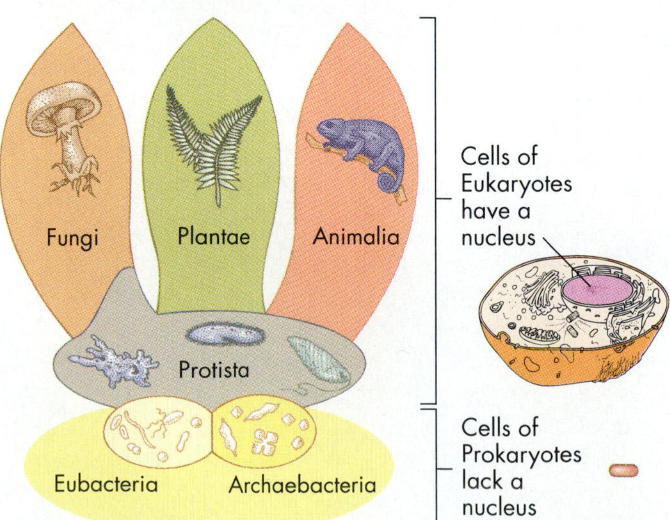

FIGURE 3.7

Prokaryotic and Eukaryotic Cells, and the Kingdoms of Life.

Eukaryotic cells are usually 10 times bigger than the simpler prokaryotic cells, and make up organisms in the fungal, plant, animal, and protistan kingdoms. Prokaryotic cells lack a nucleus and make up the eubacterial and archaebacterial kingdoms.

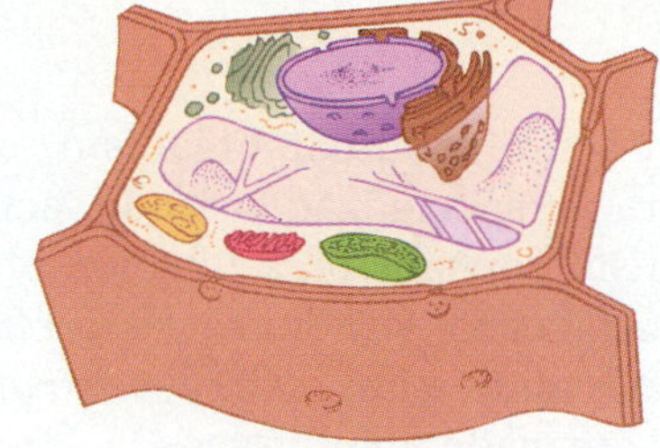

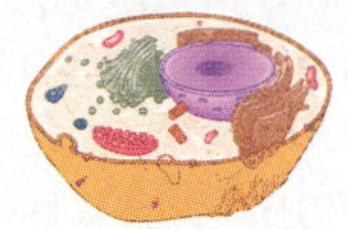

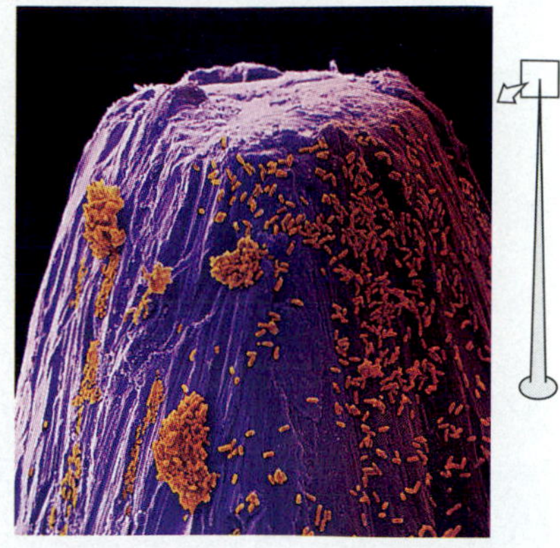

FIGURE 3.8

Cell Size Can Differ Widely.

If a typical bacterium such as *E. coli* were the size of the drawing of the animal cell in this figure, then the animal and plant cells would be considerably larger: In the scanning electron micrograph, *E. coli* cells are dwarfed by the point of a pin.

dreds of *E. coli* cells can fit on the tip of a sharp pin [FIGURE 3.8]. Prokaryotes exploit environments inaccessible to eukaryotes, with their more complex cells. Some Archaebacteria, for example, thrive in the boiling waters of Yellowstone's hot springs, while bright orange eubacteria grow in a slimy mat at the cooler but still steamy edges of the spring [FIGURE 3.9].

FIGURE 3.9

Environmental Versatility.

Prokaryotic cells can utilize a much broader spectrum of energy sources and environments than can eukaryotic cells. Archaebacteria thrive in the boiling waters of Yellowstone National Park's Grand Prismatic Spring, while photosynthetic eubacterial cells grow in a slimy orange mat at the spring's cooler edges.

EUKARYOTIC CELLS

The cells of most familiar organisms are **eukaryotic** (yoo-kare-ee-OTT-ik; "true kernel," or nucleus). Eukaryotic cells contain a large, membrane-bound nucleus [FIGURE 3.6B], which directs the cell's activities and contains DNA, the molecular repository of the cell's hereditary information. A common trait of eukaryotic cells is the separation of the cell into many compartments. This separation is generally accomplished by the cell organelles (important structures with specialized functions), which concentrate the cell machinery needed for each task and isolate potentially interfering reactions from each other. Organisms made up of eukaryotic cells are called **eukaryotes**. All plants, animals, fungi, and protists are eukaryotes.

Eukaryotic cells are generally much larger than prokaryotic cells. An average-sized animal cell is about 20 μm in length, and as we saw earlier, about five animal cells could be lined up across the thickness of a sheet of paper. A typical plant cell (also a eukaryote) is a bit larger at about 35 μm across. A typical bacterial cell, a prokaryote, is about 2 μm long or one-tenth the length of the typical eukaryotic cell.

Levels of Organization in Eukaryotes While prokaryotes are basically unicellular (one-celled), eukaryotes may be unicellular, colonial, or multicellular (many-celled).

A eukaryotic **colonial organism** consists of a group of cells that are capable of living independently as single-celled organisms, but that tend to live together in clusters, with certain cells specialized for reproduction. *Volvox*, a relative of the protist *Euglena*, is an example of a colonial form [FIGURES 3.10 and 3.11]. The evolution of cellular cooperation and coexistence probably began in colonial organisms, but reached its zenith in multicellular organisms.

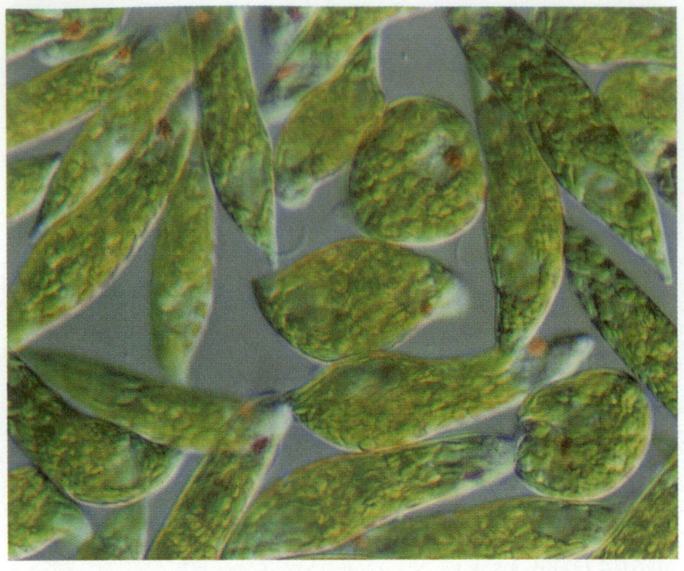

FIGURE 3.10

Complex Eukaryotic Cells.

These living *Euglena* cells, common inhabitants of pond water, contain many specialized internal organelles; for example the green chloroplasts collect energy, the orange spot detects the direction of light, and the nucleus directs cell activities.

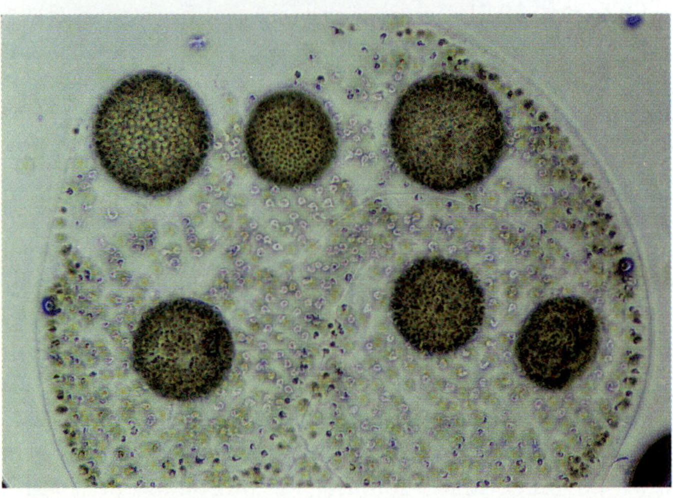

FIGURE 3.11

Colonial Organisms: Simple Division of Labor.

Volvox is a type of simple green alga that forms jewel-like spherical colonies. Each colony has dozens of regular colony cells as well as a few larger reproductive cells. The reproductive cells are capable of generating new cells within the colony or leaving to start a new colony elsewhere. Each small green cluster is a single cell that looks much like the *Euglena* in FIGURE 3.10, and is suspended in a jellylike mass.

Within a multicellular organism, eukaryotic cells usually perform specialized tasks. In fact, each multicellular organism depends on a division of labor in which groups of cells carry out their specific functions, generally in tissues and organs. (Tissues are made up of cells, and organs are made up of tissues.)

The following sections of the chapter concentrate on the specific structures and functions of eukaryotic cells, starting at the cell's outer boundary, the plasma membrane, and then moving inward through the cytoplasm to examine the nucleus and the many other organelles within the cytoplasm. In an effort to simplify the structures of cells, FIGURE 3.12 shows generalized animal and plant cells, with their internal organelles, and TABLE 3.1 describes the many cell components and their functions.

CONCEPT CHALLENGE

Let's say that you are a family physician, and you have noticed that during the last few days many more patients have complained of stomach cramps and intestinal "flu" than you would expect for this time of year. You take samples from these patients and isolate a single-celled organism that has a nucleus, but few other membrane-bound bodies within its cytoplasm. To treat these patients, will you prescribe drugs effective against prokaryotic or eukaryotic organisms? Defend your decision to one of your patients.

The Plasma Membrane

A cell's contents are separated from the surrounding environment by a flexible sheet of fatty material called the **plasma membrane** [FIGURE 3.13]. This outer boundary keeps most of the unnecessary or harmful materials out of the cell, while keeping most of the useful substances within. The plasma membrane is not, however, a tight seal around the cell. On the contrary, it regulates a constant flow of materials into and out of the cell, allowing water, ions, and certain organic molecules to pass through and enter the cell, while allowing toxic or useless by-products of cellular metabolism to exit the cell.

In a sense, the plasma membrane is like a gate and gatekeeper combined. In addition to its barrier activities, the plasma membrane also receives and generates signals that are an important part of the communication and coordination between cells.

THE FLUID-MOSAIC MODEL OF MEMBRANE STRUCTURE

The plasma membrane is only 0.1 μm thick, about one-thousandth the thickness of a sheet of paper. FIGURE 2.33

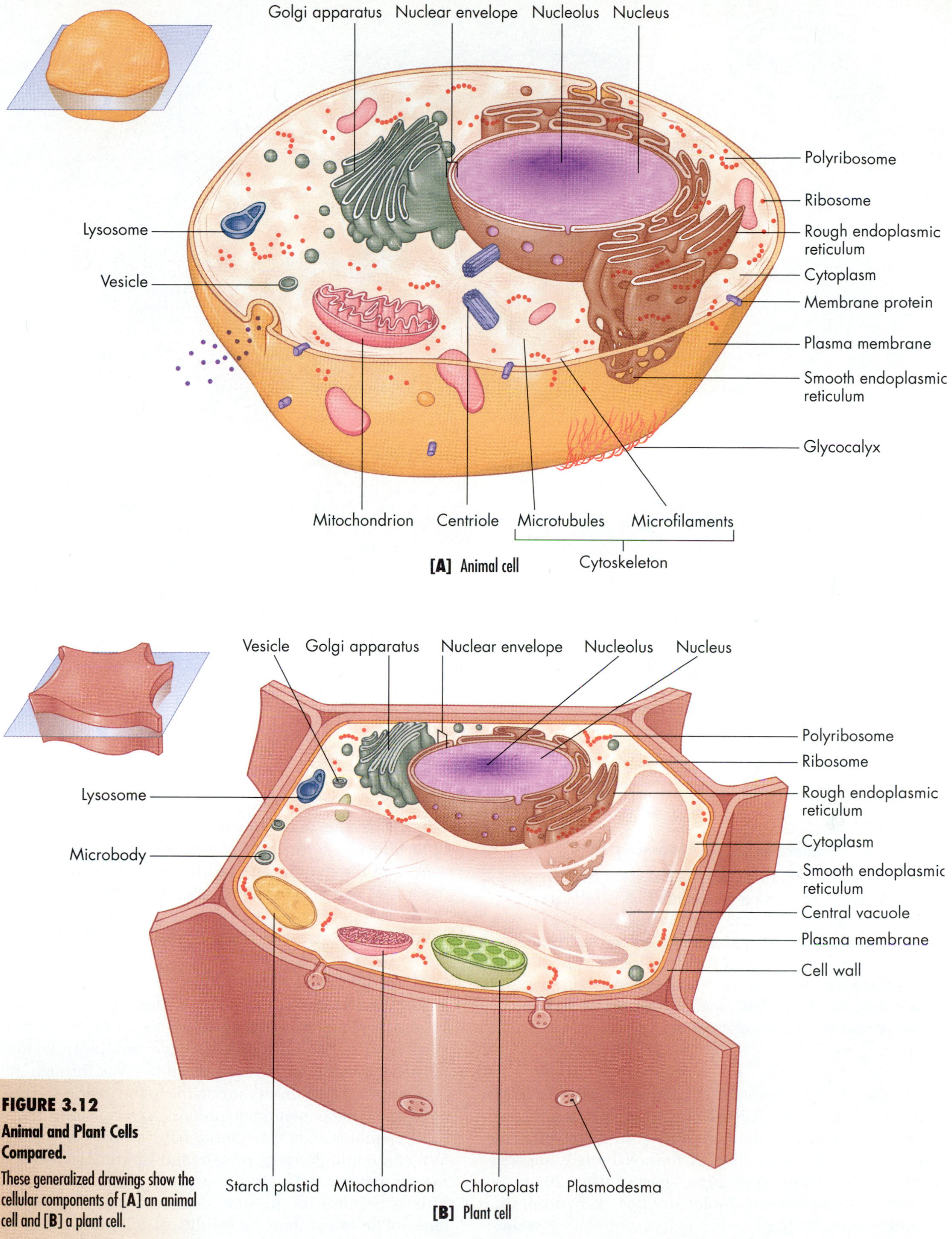

[A] Animal cell

[B] Plant cell

FIGURE 3.12

Animal and Plant Cells Compared.

These generalized drawings show the cellular components of **[A]** an animal cell and **[B]** a plant cell.

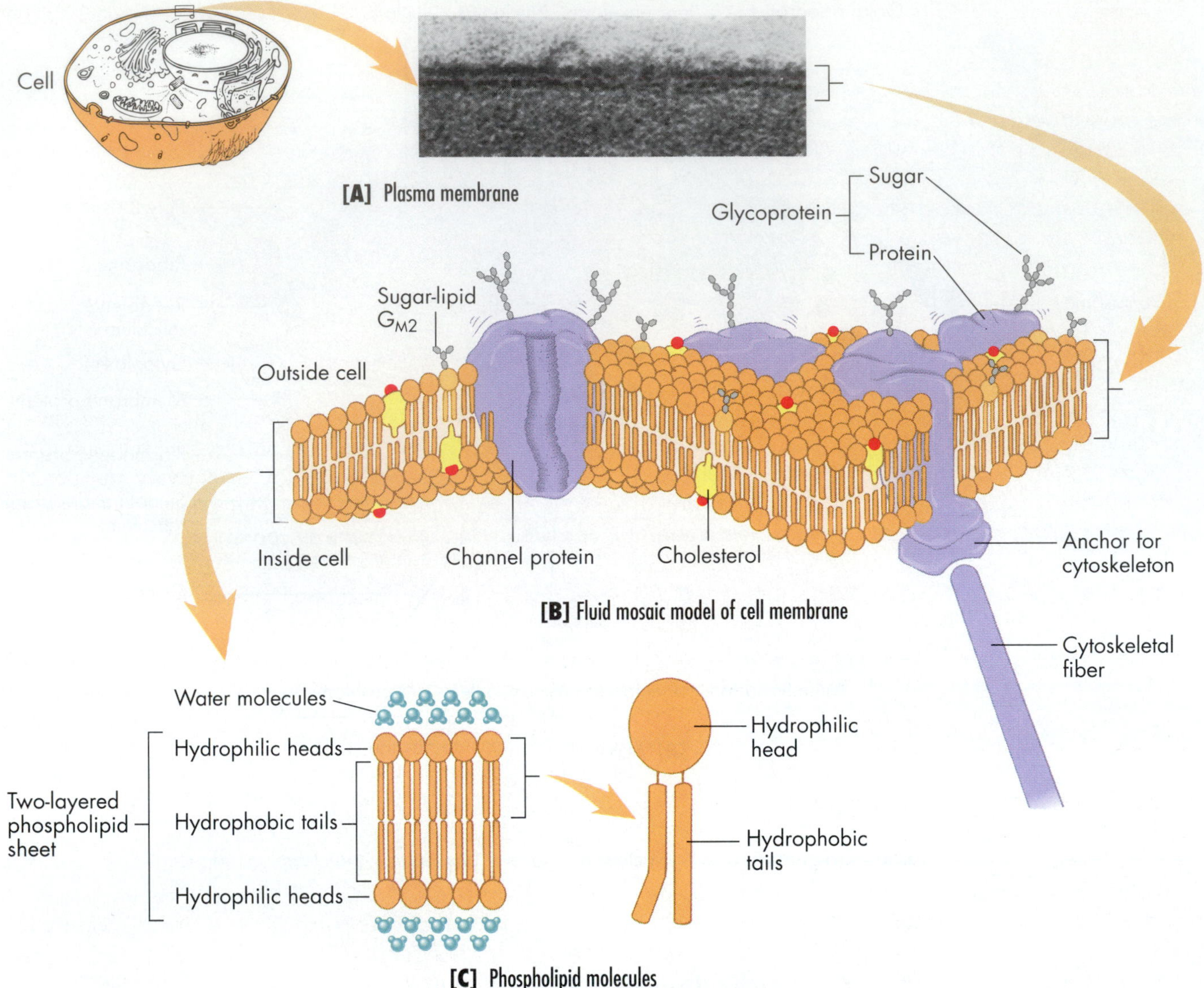

FIGURE 3.13

The Plasma Membrane: The Cell's Outer Boundary.

[A] The plasma membrane is a lipid bilayer, two tiers of phospholipid molecules with hydrophilic heads on the membrane's surface, and hydrophobic tails oriented inward toward the middle of the membrane. [B] Fluid-mosaic model. The plasma membrane is pictured as a fluid plane with floating "icebergs"—proteins that can extend through the membrane and project from either or both sides. *Glycoproteins* act as cellular labels and as receptors for incoming chemical signals. Some membrane proteins attach to cytoskeletal fibers, as shown here. [C] Phospholipid molecules.

gives a sense of a membrane's thickness relative to other biological molecules. The membrane is both flexible and dynamic, with about the consistency of motor oil. The bulk of the membrane sheet is formed by fatty organic compounds called *phospholipids* [see FIGURE 2.26].

Each phospholipid molecule has a hydrophilic ("water-loving") head and a hydrophobic ("water-fearing") tail. When phospholipids are surrounded by water, they align in a characteristic two-layered sheet—a **lipid bilayer**—with the hydrophobic tails pointing inward and the hydrophilic heads pointing outward [FIGURE 3.13A]. Water is excluded from the middle. Phospholipids assume this two-tiered configuration because the cytoplasm is watery, and cells are bathed in fluid on the outside.

In addition to their lipid content, plasma membranes contain substantial quantities of proteins, which are large, irregularly shaped bodies [FIGURE 3.13B]. Some proteins are embedded in the lipid bilayer and protrude into the cell's interior, and some pass all the way through the plasma membrane, from the outside to the inside. Cell biologists envision membrane proteins as suspended in the lipid membrane the way empty bottles float on the surface of a pond. Both the protein and lipid molecules are free to move about within the plane of the membrane. This is because the plasma membrane has the consistency of oil rather than the solidity of cold lard.

This depiction of the mobile plasma membrane is called the **fluid-mosaic model** [FIGURE 3.13B], the fluid being the lipid bilayer and the mosaic pieces being the suspended proteins.

The fluid-mosaic model of membrane structure applies not only to the plasma membrane at a cell's outer surface, but also to the internal membranes that wall off specialized compartments inside cells. These internal compartments, or *organelles* ("little organs"), carry out separate functions such as synthesizing various cellular components, harvesting energy from organic molecules in food, and transporting materials within the cell. The components of the organelle membranes may differ from the particular proteins and lipids making up the plasma membrane, but the same fluidity and suspension of proteins characterizes both internal and external membrane systems.

The Fluidity of Lipids Fluidity is important to the way a membrane functions, and cells involved in very different tasks and subjected to widely varying conditions contain differing sets of membrane lipids that keep the outer cell barrier fluid. For example, a penguin or skua may stand around all day on the ice and snow with its bare feet. Nevertheless, the membranes surrounding the cells in the birds' feet remain fluid. Likewise, a flamingo standing in the warm mud of a tropical lake has fluid membranes in the cells of its feet. Each bird has special lipids that remain fluid at the temperatures to which its feet are normally exposed. In a similar way, a motorist will use a motor oil with one viscosity for driving in the hot Arizona desert and an oil with a very different viscosity when traveling across the frigid tundras of Alaska and northern Canada. Membrane fluidity can also be affected by cholesterol molecules inserted into the membranes.

Surface Carbohydrates Membrane proteins and lipids are sometimes festooned with carbohydrate molecules that stick out from the cell surface into the surrounding medium. Certain kinds of carbohydrates project from the plasma membranes of red blood cells, for instance, and the variations between them account for the differences in blood types. People with blood type A, for example, have a different cell-surface carbohydrate than people with blood types B or O. Such sugar-lipid compounds on many cells act as cell-surface receptors that can bind to substances outside the cell and transmit information from the cell's outside environment to the cytoplasm inside the cell.

In the gray matter of the human brain, about 6 percent of the lipid molecules in the cells' plasma membrane are a compound called *ganglioside* G_{M2}. In it, a lipid portion is attached to four simple sugars that project outward from the brain cell [FIGURE 3.13B]. Children with the Tay-Sachs disease we discussed earlier can have up to 300 times the normal amount of G_{M2} in their brain cells. With this in mind, look again at FIGURE 3.1 and see if you can propose a hypothesis for the chemical nature of the cytoplasmic membrane whorls.

CROSSING PLASMA MEMBRANES

The plasma membrane defines the boundaries of a living cell, and it tends to keep the cell contents in and the external environment out. Some substances, however, such as nutrients, must pass into cells, and others, such as waste products, must pass out. Clearly, the plasma membrane is permeable to certain substances, but not all; in other words, the membrane is **selectively permeable**. How do substances pass through plasma membranes, and what accounts for the selectivity?

Fat-soluble and Water-soluble Molecules Very small inorganic molecules such as oxygen, carbon dioxide, and water can pass through plasma membranes quite readily by simple diffusion (the spontaneous movement of a substance from a region where it is highly concentrated to one where it is more sparsely concentrated). Organic molecules pass through in different ways, depending on their solubility in fat or water.

Some organic substances are *fat-soluble*—they can dissolve in fats, and thus they can pass directly through the plasma membrane, with its matrix of lipids (fats). Examples are vitamin E and the phospholipid lecithin (you can find this among the ingredients listed on a candy bar label). Most organic molecules, however, are water-soluble, including nutrients such as sugars and amino acids, and cellular wastes such as urea. A plasma membrane that was pure lipid would impede the passage of these water-soluble materials, and the cell would starve or be poisoned.

Instead, small water-soluble organic compounds pass through the plasma membrane via the proteins floating in the fluid-mosaic membrane. Some of the membrane proteins are very specific, allowing only particular sugars, ions, or amino acids to pass. There is dramatic proof that these proteins are crucial for cellular function: some people are born with a condition called cystic fibrosis, which alters a membrane protein that would normally allow chloride ions to pass through. Without a functional version of this protein, the salt content inside and outside of cells becomes seriously disrupted (see CHAPTER 10 for details). Cystic fibrosis kills most of its victims before they reach the age of 30.

While water and oxygen can simply diffuse through the membrane without any energy expenditure by the cell, the passage of many other substances does require an energy outlay, as the materials are pumped from one side of the membrane to the other, a process called *active transport*. We will discuss the processes of diffusion and active transport in more detail in CHAPTER 4.

Endocytosis While membrane proteins allow certain small molecules to pass through plasma membranes, large molecules generally do not pass directly through. Proteins, or even larger aggregates, such as debris left over from a dying cell, are usually taken into cells by means of a process called **endocytosis** (*endo*, "into," plus *cyto*, "cell"). Endocytosis involves the engulfment of material by the inpocketing of the plasma membrane, with the subsequent formation of a vesicle within the cell [FIGURE 3.14]. The endocytosis of liquid material is called *pinocytosis* ("cell drinking", and the endocytosis of solids is called *phagocytosis* ("cell eating").

Cells normally take in cholesterol via endocytosis, removing it from the bloodstream. One of every 20 people under the age of 60 who suffers a heart attack has an inherited condition in which cells fail to remove sufficient amounts of cholesterol from the blood. The fatty material thus builds up outside the cells, clogs the blood vessels, and leads to heart attacks. Endocytosis plays a role in Tay-Sachs disease, as well. Notice in FIGURE 3.14B that the inpocketing action of endocytosis brings parts of the plasma membrane into the cell's interior. Part of the engulfed membrane contains attached molecules, such as specific membrane proteins or the sugar-lipid G_{M2}. Once inside a normal cell, these membrane molecules can be destroyed or recycled. But as you will see later in this

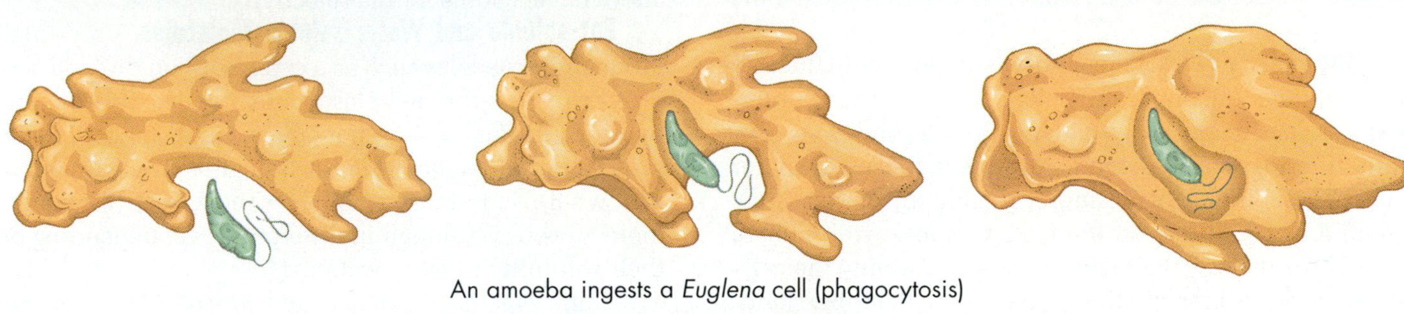

[A] Endocytosis of solid material

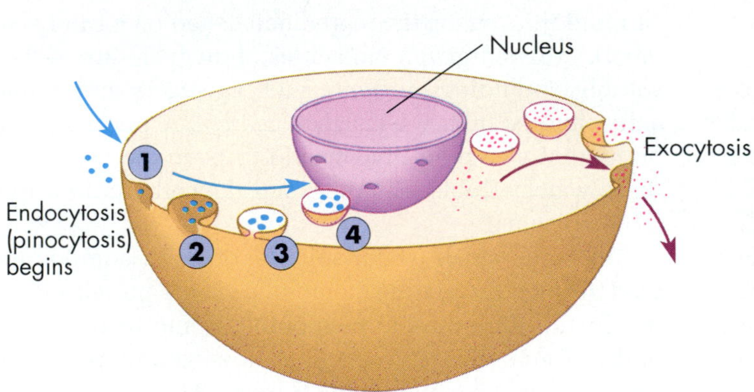

[B] Endocytosis and exocytosis of liquid material

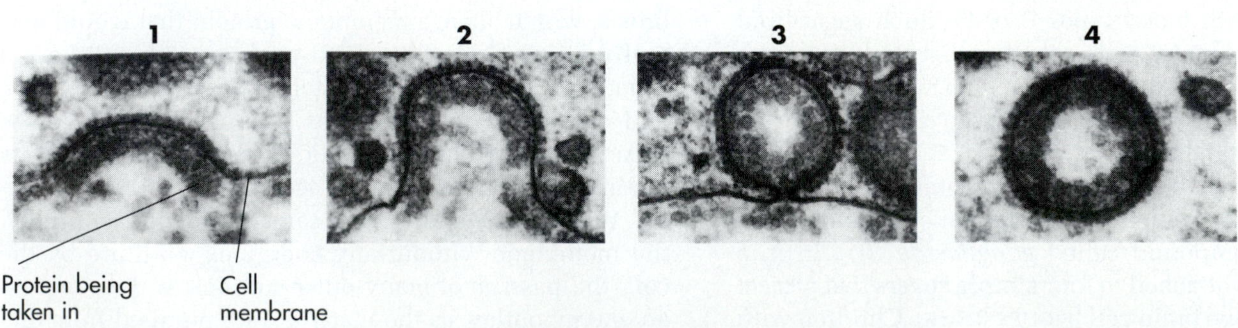

[C] Electron micrographs of endocytosis

FIGURE 3.14

Endocytosis and Exocytosis.

[A] The cell membrane forms a pocket around a food particle and engulfs it in a type of endocytosis called phagocytosis. Here, an amoeba ingests a *Euglena* cell by phagocytosis. **[B]** In the endocytosis of fluids, called pinocytosis, the membrane indents and traps a fluid (designated by blue dots) in a vesicle within the cell's interior. **[C]** A hen's egg cell accumulates yolk protein by *receptor-mediated endocytosis.* The protein to be taken up (yolk protein in these micrographs) binds to a receptor located on the cell membrane (Step 1). The receptor-yolk protein aggregate becomes concentrated in a pit (Step 2), which folds into a vesicle (Step 3). When the vesicle completely pinches off (Step 4), the yolk protein has been taken inside the cell, surrounded by a membrane. These steps are indicated in **[B]**.

FIGURE 3.15

The Nucleus: The Cell's Central Control.

Most of a cell's daily activities are directed by information molecules that issue from the nucleus. **[A]** A cross section of a nucleus reveals a large area of decondensed chromosomes and a dark-staining area, the nucleolus, which is the site of ribosome production. **[B]** The surface of the nucleus looks a bit like a golf ball because the nuclear envelope is perforated by dozens of pores that regulate the passage of RNA and other materials. The nuclear envelope is composed of two lipid bilayers with a space in between. **[C]** During cell division, the chromosomes condense into visible rods, and protein fibers appear that move the chromosomes into the forming daughter cells.

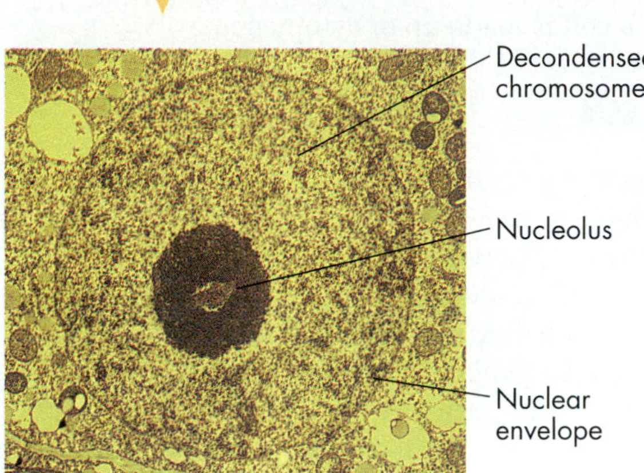

[A] Cross section of nucleus

Nuclear pores
Nuclear side
Two lipid bilayers
Cytoplasmic side

[B] Nuclear envelope

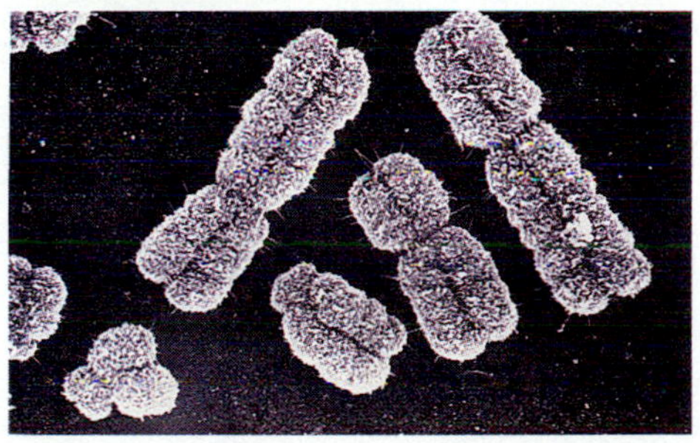

[C] Chromosomes in a dividing cell

chapter, neither happens in a Tay-Sachs cell, and the sugar-lipid molecules build up abnormally.

In a special form of endocytosis called **receptor-mediated endocytosis** [see FIGURE 3.14C], proteins being brought into the cell must first bind with a receptor, a specific cell-surface protein, before being engulfed within a vesicle and drawn into the cell.

Exocytosis The process of **exocytosis** (*exo-*, "outside") is the opposite of endocytosis. Vesicles from the cytoplasm fuse with the plasma membrane like two soap bubbles joining together. As a result, the contents of the vesicle are expelled into the surrounded extracellular fluid, and the plasma membrane of the vesicle becomes incorporated into cell's plasma membrane. Some cells discharge wastes this way, or secrete proteins, such as hormones or digestive enzymes, into the surrounding fluid.

CONCEPT CHALLENGE

The head of your local blood bank decides that blood supplies would be much more versatile if researchers could change blood types A and B into blood type O, the universal donor. You decide to take on the challenge and to develop a way to change blood cell types. What part of the cell and what type of molecule would you alter?

The Nucleus

In a eukaryotic cell viewed through a microscope the most conspicuous organelle is often the **nucleus** [FIGURE 3.15]. This roughly spherical structure contains genetic information in the form of DNA, which controls a cell's activities. The nucleus also makes and exports various forms of RNA, the nucleic acids that function in protein synthesis. RNA carries "patterns" for proteins from the nucleus to the cytoplasm. The proteins made in the cytoplasm might become part of an organelle, or might begin to carry out cell functions, as do the class of proteins called enzymes. Enzymes break down materials for the cell, build needed cell parts, and facilitate other vital tasks.

Information thus flows from the DNA in the nucleus to RNA, and then to proteins, which carry out the work of the cell. This flow of information can be diagrammed:

$$\text{DNA} \rightarrow \text{RNA} \rightarrow \text{Protein}$$

Through its RNA messengers to the cytoplasm, the nucleus also controls the rate of cell growth, cell division, and energy acquisition and use. In other words, information molecules from the nucleus direct most of a cell's moment-to-moment and day-to-day activities.

Since the nucleus contains the hereditary material, and since a disease like Tay-Sachs is inherited, it is clear that some defect in the nucleus leads to Tay-Sachs disease. Some flaw in DNA, which causes an abnormal RNA to be formed, directs the cell to produce an altered and nonfunctional protein. As we will see later in this chapter, this nonfunctional protein is an enzyme.

NUCLEAR ENVELOPE

The nucleus is surrounded by a **nuclear envelope** [FIGURE 3.15B], which is made up of *two* lipid bilayer membranes separated by a thin space. The entire envelope is perforated at dozens of points by *nuclear pores*, giving the nucleus the appearance of a dimpled golf ball. Each pore is a cluster of proteins that form a channel that regulates the passage of RNA from the nucleus to the cytoplasm.

CHROMOSOMES

Within the nucleus of eukaryotic cells, proteins associate with DNA and form long fibrils, or threads. Whenever a cell is not actively dividing, the nucleus appears grainy because of the loose, extended nature of these threads. At the start of nuclear division, however, the DNA-protein threads compact into chromosomes (literally, "colored bodies," because of the ease with which they stain with certain dyes). Chromosomes are microscopic structures that carry hereditary information [FIGURE 3.15C].

NUCLEOLUS

Within the nucleus there may be one or more dark-staining regions called **nucleoli** (new-KLEE-oh-lie; "little nuclei," singular, *nucleolus*) [FIGURE 3.15A]. Each nucleolus makes a special type of RNA (called ribosomal RNA) necessary for the synthesis of proteins. Nucleoli, therefore, help set the pace for the productivity and growth of a cell. The nucleus of a very active cell, such as a growing egg cell primed for the rapid development of an embryo, can have as many as 1000 nucleoli. In a cell with little or no protein synthesis, such as a mature sperm cell, there is no nucleolus.

To summarize our discussion of the nucleus, this central organelle with its store of DNA directs cell activities by ordering the synthesis of specific RNAs and hence specific proteins. Each protein then carries out a particular job in the cell. Many of those tasks take place within individual organelles in the cytoplasm.

Cytoplasm and Organelles

Most of a cell is made up of cytoplasm.

CYTOPLASM

Cytoplasm is a semifluid, highly organized ground substance that acts as a pool of raw materials. About 70 percent of the fluid portion of cytoplasm (*cytosol*) consists of water molecules, and about 15 to 20 percent is made up of about 10,000 different kinds of protein molecules, with 10 billion or so protein molecules occurring in an average cell. Many of the proteins are enzymes that speed biochemical reactions, while others are structural proteins that can be assembled into various cell components. Suspended in the cytoplasm are numerous fibers, particles, and functional structures—the **organelles**—some of which are described in the following sections, beginning with ribosomes.

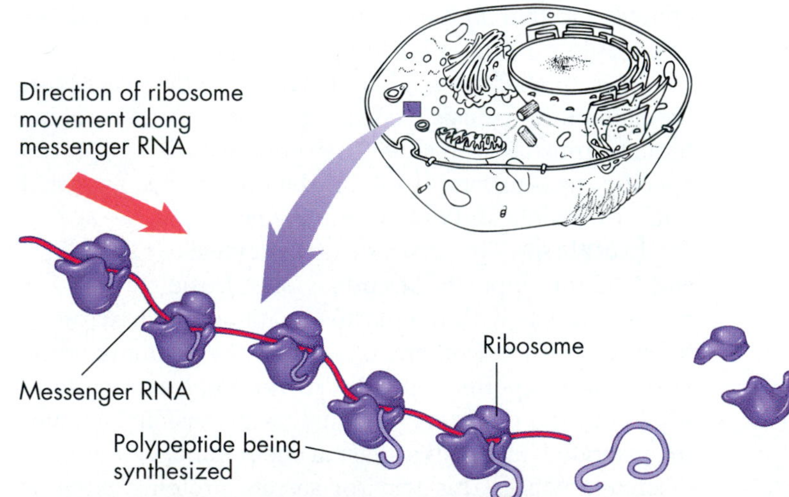

FIGURE 3.16

Ribosomes: Sites of Protein Manufacture.

Ribosomes are dense, beadlike clusters of RNA and certain proteins. Ribosomes can be seen throughout the cell cytoplasm, associated with various membrane surfaces, or in chains called polyribosomes. In a polyribosome, several ribosomes become associated with a molecule of messenger RNA like beads on a string. From information in the RNA, each ribosome builds a protein.

RIBOSOMES

Thousands of beadlike clusters called **ribosomes** are embedded throughout the cytoplasm. They serve as the workbench upon which proteins are synthesized [FIGURE 3.16]. Since there are no functional ribosomes in the nucleus, protein synthesis occurs only in the cytoplasm, outside the nucleus. Ribosomes are made up of a number of special proteins and three or four specific RNA molecules, called *ribosomal RNAs*, which are produced in the nucleolus. The mechanisms of protein synthesis are described in CHAPTER 10.

CYTOSKELETON AND CENTRIOLES

The electron microscope reveals another major feature of eukaryotic cells: the **cytoskeleton**. This three-dimensional structure of minute protein fibers forms a lattice throughout the cytoplasm, suspending the organelles in proper spatial relationships to each other and allowing precisely regulated movement of cell parts. Protein fibers in the cytoskeleton also act as internal girders and cables that help maintain a cell's shape and transport cell organelles, much as skiers are pulled along on moving rope tows. FIGURE 3.17 shows the elements of the cytoskeleton—microfilaments, microtubules, and intermediate filaments.

FIGURE 3.17

The Cytoskeleton.

The cytoskeleton acts as an internal skeleton, supporting the cell's shape and activities. Three types of protein fibers make up the cytoskeleton: **[A]** microfilaments; the alga has giant cells that rely on actin filaments to move organelles from one end of the cell to the other; **[B]** microtubules, which move colored pigment granules around in the skin cells of reptiles and fish, allowing chameleons, for example, to blend in with the color of their environment; and **[C]** intermediate filaments which provide reinforcement that makes cells tough, providing animals such as the peacock with protection, camouflage, and sexual attraction. The cell photos are analogous to X-rays of bones; they show only one particular cytoskeletal element at a time. However, all three types occur simultaneously in many cells.

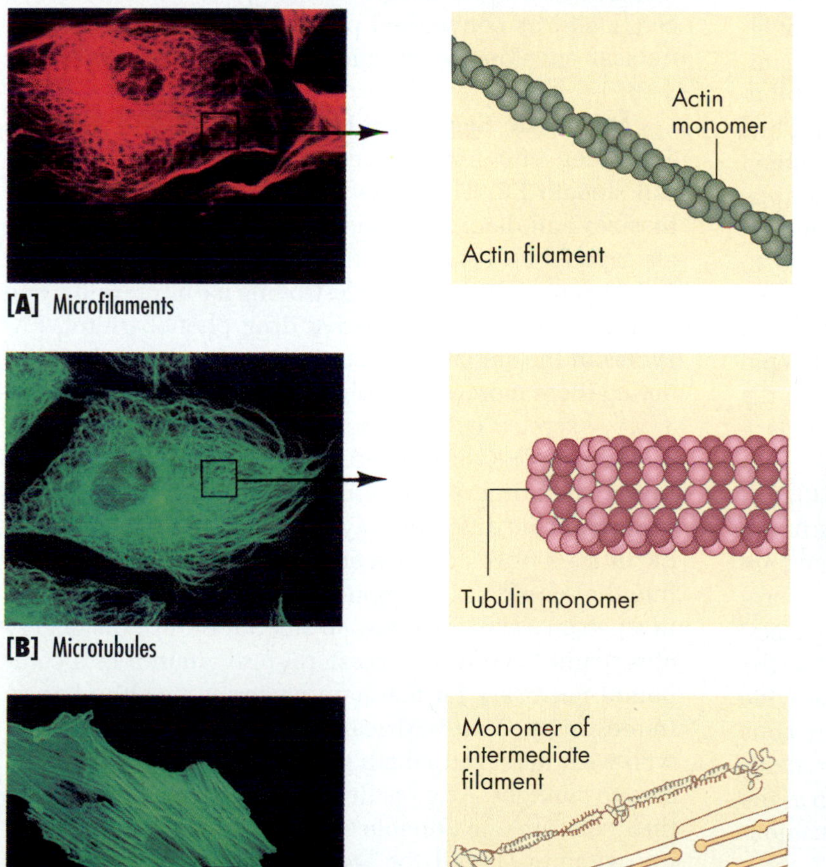

[A] Microfilaments

[B] Microtubules

[C] Intermediate filaments

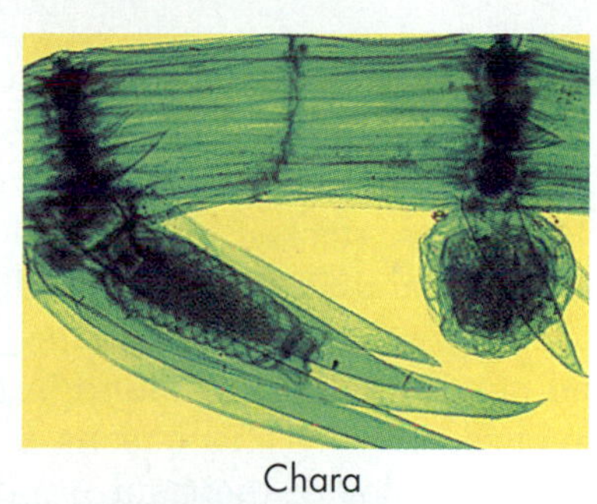

Chara

Chameleon

Peacock

Microfilaments Microfilaments are cytoskeletal elements about 3 to 6 nanometers* in diameter, made mostly of a protein called *actin*. One end of the fiber is usually anchored to the inner surface of the plasma membrane [review FIGURE 3.13], and along its length, it may associate with other proteins and cause a contracting force. Actin molecules can also rapidly assemble to or disassemble from the filament, and this dynamic activity causes the filament to lengthen or shorten. Actin filaments can pinch an animal cell in two via a drawstring-like action during cell division, and they are also critical elements in the contraction of muscles. If actin from a rabbit's muscle is mixed with a protein from a funguslike organism called a slime mold, the filaments contract forcefully, just as if they were in a muscle cell. Thus, actin and the related molecule myosin are widespread in life, suggesting a very ancient and common origin. Actin filaments can move organelles around inside of cells. The alga Chara [FIGURE 3.17] has giant cells that rely on actin filaments to move organelles from one end of the cell to the other.

Microtubules Microtubules, a second cytoskeletal element, are hollow cylindrical tubes 20 to 25 nm in diameter composed of subunits of the globular protein *tubulin*. These subunits are stacked in a spiral form, and like actin, they assemble and disassemble rapidly to lengthen or shorten the microtubules. Microtubules, for example, move colored pigment granules around in the skin cells of reptiles and fish, allowing chameleons, for example, to blend in with the color of their surroundings [FIGURE 3.17]. Microtubules also help establish and maintain the cell's shape, make up the *spindle* apparatus that separates the chromosomes during cell division [see CHAPTER 7], and are instrumental in generating cell movement [see FIGURE 3.25].

Intermediate Filaments A third component of the cytoskeleton, the intermediate filaments (about 10 nm across), are made up of *keratin* and other proteins and are midway in size between microfilaments and microtubules. These filaments help to maintain cell shape, acting like girders or tension-bearing cables that stabilize the cell's perimeter. Intermediate filament proteins provide the reinforcement that makes an animal's skin cells tough, and they make up most of the matter in skin, nails, hair, and feathers. They thus provide animals, such as the male peacock in FIGURE 3.17, with protection, camouflage, and sexual attraction.

Centrioles A pair of rod-shaped organelles called **centrioles** [see FIGURE 3.12] organize certain cytoskeletal fibers and are instrumental in the movement of cellular parts during cell division.

* nm (one-billionth of a meter).

ENDOPLASMIC RETICULUM

Cells take energy from the environment, expend it during the construction of proteins and other molecules, then transport those molecules to the proper regions of the cell. There, the materials can be stored or incorporated into new cell parts for repair, growth, or reproduction. In eukaryotic cells, these functions are accomplished by an interconnected system of membranous organelles, including the smooth endoplasmic reticulum, the rough endoplasmic reticulum, the Golgi apparatus, and transport vesicles. Both smooth and rough endoplasmic reticulum synthesize materials, and the Golgi apparatus modifies these materials and releases them in transport vesicles that carry them to their final location in the cell [FIGURE 3.18].

A cell's internal membrane system is a network of flattened, hollow tubules, sheets, and cisterns (reservoirs) that form an interconnected set of channels throughout the cell. That system of channels is called the **endoplasmic reticulum**, or **ER** (literally, "network within the cell"), and its convoluted passageways extend from the nuclear envelope to the plasma membrane [see FIGURE 3.18].

Smooth ER Some areas of the endoplasmic reticulum are folded into smooth sheets and tubes that biologists call **smooth ER**. The smooth ER [FIGURE 3.18] both synthesizes and detoxifies lipid-soluble substances (materials capable of dissolving in lipids). Smooth ER in liver cells is especially active in destroying harmful lipids; one such substance is the sedative drug phenobarbitol. Enzymes in the smooth ER act on the drug molecules, rendering them more water-soluble so that they can eventually be excreted in the urine.

Smooth ER also manufactures different types of lipids, including cholesterol. Various cells modify the cholesterol in different ways. For example, the smooth ER in cells of the ovaries and testes modify cholesterol into the steroid sex hormones estrogen and testosterone. In a person's skin, enzymes in smooth ER use the energy of sunlight to convert cholesterol into vitamin D, a compound necessary for maintaining strong, healthy bones. Interestingly, North African women of Bedouin tribes who wear dark, full-length garments get very little exposure to sunlight. As a result, the regions of smooth ER in their cells are often unable to convert enough cholesterol into vitamin D, and the women's bones grow soft and weak [FIGURE 3.19].

In human brain cells, smooth ER makes the G_{M2} sugar-lipid involved in Tay-Sachs disease. Specifically, smooth ER makes the lipid portion of G_{M2}—the red pathway in FIGURE 3.18—then it becomes inserted into the membrane. Little membranous spheres, called **transport**

FIGURE 3.18

The Internal Membrane System.

The organelles of the *cytomembrane system* are: **[A]** smooth endoplasmic reticulum, **[B]** rough endoplasmic reticulum, **[C]** Golgi apparatus, and **[D]** lysosomes. Parts are shown here in transmission electron micrographs and in fluorescently labeled whole cells. The red pathway is the one followed by G_{M2} and other membrane lipids. The blue pathway is the one followed by lysosomal enzymes such as hexoseaminidase. The yellow pathway is followed by secreted proteins, such as insulin hormone or growth hormone, or the digestive enzymes in your gut. The green pathway is the route from the cell surface to the lysosome followed by membrane components such as G_{M2} and substances such as cellular debris ingested via endocytosis.

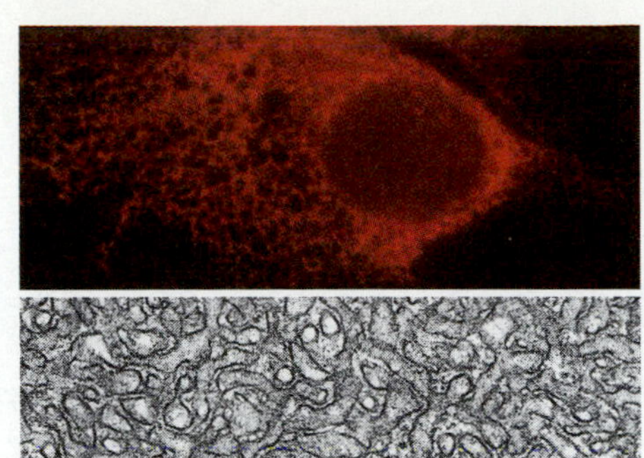

[A] Smooth endoplasmic reticulum

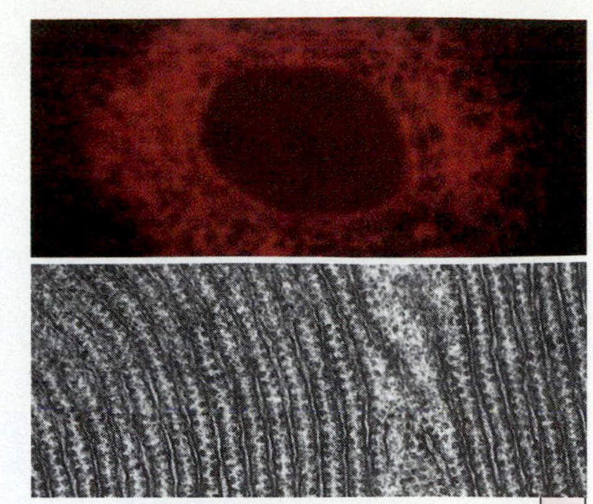

[B] Rough endoplasmic reticulum

Smooth Endoplasmic Reticulum synthesizes lipids and detoxifies many poisons.

Golgi apparatus further modifies lipids and proteins from ER, then sorts them and ships them out to plasma membrane or other organelles.

Transport vesicles shuttle lipids, proteins, and other materials to and from various organelles.

Exocytic vesicles transport materials to the cell surface; lipids for inclusion in plasma membrane, and proteins for secretion.

Plasma membrane

Newly synthesized protein

Newly synthesized lipid

Sugars added

Nucleus

Secretory protein pathway

Lysosome protein pathway

Sugars added

Membrane lipid pathway

Endocytic pathway

Membrane lipid like G_{M2}

Debris

Rough endoplasmic reticulum synthesizes proteins that are secreted from the cell, or inserted in the plasma membrane, or enter various organelles, such as the lysosome.

Lysosomes contain digestive enzymes that recycle worn out cell parts or digest materials taken into the cell.

Endocytic vesicles bud off the plasma membrane and carry the engulfed liquid or solid materials to lysosomes, where they are digested and recycled.

[C] Golgi apparatus

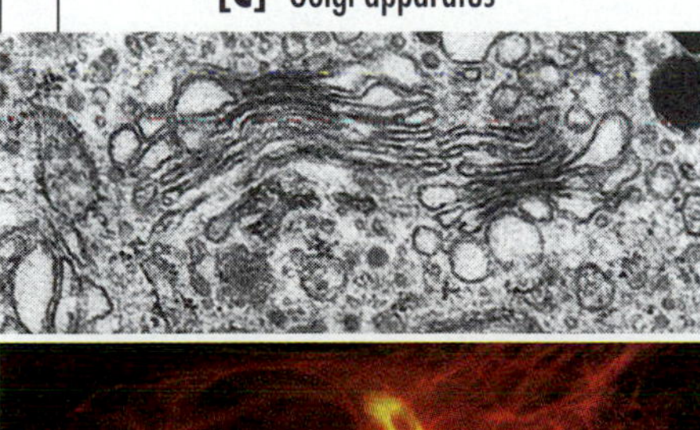

[D] Lysosome

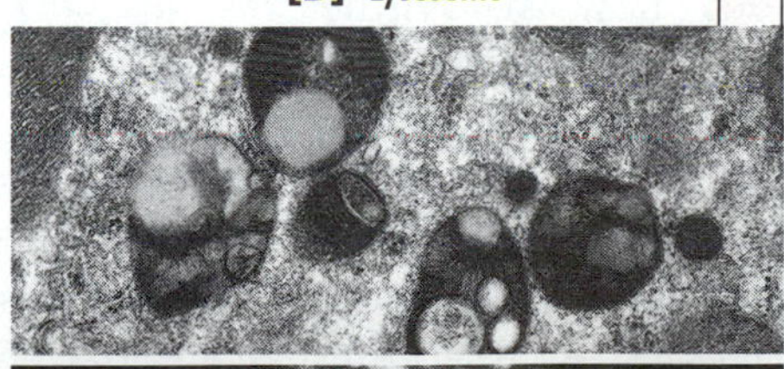

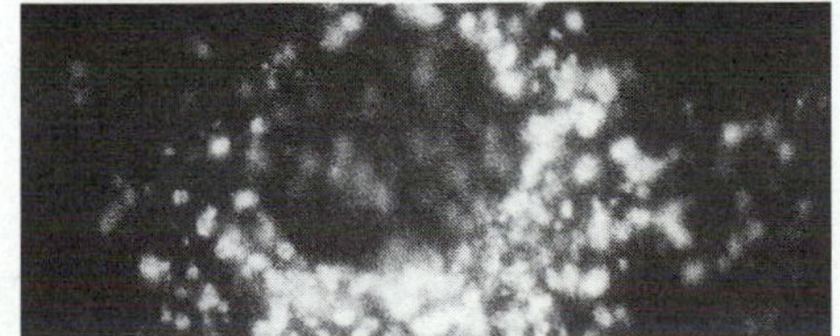

FIGURE 3.19

Some Bedouin Women Have a Smooth ER Problem.

Because this woman's clothing leaves little or no skin exposed to sunlight, her smooth ER may not be able to make enough of the vitamin D necessary to maintain strong, healthy bones.

vesicles, pinch off from the smooth ER and carry membrane lipids such as G_{M2} to the Golgi apparatus. There, simple sugar molecules are added to the G_{M2}. This sugar-lipid is then shunted to vesicles that move to the plasma membrane and become part of it. Once the G_{M2} molecule is in the plasma membrane, it can assist the cell in recognizing substances from the surrounding environment.

Rough ER Another part of the endoplasmic reticulum, called **rough ER**, is studded with ribosomes and looks rough under a microscope [FIGURE 3.18B]. This region is involved in the synthesis of many proteins that are not dissolved in the fluid portion of the cytoplasm. For example, the rough ER helps produce the enzymes that digest most of the foods you eat. Tracing the production and transport of an enzyme through this membrane system helps us understand how the ER functions. Let's focus on the rough ER in a normal person's brain cells, and how it synthesizes an enzyme called *hexoseaminidase*. This protein will prove important to our understanding of Tay-Sachs disease.

In normal brain cells, ribosomes attach to the rough endoplasmic reticulum [review FIGURE 3.18] and synthesize the enzyme hexoseaminidase. As this protein is assembled, it passes through a pore in the rough ER membrane and enters the cavity inside the ER's membranous channels. (You can trace the protein's transport route by following the blue pathway in FIGURE 3.18.) Once the hexoseaminidase molecules are inside the ER cavity, sugar molecules are added to the proteins as they move through the channels. Eventually, the newly synthesized proteins enter transport vesicles, which pinch off from the rough ER. Most of the sacs pinched off from the endoplasmic reticulum fuse with membranes of the Golgi apparatus.

GOLGI APPARATUS

Named for the Italian cell biologist who first described it, a **Golgi apparatus** (GOAL-gee) is a series of flattened membranous sacs resembling a pile of nearly empty hot-water bottles [FIGURE 3.18C]. Some cells have just one Golgi apparatus, but others may have hundreds. In a brain cell, the Golgi apparatus can modify the sugars previously added to hexoseaminidase or other proteins. These modified sugars can act like the tags an airline ticket agent sticks on your luggage, intended to direct your bags to the correct destination.

A Golgi apparatus directs newly synthesized proteins to enter vesicles that move to various parts of the cell. Some proteins, like hexoseaminidase, are conveyed from vesicles to other organelles. Other proteins, like the one involved in the disease cystic fibrosis, enter vesicles that place them within the plasma membrane on the cell's surface. Still other proteins, including the enzymes that digest your food, are directed into vesicles that transport them into the space outside the cell (yellow pathway in FIGURE 3.18).

A Golgi apparatus directs the traffic not only of proteins such as hexoseaminidase and lipids such as the sugar-lipid G_{M2}, but also of carbohydrates. In plants, the Golgi apparatus, which is called a *dictyosome* by some botanists, packages for export the cellulose that forms the rigid outer cell wall.

LYSOSOMES AND MICROBODIES

You have seen that some transport vesicles pinch off from a Golgi apparatus and move to the cell surface. Other vesicles, however, are permanent residents of the cytoplasm. Such vesicles include lysosomes and microbodies.

Lysosomes **Lysosomes** ("loosening bodies") are spherical vesicles that contain powerful digestive enzymes [FIGURE 3.18D]. These enzymes can help to break down worn-out cell constituents or to digest debris that the cell has taken in by endocytosis. Lysosomes are involved when you get a splinter in your finger and the finger becomes red, inflamed, and infected. When this happens, many of your white blood cells move to the site of the injury and engulf the bacteria or debris that has built up in the wound (green pathway in FIGURE 3.18). Once inside a white blood cell, the membrane-encapsulated "meal" fuses with a lysosome. Digestive enzymes in the lysosome generally break down the bacteria or debris, into smaller molecules. The small molecules can pass through the lysosomal membrane and enter the fluid portion of the cytoplasm, where they are available for reuse by the cell. In this way, lysosomes halt the infection.

Lysosomes also sometimes act as "suicide bags," breaking open and spilling their contents, and literally digesting entire damaged or aged cells from the inside out. This process is especially dramatic in insects that are transformed within cocoons from caterpillars to moths or butterflies. One by one, the caterpillar cells die and are digested, making way for the new adult moth or butterfly cells.

It is still not clear why the enzymes in the suicide bags don't digest the bags themselves, but clearly the enzymes cannot break down every type of material. In fact, aging eukaryotic organisms often accumulate brown-pigmented granules in their lysosomes. Biologists think that these granules are undigested cellular "refuse" that must be stored until the cell dies. In an 80-year-old person, the refuse material can occupy 10 percent of the volume of each heart cell. Elderly people often develop brownish "age spots" on their skin from the same pigments.

How does a lysosome recycle worn-out cell parts? If we consider a normal human brain cell, parts may become damaged, including molecules of the sugar-lipid G_{M2} on the cell's surface. Patches of surface membrane containing the damaged sugar-lipid will then pinch off and form vesicles inside the cell (green pathway in FIGURE 3.18). The vesicles can then fuse with lysosomes, and digestive enzymes inside the lysosomes can degrade the defective sugar-lipid and recycle its components. The enzyme that carries out this degradation and recycling is hexoseaminidase. What would happen to the G_{M2} in the lysosomes if those baglike organelles lacked the digestive enzyme hexoseaminidase? The answer is: whorls. It turns out that people with Tay-Sachs disease do, in fact, have lysosomes missing the hexoseaminidase enzyme. As a result, they cannot digest G_{M2} from worn-out patches of cell membranes, and the G_{M2} sugar-lipid accumulates, forming the whorls visible in the brain cells of Tay-Sachs patients [review FIGURE 3.1].

Microbodies In addition to lysosomes, cells have another class of membranous vesicles residing permanently in the cytoplasm: **microbodies**. One type of microbody, called a *peroxisome*, contains the enzyme catalase, which breaks down the corrosive compound peroxide that forms as a by-product of a cell's metabolic functions. If you have ever used hydrogen peroxide to cleanse a cut on your finger, you have seen catalase at work. The bubbling results from the action of catalase from your cells as this enzyme breaks down the hydrogen peroxide into bubbles of oxygen and water. Peroxisomes within liver and kidney cells break down and detoxify about half the alcohol a person drinks, and other peroxisomes can produce a small amount of energy for the cell. However, such microbodies provide only a small percentage of the energy compounds used by a cell. A cell's real power plant is the mitochondrion.

MITOCHONDRIA

The organelles called **mitochondria** (singular, *mitochondrion*) provide chemical fuel for cellular activities that go on continually, including the production of proteins, the replication of DNA, and cell movement. Mitochondria operate by breaking down certain carbon-containing molecules into carbon dioxide and water. During the breakdown process, energy is released which is then stored in high-energy molecules of ATP, a nucleotide [review FIGURE 2.32]. These "energy packets" quickly diffuse throughout the cell and fuel the biochemical transactions of life's processes. The chemical conversions that go on inside the mitochondria require oxygen and are collectively called *aerobic respiration*. The details of these important chemical reactions are discussed in CHAPTER 5.

The term *mitochondrion* comes from the Greek words for "thread" and "grain," and a mitochondrion can indeed look like a tiny grain of rice or, at low magnification, threads [FIGURE 3.20]. The shape can vary from small spheres to long, sausagelike bodies roughly the size of a bacterium.

While nearly all eukaryotic cells have mitochondria, the characteristics and abundance of the organelles depend on the cell's activity level. Cells that expend a great deal of energy have larger, more numerous mitochondria than relatively inactive cells. A heart muscle fiber, for example, can have thousands of large mitochondria. Also, mitochondria are often packed most densely in regions of a cell with the highest energy demands. The mitochondria in a human sperm cell, for instance, are wrapped around the base of the whiplike tail that propels the cell.

Mitochondria have a semiautonomous existence in a cell: They have their own DNA that directs the production of some, but not all, of their component proteins. Information to form the rest of the mitochondrion is encoded in the nuclear DNA. Mitochondria also have their own ribosomes, and they can divide in half, thus reproducing independently of the cell's normal division cycle. They are bacterialike in size and shape and carry out energy-conversion reactions on the inner surfaces of their membranes, as do bacteria. These facts have led some biologists to suggest that mitochondria originated over 2 billion years ago when primitive bacteria were engulfed by larger cells. This hypothesis is called *endosymbiosis*, and we will return to it in CHAPTER 17.

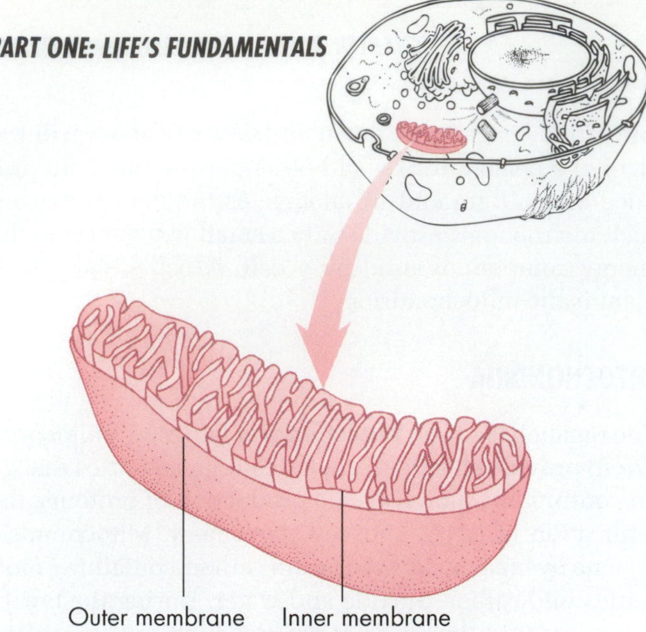

[A] Mitochondrion

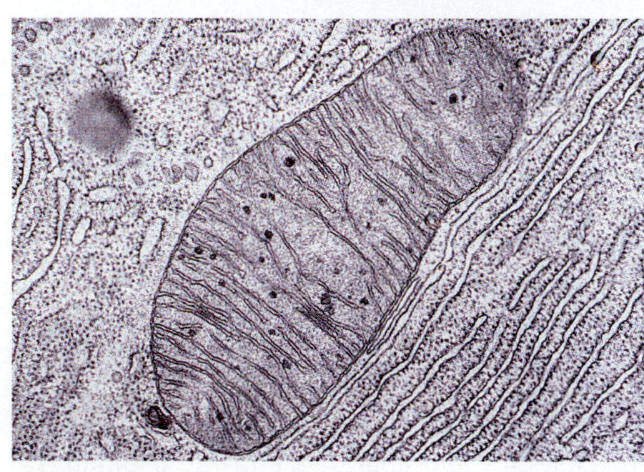

[B]

FIGURE 3.20

Mitochondria: Power for the Cell.

[A] Mitochondria help eukaryotic cells harvest energy from foods. Mitochondria have a smooth *outer membrane* and an *inner membrane* folded into a complex of sheets. The space between the two membranes is crucial for the generation of ATP, the cellular energy currency. Inside the inner membrane, is a compartment that contains the genetic material DNA, ribosomes, and many enzymes involved in producing ATP. [B] The electron micrograph shows a mitochondrion from a pancreas cell.

Not only do mitochondria have their own DNA, but it is passed to offspring only through the mother, since mitochondria are present in eggs but not the part of the sperm that enters the egg at fertilization. For this reason, people can trace all the mitochondria in their cells back to their mothers, grandmothers, great-grandmothers, and so on. Analysis of mitochondrial DNA from people all over the world has led some researchers to claim that all people alive today are members of a lineage that began with one woman or a small group of women who probably lived in Africa about 200,000 years ago. Controversy still surrounds the exact time the woman or women lived, and whether or not they were anatomically modern. Nevertheless, this idea has caught the popular imagination as "mitochondrial Eve," the woman from whom all people now alive supposedly descended.

While a time-lapse movie may give the impression that mitochondria lead an independent life in a cell, the fact is that when mitochondria are removed from a cell, they are just bags of biochemicals. It is the simultaneous, coordinated activities of mitochondria organized in a cell, along with the nucleus, endoplasmic reticulum, and other organelles, that add up to life. Separately, organelles are marvels of biochemical organization, but they are not alive.

CONCEPT CHALLENGE

I-cell disease is a rare inherited disorder that causes a baby's cells to accumulate abnormal granular sacs—bags of grainy material. The substances in these sacs would normally be digested by lysosomal enzymes. These enzymes are missing in the sacs, but they are present in the baby's blood. Apparently, the patient's cells synthesize the enzyme normally, but the molecules are misdirected to the cell surface and are secreted into the bloodstream rather than deposited and stored in lysosomes. What cell organelle could be malfunctioning in I-cell disease? What kind of mistake is the organelle making?

Specialized Cell Structures

The structures we have discussed so far are found in most eukaryotic cells. A number of specialized organelles, however, are found only in certain cell types. These specialized structures include plastids, vacuoles, cilia, and flagella.

PLASTIDS: ENERGY CAPTURE AND STORAGE

Plants and some protists have several types of oval, membrane-bound organelles called **plastids**, which harvest solar energy, manufacture nutrient molecules, and store materials. Plastids called **chloroplasts** ("green plastids") are membranous organelles that trap the energy of sunlight in a chemical form—generally, sugar molecules—in the process of *photosynthesis* [CHAPTER 6]. All animals and fungi, and most other organisms on earth, ultimately

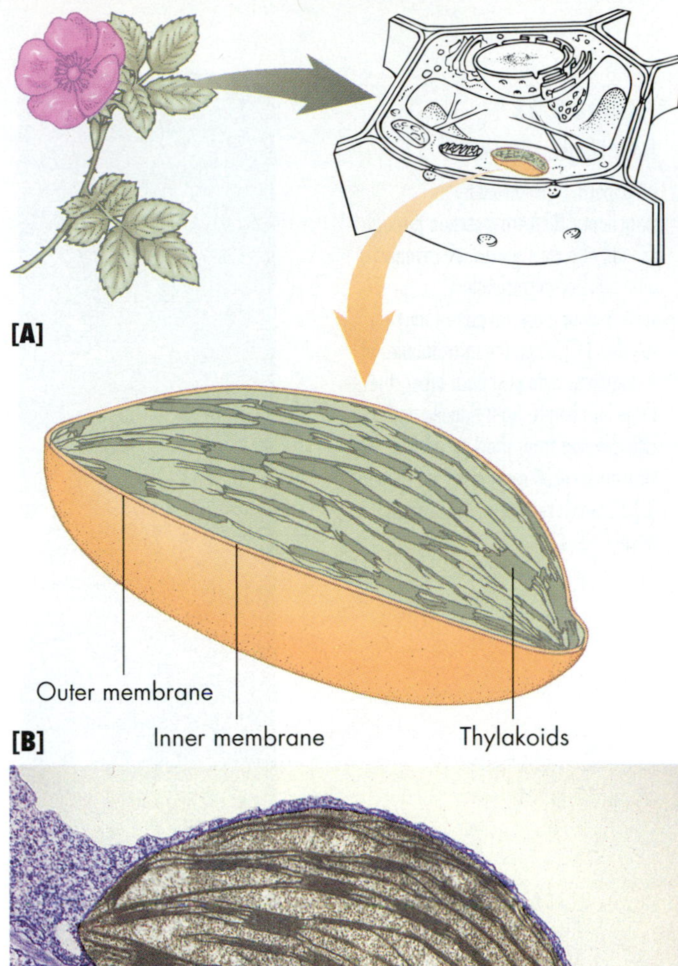

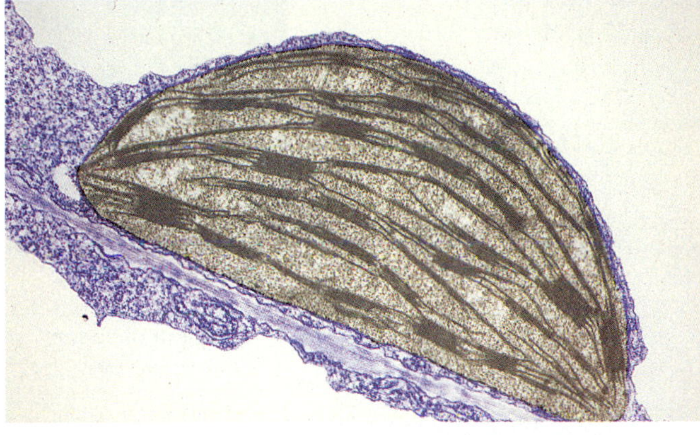

FIGURE 3.21

Chloroplasts: Organelles of Photosynthesis.

[A] The chloroplasts in plant cells are elongated organelles that contain green-colored pigments that trap the energy from sunlight. [B] Chloroplast pigments are embedded in membranes that arise from the inner membrane and form a stack of flattened disk-like sacs called the *thylakoids*. These thylakoid membranes trap sunlight and transfer the energy to energy-carrying molecules that fuel the production of sugar and starch [see CHAPTER 6] [C] The electron micrograph of a chloroplast is magnified 40,000 times.

depend on this critical process of energy conversion to supply the nutrients, materials, and energy compounds they need for their own survival.

FIGURE 3.21 depicts the internal structure of chloroplasts. Membranes within a chloroplast contain the green, light-absorbing pigment *chlorophyll*. Chloroplasts house their own circular DNA molecules, which encode some of the chloroplast's proteins. This separate DNA suggests to biologists that chloroplasts, like mitochondria, evolved from primordial bacteria after they came to inhabit larger eukaryotic cells.

Two other types of plastids are common in plant cells. *Chromoplasts* ("colored plastids") store yellow, orange, and red energy-trapping pigments, and give color to many fruits and flower petals. *Leucoplasts* ("white plastids"), in contrast, are colorless organelles. Some leucoplasts store starch granules, while others store proteins or lipids. Potatoes, for example, contain billions of starch-storing leucoplasts.

VACUOLES

Seen through a microscope, most plant cells look strangely hollow, with what appears to be empty space filling most of the cell. Actually, the "space" is an important organelle called the **central vacuole** (from the Latin *vacuus*, "empty"). This organelle contains water and various storage products, has a single membrane, and can occupy from 5 to 95 percent of the total cell volume. The orange cell in FIGURE 3.4 consists mostly of a vacuole filled with orange juice.

Far from empty, the central vacuole's main role is to take up space. It fills up a plant cell with what amounts to a bag of water that presses a small amount of cytoplasm and all the cell's organelles into a thin layer just below the plasma membrane [see FIGURE 3.12B]. This makes the surface-to-volume ratio quite favorable. When plenty of water is available, a plant's many vacuoles store the water, keeping the cells plump and giving firm shape to the leaves, stems, and other plant parts. The next time you forget to water one of your houseplants and it wilts, you'll know that its vacuoles need refilling.

Although the central vacuole contains mostly water, it can also store sugars, proteins, ions such as Na^+, pigments, and other materials, including orange juice. Vacuoles in some plant cells harbor poisonous substances that protect plants from hungry animals, or they may contain specialized products such as rubber or opium.

Single-celled organisms that live in fresh water, such as the protists *Euglena* and *Paramecium*, have another type of vacuole that eliminates the water that tends to accumulate in these aquatic cells. This "bailing out" is done by **contractile vacuoles** that collect water and pump it out. *Paramecium* has a contractile vacuole at each end of the cell [FIGURE 3.22]. Water collects quickly in the vacuole and is squeezed out through the discharge pore every 15 to 60 seconds by contractions based on the activity of the cytoskeleton. In this way, the contractile

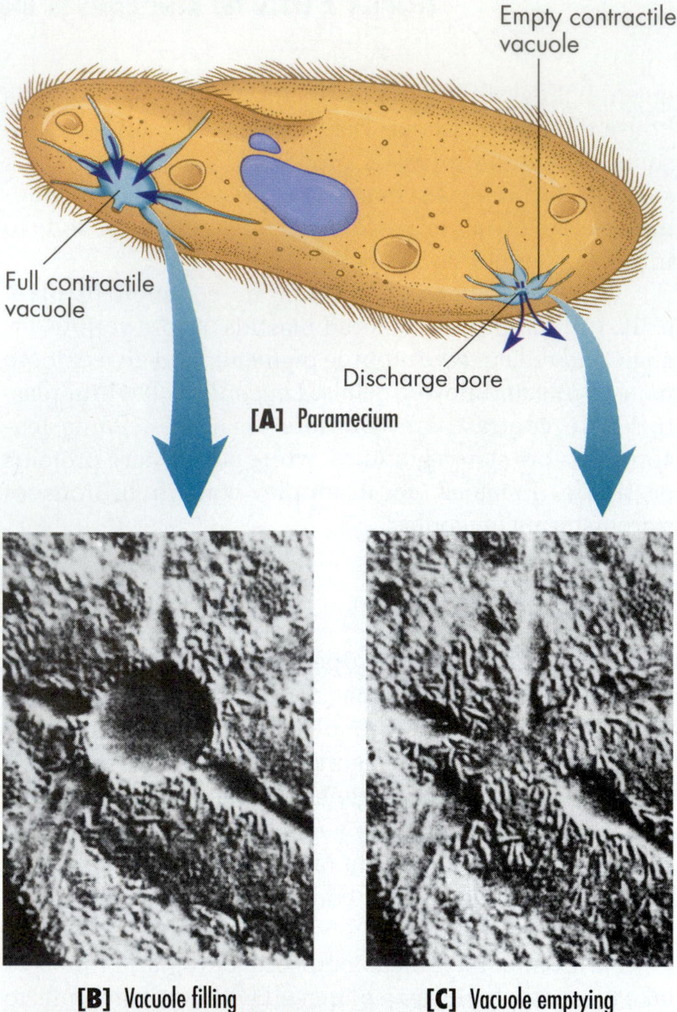

[A] Paramecium

[B] Vacuole filling

[C] Vacuole emptying

FIGURE 3.22

The Contractile Vacuole: Wastewater Management.

[A] *Euglena, Paramecium,* and many other large, single-celled organisms living in fresh water have contractile vacuoles, an organelle that collects and pumps out water. In *Paramecium*, the contractile vacuoles are near the cell's surface, expelling water one or more times per minute. Water then exits via the discharge pore. [B] Tiny ducts lead water into the nearly filled contractile vacuole of a paramecium. [C] When the vacuole contracts, the fluid is forced out.

FIGURE 3.23

How Flagella Move.

[A] Sperm is propelled by its flagellum. [B] A cross section through the tail of a rat's sperm cell reveals nine pairs of microtubules surrounding a central pair of microtubules. [C] When the microtubules of a flagellum slide past each other, the flagellum bends. Both flagella and cilia develop from short cylindrical structures called centrioles [see FIGURE 3.12], which remain at the base as *basal bodies.*

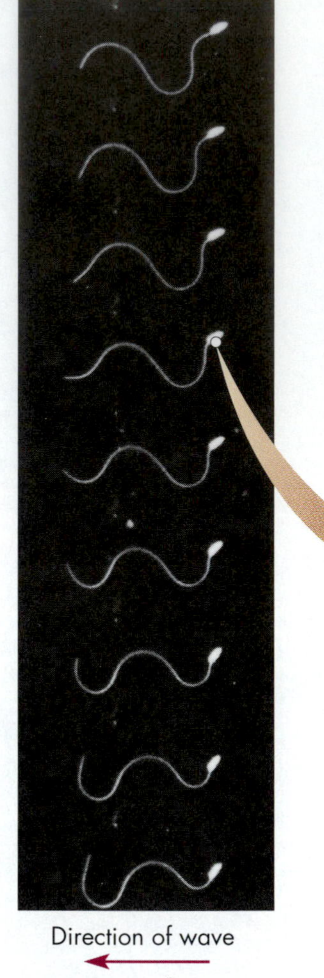

[A] Waves moving down a flagellum propel a sperm

vacuole prevents excess water from accumulating and perhaps, eventually bursting and killing the cell.

ORGANELLES OF CELL MOVEMENT

For cells such as *Euglena* that pursue sunlight, or for a mammal's white blood cells that chase invading microorganisms, movement is an important function, and it is made possible for these and most other mobile cells by the contraction of certain cytoskeletal fibers [see FIGURE 3.17]. Many cells possess either of two specialized organelles of movement: flagella or cilia, both of which are made from tiny, tube-shaped elements of the cytoskeleton called microtubules. Some free-living cells like *Euglena*, as well as the sperm cells of many animals and some plants, have a fine, whiplike organelle called a **flagellum** (Latin, "small whip"; plural, *flagella*) that extends from the cell surface and undulates, pulling or pushing the cell through its liquid medium [FIGURE 3.23]. Without this propulsion, a human sperm cannot bring about fertilization. Some Maori tribesmen from New Zealand, for example, are sterile because of a genetic defect that renders their sperms' flagella nonfunctional.

Certain single-celled protists bear not one flagellum, but thousands of shorter projections called **cilia** (Latin,

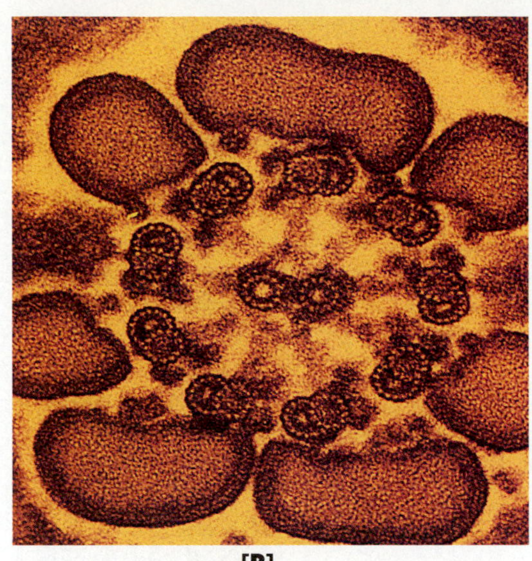

[B]

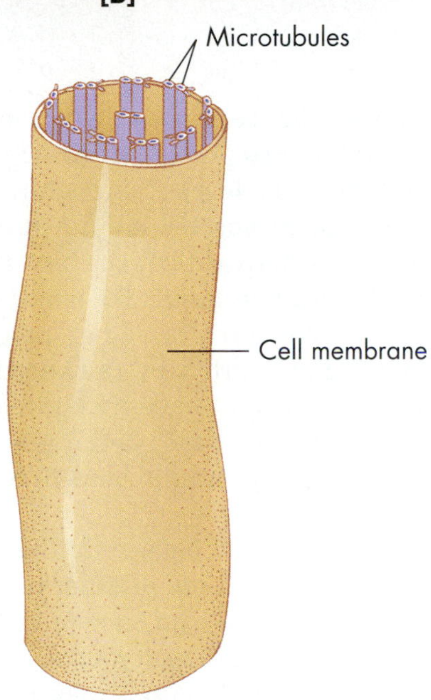

[C] A drawing of a section of a flagellum reveals its microtubules surrounded by a cell membrane

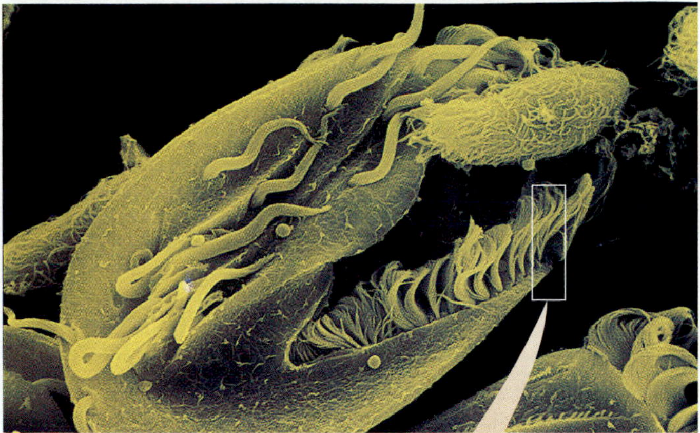

[A] Large ciliated protist devours smaller one

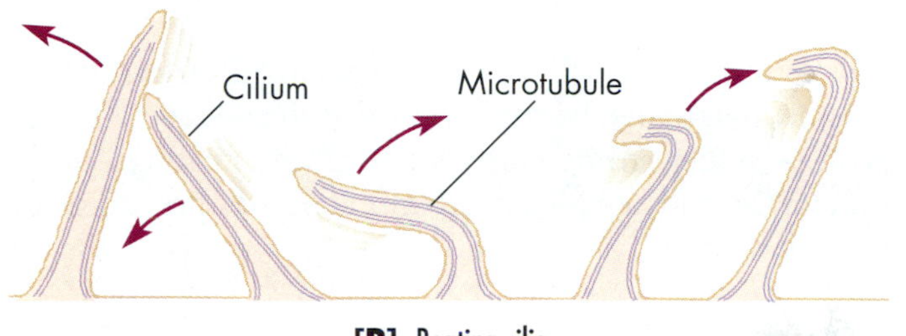

[B] Beating cilia

FIGURE 3.24

Cilia: Hairline Projections That Beat in Concert.

Cilia tend to occur in vast numbers, like short, movable hairs over the cell surface. **[A]** The slipper-shaped *Paramecium* cell in the upper right corner can swim quickly forward or backward by the wavelike motions that sweep continuously and sequentially across its thousands of surface cilia. The much larger *Euplotes* cell, with its own fringe of waving cilia, is attempting to consume the *Paramecium*. **[B]** As in flagella, microtubules span the interior length of each cilium and enable it to flex.

"eyelashes"; singular, *cilium*) all over the cell's surface [FIGURE 3.24]. The cilia beat in concert like the oars of a medieval galley ship, allowing the cell to swim quickly. In some protozoa, and in the cells of certain multicellular organisms, cilia can serve a different function, sweeping fluid and particles across the stationary cell. In cells that line the human breathing passages, for example, cilia sweep dust particles out toward the mouth and nose, where they are eventually expelled in mucus or swallowed. These cilia are often damaged in smokers, contributing to "smoker's cough" and greatly increasing smokers' chances of developing lung cancer. This is because damaged, inactivated cilia cannot move mucus and harmful particles from the lungs.

A kind of cell movement that involves neither flagella nor cilia causes the devastating migration of cells when a malignant cancer begins to spread, or *metastasize* (from the Greek word "to change"). These creeping movements of animal cells rely on microtubules and other components of the cytoskeleton, just as a running squirrel depends on its bony skeleton [FIGURE 3.25]. Using magnification, you can see that the rear of the cell attaches to a surface and that the front edge of the cell protrudes forward in sheetlike ruffles that appear to flutter. The ruffling edge thrusts forward, and the lower cell surface sticks to the substrate. Eventually, the rear end snaps free of its surface attachment, and the cell moves forward.

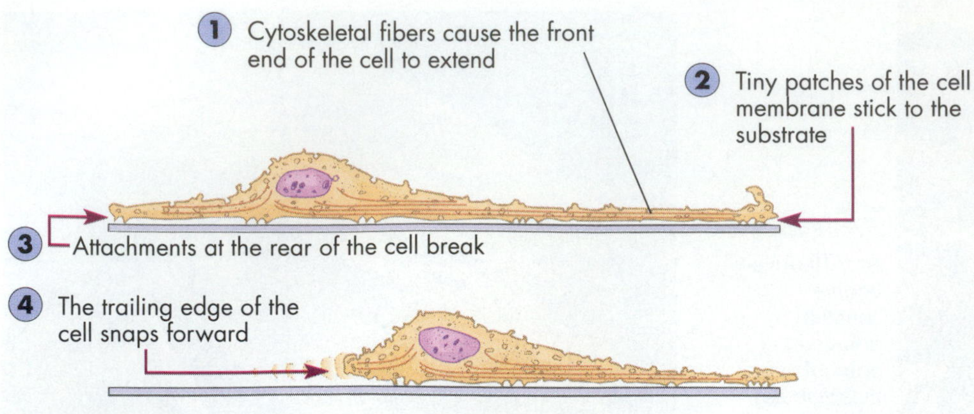

FIGURE 3.25
Cell Movements.
The changeable shape of an animal cell, as well as the ability to creep and glide forward, is maintained by the dynamic cytoskeleton. As an animal cell in a Petri dish creeps forward, the front end flattens like a ruffle blowing in the wind as microfilaments rapidly extend (1) and contract. As the leading edge stretches forward, sticky areas (adhesion plaques) on the lower surface cling to the substrate below (2). Eventually, the cell contracts (3), the trailing edge snaps forward in inchworm fashion, and the cell moves toward the ruffling edge.

Cell Coverings

The boundary of a cell cannot be said to fall strictly at the plasma membrane, because virtually all cells secrete coverings—either strong cell walls or fluffy coatings—that protect the delicate plasma membrane and offer other advantages as well.

CELL WALLS

Cell walls, composed largely of cellulose, surround plant cells on all sides [FIGURE 3.26]. First the cells lay down a *primary cell wall* immediately outside the plasma membrane. The primary cell wall remains stretchable and flexible until the cell has stopped growing and has begun to mature. The wall is also porous, allowing water, gases, and some solid materials to pass through to the plasma membrane. Many plant species lay down a *secondary cell wall* inside the primary wall; it can be very rigid from stiffening materials such as *lignin*, the major component of woody tissue. If you are working at a wooden desk, your work is being supported by thickened, dead cell walls—all that remains of a once-living tree. If you are dressed in a cotton garment today, you are wearing the spun and woven cell walls of little hairs that grew from a cotton plant's seeds.

Some fungi and protists have cellulose in their cell walls, as well as a wide variety of other molecules. Eubacteria and archaebacteria also secrete cell walls, but these are made up of complex materials that include sugars, lipids, and amino acids rather than cellulose. The cell walls of archaebacteria contain unusual polysaccharides, which is one reason why they are classified as a kingdom distinct from Eubacteria.

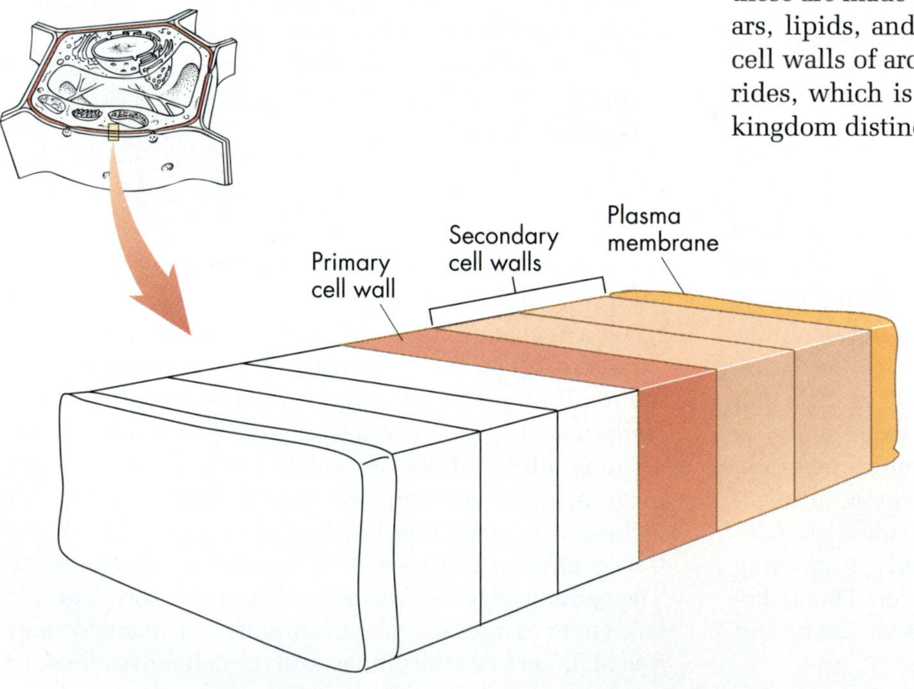

FIGURE 3.26
Plant Cell Walls.
Plant cells are surrounded by walls made up mostly of cellulose, the most abundant organic molecule on earth. As a newly divided plant cell matures, it lays down a porous primary cell wall just outside the plasma membrane that remains flexible until a secondary cell wall, usually more rigid than the first, is deposited inside the primary wall, close to the plasma membrane.

EXTRACELLULAR MATRIX

Most animal cells secrete an **extracellular matrix**, an external meshwork of fibrous proteins linked to polysaccharides. The matrix surrounds and supports the cell and glues it to adjacent cells [FIGURE 3.27]. The most common of these fibrous proteins is **collagen** (KAHL-uh-gen), which, as you saw in CHAPTER 2, has stiff, ropelike polypeptide chains wound around each other into fibrils. Collagen molecules constitute 25 percent of all the protein in a typical mammal, and most of the material in tendons, the cables that enable muscles to move bones. One can usually feel the cords that run behind the knee by bending the leg and pulling the heel backward. Those cords—the tendons—are almost entirely made up of proteins in the extracellular matrix.

Bone is also largely composed of extracellular matrix. One of the symptoms of a child with Tay-Sachs disease is a greatly enlarged skull by 18 months of age. This enlargement is caused by an increased deposition of extracellular matrix.

Another protein in the extracellular matrix is *fibronectin*, which abundantly surrounds normal cells, but is absent from cancer cells and is a factor in their ability to spread throughout the body. This shows that proteins of the extracellular matrix can dramatically affect cell behavior as well as cell shape.

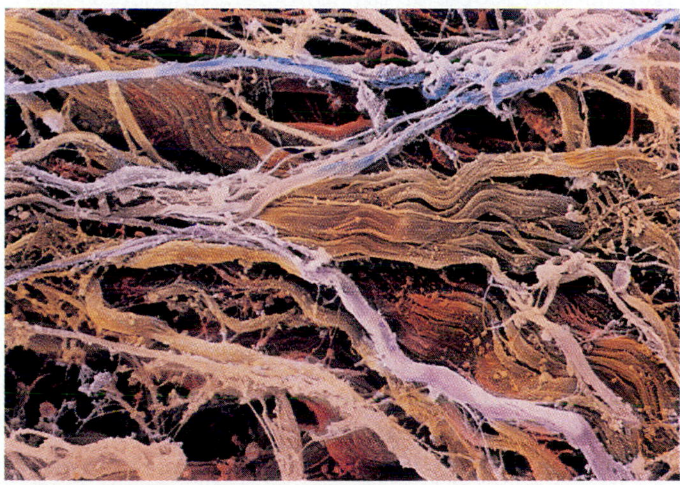

FIGURE 3.27

Extracellular Matrix: An Exterior Meshwork.

Most animal cells are surrounded by a meshwork of protein and polysaccharide molecules. This extracellular matrix cushions the cell, strengthens the tissue the cell is part of, and helps establish and maintain the cell's shape. The most common extracellular material in animals is collagen. This false-color scanning electron micrograph reveals bundles of collagen and elastic fibers surrounding an animal cell.

Links Between Cells

Cells in multicellular organisms are not only embedded in an extracellular matrix, they are also attached to neighboring cells by physical linkages called **intercellular junctions**. Like the matrix materials, these junctions help weld cells together into functional tissues and organs, but they also allow the cells to communicate freely, and they enable the activities of various cell types to be coordinated. While some linkages allow materials to flow between cells, others, in organs such as the urinary bladder, prevent leaks between cells. Still other junctions, called **adhering junctions**, hold cells together. People with genetically defective adhering junctions often lose large patches of skin from the slightest scrape. FIGURES 3.28 and 3.29 describe the various cell junctions.

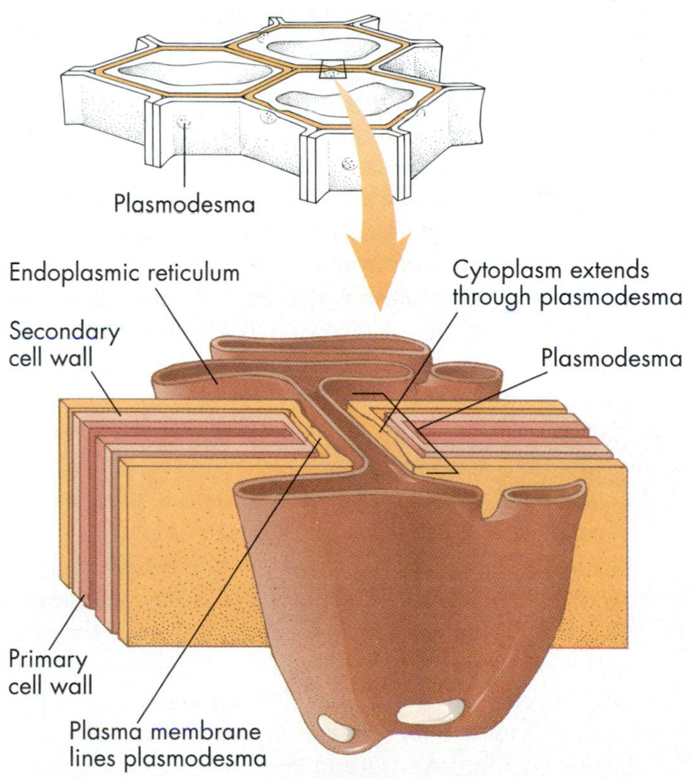

FIGURE 3.28

Communicating Links between Plant Cells.

In plants, encasement by the cell wall would isolate and block intercellular coordination if it weren't for special links. Called *plasmodesmata*, these plant cell connectors are delicate bridges of cytoplasm that pass through the walls of adjacent cells and link them, allowing the rapid exchange of materials and sometimes electrical signals.

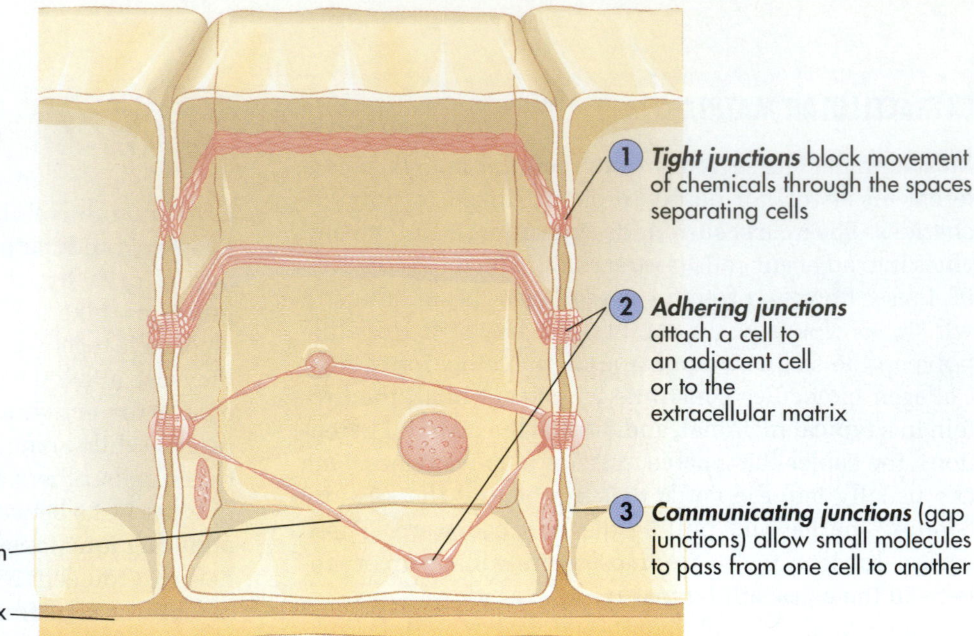

FIGURE 3.29

Junctions between Cells.

Numerous junctions allow linkage and communication between cells. A *tight junction* is a band around the cell that forms a tight seal and prevents molecules from passing between it and the next cell. A *belt desmosome* is a ring of actin filaments that encircles a cell and helps it adhere to adjacent cells. A *spot desmosome* is like a tiny spot weld that joins two cells and anchors intermediate filaments. A *gap junction* is a protein-lined pore that connects adjacent cells and allows communication.

Tay-Sachs Disease

In this chapter you had the opportunity to apply the general principles of cell biology to a specific medical problem and to use your critical thinking skills to propose a solution. Now let's pull all the clues and hypotheses together to reach an understanding of Tay-Sachs disease.

The problems of a child with Tay-Sachs disease seem related to the accumulation of membranous whorls in brain cells [see FIGURE 3.1]. Those whorls, as it turns out, disrupt brain cell function, accounting for the child's stiff response when startled, and preventing the child from learning to speak or even to sit up. The whorls also increase the size of the brain cells, and in turn, the whole brain. This causes cells in the skull to secrete more extracellular matrix and causes skull bones to enlarge.

The whorls consist mainly of the sugar-lipid G_{M2}, a normal component of a brain cell's plasma membrane. Old G_{M2} molecules are normally removed from the cell surface by endocytosis and delivered to lysosomes where they are broken down and recycled. In a Tay-Sachs patient, the G_{M2} molecules are not broken down, and as a result they accumulate. We can conclude that something must be wrong with the lysosomes, and children with Tay-Sachs disease are missing the activity of the lysosomal enzyme hexoseaminidase, which normally removes part of the sugar from G_{M2}. Without the enzyme, G_{M2} accumulates, is stored in whorled membranes, and causes the symptoms we have discussed. Since this disease is genetic, the basic defect must lie in DNA present in all of the child's cells. The cells of Tay-Sachs patients lack information for making functional hexoseaminidase protein. Since brain cells are especially dependent on this enzyme, brain function is disrupted in its absence.

Physicians still do not know how to prevent the damage caused by the absence of normal hexoseaminidase in brain cells. Knowing that Tay-Sachs disease is genetic, and therefore inherited, scientists have been able to identify couples who might produce children with the disease. If both the husband and wife show hexoseaminidase concentrations about half as large as the average, their chances are about 1 in 4 of producing a child with Tay-Sachs disease. In the last 20 years or so, many Jewish people of Eastern European ancestry have chosen to have their hexoseaminidase levels tested before having children. On the basis of results from those tests, the couples have been able to plan their families. As a result, the number of babies who suffer from Tay-Sachs disease has decreased dramatically.

Connections

This chapter investigated the fundamental nature of cells. To understand how cells function, we drew on our knowledge of the molecules of life from CHAPTER 2. An example was the way lipids form cell membranes. To understand the diversity of cell types, we had to recall the kingdoms of life from CHAPTER 1, recognizing that the two prokaryotic kingdoms have much simpler cells than the four eukaryotic kingdoms, and recognizing the special organelles that occur, for instance, in cells of the plant kingdom.

The principles of cell biology apply directly to human life, and the example of Tay-Sachs disease we discussed in this chapter is but one instance. During your lifetime, you will have many opportunities to apply the knowledge of cell structure you learned in this chapter.

KEY TERMS

cell theory, 68
cell wall, 90
central vacuole, 87
chloroplast, 86
chromosome, 80
cilium, 88
collagen, 91
contractile vacuole, 87
cytoplasm, 80
cytoskeleton, 81
electron microscope (EM), 69
endocytosis, 78
endoplasmic reticulum, 82
eukaryotic cell, 73
exocytosis, 79
extracellular matrix, 91
flagellum, 88
fluid-mosaic model, 77
Golgi apparatus, 84
intercellular junction, 91
light microscope, 68
lipid bilayer, 76
lysosome, 84
microbody, 85
mitochondrion, 85
nuclear envelope, 80
nucleolus, 80
nucleus, 79
plasma membrane, 74
prokaryotic cell, 72
ribosome, 81
rough ER, 84
scanning electron microscope (SEM), 69
selectively permeable membrane, 77
smooth ER, 82
surface-to-volume ratio, 67
transmission electron microscope (TEM), 69
transport vesicle, 82

HIGHLIGHTS IN REVIEW

1 Cells are the fundamental units of life, the smallest units that can independently acquire energy, metabolize, develop, and reproduce.

a] Cells vary in shape, size, and internal structure, depending on the function.

b] The parts of a cell do not in themselves make up a living unit because life is an emergent property based on the physical organization and integrated activities of all cell parts.

c] Most cells are extremely small: About five average animal cells can fit across the thickness of a piece of paper. Cell size is limited by the surface-to-volume ratio, because the larger a cell's volume, the greater are its needs for material exchange at the cell surface.

d] Two centuries after the discovery of cells, German scientists developed the cell theory: (1) All living things are composed of one or more cells; (2) cells are the basic units of life and all of life's chemical reactions take place inside of cells; (3) all cells arise from preexisting cells.

2 A lipid-containing plasma membrane surrounds all cells and separates the cell's interior from its surroundings. Within the cell is a region packed with DNA, the hereditary material. The space between the plasma membrane and the DNA is filled with cytoplasm, which contains structures that carry out life processes.

a] A plasma membrane keeps harmful substances out, sequesters useful substances within, and regulates the traffic of raw materials and wastes into and out of cells. The plasma membrane is a semifluid bilayer of phospholipid molecules studded within protein molecules that accomplish the actual "traffic control."

b] Some materials enter or exit the cell via inpocketing and outpocketing of the cell membrane. This is how cholesterol is cleared from the blood.

c] Most of the cell contains cytoplasm, a gel-like substance that is mostly water with dissolved proteins, small molecules, and suspended particles. The cytoplasm acts as a pool of raw materials and contains a latticework, the cytoskeleton, that suspends the organelles.

d] Ribosomes are small spheres of RNA and protein that exist either free in the cytoplasm or attached to the endoplasmic reticulum. Ribosomes are the workbenches on which proteins are made.

3 Eukaryotic cells contain a membrane-bound nucleus, as well as many other membrane-enclosed organelles. Prokaryotic cells lack a nucleus and other membrane-bound organelles.

a] The kingdoms Eubacteria and Archaebacteria contain only species made up of single prokaryotic cells. The kingdoms Plantae, Animalia, Fungi, and Protista contain only species made up of one or more eukaryotic cells.

b] Most of a cell's day-to-day activities are determined by information molecules issuing from DNA in the cell nucleus. In a eukaryotic cell, the nucleus is often the largest organelle. The nucleus is surrounded by a nuclear envelope, and it contains a dark-staining region, the nucleolus, and produces ribosomal RNA.

c] A eukaryotic cell has a system of internal membranes and organelles through which materials move toward various destinations.

d] The endoplasmic reticulum is a network of flattened, hollow tubules and sheets extending throughout the cell and forming interconnected channels. Rough ER is studded with ribosomes and is involved in the production and transport of proteins such as the enzymes that digest a person's food. Smooth ER lacks ribosomes and produces, stores, and secretes steroid hormones and other lipids, including vitamin D, which is necessary for strong bones.

e] Materials made in the ER can be stored in vesicles or enter the Golgi apparatus—a stack of saucer-shaped, baglike membranes that modify the molecules produced in rough and smooth ER and package them for export from the cell or for storage in the cytoplasm.

f] Lysosomes are spheres containing digestive enzymes that recycle worn-out cell parts. The lack of even one of these enzymes can and does cause Tay-Sachs disease.

g] Cells of many-celled organisms (plants, animals, and fungi) have evolved specialized structures and activities that permit the coordinated functioning of whole organism.

h] Mitochondria are small, sausage-shaped bodies that harvest energy from certain organic molecules and produce the chemical fuel ATP that powers cell activities. During that harvest, mitochondria use up oxygen.

i] Plant cells and some protists contain plastids, which

are small bodies with double membranes. Chloroplasts are green plastids that capture the energy of sunlight by photosynthesis. Chromoplasts store yellow, orange, and red pigments in flower petals and other structures, while leucoplasts store starch granules, for example, in a potato.

j] A central vacuole filled with water and some stored molecules takes up most of the space in a plant cell. A wilted plant's vacuoles are not filled with water. Contractile vacuoles pump excess water from cells that live in fresh water.

k] Organelles of movement, the flagella and cilia, have a similar array of microtubules and can move entire cells, for example, sperm, or can move substances past cells, for example, clearing the lungs and airways of debris.

l] Plant cells have outer walls made mostly of cellulose. Animal cells have an extracellular matrix that surrounds them in a supportive meshwork and that acts like intercellular glue. This meshwork is the main component of tendons.

m] Cells often have physical links to their neighbors that allow passage of materials and signals between cells and that prevent leakage from one side of a cell to the other, among other functions.

UNDERSTANDING THE FACTS AND CONCEPTS For Questions 1–5, match each of the descriptions with the one item from the following list that is most appropriate. Any item may be used once, more than once, or not at all.

a] plasma membrane
b] cell wall
c] nuclear envelope
d] nucleoli
e] chromosomes

1 A specialized cell that makes little or no protein (such as a sperm cell) might not have this structure.

2 There are four layers of phospholipids in this structure.

3 This threadlike structure is not made from lipids or carbohydrates, it is not membranous, and its number is constant except during cell division.

4 Some of the lipids in this dynamic structure may be joined to sugar molecules.

5 This structure is composed of carbohydrates in many organisms, or of lipids and/or amino acids in others. It is freely permeable to watery liquids, gases, and even to some solids.

As above, for Questions 6–10, match the descriptions with the appropriate item from the following list.

a] ribosomes
b] cytoskeleton
c] smooth endoplasmic reticulum
d] rough endoplasmic reticulum
e] Golgi apparatus

6 This structure synthesizes some lipids and vitamin D.

7 This protein component of cells moves several other organelles around in the cell and also helps cells to maintain their shape.

8 This structure is comprised of a system of intracellular channels through which proteins travel and often are modified as they move.

9 These beadlike structures are composed of nucleic acids and proteins and are the site of protein synthesis.

10 This structure modifies, "packages," and releases for distribution molecules of protein, lipid, and/or carbohydrate.

As above, for Questions 11–15, match the descriptions with the appropriate item from the following list.

a] mitochondria
b] cell membrane
c] nucleoid
d] cell wall
e] chloroplast

11 A structure found in all prokaryotic cells but not in any eukaryotic cells.

12 A structure found in all prokaryotic and eukaryotic cells.

13 Found in virtually all eukaryotic cells but not present in any prokaryotic cells.

14 Found in most prokaryotic cells and in some eukaryotic cells.

15 Not found in prokaryotic cells, and only in certain eukaryotic cells.

As above, for Questions 16–20, match the descriptions with the appropriate items from the following list.

a] nucleolus
b] lysosome
c] mitochondrion
d] cilium
e] collagen

16 The most abundant fibrous protein in animals; produced intracellularly but found within the extracellular matrix.

17 Contains a special type of RNA that is required for cellular synthesis of proteins.

18 Permanent cytoplasmic vesicles that contain digestive enzymes used to destroy intracellular debris.

19 Cellular projections that either help to move a cell through its environment or to move the environment over the cell.

20 Makes some of its own protein, is passed down from mother to offspring, and is the major site of aerobic metabolism in cells.

21 Label the phospholipids, proteins, glycoproteins, and cholesterol in the figure on next page.

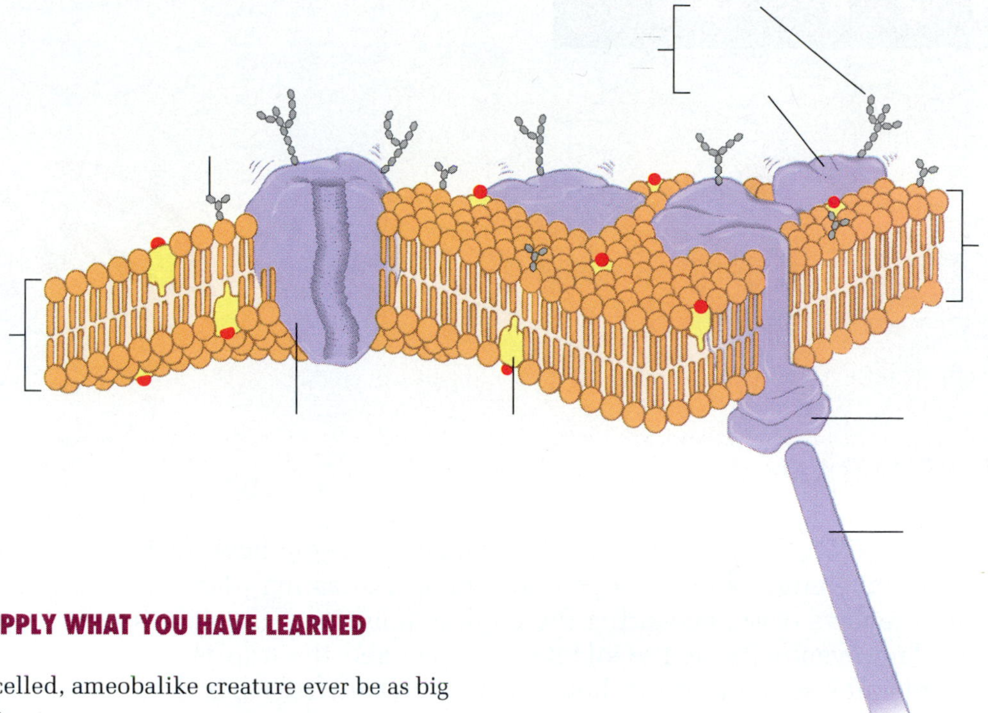

INTEGRATE AND APPLY WHAT YOU HAVE LEARNED

1 Could a single-celled, ameobalike creature ever be as big as a truck? Explain.

2 What are the *minimum* requirements for something to qualify as a cell?

3 Are the substances that are contained within endocytotic vesicles really inside the cell?

4 Compare the coverings around prokaryotic and eukaryotic cells.

ANALYSIS

1 Assuming that all the following shapes of cells contain the same volume of cytoplasm, which would have the lowest ratio of surface-to-volume?

- **a]** Cube
- **b]** Pyramid
- **c]** Cylinder
- **d]** Sphere
- **e]** Elongated thread

2 If you knew that the disease cystic fibrosis was characterized by a disrupted balance of salts inside and outside of cells, you might guess that ______ molecules in the cell membrane were faulty because ______.

- **a]** Lipid; phospholipids are freely permeable to salts
- **b]** Lipid; the lipid pores must not be functioning right
- **c]** Lipid; they regulate the passage of ions and salts
- **d]** Protein; they regulate the passage of ions and salts
- **e]** Protein and lipid; together they regulate the permeability of the membrane to ions and salts.

3 Image an arctic bird standing on ice and a flamingo (a tropical bird) standing in a muddy lake. Suppose you know that the cell membranes of the cells in their feet are equally fluid in the two birds and that their body temperatures are virtually the same. Which of the following would you logically conclude from the material that you learned in this chapter?

- **a]** The cells in the arctic bird must be larger because they must generate more metabolic heat.
- **b]** The proteins in the cell membranes of the two birds must differ because they regulate the consistency of the membrane.
- **c]** The lipids must be different because they regulate the fluidity characteristics of the cell membrane.
- **d]** The cell membranes of the two kinds of birds must be fundamentally different.
- **e]** The arctic bird's cells must be covered by more layers of cell membrane in order to insulate them.

4 You can distinguish the outer surface of the cell membrane from the inner surface because only the outer surface ______.

- **a]** Contains phospholipids
- **b]** Contains carbohydrate receptors
- **c]** Is attached to cytoskeletal proteins
- **d]** Is hydrophobic
- **e]** Is penetrated by proteins

5 If you wished to observe as much detail as possible of the surface of a mitochondrion, which of the following microscopes would you be likely to choose?

- **a]** Light microscope.
- **b]** TEM.
- **c]** SEM.
- **d]** Differential contrast microscope.
- **e]** It wouldn't matter as long as the resolving power was sufficient.

CHAPTER 4

The Dynamic Cell

THE VENUS FLYTRAP

It could be a scene from *Little Shop of Horrors*: An animal exploring new territory is lured by an enticing aroma into the open jaws of a meat-eating plant. Silently and quickly the jaws close, ensnaring the hapless prey. Digestive juices begin to dissolve it. And eventually, as the silent jaws open and the trap is set once again, nothing remains of the intruder but a skeleton. One writer has called the feeding of carnivorous plants "life imitating Edgar Allan Poe." But it is neither macabre short story nor movie plot.

Perhaps the quintessential carnivorous plant is the Venus flytrap (*Dionaea muscipula*). In a place like Croatan National Forest, a piney preserve near North Carolina's Atlantic seaboard, these plants grow freely in the waterlogged, acidic soil. Each of the plant's bizarre leaves is made up of two green lobes rimmed with stiff, whitish spines and sporting three trigger hairs at the lobe's center [FIGURE 4.1].

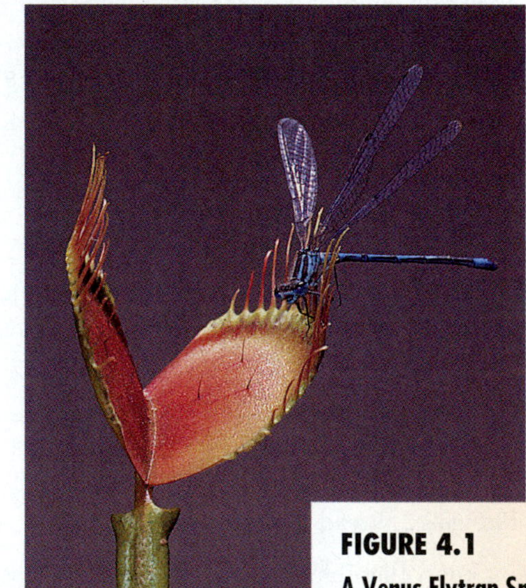

FIGURE 4.1

A Venus Flytrap Snares a Meal.

When an insect touches two sensitive hairs or one in quick succession, dynamic trigger cells at the base of the hair initiate electrical signals that cause other leaf cells to swell. Within half a second, the trap springs shut, snaring the insect. Digestive enzymes secreted by other cells digest the insect's body, and the plant assimilates most of the animal's chemical constituents.

At the base of these hairs are cells with special abilities. When an ant or dragonfly lands and bends the hair twice in quick succession, *trigger cells* at the foot of the hair are deformed, as if pried by a lever. Stimulated by the stress, trigger cells generate an electric signal that flows from cell to cell through the leaf. Specialized *motor cells* receive the signal, change shape, and cause the trap to close.

As the doomed animal struggles, the leaf closes tighter. The thrashing breaks open cells that have synthesized huge quantities of digestive enzymes, proteins that begin to dismantle the animal's body molecule by molecule. Within two days, nothing remains of the prey but the indigestible husk, the animal's other tissues having been broken down and absorbed by the plant. At this point, the leaf-mouth begins to open once again, and the smell of sweet plant secretions will probably lure another insect within minutes or hours.

The shutting of the flytrap depends on the activities of individual cells within each leaf—trigger cells, motor cells, and digestive cells. That makes carnivorous plants a good central example for our discussion in this chapter of the life-support activities within a living cell. These include chemical tasks such as breaking down nutrients, transport tasks such as moving materials into and out of cells, and mechanical tasks such as the expansion and contraction that opens and closes a Venus flytrap.

The precise activities within a cell depend on a continual flow of energy from the environment. In a Venus flytrap, as in nearly all plants, this energy originates from sunlight trapped in the plant's leaf cells in the process of photosynthesis. The light energy is transformed into chemical energy and stored in the bonds of sugar and starch molecules. The complex process of photosynthesis requires special substances called pigments which capture light energy; protein catalysts called enzymes which speed up chemical reactions; and the energy-transfer molecule ATP. Cells also depend on a flow of materials—mostly water, ions, minerals, nutrients, and wastes—which are used during the maintenance of order and activity.

Some materials can diffuse into or out of cells through the plasma membranes, but others require special transport proteins that may or may not expend ATP energy. This chapter explains how the universal laws of energy apply to living cells, why a cell requires energy, how energy is used, and how enzymes can facilitate most of the cell's chemical, transport, and mechanical tasks. Understanding these topics is crucial to our exploration of biology because all organisms are made up of dynamic cells carrying out the necessary tasks for sustaining order, living day to day, and reproducing.

The special cells that open and close a Venus flytrap are typical in many ways, but they are specialized, too, in ways that facilitate the plant's meat-eating habit—and in turn its survival in its environment. Carnivory among plants is an evolutionary innovation that allows the flytrap to exploit a boggy environment where most other plants would fail to flourish. There are 350 or so carnivorous (meat-eating) plant species around the world, and Croatan, North Carolina, has the widest selection of any region its size in the United States. Plant ecologists have determined that carnivorous plants generally share poor environments that lack sufficient absorbable minerals or other nutrients in the soil to support the life of most types of flowering plants. Evidence suggests that digested insects help the plants procure these missing nutrients, but at a cost: the trigger cells use energy to pump ions rapidly in and out to generate an electrical signal; the motor cells expend energy to open and close the leaves; and the digestive cells spend energy producing digestive enzymes that break down the prey.

As we investigate the dynamic activities of cells in this chapter, you will drop your textbook, float a toy boat, or mix cream in coffee for science; see how much energy a cell burns in a second; learn how a hen's egg protects itself; and find out what your red blood cells are up to. So let's begin with the ultimate power trip: a tour of the universal laws of energy. ❑

MESSAGES

1 All organisms require a source of energy. While energy cannot be created or destroyed, it can change from one form to another, say from chemical energy to electrical energy.

2 In energy conversions, some energy is inevitably lost as heat, the random, disordered motion of atoms and molecules; therefore, without the input of energy, everything tends to become more disorganized over time.

3 Enzymes are proteins that speed up the rates of chemical reactions.

4 Some substances enter cells by diffusion, but other substances cannot move into or out of cells without special protein in the cells' plasma membrane.

Cells and the Universal Energy Laws

A living cell such as a trigger cell in a Venus flytrap [FIGURE 4.2], or a lung cell in your body, is like a bustling metropolis that runs on energy. Materials stream in and out through the "city walls"—the plasma membrane. Instructions issue from the "city government"—the DNA housed in the nucleus—for cell maintenance and the synthesis of materials. Ribosomes, like microscopic factories, churn out proteins, and the endoplasmic reticulum and Golgi apparatus can process and ship these materials to other "cities." Lysosomes break down wastes and recycle useful raw materials. The plastids of plant cells produce and store carbohydrates like miniature gardens and warehouses. And the subcellular power plants called mitochondria "burn" organic molecules and capture the energy in a form more useful for powering the cell.

FIGURE 4.2

Trigger Cells and Digestive Cells in a Venus Flytrap.

A scanning electron micrograph of a sensitive hair reveals the trigger cells that generate an electric signal after becoming deformed when an insect bends the hair. The digestive cells are also present as elements of a digestive gland.

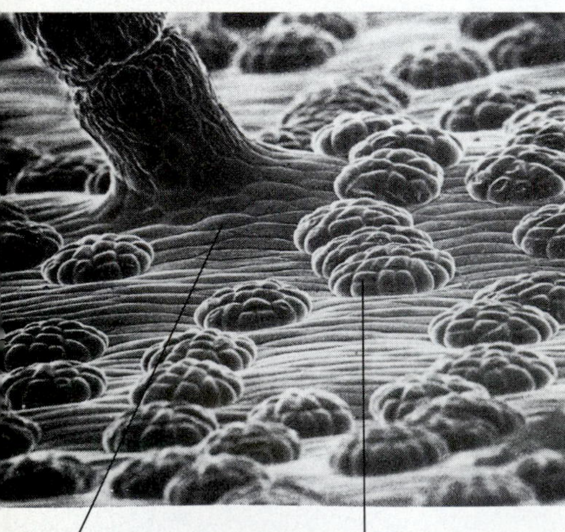

Just like cities, automobiles, and airplanes, cells need a constant flow of energy to carry out their tasks, and the energy flow proceeds according to certain universal laws. You saw in CHAPTERS 2 and 3 that cells are made up of organelles, that these cell parts are in turn constructed of atoms and molecules, and that these constituents act according to the chemical and physical laws of the universe. A great deal can therefore be understood about the behavior of cells by learning about the universal laws of energy.

THE LAWS OF ENERGY CONVERSIONS

Let's try a simple energy demonstration. Stand up, close this book, hold it at about chin level parallel to the floor and drop it. Unless you have a very thick carpet, the book probably landed with a bang. What caused this loud noise? A physicist might describe it as a series of energy conversions. **Energy** is the ability to perform work or to produce change, and you can probably think of several energy conversions that took place after you dropped the book.

States of Energy As you lifted the book from the desk to chin height, you stored energy in the book in the form of **potential energy**, energy that is available to do work [FIGURE 4.3A]. Likewise, a huge boulder poised at the edge of a cliff, the water saved up behind Grand Coulee Dam, and the chemical bonds inside a piece of firewood all contain stored energy. This potential energy is capable of being released and accomplishing work, such as smashing a hole in the road at the bottom of the cliff, turning the turbines of a hydroelectric generator, or heating pancakes on a griddle. When you dropped the book, it began falling, and as the book changed position in space, the stored potential energy was released [FIGURE 4.3B] and transformed into **kinetic energy**, the energy of motion. These two forms of energy, potential and kinetic, are the two major states of energy in the universe, and each can be converted into the other.

These two energy states can both assume a number of different forms. As your textbook slammed against the floor, the floor and book both warmed a bit. Here some of the kinetic energy was transformed into heat energy. When the book hit the floor, some of its kinetic energy was transformed into sound energy. Sound waves reaching your eardrum caused it, and other, internal, parts of the ear to vibrate (kinetic energy) until eventually, the kinetic energy was transduced (changed) into a signal of electrical and chemical energy that traveled to your brain, where it was interpreted as a loud slam.

Conservation of Energy The conversion of energy from one form into another, say, from chemical energy into electric energy, is addressed by the **first law of ther-**

[A] Potential energy

Energy of movement

[B] Kinetic energy

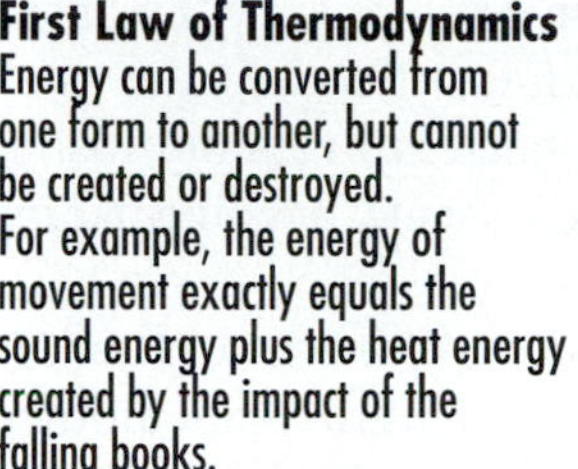

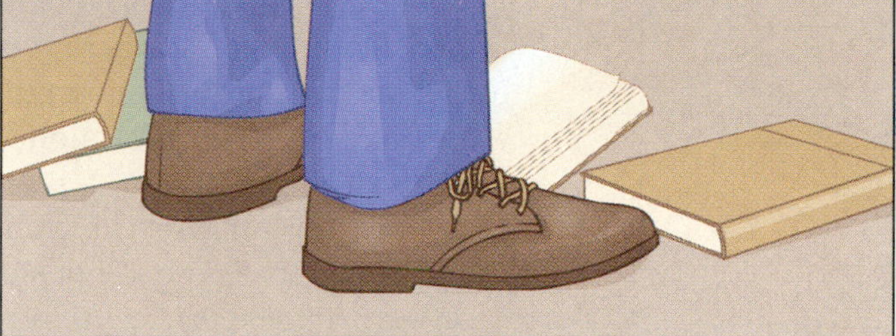

Second Law of Thermodynamics
Energy conversions are inefficient, and the disorder (entropy) in the universe continually increases spontaneously. For example, the energy of motion of the book was not completely converted to sound: some was lost as heat.
The books lying on the floor make your room look messy and disordered. You must expend energy to file them back on your bookshelf. Over time, all systems become disordered without continual input of energy.

[C] Laws of Thermodynamics

FIGURE 4.3

The Universal Laws of Energy Transformation.

[A] Potential energy (stored energy) can be as simple as the energy stored in books being raised to shoulder height. [B] As a book falls, potential energy converts to kinetic energy (energy of motion). [C] When the book hits the floor, further energy transformations occur that illustrate the laws of thermodynamics given in the figure.

modynamics (*thermo-*, "heat," and *dynamics*, "change"). This principle says that energy can be changed from one form to another, but during these conversions, it is *conserved*; that is, it is neither created anew nor destroyed [FIGURE 4.3C]. When the falling textbook, for example, abruptly stopped moving upon hitting the floor, the amount of energy released as sound waves in the air, plus the amount of energy that warmed the floor, book, and surroundings, would have exactly equaled the kinetic energy in the moving book. No new energy would have been produced, and none destroyed. If new energy can't be made, and existing energy can't be destroyed, what does this say about the total amount of energy in the universe? Evidently, the amount of energy in the universe must be constant.

Energy Is Continually Lost as Heat The amount of energy is the same before and after an energy conversion (the first law), but *the amount available to do useful work always decreases during the change.* This principle is the **second law of thermodynamics**: Energy conversions are never 100 percent efficient. For example, energy in the sound waves from the book slapping the floor is less than the total kinetic energy of the falling book because some energy was lost as heat. During every type of energy conversion, some energy is inevitably lost to the surrounding environment as **heat**, which scientists define as the random movement of atoms and molecules.

Since randomness is the opposite of order, the inefficient conversion of energy from one form into another (with some energy lost in a form that can no longer do work) results in increasing disorder in the system. Scientists use the term **entropy** (EN-tro-pee) as a measure of the disorder or randomness in a system. The more disorder, the greater the entropy. Entropy accounts for the energy that escapes from the system during transformations and is no longer usable. Anyone who has ever tried to keep a college dorm room tidy is already quite familiar with the concept of entropy: Without a constant input of energy to clean up dust and dirt, and put clothes and books away, entropy (disorder) tends to increase, and things go downhill fast.

ENVIRONMENTAL CONSEQUENCES OF THE SECOND LAW

The second law of thermodynamics has profound consequences for cells, organisms, and ecosystems. Consider the chain of energy conversions from the sun at the center of our solar system to a plant growing on the ground and to an animal nibbling on that plant. Nuclear reactions taking place in the sun release not only light, but also massive amounts of heat, which are forever lost in

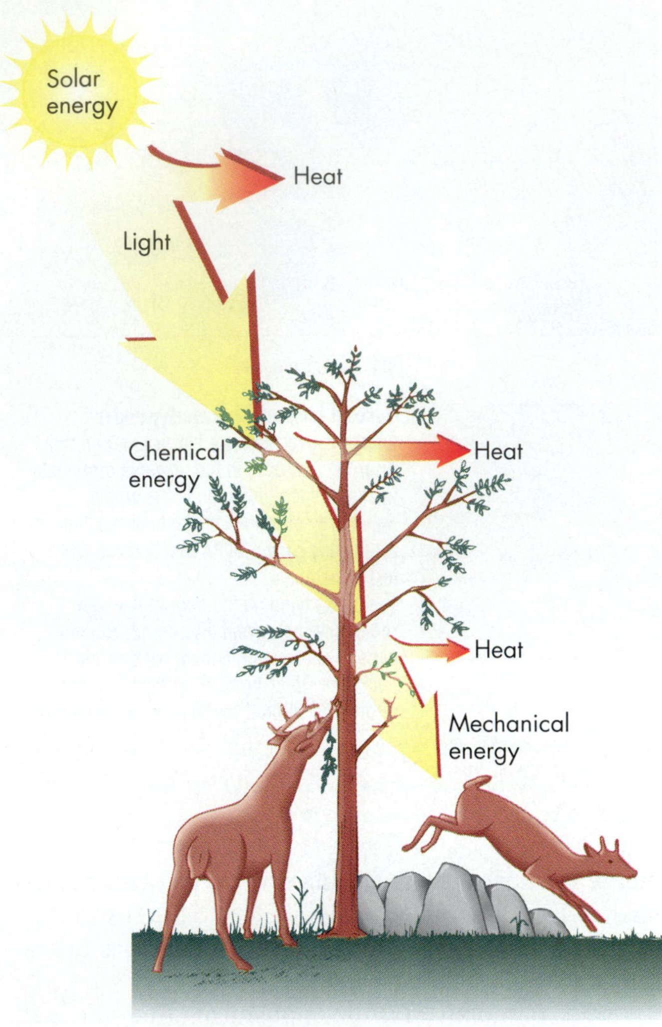

FIGURE 4.4

Energy Flow in the Environment.

The flow of energy through an ecosystem is predicted by the laws of thermodynamics. The first law of thermodynamics states that energy can be changed from one form to another but is neither created nor destroyed. Here, nuclear energy from atomic fusions taking place in the sun is converted to light, to chemical energy in the plant's tissues, and to mechanical energy in the animal's tissues. The second law of thermodynamics states that all such interconversions are inefficient to some degree. Thus, with each interconversion in this chain, some energy is lost as heat, diffuses away, and becomes uselessly disorganized.

space [FIGURE 4.4]. Sunlight enters the earth's atmosphere and strikes the leaves of an alder tree. Most of the light is reflected as heat and once again lost, simply radiating away into the atmosphere in a low-quality, unusable form. (Gathering up such heat energy for use would require the expenditure of more energy than you could recover!) Some of the light is reflected as the green light we see in the plant's color. A small amount of light energy is absorbed by substances in the leaf and converted into chemical energy and stored as starch and other molecules; but during this process, still more heat is lost. Finally, a deer grazes on the alder's leaves, and its digestive system breaks down the leaves' complex molecules into simpler ones, releasing energy that is used to fuel ongoing life processes. As always, the conversion is inefficient, and energy radiates into the air as body heat from the warm-blooded animal.

Only a tiny fraction of the original solar energy is saved in a chemical form that can be expended to maintain order—in this case, precise movements by the deer's muscles as it bounds across the forest floor. At the same time, a great deal of energy has been irreversibly lost as the random jostling of molecules (heat), accomplishing no work.

We can interpret the grand consequences of the second law of thermodynamics in a very immediate biological way. It takes many pounds of dry matter in a plant to make one pound of dry matter in a deer because most of the energy consumed by the deer warms the deer's body and is eventually lost as heat to the surrounding air. The same holds true when a cougar eats the deer. That is why a forest can support many more pounds of plants than of deer, and many more pounds of deer than of cougars. We can also apply the consequences of the second law of thermodynamics directly to the human diet: Which would be a more efficient transformation of sunlight energy into human tissue and activity, a meal of corn or a meal of corn-fed beef? On our overpopulated planet, will a mostly vegetarian diet or a mostly meat-based diet provide a more efficient sustenance of earth's people?

CELLS AND ENTROPY

Animals and plants—whether cougars, deer, alder trees, or Venus flytraps—are made of cells, and within these cells, all significant energy transformations take place. Therefore, the second law of thermodynamics, the tendency to become disordered, affects organisms most directly at the cellular level. Cells remain healthy and well-ordered only if they obtain enough energy to fuel all their synthesis and maintenance activities, despite the loss of energy dictated by the second law—the inevitable loss of heat to the surrounding environment. A living cell, in a real sense, is a temporary repository of order purchased at the cost of a constant flow of energy. If that flow of energy is impeded, order quickly fades, disorder reigns, and the cell dies. Since enzymes and other proteins are required to maintain that energy flow, the synthesis of new proteins is necessary to stave off disorder. The need for continued protein synthesis is one reason Venus flytraps evolved into insect eaters. Bogs have less nitrogen, molybdenum, and other nutrients than forest soils. The body of an ant, spider, or fly, however, contains enough of these elements to provide the plant's nutritional requirements. Clearly, the nonstop organizational activity of a living cell bears a steep price tag in terms of energy requirements. Our next topic is the ways in which cells use that energy, thus deferring entropy.

CONCEPT CHALLENGE

Some biologists sum up the laws of thermodynamics with the old saying, "There's no such thing as a free lunch." What do you think they mean by that with reference to living systems?

Chemical Reactions and Energy Flow in Living Things

Inside cells, chemical energy fuels the maintenance of order through **chemical reactions**. During a chemical reaction, one set of molecules is transformed into other kinds of molecules, and energy is shifted from the bonds of one set of molecules to the bonds of the other.

CHEMICAL REACTIONS: MOLECULAR TRANSFORMATIONS

At some point, you may have watched a child play with a toy boat powered by vinegar reacting with baking soda. The mixing of the two common kitchen ingredients causes vigorous bubbling that propels the boat forward. This chemical combination—which you could also try in a jar on your kitchen counter—sets up a chemical reaction. During a chemical reaction, the starting substances, or **reactants**, interact with each other to form new substances, the **products**. There are often two reactants and two products, but two reactants can be transformed into one product, and one reactant can become one or two products. This general relationship can be represented by the equation

$$\underset{\text{(Reactants)}}{A + B} \rightarrow \underset{\text{(Products)}}{C + D}$$

where A and B represent reactants and C and D stand for products. In the case of the baking soda and vinegar, the reactants are hydrogen ions (H^+) from the acetic acid in the vinegar and bicarbonate ions (HCO_3^-) from the baking soda. As FIGURE 4.5 shows, an OH^- cleaved

[A] Mixing reactants

FIGURE 4.5

Energy-Releasing Reactions.

[A] Vinegar and baking soda react with each other in an exergonic, or energy-yielding, reaction, manifested by vigorous bubbling. **[B]** During the reaction, the reactants are transformed into products. A hydrogen ion (H^+) from the acetic acid in the vinegar joins with an OH group from the baking soda's bicarbonate ion, and water (H_2O) forms. Taking the OH from bicarbonate leaves CO_2, the gas that bubbles away; this releases energy, and the gaseous products are more disorganized than the reactants. **[C]** In general, in an exergonic reaction, reactants change to products with the release of energy.

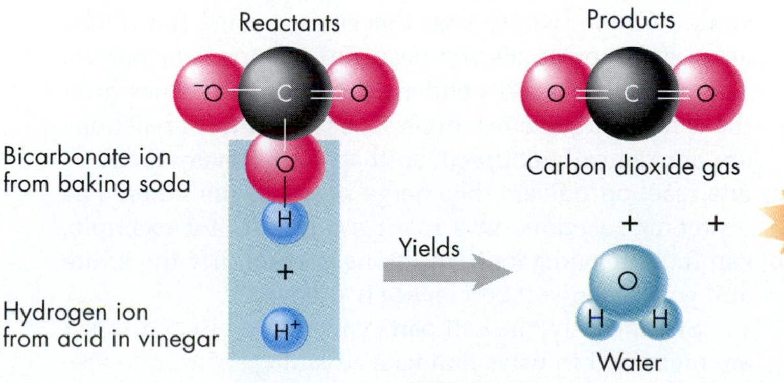

[B] The reaction

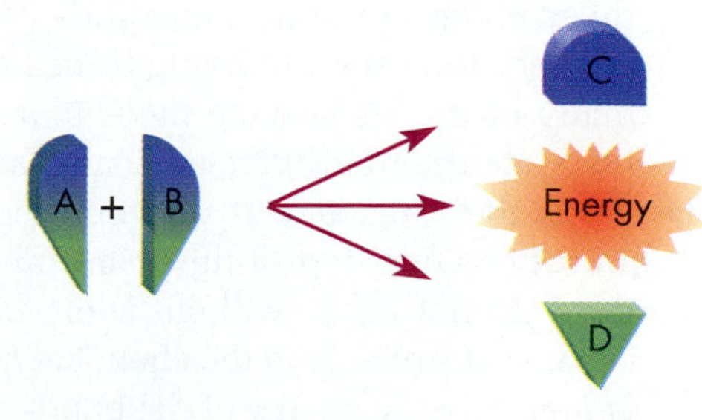

[C] Energy-releasing reaction (exergonic)

from a bicarbonate ion combines with a hydrogen ion. As a result, H_2O forms, along with CO_2, or carbon dioxide gas, the cause of the bubbling. This reaction takes place spontaneously, as soon as the ingredients are combined. There is no need to add heat, electricity, or any other form of energy to make it go.

Energy-Releasing Reactions Reactions that release energy, such as the one between vinegar and baking soda, are called **exergonic** ("energy out") [FIGURE 4.5A]. Heat is released during this reaction because of the inefficiency of energy conversions. As new products form from the reactants, the energy-containing bonds between the carbon, hydrogen, and oxygen atoms become rearranged, and during this conversion, some energy is released as heat, and the entropy (disorder) of the system increases as the carbon dioxide molecules bubble off into the air randomly. This reaction satisfies the second law of thermodynamics: The products contain less energy and are more disordered than the reactants. Likewise, the reactions that digest an insect caught in a Venus flytrap release energy, and the digested animal is more disordered than it was in life.

Some exergonic reactions are like a rock poised at the top of a hill: They need some energy input—a shove by a hiker, let's say—before they will proceed. Nevertheless, even though a little bit of energy is needed to get the reaction started, the overall result is a *release* of energy. A great number of cellular activities involve exergonic reactions and give off heat. A striking example is the contraction of muscle cells, which accounts for the warmth you feel in your muscles after a few minutes of running, aerobic dancing, or lifting weights.

Energy-Absorbing Reactions In addition to the energy-releasing reactions just discussed, there is another category of chemical reactions that do not proceed spontaneously and do not give off heat. These are called **endergonic** ("energy in") reactions. A familiar endergonic reaction takes place when you cook an egg [FIGURE 4.6]; the added energy (heat) causes the egg white proteins to form new chemical bonds and solidify. Endergonic reactions also take place when the chlorophyll pigments in a Venus flytrap's leaf absorb energy in sunlight and synthesize carbohydrate molecules. Endergonic reactions are very important to living things because they include many of the underlying molecular transformations that bring about order in the cell, such as the building of proteins, the replacement of worn sections of membrane, and the generation of new ribosomes. Endergonic reactions do not occur without some form of added energy because the energy of the chemical bonds in the reactants is *less than* the energy of the bonds in the products.

The Energy Source for a Cell's Endergonic Reactions If the order and activity of the dynamic cell depend on an appropriate source of energy for each and every endergonic reaction, where does that energy come from? The answer demonstrates the beautiful economy of nature: The energy for a cell's endergonic reactions comes from the cell's exergonic reactions. The two kinds of reactions are energetically *coupled*, so that the leftover energy of one reaction powers the energy needs of the other. The exergonic reactions in a roaring campfire, for example, can fuel the endergonic reactions that solidify the white and yolk when you boil an egg [FIGURE 4.7].

Significantly, the cell parts you studied in CHAPTER 3 are organized in ways that take advantage of coupled reactions. Ribosomes, for example, are constructed in such a way that the energy-producing reactions that provide

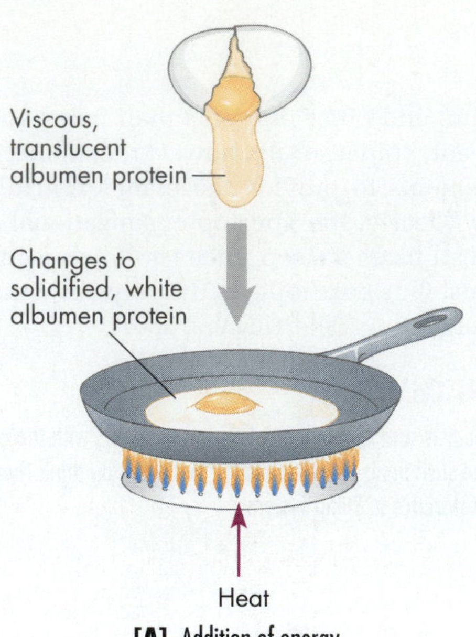

[A] Addition of energy

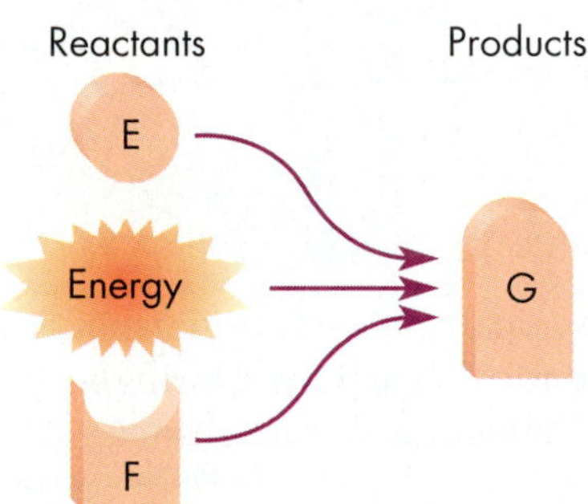

[B] Energy-absorbing (endergonic) reaction

FIGURE 4.6

Energy-Requiring Reactions.

It's strange to think that frying an egg can induce an endergonic, or energy-requiring, reaction—but it does. The addition of heat energy alters the hydrogen bonds in the proteins in egg white. This changes the proteins' shape, and we can see the result: A clear, viscous solution of the proteins becomes opaque, white, and solid.

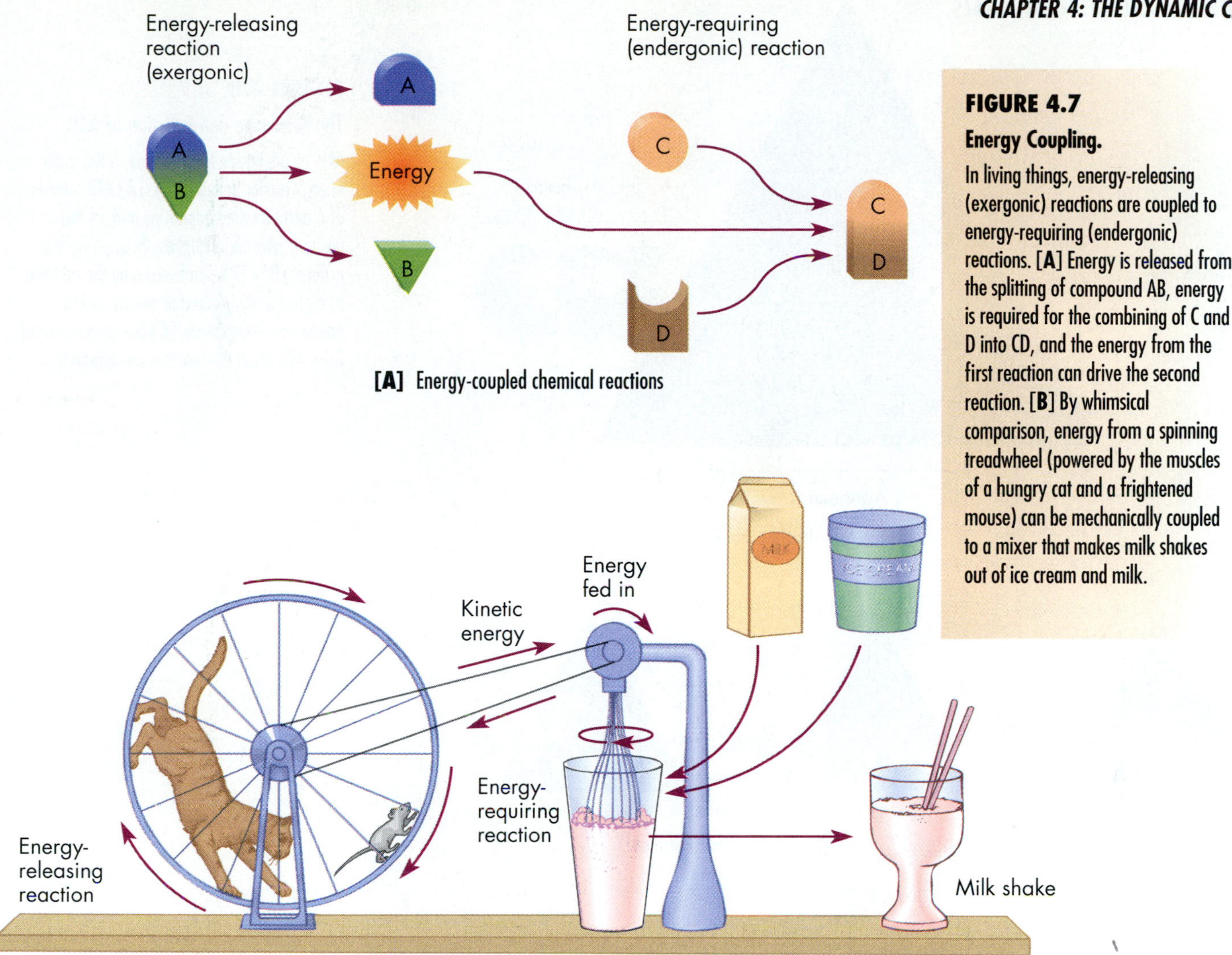

[A] Energy-coupled chemical reactions

[B] Energy-coupled mechanical reactions

FIGURE 4.7

Energy Coupling.

In living things, energy-releasing (exergonic) reactions are coupled to energy-requiring (endergonic) reactions. **[A]** Energy is released from the splitting of compound AB, energy is required for the combining of C and D into CD, and the energy from the first reaction can drive the second reaction. **[B]** By whimsical comparison, energy from a spinning treadwheel (powered by the muscles of a hungry cat and a frightened mouse) can be mechanically coupled to a mixer that makes milk shakes out of ice cream and milk.

power for building new proteins take place right next to the energy-requiring reactions that add new amino acids onto a growing protein chain.

ATP: THE CELL'S MAIN ENERGY CARRIER

If the energy from one type of reaction in a living thing can drive another such reaction, what form does that energy take? Light? Electricity? Heat? Nuclear power? The logical assumption might be heat, since we saw that exergonic reactions often give off heat. But the heat given off in a cell dissipates, and there is no mechanism in cells for gathering it up and putting it to good use. Instead, most of the energy released during an exergonic reaction is quickly trapped in the chemical bonds of an energy-carrier compound that can transfer energy from one molecule to another. The most common energy-carrying molecule is ATP [FIGURE 4.8A]. The amount of energy stored in special high-energy bonds of ATP is about the amount needed for many metabolic reactions.

When energy is released from an exergonic reaction, it can be used to alter a molecule called ADP (adenosine diphosphate), which has two phosphate groups and lower energy content, into ATP (adenosine triphosphate), which has three phosphate groups and higher energy content [FIGURE 4.8B]. Likewise, when ATP is cleaved, it releases ADP and a small amount of energy [FIGURE 4.8C]. The significance of ADP and ATP is that they function as intermediates, or links, between energy exchanges in cells.

Energy intermediates are important because for energy to flow from one chemical reaction to another, the reactions must have some chemical compound in common—that is, a compound that appears among the products of one reaction and the reactants of the next. ATP often appears in the equations for both exergonic and endergonic reactions [FIGURE 4.9].

An analogy for ATP is money. When a farmer sells his or her crops and then uses the money to buy new tools, the intermediate for both exchanges is money. ATP

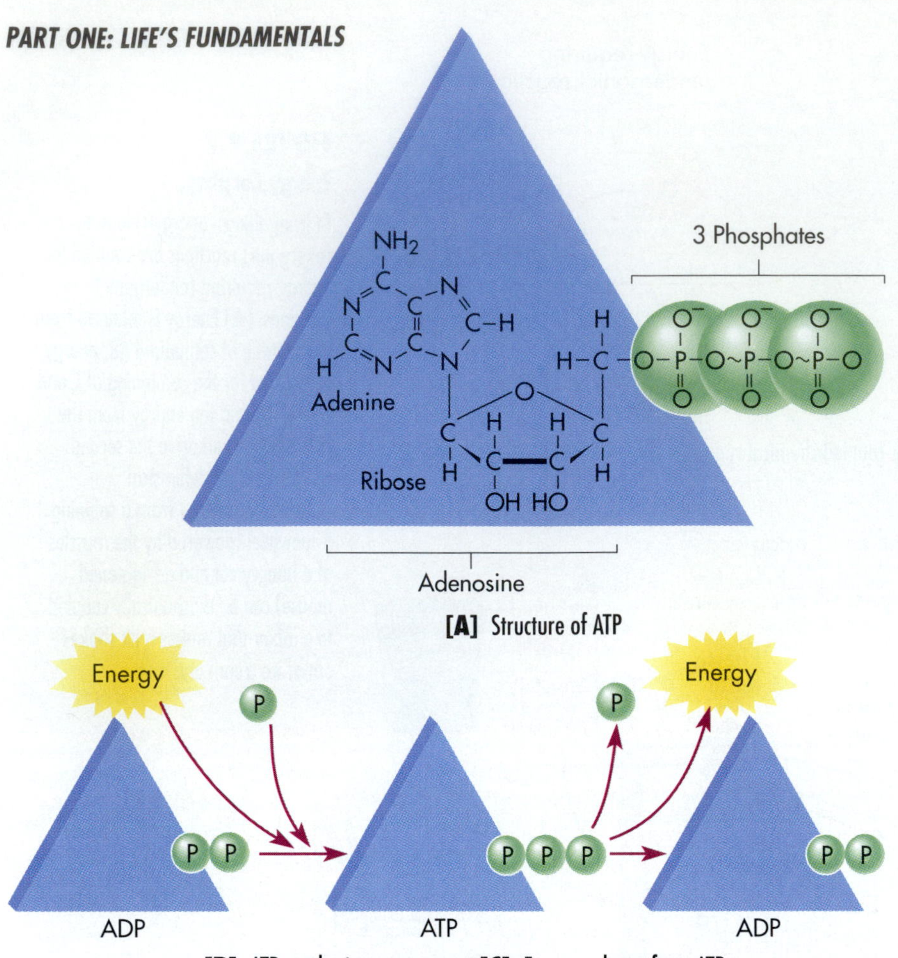

FIGURE 4.8

The Structure and Function of ATP.

ATP, or adenosine triphosphate, is the main energy carrier in living cells. [A] ATP consists of a portion called adenosine and, as the name *triphosphate* implies, three phosphate groups. [B] ATP is synthesized by the addition of a phosphate group and energy to ADP, adenosine diphosphate. [C] Energy is released from ATP when the terminal phosphate is

is sometimes considered a form of currency in the cell—the chemical coin of the realm—that is "earned" during energy-yielding exergonic reactions and "spent" during energy-absorbing reactions.

The ATP "budget" of a single cell reads like the annual financial ledger of a small nation. Consider the bacterium *Escherichia coli*, which resides (usually harmlessly) in the human intestine. An *E. coli* cell contains about 5 million ATP molecules, but it spends 2.5 million *per second* to fuel the work of building DNA, RNA, protein, and lipid molecules. Thus, its energy reserves can "buy" just 2 seconds' worth of maintenance and growth. The cell must continually break down food molecules, store the energy in ATPs, and stockpile its currency in the cytoplasm to keep up with the staggering energy costs of resisting entropy and staying alive. The human body—with its trillions of cells—cycles an estimated 40 kg (about 88 lb) of ATP molecules into ADP and back again each day, representing quadrillions of chemical reactions and energy exchanges daily. Experiments show that when the leaf of a Venus flytrap snaps shut, about 30 percent of the ATP in the moving leaf is lost.

The fact that ATP is the major energy storage and transfer compound of all living things reinforces the idea that all living organisms share a common ancestry.

In summary, chemical reactions involve energy exchanges between molecules, some releasing energy and some absorbing energy. Energy-absorbing reactions are possible because they are linked to energy-releasing reactions. Energy is often shuttled from an energy-releasing reaction to an energy-requiring reaction by the energy carrier molecule ATP. As the next section shows, these reactions accelerate in the presence of a special kind of protein called enzymes.

CONCEPT CHALLENGE

We have used the analogy of currency—cold cash—for ATP. Using this same analogy, what substance(s) in living cells would a bank savings account represent?

METABOLISM: CHAINS OF REACTIONS

As FIGURE 4.7 showed rather whimsically, two reactions can be linked by energy. They can also, however, be linked by substances. Frequently the chemical product of

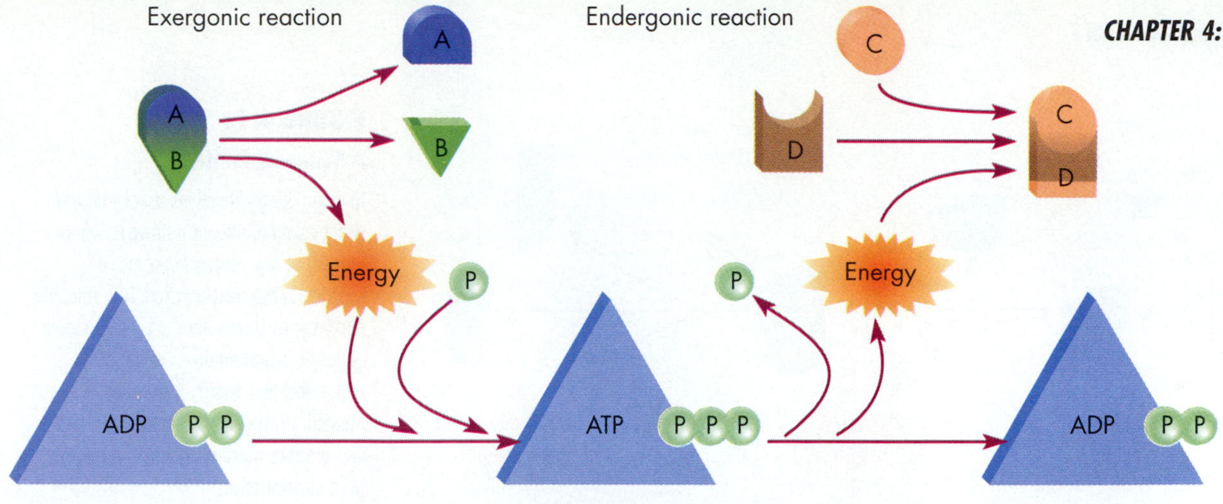

[A] ATP serves as an energy carrier

[B] The flashy results of an ATP-mediated reaction

FIGURE 4.9

How ATP Serves as an Energy Carrier.

[A] Energy released from an exergonic reaction can make possible the synthesis of ATP from ADP and phosphate. Energy now trapped in ATP can be liberated when a phosphate is cleaved, and it can power an endergonic reaction. **[B]** The glowing of this bioluminescent fish, *Kryptophanaron alfredi*, in the inky recesses of the Caribbean Sea, is fueled by ATP-coupled reactions such as those in part **[A]**. The fish consumes food, and this is broken down into small molecules. Bacteria within special pouches under the fish's eyes and along its sides extract chemical energy from the small molecules and store much of the energy in the form of ATP. Later, the cells break down the ATP to ADP, and the liberated energy helps to form a chemical that releases energy as light. Chemical energy is thus converted to light energy, and ATP is the intermediary.

one reaction becomes a reactant in the next reaction, like an automotive part passing down an assembly line for one modification after another. Together, all the reactions of the body that sustain life constitute **metabolism**. The reactions of metabolism break down molecules and build up molecules, give off energy and use energy. FIGURE 4.10 illustrates the concept of interlinked chains of reactions within cells, or **metabolic pathways**. Metabolic pathways allow living cells to subdivide a big chemical change—one that might, for example, release a tremendous amount of heat—into a number of smaller steps, liberating energy in packets small enough for the cell to use efficiently.

Some metabolic pathways break down complex molecules to simpler products, with an accompanying release of energy. Such pathways are called **catabolic pathways**. Other metabolic pathways bring about the synthesis of complex molecules from simpler precursors, and usually require an input of energy. Such series of reactions are called **anabolic pathways**. Metabolism is so important to life that we devote all of CHAPTERS 5 and 6 to its key features.

How Enzymes Speed Up Chemical Reactions

When an insect becomes snared in a Venus flytrap, it is rapidly broken down by digestive enzymes. Clearly, enzymes can facilitate chemical transformations. Let's investigate how these biological catalysts accomplish such an important activity.

Most chemical transformations in living cells are facilitated by **enzymes**. Enzymes are proteins that function as **catalysts**, agents that speed up specific chemical reactions without being permanently changed by the reaction. We know that enzymes are important because many

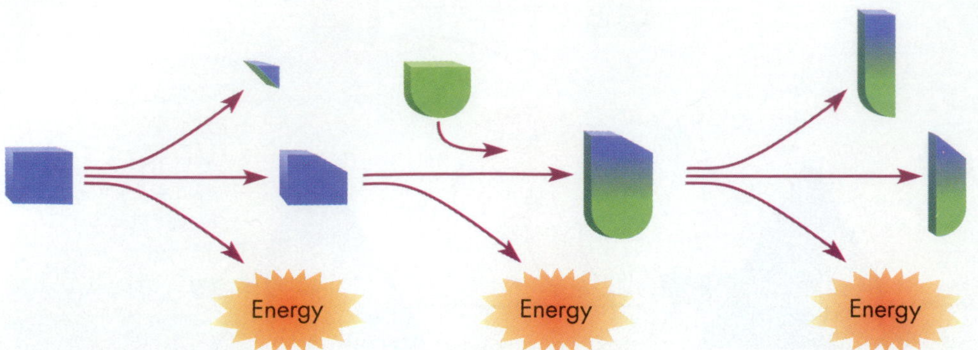

FIGURE 4.10

A Metabolic Pathway.

In living things, reactions usually occur in series called metabolic pathways, where a product of one reaction becomes a reactant in the next reaction. In a catabolic pathway, as shown here, a more complex molecule is broken down to simpler molecules, and energy is released. Anabolic pathways require energy input and produce more complicated molecules from simpler ones.

human diseases are caused by a single defective enzyme, even though all the thousands of other enzymes in the body function perfectly. Some children, for example, are born without the ability to synthesize an enzyme needed to change an inactive form of vitamin D to the active form. Without active vitamin D, the children cannot generate strong, healthy bones, and they develop the disease called *rickets*, with its characteristic curved legs [FIGURE 4.11]. Fortunately, if these children get enough of the active form of vitamin D in their diets, they can bypass the need for the enzyme and develop and grow normally.

FIGURE 4.11

Rickets Reveal the Importance of Enzymes in the Body.

A specific enzyme with a complicated name (25-hydroxy-cholecalciferol-1-hydroxylase) is just one of thousands of enzymes in human cells. If this single enzyme functions improperly, however, a child's body cannot carry out a certain chemical reaction that converts vitamin D to its active form. Lacking the vitamin, the child cannot build strong bones. More normal growth and development is possible if the child gets enough active vitamin D in the diet.

Why do human cells require a specific enzyme to convert vitamin D to an active form? Why can't that conversion just occur spontaneously? The answer is that vitamin D *does* change spontaneously from an inactive to an active form, but the conversion happens very slowly—so slowly that too little of the active form would be available to ensure strong bones and normal growth. When a specific enzyme is present, however, the reaction occurs far faster than without the enzyme. Likewise, the digestive enzymes secreted by the leaf gland cells of a Venus flytrap greatly speed the release of nutrients from an insect's body. How, then, do enzymes speed up chemical reactions? To understand the answer, we must consider a concept that biochemists call activation energy.

ENZYMES LOWER ACTIVATION ENERGY

In any chemical reaction, whether in a cell or in a test tube, there is a barrier, or hill, separating the reactants and products. For vitamin D to be converted from inac-

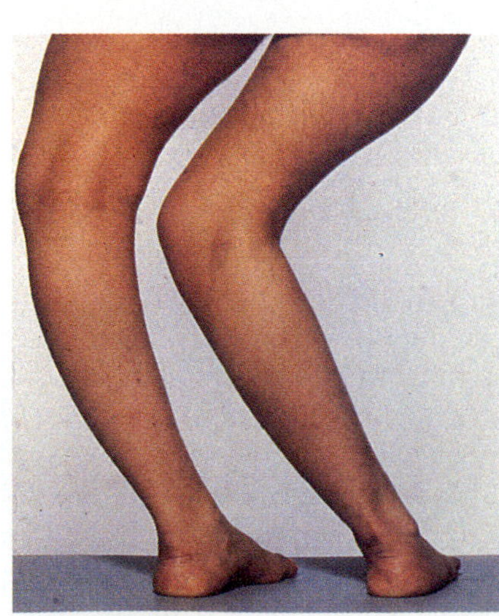

[A] Legs of a child with rickets

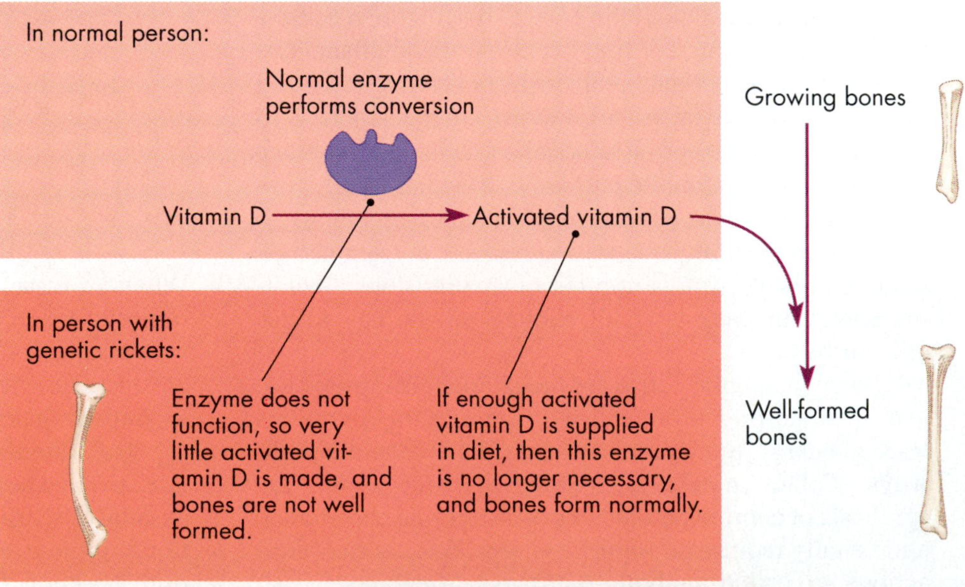

[B] Role of an enzyme in rickets

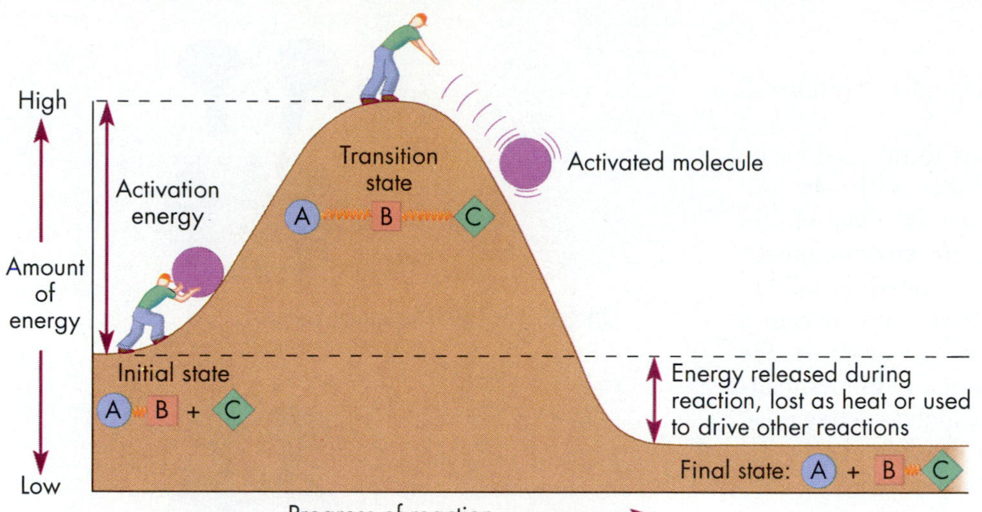

FIGURE 4.12

Activation Energy.

Reacting molecules need to collide with a certain minimum amount of energy before they can reach the springlike transition state. The amount of energy needed to reach the transition state, the activation energy, is much higher than the initial energy state, and that energy resembles a hill, or barrier, when graphed. Once activated molecules reach and then pass the transition state, they proceed to the final state, the new products.

tive to active form, or for a pair of imaginary reactants, which we'll call AB and C, to be converted into A and BC, the reactants must collide with each other [see FIGURE 4.12]. That collision cannot be a gentle bump, but instead must be a solid, energetic thump, so that the chemical bonds in the reactants can be broken and the new bonds in the products formed. For a fleeting instant during the conversion, chemical bonds in the reactants are distorted like springs being stretched, and this fleeting intermediate state, A~B~C, cannot be reached without a very energetic collision between the molecules. The intermediate springlike state is the **transition state**, and because energy input is required to achieve it, it is often characterized as a barrier, or hill, separating reactants and products [FIGURE 4.12].

Most molecules jostling about and colliding randomly do not have enough energy of motion (kinetic energy) to achieve and overcome the energy barrier of the transition state; they simply bounce off each other. Some individual molecules, however, do jostle about with enough kinetic energy so that they collide in a **productive collision** [see FIGURE 4.12], an impact that generates the springlike transition state A~B~C, and AB and C can be converted into A and BC. The energy of impact great enough to cause molecules to cross the energy barrier is called the **activation energy** [FIGURE 4.12].

In a living cell, most molecules do not on their own collide with sufficient energy to cross the barrier and react—at least not fast enough to keep pace with the cell's needs for energy and materials. In principle, one way the reactions in a cell could be speeded up would be for the speed of colliding molecules to increase. But this increase of speed would require an increase of temperature, and high temperatures would speed all the reactions in a cell, not just those needing a boost. This overall increase would cause the cell to become disorganized, and it would die. Instead, enzymes play by far the most important role in speeding reactions in living cells. Enzymes act by lowering the activation energy needed for biochemical reactions to take place [FIGURE 4.13]. The action of an enzyme, in effect, bores a tunnel through the energy barrier, allowing the reactants to react and become prod-

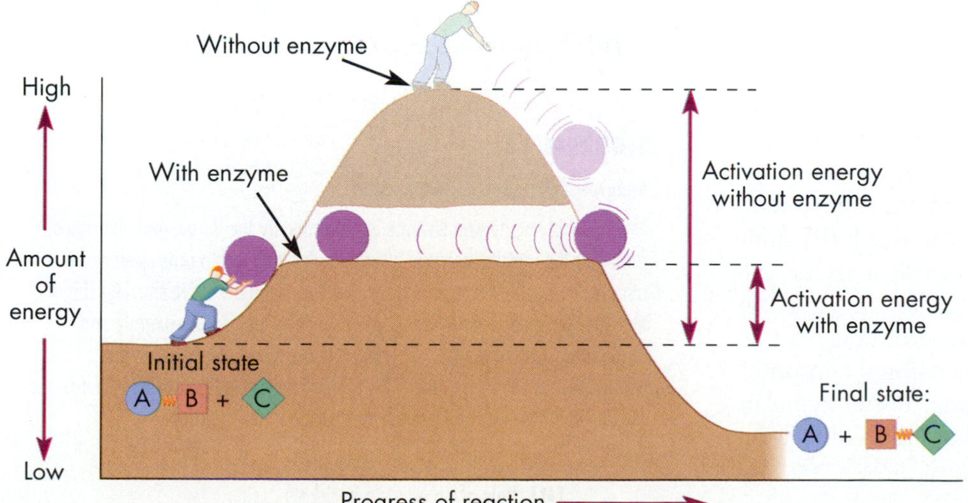

FIGURE 4.13

Enzymes Lower the Energy of Activation.

An enzyme can lower the activation energy—the barrier that must be crossed by reacting molecules—without itself being changed or used up. It's as if the enzyme "bores a tunnel" through the activation energy "hill." In this way, enzymes speed up biological reactions.

ucts without the need to overcome the high activation energy barrier.

Enzymes not only allow reactions to proceed at the relatively low temperatures compatible with life processes, but they also allow only specific reactions to take place. As you have seen, one specific enzyme speeds up the conversion of vitamin D into its active form. Another enzyme, called *carbonic anhydrase* (enzyme names often end in the suffix *-ase*), catalyzes, or speeds up, the reaction shown in FIGURE 4.5B and helps red blood cells transport carbon dioxide from tissues such as active muscles to the lungs, where the carbon dioxide can be exhaled. Each single molecule of carbonic anhydrase facilitates this transport at the amazing speed of 600,000 times per second! This enables the conversion to proceed nearly a million times faster than it would without a catalyst.

Let's take a closer look now, at *how* enzymes carry out their work.

LOCK-AND-KEY MODEL OF ENZYME ACTION

Enzymes are like virtually all other biological molecules in one important way: Their functioning is directly related to their form. Certain RNA molecules called *ribozymes* have recently been shown to catalyze some reactions among other RNAs. However, protein enzymes are by far the most versatile biological catalysts.

Like other proteins, each enzyme has its own unique three-dimensional shape, but most share an important feature: a deep groove, or pocket, on their surface called the **active site** [FIGURE 4.14A and review FIGURE 2.29]. The shape of the enzyme's active site fits a specific reactant or set of reactants. These specific reactants, or **substrates**, fit into the groove rather like a key in a lock, and once locked in, they are acted on by the enzyme.

The enzyme holds, or *binds*, the substrates in place, forming an **enzyme-substrate complex** [FIGURE 4.14B]. Digestive enzymes within a Venus flytrap, for example, bind to molecules in a trapped insect's body, forming enzyme-substrate complexes. The active site of the enzyme can change shape as substrates bind to it, improving the "fit." This in turn orients the substrates near to each other and then subtly changes the shape of the substrates, straining their bonds and helping them to reach the transition state [FIGURE 4.14C]. By means of this activity, the enzyme effectively lowers the activation energy, making it easier for substrates to react and form products [FIGURE 4.14D]. FIGURE 4.15 shows a three-dimensional computer graphic of an actual enzyme, lysozyme, the enzyme in egg white that destroys any bacteria that might otherwise infect and kill the incubating embryo.

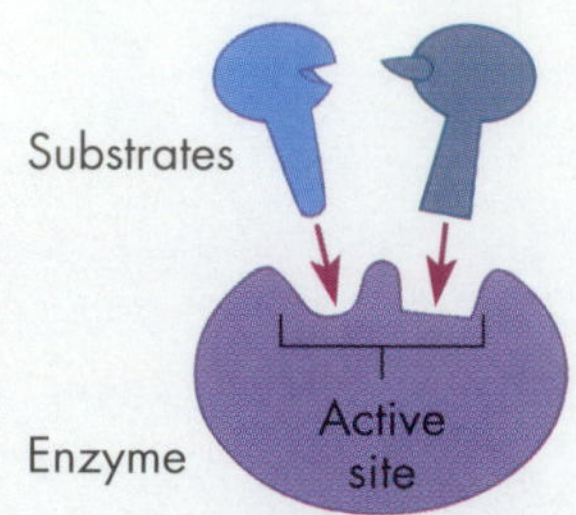

[A] Substrates bind to the enzyme's active site

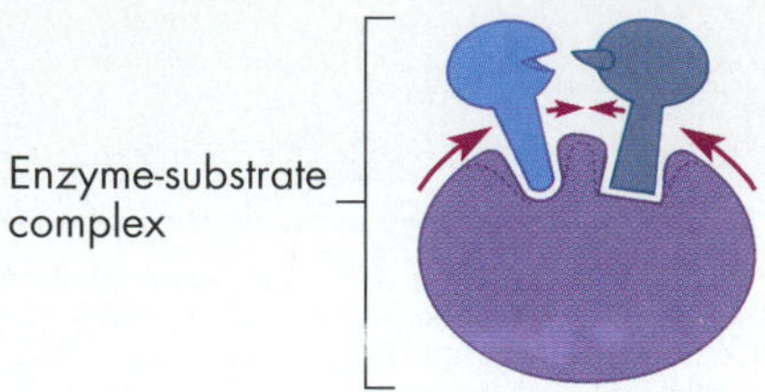

[B] The enzyme's active site changes shape, improving fit and orienting the substrates

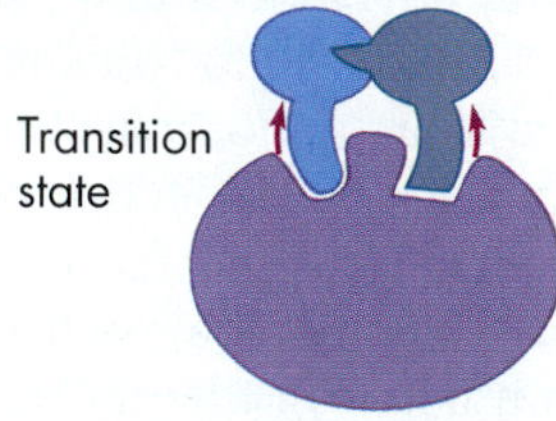

[C] The enzyme strains the shape of the substrates, facilitating a productive collision

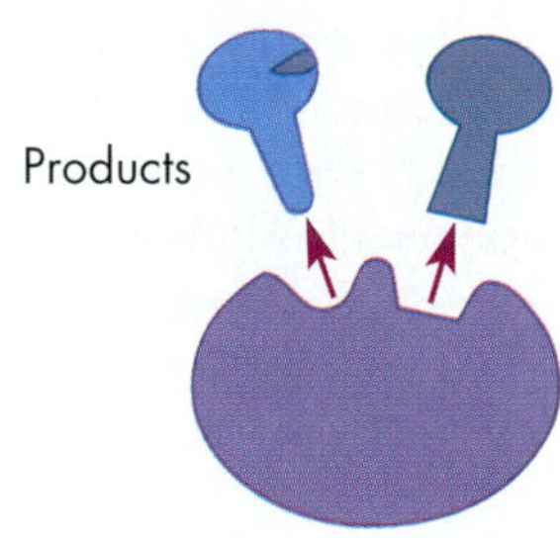

[D] The reaction finishes, and the enzyme returns to normal

FIGURE 4.14

Enzymes: Structure and Function.

This figure shows how an enzyme, represented by the purple globular shape, facilitates the reaction between two substrates, one with a protrusion and the other without. **[A]** The substrates bind to the enzyme's active site. **[B]** This enzyme-substrate complex brings the substrates into close proximity and orients them appropriately. **[C]** These changes strain the bonds of the substrates and help them react. **[D]** When the reaction is complete, the enzyme regains its initial state and is ready to catalyze a new reaction.

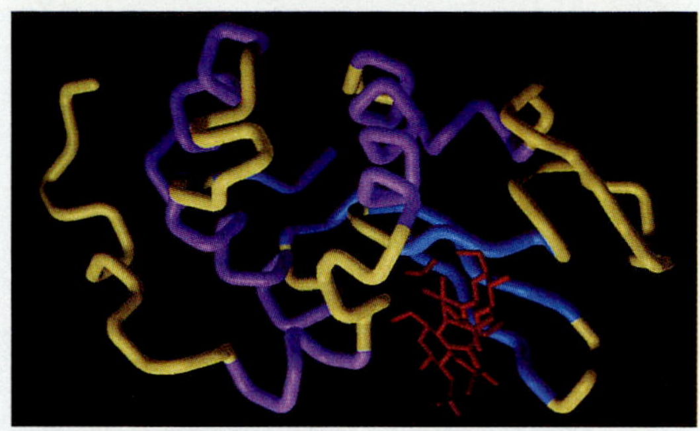

FIGURE 4.15

Lysozyme, the Major Enzyme in Egg White.

Lysozyme protects a chick embryo from bacterial infection. The enzyme binds substrate (shown in red)—the bacterial cell wall—in its central V-shaped cleft, which is visible near the center of this three-dimensional computer-generated model of lysozyme. Can you identify regions of α helix, β sheet, and disordered loop in this molecule?

The molecule shown in red in FIGURE 4.15 is the enzyme's substrate. In this case a portion of the bacterium's cell wall bound to the deep groove that is the enzyme's active site. When the enzyme disrupts many of these substrate molecules in a bacterium's cell wall, the bacterium bursts and dies. When all cells have died, the infection is overcome.

BOX 4.1 on page 110 describes a newly developed enzyme drug for a painful storage disease and explains how the drug works—that is, if the patient can afford to buy it.

In summary, enzymes speed up chemical reactions by lowering the activation energy of the reaction and by binding substrates and increasing their local concentrations so tremendously that it is much more likely that reactants will collide and react at normal body temperatures. Each of your cells has thousands of enzymes, each increasing the rate of only one chemical reaction. The lack of even one enzyme can cause abnormal development or even death.

CONCEPT CHALLENGE

A certain protein-digesting enzyme encoded in genes of the AIDS-causing virus is necessary for the virus to complete its replication in an infected person. Researchers have designed drugs that irreversibly bind this enzyme's active site. Use your knowledge of enzyme function to explain how this drug might provide a useful therapy for AIDS patients.

Cell Transport

For enzymes to do their jobs, they must have access to their substrate molecules. Some of those molecules are provided by mechanisms that transport materials into cells.

A cell's use of energy postpones the inevitable disorganization dictated by the second law of thermodynamics. That energy comes from a series of enzyme-catalyzed reactions that produce ATP molecules from the stepwise breakdown of sugars and other energy-rich molecules. Since digestive enzymes are located inside cells, a sugar molecule can become useful only after it enters the cell. A logical question, therefore, is, How do molecules get into cells? Let's start by examining the fluids through which sugars and other materials move as they enter living cells.

FLUID COMPARTMENTS OF CELLS

Cells are minute pools of liquid surrounded by lipid bilayer membranes [see CHAPTER 3]. Most of a cell's internal activities take place surrounded by water molecules, and the outside of the plasma membrane remains wet. Evidence suggests that life arose in the sea, and living things have retained an "inner sea" ever since. The pool inside each cell is a bit like seawater; it is a rich solution of ions and dissolved gases, but it also contains sugars, amino acids, and other raw materials, usually with a distinctly different composition than the fluid outside the cell.

The cell's inner pool is called the **intracellular fluid**, and the materials for maintaining that pool are taken up from the fluid outside the cell [FIGURE 4.16]. A single-celled organism is almost always surrounded by fresh water or seawater or by the body fluids of another organism that it inhabits. In a multicellular organism, the fluid outside the cell is usually **extracellular fluid**, which fills all the spaces around and between cells. As we shall see later, such fluid plays a role in the shutting of a Venus flytrap.

Some animals (for example, jellyfish and snails) have a single extracellular fluid, while other animals (for example, goldfish and armadillos) have two types: the clear extracellular tissue fluid in the spaces between cells and the blood that flows through the animal's vessels. Materials can move back and forth between the intracellular fluid, extracellular fluid, and blood across the membrane barriers that separate them.

Movements between fluid compartments are governed in part by **gradients**, differences in the concentration or pressure of materials in one part of a defined area compared to those in another. Just as gradients are central to the transport of substances into and out of cells, a house mover thinks about gradients when moving a piano into or out of a house. Cellular substances can often move down a concentration gradient passively—much as a piano will roll down a ramp. For a substance to move up its concentration gradient, however, an input of energy is required, just as it is when movers must push a

box 4.1
Human Impact

An Enzyme More Precious Than Diamonds

Have you ever paid for a prescription drug and been shocked by the price? "This little bottle," you say, "is $47.50? You must be kidding!" Well, what if you needed a drug that cost $14,700 per week, or $765,000 per year? Would that strain your budget? There actually *is* such a medicine, a modified enzyme that *The New York Times* and the *Wall Street Journal* have called the world's most expensive drug. Ironically, this substance is the main treatment for the most common inherited storage disease in America. If this sounds like a recipe for economic disaster as well as a complex social issue, it is.

Recall that in CHAPTER 3, we already encountered one genetic storage disease called Tay-Sachs. Patients lack the enzyme hexoseaminidase that is normally present in the lysosomes of brain cells. Missing that digestive enzyme from the cell's refuse organelles, the brain cells accumulate whorls of the sugar-lipid G_{M2}. This abnormal storage, in turn, interferes with hearing, vision, learning, movement, and other nervous system activities.

An even more common cellular storage problem is Gaucher's disease (pronounced Go-SHAYZE). Like Tay-Sachs, it is most common in Jewish people from Eastern Europe or their descendants. Although there are different forms, most people with Gaucher's lack an enzyme called glucocerebrosidase. Like the Tay-Sachs enzyme, this one normally breaks down a sugar-lipid, but in this case it's called glucocerebroside. Since a Gaucher's patient lacks the enzyme that digests the sugar-lipid, it builds up, but this time not in brain cells. Instead, the excess material is gobbled by and stockpiled in a type of white blood cell called a macrophage [FIGURE 1]. These blood cells, in turn, collect in the liver and spleen, which can greatly enlarge. As a result, the patient can suffer abdominal pain, digestive problems, and anemia. Since macrophages are produced in the marrow—the central reddish matter—of the bones, Gaucher's patients can also have painful or deformed bones and joints, and must often undergo repair or replacement surgery.

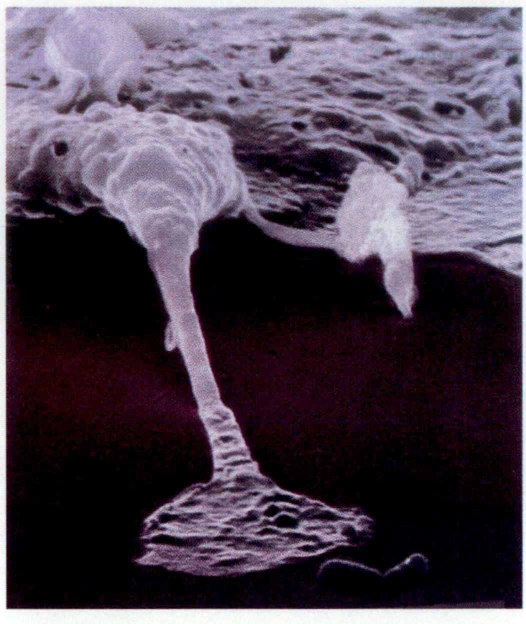

FIGURE 1
Macrophage Approaching Bacteria

During the 1970s, researchers tried extracting the missing glucocerebrosidase enzyme from human placentas discarded after childbirth. They would painstakingly purify the enzyme and infuse it into Gaucher's patients, but the results were disappointing. Until, that is, one group discovered that macrophages have a mannose receptor—something akin to a "mailbox" for receiving enzymes tagged with a mannose "address" label. By adding mannose (a type of sugar) to the glucocerebrosidase enzyme extracted from placentas, they were able to create a modified enzyme drug called aglucerase. This modified enzyme, with its added "address

piano up a ramp. The next two sections consider both types of transport.

PASSIVE TRANSPORT, DIFFUSION, AND THE SECOND LAW OF THERMODYNAMICS

Materials—sugars, ions, water, amino acids, and so on—tend to move *down concentration gradients*, that is, from regions of higher concentration to regions of lower concentration. When a substance enters or leaves a cell by moving down such a gradient, the process is called **passive transport**: The material moves without energy expenditure by the cell.

Cells need a constant supply of materials for maintenance activities and building new cell parts. Cells must also get rid of excess ions, organic wastes, and other materials. Many of the substances that enter or leave cells via passive transport follow gradients in or out. The tendency of materials to move from areas of higher concentration to areas of lower concentration is called **diffusion**. You can witness the results of diffusion if you very carefully place a small dollop of whipped cream in a cup of label," goes straight to the white blood cells and can reduce the build-up of sugar-lipids stored there. As a result, it can reverse some of the liver, spleen, and bone symptoms in Gaucher's patients.

The only trouble with aglucerase has been the large quantity many patients needed weekly, plus the astonishing price tag it carries. One researcher has estimated that if 5000 of the most seriously affected Gaucher's patients in the United States received weekly treatments of aglucerase, the medical costs would come to $1 billion per year! Insurance companies balk at paying bills that big, and in fact, many Gaucher's patients do not receive the modified enzyme drug they need.

Gene researchers are working hard at two alternative approaches to this dilemma: One involves inserting healthy human genes for glucocerebrosidase enzyme into yeast, bacteria, or other easy-to-grow microbes. Their hope is that the cells will produce great quantities of the precious enzyme for commercial harvest. The other involves implanting functional genes for the enzyme first into laboratory animals then eventually into patients, so they can produce the enzyme in their own bodies.

Investigators have seen some promising results with the latter approach, also known as gene therapy [see CHAPTER 11]. So far, however, the small victories have been in mice, not people. In the meantime, many Gaucher's patients are left to battle with their insurance companies and to hope for some kind of national health insurance and for further research breakthroughs. What do you think about their situation? Should society pick up the tab where insurance companies refuse? Would you feel the same if you or a loved one had Gaucher's disease?

It's all a matter of enzymes. And money.

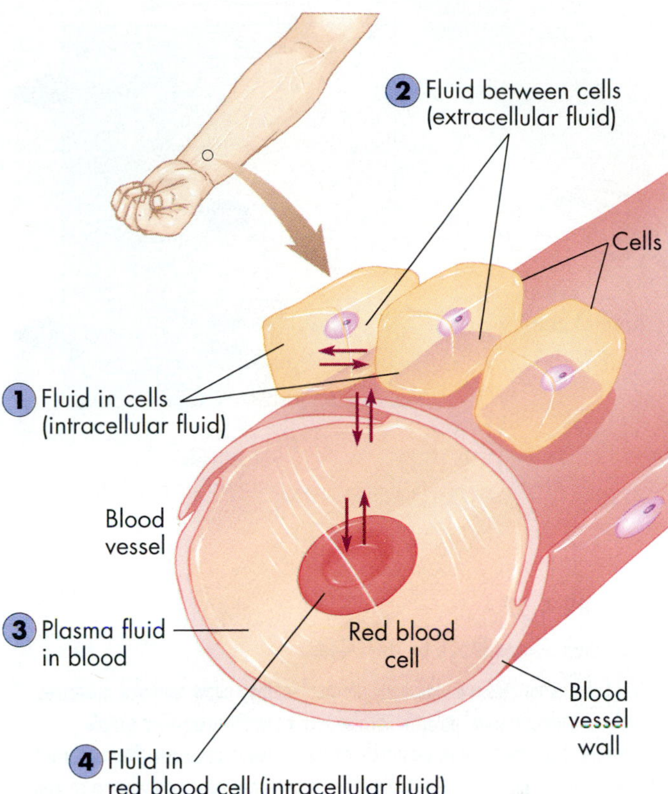

FIGURE 4.16

Biological Fluids and Their Movements.

A small blood vessel from a person's arm illustrates the fluid compartments in and around cells. There is intracellular fluid inside the cells that surround the blood vessel (1) and also inside the red blood cells (4). There is extracellular fluid between the cells (2). And there is fluid called plasma in the blood (3). Water and materials from these various fluids can move back and forth across cell membranes.

black coffee. Billions of collisions send water, cream, and coffee molecules bouncing off each other in random directions like billiard balls, until eventually, the cream molecules have spread evenly throughout the cup.

This process, however, will take a long time. If, on the other hand, you pour in a quantity of liquid cream rapidly so that it plunges to the bottom and mixes with the coffee, you will be witnessing, not diffusion, but *bulk flow*, the movement of a stream of fluid molecules all in the same direction. This is the more typical pattern when adding cream to coffee or milk to tea.

A popular classroom experiment is to build a diffusion system that acts something like the plasma membrane of a red blood cell or other cell. This usually involves a container separated in half by a cellophane sheet perforated with tiny holes through which water and other molecules can move. Both the dissolved molecules and water will migrate back and forth passively, with the net migration always from higher concentration to lower. Although molecules move in both directions across the sheet, more move from the higher-concentration side to the lower-concentration side. When the concentrations on both sides are equal, *equilibrium* is reached. At equilibrium, the same number of molecules move in both directions at the same rate, and so no further net change occurs. Most materials enter or leave cells via this process of *simple diffusion* [FIGURE 4.17A].

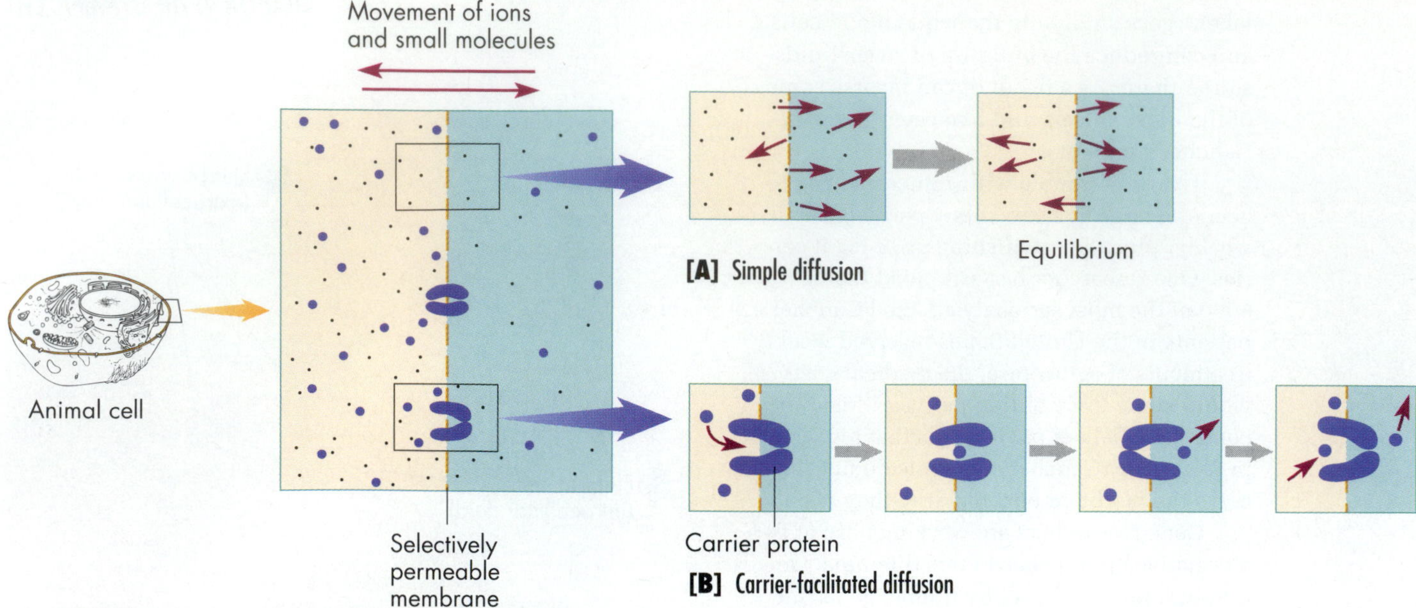

FIGURE 4.17

How Substances Move by Diffusion.

[A] Cell membranes are selectively permeable; they allow nonpolar molecules such as lipids and small polar molecules such as water to pass via *simple diffusion.* Note that a constant number of molecules pass across the membrane in a given amount of time, but that there is *net* movement only when a concentration gradient is present. [B] Other somewhat larger molecules, such as glucose or the nitrogenous waste product urea, cannot diffuse through the membrane without the action of special carriers in a process called *carrier-facilitated diffusion.* Specific carrier proteins span the membrane and change shape to allow the solute to cross the bilayer, then spontaneously change back again, ready for the next incoming molecule. Because the solute always moves down a concentration gradient, this is true diffusion, with no energy expenditure by the cell. But carrier-facilitated diffusion can allow substances to diffuse across a membrane that would otherwise block their passage, thus enabling the cell to take in or get rid of substances quickly enough to keep pace with metabolism. In a red blood cell, both simple and facilitated diffusion take place simultaneously across the plasma membrane.

The lipid bilayer surrounding a cell is permeable to some very small molecules and lipid-soluble molecules. It does not, however, allow large water-soluble or electrically charged molecules, including ions, to pass: It is a *selectively permeable membrane.* How then, do larger or charged molecules pass in and out? The selectively permeable membrane contains *pores* or *channels* that allow charged molecules such as ions, water, and certain other small polar molecules to pass freely. Some molecules, however, are so large or so charged that they can penetrate neither the membrane nor the pores, and thus cannot enter or leave the cell by diffusion.

For some large or electrically charged molecules, this problem is overcome by means of a special type of diffusion called **carrier-facilitated diffusion**. This process still requires no energy expenditure by the cell, but it does involve special *carrier proteins* embedded in the plasma membrane [FIGURE 4.17B]. You can think of carrier proteins as little trap doors that snap open and shut, allowing a molecule which happens to bump into it to diffuse into the carrier on one side of the membrane and out of the carrier on the other, always tending to move down the concentration gradient. The exchange of ions across cell membranes that results when the trigger cells in a Venus flytrap are tripped probably relies on carrier-facilitated diffusion. Carrier-facilitated diffusion allows substances such as glucose and the waste product urea to pass through plasma membranes.

PASSIVE TRANSPORT AND THE MOVEMENT OF WATER

Since a cell contains mostly water, and since its delicate outer plasma membrane must be immersed in fluid, water itself is probably the most important substance to enter and leave the cell, and it does so by diffusion. The movement (diffusion) of water through a selectively permeable membrane (such as a cell's plasma membrane) in response to gradients of concentration or pressure has its own name: **osmosis**. The principles involved in osmosis are the same as those governing the diffusion of all materials, and in fact, the flow of water molecules is directly related to the concentration of other materials.

Water moves from a region of higher water concentration to a region of lower water concentration. In this case, higher and lower concentration refer to the relative number of water molecules, not to sugar or salt molecules or some other solute in the solution. Therefore, pure distilled water has a higher concentration of H_2O molecules than water containing a solute, such as salt.

The red blood cell is a good example of the impor-

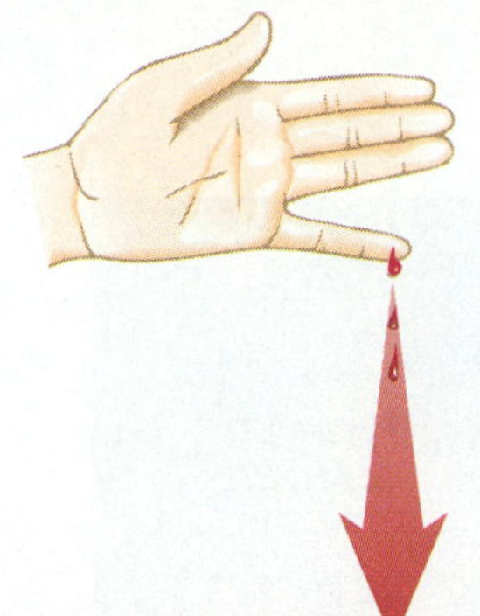

FIGURE 4.18

Osmosis and Red Blood Cells.

Water moves across a selectively permeable membrane such as the plasma membrane of a red blood cell by osmosis, that is, passively from regions where water is most concentrated (i.e., where the solute concentration is low) to regions where water is least concentrated (where the solute concentration is high). **[A]** When a cell is placed in a solution that is highly concentrated in ions or other solutes (a *hypertonic* solution), water moves out by osmosis, leaving the red blood cell shriveled. **[B]** In a solution with solute concentrations similar to those of the cell interior (an *isotonic* solution), water moves in and out at equal rates, and the cell maintains its normal disk-like shape. **[C]** When a cell is placed in pure water, a solution lacking solutes (a *hypotonic* solution), water rushes into the cell, and the cell expands rapidly, becoming spherical and sometimes bursting. The red blood cells in these photographs are magnified about 400 times.

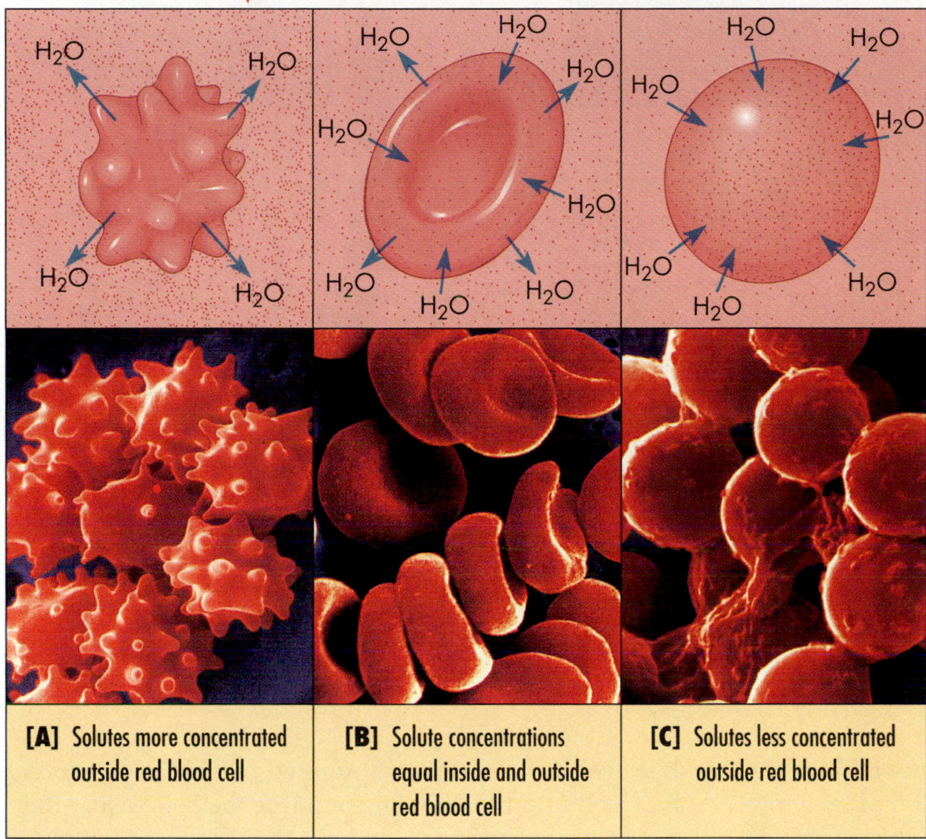

[A] Solutes more concentrated outside red blood cell

[B] Solute concentrations equal inside and outside red blood cell

[C] Solutes less concentrated outside red blood cell

tance of diffusion and osmosis to living cells. As it passes through the blood vessels, a red blood cell is bathed in a yellowish fluid called *blood plasma*, and significantly, the total concentration of salt ions, sugars, and other solute molecules in this plasma is the same as that inside the cell. As a result, the extracellular plasma is said to be **isotonic** relative to the fluid in the cell. (*Iso-* means "same," and *tonic* refers to osmotic properties.) Thus, an isotonic solution outside the cell has the same osmotic properties as the fluid in the cell's interior [FIGURE 4.18B].

Surprisingly, 100 times the total volume of water in a red blood cell moves rapidly back and forth across the cell's plasma membrane every second. Yet, there is no *net* water movement in or out because the concentration of water molecules is similar on both sides. (This is something like a busy urban department store where many people enter and exit through revolving doors every minute, but the total number of shoppers inside remains the same.) What happens if you place red blood cells in a solution containing a higher concentration of salts, sugars, and proteins (and hence a lower concentration of water) than the fluid inside the cell? With a lower concentration of water molecules outside the cell than inside, water would move down its concentration gradient and out of the cell. A cell containing less water would have a shrinking volume, and its membrane would crinkle like a partially deflated balloon [FIGURE 4.18A]. Such a solution is called **hypertonic**. Finally, what happens when you place blood cells in pure water? The pure water contains a lower quantity of dissolved materials than the fluid inside the cells, and a higher concentration of water molecules. Such a solution is called **hypotonic**. There would be a net movement of water into the cells by osmosis, and the cells would swell and eventually burst [FIGURE 4.18C].

The day after you eat a big helping of a very salty or sweet food—say, a large bag of potato chips or a quart of

FIGURE 4.19

Osmosis and Plant Cells.

Water in the central vacuoles of plant cells exerts an outward force called turgor pressure that keeps the plant tissue firm. **[A]** In the sensitive plant, the appearance of the entire leaf depends on the turgor pressure of a single cell in the junction between the leaf's stalk and the plant's main stem. When turgor pressure in that cell is high, the leaf stands erect. **[B]** If you touch the plant, this cell pumps ions out, and water follows by osmosis. This lowers the turgor pressure, and the leaf looks wilted.

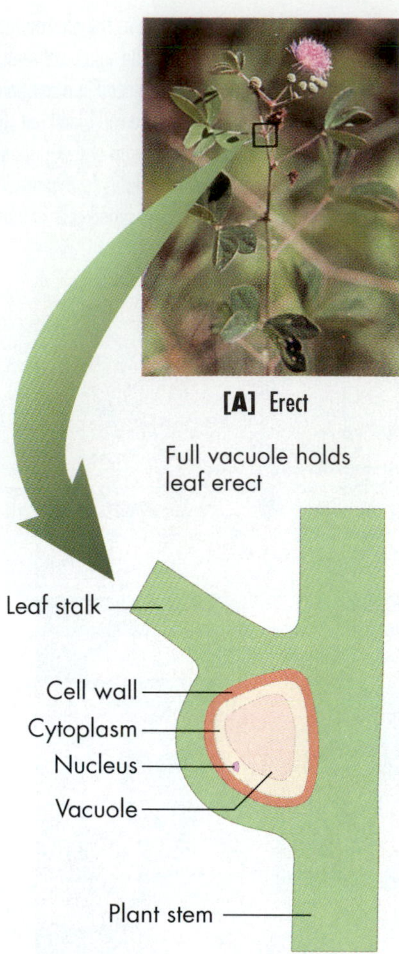

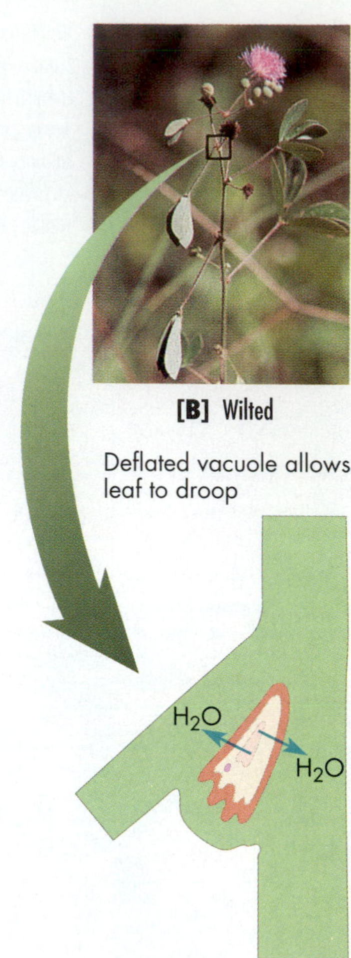

ice cream—your weight may go up a few pounds. The reason is that the ion or sugar concentration of body fluids increases, and so you retain more of the water you drink as it moves into the fluids and offsets the extra molecules. Near the end of a woman's menstrual cycle, hormones can also cause the solute concentration of cells to rise or fall, often bringing about water retention and the bloated feeling some women experience at that time.

The concentration of substances inside the cells of most plants is usually higher than that in the soil water around its roots, and therefore water tends to enter the plant and its cells by osmosis. The cells do not burst, however, because they are contained by their strong cell walls. The water fills the vacuole within the cell and exerts an outward force against the cell wall called **turgor pressure** that plumps out the cells.

By blowing up a balloon inside a box so that air pressure is exerted against the inner box cells, you can create a condition analogous to turgor pressure. In an interesting tropical plant called the sensitive plant, cells with large, plump vacuoles at the base of the leaf stalk hold the leaves erect [FIGURE 4.19A]. When the plant is touched, however, certain ions are pumped out of each such leaf base cell into the fluid surrounding the cell. Water then leaves the vacuoles by osmosis, the vacuoles and whole cells shrink in size, and the leaves droop [FIGURE 4.19B].

A similar mechanism accounts for the closing of the Venus flytrap: Cells on one side of the leaf elongate, causing the leaf to bend and the teeth to intermesh like the bars on a prison gate. Botanists agree that once an animal disturbs the leaf's trigger hairs, the trap springs shut due to turgor pressure in expanding leaf cells. Plant scientists still disagree, however, about the mechanism that prevents the cells from expanding when the leaf traps are open. Some plant physiologists have noted that ATP disappears as the trap closes. As a result, they have suggested that before a flytrap can snap shut, certain substances must be transported. This movement would require energy expenditure by the cell—our next topic.

ACTIVE TRANSPORT: ENERGY-ASSISTED PASSAGE

While many substances are transported passively down concentration gradients within cells, certain materials

active learning: How do leaves of a flytrap snap shut?

The trapping mechanism of a Venus flytrap is a marvel of evolutionary engineering, but how does it work? We can devise at least two models for the closing mechanism.

Hypothesis 1: The two lobes of the leaf close like a slamming book or a clam shell.

Hypothesis 2: The lobes of the leaf resemble your two hands with your fingers extended but with the palms of your hands joined at the wrist; closing involves the fingers simply curling together.

To test the two hypotheses, imagine that you cut a flytrap leaf in two and looked at a cross section of the open leaf as shown.

[A]

[B]

Now let's stimulate the sensitive trigger hair on an open leaf and let the two lobes snap shut. If we cut the closed trap in half, we could observe the cross section as shown.

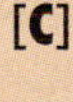

[C]

Which of the models above is ruled out by the results?

. . . .

Write a sentence that supports your reasoning.

. . . .

. . . .

Next, let's extend the analysis to the cellular level. Imagine that the cells in the indicated area of an open leaf are arranged as shown.

[D]

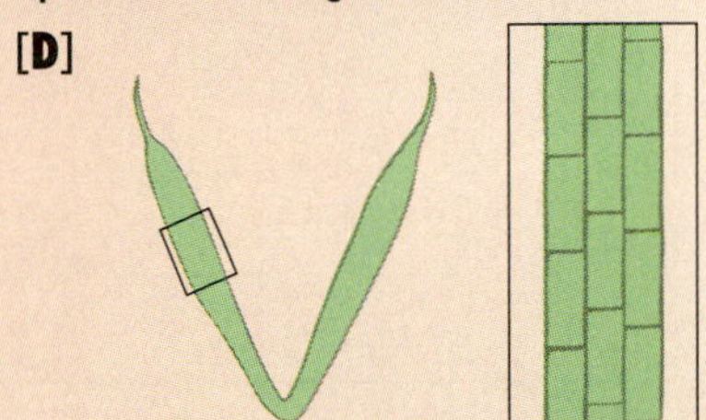

We have seen that a closed leaf curls like fingers curving together around a ball. Therefore, we can reason that changes in the size or number of the cells in this curling region may actually *cause* the trap to shut. In the figure below, draw in cells to show how changes in either cell size or cell number could cause the bend.

[E]

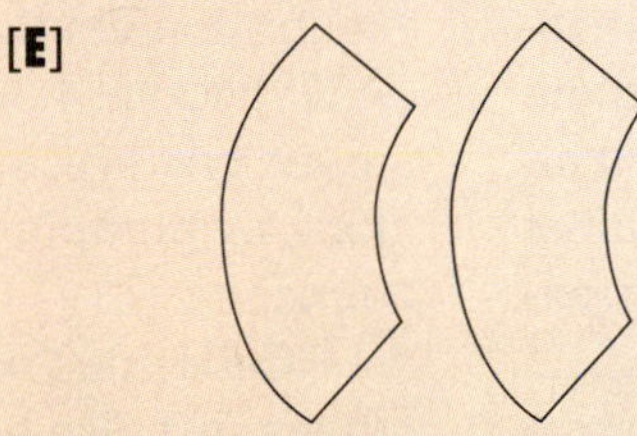

Microscopic observation revealed that the cells in the closed Venus flytrap leaf actually look like this:

[F]

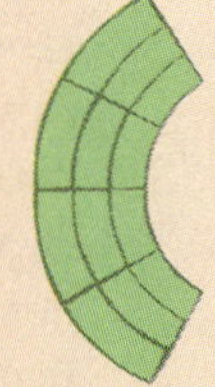

Which prediction is confirmed by these results?

. . . .

How could such changes in cell shape cause the trap to close?

. . . .

. . . .

The experiment showed that the cells on the outside of the bend elongated, while those on the inside of the bend remained the same size, and this asymmetric lengthening caused a bend in the opposite direction. Other experiments revealed that as the trap closed, the amount of ATP dropped drastically. Review your knowledge of osmosis, membrane pumps, and active transport (Figure 4.20), as well as the mechanism that causes size changes in red blood cells (Figure 4.18) and sensitive plant cells (Figure 4.19). On the basis of what you know, write down a possible mechanism for the enlargement of the outside cells before you read on.

. . . .

. . . .

. . . .

Plant physiologists suspect that cells on the outside of the bend actively pump ions across the cell membrane. The ion transport allows turgor pressure to expand these cells and it also accounts for the observation that ATP is used up as the leaf closes. Plant physiologists are still arguing, however, about which ions cross the membrane and how they might affect turgor pressure, soften cell walls, and allow the cells to lengthen on one side. Perhaps one of you will one day perform the experiments to settle this mystery of how Venus flytraps snap shut!

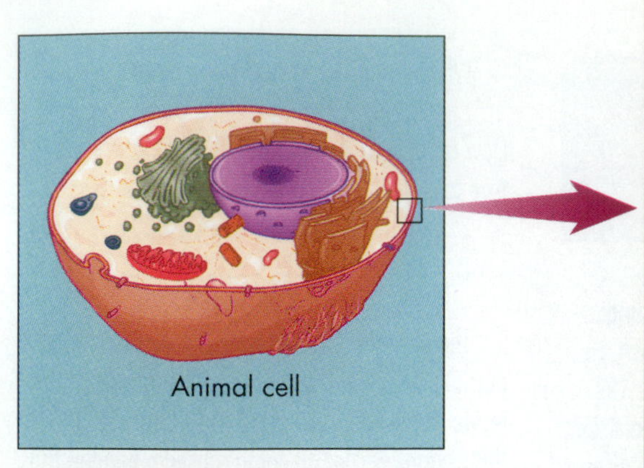

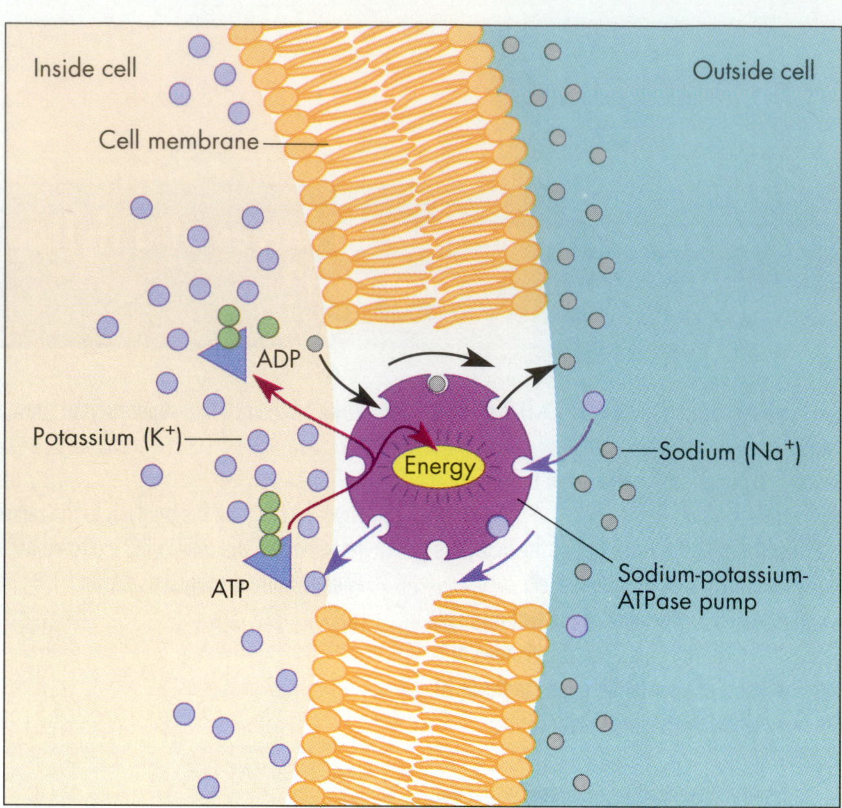

FIGURE 4.20

Active Transport by Membrane Pumps.

Specific membrane proteins actively pump ions into and out of the cell against the ion's concentration gradients using energy in ATP. As you rest quietly, most of the ATP being used by your body forces sodium out and potassium into your cells against their gradients.

are also transported in or out *against* gradients; this means that substances can become less randomly distributed within cells. This happens, for example, when there is already more of a certain substance inside the cell than outside, but the cell needs even more. Movement against gradients, like rolling a piano uphill, requires energy expenditure by the cell, and is called **active transport**.

Cells contain such an array of raw materials relative to extracellular fluid that most cells must contend with a continuous tendency for water to enter the cell and cause swelling. Cells must also accumulate certain materials for building, maintenance, and energy needs that are too large to enter the cell by passive transport. Thus the dilemma: To prevent swelling, the cell must move certain materials *out* (with water following osmotically), but to continue functioning, it must move certain materials *in* against their concentration gradients. Moving a substance against a concentration gradient involves the energy-costly process of active transport. In many cells, active transport is accomplished by a set of proteins in the plasma membrane that act like pumps, pushing ions into and out of the cell.

One type of pump, the **sodium-potassium pump**, maintains osmotic balance in the cell at a high energy cost—30 to 70 percent of a cell's entire energy budget. Cells usually contain a high concentration of K^+ but a low concentration of Na^+, while the fluid outside the cell contains the reverse. The special plasma membrane proteins in the pump actively move two potassium ions in for every three sodium ions moved out against their concentration gradients by expending energy in the form of ATP [FIGURE 4.20]. This pumping results in a proper osmotic balance, since water follows the Na^+ out. Thus, so long as pumping continues, the cell does not shrink or burst.

Cells must have the proper concentration of ions, but they must also have glucose, amino acids, and other raw materials that are building blocks for cell parts. Many of these valuable commodities, however, occur in greater concentration inside cells than outside, and thus would tend to diffuse out rather than remain inside. This problem is overcome by *membrane transport proteins* that actively bring in the needed molecules, enabling the cell to stockpile these commodities as fast as it uses them. The raw material diffuses to the cell and binds to a transport protein that spans the plasma membrane. The membrane protein then changes shape (for some compounds, at a cost of ATP energy), and in so doing, acts as a carrier that transports the passenger molecule, releasing it on the other side of the plasma membrane. Interestingly, no molecular pump directly transports water into a cell; instead, cells actively transport ions, and water follows by osmosis.

The inpocketing and outpocketing of cell membranes (endocytosis and exocytosis) also help many cells take in or expel particles of food, water, or wastes and thus accomplish their transport work to keep up with the staggering need for raw materials [review FIGURE 3.14]. Once inside the cell, however, the materials from all routes of entry are used in the cell's crucial activities, enabling it (or the organism the cell is part of) to pursue food, to divide and reproduce, and to carry out chemical and transport work. This is just one more reminder that all the work of the dynamic cell is inextricably intertwined into one ongoing process—life.

CONCEPT CHALLENGE

Endurance athletes obtain both water and sugars in a sports drink such as Gatorade; the sugars leave the stomach and enter the blood rapidly to replace lost fuel, and water replaces moisture lost in sweat. Consuming a sports drink with a concentration of sugar higher than the concentration of substances inside the athlete's cells actually causes water to leave the cells and enter the stomach. How can this be explained on the basis of osmosis? What would you recommend to an endurance athlete about the concentration of sugars in his or her choice of sports drink?

Connections

In the time it takes you to blink, sneeze, yawn, or swallow, nearly every cell in your body will have performed thousands of individual activities. Membrane pumps will have transported billions of ions into and out of your cells accompanied by moving water molecules, back and forth, over and over. Enzymes will have sped up hundreds of metabolic reactions—with interconnected metabolic pathways like the strands of a web—that break down food and build up needed materials.

The energy costs for these cellular activities are enormous, and yet, collectively, life processes are energy-releasing. Cells are temporary repositories of chemical energy; the chemical energy in food molecules fuels the processes needed for the cells' internal organization. They give off heat that disorders the environment ever so slightly by causing random jostling motions of the air or water around them. Efficient energy collection and use are so critical to survival that they have been a prime cutting edge for evolutionary processes. Natural selection rewards those organisms that can grab the most energy and use it efficiently, while the less efficient life forms tend to die out.

Evolutionary pressures like these have allowed the Venus flytrap to evolve animal-snaring adaptations. These adaptations provide needed nutrients and allow the plant to invade boggy environments that are inhospitable to most plants. With its adaptations for supplemental feeding, this small plant can survive in nutrient-poor ecosystems, and thereby gather sunlight energy that would be unavailable to most other plants.

In the next two chapters, we will trace the energy flow through cells. As we do, try to envision the bustling metropolis—the dynamic cell—that surrounds and is powered by that flowing energy.

CONCEPT MAPPING

Draw a concept map relating the following terms. Link the concepts with arrows and a connecting word or short phrase.

activation energy
active site
energy
enzyme
second law of thermodynamics
substrate
product

KEY TERMS

activation energy, 107
active site, 108
active transport, 116
ATP, 103
carrier-facilitated diffusion, 112
chemical reaction, 101
diffusion, 110
energy, 98
entropy, 99
enzyme, 105
first law of thermodynamics, 98
gradient, 109
heat, 99
hypertonic, 113
hypotonic, 113
intracellular fluid, 109
isotonic, 113
kinetic energy, 98
metabolic pathway, 105
osmosis, 112
passive transport, 110
potential energy, 98
product, 101
reactant, 101
second law of thermodynamics, 99
sodium-potassium pump, 116
substrate, 108
transition state, 107
turgor pressure, 114

HIGHLIGHTS IN REVIEW

1 All organisms require a source of energy. While energy cannot be created or destroyed, it can be change from one form to another say, from chemical energy to electrical energy.

2 In energy conversions, some energy is inevitably lost as heat, the random disordered motion of atoms and molecules; therefore, without the input of energy, everything tends to become more disorganized over time.

a] Energy, the capacity to do work, exists in two states: potential energy (stored energy) and kinetic energy (the energy of motion). These states, as well as different forms of energy, can be interconverted, but according to the first law of thermodynamics, energy is neither created nor destroyed in the process.

b] Some energy is released as heat during any energy interconversion. The release, according to the second law of thermodynamics, adds to the entropy—the increasing disorder of the system.

c] Cells are subject to the universal laws of energy, but the chain—sun → organic molecules → life processes—makes energy available in chemical form, which the cell can then expend to "purchase" order through the construction and maintenance of cell parts and processes.

d] Living cells can perform chemical reactions that absorb energy because they are coupled to chemical reactions that release energy. For example, the reactions involved in building a protein are coupled to the reactions involved in breaking down sugar molecules.

e] Chemical reactions, during which reactants are converted to products, underlie virtually all cellular energy conversions. Exergonic reactions proceed spontaneously and give off heat while endergonic reactions absorb energy.

f] In cells, energy-releasing and energy-requiring reactions are coupled so that the energy for endergonic reactions is generated during exergonic reactions. The energy is usually trapped in a carrier called ATP, which serves as a chemical intermediate, linking all the energy exchange reactions collectively called metabolism.

3 Enzymes are proteins that speed up the rate of chemical reactions.

a] Before reacting molecules can be transformed into products, they must have sufficient energy to undergo a productive collision and reach a very brief intermediate transition state. The minimum amount of energy they need for a reaction, the activation energy, forms a barrier between reactants and products and would prevent most reactions in cells from occurring quickly enough if there weren't catalysts called enzymes which facilitate and speed up reactions.

b] Enzymes lower the activation energy needed for biochemical reactions by binding reactants in a groove in the enzyme called the active site. This orients the substrates in such a way that appropriate regions of the molecules will collide, and changes the shapes of the substrates, which helps them reach the transition state. Enzymes can speed up reactions a million-fold.

4 Some substances enter cells by passive diffusion, but other substances cannot move into or out of cells without special protein in the cells' plasma membrane.

a] A major task of the dynamic cell is to maintain the appropriate mix of ions, molecules, and gases in the intracellular fluid so that materials are always on hand for the building of new cell parts. Cells accomplish this by passive and active transport.

b] Materials naturally move down gradients from areas of high concentration to areas of low concentration, and this simple process, called diffusion, accounts for most of the passage of materials through the cell membrane.

c] Water also moves down gradients from areas of higher water concentration (purer water with fewer solutes) to areas of lower concentration (solutions with more solutes) in a process called osmosis. If the fluid inside and outside the cell have the same concentration of dissolved substances (an isotonic solution), there will be no net movement of water due to osmosis. However, if the extracellular fluid has a lower concentration of dissolved materials than the cell (a hypotonic solution), the cell will imbibe water and swell or burst. If the bathing solution has a higher concentration of dissolved materials (it is hypertonic), the cell will shrink from water loss.

d] Cells expend a great deal of energy pumping ions in and out, helping maintain the correct internal concentration of each substance.

e] Cells carry out additional active transport work: Carrier proteins embedded in the plasma membrane ferry large molecules across, and the entire membrane can inpocket or outpocket, thus taking in or expelling water or materials.

UNDERSTANDING THE FACTS AND CONCEPTS

For Questions 1–5 match each of the descriptions with the one item from the following list of terms that is most appropriate. Any item may be used once, more than once, or not at all.

a] potential energy
b] kinetic energy
c] activation energy
d] enzyme
e] entropy

1 Heat-caused random atomic or molecular motion that is not available to do useful work.

2 A stored form of energy, usually as a result of position or tension, as in a chemical bond or a boulder on the top of a hill.

3 Protein molecules of specific shape that form a temporary union with one or more specific reactants.

4 The energy of motion, released by the movement of atoms, molecules, or other objects of mass.

5 Energy required to make the collision between reactants sufficiently forceful to effect the transition state.

As above, for Questions 6–10, match the descriptions with the appropriate item from the following list:

a] exergonic
b] endergonic
c] anabolism
d] catabolism
e] coupled reactions

6 Energy-yielding metabolic reactions in which larger molecules are broken down to smaller ones.

7 Metabolic reactions in which larger molecules are produced from smaller ones.

8 The combination of an energy-yielding reaction and an energy-requiring reaction.

9 A chemical reaction in which the reactants have more potential energy than the product(s).

10 A chemical reaction that depends upon another, concomitant reaction for the energy needed for completion.

As above, answer Questions 11–15 using the following list:

a] productive collision
b] transition state
c] active site
d] substrate
e] ATP

11 Reactants that participate in an enzymatically regulated reaction.

12 The participant in an exergonic reaction that is widely used as a coupling agent with anabolic or endergonic reactions.

13 The specific site within an enzyme to which the correspondingly specific site of a particular substrate molecule can attach by chemical bonding.

14 The general term that includes enzyme-substrate complexes and other energetically unstable combinations of atoms or molecules.

15 An energetic interaction between two or more individual atoms or molecules that results in a chemical reaction.

As above, for Questions 16–20 match the descriptions with the appropriate item from the following list:

a] diffusion
b] osmosis
c] bulk flow
d] carrier-facilitated diffusion
e] active transport

16 This caused you to awaken to the seductive odors of breakfast.

17 The movement of individual molecules against their concentration gradient, usually requiring an external energy source, such as ATP.

18 A relatively rapid and forceful movement of a fluid stream, such as an oceanic current.

19 The transport of large hydrophilic or polar molecules through the cell membrane from a region of high concentration to low (external energy source usually not required).

20 The term that applies only to the passive movement of water through a selectively permeable membrane.

INTEGRATE AND APPLY WHAT YOU HAVE LEARNED

1 Is the growth of a child into an adult mainly an endergonic or an exergonic process? Explain.

2 Living, growing organisms defer the rise in entropy by using metabolic energy to construct more highly ordered cellular material. How can this situation be reconciled with the second law of thermodynamics, which holds that the amount of energy available to do useful work always decreases during an energy exchange?

3 Why is ATP not considered to be an enzyme?

4 Our planet can support a larger number of vegetarians than meat eaters. Why?

5 How does the sodium-potassium pump help to maintain osmotic balance in a cell?

ANALYSIS

1 Chemical energy, the energy contained within chemical bonds, is an example of:

a] Potential energy
b] Kinetic energy
c] Entropy
d] Activation energy
e] Heat

2 Motion, including passive transport, involves the expenditure of energy. What is the source of this energy for simple diffusion and osmosis, which are examples of passive transport?

a] Attractive forces between molecules.
b] Pressure from surrounding molecules.
c] The first molecules to move act as magnets, pulling along the next in line, and so on, like a train.
d] The random, jostling movement of the molecules themselves.
e] The potential energy within the surrounding medium, whether gaseous or liquid.

CHAPTER 5

How Living Things Harvest Energy

SPLITTING SUGAR FOR MUSCLE POWER

Suppose you and a friend decide one day to hike up a steep trail that winds alongside a tumbling stream—a trail, let's say, like the many criss-crossing the Green Mountains of Vermont. Fortified by a hearty breakfast of pancakes and maple syrup, you start up the trail in the brisk morning air. As you ascend into the fragrant pine forest, the steepness of the terrain begins to tax you. Your heart starts to pound; your chest heaves, and your leg muscles burn. Eventually, adopting a slower, steadier pace, you reach the summit and take in a panoramic view of blue-green forested ridges that gently fade to gray in the distance.

To carry out this simple weekend activity, the human body must rely on many fundamental life processes. Our subject here is one of the most basic: how cells extract energy from nutrient molecules, and how that energy is expended to fuel cellular activities. The sugar in maple syrup, the flour in pancakes, and the butter melting on top contain stored chemical energy. Cells in the maple tree, the wheat plant, or the hikers' muscles, lungs, and other organs can harvest that energy. The energy, in turn, fuels the growth of new leaves on the plants, and powers the walking, breathing, and other activities that carry two hikers up a mountain trail. To understand how muscle and other cells harvest the energy stored in nutrients, we need first to follow the pathway that energy takes in the physical and biological world. Energy leaves the sun as light, is captured by living cells in chemical form, and eventually passes back to the nonliving environment as heat.

We saw in CHAPTER 4 that ATP is the cell's energy currency, the form in which most energy actually moves around inside a cell. Once we discuss how a simple molecule like ATP stores and releases energy, we can see how living cells get that energy from the nutrients they take in. Biochemists have identified two main metabolic pathways for energy release within cells: (1) The major pathway requires that oxygen be present as nutrients are broken down, and it results in a prodigious amount of energy stored in molecules of ATP and other storage compounds. (2) The second pathway can go on even if the cell is deprived of oxygen, but the cell harvests less energy from the nutrient breakdown. It turns out that your muscle cells, as well as yeast cells and a few kinds of bacteria, can use both pathways of energy harvest, depending on the amount of oxygen present in the cell's environment. We'll discuss the different ways human muscle cells extract energy from sugar molecules when oxygen is plentiful and when it is scarce. And we'll see how athletes trained for an event like sprinting use one pathway, while those involved in an activity like marathon running use another.

The subject of energy harvest may seem like one of the more detailed and challenging in your study of biology. However, it is also central to all that goes on in cells and organisms, and that's why it is crucial to understanding the nature of life. The subject of energy harvest is also as personal as what you will consume in your next meal and how your body is currently utilizing what you ate for your last. The energy pathways within the cell are also a unifying concept in biology. Most of the metabolic reactions operating in human cells are also at work in other animals, plants, fungi, protists, and bacteria. This is a powerful reminder of life's unity, its evolutionary descent from common ancestors. On a larger scale, the subject of cellular energy harvest is also a key to understanding ecology, including the ways in which energy flows in the environment, as one organism feeds on another. We'll look at energy flow in the environment in detail in Chapter 40.

We will begin this chapter by discussing how ATP and electrons transfer energy in cells, and how much energy cells harvest in the presence and absence of oxygen. Next, we'll outline the individual steps of energy harvest. We'll see how energy-harvesting processes are linked to the acquisition of materials in the cell. Finally, we'll talk about several different forms of exercise, and how they emphasize different portions of the energy-harvesting pathways.

Along the way, we will see why a person's diet must always include at least 500 calories per day; what vitamins do within the cell; and how our bodies fuel fast, moderate, and slow exercise. ❑

MESSAGES

1 Energy flows through ecosystems. Plants capture energy from sunlight and store it in carbohydrates. This energy can then be harvested by the plant's cells or by the cells of animals and other organisms that consume plant matter. Eventually the energy returns to the environment as heat.

2 The aerobic pathway of energy harvest yields prodigious amounts of ATP in three stages—glycolysis, the Krebs cycle, and the electron transport chain.

3 The anaerobic pathway of energy harvest produces only about 5 percent as much ATP as the aerobic pathway. The splitting of glucose harvests some ATP, and the incompletely dismantled carbon products are discarded as wastes.

4 Muscle cells can use the aerobic pathway for a sustained energy harvest if sufficient oxygen is present; when oxygen is in short supply, they can use the anaerobic pathway, which produces toxic waste products, for short bursts of activity.

Sources of Energy

All cells need energy to live, whether they are muscle cells in a hiker's leg, budding yeast cells turning a tank of grape juice into wine, or cells in a maple leaf fluttering in the wind. Regardless of type, however, no cell can "make" energy. Recall from CHAPTER 4 that energy can neither be created nor destroyed, but can merely be transformed from one state to another [review FIGURE 4.3]. Each cell must therefore obtain its energy from some outside environmental source, and cells do so in a range of ways. Yeast cells growing on a grape skin obtain energy from nutrient molecules originally synthesized within the grape leaves and stored within the fruit. Humans obtain energy from molecules present in the plant and animal matter they eat and digest. Maple leaf cells obtain their energy directly from sunlight, which they trap and convert into a chemical form [FIGURE 5.1].

Let's follow the flow of energy through the Vermont ecosystem that includes our weekend hikers. Sunlight strikes the leaves of maple trees, and the leaves carry out *photosynthesis*—they trap solar energy and store it in the form of sugars. (The next chapter discusses in more detail exactly how energy capture and sugar production take place in leaves.) In the spring, a Vermont farmer may tap the maple tree to obtain the energy-rich sugar molecules from the tree's sap. Once this sap is boiled down into maple syrup, it contains a concentrated supply of sugar, which the hikers can pour onto their pancakes. Inside their bodies, the glucose molecules leave the stomach

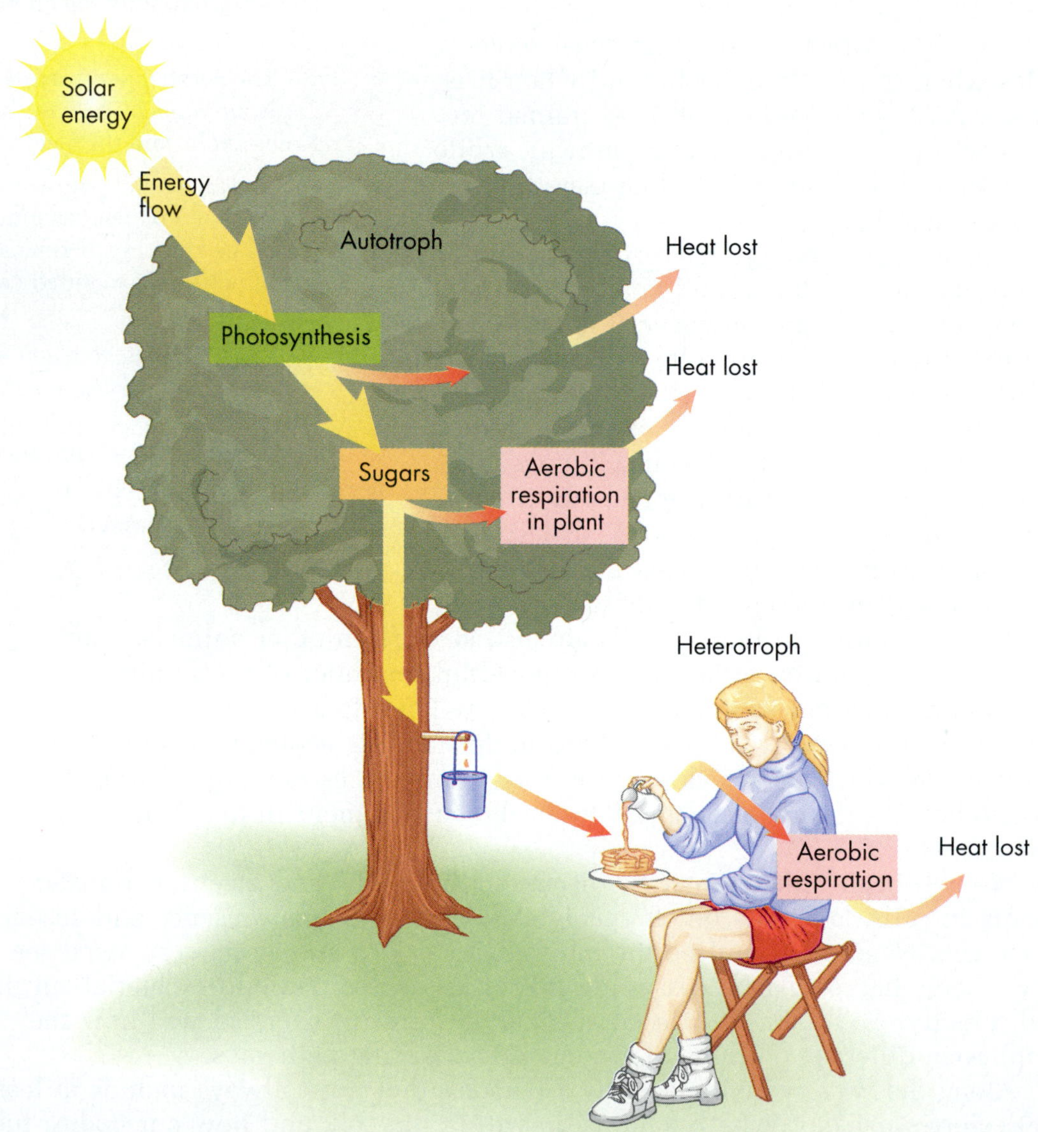

FIGURE 5.1

Energy Flow Through Ecosystems.

Energy usually enters as sunlight and is eventually lost as heat. During photosynthesis, light energy is converted to chemical energy and stored in sugars. During aerobic respiration in plant and animal cells, the energy stored in sugars is harvested, and available to fuel cellular processes. At each energy conversion, some energy is lost as heat to the environment.

and intestines, flow through the bloodstream, and enter muscle, brain, lung, and other tissue cells. There, the glucose molecules are gradually broken down, and some of the energy from sunlight that the maple leaves trapped is released. That energy is now available to power the hikers' vision, thinking, breathing, movement, and so on. Since, however, all energy transformations inevitably lose some energy as heat [review FIGURE 4.3], the hikers warm up as they climb, and the excess body heat is lost to the cool mountain air around them.

In summary, energy flows through ecosystems. Most of it originates in the environment in the form of sunlight and flows through plants where it is captured and stored in carbohydrates. Plants can then use those carbohydrates for their growth or other activities, or energy in the carbohydrates can move to the animals that consume the plants. Some of the energy is lost directly from the animals' bodies as heat, and, for practical purposes, the rest is lost as heat when the animals die and their bodies are broken down by bacteria and fungi. During its temporary sojourn through an organism, the energy is converted into a chemical form that can help maintain the order of cell constituents and other activities; that chemical form is the energy-rich molecule ATP.

CONCEPT CHALLENGE

If all the animals on the earth died, what would be the effect on the trees, flowers, and other plants? If all the plants in the world died, what would be the effect on people and the other animals?

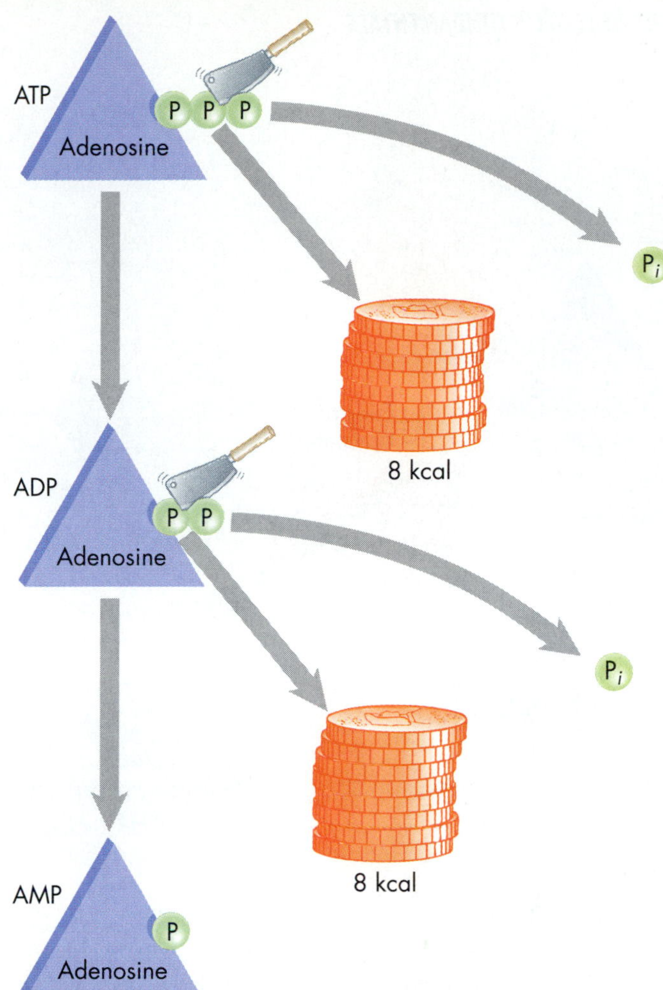

FIGURE 5.2

Comparison of ATP, ADP, and AMP.

ATP, or adenosine triphosphate, has a long, three-phosphate tail, and the molecule contains a substantial amount of energy. ADP has a shorter tail with just two phosphate groups and contains less energy. AMP has just a one-phosphate tail and considerably less energy than either ADP or ATP. Cleaving one phosphate group from ATP yields 8 kcal of energy, one inorganic phosphate ion, and ADP. Cleaving another phosphate group yields an additional 8 kcal of energy, another inorganic phosphate ion, and AMP.

Energy Transfer in Cells

For all living cells—whether in a hiker's complex muscle cells or in the simpler cells of bacteria living on the hiker's skin—two energy-carrying molecules link energy harvest and energy use: (1) ATP and (2) its lower-energy partner ADP. To understand how cells harvest energy from nutrients, we must therefore look first at the structure of ATP and how it stores energy.

STRUCTURE OF ATP

ATP, or adenosine triphosphate, is a molecule with a long tail of three phosphate groups, and this structure explains its utility to the living cell [review FIGURE 4.8]. ADP, or adenosine diphosphate, contains two phosphate groups, and AMP, or adenosine monophosphate, just one [FIGURE 5.2]. When a phosphate bond is broken, a large amount of useful energy contained in the whole molecule is readily released. This release of energy from a chemical bond is similar to pulling the plug at the bottom of a rain barrel. The energy released in the outflow was stored in the total water contents, not simply in the rubber plug.

Phosphorylation When one of the three phosphate groups is cleaved from the energetic tail of ATP, the phosphate can be transferred to another molecule along with some of the released energy. This transfer is called *phosphorylation*. For example, ATP can transfer a phosphate group to glucose, making glucose phosphate and ADP [FIGURE 5.3]. This energizes the glucose and allows it to participate in reactions that could not otherwise take place.

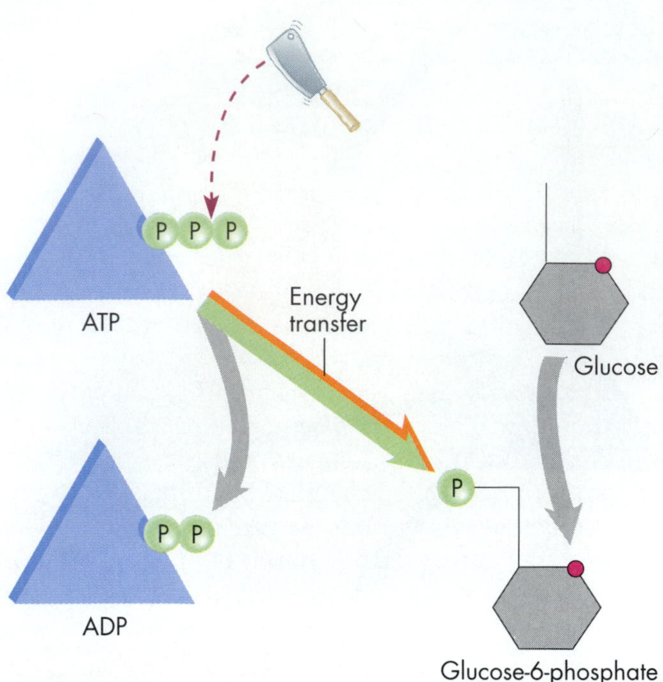

FIGURE 5.3

Phosphorylation of Glucose.

Glucose can be "supercharged" by the cleaving of ATP and the addition of the phosphate group to the glucose molecule. This glucose-6-phosphate can then participate in reactions in which a regular glucose molecule cannot take part.

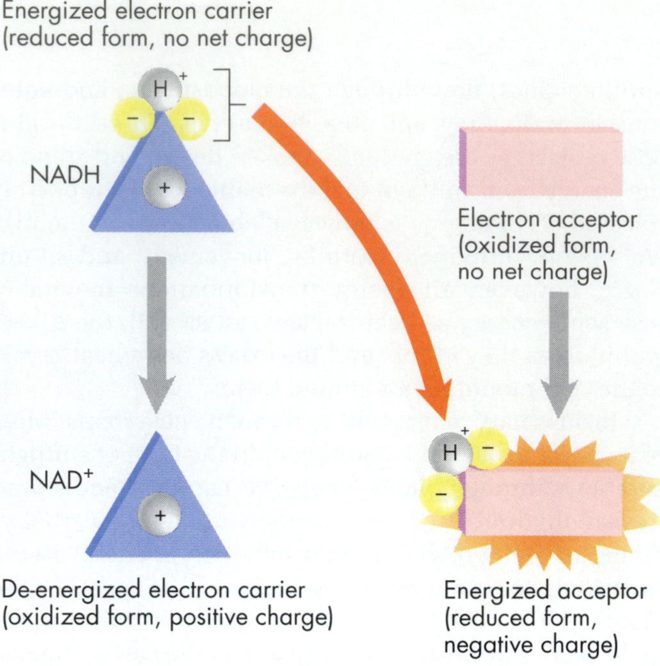

FIGURE 5.4

Electron Exchange Reactions and Electron Carriers.

The electron carrier NADH can transfer electrons, along with a hydrogen ion, to an electron acceptor. The NADH loses electrons, forming NAD^+ and releasing energy. Within mitochondria, some of the energy released in the electron transfer can be trapped and stored in ATP molecules.

ELECTRON FLOW AND ENERGY TRANSFER

Phosphate is not the only substance that transfers energy from one molecule to another—electrons can also transfer energy. Recall from CHAPTER 2 that the formation of chemical bonds involves the sharing or transfer of electrons between atoms [review FIGURES 2.8, 2.11]. When electrons move from one molecule to another, they can also transfer energy.*

In biological systems, pairs of electrons along with a hydrogen ion (H^+) often move between special molecules called *electron carriers*. One such carrier, called **NADH** (nicotinamide adenine dinucleotide), can give up two electrons and a hydrogen ion and become NAD^+. Simultaneously, one of a large number of other compounds traps the electrons and hydrogen ion [see FIGURE 5.4]. NAD^+ helps the alcohol-degrading enzyme we discussed in CHAPTER 2 (called ADH) metabolize the ethanol in wine or beer by the transfer of electrons and hydrogen ions [review FIGURE 2.29]. NAD^+ and other enzyme-helping molecules (called *cofactors*) are made from vitamins. NAD^+ is made from vitamin B_3 (or *niacin*), which you will probably find listed among the ingredients on your breakfast cereal box.

As electrons are passed from one electronic carrier to another, energy is released. This energy can be channeled into ATP formation. Thus, during the overall breakdown of a nutrient molecule—a molecule of glucose in a muscle cell, for example—electrons are removed from the sugar and transferred to an end product. In the process, energy is channeled into the production of ATP. Now let's look at that overall breakdown.

*The transfer of electrons from one molecule to another is called an **oxidation-reduction reaction**. The molecule that loses electrons is said to be *oxidized*, and the molecule that gains electrons is *reduced*. The molecule that is oxidized (loses electrons) also loses energy, while the molecule that is reduced (gains electrons) also gains energy.

An Overview of Energy Harvest

Recall that as our hikers moved slowly up a gradual hill, they breathed in deeply and regularly. When they moved rapidly up a particularly steep slope, however, their muscles started to ache and they began to gasp for air. These two states of our hikers reflect two pathways of energy harvest taking place in their muscle cells—energy harvest with oxygen, the *aerobic pathway* and energy harvest without oxygen, the *anaerobic pathway*. Human muscle cells and yeasts are somewhat special among eukaryotic cells in having the ability to obtain energy from either pathway depending on the availability of oxygen.

During an easy hike, the lungs and blood circulation provide enough oxygen to muscles so that muscle cells can use the aerobic pathway (the orange arrows in FIGURE 5.5). During a more strenuous climb, however, the body systems cannot provide oxygen as fast as muscle cells require it. For this reason, the muscle cells switch to the alternative anaerobic pathway (the broad yellow arrows in FIGURE 5.5). Some cells, such as brain cells, are limited to the aerobic pathway, while other cells, such as the bacterium that causes tetanus, can only use the anaerobic pathway. Let's survey the broad outlines of the two main energy-releasing pathways now, and then in later sections of the chapter, look at them in more detail.

THE AEROBIC PATHWAY OF ENERGY HARVEST

The aerobic pathway involves three main processes, or series of reactions. These are called (1) glycolysis, (2) the Krebs cycle, and (3) the electron transport chain, and are shown in the broad orange arrows in FIGURE 5.5. These processes occur in the cells of plants and animals, and in many cells of the other kingdoms as well.

Glycolysis In brief, energy harvest begins with the breakdown of glucose in a sequence of reaction steps called **glycolysis** (*glyco-*, "sugar," and *lysis*, "splitting"). The energy stored in the chemical bonds of glucose molecules originally came from the sun, and it was trapped in molecular form by photosynthesis taking place in green plants, algae, and certain kinds of bacteria.

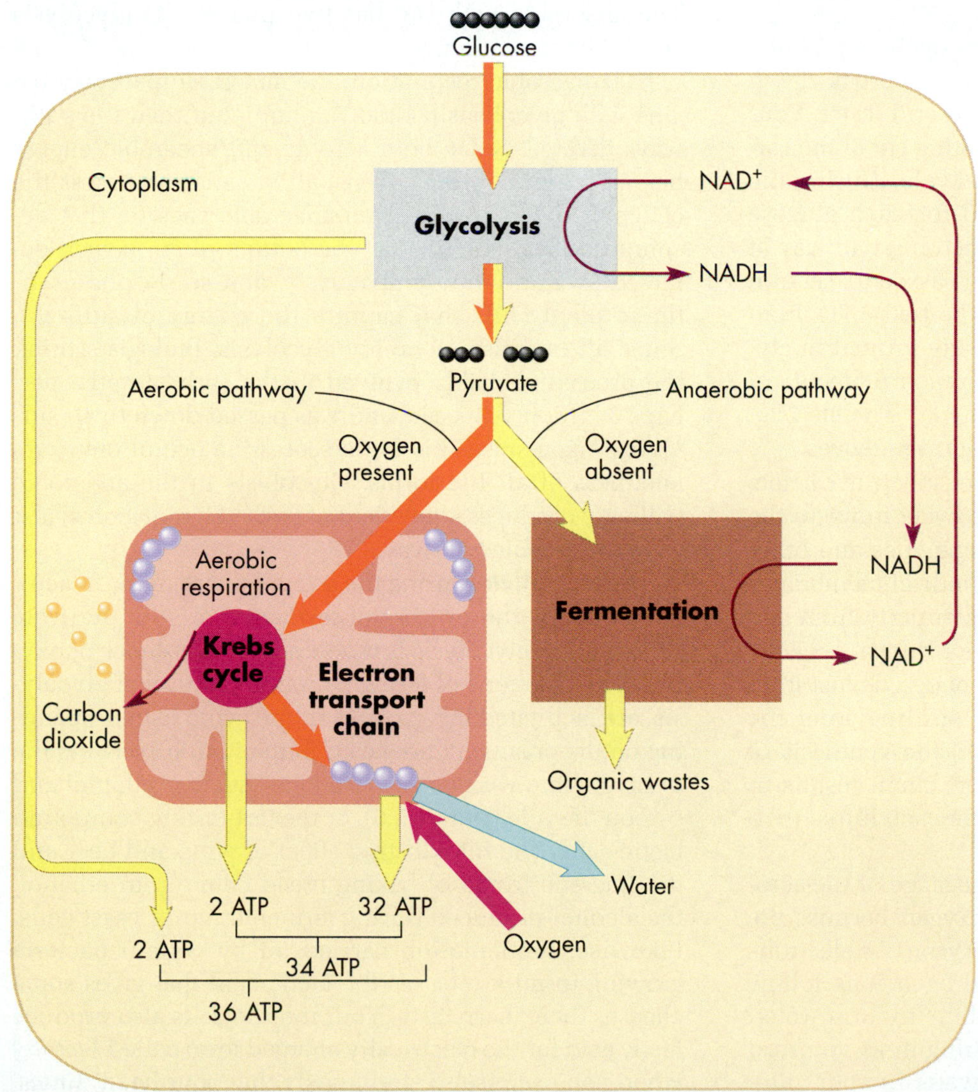

FIGURE 5.5

Energy Harvest.

The aerobic pathway shown by the orange arrows includes glycolysis in the fluid portion of the cytoplasm as well as the Krebs cycle and the electron transport chain in the mitochondrion. The aerobic pathway breaks down glucose to carbon dioxide and water, consumes oxygen as a final electron acceptor, and produces a total of about 36 ATP per molecule of glucose. The anaerobic pathway shown by the yellow arrows includes glycolysis and fermentation, both of which occur on enzymes in the fluid portion of the cytoplasm. This pathway leaves organic wastes like alcohol and carbon dioxide or lactic acid, and liberates only two ATP per initial molecule of glucose.

The reactions of glycolysis split the six-carbon sugar glucose [FIGURE 5.5] into two molecules of the three-carbon compound *pyruvate.* The steps of glycolysis take place in the cell's cytoplasm, rather than in a membrane-bound organelle, and they are facilitated by enzymes dissolved in a cell's watery cytoplasmic solution. For each glucose molecule split during glycolysis, there is a net gain of two ATPs.

The three-carbon pyruvate molecules produced in glycolysis leave the cytoplasm and enter a mitochondrion, where they serve as intermediates—molecular raw materials—for **aerobic cellular respiration**, the breakdown of nutrients and the production of ATP energy in the presence of oxygen [see FIGURE 5.5]. Aerobic respiration in mitochondria consumes oxygen and yields carbon dioxide and water plus a large harvest of ATP molecules in addition to the two ATP from glycolysis. That energy harvest accounts for the mitochondrion's reputation as a cellular powerhouse. The two phases of aerobic cellular respiration are called the Krebs cycle and the electron transport chain.

The Krebs Cycle In brief, the **Krebs cycle** is part of a series of enzyme-catalyzed reactions that break down pyruvate completely into carbon dioxide and water. Your exhaled breath is the body's way of getting rid of the carbon dioxide produced in the Krebs cycle. During the Krebs cycle, two ATPs are produced for each glucose molecule that originally entered the aerobic pathway at glycolysis, and numerous electrons are passed to electron carriers such as NADH and $FADH_2$. The two ATPs from the Krebs cycle added to the two already formed in glycolysis create a total of four ATPs for each molecule of glucose reaching this stage in the pathway. The electron carriers now undergo additional reaction sequences.

The Electron Transport Chain In brief, electron carriers loaded with electrons from the Krebs cycle move to the **electron transport chain**, a group of enzymes and other electron carriers embedded in mitochondrial membranes [FIGURE 5.5]. Electrons are passed sequentially from one member of the chain to another, like a bucket brigade. During each electron transfer, a bit of energy is lost from the electron (like a splash of water spilling from the bucket); this released energy goes into the synthesis of ATP from ADP. The electron transport chain results in the synthesis of a whopping 32 ATPs for each initial molecule of glucose.

The electron transport chain is the stage of the aerobic pathway that actually uses the oxygen because the last electron acceptor in the chain is oxygen. As electrons are added to oxygen atoms, and hydrogen ions follow along, the hydrogen and oxygen combine to form water, H_2O. Thus, the oxygen you are breathing in as you read these sentences will be converted to water in your mitochondria as oxygen accepts electrons and aerobic respiration takes place.

An ATP Tally Let's pause and tally up the ATP yield from the aerobic pathway (see the bottom of FIGURE 5.5). From each initial glucose molecule metabolized in the cell, the net yield from glycolysis is 2 ATPs, the Krebs cycle produces 2 more, and the electron transport system produces 32. This gives a total net yield of 36 ATP molecules per glucose molecule. Now let's compare this yield to that of the anaerobic pathway energy harvest when oxygen is absent.

THE ANAEROBIC PATHWAY OF ENERGY HARVEST

We return to our hikers, now out of breath as they rush up a steep slope. Muscle cells in their tired legs are using the aerobic pathway to its maximum, but still there is not enough ATP to fuel the rapid activity. Their muscle cells, in addition, then begin to utilize the anaerobic pathway of energy harvest (the broad yellow arrows in FIGURE 5.5). The anaerobic pathway has two phases: (1) glycolysis and (2) fermentation.

Like aerobic respiration, the anaerobic pathway begins with glycolysis [FIGURE 5.5, top], but then the pathways diverge. Since both aerobic and anaerobic energy harvest begin with glycolysis, all organisms possess the enzymes of glycolysis. Even anaerobic bacteria that decompose organic sludge at the bottom of an oxygen-depleted swamp employ glycolysis, and so do photosynthetic plant cells that capture the energy of sunlight. Since all organisms employ glycolysis, biologists think the process probably evolved in the earliest cells, perhaps 3 billion years ago, and was passed down to all surviving organisms. This underscores the evolutionary relatedness of all life forms. Glycolysis in the anaerobic pathway produces a net yield of two ATP molecules, just as in the aerobic pathway.

Fermentation During the second phase of the anaerobic pathway, the process of **fermentation**, the pyruvate generated by glycolysis is modified in the absence of oxygen. The reactions of fermentation, like those of glycolysis, are facilitated by enzymes in the cytoplasm. Depending on the organism, however, fermentation converts the pyruvate into various end products, such as ethanol and carbon dioxide [FIGURE 5.6], or the tart-tasting compound lactic acid. The intoxicating effect of wine and beer and the fragrant aroma of baking bread come from ethanol, the alcohol produced during fermentation in yeast cells. Likewise, fermentation carried on by certain bacteria growing in milk releases the lactic acid that gives some cheeses their sharp taste. Your muscle cells also produce lactic acid (or the electrically charged form called *lactate*) when you exercise so strenuously that your heart, lungs,

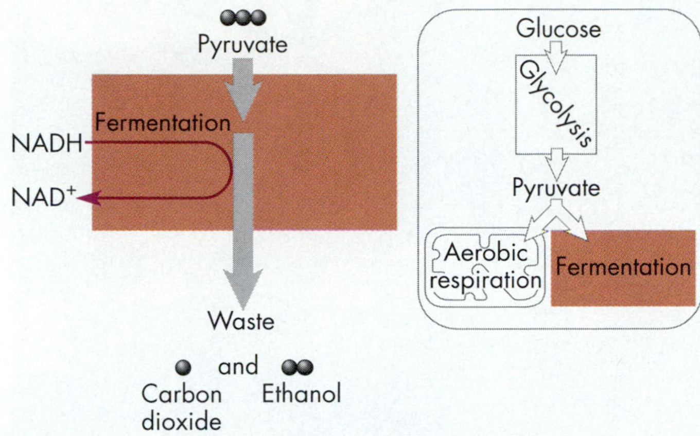

[A] The fermentation reaction

[B] Baking and alcoholic fermentation

FIGURE 5.6

Fermentation and the Aroma of Baking Bread.

Yeast cells growing in a lump of dough carry out glycolysis, during which glucose from the starch in flour is broken down into pyruvate molecules. In the process, NADH and ATP are produced and these can power cellular activities. For glycolysis to continue, the cells require a steady supply of NAD^+; this is regenerated during fermentation from the NADH by enzymes that split pyruvate into ethanol and carbon dioxide. The warm carbon dioxide gas expands and causes the bread to rise; at the same time, the ethanol evaporates and produces the fragrant aroma.

and blood vessels cannot supply oxygen fast enough for them to provide sufficient energy solely by aerobic cellular respiration. The "burn" aerobics instructors like to talk about comes from the liberation of lactate as you work a muscle very hard.

The process of fermentation yields no ATP in addition to the two produced in glycolysis. The wastes, in fact, contain all the energy that would be captured in ATP if the cell were using the aerobic pathway. You might wonder, then, why cells would carry out fermentation and produce waste substances such as ethanol and lactate, which are toxic to the cell? The answer is that fermentation reactions recycle the electron carrier NAD^+, which, as we saw, is required for glycolysis to continue (notice the purple arrows in FIGURE 5.5). Without this recycling of NAD^+, the cell would run out of electron carriers, and glycolysis would cease.

AN ENERGY TALLY

We can compare the two pathways of energy harvest now:

1. The aerobic pathway of cellular respiration generally yields 36 molecules of ATP for each molecule of glucose dismantled (a net of 2 from glycolysis, 2 from the Krebs cycle, and 32 from the electron transport chain) [TABLE 5.1]. In the process, the six carbons from the original glucose molecule end up in carbon dioxide molecules, and the hydrogens and numerous electrons also released from glucose are transferred to oxygen, resulting in the formation of water molecules.

2. The net yield from the anaerobic pathway is only the two molecules of ATP from the reactions of glycolysis and the wastes, such as ethanol and carbon dioxide or lactate [TABLE 5.2].

Clearly, the aerobic pathway is more efficient than the anaerobic pathway—in fact, 18 times more efficient—for producing ATP from glucose. This explains why a person can exercise aerobically for long periods, but can carry on very strenuous anaerobic exercise for only a short time before running out of energy.

The energy yield of the anaerobic pathway may seem meager—just two ATPs per glucose. However, anaerobic metabolism is crucial to the global recycling of carbon and the stability of the environment. Organic matter from dead leaves, dead microorganisms, and other sources often sinks into an environment devoid of oxygen, such as the soft layers at the bottom of lakes or oceans [FIGURE 5.7]. If it weren't for anaerobic decomposers—organisms capable of breaking down organic matter via anaerobic metabolism—most of the world's carbon would eventually be locked up in undecomposed organic material in these oxygen-poor environments, as deeper and deeper layers built up. As a result, there would be too little carbon dioxide available as a raw material for photosynthesis, plants would be unable to generate new glucose molecules, and neither plants nor animals would survive.

FIGURE 5.7

Environmental Carbon Cycling: Anaerobic Respiration Plays a Key Role.

Hikers pausing for a rest by a clear mountain lake see only the sparkling upper layers of water, where photosynthesis in the tissues of algae and green plants fixes carbon and releases oxygen into the water. When water plants and oxygen-breathing fish die, their bodies sink to the lake bottom. These deep layers are colder and soon become depleted of oxygen, as oxygen-utilizing microorganisms begin to decompose the debris. Once the oxygen is fully depleted, the microorganisms switch to anaerobic metabolism and complete the decomposition of the organic matter to carbon dioxide and often methane gas, which hikers can sometimes see bubbling up to the surface. Without this anaerobic recycling, most of the carbon would be tied up in dead plants and animals.

TABLE 5.1 Summary of the Aerobic Pathway of Energy Harvest

Process	Raw Materials	Requires Oxygen?	Number of ATPs Used per Glucose	Number of ATPs Produced per Glucose	End Products
Glycolysis	Glucose, NAD^+	No	2	4	Pyruvate NADH
Krebs cycle (plus acetyl-CoA step)	Pyruvate, NAD^+, FAD	No	0	2	Carbon dioxide NADH $FADH_2$
Electron transport chain	NADH, $FADH_2$	Yes	0	32	Water NAD^+ FAD

Total = 4 + 2 + 32 − 2 = 36

TABLE 5.2 Summary of the Anaerobic Pathway of Energy Harvest

Process	Raw Materials	Requires Oxygen?	Number of ATPs Used per Glucose	Number of ATPs Produced per Glucose	End Products
Glycolysis	Glucose, NAD^+	No	2	4	Pyruvate NADH
Fermentation	Pyruvate, NADH	No	0	0	NAD^+ and either ethanol or lactate and water

Total = 4 − 2 = 2

This section dealt with the core concepts of energy harvest. Knowing these basic principles allows us to understand other important biological principles, such as how energy flows through ecosystems, how feeding strategies evolve in different organisms, and how gases are exchanged in animal tissues. In the next section, students curious about the individual phases of energy harvest can learn about them in more detail.

CONCEPT CHALLENGE

A geneticist gives you test tubes containing one of two types of yeast cells that are the same in every way except that one can carry out only the aerobic pathway of energy harvest and the other can carry out only the anaerobic pathway. Unfortunately, the tubes are labeled only A and B, but they do display a difference: Yeast from tube A grow rapidly, while yeast from tube B grow slowly. Which tube contains the cells capable of performing only the aerobic pathway? How did you make your choice?

Energy Harvest: A Deeper View

The energy stored in a nutrient molecule is not released all at once as in the burning of a match; instead, molecular bonds are broken and rearranged in numerous small steps, and the sequentially released energy is stored for future use. Since the transfer of energy is always inefficient, some heat is lost at each step. Despite this inevitable waste, living cells harvest enough energy in the form of ATP to fuel their growth, maintenance, and reproduction. Let's look at the specific steps that take place as the energy is gradually released from glucose inside the cell. We'll begin with glycolysis and fermentation, then investigate the Krebs cycle and the electron transport chain.

GLYCOLYSIS: THE UNIVERSAL FIRST STEP

At this moment, practically every cell on earth, including those of plants, animals, and microorganisms, is burning the six-carbon sugar glucose via glycolysis, with its nine sequential reaction steps. As the reactions of this pathway proceed in the cell's cytoplasm, each glucose molecule is split into two molecules of the three-carbon compound pyruvate, and two ATPs are harvested. The overall net reaction for glycolysis is:

$$\underset{\text{Glucose}}{C_6H_{12}O_6} + 2\ ADP + \underset{\text{Phosphate}}{2P} + 2\ NAD^+ \rightarrow$$

$$2\ \underset{\text{Pyruvate}}{C_3H_4O_3} + 2\ ATP + 2\ NADH + 2\ H^+ + 2\ H_2O.$$

But this change does not happen in one big energy-releasing bang. Instead, there are four stages, or groups of steps. The significant outcome of each stage is illustrated in FIGURE 5.8 and summarized here:

Stage 1: Energizing the Sugar In two separate reactions, enzymes remove a phosphate group from each of two ATP molecules and transfer them to the six-carbon sugar molecule [review FIGURE 5.3]. This charges the sugar with high-energy phosphates, but expends cellular energy rather than harvesting it. This addition of phosphates to the sugar costs the cell some energy, since ATP must be cleaved to provide the phosphate group and form the new bond. The expenditure, however, is like priming a pump or spending money to make money; as the reactions proceed further, that original cost is paid back with interest—the harvesting of additional ATP. The result of the reactions in Step 1 is an energized sugar called *fructose 1,6-bisphosphate*, with two phosphates (the numbers 1 and 6 refer to which carbon in the sugar bears the phosphate; Stage 1 in FIGURE 5.8).

Stage 2: Splitting the Sugar The high-energy sugar formed in the initial reactions is split into two molecules of the compound G-3-P (glyceraldehyde 3-phosphate), each of which has three carbons and one phosphate group (Stage 2 in FIGURE 5.8). Each of the remaining steps takes place twice, once for each of the three-carbon molecules formed.

Stage 3: Exchange of Electrons and Hydrogen Ions Electrons and hydrogens are removed from the phosphorylated three-carbon G-3-P. The electrons and hydrogens are transferred to NAD^+, producing NADH. In addition, phosphate groups are added to the three-carbon G-3-P, forming a new three-carbon compound with two phosphates called 1,3-BPG (1,3-bisphosphoglycerate; see Stage 3 in FIGURE 5.8).

Stage 4: Energy Harvest The 1,3-BPG, with its three carbons and two phosphates, is next shaped and reshaped into a series of related compounds (summarized as Stage 4 in FIGURE 5.8). Enzymes remove phosphates one at a time and transfer them to ADP, generating a total of four new ATPs for each molecule of glucose dismantled via glycolysis.

The reactions of glycolysis can be summarized as follows: Each six-carbon glucose molecule is boosted to a higher energy state with the addition of two phosphates and rearranged. The six-carbon sugar-phosphate is then split into 2 three-carbon compounds. Each of the three-carbon molecules is rearranged and a phosphate removed and used to form ATP, leaving 2 three-carbon pyruvate molecules.

For each initial glucose, two ATPs are spent in Stage 1 [see FIGURE 5.8], and four ATPs form in Stage 4, leaving

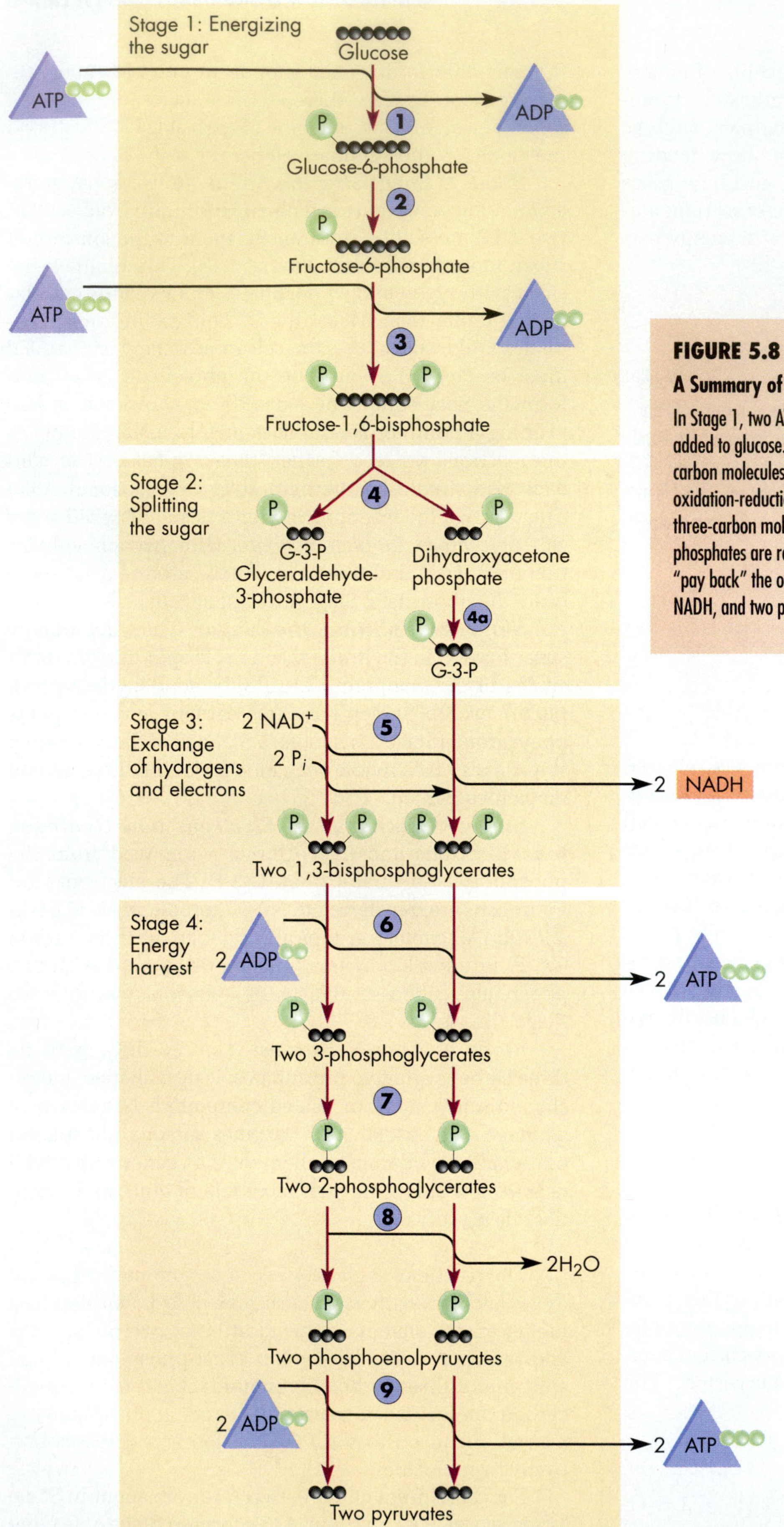

FIGURE 5.8

A Summary of Glycolysis.

In Stage 1, two ATP molecules are spent, and two released phosphates are added to glucose. In Stage 2, this charged glucose is split into 2 three-carbon molecules, each with one phosphate (G-3-P). In Stage 3, an oxidation-reduction reaction and the addition of phosphates result in 2 three-carbon molecules, each with two phosphates. In Stage 4, those phosphates are removed and energy is captured in four ATPs, two of which "pay back" the original investment. The net yield, then, is two ATP, two NADH, and two pyruvate molecules.

a net gain of two ATP molecules. These ATPs can power cell activities, and the pyruvate and NADH can participate in the other two phases of the aerobic pathway, the Krebs cycle and the electron transport chain, respectively. Since both of these additional processes take place in mitochondria, let's look more closely at why a mitochondrion's structure is so integral to its function of providing energy for a eukaryotic cell.

MITOCHONDRIA: SITES OF ATP PRODUCTION

Highly active cells contain large numbers of sausage-shaped ATP-generating organelles called **mitochondria**. A liver cell or a large plant cell can contain more than a thousand mitochondria, and they can occupy nearly a fifth of the cell's volume [review FIGURE 3.20]. Mitochondria consist of two membranes, like a large lightly inflated balloon folded up inside a smaller, inflated balloon [FIGURE 5.9]. A mitochondrion's **outer membrane** is surrounded by the cell's cytoplasm. Large protein-bound pores perforate the membrane, and molecules up to the size of small proteins can pass through the openings. Molecules can thereby gain ready entrance to the gap between the two mitochondrial membranes, the **intermembrane space**.

A mitochondrion's **inner membrane** is thrown into folds called **cristae**, which are studded with the enzymes of the electron transport chain and an enzyme that synthesizes ATP [FIGURE 5.9B, C]. The inner membrane is much less permeable than the outer membrane; hence, the area enclosed by the inner membrane, the **matrix**, is a compartment well isolated from the rest of the cell. The matrix contains several copies of the circular mitochondrial DNA molecule, many mitochondrial ribosomes, and hundreds of enzyme molecules, including those that carry out the reactions of the Krebs cycle.

THE KREBS CYCLE PRODUCES CO_2 AND CHARGED ELECTRON CARRIERS

The three-carbon pyruvate molecules produced in the cytosol by glycolysis can pass through both mitochondrial membranes and enter the mitochondrial matrix. In the matrix, enzymes strip high-energy electrons and hydrogen ions from the three-carbon skeleton of each pyruvate and transfer them to electron carriers. At the same time, carbon dioxide is released as a waste product. This dismantling takes place in three stages [FIGURE 5.10]:

Stage 1: Bridge to the Krebs Cycle In the mitochondrial matrix, one of the three carbons of pyruvate is removed and converted to carbon dioxide. The two remaining carbon atoms (an acetyl group) are attached to a carrier molecule called coenzyme A (or CoA), forming a compound called *acetyl-CoA* [FIGURE 5.10, Stage 1]. In the process, one NADH molecule is produced.

Stage 2: Acetyl-CoA Enters the Krebs Cycle Acetyl-CoA binds to the first of a series of eight enzymes that are integral to the Krebs cycle. The first enzyme in the cycle joins two carbons from acetyl-CoA to a four-carbon compound, generating a six-carbon compound called *citric acid* (or, the ionized form, *citrate*). Chemists first discovered citric acid more than 200 years ago in lemon juice, a citrus fruit, hence the name. The Krebs cycle is often called the *citric acid cycle* because of this compound.

Stage 3: Carbon Dioxide Formation and Electron Carriers From the six-carbon compound citric acid, two carbons are stripped off one at a time, and released as two carbon dioxide molecules, leaving a four-carbon compound. Additional electrons and hydrogens are also stripped off, and picked up by the electron carrier NAD^+ to form NADH [see FIGURE 5.10]. The resulting four-carbon compound is further rearranged to the original four-carbon compound of Stage 2 (see above), thus completing the cycle. In the process, another electron carrier, *$FADH_2$*, captures more electrons and hydrogens and energy is released that is eventually captured in an ATP.

In total, for each pair of pyruvate molecules and each two turns of the cycle, 2 ATP molecules and 10 charged carriers form (8 of NADH and 2 of $FADH_2$) and 4 carbon dioxide molecules (representing the rest of the carbon atoms in the original glucose) are released.

The ATP harvest from the Krebs cycle is the same as from glycolysis—two ATPs per glucose molecule. However, the 10 pairs of high-energy electrons in NADH and $FADH_2$ represent a treasure trove of stored energy that will be released, bit by bit, by the components of the electron transport chain.

The six carbon dioxide molecules produced by the complete breakdown of the six-carbon glucose molecule eventually diffuse out of the mitochondrion. In a plant cell, these can be used immediately for photosynthesis [see CHAPTER 6], but in an animal cell, they are released as a waste product. In our hikers puffing along up the mountain trail, these carbon dioxide molecules are exhaled in their labored breathing.

The Krebs Cycle as a Metabolic Clearinghouse The Krebs cycle is an important intermediate phase in aerobic respiration. It not only completes the breakdown of glucose begun in glycolysis, but it also functions as a metabolic clearinghouse. Organic nutrients other than glucose can enter the pathway at various points in the Krebs cycle [FIGURE 5.11]. If its supply of sugar falls, an animal or plant cell begins to break down lipids or proteins. Some of the subunits from these reactions can then be con-

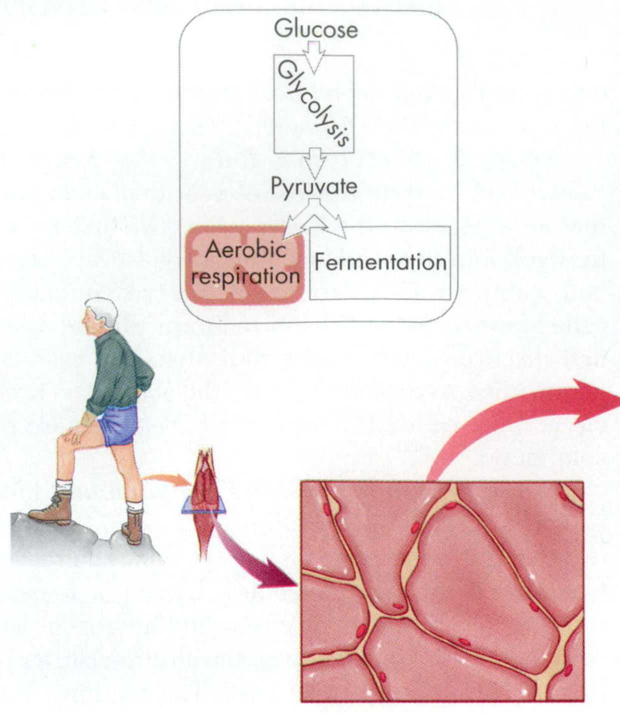

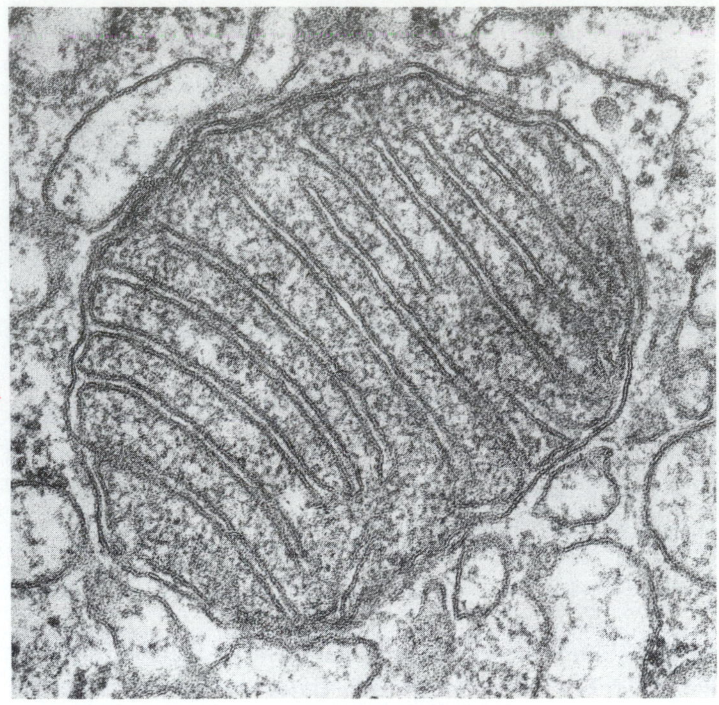

[A] Electron micrograph of mitochondrion.

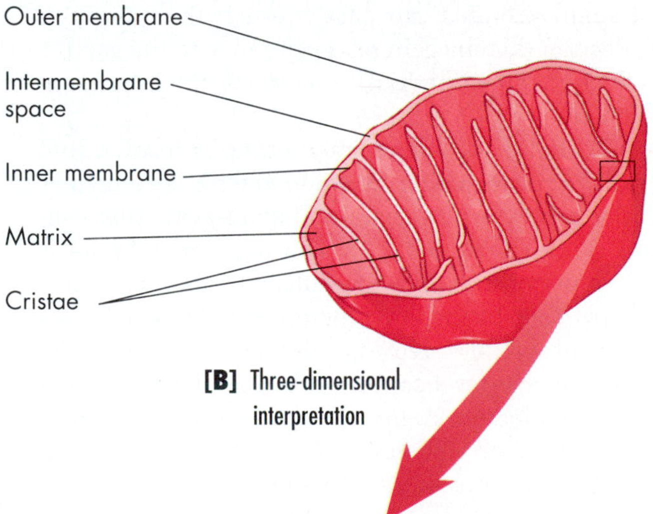

[B] Three-dimensional interpretation

FIGURE 5.9

Mitochondrial Function Depends on Structure.

[A] An electron micrograph of a mitochondrion reveals the membranes and compartments that function in energy metabolism. **[B]** The cytoplasm surrounding the mitochondrion contains the enzymes involved in glycolysis. The mitochondrial outer membrane is perforated by protein-lined pores that connect to the intermembrane space between the outer and inner mitochondrial membranes. The inner membrane is studded with proteins of the electron transport chain as well as the ATP-synthesizing enzyme. In the matrix, the space surrounded by the inner membrane, the enzymes of the Krebs cycle dismantle carbon compounds to carbon dioxide.

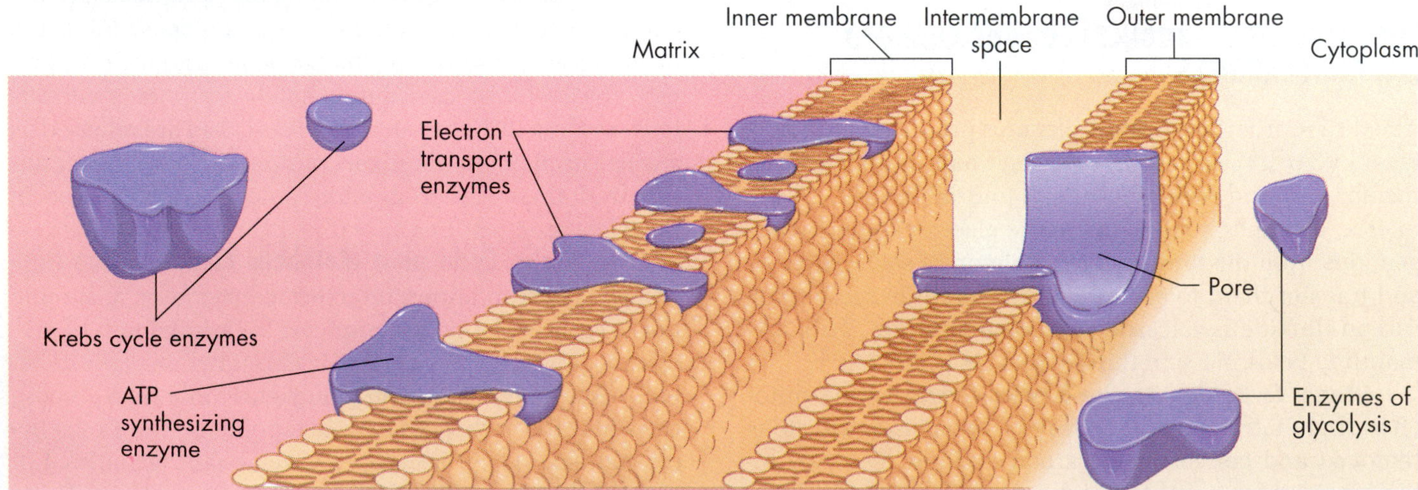

[C] Mitochondrial membranes

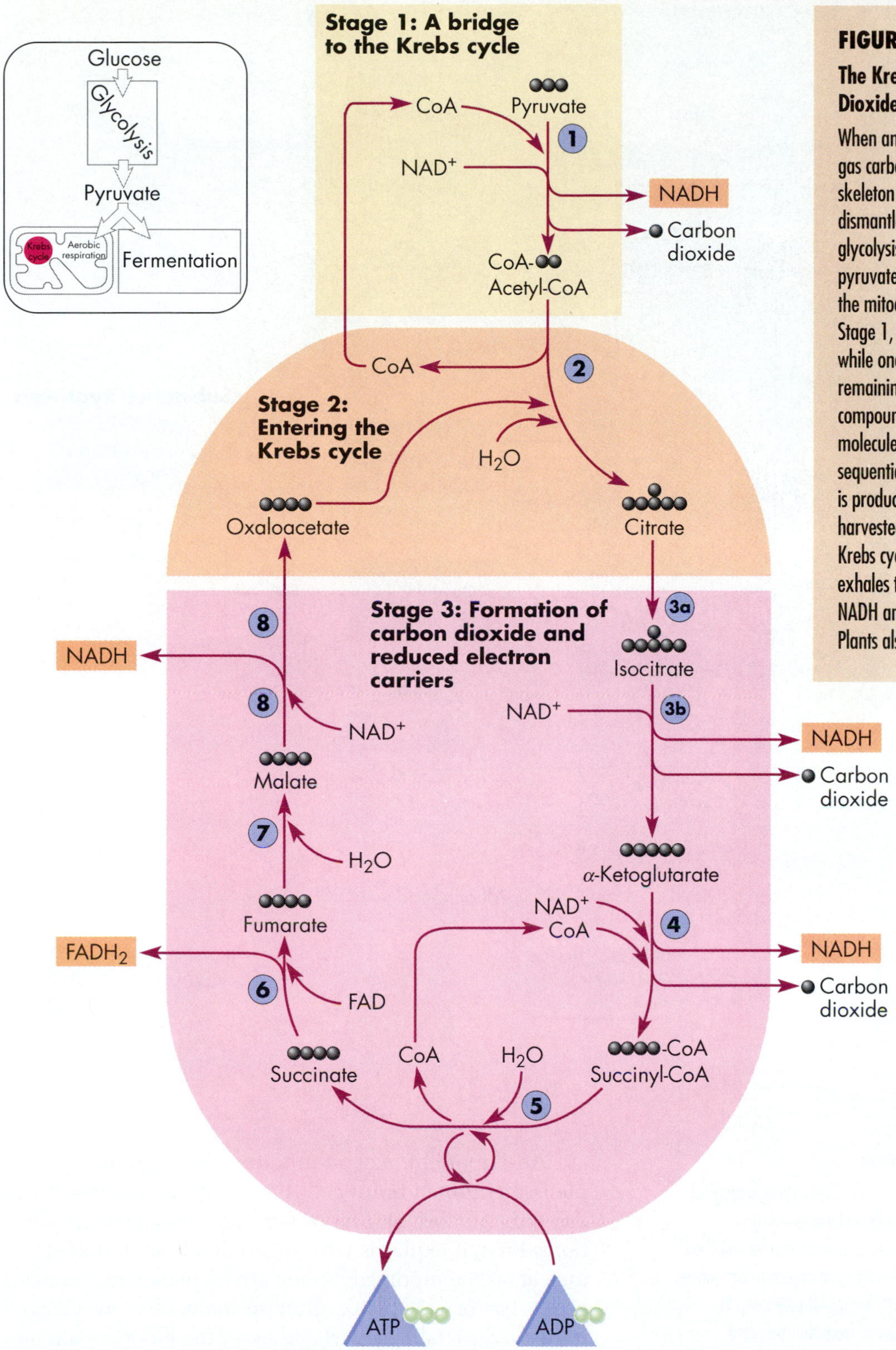

FIGURE 5.10

The Krebs Cycle: Source of Exhaled Carbon Dioxide.

When an animal exhales, it is breathing out the waste gas carbon dioxide. That gas arises as the six-carbon skeleton of a glucose molecule is completely dismantled. Prior to this phase of the aerobic pathway, glycolysis had split glucose into 2 three-carbon pyruvate molecules in the cytoplasm. Pyruvate enters the mitochondrial matrix (see FIGURE 5.9) and, in Stage 1, one carbon is removed as carbon dioxide, while one NADH is generated. In Stage 2, the remaining two carbons join to a four-carbon compound (oxaloacetate) and form the six-carbon molecule citrate. In Stage 3, two carbon dioxides are sequentially cleaved from this citrate molecule, an ATP is produced, and four reduced electron carriers are harvested. The original four-carbon compound of the Krebs cycle, oxaloacetate, is regenerated. The animal exhales the carbon dioxide, and its cells can utilize the NADH and $FADH_2$ in the electron transport chain. Plants also carry out the Krebs cycle.

verted into pyruvate, acetyl-CoA, or Krebs cycle intermediates and be shunted directly into the pathway at the appropriate stage. Thus they can be dismantled for energy harvest.

The clearinghouse concept extends still further because acetyl-CoA and other raw materials from the Krebs cycle can also be shunted *out* of the pathway and converted into building blocks for the synthesis of new fats, proteins, and carbohydrates.

Ironically, the Krebs cycle clearinghouse also contributes to the physical damage in cases of extreme malnutrition. If no carbohydrates or lipids are available for ATP production, the body breaks down proteins into amino acids. Carbon skeletons of some amino acids can

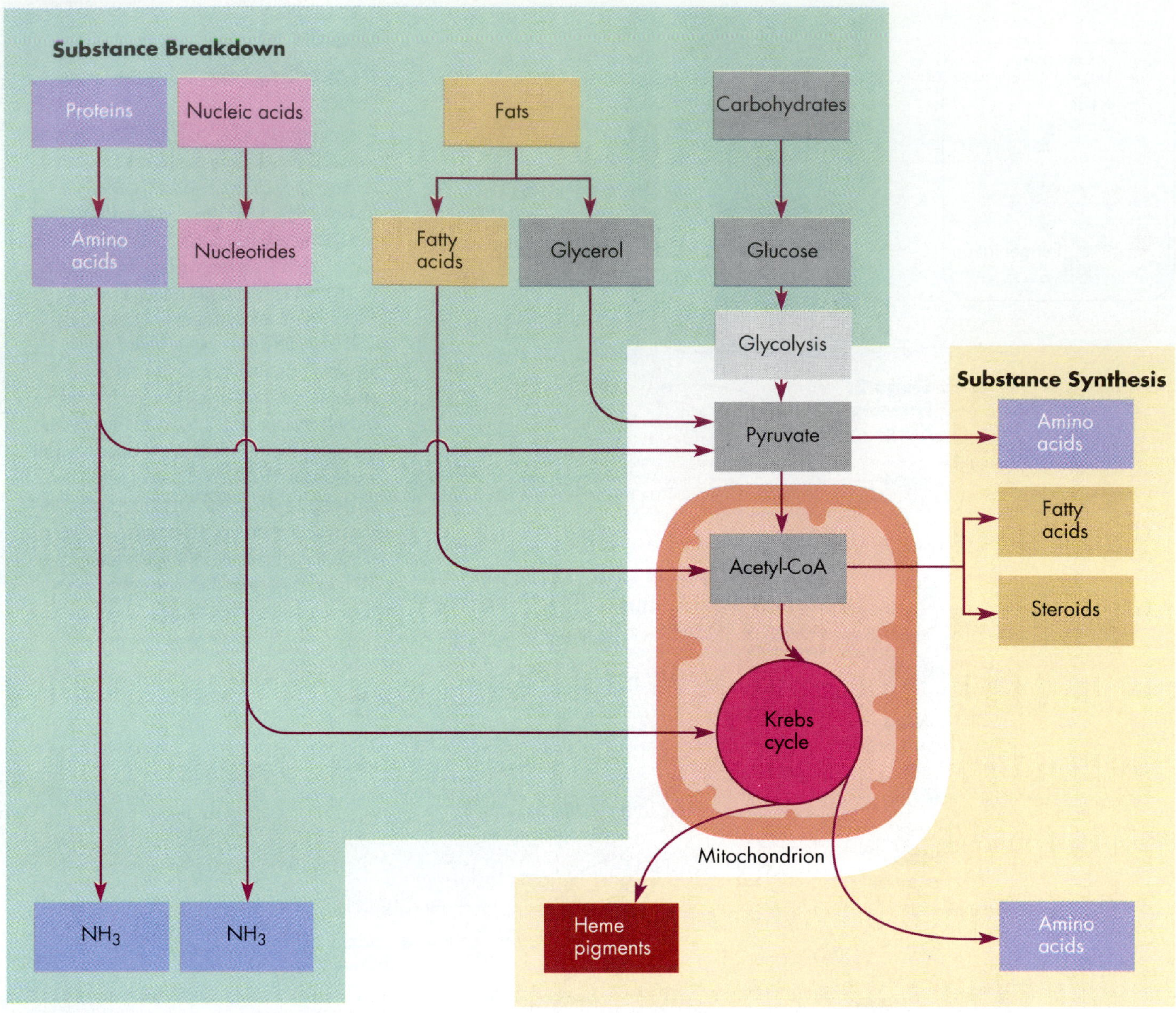

FIGURE 5.11

The Krebs Cycle: A Metabolic Clearinghouse.

While only certain molecules can feed directly into the Krebs cycle, biological polymers—proteins, nucleic acids, fats, and polysaccharides—can themselves be broken down into constituent parts, and these can be modified into intermediates that feed into the cycle. In this way, an organism can harvest energy not just from glucose but from any of the biological polymers. In addition, Krebs cycle intermediates can be removed from the cycle and modified into new materials for the cell, including amino acids, fatty acids, steroids, and iron-containing heme pigments.

enter the Krebs cycle and provide enough energy to keep the body alive; but meanwhile, the body is digesting itself to provide energy, with the Krebs cycle making it possible.

An important exception to the clearinghouse principle is the human brain cell. For complex reasons, it can use only glucose as fuel, a fact that has major implications. First, it explains why sugary foods are mood elevators, at least temporarily. Soon after a person consumes a candy bar or soft drink, glucose molecules are cleaved from sucrose (table sugar) and enter the bloodstream and brain. Infused with their favorite fuel, the brain cells can function at peak efficiency, leading one to feel happier, smarter, and livelier—at least for a time. Second, this quirk of the brain cell explains why dieters are warned to consume at the very least 500 calories in carbohydrates per day and to avoid liquid or powdered protein diets: The brain alone needs at least that many calories of glucose for normal functioning, and without it, a person can grow faint and even lapse into unconsciousness.

ELECTRON TRANSPORT CHAIN: ENERGY BUCKET BRIGADE

The final stage of aerobic respiration is where most of the energy of glucose is turned into ATP; where oxygen comes into play, making this respiration *aerobic*; and where mitochondrial structure is so crucial to the survival of eukaryotes.

The electron transport chain is both a series of structures and a series of events. To follow these structures and events, we must return to the detailed architecture of the mitochondrion. Recall that the inner mitochondrial membrane separates two fluid-filled compartments within the organelle (the inner compartment, or matrix, and the intermembrane space or outer compartment) and that the Krebs cycle occurs within the inner compartment [FIGURE 5.12; review FIGURE 5.9]. Embedded within the inner membrane, however, are the proteins of the electron transport chain, which protrude from and sometimes span the entire membrane.

Electrons stored in NADH or $FADH_2$ from glycolysis and the Krebs cycle are passed from one electron carrier to the next in a flow down an energy gradient. One can think of this as an electron bucket brigade during which small amounts of energy are released with each downward step. This energy goes into the synthesis of ATP, as shown in FIGURE 5.12.

The events of the electron transport chain represent a series of electron transfer reactions. During these reactions, the electron carriers NADH and $FADH_2$ lose electrons and hydrogen ions. Ultimately, oxygen atoms gain the electrons and hydrogen ions and become water (H_2O). Oxygen thus serves as the final acceptor of electrons and hydrogen ions for the entire aerobic respiration pathway. This explains why the pathway is called "aerobic." In animals, the oxygen for the electron transport chain must come from the atmosphere, for example by breathing. In plants, the oxygen can come directly from the plants' own photosynthesis.

If oxygen is not present to accept the electrons and protons once they have passed down the electron transport chain, the entire process quickly stops. The organism or cell, if strictly aerobic, then dies. That's how cyanide kills living things. The C≡N chemical group

FIGURE 5.12

The Electron Transport Chain and ATP Harvest.

Pyruvate molecules generated during glycolysis in the cytosol (1) are transported to the inner compartment, or matrix (2). Within the matrix, enzymes dismantle pyruvate to carbon dioxide and produce reduced electron carriers such as NADH (3). Electrons from NADH are passed down the electron transport chain. With each transfer, this energy is released, and used to pump hydrogen ions into the intermembrane space (4). Hydrogen ions build up in this space (5), and flow back into the matrix through an enzyme that synthesizes ATP (6). The electron transport chain finally passes electrons to oxygen along with hydrogen ions; this generates water (7). This process completely metabolizes glucose to carbon dioxide and water and harvests large quantities of ATP.

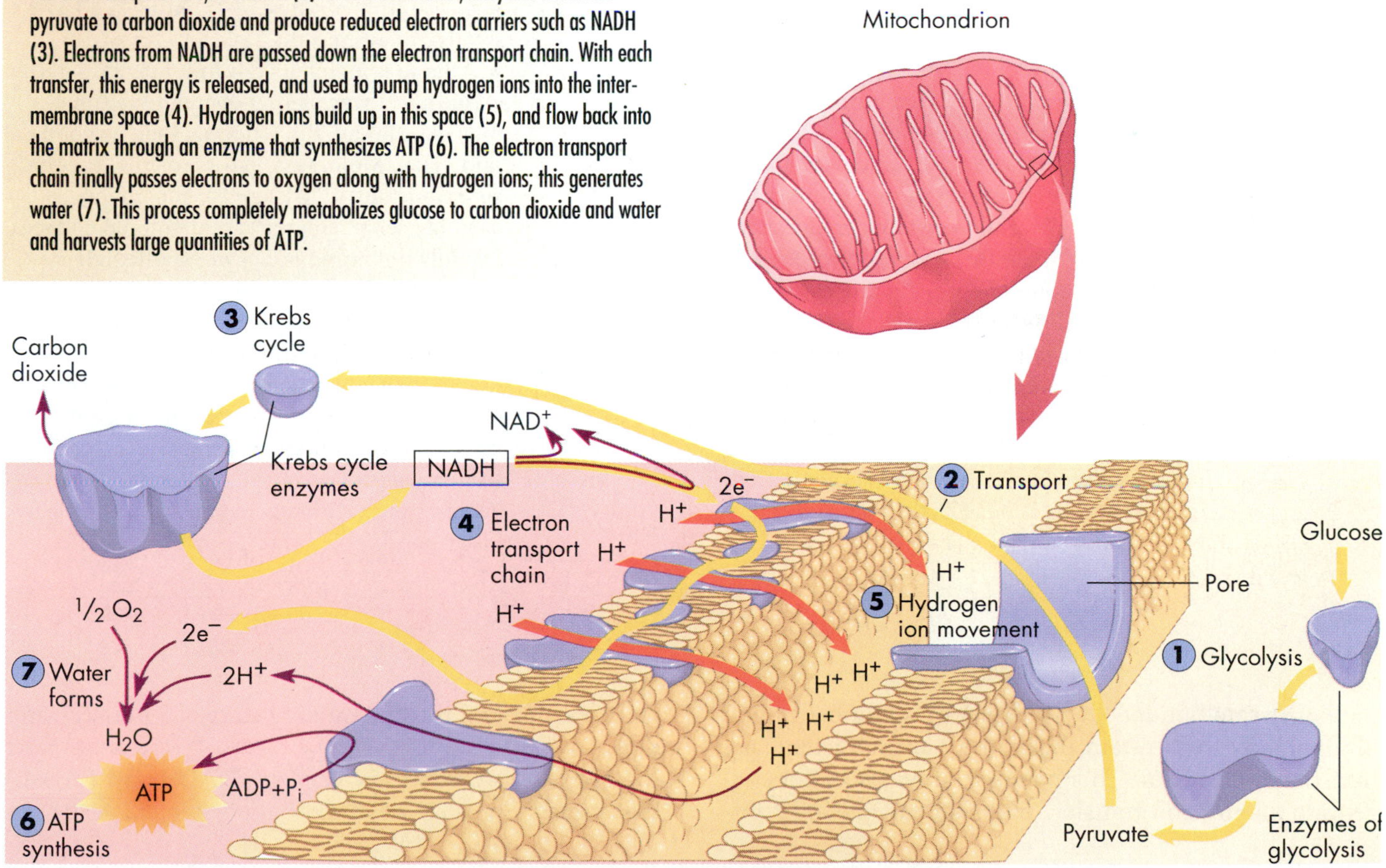

Human Impact

Sarah's Mitochondria

One day in 1974, while walking in Manhattan with her mother, 8-year-old Sarah T. (not real name) suddenly felt like her legs were being pricked by tiny needles. The symptoms continued and worsened, and over the next few months and years, Sarah could no longer walk a city block without her muscles burning and her heart racing. At age 16, she was attending her high school classes in Washington state in a motorized wheelchair. Sarah's intensifying fatigue and burning muscles baffled her doctors; the only clue to the problem was a high level of acidity in her blood. After intense study, a team of medical geneticists, biochemists, and physicians working at universities in Oregon and Pennsylvania finally recognized that Sarah's exhaustion was not based on problems with her lungs or heart, but rather with the power supply to her muscles.

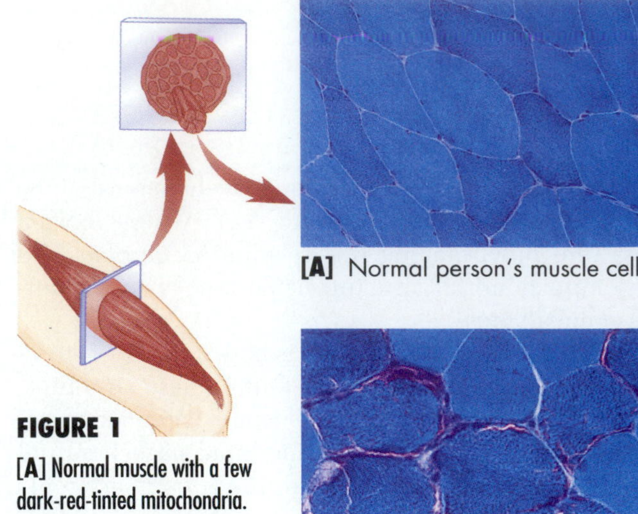

[A] Normal person's muscle cells

[B] Sarah's muscle cells

FIGURE 1

[A] Normal muscle with a few dark-red-tinted mitochondria. **[B]** Sarah's muscle fibers with masses of aberrant mitochondria encircling many fibers.

Like everyone else's muscle cells [FIGURE 1A], Sarah's generated lactic acid during physical activity. Hers, however [FIGURE 1B], made as much while strolling slowly or taking a shower as other people's do during strenuous exercise. Researchers deduced that Sarah's problem lay in her system of energy production—either in glycolysis or fermentation, or in the Krebs cycle and the electron transport system. But which of these was altered? And how could the researchers help her?

Experiments ruled out one respiratory pathway after another until biochemists traced Sarah's problem to her mitochondria. Researchers discovered that because of an altered protein in the electron transport chain (embedded in the mitochondrial membranes), her muscle cells could not transfer electrons to oxygen. Without this electron transfer, the electron transport chain—the muscles' major energy provider—could not make ATP. Instead, Sarah's muscles generated huge numbers of mitochondria, but still had to rely solely on the small ATP harvest provided by glycolysis and the fermentation that accompanies it. As we saw, however, fermentation in muscle cells creates lactic acid, and this was causing Sarah's blood acidity levels to climb so high that she occasionally blacked out.

Having identified the cause of her tremendous fatigue, researchers could look for a solution, a substitute acceptor that would allow electrons to cross the gap in the electron transport chain created by the abnormal protein in her mitochondrial membranes. That substitute electron acceptor turned out to be a simple mixture of vitamin C and a derivative of vitamin K.

Sarah remembers her first injection of the vitamins. "I was feeling dead all the time, then after taking the vitamins I felt so good I said, 'Hey, Mom, let's jog.' I was that much better, instantly." By age 20, she no longer used her wheelchair and had entered college. Sarah is not cured of her metabolic defect—the abnormal electron acceptor still fails to work in her mitochondria. But because a team of researchers understood cellular respiration, they could determine her problem and supply the appropriate substitute acceptor, so she can now generate much less lactic acid and enough ATP to lead a normal life. [Quote from *Old Oregon*, Fall 1986; article by Lisa Cohn.]

(carbon triple-bonded to nitrogen) in sodium cyanide binds tightly to a certain pigment molecule in the electron transport chain. This bonding prevents it from accepting an electron transfer. The result is that ATP formation halts, and the cell quickly starves for energy and dies.

For each set of energy carriers (NADH and $FADH_2$) that comes from a single glucose molecule, 32 ATPs can

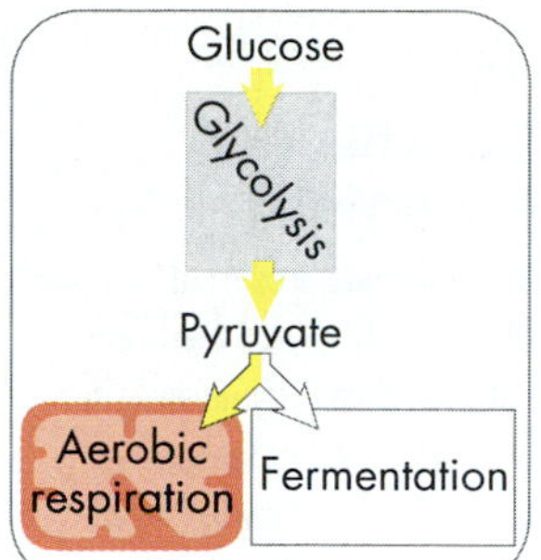

FIGURE 5.13

Aerobic Respiration: An Overview and Energy Tally.

For every glucose molecule broken down through aerobic respiration, two ATP molecules are harvested during the glycolysis phase, and two more ATPs are gained during the Krebs cycle. The Krebs cycle also produces a set of electron carrier molecules ($FADH_2$ and NADH), and these are processed by the electron transport chain with the accumulation of 32 more ATP molecules. (In some cells, notably those of the liver and heart, the ATP yield can be slightly higher—34 ATPs per glucose.) Altogether, during aerobic respiration, 36 ATPs are harvested per glucose molecule, 6 carbon dioxide and 6 water molecules are released as inorganic waste products (12 water molecules are produced, but 6 of these are used up for a net release of 6), and 6 oxygen molecules are consumed.

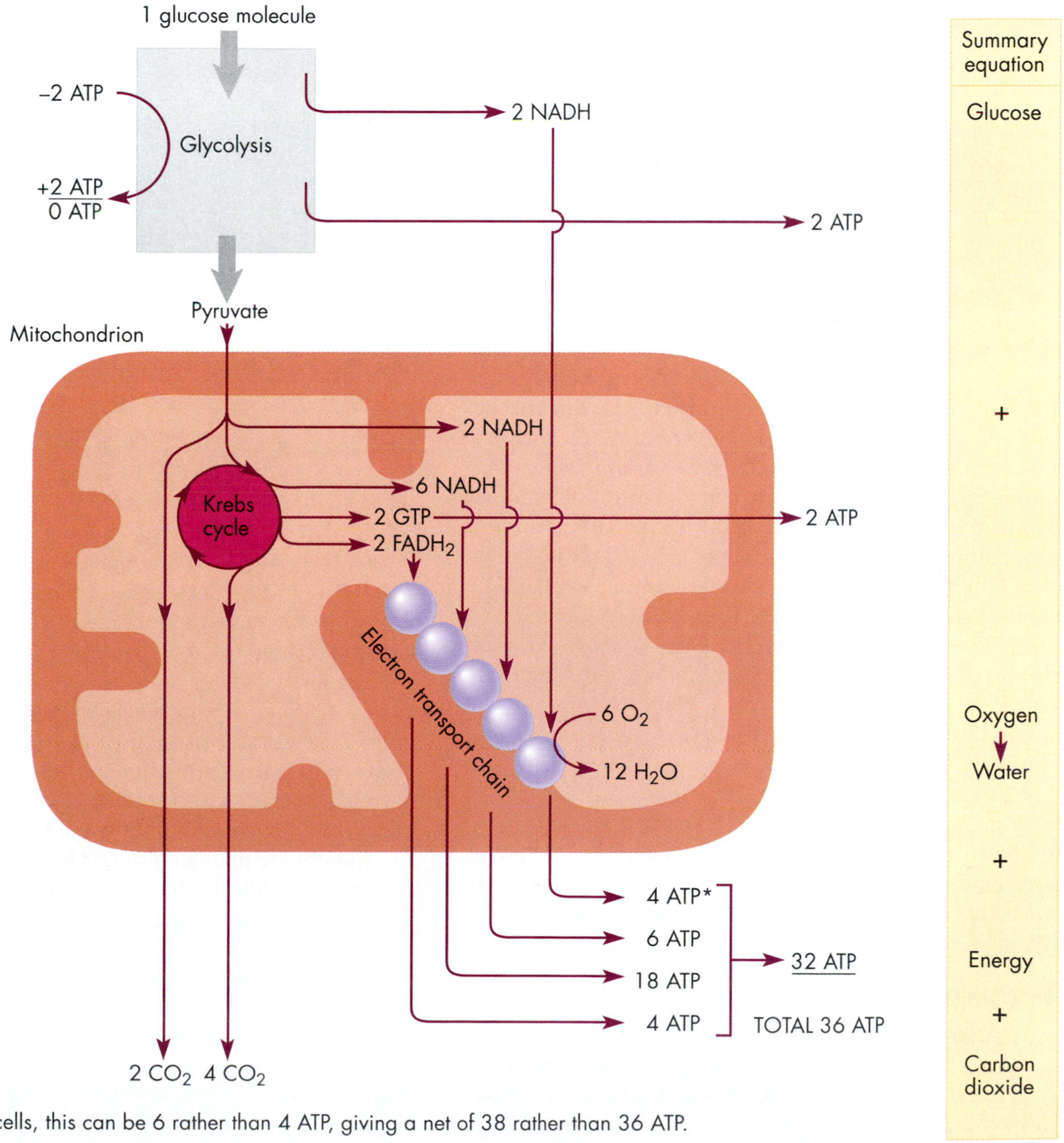

*In heart and liver cells, this can be 6 rather than 4 ATP, giving a net of 38 rather than 36 ATP.

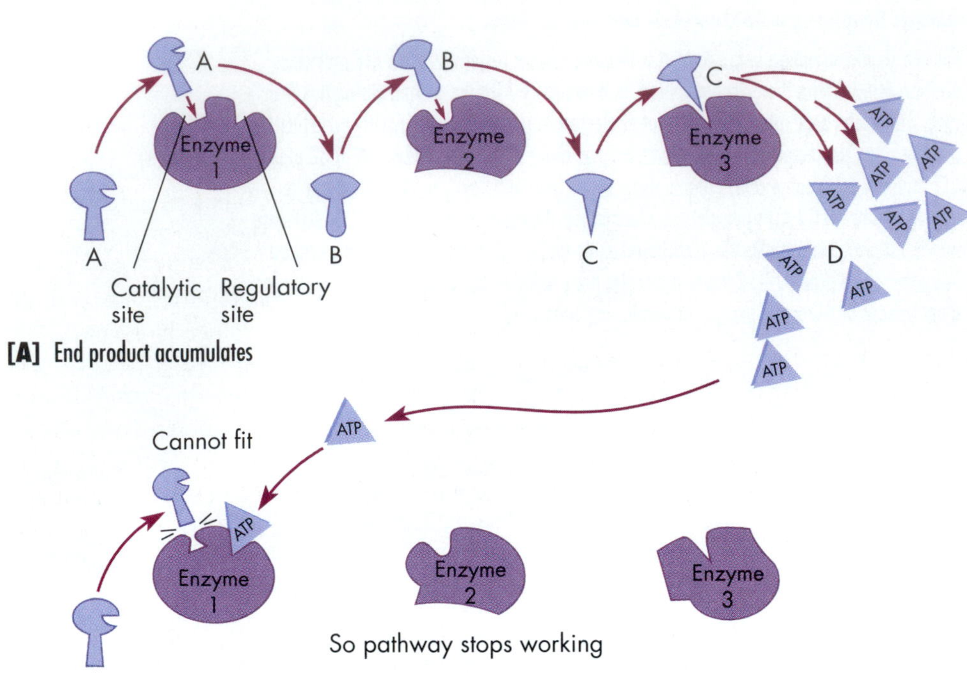

FIGURE 5.14

Feedback Inhibition: How Cells Regulate Energy Metabolism.

The accumulation of the end product of energy metabolism—ATP—can feed back and inhibit the activity of an enzyme early in the pathway of glycolysis and hence inhibit the entire pathway. **[A]** The pathway's end product (triangular molecules representing ATP) accumulates and **[B]** fills the regulatory site (or allosteric site) of enzyme 1. This alters the shape of the enzyme so that it cannot bind substrate, thus inactivating it and turning off production of the pathway leading to more ATP. With glycolysis temporarily halted, no pyruvate is made, there is thus no fuel for the Krebs cycle, and aerobic respiration comes to a halt. When the excess ATP is used up, the regulatory site on enzyme 1 opens up, the enzyme is once more active, and the metabolic energy harvest can once again commence.

form during the electron transport phase. This is nearly 90 percent of the ATP produced by the breakdown of a glucose molecule via aerobic respiration. Adding this 32 to the 2 ATPs from glycolysis and the 2 ATPs from the Krebs cycle, we get a final tally for the entire three-part aerobic pathway of 36 ATPs per glucose molecule, as itemized in FIGURE 5.13. Uncertainties remain in our understanding of aerobic respiration. Nevertheless, biologists are quite sure that the unique internal structure of the mitochondrion makes possible the high efficiency of aerobic respiration in eukaryotes. One bit of evidence for this involves aerobic bacteria, which lack mitochondria. These cells can still carry out aerobic respiration via transport proteins embedded in their plasma membranes; however, their ATP harvest is generally less than half that of eukaryotic cells. We active multicellular organisms need huge amounts of ATP energy, and our hundreds of trillions of mitochondria provide it. Any impairment of mitochondrial function can have severe consequences for health and vitality, as BOX 5.1 on page 136 explains.

CONCEPT CHALLENGE

A student friend of yours claims that plants cannot be poisoned by cyanide because plants do not employ mitochondria and the electron transport chain to harvest ATP, but instead obtain ATP by photosynthesis. Analyze this argument.

Control of Metabolism

Our hikers wandering through the Green Mountains can burn carbohydrates either by the aerobic or the anaerobic pathway. They can even burn lipids for energy if they run out of carbohydrates due to the Krebs cycle clearinghouse. If they stop for a moment to enjoy the view, eat a powerbar, and perhaps take a photograph, their systems will slow down the production of ATP. These intricacies of metabolism raise new questions: What makes a cell begin burning its stores of lipids or proteins when glucose runs out? When the supplies of glucose return, what then stops the cell from burning all its lipids or proteins—literally eating itself up from the inside and thereby destroying its cellular structures? Finally, what triggers a cell to build molecules only when they are needed?

Clearly, there must be a great deal of internal coordination and control over the cell's harvesting of energy and building of needed materials. But what form does it take?

We can create a simple analogy for the way that glycolysis and the breakdown of glucose is controlled in the cell. Imagine a shoe factory in which shoes pile up so high that they topple over and stop the assembly line; it is only after the shoes are removed and shipped out that the line is free to start up again and make more shoes.

Likewise, such high levels of ATP can build up in a cell that the cell requires no more of the molecular fuel [FIGURE 5.14A]. When this occurs, the ATP binds to a special regulatory site on the enzyme called PFK that catalyzes a reaction in stage 1 of glycolysis. This binding shuts down enzyme activity [FIGURE 5.14B]. This, in turn, switches off the entire glycolytic pathway: If an early reaction does not function, there is no substrate for subsequent reactions. The regulatory advantages of this system are obvious: When the cell already has high levels of ATP, the presence of the ATP molecule itself serves as a control to turn off its own production. Then, when ATP levels drop, the excess ATP is pulled from the enzyme "machinery," and the glycolysis "assembly line" can once more resume. The buildup of a metabolic product and its inhibitory effect on a special enzyme such as PFK is called **feedback inhibition**. This term is appropriate since the accumulation of the product in effect feeds back and turns off its own production.

Feedback inhibition works because the special enzyme has two binding sites, the catalytic (or catalytically active) site and the regulatory (or *allosteric*) site. The action begins when a molecule—usually the end product of the pathway the enzyme is part of—binds to the enzyme's regulatory site. This binding causes the enzyme to change shape, and as a result, its catalytic site stops binding the normal substrate and the enzyme's activity is turned off.

Through feedback inhibition and other forms of control, the activity of a cell's metabolic enzymes is turned on or off so that the cell burns glucose when it is available and lipids or proteins when glucose is lacking, and builds the appropriate biological raw materials just when they are needed for growth or maintenance activities. The control mechanisms within cells are truly elegant—efficient yet highly conservative. In addition, they occur in the smallest bacteria, the large muscle cells in a hiker's leg, and even the baker's yeast. The similarity of control mechanisms reminds us that all types of cells are related in a fundamental way by their metabolism. Such identical forms of control and regulation simply ensure that order rather than metabolic chaos reigns within the organism.

Energy for Exercise

Like any other aerobic organism, an exercising person relies on the energy-harvesting pathways discussed in this chapter. This is true whether the energy is needed for the powerful lunge of a center guard in football, for one leg of a relay event, which a swimmer will quickly churn through, or for a hiker's long steady climb up a mountain trail. Energy for different forms of exercise, however, comes from different parts of the metabolic pathway.

Exercise physiologists have determined that there are three energy systems: the immediate system, the glycolytic system, and the oxidative system—that supply energy to a person's muscles during exercise. The duration of physical activity dictates which system the body uses. The *immediate energy system* is instantly available for a brief explosive action, such as one football tackle, one bench press, or one ballet leap, and the system has two components [FIGURE 5.15A]. One component is the small amount of ATP stored in muscle cells, immediately useful like the few coins you carry around in your purse or pocket. This stored ATP, however, runs out after only half a second—barely enough time to heave a shot or return a tennis serve, let alone run to catch a bus. The second component of the immediate energy system is a high-energy compound called *creatine phosphate*, which muscle cells store in larger amounts than ATP. Creatine phosphate is more like a handful of dollar bills than a few coins. When the ATP in a muscle cell is depleted, creatine phosphate transfers phosphate to ADP; this regenerates ATP, and ATP can then fuel the muscle cell to contract and move the body. Even the cell's store of creatine phosphate, however, becomes depleted after only about a minute of strenuous work; thus, muscles must rely on more robust systems to power longer-term activities.

The *glycolytic energy system*, which depends on splitting glucose by glycolysis in the muscles, fuels activities lasting from about 1 to 3 minutes, such as an 800-m run or a 200-m swim [FIGURE 5.15B]. This storage form is more like an account at the bank than like dollars in your wallet, and it "purchases" more activity. Glycolysis in the muscle cell cytoplasm can cleave glucose in the absence of oxygen and generate a few ATPs. Fermentation takes the product of glycolysis a step further and makes the waste product lactic acid. Recall that there is a net yield of only two ATP molecules for each molecule of glucose from glycolysis. This is why the glycolytic energy system can sustain heavy exercise for only about 3 minutes and why lactic acid begins to build up in muscles by then.

Activities lasting longer than about 3 minutes—a jog around the neighborhood, an aerobic dance session, or a long uphill hike—require oxygen and employ the *oxidative energy system* [FIGURE 5.15C]. This system can supply energy for activity of moderate intensity and long duration and is like money invested in stocks and bonds that give long-term steady income. The oxidative energy system is based on cellular respiration and includes the Krebs cycle and oxidative phosphorylation via the elec-

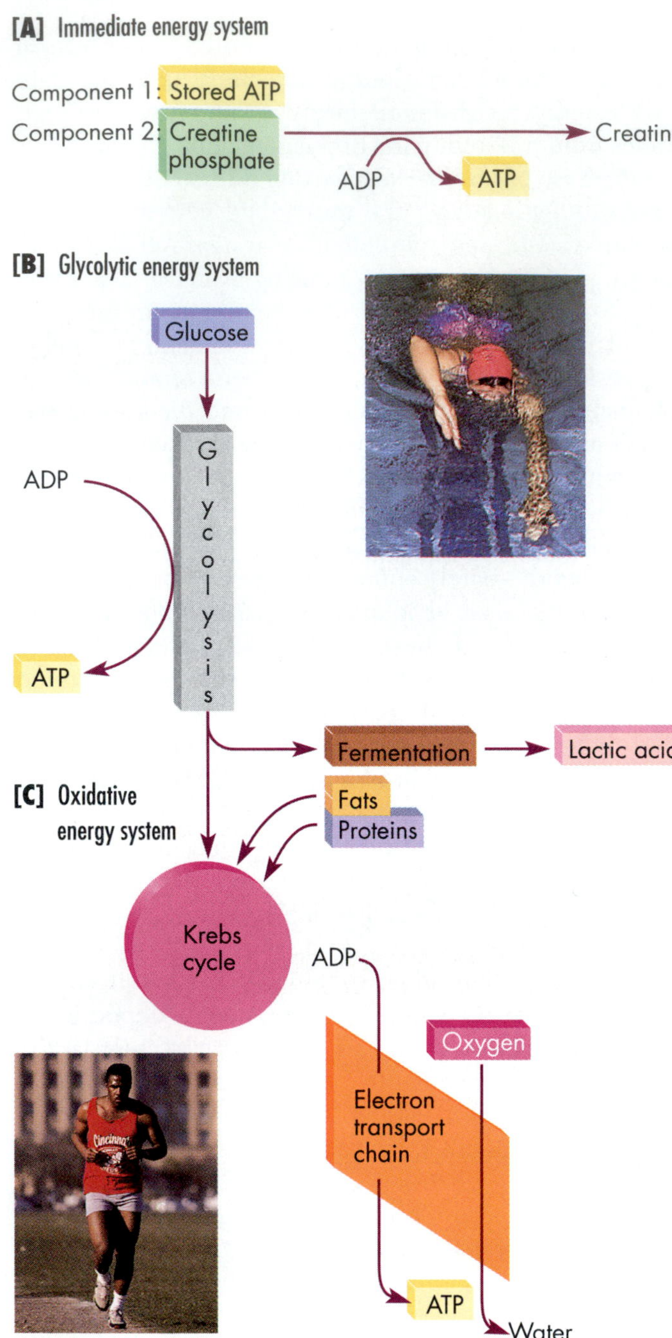

FIGURE 5.15

Three Energy Systems for Fueling Muscular Motion.

[A] The immediate energy system, based on ATP and creatine phosphate stored in muscle cells, can fuel a brief explosive activity. [B] The glycolytic energy system, based on glycolysis (anaerobic breakdown) of glucose in the muscle cells, can fuel short-duration, high-intensity activities, such as a 100-m swim. [C] The oxidative energy system, based on cellular respiration (aerobic breakdown) of glucose, glycogen, fatty acids, or amino acids stored elsewhere in the body and carried to the muscles through the bloodstream, can fuel long, sustained activities, such as a 20-km run.

tron transport chain. It uses oxygen as the final electron acceptor and generates many ATP molecules per glucose molecule burned. After an exercise session, ATPs start to build up again. The form of metabolic regulation called feedback inhibition takes over, slowing glycolysis as well as the rest of the aerobic respiration pathway.

The glycolytic energy system can make ATP only from glucose or glycogen (cleaved to release glucose), but the oxidative energy system can produce energy by breaking down carbohydrates, fatty acids, and amino acids mobilized from other parts of the body and transported to the muscle cells by way of the bloodstream [review FIGURE 5.11]. Clearly, anyone interested in melting away body fat should engage in aerobic (oxygen-utilizing) activities like jogging, swimming, bicycling, or strenuous hiking which rely primarily on the oxidative energy supply system and its ability to use fats as fuel.

Connections

Studying energy metabolism can give you a deeper understanding of two of life's simplest but most profound phenomena: breathing in and breathing out. Breathing in supplies the oxygen necessary as the last acceptor of electrons in the electron transport chain. This in turn allows cells to harvest a large amount of ATP from the energy stored in glucose. Breathing out rids the body of carbon dioxide, which arises when Krebs cycle enzymes in the mitochondria chop carbon atoms from glucose breakdown products. The plants around you—house plants, for instance—can then take up the carbon dioxide you exhale (plus the carbon dioxide the plant produces in its own mitochondria) and, with energy from sunlight, incorporate it into new carbohydrate molecules. The cycle of inhaling and exhaling keeps you alive, and the cycle of aerobic respiration and photosynthesis maintains a big part of the energy balance in the world as we know it. Our next chapter investigates photosynthesis in more detail.

CONCEPT MAPPING

Draw a concept map relating the following terms. Link the concepts with arrows and a connecting word or short phrase.

glycolysis	**lipids**
aerobic pathway	**proteins**
anaerobic pathway	**ATP**
Krebs cycle	**plant cells**
electron transport chain	**animal cells**
fermentation	**mitochondria**
glucose	

KEY TERMS

aerobic pathway, 125
anaerobic pathway, 126
electron transport chain, 126
fermentation, 126
glycolysis, 125
Krebs cycle, 126

HIGHLIGHTS IN REVIEW

1 Energy flows through ecosystems. Plants capture energy from sunlight and store it in carbohydrates. This energy can then be harvested either by the plant's cells or the cells of animals and other organisms that consume plant matter. Eventually the energy returns to the environment as heat.

2 The aerobic pathway of energy harvest yields prodigious amounts of ATP in three stages—glycolysis, the Krebs cycle, and the electron transport chain.

- **a]** The energy storage molecule ATP links energy exchanges in cells. The cleaving of a phosphate group from ATP, which forms ADP and phosphate, releases energy that can be used in cellular processes. The transfer of a phosphate from ATP to another molecule can allow the phosphorylated molecule to participate in reactions that would otherwise be energetically unfavorable.
- **b]** Energy transfer in cells involves a flow of energized electrons from one molecule to another. Carbohydrates lose electrons and hydrogen ions to electron carriers, including NAD^+, which becomes NADH. Mitochondrial enzymes can extract energy from the electrons in NADH, and these can be used to produce ATP.
- **c]** Glycolysis means sugar splitting; during this pathway, a six-carbon glucose molecule is split into 2 three-carbon pyruvate molecules with the net formation of two ATP molecules.
- **d]** During glycolysis in the cytosol, an initial energy investment of two ATPs activates glucose, and then the six-carbon molecule is split into 2 three-carbon molecules. Breakdown of the three-carbon molecules produces NADH from NAD^+, and finally, four ATPs are harvested as the three-carbon molecule pyruvate is generated.
- **e]** During aerobic cellular respiration, each glucose molecule is fully oxidized to carbon dioxide and water, and a large harvest of ATP occurs.
- **f]** Enzymes of the Krebs cycle located in the mitochondrial matrix strip carbons from derivatives of pyruvate and release them as carbon dioxide. In the process, two ATP molecules form, and a considerable amount of energy is stored in the electron carriers NADH and $FADH_2$. The Krebs cycle is a metabolic clearinghouse because breakdown products of proteins and lipids can enter the cycle, and Krebs cycle compounds can be used for synthesis of other biological molecules.
- **g]** The electron transport chain consists of electron transport proteins embedded in the inner mitochondrial membrane. Some of these proteins strip electrons from the carrier molecules NADH and $FADH_2$ and pass the electrons along to other proteins, releasing energy bit by bit. This energy fuels the pumping of hydrogen ions from the mitochondrion's matrix to its inner membrane space. When the hydrogen ions flow back across the inner mitochondrial membrane to the matrix through a special protein, they drive the synthesis of ATP.
- **h]** The final tally for aerobic respiration in most eukaryotic cells is 36 ATPs per initial glucose molecule: 2 ATPs generated during glycolysis, 2 ATPs during the Krebs cycle, and 32 during the electron transport chain.

3 The anaerobic pathway of energy harvest produces only about 5 percent as much ATP as the aerobic pathway. The splitting of glucose harvests some ATP, and the incompletely dismantled carbon products are discarded as wastes.

- **a]** Cells can harvest the energy stored in glucose by one of two major pathways, the aerobic pathway, which utilizes oxygen, or the anaerobic pathway, which does not use oxygen. Some kinds of cells, like yeasts and the muscle cells that move the human skeleton, can use both pathways, while other types of cells can use just one.
- **b]** In the anaerobic pathway, pyruvate, the product of glycolysis, can be fermented into waste products such as ethanol and carbon dioxide or lactic acid. While no additional ATP forms in this process, the NADH from glycolysis is recycled to NAD^+, and this allows glycolysis to continue. Production of bread, wine, and cheese all depend upon fermentation by microorganisms growing in dough, grape juice, and milk.

4 Muscle cells can use the aerobic pathway for a sustained energy harvest if sufficient oxygen is present; when oxygen is in short supply, they can use the anaerobic pathway, which produces toxic waste products, for short bursts of activity.

- **a]** The appropriate use of organic nutrients and the timely production of biological molecules by the cell requires metabolic controls. The feedback inhibition of allosteric enzymes such as PFK is one such control mechanism.
- **b]** The immediate system, the glycolytic system, and the oxidative system provide energy to muscles for short-term (1 second), intermediate (1 to 3 minutes), or long-term (several minutes to hours) muscular activities.

UNDERSTANDING THE FACTS AND CONCEPTS For Questions 1–3, match each of the descriptions with the most appropriate item from the following list of terms. Any item may be used once, more than once, or not at all.

a] coenzymes **b]** anaerobes **c]** vitamins

1 Nonprotein molecules that are essential components of certain functional proteins.

2 Organisms that function in an oxygen-free environment.

3 Compounds that are often chemical precursors of such molecules as NADH and other electron carriers.

As above, for Questions 4–8, match the descriptions with the appropriate item from the following list:

a] ADP **c]** NAD^+ **e]** acetyl-CoA
b] ATP **d]** pyruvate

4 A three-carbon compound formed as an end product of glycolysis.

5 A high-energy nucleotide that is a by-product of glycolysis.

6 A dinucleotide that acts as an electron carrier in the process of fermentation.

7 The two-carbon molecule and its carrier molecule that enter the Krebs cycle.

8 A nucleotide having two PO_4 groups.

As above, for Questions 9–13, match the descriptions with the appropriate item from the following list:

- **a]** outer mitochondrial membrane
- **b]** inner mitochondrial membrane
- **c]** mitochondrial cristae
- **d]** mitochondrial matrix space
- **e]** both **b** and **c**, above

9 Location of the Krebs cycle enzymes.

10 Location of the enzymes of the electron transport chain.

11 Location of mitochondrial DNA and ribosomes.

12 Membrane convolutions or folds.

13 Contains large, protein-bound pores.

As above, for Questions 14–18, match the descriptions with the appropriate item from the following list:

- **a]** aerobic pathway
- **b]** anaerobic pathway
- **c]** feedback inhibition
- **d]** creatine phosphate
- **e]** oxidative phosphorylation

14 The switching off of the glycolytic pathway by PFK that occurs when ATP levels become excessive is an example of this.

15 Driver of the electron transport chain.

16 The set of reactions that culminates in the production of pyruvate.

17 The set of reactions culminating in the production of carbon dioxide and water.

18 A high energy compound that is stored in large amounts in muscle cells.

INTEGRATE AND APPLY WHAT YOU HAVE LEARNED

1 During what stage or stages of the breakdown of sugar to carbon dioxide and water do the carrier molecules NAD^+ and FAD appear?

2 Describe the connection between the electron transport chain and the actual synthesis of ATP.

3 When bread dough rises, it produces a characteristic aroma. What is the relationship, if any, between the rising and the aroma?

4 If one wishes to lose weight, why is it better to concentrate on endurance-style exercise than sprint-style?

5 In the case study of Sarah's mitochondria [BOX 5.1], the metabolic defect was found to be the failure of electrons to be transferred to oxygen. Why do you think that this defect resulted in muscular pain?

ANALYSIS

1 Bread dough usually contains yeast enzymes, which cause the dough to expand, or rise. Cakes are made to rise by the release of carbon dioxide from baking soda (sodium bicarbonate). If you made a cake and a loaf of bread using identical amounts of sugar and starch in both, which would end up tasting sweeter? Choose your answer from below:

- **a]** The cake would be sweeter because its sweetness would not be obscured by the flavor of yeast.
- **b]** The cake would be sweeter because its sugar would not be broken down by the action of yeast into other molecules, such as CO_2.
- **c]** The bread and cake would be equally sweet.
- **d]** The bread would be sweeter because during fermentation yeasts produce sugar.
- **e]** The bread would be sweeter because carbon dioxide is produced organically rather than inorganically.

2 Compare the energy available in one molecule of sugar with that contained in 36 molecules of ATP. Which of the following statements is correct?

- **a]** The amount of energy in the sugar is greater than the amount needed to convert 36 molecules of ADP to ATP.
- **b]** There is less energy in the sugar than the amount needed to convert 36 molecules of ADP to ATP.
- **c]** There is more energy in sugar than the total amount in the 36 molecules of ATP.
- **d]** The energy in the sugar must equal that in the 36 ATP molecules.
- **e]** The energy in the sugar must equal the total amount that is needed to convert 36 ADP to 36 ATP molecules.

3 The water that appears as a by-product when sugar is burned:

- **a]** Still contains much of the energy that was in the sugar before it was burned.
- **b]** Is composed of hydrogen and oxygen, both of which were obtained from the sugar.
- **c]** Is composed of hydrogen and oxygen, both of which were obtained directly from the air used for burning.
- **d]** Is a low-energy molecule because most of the energy of the sugar has been transferred to other molecules in the burning process.
- **e]** Is a by-product of glycolysis.

4 Substrate-level phosphorylation to make ATP:

a] Can occur only within mitochondria.
b] Can take place only on mitochondrial membranes.
c] Combines a phosphorylation reaction with the transfer of electrons between carriers, such as NAD^+ and FAD.
d] Can occur only if there is a phosphate donor that has more energy than ATP.
e] Requires the presence of both mitochondria and oxygen.

5 What is the most immediate result of the energy that is released by electrons as they traverse the electron transport pathway?

a] It is used directly to make ATP from ADP + P.
b] It is used to pump hydrogen ions from the matrix space to the intermembrane space.
c] It is used to move oxygen into the mitochondrion.
d] It is used primarily to make more NAD^+ and FAD.
e] It is mostly dissipated as heat.

CHAPTER 6

Photosynthesis: Trapping Sunlight and Carbon

ATMOSPHERIC CARBON DIOXIDE: AN INTERESTING DILEMMA

An old Chinese adage says: "May you live in interesting times." As most people would surely agree, we do. Consider this simple fact: Since the Industrial Revolution began about 200 years ago, people have been burning huge amounts of fossil fuels (oil, coal, natural gas, gasoline), as well as cutting and burning forests all over the globe. The result of this combustion has been a dramatic rise in the amount of carbon dioxide (CO_2) in the atmosphere, and some experts expect the levels of carbon dioxide to keep rising all during our lifetimes. The prospects are unsettling—and interesting.

[A]

[B]

FIGURE 6.1

Atmospheric Fertilizer: Carbon Dioxide, Photosynthesis, and Plant Growth.

Researchers wanted to test how changing atmospheric levels of carbon dioxide affect plant growth. They grew phlox (pink flowers), jimsonweed (red flowers), and other plants in two test atmospheres. **[A]** One contained levels of carbon dioxide similar to those in the atmosphere early in this century (300 parts per million, left). **[B]** The other contained carbon dioxide levels three times higher (900 parts per million, right). What do these photos reveal about the effects of elevated carbon dioxide on flowering, plant height, and aging (as measured by yellowing lower leaves)?

Why should we worry about rising carbon dioxide levels? Right now, carbon dioxide represents only about 3 molecules for every 10,000 molecules in the air, and during the next 50 years or so, atmospheric scientists foresee a rise to just 7 molecules per 10,000.

The most publicized reason has been the greenhouse effect and global warming, which are discussed in CHAPTER 40. But an equally profound reason cuts straight to the heart of energy, metabolism, and life on earth: Plants take carbon dioxide from the air and use it in the synthesis of the carbohydrate molecules that directly or indirectly fuel the cellular activity of virtually all life forms.

How will increased carbon dioxide levels affect plant growth? To answer that, Harvard ecologists grew red-flowered jimsomweed, pink-flowered phlox, and other plant species in sealed chambers containing various levels of carbon dioxide. The results in FIGURE 6.1A and B show that the plants supplied with more CO_2 grew faster, flowered more abundantly, and in some cases turned yellow with age earlier than the plants with less CO_2.

On a planet where millions of people face starvation each year, wouldn't faster plant growth be a desirable thing? The answer depends on which plants grow faster—crops or weeds; on the amount of rainfall, soil nutrients, and other growth-limiting factors; and on how the plants interact and compete in their natural environments. We will address those issues in this chapter as we learn how plants use carbon dioxide during photosynthesis, our main subject here.

Photosynthesis is the metabolic process by which plants and many microorganisms trap solar energy, convert it to chemical energy, and store it in the bonds of organic nutrient molecules such as sugar-phosphates and other carbohydrates. CHAPTER 5 discussed the metabolic pathways that break down glucose and release energy in plants, animals, and other organisms. Here we find out how sugars such as glucose are produced in nature. Since photosynthesis traps light energy, we examine the nature of light energy early in the chapter. After that, we consider the nature of pigments that capture energy from light, and two main phases of photosynthesis. One phase traps light energy and converts it to ATP and energized electron carriers. A second phase then transfers this chemical energy to carbon dioxide from the air and in the process, synthesizes sugars and other high-energy carbohydrates. This marvelous metabolic feat accounts for the cellulose that constitutes most of the plant bodies standing all around us. It also supplies the nutrients that power all plant, animal, fungal, and most microbial activity. Photosynthesis is truly central to life on earth, and thus it has a significant early spot in our discussions of biology.

MESSAGES

1 Green plants trap light energy and synthesize carbohydrate molecules via the reactions of photosynthesis. Plant cells use the raw materials carbon dioxide and water, and release oxygen as a waste product.

2 Virtually all familiar life forms depend on photosynthesis for energy and organic carbon. Plants do so directly; organisms that feed on plants or on plant-consuming organisms do so indirectly.

3 In plants, photosynthetic reactions take place in green cellular organelles called chloroplasts. One set of reactions traps light energy and releases oxygen from water. Another set removes carbon dioxide from the air and combines it with hydrogen in the biologically useful form of carbohydrates. This reaction is reciprocal to aerobic respiration in plants and other organisms.

4 Carbon dioxide levels rise in the atmosphere as people burn fossil fuels. This may disrupt ecosystems, as well as contribute to greenhouse warming of the earth.

With this grounding in photosynthesis, we can consider why plants that have efficient systems for using carbon dioxide may not benefit much from a CO_2-laden atmosphere, whereas less efficient plants may grow more luxuriantly and perhaps out-compete the first group. Some ecologists fear that this may change our farm environments and natural landscapes in ways that are unpredictable, undesirable, or, likely, both. Finally, we'll see how carbon atoms cycle between organisms and the nonliving world in a massive global exchange.

Along the way, you will learn why crabgrass grows so aggravatingly well in the summer, why the oxygen you breathe came from a plant just a short time ago, and how CO_2 levels affect sugarcane, corn, grasses, and other kinds of plants. ❑

Light and Pigments

Think, for a moment, about a jimsomweed and a phlox plant, and how differently they obtain energy from the way you do yourself. The plants obtain their energy from the nonliving environment in the form of sunlight and inorganic raw materials, while you must obtain yours by eating plants, or by eating other animals that consumed plants. Biologists classify organisms that take in preformed nutrient molecules from the environment as **heterotrophs** (*hetero-*, "other", and *troph*, "feeder"). Heterotrophs include many prokaryotes, most protists, yeasts and other fungi, and humans and all other animals. In contrast to heterotrophs, **autotrophs** (*auto-*, "self") are organisms that take energy directly from the nonliving environment and use it to synthesize nutrient molecules. Autotrophs include photosynthetic organisms—green plants and certain protists and prokaryotes that obtain energy from sunlight—and chemosynthetic organisms—a few kinds of prokaryotes that extract energy from inorganic chemicals such as hydrogen gas and hydrogen sulfide with its rotten-egg smell. We can see that, ultimately, autotrophs are the source of all energy that flows through living systems since heterotrophs obtain nutrients either from autotrophs or from heterotrophs that once consumed autotrophs. The basic source of energy flowing through organisms begins with the sun's light, so before moving on to the details of photosynthesis, let's take a look at the nature of light and of solar energy.

SOME PHYSICAL CHARACTERISTICS OF LIGHT

Visible light—the white sunlight we can separate into a rainbow of colors through a prism—is just a small part of the *electromagnetic spectrum*, which is the full range of electromagnetic radiation in the universe, from highly energetic gamma rays to very low-energy radio waves [FIGURE 6.2]. All such radiation travels through space behaving both as vibrating particles called **photons** and as waves. The amount of energy in a photon determines its *wavelength*—the distance it travels during one complete vibration. The more energetic the photon, the shorter its wavelength.

Photons of visible light have wavelengths in the narrow range of 380 nm (violet light) to 750 nm (red light). It is no coincidence that living things can absorb and use light within this restricted wavelength range. Gamma rays, with their shorter wavelengths, are so energetic that they disrupt and destroy biological molecules they strike, while radio waves are so low in energy that they cannot excite biological molecules at all. We can separate and perceive the colors in sunlight because the pigments in the retinas of our eyes are excited by different vibrational frequencies within the visible range, and nerve impulses fired off to the brain are interpreted as red, blue, violet, and other colors.

Our earth is constantly bathed in light radiating from the sun. Each minute, the solar energy striking a square meter of the earth's surface could raise the temperature of half a liter (about a pint) of water by 25°C. Plants, algae, and other autotrophs use less than 1 percent of this energy in photosynthesis, but they are so numerous on our planet that they collectively produce between 150 and 500 billion tons of carbohydrates every year—enough to build a giant mountain of cellulose and starch. Light travels so quickly and the photosynthetic process takes place so rapidly that you can practically eat sunlight in a fresh-picked leaf: If you pluck and immediately chew a growing lettuce leaf, you are consuming energy that left the sun just 8 minutes earlier and was converted to the chemical energy in carbohydrate molecules almost instantly

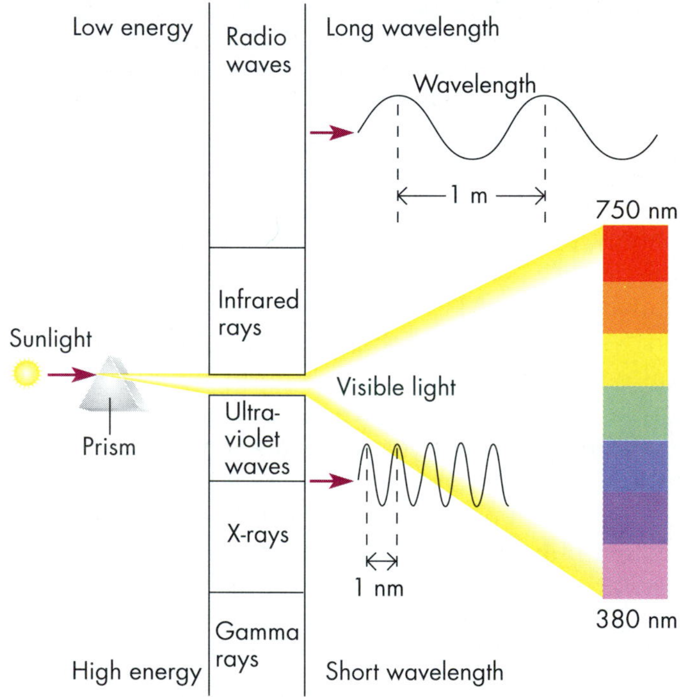

FIGURE 6.2

Visible Light and the Electromagnetic Spectrum.

We are constantly bathed by electromagnetic energy, from radio waves to infrared rays (heat), X-rays, and gamma rays. Each category has an energy range measured in oscillating waves of specific lengths (wavelengths). Only the small portion of the entire electromagnetic spectrum with wavelengths in the range of 380 to 750 nm is visible to us as white light that can be broken by a prism into colored light, each color with different wavelengths, as in a rainbow.

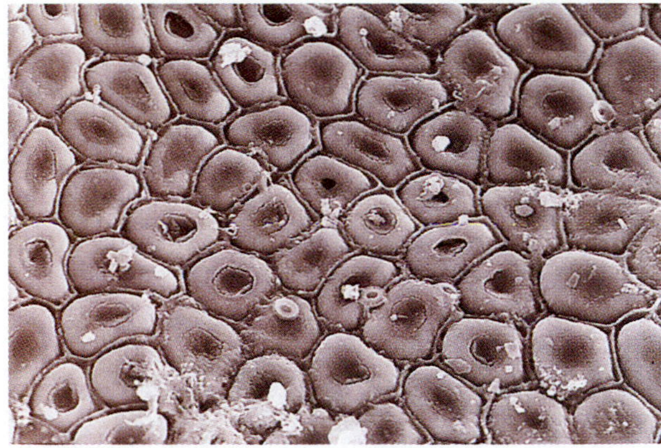

FIGURE 6.3

Red Algae that Grow in the Dark.

Botanists from the Smithsonian Institution discovered a type of red alga growing in virtual darkness some 260 m (about 880 ft) below the ocean surface on the slopes of a seamount off the Bahamas. These algae are a purplish color and grow in hard, flat colonies. The cells are shown here from the top, and are magnified 1600 times. Botanists thought that plants would be unable to grow at depths below 215 m (about 705 ft), because light levels are too low. These red algae, however, are about 100 times more efficient at capturing sunlight and converting it to carbohydrates than are red algae growing near the ocean's surface.

upon striking the plant. Photosynthesis is not only rapid but amazingly efficient. FIGURE 6.3 shows a red alga that can gather enough light for photosynthesis in the inky recesses of the deep ocean, where very little light penetrates.

CHLOROPHYLL AND OTHER PIGMENTS ABSORB LIGHT

Objects like red apples, black panthers, or green leaves look colored because they absorb and reflect light [FIGURE 6.4]. An apple looks red because pigment molecules in the skin cells absorb light from various parts of the visible spectrum and reflect only red light. A panther looks black because pigments in the animal's coat absorb all the wavelengths of light that strike them, reflecting none. A phlox leaf looks green because its green chlorophyll pigments absorb violet, blue, yellow, and red light and reflect only green. This range of absorbed light is called the *absorption spectrum*, and is unique for each pigment. There is an essential difference, however, between chlorophyll molecules and the pigments in a panther's coat: Chlorophyll participates in photosynthesis as well as giving a green leaf its color.

Two main types of chlorophyll take part in photosynthesis: *Chlorophyll a* occurs in all photosynthetic

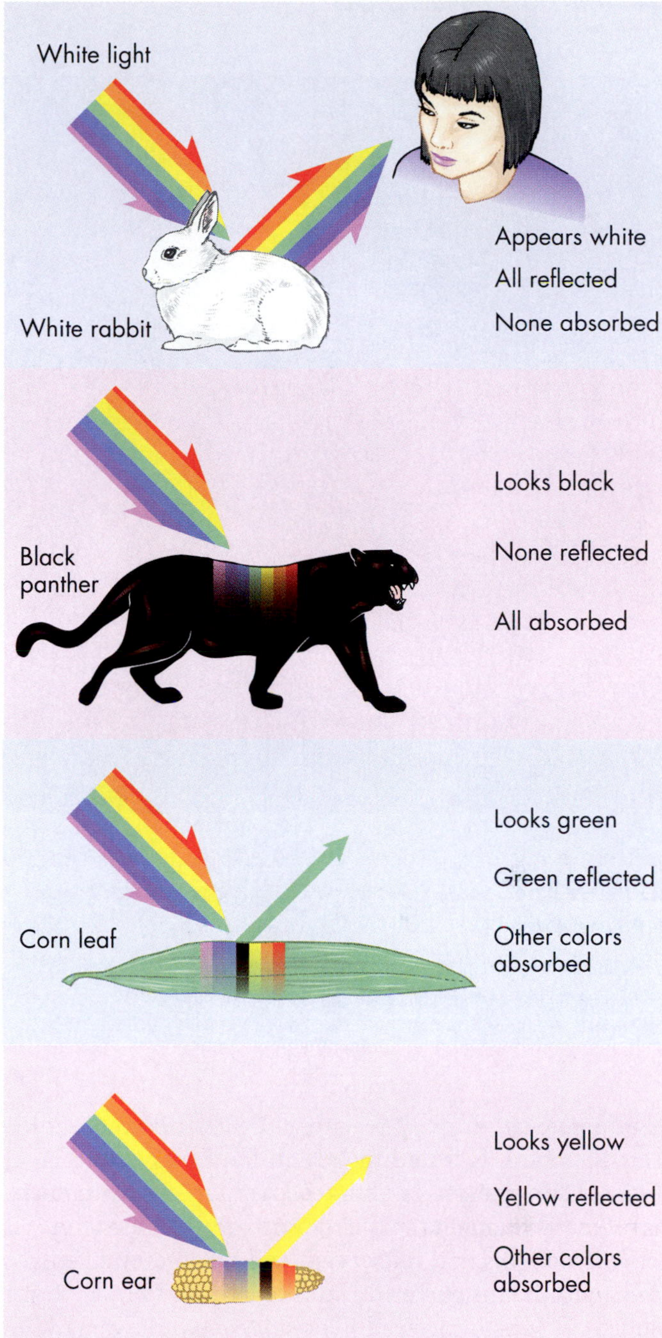

FIGURE 6.4

Pigments and Absorption Spectra.

The colors we see in a given object depend on which wavelengths of light the pigments in that object absorb and which are reflected back to our eyes. A white rabbit looks white because all wavelengths are reflected. Black pigment in a panther's coat absorbs light throughout the whole color spectrum and reflects nothing; thus, it looks black to us. Chlorophyll in a corn leaf absorbs many wavelengths of light, but green light is most strongly reflected, and thus the leaf looks green. Corn ears contain carotenoid pigments, which absorb wavelengths in the blue range and reflect back red, yellow, and orange light. Because a leaf often contains both chlorophyll and carotenoids, its pigments can absorb and use for photosynthesis most of the wavelengths of visible light that strike it.

FIGURE 6.5

Autumn Glory.

When chlorophyll breaks down in autumn, carotenoids can shine through, painting the forest with orange, gold, and scarlet hues.

organisms; *chlorophyll b*, with a slightly different molecular structure, is found mostly in land plants and green algae. **Chlorophyll** is often accompanied by colorful **carotenoid pigments**, which absorb green, blue, and violet wavelengths and reflect red, yellow, and orange light. Carotenoids are generally masked by chlorophyll and thus tend to be unnoticed in green leaves. However, they give bright and obvious color to many nonphotosynthetic plant structures, such as roots (carrots), flowers (daffodils), fruits (tomatoes), and seeds (corn kernels). Carotenoids are behind the glorious colors of autumn. As summer ends and the nights grow cool, chlorophyll begins to break down, allowing the gold and red carotenoids to show through and emblazon maple, oak, sumac, and other deciduous trees and vines [FIGURE 6.5]. Chlorophyll absorbs all but green wavelengths of light, and carotenoids all but the red, orange, and yellow wavelengths. Functioning together in pigment complexes, chlorophylls and carotenoids can absorb most of the available energy in visible light. The importance of these pigments, of course, is not just that they absorb light, but what becomes of that captured energy.

The Chloroplast: Solar Cell and Sugar Factory

The photosynthetic pigments in the leaf of a plant such as phlox are concentrated in layers of cells that are green and carry out photosynthesis [FIGURE 6.6A]. These cells contain **chloroplasts**, green organelles in which both the energy-trapping and carbon-fixing reactions of photosynthesis take place [FIGURE 6.6B]. Each leaf cell may contain about 50 chloroplasts, and each square millimeter of leaf surface more than half a million of the green organelles. Chloroplasts are analogous to mitochondria in several ways: Both are elongated organelles with complex membranous structures [see FIGURE 6.6C]; both carry out energy-related tasks in the cell; and both have their own DNA in circular chromosomes. However, while mitochondria are "powerhouses" that generate ATP, chloroplasts are both solar cells and sugar factories that capture sunlight and generate sugar-phosphates and other carbohydrates. BOX 6.1 on page 150 describes the possible evolution of chloroplasts.

CHLOROPLAST MEMBRANES

Chloroplasts have an outer membrane and inner membrane that lie side by side and that collectively enclose a space filled with a watery solution, the *stroma* [see FIGURE 6.6C]. A third membrane system lies within the inner membrane and forms a complicated network of stacked, disklike sacs, the **thylakoids**, interconnected by flattened channels. Each stack of thylakoids is called a *granum* (plural, *grana*), and each individual thylakoid disk has an internal compartment, the *thylakoid space* [FIGURE 6.6D]. Chlorophyll and other colored pigments are embedded in the thylakoid membrane and make this membrane the only part of an entire plant that is truly green. In addition to pigments, the thylakoid membrane contains members of an electron transport chain and, in some areas, many copies of an ATP-synthesizing enzyme [FIGURE 6.6E].

The energy-trapping reactions of photosynthesis take place in the thylakoid membrane and thylakoid space. These reactions provide the chloroplast's solar cell activity. The carbon-fixing reactions, which begin in the stroma and continue in the cytoplasm, collectively produce sugar phosphate molecules. The sugar phosphates produced by photosynthesis leave the chloroplast and are either (1) linked into starch molecules and stored in plastids, (2) used as subunits in the cellulose of the plant cell wall, or (3) broken down in the plant cell's own cytoplasm and mitochondria (via glycolysis and aerobic respiration, see CHAPTER 5), where they power cellular activity.

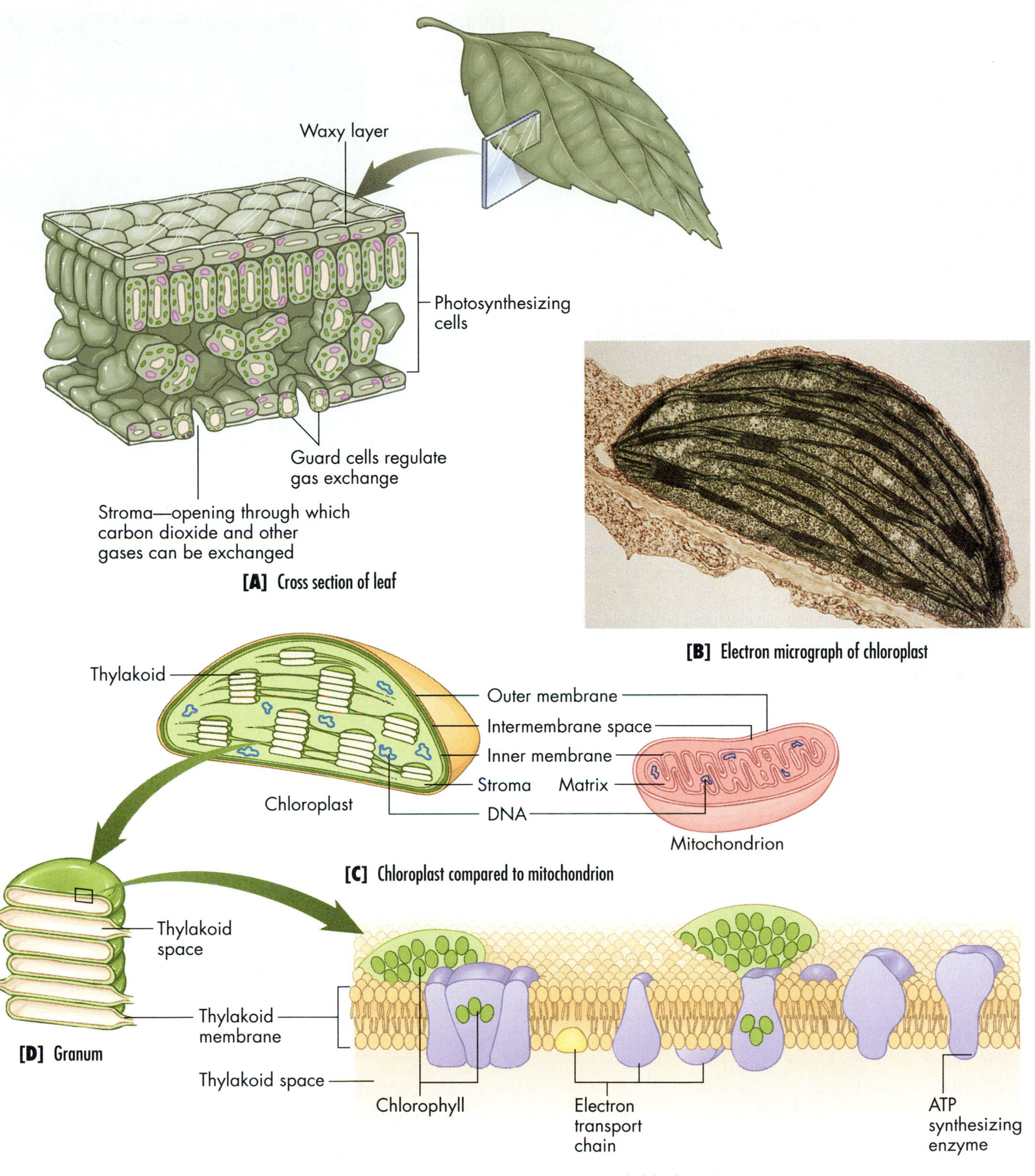

FIGURE 6.6
Chloroplasts: Photosynthetic Organelles.

[A] Below a leaf's waxy surface is a sheet of cells lacking green chloroplasts. The interior of the leaf, however, contains layers of cells that do contain chloroplasts and do conduct photosynthesis. The lower surface of the leaf contains tiny cell-lined pores called stomata through which atmospheric gases and water vapor are exchanged. **[B]** Chloroplasts have outer and inner membranes and an intermembrane space between them; a third set of membranes forms disklike sacs called thylakoids. **[C]** The stroma, or matrix, surrounds the thylakoid stacks. A mitochondrion is shown for size and shape comparison. **[D]** Each thylakoid has its own membrane and internal space, and several sacs are often stacked together in a granum. **[E]** The thylakoid membrane contains chlorophyll, the proteins of the electron transport chains, and the ATP-synthesizing enzymes.

box 6.1
Focus on the Environment

When You See Green, Think of a Green Sea

Look out across the emerald green grass in a park or at the deep verdant hues in a pine forest, and you are seeing a piece of evolutionary history. Those oceans of color come from chlorophyll pigments bound up in chloroplasts. Many plant biologists think that these banana-shaped organelles arose after ancestral photosynthetic bacteria took up residence in larger cells. Algae may have evolved from these ancient cellular communes and in turn given rise to land plants like grass and trees. While this *endosymbiotic* ("coming to live inside in harmony") theory makes good sense, skeptics want evidence, and recently, marine biologists stumbled onto an interesting bit of it.

In 1988, Sallie W. Chisholm and a team of colleagues from Harvard University and Woods Hole Oceanographic Institution reported their discovery of a strange new kind of photosynthetic cell in six different ocean areas from the tropics to the North Atlantic. The cells are prokaryotic (have no nucleus) and are so tiny that the team could see them only with an electron microscope. They were also incredibly numerous, with up to 100,000 cells in every teaspoonful of seawater sampled. The cells have a kind of chlorophyll *a* never seen before as well as chlorophyll *b*, alpha-carotene rather than beta-carotene, and traces of other pigments.

The team places the new cells in the phylum Prochlorophyta, which contains species generally thought to be living relatives of the very early cells that gave rise to chloroplasts. One type of prochlorophyte called *Prochloron*, discovered in 1976, grows inside the bodies of sea squirts—and other little marine animals [FIGURE 1]—trading photosynthetic products for room and board. Some biologists consider *Prochloron*'s symbiosis, its ability to coexist within other organisms, to be significant, since some similar photosynthetic bacterium probably entered into communal living with ancestors of plant cells.

FIGURE 1
***Prochloron* and a Sea Cucumber.**
The green tinge on this animal, a type of sea cucumber, comes from the symbiotic alga *Prochloron.*

In 1986, researchers discovered a second type of prochlorophyte—this one free-living—in shallow lakes in the Netherlands. The new type discovered by the Chisholm team in 1988 is also free-living, but is perhaps more significant for three reasons: (1) It contains both chlorophylls *a* and *b*, the same pigments in the chloroplasts of virtually all land plants, a striking evolutionary coincidence; (2) the internal structure of the new species is quite similar to the Prochloron cells that live within sea squirts and may, in fact, be a free-living counterpart resembling the ancient marine cells that existed independently before coming to inhabit plant cells; and (3) the newly discovered green cells are so abundant that they must have great ecological significance as fixers of carbon and producers of oxygen in the open oceans.

It is always exciting when biologists discover something so potentially important right beneath our collective noses. In this case, the discovery may also help explain the sea of green right outside our windows.

An Overview of Photosynthesis

There is a beautiful symmetry to the metabolic processes of respiration and photosynthesis that is revealed by their nearly opposite overall equations [FIGURE 6.7].

(1) Aerobic cellular respiration:

$$C_6H_{12}O_6 + 6\ O_2 \rightarrow 6\ CO_2 + 6\ H_2O + \text{chemical energy}$$

glucose + oxygen → carbon dioxide + water + ATP + NADH

(2) Photosynthesis:

$$6\ CO_2 + 6\ H_2O + \text{light energy} \rightarrow C_6H_{12}O_6 + 6\ O_2$$

carbon dioxide + water + light energy → glucose + oxygen

In CHAPTER 5 [review FIGURE 5.5], we saw that when oxygen is present, aerobic respiration in mitochondria allows cells to break down glucose into carbon dioxide and water and to release chemical energy. In photosynthesis, nearly the reverse takes place in the chloroplast. Light energy is trapped and transformed into chemical energy, which is then used to convert carbon dioxide and water into sugars, and oxygen is released as a waste product.

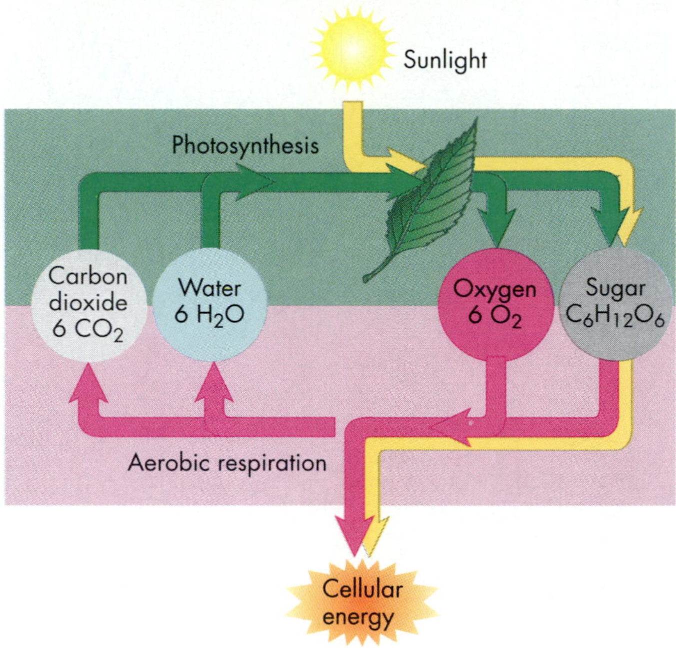

FIGURE 6.7

Comparison of Photosynthesis and Aerobic Respiration.

Photosynthesis uses carbon dioxide and water, generates oxygen, and traps solar energy in the chemical bonds of carbohydrate molecules. Aerobic respiration uses the products of photosynthesis—carbohydrates and oxygen—and produces carbon dioxide and water, while liberating energy from the bonds of organic molecules.

(Sugar phosphates related to glucose are the actual products of photosynthesis, as we will see later, but here we use glucose for simplicity.)

Clearly, living things must have both a source of energy as well as a means of releasing it, and for green plants and most other autotrophs, the direct energy source is sunlight. To understand the process of photosynthesis, we must follow the path of light and of the electrons whose energy level is boosted by sunlight. There are two main phases to this pathway. First, sunlight boosts the energy level of electrons from chlorophyll; this energy is then trapped as ATP and in electron carriers, while water is split, releasing oxygen and hydrogen ions [FIGURE 6.8, left]. Second, these high-energy electrons and hydrogen ions are added to carbon dioxide; as this occurs, their energy is transferred to the bonds of carbohydrate molecules [FIGURE 6.8, right].

THE LIGHT-TRAPPING PHASE OF PHOTOSYNTHESIS

When sunlight strikes green chlorophyll or other colored pigments in the chloroplast of a phlox leaf, let's say, some of the solar energy becomes trapped as it boosts electrons in the pigment molecules to higher energy levels. Then, before the electrons drop back to their original energy

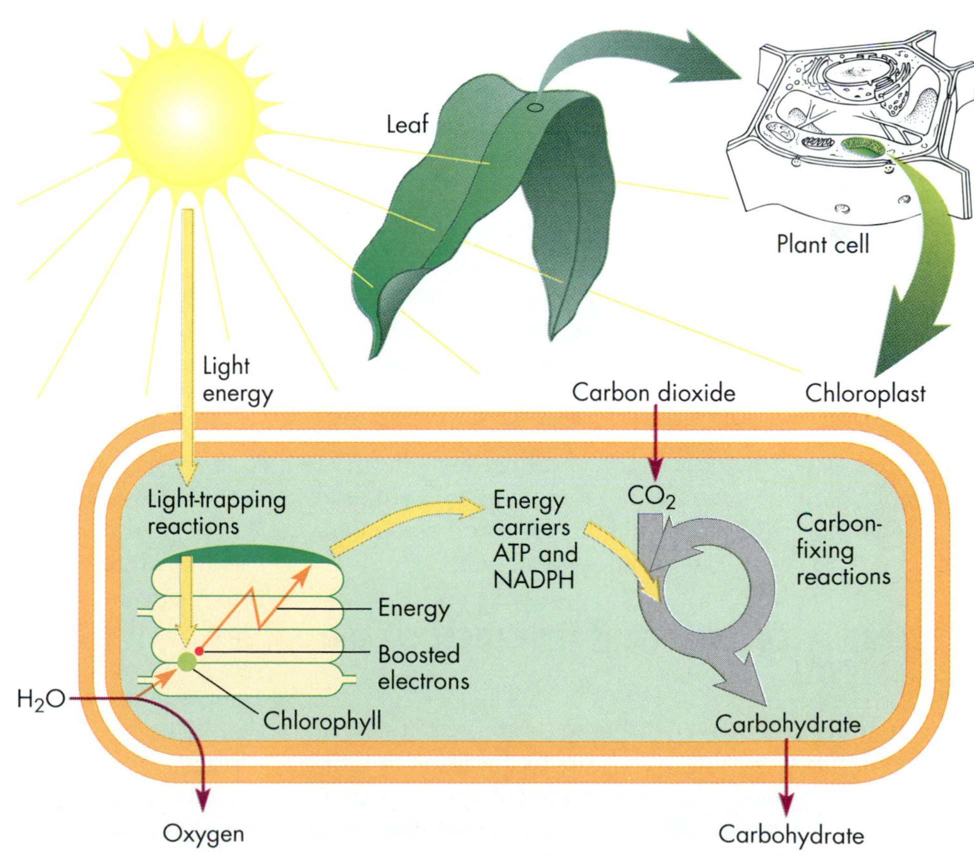

FIGURE 6.8

Overview of Photosynthesis.

Photosynthesis includes two phases, a set of light-trapping reactions and a set of carbon-fixing reactions, both of which occur in chloroplasts. During the first reactions, light energy is trapped by the green pigment chlorophyll, and this process boosts certain electrons to higher energy levels. As these electrons pass from one electron acceptor to another, energy is released and stored in chemical form in the high-energy carriers NADPH and ATP. During the carbon-fixing reactions, the high-energy carriers supply energy to a cycle of reactions that fix carbon into biological molecules such as sugar phosphates and starch.

levels, they leave the chlorophyll and pass down an electron transport chain much like the one in the mitochondrial membrane. The "hole" in the chlorophyll left by the energized electrons is filled by electrons stripped from a water molecule (H_2O). The hydrogens from the water molecule stay in the chloroplast, but the oxygen is released to the atmosphere, where humans and other organisms, including plants, can use it in aerobic respiration.

The solar energy transferred to chlorophyll electrons is converted to chemical energy in the chloroplast. The energy is released from the light-energized electrons bit by bit and stored in the chemical bonds of ATP and the energized electron carrier NADPH. (NADPH is similar to NADH in that it carries electrons and an extra hydrogen. It differs, however, by the presence of an extra phosphate group, hence the "P" in NADPH.) These events make up the first phase of photosynthesis, the **light-dependent**, or **energy-trapping**, **reactions**. The reactions are driven by light energy and can take place only when light is available, and they produce oxygen, ATP, and energized electron carriers [see FIGURE 6.8].

THE CARBON-FIXING PHASE OF PHOTOSYNTHESIS

The ATP and NADPH produced by the energy-trapping reactions supply the energy needed for the second phase of photosynthesis, the **carbon-fixing reactions** (also called the **Calvin-Benson cycle**, or **light-independent reactions**). Carbon fixation refers to the fixing of inorganic carbon from the air into a biological molecule. During the carbon-fixing reactions, energy stored in the bonds of ATP and NADPH is transferred to the bonds of carbohydrate molecules [see FIGURE 6.8, right]. The carbon-fixing reactions do not directly require light energy, they only require the energy carriers ATP and NADPH produced by the light-dependent reactions. During the carbon-fixing reactions, NADPH is stripped of some electrons and hydrogen atoms. These are added to carbon dioxide, eventually producing carbohydrates, such as the phosphorylated version of fructose (fruit sugar) and sucrose (table sugar). These compounds can then diffuse away from the chloroplast and be used in the same plant cell, either for the energy-harvesting steps of glycolysis and aerobic respiration, or to synthesize cellulose or starch.

THE CARBON-FIXING REACTIONS AND GLOBAL CARBON DIOXIDE LEVELS

We can apply what we've discussed so far about the carbon-fixing phase of photosynthesis to the issue of rising levels of atmospheric carbon dioxide between the 1950s and the 1990s [FIGURE 6.9]. The most obvious change, as

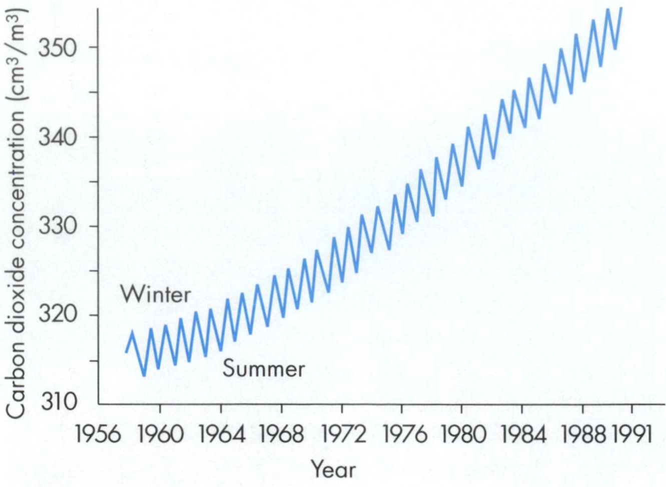

FIGURE 6.9

Carbon Dioxide in the Air.

Measurements of carbon dioxide levels in the atmosphere have showed steady increases for many decades. Superimposed on that rise is a seasonal up-and-down fluctuation, high in the winter and low in the summer. What causes these changes?

you can see in the graph, is a steady rise in carbon dioxide levels, largely due to the burning of fossil fuels (coal, oil, and natural gas) and rain forests. Superimposed on this, however, is a yearly fluctuation of CO_2 levels—high in the winter and low in the summer. What do you suppose could account for the decrease in carbon dioxide each summer? Here's the reason: Since plants photosynthesize more in the spring and summer, they take more carbon dioxide from the air in those seasons, thus lowering the atmospheric level of the gas. Conversely, in the winter, photosynthesis occurs at a slower rate because temperatures are lower and because there is less sunlight shining on the northern hemisphere. Consequently, the amount of carbon dioxide in the atmosphere increases. Such data show quite convincingly how great an effect living organisms have on the nature of the earth and its atmosphere.

Plant Growth and Rising Carbon Dioxide Levels Rising levels of carbon dioxide in the atmosphere exert their effect on plant growth during the carbon-fixing reactions of photosynthesis. In the greenhouse used by plant scientists to grow phlox and jimsonweed [see FIGURE 6.1], the extra carbon dioxide they supplied sped up the carbon-fixing cycle just as supplying more gas by pressing a car's gas pedal speeds up the engine. This helps explain the additional photosynthesis and extra plant growth observed under these conditions.

The structure of the leaf in FIGURE 6.6A reveals one reason why scientists could have expected a plant like phlox to grow better when they supplied extra carbon dioxide: Most plant leaves are coated with a waxy layer that blocks the passage of gases to the leaf interior where the photosynthesizing cells are located. Gases can exchange with the leaf interior only through *stomata* (singular, *stoma*), cell-lined pores in the leaf's lower surface. If the stomata are opened widely, more carbon dioxide can diffuse in, but more water can also diffuse out, and this can lead to drying out. In fact, for each carbon dioxide molecule that passes into the leaf through the stomata, 200 molecules of water can evaporate out. When carbon dioxide concentrations are higher in the air, the stomata can be partially closed and still admit the same amount of carbon dioxide. Partial closure has the advantage of retarding water loss and hence increasing the efficiency of the plant's water use. In regions where water is scarce and is a limiting factor in plant growth, high carbon dioxide levels could make the difference between luxuriant and scraggly growth. Recent experiments with oats and wheat have in fact shown that carbon dioxide improves water use, and hence plant growth. Just because a few types of plants grow better with higher levels of carbon dioxide, however, does not mean that all plants will do equally well, and the result could be ecological change.

The principles of photosynthesis presented above form a basis for understanding the life styles of plants, which we'll discuss in PART SIX of this book, and ecological relationships among plants. Many curious students, however, will want a deeper understanding of the mechanisms of photosynthesis, and we provide that in the next section.

CONCEPT CHALLENGE

Suppose that plant scientists grow one group of plant cells in water containing a radioactively labeled oxygen atom ($H_2{}^{18}O$) and grow a second group of plant cells in carbon dioxide containing two radioactively labeled oxygen atoms ($C^{18}O_2$). Next, they shine light on the cells and capture and measure the oxygen released from each. Which group of cells releases radioactive oxygen gas ($^{18}O_2$)?

Photosynthesis: A Closer Look

This section provides a more detailed step-by-step explanation of the two phases of photosynthesis. However, before we get into those steps, we need to examine the arrangements of pigments and proteins in the thylakoid membranes of chloroplasts, since this is where the steps take place.

THE LIGHT-TRAPPING REACTIONS OF PHOTOSYNTHESIS

When sunlight shines on a phlox leaf, the energy-trapping reactions of photosynthesis proceed almost instantaneously, converting solar energy to chemical energy. These reactions take place in the thylakoid membranes of the chloroplasts and inside the thylakoid disks. This phase of photosynthesis involves three sets of reactions: First, light excites certain electrons in chlorophyll to higher energy levels; second, energy from these excited electrons is stored in stable energy carriers; and third, water molecules are split; some of the water's low-energy electrons replace the electrons lost from chlorophyll, while its oxygen is released. Let's consider these three sets of reactions in more detail.

Stage 1: Light Excites Electrons Embedded in the thylakoid membranes of each chloroplast are so-called **antenna complexes** [FIGURE 6.10A]. These are clusters of about 250 chlorophyll, carotenoid, and other pigment molecules that act in concert during the harvest of light energy. The pigment molecules are arranged around a central chlorophyll, the *reaction center chlorophyll*. That central chlorophyll is associated with a special set of proteins and other molecules called a **photosystem**. There are two photosystems, *photosystem I* and *photosystem II*, which differ slightly in the wavelength of light they absorb.

The reaction center chlorophyll in a photosystem absorbs slightly lower-energy light than the other pigment molecules. When the other pigment molecules are struck by photons of light, they pass the energy they absorb to the reaction center, which actually participates in photosynthesis [see FIGURE 6.10A]. Energy bounces from pigment to pigment within the antenna complex at high speeds: each energy transfer takes only 10^{-12} second. Some biologists compare the passage of energy in the antenna complex to the movement of energy between metal atoms in a radio wave antenna. But the antenna complexes are biological molecules, and the funneling of light energy among them results in an excited chlorophyll molecule. This molecule possesses—but fleetingly—a pair of highly energized electrons.

Stage 2: Energy Is Stored in Energy Carriers The light-excited chlorophyll of photosystem II has a very short lifetime. It quickly passes its energized electrons to a series of electron acceptors, each of which passes them to the next acceptor in the series [FIGURE 6.10]. The result of this electron bucket brigade—an *electron transport chain*—is the storing of energy in two kinds of molecules that can accomplish useful work in the cell: the energy carriers ATP and NADPH.

As the electrons pass from one acceptor to the next,

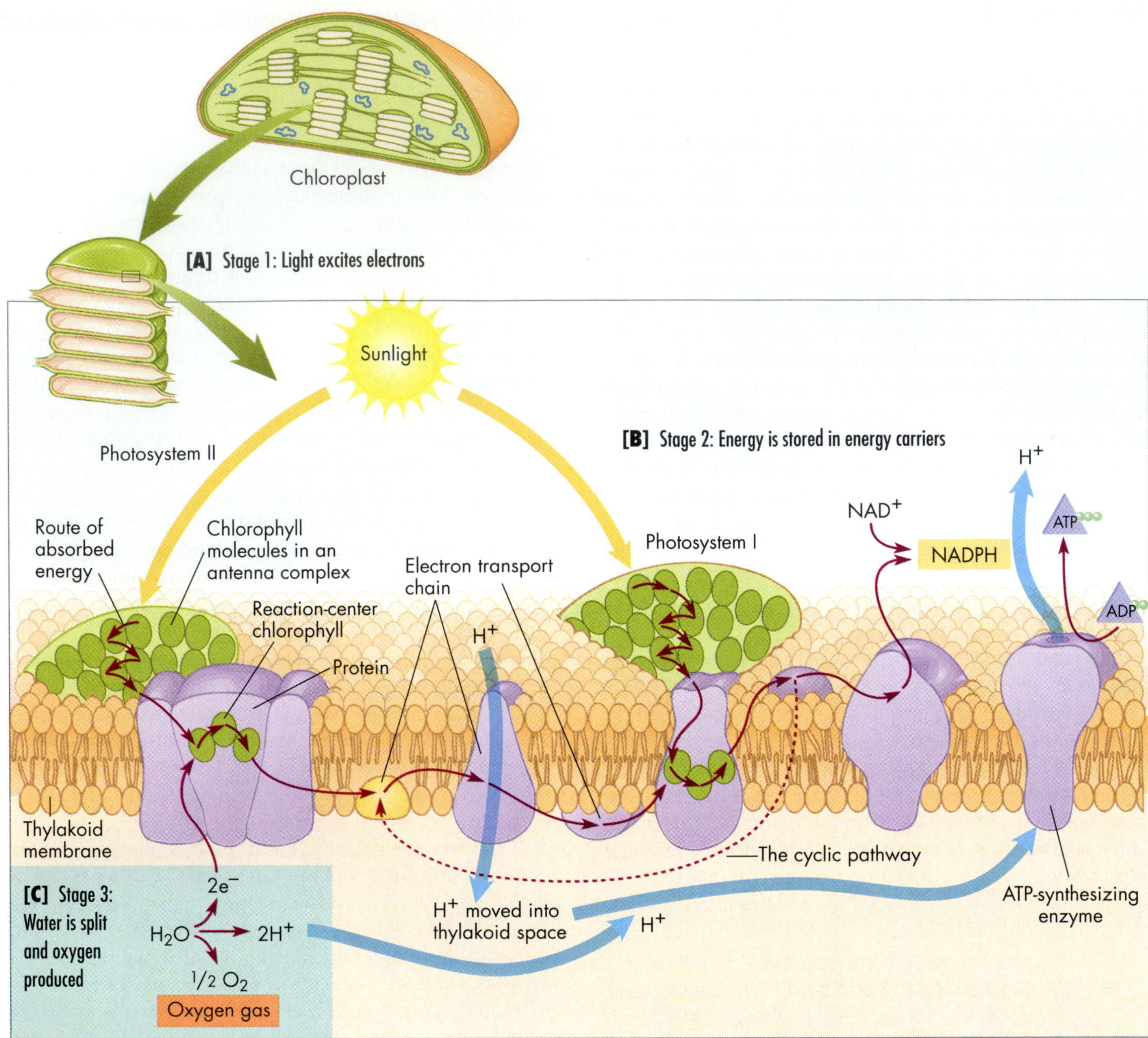

FIGURE 6.10

The Light-Trapping Reactions of Photosynthesis.

[A] In Stage 1, light energy excites electrons to higher energy levels. **[B]** In Stage 2, electrons are passed down an electron transport chain and reboosted with more light; the energy is stored in the energy carriers ATP and NADPH. **[C]** In Stage 3, electrons boosted from chlorophyll in Stage 1 are replaced by electrons stripped from water, and oxygen gas is released.

they lose a bit of their energy. That energy becomes stored in ATP molecules by a process similar to the one that takes place in mitochondria—the flow of hydrogen ions through a membrane protein that synthesizes ATP [FIGURE 6.10B].

At the end of the electron transport chain, the electrons join a second photosystem, photosystem I, and are again boosted to higher energy levels by light [see FIGURE 6.10B]. This boost-drop-boost pathway of energy gain and release can be plotted as a zigzag path on a graph of energy content and hence is called the *Z scheme* of photosynthesis. Finally, the energetic electrons are transferred one last time, and NADPH forms. This electron carrier stores seven times more energy than an ATP molecule and can move around in the chloroplast and the cell, where the energy is released once again and can perform useful work. (In an ancient type of photosynthesis involving only photosystem I, electrons travel the dashed circular path shown in FIGURE 6.10B that excludes photosystem II. Called *cyclic photophosphorylation*, this pathway provides a small amount of ATP and no NADPH. In contrast, the zigzag Z scheme is called *noncyclic photophosphorylation*.)

With the production of the stable energy carriers ATP and NADPH, the energy-trapping reaction has trapped light energy into useful chemical energy. But it has removed an electron pair from the reaction center chlorophyll in photosystem II, leaving a hole that must be filled with another electron pair. That pair comes from the splitting of a water molecule.

Stage 3: Water Is Split and Oxygen Produced The ecopolitical slogan "Have you thanked a plant today?" is apt indeed, for plants produce oxygen, a necessity for our lives. This oxygen comes from water molecules broken down by the process of photosynthesis.

The photosynthetic pathway inside plant cells strips electrons and hydrogens from water molecules [FIGURE 6.10C]. These electrons replace the electrons that were boosted from chlorophyll by incoming light. The oxygen left from the breakdown of water (H_2O) is then discarded as a metabolic waste product. The hydrogen atoms left over when water is split are also used in the plant cell's chloroplasts during the production of ATP, much as hydrogen ions are used in mitochondria [compare FIGURES 5.12 and 6.10B].

THE CARBON-FIXING REACTIONS OF PHOTOSYNTHESIS

A green leaf is a far better solar collector than any mechanical solar device yet built. Solar devices usually require expensive silicon crystals and rare metals, moving parts, and constant upkeep. A leaf, on the other hand, can track the sun's path without wheels and gears; it is mostly organic and biodegradable; and it uses materials that are abundant in the environment. Moreover, a leaf does something no solar device can do: It builds organic nutrients in a cyclic series of enzyme-catalyzed reactions. This cycle of reactions, called the *carbon-fixing reactions* of photosynthesis, or the **Calvin-Benson cycle** (after the biologists who figured out its steps), takes place on the outer surface of the thylakoid membrane, facing the stroma.

The Calvin-Benson reactions form a cycle with three main stages. In the first, carbon is fixed to a biological molecule. In the second, carbohydrates are synthesized as stable energy-storage molecules. In the third, the starting compound for the cycle is regenerated [FIGURE 6.11].

Stage 1: Carbon Fixation In carbon fixation, carbon dioxide from the air becomes attached to a biological molecule, which makes carbon available for carbohydrate synthesis [FIGURE 6.11]. The acceptor of carbon dioxide is a five-carbon molecule called *ribulose bisphosphate*. When the single carbon of carbon dioxide is joined to this five-carbon molecule, the result is an unstable six-carbon molecule that immediately breaks down to 2 three-carbon molecules. The addition of carbon to a biological molecule is accomplished by an enzyme nicknamed **rubisco** (for ribulose bisphosphate carboxylase oxygenase). There are so many rubisco molecules in plants that botanists have calculated that rubisco may be nature's most common protein—a fact that graphically demonstrates the importance of photosynthesis to life on earth.

The net result of carbon fixation, the first step in the Calvin-Benson cycle, is the linkage of inorganic carbon dioxide to a biological molecule within the plant cell. This fixed carbon is the molecular starting point from which the plant will build proteins, DNA, and cellulose (cell-wall material). These biological molecules may end up as wood in your house or a cotton garment on your back or may be consumed and end up as animal or fungal tissue.

Stage 2: Carbohydrate Manufacture The second stage of the carbon-fixing reactions traps energy into stable storage molecules [FIGURE 6.11]. Specifically, energy from ATP, and electrons and hydrogens from the NADPH generated during the energy-trapping reactions, are added to the newly fixed carbon, now in the form of three-carbon molecules. Roughly speaking, the carbon goes from CO_2 (carbon dioxide gas) to CH_2O (carbohydrate) through the addition of hydrogens and electrons. The three-carbon carbohydrates formed can be joined together, forming six-carbon carbohydrates, such as fructose phosphate and glucose phosphate or eventually starch, fats, and amino acids. Since one glucose subunit of a starch molecule contains six carbons, six carbon dioxides must be fixed, one at a time, in six turns of the cycle, before one starch subunit can be produced.

Stage 3: Regenerating the Starting Compound So far, the first two stages of the carbon-fixing reaction have fixed both carbon and energy into useful storage forms such as sugar phosphates and starch. If the cycle is to continue, however, the five-carbon compound that began the cycle must be regenerated, a process requiring some ATP generated during the energy-trapping reactions [FIGURE 6.11]. Once this carbon dioxide acceptor is rebuilt, the enzyme rubisco can initiate a new cycle and the formation of more carbohydrate.

WHAT MAKES CARBOHYDRATES BETTER ENERGY STORES THAN ATP?

You may wonder why a plant cell doesn't simply use the ATP and NADPH from the energy-trapping reactions directly to fuel cellular activity instead of converting it in a second reaction series to yet another form like sugar or starch. The answer is that carbohydrates are far more compact and practical forms for storing chemical energy than are the high-energy carriers. The glucose molecule, for instance, is smaller than a single ATP, yet stores in its bonds the energy equivalent of 36 molecules of ATP. Glucose is also less reactive than ATP and NADPH in cells—in fact, you can store it for years in a jar on the shelf in your kitchen without its breaking down—and it can

FIGURE 6.11

The Carbon-Fixing Reactions of Photosynthesis.

These reactions take carbon molecules from the air and fix them into biological molecules. In Stage 1, carbon is fixed. Three CO_2 molecules from the atmosphere are added to three molecules of ribulose bisphosphate by the enzyme rubisco. This forms three molecules of an unstable six-carbon compound, which breaks down rapidly to six molecules of a three-carbon compound; thus fixing the carbon. In Stage 2, carbohydrate is manufactured. Enzymes act on energy, electrons, and hydrogens from the NADPH and ATP made in the light-dependent reactions, altering the fixed carbon into six molecules of the three-carbon compound G-3-P (glyceraldehyde 3-phosphate). One of these G-3-P molecules can be siphoned from the cycle and joined to others, generating sugars. The other five G-3-P molecules are left for the final step. In Stage 3, the cycle is regenerated. Energy from ATP generated in the light-dependent reactions modifies the 5 three-carbon G-3-P molecules into 3 five-carbon ribulose bisphosphate molecules. With this starting compound regenerated, the cycle can continue.

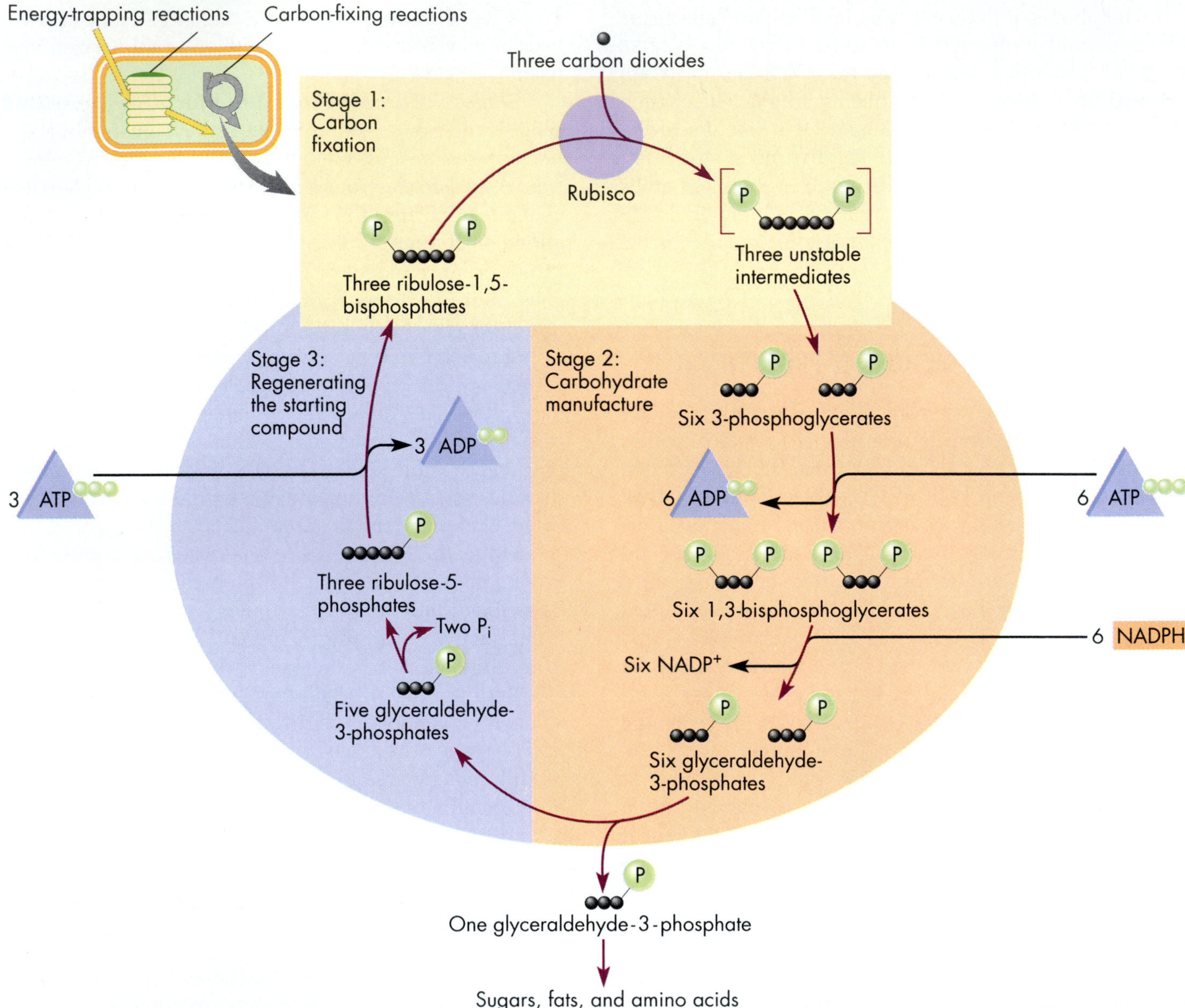

travel easily from place to place in an organism. Carbohydrates can be easily converted to other biological molecules [see FIGURE 5.11], and glucose and other sugars can be easily linked into huge inert storage molecules such as starch or glycogen (to be tapped only when needed) or into structural molecules like the cellulose in plant cell walls [see CHAPTER 3].

The carbon-fixing reactions of photosynthesis ultimately provide all the food energy you eat. These reactions are also the ones which will be affected by increasing levels of atmospheric carbon dioxide, as we discussed in the introduction. The carbon-fixing reactions of some plants will speed up more than others, however, due to a special type of carbon fixation.

Adaptations for Hot, Dry Environments

Each summer, as the hot sun dries up unwatered suburban lawns, annoyingly luxuriant tufts of green crabgrass flourish among an otherwise brown sea. Midwestern farmers often note that during a hot, dry spell, a field of corn [FIGURE 6.12A] can be green next to a dying field of beans. And motorists may also notice that while wildflowers shrivel and die back along the roadside in the summer, ice plants remain plump and healthy [FIGURE 6.12B]. Interestingly, many plants that evolved in hot, dry climates display a variation on the principles of photosynthesis we just learned—a variation that enables them to continue to photosynthesize at times when other plants cannot.

When the weather is hot and dry, the tiny openings (or stomata, see FIGURE 6.6) on leaf surfaces close, preventing water from escaping. This closure, however, also shuts out carbon dioxide and shuts in oxygen. Thus, in hot, dry weather, oxygen from the energy-trapping reactions builds up in the leaf, and carbon dioxide decreases. Surprisingly, oxygen competes with carbon dioxide for the rubisco enzyme. When levels of oxygen rise inside the leaf, the rubisco enzyme carries out reactions that result in the *loss* of fixed carbon rather than the *gain* of carbon. Plant biologists call this wasteful competing reaction **photorespiration**.

Some plants, such as corn and crabgrass, have evolved a solution to this seeming defect in rubisco: Carbon dioxide is supplied to cells that specialize in the carbon-fixing reactions by a special *carbon dioxide pumping cycle*. This biochemical cycle increases carbon dioxide concentrations inside the photosynthesizing cells in the leaf and thus decreases wasteful photorespiration. The plants can then grow with partially closed stomata even when the weather is hot and dry.

The essentials of a carbon dioxide pumping cycle are shown in FIGURE 6.12C. Carbon dioxide from the air is added to a three-carbon compound, making a four-carbon compound. When the four-carbon compound breaks down, it delivers carbon dioxide to the carbon-fixing reactions of the Calvin-Benson cycle directly and regenerates the original three-carbon compound. With the concentrated delivery of carbon dioxide directly to the carbon-fixing reactions, photosynthesis can continue and the plant remains green and healthy.

There are two different types of carbon dioxide pumping cycles in different groups of plants. Corn, crabgrass, sugarcane, and many other tropical plants are called **C_4 plants** because the first stable compound after carbon fixation has four carbons [see FIGURE 6.12C]. In

[A] Corn: A C_4 plant

[B] Ice plant: A CAM plant

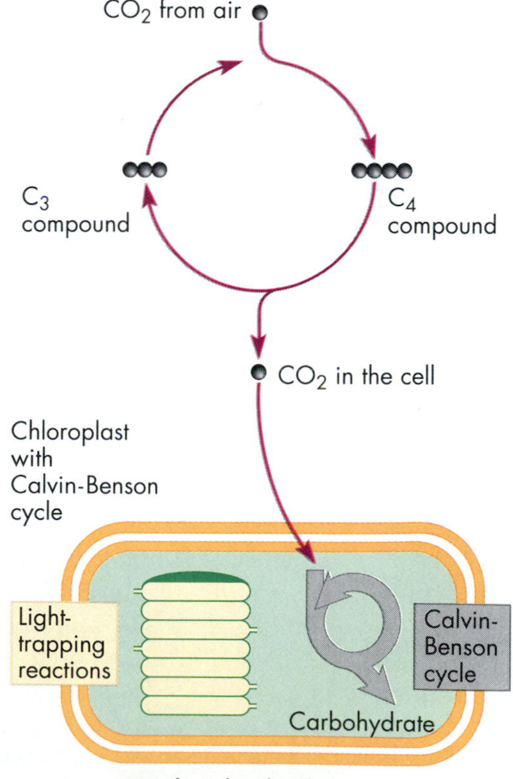

[C] The carbon dioxide pumping cycle

FIGURE 6.12

Photosynthesis in All Seasons.

Some plants adapted to hot, dry environments have special methods for fixing carbon. Corn **[A]**, which thrives in the hot summers of Iowa and Kansas, and ice plant **[B]**, which flourishes along dry berms of California freeways, have special ways of fixing carbon that other plants lack. Unlike C_3 plants, corn and the ice plant initially fix carbon into a compound with four carbons by a carbon dioxide pumping cycle **[C]**. Hence, corn is called a C_4 plant, and ice plant is called a CAM plant based on the special type of carbon compound its carbon fixation involves. In C_4 plants, the carbon dioxide pumping cycle is in one type of cell, and the light-independent reactions of the Calvin-Benson cycle are in another type of cell. In CAM plants, the carbon-dioxide pump and the light-independent reactions are in the same cells, but they occur at different times during the day.

contrast, as we saw in FIGURE 6.11, most plants are **C_3 plants**, with a three-carbon compound being the first stable product after carbon fixation. Succulents such as ice plant, jade plant, and other plants with water stored in thick leaves, as well as some cacti, have a slightly different carbon dioxide pumping cycle and are called **CAM plants** (CAM stands for *crassulacean acid metabolism*). Interestingly, in CAM plants, the carbon dioxide pump fixes carbon at night, when temperatures are lower, humidity is higher, and the dangers of dessication are reduced. Then, during the daytime, when the sun comes up and the risk of drying out rises, the stored carbon dioxide can be delivered to the Calvin-Benson cycle without a need for the stomata to open. Because of their carbon dioxide pumping cycles, C_4 and CAM plants have a significant advantage over the more common C_3 plants—and this advantage accounts for the exasperating luxuriance of crabgrass in hot, dry conditions.

C_3 AND C_4 PLANTS IN A CARBON DIOXIDE–RICH WORLD

Let's consider, for a moment, how C_3 and C_4 plants might fare in our polluted world of high carbon dioxide. As levels of carbon dioxide rise in the atmosphere, the carbon-fixing reactions of photosynthesis should be favored over the carbon-loss reactions of photorespiration. In a carbon dioxide–rich environment, then, the photosynthetic efficiency of many plants should increase rather than decrease. We just saw that C_4 plants already have special mechanisms enabling them to carry on photosynthesis efficiently despite hot, dry conditions, and one might predict that they might not be stimulated as much as C_3 plants. Some plant ecologists, therefore, have predicted that in a carbon–rich world, C_3 species may grow disproportionately well and create an ecological imbalance.

To test this hypothesis, ecologists set up carefully controlled growing chambers—mini-greenhouses—on salt marshes in Chesapeake Bay [FIGURE 6.13] and on the tundra in north central Alaska. They found that in some cases, C_3 plants do indeed grow faster and larger when carbon dioxide levels are elevated. This has caused some plant biologists to worry that as carbon dioxide levels rise due to human activity, important C_4 crops such as corn and sugarcane may suffer stiffer competition from C_3 weeds. Some fear that climbing carbon dioxide levels will also disrupt natural environments. Ecologists found, for example, that a particular type of bromegrass outcompeted some other grasses when carbon dioxide levels were raised experimentally. Since a prairie dominated by bromegrass is more likely to burn, wild fires could become more common in rangeland communities.

The message here is that as people burn more fossil fuels and cut and burn down more forests, atmospheric

FIGURE 6.13

Ecologists Use Mini-Greenhouses to Test CO_2 and Plant Growth. Ecologists mimic the high carbon dioxide atmosphere of the near future to determine its effects on plant growth, in a salt marsh in Chesapeake Bay.

CO_2 levels will continue to rise. Those increases may stimulate the efficiency of some plants, but this is not the unmitigated blessing it may at first seem. To successfully preserve healthy world environments, biologists, policy makers, and the public will need a deeper understanding of photosynthesis and the complicated ways in which organisms interact in the environment.

Photosynthesis and the Global Environment

Look at your hands. The carbon and oxygen atoms that make up the skin, tendons, and bone of your fingers were once physically present in the tissues of plants. Now look at a nearby houseplant, or out the window at a tree. Those green organisms contain carbons and oxygens that were once present in your body and the cells of other animals, fungi, or microorganisms. Aerobic respiration and photosynthesis are not simply chemical processes with reciprocal equations. They are metabolic engines that drive a global **carbon cycle**, a flow of atoms from autotrophs to heterotrophs, to the atmosphere, soil, and water, and back to autotrophs once again [FIGURE 6.14]. Most autotrophs fix carbon dioxide into organic compounds and release oxygen. Most heterotrophs break down those compounds and in the process consume oxygen and release carbon dioxide. Autotrophs fix the released carbon dioxide once again, release new oxygen, and the cycle continues.

These cyclic events ultimately involve all of the trillions of tons of carbon on our planet, whether it is currently tied up in wood, leaves, animal tissues, microorganisms, or rotting soil particles, and whether it is

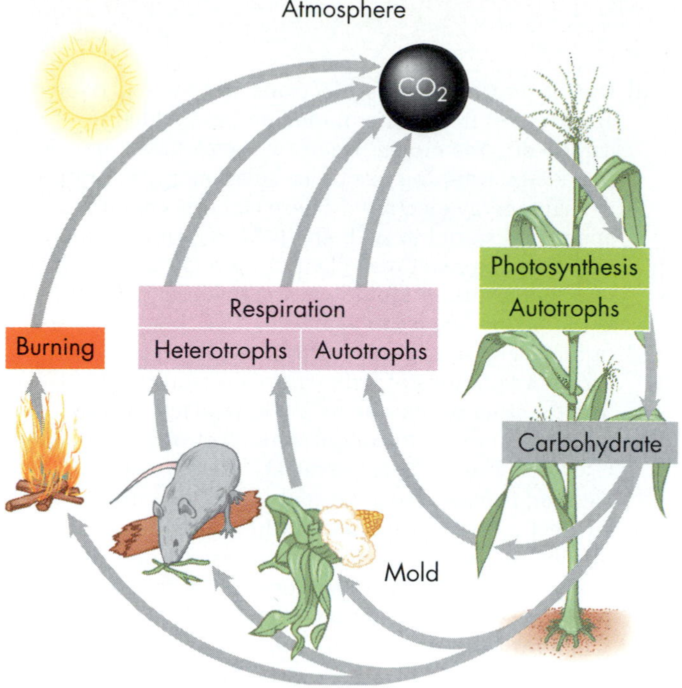

FIGURE 6.14

The Global Carbon Cycle.

Vast amounts of carbon cycle through the air, soil, and water as autotrophs fix carbon dioxide into organic compounds, and heterotrophs (along with nonliving combustion processes such as burning) break down those compounds and once again release carbon dioxide.

floating in the atmosphere as carbon dioxide gas, dissolved in seawater, or tied up as carbonates in rocks.

If all autotrophs suddenly stopped functioning today, biologists estimate that all organic compounds would be metabolized to carbon dioxide within 300 years, and no free oxygen would remain in the oceans and atmosphere after just 2000 years. This is an astonishingly short time, considering that it took many millions of years for oxygen to accumulate to its current levels in air.

FIGURE 6.14 shows the global carbon cycle and the relationships between autotrophs and heterotrophs and between photosynthesis and aerobic respiration. Notice that the burning of fossil fuels (a form of fixed carbon) makes a significant contribution to the pool of carbon dioxide available for fixation by autotrophs. The burning of fossil fuels and of our rain forests is contributing to a buildup of atmospheric carbon dioxide gas. Besides the carbon dioxide fertilizer effect we have discussed throughout the chapter, this buildup contributes to a phenomenon called the **greenhouse effect**. In the greenhouse effect, gases trap heat from sunlight beneath the atmospheric layers, just the way the glass windows of a greenhouse trap heat inside the greenhouse. This trapped heat may in turn be leading to increased global temperatures, changing climate patterns, additional photosynthesis by autotrophs, and a slight melting of the polar ice caps.

Just as the metabolic processes in individual cells have a massive collective effect in the global carbon cycle, worldwide disturbances of that cycle caused by human activity will have their own pervasive effects. Those effects, if we allow them to occur, could disrupt biochemical processes such as photosynthesis and respiration to the point of exterminating many of the complex organisms we take for granted today.

CONCEPT CHALLENGE

CAM plants, including ice plants and many cacti, are efficient photosynthesizers under arid conditions because their stomata close during the day, allowing greater water conservation. Explain how in moister environments, C_3 plants might out-compete CAM plants.

Connections

Photosynthesis is far from universal. Of the 3 to 5 million living species cataloged so far, heterotrophic species outnumber autotrophic species by a wide margin. While nearly all living things break down organic nutrients for energy, only about half a million species can build those nutrients through solar-powered carbon fixation. Nevertheless, those half million support millions of other living species. If photosynthesis stopped today, virtually all heterotrophs and most autotrophs alike would die from lack of energy in a surprisingly short time. Even a seemingly minor perturbation in photosynthesis, caused by increasing carbon dioxide concentrations in the atmosphere, could potentially change large ecosystems in ways that we can scarcely imagine or predict.

This chapter ends the unit on cells and cellular chemistry. We have come a long way—from atoms and molecules to cell structure, mechanisms of water and solute transport, and finally to the flow of energy from the sun and other sources through the dynamic systems of living cells. In BOX 6.2, you can apply what you learned in this section on atoms, molecules, and cells to a very real problem, Americans' use of products from the tobacco plant. This will help you make informed decisions as consumers and citizens.

Our next chapter deals with another of life's characteristics: reproduction. You will see, however, that cell structure, chemistry, and energetics are encountered again in this new topic, as well as throughout the rest of the book. The cell is the fundamental living unit, and its properties are closely reflected in all larger and more complex living things.

KEY TERMS

antenna complex, 153
autotroph, 146
C_3 plant, 158
C_4 plant, 157
CAM plant, 158
carbon cycle, 158
carbon fixation, 155
carbon-fixing (light-independent, or Calvin-Benson reactions), 152
carotenoid pigment, 148
chlorophyll, 148
chloroplast, 148
energy-trapping (light-dependent) reactions, 152
greenhouse effect, 159
heterotroph, 146
photon, 146
photorespiration, 157
photosynthesis, 145
photosystem, 153
reaction center, 153
rubisco, 155
thylakoid, 148

HIGHLIGHTS IN REVIEW

1 Green plants trap light energy and synthesize carbohydrate molecules via the reactions of photosynthesis. Plant cells use the raw materials carbon dioxide and water, and release oxygen as a waste product.

a] Photosynthesis is the metabolic process by which solar energy is trapped, converted to chemical energy, and stored in the bonds of carbohydrates. Nearly all living organisms depend on photosynthesis, either directly, by photosynthesizing themselves, or indirectly, by consuming photosynthesizers or their products.

b] The reactants and products of photosynthesis and aerobic respiration are reciprocal. The photosynthetic process uses carbon dioxide and water while capturing energy and producing sugars and oxygen. The respiration process breaks down sugars and uses oxygen while harvesting energy and releasing carbon dioxide and water.

2 Virtually all familiar life forms depend on photosynthesis for energy and organic carbon. Plants do so directly; organisms that feed on plants or on plant-consuming organisms do so indirectly.

3 In plants, photosynthetic reactions take place in green cellular organelles called chloroplasts. One set of reactions traps light energy and releases oxygen from water. Another set removes carbon dioxide from the air and combines it with hydrogen in the biologically useful form of carbohydrates. This reaction is reciprocal to aerobic respiration in plants and other organisms.

a] Visible light is just a small part of the electromagnetic spectrum, which also includes radio waves and X-rays.

b] Chlorophyll molecules absorb violet, blue, yellow, and red wavelengths of light, reflecting only green wavelengths. Carotenoids are red, yellow, and orange pigments that function as accessory pigments to chlorophyll in the leaf and give many fruits, seeds, roots, and autumn leaves their color.

c] Chloroplasts are the sites of photosynthesis in eukaryotic autotrophs. Chlorophyll molecules are embedded in the membrane of the thylakoid disk.

d] During photosynthesis, electrons and hydrogen ions are stripped from water molecules and added to carbon dioxide molecules in two sets of reactions. In the energy-trapping reactions, solar energy is captured in high-energy electrons. These electrons and their energy are stored in ATP and NADPH. During the carbon-fixing reactions, chemical energy from ATP is used to fix carbon dioxide from the air, and the energy is trapped in sugar molecules.

e] In the energy-trapping reactions of photosynthesis, incoming photons of light cause electrons to be boosted from photosystem II. These electrons are replaced in the chlorophyll with electrons donated by H_2O, which is split, releasing O_2. Each boosted electron is passed to an acceptor and then moves down an electron transport chain in the thylakoid membrane. As the electrons are passed along, some of the energy moves hydrogen ions across the thylakoid membrane, and the hydrogen ions flow back, driving ATP synthesis. At the end of the electron transport chain, the electrons are reboosted by light striking photosystem I. The boosted electrons are transferred to other electron acceptors, and finally reduce $NADP^+$ to NADPH.

f] In the carbon-fixing reactions (Calvin-Benson cycle), the enzyme nicknamed rubisco adds a carbon dioxide molecule to the five-carbon compound ribulose bisphosphate, generating a short-lived six-carbon unit that rapidly splits into 2 three-carbon molecules; this fixes carbon in a biological form.

g] Energy, electrons, and hydrogens from ATP and NADPH are added to the three-carbon compounds, forming carbohydrates such as sugar phosphates and starch. The final stage of the carbon-fixing reactions regenerates ribulose bisphosphate.

h] If the weather is hot and dry, leaf stomata close and photorespiration occurs in leaves of C_3 plants, causing a loss of carbon. In a C_4 plant such as corn, or a CAM plant such as ice plant, a carbon dioxide pumping mechanism allows photosynthesis to continue even in hot, dry weather.

4 Carbon dioxide levels rise in the atmosphere as people burn fossil fuels. This may disrupt ecosystems, as well as contribute to greenhouse warming of the earth.

a] The reciprocal metabolic processes of photosynthesis and aerobic respiration drive a global carbon cycle—a flow of carbon from the atmosphere into organisms, and back into the atmosphere.

b] Carbon dioxide levels in the atmosphere have risen substantially since 1800 due to human activities, such as combustion of fossil fuels and the cutting and burning of tropical forests. Increasing carbon dioxide concentrations may boost photosynthesis in C_3 plants due to a decrease in water loss and photorespiration. It may not, however, benefit many C_4 plants as much. This imbalance could potentially alter human agriculture as well as natural ecosystems in ways that remain to be fully researched.

UNDERSTANDING THE FACTS AND CONCEPTS For Questions 1–5, match each of the descriptions with the most appropriate location from the following list.

a] reaction center
b] thylakoid membrane
c] thylakoid space
d] stroma
e] granum

1 The only structural layer in a green plant that is truly green.

2 A stack of sacs.

3 Light-absorbing pigment molecules within an antenna complex.

4 The region inside a granum in which the light-dependent reactions occur.

5 The region outside a granum but inside a chloroplast where carbon fixation takes place.

As above, for Questions 6–15, match the descriptions with the appropriate item from the following list. Any item may be used once, more than once, or not at all.

a] photosystem I
b] photosystem II
c] both photosystems I and II
d] electron transport chain
e] Calvin-Benson cycle

6 Water is split and gaseous oxygen is released.

7 Light-absorbing pigments.

8 Carbon dioxide is first incorporated into organic form in C_3 plants.

9 ATP is synthesized.

10 ATP is broken down into ADP and phosphate.

11 Referred to as the Z scheme.

12 NADPH is formed.

13 NADPH is oxidized.

14 Found in ancient forms as a cyclic way to make ATP.

15 Noncyclic photophosphorylation.

As above, for Questions 16–20, match the descriptions with the appropriate item from the following list.

a] ribulose bisphosphate
b] rubisco
c] photorespiration
d] C_3 plant
e] C_4 plant

16 The process by which O_2 rather than CO_2 is utilized in the Calvin-Benson cycle.

17 A plant that fixes CO_2 directly in the Calvin-Benson cycle.

18 An enzyme that can fix either O_2 or CO_2, depending upon their relative concentrations.

19 When CO_2 is fixed in C_3 plants, the molecule that is its first substrate acceptor.

20 A plant that first fixes CO_2 in a reaction that precedes the Calvin-Benson cycle.

INTEGRATE AND APPLY WHAT YOU HAVE LEARNED

1 In what ways are the process of photosynthesis and the harvesting of energy dependent upon each other?

2 In what ways are the light-dependent and light-independent reactions of photosynthesis coupled?

3 Explain why photorespiration makes photosynthesis less efficient.

4 Is the energy contained in the green part of the visible spectrum used at all for photosynthesis?

5 Why is it an advantage for a desert plant to be a C_4 plant?

ANALYSIS

1 The concentration of certain gases in the atmosphere fluctuate with the seasons. For example, the concentration of carbon dioxide is lowest in the summer. When would you logically expect that atmospheric oxygen would be lowest?

a] Spring
b] Summer
c] Fall
d] Winter
e] None of these; no seasonal changes for O_2

2 Refer to page 153, where it is stated that for each molecule of carbon dioxide that enters a leaf through a stoma, 200 molecules of water leave by evaporation. Which of the following explanations seems to be the most logical?

a] The difference between the concentration of CO_2 from inside to outside the leaf is greater than it is for water.
b] The concentration gradient for water from inside to outside is steeper than the gradient for CO_2 from outside to inside.
c] The water molecules, being smaller, can more easily slip out through the stoma.
d] Carbon dioxide is actively transported, whereas water moves by the passive diffusion of osmosis.
e] Water molecules, being polar, have more diffusion energy than carbon dioxide molecules have.

3 The pigment chlorophyll differs from the pigments of animal fur in this respect:

a] Only chlorophyll can absorb light; animal pigments cannot.
b] Only chlorophyll can reflect light; animal pigments cannot.
c] Animal fur pigments absorb light; chlorophyll does not.
d] Animal fur pigments reflect light; chlorophyll does not.
e] Whereas both chlorophyll and animal pigments can reflect and/or absorb light, chlorophyll can be reversibly oxidized; animal fur pigments cannot be reversibly oxidized.

4 At the start of the light-dependent reactions, when chlorophyll absorbs light photons it becomes:

a] Oxidized
b] Reduced
c] Phosphorylated
d] Synthesized
e] Bleached

5 During the light-dependent reactions, water is:

a] Oxidized
b] Reduced
c] Phosphorylated
d] Ionized
e] Evaporated

PART ONE: APPLY AND DECIDE

Tobacco Use: A Cellular Assault

Imagine the headlines—and the general sense of panic—if three separate jumbo jets crashed on the same day and all 1475 passengers were killed. Imagine, further, that three jets crashed every day for a year. The daily crash story might be relegated to the back page after a while, but you can bet that few travelers, if any, would willingly climb aboard one of the "death planes." How ironic, then, that smoking claims an average of 1475 lives every day, that nearly all Americans know about the risks involved in tobacco use, and yet 50 million people still smoke or chew tobacco.

The cigarettes, cigars, pipe and chewing tobacco made from the tobacco plant seem to be uniquely addictive. This is because the nicotine in tobacco leaves causes a strong physiological addiction and because many people become psychologically dependent on the taste and feel of tobacco.

Researchers are anticipating a future epidemic of lung, mouth, throat, and other cancers based on the smoking and chewing habits of women and men of college age today. For this reason, it is worth examining the biological costs of tobacco use, focusing on the level of the cell, where the damage begins. You will then be better prepared to make personal decisions about smoking or chewing tobacco, to communicate to friends and family, and to vote on public issues such as cigarette taxes and smoking bans.

Tobacco (*Nicotiana tabacum*) is a hardy C_3 plant that grows up to 2 m tall and produces very large leaves and spikes of pretty, usually pink, flowers. When Christopher Columbus arrived in the New World, he found Native Americans cultivating and smoking tobacco for what they believed to be medicinal properties. Columbus and other explorers carried tobacco to Europe and eventually the rest of the world. Today, 500 years later, the plant is still widely used.

As a natural defense, tobacco leaves and stems produce various compounds that discourage insects and other predators. Among them is a bitter-tasting, nitrogen-containing compound, the alkaloid called nicotine. Smokers and chewers become physically hooked on this compound, which is toxic to the body but has profound and in some ways pleasant psychoactive properties. Researchers have found that after regular smoking or chewing for a year, a person has only one chance in five of quitting successfully on a given attempt. It takes most people three or four serious efforts to stop—and some are never able to.

Nicotine addiction is only part of the problem of tobacco use. Analysis of smoke from burning tobacco reveals over 4000 separate compounds, including DDT, arsenic, nitrosamines, and formaldehyde, all known carcinogens (cancer-causing agents). Similar tests of chewing tobacco reveal traces of three additional carcinogens: cadmium, uranium, and polonium. One of the most damaging compounds in tobacco smoke is the poisonous gas carbon monoxide—the culprit when someone dies from the fumes of a running car engine in a closed garage.

What do the free radicals, nicotine, and other tobacco compounds actually do to human cells?

1. Paralyzed Cilia Tobacco smoke can paralyze the cilia, the microscopic hairlike projections from cells lining the airways of the human respiratory tract [review FIGURE 3.24]. Without these continuously beating cilia, germs and particles of foreign matter can enter the lungs and cause irritation and infection. The lungs and respiratory passageways compensate by producing more mucus, which is expelled in a smoker's characteristic cough. Perennial coughing can weaken the lungs and lead to chronic bronchitis.

2. Lung Cell Changes Cadmium, nitrogen dioxide, and other substances in cigarette smoke can rupture cells in the lungs' tiny, balloonlike air sacs, or alveoli. They can also prevent a cell's smooth endoplasmic reticulum from producing normal amounts of surfactant [FIGURE 1]. Both of these changes can contribute to permanent shortness of breath and to lung diseases such as emphysema.

3. Disturbed Mitochondria Cell biologists have exposed eukaryotic cells to cigarette smoke, then viewed their mitochondrial membranes with an electron microscope. The smoke destroyed the mitochondria's normal internal structure, and with it, their ability to carry out the reactions of the Krebs cycle and the electron transport chain. Thus the cell is starved for ATP energy, and eventually dies.

4. DNA Damage Many of the toxic compounds in tobacco can attack and damage DNA. Repair enzymes in the nucleus attempt to fix these broken strands, and correct bungled base pairs. However, continual exposure to the toxins can lead to an accumulation of

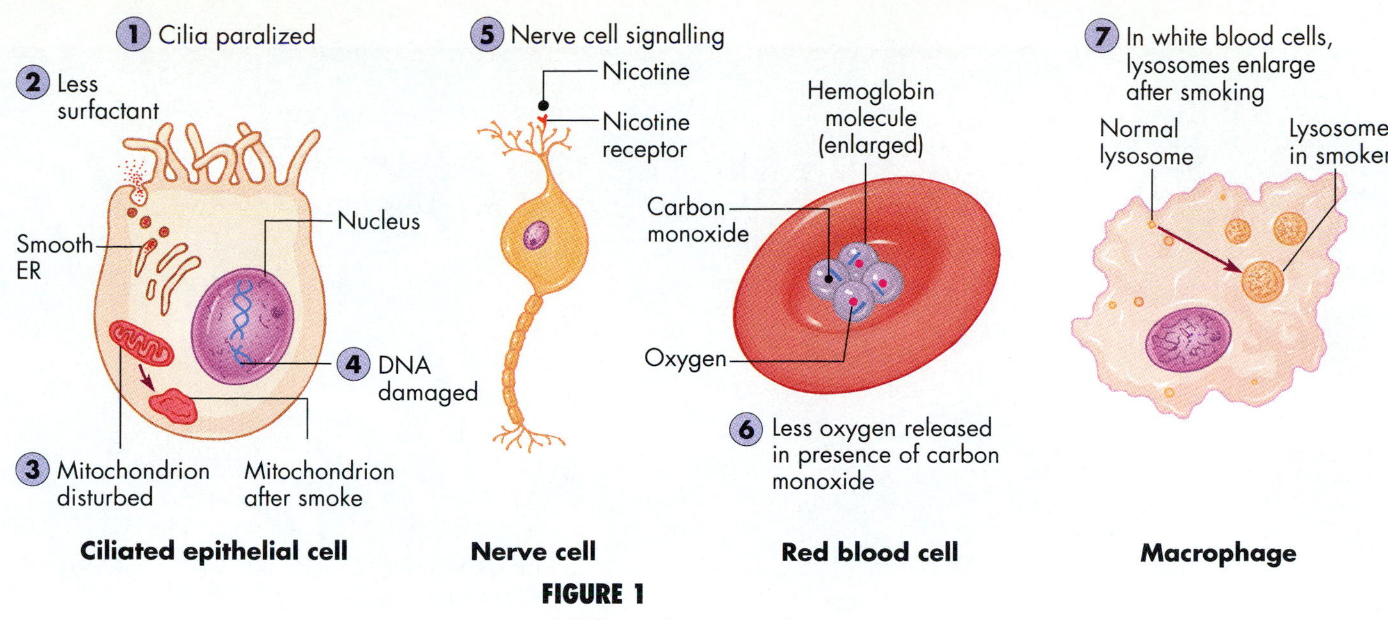

FIGURE 1

The Effects of Tobacco Smoke on Cells.

errors, which are implicated in the formation of cancerous tumors.

5. Nicotine and the Cell Certain cells in the nervous and muscular systems have receptor proteins in their membranes that bind nicotine. After binding the drug, these cells respond a bit differently to normal nerve signals, explaining how nicotine acts as a stimulant.

6. Increased Carbon Monoxide Smokers have elevated levels of CO in their bloodstreams. When CO combines with hemoglobin, the pigment delivers less oxygen to the body's tissues, including the brain, and the person's ability to think clearly is reduced.

7. Immune Cell Changes Researchers studying the white blood cells called macrophages that patrol and protect the airways and lungs have found an increase in the size and numbers of lysosomes within the cells and a decrease in protein synthesis. The cells are apparently so busy ingesting foreign particles and the debris of damaged cells that they can't grow and function properly. Immune cell disruption helps explain why smokers catch colds, flu, and pneumonia more easily than nonsmokers, as well as experience increased cancer rates.

8. Consequences, Immediate and Delayed Some cellular changes from smoking produce immediate effects. After just a few days or weeks of smoking, most people have shortness of breath and decreased ability to perform aerobic exercise. Tobacco users have lowered resistance to colds and flu, and slower healing of broken bones and other wounds. Female smokers tend to have more miscarriages and babies of lower birth weight. Even in their teens and twenties, tobacco users tend to develop periodontal (gum) disease four times as often as nonusers. And many recent studies show that tobacco use interferes with complex tasks involving memory and reasoning, and addicts are more likely to have automobile or industrial accidents.

Some of the more life-threatening health consequences of smoking and chewing include lung diseases, heart disease, thyroid disease, and cancers of the mouth, throat, lungs, bladder, and other sites. These can take 20 or 30 years to develop—hence the factor of deferred disease in tobacco addiction. Users have twice the overall cancer risk of nonusers, 15 times the risk of oral and lung cancer, twice the probability of heart attack or stroke, and a much greater chance of lung damage. In all, smokers die an average of eight years earlier than nonsmokers, and are less healthy and less active during the intervening years.

While one-quarter of American adults smoke or chew tobacco, the problem affects everyone. In 1993, physicians who looked at nicotine levels in Americans' blood were surprised to report that *every person they screened* had nicotine in their blood. Other measures of the pervasiveness of tobacco smoke in the environment are the statistics released in 1992 by the Environmental Protection Agency. These blamed 2500 to 3300 lung cancer deaths each year among nonsmokers on so-called second hand smoke (fumes exhaled from smoker's lungs or rising from their burning smokes). The EPA also attributed 150,000 to 300,000 respiratory tract infections per year among children to adults' smoking.

In an attempt to reduce the exposure of nonsmokers, many municipalities have banned smoking in restaurants, office buildings, and other public places. In addition, a professional society of 9000 cancer specialists has called upon the federal government to levy an excise tax on cigarettes of about $3 per pack, compared to the current 16 cents. They cite plummeting tobacco consumption in countries like Canada that have imposed such stiff taxes, and note that budget-conscious young people are among the first to quit when tobacco prices skyrocket.

We hope, in the name of good health, you will apply these findings on the cell biology of smoking to decisions you make about your own life. ❑

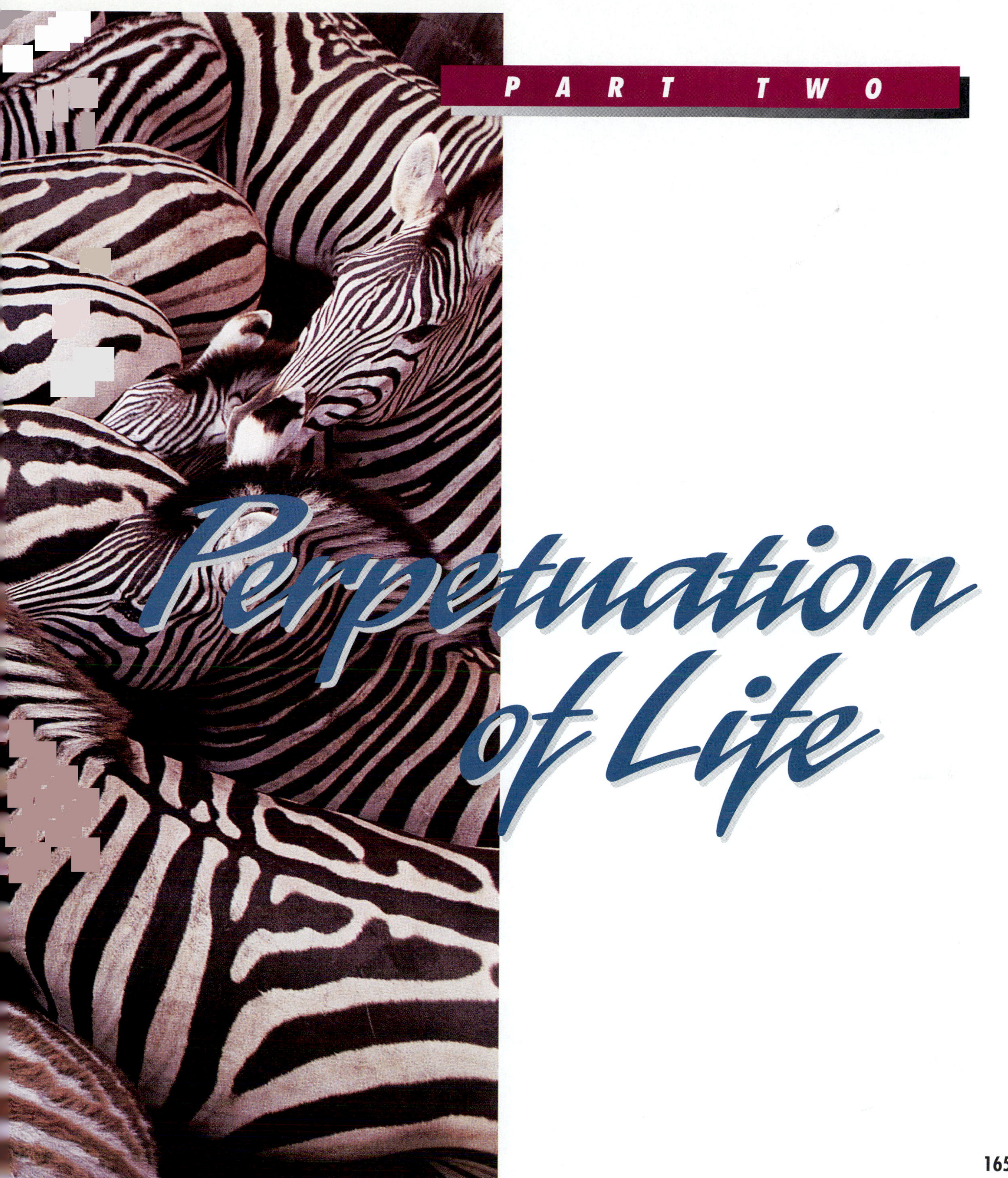

PART TWO

Perpetuation of Life

CHAPTER 7

Cell Cycles and Life Cycles

CALLUSES, SCARS, AND CELL DIVISION

Once a prevalent symbol of the American West, cowboys are, today, a relative rarity. A few thousand ranch hands, mostly men, still ride horseback to round up and drive cattle where mechanized vehicles can't maneuver easily. One reason is that ranch hands live a very hard life. Years of exposure to sun and wind, blowing sand, snapping tree branches, and slipping hand tools carve a record of fine lines, deep wrinkles, callused areas, pigmented blotches, and small scars in the skin of hands and face [FIGURE 7.1].

FIGURE 7.1 Portrait of a rancher: A living record of exposure to the elements.

Calluses and scars are evidence of the skin's self-protective mechanisms, whether on a rancher's leathery skin or on your own elbows, knees, or other hardened zones. These areas are based on an increase in the number of skin cells. This increase produces tough, hardened regions at places of continual abrasion and very strong bands of tissue where the skin was torn or gashed. The addition of new skin cells takes place in a deep layer below the dry surface zones. Individual cells in this deep layer increase in size and then split, forming two new daughter cells. The daughter cells can themselves increase in size and then split. This repeating cycle of cell growth followed by cell splitting is called the cell cycle, and it is the subject of this chapter.

At the most fundamental level, the repair of a wound requires cell division, the splitting of one cell into two. So, in fact, does the perpetuation of any living thing, both by means of day-to-day maintenance or by reproduction. For a one-celled organism, cell division and reproduction can be the same thing. The single-celled organism simply grows larger, and then divides in half, forming two smaller daughter cells. For a multicellular organism, however, reproduction is usually more complex, involving specialized reproductive cells and the organs that produce them. Reproductive cells are often large, relatively immobile cells called eggs and small, highly motile cells called sperm. When an egg and sperm unite, they form a single fertilized egg cell which then divides repeatedly and develops into the same type of multicellular organism as its parents.

Understanding how cells divide helps us to explore the broader subjects of genetics and development, our major concerns in this new section. In them, we learn how inherited traits such as eye color, height, or human genetic diseases (for example, cystic fibrosis) are passed down from parents to offspring, and how a single fertilized egg can become a frog or a tree.

Before most cells divide, a copy is made of the genetic material. One copy is then apportioned to each of the two daughter cells during division. The events that take place inside the cell from one division to the next are collectively called the cell cycle, and they result in cellular reproduction—one cell becoming two. Each organism also has a life-cycle, the events that take place between its initiation and its death, or sometimes between the reproduction of one generation and the reproduction of the next.

Cell division is usually precisely controlled, and the importance of this regulation is a theme we will encounter again and again throughout this chapter. If a ranch hand suffers a cut on the cheek, then the measured pace of cell divisions that normally replace worn-out skin is stepped up temporarily, and the additional skin cells fill in the wound and perhaps form a scar. After years of exposure to harsh sunlight, however, a rancher might develop a wartlike lump or a white crusty circle of renegade cells on the forehead or cheek—a skin cancer. If left untreated, these altered cells would continue to pass through the cell cycle, dividing and redividing into a larger and larger lump or crusty patch. Eventually, renegade malignant cells could leave the skin cancer site, invade other parts of the body, divide there in an uncontrolled way, and threaten the rancher's life. The same general sequence claims millions of Americans every year.

Our study of cell growth, cell division, and growth controls will address several major topics. We'll begin by seeing how scientists traced the location of the cell's hereditary information to the chromosomes. Next, we'll see details of the cycle of cell growth and division, and how it is controlled. We'll follow the movement of chromosomes during mitosis, the division of normal body cells. We'll discuss the special cell division called meiosis that produces reproductive cells. Finally, we'll see what roles the two different kinds of cell division play in evolution.

Along the way, you will see exactly how radiation damages cells; you will learn what causes Down syndrome; and we will discuss new medical techniques for controlling cell division to speed the healing of wounds. ❑

MESSAGES

1 A cell's hereditary information resides in the DNA molecules of its chromosomes, which are housed in the cell's nucleus.

2 The cell cycle includes a period of growth and a period of division. In the growth phase, chromosomes and other cell constituents replicate. During the division phase, replicated parts are distributed to offspring cells.

3 Regulation of the cell cycle is crucial for controlled cell division and for wound healing; inappropriate regulation can lead to cancers and other tumors.

4 Cell division by mitosis produces two genetically identical offspring cells.

5 Sexual reproduction involves meiosis—two consecutive cell divisions. When a diploid cell with two copies of each chromosome divides by meiosis, four haploid cells result, each cell bearing only one copy of each chromosome. Haploid cells can fuse in fertilization, forming a new diploid cell.

Chromosomes Store Hereditary Information

The skin cells on a cowboy's face share a general strategy of cell reproduction with other living cells: They take in nutrients, increase in size, and then divide in two. But what information directs the way a skin cell uses nutrients, grows, or divides? The answer is that **chromosomes** in the cell nucleus contain the genetic information that controls what a cell looks like and how it performs.

THE NUCLEUS: A STOREHOUSE OF INFORMATION

In the 1930s, the German biologist Joachim Hämmerling determined that the cell nucleus rather than the cell cytoplasm governs cell growth. He arrived at this conclusion by studying *Acetabularia*, an alga that inhabits warm seas and looks like a slender green toadstool. Each individual of *Acetabularia* is a huge single cell about 6 cm (2.4 in) long, easily visible to the naked eye [FIGURE 7.2A]. The cell has three main parts: a *foot*, which attaches the cell to a rock and contains the nucleus; a *stalk*, which grows upward from the foot; and a photosynthetic *cap*, which grows on the stalk and resembles an inverted umbrella in some species and a daisy in others [FIGURE 7.2B]. If the cap is cut off, the cell will regenerate another one identical to the first. But where do the directions for regenerating the new cap come from? From the cytoplasm in the stalk from which the new cap grows? Or from the nucleus in the foot?

To find out, Hämmerling took the stalk of an umbrella-cap cell (lacking both foot and cap) and transplanted it to the foot of a daisy cell (lacking both stalk and cap). This composite cell generated a new cap intermediate in shape between umbrella and daisy, suggesting that some information came from the stalk and some from the foot. When the researcher removed this regenerated cap, another grew—this time, daisy-shaped [see FIGURE 7.2B]. Clearly, some material in the foot of the composite cell (originally a daisy-cap species) must have directed the construction of the new cap. From this and other experiments, biologists concluded that the nucleus of *Acetabularia* cells, and, in fact, the nucleus of all eukaryotic cells, is the central repository of information for constructing new parts of each new cell.

FIGURE 7.2

The Genetic Information Lies in the Nucleus.

A classical experiment using the green alga *Acetabularia* revealed the importance of the nucleus as the repository of hereditary information. **[A]** Species of *Acetabularia* have large single cells with either an umbrella-shaped cap or a daisy-shaped cap that will regenerate if removed. **[B]** Transplanting the stalk of the umbrella species to the foot of the daisy species eventually results in the regrowth of a daisy-shaped cap, proving that genetic instructions for cap shape lie in the cell nucleus in the foot.

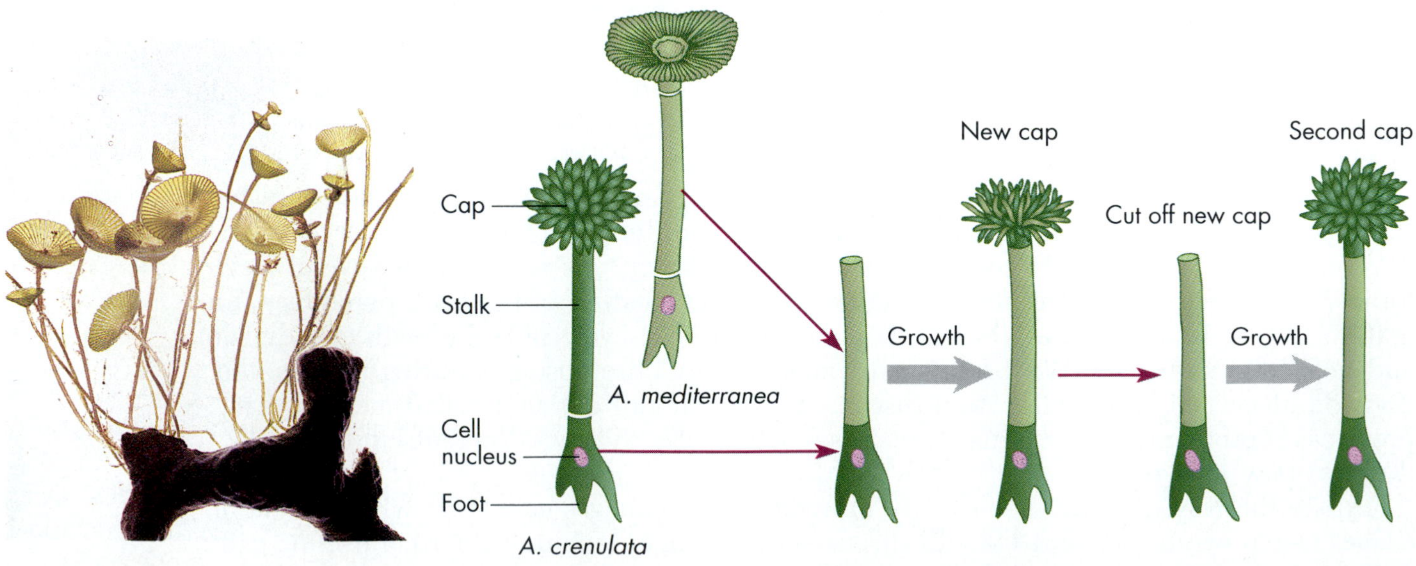

[A] *Acetabularia mediterranea*

[B] The transplant experiment

GENETIC INFORMATION: STORED IN CHROMOSOMES

What substance within the nucleus carries the information to build a cell? Biologists knew that an individual's mother and father both donate hereditary information to the offspring about equally. They speculated that hereditary information must therefore lie in some cell part that is donated equally by egg cell and sperm cell despite the fact that egg cells are much bigger than sperm cells. Re-

(a) Fertilization

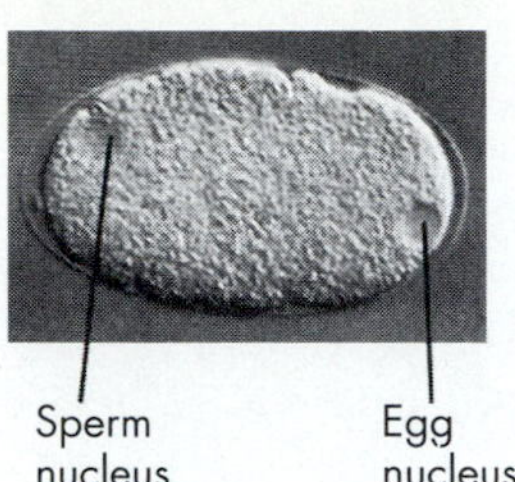

(b) Nuclear fusion

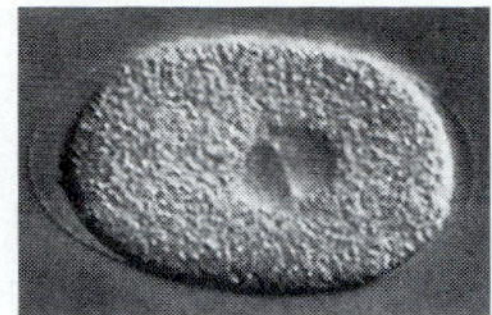

(c) Fertilized egg (zygote)

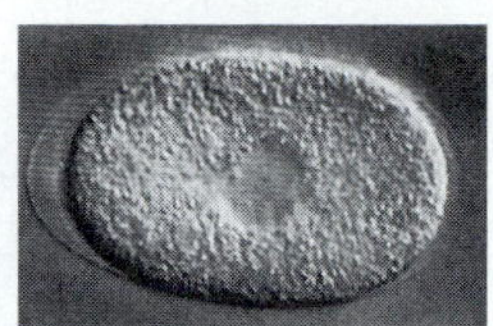

(d) Cell division by mitosis

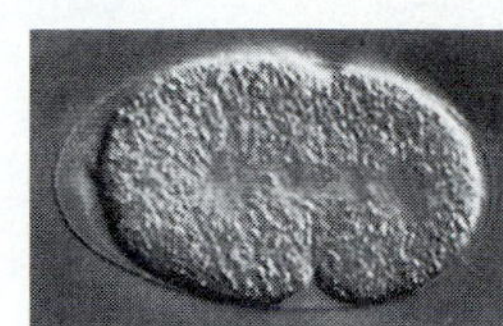

(e) 2-cell stage

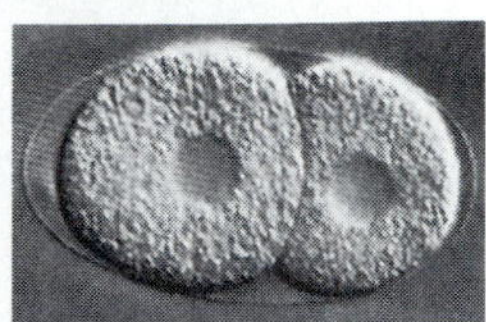

FIGURE 7.3

How Do We Know That the Nucleus Contains Hereditary Information?

While studying a type of roundworm or nematode, nineteenth-century biologists observed that during fertilization of the egg cell, the nucleus is the only component of the sperm that actually enters the egg. Since the offspring show hereditary traits from the father, they reasoned that the nucleus contains hereditary information. These photographs, taken through a modern microscope, show the sperm and egg nuclei at fertilization [**A**], during fusion [**B**], and forming a zygote, or fertilized egg [**C**]. When the zygote divides by mitosis [**D**], it produces a two-cell embryo [**E**], which continues division. Eventually many cells result and form the adult worm.

searchers identified this structure in 1883 in experiments with a roundworm called *Ascaris*, a parasite that infects some domestic dogs and other animals. Researchers found that the nucleus of each egg and sperm cell contained two bodies that took up stain and looked bright when viewed through a microscope. These two stained structures were named *chromosomes* ("colored bodies").

By watching fertilization under a microscope, biologists could see that chromosomes are the only structures that egg and sperm cells contribute in equal amounts to the fertilized egg and the new individual that grows from it [FIGURE 7.3]. The conclusion: Hereditary information lies in the chromosomes. So to understand cellular aspects of heredity, we have to know about chromosomes.

The Structure of Chromosomes Biologists know today that all organisms contain chromosomes, although the size, shape, and number of these structures differ from species to species. They also know that chromosomes contain DNA [see FIGURE 2.32], and experiments that we'll discuss in detail in CHAPTER 9 show that this DNA carries hereditary information. Each prokaryotic cell has a single long DNA molecule whose ends are joined into a circle [FIGURE 7.5A].

Eukaryotic cells have two or more chromosomes: the nucleus of the roundworm mentioned earlier contains 4 chromosomes; a giant sequoia nucleus, 22; a goldfish nucleus, 104; and the nucleus in a human skin cell, 46 [FIGURE 7.4A]. Each chromosome from a eukaryotic cell that has just finished dividing contains just one long DNA molecule, but the molecule is linear, like a strand of spaghetti, not circular, like a rubber band. Within the chromosome, this long DNA molecule winds around proteins like string wrapped around thousands of tiny spools. The spools are called *nucleosomes*, and this packaging helps to decrease tangling. Electron micrographs reveal great loops of the DNA-protein thread extending out from the axis of a condensed chromosome [FIGURE 7.4B].

In the early stages of nuclear division, a cell's chromosomes have already replicated (made copies of themselves). They now consist of two identical rods, called sister **chromatids**, held together at a single point, the **centromere** [FIGURE 7.4B]. Each chromatid has just one DNA molecule, and the DNA molecules in two sister chromatids are identical [FIGURE 7.4C]. The relationship of chromosomes to chromatids is always difficult the first time you encounter the concept. Just keep in mind that each chromatid is made of a single long DNA molecule [FIGURE 7.4C and D].For this reason, a replicated chromosome has two chromatids attached to the same centromere and an unreplicated chromosome has just one DNA molecule in one chromatid and one centromere.

When a cell divides, each new offspring cell receives its own set of chromosomes containing a copy of the hereditary material. For this reason, the duplication and distribution of the chromosomes are central activities in the cell cycle.

CONCEPT CHALLENGE

How many DNA molecules are contained in the nucleus of one of your skin cells just after it has completed cell division? Justify your answer.

The Cell Cycle

Just as it takes two very different sets of blueprints to build a suburban ranch house and an urban skyscraper, each living species has its own unique genetic blueprint. That genetic information is contained in chromosomes located in the nucleus of a cell. Normally, when a cell doubles in size and then divides, the chromosomes also double in number and then separate, so that each of the

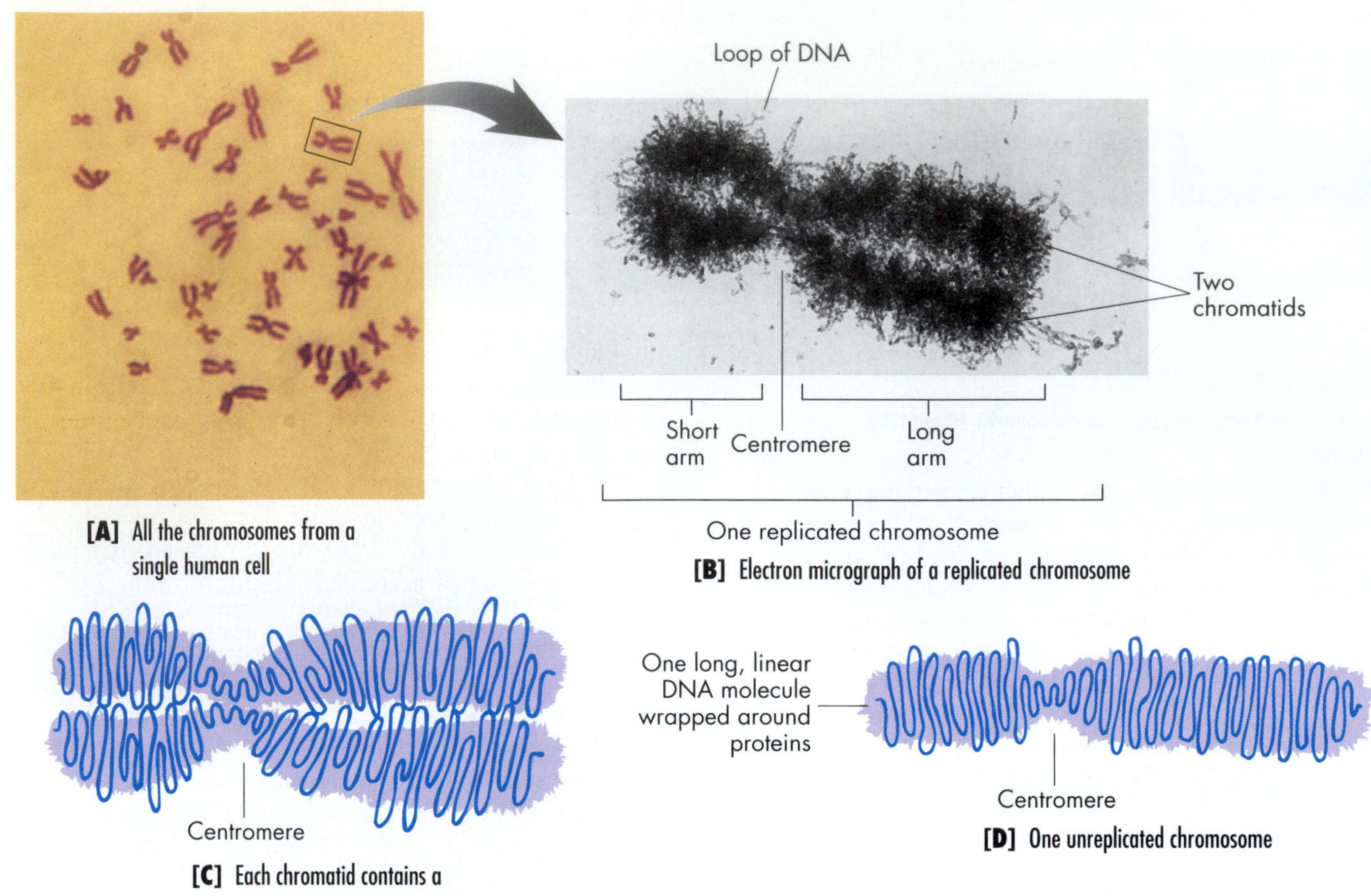

[A] All the chromosomes from a single human cell

[B] Electron micrograph of a replicated chromosome

[C] Each chromatid contains a single DNA molecule

[D] One unreplicated chromosome

FIGURE 7.4

Chromosomes in the Nucleus.

[A] Just before cell division, dark-staining structures called chromosomes become visible in the cell. A person has 46 chromosomes in each body cell nucleus. [B] In an electron microscope, a chromosome during cell division appears as two fuzzy rods connected at a point. Biologists call the entire structure a chromosome, and each of the two rods a chromatid. [C] A single very long DNA molecule wrapped around little balls of protein makes up each chromatid. You can see these DNA-protein fibers flaring out from the edges of the chromosome. Part [D] shows an unreplicated chromosome.

two new daughter cells receives an identical set of chromosomes. This chromosomal duplication is an essential part of the **cell cycle** of growth, duplication, and division, which is apparent both in single-celled prokaryotic organisms like bacteria, which lack a cell nucleus, and in multicellular eukaryotes like human beings whose cells each contain a nucleus.

THE CELL CYCLE IN PROKARYOTES: SIMPLE DIVISION

The cell cycle in a typical bacterium such as the intestinal inhabitant *Escherichia coli* alternates between periods of cell growth and cell division. The rod-shaped *E. coli* cell doubles in length, forms a partition in the middle, and then separates into two cells [FIGURE 7.5]. This prokaryotic process of cell division is called **binary fission** (literally, "two-way splitting"). The two new daughter cells increase in size, and under good conditions can divide again in about 30 minutes.

During binary fission, each offspring cell receives an identical set of genetic information. Recall that a prokaryote's hereditary information resides in a single circular molecule of DNA. Biologists generally call this DNA circle a chromosome even though it is quite different in shape from the linear chromosomes of eukaryotes [review FIGURE 7.4]. The prokaryote's circular DNA molecule is found in the dense-looking center of the cell rather than in a true membrane-bound nucleus [FIGURE 7.5A]. One specific point on the circular bacterial DNA is attached to the cell membrane, and as the cell doubles in size, the DNA duplicates, or *replicates* [FIGURE 7.5B]. The result is two side-by-side circles, each with its own point of attachment to the membrane [FIGURE 7.5C]. A central partition then forms between the two circles, cleaving the original cell into two identical daughter cells, each with a complete DNA circle [FIGURE 7.5D and E].

Ribosomes and other cell structures, which have also doubled in number during the growth phase, are distributed equally during cell division. The two identical offspring cells that result from binary fission then grow and repeat the cell cycle.

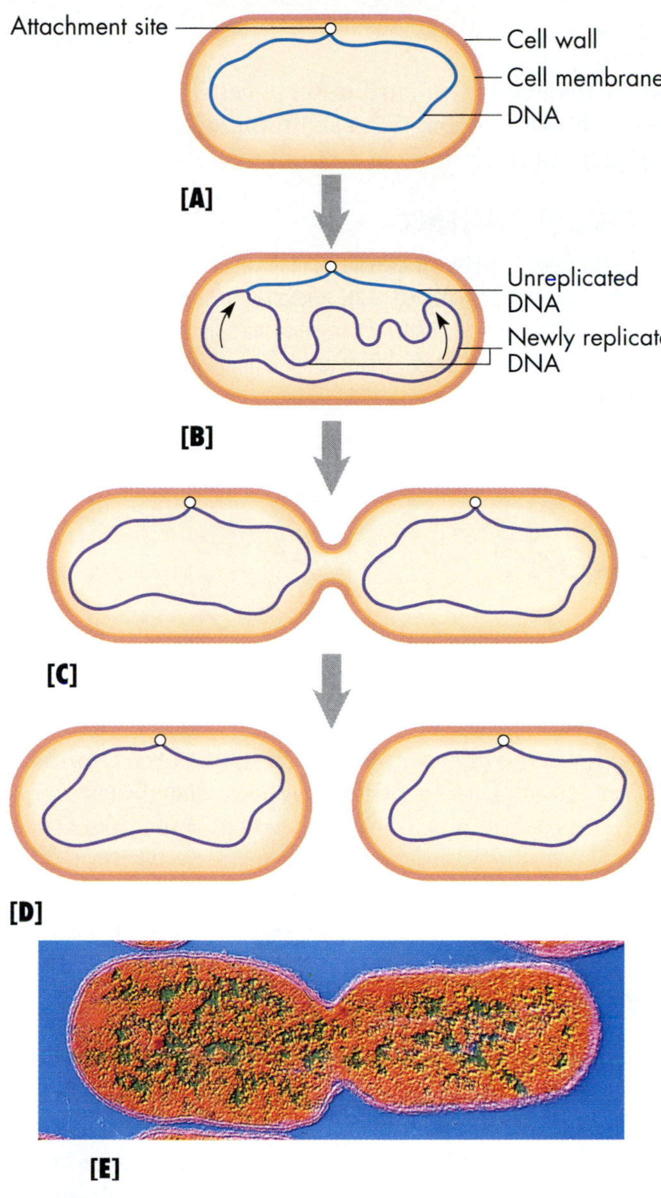

FIGURE 7.5

Cell Division in Prokaryotes.

[A] *E. coli* and other bacteria reproduce by binary fission; that is, [B] DNA doubles, and [C] each cell elongates and [D] divides in two. [E] Electron micrograph of a dividing bacterium, a species of *Shigella*, responsible for dysentery.

THE CELL CYCLE IN EUKARYOTES: PHASES OF GROWTH AND DIVISION

The cell cycle in eukaryotes differs from that of prokaryotes in several ways due to differences in cell complexity. Biologists divide the eukaryotic cycle into four phases, three involving cell growth and one resulting in actual cell division. The overall growth period of a eukaryotic cell is called **interphase** because it intervenes between two division periods [FIGURE 7.6]. The three parts of interphase are called G_1 (gap 1), S (DNA synthesis), and G_2 (gap 2). (G_1 and G_2 are called *gaps* because they are periods in the cell cycle that come between the active division phase and the DNA synthesis phase.) The remaining fourth phase, the division period, is called M (mitosis).

Different cells require different amounts of time to pass through the entire cell cycle. Cells in your bone marrow replace worn-out blood cells by dividing every 18 hours or so. Cells that line your stomach divide in a cell cycle about 24 hours long. If the ranch hand in our chapter opener developed a type of skin cancer called a *basal cell carcinoma*, the cells in this growth would undergo new rounds of the cell cycle every 67 hours or so. For each of these cell types, the time it takes for DNA synthesis, G_2, and mitosis is about the same. The different cell types vary most in the time it takes them to complete G_1.

G_1: An Active Growth Phase G_1 and G_2 are important growth phases during which the cell generates new components in preparation for cell division. During the G_1

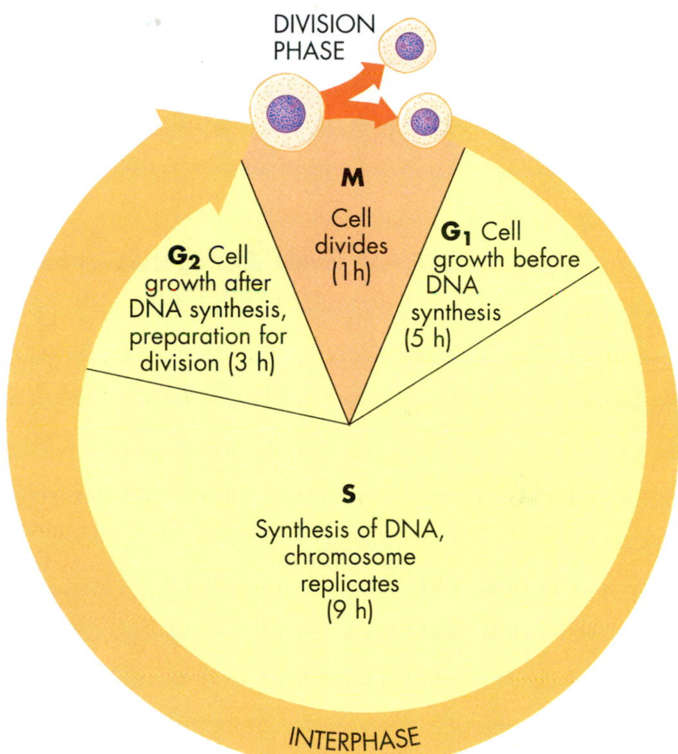

FIGURE 7.6

Four Phases of the Cell Cycle: M, G_1, S, G_2.

In a eukaryotic cell, growth (interphase) and division take place in four phases: an active growth phase (G_1), of varying length (here, about 1 hour); a long phase of DNA synthesis (9 hours or so), called S; a preparation for cell division (3 hours or so), called G_2; and then the rapid division phase (about 1 hour), called M for mitosis.

phase, new proteins, ribosomes, mitochondria, and other cell components are formed in preparation for DNA synthesis and cell division.

The length of the G_1 phase determines the length of the entire cell cycle. The G_1 phase can be quite short or very long, depending on the type of cell, its role in the organism, and conditions in its environment. For example, skin cells normally have a long cell cycle and a long G_1 (a few days). But that can change. If a wound removes some of those cells, the G_1 phase may shorten in the remaining skin cells at the edge of the wound, speeding up growth and division, and enabling the wound to heal rapidly. If the bark of an aspen tree is nibbled away by an animal, the bark-forming cells can enter a shortened G_1 phase, rapidly producing new bark, which protects the damaged area. Following G_1 eukaryotic cells enter the synthesis phase, S.

S: A Period of DNA Replication When cells complete the G_1 phase, they enter the **S phase**, during which the double-stranded DNA molecule in each chromosome is duplicated and certain proteins associated with chromosomes are synthesized in the cytoplasm [see FIGURE 7.6]. When the S phase ends, each chromosome is made up of two identical and parallel DNA molecules. Every stretch of DNA in a chromosome is copied only once during each S phase. After this copying is complete, the cell enters G_2.

G_2: The Cell Prepares to Divide During the **G_2 phase**, the cell continues to synthesize many proteins. If a researcher artificially blocks this synthesis, the cell fails to divide, suggesting that some proteins synthesized during G_2 promote mitosis. When all the necessary proteins have been synthesized, the cell leaves the final growth phase and begins to divide through mitosis.

M: The Cell Divides The **M phase** generally consists of two main events: mitosis, the division of the nuclear material, and cytokinesis, the division of the cytoplasm [FIGURE 7.7]. Both mitosis and cytokinesis are so important for the equitable distribution of genetic material and other cell components that each topic will be discussed separately in the following sections.

CONCEPT CHALLENGE

An oncologist (a physician specializing in the treatment of cancer) compares two tumors by looking at the fraction of cells found in the M phase of the cell cycle. In one tumor 5 percent of the cells are in M, and in the other, 1 percent in M phase. Which tumor is growing more rapidly? Defend your choice.

Mitosis: The Nucleus Divides

Viewed through a microscope, the G_1, G_2, and S phases of the cell cycle do not seem very active. In contrast, mitosis is spectacular. During **mitosis**, which distributes DNA equally to each new cell, the two sets of threadlike chromosomes produced during the S phase are distributed to opposite ends of the cell, and two new daughter nuclei form, each with its own complete set of chromosomes.

THE ACTIVITY OF CHROMOSOMES DURING MITOSIS

The cell watcher equipped with a light microscope will notice that the chromosomes, which are invisible during interphase, become visible during the M phase, making it possible to witness the movements of the chromosomes. The chromosomes become visible during the M phase because during this period the DNA becomes wrapped around protein spools much as a kite string is wrapped around a holder. By contrast, during interphase, DNA is loose and jumbled like a pile of string lying on the floor.

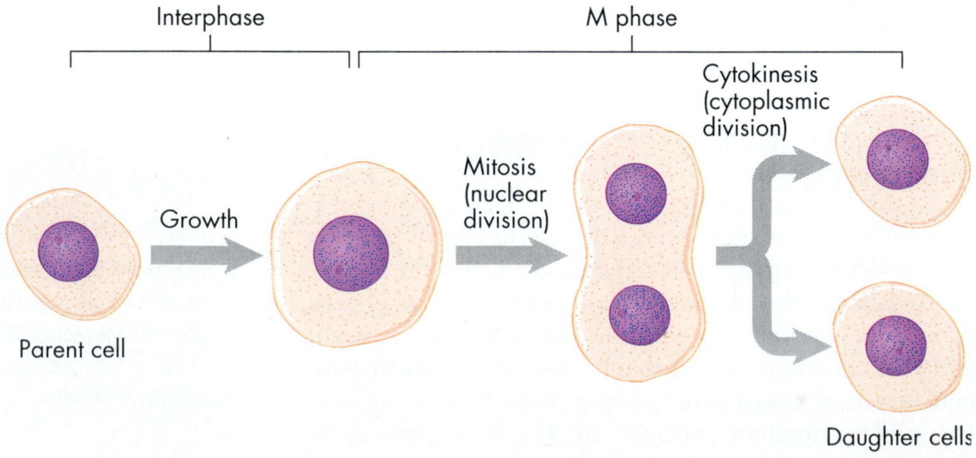

FIGURE 7.7

Events of the M Phase: Division of the Nucleus and Cytoplasm.

A cell grows during interphase and then enters the M phase. In the M phase, the nucleus divides during mitosis, and the cytoplasm divides during cytokinesis. Division results in two daughter cells identical to the original parent cell.

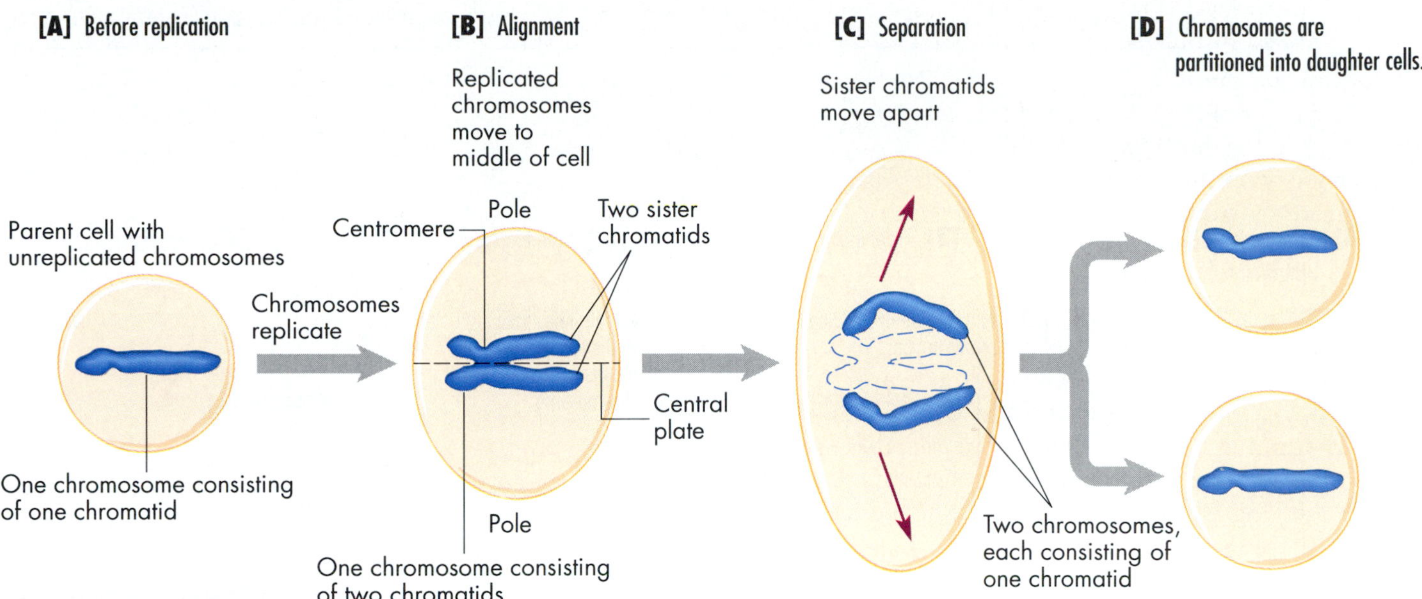

FIGURE 7.8

Chromosome Movement During Cell Division.

Each chromosome in the parent cell originally contains a single chromatid. [A] During interphase, each chromosome replicates and comes to contain two chromatids. During the rapid M phase, each chromosome carries out a series of movements. [B] First, the duplicated chromosome aligns along the cell's midline. [C] Then the chromatids separate and move toward opposite poles of the cell. [D] Finally, the chromatids—now individual chromosomes—become partitioned into separate daughter cell nuclei.

To understand mitosis, we must focus on three key processes: chromosome replication, chromosome alignment, and separation of the chromosome sets.

After a cell has divided, each chromosome is unreplicated; it consists of a single chromatid with one DNA molecule [FIGURE 7.4D]. During interphase, the DNA replicates, or doubles, and when chromosomes first appear in the M phase, they consist of two sister chromatids each with its own DNA molecule held together at the centromere [FIGURE 7.4B]. Thus, a chromosome at the beginning of mitosis consists of two chromatids, each containing a single DNA molecule. The two DNA molecules in a replicated chromosome are identical.

The replicated chromosomes undergo a series of movements that cause them to *align* in the middle of the cell [FIGURE 7.8B]. The sister chromatids then *separate* and move to opposite ends, or *poles*, of the parent cell [FIGURE 7.8C], where they can be evenly distributed to the two new daughter cells [FIGURE 7.8D]. The daughter cells are genetically identical to each other, and to the original parent cell. (Compare the parent cell in FIGURE 7.8A to the products of mitosis in FIGURE 7.8D.)

There is a key difference between basic cell replication and the reproduction of a person or a rose. When these organisms reproduce, the parent and offspring both generally continue to exist. In contrast, during cell reproduction, the parent cell [FIGURE 7.8A] ceases to exist as an entity and its parts are distributed to the two offspring cells [FIGURE 7.8D].

THE PHASES OF MITOSIS

Although mitosis is a continuous process, biologists generally divide it into distinct phases to simplify its description. The events of these phases are summarized here and illustrated and explained in more detail in FIGURE 7.9. Recall that chromosome replication has already occurred during the S phase, before mitosis begins.

Prophase In **prophase** (*pro-*, "before"), the chromosomes condense, the nucleolus disappears, and a *mitotic spindle* forms. The mitotic spindle is a weblike structure of microtubules that suspends and moves the chromosomes [FIGURE 7.9A and B]. In late prophase a stage cell biologists call **prometaphase** (*meta-*, "middle"), the nuclear envelope disappears, the spindle enters the nuclear region, and the chromosomes attach to the spindle at the centromere [FIGURE 7.9C]. Individual chromosomes jostle back and forth, as if they were involved in a tug-of-war between the two poles.

Metaphase In **metaphase**, the spindle microtubules align the chromosomes in the middle of the spindle, each chromosome lined up independently of the others along a single plane, called the metaphase plate, in the middle of the spindle [FIGURE 7.9D]. The metaphase plate is a bit like the cut surface of a grapefruit that has been sliced in half.

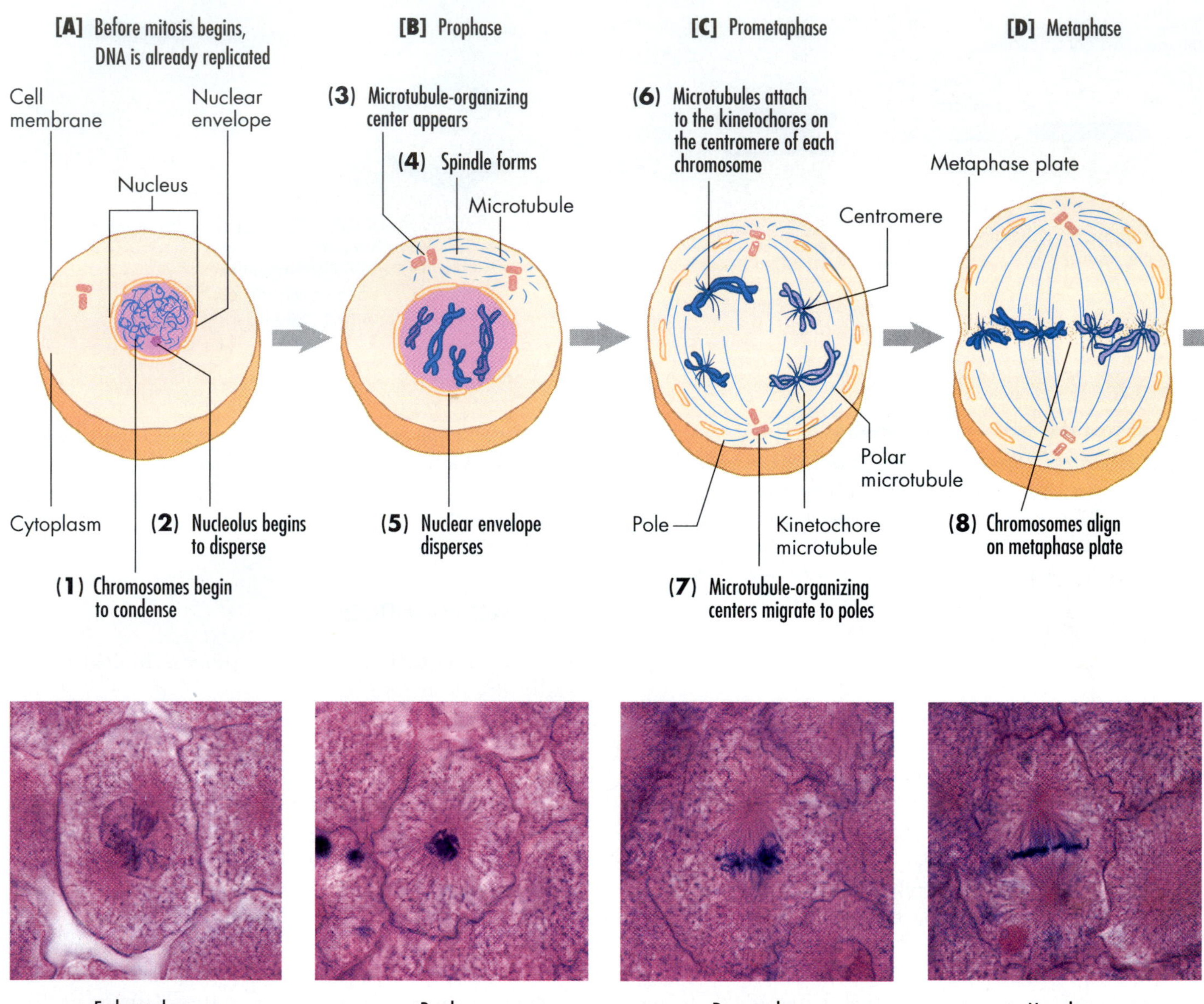

FIGURE 7.9

The Phases of Mitosis.

The cell's DNA has already replicated during interphase. As the cell enters the first part of mitosis, called prophase, the DNA changes from its diffused and tangled state during interphase to become more tightly packaged (1). Also, the nucleolus disperses (2). The microtubule-organizing centers duplicate and spin out microtubules (3). Animal cells have centrioles, or short cylinders of microtubules, at the centers of their microtubule-organizing centers, but plant cells lack centrioles. The microtubule-organizing centers separate and move toward opposite ends, or poles, of the cell, spinning out the mitotic spindle, a gossamer structure that suspends and moves the chromosomes (4). At the beginning of prometaphase, the spindle invades the nuclear region, and the nuclear envelope disperses (5). Microtubules attach to chromosomes by kinetochores, special sites located at the centromere of each chromosome (6). The chromosomes then jostle back and forth as the polar microtubules interact with the kinetochore microtubules, and the microtubule-organizing centers complete

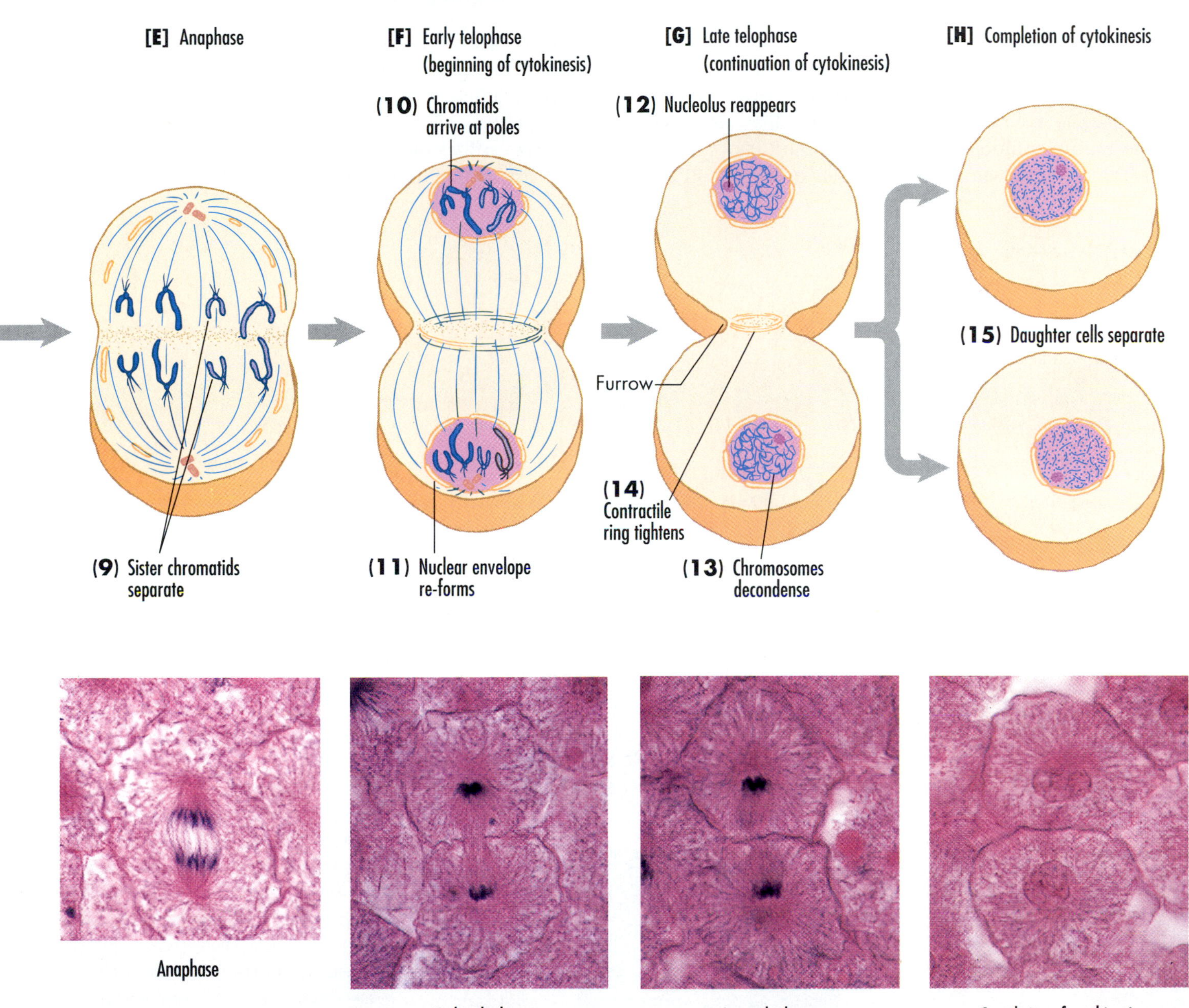

their migration to the poles (7). (Also see FIGURE 7.10). During metaphase, the chromosomes become aligned in a plane on the metaphase plate (8), a plane lying halfway between each pole, like the plane a knife makes when cutting a grapefruit in half. Next, during anaphase, the kinetochores separate, and microtubules pull sister chromatids apart, toward opposite poles (9). Early telophase marks the beginning of cytokinesis. The daughter chromatids (now independent chromosomes) arrive at each pole (10), after which the nuclear membrane re-forms around the chromosomes (11). Then the nucleolus reappears (12), the spindle dissolves, and the chromosomes reel out again into a tangled mass of DNA and protein (13). In addition, in late telophase in animal cells, a contractile ring tightens around the cell's midline where the metaphase plate had been, creating a furrow (14). Separation of the daughter cells completes cytokinesis (15). The micrographs show cells of the whitefish, magnified 450 times.

Anaphase In **anaphase** (*ana*-, "against", "opposed"), the centromeres divide and the spindle microtubules separate the chromatids (now called chromosomes) and pull them toward opposite poles [FIGURE 7.9E]. The chromosomes appear to be dragged through a viscous fluid by spindle fibers attached to the centromere. FIGURE 7.10 describes the precise role of the spindle in separating the chromatids during anaphase.

Telophase In **telophase** (*telo*-, "goal"), the chromosomes arrive at opposite poles of the cell, and the preparatory events are reversed: the nuclear envelope reappears, the spindle dissolves, and so on [FIGURE 7.9F and G]. Once telophase is over, the division of the cell nucleus (mitosis) is complete. Now the cell has two nuclei carrying identical sets of chromosomes. The M phase continues, however, as cytokinesis begins [FIGURE 7.9H].

THE CONCEPTS OF MITOSIS APPLIED: THE GENETIC EFFECTS OF RADIATION

Have you known someone with cancer who received radiation therapy? During this treatment, technicians aimed a radiation beam precisely at the patient's tumor and shielded other parts of the patient's body from the beam. How does radiation therapy work and why was it necessary to shield the patient's body?

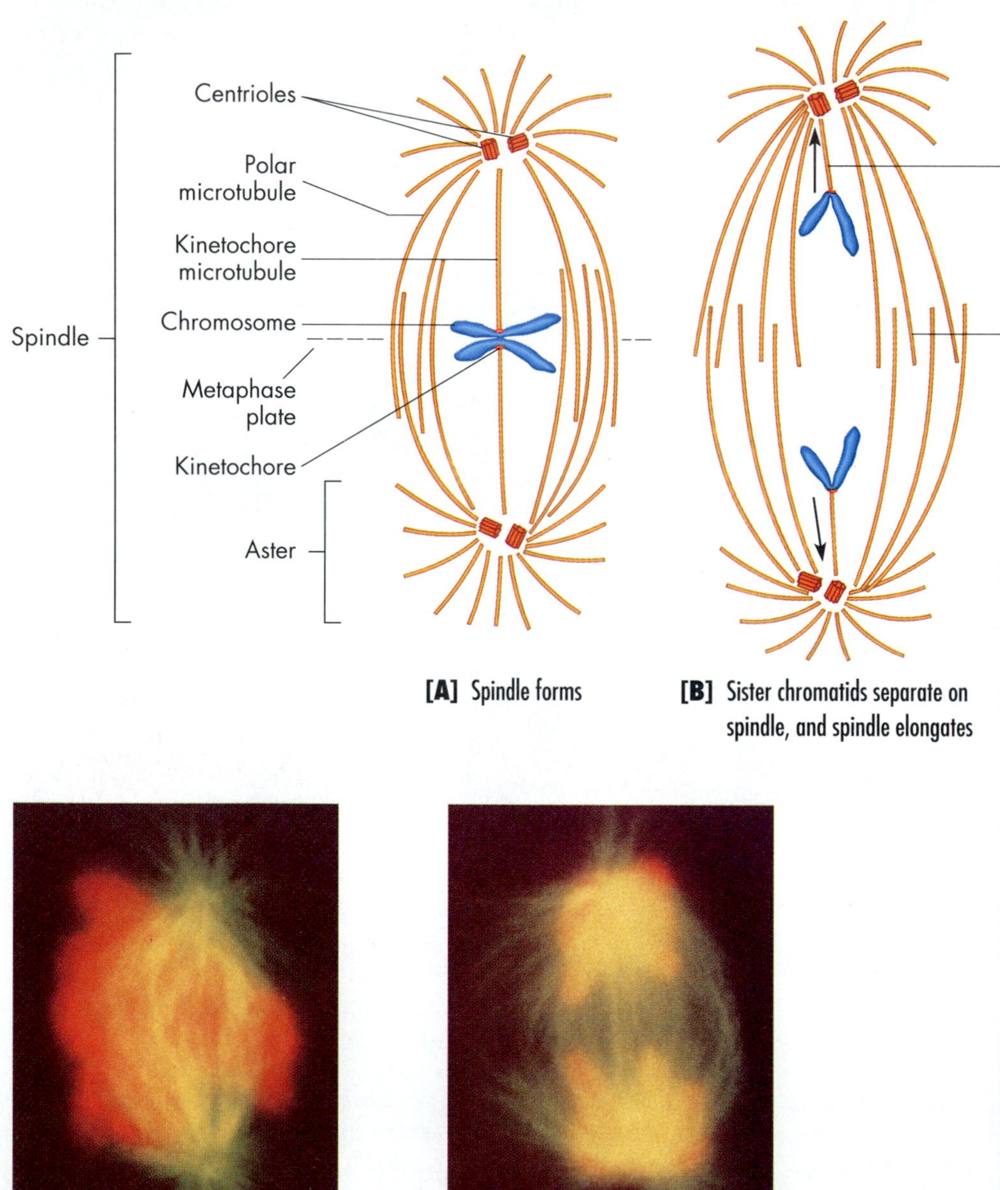

[A] Spindle forms

[B] Sister chromatids separate on spindle, and spindle elongates

[C] Metaphase

[D] Anaphase

FIGURE 7.10

What Moves the Chromosomes: Structure and Function of the Spindle.

[A] As a spindle forms in animal cells, microtubules radiate around the centrioles, the cell's microtubule-organizing center, like a star, or *aster*. Longer *polar microtubules* extend from each pole toward the other and overlap at the metaphase plate in the middle of the cell. Another set of microtubules, the *kinetochore microtubules,* lead from the centrioles at the poles to each chromatid, attaching at the kinetochore, a special group of proteins at the chromosome's centromere. **[B]** During anaphase, sister chromatids move toward the poles, and the cell lengthens. The chromosome's centromere splits, and the sister chromatids separate, moving toward opposite poles along the kinetochore microtubules. A tiny protein-based motor in the kinetochore probably runs along the microtubule like a locomotive along a train track. The microtubule then dissolves after the "locomotive" passes by. The overlapping polar microtubules slide apart, generating the pushing force that lengthens the cell in preparation for division into two cells. **[C]** This photograph of a cell in metaphase shows the spindle microtubules stained green and the chromosomes stained red and lined up at the metaphase plate. **[D]** A cell in anaphase is longer and has chromosomes much nearer the poles.

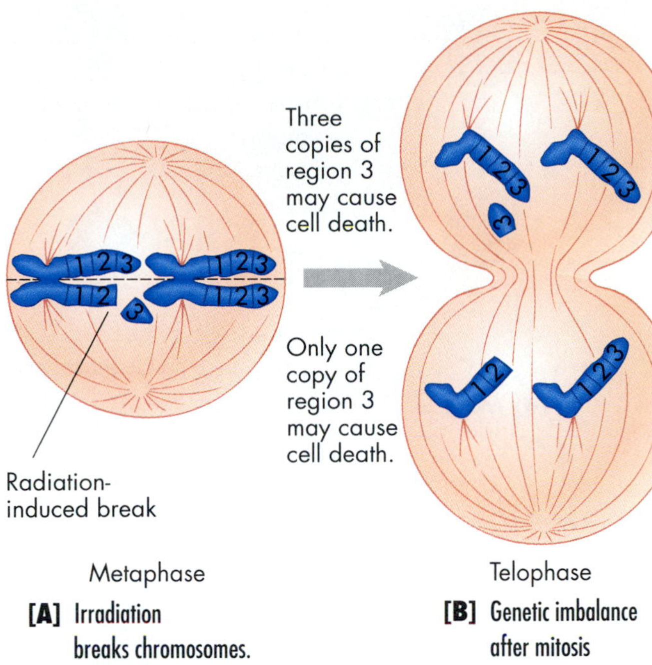

FIGURE 7.11

Mitosis in an Irradiated Cell with a Broken Chromosome Can Lead to Genetic Imbalance.

Biologists learned many decades ago that radiation can break chromosomes [FIGURE 7.11A]. If such breakage occurs in a cell that does not divide, the cell will still be able to carry on in a normal fashion because most body cells contain two copies of each chromosome, one copy from the individual's mother, the other from the father. Even if one chromosome is damaged at a single point by radiation, the intact copy will usually compensate for this.

If chromosome breakage takes place in a cell that later divides—a skin cell, for example, or a tumor cell—the consequences are often quite different, and your knowledge of mitosis can help explain why. During cell division in an irradiated cell, a broken piece of a chromosome that is no longer attached to a centromere will not be distributed normally to the daughter cells. This is because the centromere is the only part of the chromosome that is directly pulled to the cell's opposite poles during mitosis [FIGURE 7.11B]. This unequal distribution of chromosomal material can cause cells to have too few or too many chromosome parts, and this genetic imbalance can lead to the cell's death.

As we saw earlier, the most rapid cell division taking place in a cancer patient often involves the cancer cells themselves. Therefore, well-aimed radiation therapy damages those cells more than it hurts the patient's surrounding cells. Like all humans, cancer patients do have other rapidly dividing cells. These include the precursors of red and white blood cells, skin cells, and cells of the intestinal lining. Not surprisingly, these cell types can be damaged during radiation therapy, and their disruption explains some of the side effects of radiation therapy: anemia (due to too few red blood cells), susceptibility to infection (due to too few white blood cells), hair loss (due to damaged skin cells), and nausea (due to damaged intestinal cells). The disruption of these cell types also explains the effects of uncontrolled exposure to high levels of radiation, as in an accident at a nuclear power plant. These facts underscore the devastating consequences nuclear war or nuclear accidents could have for people and most other life forms on Earth. It also highlights the restorative powers of mitosis, which keep most of us healthy most of the time, as well as the application of basic biology in the treatment of serious diseases.

CONCEPT CHALLENGE

Biologists have isolated a compound from Pacific yew trees that physicians have been able to use as an effective treatment for ovarian and breast tumors. Biologists have also isolated a compound from the Madagascar periwinkle that is effective against childhood leukemias, which are forms of cancer. In both cases, the molecules act by binding to the proteins of the spindle and interfering with the spindle's separation of chromosomes in mitosis. Why would these compounds be more harmful for cancer cells than normal cells?

Cytokinesis: The Cytoplasm Divides

Toward the end of mitosis, the cytoplasm of most plant and animal cells begins to divide by means of **cytokinesis** (literally, "cell movement") [FIGURE 7.9F to H]. In both animal and plant cells, new plasma membranes form at or near the place once occupied by the chromosomes during metaphase [FIGURE 7.12] and separate the two nuclei into the two new cells. The details of cytokinesis vary because animal cells have a pliable outer surface, while plant cells have a rigid cell wall.

CYTOKINESIS IN ANIMAL CELLS

Animal cells divide from the outside in, as a circle of microfilaments called a **contractile ring** pinches each cell in two. During anaphase the constriction of the ring, which contains the contractile protein actin, creates a dent, or *furrow*, in the cell surface in much the same way that a

FIGURE 7.12

Cytokinesis in Animal and Plant Cells.

[A] Animal cells are pinched in two from near the cell surface by a contractile ring that creates a furrow and eventually separates the daughter cells. [B] A cell plate partitions a plant cell from the inside out, forming two daughters.

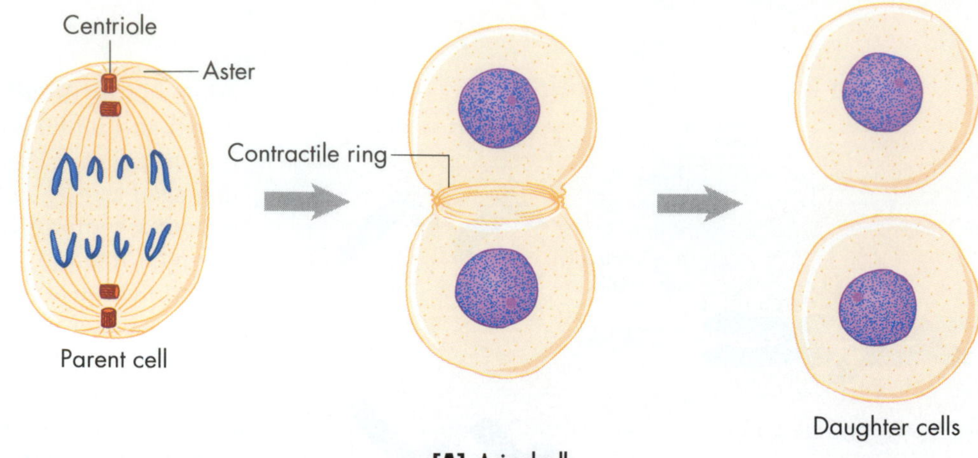

[A] Animal cell

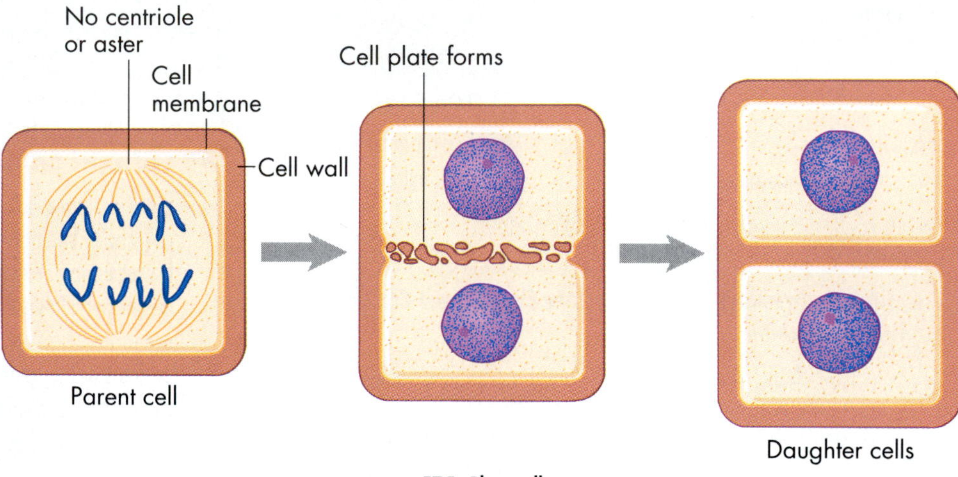

[B] Plant cell

purse string tightens around the neck of a purse [FIGURE 7.12A]. The anaphase furrow deepens, and eventually squeezes the cell in two.

CYTOKINESIS IN PLANT CELLS

Plant cells, with their rigid cell walls [review FIGURE 3.26], retain their shape throughout the cell cycle, dividing from the inside out [FIGURE 7.12B]. During telophase, vesicles filled with cell-wall precursors collect in the center of the cell, where the chromosomes had aligned during metaphase. The many separate vesicles gradually fuse, forming a central partition, or **cell plate**, made of cell-wall material sandwiched between plasma membranes. This fusion completes the central partition and divides the plant cell into two identical daughter cells, which remain connected [FIGURE 7.12B]. Each cell now has its own nucleus and is ready to begin interphase.

CONCEPT CHALLENGE

Consider a cell within a maturing pollen grain of a flowering plant. This cell undergoes not one but two divisions of the nucleus by mitosis without cytokinesis. How many offspring cells will result and how many nuclei will each contain?

Regulating the Cell Cycle

Viewed collectively, the events of the cell cycle ensure that a cell grows and that its hereditary material is distributed equally into two offspring cells. Some cells, however, such as normal liver or bark cells, cycle slowly. Other cells, like the cells that line a person's intestines or the cells at the tip of a plant's growing roots, cycle rapidly. And some cells do not cycle at all, including most cells within an adult person's brain or the cells within a tree trunk that transport sugar. Sometimes, as in a benign or cancerous tumor, the cells continue to divide out of control. Clearly, the cell cycle must be regulated—and it sometimes escapes that normal regulation. Biologists have performed numerous experiments to learn how regulation takes place, and how it can go awry.

EXTERNAL FACTORS REGULATING THE CELL CYCLE

The cowboy we saw at the beginning of the chapter is constantly suffering very small injuries to his face and hands—dry chapped skin from wind burn; scrapes from

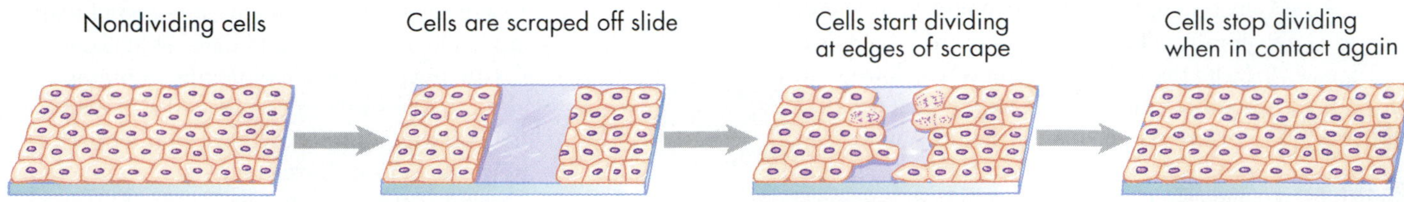

[A] Normal cells: Contact inhibition of division and movement

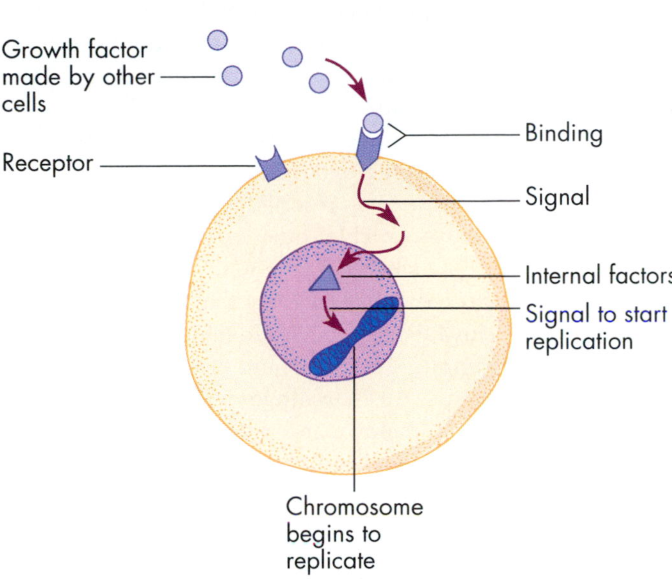

[B] External and internal factors regulate the cell cycle

[C] Wound without growth factor

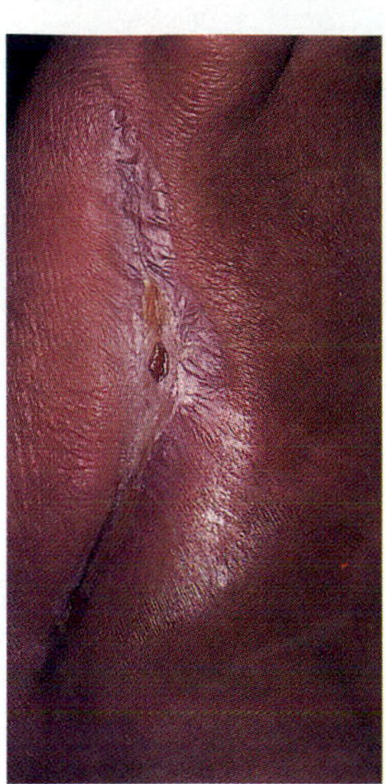

[D] Wound with growth factor

FIGURE 7.13

Contact Inhibition: Cells Stop Growing When They Contact Other Cells.

[A] Cells will grow and divide only until they cover a slide or dish with a layer one cell thick. The contact of other cells at their edges inhibits further division. If a swath is cut through a layer of nondividing cells (as during a wound), cells along the edge will grow and divide until they are once more in contact with their neighbors. **[B]** External factors (including growth factors made by other cells) can stimulate internal factors within the cell's cytoplasm and nucleus. This can trigger DNA replication, which leads to another round of cell division. **[C, D]** Growth factors can stimulate cell divisions and wound healing. The patient in these photos suffered chemical burns that remained unhealed for several months. Treatment with epidermal growth factor brought about rapid healing.

leather straps; minor cuts from barbed wire and badly aimed nails. Following each incident, his skin cells divide, repairing the damaged tissue. But once the wound is healed, the skin cells stop dividing, and this prevents the proliferation of an unorganized mass of skin tissue. These starts and stops of cell division are controlled by factors in the cell's outside environment interacting with factors inside the cell. Let's see how.

Cell Contact and Cell Division Dividing cells at the edge of a wound behave somewhat like people entering an elevator: Just as the riders form a layer one person high and do not climb onto each other's shoulders, cells at the edge of a wound divide rapidly, grow inward from all directions, close the gap, and stop dividing when they touch other cells at the center of the wound [FIGURE 7.13A]. These cells are exhibiting **contact inhibition** of cell division and cell migration, which means that the cells at the edges of a wound stop growing and migrating when they contact other cells and close the gap caused by the wound. BOX 7.1 on page 180 describes one life-saving application of this concept, using contact inhibition to create artificial skin. Biologists do not yet know exactly how physical contact with other cells inhibits cell division, but they do know that the rate of cell division is influenced both by growth factors external to the cell and internal regulatory factors.

Growth Factors and Cell Division Just as the external physical factor of cell contact can control when and if a cell divides, external chemical factors can also affect cell division. Proteins in a cell's environment called **growth factors** can enhance the growth and division of specific cell types [FIGURE 7.13B]. At the site of a cut or other

box 7.1
Biology Applied

Cell Cultures and Artificial Skin

Being burned is among the greatest traumas a person can experience. And it is woefully common: More than 150,000 Americans are admitted to hospital burn units each year. More than half have injuries so severe they require skin grafts or transplants from other surface regions. Unfortunately, more than one-third of the patients who require grafts have such extensive wounds that the unburned areas are too small to harvest grafts for the burned parts. Until now, doctors have had few alternatives for treating these patients, and in many cases, they have been left with disfiguring scars and restricted movement.

Researchers, mostly at private corporations, have taken on the challenge of growing artificial skin for burn patients by applying their knowledge of cell division and techniques for growing cells in large batches. Their new products, currently being tested in hospital trials, hold real promise for reducing the pain, the costly multiple surgeries, and the terrible scarring of serious burns [FIGURE 1].

Our skin covers nearly 2 square meters (18 square feet) of body surface, and is our largest organ. Yet its structure is simple—basically layers of cells with a few embedded glands and hair follicles, serviced by nerve endings and blood vessels. As FIGURE 2 shows, the *epidermis* is made up of four or five layers of flattened cells (depending on the skin's location on the body). The outermost layers are dead and *keratinized* or horny and dry, and the inner layers are still dividing and pushing new cells toward the surface.

The *dermis* is a strong, flexible meshwork containing three main types of fibers secreted by *fibroblast* cells. Some of the secreted fibers are made of collagen, and make skin tough. Elastic fibers are made of flexible proteins and provide stretchiness. So-called reticular fibers are more rigid and form a supporting network. Serious burns penetrate to and often destroy expanses of dermis. Without this underlying zone, as a burn heals, the area usually contracts tightly and this leads to scarring and immobility of the burned areas. When a physician can graft dermis and overlying epidermis from a patient's unburned skin, the wound can heal more normally. But when the burns are too extensive, doctors can't find enough on other parts of the body. Then, in order to cover the wound, physicians have often resorted to grafting skin from the only available sources—cadavers or fetal pigs—to cover the wound. These are better than no dermis at all, but they have several disadvantages: They pose the risk of triggering immune rejection and of introducing hepatitis, AIDS, or other viruses; they require eventual removal and repeat surgeries; and pigmentation can be missing (in fetal pig tissue) or different from the patient's.

Ideally, physicians would be able to do something very different: They would remove a small patch of a burn patient's undisturbed dermis and epidermis; they would grow big, healthy sheets of a person's own skin cells quickly; then they would graft it onto the burned areas permanently. This best of all possible solutions lies many years—perhaps

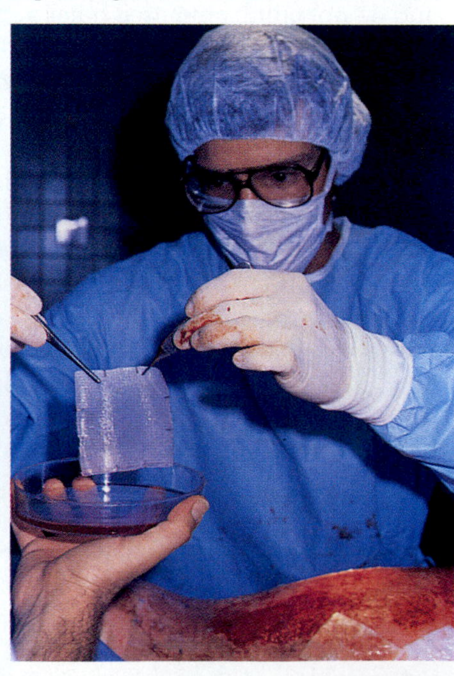

FIGURE 1
Surgeon holds square of human skin cultured on an artificial meshwork, and prepares to apply it to the burned flesh of a fire victim.

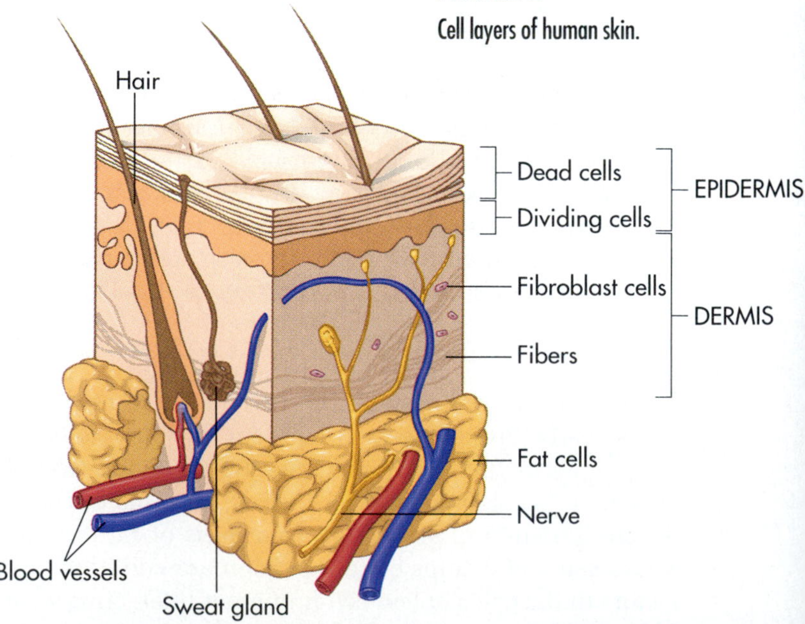

FIGURE 2
Cell layers of human skin.

decades—away. But in the meantime, researchers are developing some impressive substitutes that employ a knowledge of cell division as well as modern medical materials.

As we saw on page 179, cells exhibit contact inhibition: cells growing on a flat surface stop dividing when their outer margins contact other adjacent cells. They thus form a sheet just one cell thick—generally too thin and fragile to pick up, manipulate, and transplant. Knowing this, researchers at MarrowTech, Inc., of La Jolla, California, devised an ingenious meshwork made up of the same materials used in surgical suturing threads. They obtain human dermal fibroblasts from the foreskins of newborn baby boys who have undergone circumcision, and then culture these fibroblast cells upon the meshwork. The cells attach to the meshwork, grow in a layer 15 to 20 cells thick, and secrete their own supportive matrix of collagen that helps cement the entire culture into a solid tissue. Because a burn patient's body does not reject dermal cells from newborns, physicians can graft this artificial human dermis (DermaGraft) onto a patient's skin wound without fear of rejection or infection [FIGURE 1].

Devising a replacement for the outer skin, the epidermis, has been more complicated because a host's body will readily reject epidermis transplanted from a donor. To circumvent this problem, researchers at Biosurface Technology, Inc., in Cambridge, Massachusetts, have learned to culture a burn patient's own epidermal cells; in about 3 weeks enough epidermis to cover the whole body is generated.

Using these two new techniques together, a physician could first graft on artificial dermis, then, after a few weeks, overlay the burn with the patient's own cultured epidermis. In theory, the patient's body would accept the combination, which would not have to be replaced later, thus avoiding the pain and expense of repeat surgeries. Equally important, the transplanted human dermal layer would protect the wound, give an anchor for the growth of new epidermis, and help prevent contraction, severe scarring, and inflexibility of the burned area. Hospital tests on both products are under way, and the results look positive. Clearly, the modern biologist's knowledge of cell division and cell culture is helping save skins and lives.

wound, growth factors are released from broken cell parts. The factors diffuse locally and stimulate division in the cells that line damaged blood vessels and in other nearby cells. At the wound site, growth factors bind to special receptors embedded in the plasma membrane, triggering cellular events that cause the targeted cells to divide. As a result, the tear in the skin is eventually filled in with new cells.

Although growth factors are normally released in very small amounts, researchers have used biotechnology to produce large quantities, which physicians can use to treat injured patients. In Oklahoma, for example, a man who was repairing an industrial refrigeration system accidentally splashed ammonia into his eye. After three weeks of conventional therapy, his cornea (the transparent protective covering of the eye) had not healed, and he could see only blurs and shadows with the injured eye. Then clinical molecular biologists at the University of Oklahoma treated his eye with drops containing epidermal growth factor. In just 4 days, the cornea had healed, and a few weeks later his vision had returned to nearly normal. Intensive research continues in this fascinating field [see FIGURE 7.13C].

INTERNAL FACTORS REGULATING THE CELL CYCLE

While cell contact and growth factors are external factors that influence cell division, other internal regulatory factors function from within the cell itself. Recent studies have shown that the cell cycle is regulated by at least two internal proteins. Apparently, each step in the cycle causes the next step to occur, and interestingly, the same proteins act in the cells of many different kinds of organisms, a reflection of their common evolutionary origin.

In the early 1990s a wave of excitement spread throughout the community of cell-cycle researchers when they realized that the same two types of internal regulatory proteins appear to control the cell cycle in eukaryotes as different as yeasts, fruit flies, and frog embryos. The researchers knew that if they could understand how these internal proteins regulate the cell cycle, they might be able to design new therapies to help heal patients' wounds.

Cell biologists are still working out the details of cell-cycle control, but it is already evident that external regulatory factors—cell contact and growth factors—act on the cell's internal mechanisms, triggering events inside the cell that influence cell division. When these fine-tuned controls break down, as they do occasionally, the result can be cancer.

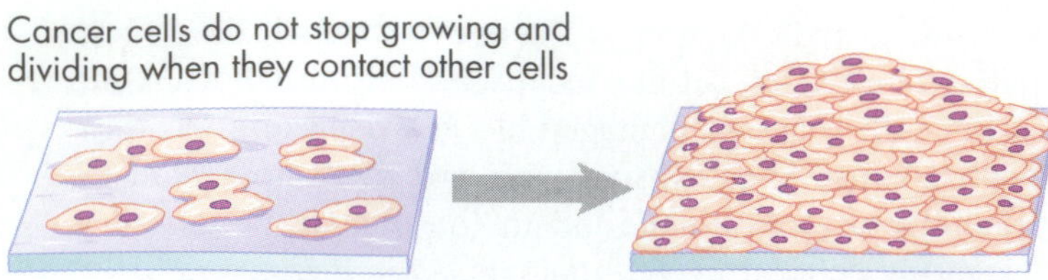

FIGURE 7.14

Cancer Cells Do Not Respond to Contact Inhibition.

In cancer cells, the contact of neighboring cells does not inhibit growth; hence, cancer cells can form masses (tumors) that can invade healthy tissue.

CANCER: CELL-CYCLE REGULATION GONE AWRY

Through the years, and especially since World War II, the life-threatening group of diseases called **cancer** has been on the rise. Today, cancer causes about one of every three deaths in the United States. Cancer occurs when the strict regulation of the cell cycle goes out of control for some reason. Cancer cells don't necessarily grow and divide faster than normal cells. Instead, they grow and divide unaffected by the usual regulatory factors [FIGURE 7.14]. Unlike the normal cells we saw earlier, cancer cells do not stop moving when they contact other cells, and as a result, they invade healthy tissues and multiply into masses called **tumors**.

Recent evidence indicates that most cancers are related to changes in a cell's DNA. Many such changes are caused by substances in the environment, such as chemicals, radiation, the tar in cigarettes, or viruses that alter the structure of DNA. Some of these changes to DNA can affect either growth factors or receptors for growth factors on the plasma membrane. Some cancer cells may produce both the growth factor and its receptor so that the cells are constantly stimulating themselves to divide. Some cancer-causing genes called **oncogenes** may disrupt the normal activities of regulatory proteins that work inside cells. There is still much uncertainty about the causes of cancer, but it is clear that the answers will be found in the regulation of the cell cycle.

CONCEPT CHALLENGE

Researchers have determined that in some types of breast cancer, the hormone called estrogen binds to special receptors on the plasma membranes of the cancerous cells, and the estrogen stimulates factors within the cells that regulate cell division. Reexamine FIGURE 7.13B and label the components that correspond to estrogen and the estrogen receptor. Suggest a treatment for this type of cancer.

Life Cycles of Multicellular Organisms

For many single-celled organisms, the cell cycle is the same as the life cycle: Each cell division normally reproduces the entire organism, producing genetically identical offspring. Most multicellular organisms, however, are too complex for that. The male and female parents can't simply split down the middle and somehow reorganize their cells into half-size versions of themselves. Instead, the parents provide specialized sex cells that unite into a single new cell that begins to grow and divide repeatedly. The cells in the dividing cluster soon take on various specialized roles, finally developing into a new individual.

Life cycles are more complicated than cell cycles, and the reproductive stage of a life cycle can follow one of two very different strategies: asexual or sexual reproduction.

ASEXUAL REPRODUCTION: IDENTICAL OFFSPRING FROM ONE PARENT

If you have ever left a potato in the pantry too long, you have undoubtedly witnessed the beginnings of asexual reproduction. Cells in the potato's eyes (buds of tan tissue on the surface of the potato) begin to reproduce by mitosis, drawing energy from the starch inside the potato and developing stems and leaves. If you remove one of the growing eyes along with a chunk of the potato and put it in soil, and conditions are right, your plant will photosynthesize, grow, and become a plant genetically identical to the ones you left in the pantry. Potato farmers, in fact, usually plant sprouted eyes instead of potato seeds.

In such **asexual reproduction** (also called *vegetative reproduction*), a new, genetically identical offspring grows directly from one or more body cells of a single parent. Farmers and gardeners often induce asexual reproduction artificially to duplicate a prize-winning specimen. Using the technique described in FIGURE 7.15, a gardener can make a cutting that will grow into a new plant with features identical to the parent plant. More intricate methods enable biologists to grow laboratory forests of identical miniature trees.

Asexual reproduction is by no means limited to experiments in the laboratory or the field. It is a common survival strategy in the natural world. Strawberries, for example, send out **runners**, special arching stems that reach the ground a short distance from the parent plant, take root, and develop stems and leaves. In Southeast Asia, all the bamboo plants in a forest are genetically

[A] 

 [B]

FIGURE 7.15

Asexual Reproduction in Plants.

To make a cutting from a prize fuchsia, a gardener cuts off the new growth at the end of a young branch, dips the cut stem in a plant hormone that encourages rapid mitosis of cells at the cut surface [**A**], and sticks the stem into potting soil [**B**]. In a few weeks, roots develop from the stem and establish the new plant. Since the cutting is genetically identical to the original plant, the eventual bush will have flowers the same showy size and color as the favorite original fuchsia.

FIGURE 7.16

Asexual Reproduction in Animals.

The large orange arm of this sea star is regenerating an entire new individual, including the six smaller orange arms.

identical, all derived asexually from a single founder plant. This is a helpful evolutionary strategy, since an area that is suitable for the parent is probably also suitable for its identical offspring.

Certain animals also reproduce asexually. The hydra, a relative of the jellyfish, reproduces by **budding**. Cells in special regions of the body undergo rapid mitosis and become organized into new hydras that bud off from the parent. Another method of asexual reproduction is **regeneration**. For example, if a sea star like the orange one in FIGURE 7.16 is cut in half, it can regenerate the missing portion from each half, producing two sea stars in place of the single parent. Many other animals are also capable of regeneration, but many plants and animals that usually reproduce asexually can switch to sexual reproduction under specific environmental conditions.

SEXUAL REPRODUCTION: OFFSPRING FROM FUSED GAMETES

Although asexual reproduction can produce new individuals, the life cycles of most eukaryotes involve **sexual reproduction**, in which parents (usually two, but sometimes one) generate specialized sex cells called **gametes** [FIGURE 7.17]. When gametes from mates (usually male and female) fuse, they set into motion the life of a new individual.

Consider sexual reproduction and the life cycle of the roundworm *Ascaris* [FIGURE 7.17]. As with most plants and animals, its female gametes are large, immobile cells called **eggs**, while the male gametes are small motile cells called **sperm**, which can move or be carried from the male to the egg [FIGURE 7.17A]. Eggs and sperm are usually produced in specialized organs. In flowering plants, they are produced by structures in the flowers; in animals, eggs are made by the *ovaries* and sperm are made by the *testes*.

In many organisms, including human beings and ginkgo trees, each individual produces just one kind of gamete, either egg or sperm. However, in pear trees, roundworms, and many other species, each adult individual can produce both types of gametes.

The fusion of egg and sperm, or **fertilization**, results in a single cell called the **zygote** [FIGURE 7.17B]. In the zygote, the hereditary information from both parents unites, creating a genetically unique combination of genes and chromosomes. The single-celled zygote undergoes *development*, usually a period of rapid mitotic division accompanied by cellular specialization [FIGURE 7.17C]. During this early developmental period, an immature form emerges, continues to grow, and eventually changes into a mature adult [FIGURE 7.17D].

In animals, eggs and sperm come from specialized cells called **germ cells** [FIGURE 7.17D]. As you have seen, the body cells, or **somatic cells**, of a multicellular organism eventually die. However, if an organism's egg or sperm cells unite with those of another individual of the same species, then at least some portion of the organism's genetic material lives on in the offspring even though the somatic cells die. An organism completes its life cycle by undergoing a special type of cell division called meiosis, which produces reproductive cells that may lead to a new generation [FIGURE 7.17E].

CONCEPT CHALLENGE

Where in your body are cells dividing right now by mitosis? Where in your body are cells dividing by meiosis?

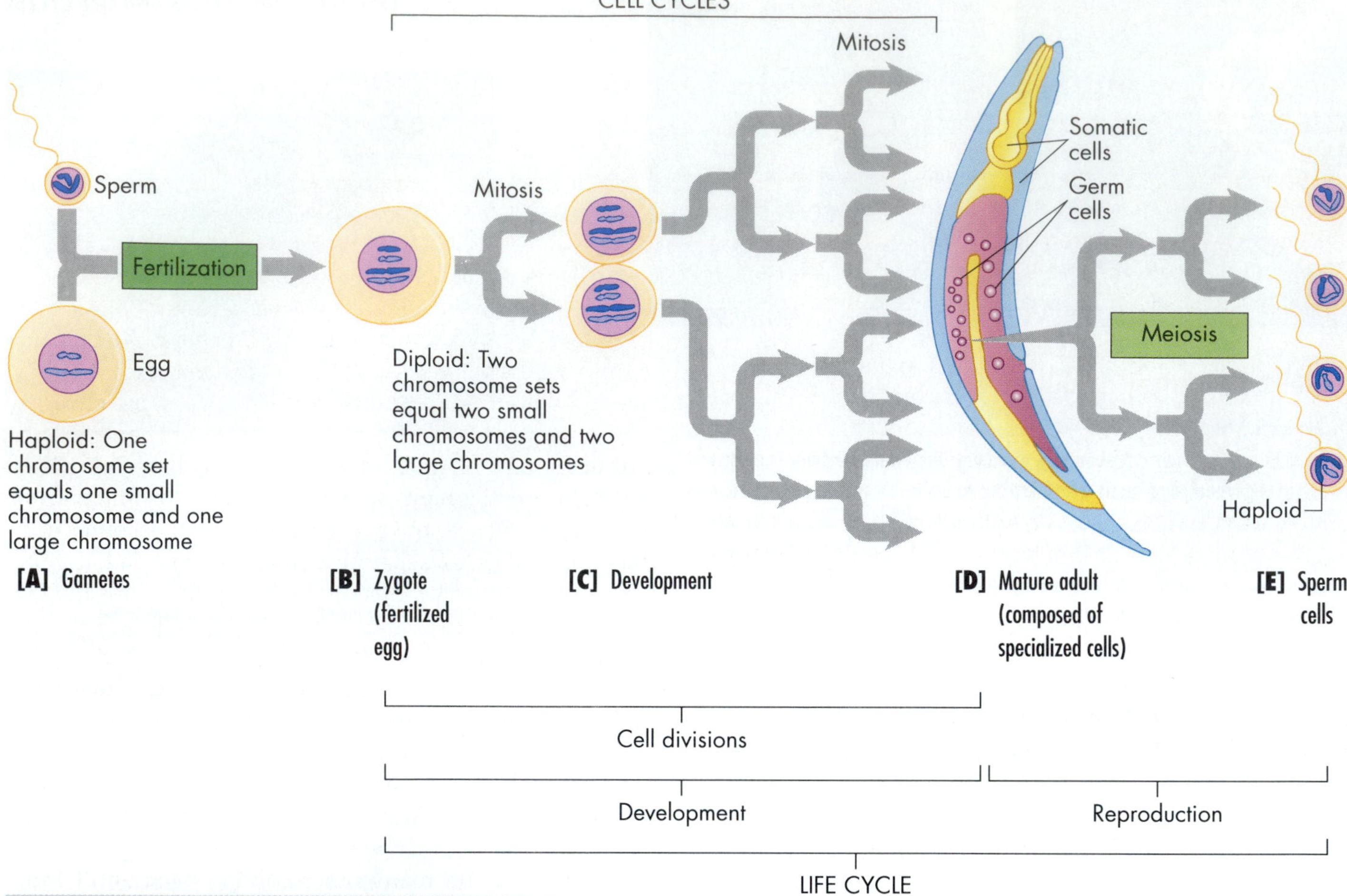

FIGURE 7.17

Chromosomes and the Life Cycle: Changes in Chromosome Number During Fertilization and Meiosis.

[A] Gametes (egg and sperm) are haploid, with just a single set of chromosomes, two in this example. [B] A zygote, formed at fertilization by the union of egg and sperm, represents a new genetic combination with the normal diploid (2*n*) chromosome number. [C] Cell division by mitosis gives rise to an immature form. [D] Further mitotic divisions and cell specializations lead to the mature adult—here, the roundworm *Ascaris*, a parasite that causes diarrhea and other digestive problems in humans. [E] Meiosis takes place in gamete-forming organs and results in the generation of eggs or sperm with the haploid (1*n*) chromosome number.

Meiosis: Halving the Chromosome Number

Why would an organism need a special type of cell division—meiosis—before producing gametes? A good case study is the parasitic roundworm *Ascaris*. Each *Ascaris* egg contains two chromosomes, and each sperm also contains two chromosomes [FIGURE 7.17A]. Therefore, zygotes formed when sperm fertilize eggs will have four chromosomes, and as the organism develops, all of its cells will contain four chromosomes [FIGURE 7.17B and C]. Now if these individuals with their four-chromosome cells made eggs or sperm that contained four, not two chromosomes, what would happen during reproduction? The next generation would have cells containing 8 chromosomes, the next 16, and so on. As we have seen, however, the number of chromosomes in a species is constant and does not increase over time. A special type of cell division called meiosis prohibits this runaway increase in chromosome number. Meiosis ensures that gametes contain half as many chromosomes as normal body cells. Thus, when the gametes unite at fertilization, the original parental chromosome number is restored, not doubled.

Meiosis (literally, "to make smaller") is a special type of cell division that produces gametes or spores, specialized cells whose chromosome number is half that of other body cells [see FIGURE 7.17D and E].

In terms of chromosome number, fertilization and meiosis play opposite roles in a life cycle involving sexual reproduction. Fertilization doubles the chromosome number, while meiosis divides it in *half*. As we will see later in the chapter, meiosis also increases genetic variation, and this increase is a precondition for evolution.

CHROMOSOME NUMBER AND HOMOLOGOUS PAIRS

Before we can investigate how meiosis reduces the number of chromosomes, we must understand how many chromosomes each cell contains at different phases of the life cycle. As we have seen, the body cells of each species have a characteristic number of chromosomes: human beings have 46, chimpanzees 48, cows 60, houseflies 12, onion plants 16, cotton plants 52, tomato plants 24, potato plants 48. What do all these chromosome numbers have in common? They are all even numbers. This is because in the body cells of eukaryotes, chromosomes are present in pairs. Such chromosome pairs are called **homologous chromosomes**. A cell that has just *one member* of each homologous pair in its nucleus is **haploid** [FIGURE 7.17A], while a cell that has two members of each homologous pair is **diploid** [FIGURE 7.17B].

Thus, a gamete from *Ascaris* with unpaired chromosomes (one long, one short) is haploid [FIGURE 7.17A], whereas body cells in which the chromosomes are paired are diploid [FIGURE 7.17B]. In an individual with diploid cells, one member of each homologous chromosome pair comes from the female parent (light blue in FIGURE 7.17), while the other member comes from the male parent (dark blue). A human egg or sperm has 23 chromosomes (the *haploid* number); a human body cell has 46 chromosomes (the *diploid* number). Haploid cells and diploid cells can both divide by mitosis, but only diploid cells can divide by meiosis. A person conceived without exactly 46 chromosomes can have very serious problems. One such consequence is Down syndrome, which results from an error in meiosis [see BOX 7.2 on page 186].

MEIOSIS I

The change from diploid to haploid involves two sequential cell divisions called meiosis I and meiosis II. In **meiosis I**, a diploid parent cell divides, forming two haploid daughters. In **meiosis II**, those two daughter cells divide again, resulting in four haploid cells. As in mitosis, chromosome *replication*, *alignment*, and *separation* are our keys to the understanding of meiosis. FIGURE 7.18 focuses on chromosome movements during meiosis. Be sure that you thoroughly understand how chromosomes move during meiosis before you investigate FIGURE 7.19, which illustrates the cellular details of meiosis.

For purposes of illustration, let's look at FIGURE 7.18 to track chromosome movements during meiosis of a diploid cell containing three pairs of chromosomes—one long pair, one intermediate pair, and one short pair [FIGURE 7.18A]. Each pair consists of two homologous chromosomes; for example, the two long chromosomes with a centromere in the middle in FIGURE 7.18A are homologous chromosomes. In the interphase that precedes meiosis I, the chromosomes replicate, as they do in mitosis. Thus, when the chromosomes become visible during prophase, they each reveal two chromatids [FIGURE 7.18B and C].

The pattern of chromosome alignment in meiosis I is striking and shows clearly how it differs from mitosis. In meiosis I, the two chromosomes of each homologous pair come together, with the two long chromosomes, the two intermediate chromosomes, and the two short chromosomes each side by side [FIGURE 7.18C]. Then pairs of chromosomes align independently on the same plane within the cell [FIGURE 7.18D]. This chromosome pairing and alignment are key to how a diploid cell becomes haploid because in the separation phase of meiosis I, homologous chromosomes separate and enter the two daughter cells [FIGURE 7.18E].

Compare the chromosome number of each daughter cell in FIGURE 7.18E with that of the parent cell after chromosome replication in FIGURE 7.18B. Each daughter cell contains three chromosomes, each chromosome consisting of two chromatids joined at the centromere. (Remember, a chromosome has a single centromere.) The parent cell after replication, however, had six chromosomes, each consisting of two chromatids [FIGURE 7.18B]. Clearly, meiosis I has reduced the number of chromosomes within a cell from the diploid number (6) to the haploid number (3). Although the chromosome number is now haploid, each chromosome is still made up of two chromatids. In meiosis II, a further division results in each chromosome containing just a single chromatid.

MEIOSIS II

In the interphase following meiosis I and preceding meiosis II, there is no chromosome replication [FIGURE 7.18F]. That is why all the chromosomes remain as they were after meiosis I, with two chromatids each. When the chromosomes align in meiosis II, they line up independently on the central cell plate, as they do in mitosis [FIGURE 7.18G]. During the separation phase of meiosis II, sister chromatids separate, producing four haploid cells, each with three single-chromatid chromosomes [FIGURE 7.18H]. These cells can then become eggs or sperm, or may continue dividing by mitosis, depending upon the species.

To summarize the key features of meiosis, focus on replication, alignment, and separation in FIGURE 7.18. In meiosis I, chromosomes have already replicated; during alignment, homologous chromosomes pair up; and finally, homologous chromosomes separate, yielding two haploid cells containing replicated (or two-chromatid)

box 7.2
Human Impact

Down Syndrome, Meiosis, and Development

Like 1 in every 1000 children born in the United States, the boy in FIGURE 1 has Down syndrome, a condition characterized by lower than average IQ; heart malformation; eyelid folds; short stature; foreshortened hands and feet; and palm prints with a unique crease. In searching for the cause of Down syndrome, researchers observed that a woman over 45 years old is 100 times more likely to have a baby with Down syndrome than a woman of 19.

When researchers closely examined cells from children with Down syndrome, they found 47 chromosomes in the nucleus of each cell rather than the normal human complement of 46. The cells had three copies of chromosome 21; for this reason, the syndrome is sometimes called *trisomy 21* (literally, "three copies of chromosome 21"). Geneticists traced this extra chromosome to an occurrence during meiosis that could be linked to the mother's age.

Long before a woman reaches maturity—in fact, at the time of her own birth—she possesses reproductive organs (ovaries) that contain all the egg cell precursors (oocytes) she will ever have. Each precursor is arrested in prophase I, the first phase of meiosis, and can remain that way for many decades until a hormone signal causes the egg to resume meiosis in preparation for fertilization. Somehow, during that long stage of arrested meiosis, an egg is occasionally damaged in a way that prevents meiosis from occurring properly when it resumes. The older a woman gets, the more likely this damage becomes.

One possible result of the damage is failure of the homologues to separate, or disjoin, an occurrence known as **nondisjunction**. The egg would then contain two copies of chromosome 21. When fertilized by a normal sperm with one copy of chromosome 21, the zygote would then have three copies of chromosome 21 (trisomy 21). The embryo with trisomy 21 has a gene imbalance in each cell, and this alters normal development and leads to the characteristic birth defects seen in Down syndrome.

Unraveling the cause of Down syndrome was an important scientific achievement. Once researchers revealed the age connection, women could decrease their likelihood of producing a Down syndrome child in two ways: by planning to complete their families before age 40 and by seeking specific tests during pregnancy that reveal if a fetus has chromosomal defects [see CHAPTER 12].

FIGURE 1
A child with Down Syndrome.

FIGURE 7.18
Summary of Major Events in Meiosis.
During meiosis I, chromosomes replicate, homologous chromosomes pair and cross over, and the homologues separate (**A–E**). During meiosis II, a chromosome's two chromatids separate, and each gamete (germ cell) receives a haploid set (**F–H**).

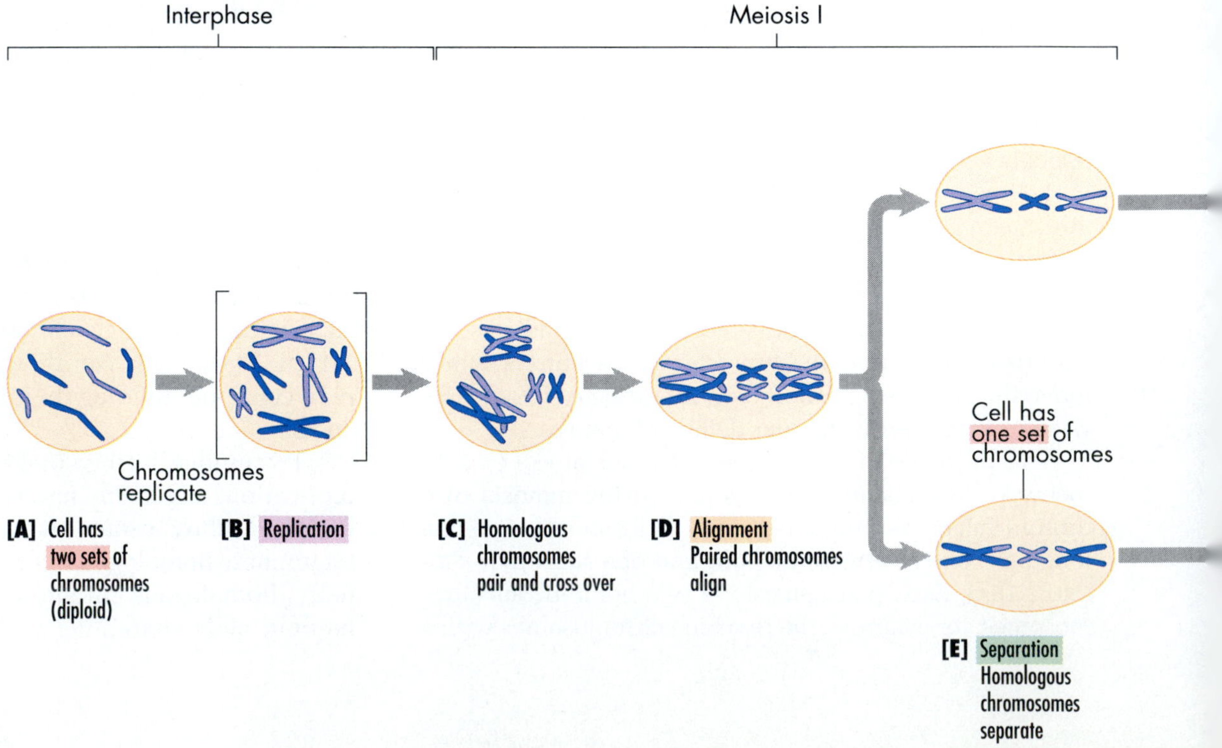

Interphase

Meiosis I

Chromosomes replicate

[A] Cell has two sets of chromosomes (diploid)

[B] Replication

[C] Homologous chromosomes pair and cross over

[D] Alignment
Paired chromosomes align

Cell has one set of chromosomes

[E] Separation
Homologous chromosomes separate

chromosomes. In meiosis II, there is *no* chromosome replication; individual chromosomes align independently; and, finally, chromatids separate, yielding four haploid cells containing unreplicated chromosomes (with only one chromatid). Cellular aspects of meiosis are shown and described in more detail in FIGURE 7.19.

ORIGIN OF GENETIC VARIATION DURING MEIOSIS

The cell division of a body cell by mitosis leads to two genetically identical daughter cells. However, cell division by meiosis of a gamete-producing cell leads to four eggs or sperm that are only genetically *similar*, not identical. To get the idea, think of one set of identical twins and one set of fraternal twins that you may be acquainted with. The identical twins arose from two cells that were the mitotic offspring of a single cell, the zygote. In contrast, the fraternal twins arose from two different cells that resulted from division by meiosis, with its shuffling of genetic information. The important message here is that a *mitotic* division, which can occur in either a diploid or a haploid cell, leads to two cells that are genetically identical to each other and to the previously existing parental cell. In contrast, a *meiotic* division converts a diploid cell to four genetically similar but not identical haploid cells.

Recall that each of your body cells contains two copies of 23 different chromosomes, making a total of 46. Each chromosome contains different hereditary information. Take chromosome 8, for example. You have a pair of these homologous chromosomes, one chromosome from your mother and one from your father. The two homologous chromosomes have similar but not identical genetic information. Both, for example, contain a hereditary factor that controls your blood type. The copy of chromosome 8 from your mother might bear the factor for blood type A, while chromosome 8 from your father might bear the factor for blood type B (making your blood type AB). Similar differences could exist for many of the other hereditary factors on chromosome 8. During meiosis, a reshuffling of the maternal and paternal chromosomes leads to new genetic combinations. The reshuffling is called **genetic recombination**, and it occurs in two ways: crossing over and independent assortment.

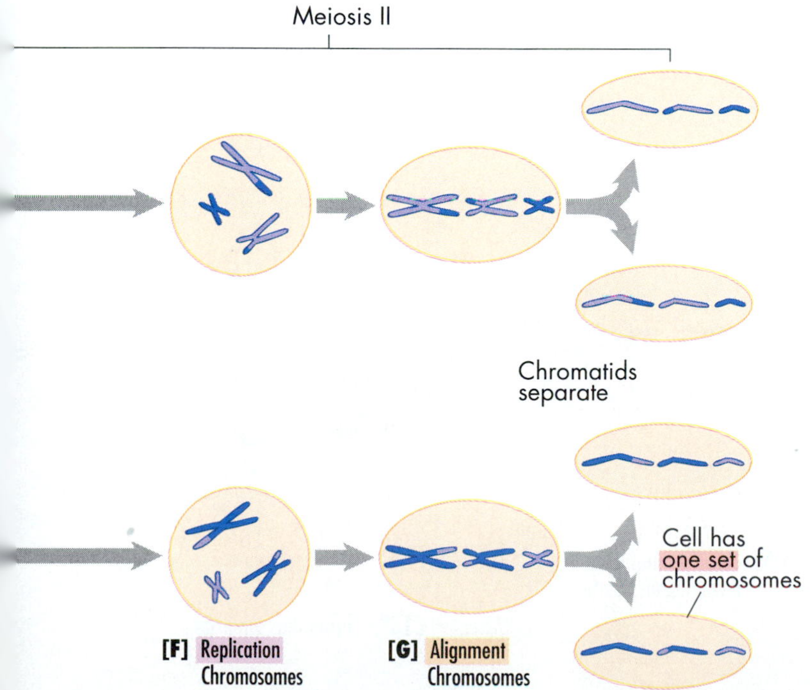

Crossing Over: An Exchange of Chromosome Parts

Crossing over occurs during meiosis I, when homologous maternal and paternal chromosomes pair [FIGURE 7.20A]. While the chromosomes are paired, enzymes randomly break the DNA molecule in each homologue, switch corresponding regions of each chromosome (the actual crossing over), and then attach them to the new chromosome [FIGURE 7.20B]. After meiosis II, individual haploid cells can have chromosomes of mixed ancestry, as the dark and light blue colors in FIGURE 7.20C indicate. Note that in FIGURE 7.20C each haploid cell is genetically unique.

Independent Assortment: Chromosome Pairs Align Randomly A second meiotic mechanism contributes to genetic variety in the egg or sperm, depending on how the paired chromosomes align during meiosis I. The chromosomes in FIGURE 7.21A are arranged with all of the paternal copies of the chromosomes oriented toward the same pole of the cell. Because of this alignment, each of the haploid products of meiosis will contain either all paternal copies or all maternal copies. Another possible arrangement is shown in FIGURE 7.21B, with one of the paternal chromosome copies oriented toward one pole, and the other toward the opposite pole. With this arrangement, each meiotic product will have one paternal and one maternal chromosome.

Take a moment to look at the set of eight gametes in FIGURE 7.21A and B. Try to determine how many different types of gametes have been formed. Each gamete pictured has a haploid set of chromosomes (one short and one long). A total of four combinations is possible, however, based on the parental origin: (1) both paternal, (2) both maternal, (3) long paternal and short maternal, and (4) long maternal and short paternal. These four combinations differ not only from each other, but also from the parent cells.

Meiosis II

Chromatids separate

Cell has one set of chromosomes

[F] Replication
Chromosomes do not replicate

[G] Alignment
Chromosomes align

[H] Separation
Four nonidentical haploid cells

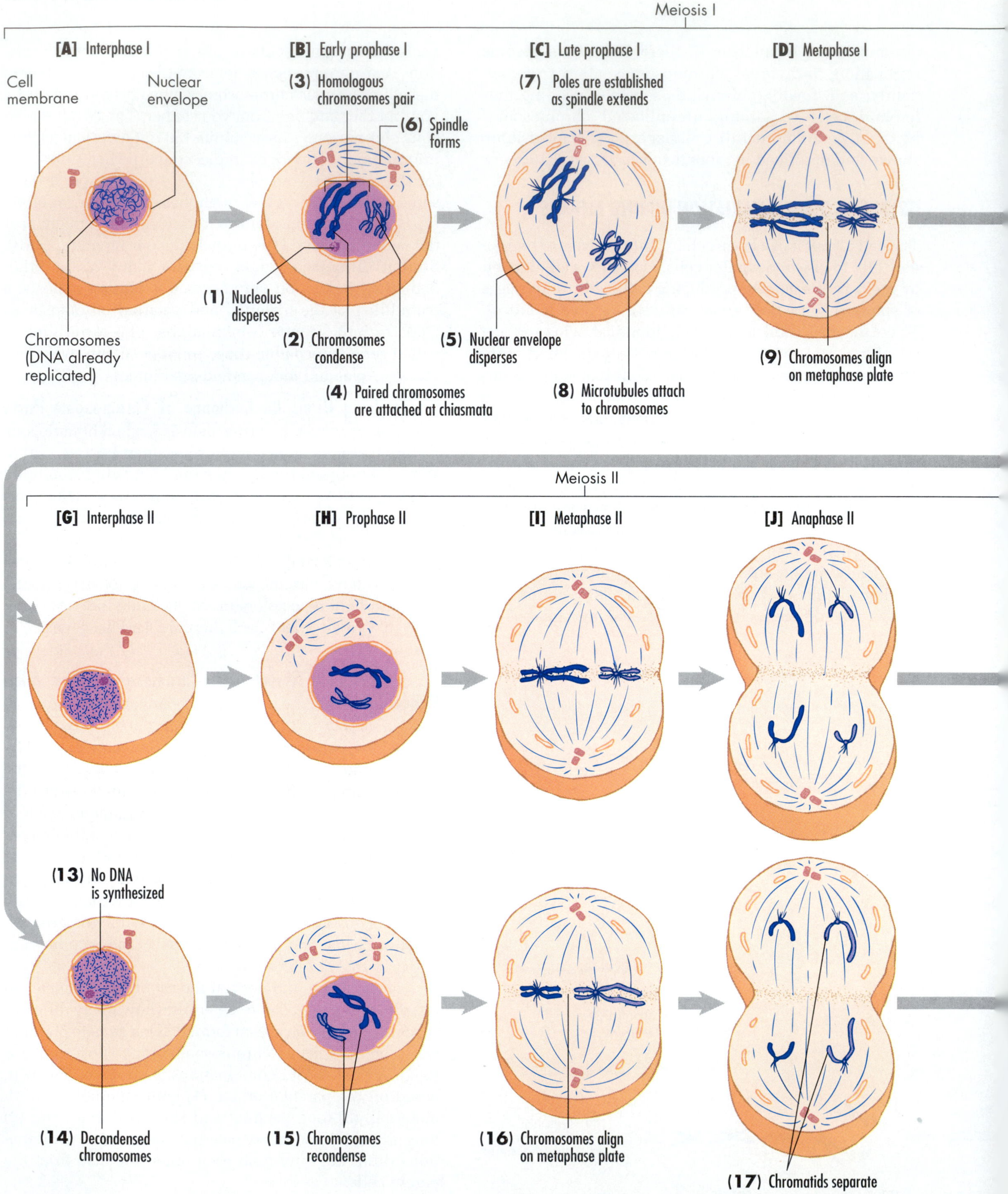

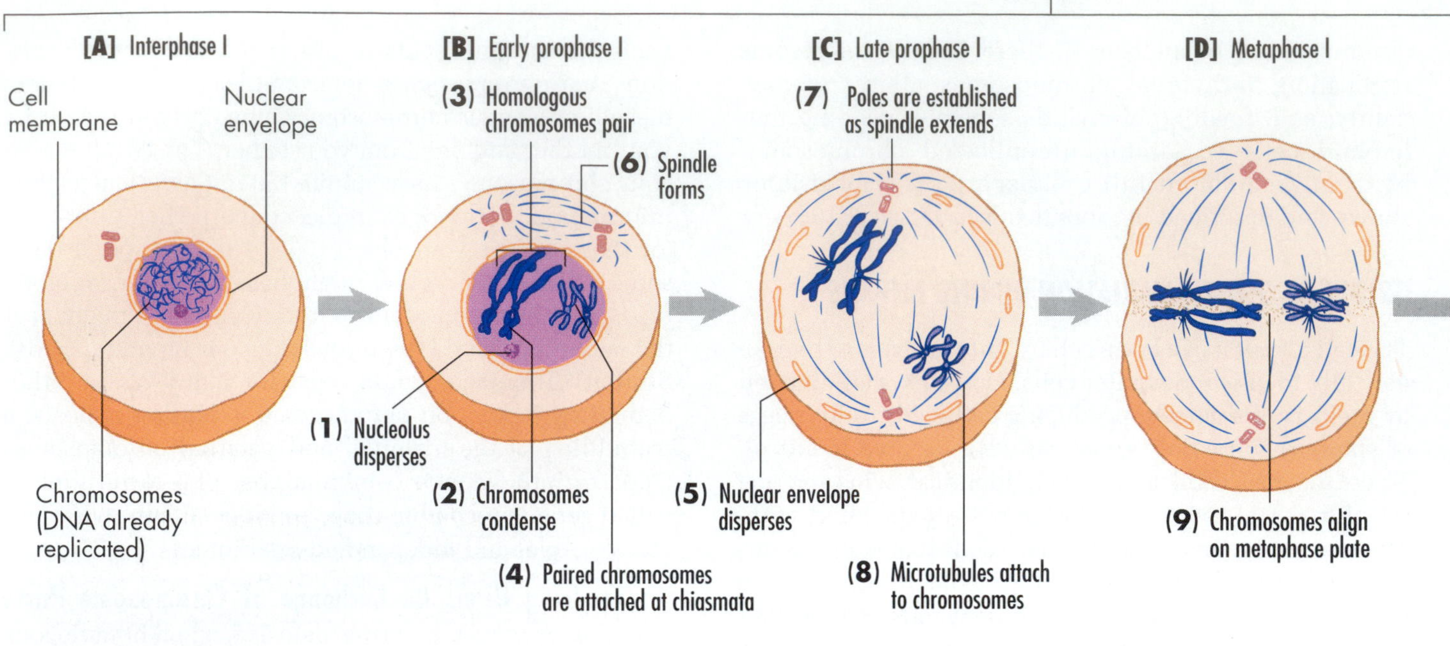

Meiosis I
[A] Interphase I
Cell membrane
Nuclear envelope
Chromosomes (DNA already replicated)
[B] Early prophase I
(3) Homologous chromosomes pair
(6) Spindle forms
(1) Nucleolus disperses
(2) Chromosomes condense
(4) Paired chromosomes are attached at chiasmata
[C] Late prophase I
(7) Poles are established as spindle extends
(5) Nuclear envelope disperses
(8) Microtubules attach to chromosomes
[D] Metaphase I
(9) Chromosomes align on metaphase plate
Meiosis II
[G] Interphase II
[H] Prophase II
[I] Metaphase II
[J] Anaphase II
(13) No DNA is synthesized
(14) Decondensed chromosomes
(15) Chromosomes recondense
(16) Chromosomes align on metaphase plate
(17) Chromatids separate

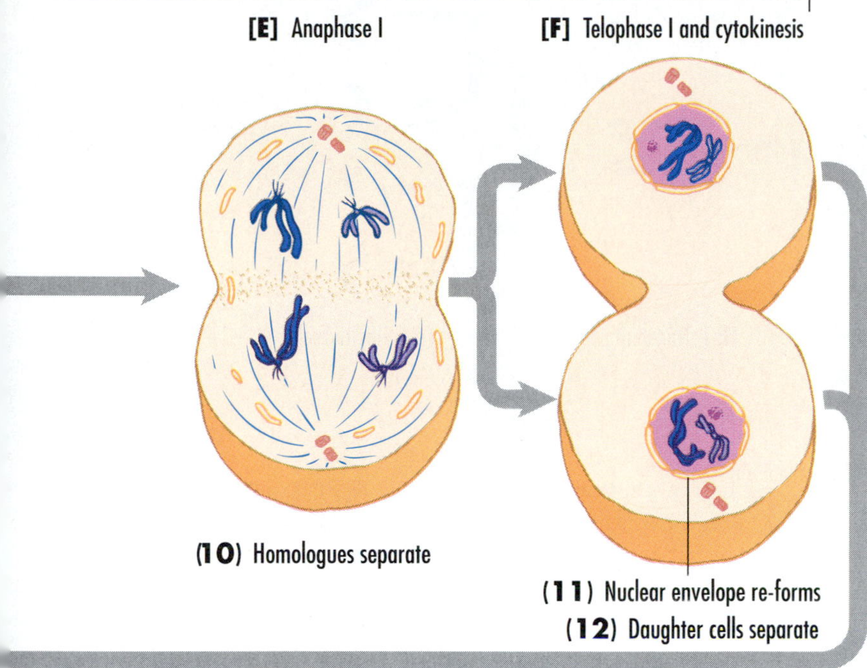

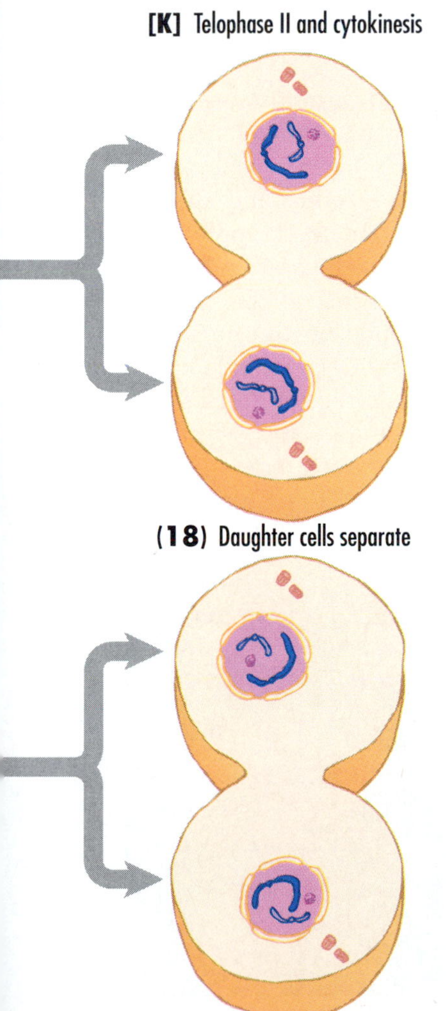

FIGURE 7.19

The Phases of Meiosis.

During the S phase of interphase I, DNA is synthesized, as in a normal diploid cell at the corresponding phase of a conventional mitotic cell division. (1) In early prophase I, just as in mitosis, the nucleolus disperses. (2) The duplicated chromosomes condense and become visible through the light microscope. (3) Homologous chromosomes line up together. In the cells shown here, the two long chromosomes pair with each other, and the two short chromosomes pair side by side. The pairing of homologous chromosomes is the most significant event of prophase I and does not occur in mitosis. (4) Toward the end of early prophase I, homologous chromosomes separate slightly, but remain attached to each other at one or two places called the *chiasmata.* In these regions, one chromosome appears to cross over another. (5) In late prophase I, the nuclear envelope breaks down. (6) The spindle forms. (7) The spindle poles are established. (8) Chromosomes are attached to kinetochore microtubules. (9) During metaphase I, paired homologous chromosomes align at the metaphase plate. Both chromatids of each homologue orient toward the same pole, while the chromatids of the other homologue orient toward the opposite pole. (In mitosis, since homologous chromosomes do not pair, the two chromatids of each chromosome orient toward opposite poles instead of toward the same pole.) (10) During anaphase I, the homologues separate from each other, but sister chromatids remain linked at their centromeres and do not separate from each other as they do in mitosis. (11) Nuclear envelopes re-form during telophase I, and the cytokinesis that follows produces two daughter cells. (12) Each of the daughters has two chromosomes consisting of two sister chromatids attached at their centromeres, with the haploid number of chromosomes. (13) During interphase of meiosis II, no DNA synthesis occurs, in contrast to mitosis and meiosis I. (14) The chromosomes decondense into stringlike masses. Since there is no DNA synthesis, the two daughter cells that have passed through meiosis I continue to have the haploid number of chromosomes but the diploid amount of DNA as they enter prophase II. (15) During prophase II, individual chromosomes become visible in the light microscope, the nuclear envelope breaks down, and the spindle forms. (16) During metaphase II, the chromosomes line up on the metaphase plate, and this time the two chromatids orient to opposite poles, as in mitosis.
(17) During anaphase II, the two sister chromatids separate and become individual chromosomes. (18) After the nuclear envelope re-forms during telophase II, cytokinesis takes place and four cells result. Each of the four daughter cells has the haploid number of chromosomes and the haploid amount of DNA. The events of meiosis result, overall, in a single diploid cell giving rise to four haploid cells.

FIGURE 7.20

Genetic Recombination Through Crossing Over: Chromosomes Exchange DNA.

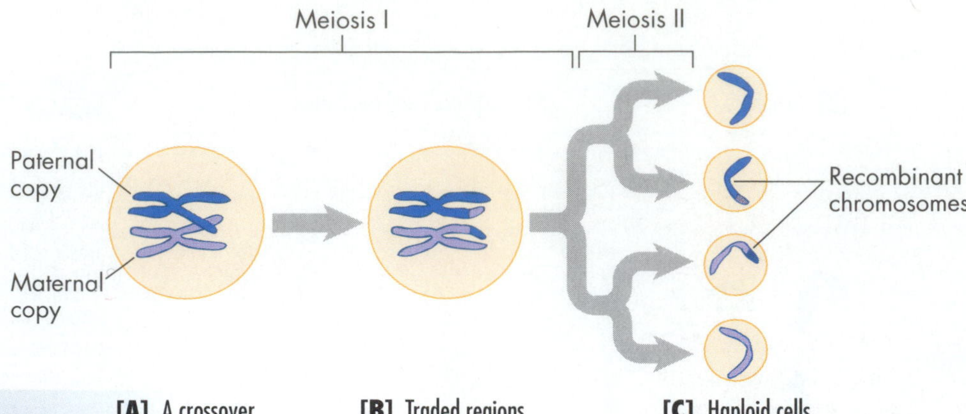

[A] A crossover **[B]** Traded regions **[C]** Haploid cells

FIGURE 7.21

Meiosis, Genetic Recombination, and Genetic Variation.

[A, B] Independent assortment during meiosis can help provide the genetic variability evident in [C] puppy siblings of different sizes, colors, and fur patterns. In one of the mother dog's egg-forming cells, two chromosomes (dark blue) inherited from her father (the maternal grandfather of the puppies) lie on the same side of the metaphase plate during meiosis I [A], while in a different egg-forming cell, two chromosomes inherited from her father lie on opposite sides of the metaphase plate [B]. These two arrangements give different combinations of chromosomes in the gametes, the eggs, as the diagram shows on the right. Since different puppies inherit different combinations of chromosomes from their maternal grandfather and grandmother, they can be genetically quite variable [C].

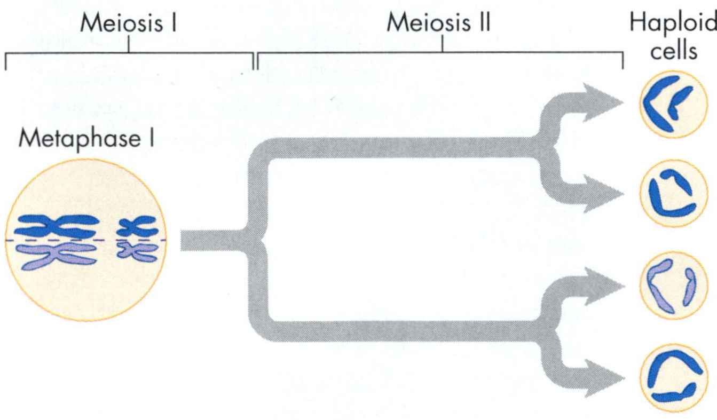

[A] One possible chromosome arrangement

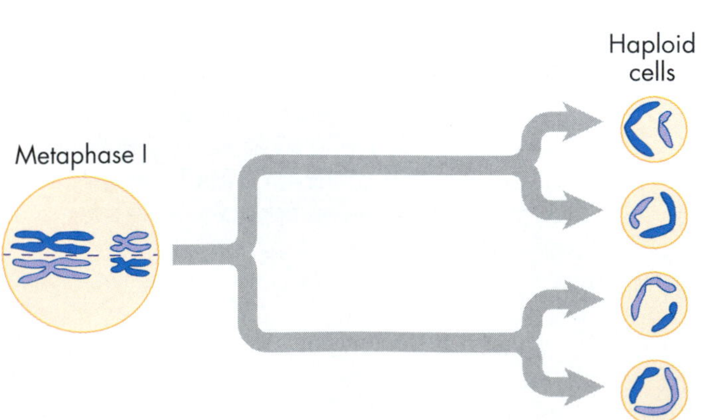

[B] Another possible arrangement

Clearly, new genetic combinations can arise during meiosis, since maternal and paternal chromosomes are slightly different genetically. All four combinations occur with equal frequency, because it is totally random whether a chromosome will face one pole or the other and because the chromosome orientations occur independently from what is happening with nearby chromosomes. This random distribution process is called **independent assortment**, and it will be discussed further in the next chapter.

The independent assortment of chromosomes during meiosis can be compared to choosing from the menu of a restaurant where there are two choices for each course. Each course represents a chromosome, and the two choices—duck or veal, Brie or Camembert, apple tart or crème caramel, and so on—represent the two homologous copies of each chromosome. Just as the potential number of different meals depends on the number of courses, the potential number of unique genetic combinations in the gametes depends on the number of chromosomes. From a menu with six courses and two choices per course, you could make up 2^6 ($2 \times 2 \times 2 \times 2 \times 2 \times 2$), or 64, different meals. For gametes, that number is 2^n, where n is the haploid chromosome number. For the

[C] Puppies with different phenotypes

active learning: Mitosis and meiosis

To help understand how chromosomes move during cell division, compare mitosis and meiosis. Focus on the processes of replication, alignment, and separation as you draw chromosomes in the empty cells. First do mitosis (cells 1–13); after studying meiosis, come back and fill in cells 21–33.

Mitosis:

1 How many chromosomes are in cell 1? . . . . .

2 Is cell 1 haploid or diploid? . . . .

3 How many chromatids does each chromosome have in cell 2? . . . .

4 Is cell 2 haploid or diploid? . . . .

5 In which phase of mitosis is cell 3? . . . .

6 How similar genetically is cell 4 to cell 5?
. . . .

7 How similar genetically is cell 13 to cell 1?
. . . .

Meiosis:

1 Is cell 22 haploid or diploid? . . . .

2 What is the key difference in chromosome alignment between cell 23 and cell 3? . . . .

3 Compare separation of homologous chromosomes and sister chromatids in the offspring of cell 3 and cell 23. . . . .
. . . .

4 Is cell 24 haploid or diploid? . . . .

5 How many chromatids does each chromosome have in cell 27? . . . .

6 Compare the replication step between cells 5 and 7 to that between cells 25 and 27. . . . .
. . . .

7 How similar genetically is cell 32 to cell 21?
. . . .

8 How similar genetically is cell 32 to cell 30?
. . . .

9 Is cell 32 haploid or diploid? . . . .

10 Crossing-over is not shown in this figure; if it were, in which cell would it occur? . . . .

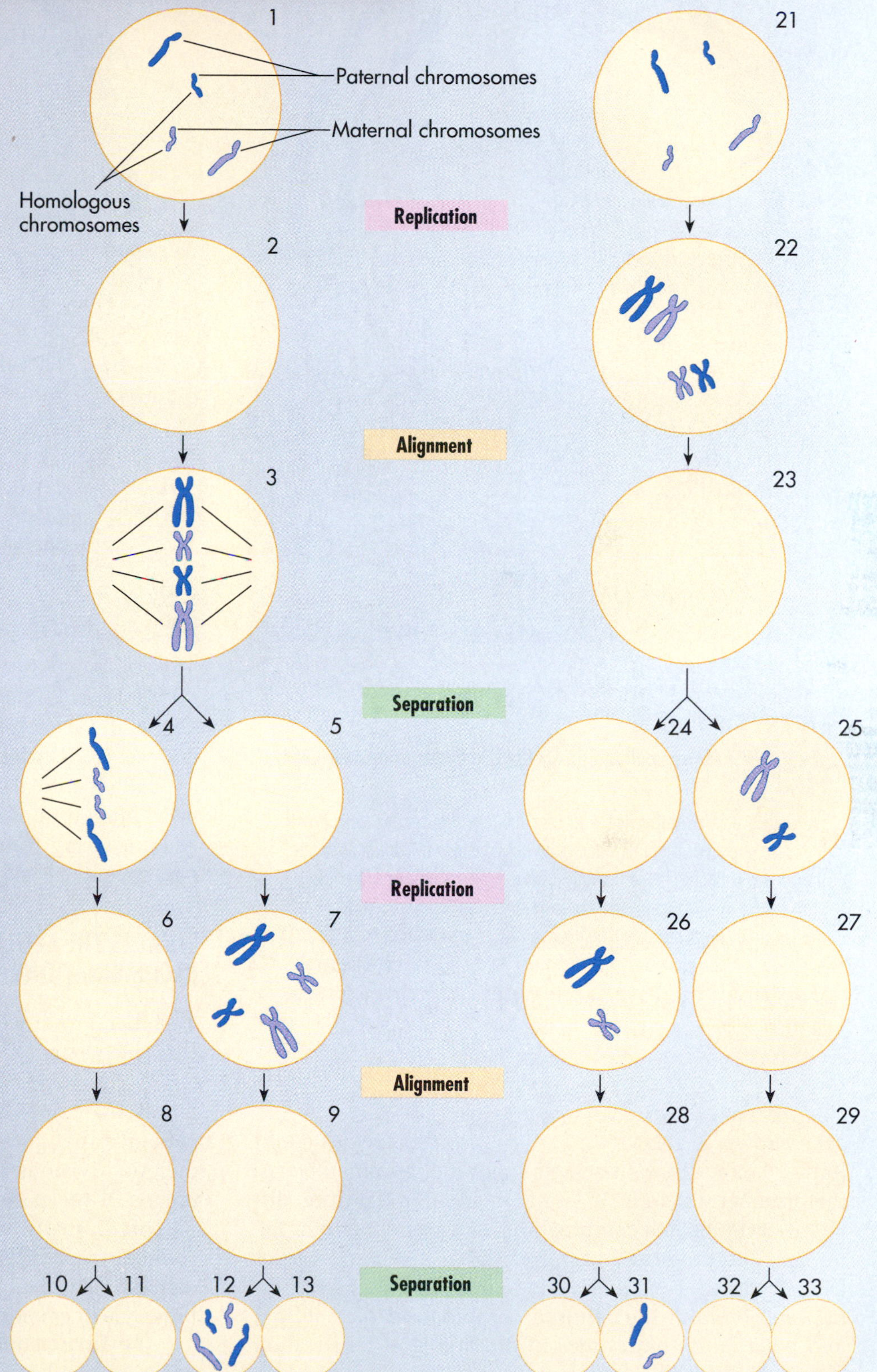

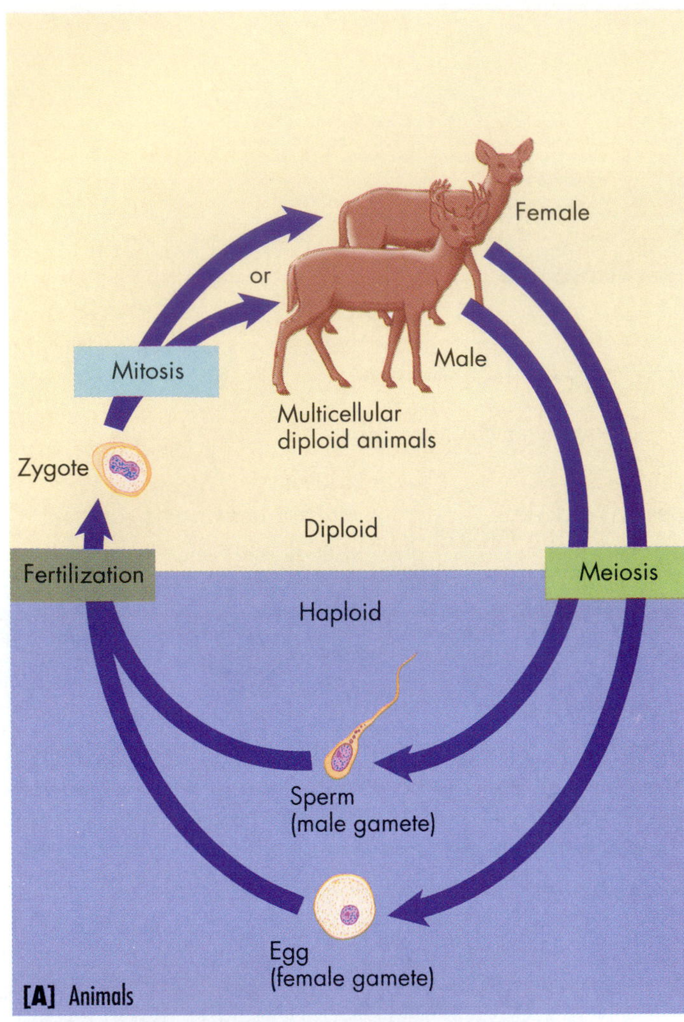

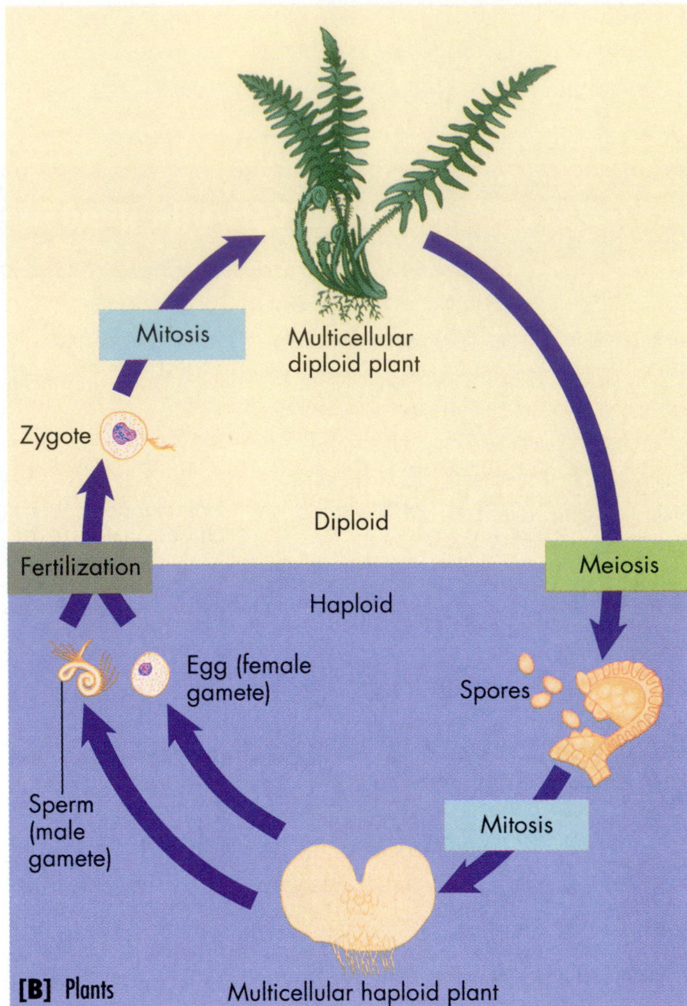

FIGURE 7.22

The Roles of Mitosis and Meiosis in the Life Cycles of Animals and Plants.

[A] In most animals, each adult produces either egg or sperm via meiosis, and the union of gametes at fertilization yields a zygote that divides mitotically to form a new adult. **[B]** In many plants, meiosis produces haploid spores that grow mitotically into multicellular haploid plants, sometimes free-living, as in this fern, sometimes dependent on the adult plant, as in pines or tulips. Each haploid plant then produces gametes through mitotic divisions, and the union of haploid gametes creates a zygote, restoring the diploid chromosome number in the future adult.

cells shown in FIGURE 7.21, $n = 2$, so the species could form 2^2, or 4, different gamete types. For a human being, that number would be 2^{23}, or a potential of 8 million different chromosome combinations of eggs or sperm in the nuclei. The crossing over process we discussed earlier [review FIGURE 7.20] gives rise to even more new combinations. Finally, the random combination of maternal and paternal chromosomes in the zygote at fertilization further increases the number of genetic combinations.

The incredible genetic diversity that results from independent assortment helps explain why an organism with several chromosomes is unlikely to produce two genetically identical gametes, and why, in turn, cucumbers, prize bulls, puppies [FIGURE 7.21C], and all other organisms resulting from sexual reproduction resemble their parents but are never exactly like either parent.

WHAT IS THE EVOLUTIONARY SIGNIFICANCE OF GENETIC RECOMBINATION?

Together, crossing over and independent assortment ensure that sexually produced offspring are genetically different from their parents. A litter of puppies, all with different fur color, fur pattern, and body size, is living evidence that genetic recombination took place during meiotic divisions in the gamete-producing reproductive organs of the parent dogs. The puppies resemble their parents in many ways, but differ in at least a few, and some of these differences may give the puppies a better chance of survival. And therein lies the evolutionary significance of genetic recombination during meiosis.

The Environment, Evolution, and Meiosis Why does genetic recombination occur in the first place? If a parent has a favorable genetic combination that allows it to survive in an unchanging environment—say the cold, dark, mud at the bottom of the deepest oceans—one might argue that asexual reproduction would best retain the parent's genetic advantage passed directly to its offspring. One could argue differently, however, if the environment were a changing one, such as salt flats near the ocean shore, which are subject to periodic drying, flooding, and changing salt levels. In such a variable setting, one genetic combination might be advantageous during dry conditions but disadvantageous during wet conditions. Here, asexual reproduction might doom an offspring with its parent's disadvantageous gene combination. In contrast, sexual reproduction—with its reshuffling of genes during meiosis—might deal a few of the offspring a winning hand in surviving unpredictable conditions.

MITOSIS AND MEIOSIS COMPARED

Because it is relatively easy to confuse the details of mitosis and meiosis, let's compare the two [FIGURE 7.22]. The main distinction between mitosis and meiosis is that a mitotic division can occur in either a haploid or diploid body cell, producing two cells genetically identical to each other and to the parent cell. Meiosis, however, can occur only in special reproductive cells, such as ovaries and testes in human beings, the flowers of flowering plants, and special reproductive organs in haploid fern plants [FIGURE 7.22B]. The products of meiosis are unique. They are haploid and can undergo fertilization and produce a new diploid individual.

Mitosis accomplishes two things: (1) The reproduction of cells, and (2) the equal distribution of DNA to each new daughter cell. Meiosis accomplishes three things: (1) By reducing the chromosome number from diploid to haploid, meiosis prevents an increase in chromosome number which otherwise would occur at fertilization. (2) Crossing over during meiosis permits new combinations of maternal and paternal hereditary traits. (3) Independent assortment allows for the further random combination of maternal and paternal chromosomes.

Differences in chromosome movements generated by differences in replication, alignment, and separation are summarized in ACTIVE LEARNING BOX 7.1.

CONCEPT CHALLENGE

A person with Down syndrome has three copies of chromosome 21 in each body cell, but two copies of all the other chromosomes. Modify FIGURE 7.18 to show how chromosomes 21 and 22 might behave during metaphase of meiosis I in a person with Down syndrome. How many of the four haploid cells arising from the completion of this meiotic division would be normal haploid cells?

Connections

The principles of cell division we discussed in this chapter are relevant to our individual lives and to life in general. We have all witnessed the normal control of cell division after accidentally cutting a finger: Surviving cells are stimulated to divide and then division slows down when the wound is healed. Most of us also know someone who developed a group of renegade cells that multiplied into a benign or malignant tumor, outside normal regulatory controls. Regardless of outcome—normal wound healing or tumor formation—cell division was at work.

The principles of cell division help us in other ways as well. By understanding how chromosomes are distributed during mitosis, we can appreciate why radiation would do more damage to dividing cells than to cells that do not undergo division. Studying the special division of meiosis allows us to understand how chromosomal irregularities arise, such as having an extra chromosome 21, the cause of Down syndrome. Finally, the independent assortment and crossing over that occur in meiosis provide the cellular bases for the origin of genetically diverse gametes. Since genetic diversity is a prerequisite for evolution by natural selection, as we saw in CHAPTER 1, these cellular mechanisms underlie the broader theme of evolution—species change over time. The topic of the next chapter, how genes are passed from parent to offspring, sets the stage for understanding the production and development of the individual, and the continuation of the species, as well as evolutionary changes.

CONCEPT MAPPING

Draw a concept map relating the following terms. Link the concepts with arrows and a connecting word or short phrase.

asexual reproduction
cancer
cell cycle
cytokinesis
diploid
haploid
homologous chromosomes
meiosis
mitosis
sexual reproduction

KEY TERMS

anaphase, 176
asexual reproduction, 182
binary fission, 170
budding, 183
cancer, 182
cell cycle, 170
cell division, 171
cell plate, 178
centromere, 169
chromatid, 169
contact inhibition, 179
contractile ring, 177
crossing over, 187
cytokinesis, 177
diploid, 185
Down syndrome, 186
fertilization, 183
furrow, 177
gamete, 183
genetic recombination, 187
germ cell, 183
growth factor, 179
haploid, 185
homologous chromosomes, 185
independent assortment, 190
interphase, 171
life cycle, 182
meiosis, 184
metaphase, 173
mitosis, 172
mitotic spindle, 173
pole, 173
prometaphase, 173
prophase, 173
regeneration, 183
sexual reproduction, 183
somatic cell, 183
telophase, 176
tumor, 182
zygote, 183

HIGHLIGHTS IN REVIEW

1 A cell's hereditary information resides in the DNA molecules of its chromosomes, which are housed within the cell's nucleus.

a] Studies with a single-celled alga revealed that the information for constructing each part of each new cell resides in the nucleus.

b] Hereditary information lies in the chromosomes. Prokaryotes have a single circular chromosome, while eukaryotes have at least two separate linear chromosomes. The duplication and distribution of the chromosomes are central to the cell cycle.

2 The cell cycle includes a period of growth and a period of division. In the growth phase, chromosomes and other cell constituents replicate. During the division phase, replicated parts are distributed to offspring cells.

a] Prokaryotic cells increase in size and then undergo binary fission. During this process, the circular chromosome is duplicated, each copy is localized at one end of the double-sized cell, and a partition forms, separating the two cell halves.

b] Eukaryotic cells have a four-phase cell cycle. Three phases are devoted to growth and DNA duplication, and one to cell division. The three-phase growth period is called interphase, and the division period, mitosis and cytokinesis.

c] Interphase has three phases: G_1 (gap 1), during which the cell produces new proteins and cell parts; S (synthesis), during which DNA molecules are copied; and G_2 (gap 2), during which the cell makes more protein and prepares to divide.

3 Regulation of the cell cycle is crucial for controlled cell division and wound healing; inappropriate regulation can lead to cancers and other tumors.

a] Contact inhibition of cell division is an external control over the cell cycle. Growth factors also affect the timing of cell division.

b] Cancer cells do not stop growing or migrating upon contact with other cells and hence invade healthy tissue and form tumors.

4 Cell division by mitosis produces two genetically identical offspring cells.

a] During mitosis, or the M phase, duplicated chromosomes are apportioned equally to new daughter cells, and cytokinesis creates two new cells.

b] Mitotic division has five parts. In prophase, the chromosomes condense, the nucleolus disappears, and the mitotic spindle forms. In prometaphase, the nuclear envelope disappears and the chromosomes attach to the spindle. In metaphase, the chromosomes become aligned in the middle of the cell. In anaphase, the chromatids are pulled to opposite poles. In telophase, the events of prophase and prometaphase reverse in preparation for cytokinesis. The cell now has two identical nuclei.

c] During cytokinesis in animal cells, a contractile ring of microfilaments makes a furrow in the pliable cell surface and eventually squeezes the cell into two daughters, each with an identical nucleus formed during mitosis. In plant cells, a cell plate made of cell-wall and cell-membrane material forms down the cell's midline and partitions the cell into two identical daughters.

5 Sexual reproduction involves meiosis—two consecutive cell divisions together. When a diploid cell with two copies of each chromosome divides by meiosis, four haploid cells result, each cell bearing only one copy of each chromosome. Haploid cells can fuse in fertilization, forming a new diploid cell.

a] While the cell cycle is an alternation between cell division and cell growth, the life cycle involves formation of the zygote, development of the immature stage, continued development to a mature stage, and reproduction.

b] In multicellular organisms, only the sperm nucleus enters and merges with the egg nucleus during fertilization. The nucleus therefore directs the life cycle as well as the cell cycle.

c] Reproduction can be asexual and involve mitotic divisions alone, or it can be sexual and involve both meiosis and mitosis.

d] In asexual reproduction, a few body cells of a single parent divide mitotically, producing an offspring genetically identical to the parent.

e] During sexual reproduction, gametes form after meiotic divisions. Meiotic divisions halve the chromosome number, making the gamete haploid; fertilization then restores the diploid chromosome number in the zygote.

f] Meiosis has two parts, meiosis I and meiosis II. In meiosis I, homologous chromosomes pair and then separate to opposite poles, halving the number of chromosomes in the two daughter cells. In these cells, meiosis II proceeds without DNA synthesis, and chromatids separate to opposite poles, yielding four haploid cells.

g] The four haploid daughter cells produced through meiotic divisions are similar but not identical to the parent cell. Independent assortment causes a random distribution of parental homologous chromosomes in the gametes, and crossing over results in a switching of parts of homologous chromosomes.

UNDERSTANDING THE FACTS AND CONCEPTS For Questions 1–5, match each of the descriptions with the most appropriate item from the following list of terms. A given item may be used once, more than once, or not at all.

a] chromatid
b] centromere
c] chromosome
d] microtubules
e] cytokinesis

1 The mitotic spindle.

2 During G_1, each chromosome is composed of one of these structures.

3 During prometaphase this structure attaches directly to a spindle fiber.

4 To determine the number of these structures, just count the number of centromeres.

5 Final stage of the M phase.

As above, for Questions 6–10, match the descriptions with the appropriate item from the following list.

a] S b] G_1 c] G_2 d] M
e] more than one of the preceding

6 Included in interphase.

7 Primary stage of cell growth and metabolic function.

8 Replication of DNA occurs during this stage.

9 The chromosomes are composed of two chromatids during the first part of this stage.

10 The stage during which sister chromatids separate.

As above, for Questions 11–15, match the descriptions with the appropriate item from the following list.

a] binary fission
b] mitosis
c] meiosis
d] cell cycle
e] life cycle

11 G_1, S, G_2, M.

12 Diploid, haploid.

13 Immature, mature, reproduction, immature, etc.

14 Prophase, metaphase, anaphase, telophase, G_1.

15 DNA replication, cell elongation, no spindle, cell splitting.

As above, for Questions 16–20, match the descriptions with the appropriate item from the following list.

a] prophase
b] metaphase
c] anaphase
d] telophase
e] prometaphase

16 Spindle forms, chromosomes coil and shorten.

17 Alignment phase.

18 Nuclear membrane disperses, centromeres attach to spindle fibers.

19 Chromatids separate.

20 Stage in which a cell has two nuclei.

INTEGRATE AND APPLY WHAT YOU HAVE LEARNED

1 What is the difference between mitosis and cytokinesis?

2 How does cytokinesis differ in plant and animal cells?

3 What is the unique and critical event in meiosis that promotes an increase in genetic variation by means of independent assortment?

4 What is the critical event in meiosis that leads to halving the number of chromosomes in a cell?

5 Would there be an increase in genetic variability if pieces of sister chromatids were exchanged during meiosis?

ANALYSIS Select the best answer to each of the following questions from the choices provided.

1 What distinguishes cyclins from growth factors?
a] Nothing—they are the same.
b] Cyclins regulate binary fission only; growth factors regulate mitosis.
c] Cyclins are used by the cell that produces them; growth factors are used by different cells.
d] Cyclins act extracellularly; growth factors act intracellularly.
e] Cyclins combine with membrane receptors; growth factors do not.

2 Suppose that a cell has a mutation that results in the complete absence of growth factor receptors. What would be the most likely consequence?
a] No change in the cell or in its cell cycle.
b] Cell division could not occur.
c] Cell division would occur too often and might be out of control.

3 Suppose that a cell had a mutation that resulted in a great excess of growth factor receptors in its membrane. What would be the most likely result?
a] No change in the cell or its cell cycle.
b] Cell division would cease.
c] Cell division would occur too often and might become out of control.

CHAPTER 8

Mendelian Genetics

WHITE TIGERS AND FAMILY PEDIGREES

In the dry season of 1951, hunters found four tiger cubs playing in the hunting preserve of the Maharajah of Rewa, in central India. One of these cubs caused great excitement: Instead of the normal orange-and-black-striped coat, it had a pure white pelt! Lured into a cage by a bowl of water, the young tiger was brought to the maharajah's palace and named Mohan, meaning "Enchanter." Mohan grew into a magnificent creature, larger than most tigers, strong, and healthy. But he lacked the usual rich orange and jet black pigments of normal tigers; his coat was nearly pure white, with ashen gray stripes. In addition, his nose and paw pads were pink instead of black, and his eyes were ice blue with a tendency to cross [FIGURE 8.1].

FIGURE 8.1

Mohan: A White Tiger with Blue Eyes That Tend to Cross.

How can we understand the inheritance of a pale pelt in tigers?

Was Mohan just a unique occurrence, or were his special traits *hereditary*—that is, capable of being passed to offspring and through them to future generations? Mohan was allowed to mate with Begum, an orange and black tiger, but the results disappointed the maharajah: None of the cubs had Mohan's sparkling white coat. The royal tiger keepers continued to house Mohan in a cage with one of his daughters, Radha, and you can imagine their surprise when they saw the offspring from a mating of the two: Some of the cubs in the second generation had beautiful white coats—the trait could be passed on!
Although the white trait appeared to have skipped Radha's generation, it had remained intact (though hidden) and had reappeared in the next generation. What could cause a hereditary trait to behave in this hit-or-miss

fashion? And what was the relationship of the white pelt to the crossed blue eyes?

We will see the answers to these questions unfold as we investigate **heredity**, the transmission of physical, biochemical, and behavioral traits from parent to offspring. The biological field that deals with these subjects is genetics, and it will take the next five chapters to explore this important and relevant field. We begin with the history of genetics, and how the monk Gregor Mendel working in the mid-nineteenth century deduced the existence of hereditary factors we now call *genes*. Observable traits, such as a tiger's coat color and eye color, are controlled by genes, as are the shades of human hair and eyes. We follow a historical approach in this chapter because the pattern of reasoning that guided Mendel's analysis is a model of scientific investigation and clear thinking. We will explore the rules of heredity, which describe how genes and the traits they determine are passed from generation to generation. The rules follow from the process of meiosis, the special cell division that occurs during the formation of eggs and sperm [review FIGURE 7.19]. They hold in principle for all sexually reproducing organisms, from the single-celled paramecium to peas and tigers. Finally, we will analyze how genes interact with each other and with the environment to control an organism's appearance, function, and behavior.

An understanding of genetics is pivotal in many ways to our exploration of biology. First, it helps reveal the workings of evolution by natural selection. Genes that underlie an unfavorable trait tend to occur less frequently over the course of several generations because fewer individuals carrying that trait will survive and reproduce. For example, crossed eyes in a predator like the tiger would be selected against, since the animal relies on split-second coordination to fell fast-moving prey.

Second, the principles of genetics are important for understanding how embryos develop. By studying the ways genes interact, we can begin to understand how individuals develop different physical or behavioral traits that, in turn, may affect their ability to survive and reproduce. Third, a study of genes helps us understand *physiology*—how organisms function. For example, by studying mice with a hereditary eye condition similar to Mohan's, researchers can investigate what factors are necessary for eye muscles to develop correctly and how nerves arise and link the eye's visual field to the visual centers of the brain.

In the last of our genetics chapters, CHAPTER 12, you will see dozens of human applications of the principles that Mendel discovered and that we discuss in this chapter. By learning about the way genes act, you can gain a deeper appreciation for the study of genetics and the impact genes have on a person's daily life.

Our major topics in this chapter pertain to eukaryotic organisms in general. They include how the monk Gregor Mendel discovered and analyzed genes in his quiet abbey garden; the rules governing the inheritance of two genes that control two different traits; how genes are organized on chromosomes and how new gene combinations and alternative gene forms arise; and how genes interact with one another and with the environment to govern the way an organism looks, functions, and behaves.

Along the way, we will see how Mendel's contemporaries received his new ideas; why Siamese cats have dark ears, face, tail, and feet; how curly-eared cats inherit their special trait; why some snapdragons are pink; and how you inherited your blood type. ❑

MESSAGES

1 Heredity traits are controlled by genes, which occupy specific positions on the chromosomes.

2 According to Mendel's principle of segregation, the two copies of a gene separate into different cells during meiosis. During fertilization, egg and sperm combine at random with respect to their genetic content.

3 Genes on the same chromosomes assort independently of each other during meiosis.

4 Genes on the same chromosome tend to be inherited together. Based on this tendency, geneticists have learned to map the locations of genes on chromosomes.

5 Genes interact with each other and with the environment and determine the eventual physical makeup of the individual.

Genetics in the Abbey

Since ancient times, gardeners and farmers have realized that many characteristics of domesticated plants and animals pass from parent to offspring. They did not understand, however, exactly how the traits were transmitted. Many observers thought individuals had a nonspecific mixture of their parents' traits.

Looking at certain organisms in nature, it is not hard to see why people believed that offspring were intermediates between their parents. Consider, for example, two hibiscus plants whose flowers have petals of different color, orientation, and length, and sepals (leafy green extensions below the petals) of different orientation [FIGURE 8.2]. When plants with these two flower types are mated, the result is a **hybrid**: the offspring of two individuals with differing forms of a given trait. In this case, the hybrid's flowers have the petal length of one parent, the sepal orientation of the other, and a petal color and orientation intermediate between those of the two parents. Such observations made people think that the hereditary "stuff" of a mother and father *blend* to produce the characteristics found in the offspring, just as cream mixes with black coffee to produce the tan-colored café au lait. This idea became known as the *blending hypothesis of heredity*. A monk named Gregor Mendel, however, eventually used the process of science to prove the blending hypothesis to be incorrect.

FIGURE 8.2

Casual Observations of Traits Such as Flower Color or Shape Led to the Now-Disproven Blending Hypothesis of Inheritance.

Petal color in hibiscus appears to be a blended trait: Red and yellow parents give rise to orange offspring. However, most traits, even in hibiscus flowers, do not appear to blend. A parent with long petals and a parent with short petals, for example, give rise to offspring with long petals.

GREGOR MENDEL

Mendel was the son of a farmer in what is now Slovakia. Nature fascinated the child, as did science, mathematics, horticulture, and beekeeping. A teacher at the local elementary school recognized Mendel's talent and urged his parents to continue his education. Through the financial sacrifices of his parents and sisters, he was able to attend school at a very special monastery in old Brno. There Mendel came into contact with many of the prominent philosophers and scholars of the time. He was so successful in his studies that the school administrators sent him to the University of Vienna to take a qualifying exam to become a certified teacher of natural history and physics.

Mendel flunked the examination twice, and developed a chronic illness that forced him to return to the monastery. There, he continued his teaching and began quietly cultivating the abbey garden [FIGURE 8.3]—an activity that would ultimately change the study of biology.

While at the University of Vienna, Mendel had learned that all matter is made up of discrete atoms and molecules. He wondered if heredity could also be governed by particles that retain their identity from generation to generation. He suspected that some of the particles could be hidden in a hybrid. Mendel put his new *particulate hypothesis of heredity* to the test in a carefully devised long-term study that cataloged the number and types of pea plants in successive generations.

FIGURE 8.3

Gregor Mendel (1822–1884) with His Beloved Pea Plants.

FIGURE 8.4

The Advantages of Experimenting on Pea Plants.

[A] Pea plants not only have attractive flowers, but their petal arrangement protects one pea flower from fertilization by the pollen of another pea flower. As a result, each flower fertilizes itself in nature. **[B]** An experimenter can cross-fertilize pea flowers precisely and deliberately and thus decide which will be the male parent and which the female parent of a new generation. Mendel was able to gently tease back the petals from the flower he chose to be the female parent, reach in with a pair of scissors, and sever the anthers, the source of pollen. Thus, he could eliminate the male gametes from that flower. He could then dust the stigma (the pollen-receiving structure of the female flower) with pollen from another pea plant he chose to be the male parent. **[C]** Finally, Mendel could collect seeds from the mated plant, grow them, and observe the characteristics of the offspring's flowers.

[A]

MENDEL'S EXPERIMENTS

Whereas the blending hypothesis predicted that each hereditary factor would be permanently diluted in the hybrid, Mendel's particulate hypothesis predicted that each hereditary factor would remain unchanged in a hybrid. Mendel realized that he could disprove one of these two predictions by checking the offspring of a hybrid generation: If the original parental forms reappeared in the second generation, this would show that the hereditary factors had passed through the hybrids unchanged and particulate. If the original forms *failed* to reappear in the hybrid's offspring, however, then the factors would appear to have been blended.

Peas Were a Good Choice Mendel chose the garden pea as his test subject because it has several advantages [FIGURE 8.4A]. First, he could purchase pea strains that showed clear alternative forms for single traits, such as short stems versus long stems or white flowers versus purple flowers. By selecting strains that differed in only one trait, he could study inheritance of one feature unconfused by all other variations. Second, Mendel could easily control which peas mated with which because peas normally *self-fertilize*—an individual plant mates with itself. Self-fertilization occurs because, in a single

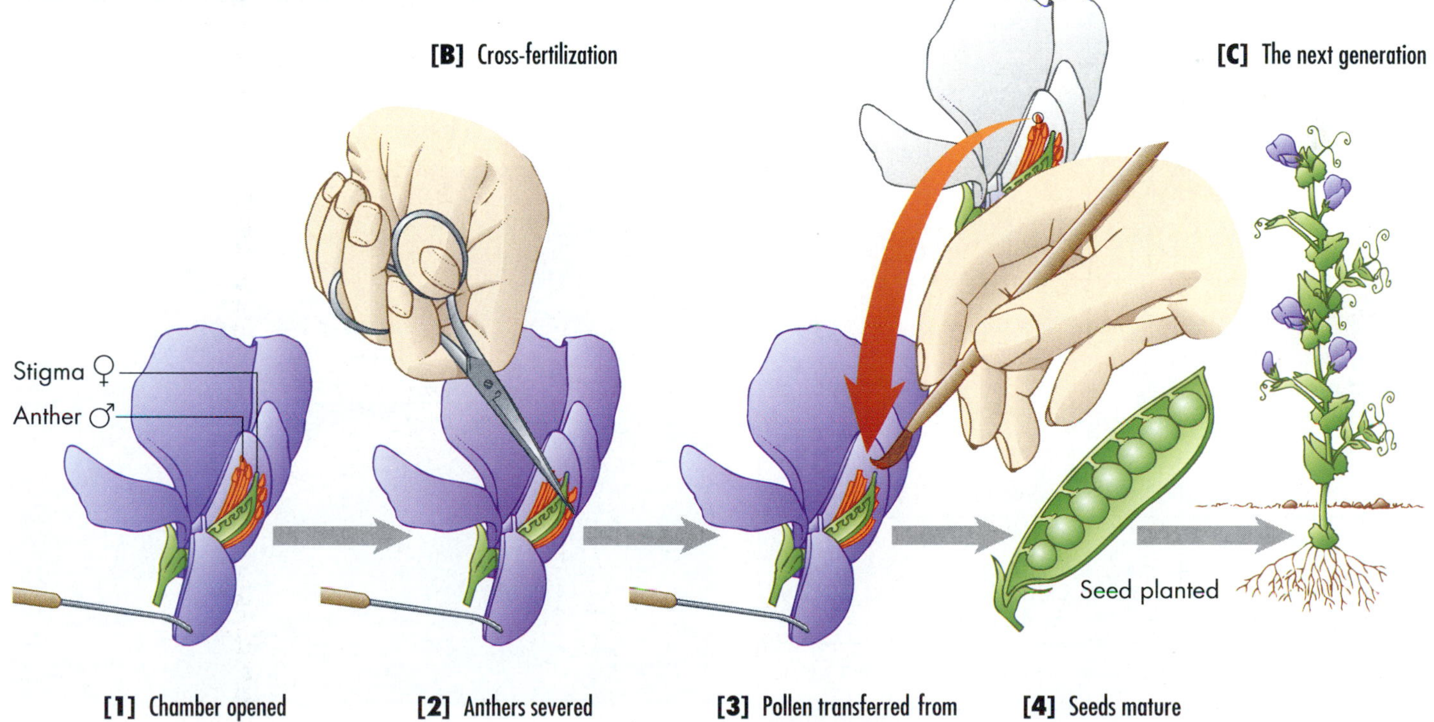

pea flower, a chamber of petals encloses both male and female sex organs. This insures that the egg and the pollen (the source of the sperm) come from the same individual flower. A third advantage of pea plants was that Mendel could *cross-fertilize* plants. He simply clipped off the pollen-producing organ from one flower (for example, a purple flower) before its maturation and dusted the egg-containing organ in that flower with pollen from another selected plant (for example, a white flower) [FIGURE 8.4B]. From the seeds of this cross, Mendel could grow a new generation of pea plants [FIGURE 8.4C].

With clearly stated hypotheses to be tested, and a well-chosen experimental system to test them, Mendel was now ready to perform some scientific tests.

Mendel Discovers Some Rules of Genetics

Mendel analyzed the inheritance of clear-cut alternatives for seven pea traits, including stem length (long versus short), flower color (purple versus white), and seed shape (round versus wrinkled) [FIGURES 8.5A and B]. He began by demonstrating that he had 14 strains of **pure-breeding** peas, pea strains that when self-fertilized, always produced offspring like themselves (for example, short-stemmed plants always produced more short-stemmed plants when allowed to self-fertilize).

MENDEL DISPROVES THE BLENDING HYPOTHESIS

To test the blending hypothesis, Mendel carried out matings between individuals that differed in only one trait, or *monohybrid crosses*. In one such monohybrid cross, he planted long-stem and short-stem seeds early one spring and let them grow into the **parental (*P*) generation** [FIGURE 8.6A]. Later that spring, when the parental plants had flowered, Mendel cross-fertilized long-stemmed plants with pollen from the short-stemmed plants. In the summer, when the pods became swollen with plump peas, he collected the seeds. These seeds would produce the next generation, called the **first filial (F_1) generation**, meaning the first generation in the line of descent. Planted in the spring of the second year, the F_1 seeds of the long-stem/short-stem cross all grew into long-stemmed plants [FIGURE 8.6B]. Mendel found that for each of the contrasting traits he was studying, only one appeared in the F_1 generation. The trait that appears in the

FIGURE 8.5

Mendel Studied Several Pairs of Traits in Pea Plants.

Each of the traits Mendel examined (stem length, flower color, seed shape, and four others) can appear in two forms: a dominant form and a recessive form. **[A]** Crosses for three traits. The forms are long versus short stems, purple versus white flowers, and round versus wrinkled seeds. **[B]** In 1990, researchers discovered that the normal dominant allele for the trait of seed shape causes starch to be stored in the seed, and this makes it plump and round. Seeds homozygous for the recessive allele do not make or store this starch, and so they look shriveled and wrinkled.

	Stem length	Flower color	Seed shape
Dominant characteristic (dominant allele)	Long	Purple	Round
Recessive characteristic (recessive allele)	Short	White	Wrinkled
Phenotypic ratio in F_2	2.84 long: 1.00 short	3.15 purple: 1.00 white	2.96 round: 1.00 wrinkled
Total counted	1064	929	7,324

[A]

[B]

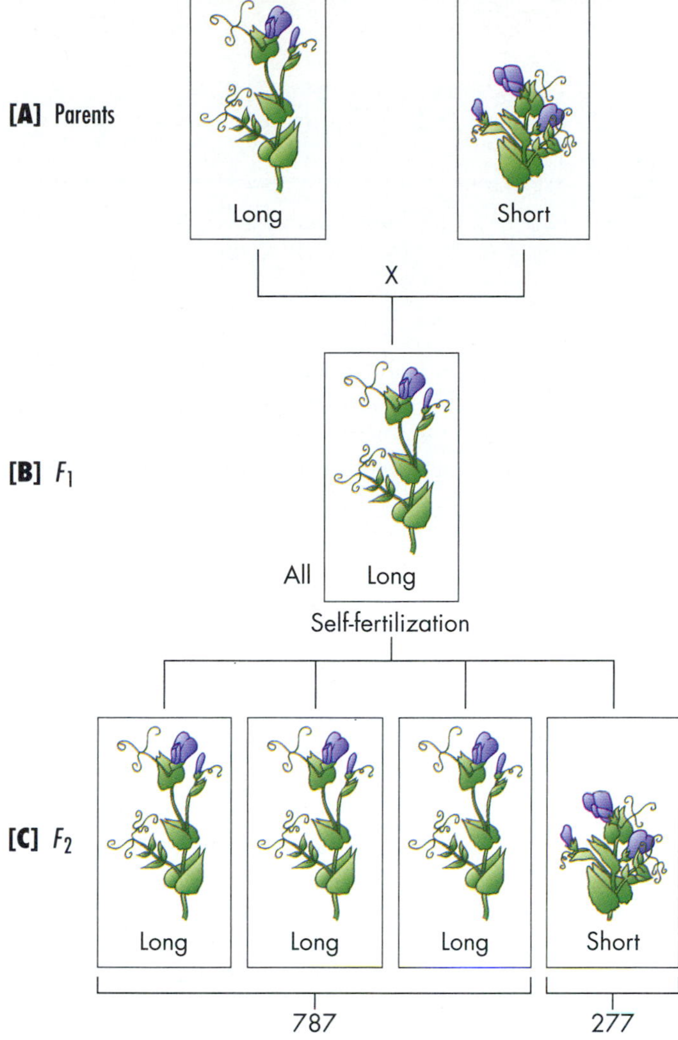

FIGURE 8.6

Mendel's Evidence for the Particulate Theory of Heredity.

When Mendel crossed long- and short-stem pea plants [A], he got all long-stem progeny in the F_1 generation [B], but in the F_2 generation [C], about one-quarter 26% of the progeny were short-stemmed and about three-quarters 74% were long-stemmed.

F_1 hybrid, such as long stems in peas, is said to be dominant, while the trait that does not appear (such as short stems in peas) is referred to as recessive.

What happens to the recessive characteristic? Does it disappear completely? Does it blend with the dominant characteristic? Or does it remain intact but hidden in the F_1 generation? Mendel knew exactly how to learn the answers to these questions. He allowed the long-stemmed F_1 hybrid plants to self-fertilize, and the next spring he planted the seeds of the **second filial (F_2) generation**. When the second generation of pea plants grew up, most of them had long stems, but significantly, some plants had short stems. Again, there were no stems of intermediate length [FIGURE 8.6C]. The reappearance of pure short stems among the offspring of long-stemmed hybrids and the absence of any intermediates were the results predicted by the particulate hypothesis and dramatic disproof of the blending hypothesis of heredity.

RESULTS OF A CROSS

Mendel was not satisfied with just saying that "some" of the F_2 plants had short stems. He counted the plants and found that 787 of the F_2 plants had long stems and 277 had short stems. These numbers showed about a 3:1 ratio of long-stemmed to short-stemmed plants in the F_2 generation. (A perfect 3:1 ratio would be 798:266, not much different from the 787:277 actually observed.) When Mendel examined the offspring of crosses for all seven traits individually, he obtained similar results. In each case, all the F_1 hybrid plants were identical, showing only the dominant form of two alternatives, and there were no plants with intermediate features. Self-fertilization of the F_1 generation produced an F_2 generation in which the recessive form of the trait in question reappeared in about one-quarter of the plants, while the other three-quarters showed the dominant form. Clearly, the unit of heredity that produced short-stemmed plants in the parental generation had been passed along to the F_1 generation, although it remained hidden.

Genes and Alleles Mendel reasoned that since short stems reappeared in the F_2 plants, the hereditary factor that causes short stems had to be an individual unit, like a particle, and not like a liquid that could be mixed with another liquid of a different color. This particulate factor of inheritance is now called a **gene**. While Mendel did not use that term, we will use it in the following discussion for clarity.

A gene influences a specific trait in an organism, such as the length of a pea stem or the color of a tiger's coat. The gene is not the trait itself, but a factor that causes the organism to develop a specific trait.

Mendel's results showed that genes can come in different forms, which we now call alleles. An **allele** (AL-eel) is an alternative form of a gene. In pea plants, the gene for stem length has two alleles, one causing long stems and one causing short stems. Likewise, the gene for coat color in tigers has two alleles, one for orange fur and one for white fur.

We now know that a gene is a portion of a DNA molecule in a chromosome [FIGURE 8.7 and TABLE 8.1]. Although an individual chromosome may contain thousands of genes controlling hundreds of different traits, each chromosome will have just one allele (alternative form) for any individual gene.

Dominant and Recessive Alleles Mendel realized that the reappearance of short-stemmed plants in the F_2 gen-

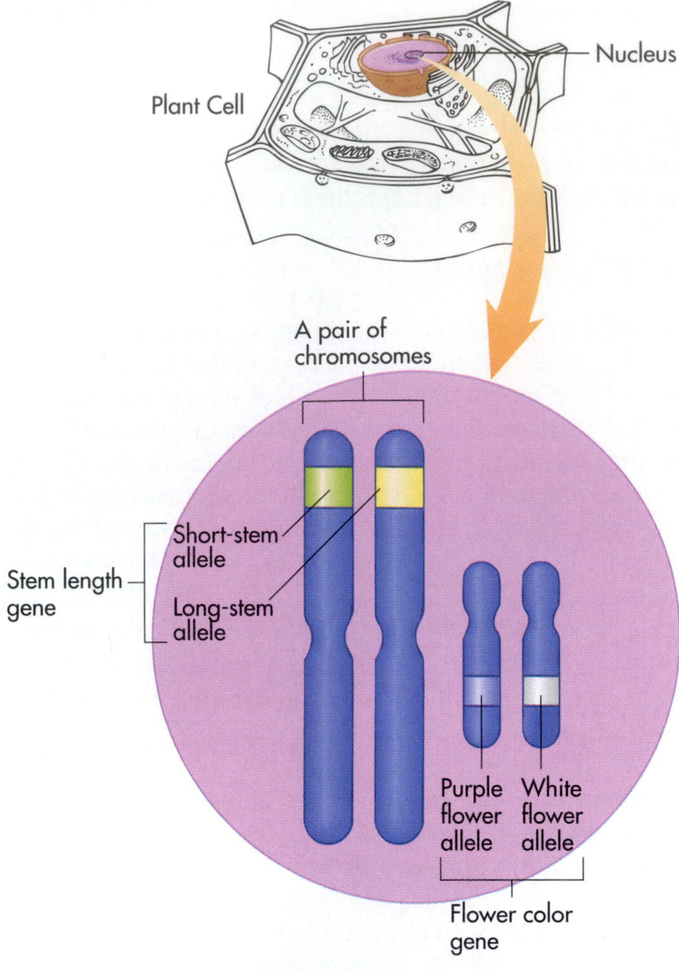

FIGURE 8.7

Genes, Alleles, and Chromosomes.

Although Mendel discovered the entities of inheritance we now call genes, he was not aware of chromosomes, cell organelles containing the genes. Today we know that most cells in a multicellular eukaryotic organism contain two copies of each chromosome. These two copies are a pair of homologous chromosomes. Each specific gene lies at a specific place on one of the chromosomes. For example, the stem length gene is at a specific location on a certain chromosome in peas. The two different copies of that long chromosome might have different alleles, or alternative forms, of the stem length gene, for example, long stem length or short stem length. A shorter chromosome might have a certain flower color gene. The two copies of that shorter chromosome likewise might have two different alleles—white versus purple color.

TABLE 8.1 Principles of Heredity

1. A hereditary trait is governed by a sequence of DNA (or in some viruses, RNA) called a *gene*.
2. Genes reside on chromosomes.
3. The gene for each trait can exist in two or more alternative forms. Called *alleles*, these forms, interacting with the environment, help determine an organism's external appearance, biochemical functioning, and behavior.
4. Most higher organisms have two copies of each gene in body cells (they are *diploid*). Gametes (egg or sperm), however, are *haploid* (have one copy of each gene).
5. *Homologous chromosomes* are two chromosomes that are similar in size, shape, and genetic content.
6. A *homozygote* has two identical alleles of a gene; a *heterozygote* has two different alleles of a gene.
7. A heterozygote may have visible traits dictated by only one of the alleles, called the *dominant* allele. The hidden allele is called the *recessive* allele.
8. *Phenotype* is an individual's physical makeup, the way it looks and functions; *genotype* is an organism's genetic makeup.
9. Pairs of alleles separate, or *segregate*, before egg and sperm formation, and each gamete is *haploid* (has one copy of each gene). At fertilization, sperm and egg combine randomly with respect to the alleles they contain, and the resulting zygote is diploid.
10. According to the principle of *independent assortment*, genes on different chromosomes assort into gametes independently of each other.
11. Linked genes lie on the same chromosome and tend to be packaged into gametes together.

eration meant that the short-stem allele was present but invisible in the hybrid, F_1 plants. If the short-stem allele had not been present in the hybrid, it could not have been passed on to the F_2 offspring. Since the hybrids showed the long-stem trait, Mendel knew that the long-stem allele was also present in the hybrid generation. For this reason, Mendel concluded that a hybrid contained two copies, or alleles, of each gene, one visible allele and one invisible allele.

The allele whose trait shows in a hybrid is said to be **dominant**. The allele that is overshadowed each time it is paired with a dominant allele is said to be **recessive**. This explains the coat color in a tiger like Mohan's hybrid daughter Radha. She had one dominant allele for an orange coat and one recessive allele for a white coat, but looked just like a pure-breeding orange tiger with two dominant alleles for orange coat. Likewise, the short-stem allele in pea plants is recessive to the dominant long-stem allele.

In his work with pea plants, Mendel reasoned that since the hybrid has two alleles of each gene, the pure-breeding parents must have two copies of each gene as well. But in the case of the pure-breeding parents, both alleles are identical. About 40 years after Mendel pub-

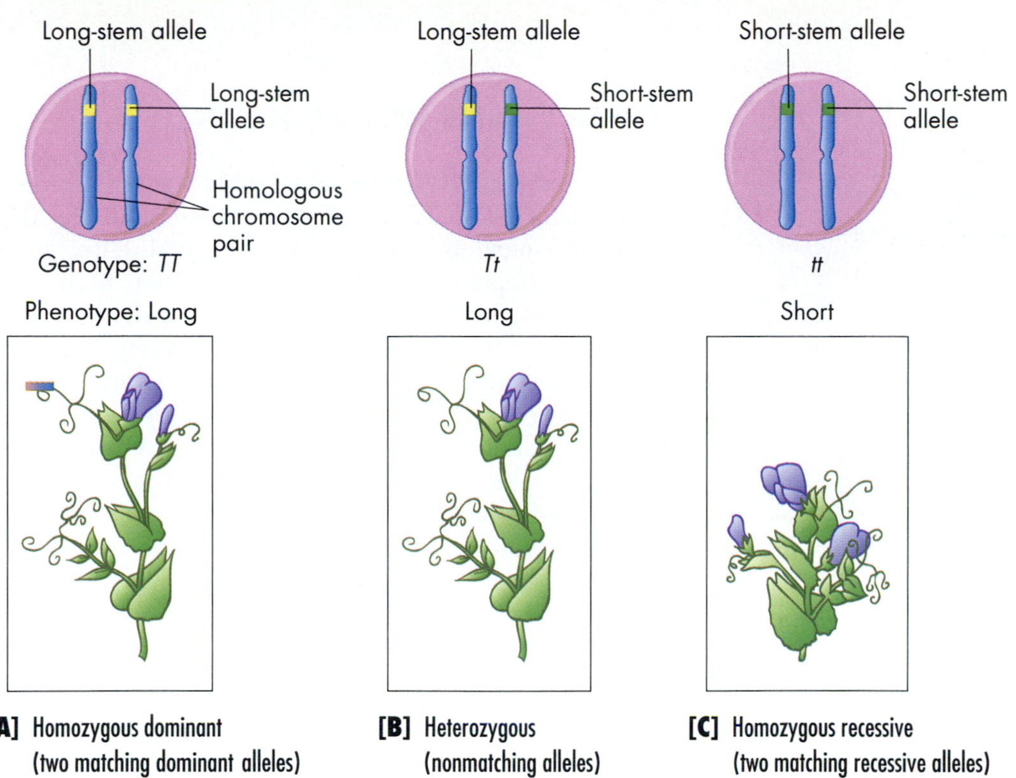

[A] Homozygous dominant (two matching dominant alleles)

[B] Heterozygous (nonmatching alleles)

[C] Homozygous recessive (two matching recessive alleles)

FIGURE 8.8

Mendel's Synthesis.

Underlying an organism's visible, physical appearances (its phenotype) is a pair of alleles for each gene (the genotype) that determine the phenotype. Because a dominant allele such as the one for long stems can hide a recessive allele such as the one for short stems, a homozygote with two dominant long-stem alleles **[A]** will have the same phenotype as a heterozygote **[B]** with one copy of each allele, even though it has a different genotype. Only the homozygous recessive **[C]**, with two matching short-term alleles, will have the short-stem phenotype.

lished his work, biologists confirmed that most large familiar organisms possess two alleles for each gene because each cell contains two copies of each chromosome [see FIGURE 8.7]. One chromosome of each pair might bear one allele, for example the short-stem allele of the stem-length gene. At the same time, the other chromosome of the pair could bear the other allele (the long-stem allele).

Genotype and Phenotype Mendel realized that the long-stemmed F_1 hybrid plants looked like the long-stemmed pure-breeding plants of the parent generation but they were genetically different. The physical characteristics of an organism, such as its stem length or fur color, is its **phenotype**. The genetic makeup of that individual is its **genotype**. In the case of a long-stemmed phenotype, the genotype could be either two dominant long-stem alleles or one long- and one short-stem allele [FIGURE 8.8A and B]. Organisms with two different types of alleles for a given trait are said to be **heterozygous** for that trait [FIGURE 8.8B]; pure-breeding organisms, with a pair of identical alleles for a given trait, are **homozygous** for that trait [FIGURE 8.8A and C]. (A heterozygous individual is called a **heterozygote**, and a homozygous individual is called a **homozygote**.) Thus, the pure-breeding long-stemmed and short-stemmed parents were homozygotes, while their hybrid offspring were heterozygotes.

CONCEPT CHALLENGE

Can you always tell an organism's genotype by looking at its phenotype? Why or why not?

Mendel's Segregation Principle

You have seen that a heterozygote for a particular gene has two different alleles of that gene, for example, one long-stem allele and one short-stem allele. Mendel suggested that an offspring receives one allele from each parent. This implies that the two alleles from a parent separate, or *segregate*, from each other before gamete (sex cell) formation, producing eggs or sperm that have only one allele apiece.

Let's say a parent plant's genotype includes two different alleles for the stem-length gene: one long-stem allele and one short-stem allele [FIGURE 8.8B]. From what we have learned about meiosis we can understand that an individual egg cell or an individual sperm cell will bear either one allele or the other, but not both [FIGURE 8.9B]. A cell containing just one allele of each gene is said to be *haploid* and a cell with two alleles of each gene is *diploid*. Eggs and sperm are haploid, but parent pea plants are diploid.

If we generalize from Mendel's pea experiments, we can define his **segregation principle**: Sexually reproducing diploid organisms have two copies of each gene, which segregate from each other during meiosis *without blending* or being altered in any way. Thus, when gametes form, they each contain only one copy of each gene.

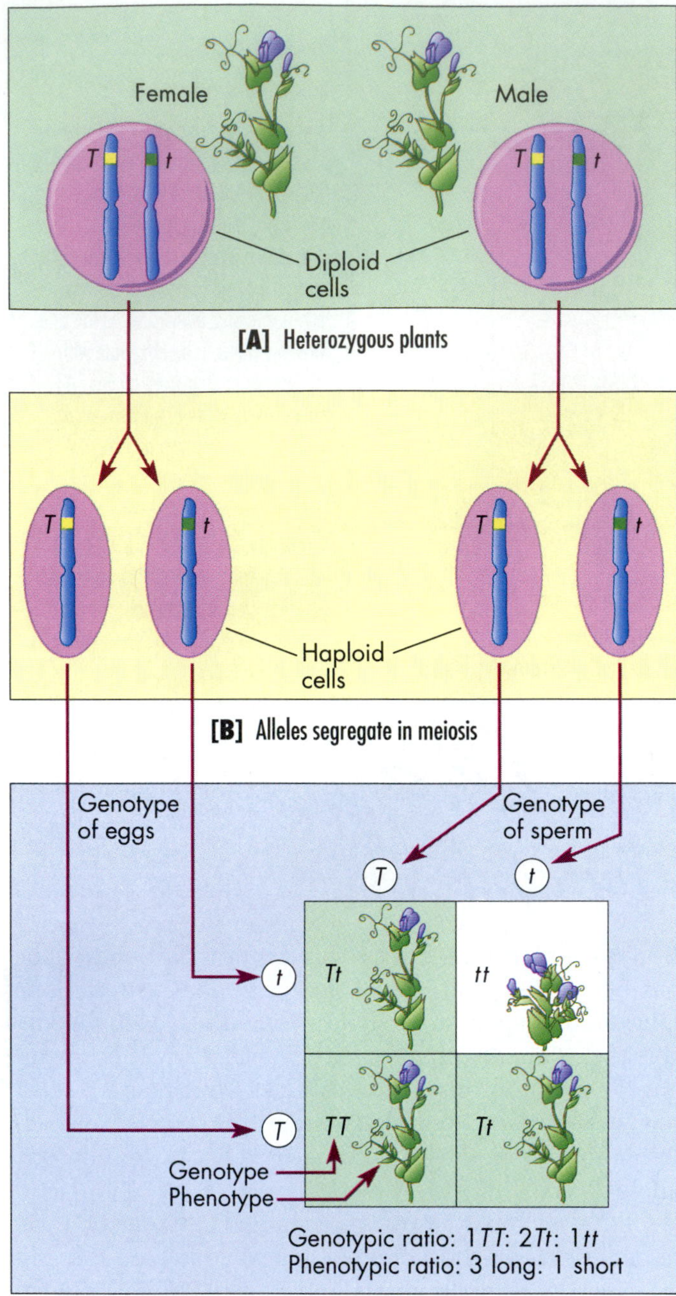

FIGURE 8.9

Meiosis and Mendel's Principle of Segregation.

Mendel's principle of segregation follows from the process of meiosis. **[A]** These individuals heterozygous for the long (*T*) and short (*t*) alleles of the stem length gene are like the F_1 of FIGURE 8.6. **[B]** After chromosome replication, cells pass through meiosis I and II, resulting in haploid cells. Each of these cells has a single allele of the stem-length gene, either *T* or *t*. Thus, the *T* and *t* alleles have segregated away from each other. **[C]** During fertilization, alleles recombine at random. This can be sketched by using a Punnett square. The offspring have the genotypic ratio of 1 *TT* to 2 *Tt* to 1 *tt*, and a phenotypic ratio of 3 long to 1 short.

GENETIC SYMBOLS AND PUNNETT SQUARES

The segregation principle is probably easiest to understand using symbols to represent genes and alleles [FIGURE 8.9]. Geneticists use uppercase (capital) letters to indicate dominant alleles, and lowercase letters to indicate recessive alleles. If *T* represents the long-stem allele of the gene for stem length, then *t* represents the short-stem allele of that gene. The heterozygous F_1 generation is then shown as *Tt* [see FIGURE 8.9A].

During the meiotic cell divisions that occur prior to gamete formation in the *Tt* heterozygote [FIGURE 8.9A], the alleles (one on each chromosome of a homologous pair) segregate (separate). As a result, half the gametes end up with the *T* allele and the other half with the *t* allele [FIGURE 8.9B]. Mendel then predicted *random fertilization*, that is, that eggs and sperm come together totally at random with respect to the allele they carry [FIGURE 8.9C]. In other words, an egg cell with a *t* allele is just as likely to be fertilized by a sperm cell with a *t* allele as it is to be fertilized by a sperm carrying a *T* allele.

A good way to visualize the consequences of random fertilization is to draw a diagram called a **Punnett square**, named after the British mathematician who first devised it. To construct a Punnett square for the mating of two pea plants that are heterozygous for stem lengths, draw a large square made up of four smaller squares. Write the two pollen types produced by meiosis along one side of the square (*T* or *t*) and the two egg types along the top (*T* or *t*) [FIGURE 8.9C]. Then fill in the empty boxes with the genotypes of the offspring that result from the fertilization of each egg type with each pollen type. For a tutorial in the use of Punnett squares, work through the Active Learning Box on page 205.

FIGURE 8.9C shows four F_2 genotypes: *TT*, *Tt*, *tT* and *tt*. Since the order of alleles is not important, *Tt* and *tT* are equivalent, and so there are really only three genotypes, found in the ratio 1*TT*:2*Tt*:1*tt*. If we look at the physical characteristics of the plants themselves, however, we find that the 1:2:1 genotypic ratio produces a 3:1 phenotypic ratio (3 long stem to 1 short stem). The reason is that the single *TT* genotype and both *Tt* genotypes have the same long-stem phenotype because *T* is dominant over *t*. So the phenotypic ratio expected from the segregation principle and random fertilization is (1 + 2):1, or three long-stemmed plants to one short-stemmed plant, a result close to what Mendel actually counted.

Mendel's segregation principle explains one of the mysteries concerning the white-coated tiger Mohan. Mohan's orange F_1 offspring, including Radha, were hybrids bearing one white-coat allele received from the white-furred Mohan, and one orange-coat allele from their normal, orange-furred mother. Owing to the fact that the F_1

active learning: How to use a Punnett square

The purple-flower allele (*P*) is dominant to the white-flower allele (*p*) in peas. What are the genotypic and phenotypic ratios in the offspring of two heterozygous (*Pp*) purple flowers? (Enter your responses in the red boxes.)

1. Determine the genotype of the gametes.

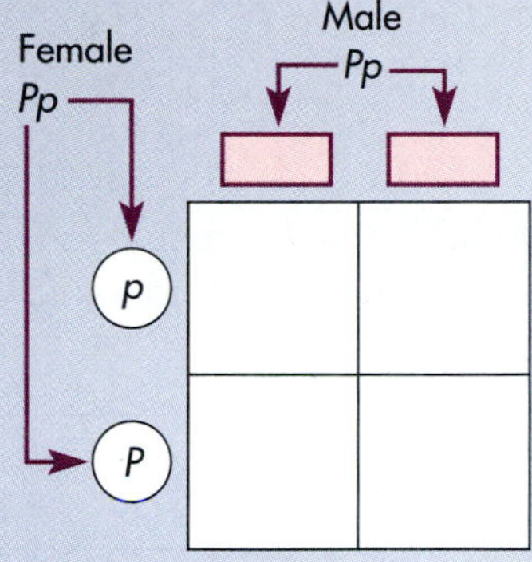

2. Combine gametes.

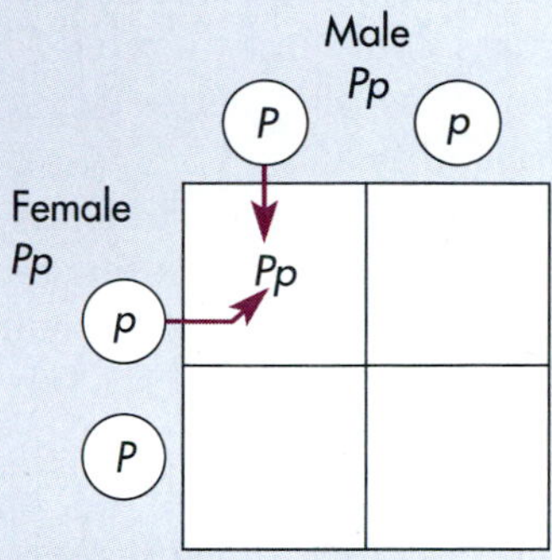

3. Continue combining gametes.

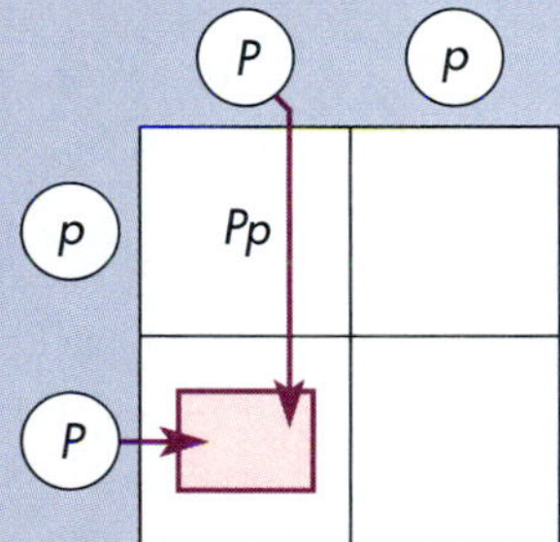

4. Finish filling in box.

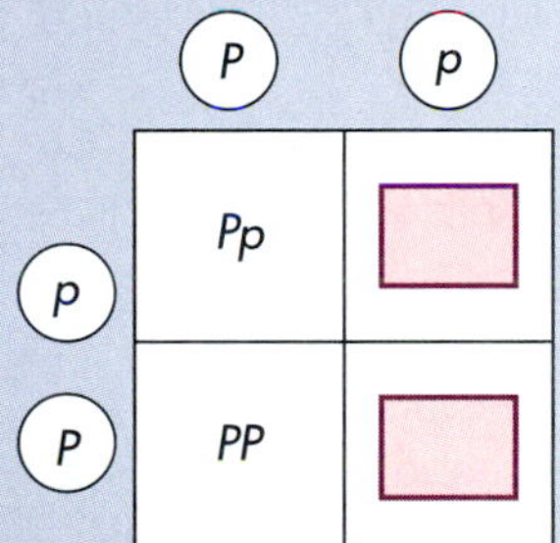

5. Determine phenotypes.

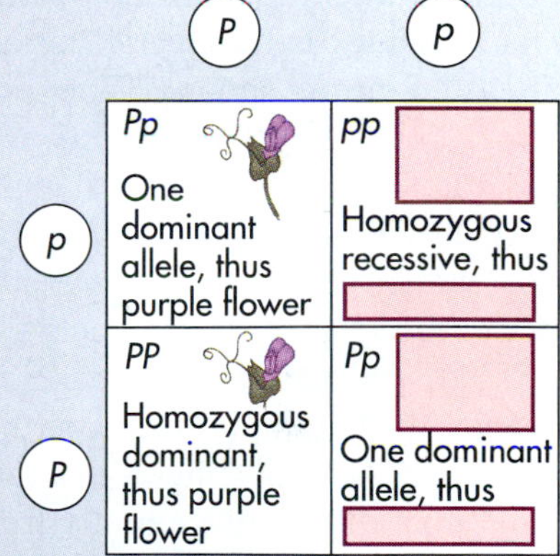

6. Determine genotypic ratio.

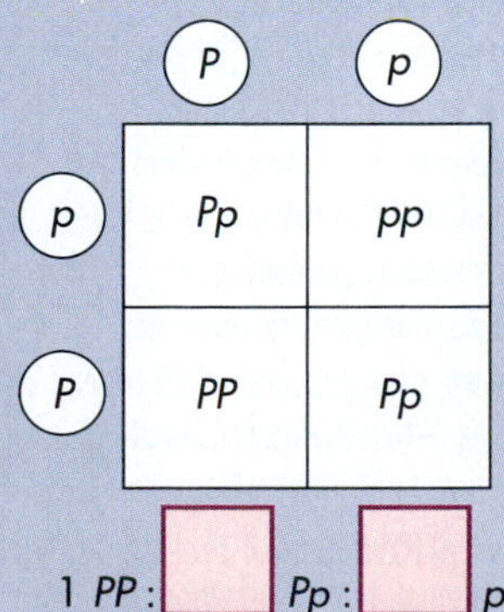

7. Determine phenotypic ratio.

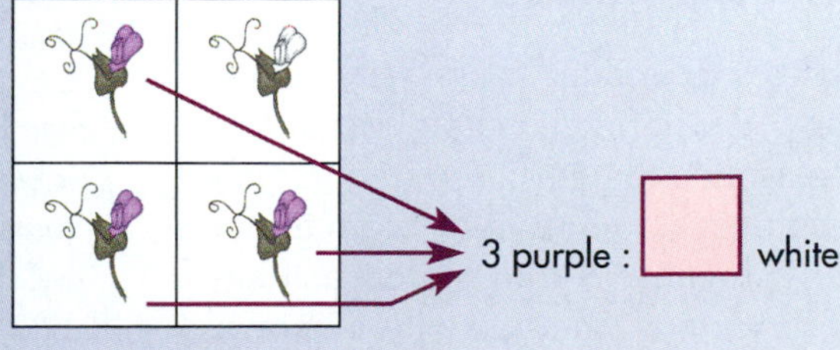

heterozygotes had orange fur shows that the white-coat allele is recessive and the orange-coat allele dominant.

PEAS AND PROBABILITIES: PREDICTING PHENOTYPIC RATIOS

Mendel's segregation principle predicts a 3:1 ratio in the offspring from a mating of two heterozygotes. But he actually found 2.84:1 for the *Tt* × *Tt* mating, and 3.15:1 and 2.96:1 for other matings [see FIGURE 8.5]. Why don't the figures come out to exactly 3:1? The answer is that the principles of genetics rely on the laws of chance and probability.

You can demonstrate the probability of obtaining the 3:1 relationship by tossing two different coins simultaneously. Let a penny represent sperm from a pollen grain, and let a nickel represent an egg. The head of each coin represents the *T* allele, and the tails represent the *t* allele, so each coin has an equal number of *T* and *t* alleles, just like the population of gametes from a heterozygote.

To represent fertilization, flip both coins at the same time and record whether they land heads up or tails up. If both are heads, the genotype is *TT*; two tails represent a *tt* genotype; and if one coin is heads and the other is tails, the "offspring" will be heterozygous (*Tt*). Flip the coins 20 times. Is the overall ratio exactly what you got using the Punnett square? (You would expect $\frac{1}{4} \times 20 = 5$ heads/heads; $\frac{1}{2} \times 20 = 10$ heads/tails; and $\frac{1}{4} \times 20 = 5$ tails/tails.) If not, how can you explain the discrepancy?

What would be the probable result if you tossed the coins many more times than 20, say 7324 times, as Mendel did when he was experimenting with round and wrinkled seed shape? (Review FIGURE 8.5.) Like the toss of a coin, the combination of sperm and egg alleles is governed by the laws of chance. In a low number of trials, the results may differ substantially from those predicted for random tossing, but as the number of trials increases, the results will come closer to the mathematically predicted values.

A TESTCROSS CAN DISTINGUISH GENOTYPES

When an organism shows a dominant characteristic, we cannot know by looking at it whether it is homozygous or heterozygous. Mendel reasoned that he could distinguish a heterozygote from a homozygote only if he were to observe the traits of their progeny. So he mated a long-stemmed individual of unknown genotype with a known homozygous recessive short plant. He reasoned that if the long-stemmed parent was homozygous *TT*, then all the offspring would be *Tt* and have long stems [FIGURE 8.10A]. If, on the other hand, the unknown parent was

You can distinguish a homozygous dominant individual from a heterozygous individual by mating it to a known homozygous recessive

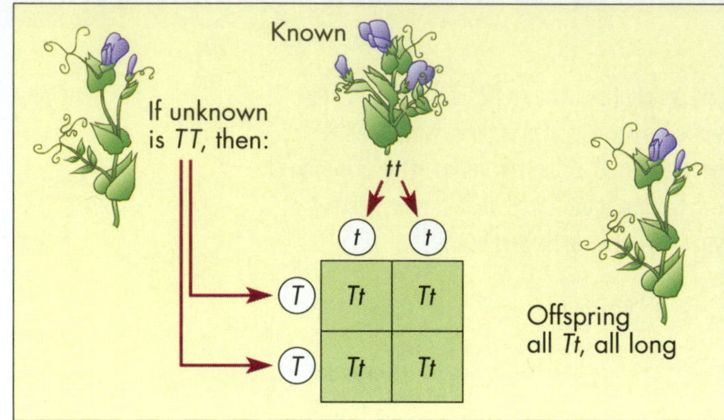

[A] If unknown is homozygous dominant

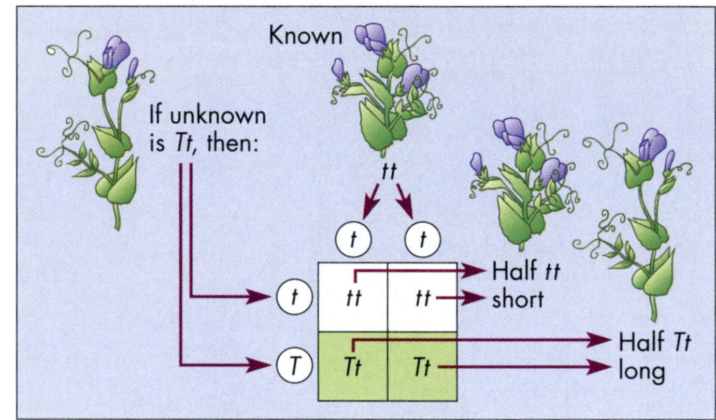

[B] If unknown is heterozygous

FIGURE 8.10

The Testcross: A Way to Determine Genotypes.

A homozygous dominant individual has the same phenotype as a heterozygote. So how can a geneticist tell them apart? The trick is to mate the unknown individual to a known homozygous recessive. If the unknown is homozygous dominant [**A**], then all of its gametes will have the dominant allele, in the case of the stem-length gene of peas, *T*. When these *T* alleles join with the recessive *t* alleles from the homozygous recessive known tester individual at fertilization, all the offspring will be *Tt* and have the dominant phenotype, in this case long stems. If the unknown is heterozygous *Tt* [**B**], then half of the gametes will have the recessive *t* allele. When these combine at fertilization with the *t* alleles of the known recessive tester, they will produce homozygous recessive offspring. The difference is clear: In a testcross, the offspring of a homozygous dominant are all dominant, but the offspring of a heterozygote are half recessive and half dominant.

FIGURE 8.11

Curly-Eared Cats.

In 1981, a stray cat with oddly curled ears moved in with a family in Lakewood, California. Today, the animal's descendants are exciting feline fanciers all over the world. When that curly-eared cat (which resembled this black male) mated with a normal cat, about half the kittens had curly ears and the rest were normal. To explain these results, geneticists saw this as a testcross between the curly-eared cat, which had to be heterozygous for a dominant allele causing the curly ears, and a normal cat that was homozygous for the recessive normal allele.

heterozygous *Tt*, then a mating with a homozygous recessive *tt* plant would generate offspring that were half long-stemmed (*Tt*) plants and half pure-breeding short-stemmed (*tt*) plants [FIGURE 8.10B]. Since the two different long-stem genotypes produced different kinds of progeny in this cross, Mendel could distinguish the parental genotype. This technique, mating a phenotypically dominant individual with a known homozygous recessive, has come to be known as a **testcross**.

Testcrosses have proven useful to geneticists for over a century. A recent instance was the curly-eared cat discovered in 1981 [FIGURE 8.11].

Let's apply the principle of a testcross to the mating between Mohan and his hybrid daughter Radha. We can consider the mating as a testcross between the homozygous recessive white Mohan (whose genotype could be written as *oo*, for white, or nonorange) and his daughter Radha, who must have been heterozygous (genotype *Oo*). Viewed in this way, the genetic basis for the appearance of the second-generation cubs becomes clear: some were homozygous white (*oo*) and others were heterozygous orange (*Oo*).

CONCEPT CHALLENGE

A zebrafish bearing long fins mates with another long-finned zebrafish, and their offspring show a ratio of 3 long-finned fish to 1 short-finned fish. Are the two parents heterozygotes or homozygotes for the long-fin allele? Are long fins dominant or recessive to short fins?

Mendel's Principle of Independent Assortment

At first, Mendel studied only one trait at a time. When he observed stem length, for instance, he did not also record the incidence of flower color or any other trait. After using these **monohybrid crosses** to establish the concepts of dominance, recessiveness, and segregation, Mendel was ready to find out what would happen if he studied *two* traits at a time—a **dihybrid cross** [FIGURE 8.12].*

THE RATIOS REVEAL INDEPENDENT ASSORTMENT

To study the inheritance of two different genes in the same cross, Mendel chose short-stemmed plants with white flowers (*ttpp*) and long-stemmed plants with purple flowers (*TTPP*). (Purple is dominant over white, so we represent the purple allele by *P* and the white allele by *p* [FIGURE 8.12A].)

In the F_1 generation produced by mating *TTPP* with *ttpp*, all the offspring had purple flowers at the ends of long stems [FIGURE 8.12B]; in other words, the dominant forms for both stem length and flower color appeared in the F_1 generation.

Mendel then allowed the F_1 plants to self-fertilize and produce the F_2 generation. The outcome of this mating was not so easy to predict. Would purple flowers always be found with long stems, as in the original parents? (Offspring showing the same traits as the parents are referred to as **parental types**.) Or would some purple flowers appear on short stems? (Offspring with combinations of traits different from the parents are called **recombinant types**.)

When Mendel looked at his field of F_2 pea plants, he saw that indeed new combinations had formed [FIGURE 8.12C]. Some of the plants had short stems and purple flowers, while others had long stems and white flowers. Evidently, the alleles had reassorted during the formation of gametes. When Mendel counted the number of each type of plant, he found approximately the following ratio: $9/16$ long-stemmed purple to $3/16$ long-stemmed white to $3/16$ short-stemmed purple to $1/16$ short-stemmed white—a 9:3:3:1 ratio [FIGURE 8.12D]. If, however, he looked at just flower color or just stem length, he found a phenotypic ratio of 3:1. For instance, there were $9 + 3 = 12$ purple-

* Note that when writing of monohybrid or dihybrid crosses, a geneticist means that attention is being paid to only one or two pairs of alleles, respectively. It does *not* mean that the organisms differ in only one or two features. In reality, it is unlikely that a pair of organisms would differ in only one or two features.

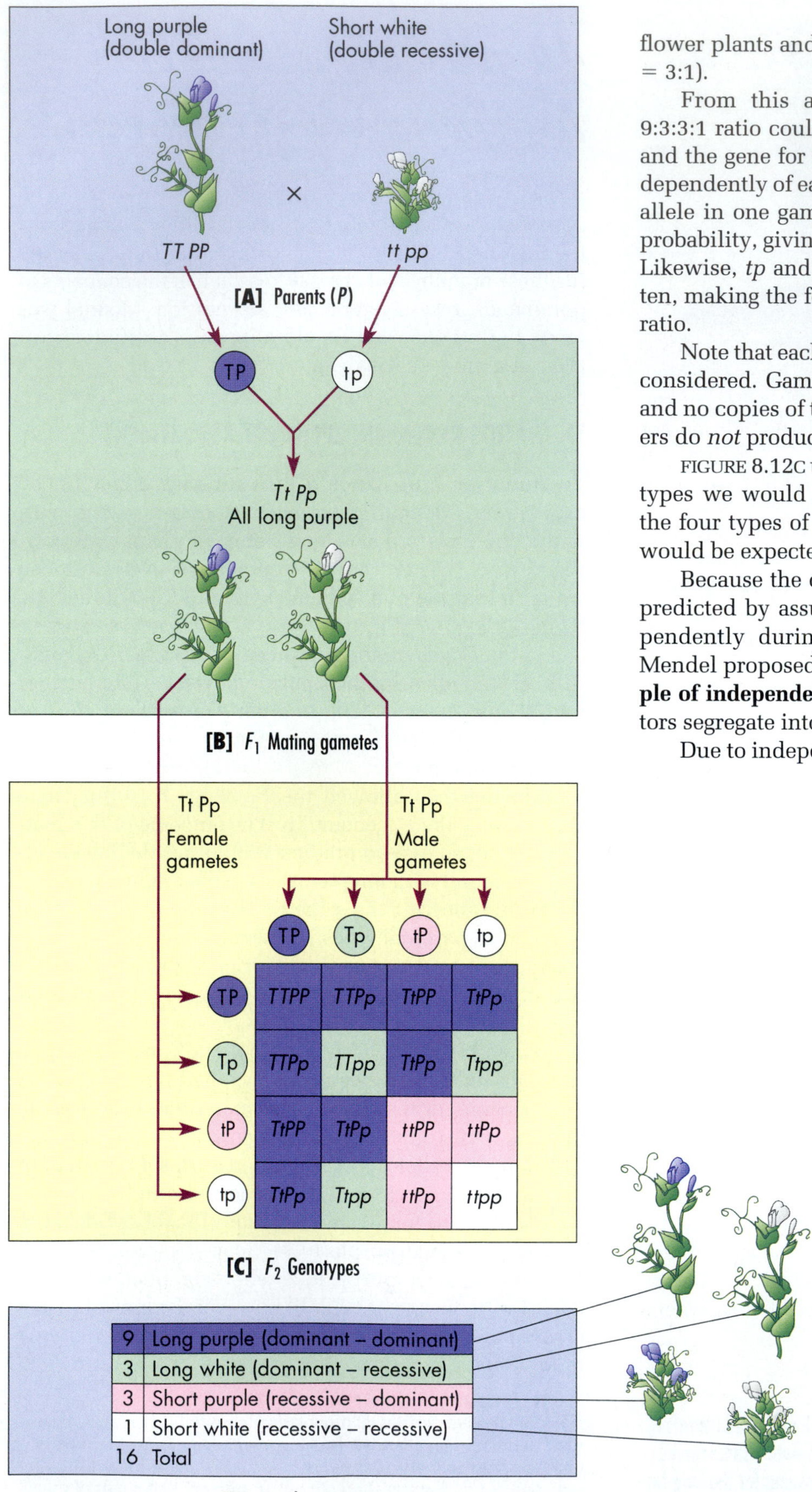

flower plants and 3 + 1 = 4 white-flower plants (12 to 4 = 3:1).

From this analysis, Mendel concluded that the 9:3:3:1 ratio could arise only if the gene for stem length and the gene for flower color segregated, or assorted, independently of each other into four types of gametes. A *T* allele in one gamete was found with *p* or *P* with equal probability, giving equal numbers of *Tp* and *TP* gametes. Likewise, *tp* and *tP* gametes would occur equally as often, making the four types *TP*, *Tp*, *tP*, and *tp* in a 1:1:1:1 ratio.

Note that each gamete has only one copy of each gene considered. Gametes with two copies of the same gene and no copies of the other gene are not found—i.e., flowers do *not* produce gametes of the type *TT* or *Pp*.

FIGURE 8.12C uses the Punnett square to find the genotypes we would expect from a random combination of the four types of gametes. Sixteen (4 × 4) combinations would be expected with a phenotypic ratio of 9:3:3:1.

Because the outcome of the dihybrid cross could be predicted by assuming that two gene pairs assort independently during the formation of eggs and sperm, Mendel proposed his second rule of heredity, the **principle of independent assortment**: Different hereditary factors segregate into gametes independently of each other.

Due to independent assortment [FIGURE 8.12] some of

FIGURE 8.12

Results of a Dihybrid Cross.

[A] A double-dominant long-stemmed purple-flowered plant *TTPP* mated to a double-recessive short-stemmed white-flowered plant *ttpp* gives **[B]** F_1 progeny that are all double heterozygous *TtPp* with long stems and purple flowers. **[C]** When the F_1 generation self-fertilizes, some of the F_2 generation represent new combinations: plants with long stems and white flowers and other plants with short stems and purple flowers. **[D]** These results can be explained by Mendel's principle of independent assortment. The Punnett square in **[C]** models the results. The double heterozygous plants make four types of gametes. When these combine randomly at fertilization, 16 boxes are filled with various genotypes. These genotypes provide a phenotypic ratio of 9:3:3:1—9 long purple to 3 long white to 3 short purple to 1 short white.

the offspring were unlike either parent. Neither parent showed both short stature and purple flowers but some of the offspring did. How was this possible? **Recombination**, which produces new genetic combinations, had occurred during meiosis. Recombination provides the opportunity for evolution to proceed in a new direction, offering nature something new to "select."

As an example, let's consider what might happen if the pea plants were growing in an environment that suddenly changed so that the wind speed increased and the levels of damaging ultraviolet light became much stronger. (Both of these changes, by the way, are easily possible in our present world, as human activities have brought about global warming and the thinning of the protective layer of ozone which screens out ultraviolet light.) In such an altered environment, pea plants with long stems would be at a disadvantage because the wind might blow them over before they could reproduce. Likewise, plants with white flowers (and thus no dark protective pigment) might be at a disadvantage because the strong ultraviolet irradiation might damage the flower's petals. A plant with short stems, however, would be less likely to be blown over, and one with deep purple pigments in petals would receive a measure of protection from ultraviolet light. Thus, short plants with purple flowers would be able to survive and reproduce better than plants with any other combination of these traits. Recombination, occurring through meiosis and due to independent assortment, would provide plants with new genetic combinations, some of which, like the short, purple pea plants, might be selected for by natural selection, altering an evolutionary pathway.

CONCEPT CHALLENGE

A plant geneticist mates a corn plant with smooth yellow seeds to a corn plant with wrinkled purple seeds. She then mates the F_1 plants to each other to produce the F_2. In the F_2, the geneticist finds a phenotypic ratio of 9 purple smooth to 3 purple wrinkled to 3 yellow smooth to 1 yellow wrinkled. For the seed-color gene, is the yellow or purple allele dominant? For the seed-shape gene, is the wrinkled or smooth allele dominant?

MENDEL'S RESULTS IGNORED AND REDISCOVERED

Biologists of Mendel's time paid no attention to the monk's analyses of inherited characteristics. Mendel published his results in 1865 in a journal that was circulated in about a hundred scientific libraries, but there was practically no response to the monk's paper. The few plant breeders who cited Mendel's findings indicated by their remarks that they did not understand the importance of his three main concepts:

1. Hereditary factors (genes) are inherited as distinct units.

2. Alleles separate before the formation of gametes (sex cells) and join together randomly at fertilization.

3. Two genes assort independently during sexual reproduction.

Perhaps Mendel's ideas were not appreciated because he was a monk, and not a "scientist" who was part of a scientific community, or because his theories were essentially mathematical, and at that time mathematicians and plant breeders worked in completely different realms. At any rate, the original work of Mendel lay practically unnoticed. Discouraged, Mendel eventually gave up his breeding experiments and became abbot of his monastery.

Finally, in 1900, Mendel's ideas were "rediscovered" by three European botanists who performed plant experiments similar to those of Mendel. They did not know of Mendel's work at the time, and were later surprised at the similarity, but the most important thing was that they recognized the importance of Mendel's findings. The farmer's son who had seen a connection between mathematics and agriculture had launched a new era of genetics.

Genes Are Located on Chromosomes

While Mendel's report was gathering dust in European libraries, other biologists were advancing our understanding of cell structure. Particularly important were microscopic observations that showed how chromosomes move during mitosis and meiosis [FIGURE 8.13; CHAPTER 7]. Shortly after the rediscovery of Mendel's principles, biologists realized that there are several parallels between the inheritance of genes and the distribution of chromosomes during meiosis:

1. Two copies of each gene and two copies of each chromosome exist in each body cell [FIGURE 8.13A].

2. Pairs of alleles and pairs of homologous chromosomes both segregate during gamete formation [FIGURE 8.13A].

3. Genes for different traits and nonhomologous chromosomes both assort independently when egg and sperm are formed [FIGURE 8.13B].

These correlations suggested that genes were physically linked to chromosomes. To test that suggestion, in-

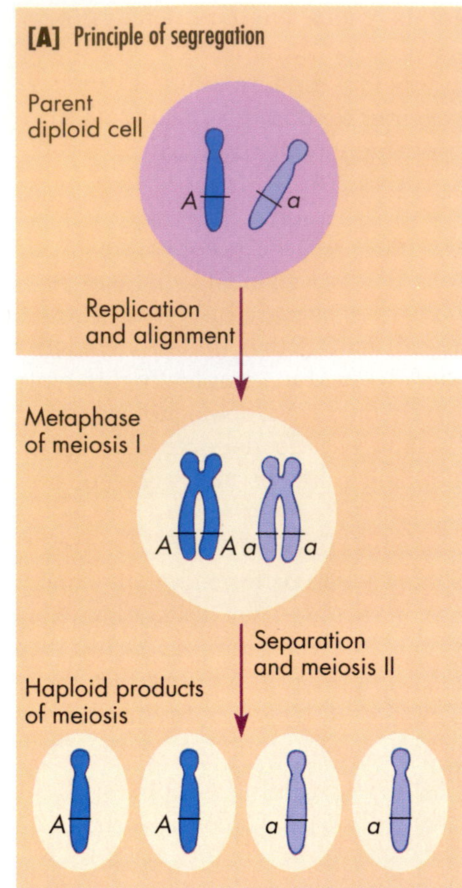

Alleles segregate into different gametes and therefore into different offspring, i.e., Alleles *A* and *a* separate during meiosis and appear in different haploid cells.

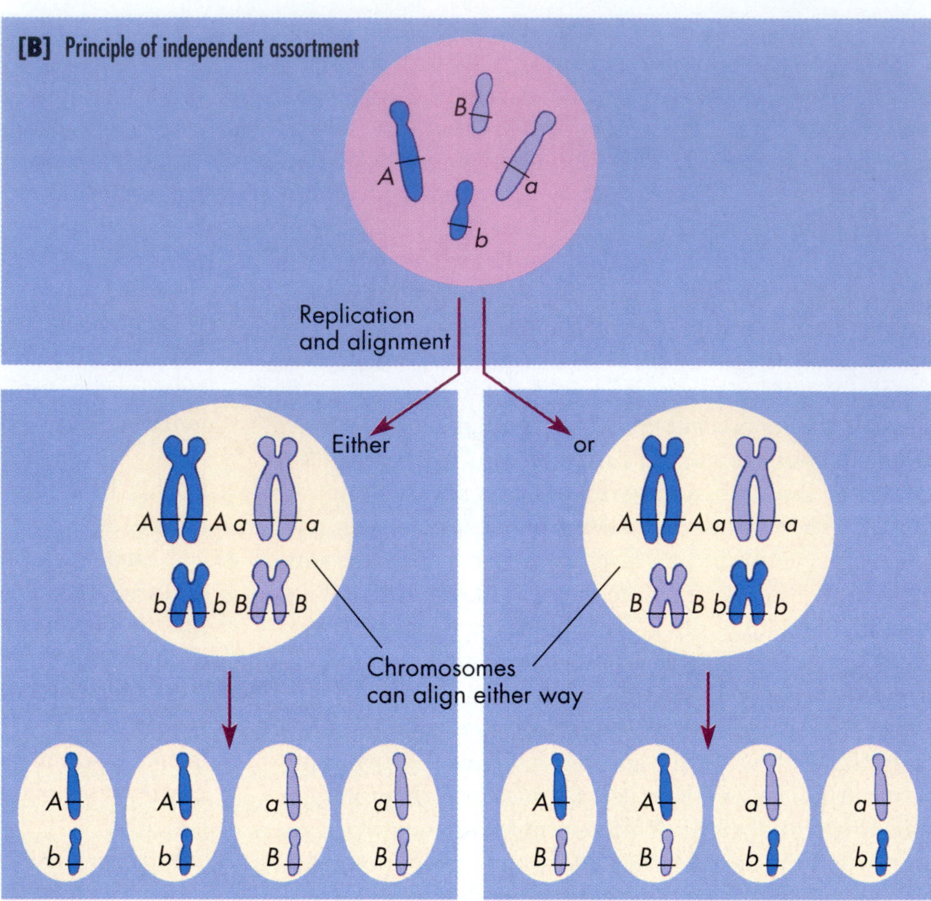

Alleles of genes on different chromosomes assort independently into haploid meiotic products, and fertilization occurs at random, i.e., *Ab* and *aB* gametes are equal in frequency to *AB* and *ab*.

FIGURE 8.13

Mendel and Meiosis.

An understanding of meiosis explains Mendel's principles. [A] Segregation occurs because homologous chromosomes separate from each other in meiosis. [B] Independent assortment occurs because non-homologous chromosomes can align independently during meiosis I, sometimes aligning as on the left, sometimes aligning as on the right.

vestigators would have to identify individual chromosomes and show that a specific trait is always transmitted with a specific chromosome. That became possible by investigations of chromosomes associated with gender.

SEX CHROMOSOMES

Have you ever wondered why there are roughly as many boy babies born as girl babies (the actual ratio is about 106 boys to 100 girls)? The reason became apparent from chromosomal studies. In 22 of the 23 pairs of human chromosomes, the two members of the pair are identical in size and shape; in the other chromosome pair, however, males and females differ. Biologists call chromosome pairs in which both chromosomes look the same in both sexes **autosomes**, while they call chromosome pairs with dissimilar members in males and females **sex chromosomes.** FIGURE 8.14 demonstrates this concept for fruit flies. Humans have 22 pairs of autosomes and one pair of sex chromosomes. In fruit flies, tigers, and people, females have two identical sex chromosomes, called ***X* chromosomes**, and males have one *X* chromosome and another, often smaller chromosome called a ***Y* chromosome**. Although sex chromosomes are common in animals, they are rarely found in plants, fungi, or protists.

The distribution of sex chromosomes during meiosis can account for the appearance of about equal numbers of males and females. An *XY* male is like the heterozygous parent, and an *XX* female is like the homozygous recessive in a testcross [FIGURE 8.15]. In the male, the *X* and *Y*

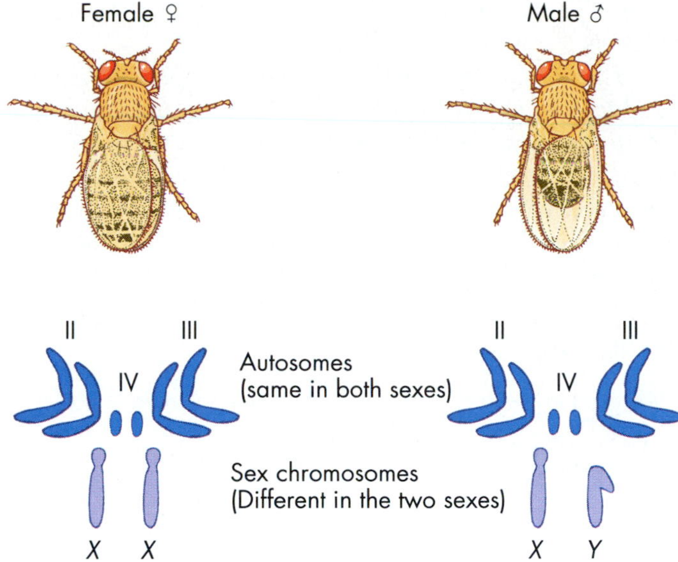

FIGURE 8.14

Sex and Chromosomes.

In fruit flies and many other animals, males and females have identical sets of autosomes but differently shaped sex chromosomes. Fruit flies have four chromosome pairs, three identical sets of autosomes and one pair of sex chromosomes. Males have one *X* and one *Y* sex chromosome, while females have two *X* chromosomes.

segregate during meiosis, and as a result, half the sperm contain a *Y* chromosome and half an *X* chromosome. In the female, the two *X* chromosomes segregate during meiosis; as a result, each egg contains one *X* chromosome. If the *X* and *Y* sperm randomly fertilize a group of eggs, then half of the zygotes formed will be male (*XY*) and half female (*XX*). Note that a male's single *X* chromosome must be inherited from his mother. Since males and females have different chromosomes, we know that at least one trait—gender—is regulated by chromosomes. But are there any others?

SEX-LINKED TRAITS

In 1910, Thomas Hunt Morgan and his associates at Columbia University began a series of experiments. Morgan wanted to find out about genes and chromosomes but did not have Mendel's monastic patience. Because Morgan didn't want to wait a year between generations, as Mendel had, he chose the fast-breeding fruit fly, *Drosophila melanogaster*. Although fruit flies are pests in the kitchen, they are ideal subjects for microscopic inspection, with brick-red eyes, tan abdominal stripes on a gray body, and glistening black bristles [FIGURE 8.14]. No bigger than an "l" in this sentence, fruit flies are easy to raise and breed, and they develop quickly. In just 12 days, an egg becomes a reproductive adult ready to produce hundreds of offspring.

One day, Morgan was observing some of his fruit flies under the microscope when he noticed a fly with white eyes instead of the usual red [FIGURE 8.16]. A *mutation*—a permanent change in the genetic material—had altered a gene for eye color from the normal red-eye allele (symbolized by *Drosophila* geneticists as w^+) to the mutant white-eye allele (w).

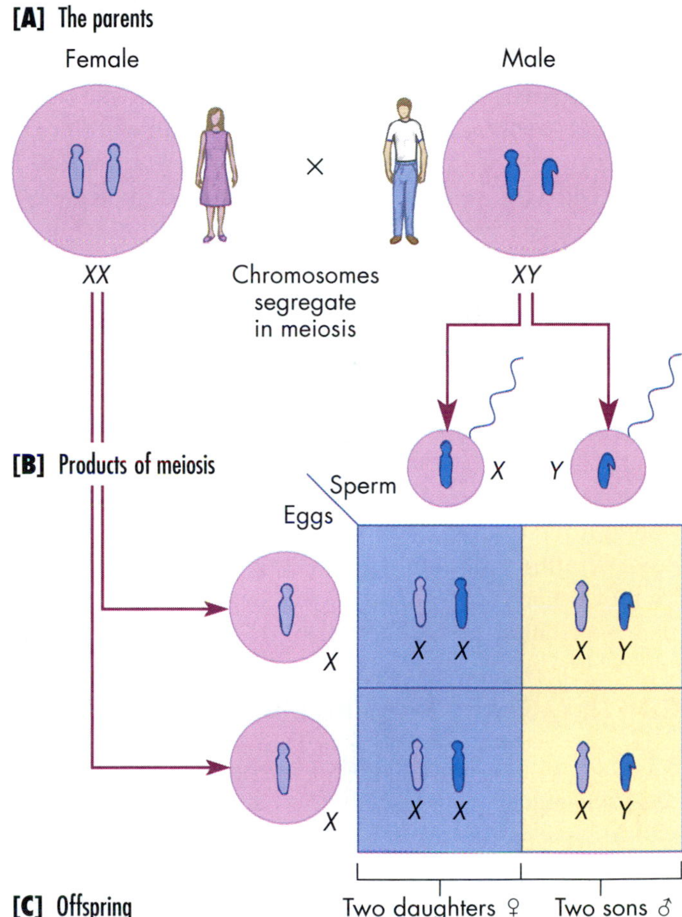

FIGURE 8.15

Inheritance of the *X* Chromosome.

[A] Women have two *X* chromosomes, but men have one *X* and one *Y* chromosome. [B] Sex chromosomes separate during the meiotic divisions that produce eggs and sperm. [C] Daughters inherit one of their *X* chromosomes from their father, while sons inherit their only *X* chromosome from their mother. Sons inherit their *Y* chromosome from their father.

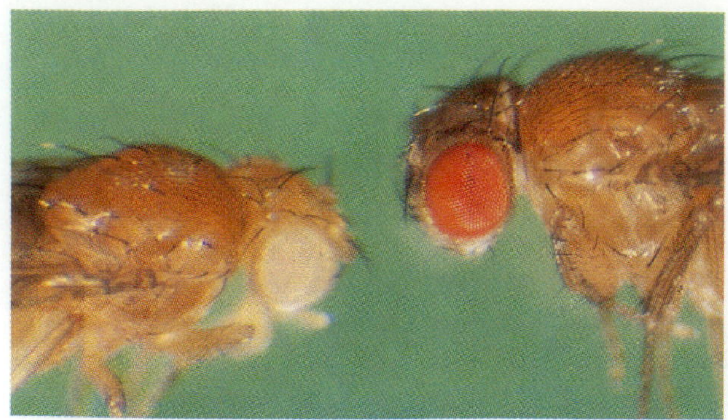

FIGURE 8.16

The White-Eye Mutation in *Drosophila.*

The fly on the right with red eyes is normal, or wild-type, in color, while the fly on the left is a white-eyed mutant (magnification 60×).

From a series of crosses, one of which is shown in FIGURE 8.17, Morgan realized that the gene for eye color is carried on the *X* chromosome. That gene is therefore a sex-linked gene, or more specifically, an **X-linked gene**.

After discovering and experimenting with several *X*-linked genes, Morgan and his coworkers drew an important conclusion: The *Y* chromosome carries no allele of the gene for eye color, or for most of the other *X*-linked genes. From this series of experiments, Morgan concluded that (1) genes are located on chromosomes, and (2) each chromosome carries many different genes.

Later work with human family pedigrees [review FIGURE 8.15] showed that the genes for muscular dystrophy, fragile-*X* retardation, hemophilia (bleeder's disease), and color blindness are among the 2000 or so genes located on the human *X* chromosome. We will discuss the inheritance of such genes in CHAPTER 11.

GENE LINKAGE AND CROSSING OVER

Morgan and his research group wondered how the many different genes located together on the same chromosome would be inherited. Would they assort independently as predicted by Mendel's second principle, or would they always be inherited together? Experiments revealed that two genes on the same chromosome tended to be inherited together in the original parental combinations, but that a few new combinations were found. The tendency of genes located on the same chromosome to be inherited together is called **linkage**.

To explain linkage, geneticists suggested that different genes residing on the same chromosome are physically linked, like beads on a string. However, if genes strung along the same chromosome tend to be inherited together, how can some offspring receive new combinations of alleles? The solution to this problem lies in the process of crossing over in meiosis, which we studied in CHAPTER 7 [review FIGURE 7.20].

If meiosis occurs without crossing over, the haploid products of meiosis will have chromosomes with the original parental genetic combinations [FIGURE 8.18A]. If, however, crossing over occurs, some of the resulting haploid cells will have new genetic combinations [FIGURE 8.18B]. If meiosis like that shown in FIGURE 8.18B with crossing over between two specific genes is a rare cir-

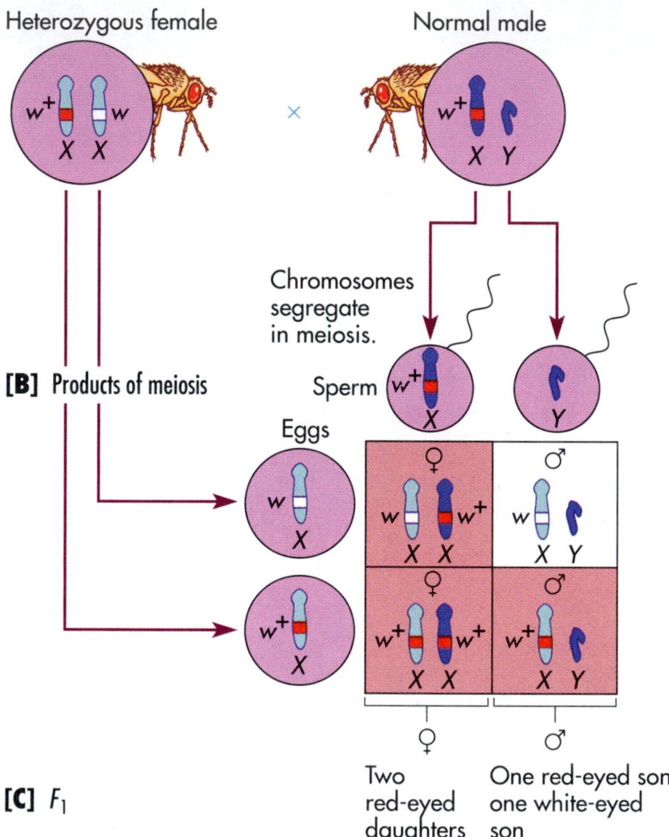

FIGURE 8.17

One of Morgan's Experiments with the White-Eye Color Gene.

[A] A female heterozygous for the white allele (w) and the dominant red allele (w^+) will have red eyes. **[B]** During meiosis, she will produce some eggs with a w allele and others with a w^+ allele. Her normal mate will produce some sperm with an *X* chromosome bearing the w^+ allele and others with a *Y* chromosome bearing no allele of the eye color gene. **[C]** In the next generation, white eyes reappear in one-quarter of the offspring—but always in males. Half the males are white-eyed. A Punnett square for these crosses reveals that daughters receive an *X* chromosome from their father as well as from their mother, while sons receive an *X* chromosome only from the mother. Therefore, a son's phenotype will reflect whichever allele is carried on the *X* chromosome he got from his mother.

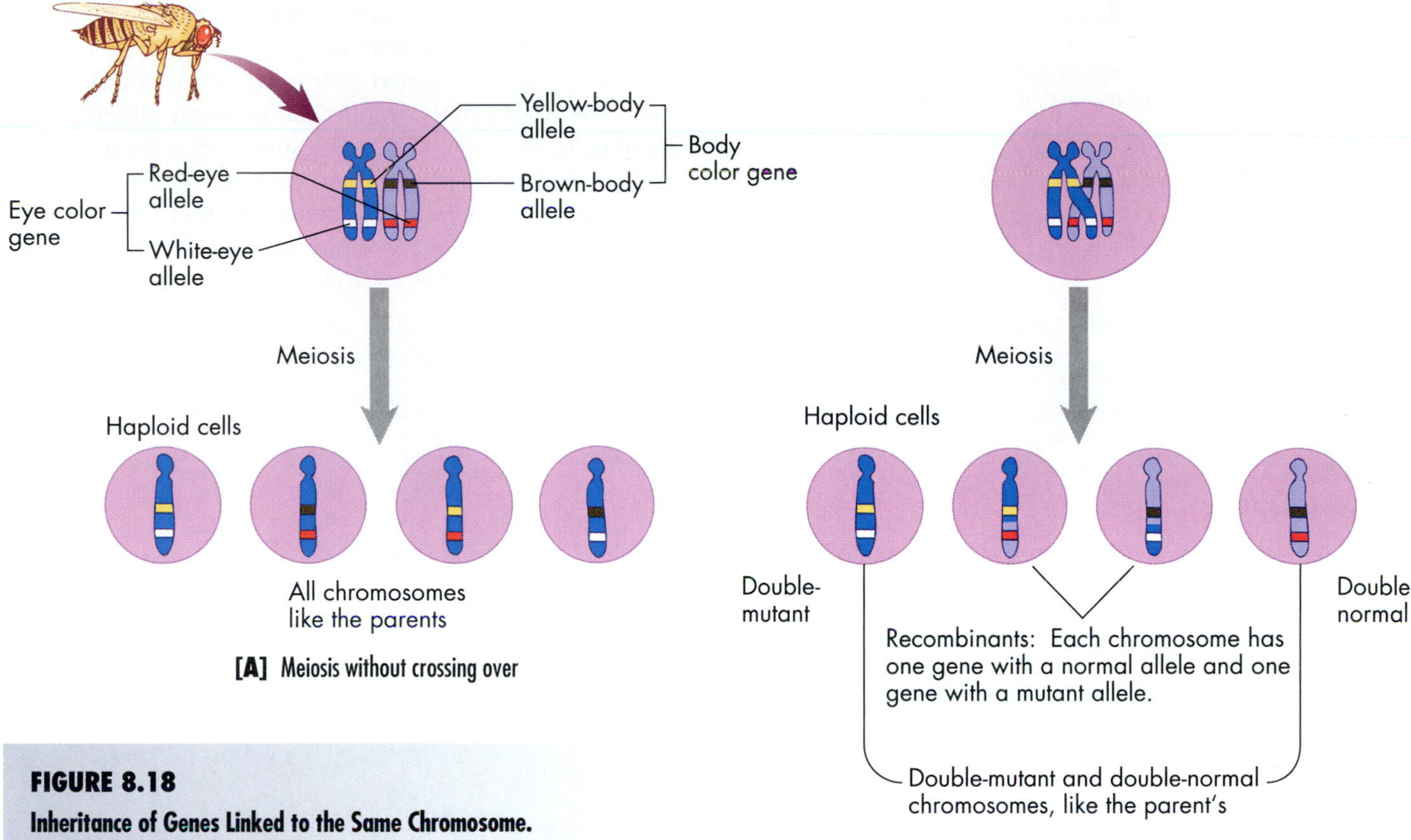

FIGURE 8.18

Inheritance of Genes Linked to the Same Chromosome.

[A] A body color gene (with yellow-body versus brown-body alleles) and an eye color gene (with white-eye versus red-eye alleles) are both linked on the *X* chromosome. If one copy of the *X* carries both recessives (alleles for yellow body and white eyes) and the other *X* chromosome carries both dominants (alleles for brown body and red eyes), then the eggs will have the double-recessive or the double-dominant arrangement of alleles if no crossing-over occurs. [B] If crossing-over occurs in meiosis, then recombinant chromosomes will appear in the haploid cells resulting from meiosis. Some recombinants will have the white-eye allele linked to the brown-body allele, and others will have the red-eye allele linked to the yellow-body allele.

cumstance, then recombined chromosomes will be rare among the population of haploid cells resulting from meiosis. Some of these new genetic combinations may provide an organism with different physical structures or abilities that might cause it to be favored by natural selection, thus contributing to evolutionary change.

GENETIC MAPPING OF CHROMOSOMES

Crossing over is important for our understanding of genetics for two reasons: (1) Crossing over allows for the reshuffling of alleles on the same chromosome, which, as we just saw, is important for providing new genetic variation upon which natural selection can act, and (2) it allows geneticists to determine the order of genes along a chromosome, a process called **genetic mapping**.

Genetic mapping is possible because the closer together two genes lie along a chromosome, the less the chance that a crossing-over event will occur between them. Conversely, as the distance increases between two genes on a chromosome, the chances of a crossing-over event between them increases, as does the chance of genetic recombination taking place. This is similar to saying that when you drive your car on a 100-mile trip you are likely to change lanes more often than when you drive to your local market. In a similar way, comparing the amount of recombination between different genes gives a measure of how far apart they are on the chromosome.

Geneticists drew two important conclusions from linkage maps:

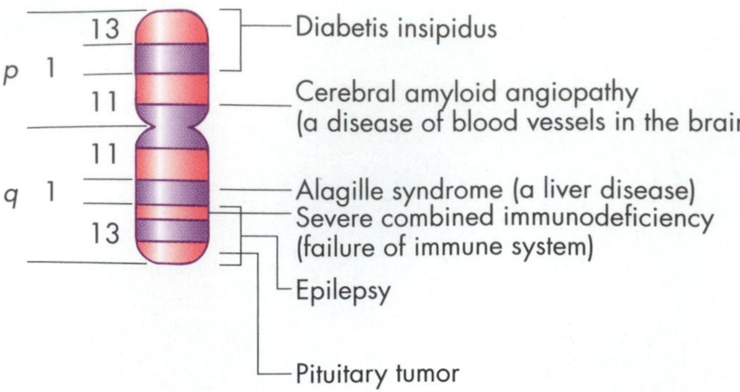

FIGURE 8.19

Map of Human Chromosome 20.

This map shows the location of six different genes on chromosome 20. People with the normal alleles of these genes do not have the conditions that are listed in the figure. The numbers to the left of the chromosome indicate various bands that appear on the chromosome.

1. Each gene is located on a chromosome at a specific location called a *locus*. Thus, in all humans, the gene for diabetes insipidus is located in the same specific place on chromosome 20 [FIGURE 8.19].

2. A chromosome is a linear array of genes, like beads on a string. For example, FIGURE 8.19 shows separate genes for diabetes insipidus, a blood vessel disease, a liver disease, epilepsy, and a certain kind of tumor lined up along the human chromosome 20, one of our smallest chromosomes. When we discuss human genetic diseases in CHAPTER 11, you will see that the ability to locate a gene has enabled medical geneticists to isolate many genes that cause human diseases, including genes for cystic fibrosis, muscular dystrophy, and certain types of cancer. The isolation of these genes may someday lead to effective therapies.

CONCEPT CHALLENGE

Explain how it happens that the father, rather than the mother, determines the gender of a baby.

Gene Interactions

The series of experiments leading from Mendel to Morgan established fundamental principles about genes, and later experiments have shown that a gene does not function in a vacuum. Instead, it interacts with other genes and with the environment in specifying how an organism develops, functions, and lives.

As biologists studied more and more about genes, they discovered that interactions between alleles of the same gene, between different genes, and between genes and the environment are often unexpected. On closer inspection, however, the unforeseen genetic behavior can usually be explained by slight extensions of Mendelian principles, as you will see in the following sections.

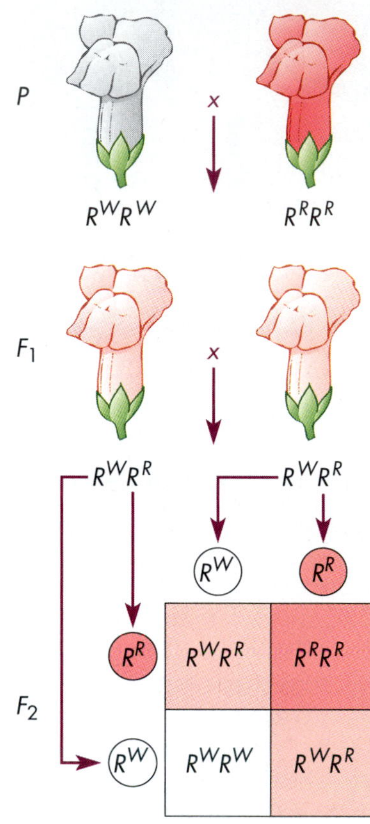

FIGURE 8.20

Flower Color in Snapdragons: Incomplete Dominance.

A heterozygous plant with one red-flower and one white-flower allele (R^WR^R) has pink flowers because a single red-flower allele in a heterozygote makes less pigment than the two red-flower alleles in a homozygous R^RR^R plant. Since less pigment is made, the heterozygote appears pink.

INTERACTIONS BETWEEN ALLELES

According to the results of Mendel's experiments, each gene has two alleles, which are either dominant or recessive. But as later geneticists found, the alleles of some genes fail to fall clearly into either category, and some genes have many more than two alleles.

Incomplete Dominance The flowers called snapdragons can be white or red [FIGURE 8.20]. When snapdragons are self-pollinated, white flowers breed true and so do red flowers. But if pollen from a white flower fertilizes the eggs in a red flower, the heterozygous offspring have *pink* flowers, not white or red as Mendel's principle of dominance would have predicted. Since the phenotype of the heterozygote is intermediate between pure-breeding red flowers and pure-breeding white flowers, this phenomenon is called **incomplete dominance**.

It's important to note that the intermediate pink phenotype does not indicate a blending inheritance because when two heterozygous pink flowers mate, the offspring have the expected Mendelian ratio of 1:2:1, or 1 homozygous red to 2 heterozygous pink to 1 homozygous white [FIGURE 8.20]. The alleles emerge from the hybrid unchanged.

Look back at FIGURE 8.2. Can you explain the results with the hibiscus flowers in terms of the particulate theory and dominant, recessive, and incompletely dominant genes?

Codominance When two alternative alleles are *fully apparent* in a hybrid, with both phenotypes showing in the organism, we say that the hybrids exhibit **codominance**. A familiar example of codominance is the blood-type gene called *ABO* [FIGURE 8.21]. Allele *A* causes a certain sugar called type A to appear on the surface of red blood cells, and a person with this allele is said to have blood type A. Allele *B* causes a different sugar, called

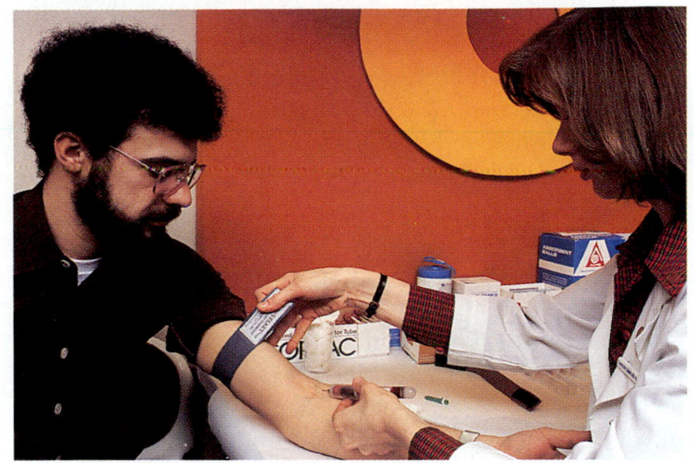

[A]

Phenotype		Genotype
Blood type	Cell surface molecule	
A	Red blood cell	*AA* or *AO*
B		*BB* or *BO*
AB		*AB*
O	(Neither A nor B)	*OO*

[B] Blood types

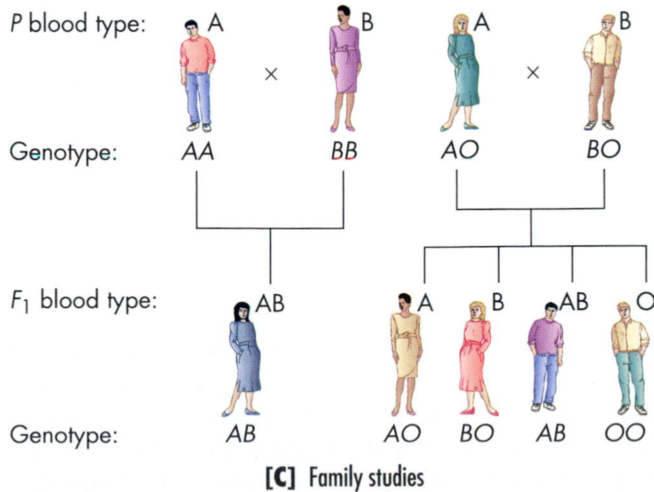

[C] Family studies

FIGURE 8.21

Blood Types Illustrate Codominance and Multiple Alleles.

[A] Human blood is a convenient source of cells for the investigation of cell surface molecules. One group of sugar molecules on the surface of red blood cells is controlled by the *ABO* gene. [B] Red blood cells from people with the four different blood types have different combinations of sugars on their surfaces. These phenotypes are given by the genotypes shown in the table. There are three alleles at this locus, *A*, *B*, and *O*, but any individual has only two copies of this gene, and can thus be *AB* or *BO*, among others, but never *ABO*. Since *A* and *B* both can show in the phenotype (blood type) of a person with the *AB* genotype, this gene displays codominance. [C] Since people with an A or B blood type can have different genotypes, the blood types of families can differ. Use the data in this figure to help solve ACTIVE LEARNING BOX on page 217.

type B, to appear on the surface of red blood cells, giving blood type B. Someone who has two *A* alleles has only the A sugar, and a person with two *B* alleles has only the B sugar. But a heterozygote with one *A* allele and one *B* allele has both A and B sugars on the surface of red blood cells. That person has blood type AB, a codominant phenotype.

Students are sometimes confused by the difference between codominance and incomplete dominance. In codominance, both alleles are *fully* expressed in the heterozygote, while in incomplete dominance, the phenotype is intermediate. AB blood type is not intermediate between A and B—it is fully A *and* B. But pink snapdragons cannot be fully red and fully white, and thus incomplete dominance is at work. In general, codominance occurs in biochemical traits.

Multiple Alleles Although Mendel studied only two alleles for each of his seven pea plant genes, some genes have more than two alleles, as shown with human *ABO* blood groups. In addition to the two codominant alleles *A* and *B*, one blood-group gene has a third allele that is fully recessive to both *A* and *B*. This recessive allele is called *O*. A person with two doses of *O* has neither the A nor B sugar marker and has blood type O. Since *O* is recessive, an *AO* heterozygote has blood type A, and a *BO* heterozygote has blood type B. Although there are three alleles of the human *ABO* gene, no one person can have all three at once, because as you saw in CHAPTER 7, each child gets only *one* allele of a gene from each parent, for a total of two copies of any gene in a given individual.

Genes with codominant alleles are especially useful for paternity disputes. To see how this works, try the Active Learning Box on page 217.

The *ABO* gene has three alleles, but some genes have even more than that. The genes controlling tissue rejection in organ transplantations all have many alleles. As part of the *major histocompatibility complex* (*MHC*), such genes encode certain proteins on a cell's surface. These substances serve as identification markers that help the body distinguish its own cells from other foreign substances like bacteria, viruses, or parasites that might otherwise successfully invade the body and cause disease.

Because there are several genes in the major histocompatibility complex, each with several alleles, it is very unlikely that two unrelated (or even related) persons will have precisely the same combination of alleles at all loci. That is why a person with kidney or liver disease must often wait a long time before being matched with a suitable donor. If there are too many allelic differences between the tissues of donor and recipient, the immune-system cells of the recipient will kill the cells of the donated organ. Even with immunosuppressant drugs that help stop tissue rejection, the multiple alleles for tissue types constitute a major obstacle to many life-saving transplants.

PLEIOTROPY: MULTIPLE EFFECTS OF A SINGLE GENE

A single gene may determine several different phenotypes in a phenomenon known as **pleiotropy** (*pleio-*, "more"). The white tiger Mohan shows the action of a pleiotropic gene. The primary effect of the white-coat allele is an absence of *melanin*, the dark pigment in the hair, skin, feathers, and scales of vertebrates. However, the same allele also seems to give rise to crossed eyes, not only in tigers, but in other species that have a similar mutation—for instance, in some albino people [FIGURE 8.22]. We do not yet know why this one allele produces two apparently unrelated phenotypes. However, it is easy to see how the process of natural selection might keep the frequency of the white-coat allele very low among tigers, since white tigers have little protective camouflage. Also, crossed eyes probably makes it more difficult for a tiger to hunt successfully. Thus, homozygous white tigers might have a difficult time surviving to sexual maturity. Over the course of several generations, nature would select against the white-coat allele.

FIGURE 8.22

Albinos Illustrate Genes with Many Effects.

A mutant allele of a single gene for melanin production in humans can affect several phenotypic traits, including the pigmentation of eyes, hair, and skin and the occurrence of crossed eyes.

active learning: Solving a paternity case

Olivia and Abner have a baby in a hospital, but suspect that the nurses have accidentally switched their baby with another. Abner is blood type AB and Olivia is blood type O. The newborn the staff brought to them is blood type AB. The nurse on duty contends that this is their child because its blood type is just like the father's. The nurse says that the only other baby on the ward at the time is blood type A, and since this second baby is different from both Olivia and Abner, it cannot be their child.

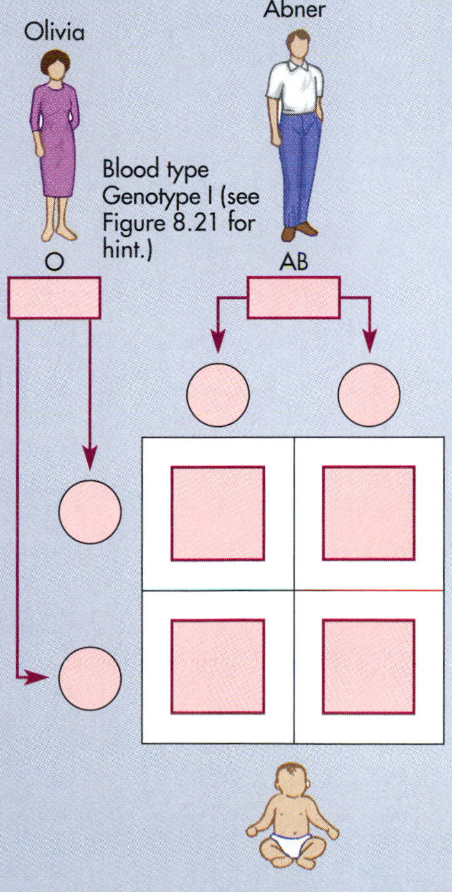

Do you agree with the nurse's reasoning? Use a Punnett square to determine the truth.

. . . .

From examining your Punnett square, is it possible that the baby with blood type AB could come from Abner and Olivia?

. . . .

Is it possible that the baby with blood type A could come from Olivia and Abner?

. . . .

Does the Punnett square *prove* that the baby with blood type A really is the child of Olivia and Abner? Explain.

. . . .

box 8.1
Human Impact

Hinnies, Mules, and Mendel's Rules

It's a good thing Gregor Mendel didn't ride a mule; the principles of inheritance might never have been the same.

When Mendel crossbred plants that produced only round peas with others producing only wrinkled peas, all the offspring had round seeds. This was true whether the male gametes came from the round-seeded plants and the female gametes came from the wrinkled-seeded plants or vice versa. The gene for seed shape, in other words, acts the same in the offspring regardless of which parent donates it. From such simple observations, Mendel derived his straightforward laws of heredity.

Imagine, though, how scientific history might have been changed had Mendel trotted about Brno on a big brown mule, coarse-woven robes flapping in the wind. Had he fancied mules, Mendel would certainly have known that breeders produce the hybrids by crossing a male donkey with a female horse. And he would have known that while a mule can be as tall as its mother, it has the ears, legs, hooves, tail, and stubbly mane of the donkey father [FIGURE 1A]. Mendel would also surely have been aware that the reciprocal cross of a female donkey and a male horse produces not a mule but a hinny. And as all mule lovers know, a hinny looks basically like its equine father and resembles its donkey mother much less [FIGURE 1B].

Mules and hinnies are obvious exceptions to Mendel's discovery that in pea plants, genes are inherited in the same way from the male and female parent. Enough examples like this, and the good monk might have gotten discouraged and gone back to mathematics!

Modern geneticists have studied the curious differences between mules and hinnies, along with a collection of similar genetic oddities that result from what they call "genomic imprinting." This phrase refers to a hypothetical imprinting process that takes place during gamete formation. By some mechanism, the process causes a gene to be expressed one way in the offspring if the male parent donates it, and another way if the female parent donates it.

Several human diseases show evidence of genomic imprinting, including Huntington's disease and the Prader-Willi and Angelman syndromes. Huntington's is a severe degenerative condition that is inherited as a dominant trait and tends to onset in middle age (details in CHAPTER 12). About 10 percent of the time, however, the disease onsets in children, and of those young victims, 9 out of 10 inherit the disease from their father. A child with Prader-Willi syndrome is usually short, obese, and mentally retarded and tends to have inherited both copies of chromosome 15 from the mother. A child with Angelman syndrome is usually retarded, laughs excessively, and moves in a jerky, puppetlike manner. Research shows that many such children inherit a complete chromosome 15 from the father, but lack pieces of the maternal chromosome 15.

Geneticists are not certain how genomic imprinting occurs, although some think it may involve DNA methylation—the attachment of methyl groups to parts of the DNA molecule—and with that, the turning on or off of certain genes. They do know, however, that since genomic imprinting results in a gene's being expressed differently when inherited from the mother or father, it represents an important exception to Mendel's basic laws of equivalent inheritance. And many suspect that imprinting is playing unknown but significant roles in normal development and genetic diseases—not to mention in phenotypic patterns as dramatic as the hinny and the mule.

FIGURE 1
A Mule [A] and a Hinny [B].

[A] Mule: Offspring of a female horse and a male donkey

[B] Hinny: Offspring of a female donkey and a male horse

ENVIRONMENTAL EFFECTS ON GENE EXPRESSION

If you go to a cat show or a feline fanciers meeting, you may see some of the owners brushing ice over the ears and tails of their prize Siamese cats. Queried about the practice, a proud owner will often explain that this practice causes the points—the tips of the ears, tail, and feet—to become darker [FIGURE 8.23]. This curious practice indicates that the expression of a gene can be altered not only by other genes, but also by the environment. An enzyme catalyzes the production of the dark pigment melanin in Siamese cats, and this enzyme is genetically changed in such a way that it is unable to work at normal body temperature. At the lower temperatures found in a cat's extremities, however, the enzyme is catalytically active and can produce melanin that darkens the ears, paws, and tail. The ice packs at the cat show, if continued over a long period of time, just augment the production of the pigment.

FIGURE 8.23

Siamese Cats and Environmental Effects on Gene Expression.

Siamese cats have dark ears, nose, paws, and tail because those extremities are cooler than the rest of the body, and the enzyme involved in producing dark coat pigment can only function at this lower temperature. Thus, a gene's environment can sometimes influence its expression.

The environment of a gene can also be its immediate cellular surroundings, and these, too, can affect the gene's expression. BOX 8.1 explains the strange case of male and female mules, and how some genes are expressed differently depending on whether they are inherited from the mother or the father.

CONCEPT CHALLENGE

A horse breeder consults pedigree books and finds that the mating of palomino (tan) stallions to palomino mares can create foals of several coat colors. The foals can be chestnut (dark brown), palomino, and cremello (cream) in a 1:2:1 ratio. Set up a Punnett square to explain these results. Pay special attention to describing dominance relationships between the alleles of this coat-color gene.

Connections

The story of Mendel's experiments is a remarkable illustration of a scientist's ability to form concrete conclusions from indirect data. Mendel did not know about chromosomes or meiosis, but from his careful analysis of stem length and flower color in peas, he was able to infer the existence and behavior of factors now called genes—tiny objects he would never see.

Mendel's insights also had tremendous theoretical impact once they were rediscovered. By demystifying how alleles pass intact through a hybrid, he gave evolutionary theorists the kind of mechanism they needed to explain the origins and maintenance of biological diversity.

Research has shown that many of the observable traits we have discussed are caused by some chemical in the organism's body. Color in the hair or skin of tigers, dogs, or people is caused by the dark pigment melanin; the color of pea flowers is caused by a red pigment; and cell surface substances that thwart organ transplants differ because they are composed of different protein-carbohydrate complexes. Somehow, genes control these chemical differences. Exactly how they achieve that control will be the subject of the next chapter.

KEY TERMS

allele, 201
autosome, 210
crossing-over, 212
dominant, 202
first filial (F_1) generation, 200
gene, 201
genetic mapping, 213
genotype, 203
heredity, 197
heterozygous, 203
homozygous, 203
incomplete dominance, 215
linkage, 212
parental (*P*) generation, 200
parental type, 207
phenotype, 203
principle of independent assortment, 208
principle of segregation, 203
Punnett square, 204
pure-breeding, 200
recessive, 202
recombinant type, 207
second filial (F_2) generation, 201
sex chromosome, 210
testcross, 207
***X* chromosome,** 210
***X*-linked,** 212
***Y* chromosome,** 210

HIGHLIGHTS IN REVIEW

1 Hereditary traits are controlled by genes, which occupy specific positions on the chromosomes.

a] Inherited traits, from pelt or petal color to stem length and even certain behaviors, are controlled by one or more genes, discrete stretches along a DNA molecule. An organism's physical makeup, the way it looks or acts, is called its phenotype, while the underlying genetic makeup, its complement of genes, is its genotype.

2 According to Mendel's principle of segregation, the two copies of a gene separate into different cells during meiosis. During fertilization, egg and sperm combine at random with respect to their genetic content.

a] The genes for each trait can exist in two or more alternative forms called alleles. At conception, every individual receives one allele of each gene from each parent. If the two alleles are identical, the individual is said to be homozygous for that trait. If the two alleles are different, the individual is heterozygous for that trait.

b] In heterozygotes, some alleles hide the expression of other alleles. The hidden allele is the recessive allele; the apparent allele is the dominant allele. Without further analysis, it is often impossible to distinguish whether an individual with a dominant phenotype is heterozygous or is homozygous for the dominant allele.

c] The two alleles for a given gene in each cell are carried on two chromosomes that are similar in size, shape, and genetic content; these are called homologous chromosomes. Each chromosome carries many genes.

d] Mendel's first law, the principle of segregation, states that when gametes form, homologous chromosomes—and hence the gene copies they contain—separate, giving each sperm or egg cell only one copy of each gene. Furthermore, when fertilization occurs, sperm and egg fuse randomly with respect to the alleles those cells contain.

3 Genes on the same chromosomes assort independently of each other during meiosis.

a] Mendel's second law, the principle of independent assortment, explains that the genes on different chromosomes are distributed into gametes independently of each other. However, genes that are on the same chromosome tend to be packaged into gametes together as linked genes.

b] Most chromosome pairs (the autosomes) are identical in both sexes. But one pair, the sex chromosomes, are different in shape and genetic content in the two sexes of many animal species. Human males, for example, have an *X* and *Y* chromosome, while females have two *X*'s. Genes that reside on the *X* chromosome are said to be *X*-linked.

4 Genes on the same chromosome tend to be inherited together. Based on this tendency, geneticists have learned to map the locations of genes on chromosomes.

a] Genes reside on chromosomes, which explains the close relationship between the rules of heredity and the principles of meiosis. Each chromosome contains hundreds of genes strung along end to end like beads on a string.

b] Occasionally, homologous chromosomes exchange parts during meiosis (crossing-over). This generates new combinations of alleles different from their arrangements in the parents, a phenomenon called recombination. The rates of recombination between two genes serves as a way of mapping the locations of genes relative to each other.

5 Genes interact with each other and with the environment and determine the eventual physical makeup of the individual.

a] Sometimes alleles are neither clearly dominant nor clearly recessive. In incomplete dominance, the heterozygote's phenotype is unlike either parent but is intermediate between those of the homozygous parents, like the pink heterozygous offspring of a red and a white snapdragon. In codominance, the characteristics of *both* phenotypes show up fully in the heterozygote, for example, the AB blood type occurs in a heterozygote for the *A* and the *B* allele.

b] Interactions with the environment or with other genes can mask the presence of a particular allele. A single gene may influence several different traits, in a phenomenon known as pleiotropy.

UNDERSTANDING THE FACTS AND CONCEPTS

For Questions 1–5 choose from the following list of terms the one that best fits the description or situation given.

a] dominant allele
b] recessive allele
c] homozygous
d] heterozygous
e] testcross

1 An organism is bred to a homozygous recessive individual in order to determine whether it is homozygous or heterozygous for a gene under study.

2 Alleles that are expressed phenotypically only when homozygous.

3 The presence of two identical forms of a gene on a pair of homologous chromosomes.

4 An allele that is expressed phenotypically whether homozygous or heterozygous.

5 Term describing an organism with different allelic forms of a gene on a pair of homologous chromosomes.

As above, for Questions 6–10, choose from the following list of terms the one that best fits the description or situation.

a] phenotype
b] genotype
c] hybrid
d] pure-breeding
e] F_1

6 The physical manifestation of a genetic trait, usually expressed in words.

7 A synonym for heterozygous, usually referring to the progeny of individuals of different homozygous strains.

8 The symbolic representation of the allelic composition of an individual.

9 In a formal genetic cross, the heterozygous offspring of two contrasting, homozygous parents are members of the ______.

10 A synonym for homozygous.

As above, answer Questions 11–14 using the following list of terms:

a] locus
b] multiple alleles
c] pleiotropy
d] polygenic trait

11 More than two forms of a given gene exist.

12 The position of a gene on a chromosome.

13 A phenotype that is determined by many different genes.

14 Multiple and apparently unrelated effects of an allele.

As above, answer Questions 15–19, using this list:

a] principle of segregation
b] independent assortment
c] random fertilization
d] crossing-over
e] codominance

15 The phenotypic and simultaneous expression of both alleles in a heterozygote.

16 The physical event that enables genes to form new relationships within the same linkage group.

17 The alignment of homologous chromosomes in metaphase I of meiosis provides the physical basis for this phenomenon.

18 Assumes the absence of any genetic factors that would favor the union of a particular egg with a particular sperm cell.

19 The separation of homologous chromosomes during meiotic anaphase I.

INTEGRATE AND APPLY WHAT YOU HAVE LEARNED

1 Show all the possible genotypes for a pea plant with long stems (dominant) and purple flowers (dominant). Use the letters *L* or *l* to symbolize the stem-length alleles and *P* or *p* for the color alleles.

2 Show all the genotypes of the gametes produced by the hybrid pea plant *LlPp*.

3 In guinea pigs, black fur (*B*) is dominant to white fur (*b*). How could an animal breeder test whether a given black guinea pig is homozygous or heterozygous for this color gene?

4 How are the sex chromosomes of sperm and eggs of mammals alike? How are they different?

5 Do closely linked genes—say, two genes right next to each other—ever exhibit independent assortment? Explain.

ANALYSIS Complete the following multiple-choice problems:

1 There are many more color-blind men than color-blind women. Color-blind women must have color-blind fathers. Using this information, you would deduce that the allele for color-blindness is:

a] An autosomal dominant
b] An autosomal recessive
c] A sex-linked dominant
d] A sex-linked recessive
e] Incompletely dominant and autosomal

2 In the Smith family, all the children have small, tightly connected earlobes, commonly called attached lobes. Their parents both have large, pendulous earlobes. What is the most likely explanation of the children's lobes?

a] They all developed mutations for attached earlobes.
b] Earlobe phenotype is determined by multiple alleles.
c] The allele for attached earlobes is recessive; the parents are heterozygous.
d] The allele for attached earlobes is dominant; the parents are heterozygous.

3 The Lee children and one of their parents are all blood group B. The other parent must be:

a] Blood group A.
b] Blood group B.
c] Blood group AB.
d] Blood group O.
e] Any of the above.

CHAPTER 9

DNA: The Thread of Life

DNA AND A DEADLY REPAST

In January 1993, newspapers all over the United States carried a shocking story: two children had died, four youngsters were in critical condition, and 500 people of various ages had grown ill from tainted food. All had eaten hamburgers at restaurants in the Pacific Northwest. In each case, the meat came from the same distributor, it was undercooked, and it caused a severe form of diarrhea that damaged or killed the people's intestinal cells.

This food poisoning raised public awareness about food safety issues such as insufficient cooking of meat, and lax hygiene at meat-

FIGURE 9.1
DNA: A Twisted Ladder.
A computer-generated image of a DNA molecule.

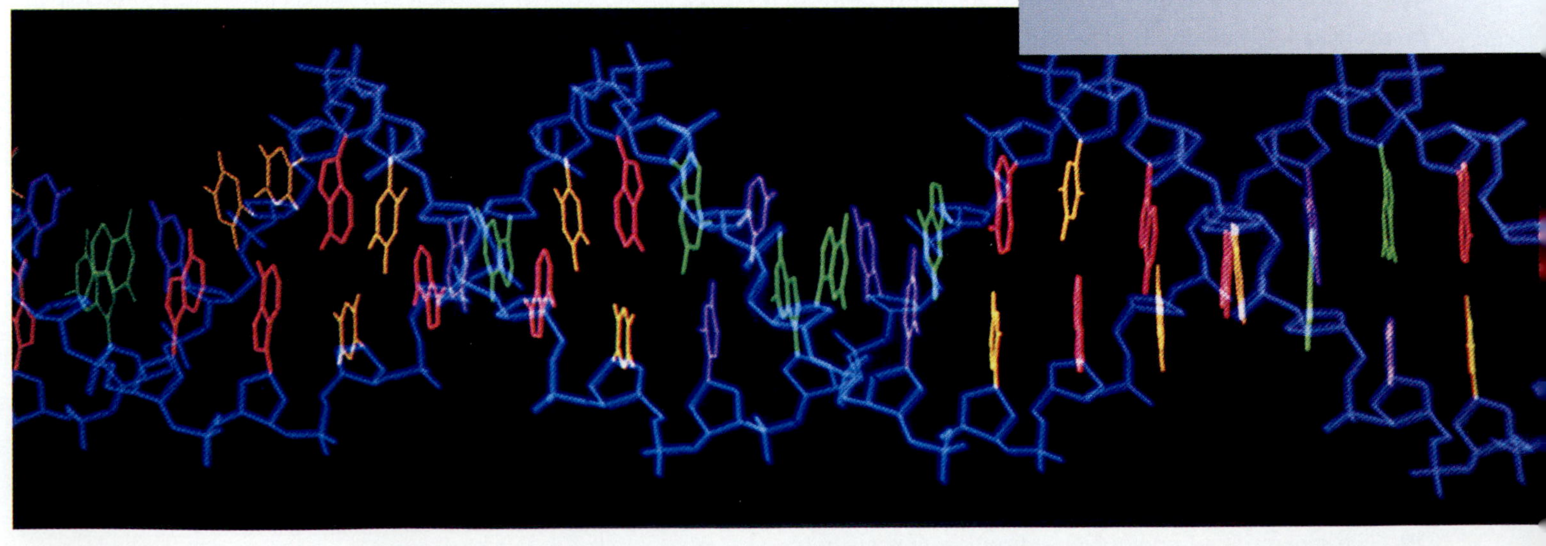

packing plants. It also drew the nation's attention—if only briefly—to a particularly harmful strain of the common bacterium *Escherichia coli*. This microbe inhabits the human intestine, and is normally harmless. The tainted hamburgers, however, contained cells of a dangerous strain of *E. coli* called O157:H7. Why can one form of *E. coli* exist in the human gut without injuring it, while another can kill intestinal cells and in extreme cases, even the host? The answer goes straight to the heart of the genetic repository inside *E. coli* and every other cell: DNA.

The fine, threadlike DNA molecule carries hereditary information and guides the daily functioning of the cell [FIGURE 9.1]. Its elegant structure is the key to how DNA stores genetic information and directs cellular activity and is the subject of this chapter.

DNA structure is crucial to the ability of *E. coli* O157:H7 to cause food poisoning. The reason is that this particular strain of bacteria is infected with a virus. Free in the environment, this bacterial virus consists of nothing more than a single DNA molecule plus a few proteins. It has no metabolism, and cannot replicate on its own, so it's not alive; it's just a particle of chemicals. When the viral DNA enters the bacterial cell, however, it inserts itself into the bacteria's own DNA molecule. Whenever the bacterium reproduces, the offspring cells get exact copies of the viral DNA. The DNA of the virus causes the bacterium to make a toxin that can kill the intestinal cells of a person unfortunate enough to eat food that has not been heated sufficiently to destroy the bacteria. To understand how a DNA molecule can be copied exactly, be passed to offspring cells, and can direct the synthesis of a toxin, we must first understand the structure of DNA.

MESSAGES

1 Genes are made of DNA.

2 A DNA molecule consists of two strands of nucleotides oriented in opposite directions and twisted into a double helix.

3 The structure of DNA explains its ability to replicate and to control an organism's physical makeup and abilities.

4 During DNA replication, the two strands separate and enzymes synthesize a complementary copy of each one; this yields two daughter molecules identical to the original molecule.

Within all living cells, DNA is the hereditary material. Its structure and function are keys to many core concepts in biology, including how cells replicate; how genetic changes occur and how they lead to cancer or genetic diseases, such as hemophilia, sickle-cell anemia, and Tay-Sachs disease; and how modern groups of organisms are descended from evolutionary progenitors. The structure of DNA even has an important place in the identification of criminals, as you will see in a later chapter.

We begin this chapter with the pivotal experiments that showed that genes are made of DNA. These experiments relied heavily on bacterial viruses like the one that renders *E. coli* O157:H7 so deadly. We go on to see how researchers determined the precise structure of DNA, and then we examine that elegant structure. The chapter ends by discussing how the structure of DNA allows it to replicate, so that like can beget like. We will also see how a stretch of DNA can determine the structure of proteins, including the toxin made by O157:H7.

Along the way, we will see how anti-AIDS drugs interfere with DNA replication and slow the progress of this deadly disease, and how the misuse of antibiotics may allow epidemics of tuberculosis and other infectious diseases to break out and continue unchecked. ❑

Identifying the Hereditary Material

Mendel's experiments showed that the discrete hereditary factors we now call genes determine a pea plant's physical appearance, including the color of its flowers, the length of its stem, and the shape of its seeds. But Mendel had no idea what these "factors" were or how they worked. To understand how Mendel's hereditary factors work, we must first determine their chemical makeup and then how they function. The first step toward our present knowledge of genes was the understanding that genes are located on chromosomes. Since chromosomes in eukaryotes contain mainly protein, RNA, and DNA, one of those substances was presumed to be the genetic material. But which one?

Geneticists first thought that proteins were the only substances versatile enough to carry the complex information in genes. This was a logical assumption, since proteins are constructed from 20 different subunits (amino acids) [see FIGURE 2.28], while DNA has only four: the nucleotide bases adenine (A), thymine (T), guanine (G), and cytosine (C) [see FIGURE 2.32]. Clearly, 20 amino acids can form many more combinations—and hence carry much more information—than four bases, just as you can form more words from an alphabet of 20 letters than from an alphabet of 4 letters. In the mid-twentieth century geneticists began to test whether genes are made of protein or nucleic acid.

EVIDENCE FOR DNA: BACTERIAL TRANSFORMATION

In the 1930s, researchers found that they could transfer an inherited characteristic—the ability to cause pneumonia—from one strain of bacteria to another. To do so, they simply exposed a harmless bacterial strain to cells of a disease-causing strain that had been killed by heat. The formerly harmless bacteria acquired the inherited ability to cause disease from some substance in the heat-killed cells. Clearly, whatever that substance was, it could carry hereditary information.

Researchers wanted to determine which chemical component was capable of carrying hereditary information. To do so, they extracted the carbohydrates, proteins, and DNA from the *pathogenic* ("disease-causing") strain [FIGURE 9.2A to C]. They observed that the harmless strain of bacteria treated with either carbohydrate or protein remained harmless. In contrast, the harmless bacteria treated with pathogenic DNA became pathogenic and passed that trait to their progeny [FIGURE 9.2D to F]. Researchers call the transfer of an inherited trait by the uptake of pure DNA *transformation*. In some medical cases, transformation results in the transfer of circles of pure DNA called plasmids [FIGURE 9.3] moving from one bacterial cell to another.

CONFIRMATION THAT GENES ARE MADE OF DNA

Despite the transformation experiments, biologists did not totally accept DNA as the genetic material. Important confirmation came in 1952, when Alfred D. Hershey and Martha Chase performed a series of experiments with **bacteriophages** ("bacteria eaters") or simply *phages*, viruses that infect and reproduce in bacteria such as the virus that turns normal *E. coli* into the killer bacterium *E. coli* O157:H7.

Several features of bacterial viruses make them convenient subjects for genetic study. First, these viruses are easy and inexpensive to maintain. Second, they can produce new viruses rapidly, spawning exact copies about 25 minutes after infecting a host cell. Third, and most important, these bacterial viruses consist simply of a core of DNA surrounded by a protein coat [FIGURE 9.4A and B]. Clearly, here was a chance to show which component was responsible for heredity.

Bacterial viruses infect bacterial cells by attaching themselves to the host and injecting their single molecule of DNA into the bacterium [FIGURE 9.4C and D]. Inside the bacterium, the viral DNA can either join to the bacterium's DNA and be carried along with it, as happens with O157:H7, or it can immediately start multiplying and provide instructions for making new viruses [FIGURE 9.4E and F]. The new viruses soon burst from the host cell, able to infect other bacteria [FIGURE 9.4G].

Hershey and Chase reasoned that if newly created viruses were identical to their parent, a hereditary substance must be at work. If only viral DNA was injected into the host cell, then it was sensible to assume that DNA was the genetic material. Hershey and Chase knew that proteins contain sulfur but do not contain phosphorus, and that the content of DNA is the opposite—it contains phosphorus but no sulfur. By labeling the proteins and DNA with radioactive isotopes of sulfur and phosphorus, they were able to determine which substance entered the host cell and thus directed the genetic activity. The results showed that phosphorus was on the inside and sulfur on the outside, demonstrating that DNA alone had entered the cell. They concluded that DNA, not protein, was responsible for directing the genetic activity of the virus [FIGURE 9.2].

While these experiments showed that DNA is the hereditary material in a virus, some observers have wondered whether the traits of a simple bacterial virus can be

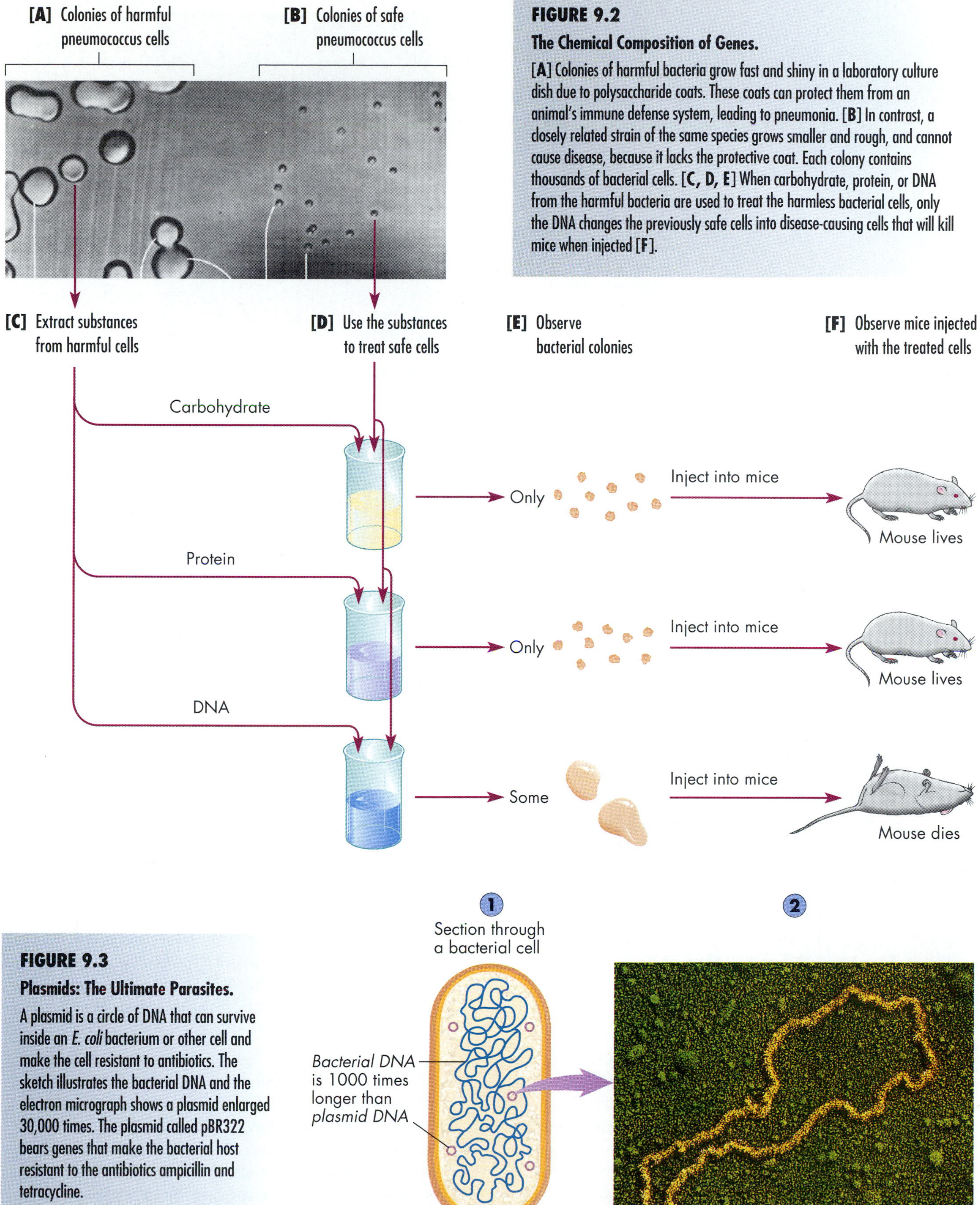

FIGURE 9.2

The Chemical Composition of Genes.

[**A**] Colonies of harmful bacteria grow fast and shiny in a laboratory culture dish due to polysaccharide coats. These coats can protect them from an animal's immune defense system, leading to pneumonia. [**B**] In contrast, a closely related strain of the same species grows smaller and rough, and cannot cause disease, because it lacks the protective coat. Each colony contains thousands of bacterial cells. [**C, D, E**] When carbohydrate, protein, or DNA from the harmful bacteria are used to treat the harmless bacterial cells, only the DNA changes the previously safe cells into disease-causing cells that will kill mice when injected [**F**].

FIGURE 9.3

Plasmids: The Ultimate Parasites.

A plasmid is a circle of DNA that can survive inside an *E. coli* bacterium or other cell and make the cell resistant to antibiotics. The sketch illustrates the bacterial DNA and the electron micrograph shows a plasmid enlarged 30,000 times. The plasmid called pBR322 bears genes that make the bacterial host resistant to the antibiotics ampicillin and tetracycline.

Head — Protein shell, DNA, Collar

Tail sheath

Tail fiber

[A] Anatomy of a phage, a bacterial virus

DNA

[B] DNA bursting from the head of a phage

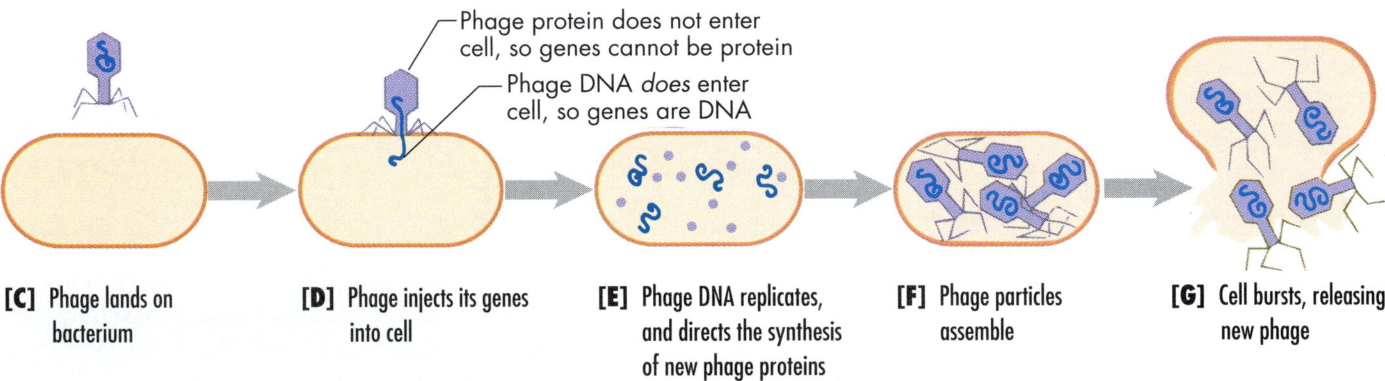

[C] Phage lands on bacterium

[D] Phage injects its genes into cell

[E] Phage DNA replicates, and directs the synthesis of new phage proteins

[F] Phage particles assemble

[G] Cell bursts, releasing new phage

FIGURE 9.4

Life Cycle of a Bacterial Virus.

[A] The bacterial virus phage T4 has a head with a protein shell surrounding DNA, a columnar tail, and leglike protein tail fibers. **[B]** The DNA from the virus has spewed out of the head. If this DNA were stretched out, it could wind around the perimeter of the photo six times. **[C]** In the virus life cycle, the phage lands on a bacterial cell and injects its genes **[D]**. Inside the cell, **[E]** the genes direct the cell to make new virus particles **[F]**, which burst out and infect additional cells **[G]**. Since virus DNA enters the cell, but virus protein does not, Hershey and Chase concluded that genes are made of DNA.

generalized to more complex living organisms. The answer is yes—or almost always yes. Subsequent research has confirmed that DNA is the genetic material in all cells and many viruses. In other viruses, such as the virus that causes AIDS (acquired immune deficiency syndrome), the genes are made of RNA.

The experiments described here changed our concept of the gene from Mendel's abstract particle that determines heredity to a tangible chemical that biologists could see and manipulate. Nowadays, molecular biologists break open cells, separate the DNA from the proteins, add ice-cold ethanol to the DNA, and put it in the freezer. The next morning, at the bottom of the test tube is a stringy white precipitate a bit like powdered sugar—but it is pure DNA, pure genes [FIGURE 9.5].

Once it was known that DNA is the genetic material, researchers began their attempt to discover DNA's *three-dimensional* structure.

CONCEPT CHALLENGE

HIV, the virus that causes AIDS, consists of a lipid membrane surrounding a layer of protein that encloses two molecules of RNA (a nucleic acid related to DNA). When HIV infects a human cell, it replicates and kills the cell, and in the process, more virus particles are released. Design an experiment to determine which substance in HIV carries the hereditary material of the virus—lipid, protein, or RNA.

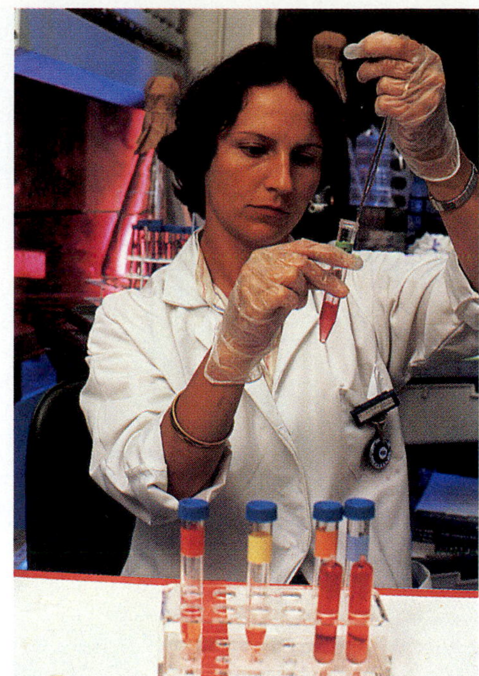

FIGURE 9.5
Manipulating DNA in the Lab.
In a laboratory, researchers isolate DNA from the cells that carried it. Here, a biochemist is separating DNA from solution.

DNA: The Twisted Ladder

A delicate butterfly and the plant on which it rests both have genes encoded in molecules of DNA. DNA can replicate so perfectly that the offspring of two swallowtail butterflies look very much like the parents. The genetic material can still contain enough variation, however, so that some swallowtails are black rather than yellow [FIGURE 9.6]. The variability inherent in DNA is the basis of natural selection, the key to evolution, and the explanation for life's wonderful variety. How does the structure of DNA account for both the unity of life—the shared traits and common descent of living things—and also its stunning diversity?

FIGURE 9.6
DNA: The Universal Hereditary Material.
How can the hereditary material be constant enough so that yellow swallowtail butterflies have yellow offspring, but variable enough so that black swallowtails are also found?

In the early 1950s, James D. Watson, an American postdoctoral student, and Francis H. C. Crick, a British researcher, met in Cambridge, England, and began their study of DNA [FIGURE 9.7A]. Like most of the scientists at that time, Watson and Crick believed that DNA would be shaped like a *helix* (similar to the winding of a spiral staircase).

Watson and Crick knew from published experiments that DNA is a linear molecule; that is, it is a very long thread. To imagine how long DNA is, look at the length of the thread in a simple bacterial virus like the one that infects *E. coli* O157:H7 [review FIGURE 9.4B]. The host bacterium's own DNA molecule is 100 times longer than the virus's DNA molecule. And the lengthiest DNA in your cells can be 5000 times as long!

Watson and Crick also knew that DNA was a chain of nucleotides of four types [FIGURE 9.8]. The four kinds of nucleotides are identical except for the bases they contain—either *adenine* (A), *cytosine* (C), *guanine* (G), or *thymine* (T). The four bases dangle like charms from a charm bracelet whose chain is made of alternating sugars and phosphates. Thus we often describe a section of DNA by the sequence of bases in the chain. For example, a 15-nucleotide portion of a single-stranded DNA chain might have bases in the sequence ACCCGTCCGTGTTAG.

Watson and Crick also knew that in any given molecule of DNA, A and T appear in equal amounts, and C and G are also present in equal amounts. But the two scientists needed more information to build an accurate model of a DNA molecule. This information came from the British biophysicist Rosalind Franklin, who was working in the laboratory of Maurice H. F. Wilkins at King's College in London [FIGURE 9.7B]. Franklin took some particularly good pictures of DNA in an attempt to see how the four subunits—A, T, G, and C—were arranged [FIGURE 9.7C]. The pictures were made using *X-ray diffraction*, a process in which a beam of X-rays passes through a fiber made of many parallel strands of pure DNA. Within the DNA fiber, similar structures are repeated, like the pattern of tiles on a floor, causing the X-rays to expose film in a spot pattern. By analyzing these patterns biophysicists can map the relative positions of atoms in a molecule.

When Watson saw the X-ray photos he noticed a certain symmetry suggesting that the molecule might consist of *two* connected strands of DNA. The pictures showed that DNA was most likely a coiled structure, with 10 nucleotides in one loop of the coil. Now the young researchers had to find out how the bases and sugar-phosphate backbones were arranged in three-dimensional space before they could understand how the molecule could replicate itself.

Excited by Rosalind Franklin's stunning data, Wat-

FIGURE 9.7

Elucidation of the Structure of DNA.

[A] James Watson and Francis Crick worked out the structure of DNA in 1953 by building scale models of the molecules involved and trying to arrange them in ways that were consistent with all the known data. [B] Rosalind Franklin working with Maurice Wilkins performed X-ray diffraction experiments that provided the key data—the pattern of symmetry shown in the image in [C]—required to completely determine the structure.

[A]

[B]

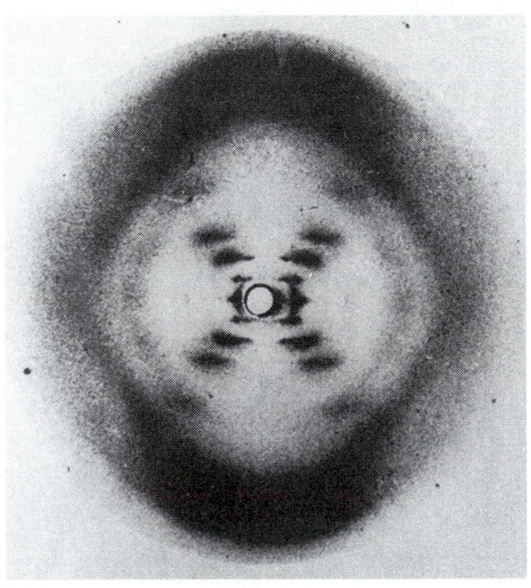

[C]

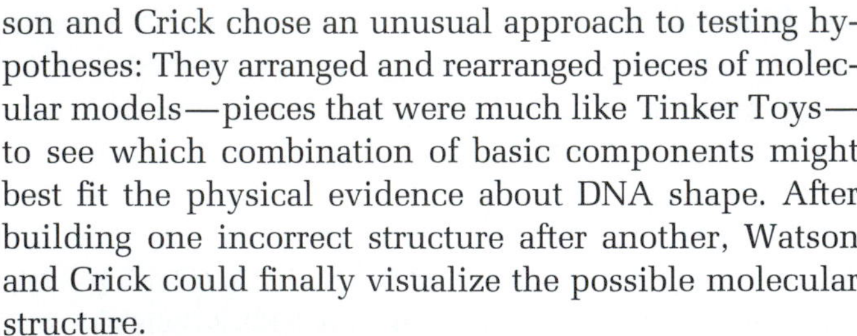

son and Crick chose an unusual approach to testing hypotheses: They arranged and rearranged pieces of molecular models—pieces that were much like Tinker Toys—to see which combination of basic components might best fit the physical evidence about DNA shape. After building one incorrect structure after another, Watson and Crick could finally visualize the possible molecular structure.

The structure was a **double helix**, which resembles a twisted ladder [FIGURE 9.9]. Watson and Crick had won the race. In 1953, in the international journal *Nature*, Watson and Crick published their ideas about the DNA molecule. In a classic understatement, they wrote that it had "not escaped our notice" that the model immediately suggested ways in which DNA could fulfill its two major biological functions: replicating and storing information. The model's simplicity plus its enormous power to explain observations of nature led to rapid acceptance by the scientific world.

In 1962, Watson, Crick, and Wilkins were awarded the Nobel prize for their assessment of the structure of DNA. Franklin had died four years earlier of breast cancer at the age of 37, and most observers agree that she had played a crucial, but underappreciated, role in the elucidation of the structure of DNA.

The Structure of DNA

Having explored how the structure of DNA was discovered, let's investigate the structure itself.

1. A DNA molecule is composed of two nucleotide chains oriented in opposite directions, like the northbound and southbound lanes of a highway [FIGURE 9.9,

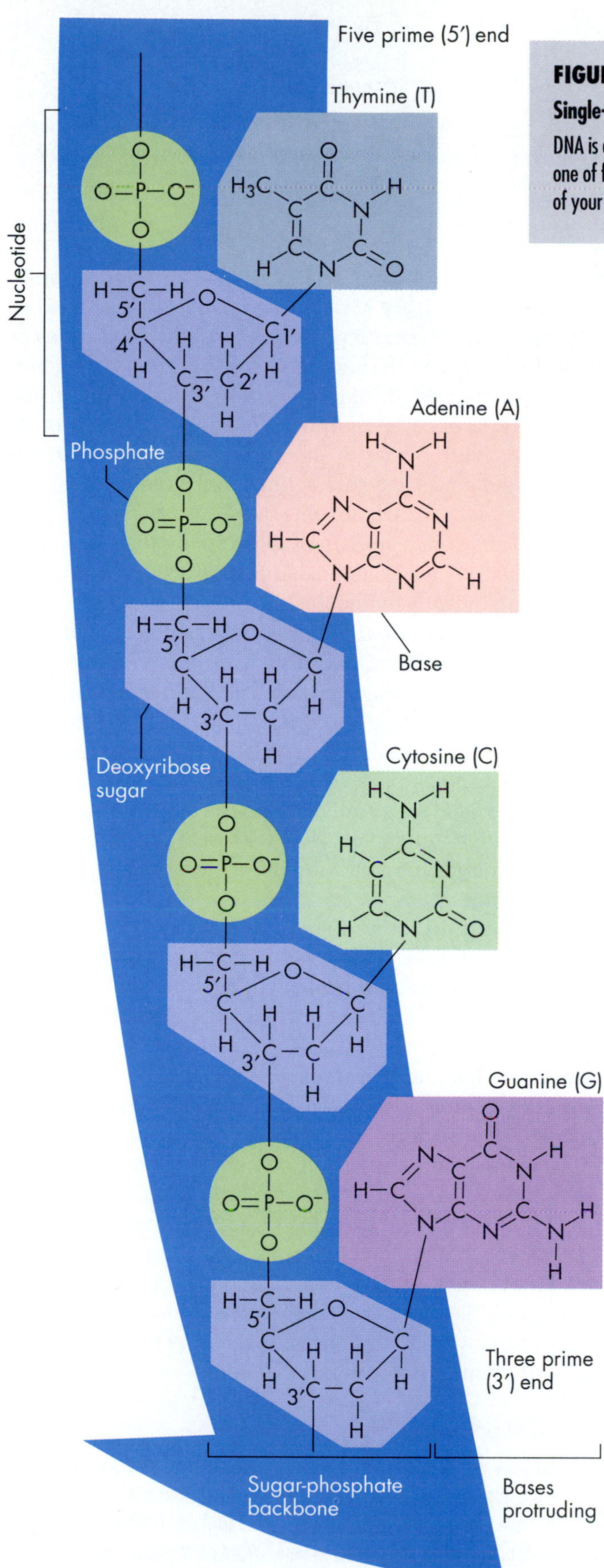

FIGURE 9.8

Single-Stranded DNA.

DNA is a chain of nucleotides. Each nucleotide consists of a phosphate attached to a sugar attached to one of four bases — A, T, C, or G. The chain here has only four nucleotides. A DNA in a nucleus in one of your cells contains on average 130 million nucleotides.

Point 1]. To emphasize the opposite orientation of the two strands, geneticists refer to them as *antiparallel*.

2. The two sugar-phosphate backbones face the outside of the molecule, while the bases attached to the backbones face inward, connecting in the middle like the rungs of a ladder [FIGURE 9.9, Point 2]. The sugar-phosphate backbones form the uprights of the ladder.

3. The bases A and T pair with each other, and the C and G bases also pair; that is, A *complements* T, and C *complements* G—the two fit together like adjacent puzzle pieces. Bases are held together by hydrogen bonds [review FIGURE 2.32], hydrogen atoms shared between an A and a T base or a C and a G base [FIGURE 9.9, Point 3]. This base pairing by means of hydrogen bonding is significant in three ways. First, it provides the force that holds the two single strands of DNA together into a double-stranded molecule. Second, this **complementary base pairing** explains why DNA always has equal amounts of A and T and equal amounts of C and G. Third, complementary base pairing is not only a fundamental feature of DNA's shape, but also of its biological activities: replication and the storage of genetic information. Base pairing is also crucial for gene splicing conducted by molecular biologists, as you will see later.

4. Finally, the two DNA strands are twisted together to form intertwined helices, or the double helix [FIGURE 9.9, Point 4]. In short, DNA looks like a twisted ladder.

PACKAGING DNA IN CHROMOSOMES

Watson and Crick derived the twisted-ladder structure of DNA from linear DNA, but DNA can also be circular. The DNA of bacterial cells loops back on itself to form a circle. In addition, the DNA of plasmids [review FIGURE 9.3] and of many viruses that infect mammalian cells is circular, as is the DNA of chloroplasts and mitochondria in eukaryotic cells.

In contrast, the nuclear DNA of eukaryotes and the DNA of many viruses is linear. In these organisms, single molecules of DNA are extraordinarily long and are organized into chromosomes. In a eukaryotic cell's nucleus, each chromosome consists of a single, long, tightly wound DNA molecule. For example, the actual length of

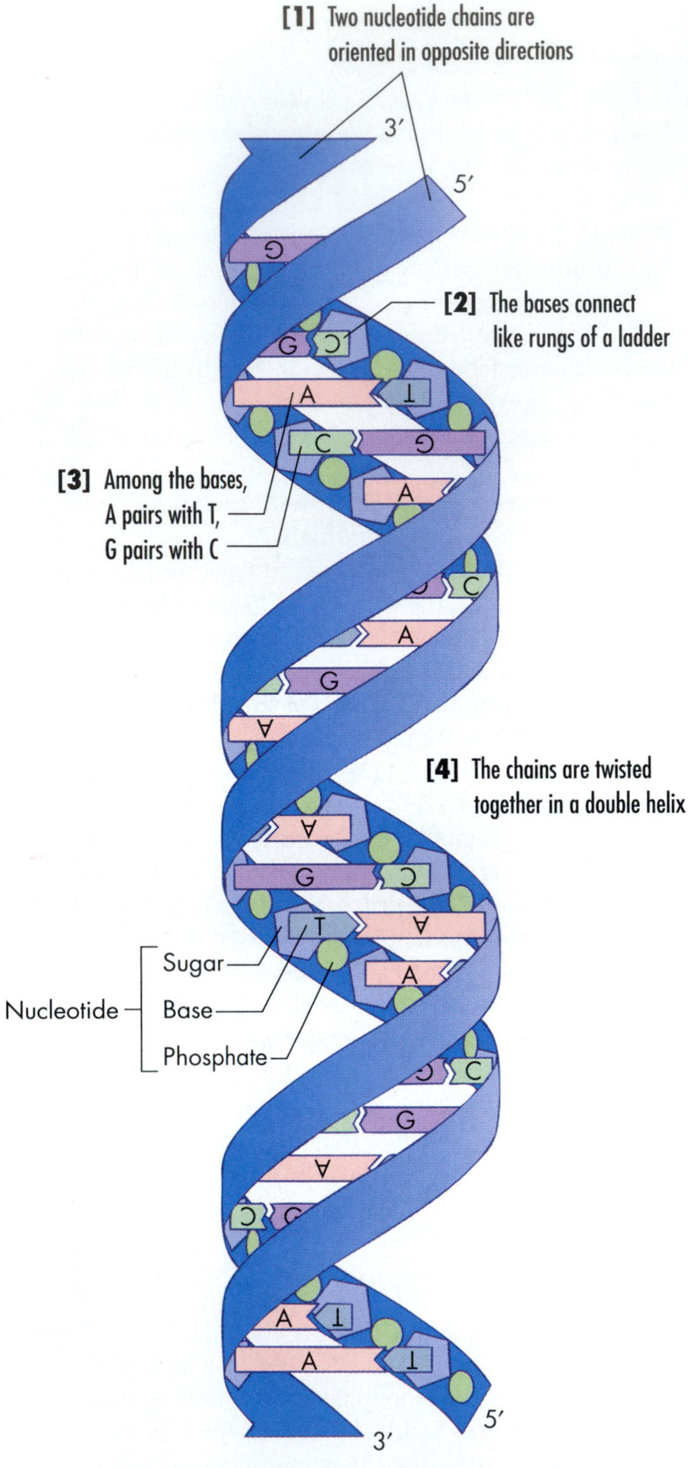

FIGURE 9.9

Double-Stranded DNA.

DNA consists of two sugar-phosphate ribbons from which bases protrude toward the middle holding the two strands together into a single molecule; A links T and G with C like rungs on a twisted double ladder.

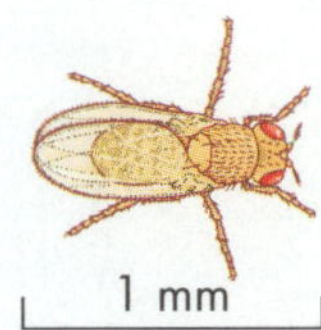

FIGURE 9.10

The DNA Molecule of a *Drosophila* Chromosome.

If stretched out, the DNA in just one fruit fly chromosome would be 12 times longer than the insect itself!

the DNA molecule from the largest of the four chromosomes of the fruit fly *Drosophila* is about three times as long as the word *genetics* printed here (that's 12 times as long as the fly itself), and carries a few thousand genes [FIGURE 9.10]. Despite the molecule's length, each of the fly's millions of cells contains two copies of this chromosome, as well as pairs of the other three chromosomes. How do cells package such a huge molecule into a structure as small as a chromosome?

The enormous length of DNA in a eukaryotic cell cannot just be wadded up haphazardly. If it were, the separation of DNA molecules during cell division would be as difficult as untangling two kite strings. What happens instead is that DNA, like a proper kite string, is wound around spools of proteins called *histones* [FIGURE 9.11D and E]. A single spool wrapped with two loops of DNA (140 base pairs long) is called a *nucleosome*. Adjacent nucleosomes pack closely together to form a larger coil, somewhat like a coiled telephone cord. This cord, in turn, is looped and packaged with scaffolding proteins into *chromatin*, the substance of a chromosome [FIGURE 9.11A and B].

Compare this new knowledge about DNA packing into chromosomes with what you already know about chromosome activity in mitosis. Recall that in the interphase portion of mitosis, chromosomes are not individually visible [review FIGURE 7.9A]. This is because the spools of DNA and protein are not packed closely together, allowing the chromatin to spread out diffusely through the nucleus. In contrast, during prophase [review FIGURE 7.9B], the spools compact together as shown in FIGURE 9.11D, leading ultimately to a visible chromosome [FIGURE 9.11A]. It is this orderly packaging of DNA that prevents massive tangles during cell division.

From base pair to double helix to DNA-wrapped histone spool, the structure of DNA is a boundless biological resource. How that structure allows DNA to create diversity out of unity is a marvel of biology.

➤ CONCEPT CHALLENGE

Let's say you decide to test whether or not the usual rules for base pairing apply to sea urchins. You remove an urchin from a tide pool and determine that 33 percent of the bases in its DNA are T. What percentage of its bases would you predict to be A? What percentage should be left over for the other two bases? What percentage of C should you expect?

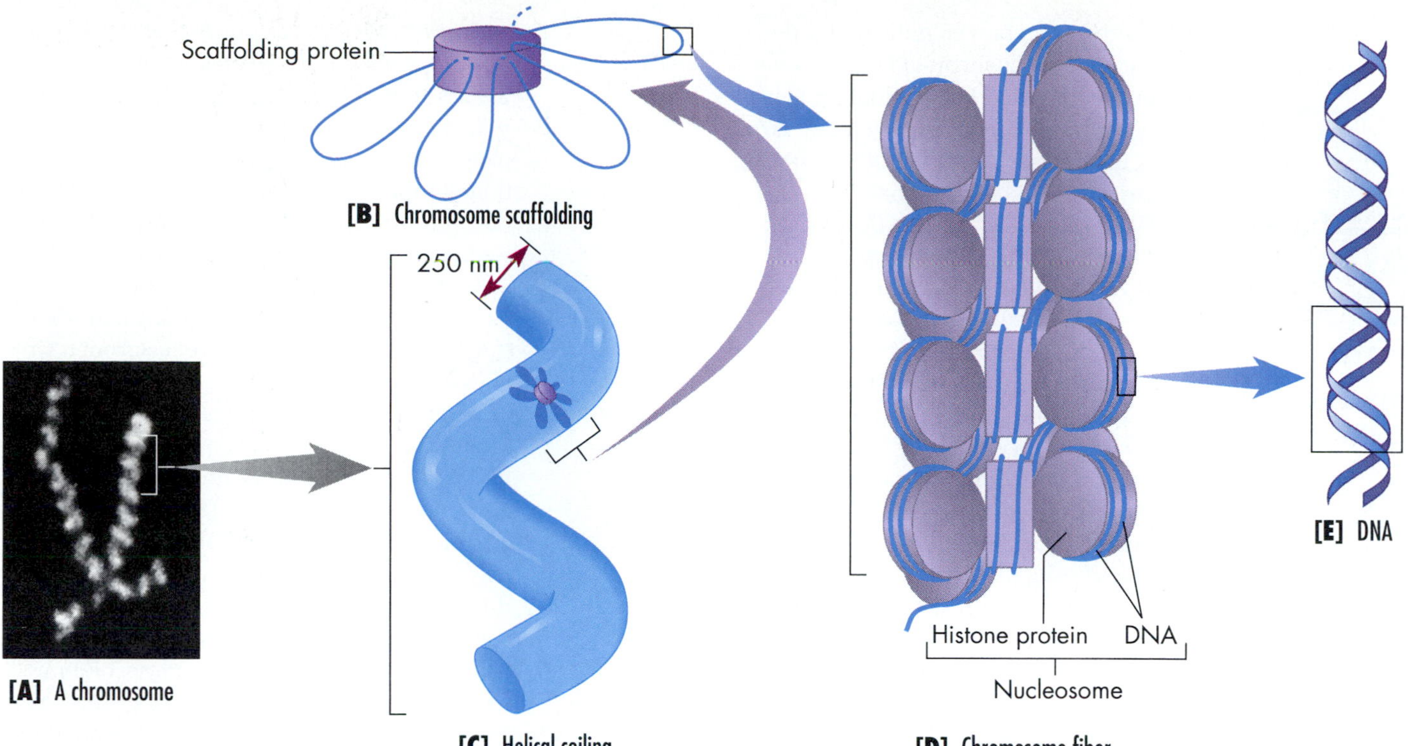

FIGURE 9.11

How DNA Is Packaged into Chromosomes.

DNA is packaged in a way that allows its enormous length to be contained in a chromosome without tangling. The DNA double helix [E] wraps around protein spools, forming nucleosomes [D]. Nucleosomes interact with each other, probably forming a coil in a chromosome fiber. [C] The chromosome fiber loops in and out of a central core of scaffolding proteins [B] that also gently coil. This coiling is visible in an intact chromosome [A].

DNA Replication

Watson and Crick first publicly presented the double helix at a meeting of geneticists in Cold Spring Harbor, New York, in the spring of 1953. Even before they began to speak, they knew it would generate excitement, for it suggested ways that DNA might replicate and store information. But even Watson and Crick were startled by the enthusiastic reception their peers gave the double helix model. In the eyes of a scientist, a hypothesis takes on power and beauty in proportion to how well it explains diverse facts simply and coherently; and the scientific community agreed that the double helix model was a powerful tool.

Returning from the meeting, geneticists put on their lab coats, went to their benches, and quickly confirmed the Watson and Crick model. In various ways, they found that complementary base pairing and linear variation in base order do indeed account for a gene's ability to replicate itself and store information. This section explores how DNA replicates, and the next chapter explores how DNA stores hereditary information.

People have always recognized that like begets like. It is obvious at some family reunions, for instance, that red hair, a big nose, a bald head, or short stature has been passed along for several generations [FIGURE 9.12]. For traits to be conserved this way, genes must produce exact copies of themselves, and those copies must be transmitted from parent to child. The key to this gene copying is

FIGURE 9.12

Family Features.

The inheritance of physical traits, passed down through generations, depends on the precise copying of DNA molecules.

box 9.1
Biology Applied

DNA Synthesis, AZT, and Treatment of AIDS

Thanks to researchers who understood the basic biological mechanism of DNA synthesis, there is a drug that helps people infected with HIV, or human immunodeficiency virus, the virus that causes AIDS. In 1985, medical researchers at Duke University showed that a drug called zidovudine (AZT) effectively slows the progression of acquired immune deficiency syndrome, a virus-caused illness that destroys the immune system and leaves the body susceptible to life-threatening infections and cancers [details in CHAPTER 26]. These studies were preliminary, however, and AZT did not receive approval from the U.S. Food and Drug Administration for two more years. This delay triggered much anger and protest from AIDS patients, hundreds of whom died during the interim period. Why was the U.S. government initially hesitant about legalizing the drug? And how, in the first place, does AZT therapy work?

AIDS is caused by the HIV virus, which has about nine genes stored in an RNA molecule—a single-stranded molecule of heredity related to the double-stranded DNA. For the HIV virus to successfully infect a cell of the human immune system (a white blood cell), the virus's RNA genes must be copied into the matching DNA form. This RNA-to-DNA copying is carried out by a viral enzyme (called *reverse transcriptase*) that synthesizes DNA from the viral RNA template much as the normal DNA-synthesizing enzyme (*DNA polymerase*) makes DNA from a DNA template [see FIGURE 9.13]. The researchers thought that by blocking this synthesis of viral DNA they might be able to block virus replication in an infected person and hence slow the progression of the disease.

In both DNA-to-DNA copying and RNA-to-DNA copying, synthesis proceeds in a single direction and accomplishes a hooking up of two nucleotides, with the phosphate of one nucleotide connected by an oxygen atom to the sugar of the previous nucleotide in the chain [FIGURE 1A, Step 1]. The Duke University biologists knew that DNA polymerase acts by bringing an OH group that hangs down from the sugar of one nucleotide together with another OH group that protrudes upward from the phosphate of a second nucleotide [FIGURE 1A, Step 2]. As a result, an H from one joins an OH from the other, making H_2O, or water (Step 2). In the process, the enzyme leaves the remaining oxygen stuck as a bridge between the two nucleotides, and the newly forming DNA chain grows one nucleotide longer.

The AIDS researchers at Duke knew two other important things: (1) AZT is a nu-

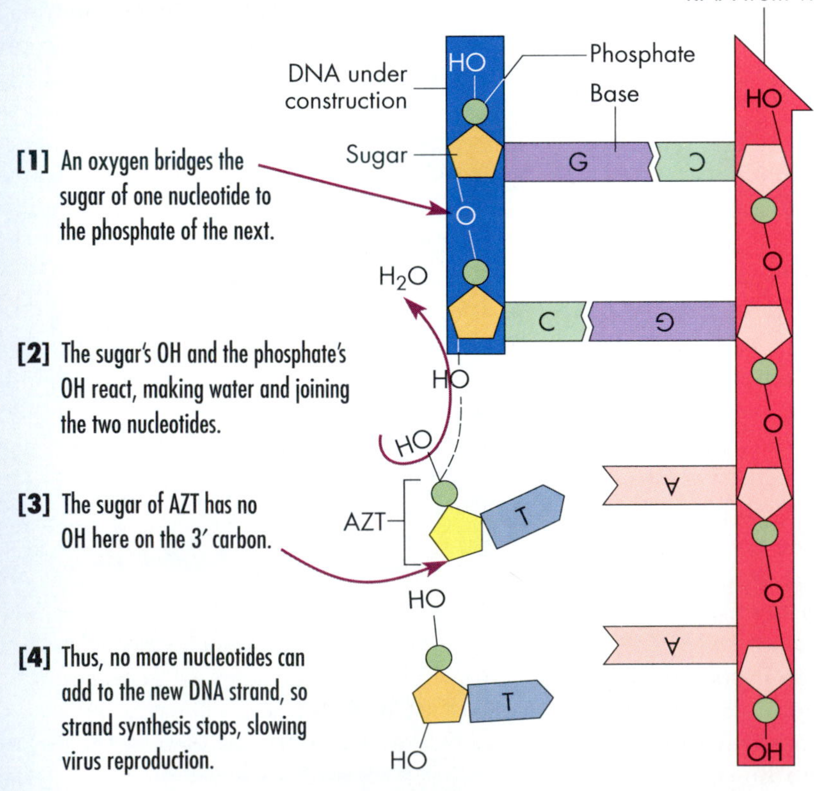

[A] DNA action

[B] Social action

FIGURE 1
[A] How AZT blocks the synthesis of AIDS virus DNA. With the synthesis of viral DNA slowed, the virus cannot replicate rapidly inside an infected person, and the person with AIDS retains health longer. **[B]** Political action taken to support AIDS research.

cleotide very similar to the normal T nucleotide in shape. As a consequence, both the cell's DNA polymerase and the virus's reverse transcriptase enzymes can pick up the drug AZT instead of the normal nucleotide containing T and incorporate it into growing DNA [FIGURE 1A, Step 3]. (2) Because AZT is missing the critical OH that forms the link to another nucleotide in DNA, synthesis of a new DNA chain halts at the point where AZT is inserted [FIGURE 1A, Step 4]. Researchers reasoned that if AZT is incorporated in this way and prevents the viral RNA from being copied into DNA, the virus could not replicate and infect additional immune system cells; thus, the spread of AIDS through the body would be slowed, and the patient's life would be extended.

In July of 1985, AZT was first administered to AIDS patients at Duke University Medical Center. During this initial study, AIDS patients improved considerably from the treatment. After reviewing this work, however, scientists and physicians from the FDA insisted that they could not approve and release the drug nationally until additional studies were carried out. This was their reason: It was clear that AZT blocks synthesis of DNA from the viral RNA template. It was not yet clear, though, how much the drug also blocks necessary DNA synthesis in the patient's own cells. If this blockage was too pronounced, it could lead to a condition similar to radiation sickness [see CHAPTER 7, page 177] or to cancer and would be, in the long run, worse for patients than other AIDS therapies already available.

As these additional, longer-term tests were going on, many AIDS patients protested and demonstrated, demanding faster access to AZT [FIGURE 1B]. Fortunately, the follow-up studies confirmed that AZT therapy does block viral replication more effectively than it blocks DNA synthesis in the patient's own cells and that while there are significant side effects from this cellular blocking, such as headaches, mood swings, anemia, and digestive disturbances, in general AZT improves AIDS patients' quality of life. In March of 1987, the FDA approved AZT as a prescription drug for AIDS patients. Today, along with antibiotics and other drugs we will discuss in later chapters, AZT is helping to keep thousands of people alive as the search for better treatments and prevention goes on.

the extraordinary accuracy with which DNA molecules are copied in cells.

Let's link back to CHAPTER 7 again to place the biochemical process of DNA replication into the familiar context of the cell cycle. Recall that cells cycle through a period of growth (interphase) and a period of division (M phase) [review FIGURE 7.6]. Recall that interphase has three parts: in G_1, cells have a single copy of each nuclear DNA molecule; in the S phase, DNA is synthesized, or replicated; and in G_2, there are two copies of each DNA molecule. Finally, in mitosis, those two DNA copies are separated into the two offspring nuclei. Let's now focus on the events of the S phase, the copying of DNA.

STEPS IN REPLICATION

The replication of DNA follows directly from the principle of complementary base pairing and can be divided into three steps: (1) strand separation, (2) complementary base pairing, and (3) joining.

1. Separation. For replication to begin, the two strands of the double helix must first unwind and separate from each other (FIGURE 9.13A; the figure does not attempt to show the unwinding and rewinding of DNA). The unwinding and separation of the strands are catalyzed by enzymes that break the "rungs" of the ladder. Those rungs, remember, are the bases on each strand that are bound together by weak hydrogen bonds. The separated base pairs are temporarily unpaired.

2. Complementary base pairing. The unpaired bases form new hydrogen bonds with free nucleotides that drift into the area, but the pairing is anything but random. An A base on one DNA strand pairs only with a free T base (complete with its sugar-phosphate backbone), and an attached C pairs only with a free G. (Likewise, attached Ts bond only to free As, and attached Gs bond only to free Cs.) Thus, the sequence of bases in the original strand specifies the same sequence in the new strand according to the rules of complementary pairing [FIGURE 9.13B].

3. Joining. The joining together of the newly added bases forms a new strand that is complementary to the parent strand, forming two new double helices that are identical to the original double helix [FIGURE 9.13C]. The joining, or *polymerization*, of the new double helices is catalyzed by an enzyme called *DNA polymerase*. Because this enzyme can synthesize DNA in only one direction, there are differences in the way the two newly made DNA strands are synthesized; for the curious student the details are shown and described in FIGURE 9.14. The drug AZT, which physicians have used for several years to help prolong the lives of people infected with the AIDS-causing virus, prevents the normal joining of bases during DNA synthesis, as BOX 9.1 explains.

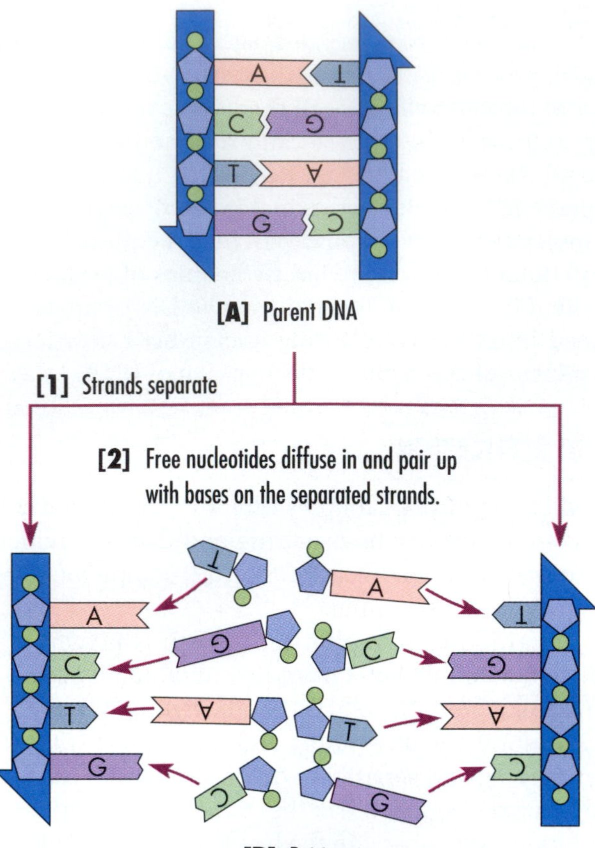

[3] Each new row of bases is linked into a continuous strand.

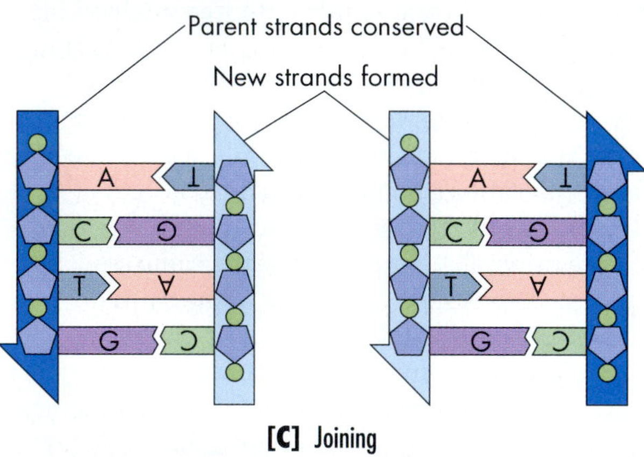

[C] Joining

FIGURE 9.13
DNA Replication.

When DNA replicates, the strands separate [A] and previously unattached nucleotides that have accumulated in the nucleus diffuse in and pair up with appropriate unpaired bases [B]. The new bases link together [C], forming new strands. The two new double-stranded molecules are identical to the parent DNA molecule, and in each, one strand is inherited intact from the parent and one is newly formed.

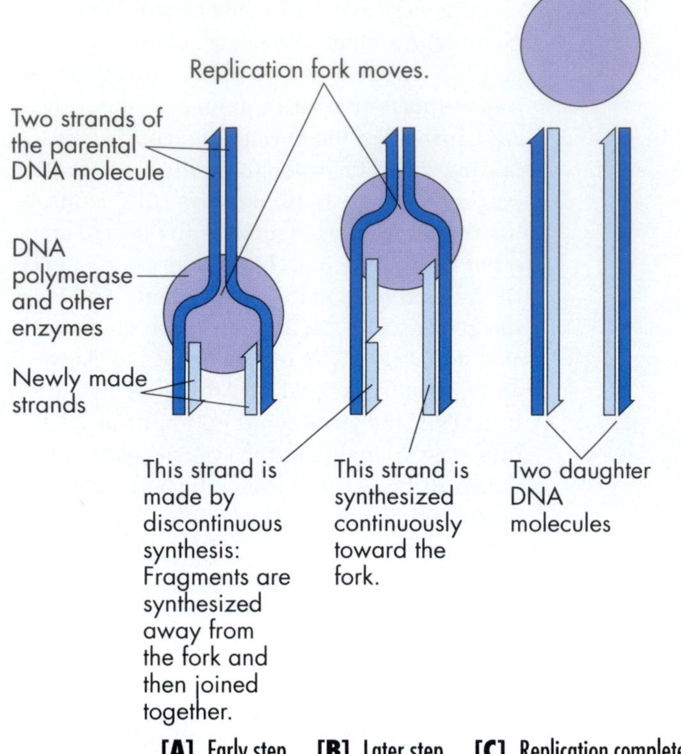

[A] Early step [B] Later step [C] Replication complete

FIGURE 9.14
Action at the Fork.

[A] The enzyme DNA polymerase can add nucleotides to a newly forming strand of DNA in only one direction. Since the two strands of a DNA molecule are oriented in opposite directions, this causes a problem. [B] When the replication fork, the point where the two strands separate, has moved forward, the strand made in the direction of the fork can be synthesized continuously. The other newly made strand, however, is made discontinuously in short fragments, starting near the fork and moving away from it. [C] Enzymes then join fragments together forming a new intact DNA strand.

SEMICONSERVATIVE REPLICATION

The continuous repetition of strand separation, base pairing, and joining along the length of the DNA molecule produces two double strands of DNA from one. Each new double-stranded DNA has a base sequence identical to the base sequence of the original. FIGURE 9.14 shows that each of the two daughter DNA molecules has one strand intact from the original parent, while the other strand is completely new. Because only one of the two strands in the daughter molecule is inherited intact—or conserved—from the parent molecule, this type of replication is called **semiconservative replication**.

FIGURE 9.15 depicts experiments that proved that DNA replication is semiconservative in higher organisms. Apparently, all living creatures share this mode of replication. Note that the semiconservative copying of DNA is very different from copying a piece of paper on a

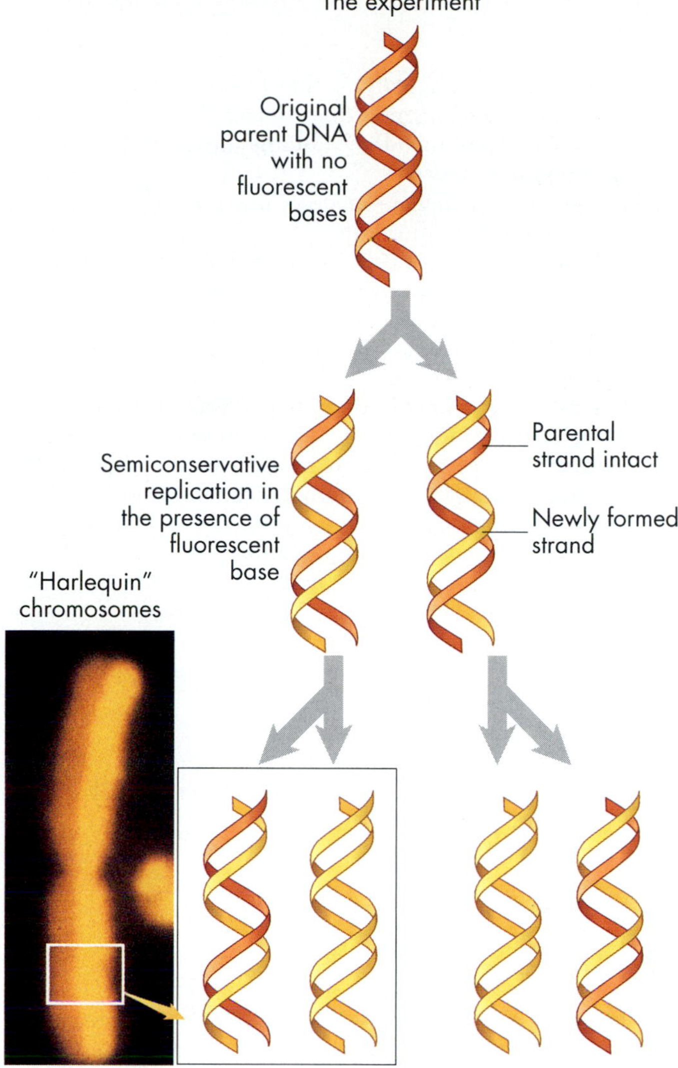

FIGURE 9.15

DNA Replication Is Semiconservative.

This hypothesis predicts that each new daughter DNA molecule will contain one strand from the parent double helix (brown), and the other strand will be newly formed (yellow). Follow the path of one original (brown) strand through two rounds of DNA synthesis. To see if the semiconservative path predicted by the hypothesis is true, researchers grew cells from a Chinese hamster ovary for two rounds of DNA synthesis—two cell cycles—in the presence of a compound that fluoresces bright yellow. All newly made strands will therefore show up bright yellow in the chromosomes. DNA molecules with one newly made strand and one parental strand will have half as much fluorescent material and will appear brown. The results, as shown in the photograph, are consistent with the hypothesis.

photocopying machine, which produces a totally new copy but conserves the original fully intact. In contrast, with DNA replication, the original molecule ceases to exist, but half of it becomes part of one offspring molecule and the other half becomes part of the other offspring molecule.

ACCURACY OF DNA REPLICATION

It is a wonder that DNA molecules are ever copied correctly, considering the complexities of unwinding, separation, base pairing, and polymerization required for semiconservative replication. Yet DNA synthesis is incredibly accurate: An error is made about once in every 10^9 bases. To approach that level of accuracy you would have to type this entire book a thousand times with only one typing error!

For many organisms, survival requires extreme accuracy of DNA replication. For example, the human **genome** (the total of all the genes in a single haploid cell) contains about 3×10^9 base pairs. On the average, then, each egg or sperm will have about three new errors. If the rate were much higher, the genetic information would be so altered that the organism could not function.

➤ CONCEPT CHALLENGE

Many agents used during chemotherapy to treat cancer act by disrupting the base-pairing step of DNA replication. Why would cancer cells be more susceptible to those agents than would very slowly dividing cells?

Connections

The experiments we discussed in this chapter revolutionized the geneticist's view of the gene: They gave Mendel's abstract theory a physical reality in the form of DNA. As genes became tangible, geneticists learned to manipulate them so that they could see how they worked. Knowing the structure and function of DNA allows us to appreciate that in *E. coli* O157:H7, the DNA of a virus that infects the bacterium is replicated along with the bacterium's DNA, and causes a toxic protein to be made.

The conceptually simple DNA molecule—just a couple of strings of nucleotides twisted around each other—unifies all life. DNA also is the source of life's diversity. Spinach leaves are green and cardinal feathers red because the DNA in the pigment-forming cells differs in base sequence.

Ultimately, this central molecule of life provides each species with a means of obtaining energy and reproducing—of passing along exact copies of the genes to offspring via the remarkable fidelity of DNA's template mechanism during replication. How DNA blueprints are actually decoded to form the proteins that carry out cell function is our subject in CHAPTER 10.

KEY TERMS

complementary base pairing, 229
double helix, 228
genome, 235
semiconservative replication, 234

HIGHLIGHTS IN REVIEW

1 Genes are made of DNA.

- **a]** The ability of a bacterial virus (a molecule of DNA surrounded by protein) to cause disease by encoding a protein that kills human cells illustrates gene function.
- **b]** Classic experiments of the mid-twentieth century showed that genes are made of DNA, not protein.

2 A DNA molecule consists of two strands of nucleotides oriented in opposite directions and twisted into a double helix.

- **a]** Each DNA nucleotide is made of a sugar group, a phosphate group, and one of four nitrogen-containing bases. Thousands of nucleotides are connected in a chain—sugar to phosphate to sugar—with the bases hanging off each of the sugars.
- **b]** The four types of bases can be in any order, so that each gene has a different—but specific—base sequence. From this variability comes DNA's remarkable capacity to store almost infinite amounts of information.
- **c]** The two chains in each DNA molecule are oriented in opposite directions.
- **d]** Each of the four bases tends to form weak hydrogen bonds with only one of the other three bases. A joins with T, and C joins with G. This phenomenon is called complementary base pairing.
- **e]** Because of complementary base pairing, the bases on the two polynucleotide chains bond to each other, holding the two chains together, with the bases facing inward and the sugar-phosphate backbones facing outward.
- **f]** DNA in eukaryotes is extremely long; it is wound around spools of protein, and then it coils upon itself repeatedly. This organization avoids tangling during cell division.

3 The structure of DNA explains its ability to replicate and to control an organism's physical makeup and abilities.

- **a]** The key to how genes function—both to specify polypeptides and to replicate—lies in the unique double helix structure of DNA.
- **b]** Genes have two major functions: They replicate themselves, and they carry information that specifies amino acid sequences in the proteins of each organism.
- **c]** Genes look and act fundamentally the same in all living things.
- **d]** Information is encoded in DNA in a linear fashion and can be read in only one direction.

4 During DNA replication, the two strands separate and enzymes synthesize a complementary copy of each one; this yields two daughter molecules identical to the original molecule.

- **a]** Genes work by specifying the structure of proteins, including the enzymes that control chemical reactions in the cell and the other proteins that make the structure of the cell.
- **b]** In general, each gene specifies one polypeptide, that is, one chain of amino acids.
- **c]** Because of base pairing, each chain acts as a template for the creation of a new copy of the other chain. Thus, two new strands are created, one from each half of the old chain in a process called semiconservative replication.
- **d]** Replication occurs in three stages: Enzymes enable the two chains of the double helix to unwind from each other; nucleotides float in and pair with their complements, forming a new chain; and an enzyme called DNA polymerase joins the nucleotides in the new chain together—sugar to phosphate to sugar.
- **e]** DNA replication is incredibly accurate, with less than one error in every 10^9 bases. This accuracy is due to DNA polymerase, which removes incorrect (i.e., unpaired) bases from the end of the newly forming nucleotide chain.

UNDERSTANDING THE FACTS AND CONCEPTS

For Questions 1–5 choose the term that correctly completes the statement.

- **a]** double helix
- **b]** cytosine
- **c]** adenine
- **d]** hydrogen
- **e]** covalent

1 Thymine in one strand of DNA pairs with ______ in the other.

2 The pairing of complementary bases to form "rungs" of the DNA ladder occurs by means of ______ bonding.

3 A strand of DNA is composed of a linear array of nucleotides joined to each other by ______ bonds.

4 Guanine in one strand of DNA pairs with ______ in the other.

5 DNA normally exists as a double-stranded molecule called a ______.

For Questions 6–10, match each of the descriptions with the most appropriate item from the following list of terms.

- **a]** plasmid
- **b]** bacteriophage
- **c]** helix
- **d]** antiparallel
- **e]** backbone

6 A virus that infects bacteria.

7 The term that describes the opposite orientation of the strands of DNA in a double helix.

8 A linear stack of the sugar and phosphate components of nucleotides in a strand of DNA.

9 Consists of a small, extrachromosomal circle of DNA.

10 A spiral of constant diameter.

For Questions 11–15 match each of the descriptions with the appropriate item from the following list.

a] histone
b] nucleosome
c] chromatin
d] chromosome
e] circular DNA

11 A length of DNA wrapped around a protein spool.

12 The hereditary material of bacteria.

13 An organized collection of DNA, histone, and other proteins.

14 A nuclear organelle made of chromatin.

15 A chromosomal protein that acts as a support for DNA.

For Questions 16–20 match the hypothesis or process cited in the following list with the experimental evidence described.

a] semiconservative replication
b] bacterial transformation
c] production of sequential generations of phage
d] DNA is the transforming principle
e] DNA is a double helix

16 Experiments that isolated all of the molecular constituents of S-strain pneumococci.

17 Experimental evidence showed that one strand in a DNA duplex is inherited intact while the other strand is newly synthesized one nucleotide at a time.

18 Regularities in X-ray diffraction photographs appeared.

19 The deliberate or accidental uptake of pure DNA by a cell can change that cell's phenotype.

20 The primary function of bacteriophage infection.

INTEGRATE AND APPLY WHAT YOU HAVE LEARNED

1 Contrast the meaning of the terms *complementary bases* and *semiconservative.*

2 Explain how experiments with viruses demonstrated that DNA, rather than protein, is the carrier of hereditary information.

3 Why was the structure of DNA considered to be so important that several research groups engaged in a race to find a model?

4 Suppose that during replication of a molecule of DNA an adenine molecule paired with a cytosine molecule. Further, suppose that this error was not detected and the DNA completed its replication. Trace the consequences through two or more rounds of replication.

5 Why were geneticists reluctant at first to accept the idea that genes are composed of DNA rather than protein? Can you find any flaw in their reasoning?

ANALYSIS

1 A botanist discovers a new plant in the Amazonian rain forest. This plant's DNA bases are 18 percent adenine. What is its percentage of cytosine nucleotides?

a] 18 b] 32 c] 36 d] 64 e] 82

2 Suppose that life is discovered on another planet, but the replication of its genetic material results in both new strands of a double-stranded molecule being produced continuously. In other words, there are no lagging strands that occur during replication. What is the *most likely* hypothesis to explain this finding?

a] The genetic material was not replicated semiconservatively.
b] The genetic material was not antiparallel.
c] The genetic material was not in the form of a helix.
d] The genetic material was not replicated by means of an enzyme.
e] The genetic material was mutated DNA.

3 Pretend for a moment that DNA does not replicate semiconservatively. Rather, the strands in the DNA duplex separate completely and temporarily, while each replicates a new strand by complementary base pairing. Then the two old strands snap back together again, and the two new ones also join together. If this were to happen, what would you expect to see in the photograph of chromosomes in FIGURE 9.15?

a] All chromosomes would be completely brown.
b] All would be completely yellow.
c] All would be half brown and half yellow.
d] Some chromosomes would be completely brown and some would be completely yellow.
e] Some would be all brown, some all yellow, and some both brown and yellow.

4 Refer to the experiment illustrated in FIGURE 9.2. What would you expect to happen to the mouse after being injected with heat-killed R cells and live S cells? The mouse would:

a] Die because the R cells would transform the S cells.
b] Live because the R cells would transform the S cells.
c] Die because of the virulence of untransformed S cells.
d] Live because of lack of virulence of the R cells.
e] Die because the S cells would transform the R cells.

5 AZT slows the progression of AIDS primarily because it interferes with the:

a] Unwinding of the viral DNA.
b] Elongation of the viral RNA.
c] RNA-to-RNA replication of the virus.
d] Elongation of DNA in the RNA-to-DNA copying process.
e] Normal base pairing during DNA replication.

CHAPTER 10

How Genes Work

CYSTIC FIBROSIS: A CASE STUDY IN GENE ACTION

Jeff Pinard, a microbiology major at the University of Michigan, set out in the summer of 1989 to research the genetic causes of his own disease: *cystic fibrosis*, the most common lethal genetic disease among Caucasians. Of the thousands of Americans born each year with cystic fibrosis, only 10 percent survive to their thirties. Both of Jeff's parents are carriers for the cystic fibrosis trait: like 1 in 20 people of northern European ancestry, his mother and his father each carry one copy of the recessive mutation that leads to the disease. Like other carriers, neither parent showed symptoms of the disease. Having inherited one copy of the allele for cystic fibrosis from each parent, Jeff, however, did show disease symptoms.

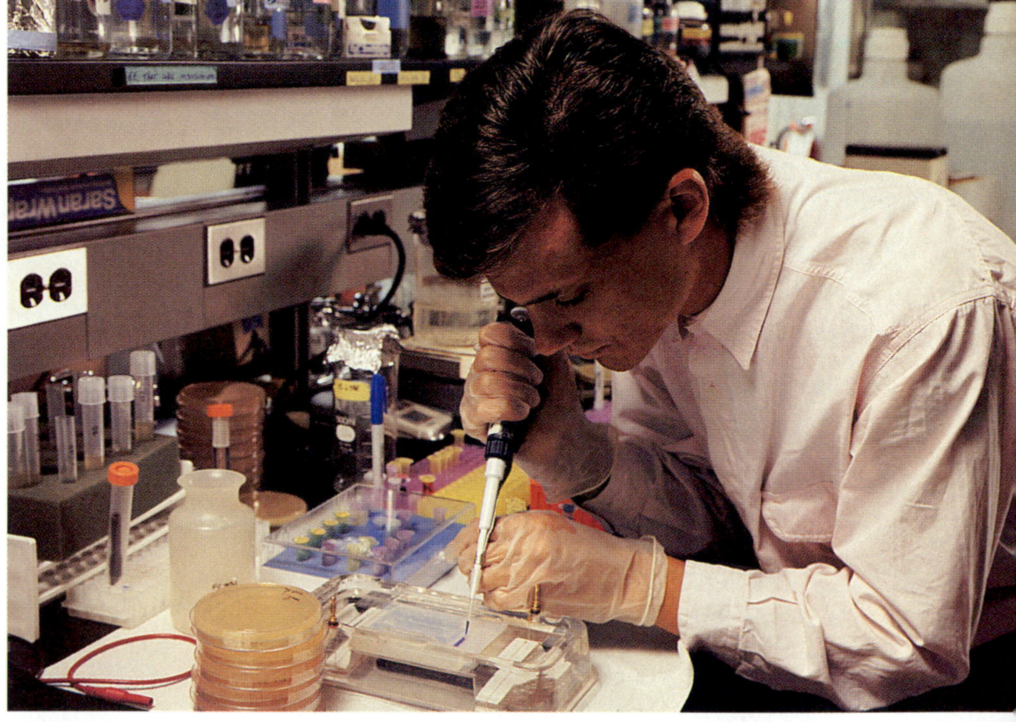

FIGURE 10.1
Jeff Pinard Researches His Own Disease.
An undergraduate at the University of Michigan, Pinard is analyzing DNA samples here as part of a study to find the mutation responsible for his type of cystic fibrosis disease.

Cystic fibrosis is characterized by clogged ducts in many parts of the body. The recessive allele encodes a faulty protein, and this protein leads to a buildup of sticky mucus in lung passages and to blockages of the pancreas ducts, the sweat glands, and in males, the sperm ducts. The impaired lung passages lead to breathing difficulties and recurrent lung infections; the blocked pancreatic ducts to digestive problems; the impeded sweat glands to salty-tasting skin; and the blocked sperm ducts to sterility. By the time Jeff Pinard was 6 years old, he had contracted pneumonia twice, he weighed only 30 pounds, and he had severe stomachaches because his body could not digest food properly.

The saltiness of sweat suggested to physicians that Jeff might have cystic fibrosis, and a test that measures the salt content of his sweat confirmed the diagnosis. This knowledge enabled his family to administer treatments under medical supervision. Several times each day, his parents thumped him firmly on the back and chest for half-hour-long "percussion" sessions to dislodge gummy mucus and to allow easier breathing for a few hours. His parents gave him digestive enzymes in pill form with every meal—or even with a single M&M candy—to supply the enzymes that could not reach his digestive system through his plugged pancreatic ducts. His doctor prescribed powerful antibiotics to fight the lung infections, and with all these treatments, Jeff's condition improved.

Jeff had always been interested in science, and when he entered the University of Michigan as a freshman, he volunteered to help out in the laboratory of Dr. Francis Collins [FIGURE 10.1]. The previous year, Collins' group and a second set of researchers in Toronto had isolated the cystic fibrosis gene, and Jeff wondered about the alleles he had inherited. He knew that he had a relatively mild form of the disease because several of the friends he had made at a summer camp in Michigan for children with cystic fibrosis had died of the disease while Jeff had remained comparatively healthy. Could he have inherited two not-so-dangerous alleles for cystic fibrosis? Jeff learned laboratory procedures to help find out.*

MESSAGES

1 In general, a gene affects an organism's form and function by specifying the structure of a certain protein.

2 Genetic information is stored in DNA in the order of its base pairs.

3 The flow of genetic information proceeds in two steps. (1) Enzymes copy a portion of a DNA molecule into an RNA molecule. (2) The order of bases in RNA specifies the order of amino acids in a protein.

4 In a eukaryotic cell, DNA is copied to RNA in the nucleus, and proteins are made in the cytoplasm on ribosomes.

5 A change in the order of bases in DNA (a mutation) can change the order of amino acids in a protein, and this may disrupt the protein's function.

6 In general, all cells in a multicellular organism contain the same genes, but different cells utilize different genes.

7 An understanding of gene action can help biologists design therapies to improve the lives of people with genetic diseases.

Jeff's condition focuses our attention on the subject of this chapter, the principles of gene action: How do cells use the information stored in DNA to control the way an organism looks and performs? Since an organism's appearance and behavior depend on its proteins, this question becomes a more specific one: How does DNA control protein structure?

Each of an organism's proteins is formed when the information from a portion of a DNA molecule is copied into a special form of an RNA molecule called messenger RNA. The information coded in that RNA then specifies the structure of an individual protein. A mutation—a DNA alteration such as the one that leads to cystic fibrosis—brings about changes in the structure of the specific protein that the DNA encodes. For example, people with cystic fibrosis have an abnormal transport protein in cells lining the body's airways, pancreatic ducts, sweat ducts, and sperm ducts.

Studying protein synthesis helps us understand how cells and organisms operate day to day, and how a many-celled organism can arise from a single fertilized egg cell. It also clarifies some of the fundamentals of Mendelian genetics, such as the basis for dominant and recessive alleles. The mechanisms of gene action, being identical in all living cells, provide evidence that organisms arose from a common origin. Finally, the design of revolutionary therapies for diseases such as cystic fibrosis is centered around manipulating genes and the proteins they encode.

The topics in this chapter include how information flows from DNA to RNA to proteins; how alterations in DNA lead to diseases such as cystic fibrosis; and what factors determine when and where a cell uses specific genetic information. Along the way, we will see how certain life-style changes can help to protect our cells from some environmentally caused DNA damage; how the deadly by-products of some fungi and bacteria disrupt human cells; and how medical researchers are applying the principles of protein synthesis to curing human genetic diseases. ❑

* If you wish to read more about Jeff Pinard's cystic fibrosis research, see the article "Stalking a Lethal Gene" by writer Maya Pines. It appears in *Blazing a Genetic Trail*, a report you can obtain from the Howard Hughes Medical Institute, 6701 Rockledge Drive, Bethesda MD 20817.

Using Mutations to Learn How Genes Work

A mutation like the one that causes cystic fibrosis proves that the information content of DNA is important. Nevertheless, biologists wanted to know exactly *how* DNA stores information in genes. A view of what genes do, particularly the function of information storage, emerged from studies of a harmless but startling defect in human babies.

WRONG GENE, WRONG ENZYME

Much to the consternation of their parents, some otherwise normal babies produce jet black urine. This rare condition is called *alkaptonuria*. Around the time that Mendel's laws were discovered, almost 100 years ago, the English physician Archibald Garrod suggested that the substance excreted by these infants is an ordinary product of metabolism that is usually broken down by an enzyme [FIGURE 10.2A]. Garrod speculated that this enzyme does not function in people with alkaptonuria. As a result, the product is not broken down and accumulates to excess [FIGURE 10.2B]. The excess substance is excreted in urine, eventually turning black when it is exposed to air. Garrod also observed that the black substance accumulates in dark spots on the person's outer ear.

In the first application of Mendel's laws to humans, Garrod studied the families of affected children, noting that alkaptonuria is an "inborn error of metabolism," an inherited mutation that results in an abnormal recessive allele. Garrod also reasoned that if a mutant allele causes the *absence* of enzyme function, then the normal allele is responsible for the *presence* of enzyme function. In other words, a gene functions by allowing a specific enzyme reaction to occur.

THE ONE GENE–ONE ENZYME HYPOTHESIS

As with Mendel's ideas, the significance of Garrod's work went unrecognized for many years. Then, in the 1940s, geneticists George Beadle and Edward Tatum devised a series of sophisticated laboratory experiments to deter-

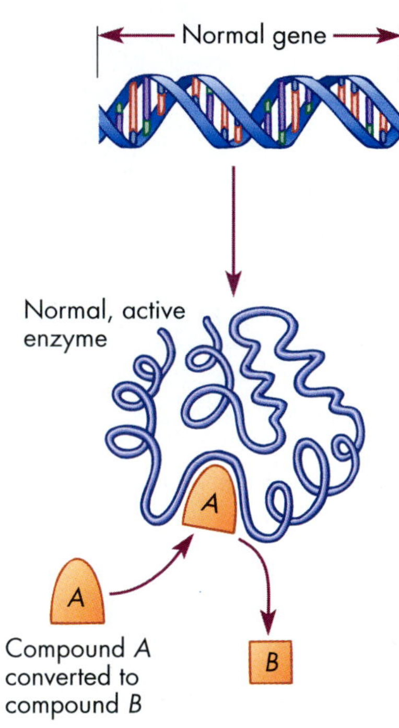

FIGURE 10.2

The Link between Genes and Enzymes.

Analysis of normal infants [A] and infants with the hereditary condition alkaptonuria [B] led to the realization that genes can specify the activity of enzymes.

FIGURE 10.3

Mutant *Neurospora* and the One Gene–One Enzyme Hypothesis.

Beadle and Tatum irradiated *Neurospora* spores; some spores remained unchanged, but in others, genes were altered in ways that blocked the mold's ability to make needed substances. To distinguish between the normal and mutant molds, Beadle and Tatum grew irradiated spores on different kinds of nutrient media. Complete medium provided all the complex molecules *Neurospora* needs to grow [**A**]. Minimal medium provided only small molecules from which normal, but not mutant, mold can synthesize their own complex nutrients [**B**]. The third medium is minimal medium supplemented with just one complex nutrient, here vitamin B_1 [**C**]. Work through the figure to see how Beadle and Tatum came up with the one gene–one enzyme hypothesis.

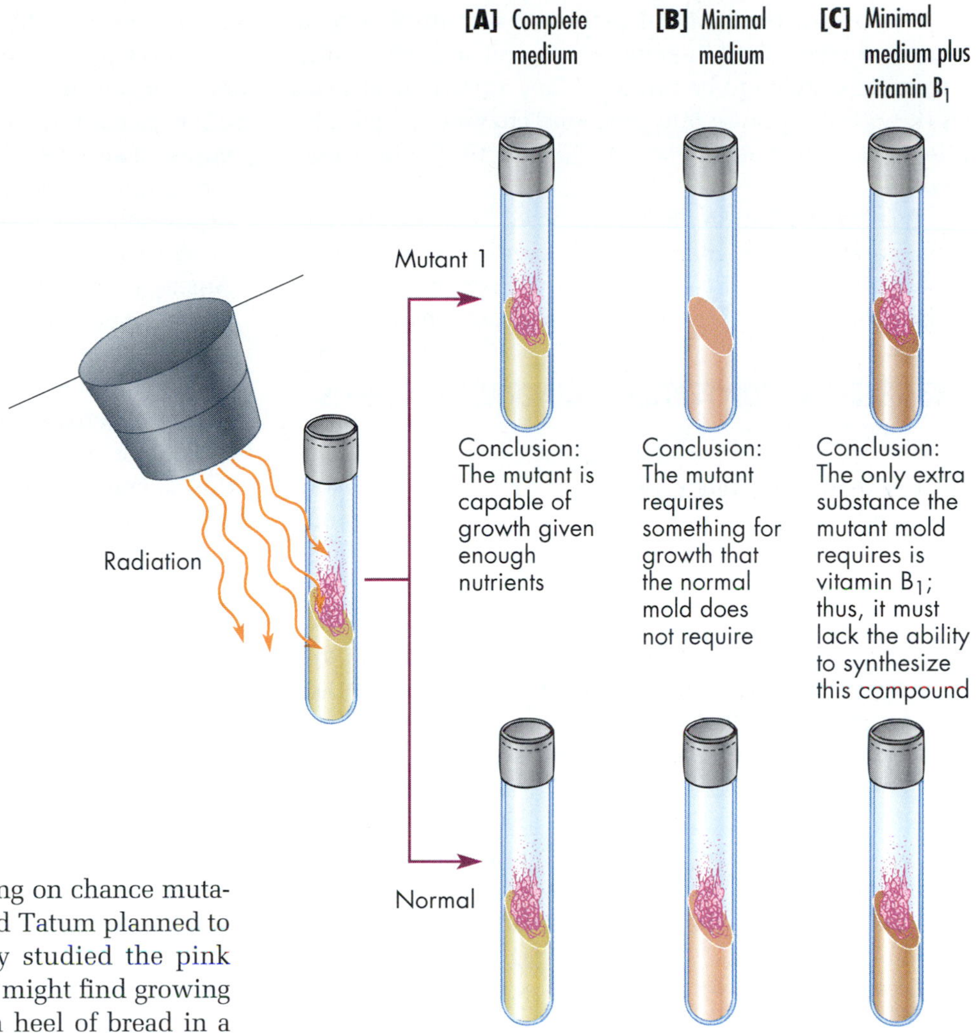

mine what genes do. Instead of relying on chance mutations, as Garrod had done, Beadle and Tatum planned to generate their own mutations. They studied the pink mold *Neurospora crassa*, which you might find growing as a fuzzy pink patch on a forgotten heel of bread in a kitchen cabinet.

Although structurally simple, *N. crassa* can make nearly every compound it needs for growth and reproduction from sugar, salts, and a single vitamin, biotin. *Neurospora* can be grown in the laboratory on a synthetic medium containing inorganic salts, nitrogen, a source of energy (such as sucrose), and biotin. This medium is called a *minimal medium*. Beadle and Tatum wanted to produce, detect, and study mutants that could no longer synthesize all of their necessary amino acids and vitamins.

Follow FIGURE 10.3 to see how the researchers found a mutant form of the bread mold that was unable to synthesize vitamin B_1 (thiamine). First, they irradiated the mold with ultraviolet light, then picked out spores (reproductive cells) and placed them individually into test tubes containing *complete medium*, a substance that contains all the amino acids and vitamins that *Neurospora* needs to grow. You can see two such test tubes in FIGURE 10.3A. Both normal and mutant cells grow on the complete medium because it supplies the needed vitamin even to cells that can't generate it themselves. Next, the researchers transferred cells from the complete medium to a minimal medium, which lacks the needed vitamin [FIGURE 10.3B]. On this medium, the mutant cells failed to grow because they could not generate the missing vitamin themselves. On that same medium, the normal strain could grow well because the cells synthesized their own vitamins. Finally, the researchers mixed up a batch of minimal medium to which they specifically added vitamin B_1. Then they transferred the cells that could not grow on the minimal medium to this supplemented medium. The mutant cells grew under this regime, showing that vitamin B_1 was the only substance they could not make on their own [FIGURE 10.3C].

Beadle and Tatum collected several different mutants that had to be supplemented with vitamin B_1 and many more that required the researchers to add other vitamins or particular amino acids before they would grow. They subjected their collection of mutants to genetic analysis and found that different genes controlled the inheritance of each nutritional requirement.

Beadle and Tatum concluded that (1) each step in a biochemical pathway is controlled by a specific gene, (2) each step in a biochemical pathway is controlled by a specific enzyme, and therefore (3) a gene acts by causing a specific enzyme to be formed. They summarized their work with the phrase **one gene–one enzyme**, which implies that each gene regulates the production of only one enzyme.

Beadle and Tatum further proposed that because enzymes are proteins, genes actually control the synthesis of specific proteins. Beadle and Tatum were awarded the Nobel prize in 1958 for their work with *Neurospora*.

ONE GENE–ONE POLYPEPTIDE: A GENERAL HYPOTHESIS

Beadle and Tatum showed that genes determine the presence of an enzyme, but a crucial question remained: *How* does a gene cause an enzyme or other protein to be produced? The answer came from studies of young African American patients with an inherited blood disease.

In 1905, a young African American experiencing pains in his joints and abdomen, chronic fatigue, and shortness of breath consulted a Chicago physician. A blood test showed that the man's blood had too few red blood cells and that many of the cells were shaped like crescents, or sickles, instead of the normal biconcave disks [FIGURE 10.4]. Later it was discovered that the sickle-shaped blood cells tended to clump together, lodging in very small blood vessels—especially in the joints and abdomen—blocking blood flow and causing pain. The sickle-shaped cells were also fragile and easily destroyed, causing a shortage of red blood cells, a condition called *anemia*.

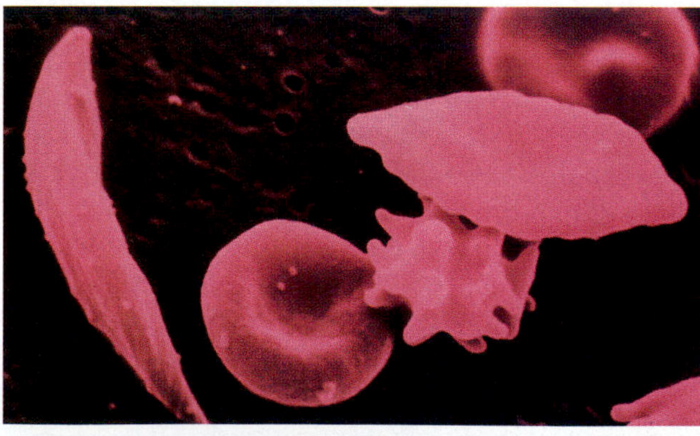

FIGURE 10.4
Normal and Sickled Blood Cells.
This scanning electron micrograph of blood from a child with sickle-cell anemia shows a sickled cell (left), then a normal cell, and then three more abnormally shaped red blood cells (right).

Because red blood cells carry oxygen from the lungs to the rest of the body, a deficiency of these cells results in a chronically inadequate oxygen supply and a feeling of breathlessness and fatigue. The patient had what we now call *sickle-cell anemia*, a genetic condition that currently affects about 60,000 people in the United States. Sickle-cell anemia is found in about 1 out of every 50 West Africans, and in about 1 out of every 400 African Americans. In many other populations of different origins, the condition is much less frequent. (You may be curious about why sickle-cell anemia is distributed this way in different populations and localities. We'll discuss the evolutionary explanation for this in CHAPTER 16.)

Sickle-cell anemia is inherited as a recessive mutation in simple Mendelian fashion: In order to suffer from sickle-cell anemia, a child must inherit a sickle-cell allele from each parent.

Scientists knew that red blood cells contain the protein hemoglobin, which carries oxygen from the lungs to the rest of the body. Recall that a single molecule of the hemoglobin protein consists of two pairs of polypeptide chains: two α chains and two β chains (a polypeptide is a chain of amino acids; see FIGURE 2.31). When researchers compared the molecular structure of hemoglobin taken from well people with that of hemoglobin from sickle-cell patients, they found a subtle difference. The sickle-cell beta protein chains differ from normal beta chains by just 1 amino acid out of the 146 present in each chain. (The alpha chains are identical in sickle-cell hemoglobin and normal hemoglobin.) In normal beta chains, the sixth amino acid is glutamic acid, but in sickle-cell beta chains it is a valine [FIGURE 10.5]. This small change is enough to distort the shape of the hemoglobin molecule, and the distortion causes groups of hemoglobin molecules to join together in long fibers, instead of remaining separate. The fibers, in turn, alter the shape of the red blood cells and cause sickle-cell anemia.

The work with sickle-cell anemia was important for three reasons. First, this work and later studies based on it enabled physicians to diagnose the disease earlier and more accurately and to work on treatments that help patients live longer lives with less pain. Second, it showed that a single gene does not necessarily specify the structure of an entire protein. For proteins that consist of several polypeptide chains, the synthesis of each polypeptide chain is regulated by a different gene. Thus the principle of one gene–one enzyme had to be amended to **one gene–one polypeptide**. Third, it revealed how a simple genetic mutation can alter the functioning of an entire organism. A mutant gene alters the shape and function of a specific polypeptide; cells carrying the mutant polypeptide may not function properly, and the whole organism can suffer as a result.

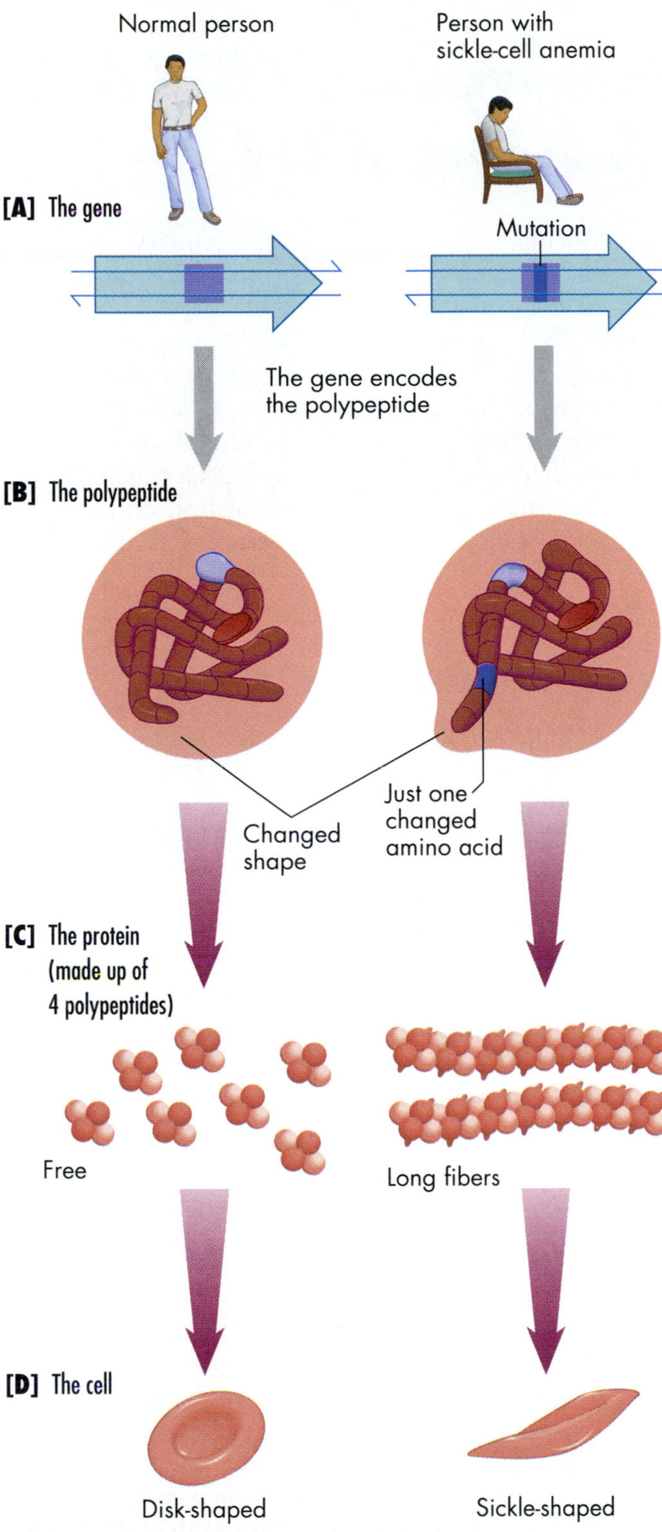

FIGURE 10.5

Mutation in a Single Gene Causes Sickle-Cell Anemia.

As the figure shows and the text describes, a changed gene leads to an altered hemoglobin protein, changed blood cell shape, and widespread effects on the body.

We can apply the one gene–one polypeptide principle to cystic fibrosis. The normal allele of the cystic fibrosis gene encodes a polypeptide that apparently functions in some way to prevent mucus buildup in lungs. In contrast, in a patient like Jeff Pinard, the disease allele fails to make a polypeptide that functions normally, and as a result, mucus accumulates.

➤ CONCEPT CHALLENGE

How would you interpret the difference between normal orange tigers and genetically white tigers in terms of the one gene–one polypeptide principle?

How Do Genes Control Life Activities?

The synthesis of proteins is one of a cell's most important tasks, since the activities of proteins underlie nearly all cellular functions. For muscle cells to flap the wings of a honeybee, for seaweed to synthesize chlorophyll, or for ions to be transported properly into and out of cells, proteins—including many enzymes—must be constructed to carry out highly specialized roles.

Protein synthesis provides each cell with exactly the right amount and types of enzymes and structural proteins it needs to perform its function, maintain an inner stability, grow, and reproduce. Thus, genes control life activities by controlling protein synthesis.

The experiments we have discussed already uncovered two basic secrets of a gene: (1) Genes are made of DNA [review FIGURES 9.2 and 9.4]. (2) A gene specifies the amino acid sequence of a polypeptide chain [review FIGURES 10.2, 10.3, and 10.5]. The next question was simple but profound: Exactly how does the information in DNA become decoded and translated into protein structure?

The answer unfolded over decades of research and can be summarized this way: *Genetic information in cells generally flows from DNA to RNA to protein.* Stated another way: DNA codes for RNA, which codes for the composition and structure of a specific protein.*

The information flow from DNA to protein is a two-step process [FIGURE 10.6]. In the first step, information in

* Note that in some instances information from RNA can flow back to DNA. This happens, for example, when the AIDS-causing virus HIV infects a human cell. In the virus particle, the genetic information is stored in RNA. When the virus infects a cell, an enzyme called *reverse transcriptase* copies the genetic information in RNA into DNA. After that step, information flows in the usual way, from viral DNA (now in the cell) to new viral RNA, to viral proteins, eventually causing AIDS.

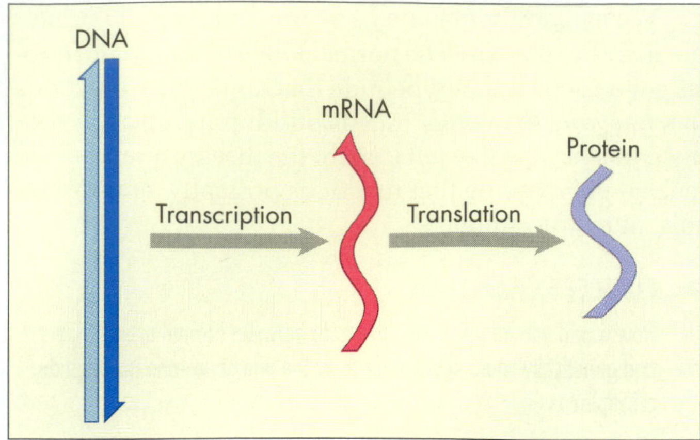

FIGURE 10.6

Information Flows from DNA to RNA to Protein.

Enzymes transcribe DNA into a messenger RNA molecule; ribosomes translate the mRNA into the amino acid chain in a protein.

a portion of the DNA molecule is copied, or *transcribed*, into an RNA molecule. RNA is an information-storing molecule with a structure somewhat similar to DNA. This first step, an RNA copy of a portion of a DNA molecule, is called *transcription*, a word that implies that information in one dialect (the base sequence of DNA) is copied into a different dialect (the base sequence of RNA). In the second step, the information in one type of RNA molecule is *translated* by ribosomes into specific sequences of amino acids that make up polypeptide chains. This step is called *translation* because genetic information is translated from the language of nucleic acids into the language of proteins.

Since transcription copies DNA to RNA, it must occur where DNA is located in the cell; in a eukaryote, that's the nucleus. Likewise, since translation occurs on ribosomes, it must take place where the ribosomes are located in the cell; that's the cytoplasm. After an RNA is transcribed in the nucleus, it moves through the pores in the nuclear envelope to the cytoplasm [review FIGURE 3.15]. There the RNA binds to ribosomes and is translated into protein molecules. The proteins can then be transported to their place of action in the cell via the routes we discussed in CHAPTER 3. In the case of the cystic fibrosis protein, for example, once synthesized, the molecule moves to the cell surface, where it regulates the exchange of ions between the cell and its surroundings.

Now let's look more closely at the process of transcription, which allows the hereditary information in nuclear DNA to reach the cytoplasm, where protein synthesis occurs.

Transcription: DNA Is Copied into RNA

While the one gene–one polypeptide principle is a good generalization, this phrase implies that genes make proteins directly. In fact, they do not. Instead, genes act through RNA intermediates [review FIGURE 10.6]. Many biologists today would define a gene as a portion of a DNA molecule that is copied or transcribed into an RNA molecule. Biologists call the synthesis of a single-stranded RNA under the direction of DNA **transcription**. Before we investigate the process of transcription, we must understand the differences in structure between DNA and RNA.

COMPARISON OF DNA AND RNA

Like DNA, RNA consists of a long string of nucleotides linked by sugar-phosphate backbones [FIGURE 10.7A]. Unlike DNA, RNA can move from the nucleus to the cytoplasm, which we will see later is an important factor in protein synthesis. RNA also differs from DNA in four other ways.

1. RNA nucleotides contain the sugar ribose instead of deoxyribose, the sugar in DNA. As its name suggests, deoxyribose has one less oxygen atom than ribose. Although this is the only structural difference between the two sugars, many functional differences arise from it.

2. RNA contains the base uracil (U) instead of thymine (T), which is found in DNA. Uracil pairs with adenine on the DNA template during transcription.

3. RNA usually consists of a single strand of nucleotides, whereas DNA usually consists of two strands. In some kinds of RNA, however, a single RNA molecule folds back on itself and forms short, double-stranded regions connected by complementary base pairs [FIGURE 10.7b].

4. RNA molecules are much shorter than the DNA molecules that make up chromosomes. Each DNA molecule carries hundreds or thousands of genes, but an RNA molecule usually contains information from only one gene.

With these points about the structure of RNA in mind, we can now understand how DNA is transcribed into RNA.

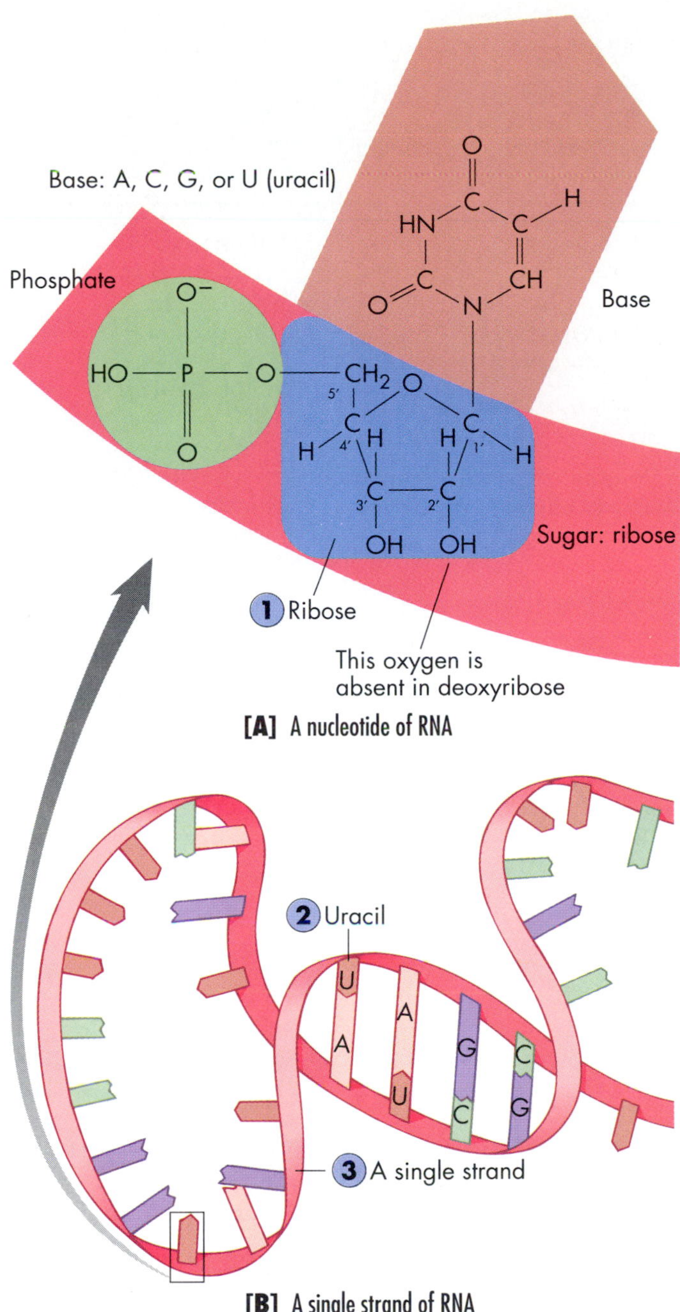

FIGURE 10.7

The Structure of RNA.

[A] An RNA nucleotide consists of a base (A, C, G, or U), the sugar ribose, and a phosphate. The base shown here is U (uracil), which is found in RNA but not DNA. [B] Many RNA nucleotides are joined together in an RNA molecule that can loop back and form complementary base pairs with itself. A pairs with U, and C pairs with G.

THE PROCESS OF TRANSCRIPTION

The process of transcription involves three basic steps—DNA strand separation, complementary base pairing, and nucleotide joining.

1. Strand separation. In the nucleus, at the start of transcription, enzymes cause portions of the DNA double helix to unwind and separate [FIGURE 10.8, Step 1]. For any given gene, only one of the two separated DNA strands serves as a template (or model) for the synthesis of a complementary single strand of RNA. The other DNA strand simply stays out of the way.

2. Complementary base pairing. As nucleotides containing the sugar ribose diffuse in, they pair with complementary nucleotides on one DNA strand [FIGURE 10.8, Step 2]. The base A on the DNA pairs with a U RNA nucleotide, while T on the DNA pairs with an A RNA nucleotide. As in DNA replication, C pairs with G.

3. Nucleotide joining. The enzyme RNA polymerase joins two adjacent RNA nucleotides together [FIGURE 10.8, Step 3]. Other RNA nucleotides diffuse in and pair with their complementary bases in the DNA strand, and the enzyme joins them together one by one, forming a single strand of RNA. About 30 nucleotides per second are added to the new RNA molecule until the action of RNA polymerase is halted by a stop signal, a specific nucleotide sequence in DNA. The parental DNA rewinds as transcription is completed. The DNA is now available to produce more copies of the same RNA. The newly made RNA may be chemically modified, and then it passes out of the nucleus into the cytoplasm.

COMPARISON OF TRANSCRIPTION AND DNA REPLICATION

The transcription of DNA into RNA is somewhat similar to DNA replication. In both cases, the two DNA strands unwind and separate, nucleotides diffuse to the site and line up by base pairing, and a polymerase enzyme joins the new nucleotides into a nucleic acid strand. In addition, the RNA polymerase enzyme is similar to DNA polymerase in that both copy in the same direction along a single DNA strand.

Transcription differs from DNA replication in the following ways [FIGURE 10.9 and TABLE 10.1]:

1. The entire DNA molecule is copied during DNA replication, but during transcription, a limited portion of the DNA is transcribed at any one time. Most transcribed regions are between 100 and 10,000 nucleotides long, and a single DNA molecule can have several thousand transcribed regions.

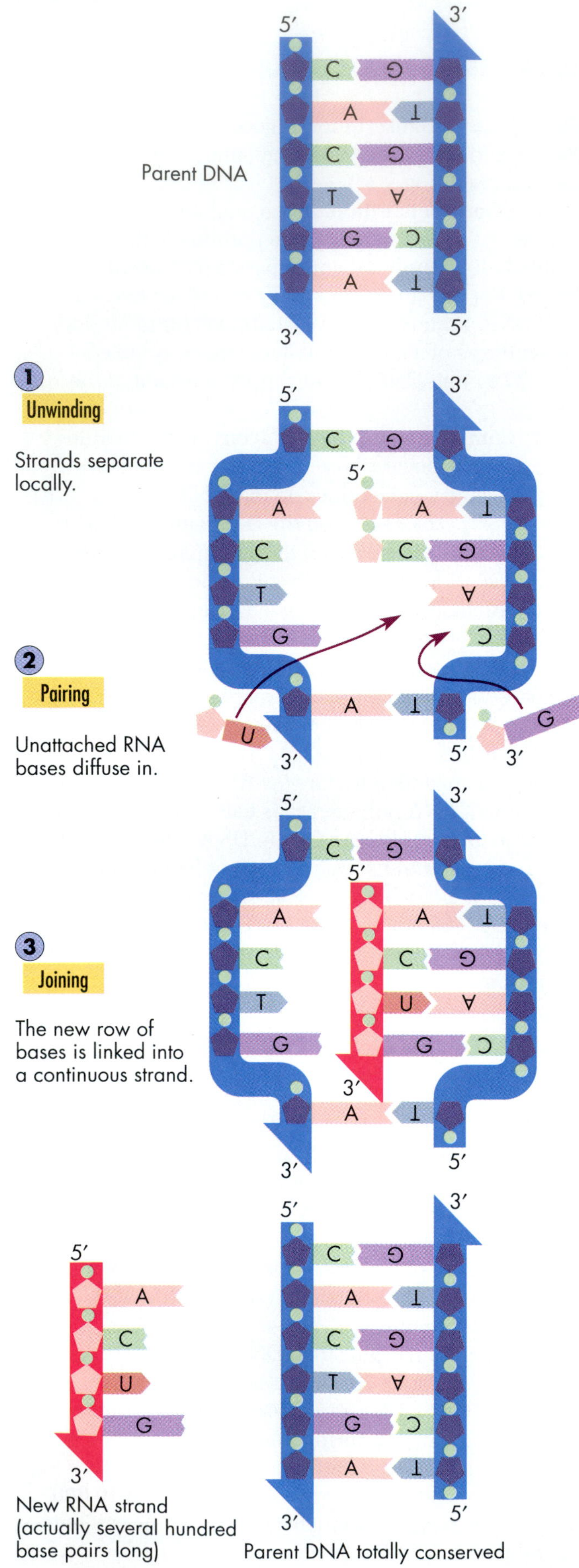

TABLE 10.1 A Comparison of the Syntheses of DNA, RNA, and Proteins

Type of Synthesis	Description
DNA Synthesis (replication)	The copying of information from parental DNA to two daughter DNA molecules during DNA replication.
RNA Synthesis (transcription)	The transfer of information from one strand of the DNA double helix molecule to a complementary single-stranded RNA molecule.
Protein Synthesis (translation)	Information encoded in the specific sequence of mRNA nucleotides determines the specific sequence of amino acids in a specific protein. This process translates the four-letter alphabet of nucleotides into the 64-word vocabulary signifying amino acids, "start," and "stop" signals.

FIGURE 10.8
How DNA Is Transcribed into RNA.

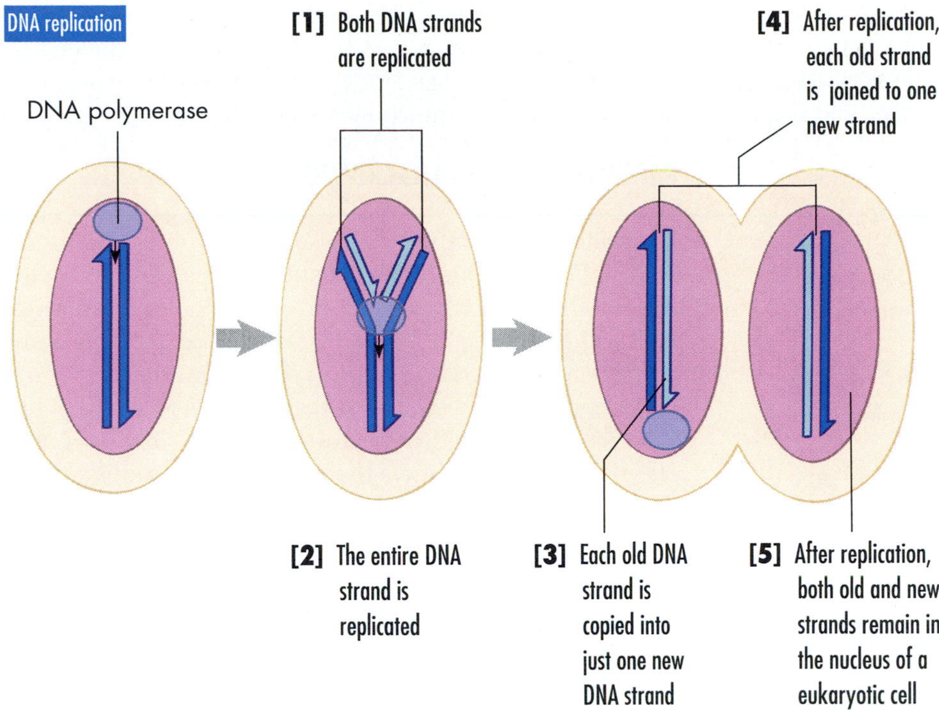

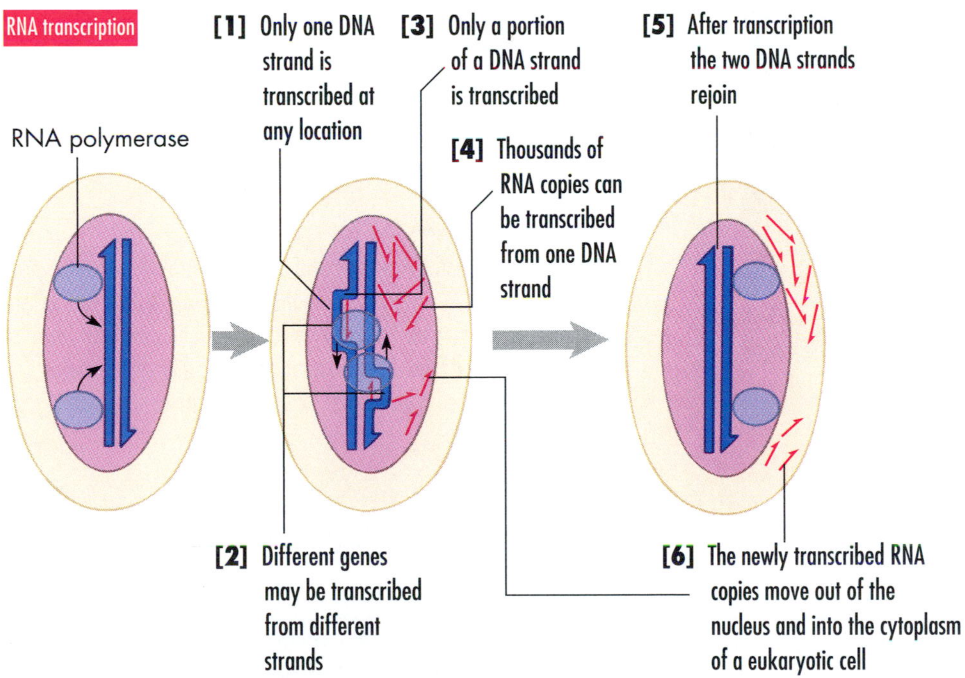

FIGURE 10.9
DNA Replication and Transcription: Similarities and Differences.

2. During DNA replication, both strands of a DNA molecule are copied. In the transcription of a gene, only one of the two separated DNA strands is transcribed into RNA. However, different genes may be transcribed from different DNA strands.

3. During DNA replication only one copy of each gene is made, but in transcription, a single gene may be transcribed thousands of times. Thus, a cell can make thousands of copies of an individual RNA molecule, each bearing the encoded message from a single gene.

4. After DNA replication, each of the old strands of DNA is joined to a new strand; after transcription, the two original DNA strands rejoin.

5. After DNA replication takes place in a eukaryotic cell, both old and new strands remain in the nucleus. After transcription, however, the newly formed strands of RNA separates from the DNA and moves from the nucleus into the cytoplasm, where the RNA performs its function.

FIGURE 10.10

Caesar Experiments with RNA Synthesis.

How long can a person live without transcribing DNA? The Roman emperor Claudius Caesar provided an answer in A.D. 54 with an experiment he unwittingly—and no doubt unwillingly—performed with his wife Agrippina. She mixed into Caesar's favorite dish of edible mushrooms (*Amanita caesarea*) a few of the poisonous species *A. phalloides.* These poisonous mushrooms contain a substance called *alpha-amanitin,* which blocks the activity of RNA polymerase, and hence, the synthesis of messenger RNA. As Caesar digested the fungus, the alpha-amanitin entered his bloodstream and was absorbed by the liver and kidneys, where it began to block transcription. About 15 hours after Caesar's repast, with no new mRNA to make new proteins, his liver cells stopped functioning, and nausea, diarrhea, and delirium began to hit him. Two days later, he died of liver failure. It is highly doubtful that Caesar learned to appreciate the valuable role of RNA polymerase in DNA transcription. But perhaps, in a general way, Agrippina did.

6. Finally, as we discussed in CHAPTER 7, nuclear DNA replication occurs only in the S phase of the eukaryotic cell cycle [see FIGURE 7.6], while transcription occurs throughout interphase, in G_1, S, and G_2.

IMPORTANCE OF TRANSCRIPTION

Accurate transcription is essential to life and good health. For example, if the cystic fibrosis gene in a person's cells is not transcribed properly due to a mutation in that gene, then no CFTR (cystic fibrosis transmembrane regular) protein is made, cells do not transport ions correctly, ducts become clogged, and people like Jeff Pinard have difficulties breathing and digesting food. Another example that shows the importance of transcription involves mushrooms. Although some kinds of mushrooms are a diner's delicacy, many others produce chemicals that block the action of the enzyme RNA polymerase [FIGURE 10.10]. After a person eats just a small serving of mushrooms containing these substances, the poisons enter the blood and are transported to the liver, where they block transcription. Liver cells, deprived of new RNA molecules and hence the synthesis of new proteins, begin to die. Denied the activities of the liver, the mushroom eater dies within a couple of days.

The process of transcription makes several types of RNA molecules, three of which are involved in *translation,* the synthesis of polypeptide chains. In the following section, we will investigate those three types of RNA.

CONCEPT CHALLENGE

RNA molecules are rather unstable and tend to be broken down rapidly in cells. Let's say you want to compare the stability of RNAs in two different types of cells. You treat the cells with a transcription inhibitor from mushrooms. One cell type dies in 1 hour, and the other cell type dies in 12 hours. Which cell type is more likely to have more stable RNA molecules?

Types of RNA

Three specific types of RNA molecules function in protein synthesis. These are called messenger RNA, transfer RNA, and ribosomal RNA. Remember that all three types of RNA are made in the nucleus by the transcription of specific genes and then move to the cytoplasm where they function in translation.

MESSENGER RNA

Messenger RNA, or **mRNA**, carries genetic information from DNA in the nucleus to the ribosomes in the cytoplasm, where it directs the synthesis of proteins. The base sequence in mRNA is complementary to the sequence of bases in a gene on the DNA molecule, and it therefore carries the same genetic information in complementary form. Messenger RNA molecules are generally from about 1000 to 10,000 nucleotides long and are the least abundant type of RNA (5 percent of the total RNA).

TRANSFER RNA

The function of **transfer RNA (tRNA)** is to pick up amino acids in the cytoplasm and to align them in the order specified by mRNA on the ribosome. To do that, a molecule of tRNA, which is about 75 nucleotides long, must be able to do two things: (1) It must recognize and attach to a particular amino acid, and (2) it must find and temporarily bind to the appropriate place on the mRNA.

Each tRNA molecule carries only one specific kind of amino acid. Since there are 20 different amino acids, there must be at least 20 different tRNAs (some amino acids are transported by more than one kind of tRNA). The two functions of tRNA are carried out by opposite ends of the molecule. At one end of each tRNA molecule is a site that can attach to one of the 20 amino acids [FIGURE 10.11A]. Specific enzymes attach the right amino acid to the right tRNA (a tRNA with the amino acid serine attached to one end is shown in FIGURE 10.11A).

Codons and Anticodons Sequences of three adjacent bases on mRNA specify the insertion of a particular amino acid during the synthesis of a polypeptide chain. These mRNA sequences are called **codons**. The complementary sequences on tRNA that recognize the codons are called **anticodons**. For example, one of the six tRNAs that carries the amino acid serine has the anticodon UCG. The UCG anticodon pairs up with the AGC codon on an mRNA:

UCG
AGC

Thus, AGC is a codon for serine [FIGURE 10.11B]. The codon-anticodon pairing is the key to how amino acids line up in proper sequence for polypeptide synthesis. The order of the codons in mRNA specifies the order in which specific amino acids carried by tRNAs will line up. To join the lineup of amino acids to each other is the job of the ribosome.

RIBOSOMAL RNA AND RIBOSOMES

The third type of RNA is called **ribosomal RNA (rRNA)** because it eventually becomes parts of ribosomes, the actual sites of protein synthesis. By weight, eukaryotic ribosomes are about half protein and half rRNA. Ribosomal RNA is the most abundant type of RNA (80 percent).

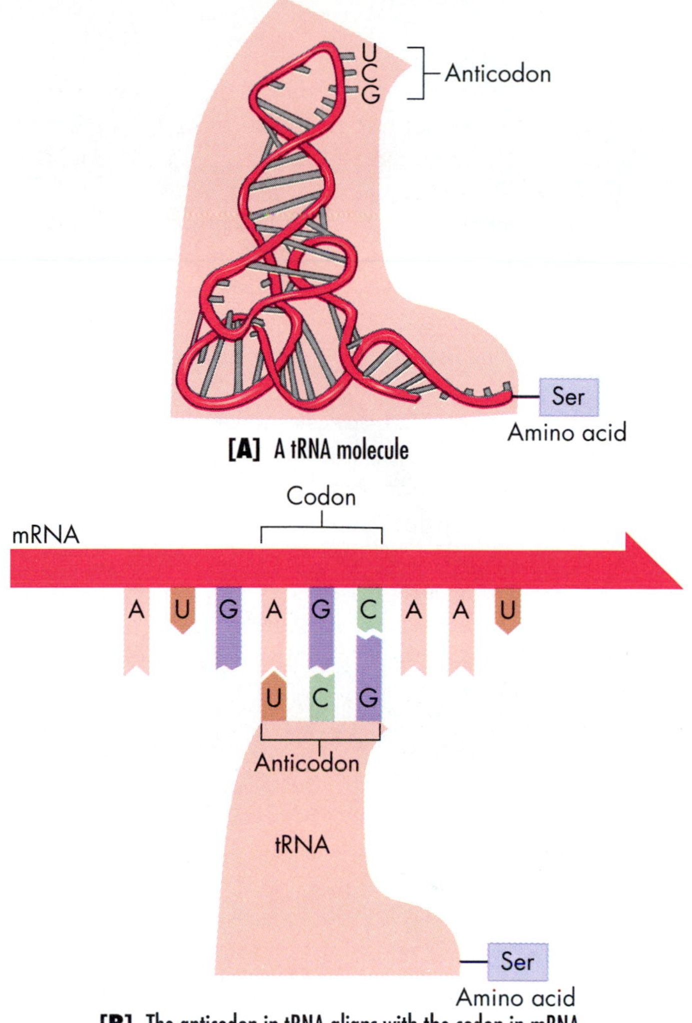

[A] A tRNA molecule

[B] The anticodon in tRNA aligns with the codon in mRNA

FIGURE 10.11

Transfer RNA: Amino Acid Carriers.

[A] Each transfer RNA is a ribbon about 75 bases long that is looped back on itself and folded in space by complementary base pairing, which forms a bootlike shape. At one end of each tRNA is its specific anticodon, in this case, UCG; at the other end is attached the amino acid corresponding to this anticodon, here, Ser (serine). [B] The tRNA's anticodon forms base pairs with a codon in mRNA.

Ribosomes support mRNAs and tRNAs and help to link amino acids to each other in a growing polypeptide chain. In eukaryotes, rRNAs come in four different sizes, and each ribosome has one molecule of each size. Approximately 75 proteins are wrapped around these rRNAs, forming the beadlike structures.

Each ribosome consists of one large and one small subunit separated by a groove [FIGURE 10.12A]. An mRNA strand "threads" through the groove, and the ribosome

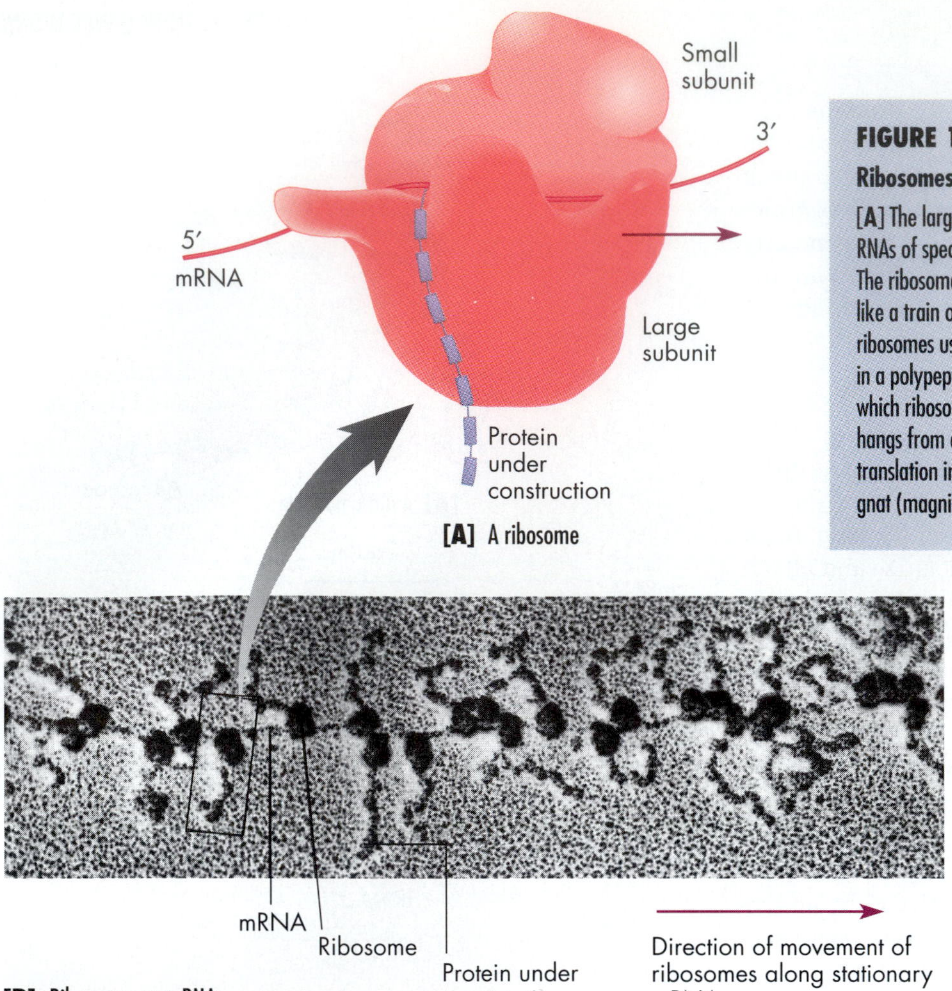

[A] A ribosome

[B] Ribosomes on mRNA

FIGURE 10.12

Ribosomes: Sites of Protein Synthesis.

[A] The large and small ribosomal subunits are made up of a few RNAs of specific lengths as well as many kinds of associated proteins. The ribosomal subunits join to the mRNA, then slide along the mRNA like a train on a track or a pulley on a rope. As they slide, the ribosomes use the order of bases in the mRNA to order the amino acids in a polypeptide. **[B]** The electron micrograph shows an mRNA along which ribosomes are sliding. An ever-elongating polypeptide chain hangs from each ribosome. Researchers photographed this example of translation in cells from the salivary glands of *Chironomus*, a kind of gnat (magnification 17,500×).

moves along the mRNA like a pulley on a rope. As it moves, the ribosome joins amino acids to the growing polypeptide chain one by one in the order specified by the mRNA. Recent evidence suggests that rRNA may help to catalyze the addition of one amino acid to the next. Usually, a linear cluster of ribosomes, called a **polyribosome**, moves together along each mRNA strand like cars in a caravan, producing multiple copies of the same polypeptide [FIGURE 10.12B].

Cells contain tens of thousands of ribosomes, located wherever proteins are being synthesized—in the cytoplasm of eukaryotic cells or in the DNA-containing region of a prokaryotic cell. Working together, then, the ribosomal, messenger, and transfer RNAs carry out protein synthesis.

Translation: Protein Synthesis

Protein synthesis is the assembly of amino acids into a polypeptide. After mRNA carries the encoded genetic message from DNA in the nucleus to the cytoplasm, the message must be read by a ribosome and tRNAs to produce the polypeptide chains that make up proteins. The formation of a polypeptide is called **translation** because the genetic information in one class of chemicals, nucleic acids, is translated into a different class of chemicals, proteins. The sequence of nucleotides in DNA and RNA determines the sequence of amino acids in proteins.

STAGES OF PROTEIN SYNTHESIS

The assembly of amino acids into a polypeptide has three main stages: initiation, elongation, and termination. These stages are pictured and described in detail in FIGURE 10.13. During *initiation*, the necessary machinery for protein synthesis is assembled. The two ribosomal subunits come together on the mRNA, and a tRNA attached to the amino acid methionine, which is the first amino acid of most polypeptides, binds to its spot on the mRNA. During *elongation*, the polypeptide grows as amino acids are bonded together by peptide bonds to form the growing polypeptide chain. *Termination* occurs when the ribosome reaches a stop signal in mRNA, for example, the three nucleotides UAG in a row. At that point the growth of the polypeptide chain comes to a halt, and the newly formed polypeptide falls away from the ribosome.

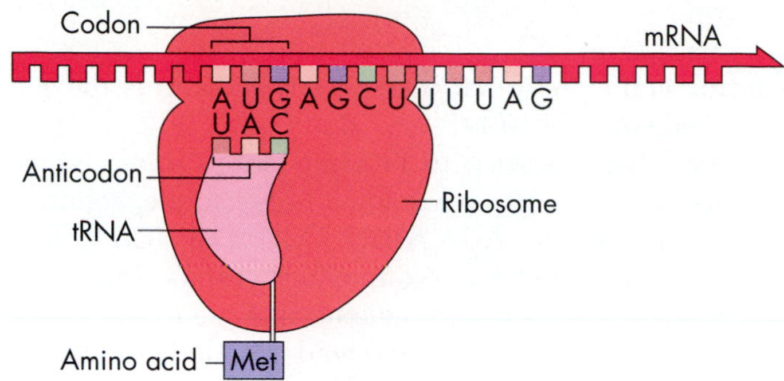

[A] Initiation

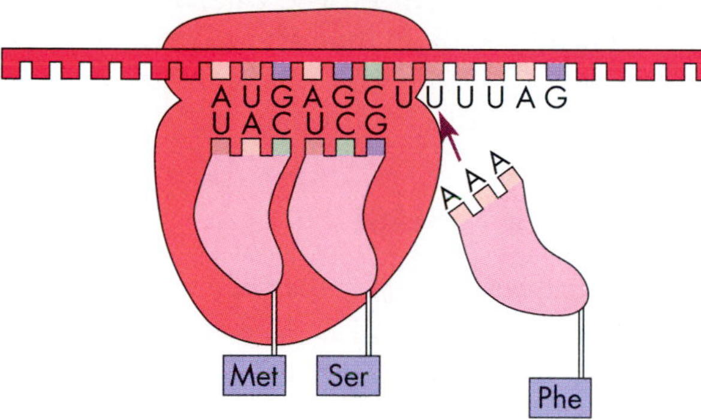

[B] Elongation

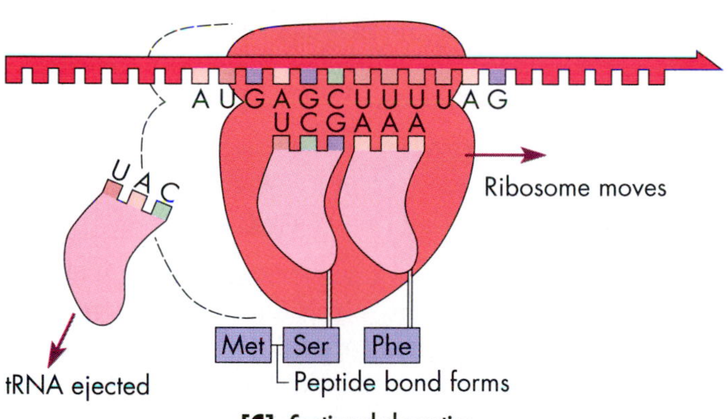

[C] Continued elongation

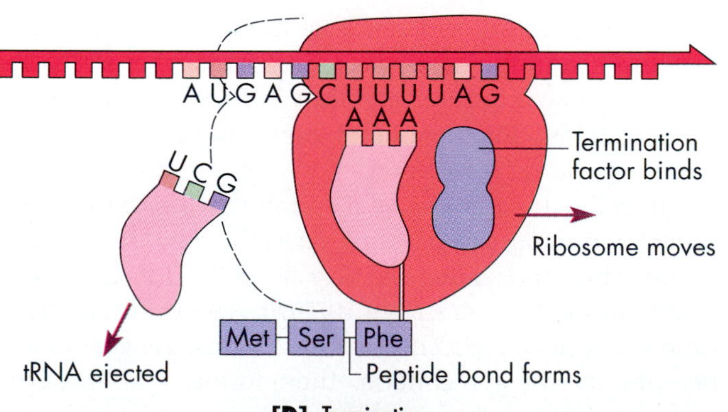

[D] Termination

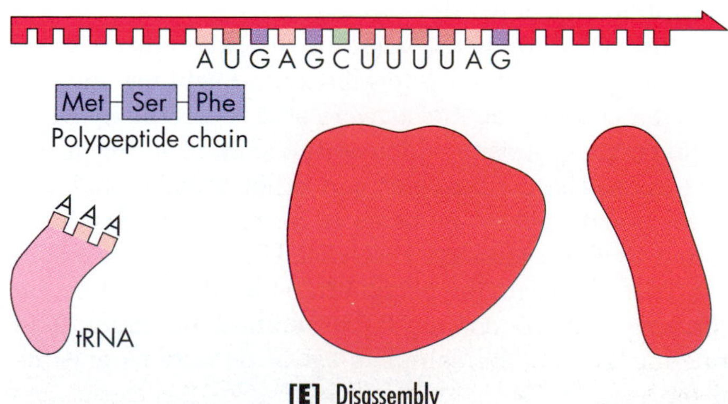

[E] Disassembly

FIGURE 10.13
The Main Events of Protein Synthesis.

[A] ***Initiation.*** Several components assemble before translation begins. A special initiator tRNA molecule charged with the amino acid methionine binds to the small ribosomal subunit, and then the small and large subunits and the mRNA come together. The anticodon of the initiator tRNA binds to the AUG start codon on the mRNA. **[B]** ***Elongation.*** A second charged tRNA forms base pairs between its anticodon and the second codon in the mRNA, thus aligning the first and second amino acids, each still attached to its tRNA. **[C]** ***Continued elongation.*** During peptide bond formation, the bond between the first tRNA and its amino acid is broken, and the ribosome forges a peptide bond between the first and second amino acids. The first tRNA, now without its amino acid, is ejected from the mRNA and recycled; the ribosome moves down the mRNA one codon. Now a dipeptide (Met-Ser in this case) is hooked to the second tRNA. To continue peptide elongation, the foregoing steps are simply reiterated many times until the complete polypeptide is formed. A charged tRNA binds to the third codon. A peptide bond is formed between the dipeptide and the newly arrived amino acid, forming a tripeptide; the tRNA from the second amino acid is ejected; and the ribosome moves once more. **[D]** ***Termination.*** At the end of the message, the ribosome reaches a stop codon along the mRNA such as UAG, a codon for which there is no tRNA. At this point, a termination factor, a certain protein, binds in place of tRNA and translation halts. **[E]** ***Disassembly.*** The two ribosomal subunits separate from each other and release the mRNA, and the polypeptide is cleaved from the last tRNA molecule. The polypeptide is then free to do some job in the cell, and the translation machinery can reassemble and initiate the synthesis of another polypeptide molecule.

As you can tell from FIGURE 10.13, the key feature of translation is the way in which the mRNA base sequence specifies the alignment of tRNAs, each carrying one particular kind of amino acid. The sequence of bases in mRNA, which is complementary to the base sequence of a gene in DNA, contains the genetic code.

The Genetic Code

The **genetic code** is a lexicon that shows the amino acid "definition" of an individual series of bases in mRNA. The genetic code determines which amino acids the cell will translate from each base sequence. The language of ribonucleic acids has an alphabet of only 4 letters (the 4 bases A, C, G, and U), while the language of proteins has 20 (the 20 amino acids). How can these four nucleotides code for the 20 different types of amino acids found in proteins?

In working out the genetic code, scientists reasoned that if one nucleotide coded for one amino acid, four nucleotides could code for only four different amino acids instead of the 20 actually found. Alternatively, one might guess that two bases could encode an amino acid. For example, AU might encode one amino acid, CA another, and so on. But that wouldn't work either, for there are only 16 (4×4) ways to combine the four bases into pairs:

AA	AC	AG	AU
CA	CC	CG	CU
GA	GC	GG	GU
UA	UC	UG	UU

Clearly, 16 is still not enough "words" to encode the 20 amino acids found in most proteins. Three nucleotides at a time, however, with 64 possible combinations ($4 \times 4 \times 4$) are more than enough to code for the 20 required amino acids. We can say that the language of RNA contains only four letters, from which 64 three-letter words may be written. (In the same way, the 26 letters of our alphabet can be combined in various ways to make an entire dictionary of words.)

In a DNA or RNA, a group of three adjacent nucleotides that code for a specific amino acid together are called a codon. Geneticists were able to decipher which codon corresponds to which amino acid by adding artificial messages to ribosomes and other translational machinery which they extracted from bacterial cells. They found that the message UUUUUUUUUUUU, for example, translated into the polypeptide Phe-Phe-Phe-Phe, indicating that the codon UUU specifies the amino acid phenylalanine. The complete set of 64 codons and the specific amino acid that each codon encodes is the genetic code [FIGURE 10.14].

Note that in FIGURE 10.14 some amino acids can be encoded by more than one codon. For example, arginine can be encoded by AGA, AGG, CGU, CGC, CGA, and CGG. As a result of this redundancy of the code, some mutations (changes in the sequences of nucleotides in a gene) do not change the amino acid encoded. For example, the mutation of a CGC codon to CGG would not change the amino acid encoded—in both cases it would be arginine—and so this mutation would likely have no effect on the organism. The genetic code is redundant, but if you examine FIGURE 10.14, you can see that while an amino acid can have more than one codon, there are no codons that encode more than one amino acid: each codon uniquely signifies a single amino acid.

READING THE GENETIC MESSAGE

When researchers tested a genetic message with a repeating triplet motif such as AGCAGCAGCAGCAGC, they obtained an unexpected result. This message translated into not one but three different polypeptides, each consisting of only one kind of amino acid: either all serine (codon AGC), all alanine (codon GCA), or all glutamine (codon CAG). We can understand how this happens if we assume that in this test-tube experiment, translation can start at any place along the message. Thus, we can write the message in any of the following ways, with imaginary breaks separating the codons:

. . . AGC AGC AGC AGC AGC . . .
. . . GCA GCA GCA GCA GCA . . .
. . . CAG CAG CAG CAG CAG . . .

This result shows that once translation starts at one point, it then reads off bases in groups of three. Codons do not overlap, and no bases are skipped. The result also shows that the place where translation begins determines the meaning of the message.

The genetic message given above can be divided into codons in three ways. Each different way is called a *reading frame*.

Differences in the reading frame allow different meanings from the same stretch of messenger RNA. To appreciate the importance of the reading frame in a message, consider for a moment howc hangesi nreadingf ramec ana lterthism essage.

In cells, there is a special codon that signifies where translation should start; thus it establishes the reading frame. This **start codon**, AUG, also codes for the amino acid methionine [see FIGURE 10.13A]. As a result, methionine starts nearly all polypeptide chains. As geneticists deciphered the 64 codons, they found that three of

Codon			Amino acid
U	U	U	Phe
U	U	C	Phe
U	U	A	Leu
U	U	G	Leu
U	C	U	Ser
U	C	C	Ser
U	C	A	Ser
U	C	G	Ser
U	A	U	Tyr
U	A	C	Tyr
U	A	A	STOP
U	A	G	STOP
U	G	U	Cys
U	G	C	Cys
U	G	A	STOP
U	G	G	Trp

Codon			Amino acid
A	U	U	Ile
A	U	C	Ile
A	U	A	Ile
A	U	G	Met (START)
A	C	U	Thr
A	C	C	Thr
A	C	A	Thr
A	C	G	Thr
A	A	U	Asn
A	A	C	Asn
A	A	A	Lys
A	A	G	Lys
A	G	U	Ser
A	G	C	Ser
A	G	A	Arg
A	G	G	Arg

Codon			Amino acid
C	U	U	Leu
C	U	C	Leu
C	U	A	Leu
C	U	G	Leu
C	C	U	Pro
C	C	C	Pro
C	C	A	Pro
C	C	G	Pro
C	A	U	His
C	A	C	His
C	A	A	Gln
C	A	G	Gln
C	G	U	Arg
C	G	C	Arg
C	G	A	Arg
C	G	G	Arg

Codon			Amino acid
G	U	U	Val
G	U	C	Val
G	U	A	Val
G	U	G	Val
G	C	U	Ala
G	C	C	Ala
G	C	A	Ala
G	C	G	Ala
G	A	U	Asp
G	A	C	Asp
G	A	A	Glu
G	A	G	Glu
G	G	U	Gly
G	G	C	Gly
G	G	A	Gly
G	G	G	Gly

FIGURE 10.14

The Genetic Code.

This dictionary shows how just four ribonucleotide bases (U, C, A, and G) are combined into 64 possible condons, and the genetic meaning—either a specific amino acid or a stop signal—for each one. For example, U in the first position, A in the second position, and C in the third position (UAC) equals Tyr, or tyrosine. AUG encodes Met, and sets the reading frame; three codons stop translation. (Abbreviations of amino acids: Phe, phenylalanine; Leu, leucine; Ser, serine; Tyr, tyrosine; stop, stop codon; Cys, cysteine; Trp, tryptophan; Ile, isoleucine; Met, methionine; Thr, threonine; Asn, asparagine; Lys, lysine; Arg, arginine; Pro, proline; His, histidine; Gln, glutamine; Val, valine; Ala, alanine; Asp, aspartic acid; Glu, glutamic acid; Gly, glycine.)

them—UAA, UAG, and UGA—do not code for any amino acid. Instead, these codons are called **stop codons** because whenever any of these codons appears in a message, translation stops and the completed polypeptide is released from the ribosome.

Now let's see how the genetic code is used to translate a portion of an mRNA into a string of amino acids. Assume the mRNA had a sequence

GAAUGAAAGUUGAC

Since the start codon AUG sets the reading frame, the polypeptide starts with methionine (Met).

GA AUG AAA AGU UGA C
Met

Since codons do not overlap, the next codon is AAA, or lysine (Lys), then AGU, or serine (Ser), and then UGA, or stop [review FIGURE 10.14].

GA AUG AAA AGU UGA C
Met Lys Ser stop

Thus, the unrealistically short amino acid string in this example would be Met-Lys-Ser.

THE GENETIC CODE IS ALMOST UNIVERSAL

The scientists who broke the genetic code were studying the protein-making apparatus of bacteria, but experiments with other life forms have shown that nearly all organisms use the same genetic code. The only exceptions so far are certain mitochondria and some protists, in which a few codons differ from the norm. It is an amazing testament to the unity of all living things that human cells and bacterial cells "speak" exactly the same genetic language. A gene from a human cell translates into the very same amino acid sequence, whether the gene resides in the nucleus of that human cell or whether a molecular biologist transfers that gene to a bacterial cell. The near-universal nature of the genetic code is strong evidence that all living things derive from ancestral cells with the same origin in the earth's history. Developed at the dawn of life, this ancient code has apparently been inherited intact for billions of years.

It has taken you much longer to read about protein synthesis than the process itself takes. The average eukaryotic protein is made up of about 400 amino acids and can be synthesized in about 20 seconds. Despite this speed, very few errors occur. Although errors happen more frequently during protein synthesis than during DNA replication, they are usually not as serious, because more than enough normal protein is available. In contrast, a single error in a sex cell during DNA replication may affect the offspring dramatically.

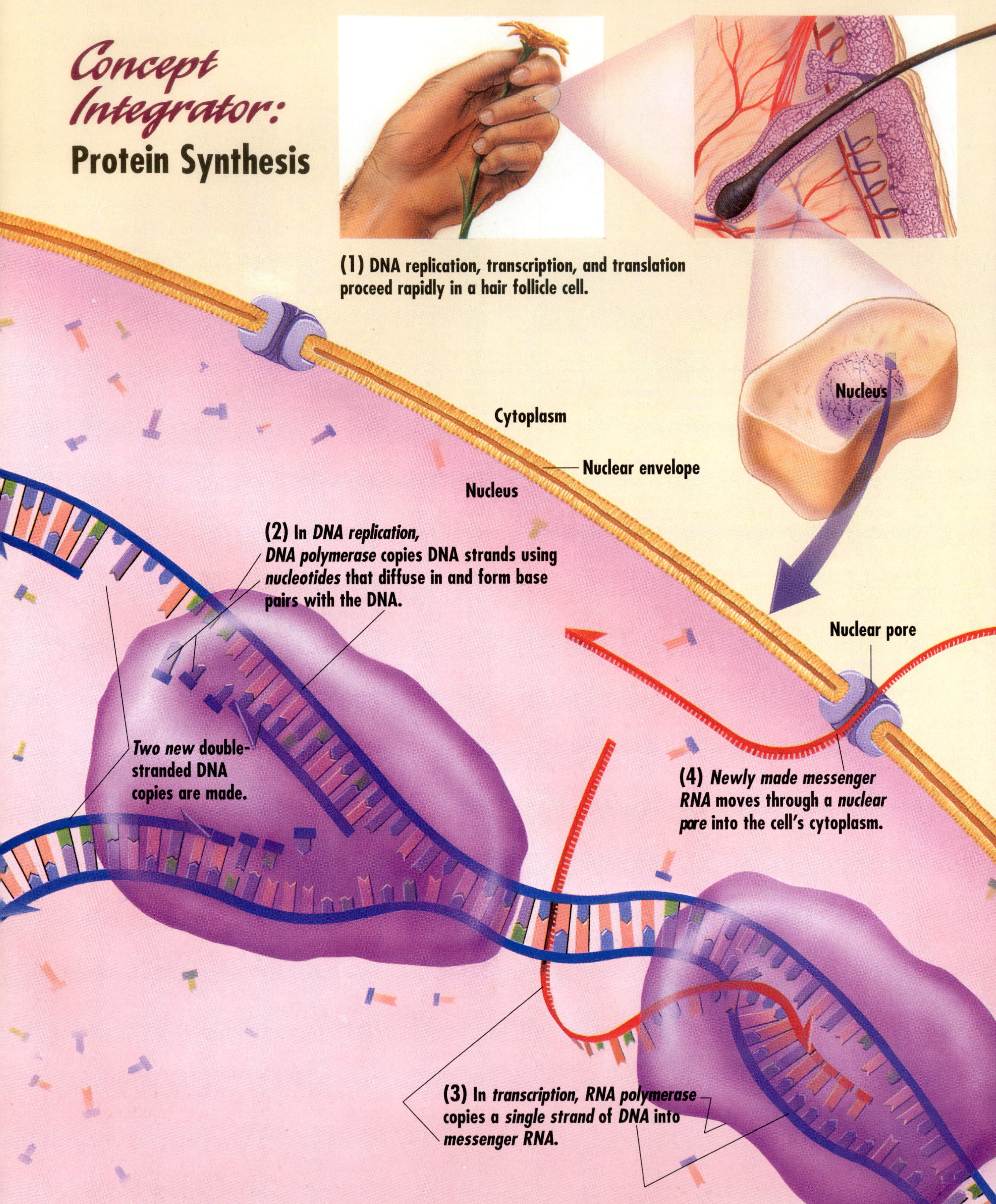
Concept Integrator:
Protein Synthesis
(1) DNA replication, transcription, and translation proceed rapidly in a hair follicle cell.
Nucleus
Cytoplasm
Nuclear envelope
Nucleus
(2) In DNA replication, DNA polymerase copies DNA strands using nucleotides that diffuse in and form base pairs with the DNA.
Nuclear pore
Two new double-stranded DNA copies are made.
(4) Newly made messenger RNA moves through a nuclear pore into the cell's cytoplasm.
(3) In transcription, RNA polymerase copies a single strand of DNA into messenger RNA.

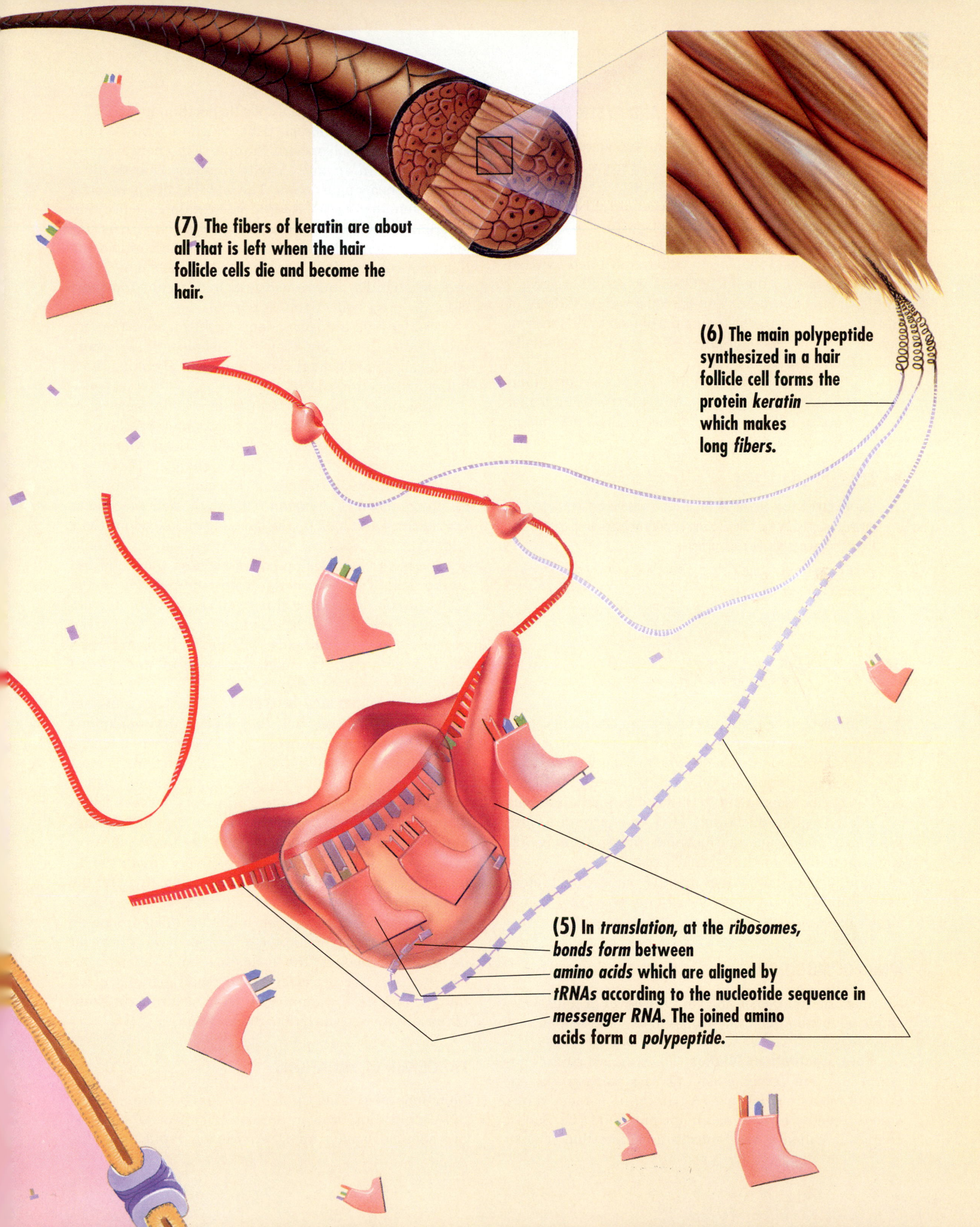
(7) The fibers of keratin are about all that is left when the hair follicle cells die and become the hair.
(6) The main polypeptide synthesized in a hair follicle cell forms the protein *keratin* which makes long *fibers*.
(5) In *translation*, at the *ribosomes*, *bonds form* between *amino acids* which are aligned by *tRNAs* according to the nucleotide sequence in *messenger RNA*. The joined amino acids form a *polypeptide*.

TRANSLATION, THE GENETIC CODE, AND CYSTIC FIBROSIS

We can apply what we learned about translation and the genetic code to the genetic disease cystic fibrosis. The cystic fibrosis protein CFTR consists of 1480 amino acids. How many nucleotides must the mRNA for this gene have, at a minimum? Since it takes three bases to encode one amino acid, the minimum size would be 4440 nucleotides—the 1480 amino acids multiplied by 3. In addition to this minimal length, messenger RNAs generally have tens to hundreds of bases tacked on in front of and behind their protein coding portions. For this reason, the mRNA that encodes the CFTR protein is actually longer than 4440 nucleotides.

In the next section, we will concentrate on mutations, changes in the genetic code and translation that produce nonfunctional proteins, such as the one found in people with cystic fibrosis.

➤ CONCEPT CHALLENGE

Use the genetic code [FIGURE 10.14] to determine the amino acid sequence encoded by the following short stretch of bases in an mRNA. Use the start codon to set the reading frame.

CGAUGGUAUAGUCCCUGGUAGAGA

Gene Mutation

A **mutation** is a change in the base sequence of an organism's DNA. A mutation may have no effect, or it may cause a massive, often deleterious, shift in the way an organism looks or functions. Geneticists recognize two general categories of mutations: (1) **Chromosomal mutations** affect large regions of chromosomes, or even entire chromosomes, and hence the locations of many genes. (2) **Gene mutations** alter individual genes.

Chromosomal mutations include changes in chromosome number (as in Down syndrome; see BOX 7.1) and structure (see CHAPTER 11).

KINDS OF MUTATIONS

This section focuses on the simplest level of mutations: single-gene mutations, or changes in the base sequence of a single gene.

Base Substitution Mutation The simplest sort of mutation in a gene occurs when one base pair replaces another, for example, when a TA pair in the parental molecule is replaced by a GC pair in the mutated DNA. Such a change is called a **base substitution mutation** [FIGURE 10.15A and B]. It occurs at a specific position in the CFTR gene of some cystic fibrosis patients whose families originally came from Northern Europe. The change in the DNA results in an mRNA with one base substitution, which alters a single codon. The altered mRNA encodes one protein that has an aspartic acid residue in a position where the normal protein would have glycine. This single replacement—a change in 1 of the 1480 amino acids that make up the CFTR protein—alters the shape of the protein slightly, and as a result, it does not transport chloride ions as effectively as normal.

Many cystic fibrosis patients with this mutation are less ill than most, but they still usually die as young adults, all due to a single base pair change among the 3 billion in the child's DNA.

Sometimes a base substitution in a gene creates a severe change in the structure and function of the polypeptide it encodes. For example, some African Americans with cystic fibrosis have a GC to TA change in DNA at a specific position in the gene. This change transforms a GGA codon to a UGA codon. Turn to the genetic code in FIGURE 10.14 and look up the meaning of the codons GGA and UGA. As you see, GGA encodes the amino acid glycine but UGA is a stop codon because it encodes no amino acid. This stop codon causes translation of the CFTR protein to cease at this point. As a result, the abnormal CFTR protein has 552 amino acids instead of the normal 1480. The patient's cells probably destroy this short protein rapidly, so it has none of the normal functions.

Base Deletions and Insertions While base substitutions change one base pair into another, **base deletions** and **base insertions** remove or add base pairs to the gene [FIGURE 10.15C and D]. In the form of cystic fibrosis most common in North America, three adjacent base pairs are deleted from the gene, and this brings about the deletion of a single amino acid from the protein. This altered protein has 1479 amino acids rather than the 1480 in the normal protein. Although the change is very small, the altered CFTR protein does not function normally, and the consequence is the disease of cystic fibrosis. In less common forms, a single base pair might be deleted, making the gene one base pair shorter, or a single base pair might be inserted, making the gene one base pair longer. Changes like these can alter the reading frame of an mRNA, garbling the amino acid sequence from the point of the mutation to the end of the protein.

THE ORIGIN OF MUTATIONS

Mutations arise either through spontaneous errors during DNA replication or later, through the effects of physical or chemical agents on DNA. During replication, the

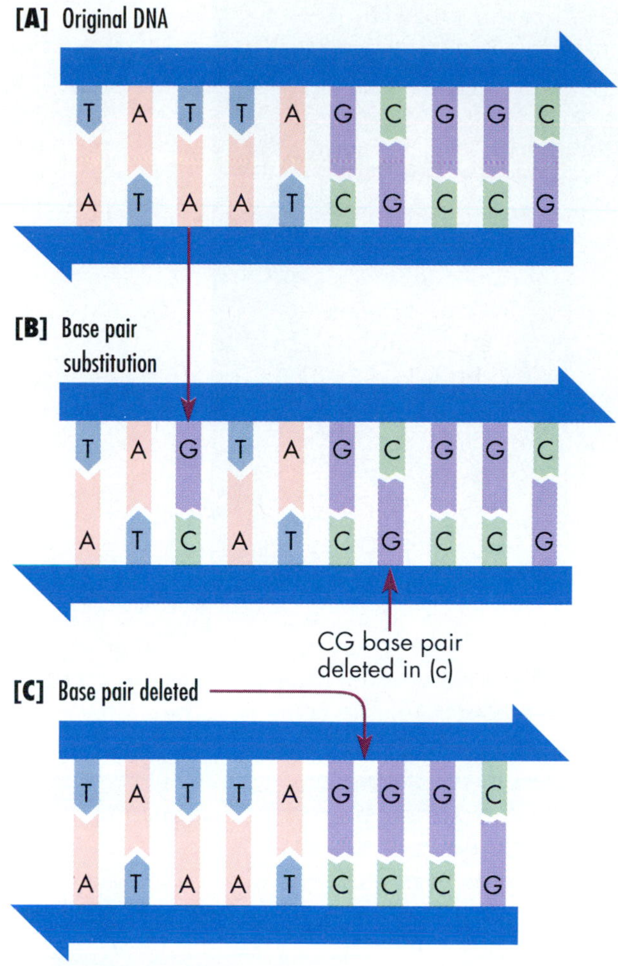

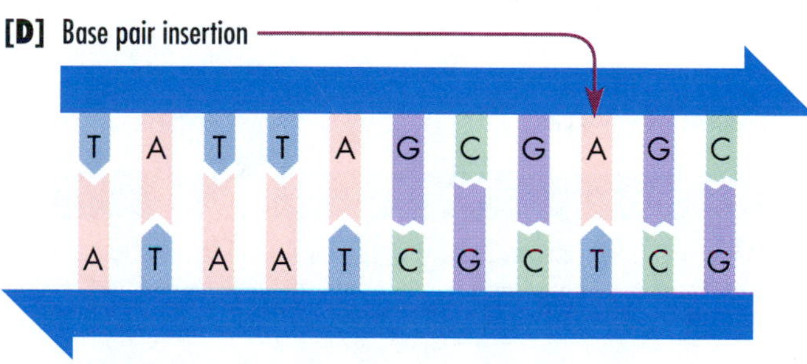

FIGURE 10.15

Mutations in Individual Genes.

Changes in the nucleotide base sequence of an original DNA **[A]** can occur in several ways. **[B]** In a base substitution, one base pair (a TA pair in part **[A]**) can be replaced by another (here CG). **[C]** In a base deletion, a base pair is removed from the DNA. **[D]** In a base insertion, a base pair is added to the DNA. In each case, the sequence is altered so that the gene is mutated and may function very differently, if at all. Sometimes several adjacent bases many be deleted or inserted at once. The most common mutation causing cystic fibrosis is the deletion of three adjacent base pairs.

wrong base can be inserted into a DNA molecule. Usually, this kind of replication error is edited out by the proofreading of DNA polymerase. More commonly, physical and chemical agents called *mutagens* change DNA structure. The many mutagens include ultraviolet rays from the sun, chemicals in cigarette smoke, and even many natural substances from plants or fungi. For example, aflatoxin, a natural compound found in moldy peanuts, is an extremely potent mutagen.

Despite the many mutagens in our environment, a mutation is a relatively rare event in most cells. But when mutations are common, the consequences can be dire. For example, people with the inherited skin disease *xeroderma pigmentosum* accumulate mutations in their cells rapidly. The girl shown in FIGURE 10.16A suffers from this condition and has dozens of skin cancers where ultraviolet light from the sun has mutated the DNA in her skin cells.

In most people, DNA repair enzymes constantly patrol the DNA, moving up and down the long strands, detecting and fixing damaged or unpaired bases [FIGURE 10.16B to D]. People with xeroderma pigmentosum, however, have defective DNA repair enzymes. As a result, the few mutations induced by even a low exposure to sunlight accumulate in cells and cause numerous skin cancers.

Cancer-causing substances are known as *carcinogens*; they often act by generating mutations. A mutagen has the potential to cause cancer if it happens to affect a gene that regulates cell growth. Because many agents that cause mutations in bacteria are the same as those that damage human DNA, nearly anything that increases the mutation rate in bacteria also increases the mutation rate (and potentially the cancer rate) in people. Thus, researchers can test the carcinogenicity (cancer-causing ability) of chemical and physical agents by exposing certain bacteria to suspected carcinogens and measuring the bacteria's mutation rates. With this test, called the *Ames test*, researchers have generated an extensive catalog of mutagenic substances that are potentially carcinogenic. We may encounter many of these carcinogens in the environment, including asbestos, ultraviolet light, nitrates in foods that become carcinogenic during cooking, and some dietary contaminants. Individuals who expose themselves to known mutagens, such as excessive sunlight or cigarette smoke, place a heavy burden on their DNA repair enzymes, greatly increasing the risk of cancer.

An individual with a defective DNA repair enzyme has an increased susceptibility to cancer in many different organs. In contrast, a person with cystic fibrosis has abnormalities in only certain organs especially in the ducts of the lungs and pancreas. Why is it that some mu-

FIGURE 10.16

How Enzymes Repair Mutations.

[A] This Navajo child suffers from the skin disease xeroderma pigmentosum. She has developed hundreds of skin cancers, visible as small, dark spots, on her facial skin, which has been exposed to the sun, but not on her arms and body, which are protected by clothing. Deficient DNA repair leaves the skin cells vulnerable to the mutagenic effects of ultraviolet light. When ultraviolet light from the sun strikes and damages DNA in a normal person, [B] enzymes find the damage, [C] cut it out, and [D] make a new reciprocal copy of the old strand and ligate, or join, the two strands, leaving the DNA as good as new.

[A] Skin cancer from unrepaired DNA damage

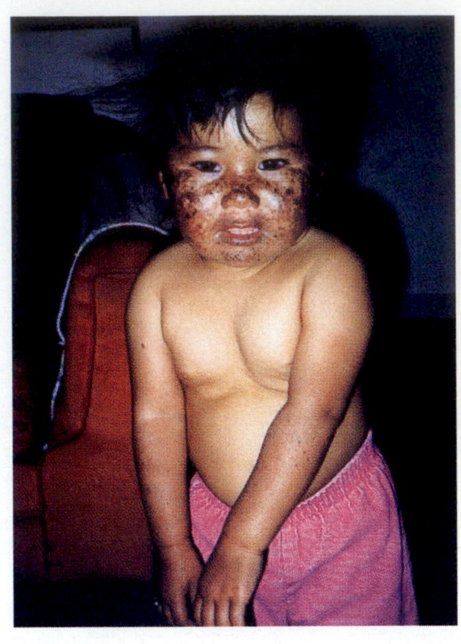

[B] Damaged DNA

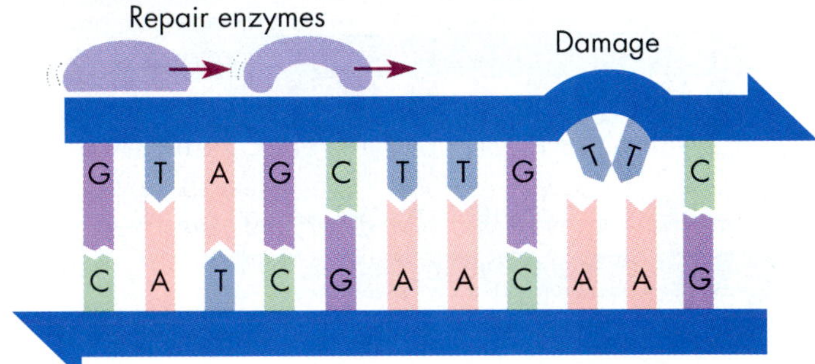

[C] Damaged DNA excised

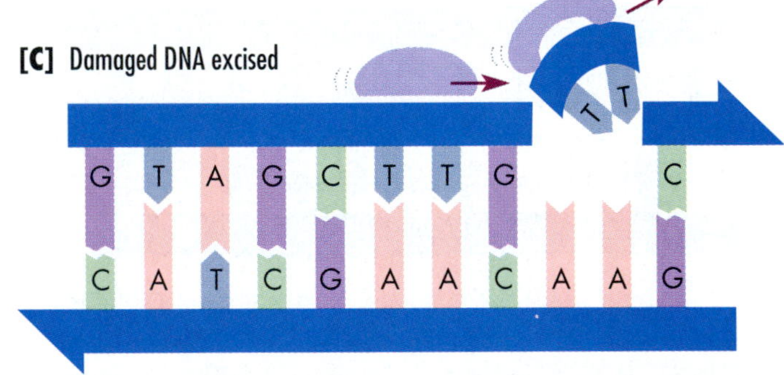

[D] Excised DNA filled in and repaired

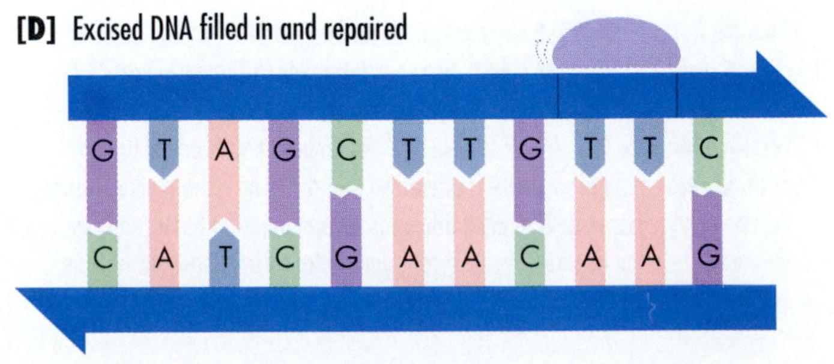

tations affect many organs while others affect only certain organs? The answer lies in gene regulation.

➤ CONCEPT CHALLENGE

An insertion mutation occurs in the DNA that encodes the RNA which you translated in the previous CONCEPT CHALLENGE. The result of this mutation is that a G becomes inserted after the second U in the mRNA. Make the insertion in the space provided and translate the mRNA. What is the net result of the mutation?

CGAUGU   AUAGUCCCUGGUAGAGA

Regulation of Gene Activity

Different cells have different constellations of proteins. While all cell types produce the enzymes that repair DNA, only red blood cells make hemoglobin, only pancreas cells make the digestive enzyme trypsin, and only cells lining the ducts of organs such as the lungs, pancreas, and sweat glands make the CFTR protein. Clearly, each cell regulates the processes of transcription and translation in such a way that it produces the correct amounts of specific proteins. The process that controls each cell's gene activity is called **gene regulation**.

GENE REGULATION IN PROKARYOTES

Even though bacterial cells have a simpler structure than eukaryotic cells, bacteria have streamlined and highly sophisticated mechanisms of gene control. Bacterial cells can change the kinds of proteins they make quickly and frequently. Consider, for example, the *Escherichia coli* bacteria growing in your intestine. After you have eaten a big delicious bowl of ice cream, the bacteria in your intestine are floating in a sea of lactose, the main sugar in milk. The cells quickly make enzymes that help break down lactose inside the bacterial cell.

A bacterial cell that makes unnecessary proteins expends energy and materials that could be used to prepare the cell for reproduction; thus, it grows and divides more slowly than a bacterium using its resources more efficiently. Bacteria have a gene regulation system called an *operon*, which prevents the production of unnecessary proteins.

The best-studied operon regulates lactose use. Scientists in the mid-1900s knew that the enzyme beta-galactosidase breaks down lactose into the two simple sugars glucose and galactose [FIGURE 10.17]. *E. coli* cells growing in the absence of lactose contain none of this enzyme. However, if lactose is supplied, each cell produces thousands of beta-galactosidase molecules in just 10 to 20 minutes. Somehow, lactose rapidly *induces* the formation of the sugar-digesting enzyme and two other functionally related proteins.

In 1960, through some elegant experiments, French geneticists François Jacob and Jacques Monod determined how *E. coli* bacteria regulate the synthesis of beta-galactosidase in response to the presence or absence of lactose—the enzyme's substrate. They found that a certain protein in the *E. coli* cell can sense the absence or presence of lactose. If lactose is absent, the protein signals the DNA not to make mRNA for the beta-galactosidase enzyme [FIGURE 10.18A]. Conversely, if lactose is present, the sensor protein allows the DNA to make mRNA for the enzyme [FIGURE 10.19B]. Careful experiments showed that the sensor protein (which Jacob and Monod called the *repressor*, since it represses transcription) acts by binding to DNA. Specifically, it binds to a spot they called the *operator*, which is near the gene that encodes the lactose-digesting enzyme [FIGURE 10.19B]. When the repressor binds to the operator DNA, it blocks transcription of the gene. With no transcription, there can be no mRNA, and hence no lactose-digesting enzyme is formed. Jacob and Monod gave the name **operon** to the protein signaler, the operator DNA sequence to which it binds, and the group of genes that encode specific proteins all related to a single function.

An important feature of gene regulation in bacteria is *transcription-level control*, whereby the cell regulates the amount of enzyme by regulating the transcription of mRNA. Another important feature is *simultaneous transcription and translation* [FIGURE 10.19A]. Because bacteria lack a nuclear envelope, translation of an mRNA molecule can start at one end of the message even before transcription of DNA into mRNA has finished at the other. This simultaneous activity increases the speed with which bacteria can respond to changing levels of materials in their environment.

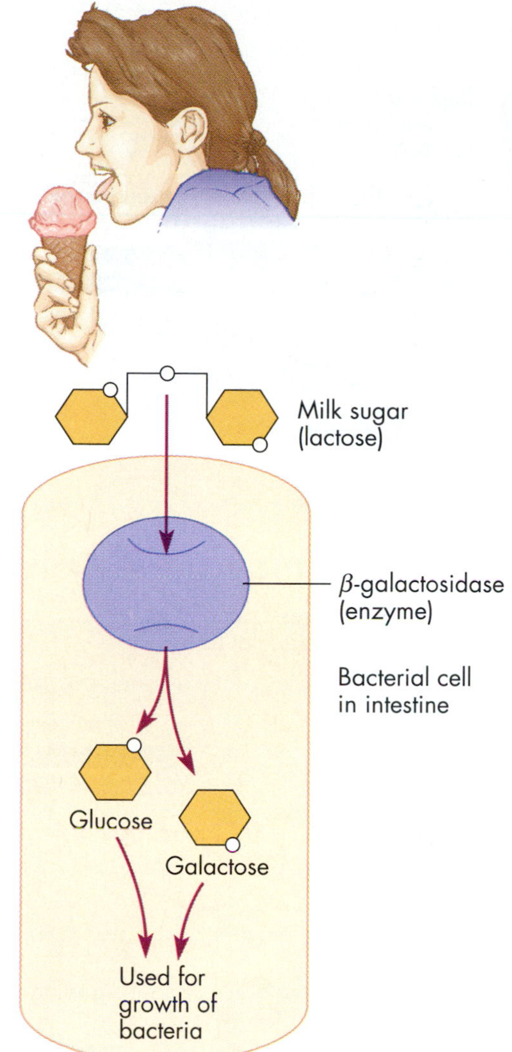

FIGURE 10.17

An Enzyme Breaks Lactose into Simple Sugars in a Bacterial Cell. The enzyme beta-galactosidase cleaves lactose (milk sugar) into the simple sugars glucose and galactose. If you eat ice cream, bacteria in your intestine break down the lactose and use it for their own growth. A regulator protein can detect the presence of the lactose, turn on the beta-galactosidase gene, and cause more enzyme to be made. Thus, when the milk sugar is present, the bacteria have the enzymes they need to quickly use it.

The last important feature of bacterial gene regulation is *short-lived mRNAs*. The *E. coli* cell synthesizes an mRNA message in four minutes, but destroys it in three. Thus, when the cell no longer needs to make a particular enzyme, it rapidly dismantles the machinery for making it. Again, this allows a rapid response to changing metabolic needs.

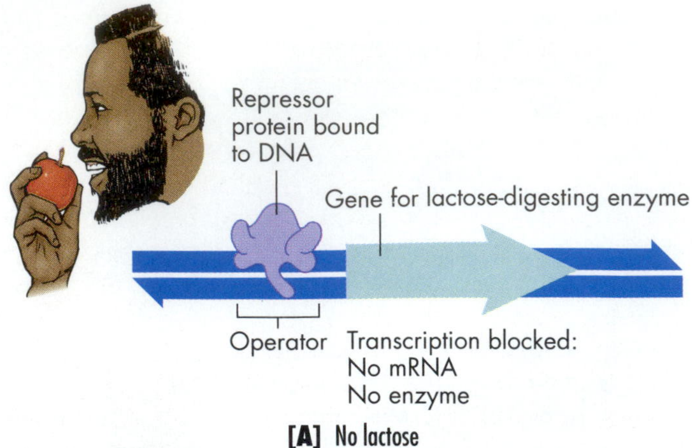

[A] No lactose

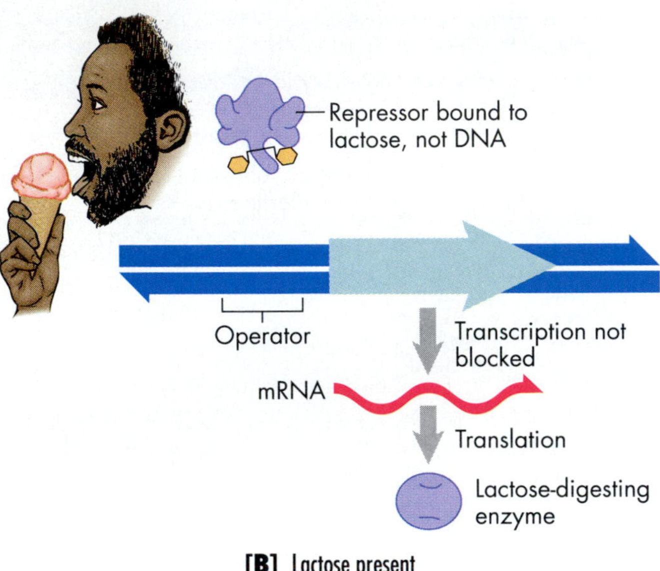

[B] Lactose present

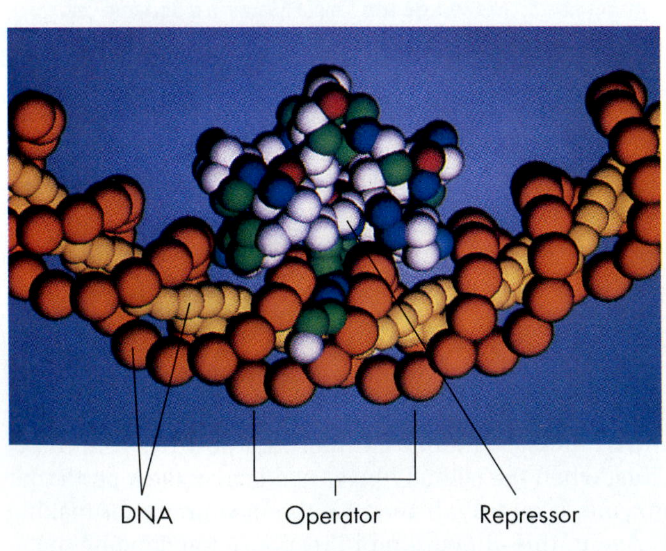

[C] Repressor protein bound to DNA

FIGURE 10.18

How a Bacterial Cell Makes an Enzyme Only When It Is Needed.

A normal bacterial cell makes enzymes for digesting the milk sugar lactose only when that sugar is present in the cell's environment. This is due to the operon mechanism of gene regulation. **[A]** In the absence of lactose, a repressor protein binds to DNA at the operator, a region of DNA near the gene for the digestive enzyme beta-galactosidase. The presence of the repressor physically blocks the ability of RNA polymerase, the RNA-making enzyme, from transcribing the gene. When transcription is blocked, no mRNA is formed, so there can be no translation of the enzyme protein. **[B]** When lactose is present, it binds to the repressor, and the repressor is then unable to bind to DNA. Now RNA polymerase has access to the gene and transcribes it into mRNA. This mRNA is translated, the enzyme is synthesized, and it can digest the sugar lactose, providing energy and materials for the bacterium. **[C]** A computer image shows how the repressor wraps an arm around one of the grooves in DNA and hence blocks transcription. DNA is shown in rust and gold, and protein in white, blue, red, and green.

GENE REGULATION IN EUKARYOTES

Eukaryotes and prokaryotes share many mechanisms of gene regulation, such as controls at the transcriptional level and control by regulatory proteins. But there are variations, too, based on the fundamental differences between single-celled and multicelled organisms.

How is it that different genes are turned on in the different cell types of a multicellular organism? The short answer is that all cells have the same genes, but in some cells, certain genes are permanently unavailable for transcription. Compare, for instance, the regulation of a lactose-digesting-enzyme gene in humans with the lactose operon in bacteria [review FIGURE 10.18].

Baby mammals produce a lactose-digesting enzyme called *lactase* in their intestinal cells, and the enzyme allows them to digest milk sugar. But as the babies mature, they permanently lose the ability to make the lactase, and along with it the means of digesting milk sugar. The mammalian regulation of lactase production—continuously on, then permanently off—seems crude compared to the environmentally sensitive bacterial lactose operon. But then, most mammals never drink milk as adults, so losing this digestive capacity is actually an efficient physiological response.

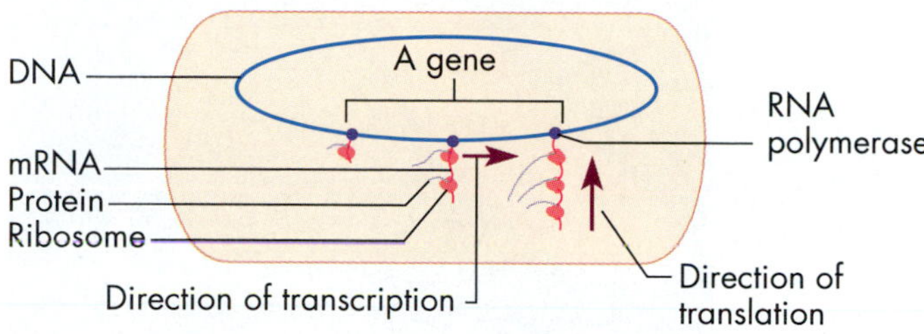

[A] Prokaryote: Simultaneous transcription and translation

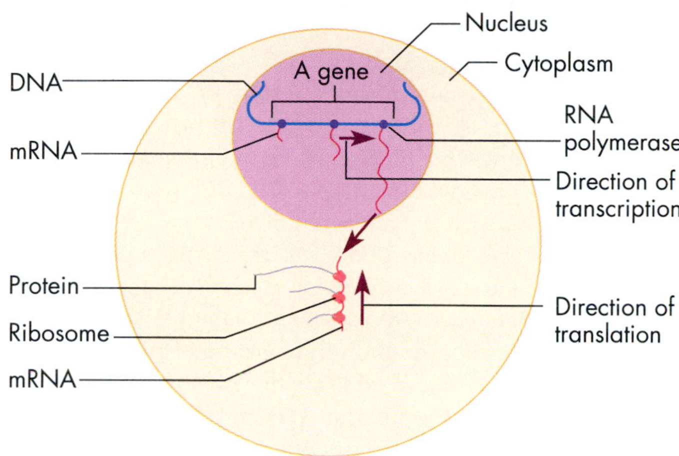

[B] Eukaryote: Separation of transcription and translation

FIGURE 10.19

Where Transcription and Translation Take Place in Cells.

[A] In a bacterial cell, transcription and translation occur simultaneously along the naked, circular chromosome. [B] In a eukaryotic cell, transcription takes place in the nucleus, and the newly made RNA is transported to the cytoplasm, where translation occurs.

Like other mammals, most adult humans lose the ability to digest lactose. After early childhood, drinking milk gives most people abdominal cramps. Unfortunately, many adults without lactase have a taste for ice cream and other milk products. If a lactose-sensitive person takes a Lactaid or Dairy-Ease pill containing lactose-digesting enzymes made by microorganisms just before eating a giant bowl of mocha almond fudge ice cream, the enzyme digests the lactose before it can reach the intestine and cause discomfort.

The transcribing of certain genes only in specific cells during a limited period of development is a general strategy in multicellular eukaryotes. It results in different cells expressing different genes [FIGURE 10.20].

Knowing that different cell types express different genes, we can now understand the range of symptoms experienced by a person with cystic fibrosis. Recent experiments show that the mRNA from the cystic fibrosis gene is found in duct cells (air ducts, pancreatic ducts, sweat ducts, and sperm ducts) but is not found in brain cells or many other cell types. Apparently, duct cells are the only cells of the body that transcribe the cystic fibrosis gene, as well as the only cells that require the CFTR protein to function normally. Thus, when a mutation changes the cystic fibrosis gene and hence the CFTR protein, duct cells are the only ones affected.

What are the genetic regulatory mechanisms that ensure that only duct cells transcribe the cystic fibrosis gene both in patients and in unaffected people?

Cell types:	Genes: Hemoglobin	Skin protein	Digestive enzyme
Red blood cells	RNA		DNA
Skin cells		RNA	DNA
Stomach cells			RNA DNA

Active gene   Inactive gene

FIGURE 10.20

Gene Activity in Multicellular Organisms.

Multicellular organisms use different genes in different cells at different times during development. In general, each cell in the body has a complete set of genetic information. For example, each cell has the gene for the oxygen-carrying protein hemoglobin normally found only in red blood cells and for digestive enzymes made only by cells of the digestive tract. Each cell has about 100,000 genes for proteins made by other groups of cells. The reason that red blood cells accumulate hemoglobin but not digestive enzymes is that immature red blood cells transcribe the hemoglobin gene into RNA, but they do not transcribe the genes for digestive enzymes. Likewise, digestive tract cells transcribe digestive enzyme genes but not genes for hemoglobin.

box 10.1
Human Impact

The Science of Deadly Diarrhea

In many of the world's developing countries, cholera and other diarrheal diseases kill one quarter of the infants who die each year—4 million of the 14 million annual deaths. This death rate is not only tragic but it helps foster the high birth rates that are leading to rapid population growth in many poor nations. Intensive research efforts into the causes of cholera have revealed a lesson in posttranslational control that illustrates one of this chapter's important topics. Recent evolutionary findings may also help explain why another deadly illness we encountered in this chapter is so relatively common. Finally, work done in the 1970s pointed out a simple, inexpensive therapy for cholera that has, since the early 1980s, saved nearly a million children each year.

Epidemiologists (researchers who study disease outbreaks) believe that cholera originated in India centuries ago. While it took a heavy annual toll of human life, it remained isolated on that subcontinent until improvements in transportation allowed greater contact with India, and ship and rail passengers inadvertently spread cholera to Europe and beyond. Since this happened in the early 1800s, there have been seven *pandemics*, or major epidemics, that began in one country and spread rapidly. The most recent outbreak, for example, started in Peru in early 1991, and moved throughout much of South America within three months, causing over 300,000 cases of cholera.

Physicians and scientists of the nineteenth century could only speculate about the etiology (cause) of cholera until in 1883, German microbiologist Robert Koch found the culprit through his microscope: the distinctive, comma-shaped bacterium *Vibrio cholerae* [FIGURE 1]. It took researchers the better part of the twentieth century to ferret out exactly how these bacteria trigger massive, life-threatening diarrhea in human hosts, but the details are now fairly clear.

Like many pathogenic (disease-causing) bacteria, *V. cholerae* harms its host by secreting a toxin as it grows. This substance is an *enterotoxin*, meaning that it acts on the small intestine. *V. cholerae* shares this habit with several other dastardly microorganisms, including the causes of gas gangrene, salmonella food poisoning, garden-variety "turista," and more serious forms of *Escherichia coli* food poisoning [review CHAPTER 9]. The *V. cholerae* enterotoxin is a protein made up of three kinds of polypeptide chains linked into an A subunit and a B subunit. The latter sneaks the toxin past the intestinal cell's outer membrane, while the former turns on the flow of diarrheal fluid.

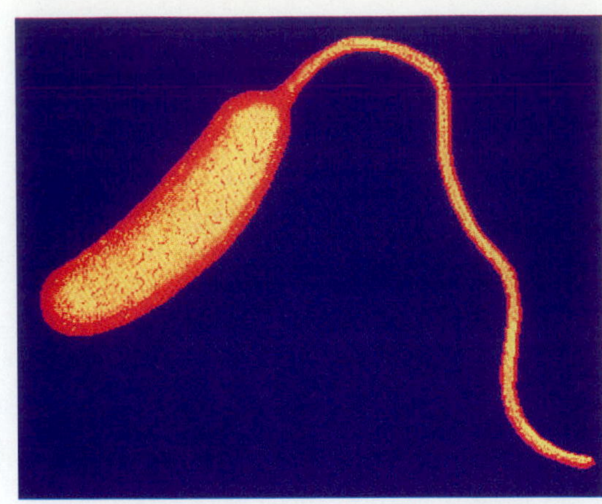

FIGURE 1
The bacterium *Vibrio cholerae*, the cause of cholera in humans.

At the molecular level, the A subunit of the cholera toxin acts by causing a posttranslational modification; specifically, it activates molecules of an enzyme, adenyl cyclase, already present in the intestinal cell. This enzyme's normal job is to convert molecules of the energy carrier ATP into molecules of the regulatory compound cyclic AMP, which, among other things, helps govern the traffic of ions, water, and other materials into and out of intestinal cells. Once the toxin modifies the enzyme, it converts large quantities of ATP to cAMP. The extra cAMP molecules in turn cause the intestinal cells to pump abnormally large numbers of chloride and bicarbonate ions into the lumen, the internal cavity of the intestine. As one would expect, water follows osmotically [FIGURE 2, and review FIGURE 4.18], and this becomes diarrheal fluid. Collectively, the cells of the small intestine can pour out so much fluid that the large intestine can't absorb it all, and the cholera victim loses shocking amounts of liquid—from 2 to 5 *gallons* per day!

If untreated, this outpouring can kill a person within 24 to 72 hours, due to dehydration, low blood pressure, and organ failures. But 99 percent of the time, the patient will survive if he or she ingests an equivalent amount of water laced with precise amounts of salts and sugars so that the body can survive until the immune system fights off the *V. cholerae* infection.

This so-called oral rehydration therapy was the result of basic research into the mechanisms of normal digestion and the impaired function during diarrhea, as well as

FIGURE 2
How the bacterium causes the disease cholera.

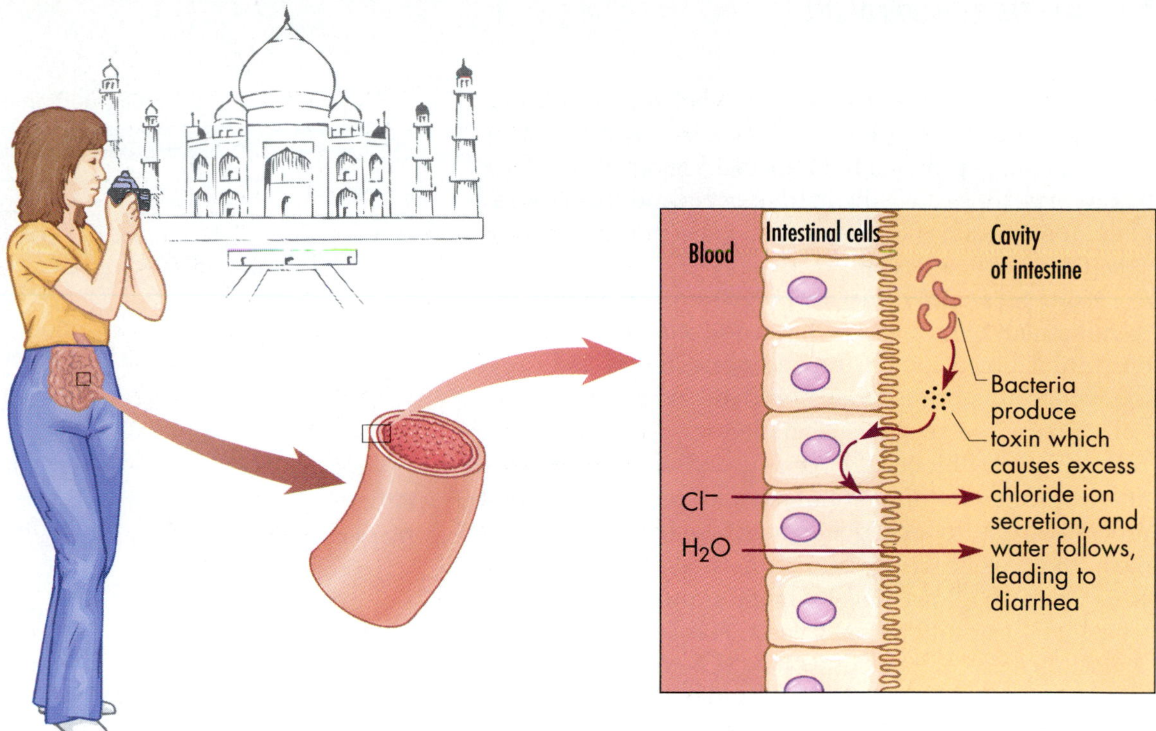

specific attempts to develop simple, inexpensive treatments that could be used widely in developing countries. Norbert Hirschhorn and others working in Bangladesh in the early 1970s discovered the exact combination of salt, potassium chloride, trisodium citrate, glucose, and water that can keep a person hydrated during a bout of cholera and provide sufficient salts and food calories to tide them over until they can receive antibiotics or until the body's own defenses take over and fight the disease. This strategy costs pennies, and yet has saved millions of lives.

Two aspects of the cholera story illuminate subjects in this chapter. First, the cholera toxin is another example of posttranslational regulation of gene activity. By binding to the adenyl cyclase enzyme, the toxin causes a modification, a shape change, that activates the protein—turns it on—and thereby leads to the host's diarrhea. This may seem like a ghoulish joke, but it has real evolutionary significance for the *V. cholerae* bacterium. This organism has no means of its own to disperse to new hosts, but when billions of the cells are flushed out in gallons of body fluid that can contaminate nearby streams and wells, the chances of reaching a new host through drinking water, food, or body contact are greatly increased.

Curiously, human evolution has provided a countermeasure that is the second topic of note: Some biologists now believe that the gene for cystic fibrosis may have persisted in Caucasian populations because it confers some protection against cholera. Recall from CHAPTER 8 that when the gene for sickle-cell anemia is inherited in two doses (one from each parent) it leads to a painful genetic disease. When inherited in just one dose, however, the carrier is free of anemia and also has natural protection against the malarial parasite. In a similar way, two doses of the disease-causing allele of the CFTR gene [see page 238] lead to a life-threatening condition, cystic fibrosis. However, one dose of the mutant allele may confer partial immunity to diarrhea from enterotoxin. This is because, as we saw, the membrane protein responsible for chloride secretion is abnormal in cystic fibrosis, and can't be turned on with the subsequent outpouring of water. Now, for every cystic fibrosis sufferer, there are 100 carriers with a single dose of the CFTR gene. According to current speculation, those heterozygotes would have partial protection from enterotoxins because the "diarrhea switch" could only be turned on part way. Presumably, they would lose less fluid during a bout of cholera. This theory has yet to be confirmed, but it makes good sense. In the meantime, it is heartening to know that with oral rehydration therapy, lives are saved regardless of genetic predisposition.

In eukaryotes, these regulatory mechanisms follow a strategy similar to those we already discussed for bacteria: Regulatory proteins bind special regions of DNA (in eukaryotes, these are called *enhancers*), and this binding alters the transcription rate of the gene in a way that is specific to each cell type.

Now, let's apply this idea of gene regulation to the development of an embryo. As a one-celled zygote develops into a larger, more complex organism, the multiplying cell groups begin to take on their various functions. Cells that are at first unspecialized soon become nerve cells, skin cells, blood cells, or duct cells as a result of the process of *cellular differentiation*, and this specialization process is due to selective gene expression. In eukaryotes, then, gene regulation underlies differentiation, which in turn allows development to proceed, changing a fertilized egg into an embryonic mouse, plant, or person. Biologists don't yet know how different genes are selected for expression in different cells or how the selection, once made, is stabilized.

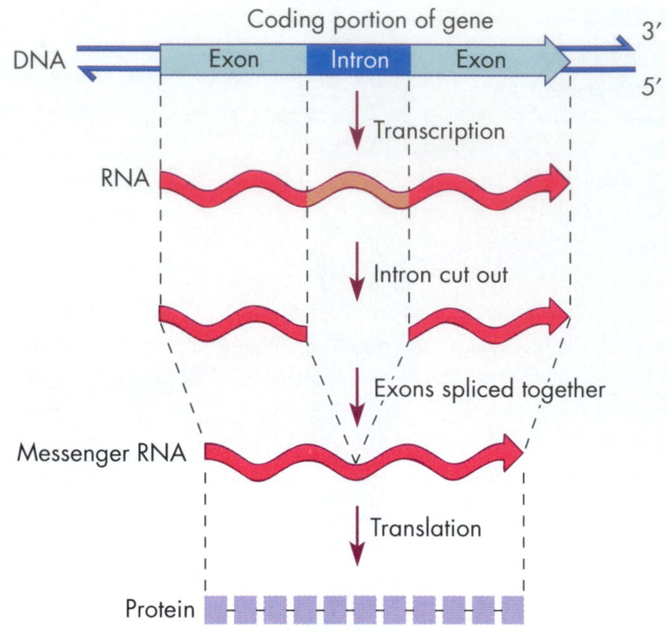

FIGURE 10.21

Introns and Exons: Structure of a Eukaryotic Gene.

In eukaryotes, only certain parts of a gene actively encode proteins. The whole coding portion of a gene is transcribed into an RNA, but regions called introns are spliced out before the mature mRNA forms. The figure shows one intron, but the cystic fibrosis gene has 24, and other genes can have as many as 50.

REGULATION, DEVELOPMENT, AND GENETIC DISEASE

Biologists have been working hard to learn how genes are regulated during embryonic development, since understanding this process will ultimately help physicians treat genetic diseases such as sickle-cell anemia.

MULTIGENE FAMILIES AND SICKLE-CELL ANEMIA

Recall that people with sickle-cell anemia have defective genes for hemoglobin [review FIGURE 10.5]. Researchers discovered some time ago that an adult mammal expresses a different gene for hemoglobin than does a fetus enclosed in the mother's womb. People and other mammals, in fact, have a *multigene family* for hemoglobin, a group of genes related in structure and function that code for closely related blood pigment molecules.

In people with sickle-cell anemia, a gene for the adult form of hemoglobin is defective, but a gene for the fetal form of hemoglobin is normal. If biologists could fully understand gene regulation during development, perhaps they could prevent the fetal hemoglobin gene from turning off during development. If that fetal gene were continually transcribed, even in the child and adult, the individual would have sufficient amounts of normal hemoglobin to prevent the symptoms of sickle-cell anemia. This would be true even though the mutant gene for the adult form of hemoglobin remained in the cells unexpressed. Medical researchers are trying to learn how to manipulate the expression of these genes during development. A few research teams have shown that the compound *hydroxyurea* does cause an adult to produce fetal hemoglobin and that this can in fact help some patients with sickle-cell anemia.

CYSTIC FIBROSIS, INTRONS, AND EXONS

Recent experiments suggest that researchers may be able to develop a therapy for cystic fibrosis based on the principles of transcription and translation within a few years' time, rather than the decades many thought it might require. When geneticists isolated the cystic fibrosis gene, they discovered that, like most other eukaryotic genes, it was broken into many different parts, in this case scattered along a portion of chromosome 7. In general, eukaryotic genes differ from prokaryotic genes in their possession of *introns*, stretches of DNA that *intr*ude into the gene but do not appear in the final mRNA and hence are not translated into protein. The remaining portions of a gene, which *do appear* in the mRNA, are called *exons*, since they are *ex*pressed. The cystic fibrosis gene has 24 exons. When DNA is transcribed, the product of transcription includes an RNA copy of both the introns and exons [FIGURE 10.21]. The RNA is then processed or edited by cleaving out the introns and splicing the remaining exons together. Surprisingly, some introns act as a nonprotein enzyme (or *ribozyme*) that splices itself out. The result is an mRNA molecule ready to be exported from the nucleus, now bearing the actual exon sequences that will be translated into protein.

Cystic fibrosis researchers have studied the details of exons and introns in the CFTR gene. As a result, they have recently learned to produce mice with a CFTR gene that is disrupted by an extra intron and exon. These so called *knockout mice* show many of the symptoms of a human with cystic fibrosis. With the new laboratory mice, researchers should be able to develop therapies much more rapidly, inexpensively, and safely than they have been able to with human volunteers.

In 1990, researchers used their knowledge of introns and exons to build a nonmutated copy of the cystic fibrosis gene possessing only exons (expressed regions) and no introns. They inserted a synthetic gene into airway cells and pancreas duct cells that they had removed from a person with cystic fibrosis and grown in the laboratory. Wonderfully, the inserted gene was transcribed into an mRNA, and the mRNA was translated into properly functioning CFTR proteins that worked to pump chloride ions normally: The diseased cells had been cured! Officials of the Cystic Fibrosis Foundation now hope that researchers will be able to design a spray that can deliver the engineered gene to airway cells in a living person, and that transcription and translation of the CFTR gene will allow affected cells and respiratory systems, to function normally. (No genetic change in airway cells would be transmitted to a person's children, of course, since airway cells do not give rise to sex cells.)

Formidable obstacles remain in designing effective delivery systems for engineered genes, but before long, medical workers will be applying the basic biology of transcription, translation, and gene regulation described in this chapter to diseases like cystic fibrosis and sickle-cell anemia and helping to improve the quality of life for millions of sufferers around the world. The next chapter examines human genetic diseases and other aspects of human genetics. CHAPTER 12 then explores how researchers are using the new techniques of biotechnology to help overcome biological problems including human diseases.

CONCEPT CHALLENGE

One type of mutation that causes cystic fibrosis interferes with the removal of an intron. Will the mRNA from the cystic fibrosis gene be longer or shorter than normal, or will there be no change? What is likely to happen to the CFTR protein coming from such a mutant gene? If a person had two copies of such a gene, would the person be quite ill with cystic fibrosis or have very few symptoms?

Connections

In a very real sense, the possibility of helping people with cystic fibrosis and other genetic diseases stems directly from life's basic unity at the level of genes and molecules. Mendel initially discovered genes in plants; later experimenters learned that genes are made of DNA by studying viruses; research on fungi supported the one gene–one enzyme concept; and studies with bacteria elucidated the mechanisms of transcription, translation, and gene regulation. All these principles are being applied now to helping people with genetic diseases, and the future uses of this knowledge are potentially limitless.

CONCEPT MAPPING

Draw a concept map relating the following terms. Link the concepts with arrows and a connecting word or short phrase.

base substitution
cystic fibrosis
cytoplasm
DNA
enzyme
gene
mRNA mutation
nucleus
phenotype
polypeptide
ribosome
transcription
translation

KEY TERMS

anticodon, 249
chromosomal mutation, 256
codon, 249
gene mutation, 256
gene regulation, 258
genetic code, 252
messenger RNA (mRNA), 248
mutation, 256
operon, 259
protein synthesis, 250
ribosomal RNA (rRNA), 249
RNA polymerase, 245
transcription, 244
transfer RNA (tRNA), 249
translation, 250

HIGHLIGHTS IN REVIEW

1 In general, a gene affects an organism's form and function by specifying the structure of a certain protein.

2 Genetic information is stored in DNA in the order of its base pairs.

3 The flow of genetic information proceeds in two steps. (1) Enzymes copy a portion of a DNA molecule into an RNA molecule. (2) The order of bases in RNA specifies the order of amino acids in a protein.

- **a]** Three kinds of RNA—messenger RNA (mRNA), transfer RNA (tRNA), and ribosomal RNA (rRNA)—are involved in protein synthesis. Nucleotides in RNA differ from those in DNA by the sugar—ribose versus deoxyribose. RNA consists of only a single strand of nucleotides, not a double strand, as in DNA; and the RNA base uracil replaces the DNA base thymine. RNA molecules are also generally much shorter than DNA molecules because they tend to code for only one or a few genes rather than for the whole genome.
- **b]** Like DNA replication, transcription is directed by base pairing and carried out by a polymerase enzyme. During transcription, a few genes may be copied thousands of times, while during replication, the whole genome is copied just once. The newly formed RNA strand separates from the DNA molecule in transcription, instead of joining with it as does a newly replicated DNA strand.
- **c]** The genetic code, tRNA, and ribosomes together translate an mRNA into a protein. The genetic code is identical in nearly all organisms. A codon is a group of three bases in RNA that specifies an individual amino acid in a protein. Specific condons indicate where on the mRNA molecule translation begins and ends. Several different codons may specify the same amino acid, but no codon specifies more than one amino acid.
- **d]** The tRNAs translate the mRNA codons into an amino acid sequence. At one end of each tRNA molecule is an anticodon; at the other end is a specific amino acid. The anticodon pairs up with the mRNA codon, so that the amino acids are ordered according to the mRNA codon sequence.
- **e]** Ribosomes hold mRNAs, tRNAs, and amino acids in place and join the amino acids into a polypeptide.

4 In a eukaryotic cell, DNA is copied to RNA in the nucleus, and proteins are made in the cytoplasm on ribosomes.

5 A change in the order of bases in DNA (a mutation) can change the order of amino acids in a protein, and this may disrupt the protein's function.

- **a]** A change in the base sequence of DNA, a mutation, may consist of a base substitution, insertion, or deletion. Usually, DNA repair enzymes detect and fix altered DNA; if these enzymes fail, a permanent mutation occurs. A change in the sequence of DNA bases results in a change in the sequence of bases in mRNA, which can result in a change in the sequence of amino acids in a polypeptide. These changes can frequently destroy the polypeptide's function, which can alter the way an organism looks or acts.
- **b]** Mutations in the sequence of base pairs in DNA can result in genetic defects or cancer.
- **c]** Cells limit the production of unnecessary proteins by regulating gene activity. They can accomplish this by regulating transcription, modifying the mRNA after transcription, regulating translation, modifying the polypeptides after translation, and adjusting protein activity itself.
- **d]** A classic example of transcription-level control in prokaryotes is the lactose operon of *Escherichia coli.* In the absence of lactose, a repressor protein binds to an operator sequence (a special site on the DNA) and so prevents transcription of genes for lactose-digesting enzymes. In the presence of lactose, the repressor protein binds to the lactose instead of to DNA. RNA polymerase then transcribes the structural genes that code for enzymes that digest lactose.
- **e]** Prokaryotes can change the kinds of proteins they make often and rapidly, which accommodates changes in their environment. They achieve flexible gene regulation dependent on environmental conditions by using transcription-level control, by transcribing and translating simultaneously, and by destroying mRNAs almost as fast as they are made.

6 In general, all cells in a multicellular organism contain the same genes, but different cells utilize different genes.

- **a]** In a eukaryotic organism, all cells generally have the same genes, but some cells use one set of genes while others use another set. These differences arise through the developmental process called cell differentiation. How eukaryotes achieve this cell differentiation is one of the great mysteries of biology.
- **b]** Unlike prokaryotic genes, eukaryotic genes that are regulated simultaneously are not necessarily located near each other; they may even be on different chromosomes. Like prokaryotic genes, eukaryotic genes may be flanked by regulatory sequences of DNA that are responsive to gene regulators made of protein.
- **c]** In eukaryotes, genes are sometimes clustered according to their historical relatedness in multigene families. The hemoglobin gene family in humans encodes several structurally similar globins, some of which are expressed at different times in development.

7 An understanding of gene action can help biologists design therapies to improve the lives of people with genetic diseases.

UNDERSTANDING THE FACTS AND CONCEPTS For Questions 1–5, match each of the descriptions with the most appropriate item for the following list of terms.

a] mRNA c] rRNA e] anticodon
b] tRNA d] codon

1 A sequence of three adjacent nucleotides, usually in linear series with other similar, though different, sequences, that carries the genetic message from DNA to a ribosome.

2 A nucleic acid that is found associated with several protein molecules and is essential to the construction of all kinds of cellular proteins.

3 A sequence of three adjacent nucleotides that is part of a molecule that is capable of bonding to an individual amino acid.

4 A nucleic acid of complex shape that includes nucleotides, some of which pair internally with other nucleotides of the same molecule and others that pair externally with another molecule.

5 A linear nucleic acid that is sometimes found paired with DNA and at other times with numerous smaller molecules of RNA.

For questions 6–10, complete each of the statements below by choosing one of the terms from the following list.

a] nonsense d] redundant
b] base substitution e] stop
c] reading frame

6 A protein that is changed by the inclusion of one wrong amino acid is a result of a ______ mutation.

7 The genetic code is said to be ______ because there may be several other tRNA molecules that, although bearing a different anticodon, may transport the same amino acid to the ribosome.

8 If a mutation results in the production of a protein that is significantly shorter than normal, a ______ mutation may have caused the change.

9 Deletion or insertion of one or two bases in DNA will most likely cause a change in the ______.

10 A sequence of three adjacent bases in mRNA for which no complementary tRNA pairing can occur is called a ______ codon.

As above, complete Questions 11–15 by choosing the most appropriate term from the following list.

a] transcriptional regulation
b] posttranscriptional regulation
c] translational regulation
d] posttranslational regulation
e] protein activity regulation

11 A mutation that renders the ribosome nonfunctional would constitute ______.

12 The production of lactose-digesting enzymes in bacteria is an example of ______.

13 If excess product of a metabolic pathway inhibits one of the enzymes that act early in that pathway, ______ is said to be working.

14 The removal of introns is an example of ______.

15 The modification of a polypeptide after its synthesis is complete is known as ______.

For Questions 16–20, match the descriptions with the appropriate item from the following list of terms.

a] operon d] intron
b] repressor e] multigene family
c] exon

16 A group of related genes in eukaryotes that may operate sequentially at different times in the life cycle.

17 A group of functionally different genes in bacteria that always includes both a gene that acts as an on/off switch and genes that code for enzyme synthesis.

18 A length of DNA within a gene that codes for a section of mRNA that is removed after transcription and before translation.

19 A protein that has the ability to bind to a specific gene, thereby blocking the access of transcriptional enzymes to that gene.

20 A length of DNA within a gene that is transcribed into mRNA and remains intact and functional during translation.

INTEGRATE AND APPLY WHAT YOU HAVE LEARNED

1 What are the most important differences between DNA replication and RNA transcription?

2 How does RNA differ from DNA, and how do the three different kinds of RNA differ from each other?

3 Assume that there is an AGU codon on a molecule of mRNA. Which tRNA anticodon will pair with this codon? Which amino acid will be positioned? What was the order of bases on DNA that coded for this mRNA codon?

4 Why might cancer-causing agents (carcinogens) also be mutation-causing agents (mutagens)?

ANALYSIS

1 Which of the following is *not* different in the processes of DNA replication and transcription?

a] The particular enzyme that assists polymerization.
b] The particular bases involved in the processes.
c] The antiparallel nature of the nucleic acid strands in the duplex.
d] The sugar portion of each of the nucleotides involved.
e] The length of the newly synthesized strand.

2 Suppose for a moment that each amino acid requires a code consisting of four sequential nucleotides, rather than three. How many different codons could be composed?

a] 32 b] 64 c] 128 d] 256 e] 1024

C H A P T E R 1 1

Human Genetics

PHENYLKETONURIA

Have you noticed warning labels on cans of diet soda that read "Phenylketonurics: Contains phenylalanine" and wondered why they were there? The answer is that the amino acid phenylalanine can cause mental retardation in the tiny percentage of children who are born with the genetic defect phenylketonuria. Fortunately, advances in the field of human genetics—the subject of this chapter—allow physicians to detect this condition even before a child is born, thus preventing its devastating effects [FIGURE 11.1].

Most people can digest the low-calorie sweetener called *aspartame* used to flavor many diet soft drinks and other foods. The discovery of aspartame was hailed as a major advance, since unlike some earlier artificial sweeteners, it was less toxic and quite safe in moderate doses. Aspartame is intensely sweet, and

FIGURE 11.1

Some Genetic Diseases, Such as Phenylketonuria, Can Be Detected Early and Controlled.

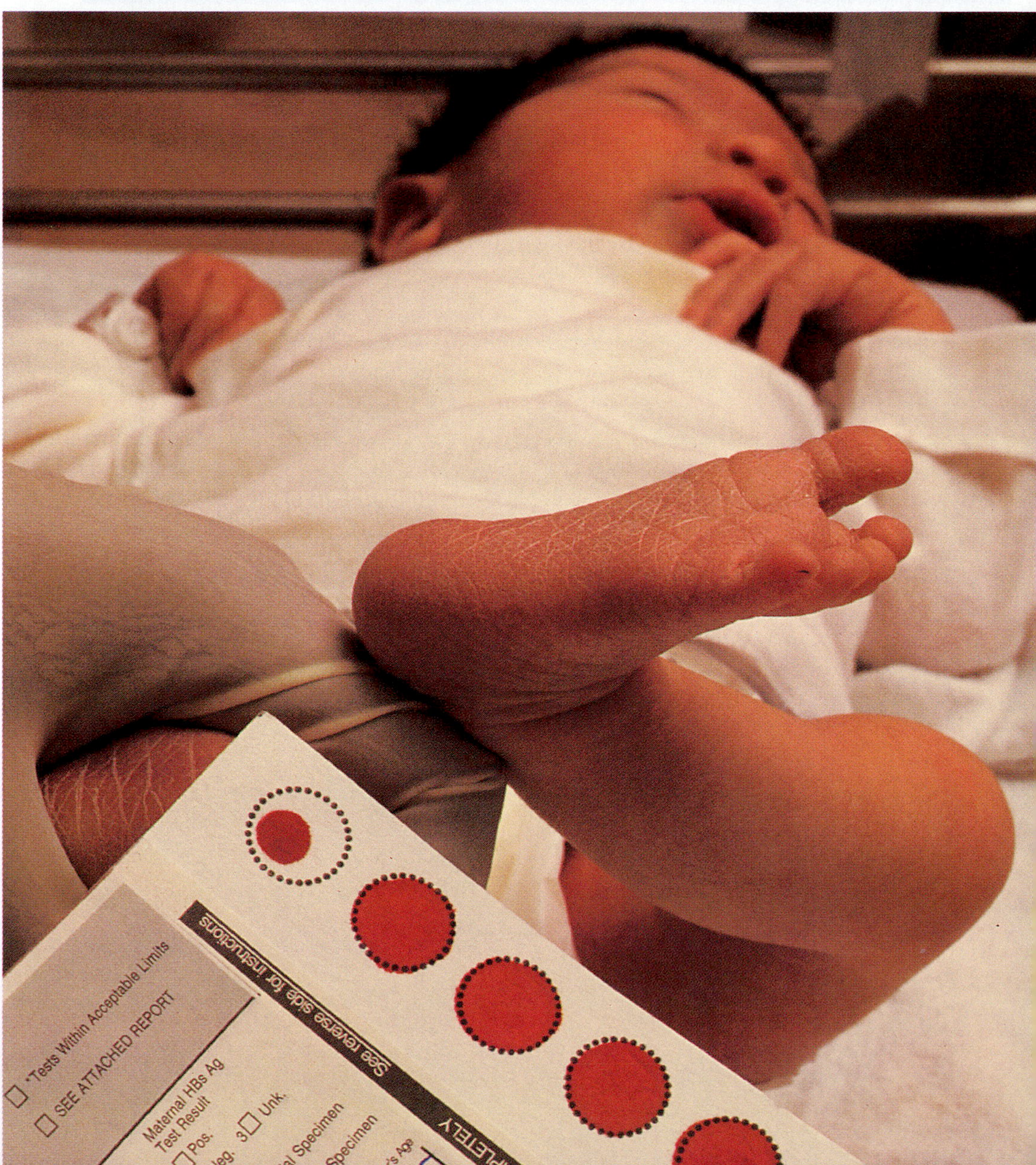

being naturally low in calories, it can help prevent obesity. Nevertheless, it contains large amounts of phenylalanine, an amino acid found in virtually all proteins.

Most people have a dominant allele of a gene that encodes the enzyme *phenylalanine hydroxylase* (*PAH*), which breaks down phenylalanine by converting it into the amino acid tyrosine. Unfortunately, a small percentage of people lack PAH. These people inherit two mutant alleles of the PAH gene, and as a result they do not produce the enzyme. Such people are called *phenylketonurics,* and their disease is called **phenylketonuria**, or **PKU**. For people with PKU, the phenylalanine they ingest in foods builds up in the body and is transformed, not to tyrosine, but to compounds known as phenylketones. These mousy-smelling substances are excreted in urine, giving it a telltale odor.

Odd-smelling urine is of little medical importance in itself, but the production of phenylketones carries an additional, more serious consequence: Phenylketones cause nerve cells in the brain to develop abnormally. If a child born with PKU drinks breast milk, diet soda, or any other food containing phenylalanine, the child's brain development will be stunted. The afflicted child will start having seizures at about the age of 6 months, and will probably not advance beyond a mental age of 2.

In the past, 1 out of every 100 mentally retarded children suffered from PKU. But today, thanks to the techniques of modern human genetics, the recessive alleles that cause PKU can be detected before the child is born. Then, after birth, the parents can control the child's diet to exclude phenylalanine, thus preventing mental retardation.

Geneticists have come to understand human inheritance and genetic traits by building on the studies of more classic laboratory subjects, such as fruit flies and bacteria. Scientists who study human genetics apply their basic knowledge of the cellular and molecular biology we have discussed earlier to a special animal, *Homo sapiens*. This chapter investigates the fundamentals of human genetics, and how biologists have studied thousands of inherited human traits, including phenylketonuria. Our discussion of human genetics—how people inherit family resemblances, particular metabolic traits, and other characteristics—deals with a number of issues. First we'll discuss how researchers use standard genetic techniques to study the inheritance of human traits. Then we'll see how new molecular techniques have allowed scientists to map human genes. We'll consider how scientists and physicians are currently treating human genetic diseases. Finally, we'll see how geneticists counsel prospective parents and help to prevent new cases of genetic diseases by identifying adult carriers and detecting defective genes in the fetus.

Along the way, we will see how our collective knowledge of human genes could lead to new disease cures, but may create other societal problems as yet unimagined. ❑

MESSAGES

1 The genetic principles that apply to peas and fruit flies also apply to humans.

2 By analyzing pedigrees, geneticists can often determine whether a trait is *X*-linked or autosomal, and whether it is dominant or recessive.

3 Changes in chromosome number and structure can alter a person's phenotype.

4 Geneticists can now use variations in DNA structure to map human genes. Knowing a gene's location on a chromosome may help a geneticist isolate pure copies of the gene.

5 Analyzing cloned genes can help a geneticist determine the gene's function and lead to improved medical diagnoses and treatments.

6 Ethical dilemmas surround our new ability to detect various human genetic conditions.

Studying Human Genetic Conditions

Human beings are wonderfully complex organisms. We come in a nearly infinite variety of shapes, sizes, and colors, and have a huge range of individual traits. Many of our traits are *polygenic* [review FIGURE 8.22], which means that many genes work together to determine individual characteristics such as how well our livers function, how long our limbs are, whether our skin is clear or dotted with acne, and whether we are susceptible to a few types of cancer, diabetes, allergies, heart attacks, or early senility (Alzheimer's disease).

While a person's health does have a strong genetic component, not all human diseases have genetic causes. Many are due to infection by microorganisms (malaria, measles, colds, AIDS), others are caused by substances in the environment (many cancers, emphysema, lead poisoning). Even if a disease does have a genetic component, simply inheriting the gene does not necessarily mean that the person will develop the disease. For example, there are 180,000 cases of breast cancer diagnosed in the United States each year, but of them, only 10 percent have been shown to be hereditary. Of those 18,000 cases, half are due to a specific gene called BRCA1 (for Breast Cancer 1). In women with the disease-causing allele of the breast cancer gene, 60 percent will develop breast cancer by the age of 50 (instead of the usual 2 percent), and 80 percent by age 65. Clearly, the allele dramatically increases the likelihood of developing the disease, but other environmental or genetic factors must also be involved, and unless all factors are present, some women with the allele will be spared the disease.

This chapter surveys what we know about human genetics, and since we consider many genetic diseases, it may seem that all that human genes do is code for diseases. Bear in mind, however, that geneticists classically identify genes by detecting hereditary differences among organisms. Since most genes encode necessary functions in the body, when mutations alter these genes, the functions don't occur, and disease can be the outcome. The focus of many geneticists, therefore, becomes the genes that cause variations, and many of these are medically important. The tens of thousands of human genes that do not cause difficulty simply come to the geneticist's attention less often.

At the same time, geneticists have discovered more than 3000 human traits determined by single genes in simple Mendelian fashion. These traits include hair on the middle joint of the fingers; attached or unattached earlobes [FIGURE 11.2A and B]; the widow's peak hairline; ear pits [FIGURE 11.2C]; tufts of hair on the ears [FIGURE 11.2D]; extra fingers or toes (*polydactyly*); baldness; hemophilia; color blindness; counterclockwise cowlicks in the hair; uncombable hair syndrome; dry, brittle ear wax; albino skin and hair coloring; and PKU. TABLE 11.1 lists some human traits that are inherited in simple Mendelian fashion.

A list like the one in TABLE 11.1 represents a stunning achievement because humans are uniquely difficult genetic subjects, for the following reasons: (1) People don't choose mates and produce offspring to satisfy a geneticist's curiosity. The investigator must search for existing subjects and matings that express traits of interest.

FIGURE 11.2

Alternative Ear Characteristics May Be Controlled by Single Genes.

Attached [**A**] versus unattached [**B**] earlobes may be due to alternative forms of a single gene. Other genes may cause ear pits [**C**] or hairy ears [**D**].

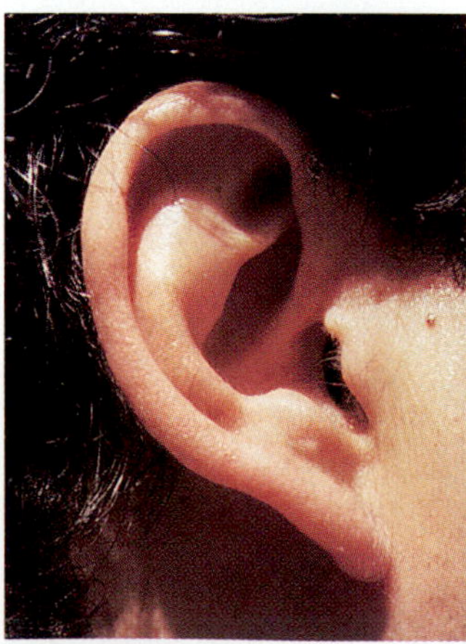

[**A**] Attached earlobe

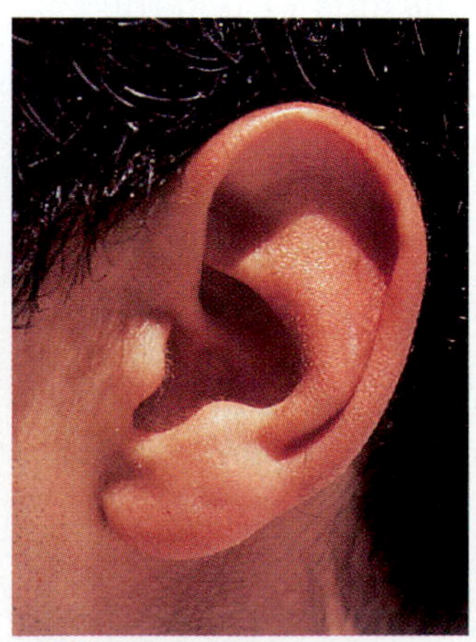

[**B**] Unattached earlobe

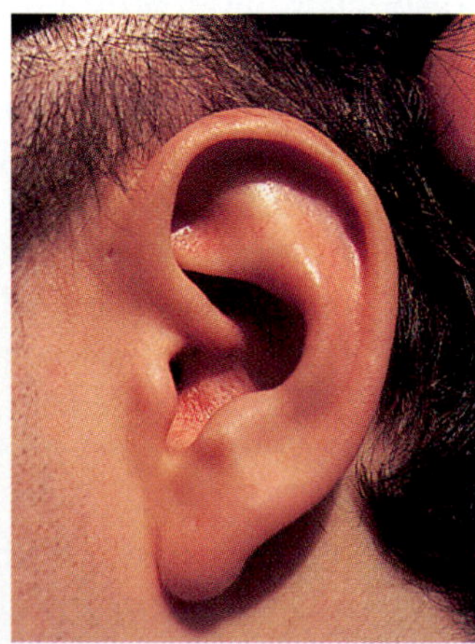

[**C**] Ear pit

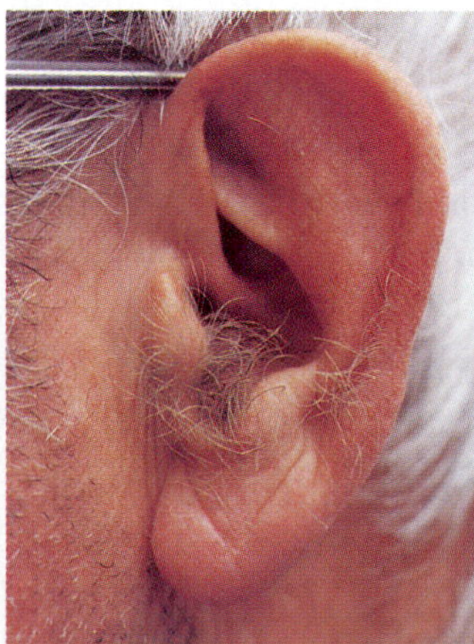

[**D**] Hairy ear

TABLE 11.1 Some Human Genetic Conditions

Disease	Effect
Recessive Allele on Autosomal Chromosome	
Albinism (chromosome 11)	Missing enzyme. Unpigmented skin, hair, and eyes
Cystic fibrosis (chromosome 7)	Defective membrane protein. Excessive mucus production; digestive and respiratory failure
Phenylketonuria (PKU) (chromosome 12)	Missing enzyme. Mental deficiency
Severe combined immunodeficiency (chromosome 20)	Missing enzyme. No immune response
Sickle-cell anemia (chromosome 11)	Abnormal (mutant) hemoglobin. Sickle-shaped red cells; anemia; blocked circulation
Tay-Sachs disease (chromosome 15)	Missing enzyme. Buildup of fatty deposits in brain; no mental development
Thalassemia (chromosome 16 or 11)	Abnormal hemoglobin. Anemia; bone and spleen enlargement
Recessive Allele on X Chromosome	
Duchenne muscular dystrophy	Missing membrane protein. Muscle cells die
Green weakness (color blindness)	Abnormal green-sensitive pigment in eyes. Inability to see the color green
Fragile *X* syndrome	Easily broken *X* chromosome. Mental retardation
Hemophilia	Missing clotting factor(s). Uncontrolled bleeding from wounds
Dominant Allele on Autosomal Chromosome	
Hypercholesterolemia (chromosome 5)	Missing protein that takes cholesterol from the blood. Heart attack by age 50
Huntington's disease (chromosome 4)	Progressive mental and neurological damage
Conditions Due to Variations in Chromosome Number	
Down syndrome (an extra chromosome 21)	Mental retardation. Heart abnormalities
Klinefelter syndrome (*XXY*)	Defect in sexual differentiation
Turner syndrome (*XO*)	Webbed neck. Sterility

(2) There is almost never a true F_2 generation available for study because brothers and sisters rarely mate. (Recall that the F_2 Mendel used was the result of crossing one member of the F_1 generation with another F_1—for example, a brother mated to a sister.) (3) A given couple seldom produces more than 10 children, and usually fewer than three. Therefore, sampling populations are too small to accomplish meaningful statistical analyses. (4) It could take more than half of a geneticist's career to follow the traits in even a single human generation, not to mention the many generations needed for substantive analysis.

To determine whether an inherited condition is caused by one gene, many genes, or by a change in chromosome structure, human geneticists have had to rely on two traditional tools borrowed from genetic researchers who study other animals, plants, or microbes. The first tool is the analysis of family histories. The second is the study of chromosome variations. Let's look at each tool and the kinds of conditions it can uncover.

PEDIGREES: FAMILY GENETIC HISTORIES

Because the human life cycle is so long and the number of offspring in the average family is so small, a geneticist must follow the inheritance of a trait through all the members of an extended family. Fortunately, some families have kept detailed records for several generations, enabling a geneticist to gather and study past medical records. From such records, the investigator draws up **pedigrees**, orderly diagrams that show family relationships, birth order, gender, phenotype, and sometimes genotype of each family member. The researcher assigns each person in the pedigree an identifying number consisting of a roman numeral and an arabic number. The roman numeral indicates the person's generation in the extended family tree [FIGURE 11.3]; the arabic number indicates his or her position in that generation (the rows in the diagram).

FIGURE 11.3 diagrams a family with PKU expressed in

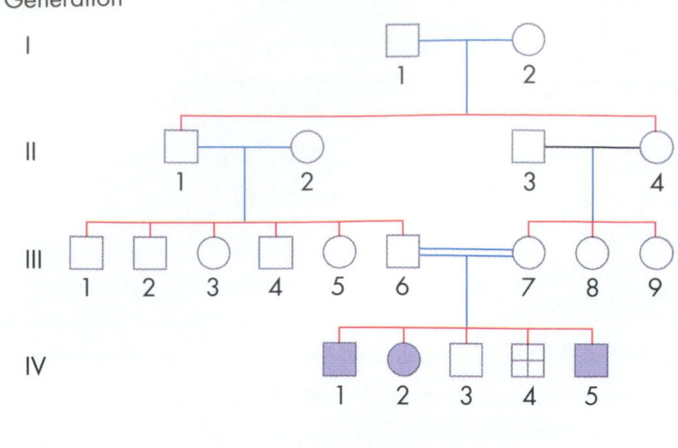

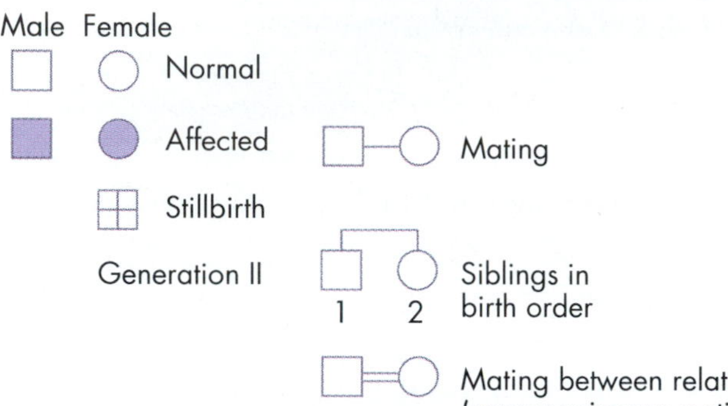

FIGURE 11.3

Pedigree of a Family with PKU.

A single horizontal line connecting a male and female represents a mating; a double line indicates a mating between relatives. A horizontal line above a series of symbols designates the siblings of one family in birth order from left to right. Vertical lines connect parents to offspring. Generations are labeled with roman numerals, and each individual in the pedigree is numbered. In this family, children with PKU (IV1, IV2, and IV5) had parents (III6 and III7) who were first cousins and phenotypically normal. Both parents were heterozygous for a recessive PKU allele inherited from I1 or I2; this would explain why some offspring were homozygous for the disease.

certain members. The only female with PKU in this pedigree is individual IV2 (generation IV, second person listed). Geneticists say that a person like IV2, who shows the trait in question, is *affected*; a person showing the more common normal trait is *nonaffected*. Analysis of pedigrees such as this one can tell a geneticist whether a given trait is dominant or recessive, whether a gene lies on a sex chromosome or on an autosome, and sometimes whether two genes are linked. (Recall from CHAPTER 8 that in a heterozygote, an individual with two different alleles of a particular gene, the dominant allele is the one that shows up in the phenotype—the organism's physical makeup. The recessive allele is hidden. Recall, too, that an autosome is a chromosome other than a sex chromosome—an *X* or a *Y* chromosome.)

INHERITANCE PATTERNS OF SOME GENETIC DISEASES

A pedigree can look rather formidable, with its rows of grandparents, parents, aunts, uncles, brothers, and sisters. Nevertheless, the rules for analyzing a pedigree follow Mendel's principles. FIGURE 11.4 depicts these simple rules as they apply to human traits. The following sections describe three such traits, with each trait showing a different pattern of Mendelian inheritance.

PKU: An Autosomal Recessive Most genetic conditions are caused by recessive alleles on autosomal chromosomes, and PKU is a classic example. Let's analyze the pedigree in FIGURE 11.3 to see how it represents this inheritance pattern. The pedigree shows that two brothers and one sister in the same family inherited PKU, but that neither parent showed the trait. Because the PKU trait is inherited, it must come from one or both of the parents, but since neither parent is affected, the PKU allele must have been hidden in one or both of the parents. By definition, then, a hidden trait like PKU must be recessive [review TABLE 8.1]. In general, if an offspring inherits a condition but neither parent shows the condition, the trait is recessive.

Now let's see if PKU is *X*-linked. A girl who is homozygous for a recessive condition like PKU must get one PKU allele from her father and one from her mother. If PKU were on the *X* chromosomes, then the *X* chromosome the girl got from her father would have to bear the allele that causes PKU. Thus, her father's only *X* chromosome (he is *XY*, remember) would bear the PKU-causing allele, and, significantly, *he would show the disease.* Since, as you just saw, neither parent shows the disease, the PKU-causing allele cannot reside on the *X* chromosome. Therefore it must be autosomal. In general, for a recessive condition, if an affected *girl's* father does not show the condition, then the gene for the condition lies on an autosomal chromosome. In a pedigree like this, one cannot rule out the hypothesis that an allele is *X*-linked by looking only at affected boys. This is because affected boys could arise from unaffected parents if the trait is autosomal and both parents are heterozygous for the condition *or* if the trait is *X*-linked and the mother is heterozygous.

Because the PKU allele is recessive, the affected offspring must be homozygotes to show the trait. Since the parents pass on the gene but do *not* show its effects, they must be heterozygotes, or **carriers**. In carriers, the dominant normal allele masks their recessive mutant allele. Another prominent example of a condition inherited as

FIGURE 11.4

Inheritance of Human Traits Determined by a Single Gene.

[A] With dominant genes on an autosomal chromosome, all affected offspring have at least one affected parent. With one normal and one affected parent, half the offspring will be affected, half will be normal (1). For recessive genes on autosomal chromosomes, affected children usually have two unaffected heterozygous parents, and the ratio of offspring will be one normal to two carriers (normal phenotype) to one affected (2). **[B]** With recessive genes on an *X* chromosome, males are affected more often than females, and inheritance is from mother to son, not father to son. With a normal father and a carrier mother, half the daughters will be normal and half will be carriers, while half the sons will be normal and half will be affected.

[A] Is a trait dominant or recessive?

1. If a child has a rare trait and one parent is affected, the trait is usually dominant.

2. If a child has a rare trait but neither parent is affected, the trait is usually recessive.

[B] Is a trait *X*-linked or autosomal?

3. If a son inherits the trait from his father, it cannot be *X*-linked.

4. If all or nearly all the affected people are males, the trait may be *X*-linked.

5. If both males and females are affected about equally, the trait is probably autosomal.

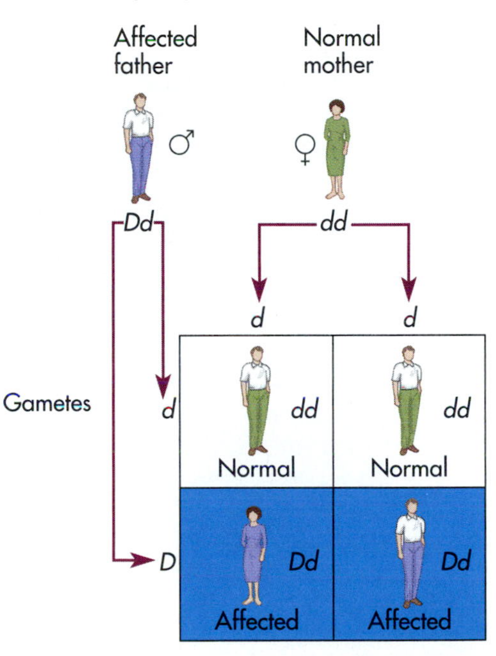

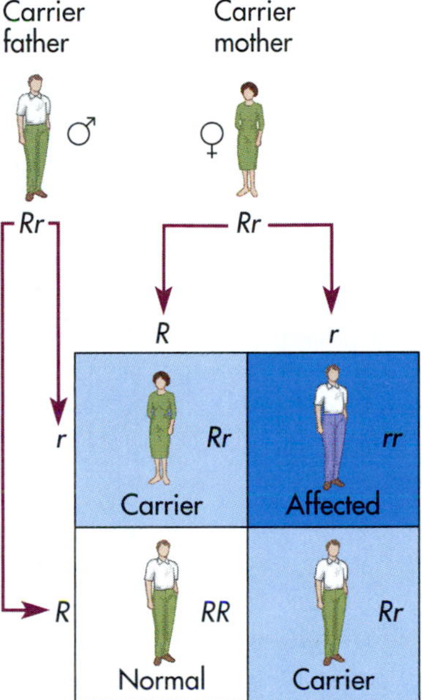

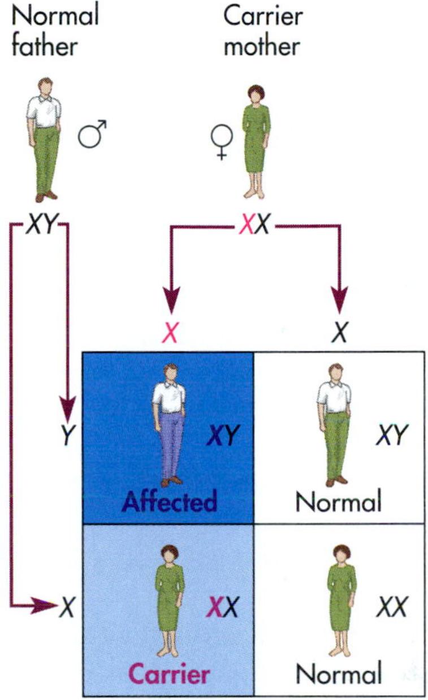

Offspring

Ratios: 1 normal : 1 affected — ♂ or ♀, ♂ or ♀

1 normal : 2 carriers : 1 affected — ♂ or ♀, ♂ or ♀, ♂ or ♀

1 normal : 1 carrier : 1 normal : 1 affected — ♀, ♀, ♂, ♂

an autosomal recessive is **albinism**, a genetic condition in which the homozygote's skin cells and hair follicles fail to produce the dark pigment melanin.

Duchenne Muscular Dystrophy: An *X*-Linked Recessive

An *X*-linked genetic disease is a condition in which the causative allele is on the *X* chromosome. The most common *X*-linked recessive genetic disease is **Duchenne muscular dystrophy** (**DMD**), a degenerative muscle condition that strikes 1 out of every 3500 boys.

The DMD gene is the largest one yet discovered, having over 2 million base pairs, due to its very long introns [see FIGURE 10.22]. Its immense size probably explains why this gene mutates more frequently perhaps than any other human gene, with a new mutation occurring about once every 20,000 births. The muscle cells of an affected boy die slowly, beginning in the legs. By age 5 the child cannot stand up easily, and he rises in a characteristic way [FIGURE 11.5]. At about age 20, the diaphragm muscles degenerate, and the affected person can no longer breathe.

Research based on recombinant DNA techniques [see CHAPTER 12] has shown that the normal allele of the DMD gene encodes a protein located on the plasma membrane of muscle cells. The protein is involved in regulating the entry of calcium ions into muscle cells. DMD patients lack this protein, and as a result, high levels of calcium

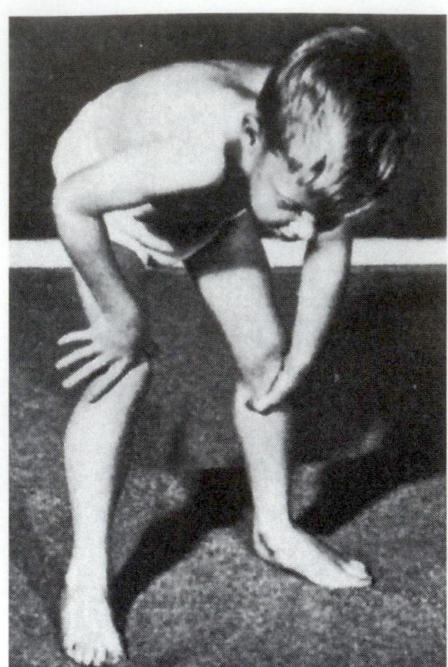

[A] How a boy with DMD rises

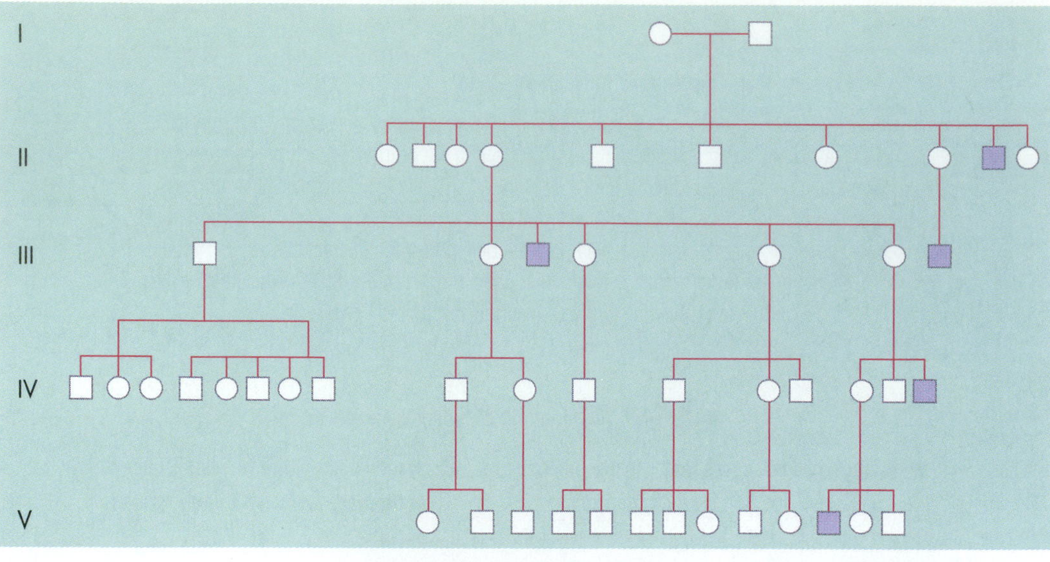

[B] Pedigree of DMD, an *X*-linked gene

FIGURE 11.5

Duchenne Muscular Dystrophy: Caused by a Recessive Allele on the *X* Chromosome.

[A] To rise from the floor, boys with this disease "climb up themselves," as this 5-year-old is doing, because their lower limbs grow weak before their upper bodies do. [B] A pedigree of a family with Duchenne muscular dystrophy shows the characteristic inheritance of an *X*-linked gene: Nearly all the affected people are male, and a son never inherits the condition from his father. To simplify the pedigree, the geneticist has not drawn in unaffected unrelated spouses.

ions enter muscle cells and cause them to degenerate. Based on this finding, physicians are planning to test drugs that might slow the entry of calcium ions into muscle cells, perhaps also slowing or preventing muscle degeneration.

In another approach in the fight against DMD, physicians took immature muscle cells from normal donors and injected them into the feet of three boys with the disease. These normal muscle cells fused with the patients' own defective muscle fibers, and the strength in their foot muscles increased. It is hoped that through drug therapy, muscle-cell injections, and other future discoveries, successful treatments will emerge for this debilitating recessive condition.

Because Duchenne muscular dystrophy lies on the *X* chromosome, it is passed to boys from their mothers, who are heterozygous carriers—the mothers transmit the trait but they don't show it themselves. Another *X*-linked recessive allele causes *hemophilia*, or bleeder's disease, in 1 out of every 5000 male births [FIGURE 11.6].

FIGURE 11.6

Hemophilia and the Royal Houses of Europe.

Nicholas, Czar of Russia (right), married Alexandra (left), granddaughter of Queen Victoria of England (1819–1901). Like Victoria, Alexandra was a heterozygote for hemophilia, or bleeder's disease, a disorder in which a person cannot produce a protein needed to bring about rapid blood clotting. Their son and heir to the throne, Alexis, inherited the hemophilia allele and suffered serious bleeding from even small bruises that would go unnoticed in a normal child. Desperate, Alexandra turned to Rasputin, a self-styled religious mystic who seemed able to put her son Alexis into a trance. The lack of movement may have helped the hemorrhages to heal, and Alexandra was convinced that Rasputin had wrought a miracle. Through Alexandra, the mystic gained influence over Czar Nicholas. The man who became prime minister after the czar's overthrow said, "If there had been no Rasputin, there would have been no Lenin." (Quote from R.K. Massie, *Nicholas and Alexandra*, New York: Dell, 1967, pp. 529–530.) A geneticist might add, "If there had been no mutation in Queen Victoria, there would have been no need for Rasputin."

Huntington's Disease: An Autosomal Dominant **Huntington's disease** is the result of a dominant allele on autosomal chromosome 4. In contrast to PKU and DMD, which appear in infants or young children, Huntington's disease does not manifest itself until ages 35 to 50. Researchers do not understand the physical basis for the disease or its timing, but they have chronicled a set of distressing symptoms that include progressive degeneration and death of nerve cells, irregular and jerky movements, intellectual deterioration, and often severe depression.

Pedigrees have shown that Huntington's disease is inherited as a dominant allele on an autosomal chromosome. For this reason, each child who will eventually be affected inherits the disease-causing allele directly from a parent who is, or will become, affected [FIGURE 11.7]. Whether afflicted or not, all the victim's children suffer years of anxiety because they have a 50/50 chance of inheriting the disease and because they worry about passing it on to their children. To make it worse, until recently, the children of an affected parent had to wait until middle age—usually *after* they had reproduced—to learn their fate. In the mid-1980s, however, molecular geneticists developed a way to detect the gene earlier in the life of a person whose parent has Huntington's disease—if, that is, they choose to know. We will discuss this difficult dilemma later in the chapter.

[A] Woody Guthrie

[B] Arlo Guthrie

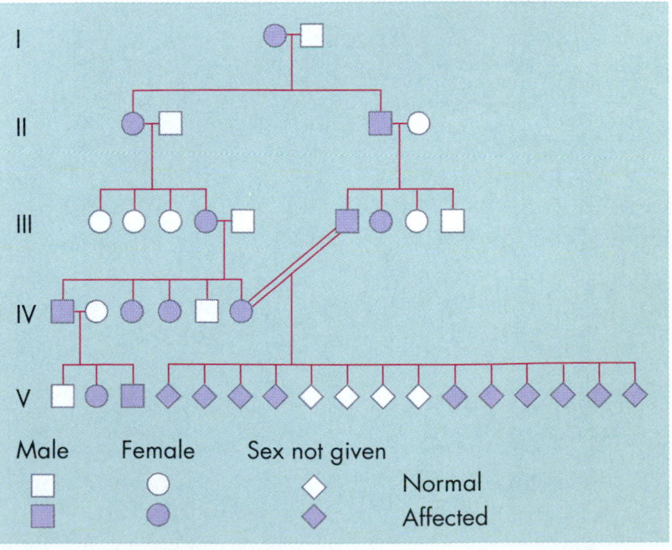

[C] Pedigree of a family with Huntington's disease

FIGURE 11.7

Huntington's Disease: Caused by a Dominant Allele on an Autosomal Chromosome.

[A] Woody Guthrie, singer and writer of "This Land Is Your Land," "So Long, It's Been Good to Know Ya," and other popular songs, died of Huntington's disease. **[B]** His son, Arlo Guthrie, a popular singer in the 1970s ("You Can Get Anything You Want at Alice's Restaurant" and "The Train They Call the City of New Orleans") had a 50/50 chance of inheriting the condition. **[C]** This pedigree of a family with Huntington's disease was collected by Nancy Wexler from a family in Venezuela in 1987. A quick look at the pedigree shows that every person with the trait has at least one parent with the trait. The 14 children of the marriage between cousins are shown as diamonds. For confidentiality, there is no indication of sex or birth order.

After years of work, researchers recently identified the portion of human DNA containing the Huntington's gene. Geneticist Marcy MacDonald worked with James Gusella, and both were amazed to find that the base sequence CAG was repeated over and over near the start of the gene. If you look in the genetic code in FIGURE 10.14 you will see that this repeat encodes the amino acid glutamine. Their unexpected finding was that normal people have from 11 to 34 repeats of this short sequence, but Huntington's patients have between 37 and 86 repeats. Stranger still, the greater the number of repeats, the earlier the disease strikes and the more severe it becomes. The number of repeats can increase still further when passed from father to offspring, and often the disease becomes more debilitating over the generations.

The researchers and the medical community in general hope this new knowledge of gene structure will soon lead to better treatments for Huntington's disease. Currently, however, there is no effective treatment for the disease, and the patient dies a protracted death. Physicians have tried one experimental treatment recently with *nerve growth factor*, a protein that stimulates cell division in nerve cells [see FIGURE 13.20]. The administration of nerve growth factor seems to prevent the typical brain cell deterioration seen in Huntington's. Researchers hope that nerve growth factor will continue to be accepted as an effective treatment for the disease.

CHROMOSOMAL ABNORMALITIES

Pedigrees provide the geneticist's first tool in studying human genetic disease. Another technique is the preparation and study of chromosomes using special dyes and labeling techniques. Called *karyotyping* (after the same root word for kernel, or nucleus, that occurs in eu*kary*ote), this technique enables geneticists to probe conditions caused not by single-gene mutations but by changes in chromosome number or structure.

Changes in Chromosome Number As you saw in CHAPTER 7, people normally have 46 chromosomes: 22 pairs of autosomes and 1 pair of sex chromosomes—*XX* in females and *XY* in males [FIGURE 11.8]. If a chromosome set deviates from that pattern, development is generally abnormal. Errors in the distribution of chromosomes during meiosis cause changes in the chromosome number. Homologous chromosomes normally separate during meiosis, but occasionally some chromosomes fail to separate. When the sex chromosomes are involved, the result can be that an egg, instead of containing the usual one *X* chromosome, contains two *X*s or none. This failure of paired chromosomes to separate during meiosis is called **nondisjunction**. Recall that three copies of autosomal chromosome 21 causes Down syndrome [see BOX 7.2, page 186].

People can inherit too many or too few sex chromosomes, with dramatic consequences. A person with one *X* and no *Y* chromosome (*XO*) is a sterile female with **Turner syndrome**. This condition is characterized by folds of skin along the neck, a low hairline at the nape of the neck, a shield-shaped chest, and failure to develop adult sexual characteristics at puberty.

About 1 newborn male in 1000 has two *X* chromosomes and one *Y* chromosome (*XXY*), a condition called **Klinefelter syndrome**. Affected people develop as sterile males with small testes, long legs and arms, and some-

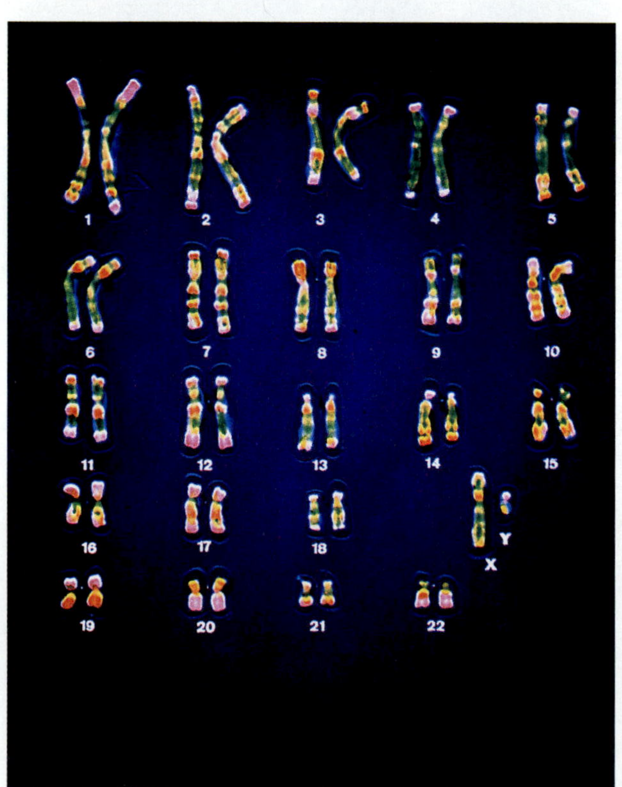

FIGURE 11.8

The Human Genome: A Karyotype.

To view human chromosomes, geneticists remove white blood cells, stain and photograph their nuclei, then cut the chromosomes from the photo with scissors and arrange them in pairs in decreasing size, as shown here. Geneticists have been able to localize many genetic abnormalities to specific chromosomes. Examples include Huntington's disease, chromosome 4; cystic fibrosis, chromosome 7; interferon deficiency, 9; albinism, sickle-cell anemia, beta-thalassemia, 11; phenylketonuria, 12; Prader-Willi syndrome, Tay-Sachs disease, 15; alpha-thalassemia, 16; severe combined immunodeficiency, 20; Alzheimer's disease, 21; color blindness, Duchenne muscular dystrophy, hemophilia, *X*; testis determining factor, *Y*.

what diminished verbal skills, although their IQ scores are near normal. Most men with Klinefelter syndrome manage well in society, and many are unaware of their chromosomal abnormality until they marry and are unable to father a child. If identified early, a Klinefelter male can be treated with male hormones in early adolescence to enhance his eventual sexual performance, prevent osteoporosis, and improve his mood and general sense of well-being.

Many boys who inherit one *X* and two *Y* chromosomes (*XYY*) grow to be over 6 feet tall and have below-average intelligence.

Curiously, only changes in the number of sex chromosomes and changes in the smallest autosomal chromosome (chromosome number 21) show up in relatively healthy human infants. Embryos that receive too many or too few of the other chromosomes have such disturbed development due to gene imbalance that they die in the womb or soon after birth. A child with trisomy 21 (Down syndrome) survives presumably because cells are not very sensitive to the exact amount of the products that come from the few genes on that small autosome.

Translocation: Exchange of Chromosome Parts Researchers have found that over 30 types of cancer are caused by chromosome **translocation**, instances in which a part of one chromosome moves to a new location on a different chromosome due either to an environmental factor, such as irradiation, or to unknown causes. In one such translocation common among Central Africans, the long arms of chromosomes 8 and 14 break in specific places, and the free ends exchange places [FIGURE 11.9A]. If the 8/14 translocation occurs in a certain type of white blood cell at any time during a person's life, it can lead to a cancer called *Burkitt's lymphoma* [FIGURE 11.9B]. This cancer can grow into an enormous jaw tumor within a matter of days. Fortunately, in about half the cases treatments with anticancer drugs rapidly shrink the tumor and lead to long-term survival.

When a translocation occurs between chromosomes 8 and 14, a cancer gene, or **oncogene**, called *myc* is relocated. In its normal spot on chromosome 8, *myc* encodes normal amounts of a protein that regulates cell growth. But in its new position on chromosome 14, it is influenced by its new neighbor, a gene rapidly transcribed in normal white blood cells. This influence causes *myc* itself to become rapidly transcribed, somehow causing the white blood cells to grow and reproduce out of control. This proliferation leads to cancer. Survivors of Burkitt's lymphoma can usually reproduce without fear of passing on the chromosome translocation to their offspring since the condition normally occurs in a white blood cell rather than in a germ cell (egg or sperm).

Fragile *X* Chromosome In the mid-1980s, researchers discovered a chromosome alteration that is second only to Down syndrome as a genetic cause of mental retardation. The condition is called **fragile *X* syndrome**, and a person with this condition has one or two *X* chromosomes whose tips break off easily at a specific place [FIGURE 11.10].

People with fragile *X* syndrome often have large, protruding ears and long faces, and as babies are hyperactive and make eye contact poorly. Geneticists were initially mystified by the inheritance of the fragile *X* syndrome because the mental retardation it causes seems to get worse as the generations go by. They recently discovered that a sequence of three bases (CGG) near the fragile site is normally repeated between 6 and 50 times, but in affected people the repetitions can increase to between 50 and 200. This fragile chromosome syndrome underscores the importance of regular chromosome structure for normal embryonic development.

FIGURE 11.9

Chromosome Translocation Can Turn On a Cancer Gene.

[A] In a reciprocal chromosome translocation, the tips of two chromosomes can exchange places. If the tip of chromosome 8 and the tip of chromosome 14 exchange in a white blood cell precursor, a growth control gene (an oncogene) from chromosome 8 comes to lie near a DNA sequence that specifies a high level of gene expression. This can cause the growth control gene to be expressed improperly, and the white blood cells divide without ceasing, causing a tumor called Burkitt's lymphoma. **[B]** A jaw tumor like this child's is a common symptom of Burkitt's, although such a tumor can also be triggered by a viral infection.

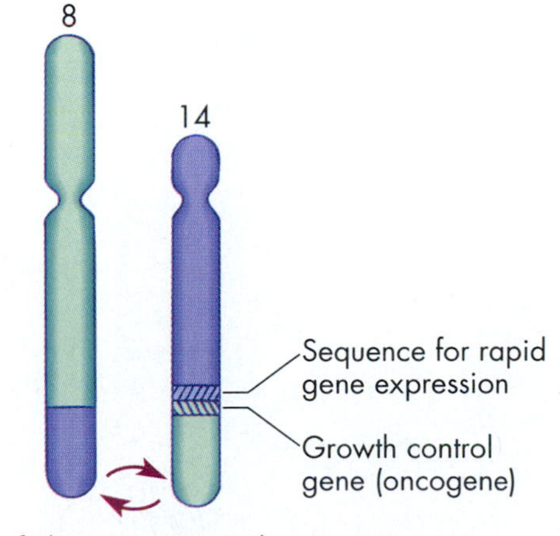

[A] Translocation between chromosomes 8 and 14

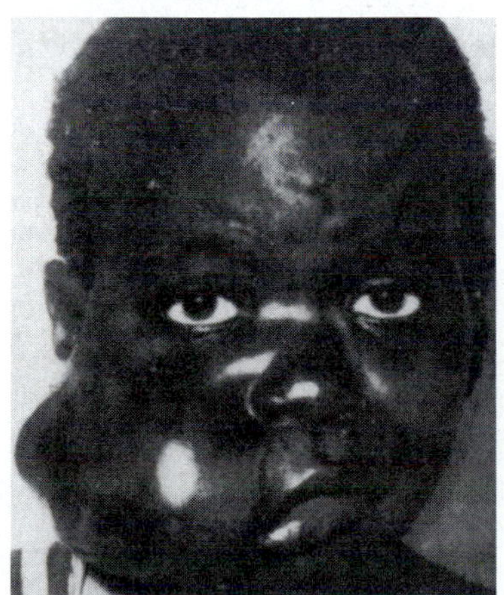

[B] A child with Burkitt's lymphoma

[A] Men with fragile *X* syndrome

[B] A broken *X* chromosome

FIGURE 11.10

Fragile *X* Chromosome and Mental Retardation.

[A] Three brothers who have fragile *X* syndrome each display the elongated face, jutting chin, squarish forehead, and large ears that characterize the condition. [B] The lower tip has broken off the right chromatid of this fragile *X* chromosome from a mentally impaired male.

FIGURE 11.11

Calico Cats and Mosaic Women.

A female mammal is a mosaic of cells containing maternal or paternal active *X* chromosomes. [A] Mosaicism in a calico cat's tricolored coat, with its patches of black and orange fur. [B] Women who are heterozygous for a *X*-linked gene that causes a condition called anhidrotic dysplasia have a mutant allele that prevents the development of sweat glands; thus, heterozygous females have some patches of skin that lack sweat glands and other patches that produce sweat glands normally. Different skin patterns occur even in identical twins and depend on the random activation and inactivation of *X* chromosomes that took place in the embryo.

[A] Calico cat

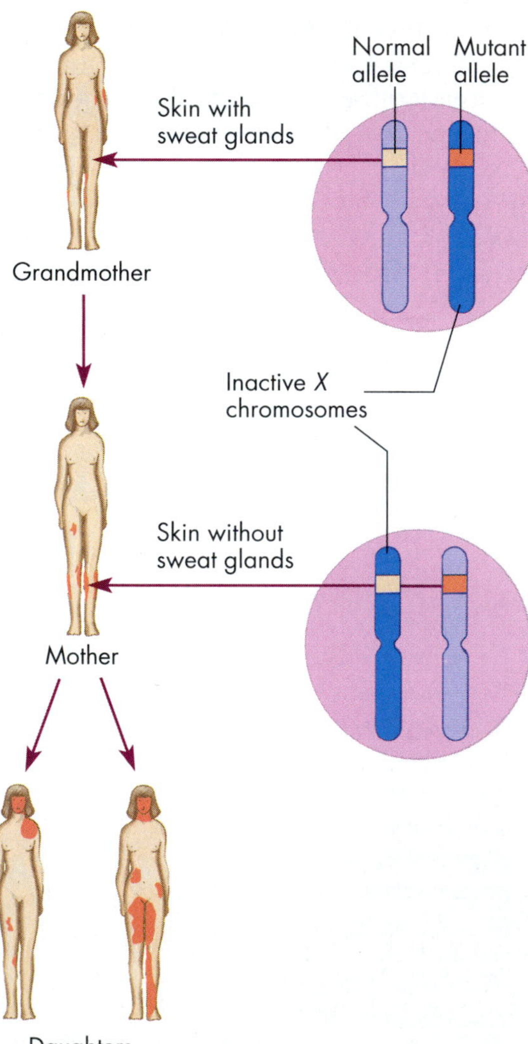

[B] Heterozygous people

X CHROMOSOME INACTIVATION

Human cells can tolerate extra *X* chromosomes because only one *X* chromosome is actually functional in any given human cell. As discovered by geneticist Mary Lyon, when a female embryo of a human or other mammal is just 2 weeks old and consists of only about 500 to 1000 cells, one of the *X* chromosomes in each cell becomes genetically inactive. By *genetically inactive*, Lyon means that this chromosome does not transcribe genes into RNA, and hence the genes on that particular chromosome can have no effect on the phenotype.

In any given female human cell at this early stage, either the mother's *X* chromosome or the father's *X* becomes inactivated—the other *X* remains genetically active. Once an *X* is inactivated in an embryonic cell, however, all of the thousands or millions of daughter cells derived from it will have the same inactive *X*. Thus, a female mammal is a mosaic of cells containing active *X* chromosomes of maternal or paternal origin. You can see this mosaicism in the patches of black and orange fur in a calico cat [FIGURE 11.11A], as well as in women with a certain skin condition [FIGURE 11.11B].

Women have similar mosaic patterns for various traits in each of their organs. A woman whose father carries an *X*-linked allele for color blindness, for example, will have patches of eye cells that see color normally and patches that are color-blind, depending on which *X* chromosome is inactivated in each group of cells.

Because of the phenomenon of *X* chromosome inactivation, the cells of both adult men and adult women have just one active *X* chromosome. Likewise, a person with Turner's syndrome (one *X*) has a single active *X*, and even a person with three *X* chromosomes in every body cell has just a single active *X* in any particular cell, and hence develops normally.

In a female with two *X*s (the typical number), a geneticist can see the inactive *X* chromosome by scraping cells from the insides of the mouth and looking for a small, dark spot on the edge of a nucleus. *XY* males, who have no inactive *X*, lack this dark spot in the nucleus. For several Olympic games before 1992, this procedure was used as a sex test for female athletes.

➤ CONCEPT CHALLENGE

Some members of a family have a certain very rare inherited condition. The father and three daughters are affected, but the mother and three sons are not affected. What modes of inheritance (dominant versus recessive, *X*-linked versus autosomal) can you rule out and why? With which modes are the data consistent? What is the most likely mode of inheritance?

Mapping Human Genes

Translocations and broken chromosomes exert their damaging effects partly because they rearrange the order of genes. Because a gene's location is so important, geneticists have tried to assign as many human genes as possible to precise locations along the 23 human chromosome pairs. To do this researchers have applied new methods first developed on nonhuman subjects to map, detect, and treat a much broader range of serious human genetic diseases than ever before.

In 1911, geneticists noticed the striking form of color blindness called *green weakness*, a condition in which males inherit a visual defect that prevents them from seeing greens, but allows them to see other colors normally [FIGURE 11.12]. Pedigrees showed that a man passes color blindness through his daughters to his grandsons. On the basis of these experiments, geneticists were able to assign green color blindness to the human *X* chromosome.

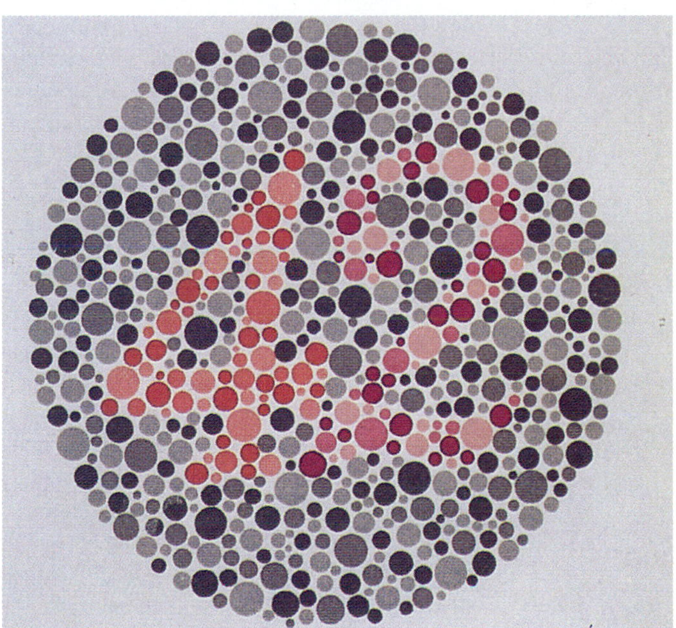

FIGURE 11.12

Color Blindness Test: What Number Do You See?

The first human gene that geneticists ever located was the gene for green weakness, which lies on the *X* chromosome and causes a partial color blindness. About 5 percent of males of northern European descent have this type of color blindness. The gene causes a decrease in the amount of a light-sensitive pigment in the eyes, and affected people see reds as reddish brown, bright greens as tan, and olive greens as brown. If you look at this plate and see only the number 4, you are green-blind. If you see only the number 2, you are red-blind. If you see the number 42, you have normal color vision. If the 2 is not very clear, you are the intermediate called green-weak. If the 4 is not clear, you are a red-weak intermediate.

box 11.1
Biology Applied

The Human Genome Project

In October of 1990, the journal *Science* included a large foldout wall chart printed with figures that resembled military decorations and candy-striped worms. The poster was, at that time, the most up-to-date map of the human genome. The bands of yellow, orange, blue, magenta, and white represented regions on blown-up, diagrammatic versions of the human's 22 chromosomes plus *X* and *Y* [FIGURE 1]. Within these bands, gene mappers had indicated the correct locations of hundreds of human genes, as well as some other genetic data then available.

In October 1992, *Science* printed another foldout genome map, this one less colorful but with many fewer blank spaces between charted areas. In just two years, researchers had mapped the major segments of two entire chromosomes: chromosome 21 and the *Y* chromosome. They had also pinpointed the exact locations of 2372 genes, one quarter of which are associated with specific human genetic diseases. Progress has been far faster than geneticists had predicted 10 years ago when the project began. This improved pace may help to quell the criticisms that surround the Human Genome Project, the largest coordinated scientific effort ever undertaken.

First suggested in 1985, the Human Genome Project, will cost more and take longer than building the atom bomb or landing people on the moon. By now, hundreds of researchers are laboring simultaneously in labs all over the world to do three things: They are trying to place the small percentage of currently known human genes in the proper positions on the 23 pairs of chromosomes. They are working to locate and identify the 50,000 to 100,000 additional genes (some 4400 of which occur on an average chromosome). And they are hoping to determine the nucleotide sequences of all 3 billion bases in the human DNA—even sequences that don't code for proteins or that simply repeat each other. Some observers have labeled the Human Genome Project the ultimate measure of humankind—an effort that could revolutionize medicine, biology, and psychology. But critics have wondered whether the benefits to society and individuals can possibly outweigh the tremendous costs in research funding and ethical dilemmas.

Scientific funding agencies will provide at least $3 billion for the mapping effort, and participants hope the task will be completed by the year 2005. The result will be a 500-volume encyclopedia of human genetics spelling out the detailed instructions for every protein in the body. This will provide the most complete view ever assembled of the amazing diversity of alleles within the human genome. It will also help researchers understand how each of our proteins functions routinely day to day and perhaps how each participates in mental and physical health or illness. While the view of genetic diversity and of normal protein functioning is profoundly important to our basic understanding of the human body, researchers and funding agencies have often focused on the explication of human genetic diseases as justification for the genome project's enormous price tag. Collectively, humans have more than 4000 hereditary diseases, including sickle-cell anemia, cystic fibrosis, and Huntington's disease. These three and a few hundred others can be traced to single-gene mutations. Most diseases, however, including cancer, diabetes, and heart disease, probably have roots in multiple genes, and the interactions between them will be challenging to work out, even with a genome map.

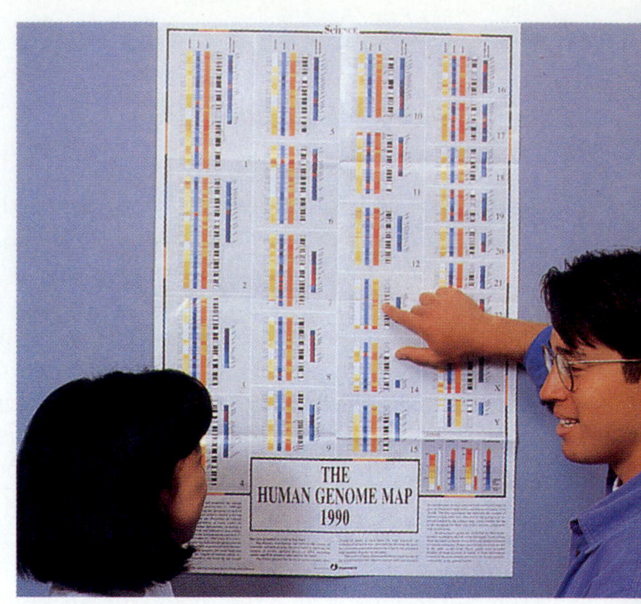

FIGURE 1
The Human Genome Map in 1990.

The physical mapping of chromosomes *Y* and 21 is a milepost, even though the two are the smallest human chromosomes and account for only 2 percent of the total genome. With them, gene mappers have successfully determined the order of large overlapping sections of the DNA that makes up each chromosome. They have also pinpointed 18 specific genes within the segments of chromosome *Y*, and 39 genes within the segments of chromosome 21, including those that contribute to Lou Gehrig's disease and to forms of Alzheimer's disease and epilepsy.

Some geneticists compare these segments to books on a library shelf, and for 21 and *Y*, all the books now shelved are in correct order. Researchers still can't decipher the locations or meanings of most phrases (genes) and letters (nucleotide sequences) within those volumes. At least, though, once an experimenter does discover which large segment contains a gene or base sequence of interest, he or she can get it straight off the "library shelf" (test tubes of refrigerated DNA, arranged in ordered segments).

To some critics of the Human Genome Project, a high-resolution view of our species' genome would be a mixed blessing. Knowing the complete nucleotide sequence, wrote Bernard Davis and colleagues, would be "like viewing a painting through a microscope," with researchers having to "plow through 1 to 2 million 'junk' bases before encountering an interesting sequence" and then discover its unknown function. Everyone from the harshest critics to Nobelist James Watson, who has been a leader of American gene mapping efforts from the outset, agrees that only 3 percent of human DNA codes for proteins of any kind, and most genetic problems will be concentrated in and near that 3 percent.

Other major criticisms of the giant project are financial and ethical. The last 15 years has seen major cuts in American biomedical research funding. Money for genome project work in any single year could provide sizable ($200,000) research grants to over 750 investigators. With a fully mapped genome, doctors, insurance companies, and employers would be able to "read" your inherited tendencies toward heart disease, cancer, or dementia, even though you have no symptoms now and may never develop them.

Despite these legitimate concerns, many scientists are convinced that the benefits of mapping our genome will far exceed the risks and point to the wealth of basic information already unlocked from DNA sequencing in viruses, bacteria, plants, and other animals. James Watson, probably the genome project's most eloquent spokesperson, sees it this way: "When finally interpreted, the genetic message encoded within our DNA molecules will provide the ultimate answers to the chemical underpinnings of human existence."

The traditional tools of pedigrees and karyotyping are so difficult, however, that it took another half century before geneticists could assign any gene to a specific autosomal chromosome.

Since 1970, thanks largely to three new techniques, geneticists have traced about 1000 human genes to specific places on sex and autosomal chromosomes. One of the methods used actually bypasses sexual reproduction, and the other uses DNA fragments called restriction fragment length polymorphisms, or RFLPs. Both techniques would have amazed Gregor Mendel, and collectively they have been called the "new genetics." Let's look at two exciting new tools and see how the gene mapping they make possible is revolutionizing the study and treatment of genetic diseases.

MAPPING VARIATIONS IN DNA STRUCTURE

A new phase of research begins after a geneticist discovers a new mutation and determines whether it is dominant or recessive, autosomal or *X*-linked. The next thing to be learned is *where the mutant gene occurs on a particular chromosome.* Locating a gene on a map of the human genome [see BOX 11.1 on page 280] is like locating yourself on a road map. When driving through New England with its many towns and villages, it is easy to find your position precisely on a road map; you just identify your location relative to the nearest town. It is much harder to do that in a state like Nevada, however, with so few towns and such great distances between them. Until the mid-1980s, human geneticists were "touring Nevada": They had located very few *genetic markers*, or identifiable sites such as genes for particular diseases or for individual blood types, as reference points for placing new genes on the map.

To make a really good map, geneticists need a large number of genetically identifiable sites, and they found these in the structure of DNA itself. They knew that the order of bases is extremely similar in DNA from corresponding places on homologous chromosomes. An example would be the identical region on chromosome 13 from two different people. These sequences, however, occasionally differ in a single base pair, and geneticists figured out a way to use these differences as identified sites to map DNA. As you will see in greater detail in the next chapter, specialized enzymes called *restriction enzymes* recognize and cleave DNA molecules at specific base sequences. When the enzymes make these cuts in DNA, they produce DNA fragments, which are called *restriction fragments.* One person's DNA may be cut at a specific place on a given chromosome by a certain enzyme, but another person's may not be, due to a difference in base sequence in the two DNAs. [FIGURE 11.13].

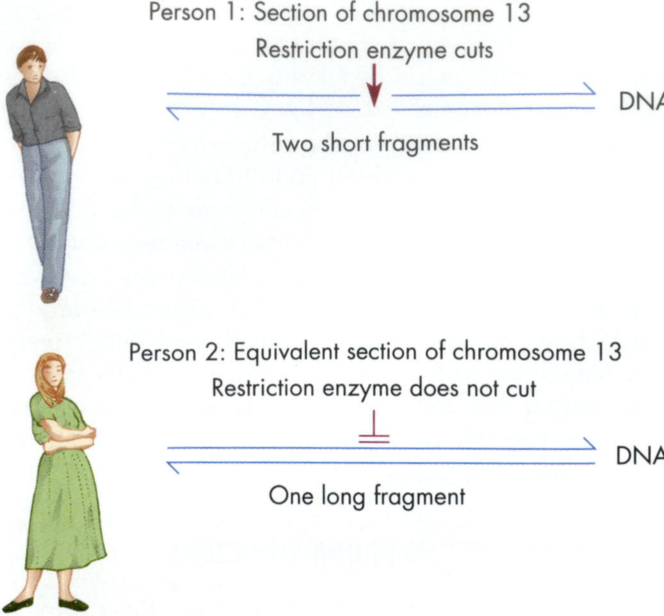

FIGURE 11.13

The Principle of RFLPs.

RFLPs (restriction fragment length polymorphisms) get their name in the following way: Enzymes called *restriction enzymes* cleave long strands of DNA into *fragments* by cutting at specific sites in the DNA's base sequence. Because of variations in DNA sequence, different people's fragments can be of different sizes, and geneticists call these size differences *length polymorphisms*. RFLPs are due to mutations in the DNA sites that are cut by restriction enzymes. These different fragment sizes can be used to help detect or predict diseases caused by mutated genes that contain cutting sites or that lie near cutting sites.

A genetic difference such as this one, where an enzyme cuts the chromosomes at different sites, is called a *polymorphism* (*poly*, "many"; *morph*, "form" or "shape"). This change in cutting sites in DNAs is a bit like cutting a pair of 1-foot-long strings at different places: You get fragments of different lengths. These fragments are inherited according to Mendel's rules and can be used as genetic markers. Such DNA variations are called **restriction fragment length polymorphisms**, or **RFLPs** (pronounced "riflips"), and they are quite frequent in human populations. RFLPs greatly expanded the number of genetic markers for establishing reference points ("familiar towns") on the human gene map.

The more precise mapping made possible by using RFLPs has been significant in two ways. First, it allows a genetic disease like PKU to be located on the map near a particular RFLP. Since researchers can identify a RFLP in the tiny bit of DNA obtained from just a few cells, they can tell with some certainty whether a particular family member is at risk for PKU. With this detection system, even unborn fetuses can be identified as either carriers of the gene or potential victims of the disease. (A technique similar to RFLPs has also been used to take DNA "fingerprints" and to solve crimes and cases of disputed paternity.)

A well-defined genetic map has a second significance: It helps a molecular geneticist to clone a desired gene—that is, to use recombinant DNA methods [see CHAPTER 12] to isolate, purify, make copies of, and analyze a gene for a given disease. By knowing exactly where the gene lies on a particular chromosome, researchers can isolate it more easily in a test tube. Once they analyze the isolated disease-causing gene, medical researchers can devise new treatments to help victims of the disease. Recall that researchers studying muscular dystrophy and cystic fibrosis were able to isolate and analyze the responsible genes. This allowed them to determine the structure and functions of the abnormal proteins and to design possible new therapies. FIGURE 11.14 shows a map of the *X* chromosome pinpointing the location of various human diseases, including PKU, hemophilia, Duchenne muscular dystrophy, dyslexia, cystic fibrosis, Lesch-Nyhan syndrome (a movement disorder that includes uncontrollable arm and leg spasms), a certain type of adult-onset diabetes, and other genetic conditions. As BOX 11.1 (page 280) explains, geneticists have already located more than 2300 functional genes, and they plan to finish mapping all of our 50,000 to 100,000 human genes within the next decade.

MAPPING GENES ON CHROMOSOMES

By cloning a specific RFLP, a geneticist can generate a test tube full of DNA fragments with a defined base sequence lying within or near a gene of interest. Researchers have developed techniques for directly viewing where that gene might reside on a chromosome. They can label the stretch of DNA in the test tube with a fluorescent dye to create a *probe*, a colored tag for a specific gene. If an experimenter labels a DNA stretch in this way, and then incubates the probe with a set of chromosomes, the probe will stick (or *hybridize*) only to the gene that originally gave rise to the fragment in the test tube, since that's the only DNA with which it can readily form complementary base pairs. Geneticists call the use of a probe to find an interesting gene's location on a chromosome an *in situ hybridization* experiment (*situ* means "site" or "location"). In FIGURE 11.15, you can see the fluorescent signal from the probe showing up as a pair of brilliant yellow spots, one spot for each sister chromatid [review FIGURE 7.8]. In a recent in situ hybridization experiment, geneticists showed that a fragment of DNA that no one

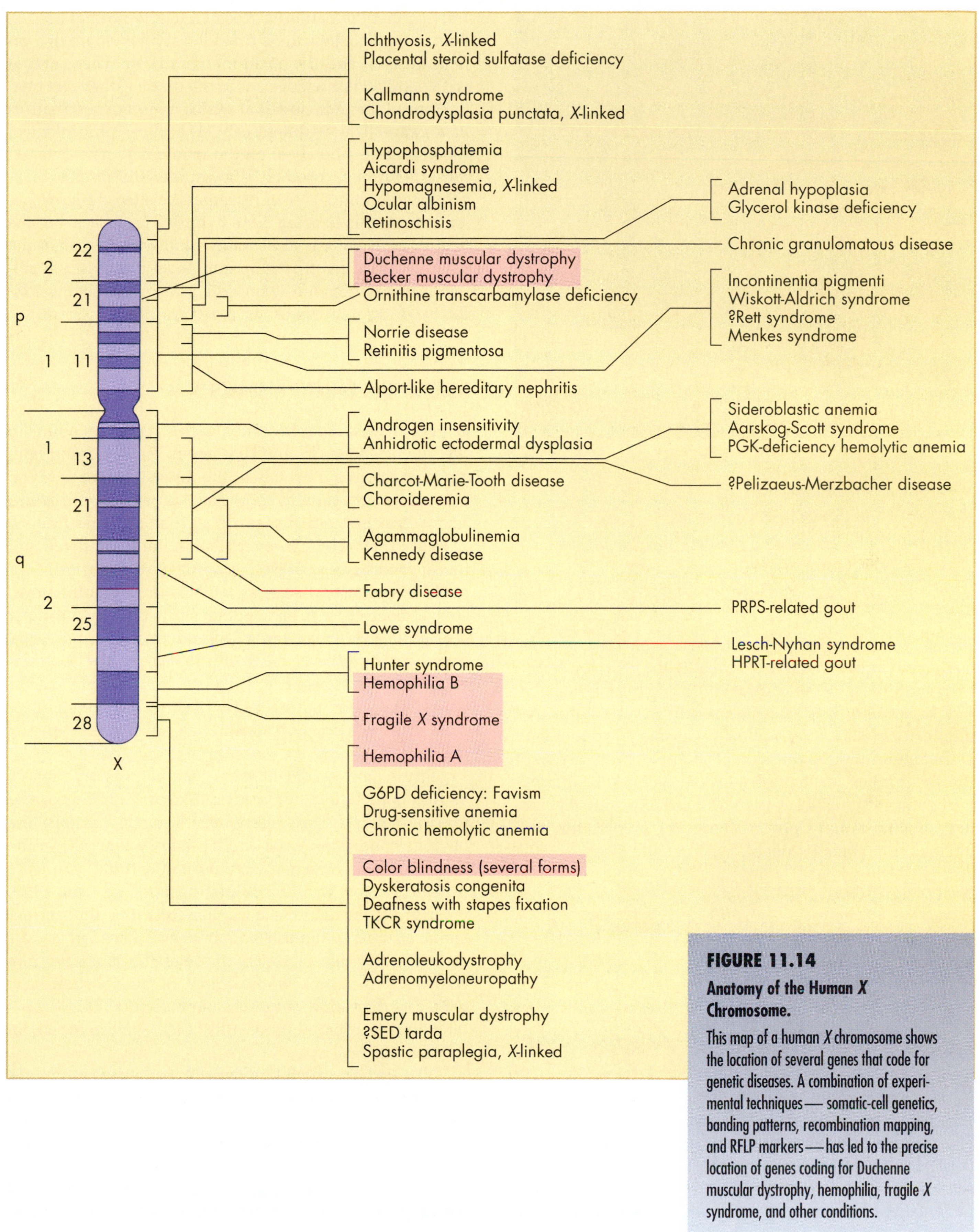

FIGURE 11.14

Anatomy of the Human *X* Chromosome.

This map of a human *X* chromosome shows the location of several genes that code for genetic diseases. A combination of experimental techniques—somatic-cell genetics, banding patterns, recombination mapping, and RFLP markers—has led to the precise location of genes coding for Duchenne muscular dystrophy, hemophilia, fragile *X* syndrome, and other conditions.

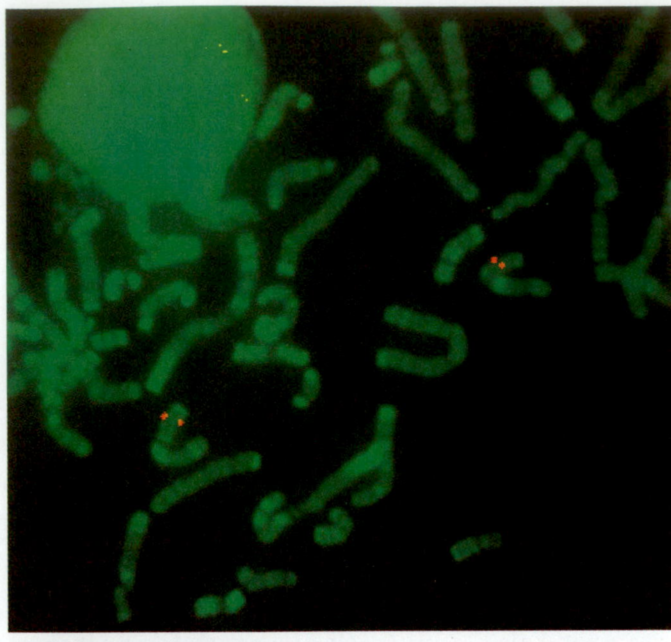

FIGURE 11.15

Mapping a Gene Directly on Its Chromosome.

The two bright yellow spots reveal that a gene for a certain immune system protein resides on human chromosome 19.

could previously identify is probably the gene for Angelman syndrome. We encountered this condition once before in BOX 8.1; it is characterized by severe motor and intellectual retardation, epilepsy, absence of speech, and an open-mouthed expression. The application of two of the techniques we have been discussing—in situ hybridization and RFLPs—is revolutionizing human genetics. Geneticists have never been better prepared to provide real therapies for people with genetic disease.

➤ CONCEPT CHALLENGE

Why do human geneticists find mapping genes to be useful?

Treating Genetic Diseases

Gene mapping and sophisticated tools allow modern physicians to do two things doctors only dreamed of a few decades ago: First, they can understand the entire pathway of a genetic disease—from mutant genes to absent or defective proteins to physiological effects on the patients. And second, they can often intervene successfully so that the patient can lead a nearly normal life. To help, the physicians must first identify the victim of a genetic disease and diagnose the condition. Then, armed with detailed knowledge about the disease, they can treat it at one of the three levels at which a mutant gene causes its damage. The treatment can: (1) address physiological repercussions (the effect on the whole patient); (2) target the mutant gene product (the protein that is absent or produced improperly); or (3) target the altered gene itself. At present, this is being done in only a few cases, but the technique is certain to become more widespread in the future. Physicians treat different genetic diseases at different levels, depending on which technique is most effective or, in some cases, currently available to them.

DIAGNOSING GENETIC DISEASES

Genetic diseases manifest themselves at various points in the human life cycle, and the time of onset is important, since physicians must detect the disease before they can treat it. Several genetic conditions appear only in adulthood: Huntington's disease is an obvious one [review page 275]. The list, however, also includes 10 percent of breast cancer cases, as well as adult-onset diabetes based on inherited tendencies, as well as less harmful traits, such as acne or hairy ears, which show up only during or after puberty. The majority of genetic traits, however, become noticeable in early childhood. As the body develops, the child begins to walk and talk, and physical or behavioral abnormalities begin to show. Duchenne muscular dystrophy is an example, as are certain types of mental retardation.

Physicians can detect some genetic traits as soon as a baby is born. Albinism, for example, or extra fingers and toes (polydactyly) are easily observed. And certain less obvious conditions, such as PKU can be detected in the hospital nursery. (If you were born after 1962, you were probably tested for PKU.) Blood from a baby with PKU, taken from the infant's heel, has 20 times the normal about of phenylalanine and can be identified as described in [FIGURE 11.16]. Early detection is crucial in a condition like PKU, where brain damage accumulates with each ingestion of more phenylalanine. This simple blood test has spared thousands of children from mental retardation.

The earliest possible diagnosis for other conditions, whether their onset is at birth, during childhood, or in adulthood, is critical. This enables the physician to prevent as much damage to the body as possible. Early detection is beneficial whether the effects manifest themselves as cancer, muscle or nerve deterioration, uncontrolled bleeding, or diabetes. The next step, of course, is treatment.

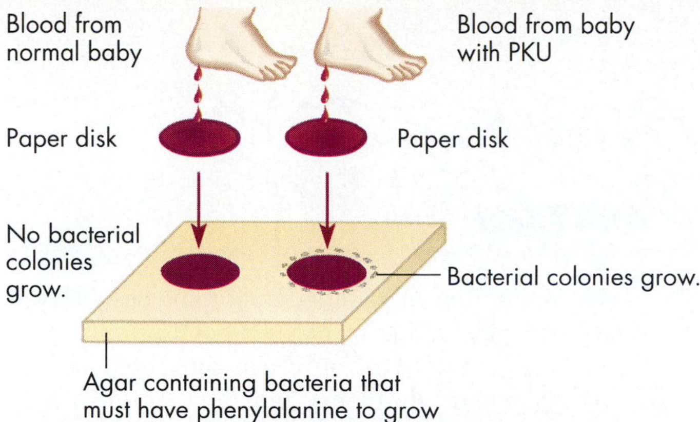

FIGURE 11.16

Test for Detecting PKU in Newborns.

A technician collects a tiny drop of blood from the newborn's heel, drops it onto a disk of filter paper, and places the disk on the surface of a block of agar containing bacteria that must have high levels of phenylalanine to grow. No colonies grow around a disk containing a normal baby's blood because of its low amount of phenylalanine; but because the blood of a baby with PKU has 20 times more phenylalanine than normal, bacterial colonies will grow in a halo around the disk.

PHYSIOLOGICAL THERAPY

A mutant gene can adversely affect the patient's entire physical health, and thus, many treatments intervene at the level of physiology. Doctors who identify a baby with two copies of the PKU allele, for example, prescribe a diet that prevents the buildup of phenylketones and hence protects the infant's brain. The diet is a monotonous but nutritious mush of dried milk powder (treated to remove phenylalanine), cornstarch, butter, fresh fruits, and vegetables (except beans and peas, which are high in phenylalanine). The baby and its parents walk a dietary tightrope, because the infant's caloric intake must be constantly high enough to prevent its body from breaking down its own proteins and thereby releasing phenylalanine from within. This high-calorie diet must be initiated within the first 2 or 3 months of life if the baby with PKU is to develop nearly normal intelligence. If treatment is delayed past 6 months, it is ineffective at preventing brain damage. The child must avoid phenylalanine completely until age 6 and to a lesser degree throughout life; hence the warning notice on cans of diet soft drinks that contain phenylalanine.

Another example of a physiological treatment for a genetic disease is giving cystic fibrosis patients an enzyme that loosens up the mucus that accumulates in their lungs [see CHAPTER 10].

PROTEIN THERAPY

Some genetic diseases can be treated by giving patients the protein they cannot make. In hemophilia, for example, the gene mutation blocks the synthesis of a clotting protein; thus, the blood fails to clot normally [review FIGURE 11.6]. However, after receiving blood transfusions that contain this clotting factor, a hemophiliac is able to form blood clots and is relatively safe from bleeding to death. A pituitary dwarf remains short because he or she lacks the gene for the protein growth hormone. Injections of that hormone will stimulate growth and allow the abnormally short child to partially "catch up" with his or her classmates. Likewise, some diabetics lack the gene for normal insulin production but can respond to injections of insulin. Protein therapies such as these can work as long as the needed protein is carried by the blood, but, unfortunately, this is true for only a few diseases. Most proteins must be positioned in a specific organ—for example, the PAH enzyme in cell membranes in the liver—and so injecting PAH into a PKU victim's blood serves no purpose. Ultimately, gene therapy may be the answer.

GENE THERAPY

Treating a genetic disease at the physiological or protein level may help the patient's symptoms, but it leaves the defective gene unchanged. With gene therapy, the physician could substitute a normal gene for a defective one. The normal gene would encode the normal protein, thereby allowing the person's body to function in a healthy way.

Physicians first used gene therapy in 1990 to cure a 4-year-old girl. "Linda" (not her real name) has an inherited condition called *severe combined immunodeficiency disease* (*SCID*). Patients with SCID cannot make a specific enzyme called adenosine deaminase (ADA) needed by white blood cells to mount an effective immune response and thus protect the body from bacterial and viral infections. Children who lack the ADA enzyme usually contract and die from pneumonia or influenza, and can survive for extended periods only if raised in a sterile chamber, fed sterilized food, and protected from physical contact with everyone, even their parents.

To help Linda, geneticists from the National Heart, Lung, and Blood Institute withdrew some of the child's genetically defective white blood cells. Into these cells, they inserted a normal copy of the gene for ADA. Just 4 hours after governmental regulatory agencies had approved their petition to perform the first gene therapy, the physicians were slowly transfusing the genetically engineered cells into Linda's blood. For her part, the girl sat watching with a mixture of hope and fascination as a

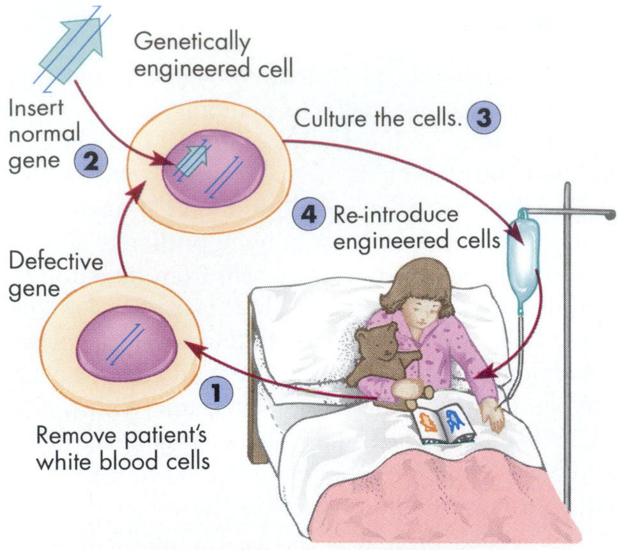

FIGURE 11.17

Immune Deficiency and the First Trial of Human Gene Therapy.

This sketch of a 4-year-old little girl traces the steps of the first attempt to use gene therapy to treat an inherited disease. The child cannot fight off infections because her white blood cells lack a certain enzyme. Researchers at the National Institutes of Health in Bethesda, Maryland, took special white blood cells from the little girl (Step 1) and inserted into them the normal gene that codes for the enzyme she lacks (Step 2). Next they cultured the genetically engineered cells and raised billions of them (Step 3). They then reinserted these billions of engineered cells into the little girl (Step 4). The cells do in fact make the missing enzyme and two years later, the little girl is able to carry on a nearly normal life.

bluish liquid drained from a plastic bag through a tube into her left hand [FIGURE 11.17]. In less than a week, she was able to leave the hospital. She is 6 years old now, and no longer homebound. Linda was able to finish kindergarten in a public school, and can take dancing and swimming lessons and go swimming with her friends. Her life is nearly normal now, except for occasional trips to the National Institutes of Health in Bethesda, Maryland, where Linda receives additional regular infusions of genetically modified cells. She must have repeated treatments because white blood cells normally have a lifetime of only a few months. Linda's case has greatly encouraged the families of children with genetic diseases, as well as the scientific community at large. The kinds of techniques the researchers tried may eventually lead to effective gene therapy to correct many more genetic disorders at a fundamental level.

Preventing Genetic Diseases

The recent revolution in genetic techniques has allowed scientists and physicians to carry detection and treatment one step further: Not only can they help people with existing genetic problems, but they can also help prevent the conception and birth of children doomed to a lifetime of suffering. They do this by detecting deleterious alleles in potential parents and in the unborn and by counseling the parents as to their options.

GENETIC COUNSELING

Genetic counselors provide information to families at risk for transmitting genetic defects. They acquaint the prospective parents with the manifestations of the disease they may pass along, with the potential suffering of an affected child, and with the emotional and financial impact on parents and siblings. Counselors can assess the couple's probability of having an afflicted child, help them consider the available options, and assist in their acceptance of the difficult reproductive decisions they must make.

DETECTING CARRIERS

For decades, geneticists have been able to identify carriers of defective alleles through pedigrees: Children who are homozygous recessive for a trait like PKU have parents who are heterozygous carriers. This approach, however, has predictive value only for the additional children of those same sets of parents. Sensitive biochemical analysis can now reveal the presence of a defective enzyme in unaffected carriers of some conditions. A heterozygote for PKU or Tay-Sachs disease, for example, has just one allele, rather than two, making the normal enzyme. A heterozygote, therefore, has less enzyme than a person with two normal alleles. In this way, the person can be shown to be a carrier even if he or she hasn't yet produced children. These techniques won't work for some genetic ailments, particularly those whose defective protein has yet to be identified. However, RFLPs closely linked to the gene of interest can sometimes be used instead; carriers for Duchenne muscular dystrophy and Huntington's, for example, are detected this way.

PRENATAL DIAGNOSIS

Tremendous progress has been made in the area of diagnosing genetic disorders in fetal cells before birth.

Amniocentesis In the mid-1960s, geneticists pioneered a method for collecting fetal cells called **amniocentesis** (AM-nee-oh-sen-TEE-siss). Today, this procedure is frequently used for most pregnant women over 35 or for those with family histories of genetic disease [FIGURE 11.18A]. In amniocentesis, the physician inserts a needle through the mother's abdominal and uterine walls and removes a few milliliters of the amniotic fluid that surrounds and cushions the fetus. This fluid contains sloughed-off fetal skin cells, which are collected and grown in the laboratory, then tested for defective chromosomes or genes. In these cells, certain defective genes can be detected by the absence of normal enzymes, such as the enzyme needed to prevent Tay-Sachs disease [see TABLE 11.1]. Over 100 genetic abnormalities can be identified through amniocentesis, but the technique does have a drawback: It cannot be accomplished before 14 to 20 weeks of pregnancy. By that time, the mother may have already felt the fetus move and may find it emotionally difficult to consider an abortion, even if test results reveal massive genetic defects.

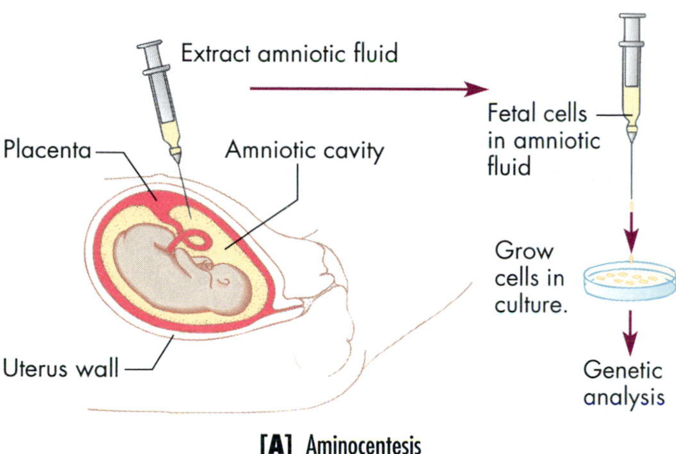

[A] Aminocentesis

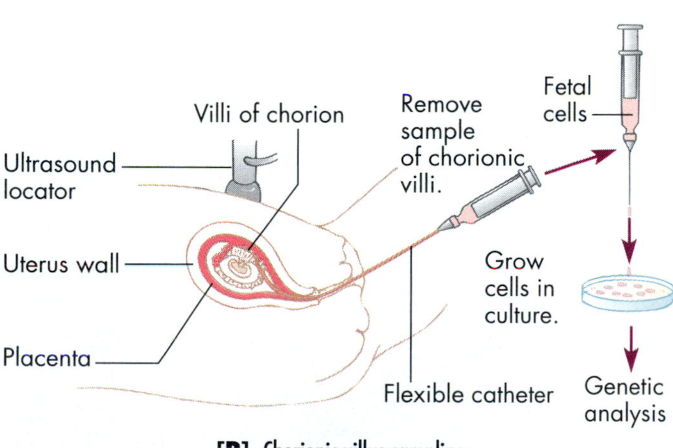

[B] Chorionic villus sampling

FIGURE 11.18

Prenatal Diagnosis of Defective Genes and Chromosomes.

As the text explains, the differences between **[A]** amniocentesis and **[B]** chorionic villus sampling lie in when and how the fetal cells are removed. In both techniques, the cells are cultured and the chromosomes analyzed for genetic abnormalities.

Chorionic Villus Sampling In a newer procedure called **chorionic villus sampling**, the physician removes fetal cells from the developing placenta, the organ that nourishes the fetus [FIGURE 11.18B]. The physician can then examine the fetal chromosomes and study the DNA directly. The mutation that causes sickle-cell disease, for example, eliminates a restriction-enzyme cutting site present in normal DNA. Thus, a researcher can detect the sickle-cell allele through direct DNA analysis with 100 percent accuracy. This technique can also detect fetuses with PKU or Huntington's genes, but with somewhat less accuracy. Geneticists predict that soon they will be able to detect most common genetic diseases in the unborn on the basis of RFLPs. Unfortunately, treating fetuses for these gene defects is a more distant prospect. It may someday be possible to replace defective fetal genes with normal DNA sequences and thus cure the baby's condition before birth. For now, however, the options are limited mainly to detecting carriers and providing genetic counseling.

Choosing Disease-Free Gametes As you know from CHAPTER 8, heterozygous carriers do not themselves exhibit disease symptoms because the normal gene copy masks the defective copy. If, during conception, carriers could somehow donate only eggs or sperm containing the normal gene copy (not the defective one), they would not transmit the disease to their offspring. Recent experiments suggest that this selectivity may one day be possible, beginning with female carriers. A newly ovulated egg has completed only the first meiotic division of meiosis [review FIGURE 7.19] and the huge egg remains attached to its small sister cell (called the *polar body*). Because homologous chromosomes separate during the first meiotic division, only one of the two cells—either the egg or the polar body—gets the homologue with the defective gene (except for infrequent cases of crossing-over). Thus, if the polar body contains that defective copy, then the egg must have the normal copy. Researchers in Chicago devised a way to remove the polar body without injuring the egg and to test its DNA. They performed this test on eight egg–polar body pairs from a female carrier of a genetic disease and determined that two of her eggs were free of the defective gene. They combined these with her husband's sperm and reimplanted the eggs into her uterus.

If this sort of procedure becomes more generally available to both men and women, it could allow the healthy carriers of genetic disease to produce disease-free children, For now, however, prospective parents who know they carry defective genes face hard choices. So do the parents of a fetus revealed as defective through prenatal diagnostic techniques. Genetic counselors try to assist people with these difficult dilemmas, including helping prospective parents prepare emotionally and financially to support a child with special needs.

NEW DILEMMAS

The recent revolution in genetic techniques has given us a newfound ability to map human genes, to find markers for specific diseases, and to screen unborn fetuses, children, and adults for various genetic conditions. While there is great enthusiasm in many quarters over the potential these techniques present for studying the genetic roots of many diseases, there is also considerable caution.

People at risk for Huntington's disease, for example, can now be tested to learn whether they carry the dominant harmful allele (and will inevitably develop the degenerative condition, since there is no cure). Yet many of them have little desire to know their fate in advance. For example, Nancy Wexler [FIGURE 11.19], a prominent Huntington's researcher who is herself the daughter of a woman who died from Huntington's, has a 50 percent chance of having the disease. She prefers, however, to live life without knowing her genetic status. As she explains it, without this knowledge, she doesn't have to worry that the disease is coming on "every time I knock over a cup of tea." Turning the question around on the listener, she asks, "Do you want to know how and when you are going to die, especially if you have no power to change the outcome?"

Some bioethicists point to other potential problems stemming from the new techniques for genetic screening and prediction: They fear that insurance companies and employers will require in-depth genetic screening for new customers or job seekers and may reject anyone with a gene that merely raises the *probability* of *someday* developing heart disease, cancer, or other conditions. Some activists worry that industrial firms will select for (even someday engineer people for) the ability to withstand occupational exposures rather than cleaning up their factories. Some social and religious groups are concerned that

FIGURE 11.19

Charting a Genetic Killer.

Nancy Wexler studies the extensive pedigree of a family with Huntington's disease. If you were in a family at risk for the disease, would you want to know if you had it?

detailed fetal screening will result in more abortions, since at present, most genetic defects that can be detected cannot be corrected or, in some cases, even treated effectively.

Some observers of the revolution in genetic screening have proposed the creation of new panels, laws, and rules. These, they say, should govern and probably restrict the speed and extent to which insurance companies, employers, and individuals can make use of detailed genetic information. Some geneticists, however, feel that existing bodies, such as the Recombinant DNA Advisory Committee of the National Institutes of Health, are already sufficient to meet the ethical and legal concerns posed by the rapid expansion of genetic screening. People on all sides of the issue seem to agree that widespread public awareness of the issues is essential to prevent abuses.

➤ CONCEPT CHALLENGE

Let's say that one of your parents had a dominant allele that gives a 75 percent chance of developing a certain type of cancer between the ages of 35 and 65. What is the likelihood you have inherited the cancer allele? If a DNA-based test became available tomorrow that could tell with 97 percent certainty whether or not you have the cancer allele, would you have the test? Why or why not? If you had the test and were found to have the cancer allele, would you tell your boyfriend or girlfriend? Would you tell your insurance company? If you ran a business that paid for your employees' health costs, would you want to screen potential employees for this allele before hiring them? How do you think our society should handle questions like these?

Connections

The principles of human genetics, initially founded on our understanding of bacteria, fruit flies, and other simple organisms, have modified the study of our own species and have opened a new era in genetics. They have allowed scientists to detect several genetic disorders early enough to prevent organic damage and to predict several others so that if parents wish, they can prevent the birth of an infant doomed to suffer and perhaps die prematurely. A solution favored by many would be to eliminate gametes bearing defective genes by screening and removing them so that they can't participate in fertilization—an approach now in the experimental stage. Perhaps before such manipulation becomes possible, however, geneticists may learn how to alter the genotype of a defective fetus in the womb or to influence the way its phenotype unfolds so as to prevent the mutant allele from exerting deleterious effects. Some of this knowledge may come from the young field of biotechnology—the application of the principles of genetics and molecular and cellular biology to solving human problems. Our next chapter discusses this fast-moving area of biological science.

CONCEPT MAPPING

Draw a concept map relating the following terms. Link the concepts with arrows and a connecting word or short phrase.

amniocentesis
chromosome translocation
dominant gene
genetic counselor
genetic mapping
identifying carriers
karyotype
pedigree
recessive gene

KEY TERMS

albinism, 273
amniocentesis, 287
carrier, 272
chorionic villus sampling, 287
chromosome translocation, 277
fragile *X* syndrome, 277
Klinefelter syndrome, 276
oncogene, 277
pedigree, 271
RFLP, 282
Turner syndrome, 276

HIGHLIGHTS IN REVIEW

1 The genetic principles that apply to peas and fruit flies also apply to humans.

2 By analyzing pedigrees, geneticists can often determine whether a trait is *X*-linked or autosomal, and whether it is dominant or recessive.

- a] PKU and albinism are due to recessive alleles on autosomal chromosomes. Affected people are homozygotes, and their parents are heterozygous carriers.
- b] Duchenne muscular dystrophy and hemophilia are both caused by recessive alleles on the *X* chromosome. *X*-linked genes are passed to a son from a mother.
- c] Huntington's disease is due to an autosomal dominant allele; even a heterozygote with just one dose of the mutant allele shows symptoms of nerve degeneration in middle age.

3 Changes in chromosome number and structure can alter a person's phenotype.

4 Geneticists can now use variations in DNA structure to map human genes. Knowing a gene's location on a chromosome may help a geneticist isolate pure copies of the gene.

5 Analyzing cloned genes can help a geneticist determine the gene's function, and lead to improved medical diagnoses and treatments.

- a] At a given location, the same fragment of DNA from two different people may have slightly different base sequences. This difference can cause a restriction enzyme to cut one of the DNAs but not the other, leading to an RFLP. RFLPs can serve as markers that help locate and map human genes. The Huntington's disease gene was mapped this way, and RFLPs are being applied to the study of PKU, hemophilia, and other diseases.
- b] Some genetic diseases, like Huntington's, don't show up until adulthood. Others, like Duchenne muscular dystrophy, are apparent by early childhood and PKU can be detected at birth with a blood test.
- c] Early detection allows the physician to treat a genetic disease by physiological means (e.g., prescribing a strict phenylalanine-free diet for PKU victims), by protein therapy (e.g., providing blood-clotting factor to hemophiliacs), or by implanting normal genes into the cells of a person with genetic disease.

6 Ethical dilemmas surround our new ability to detect various human genetic conditions.

UNDERSTANDING THE FACTS AND CONCEPTS

For questions 1–5, match each of the descriptions below with the most appropriate item in the following list. In this group as well as in those that follow, any item may be used once, more than once, or not at all.

- a] Burkitt's lymphoma
- b] Duchenne muscular dystrophy
- c] PKU
- d] Huntington's disease
- e] Klinefelter syndrome

1 A fatal disease, most common in males, caused by an *X*-linked recessive allele that leads to elevated calcium concentration in muscle cells due to an abnormal protein in the cell membrane. Muscle cell death results.

2 Newborns routinely tested for this. Caused by a recessive allele that leads to the accumulation of a certain metabolite in body cells, especially in brain cells, resulting in mental retardation.

3 An autosomal dominant lethal trait with late age of onset. Brain cell degeneration, severe depression, jerky muscle movements, and nerve cell death are major symptoms.

4 Karyotypes show two *X* chromosomes and one *Y*. Restricted to males, who are sterile and long-limbed.

5 If pieces of chromosomes 8 and 14 undergo reciprocal translocation in a white blood cell precursor, any nearby oncogene will be stimulated to enhanced activity, resulting in uncontrolled white blood cell proliferation and cancer.

As above, for Questions 6–10, match the descriptions with the appropriate item from the following list.

- a] carrier
- b] karyotype
- c] amniocentesis
- d] restriction fragment
- e] RFLP

6 A procedure by which fetal cells are obtained from amniotic fluid.

7 An individual who is heterozygous for a recessive allele which may be passed on to another generation.

8 Homologous sections of DNA that differ in length between individuals who carry different alleles for the same particular gene.

9 A representation of a single cell's chromosome complement, usually arranged in homologous pairs.

10 A length of DNA from a chromosome that has been treated with one or more restriction enzymes.

As above, for Questions 11–14, match the descriptions with the appropriate item from the following list:

- a] nondisjunction
- b] reciprocal translocation
- c] chorionic villus sampling
- d] polygenic trait

11 A meiotic error in which a pair of homologous chromosomes fails to properly separate.

12 A technique whereby cells are removed from the placenta so that chromosomal and genetic analysis can be made.

13 Heritable traits, such as skin pigmentation and height, that are determined by several sets of alleles.

14 A process of mutual exchange of chromosomal segments between nonhomologous chromosomes.

As above, for Questions 15–19, match the description with the appropriate item from the following list of terms.

- a] hypervariable region
- b] DNA fingerprint
- c] probe
- d] oncogene
- e] in situ hybridization

15 A cancer-causing gene.

16 The pinpointing of a gene by a probe while the gene is within an intact chromosome.

17 Short regions of DNA consisting of base sequences that are repeated numerous times.

18 A single-strand length of nucleic acid with a known base sequence that is used to pinpoint the location of specific genes.

19 A pattern of restriction fragments obtained by electrophoresis that compares the hypervariable regions of the DNA from two individuals.

INTEGRATE AND APPLY WHAT YOU HAVE LEARNED

1 Distinguish between the uses and procedures of pedigree analysis, amniocentesis, and karyotyping. Which tests reveal definitive information, and which indicate only the probability of either having or of being a carrier for some specific mutation or heritable trait?

2 If individuals who carry an autosomal dominant lethal gene die as a consequence, how do you account for the fact that such lethal genes, like that for Huntington's disease, are perpetuated in a population?

3 How can a chromosomal translocation cause cancer?

4 What advantages and what pitfalls are likely to characterize the use of DNA fingerprinting as definitive evidence in courts of law?

5 Outline the major benefits and risks associated with the Human Genome Project. What is your stand on the issues that you present?

ANALYSIS

Answer the following multiple-choice questions.

1 What describes the genetic nature of the trait characterized in the pedigree shown in the accompanying figure?

- a] *X*-linked recessive
- b] *X*-linked dominant
- c] *Y*-linked gene
- d] Autosomal recessive
- e] Autosomal dominant

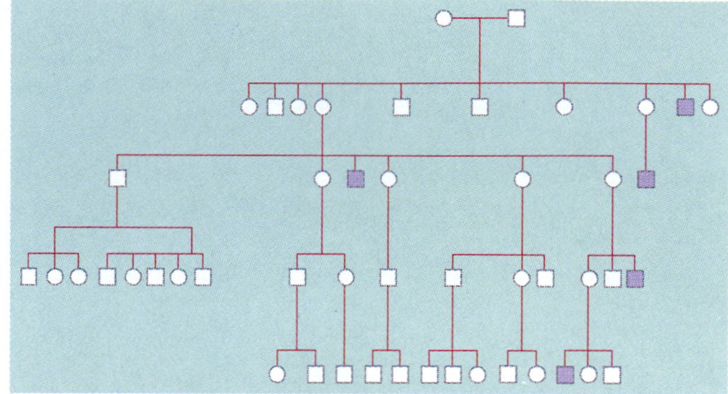

2 Which is the most likely statement about the two people in generation I of the pedigree in the illustration?

- a] They are both carriers.
- b] One is a carrier, the other may or may not be.
- c] Both are normal because the affected individuals could have mutations that arose spontaneously.
- d] Both are homozygous for the trait but for some reason neither shows it.
- e] The female is a carrier, the male is not.

3 Which of the following statements is correct?

- a] DNA fingerprints reflect natural differences in the DNA between individuals: RFLPs do not.
- b] DNA fingerprints are obtained by the use of restriction enzymes, whereas RFLPs do not require their use.
- c] Both an individual DNA fingerprint and RFLPs include the base sequence in all of the chromosomes within one cell.
- d] Both fingerprinting and mapping by means of RFLPs utilize maps of short sections of the DNA that vary greatly throughout the population.
- e] DNA fingerprinting compares the base sequence between two different samples of DNA, whereas mapping a mutant gene with RFLP involves correlating data from pedigrees with a particular RFLP pattern.

4 Which of the following abnormalities could not be detected by karyotyping? Give the reason for your answer.

- a] Down syndrome
- b] Turner syndrome
- c] Klinefelter syndrome
- d] Phenylketonuria
- e] Burkitt's lymphoma

CHAPTER 12

Biotechnology and Recombinant DNA

A 55-year-old man sees his bus pulling away from the curb and starts running after it. Suddenly he feels sharp pains in his chest, his fingers start to tingle, and his breathing becomes labored. He slumps to the ground, in the midst of a classic heart attack.

A tiny blood clot, lodged in a nearby blood vessel, has broken free and traveled to a vessel that supplies blood to the heart muscle. The clot has clogged the heart vessel so completely that muscle cells downstream from the clot no longer receive the oxygen they need, and a portion of the heart muscle begins to die.

Bystanders call an ambulance, and soon the man is en route to the hospital. What the paramedics and emergency room physicians need is a quick way to dissolve the clot before too much damage has been done. Fortunately, such a drug is available—an enzyme called *tissue plasminogen activator*, or *tPA*—and a medic administers it within minutes.

FIGURE 12.1
Molecular Geneticist Karl Ebert with Transgenic Goats.

Researchers first discovered tPA in human blood, but in such small quantities that it was extremely difficult and expensive to isolate usable amounts from human donors. In the early 1980s, molecular geneticists at a biotechnology firm near San Francisco isolated the gene for human tPA and inserted it into bacteria. The researchers then grew billions of *transgenic cells*—cells into which researchers have transferred foreign genes—in huge fermentation vats. This genetic engineering process has increased the availability of tPA. Nevertheless, the production processes are so delicate and the research and operating costs are so high that commercial tPA costs $22,000 per gram.

Clearly, an alternative solution was needed, and biologists suggested another way to produce large quantities of tPA. The idea was to splice the human gene for the tPA protein into the chromosomes of a large mammal. Then the scientists would trigger the secretion of the foreign tPA protein into the animal's milk. Finally, they would milk the animal and collect the tPA. Before long, the process—dubbed "pharming" by the press—was actually happening, as you will see in the following paragraphs.

Molecular geneticists took the human gene for tPA and attached it to DNA sequences that regulate how a goat's cells make a particular milk protein. This regulatory DNA ensures that the spliced-in gene for tPA is expressed in only one type of cell: milk-producing cells in the goat's mammary glands. With this accomplished, the researchers harvested ripe eggs from female goats [FIGURE 12.1] and added sperm to fertilize the eggs. Then they injected the engineered-tPA DNA into each fertilized egg. After the egg cells had divided several times and formed embryos of 8 to 16 cells, the researchers implanted 200 of the embryos into 36 female goats.

The goats gave birth to 29 kids, and two of them, one male and one female, contained the human tPA gene. The female kid was raised and bred to initiate pregnancy and lactation, and her milk indeed contained the human protein. The gene was also expressed in one of her five kids, and in later experiments, a female produced 3 milligrams of tPA per milliliter of milk. This one goat was producing 9 to 12 grams of tPA each day—or over $200,000 worth of the drug—much more than from a bioengineering firm.

MESSAGES

1 The movement of DNA sequences from one chromosome to another occurs frequently in nature, and gives rise to new genetic combinations.

2 Certain enzymes can cut DNA molecules at specific locations, while other enzymes can join DNAs of separate origin. Once genetic engineers create recombinant DNAs using these enzymes, they can replicate, or clone, the DNA in bacteria and produce millions of copies of a single DNA.

3 Researchers can design DNA molecules that encode proteins useful in treating diseases or in improving the qualities of certain crop species. Researchers can also modify human cells with recombinant DNAs, and while this can help cure diseases, it also raises serious ethical questions.

The tPA protein made by goats must be tested for safety and effectiveness, but researchers predict that it will be ready for commercial distribution before the end of the century. The use of tPA should give a person who suffers a heart attack an excellent chance for recovery.

The techniques of genetic engineering we discuss in this chapter pull together aspects of many of the chapters you have studied so far. These topics include the structure of proteins and cells [CHAPTERS 2 and 3]; the mechanisms of inheritance and chromosomes [CHAPTERS 7 and 8]; the structure, transcription, and mutation of DNA [CHAPTER 9]; the processes of translation and gene regulation [CHAPTER 10]; and the principles of human genetics [CHAPTER 11]. The ethical questions raised by the new technology will be discussed here and throughout the book, especially in the chapters on ecology and on human embryology and development.

Along the way, we will see how researchers engineered giant mice and helped a little girl grow taller; what a clone is; and why some agriculturalists are opposed to pharming. ❑

Genetic Recombination in Nature

Genetic recombination creates a DNA molecule with a new arrangement of genes. Before discussing how molecular biologists create new combinations of genes from different species in the laboratory—**recombinant DNA**—we must reexamine details of how genetic recombination can occur within a single species [review FIGURES 7.20 and 8.18]. As you saw earlier, during meiotic cell division in a eukaryotic cell, chromosomes can cross over and produce new genetic arrangements and combinations. The phenotype of an organism carrying these recombined genes will sometimes differ from the phenotype of either parent, or, for that matter, from the phenotype of any ancestor. In fact, the recombined genes may give the organism an advantage in dealing with its environment, and for this reason, recombination plays an important role in evolution by natural selection.

EVIDENCE OF RECOMBINATION BETWEEN SPECIES

The kind of genetic exchange you have already encountered is a function of sexual reproduction between members of the same species. In contrast, today's molecular geneticists are producing genetic exchanges between *different* species. We just saw an example: inserting a human gene for the tPA protein into goat chromosomes. Do genetic exchanges between species also occur in nature, or are scientists doing something that has never happened before?

Recent evidence suggests that different species apparently can and do exchange bits of DNA. For example, the bacterium *Escherichia coli*, which inhabits human and other animal intestines, has two genes for an enzyme that helps catalyze the energy-harvesting pathway of glycolysis. One of these genes is structurally similar to the same gene in a related bacterial species. The other gene, however, is similar to the corresponding gene in a *eukaryotic* cell. A possible explanation is that the bacterium acquired a second gene for this enzyme by exchanging DNA with a eukaryote.

Evidence that DNA can move from one animal species to another comes from observations of two very different species of fruit flies. Until the 1950s, a piece of DNA called the *P element* was absent from one species of fruit fly. Researchers later found evidence that this P-element DNA had evolved in a different species of fruit fly. How did the P element get from one species to another? Researchers recently found a mite, a small spiderlike parasite, that sucks juices from fly eggs [FIGURE 12.2]. Several lines of evidence suggest that the mite sucked P-element DNA from the eggs of one fruit fly species and transferred it to the eggs of the other some time in the first half of this century. Since the P element has moved between species, biologists think that other DNAs may also move, and some classic research has provided underpinning to this idea.

TRANSPOSABLE GENES

The P element is a special DNA because it contains genetic instructions that allow it to move from one part of the genome to another, moving in and out of chromosomes—a mobile DNA.

FIGURE 12.2

Gene Transfer Between Species.

Evidence suggests that the mite *Proctolaelaps regalis* sucked fluid containing DNA from the eggs of one species of fruit fly and transferred it to the eggs of another; from there, the DNA hopped into the chromosomes of the second species. Evidence such as this led researchers to conclude that during evolution, genes can occasionally be exchanged between species.

Egg of one species
P-element in chromosome
Egg of the other species
Before 1950
After 1950

[A]

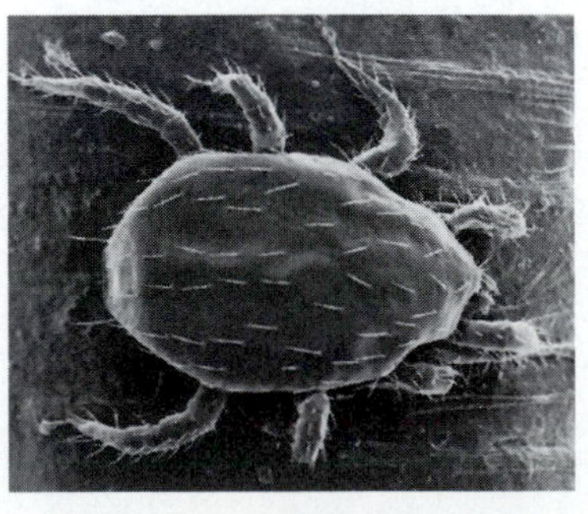

[B]

FIGURE 12.3
The Late Barbara McClintock, Nobel Prize–Winning Geneticist, Studying Jumping Genes in Corn.

Geneticist Barbara McClintock identified mobile genetic elements in corn as early as the 1940s, providing the first evidence that some DNA sequences can move around in chromosomes [FIGURE 12.3]. McClintock made a startling discovery in the mid-1940s that explained why Indian corn can have brilliantly colored kernels. She knew that some corn strains grow pale seeds with bright, irregular patches of purple. When she tried to map the gene responsible for regulating the purple pigment, however, she made the unprecedented observation that the gene in question jumped around from one chromosome to another. This was surprising, to say the least, because in all previous experiments, each gene had been shown to occupy a constant location on a specific chromosome. The "jumping gene" phenomenon was due to a specific class of genetic elements that came to be called **transposons**, which can move about in the genome and alter the action of the genes they enter. Transposons were first discovered in corn, then later in bacteria, fruit flies, and even human beings.

The mechanisms McClintock described were so complicated that her colleagues largely misunderstood the significance of her work. She belatedly received a Nobel prize in 1983 for her ground-breaking experiments.

After biologists learned that transposons move genetic material from one chromosome to another in nature, they wondered whether they could somehow use jumping genes to make genetic transfers in a controlled way. Using sophisticated instruments, specific enzymes, and other materials, geneticists can now create new gene combinations—recombinant DNAs—tailored to serve specific human purposes. They can then transfer these new gene combinations into chromosomes using transposons and other methods. The remainder of this chapter describes some of the exciting techniques and applications of recombinant DNA.

Recombinant DNA Research

Because mutation and recombination occur randomly in nature, the production of new genes, proteins, and phenotypes is extremely slow. Today, with revolutionary new techniques, genetic engineers can design and construct new genes in just a few weeks.

CONSTRUCTING RECOMBINANT DNA

The revolutionary microtechniques used to make recombinant DNA include the capacity to cut DNA molecules in specific places, to paste selected fragments together to construct a specially designed, recombined DNA molecule, and to allow the new DNA to enter a cell, where it can become part of the cell's hereditary material.

Cutting the DNA Researchers obtain proteins called **restriction enzymes** from bacteria and use them as scissors for cutting DNA. The bacteria produce restriction enzymes naturally, and use them to cut certain viral DNAs into pieces, thus restricting the types of viruses that can infect them. A restriction enzyme recognizes a specific sequence of a few base pairs wherever it occurs in a DNA molecule. The enzyme then cleaves the DNA at a consistent place in or near the same sequence. Some restriction enzymes cut the DNA in a staggered fashion, so that one strand of the double helix sticks out beyond the other. Other restriction enzymes, however, cut the DNA strands straight across. One restriction enzyme called *Eco*RI, for example, recognizes the sequence

GAATTC
CTTAAG

and cleaves the DNA between the G and A on both strands as shown in FIGURE 12.4B. When the weak hydrogen bonds between the strands break, the two DNA fragments can separate, leaving single-stranded ends protruding [FIGURE 12.4C].

Joining the DNA The unpaired sequences protruding from a staggered cut are called *sticky ends* by biologists. They can easily re-form hydrogen bonds with comple-

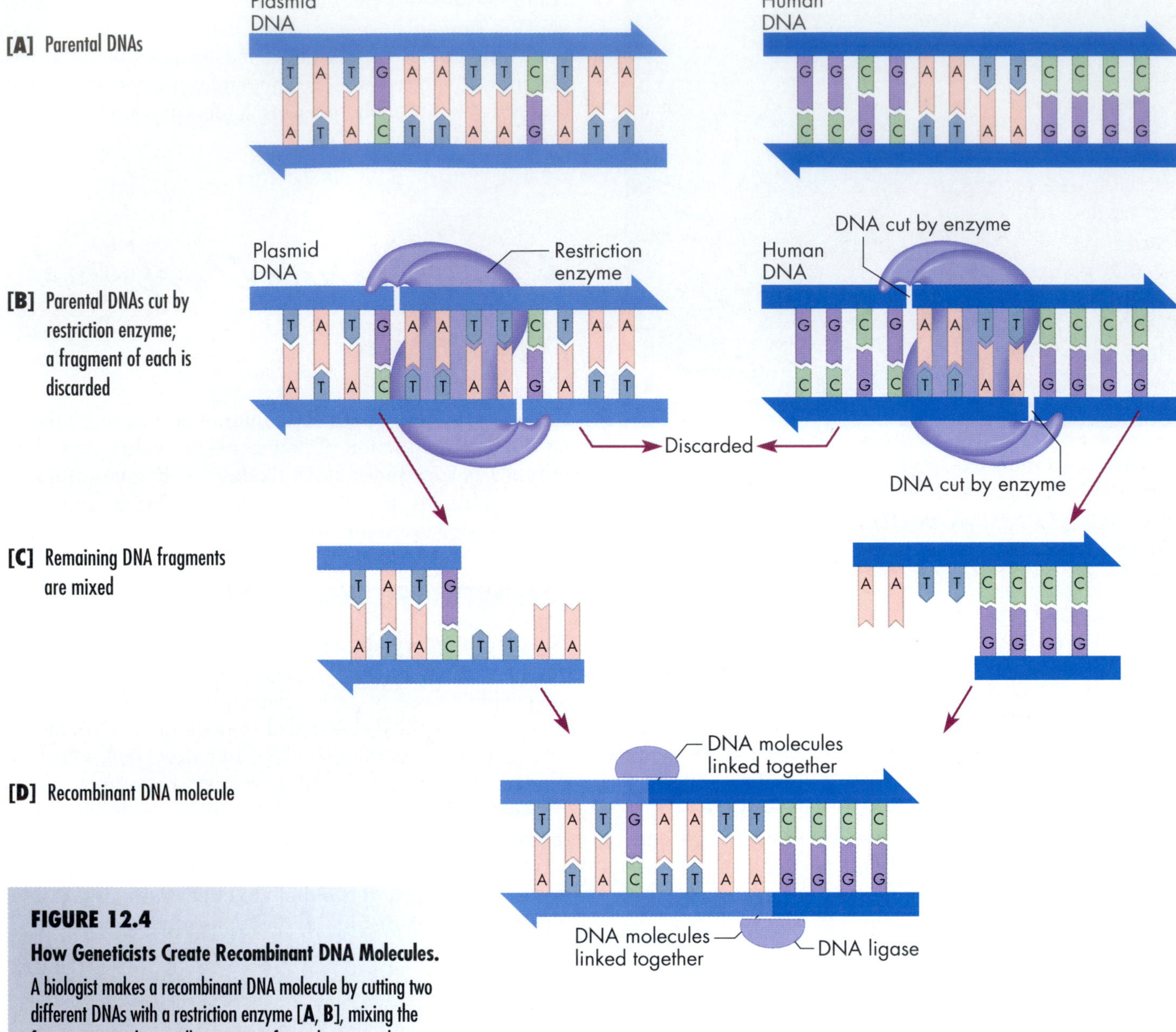

FIGURE 12.4

How Geneticists Create Recombinant DNA Molecules. A biologist makes a recombinant DNA molecule by cutting two different DNAs with a restriction enzyme [**A**, **B**], mixing the fragments together to allow pairing of complementary bases at the ends of the molecules [**C**] and then joining them together with ligase enzyme to create a recombinant DNA [**D**].

mentary unpaired sequences on the ends of other DNA molecules. In fact, any two DNA molecules with complementary protruding sequences can join together, no matter how unrelated the rest of the two DNA molecules may be. This tendency for sticky ends to join together through complementary base pairing turns out to be a boon to genetic engineers. It allows them to recombine DNA molecules from organisms as different as humans and bacteria. For instance, in FIGURE 12.4, the DNA shown in dark blue could represent a small portion of the human tPA gene cut by the *Eco*RI restriction enzyme, while the pale blue DNA could represent a similarly cut fragment of DNA from any other source.

Hydrogen bonds between the AATT of the human DNA and the TTAA of the other DNA can hold the two fragments together [FIGURE 12.4D]. Once this happens, the enzyme **DNA ligase** can make strong covalent bonds between the opposing ends of the two molecules [FIGURE 12.4D]. The covalent bonds join the two fragments at the designated sites. (In nature, DNA ligase repairs damaged DNA; review FIGURE 10.16.)

Carrying the DNA For recombinant DNA molecules to be useful, the experimenter must make many identical copies, a process called cloning. To clone a desired piece of DNA, researchers use restriction enzymes and DNA ligase to insert the DNA into the circular DNA of a *vector*, usually a virus or plasmid that occurs in nature [review FIGURE 9.3]. Plasmids have three useful properties: (1) They contain genes that make a bacterial cell resistant to certain antibiotics, such as streptomycin. (2) They can replicate inside bacterial cells. (3) They can move readily from one bacterial cell to another. While these properties make plasmids useful in the genetics lab, they make plasmids a disaster in a hospital because they can carry genes that confer antibiotic resistance from one bacterial species to another. As resistance spreads from one type of disease-causing bacterium to another, antibiotics can very quickly become ineffective at preventing disease. When plasmids are harnessed as carriers of recombinant DNA, however, they can be both safe and useful.

HOW TO CLONE A HUMAN GENE

A researcher interested in cloning a human gene might carry out the following experiment: The researcher splices a DNA fragment such as the gene for human tPA into a plasmid and then makes many copies of the plasmid plus tPA gene insert. Eventually, the geneticist isolates the tPA genes from the plasmids. This process is a bit like scribbling a short note in the middle of a typed letter and then making 100 photocopies of the letter; both the letter and the note get copied.

FIGURE 12.5 traces the steps that researchers follow to **clone** a DNA fragment, that is, to select a desired DNA sequence and reproduce many copies of it. The following account is numbered to coincide with the steps in FIGURE 12.5:

1. The circular DNA of plasmid molecules is cut with a restriction enzyme, opening the plasmids and forming two sticky ends on each plasmid. The DNA containing the gene to be cloned is also cut with the same restriction enzyme; each of the thousands or millions of fragments has sticky ends.
2. The sticky ends of a plasmid DNA molecule and the ends of a fragment of human DNA adhere by base pairing. Each plasmid will contain only one of thousands of different fragments of human DNA. Treatment with DNA ligase re-forms the circle.
3. The plasmid is transferred, along with the hitchhiking human DNA, into a bacterial cell. The process is called *transformation* [see CHAPTER 9].
4. The bacterial cell (which now contains the plasmid-bearing human DNA) is cultured. The cell divides repeatedly, producing a clone of cells—a group of identical offspring cells, all descended from the same founder cell. An antibiotic is added to the bacterial growth medium to kill any bacterial cell that does not contain the plasmid with its antibiotic-resistant gene.

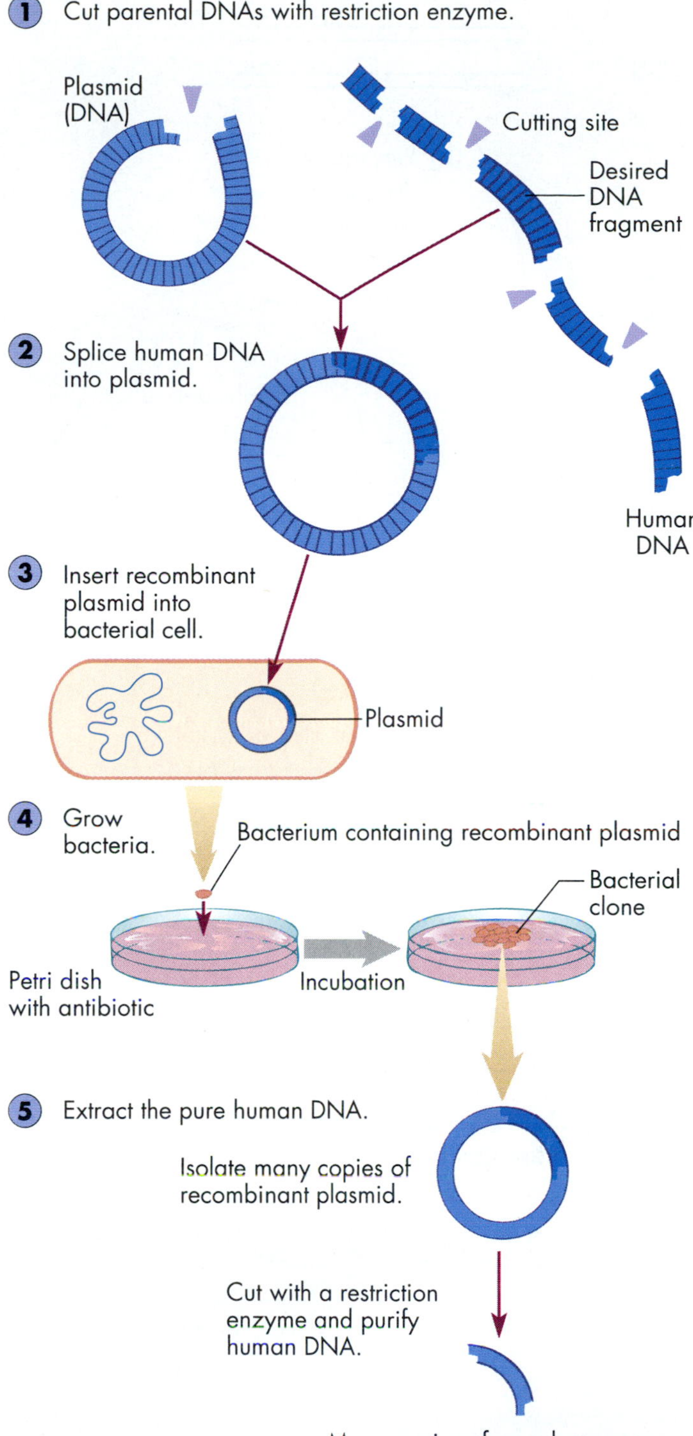

FIGURE 12.5
How to Clone a Human Gene.
Follow the steps described in the text to see how molecular biologists clone human genes.

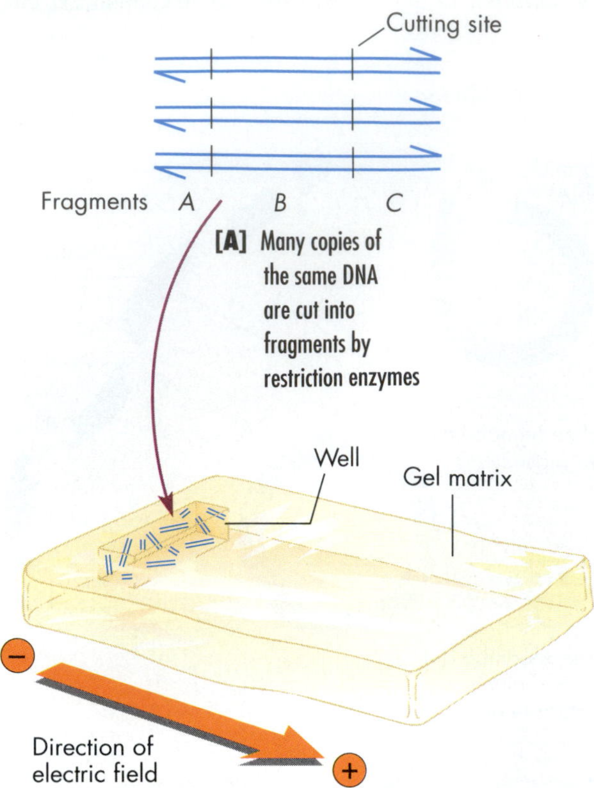

[A] Many copies of the same DNA are cut into fragments by restriction enzymes

[B] The fragment mixture is put into a well cut in a gel and an electric field is applied across the gel

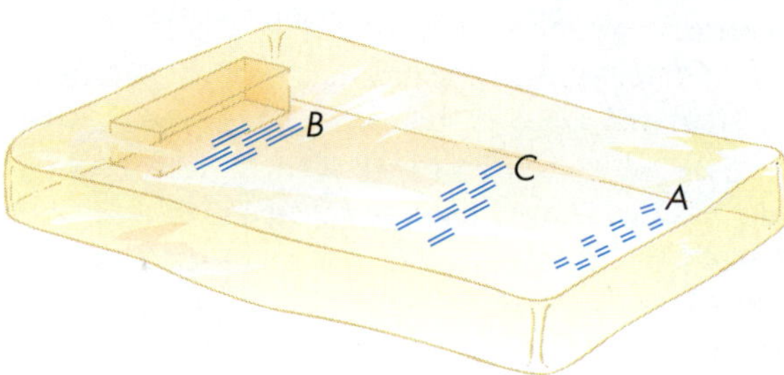

[C] The smaller fragments move faster than the larger fragments and, in time, separate into distinct bands

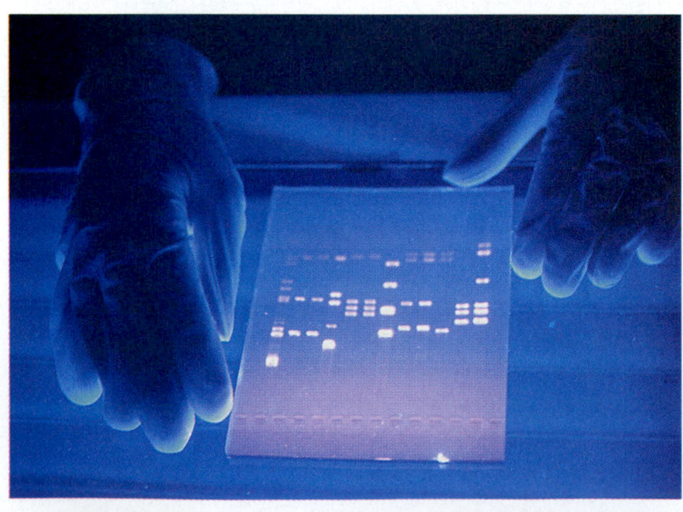

[D] The DNA bands shine a bright orange-pink in UV light

FIGURE 12.6

The Process of Gel Electrophoresis Separates DNA Fragments According to Their Size.

The steps in **[A-C]** allow geneticists to identify and work with individual DNA molecules. In **[D]**, a molecular geneticist examines under ultraviolet light DNA fragments of various sizes separated by electrophoresis.

Now comes a problem: Which one of the bacterial colonies contains bacteria bearing the desired gene, and how can the researchers find it? Biologists identify the bacterial colony containing the gene they wish to isolate by using a *probe*, a marker molecule that binds specifically to the desired gene by complementary base pairing (recall A pairs with T and C pairs with G in DNA). A probe for the human tPA gene, for example, could be tPA messenger RNA that the researcher has marked with a radioactive label. Because this radioactive probe contains a sequence of bases complementary to the tPA gene, any bacterial colony that becomes radioactive after exposure to the probe must contain the human tPA gene.

5. The rare colony with its desired gene is identified, and the plasmids—still carrying the human tPA gene—are isolated and separated from the cloned bacterial cells. The plasmids are cut with a restriction enzyme, which releases the human gene from the plasmid. The researcher can then purify the human gene from the plasmid using the process of **electrophoresis** (see FIGURE 12.6 for an explanation). In this way, the human tPA gene can be mass produced through cloning.

BOX 12.1 explains another useful technique for making multiple copies of DNA segments: polymerase chain reaction.

➤ CONCEPT CHALLENGE

Explain two ways geneticists utilize the property of complementary base pairing in recombinant DNA research.

Promises and Problems of Recombinant DNA

Genetic engineering enthusiasts point out that the technology holds tremendous promise for humankind. Imagine a drug, available in great supply for only a few dollars, that can kill the viruses that cause the common cold.

Also, an enzyme that can arrest a heart attack already in progress; bacteria that can make automobile fuel from discarded corn stalks; other bacteria that can protect crops from late spring freezes; corn that contains proteins with the nutritional value of beef proteins; gene replacements to cure people and animals of crippling genetic diseases; engineered mice that can be used to screen for and identify cancer-causing agents.

All of these sound helpful—even life-saving. But now consider the potentially less favorable side to genetic engineering. For example, bacteria could be engineered to break down oil and then used to disperse oil slicks, but what if these hypothetical bacteria moved beyond oil slicks and began to devour the world's dwindling oil supplies? What would happen if it became possible for people to genetically engineer children with traits the parents or society deemed desirable? Many feel that recombinant DNA research has the potential to make certain dreams come true, but worry about the social and ecological impacts that might result. Let's look at the promise, the present performance, and the nagging questions that surround recombinant DNA research.

FIGURE 12.7 A Growth Spurt Thanks to Genetic Engineering. As the marks on the door frame show, this child grew several inches during treatment with growth hormone made through genetic engineering techniques.

HOW GENETIC ENGINEERING MAY RESHAPE LIFE

Scientists have three approaches for exploiting custom genes for their own purposes. First, they can alter bacteria so that the bacteria produce a specific protein, such as human insulin, that can be used to treat a disease (diabetes). Second, scientists can alter an organism to make it more useful; for example, they could engineer corn or cattle to grow faster and resist diseases better. Third, they could alter the human genome itself to cure disease.

Mass-Producing Proteins The 10-year-old girl in FIGURE 12.7 was born unable to produce growth hormone, and she seemed destined to live out her life as an abnormally short individual. But in just one year of treatment with human growth hormone the girl grew about 12.6 cm (5 in). The human growth hormone was mass-produced by genetically engineered bacteria containing the human growth hormone gene. Using the standard cloning techniques, scientists isolated the gene for human growth hormone and then inserted it into bacteria. Grown in an enormous vat, the bacteria then produced commercial quantities of human growth hormone, enough for treating the thousands of children in the United States who make too little of the hormone to grow to normal height. The precious substance is now available in pharmacies by prescription, and physicians hope someday to also use it to treat burns, slow-healing fractures, and bone-loss diseases.

Biotechnologists have also succeeded in cloning the gene for growth hormone of cows, producing large amounts in bacterial cells and harvesting the protein for commercial use. When injected into dairy cows, this hormone can increase milk production by 10 to 20 percent. The U.S. Department of Agriculture has certified that milk from treated cows is safe to drink, but some observers note that it requires more antibiotics to keep treated cows healthy, and some of the antibiotics as well as some of the hormone itself may end up in the milk. This advance might seem like a boon to the dairy industry, but using a product of recombinant DNA may have a significant economic downside. Agricultural economists have pointed out that farmers already produce a surplus of milk products. Given this surplus, competition from large milk-producing operations, which can more easily afford to use the expensive drug, may drive small family-owned dairies out of business.

Growth hormone is just one type of protein now mass-produced through engineered bacteria. Bacteria have also been fitted with genes for insulin hormone that is chemically identical to the insulin produced by the human pancreas. Until recently, diabetics had to rely on insulin obtained from pigs or cows. This animal insulin differs slightly from human insulin, and as a result, it can cause an allergic response in some people.

Genetically altered cells are also producing a number of other medically useful proteins. These include: (1) *alpha-1-antitrypsin*, a protein that can help patients with the lung-destroying disease emphysema; (2) *transforming growth factor beta* (TGF β), a protein that can help heal previously untreatable degenerative eye disease; (3)

box 12.1
People in Biology

Kary Mullis and Copies of Copies of Copies

As Kary Mullis tells it, he invented one of the most powerful tools of modern molecular biology while driving down a dark country road one night in April of 1983. Mullis had been working for the Cetus Corporation in Emeryville, California, synthesizing nucleic acid probes with which to identify specific base sequences in DNA. If a researcher wanted to study a particular sequence further, he or she usually needed to generate millions or billions more copies. In the early 1980s, it could take weeks of lab work to amass such a stockpile—still no more than a small test tube of DNA in a clear solution.

Working with probes and the subsequent tortuous copying procedures was not only difficult but somewhat repetitious and tedious, and it left Mullis lots of time to brainstorm easier approaches to manipulating DNA. That spring night, he was musing about new ways to create a molecular photocopying machine that could churn out unlimited copies of a desired piece of DNA. By the time Mullis reached his cabin near Mendocino, he had solved the problem in his head, and by the following Monday, had tested it in the lab. The result: the *polymerase chain reaction*, or PCR.

The technique is so fundamentally simple, writes Mullis, that other DNA workers wonder, Why didn't I think of that? Starting with a long double-stranded DNA molecule containing a short sequence he wished to copy and study, he heated the long molecule to separate the strands. Next he added two different primers, short pieces of DNA that bracket the desired DNA sequence. One primer attaches to one strand of the DNA at one end of the desired sequence, while the second primer attaches to the other strand at the other end. The primer tells the DNA-synthesizing enzyme DNA polymerase to "start copying DNA here." Mullis then added DNA polymerase to the mixture and waited for it to generate a copy of the bracketed DNA segment. As long as (1) heat is supplied at the right time to separate the DNA strands, (2) primers are present to bracket the desired piece of DNA, and (3) the polymerase enzyme is available to copy the designated segment, cycle after cycle of replication can take place with the copied strands serving as templates for the next round of copying. In this chain reaction, 2 copies become 4, 4 become 8, then 16, 32, 64, 128, and so on, up to a million copies in just 20 relatively rapid cycles.

Mullis had to make some adjustments to the process, including using DNA polymerase from a hot springs bacterium that can withstand the near-boiling strand-separating heat jolt needed for each duplication cycle. Eventually, he was able to automate the entire process and enclose it in a machine housing. He had invented a DNA copying machine that has since become standard equipment in most DNA laboratories.

Like a paper copier, a PCR machine has innumerable uses, some of them ingenious. For example, by employing PCR, crime investigators can take DNA fingerprints from the cells in a tiny speck of dried blood or at the base of a single human hair in their search for a criminal's identity. With conventional techniques, the sleuth needs 1000 times more blood (a sizable stain) or 1000 strands of hair (a yanked-out handful!) to collect enough DNA to study.

PCR is also allowing physicians to diagnose genetic defects in very early embryos by removing the DNA from a few sloughed-off or collected cells and amplifying it. They are also using PCR to detect the presence or absence of AIDS virus in the infants of mothers with the immune system disease. And finally, biologists can now study the tiny stands of DNA they retrieve from ancient specimens—Egyptian mummies, woolly mammoths, even human brains preserved in old bogs—by copying the genetic material via PCR.

Kary Mullis's spring evening brainstorm won him a Nobel Prize in 1993 and made him a wealthy man. Most importantly, it enriched the study of biology immeasurably.

granulocyte colony stimulating factor, which can stimulate the body to form new white blood cells that help fight infection; and (4) *interferon gamma*, which can protect laboratory-grown human cells from infection by hepatitis and herpes viruses. In laboratory tests, interferon can also stimulate the growth of tumor-killing cells. This partial list of early successes only hints at the many useful products that genetically engineered microbes may ultimately provide for us.

While genetic engineers have used bacteria and other cultured cells to produce useful proteins for humans, they have also been able to genetically alter farm animals. FIGURE 12.8 explains how researchers produced the goats described in the chapter introduction that make tPA.

Molecular geneticists are also engineering plants to produce useful substances. European researchers, for example, induced plants to make a plastic that manufacturers are now using in biodegradable shampoo bottles [FIGURE 12.9]. Researchers are engineering other plants as well, to make vaccines that protect farm animals from disease.

Improving Plant and Animal Stocks Recombinant DNA can accelerate genetic manipulation of agricultural animals. The alteration of farm animals originated with ancient human societies more than 10,000 years ago, as they applied natural breeding and artificial selection to the domestication of cattle, sheep, goats, and other animals. In the past, breeders selected individuals with desirable traits—traits that arose by mutation and naturally occurring DNA recombination—to be parents of the next generation. Today, geneticists can insert into livestock specific genes, for example, genes that increase bone and tissue growth or milk production, just as researchers inserted the human tPA gene into a goat. The greatest impact on human lives, however, probably will be made through advances in the genetic engineering of plants.

Today, only 30 plant species make up nearly 93 percent of the human diet, and our dependence on these 30 species will only increase in the future. The global hu-

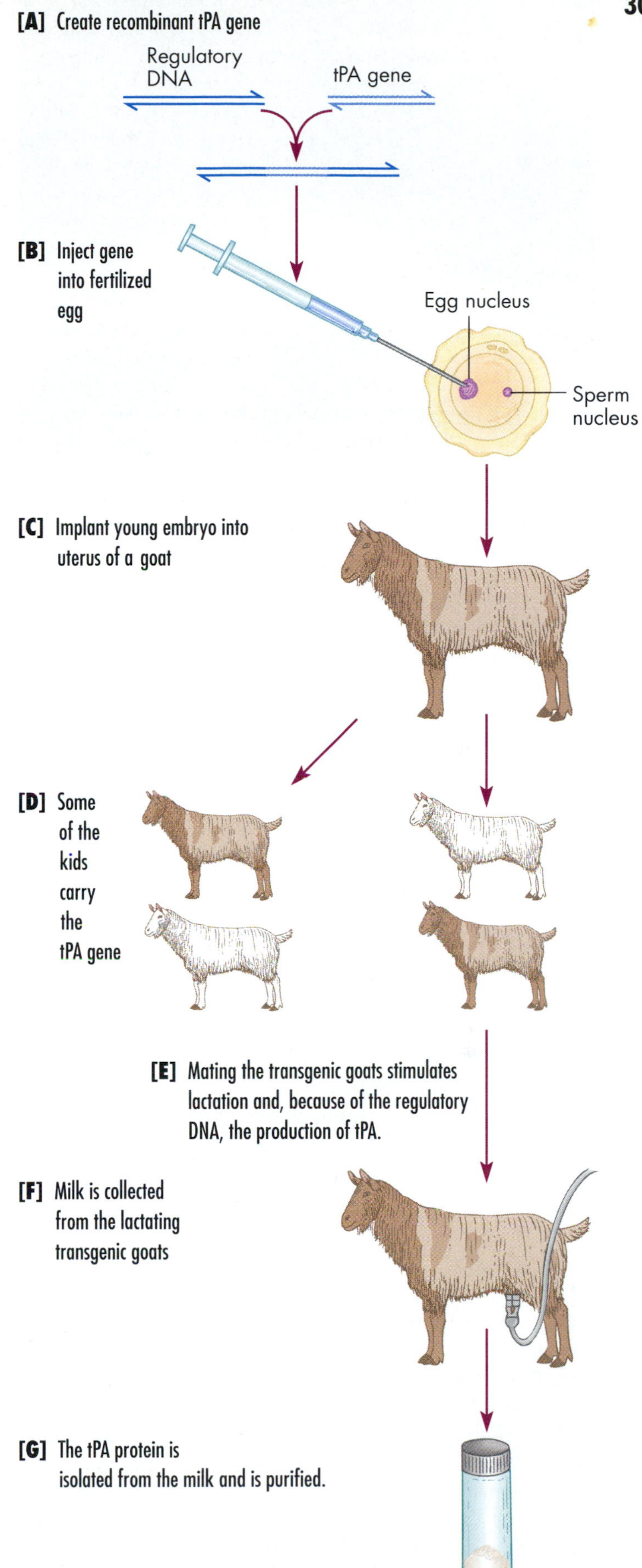

FIGURE 12.8

Creating a Transgenic Goat.

[A] Geneticists use recombinant DNA techniques to place the tPA gene adjacent to DNA that causes tPA to be secreted into milk. [B] They inject the recombinant gene into a fertilized goat egg, and then [C] implant it into the uterus of a nanny goat. There it develops into an embryonic kid that may carry the tPA gene. [D]. When a female transgenic kid is born, matures, and produces milk [E], technicians can collect and purify tPA from the milk [F,G] and use it to treat human heart attacks and strokes.

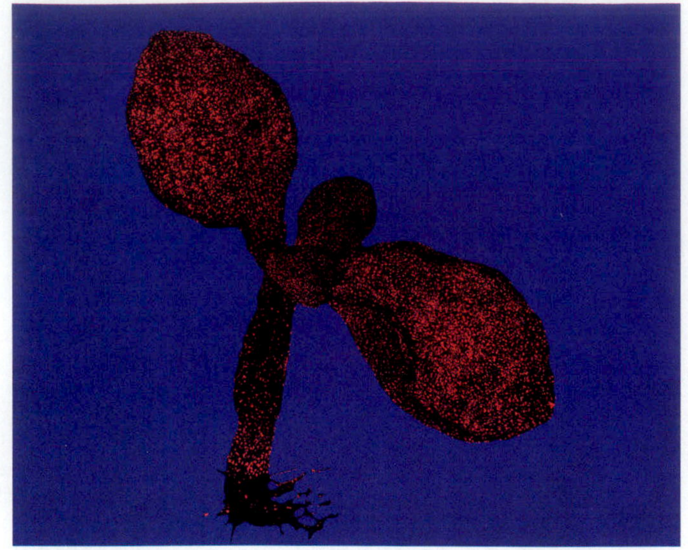

FIGURE 12.9

A Plant Engineered to Make Plastic.

In this seedling of the common weed *Arabidopsis*, the red dots are cell vesicles filled with a plastic used in making shampoo bottles.

man population continues to explode, even as agricultural land gives way to encroaching cities and deserts, and as energy for producing fertilizers and pesticides becomes increasingly scarce and expensive. For decades, botanists have been breeding better crop plants to help meet the growing worldwide demand for food. With the current capacity to insert genes directly into individual plants and animals, botanists can begin to circumvent these time-consuming breeding programs and use genetic engineering techniques to improve the efficiency of photosynthesis; decrease crop dependence on fertilizers; improve the nutritional quality of seeds, grains, and vegetables; and increase plant resistance to pests, salt, drought, and extreme temperatures.

For example, researchers have introduced a gene into tomato plants that renders them resistant to some plant viruses. The protective gene, obtained from a particular virus, prevents this virus and certain other disease-causing viruses from infecting plant cells. Normal tomato plants become stunted when infected with disease viruses, but the *transgenic* plants (plants carrying a foreign gene in their chromosomes) resist viral infection. Researchers are also using gene-transfer technology to design new varieties of oranges, potatoes [FIGURE 12.10] wheat, and rice that need far fewer chemicals to resist insects and disease-causing fungi and that use nutrients more efficiently. Some biologists predict that these genetic improvements should lower the farmer's costs in producing these and other foods, as well as decreasing environmental pollution from pesticides and fertilizers. Plant researchers have also developed herbicide-resistant crop varieties using recombinant DNA methodologies. While these plant scientists are striving to decrease the amounts of herbicides needed to control weeds, some critics argue that planting herbicide-resistant crops will only increase our dependence on the use of chemicals for agriculture.

A natural question arises from achievements in genetics research: If beneficial genes can be inserted into plants, animals, and livestock, then why not into people?

Human Gene Therapy Altering a person's genes to combat disease, a process called gene therapy, is a field still in its infancy. In theory, gene therapy could be applied in either of two ways: (1) Genes could be inserted into the somatic (body) cells, or (2) into the germ cells (the cells that give rise to the sperm and eggs). The first procedure, *somatic-cell gene therapy*, is the straightforward treatment of an individual's disease—a simple extension of current medical practice. But the second kind, *germ-line gene therapy*, would also affect the genetic makeup of the treated individual's offspring.

In 1990, physicians carried out the first instance of somatic-cell gene therapy—the introduction of genetically engineered cells into a human being to cure disease. In this case, a 52-year-old woman was terminally ill with melanoma, a fast-growing cancer that often begins with a dark mole and then enlarges rapidly. The patient's cancer

FIGURE 12.10

Potato Plants With Genetic Resistance to Insect Pests.

The luxuriant plants in the center row contain a gene from a bacterium that kills certain insect pests, while the devastated plants in the flanking rows lack this gene.

had spread, and she had tumors in her chest wall, abdomen, and limbs, which had failed to respond to conventional cancer therapies. Physicians at the National Cancer Institute removed a portion of her tumor and extracted from it a type of white blood cell called *tumor-infiltrating lymphocytes*, or *TILs*. These cells naturally home in on a tumor and crawl into the mass of tumor cells. Previous experiments had shown that specially treated forms of these cells can slow tumor growth. [FIGURE 12.11].

Physicians then inserted into the patient's own TIL cells a gene coding for *tumor necrosis factor* (*TNF*), a protein that kills cancer cells in mice. Within nine months of the first treatment, the multiple tumors had either regressed or disappeared. After 13 months (the latest data available), regression was continuing. While researchers will have to apply the techniques to additional patients, these initial results were striking, and provide the hope that somatic-cell gene therapy of this type might help many kinds of cancer patients.

Since the gene therapy in this case altered white blood cells (which are somatic cells) but not sex cells, the altered genes could not be passed on to the patient's children. For most people, this sort of gene therapy presents no new ethical problems: As in conventional surgery or drug therapy, a single individual is treated, he or she benefits, and the defective genes remain to be perpetuated in the human population.

Gene therapy of germ-line cells is another matter, since it constitutes a totally new approach to medicine, raising complex issues of safety and ethics. In germ-line gene therapy, recombinant DNA would be inserted into human sex cells via techniques similar to those used to make the transgenic goats [see FIGURE 12.8]. Not only would the treated individual be affected, but so would all

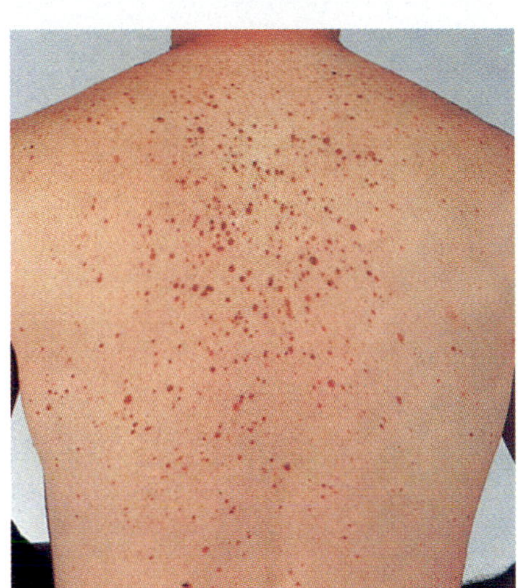

[A] Before treatment

FIGURE 12.11

Using Gene Therapy to Fight Cancer.

[A] A patient with dozens of potentially lethal melanoma skin cancers. **[B]** White blood cells called tumor-infiltrating lymphocytes (TILs) were isolated from the patient and treated to increase their production of a certain anticancer protein. When put back into the patient, the TILs caused many of the melanomas to regress. In the first trial of gene therapy, physicians removed TIL cells from a different group of skin cancer patients (Step 1). Next, they inserted into the TIL cells the gene for a tumor-killing protein called tumor necrosis factor (TNF) (Step 2). These engineered TIL cells, reintroduced into the patients, invaded the tumors (Step 3). TNF produced by the genetically engineered cells should destroy the skin cancer cells (Step 4) and allow patients to survive. In **[C]**, a patient's melanomas have regressed.

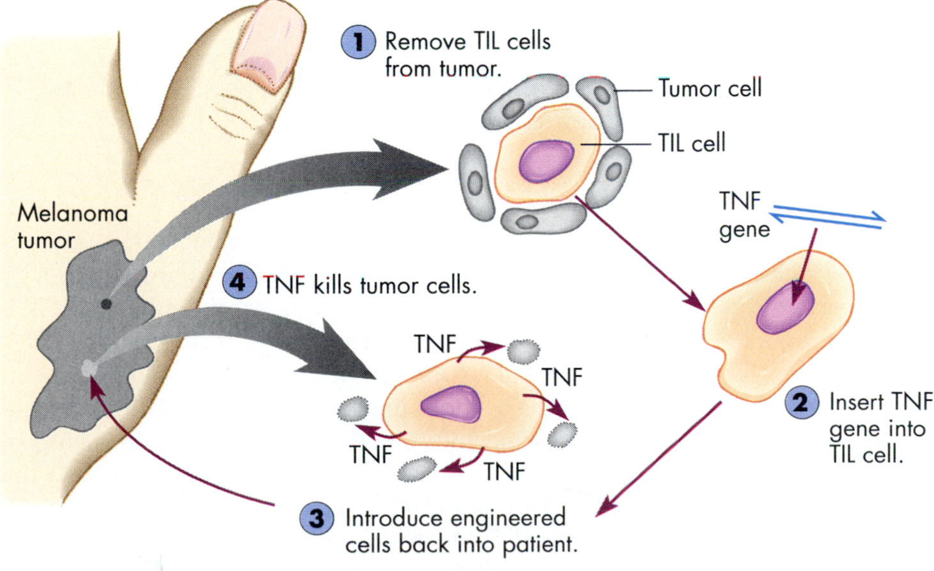

[B]

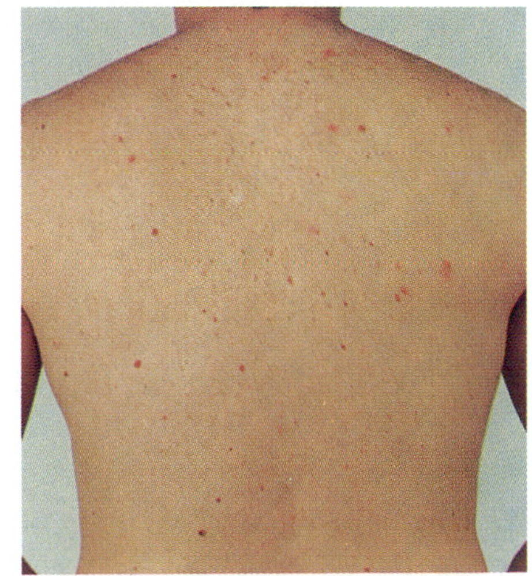

[C] After treatment

the individual's descendants. While researchers have had some success with techniques for this kind of gene transfer in mice and goats, they must still solve many technical and ethical problems before treating human sex cells. One problem is that the success rate is low—only 6 successes out of 300 injected eggs in a typical experiment with mice. Another problem is that the inserted DNA sometimes causes a new mutation, possibly creating a new defect as serious as the disease the therapy was meant to cure. Finally, the inserted normal gene usually does not replace the native defective gene, but is simply added to the genome, so that the defective gene can still be inherited by some of the children of the treated individual. At present, germ-line therapy seems out of the question for purely technical reasons; however, it will someday be possible and will pose significant long-term ethical questions. Since its application would alter the genetic makeup of children not yet born, it could change the course of human evolution. Would a treatment that causes a heritable change infringe on the rights of future generations? Are the expected benefits of germ-line gene therapy worth the unknown risks? Could a problem be solved in more traditional ways, whose risks are understood and whose benefits are clear? These and other questions have led scientists and lay observers to debate—at times heatedly—the potential misuse, accidental or deliberate, of recombinant DNA techniques.

FIGURE 12.12

The Introduction of Novel Organisms Can Cause Unexpected Environmental Consequences.

The kudzu vine (which is not genetically engineered) was brought to the American South to control erosion. But the hardy plant competed with native vegetation and is now destroying some forests.

RECOMBINANT DNA: ENVIRONMENTAL RISKS

Some critics have suggested that introducing organisms altered by recombinant DNA methodologies into the environment poses significant risks. Professional ecologists (scientists who study the interactions of organisms with each other and with their physical surroundings) agree that some dangers do exist. A dangerous new weed could conceivably be created if, say, a new variety of rice engineered with genes for salt tolerance escaped from cultivated fields and invaded the brackish water in the mouths of rivers. A weed might become a worse pest if, for example, it interbred with a genetically engineered domestic crop and gained a gene for herbicide resistance.

There have, of course, been many ecological disruptions with natural, unengineered organisms; a prime example is the devastating growth of kudzu, a fast-growing, large-leafed Japanese vine that was introduced into the southeastern states to control soil erosion but has ended up choking tens of thousands of acres of native and commercial forests [FIGURE 12.12]. The Ecological Society of America, the professional society of scientists studying the environment, agrees that the modern methodologies of recombinant DNA and genetic engineering can produce genetically novel organisms that can be used for agricultural advances, in waste management, and for detoxification of chemicals. With this in mind, while the Ecological Society of America does recognize the risk in introducing any nonnative species, it lent its support in 1989 to the development of genetically modified organisms, as long as those products of biotechnology are carefully evaluated by regulatory agencies before being released into the environment. The society contends that any organism proposed for environmental release should be evaluated according to its biological properties, and not simply according to whether developers used recombinant genetics or traditional breeding techniques.

While ecologists have been concerned with the environmental effects of genetically novel organisms, others have voiced concern about the consequences for human health, as well as the ethics of gene manipulation.

RECOMBINANT DNA: NOVEL PROBLEMS OF SAFETY AND ETHICS

Questions about both the safety and morality of recombinant DNA research have cast a shadow over its promise of a better future for the human race. What, for example, would happen if bacteria transformed with a cancer-causing gene escaped from a laboratory and infected people? Would a cancer epidemic decimate the population? This possibility so horrified scientists in the mid-1970s that molecular biologists agreed among themselves to stop research on recombinant DNA for several months as a small group of researchers conducted a potentially dangerous experiment—a worst-case scenario.

In a laboratory specially designed to contain dangerous microorganisms, they spliced cancer-causing genes into the DNA of bacterial plasmids and then exposed mice to potentially cancer-causing bacteria. As it turned out, the mice developed no more tumors than did unexposed mice. The public's worst fears were allayed, and the National Institutes of Health allowed recombinant DNA researchers to resume their work under strict guidelines designed to prevent the escape of recombinant organisms into the outside world.

Many other questions about the applications of recombinant DNA remain to be resolved. Significant disagreements arise over using gene therapy to "improve" a healthy embryo. For example, inserting a human growth hormone gene into mice causes them to grow larger [FIGURE 12.13]. Might not prospective parents ask physicians to insert additional growth hormone genes into their newly conceived embryo so they could produce a potential linebacker, or a center forward?

Other parents might want a child with greater intelligence or a more winning personality. From a scientific standpoint, such applications of recombinant DNA research now seem only a remote possibility. Untold numbers of genes control each of these complex traits in unknown ways. Geneticists may never be able to discover and clone all the relevant genes and introduce them into a person with any degree of control. However, society should consider these issues before biotechnologists achieve such breakthroughs. And if history repeats itself, the potential for human gene therapy will arrive more quickly then we expect.

Only through free and open discussion will potential problems be solved and appropriate guidelines be devised to ensure that recombinant DNA technology can help produce higher-yield crops, better livestock, and healthier humans while minimizing risk.

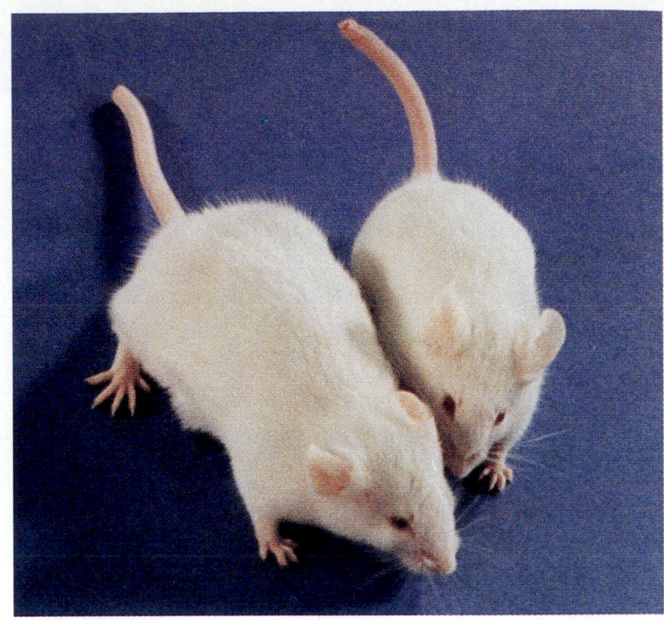

FIGURE 12.13

An Example of Germ-Line Gene Therapy.

Genetic engineers produced a giant mouse twice as big as its normal sibling by inserting a gene from human growth hormone into one of its chromosomes. Applications of such techniques to human embryos would raise serious ethical questions.

➤ CONCEPT CHALLENGE

Describe an important ethical issue in recombinant DNA research. List arguments on both sides of that issue.

Connections

The last few chapters have taken us on a genetic journey. The journey started with the simple statement that genes exist and control the way organisms look and act. It continued through an exploration of gene structure and how that structure enables genes to determine an organism's traits. And it included the discovery that geneticists can isolate individual genes, alter them, and insert them into the chromosomes of organisms as different as bacteria, tomatoes, and humans, and by so doing change the organisms' traits.

In the remaining two chapters of this section, we look at how the genetic blueprints of life encode and control the building of a many-celled organism during embryonic development. We will address the question: How does an egg become an organism? The answers, as you will see, are fascinating indeed.

KEY TERMS

clone, 297
DNA ligase, 296
electrophoresis, 298
recombinant DNA, 294
restriction enzyme, 295
transposon, 295

HIGHLIGHTS IN REVIEW

1 The movement of DNA sequences from one chromosome to another occurs frequently in nature, and gives rise to new genetic combinations.

a] During genetic recombination in nature, two different DNA molecules may break and rejoin. This can create novel phenotypes of evolutionary significance.

b] Genetic recombination within a species is a natural process that occurs in all sexually reproducing organisms. In some cases, genes have apparently been transferred from one species to another in nature.

2 Certain enzymes can cut DNA molecules at specific locations, while other enzymes can join DNAs of separate origin. Once genetic engineers create recombinant DNAs using these enzymes, they can replicate, or clone, the DNA in bacteria and produce millions of copies of a single DNA.

a] Restriction enzymes cut DNA molecules at specific places and leave protruding ends. When the ends of two DNA molecules have paired, DNA ligase enzyme can be used to link the two molecules together.

b] Circles of DNA called plasmids can ferry recombinant DNA molecules into bacteria cells, where the DNA is copied.

3 Researchers can design DNA molecules that encode proteins useful in treating diseases or in improving the qualities of certain crop species. Researchers can also modify human cells with recombinant DNAs, and while this can help cure diseases, it also raises serious ethical questions.

a] Foreign genes can be inserted into bacterial, plant, or animal genomes. Even in this strange new environment, the foreign gene can often be expressed to make its characteristic protein.

b] Genes designed and made by molecular geneticists can be used to cure human disease and to improve crop species.

c] Society must confront the questions of safety and ethics that recombinant DNA technology poses.

UNDERSTANDING THE FACTS AND CONCEPTS

For Questions 1–5, match each of the descriptions with the most appropriate item in the following list. Any item may be used once, more than once, or not at all.

a] restriction enzyme
b] DNA ligase
c] transposon
d] plasmid
e] clone

1 A group of identical items, all of which have been derived from a single, common source.

2 The P element.

3 A protein that acts as scissors for cutting DNA.

4 A molecular glue that mends breaks in the helix.

5 An extracellular polynucleotide ring that can infect a cell and become integrated in the cell's chromosome.

As above, for Questions 6–10, match the descriptions with the appropriate item from the following list.

a] *Eco*RI
b] plasmid
c] transformation
d] electrophoresis
e] tPA

6 The process of incorporating a foreign piece of DNA into a living cell.

7 An agent that has been responsible for imparting antibiotic resistance to pathogenic bacterial strains.

8 A process whereby DNA fragments can be sorted according to their length.

9 A restriction enzyme.

10 A human protein that is produced by transgenic goats.

As above, for Questions 11–15, match the descriptions with the appropriate item from the following list of terms.

a] transgenic organism
b] somatic-cell gene therapy
c] germ-line gene therapy
d] interferon gamma
e] plasmid

11 A blood component that is used to protect human cells in laboratory culture from infection by hepatitis and herpes viruses.

12 Treating a disease or deficiency condition by inserting a correcting gene into somatic cells.

13 Treating a disease or deficiency condition by inserting a correcting gene into reproductive cells.

14 A term that applies to an individual of one species having a gene from another species in its genome.

15 A vehicle for moving desired DNA fragments into or out of cells.

For Questions 16–20 indicate whether each statement is true or false. Where you have said that a statement is false, indicate why it is false and revise it to make it correct, using a separate piece of paper for your answers.

16 Domestic animals can be made to synthesize certain human proteins by recombinant DNA technology.

17 Human milk can be harvested from a cow if her mammary glands have been inoculated with human milk protein.

18 Natural genetic recombination normally takes place during meiosis.

19 The polymerase chain reaction makes it possible to generate large numbers of identical DNA copies from a very small initial sample.

20 Electrophoresis is a method for sequencing genes.

INTEGRATE AND APPLY WHAT YOU HAVE LEARNED

1 Does naturally occurring genetic recombination always require meiosis and homologous chromosomes? What about artificially engineered recombination?

2 Contrast the effects and implications of germ-line gene therapy with somatic-cell gene therapy.

3 Describe the polymerase chain reaction technique and comment on its usefulness.

4 Explain how a genetically engineered plant might pose unacceptable ecological risk.

5 Cite and describe three examples of human diseases that have been either cured or significantly alleviated through the use of genetic engineering technology.

ANALYSIS

Answer the following multiple-choice questions.

1 Refer to FIGURE 12.11 concerning the regression of lethal tumors in the skin after TILs were genetically engineered, and answer the following question. Why is it preferable to go to the trouble of inserting the TNF gene into the TILs rather than simply treating the patient with the TNF directly?

- **a]** By genetically engineering the TILs the medication can be localized more specifically.
- **b]** The genetically engineered cells could probably produce more TNF than could be produced by other means.
- **c]** The TNF would be more "human" than if it were produced by a bacterial clone or a transgenic goat.
- **d]** Presumably the genetically altered TILs will form an ever-increasing clone that will attack tumors even before they are visible or dangerous.
- **e]** In order to produce TNF commercially, the gene for TNF must be isolated. This step is avoided by using genetically engineered white blood cells.

2 What is the advantage of using a restriction enzyme that produces "sticky" ends?

- **a]** It makes it unnecessary to use DNA ligase molecular glue.
- **b]** Sticky ends form recognition sites for DNA ligase.
- **c]** Diagonally cut ends allow hydrogen bonding of base pairs to occur and thus position the strands more firmly during repair.
- **d]** Diagonally cut ends result in more exposed surface for recognition by DNA ligase.
- **e]** Restriction enzymes with sticky ends excise a larger seg-ment than restriction enzymes that cut straight across.

3 Why is an antibiotic-resistant gene included in a plasmid that is to be used for gene transfer?

- **a]** It is primarily to keep the plasmid alive during its use.
- **b]** It keeps the target cell alive.
- **c]** It functions as a marker for the cells that have been successfully transformed.
- **d]** It prevents the target cells from becoming infected after being transformed.
- **e]** All the foregoing are correct.

4 In the procedure described in the chapter introduction, why was the human tPA gene inserted in the field of influence of the goat's lactation genes rather than in a randomly selected gene area?

- **a]** To effect recombination, the insertion site must be identifiable by a landmark.
- **b]** So that tPA would be synthesized and secreted only with the milk.
- **c]** Because more tPA would be produced if the tPA gene were under the influence of a large-volume product, rather than a low-volume one.
- **d]** Both **b** and **c** are correct.
- **e]** None of the foregoing is correct.

CHAPTER 13

Animal Reproduction and Development

WHEN RIBS ARISE

Run your hand down your side and feel your ribs beneath the skin. These 12 pairs of evenly spaced, gently curving staves make a barrel that surrounds the heart, lungs, and other organs inside your chest. When muscles between your ribs contract, they bend your trunk to the side. How did ribs and muscles come to form in such an orderly fashion? What processes controlled the way they grew while you were an embryo in your mother's womb? Developmental geneticists Marnie Halpern and Charles Kimmel at the University of Oregon wondered about questions like this and decided to search systematically for mutations in early embryos that would block fundamental processes like the organization of ribs and the development of the brain. They couldn't, of course, induce deliberate mutations in human embryos; that would be unethical. Besides this, human embryos like the embryos of mice and other mammals, are difficult to study because they develop inside the uterus. Kimmel and Halpern decided

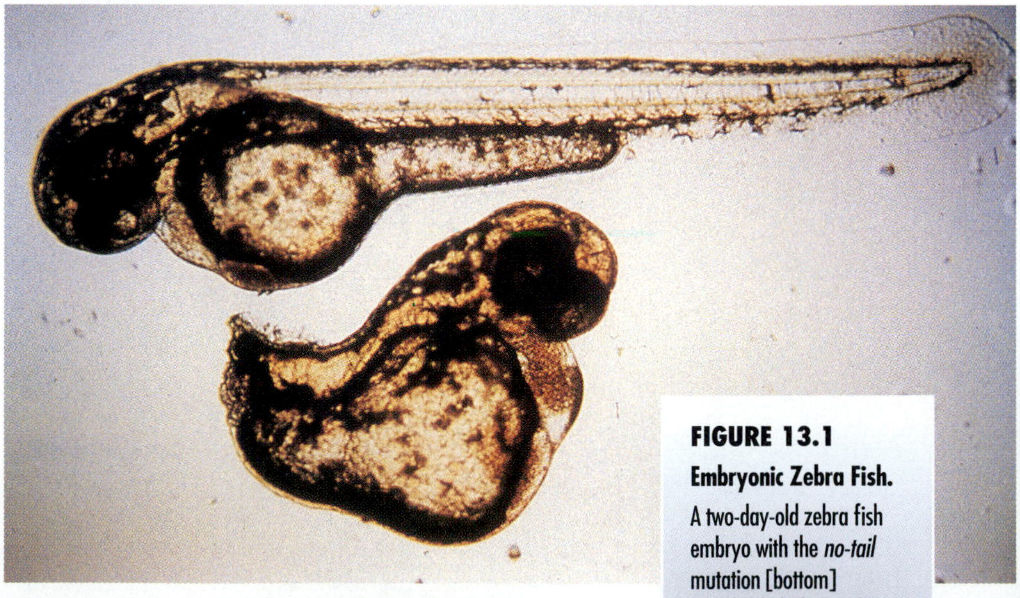

FIGURE 13.1
Embryonic Zebra Fish.
A two-day-old zebra fish embryo with the *no-tail* mutation [bottom] completely lacks a tail, while a wild-type embryo of equivalent age [top] has a long, straight tail.

to look instead for mutations in a small fish species called the zebra fish. They knew that the early steps of development in vertebrates (animals with backbones) are very similar in zebra fish and humans and that the rules governing these steps are almost certainly the same or very similar in the two groups of animals.

Halpern and Kimmel found a mutation affecting the *embryo*, or developing immature organism, which they called *no tail* [FIGURE 13.1A]. As you can see from the photograph, mutant embryos with this alteration lack the long, straight tail that is normal in zebra fish. As the biologists studied the mutant embryo more closely, they realized that it had several other defects. As you can see from the photograph, the blocks of tissue on the flanks of the mutant animal are misshapen. These tissues, called *somites*, will form the fish's ribs and flank muscles. Instead of being shaped like chevrons (>>>>>>>) as in the wild type [FIGURE 13.1B], they are shaped like rectangular blocks (|||||||||). In addition, the mutant zebra fish embryo was missing the stiff rod that marks the future location of the vertebrae in the backbone. The researchers knew that this mutation signified a defect in a gene that is necessary for embryos to form normal tail, ribs, and spinal cord, and they wondered how the gene does its job.

In this chapter, we explore the mysteries of **development**: how a single cell, the fertilized egg, divides into a many-celled embryo and then a fully functioning organism, with ribs, vertebrae, spinal cord, and other organs arranged in the proper fashion. We will also apply our understanding of development to cancer, since cancer cells continue to act like embryonic cells in many ways and divide and invade other tissues inappropriately. Some developmental mechanisms are universal, and some are specialized in different types of animals and plants. In this chapter, we concentrate on the development of animals with backbones. We will discuss the special mechanisms of human development in CHAPTER 14; of other animal groups in CHAPTERS 19, 21, and 22; and of plants in CHAPTERS 36 and 37.

The mechanisms of development are integral to understanding several aspects of biology. First, the genetic mechanisms we discussed in CHAPTERS 7 through 12 bring about all phases of development. Genetic mutations, such as those that cause PKU [CHAPTER 12] and Tay-Sachs [CHAPTER 3] alter the course of development—in these two cases, changing the way a person's brain develops. Development is also closely related to the study of anatomy (organismal structure) and physiology (organismal function), since developmental processes bring about the organism's shape and provide the proteins and enzymes that carry out day-to-day functioning. Finally, development helps us understand evolution. This is because much evolutionary change begins with alterations in the genes that control development [see, for example, FIGURE 16.3]. Some of Darwin's major arguments in defense of evolution by natural selection turned on the comparison of different kinds of embryos. As we go through the chapter, we will see how embryologists studied the no-tail mutation in the zebra fish embryo, since their research is typical of the ways biologists have learned the known details of animal development. ❑

MESSAGES

1 Sexual reproduction involves behaviors that promote the joining of eggs and sperm in fertilization, as well as the development of the fertilized egg into a new individual organism.

2 Animal development involves a series of mitotic cell divisions (cleavage) that results in many cells arising from the single fertilized egg. It also involves organization of three main cell layers (gastrulation); formation of body organs (neurulation and organogenesis); specialization of cells for particular functions (differentiation); increase in body size (growth); and sexual maturation. Meiosis in the new individual results in the production of eggs or sperm (gametogenesis).

3 Development occurs in an ordered sequence due to gene-regulatory substances (developmental determinants) in the egg and to interactions between developing cells.

4 In many ways, cancer cells behave like embryonic cells as a result of mutations in genes that normally regulate cell growth and differentiation.

Sexual Reproduction — Mating and Fertilization

We saw in CHAPTER 7 that sexual reproduction begins with *meiosis*, the process that produces genetically distinct haploid cells that can form *gametes*, the egg and sperm. **Fertilization**, the fusion of egg and sperm, produces the one-celled **zygote**, the first cell of the new individual. In animals as different as zebras and zebra fish, the basic elements of fertilization are the same: A tiny, mobile sperm cell fuses with a relatively huge ovum, or egg cell. The ovum contributes a haploid set of chromosomes, a storehouse of nutrients, and information that controls development. The sperm furnishes a second haploid set of chromosomes (thereby forming a diploid zygote) and acts as a trigger for the development of the new individual.

The meeting of egg and sperm requires events and processes at both macro- and microlevels. At the macrolevel, organisms **mate**; that is, they behave in a way particular to their species that brings egg and sperm together at the appropriate time for fertilization to occur [FIGURE 13.2]. At the microlevel, egg and sperm make contact and fuse through a series of precise cellular events that depend on the inherent structure and behavior of the gametes themselves.

In the following sections we first consider the range of common mating behaviors in animals, and then examine the specific microevents of fertilization.

MATING STRATEGIES

Cooperation and timing are the keys to mating, since in most animal species individuals of each sex produce only one type of gamete, and some degree of cooperation is necessary to ensure that when an egg is ready to be fertilized, sperm are available. Even in hermaphroditic species like earthworms and sea slugs, in which each individual produces both egg and sperm, reproductive cooperation often occurs. Hermaphrodites usually pair up and reciprocally exchange sperm, thereby fertilizing each other's eggs. Most animals are not hermaphrodites, however, and for mating to occur, individuals of both sexes in a species must first recognize each other, and then synchronize the release of gametes.

Mate Recognition Animal mates recognize each other mainly through visual, auditory, and olfactory cues. Female birds are often attracted to splendidly colored male birds. The peacock's tail, the indigo bunting's

[A] External fertilization in frogs

FIGURE 13.2

Mating Brings Egg and Sperm Together.

[A] External fertilization. The male golden toad (*Bufo periglenes*) clasps the larger female, and they release gametes simultaneously. [B] Internal fertilization. Mating behavior brings animals together, and the male releases sperm inside the female's reproductive tract. Praying mantises can continue to mate even if the female bites off the male's head.

[B] Internal fertilization in mantises

vivid blue plumage [FIGURE 13.3], the male frigate bird's expandable red throat sac, and the male widowbird's long, fluttering tail all stimulate the female's sexual interest at appropriate times of the year. Such visual cues abound throughout the animal kingdom: The lion's mane, the elephant seal's proboscis, and the stickleback fish's red stripe are all examples.

Auditory cues can be heard everywhere, especially on a summer night. Crickets respond to whirring sounds made when their bristled back legs are rubbed together, cats howl, and birds sing songs that attract members of the opposite sex of their own species. Olfactory cues include the odor signals that mammals and insects generate through the production of *pheromones* (FERE-o-moanz), chemical compounds that can attract members of the same species. Male pigs, for example, exude a strong-smelling pheromone called *androstenone* (also known as boar "taint"). If a sow in the receptive phase of her sexual cycle detects even a faint whiff of boar taint, she will assume a sexually receptive stance and wait to be mated.

Synchronization of Gamete Production and Release Within a given species, eggs and sperm must reach maturity at the same time for fertilization to occur. Some species mate only at specific times of the year (most often in spring), during which all adult individuals reach sexual readiness. For this reason, hens' eggs are more plentiful, and thus less expensive to buy, in springtime. Environmental cues, such as gradual changes in day length, often trigger the synchronized maturation of gametes. The release of gametes is often synchronized as well, and is usually stimulated by recognizable behavior by a member of the opposite sex. For example, at the time of mating, a male zebra fish rams a female's abdomen, and this action stimulates both fish to release gametes.

External and Internal Fertilization The synchronized release of gametes into the environment is a prelude to external fertilization, which is found in many water-dwelling species. In **external fertilization**, eggs and sperm are deposited directly into the surrounding environment (usually water), where some gametes meet by chance and fuse. External fertilization can occur with little or no contact between the mating adults, as in sea urchins. In other species, such as frogs and salamanders, the male clasps the female firmly for prolonged periods, and this touching stimulates the simultaneous release of gametes. Because unfertilized eggs survive only a short time, this synchronized release greatly increases the likelihood that an ovum will still be healthy when encountered by sperm.

Land-dwelling animals, including most mammals, birds, reptiles, insects, and snails, use **internal fertilization**, in which the male deposits sperm directly into the female's genital opening, and the gametes meet in a chamber (such as the uterus) or a tube. The synchronized sexual meeting of male and female animals is based on internal chemical signals, or *hormones*, as well as on behavioral cues. (Unlike most other animals, human beings are more or less continually sexually receptive.) Behavior resulting in sperm deposition within the female's body is called **copulation**. It has a distinct advantage over external fertilization in that sperm can be concentrated and protected within the female's body for a short time, until viable eggs are available for internal fertilization.

FIGURE 13.3

Flamboyant Displays Increase Male Mating Success of Some Birds.

The brilliant blue plumage of the indigo bunting [A] and the bright red throat pouch of the frigate bird [B] attract females.

[A]

[B]

[A]

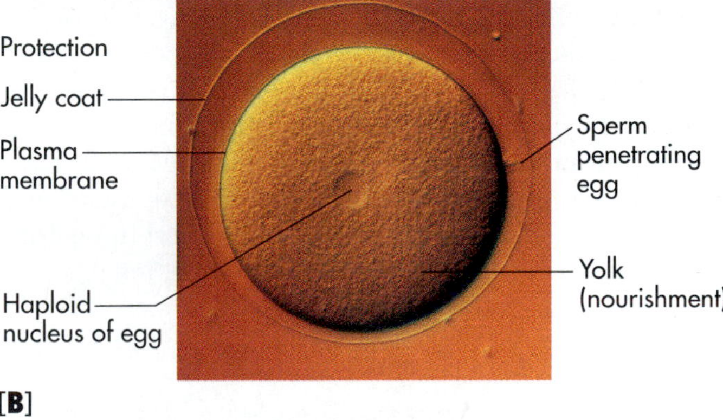

[B]

FIGURE 13.4

The Egg.

[A] The largest single living cell is the egg cell (the yolk) of an ostrich. Hunter-gatherer people of the Kalahari Desert in Namibia are about to cook one here. [B] The egg cell of a sea urchin protects the new individual, instructs its development, and contributes half the chromosomes. Caught at the moment of fertilization, a sperm penetrates the egg from the right. Because this egg warehouses nourishment for the embryo, it is 1000 times the volume of the sperm, which just donates a haploid set of chromosomes.

STRUCTURE AND FUNCTION OF EGGS AND SPERM

While mating brings the eggs and sperm into close contact at appropriate times, fertilization, the actual *fusion* of egg and sperm, results from a series of events that depend on the structures and activities of the gametes themselves.

Eggs and sperm are strikingly different. An egg is usually the largest single cell in a female's body. The egg of an ostrich is the largest animal cell on earth [FIGURE 13.4A]. The baseball-sized yolk of an ostrich egg is actually an **ovum** (an egg cell after it leaves the ovary), plus a huge store of food for the future embryo. The sea urchin egg in FIGURE 13.4B, while far smaller, contains ample nutrients, as well. An ovum performs many jobs:

1. It donates a haploid nucleus to the new embryo.

2. In some animals, it protects the developing embryo inside jellylike protein coatings, strong membranes, sacs of fluid, or hard or leathery shells.

3. It nourishes the embryo with **yolk**, which contains rich stores of lipids, carbohydrates, and special proteins and often, all the new embryonic mitochondria.

4. It provides the machinery for protein synthesis in the form of mRNAs, tRNAs, and ribosomes, enabling the zygote to undergo rapid cell division without slowing down for gene transcription.

5. Special substances in its cytoplasm that control the expression of genes direct the development of the early embryo. An ovum is like a computer loaded with a program, ready to play out a logical sequence of actions when the sperm's entry and fusion cause the program to run.

In contrast to an egg, a **sperm** is one of the smallest cells in the male body, stripped down to just a few elements [FIGURE 13.5]: (1) a compact haploid nucleus, (2) several mitochondria, which provide energy, (3) a long flagellum (tail) that propels the sperm, and (4) a sac of enzymes, the *acrosome*, that digests a path through the egg's protective outer coatings.

A sperm's streamlined size and shape reflect its narrow function: Reach the egg, penetrate its coating, and deliver a haploid nucleus into the egg's cytoplasm. This penetration and delivery of the nucleus, together with the activation of the resting egg, make up the events of fertilization.

EVENTS OF FERTILIZATION

The events of fertilization in sea urchins are described below and shown with corresponding numerals in FIGURE 13.6:

1. As its lashing flagellum propels a sperm headlong into an egg's jellylike coating, a sack at the tip of the sperm head releases enzymes that allow the sperm to penetrate the coatings on the egg's surface. Specific protein molecules on the surface of the now-exploded sack bind to other proteins, called *sperm receptors*, on the surface of the egg. Like a key fits a lock, the binding of the sperm protein to its receptor ensures that an egg

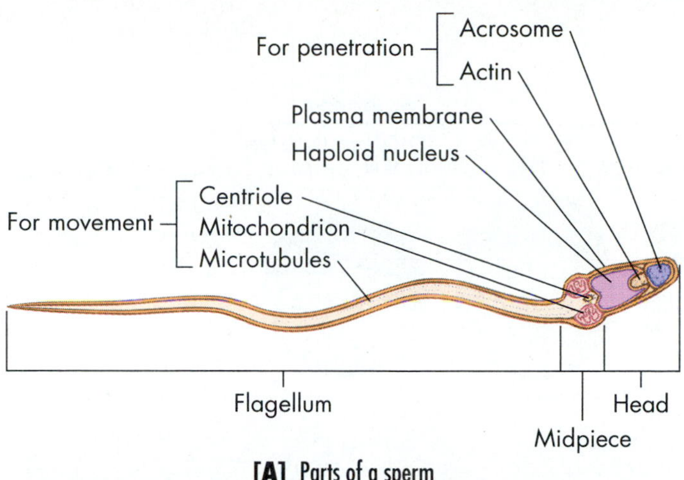

[A] Parts of a sperm

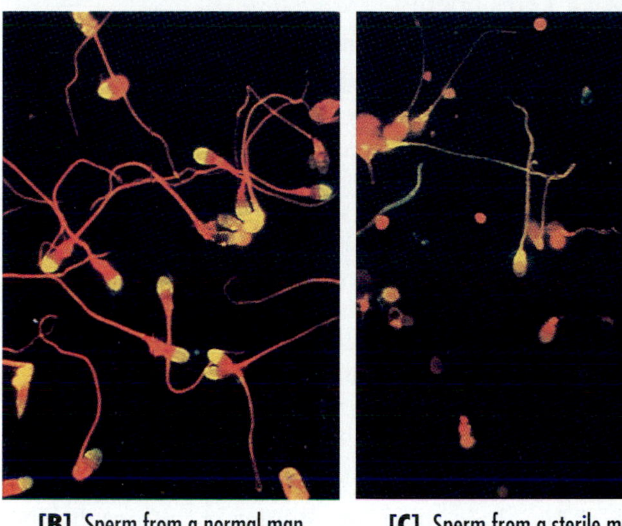

[B] Sperm from a normal man **[C]** Sperm from a sterile man

FIGURE 13.5

The Sperm.

This mobile cell contributes half the chromosomes during fertilization and triggers development in the zygote. **[A]** The streamlined sperm cell contains its payload, the haploid nucleus, plus factors that allow it to recognize and penetrate the egg, as well as organelles that power the lashing tail and allow it to swim. **[B, C]** Some men may be infertile because their sperm cells lack the normal form of a protein necessary for recognition and penetration of the egg. This protein (here stained bright yellow) is located around the front tip of a normal man's sperm. The sperm of some sterile men have less of the protein, and it is distributed abnormally. This evidence supports the hypothesis that the protein is crucial for fertilization.

FIGURE 13.6

Fertilization: Egg and Sperm Fuse, Forming a Genetically Unique Zygote.

The scanning electron micrograph (insert) shows a flotilla of sperm cells surrounding a sea urchin egg. The events of fertilization are discussed in the text. Note that the mitochondria of the sperm do not enter the egg—all of your mitochondria came from your mother.

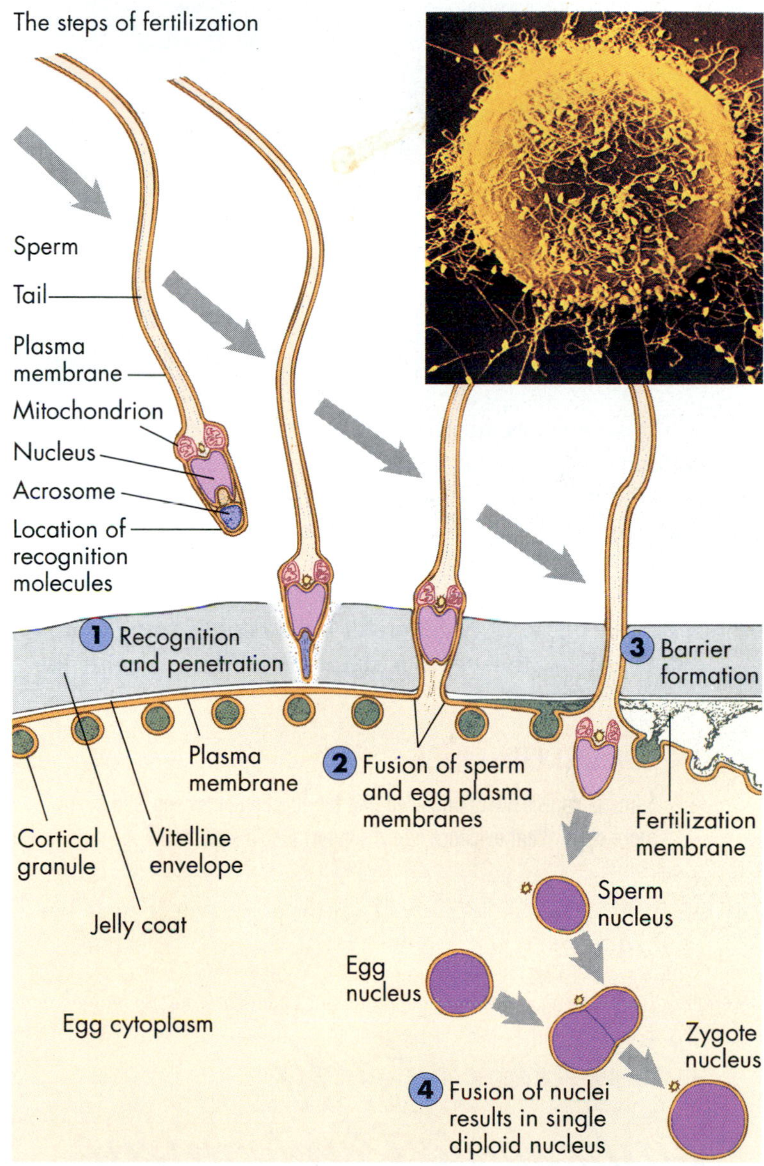

will be fertilized by a sperm only from its own species. Men whose sperm lack these recognition proteins are usually infertile [see FIGURE 13.5B and C].

2. The plasma membranes of the egg and sperm fuse, and the sperm head, including the nucleus, plunges into the egg's cytoplasm. The tail of the sperm remains outside the egg.

3. A sweeping wave of chemical reactions causes the breakdown of the granules beneath the ovum's plasma membrane to release substances that generate the *fertilization membrane*, a barrier that prevents additional sperm from entering the egg.

4. The sperm and egg nuclei fuse, and a novel genetic combination is established within the zygote's newly formed diploid nucleus.

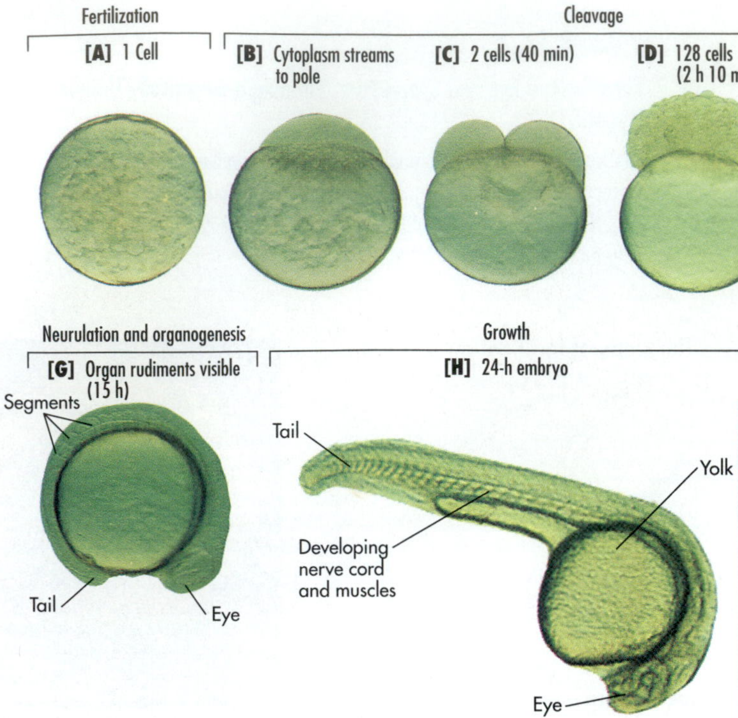

FIGURE 13.7

How a Fertilized Egg Develops into a Zebra Fish.

A newly fertilized egg [**A**] rapidly separates into a clear cap of cytoplasm above a denser yolk mass [**B**]. Within 40 minutes, the zygote divides into two cells, beginning the process of cleavage [**C**]. About two hours after fertilization, there are 128 cells in the cluster [**D**], and at three hours, the embryo consists of about 1000 cells [**E**]. By six hours, some of these cells have started to migrate downward, around the yolk, and have started to form cell layers in the process of gastrulation [**F**]. At 15 hours, organs have started to form in the process of organogenesis [**G**]. Here, the thickened part of the embryo has developed an eye, the rudiments of vertebrae along the back, and a tail. Finally, just 24 hours after fertilization [**H**], the fertilized egg has become a fish larva, almost fully developed and ready to hatch.

In some eggs, fertilization also triggers dramatic relocations of cytoplasm. For example, in zebra fish a clear cap of cytoplasm collects at the top of the egg above the yolk [FIGURE 13.7A and B]. In frogs, the entry of a sperm into a frog egg sets off a "shudder" of activity within the cytoplasm that sweeps around the egg just beneath its surface. This activity generates a crescent of gray-colored cytoplasm at the egg's equator, directly opposite the sperm's entry point.

➤ **CONCEPT CHALLENGE**

Some biologists have suggested that the function of an egg is to make more eggs. What evidence would support such a statement?

Patterns of Early Embryonic Development

In zebra fish, just a dozen hours or so after fertilization, a multicelled embryo has begun to take visible shape. It shows distinct head and tail regions, definite right and left sides, and the signs of a future backbone along the dorsal surface. After the fertilized egg cell divides into many cells [cleavage, FIGURE 13.7B to E], the cells become rearranged into three layers of cells [gastrulation, FIGURE 13.7F]. Then these three layers interact with each other and develop into individual organs such as the brain, gut, and muscles via the processes of neurulation and organogenesis [FIGURE 13.7G]. Eventually, the fully developed embryo [FIGURE 13.7H] grows larger and matures sexually. Table 13.1 summarizes the key events of animal development. The following sections examine each developmental event.

CLEAVAGE: FORMING A MULTICELLULAR EMBRYO

After fertilization, the newly formed zygote (fertilized egg) undergoes a series of rapid, synchronous mitotic divisions called **cleavage**, which results in a ball of many cells [FIGURE 13.8]. In contrast to cell division in later life [review FIGURE 7.7], cleavage divisions are not interrupted by growth periods. As a result, as the egg is cleaved in half, then quarters, eighths, sixteenths, and so on, the cells get smaller and smaller, forming a ball of cells. Because the cells have not yet begun to grow, this ball of cells is often about the same size as the original fertilized ovum [see FIGURES 13.7D and 13.8C].

During late cleavage in many animals, a fluid-filled cavity (a *blastocoel*) forms in the ball of cells, transforming it into a hollow ball called a **blastula** [FIGURE 13.8D].

Cleavage Patterns and Yolk Size The amount of yolk in an egg cell affects the pattern of cleavage divisions. Eggs with a small yolk divide completely through. For example, the eggs of sea urchins, zebra fish, and humans and other mammals divide repeatedly until they form what looks like a tightly clustered ball of grapes. In contrast, the eggs of birds and fish have so much yolk that cleavage furrows that usually divide the egg cannot pass all the way through [see FIGURE 13.7C]. Therefore, in the blastula stage, embryo cells collect in a disk at the top of the egg [see FIGURE 13.7D and E].

Regardless of the pattern of division, individual cells are no longer alike after cleavage. Cells have begun to specialize and have different developmental capabilities.

Developmental Determinants An egg cell contains not only nutrients, but also developmental information. Egg cytoplasm contains chemicals called **developmental determinants** that provide instructions for the developing embryo. In many species, these developmental determinants become localized in particular regions of the egg. After cleavage, certain chemicals and their instructions end up only in certain cells. The determinants activate the expression of specific genes in specific cells, causing various cell groups to develop differently.

For example, in a fertilized frog egg, a sliver of gray cytoplasm called the **gray crescent** develops opposite the point of sperm entry [see FIGURE 13.8B]. After cleavage only certain cells contain the gray crescent [see FIGURE 13.8C]. Cells containing the gray crescent material eventually become the embryo's dorsal (back) region [FIGURE 13.8D and E]. If a frog egg is manipulated in the laboratory

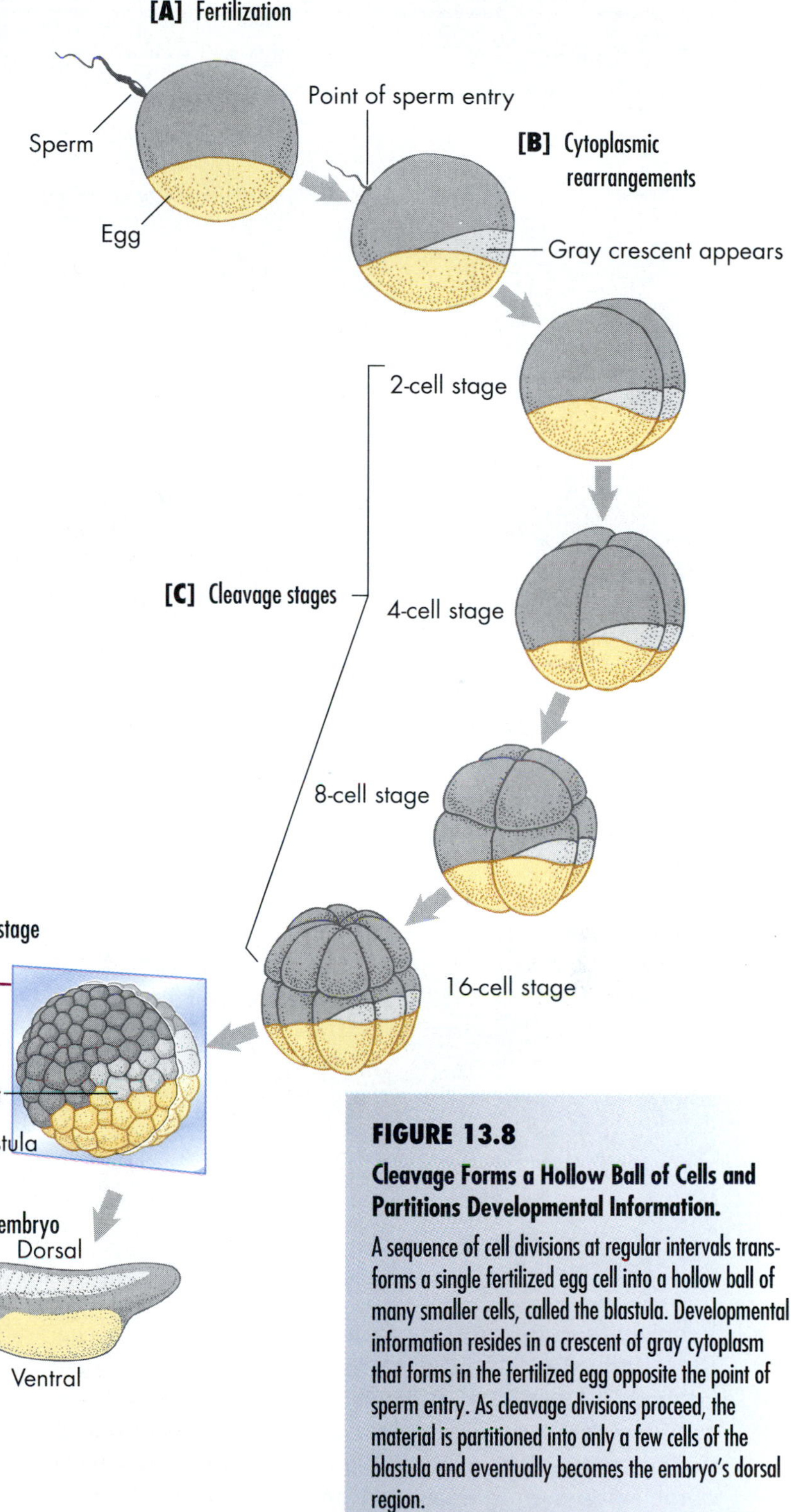

FIGURE 13.8

Cleavage Forms a Hollow Ball of Cells and Partitions Developmental Information.

A sequence of cell divisions at regular intervals transforms a single fertilized egg cell into a hollow ball of many smaller cells, called the blastula. Developmental information resides in a crescent of gray cytoplasm that forms in the fertilized egg opposite the point of sperm entry. As cleavage divisions proceed, the material is partitioned into only a few cells of the blastula and eventually becomes the embryo's dorsal region.

TABLE 13.1 Key Events of Development

Event	What Happens	Consequences
Fertilization	Egg and sperm fuse	A unique genetic combination is created
Cleavage	Fertilized egg subdivides into a ball with many cells	Developmental determinants are partitioned into different cells
Gastrulation	Cells rearrange and produce an embryo with three cell layers	Precursors of internal organs migrate to inside of embryo; cells interact with new neighbors
Neurulation	Neural tube forms	Conditions are established that will generate other body organs
Organogenesis	Body organs form	Cells interact, differentiate, change shape, proliferate, die, and migrate
Growth	Organs and whole organism increase in size	Adult body form is attained
Gametogenesis	Eggs and sperm develop	Reproduction becomes possible

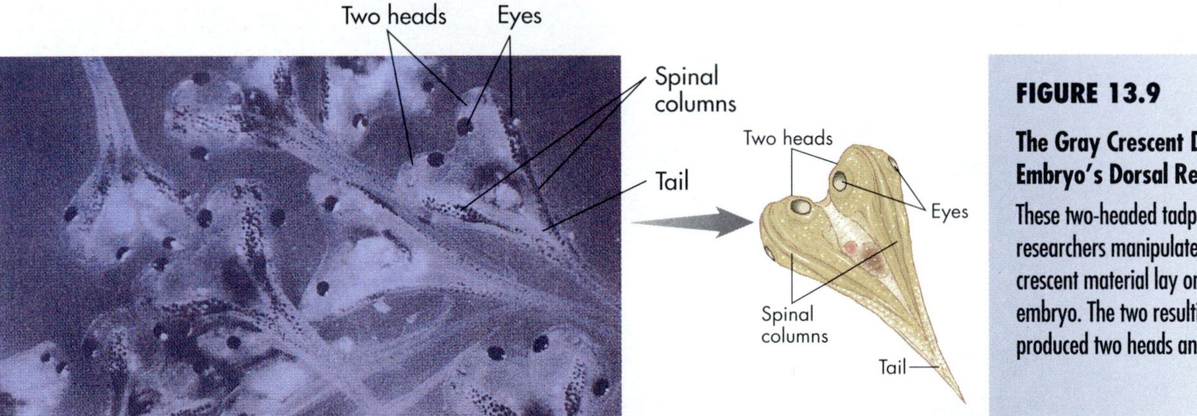

FIGURE 13.9

The Gray Crescent Determines the Embryo's Dorsal Region.

These two-headed tadpoles formed after researchers manipulated eggs so that gray crescent material lay on two sides of the embryo. The two resulting dorsal regions produced two heads and two spinal columns.

to create *two* gray crescents, one on each side of the egg, the result is a tadpole with two heads and two spinal columns [FIGURE 13.9]. Biologists concluded from this experiment that the gray crescent contains a developmental determinant that instructs cells to become specific parts of the embryo, in this case, the head and back.

While determinants may help direct certain developmental events in frogs and insects, other events in these animals, and most events in mammals, rely on a different key to early development: the position of cells relative to one another.

Cell Position and Specialization At the four- and eight-cell stages of cleavage, all the cells of a mammalian embryo are developmentally the same [FIGURE 13.10A]. We know this because when a researcher separates all four cells of an early mouse embryo, each cell can develop into a complete mouse instead of into various parts of a single mouse. Likewise, if a human embryo splits into

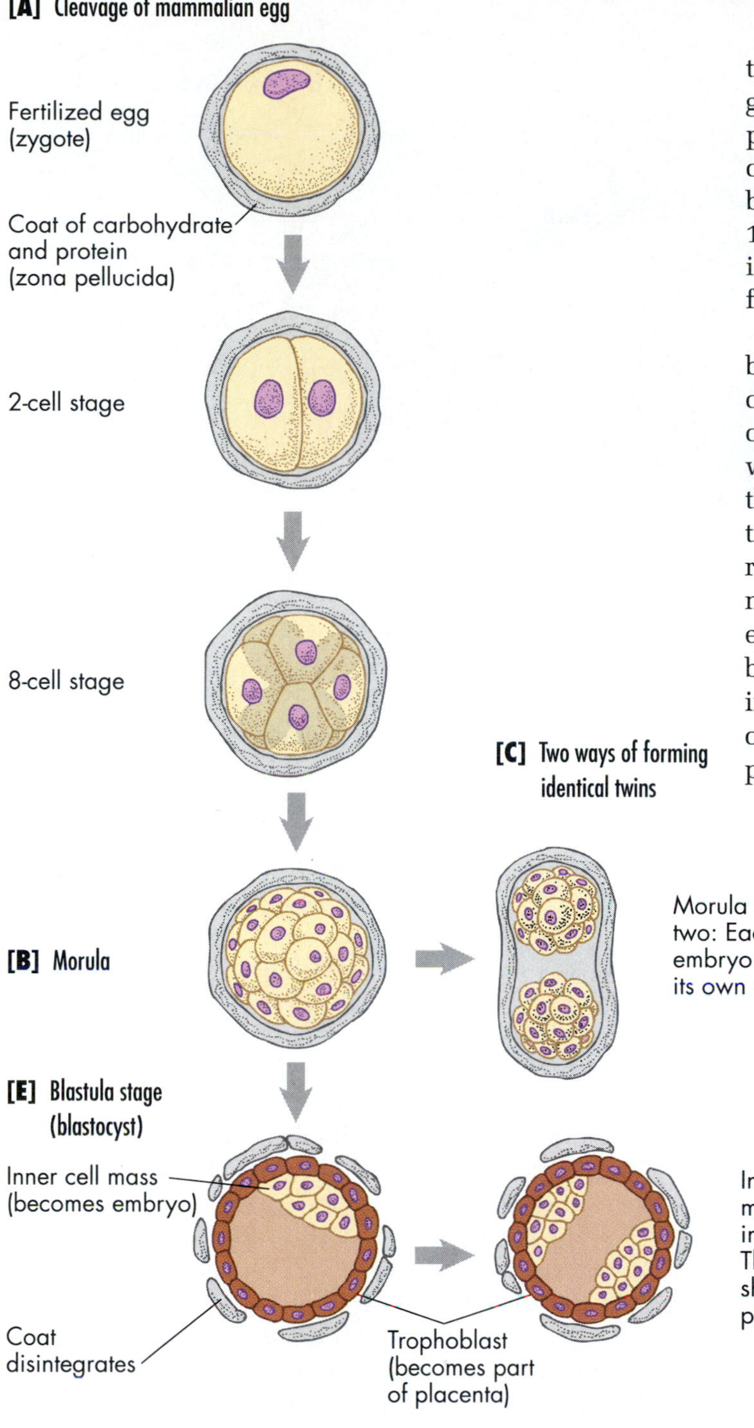

two groups of cells just prior to the blastula stage, each group can grow into a complete human being. This proves that at the time of the split, cells had made no unchangeable commitment to making just one part of the body. Identical twins develop from such splits [FIGURE 13.10D]. (In contrast, *fraternal* twins arise from the fertilization of two separate eggs by two separate sperm, not from the splitting of one embryo into two.)

With the further development of a mammalian embryo, some cells migrate inside the blastula and other cells remain on the surface. The cells on the inside, called the *inner cell mass*, become the living embryo, while the outer cells become the *trophoblast*, a structure that helps nourish the early embryo [FIGURE 13.10E]. The trophoblast develops into part of the **placenta**, the blood-rich organ that supplies maternal nourishment to the mammalian embryo. A cell's fate—becoming part of an embryo or part of a nutritional structure—depends upon being at the right place at the right time. The cells of the inner cell mass express genes appropriate to one developmental pathway, while the cells of the trophoblast express other genes.

[D] Identical twins

FIGURE 13.10

Cleavage Patterns in a Mammalian Embryo.

[A] In a mammal, early cleavage divisions create a solid cell cluster, or *morula* **[B]**. If the morula splits into two groups of cells **[C]**, identical twins, each with its own placenta, can develop **[D]**. Cells within the morula become rearranged and create a hollow blastula stage called the *blastocyst*. This has a trophoblast on the outside (destined to become part of the placenta, the embryo's source of nourishment) and a thick group of cells called the *inner cell mass* (destined to become the mammalian embryo) **[E]**.

FIGURE 13.11

Simplified Body Plan of an Adult Animal: Three Concentric Tubes.

Snakes, like most animals, have three distinct body layers established during early development. The embryonic endoderm gives rise to the gut, lungs, liver, and pancreas; the mesoderm generates muscles, bones, blood, kidney, and gonads; and the ectoderm leads to the skin and nervous system.

As you just saw, a cell's early fate can be determined by developmental determinants in the cytoplasm or by the cell's position relative to other cells. In either case, cleavage accomplishes two things: (1) It divides one large fertilized egg cell into many small cells, and (2) it establishes developmental differences among the small cells. These differences result in the expression of different genes in different embryonic cells, which leads to a dramatic series of migrations—gastrulation—that establish the animal's body plan.

GASTRULATION: FORMING THREE BODY LAYERS

Gastrulation is the process that transforms the hollow ball of cells generated during cleavage into a three-layered structure, eventually forming the basic elements of an animal, with front and back ends, right and left sides, a future skin outside, and a future gut inside. (*Gastrulation* literally means "gut formation.")

Formation of Germ Layers The bodies of most types of animals consist of three concentric tubes or layers of cells [FIGURE 13.11]: an inner tube, which forms the gut; a middle layer, including bones and muscles, which surrounds the gut; and an outer tube, which consists of skin with its sense organs. These three cell layers, or **germ layers**, are established by gastrulation and give rise to the organs and organ systems of the body:

1. The inner layer, or **endoderm** ("inner skin"), gives rise to the inner linings of the gut and the organs that branch from it, including the lungs, liver, and pancreas.

2. The middle layer, or **mesoderm** ("middle skin"), develops into muscles, bones, connective tissue, blood, and reproductive and excretory organs. These muscles and ribs come from mesoderm in blocks called **somites**.

3. The outer layer, or **ectoderm** ("outer skin"), becomes the skin and nervous system.

FIGURE 13.12 shows how all the parts and cells of an embryo (except the gametes) can be traced back to the three germ layers.

The essential feature of gastrulation is that cells move from the outside of the embryo to the inside, leading to the creation of an embryo with three tissue layers. While the details of gastrulation depend on the type of egg an animal forms, sea urchin embryos illustrate the main features of the process [FIGURE 13.13]. In Stages 1 and 2, future mesodermal cells leave the surface of the blastula and enter the cavity of the hollow ball [FIGURE 13.13A]. In Stages 2 and 3, a dent appears in the lower part of the embryo and deepens, causing the cells to move inward, like a finger pushing into a balloon. The indentation produces the endoderm, and proceeds toward the top of the embryo, eventually breaking through and forming the mouth (Stage 4). The original opening becomes the anus, and the endoderm develops into the digestive tube from the mouth to the anus. The cells remaining on the outside become the skin and nervous system.

The movement of mesoderm and then endoderm into the blastula thus generates a three-layered embryo, but it also establishes body axes—top and bottom, front and rear. In frogs, mesodermal cells begin to migrate into the embryo at the gray crescent [FIGURE 13.14A to G], establishing the embryo's dorsal (back) and ventral (belly)

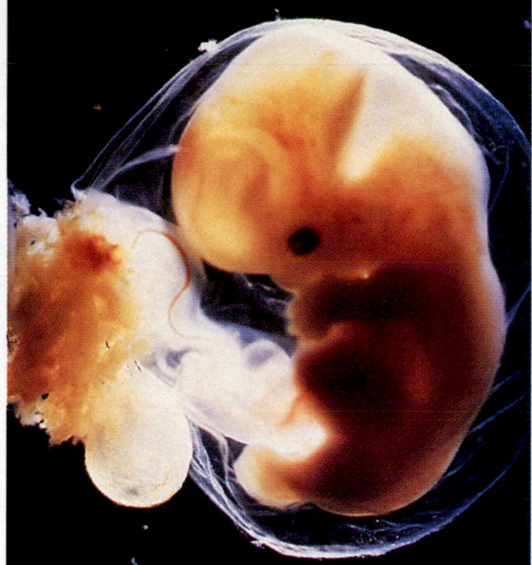

FIGURE 13.12

Origin of a Mammal's Body Parts.

This diagram shows the origin of the liver, brain, kidney, and other organs and tissues from the three embryonic cell layers: the endoderm (yellow), mesoderm (red), and ectoderm (blue). For example, mesoderm blocks off into *somites*, cubes of tissue that form the vertebrae of the spine and the muscles of the trunk.

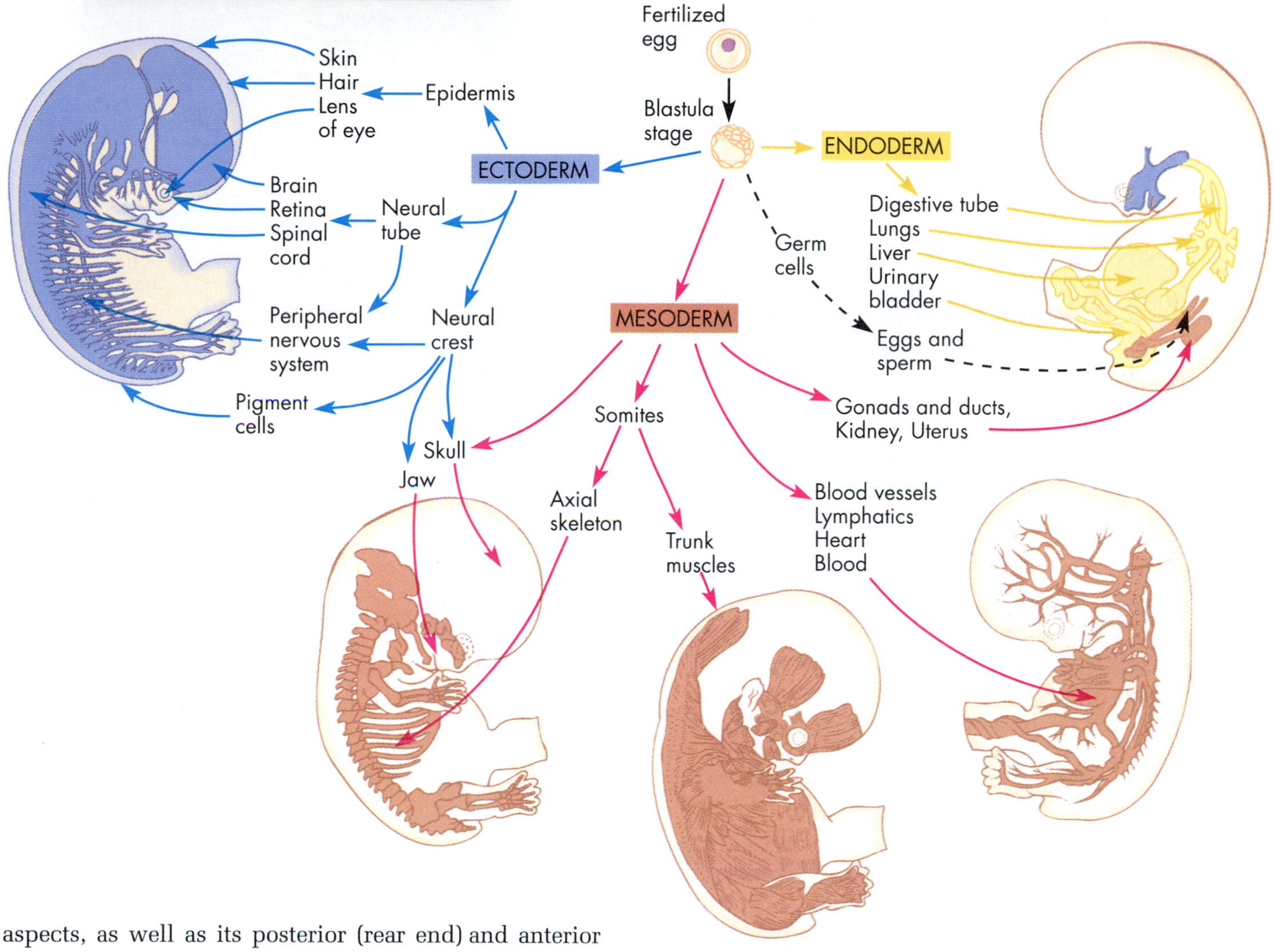

aspects, as well as its posterior (rear end) and anterior (front end), and creating a left and right side [FIGURE 13.14H and I]. Thus the gray crescent, whose own position was determined by the sperm's point of entry into the egg, plays a primary role in orienting the axes of a frog's body.

➤ CONCEPT CHALLENGE

If Marnie Halpern and Charles Kimmel found a mutation that blocked the process of gastrulation in the zebra fish, what do you think the mutant embryos would look like? (Consult FIGURE 13.7.)

NEURULATION: FORMING THE NERVOUS SYSTEM

Gastrulation sets up the conditions for the next developmental stage, **neurulation**, the development of the **neural tube**, the hollow cylinder that differentiates into the brain and spinal cord.

As gastrulation is completed in a vertebrate embryo, an arch of ectoderm, mesoderm, and endoderm is formed above the yolk, where the embryo's back will form [FIG-

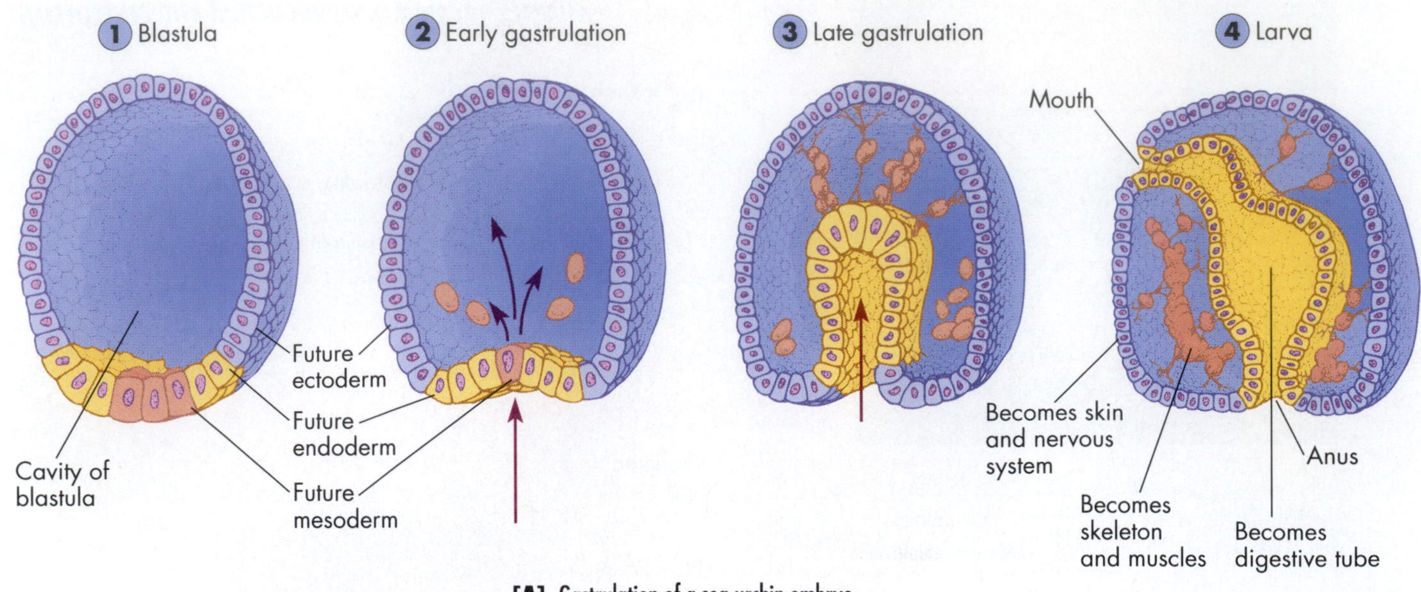

[A] Gastrulation of a sea urchin embryo

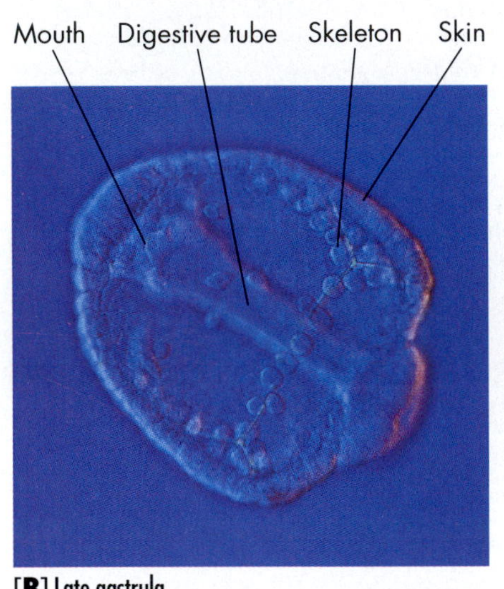

[B] Late gastrula

[C] Adult sea urchin

FIGURE 13.13

Forming Three Body Layers: Gastrulation in a Sea Urchin Embryo.

[A] The hollow ball of the blastula (1) begins to indent at the bottom, and individual mesoderm cells (red) pop from the wall of the blastula into the cavity (2). The indentation of the endoderm (yellow) continues, pulled along by some mesoderm cells (3), until it reaches the ectoderm (blue), with which it fuses and forms a complete digestive tube through the embryo (4). The mesoderm forms the skeleton and muscles. **[B]** Photomicrograph of a gastrulating sea urchin embryo. **[C]** An adult sea urchin.

URE 13.15A and B]. Along the embryo's midline, the mesoderm becomes organized into a rod called the *notochord*, which marks the future site of the backbone, or spinal column. (In the *no-tail* embryo described in the chapter introduction and shown in FIGURE 13.1, the notochord is missing, which is why the tail fails to elongate.) The backbone is the structure that distinguishes vertebrates from invertebrates, animals without backbones, which we discuss in CHAPTER 21.

The neural tube forms when the sheet of ectodermal cells rolls up. The surface layer of cells folds up, forced from the sides like two wrinkles in a bed sheet [FIGURE 13.15D]. The two folds fuse and bud off to form the neural tube [FIGURE 13.15F and G]. Cells forming the neural tube assume a wedgelike shape [FIGURE 13.15H], helping the tube roll up.

Spina Bifida: A Developmental Error Unfortunately, mistakes can occur during neurulation. In about 1 out of every 200 live human births, for example, the neural tube does not roll up completely and fails to close, producing **spina bifida**, the most common birth defect among babies in the United States. This condition leaves the spinal cord open to the outside, often causing paralysis. Although spina bifida is not preventable at this time, appropriate doses of some vitamins taken by the mother during pregnancy may help reduce its frequency. The condition can be detected early in the pregnancy, and can usually be treated surgically after the baby is born.

FIGURE 13.14

Gastrulation in a Frog Embryo.

Recall that fertilization and cleavage create the blastula, a hollow ball of cells [A, B]. On the surface of this ball [C, D], embryologists have mapped the location of cells that will eventually form the three germ layers—the outer ectoderm (blue), the intermediate mesoderm (red), and the inner endoderm (yellow). As gastrulation begins [E], the mesodermal cells of the gray crescent leave the surface of the blastula and enter the hollow ball's cavity. The endoderm follows. During midgastrulation [F], the future gut cavity enlarges and the blastula cavity becomes smaller. When gastrulation is complete [G] a three-layered embryo has developed, and an exterior view [H] shows an embryo with the yolk plug marking the posterior, the site of the future anus, and the anterior, dorsal (back), and ventral (belly) sides apparent. For comparison, [I] shows a newly hatched larva in the same orientation.

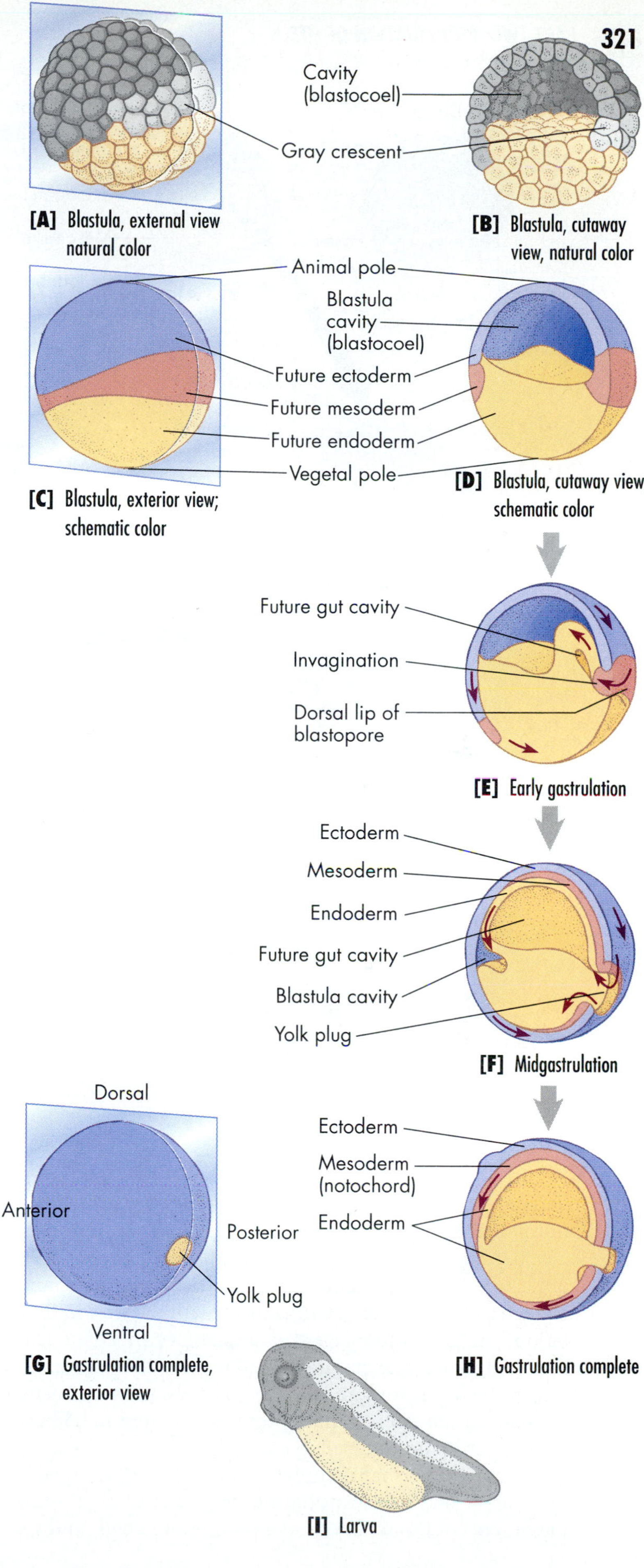

Embryonic Induction In 1924, as part of their investigation of neural tube formation, the embryologists Hans Spemann and Hilde Mangold removed cells that would normally make the notochord, a dorsal (back) structure, from the gray crescent area of a newt embryo. They transplanted these cells into a host newt embryo in a region that would normally form belly skin [FIGURE 13.16]. The implant developed into a notochord, and, remarkably, caused a second neural tube to form on the belly of the host embryo, which developed into a bizarre set of Siamese twins joined belly to belly. (This is another example of how a cell's position affects the type of structure it will form; future skin cells located near the notochord are caused to develop into neural cells.)

Spemann and Mangold concluded that the transplanted cells, which contained gray crescent cytoplasm, had liberated a developmental signal that *induced* the host's ectoderm to form a neural tube, notochord, muscles, and eventually an entirely new embryo. In general, **embryonic induction** is the interaction of two adjacent cells or groups of cells such that the developmental fate of one or both cell groups is altered.

The *no-tail* mutation in zebra fish [see FIGURE 13.1] gave biologists an opportunity to investigate the process of induction further. Researchers wondered why there were two defects in the mutant: the failure of a notochord to form and the failure of the somite blocks to assume their normal shape. One hypothesis, referred to as the *independent hypothesis*, was that the two abnormalities were independent of each other. To test that hypothesis, researchers transplanted normal notochord cells into the mutant embryo. They found that genetically normal notochord cells "cured" the defect in the adjacent genetically mutant somite blocks. They concluded that the in-

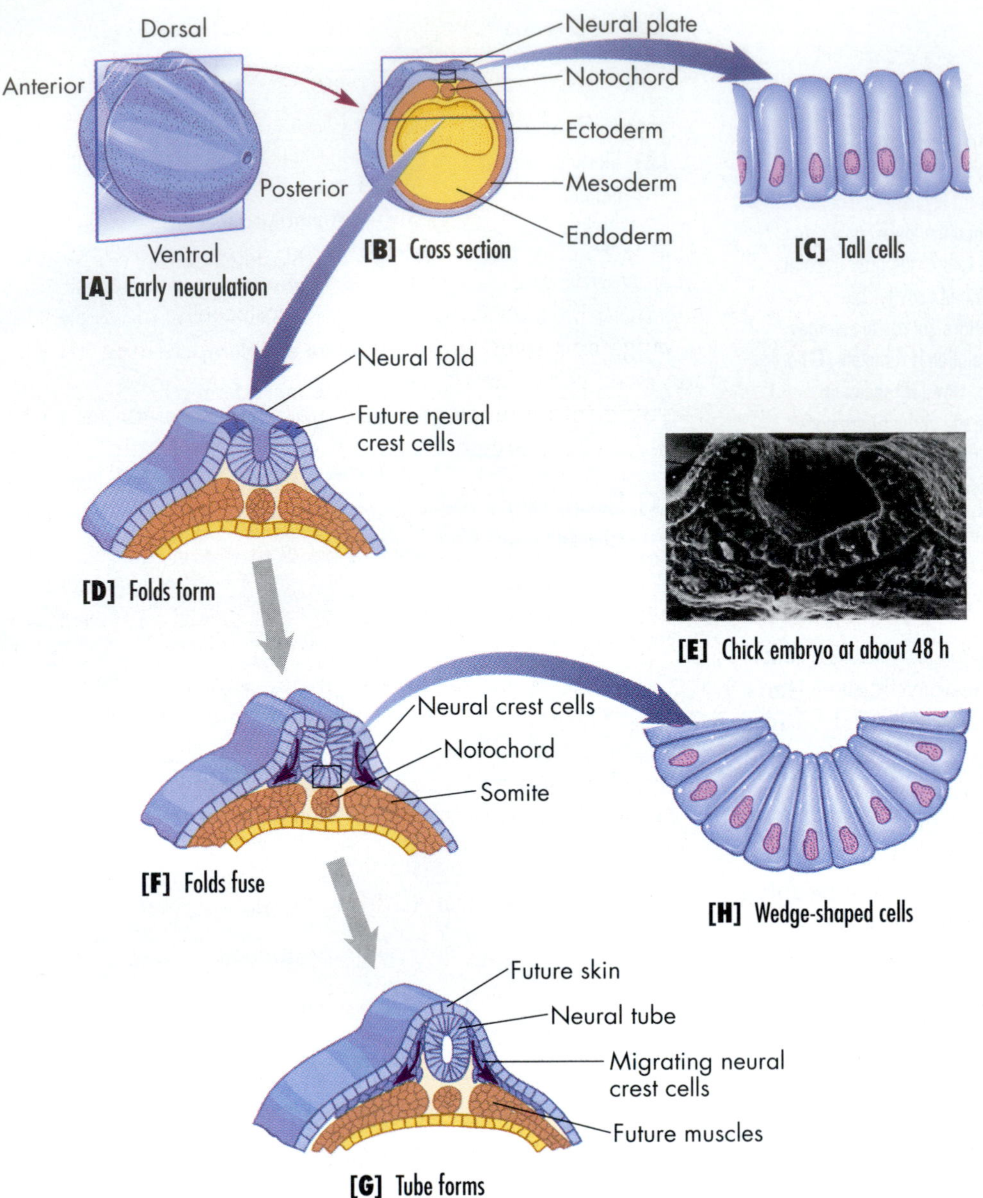

FIGURE 13.15

Neurulation: Formation of the Future Nervous System in a Frog Embryo.

The events of neurulation, or neural tube formation, are easier to envision by considering only the upper surface of this frog embryo cut in half. [**A**, **B**] The neural plate, a region of tall ectoderm cells [**C**], takes shape on the embryo's dorsal (back) surface. [**D**] Folds form as cells in the neural plate change shape and cells from the side push in, causing a wrinkle in the sheet. The insert [**E**] shows a scanning electron micrograph of a chick embryo at this stage of development. [**F**] When the neural folds come together, the neural tube pinches off, the surface ectoderm (future skin) becomes a continuous sheet, and the neural crest (the cells at the crest of the neural fold) begins to disperse. [**G**] By comparing [**G**] to [**H**], you can see how changes in cell shape from columns to wedges help the neural tube form.

dependent hypothesis was wrong. They decided that the normal notochord must send a signal to the somites, causing *them* to develop their typical chevron shape. In other words, the notochord normally induces the shape change in the somites. In an embryo with the *no-tail* mutation, that inductive signal is apparently missing. The development of the somites—and of your ribs and muscles, as well—is thus an example of how embryonic cells depend on the environment provided by their neighboring cells in order to develop correctly.

With the normal completion of early animal development—cleavage, gastrulation, and neurulation—the major axes of the embryo have been established, and its cells are in position to form body organs. But the embryo is far from being a finished animal. The next developmental priority is **organogenesis**, the establishment of its body organs.

➤ CONCEPT CHALLENGE

An embryologist isolates and grows the lower third of an embryo (the yellow cells from FIGURE 13.8D) and finds that they form only endoderm. When she isolates the upper third of cells (the gray cap), she finds they form only ectoderm. Yet when she isolates the lower cells and upper cells and grows them together, all three layers form, including mesoderm. What kinds of cell interactions are taking place in her experiment?

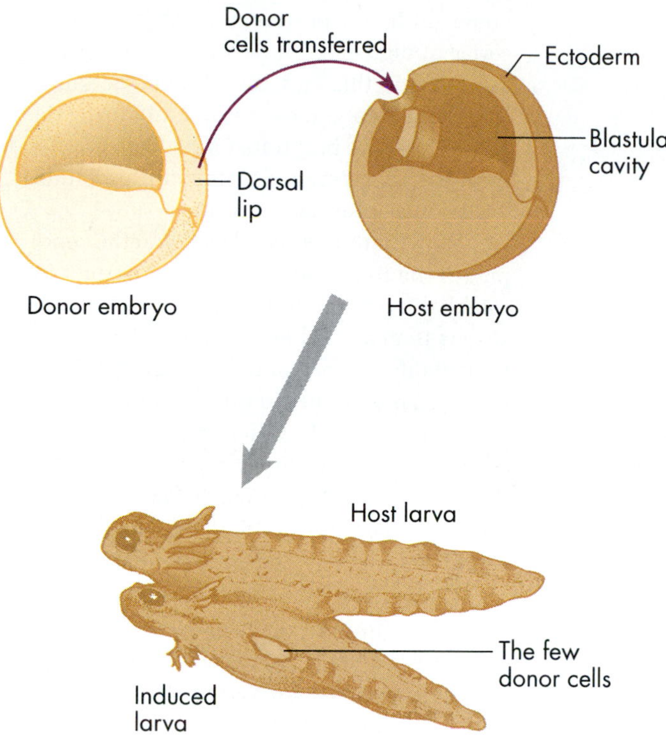

FIGURE 13.16

What Establishes the Embryo's Head-Tail Axis?

Experimenters transplanted cells from the gray crescent area (dorsal portion) of one newt embryo to the future belly of another newt embryo and watched as neurulation took place on both sides of the embryo. Two animals emerged, joined belly to belly. Gray crescent cells induced the host's ectoderm layer to generate a new head-tail axis and eventually a new embryo.

Organogenesis: Development of Body Organs

Suppose you collected some zebra fish embryos from the bottom of the Ganges River and observed their development under a microscope for three days at six-hour intervals. As you watched each tiny fish develop still enclosed in its clear, jellylike globe, you would see various organs take shape in a translucent embryo: eyes, fins, kidneys, muscles, and other organs. This organ formation involves changes in the shape of cells and cell groups, or **morphogenesis**, and changes in the types of proteins made by individual cells, or *differentiation*.

MORPHOGENESIS: DEVELOPMENT OF ORGAN SHAPE

The three germ layers formed during gastrulation make the rudiments of organs, but the cells of those layers undergo a procession of changes during morphogenesis. Some cells change shape, some divide rapidly, others die, and still others move to new places as the organ develops. This cavalcade of proliferating, disappearing, and migrating cells goes on continuously, and the results include endoderm bulging from the gut and becoming lungs and liver, a pocket of ectoderm pinching off and forming the future ear, and loose mesodermal cells aggregating into a block of future muscle tissue. FIGURE 13.17 shows how programmed cell death helps sculpt the developing hand, and BOX 13.1 [page 324] explains how correct axes form in a developing arm.

Cell Migration Recall that the neural tube arises as a sheet of ectodermal cells that curls up, thereby creating folds that fuse together [see FIGURE 13.15]. Where these folds meet, a group of cells called the **neural crest** begins to migrate along defined pathways around the embryo, giving rise to cells and organs in distant locations. Some neural crest cells stay near the epithelium, migrate into the skin, and form pigment cells. If you can see brown moles, freckles, or an overall dark skin pigment on your

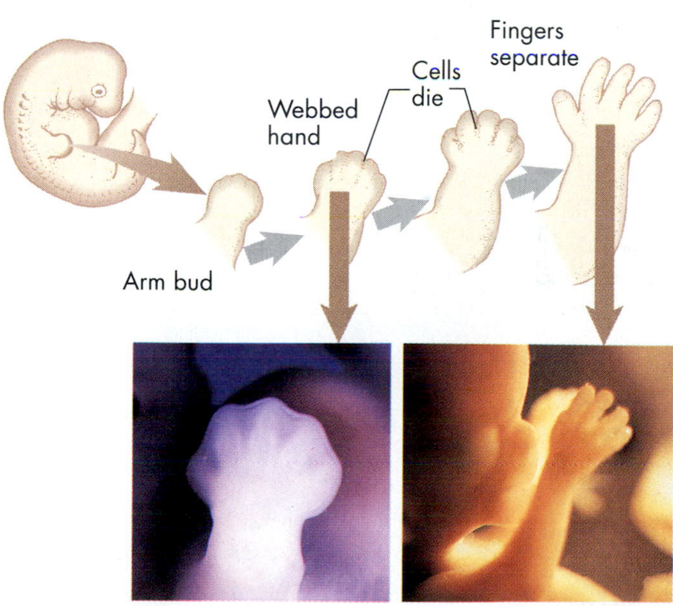

FIGURE 13.17

Cell Death Generates the Shape of Your Hands.

Your hand first forms as a paddle. The fingers emerge when four rows of cells between the fingers die away. In ducks cells between the digits don't die, leaving the webbing which helps them paddle across a pond.

box 13.1
Biology Applied

How Does a Cell "Know" Where It Is?

Sometimes, as a human embryo develops, one of its limbs gets "confused." Instead of forming a normal hand with a thumb on one side, a little finger on the other, and three digits in between [FIGURE 1A], it generates a hand lacking a thumb but possessing two little fingers that flank five middle digits [FIGURE 1B]. A machinist who lived in Boston about 100 years ago had such a hand and reported that it "not only helped him do his job, but also gave him certain advantages in playing the piano."

An unusual appendage like this raises a special question: What makes a little finger form on one side of the hand but not on the other? Or, more generally: How does a cell in an embryo "know" where it is so it carries out its functions in the appropriate place? Over 20 years ago, Lewis Wolpert suggested that such cellular localization was based on *positional information*—spatial cues that tell cells where they are in the embryo, perhaps via diffusible molecules. Experiments on chicken wings and other vertebrate limbs have revealed much about how positional signaling takes place in embryos.

A chicken's wing develops much like a human arm [FIGURE 1C]. It has a single large bone attached to the shoulder, two smaller bones in the lower limb, a wrist joint, and a series of digits, and it develops from a tiny bud. Within this bud, a shell of ectoderm overlies a core of mesoderm, and during development, the bud transforms from a rounded protuberance to an elongated limb with bones, muscles, and skin.

W. J. Saunders wondered whether one part of the limb bud was more important in establishing the thumb-to-pinky series of positions than any other. He found that if he transplanted a special region (called the *polarizing zone*) from the posterior portion of one wing bud to the anterior portion of another, the limb produced a wing similar to the hand of the Boston machinist, with two mirror-image copies of digits 4 and 3 and a single central copy of digit 2 [FIGURE 1D].

It was later found that researchers can produce a similar mirror-image wing by applications of a substance called retinoic acid. (Retinoic acid is a derivative of vitamin A, which is chemically related to the bright

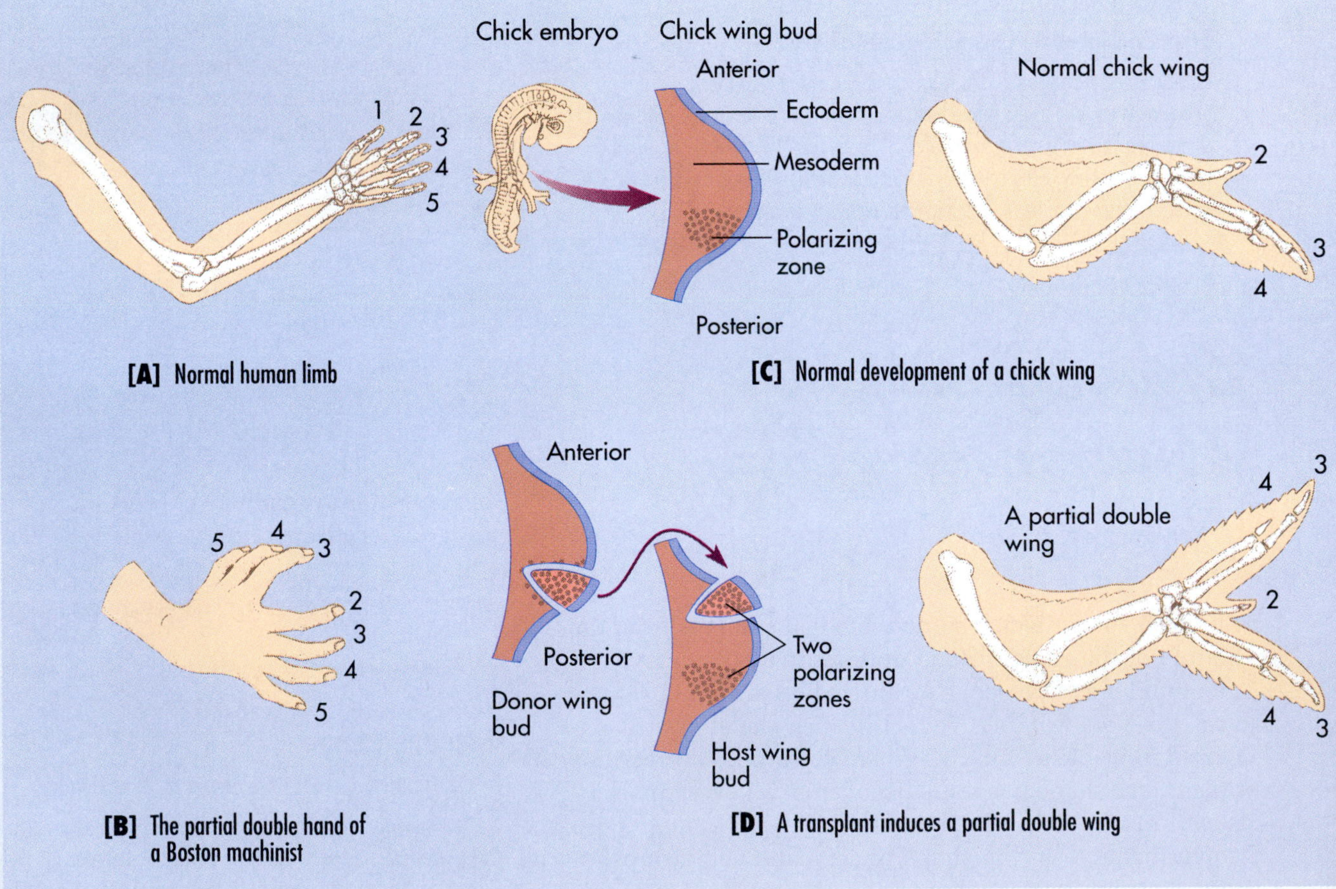

FIGURE 1
Development of the Hand.
[A] A normal human hand contrasts with **[B]** that of a machinist from Boston. This man's hand had a double set of digits. **[C]** A normal wing bud in the embryonic chick leads to normal bones in the adult chicken wing. **[D]** When the posterior part of a limb bud is removed from a donor and transplanted to the anterior portion of a host wing bud, the result is a limb with two posterior axes and two mirror-image sets of digits.

orange pigment in carrots.) In the late 1980s, it was shown that retinoic acid is distributed in a gradient across the limb, with a high concentration at the limb bud's posterior and diminishing concentrations toward the anterior edge. It was suggested that this concentration gradient directs the pattern of digit formation, allocating digits to final locations on the limb and specifying the type each will develop into (thumb, middle finger, pinky, and so on). Limb cells can "read" the amount of retinoic acid around them because it binds to a receptor on the cell surface for that particular substance. The binding of retinoic acid to the cell-surface receptor causes new sets of genes to be expressed, and scientists think these genes, in turn, cause a group of cells to develop as a thumb or as a little finger (or equivalent digits).

Exposure to developmental regulatory molecules similar to retinoic acid can have dramatic and often tragic consequences to a developing human. A Swiss pharmaceutical company released a drug in 1982 called Accutane that can significantly help—even permanently cure—people with severe acne. But there was a catch: the drug turned out to cause birth defects. Not coincidentally, Accutane is a derivative of vitamin A called isotretinoin with a structure similar to the molecule retinoic acid. Like retinoic acid, Accutane exerts profound influences on a developing fetus. If a woman becomes pregnant while taking Accutane, her baby may be born with a variety of defects, including stunted growth of the heart muscle, mental retardation, abnormally small jaws, and ears growing below chin level. Despite strong warning labels and the advice of physicians and pharmacists, hundreds of women have taken Accutane during pregnancy, and the U.S. Food and Drug Administration estimates that between 1982 and 1988 nearly 600 babies were born with Accutane-induced birth defects.

Researchers are continuing to study the mechanisms by which Accutane alters fetal development and to search for safer alternatives. In the meantime, the FDA has directed the drug's manufacturers to print photographs of deformed newborns on the medicine's label. Only continued public education about the teratogenic (inducing birth defects) nature of isotretinoin will prevent tragic deformities and the loss of an acne drug that is changing some people's lives for the better.

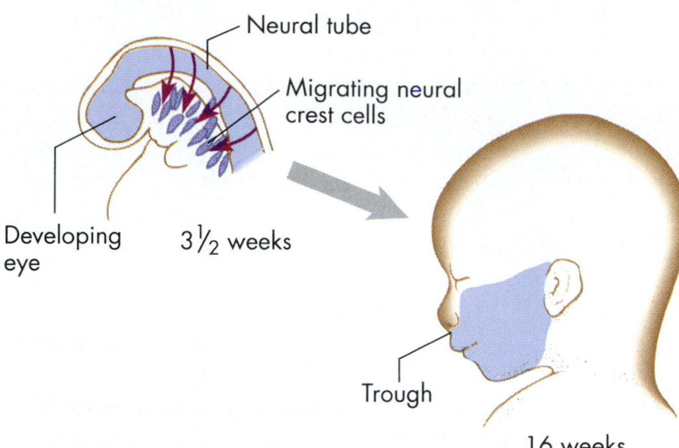

FIGURE 13.18

Migrating Neural Crest Cells Help Form the Face.

Neural cells move from the neural folds around the sides of the neural tube and embryonic head and help form the lower part of the face and jaw.

arm, you are observing the results of embryonic pigment cells that migrated from the neural crest near your spinal cord across your back and down your arm.

Other neural crest cells migrate from the back of the embryonic head and around both sides of the developing mouth, forming the teeth, as well as some of the muscles, bones, and cartilage of the lower face [FIGURE 13.18]. As the migrating cells approach the front of the face, they grow together at the "trough" between the nose and upper lip. If the mass of cells fail to migrate and grow together, they leave a space between the nose and mouth, or they leave a gap in the roof of the mouth. These birth defects, respectively called *cleft lip* and *cleft palate*, can usually be corrected by surgery.

While cell migration, cell death, and the folding of cell sheets can shape organs, another set of changes—this time at the protein level—occurs before an organ can carry out its specific functions.

DIFFERENTIATION: DEVELOPMENT OF ORGAN FUNCTION

Each type of cell in an animal's body performs a specific function: muscle cells move body parts, retinal cells in the eye detect light, red blood cells carry oxygen. A cell can carry out its particular functions in large part because of the distinctive proteins it contains. Some proteins break chemical bonds in food molecules, others cause a cell to contract, still others change form when struck by light. Each cell type that has its own form and contains its own type of proteins is called a *differentiated*

TABLE 13.2 Processes That Bring About Development

Process	What Happens
Formation and storage of cytoplasmic determinants	Chemical factors stockpiled in the egg (like gray crescent cytoplasm) become partitioned into different cells during cleavage and cause cells to select and express a specific set of genes
Induction	One cell group causes another group to start developing into a different cell type; e.g., the notochord induces nearby ectoderm to form the neural tube
Morphogenesis	Organs form via changes in cell shape, cell proliferation, cell death, and cell migration; e.g., neural tube rolls up; digits are carved; neural crest cells migrate
Differentiation	Specialized cell types develop; includes the production of specific proteins that allow a cell to carry out its particular role; e.g., skin cells make fibrous proteins; intestinal cells make digestive enzymes

cell. Cells become specialized through the process of **differentiation**.

Different cell types contain different proteins because specific genes are active in those cells. In muscle cells, for example, genes encoding contractile proteins are turned on, while in cells in the retina of the eye, genes and mRNAs that encode light-sensitive proteins are active while many other genes are inactive. One of the hottest research topics in biology today is the mechanisms that cause the expression of specific genes.

REGULATION OF DEVELOPMENT

So far, we have followed a developing embryo through a series of regulated events, beginning with a single fertilized cell and progressing to the formation of a ball of cells, the rearrangement of the cells into a three-layered embryo, and finally, the molding of these layers into organs that have a specific shape, make specific proteins, and function in a specific way. Developmental biologists are still trying to understand one of the great mysteries of modern biology: How do the regulatory mechanisms actually turn on different genes in different cell types to cause these events? Although the full answer is not known, enough information is available to discuss a general model of developmental regulation [TABLE 13.2].

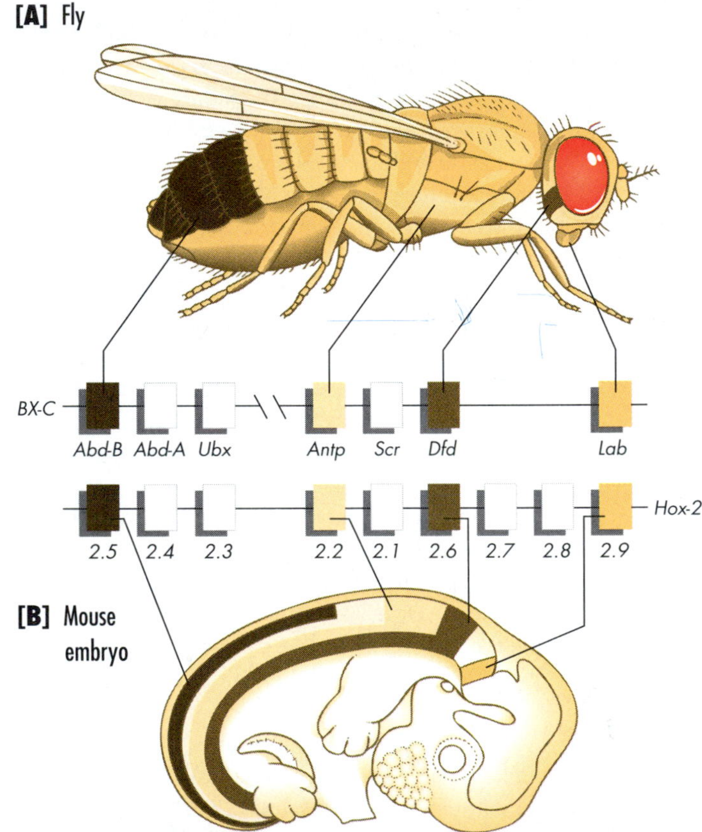

FIGURE 13.19

Similar Genes Control Development in Very Different Animals. The body of a fly [A] and a mouse [B] seem to be very different, but very similar homeotic genes are expressed in both animals at specific times during their development. Corresponding genes are expressed in the same order from the front to the rear of both animals, and the genes are even in similar orders on the chromosomes. The fact that mutations in these genes cause defects in development supports the notion that they play an important regulatory role. This evidence suggests that, due to the process of evolution, very similar rules program development in all animals.

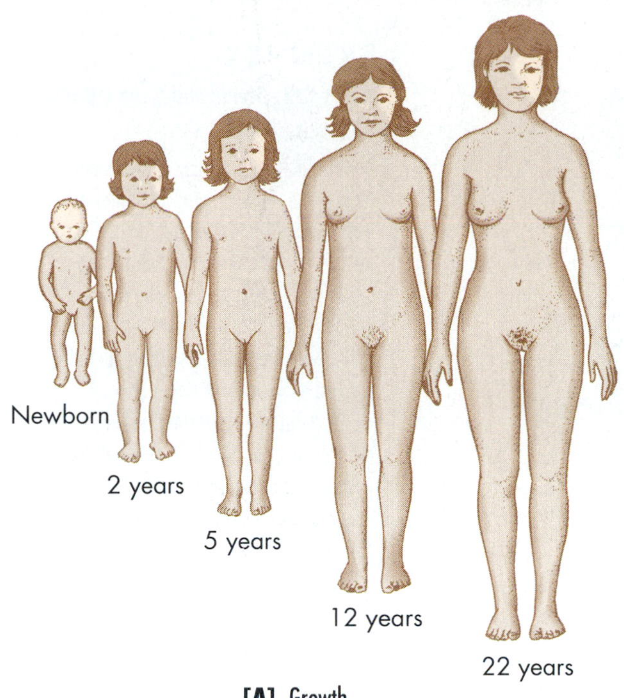

[A] Growth

FIGURE 13.20

Growth: Weight Quadruples and Proportions Change between Birth and Adulthood.

[A] As this girl grows into a woman, her weight multiplies from 7 lb to 14, 28, 56, and 112 lb. Also, her head grows proportionately smaller, her arms and legs longer, and her trunk less dominant. [B] X-rays of human hands reveal that the finger bones lengthen and enlarge and the knuckles become more and more prominent throughout life.

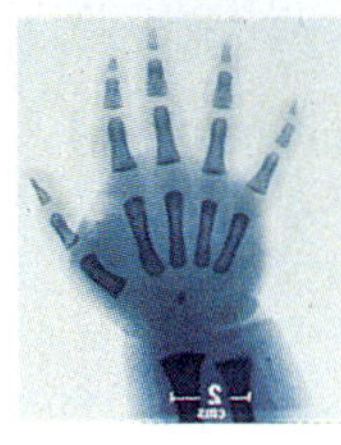

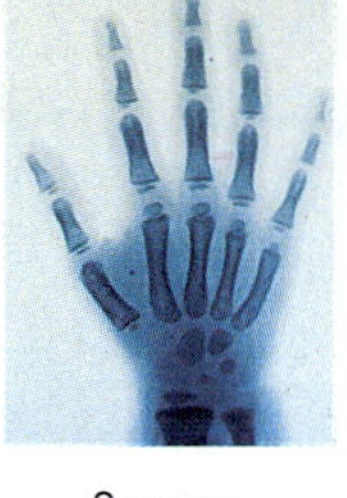

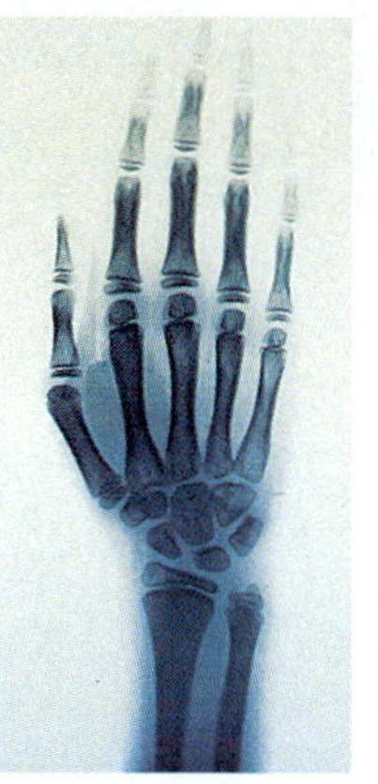

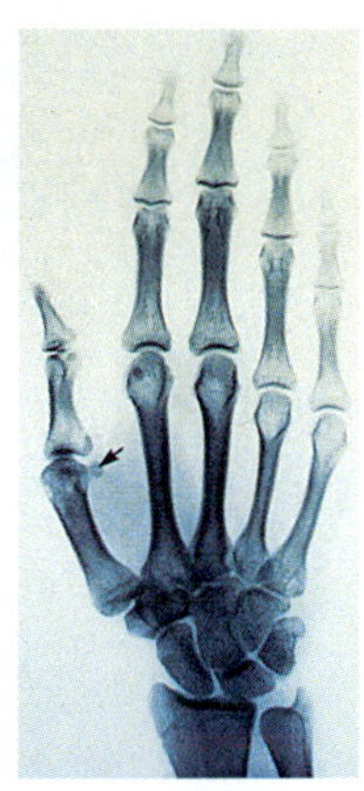

[B] Maturation of the hand bones

HOMEOTIC GENES

Although it would appear that your body and the body of an insect are organized very differently, researchers have learned recently that nearly identical groups of genes control the overall body plan of various animals, including humans and fruit flies. These regulatory genes, called *homeotic genes* (*homeo-* means "the same"), were first identified in fruit flies. Homeotic genes each contain a very similar DNA sequence of about 180 nucleotides. This sequence is called a **homeobox**.

A mutation in one of the homeotic genes of a fruit fly can change an antenna into a leg; the mutant fly has two complete legs growing from its head. It seems clear that the homeotic gene in question is important in determining the developmental difference between a leg and an antenna. Other homeotic genes control the development of other parts of the body [FIGURE 13.19A].

Geneticists were surprised to find that nearly identical homeotic genes direct embryonic development in animals as different as fruit flies, zebra fish, mice, and human beings [FIGURE 13.19B]. Even more remarkable, however, was the finding that the homeotic genes of different species are turned on in the same anterior-to-posterior order. This extraordinary conservation of gene structure and function is consistent with the idea that fruit flies, zebra fish, mice, and human beings all arose from a single common ancestor.

➤ CONCEPT CHALLENGE

A baby born with a particular mutant allele has webs of skin between its fingers and toes, a condition called syndactyly. What aspect of morphogenesis is altered in these infants? What does the normal allele of this gene do to cause the development of hands without webs?

Development Continues Throughout Life

An individual's development continues after hatching or birth and helps both the individual and the species survive by allowing the organism to grow larger and to mature sexually.

CONTINUAL GROWTH AND CHANGE

If you've ever adopted a young pet, then you know that **growth** is the most apparent aspect of postembryonic development. The animal's weight doubles several times before stabilizing in adulthood. The same is true for a newborn child [FIGURE 13.20]. Weight gain comes largely from an increase in cell number due to cell division throughout the body. Thus, a newborn with billions of cells will have trillions as an adult.

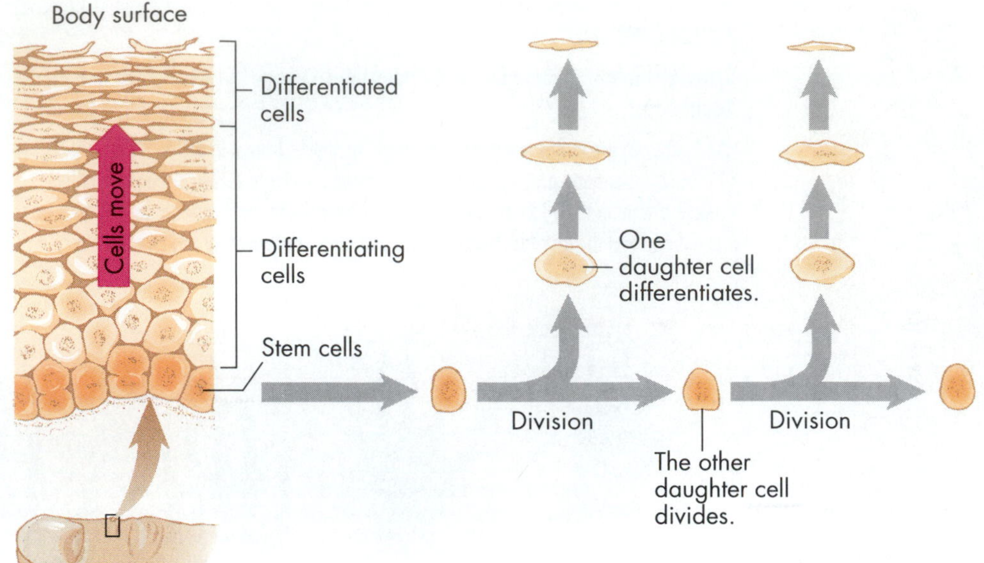

FIGURE 13.21

Stem Cells Divide and Renew Body Parts.

In the skin on your index finger, stem cells in the dividing layer continuously reproduce as time goes by. As each stem cell divides, one daughter remains an undifferentiated stem cell with the potential for future divisions, while the other daughter becomes differentiated into an epidermal cell and is pushed outward, becoming part of the finger's protective layer. The differentiated skin cell can no longer divide and eventually dies and flakes off as dry skin while younger cells push up from below.

Most organs stop enlarging after an animal reaches its adult size, and development continues mainly by the replacement of dying cells. Cell replacement usually involves **stem cells**, which are generative cells that retain the ability to divide throughout life [FIGURE 13.21]. Stem cells in the skin, bone marrow, intestines, and gonads are especially active. When a stem cell divides, one offspring cell differentiates into the specialized cell type (for example, a blood cell in the bone marrow or a gamete in the gonad), while the second daughter becomes another stem cell.

Natural substances in the body called *growth factors* regulate the division and differentiation of stem cells so that growth and change occur only where and when needed. Sometimes, however, something goes terribly wrong: Growth regulation fails and tissues grow unchecked, resulting in a massive tumor or invasive cancer.

CANCER: DEVELOPMENT RUNNING AMOK

A *tumor* is, in a sense, an inappropriate continuation of development. A *benign* tumor is a clump of cells that continues to grow unchecked, but stays localized in a tightly packed group and thus can usually be removed surgically. A cancerous, or *malignant*, tumor, however, grows without ceasing and also spreads throughout the body, a process called **metastasis**, invading healthy organs and destroying them [FIGURE 13.22A]. In insidious ways, cancer cells mimic normal embryonic cells.

Cancer cells appear to be mutant stem-cell derivatives that normally would differentiate and stop dividing. Cancer cells assume the rounded shape of embryonic cells [FIGURE 13.22B] and proliferate as if they were still part of a rapidly growing embryo. This happens because mutations occur in genes that usually encode growth factors or a cell's ability to detect or respond to growth factors. As a consequence, cells proliferate out of control. The mutant forms of these growth control genes are called *oncogenes*. Women whose breast cancer cells have extra copies of a specific oncogene that encodes a growth factor receptor are much more likely to die than patients whose cancer cells lack the oncogene. In an exciting new approach to cancer therapy, researchers have found, for example, that they can prevent the growth of certain human breast cancer cells by blocking the mutant receptor protein. This prevents the cell from receiving a signal to grow and thus inhibits the growth of the breast tumor.

Cancer cells mimic not only the continual divisions of embryonic cells, but also their ability to migrate. In metastasis, tumor cells invade healthy tissues. In a similar way, early embryonic cells invade the wall of the uterus in a limited way and secure a nutrient supply. Normal embryonic cells, such as neural crest cells [review FIGURE 13.18], flatten out and migrate along trails of a special protein (fibronectin) laid down in the embryo. Cancer cells, however, have very little of this protein, and researchers speculate that this may partially explain why the cells are round, not flat, and why they migrate between

[A] Mouse with a mammary tumor

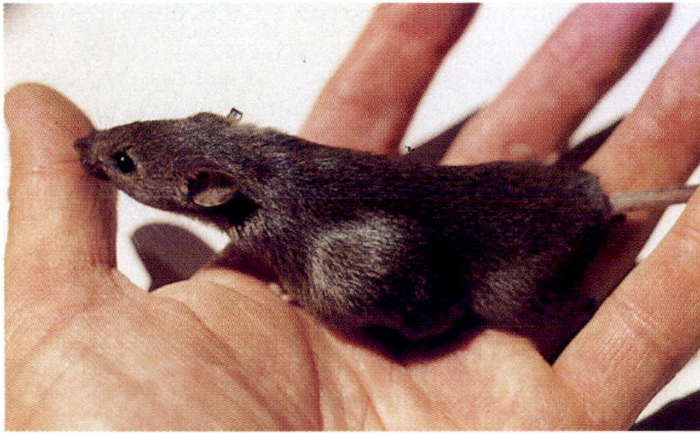

[B] Cancer cells have rounded shapes

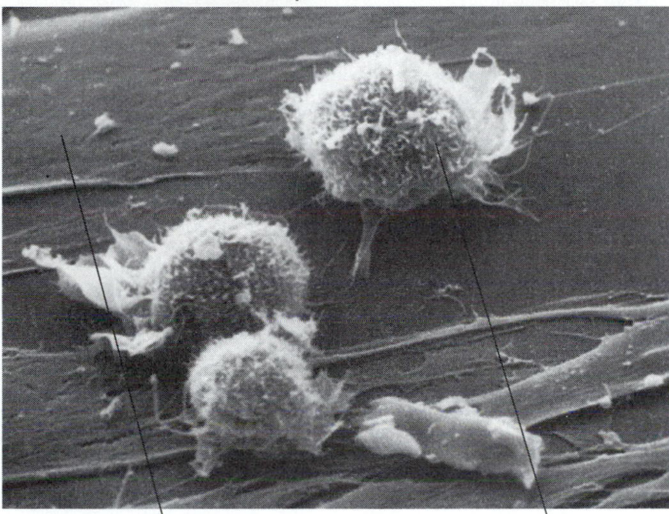

FIGURE 13.22

Cancer: Development Gone Awry.

[A] A mammary tumor grows out of control in a mouse. **[B]** Cells in such a tumor have a different shape from normal differentiated cells. This scanning electron micrograph shows how the flattened growth characteristics of normal cells contrast with the rounded shape of cancer cells.

nearby cells. If they escape into the blood or lymph, these migrating cells can spread far from the original tumor. If the cells are able to acquire an adequate blood supply in their new location, they may disrupt the structure of tissues enough to lead to the person's death.

With cancer, as with all diseases, an ounce of prevention is worth a pound of cure. Since we now know that cancers are due to mutations in genes that control normal growth and differentiation and that several mutations will usually accumulate before a cancer metastasizes, we must avoid repeated exposure to cancer-causing agents such as sunlight, cigarette smoke, and high-fat, low-fiber diets.

➤ CONCEPT CHALLENGE

What evidence do biologists have that cancer is a genetic disease and that cancer cells are permanently stuck in a developmentally immature state?

Connections

The story of animal development carries us full circle from a single fertilized egg, to an embryo, a young animal, an adult organism, and finally to the production of new eggs or sperm. A description of animal development shows the continuity of hereditary information in living things. An organism's physical structures and life cycles have precursors in previous parts of the cycle, are constrained by prior structures and events, and develop as a result of chemical instructions or interactions among previously existing cells.

Embryology is a pivotal discipline in the biology of multicellular organisms. The actions of genes and proteins at the molecular level cause cells to become committed to a particular developmental fate and to differentiate, divide, and specialize into the organism's body plan and organs. These embryonic organs become the functioning anatomical parts of organisms that we will study in later chapters of the book.

While this chapter has laid out the general principles of animal reproduction and development, there is a great deal more to say about how these principles apply to our own species and how we progress through the life cycle from fertilized egg to fertilized egg. That is the subject of the next chapter.

KEY TERMS

benign, 328
blastula, 314
cleavage, 314
copulation, 311
development, 309
cytoplasmic determinant, 315
differentiation, 326
ectoderm, 318
endoderm, 318
external fertilization, 311
gastrulation, 318
germ layer, 318
growth, 327
homeobox, 327
homeotic genes, 327
induction, 321
internal fertilization, 311
malignant, 328
mate, 310
mesoderm, 318
metastasis, 328
neural tube, 319
neurulation, 319
organogenesis, 322
ovum, 312
stem cell, 328
yolk, 312

HIGHLIGHTS IN REVIEW

1 Sexual reproduction in animals involves behaviors that promote the joining of eggs and sperm in fertilization, the development of the fertilized egg into a new individual as well as organism.

a] Mating between animals usually begins with mate recognition, often based on visual or olfactory cues. Hormonal or behavioral signals then help to synchronize the production and release of gametes—the eggs and sperm.

b] Eggs are storehouses for nutrients and developmental instructions. Sperm, small cells with few organelles, are specialized for rapid swimming and delivery of a haploid nucleus into the egg's cytoplasm.

c] During fertilization, the plasma membranes of sperm and egg fuse, and the sperm nucleus enters the egg cytoplasm and fuses with the egg's nucleus, creating a diploid nucleus for what is now the zygote, the first cell of the new individual.

2 Animal development involves a series of mitotic cell divisions (cleavage) that results in many cells arising from the single fertilized egg. It also involves organization of three main cell layers (gastrulation); formation of body organs (neurulation and organogenesis); specialization of cells for particular functions (differentiation); increase in body size (growth); and sexual maturation. Meiosis in the new individual results in the production of eggs or sperm (gametogenesis).

a] Cleavage is a series of rapid, synchronous mitotic divisions that in many species transforms the zygote into a hollow ball of cells called the blastula.

b] During cleavage, instructional chemicals called cytoplasmic determinants become localized into individual cells.

c] During gastrulation, cells move from the outside of the blastula to the inside and form three layers—the ectoderm, mesoderm, and endoderm. These three germ layers give rise to all other cell types (except the germ cells).

d] During neurulation, the ectoderm layer is induced by the underlying mesodermal notochord cells. The layer rolls up and forms the neural tube, the future nervous system.

e] During organogenesis, the body's organs assume their form and function. The development of organ form, or morphogenesis, involves changes in cell shape, cell proliferation, programmed cell death, and cell migration. The development of organ function, or differentiation, involves the selective expression of genes through the actions of regulatory proteins.

f] An animal's development continues after hatching or birth. The animal grows and takes on adult characteristics that enable it to collect food, defend itself and its territory, and find mates. In adulthood, cell division usually involves stem cells and is regulated by growth factors and other substances.

3 Development occurs in an ordered sequence due to gene-regulatory substances (developmental determinants) in the egg and to interactions between developing cells.

a] In frog embryos, the gray crescent contains cytoplasmic determinants that specify the dorsal side of the embryo, resulting in the formation of the notochord.

b] The notochord sends an inductive signal to the overlying ectoderm, resulting in the formation of the neural tube.

4 In many ways, cancer cells behave like embryonic cells as a result of mutations in genes that normally regulate cell growth and differentiation.

a] Benign or malignant tumors form as normal growth controls break down and cells proliferate unchecked; malignant tumors spread, or metastasize, to other areas of the body.

b] Cancer is a multistep process in which cells accumulate several genetic changes as they become more malignant.

UNDERSTANDING THE FACTS AND CONCEPTS

For Questions 1–4, match each of the descriptions with the most appropriate item from the following list of terms. In this group, as well as those that follow, any item may be used once, more than once, or not at all.

a] copulation
b] fertilization
c] mating
d] hermaphroditism

1 The condition in which an animal contains both male and female sex organs and produces both eggs and sperm.

2 Behavior that has evolved to assure fertilization of the egg by the sperm.

3 The act of behavior that results in fertilization of ova internally.

4 The process whereby a sperm cell recognizes and penetrates the cell membrane of an ovum and delivers a haploid set of chromosomes to that ovum.

As above, for Questions 9–10, match the descriptions with the appropriate item from the following list.

a] trophoblast
b] inner cell mass
c] germ layers
d] neural crest
e] gray crescent

5 A region within the fertilized frog egg that contains cytoplasmic determinants that induce the formation of the head and back.

6 Three concentric tubes of cells that arise during gastrulation; most of the organs of the body are derived from one or another of these sources.

7 The cells of a very young mammalian embryo that will be the source of all the cells of all the body parts as development proceeds.

8 Cells of the early mammalian embryo that will contribute to the formation of the placenta, but not to the embryo proper.

9 Cells that wander far from their site of origin and give rise to certain structures in the face and nervous system and to the pigmented cells of the skin.

As above, for Questions 10–14, match the descriptions with one of the terms from the following list:

a] differentiation
b] morphogenesis
c] gastrulation
d] cleavage
e] neurulation

10 The process by which ectoderm rolls up to form a tube along the dorsal side of the embryo.

11 The acquisition of a specialized cellular structure and function due to the synthesis or particular cellular proteins.

12 The processes whereby an embryo, or part of an embryo, acquires a particular form or three-dimensional shape.

13 Rapid cycles of cell division that transform a zygote into a blastula.

14 The differential movement of cells in the embryo that results in the formation of three embryonic tissue layers and the establishment of the basic body plan.

As above, for Questions 15–19, match the descriptions with the appropriate item from the following list.

a] cytoplasmic determinants
b] stem cells
c] germ cells
d] homeotic genes
e] embryonic induction

15 A process whereby one group of cells alters the developmental fate of another group of cells.

16 A general name for any relatively undifferentiated cells that can give rise to more specialized types of cells throughout the life of the organism.

17 The relatively undifferentiated cells in an embryo that ultimately give rise to the gametes.

18 Instructional chemical substances that are unequally distributed in the egg. As a result of cleavage they are distributed in differing amounts in different cells and serve to regulate cellular differentiation.

19 Control genes found in various organisms. Their products bind to DNA and regulate the expression of other genes that govern the sequence of anatomical structures in the head-to-tail axis.

INTEGRATE AND APPLY WHAT YOU HAVE LEARNED

Answer the following questions on a separate sheet of paper.

1 Describe the events that occur during and immediately following fertilization.

2 In addition to subdividing a fertilized egg into many cells, how does cleavage assist in the further development of the embryo?

3 When are the germ layers formed, and what does each become?

4 What is the difference between differentiation and morphogenesis? When in embryonic development does each begin?

5 How is differentiation as a result of embryonic induction different from that resulting from cytoplasmic determinants.

ANALYSIS

Answer the following multiple-choice questions.

1 How could you determine whether a mammalian trophoblast cell develops into placenta because of its position or because of intrinsic developmental determinants?

a] Remove the cell and transplant it into the trophoblast of another embryo with recognizably different cells.
b] Remove the cell and see if the placenta develops despite the cellular loss.
c] Remove the cell and insert it into the inner cell mass of the same embryo.
d] Remove the cell and insert it into the inner cell mass of another embryo that has recognizably different cells.
e] Transfer the cell to tissue culture medium in a test tube and see what it develops into.

2 Into which of the following categories would the product of homeotic genes most logically fall?

a] Structural protein.
b] Enzymatic protein.
c] Gene regulator protein.
d] Operons.
e] Enzyme inhibitor.

3 Which of the following is *not* a stem cell?

a] Generative cells in the skin.
b] Germ cell.
c] Inner cell mass cell.
d] Unspecialized cell in the bone marrow.
e] Trophoblast cell.

4 Refer to FIGURE 1B in BOX 13.1. Recall that the polarizing zone (PZ) produces and releases retinoic acid, which then diffuses outward from the PZ. How do you account for the fact that the Boston machinist had no thumb?

a] The two PZs must have been too close together, leaving no room for a thumb to develop.
b] The two PZs must have been too far apart, so that the retinoic acid was too diluted to induce a thumb.
c] There was no region in the limb bud with low enough retinoic acid concentration.
d] There was no region in the limb bud with high enough concentration of retinoic acid.
e] Perhaps the limb bud was so disorganized that nothing could develop normally.

CHAPTER 14

The Human Life Cycle

A SCOTTISH TEST OF A FRENCH PILL

Over the course of a few months, 800 frightened women between 16 and 45 years old came to the Dean Terrace Health Clinic or to the emergency room of Edinburgh's Royal Infirmary. All these women had the same problem: Each had had sexual intercourse within the previous 72 hours without using any form of birth control, thus risking pregnancy, and each felt unable to provide the care every child deserves. Each sought treatment in the form of a "morning after" pill to prevent pregnancy after sexual intercourse. For several years, such a pill, containing high doses of the female hormones estrogen and progesterone, had been available to European women. During 1991, however, a team of physicians at the two health clinics led by Dr. Anna Glasier wanted to compare this traditional "morning after" pill to the safety and usefulness of a newer drug, mifepristone, or RU 486.

Dr. Glasier knew that mifepristone acts by blocking the action of the hormone progesterone, a substance that maintains the nutritive lining of the female reproductive tract [FIGURE 14.1]. Without progesterone activity, pregnancy, the implantation of an embryo into the womb, cannot occur. It was unclear, however, whether mifepristone would prevent pregnancy more safely and effectively than the traditional drug designed for that purpose. Hence the Scottish study.

FIGURE 14.1

Hormones and Fertility. Progesterone [A] maintains the nutritive lining of the womb, but the similarly shaped molecule RU 486 [B] can block progesterone activity.

[A]

The Glasier team administered mifepristone to 402 of the 800, the first test group, and the traditional pill to the other 398, the second test group. Of the 402 in the mifepristone group, none became pregnant, while four pregnancies resulted among the 398 in the traditional group. Glasier's team also found that mifepristone caused far less nausea and vomiting than the traditional pill. Some users of mifepristone, however, did experience a delay in their menstrual cycle.

Professionals concerned about the physical and emotional health of women and their babies have suggested that the use of mifepristone after unprotected intercourse could reduce the number of induced abortions, a goal that nearly everyone would agree is important. Despite the promise of mifepristone to decrease the unnecessarily high rate of abortions in the United States, where the rate of abortions is among the highest in the world, mifepristone and related drugs may not become available because of conflicting political opinions.

The use of mifepristone to block pregnancy focuses our attention on the central topic of this chapter, human reproduction and development. The chapter explores the steps leading up to pregnancy, including male and female sexual anatomy and function and the processes of fertilization and implantation into the womb. We then turn to how a ball of cells changes into a human embryo, and then a baby. We end the chapter by briefly looking at the life cycle after birth. These topics have direct relevance for each of us, whatever our stage of the life cycle and our reproductive status because we are confronted daily with information and choices regarding important reproductive issues: sexually transmitted diseases; birth control; overpopulation; infertility; delivery and birth; and aging.

As biology students, you will also find that our current topics are closely connected to other chapters in the book. This will be especially true when we discuss hormones in CHAPTER 29, human population trends in CHAPTER 37, and the evidence for evolution found in the similarities of human embryos to those of chickens, frogs, and fish. As citizens of the 1990s, you might find various facts in this chapter of special interest, including how home pregnancy test kits work; what are the newest types of birth control; and how to avoid sexually transmitted diseases. ❑

MESSAGES

1 In both males and females, the same brain hormones regulate (a) the gonads (testes or ovaries), which produce gametes (sperm or eggs), and (b) steroid hormones that control development of gender and the functioning of mature sexual organs.

2 The management of human fertility is a pressing issue worldwide. A detailed knowledge of reproductive physiology allows us to promote or prevent human reproduction.

3 Human embryos pass through the same developmental stages that unfold in other vertebrate animals, modified slightly because human embryos, like most other mammals, obtain nourishment through the placenta.

4 Growth, maturation, and aging continue to occur throughout life. Biologists do not yet understand the causes of aging.

[B]

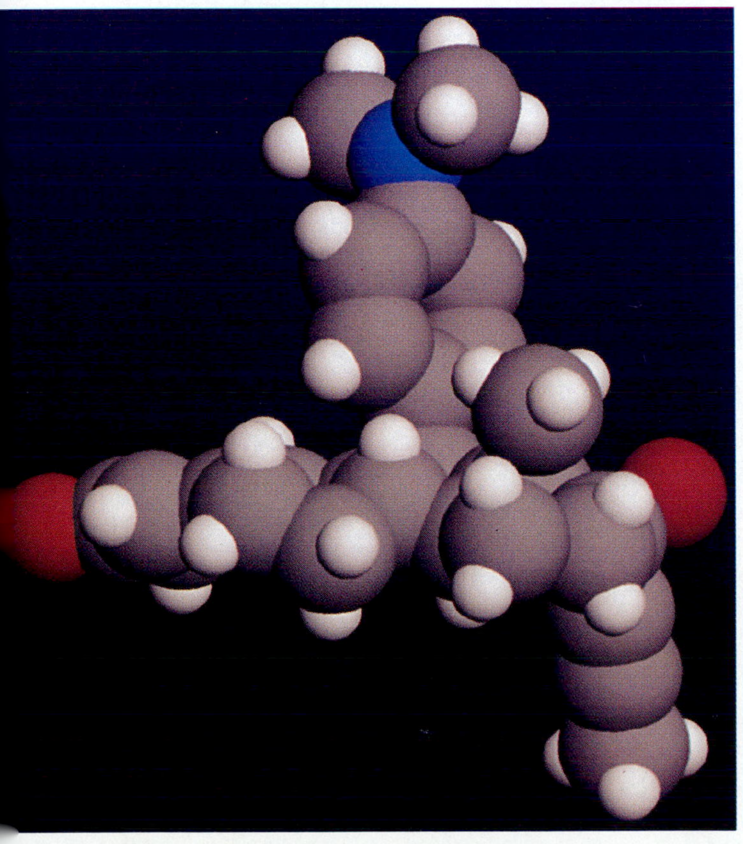

Male and Female Sexual Characteristics

Every human being has a complex system of primary and secondary sexual characteristics involved in sexuality and reproduction. **Primary sexual characteristics** are reproductive organs capable of passing along parts of a parent's genome to the next generation by sexual reproduction. In contrast, **secondary sexual characteristics** are external features, such as chest hair in males and milk-producing breasts in females, that are not directly involved in sexual intercourse, but are significant in reproductive behavior [FIGURE 14.2].

Male and female primary sexual characteristics are different but parallel in structure and function. Each sex has a set of *gonads*, or reproductive organs: testes in males and ovaries in females. Gonads produce sex cells by meiotic divisions (sperm in males and eggs in females), and they also produce sex hormones. Both sexes also have *accessory reproductive structures*: various ducts, chambers, and glands that transport and store sex cells, and, in the female, that nurture the developing embryo.

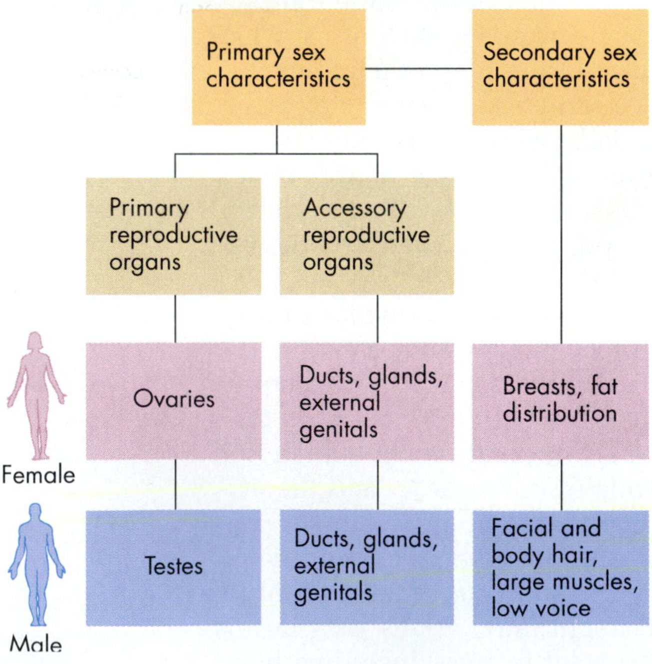

FIGURE 14.2
Human Sex Characteristics.

Male Reproductive System

The functions of the male reproductive organs are the production of sperm and the transfer of the sperm into the female reproductive tract. The organs that produce the sperm cells, the testes, also produce the primary male hormone, testosterone, which aids sperm production.

TESTES: SPERM-PRODUCING ORGANS

The primary male reproductive organs are the paired **testes** (singular, *testis*). They are formed inside the body cavity, but they descend shortly before birth into an external sac between the thighs, the **scrotum** [FIGURE 14.3A and TABLE 14.1]. The lower temperature in the scrotum, *outside* the body, is necessary for active sperm production; sperm remain infertile if they are retained inside the body. In some men, wearing tight pants or underwear, exercising too hard, or sitting in a hot tub can elevate the

TABLE 14.1 Male Reproductive System

Structure	Function
Parts of Testis	
Seminiferous tubules	Produce and transport sperm
Interstitial cells (Leydig cells)	Produce testosterone
Sustentacular cells (Sertoli cells)	Sustain developing sperm
Accessory Ducts	
Epididymis	Stores and matures sperm
Vas deferens	Conducts sperm
Ejaculatory duct	Transports sperm
Urethra	Transports sperm and urine
Accessory Glands	
Seminal vesicles	Secrete fluids that aid sperm function
Prostate gland	Secretes fluids that aid sperm function
Bulbourethral gland	Secrete lubricants
External Genitals	
Scrotum	Contains testes
Penis	Organ of sexual intercourse; contains urethra

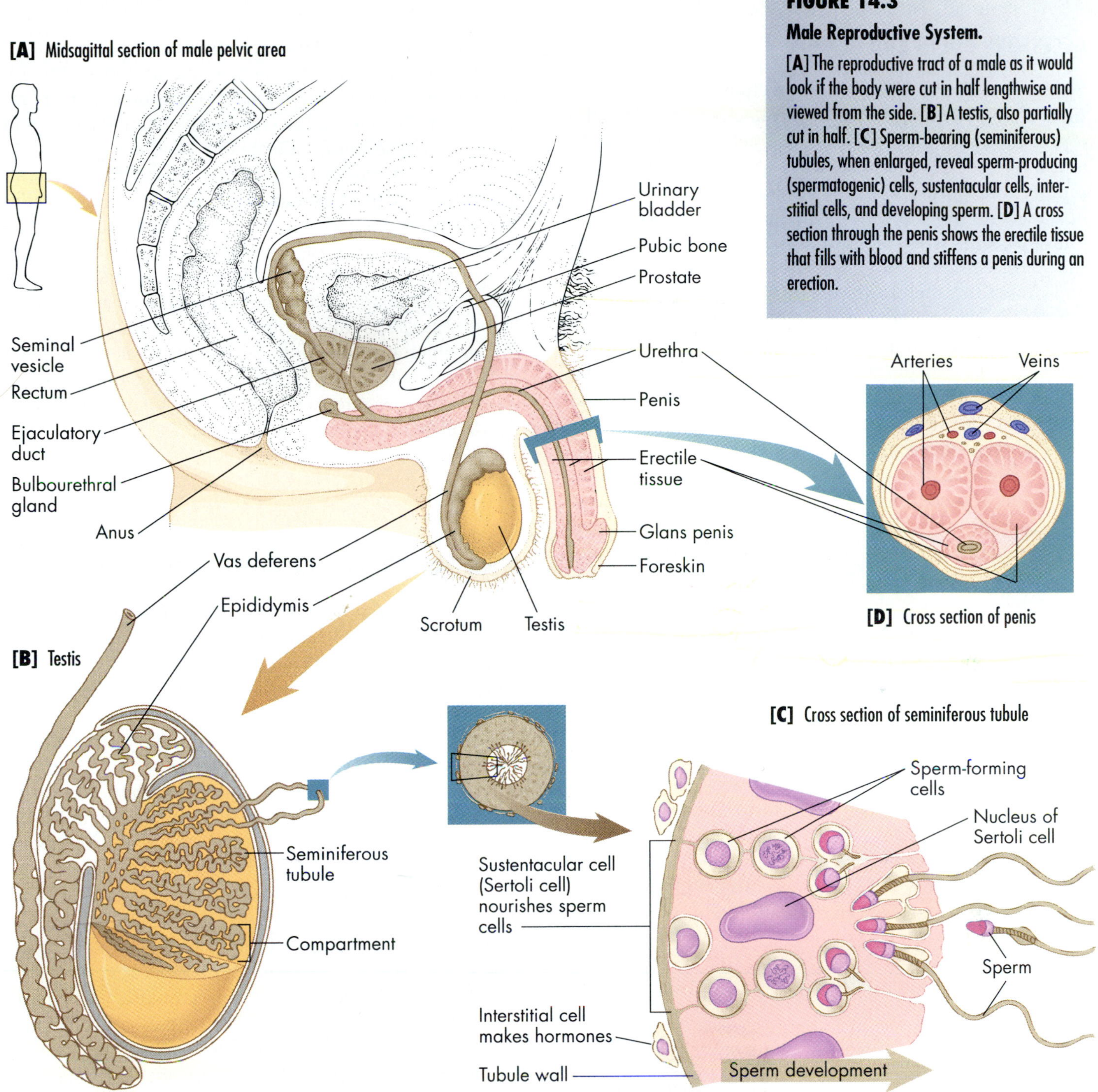

FIGURE 14.3

Male Reproductive System.

[A] The reproductive tract of a male as it would look if the body were cut in half lengthwise and viewed from the side. **[B]** A testis, also partially cut in half. **[C]** Sperm-bearing (seminiferous) tubules, when enlarged, reveal sperm-producing (spermatogenic) cells, sustentacular cells, interstitial cells, and developing sperm. **[D]** A cross section through the penis shows the erectile tissue that fills with blood and stiffens a penis during an erection.

temperature of the scrotum and testes to a point where sperm development stops temporarily. A fever can kill hundreds of thousands of sperm cells.

Each testis is an oval structure about 4 cm (1.5 in) long and is subdivided into about 250 compartments [FIGURE 14.3B]. Each compartment contains up to four highly coiled, hollow tubes called **seminiferous tubules**, each of which is about 70 cm (28 in) long. The combined length of the seminiferous tubules in both testes is about 225 m (750 ft). Sperm-forming cells, or **spermatogenic cells**, are produced in the walls of the seminiferous tubules, where they undergo meiosis. The resulting haploid cells develop into *sperm*.

While sperm cells are developing in the seminiferous tubules, they are sustained, surrounded, and nourished by **sustentacular cells**, also called *Sertoli cells* [FIGURE 14.3C]. Also present in the tissue that encases the seminiferous tubules are **interstitial cells**, or *Leydig cells*, which produce the male steroid hormone **testosterone**, which is chemically related to cholesterol.

ACCESSORY DUCTS AND GLANDS

Sperm travel from the seminiferous tubules in each testis into the **epididymis**, a coiled tube attached directly to the top of the testis where the sperm mature [FIGURE 14.3A and B]. When a male is sexually stimulated, sperm are rapidly transported from the epididymis down a system of ducts and are forcefully released from the body. The journey begins when smooth muscle cells in the walls of the epididymis adjacent to each testis contract repeatedly and propel the sperm into the **vas deferens** (plural, vasa deferentia, also called *ductus deferens* or sperm duct), a connecting tube that is 45 cm (18 in) long with muscular walls that contract and continue to propel the sperm [see FIGURE 14.3A and B]. The vas deferens leads back into the body from the testis. Near the urinary bladder, the left and right vas deferens merge into the **ejaculatory duct**. Secretions from the paired **seminal vesicles** are added and bathe the sperm in a fluid that contains the sugar fructose and other nutrients as well as substances that regulate pH and stimulate muscular contractions in the female reproductive tract. The ejaculatory duct passes through the **prostate gland**, a chestnut-shaped gland that secretes a milky alkaline fluid that mingles with the sperm and the fluid of the seminal vesicles. The added secretions enlarge the volume of sperm and lower the mixture's pH to about 7.5. This pH helps neutralize the acidity of the female reproductive tract. This natural acidity protects her delicate tissues from microorganisms but also tends to inhibit sperm motility. The **bulbourethal glands** also add alkaline fluid to the sperm in the urethra.

Once surrounded by milky fluid, the sperm passes from the ejaculatory duct to the **urethra**, a tube that runs through the penis. The mixture of sperm and fluid is called **semen**.

THE PENIS

The **penis** has a dual purpose: It carries urine through the urethra to the outside, and it transports semen through the urethra during ejaculation [see FIGURE 14.3D]. In addition to the urethra, the penis contains three cylindrical stands of *erectile tissue*. During erotic stimulation, the spaces of the erectile tissue become engorged with blood under high pressure. As part of the overall response to stimuli from the nervous system, the arteries leading to the penis are dilated (enlarged), and the veins leading away are constricted. This dual action prevents blood from escaping, and the penis becomes enlarged and firm in an *erection*. The erect penis can be inserted into the vagina during sexual intercourse.

The average ejaculation produces about 3 or 4 mL of semen (about a teaspoonful) and contains 300 to 500 million sperm. If fewer than about 150 million sperm are released per ejaculation, the semen may be ineffective, and the man may be *sterile*, unable to reproduce.

Nearly 10 million men in the United States are affected to one degree or another by *impotence*, the inability to sustain an erection until ejaculation. Recent advances in the study of the erection process show that tissue within the penis (probably nerve tissue) produces the gas nitric oxide, which in turn triggers changes in the blood vessels that allow blood to accumulate in the penis, producing an erection. Researchers have found that many men with severe impotence produce much less nitrous oxide than normal, and future treatment may be based on this discovery.

HORMONAL CONTROL OF SPERM PRODUCTION

Hormones from the brain and testes work together in an interconnecting feedback loop that regulates the timing of sperm production. The basic scheme is relatively simple: If sperm production falls below a specific level, hormones are released from the brain [FIGURE 14.4 step 1] and transported in the bloodstream to the testes (step 2), where they cause increased production of both sperm and testosterone (steps 3 and 4). When the testosterone concentration rises above a certain level, it acts on the brain, blocking further release of the brain hormones (step 5). With a lower concentration of brain hormones, the testes' production of sperm and testosterone decreases. When sperm is released during ejaculation, the hormone levels fall below the set point, and the cycle begins again. This is an example of a *negative feedback loop*, a cyclic series of processes that tends to maintain conditions in a constant state. The loop is called *negative* because it operates by opposition: When the level of testosterone gets too high, the feedback mechanism brings it back down, and when the level gets too low, the feedback brings it back up.

Testosterone has other functions besides triggering the release of brain hormones; it causes a male embryo's sexual organs to differentiate and it induces the development of male secondary sexual characteristics at puberty [see FIGURE 14.2]. These characteristics include the development of mature genitals, male facial and chest hair, and muscles; the stimulation of sperm production by sperm-producing cells; an increased interest in sex, and in many animals and often in humans, the tendency toward expression of aggressive behavior.

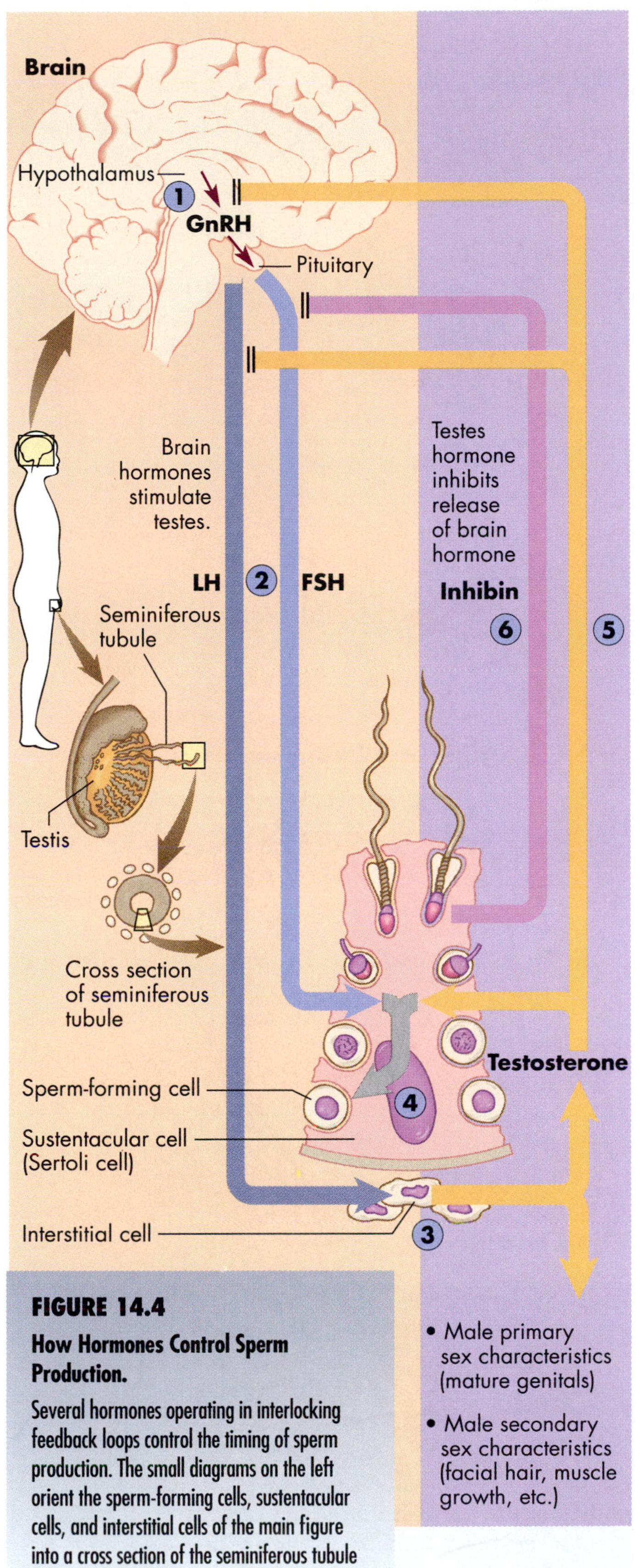

FIGURE 14.4

How Hormones Control Sperm Production.

Several hormones operating in interlocking feedback loops control the timing of sperm production. The small diagrams on the left orient the sperm-forming cells, sustentacular cells, and interstitial cells of the main figure into a cross section of the seminiferous tubule drawn out from the testis, as in FIGURE 14.3.

Female Reproductive System

Reproductive functions in females involve more steps than do equivalent processes in males and include many additional roles. Females produce their gametes and hormones in a monthly cycle, rather than continuously as in males, and after fertilization they nourish, carry, and protect the developing embryo. They also nourish the baby for a time after it is born. FIGURE 14.5 and TABLE 14.2 summarize the structures and functions of the female reproductive system.

PRODUCTION AND PATHWAY OF EGGS

Women produce eggs within two solid almond-shaped organs called **ovaries**, which lie inside the body cavity just below the waistline [FIGURE 14.5A and B]. Each ovary contains (1) immature cells called **oocytes**, which develop into mature eggs, and (2) cells called *follicular cells*

TABLE 14.2 Female Reproductive System

Structure	Function
Parts of Ovary	
Oocytes	Develop into eggs
Follicle cells	Support oocytes; produce estrogen and progesterone
Accessory Ducts	
Fallopian tube (oviduct)	Transports ovum; site of fertilization
Uterus	Sustains embryo throughout pregnancy
Cervix	Secretes mucus; reduces bacterial contamination
Vagina	Organ of sexual intercourse; acts as birth canal
Accessory Glands	
Bartholin's glands	Secrete lubricant
External Genitals	
Labium major	Protects other reproductive organs
Labium minor	Protects other reproductive organs
Clitoris	Sensitive to sexual stimulation

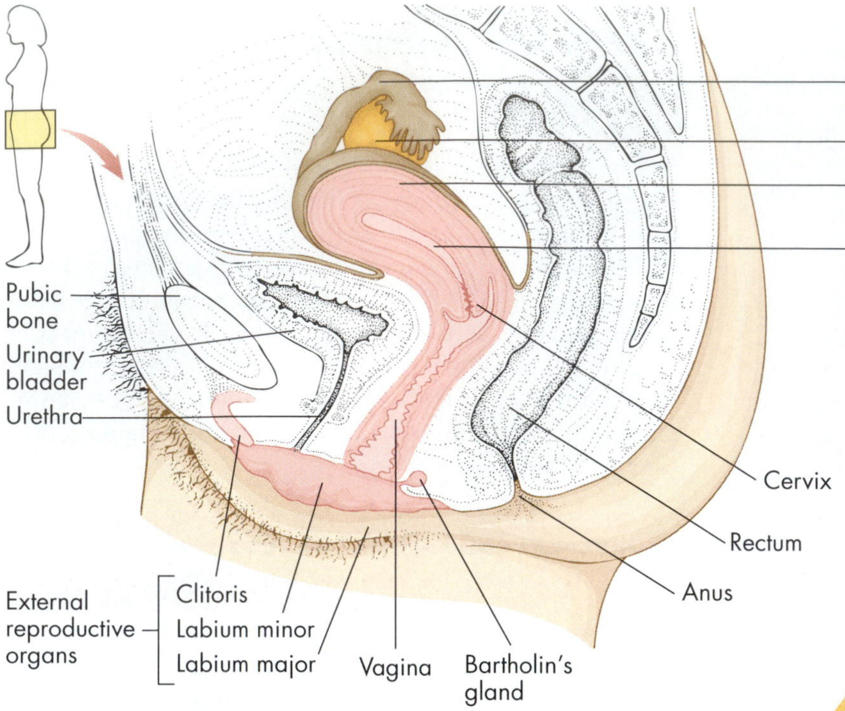

Fallopian tube (oviduct)
Ovary
Uterine wall
Uterine cavity
Pubic bone
Urinary bladder
Urethra
Cervix
Rectum
Anus
External reproductive organs
Clitoris
Labium minor
Labium major
Vagina
Bartholin's gland

[A] Midsagittal section of female pelvic area

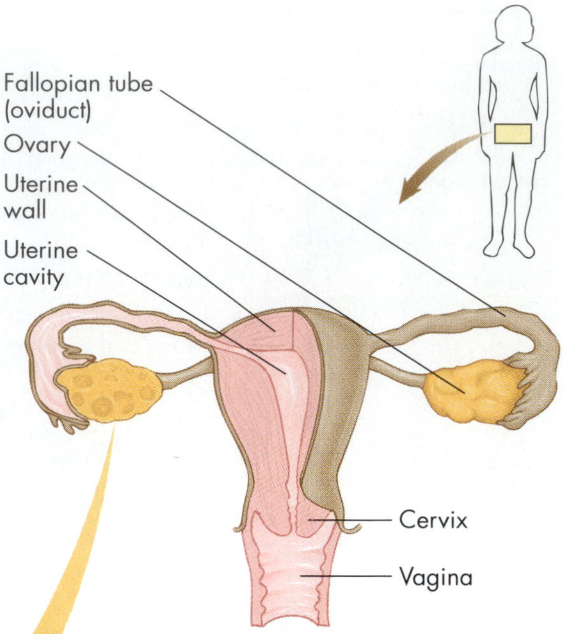

Cervix
Vagina

[B] Anterior view

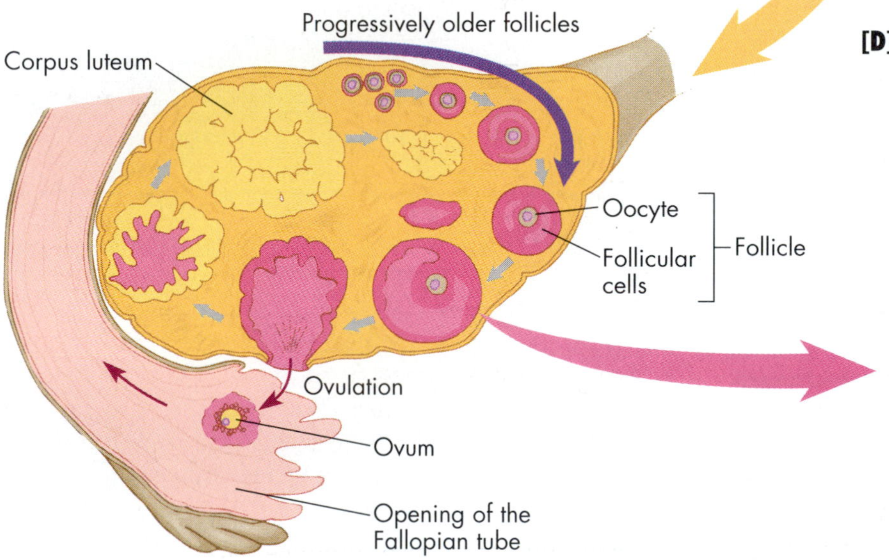

Progressively older follicles
Corpus luteum
Oocyte
Follicular cells
Follicle
Ovulation
Ovum
Opening of the Fallopian tube

[C] Cross section of ovary

[D] Cross section through an ovarian follicle

Nucleus
Nucleolus
Oocyte
Follicular cell

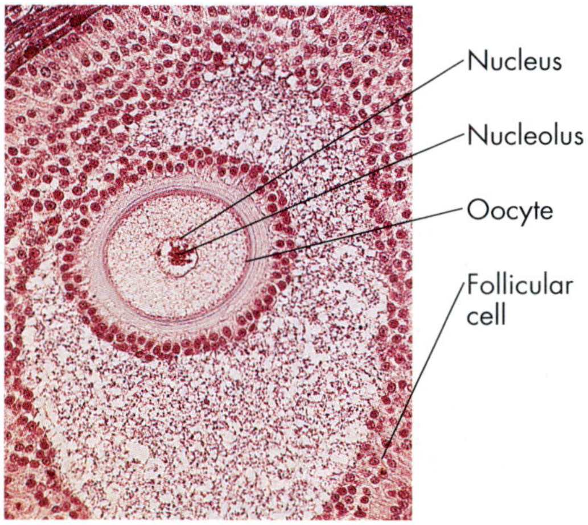

[E] A human egg.

FIGURE 14.5

Female Reproductive System.

The female reproductive system as it would look if it were cut in half lengthwise and viewed from the side [A], or viewed from the front [B]. [C] An enlarged view of one ovary and fringed fallopian tube opening. [D] Cross section through an ovarian follicle. [E] A human egg, surrounded by a jellylike coat and many follicular cells.

that surround and support the oocytes [FIGURE 14.5C and D]. An oocyte surrounded by its follicular cells constitutes a functional unit called a **follicle** [FIGURE 14.5D]. At birth, each ovary in a typical baby girl contains about 2 million follicles with oocytes arrested in prophase of meiosis I [review FIGURE 7.19].

When a female reaches puberty, about the age of 12 or 13, a process called **ovulation** takes place every 28 days or so. Usually, a single follicle in one of the ovaries enlarges, its oocyte matures, the follicle ruptures, and a mature egg, now called an ovum, is released into the body cavity [FIGURE 14.5C]. (An ovum is the largest human cell; it is about the size of the dot over this *i*.)

After being released from the ovary, and before it can become lost in the body cavity, the ovum moves toward the fringed opening of one of two 10-cm-long (4-in-long) tubes called **fallopian tubes**, or **oviducts** [see FIGURE 14.5B and C]. Both fallopian tubes are lined with millions of cilia that act like paddles, sweeping the ovum down the tube toward the uterus. Fertilization usually takes place in the fallopian tube as an up-swimming sperm meets and penetrates a down-moving ovum. Eventually, the ovum, whether or not it has been fertilized, is swept into the thick-walled, pear-shaped **uterus**, which is contiguous with the fallopian tubes. If the ovum has been fertilized, it will begin cell division and become implanted into the uterine wall, where it is nourished and where it develops into an embryo and fetus.

During most months, the ovum is not fertilized, and instead of being implanted into the uterine wall it degenerates and may be discharged from the uterus through the **cervix**, the neck of the uterus. Just beneath the cervix lies the **vagina**, a hollow, muscular tube that receives the penis during intercourse, conveys uterine secretions (including the monthly menstrual flow), and serves as the stretchable birth canal that allows the passage of the fetus during childbirth.

The vaginal opening is surrounded by external reproductive structures that include protective tissues called the *labia minor* and *major*, and tissues sensitive to sexual stimulation, such as the **clitoris**. Also included are the lubricating **Bartholin's glands** [see FIGURE 14.5A].

HORMONAL CONTROL OF THE OVARIES AND UTERUS: THE MENSTRUAL CYCLE

Most women have a reproductive cycle, or **menstrual cycle**, that fluctuates around 28 days. This cycle is coordinated by several hormones [see FIGURE 14.6]. During each menstrual cycle, the inner lining of the uterus, the **endometrium**, builds up in preparation for pregnancy just before an ovum is released from an ovary. If the ovulated ovum is not fertilized, the uterine lining sloughs off, and menstrual bleeding occurs, beginning a new 28-day cycle.

During the menstrual cycle the ovaries produce the female steroid hormones **estrogen** and **progesterone**. The pituitary gland also synthesizes LH and FSH, identical to the brain hormones produced in the male. Together, these four hormones drive a feedback loop that produces a fairly regular monthly cycle.

The menstrual cycle is usually counted from day 1, when the menstrual flow begins. See the numbered steps in FIGURE 14.6:

Step 1: At the beginning of the menstrual cycle, estrogen and progesterone are at their lowest levels in the woman's bloodstream. As in males, the drop in gonadal hormones triggers the hypothalamus to secrete GnRH.

Step 2: The releasing hormone stimulates the pituitary gland to produce FSH and LH, which travel through the bloodstream to the ovary.

Step 3: As its name suggests, FSH (follicle-stimulating hormone) stimulates the ovarian follicles to grow, but usually only one follicle with its oocyte matures each month.

Step 4: The maturing follicle grows rapidly, and secretes increasing amounts of estrogen.

Step 5: The increased estrogen causes the uterine lining to become thicker and more heavily supplied with blood.

Step 6: On about the fourteenth day of the 28-day cycle, the pituitary gland secretes a large pulse of LH and additional FSH, and these trigger the oocyte to break its meiotic arrest and complete its first meiotic division.

Step 7: The developing follicle ruptures and releases the ovum; this is ovulation.

Step 8: Once the ovum has left the ovary and begins its journey down the fallopian tube, the follicle cells left behind in the ovary enlarge and form a new gland, the **corpus luteum** ("yellow body").

Step 9: Corpus luteum cells continue to secrete estrogen, but they also produce large quantities of progesterone.

Step 10: Together, estrogen and progesterone promote the buildup of the endometrium.

Step 11: Estrogen and progesterone also inhibit the hypothalamus from making releasing factors and the pituitary gland from releasing LH and FSH. The drug we discussed earlier, RU 486, mifepristone, blocks the action of progesterone. It can therefore prevent the buildup of the endometrium. Even if an ovum becomes fertilized, completes meiosis II, and starts to divide, the thinness of the endometrium will prevent the successful implantation of an early-stage embryo, which would ordinarily mark the beginning of *pregnancy*.

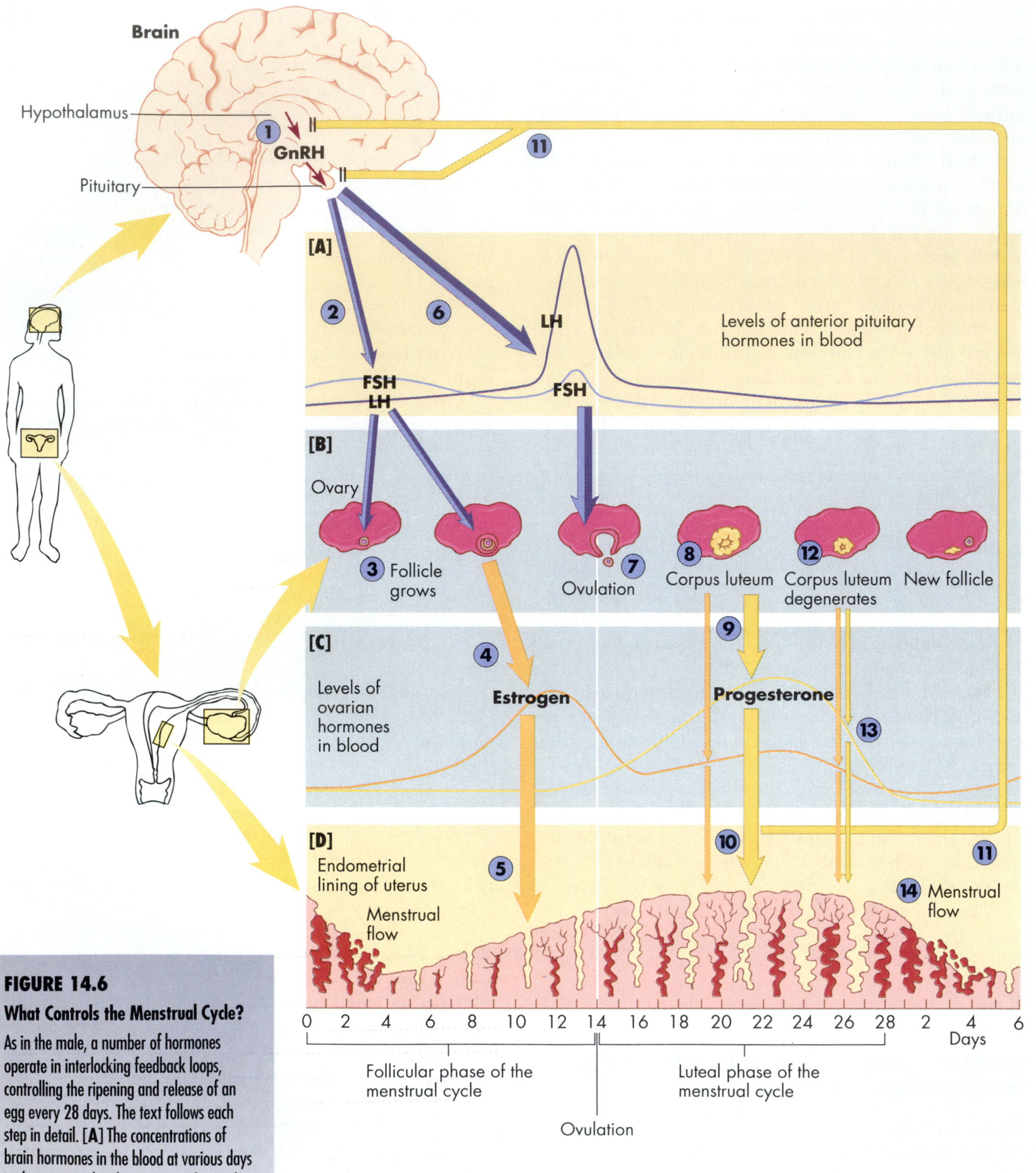

FIGURE 14.6

What Controls the Menstrual Cycle?

As in the male, a number of hormones operate in interlocking feedback loops, controlling the ripening and release of an egg every 28 days. The text follows each step in detail. **[A]** The concentrations of brain hormones in the blood at various days in the menstrual cycle starting at day 1, the first day of menstrual flow. **[B]** Events in the ovary on different days. **[C]** The levels of ovarian hormones in the blood. **[D]** The condition of the uterine lining in response to the cycling hormones.

Step 12: If the ovum in the fallopian tube is not fertilized by a sperm, diminishing levels of LH and FSH cause the corpus luteum to degenerate on day 24 of the cycle.

Step 13: As a result of the degeneration of the corpus luteum, less and less estrogen and progesterone are released.

Step 14: As estrogen and progesterone supplies diminish, the endometrium begins to slough off, and an approximately five-day-long period of menstrual flow starts again, marking the beginning of a new menstrual cycle.

The menstrual cycle becomes irregular or stops altogether in some women who exercise strenuously and achieve a low level of body fat. Apparently, a certain amount of body fat is necessary to maintain the production of GnRH. Once these women begin to exercise less and accumulate more body fat, they start producing sufficient amounts of GnRH to allow their menstrual periods to begin again.

The female's hormonal feedback loops are similar to the male's in several ways: (1) They both involve the same hormones from the brain, GnRH, LH, and FSH. (2) Hormone concentrations are self-regulating in that they feed back and turn off their own production. (3) Reproductive hormones in both males and females ensure the continuation of the species by making gametes available, continuously in the male and cyclically in the female. (4) Estrogen has developmental effects in the adolescent female that correspond to those of testosterone in the male.

➤ CONCEPT CHALLENGE

Suppose that a chemist produced a substance that blocked the action of the hormone LH. How would this discovery affect the reproductive cycles and/or secondary sexual characteristics of men and women? Can you think of a practical use for such a substance?

Human Mating, Conception, and Pregnancy

After an ovum is released from an ovary, it is receptive to sperm, and thus *fertile*, for only a day or so. If fertilization (conception) takes place, the fertilized ovum is able to implant in the uterine wall, and pregnancy begins.

HUMAN MATING

For fertilization to occur in human beings, the male and female must engage in **coitus**, or sexual intercourse, within a few days of the time of ovulation.

Male and female human bodies are admirably suited to help sperm and egg converge. Nerve impulses generated during sexual stimulation increase blood flow to the penis and cause an erection in males. In the female, the clitoris, tissues around the vagina, and the nipples of the breasts also swell and grow more sensitive to sexual stimulation. Lubricants flow from the male bulbourethral glands and the female Bartholin's glands and vaginal wall, easing the penetration by the penis into the vagina, and making the vagina more hospitable to sperm.

After penetration, further clitoral stimulation and pelvic thrusts build sexual excitement in both partners, and the excitement often peaks with **orgasm**, involuntary muscular contractions of the vaginal and uterine walls in the female, and in the male, contractions of the muscles lining the seminal vesicles, urethra, and other structures. The contractions accompanying orgasm are highly pleasurable in both sexes, and in the male cause the high-pressure ejaculation of sperm.

CONCEPTION AND PREGNANCY

Ejaculation during sexual intercourse sends more than 400 million living sperm racing toward the ovum at 1 cm (0.39 in) per minute. Of these millions of sperm, only about 50,000 navigate the cervix successfully and enter the uterus, and fewer still encounter the ovum in the fallopian tube, where fertilization usually takes place. Over 100 sperm may reach the egg, with the tip of each sperm releasing enzymes that digest their way through the ovum's protective coats [review FIGURE 13.6]. The race ends when one sperm enters the egg [FIGURE 14.7], and the entry triggers the ovum to erect barriers to further sperm penetration. At this point, the ovum completes its second meiotic division. (See FIGURE 7.19 for a review of meiosis I and meiosis II.) (LH triggered the completion of the first meiotic division at ovulation.)

After the single sperm has entered the ovum, its haploid nucleus moves toward the ovum's haploid nucleus, and the two nuclei fuse in the cytoplasm of the ovum, forming a diploid zygote with a unique genotype.

The fertilized egg continues its journey through the fallopian tube until it reaches the uterus four or five days later. The process of ovulation has already caused the corpus luteum to complete the monthly development of the uterine walls [review FIGURE 14.6, Steps 7–10] so the fertilized egg can implant in the uterus.

The fertilized egg has been dividing while it has been moving toward the uterus. By day 4, the berrylike cell mass increases to about 30 cells, and a cavity forms in its

FIGURE 14.7
Human Fertilization.
This photo documents the brief moment of fertilization, during which a sperm enters the egg (here, near center of photo). The sperm nucleus will move into the egg cytoplasm, and the two nuclei will fuse, forming the unique genotype of a new individual.

middle. The embryo is now called a **blastocyst** [CONCEPT INTEGRATOR, p. 348, step 5], which consists of two cell groups: (1) the *inner cell mass*, which will continue to develop as the embryo proper, and (2) the *trophoblast*, which will form nutritive tissue.

About six days after fertilization, when the blastocyst consists of about 100 cells, it attaches to the uterine wall, secretes enzymes that break down a small portion of the endometrium, and burrows in, establishing the first physical connection between mother and young. This connection process is **implantation** [CONCEPT INTEGRATOR, step 6] and is the point at which **pregnancy**, the development of an embryo in the uterus, begins.

FETAL MEMBRANES AND THE PLACENTA

Trophoblast cells contribute to the **chorion**, a fluid-filled sac that surrounds the embryo [CONCEPT INTEGRATOR, steps 7 and 8]. The chorion is the embryo's three-way ticket to survival: (1) It absorbs nutrients from the mother's blood and passes them on to the rapidly dividing embryo, (2) it develops into the larger placenta, which sustains the embryo throughout the nine months of pregnancy, and (3) it produces a hormone called **hCG (human chorionic gonadotropin)**. This hormone prevents the onset of a new menstrual cycle, which would flush the implanted embryo from the uterus. The production of hCG, the first major biochemical event of pregnancy, signals the embryo's presence and positively confirms pregnancy. Home pregnancy tests use a simple but ultrasensitive system for detecting hCG, and 97 percent of the time these tests accurately reveal a pregnancy just a week old.

As hCG enters the mother's bloodstream (starting on day 10 after fertilization), the corpus luteum produces estrogen and progesterone, and these hormones maintain the uterine lining, prevent menstruation, and allow pregnancy to continue for the first two months. This step—the implantation of the embryo into the uterus—was blocked in the women attending the Scottish clinic who wished to prevent pregnancy. The mifepristone they took blocks the action of progesterone, and so the uterine lining sloughs off, preventing pregnancy.

The chorionic membrane contains many fingerlike projections called *villi* (singular, *villus*) that allow the exchange of nutrients, gases, and metabolic wastes between mother and embryo [CONCEPT INTEGRATOR, step 8].

Embryonic cells also grow upward and form a second important fetal membrane, a hollow ball called the **amnion** [CONCEPT INTEGRATOR]. The space within this ball becomes the **amniotic cavity**, which encloses a watery, salty fluid that enables the developing embryo to float suspended in a relatively injury-free environment during pregnancy [CONCEPT INTEGRATOR, step 9]. The embryo sloughs off cells into this amniotic fluid. Much later these cells can be collected and analyzed for possible genetic abnormalities during amniocentesis [see FIGURE 11.18].

The chorion grows and becomes enmeshed with maternal tissue, forming the dark red, spongy **placenta**, an exchange site where a thick tangle of embryonic blood vessels encounters blood-filled spaces in the uterine lining, forming the vital link between mother and embryo [FIGURE 14.8]. The **umbilical cord** forms a lifeline connecting the embryo to the placenta.

Embryonic and maternal bloods do not mingle, but nutrients and oxygen pass from the mother's blood across embryonic vessel walls and into the embryo's blood. In the reverse direction, carbon dioxide and other wastes pass from the embryo into the placenta. After a few weeks, the placenta begins to produce estrogen and progesterone, which maintain the uterine lining and block the production of LH and FSH, thus preventing menstruation and the termination of pregnancy. The corpus luteum then slowly degenerates.

MOTHER'S CONTRIBUTION TO THE FETAL ENVIRONMENT

Because the placenta forms such an intimate physical connection between mother and offspring, the foods, drugs, and other chemicals the mother takes into her body profoundly affect the baby. She must, therefore, be especially careful during pregnancy.

For proper growth of muscles and bones, the fetus requires a constant supply of protein and calcium, as well as fatty acids, carbohydrates for energy, vitamins, and minerals. Because the mother's diet must provide these,

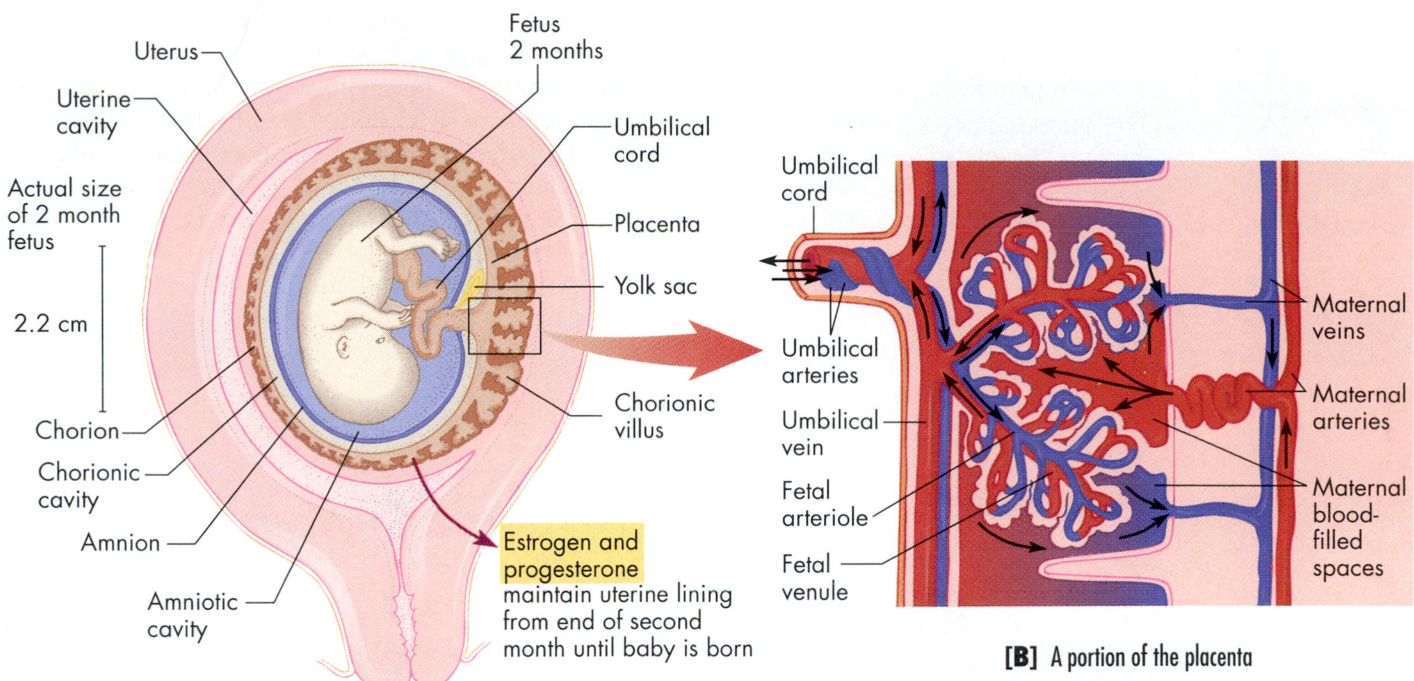

FIGURE 14.8

Development of Fetal Membranes and the Placenta.

After 60 days, the placenta is well established and produces the hormones that prevent menstruation for the remainder of the embryo's nine-month gestation. Narrow fingers of tissue, or chorionic villi, project from the chorion, and each eventually houses a tiny blood vessel. The maternal blood fills the spaces around the villi, and exchange of gases and materials takes place across the delicate layer separating the maternal and fetal blood supplies. (FIGURE 11.18 shows how chorionic villi are sampled for various genetic conditions.)

obstetricians usually advise pregnant women to drink lots of milk, eat plenty of protein, and take vitamin and mineral supplements. Protein intake is especially important in the final trimester, when the fetus experiences the greatest expansion in brain size. A maternal weight gain of 11 kg (25 lb) or so is now believed appropriate for most women to help prevent premature or underweight infants with their greater susceptibility to infections and breathing problems.

A mother's life-style habits, such as smoking, drinking, or drug use, can have severe repercussions for the fetus. Two such consequences are *fetal alcohol syndrome* and *fetal tobacco syndrome*. Babies born to women who drink substantial amounts of alcohol during pregnancy show greater incidence of mental retardation, emotional abnormalities, cleft palates, underdeveloped hearts, slow growth, and facial anomalies. Many obstetricians suggest that their patients avoid alcohol altogether during pregnancy. Miscarriage (premature expulsion of the fetus) is much more likely in smoking mothers, and babies are more likely to have a low birth weight and thus greater susceptibility to respiratory disease and *sudden infant death syndrome* (suffocation during sleep). Overall, infants with fetal tobacco syndrome have a death rate 10 to 100 percent higher than children from nonsmoking mothers. Mothers who take amphetamines or cocaine risk having infants with neurological defects, and mothers who take heroin or other narcotics often give birth to addicted infants.

Some of the saddest chapters in modern medicine concern the fetal damage that can be caused by certain prescription drugs. The antinausea drug thalidomide calms adult nerves, but it alters the embryonic nerves that help direct proper limb development and growth. Some pregnant women who took thalidomide in the early 1960s had babies that were mentally and emotionally normal but had shortened arms with deformed hands. Ironically, thalidomide does not cause these effects on mice or rats, and so its devastating effects were not discovered during premarketing laboratory tests.

Premarket drug testing has become much more stringent in recent years, but pregnant women must still be wary of exposure to drugs, other chemicals, and certain viral diseases such as German measles (rubella). As late as 1988, about 100 women taking the prescription drug Accutane for severe acne ignored the manufacturer's warning and became pregnant while on the medication [see BOX 13.1, page 324]. As a result, they gave birth to mentally retarded babies with abnormally small jaws and very low-set ears. It is not surprising that Accutane, a chemical relative of retinoic acid, the developmental signaling molecule we discussed in CHAPTER 13, affect a developing embryo or fetus. Pregnancy is clearly a time when a woman must take special care of her general health and nutrition—and, indirectly, her baby's.

FIGURE 14.9

Development of the Neural Tube and Spina Bifida.

[A] In a 23-day-old human embryo, roundish blocks of cells (somites) that will form the backbone and trunk muscles are being added at the growing embryo's posterior end. The neural tube has folded up in the region of the somites, but folding is not yet complete at the ends of the embryo. **[B]** If the neural tube does not close completely, a common birth defect called spina bifida may result. The embryo in this photo has a mild case, which was corrected by surgery after the baby was born.

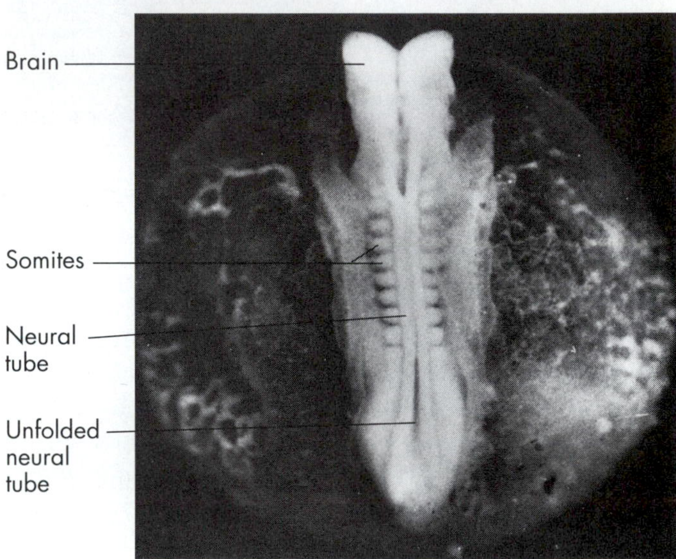

[B] Spina bifida

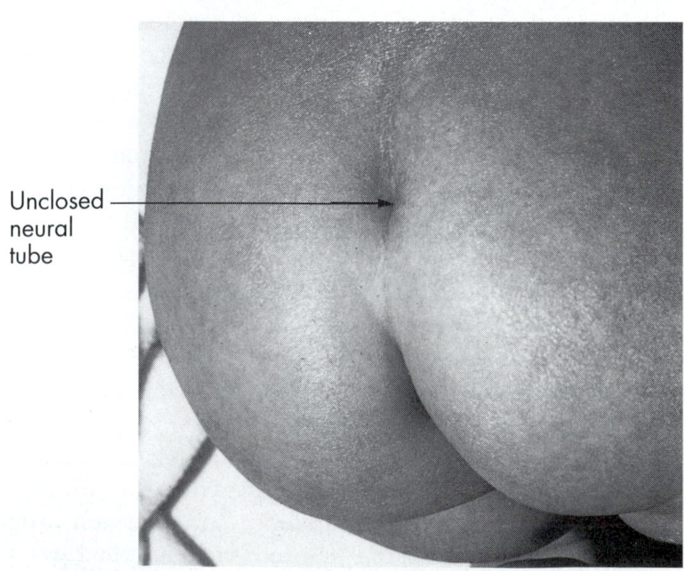

Human Development, Birth, and Lactation

Even before a woman suspects that she may be pregnant, the embryo has embarked on the early stages of development.

DEVELOPMENTAL STAGES IN A HUMAN EMBRYO

In the third week of pregnancy, when the embryo is still smaller than the length of this *l*, gastrulation establishes the embryo's three-layered body plan. During the next eight weeks, the nervous system and other organs form, transforming the tiny mass into an organism with a characteristic human shape.

The nine months of a human pregnancy are divided into three 3-month periods, or **trimesters**. The main developmental events of each trimester are described in the following sections.

The First Trimester The formation of body organs begins as the neural tube rolls up midway through the *third week* of pregnancy [FIGURE 14.9]. The neural tube will eventually develop into the nervous system. Primitive blood cells and blood vessels have formed, and during the *fourth week* the U-shaped heart begins to pump blood. The limb buds are visible, and the intestine is a simple tube. The liver, gallbladder, pancreas, lungs, eyes, nose, and brain begin to form. Although the embryo is only about 5 mm (less than ¼ in) from crown to rump, it is already about 10,000 times larger than the fertilized egg.

When the embryo reaches its *eighth week* it is called a **fetus** because most organs have already formed, and during the rest of gestation—the fetal period—they simply enlarge and mature. The fingers and toes are well formed, and the head lifts away from the chest. The first indications of skeletal bone formation can be seen during the *ninth week*. The fetus is about 23 mm (1 in) from crown to rump. The first trimester ends with the *tenth to twelfth weeks*; by which time the fetal pulse is detectable, and the fetus is about 40 mm (1.5 in) long.

Sex Differentiation during the First Trimester The primary sex organs have an interesting, if somewhat startling, course of development. The embryo produces gonads, sex ducts, and external genitals before the eight-week mark, but you can't tell a boy from a girl at that stage because sexual structures are still **indifferent**—they have the potential to form structures of either sex. As FIGURE 14.10A shows, embryos at this stage have two sets of sex ducts. In female embryos, one set of ducts, called the *Müllerian ducts*, develops into the fallopian tubes and uterus while the other disappears [FIGURE 14.10B]. In the male, the Müllerian ducts disappear, while the second set becomes the vas deferens [FIGURE 14.10C].

A fascinating thing about sex differentiation is that in the absence of a specific biochemical signal from the *Y* chromosome, these indifferent structures become female in type. Without a *Y* chromosome, gonads become ovaries, the pre-vas deferens cells die, and the sex ducts develop into the fallopian tubes and uterus. The external genitals follow suit [FIGURE 14.12B].

Nature, however, has ensured an alternative means of sexual differentiation by providing a set of signals that can turn indifferent gonads into male instead of female organs. The signals begin with a specific gene on the *Y* chromosome called the **testis-determining factor (TDF)** gene. Researchers confirmed the presence of such a gene by studying four men who suffered from infertility. Each man turned out to have two *X* chromosomes—the female chromosomal constitution—yet they were phenotypical males. When these mens' chromosomes were examined at a molecular level, they were found to have a short stretch of *Y* chromosome DNA inserted into another chromosome. Since this *Y* chromosome DNA was sufficient to direct the development of testes, geneticists realized that it must contain a gene for male development. Eventually, researchers isolated the gene from this DNA. The gene's structure suggests that its action is to regulate the activity of other genes, causing an indifferent gonad to develop into a testis.

Once testes differentiate, they release two substances that complete male differentiation, testosterone and a protein called *Müllerian-inhibiting hormone*. Most of the testosterone maintains the male duct system, while an enzyme converts the rest into a second hormone, a derivative of testosterone called *5DHT* (*5-dihydrotestosterone*), which causes the external genital tissue to develop into the penis and scrotum [FIGURE 14.10C]. The Müllerian-inhibiting hormone blocks the growth of the cells that form the female reproductive ducts, leaving the other duct to become the vas deferens. Without this Müllerian-inhibiting hormone, the male would retain the ducts that become the fallopian tube and the uterus. Without the testosterone and its derivative, the male would retain female external genitalia. Since Müllerian-inhibiting hormone blocks the growth of future uterine cells, biotechnologists are hoping they can manufacture

[A] External genitals in a 7-week embryo

Genital tubercle
Genital fold
Membrane covering urogenital opening
Labio-scrotal swelling
Anal membrane

If no *Y* chromosome
If *Y* chromosome present
No testosterone
Female organs develop
Testosterone→5DHT
Male organs develop

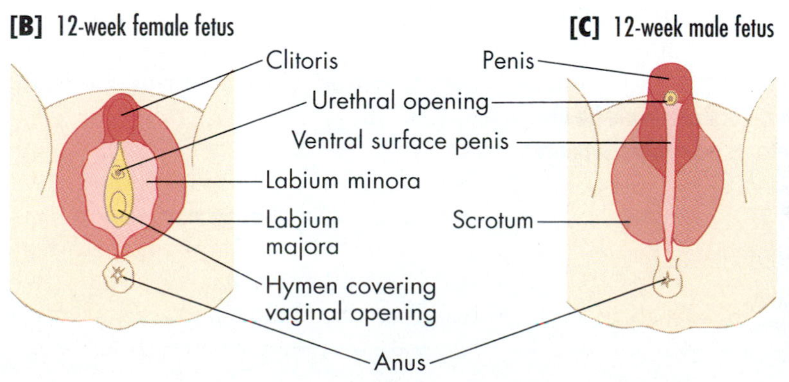

FIGURE 14.10

Development of the Reproductive System.

[A] Until seven weeks of development, the external genitals of male and female embryos look alike, and two systems of reproductive ducts have developed. **[B]** In the absence of a *Y* chromosome, the gonad becomes an ovary and one of the duct systems dies while the other (the Müllerian duct) becomes the fallopian tube and uterus. **[C]** In the presence of a *Y* chromosome, the gonad becomes a testis and secretes testosterone and Müllerian-inhibiting hormone. The Müllerian-inhibiting hormone kills the female duct system. The testosterone is converted to a related hormone (5DHT) that induces the development of male genitals.

active learning: Male Chromosomes, Female Gender

In 1991, the female hurdles champion from Spain flew to Japan to compete in the World University Games. Soon after she arrived, officials asked her to undergo a second test of her sex chromosomes. Although she had never menstruated, Carmen had passed a similar test once, and was not worried about this retest. The new examination, however, showed that Carmen had just one *X* chromosome, not two, and one *Y* chromosome—in other words, the male sex constitution. She was shocked because she'd never had cause to doubt that she was a female. Spanish sports officials told her to fake an injury as she warmed up for her race, sent her home, and asked her to resign from the national team.

Let's apply what you learned on page 345 about sex determination to Carmen's condition, and try to understand how she could have a male chromosomal constitution but the breasts, body contours, sexual behavior, and external genitalia of a female.

Medical examinations showed that, while Carmen had a vagina, she lacked female sex ducts—uterus and fallopian tubes. She also lacked male sex ducts—no vasa deferentia. What's more, instead of ovaries, she had testes.

Did Carmen have the TDF (testis-determining factor) gene? ....
What evidence from chromosomes and phenotype supports your answer?......
....

Since the TDF gene lies on the *Y* chromosome, it makes sense that Carmen had this gene and that it led to the formation of testes. We know further that developing testes produce two substances that affect the development of sex ducts and genitalia.

What are they?.....................
What effect does Müllerian-inhibiting factor have on female sex ducts?.........
What effect does testosterone have on male sex ducts?....................
Did Carmen's gonads produce Müllerian-inhibiting factor?..................
What evidence supports your answer?...
....

The absence of female sex ducts proves that Müllerian-inhibiting factor was present in Carmen's body.

What, however, does the absence of male sex ducts and the presence of female instead of male genitalia suggest about testosterone? ....

Since testosterone is necessary to maintain male sex ducts and to form male genitalia, Carmen's sex ducts and genitalia acted as if no testosterone were present.

We could understand Carmen's condition if testosterone was absent from her body. However, steroid tests have revealed that she does indeed produce testosterone, and it comes from her testes. What is actually lacking is the ability of Carmen's cells to respond to testosterone. Without a response to testosterone, her male sex ducts died and her genitalia and secondary sex characteristics remained female. With the presence of Müllerian-inhibiting factor, her female sex ducts died, leaving her with no uterus. The overall result was a female body with male sex chromosomes and sex hormones.

Do you think that Carmen should be allowed to compete as a female? Support your opinion. ....
....

enough of it to treat women suffering from cancers of the reproductive tract. The hormone should stop abnormal cells from growing and dividing.

The Second Trimester The second and third trimesters of pregnancy are devoted to the increase in size and maturation of the organs developed during the first trimester. External genitalia attain distinctive features during the *twelfth week*, and the mother feels her uterus enlarging. By the end of the *fifteenth week*, the fetus measures about 56 mm (2.25 in) from crown to rump.

By the *sixteenth week* [CONCEPT INTEGRATOR, step 14], the face looks "human," and stretching movements begin to occur. Many developmental changes are evident from the *twentieth* to *twenty-fourth weeks*. The body is now covered by a downy hair called *lanugo*, and fetal heart sounds can be heard with a stethoscope. The lungs are formed, but do not function yet. The gripping reflex begins to develop, and kicking movements and hiccuping may be felt by the mother.

The Third Trimester At the beginning of the third trimester [CONCEPT INTEGRATOR, step 15], the fetus's eyelids open, eyebrows and eyelashes form, and light can be perceived. At this stage, with almost two months to go, a baby born prematurely would have at least a 10 percent chance of survival.

During the final couple of weeks, the lack of space in the mother's uterus causes a decrease in fetal activity, and the fetus usually turns to a head-down position. The fetus is said to have come to *term*, about 280 days after the beginning of the last menstruation before conception.

BIRTH AND LACTATION

As a pregnant woman waits, the fetus prepares itself and its mother for impending labor and delivery. Special brown fat, stored around the neck and down the fetus's back, will produce heat after the baby is expelled from the warm uterus. Special reserves of carbohydrates are laid down in the heart and liver that nourish the baby until it can suckle milk. The placenta makes and secretes a hormone that prepares the mother's mammary glands to produce milk. The maternal uterine muscles become more excitable, contracting periodically in "false labor" and building strength as the time of birth approaches.

During the last trimester, the uterus's sensitivity to the hormone **oxytocin** increases a hundredfold [FIGURE 14.11]. Oxytocin is produced in the hypothalamus of the brain and is released from the pituitary gland. When the sensitivity to oxytocin reaches a threshold, **labor**, the physical effort and uterine contractions of childbirth, begins. Nerves from the cervix signal the brain and pituitary to release more oxytocin, causing the uterine muscles to contract even more. Oxytocin also stimulates the uterine wall to release **prostaglandins**, hormones that stimulate uterine contractions still further. This causes the release of even more oxytocin, in a continuing *positive feedback loop*. This is a *positive* feedback loop because an increase in oxytocin causes still more oxytocin to be released. In contrast, in the *negative* feedback loop in a male, an

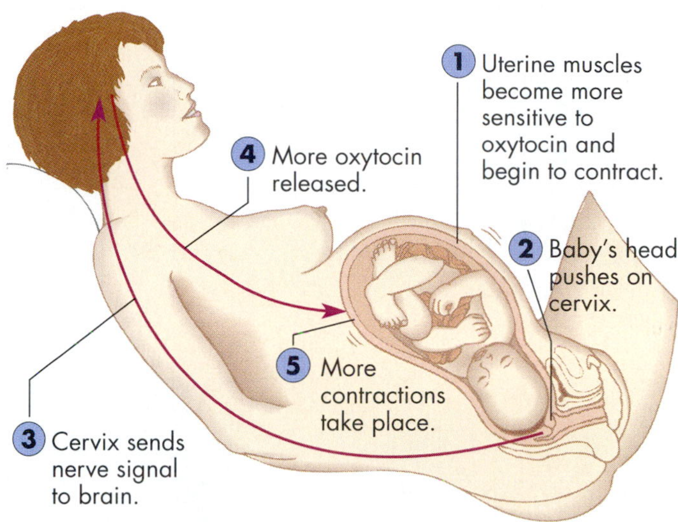

[A] A positive feedback loop propels labor

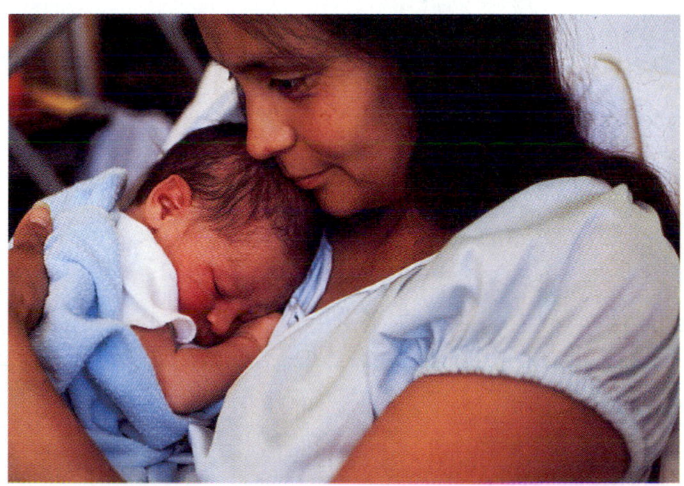

[B] Shortly after birth

FIGURE 14.11

Birth: Hormones Trigger Contractions and Delivery.

[A] As birth approaches, uterine muscles become more sensitive to oxytocin (1), and they begin to contract, pushing the baby's head into the mother's cervix and causing it to widen (2). The stretched cervix sends a signal via nerves to the brain (3), which causes more oxytocin to be released (4). This increase in oxytocin causes more uterine contractions, and the cycle escalates (5). **[B]** The cycle continues until the baby is expelled through the birth canal. Finally, the mother and newborn can begin to recover from the arduous process.

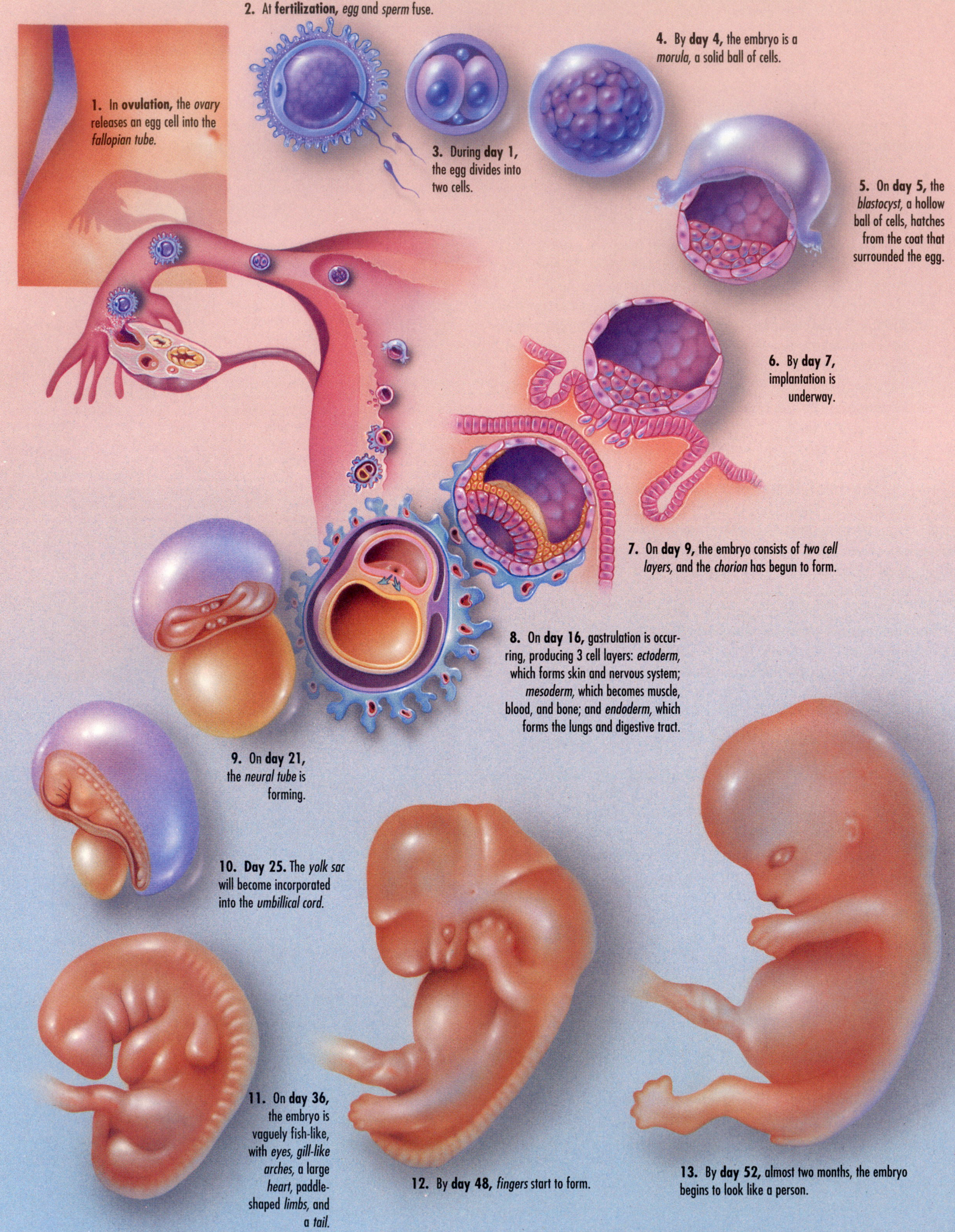

2. At **fertilization,** *egg* and *sperm* fuse.
1. In **ovulation,** the *ovary* releases an egg cell into the *fallopian tube.*
3. During **day 1,** the egg divides into two cells.
4. By **day 4,** the embryo is a *morula,* a solid ball of cells.
5. On **day 5,** the *blastocyst,* a hollow ball of cells, hatches from the coat that surrounded the egg.
6. By **day 7,** implantation is underway.
7. On **day 9,** the embryo consists of *two cell layers,* and the *chorion* has begun to form.
8. On **day 16,** gastrulation is occurring, producing 3 cell layers: *ectoderm,* which forms skin and nervous system; *mesoderm,* which becomes muscle, blood, and bone; and *endoderm,* which forms the lungs and digestive tract.
9. On **day 21,** the *neural tube* is forming.
10. **Day 25.** The *yolk sac* will become incorporated into the *umbillical cord.*
11. On **day 36,** the embryo is vaguely fish-like, with *eyes, gill-like arches,* a large *heart,* paddle-shaped *limbs,* and a *tail.*
12. By **day 48,** *fingers* start to form.
13. By **day 52,** almost two months, the embryo begins to look like a person.

Concept Integrator:
The Marvel of Human Development

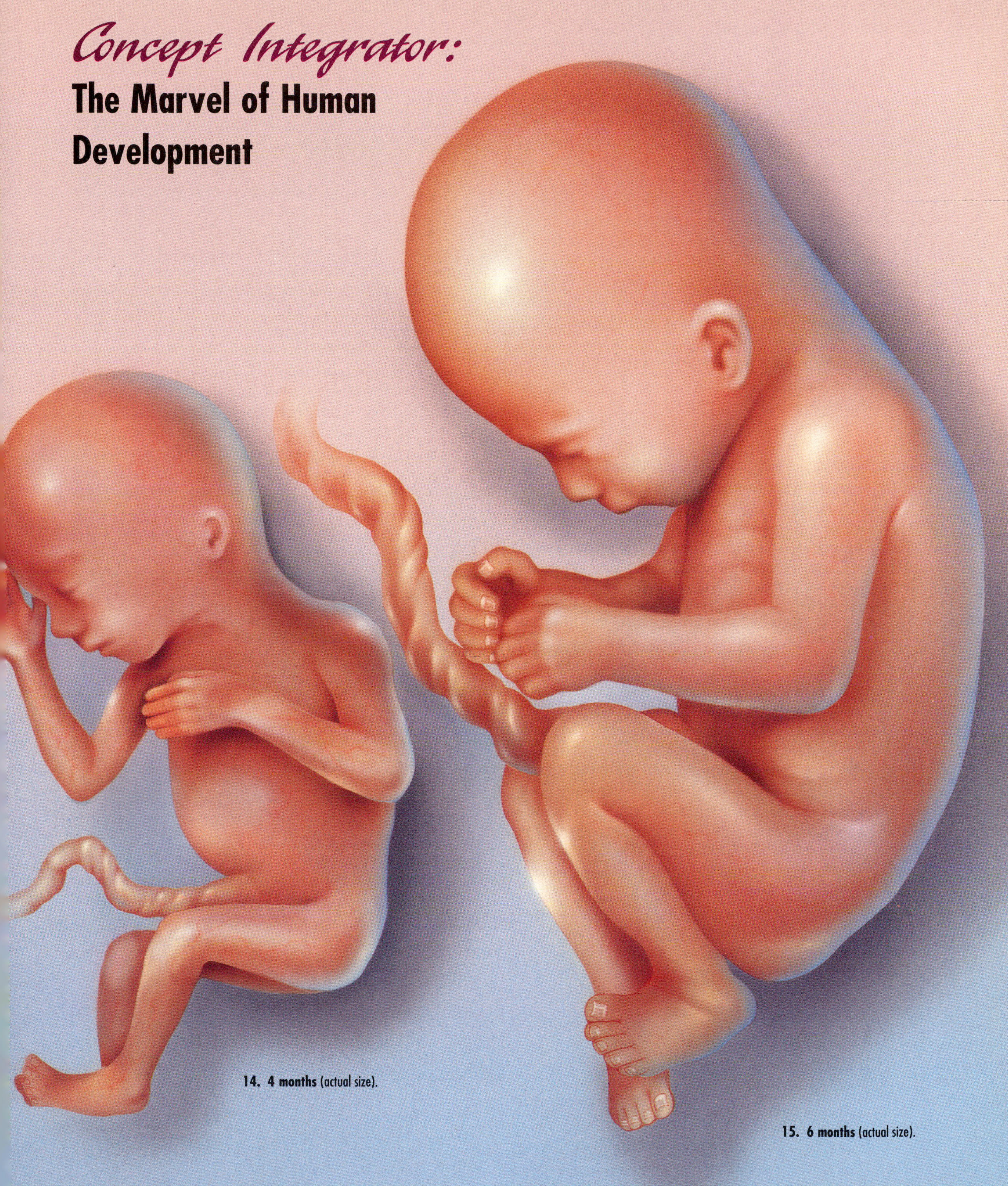

14. 4 months (actual size).

15. 6 months (actual size).

increase in testosterone causes *less* testosterone to be produced.

As the uterine contractions grow longer, stronger, and more regular, the cervix widens. Each contraction starts at the upper end of the uterus and moves downward toward the cervix, which the baby's head pushes further and further open as it moves toward the vagina, or birth canal. The squeezing eventually causes the amniotic sac to burst (the water breaks), and when the cervical opening reaches a width of about 10 cm (4 in), delivery is usually only minutes away. With considerable pushing of her abdominal muscles, the mother is able to force the baby's head past the pelvic opening, allowing the baby's shoulders and hips to emerge.

The newborn baby, still attached to the placenta by the umbilical cord, takes its first gasp of air. Soon the blood vessels in the cord cease pulsing, and attendants sever it. Maternal blood vessels that supply the uterus shut down, which prevents excessive blood loss, and with a few more uterine contractions the placenta (now called the afterbirth) is expelled.

Recent studies show that the heavy stress of being born is actually beneficial in that it causes stress hormones (adrenaline and noradrenaline) to surge through the infant's body, clearing its lungs, promoting normal breathing, mobilizing stored energy, and sending extra blood to the heart and brain. Although the transition from semiaquatic fetus to air-breathing land animal is difficult, most babies make it quite well.

MAMMARY GLANDS

The **mammary glands** within the paired breasts of a woman produce and secrete milk. Breasts are present in an undeveloped form in children and men, but during puberty in the adolescent female the breasts begin their development as milk-producing organs. During pregnancy, secretions of estrogen and progesterone cause the milk-secreting glands within the breasts to develop; and *prolactin*, a hormone from the pituitary gland, causes milk production after the woman gives birth.

The breasts contain an extensive drainage system made up of many lymph nodes. These nodes may be susceptible to breast cancer (the most common form of cancer in females), and frequent self-examination of the breasts for lumps and regular physical checkups should be routine practices.

➤ CONCEPT CHALLENGE

What problems might occur if the umbilical cord became knotted up during a prolonged delivery?

Growth, Maturation, and Aging

Although birth marks the end of fetal development, it is not the end of growth and change. People continue to cross a series of more gradual thresholds as they move from infancy through childhood, puberty, adulthood (the longest stage in the human life cycle), and old age.

Human growth follows interesting patterns. Recent experiments show that infants do not grow larger gradually, but rather they go for periods of 2 to 15 days with no increase in size at all, and then, in less than 24 hours, they grow from 0.5 to 1.65 cm (about ¼ in to ¾ in) [FIGURE 14.12]. Growth is fastest in the head and body center, and slowest toward the periphery. At birth, the brain is the most developed organ, and is 25 percent of its final adult weight. By 6 months of age, the brain has reached 50 percent of adult weight, and by age 10, 90 percent. Body weight lags far behind brain weight, advancing from 5 percent of adult weight at birth to only 50 percent at age 10. The child's trunk grows faster than its arms and legs, and it gains coordination over gross limb movements before finger and toe movements.

Adolescence is marked by the most dramatic physical changes since fetal life. **Puberty**, maturation of the reproductive system and development of secondary sexual characteristics, usually takes place in girls between the ages of 11 and 13, and in boys between 13 and 15.

With modern nutrition, sanitation, and health care, physical maturity can begin at the end of adolescence and last for 50 to 75 years or more. The peak performance of all the body's organ systems usually occurs between the early 20s and the early 30s. Sometime after age 30, the process of **aging**, a progressive decline in the maximum functional level of individual cells and entire organs, begins to accelerate. About age 50, females experience **menopause**, a cessation of the menstrual cycle, while males may lose some *potency*, or the ability to maintain an erection. Sometime after age 60, people reach **senescence**, or old age, when the decline in cell and organ function is less gradual and more profound.

WHAT CAUSES AGING?

Many researchers have proposed hypotheses to explain the causes of aging. Most hypotheses can be grouped into two categories: (1) genetic clock hypotheses, and (2) wear-and-tear hypotheses.

Many biologists support the idea of a genetic clock, arguing that there is a genetically specified timetable for aging and death. Just as genes regulate the timing of organ

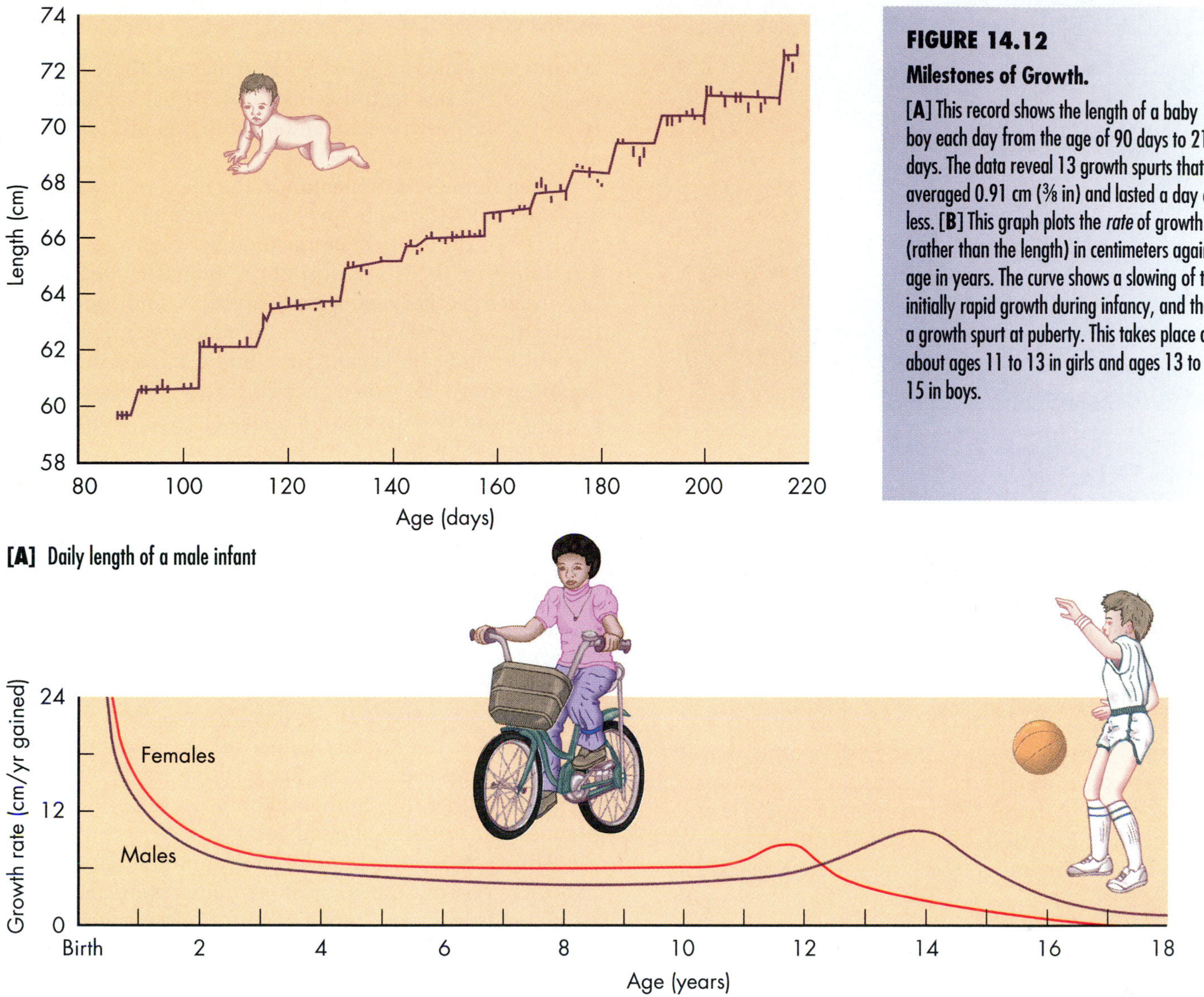

[A] Daily length of a male infant

[B] Rate of growth

FIGURE 14.12

Milestones of Growth.

[A] This record shows the length of a baby boy each day from the age of 90 days to 218 days. The data reveal 13 growth spurts that averaged 0.91 cm (3⁄8 in) and lasted a day or less. **[B]** This graph plots the *rate* of growth (rather than the length) in centimeters against age in years. The curve shows a slowing of the initially rapid growth during infancy, and then a growth spurt at puberty. This takes place at about ages 11 to 13 in girls and ages 13 to 15 in boys.

formation, cell death in the formation of fingers or toes [review FIGURE 13.17], and sexual maturation, they also regulate when our organs and systems will cease functioning. There is tantalizing evidence for genetic clock hypotheses. First, cells seem to have preset limits to the number of times they can divide. For example, skin cells taken from a human infant and grown in the laboratory divide about 50 times and then stop, while similar cells from a 90-year-old divide only a few times in the lab. Second, organs seem to age in a preprogrammed way. Each year, starting at about age 30, we experience a 1 percent decline in the maximum function of various organs. This includes lung capacity, the amount of blood the heart pumps, the strength and coordination of muscles, the filtering power of the kidneys, and the highest-pitched sounds we can hear.

Third, certain genetic conditions can bring about symptoms of aging. Children with *progeria* (early aging) begin to lose their hair, show wrinkles, and experience arthritis and heart attacks by age 5 or 6 [FIGURE 14.13]. These children, however, do not suffer the whole spectrum of age-related illnesses, which includes cataracts, diabetes, and cancer. Certain other people inherit a dominant mutation that appears to cause Alzheimer's disease, a kind of senility. The brains of Alzheimer's patients shrink and contain far greater numbers of tangles and knots, called senile plaques, than do the brains of normal older people. While the complete genetic basis for diseases like progeria and senility remains unsolved, the conditions do seem to support the genetic clock concept.

Fourth, researchers have discovered genetically dis-

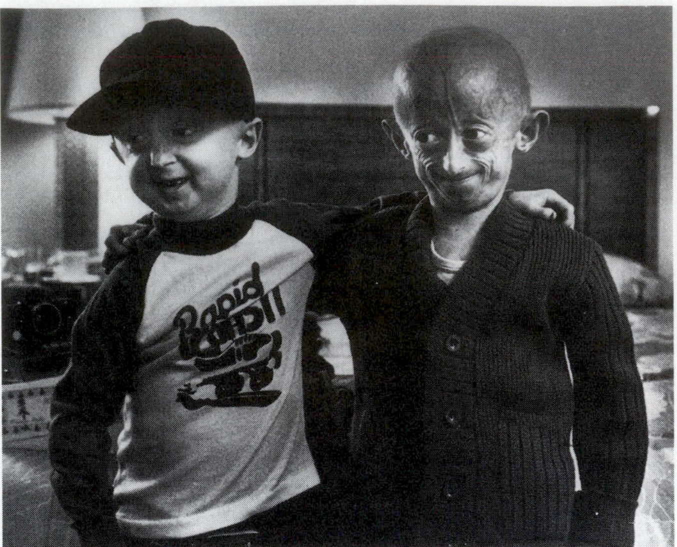

FIGURE 14.13

Premature Aging: A Genetic Basis?

Some children have a genetic condition called progeria, in which certain signs of aging develop by age five or six. The boy on the left (age 9) from Texas and his new acquaintance (age 8) from South Africa both have progeria. They are pictured about to tour Disneyland.

tinct strains of fruit flies and nematode worms that live twice as long as normal—equivalent, in fact, to a person living 150 years or more. This suggests that aging is under genetic control. Exactly how the "Methuselah" genes extend the lives of flies or worms, however, will require further analysis.

Evidence suggests to other biologists that our preprogrammed genes are less likely to cause aging than is *wear and tear*: the accumulation of random errors in the replication and use of DNA and in the synthesis of proteins; or the accumulation of metabolic by-products that disable enzymes, other proteins, and lipids. Informational errors may be due to environmental insults such as sunlight, radiation, or chemicals. One class of culprits may be the toxic, chemically active, oxygen-containing molecules called free radicals [review CHAPTER 2] that arise naturally during energy metabolism. The relentless piling up of metabolic garbage probably interferes with normal cell function, including the flow of information from DNA to RNA to protein, and perhaps contributes to the 1 percent per year decline in organ function we all experience.

Evidence on the long-lived flies and worms we just discussed reveals that these animals have increased levels of one particular enzyme called superoxide dismutase, which breaks down free radicals of oxygen. This finding is extremely provocative, in that it may lead to a theory unifying the genetic and wear-and-tear hypotheses. Researchers are still a long way, however, from understanding the basic mechanisms of aging.

AGING GRACEFULLY

Whatever the major causes of aging may be, there are several good reasons to think that most of us can and will live long, healthy lives despite the inevitabilities of decline and mortality.

First, human **life expectancy**, the maximum probable age a person will reach, has never been higher. For most of human history, most people died between ages 20 and 40. "Life," as English philosopher Thomas Hobbs wrote, was indeed "solitary, poor, nasty, brutish, and short." But thanks to advances in agriculture, emergency medicine, obstetrics, and public sanitation and immunization programs, average life expectancy in North America has increased steadily to its current levels of 78 for women and 71 for men. The oldest verified age is just over 120 years.

Second, studies show that adhering to certain health practices can increase life expectancy—as well as the *quality* of those extra years—significantly. Researchers have found that a man who eats regular meals (including breakfast), exercises regularly, sleeps an adequate amount, maintains an ideal weight, does not smoke, and limits alcohol consumption can live an average 11 years longer than one who follows three or fewer of those practices. A woman who observes all six healthy practices can add seven years to her already longer life expectancy. Moreover, those observing these six positive life-style habits remain healthier during their extra years than other people their age with fewer good habits.

Third, there is a known environmental parameter that increases the life span of laboratory animals: a diet restricted in calories. Rats given a calorie-restricted diet *with adequate nutrition* live about 30 percent longer than rats allowed to feed freely. There is some collateral evidence that caloric restriction can slow human aging, perhaps, biologists speculate, by slowing the genetic clock, by decreasing metabolic rate and hence the generation of free radicals, or both.

Finally, the dread many people feel over the prospects of growing old may be the result of misconceptions and negative stereotyping. A recent study revealed that 95 percent of the elderly in the United States live independently, not in institutions; most are in regular contact with their families, not isolated or lonely; most are vigorous and active, not frail and sedentary; and most are financially secure, well educated, and integrated into their communities. Clearly, the last decades of life can hold great satisfactions rather than dependence and illness. It all seems to depend on the habits and support systems one establishes much earlier in life.

➤ CONCEPT CHALLENGE

What is the evidence that aging is a genetically programmed step in normal development?

Fertility Management and Overpopulation

The drama of human fertilization happens quite naturally, and, many think, all too often. Every minute, 230 babies are born throughout the world, but only 90 people die during that same time, leaving a net increase of 140 people per minute, or 1.4 million additional people every week.

Many eminent ecologists, including Paul and Anne Ehrlich, believe that the exploding human population is the single most serious problem in the world. In some areas unchecked population growth has led to severe food shortages, overclearing of forestland, and excess burning of fossil fuels. These practices can contribute to the extinction of valuable species and even the ruin of entire ecosystems, and are partially responsible for what seems to be a worldwide global warming trend. Even in industrialized countries like the United States, where the birthrate is relatively low, overpopulation is still a serious problem because American parents use about 30 times more energy, food, and other resources in raising a child than parents in poorer countries like Venezuela or Bangladesh. And so for personal reasons as well as environmental ones, many people around the world seek to avoid unwanted pregnancies.

BIRTH CONTROL

Birth can be prevented by blocking fertilization, or *contraception*; by interfering with implantation; and by terminating pregnancy, or *abortion*. Table 14.3 outlines different types of birth control.

All techniques for blocking fertilization share the same goal—to prevent pregnancy—but they vary widely in approach. Behavioral techniques include abstaining from intercourse totally, abstaining during a woman's fertile period—the so-called rhythm method, natural family planning, or fertility awareness method—and withdrawal of the penis before ejaculation. None of these three approaches is very reliable.

A widely used technique for avoiding fertilization is natural family planning. This requires the couple to abstain from intercourse for at least two days before ovulation, since sperm can live in the female's reproductive tract for two days, and for three days after ovulation, since the egg remains fertile for three days. In practice, the period of abstinence must be much longer because of the difficulty in predicting and detecting the time of ovulation in most women. Some women, however, who are highly motivated, have regular menstrual cycles, and have adequate training, can determine relatively accurately the period of time during which fertilization can occur by examining mucus and monitoring their body temperature as well as by watching the calendar. For these women, the method works relatively well.

Several companies are planning to market a more accurate system for predicting the time of ovulation, based on the timing of hormone peaks in a woman's blood. As you can see from FIGURE 14.6, the amount of estrogen in a woman's blood peaks just before she ovulates, the amount of LH peaks at ovulation, and the amount of progesterone rises just after ovulation. These peaks, in effect, provide the red light and green light for sexual intercourse. Pharmaceutical companies have developed sensitive "dipstick" tests based on monoclonal antibodies (immune system molecules described in CHAPTER 26). One type now sold in drugstores can be dipped in a few drops of urine, and a color change indicates the increase in progesterone just after ovulation. A second type, still being developed, would work the same way but detect the increase in estrogen and hence predict ovulation so that couples wishing to avoid pregnancy could abstain from intercourse between the estrogen peak and the progesterone peak. A dipstick test for LH, on the other hand, is valuable for couples wishing to conceive, so that sexual intercourse can occur during maximum fertility.

Besides behavioral techniques, a number of chemical, physical, and surgical methods prevent sperm from fusing with eggs. Chemical techniques for preventing conception, such as douching (rinsing the vagina) or applying spermicidal (sperm-killing) creams, foams, or jellies, are about as effective as the rhythm method or withdrawal. Physical barriers include the male rubber condom (a sheath worn over the penis), the diaphragm (a dome covering the cervix), or the female condom, a clear plastic sheath attached to two flexible rings which inserts into the vagina like a diaphragm, and covers both the woman's internal tissues and her external genitals as well. Physical barriers are much more effective if combined with spermicides. Male and female condoms can also help in preventing the spread of AIDS and other sexually transmitted diseases [see BOX 14.1, page 354]. Sterilization, a permanent barrier to conception, is achieved when a surgeon ties off a woman's fallopian tubes (in a *tubal ligation*). This prevents eggs from entering the uterus. Male sterilization (*vasectomy*) severs a man's vasa deferentia, thus blocking the exit of sperm from the body.

The widely used oral contraceptive known as "the pill" is a chemical strategy for blocking the production of eggs or sperm. The female pill is a mixture of synthetic hormones that block FSH and LH production and hence maturation and ovulation of eggs. Pills have side effects

box 14.1
Biology Applied

Sexually Transmitted Disease: A Growing Concern

One consequence of America's sexual revolution during the 1970s and 1980s was a dramatic rise in the incidence of *venereal diseases*—diseases of the genital tract and reproductive organs caused by microorganisms. These diseases are part of the broader category of *sexually transmitted diseases* (STDs)—infectious diseases of any body region that can be passed to a partner through sexual contact. Regardless of terminology, however, the problem remains: A person with multiple sex partners has a substantial risk of contracting a sexually transmitted disease. Over 10 million cases are treated each year in the United States, and a study released in 1993 showed that 1 American in 5 is infected with an STD. If undetected or untreated, such diseases can lead to severe complications. Thus it is important to be familiar with the indicators of the most common STDs.

The most common STD is *chlamydia*, an infection by the bacterium *Chlamydia trachomatis* picked up through sexual contact with an already infected person [FIGURE 1]. A woman may have no symptoms of chlamydia at all, or she may experience pelvic pain, painful urination, vaginal discharge, fever, and swollen glands near the groin; a man may have a discharge from the penis or painful urination. The infection can be simply and effectively treated with antibiotics, but if undetected or untreated, it can lead to severe infection of the reproductive organs and even sterility. Because a person with chlamydia can be symptom-free, pregnant women can unknowingly pass the infection on to their newborns. This STD, in fact, is the most common infection in newborns, with 100,000 cases per year.

The second most common STD is *herpes genitalis*, caused by the herpes simplex virus. At least 20 million Americans have herpes, and 300,000 to 500,000 new cases arise annually. The virus causes watery blisters to form around the genitals; these break and form painful open sores that eventually heal. The virus can lie dormant for weeks, months, or years. Then, due to sunlight, emotional stress, or other triggers, the virus can break out once again and cause a new cycle of pustules and sores. Right now, there is no cure for herpes, and only partially effective antiviral drugs are available. If birth coincides with an active herpes phase in the mother, the newborn can suffer damage to the brain, liver, or other organs.

The fastest-spreading, and in many ways most worrisome, STD is *venereal warts*, caused by the papilloma virus. These small, painless, cauliflowerlike bumps grow around the sex organs, rectum, or mouth and are passed through skin-to-skin contact. Getting rid of the warts is usually no problem; they can be burned or frozen off or surgically removed. More ominously, however, researchers are finding that the papilloma virus (which can remain in the body even after the warts are removed) has been present in 90 percent of the cervical cancer tissues studied so far.

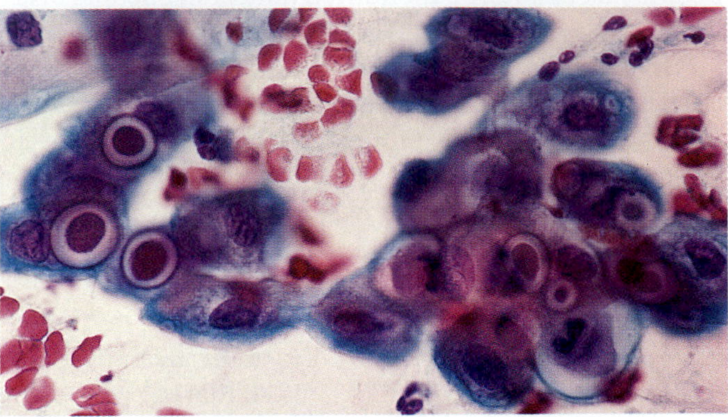

FIGURE 1 ***Chlamydia trachomatis*** **Causes the Most Common Sexually Transmitted Disease.**

Perhaps the best-known STDs are *gonorrhea* and *syphilis*. Both are caused by microorganisms; both are contracted through sexual activity with an infected person; both can have mild initial symptoms (a discharge or painless sore) or no symptoms at all; and both can usually be treated successfully with antibiotics. As in chlamydia, failure to find and treat gonorrhea at an early stage can lead to severe infection of reproductive organs or sterility. Failure to treat syphilis can also result in widespread damage to the heart, eyes, and brain and can result in severe damage or death to an unborn child.

Other STDs include pubic lice (crabs); scabies (parasites that burrow under the skin); certain types of vaginal yeast infections; trichomonas (flagellated protists that infect the vagina or penis); and acquired immune deficiency syndrome (AIDS; see details in CHAPTER 22).

Clearly, sexually transmitted diseases have become a serious threat. Planned Parenthood recommends that sexually active adults (particularly those with numerous sex partners) use condoms to prevent the passing of infections; be alert to sores, bumps, discharges, painful urination, or pelvic pain; and get tested regularly for STDs. This is especially important before or during pregnancy.

TABLE 14.3 Birth Control Methods

Method	How It Works	Percent Accidental Pregnancies in First Year	Side Effects or Other Problems
Behavioral Techniques			
No method	Chance	Up to 80 or more	Unwanted pregnancies and risks associated with childbirth
Rhythm method	No intercourse during woman's fertile period (days 12–20)	20	Often ineffective because eggs can be released before or after this period
Withdrawal (coitus interruptus)	Penis is withdrawn before ejaculation	20	Depends on willpower and timing; small amounts of semen are often released before ejaculation
Chemical Techniques			
Spermicides (foams, creams, vaginal rinses)	Chemicals introduced to vagina before or after intercourse kill sperm	21	Applications often too little, too late, or not done at all, and sperm can survive
Physical Barriers			
Male condom	Rubber sheath on erect penis catches ejaculated semen	12	Effectiveness depends on how carefully used; condom can slip and allow semen to leak; some people are allergic to condom material or coatings
Condom plus spermicidal foam	Same as above plus sperm-killing action of contraceptive foam	0–2	Effectiveness depends on how carefully method is used; helps protect against AIDS and other STDs
Female condom	Plastic pouch inserted into vagina catches semen	24	
Diaphragm plus spermicidal jelly or cream	Round rubber dome covers cervix and blocks sperm entry; chemicals kill sperm	2–10	Effectiveness depends on how carefully both are used; some people are allergic to diaphragm or chemicals
Contraceptive sponge	Disposable sponge containing spermicide blocks cervix, absorbs sperm, and kills them	18	Effectiveness depends on how carefully used; some people are allergic to the sponge or chemicals
Sterilization			
Tubal ligation	Doctor ties off a woman's fallopian tubes, permanently blocking sperm passage	0.4	Slight chance operation will not completely block tubes; requires surgery
Vasectomy	Doctor severs and ties off a man's vasa deferentia, permanently blocking sperm release	0.15	Very slight chance procedure will not completely block vasa deferentia; simple procedure; can be performed in doctor's office; possible immune system reaction
Hormonal Strategies			
Birth control pills, implants	Synthetic estrogens and progesterones prevent normal menstrual cycle from occurring, so eggs are not released	3	Effectiveness depends on how conscientiously pills are taken; some women experience nausea, missed periods, other side effects; pills increase the risk of blood clots and strokes
Implantation Blockers			
Intrauterine devices (IUDs)	Prevent fertilized egg from implanting in uterine wall	6	Most no longer sold in U.S. because manufacturers fear lawsuits due to infections, scarring, and sterility risk
RU 486	Blocks hormone that maintains thick utrine wall	Less than 1%	Abdominal cramps; late period

box 14.2
Human Impact

Overcoming Infertility

Limiting human fertility is, for many people, a personally and environmentally important goal. For many others, however, life without children is unthinkable and infertility can seem a personal tragedy. One in six couples cannot conceive without medical treatment, and that number is growing, partly because of the increase in venereal diseases [see BOX 14.1 on page 354] and the use of IUDs and partly because of the common practice of deferring parenthood to the 30s, when fertility naturally declines. About half the time the woman is the infertile partner, and in most of these cases, her fallopian tubes or uterus are blocked or scarred. When the male is the infertile partner, he usually has a low sperm count, or his sperm lack normal motility. Today, physicians can successfully treat about 70 percent of infertile men and women through therapy with synthetic hormones, or corrective surgery to unblock passages.

A relatively recent solution to infertility is **in vitro fertilization**, fertilization in a laboratory dish. To accomplish in vitro fertilization, a medical team monitors the ripening of an egg cell while it is still inside one of the ovaries of the mother-to-be. The medical workers then insert a pencil-thin viewing tube (a laparoscope) near the cell, and carefully suck the ripe egg into a thin, hollow needle. They place the egg in a laboratory culture dish and mix it with sperm from the future father [FIGURE 1]. The fertilized egg undergoes cleavage in the laboratory dish, and when it reaches the eight-cell stage, the team draws the microscopic cluster into a flexible plastic tube and transfers the living cell group into the mother's uterus. There it burrows into the lining of the uterine wall and can develop into an embryo, fetus, and finally a baby. Each attempt to become pregnant by in vitro fertilization has only about a 15 percent change of succeeding, and costs about $10,000. Despite the expense and low success rate, many couples are willing to pay considerable sums in the hopes of bearing their own biological children, and there are now 250 in vitro fertilization clinics in the United States.

In some cases, when a man's sperm lack the ability to penetrate an egg, a medical technician can drill a hole in the coverings that surround the egg, then inject sperm directly onto the egg. In another variant, called GIFT (gamete intrafallopian transfer), medical workers implant the egg and sperm together in the woman's fallopian tube rather than mixing them in the test tube. In ZIFT (zygote intrafallopian transfer), doctors insert a fertilized egg into the fallopian tube rather than depositing an eight-celled embryo into the uterus. Some women are unable to carry a baby to term, and in such cases, a physician can carry out an in vitro fertilization with her egg and her partner's sperm, then insert the early-stage embryo into another woman's uterus. This *surrogate mother* then carries and gives birth to a baby. In one recent case, a woman served as a surrogate mother for her daughter's and son-in-law's embryo, and thus gave birth to her own grandchild!

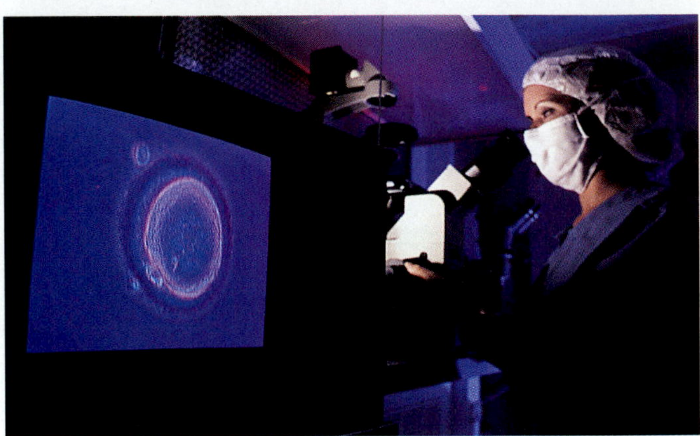

FIGURE 1

Test-Tube Fertilization.

A technician examines the projected image of a ripe egg removed from a woman's ovary in preparation for in vitro fertilization—literally, fertilization "in glass," meaning in a glass laboratory dish—instead of in vivo, meaning in a living organism.

Researchers have devised ways to freeze embryos. If the first attempt to artificially implant an embryo fails, physicians can retrieve another from cold storage and try again without having to repeat all the prior steps of in vitro fertilization. In 1990, a couple produced seven embryos by in vitro fertilization, and medical workers froze them for future implantation. Meanwhile, the couple divorced, and neither parent wanted the other to implant or destroy the embryos. The court finally gave both parents joint custody of the frozen embryos.

With each of these methods of in vitro fertilization, the goal is to produce a baby. The next section of the chapter follows the steps of human development from fertilization to the birth of a child.

some people find unacceptable; blood clots are a particular risk to women who are over 35 or who smoke. In a recent innovation called Norplant, an implant is placed under the skin of a woman's upper arm. It slowly releases the same hormones used in oral contraceptive pills and can prevent conception (but not menstruation) for up to five years.

Even if these means of inhibiting fertilization fail, a woman still does not become pregnant for several days *after* fertilization. During this time, the fertilized egg moves down the fallopian tube and begins to divide into a group of cells. Pregnancy begins with implantation, when the fertilized egg becomes embedded in the thickened wall of the uterus. Medical researchers have discovered several ways to block implantation. An *intrauterine device* (IUD), a small plastic or metal insert worn in the uterus, can interfere mechanically with implantation of the embryo. While very effective, these devices are associated with increased risk of pelvic infections, potential scarring, and in some cases, future infertility. The morning-after pill, with its high doses of estrogen and progesterone, also interferes with implantation, as does mifepristone. As the experiments of Anna Glasier showed, the latter drug is apparently safer and more effective than either of these other alternatives.

Abortion is the expulsion of a fetus, either spontaneously or deliberately induced. The Supreme Court ruled more than a decade ago that the government cannot forbid deliberate abortions early in pregnancy. Many people, however, have strong moral objections to abortion, and most people would agree that safe and effective contraception is medically and emotionally superior to abortion. Still, there are several medically supervised abortion procedures currently available in the United States and these procedures, if properly carried out, are statistically safer for the mother than carrying a baby to term and delivering it.

While mifepristone can be used to prevent pregnancy if given within 72 hours of sexual intercourse, the drug became known as the "French abortion pill" because this progesterone-blocking agent can induce abortions after implantation has occurred. Progesterone promotes the growth of the uterine lining, the relaxation of the uterine muscle, the firming of the cervix, and the formation of a mucus plug, all crucial to the continuation of pregnancy. If an agent such as mifepristone blocked the action of progesterone in the first 9 weeks of pregnancy, the uterine lining would begin to slough off, the uterine muscles would begin to contract, the cervix would soften and dilate, and eventually the embryo would be expelled from the womb. In France, Great Britain, Sweden, and China, hundreds of thousands of women have terminated pregnancies by taking mifepristone together with prostaglandin, which stimulates uterine contractions. Mifepristone, when used to induce abortions, can cause strong abdominal cramps and heavy bleeding. Nevertheless, observers believe that, on the whole, this method of abortion offers fewer medical drawbacks than surgical abortions, and has the additional advantage that physicians not associated with designated abortion clinics can safely administer and monitor the therapy. Mifepristone is not currently available in the United States, but talks are under way between U.S. health officials and the French drug company that manufactures most of the mifepristone, and human trials are planned in at least one state.

While preventing fertilization is a major goal for many couples around the world, others face an opposite problem: they cannot conceive a baby. BOX 14.2 discusses this problem.

➤ CONCEPT CHALLENGE

Which do you think is a bigger problem in the world, infertility or overpopulation? Support your answer. Why do you think there are more U.S. businesses involved in infertility research than in contraception research?

Connections

We followed the human life cycle from sperm and egg production in the human reproductive systems, through mating, fetal development, childhood, and adulthood. We saw that the developing embryo undergoes remarkable changes during the first few weeks of life.

Reproduction solves a central problem for the human species. Our bodies age, wear out, and die. Maturation of sex organs and blossoming of secondary sex characteristics in adolescence make it possible for humans to start new life cycles through their children. Reproduction is our only means of continued existence as well as our anchor to parents, ancestors, and progenitors. How far back does the unbroken chain extend? And where did it start? We may never know the answers, but the laws of nature and genetics, operating today as they did on earth billions of years ago, allow us to make testable hypotheses about how natural laws could have resulted in life and its diversity. The four chapters in our next section analyze the processes of evolution, and how the organisms today may have arisen, diverged, and flourished.

KEY TERMS

aging, 350
amniotic cavity, 342
Bartholin's gland, 339
bulbourethral gland, 336
cervix, 339
chorion, 342
clitoris, 339
coitus, 341
contraception, 355
corpus luteum, 339
ejaculation, 336
ejaculatory duct, 336
endometrium, 339
epididymis, 336
estrogen, 339
fallopian tube, 339
follicle, 339
human chorionic gonadotropin (hCG), 342
implantation, 342
interstitial cell, 335
in vitro fertilization, 356
life expectancy, 352
menopause, 350
menstrual cycle, 339
oocyte, 337
orgasm, 341
ovary, 337
oviduct, 339
ovulation, 339
oxytocin, 347
penis, 336
pregnancy, 342
primary sexual characteristics, 334
progesterone, 339
prostaglandin, 347
prostate gland, 336
scrotum, 334
secondary sexual characteristics, 334
semen, 336
seminal vesicle, 336
seminiferous tubule, 335
senescence, 350
spermatogenic cell, 335
sustentacular cell (Sertoli cell), 335
testes, 334
testis-determining factor, 345
testosterone, 335
trimester, 344
umbilical cord, 342
urethra, 336
uterus, 339
vagina, 339
vas deferens, 336

HIGHLIGHTS IN REVIEW

1 In both males and females, the same brain hormones regulate (a) the gonads (testes or ovaries), which produce gametes (sperm or eggs), and (b) steroid hormones that control development of gender and the functioning of mature sexual organs.

a] Male reproductive structures make sperm and transfer them into the female's reproductive tract, where the sperm encounter and fertilize eggs.

b] A negative feedback loop maintains a continual sperm supply.

c] Female reproductive structures make and transport eggs, prepare and nourish developing embryos, and deliver babies.

d] Hormonal feedback loops bring about the cyclic production and release of eggs and the preparation of the uterus for an embryo (the menstrual cycle). The releasing hormone GnRH and LH and FSH in females cause the ovarian follicle to produce estrogen and progesterone.

e] After ovulation, the ovarian follicle becomes the corpus luteum which produces less estrogen and more progesterone, continuing to promote the buildup of the uterine lining. If an egg is fertilized, this lining is maintained throughout pregnancy. If the egg is not fertilized, the corpus luteum, degenerates and the lining is sloughed off in menstrual flow.

2 The management of human fertility is a pressing issue worldwide. A detailed knowledge of reproductive physiology allows us to promote or prevent human reproduction.

a] Sexual intercourse can bring egg and sperm together in the female's fallopian tubes, followed by fertilization. After a few cell divisions, the new ball of cells burrows into the uterine wall (implantation).

b] One of the world's major problems is overpopulation of the human species. Birth control can be achieved by preventing fertilization or implantation, or by terminating pregnancy.

c] To overcome infertility, men or women may need hormonelike drugs, corrective surgery, in vitro fertilization, insemination with donor gametes, or the help of a surrogate mother.

3 Human embryos pass through the same developmental stages that unfold in other vertebrate animals, modified slightly because human embryos, like most other mammals, obtain nourishment through the placenta.

a] Pregnancy is a partnership between embryo and mother. After the embryo implants in the uterine lining, it develops a surrounding chorion, which produces a hormone (hCG) that allows the corpus luteum to continue producing progesterone; this hormone prevents the uterine lining from being sloughed off.

b] The chorion grows and enmeshes with maternal tissue, thus forming the placenta, a spongy organ that nourishes the embryo, removes wastes, and screens out bacteria and toxins.

c] Sexual differentiation in the fetus begins with an indifferent embryo that is neither male nor female. In the absence of a signal from the *Y* chromosome, the gonads, sex ducts, and external genitals express genes causing them to develop as female structures.

d] The mother's nutrition, life-style habits, and taking of prescription drugs can affect fetal development.

e] The fetus programs itself and its mother for delivery. Among other changes, it stores special fat and carbohydrate reserves, and it secretes hormones that prepare the mother's breasts for milk production and cause her body to release the signals (prostaglandins and oxytocin) that initiate labor and delivery.

4 Growth, maturation, and aging continue to occur throughout life. Biologists do not yet understand the causes of aging.

a] Infancy and childhood are periods of phenomenal physical growth and mental development. Puberty is marked by an additional growth spurt and by the maturation of sexual organs and the development of secondary sexual characteristics. During adulthood, the body reaches its peak physical performance, then gradually declines until senescence sets in, when the body ages more rapidly.

b] Aging may be caused by genetic programming, by wear and tear, or by other factors. Observing good health habits throughout life increases the likelihood that the last years of life will be spent in good health.

UNDERSTANDING THE FACTS AND CONCEPTS For Questions 1–10, match each of the descriptions with the most appropriate item or items from the following list of terms. In this group, as well as those that follow, any item may be used once, more than once, or not at all.

a] testis
b] hypothalamus
c] pituitary
d] follicle
e] corpus luteum
f] chorion

1 Site of interstitial cells.

2 GnRH is made here.

3 Secretes estrogen and progesterone.

4 Secretes LH and FSH.

5 This temporary structure develops after ovulation; if pregnancy ensues, its function will be taken over by the placenta.

6 Sustentacular cells produce inhibin within this structure.

7 Place where gonadotropins are made in males and females.

8 Structure that surrounds a maturing oocyte.

9 Structure(s) that secrete hormones that block gonadotropins.

10 Secretes testosterone.

As above, for Questions 11–15, match the descriptions with the appropriate item from the following list.

a] epididymis
b] vas deferens
c] urethra
d] oviduct
e] endometrium

11 This structure cyclically changes in thickness.

12 A structure that carries urine and sperm.

13 Located adjacent to the testis; site of sperm maturation and storage.

14 A long sperm duct that enters the abdomen from the scrotal sac.

15 Ciliated passageway that is the place where fertilization occurs.

As above, for Questions 16–20, match the descriptions with the appropriate item from the following list.

a] inner cell mass
b] trophoblast
c] chorion
d] amniotic cavity
e] neural tube

16 The development of body organs begins with the formation of this structure.

17 Appears at the blastocyst stage, and will give rise to the body of the embryo.

18 Contributes directly to the placenta and also secretes the hormone that home pregnancy kits measure.

19 Develops at the blastocyst stage and will contribute to the chorion.

20 A fluid-filled enclosure that protects the embryo.

INTEGRATE AND APPLY WHAT YOU HAVE LEARNED

1 Contrast the action of mifepristone (RU 486) as a morning-after contraceptive and as an agent that induces abortion.

2 Outline the feedback loop that maintains the sperm supply.

3 Explain how pharmaceutical tests that could indicate peaks in LH, estrogen, and progesterone might be used in family planning and birth control.

4 Does the embryo develop within the cavity of the uterus? Explain.

5 Contrast the feedback loop between oxytocin and the uterus with the feedback loop that occurs between the chorion and the pituitary.

ANALYSIS

1 One difference between the production of a mature sperm and that of a mature egg is:

a] Sperm production requires the action of the hypothalamus, but egg production does not.
b] Egg cells are supported and maintained by accessory cells, but sperm cells are not.
c] Sperm always travel within closed tubes, but eggs are released into the body cavity.
d] Only egg cells require the action of follicle-stimulating hormone.
e] Egg production is regulated by negative feedback, but sperm production is regulated by positive feedback.

2 A testosteronelike drug (an anabolic steroid) that athletes are banned from taking was found in the urine of an Olympic sprinter. He claimed that someone must have added the drug to his urine sample. The laboratory technicians did not believe this because the amount of natural testosterone in his urine was lower than normal. Why was a low level of urinary testosterone a clue that the athlete was lying and probably had taken the drug?

a] Because it resulted from drug-induced elevation of GnRH.
b] Because it resulted from drug-induced elevation of LH.
c] Because it resulted from low levels of inhibin.
d] Because it resulted from depressed LH.
e] Because it resulted from enhancement of interstitial-cell activity.

3 Recall that the TDF gene was discovered by studying a few infertile men who had two *X* chromosomes and a small piece of *Y* chromosome attached to another chromosome. This discovery was important because it showed that:

a] The men were infertile because they lacked two complete *Y* chromosomes.
b] The gene that signals the indifferent gonad to become a testis must have been on the piece of *Y* chromosome that they possessed.
c] The gene that signals the indifferent gonad to become a testis must have been on the *X* chromosome and it was not inhibited because they did not have an intact *Y* chromosome.
d] The patients were infertile because as embryos their testes were not producing any testosterone.
e] The men were infertile because they had no 5DHT.

Should We Patent Life Forms, Genes, and DNA?

The gene splicing revolution began in the mid 1970s, and within a decade, there had been a legal revolution, as well: The U.S. Patent and Trademark Office had issued its first patent for a living organism. Researchers had genetically engineered an oil-eating bacterium and wanted to market this "product" for use during oil spills and similar applications. At the same time, they wanted to cover their costs for "inventing" this environmentally useful organism, and so desired assurance that no one could collect a sample of the modified cells, grow large quantities of them, and sell them without permission. The patent conferred just that protection, and since that first issuance, the U.S. government has granted patents for cotton, potato, and tobacco plants engineered to resist pests or pesticides. For example, farmers who planted cotton engineered for resistance to an herbicide can spray their fields with the herbicide, and only the weeds will die, leaving the cotton crop healthy and weed-free [FIGURE 1]. In 1988, the patent office even granted a patent for an animal, the "Harvard oncomouse," a laboratory mouse containing a human cancer gene. It was just a matter of time until someone would try to patent human genes. And that, in fact, is just what happened, creating a storm of controversy.

Do you think the government should grant individual researchers, universities, or private companies the exclusive rights to sell certain living organisms? Should that right extend to our own DNA and proteins? Let's apply some of the principles we've discussed in this section on genetics and development to the questions, so you can decide for yourself how you stand.

It's important, first of all, to understand a few things about patents. These legal grants are designed to protect someone who creates a novel and useful invention against imitation by someone who would steal or copy the idea and profit from it unfairly. Patents are intended to encourage innovation, and the heavy investment of time and money it requires, by providing a limited monopoly for the seller. To merit this right, a patentable object or organism must display three criteria: *novelty* (it must be truly a unique creation); *utility* (it must have a specific use); and *nonobviousness* (it must be an invention, not a discovery of something preexisting). In addition, in Europe, it must not be "contrary to public order or morality." Let's look at a series of cases and how they reflect these criteria and the problems they raise.

Many genes of interest—genes for resistance to particular insect pests, viruses, bacteria, drought, fungal diseases, or pesticides, for example—may already exist in wild plants or animals in nature. What if a researcher applies what we learned about the structure of DNA in CHAPTER 9 and the recombinant DNA methods discussed in FIGURE 12.5 to transfer a resistance gene from a wild relative of corn to an agricultural variety of corn? Would this organism be novel enough to patent? The U.S. and British patent offices agree on this one: No. The wild corn and the domestic corn could conceivably interbreed and the resistance gene could be transferred naturally. But what about a corporation's time and money in making the wild gene immediately useful? Shouldn't the corporation be allowed a patent in order to recoup its investment in producing the improved corn?

Another issue revolves around the exploitation of wild genes from plants or animals in economically developing countries by researchers from developed nations. Should a patent go to the researchers who splice the useful wild gene into a novel organism? Or should the patent go to the country where the native plant or animal grows, even though scientists there were not the first to use the gene?

The "novelty" and "nonobviousness" criteria have ramifications for the free flow of ideas in science. What if a molecular geneticist studies a particular gene and its protein product, say for cystic fibrosis [FIGURE 10.1]. Suppose he or she publishes this work, and then later goes on to tinker with the nucleotide sequences, modifying the protein in such a way that it is now useful as a new drug that helps cystic fibrosis patients breathe easier. Since the information has already been published in the scientific literature, it may not be considered novel and nonobvious later by a patent official. Will researchers hold back on sharing results that could help other experiments and thus speed the advance of knowledge in a given field? This holding back is already standard in many corporate laboratories, but a new wrinkle seems to have arisen just recently, and touched off an outcry over patenting human DNA.

Consider the Harvard oncomouse. It contains a human oncogene, a gene that regulates the cell cycle [CHAPTER 7] and causes cancer [see FIGURE 13.23]. This oncogene frequently causes the oncomouse to develop breast tumors spontaneously during the hormonal changes of pregnancy [see FIGURE 14.8], and at an even greater rate when treated with cancer-causing agents. These properties make the oncomouse a model for understanding the causes of human breast cancers. German and Swiss groups charge that the mouse cannot be patented in Europe because this engineered mouse is designed purposefully to suffer, and hence is inherently immoral. The European Patent Office ruled in 1993, however, that the invention's usefulness for solving the

FIGURE 1

A Patented Plant.

Normal cotton (right) wilts and dies when sprayed with a herbicide. Genetically engineered cotton (left) survives the spray, allowing farmers to rid their fields of herbicide-sensitive weeds, but not their crop.

growing problem of women dying of breast cancer does in fact make it moral. Which side do you believe is right, and why?

An unanticipated patent problem arose in the early 1990s, as a result of the Human Genome Project [see BOX 11.1, page 280]. The goal of this massive effort is to determine the location of every gene on each of our 23 pairs of chromosomes, as well as to determine the nucleotide sequences of the entire complement of human DNA. People had expected this wealth of genetic data to lead to new drugs and enzyme therapies, for example, therapies to treat diseases such as melanoma cancer, pituitary dwarfism, and severe combined immunodeficiency [FIGURES 11.16, 12.7, and 12.11]. People also expected researchers eventually to file patent applications, after such drugs and proteins neared commercial application. But observers were utterly shocked when in 1991 a scientist then at the National Institutes of Health, Craig Venter, published the sequences for about 340 fragments of DNA from genes used in the human brain and simultaneously applied to patent them all!

Two camps immediately formed. Venter and some NIH officials felt it was in the interests of the American public, who funded most of the work, to seek patent protection for any future uses of that DNA. Many biotechnology corporations and many researchers in other countries, however, immediately cried foul. They pointed out that Venter's group had sequenced the DNA fragments but had not examined their functions. Thus, they argued, Venter had no idea what the genes might do, or what proteins they might encode, and hence the case failed the patent test of utility. Furthermore, they predicted that if the practice were sanctioned and patents were granted, a few large, highly automated laboratories could "own" the entire human genome within a few years, even though the exact locations and functions of most of the genes were still a mystery. (Venter later left the NIH to head a private biotechnology company to continue applying the methods he developed.) And they feared that whoever won the patents for such fragments would charge licensing fees to any research team, corporation, or foreign government trying to generate new drugs, enzymes, or novel organisms based on the human genome, and hence hinder the advance of science.

As it turned out, the U.S. patent office denied patents for the human DNA fragments in 1992, under the criterion of utility, since no one knew the functions of the genes, and novelty, because the gene came from publicly available DNA libraries. NIH officials appealed this decision in 1993, however, and at least one American company, Incyte Pharmaceuticals, Inc., of Palo Alto, California, filed similar applications for 100,000 fragments in 1993, and expects to file for the entire human genome by 1995.

Prior publication of genetic data such as the nucleotide sequences can invalidate a future patent claim. Do you think government and corporate labs should be allowed to patent DNA fragments of unknown utility? If not, what incentive will such labs have for the enormous expenditures required to develop new drugs or therapeutic proteins? How do you think British, Japanese, and other international scientists might feel about the new American "gold rush" to patent human DNA? And do you see a difference between patenting a genetically engineered mouse and patenting DNA sequences that might eventually provide gene therapy to a person with a disease like cystic fibrosis or sickle-cell anemia or Tay-Sachs [CHAPTER 3]? These questions are still open. You decide... ❏

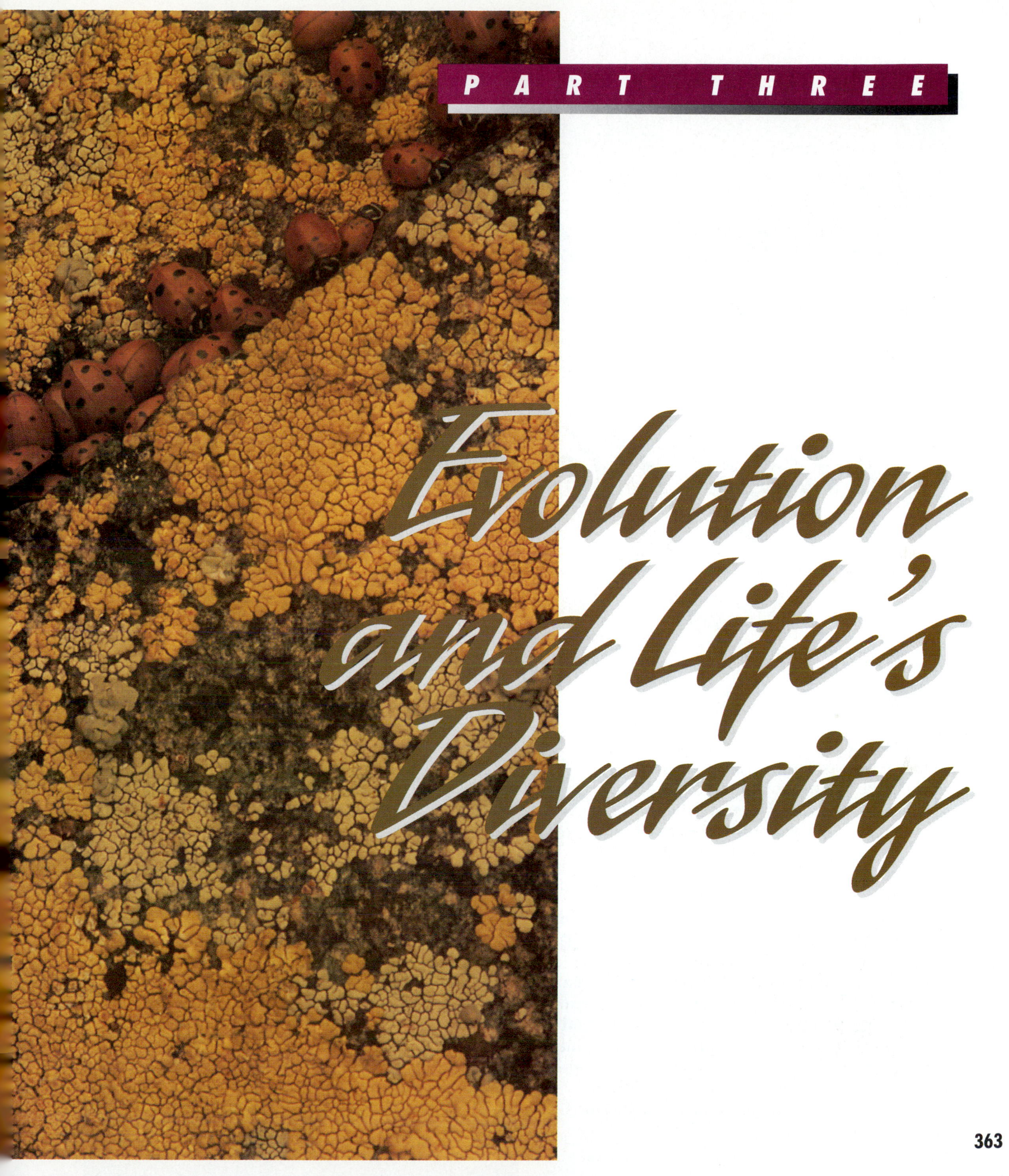

PART THREE

Evolution and Life's Diversity

CHAPTER 15

Nature's Evidence for the Evolution of Living Things

FIGURE 15.1
Hallucigenia.
This extinct marine worm lived about 600 million years ago.

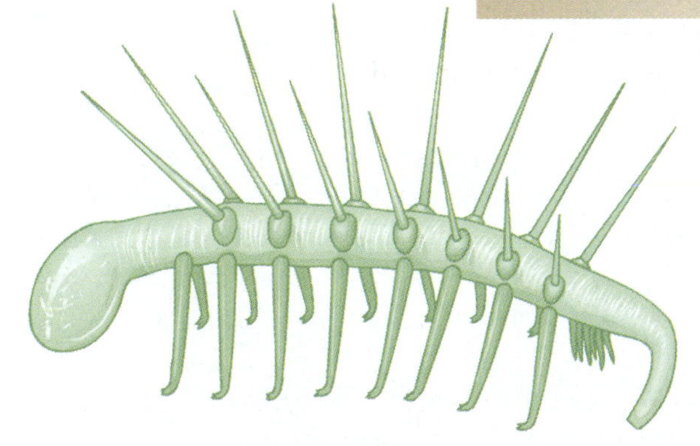

HALLUCIGENIA: A FOSSIL NIGHTMARE

More than half a billion years ago, immense seas covered what is now western Canada. Weird-looking animals swam, crawled, and burrowed beneath these ancient oceans, and in 1977 researchers discovered evidence of a strange wormlike organism from this long-extinct menagerie. They dubbed their find *Hallucigenia sparsa* because of its bizarre appearance. The fossil specimens of *Hallucigenia* revealed a grotesque organism just 1 in (2.5 cm) long. It had a wormlike body, that resembled a head at one end, but lacking eyes, antennae, or other obvious sense organs. Jutting from its trunk were seven pairs of formidable spines [FIGURE 15.1]. Opposite these spines along the body tube's other surface were a series of flexible protrusions with split tips. Behind them, on the same side, sprouted a cluster of six short tentacles.

Some biologists speculated that *Hallucigenia* might have been an appendage broken off from a larger and even more fantastic animal. Others thought it was a free-living worm that walked across the muddy ocean bottom on its spines, like a centipede on stilts. The major question facing biologists was not how the creature moved, or even lived, but how it came to appear suddenly in the fossil record without apparent ancestors, and why it seemed to be totally different from any other known organisms.

Scientists who study **fossils**, the remains or traces of organisms that lived in the past, are called *paleontologists*. These experts could use *Hallucigenia* to test their notions of **evolution**, the process whereby genetic changes in populations of organisms lead to branching pathways of descent over time. The spiny, wormlike *Hallucigenia* did not fit neatly into scientists' ideas in 1977, but the scientific sleuths nonetheless used clues such as the appearance and disappearance of *Hallucigenia* to reconstruct events that occurred millions of years ago. The evidence they and other scientists have amassed so far supports the idea that the full diversity of living forms on earth evolved from common ancestors who lived in the distant past.

In this chapter we will trace some of the evidence for evolution and for the strong scientific conviction that evolution occurs and can be documented. We'll begin with a brief look at how the idea of evolution first arose, and then consider in detail the mechanism for evolution that Darwin and Wallace hypothesized: natural selection.

MESSAGES

1 Fossils help document life's history on earth. They show that today's living organisms are different from, and often more complex than, those that lived in the past. Fossils provide firm physical evidence that life has evolved.

2 Natural selection is the differential survival and reproduction of a small fraction of a population due to their physical, biochemical, or behavioral traits.

3 Evidence for evolution comes from genes comparative anatomy, and from ecology.

4 Macroevolution, the large-scale evolutionary changes in groups of species over geologic time, can be explained by microevolution, genetic changes in populations of individual species.

5 Extinctions of some species followed by adaptive radiations by others have marked the course of evolution. The tempo of evolutionary change is not constant; bursts of evolutionary change are often followed by long periods with little change.

After natural selection we'll examine evidence for evolution. Some of the clues that helped biologists reach their conclusions are literally hard evidence—information inscribed in stone, such as the fossils of *Hallucigenia*. Other evidence is more circumstantial, such as the presence of unused structures still found in living organisms because of unchanged DNA sequences passed on to living species. Still other clues lie in anatomical comparisons of living groups to each other and to fossils, and in comparisons of genes and proteins among closely and distantly related organisms. Researchers are devising exciting new tests of evolutionary theory based on a direct analysis of DNA from fossilized plants and insects, and perhaps even dinosaurs. As we will see, some evidence for evolution comes from the study of where and how organisms live in their physical settings. As the chapter ends, we will also look at pathways of divergence during evolution, including the rates of evolutionary change, and the effect extinction and other evolutionary events can have on life's history.

Subsequent chapters in this section discuss the genetic basis for evolution, the origin and history of life, and finally, how biologists classify the broad diversity of life forms that evolution has produced.

Let's return, for now, to the long-extinct *Hallucigenia*, and to other evolutionary tidbits, as well, including the gill slits of human embryos, the snake's obsolete pelvis, flying mammals called sugar gliders, and the common, ubiquitous relatives of dinosaurs still alive today. ❑

Emergence of Evolutionary Thought

A nature lover can fully enjoy the diversity of life while walking through a meadow filled with wildflowers, birds, butterflies, and other plants and animals. Biologists, too, can appreciate this diversity for its own sake, but in addition, they want to learn how it came about. What are the mechanisms that caused so many kinds of organisms to appear on our planet? The dominant view of Europeans in the early nineteenth century was that species were individually created at one time and place in the past, and have continued unchanged to the present [FIGURE 15.2A]. Robins and daisies, for example, were thought to have been created at the beginning of the world and to have remained exactly the same ever since.

By the early nineteenth century, however, scientific exploration of the natural world had begun to provide facts that contradicted the notion of a single event of special creation. Since the sixteenth century, Europeans had been discovering strange plants and animals in distant lands, such as the platypus in Australia [see FIGURE 18.1]. If all types of organisms were created at a single place at one time, then *why were organisms different in different regions of the earth*? Geologists had discovered fossils that were quite different from today's organisms. If all organisms were created at one time, then *why would ancient organisms be different from those that live now*? Furthermore, anatomists had found that the limbs of various mammals, such as whales, bats, and people, contain bones that are clearly modifications of a single basic plan, even though they perform very different functions, such as a paddle, a wing, or a hand. If each species was specially created, then *how could the same fundamental bone plan be the best design for swimming, flying, and fine manipulation*? In the face of such questions, a few naturalists began to suggest that different types of organisms might have changed, or evolved, over time.

LAMARCK'S DISPROVEN THEORY OF EVOLUTION

French naturalist Jean Baptiste Lamarck (1744–1829) agreed with the prevailing notion that organisms were individually created, but suggested that they may change over the course of time [FIGURE 15.2B]. Furthermore, Lamarck proposed a hypothetical mechanism that could, in principle, cause this change to come about.

Consider, for instance, the graceful, towering giraffe and its extremely long neck and legs. Early naturalists noted that a its long neck allows it to eat leaves from high branches inaccessible to wildebeests, zebras, elephants, and others of the African savanna. In 1809 (the year Charles Darwin was born), Lamarck suggested that an animal's physical needs determine how its body will develop. For example, he proposed that early giraffes must have stretched their necks trying to graze on the leaves of high branches, and that the longer neck that they acquired through such stretching was passed on to their offspring. Experiments eventually showed that Lamarck's hypothesis, the **inheritance of acquired characteristics**, was wrong and not the mechanism underlying evolution.

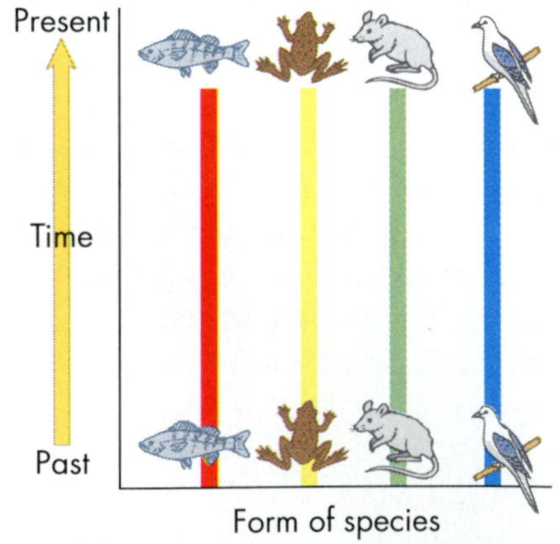

[A] Special creation, fixed species

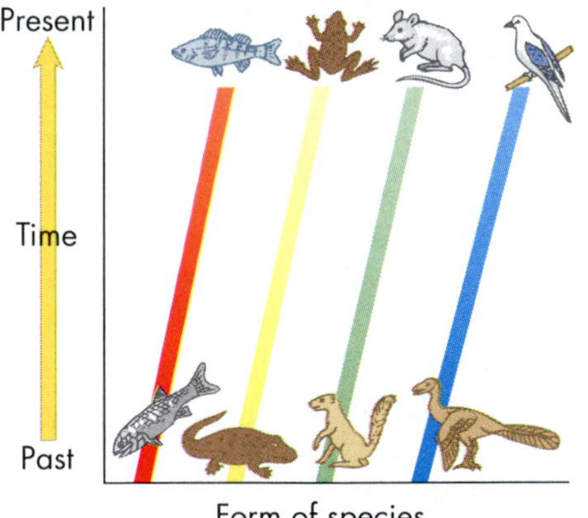

[B] Special creation, evolution

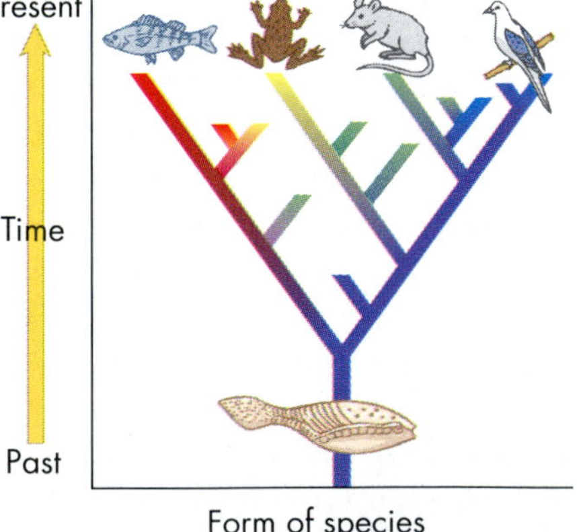

[C] Descent with modification

FIGURE 15.2
Evolutionary Hypotheses.
A diagram representing three different hypotheses for the origin of present-day species.

Consider an "experiment" that clearly disproves Lamarck's idea: Many human families have for thousands of years and hundreds of generations removed the foreskins from the penises of infant males in the process of circumcision. Despite this deliberate alteration for generation after generation, every baby boy continues to be born with a foreskin—clear evidence that the acquired characteristic is not inherited.

Having studied the mechanisms of genetics and development in CHAPTERS 7 to 14, you can appreciate that the theory of the inheritance of acquired characteristics has no basis in biology: The stretching of a giraffe's neck or the circumcision of a baby boy does not change the genes that control neck length or foreskin development within egg or sperm cells.

It remained for two English naturalists working in the mid-1800s—Charles Darwin and Alfred Russel Wallace—to devise an alternative explanation for how living things evolve.

DARWIN'S VOYAGE OF DISCOVERY

In the 1830s, young Charles Darwin sailed around the world, investigating nature's diversity. (An account of his voyage and a description of Wallace's very similar scientific work appear in BOX 1.1.) By observing the diversity of organisms on a series of isolated islands, Darwin became convinced that this diversity was the result of evolution. His concept of *how* evolution took place, however, differed from that of Lamarck and others before him in two highly significant ways. First, he proposed **descent with modification**—the idea that all organisms are descended from common ancestors with modifications. Second, he proposed a biologically feasible mechanism for the modification: natural selection working on individual heritable variation.

Descent with Modification Darwin suggested that the evolutionary pathways of organisms resembled a branching tree with living organisms at the tips of the outer branches and extinct ones at the tips of short branches [FIGURE 15.2C]. This contrasts sharply with the form predicted by special creation or Lamarckianism [FIGURE 15.2A and B]. It also contradicts the still popular misconception that evolution is a ladder from lower forms, like worms and fish, to higher forms, like birds, dogs, and humans. No species that exists today was an ancestor to any other living species. Darwin's true genius, however, was to propose a mechanism for how this descent with modification could occur.

Evolution by Natural Selection Darwin's experiences in South America and in the Galápagos Islands convinced him that evolution had occurred. But on his return to England, his mind still searched for some mechanism that could cause evolution. He eventually drew together two indisputable facts based on his own observations, and from these, he synthesized a far-reaching conclusion:

Fact 1 Individuals in a population vary in many ways, and some of these variations are inheritable.

Fact 2 Populations have the inherent ability to produce many more offspring than the environment's food, space, and other assets can possibly support. As a consequence, individuals of the same population compete with each other for limited resources.

Darwin's Conclusion Individuals whose hereditary traits allow them to cope efficiently with the local environment leave more offspring than individuals without those traits. As a result, certain heritable variations become more common in succeeding generations.

THE MECHANISM OF NATURAL SELECTION

Darwin used the term **natural selection** to describe the greater reproductive success displayed by those individuals with certain favorable characteristics, as compared to members of the same species lacking those traits. He chose this term because nature "selects" the favorable traits that are passed on to the next generation. The favorable traits, or **adaptations**, include body parts, behaviors, or physiological processes that are modified or maintained as a result of natural selection because they improve an organism's chance of surviving and reproducing successfully.

The principle of natural selection, now widely accepted as a main mechanism behind evolution in nature, can explain adaptations such as the long necks of giraffes. If we begin with a population of giraffes browsing on trees in the savanna many thousands of years ago [FIGURE 15.3], we can envision that some of the giraffes would have longer necks than others, just as some people have longer necks than others. Experiments have shown that genes help determine neck length in various animals.

Now, if on the savanna there were too few low-hanging leaves to feed all the giraffes in the population, then the longer-necked giraffes could reach more food, harvest more energy and materials, and survive and produce more offspring. Many of these offspring, like their parents, would possess the genes for long necks. With proportionately more long-neck genes around, the average neck length in the giraffe population would increase over time to its present-day length.

It is important to note, however, that natural selection is not the *cause* of variations within a population. Short necks, long necks, and other variations preexist in the population as a result of random changes in genes, or

FIGURE 15.3
Natural Selection and Neck Length in Giraffes.
Amid tall trees and long-necked neighbors, short-necked giraffes would have died and left few offspring, while long-necked ones would have survived, reproduced more often, and left more long-neck genes in the population.

mutations. Natural selection simply results in the best-adapted, best-competing individuals surviving and producing offspring in the next generation.

We will return to a more detailed, genetic explanation of natural selection in the next chapter. But first, let's examine some of the evidence amassed by Darwin and subsequent generations of biologists for the fact of evolution.

Evidence for Evolution

Evolution has such tremendous power to explain so many different aspects of biology that most biologists accept it as a fact. Evidence to support evolution comes from the fossil record, from anatomy, from biochemistry, and from the geographical distribution of organisms.

THE FOSSIL RECORD

About 4.5 billion years have passed since the earth formed. During that immense span of time, the face of the planet has changed dramatically. Many millions of species have appeared and survived for a period; then, ultimately, most of them (about 99 percent) have become extinct. The evidence for this claim lies in the fossil record, which extends back at least 3.45 billion years. Fossils provide scientists with tangible evidence of what past organisms looked like, and when and where they lived.

Types of Fossils A fossil can be nothing more than the trail, preserved in rock, of an animal that once slithered across some ancient ooze. But the fossils more familiar to us form from hard, decay-resistant structures such as bones and teeth. Others form from leaf prints, footprints, casts, outlines, or body parts preserved in rock after the perishable organic matter is removed and replaced with minerals. FIGURE 15.4 shows the fossilized skull and teeth of *Albertosaurus*, a savage meat-eating reptile that lived about 75 million years ago.

Fossils and Sedimentary Rock Although some fossils may be found in ice, peat bogs, or tar pits, most fossils are found in *sedimentary rock*, a type of layered rock formed by the accumulation of layers of sand and dirt particles over long periods of time. These materials may either be transported by the wind, water, or ice to the site of deposition, or may form chemically at the site. Over time, layer after layer accumulates, with newer layers pressing down on older ones. Pressure and heat from overlying layers cement the particles together, gradually changing them into rock.

The Burgess Shale, the rock formation in Canada where the fossil remains of *Hallucigenia* were discovered, formed about 520 million years ago. The shale developed at the foot of the continental shelf below many meters of seawater. *Hallucigenia* and neighboring species

FIGURE 15.4
Albertosaurus.
Fossilized remains of animals such as *Albertosaurus* reveal that fierce meat-eating dinosaurs roamed the earth 75 million years ago.

FIGURE 15.5

Layers in Time.

Dozens of layers of sedimentary rocks lie exposed by erosion in Badlands National Park, South Dakota. Within the layers lie fossils from a procession of different organisms that arose, adapted, and died out as the environment changed, over the millennia, from barren rock plateau to ocean to swamp to ice fields to dry, exposed hills. Shelled, squidlike ammonites lie entombed in the oldest (deepest) strata, while later layers contain huge titanotheres, small, early ancestors of the horse (mesohippus), and sheeplike oreodonts.

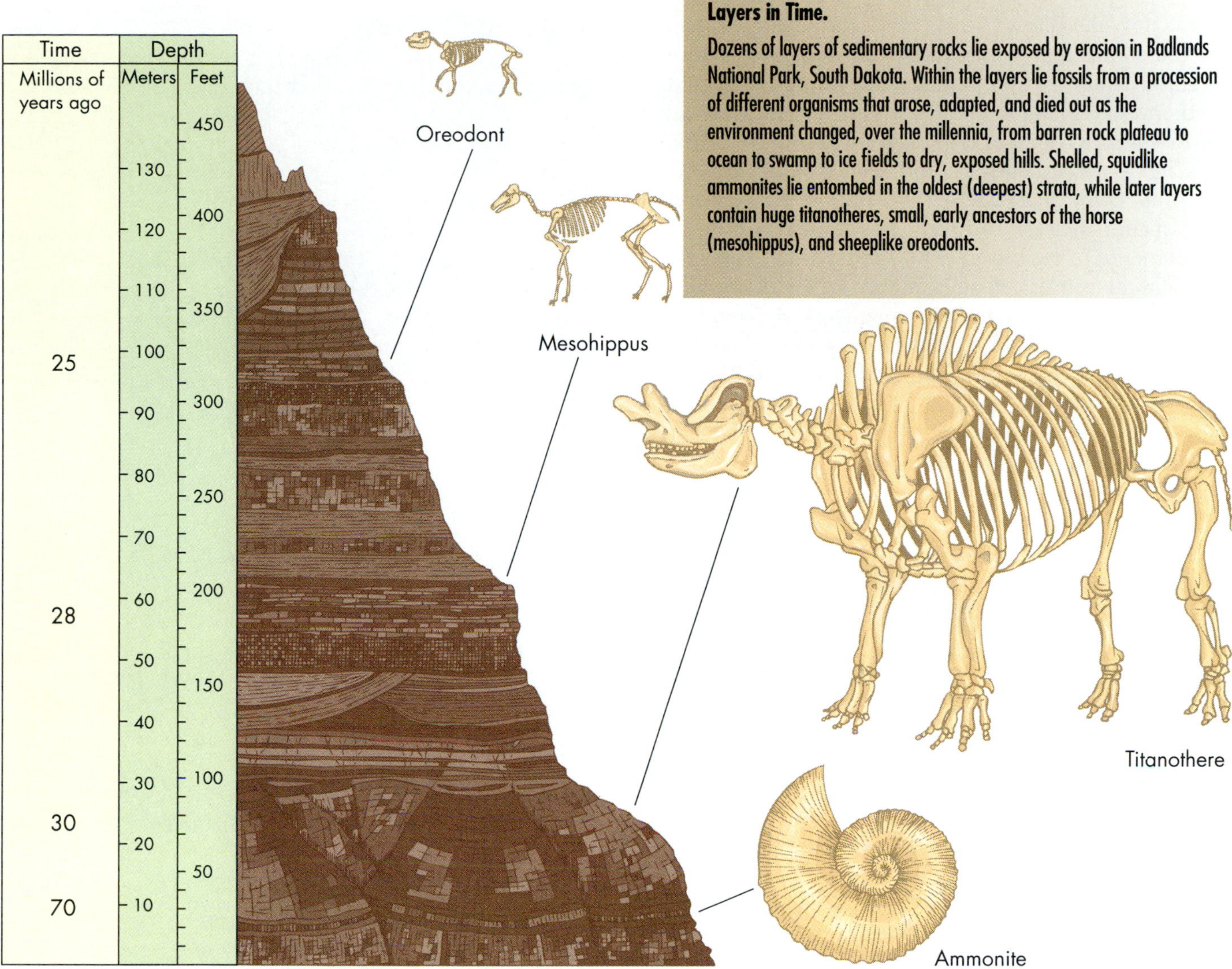

probably lived on the sea floor at the edge of the continental shelf. Occasionally they were hurled off the edge of the shelf and buried by sudden mudslides that created a muddy anaerobic grave, safe from scavenging animals and rapid decay. Over millions of years, the many layers of sediment deposited at the foot of the continental shelf became compressed, one below the other, into sedimentary rocks.

Because fossil formation is a sporadic process requiring just the right conditions, the fossil record shows only a small sample of the organisms that have ever existed. Even where sedimentation does take place, it is rare for fossils to form when organisms lack hard parts such as bones, shells, or teeth. The preservation of the soft-bodied *Hallucigenia* in the anaerobic mud of the Burgess Shale is one of these lucky rarities.

Some Conclusions from the Fossil Record Despite gaps in the fossil record (we do not have all of life's evolution preserved in rocks) paleontologists can still piece together an overall picture of how living things evolved and diverged. This portrait depends on one simple fact: in undisturbed rock layers, newer sediments are deposited on top older layers of sediment [FIGURE 15.5]. This stratification, or layering, often yields important information about when an organism lived and died. To establish dates for the sequence of events, paleontologists determine how much radioactive decay has occurred in certain radioactive isotopes deposited in nonsedimentary rock associated with the fossil rocks.

Paleontologists discovered one of the richest fossil beds in North America in the Badlands near the border of North and South Dakota. This area has provided priceless lessons about the evolutionary process [see FIGURE 15.5]. About 70 million years ago, the area lay beneath a shallow sea and was densely populated by *ammonites*, extinct shelled relatives of squids and octopuses. By about 30 million years ago, the seas had retreated, and huge rhinoceroslike animals, the *titanotheres*, roamed what is now the Black Hills of South Dakota. A few million years after that, a three-toed, knee-high ancestor of the horse named *Mesohippus* grazed where the larger titanotheres had roamed. Finally, fossils reveal that sheeplike *oreodonts* became abundant by about 25 million years ago. Paleontologists found each type of fossil at a particular level, suggesting that each of the different kinds of animals lived during a well-defined period.

A student of evolution would predict that paleontologists might unearth fossils intermediate in form between major groups of today's organisms. A famous instance involves the 150-million-year-old crow-sized *Archaeopteryx*. This animal looked like a small dinosaur in nearly all respects, but it had feathers like a bird [FIGURE 15.6]. Some paleontologists even claim that birds are, in fact, a type of dinosaur that did not become extinct. The many intermediate forms of the horse are another classic example of fossil evidence of evolution [see FIGURE 15.18].

These and other sets of fossils are what would be predicted by evolution: (1) different organisms lived at different times, (2) organisms of the past were different from those alive today, (3) fossils in adjacent layers are more similar to each other than to fossils from distant layers, (4) intermediate forms are sometimes found, and (5) in general, older rocks contain simpler forms, and younger rocks contain more complex ones. Although that pattern is the most common one, there is no evolutionary rule that says that living things proceed from the simple to the complex. Instead, it is understood that living things have proceeded from the well suited to the better suited in relation to their environment.

FIGURE 15.6

Archaeopteryx.

Fossils of this extinct animal reveal a skeleton like a 200-million-year-old reptile, but feathers like a modern-day bird. Birds are the closest living relatives to the dinosaurs.

➤ CONCEPT CHALLENGE

What sort of fossil evidence would it take to contradict the major predictions of evolutionary theory?

EVIDENCE FROM COMPARATIVE ANATOMY

Patterns of evolutionary descent can be found not only from fossil evidence but also from the anatomical features of living and extinct species. Telling evidence comes from homologous structures, from vestigial organs, and from the embryos of related organisms.

Homology: Organs with Similar Origins Natural selection can only work on genetic variation that is already present in a population, gradually modifying preexisting structures into structures with a new form or function. This principle applies to cases of **homologous** (*homo-* means "same") **structures**, similar structures in related organisms. These structures may have different functions, but they share the same embryonic origin and almost certainly arose from the same structure in a common ancestor.

The forelimbs of various mammals exemplify homology. Forelimbs can perform many different functions: Human hands can manipulate different-sized objects, a cheetah's forelimb permits rapid running, a whale's flippers allow efficient swimming, a bat's wings enable it to fly. Yet each type of limb is composed of the same skeletal elements [FIGURE 15.7]. Fossil evidence supports the hypothesis that the forelimbs, including the digits, of various mammals arose from the forelegs of ancestral five-fingered reptiles and became modified by natural selection during the descent of these animals in ways that facilitated different tasks. The alternative hypothesis, that these exact bones—one in the upper part of the limb, two in the lower part, and five digits—are the very best

FIGURE 15.7

Comparative Anatomy Suggests Descent with Modification.

Natural selection has modified the same set of skeletal elements in the forelimbs of a human, a cheetah, a whale, and a bat for different functions: manipulation, running, swimming, and flying.

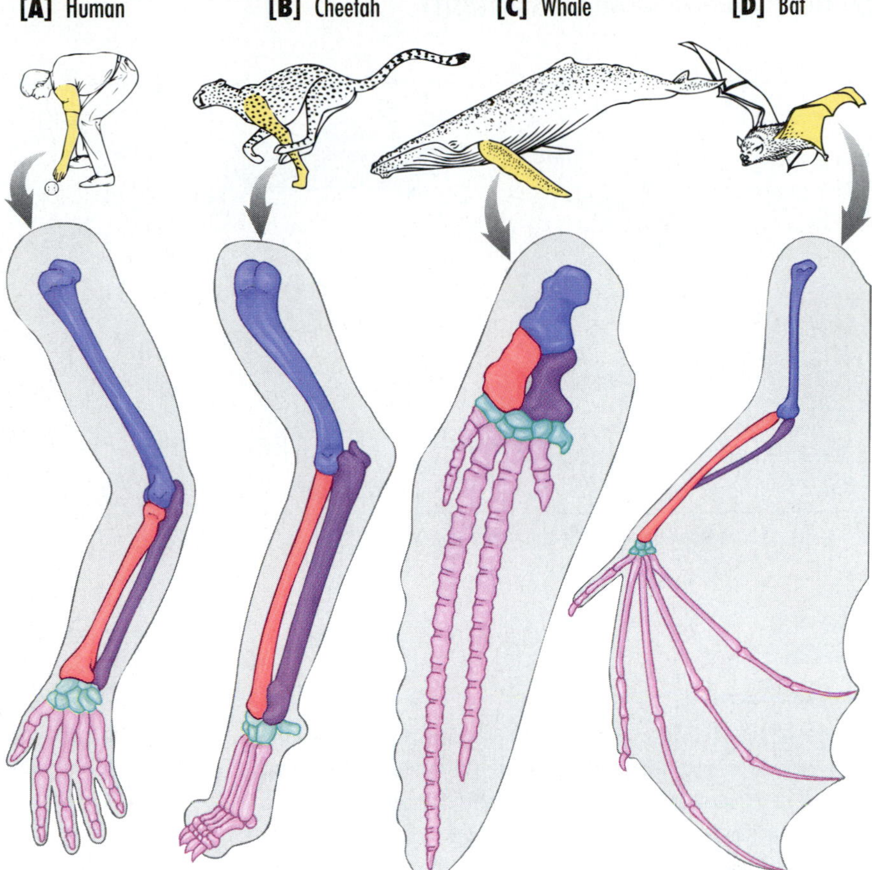

FIGURE 15.8

A New Chinese Fossil Helps Explain *Hallucigenia*.

Chinese paleontologists discovered a fossilized "armored lopopod"—an animal with lobe-shaped legs covered by armorlike scales [**A**]. It shares many fundamental features with the extinct *Hallucigenia*, as well as with species of the living velvet worm, *Peripatus* (phylum Onychophora), which inhabit the leafy forest floor in wet tropical areas [**B**].

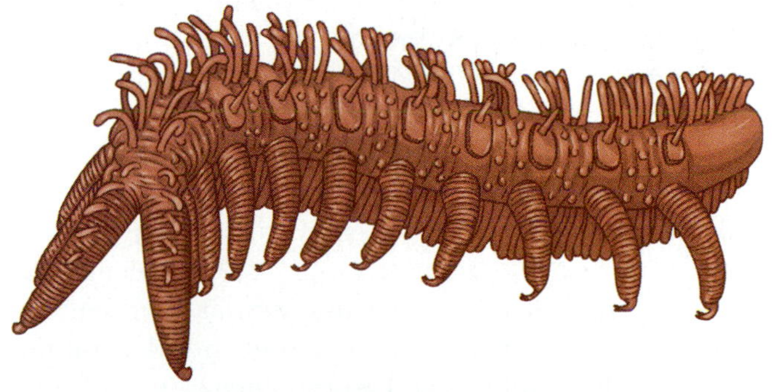

[A] Reconstruction of fossil armored lopopod

[B] *Peripatus*

for manipulation *and* running *and* swimming *and* flying seems unreasonable to most anatomists.

By carefully investigating homologous organs, researchers have begun to understand the problematic *Hallucigenia*. In 1991, paleontologists found a new fossil organism in sedimentary rock beds in China that dated to roughly the same era as *Hallucigenia*. The Chinese fossil showed several features the researchers interpreted as homologous to those in its Canadian counterpart. These features include rows of stiff spikes, soft appendages with claws, and tentacles sprouting between the appendages [FIGURE 15.8].

These very similar structures place both *Hallucigenia* and the Chinese fossil in the animal phylum of velvet worms (Onychophora) [FIGURE 15.8B]. The few surviving members of this phylum live in tropical Central and South American forests, and one observer has described them as a cross between a centipede and the Michelin tire man [see FIGURE 21.26]. Homologous organs like those appearimg in various fossils and living forms of velvet worms serve as strong evidence for evolution.

Vestigial Organs We can see an especially telling type of homology in **vestigial organs** (veh-STIHJ-uhl; Latin *vestigium*, "footprint," "trace"), rudimentary structures with no apparent use in the organism, but with a strong

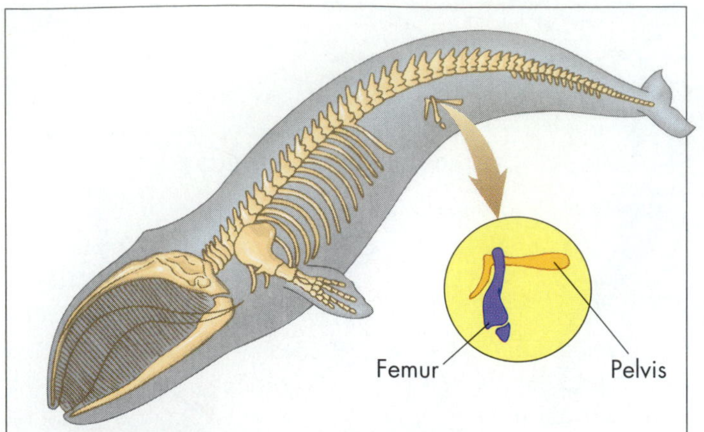

[A] Vestigial pelvis and femur of a whale

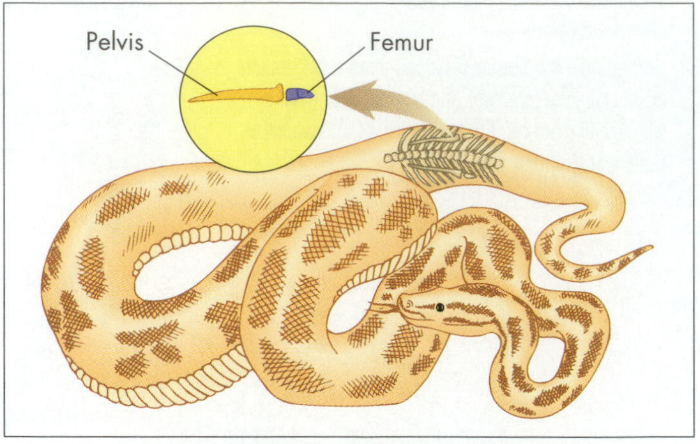

[B] Vestigial pelvis and femur of a python

FIGURE 15.9

Vestigial Organs.

A whale **[A]** and a python snake **[B]** both have a rudimentary pelvis and femur bones, even though these structures provide no apparent function. They were inherited from ancestors and modified, and they remained, even though they don't contribute to the new modes of locomotion.

resemblance to functional structures in probable ancestors. Humans, for example, still have the vertebrae (spinal bones) for a tail, muscles that can wiggle the ears, an appendix, and extra back teeth (wisdom teeth) none of which has survival functions today. In the same way, some snakes and whales still have a pelvis (hipbones) and femur (thigh) bones, although these bones are not functional [FIGURE 15.9].

Various forms of evidence suggest that while the ancestors of humans, whales, and snakes had well-developed examples of these vestigial structures, changing environments selected against the genes needed for the structures to be fully developed in today's organisms.

➤ CONCEPT CHALLENGE

In the fifth week of development, human embryos form a small tail that later disappears [see FIGURE 14.11]. What is the evolutionary significance of this embryonic tail?

EVIDENCE FROM COMPARATIVE EMBRYOLOGY

Evidence for evolution can be seen in embryonic structures as well as in homologous and vestigial organs. The adult forms of related organisms may be quite different, while their embryonic forms may show striking similarities. Adult vertebrates, for example, can be as different as fish, birds, and reptiles, yet their early embryos all resemble each other strongly [review CHAPTER 13].

Most vertebrates, including humans, breathe through lungs, not gills (as fish do), but all vertebrate embryos pass through fishlike stages in which their neck region forms pouches like gill slits, which, like gills, are surrounded by arches containing blood vessels [FIGURE 15.10]. One embryonic gill pouch in humans becomes the eustachian tube leading from the mouth to the ear, while tissues surrounding other gill pouches become the thymus gland and tonsils. In some people, a gill pouch does not change or disappear during development, and these people are born with a tube leading from inside the mouth to the outside of the neck. This abnormality, which is evidence of evolution, can be fixed easily by surgery.

Evolutionary theory can account readily for the fishlike appearances of vertebrate embryos. Organisms that share a common descent often use similar fundamental embryological patterns on which they build different species-specific adult forms. The evidence from embryology is consistent with the idea that structures evolve by genetic modification of features originally inherited from ancestral forms.

EVIDENCE FROM BIOCHEMISTRY

You saw in several earlier chapters that all living organisms share basic features at the molecular level, including the same four bases in DNA, the same 20 amino acids in proteins, similar ribosomal RNA molecules, and, with rare deviations, a single genetic code [review FIGURE 10.14]. There is no known chemical reason why these features are the only ones compatible with life. However, this is exactly what would be predicted if all life evolved from a common ancestor that happened to have these biochemical features.

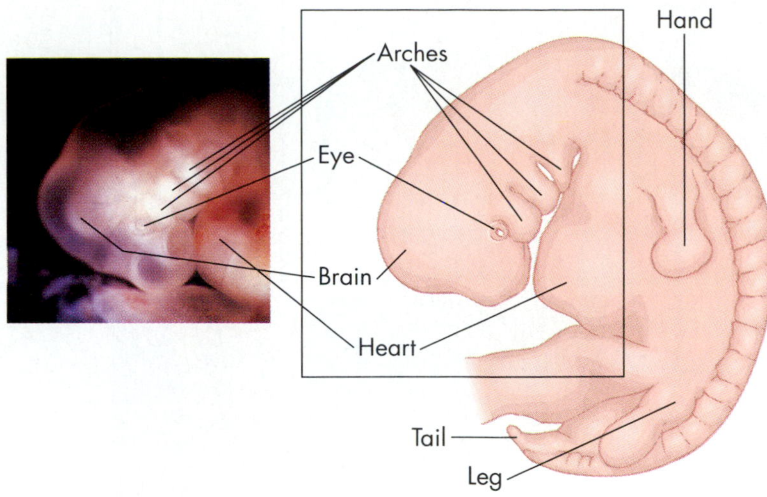

FIGURE 15.10

Human Embryos Have "Gills."

For a short time about a month after conception, each human embryo has openings from the inside of the mouth to the outside, like a fish's gills. This surprising commonality makes sense in evolutionary terms, since fish and humans shared an ancient gilled ancestor.

Molecular biologists uncovered one of the strongest proofs of evolution by comparing the structure of specific genes and proteins found in different living species. By comparing many genes or proteins from dozens of species, molecular geneticists can construct family trees for the proteins [FIGURE 15.11]. On the family tree, very similar molecules are placed at the tips of two branches that split recently in time, and less similar proteins are placed on branches that split in the more-distant past.

In general, a family tree derived from molecular data is very similar to a family tree derived from fossil data. Strong support for evolutionary theory comes from the fact that researchers can derive molecular and fossil data quite independently, and yet both data sets show a common pattern of relationships between living and extinct species, exactly as predicted by evolutionary theory.

Recently, scientists gave an exciting boost to the use of DNA as a probe for investigating evolutionary relationships. Teams of researchers succeeded in extracting DNA from 30-million-year-old termites trapped in amber. The DNA from the fossil termite was very similar to DNA from anatomically similar living termites. The ancient DNA was less similar to DNA from living termites with more divergent physical features. These results are exactly what evolutionary biologists would predict. Other teams of investigators are attempting to isolate DNA from 200-million-year-old fishes, and even from a dinosaur or two.

➤ CONCEPT CHALLENGE

The alpha chain of the blood protein hemoglobin is identical in chimpanzees and humans, while the alpha chain from gorillas differs by one amino acid, and the alpha chain from rhesus monkeys differs by three. How would an evolutionary biologist interpret these data?

EVIDENCE FROM BIOGEOGRAPHY

The study of the geographic distribution of living organisms is called **biogeography**. When Darwin sailed around the world aboard the *Beagle*, the field data he collected about where organisms live were probably his first clue that evolution occurs. Darwin was especially astonished by the unusual plants and animals of the Galápagos Islands. On these barren islands, he saw a rare instance of tool use outside the apes and humans: finchlike birds using cactus spines as if they were miniature picks to pry insect larvae out of dead wood. Darwin also found other interesting animals, including huge tortoises that varied in form from island to island, and iguanas that swam in the sea instead of living exclusively on land. Although these organisms were not identical to any others, they were very similar to South American forms.

The organisms Darwin saw on the Galápagos made him reconsider the plants and animals he had observed in the Cape Verde Islands off the coast of Africa. These islands are volcanic and mountainous, and lie near the equator, just as the Galápagos Islands do. Despite this similarity, Cape Verde organisms were not closely related to Galápagos wildlife—the species, genera, and families were very different. Instead, each island's organisms were more closely related to those of the nearby mainland, even if the mainland was very different ecologically.

At this point, Darwin was in a quandary. The idea of special creation would suggest that each individual island had had its own individual creation event. Darwin thought it much more likely that organisms must have flown, swam, or rafted on floating vegetation from the nearest continental shore to a newly formed island. Darwin reasoned that because these traveling colonists encountered no competition on the young islands, they might then diverge into new species as they adapted to different environments through natural selection. Darwin had begun his voyage believing that every species was created in its special place, but this evidence from biogeography argued against unchanging species.

While the organisms of the Galápagos and Cape Verde Islands were not closely related, they did share some superficial similarities related to their similar environments. In fact, wherever Darwin traveled, he found

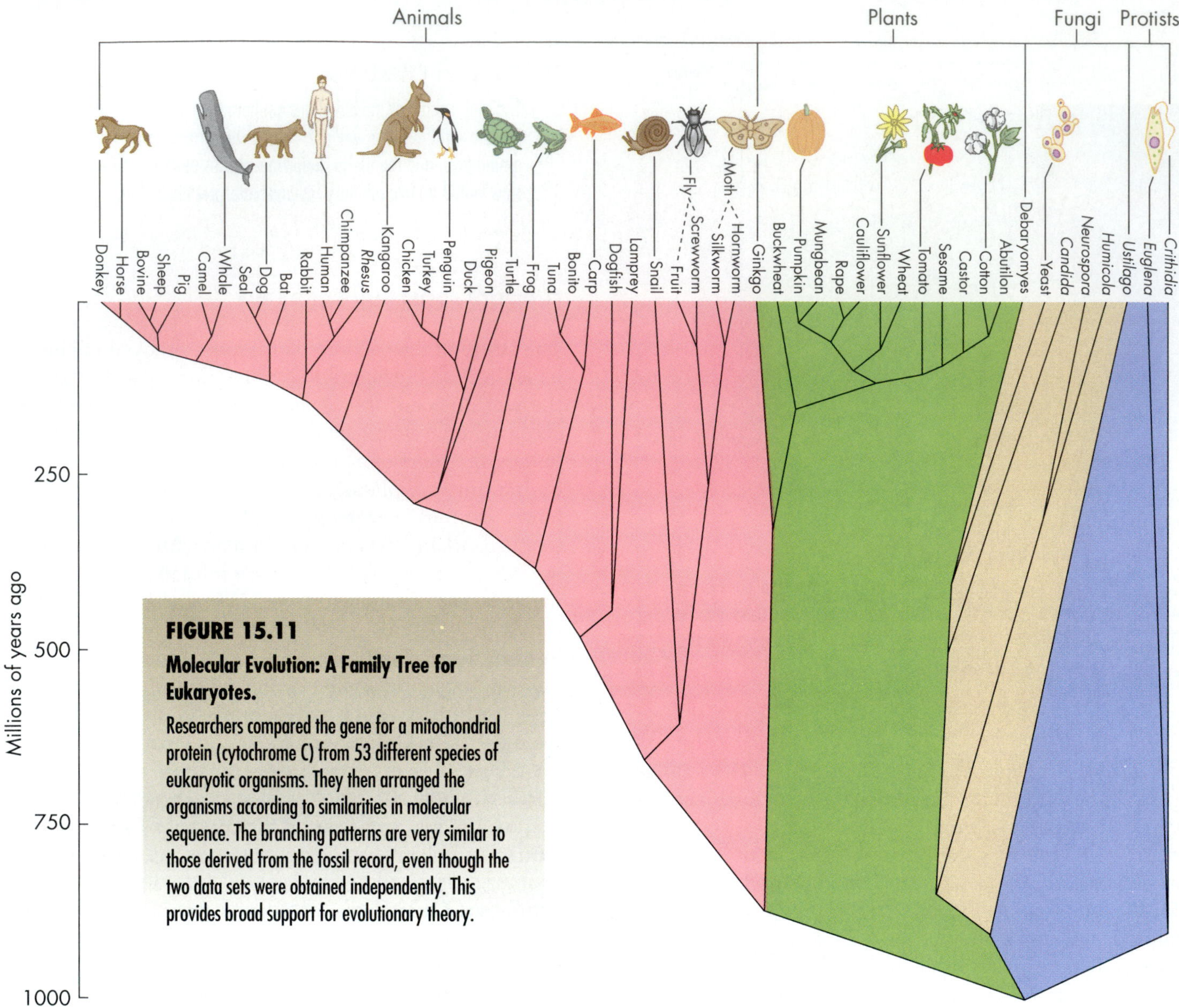

FIGURE 15.11

Molecular Evolution: A Family Tree for Eukaryotes.

Researchers compared the gene for a mitochondrial protein (cytochrome C) from 53 different species of eukaryotic organisms. They then arranged the organisms according to similarities in molecular sequence. The branching patterns are very similar to those derived from the fossil record, even though the two data sets were obtained independently. This provides broad support for evolutionary theory.

that unrelated organisms display similar characteristics in response to common environmental conditions. This is the phenomenon that modern biologists call *convergent evolution*. Darwin cited as examples of convergent evolution the superficial physical similarities shared between marsupial (pouched) mammals and placental mammals that inhabit similar environments on different continents. Convergent evolution is discussed in more detail with other patterns of evolution on page 376.

Data from biogeography underscore Darwin's idea of descent with modification, and examples of convergent evolution suggest that at least some of the modifications can occur by natural selection.

EVOLUTION BY NATURAL SELECTION

Have you ever wondered how an evolutionary biologist could ever think that the eye, a complex organ beautifully adapted for detecting light, could arise "by chance"? This type of question often confounds nonbiologists. The key to solving this problem is to realize that it is *mutation and genetic recombination* that occur by chance, not the appearance of a complete organ.

Natural selection works on undirected genetic changes that in some cases bring about genotypes (and in turn, phenotypes) that make organisms better suited to their environments. These genetic combinations allow the organisms to survive and reproduce a little bit more effectively than individuals without these genotypes, and hence the genes for beneficial combinations increase in frequency. Let's apply those principles to the evolution of eyes.

Eye evolution probably began with small groups of light-sensitive cells like those in limpets [FIGURE 15.12A]. Mutations that caused such cells to sink into cuplike depressions benefited their bearer by enabling it to detect

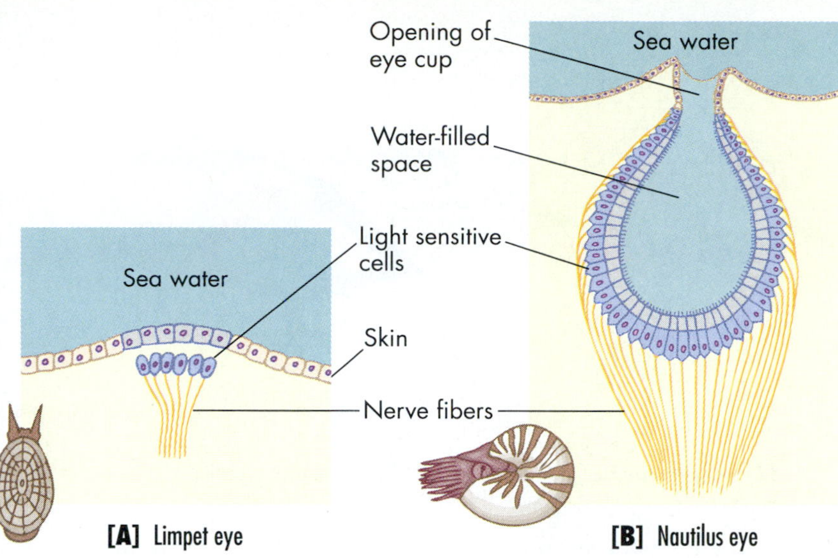

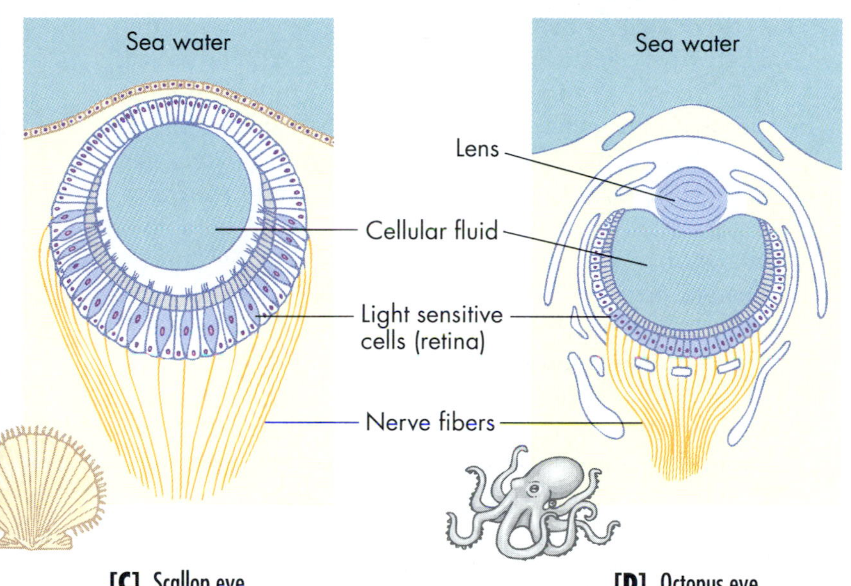

FIGURE 15.12

In the Eye of the Octopus: Analogies for Intermediates.

Natural selection can explain the origin of complex organs like the eye. Mutations which change eye shape in a way that provides an advantage for survival and reproduction will be maintained in a population, as in modern (**A**, limpet; **B**, nautilus; **C**, scallop; and **D**, octopus).

the direction of light sources and hence to find food and avoid predators more efficiently. Today's nautilus, a mollusk [see FIGURE 21.18], possesses such an eye [FIGURE 15.12B].

Further random mutations could cause the edges of the cup to join and a cellular fluid to accumulate in the cavity. If the fluid focused light on the sensitive cells, it would be easier for the organism to detect specific images. Modern scallops have this type of eye [FIGURE 15.12C]. Other sorts of mutations could arise, but if they resulted in a less-effective eye, natural selection would weed them out (unless the animal inhabited an environment where light detection provides no advantage, as in cave-dwelling fish).

Finally, if mutations arose that caused skin to grow over the eye, the cellular fluid to harden into an effective lens, and become connected to eye muscles, a complex eye capable of forming tightly focused images would have evolved. Some modern complex mollusks—the squid and octopus—have such eyes, which are very similar to our own, a case of convergent evolution [FIGURE 15.12D; see also FIGURE 15.13D].

The eyes shown in FIGURE 15.12 do not form an evolutionary series because all the organisms pictured are alive today and none was a descendant of another. What the group of eyes does show is that an organ doesn't need to be totally evolved to work well for an organism in its environment. In addition, the series shows that eyes exist today that are predicted by the hypothesis of a stepwise accumulation of mutations occurring gradually over many generations altering eye shape and visual acuity. The hypothesis involves no "forethought," no structural "goal" toward which evolution points. The idea of natural selection simply suggests that individuals who by chance have genes that provide better adaptation to the environment—however slight—will on average reproduce more successfully than individuals without this advantage.

Pathways of Descent

You have seen in previous sections how evidence from fields as different as paleontology, anatomy, genetics, embryology, and biogeography supports the principles of

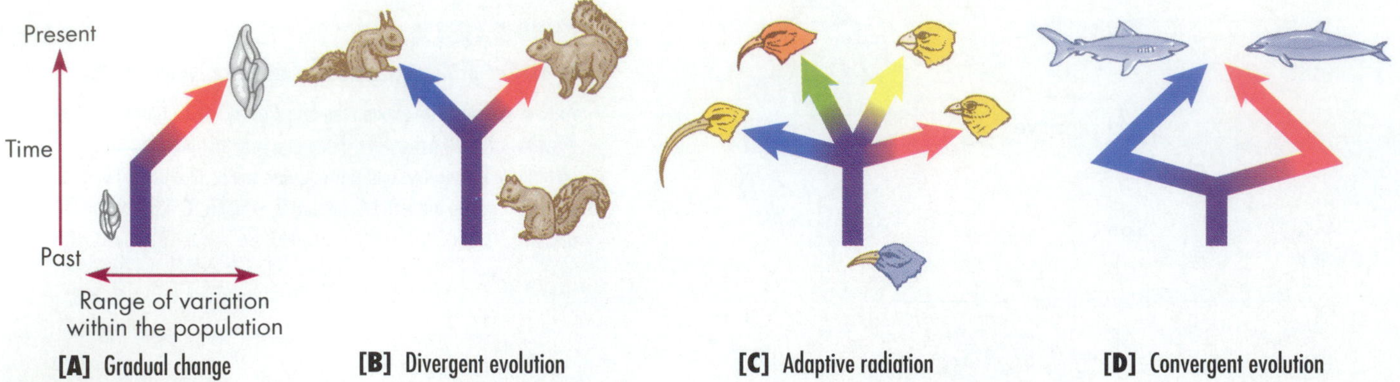

FIGURE 15.13

Patterns of Descent in Evolution.

Over time, [A] a population's gene pool may gradually change in a certain direction, as when cheetahs became smaller, or [B] the population may split into two populations, each changing in a different way as time progresses, as when two species of squirrels evolved from one original species when the Grand Canyon formed and blocked reproduction between the two populations. [C] Adaptive radiation results in several new species arising from a single common ancestor, as with the small honeycreepers of Hawaiian forests. [D] In convergent evolution, separate lines of distantly related organisms evolve similar forms that allow them to exploit similar environments, as with sharks and dolphins.

Darwinian evolution. This section focuses on how that evolutionary change occurs over time. We'll look first at various patterns of descent, then investigate how rapidly organisms may change over vast periods of geological time, and finally examine how extinctions can affect the pathways of evolution.

PATTERNS OF EVOLUTION

Evolutionary change can follow several patterns, as related species become more dissimilar or unrelated species become superficially more alike over time. We consider here four different patterns: gradual change, divergent evolution, adaptive radiation, and convergent evolution.

Gradual Change In the course of **gradual change**, genetic changes in a population occur slowly over many generations [FIGURE 15.13A]. One species gradually changes into a new species. As an example, consider a single lineage of tiny marine protists called foraminifera. From about 10 million to 5.6 million years ago, there were no significant changes in the shape of the shells secreted by these organisms. But then gradually over a period of 600,000 years they underwent a physical change into what experts consider a new species, which remains alive and unchanged in today's Indian Ocean.

Divergent Evolution In another pattern of evolution, called **divergent evolution**, a single population is split into two or more populations followed by a gradual accumulation of genetic differences in the isolated populations. Before the Colorado River cut a mile-deep gash through an Arizona plateau, a single species of squirrel occupied the area. When the Grand Canyon became an uncrossable barrier, the squirrels were divided into two separate populations. Subsequent genetic changes produced the two species of squirrels, which now inhabit opposite rims of the Grand Canyon [FIGURE 15.13B].

Adaptive Radiation Sometimes, divergence occurs simultaneously among a number of populations of a single species in ways that produce a variety of phenotypes. This fan-shaped branching pattern of evolution is called **adaptive radiation** [FIGURE 15.13C]. Adaptive radiation may be fostered when an ancestral species invades new territories that allow a variety of new and different ways of life. An example is the colonization of the Hawaiian Islands by colorful birds called honeycreepers, which led to about 20 different species adapted to survive on different foods.

Convergent Evolution Still another evolutionary pattern occurs when two or more dissimilar and distantly related lineages evolve in ways that make them appear superficially similar. As you saw earlier, this phenomenon is called **convergent evolution** [FIGURE 15.13D].

An example of convergent evolution can be seen in the squirrel-like sugar gliders of Australian forests and the flying squirrels of Europe, Asia, and North America. Bushy-tailed sugar gliders soar through Australian forests on thin folds of skin that extend along both sides of the body from wrist to ankle [FIGURE 15.14A]. Like all of Australia's original mammals, sugar gliders are marsupials, nurtured in the mother's marsupium, or pouch. Sugar gliders, however, look and act very much like the placental mammals called flying squirrels [FIGURE 15.14B]. Throughout much of the northern hemisphere, flying

[A] Sugar glider

[B] Flying squirrel

FIGURE 15.14

Convergent Evolution.

Organisms adapted by natural selection to exploit similar environments evolve similar adaptations even though they are only distantly related. The sugar glider **[A]** and the flying squirrel **[B]** belong to two very different groups of mammals, but they have both evolved bushy tails and flaps of skin that help them exploit their forest environments more efficiently.

squirrels inhabit forest environments similar in many ways to those of the sugar gliders, which evolved independently thousands of miles away. Marsupial counterparts of placental wolves, mice, moles, and anteaters also inhabit parts of Australia. How could such remarkable similarities occur?

Geologists agree that the Australian continent separated from the rest of the earth's continents more than 50 million years ago [see CHAPTER 17], and that few if any placental mammals had reached this landmass at the time of the separation. Without competition from placental mammals, Australian marsupials diverged into a great number of species, most of which came to perform ecological roles very similar to roles filled by placental mammals on other continents. Similar physical forms and behaviors evolved in the two groups of animals when they inhabited similar environments.

➤ CONCEPT CHALLENGE

Echidnas and anteaters both eat ants, and have long snouts and tongues, and sharp, powerful claws. These features evolved independently in the two groups. Which of the four patterns of evolution just discussed does this exemplify and why?

THE TEMPO OF EVOLUTION

The patterns of evolution we have just discussed seem to suggest that all lines change at about the same constant rate over time, at least as drawn in FIGURE 15.13, an idea called phyletic gradualism. More recent analysis, however, suggests that instead, structural changes usually occur in fits and starts, an idea called punctuated equilibrium.

Phyletic Gradualism Traditionally, evolutionary biologists have agreed that after a population splits into two, as when one population of squirrels becomes separated by a river, the new subgroups diverge from each other at about equal and constant rates as natural selection and random genetic changes cause modifications in each line. This view is called **phyletic gradualism** [FIGURE 15.15A]. Adherents of phyletic gradualism suggest that over time, a new species arises as changes accumulate within a population. The population may or may not split into two different lines.

Punctuated Equilibrium If new species form as one species gradually transforms into another, then the fossil record should show numerous intermediate species. While Darwin advocated this view, there was little fossil evidence to support it in his time, and despite a century and a half of additional fossil collecting, paleontologists still have few records of gradually transformed lineages with numerous intermediate species. Scientists traditionally attribute such gaps in fossil lineages to gaps in the fossil record, explaining that intermediate fossils are missing because the rocks that contained them have eroded away, are still undiscovered, or were never formed in the first place.

In 1972, professors Niles Eldredge and Stephen Jay Gould suggested that instead of making the data fit our ideas, we should make our ideas fit the data. They proposed that the gaps themselves might be telling us how speciation takes place. They pointed out that according to the fossil record, the typical species is *not* in a constant process of change; instead, species generally exist for millions of years without significant alteration. These long periods of phenotypic equilibrium are interrupted, or "punctuated," only rarely by great phenotypic changes that result in new species.

This alternative to the phyletic gradualism model is called **punctuated equilibrium** [FIGURE 15.15B]. It has three tenets:

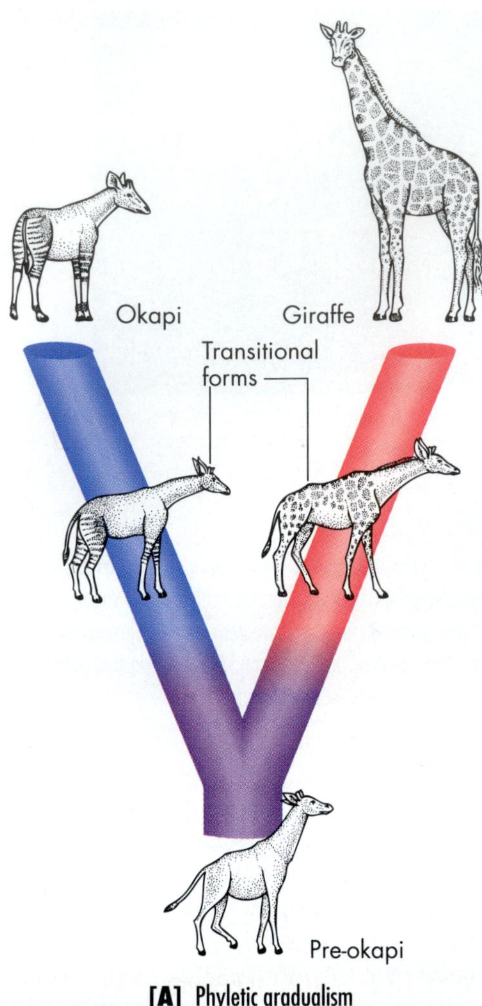

[A] Phyletic gradualism

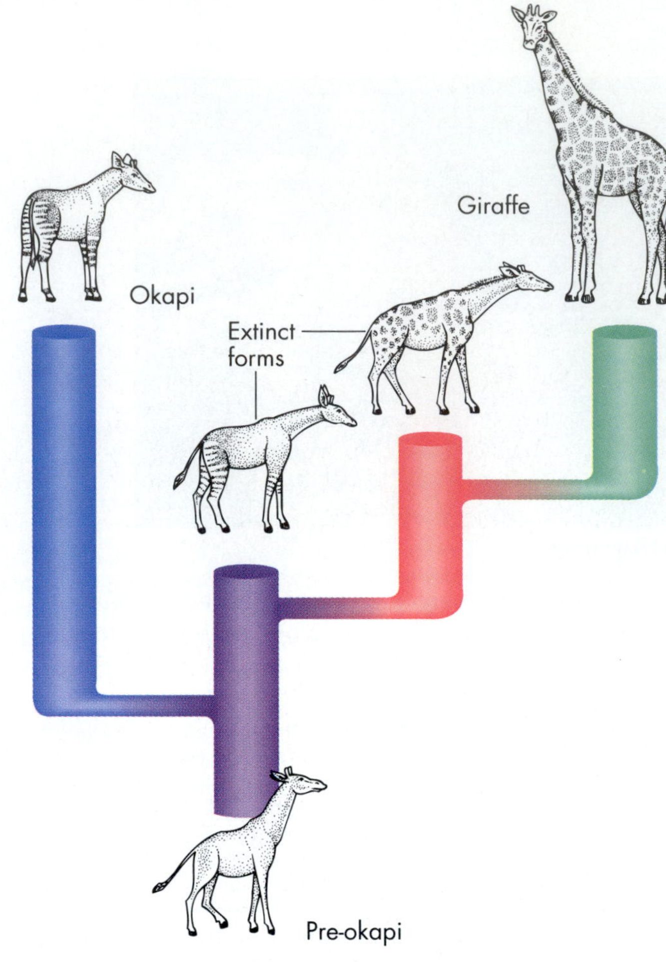

[B] Punctuated equilibrium

FIGURE 15.15

Two Hypotheses for the Descent of Okapis and Giraffes.

[A] The evolutionary pattern of phyletic gradualism portrays okapi and giraffe evolution as two equal, slowly diverging lines from a common ancestor. **[B]** The punctuated equilibrium hypothesis suggests that a small population budding off from the ancestral pre-okapi group would lead to modern okapis, while a number of more giraffelike species would arise and became extinct as time passed.

1. Alterations in body form evolve rapidly, in evolutionary terms.

2. During **speciation** (the formation of a new species), changes in form occur almost exclusively in small populations, and the resulting new species are quite different from their ancestral species.

3. After the burst of change that results in speciation, species retain much the same form until they become extinct, perhaps millions of years later.

To get a feel for the different patterns of descent predicted by punctuated equilibrium and phyletic gradualism, consider two contrasting family trees for the okapi and giraffe: (1) Phyletic gradualism predicts that the two evolutionary lines began to diverge slowly. Gradually, changes accumulated that eventually prevented the two groups from interbreeding. They are now separate species. Modifications in form continued to occur gradually in both species until the present-day okapi and giraffe emerged. (2) In contrast, punctuated equilibrium predicts that offshoots of the evolutionary lines emerged repeatedly as small, isolated populations that split from the main group, underwent bursts of evolutionary change, and then maintained their new features for a long or short period before the line died out.

In fact, it seems quite likely that *both* processes took place. Evolution may often occur when a small population becomes isolated geographically or reproductively from other members of its species. Gradually, over many generations, but nonetheless a very short stretch of geologic time, the isolated population accumulates genetic changes. Once sufficient change has occurred, and the new species is fitted to its environment, very little additional change may take place.

SPECIES SELECTION, EXTINCTION, AND EVOLUTIONARY PATHWAYS

The evolutionary patterns discussed so far have mainly documented changes in lineages of organisms over relatively short periods of time, as one species splits into two or more, such as the Grand Canyon squirrels or the Hawaiian honeycreepers. Such changes at the level of the population and species are called **microevolution**, and we will discuss them from a genetic viewpoint in more detail in the next chapter. In contrast, grander evolutionary changes above the level of the species are called **macroevolution**. Examples of macroevolution include the sudden appearance of *Hallucigenia* in the fossil record or the evolution of mammals from reptilian ancestors. Evolutionary biologists have been trying to determine whether microevolutionary mechanisms—small, gradual transformations brought about largely by natural selection and unpredictable genetic changes—can account for the origin of new groups above the species level, or whether different kinds of evolutionary mechanisms are required to produce such major differences.

Major Evolutionary Transitions The origin of mammals from reptilian ancestors is so well documented by an extensive fossil record that it provides a good model for understanding major evolutionary transitions [FIGURE 15.16]. The fossil record shows that in the course of evolution, the orientation of the reptile legs, the complexity of the jaw muscles, and the variety of tooth types became gradually more mammal-like. The record does not show a single line of evolution from reptile to mammal. Rather, at each stage, there was a radiation into several evolutionary lines, most of which died out without giving rise to further species. In some lines, there was evolution away from the mammalian condition, but many of these lineages also became extinct.

The fossil record shows that (1) mammalian evolution occurred as a result of many small changes, and (2) these changes, like those related to locomotion and feeding, involved adaptations to environmental conditions and presumably occurred by natural selection. From such evidence, biologists concluded that macroevolution is just microevolution extended over very long time periods.

How can the gradual, stepwise evolution of new features occur in macroevolution? An answer comes from huge, extinct, rhinoceroslike animals called titanotheres. Titanotheres had large horns [FIGURE 15.17], but their smaller ancestors did not. Fossil evidence shows that the huge horns arose gradually in stages from small bumps already present on the heads of earlier species. To generalize from this finding, it can be said that new structures often evolve by a change in size, shape, or function of a preexisting body part. The original structure did not appear in anticipation of a future need, but once the structure was present it could be modified by natural selection to perform a new function. One important example of this, the panda's "thumb," is described in BOX 15.1

Species Selection and Evolutionary Trends So far, we have viewed evolution mainly as a result of differences in reproductive success among individuals within a population. Superimposed on this selection within a species is competition among members of different species that have similar ecological needs, including the same foods and living sites. This process is called **species selection**, or *species replacement.*

Species selection among closely related species may lead to apparent *evolutionary trends*, a series of morphological changes in a given direction over a long period of time. The evolution of the horse exemplifies how species selection might lead to perceived evolutionary trends [FIGURE 15.18].

Modern horses are large animals with long legs, and hoofs bearing one toe apiece; they also have high-crowned cheek teeth with complex grinding surfaces. An early ancestor of the modern horse was the size of a German shepherd dog. Its legs were short and bore four toes, and its teeth were small and lacked grinding surfaces. The evolutionary changes in horses did not occur smoothly in a single lineage, from small to large body size, four toes to one, and small to large grinding teeth, as if directed toward a goal. Rather, this perceived trend unfolded in fits and starts, with different species displaying various combinations of toe number, body size, and tooth shape, and with smaller species occasionally replacing larger ones. Each evolutionary line proceeded separately and independently of purpose, subject to changes in the environment and competition with other similar species.

Species selection can explain the trends in horse evolution in two different ways: (1) Smaller-sized horse species may have tended to become extinct more rapidly than larger horse species, or (2) larger horse species may have tended to form new species more rapidly than species with smaller individuals. It is possible that a combination of the two hypotheses was operating during horse evolution. Ultimately, with the passage of time, evolution in all lineages typically leads to the great equalizer—extinction.

Adaptive Radiation, Extinction, and Modern Species Modern organisms tend to fall into a few large taxonomic groups called **phyla** (singular, *phylum*) for animals, and **divisions** for fungi and plants. Some major animal phyla are mollusks (snails, clams, squid), arthropods (crabs,

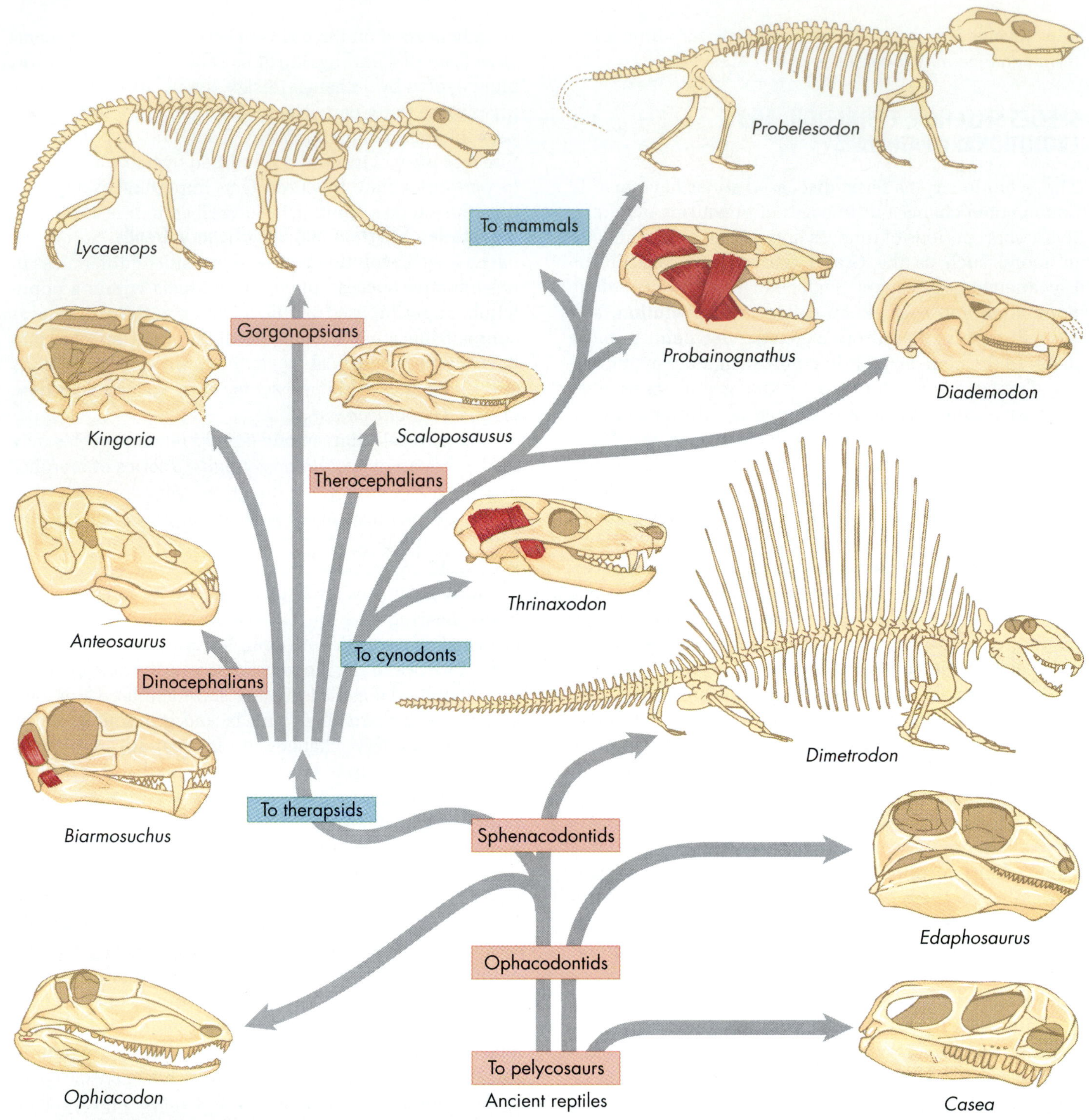

FIGURE 15.16

Reptiles to Mammals: How Major Evolutionary Transitions Can Occur.

In the mammal-like reptiles several great adaptive radiations occurred. The changes occurred gradually in many small steps, and that they occurred in adaptive traits, those involved in locomotion and feeding. They show that macroevolutionary changes can occur by microevolutionary principles applied over very long time periods.

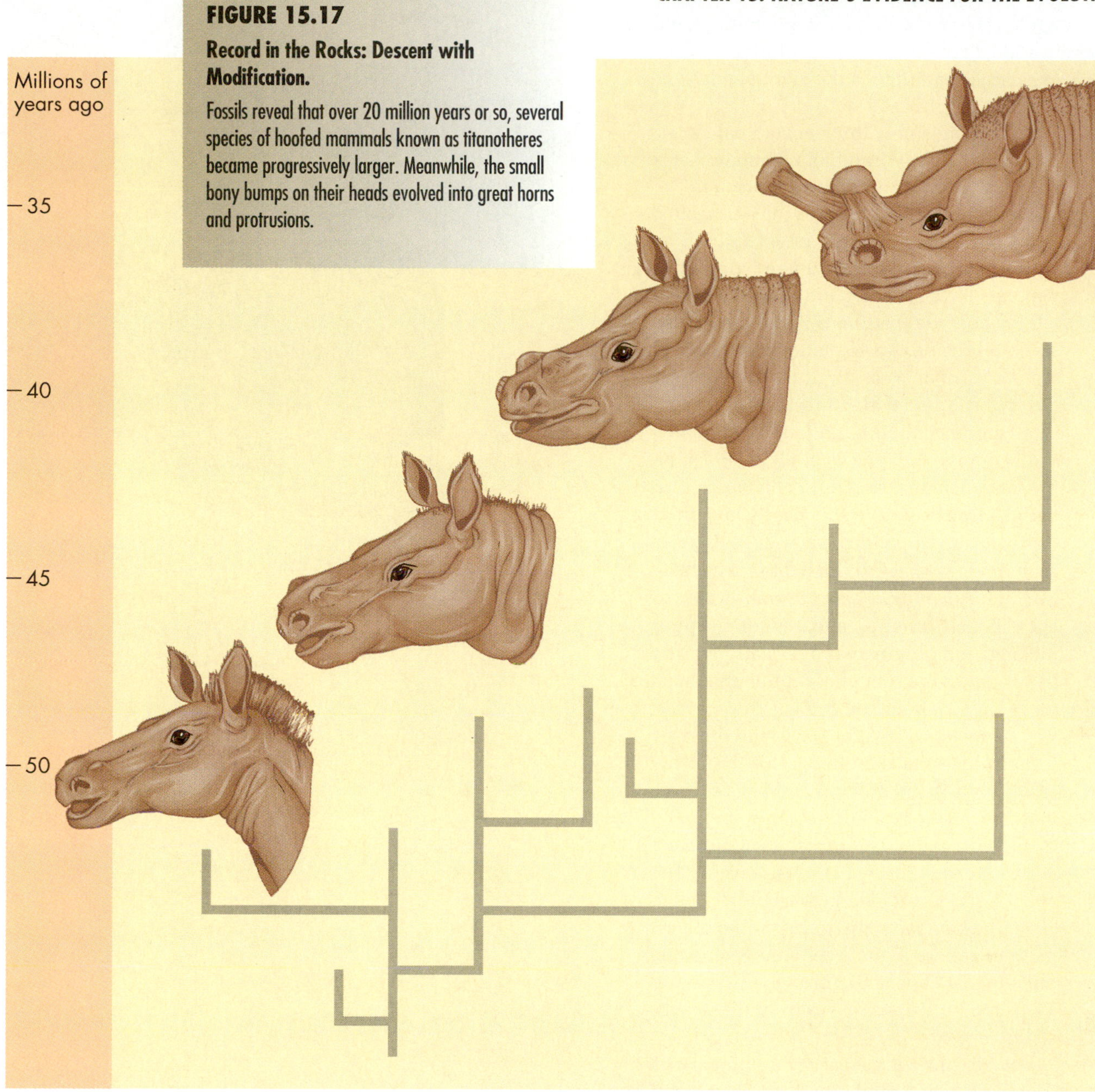

FIGURE 15.17

Record in the Rocks: Descent with Modification.

Fossils reveal that over 20 million years or so, several species of hoofed mammals known as titanotheres became progressively larger. Meanwhile, the small bony bumps on their heads evolved into great horns and protrusions.

spiders, insects), and chordates (amphibians, fish, reptiles, birds, mammals).

Could the origins of these phyla—a classic example of macroevolution—be explained by the mechanisms of microevolution? Part of the answer may come from *Hallucigenia* and the other curious fossilized organisms discovered in the Burgess Shale and similar deposits. These sediments were laid down over 500 million years ago, as the Cambrian time period began (see the geologic timetable on the inside front cover of the book). The major animal phyla appeared rapidly during the Cambrian, and in the hundreds of millions of years since, no new body plans have appeared. For this reason, paleontologists call the events of this time period the **Cambrian explosion**.

Fossil evidence confirms that during the early Cambrian period, the range of body plans was far greater than it is today. While *Hallucigenia* can be placed in today's phylum Onycophora [review FIGURE 15.8], geologists assign other strange cohorts to extinct phyla. *Nectocaris*, for example, had a front half similar to an insect, but a tail end vaguely like an embryonic fish, and *Dinomischus* resembled a long-stemmed goblet. Neither seem to fit in any existing phylum.

Many branches of the Cambrian period's bushy family tree were subsequently pruned by extinction, and

box 15.1
Biology Applied

Evolutionary Oddities Prove the Rule

"Odd arrangements and funny solutions," wrote Harvard professor Stephen Jay Gould in 1980, "are the proof of evolution."* And one classic example is the giant panda's thumb.

The handsome, thick-coated panda [FIGURE 1A] lives only in remote, remnant bamboo forests along the Tibetan plateau in eastern China. Giant pandas once inhabited much of China, but pelt hunters and the steady encroachment of human farms and settlements have caused panda populations to fall to about 1000 animals surviving in just six small regions with a total land area less than the size of West Virginia. Although pandas will eat pine bark, horsetails, and even small animals, their main food is bamboo stalks—up to 650 pieces or 88 lb (40 kg) of the tender shoots each day. The panda's most curious trait relates to this hearty appetite for bamboo shoots: pandas have seven digits on each front paw. The animal's five true "fingers" plus one opposable "thumb" and a seventh "finger" are fully adapted for grasping, picking, and holding bamboo stalks while the heavy omnivore munches on the plant matter for at least 16 hours each day.

Biologist D. Dwight Davis and others have studied the anatomy of this thumb and extra finger in great detail and have found that they are not true digits. They are, instead, enlarged wrist bones (sesamoid bones) with attached muscles and tendons [FIGURE 1B]. This arrangement enables the thumb to grasp the bamboo tightly against the other, more fingerlike digits. When the panda then pulls the bamboo stalk through this vise, it strips the nutritious leaves from the woody stalk.

Some time in the early evolutionary history of the giant panda, mutations occurred in the forearms. These gave some individuals enlarged wrist bones along with muscles and tendons connected in such a way that the new digits had opposable movement. These mutant pandas with their ersatz thumbs must have been able to gather more bamboo, survive in greater numbers than their five-digit cohorts, and leave more offspring. Biologists think this because today, all surviving pandas have the opposable thumb. This odd arrangement, these extra digits, is far from the kind of graceful innovation some skilled bioengineer might have designed. But it functions like a thumb, and its origin from a preexisting structure is a good demonstration of how evolution generally proceeds in the natural world.

[A]

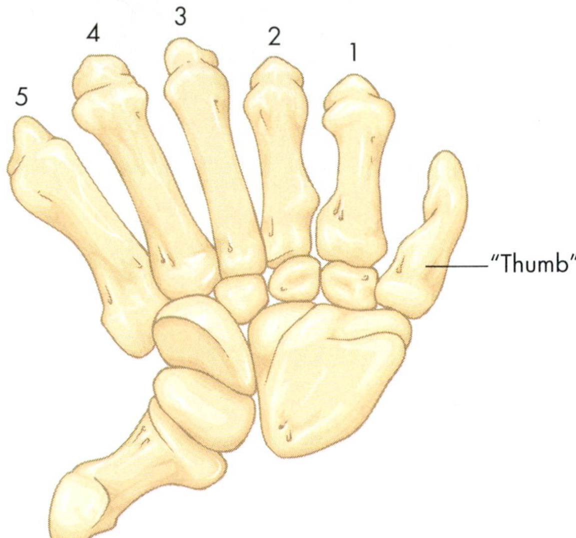

[B] Bones of the panda's hand

FIGURE 1 **[A]** The giant panda uses its hand to strip bark and leaves from bamboo, trapping the plant between its five fingers and a "thumb" evolved from an enlarged wrist bone. **[B]** Bones of the panda's hand include those of the normal five digits of most mammals (only bones at the base of the digits are shown here, colored yellow and labeled 1–5) plus modified wrist bones that form an extra finger and a functional extra "thumb."

*Stephen Jay Gould, *The Panda's Thumb*, New York: Norton, 1980, pp. 20–21.

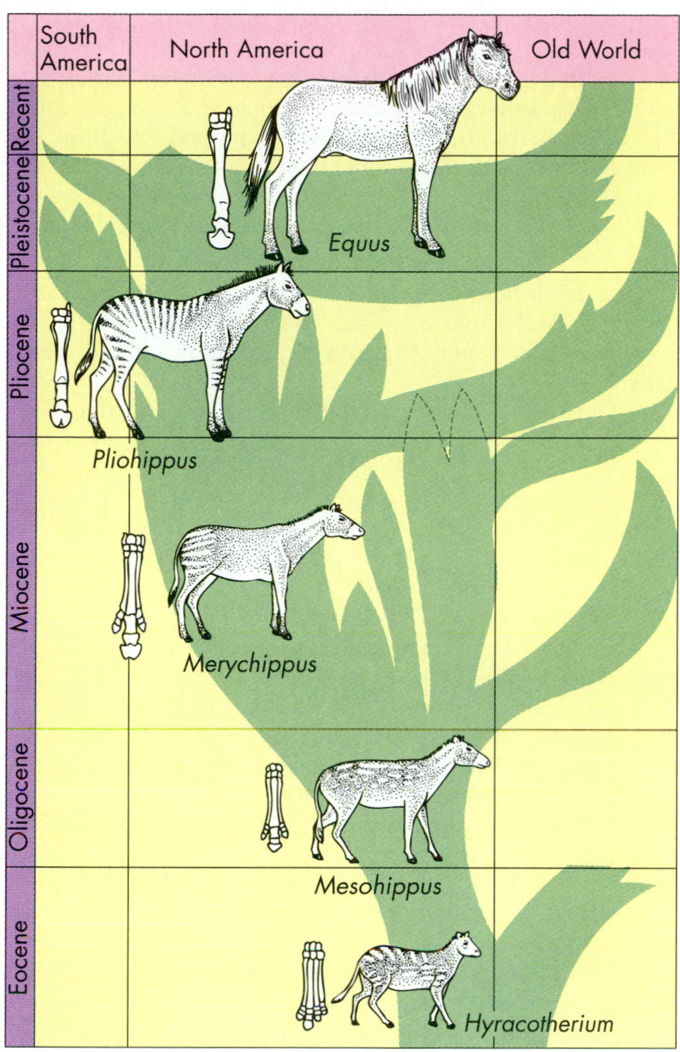

FIGURE 15.18

Species Selection, Extinction, and Evolutionary Trends.

The tiny, dog-sized progenitor of modern horses, *Hyracotherium*, had four toes, but modern *Equus* has one toe. The lineage is a bush with many branches, all of which have died out except modern horses.

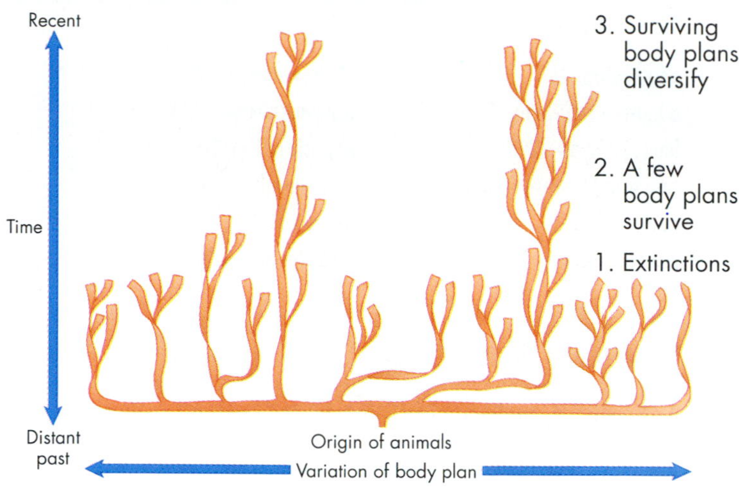

FIGURE 15.19

Radiation, Decimation, Diversification.

During the Cambrian explosion, animals radiated into diverse body plans. Paleontologists think these divergent lines were decimated by extinction, and that only a few lines persisted in time. These few subsequently diverged once more, but in all lines, the basic body plans that evolved in the Cambrian are retained.

only organisms with the basic body plans we know today survived [FIGURE 15.19]. After these lines were well established, they diversified, but always within the constraints of their characteristic general forms. Most animals alive today have a head, a nerve cord, and other features because this basic pattern arose in organisms like *Hallucigenia* crawling across the sea floor nearly 600 million years ago.

➤ CONCEPT CHALLENGE

Explain how the evolution of horses illustrates the following principles: species selection, punctuated equilibrium, divergent evolution, and evolutionary trends.

Connections

While many questions remain about the detailed mechanisms of evolution, there is overwhelming evidence that descent with modification has indeed occurred and that natural selection is a potent force that drives the change in organisms over time. Evolution provides a powerful explanation for diverse biological phenomena, from the appearance of fossils such as *Hallucigenia*, to a snake's pelvis, to the gill slits in human embryos. In the following chapter you will learn more about microevolution and the genetics of evolution, and in CHAPTER 17 you will investigate how evolutionary principles can account for the emergence of life itself and for the major events in life's early history on earth.

KEY TERMS

adaptation, 367
adaptive radiation, 376
biogeography, 373
convergent evolution, 376
homology, 370
macroevolution, 379
microevolution, 379
natural selection, 367
phyletic gradualism, 377
punctuated equilibrium, 377
speciation, 378
vestigial organ, 371

HIGHLIGHTS IN REVIEW

1 Fossils help document life's history on earth. They show that today's living organisms are different from, and often more complex than, those that lived in the past. Fossils provide firm physical evidence that life has evolved.

a] Fossils are the remains of prehistoric organisms, often mineralized and preserved in sedimentary layers. By studying fossils, biologists have confirmed that organisms have changed over time, that organisms display a general trend toward greater complexity, and that intermediates once existed between major groups of organisms.

2 Natural selection is the differential survival and reproduction of a small fraction of a population due to their physical, biochemical, or behavioral traits.

a] Natural selection is often the most important agent of evolution. It results in adaptation, evolved genetic traits that allow an organism to function better in its environment. Well-adapted organisms have high evolutionary fitness, the relative ability of an individual to survive and reproduce successfully.

b] Directional selection is a form of natural selection in which alleles that cause individuals to exhibit one extreme form of a trait become more frequent. Stabilizing selection results in a high frequency of alleles specifying the average expression of a trait. Disruptive selection results in a high frequency of alleles specifying two extreme forms of a trait.

3 Evidence for evolution comes from genes, comparative anatomy, and from ecology.

a] The structure of genes and proteins in living organisms reveals the degree to which modern organisms are genetically related. Family trees drawn from such data are broadly similar to family trees derived from fossil data. Since the two types of data are independent, this is strong evidence for evolution by descent with modification.

b] Homologous organs (organs with a similar origin but different functions), such as the limbs of various vertebrates, suggest that evolution works by modifying a basic plan.

c] Vestigial organs (structures with no apparent utility in a living organism) like the snake's pelvis, provide evidence of descent with variation.

d] Evolutionary principles can readily explain the passage of embryos through stages that strongly resemble those of different but related organisms.

e] Biogeography, the study of where on earth organisms live, reveals that island species are different from, but related to, species on the nearby mainland. Evidence suggests that mainland forms invaded the islands and evolved into the island forms via natural selection.

f] In convergent evolution, organisms that are not closely related become adapted to similar environments by natural selection and display similarities in body form or other characteristics.

4 Macroevolution, the large-scale evolutionary changes in groups of species over geologic time, can be explained by microevolution, genetic changes in populations of individual species.

a] Organisms can evolve by gradual change, by divergent evolution, and by adaptive radiation. Each pattern depends on the organism's evolutionary history and its physical environment.

b] Phyletic gradualism is the notion that evolutionary change is evenly paced and based on the processes of microevolution.

c] The punctuated equilibrium model suggests that in small, isolated populations, evolution can occur very rapidly over geologic time. After the burst of change, new species tend to retain the same form until they become extinct, perhaps millions of years later.

d] Species selection occurs when various species compete in the same ecological setting, and one species drives another to extinction. Species selection can help explain evolutionary trends, if different rates of extinction or speciation occur in species with phenotypes at one end of the trend (for example, larger species).

e] The microevolutionary principles of adaptive radiation followed by extinction seem sufficient to explain the origin of new groups above the species level, such as the evolution of animal phyla.

5 Extinctions of some species followed by adaptive radiations by others have marked the course of evolution. The tempo of evolutionary change is not constant; bursts of evolutionary change are often followed by long periods with little change.

UNDERSTANDING THE FACTS AND CONCEPTS

For Questions 1–5 match each of the descriptions with the most appropriate item from the following list of terms. Any item may be used once, more than once, or not at all.

a] convergent evolution
b] divergent evolution
c] adaptation
d] species selection
e] radiation

1 Divergence of an ancestral species into many species that are well adapted to slightly different parts of a new territory.

2 A result of competition between members of different species for the same habitat.

3 Two or more groups of organisms that live in similar environments are superficially similar but are descended from different ancestors.

4 Body parts, behavior, or physiological processes that improve an organism's chance of survival and/or reproduction.

5 Two or more species that have recently split from a common ancestral species that may now be extinct.

As above, for Questions 6–10, match the name with the description.

- **a]** paleontologist
- **b]** macroevolution
- **c]** microevolution
- **d]** *Hallucigenia*
- **e]** Cambrian

6 A bizarre, extinct marine worm.

7 A period of time when most animal phyla appeared.

8 Evolution of phyla of organisms.

9 A biologist who studies ancient life through fossils.

10 Evolution of one species from another.

For Questions 11–15, match each of the descriptions with the most appropriate item from the following list.

- **a]** homologous organ
- **b]** vestigial organ
- **c]** phyletic gradualism
- **d]** punctuated equilibrium
- **e]** convergent evolution

11 An example of evolution that is based on firm ecological and biogeographic evidence.

12 Rapid changes in small, offshoot populations, followed by long periods of stability.

13 Slow but steady descent with modifications consistent with natural selection in ever-changing environments.

14 Two structures that may have different functions but are descended from a common ancestral structure.

15 A form of homology in which a rudimentary structure in one species is believed to share a common evolutionary ancestry with a functional structure in another species.

For Questions 16–20, match the organisms with one of the patterns of evolution from the following list.

- **a]** gradual change
- **b]** divergent evolution
- **c]** convergent evolution
- **d]** adaptive radiation
- **e]** species selection

16 Honeycreeper birds of Hawaii.

17 The horse.

18 Squirrels on the north and south rims of the Grand Canyon.

19 Foraminiferans in the Indian Ocean.

20 Flying squirrels (placental mammals) and marsupial sugar gliders, which both glide from tree to tree.

INTEGRATE AND APPLY WHAT YOU HAVE LEARNED

1 What is the role of the environment in the evolutionary theory of Lamarck as compared to that of Darwin?

2 Despite the similar ecology of the Galápagos Islands and the Cape Verde Islands, different plants and animals live on them. How did these observations influence Darwin?

3 Opponents of evolution point to the theory of punctuated equilibrium as evidence that even scientists doubt the validity of Darwin's ideas. Do you agree? Explain.

4 Proponents of evolution point to examples of convergent evolution as an argument against special creation of unchanging species. Do you agree? Explain.

5 Why is extinction the likely fate of any species?

ANALYSIS

1 Suppose you are observing a group of five species of "smurfs," each of which lives in the tree canopy in a jungle. Each species has a grasping, prehensile tail. They have numerous small physical differences, eat different diets, and prefer to remain at different levels in the canopy. The molecular data strongly suggest that they evolved from a common ancestor after the ancestral species populated the jungle. Which of the following terms best describes this evolutionary situation?

- **a]** Parallel evolution.
- **b]** Convergent evolution.
- **c]** Adaptive radiation.
- **d]** Divergent evolution.
- **e]** Coevolution.

2 Why was the existence of *Hallucigenia* fossils hard to explain?

- **a]** The animal's structure is strange.
- **b]** Biologists did not see how it could move or eat.
- **c]** It is more bizarre than any other organism known to have lived at the same time.
- **d]** Its location in western Canada was hard to account for.
- **e]** There were no known fossil traces of ancestral types.

3 If you were an avid fossil hunter, which of the following locations would appeal most to you?

- **a]** Inside an active volcano.
- **b]** Sedimentary rock that had once been a lake bottom.
- **c]** An area that had always been a part of the floor of the open ocean.
- **d]** Hot sulfur springs.
- **e]** Land that had always been on the top of a mountain range.

4 Which of the following theories or observations *requires* one to assume that the fossil record is incomplete with respect to extinct intermediate species?

- **a]** Punctuated equilibrium.
- **b]** Homologous organs.
- **c]** Convergent evolution.
- **d]** Vestigial organs.
- **e]** All of the above.

CHAPTER 16

The Genetic Basis for Evolution

CHEETAHS, SPRINTING TOWARD EXTINCTION

Cheetahs of the African savanna are lithe, fast-running predators. When hunting, they climb to elevated lookouts and then search for prey. If a gazelle or wart hog nears the cat's perch, the cheetah descends noiselessly and approaches the prey, its buff and brown-spotted coat camouflaged in the long, uneven grasses. The cheetah crouches, then springs toward the startled animal. In one second and two bounding strides, the cheetah is running more than 70 km per hour (45 mph), closely pursuing the terrified prey [FIGURE 16.1]. If the feline hunter is lucky, it will soon trip the prey with a powerful swipe of a forepaw, then strangle it with a crushing bite to the throat.

FIGURE 16.1

The World's Fastest Runner. The cheetah (*Acinonyx jubatus*) evolved with a slender body, camouflage markings, and a high-speed gait. Here a cheetah pursues a wart hog.

The cheetah is the world's fastest-running animal, and many of its features contribute to the single task of sprinting. The lungs, blood vessels, and heart are huge for an animal its size; the legs are long and slender; the claws are partially extended like spikes on running shoes; and the backbone is capable of extreme flexing which allows it to bound in strides up to 7m (23 ft) long.

Despite the cheetah's prowess as a sprinter and hunter, the species is dangerously close to extinction. While the animals once enjoyed a worldwide distribution, the approximately 20,000 remaining animals are now limited to a few sites in southern and eastern Africa. These relatively few remaining cats have inherited many genetic defects that hinder day-to-day survival and reproduction. They are highly susceptible to diseases such as cat distemper; many male cheetahs produce defective sperm that cannot fertilize eggs; and cubs tend to be sickly. These genetic problems surface frequently because, among the remaining

cheetahs, every individual is genetically very similar to every other. The result is the same depression of genetic vitality and expression of undesirable recessive traits that would be expected if humans married their cousins for generation after generation.

How cheetahs became such beautifully adapted hunters and yet have reached their current state of low genetic diversity and near extinction serves as a case study for this chapter's topic, the genetic basis of evolution. This topic was introduced in broad terms in CHAPTERS 1 and 15: Recall that Charles Darwin was amazed by the diversity he found among the plants and animals he observed in his voyage on the *Beagle*. He also noted that each species has the potential to produce many more individuals each generation than actually survive and reproduce. From these facts he reasoned that some individuals in a population might carry hereditary variations that would allow them to cope with their environment and to reproduce more successfully than other members of the population. He concluded that this process, which he called natural selection, would cause some hereditary variations to become more frequent in subsequent generations—in other words, evolution would occur.

This chapter discusses these arguments in more detail and explores their genetic basis. It begins by investigating the extent of genetic diversity within various species, then moves to how that variation arises and is maintained in populations, including why some species, like cheetahs, have such a limited genetic diversity today. Next, the chapter examines the consequences of genetic diversity and how various factors, such as migration, small population size, and natural selection act on preexisting genetic diversity to change the frequencies of various genes and alleles. We move on to the mechanisms of natural selection, analyzing them in ways that were not available to Darwin and Wallace because they were unfamiliar with Mendel's principles of genetics. The chapter closes with an up-to-date view of the origin of species based on the latest laboratory and field research.

MESSAGES

1 Genetic variation—hereditary differences between individuals—provides the diversity of appearances, capabilities, and behaviors from which the environment selects the parents for the next generation.

2 The frequencies of genes and alleles in a population tend to remain the same from generation to generation unless the population is disturbed. Perturbations can include mutation, migration, nonrandom mating, genetic drift, and natural selection.

3 New species arise when genetic barriers divide one population into two populations that do not interbreed in nature and that follow separate evolutionary pathways.

This chapter continues a three-chapter section on evolution. The previous chapter examined evidence for Darwin's thesis of descent with modification. This chapter concentrates on genetic mechanisms of evolution and how natural selection and genetic change affect species of organisms. CHAPTER 17 looks back to the origins of life and surveys the evolutionary history of life on earth.

Recall that Darwin's amazement at the diversity of life forms he found in his travels initially set him on the track that led to his theory of evolution. Human activities are currently diminishing this diversity at a rate perhaps never before experienced on our planet. Cheetahs, likely, will be one of those species to become extinct in the near future.

As we explore the genetic basis for evolution, we will encounter a host of examples that help explain evolutionary principles and processes. These include human skin color, foreshortened fingers among the Amish in Pennsylvania, St. Bernards, and peppered moths. ❑

Genetic Variation: The Raw Material of Evolution

The principles of evolution are founded on **genetic variation**—genetic differences among individuals of the same species that cause the individuals to vary in inheritable traits, such as weight or running ability or leaf shape or petal color. Charles Darwin devoted the first two chapters of his book *On the Origin of Species* to this topic. Darwin was enormously impressed with variation in domestic species, such as pigeons and dogs [FIGURE 16.2]. On his voyages he noted that similar variations in form occur in nature. But Darwin did not know about genes and could not understand how the mechanisms of heredity could produce this variation, or how genetic variants could be passed from parent to offspring. Mendel published his genetics experiments [CHAPTER 8] six years after Darwin and Wallace first published their work on evolution, and if either of the two naturalists read Mendel's papers, they did not understand their significance.

It took about 70 years for evolutionary biologists to combine Darwin's original theory with Mendelian genetics. The current *synthetic theory of evolution* explains natural selection in genetic terms. It suggests that (1) gene mutations occur in reproductive cells at frequencies that are high enough to have an evolutionary impact, (2) gene mutations occur in random directions unrelated to any need the organism may have in the environment—for example, one mutation may make a cheetah run slower, while another enables a cheetah to run faster, and (3) natural selection acts on the genetic diversity introduced through such random mutations. To understand the modern synthetic view of evolution, we must understand the source and extent of genetic variation. We must also learn how genetic variation is maintained in a **population**, a local group of interbreeding members of a species. (We will discuss populations later in the chapter).

FIGURE 16.2

Domestic Variations.

The varied physical traits within a species, such as the domestic dog, fascinated Darwin, as well as dog breeders and modern geneticists. The traits of a golden Labrador retriever, a huge Scottish deerhound, a diminutive King Charles Cavalier spaniel, and a wirehaired mixed-breed terrier demonstrate the range of genetic variation in just one species.

SOURCES OF GENETIC VARIATION

Genetic variation can increase in a population by single-gene mutation, gene duplication, exon shuffling, and recombination.

Single-Gene Mutation The first source of genetic variation, a **single-gene mutation**, changes the DNA sequence of a single gene in a way that may or may not alter the gene's function [FIGURE 16.3A]. Some mutations are neutral; they leave the gene's function basically intact, neither harming nor helping the organism. Other mutations have favorable effects on the phenotype. Favorable mutations, for example, increased the size of a cheetah's heart over a long period of evolutionary time. In more recent history, favorable mutations increased the number of corn kernels a corn plant produces. But many mutations, are either harmful or lethal; they reduce, modify, or destroy the function of a gene necessary for survival.

Gene Duplication **Gene duplication** can produce new genes without destroying the original function [FIGURE 16.3B]. A chance error in DNA replication or recombination can result in two identical copies of a gene, arranged one after the other on the same DNA molecule. A mutation in one of these copies would not affect the other copy; thus, the nucleotide sequences and functions of the two copies can differ. One of the copies might maintain the original gene function, while the adjacent gene might mutate to a new form that encodes a different but related function. Favorable gene duplication has resulted, for example, in several different human hemoglobin genes. Each of these hemoglobin genes is active at a specific time in human development—the embryo, fetus, child, or adult. The function of each hemoglobin gene relates to the specific oxygen availability and use at each developmental stage [see FIGURES 10.5 and 10.21].

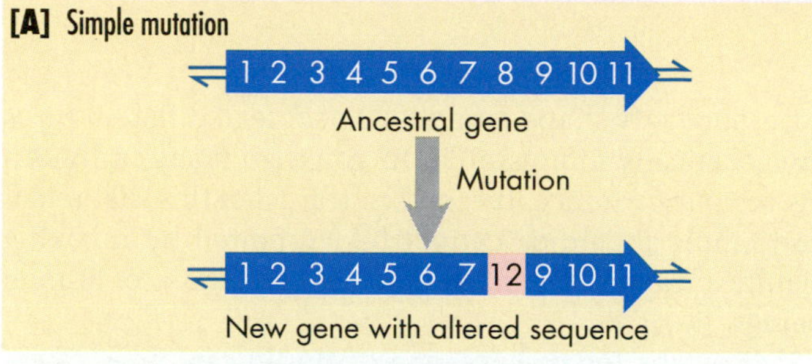

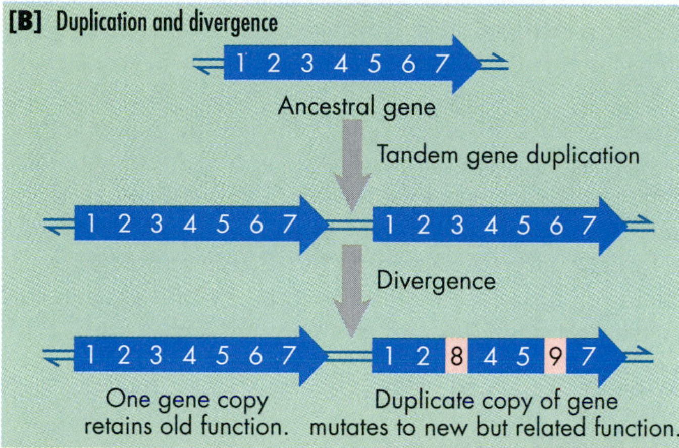

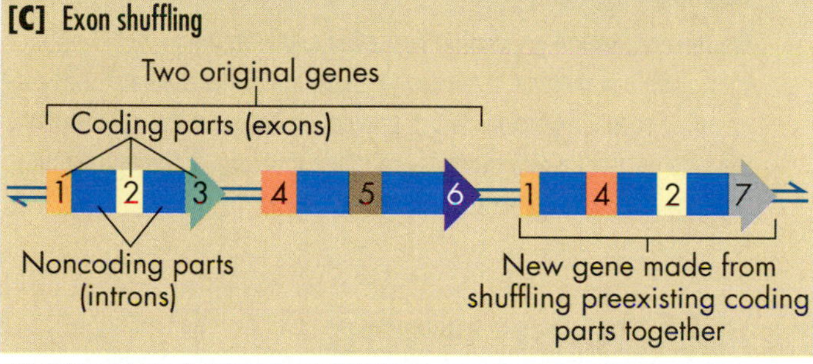

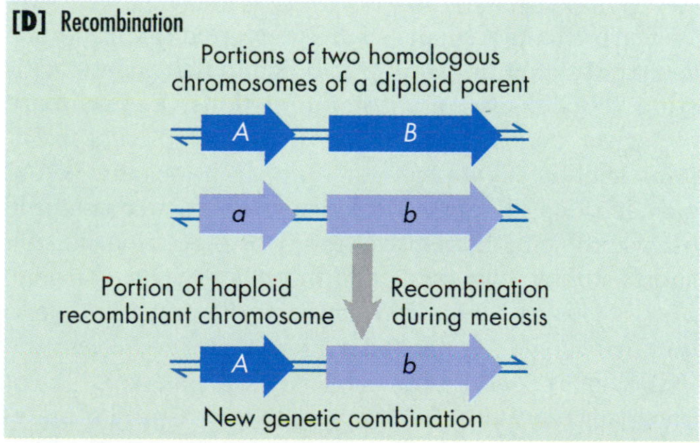

FIGURE 16.3

The Sources of Genetic Variation.

[A] In a simple mutation, a nucleotide (here labeled 8) in a gene can be replaced with an entirely different one (here labeled 12). [B] In duplication and divergence, an ancestral gene doubles, resulting in two copies of the gene in tandem on the same chromosome. One copy can remain unchanged (left side), while the other can diverge in sequence by simple mutation (bottom right) and perhaps eventually assume a new function. [C] In exon shuffling, separate pieces of a gene (here 1–6) get shuffled into a different set and sequence (1, 4, 2, 7) with a new function. [D] In recombination, random assortment and crossing-over lead to chromosomal changes and hence new genetic combinations.

Exon Shuffling **Exon shuffling** is a third source of new genes [FIGURE 16.3C]. Recall that eukaryotic genes often consist of *exons*, which can be expressed in proteins, and *introns*, which interrupt the exons and are not translated into protein [review FIGURE 10.22]. As a result of chromosome rearrangements that occur infrequently but spontaneously, an exon with one function can become positioned by chance next to a different exon. The new combination might evolve into a new gene encoding for a novel protein. For example, the gene for the protein that removes excess cholesterol from the blood, thus protecting against heart attacks, contains an exon from an immune-system gene, an exon from a growth-factor gene, and exons from genes for blood-clotting proteins.

Recombination Single-gene mutation, duplication, and exon shuffling can create new alleles of old genes, or new genes with new functions, thus increasing genetic variability. Once these processes produce a large number of alleles, **recombination** in sexually reproducing species can rearrange those alleles further, providing the fourth source of genetic variation [FIGURE 16.3D; review also FIGURE 7.20].

It is important to keep in mind that recombination shuffles existing alleles into new combinations, but does nothing to change the frequency (commonness) of alleles in a population's **gene pool** (all the genes in a population at any given time). By analogy, you could shuffle and deal the 52 cards in a deck into millions of new combinations (hands), but the deck would always have the same frequencies of cards—four aces, four queens, and so forth. Only if a mutation changed an ace into a queen, or if duplication produced two jacks of diamonds, for example, would the frequencies change.

EXTENT OF GENETIC VARIATION

You have seen how single-gene mutations, gene duplications, exon shuffling, and recombination can generate new genes and new gene combinations. Now we can pose and answer another question: How much genetic variation actually exists within a natural population?

Individuals who are heterozygous for any given trait have two different alleles for a certain gene, and often one allele is dominant and the other is recessive. This means that a large degree of genetic variation is *masked*, transmitted from generation to generation as hidden recessive alleles when paired with dominant alleles.

In an effort to probe the potential genetic diversity hidden by dominant alleles, molecular biologists developed simple yet sensitive techniques, including gel electrophoresis [see FIGURE 12.6]. These tools enabled them to detect many alleles that would otherwise be hidden, and revealed a great amount of genetic variation in most

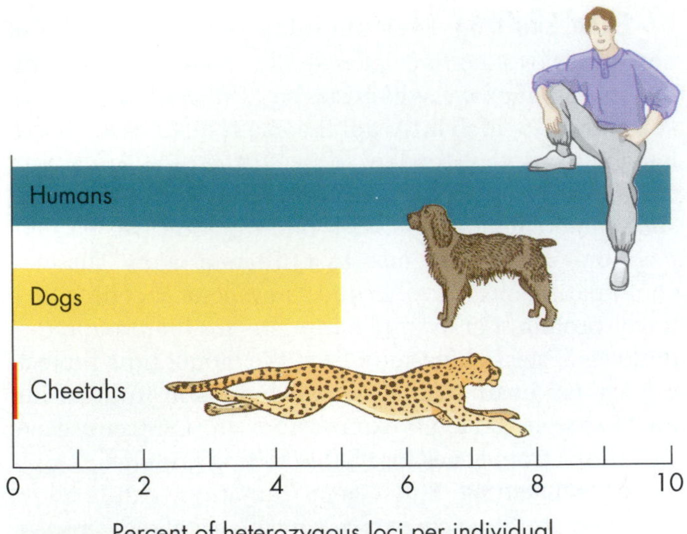

FIGURE 16.4

Humans and Dogs Have More Genetic Variation than Cheetahs. Each person is heterozygous (has two different alleles) for 10 percent of their 100,000 genes. Dogs are heterozygous for about 5 percent of their genes. By contrast, cheetahs are heterozygous for only about 0.07 percent of their genes.

populations. For example, the results of electrophoresis showed that each person is heterozygous for about 10 percent of their genes. In other words, for 10 percent of a person's genes, the allele the person inherited from the mother differs from the allele inherited from the father [FIGURE 16.4].

Humans have about 100,000 genes, and since each person is heterozygous for about 10 percent of their genes, each person has two different alleles for about 10,000 genes. When biologists compare the distribution of these heterozygous genes *within* a given racial group to the distribution among all racial groups, they find that about 90 percent of the genetic variation that exists in the human species for traits like height, weight, and the amino acid sequences of individual proteins [FIGURE 16.5], exists among individuals of a single racial group. If a catastrophe destroyed everyone on earth except a few thousand Native Alaskans, Bora Bora Islanders, or African Bantus, about 85 percent of all presently existing genetic variation would be represented in the survivors.

In contrast, only about 10 percent of the total genetic variation accounts for all the racial differences between European, African, Indian, East Asian, New World, and Oceanic peoples [see FIGURE 16.5]. Just 2 to 5 of our 100,000 genes are responsible for the differences in skin color of Africans and Europeans. In other words, except for superficial traits like skin color, facial features, hair type, and body shape, numerous people of different races are genetically more similar to you than many individuals of your own race. It is quite clear from this that racial and ethnic prejudice cannot be supported by invoking significant genetic demarcations among groups of human beings.

Cheetahs today have one of the lowest rates of genetic diversity among mammals. Cheetahs are heterozygous for only 0.07 percent of their genes [see FIGURE 16.4]. Thus, they have about 100 times less genetic variability than humans do, even less than certain highly inbred strains of livestock. This genetic uniformity may help to propel cheetahs toward extinction. For example, a single new virus to which all cheetahs were vulnerable could wipe out their entire population. Furthermore, since genetic variation is a precondition for evolution, cheetahs, with their small amount of genetic diversity, have little chance for evolutionary change.

➤ CONCEPT CHALLENGE

The current cheetah population has little genetic variation, but Bengal tigers, with a population size roughly the same as the cheetah's, has considerable genetic variation. Which population has a greater capacity to adjust to environmental changes brought about largely through human activities in their range? Support your answer.

THE HARDY-WEINBERG PRINCIPLE

We have seen that agents such as mutation can introduce genetic variation into a population and that genetic variation is extensive in most populations. Evolutionary biologists, pondering these facts, wondered what maintains genetic diversity over time. In 1908, the young geneticist R. C. Punnett mulled over a particular example of this problem: Punnett was aware of a mutation causing short, stubby fingers, a condition known as *brachydactyly* [FIGURE 16.6]. He knew that the mutant allele is dominant over the allele for normal fingers. Since the mutation is dominant, why doesn't everyone have brachydactyly within a few generations after the mutation arises?

One day, Punnett posed the problem to his friend G. H. Hardy, a famous British mathematician. Hardy thought about it for a minute and then wrote two simple mathematical equations on a napkin to illustrate what he thought was happening genetically in the case of brachydactyly. To simplify the problem, Hardy assumed that there would be no significant outside influences on a population of organisms. Then he reasoned that this freedom from external pressures would have two consequences: (1) The frequencies of *alleles* in the population

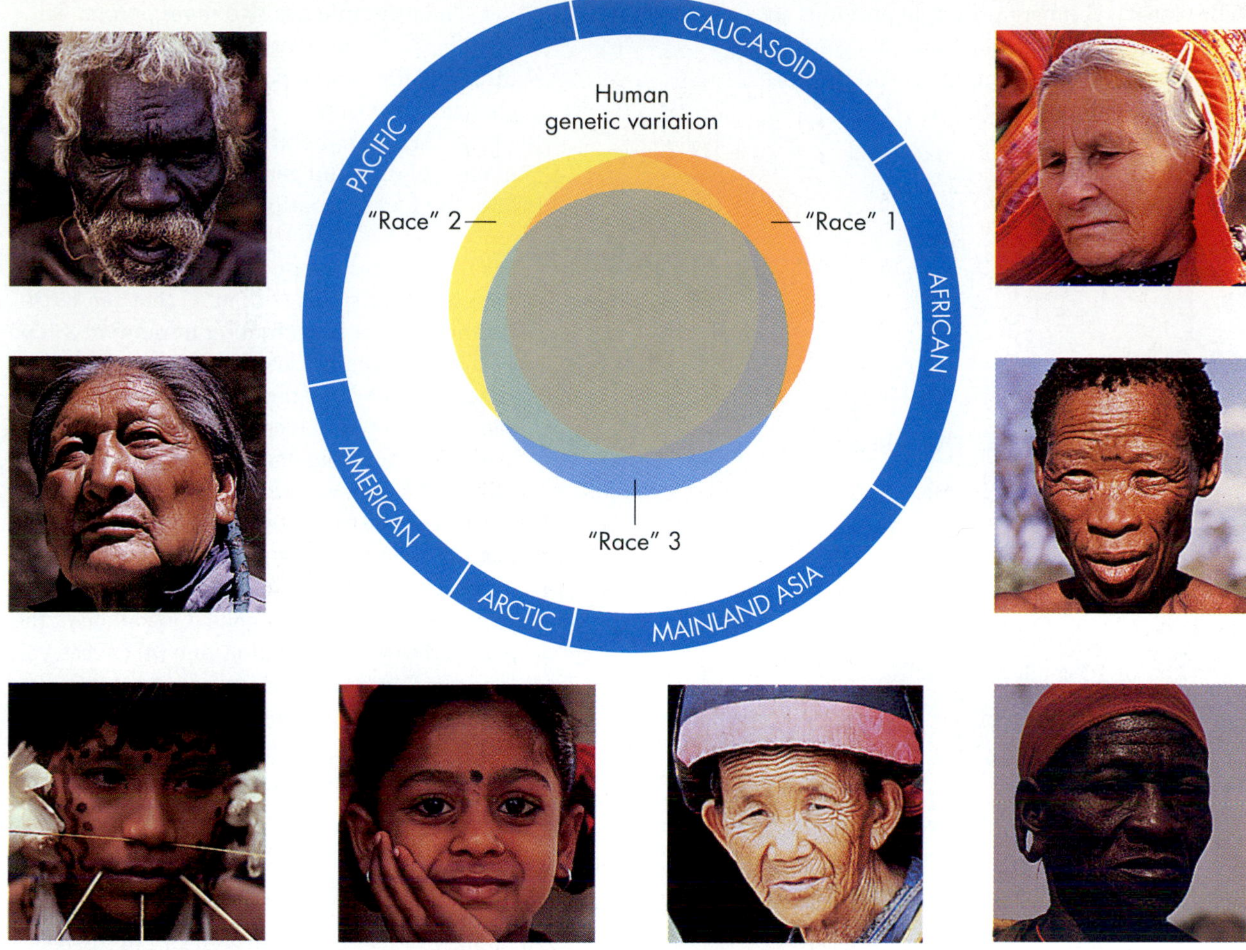

FIGURE 16.5

Genetic Variation and the Races of Humankind.

Genetic variation among individuals within each human race is far greater than the variation between races. Each circle represents the genetic variation encompassed by a human race. The variations are broadly overlapping, with very few genetic differences confined to a single race.

would not change over the generations, and (2) the relative frequency of various *genotypes* (heterozygotes and homozygotes) in the population would stay the same after the first generation. Biologists still call the mathematical model describing the genetic behavior of populations the **Hardy-Weinberg principle**, after Hardy and E. Weinberg, a German physician who discovered it independently [see BOX 16.1].

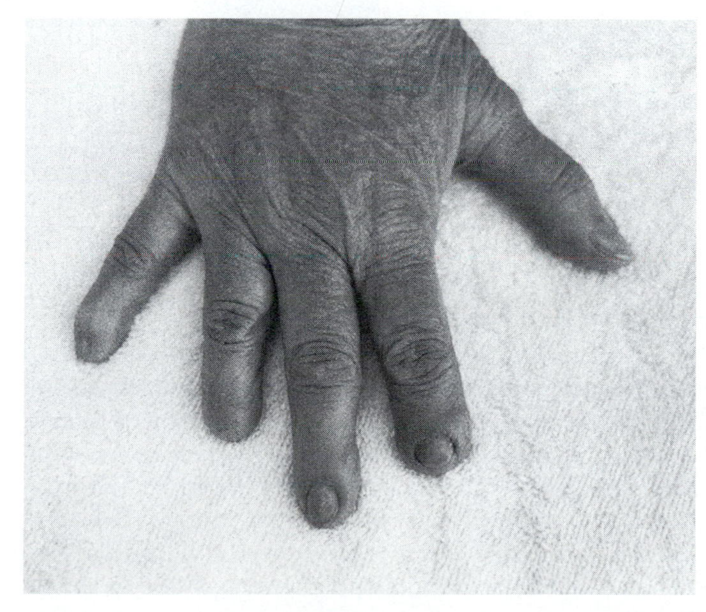

FIGURE 16.6

Are Dominant Alleles the Most Common Alleles in a Population?

The dominant allele of the brachydactyly gene causes the end bone of the fingers to be short or missing, while the recessive allele causes fingers of normal length, with three joints. Why don't most people have the dominant allele of this gene?

box 16.1 Biology Applied

The Hardy-Weinberg Principle

The Hardy-Weinberg principle provides an idealized standard against which a geneticist can compare a real population and thus detect evolutionary changes.

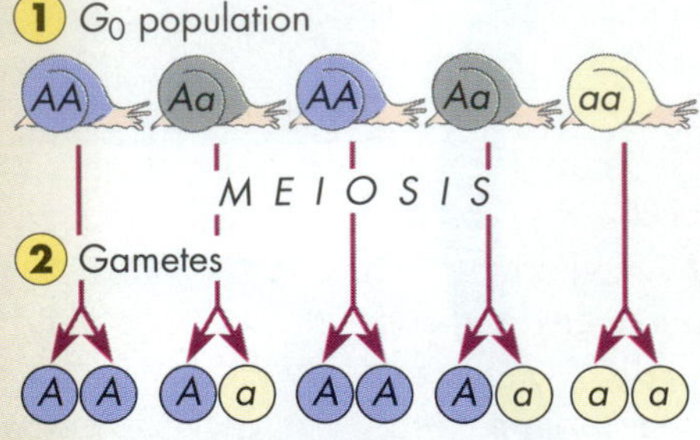

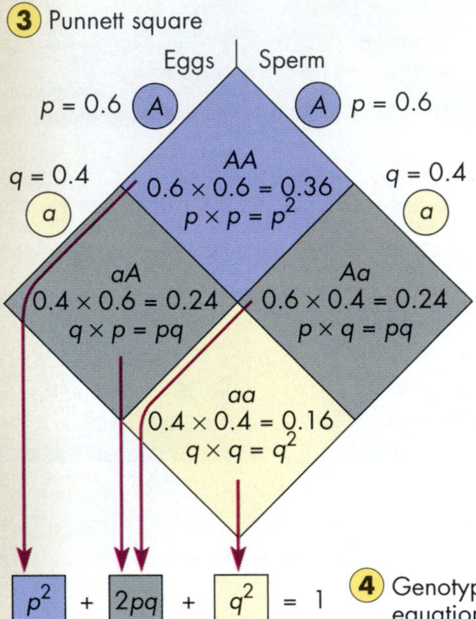

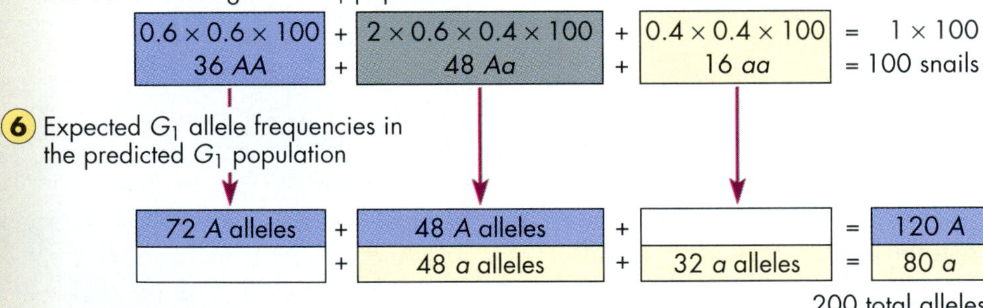

The principle has two main points:

1. If left undisturbed, the frequency of different alleles in a population remain unchanged over time.
2. With no disturbing factors, the frequency of different genotypes will not change after the first generation.

We understand the principle by considering a population of snails [FIGURE 1] that can fertilize themselves or each other at random [Step 1]. These snails possess a partially dominant gene for shell color that determines whether each animal will have a blue, yellow, or green shell. By analyzing changes in frequency of this color gene with the Hardy-Weinberg equations, we can determine whether the snail population is evolving.

Each of the five snails is a diploid with two copies of the color gene. One allele of the gene (*A*) causes blue color, one (*a*) causes yellow color, and the heterozygote (*Aa*) is green. In this population's gene pool, there are ten alleles: six *A* alleles and four *a* alleles. If the symbol p represents the fraction of *A* alleles, then $p = 6/10$, or 0.6. If the symbol q represents the fraction of *a* alleles, then $q = 4/10$, or 0.4.

Since the number of *A* alleles plus the number of *a* alleles represents all of the alleles of this gene in this snail population, $0.6 + 0.4 = 1$, or $p + q = 1$. This is the *allele pool equation.*

To see if allele frequencies change and evolution occurs, we must examine what

Punnett and other contemporary geneticists in the Mendelian tradition were puzzled by the brachydactyly problem because they confused the concepts of genotype and phenotype. They were more concerned with the number of stubby-finger phenotypes than with the frequency of the allele that caused the condition. This remains a common misconception among students today. For example, if a person homozygous for the dominant trait of brachydactyly (*BB*) marries a person with the recessive trait for normal fingers (*bb*), then all their first-generation offspring will be heterozygous (*Bb*) and all will show the dominant stubby-finger phenotype.

By considering only phenotype, we might conclude that the dominant trait (allele) has increased, and the re-happens when the snails reproduce, keeping in mind that alleles separate when egg and sperm are formed. Notice that despite meiosis, the frequency of A and a alleles is the same in the gametes as it was in the original population, 6/10 A and 4/10 a [Step 2].

Now what happens to allele frequencies at the time of fertilization? If we assume random mating, then we can write the frequencies into a Punnett square [see CHAPTER 8 and Step 3].

We can call $p^2 + 2pq + q^2 = 1$ the *genotype equation* [Step 4]. This equation says that the sum of the individuals with AA and Aa and aa genotypes adds up to the entire population (the 1 in the equation).

To determine whether evolution has occurred in the snail population, we must look for a change in allele frequencies between generations. If five snails of the parental G_0 generation produce 100 snails in the G_1 generation, then we can *expect* the genotype frequencies and resulting genotype numbers shown in Step 5, barring outside influences. (Thus, the genotype equation predicts the number of each of the three different genotypes in the population.) What, then, are the frequencies of alleles in this new generation? Have they changed?

You can see from Step 6 that the frequency of A alleles is 120/200, or 0.6, and the frequency of a is 80/200, or 0.4—the same as in the original generation. From this generation on, the allele and genotype frequencies will remain the same in the absence of outside influences, as you can prove for yourself by making another Punnett square and filling it in with the new data.

Applying the Hardy-Weinberg principle this way, we can predict that in populations without some outside influence, no evolution will occur over the generations. The Hardy-Weinberg equations provide a benchmark, a point of comparison for measuring any gene changes—or evolution—that might occur.

cessive trait (allele) has disappeared. But by considering the genotype, we can see that the parents ($BB \times bb$) have four alleles, half of which are B and half of which are b. Because the children's genotypes are all Bb (again, half B, half b), the allele frequency had not changed. This was Hardy's *first consequence*: allele frequencies remain constant in the absence of outside influences.

The *second consequence* begins to show up in the next generation. Suppose that two people from the first generation, each one possessing the genotype Bb, mated with each other. Their offspring (the second generation) would have the genotypes $1BB : 2Bb : 1bb$ and the phenotypes 3 short fingers to 1 normal finger as predicted by Mendelian genetics. Hardy's equations [BOX 16.1] show that this genotype ratio would remain stable throughout further generations as long as outside perturbing influences remained absent. Biologists say that a population is in **Hardy-Weinberg equilibrium** when it has both stable allele frequencies and stable genotype frequencies over many generations due to the absence of external pressure.

How Is the Hardy-Weinberg Principle Used? The Hardy-Weinberg principle is a useful model for predicting allele frequencies in populations, and therefore for determining whether or not a population is evolving. The allele frequencies remain unchanged, and hence the population does not evolve as long as the population is free of outside influences. What are those outside influences? Five conditions must hold for allele frequencies to remain constant over generations and for the Hardy-Weinberg principle to predict the distribution of genotypes in a population: (1) no mutation, (2) no migration into or out of the population, (3) large population size, (4) random mating, and (5) no selection. As long as all five conditions are met, allele frequencies in a population will remain unchanged, and the Hardy-Weinberg principle accurately predicts these frequencies from one generation to the next.

Thus a population in equilibrium would show no net change. Populations *do* evolve, however, and in fact, one definition of evolution is changes in allele frequencies within a population over time. This is also called *microevolution* if the time scale is short from a geological perspective and the changes are occurring at the level of the population and species. The Hardy-Weinberg predictions provide a theoretical standard against which to compare all changes occurring in real populations that are interacting and evolving in real environments.

In the next section you will see how conditions in nature, which deviate from the theoretical conditions of the Hardy-Weinberg principle, affect populations. In so doing, you will also see the actual agents of evolution at work.

➤ CONCEPT CHALLENGE

Five percent of the individuals in a population of foxglove plants have dark magenta flowers and 95 percent have white flowers. Crosses reveal that the magenta allele is dominant over the white allele of this flower color gene. If you were to observe this population in 60 years, what changes in the relative proportions of individuals with the two flower colors would you expect? What assumptions did you make in your estimation?

The Agents of Evolution

Cheetahs and other living cats evolved several million years ago from an ancient population of cats with considerable genetic variation. But while such variation is the raw material of evolution, it could no more change an ancient cat into a cheetah, or an ancestral dog into a St. Bernard, than flour could make itself into a bran muffin. Something has to act on the raw materials first. What, then, are the *agents of evolution*? They are simply the outside factors that can perturb the Hardy-Weinberg equilibrium: mutation, migration, small population size, nonrandom mating, and selection. Let's see how they work.

MUTATION AS AN AGENT OF EVOLUTION

Mutation can alter allele frequencies within a population by changing one allele into a different allele. Consider, for example, the most common eye tumor in children, retinoblastoma, which in about 30 percent of all cases acts as if it were due to a newly arisen dominant mutation. With each new mutation, one normal allele is removed from the gene pool and replaced with the tumor-causing allele. The allele frequency changes by a tiny amount, and the Hardy-Weinberg equilibrium is disturbed slightly. The primary importance of mutation to evolution, however, is not this small change in allele frequency, but rather the new phenotype the new mutation may provide, and how it is acted on by natural selection.

GENE FLOW: MIGRATION AND ALLELE FREQUENCY

If individuals migrate from one population to an adjacent population, they may remove alleles from one group and introduce them into a second group. The change in allele frequencies due to immigration or emigration is called **gene flow**, and few populations are so isolated that they escape it entirely.

Interestingly, archaeologists have been able to apply the principles of gene flow and the Hardy-Weinberg model to questions about human history. A certain type of decorated beaker began to show up in archaeological digs at sites of human habitation dated about 9000 years ago. The artifacts seem to have been made originally in the Middle East, but over several thousand years, their production and use spread across the European continent. Archaeologists wondered whether it was the knowledge of how to make and decorate this new pottery that spread gradually, or whether it was the Middle Eastern farmers who invented the pottery style who themselves migrated slowly across the continent. Evidence suggests that the artful farmers moved their farms northwestward at an average of about 1 km (0.6 mi) per year, beginning about 9000 years ago. As they founded new farms, the migrating artisans intermarried with the local Europeans. The further the farmers migrated from their Middle Eastern homeland, the more they intermixed with the local populations. This migration produced a genetic gradient still evident in allele frequencies as one travels from southeast to northwest [FIGURE 16.7]. A single holdout tribe—the Basques living at the border of what is now Spain and France—resisted intermarriage with the migrating Middle Easterners, and today they alone retain the language and genetic makeup of the prehistoric Europeans.

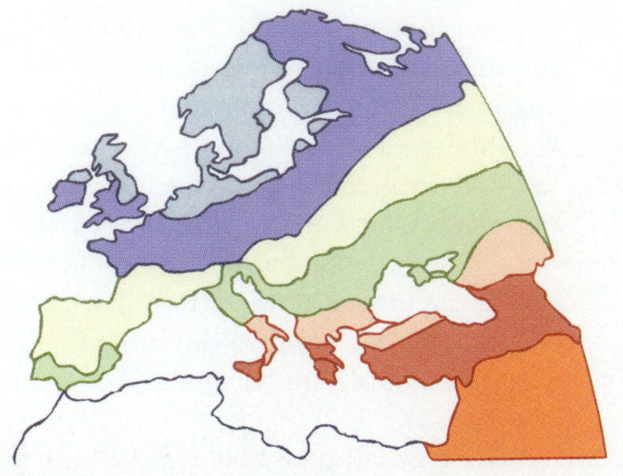

FIGURE 16.7

Spreading Culture and Human History in Europe.

People arising from regions of Europe shown in the same color are more genetically similar for a group of 95 different genes than they are to people living in other regions. The gradient of genetic similarity suggests a wave of migration across Europe from the south and west.

GENETIC DRIFT: CHANGES DUE TO CHANCE

The term **genetic drift** refers to chance, unpredictable fluctuations in allele frequency. While genetic drift can occur in populations of any size, we can see its effects most clearly in small populations. To understand why, suppose that an allele *r* exists in one-tenth of the individuals in both a large and a small population of cherry trees. If a chance occurrence like a severe spring ice storm struck a large population with 1 million cherry trees and half died, 500,000 would still survive. It is quite likely that one-tenth of the survivors, or about 50,000, would still bear the *r* allele. If, however, that same storm hit a small population of just 10 trees, only one of which had the *r* allele, the results could be very different. Let's say that of the 10 trees, 5 survived at random. There is just a 50 percent chance that the single tree bearing the *r* allele would be among the survivors. If none of the 5 bore this gene, the population would lose one allele completely, and the allele frequency would have changed in an unpredictable way.

Bottleneck Effect Let's look at how genetic drift can affect some real-world organisms in small populations. A natural disaster like our hypothetical storm in the cherry orchard can cause a type of genetic drift called a **bottleneck effect**. When a large, genetically varied population suddenly decreases in size, the gene pool might well change even if the population expands again after the decrease. This is because genetically, the few survivors were not necessarily a representative sample of the original large population.

Cheetahs probably experienced a population bottleneck due to disease, drought, or overhunting by humans about 10,000 years ago [FIGURE 16.8]. At that time, about 75 percent of existing large mammals became extinct, including mammoths, cave bears, and saber-toothed tigers. A bottleneck event such as this would probably have had two consequences for the cheetah. First, it could have reduced the genetic variation in cheetah populations to its present low level [review FIGURE 16.4]. Second, the small population of survivors might, by chance, have included alleles for susceptibility to disease and poor reproductive rates. These alleles could then have been passed down to virtually all the members of the current populations.

Because of human activity, more and more species are experiencing genetic bottlenecks like the one that has endangered the cheetah. The destruction of conifer forests in America's Pacific Northwest, tropical forests in Brazil and Borneo, and grasslands in North Africa are devastating the natural habitats on which many species depend. Some biologists, including Harvard ecologist E. O. Wilson, warn that the majority of species on earth will become extinct during the next 50 to 100 years—an extinction crisis of unprecedented proportions. To help protect endangered species, biologists apply the principles of population genetics and evolution to estimate the *minimum viable population*, the smallest population with enough genetic variability to ensure continued species survival. Armed with this information, conservation biologists can then help citizen groups and government agencies to set aside appropriate tracts of natural habitat to sustain populations above their minimum viable levels. This, in turn, can help prevent the chance effects of genetic drift within small populations from leading those groups toward extinction.

Founder Effect Besides its significance to population bottlenecks, genetic drift can also be involved in the **founder effect**, which takes place when a few individuals separate from a large population and establish a new one, and both populations continue to exist. Since the tiny group of founders bears only a small fraction of the alleles from the original large population, it may represent an atypical genetic sample. For example, the 14,000 Amish people who live in thriving communities in eastern Pennsylvania are all descendants of about 200 individuals who emigrated from Switzerland beginning about 1720. One of these founders—either Mr. or Mrs. Samuel King, who emigrated in 1744—had a recessive allele that, when homozygous, causes a number of dis-

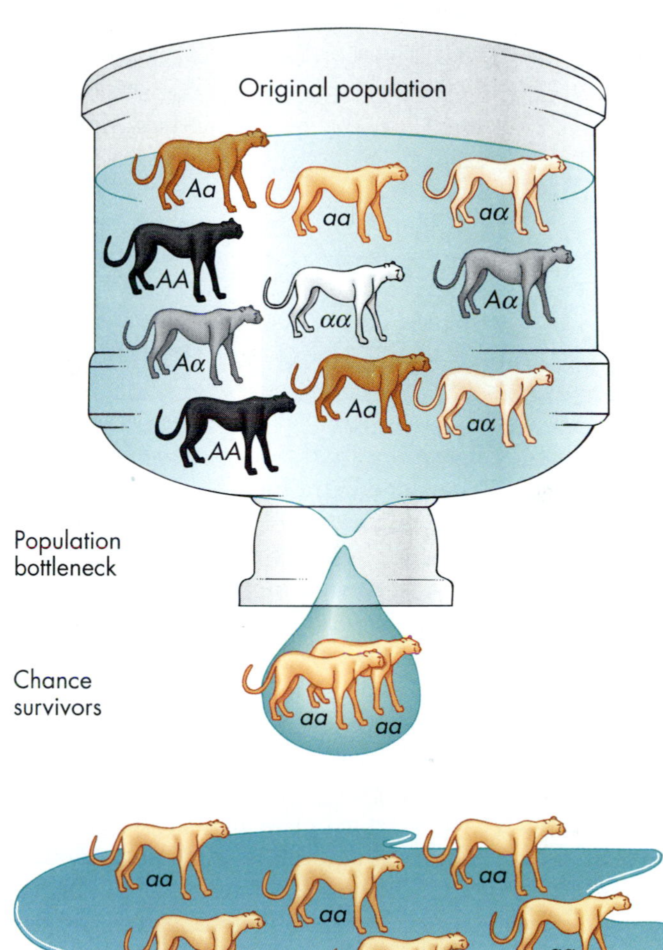

FIGURE 16.8
Bottleneck Effect.
A dramatic reduction in population size leads to reduced genetic variability. Ancestral cheetahs probably had normal levels of genetic variability, represented in the bottle by the three alleles *A*, *a*, and *alpha*, and the six genotypes *AA*, *Aa*, *A alpha*, *a alpha*, and *alpha alpha*. The few individuals surviving after some apparent catastrophe about 10,000 years ago contained just a fraction of the original population's genetic diversity. The survivors of this bottleneck effect established a lineage with greatly reduced genetic variation.

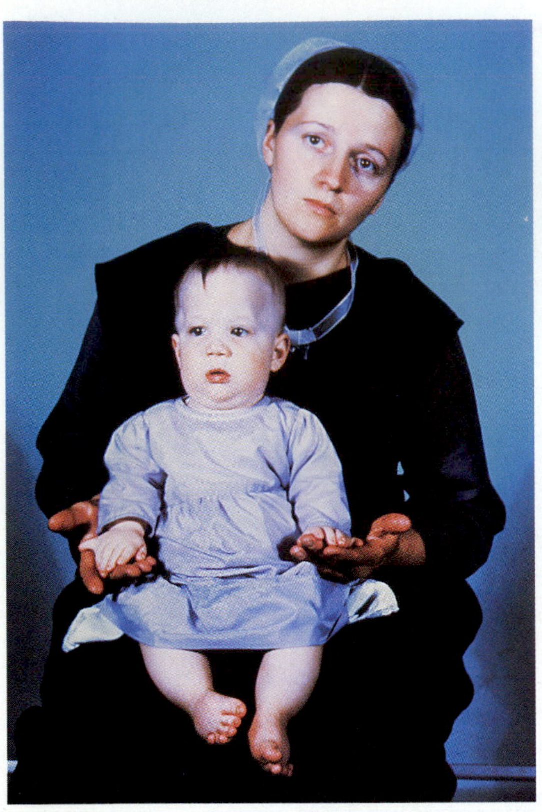

FIGURE 16.9

The Founder Effect.

One of the original members of the Amish community in Pennsylvania happened by chance to carry a recessive allele for a rare kind of dwarfism that causes shortened forearms and lower legs, an altered upper lip, and, often, extra fingers and toes. Note the six fingers on this child's right hand. Since there were few founders, this allele occurs more frequently among the Amish than in the general population.

FIGURE 16.10

The Javan Rhinoceros.

The effective population size for this forest dweller in Indonesia is approaching the minimum for continuation of the species.

tinctive changes, including short forearms and lower legs as well as extra fingers and extra toes [FIGURE 16.9]. As a result, in the Amish living around Lancaster, Pennsylvania, the frequency of this allele is 1 per 14, instead of the 1 per 1000 found in other populations. Later in this chapter, you will see how the founder effect can help give rise to entirely new species, proving how important genetic drift is to the process of evolution.

NONRANDOM MATING

The processes we've investigated so far—mutation, gene flow, and genetic drift—alter allele frequencies. Another process, a population's mating tendencies, can alter the genotype frequencies—the ratio of heterozygotes and homozygotes predicted by the Hardy-Weinberg principle. In making his analysis, Hardy assumed that every individual in a population has an equal chance of mating with any other member of that population. Often, however, it doesn't turn out that way.

A common type of nonrandom mating is **inbreeding**, a situation in which relatives are more likely to mate with each other than with unrelated individuals. In many species, individuals tend to mate with other individuals that were born nearby and thus may be related, closely or distantly. Since related individuals are more likely to have identical alleles than nonrelatives, inbreeding increases the frequency of homozygotes and decreases the frequency of heterozygotes. Inbreeding therefore alters the relative frequencies of genotypes, but not the frequencies of the various alleles. Since all living cheetahs are very closely related genetically, every mating that takes place today is an example of inbreeding. This can explain why so many cheetahs are homozygous for alleles that result in high percentages of defective sperm and in susceptibility to disease.

As the population of a species shrinks, biologists may attempt to save the species by applying the principles of population genetics to offset the effect of inbreeding. The rarest large mammal on earth is the Javan rhinoceros. Although this huge grazer once ranged throughout all of Southeast Asia [FIGURE 16.10], the current population has been whittled down to about 60 individuals. Is 60 enough to prevent the deleterious effects of inbreeding? Geneticists calculate that it would take a population of about 500 rhinos breeding randomly — a so-called **effective population** — to retain about 90 percent of the current genetic variation for the next century. If these calculations are correct, the current population seems much too small to ensure the continuance of the species. What should we do to preserve Javan rhinos? Should we capture the few remaining wild rhinos and try to breed them in captivity in ways that would maximize genetic diversity in the offspring? Or should our priority be more stringent protection of wild populations? Biologists and other citizens will face important and difficult choices like this with more and more species as human activities continue to precipitate extinctions.

We've seen that with inbreeding, a shift of genotype frequencies occurs away from Hardy-Weinberg equilibrium. This shift also occurs with assortative mating. In **assortative mating**, males of a particular type tend to mate with females of a particular type. Consider a human population in which short adults tended to mate only with other short adults, and tall adults with other tall adults. If height were controlled by two separate alleles, one for tall stature and one for short, then such nonrandom mating would lead to few heterozygotes—far fewer than predicted by the Hardy-Weinberg principle (the 2 *pq* of BOX 16.1). The proportion of the two alleles would remain the same, but the frequencies of the three genotypes in the population would change. If assortative mating of this type were complete—that is, if no tall people ever mated with short people—then the two phenotypes would constitute different species.

NATURAL SELECTION: HOW POPULATIONS BECOME BETTER ADAPTED TO THEIR ENVIRONMENTS

In natural selection, the environment plays the role of the breeder, selecting as parents those individuals who can survive, exploit available resources, and reproduce most effectively in their specific environment. An analogy to natural selection is **artificial selection**, wherein humans choose individuals with certain phenotypes to become the parents of succeeding generations. Swiss dog breeders, for example, selected and mated the largest, strongest animals for many consecutive generations to create the St. Bernard, a breed that historically has helped save skiers lost around Greater St. Bernard Pass in the Swiss Alps. Regardless of the agents of selection, an individual's ability to survive and reproduce in a particular environment relative to other members of its species is its **evolutionary fitness**.

Cheetahs, for example, may act as agents of selection upon gazelles. The big cats prey on the slowest individuals among the gazelles, leaving the more fit animals—those who can escape, exploit available resources, and reproduce most effectively in their specific environment.

Let's consider another example of natural selection, one of a very few that has actually been documented as it was in progress. Darwin noticed that the Galápagos Islands are home to several finch species, each with a distinctive body size, bill shape, choice of foods, and habits for food gathering. Recent studies have shown that even a single species of finch possesses significant genetic variability, the result, presumably, of mutation and recombination. Scientific observations of certain finch species show how natural conditions can act on this genetic variability to alter the populations. In 1977, the Galápagos Islands, an already dry and largely barren archipelago, suffered a severe drought during which plants produced few leaves and seeds [FIGURE 16.11]. For several years before and after the drought, a team of evolutionary biologists led by Peter Grant of Princeton University remained on the islands and leg-banded and monitored populations of a ground finch species, the medium ground finch, *Geospiza fortis*. The biologists determined that during 1977, the medium ground finch failed even to breed and 85 percent of the species' individuals disappeared. Both before and after the killing drought, Grant's team measured the sizes of the birds' bodies and beaks. They found that individuals with larger bodies and beaks had a much greater likelihood of surviving than smaller birds. In some ways, this runs counter to common sense because bigger birds need to find and eat more food than smaller birds. So how could it be true?

[A] Wet year

[B] Dry year

FIGURE 16.11

Natural Selection Observed in a Galápagos Ground Finch.

In a wet year **[A]**, small and large seeds provide plentiful food for birds (*Geospiza fortis*) with either large or small beaks. But in a dry year **[B]**, small seeds are more scarce than large seeds. Birds with beaks that are too small to crack large seeds usually die, leaving birds with larger beaks to be the parents of the next generation. The frequency of alleles causing larger beaks therefore increases due to natural selection.

The team's long-term evidence provided an explanation. During a normal, wetter year, herbs and grasses grow abundantly and produce small, soft seeds in great quantity. In such a year, only a few kinds of plants produce large seeds, and these grow in much more limited numbers. Apparently, birds eat the small seeds first, and then turn to the large seeds. Only the bigger birds with stouter beaks, however, can crack large seeds and devour the nutritious kernels. During a drought, there are few small seeds, so little birds, unable to make use of the larger seeds, are more likely to die, whereas bigger birds can utilize both types of seeds. Since experiments had revealed that in this finch species, heredity largely determines beak and body size, Grant was able to confirm changes in the frequency of the alleles that control these characters. He and his colleagues showed that climate fluctuations can act very quickly on an inheritable characteristic such as body size, leaving only certain individuals to become parents for the next generation. *This is the essence of natural selection.*

While all five agents of evolution (mutation, migration, genetic drift, nonrandom mating, and selection) can nudge a population away from Hardy-Weinberg equilibrium, evolutionary biologists have been particularly interested, in recent years, in determining what role each agent plays in *maintaining* genetic variation.

MAINTENANCE OF GENETIC VARIATION

We have seen in this chapter that aside from rare exceptions such as the cheetah, natural populations have a large amount of genetic variation. Some biologists, called *selectionists*, suggest that natural selection maintains the genetic diversity found in most populations. Selectionists contend that different alleles of a gene affect an individual's health, survival, or fertility differently, depending on the environment. They also believe that natural selection acts on these differences, and thereby leaves only certain individuals to become parents for the next generation.

Neutralists, on the other hand, claim that most variation isn't linked to an organism's survival and reproduction. Neutralists point out that even organisms living in very different environments generally accumulate mutations in a given gene at quite similar rates suggesting that many genes mutate at a given rate regardless of environmental factors. In each of the lineages that gave rise to humans and goldfish, for example, amino acids changed in a certain protein only once every 7 million years on average. This seemingly steady tempo of change has been called a **molecular clock** because it seems to allow a new, mutant allele to replace all the alleles of a given gene in the population over a constant but very long period of time. Neutralists suggest that most alleles are equivalent and that most mutations are neutral, conferring neither advantages nor disadvantages on an organism's fitness. If a mutation leads to a slightly different protein, that protein need not be identical to the original as long as it works about as well. Neutralists maintain that most genetic change is the result of neutral mutations, unpredictable genetic drift, and gene flow, while selectionists contend that genetic change is due to the various forms of natural selection.

Many biologists take a view that combines the best arguments of the selectionists with the soundest observations of the neutralists. The intermediate view is that even if many mutations are neutral, natural selection can still act on plenty of nonneutral variation in every living generation and thereby affect the course of evolution. The next section examines natural selection in action, focusing on how environmental pressures help bring about adaptive genetic change.

➤ CONCEPT CHALLENGE

For many generations, an isolated population of about 50 baboons has inhabited a seasonally dry gulley in Africa's arid Namibian Desert. Infant mortality is high, and the animals must go with very little water for many weeks at a time. What may be happening to the genetics and evolution of this population?

Natural Selection in Action

Question: What do cactuses, stone plants, poison ivy, and puffer fish have in common? Answer: All have adaptations that help prevent them from being eaten. The cactus has thorns; a stonelike appearance camouflages the stone plant [FIGURE 16.12A]; poison ivy leaves produce a toxic alcohol that poisons and repels most hungry animals; and a puffer fish enlarges itself beyond the grasping capacity of most predators. [FIGURE 16.12B] In each case, the harder-to-eat organisms—the genetically thornier, better camouflaged, more poisonous, and faster expanding individuals—survived to reproductive age more frequently and left more offspring. This exemplifies natural selection, the changing of allele frequencies by differential reproduction and survival.

Let's look at an example of how humans have become adapted by natural selection to live in particular environments, then examine various mechanisms by which natural selection can act.

[A]

FIGURE 16.12

Strategies to Avoid Being Eaten.

[A] The stone plant (*Lithops divergens*) is so well camouflaged that an animal is unlikely to find and consume it. [B] This guinea fowl puffer fish in Hawaii inflates itself, making it too difficult for most predators to eat.

[B]

SICKLE-CELL ANEMIA AND NATURAL SELECTION

The inherited human disease sickle-cell anemia is an excellent example of natural selection at work. In a person with this condition, red blood cells change shape, clog up capillaries, and are removed rapidly from circulation by the spleen, resulting in anemia (too few red blood cells) [review FIGURE 10.4]. Before the days of modern medicine, the clogged capillaries, enlarged spleen, and other symptoms usually killed victims of sickle-cell anemia before they reached reproductive age.

Affected red blood cells get their sickled shape because of an altered hemoglobin, the protein in red blood cells that transports oxygen. People with sickle-cell anemia have two copies of the sickle-cell allele for hemoglobin, while people with one normal allele and one sickle allele have *sickle-cell trait*, a usually harmless condition in which cells sickle only under extreme conditions.

In some areas of Africa, 40 percent of the alleles of this hemoglobin gene in the local human population are the sickle allele. Among African-Americans, the figure is 5 percent, and among European-Americans, it is only 0.1 percent. What could produce such a high frequency of a harmful allele in specific populations? Biologists began to understand this puzzle when they discovered that the sickle allele was most frequent in areas with a high incidence of malaria, an often fatal disease [FIGURE 16.13]. They later determined that heterozygotes with one normal and one sickle allele for hemoglobin are less likely to die of malaria than people with two copies of the normal allele. Individuals with one copy of each allele benefit from **heterozygote advantage**, a situation in which the heterozygote has a greater fitness than either homozygote. When malarial parasites cause a heterozygote's red blood cells to sickle, the spleen destroys the resident parasites along with the sickled red blood cells, and this limits the infection.

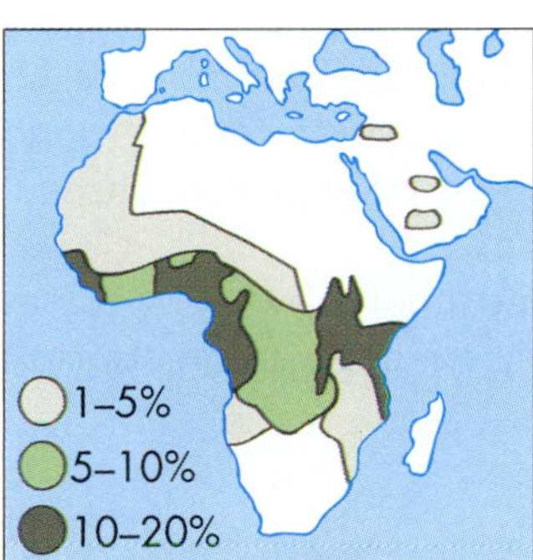

[A] Frequency of sickle-cell allele

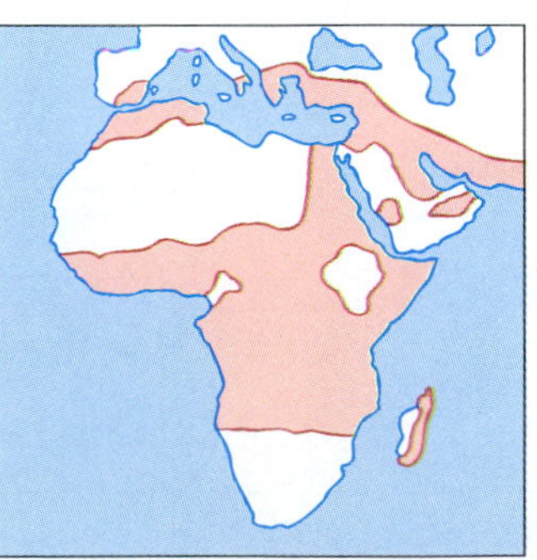
[B] Distribution of malaria

FIGURE 16.13

Sickle-cell Anemia and Malaria: Natural Selection in Action.

[A] The sickle-cell allele that encodes the beta chain of hemoglobin is found at surprisingly high frequencies in certain parts of Africa, the Mediterranean, and the Middle East. [B] The map highlights regions where the malarial parasite lives. The similar distribution of the sickle hemoglobin allele and malaria provided the first clue that malaria might be an agent of natural selection.

In malaria-ridden environments, people with two copies of the sickle allele for hemoglobin die of sickle-cell anemia, while people with two copies of the normal allele often die of malaria. In those same areas, however, heterozygotes, people with one copy of each allele, usually survive both conditions. These surviving heterozygotes reproduce and pass on sickle hemoglobin alleles to the next generation.

Natural selection thus caused the frequency of the sickle allele to increase in these populations. On the other hand, the allele did not increase in frequency in malaria-free regions, where it did not confer any advantage.

➤ CONCEPT CHALLENGE

Suppose that in a malaria-infested region of the globe, all children with sickle-cell anemia died, and all people with no sickle-cell genes died of malaria before reproducing. What genotypes would be left as parents for the next generation? What would be the expected frequency of the sickle allele for hemoglobin in this population? Could the frequency of the sickle hemoglobin allele ever increase beyond this level?

SOME MODES OF SELECTION

Natural selection can affect populations in three ways, which biologists refer to as *directional*, *stabilizing*, and *disruptive selection*.

Directional Selection **Directional selection** shifts a population toward one extreme form of a trait [FIGURE 16.14]. For example, when the first cheetahs appeared about 4 million years ago, they weighed more than twice as much as modern cheetahs. But over the years, light, fast-running animals reproduced more successfully, and thus natural selection favored alleles that pushed cheetah weight in one direction — downward.

Another classic case of directional selection involves a speckled insect called the English peppered moth. Ancient mutations resulted in two alleles for one of the moth's color genes: a dominant allele for black coloration and a recessive allele for pale gray. During the early 1800s, before the Industrial Revolution in England, genetically light-colored moths often rested on tree trunks that were gray because they were encrusted with lichens. These light-colored moths were so well camouflaged that they often escaped being eaten by birds [FIGURE 16.15]. After the Industrial Revolution began, soot darkened many tree trunks so that the light-colored moths were conspicuous and became easy targets for birds. Moths bearing the allele for black coloration, on the other hand, were well camouflaged on the sooty trees and survived in large numbers, contributing their alleles for dark color to succeeding generations. In urban areas, therefore, the frequency of alleles for the extreme trait of dark color increased. At the same time, the frequency of alleles for light color decreased, and the moth population continued to evolve.

Directional selection is also responsible for the resistance that insects and microorganisms acquire to pesticides and antibiotics. A new insecticide may kill all but a few of the insects in a population. The few survivors persist because, by chance, they already contain mutated genes that render them resistant to a certain concentration of the poison. These organisms pass on the resistance allele to their offspring, and eventually, a large population of resistant individuals arises. Continued random mutations and the application of higher concentrations of pesticides has brought about an escalation in resistance. Among the microorganisms that cause syphilis and gonorrhea, a similar process has resulted in resistance to ever-increasing concentrations of penicillin. It takes 10 times as much penicillin to fight these infections today as it did just 30 years ago.

Stabilizing Selection While directional selection results in the evolution of an extreme phenotype **stabiliz-**

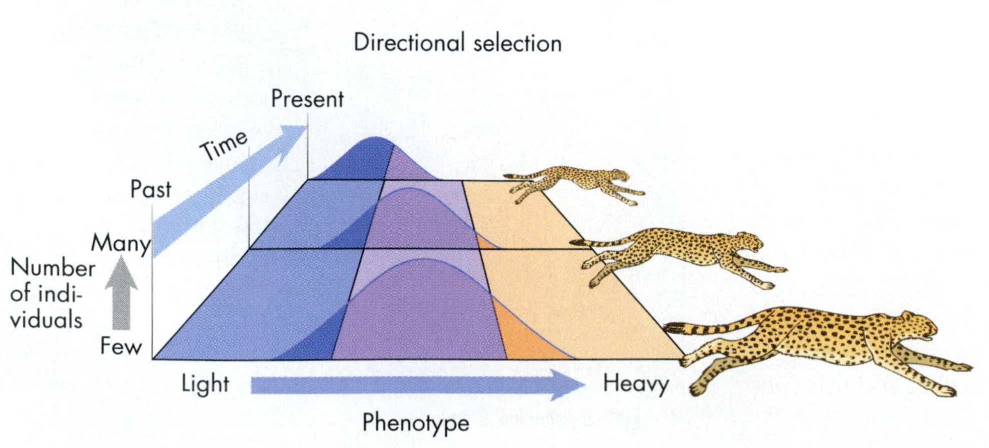

FIGURE 16.14
Directional Selection.
In the original cheetah population, individuals expressed a range of weights, but lighter cheetahs reproduced more successfully and passed to their offspring alleles causing lighter weight; thus the percentage of light cheetahs increased in the population.

[A]

[B]

FIGURE 16.15

The Peppered Moth: Directional Selection in Action.

[A] The light moth on this gray, lichen-covered tree nearly escapes detection, while the black form stands out plainly. [B] The light form of the moth is more conspicuous to humans—and birds—on a dark, soot-covered tree trunk. Experiments showed that birds preferentially eat poorly camouflaged moths. Selection was thus directional, toward the color of tree trunks in a given locale.

ing selection results in individuals with intermediate phenotypes [FIGURE 16.16A]. For example, far more human babies are born weighing about 3.2 kg (7 lb) than any other weight. Very heavy or very light babies have less chance of surviving, since heavy babies complicate delivery, while lightweight babies tend to be incompletely developed and less ready for life outside the womb. Height in adult humans is a similarly stabilized trait, with most people falling toward the center of the bell curve for height [FIGURE 16.16B]. Apparently in the past, very large or very small people left fewer progeny than those with intermediate heights.

Stabilizing selection can explain why certain organisms—so called living fossils—have persisted for millions of years with little or no outward change in form. Examples include coelacanths [see FIGURE 22.9A], which date back about 300 million years; scorpions, which have looked very similar for 350 million years; and redwood trees, which first appeared about 50 million years ago. Each organism probably arose in an environment that remained little changed for immense time spans, such as deep oceans or dense rain forests. Alleles causing a particular phenotype especially suited to the stable surroundings would have been selected for, while other alleles would have eventually disappeared.

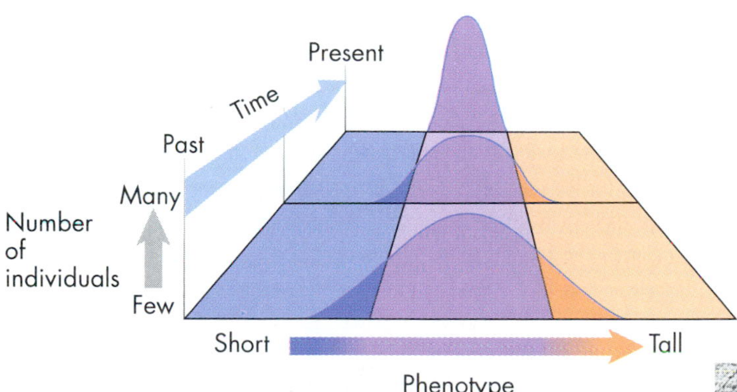

[A] Stabilizing selection

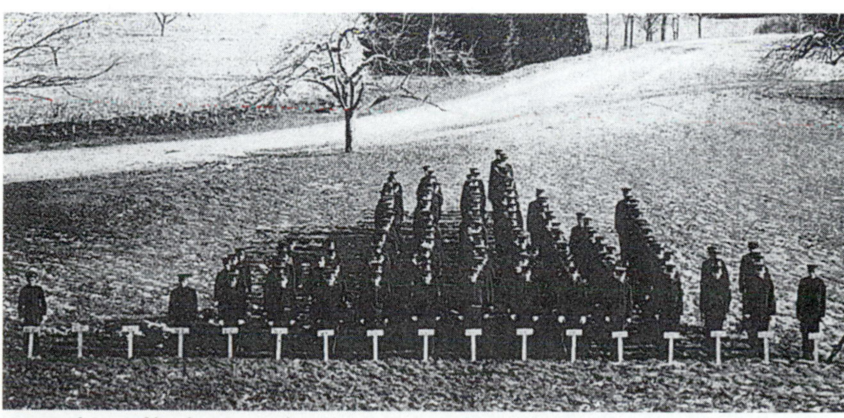
[B] Distribution of heights in a population

FIGURE 16.16

Stabilizing Selection.

[A] This form of natural selection favors the middle ground between extreme expression of a trait. [B] People are more likely to be of average height, and less likely to be very tall or very short, as illustrated by this historical photograph of cadets from a military academy. This distribution results because very short or very tall people have fewer children on average than people of medium height.

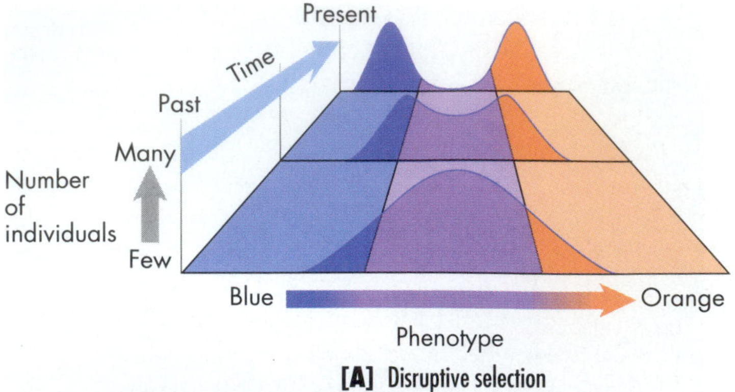

[A] Disruptive selection

FIGURE 16.17

Disruptive Selection.

[A] This form of natural selection favors two extreme expressions of a trait. For example, the butterfly *Pseudacraea eurytus* is found in two forms, one with blue fans [B], and the other with orange fans [C]. These two edible forms mimic two other foul-tasting species, the blue *Bematistes epaea* [D], and the orange *B. macaria* [E]. Because *P. eurytus* butterflies that are intermediate in color pattern mimic neither model very well, they are selected against by predators.

[B] **[D]**

[C] **[E]**

Disruptive Selection Whereas in stabilizing selection the two extreme forms of a trait become *less* frequent in the population, in **disruptive selection** the two extreme phenotypes become *more* frequent [FIGURE 16.17A]. In parts of Africa, females of the butterfly *Pseudacraea eurytus*, for example, occur in either a blue or an orange form [FIGURE 16.17B]. The blue form looks like one foul-tasting butterfly species (*Bematistes epaea*), while the orange form looks like another species (*B. macaria*) that also tastes offensive to birds [FIGURE 16.17C]. *P. eurytus* individuals that are mimics of either foul-tasting model species are likely to escape being eaten. In contrast, butterflies with the alleles that produce an appearance intermediate between the two extremes mimic neither of the bad-tasting species and are more likely to be eaten by predators. Thus they are less likely to serve as parents of the next generation, and alleles for the intermediate phenotype will decrease in frequency.

Directional, stabilizing, and disruptive selection can clearly change gene frequencies within populations, but are these changes sufficient to account for actual speciation—for the separation of a population into two separate species? Many kinds of evidence suggest that indeed they are.

➤ CONCEPT CHALLENGE

In humans, women of childbearing age, due to menstrual bleeding, have a much greater need for iron than men. At the same time, an excessive accumulation of iron in the body causes disease, and this occurs with a much greater frequency in men than in women. Can you propose a hypothesis for the ways natural selection might act on genes that promote the accumulation of iron in humans? Will natural selection act the same way on the fitness of men and women in this regard? Which mode of selection might be involved in each case: directional, stabilizing, or disruptive?

What Is a Species?

Every zoo-goer knows that cheetahs, tigers, and lions are three different kinds of cat: one is slender and spotted, one is powerful and striped, and one is large and tan with a dark mane in males. Biologists who have tested these three cats have also found that each has its own unique genetic makeup, and in the wild, each breeds and reproduces only with others of its kind. Thus, not only are these three kinds of cats phenotypically different, but they are also genetically different. These criteria place them unambiguously in three different species. There are no half tigers/half cheetahs or half cheetahs/half lions, and hybrids between tigers and lions are produced only in zoos, not in nature.

According to modern evolutionary theory, *species* are groups of populations that interbreed with each other in nature and produce healthy and fertile offspring. Each species, in short, is *reproductively isolated* from (fails to generate fertile progeny with) every other species in nature. Species that to a human observer look identical can live side by side and not interbreed. Conversely, if different-looking organisms—like the blue butterflies and orange butterflies of FIGURE 16.17—can interbreed in nature, they are considered *races*, or varieties, of a single species.

The definition of a species as a shared, reproductively isolated gene pool has been the most satisfactory to date, but even it has limitations. One drawback is that it applies only to sexually reproducing organisms. Asexual organisms, such as some bacteria that reproduce only by cell division, cannot be grouped in species according to this criterion. This definition also makes it harder to assign a fossil organism to a particular species because long-dead species can't be tested for reproductive isolation. In both cases, different species must be identified on the basis of morphologic, ecologic, evolutionary, or biochemical traits, rather than by the potential for interbreeding.

How Do New Species Arise?

To see how a species arises, we must first consider the causes of reproductive isolation. In other words, what factors keep separate species like lions and tigers or sets of identical-looking tropical frogs from interbreeding and producing fertile offspring in nature?

REPRODUCTIVE ISOLATING MECHANISMS

Any biological feature that prevents the members of one species from successfully breeding with those of another is considered a **reproductive isolating mechanism**. For example, dozens of frog species may inhabit the same area, but one species may mate only in ponds, another only in flowing streams, and a third in shallow pools and puddles. Or some may mate only in the spring and others only in the fall. Specific physical structures can also hinder mating. In many spider mites, for example, the male genitals are shaped like a "key" that opens the genital plates of females of their species; the "key" of one species cannot open the "lock" of another.

Sometimes, the sperm of one species can contact the eggs of another species, but reproductive isolating mechanisms prevent them from producing fertile offspring. For example, the gametes of two species can be so different that fertilization is impossible, as with sea urchin eggs that are surrounded by a specific chemical that binds only to sperm from the same species.

Sometimes, matings are partially successful and fertilization does take place, but the **hybrids**, the offspring of the two cross-breeding species, are not viable individuals. With many American frogs, genetic differences between species lead to abnormal hybrid zygotes and freakish individuals.

Occasionally, the cross-mating of two species produces extremely healthy offspring (a result known as *hybrid vigor*), but these offspring may themselves be nonreproductive (they experience *hybrid sterility*). Breeders produce mules [see BOX 8.1 on page 218] by crossing a male donkey (31 chromosomes in each sperm) with a female horse (32 chromosomes in each egg). Although mules (with their 63 chromosomes per somatic cell) are noted for strength and endurance, they cannot produce further offspring with other mules, with horses, or with donkeys because the odd-numbered chromosomes fail to pair normally during meiosis.

Since reproductive isolating mechanisms separate the gene pools of two related species, we can begin to understand **speciation**, the emergence of new species, by learning how reproductive isolating mechanisms develop.

THE ORIGIN OF NEW SPECIES

Most biologists believe that the majority of species arise after populations become geographically isolated and evolve in separate ways.

Allopatric Speciation A physical barrier, such as a river, a desert, different vegetation belts, or even a new highway or pipeline, can separate populations of a single species and prevent gene flow between the groups. The split populations slowly diverge as mutation, genetic drift, and adaptation cause different sets of characteristics to accumulate. Eventually, barriers to reproduction emerge and prevent matings even if, in the future, the two populations once again come into contact. This mechanism is called geographical, or **allopatric speciation** (*allo-* means "different," and *patric* is from the Greek *patra*, "native land") [FIGURE 16.18].

For example, about 7 million years ago, the Colorado

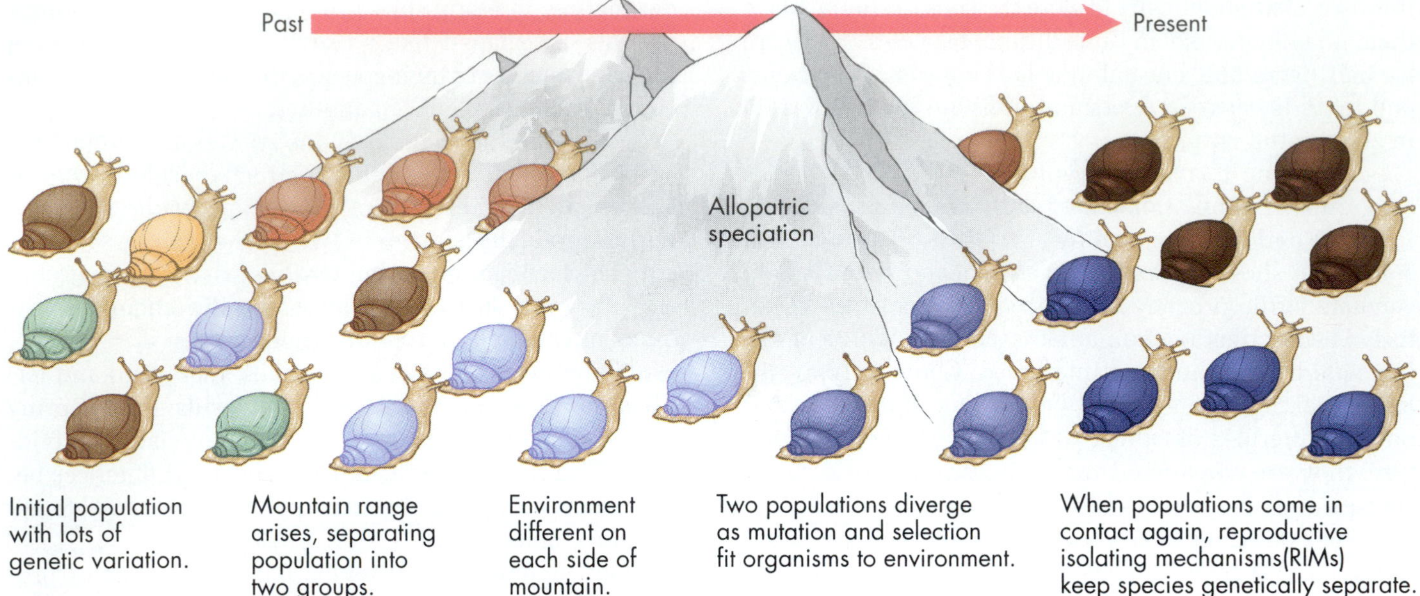

FIGURE 16.18

Geographical Barriers and Allopatric Speciation.

After a mountain range or other geographical feature separates an initial population (here, snails of various colors) into two noninterbreeding subpopulations, the processes of natural selection and genetic drift may genetically differentiate the two groups (as seen by their changing colors). If the mountain range erodes and the two groups eventually come into contact again, they may be so dissimilar genetically that they can no longer interbreed, and they will have become two different species (here brown and blue snails).

River began carving out the Grand Canyon in an area inhabited by one squirrel species. Today, the Kaibab squirrel occupies the north rim of the canyon, while the closely related Abert's squirrel inhabits the south rim. The two species most likely arose through allopatric speciation, gradually diverging after the canyon became an uncrossable barrier [see FIGURE 15.13].

The founder effect can bring about a second type of allopatric speciation. Sometimes, a few founding individuals migrate long distances and settle into a new geographical area. Since the founders can be few in number—even a single pregnant female can suffice—the absence of gene flow plus the effects of genetic drift and selection can result in rapid divergence. For example, a species of pygmy mammoths, averaging only 1 m (3.28 ft) in height, once lived on Santa Catalina Island off the coast of southern California. Biologists suspect that this species of diminutive pachyderms descended from a few 5-ton mammoths that reached the island some 50,000 years ago. Since there was no contact between the island mammoths and their mainland cousins, they split into two species by allopatric speciation.

Parapatric Speciation Sometimes, two populations of a species occupying separate but adjacent territories may interbreed at one edge of their range but still diverge anyway—a phenomenon called **parapatric speciation** (meaning "parallel native land"). For example, the hooded crow, a gray bird with a black head and tail (*Corvus corone*) inhabits western Europe, while the completely black carrion crow (*C. cornix*) inhabits eastern Europe. The two species meet in a line a few tens of miles wide running across central Europe, and they interbreed. Despite this intermixture, the two populations of crows have diverged sufficiently so that bird specialists consider them to be separate species.

Sympatric Speciation We've seen so far that allopatric speciation occurs when two populations occupy totally *separate* territories, and parapatric speciation occurs when two populations occupy territories that are *adjacent*. **Sympatric speciation** ("same native land") occurs when two species evolve from a single species without any geographical separation of the populations. Controversy surrounds most models of sympatric speciation, but many biologists agree that speciation can occur in just one generation as a result of *polyploidy* in plants, an increase in the number of chromosome sets in a cell. A good example of polyploidy is seen in a genus of pretty pink wildflowers called *Clarkia*, first collected in Idaho by members of the Lewis and Clark expedition [FIGURE 16.19]. *Clarkia concinna* has 14 chromosomes in a diploid set, while *C. virgata* has 10. If these two species mate, the hybrid progeny gets 7 haploid chromosomes from one parent and 5 from the other, making a total of 12. The diploid hybrid is sterile because the two different sets of chromosomes do not pair and distribute normally

[A]

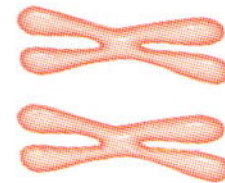
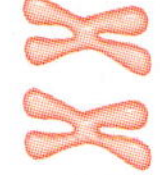

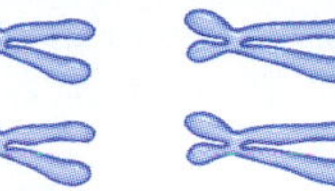

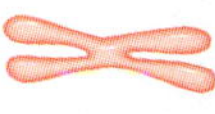

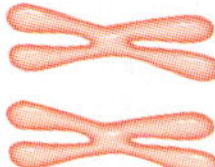
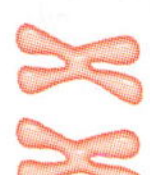
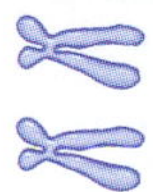
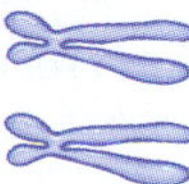

[B] *Clarkia* family tree

FIGURE 16.19

Genetic Barriers and Sympatric Speciation.

[A] The diploid western wildflower *Clarkia concinna* has two sets of seven chromosomes, while **[B]** the related diploid species *C. virgata* has two sets of five chromosomes. A diploid hybrid of these two species has one set of five chromosomes and one set of seven chromosomes and is sterile because of mismatched chromosomes during meiosis. But if the chromosome sets double in this diploid hybrid, then a tetraploid results with two sets of five chromosomes and two sets of seven chromosomes. Meiosis occurs normally in this tetraploid *C. pulchella*, because each chromosome can pair with a homologous chromosome. Since the newly formed tetraploid cannot successfully mate with either parental species, and since it arose from the parental populations without a geographical barrier, it represents a case of sympatric speciation.

during meiosis and gamete formation. At some time in the past, however, a tetraploid hybrid plant arose with four sets of haploid chromosomes. In this plant, the chromosome count had doubled from 12 to 24 (two sets of 7 and two sets of 5). In this tetraploid plant, every chromosome had a partner, and so meiosis could occur normally. The new tetraploid hybrid, called *Clarkia pulchella*, is self-fertile, but cannot produce fertile hybrids with either diploid species. A new species was therefore created in one step within one original population without a geographic barrier. Sympatric speciation like this is relatively common in plants, and has led to some of our most important crops, including wheat and cotton.

➤ CONCEPT CHALLENGE

Chihuahuas and Great Dane dogs do not interbreed. Why do we consider them to be members of the same species? How do the concepts of biological species, reproductive isolating mechanisms, and gene flow enter into the answer?

Connections

About ten thousand years ago, cheetahs were found in many parts of the globe. Today, however, they are limited to just two areas of Africa, with one population of interbreeding cats in each region. Populations like these constitute the basic units of evolution—the units in which allele frequencies change and evolution occurs. These changes in allele frequencies happen because some individuals survive and reproduce more effectively than others. Often this takes place by natural selection. Other genetic events occur by chance and have less predictable, but still powerful, consequences. These include migration, mutation, and even environmental catastrophes, such as those that forced cheetah populations through bottlenecks that diminished their genetic variability. Many of the species now alive are currently being forced into potential environmental and genetic catastrophes by the rapid expansion of the human population. By applying the principles of population genetics and by taking political action to preserve habitats, our species might be able to prevent the extinction of half the world's living species, which is predicted to occur in the near future.

We have dealt here with microevolution, evolutionary changes occurring within and between populations over geologically short periods of time leading to new species. Our next chapter explores the history of the earth and of life itself.

KEY TERMS

allopatric speciation, 403
bottleneck effect, 395
directional selection, 400
disruptive selection, 402
effective population, 396
exon shuffling, 389
founder effect, 395
gene flow, 394
gene pool, 389
genetic drift, 394
Hardy-Weinberg principle, 391
hybrid, 403
inbreeding, 396
molecular clock, 398
parapatric speciation, 404
population, 388
reproductive isolating mechanism, 403
speciation, 403
stabilizing selection, 400
sympatric speciation, 404

HIGHLIGHTS IN REVIEW

1 Genetic variation—hereditary differences between individuals—provides the diversity of appearances, capabilities, and behaviors from which the environment selects the parents for the next generation.

a] In most sexually reproducing organisms, genetic variation is extensive.

b] Genetic variation arises from new alleles, new genes, and new genetic combinations. The sources of these genetic novelties include mutation, gene duplication and divergence, exon shuffling, and recombination.

2 The frequencies of genes and alleles in a population tend to remain the same from generation to generation unless the population is disturbed. Perturbations can include mutation, migration, nonrandom mating, genetic drift, and natural selection.

a] Evolution is the change in allele frequencies in a population over generations.

b] The Hardy-Weinberg principle predicts that in a population, the frequency of alleles will remain unchanged over generations, and after two generations, the frequency of genotypes (homozygotes and heterozygotes) will not change. This principle holds only if there is no mutation, no migration, large population size, random mating, and no selection.

c] Unpredictable fluctuations in allele frequency can occur by chance; such changes are called genetic drift, and are more apparent in small populations. The bottleneck effect and the founder effect are examples of genetic drift.

d] Nonrandom mating, including inbreeding between relatives, and assortative mating between individuals with similar phenotypes, can alter genotype frequencies without altering allele frequencies.

3 New species arise when genetic barriers divide one population into two populations that do not interbreed in nature and that follow separate evolutionary pathways.

a] A sexually reproducing species is a group of populations that can interbreed and produce healthy, fertile offspring.

b] Reproductive isolating mechanisms may prevent crosses between species or make those crosses that do occur unsuccessful at producing healthy, fertile young.

c] In allopatric speciation, populations become geographically separated and then diverge along different evolutionary paths. Parapatric speciation occurs in populations that overlap at their edges. Sympatric speciation occurs in populations that overlap for most of their ranges.

UNDERSTANDING THE FACTS AND CONCEPTS

In Questions 1–5, match each of the descriptions with the most appropriate item or items from the following list.

a] mutation
b] gene duplication
c] exon shuffling
d] recombination
e] more than one of these apply

1 This results in changes in allele frequency in a population.

2 Several alleles for one gene can be produced by this process.

3 Several different human hemoglobin genes have been produced by this process.

4 Although new alleles are not created, new combinations may result in new phenotypes.

5 A new gene may be created by putting together functional parts of other genes by this process.

For Questions 6–10; match the descriptions with the most appropriate item from the following list.

a] population
b] gene pool
c] heterozygosity
d] gene flow
e] fitness

6 The relative ability to survive and reproduce in a particular environment.

7 Sum of all the alleles in a population.

8 Relative proportion of genes in a population for which there is more than one allele.

9 Change in allele frequency caused by individuals entering or leaving a population.

10 A group of interacting individuals of the same species inhabiting a circumscribed area.

For Questions 11–15, match the descriptions with the most appropriate item from the following list.

a] directional selection
b] stabilizing selection
c] disruptive selection
d] more than one of these
e] none of these

11 Occurs when a new mutation is neither harmful nor beneficial.

12 A gradual change of a species toward a more distinctly different phenotype.

13 Generally produces two new species from one parental species.

14 Unlikely to lead to the formation of a new species.

15 Often occurs when individuals are not highly adapted to a stable environment.

For Questions 16–20, match the descriptions with the most appropriate item from the following list.

a] allopatric speciation
b] parapatric speciation
c] sympatric speciation
d] more than one of these
e] none of these

16 Method of speciation begun by the founder effect.

17 Has resulted in speciation in plants within a single generation.

18 Reproductive isolating mechanisms that evolve by chance alone.

19 Strong reproductive isolating mechanisms that evolve at one end of a geographic range.

20 Speciation without the evolution of a reproductive isolating mechanism.

INTEGRATE AND APPLY WHAT YOU HAVE LEARNED

1 What aspects of cheetahs' heredity constitute the bases of their status as an endangered species?

2 The Hardy-Weinberg equilibrium is relevant under only such special conditions that it does not apply to most natural populations. What, then, is its value?

3 Explain what each of the terms in the Hardy-Weinberg equilibrium equation ($p^2 + 2pq + q^2 = 1$) represents. Which of these terms is usually most easily identified in a population by simple observation?

4 Contrast inbreeding and assortative mating.

5 The northern spotted owl is threatened with extinction in the Pacific Northwest. Some people argue that a minimum of several hundred reproducing individuals would be necessary in order to maintain a viable population. Preserving the habitat for this number of breeding pairs would prevent logging in much of the remaining old-growth forest. Other people argue that we can log the rest of the old growth while keeping a few breeding pairs of owls in captivity, where they could reproduce and avoid extinction. Compare the underlying assumptions of these two arguments, discussing any ethical principles that you see.

ANALYSIS

1 Review the material on sickle-cell anemia, pages 399 and 400, and then decide which of the following statements correctly follows from the information given, and which does not.

a] In certain areas of Africa the presence of malaria acts as an agent of natural selection.
b] In the United States, malaria acts as an agent of natural selection.
c] The degree of heterozygosity for the sickle-cell allele is greater in the United States than in the malarial areas of Africa.
d] If malaria were totally eradicated in Africa, the frequency of the sickle allele would decrease.
e] If malaria were eradicated, individuals who were heterozygous for the sickle allele would be less fit than either homozygote.

2 Review the study of finches in text pages 397 and 398. Suppose that Grant had found that the survival rates among the small birds was greater than among the larger birds. Which of the following statements would then be the most logical deduction?

a] During the drought, all of the different kinds of plants must have produced small seeds.
b] Body size in finches is not controlled by genes but rather by food supply. During the drought few birds had enough food to grow large.
c] The meager food supply that remained during the drought was equally suitable for large and small birds. Therefore the smaller birds, requiring less food, showed a better survival rate.
d] The large birds must have succumbed to disease.
e] Natural selection was not involved—some other force or factor was at work.

3 Refer to the puzzle of the pottery and the farmers that appears on text page 394. If knowledge of how to make this sort of pottery spread by means other than farmers migrating, which of the following would most likely have been observed with respect to the alleles studied? (More than one may be correct.)

a] A gradient of frequencies of Middle Eastern alleles, with a high point in western Europe.
b] A gradient of frequencies of Middle Eastern alleles, with a high point in the Middle East.
c] A loss of the alleles in all of the populations.
d] An isolated pocket of high frequency in the Middle East; low frequency in western Europe; no gradients.
e] A random scattering of frequencies between the Middle East and western Europe.

4 Which of the following best explains why genetic bottlenecks are the major cause of seriously endangered species?

a] The population size is reduced below effective size.
b] The chance survivors may not be representative of the original population in terms of variability.
c] The chance survivors may be more homozygous than their ancestors.
d] The chance survivors may harbor a higher frequency of harmful genes than their ancestors did.
e] All of the choices above are correct.

CHAPTER 17

Origin and History of Life

STROMATOLITES

Visitors to Hamelin Pool, a tiny inlet along the central part of Australia's western coast, can stand barefoot on the white sand and feel the warm waves of the Indian Ocean lapping against their toes. Observers can also see—and smell—acres and acres of rock mounds rising from the sand, each shaped like a knee-high, free-form footstool and covered with a greenish glaze that smells a bit rotten [FIGURE 17.1]. Biologists studying Hamelin Pool and a handful of other warm bays in other parts of the world have discovered that these slimy green mounds are the most enduring forms of life on our planet!

The rocky structures, called **stromatolites**, are like very stale, mineralized layer cakes. At the surface of each mound, pigmented microorganisms called **cyanobacteria** thrive in matlike communities, along with a few other kinds of cells. Cyanobacteria (formerly known as blue-green algae) are prokaryotes with simple nutritional requirements. They use energy from sunlight, carbon and nitrogen from the air, and water from their surroundings to produce their own organic compounds via photosynthesis. Their metabolic activities cause fine layers of minerals to be deposited on the stromatolites. It takes, on average, more than 50 years for a stromatolite to grow 1 in (2.54 cm) taller, and some of the mounds are 1000 years old. While this is quite elderly for a structure of biological origin, stromatolites have an even older heritage that gives them their great importance to evolutionary biology.

A few hundred miles inland from Hamelin Pool, a town with the unlikely name of North Pole sits baking in the red desert of Australia's western Outback. Near the town, geologists have found outcroppings more than 3.5 billion years old. These are

FIGURE 17.1
Today's Stromatolites: Evidence of Life's Early History. These pillow-shaped mounds were built by large colonies of cyanobacteria, photosynthetic prokaryotes formerly known as blue-green algae, in Hamelin Pool along Australia's west coast. Fossils of similar structures built 3.5 billion years ago may be evidence of some of earth's earliest cells.

FIGURE 17.2
Fossil Stromatolites. This South African reef of fossil stromatolites, strikingly similar to those living today, is about 2.3 billion years old. Similar fossils from North Pole, Australia, are 3.5 billion years old, and are among life's earliest traces on earth.

some of earth's oldest rocks. Embedded within the rocks, researchers found a set of layered mounds remarkably similar to the stromatolites of Hamelin Pool [FIGURE 17.2]. These mounds must be at least as old as the rock in which they are found. In sedimentary rocks near the layered fossils, researchers found mineralized structures that look nearly identical to the filamentous and spherical cyanobacteria that live today in stromatolites.

Scientists know from various kinds of evidence that earth formed about 4.5 billion years ago, and as we just saw, cells with shapes and growth patterns similar to some of today's cells had already appeared by 3.5 billion years ago. It is difficult to comprehend such vast reaches of time, but we can use a few comparisons to put the early cells into a time perspective. Depending on your age, your grandparents or great-grandparents may have been born 100 years ago—a long time by human standards. Now think of 10 times that long—a thousand years, and perhaps 40 or 50 generations. A thousand thousand years—a million—is still very recent in terms of a planet that is 4500 million, or 4.5 billion, years old. If we represent the time span from earth's formation to the present as a single year, then the earth was formed on January 1, cells had already evolved before March 25, and all of human history took place at the end of December.

MESSAGES

1 Earth's size, composition, and distance from the sun provided the physical and chemical factors necessary for the origin of life.

2 While experiments can show how life *might* have arisen according to natural laws, scientists may never show exactly how it *did* originate.

3 The evolution of life has depended on the evolution of earth's geology. Reciprocally, life has altered the atmosphere and surface of the planet.

The early cyanobacterial cells, so similar to those in Australia's Hamelin Pool, bring us to the chapter's central question: How did life evolve on earth? The origin of life is an exciting research area within the larger field of evolutionary biology. In such studies, scientists try to piece together evidence that reveals the conditions on the newly formed earth, what kinds of raw materials were present, and how these may have coalesced in the first living cells. This field also addresses the early evolution of single-celled and multicellular living things once the earliest primitive cells had appeared.

We will begin our overview of this interesting field by investigating how our planet formed, and what made it conducive to life. We will see that biologists do not know exactly how life originated. All the physical evidence for life's origins, however, is consistent with the universal laws of chemistry and physics in operation today, but acting in an environment that differed significantly from that of the modern earth. Finally, the chapter presents evidence that the histories of earth and life are inextricably intertwined. The chemical composition of the early earth determined the properties of the first cells, the biological activities of the early organisms caused massive changes in the earth's atmosphere and land, and geological changes in the earth's crust altered the pathway of life's evolution.

As we explore life's origins and history, we'll encounter much of interest besides stromatolites, including a world with belching volcanoes, flaming meteors, and a choking atmosphere. Let's dig in, now, to the scientific study of one of humankind's oldest questions: How did life begin? ❑

Formation of the Universe and Solar System

The exploration of space in the last few decades has provided tremendous new knowledge about the sun and nine planets of our solar system. The more we explore and learn, the more it appears that life on earth is the only life in the solar system. Why did living things arise, flourish, and survive only on this planet and not on the others?

Evidence suggests that the universe began with the so-called Big Bang, an event about 13 to 18 billion years ago, during which all matter in the universe, which had previously been concentrated into a single mass, was scattered in an enormous explosion [FIGURE 17.3]. About 6 billion years ago, a giant disk made up primarily of hydrogen and helium, the lightest elements, spun in our corner of the universe. Gravitational attraction among atoms led to an accumulation of matter at the disk's center. The resulting compaction created a new star, our sun, and increased its internal temperature by millions of degrees. Thermonuclear reactions ignited, which still continue to increase in intensity.

At various distances from the center of the spinning disk, planets condensed from the cold gas and interstellar dust orbiting far from the pale young sun. Earth, the third planet from the sun, formed about 4.5 billion years ago.

The Early Earth

Current evidence suggests that the newly formed earth was nothing but a huge, frigid ball of ice and rock, devoid of oceans and atmosphere. Several factors, however, began to warm the planet. Deep within the earth, radioactive elements [see page 33] were decaying and releasing energy, including heat. The enormous, crushing pressure caused by the force of gravity on the rock layers also tended to release heat deep in the earth. In addition, a huge meteor apparently struck the earth, and the impact heated the rock. (This event probably also created the moon from what is now the Pacific basin.) These factors heated the earth into a molten mass that did not cool for several hundred million years (mid-February on our time scale) [FIGURE 17.4A].

As the earth cooled, a thin crust of rock formed at the surface. Molten rock (lava) frequently erupted through the crust, and clouds of gases issued from the planet's interior. The earth's first atmosphere began to form from these vapor clouds; this atmosphere is thought to have consisted of carbon dioxide (CO_2), nitrogen (N_2), water vapor (H_2O), and hydrogen sulfide (H_2S), with traces of ammonia (NH_3), and methane (CH_4).

FIGURE 17.3

Formation of the Solar System.

Billions of years after the Big Bang, a monstrous cloud of gases and dust condensed to form the sun and planets of our solar system, including the earth. This figure shows a cosmic time frame, from the Big Bang to the first cells.

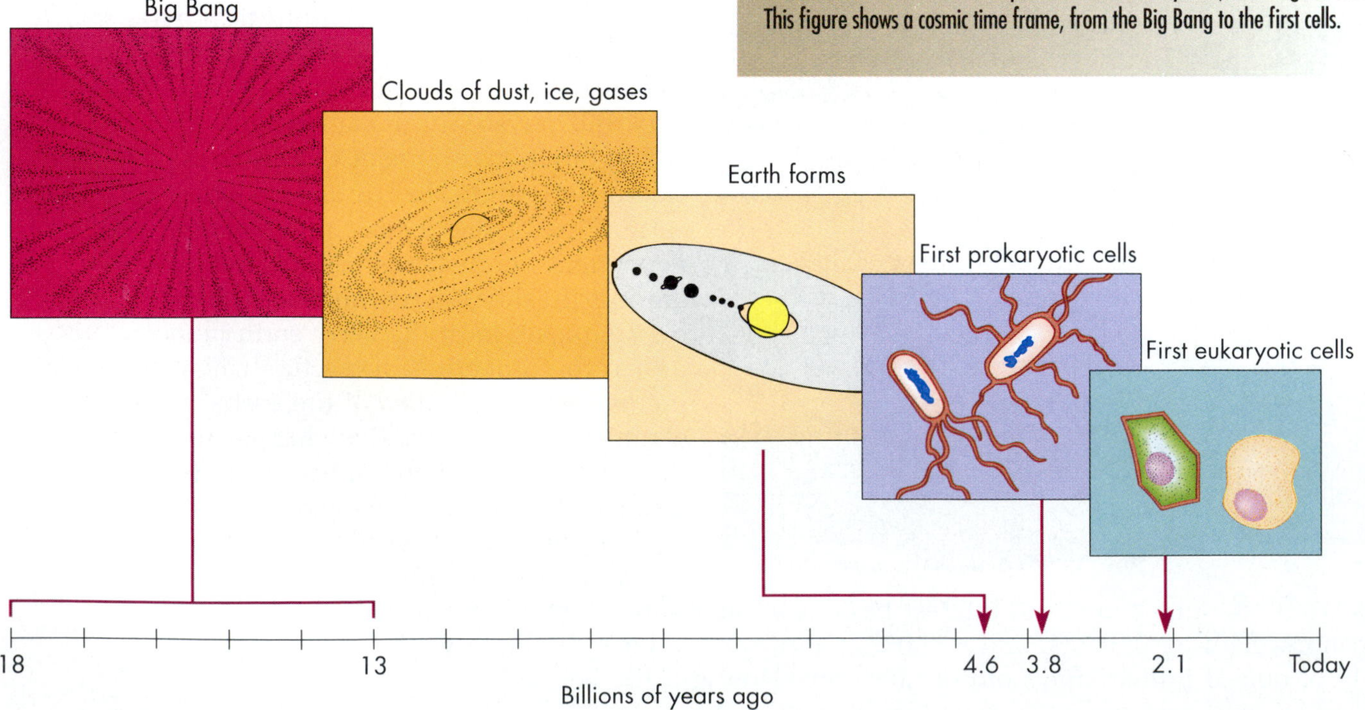

[A] Earth: 4 billion years ago

[B] Earth: 3.5 billion years ago

FIGURE 17.4

The Primordial Earth.

[A] A molten scene about 4 billion years ago as the forces of gravity and radioactive decay melted the initial ball of rock and ice and created a red inferno of lava that boiled and smoked for millions of years. **[B]** By about 3.5 billion years ago, the planet had cooled, and the landscape was nothing but somber shades of black and blue. (Artist's conceptions.)

As the earth's surface and surrounding atmosphere cooled, water vapor in the atmosphere condensed and fell to the surface as rain. The water then evaporated and fell again, eventually cooling the surface and creating a shallow ocean that covered the entire globe except for sharp volcanic cones that spewed out more lava and gases.

Because of the volcanic ash in the air, the sky was a much grayer blue than it is today; it was probably also dotted with clouds, and frequently streaked by flaming meteors and asteroids—small hunks of primordial rock hurtling in from space. The sun was about 25 percent weaker than it is now, and looked pale in the sky. It is in this watery environment, with its air free of molecular oxygen (O_2), that life is thought to have arisen [FIGURE 17.4B].

Life as we know it—based on carbon compounds and liquid water—could have arisen only on a planet with sufficient amounts of both. Mercury and Venus, the sun's closest satellites, are blisteringly hot, with daytime surface temperatures of about 325°C and 480°C (520°F and 900°F), respectively. Under such conditions, water occurs only as a gas, and all carbon compounds occur as inorganic gases. In contrast, Mars may once have had flowing water and living things, but now it is barren, with all of its water frozen and most of its carbon trapped in rocks. Of all the planets in our solar system, only earth had the combination of composition, geological activity, size, and distance from the sun necessary for life to arise and survive.

Because of the earth's size, its gravity held gases escaping from the molten core like an encircling blanket. This atmospheric blanket, in turn, trapped enough to the sun's heat energy to keep the earth's surface temperature moderate and to retain most of the liquid water. Our distance from the sun also plays a part in the relatively temperate conditions over much of the globe. All these factors contributed to the emergence of life on earth.

➤ CONCEPT CHALLENGE

If the earth's orbit were to shift suddenly toward that of Mars, what would be the likely effect on life on earth? How about toward Venus? Cite evidence to support your answers.

The Emergence of Life

We have seen that the earth's size and distance from the sun set up two necessary physical and chemical conditions for life: (1) temperatures at which water is liquid, and (2) an atmosphere containing the elements of living things, including hydrogen, oxygen, carbon, and nitrogen. (The oxygen was not the molecular oxygen, O_2, present in the modern atmosphere, but was combined with other elements.) But how did these elements become organized into living systems? Scientists investigating the origin of life have broken down this question into several smaller problems, namely: (1) When did life arise? (2) How did the organic molecules of life originate? (3) What were the first self-replicating systems? (4) What were the earliest cells like?

Although no firm consensus exists among biologists about the answers to these questions, scientists do share one underlying assumption: Life originated in a succession of many small steps, and each step was a likely event that followed the laws of chemistry and physics. The challenge to biologists is to identify those small, likely steps and to demonstrate how they could have occurred on the earth billions of years ago.

WHEN DID LIFE EMERGE?

Try as they will, researchers will probably never find direct physical evidence of the earliest forms of life. A record of the chemical events leading to life's emergence may well have been laid down in sediments that solidified into rocks about 4 billion years ago (mid-February on our time line). However, that fossilized sequence was almost certainly obliterated: giant meteors flamed into the earth's atmosphere with great frequency until about 3.8 billion years ago (March 1 on our time line) [FIGURE 17.5]. Every square kilometer of the earth's initial crust must have melted from the force of the meteor impacts, thereby destroying any living systems that might have evolved and any evidence of how they had emerged. You can observe direct evidence for such meteor bombardment by looking at the surface of the moon through a pair of binoculars or a telescope. The meteor showers that pelted earth also pounded the moon, leaving thousands of craters you can see on a clear night.

Living organisms must have evolved or reappeared* shortly after the intense meteor bombardment ceased because scientists have discovered fossils of simple cells in rocks about 3.6 billion years old (March 25 on our time line). These fossils are, in fact, the layered fossils of North Pole, Australia, that so closely resemble today's stromatolites. We will discuss the primitive cells that generate these rock mounds in more detail later. For now, it is enough to remember that the evolution of living things from nonliving materials must have taken place relatively quickly—within a few hundred million years at most—when our planet was young. By human standards, of course, a few hundred million years is an eternity. But viewed in another way, the evolution of cats, kangaroos, and humans from reptilian ancestors took about as long as it did for the earliest cells to emerge from lifeless molecules—a few weeks in terms of our yearlong analogy.

* Some biologists think that life may have arisen many times and in many places on earth.

The obvious question is: What happened during those first few hundred million years after the streams of meteors stopped falling? The first part of the answer seems to be that inorganic compounds formed the molecules of life [TABLE 17.1].

ORIGINS OF SIMPLE ORGANIC MOLECULES

There are two parts to the story of prebiotic ("before life") chemistry: (1) the origin of small molecular building blocks, such as simple sugars and amino acids, and (2) how these building blocks joined to make the large molecules (macromolecules) that characterize living organisms. In the 1950s, a 23-year-old chemistry student named Stanley Miller pondered the physical conditions on the early earth and the chemical constituents of earth's first atmosphere. Miller wondered if something in that setting could have produced the amino acids, sugars, and nitrogen-containing bases necessary for forming the proteins, carbohydrates, and nucleic acids in living things [FIGURE 17.6].

FIGURE 17.5

Meteor Crater Near Flagstaff, Arizona.

This crater, which is about 9.5 km (6 mi) wide, is just a few thousand years old. It is similar to craters created during the long period early in the planet's history when meteors constantly pounded earth's surface.

TABLE 17.1 Steps in the Evolution of Life

Event	Billions of Years Ago	Consequences
Full diversity of life forms present	0.5	Complete ozone screen; atmosphere same as today; 20% oxygen; large, active fishes, land plants
Shelled animals and early land plants appear	0.55	Diversity evident in fossil record; 10% oxygen in atmosphere
Many-celled organisms appear	0.67	Fossils and tracks made; oxygen and ozone accumulate in atmosphere to about 7%
First eukaryotic cells appear	2.1	Mitosis, meiosis, genetic recombination, and aerobic respiration occur; 2% oxygen in atmosphere
Oxygen-tolerating blue-green algae appear	2.0	Ozone screen begins to form; iron deposits appear on earth's surface
Strong evidence of photosynthetic organisms	2.8	Oxygen is given off into atmosphere, but still is less than 1%
Autotrophs, methane-generating bacteria, and sulfur bacteria appear; suggestive evidence of photosynthesis	3.6	Little change in atmosphere
The origin of life	3.8	Primordial atmosphere lacks oxygen

In a seminal experiment, Miller constructed an apparatus featuring a large glass beaker filled with what were thought to be the raw materials of the primitive atmosphere: water (H_2O), traces of ammonia (NH_3), and methane (CH_4) or carbon dioxide (CO_2). For about a week, Miller passed electrical discharges through this system to simulate lightning. Amazingly, the pinkish sludge that collected in the beaker contained large quantities of a few types of amino acids, plus a few other compounds, most with biological significance. In later experiments, Miller and others showed that by altering the conditions somewhat, they could also generate the bases

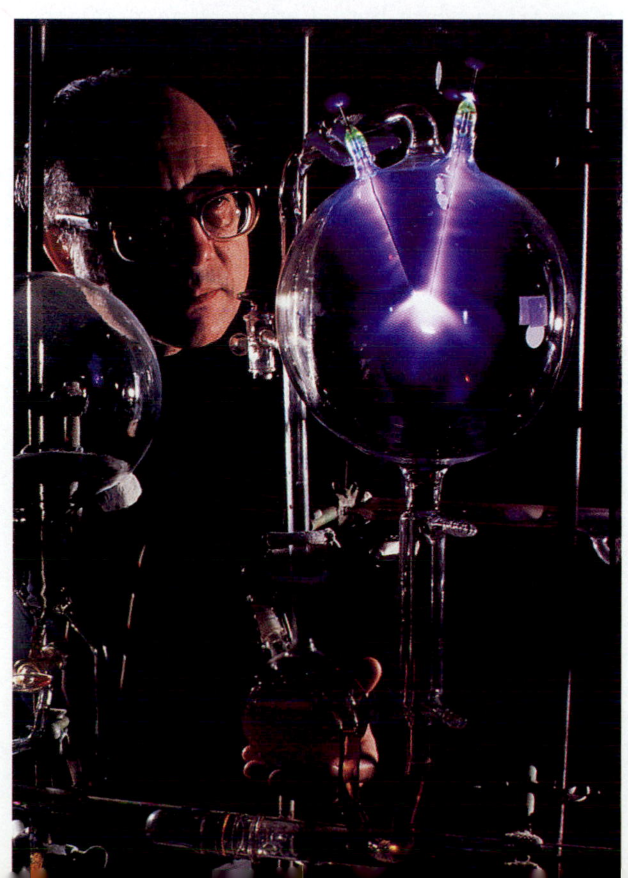

FIGURE 17.6

Biological Subunits Form in Simulations of Earth's Primordial Conditions.

Stanley L. Miller (pictured) and Harold C. Urey, working at the University of Chicago in 1955, designed a simple system to re-create the supposed conditions on the early earth.

and sugars that are components of nucleic acids. Although these substances did not readily join into nucleotides, the subunits of RNA and DNA, their presence made it clear that under the physical and chemical conditions thought to be present on the early earth, the formation of the organic building blocks of proteins and nucleic acids would have occurred spontaneously.

Somewhat surprisingly, scientists have found that outer space is another ready source of organic compounds. Astronomers have recently discovered enormous interstellar clouds containing organic molecules [FIGURE 17.7]. This finding suggests that some organic molecules are easily synthesized in the absence of living cells and in the vacuum of space. Still more evidence comes from a class of meteorites called *carbonaceous chondrites*. In 1969, one of these dense, charred meteorites fell in Australia. Chemists immediately examined it using reliable analytical methods that protected the specimen against contamination by earthly substances. Intriguingly, this object from space contained organic compounds that closely matched those in the pinkish sludge Miller produced in his electrical-discharge experiments. Analysis of the carbonaceous chondrite showed that it had broken off from a larger body, and that important biological molecules were synthesized on this parent body. This does not automatically prove that organic compounds formed on the early earth, but it does show that formation actually did take place in other parts of space and that, by the common laws of physics and chemistry governing the universe, the same production could have taken place here. Carbonaceous chondrites and other meteorites may even have carried substantial amounts of organic material to earth's surface during its long period of intense meteor bombardment.

FIGURE 17.7

Planetary Formations.

Clouds of organic molecules such as those aglow in the light of Orion nebula M42 take part in the formation of planets. Such organic molecules from space may have contributed organic materials for early life, although they were not of biological origin.

FORMATION OF MACROMOLECULES

Scientists attempting to reconstruct the events in the origin of life suggest that the next step after small organic molecules would have been macromolecules. They think that small molecules would have joined, forming larger polymers (proteins and nucleic acids). One obstacle to this hypothetical formation is that fact that water tends to destroy long organic molecules. Chemists have shown, however, that when they freeze solutions of amino acids or heat them to drive off water molecules, small polypeptides (chains of amino acids) form spontaneously. Perhaps a "primordial soup"—seawater containing organic precursors—collected in ancient tide pools, became concentrated via solar evaporation, and finally dried in the hot sun or froze at night. These conditions could have forced out water and provided the energy to form polypeptides and other macromolecules. Experimenters have also shown that under certain conditions, organic molecules can become trapped on the surfaces of clay particles. When molecules are arrayed on that substrate, light or heat energy can promote certain chemical reactions between them. Organic molecules could thus have accumulated on the surfaces of inorganic clays, or minerals such as iron pyrites (fool's gold). The destructive role of water could thus have been neutralized, and polymerization reactions could have taken place. The result would have been polypeptides, polynucleotides, polysaccharides, and perhaps lipids—the raw materials of life.

SELF-REPLICATING SYSTEMS

It's one thing to explain how small organic molecules might have accumulated and linked up into proteins and nucleic acids. It is another entirely to understand how these primitive organic chains could have begun making copies of themselves. Since self-replication is a fundamental aspect of life's chemistry, scientists needed an explanation for how it could have happened.

The Origin of Self-Replication Researchers studying the origins of life realized that describing the evolution of self-copying molecules is a chicken-or-egg problem. In today's cells, nucleic acids require protein enzymes for their replication, but proteins require nucleic acids for *their* synthesis. Which, then, came first, proteins or nucleic acids?

RNA Catalysts Many biologists are convinced that the first replicating systems were based on the nucleic acid RNA [FIGURE 17.8A]. Experiments conducted in the 1980s showed that certain types of RNA can act as catalysts. These RNA catalysts, or **ribozymes** can snip themselves in two, and they can join together nucleotides or RNA molecules [FIGURE 17.8B]. In more recent experiments, researchers have even shown that the ribosomal RNA in today's cells can catalyze the assembly of amino acids into protein [see BOX 17.1]. These studies suggest a stage in the evolution of life some have called the "RNA world." In it, RNA molecules would have catalyzed both their own replication and the joining of amino acids into protein.

While the RNA-world hypothesis seems attractive to many biologists, it does have weak points. First, RNA is a very unstable compound, and it would have degraded quickly at the high surface temperatures of the early earth. Second, while RNA molecules can replicate themselves, they don't do it very well in a laboratory setting simulating the earth's harsh early conditions. By adding certain types of energy-transfer molecules to the system, researchers can apparently improve RNA replication. For this reason, one Nobel Prize–winning biologist, Christian de Duve, suggests that sulfur-containing compounds called *thioesters* may have acted as such molecules.

Natural Selection of Self-Replicating Molecules For life to evolve, natural selection must have come into play. Indeed, little sets of self-replicating RNA molecules and proteins would have been subject to Darwinian selection and differential survival. Some of the daughter RNA and protein molecules would no doubt have developed mutations. Some of the mutated molecules might have been more stable or more easily copied than others, and these would have remained in existence longer and/or replicated faster. Such RNAs would then have accumulated more quickly than others in the primordial soup. In this way, over a great many years, RNA molecules would have become tuned by natural selection for stability and rapid replication.

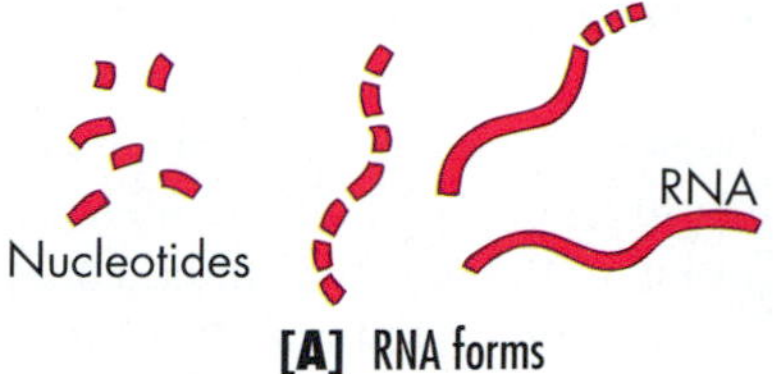

[A] RNA forms

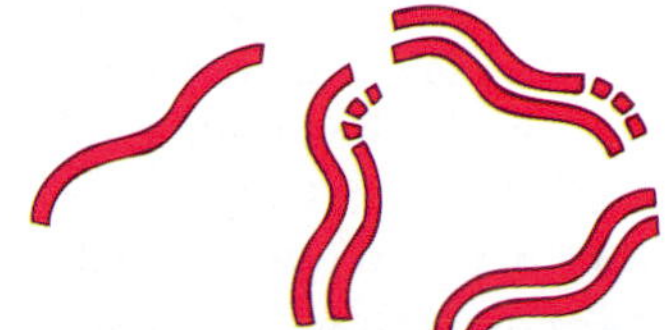

[B] Ribozymes catalyze RNA replication

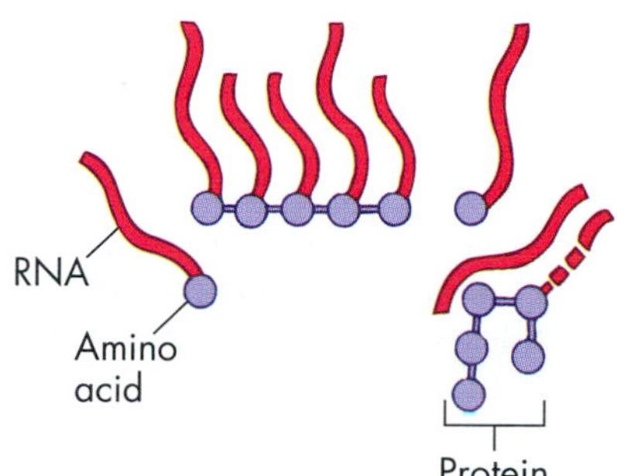

[C] RNA catalyzes protein synthesis

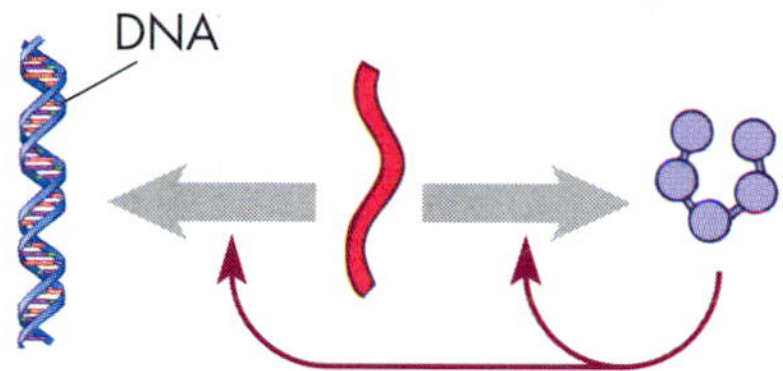

[D] RNA encodes both DNA and protein

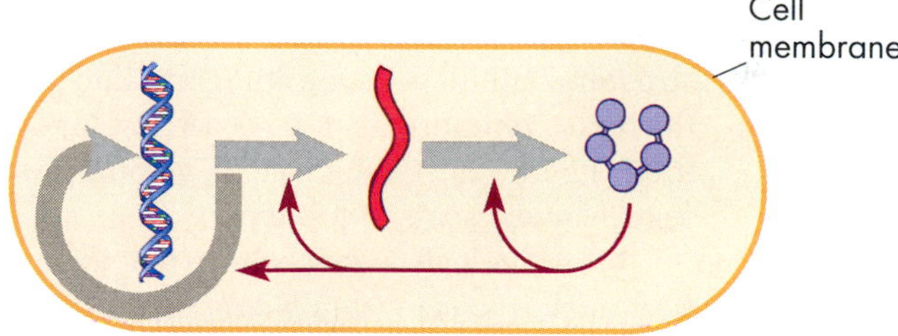

[E] Proteins catalyze cell activities

FIGURE 17.8

The Origins of Life in a Proposed RNA World.

[A] Organic subunits could have combined and formed RNA molecules. **[B]** RNA molecules could have acted as ribozymes, catalyzing their own replication. **[C]** RNA molecules could also have catalyzed the synthesis of protein, which in turn stabilized RNA molecules and catalyzed RNA replication. **[D]** DNAs could have been copied from RNA molecules, and at some point, proteins may have begun to catalyze the synthesis of more proteins from information in RNA. **[E]** DNA assumed an information storage role, while RNA continued to be involved in protein synthesis; a cell membrane also appeared.

The First Cells

Self-replication is a fundamental feature of life's biochemistry, and for this reason, the RNA world and its evolution are convincing ideas. But true cells, as you have seen throughout this book, have genes that encode proteins, are surrounded by a plasma membrane, and have complex metabolic pathways. How did all of these features evolve?

box 17.1
Biology Applied

Attempts at Artificial Life

Some observers call it "playing God." Some call it "irrelevant." Most simply think the pursuit of artificial life in the laboratory (AL) is interesting—if slightly abstract—research. AL experimenters themselves are in that last group, of course, and defend their work by pointing up the practical and theoretical offshoots already accumulating as they attempt to simulate life from scratch. Moreover, they feel that given enough time, a successful spontaneous self-organization in the laboratory will be like the origin of life on earth—inevitable.

Most of the AL experiments under way in dozens of teaching and research institutions around the world are so-called *dry labs*. Scientists create computer programs for cells, larger organisms, and communities of living things. These programs generate electronic species that undergo genetic change through mating and mutation—in other words, they evolve: whole lineages can emerge, evolve, and become extinct within minutes or hours. Nevertheless, many biologists consider dry labs like this to be irrelevant to origins of life research.

By contrast, many observers of AL experiments see *wet labs*, which simulate living structures and processes in the test tube, as more relevant than theoretical computer programs to origins of life research. Among the earliest AL wet-lab research efforts were the Urey-Miller experiments on the chemistry of the early atmosphere, and Sidney Fox's work on proteinoid microspheres, which attempted to recreate the possible structures of early protocells. Like these earlier experimenters, researchers now working on biochemical and structural simulations are still grappling with the basic conundrums of life's earliest history: Which came first, metabolism or replication? Nucleic acids or proteins?

Many biologists now believe that the earliest nucleic acids were RNAs; indeed, Thomas Cech's Nobel Prize–winning work showed that certain RNA molecules or ribozymes can carry out simple catalyses without the assistance of enzymatic proteins. For this reason, recent wet lab AL experiments with RNA have been getting a warm reception.

One group of Boston researchers recently figured out that most of the time, ribozyme (RNA) molecules floating about in the test tube are chopped into three pieces, each of which can act as a template for replication of new ribozyme segments. Significantly, these segments can also self-assemble into a complete RNA that can act like an enzyme. The three-piece-then-one-piece RNAs therefore carry out both DNA-like self-replication, and proteinlike catalytic activities. The workers have also tried surrounding the ribozyme pieces with little phospholipid bubbles (the main biochemicals in cell membranes; see CHAPTERS 2 and 3). These artificial protocells can "grow" by merging with other bubbles, and "divide" if squeezed through a fine mesh.

Other researchers have found that assembled ribozymes can catalyze two key reactions in protein synthesis [see CHAPTER 10]. These steps include the joining of amino acids to a growing polypeptide chain, and the step that comes before it during protein synthesis: the release of an amino acid from a transfer RNA [see FIGURE 10.13].

These findings topple the old view of RNAs in a cell's ribosomes as mere scaffolding material upon which protein enzymes carry out the steps of protein synthesis. They also bolster the view that RNAs existed on the primordial earth before cells, and were capable of self-replication, self-assembly, and catalysis of biochemical reactions. Such molecules would have obviated the chicken-or-egg problems of metabolism versus replication or DNA versus proteins.

A consensus view has emerged from the discussions and brainstorming of AL researchers in recent years: It is a conviction that spontaneous self-organization—the automatic linking of order and organization within a dynamic system—is not a rare and highly unlikely event that defies the second law of thermodynamics [see CHAPTER 2]. It is, instead, ubiquitous in nature, as evidenced by a variety of phenomena from snowflakes to organized weather patterns, sunspot cycles, periodic mass extinctions, and even the sudden spurts of evolutionary change called punctuated equilibrium [see CHAPTER 16]. Granted, this consensus is largely based on the same kinds of models, computer algorithms, and lab experiments that some observers find irrelevant. But AL researchers predict that within the next century, we will see lab-designed organisms surviving, reproducing, and evolving alongside natural living things.

Scientists don't really understand how the process of translation—the sequence of events that translates the genetic information in DNA into polypeptides—evolved on earth [FIGURE 17.8C]. The presence of ribozymes does not explain how messenger RNA evolved, or how stable, double-stranded DNA took over the job of storing genetic information from primordial RNA. But within every living cell lies evidence that these evolutionary steps did take place.

EVOLUTION OF MEMBRANES

The compartmentalization of self-replicating systems inside membranes was an important step in the origin of life. Let's consider the advantages of a plasma membrane and then think about how membranes may have formed.

The big advantage of membranes is that the concentrations of substances inside the membrane-bound compartment can be very different from those outside. Harmful substances can be excluded, and useful chemicals retained within instead of diffusing away. For example, consider a protein that can speed the replication of RNA molecules. If this protein was encoded by a membrane-enclosed RNA, it would stay close to that particular RNA molecule. But if the replication-enhancing protein was encoded by a *free* RNA molecule, it could diffuse away and enhance the replication of other RNA molecules rather than the RNA that encoded this useful protein.

While membranes are clearly advantageous, how could they have formed in the absence of life? Experiments show that membranelike sheets can form spontaneously when polypeptides and phospholipids are heated together [FIGURE 17.9]. The membrane-enclosed compartments can even divide under certain conditions. Substances in the surrounding medium can become trapped inside these membranes. If self-replicating RNAs became trapped in such a membrane, they might replicate more rapidly than unenclosed RNAs due to the advantages discussed above.

FIGURE 17.9

Cell-Like Lipid-Coated Droplets Can Form Spontaneously in the Laboratory.

Under specific conditions of temperature, moisture, and acidity, tiny spheres will form in solutions that contain amino acids and phospholipids. These lipid-coated droplets (magnification about 10,000×) form as the solution of phospholipids cools. Perhaps the earliest cell membranes formed spontaneously under similar circumstances.

ORIGIN OF METABOLIC PATHWAYS

Biologists also suggest that natural selection could have led to complex metabolic pathways within membrane-bound units. Initially, all materials needed for replication were present in the primordial soup. But some substances may have become depleted quickly. Any membrane-enclosed RNA that by chance encoded a protein that could change a more abundant material, say substance Y, into a less abundant material, say substance X, would have a selective advantage. Then when substance Y became depleted, another round of mutation and natural selection could result in an enzyme that might change substance Z into substance Y. In this way, metabolic pathways could have evolved; in this case, Z → Y → X. At some point, the coordinated activities—replication, protein synthesis, and metabolism—became so closely integrated that the membrane-bound compartments were indistinguishable from true biological units—living cells—with all the fundamental characteristics described in CHAPTER 1.

CHARACTERISTICS OF THE FIRST CELLS

Biologists will probably never find fossils of the earliest living cells because, as we have seen, the rocks that may have contained them almost certainly melted under a barrage of meteors billions of years ago. Nevertheless, biologists can piece together a fairly solid picture of these earliest cells by studying cellular genetics and certain very primitive living cells. Geneticists have discovered, for example, that DNA is the hereditary material in all surviving cells and that certain genes, including those for ribosomal RNA, occur in all modern organisms. By comparing the base sequence of these common genes and finding them to be startlingly similar, geneticists have become convinced that all organisms now alive must have descended from a single common ancestral group [FIGURE 17.10].

Given this information, it makes sense to look for a living organism with traits as close to the ancestral cells as we can find. Gene-sequencing studies show that certain archaebacteria found in hot, sulfurous pools in geothermal areas have evolved especially slowly, which means that they are as similar to the ancient progenitor cells as any we may ever find. A prime example is *Pyrodictium*, a dish-shaped cell that inhabits nearly boiling submarine sulfur springs off the Italian coast near a town called Volcano [FIGURE 17.11].

The traits of organisms like *Pyrodictium* suggest that the very first cells and the common ancestors of all life forms probably tolerated or required temperatures near the boiling point of water in order to grow. They probably used sulfur and hydrogen gas as energy sources (rather than light and water as in photosynthesis). As a carbon source, they probably used carbon dioxide, and they could survive only in oxygen-free (anaerobic) environments. In addition, they probably lived submerged in oceans, lakes, and hot springs, deep enough so that the overlying water screened out the highly energetic ultraviolet light that can destroy complex biological molecules.

If you think about the pathway from the formation of organic molecules to the probable first cells in their hot, sulfurous pools, the way was long and surely involved millions of small evolutionary steps. Nevertheless, each of these small steps had a high probability of occurring given the laws of chemistry and physics and the conditions existing on the earth 4 billion years ago. In fact, many biologists feel that once these rare conditions existed, life was bound to emerge. Cell biologist Christian de Duve expressed the sentiment this way: "The universe was pregnant with life."

Life Alters the Earth

Based on the evidence they have collected, biologists have suggested a cycle in which earlier organisms altered the physical conditions on earth and paved the way for later organisms [CONCEPT INTEGRATOR page 420]. Let's look at their evidence.

Geologists have discovered ancient rocks in southern Greenland that contain deposits of organic carbon—deposits that provide some of the earliest evidence of life. These rocks formed in ocean sediments about 3.8 billion years ago, shortly after meteorites stopped bombarding the earth and obliterating all traces of earlier events. Like today's biological molecules, these carbon deposits are enriched in the lighter isotope of the carbon atom (^{12}C) instead of the heavier carbon isotope (^{13}C) characteristic of inorganic deposits.

This isotope ratio makes sense if the deposits had a biological origin. Some scientists, however, still argue that the deposits had a purely geochemical origin on certain technical grounds. Although biologists do not know what kinds of living systems generated the carbon compounds, they were probably anaerobic cells, since the primordial atmosphere lacked oxygen. These earliest living systems were probably heterotrophs that obtained their raw materials from the carbon-containing organic

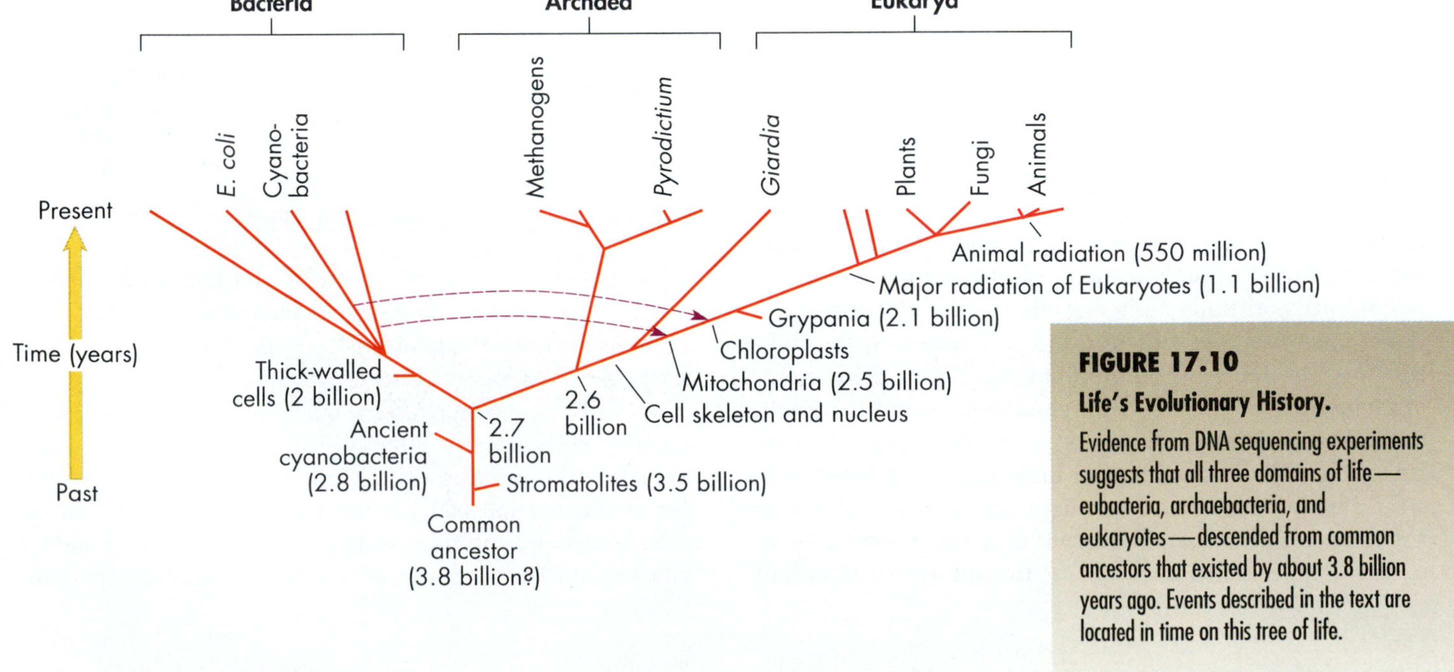

FIGURE 17.10

Life's Evolutionary History.

Evidence from DNA sequencing experiments suggests that all three domains of life—eubacteria, archaebacteria, and eukaryotes—descended from common ancestors that existed by about 3.8 billion years ago. Events described in the text are located in time on this tree of life.

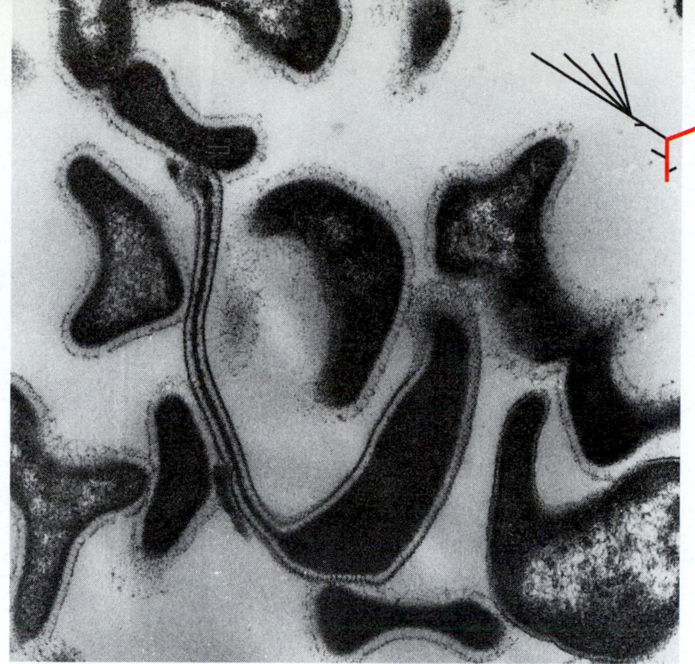

FIGURE 17.11

Hot Springs Archaebacteria.

Sulfurous submarine hot springs heated well above the boiling point of water may simulate environments in which the earliest cells lived. Inhabitants of these hellish environments include *Pyrodictium,* an archaebacterium whose genes have changed little from life's common ancestor.

molecules that fell to earth as meteors or were generated when lightning shot through the primitive atmosphere.

A time would have come, however, when so many heterotrophs were removing organic molecules from the environment that the supply did not match the demand. Eventually, all the free organic compounds would have been incorporated into the molecules of living systems. For the first time, organisms would have changed their environment.

➤ CONCEPT CHALLENGE

The investigation of the origins of life differs from many other scientific investigations because the event occurred in the distant past and conditions have changed so dramatically since then. How has this fact affected the way scientists study life's origins?

Evolution of Autotrophs

As the primordial oceans became depleted of organic materials, the first heterotrophic organisms would have run short of carbon and energy supplies. Natural selection would have favored autotrophs, organisms that could produce their own organic molecules from simpler inorganic precursors, such as methane and carbon dioxide.

The fossilized remains of stromatolites from South Africa [see FIGURE 17.2], which so resemble today's stromatolites still sprouting from Hamelin Pool, hint at the results of that natural selection: Cells able to make their own organic molecules via photosynthesis had evolved by 3.6 billion years ago. The prokaryotic cells that had generated the fossil stromatolites decayed without leaving any traces. They did, however, leave finely layered stone remains that suggest the presence of cells such as those in today's stromatolites. Just as significantly, scientists have uncovered fossils of cells much like cyanobacteria in 3.5-billion-year-old sedimentary rocks quite close to the sites of the fossil stromatolites [FIGURE 17.12].

EARLY PHOTOSYNTHESIS

While the photosynthetic cyanobacteria may have used hydrogen sulfide and hydrogen gas as electron donors in their energy metabolism, these raw materials were probably limited. The supply of water, however, was virtually unlimited, and any organism that could use water as a source of electrons in energy metabolism could invade

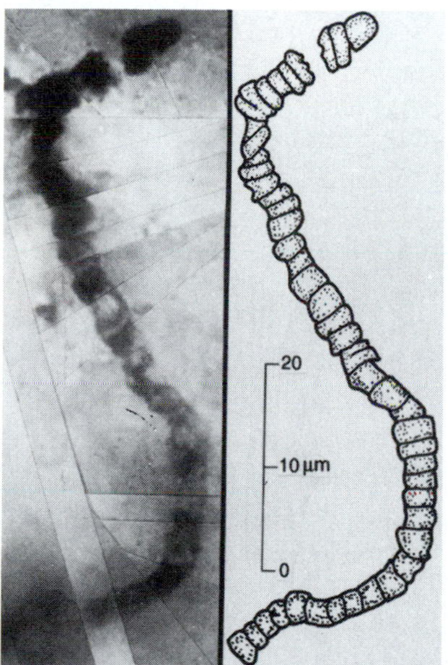

FIGURE 17.12

Fossils of 3.5-Billion-Year-Old Prokaryotic Cells.

These fossils from near North Pole, Australia, are the earliest remains of distinct cells yet found.

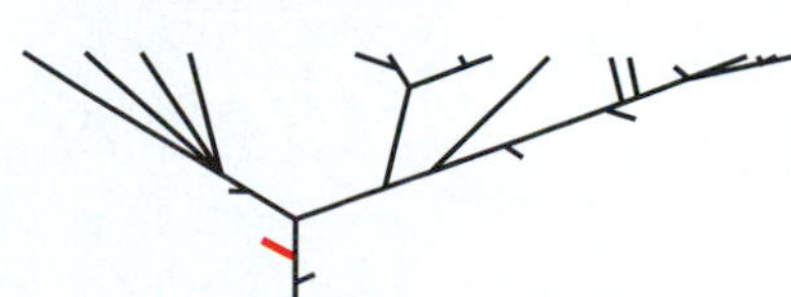

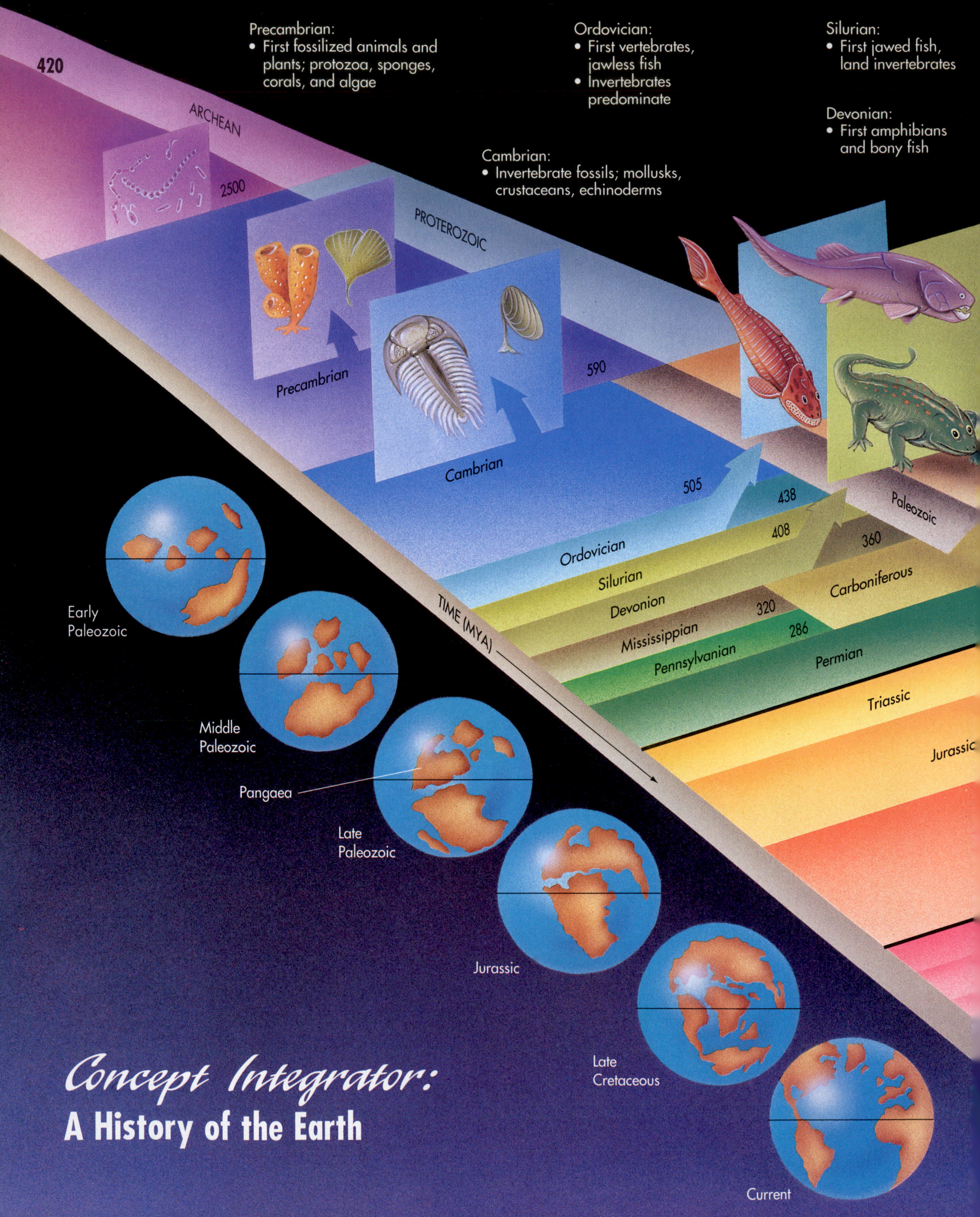

Concept Integrator: A History of the Earth

Carboniferous:
- First conifers, insects, and reptiles

Permian:
- World-wide mass extinctions

Triassic:
- Conifers predominate
- Dinosaurs appear

Jurassic:
- First mammals and birds
- Rise of giant dinosaurs

Cretaceous:
- Mass extinction at end of Cretaceous
- First flowering plants early Cretaceous
- First placental mammals

Tertiary:
- Bony fishes proliferate
- Radiation of mammals
- First monkeys and apes
- Flowering plants predominate

many more environments. Today's cyanobacteria do obtain electrons from water in photosynthesis and release oxygen as a waste product. Significantly, paleontologists have discovered filaments and cell fossils strongly resembling today's cyanobacteria in Australian rocks dating from about 2.8 billion years ago [review FIGURE 17.10]. From this evidence, biologists have concluded that photosynthesizing cells capable of splitting water molecules during photosynthesis and or releasing oxygen gas into the environment had evolved by 2.8 billion years ago. With this development, life had begun to modify the planet's atmosphere in a major way.

Oxygen Tolerance and Aerobic Cells For over a billion years after the evolution of oxygen-producing photosynthetic cells (about 3 months on our year-long analogy), there was little or no accumulation of oxygen in the atmosphere. Instead, iron compounds in seawater reacted with the free oxygen, forming iron oxides, which fell to the ocean floor by the ton. Geologists have found iron deposits dating from this time lying in horizontal bands in deep rock strata at sites all over the earth. We have tapped some of these deposits, in fact, in the iron mines of Michigan and Minnesota. After most of the iron in seawater was bound in this way, oxygen probably began to build up in the atmosphere. Ironically, since oxygen is poisonous to anaerobic cells, this initial buildup would have been harmful to both the early heterotrophs and to the photosynthesizers themselves.

About 2 billion years ago (July 24), as oxygen levels continued to build, cells became fossilized in the Lake Superior area—filaments and spheres, and occasionally thick-walled cells that were probably resistant to the harmful effects of oxygen [FIGURE 17.13]. Such fossils of thick-walled cells suggest that oxygen given off by cyanobacteria all over the earth had begun to accumulate in the atmosphere. Concentrations may have reached as much as 1 percent of today's atmospheric O_2 concentrations (about 0.2 percent). The fossils also suggest that mechanisms that allow a cell to avoid oxygen poisoning had also begun to appear.

The accumulation of oxygen would have had two other consequences. First, it would have allowed the evolution of cells that obtain their energy by aerobic respiration, with oxygen as the final electron acceptor [review FIGURE 5.13]. These cells might have been similar to the common bacterium *Escherichia coli*. Second, the increase in oxygen would have had a significant impact on the global environment with the formation of an ozone screen. Sunlight striking gaseous oxygen (O_2) creates ozone (O_3), which in turn absorbs ultraviolet light. The accumulating ozone would have protected living things from some of the damage that ultraviolet light inflicts on DNA. Because of this newly developing ozone shield, cells could begin to live closer to the planet's surface.

Evolution of Eukaryotic Cells

The earth has probably been inhabited by living things for more than 80 percent of its history, or about 3.6 billion years (March 25 on our time line). For more than half that time, the most complex life forms were prokaryotic cells, such as the ones that build the stromatolites of Hamelin Pool. About midway through life's history on earth (about 2.1 billion years ago, or July 15 on our time scale), eukaryotic cells appeared. Multicellular organisms—animals, plants, and most fungi—began to show up only about 700 million years ago (November 5). Biologists have carefully examined the stepwise bursts of evolution that led first to the emergence of eukaryotic cells and their diversification and then to the appearance of multicellular eukaryotic organisms.

Biologists were surprised in 1992 to find eukaryotic cells (large cells with a membrane-bound nucleus) in a Michigan iron mine. These large, corkscrew-shaped cells, called *Grypania*, were about the diameter of a strand of spaghetti [FIGURE 17.14] and lived *before* the

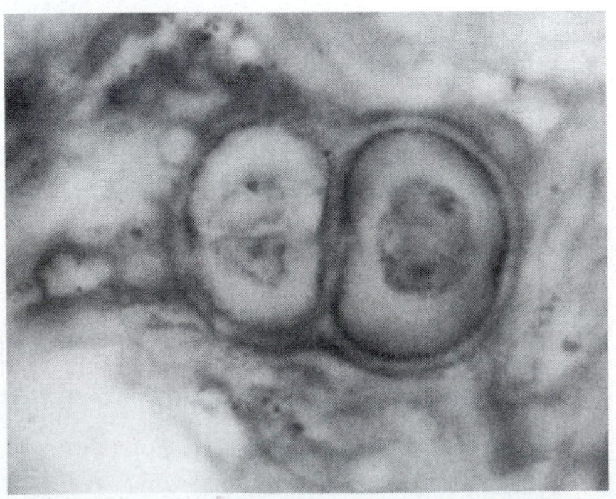

FIGURE 17.13

Thick-Walled Cells Signaled an Environmental Change.

Fossilized cells dated to approximately 1.5 billion years old reveal thick walls. These may have protected the cells from the poisonous gas, oxygen.

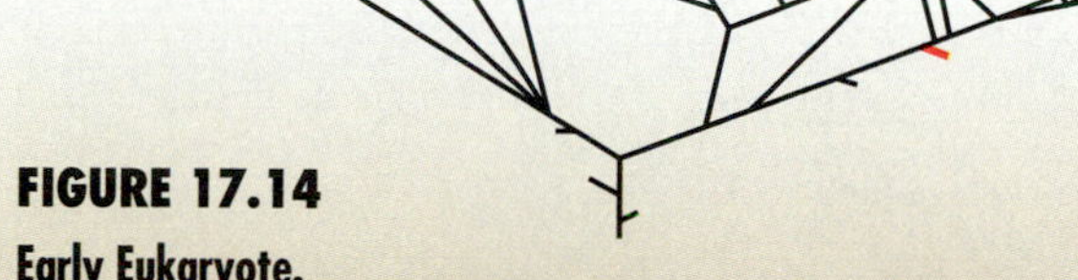

FIGURE 17.14

Early Eukaryote.

The size of this 2.1-billion-year-old corkscrew-like organism called *Grypania* suggests that it is a eukaryote that contains mitochondria and perhaps chloroplasts.

main phase of iron deposition. Hence they would have flourished before the major buildup of oxygen in the atmosphere and would have harvested energy anaerobically. Their discovery pushed back the appearance of the earliest known eukaryote to 2.1 billion years ago (July 15 on our time line). This fossil evidence, along with molecular genetic data, shows that, while the eukaryotes evolved after the bacteria and archaebacteria, they are nevertheless a very ancient group.

Some biologists hypothesize that the earliest eukaryotic cells had lost the ability to make tough cell walls, but had developed nuclear envelopes and a cytoskeleton that provided rigidity and toughness [see FIGURE 3.17]. Modifications like these would have allowed nourishment by means of phagocytosis—the engulfment of food particles in pockets of the plasma membrane [review FIGURE 3.14]. Nuclear envelopes may have originated as invaginations of the plasma membrane, which came to surround and protect the host cell's naked DNA about 2.6 billion years ago [FIGURE 17.15A and B].

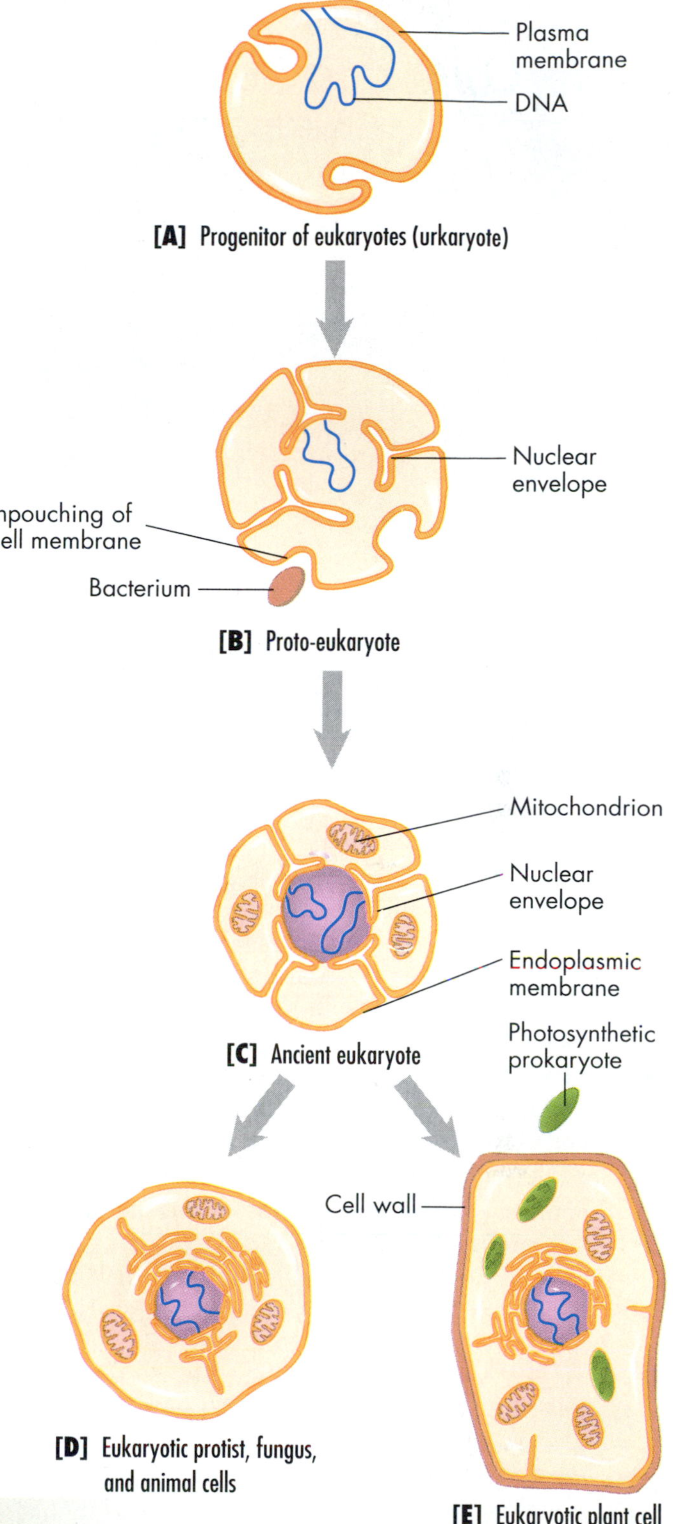

FIGURE 17.15

The Endosymbiont Hypothesis.

A prokaryotic host cell lacking a cell wall and specific internal organelles, but capable of inpouching its cell membrane [A], may have merged with symbiotic bacteria that could perform respiration in the presence of oxygen [B]. These bacteria could then begin to survive in the hosts cell's cytoplasm. The inpouching may have given rise to the endoplasmic reticulum and the nuclear membrane in this ancient eukaryote [C]. The nonphotosynthetic symbionts could have evolved into mitochondria in the ancient eukaryote [D]. Photosynthetic prokaryotes engulfed by an ancient eukaryote could have given rise to chloroplasts [E].

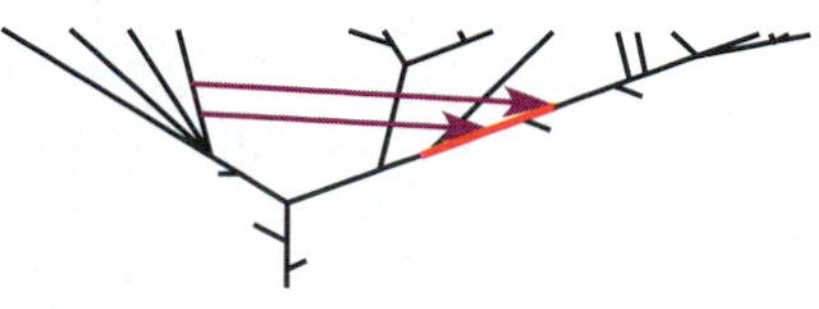

Concept Integrator: The Pre-Cambrian Sea

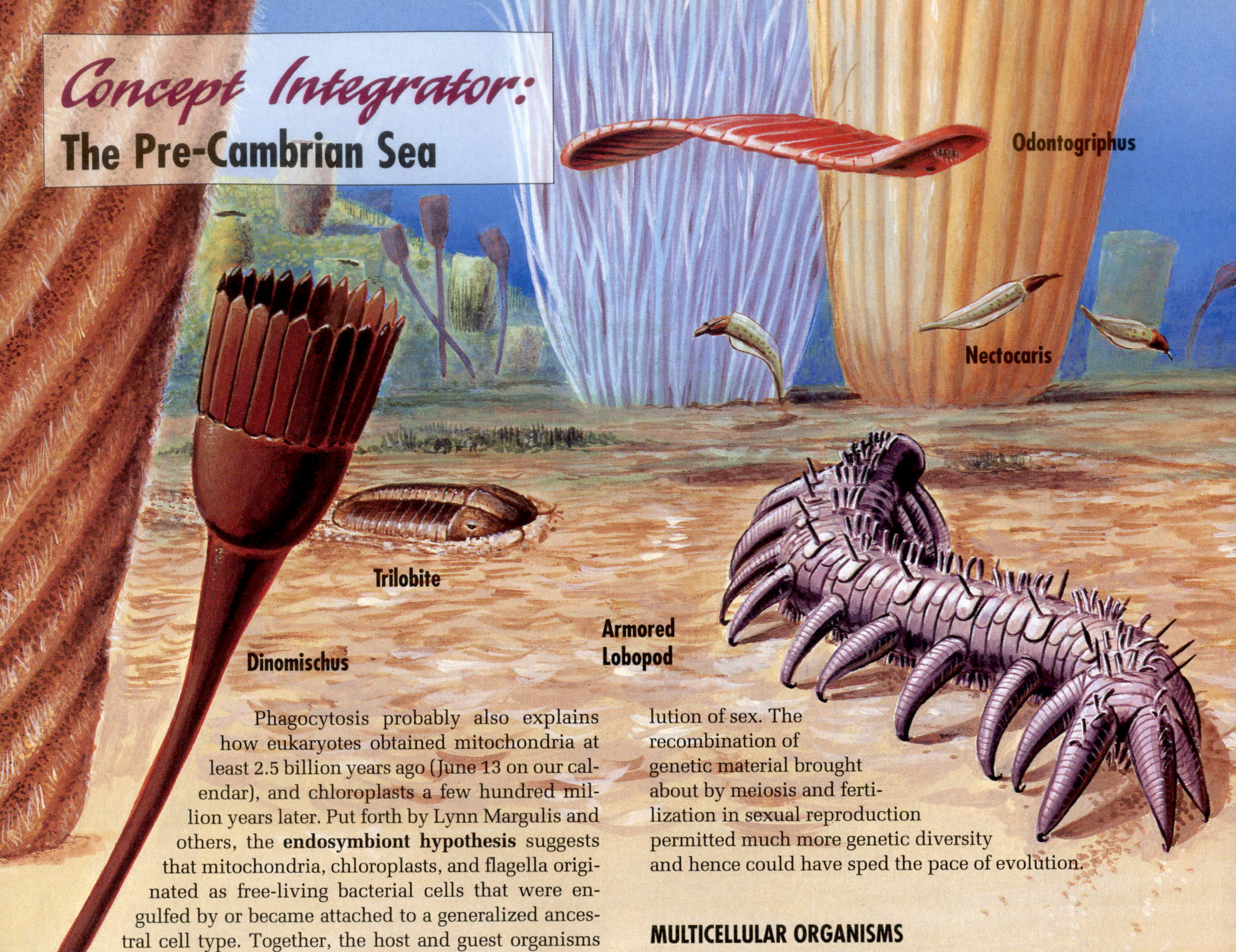

Phagocytosis probably also explains how eukaryotes obtained mitochondria at least 2.5 billion years ago (June 13 on our calendar), and chloroplasts a few hundred million years later. Put forth by Lynn Margulis and others, the **endosymbiont hypothesis** suggests that mitochondria, chloroplasts, and flagella originated as free-living bacterial cells that were engulfed by or became attached to a generalized ancestral cell type. Together, the host and guest organisms formed cellular communes, viable single organisms with each member adapted to the group arrangement and deriving benefit from it.

Evidence for this hypothesis includes the facts that mitochondria are remarkably similar in size and shape to today's aerobic bacteria and that they have their own DNA, which replicates independently of the cell's genetic material. Chloroplasts are similar in size and shape to certain photosynthetic prokaryotes [review BOX 6.1 on page 150]; like mitochondria, they have separate, self-replicating DNA with bacterialike genes. The inclusion of energy-processing organelles such as these would have been metabolically advantageous for oxygen-tolerant cells large enough to accommodate them.

While cells contained mitochondria and chloroplasts by about 1.9 billion years ago (August 1), eukaryotes did not start to radiate into numerous groups before about 1 billion years ago (October 12). Some biologists suggest that the burst in diversity at that time was due to the evolution of sex. The recombination of genetic material brought about by meiosis and fertilization in sexual reproduction permitted much more genetic diversity and hence could have sped the pace of evolution.

MULTICELLULAR ORGANISMS

Many-celled organisms—animals, plants, and certain fungi—probably arose from ancestors that existed as independent cells in a loose network or colony [FIGURE 17.16]. About 900 million years ago (October 20), the first organisms appeared that seem to organize as cell populations in response to chemical signals. These yeastlike eukaryotic cells lived in colonies with cells organized in rows, probably because they responded to each other's chemical signals. Tight packing of such cells and the emergence of a division of labor wherein some cells in the colony primarily fed, others primarily reproduced, and so on, could have led to the evolution of multicellular organisms.

THE EMERGENCE OF ANIMALS

About 580 million years ago (November 15), simple animals first appeared in the fossil record. This burst of ani-

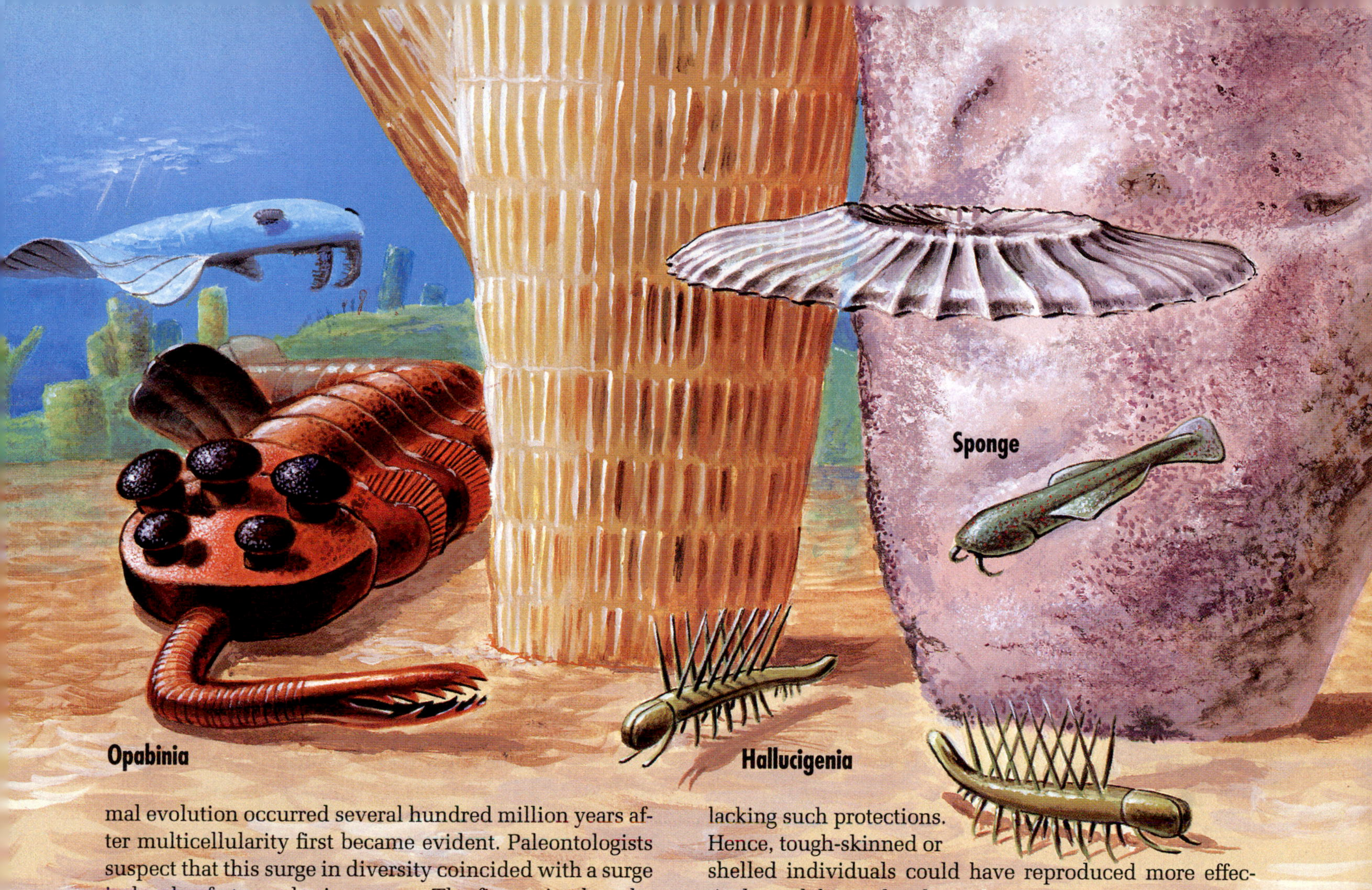

mal evolution occurred several hundred million years after multicellularity first became evident. Paleontologists suspect that this surge in diversity coincided with a surge in levels of atmospheric oxygen. The first animals to be found as fossils were soft-bodied, and are known as the Ediacaran fauna [see the Concept Integrator on pages 426 and 427]. Some of these animals looked like jellyfish, and others resembled annelid worms [see FIGURE 21.20], although they may not actually be related to these living animals. Biologists wonder why soft-bodied animals became preserved as fossils 580 million years ago whereas the soft body parts of later organisms tended not to become fossilized. Some paleontologists suspect that there were no scavenging animals during Ediacaran times, and so dead animals remained on the ocean floor longer than they would today—a process that would have increased the likelihood of fossilization. In contrast, 40 million years later, animals with hard body parts, including the prickly spined *Hallucigenia* [see FIGURE 16.1] and the trilobites, emerged in the Cambrian explosion and became preserved in the Burgess Shale [see FIGURE 16.18]. The armoring of these animals suggests that predators had begun to roam the seas. Within a population, those individuals with tougher exteriors would have been devoured less often by predators than organisms lacking such protections. Hence, tough-skinned or shelled individuals could have reproduced more effectively, and donated to the next generation a disproportionate share of the population's genes. Holes punctured in the bodies of some fossil trilobites offers silent evidence that indeed, survival had become much more tenuous.

Large Plants and Animals By 400 million years ago (November 29 on the time line), the atmosphere was essentially like today's, with 20 percent oxygen content. The ozone screen was in place, and very large, complex life forms were appearing in profusion. Large fishes swam in the ancient seas, and by 500 million years ago, primitive land plants grew along moist shores. Within another 200 million years, amphibians, reptiles, birds, and mammals would move about the continents, and large stands of conifer trees and flowering plants would grow abundantly.

Effects of Geologic Change on Life

Just as the evolution of living things affected the earth, the earth, too, has changed geologically and influenced the evolution and diversification of living things. Geologists describing these events have divided time into eons,

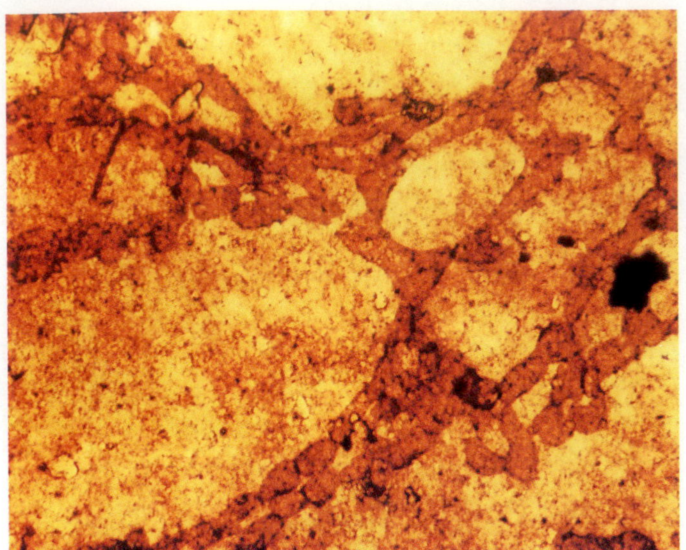

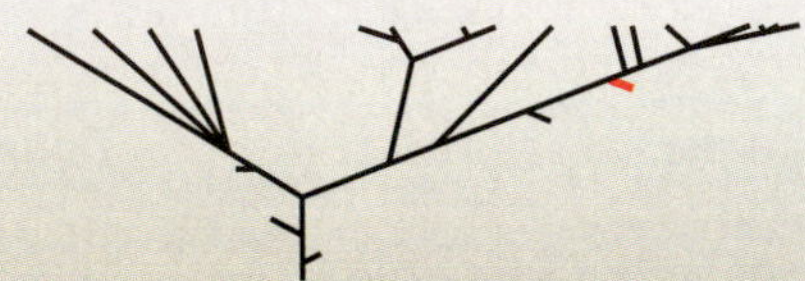

FIGURE 17.16

Colonial Cells and the Origins of Multicellularity.

These eukaryotic yeast-like organisms (*Eosaccharomyces ramosus*, the red ovals) lived in colonies of oriented cells, suggesting that they responded to each other's chemical signals in some way. Further evolution of intercellular communication and division of labor may have led to the appearance of true multicellularity.

eras, periods, and epochs, based on sudden changes in the fossil record. For our purposes, however, the names of just five time periods are useful to remember: (1) The most recent is the Cenozoic ("recent life"), preceded by (2) the Mesozoic ("middle life"), (3) Paleozoic ("old life"), and (4) Proterozoic ("earlier life"). Finally, (5) the Archean ("ancient") reaches back to the formation of the oldest known rocks, about 4 billion years ago.

ARCHEAN EON AND PROTEROZOIC ERA: LIFE EVOLVES AND SINGLE-CELLED ORGANISMS DIVERSIFY

In the 500 million years or so between the origin of the earth (January 1) and the beginning of the **Archean eon** (February 13), meteor impacts and volcanos probably prevented life from emerging. Compared to that hellish early environment (which some scientists, in fact, refer to as Hadean), the Archean was a peaceful time, with black volcanic islands jutting upward through a blue ocean of global dimensions. As lava from nearby volcanos accumulated, some areas of high, flat terrain—the first continents—rose from the seas. Environments had become stable enough for life to evolve, and stromatolites appeared early in the Archean. During the **Proterozoic era**, beginning about 2.5 billion years ago, there were shallow seas, a phase of mountain building began, and the organisms that had evolved earlier began to diversify.

PALEOZOIC ERA: AN EXPLOSION IN ANIMAL DIVERSITY

The **Paleozoic era** started about 580 million years ago (about 6 weeks before year's end on our time line) with the *Cambrian explosion*, the sudden appearance of all major easily fossilized animal phyla in the fossil record. The Cambrian is the earliest period of the Paleozoic.

During the Paleozoic, the landmasses of the earth's crust began to move about and assume new arrangements that greatly affected life's evolution. The mechanism that explains why and how the earth's landmasses move is known as **plate tectonics**. Continents are now known to rest atop huge plates of the earth's crust, and the plates are slowly moved by the circulation of underlying molten material. Thus, the continents are carried by the plates they rest on [FIGURE 17.17]. These movements explain many previously poorly understood phenomena, including where and why most earthquakes occur.

Before the end of the Paleozoic, about 245 million years ago (December 11), the earth's landmasses converged into a single super landmass called *Pangaea* ("all earth"). As the plates collided, the continental landmasses on top of them buckled and tilted, huge mountain ranges were uplifted, warm, shallow inland seas were drained, and the size and position of ocean basins were altered. These physical changes and associated changes in climate were a calamity for both sea and land animals. It is estimated that three-quarters of the families of amphibians (salamanders and relatives), 80 percent of the reptiles (lizards, snakes, and relatives), and 96 percent of all marine species became extinct. This mass extinction, the greatest we know about in the history of the earth, is called the *end-Permian extinction*, since it came at the end of the Permian period in the Paleozoic Era.

MESOZOIC ERA: THE AGE OF DINOSAURS

The convergence of the continents in the late Paleozoic was reversed in the **Mesozoic era** (between 245 and 65 million years ago; about December 20). During this period, the single landmass of Pangaea broke apart, separating populations of organisms and providing great opportunities for evolutionary change.

The island continent of Australia provides a good example of how this breakup affected the evolution of plants and animals. At the time Pangaea existed, mammals had not yet evolved the placenta, which nourishes

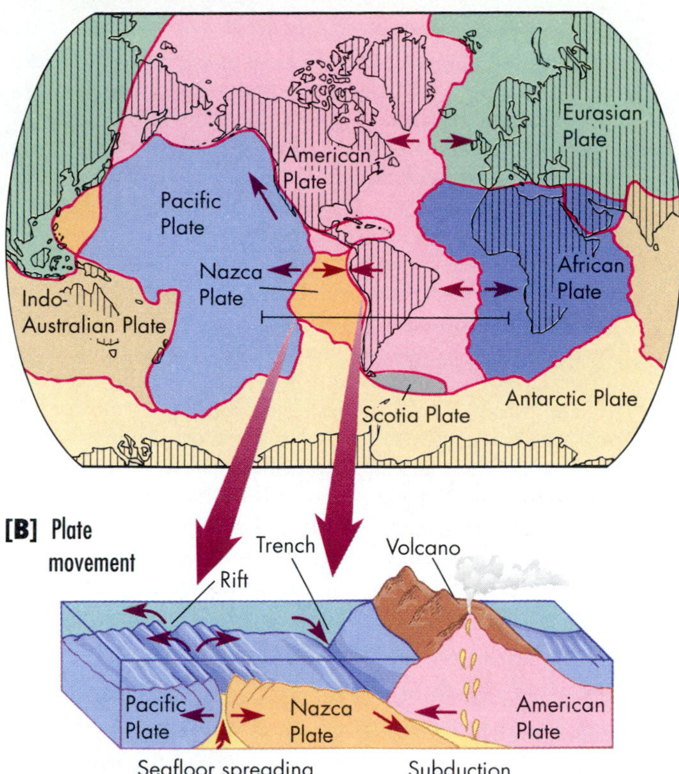

FIGURE 17.17

Plate Tectonics: Continents Colliding.

[A] The continents and oceans ride on massive crustal plates. [B] At the junctions (rifts) between certain plates, new crustal material oozes up from the earth's core and pushes the plates apart. Where the leading edges of plates collide, one pushes below the other, causing a deep trench, in a process called subduction. Here, the Nazca plate dives beneath the American plate. This process causes most of the earth's violent geological activity.

the young in the uterus. Instead, offspring were born at a very early developmental stage, found their way to their mother's pouch, or *marsupium* (Latin "purse"), where they were nourished by milk. These small primitive mammals, called *marsupials*, thrived all across Pangaea, as well as on the giant landmass of Australia after it broke apart and drifted away. Thousands of species of mammals with placentas evolved on other continents, but not on Australia. Without competition from placental mammals, the marsupials radiated into a rich collection of unique animals, including kangaroos, koalas, and wombats. Except for a few species, like the opossum, most marsupials died out in the rest of the world.

Along with the breakup of Pangaea came the emergence and diversification of the dinosaurs, as well as the growth around the globe of great swampy forests containing tropical plants and trees, and the first appearance of flowering plants, birds, and mammals.

CENOZOIC ERA: THE AGE OF MAMMALS AND FLOWERING PLANTS

A catastrophic event ended the Mesozoic era and ushered in the current era, the **Cenozoic**, about 65 million years ago (December 26). In another mass-extinction event, half of the world's species, including the dinosaurs, died out. Current research suggests that a meteor crashed into the gulf coast of Mexico at that time. The resulting airborne debris clouded the sky, and could have changed climates dramatically, leading to the extinctions (see CHAPTER 18 for more details).

In the altered environments of the Cenozoic, birds, mammals, and flowering plants continued diverging into large and varied groups of species. Extensive crusted plate movement in the Cenozoic brought the continents to their present configuration, separating groups of organisms, rafting them to entirely new locations and conditions, and further encouraging the diversity of life.

➤ CONCEPT CHALLENGE

What would the physical condition of today's earth be if life had never evolved?

Connections

Understanding the origins of life is a profound scientific challenge. The drama unfolded on a stage that no longer exists and under conditions that changed forever after living organisms appeared. Nevertheless, by knowing both the characteristic behavior of matter—particularly organic molecules—and the attributes of all living things, biologists have reconstructed a plausible sequence of the prebiotic events that led to the emergence of the first cells. While there are no cell-like fossils from 4 billion years ago to suggest the form of the universal ancestors, we can study "living fossils" like the stromatolites [FIGURE 17.1] and *Pyrodictium* [FIGURE 17.11] which may be similar to the earliest cells.

Perhaps someday we will find, near some abyssal deep-sea vent or on a distant planet, more specific evidence for how life evolved. In the meantime, the biochemical history laid down in living cells is a helpful substitute. We cannot fully appreciate living things or understand their evolutionary relationships, however, without first comprehending the intertwined pasts of life and the planet on which it arose by natural processes.

KEY TERMS

Archean eon, 426
Cenozoic era, 427
cyanobacteria, 408
endosymbiont hypothesis, 424
Mesozoic era, 426
Paleozoic era, 426
plate tectonics, 426
Proterozoic eon, 426
ribozyme, 415
stromatolite, 408

HIGHLIGHTS IN REVIEW

1 Earth's size, composition, and distance from the sun provided the physical and chemical factors necessary for the origin of life.

a] Evidence from astronomy suggests that the universe began in a huge explosion 13 to 18 billion years ago. The sun and its planets coalesced from clouds of hydrogen, helium, and heavier elements. By 4 billion years ago, the basic structure of the earth was set: a core of heavy metals surrounded by a liquid mantle of hot rock, a hard crust of cool black rock on the surface covered by blue oceans, and an atmosphere lacking molecular oxygen.

b] The earth's advantageous combination of composition, geological activity, size, and distance from the sun provided an environment favorable to the chemistry of organic molecules, and thus life was able to evolve and prosper here but not on neighboring planets.

2 While experiments can show how life *might* have arisen according to natural laws, scientists may never show exactly how it *did* originate.

a] Laboratory experiments simulating the supposed environment of the early earth show that organic molecules such as amino acids and bases of nucleic acids could probably have formed by natural processes from small molecules present in earth's air, water, and rocks. Organic molecules may also have been brought to earth on meteors from space.

b] Some biologists speculate that life might have originated in an RNA world, a world in which RNA molecules catalyzed their own replication and the joining of amino acids into proteins. Eventually, the more stable molecule DNA took over the information storage role of RNA, but RNA stayed involved in protein synthesis.

c] Some biologists suggest that life arose in underwater hot springs, and others propose that the formation of self-replicating molecules was catalyzed by clays.

d] Lipids and polypeptides can form membranelike spheres spontaneously. If they came to surround nucleic acids that could encode polypeptides that promoted nucleic acid replication, the combination would have had some of the properties of life.

3 The evolution of life has depended on the evolution of earth's geology. Reciprocally, life has altered the atmosphere and surface of the planet.

a] The earth formed about 4.5 billion years ago, but the oldest rocks yet found are only 4 billion years old due to volcanic activity and meteor bombardment, which probably destroyed older rocks. The earliest evidence of life is from 3.6 billion years ago; this leaves only a few hundred million years for the evolution of life.

b] The earliest living systems altered earth by consuming preexisting organic molecules. Early cells may have resembled some archaebacterial cells and existed in hot submarine sulfur springs. There, they might have obtained carbon from carbon dioxide, and energy in the absence of oxygen from hydrogen gas and hydrogen sulfide.

c] Stromatolites (toadstool-shaped fossils deposited layer upon layer by prokaryotic cells) have been found dating from 3.5 billion years ago. Similar formations are still found in warm bays in the world today.

d] By 2.8 billion years ago, cyanobacteria were modifying the planet by releasing oxygen into the environment. The oxidation of the iron caused it to precipitate and fall to the ocean bottom in rusty red layers.

e] Eukaryotes arose before 2.1 million years ago from a prokaryotic precursor similar to today's archaebacteria. Subsequently, according to the endosymbiotic hypothesis, aerobic and photosynthetic prokaryotic cells entered the early eukaryote and evolved into mitochondria and chloroplasts. After sexual reproduction evolved in eukaryotes, they rapidly diversified about 1 billion years ago.

f] Soft-bodied animals arose about 600 million years ago. Land plants probably appeared about 500 million years ago.

g] The Archean eon (4 to 2.5 billion years ago) lasted from the formation of the earliest known rocks until a time of massive mountain building. During the Proterozoic eon (2.5 billion years ago until 570 million years ago), life diversified immensely. The Paleozoic era (570 to 245 million years ago) began with the Cambrian explosion, the sudden appearance of all major easily fossilizable animal phyla and ended 245 million years ago with the greatest mass extinction in the history of the earth, a time when over 90 percent of all earth's species became extinct. The Mesozoic era (245 to 65 million years ago) ended with another mass extinction. Probably caused by a meteor impact, this extinction wiped out the dinosaurs and many other groups. During the Cenozoic era (65 million years ago until the present), continents were rearranged and life evolved to its present state.

UNDERSTANDING THE FACTS AND CONCEPTS

For Questions 1–5, match each of the descriptions with the most appropriate item from the following list of terms.

a] stromatolite
b] *Pyrodictium*, a bacterium found in hot, sulfurous springs
c] cyanobacterium
d] eukaryotic cell *Grypania*
e] ribozymes

1 A molecule that is both self-replicating and catalytic.

2 A cell that may be similar to the first cell.

3 Had a nuclear membrane and was an anaerobic cell.

4 A cell that confirms the early evolution of autotrophy.

5 A mineralized mound covered by a mat of bacteria.

To which of the following are the events named in Questions 6–10 causally related?
- a] evolution of autotrophy
- b] evolution of photosynthesis
- c] rise in atmospheric oxygen levels
- d] adaptive radiation of eukaryotes
- e] evolution of large plants and animals

6 Evolution of sex.

7 Surge in atmospheric oxygen to 20 percent.

8 Evolution of photosynthetic metabolism.

9 Depletion of inorganic carbon and inorganic high-energy molecules.

10 Depletion of gaseous H_2S and H_2 as electron donors.

For Questions 11–20, match the events with the period of time during which they occurred.
- a] Archean eon
- b] Proterozoic eon
- c] Paleozoic era
- d] Mesozoic era
- e] Cenozoic era

11 Appearance of first heterotrophic cells.

12 First flowering plants appeared.

13 Pangaea formed.

14 Pangaea broke apart.

15 Cambrian explosion of life forms.

16 Continents moved to present locations.

17 Adaptive radiation of birds and mammals.

18 Ended with a mass extinction that eliminated the dinosaurs.

19 Age of *Homo sapiens*.

20 Appearance of autotrophic bacteria.

INTEGRATE AND APPLY WHAT YOU HAVE LEARNED

1 What paradox is solved by postulating an "RNA world"?

2 Explain why modern marsupials are limited primarily to Australia despite having been widespread throughout Pangaea.

3 What positive function is served by mass extinctions?

4 Autotrophic bacteria have been discovered thriving in deep hydrothermal vents in the ocean. Would you expect these organisms to be photosynthetic autotrophs?

5 You are engaged in a debate with someone who argues that if cells arose spontaneously from nonliving matter once in the history of the earth, the same or similar events should be able to occur again at any time. What is your response?

ANALYSIS

For Questions 1–3, select the best answer.

1 Why are the particular stromatolites in Hamelin Pond considered to be so important?
- a] They house the most ancient organisms on earth.
- b] They are very similar to fossilized mounds that are among the earth's most ancient rocks.
- c] They contain parts of carbonaceous chondrites.
- d] They are an important source of oxygen in the southern hemisphere.
- e] The organisms that they contain are similar to the ancient organisms that grow in hot, sulfurous springs.

2 If the first self-replicating molecules were made of RNA, why don't present-day cells have genes made of RNA?
- a] It is more efficient to use two different informational molecules for protein synthesis processes.
- b] RNA cannot mutate, and thus it cannot provide genetic diversity.
- c] Only DNA can be protected within a membrane.
- d] Since DNA is double-stranded and more stable, it is better suited to the role of being a gene. Thus if it arose by chance it would have been selected for.
- e] DNA has been shown to have enzymatic properties, whereas RNA does not.

3 Why are cyanobacteria so self-sufficient?
- a] They obtain all their nutritional and energy needs from molecular water.
- b] They obtain both nitrogen and carbon from the atmosphere and can use solar energy.
- c] They obtain both nitrogen and carbon from the soil.
- d] They can live on sunlight alone.
- e] They are exceptionally good scavengers.

For Questions 4–8, refer to the Miller experiment on page 413 and to the discussion of carbonaceous chondrites on page 414, and then determine whether each of the stated conclusions is logical (yes) or not (no).

4 Organic material could not have formed spontaneously on the prebiotic earth.

5 Organic material has fallen to earth from space.

6 Organic matter can form without the activity of living cells.

7 Macromolecules probably formed in space before forming on earth.

8 If organic molecules can form within rocks in the solar system, then they could also have formed on the prebiotic earth.

CHAPTER 18

Biodiversity and Evolution

THE PUZZLING PLATYPUS

Eighteenth-century British seamen would occasionally sail home from the Far East with stuffed "mermaids" aboard ship. These specimens, featuring a monkey's head and chest stitched to a fish's tail, were nothing more than playful deceptions from Chinese taxidermists. No wonder, then, that when British naturalist George Shaw first saw an authentic platypus skin in 1799, brought back from Australia by a Navy captain, he thought the specimen a fake, sewed from a beaver's pelt and a duck's bill [FIGURE 18.1].

FIGURE 18.1
The Patchwork Platypus. With fur like a mammal, a reproductive system like a reptile, and bill like a duck, the platypus posed a problem for the first scientists who saw it: How should this hodge-podge animal be classified?

The platypus is a warm-blooded, fur-bearing, milk-producing mammal. It has, however, at least five reptilelike traits: the females lay leathery eggs; the males have venomous poison spurs; and both sexes have hair growing in swirled patterns like reptilian scales, reptilelike shoulder bones, and a cloaca (klo-AY-kuh)—a single opening used in both reproduction and excretion. Birds have cloacas, too. The platypus has another trait originally thought to be birdlike—its sensitive duck bill—plus one fishlike trait, an electrosensory system that detects weak electric currents.

After studying this odd collection of characteristics, many researchers have called the platypus "undeveloped," "inefficient," or "primitive." Yet it was paddling about in Australian streams and pools with its unique set of characteristics for 35 million years before the dinosaurs died out, and has continued for 65 million years since!

These characteristics suit the aquatic animal perfectly to its habitat and way of life. Each diminutive platypus, about half the size of an adult cat, spends the day in an earthen burrow carved into the muddy bank of a pond or stream. A female in her burrow will eventually lay two small, leathery eggs less than 2.5 cm (1 in) in diameter. After wriggling around the shell, whitish infants chip their way out with an egg tooth resembling a baby alligator's, and the mother nurses the offspring with milk from her two nipples. Both the female and male platypus are secretive and leave the burrow only after nightfall, slipping quickly into the water. There the animals may play, mate, or fight with other platypuses, but they primarily hunt, paddling webbed feet twice as fast as a duck. The platypus's soft leathery bill has 50 times more nerve fibers running to the brain than the skin surface of a person's hand, and these nerves enable the animal to locate and nab prey with amazing accuracy.

The platypus's fascinating physical and behavioral adaptations came to light little by little, and **taxonomists** (biologists who describe, identify, name, and classify organisms) tried to squeeze the animals into preexisting categories. But to which class of vertebrates did the platypus belong? Should its shoulder bones, eggs, poison glands, and cloaca assign it to the reptiles? Or should its fur, milk production, and warm blood place it with the mammals? Is it a primitive holdout, retaining premammalian traits? Is it a sophisticated modern animal, well tuned to its environment with electric field detectors and fine, warm fur? Or is it an intermediate?

The puzzling platypus brings us to this chapter's central concerns: First, how do biologists classify the diversity of living things? Second, how can we use our understanding of evolution and taxonomy to prevent the *biodiversity crisis*: the current widespread and accelerating loss of species due largely to human population growth?

MESSAGES

1 Systematics is the scientific study of the diversity of organisms and how different organisms are related.

2 Biologists often group together organisms sharing a set of similar traits which they inherited from a common ancestor (homologies). Organisms can also share similar traits due to shared environments rather than a shared recent ancestor (analogies).

3 Comparisons based on gene and protein sequences suggest that there are three major domains of life: the Bacteria, the Archaea, and the Eukarya.

4 Our planet's diversity of life is immense; there are perhaps as many as 50 million species. Human activity is threatening to destroy much of this diversity in our lifetimes. Prompt action may prevent the irretrievable destruction of the living products of millions of years of evolution.

The study of the diversity of organisms and their evolutionary relationships involves classifying living things into groups of like organisms. As is common in many areas of science, biologists sometimes disagree about what those rules should be. Biologists can classify organisms by appearance and other physical traits; by genetic evidence of common ancestry; and by other methods that we will discuss in this chapter. In one well-respected system (which we first saw in CHAPTER 1) all living things can be divided into six kingdoms: true bacteria, archaebacteria, protists, fungi, plants, and animals. This classification system helps biologists sort out life's immense diversity—as many as 50 million species.

This chapter describes the specific ways in which biologists classify living organisms. It will focus on a method of classification called *cladistics*, which is based on evolutionary history. We will see that recent molecular genetic investigations have given biologists a revolutionary way of seeing the relationships between organisms in the various living kingdoms. Finally, we will discuss the biodiversity crisis, which humankind is largely creating, but also attempting to solve.

Besides encountering the platypus in this chapter, you'll meet a monstrous toad that has gotten too successful for its own good—or anyone else's. ❑

Organizing Life's Diversity

Just thinking about familiar organisms such as dogs, frogs, oaks, grass, toadstools, and bacteria, we can see that life is diverse. The diversity is even more striking if you consider unfamiliar organisms, such as the platypus, inch-high, inverted umbrellas called moss animals (bryozoans) that carpet rocks in shallow seas [FIGURE 18.2A] or the palmlike cone-bearing plants called cycads [FIGURE 18.2B]. Despite the millions of unique life forms inhabiting earth, people have long recognized patterns of similarity among organisms and grouped similar organisms together. Even children can classify a mountain lion as a type of cat and a wolf as a type of dog, and both organisms as animals, not plants. But what about the bryozoan? Is it a plant or an animal? What about the cycad? Should we consider it a relative of the palm or the pine? And what about the patchwork platypus? For scientists, life's diversity must be cataloged systematically in ways that indicate the relationship of organisms to one another. **Systematics** is the scientific study of the diversity of organisms, and how they are related in an evolutionary context. In contrast, **taxonomy** is the science of identifying organisms, classifying them into genus, species, and so on, and naming them.

WHY STUDY DIVERSITY?

The contemporary study of organismal diversity is much more than simply making lists for their own sake. There are both scholarly and practical reasons for a vigorous push toward a catalog of all of earth's species (up to 50 million, some think) as soon as scientifically possible:

1. Much of the earth's diversity could be lost in the next 50 years due mostly to harmful human activity. Species not yet cataloged may never be known to science, and biology will be impoverished by that loss.

2. A knowledge of organisms and their life habits is crucial for identifying the causative culprits during outbreaks of disease. In April 1993, for example, dozens of Milwaukee residents were hit with a week-long bout of diarrhea. Epidemiologists suspected that a microorganism was the cause, but different drugs control different microbes, and physicians needed to know which organisms they were dealing with in order to prescribe the most effective antibiotic. Microbiologists familiar with the diversity of prokaryotes and protists were consulted, and they identified the culprit as the protist *Cryptosporidium.* With that knowledge, doctors devised appropriate treatments and were able to cure the affected Wisconsin residents.

3. Many of our current and future drugs and raw materials come from nature, and new compounds are discovered regularly. For example, researchers found a very useful treatment for childhood leukemias in the Madagascar periwinkle [review FIGURE 1.1]. Further investigations of little-known life forms in remote forests, grasslands, tide pools, and other wild places will no doubt uncover more compounds of great usefulness for saving lives, feeding people and livestock, and developing new industrial materials.

[A] Bryozoans

[B] A cycad

FIGURE 18.2 Intuitive Classification Is Not Always Right. **[A]** A colony of bryozoans looks like a carpet of moss plants, but these organisms are members of a distinct animal phylum. **[B]** A cycad resembles a palm, but because its seeds sit naked in a cone, not in a fruit, the cycad is more closely related to pine trees than palms.

CLASSIFICATION ACCORDING TO LINNAEUS

If you have examined wildflowers in a mountain meadow or a vacant lot, you may have used a field guide to help you identify which flowers are present. Some field guides organize flowering plants according to the color of the blossoms—all the plants with yellow flowers in one section, all the blues in another, and so on. This kind of organization makes it easy to enjoy nature even without expertise in plant systematics, but it is less useful for biologists wanting to classify organisms into groups that are related in fundamental biological ways, not by superficial characteristics such as flower color.

More than 200 years ago, a Swedish botanist named Carolus Linnaeus developed a classification system in which every organism then known to science was assigned to a series of increasingly more general groups. This assignment was based on the degree of structural similarity among various organisms.

Linnaeus also invented a way to name each type of organism—the system of **binomial nomenclature** (*binomial* means "two names"). Linnaeus's system of naming organisms assigns each species a two-word name, such as *Homo sapiens*, our own species' name. The first part of a species' name is a term which may include several similar species; such a term is called the *genus* (plural, *genera*). Human beings are in the genus *Homo*. The second part of a species' name is a unique term which, when combined with the genus term, refers only to one type of organism, the *species*. The second part of our species name is *sapiens*. In today's usage, the genus is a group of very similar organisms related by common descent from a recent ancestor and sharing similar physical traits. We humans, for example, are the only modern-day members of the genus *Homo*. Within the genus *Felis*, however, there are several living members, including the domestic cat and the cougar, or mountain lion.

A **species** is a unique group within a genus whose members share the same set of structural traits and can successfully reproduce in nature only with members of the same species. (Recall from CHAPTER 16 that determining species can be difficult. For example, one cannot test mating potentials between fossil organisms, and species that do not generally reproduce sexually pose a special problem. Taxonomists must apply the best techniques available given each group's particular traits and life history.) As in all species' names, *Homo sapiens* ("wise human"), begins with a capital letter, and both names are written in italics or underlined. During a written or oral discussion, scientists mention the full species name *Homo sapiens* or *Felis domestica* (the house cat) the first time they refer to it. Subsequently, they may abbreviate the genus and simply use *H. sapiens* or *F. domestica* if there is no danger of confusion with a different genus, but they would not formally refer to the species as simply *sapiens* or *domestica*. This is because more than one species may share the second part of the species name. The pesky housefly, for instance, is in the genus *Musca* but the second part of the name is *domestica*, as in the name for the house cat.

THE HIERARCHY OF CLASSIFICATION

Biologists place each organism, with its genus and species name, in a series of categories, or *taxonomic groups*, based on shared characteristics. More broadly shared characteristics define more inclusive groups, and each group is a subcategory of the next higher level. To see how this works, follow FIGURE 18.3 as you study the following few paragraphs.

Taxonomists group several species together in a genus; our genus *Homo* includes ourselves *Homo sapiens*, plus the fossil species *H. erectus* and *H. habilis* [see FIGURE 23.8]. Similar genera are grouped together in the same **family**—in our species' case, the family of great apes (Pongidae), including people and fossil members of the genus *Homo*, the chimpanzees in the genus *Pan*, the gorilla in the genus *Gorilla*, and the orangutan in the genus *Pongo*. Other families are the lesser ape family (Hylobatidae), which contains the gibbon; the cattle family (Bovidae), which contains domestic cows and the American bison; the deer family (Cervidae), which contains deer, elk, and moose; and the cat family (Felidae), which contains house cats, mountain lions, tigers, and jaguars.

Members of similar families belong to the same **order**; the great ape and lesser ape families, for instance, belong to the order primates, while the cattle and deer families belong to the order artiodactyls (animals with an even number of toes). Other orders include rodents (such as mice and rats) and carnivores (such as cats, dogs, and bears).

Members of similar orders belong to the same **class**. All the animals we have mentioned so far are grouped in the class Mammalia, animals with hair and milk glands. Other classes include Amphibia, such as frogs and salamanders, and bony fishes (Osteichthes), such as salmon.

Similar classes group together in the same **phylum** (plural, **phyla**), in the case of animals, or **division**, in the case of plants. Animals in the same phylum have the same general body plan—all the animals we have discussed so far have a rod along their back called a notochord [review FIGURE 13.15] at some point in their life cycle, and are in the phylum chordata. Other animal phyla include mollusks, such as clams, snails, and octopuses, and arthropods, containing insects and lobsters and other animals.

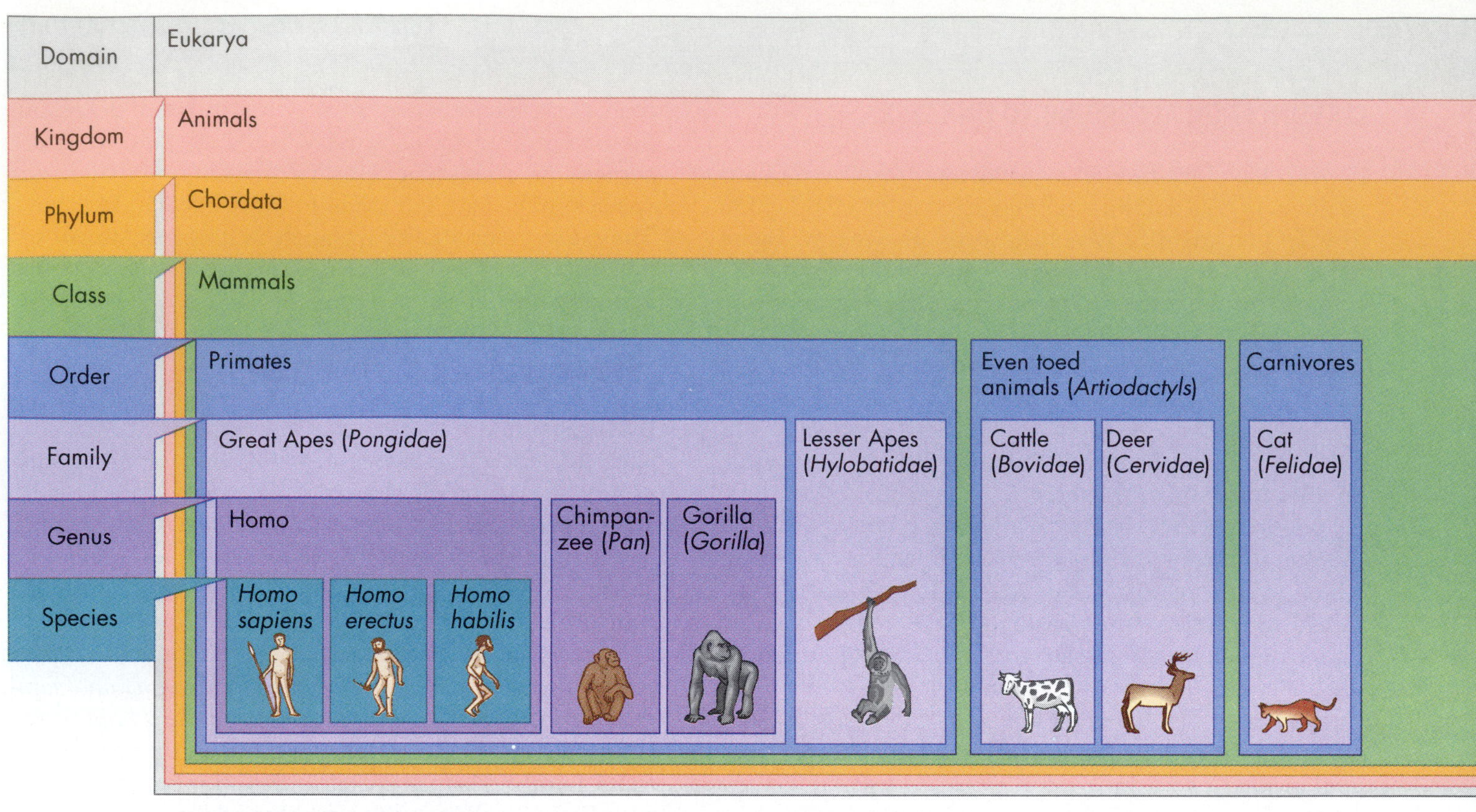

FIGURE 18.3

Hierarchies of Biological Classification.

While this diagram is far from complete, it shows how humans are related to other groups.

Taxonomists lump groups of phyla together into the same **kingdom**. All the organisms discussed here so far are in the animal kingdom, but there are a total of six kingdoms: animals, plants, fungi, protists, true bacteria, and archaebacteria. Finally, many biologists now assign similar kingdoms to the same **domain**, which we discuss in detail later. Biologists identify at least three domains, one (Eukarya) including all eukaryotes (animals, plants, fungi, protists), another (Bacteria) including the true bacteria, and a third (Archaea) including archaebacteria.

Of all the categories in this classification system, only the lowest level, the species, is *objective*; that is, biologists can test whether two different populations of individuals represent two different species by testing whether they can interbreed successfully in nature. Sometimes, however, even that definition is difficult to apply as in, for example, the fossil organisms or those that seem to lack sexual reproduction. The higher levels—genus, family, and so on—are based on the opinions of taxonomists who study particular groups of organisms. So how do taxonomists decide what to include in different genera or families?

CRITERIA FOR CLASSIFICATION

A taxonomist's job might seem easy—just group organisms according to their shared characteristics. But which characteristics should we use? Is flower color a good measure? Should we group together all plants with yellow flowers—dandelions with yellow roses, for example? Or should we group together plants with five sets of flower parts—yellow roses with white-flowered apple trees? Obviously, the characteristics taxonomists choose to lump together affect the groups they create.

Taxonomists generally choose characteristics in such a way that the groupings reflect some hierarchical principle of nature. But there's a catch: taxonomists disagree about which principle to emphasize. Some want to group organisms by the way they look or act: that is called the *phenetic approach*, because of the reliance on *phenotypes*. Other taxonomists want their hierarchy to reflect the branching pattern of evolutionary descent: this is called the *phylogenetic* (family-tree) *approach*. Since species with like physical traits usually share common

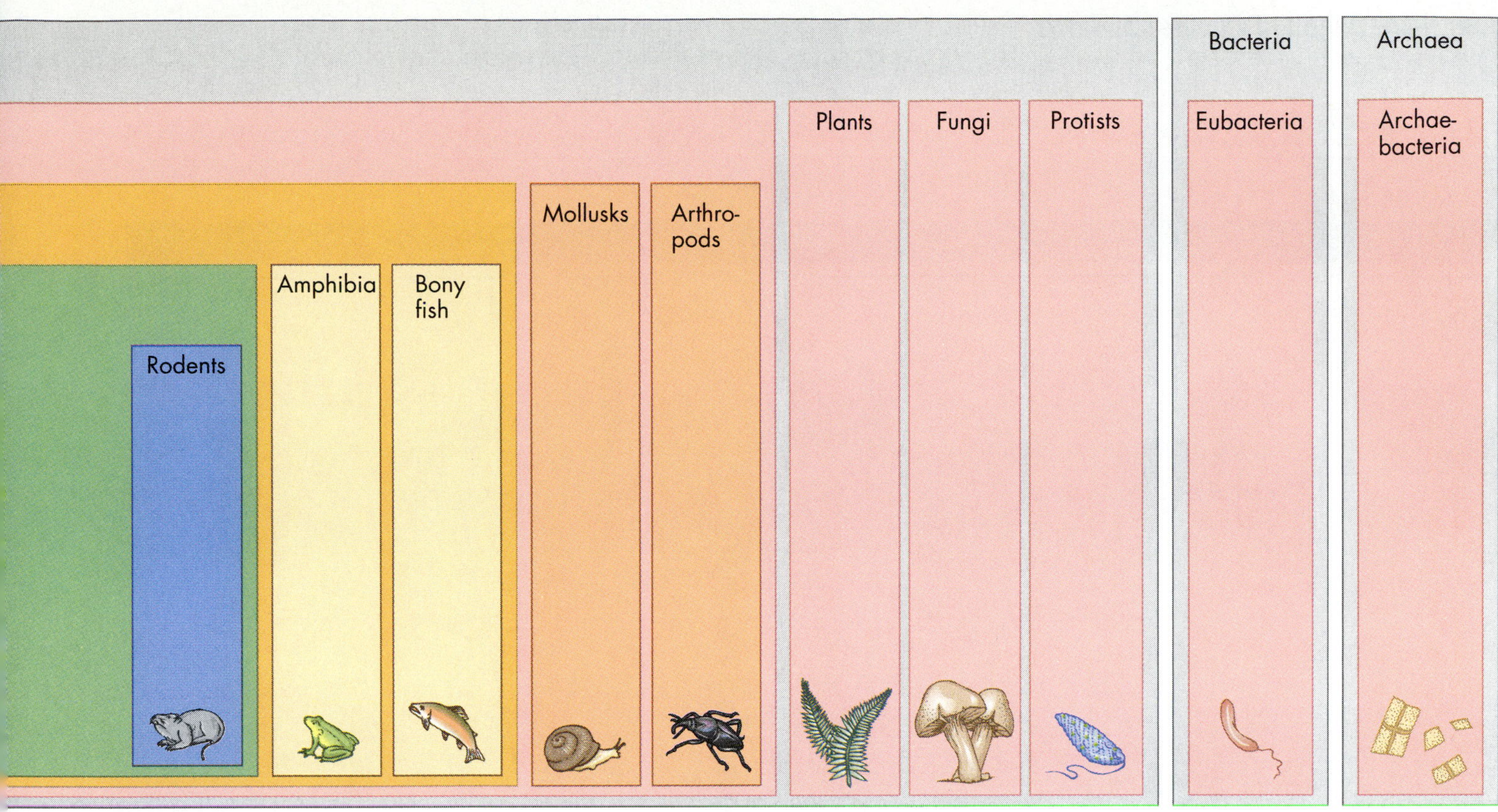

ancestors, these two schools can normally agree on how to classify organisms.

Disputes do arise, however, and in two ways: First, some organisms that look superficially similar are *not* closely related; their similarities are a result of convergent evolution [review FIGURE 15.13D]. Second, organisms that *are* closely related may look superficially different as a result of rapid evolutionary divergence [review FIGURE 15.13B]. FIGURE 18.4A depicts barnacles and limpets, two types of shelled animals that live affixed to rocks in pounding surf. Since each animal resides in a shell glued to a rock, barnacles and limpets seem more similar in appearance and life habits than either is to a crab scurrying across the ocean bottom. Actually, though, barnacles are more closely related to crabs than they are to limpets, as indicated by the lines of descent shown in FIGURE 18.4. Employing a phenotypic classification scheme, one might rely on shell shape, grouping limpets (which are mollusks) with barnacles (which are arthropods). On the other hand, a biologist using a phylogenetic classification scheme (classification based on family tree) might group barnacles with crabs (calling both arthropods), because of their jointed legs inherited from a recent common ancestor. The key here is to recognize that limpets and barnacles look alike superficially because of convergent evolution: Their similar traits arose as natural selection worked on genetic variations; as a result, adaptations evolved, suiting these organisms to their common environment, the pounding surf.

The physical-trait and family-tree systems of classification might also differ when related organisms have undergone rapid evolutionary divergence. Consider the classification of lizards, crocodiles, and birds [FIGURE 18.4B]. Birds and crocodiles share a more recent common ancestor than do crocodiles and lizards. But birds have diverged rapidly from the common ancestor because, through natural selection, many new forms have arisen and adapted to many different environments. Phenetic (phenotypic) classification might group lizards and crocodiles together because of their overall similarity. Phylogenetic classification, however, would group crocodiles with birds because of their more recent common ancestor. How can taxonomists resolve conflicting classification schemes such as these? How can they sort out the difficulties brought about by convergent evolution or rapid evolutionary divergence? While debate is sometimes quite heated on these issues, many biologists like to emphasize the phylogenetic approach.

➤ CONCEPT CHALLENGE

Certain aspects of the skeletons of birds and dinosaurs are very similar, but in exterior form, dinosaurs seem more similar to lizards. Would you classify birds in a group with dinosaurs or with lizards? If you do not have enough information to decide, what more do you need to know?

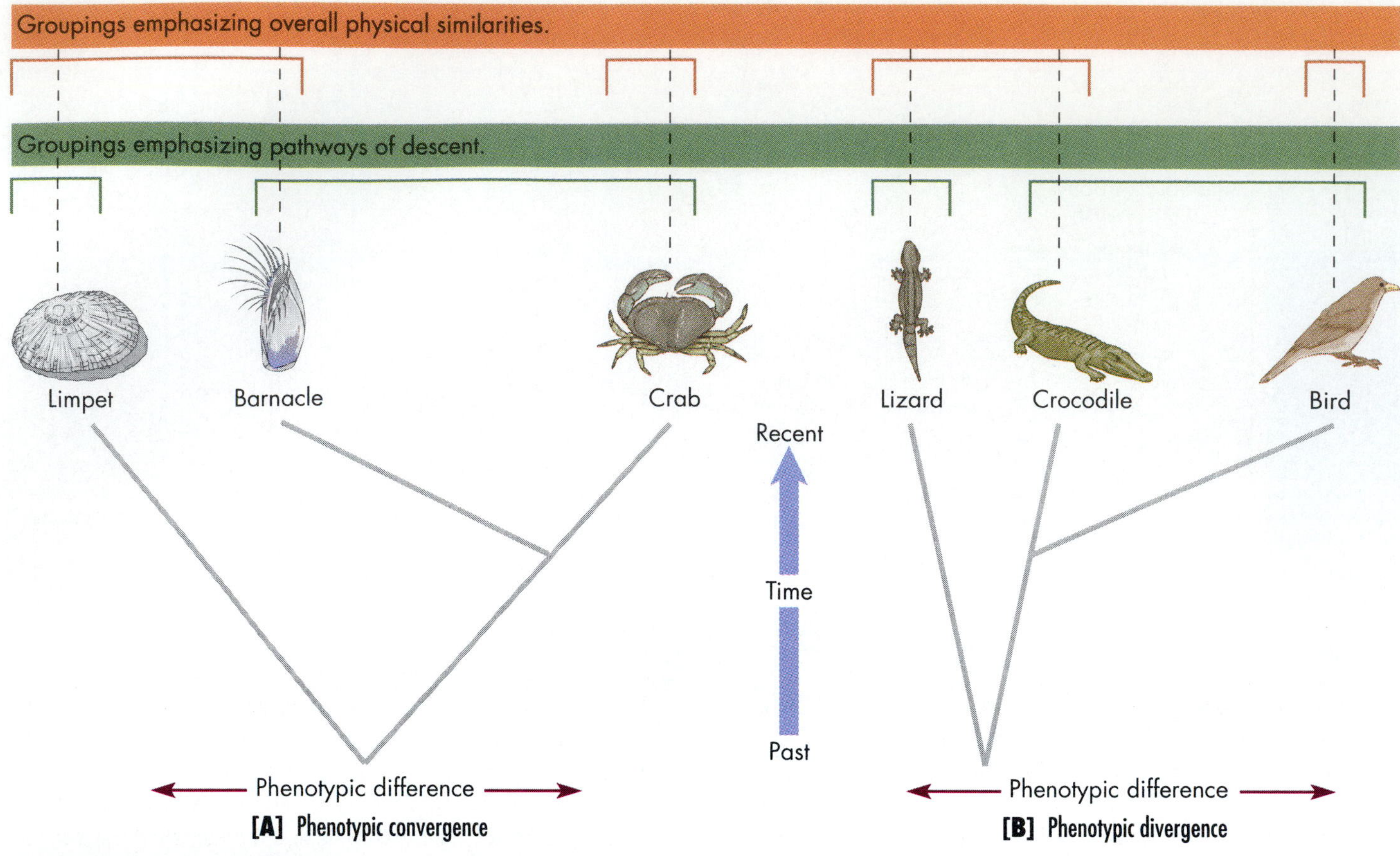

FIGURE 18.4

What Groups of Organisms Should We Classify Together?

[A] Phenotypic convergence occurs when two groups that are not closely related evolve to be like each other. [B] Phenotypic divergence occurs when an ancestral group splits, and one line evolves rapidly while the other evolves slowly. People who classify organisms by emphasizing physical similarities might categorize these animals differently than people emphasizing pathways of descent.

Cladistics: Classification by Shared Derived Traits

Many biologists today prefer to emphasize family trees as they determine taxonomic groupings, lumping organisms together by evidence of recent common ancestry rather than physical similarities. In our previous example, biologists taking this approach would group crocodiles with birds. To identify groups sharing recent common ancestry, many biologists use a procedure called **cladistics**, which attempts to reconstruct patterns of descent using traits or characteristics thought to have evolved only within the group under consideration; for example, hair evolved only in mammals.

The best way to understand cladistics is to apply it to an example. So let's classify a crocodile, a crow, a cow, and a bat [FIGURE 18.5]. A classification approach based on phenetics might group together the animals with an ability to fly: crows and bats. Fossil evidence, however, shows that the ability to fly evolved independently in ancestors of the two animals. Flight in birds and bats is an **analogous trait**, a similar trait possessed by two different groups, but not by ancestors of the groups. Another analogous trait is the streamlined body shape of porpoises and sharks. The land-living ancestors of porpoises lacked the streamlined shape, so the evolution of the porpoise's shape was independent of the evolution of the shark's shape. Cladists cannot use an analogous trait like flight in birds and bats to build a classification. They can build a classification only by using a **homologous trait**, a similarity shared by two species due to inheritance from a common ancestor. For example, hair in cows and bats is a trait possessed by the common ancestor of these two species, and hence this trait is homologous. A cladist, therefore, would use hair to group cows and bats together in the same classification category, but would not use the trait of flight to group bats and crows together.

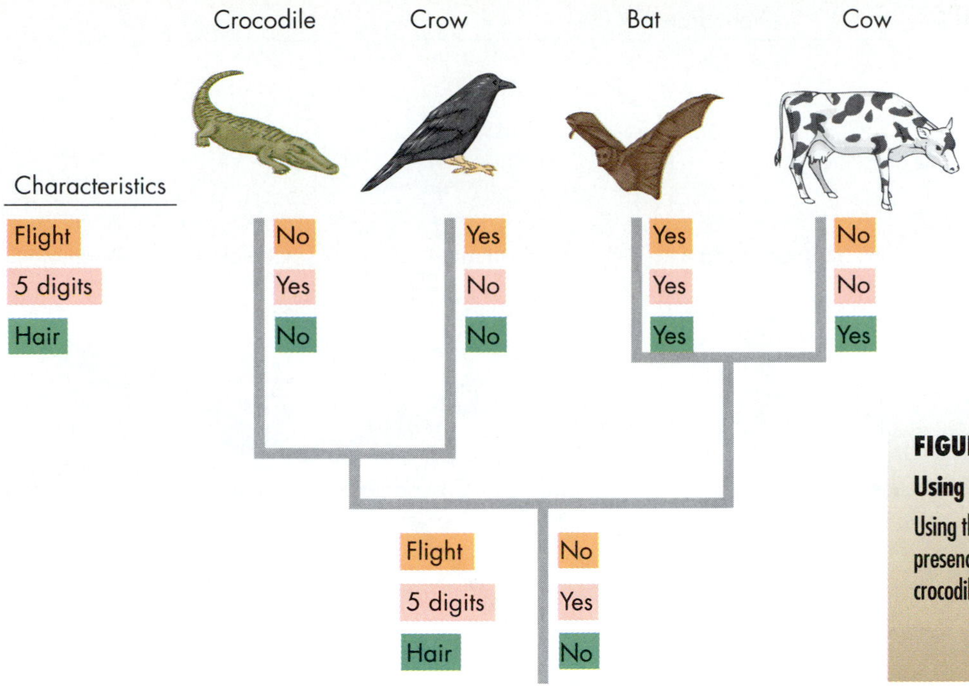

Crocodile
Crow
Bat
Cow

Characteristics
Flight
5 digits
Hair

Characteristics	Crocodile	Crow	Bat	Cow	Common ancestor
Flight	No	Yes	Yes	No	No
5 digits	Yes	No	Yes	No	Yes
Hair	No	No	Yes	Yes	No

FIGURE 18.5

Using Cladistics: Identifying Shared, Derived Traits. Using three traits (flight, five digits on each limb, and the presence of hair) we can construct a family tree for the crocodile, crow, bat, and cow.

Besides distinguishing between analogous and homologous traits, cladists must make another distinction—one between ancestral characteristics and derived characteristics.

An **ancestral characteristic** is a trait possessed by an ancient ancestor that has been modified by evolution into new characteristics in some later species. An example of an ancestral characteristic is the occurrence of five digits on each limb, as found on crocodiles and bats and other more distantly related species, such as salamanders.

In contrast to an ancestral characteristic, a **derived characteristic** is an evolutionary novelty, an inherited change of a previously existing characteristic. An example of a derived characteristic is two digits in cows (cloven hooves) and three in crows. Fossils reveal that the common ancestor of cows and crows had five digits on each appendage. However, in the evolutionary line that led to cows, three of the digits were lost, and in the line that gave rise to crows, two digits were lost. (In the line that gave rise to people, no digits were lost.) Thus, a reduced number of digits was derived from the ancestral condition of five digits. Another derived characteristic is hair. Hair was derived in the common ancestor of cows and bats from the ancestral condition of scales in ancient reptiles. Thus, the presence of hair, a shared, derived trait, would lead a cladist to cluster bats and cows together on one branch of a family tree (also called a cladogram) [FIGURE 18.5].

Let's apply cladistics to the platypus—an animal that we saw has some traits of mammals, but others of reptiles and birds. Which group should we group it with—the birds, reptiles, or mammals? We can't use the bill of the platypus and the bill of the duck to group the platypus with birds because these features are analogies; the bills are similarly shaped mouthparts that evolved independently as adaptations for shoveling small bits of food from the bottoms of streams. Likewise, we can't lump the platypus with reptiles based on the traits of reptilian shoulder girdle, reproductive system, and egg-laying habit. Although the platypus does, indeed, share these traits with reptiles, they are ancestral traits that were already present in the common ancestor of reptiles and the platypus, rather than derived traits originating in the line that gave rise solely to the platypus and its close relatives. In contrast, hair and mammary glands *are* shared, derived traits: characteristics that are shared by all mammals and only mammals and are derived from the ancestral condition (reptilian scales) via evolution. Cladistic principles emphasizing shared, derived traits place the platypus squarely with the mammals.

Using similar reasoning, cladists have developed a proposed family tree for all the vertebrates [see FIGURE 18.6]. To do this, cladists used shared, derived characteristics like the presence of hair, membranes surrounding developing embryos, and other traits.

To understand how a family tree can be used to construct a classification, consider FIGURE 18.6. The mammals that produce a placenta (such as female cattle), the mammals bearing a pouch (marsupials, such as kangaroos), and mammals that lay eggs (such as the platypus) all occupy branches from the same stem. (These are the green branches in FIGURE 18.6.) Hence they are grouped together in the class Mammalia. Likewise, dinosaurs, birds, and crocodiles all occupy a single branch (blue in the figure), and so they are grouped together in a class. If

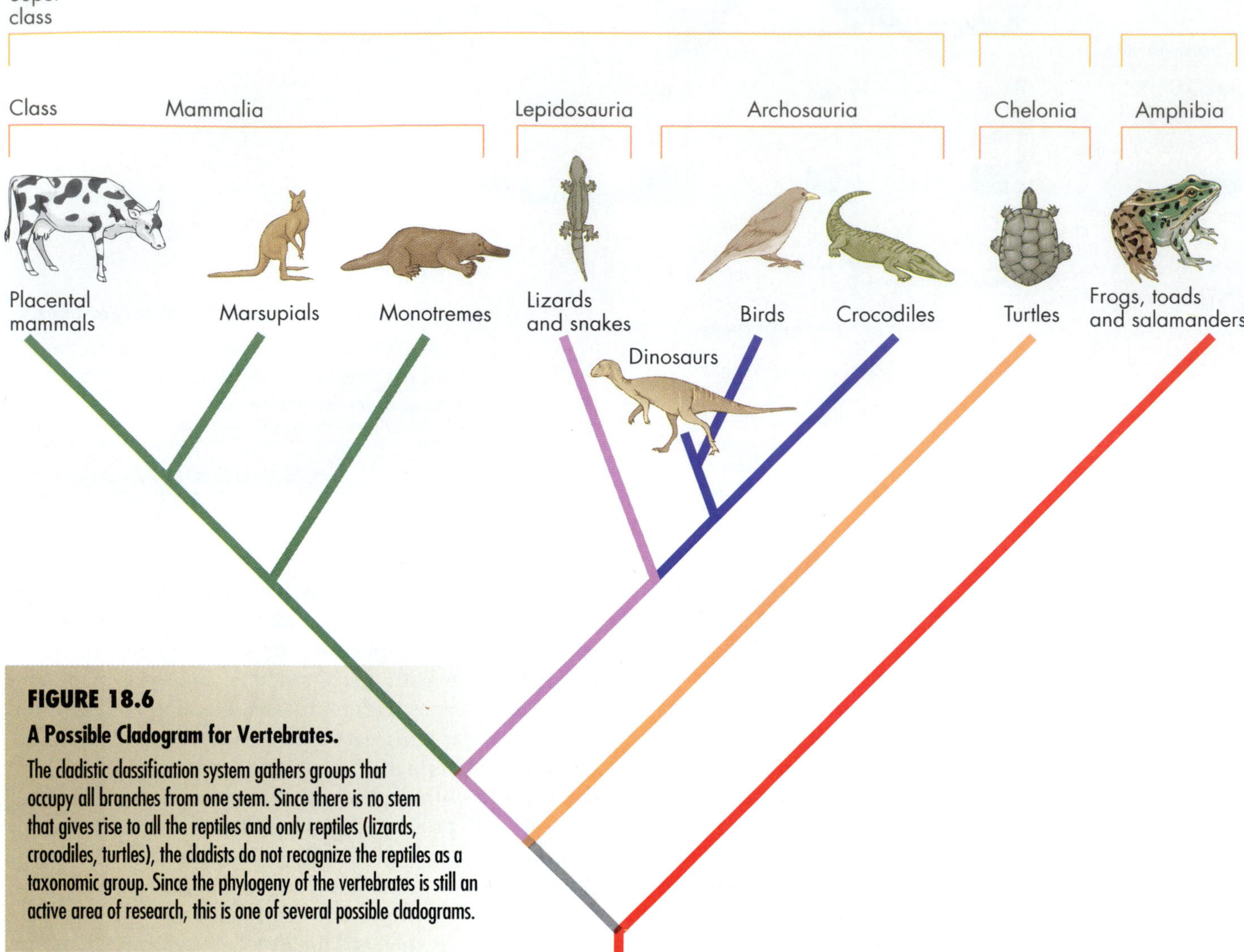

FIGURE 18.6

A Possible Cladogram for Vertebrates.

The cladistic classification system gathers groups that occupy all branches from one stem. Since there is no stem that gives rise to all the reptiles and only reptiles (lizards, crocodiles, turtles), the cladists do not recognize the reptiles as a taxonomic group. Since the phylogeny of the vertebrates is still an active area of research, this is one of several possible cladograms.

we now include the lizards (purple) with those two classes (green and blue), we have another group, which contains all the twigs coming from one branch and which is a superclass encompassing all animals whose embryos form an amnion [review FIGURE 14.9]. Finally, the turtles (orange) and amphibians (red branch) occupy their own parts of the tree, and hence are assigned to their own classes and superclasses. Note that no branch on this family tree has subbranches containing *all* reptiles and *only* reptiles (snakes, turtles, and crocodiles). Cladists, do not recognize reptiles as a distinct group.

➤ CONCEPT CHALLENGE

Consider the flippers of a whale and a seal. How would a cladist determine whether these flippers are shared, derived traits, ancestral traits, or merely analogies?

The Kingdoms of Life

During the time of Carolus Linnaeus two centuries ago, scientists considered all organisms to be either animals or plants. The more biologists learned about the cellular structure of organisms, however, the more species they discovered that did not fit neatly into these two kingdoms. Many biologists today recognize five or six main kingdoms [FIGURE 18.7]; in this book, we will use the six-kingdom scheme, Eubacteria, Archaebacteria, Protista, Fungi, Plantae, and Animalia.

The first two kingdoms are both single-celled prokaryotes, cells without membrane-bound nuclei. The **Eubacteria** (*eu* means "true") include familiar bacteria like the ones that cause teeth to decay and turn milk to yogurt. The **Archaebacteria** are also single-celled prokaryotes, but they have very distinctive cell membranes

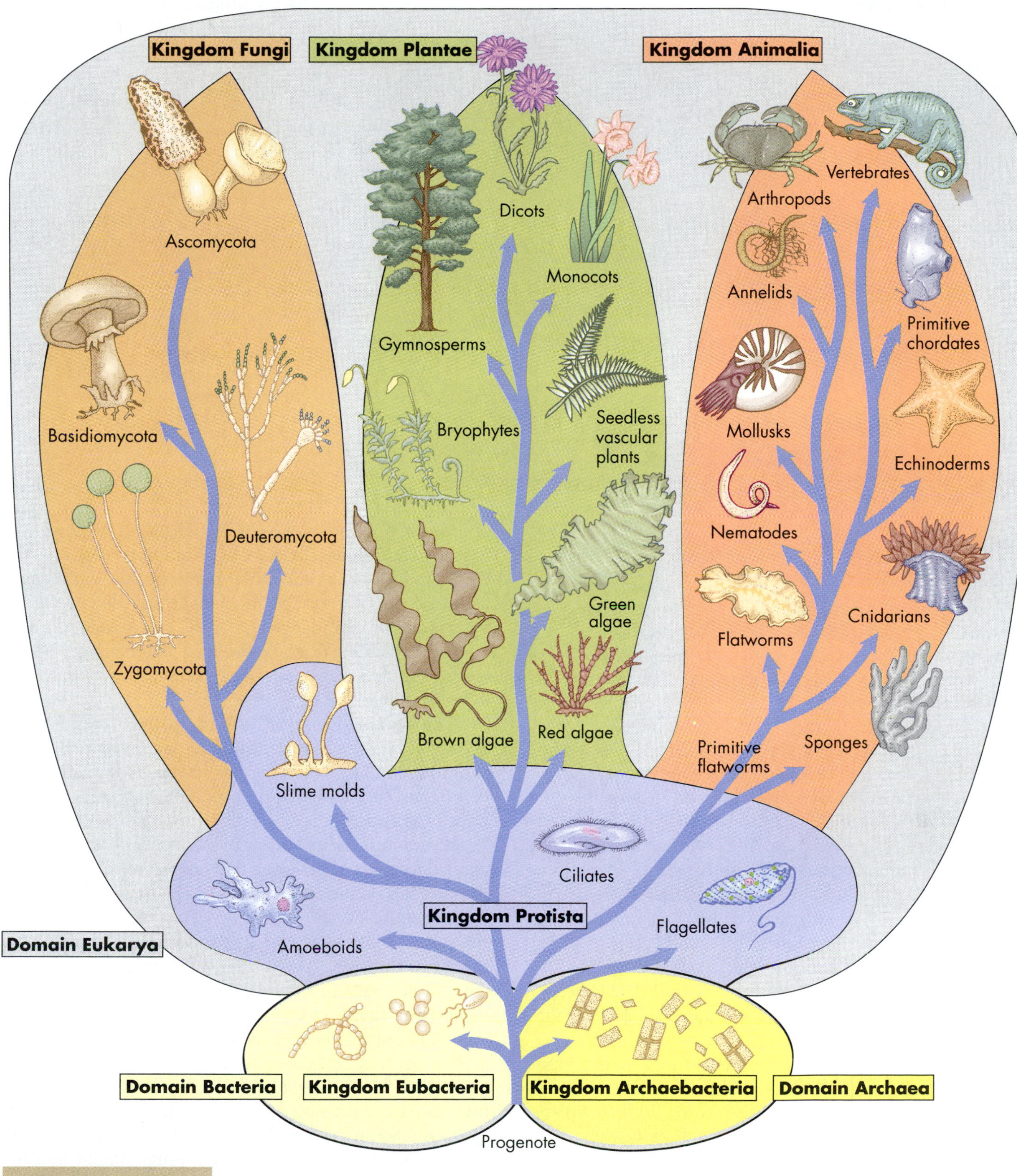

FIGURE 18.7
The Six Kingdoms of Life.

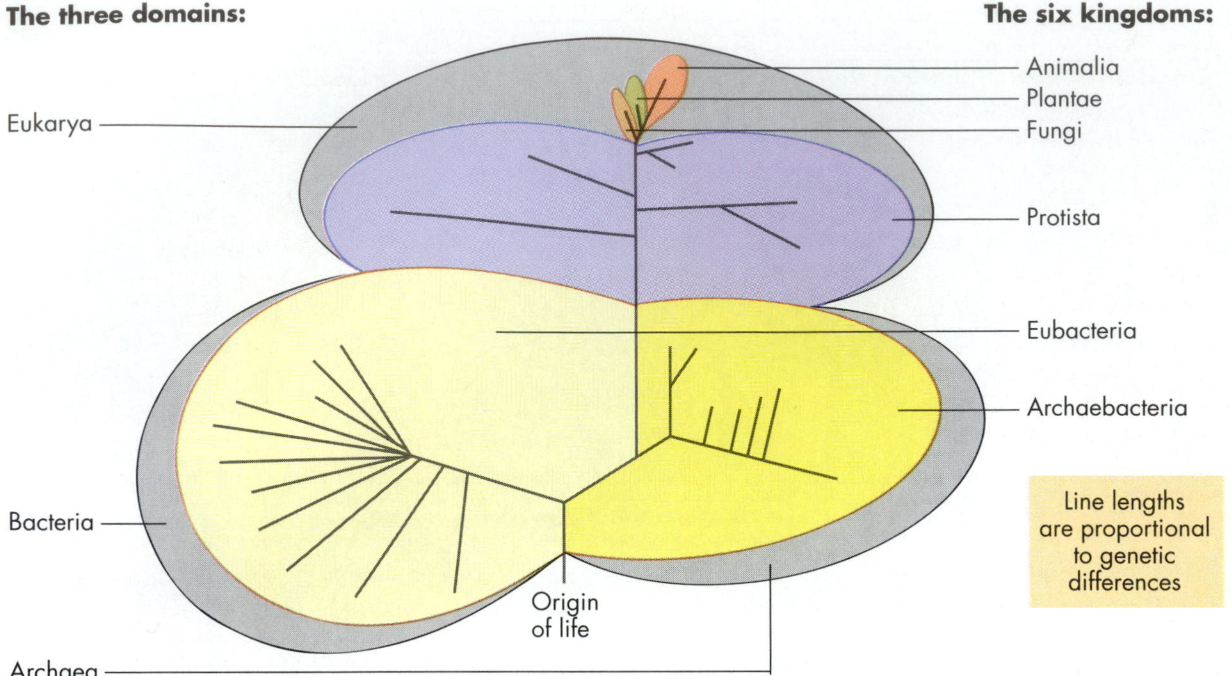

FIGURE 18.8

Domains and the Family Tree of All Life.

Based on the sequences of ribosomal RNA genes, living things fall into three domains: the Bacteria, Archaea, and Eukarya. These data revolutionized our appreciation of life's genetic diversity, since the most visible, most familiar kingdoms—fungi, plants, and animals—are just tiny branches arising from the tip of one of the three branches.

as well as other unique biochemical and genetic traits. The Archaebacteria live in harsh environments we would find unlivable, such as acidic, sulfurous hot springs, very salty lakes, and the interiors of cows' intestines! Taxonomists used to lump the Eubacteria and the Archaebacteria together in the kingdom Monera, creating the classic five kingdoms of life first proposed by R. H. Whittaker in 1969. But studies in the 1980s on the molecular biology of Archaebacteria suggested such profound differences between Eubacteria and Archaebacteria that the latter rated their own kingdom, bringing the total to six.

The last four kingdoms (protists, fungi, plants, and animals) contain eukaryotic cells, cells with envelope-enclosed nuclei. The **Protista** (also called Protoctista) are generally single-celled eukaryotes such as *Euglena* and amoebas [see FIGURE 3.14]. The **Fungi** include multicellular heterotrophs, such as mushrooms and molds, as well as unicellular yeasts. Fungi decompose other biological materials, develop from spores, and lack flagella. The **Plantae**, which are mostly multicellular autotrophs, include some algae and mosses, ferns, and flowering plants; plants generally develop from embryos. Finally, the **Animalia** are multicellular heterotrophs, which lack cellulose, usually exhibit movement and develop from embryos; they include sponges, insects, clams, birds, and mammals [FIGURE 18.7]. In the next five chapters of the book (CHAPTERS 19 to 23), we explore these six kingdoms in greater detail.

The six-kingdom classification system, then, reflects diversity in cell structure. In 1990, however, molecular evolutionist Carl Woese compared the genetic structure of different cells and revolutionized our understanding of how the six kingdoms of life relate to each other. To make genetic comparisons of very different species, Woese needed to investigate a gene that is present in all species. Because ribosomes are present in all organisms, Woese decided to look at the sequences of the genes that encode ribosomal RNA [review FIGURE 10.12]. He then assumed that organisms which are closely related by descent have more similarities in ribosomal RNA gene sequence than organisms which are only distantly related. By comparing many organisms, he could construct a family tree for all life—at least as far as this one gene is concerned.

Woese found that the six kingdoms naturally cluster into three main categories he called *domains* [FIGURE 18.8]. In Woese's system, the domain **Eukarya** includes four of the six kingdoms—protists, fungi, plants, and animals. The other two domains have just one kingdom each; the domain **Bacteria** contains the kingdom Eubacteria, and the domain **Archaea** contains the kingdom Archaebacteria.

Interestingly, the two domains of prokaryotic cells, Bacteria and Archaea, are not closely related genetically [see FIGURE 18.8]. The third domain, however, Eukarya, is related to both of the other domains. This relationship exists because the Eukarya evolved in a patchwork fashion; the cell nucleus is derived from some ancient prokaryotic cell closely related to surviving Archaea, while the mitochondria and the chloroplasts of Eukarya are derived from cells in the domain Bacteria [review FIGURE 17.15].

While the concept of three domains may seem like an unnecessary addition to the six-kingdom scheme, in fact it revolutionizes a biologist's perspective on the evolution of life. For example, the familiar organisms we see every day—squirrels, roses, robins, mushrooms, dogs, oak trees, earthworms, and flies—are just tiny branches at the tree's periphery. The vast expanse of evolutionary diversity determined through molecular analyses lies in the single-celled organisms within the domains Bacteria and Archaea, and in the kingdom Protista. The last section of this chapter further explores the vast diversity encompassed by this expansive three-domain tree of life.

➤ CONCEPT CHALLENGE

Cladists classify together groups of organisms that include an ancestral species and *all* of its descendants (a so-called *monophyletic group*). Which of the six kingdoms of life would not be such a group according to the molecular genetic data in FIGURE 18.8?

Biodiversity and Extinction

Biologists don't know how many other species are on this planet. Some think the figure could be 3 million, others upwards of 50 million. One thing is certain, species are becoming extinct faster than taxonomists can describe them.

HOW MANY SPECIES INHABIT EARTH?

Over the past 230 years, biologists have discovered, named, and recorded 1.5 million species of living organisms. This has been basically a hit-or-miss prospect, based on where naturalists have gone, what they have been interested in collecting, and how much time (and funding) they have had for studying the unknown organisms and determining whether they can in fact claim and name them as new species.

Of the 1.5 million known species, about 5 percent are single-celled prokaryotes and eukaryotes; about 22 percent are fungi and plants; and about 70 percent are animals [FIGURE 18.9]. Most of the animals are invertebrates (they lack backbones). The sea harbors the most varied groups of invertebrates, with 90 percent of all the higher taxa (phyla and classes). But these thousands and thousands of taxonomic categories contain only 20 percent of the individual animal species. By far the largest number of species are animals that inhabit land. And of these, nearly a million species so far named are insects. Ours is a planet of terrestrial insects [FIGURE 18.10], and there are probably millions of species still unnamed.

For many years, biologists estimated that an accurate number of total species would be about 3 to 5 million. They based this estimate on the fact that two-thirds of the currently known species live in temperate zones (where, coincidentally, almost all the scientists have lived). Since a long-accepted formula held that perhaps twice as many species inhabited tropical regions as temperate ones, it seemed likely that there were at least another 1.5 million species to be found and studied.

This estimate lost favor, however, when in the 1970s and 1980s biologists went out to study the "last biotic frontiers"—the seabed in the ocean deeps, and the canopy (treetop habitat) of tropical rain forests. It has been a common occurrence for researchers to collect mud from a previously untested spot of ocean bottom, or to gather the plants and animals clinging to the rarified world at the top of an individual tropical treetop, and to find that 90 percent of the organisms collected were unknown to science! For example, Terry Erwin, a researcher working in the Panamanian rain forest, set off insecticidal fog bombs in individual jungle trees and collected the organisms that dropped from the trees. He then focused on the beetles and classified them according to species.

On average, 163 species of beetles fell from each species tree. He reasoned that since there are at least 50,000 tropical tree species, beetles represent just 40 percent of animals in the jungle treetops (almost all the others are also insect species), and half again as many species live on the ground as up in the forest trees, there could easily be 30 million additional species of insects and their relatives to describe in the rain forests alone. Adding in tropical plants, seabed organisms, and all other undiscovered species in other climatic zones, the figure of 50 million new species—despite its magnitude—begins to look possible.

What is the significance of there being as many as 50 million species on the earth while only 1.5 million have been named? First, it is clear that if these species are ever to be cataloged for science before they become extinct, the world needs more professional taxonomists. Second, we do not fully understand the ecological significance of this great degree of diversity. Is a huge number of species important for ecosystems or the biosphere to function? Or are ecosystems with large numbers of species more stable and long-lived than those with fewer species? We will probe these important questions in CHAPTERS 38 to 41. Although no final answers are available, what we do know is that human activities are now causing the extinction of many species.

EXTINCTION

With perhaps 50 million species on earth, but only 1.5 million named, we clearly have a very long way to go to collect and catalog our neighbors on this planet. The prospects of this happening are dismal, however, because people are destroying habitats—tropical rain forests, coral reefs, drylands, and other natural areas—faster than scientists can survey and describe them. When habitats are destroyed, so are organisms' feeding and nesting sites, and so organisms cannot survive or reproduce normally, and more individuals die than are replaced by reproduction. When the last breeding pair of a species disappears, the species becomes extinct. Since larger areas encompass more species, we can directly translate the loss of massive tracts of habitat into staggering species extinction. Ecologists Peter Raven and E. O. Wilson estimated in 1992 that up to 20 percent of the world's species could disappear over the next 30 years due to the doubling of human populations in tropical and semitropical zones. Some biologists are skeptical about such a high projected rate of extinction. There can be no doubt, however, that species are disappearing at an alarming rate (for more details, see CHAPTER 38). The extent of the loss, in fact, begins to approach some of the great extinction episodes of the past. The main difference is that *we* are causing this one, and at a much faster rate of loss than in past events. Keep in mind, though, that humans are not the only agents of extinction. The extinction of dinosaurs and other species 65 million years ago took place long before humans were on the scene.

Historic **mass extinctions**, the extinction of many groups of organisms over relatively short spans of geologic time, can serve as models for what we can expect during the current biodiversity crisis. FIGURE 18.11 shows that the history of life has been punctuated by a series of mass-extinction events. Following each episode was a burst of evolution, with many new forms replacing the extinct ones. The radiation of the mammals following the disappearance of the dinosaurs is a good example of this type of species replacement.

Past extinctions also teach us that during a massive die-off, species are not lost entirely at random. In the extinctions that took place about 10,000 years ago after the last ice age, land animals with large bodies—mammoths, mastodons, and large cats, for example—tended to die out in greater numbers than smaller-bodied forms. This trend suggests that today's extinction crisis will claim many of our favorite animals first, including giraffes, pandas, and certain whales.

A third lesson from the past is that genera widespread in nature tend to survive mass extinctions, whereas genera localized to particular small regions are especially vulnerable. If a genus is confined to one small corner of the world, then the alteration of this small region by human or natural means will eliminate viable habitat and the organisms in the genus will probably become extinct. This suggests that over the next centuries, our world will be enriched in widespread, rapidly growing "weedy" species—rats, cockroaches, ragweed—and impoverished in the rare, locally adapted species. Ironically, these rare species have tended to be most useful to people for medicines and foods. Recent fossil evidence shows that 35 million years before the dinosaurs died out, species of the platypus ranged widely from Australia to South America. This widespread distribution may, in fact, have saved the platypus from extinction at the end of the Mesozoic [see FIGURE 18.11]. Today, however, the platypus is limited to one green strip of eastern Australia,

FIGURE 18.9

Earth's Diversity.

The sizes of the figures here represent the relative number of species currently described in each group of organism. Insects account for more species than all other kinds of organism combined. Earth is the planet of the insects.

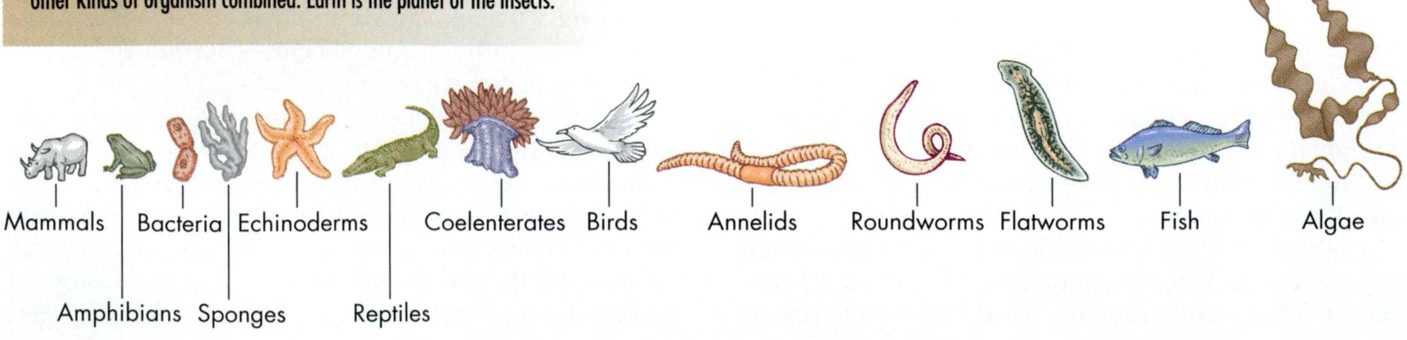

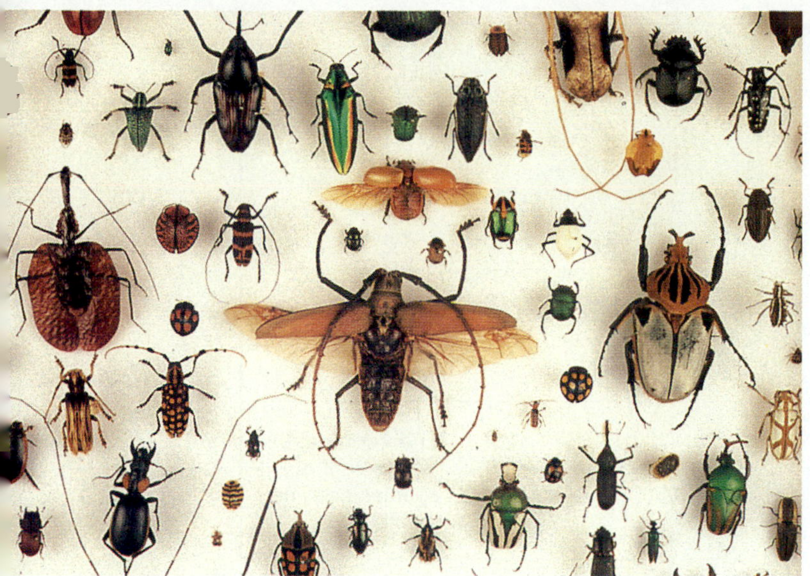

FIGURE 18.10

Radiation Run Rampant.

Nature has taken the basic insect theme and provided millions of innovative variations.

and it survives only because that nation has passed powerful laws to protect it. Without such assistance, the platypus could well be among the first to disappear of the rare and locally adapted species.

A final lesson from the fossil record of extinctions is that the permanent losses we cause in the next 50 years or so will not be recouped for many millions of years. It takes a very long time for specialized species to evolve. In fact, as long as humans continue to exist in large numbers as they do today, it is unlikely that species diversity will recover because of the dramatic way people change and manipulate their environments.

The message from a study of past extinctions is that even mild extinction crises have far-reaching biological repercussions. Many biologists conclude that the potential consequences of human-induced extinctions are so dire we must act now without hesitation or equivocation.

TAXONOMY

In recent years, taxonomy has become an exciting, high-technology science. Modern taxonomists might find themselves suspended from the tree tops of a remote rain forest [FIGURE 18.12], investigating novel ocean-bottom ecosystems in a deep-sea submersible, or working in a molecular genetics laboratory.

Taxonomy has also taken on new significance. We are changing ecosystems drastically as a result of pollution, habitat destruction, and other human activities. Our survival, and that of the millions of species that share our planet, depends on our understanding how ecosystems change as species die out and are replaced. But as we just saw, we lack the roughest estimate of how many species exist, let alone a list of them and how they live, die, and interact in the environment.

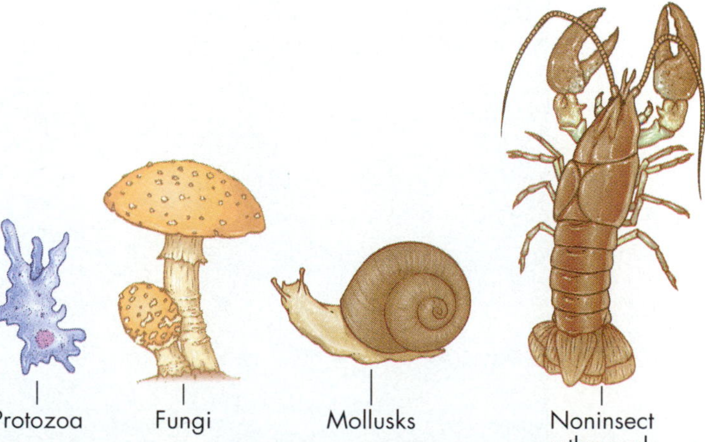

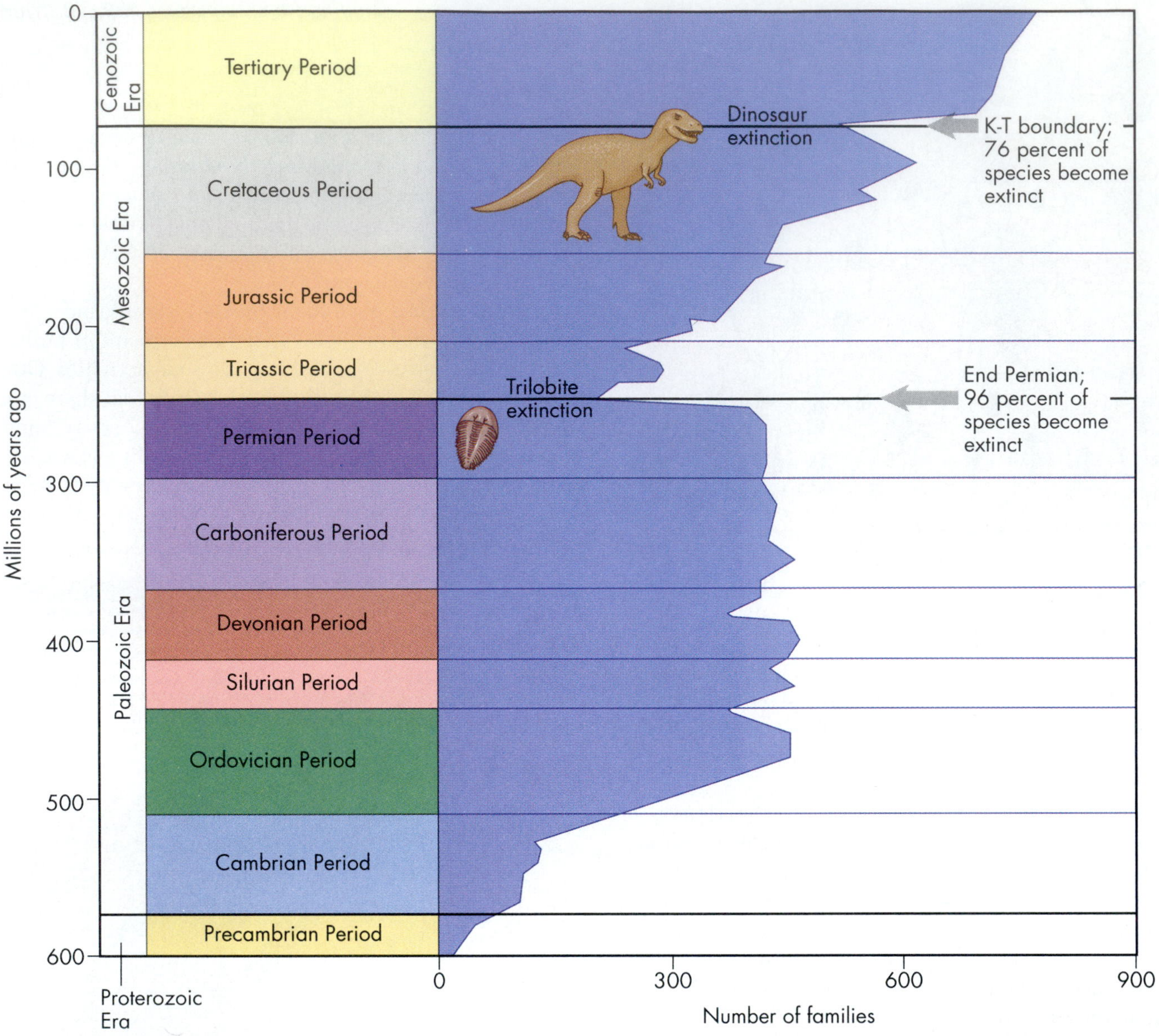

FIGURE 18.11
Extinction and Radiation.
Five major extinctions, and many smaller ones, punctuate the history of life.

FIGURE 18.12
The Costa Rican Rain Forest: A Hangout for Taxonomists.
This researcher studies the hundreds of insect species inhabiting a single tree.

Discovering, categorizing, and investigating the millions of unclassified species is important not only for our understanding of biology, but also for purely practical reasons. One reason is that researchers have developed many cancer-controlling drugs and other modern medicines from chemical constituents of rare plants. Extracts from the yew tree of the Pacific Northwest [see FIGURE 20.1] and the periwinkle flower of Madagascar [see FIGURE 1.1] are examples. Another reason is that many nutritious fruits and roots are no doubt growing in obscurity somewhere on the globe and some of those could perhaps become staple foods for small local human populations. Yet modern agriculture is teetering on a dangerously slim genetic base represented by the three or four dozen staple crops upon which our species depends for survival. We should be looking for new drug-bearing plants, new crop species, and conserving the gene pool of our heavily cultivated plants.

To combat the rapid species extinction being precipitated by burgeoning human populations, governments must commit more resources to combating overpopulation, pollution, and habitat destruction. Our government, however, may not have the will—for example, at a worldwide conference held in 1991 in Rio de Janeiro, the United States failed even to sign the global Convention on Biological Diversity designed to encourage species conservation. To catalog species before they disappear, Raven and Wilson propose a massive effort to train and deploy scientists with the skills necessary to perform the huge taxonomic tasks confronting us (see details in BOX 38.2). We hope that some portion of the audience reading this book now will become motivated to study taxonomy. It is truly a science with a deadline.

Connections

In this chapter we learned about the tens of millions of species alive today and how biologists classify and try to make sense of this vast diversity. The rest of the book will rest on the framework these topics provide. In the next chapters on life's variety we broaden our understanding of evolution by exploring the adaptations that organisms in different kingdoms have evolved and how these characteristics allow them to survive in their specific environments.

KEY TERMS

analogous trait, 436
ancestral characteristic, 437
Animalia, 440
Archaebacteria, 438
binomial nomenclature, 433
class, 433
derived characteristic, 437
domain, 434
Eubacteria, 438
family, 433
Fungi, 440
genus, 433
homologous trait, 436
kingdom, 434
order, 433
phylum, 435
Plantae, 440
Protista, 440
species, 433
systematics, 432
taxonomy, 432

HIGHLIGHTS IN REVIEW

1 Systematics is the scientific study of the diversity of organisms and how different organisms are related.

- a] A part of systematics is **taxonomy**, the theoretical basis for grouping organisms.
- b] The binomial system for naming organisms assigns each species a two-word name. The first is the name of a genus, and the second, when combined with the first, refers to a single species. A genus is a group of very similar species related by common descent from a common ancestor. For sexually reproducing organisms, a species is often defined as a group of individuals that can successfully interbreed in nature.
- c] Taxonomists group organisms in a hierarchy of ever-broader groups. Several related species belong to the same genus; several related genera are in the same family; then comes order, class, phylum (or division), kingdom, and domain.

2 Biologists often group together organisms sharing a set of similar traits which they inherited from a common ancestor (homologies). Organisms can also share similar traits due to shared environments rather than a shared recent ancestor (analogies).

- a] Some taxonomists take a phenetic approach to classification: they use shared physical traits as the basis for determining groupings. Others take a phylogenetic approach, and emphasize the branching patterns of evolutionary descent.
- b] Problems with classification arise when organisms look alike due to convergent evolution rather than common descent. Problems also arise when one group diverges rapidly in phenotype from the ancestral line, so that it looks very different from species to which it is rather closely related.
- c] Cladistics attempts to group organisms in categories related by pathways of descent. Cladists try to identify traits that are shared between two groups because the traits were *newly* derived in a common ancestor of the two groups (a so-called shared derived trait). Cladograms are diagrams that show how organisms are related based on a cladistic analysis.

3 Comparisons based on gene and protein sequences suggest that there are three major domains of life: the Bacteria, the Archaea, and the Eukarya.

- a] The domains Bacteria and Archaea include prokaryotic cells of the kingdoms Eubacteria and Archaebacteria, respectively. The Eukarya includes four of life's kingdoms: Protista, Fungi, Plantae, and Animalia.
- b] Much of life's genetic diversity lies in the domains Bacteria and Archaea. While the domain Eukarya is ancient, the familiar many-celled kingdoms (fungi, plants, and animals) are small branches at the tip of the tree.

4 Our planet's diversity of life is immense; there are perhaps as many as 50 million species. Human activity is threatening to destroy much of this diversity in our lifetimes. Prompt action may prevent the irretrievable destruction of the living products of millions of years of evolution.

- a] Extinction is the ultimate fate of all species. At several times in the history of life, a large percentage of species have become extinct over a geologically short period of time—a mass-extinction event.
- b] Following each episode of mass extinction, new forms radiated. The accelerated radiation of mammals after the extinction of dinosaurs is an example. It takes, however, a few million years for species diversity to recover from mass extinction.
- c] During mass extinctions, widespread "weedy" groups (like rats, cockroaches, and dandelions) tend to survive, and species highly adapted to local conditions (like the platypus and panda) tend to succumb.
- d] Ecologists warn that if taxonomists do not identify and classify existing species in the next decade, millions of species will perish—unknown to science—in the current mass-extinction event, which is being caused largely by the destruction of natural environments by humankind.

UNDERSTANDING THE FACTS AND CONCEPTS

For Questions 1–5, match each of the descriptions with the most appropriate item from the following list of terms. Each item in the list may be used only once, more than once, or not at all.

- a] species
- b] class
- c] genus
- d] family
- e] order

1 Only includes members of similar families.

2 A collection of closely related genera.

3 The only objective, natural, taxonomic category.

4 The group that includes similar orders.

5 The most limited group that is comprised of related species.

As above, for Questions 6–10, match the descriptions with the most appropriate items from the following list.

a] phylum
b] kingdom
c] class
d] domain
e] genus

6 A taxonomic group comprised of similar orders.

7 A taxonomic grouping that is used as part of an organism's binomial name.

8 The most limited group that includes related classes.

9 The group that is comprised of one or more phyla.

10 The most inclusive taxonomic category.

As above, for Questions 11–15, match the descriptions with the items in the following list.

a] phenetic classification
b] cladistics
c] phylogenetic taxonomy
d] homology
e] analogy

11 A system of classification based on appearance.

12 A system of classification based on family trees.

13 An anatomical similarity of two species based only on adaptation to similar environments.

14 An anatomical similarity based on inheritance from a common ancestor.

15 A method of classifying organisms based upon their shared derived characteristics.

As before, for Questions 16–20, match the descriptions with the items in the following list.

a] Archaebacteria
b] Fungi
c] Protista
d] Animalia
e] Plantae

16 Multicellular autotrophs.

17 Found living in hot, sulfurous springs and salty lakes.

18 Mostly single-celled eukaryotes.

19 Multicellular heterotrophs such as molds and yeasts.

20 Multicellular heterotrophs that lack cellulose; most are motile.

INTEGRATE AND APPLY WHAT YOU HAVE LEARNED

1 What is the binomial system of nomenclature?

2 What is the most limited taxon that includes all eukaryotic organisms? What about the prokaryotes? Explain.

3 Distinguish between the terms *analogous traits* and *homologous traits* as they are applied to the science of taxonomy.

4 Why was the platypus classified as a mammal when just as many of its characteristics are either reptilian or birdlike?

5 Why are there believed to be so many species, perhaps millions, that have yet to be discovered and described?

CHAPTER 19

Prokaryotes, Viruses, and Protists

A CELL WITHIN A CELL

Have you ever seen Russian dolls painted with red cheeks and embroidered aprons and carved to nest one inside the other from large dolls to very small ones? In an interesting analogy, termites, insects that consume and digest wood, harbor complex single-celled organisms called protists in their guts. One such protist is a particularly striking cell, *Trichonympha* [FIGURE 19.1]. It is huge for a single-celled organism, and its cell surface has thousands of lashing, whiplike flagella. Within the light-bulb-shaped *Trichonympha* live several kinds of much smaller bacteria, and these cells actually produce the wood-digesting enzymes.

This series of biological "Russian dolls" is an example of *symbiosis*—the living together of more than one species. In this case, the host termite chews wood into small pieces that move down the digestive tract and

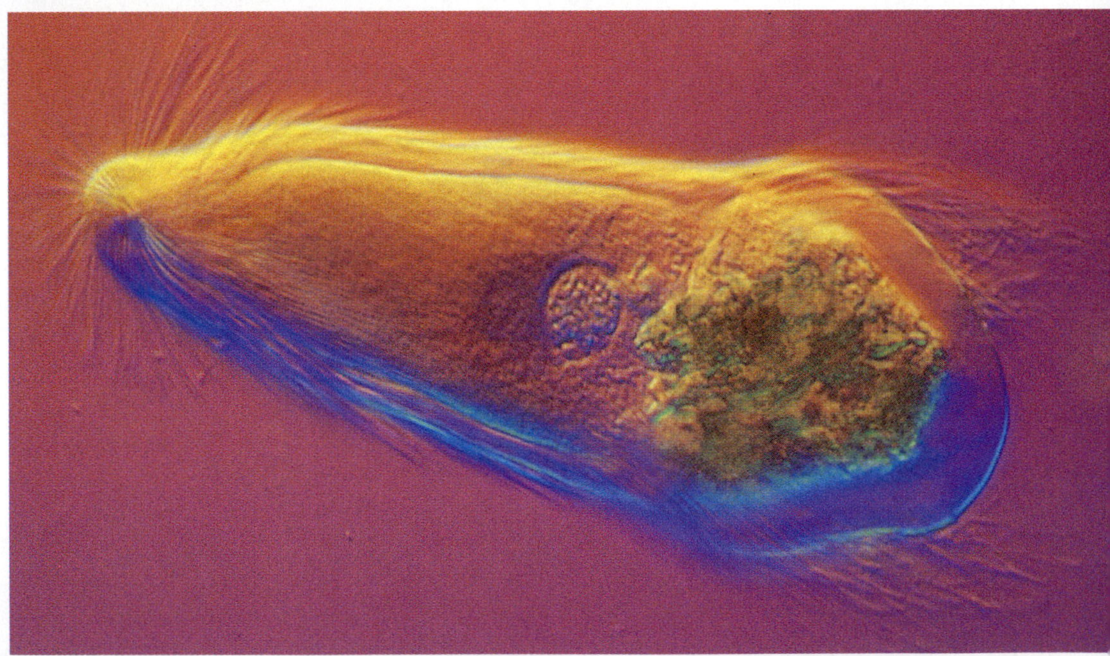

FIGURE 19.1

Cells within Cells.

Surrounded by a mane of flagella, this complicated *Trichonympha* cell has just been flushed from the gut of a termite. The central spherical, envelope-enclosed nucleus marks this organism as a eukaryote, even though it lacks mitochondria. An intracellular sac in the cell (right) is filled with tiny wood chips which the termite host rasped from a dead tree. Within the sac are bacteria that secrete digestive enzymes that break down cellulose.

form a slurry with the liquid inside the insect's gut. Each *Trichonympha* cell gliding within that slurry has a membrane at the posterior end that takes in tiny wood particles via phagocytosis [review FIGURE 3.14]. Bacteria living inside the *Trichonympha* cells produce and secrete cellulase, an enzyme that breaks the bonds in the cellulose of the wood fibers. Glucose molecules, released by the digestion of cellulose, fuel metabolism and the biosynthesis of amino acids and other macromolecules within both the bacteria and the *Trichonympha*. These raw materials are also absorbed by the termite hosts, and thus the insect can live by consuming wood, even though it actually lacks the cellulose-digesting enzymes. In a symbiosis such as this, the smaller cell provides the larger host with nutrients, while the host provides the smaller cell with a relatively protected place to live and with food.

The symbiosis among termites, protists, and bacteria brings us to our central topic in this chapter: an exploration of the diverse members of three kingdoms of mostly single-celled organisms, the Eubacteria, Archaebacteria, and Protista, as well as viruses and other noncellular forms.

We begin by investigating the lives of prokaryotes, which are among the simplest living entities. Despite their structural simplicity, prokaryotic organisms are very important in the recycling of materials in the environment, and they are also important as agents of disease. Free-living soil bacteria, for example, are involved in the breakdown of millions of tons of organic matter each year from fallen plants and animals—a breakdown that returns essential mineral nutrients to the soil.

Following this chapter's section on prokaryotes, we will discuss viruses, small particles that have no independent metabolism and that must enter a cell to reproduce, often causing disease. Finally, we enter the complex and varied world of *Trichonympha* and other members of the kingdom Protista. The cells of protists have a nucleus (clearly visible in FIGURE 19.1), and thus protists are eukaryotic organisms. Although not very elegant, the best way to describe the kingdom Protista is a group of eukaryotic organisms that are not fungi, plants, or animals. Like many prokaryotes, protists play important ecological roles in nutrient recycling and cause some of the most deadly human diseases, including malaria and schistosomiasis.

Throughout this chapter, we will see much more about the evolutionary and ecological relationships of the (mostly) single-celled kingdoms to the kingdoms of multicellular organisms—fungi, plants, and animals. We will also encounter some interesting sidelights, such as how bacteria could quickly multiply into a mass bigger than earth; what causes the common cold and how we can fight it; and what is humankind's most common infectious disease. ❑

MESSAGES

1 Single-celled organisms are ubiquitous and important: They release oxygen into the atmosphere, recycle materials, cause infectious diseases, produce medicines, and are models for studying universal biological principles.

2 A prokaryotic cell lacks a nuclear envelope and divides rapidly. All prokaryotic species belong either to the kingdom Eubacteria or the kingdom Archaebacteria. As a group, prokaryotic cells are metabolically far more diverse than eukaryotic cells.

3 Viruses are noncellular agents that enter cells to reproduce; they are often little more than a nucleic acid and a few proteins.

4 The kingdom Protista includes mostly single-celled eukaryotes, which can be quite complicated both structurally and in terms of their evolution.

The Prokaryotes

Despite their small size, prokaryotes have a profound influence on all of nature. Their primary importance is environmental, but they also have significant medical and industrial impacts. Let's first discuss the significance of prokaryotes, then explore some of their general biological properties, and finally survey the diversity of prokaryotic cells.

IMPORTANCE OF PROKARYOTES

Prokaryotes have immeasurable ecological impact. The billions of prokaryotic cells in nearly every square meter of soil, water, and air release oxygen into the atmosphere; recycle carbon, nitrogen, and other elements; and consume enormous quantities of dead animals, fungi, and plant matter; human and other animal wastes; pesticides; and pollutants that would otherwise poison the environment. The prokaryotic recycling of materials in the environment is usually gradual and unnoticed, but scientists have spent decades not only documenting this enormous natural role but also developing ways to encourage the breakdown of additional environmental pollutants (a process they call *bioremediation*). After the massive Alaskan oil spill in 1989, cleanup crews seeded the oily shoreline with a "fertilizer" containing nitrogen and phosphorus [FIGURE 19.2]. These nutrients, in combination with the carbon-containing organic compounds in the shoreline sludge, promoted the growth of local prokaryotes with a natural appetite for greasy hydrocarbons, and within two weeks, the sprayed areas had dramatically improved.

The beneficial role of prokaryotes in the environment cannot be overstated—it is the most significant single factor in the biology of this diverse group of organisms. Nevertheless, prokaryotes also have great medical import, both negative and positive. They cause hundreds of human diseases, including staph and strep infections, blood poisoning, tetanus, venereal diseases, and dental caries, as well as thousands of diseases in other animals and plants.

Despite this tremendous negative medical impact, harmful species are a small fraction of all prokaryotes, and many supply materials crucial to the health and survival of humans and other organisms. The bacterium *Escherichia coli* and other inhabitants of the human gut, for example, produce vitamins K and B_{12}, riboflavin, biotin, and other cofactors that we probably absorb and use. Usually harmless residents like *E. coli* may also blanket the intestinal walls so heavily that they block the passage of other dangerous microbes into the circulating blood.

FIGURE 19.2

Bacteria Help Clean the Environment.

In 1989, millions of barrels of oil spilled from the tanker *Exxon Valdez*, despoiling hundreds of miles of Alaskan coastline. Cleanup crews sprayed nitrogen and phosphorus fertilizer on the fouled beaches, which encouraged the growth of naturally occurring oil-consuming bacteria. The microbes destroyed oil in the fertilized plots much more rapidly than they did in the unsprayed plots.

Many plant-eating animals, including cattle, sheep, rabbits, and of course the termites we talked about in our chapter opener, would be unable to digest grasses and leaves without the cellulose-decomposing bacteria in their intestines. Biologists have confirmed this for termites by feeding them antibiotics that kill protists and bacteria, then observing that the termites begin to starve. Unknowingly, you may have performed a similar experiment on yourself. If you have ever taken large amounts of antibiotics prescribed by a doctor to cure a bacterial infection, the drug therapy may have killed off the beneficial bacteria that normally reside in your intestines and reproductive system. As a result, women may develop vaginal yeast infections and many people experience digestive difficulties and diarrhea due to the temporary absence of beneficial species and imbalances between competing species until the normal inhabitants return in large numbers.

A sizable segment of the food and chemical industries has grown up around the harnessing of prokaryotes to produce cheese, yogurt, pickles, soy sauce, chocolate, and other foods, and to generate chemical reagents, such as butanol, fructose, and lysine. As we just saw, people can also use bacteria to clean up environmental poisons and even to extract gold from low-grade ores.

Commercial applications aside, prokaryotes have been central to our understanding of biology. Much of what we know about biochemistry and molecular biology comes from the study of bacteria. The techniques of genetic engineering are largely based on bacterial chromosomes and plasmids [see CHAPTER 12]. Without prokaryotic research subjects, biology would be a century behind its current state.

Let's look, now, at the general properties of prokaryotic cells and see what makes them so important to the environment and to all living organisms, ourselves included.

GENERAL STRUCTURE OF PROKARYOTIC CELLS

Prokaryotic cells have an amazing range of habitats—from arctic snowdrifts to deep-sea hydrothermal vents and everywhere in between. Biologists recognize thousands of prokaryotic species, which they assign to the two kingdoms Eubacteria and Archaebacteria. FIGURE 19.3 shows these two separate evolutionary branches, and as you read through this chapter, you may find it helpful to refer to this diagram. Eubacteria and Archaebacteria differ in fundamental ways, but they also share several common traits, and these define the prokaryotic organism.

Prokaryotic cells have an outer cell wall that surrounds the plasma membrane [FIGURE 19.4]. This membrane, in turn, surrounds a noncompartmentalized cytoplasm dotted with ribosomes. All prokaryotic cells lack a nuclear envelope and other membrane-enclosed organelles. A prokaryote's chromosome is a circular strand of DNA that is not tightly complexed with spool-forming proteins like eukaryotic chromosomes [FIGURE 9.11].

Plasma Membrane In prokaryotic cells, adaptations of the plasma membrane serve many of the functions ac-

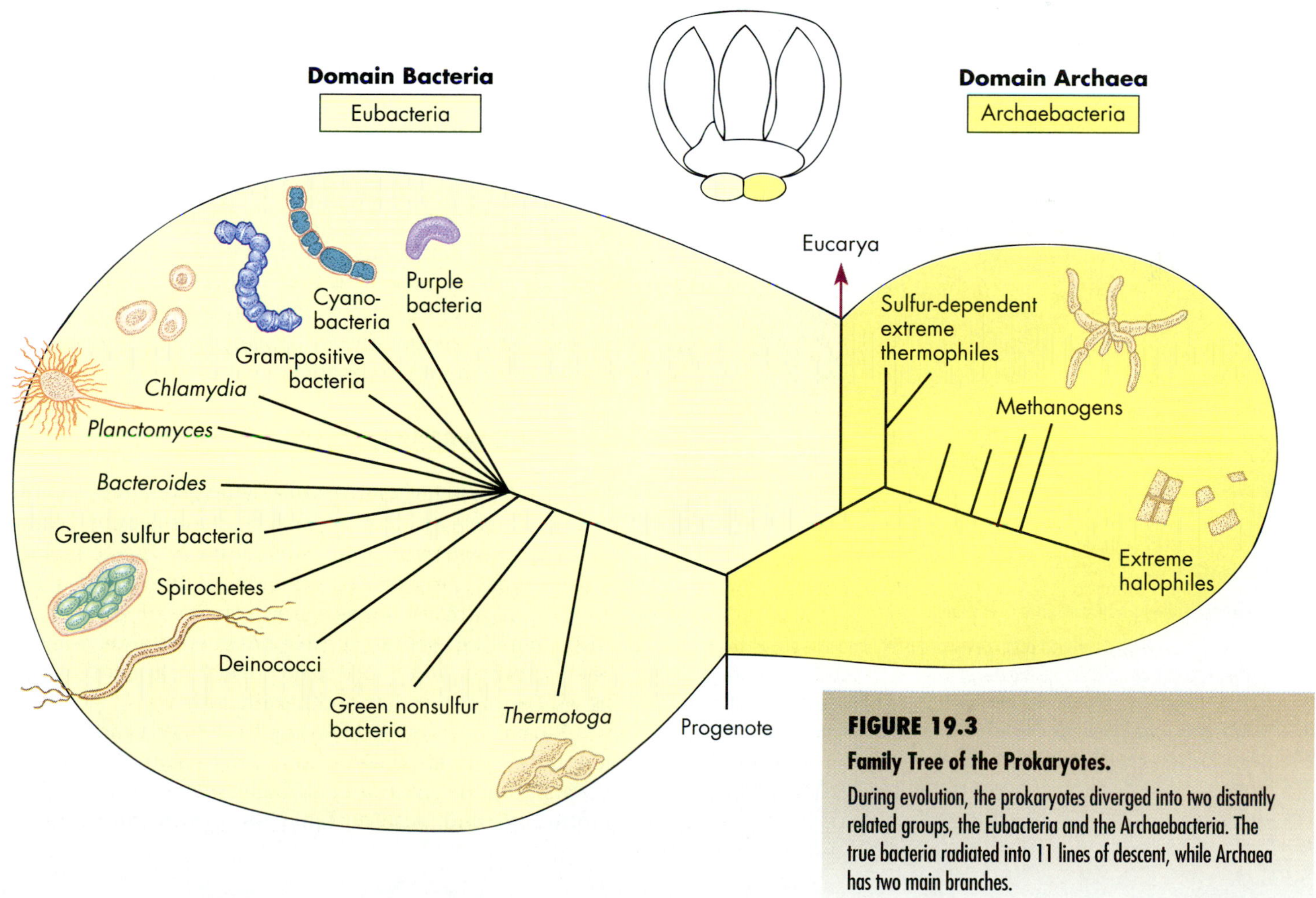

FIGURE 19.3

Family Tree of the Prokaryotes.

During evolution, the prokaryotes diverged into two distantly related groups, the Eubacteria and the Archaebacteria. The true bacteria radiated into 11 lines of descent, while Archaea has two main branches.

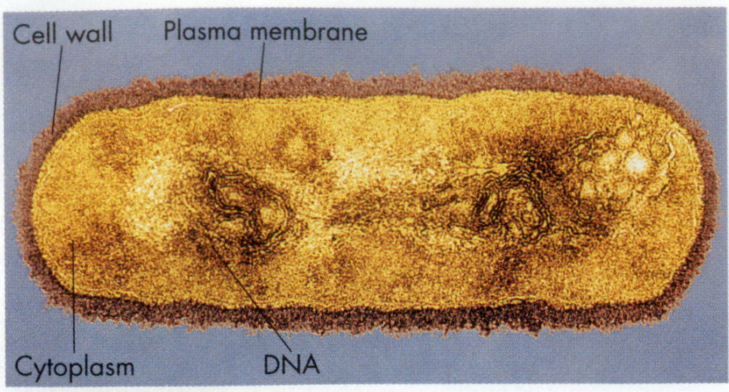

FIGURE 19.4
A Typical Prokaryotic Cell.

complished by the organelles of eukaryotic cells. Metabolic activities, such as cellular respiration and photosynthesis, take place on the plasma membrane, which sometimes folds inward into the cell's interior.

Cell Wall Support functions based in the eukaryote's cytoskeleton [review FIGURE 3.28] are centered in the prokaryote's cell wall. The outer cell wall of a eubacterial cell is a strong but flexible covering, made primarily of sugar-protein complexes called *peptidoglycans*. In so-called *gram-positive cells*, these occur in a single broad layer, while in *gram-negative cells*, the peptidoglycan layer is covered by an outer layer containing proteins and lipopolysaccharides (fat-sugar complexes).

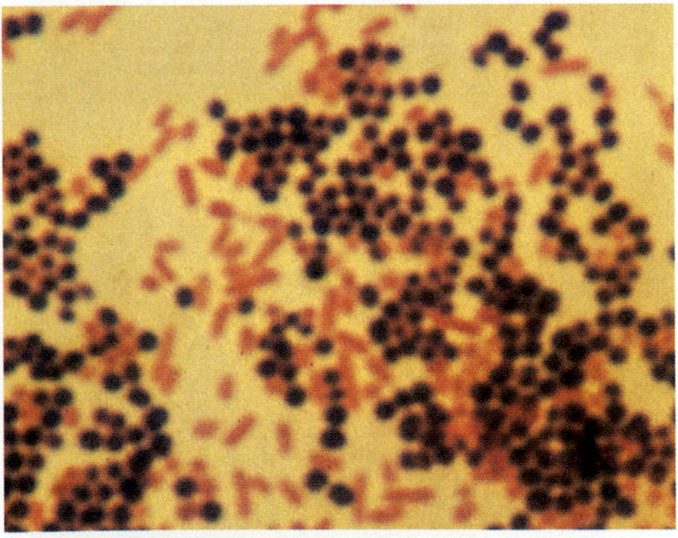

FIGURE 19.5
Gram Staining and Bacterial Cell Walls.
Danish bacteriologist Christian Gram developed a set of staining procedures that differentiate bacteria into two classes based on their ability to stain with certain dyes. Gram-positive cells, such as the *Staphylococcus aureus* shown here, stain purple; gram-negative cells, such as the *E. coli* shown here, stain red. The structural difference in their cell walls accounts both for the differential staining and for the antibiotic resistance of many gram-negative bacteria.

Gram Staining The differences in cell-wall characteristics cause gram-positive bacteria to turn purple and gram-negative bacteria to turn reddish when exposed to a special stain, called *Gram's stain* [FIGURE 19.5].

This staining technique is one of the first procedures a microbiologist will perform on a newly found or unfamiliar bacterial species. Gram staining is also important because in general, gram-positive and gram-negative bacteria are sensitive to different types of antibiotics. (An *antibiotic* is a chemical made by one microorganism that can slow the growth of or kill another microorganism.) For example, one summer many delegates at an American Legion convention in Philadelphia developed fever, chills, and pneumonia. Biologists isolated a new bacterium from the sick legionnaires and found that it was gram-negative. This suggested to physicians that penicillin would be ineffective in treating the disease. The doctors knew that penicillin kills gram-positive bacteria because it blocks the growth of the gram-positive cell wall. Conversely, penicillin does not block the growth of the gram-negative cell wall. To treat the patients, physicians tried broad-spectrum antibiotics that are effective against bacteria of both major types. One of these, erythromycin, turned out to be especially effective against Legionnaires' disease.

Forms of Bacterial Cells Although bacteria show a variety of shapes, most fall into one of the following categories: **cocci** (spheres), **bacilli** (rods), **spirilla** (spirals), and **vibrios** (comma-shaped, curved rods) [FIGURE 19.6]. The names of prokaryotes often reflect their shape. For example, *Bacillus anthracis*, the cause of anthrax, is rod-shaped, while *Streptococcus pyogenes*, the pathogen that causes strep throat, is spherical.

NUTRITION IN PROKARYOTES

Prokaryotes can be **heterotrophs**, which consume organic nutrients made by other organisms, or **autotrophs**, which make their own organic nutrients from inorganic materials in the environment. Together, these two modes of nutrition give prokaryotes as a group great ecological versatility, enabling them to survive on an enormous range of energy sources and hence to be present in air, soil, water, as well as in the bodies of other life forms.

Most prokaryotes, including the bacteria that inhabit a termite's gut, are heterotrophs: they obtain organic nutrients from the environment. Collectively, however, heterotrophic and autotrophic prokaryotes consume an enormous range of compounds, from biological materials and organic substances, such as methane, to inorganic substances, such as hydrogen sulfide; industrial chemicals, including hydraulic fluids and toxic herbicides; and cancer-causing wastes, including vinyl chloride and PCBs (polychlorinated biphenyls).

Many prokaryotes are not only heterotrophs but also **saprobes**; that is, they live on dead or dying organisms and act as decomposers. The bacteria living within *Trichonympha*, the protists inside termites, are saprobes since they obtain their nourishment from dead wood. Without the waste-recycling activity of these prokaryotic decomposers (as well as the multicellular saprobes, the fungi), the earth would quickly accumulate a thick layer of fallen leaves, dead animals, and other organic matter that could choke out living things.

Some heterotrophic prokaryotes are **symbionts** (literally, "organisms that live together"), and live on or inside other living organisms. Such organisms may be harmful (**parasites**), or they may be neutral or beneficial to the organisms with whom they live. Most disease-causing (*pathogenic*) bacteria are parasites, while beneficial symbionts include bacteria in your intestine that generate vitamins or bacteria that help digest cellulose in termites, cows, and other herbivores.

A few prokaryotes are autotrophic; these include *photoautotrophs*, which capture energy from light, and *chemoautotrophs*, which capture energy from certain chemicals. Examples of the photoautotrophic bacteria are green and purple photosynthetic bacteria and cyanobacteria. Chemoautotrophs include sulfur bacteria and methane bacteria, which can obtain the carbon they need from carbon dioxide and energy from the bonds of inorganic compounds, such as hydrogen gas, hydrogen sulfide, sulfur, iron, and certain nitrogen compounds.

REPRODUCTION IN PROKARYOTES

Prokaryotes usually reproduce by binary fission, simply dividing into identical offspring [see FIGURE 7.5]. If the dividing cells remain connected, small clusters or long filaments can form [see FIGURE 19.6C and E]. Binary fission can be incredibly fast and efficient, with division occurring as often as every 20 minutes. If a single *E. coli* bacterium and all its progeny continued dividing every 20 minutes, after just 48 hours there would be 2.2×10^{43} cells, or a mass of bacteria 4000 times the weight of the earth. This doesn't happen, of course, because the microbes have neither an unlimited source of nutrients nor ideal conditions of temperature and humidity.

Prokaryotes also have several means of genetic recombination, including *conjugation*, the direct exchange of DNA through a strand of cytoplasm that bridges two

FIGURE 19.6
Diversity of Shapes among Bacteria.
Bacteria come in a wide range of shapes and sizes, including [A] rod-shaped bacilli, such as *Bacillus subtilis*, [B] spiral-shaped spirilla, such as the spirochetes shown here, [C] round cocci, such as streptococci, and [D] comma-shaped vibrios, such as *Vibrio cholerae*, the agent that causes cholera. Note the whiplike flagella at the ends of the cells in [B] and [D]. [E] A cluster of archaebacteria (*Methanosarcina mazei*), each cell with characteristic segmentation resembling the top of a cloverleaf roll.

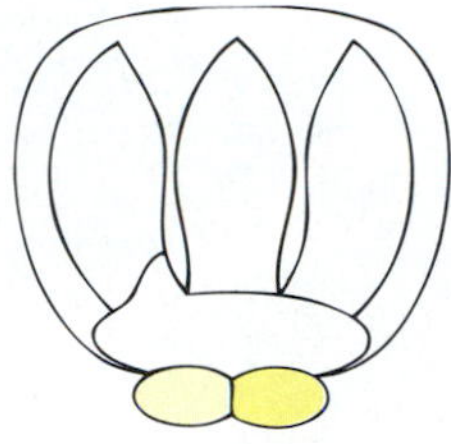

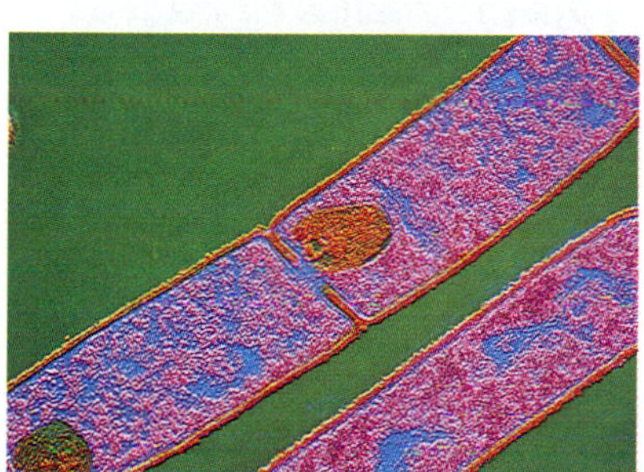

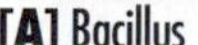

[A] Bacillus

[B] Spirilla

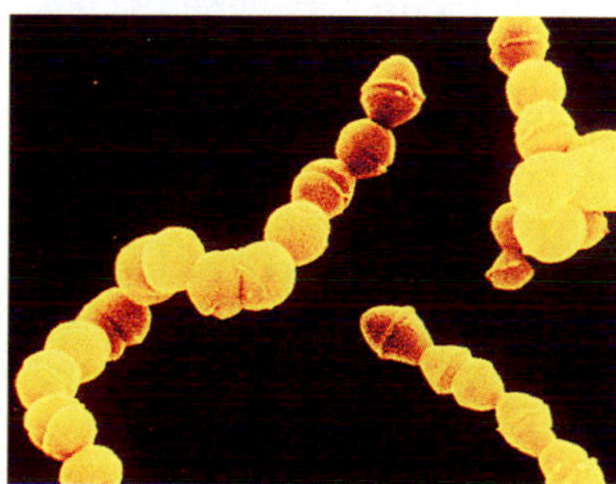

[C] Coccus

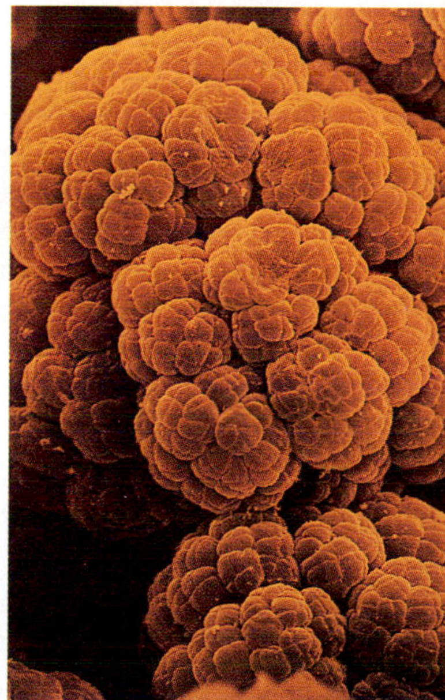

[E] Archaebacterium

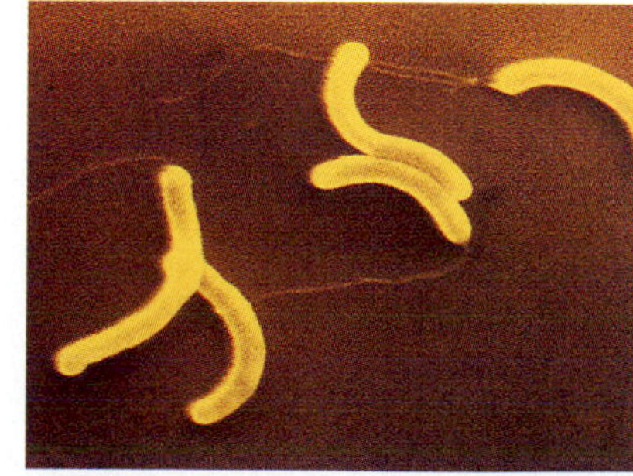

[D] Vibrio

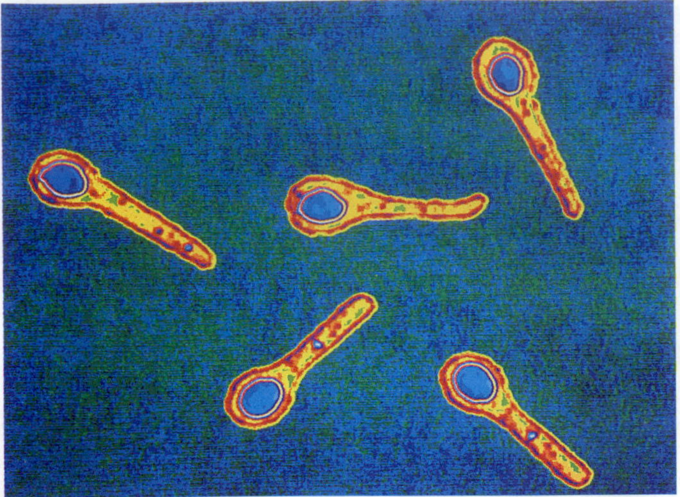

FIGURE 19.7

A Bacterial Endospore.

Certain species of bacteria can withstand extreme environmental conditions by forming tough-walled endospores, or resting cells. An endospore (false-color blue) has formed within the *Clostridium botulinum* cells shown here. Special care must be taken when canning food to use sufficient heat and pressure to kill all *C. botulinum endospores.*

cells, and indirect exchanges, such as *transduction* and *transformation. Transduction* is the transfer of genes from one bacterium to another via a virus, while *transformation* is the transfer of genes by the uptake of DNA directly from the surrounding medium [see CHAPTER 9]. As with sexual reproduction in plants and animals, these processes help to produce cells with new genetic makeups—cells which might have a survival advantage when environmental conditions change.

Many prokaryotes have yet another reproductive adaptation that enables them to survive unfavorable conditions and then divide quickly when conditions improve. Some species form small, tough-walled resting cells called **endospores** [FIGURE 19.7]. Endospore formation is triggered by worsening environmental conditions. Endospores can withstand extremes of heat, cold, drought, and even radiation for long periods. When conditions improve, the endospores grow into new bacterial cells. Because of the danger of endospore contamination, surgical instruments must be sterilized with high heat and pressure, and home-canned foods must be processed very carefully.

PROKARYOTIC BEHAVIOR

Some prokaryotes are nonmotile—they are unable to move under their own power and hence are subject to the currents and movements of their surroundings, whether that be water, blood, plant sap, or some other fluid. Other prokaryotes, however, are able to move along gradients of attractants (such as nutrients and light) or repellents (such as noxious chemicals), and thus find and exploit appropriate food sources or avoid predation. Tests show that an *E. coli* cell can detect a change in nutrient concentration of just 1 part per 10,000. That's the equivalent of a person distinguishing between two jars, one containing 10,000 marbles and the other 10,001. *E. coli* and some other bacteria have flagella (structurally different from eukaryotic flagella) that rotate and move the cell toward the higher concentration of the nutrient. Some bacteria live in puddles where the nutrients sink to the bottom. Some of these bacteria, like the one in FIGURE 19.8, contain small crystals of iron oxide (magnetite) that act like little compasses and enable the bacteria to detect up from down and to move downward into nutrient-rich sediments.

So far, we've been discussing characteristics of prokaryotes in general. Next, we consider the two taxonomic kingdoms of prokaryotes, the Archaebacteria and the Eubacteria.

ARCHAEBACTERIA

Archaebacterial cells have cell membranes, cell walls, and some aspects of protein synthesis that differ from comparable structures and processes in the Eubacteria [TABLE 19.1]. Archaebacteria also tend to inhabit harsh environments that would be inhospitable for most other organisms. One group, the **methanogens**, produces

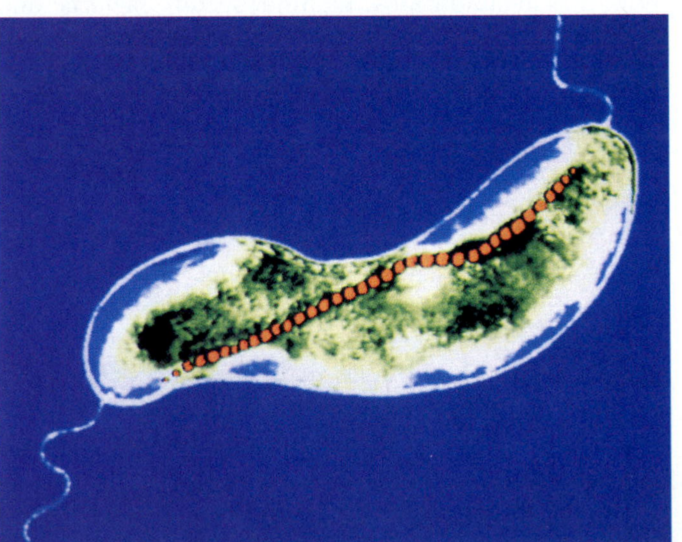

FIGURE 19.8

Iron-Containing Bacterium.

This soil bacterium contains a chain of 36 magnetic metal spheres that show up orange in this false-color electron micrograph.

TABLE 19.1 The Prokaryotes

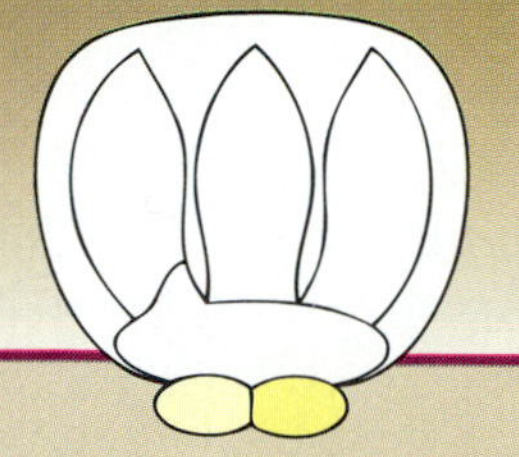

Domain and Evolutionary Group	Groups Based on Structure and Physiology	Characteristics and Significance
Archaea (Archaebacteria)		
Methanogens and relatives	Methane producers	Grow in oxygen-free environments and produce methane
	Extreme halophiles	Inhabit highly salty environments, like salt lakes and solar evaporation ponds
Sulfur-dependent extreme thermophiles	Sulfur-dependent, heat-loving archaebacteria	Inhabitants of sulfur hot springs
Bacteria (Eubacteria)		
Purple bacteria and relatives	Purple photosynthetic bacteria	Some can form a purple layer in anaerobic zone of lakes; mitochondria arose from ancient purple bacteria similar to *Rhodospirillum*
	Pseudomonads	Versatile heterotrophs in the soil; useful in cleaning up industrial pollutants
	Enteric bacteria	Inhabit animal intestinal tracts, aiding digestion (*Escherichia coli*) or causing disease (*Salmonella*)
	Myxobacteria	Gliding bacteria whose colonies make complex multicellular spore-forming stalks
	Nitrogen fixers	Free in soil or associated with plants (*Rhizobium*); fix nitrogen in biological form
	Rickettsias	Live only inside other cells; cause typhus fever and Rocky Mountain spotted fever
Cyanobacteria 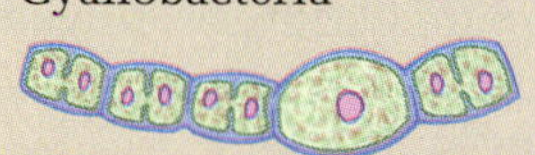	Cyanobacteria (Blue-green algae)	Photosynthetic system like a plant, and also fix nitrogen, thus most self-sufficient organisms; earliest organisms to release oxygen; precursors of chloroplasts
Gram-positive bacteria	Endospore formers	Heat-resistant spores can cause botulism or tetanus (*Clostridium* and *Bacillus*)
	Lactic acid bacteria	*Lactobacillus* makes yogurt; *Streptococcus* causes strep throat
	Gram-positive cocci	*Staphylococcus* causes toxic shock syndrome and other infections
	Actinomycetes	Cause leprosy and tuberculosis; produce antibiotics
	Mycoplasmas	Smallest of all cells; no cell wall; one cause of pneumonia
Chlamydiae	*Chlamydia*	Leading cause of blindness in humans; cause of most common sexually transmitted disease; cells live inside other cells; can't make ATP
Planctomyces and relatives	Appendaged bacteria	Cell wall consists of protein
Bacteroides and relatives	A branch of the gram-negative rods	The major anaerobic cells in the large intestine
Green sulfur bacteria	A branch of the phototrophic bacteria	Live on the warm fringes of sulfur hot springs
Spirochetes	Spirochetes	Corkscrew shape; cause syphilis and Lyme disease
Deinococcus and relatives	A branch of gram-positive cocci	Highly resistant to radiation; found near atomic reactors
Green nonsulfur bacteria	A branch of phototrophic bacteria	Grow in mats at the fringes of hot springs
Thermotoga	Anaerobic fermenters	Live in oceanic vents at highest temperature of any true bacteria

methane gas (CH_4). Methanogens include *Methanobacterium ruminantium*, archaebacteria which inhabit a cow's digestive tract. The methane gas produced by these prokaryotes and belched by the cows contributes a huge volume of methane into the atmosphere. Some biologists suggest that this gas may be contributing to the global warming of the earth via the greenhouse effect [see FIGURE 41.24].

Another group of Archaebacteria are the extreme **halophiles** (salt lovers) like *Halobacterium halobium*, which tolerate extreme salt concentrations in salt flats and places like the Great Salt Lake and the Dead Sea [FIGURE 19.9]. Other Archaebacteria are sulfur-dependent extreme **thermophiles** (heat lovers). These bacteria use sulfur in their energy metabolism and survive in hot, sulfurous springs and volcanic mud pots [review FIGURE 3.9]. One of these remarkable archaebacteria grows best at water temperatures several degrees above boiling!

EUBACTERIA

Recent analysis of eubacterial RNA suggests that there are 11 main evolutionary branches among the true bacteria [see FIGURE 19.3]. For various reasons, however, microbiologists continue to use the phenotypic classification based on bacterial form, physiology, and ecology as presented in *Bergey's Manual of Systematic Bacteriology* and other standard reference books. Our discussion of Eubacteria follows the evolutionary approach, moving through the group in FIGURE 19.3 in a counterclockwise direction. TABLE 19.1 gives the classification based on phenotypes. We discuss only the most important groups.

The largest and most diverse group of true bacteria are the *purple bacteria* such as *Rhodospirillum*. Many of these bacteria are purple and carry out photosynthesis with chlorophyll pigments that differ from those in plant cells. Some ancient member of the purple phototrophic bacteria probably gave rise to mitochondria in eukaryotes [review FIGURE 17.15]. The presence of bacterial symbionts in *Trichonympha*, the protists residing in termites' guts, suggests that such intracellular associations between cells may be frequent in nature. Other members of the group are heterotrophs, including the common intestinal bacterium *E. coli* [see FIGURE 3.8]; the bacteria in FIGURE 19.6A and D; the *Rhizobium* species that fix nitrogen in peas and beans [see FIGURE 35.11]; the bacterium that causes Legionnaires' disease; and many others. This group also includes *rickettsias*, tiny, rod-shaped parasitic bacteria that absorb nutrients from a host cell. Rickettsias are transmitted by ticks, fleas, and lice and cause typhus fever, Rocky Mountain spotted fever, and other serious human diseases.

The *cyanobacteria*, formerly called *blue-green algae*, are an environmentally important group of true bacteria. Cyanobacteria, like plants, contain chlorophyll *a*, which allows photosynthesis and the generation of carbohydrates. Many cyanobacteria can also fix nitrogen, that is, convert nitrogen gas (N_2) from the air (which is unusable for most organisms) to organic nitrogen compounds like ammonia (NH_3) for their own protein synthesis. Therefore, many cyanobacteria require only water, sunlight, a few inorganic nutrients, and the carbon dioxide and nitrogen gases readily available in air. As a result, cyanobacteria can live almost anywhere: in fresh or salt water; on damp rocks; in soil; on tree trunks; and in hot, cold, or dry climates. Because cyanobacteria are so widespread, their ecological role is tremendous; they contribute millions of tons of oxygen to the atmosphere and biologically usable nitrogen and carbon to the environments and organisms around them.

One nitrogen-fixing species of cyanobacteria (an *Anabaena* species) grows symbiotically with a water fern [FIGURE 19.10]. Before planting rice, a farmer will allow dense blooms of the water fern to grow in the paddy. When the rice plants grow large, they crowd out the water fern as well as its cyanobacterial ally. Nitrogen re-

FIGURE 19.9
Salt-Tolerating Bacteria.
These salt pans look ruddy because of the trillions of individual salt-loving (halophilic) archaebacteria that thrive in such highly saline water.

FIGURE 19.10

Cyanobacteria.

Cyanobacteria (blue-green algae) often grow as filaments, with an occasional thick-walled cell (a heterocyst) that shuts out oxygen and allows nitrogen fixation to take place. Asian rice farmers fertilize their crops naturally with cyanobacteria that grow symbiotically with a water fern. When the plant dies, it returns nitrogen to the fields.

leased from the dead ferns acts as a fertilizer for the developing rice crop and promotes high yields without the need for potentially toxic chemical fertilizers.

The fact that cyanobacteria, like plants, contain chlorophyll *a*, helped convince many biologists that an ancient cyanobacterium may have entered a eukaryotic cell hundreds of millions of years ago and become the first chloroplast [review FIGURE 17.15]. Later eukaryotes containing both this photosynthetic organelle and other membrane-bound organelles may, in turn, have given rise to the first generations of plants [see CHAPTER 17, page 428].

True bacteria include many medically important gram-positive species. *Streptococcus mutans* [FIGURE 19.11] causes tooth decay. *Streptococcus pyogenes* causes strep throat. And *Staphylococcus aureus* causes staph infections, such as *toxic shock syndrome*, a sometimes fatal vaginal infection that occurs when the bacteria multiply in and around a tampon during menstruation. Fortunately, certain other gram-positive bacteria —the *actinomycetes*—produce more than 500 different antibiotics, natural substances that fend off competing microorganisms. Antibiotics produced by these cells include streptomycin, tetracycline, and neomycin.

The smallest free-living cells are the *mycoplasmas*, simplified parasitic eubacteria that lack cell walls and live inside animals, plants, and sometimes other single-celled organisms. Only 0.2 to 0.3 μm long, mycoplasmas can cause a dangerous form of pneumonia, as well as urinary tract and other infections.

Bacterial species of the genus *Chlamydia* cause the most frequent sexually transmitted disease in North America [see BOX 14.2]. Like rickettsia, chlamydia live inside animal cells. These bacteria are even smaller than some viruses and have no way of making their own ATP; instead they must obtain energy from the cell they infect.

Another type of eubacteria, the *spirochetes*, have a spiral shape and include the agents that cause Lyme disease and syphilis [see BOX 14.2].

➤ CONCEPT CHALLENGE

Choose one group of Eubacteria and Archaebacteria, and answer these questions: How would your life be different if that group were suddenly and specifically wiped off the face of the earth? How would your life be different if that group lived inside your body?

Viruses

Some biological agents, including viruses, virions, and prions, are not organized as cells, but depend on living cells for their continued existence—and probably even evolved from them.

VIRUSES

Viruses are geometric packages of genes that are 1000 to 10,000 times smaller than most prokaryotic cells. A virus particle is, in effect, a minute package of DNA or RNA

FIGURE 19.11

Our Teeth Teem with Bacteria That Cause Decay.

Groups of rod-shaped *Streptococcus mutans*, if not brushed away, can cause cavities.

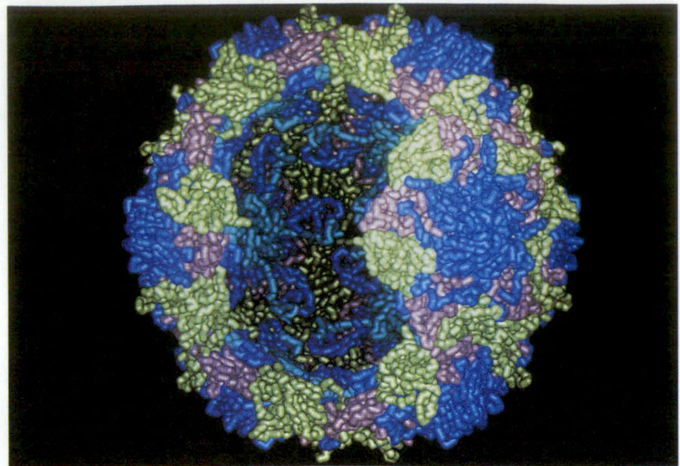

FIGURE 19.12
The Polio virus.
A protein coat surrounds the hereditary material—an RNA molecule—of the polio virus. This virus attacks nerve cells that normally stimulate muscle contraction.

surrounded by a protein coat, or *capsid*, and occasionally by other materials [FIGURE 19.12]. Viruses infect cells by attaching part of the capsid to the cell's exterior wall or plasma membrane and then either entering the cell or simply injecting their DNA or RNA into the cell, leaving the capsid outside [review FIGURE 9.4]. Once inside, the viral genes commandeer the cell's protein-synthesizing machinery, sometimes stopping production of cellular proteins entirely and preempting the machinery for the production of new virus particles complete with capsids. Eventually, the cell may burst and die, or may survive and gradually liberate thousands of viruses that can then infect other cells. Many biologists do not consider viruses to be alive because these agents lack the machinery for self-reproduction or metabolism.

There are hundreds of kinds of viruses, many of which cause plant and animal diseases such as those listed in TABLE 19.2. Some viruses, like the rhinoviruses that cause colds and the influenza viruses that cause flu, are transitory parasites. We encounter them in sneeze droplets or mucus sprayed into the air by cold or flu sufferers, or we pick them up from contaminated hands or surfaces, and they begin to multiply in our body cells. Eventually, our immune system destroys them [see CHAPTER 26].

Other types of viruses, however, like herpes simplex I and II, which cause cold sores and genital herpes, respectively, insert their DNA (or DNA copies of their RNA) into the chromosomes of nerve cells or other body cells and take up permanent residence. There they lie dormant, erupting and causing symptoms only occasionally when triggered by fever, sunlight, or other environmental stimuli. Since viruses are not cells, they are not killed by the antibiotics physicians use to block the growth of disease-causing prokaryotes and there are really no medical cures for any viral diseases. One of the biggest challenges in modern medicine is to develop drugs that can fight viruses.

VIROIDS

The **viroids** are a group of intracellular parasites that consist of an RNA molecule only about 300 nucleotides long, hundreds of times smaller than the smallest virus. Viroid RNAs lack a protein coat, and bind to RNAs in the host cell's ribosomes, thus disrupting the cell's protein synthesis. Some diseases of potatoes, cucumbers, citrus trees, and artichokes can be traced to viroids [FIGURE 19.13].

PRIONS

Finally, the **prions** are the smallest and strangest disease-causing agents that can be transmitted from one animal to another. Prions are not well understood. They appear to lack genetic material entirely, consisting of nothing but a specific protein molecule. Yet prions are implicated in serious nerve and brain diseases, including scrapie in sheep and goats; mad cow disease; Creutzfeldt-Jakob disease and kuru in humans; and possibly Alzheimer's disease, the most common form of senility and a leading cause of death. Biologists are not sure how prions reproduce or cause disease.

FIGURE 19.13
Viroids: Parasites That Infect Crop Plants.
Here, viroids—short RNA molecules—have infected artichoke leaves and damaged or killed rings of tissue in numerous places.

TABLE 19.2 Noncellular Agents of Disease

Agent	Constituents	Example	Disease
Viruses	DNA plus protein	Parvovirus	Gastroenteritis
		Herpes simplex I, II	Herpes
		Epstein-Barr virus	Mononucleosis, Burkitt's lymphoma
		Smallpox virus	Smallpox
	RNA plus protein	Paramyxovirus	Measles
		Togavirus	Rubella (German measles), yellow fever
		Rhinoviruses	Common cold
		Myxovirus	Influenza
		Poliovirus	Poliomyelitis
		Paramyxovirus	Mumps
		Rhabdovirus	Rabies
		Retroviruses	Cancer (some forms)
			AIDS
Viroids	RNA only	PST viroid	Potato spindle tuber disease
		Exocortis viroid	Citrus exocortis disease
Prions	Protein only	Various prions	Kuru (a brain disease in Borneo and elsewhere); Creutzfeldt-Jakob disease (a brain disease); scrapie (a disease of some farm animals)

THE ORIGIN OF VIRAL DISEASES

Where do new viral diseases come from? Researchers think that many new human viruses originated in non-human animals. The AIDS virus, for example, probably attacked primarily African monkeys until the continual advance of our species into formerly untouched or little-used wilderness areas brought about increased contact. The frequency of long-distance air travel was a secondary factor; once the virus infected humans, it was dispersed fairly quickly by infected people flying from Africa to other parts of the world.

There are other examples of this "old new virus" phenomenon. For example, Asian influenza A virus killed 20 to 30 million people around the world in 1918. The virus lives in ducks and other birds, and was apparently transmitted from birds to barnyard pigs and then to people. In the summer of 1993, several people in mainly southwestern states died of a mysterious respiratory virus. Researchers eventually identified it as the deadly Hanta virus, spread by contact with deer mouse feces.

COMBATING VIRAL DISEASES

Researchers have been working hard to find treatments for viral diseases such as the common cold, genital herpes, the flu, and AIDS, but is a bit like the plot of a horror movie—in which the protagonists are trying to "kill" an attacker that is not really alive. Antibiotics that kill living prokaryotes and protists leave virus particles untouched.

A drug called *interferon* is proving effective against the human cold. Interferon is a natural substance produced by the immune system in response to viral infections. Interferon released from one cell attaches to a neighboring cell and somehow prevents it from replicating viruses. With the advent of gene splicing and other techniques of bioengineering, a few companies have been able to produce large quantities of interferon, which drug researchers have incorporated into a very effective nasal spray. Clinical trials are now under way to get the necessary government approval to market both WIN 51,711 and interferon.

Yet another drug, called *acyclovir*, is proving effective against herpes infections. Acyclovir blocks the activity of certain herpes viral enzymes and thus can help control the herpes infections that cause shingles, cold sores, and genital herpes, a sexually transmitted disease affecting about 20 million Americans. Researchers think that, within a decade, physicians will have many such drugs with which to control viruses.

➤ CONCEPT CHALLENGE

Medical researchers have been less successful at identifying antiviral agents than at finding antibacterial substances. What properties of the two types of infectious agents do you think has contributed to this situation?

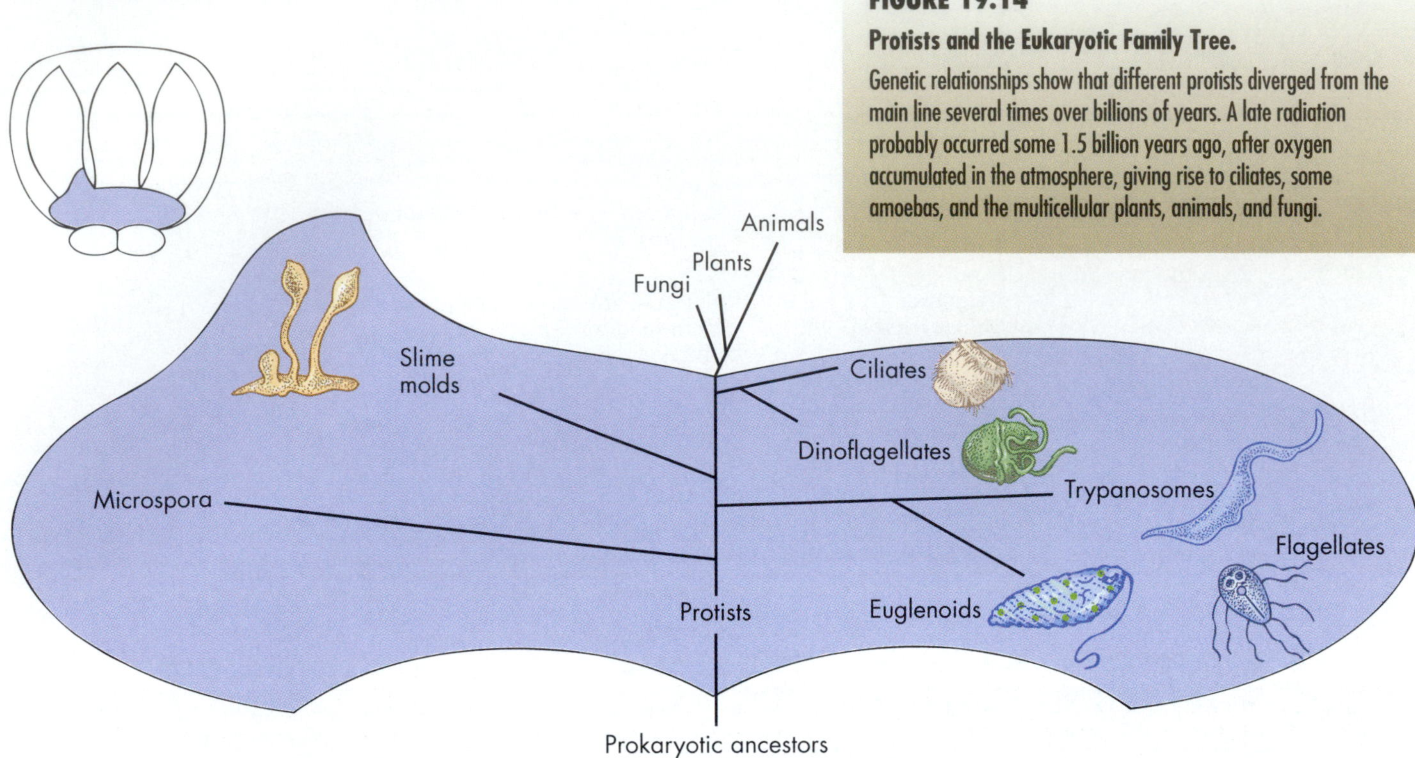

FIGURE 19.14

Protists and the Eukaryotic Family Tree.

Genetic relationships show that different protists diverged from the main line several times over billions of years. A late radiation probably occurred some 1.5 billion years ago, after oxygen accumulated in the atmosphere, giving rise to ciliates, some amoebas, and the multicellular plants, animals, and fungi.

The Protists

Our discussion turns now from the single-celled prokaryotes and the noncellular viruses to the simplest eukaryotes, the protists [FIGURE 19.14], including *Trichonympha*, the symbiont of termites discussed at the beginning of the chapter. Together, members of the kingdom Protista have a tremendous impact on health and the environment [TABLE 19.3].

CLASSIFICATION

The kingdom Protista includes a wildly diverse group of eukaryotes that, as we saw, are by definition neither plants, animals, nor fungi. Most protists exist mainly as single-cell individuals or clusters of mostly similar cells. Protists cannot be considered plants or animals because those multicellular organisms develop from embryos. Protists are not fungi because fungi develop from spores and lack flagella at any stage in their life cycle.

Biologists who study the kingdom Protista and those who study the kingdom Plantae disagree about the proper classification of various protists. Some eukaryotic algae don't seem to fit well into either kingdom. Some experts, in fact, have defined a separate kingdom they call *Protoctista*. This kingdom includes the single-cell eukaryotes, the slime molds, and the algae, including a seaweed called kelp that can be as tall as a football field is long. In this book, we take the more traditional approach and consider protists to include mostly single-celled eukaryotes. Our discussion of protists is organized around their methods of obtaining nutrition. Some protists are heterotrophs, animal-like hunters. Others are photosynthetic autotrophs, plantlike producers of carbohydrates. Still others are funguslike decomposers, and a few are strange hybrids of the different life-styles. It is important to realize that these arbitrary divisions do not reflect evolutionary relationships.

ANIMAL-LIKE PROTISTS

The hunters of the microbial world are the **protozoa** (literally, "first animals"), cells that usually consume other cells or food particles. Each of four major protozoan phyla is distinguished by its means of locomotion: by flagella, by amoeboid movement, by no means of movement, or by cilia.

Flagellates Protists of the phylum **Mastigophora** (the so-called flagellates) have flagella with the same internal structure and the same protein (tubulin) as in the tail of a human sperm cell [see FIGURE 3.23]. This evidence suggests that eukaryotic flagella have a very ancient origin.

Most flagellated protozoans live as parasites or harmless symbionts within other organisms. A common human parasite, now a scourge to campers, is the flagellate *Giardia lamblia*, which causes diarrhea [FIGURE 19.15A]. Other flagellates include members of the genus *Trichonympha*, which are the cellulose digesters that live in the

TABLE 19.3 The Kingdom Protista

Section, Phylum or Division, and Subgroup	Common Members	Characteristics and Significance
Animal-like Protists (Protozoa)		
Flagellates (Mastigophora)	Trypanosomes	Each moves by means of a flagellum; cause sleeping sickness
Amoebas (Sarcodina)	Amoebas	Move by pseudopodia; cause amoebic dysentery
	Foraminiferans, radiolarians	Shells contribute to limestone deposits
Sporozoa	*Plasmodium vivax*	Cause of malaria
Ciliates (Ciliophora)	*Paramecium*	Feeds on bacteria in ponds
	Didinium	Hunts other protists
Plantlike Protists		
Euglenophyta	*Euglena*	Moves by flagellum; contains chloroplasts
Pyrrhophyta	Dinoflagellates	Flagella in grooves; cause of red tides
Chrysophyta	Golden-brown algae, diatoms	Photosynthetic; glassy shells that serve as abrasives
Funguslike Protists		
True slime molds (Myxomycota)	*Physarum*	Many nuclei in one large cell; break down organic matter on forest floor
Cellular slime molds (Acrasiomycota)	*Dictyostelium*	Single cells converge and form slug; break down organic matter on forest floor
Water molds (Oomycota)	*Phytophthora infestans*	Cause late potato blight; thrive in dampness; damage plants

guts of termites, enabling the termites to utilize wood as food [see FIGURE 19.1]. *Trypanosoma brucei gambiense* is a flagellate that causes African sleeping sickness [FIGURE 19.15B]. Trypanosomes are dangerous pathogens because they can evade the human immune system by changing their surface coats. Trypanosomes are spread through the bite of the tsetse fly and can inflict their lethal damage on domesticated animals as well as on human beings.

Amoebas and Related Protists Members of **Sarcodina** range from creeping, bloblike **amoebas** that live in moist terrestrial habitats [see FIGURE 3.14] to delicate glassy-shelled ocean species [FIGURE 19.16A]. However, all members of this group display **pseudopodia** (literally, "false feet"), limblike cellular extensions that serve in locomotion and feeding. Pseudopodia can project outward in any direction, pulling the cell along and immobilizing prey so that the sarcodine can engulf it by phagocytosis. Several marine sarcodine species form beautiful shells around their soft cell bodies. Cells of *foraminiferans* secrete whitish, calcium-based shells that can look like spiral seashells or chambered nautiluses, while *radiolarians* produce silicon-based shells that look like miniature glass ornaments [FIGURE 19.16B]. These protozoa are so plentiful that their shells contribute to the formation of thick limestone and other rock deposits. The chalky White Cliffs of Dover formed this way.

Spore-Forming Protists A third group of protozoa, called **Sporozoa** are internal parasites residing in a wide range of animals, and they do not move under their own power like the other groups of protozoa. The sporozoa in the subgroup *Apicomplexa* have a complex of rings and tubules at the cell's apex, and a sporelike stage in which the cells lack any means of locomotion. The best-known member of this group is *Plasmodium vivax*, which causes malaria [FIGURE 19.17]. This protist is spread by tropical mosquitoes, and malaria is humankind's most prevalent infectious disease. At any one time, 100 million people are ill with malaria, and about 1 million people die of malaria each year—most of them very young African children. Vigorous attempts at eradication of mosquitoes with DDT, and treatment of the protozoan with drugs like chloroquine, have killed off susceptible protists and their susceptible mosquito hosts. However, the once-rare protist cells that, by chance, possessed mutations conferring resistance to the chemicals have now proliferated, since chemically sensitive organisms have been removed from competition. This evolution by natural selection has resulted in mosquitoes and plasmodial parasites that are harder to kill. Scientists are working to develop vaccines against this disease.

Another sporozoan subgroup is *Microspora*, parasites that live inside cells in nearly every group of animals. Microspora were one of the very earliest branches off the main line of eukaryotic evolution [review FIGURE 19.14]. Interestingly, microspora, along with *Giardia* and *Trichonympha*, lack mitochondria, so they resemble the postulated cell that became the host for the bacteria that invaded eukaryotic cells and led to the first mitochondria. More work on this little-known group of organisms could provide exciting new data for our understanding of the evolution of organisms with true nuclei.

Ciliates The fourth protozoan phylum is the aquatic **ciliates**, which are characterized by rows of cilia, whiplike cell extensions that are generally shorter and more numerous than flagella. Cilia bend with coordinated, oarlike motions, propelling the cell or helping to sweep food particles into its food-capturing organelle. [FIGURE 19.18A].

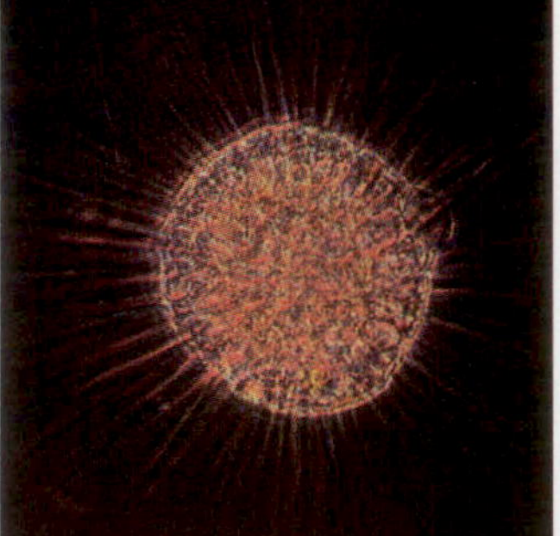

[A]

[B]

FIGURE 19.16
Sarcodina.
Sarcodines extend narrow ribbons of cytoplasm, or pseudopodia, to capture prey or pull themselves forward. **[A]** *Actinosphaerium* resembles a pincushion with hundreds of glass needles. **[B]** Radiolarians, such as this mixed group of species, have intricately patterned, glassy shells.

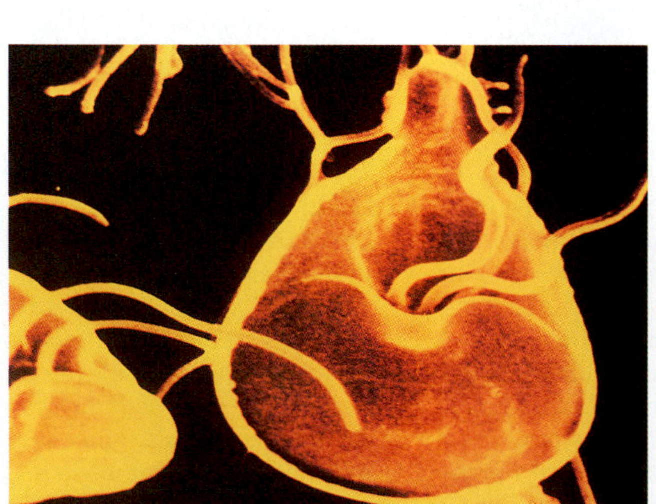

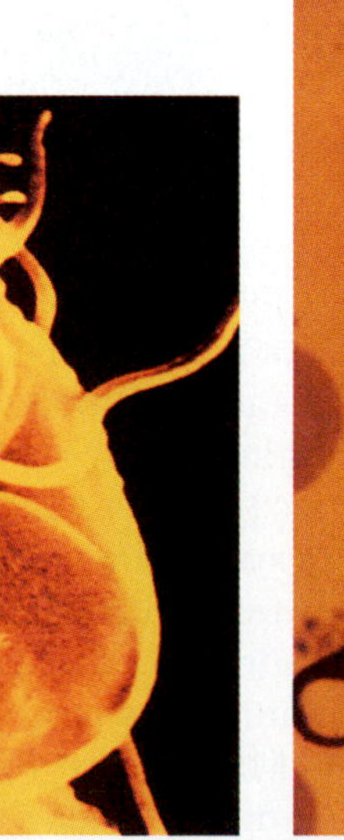

[A]

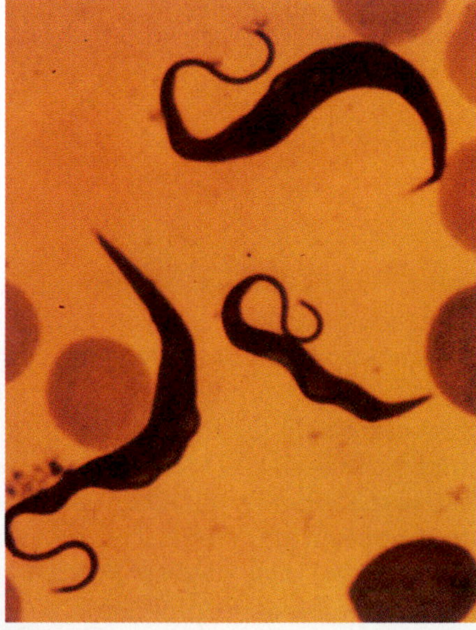

[B]

FIGURE 19.15
Some Flagellates Are Agents of Disease.
[A] A drink from a clear-looking but contaminated mountain stream can lead to infection by *Giardia*. This flagellated protist lines the inner surface of the intestines, causing severe abdominal gas, cramping, and diarrhea. **[B]** An eel-like trypanosome (magnification 5000×) wriggles among a person's lozenge-shaped red blood cells. The parasite enters the bloodstream with the bite of a tsetse fly, then invades the brain and spinal cord.

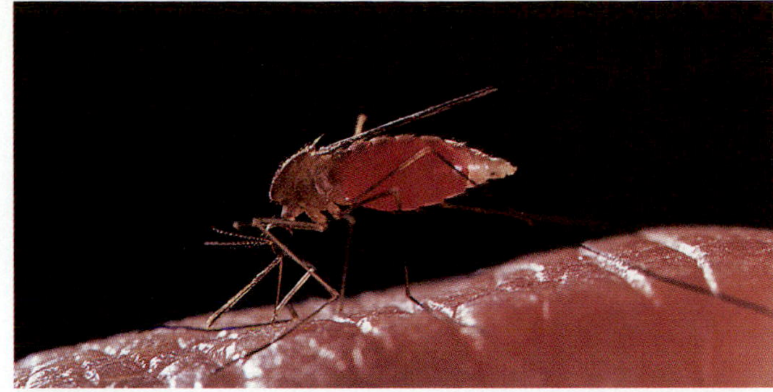

[A]

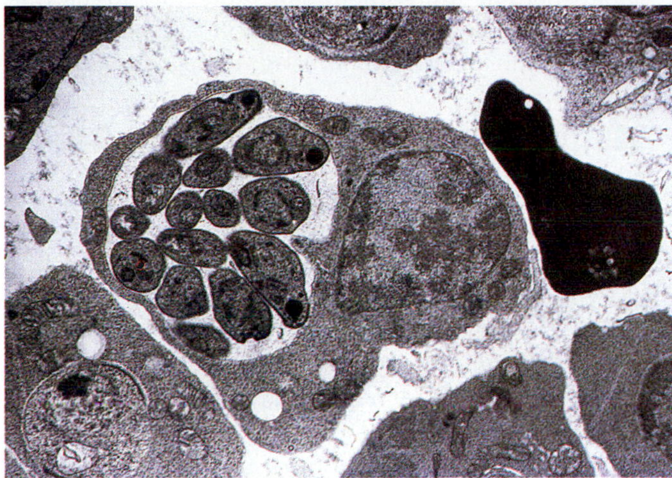

[B]

FIGURE 19.17

A Mosquito Spreads the Malarial Parasite.

[A] A tropical mosquito can inject the agent of malaria, *Plasmodium vivax* into a person's blood. [B] The parasite can enter an individual's cells and reproduce, eventually killing the cells. Unfortunately, it can also kill the person—in fact, malaria kills more people than any other single disease in the world.

Ciliates have several organelles with functions analogous to those of animal organs: anal pores that discharge wastes; microfilaments and microtubules that support and contract like bones and muscles; tiny toxic darts (*trichocysts*) used in hunting prey; and food vacuoles that are filled with enzymes and function like digestive organs. Each ciliate also contains a giant nucleus with many sets of chromosomes (a polyploid *macronucleus*) that directs cell activities, as well as one or more small diploid nuclei (*micronuclei*) that undergo meiosis and are exchanged during sexual reproduction (*conjugation*; see FIGURE 19.18B).

PLANTLIKE PROTISTS

Three divisions (or phyla) of organisms within the kingdom Protista contain chlorophyll and carry out photosynthesis: the *euglenoids*, the *dinoflagellates*, and the *golden-brown algae* and *diatoms*. These protists are part of the mass of cells called **phytoplankton** ("floating plants") that grow near the surface of fresh and marine waterways. This mass of cells has tremendous ecological significance in that it releases oxygen and forms the base of the aquatic food chain: Larger organisms graze on the phytoplankton, still larger creatures eat the grazers, and so on. Many plantlike protists swim via flagella.

Euglenoids The word **euglena** comes from the Greek for "good eye," and indeed, each of the many green, spindle-shaped euglenoids has an eyespot, or *stigma*. This structure shades a light receptor and allows the aquatic cell to swim in the direction that maximizes its photosynthetic rate. Genetic evidence suggests that the chloroplasts in Euglena arose during evolution from different photosynthetic bacteria than those that gave rise to the

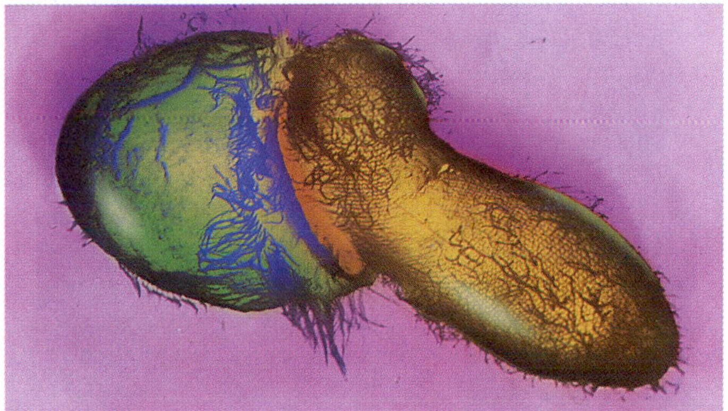

[A]

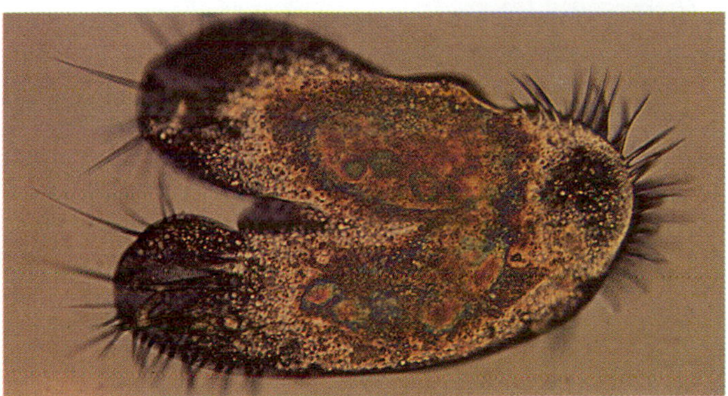

[B]

FIGURE 19.18

Ciliates Eating and Mating.

[A] A single-celled hunter and its prey. The juglike *Didinium* (left) swallowing a slipper-shaped *Paramecium* cell. [B] Conjugation in ciliates. Ciliates like the *Stylonychia* shown here exchange micronuclei.

chloroplasts of plants. The multiple evolution of chloroplasts suggests that this type of symbiotic event may occur with much more frequency in evolution than previously thought.

Dinoflagellates The **dinoflagellates** are a distinctive group that occasionally have a large economic impact. *Dinoflagellate* means "spinning cell with flagella," and indeed, these plantlike marine protists have two flagella that cause them to spin as they swim. One flagellum winds about the middle of the cell like a belt in a groove and causes the spinning motion, while another projects backward in a second groove and propels the cell [FIGURE 19.19A]. Some dinoflagellates wear a coat of armor—an elaborately embossed cellulose cell wall—while others have only a cell membrane. Dinoflagellates can cause *red tides*, dense blooms that can tint the water blood-red and produce deadly toxins that act as nerve poisons [FIGURE 19.19B]. The poisons build up in fish and shellfish, killing them and poisoning people who gather and eat them. Thus coastal areas often ban collection and consumption of shellfish during May through August, when red tides can occur.

Golden-Brown Algae and Diatoms The protists known as **golden-brown algae** and **diatoms** are the most common and perhaps the most exquisitely beautiful of the phytoplankton species. Most golden-brown algae and diatoms have a golden color as a result of a carotenoid pigment as well as their chlorophyll pigments. Their cell walls contain silica, the main component of window glass, instead of cellulose, and they store oils rather than starchy compounds. Golden-brown algae are extremely numerous in phytoplankton layers and thus major contributors to marine food chains. Diatoms get most of the popular attention, however, because of their jewel-like, glassy shells. These shells fall to the ocean floor and accumulate in crumbly white sediments called *diatomaceous earth*, which people use in toothpaste, swimming pool filters, and other applications. The golden-brown algae and diatoms are so abundant and ubiquitous that some biologists estimate they may contribute more oxygen to the atmosphere than all land plants combined.

FUNGUSLIKE PROTISTS

Among the most bizarre members of the kingdom Protista are the **slime molds**. Like mushrooms and yeasts, these glistening organisms derive food and energy by secreting digestive enzymes that break down organic matter and then absorbing the digested material back into the cell. They also form multicellular reproductive structures.

True Slime Molds A **true slime mold** exists as a flat, often fan-shaped mass that moves about slowly like a giant amoeba in damp soil, leaf litter, and fallen trees. This mass is called a *plasmodium*, and consists of a continuous cytoplasm surrounded by one plasma membrane, but containing many diploid cell nuclei.

Cellular Slime Molds **Cellular slime molds** exist as single-celled, amoebalike organisms that glide along in search of bacteria. When food is scarce, both types of slime molds send up brightly colored *fruiting bodies*, reproductive structures that look like golf balls on bent tees. Spores produced in the balls are scattered by wind or rain and germinate in new locations, dispersing the species [FIGURE 19.20].

FIGURE 19.19

Dinoflagellates: Armored Protists That Cause Red Tides.

[A] This microscopic phytoplankton has flagella in grooves along the long arms and wears a coat of cellulose armor characteristic of the dinoflagellates. **[B]** Some dinoflagellates cause dangerous red tides—dense population explosions that can tint the water blood red and produce deadly toxins, as happened during this red tide along the coast of Ethiopia.

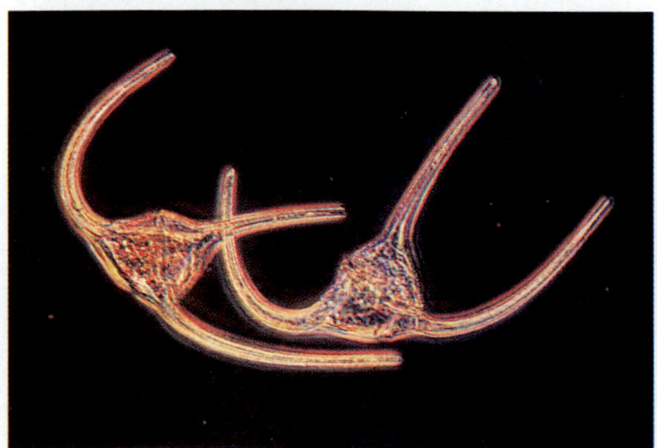

[A]

[B]

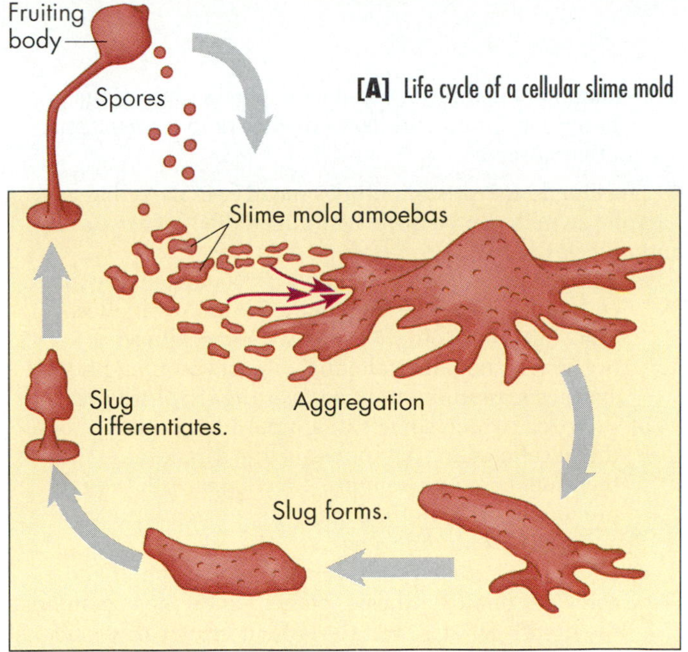

[B]

FIGURE 19.20

Life Cycle of the Cellular Slime Mold.

[A] Cellular slime molds normally move about like amoebas—as single cells. If environmental conditions become poor, they can aggregate, form a slug, and slither to some likely location. There they stop and differentiate into a fruiting body with a stalk and a spore chamber that eventually releases new, genetically recombined spores. These, in turn, germinate into new amoebalike cells that individually crawl off again in search of bacteria to consume. [B] Close-up of slime mold fruiting bodies on fallen tree leaves.

Water Molds The third group, the **water molds**, inhabit soil and water. Some are single-celled and others form fuzzy, branching filaments called *hyphae*. Like the true slime molds, water molds have several nuclei within a common cytoplasm and form large, immobile egglike cells (their formal name, in fact, is Oomycota). After fertilization, they form spores that disperse by swimming. One parasitic water mold struck Ireland in 1845–1847. This species causes *potato blight*, a disease that rapidly rots and kills growing potato vines. The Irish depended so heavily on potatoes for their daily diet that more than a million people starved, and 1.5 million more emigrated, mostly to North America.

EVOLUTIONARY RELATIONSHIPS AMONG THE PROTISTS

The survey of protists we just completed was based on phenotypes, but recent molecular data show how the groups are related genetically. Taxonomists now doubt that a common progenitor gave rise to all the protists. Instead, many separate ancestral organisms probably led to the various types of protists. Taxonomists call the protists a *polyphyletic group*—one with many different ancestors. (A *monophyletic group* includes organisms that arose from a common ancestor.)

One of the oldest lines gave rise to the intracellular parasites *Microspora* [see FIGURE 19.14]. Microspora are evolutionary relics; they contain a membrane-bound nucleus that undergoes a unique mitotic division, and they lack internal organelles such as mitochondria or chloroplasts. The microspora probably split from the main line while the earth's atmosphere was still anaerobic (about 2.5 billion years ago), and they probably resemble the host cells to the first mitochondria [see FIGURE 17.15].

Somewhat later, the flagellates diverged, and then the slime molds, and then, over a more recent and fairly short period of evolutionary time, a rapid radiation led to ciliates, some amoebas, fungi, plants, and animals. Some biologists suggest that this radiation was an evolutionary response to the accumulation of oxygen in the earth's atmosphere about 1.5 billion years ago.

➤ CONCEPT CHALLENGE

Suppose that doctors in your town suddenly started noticing many patients with severe intestinal distress and suspect that the municipal water supply is contaminated by a microorganism. You are asked to help determine which type of microorganism may be involved. How would you investigate the organism to determine whether the culprit belonged to the Eubacteria, Archaebacteria, or Protist kingdom? If it were a protist, how would you determine which group of protists was involved?

Connections

Ironically, the most numerous and in many ways the most important organisms on earth are also the smallest and the oldest. All around us, in the soil, air, and water, unseen trillions of single-celled organisms take in nutrients, metabolize them, and produce new generations in as little as 20 minutes. In contrast, the large organisms we see in our everyday lives—toadstools, grasses, robins—may have arisen from a common ancestor rather late in evolutionary history. (They are a monophyletic group.) Odd as it may seem, the fungi, plants, and animals represent recent branches on the great tree of life, and we devote the next three chapters to their characteristics and to how those branches may have appeared and diverged.

KEY TERMS

bacillus, 452
ciliate, 462
coccus, 452
flagellates, 460
parasite, 453
phytoplankton, 463
protozoan, 460
slime mold, 464
spirillum, 452
symbiont, 453
vibrio, 452
virus, 458

HIGHLIGHTS IN REVIEW

1 Single-celled organisms are ubiquitous and important: They release oxygen into the atmosphere, recycle materials, cause infectious diseases, produce medicines, and are models for studying universal biological principles.

2 A prokaryotic cell lacks a nuclear envelope and divides rapidly. All prokaryotic species belong either to the kingdom Eubacteria or the kingdom Archaebacteria. As a group, prokaryotic cells are metabolically far more diverse than eukaryotic cells.

a] Most prokaryotes are surrounded by a cell wall; they lack membrane-bound cytoplasmic organelles and have a circular strand of DNA functioning as a single chromosome. Prokaryotic cells have characteristic shapes, including bacilli (rods), cocci (spheres), vibrios (comma-shaped, curved rods), and spirilla (spirals).

b] Most prokaryotes are heterotrophs—either saprobes (consuming dead organisms) or parasites (consuming live organisms)—but some are autotrophs, obtaining energy from light (photosynthesis) or from inorganic chemicals (chemoautotrophy).

c] Most prokaryotes reproduce via binary fission; some exhibit conjugation or other forms of genetic exchange, and some form spores resistant to heat, cold, and dessication.

d] Prokaryotes can be motile or nonmotile; motile species have sensory systems that allow them to detect gradients of light, chemicals, or gravitational fields and to respond appropriately by movement.

e] Genetic evidence suggests that prokaryotes include two kingdoms. The Archaebacteria inhabit very hot or salty or sulfurous environments and tend to have unusual forms of metabolism. Cells in the kingdom Eubacteria are more common and include the gram-negative and gram-positive species, the actinomycetes, the rickettsias, the mycoplasmas (the smallest living things), and the cyanobacteria, which contain chlorophyll, have simple nutritional requirements, and show division of labor between cells.

3 Viruses are noncellular agents that enter cells to reproduce; they are often little more than a nucleic acid and a few proteins.

a] Virus particles consist of a protein capsid surrounding genes made of DNA or RNA. The genetic material enters a living cell, which then generates new virus particles.

b] Viroids are small RNA molecules devoid of a protein coat, while prions are protein particles devoid of genetic material but capable of causing infectious or hereditary brain and nerve disorders in humans and other animals.

4 The kingdom Protista includes mostly single-celled eukaryotes, which can be quite complicated both structurally and in terms of their evolution.

a] All members of the kingdom Protista are eukaryotic cells, complete with membrane-bound organelles. Many are very complex cells with specialized organelles that facilitate hunting, locomotion, photosynthesis, reproduction, and getting rid of wastes.

b] Protozoa, which some call animal-like protists, include the flagellated mastigophorans, the sarcodines with their amoeboid movement, the ciliates with their numerous oarlike cilia, and the sporozoa with their sporelike, nonmotile stage.

c] Like plants, some protists contain chlorophyll and carry out photosynthesis. They include the euglenoids, the dinoflagellates, and the golden-brown algae and diatoms.

d] The slime molds are funguslike protists that carry out extracellular digestion of organic matter.

UNDERSTANDING THE FACTS AND CONCEPTS

For Questions 1–5, match the descriptions with the most appropriate item from the following list of terms. Each item in the list may be used once, more than once, or not at all.

a] eubacteria
b] archaebacteria
c] viruses
d] viroids
e] prions

1 Genetic material is circular DNA; gram-positive or gram-negative.

2 Genetic material is unknown; made of protein only.

3 Genetic material is RNA; no capsid.

4 Genetic material is DNA or RNA; capsid present.

5 Genetic material is circular DNA; cellular; not pathogenic to humans.

As above, for Questions 6–10, match the description with the most appropriate item from the following list.

a] spirochetes
b] cyanobacteria
c] actinomycetes
d] mycoplasmas
e] none of the above

6 Source of some antibiotics.

7 Some are nitrogen-fixing autotrophs.

8 Parasitic; smallest free-living cells.

9 Include bacteria that cause syphilis and Lyme disease.

10 Include methanogens that live in a cow's stomach.

As above, for Questions 11–15, match the description with the most appropriate item from the following list.

a] parasite
b] saprobe
c] symbiosis
d] heterotroph
e] pathogen

11 The general term that includes all disease-causing biological entities.

12 Decomposers that live on dead or dying organisms.

13 The general term that refers to organisms that must obtain food from their environment.

14 A close association of two or more living species in which each species is required for the survival of the other(s).

15 A nonreciprocal relationship in which one organism lives in or on, and obtains its nourishment from, another living organism.

As above, for Questions 16–20, match the descriptions with the most appropriate item from the following list.

- **a]** phytoplankton
- **b]** flagellate
- **c]** true slime mold
- **d]** cellular slime mold
- **e]** amoeba

16 Forms a multinucleate, sluglike organism.

17 Photosynthetic marine protists.

18 Uninucleate cells that aggregate to form a fruiting body for reproduction.

19 All are heterotrophs; some are symbionts but others are parasites.

20 Protists that move by means of pseudopodia; some form shells that accumulate to form limestone or chalk cliffs.

21 Identify the organism depicted in the accompanying diagram and explain the sequence of stages of its life cycle.

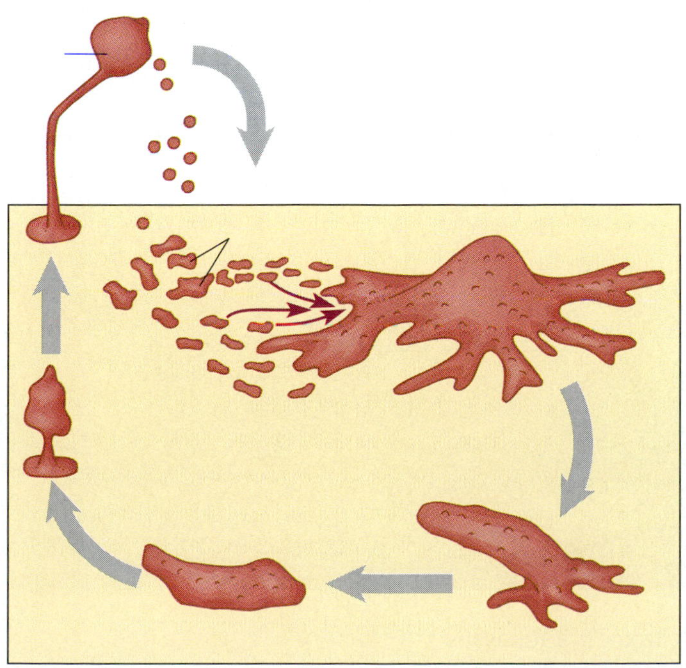

INTEGRATE AND APPLY WHAT YOU HAVE LEARNED

1 Explain the sentence, "Protists are a polyphyletic group."

2 In which of the three groups considered in this chapter (prokaryotes, viruses, protists) are there autotrophs? Specify the nature of the autotrophy in each case.

3 Why are antibiotics effective against bacteria but not against viruses?

4 Explain why cyanobacteria are among the most self-sufficient organisms.

5 In evolutionary terms, which are the more closely related pair of organisms: animal-like protists and plantlike protists, or animal-like protists and animals? Explain.

ANALYSIS

1 Which statement is correct?

- **a]** All photosynthetic bacteria fix nitrogen.
- **b]** All nitrogen fixers are photosynthetic.
- **c]** Some photosynthetic organisms are archaebacteria and fix nitrogen.
- **d]** Methanogens neither fix nitrogen nor are autotrophic.
- **e]** Some prokaryotes are saprobes but not heterotrophs.

2 Which statement is false?

- **a]** The termite obtains food material for the protist *Trichonympha*.
- **b]** *Trichonympha* obtains food for intracellular bacteria.
- **c]** *Trichonympha* produces food for the termite directly.
- **d]** Intracellular bacteria produce food for the termite.
- **e]** Intracellular bacteria produce food for *Trichonympha*.

3 Which of the following organisms has mitochondria, ribosomes, a cell wall, and chlorophyll?

- **a]** *Chlamydia*
- **b]** Microspora
- **c]** Diatoms
- **d]** Cyanobacteria
- **e]** Viruses

4 Which of the following statements is correct?

- **a]** Some viruses are symbiotic or saprobic.
- **b]** Both bacteria and protists are important in nutrient recycling.
- **c]** The first signs of multicellularity and cellular specialization did not appear until the evolution of the protists.
- **d]** Heterotrophs may be parasitic, symbiotic, or photosynthetic.
- **e]** Most protists are either harmful or not beneficial to human beings.

CHAPTER 20

Fungi and Plants: Decomposers and Producers

FIGURE 20.1

The Yew: A Taxol Producer.

A fleshy red "berry" draws birds to yew seeds. An evolutionary adaptation led to the production of taxol, a secondary metabolite, which probably defends the tree from hungry animals.

SEARCHING FOR A NEW YEW

In the 1960s, based on knowledge of common folk remedies and poisons, researchers at the U.S. National Cancer Institute collected and tested samples of the western yew tree (*Taxus brevifolia*) along with hundreds of other plant sources of potential anticancer compounds. From the shaggy, reddish-gray bark of the yew tree the researchers extracted a chemical that they named *taxol* [FIGURE 20.1]. This compound turned out to inhibit the growth of cancerous cells in test tubes, and by the 1980s, medical researchers had tested taxol in clinical trials with patients suffering from advanced and intractable ovarian cancer. Taxol caused many of the tumors to shrink. By the early 1990s, physicians were also treating thousands of breast cancer patients with taxol, and had begun testing its utility for cases of lung cancer and melanoma skin cancer.

In the summer of 1991, enter Andrea Stierle, a postdoctoral physiologist at Montana State University, searching for a research project. She was well informed not only about taxol's successes but also about its scarcity. To obtain enough taxol to treat a single cancer patient requires the bark from five to six full-grown western yews. With more than 100,000 potential recipients of the drug each year, demand would quickly exceed the forest supply. Stierle decided to search for a microorganism associated with the western yew that, like its host, generates taxol.

Stierle and her colleagues collected samples of over 200 kinds of microorganisms from yew trees in the Montana woods, and among them, discovered a new species of fungus that seemed to generate tiny quantities of taxol. Using several techniques, they confirmed that their new fungus (and not the tree tissue itself) was the source of the taxol. Dr. Stierle's university has already filed a patent application for the fungal production of taxol in industrial quantities, and she will have an important research project for years to come.

With this chapter, we move from the single-celled kingdoms to the multicellular realms of the fungi and plants.

Fungi are eukaryotes that form spores, and lack flagella and cilia at all stages of the life cycle. **Plants** are multicellular, eukaryotic organisms that are generally photosynthetic; they usually develop from embryos that arise after sexual fusion. The fungi probably arose about 900 million years ago, and the plants some 400 million years ago. Because fungi are evolutionarily much older and are, for the most part, smaller and simpler organisms, we will consider them first in this chapter.

We will see that the fungi, along with bacteria, are nature's great decomposers: they break down the remains of dead organisms and animal wastes, and as a result nutrients from the decomposed material become available to other life forms. Roles such as these give fungi vast ecological and economic importance, and show the close interdependence of fungi and plants. Many biologists think that land plants could never have evolved without the previous presence of fungi on land.

While fungi obtain nourishment by digesting organic materials in their environments, plants obtain their energy and materials from nonliving sources—solar energy and carbon dioxide in the air. Each year the earth's plants collectively produce 150 billion tons of carbohydrates, an amount that could fill the boxcars of 100 trains, each long enough to stretch from the earth to the moon! As we saw in CHAPTER 6, many microbes, most fungi, and virtually all animals live directly or indirectly on the carbohydrates that plants produce. As we survey the evolution of plants, we will see how the ancestors of the land plants moved from water to land, and how as plants evolved, a series of physical systems and physiological trends emerged that helped them meet the challenges of life on land.

In addition to investigating the evolution of plants and fungi and the characteristics of the modern groups, we'll encounter some interesting subjects in this chapter. These include the largest, tallest, and oldest living things, and some of the stranger organisms in the kingdoms of life. ❑

MESSAGES

1 Fungi, including mushrooms and yeast, are immobile eukaryotes that obtain nourishment by digesting materials produced by other organisms; hence, fungi are important in the recycling of materials in the environment. Fungi associated with the roots of many plants help the plant obtain water and mineral nutrients while the fungus receives carbohydrates from the plant.

2 Plants are generally multicellular, photosynthetic eukaryotes. They trap the energy of sunlight in organic molecules and these, in turn, directly or indirectly support nearly all other organisms.

3 The simplest plants are multicellular, photosynthetic aquatic organisms called algae. The evolution of waterproof coverings, openings through which gases are exchanged, and a system of vessels that conduct fluids allowed plants to emerge onto land. The evolution of pollen and seeds released seed plants (for example, conifers and flowering plants) from dependence on environmental water for reproduction.

4 The alternation of a haploid life form with a diploid life form is a key to understanding plant biology.

Fungi: The Decomposers

Nature works through balanced ecological cycles. Plants and photosynthetic microbes, for example, remove carbon from the air and trap it in carbohydrates. If these were the earth's only organisms, all carbon would soon be locked up in their cells and in their remains. This doesn't happen, however, because other organisms eventually decompose organic wastes and the remains of dead organisms, releasing into the environment compounds containing carbon, nitrogen, phosphorus, and other elements required for life. The earth's major decomposers are the 100,000 or more species of fungi (including the familiar mushrooms, molds, and puffballs) along with the various decomposing bacteria and protists discussed in CHAPTER 19.

NUTRITION IN FUNGI

Fungi are heterotrophs and must obtain preformed organic molecules from the environment and modify them. Fungi secrete powerful enzymes that enable the cells to digest organic matter in their environment, breaking down large molecules into smaller ones. The cells then absorb the nutrient molecules through their cell membranes. Because the actual digestive process takes place outside the organism's body, it is called **extracellular digestion**. A similar type of food breakdown takes place in an animal's hollow gut.

Most fungi are *saprobes*, organisms that decompose nonliving organic matter. Fungi consume everything from leather and cloth to paper, wood, paint, and other materials, slowly reducing old buildings, books, and shoes to crumbled ruin. During the long evolution of the fungi, however, some species changed from decomposers of nonliving tissue to *parasites*, organisms that live off other living things, without contributing to the host's survival. Today, parasitic fungi are the main cause of plant diseases, with about 5000 different fungal species attacking crops in fields, gardens, and orchards. The fungus that Andrea Stierle and her colleagues isolated from the yew tree and found to produce taxol is probably a parasite on the tree. As you will see, other types of parasitic fungi attack people and domestic animals, causing athlete's foot, ringworm, and vaginal yeast infections, to name a few.

STRUCTURE OF FUNGI

To survey the fungi, we will begin with a familiar mushroom like the common type on salads or pizzas. The body of a typical button mushroom looks solid, but is actually made up of cells joined end to end like the cars of a passenger train and forming filaments called **hyphae** (HIGH-fee; singular, hypha). Hyphae pack tightly together into a mat called a **mycelium**, rather like steel wool [FIGURE 20.2]. The aboveground portion of a mushroom is made up of densely packed hyphal filaments. Belowground, an extensive, loose mycelium spreads below the mushroom, penetrating the soil for many square meters. Whether aboveground or below, each filament or hypha has tubu-

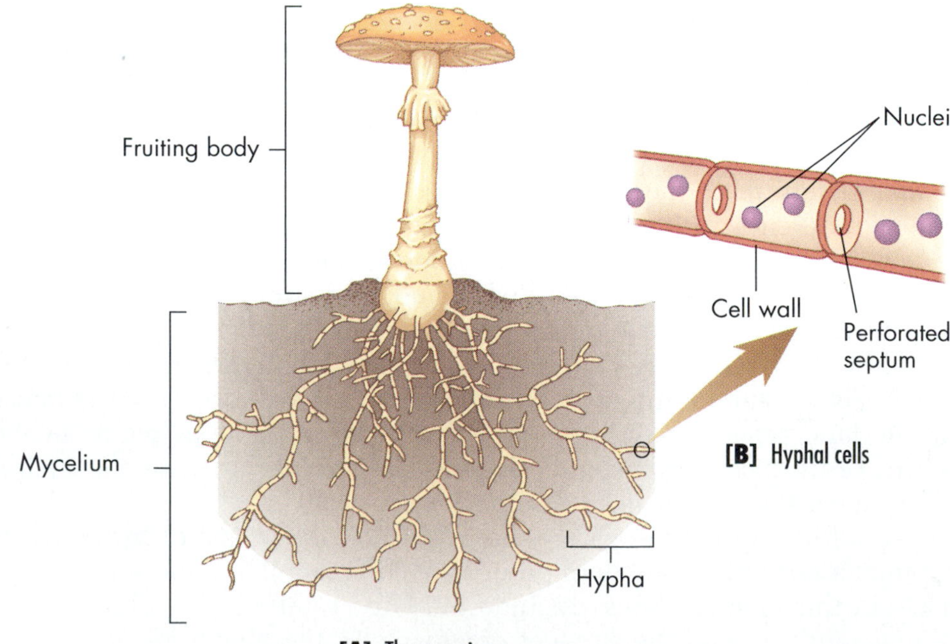

FIGURE 20.2

The General Structure of Fungi.

[A] A typical mushroom is just the reproductive portion of the fungus. Most of the fungus is a tangled mat (a mycelium) made out of filaments (hyphae), stretching below the ground surface. The mushroom, or fruiting body, produces spores that germinate and grow into new mycelial mats. [B] The fungal filaments consist of cells, each of which has two nuclei in many species. The cells have thick cell walls with perforations connecting adjacent cells.

lar side walls made mostly of *chitin* (KI-tin), a nitrogen-containing polysaccharide. The chitinous side walls surround the cell's plasma membrane [see FIGURE 20.2]. These tough walls make fungi the most resilient organisms of all the eukaryotes. The cell walls that separate adjacent cells in a filament are called *septa* (singular, *septum*), and they are perforated. These perforations in the septa permit the cytoplasm of one cell to flow into the next.

Hyphae can grow very quickly: If you added together the growth at all the hyphal tips in a tangled mycelium, just one day's new growth could easily exceed 1 km (0.62 mi). This explains how mushrooms can literally spring up overnight on damp logs or soil. This meshwork of filaments gives a fungus a large surface-to-volume ratio, which enables even a very large mushroom to digest and absorb sufficient amounts of nutrients to grow rapidly. FIGURE 20.3 shows an aerial view taken from an airplane over Montana of two huge individual fungus clones. Scientists in Michigan have found a single individual fungus covering 38 acres (15 hectares), and mycologists (fungal biologists) in Washington have found one that covers 1500 acres (about 600 hectares) and weighs well over 100 tons.

FIGURE 20.3

Is the Largest Living Thing a Fungus?

Some biologists who study fungi propose that the largest organism alive is an *Armillaria* fungus, a parasite on tree roots. This photograph taken from a plane over Montana shows a virgin forest with two centers of root disease, which has caused the trees to look a paler green. The lower center started hundreds of years ago from a single haploid spore cell, and has grown to encompass 20 acres (8 hectares). Many mycologists think that even bigger fungi are devouring trees in Michigan and Washington state.

In the early 1990s, some fungal biologists suggested that this one individual fungus may be the largest living organism in the world. As we will see later, plant biologists were quick to challenge this suggestion.

FUNGAL REPRODUCTION: WITH OR WITHOUT SEX

As we discussed in CHAPTER 7, many multicellular organisms can reproduce by either asexual or sexual means, depending on external conditions. In the fungi, asexual reproduction is the most common mode. Pieces can break off the hyphal meshwork and grow into new individuals; hyphal cells can divide or bud; or the fungus can produce asexual **spores**, cells that are dispersed and divide mitotically to form new fungi. The hyphae and the asexual spores are usually haploid, having a single copy of each chromosome, and the haploid phase dominates the fungal life cycle. Spores are adaptations for survival; they are able to withstand extreme conditions of dryness, heat, or cold, then produce a new fungus when conditions improve.

Fungi can also reproduce sexually [FIGURE 20.4]. While fungi are neither male nor female, each haploid individual is one of two mating types, *plus* or a *minus* [FIGURE 20.4, Step 1]. The haploid cells of either mating type can grow into hyphae and reproduce asexually. Eventually, a haploid cell of the plus mating type can fuse with a haploid cell of the minus mating type, forming a single cell containing two haploid nuclei (a **dikaryon**, Step 2). (Note that a dikaryon, a cell with two haploid nuclei, is different from a diploid cell containing two haploid chromosome sets in a single nucleus.) These double-nucleated cells can divide and form hyphae, and these threadlike structures can join and give rise to an organ of reproduction, the *fruiting body* (Step 3). The typical mushroom of forest and field is an aboveground fruiting body. Inside the fruiting body, the haploid nuclei can become gametes, and the nuclei of two gametes can fuse, forming a diploid zygote (a process called *sexual fusion*, Step 4). The diploid zygote usually undergoes meiosis, producing haploid cells of the two mating types, and the cycle is complete (Step 5). There are many variations of the fungal life cycle, but the essential feature is that the haploid stage dominates, and the diploid phase is often relegated to a single cell, the zygote.

FUNGAL INTERACTIONS WITH PLANTS

It is easy to imagine the earth becoming buried by a deep layer of dead plants were it not for the activity of fungi, since they are the main decomposers of cellulose and

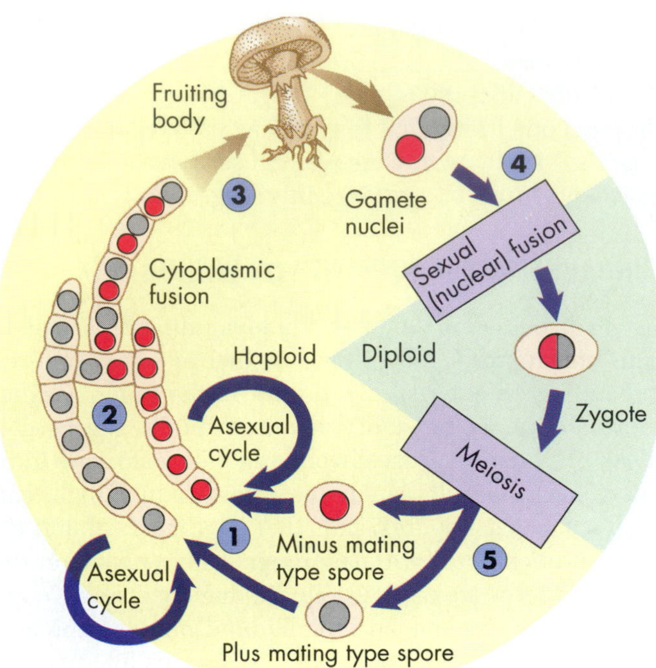

FIGURE 20.4

Generalized Life Cycle of Fungi.

The haploid phase dominates the life cycles of fungi. Spores of either the minus or plus mating type can be produced sexually and asexually, and they grow into hyphae. The cytoplasm of two cells from hyphae of opposite mating types can fuse and produce a cell with two nuclei. This cell can give rise to more hyphae, which may form a fruiting body. Within the fruiting body, the two nuclei of opposite mating types may fuse, giving a diploid zygote. The zygote undergoes meiosis and once more produces haploid spores.

lignin, the hard substances in wood and the two most abundant organic compounds on earth. But out of sight, below the soil surface, fungi and plant roots interact symbiotically in important ways. Symbiotic associations between plant roots and fungi are called **mycorrhizae** (my-CO-RYE-zee; singular, mychorrhiza, literally "fungus roots").

Biologists have discovered and named several hundred species of mycorrhizal fungi associated with the roots of perhaps 90 percent of all land plants, including the western yew we discussed earlier. In many woody plants, mycorrhizae form hairy mantles outside and between the cells of the plant's roots. These fungi expand the surface area available for the plant's uptake of water and minerals [FIGURE 20.5A]. In the majority of land plants, however, different types of mycorrhizal fungi grow *inside* the root cells, send out microscopic hyphae, and also take up from the soil large amounts of nutrients and water that benefit the plant [FIGURE 20.5B]. Infected plants can take in 10 times more phosphorus and other

[A]

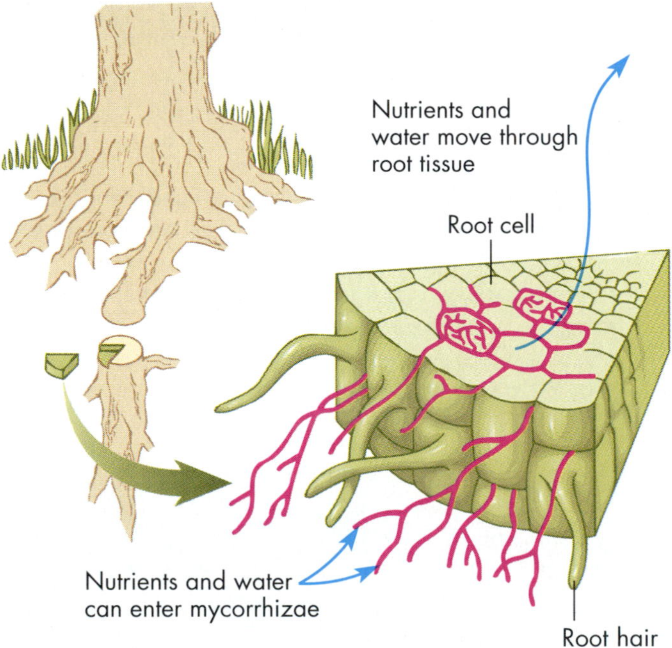

[B] Some mycorrhizae grow inside plant cells

FIGURE 20.5

Mycorrhizae: Fungus-Plant Associations.

[A] Mycorrhizal fungi can grow between and outside the cells of plant roots. Shown here are bare pine roots (left two sprigs) and pine roots associated with mycorrhizal fungi (right four sprigs). **[B]** Some species of mycorrhizae can grow inside plant root cells. Their microscopic hyphae (red) branch into the soil amid the root hairs. The fungus absorbs water and nutrients, which benefit the host, and in return, the fungus receives organic molecules from the host.

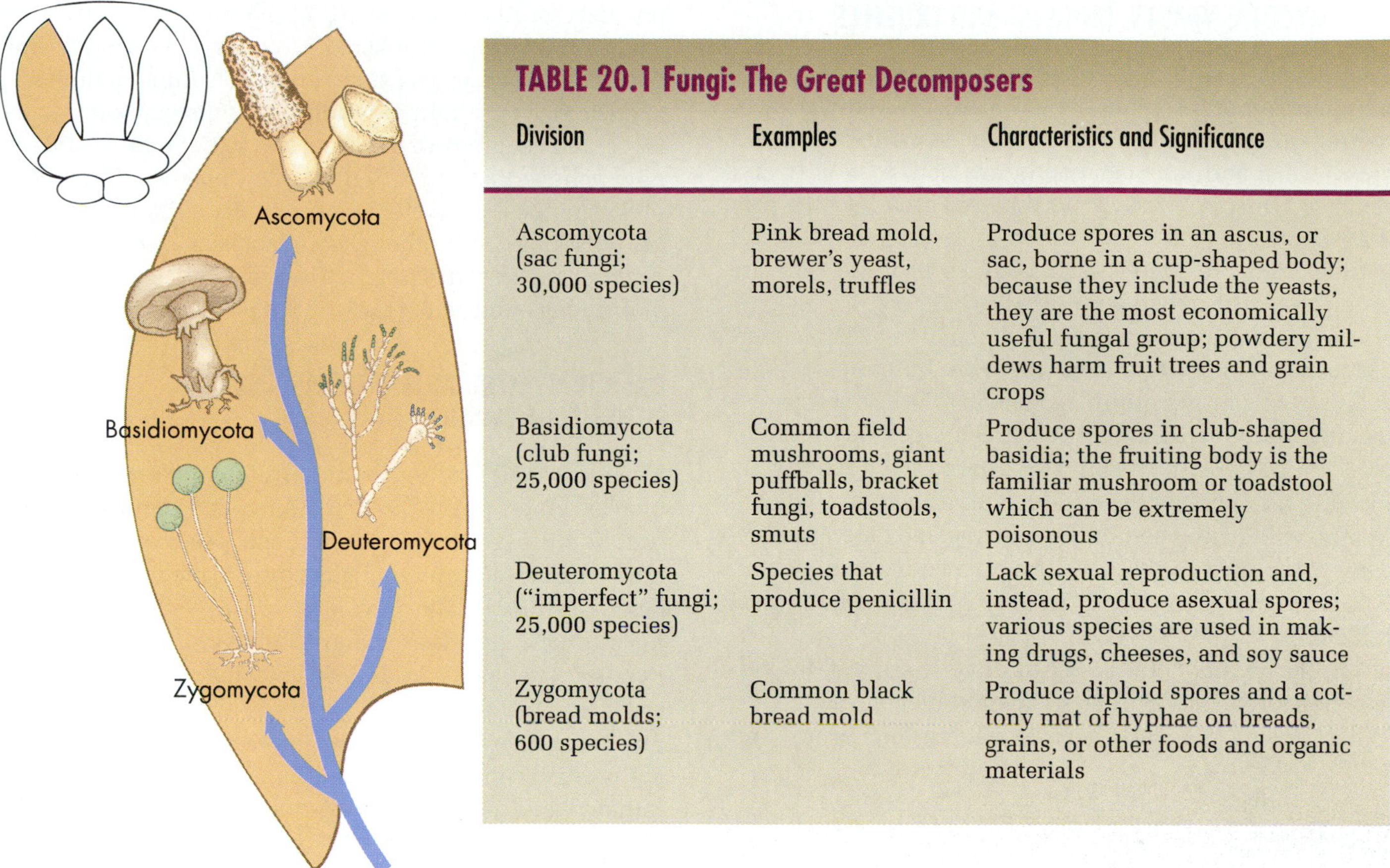

TABLE 20.1 Fungi: The Great Decomposers

Division	Examples	Characteristics and Significance
Ascomycota (sac fungi; 30,000 species)	Pink bread mold, brewer's yeast, morels, truffles	Produce spores in an ascus, or sac, borne in a cup-shaped body; because they include the yeasts, they are the most economically useful fungal group; powdery mildews harm fruit trees and grain crops
Basidiomycota (club fungi; 25,000 species)	Common field mushrooms, giant puffballs, bracket fungi, toadstools, smuts	Produce spores in club-shaped basidia; the fruiting body is the familiar mushroom or toadstool which can be extremely poisonous
Deuteromycota ("imperfect" fungi; 25,000 species)	Species that produce penicillin	Lack sexual reproduction and, instead, produce asexual spores; various species are used in making drugs, cheeses, and soy sauce
Zygomycota (bread molds; 600 species)	Common black bread mold	Produce diploid spores and a cottony mat of hyphae on breads, grains, or other foods and organic materials

nutrients than they could without the fungal symbionts. In exchange for augmenting the plant's nutrition, the fungus gets a home and probably other benefits that biologists do not yet understand.

Paleontologists have found fossilized cell clusters that look like fungal hyphae in 900-million-year-old rocks. The oldest *undisputed* fungal fossils reveal that fungi grew on land 450 to 500 million years ago. These fossils suggest that fungi apparently evolved long before plants left the oceans. Also, fossils of the oldest land plants have fossilized mycorrhizae among the roots. This evidence suggests that "fungus roots," by improving the plants' uptake of water, may have facilitated the transition of plants from water to land. Without fungus-plant interactions, plants might have evolved in very different ways.

Major Groups of Fungi

Fungi cannot move on their own and lack flagella at all stages of their life cycles. They are usually classified by the shapes of their spore-producing structures. Spores are the only means a fungus has of dispersing to new areas, and thus they are of prime importance. TABLE 20.1 describes four major divisions of fungi, focusing on how they form spores, as well as other significant characteristics. (In the fungal and plant kingdoms, a *division* is the equivalent of a *phylum* in other kingdoms.)

ZYGOMYCOTA: BREAD MOLDS AND OTHERS

Rhizopus, the common fuzzy, whitish or grayish mold that grows on bread, is a zygomycote. The typical members of this group are filamentous and grow in cottony masses, but the filaments lack cross walls between adjacent cells. Sexual fusion produces diploid zygotes that form dark, spore-forming structures; these structures may remain dormant for several months. When growth conditions are favorable, the zygote undergoes meiosis and the resulting haploid cells grow into erect stalks, each bearing an asexual spore-producing sphere that can split open and spread haploid spores. Spores can lie dormant and withstand extremes of drought and cold, germinating later and perhaps producing a new hyphal mat in another slice of bread.

ASCOMYCOTA: YEASTS, MORELS, AND TRUFFLES

This largest division of fungi includes mostly saprobes but some important plant parasites as well. Many grow in dense, fuzzy mats; they have hyphae with perforated cross walls; and they can reproduce both asexually and sexually. During sexual reproduction, ascomycotes produce spores in a little sac called an *ascus* (plural, *asci*) [FIGURE 20.6]. In many species, rows of these sacs are borne in a cup-shaped structure.

While the fungi we have discussed so far are generally multicellular organisms, there are numerous single-celled yeasts in the group Ascomycota. Although yeasts usually reproduce asexually by pinching off new smaller cells (budding), they can also reproduce sexually by forming *ascospores*, the products of a meiotic division. Yeasts, which are used in the brewing and baking industries, are probably the most economically useful fungi. Another yeast, *Candida albicans*, causes vaginal yeast infections.

Some ascomycotes are used directly as food, including the morels and truffles so favored by gourmet cooks. Other ascomycotes, however, attack our food crops, and powdery mildews parasitize apple and cherry trees as well as grain crops. Bread baked with rye infected by one particular ascomycote causes ergot poisoning in humans, characterized by hallucinations, convulsions, premature labor, and gangrene of the arms and legs. Some historians now believe that around 1690, the "possessed" witches of Salem, Massachusetts, were actually suffering from ergot poisoning. Still other ascomycotes caused *Dutch elm disease* and *chestnut blight*, diseases that have robbed us of millions of our most beautiful hardwood trees.

BASIDIOMYCOTA: MUSHROOMS AND OTHER CLUB FUNGI

The most familiar fungi—including edible mushrooms, bracket fungi, puffballs, toadstools, and smuts—are all **basidiomycotes** [FIGURE 20.7]. Each produces spores in club-shaped reproductive structures called *basidia* (singular, *basidium*). The spores are blown or carried by animals to new locations, where they germinate into new hyphae. The spore-producing abilities of the club fungi are truly prodigious. An average-sized grocery store mushroom can produce about 16 billion spores, and a giant puffball can produce 7 trillion. If all these spores germinated and developed into adult puffballs, they would collectively weigh 800 times the weight of the earth! Obviously, since we are not buried by puffballs, their spores rarely find environmental conditions favorable for germi-

FIGURE 20.6

Sexual Reproduction in Ascomycota Fungi.

[A] These fungi have a visible structure which houses a series of little sacs, or asci [B], meiosis within each ascus yields spores, and these blow about, germinate, and grow into new cup fungi.

[A]

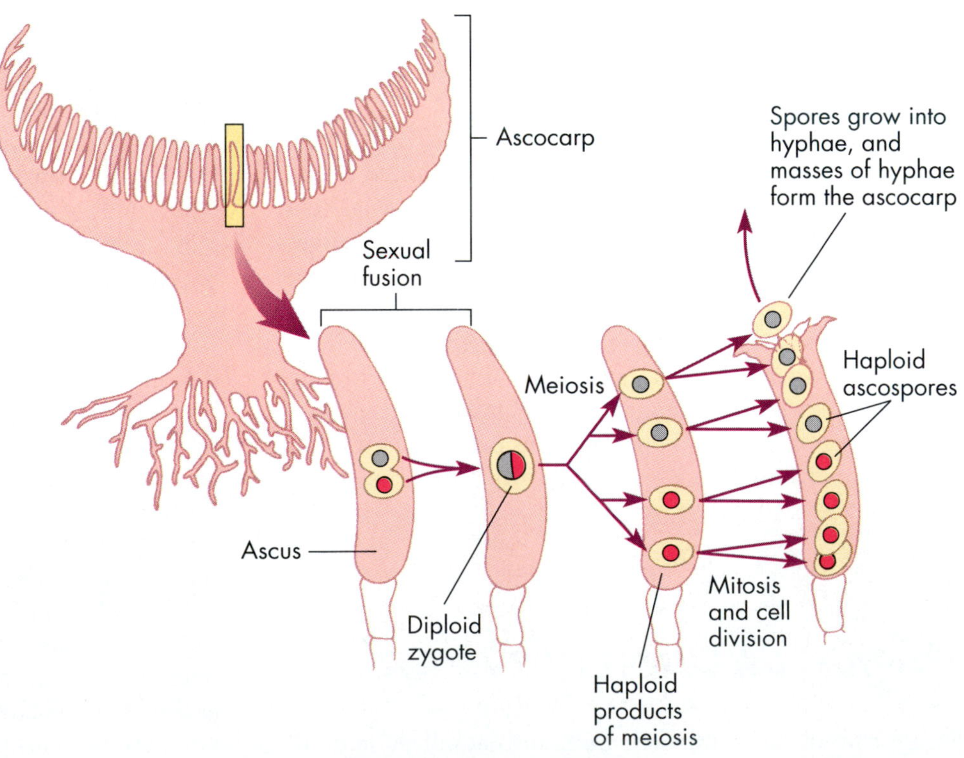

[B] Reproduction in a cup fungus

[A]

[B]

FIGURE 20.7

Common Forest Fungi.

Most familiar forest fungi are basidiomycotes. [A] The deadly poisonous *Amanita muscaria* grows in a forest in New York State. [B] A giant puffball, *Calvatia gigantica*, dwarfs this boy's hand and arms.

nation and maturation. FIGURE 20.8 shows the life cycle of a common field mushroom.

One of the most interesting relatives of the edible field mushroom is the *white rot fungus*, a basidiomycote with an appetite that would stun a teenager: This organism can digest anything from dead wood to DDT, and scientists are after its secrets in hopes of harnessing a new antipollution "device" to help clean up the environment. White rot fungus grows as small, white mycelial mats in the soil and inside rotten logs. It generates an enzyme that can degrade lignin, the hardening agent in wood. Lignin is also something of a nuisance to the paper industry, since it gives paper bags their brown color and turns newsprint yellow when exposed to air and sunlight. Papermakers add strong bleaches to combat these effects in higher-quality paper, but when these bleaches are present in industrial effluents, they cause severe pollution of lakes and rivers. Researchers are looking for ways to avoid using such bleaches by employing white rot fungus to digest lignin during papermaking.

DEUTEROMYCOTA: FUNGI IMPERFECTI

Many species of fungi seem to lack sexual reproduction (hence the name "imperfect" fungi). Most appear to be as-

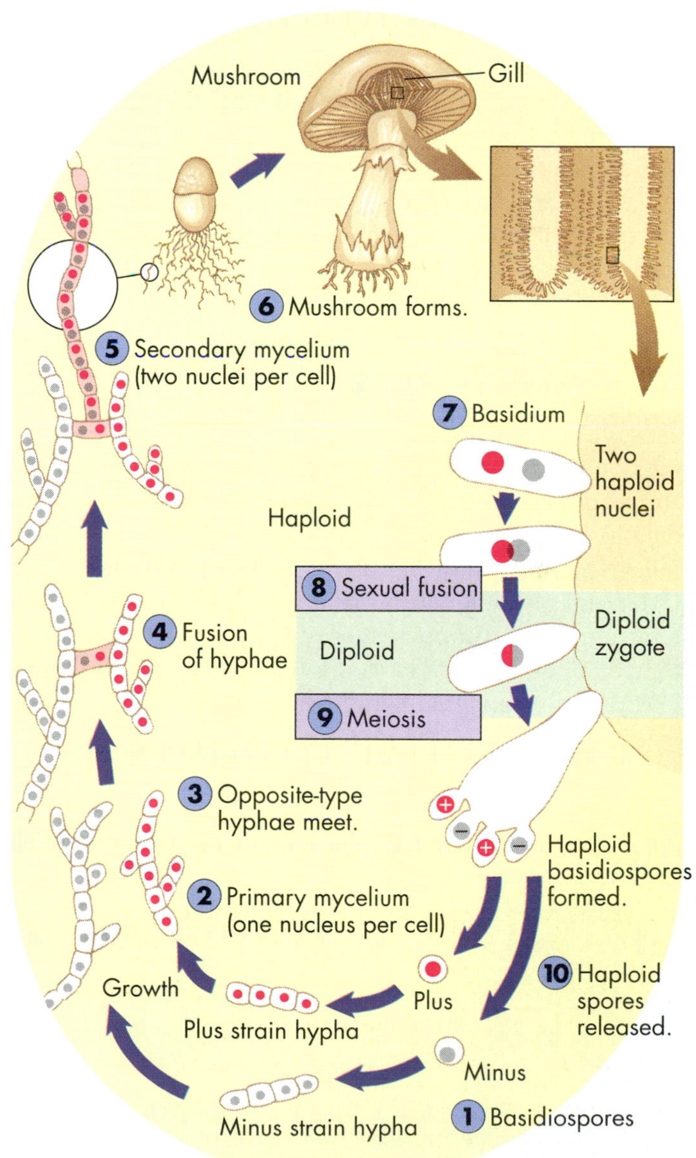

FIGURE 20.8

Life Cycle of the Common Field Mushroom.

Plus and minus haploid spores (1) grow into primary mycelia (2), which fuse, giving rise to a secondary mycelium in which each cell has two haploid nuclei (3–5). This secondary mycelium forms a mushroom, in which the two nuclei may fuse (7, 8) yielding a diploid zygote which immediately undergoes meiosis (9) to make more haploid spores (1).

comycotes that reproduce only with asexual spores called *conidia* (singular, *conidium*), which are pinched off from the tips of spore-forming hyphae. We use many kinds of Fungi Imperfecti, including *Penicillium* and *Aspergillus* species for making penicillin and other drugs, Roquefort and other blue cheeses, and citric acid, soy sauce, and *sake* (Japanese rice wine). *Aspergillus flavus*, a mold that grows on stored peanuts and various grains, produces a compound called *aflatoxin*, one of the most potent cancer-causing substances ever identified.

LICHENS: A TALE OF TWO KINGDOMS

Have you ever looked carefully at the tombstones in an old cemetery? The stones are usually encrusted with gray, orange, or greenish layers like those shown in FIGURE 20.9. These formations are **lichens** (LIE-kenz), symbiotic associations between a fungus and either a eukaryotic green alga [see FIGURE 20.14], or a prokaryotic cyanobacterium [see FIGURE 19.10]. It is widely believed that the author of the Peter Rabbit books, Beatrix Potter, was among the first—perhaps the very first—to notice this association between fungi and algae.

The fungus forms a dense hyphal mat around the algal cells—a mat that provides the photosynthetic cells with some protection and a supply of water—and lives off the alga's photosynthetic products. Lichens take almost no nutrients from the substrate on which they live, and they are often the first organisms to inhabit lava flows or other newly exposed rocks. However, they grow very slowly. You can estimate their rate of growth by measuring the diameter of the largest lichens found on tombstones of various ages, which you can calculate from the date inscribed on the stone. While lichens can survive in bleak environments, they are particularly sensitive to the industrial pollutant sulfur dioxide. Therefore, biologists have found the disappearance of lichens to be an early indicator of air pollution. Many lichens reproduce by releasing little balls that contain at least one algal cell embedded in a small network of hyphae.

➤ CONCEPT CHALLENGE

Recent evidence from molecular genetic studies suggests that fungi may be genetically more similar to animal cells than to plant cells. Which aspects of fungal biology support this hypothesis? Which contradict it?

[A]

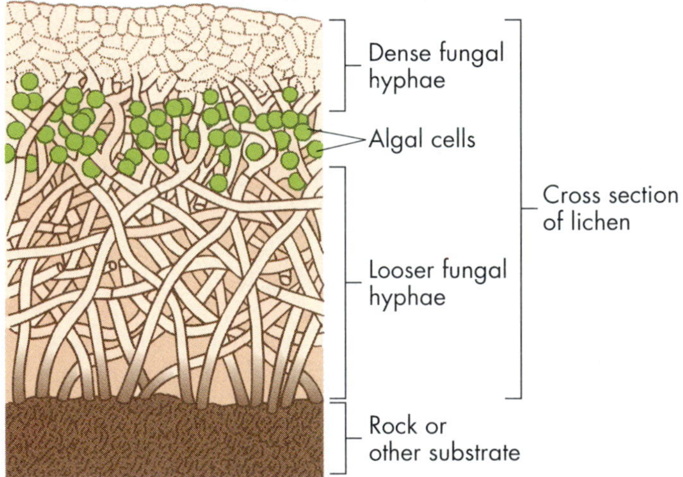

[B]

FIGURE 20.9

Lichens: Alga-Fungus Associations.

Lichens can be colorful or camouflaged [**A**], but in each type, a fungus and photosynthetic algal cell grow in such close association that they appear to be a single organism [**B**].

The Plant Kingdom

Plants can be gigantic, like the redwoods of the Pacific coast, or minute, like the duckweed on a pond's surface.

Plants can also be brilliantly colored or drably camouflaged. Regardless of differences, however, plants have certain common physical characteristics. Plants are multicellular eukaryotic organisms that generally develop from embryos or diploid multicellular young supported by maternal cells. Embryos result from sexual fusion, and most plants can reproduce by sexual means. Plant cells also generally contain chloroplasts [review FIGURE 6.6], green cytoplasmic organelles that transform the energy of sunlight into sugar-phosphates.

Plants play ecologically crucial roles. They sustain the lives of most animals, including humans, through the production of food, fibers, fuel, and shelter. In addition, plants release the oxygen that aerobic organisms, including plants, utilize during aerobic respiration. Finally, plants display an amazing array of substances called **secondary metabolites** that are not directly needed for the organism's survival and reproduction, but defend the plant against animals, fungi, and even other plants. The taxol produced by the yew tree is probably an antifungal agent. The caffeine in coffee beans and the nicotine in tobacco leaves are secondary metabolites, and may also protect the plants.

We will investigate the physiological and anatomical adaptations of plants in greater detail in CHAPTERS 34 to 37.

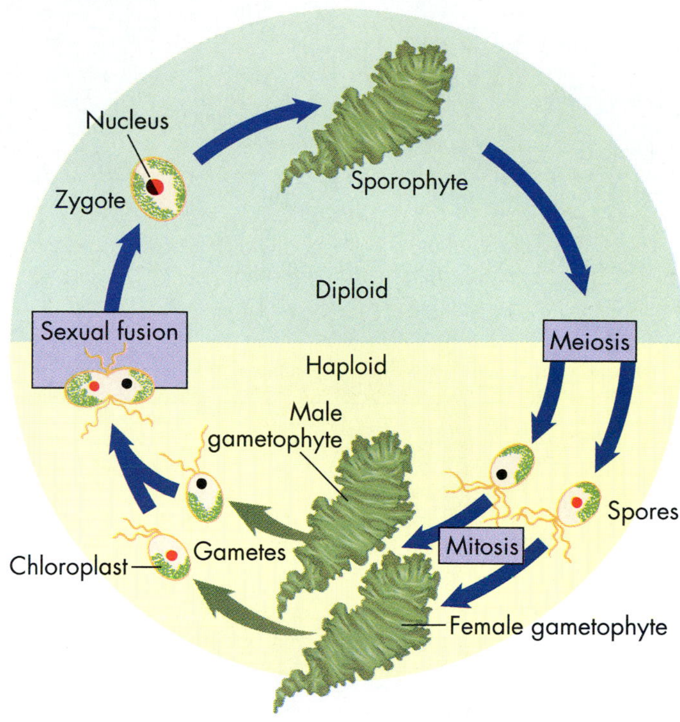

FIGURE 20.10

The Plant Life Cycle: Alternation of Generations.

Plants have an alternation between a multicellular haploid generation (a gametophyte, or gamete-making plant) and a multicellular diploid generation (a sporophyte, or spore-making plant). This figure shows the life cycle of *ulva*, an alga.

PLANT LIFE CYCLES

In their life cycles, plants have a regular alternation of multicellular haploid and diploid generations, and this **alternation of generations** is a key concept for understanding plant biology [see CHAPTER 36].

In many plant species, the haploid and diploid phases are separate, and each phase is a free-living green plant. FIGURE 20.10 shows these alternating phases in the life cycle of a green alga. The haploid phase is the **gametophyte** ("gamete-making plant"), and each gametophyte can produce male and/or female gametes. The diploid phase is the **sporophyte** ("spore-making plant"). Meiosis in the sporophyte results directly in the production of haploid spores rather than gametes. These spores divide by mitosis, and develop into multicellular, haploid gametophyte plants, which make gametes. Sexual fusion then occurs between the haploid gametes, restoring the diploid chromosome number and yielding a zygote. The zygote divides by mitosis and produces the multicellular sporophyte generation, thus completing the life cycle.

During plant evolution, there has been a gradual trend toward an increase in importance of the sporophyte generation [FIGURE 20.11]. In some algae (simple aquatic plants), the haploid phase dominates the life cycle and is the conspicuous organism we see. In a few algae and in many simple land plants, such as mosses, the diploid and haploid phases are each visible plants, as in the life cycle shown in FIGURE 20.10. Finally, in more complex land plants, like yews, coconut palms, and hibiscus, the diploid phase dominates, and is the conspicuous plant we recognize. The majority of plants fit this last description.

TRENDS IN PLANT EVOLUTION

Evidence suggests that plant life originated in water and that plants probably developed from photosynthetic protists [FIGURE 20.11]. Algae, one of the main groups of organisms traditionally included in the plant kingdom, are still aquatic. Red algae, brown algae, and green algae are actually more closely related to certain organisms in the kingdom Protista than they are to each other. Primitive green algae may have been the ancestors of all land plants [FIGURE 20.11].

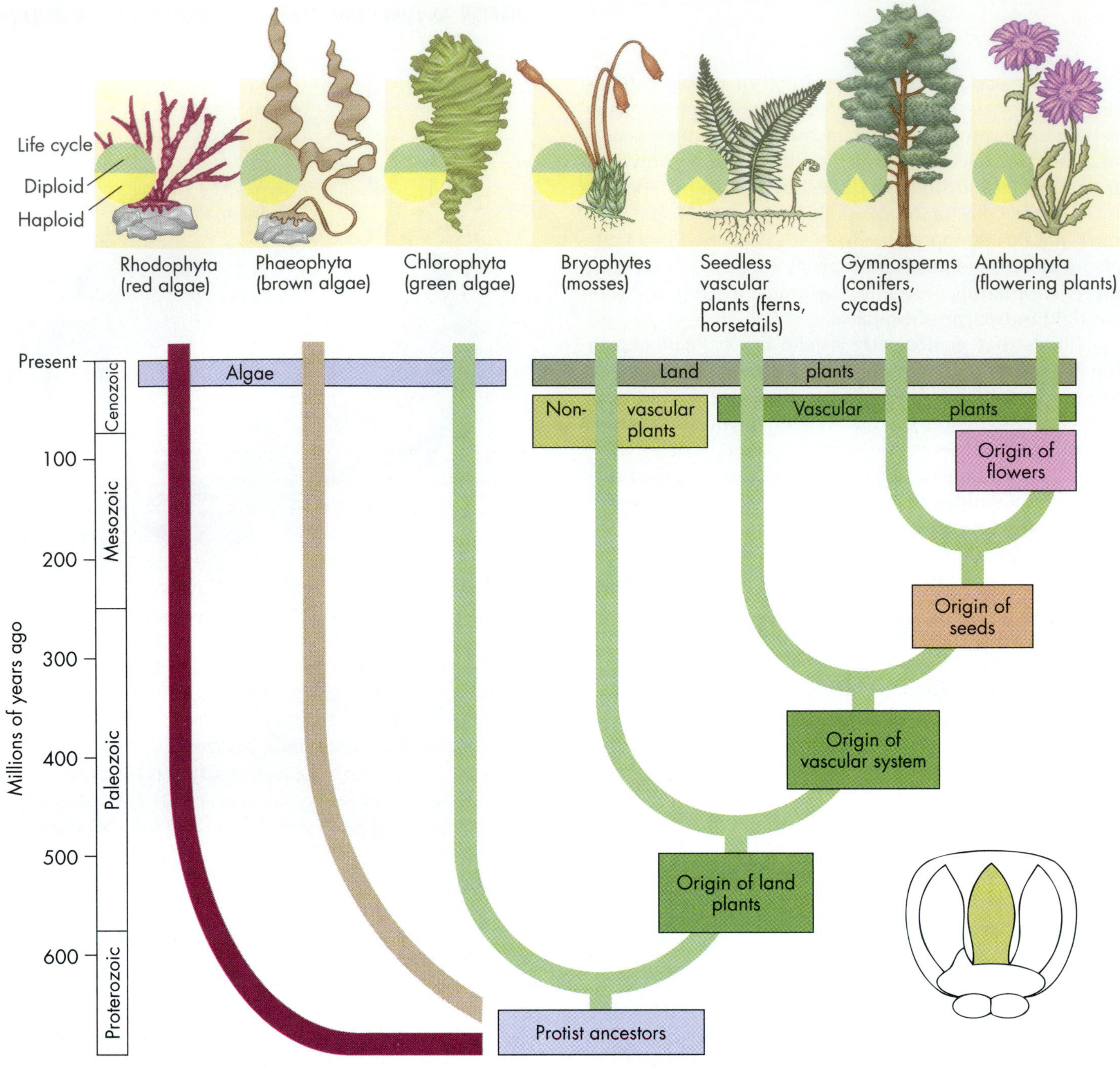

The first transitional land plants lacked specialized fluid-conducting tubes and survived on the damp fringes of ancient oceans during the Paleozoic over 450 million years ago [see TABLE 17.2]. At that time, the continents were relatively flat. Water levels probably changed dramatically as the seasons changed and during longer climatic cycles. Some individuals had combinations of alleles (through mutation and random genetic recombination) that resulted in adaptations for withstanding dessication (drying out). These individuals would have tended to reproduce more successfully in periodically dry environments than individuals lacking the alleles.

FIGURE 20.11

Trends and Mileposts in Plant Evolution.

Evidence from the structure and genetics of various organisms suggests to evolutionary biologists that the red, brown, and green algae arose independently from protist ancestors. The green algae probably gave rise to the earliest land plants, which were similar to today's bryophytes. The origins of the vascular system, seeds, and flowers were further mileposts in plant evolution. The increasing dominance of the diploid sporophyte generation, indicated by the circles in the top row of figures, was a general evolutionary trend.

Dessication-resistant adaptations include a waxy outer layer (*cuticle*) perforated by holes that allow gas exchange (*stomata*), and reproductive structures that resist drying out. Such adaptations opened up a spacious new frontier in damp areas near oceans. Sunlight was plentiful, the soil held rich mineral nutrients, and initially there were no plant-eating animals on the land.

Once plants made the transition to land, other evolutionary milestones occurred [FIGURE 20.11]. The evolution of fluid-conduction tubes, the *vascular system*, enabled the efficient distribution of water and nutrients throughout the plant. Later, the origin of **seeds** and **flowers**, the reproductive structures of angiosperms, helped improve the dispersal of young and the survival of new generations. Associated with these evolutionary innovations was a general trend toward the dominance of the diploid generation (the sporophyte) [FIGURE 20.11, top]. These innovations and trends led to the vast diversity of plants we see today.

Let's begin now to survey the diversity of the plant kingdom, starting with the algal ancestors of the land plants.

CONCEPT CHALLENGE

Some biologists have argued that plants invaded the land before animals. What kind of reasoning might lead one to this conclusion? What kind of evidence would you seek to refute or support this claim?

Algae: Ancestral Plants That Remained Aquatic

What exactly are algae? We have already encountered cyanobacteria (blue-green algae) in the kingdom Eubacteria and golden-brown algae in the kingdom Protista. Are algae monerans, protists, plants, or all three? Actually, the term **alga** (plural, *algae*), from the Latin word for "seaweed," has been used to describe simple chlorophyll-containing organisms that live in water and thus has come to include organisms in all three kingdoms. The algae of the plant kingdom are mostly multicellular aquatic organisms with simple reproductive structures. They are classified in three divisions: (1) red algae, Rhodophyta; (2) brown algae, Phaeophyta; and (3) green algae, Chlorophyta [FIGURE 20.11]. TABLE 20.2 lists the divisions of the plant kingdom.

Multicellular algae probably arose about 600 million years ago from ancestral photosynthetic protists and were the only plants "on stage" that early. They have since diversified into about 12,500 species, which vary from tiny threads to giant kelps and tangled seaweeds. Algae live in virtually all bodies of water, and sometimes on damp soils, rocks, trees, and even snowbanks. Although modern algae are generally simpler than the complex plants now dominating land habitats, they continue to serve a critical ecological role: Since aquatic habitats are so widespread (covering nearly 75 percent of the earth's surface) and algae are so abundant, the algae probably capture 90 percent of all the solar energy trapped by photosynthesis, notwithstanding the vast forests and prairies on land.

The aquatic habitat is a relatively benign and unchanging place, and its properties helped shape the organisms that live there. Because water supports the algal plant body, most algae lack rigidity, and usually undulate gently with water currents and waves. Since water surrounds the plant on all sides, individual algal cells absorb moisture and minerals directly from the surrounding water and have no need for specialized conduction tubes. Plant shape also reflects this direct contact with water: Most algae are quite flattened, which maximizes the surface area for absorbing water, minerals, and sunlight. Finally, reproduction can be asexual, involving the fragmenting of cells or body parts, or it can be sexual, with the production of eggs and sperm. The main secret to the algae's success is a range of photosynthetic pigments that can absorb the light of the different wavelengths that penetrate to varying water depths. Botanists use these same pigments to distinguish between red, brown, and green algae.

RED ALGAE: THE DEEPEST-DWELLING PLANTS

Most red algae are small, delicate organisms that occur as thin filaments or flat sheets with an ornate, fanlike appearance; some, however, are single-celled or colonial. Red algae generally live in shallow, tropical ocean waters, but a few species survive at depths of about 270 m (880 ft; see CHAPTER 6 and FIGURE 20.12). These deep-sea denizens are 100 times more efficient at capturing sunlight than are the red algae found in shallow waters. They rely on reddish accessory pigments that absorb light in the blue-green range—the only wavelengths that can penetrate to great depths. The red pigments absorb light energy, and some of it is passed to chlorophyll, which traps the energy and makes it available for the synthesis of sugars [see CHAPTER 6].

The conspicuous body of a red alga is the haploid gametophyte, which produces a gel-like protein that can be extracted commercially to make the laboratory growth

TABLE 20.2 Characteristics of the Major Plant Divisions

Division	Examples	Evolutionary Trends and Significance
Algae*		
Red algae (Rhodophyta; 4000 species) 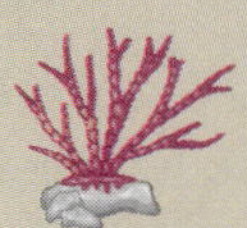	Many fanlike or filamentous types	Single-celled, colonial, or multicellular; reproduction asexual or sexual; haploid generation dominates; light-absorbing red or purple pigments can function at great water depths; used commercially for extracting thickening agents
Brown algae (Phaeophyta; 1500 species) 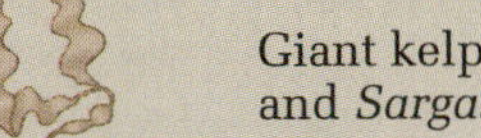	Giant kelps and *Sargassum*	The largest algae; have tubelike conducting cells, but not true vascular tissue; diploid generation dominant in many species; kelp harvested for chemicals and cattle feed
Green algae (Chlorophyta; 7000 species) 	*Ulva*, *Chlamydomonas*	Produce carotene, chlorophyll, like the land plants; many have conspicuous haploid and diploid generations; may have been ancestors to the land plants; found in fresh water
Nonvascular plants		
Bryophytes (Bryophyta; 24,000 species) 	Mosses, liverworts, hornworts	Waterproof coatings, rigid tissues for upright growth on land, rootlike rhizoids; haploid generation (gametophyte) is dominant; often the first plants to colonize an area
Vascular plants (300,000 species)		Vascular tissue for transport and vertical support on land; most have roots, stem, and leaves; diploid sporophyte usually dominant; most successful and diverse group of plants
Seedless vascular plants (Sphenophyta, Pterophyta, and Lycophyta; 13,000 species) 	Ferns, horsetails, club mosses	Vascular pipelines; rhizomes, stems, and fronds; gametophyte can be tiny, independent plant or grows from sporophyte; grow on shady forest floors in low-lying damp areas
Seed plants		Produce seeds; gametophyte reduced to a few cells, dependent on sporophyte; do not depend on water for reproduction
Cycadophyta (100 species)	Cycads	The three divisions of seed plants are cone-bearing remnants of ancient groups
Ginkgophyta (1 species)	Ginkgos	
Gnetophyta (70 species)	*Welwitschia*	
Coniferophyta (Pinophyta) (600 species) 	Pines, sequoias, firs	Naked seeds produced in cones; usually have needle-like leaves or scales; produce pollen; well-developed vascular system; true roots, stems, and leaves; sporophyte is dominant and supports gametophyte; conifers harvested in great numbers for wood products
Anthophyta (Magnoliophyta)	Flowering plants	Produce flowers, seeds, and fruits; economically useful for food, drugs, landscaping
Dicots (225,000 species)	Lilies, corn, onions, palms, daffodils	Leaves with parallel veins; seedlings have just one "seed leaf" or cotyledon; flower parts usually occur in multiples of three; seed stores much endosperm
Monocots (50,000 species)	Roses, apples, beans, daisies	Leaves with netlike veins; seedlings have two cotyledons; flower parts in multiples of four or five; seed stores little endosperm

* Some biologists consider red, brown, and green algae to be protists (or protoctists) because certain of their members are single-celled.

FIGURE 20.12

Red Algae: Deep Dwellers.

Some red algae live in cold, inky waters to depths of 268 m (884 ft) below the ocean surface due in part to the ability of their pigments to absorb the wavelengths of light that penetrate to these depths. They are also common in shallow tropical seas. These sea grapes, *Botryocladia pseudodichotoma*, grow in the cool shallow waters along California's coast.

medium agar; these algae also produce the starchy substance carrageenan, used as a stabilizing agent in ice cream, puddings, cosmetics, and paint.

BROWN ALGAE: GIANTS OF THE ALGAL WORLD

Most species of brown algae inhabit cool, offshore waters and occur as small multicellular plants. However, there are notable exceptions: The largest members of the algal world are the **kelps**, brown algae that can grow 100 m (over 325 ft) long and float vertically like tall trees [FIGURE 20.13]. Huge floating masses of the brown alga *Sargassum* thrive in the Sargasso Sea, a mass of still water that runs across the Atlantic north of the Caribbean, sometimes entangling hapless divers and ships alike.

Brown algae range from golden brown to dark brown to black. While they also contain chlorophylls *a* and *c*, it is a golden-brown carotenoid pigment that colors the plants and enables them to collect the blue and violet wavelengths of light that penetrate medium-deep water. These pigments explain why kelps can exploit an environment that many other organisms cannot.

Many brown algae possess complex structures analogous to parts of land plants: Leaflike *fronds* collect sunlight and produce sugars; the stemlike *stipe* supports the plant vertically; and the rootlike *holdfast* anchors the plant to submerged rocks. Special tubelike conducting cells carry sugars produced in the fronds to the deeper plant parts. These tubes function like the more specialized internal transport systems of land plants, but are not related tissues; both types reflect similar evolutionary solutions to the need for internal transport in a large, multicellular organism. In kelps, the diploid sporophyte dominates the life cycle, but in most smaller brown algae, the gametophyte and sporophyte generations are both free-living plants that resemble each other.

GREEN ALGAE: ANCESTORS OF LAND PLANTS

Most species of green algae live in shallow freshwater environments or on moist rocks, trees, and soil, although a few inhabit shallow ocean waters. Green algae usually occur as single cells or as multicellular, threadlike filaments, hollow balls, or wide, flat sheets.

Green algae are most notable for producing orange carotenoids and chlorophylls *a* and *b*; together these pigments absorb the sunlight penetrating air or shallow water with maximum efficiency. Green algae share this pigment combination only with the land plants. This is one of the reasons botanists think that today's green algae and land plants share a common ancestor [see FIGURE 20.11].

FIGURE 20.13

Kelps: Underwater Giants.

As sunlight slants below the ocean surface and illuminates tall, fluttering kelps, the plants tower like forest trees. Kelp can reach 100 m (more than 300 ft) in height.

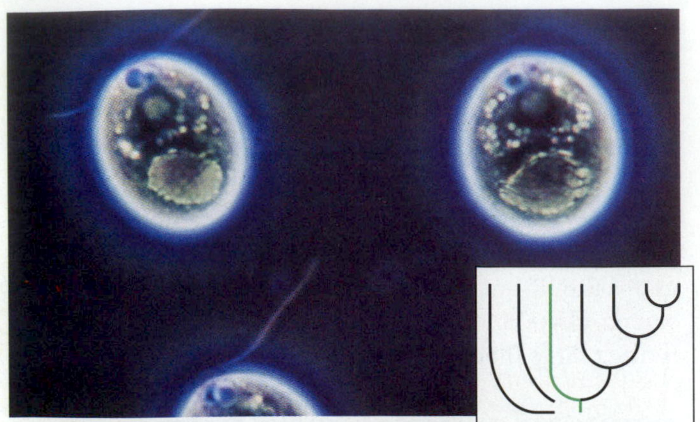

[A] *Chlamydomonas*

[B] *Ulva*

FIGURE 20.14

Green Algae: Plant Ancestors.

[A] Green algae can occur as multicellular filaments or balls, or as single cells, such as *Chlamydomonas* (shown here), a common, single-celled resident of ponds and ditches. These cells have flagella, eyespots, and a cup-shaped chloroplast. **[B]** An observer of *Ulva*, a green alga undulating in a shallow tide pool, would find it difficult to tell male or female gametophyte from sporophyte.

The unicellular green alga *Chlamydomonas* has been a favorite laboratory organism for investigating how genes control mating, the lashing movements of flagella, and the metabolic processes of photosynthesis and aerobic respiration. (Researchers affectionately refer to it as "Chlamy.") *Chlamydomonas* is an oval cell that typically lives in freshwater pools and moist soil, and is propelled by two flagella [FIGURE 20.14A]. The most prominent organelles are an eyespot that orients the cell toward light and a large, cup-shaped chloroplast that nearly fills the cell. The haploid phase dominates the life cycle of *Chlamydomonas*. The single adult cell bears a striking resemblance to a flagellated spore from the multicellular marine green alga *Ulva*, or sea lettuce, suggesting close evolutionary ties [FIGURE 20.14B].

Ulva, an inhabitant of tide pools, grows a delicate, leaflike body, or *thallus*, which resembles sheets of green cellophane and is just two cells thick [FIGURE 20.14B]. *Ulva* displays an alternation of conspicuous haploid and diploid phases, which are multicellular and (to the human observer) virtually identical [review FIGURE 20.10]. The haploid thallus, the gametophyte, produces gametes, while the identical diploid thallus, the sporophyte, produces spores.

The aquatic green algae probably gave rise to the simplest land plants, which still rely on standing water for reproduction. The next section recounts physical trends that emerged and allowed life to inhabit the land.

➤ CONCEPT CHALLENGE

Can you suggest an alternative to classifying algae in the kingdom Plantae? What are the advantages and disadvantages of the various classification schemes?

Simple Land Plants: Still Tied to Water

While algae were the major plants during the Proterozoic, it wasn't until midway into the Silurian period of the Paleozoic era that plants made the transition to land [review FIGURE 20.11]. The two groups that made the transition were the bryophytes and the seedless vascular plants. The *bryophytes*, or mosses and relatives, grow low to the ground and require standing water for reproduction. In contrast, the *seedless vascular plants*, including ferns, horsetails, and club mosses, grow taller than the bryophytes; they have an internal transport system, but still require standing water to reproduce. The seedless vascular plants were dominant during the Paleozoic, but were later surpassed by the seed-forming vascular plants, the gymnosperms and angiosperms, or Anthophyta, with their seeds and freedom from moisture-dependent reproduction. Genetic and fossil data suggest that a green alga progenitor made the transition to land and then later diversified into the bryophytes and seedless vascular plants, the latter in turn giving rise to the seed-forming vascular plants [FIGURE 20.11].

For plants to thrive on land, they needed adaptations that enabled them to overcome the following difficulties. Without surrounding water they had to (1) prevent desiccation—in particular, they needed to seal in water and minimize evaporation while still allowing for the exchange of oxygen and carbon dioxide gases; (2) support the plant body in the absence of buoyancy from surrounding water; (3) to absorb water and minerals from the new substrate—soil, mud, or sand—and transport them through the plant body; and (4) reproduce sexually without shedding gametes directly into an ocean, lake, or pond. As you will see, the relative success of each group of land plants depends, in part, on how its physical structures evolved and met these challenges.

BRYOPHYTES: PIONEERS ON LAND

The modern bryophytes include mosses, liverworts ("lobed plants"), and hornworts ("horn-shaped plants"). Second only in number of species to the flowering plants, bryophytes are small organisms that generally stand less than about 3 cm (1¼ in) tall [see FIGURE 20.15A and TABLE 20.2]. They are often the first plants to colonize a new area, and in most of the species, the diploid sporophyte grows like a miniature street lamp from the leafy haploid gametophyte. These small, simple plants have successfully met two of the main challenges of terrestrial surroundings. They have waterproof coatings that prevent desiccation, as well as tiny openings, the stomata, on their surfaces that allow gas exchange through the coatings. Bryophytes also have tissues that are rigid enough to help keep the low plants upright. The plants are small enough, however, that they do not require the kind of internal vascular system that absorbs or transports water and gives greater rigidity to other land plants. Instead of true roots, bryophytes have hairlike **rhizoids** that act only as anchors, absorbing neither water nor minerals. Water reaches the individual plant cells by slowly diffusing through the entire organism, as in a simple wick system.

[A] A moss

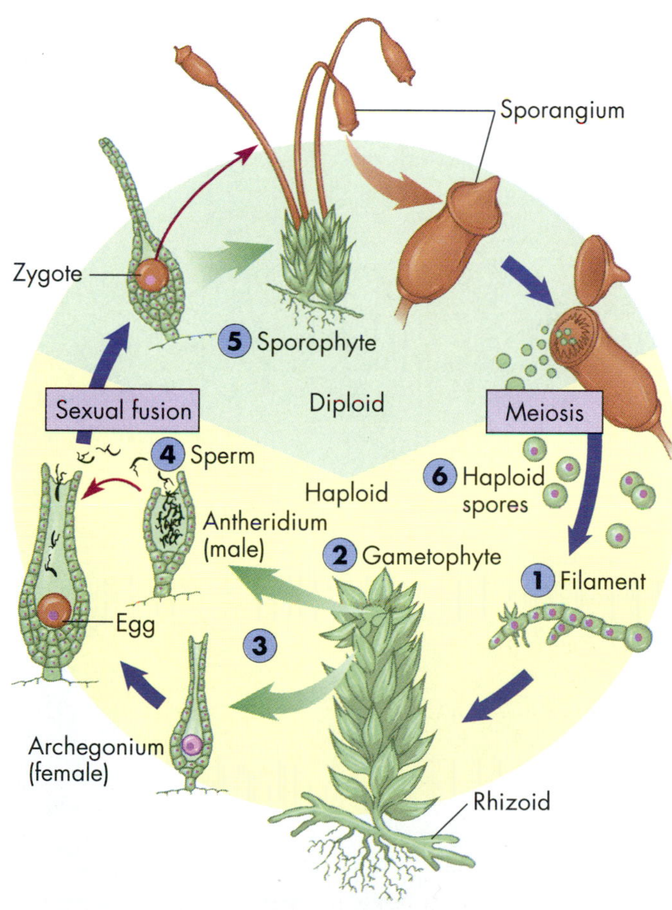

[B] Life cycle of a moss

FIGURE 20.15

Mosses: The Most Familiar Bryophytes.

[A] Although mosses grow in thick, velvety carpets on fallen trees or patches of soil, each individual moss is a leafy gametophyte (haploid) that periodically bears a slender, green-brown sporophyte (diploid). Hairlike rhizoids anchor the plant to its substrate. **[B]** As the text explains, the moss life cycle alternates between the two conspicuous phases. A spore chamber, or sporangium, produces spores, which germinate and grow into gametophytes (1, 2); and these haploid plants form egg- and sperm-producing organs (3) that make gametes (4), which in turn fuse and grow into new sporophytes (5) that once again produce spores (6).

Because of this dependence on diffusion, the plants are limited in size, and are often restricted to shady, moist places. In addition, bryophytes can reproduce only where it is at least seasonally wet. This is because they have retained swimming sperm with flagella as do the gametes and vegetative cells of so many green algae. Bryophyte sperm can reach and fertilize an egg only when the plant is drenched in water.

The life cycles of most moss species alternate between equally conspicuous gametophyte and sporophyte generations, as we first saw in algae. While many bryophyte species have survived in moist habitats, this pioneering group did not itself give rise to other more complex land plants.

VASCULAR PLANTS

During the Paleozoic, at least one more solution to the problems of life on land evolved with the vascular plants. **Vascular plants** have specialized cells that grow end to end in long internal tubes extending from the tips of root-like structures in the ground, up through the stem, to the upper parts of the plant, which carry on photosynthesis [FIGURE 20.16A]. These vascular tubes transport water, minerals, and the products of photosynthesis throughout the plant, and also lend vertical support that allows the plants to grow taller than the bryophytes—in fact to grow as tall as the General Sherman sequoia tree [see FIGURE 20.20]. The 400-million-year-old fossils of a plant called *Cooksonia*, recently found in England, show stomata and a vascular system [FIGURE 20.17]. These adaptations enabled this new group to occupy drier habitats than bryophytes and thus move farther inland into unoccupied areas.

FIGURE 20.16

The Seedless Vascular Plants.

With their vascular pipelines for support and transport of water and nutrients [A], the horsetails, ferns, and tree ferns can grow larger and taller than the mosses and relatives. [B] A horsetail, *Equisetum.* [C] A tree fern, *Dicksonia antarctica*, in a grove of eucalyptus trees.

HORSETAILS AND FERNS: THE SEEDLESS VASCULAR PLANTS

Seedless vascular plants include the horsetails (division Sphenophyta, FIGURE 20.16B); the lycopods, or club mosses and relatives (Division Lycophyta); and the familiar ferns and tree ferns (Division Pterophyta; see FIGURE 20.16C and TABLE 20.2). In the tropical climates of the Paleozoic, these first vascular plants grew in vast primordial forests, and reached sizes much larger than today's tree ferns. However, with the dawning of the Mesozoic, the climate gradually grew colder and drier, and these giants disappeared, leaving behind the more diminutive horsetails, lycopods, and ferns. In a different sort of legacy, the bodies of the fallen giants were incompletely decomposed by fungi and bacteria, and the remaining organic compounds were carbonized into the enormous coal deposits now mined for fuel.

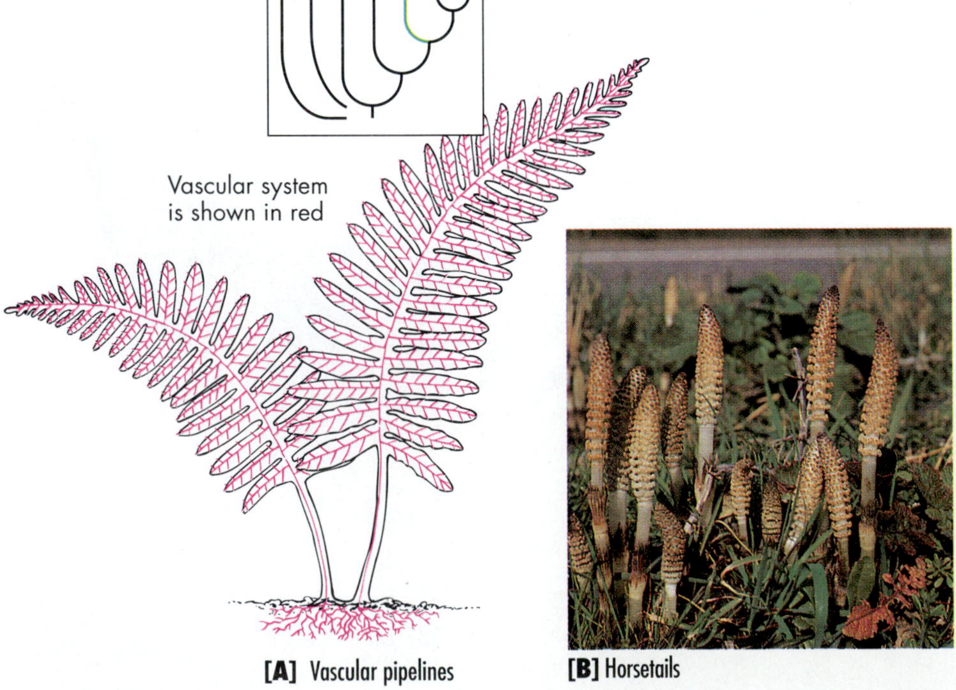

[A] Vascular pipelines

[B] Horsetails

[C] Tree ferns

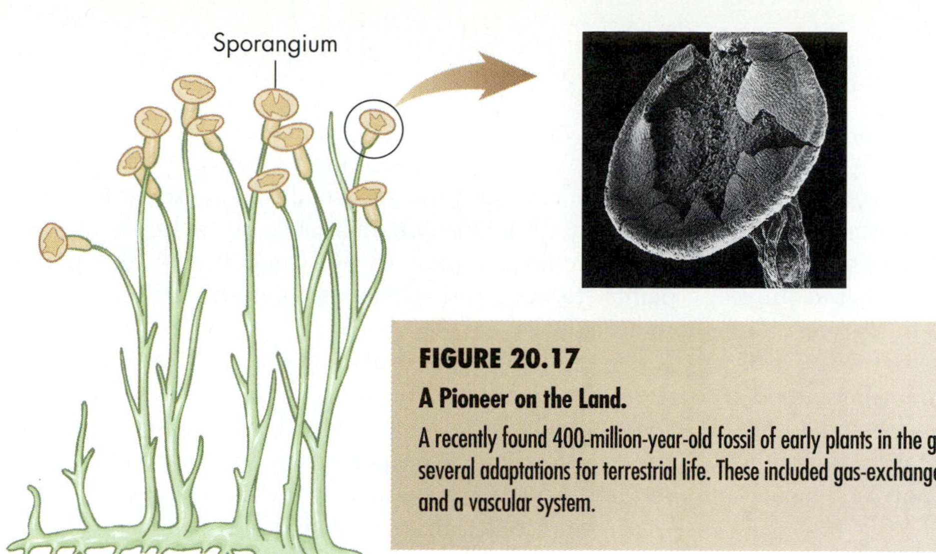

FIGURE 20.17

A Pioneer on the Land.

A recently found 400-million-year-old fossil of early plants in the genus *Cooksonia* reveals that these early plants had several adaptations for terrestrial life. These included gas-exchange ports (stomata), thick-walled, water-preserving spores, and a vascular system.

Some of the seedless vascular plants lack true roots, but have horizontal stems called **rhizomes** that grow on or just beneath the ground surface and survive from year to year. Growing up from each rhizome are many erect, leafy stalks. In horsetails and lycopods, the erect parts are true **stems**; in ferns, they are the central stalks of large leaves called *fronds*. At the end of each growing season, the stems and leaves die, and are replaced the following season by new aerial parts from new places on the slowly expanding rhizomes [FIGURE 20.18A]. This replacement is the way seedless vascular plants reproduce asexually.

In ferns, the undersides of the lacy fronds of the familiar fern adult—the diploid sporophyte—are dotted with *sori* (singular, *sorus*) [FIGURE 20.18C]. The sori on the leaves of adult sporophytes contain globes [FIGURE 20.18B, Step 1] in which meiotic divisions produce haploid spores (Step 2). Under correct conditions, the spores are catapulted into the air; they then reach land and germinate into new gametophytes (Step 3). This green gametophyte (Step 4) can produce either eggs or sperm or both (Step 5). Sperm have flagella that propel them through standing water. After fertilization (Step 6), the zygote develops into a new adult diploid sporophyte (Step 7).

[A] New fern growth.

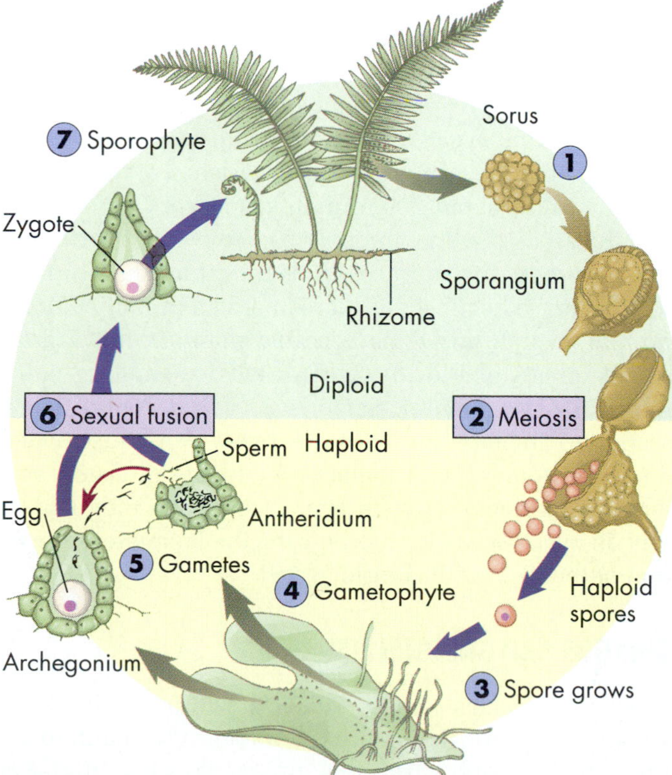

[B]

[C] Sori on fern frond.

FIGURE 20.18

Fern Life Cycle.

As the text explains, the spore chamber in the sorus produces spores, which grow into green, heart-shaped haploid gametophytes. These make eggs and/or sperm. Given sufficient water covering the gametophyte plant, the sperm can swim to the egg container, fuse with the egg, and grow into a new large, frilly diploid sporophyte.

Horsetails, lycopods, and ferns show a continuation of the trend toward dominance by the diploid generation [review FIGURE 20.10]: In these seedless vascular plants, the sporophyte is the conspicuous adult, and the gametophyte is a small, free-living green plant. The plants retain swimming sperm, like the bryophytes, and thus require standing water for sexual reproduction. In a sense, the seedless vascular plants are the botanical equivalents of the amphibians, animals such as frogs and salamanders, which can live on dry land but must have water to lay their eggs. Interestingly, giant amphibians were the dominant land animals during the warm, swampy Paleozoic.

➤ CONCEPT CHALLENGE

Explain the observation that moss sporophytes are essentially parasites on the gametophyte plant, whereas fern and horsetail sporophytes are independent organisms.

Gymnosperms: Pioneers on Dry Land

As the Mesozoic era began about 245 million years ago, Pangaea was breaking apart, the continental seas had retreated, and new masses of high, dry land became open to plant and animal colonization [see TABLE 17.2]. While the early land plants were tied to swampy lowlands, riverbanks, and coastal lagoons, new lines of plants appeared with reproductive modifications that broke the final barriers to complete colonization of land. These organisms were the **gymnosperms** ("naked-seed" plants), and they included the now-extinct seed ferns as well as the cycads, ginkgos, and conifers, such as yews, pines, firs, sequoias, and redwoods [see TABLE 20.2]. In FIGURE 20.1, you can see that the yew seed is naked in that it is not totally surrounded by the red maternal tissue. Contrast this with an apple, orange, or pea, where the seeds are surrounded by maternal cells.

The reproductive innovations of the gymnosperms include the pollen grain and the seed, as well as a further shifting toward dominance by the diploid generation. **Pollen grains** are immature male gametophytes composed of just two to five nonmotile cells plus a dry outer coat. Pollen grains are produced in huge numbers by male cones borne on the familiar conspicuous plant, the sporophyte, and are disseminated by the wind. In effect, the pollen grain "airlifts" the sperm cells to the egg cells harbored inside special chambers in female cones, also borne on a mature sporophyte. Both male and female haploid gametophytes are reduced to small nonphotosynthetic structures housed entirely by the diploid sporophyte generation; thus, they embody the trend toward a dominant diploid phase. In a sense, water needed for fertilization is provided by the moisture in the parent plants' tissues, rather than by splashing raindrops or standing water.

Another seed plant distinction is that the male and female gametophytes develop not from identical *homospores*, as in ferns, but from different-shaped *heterospores*—smaller **microspores**, which give rise to sperm, and larger **megaspores**, which give rise to eggs.

Once fertilization has taken place, the gymnosperm embryo develops inside the ovule. The **ovule**, which consists of the egg cell and its surrounding tissue layers, sits exposed on the surface of a *sporophyll*, a leaf of the female cone. (The term *naked seed* refers to the absence of an ovary surrounding the ovule). Eventually, the ovule and enclosed embryo develop into a seed, which is encased in a tough coat that houses both the young plant and a supply of food for the early stages of its growth. Seeds are the most successful dispersal devices to have evolved in the plant kingdom, and they greatly facilitated the further colonization of land. Reptiles—the dominant land animals during the Mesozoic, when gymnosperms were the ruling land plants—evolved parallel structures: eggs with tough shells that could be laid and hatched on land. These freed their reproduction from the dependence on standing water displayed by amphibians.

CYCADS

A small group of tropical plant species often mistaken for palm trees, the **cycads** (division *Cycadophyta*) are cone-bearing remnants of a group that flourished at the time of the dinosaurs, about 200 million years ago. Cycad seeds, like the seeds of all gymnosperms, are naked, carried on open reproductive surfaces in cones rather than hidden in an ovary. The most familiar American cycad is *Zamia pumila*, which grows as a native plant in the sandy-soiled forests of Florida, with leaves resembling palm fronds [FIGURE 20.19A]. Scientists recently found that a secondary metabolite in cycad seeds can cause weak muscles, tremors, and malaise. Residents of Gaum suffered an epidemic of such symptoms when they had to rely on cycad seeds for food during the Japanese occupation of their island in World War II.

GINKGOS AND GNETOPHYTES

Just a few species remain today of two formerly widespread plant divisions. *Ginkgo biloba*, the maidenhair tree, is the only surviving ginkgo species (division

[A]

[B]

FIGURE 20.19

Cycads and Ginkgos.

[A] Cycads, such as this one from Zimbabwe, resemble palms but are ancient cone bearers. **[B]** The graceful, fan-shaped leaves of the ginkgo have led to the common name maidenhair tree. In autumn, the fleshy, yellowish reproductive organs on a female tree ripen and fall, and give off a strong, rancid smell.

Ginkgophyta). However, this stately tree, with its beautiful golden autumn foliage, is still planted along many urban streets because its fan-shaped leaves are particularly resistant to smoke and air pollutants. Ginkgos produce round, fleshy cones that resemble fruits [FIGURE 20.19B]. In late summer, these cones, found only on female trees, begin to overripen and smell strongly of rancid butter. For this reason, gardeners prefer the pollen-producing male trees for ornamental plantings. Interestingly, the leaves of the ginkgo fall each year; it is **deciduous**, like many flowering trees.

The *gnetophytes* (division *Gnetophyta*) include one of the most unusual vascular plants on earth: *Welwitschia*, a low-growing, cone-producing native of southwestern Africa that, when dug up, looks like an overgrown turnip. It has a disk-shaped woody stem, long, twisting leaves that stretch across the ground for more than a meter, and a deep taproot, an evolutionary adaptation that allows the plants to survive in dry environments without rain for up to five years.

CONIFERS: FAMILIAR EVERGREENS

The cone-bearing trees called **conifers** are the most familiar and largest remaining group of gymnosperms, and they include many ecologically and economically important species. Millions of acres in mountainous regions are dominated by pine, spruce, fir, and cedar, much of which is harvested for wood, paper, and resins. Smaller conifers, such as yew, hemlock, juniper, and larch, are often used for the graceful landscaping of buildings and parks. Finally, what some botanists call the largest living things, the giant sequoias, are conifers as well as living reminders of the Mesozoic era, an age of giants [FIGURE 20.20].

FIGURE 20.20

Towering Sequoia: The General Sherman Tree.

For many years, people claimed that the sequoia called the General Sherman Tree is the largest individual organism on earth. It towers over neighboring trees and the dwarfed visitors at its base in Sequoia National Park.

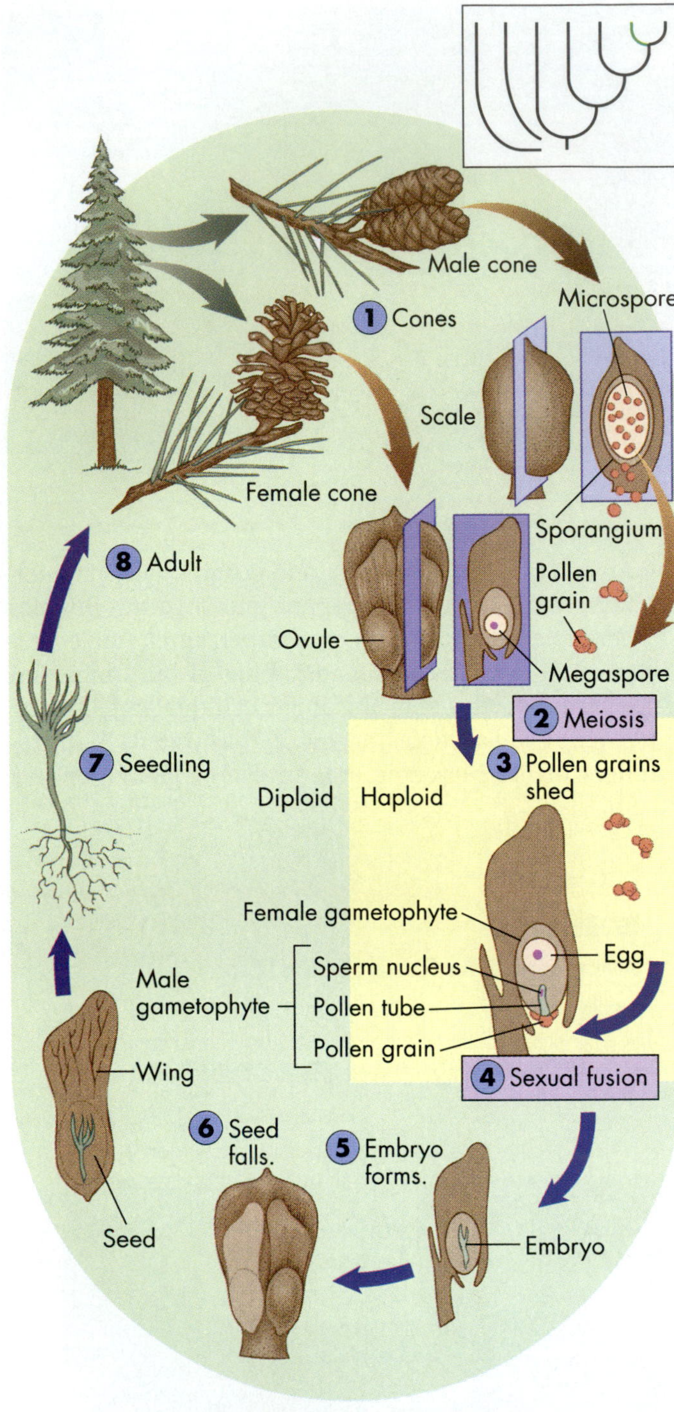

FIGURE 20.21
Life Cycle of the Pine.
The text gives a step-by-step account of the life cycle of a pine tree, a typical conifer, from the production of cones to the scattering of seeds and the sprouting of tender new seedlings.

Conifers have two distinctive characteristics: their narrow leaves and familiar woody cones. Their needle-shaped leaves are covered by a waterproof **cuticle**, or waxy layer. The needle shape resists desiccation, having little surface area relative to volume. In spring, at the start of the conifer life cycle, a large pine tree sprouts thousands of soft male cones and the more familiar large, hard female cones, each with a central axis and whorls of radiating woody scales [FIGURE 20.21, Step 1]. In spore-bearing structures of the soft male cones, diploid cells undergo meiosis (Step 2). The haploid products each develop into a *pollen grain*, the tiny male gametophyte plant. When the pollen grains are released, they shower the area around the tree with golden dust (Step 3). Borne on spring breezes, these grains occasionally encounter female cones on other trees and sift down into the open scales, becoming fixed to a sticky fluid secreted by the ovule, which houses the spore-forming tissue. When the fluid dries, it pulls the pollen inside and thereby completes pollination.

An entire year will pass before actual fertilization takes place. During this time, the pollen grows a *pollen tube*, a pathway through the wall of the female tissues toward the developing egg. Meanwhile, during the year-long period of pollen tube growth, a diploid cell (the *megaspore mother cell*) inside the ovule undergoes meiosis and forms four haploid spores. Three of these spores degenerate, and one divides mitotically into a multicellular female gametophyte containing haploid eggs. A single sperm nucleus fuses with a single egg, completing the process of fertilization (Step 4). After fertilization, the embryo begins to form while still inside the ovule's protective coat (Step 5). At the same time, this protective coat develops into the seed coat, and the rest of the female gametophyte tissue divides and forms the food supply for the expanding embryo. By late summer of the second year, the scales of the woody female cone open, and mature seeds fall to the ground (Step 6). Those that aren't consumed by hungry animals may germinate the following spring into tiny pine seedlings (Step 7) that produce the characteristic evergreen needlelike leaves and, later, the reproductive cones of an adult (Step 8).

Conifers were able to colonize higher and drier reaches of the continents during the Mesozoic because of their numerous evolutionary advances: drought-resistant leaves; protective seed coats; greatly reduced male gametophytes that depend on the sporophyte and give rise to airborne pollen grains; equally reduced female gametophytes that give rise to eggs protected inside the ovule; a well-developed vascular system that produces wood and stiffens the trunk, branches, and roots; the tendency to form mycorrhizal associations with fungi; and the ability to survive for many centuries. Because of these modifications for life on land, conifers are still a successful group, with more than 600 modern species. Nevertheless, a final set of evolutionary changes gave the flowering plants still greater success.

CONCEPT CHALLENGE

One of your friends, who knows that pines and firs are gymnosperms, looks at the photograph of a ginkgo branch in FIGURE 20.19B and says that it looks more like a cherry tree than a pine, and so couldn't be a gymnosperm. What evidence would you cite to change your friend's mind?

Flowering Plants: A Modern Success Story

As the continents rose, drier and colder conditions became more commonplace on the earth's landmasses. During the modern, or Cenozoic, era the flowering plants (formerly called *angiosperms* and now commonly referred to as either *Anthophyta* or *Magnoliophyta*) became the dominant land plants [review FIGURE 20.11]. Simultaneously, mammals became the dominant land animals. Both groups of organisms evolved with reproductive structures that protect and nourish developing embryos. Mammals evolved the placenta and the womb, while the flowering plants evolved a new reproductive structure, the *ovary*, maternal tissue that houses and protects the ovules; they also evolved flowers and fruits, which promote efficient pollination and seed dispersal. These innovations occurred along with other adaptations for life on land, such as broad leaves that collect light efficiently, and enabled the flowering plants to radiate into more species than the combined numbers of all other plant groups.

Today, flowering plants are the most common and conspicuous species in the earth's tropical and temperate regions. It is not surprising, therefore, that when fungal researchers claimed that the largest living thing is a fungal clone [review FIGURE 20.3], botanists identified and proposed a temperate zone plant for the same distinction. They located a huge stand of quaking aspens in Utah. This 106-acre, 6000-ton stand arose from a single original seed, and each shoot arises from a single interconnected root system [FIGURE 20.22] which all together weighs more than the fungal clones described earlier. Obviously, the point remains unsettled, and debate will continue—to the delight of science watchers.

From the monumentally successful flowering plants of tropical and temperate regions come virtually all of our crop plants (wheat, rice, corn, soybeans, fruits, and vegetables) and beverages (coffee, tea, colas, and fermented drinks), as well as spices, cloth, medicines, hardwoods, ornamental plantings, and, of course, flowers—symbols of beauty, affection, and renewal throughout human history. TABLE 20.2 shows the two major groups of flowering plants, the **monocots** and the **dicots**, and their main characteristics.

FIGURE 20.22

Giant Clone of Quaking Aspen.

Biologists from the University of Colorado claim that a stand of quaking aspen in Utah is the largest living individual organism. These aspen are the genetically identical offspring of a single seed and share a common, interconnected root system. The current controversy over what is the biggest organism is a standard and healthy aspect of the scientific process.

FLOWERING PLANT REPRODUCTION: A KEY TO THEIR SUCCESS

As in gymnosperms, the flowering plant life cycle displays the complete dominance of the diploid (sporophyte) generation and the complete dependence of the haploid (gametophyte) generation on the conspicuous adult plant. The older term **angiosperm** means "seed in a vessel," and indeed, seed development takes place within a structure at the base of the flower, called the **ovary**, which gives the ovules much more protection than the gymnosperm cone scale. The wall of the ovary eventually matures into a fruit that surrounds the seeds and aids in dispersal. In some cases, animals attracted to the fruit's

box 20.1
Biology Applied

The Strange Case of the Fungus Flowers

Sherlock Holmes would have liked this natural mystery. A botanist finds a stand of wild mustard plants that look normal from a distance: tall, slender, deep green, and bearing small, bright yellow flowers. Upon closer inspection, however, he sees that the plants are much too tall, they have twice too many leaves, and they are sporting what appear to be buttercup, not mustard flowers! Are they mustard plants in disguise? Is someone stealing and replacing the flowers? Has Professor Moriarty (Holmes's old enemy) taken up gardening?

A nonfictious professor, B. A. Roy of the University of California at Davis, did some botanical sleuthing, recently, and discovered the culprit behind the oddly altered mustard plants. She found that the yellow-flowering herbs (*Arabis holboellii*), growing in a mountain meadow near Gothic, Colorado, were infected with the rust fungus *Puccinia monoica* (a basidiomycote). Spores of this parasitic fungus had germinated within the mustard, and the fungus had commandeered the plant's growth controls, leading to the extra leaves and taller stature. Most significantly, the fungus had caused the mustard stem to produce clusters of yellow leaves covered with a sugary fluid that looked and smelled like buttercup flowers. The eruptions, however, were densely packed with fungal reproductive structures, and strongly attracted insects. Flies, for example, spent more time visiting and lapping sweet fluid from the "pseudoflowers" than they did from nearby mustards, buttercups, and other real flowers. The insects' movements inadvertently brought fungal hyphae of opposite mating types together, and facilitated sexual fusion and new spore formation by the *P. monoica*. These spores would then blow to new mustard plants, and some months later, new pseudoflowers would erupt.

Roy considers the *P. monoica* rust to be unique in the annals of flower mimicry (at least so far) because it triggers the construction of a counterfeit flower of a neighboring species. Since these sweet imposters lure flies, bees, and beetles away from the surrounding real flowers of many kinds of plants, the pathogenic fungus may be affecting not just their hosts, the mustard plants, but the whole plant community as well.

One can only be dazzled by the accidental brilliance of such complex coevolution between pathogens, plants, and pollinators. Holmes would, indeed, have admired this case—and its solution.

---

colorful skin or pleasing flavor eat the fruit, and then excrete the seeds along with feces in some new location. Plants have myriad dispersal mechanisms, including parachutes that loft the seeds in the wind, hooks for hitchhiking on passersby, and tasty fruits [FIGURE 20.23]. A description of the life cycle of a flowering plant is given in CHAPTER 36.

FLOWERING PLANTS AND POLLINATORS: A COEVOLUTION

The success of flowering plants can be only partly explained by the fully enclosed ovules missing in the naked-seed gymnosperms. Most bright, showy flowers attract insect, bird, or mammal pollinators that inadvertently carry loads of pollen from one plant to another as they forage for sweet nectar or for the protein-rich pollen grains. A flying or walking animal dusted with pollen and scouting for more of its favorite food will substantially increase the chances of pollination.

The mutual dependence of flowering plants on pollinators and seed dispersers and of these animal species on the same plants for nutrition is an example of *coevolution*. Plants that had genes for flowers with sweet nectar, a strong fragrance, or a bright color might have survived in greater numbers because more animals would have visited them in their constant foraging for food and carried away pollen. The animals that were attracted to the new food sources probably also survived in greater numbers, thus perpetuating the interdependence.

This coevolution has resulted in specialized physical structures in both plants and animals that help ensure the tradeoff of nutrients for pollination and seed dispersal. The mouthparts of many types of bees, butterflies, moths, birds, and even bats are precisely the right shapes for tapping the nectar or pollen of the flowers they visit. Hummingbirds, for example, which see best in the red spectrum and have long, narrow beaks and a poor sense of smell, are attracted by bright red fuchsia flowers with their long, slender shape but little fragrance [see FIGURE 20.24]. Conversely, flowers pollinated by bees tend to be yellow or blue and very fragrant, corresponding to the bees' vision and ability to perceive odors. BOX 20.1 describes one of the strangest examples of coevolution ever discovered.

[A]

[B]

[C]

FIGURE 20.23

How Plants Disperse Seeds.

[A] Milkweed (*Asclepias synaca*) blowing in a summer breeze. **[B]** Prickly seeds of the common cocklebur, clinging to a donkey's fur. **[C]** Cedar waxwing feeding berries to its young.

FIGURE 20.24

A Study in Coevolution.

A rufous hummingbird drinking from a bright-colored blossom. Adaptations evolved in the bird for collecting nectar, and adaptations evolved in the plant for attracting the hummingbirds, which assist in pollinating the flower.

➤ CONCEPT CHALLENGE

Name adaptations of the flowering plants that might account for their greater diversity than any other plant group.

Connections

This chapter has characterized the fungi and plants in terms of three major themes of this book: how organisms obtain energy, reproduce effectively, and evolve. As producers, plants capture energy, fix it, and build carbohydrates. As decomposers, fungi dismantle carbohydrates and other organic molecules, use some of the nutrients, and release more into the soil and air, where other life-forms—including plants—can use them once again.

As we consider the evolutionary diversity of the animal kingdom in the next three chapters, we will see an extension of a theme introduced here: the interrelatedness of the kingdoms. Just as plant roots need fungi for maximum absorption and nutrient recycling, and fungi need plant and animal matter for energy, so do the animals rely on all the other kingdoms in their role as mobile multicellular heterotrophs—the great consumers.

KEY TERMS

alga, 479
alternation of generations, 477
angiosperm, 489
bryophyte, 483
conifer, 487
extracellular digestion, 470
flower, 479
fungus, 469
gametophyte, 477
gymnosperm, 486
hyphae, 470
lichen, 476
mycelium, 470
mycorrhiza, 472
ovule, 486
plant, 469
pollen grain, 486
seed, 479
seedless vascular plant, 484
spore, 471
sporophyte, 477
vascular plant, 484

HIGHLIGHTS IN REVIEW

1 Fungi, including mushrooms and yeast, are immobile eukaryotes that obtain nourishment by digesting materials produced by other organisms; hence, fungi are important in the recycling of materials in the environment. Fungi associated with the roots of many plants help the plant obtain water and mineral nutrients while the fungus receives carbohydrates from the plant.

a] Fungi are the earth's major decomposers of organic matter. Some are saprobes, which break down dead tissue, and others are parasites, which attack and consume living tissues.

b] The fungal life cycle is dominated by the haploid phase. Fungi can reproduce asexually by fragmenting, budding, or producing asexual spores. They can also reproduce sexually. Hyphae are filaments of fungal cells growing end to end. Hyphae of opposite mating types can fuse, creating a dikaryon, or a cell containing two different haploid nuclei. These paired haploid nuclei can divide many times, eventually fusing and forming a diploid zygote. The zygote then undergoes meiosis and generates haploid spores, which grow into new hyphae.

c] Lichens, associations between a fungus and an alga, can obtain most of the nutrients they require from the surrounding air.

d] Fungi are multicellular heterotrophs that carry out extracellular digestion by secreting enzymes outside the fungal body and then reabsorbing digested materials. The fungal body is made up of filaments called hyphae which are packed together into a mass, the mycelium.

e] Fungi closely associated with plant roots (mycorrhizae) enable plants to absorb far more water and soil nutrients than if the fungi were not present.

2 Plants are generally multicellular, photosynthetic eukaryotes. They trap the energy of sunlight in organic molecules, and these, in turn, directly or indirectly support nearly all other organisms.

3 The simplest plants are multicellular, photosynthetic aquatic organisms called algae. The evolution of waterproof coverings, openings through which gases are exchanged, and a system of vessels that conduct fluids allowed plants to emerge onto land. The evolution of pollen and seeds released seed plants (for example, conifers and flowering plants) from dependence on environmental water for reproduction.

a] Physical trends during plant evolution include the appearance of a vascular system, the growing dominance of the diploid generation, and, in higher land plants, the production of seeds.

b] Algae probably arose from photosynthetic protists some 600 million years ago. The algal plant body is usually filamentous or flat and thin. The haploid phase generally dominates the life cycle, and reproduction involves swimming eggs and sperm.

c] Red algae have red pigments that allow some species to gather light and survive far below the ocean surface.

d] Brown algae contain a brown pigment that allows photosynthesis in deep water. Many species have plant parts called fronds, stipes, and holdfasts, analogous to leaves, stems, and roots. Some kelps grow to 100 m tall (more than 300 ft).

e] Green algae contain both chlorophylls *a* and *b*, which efficiently absorb sunlight in shallow water and in air; land plants also contain these chlorophylls. Green algae were probably the precursors to the first land plants.

f] The simplest land plants include the bryophytes (mosses, liverworts, and hornworts), which lack a vascular system or true leaves, stems, or roots. They have swimming sperm, so must have open water to reproduce.

g] The seedless vascular plants (horsetails, lycopods, and ferns) have internal pipelines for transport and vertical support, as well as horizontal stems. Fertilization still requires standing water.

h] Gymnosperms include cycads, ginkgos, and conifers. In cone-bearing plants, male cones produce pollen grains, the male gametophytes, and female cones produce ovules and, after fertilization, seeds.

i] Conifers include the familiar evergreens, such as yew, pine, fir, cedar, redwood, and sequoia. Conifers have needlelike leaves and woody female cones.

j] The pine life cycle involves airborne pollen, a slow-growing pollen tube, fertilization within the ovule's protective coat, and the maturation and dissemination of seeds.

k] In the flowering plants, the plant ovary, flowers, and fruits evolved. These structures protect the ovules and aid in pollination and seed dispersal. Flowering plants are the most diverse plant group, with more than 250,000 species, many economically important.

l] Flowering plants and animal pollinators have a long and mutually beneficial evolutionary history.

4 The alternation of a haploid life form with a diploid life form is a key to understanding plant biology.

UNDERSTANDING THE FACTS AND CONCEPTS

For Questions 1–5, match the descriptions with the most appropriate item or items from the following list of terms. Each item in the list may be used once, more than once, or not at all.

a] zygomycotes
b] ascomycotes
c] basidiomycotes
d] deuteromycotes
e] lichens

1 Consists of hyphae packed into mycelia.

2 Bread mold and most of the mycorrhizal fungi are members of this division.

3 Yeasts and many plant parasites are members of this division.

4 Does not exhibit sexual reproduction; the fungus that produces penicillin is one example.

5 The reproductive parts of these fungi are visible as mushrooms.

For Questions 6–10, match the descriptions with the most appropriate item from the following list. Each item may be used once, more than once, or not at all.

a] gametophyte
b] sporophyte
c] microspore
d] megaspore
e] sporangium
f] archegonium

6 The dominant, diploid plant in the life cycle of flowering plants.

7 In mosses, meiosis occurs within this structure, resulting in haploid cells that grow into gametophytes.

8 In flowering plants, these cells are produced within the ovary.

9 The haploid portion of a moss that is anchored to the soil by rhizoids.

10 In flowering plants, these cells develop into pollen grains.

As above, for Questions 11–15, match the description with an item from the following list.

a] red algae
b] brown algae
c] green algae
d] bryophytes
e] ferns
f] cycads

11 Contain vascular tissue, but do not produce seeds.

12 Contain chlorophyll and accessory pigments; the bacterial growth medium, agar, is extracted from one of this group.

13 The algal group that probably gave rise to land plants.

14 Looks like a palm tree, but produces seed-bearing cones.

15 Its life cycle is fairly evenly divided between haploid and diploid phases, although the diploid plant depends on the haploid plant for support and nutrition.

As above, for Questions 16–20, match the description with an item from the following list.

a] pollen grain
b] seed
c] endosperm
d] cone
e] ovule
f] ovary

16 Triploid nutritive tissue found in angiosperms.

17 Found only in angiosperms, where it protects the embryo and often develops into a fruit.

18 A protective structure of gymnosperms and angiosperms that develops from an ovule and enclosed embryo.

19 Protects the embryo in both gymnosperms and angiosperms, eventually becoming part of the seed.

20 Immature male gametophytes in the seed-producing plants.

INTEGRATE AND APPLY WHAT YOU HAVE LEARNED

1 What characteristics are shared by all of the algae that are grouped with the plants, and how do these groups of algae differ from one another?

2 In what ways are all fungi similar and how do the major groups of fungi differ from one another?

3 Outline the major trend that is apparent when comparing the life cycles of bryophytes, ferns, and seed-producing plants.

4 In surveying the representative plants in the plant kingdom, what are the primary evolutionary trends that are evident?

5 How does the presence of mycorrhizae, lichens, and pollinators contribute to the success of plants?

ANALYSIS

1 Assuming that taxol was originally the product of only the yew tree, why do you think fungi evolved that also produce taxol?

a] Such fungi would hasten the decomposition of the infected plant.
b] Taxol-producing fungi would help the plant recover from cancerous growths.
c] Such fungi would decrease the probability that the host plant would be eaten by animals.
d] The evolution of taxol-producing fungi would increase the world's supply of taxol.
e] The taxol that is produced by such fungi acts as a food source for the host plant.

2 Which of the following statements is false?

a] The primary characteristic by which fungi are different from plants is the presence of chlorophyll.
b] Without symbiotic relationships with fungi, many vascular plants would not be as large or as dominant a life-form as they currently are.
c] Both autotrophic and heterotrophic species can be found among the fungi, plantlike algae, and plants, collectively.
d] In each representative group of green plants, either the sporophyte is parasitic upon the gametophyte or the gametophyte is parasitic upon the sporophyte.
e] The absence of vascular tissue or seeds distinguishes the mosses from the other land plants.

3 Which statement is the most likely explanation for the lack of sexual reproduction in the Fungi Imperfecti?

a] Sexual cycles never developed because there is no benefit to having them.
b] Sexual cycles developed but were lost by certain ancestral fungi.
c] Sexual cycles really are present but we cannot detect them.
d] Sexual cycles evolved and then changed to become asexual.
e] Sexual cycles are primarily an animal characteristic, so we should not expect to find them in organisms that grow in the ground.

CHAPTER 21

Evolution and Diversity of Invertebrate Animals

LEARNING IN OCTOPUSES

If someone challenged you to quickly name some animals known for their intelligence, you might include chimpanzees, dolphins, and whales on your list. It's less likely that you would say "octopuses." And yet among the **invertebrates**—animals without backbones, such as worms, insects, sponges, and jellyfish—the octopus is a regular genius. Different octopus species range in length from about 5 cm (2 in.) to 5.5 m (18 ft), and have a span of up to 9 m (30 ft) between their eight outstretched, sucker-bearing arms [FIGURE 21.1]. All octopus species have complex, cameralike eyes [see FIGURE 15.12] and live as aquatic predators, hiding in holes or crevices in rocks and emerging to capture their prey. Brainpower as well as acute vision are important adaptations for their predatory lives, and researchers have recently documented some surprising evidence of this undersea invertebrate intelligence.

Until 1992, animal behavior researchers thought that only a few mammals had the ability to learn a task merely by watching another animal perform it. That year, however, Italian researchers reported an experiment in which they trained octopuses to attack a red ball, but to leave a white

FIGURE 21.1
The Octopus: A Trainable Undersea Intellect.

ball unmolested. They allowed a second group of octopuses to watch the exercise from an adjoining tank, then tossed red and white balls in with these observer animals. The experimenters found that the second group also attacked the red balls and left the white balls alone. (Other sets of octopuses easily picked up the inverse task: attacking white balls and leaving red ones untouched.)

Octopuses are typical of all animals in that: (**1**) They are multicellular heterotrophs and obtain nourishment from organic molecules originally produced by some other living thing. (**2**) They develop from the fusion of two gametes of different sizes, an egg and a sperm. (**3**) They move about under their own power at some stage of their life cycle.

The animal kingdom includes millions of species—possibly tens of millions—and many times the number of all other living species combined. It will take us three chapters to survey this unparalleled diversity. Of the millions of animal species, 95 percent are invertebrates, which, like octopuses, termites, earthworms, and oysters, lack a backbone. Because we humans, as well as most of our domesticated animals, are vertebrates, it is sometimes hard to acknowledge the truth: that we are vastly outnumbered and outranked in ecological importance by the simpler invertebrates.

In this chapter, we will trace the history of the invertebrates. As we do, we will encounter a host of colorful, strange, and sometimes fantastic organisms that will expand our notion of what animals are and how they live. Most invertebrates—in fact most species of living organisms—are land-dwelling insects. There are millions of noninsect animals, as well, however, and taxonomists place them in about 35 to 40 different major groups, or phyla. In this chapter, we will deal with only eight animal phyla: the sponges, cnidarians (jellyfish and relatives), flatworms, roundworms, mollusks (including clams, snails, and octopuses), segmented worms (such as earthworms), arthropods (including lobsters, spiders, insects), and echinoderms (sea stars and their relatives). The next two chapters then cover the phylum of the chordates: fishes, amphibians, reptiles, birds, and mammals.

Our approach here is an evolutionary one, and will apply many of the principles we learned in CHAPTERS 15 to 18. As we did with the microbes, fungi, and plants, we will examine each group of animals in light of its ecological roles, and in terms of its adaptations—how its body plan suits the organism to its environment niche. The intelligent octopuses we encounter here—along with the world's biggest jellyfish, earth's most abundant animals (the nematodes), and voracious grasshoppers that eat like a cow—are just a small part of the fascinating story of animal biology. ❑

MESSAGES

1 Animals are motile, multicellular consumers of other organisms. They arise from the fertilization of a larger egg by a smaller sperm, and pass through a ball-of-cells (blastula) stage in their embryonic development.

2 The body plans of existing animals, as well as the fossil record, and gene sequences, suggest that animals occupy a single evolutionary branch on the tree of life.

3 Themes in animal evolution include development of bilateral symmetry; a head; a one-way digestive tract; complex body cavities; and body segments.

4 Sponges are asymmetrical, and the structure of some of their cells suggest that animals evolved from certain protists. Neither the jellyfish and their relatives (the cnidarians) nor the flatworms have a body cavity, but the flatworms have bilateral symmetry and a head.

5 More complex animals belong to one of two main branches. One branch includes the roundworms, arthropods (insects and relatives), annelids (earthworms and relatives), and mollusks (octopuses, snails, and relatives). The other branch includes echinoderms (sea stars and relatives) and chordates (animals with backbones and their relatives).

An Overview of the Animal Kingdom

Animals are quite distinct from members of the other kingdoms and they share the following characteristics:

1. Unlike most prokaryotes and protists, *animals are multicellular.* In fact, some of the largest animals have trillions of cells. These cells, are organized into tissues, groups of cells performing specialized functions. Within an animal's tissues, cells are connected to each other by several types of junctions. These junctions allow communication between cells, and they hold cells to each other and to the surrounding extracellular material.

2. Unlike plants, animals are heterotrophs; they cannot manufacture their own food and instead ingest it, break it down metabolically, and obtain energy from it.

3. Unlike most fungi, which are also multicellular heterotrophs, animals can move about on their own at some point in their life cycle—usually throughout the entire cycle—to search for food and mates and to avoid danger. (Fungi and plants often produce spores which disperse the species, but these do not generally move about under their own power.

4. Finally, animals are diploid, and their primary mode of reproduction is sexual. Each individual animal originates from a zygote formed by the fusion of two unequally sized gametes (egg and sperm). The zygote becomes a blastula, which is usually a hollow ball of cells [see FIGURE 13.8], and passes through various distinct stages of development before it becomes an adult.

TRENDS IN ANIMAL EVOLUTION

As we survey the major phyla of invertebrates, we can trace the origin of the basic physical features most animals share, including a head, mouth, nervous system, heart, stomach, and appendages [FIGURE 21.2]. There were five major anatomical and physiological trends leading to these adaptations. They spanned hundreds of millions of years of evolution, and they reflected successful evolutionary solutions to the problems of survival [FIGURE 21.2]: The first was a trend away from a circular body plan (so-called *radial symmetry*) and toward a body with symmetrical right and left halves (*bilateral symmetry*). Along with this came a second trend: *cephalization*, or the development of a head, with its centralized sensory apparatus and a brain or other center of nervous integration.

Other trends were equally significant. In the third trend, invertebrate animals evolved away from a simple, saclike body with a single opening at one end toward a more complex, elongated body containing a *food-digesting tube*, the gut, with openings at both ends. A fourth trend in these animals, with a *tube-within-a-tube body plan*, was away from enclosure of the tube in solid tissues and toward suspension of the tube in a fluid-filled, tissue-lined space called a *coelom*. A fifth trend was toward *segmentation*, the development of a series of body units, each containing similar sets of muscles, blood vessels, nerves, and other structures.

Although many species of organisms with all five adaptations emerged, thousands of types, including sponges and jellyfish, evolved and survive today without one or more of these features. Clearly, while bilateral symmetry, cephalization, a one-way gut, the coelom, and

Trend	Sponges	Jellyfish	Flatworms	Roundworms	Mollusks	Segmented worms	Arthropods	Echinoderms
1	Radial symmetry		Bilateral symmetry					
2	No cephalization		Cephalization (Head)					No cephalization
3	Saclike body			Body with gut tube				
4	No body cavity			False body cavity	True body cavity			
5	No segmentation					Segmentation		No segmentation

FIGURE 21.2

Evolutionary Trends in the Invertebrates.

Trend 1: From radial to bilateral symmetry. Trend 2: Development of a head (cephalization). Trend 3: From a saclike body toward a body with gut tube. Trend 4: From no body cavity to a fully lined body cavity. Trend 5: Toward segmentation of the body.

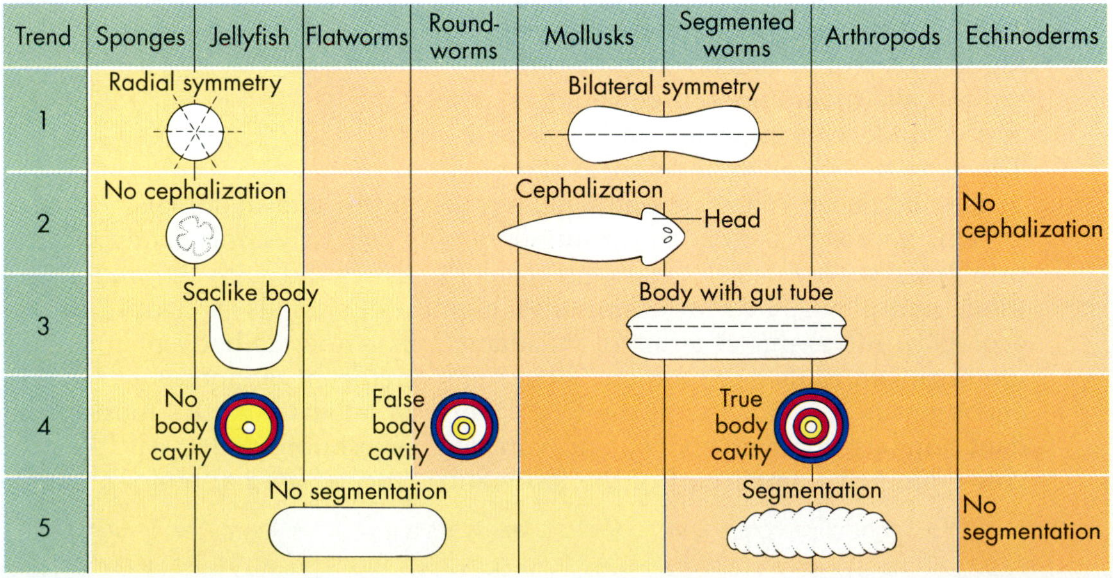

segmentation confer certain advantages, these adaptations are not prerequisites for survival.

WHEN DID ANIMALS ARISE?

As a group, animals have a long and interesting evolutionary history. The outlines of animal evolutionary pathways have emerged based on the work of paleontologists analyzing early fossils, and the efforts of molecular biologists analyzing genetic similarities between animal groups. Paleontologists have found the earliest fossils of soft-bodied animals in rocks from the Edicara region of southern Australia, and dated them to about 560 to 600 million years ago. Other fossils, however, suggest that animals already existed 800 million to 1 billion years ago—fossils of burrows, trails, and impressions in mud exist in rocks of that age. Recall one more piece of evidence from CHAPTER 17: that the rocky, shallow-water structures called *stromatolites* began to decrease in diversity about 1 billion years ago. This suggests that grazing animals were present and were feeding on the microbial communities in the stromatolites. Finally, by evidence comparing sequences of ribosomal RNA genes and globin proteins, molecular biologists have produced additional evidence that the major phyla, or body plans, of animals were already distinct lineages before 700 million years ago.

ANIMAL ORIGINS

The gene sequences and cell organelles within cells of the various animal groups are similar in many ways. Based on these similarities, many zoologists (biologists who study animals) are convinced that animals are a single lineage derived from protistan ancestors [FIGURE 21.3]. The sponges probably diverged very early from the main line leading toward the other animal groups. Recent molecular data show that a particular group of protists called *choanoflagellates* are genetically similar to animals in general and to the sponges in particular. Thus the animals may have arisen from ancient members of this group [FIGURE 21.4], and the sponges may have branched off early from these choanoflagellate progenitors. Advocates of this hypothesis point to the choanoflagellate's single flagellum—a characteristic of some cells of sponges and cnidarians (jellyfish and relatives)—as well as the tendency of some choanoflagellates to group together in colonies.

Some time after the first animals evolved, a rapid rise in the amount of oxygen in the atmosphere permitted a great evolutionary radiation of the animal kingdom—the fanning out of large numbers of animal species with a variety of forms adapted for various environments. All the animals that ever existed were products of this radiation, including invertebrates such as the first spiderlike animals to crawl on land and the giant dragonflies of primordial forests, as well as vertebrates such as the pterosaurs with their leathery wings and the lumbering mastodons and ground sloths of the last ice age. At least 35 phyla evolved (animals in a phylum share the same general body plan), most of which still have living members and include more than a million individual species.

As you read through this chapter, recall the evolutionary principles we discussed in CHAPTERS 15 to 17. Keep in mind that no organism alive today is an ancestor of any other living organism [see FIGURE 21.3]. Furthermore, remember that evolution is not a single directed path toward "bigger and better" organisms. Many branches of the animal family tree became extinct after existing for hundreds of millions of years. For reasons of brevity, this chapter focuses only on organisms alive today. Finally, as we discuss evolutionary trends, keep in mind that evolution does not tend toward any goal, but that at each point in time, natural selection works on populations of organisms in ways that allow individuals with certain inherited traits to produce more offspring. The evolutionary trends we see in retrospect were dictated simply by the genetic makeup of organisms that happened to survive to the present. In the following sections and chapters we will discuss only the most important groups of living organisms, beginning with the sponges, the simplest animals [TABLE 21.1].

➤ CONCEPT CHALLENGE

How would you test the hypothesis that all animals can be assigned to a single branch on the tree of life? (The alternative would be the independent origin of different animal phyla from different protists.)

Sponges: The Simplest Animals

The phylum **Porifera**, the **sponges**, are generally asymmetrical organisms perforated with many pores that inhabit mostly marine environments. For centuries, people considered sponges to be plants, since the adults are *sessile*, or stationary, and are permanently attached to rocks, pilings, sticks, plants, or other animals. Sponges, however, are multicellular animals that filter and consume fine food particles, bacteria, or protists from the water.

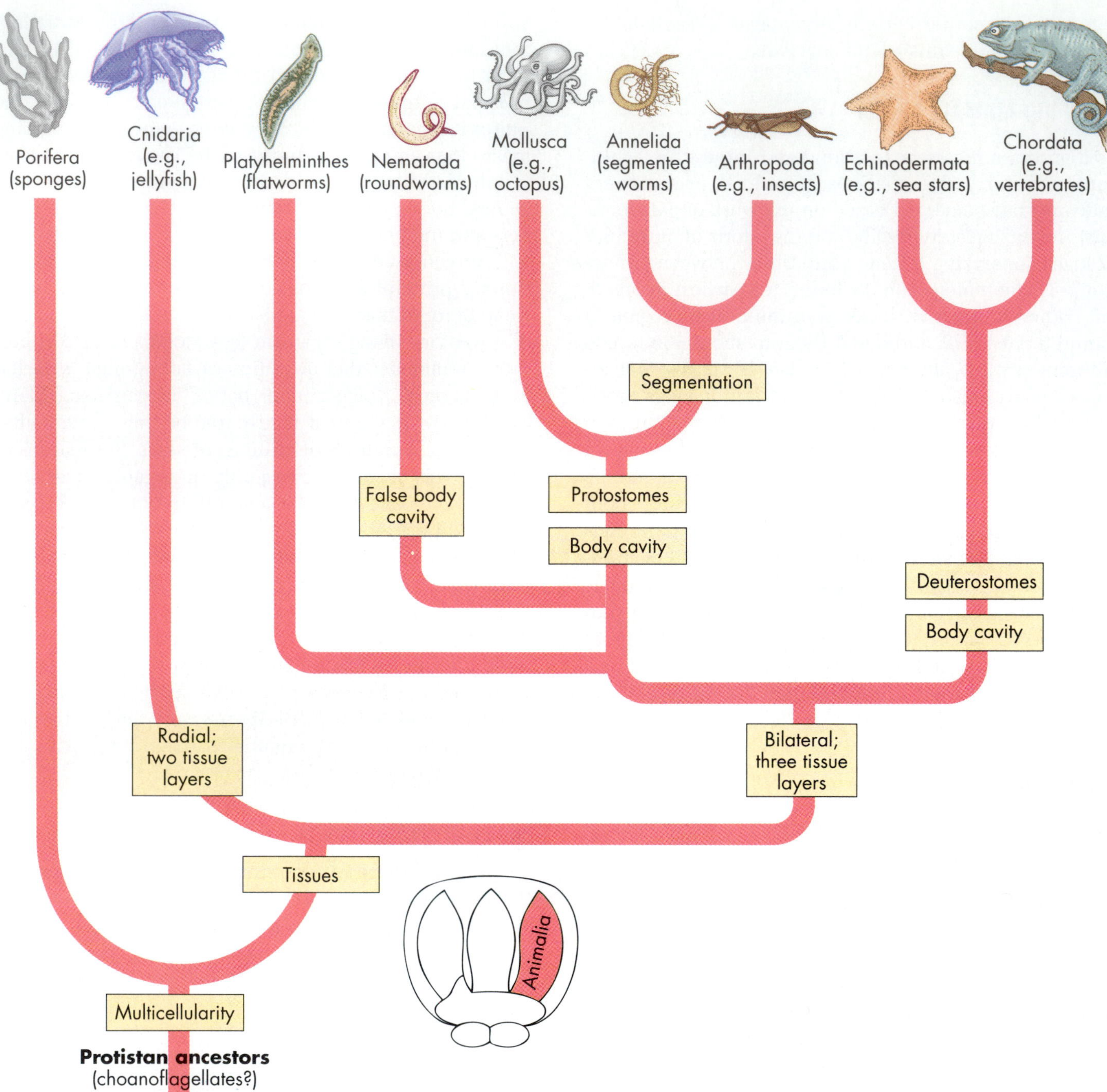

FIGURE 21.3

Evolutionary Relationships of the Major Animal Phyla.

Animals arose when true multicellularity evolved in a protistan ancestor. The sponges were an early branch off the main line. The other branches, however, have diverged into large groups, including the mollusks, arthropods, and chordates. Evolutionary biologist Simon Conway Morris suggests this hypothesis for animal relationships.

TABLE 21.1 Some Key Animal Phyla

Phylum	Examples	Number of Species	Notable Features
Porifera	Sponges	10,000	Generally asymmetrical saclike bodies; central opening; body wall perforations; spicules; no evolutionary descendants; no true tissues
Cnidaria	Jellyfish, hydras, corals, sea anemones	10,000	Radial symmetry; grow as polyps or medusae; two tissue layers; gastrovascular cavity; nerve cells, nematocysts, planula larvae; no body cavity
Platyhelminthes	Flatworms, including tapeworms, flukes	15,000	Bilaterally symmetrical with a head; three tissue layers; true organs and organ systems; life cycles often complex and include two or more hosts; no body cavity
Nematoda	Roundworms	20,000	Bilaterally symmetrical with a head and gut tube; three tissue layers with unlined (false) body cavity; hydroskeleton; extremely common in soils and as parasites on other animals and plants
Mollusca	Snails, clams, octopuses, squid, slugs	120,000	Bilaterally symmetrical with a head and gut tube; three tissue layers with a lined body cavity; gills, mantle; open circulatory system; very common marine and freshwater organisms
Annelida	Segmented worms, including earthworms, polychaete worms, leeches	18,000	Bilaterally symmetrical with a head and gut tube; lined body cavity; segmentation; hydroskeleton; move with bristles pushing against ground; crop, gizzard; closed circulatory system; earthworms occur widely and help aerate soils
Arthropoda	Spiders, mites, ticks, scorpions, millipedes, centipedes, insects, lobsters, shrimp	1–30 million	Bilaterally symmetrical with a head and gut tube; lined body cavity; segmentation; exoskeleton for support and protection; specialized segments and appendages; jointed legs; tracheae or gills; acute senses; most diverse phylum in living world
Echinodermata	Sea stars, sea urchins, sea cucumbers	7000	Gut tube and lined body cavity; a head in some larvae and adults; no segmentation; first endoskeleton; unique water vascular system for locomotion; separate evolutionary line from Mollusca, Annelida, and Arthropoda
Chordata	Invertebrates like sea squirts, and vertebrates like fish, amphibians, birds, reptiles, and mammals	50,000	Bilaterally symmetrical with head, gut tube, lined body cavity and segmentation; stiff rod of cartilage (notochord); dorsal tubular nerve cord; gill slits

STRUCTURE OF SPONGES

The simplest sponges resemble vases or irregularly shaped clusters of tubes [FIGURE 21.5A]. Whether a sponge is shaped like a branching vase or a cluster of tubes, it has a large opening called the *central cavity*, as well as hundreds of tiny perforations, or *pores*, through the body wall. If you place red dye in the water near a living sponge, you can see the colored water being drawn into the pores along with food particles. Special cells lining the central cavity remove and digest the food particles suspended in the water; the water and waste matter exit through the central opening.

Most sponges are asymmetrical and lack specialized tissues. The sponge's body wall is a thin "sandwich" with a layer of flattened cells outside, an inner lining containing *collar cells* (choanocytes), and a gelatinous middle layer containing several different kinds of wandering *amoeboid cells* [FIGURE 21.5B]. The amoeboid cells help

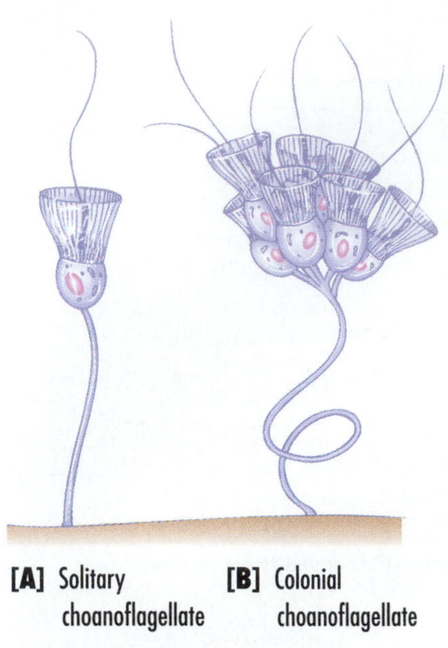

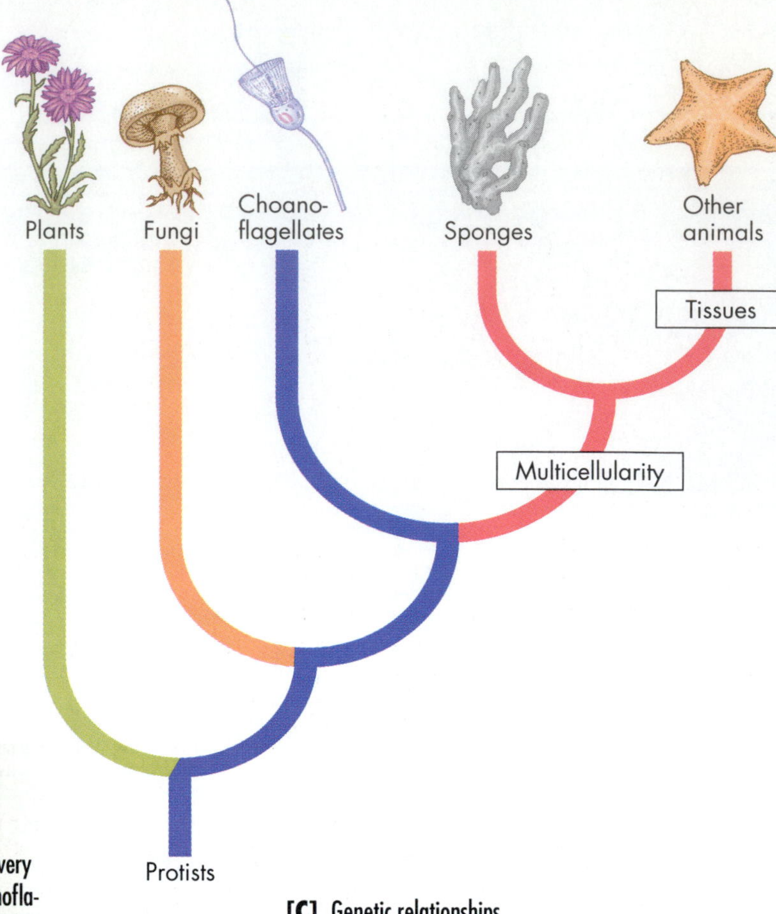

FIGURE 21.4

Animals and Certain Protists Share a Common Ancestor.

[A] Solitary protists called choanoflagellates ("collar and flagella") are very similar in structure to certain cells in sponges. [B] Some species of choanoflagellates group together into colonies. [C] Analysis of ribosomal RNA genes suggests that choanoflagellates are genetically more closely related to sponges and other animals than they are to fungi or plants.

digest food particles and distribute materials, and differentiate into gametes. Each of the collar cells in the inner layer has a long flagellum surrounded by a delicate collar of microvilli very similar to those of the choanoflagellate protists [compare FIGURES 21.4A and 21.4B]. The constant beating of the flagella draws water in through the pores and pushes it out through the central opening.

The sponge's cells are supported by an internal framework. In some sponges, this consists of fibers of a protein related to hair protein, which forms a network surrounding the various cell types. Many sponge cells also contain tiny pointed *spicules* made of silica or calcium carbonate; these act as scaffolding and also protect the sponge from predators by presenting them a spiky mouthful [see FIGURE 21.5B]. The irregular, tan bath sponges sometimes sold commercially are the tough, fibrous skeletons.

The sponges, with their primitive cell layers rather than individual cells, represent a definite advance beyond colonial protozoa in that the sponge cells function together as a single individual, rather than as a cluster of independent, single-celled organisms, as in a protozoan colony. Within such a protozoan colony, each cell retains the ability to reproduce; in sponges, however, sexual reproduction requires cell specialization.

Sponges are apparently not directly related to any other animal groups, nor did they give rise to more complex organisms.

Cnidarians: The Radial Animals

Some of the most ephemeral and beautiful invertebrates are members of the phylum **Cnidaria** (nih-DARE-ee-ah; from the Greek *knidē*, "stinging nettle".) This phylum includes the translucent hydras, the gossamer jellyfish, the sea anemones, and the colorful corals [FIGURE 21.6]. Most live in the oceans, but a few, such as the hydras, inhabit fresh water. Cnidarians have true tissues and a **radial body plan**; that is, they are circular with a central axis and structures radiating outward from the axis like the spokes of a wheel.

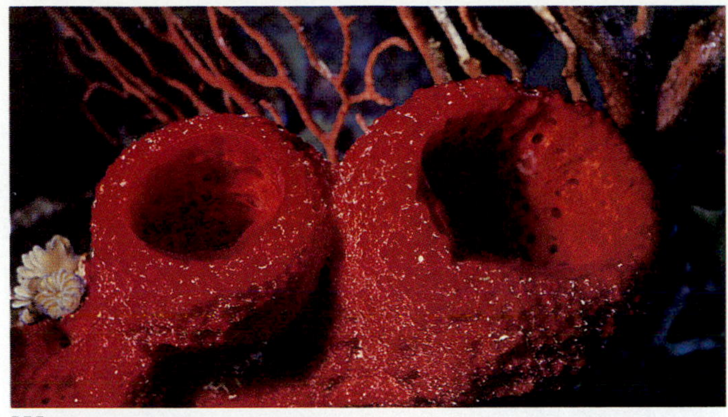

[A]

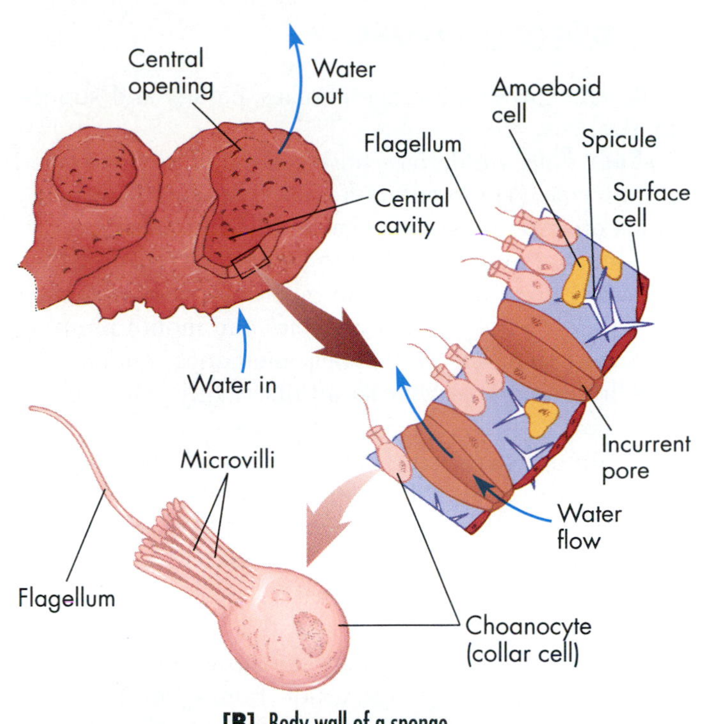

[B] Body wall of a sponge

FIGURE 21.5

The Sponges: The Simplest Animals.

[A] This red vase sponge (*Mycele* sp.) is a common denizen of coral reefs. The simple, vaselike body has a central opening and a central cavity. [B] The body wall contains collar cells (choanocytes) which look very similar to choanoflagellate cells [review FIGURE 21.4]. The body wall also contains amoeboid cells and is traversed by pores through which water flows, drawn by the flagella. Food particles collect on the microvilli of the collar cells and are brought inside the cell where they are digested.

[A] Hydra

[D] Jellyfish

[B] Sea anemone

[C] Coral

FIGURE 21.6

The Cnidarians: Ephemeral Invertebrates.

Cnidarians are radial animals, often translucent and billowing, protected by stinging weapons called nematocysts. [A] A brown hydra. [B] *Condylactis gigantea*, a pink-tipped sea anemone. [C] *Adelogorgia phyllosclera*, a coral; [D] *Aequorea victoria*, a jellyfish.

STRUCTURE OF CNIDARIANS

Cnidarians grow in a range of sizes, colors, and strange appearances.

Body Plan Cnidarians have at least one of two basic radial forms: (1) the **polyp**, a hollow, vaselike body that stands erect on a base and has a whorl of tentacles surrounding a mouth near the top [FIGURE 21.7A], or (2) the **medusa** (plural, *medusae*), an inverted umbrella-shaped version of the polyp, with tentacles and mouth pointing downward [FIGURE 21.7B]. Sea anemones, corals, and most hydras are polyps as adults, while jellyfish are medusae.

While sea anemones, hydras, and jellyfish usually live as independent individuals, thousands of coral polyps live together in huge colonies [FIGURE 21.6C]. Each individual polyp in the colony is encased in its own cup-shaped limestone skeleton. Together, the millions of tiny coral cups (the outer layers inhabited, but the inner layers empty) can form giant reefs and atolls. These include the only biological structure visible from space, the 1900-km-long (1200-mi-long) Great Barrier Reef off Australia's eastern coast, as well as islands such as Bermuda, the Bahamas, and Fiji.

Reef-building corals get their spectacular colors and obtain much of their energy from microscopic photosynthetic algae that live symbiotically inside their cells. Unfortunately, changes in the ocean are causing an epidemic of coral "bleaching"—the expulsion of the algae from the coral cells. Bleaching often leads to death of the coral [FIGURE 21.8]. Researchers fear that pollution and global warming will accelerate this trend, and perhaps destroy the largest and most impressive structures ever made by living organisms.

FIGURE 21.8

Coral Bleaching.

Pollution and temperature changes in the oceans are causing coral colonies to expel their photosynthetic algae. This "bleaching" starves the coral polyps since they get much of their nourishment from their symbiotic algae.

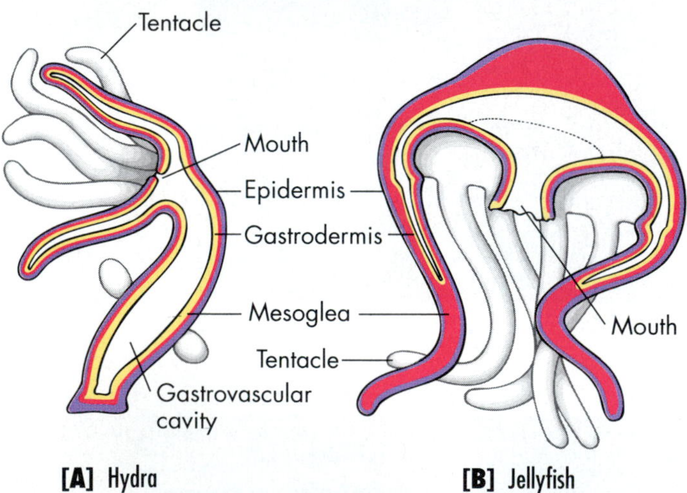

FIGURE 21.7

Cnidarians Have Two Tissue Layers.

A polyp such as this hydra [A] and a medusa, such as this jellyfish [B] both have just two layers of cells. The blue layer is the epidermis and the yellow layer is the endodermis. Sandwiched between the two tissue layers is a noncellular mass of proteins and carbohydrates, the mesoglea, indicated in red.

Body Layers Each cnidarian has a three-layered body wall with an outer skin (*epidermis*), an inner "stomach skin" (*gastrodermis*), and a jellylike substance in between called **mesoglea** (literally, "middle glue"; see FIGURE 21.7). In polyps, the mesoglea layer is thin, but in medusae, it can be thick; in fact, jellyfish resemble jelly because of the extreme thickness of the mesoglea. The mass of one monstrous North Atlantic jellyfish that is nearly 3 m (10 ft) across and weighs a ton is virtually all mesoglea.

Nematocysts Embedded in the epidermis of the tentacles are remarkable organelles called **nematocysts**, or stinging capsules. When triggered by the approach or contact of an enemy or potential prey, the tube everts like a sock turned inside out. In some types of nematocysts, this sock has a sharply pointed end that can strike the prey and release paralytic toxin. Several human deaths have been attributed to nematocysts from *Chironex*, a large tropical jellyfish.

Nerve Cells Another adaptation helps cnidarians such as jellyfish and anemones detect prey, coordinate body movements, and capture and swallow the victim. **Nerve cells**, highly elongated cells that can conduct electrical signals, are arranged in a loose network that sends information in all directions, permeating the animal's tissues and enabling it to detect stimuli and activate cells in response. Cnidarians and all more complex animals have nerve cells and nervous systems [see CHAPTER 27].

NUTRITION IN CNIDARIANS

Cnidarians use their tentacles to catch and hold their prey, which they sting and immobilize with toxin from their nematocysts. The prey is then moved into the mouth by the tentacles and digested within the body's central cavity [review FIGURE 21.7]. Some gastrodermic cells produce enzymes that begin to break down the food. This enzymatic food breakdown within a cavity is extracellular digestion; cnidarians and all more complex animals make use of it. Extracellular digestion allows animals to digest larger food pieces, and thus expand their range of food, compared with the intracellular digestion of very small food bits found in most protists and sponges.

REPRODUCTIVE CYCLES OF CNIDARIANS

Sea anemones and corals reproduce asexually by budding, or sexually by releasing eggs and sperm into the water. Adult jellyfish live as medusae and develop either egg- or sperm-producing organs (gonads), which release gametes into the water [FIGURE 21.9, Step 1]. When fertilized, the egg develops into a *planula*, a small, flat, solid larva (an immature form), which is propelled by cilia (Step 2). The planula attaches to a rock or other object, develops into a stationary *polyp* (Step 3), and then differentiates into a colony of polyps, or *strobila* (Step 4). The strobila resembles a stack of saucers, and in a type of asexual reproduction, each saucer in the stack develops tentacles and swims away as a new medusa (Step 5). At maturity, the medusa develops male or female gonads, and the cycle continues.

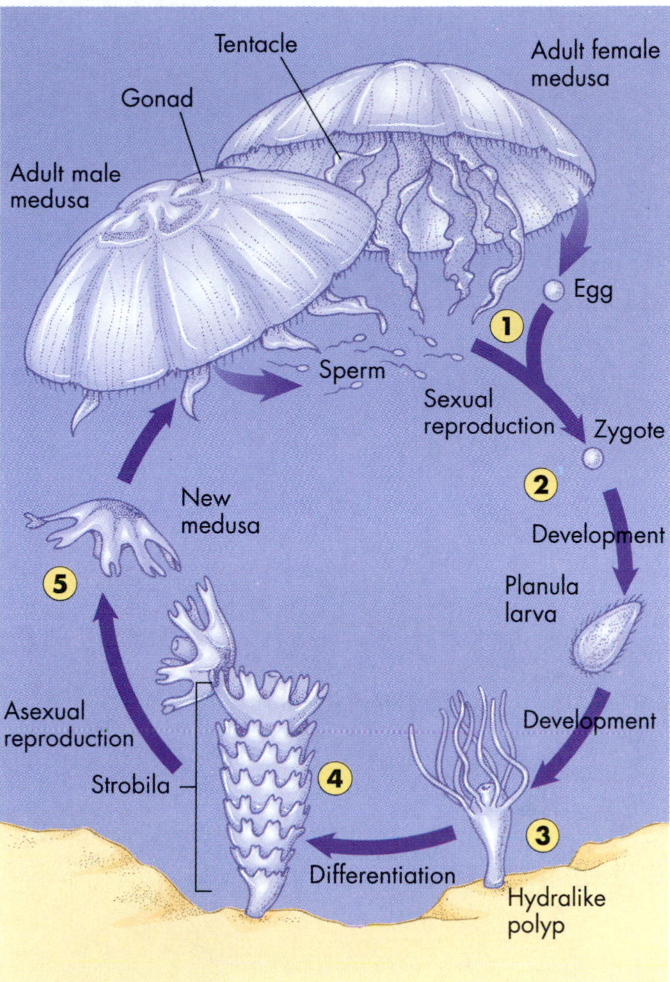

FIGURE 21.9
Life Cycle of a Jellyfish.
The text describes the jellyfish life cycle, from adult medusa to planula larva, sessile polyp, and new medusa.

➤ CONCEPT CHALLENGE

What arguments support the contention that the sponges split from the main line of animal evolution before the cnidarians arose? Consider symmetry, cell structure, and tissues.

Flatworms: Bilaterally Symmetrical

The simplest invertebrates to display bilateral symmetry and cephalization—traits seen in most animal species—

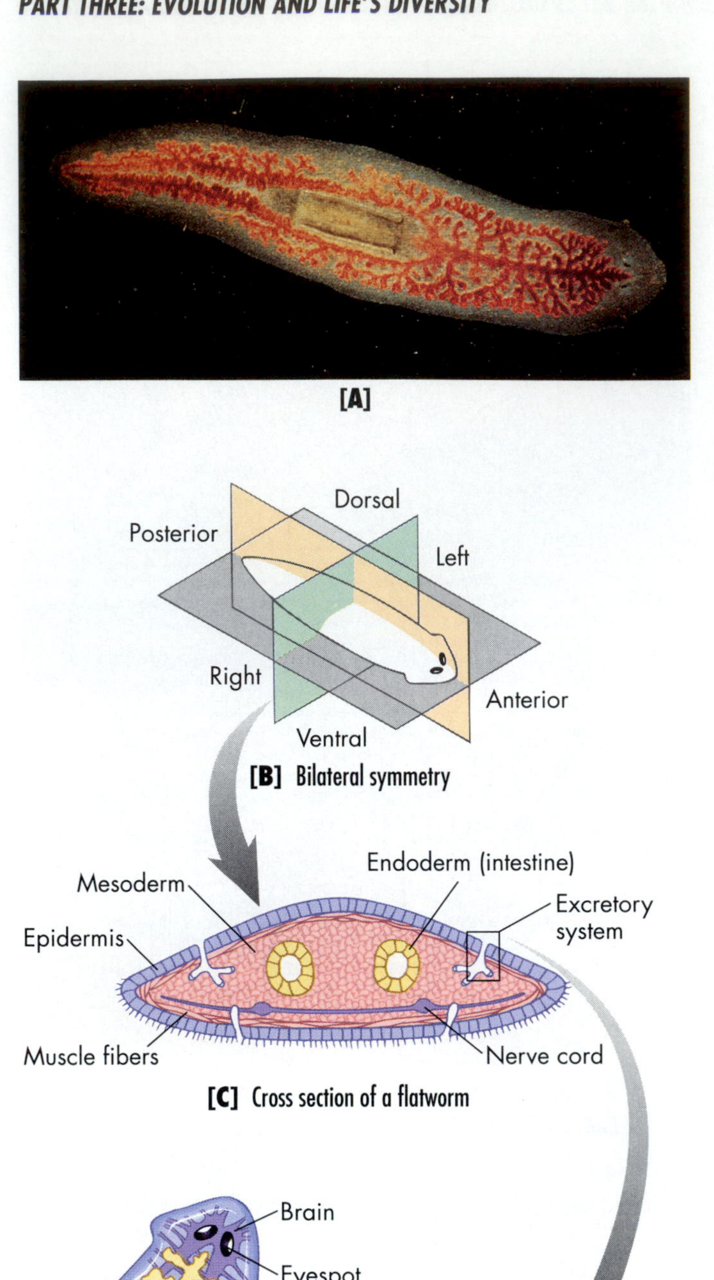

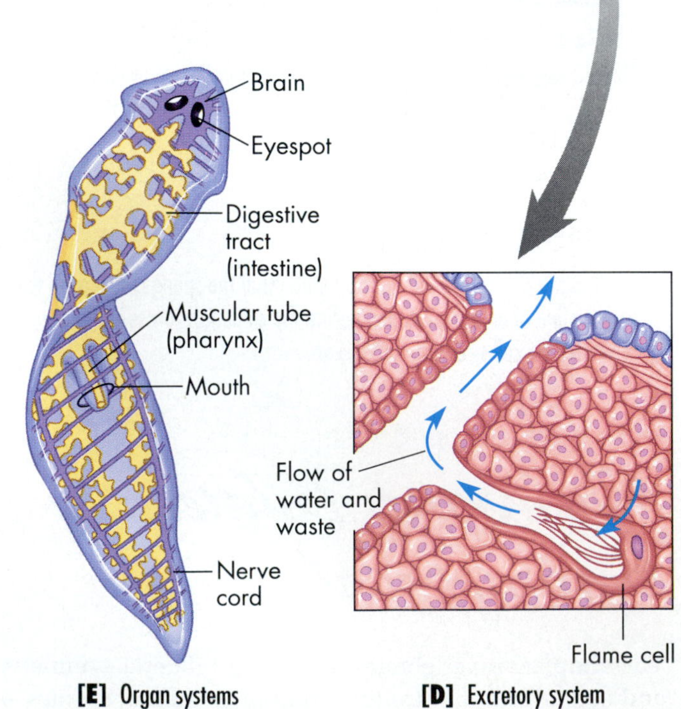

are the **flatworms** [FIGURE 21.10], members of the phylum **Platyhelminthes** (from the Greek words for "flat worms"). Free-living flatworms, such as planaria, inhabit freshwater lakes, rivers, or bodies of salt water [FIGURE 21.10]. Parasitic flatworms, such as tapeworms, live within the bodies of their hosts [FIGURE 21.11].

BILATERAL SYMMETRY AND CEPHALIZATION

Flatworms have **bilateral symmetry**—their right and left halves are mirror images. However, *anterior* and *posterior* (front and back) ends are different, and so are their *dorsal* and *ventral* (top and bottom) surfaces [FIGURE 21.10B]. In addition, flatworms display **cephalization** (from the Greek word for "head"); one end functions as a head containing both a nerve mass that serves as a brain, and specialized regions that can sense light, chemicals, and pressure. Since flatworms are bilaterally symmetrical instead of radially symmetrical, when they move forward, the head region—with its brain and sensers—encounters a new environment first. Depending on the data collected by the head region, an animal can continue forward or back up and try a different direction. This evolutionary adaptation is so successful that heads occur in almost all animals more complex than flatworms.

TISSUES, ORGANS, AND ORGAN SYSTEMS

Flatworms display two additional advances seen in all higher animals. They have three distinct tissue layers and true organs and organ systems. Recall that in the three-layered body walls of sponges and cnidarians the middle layer is a gelatinous material that contains only scattered cells. In flatworms, the middle layer, or mesoderm, is made up of living cells, not jelly, and lies between an outer cell layer, the epidermis, and an inner cell layer, the endoderm [FIGURE 21.10C]. The mesoderm is important because it gives rise to muscles, blood, and other tissues. Flatworms have organs that carry out the processes of digestion, movement, nervous responses, excretion, and reproduction.

FIGURE 21.10
Flatworms: Simple Animals with Primitive Organ Systems.
[A] The planarian *Dendrocoelum lacteum.* [B] Its elongated body shows bilateral symmetry: the right and left sides are mirror images of each other. The animal is also cephalized: the head is anterior, the tail posterior. [C] The planarian is characterized by three distinct tissue layers. [D] The flame cells of the excretory system remove excess water from the body. [E] Flatworm organ systems.

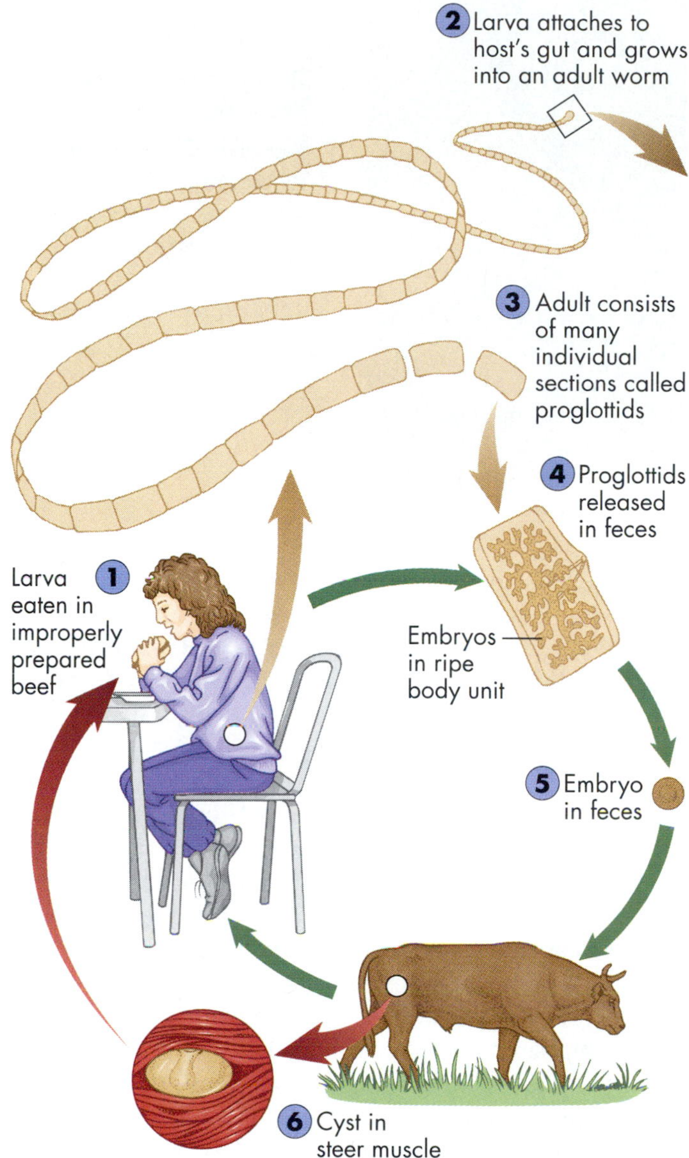

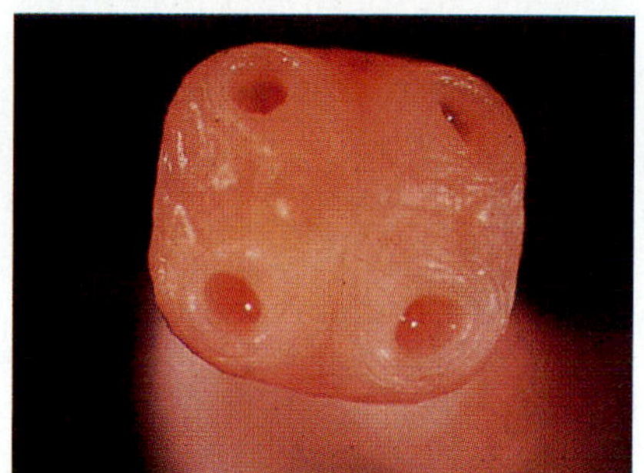

The "head," or scolex, of a tapeworm

FIGURE 21.11

Tapeworms: Parasitic Flatworms.

The beef tapeworm infects about 60 million people worldwide. It can reach 60 feet in length inside a person's intestines. A larva ingested with undercooked beef (Step 1) develops into an adult with a head (see inset) that attaches to the host's gut (Step 2), where it can cause the host to lose weight, suffer chronic indigestion, and have persistent diarrhea. Body segments (Step 3) develop reproductive organs and embryos (Step 4), detach, and are released in the person's feces (Step 5). If a cow eats food contaminated by embryo-bearing human feces, the egg hatches in the new host (Step 6) and the newly hatched organism penetrates the gut wall, passes through the bloodstream, and lodges in the cow's muscles, forming a cyst. If the cysts are not killed by cooking, they can start another cycle.

The **digestive system** of the common planarian [FIGURE 21.10E] has only one opening, like a glove; food enters and wastes exit through the same opening—the mouth.

The **excretory system** rids the body of excess water and includes a network of water-collecting tubules adjacent to *flame cells* [FIGURE 21.10D]. These saclike cells contain clusters of cilia that appear to flicker like flames; as the cilia move, they drive water into the tubules and out through pores in the body wall, thus maintaining the salt and water balance of the animal's body and ridding the body of wastes.

Planarians move by means of cilia on their external surfaces and by layers of contractile muscle cells that lie below the epidermis in a **muscular system**.

Muscles receive signals from the **nervous system**, which includes the light-detecting cells of the eyespots; the brain; longitudinal "trunk lines," or **nerve cords**; and a network of lateral nerves [see FIGURE 21.10C and E]. In addition to the eyespots, which detect light and movement, there are regions on the head that can detect food.

The **reproductive systems** of most individual flatworms have both testes and ovaries (they are hermaphroditic), and they pair up and exchange both sperm and eggs with another individual, so that both are fertilized.

A key to the flatworm's success is its flatness. Because it is so thin, oxygen and carbon dioxide diffuse directly through the thin layers to every cell, nutrients diffuse outward from the branching intestine to all body cells, and the animal needs no extra internal or external support. Free-living flatworms, such as planarians, must hunt and avoid danger to survive, and their nervous systems and senses are especially crucial.

PARASITIC FLATWORMS

In contrast, parasitic flatworms, such as flukes and tapeworms, live a sheltered life, with most of their needs provided for by a host. Parasitic flukes display a prodigious reproductive capacity. A larva of the Chinese liver fluke, for example, can infect a snail and reproduce annually, releasing a quarter of a million larvae. Each of these, in turn, can infect a fish, and if the fish is insufficiently cooked, parasitize a human being.

The blood fluke of Southeast Asia is responsible each year for 200 million cases of a human disease called *schistosomiasis*, which is second only to malaria in claiming human victims.

Parasitic tapeworms are usually quite flat, with a head, or *scolex*, that is little more than knobs with ghoulish hooks or adhesive suckers around the mouth that attach to host tissues [see FIGURE 21.11]. Their bodies consist mostly of hundreds of individual reproductive units. Flukes and tapeworms shed thousands of eggs and larvae, and these are their only means of reaching new hosts and continuing their life cycle. The tapeworms that infect a cow's intestines shed embryos that bore into its muscles and form protective cysts [see FIGURE 21.11]. If a person consumes undercooked beef bearing these cysts, the cysts can grow into new tapeworms in the person's intestines.

Flatworms have had great evolutionary importance. Biologists have reason to suspect that a small, bilaterally symmetrical creature similar to a flatworm probably gave rise to all the more complex animal groups.

Roundworms: Advances in Digestion

In sheer numbers of individuals, the most abundant animals on earth are **roundworms**, members of the phylum **Nematoda**. Roundworms, as the name implies, are round in cross section rather than flat, and most are very small; many, in fact, are microscopic [FIGURE 21.12]. A cubic meter of rich soil can contain 3 billion nematodes, and biologist A. M. Cobb once said that if all our planet's lands and seas were swept away, but the nematodes somehow stayed in place, a clear outline of the earth and its geological features would remain. Nematodes thrive in mud, sand, soil, and running and standing water—even in acidic fruit juices—and have been found in habitats from boiling hot springs to the Antarctic Ocean and the top of Pike's Peak. When the environment grows harsh, nematodes can curl up, dry out, and shut down metabolically for up to 30 years. Then, when conditions improve and water is once again present, the animals rehydrate and revive: instant nematode! While many roundworms are free-living, hundreds of the 12,000 or more nematode species are plant and animal parasites that damage crops and cause diseases.

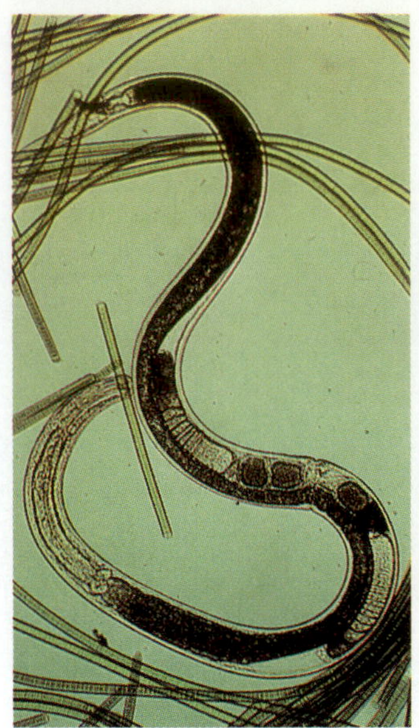

FIGURE 21.12

The Roundworm: Phylum Nematoda.

Nematodes thrive in mud, sand, soil, and running and standing water—even in acidic fruit juices. Here is the nearly transparent pinworm, *Enterobius oscillatoria*, which infects humans.

CHARACTERISTICS OF ROUNDWORMS

The roundworm's success is often attributed to two new characteristics—a fluid-filled body cavity and a functional skeleton. While roundworms share bilateral symmetry and cephalization with the flatworms, they also possess a body cavity called a **pseudocoelom** ("false body cavity"), as well as a one-way tubular gut, or digestive tract, with openings at both ends—the mouth and anus.

Sponges, cnidarians, and flatworms have no spaces between their two or three tissue layers; the layers are solidly packed together and surround the central digestive tube like a single envelope [FIGURE 21.13]. In contrast, the digestive tube of a roundworm lies in a fluid-filled space within the body. On the inside of this space is the digestive tube and on the outside are the mesoderm and skin; the space therefore separates two body-tissue layers. This type of body cavity is called a false body cavity because mesoderm lines only one side of the cavity. In contrast, octopuses, insects, and humans have a body cavity fully lined with mesoderm—a cavity called a **coelom** (SEE-lome).

The advantages of a coelom (or pseudocoelom, which works basically the same way) are fourfold: (1) Because of this space within the body cavity, the reproductive and digestive organs can evolve more complex shapes and functions. (2) In a fluid-filled chamber, the gut tube and other organs are cushioned and thus better protected. (3) Since a liquid cannot be compressed, the pseudocoelom (or true coelom, when present) can act as a **hydroskeleton** (literally, "water skeleton"), providing

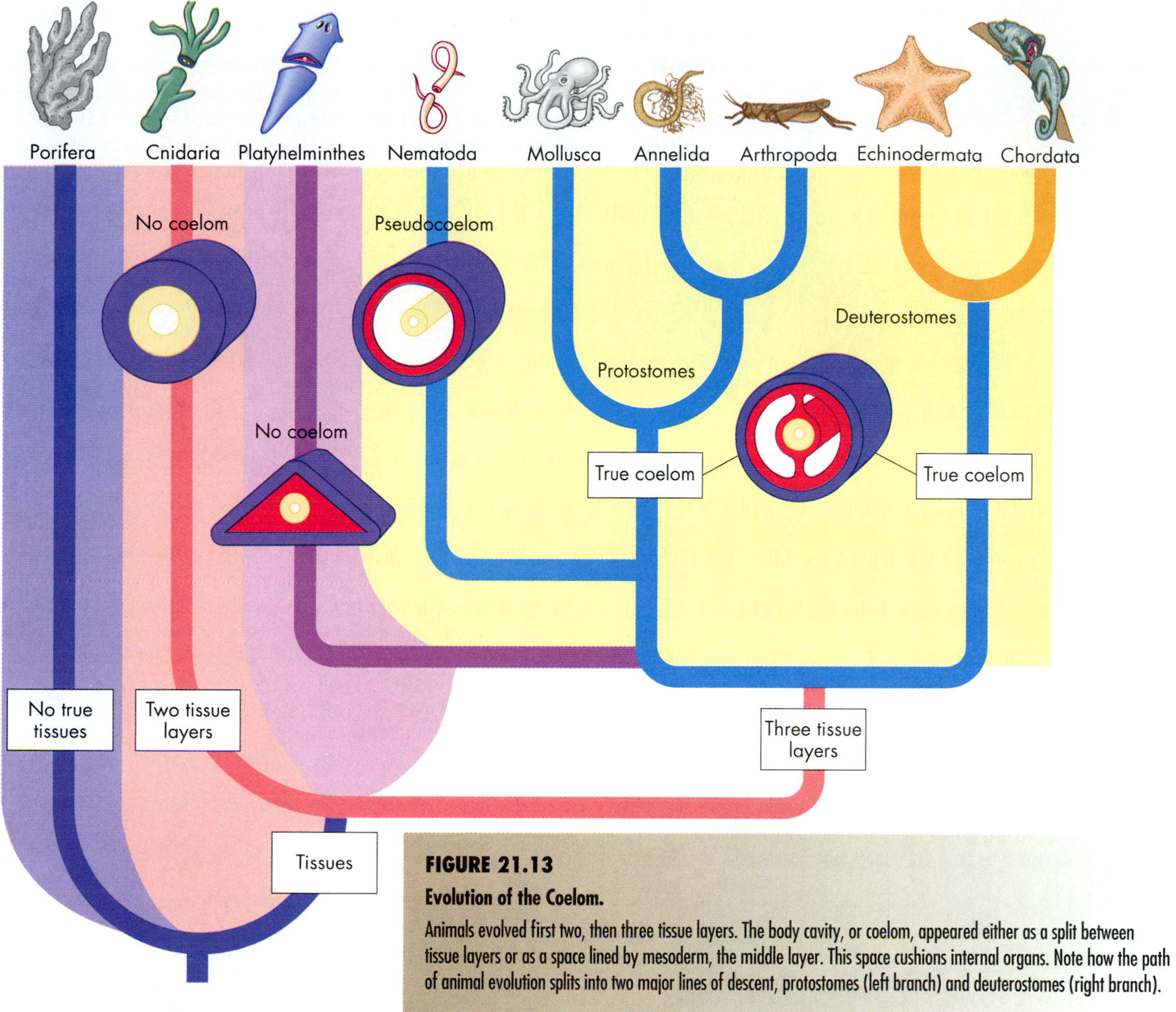

FIGURE 21.13

Evolution of the Coelom.

Animals evolved first two, then three tissue layers. The body cavity, or coelom, appeared either as a split between tissue layers or as a space lined by mesoderm, the middle layer. This space cushions internal organs. Note how the path of animal evolution splits into two major lines of descent, protostomes (left branch) and deuterostomes (right branch).

support and rigidity for the soft animal. (The same principle stiffens a man's penis during an erection.) (4) Since the gut is suspended in a cavity, its activities can take place undisturbed by the activity or inactivity of the animal's outer body wall.

One can easily see certain benefits of this architecture in the roundworms. The roundworm's gut has two separate openings, the anterior mouth and posterior anus. Food enters the mouth and moves in just one direction through the gut tube, with wastes exiting through the anus. Regions of specialized function developed along the gut for grinding food into small pieces, breaking it down with enzymes, absorbing nutrients and water, and expelling wastes. Over time, roundworms became highly efficient digesters, capable of consuming a wide variety of foods and therefore of inhabiting environments all over the earth.

ROUNDWORMS AND THE ENVIRONMENT

Many types of free-living roundworms help consume rotting plant and animal matter, and thus are ecologically important decomposers, like the bacteria and fungi. However, the parasitic roundworms get most of the fame—or infamy. At least 1000 nematode species parasitize plants, and some ecologists estimate that roundworms consume fully 10 percent of all crops annually. Nearly 50 species parasitize humans, entering in food or

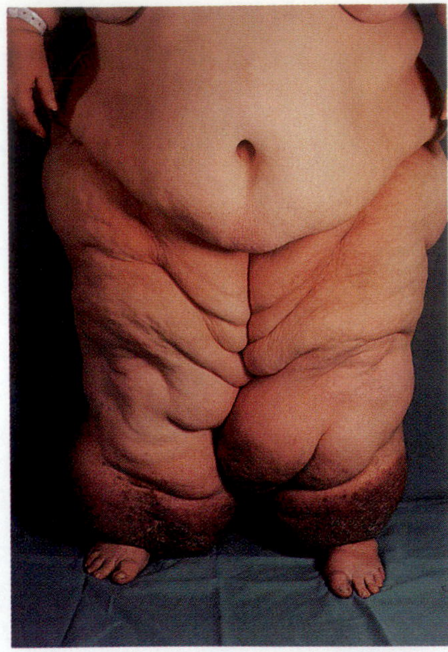

FIGURE 21.14
Elephantiasis. Grotesque limb enlargement can occur when roundworms block lymphatic vessels [see CHAPTER 25] and the tissues accumulate fluid. This woman with elephantiasis has hugely swollen legs but normal feet.

contaminated water or through bare skin. They cause many diseases, including trichinosis (from eating undercooked worm-infested pork), ascariasis (a common disease in tropical regions, characterized by lung infections and intestinal blockages due to masses of worms), hookworm (an infestation in the internal organs, also common in the tropics), and elephantiasis [FIGURE 21.14]. Clearly, much of the nematode's evolutionary success comes at the expense of other organisms, ourselves included.

➤ CONCEPT CHALLENGE

One of your classmates notes that the parasitic flatworms and roundworms lack certain sensory and digestive adaptations, and reasons that the free-living types must have evolved from parasitic forms. What evidence would you bring out to support or challenge this hypothesis?

Two Evolutionary Lines of Animals

Our discussion so far has seemed roughly linear—a history of increasing physical complexity in successive animal groups, as summarized in TABLE 21.1 on page 499. It may seem logical, therefore, to assume that evolution is also linear, with each group giving rise to the next more complex set of animals. As we discussed in CHAPTER 15, however, evolution is not linear, but rather occurs in a branching and rebranching course, rather like a bush or tree [see FIGURE 15.2]. The evidence suggests that an ancient progenitor similar to a free-living flatworm gave rise to today's flatworms, as well as roundworms and all the other bilateral phyla [review FIGURE 21.3]. Shortly after bilateral organisms arose, another split occurred, leading to two great animal lineages [see FIGURE 21.13]. One branch contains only invertebrates, including the next three phyla we will consider, the mollusks, annelids, and arthropods. The other branch includes the echinoderms and our own phylum, the chordates [see CHAPTER 19]. The two branches are distinguished by very different patterns of early embryonic development.

Recall that during animal development, the embryo forms a hollow ball (the blastula). This ball then indents, with the infolding cells becoming the digestive tract [see FIGURE 13.12]. In the evolutionary branch that contains only invertebrates, the initial indentation becomes the mouth—hence animals in this evolutionary line are called the **protostomes** ("first mouth")—and a second indentation becomes the anus. In the other evolutionary line—the one that yielded the vertebrates—the initial indentation becomes the anus, and a second opening becomes the mouth; biologists call organisms with this particular developmental pattern the **deuterostomes** ("second mouth"). The two lines of descent also show differences in the way the eggs cleave and the way the body cavity (coelom) forms. Such differences provide the kind of evidence biologists use to understand the pathways of animal evolution.

Let's look at each of the diverging evolutionary branches, beginning with the protostomes.

Mollusks: Soft-Bodied Animals

The octopus, the brainiest of all the invertebrates, is a **mollusk**, a member of the phylum **Mollusca**. This collection of more than 120,000 species of soft-bodied animals also includes snails, slugs, clams, oysters, and squid. All but the land snails and slugs are aquatic. Although mollusks are soft, some members secrete a hard shell; therefore, unlike the soft-bodied groups we've covered so far, a fairly good fossil record of the mollusks remains. The phylum probably arose from wormlike, bilaterally symmetrical ancestors some time before the early Cambrian.

CHARACTERISTICS OF MOLLUSKS

You may have the impression from eating raw oysters that mollusks are slimy blobs with no distinct shape. However, there is an identifiable body plan to all mollusks, no matter how different or soft they look inside

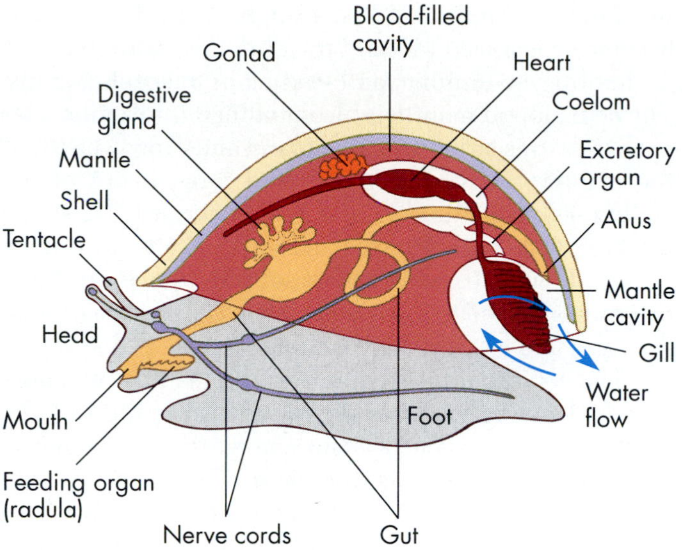

FIGURE 21.15

The Molluscan Body Plan.

This generalized creature has the foot, head, mantle, and other major molluscan organs.

their shells. As FIGURE 21.15 shows, each mollusk has a *head* housing the mouth, brain, and sense organs; a *foot*, a muscular organ used for gripping or creeping over surfaces; a *visceral mass* containing internal organs (heart, gut, sex organs, and excretory and respiratory apparatus); and a **mantle**, a thick fold of tissue that covers the visceral mass and in some mollusks secretes the calcium carbonate that makes up the hard shell.

Most members of the phylum have a special feeding organ called the *radula* (see FIGURE 21.15). This strap-shaped structure bears rows of tiny teeth and works something like a cheese grater, quickly rasping off successive layers of food.

The mollusk's gills function in the distribution of oxygen and carbon dioxide in tandem with a *circulatory system* that includes a *heart* with chambers that pump blood; *vessels*, or blood-carrying tubes, that pass through the gills; and a *blood-filled cavity* where internal organs and tissues are bathed with blood [see FIGURE 21.15]. The circulatory system is considered an *open circulatory system* because although in part of the system the blood is confined within vessels, it also flows through open spaces, where it bathes body tissues directly. As blood passes through the gills, it releases carbon dioxide and picks up oxygen; the blood then carries the oxygen from the gills to all internal cells. Specialized organs and organ systems for respiration and for circulation of blood represent evolutionary advances not found in less complex animals.

Although many mollusks are simple, some mollusks, such as the trainable octopuses we saw in the chapter opener, have well-developed nervous systems with large brains, acute senses, and a capacity for learning.

Most mollusks produce highly mobile fringed larvae (*trochophores*); their similarity to certain larvae of the next animal group, the segmented worms, suggests that as different as they look, mollusks and annelids are related by evolution.

SOME CLASSES OF MOLLUSKS

We consider here three classes of mollusks: gastropods, bivalves, and cephalopods.

Gastropods Snails, garden slugs, and sea slugs, or nudibranchs, are all **gastropods**, members of the class Gastropoda (meaning "belly foot"). Most snails are identifiable by their coiled shells, as well as by *torsion*—an internal twisting of the body mass during embryonic development. Land snails and slugs are the only terrestrial mollusks, and they move slowly along glistening *slime trails* [FIGURE 21.16A]. While most aquatic snails have

[A]

[B]

FIGURE 21.16

Gastropods: Snails, Slugs, and Nudibranchs.

[A] Gastropods can be drab and familiar pests like this common garden snail, sliding across its slime trail. **[B]** The soft, shell-less mollusks called nudibranchs are often dazzlingly colored and patterned, such as this species in the genus *Hermissenda* of California's Monterey Bay.

gills, land snails have air-breathing lungs, and they can close off their shell opening to help retain moisture during dry weather.

Some snails and most other gastropods have two antennae and two eyes. Nudibranchs often produce sour or bitter substances that repel attackers. Their bright colors serve as a warning that a bad meal lies ahead for the predator [FIGURE 21.16B].

Bivalves The word **bivalve** ("two valves") refers to the two half shells that enclose each member of this class: the oysters, clams, mussels, scallops, and relatives. Strong muscles close the valves and stretch an elastic ligament at the hinge. When their shells are closed, the animals are protected from most predators. When the muscles relax, the shell snaps open. Rapid closing of the shell in scallops produces a water jet that propels the animal. When a mollusk's shells are open, particles can enter and irritate the soft tissue. In oysters, a region of the mantle responds to injury or foreign particles by secreting layers of nacre—mother-of-pearl—around the offending object. Pearls are the result of the reddish, whitish, or black nacre deposits.

Bivalves are **filter feeders**; they have gills that, in addition to collecting oxygen and releasing carbon dioxide, can strain out and collect tiny food particles suspended in water. Beating cilia draw water across the gills, and a mucous layer traps the particles, which are then passed into the mouth. One biologist has estimated that each week, all the water in San Francisco Bay is drawn through the bivalves that lie half-buried in the silt and sand at the bottom of the bay.

Cephalopods The squid, the octopus, the chambered nautilus, and other **cephalopods** (literally, "head foot") are the most complex mollusks, and evolved as fast-swimming predators of the deep sea [FIGURE 21.17]. In these creatures, the foot is modified: It bears a circle of 8 or 10 arms, each studded with suckers, and it terminates in a funnel, or **siphon**. Thus, a single organ, the foot, has become specialized for land travel in the gastropods and for hunting, swimming, and feeding in the cephalopods. The cephalopod mantle is also modified into a muscular enclosure; this structure can expand and draw water into the mantle cavity or contract and force it out of the siphon, jet-propelling the mollusk backward. These explosive bursts can carry the animal to safety or bring its suckered tentacles within reach of prey. Once captured, the prey is moved to the mouth, which contains a pincer-like beak and a razor-sharp radula.

The coordination for hunting and feeding in cephalopods depends on acute senses, a large brain, and the most complex nervous system among the invertebrates. The largest cephalopods, the giant deep-sea squid, can weigh 450 kg (1000 lb) and reach nearly 18 m (60 ft) in length. Giant squid have the largest eyes in the animal kingdom (they can grow larger than a car's headlights), and these highly sensitive organs can form images like our own. The octopuses, with their radiating arms, may look sinister to some and dim-witted to others, but their brains are actually quite massive, and can contain more than 170 million nerve cells. Octopuses can be trained to accomplish a number of tasks, even, as we saw earlier, learning to attack a red or white ball just by watching other octopuses perform the same task. This type of learning causes biologists to suggest that octopuses are the most intelligent invertebrates.

The mollusks show many of the trends we outlined at the beginning of the chapter, such as bilateral symmetry (at least when young), a head, a one-way digestive tract, and a body cavity. They lack, however, a key structural feature displayed by the next two phyla, the annelids and the arthropods.

[A]

[B]

FIGURE 21.17
Cephalopods: Predators of the Deep.
The most complex mollusks include squid [A], octopus, and the chambered nautilus [B], which Oliver Wendell Holmes called the "ship of pearl."

Annelids: Segmented Worms

Marine sandworms, common earthworms, and leeches are all members of the phylum **Annelida**, the **segmented worms**, whose approximately 15,000 species belong to three classes. Most are members of the class Polychaeta (meaning "many-bristled") and some often colorful marine worms that burrow in the mud or sand and bear such common names as fireworms, clam worms, and feather dusters [FIGURE 21.18A]. The familiar reddish earthworms are in the class Oligochaeta ("few bristles"). These ubiquitous inhabitants of moist soils, often numbering 50,000 or more per acre, literally eat their way through dense, compacted earth, excreting the displaced material in small, dark piles. Each year, earthworms carry as much as 18 tons of rich soil per acre to the surface.

Leeches, which live mainly in fresh water, are in a separate class, Hirudinea. Many prey upon worms and mollusks, but the most infamous types parasitize large animals and suck their blood. Physicians sometimes use a substance called hirudin, which they obtain from leeches, to block undesirable blood clotting in their patients. A leech can consume three times its weight in blood and go for as long as nine months between meals [FIGURE 21.18B].

Annelid reproduction is generally sexual. Many marine annelids shed sperm and eggs into seawater, where the gametes unite and develop into larvae similar in form to mollusk larvae. Earthworms and leeches are hermaphrodites; each individual produces both sperm and eggs; when two individuals couple, they exchange sperm so that each fertilizes the other's eggs.

[A]

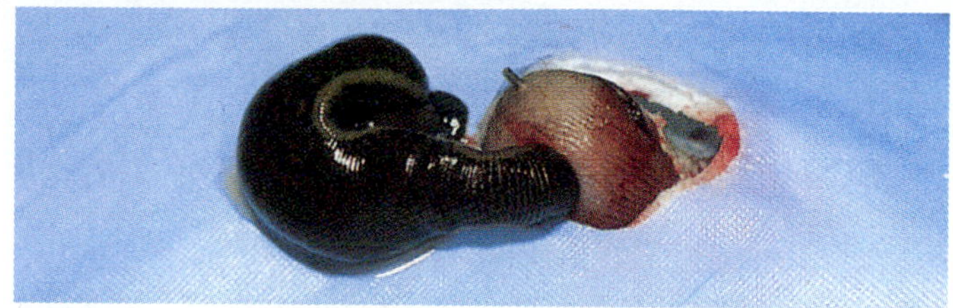

[B] A leech

FIGURE 21.18

Polychaetes and Leeches.

Segmented worms range from the delicate to the despised, including **[A]** this Caribbean feather duster worm and **[B]** this black leech. Sticking out of the blue surgical draping is a human fingertip. Surgeons reattached the severed fingertip, and medical workers then applied the leech to suck blood from tissues around the reattachment site, release anesthetics and anticoagulant, and thereby relieve pressure and decrease pain.

CHARACTERISTICS OF ANNELIDS

The term *annelid* means "tiny rings," and refers to the external segments visible on members of this phylum. Annelids are the first group possessing all five of the physical traits we have been tracing: bilateral symmetry, cephalization, a tubular gut, a coelom, and segmentation. Although more than a million animal species evolved after the annelids, no new features as basic as these emerged; all later animals, including humans, show variations on the anatomical themes that first appeared together in the annelids.

Segmentation Segmentation results in organisms of larger size. In certain environments, large size can be an advantage—for example, a large animal can more easily ingest a smaller one. The advantages of segmentation can be appreciated by analyzing earthworms—typical annelids with 100 or more body segments [FIGURE 21.19A]. Each segment is separated from the next by an internal partition, or *septum* (plural, *septa*), and each segment contains a set of internal structures. Most earthworm segments contain two excretory units called *nephridia* [FIGURE 21.19B]. Each nephridium removes excess water and wastes from the body fluids by means of a ciliated funnel (reminiscent of the flatworm's flame cell) and excretes them through a pore in the body wall.

Each segment also contains a fluid-filled compartment of the coelom and is surrounded by circular and longitudinal muscles in the body wall. As the circular muscles squeeze against the incompressible fluids in the coelom and gut, force is transmitted to adjacent segments, creating a hydroskeleton, an internal skeleton made of fluid. Contractions in sequential segments produce waves of force that propel the animal forward. *Setae*, which are not true legs but pairs of bristles attached to each segment, push against the ground with each contraction and help the animal move. In addition to nephridia, muscles, and coelom compartments, each segment contains clusters of nerve cells connected to the brain by nerve cords.

The Digestive System Organ systems for digestion have two specialized regions not seen in roundworms: a

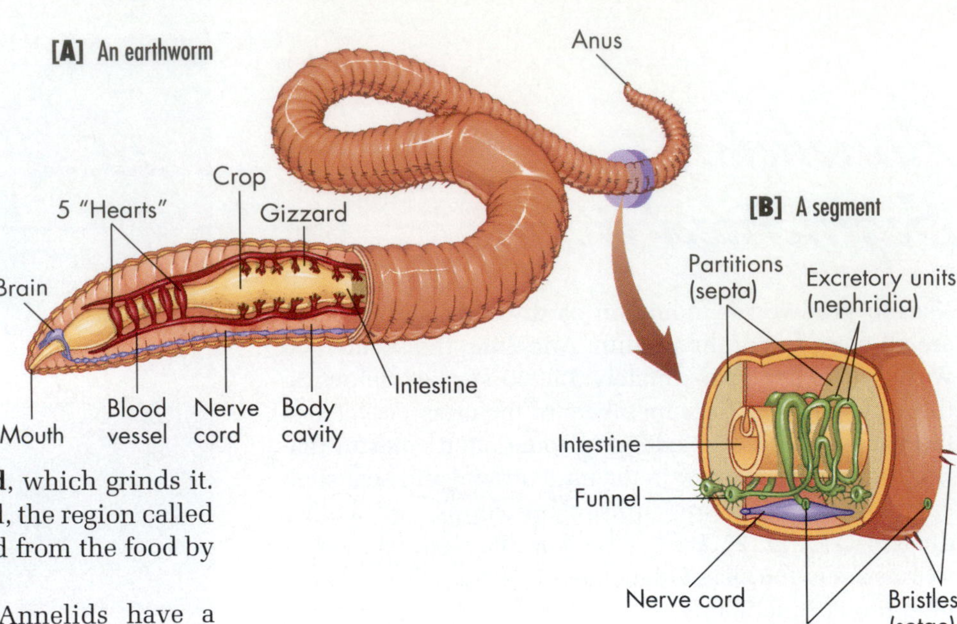

FIGURE 21.19

Earthworms: A Study in Segments.

[A] Anatomy of an earthworm, the most familiar annelid. [B] Segments are repeating modules, each containing organs of the excretory, nervous, and locomotory systems.

crop, which stores food, and a **gizzard**, which grinds it. After the gizzard has ground some food, the region called the intestine absorbs nutrients released from the food by digestive enzymes [see FIGURE 21.19A].

The Closed Circulatory System Annelids have a *closed circulatory system*: Blood is carried entirely in tubes, or vessels, and at no point bathes the body tissues directly, permitting blood to be pumped at higher pressure through the animal's body. The result is a more regular and constant delivery of supplies and a more efficient removal of wastes from cells at times of high activity.

Innovations in segmentation, digestion, and circulation allowed the annelids to grow longer and thicker than flatworms and roundworms. Segmentation also led to evolution of the arthropods.

Arthropods: Jointed Legged Animals

The largest phylum on earth is **Arthropoda**. Arthropods include fossil trilobites, spiders, mites, ticks, scorpions, centipedes, millipedes, lobsters, crabs, and insects. The insects, numbering at least 1 million species, make up the great majority of all animal species [FIGURE 21.20]. In fact, if you took a book that listed every animal species and pointed to any name at random, it would probably be an insect. As CHAPTER 18 explained, there may be many, many more insect species as yet undiscovered—perhaps 10 to 30 million more! The insects' astounding diversity and success are based on their possession of all five major features we have been tracing: Insects are bilateral; have a head; have a one-way gut; have a fully lined body cavity; and are segmented [review FIGURE 21.2]. Furthermore, there are several other characteristics shared by all arthropods: (1) an external skeleton; (2) modified, specialized segments; (3) rapid movement and metabolism due to special respiratory structures; and (4) acute sensory systems.

These arthropod features contributed to their successful invasion of the land. Fossils of the earliest land-living animals yet discovered were tiny, spiderlike organisms that lived among the stems of *Cooksonia* plants [review FIGURE 20.17] more than 405 million years ago [FIGURE 21.21]. These arthropods, called "trigs" (trigono-

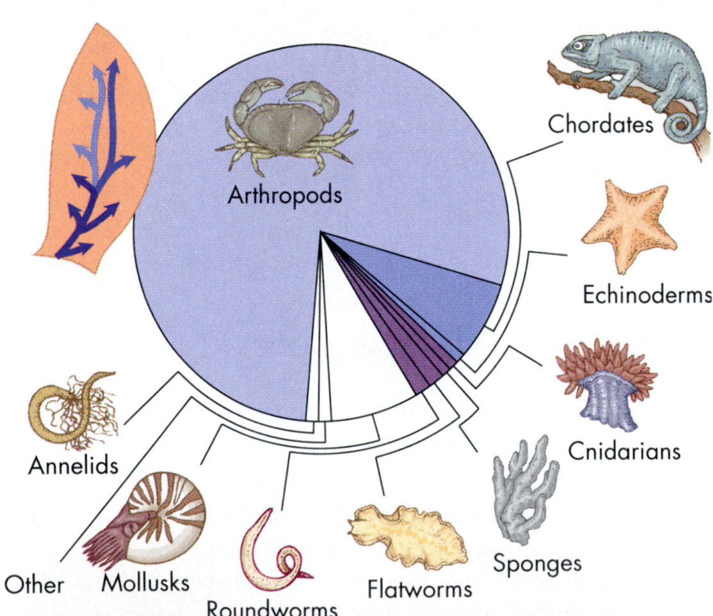

FIGURE 21.20

Arthropods: The Vast Majority of All Animal Species.

Compared to the insects and other arthropods, all other species make up a small slice of the animal kingdom. If the projections of 10 to 30 million total tropical insect species are correct, the rest of the animal kingdom is, in fact, a miniscule slice! Our era could be called the Age of the Arthropods.

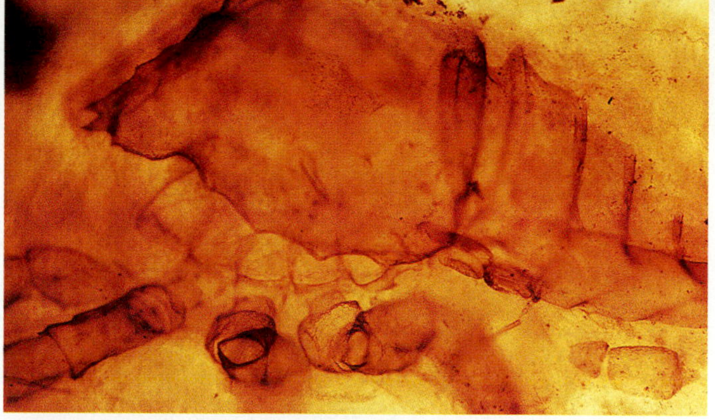

FIGURE 21.21

The Oldest Land Animal Yet Discovered.

More than 400 million years ago, "trigs" wandered through stands of the first land plants, *Cooksonia* [review FIGURE 20.17]. The fossil pictured here is 405 million years old, and shows a trigonotarbid fossil from what is now Scotland. Jointed arthropod legs are clearly visible.

tarbids), had jointed legs, armor plates on the abdomen, air-breathing book lungs like a spider, and eyes with 10 or so loosely packed lenses. The mouthparts were similar to those in modern predatory spiders.

THE ARTHROPOD EXOSKELETON

The key to the success of insects and other arthropods is the **exoskeleton**, or external skeleton, that completely surrounds the animal and provides strong support as well as rigid surfaces that muscles can pull against. In arthropods, the epidermis secretes a thick, hard cuticle that contains the polysaccharide *chitin* [see CHAPTER 2], in addition to sugar-protein complexes, waxes, and lipids that make the body covering waterproof. This mixture of substances makes a tough but somewhat flexible outer barrier for most arthropods that protects and supports the animal's soft internal organs. In crustaceans, such as crabs, lobsters, and shrimp, the exoskeleton also contains calcium carbonate crystals, which make it a hard, inflexible armor. Besides providing shieldlike protection from enemies and resistance to general wear and tear, the exoskeleton prevents internal tissues from drying out. This is extremely important, since most arthropods live on land.

In all arthropods, the exoskeleton remains thin and flexible at the *joints*, the hingelike areas of the legs and body. The term *arthropod*, in fact, means "jointed foot." The presence of jointed appendages allows arthropods to move quickly and efficiently above the ground or sea floor instead of dragging the body directly along the ground on stubby legs or bristles. Muscles attached to the inside of the exoskeleton on either side of the joints provide the leverage needed to move each appendage. Exoskeletons do, however, have a major disadvantage: The animal cannot grow larger unless it periodically sheds its constricting armor and produces a larger exoskeleton; this process is called *molting*. During the period between the shedding of one exoskeleton and the growth of the next, the animal is in a vulnerable condition.

SPECIALIZED ARTHROPOD SEGMENTS

While annelid worms have the same simple segments repeated throughout most of the body, arthropods have evolved different and highly modified segments that give them a far greater repertoire of activities. Their segments, first of all, are usually fused into a few major body regions; insects, for example, have three regions, **head**, **thorax**, and **abdomen**, whereas in spiders and crustaceans, there is an abdomen plus one region with fused head and thorax, the **cephalothorax** [FIGURE 21.22A].

[A]

[B]

FIGURE 21.22

Arthropod Appendages: Modified for Different Tasks.

[A] A brilliantly colored creature like the scarlet cleaner shrimp (*Hippolysmata grabbhami*) of the tropical western Atlantic has a fused head and thorax and a separate abdomen, as well as appendages modified for grasping, walking, mating, and swimming. [B] The appendages of the giant swallowtail (*Papilio cresphontes*) are modified for flying, walking, and sensing the environment.

During evolution, specific appendages for walking, swimming, and flying arose, with each species having its own modifications. Some arthropods also developed pincers or palps (feelers), which facilitate hunting and feeding. From the head region grew other appendages: *mouthparts* that allow chewing and sucking and *antennae* that sense odors and vibrations. The brightly colored shrimp in FIGURE 21.22A, for example, has appendages modified for eating (mouthparts), grasping (pincers), walking (legs), mating (swimmerets), and fast swimming (tail). The appendages themselves became further modified during evolution. Just compare the flapping wings of the butterfly [FIGURE 21.22B] and the small, buzzing wings of the housefly. The many modifications enabled the arthropods to exploit a wide variety of environments.

RESPIRATORY SYSTEM

Arthropods metabolize and move rapidly, and, in fact, some insects do both at the highest rates in the animal kingdom. Some tiny flies can generate enough metabolic energy to beat their wings 1000 times per second. High metabolic rates require rapid oxygen delivery, and the arthropods' efficient respiratory organs provide a large surface area for collecting oxygen and releasing carbon dioxide quickly. Insects have *tracheae* (TRAY-kee-ee; singular, *trachea*), branching networks of hollow air passages [FIGURE 21.23A]; spiders and relatives have *book lungs*, chambers with leaflike plates for exchanging gases; and finally, aquatic arthropods have *gills*, flat tissue plates that act as gas-exchange surfaces [FIGURE 21.23B]. High metabolic rates and rapid movements allow arthropods to fly, run, and swim faster than any previous animal group, and hence to escape predators with agility and to disperse over wider ranges.

SENSE ORGANS

Most arthropods have antennae capable of detecting movement, sound, or chemicals with great sensitivity. The antennae of several male moth species, for instance, can detect the *pheromones*, or odor signals, given off by adult female moths at distances of over 11 km (about 7 mi). Arthropods also have various types of organs on the head and body that detect sound and taste, and many have special compound eyes made up of 2500 or more six-sided segments called *facets* [FIGURE 21.24]. Many compound eyes are capable of color vision and can detect the slightest movements of prey, mates, or predators.

IMPORTANT ARTHROPOD CLASSES

The successful and highly divergent phylum Arthropoda is divided into a number of taxonomic classes, most of which will sound quite familiar.

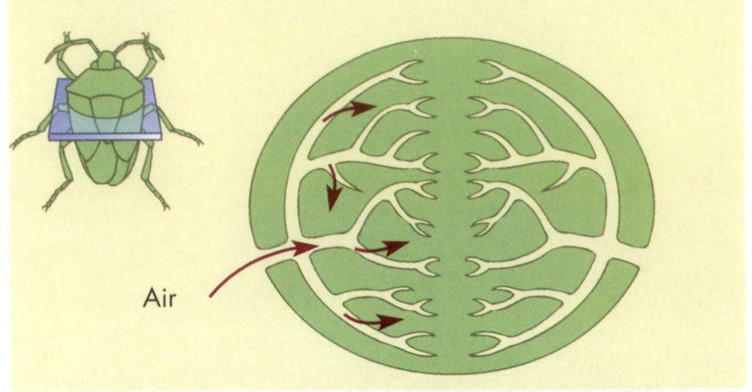

[A] Insect's tracheae

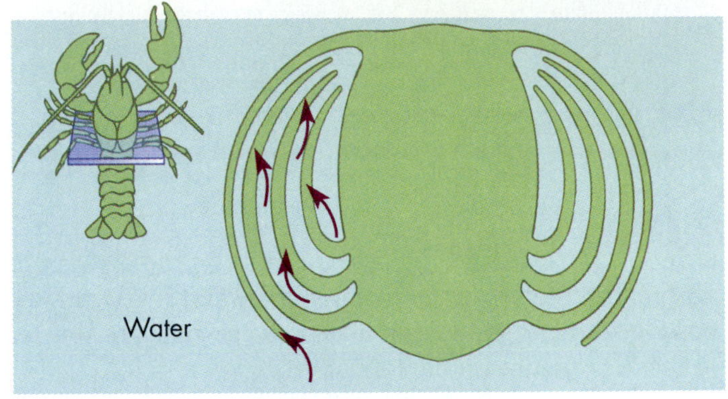

[B] Lobster's gills

FIGURE 21.23

Arthropod Respiratory Organs.

[A] An insect's tracheae are branching hollow tubes that carry air deep into the body tissues. The spider's book lungs are leaflike plates that project internally and have a large surface area for gas exchange. [B] The gills of a lobster, an aquatic crustacean, also have a large surface area, but project outward so that they are bathed in the oxygen-rich water.

Centipedes and Millipedes The 3000 or so centipede species, members of the class Chilopoda, look a lot like annelid worms, and have a series of flattened body segments. However, each segment bears a pair of true-jointed legs (not setae) that move the centipede swiftly in search of insects, worms, small mollusks, or other prey. Centipedes kill their prey with *poison claws*, modified legs on the first body segment [FIGURE 21.25]. Some huge tropical centipedes are dangerous to humans, but the common varieties that lurk in damp basements are harmless and consume insects.

Millipedes, in the class Diplopoda, are slow-moving counterparts to the centipedes. They have two pairs of much smaller legs per segment; a round, not flattened body; and a preference for decaying vegetable matter instead of live prey.

Crustaceans The crabs, shrimp, lobsters, barnacles, crayfish, and sow bugs in the class Crustacea are so different from each other that only two generalizations ap-

FIGURE 21.24

The Many Facets of an Arthropod's Eye.

The compound eye of this horsefly has hundreds of geometric facets. Each facet detects light and movement, and the insect sees a composite image created by the many individual segments.

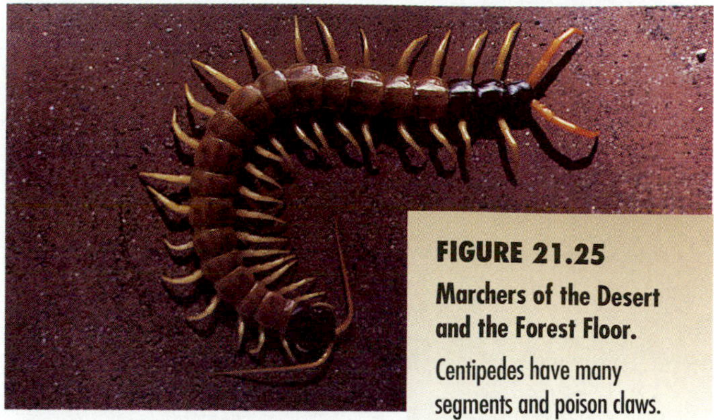

FIGURE 21.25

Marchers of the Desert and the Forest Floor.

Centipedes have many segments and poison claws.

FIGURE 21.26

The Blue Lobster: A Vividly Representative Crustacean.

The lobster *Homarus americanus*, a denizen of the Atlantic Ocean, shown in rare blue phase, can grow to 1 m (3.3 ft) in length and weigh 18 kg (40 lb). Like the shrimp [FIGURE 21.22], it has a fused head and thorax, as well as several sets of specialized appendages. Its mighty pincers help in food gathering and defense.

ply: (1) Almost all have an exoskeleton hardened with calcium salts that covers most of the animal as a protective shell, or *carapace*, and (2) all have two pairs of antennae.

The lobster, a familiar crustacean, has two major body regions: the anterior *cephalothorax* and the posterior *abdomen* [FIGURE 21.26]. Walking legs and pincers sprout from the cephalothorax, while the abdomen bears feathery swimming appendages.

Crustaceans are so numerous and so diverse that they are found in most bodies of water. There are even a few terrestrial species, like the familiar sow bugs (also called pill bugs or armadillo bugs) that scurry away when you lift a rock or flower pot sitting on the ground. Many small aquatic species, such as fairy shrimp, brine shrimp, and copepods, serve as the primary food for various fish, whales, and other species. Barnacles, an aquatic species with a contrasting life-style, cling tenaciously to tidal-zone rocks; the natural cement they secrete is strong enough to keep the animals from washing away in pounding ocean surf—a feat comparable to a person standing upright in winds of 480 to 640 km/hr (300 to 400 mph). Some biologists are studying barnacle glue in hopes of developing medical cements for fastening dentures and rejoining broken bones.

Spiders and Relatives Class Arachnida includes 55,000 species of spiders, ticks, mites, and scorpions. Arachnids belong to the subphylum Chelicerata, which also includes the extinct trilobites, the sea spiders, and the primitive-looking horseshoe crabs often seen on Atlantic coastal beaches. Arachnids lack antennae; have a cephalothorax and abdomen, each divided into segments; and usually have six pairs of appendages [FIGURE 21.27]. The first pair of appendages is modified into *chelicerae*, or poison fangs, used for killing prey or for self-defense. The second pair holds the prey while the spider injects poison or enzymes. The other four pairs are walking legs. Spiders also have silk-spinning organs called *spinnerets* at the rear of the abdomen; the spinnerets reel out threads that are, size-for-size, stronger than steel. Despite their sinister reputation, most spiders are capable of poisoning only small animals. The potent poisons of the black widow and brown recluse spiders, however, can be dangerous to humans. Present-day scorpions have a body form very similar to the huge water scorpions that lived 500 million years ago and that probably gave rise to the land scorpions and spiders. Mites and ticks are, for the most part, merely irritating biters, but some ticks carry serious diseases, including Lyme disease and Rocky Mountain spotted fever, with its fever, rash, and joint pain.

FIGURE 21.27

Arachnids: Spiders and Relatives.

This tropical *wandering* spider is fearsome-looking but harmless to humans; it weaves a broad web and preys on small insects.

[A]

[B]

FIGURE 21.28

Insects: Mostly Terrestrial and Highly Diverse.

[A] The armored head of a worker termite is adapted to gnaw through tough wood fibers, while microorganisms in its digestive tract turn the fibers into useful fuel for the animal. **[B]** A swarm of grasshoppers can denude a farmer's field in minutes. This Lubber grasshopper has a hearty appetite and chewing mouthparts.

Insects It is difficult, in a few paragraphs, to do justice to insects, the largest—and, by that measure, most successful—class of animals on earth. Most insects are terrestrial animals living in habitats from the tropics to the poles, but a few are aquatic. Many zoologists attribute insect success to the general arthropod traits we've already considered, but also to the insect's small size and the fact that many can fly. Smallness enables the insects to exploit a vast array of microhabitats—the bark of a tree, the planar landscapes on the backs of leaves, the dense thickets of an animal's fur, or the universe of midair.

Specific body parts also help account for the insects' success. Many insects have organs for smelling, touch reception, tasting, seeing, and hearing in various parts of the body. An insect's head bears one pair of antennae; its thorax bears three pairs of legs and usually one or two pairs of wings; and its abdomen is usually free of appendages. A series of modified segments, the mouthparts, enables the insect to feed efficiently. Some, like mosquitoes, have pointed stylets that pierce and suck. Others, like locusts and grasshoppers, have chewing mouthparts that can quickly decimate foliage [FIGURE 21.28B].

Insects have evolved various ways to grow, despite the confining exoskeleton, and various ways to thrive, despite the changing seasons. In some insects, like grasshoppers and cockroaches, the embryo emerges as a miniature version of the adult, but without wings or mature reproductive organs. This organism feeds, grows, and molts five or six times until it reaches adult size, then does not molt again. In most insects, however, including butterflies, flies, and beetles, the embryo develops into an immature form, or **larva**, eats voraciously, then forms a transitional stage, or **pupa**, sometimes inside a cocoon. A complete change in form, or **metamorphosis**, takes place in the body within the pupal exoskeleton. Finally, a nonmolting, reproductively mature adult emerges [see FIGURE 30.12]. In insects that metamorphose, the larvae and adults can be adapted to very different foods and environmental conditions. This successful evolutionary solution allows larvae to specialize in feeding and obtaining energy, and the adult to specialize in reproduction.

Insects have great environmental significance. To assess the relative importance of insects and humans, ask yourself how many types of organisms would become extinct if people were suddenly wiped off the face of the earth. As you can probably guess, very few would disappear. In contrast, if insects became extinct, earth would change dramatically. Thousands of plant species would die off without the help of their insect pollinators, and the animals that depend on those plants would also disappear quickly.

Insects have additional significance for medicine, agriculture, and science. Insects are vectors of many diseases, including malaria [review FIGURE 19.17] and yellow fever. Insects are major agricultural pests, consuming a substantial portion of grains and other crops each year. On the other hand, insects—the fruit fly *Drosophila melanogaster* in particular—have contributed greatly to our understanding of genetics, cell biology, and developmental biology [see FIGURES 8.16 and 13.19].

Perhaps the most fascinating of the arthropods are the **social insects**: termites [FIGURE 21.28A], ants, wasps, and bees. Most species of these insects live in large colonies with labor divided among *castes*, or subgroups, that differ in appearance and behavior. Such insect colonies are highly successful. Ants, for example, may make up one-third the weight of all the animals on earth, with 200,000 ants for every person now alive! Insect colonies are highly evolved, functioning, in a sense, as a single well-coordinated organism with the capacity to simultaneously build homes and cities, defend their own, harvest food, and reproduce. When you consider that the roots of the animal kingdom lie in Precambrian creatures no more complex than sponges, cnidarians, and flatworms, social insects like ants or termites demonstrate vividly the great evolutionary distance that invertebrates traveled.

CONCEPT CHALLENGE

Do you agree that earth should be called the Planet of the Insects?

[A]

[B]

FIGURE 21.29

Echinoderms: Spiny-Skinned Ocean Dwellers.

Echinoderms such as **[A]** this yellow feather star (*Himeroptera* sp.) in tropical Pacific waters can resemble bright flowers blooming on coral reefs (golden). **[B]** This sea cucumber (*Parastichopus californicus*) lives in the warm, shallow waters off the coast of Baja, California, and is shown here next to a blue-banded goby fish.

Echinoderms: The First Endoskeletons

An expert on invertebrates once called the echinoderms "a noble group especially designed to puzzle the zoologist." The phylum **Echinodermata** includes the sea stars, brittle stars, sea urchins, sea cucumbers, and sea lilies.

The word *echinoderm* means "spiny-skinned," and virtually all members have spines, bumps, spikes, or unappetizing projections that help to protect these slow-moving marine creatures from predators.

Echinoderms have an odd mixture of traits, some of which are innovations appearing in all higher animals and some of which are apparent regressions toward simpler forms. Besides their bizarre and interesting lifestyles and appearances [FIGURE 21.29; also see FIGURE 13.13C], the echinoderms display two new features: (1) an internal skeleton and (2) a hydraulic pressure system, unique in the animal kingdom, that facilitates locomotion. Echinoderms are also deuterostomes with a developmental pattern that sets them off from the protostomes and other invertebrates, and suggests an ancestry common to our own phylum, Chordata.

Echinoderm larvae are bilaterally symmetrical, but adults of many species are radial, headless, and brainless. Nerve trunks run along each of the adult's arms (which usually number five or multiples of five) and unite in a ring structure around the mouth [FIGURE 21.30A]. This simple system allows coordinated, but slow, movement of the limbs.

The radial body plan of echinoderms is not an advanced trait, nor is the absence of a brain. Echinoderms also lack excretory and respiratory systems, relying mainly on diffusion for all exchanges with the outside. However, they do show one important evolutionary development: an **endoskeleton**, or true internal support system. Just below the outer epidermis lie calcium-based plates that protect the internal organs and give rise to the outer spines and bumps. (We will consider other types of endoskeletons in CHAPTER 29.)

box 21.1
People in Biology

A Case of Arachnophilia

In some ways, Charles Griswold and Teresa Meikle are a very unusual married couple: they met through their academic love of spiders, their courtship included spider-collecting trips to the desert, and they have shared their passion for arachnids happily ever after. In another sense, though, their story includes some significant compromises that are fairly typical of married scientists.

Griswold and Meikle met as graduate students in entomology at the University of California, Berkeley, in the late 1970s. Both had been interested in spiders and insects since childhood. An urban kid, Griswold liked the fact that a collection of easy-to-observe wild animals existed in his own backyard. Meikle, inspired by an excellent elementary school science teacher, helped start a "bug club" and a "bug hospital" in third grade.

After Meikle finished her master's degree in 1982, she worked for several years, mostly as an entomologist for government agencies. Griswold went on to complete his doctoral degree in spider taxonomy in 1983. There being a notable dearth of jobs in taxonomy—and even fewer available to spider experts—Griswold took his first full-time research job at the Natal Museum in South Africa. There, among other projects, Griswold and Meikle made some unique observations of social nest spiders on the African veldt.

In silken orbs as large as footballs, hundreds of small Stegodyphus spiders live and work together in communities [FIGURE 1]. These communities, the researchers reported, can trap more insect prey than can solitary spiders, and can better withstand the torrential rains of autumn and the fires that sweep the grassy plains during dry season. Within these nests, the researchers identified several other spider species that pilfer food from their Stegodyphus hosts. One, *Archaeodictyna ulova* (a species name they based on the Zulu word for "loafer") will wait while the Stegodyphus spiders attack and kill insect prey entangled in the web. Then they will saunter over. If there is no room at the trough, so to speak, an *A. ulova* spider will tap a host's legs quickly with its own until the host shifts and the "loafer" can feed beside it. Griswold and Meikle never found *A. ulova* living alone or with host species other than Stegodyphus; it is a classic parasite. In total, they found spiders, lice, cockroaches, wasps, birds, and a mouse among the 25 animal species that live in or on the community spider nests.

After returning from South Africa in 1986, Meikle decided to apply her biology background to the field of library science. "I always liked asking questions, and searching for answers," she says, "and at libraries, you find out how to find out." The role that libraries can play in supporting academic research and science literacy also appeals strongly to the entomologist in her, she says. Like Meikle, many people with basic degrees in biology successfully carry their knowledge of life science into other fields such as library work, illustration, writing and editing, teaching, and marketing.

Griswold's career path, though unusually smooth for his field, has still been very difficult. At one point, he joked about his applications for grants and jobs as "a series of cliff-hangers," and himself as a "mendicant scholar," traveling from place to place to work for low postdoctoral wages because he believes in his work.

In 1992, both Griswold and Meikle landed jobs they wanted at the California Academy of Sciences in San Francisco's Golden Gate Park, he as a curator and spider researcher, she as a librarian in the institution's impressive science collection. She has since moved on to an equivalent position with a high-technology firm near San Francisco. The years of sacrificing, postponing, and relocating finally paid off in desirable jobs in their chosen fields. That probably makes them as unusual a couple as their mutual love of spiders.

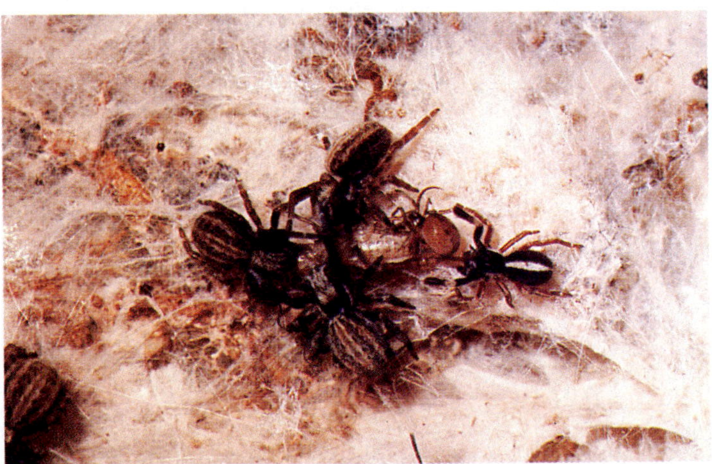

FIGURE 1
Stegodyphus spiders in silken community.

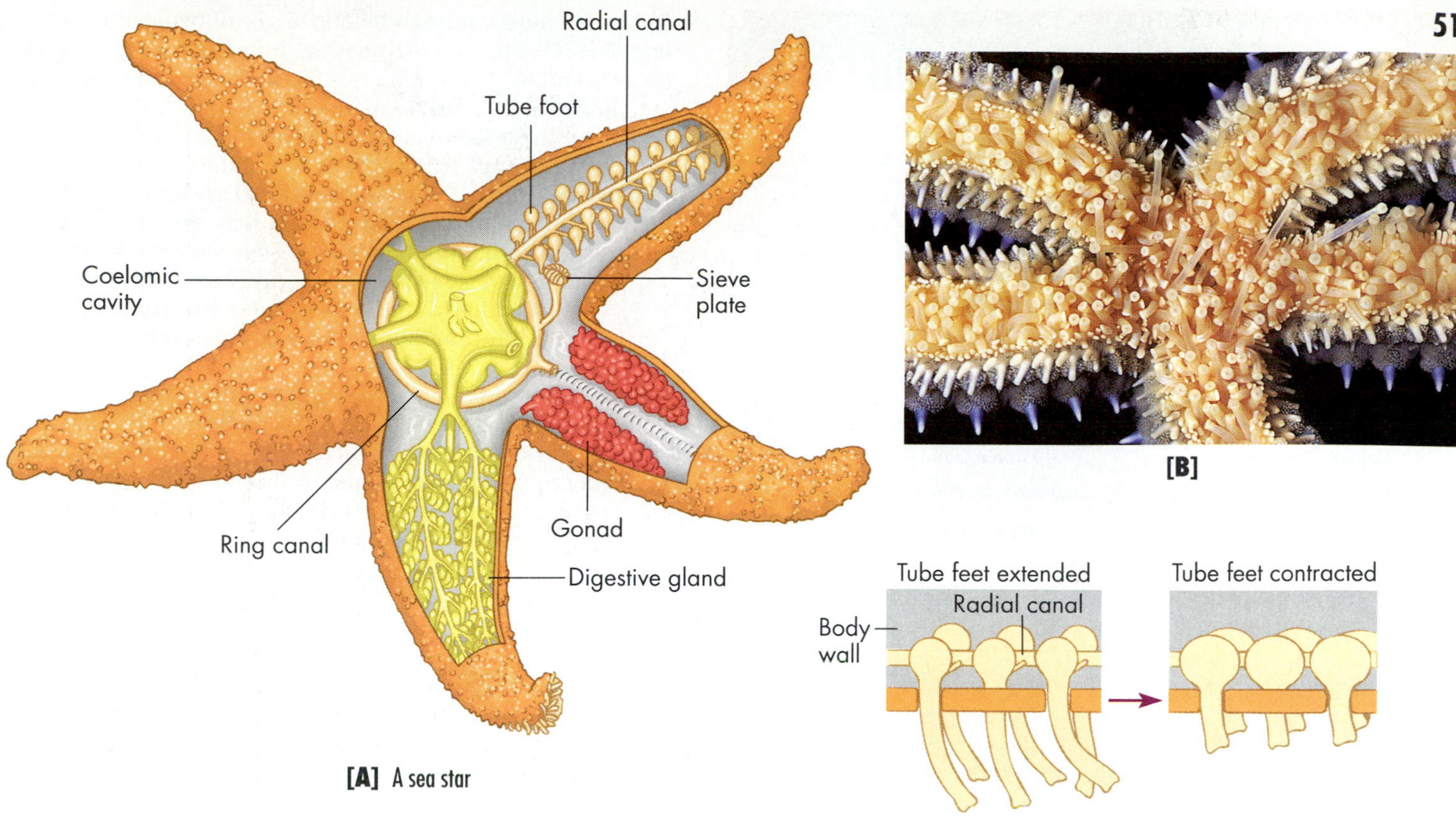

[A] A sea star

[B]

[C] Action of tube feet

FIGURE 21.30

Sea Stars and Hydraulic Tube Feet.

[A] Sea stars have radial bodies with 5 to 25 arms or more, internally branching digestive pouches, and well-developed gonads. Most remarkable are the hundreds of tube feet on the vertical surface **[B, C]** that can jut outward to an extended position under the hydraulic force of fluid from the internal water vascular system or that can contract to rounded nubbins when the pressure diminishes. By sequentially jutting and contracting these tube feet in waves, the animal can slowly move along its substrate.

Echinoderms have a large coelom, and in many species, it is divided to form the unique *water vascular system*, a set of canals filled with modified seawater [FIGURE 21.30A]. Hundreds of short branches off the canals become the tube feet that dot the ventral surface of a sea star, for example, and enable the animal to move across the sea floor [FIGURE 21.30B and C]. A bulblike portion of each tube foot can contract and force water from the canals into the foot. This pressure causes the foot to extend and a terminal sucker to grip the rock and sand below. This extension—adhesion—release of countless tube feet moves the animal along. A sea star can also force open a tightly shut clam shell with its arms and tube feet, then evert its stomach though its mouth into the open shell and digest the prey. Capable of eating more than 10 clams or oysters in a day, sea stars can be serious predators on commercial oyster beds.

Connections

The student of invertebrates enters a strange realm of worms and parasites, swarming insects, and intelligent monsters of the deep like the giant octopus. Many of these creatures are unfamiliar and live in habitats we can explore only with the help of scuba gear, shovels, or microscopes. Nevertheless, the animals without backbones make up 95 percent of all animal species. We tend to think of vertebrates as physically more complex than invertebrates, and many are. But the invertebrate world is full of organisms finely fitted by natural selection with adaptations that allow them to compete successfully in their own environments. Complicated processes like observational learning in the octopus belie any attempt to dismiss all invertebrates as "simple" or "primitive." In fact, the invertebrates, during an immense history that began more than 700 million years ago, "tested" a number of evolutionary "ideas" and came up with most of the physiological solutions employed throughout the animal kingdom in invertebrates and vertebrates alike. The next chapter examines some of these modifications in the phylum Chordata, which includes animals with backbones.

KEY TERMS

Annelida, 511
Arthropoda, 512
bilateral symmetry, 504
cephalization, 504
Cnidaria, 500
coelom, 506
deuterostome, 508
Echinodermata, 517
endoskeleton, 517
exoskeleton, 513
flatworm, 504
insect, 516
invertebrate, 494
metamorphosis, 516
mollusk, 508
Mollusca, 508
Nematoda, 506
nerve cord, 505
Platyhelminthes, 504
Porifera, 497
protostome, 508
roundworm, 506
segmented worm, 511
sponge, 497

HIGHLIGHTS IN REVIEW

1 Animals are motile, multicellular consumers of other organisms. They arise from the fertilization of a larger egg by a smaller sperm, and pass through a ball-of-cells (blastula) stage in their embryonic development.

2 The body plans of existing animals, as well as the fossil record, and gene sequences, suggest that animals occupy a single evolutionary branch on the tree of life.
- **a]** Fossil and genetic evidence suggests that animals already existed 800 million to 1 billion years ago.
- **b]** Animals may have arisen from ancient protists called choanoflagellates.

3 Themes in animal evolution include development of bilateral symmetry; a head; a one-way digestive tract; complex body cavities; and body segments.

4 Sponges are asymmetrical, and the structure of some of their cells suggest that animals evolved from certain protists. Neither the jellyfish and their relatives (the cnidarians) nor the flatworms have a body cavity, but the flatworms have bilateral symmetry and a head.
- **a]** Sponges are assigned to the phylum Porifera and are simple, sessile, saclike aquatic animals without true tissues. Sponges were the first to branch from the main line of animal evolution; they have genetic and structural similarities to certain colonial protists, whose ancestors may have given rise to all animals.
- **b]** The Cnidaria (hydras, jellyfish, corals, and sea anemones) are radially symmetrical animals with stinging tentacles used in self-defense and feeding. They can be vaselike polyps or umbrellalike medusae. They have only two tissue layers and no body cavity.
- **c]** Flatworms (phylum Platyhelminthes), including planarians, flukes, and tapeworms, are bilaterally symmetrical with a head at one end. Flatworms have three distinct tissue layers and develop true organs that accomplish digestion, nervous responses, body movement, excretion, and reproduction. They lack a body cavity.

5 More complex animals belong to one of two main branches. One branch includes the roundworms, arthropods (insects and relatives), annelids (earthworms and relatives), and mollusks (octopuses, snails, and relatives). The other branch includes echinoderms (sea stars and relatives) and chordates (animals with backbones and their relatives).
- **a]** Roundworms (phylum Nematoda) are tiny, abundant invertebrates that parasitize animals and plants and decompose rotting materials. They were the first to have a two-ended gut with mouth and anus. The gut is cushioned in a fluid-filled chamber (the pseudocoelom) surrounding the endoderm, or inner tissue layer.
- **b]** In one major animal lineage, the protostomes (mollusks, annelids, and arthropods), the indentation, or blastopore, of the developing embryo becomes the mouth, and a second opening becomes the anus. In the other major line, deuterostomes (echinoderms and chordates), the blastopore becomes the anus, and a second opening becomes the mouth.
- **c]** Mollusks (phylum Mollusca), including snails, clams, oysters, squid, and octopuses, have major body parts not found in most other invertebrates: the foot, visceral mass, mantle, mantle cavity, and gills.
- **d]** Annelids (phylum Annelida), including marine polychaetes, earthworms, and leeches, are segmented and display all five major anatomical advances. Body fluids in the coelom act as a hydroskeleton. The one-way gut has specialized regions for grinding and storing food, and the circulatory system is closed.
- **e]** Arthropods (phylum Arthropoda), including insects, crustaceans, spiders, ticks, mites, scorpions, millipedes, and centipedes, are the largest animal phylum. Each type has an exoskeleton; modified, specialized segments; a relatively high metabolic rate and rapid movement compared to other invertebrates; and acute senses.
- **f]** Echinoderms (phylum Echinodermata), including sea stars, brittle stars, sea urchins, sea cucumbers, and sea lilies, were the first to have endoskeletons. Most of the slow-moving, radially symmetrical marine animals have a water vascular system that allows locomotion.

UNDERSTANDING THE FACTS AND CONCEPTS

For Questions 1–5, match each of the descriptions with the appropriate item or items from the following list of terms. Each item in the list may be used once, more than once, or not at all.

- **a]** sponges
- **b]** cnidarians
- **c]** flatworms
- **d]** roundworms
- **e]** none of the above

1 Radial symmetry; true tissues; sessile and motile forms; beginning of coordination by means of electrical signals.

2 Extracellular digestion; parasitic and free-living forms; three tissue layers.

3 Neither radially nor bilaterally symmetrical; intracellular digestion of food molecules; beginning of cell specialization.

4 Bilaterally symmetrical; beginning of cephalization; beginning of organs organized into organ systems.

5 Extracellular digestion; one-way gut; parasitic and free-living forms.

For Questions 6–10, match each of the descriptions with the most appropriate item from the following list of terms. Each item may be used once, more than once, or not at all.

a] flatworms
b] roundworms
c] segmented worms
d] flatworms and roundworms
e] roundworms and segmented worms
f] flatworms, roundworms, and segmented worms

6 Both free-living and parasitic forms.

7 First phylum to exhibit bilateral symmetry, cephalization, a tubular gut, a coelom, and segmentation.

8 One-way gut and either a coelom or a pseudocoelom.

9 Setae, nephridia, and a hydroskeleton.

10 Tapeworms, liver flukes, and the organisms that cause schistosomiasis are members of this phylum.

For Questions 11–15, match each of the descriptions with the appropriate item or items from the following list of terms. Each item in the list may be used once, more than once, or not at all.

a] arthropods
b] echinoderms
c] annelids
d] mollusks
e] none of the above

11 Deuterostomes.

12 Protostomes.

13 Complex brain capable of observational learning; image-forming eyes; unsegmented.

14 Includes clams, crabs, and lobsters.

15 Specialization of body segments; exoskeleton made of chitin; social groups showing division of labor.

For Questions 16–20, pick the phylum in which each of the characteristics listed first appears.

a] flatworms
b] roundworms
c] annelids
d] mollusks
e] echinoderms

16 Bilateral symmetry.

17 Cephalization.

18 Segmentation.

19 True coelom.

20 Internal skeleton.

INTEGRATE AND APPLY WHAT YOU HAVE LEARNED

1 What is a coelom and why is it important?

2 Why is segmentation considered to be an important evolutionary advance?

3 When outbreaks of red tides occur, local or state boards of health prohibit the digging of clams for food. What is it about clams, in particular, that explains the prohibition.

4 Although cephalopod mollusks appear to be the "smartest" of the invertebrates, most zoologists consider the arthropods to be the most successful invertebrates. What characteristics of the arthropods account for their spectacular success?

5 In what ways are echinoderms similar to some of the more primitive invertebrates, in what ways are they similar to the chordates, and in what ways are they unique?

ANALYSIS

1 You would expect the fossil history of the mollusks and arthropods to be more fully documented than that of the nematodes and annelids because:

a] Mollusks and arthropods are more recent, and thus there is less history to document.
b] Some mollusks and arthropods live on land, which is a better site for the development of fossils.
c] Mollusks and arthropods have hard parts that can fossilize; soft-bodied animals are less likely to leave fossil traces.
d] Mollusks and arthropods are the most intelligent and/or the most numerous of the invertebrates.
e] More mollusks and arthropods are alive now or have lived in the past than other animal groups; therefore they have an increased probability of leaving fossil traces.

2 Why was the evolution of bilateral symmetry fundamental to the radiation of animals?

a] Only a bilaterally symmetrical form can have a head end and a tail end.
b] The development of bilateral symmetry enabled animals to develop a specialized nervous system, including an expanded region that functioned as a site for integration of sensory stimuli.
c] Bilateral symmetry enabled the evolution of a one-way, tubular gut.
d] Bilateral symmetry appears to be necessary for the evolution of segmentation.
e] All of the above are correct.

3 What kind of evidence would be the *most* significant in proving that sponges evolved from choanoflagellates?

a] Similarity of DNA sequences in genes for ribosomal RNA.
b] General similarity of external appearance and habitat.
c] Colonial habit of choanoflagellates.
d] Presence of a flagellum in both.
e] Similarity of appearance of choanocytes and choanoflagellates.

4 How many times did segmentation evolve in animals?

a] Once for sure.
b] Not more than twice.
c] No fewer than twice.
d] At least three times—in arthropods, annelids, and mollusks.
e] At least four times—in arthropods, annelids, mollusks, and chordates.

CHAPTER 22

Evolution of Chordates

ANGLERFISH

In biological terms, every organism has two main issues: one, procuring enough energy for day-to-day survival, and two, reproducing and thus passing on its genetic legacy. If you think about it, eating a proper diet and establishing a healthy partnership that might lead to a family are important human issues, too, from both psychological and biological perspectives. These processes are lifelong, and require considerable effort. But compared to a distant cousin like the anglerfish [FIGURE 22.1], we've got it made.

The anglerfish inhabits a remarkably challenging environment: the icy cold, inky black expanses of the deep sea, more than 1000 feet below the ocean surface. Food and mates are not only widely dispersed and in short supply, but the fish exist in a realm where even the tiniest glimpse of sunlight fails to penetrate. Despite all this, the anglerfish's adaptations—which mainly involve luring and clinging—solve the problems of eating and reproducing in the ocean abyss.

FIGURE 22.1
Chordate Characteristics Are Visible in the Anglerfish.

An anglerfish of either sex has an organic light bulb—an organ packed with bioluminescent bacteria—located between its eyes and above its gaping, sharp-toothed, proportionately gigantic mouth. Every so often, as the fish cruises slowly about in the pitch-black ocean depths, a pale blue glow emanates from this "bulb," and it may attract a small passerby, such as a hatchetfish. As the latter investigates the light source, it may enter the wide-open, motionless jaws, and with one snap, the angler has caught a fish.

The same azure glow—possibly assisted by faint but alluring odors—draws a male to a female. This time, however, the attracted party becomes a permanent appendage rather than a meal. At one-twentieth her size, the male is dwarfed by the female, and he becomes very attached to her, literally—he clamps on to her side or belly with specialized, viselike mouthparts, and after a while, his blood system actually fuses with hers. He receives all his nourishment from her; and her eggs join with sperm from his sex organs, probably when hormone surges in her body coordinate the simultaneous release of her eggs and his sperm. In some anglerfish species, the symbiosis is so complete that the male is reduced to nothing more than a pair of testes attached to the female! In FIGURE 22.1, two dwarf males are attached to the much larger translucent female. Talk about keeping one's suitors dangling. . . .

While the anglerfish has fascinating adaptations for life in the deep sea, its glassy body wall reveals other adaptations that are common to all members of the phylum Chordata, our subject in this chapter. Unlike the vast majority of animal species (invertebrates), the **chordates**, members of the phylum **Chordata**, have a rod of cartilage running down their back at some time during their life cycle. Chordates have either a stiff but flexible rod (a *notochord*) or a series of interlocking bones called the backbone that provides internal support for the body and protects a central spinal, or nerve, cord. Chordates with backbones (the great majority) are called **vertebrates** after the sets of interlocking bones, or **vertebrae**, that make up the bony column. In FIGURE 22.1, the anglerfish's backbone is visible as chevron-shaped white lines. Although there are only about 50,000 species of vertebrates compared with well over a million invertebrate species, the vertebrates include the earth's fastest runners, highest fliers, deepest divers, most agile climbers, best problem solvers, and largest and most familiar animals. The vertebrates include the fishes, amphibians, reptiles, birds, and mammals. The primates, the group that includes our own species, *Homo sapiens*, are in the class Mammalia of the subphylum Vertebrata of the phylum Chordata.

MESSAGES

1 Each chordate, at some point during its life, possesses a solid, flexible rod along its back, a dorsal, hollow nerve cord, gill slits (openings from the inside of the throat to the outside), muscle blocks, and a tail.

2 Chordates include some groups that lack vertebrae, but most chordates are vertebrates, including fishes, amphibians, reptiles, birds, and mammals.

3 Evolution of fishes involved the appearance of bone, a skull that protects the brain, jaws that facilitate feeding, fins that aid locomotion, and vertebrae that help support the body axis.

4 Specializations that evolved in amphibians and reptiles allowed vertebrates to invade and occupy land. Adaptations included air-breathing lungs, more efficient circulatory systems, and walking legs.

5 Birds have adaptations that allow full exploitation of terrestrial and arboreal habitats, including feathers and a constant body temperature. Mammals evolved hair, constant body temperature, and milk, and became the most numerous large animals on earth.

This chapter begins with the general features that distinguish chordates from other animals. Then we examine chordates that lack a backbone, small aquatic animals unfamiliar to most people. Next we explore the five familiar groups of vertebrates: fishes, amphibians, reptiles, birds, and mammals. For each group, we look at evolutionary adaptations, just as we did with the strange and interesting feeding and reproductive adaptations of the anglerfish. You'll see that each class of vertebrates possesses characteristics not present in the previous group. Keep in mind, though, that no *living* species of fish is an ancestor of today's fishes, frogs, bats, or other vertebrates. Instead, anglerfish and others on their evolutionary lineage have simply changed less from the common ancestor of all the fishes, frogs, and bats than did the lineages leading to those animals.

In this chapter, we'll not only see more details about the anglerfish and its bizarre life habits down under, we'll also encounter giant-jawed monsters of the deep; fish that walk and breathe air; and flying reptiles. ❑

Characteristics of Chordates

The diverse vertebrates—including many graceful, powerful, and fleet animals of land, sea, and air—evolved along one of the two great pathways of animal evolution, the *deuterostome line* [review FIGURE 21.13]. This path includes echinoderms (sea stars and their relatives) and acorn worms (phylum Hemichordata), marine invertebrates that share some traits with sea stars, and some with chordates. The most primitive chordates include the tunicates, or sea squirts, and fish-shaped animals called lancelets, which are small, inconspicuous, bottom-dwelling sea creatures without vertebrae. These invertebrate chordates, along with the much larger array of chordates that are vertebrates, display bilateral symmetry, cephalization, a coelom, a gut tube with mouth and anus, and body segmentation—traits that first emerged in the invertebrates [FIGURE 22.2].

NOTOCHORDS AND NERVE CORDS

The chordates almost certainly evolved from ancient aquatic invertebrates, but chordates also share a number of new structures. One of these new structures is the **notochord**, a solid, flexible rod that provides internal support and generally runs from the middle of the brain to the tip of the tail at some time during development [FIGURE 22.3]. A second new structure is the dorsal, hollow *nerve cord*, or **spinal cord**, which is a hollow tube of nerve tissue that also runs the length of the animal, just

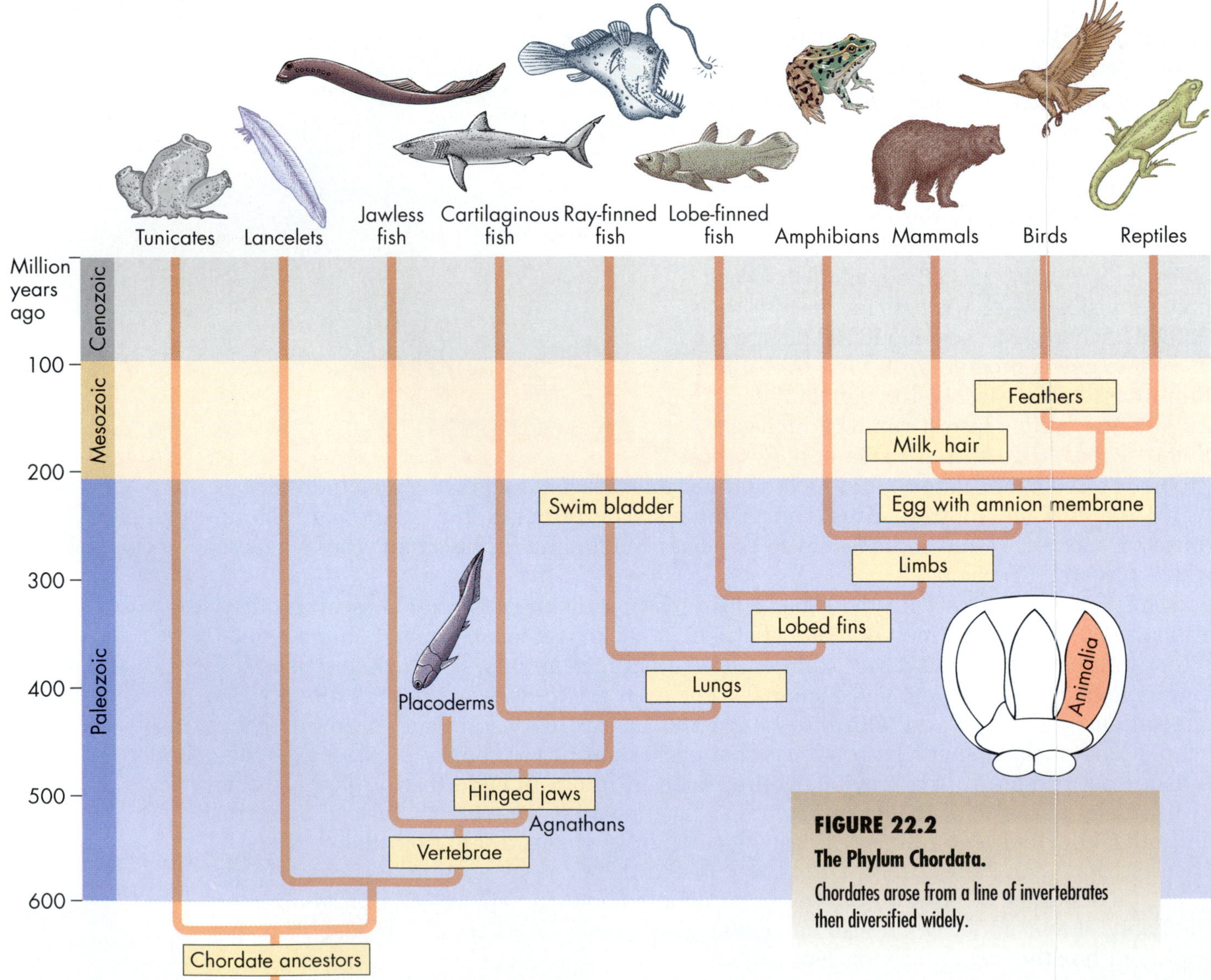

FIGURE 22.2
The Phylum Chordata.
Chordates arose from a line of invertebrates then diversified widely.

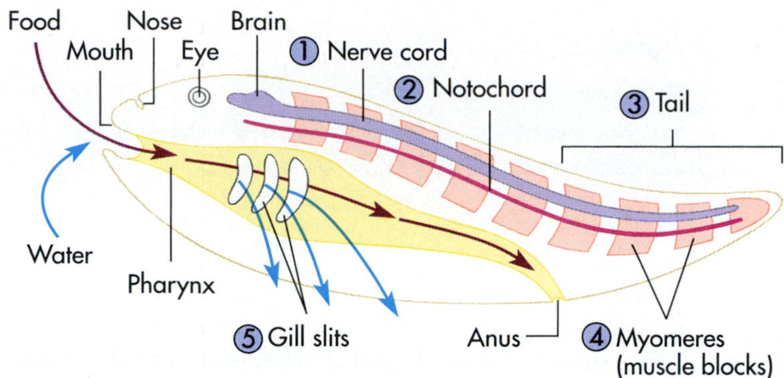

FIGURE 22.3

Generalized Chordate Body Plan.

Chordates show the major animal characteristics that emerged in the invertebrates, but five additional traits as well: the notochord, a dorsal hollow nerve cord, gill slits, muscle blocks (myomeres), and a tail with muscles and notochord behind the anus. In complex chordates, such as our own species, some of the five chordate innovations occur only briefly during early embryonic development, then disappear.

above (dorsal to) the notochord. (Recall from FIGURE 13.15 that as a vertebrate embryo forms, the nerve cord appears after induction by the notochord.) The nerve cord helps to integrate the body's movements and sensations. In most chordates, the nerve cord is present throughout embryonic and adult life.

GILL SLITS

Chordates also have **gill slits**, pairs of openings through the *pharynx*, or throat, the anterior region of the digestive tract. If you still had gill slits, they would be openings from the back of your throat, to the outside of your neck. In tunicates and lancelets, food is filtered through the gill slits. In most reptiles, birds, and mammals, gill slits are either vestigial structures found only in the embryo, or they develop into other structures. For example, a vestige of one of the gill slits is your eustachian tube, which runs from the back of your throat to your middle ear. This tube allows you to equalize the pressure in your ears with that of the environment. This is particularly important when you're in an airplane or diving in deep water.

MYOMERES

Flanking the notochord and nerve cord is a fourth type of structure and major chordate trait, muscle blocks called **myomeres**. Myomeres are actually modified body segments and the major signs of segmentation in chordates. When you eat fish, you can easily see these stacked muscle layers in the "flaky" white or pink meat.

TAILS

The notochord, nerve cord, and myomeres extend into the **tail**, the fifth chordate characteristic, a structure that protrudes beyond the anus at some point in development. Most chordates have a tail throughout life, but in some chordates, including humans, the tail appears only briefly in the embryo.

IMPORTANCE OF CHORDATE CHARACTERISTICS

The new chordate characteristics—notochord, spinal cord, gill slits, tail, and muscle blocks—had dramatic evolutionary implications [FIGURE 22.4]. The physical support of a notochord, and later of a vertebral column, allowed chordates to grow larger and larger—as large, in fact, as dinosaurs and whales. The dorsal, hollow nerve cord allowed centralized nerve coordination and control and allowed the evolution of acute senses, a wide range of behaviors, and greater intelligence than in most invertebrates. The gill slits could be used for filter feeding or for exchange of respiratory gases. The tail in its many manifestations became a major organ of locomotion (especially in fish), balance (in hopping animals), and even communication (in many birds and mammals).

Tunicates

The larvae of **tunicates**, or sea squirts, show all five chordate traits: notochord, nerve cord, gill slits, muscle blocks, and tail. Only 1 cm (⅓ in) long, tunicate larvae look like miniature tadpoles and move about by thrashing their long tails [FIGURE 22.5A]. They eventually settle down—head first—and undergo a dramatic metamorphosis. The tail, with its notochord and nerve cord, disappears, and a stationary adult develops and remains attached to a submerged rock or harbor piling [FIGURE 22.5C]. The adult has an outer *tunic*, an envelope of carbohydrate and protein, enclosing a large basket-shaped pharynx perforated by hundreds of gill slits [FIGURE 22.5B]. Cilia line the gill slits, and their beating draws water in through the mouth, or *incurrent siphon*, and sweeps it out again through the *excurrent siphon*. Food particles suspended in the feeding current are passed to the stomach. If disturbed, a tunicate will often squirt water forcibly from its siphon, and indeed the common name for tunicate is "sea squirt."

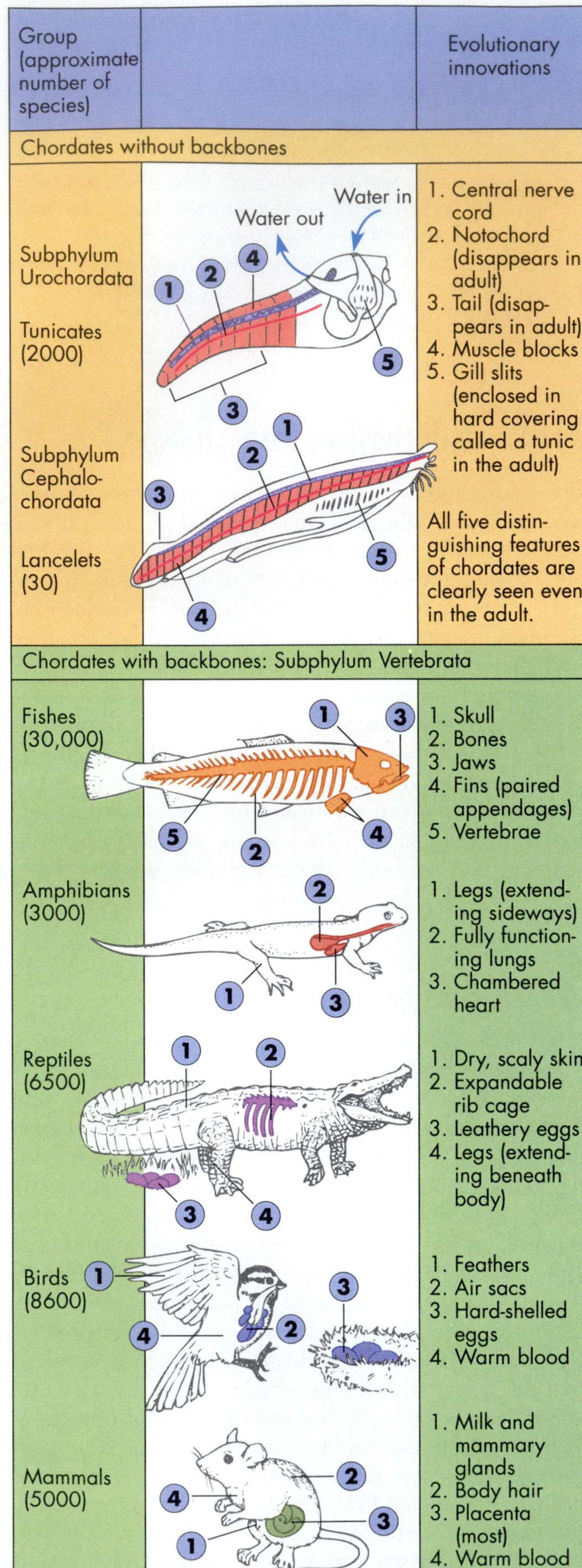

FIGURE 22.4

Major Innovations in the Chordates.

Each group of chordates shows a slightly different set of evolutionarily new characteristics, each modifying the adaptations of earlier species.

Sea squirts are small and seldom seen, but they have major significance for our study of chordates: The ancestor of fishes probably evolved from a tunicatelike larva that failed to grow up. If that ancient larva became sexually mature before the rest of its body metamorphosed into the baglike adult form, then the adult would be unnecessary in evolutionary terms.

Lancelets

Lancelets are small, streamlined marine animals about 5 cm (2 in) long that live half-buried—tail first—in the sandy bottoms of shallow saltwater bays and inlets [FIGURE 22.5D]. All the major chordate traits are visible throughout embryonic, larval, and adult life—a contrast with tunicate development. Like tunicates, however, lancelets are filter feeders; they have cilialike structures around their mouths and sticky, food-trapping mucous regions in the pharynx. If food becomes scarce, lancelets can pull up anchor and move to more fertile grounds. The ability to move as adults gives lancelets access to food supplies that are unavailable to the stationary tunicates.

➤ CONCEPT CHALLENGE

A zoologist jests that a fish is just a tunicate who never grew up. Can you present a hypothesis for the evolution of fishes from tunicatelike ancestors on the basis of natural selection, including concepts like random mutations, differential survival and reproduction, and interactions with the environment? Could fossil evidence support any of your predictions, and if so, which?

Fishes

Emerging about 550 million years ago, at the dawn of the Paleozoic era, the first fishes were streamlined filter feeders that lived in the muddy bottoms of ancient seas. About 30 cm (1 ft) long, they had fixed, circular, jawless

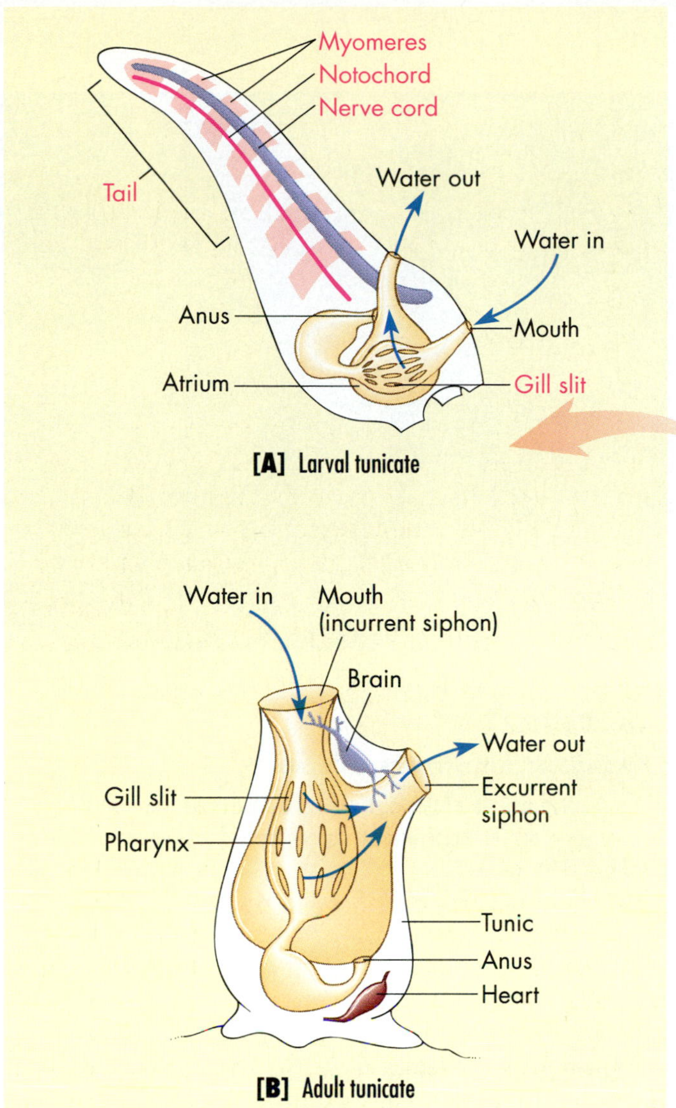

[A] Larval tunicate

[B] Adult tunicate

FIGURE 22.5

Chordates without Vertebrae: Tunicates and Lancelets.

[A] Tunicate larvae, with their streamlined body shape, resemble the generalized chordate [see FIGURE 22.3], but **[B]** adult tunicates are stationary filter feeders enclosed in a baglike tunic. **[C]** This photo shows a large group of yellow bubble tunicates. **[D]** Lancelets such as *Amphioxus* superficially seem fishlike, but they rest vertically with transparent bodies planted tail first in the sand with only the anterior end exposed for capturing tiny food morsels.

[C] Adult tunicate

[D] Adult lancelet

mouths that could filter sediments. These fishes are called *agnathans*, meaning "jawless" [see FIGURE 22.2]. Jawless fishes retained the chordate notochord, which provided internal support and induced the nerve cord during development, as well as gill slits, which facilitated feeding. New characteristics, however, were the presence of bony plates that functioned as a skull and protected the brain, and muscles, rather than tiny cilia, that powerfully drew water and the food it contained into the mouth. Muscle-powered gill slits allowed the agnathans to consume greater quantities of food than their earlier cousins and to grow 6 to 30 times larger. In addition, the gill openings were supported by gill bars made of *bone*, a meshwork of cells and proteins hardened by calcium, phosphate, and other minerals. Bony plates beneath the skin of these early fishes served as armor that probably protected them against dangerous invertebrates, such as giant sea scorpions.

These ancient jawless fishes gave rise to all modern fishes, but they also gave rise to the amphibians, reptiles, birds, and mammals. Therefore, the animals we generally lump together as fishes do not constitute *all* the descendants of a single branch on the family tree of life. Rather, there are four main lines of fish evolution, as FIGURE 22.2 depicts. These include (1) the *jawless fishes* (such as lampreys); (2) the *cartilaginous fishes* (such as sharks); (3) the *ray-finned fishes* (such as trout and goldfish); and (4) the *lobe-finned fishes* (such as lungfish and coelacanths). Frequent referral to FIGURE 22.2 will help you visualize these various fishes and their evolutionary relationships as you read this section.

Modern jawless fishes have lost the bony plates of their ancestors, and instead have very flexible internal skeletons of cartilage. The jawless lamprey is a parasite that attaches to its prey by suction, rasps through the victim's body wall with a sharp tongue, digests its tissues, and then consumes the liquified remains. At times, lampreys have been serious pests of commercially fished species in Lake Michigan.

FISH EVOLUTION

About 425 million years ago, the ancient agnathans gave rise to a second class of fishes, the *placoderms* (early jawed fishes) [see FIGURE 22.2]. The placoderms had three basic new characteristics that were so useful they appeared in nearly all the vertebrates that followed:

1. The placoderms had **hinged jaws**, derived from the bones (gill bars) that support gills in the agnathans [FIGURE 22.6]. Jaws allowed the placoderms to consume large chunks of food—kelp fronds, clams, other fishes—instead of filtering small particles. As a result, some became huge; a person can stand upright in the fossilized jaws of one placoderm.

2. The placoderms had the first **vertebrae**. They had a series of separate bones that fused with the notochord and arched over the spinal cord. The resulting vertebral column provided a site of attachment for muscles in the body wall, anchoring them as the body bent to and fro and allowing the muscles to grow larger and propel the fish more powerfully. The backbone also encased and protected the delicate spinal cord.

3. Placoderms had the first paired, lateral **fins**, appendages that provided more control over swimming direction, speed, and depth.

FIGURE 22.7

Cartilaginous Fishes: Hunters of the Sea.

A skeleton of cartilage is lightweight and flexible and makes this spotted eagle manta ray of the Caribbean resemble a predatory bird flapping beneath the waves.

MODERN FISHES

Modern fishes fall into two main groups, the *cartilaginous fishes* and the *bony fishes.*

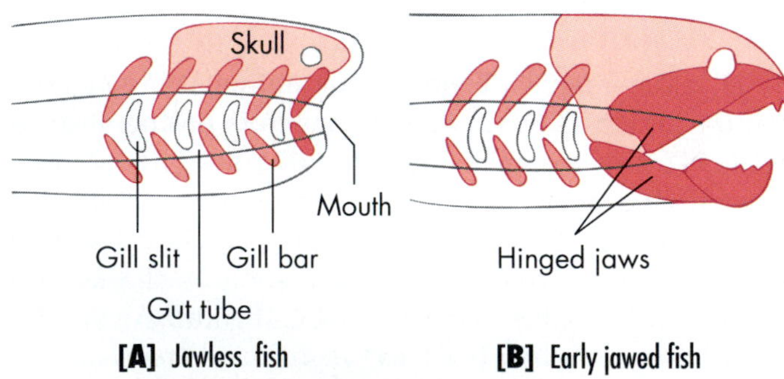

FIGURE 22.6

Evolution of the Jaw.

How the gill support bones (gill bars) of earlier jawless fishes may have evolved into the powerful jaws of the first jawed fishes, the placoderms. Jaws allowed placoderms to function as efficient grazers and hunters that preyed on kelp, fishes, and invertebrate neighbors. Their descendants include the sharks and the modern bony fishes.

Cartilaginous Fishes The placoderms were heavily armored with bony plates beneath the skin, but in their modern descendants, the **cartilaginous fishes (Chondrichthyes)**, the skull, vertebrae, and the rest of the skeleton are made entirely of cartilage, a matrix of fibrous proteins. This class includes sharks, skates, and rays. Cartilage is lighter than bone, and affords these predatory fish the speed they use to catch prey. In skates and rays, large fins provide lift, much like graceful underwater wings [FIGURE 22.7], while in sharks, stiff fins knife through the water, helping these large and common oceanic hunters to maneuver easily toward their victims.

Bony Fishes The other group derived from the ancient jawless fishes is the **bony fishes (Osteichthyes)**, which include anglerfish, bass, bluegill, minnows, trout, tuna, and virtually all of today's familiar fresh- and saltwater fishes. There are two major groups of bony fishes. The **lobe-finned fishes**, which arose about 400 million years ago, had large, muscular, lobed fins that allowed them to "walk" across the bottoms of shallow bays, and they had **lungs**, or air sacs; that allowed them to breathe air. When water levels fell, these adaptations allowed them to survive by breathing air and migrating overland to other pools or bays. Present-day lobe-finned fishes

[A]

[B]

FIGURE 22.8

Bony, Jawed Fishes: The Most Diverse Vertebrates.

The bony, jawed fishes have radiated into the largest vertebrate group, with 30,000 living species. Lobe-finned fishes had muscular appendages supporting their fins, as well as lungs that allowed air breathing during migration across land from pool to pool. **[A]** Coelacanths today inhabit the deep waters off the Comoro Islands east of India, but there may be fewer than 300 of these ancient fish left. Ray-finned fishes, including teleosts, have delicate bones in their fins, as well as a swim bladder. The modern teleosts include brilliantly colored species, for example, vertical swimmers, such as **[B]** the bizarrely lobed leafy sea dragon, a sea horse, which seems to disappear when grazing in a clump of seaweed.

include the **lungfish** and the **coelacanths** [FIGURE 22.8A]. An extinct group of lobe-finned fishes, the *rhipidistians*, were probably the first vertebrates to crawl out on land and live for extended periods.

A second group of bony, jawed fishes are the **ray-finned fishes**, including **teleosts** such as anglerfish, trout, and zebra fish. Teleosts lost their ancestors' fleshy fins, but developed much more versatile spiny fins with webs of skin over delicate rays of bone. They also lost the lungs, but developed a swim bladder, an internal balloon below the backbone that can change volume and allow the animal to adjust its swimming depth. These fishes radiated early on into a huge group. In fact, the Devonian period (about 345 to 395 million years ago) is called the Age of Fishes because the teleosts and their relatives so dominated the earth's lakes and seas.

Even today, with about 30,000 recorded species, teleosts are the largest group of vertebrates, and they owe much of their success to specializations that evolved as the group radiated, adapting them to various diets and habitats [FIGURE 22.8B]. Most fish, however, are relegated by their anatomy to life in water. The great landmasses were to be dominated by other vertebrates—descendants of the early fishes. Nevertheless, fishes evolved important new adaptations, including the skull, bone, hinged jaws, paired fins, and vertebrae [review FIGURE 22.4].

Amphibians

Fossils show that during the Devonian, about 345 to 395 million years ago, the vertebrate lineage that included air-breathing lobe-finned fishes produced the **amphibians**—vertebrates that can live both on land and in water (*amphi-* means "both"; *bios* means "life"). These ancient animals overcame the formidable problems of life on land (including more efficient means of walking, breathing, and staying moist) beginning around 375 million years ago. Modern amphibians include the approximately 3000 species of frogs, toads, salamanders, and wormlike apodes.

Early amphibians had front and hind **legs** containing strong bones and powerful muscles. Extending sideways from the body, the limbs could support the animal's weight far better than lobed fins; and with two pairs of legs, the animal could move about on land—albeit slowly and clumsily—for greater distances, even without water's buoyant support. The bones in an amphibian's limbs, and indeed, the bones in our own limbs, share the characteristic pattern we see in 380-million-year-old lobe-finned fishes [FIGURE 22.9].

Laborious walking, however, required a great deal of energy from food, and large quantities of oxygen were needed for aerobic respiration. Fossil evidence reveals that early amphibians had simple but fully functioning lungs, or air sacs, that provided a site for gas exchange. Air was probably pumped in by swallowing movements, much as it is in modern frogs and toads [FIGURE 22.10A]. Moreover, the evolution of specialized chambers in the heart and other modifications of the circulatory system tended to separate oxygenated blood en route to the body

FIGURE 22.9

The Transition to Land.

[A] Ancient lobe-finned fishes, like today's lungfish, would have risen to the surface to gulp air when the stagnant water became depleted of oxygen. From time to time they may have ventured into shallower water and to land in search of invertebrate food. Natural selection would have favored alleles that modified the limbs, over millions of years, into those of early amphibians **[B]**.

[A] Ancient lobe finned fish

[B] Early amphibian

tissues from deoxygenated blood bound for the lungs—a separation that became complete in crocodilians, mammals, and birds [see FIGURE 25.6]. Finally, an amphibian's smooth, moist skin could absorb about half the oxygen and release most of the carbon dioxide required by these animals.

These characteristics provided sufficient oxygen but did not solve the problem of water loss and drying out. Since the amphibian skin remains moist, and this facilitates gas exchange across the skin as a supplement to the lungs, the animals were restricted to life near the water's edge. In addition, amphibians lay eggs with a clear, jelly-like coating that must also stay moist, or the embryos die before the fishlike tadpoles emerge. Because their eggs and larvae require water, amphibians must return to the water to reproduce, and in this sense, they are analogous to the first land plants whose sperm required standing water to swim from one plant organ to another.

Today, frogs, toads, salamanders, and a few relatives are the only remaining members of the class Amphibia. They display the new evolutionary characteristics of the amphibian lineage: (1) legs, (2) simple but fully functioning lungs, and (3) partial separation of oxygenated and deoxygenated blood in a three-chambered heart. These generally small vertebrates are very common in freshwater environments, and show a range of interesting specializations. In some, the moist skin has become brightly colored, attracting potential mates and warning would-be predators of their poisonous glands [see FIGURE 22.10B]. In others, the skin is dry and adapted to life in dry habitats. The nervous system of many amphibians, protected by the vertebral column and bony skull, has become well developed and accompanied by keen senses of sight and hearing. In addition, rapid-fire nervous reactions enable some of these animals to catch flies by flipping out their long tongues. Amphibian limbs are often specialized, with webbed feet that allow efficient swimming and with thick muscles that enhance hopping or walking on land.

Unfortunately, amphibian populations all over the world apparently began declining about 20 years ago. Some scientists think that acid rain or global warming may be contributing to the animals' slow demise, because amphibians have porous skin that is especially susceptible to the quality and quantity of available water. BOX 22.1 explores this puzzling disappearance.

➤ CONCEPT CHALLENGE

Summarize the new evolutionary features that appeared in the amphibians. What aspects of amphibian biology inhibit them from being the dominant land vertebrates?

[A] Play-actor frogs

[B] Salamander

FIGURE 22.10

Modern Amphibians.

The brilliant gold and red pigments in the skin of **[A]** the tropical play-actor frog (*Dendrobates histrionicus*) and **[B]** the Chinese salamander (*Tylototriton verrucosus*) warn would-be predators of the amphibians' poison skin glands. **[C]** Some amphibians, such as the caecilian, burrow in the soft, moist soils of tropical rain forests in search of insects and have lost all trace of limbs.

[C] Caecilian amphibian without limbs

Reptiles

As the amphibians—which evolved far earlier—crawled about in profusion in the Carboniferous swamps, other, quite different, four-legged land animals—the earliest **reptiles**—were walking about among them [FIGURE 22.11]. Like the early seed plants, these reptilian creatures evolved important traits for life on land that freed them from dependence on moist environments.

REPTILIAN TRAITS AND TERRESTRIAL LIFE

The early reptiles, which superficially resembled crocodiles, had four innovations for terrestrial life [review FIGURE 22.4]:

1. Ancestral reptiles had dry skin with scales that provided a barrier to evaporation and sealed in body moisture. This eliminated the skin surface as a major site for gas exchange.

2. Respiratory modifications in early reptiles included lungs with a larger surface area for gas exchange than in amphibians, an expandable rib cage that could draw in large quantities of air like a bellows. Modifications in the heart and circulatory system separated oxygenated and deoxygenated blood more fully than in the amphibians, allowing more oxygen to reach body tissues.

3. Early reptiles displayed changes that eliminated reliance on open water for reproduction. Males had copulatory organs that could deliver sperm directly into the female's body (internal fertilization) rather than into the surrounding water (external fertilization). In addition, females produced so-called *amniote eggs*. These eggs essentially encased the developing embryo in a pool of fluid (the *amniotic sac*), provided a source of food (the *yolk*), and surrounded both embryo and food with membranes and a leathery shell that prevented crushing and desiccation in open air.

4. The legs of the early reptiles extended directly beneath the body rather than out to the side (as amphibians' legs did then and continue to do today); this arrangement provided better support and made walking and running easier [FIGURE 22.12].

Birds
Crocodiles
Lizards
Snakes
Million years ago
Cenozoic
Stegosaurs
Sauropods
Tyrannosaurus
Pterosaurs
100
Archaeopteryx
Mesozoic
200
Thecodonts
Paleozoic
300

FIGURE 22.11

Reptiles and Their Descendants, the Birds and Mammals.

Fossil evidence reveals that dinosaurs and birds arose from thecodonts and mammals from therapsids, both descendants of the early reptiles that walked the earth some 300 million years ago.

[A] Ancient amphibian (*Ichthyostega*)

[B] Crocodile

[C] Dinosaur

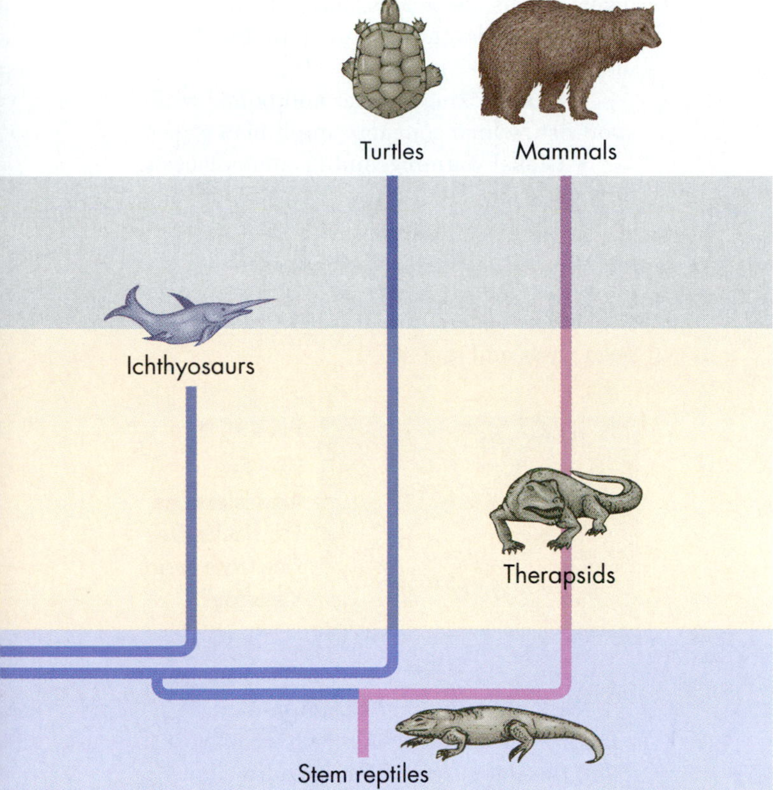

[A]

[B]

FIGURE 22.13

Today's Reptiles.

[A] This giant land tortoise lives on Isla Isabela in the Galápagos archipelago, and like all turtles, hauls about a protective shell. [B] The Komodo dragon of Indonesia can be up to 3 m (10 ft) long and weigh 115 kg (255 lb). The lizard has such a huge appetite that it will readily kill and devour other Komodo dragons and consume the dead bodies of nearly every other species of animal.

The reptiles that followed built on the legacy of these new characteristics, and in fact, were pivotal to all subsequent vertebrate history [see FIGURE 22.11]. They gave rise not only to the modern reptiles—the crocodiles, turtles, lizards, tortoises, snakes, and the tuatara—but to the birds and mammals as well. Descendants of the stem reptiles became the *thecodonts*, small lizards that ran on two legs and gave rise to the giant reptiles collectively known as the **dinosaurs**, and the *therapsids*, small, heavyset, fiercely toothed animals that led to the mammals.

The warm Cretaceous was the heyday of dinosaurs ("terrible lizards") like the 15-m-long (50-ft-long) meat-eating *Tyrannosaurus* and the 24-m-long (80-foot-long) plant-eating *Apatosaurus* (formerly *Brontosaurus*). The grand radiation of reptiles into forms that swam, flew, and lumbered across the earth lasted nearly 150 million years, and inspired the common name for the entire Mesozoic era: the Age of Reptiles. Most of the great and diverse reptiles died out, however, in a massive extinction event at the end of the Cretaceous.

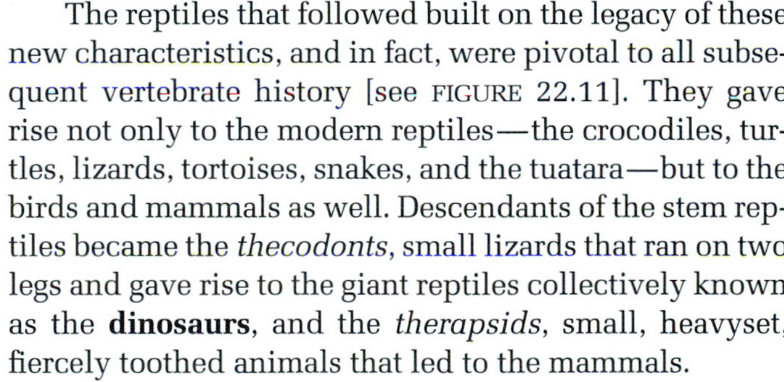

FIGURE 22.12

Reptiles Evolved Changes in Stance.

[A] Ancient amphibians had legs directed out to the sides, while reptiles, such as the crocodile [B] and the dinosaurs [C], have legs more directly underneath the body.

A few types of small reptiles weighing less than 20 kg (44 lb) survived the extinctions and radiated once again during the Cenozoic, the current geological era, beginning about 65 million years ago. Today, there are only about 6500 species in the reptile class—animals that continue to show the characteristics of dry skin with scales; expandable rib cage, improved separation of oxygenated and unoxygenated blood supplies, copulatory organs, amniote eggs with leathery shells, and usually, legs that extend directly beneath the body. Alligators, caimans, and crocodiles are all streamlined carnivores that inhabit warm climates. The tortoises and turtles have a tough and unique protective structure, the shell, or carapace [FIGURE 22.13A]. The lizards, snakes, and iguanas are elongated reptiles that inhabit wet, dry, or hot environments and sometimes reach great size. The heaviest lizards today are the Komodo dragons of Indonesia,

box 22.1
Focus on the Environment

Amphibians: Missing in Action

Mountain yellow-legged frog. Spade foot toad. Western spotted frog. Tiger salamander. Glass frog. Stomach brooding frog. Goliath frog. Golden toad. These curious-sounding species are real amphibians, and they seem to be vanishing from California, Colorado, Costa Rica, Cameroon, Haiti, and other places around the world [FIGURE 1]. A congress of experts convened recently in Corvallis, Oregon, to discuss possible explanations, and they compared notes on worldwide amphibian populations—the curious as well as the commonplace. Many came to a distressing conclusion: We are not just losing a few rare and particularly delicate animals. Instead, of the world's 3000 to 4000 amphibian species, all but a handful seem to be fading away. A group of dissenters from the University of Georgia's Savannah River Ecology Laboratory presented evidence that in their area, amphibian populations fluctuate wildly. This, they say, suggests that perhaps many amphibians are in a normal downswing right now. Nevertheless, most members of the international group of herpetologists (zoologists who study reptiles and amphibians) who gathered in Oregon considered the decline both real and marked. When they compared notes, they failed to find a single factor behind the worrisome losses. Instead, they suspected that a number of environmental insults may be playing a role, including:

- Destruction of the amphibians' natural habitat, mainly due to the harvesting of timber and the opening of new agricultural land.
- Acid rain, which alters the pH of freshwater lakes and streams and can harm amphibian eggs and larvae [see CHAPTER 1].
- Air, water, and soil pollution. (Atmospheric gases and toxic materials, such as heavy metals, pesticides, and industrial wastes, can pass through an amphibian's moist skin.)
- The stocking of lakes and ponds with sport fish, which consume amphibian eggs.
- Global warming, and in some places, unusually dry conditions.
- Thinning of the ozone layer and increased penetration of ultraviolet light.
- Overhunting, especially in tropical areas, for the exportation of frog legs to European chefs and diners.

FIGURE 1
Why Are Amphibians Like This Haitian Tree Frog (*Hyla vasta*) Vanishing?

With all their theories, amphibian experts have been unable to reach a consensus about the actual causes for dwindling amphibian populations. They did agree on one thing, however: A worldwide pattern of vanishing amphibians would be an indicator of serious global environmental degradation. As we saw in this chapter, amphibians were the earliest land vertebrates, and they have withstood innumerable climatic shifts and catastrophes. Population losses ranging over this entire taxonomic group could well portend continued losses throughout all the kingdoms of life.

which can weigh up to 115 kg (255 lb) [FIGURE 22.13B], and the longest snakes are the pythons of the same region, which can grow to 8 m (27 ft).

Many biologists no longer consider the reptiles as a formal taxonomic group because reptiles do not constitute all the animals on a single branch of the tree of life; the branch that includes the reptiles also includes the birds. Alternatively, some zoologists think of feathered and nonfeathered reptiles.

Birds

Except for domesticated animals such as dogs, sheep, and cows, the most common chordates we see daily are **birds**. Curiously, some biologists have recently suggested that, from a zoological perspective, birds are actually a type of dinosaur. The first winged vertebrates—the giant, soaring pterosaurs [FIGURE 22.11]—might seem like logical ancestors to the birds, but pterosaurs died out long before birds evolved. Instead, small, two-legged, lizardlike thecodonts appear to be the real forerunners of the birds, and important fossil evidence supports this possibility. Six skeletons of the oldest bird, the crow-sized *Archaeopteryx*, have been found at different sites in rocks dated back to the Upper Jurassic period (150 million years ago). The fossil imprints suggest that this animal was a true intermediate: It had scaly skin, curving claws, a long, jointed tail, and sharp teeth like a reptile, but it had feathered forelimbs and a tail like a bird.

Most avian evolutionary adaptations prepared the animals for efficient flight [see FIGURE 22.4]. **Feathers** are exceedingly lightweight structures made of dead cells containing the protein keratin [see CHAPTER 2]. Birds also have lightweight, hollow bones and a breastbone, or keel, enlarged into a blade-shaped anchor for the powerful pectoral muscles that raise and lower the wings. The legs are reduced to skin, bone, and tendons and can be folded up like an airplane's landing gear, reducing drag during flight.

Flight is a strenuous activity that requires plenty of oxygen for the aerobic respiration of muscle and other tissues, and birds have a third set of modifications that ensures an adequate oxygen supply. First, birds are **homeothermic** (meaning "constant-temperature"). They maintain a constant internal temperature despite environmental changes; this steady temperature helps to maintain a steady production of ATP energy during cellular respiration, which in turn fuels the activities of flight and leg muscles. In fact, the earliest feathers may have been more useful as insulation than as aids to flight. Most fish, amphibians, and reptiles have body temperatures that vary, and hence they tend to be active in warm environments but sluggish in cold ones. (See CHAPTER 24 for further discussion of warm- and cold-bloodedness.) Second, birds have a series of connected lungs and **air sacs** that exchange oxygen and carbon dioxide in an efficient one-way flow. And third, birds have a **four-chambered heart** that completely separates oxygenated and deoxygenated blood so only the former reaches body tissues.

Finally, in addition to the evolutionary features of feathers, constant body temperature, and air sacs, birds have **hard-shelled**, amniote **eggs**, which free them from dependence on open water for reproduction.

Avian innovations were so successful that in the last few million years of the Cenozoic, birds radiated into a highly diverse class with more than 8600 species specialized for distinct modes of life [FIGURE 22.14].

[A]

[B]

FIGURE 22.14
Modern Birds.
[A] The smallest bird, the bee hummingbird (*Mellisuga helenae*), can easily perch on a pencil eraser. **[B]** The largest flying bird, the Andean condor, glides in the mountains of Peru.

➤ CONCEPT CHALLENGE

Birds are probably the most common vertebrates we see in our daily lives. What factors contribute to this visibility?

Mammals

While mammals arose from the fierce, heavyset reptiles called therapsids, they developed their own distinct body traits at least 180 million years ago. The very early mammals resembled shrews, and probably scurried around the great dinosaurs' feet until the end of the Cretaceous. These mammals survived the mass extinctions of that time and radiated into 5000 modern species that give the current geological era, the Cenozoic, the name Age of Mammals [FIGURE 22.15].

CHARACTERISTICS OF MAMMALS

Like birds, **mammals** (Latin *mamma*, "breast") are warm-blooded and have a four-chambered heart, but these two traits probably evolved independently in the two groups. Mammals have two unique traits as well: (1) milk and (2) body hair or fur [see FIGURE 22.4]. A fluid rich in fats and proteins, **milk** is produced in special glands called **mammary glands** and is used to nourish newborns. Many mammals have dense carpets of hair called **fur** covering the entire body, while others, such as certain monkeys and humans, have sparse body hair, and a few, including the whales and porpoises, have virtually no hair at all and rely instead on thick layers of fat called *blubber* for insulation. Significant changes also occurred in the skull.

The majority of mammals also have a unique reproductive structure, the *placenta* [see FIGURE 14.9], which supports the growth of the embryo to a fairly complete stage of development before birth. Placental mammals descended from one branch of the earliest mammals. Two other branches led to nonplacental mammals with different ways of harboring embryos. The **monotremes**, which include only the duck-billed platypuses and spiny anteaters of Australia and New Guinea [see FIGURES 18.1 and 22.16A], lay leathery eggs and warm them until the

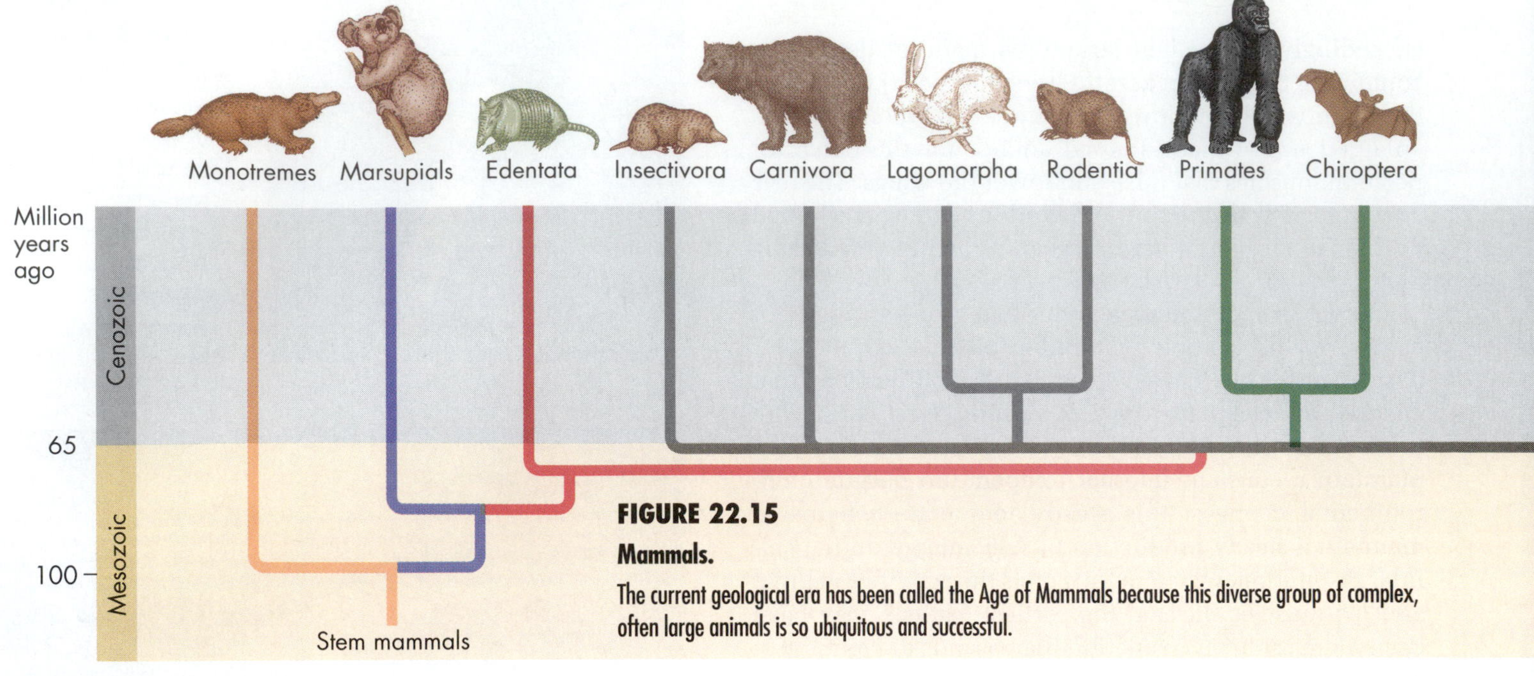

FIGURE 22.15

Mammals.

The current geological era has been called the Age of Mammals because this diverse group of complex, often large animals is so ubiquitous and successful.

young hatch. The females then nourish the young with milk. The **marsupials**, including kangaroos, opossums, koala bears, wombats, and dozens of other animals from Australia and the Americas, give birth to immature live young no bigger than a kidney bean. When the newborn emerges from the birth canal, it crawls upward into an elastic pouch of skin on the mother's ventral surface (the *marsupium*), attaches to a teat, starts to consume milk, and continues to develop inside the pouch for several months.

With their body hair and successful reproductive strategies based on the evolution of milk and the placenta, mammals can be compared to the flowering plants, which, with the evolution of flowers as reproductive organs, underwent a parallel radiation during the Cenozoic. Mammals radiated into a large and diverse class that can live in more environments and with more life-styles than any other class of animals except perhaps the birds.

FIGURE 22.16

A Collection of Modern Mammals.

Mammals are remarkably diverse in body size and appearance. **[A]** The echidnan (spiny anteater) is an egg-laying monotreme that nourishes its young with milk. **[B]** Bats are mostly beneficial insect eaters and plant pollinators. **[C]** A North American cougar spots a snowshoe hare. **[D]** The Florida manatee is a mild-mannered, herbivorous "sea cow" that can weigh 1 ton.

Temperature Regulation Various specializations of structure and behavior underlie mammalian success. Mammals can keep a constant body temperature that is warmer than the ambient air on cool days and cooler than the ambient air on hot days. This is partly because of their homeothermic metabolism, partly because of their hair or blubber, and partly because of behaviors such as hibernating and migrating.

[A]

[B]

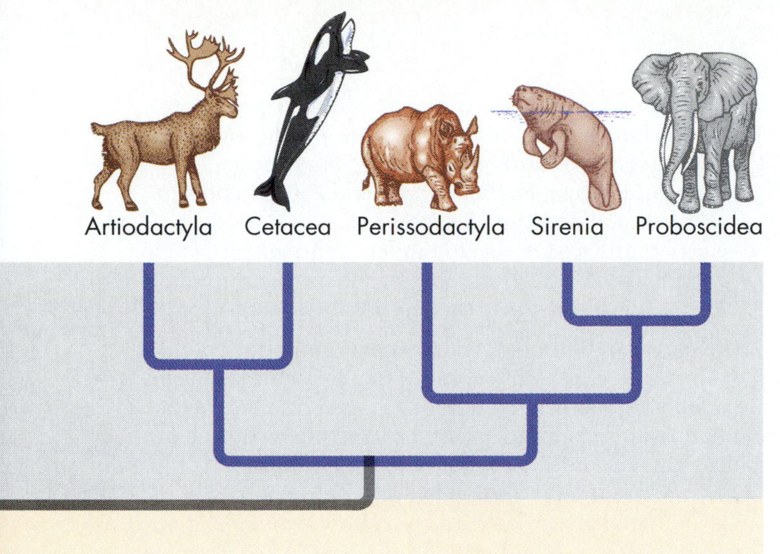

Specialized Limbs and Teeth Mammals display a range of limb types, with bones elongated, shortened, or broadened, depending on the animal's particular locomotor or food-gathering needs. In bats, the forelimb bones are light and strong, and the finger bones are greatly elongated and widely spread and support the flight membranes [see FIGURE 22.16B]. In moles and other digging mammals, the forelimbs are short and powerful, with oversized claws, while antelope forelimbs have strong, slender bones that allow swift running. Mammalian teeth are modified into chisels that gnaw wood (rats and beavers); flat molars that grind grain (cows); and sharp points that tear flesh (bears and tigers). (Humans have all three types of teeth.)

[C]

[D]

Parental Care While animals of many types care for their young, parental care is especially common and extended in mammals. Mammals not only suckle their young for days or months, depending on the species, but usually guard the new generation fiercely and teach them survival skills. The extended care of young is especially pronounced in *primates* (monkeys, apes, and humans), as we will see in CHAPTER 23.

Highly Developed Nervous Systems and Senses The keen senses of smell and hearing in most mammals, and of sight in a few, are instrumental in helping the animals find food, interact with mates and young, and avoid danger, just as these traits help birds, reptiles, and other animals, but to differing degrees. Such sensory information is integrated and often stored in memory, enabling the animals to learn from experience and to minimize the effort they expend to survive. The brains of primates and *cetaceans* (dolphins, porpoises, and whales) represent the highest level of nervous system development among all animals.

Connections

The evolutionary distance is great from a tunicate to an anglerfish to a tiger, yet each is a member of the same phylum, Chordata, and each type arose from the same marine ancestors that lived more than 500 million years ago. Innovations, radiations, and specializations within the vertebrates led to the emergence of amphibians, reptiles, birds, and finally mammals, including the biggest, fastest-running, and most intelligent animals ever to live. But that should not overshadow the continued coexistence of millions of invertebrates, which may not have all the mammals' complex physiological systems, but which have their own successful adaptations to life in modern environments.

It is interesting that only one mammal, *Homo sapiens*, is in a position to study the emergence, extinction, and adaptation of fellow living things. The enormous brain capacity and concomitant cultural development of our species were the natural outcomes of that evolution. CHAPTER 23 explores our species' biological origin in more detail.

KEY TERMS

amphibian, 529	**leg,** 529
body hair, 535	**lobe-finned fishes,** 528
blubber, 535	**lung,** 528
Chondrichthyes, 528	**mammal,** 535
chordate, 523	**milk,** 535
coelacanth, 529	**mammary gland,** 535
dinosaur, 533	**marsupial,** 536
feather, 534	**nerve cord,** 524
fin, 528	**Osteichthyes,** 528
fur, 535	**ray-finned fish,** 529
gill slit, 525	**reptile,** 531
hinged jaw, 528	**tail,** 525
homeothermic, 535	**vertebra,** 528

HIGHLIGHTS IN REVIEW

1 Each chordate, at some point during its life, possesses a solid, flexible rod along its back, a dorsal, hollow nerve cord, gill slits (openings from the inside of the throat to the outside), muscle blocks, and a tail.

2 Chordates include some groups that lack vertebrae, but most chordates are vertebrates, including fishes, amphibians, reptiles, birds, and mammals.

- **a]** Tunicates, or sea squirts, are chordates without vertebrae having a tadpolelike larva displaying the five chordate traits. When the larva metamorphoses into an adult, the notochord disappears.
- **b]** In the small chordates without vertebrae called lancelets, adults show all chordate traits.

3 Evolution of fishes involved the appearance of bone, a skull that protects the brain, jaws that facilitate feeding, fins that aid locomotion, and vertebrae that help support the body axis.

- **a]** Fishes occupy more than one branch on the family tree of life. Early jawless fishes, or agnathans, gave rise to modern jawless fishes, such as the lamprey; early jawed fishes, or placoderms, now extinct; cartilaginous fishes, such as sharks, skates, and rays; ray-finned fishes, such as goldfish; and lobe-finned fishes, such as lungfish and coelacanths.

4 Specializations that evolved in amphibians and reptiles allowed vertebrates to invade and occupy land. Adaptations included air-breathing lungs, more efficient circulatory systems, and walking legs.

- **a]** The amphibians were the first vertebrates to live on land. Amphibian innovations include front and hind legs; fully functioning lungs; a moist, scaleless skin capable of gas exchange; and a three-chambered heart.
- **b]** The amphibians radiated into a large, diverse group but now include only frogs, toads, salamanders, and wormlike apodes.
- **c]** Reptilian descendants of the amphibians have dry, scaly skin; an expandable rib cage; a male copulatory organ; eggs containing the watery amniotic sac surrounded by a leathery shell; and legs extending more directly beneath the body.
- **d]** Early reptiles gave rise to modern reptiles, birds, and mammals. About 6500 reptile species survive today, including snakes, turtles, lizards, and crocodilians.

5 Birds have adaptations that allow full exploitation of terrestrial and arboreal habitats, including feathers and a constant body temperature. Mammals evolved hair, constant body temperature, and milk, and became the most numerous large animals on earth.

- **a]** Mammals are warm-blooded and have a four-chambered heart like birds but show two additional traits: body hair and milk production. A placenta also develops in most female mammals during pregnancy. The monotremes, however, lay leathery eggs, and marsupials nurture their immature embryolike young in a pouch.
- **b]** Mammals show four major types of specializations: constant warm body temperature (homeothermy); varied shapes of limbs and teeth; prolonged parental care; and highly developed nervous systems, senses, and behaviors.

UNDERSTANDING THE FACTS AND CONCEPTS

For Questions 1–5, match the descriptions with the most appropriate item or items from the following list. Each item in the list may be used once, more than once, or not at all.

- **a]** spinal cord
- **b]** vertebral column
- **c]** gill slits
- **d]** notochord
- **e]** myomeres

1 A hollow, dorsal tube or nerve tissue found at some stage in all chordates.

2 A solid, flexible rod that is present in all chordate embryos.

3 Found in vertebrates but not in nonvertebrate chordates.

4 Bilateral, segmented blocks of tissue surrounding the vertebral column.

5 A structure originally used to filter food from seawater.

As above, for Questions 6–10, match the descriptions with the most appropriate item or items from the following list.

- **a]** tunicates
- **b]** lancelets
- **c]** lampreys
- **d]** echinoderms

6 The first adult chordate to exhibit all major chordate characteristics.

7 Sea squirts with a larval form that is of evolutionary importance.

8 A deuterostome but not a chordate.

9 Vertebrate chordate.

10 Nonvertebrate chordate.

For Questions 11–15, match the descriptions with the most appropriate item or items from the following list.

- **a]** agnathans
- **b]** placoderms
- **c]** ray-finned fishes
- **d]** Chondrichthyes
- **e]** lobe-finned fishes

11 Bony fishes with jaws that are not believed to be ancestral to land vertebrates.

12 Bony fishes with jaws and lungs; an extinct group of these fishes is believed to be ancestral to land vertebrates.

13 Modern cartilaginous fishes including sharks and rays.

14 An extinct group of fishes but the first to exhibit jaws and paired appendages.

15 Jawless fishes; an extinct group of these fishes is believed to be the first of the vertebrates.

For Questions 16–20, match the descriptions with the most appropriate item or items from the following list.

- **a]** amphibians
- **b]** reptiles
- **c]** birds
- **d]** mammals

16 Vertebrates that have lungs and a three-chambered heart but do not produce amniote eggs.

17 Vertebrates that exhibit internal fertilization and produce amniote eggs.

18 Vertebrates that gave rise to both birds and mammals.

19 Warm-blooded vertebrates that include egg layers as well as bearers of live young.

20 Homeothermic, with air sacs and a four-chambered heart.

INTEGRATE AND APPLY WHAT YOU HAVE LEARNED

1 Compare chordate characteristics with vertebrate characteristics. Are there more animals classified as chordates or as vertebrates?

2 Although lancelets look like fishes, they are not classified as such. Why?

3 Why do zoologists believe that modern groups of fishes and land vertebrates are descended from fishes that are now extinct?

4 In what ways are reptiles more suited to life on land than amphibians?

5 Trace the phylogenetic lineage of birds and mammals from the first vertebrates.

ANALYSIS

1 All of the following characteristics of amphibians except one explain why most species died out before modern times. Which one is the exception?

- **a]** Moist, absorbent skin.
- **b]** Eggs that develop in fresh water.
- **c]** Specializations of limb structures for jumping and climbing.
- **d]** A three-chambered heart.
- **e]** Production of many eggs per female.

2 Given the fact that birds and all mammals are homeothermic, which is the most logical conclusion regarding the number of times the trait of homeothermy evolved among the vertebrates?

- **a]** Once, giving rise to ancestral birds from which mammals are descended.
- **b]** Once, giving rise to stem reptiles which then gave rise to birds and mammals.
- **c]** Twice, once in the ancestral birds and once in the ancestral mammals.
- **d]** Three times—once in each of the three groups of mammals.
- **e]** Four times—once in the birds and once in each of the groups of mammals.

3 Some zoologists characterize birds as "flying dinosaurs." What do they really mean?

- **a]** Birds are descended from dinosaurs.
- **b]** Dinosaurs evolved from primitive birds.
- **c]** Modern birds and extinct dinosaurs are both descended from *Archaeopteryx*.
- **d]** Thecodont reptiles are a common ancestor for both birds and most dinosaurs.
- **e]** Birds are really dinosaurs.

4 Which of the following statements most accurately represents the fundamental difference between chordates and vertebrates?

- **a]** Chordates are fundamentally aquatic, whereas vertebrates are fundamentally land animals.
- **b]** Chordates live mostly in the oceans, whereas vertebrates are found in fresh water and marine environments.
- **c]** Adult chordates rely on their notochord more than do adult vertebrates.
- **d]** By virtue of bone, jaws, paired appendages, and a complex nervous system, vertebrates are more active animals than are chordates.
- **e]** Chordates are more streamlined in shape than are vertebrates.

5 There are more species of fishes than all other vertebrates combined. What might explain this observation?

- **a]** More of the earth is covered by water than by land.
- **b]** Water provides a more stable environment than does land and air.
- **c]** Fishes have been around for a longer time, with more opportunity for adaptive radiation.
- **d]** The abundant phytoplankton that occupy marine environments result in a proportionately greater area of ecological richness than is found on land.
- **e]** All of the above are logical explanations.

CHAPTER 23

Human Origins and Evolution

FOOTPRINTS IN TIME

In the late 1970s, anthropologist Mary Leakey led a team of geologists and other anthropologists to Laetoli, a rich fossil bed in the Serengeti plain of East Africa. The site is south of the well-known Olduvai Gorge, a steep-sided ravine where Leakey and others had found important evidence of humankind's early ancestors. One of the members, Dr. Paul Abell, found what appeared to be a heel print at Laetoli in tuff, a type of rock made of solidified volcanic ash. Soon, African team member Ndibo Mbuika discovered that he could lift small clods of matted topsoil out of the depression to reveal a distinct footprint. More excavations followed, and soon a string nearly 40 footprints long spanning 27.7 m (90 ft) emerged in the hardened ash [FIGURE 23.1A]. The footprints looked nearly human, and the team realized that they might yield important information about human ancestors, including which came first, the evolution of upright posture—that is, walking fully erect on two legs—or the evolution of a very large brain. Other basic questions about human evolution involved behavior: Did ancient human relatives live in the forest or on the grassy savanna? And did they live in social groups or as solitary individuals?

Based on the small size of the prints, their extreme age (at least 3.5 million years), and their positions in the ash, Leakey's team hypothesized that an early human ancestor such as *Australopithecus*

[A]

[B]

FIGURE 23.1

Making Tracks at Laetoli.

[A] A photo of the footprints Mary Leakey's team discovered in 1978. **[B]** This painting recreates a hypothetical scenario in which a male and a female of the species *Australopithecus afarensis* walked through loose ash in what is now Tanzania.

afarensis [FIGURE 23.1B] may have left the tracks. Evidence suggests that a male may have walked side by side with a smaller female through newly fallen ash from a nearby volcano. He may have been touching the female as they walked, and a smaller individual (probably a child) may have trailed along behind them, stepping in the male's footprints, perhaps to conceal his own, perhaps because the ash was hot, or maybe just as a game.

This momentous find helped show that bipedalism—walking erect on two feet—evolved a million years before human ancestors had enlarged brains or used stone tools, according to other fossil evidence. It also showed that ancestral humans were living in at least small social groups and walking on the open savanna 3.5 million years ago, and not just residing in forested areas as did most other primates.

The scientific implications of the footprints are important, but the tracks are also significant for the poignant hints they give about humankind's earliest social behavior. As the small group, perhaps a young family, ambled along through the soft ash, one stopped and turned slightly to the left. Did he or she hear a large animal grunt and fear an impending attack? Just at that moment, did the volcano send up new ash, steam, and lava? Did a parent turn to urge on a dawdling child? We'll never know. But as Mary Leakey writes, "This motion, so intensely human, transcends time."

The study of human origins attempts to peer through that window of time to discover how human beings evolved, one of the most intriguing subjects in modern biology. We humans have all the requisite traits for membership in the mammalian class, with our body hair, warm blood, mammary glands, and our mode of giving birth. But our unique combination of behavioral abilities—including spoken and written language, agriculture, and extensive tool use—has allowed us to dominate the environment like no other animals before us. Despite our unique abilities, there is ample evidence today, both fossil and genetic, that humans represent one recent branch of the Primate order within the mammals and that our branch separated less than 10 million years ago from the lineage leading to chimpanzees. Ethologist Desmond Morris called humans "naked apes," and so, it seems, we are subject to the same kinds of evolutionary forces we discussed in earlier chapters, including natural selection and genetic drift.

In the chapter's first half, we will see how the primate order, including the primitive monkeys or prosimians, the true monkeys, and the apes, each evolved their characteristic traits. In the chapter's second half, we'll follow the evidence for the rise of our ancestors in now-extinct genera and species, as well as for the sole surviving human species, *Homo sapiens*.

Along the way, we'll discover that certain monkeys have a "fifth hand"; we'll see how surprisingly many genes we share with the chimpanzee; how the heat of the savanna and the diet available there may have influenced our upright posture and large brains; and about some of the lively controversies in the field of anthropology, including who or what left the footprints at Laetoli. ❑

MESSAGES

1 Biologists have discovered patterns of evolution among the primates—lemurs, tarsiers, monkeys, apes, and humans—by studying evidence from fossils, from the structure of DNA and proteins, and from the anatomy and behaviors of present-day organisms.

2 Fossil evidence suggests that primates evolved in the canopy of the tropical forest. The features that evolved among earlier primates were the basis for the later evolution of humanlike forms.

3 Upright posture frees hands to manipulate tools and foods, and probably helps cool the brain. The enlarged brain supports increased dexterity. However, a fully upright posture constrains the form of the birth canal in humans and limits the size of a baby's head. Human babies are thus born in a helpless state and require intensive care for many years. Human social systems and spoken languages promote that care.

Evolution of the Primates

Because human beings are a branch of the primate evolutionary tree, the shape and size of that tree are relevant for understanding ourselves [FIGURE 23.2].

THE PRIMATE FAMILY TREE

Consider a monkey at the zoo. What features does this animal share with people that might distinguish the order Primates from other mammals? The monkey can wrap its hands around a bar or perch just as a person grasps a rope or a hammer because both humans and monkeys have an **opposable thumb**, a first digit that can touch the ends of all the other digits. In addition, the monkey is constantly turning its head, looking at zoo visitors who stare right back at it, or peering into a roommate's fur looking for parasites. The monkey is clearly very visually oriented

FIGURE 23.2

Some Living and Extinct Members of the Primate Family Tree.

Experts disagree as to whether tarsiers are prosimians or anthropoids. Lines giving rise to gorillas, chimpanzees, and humans began diverging less than 10 million years ago.

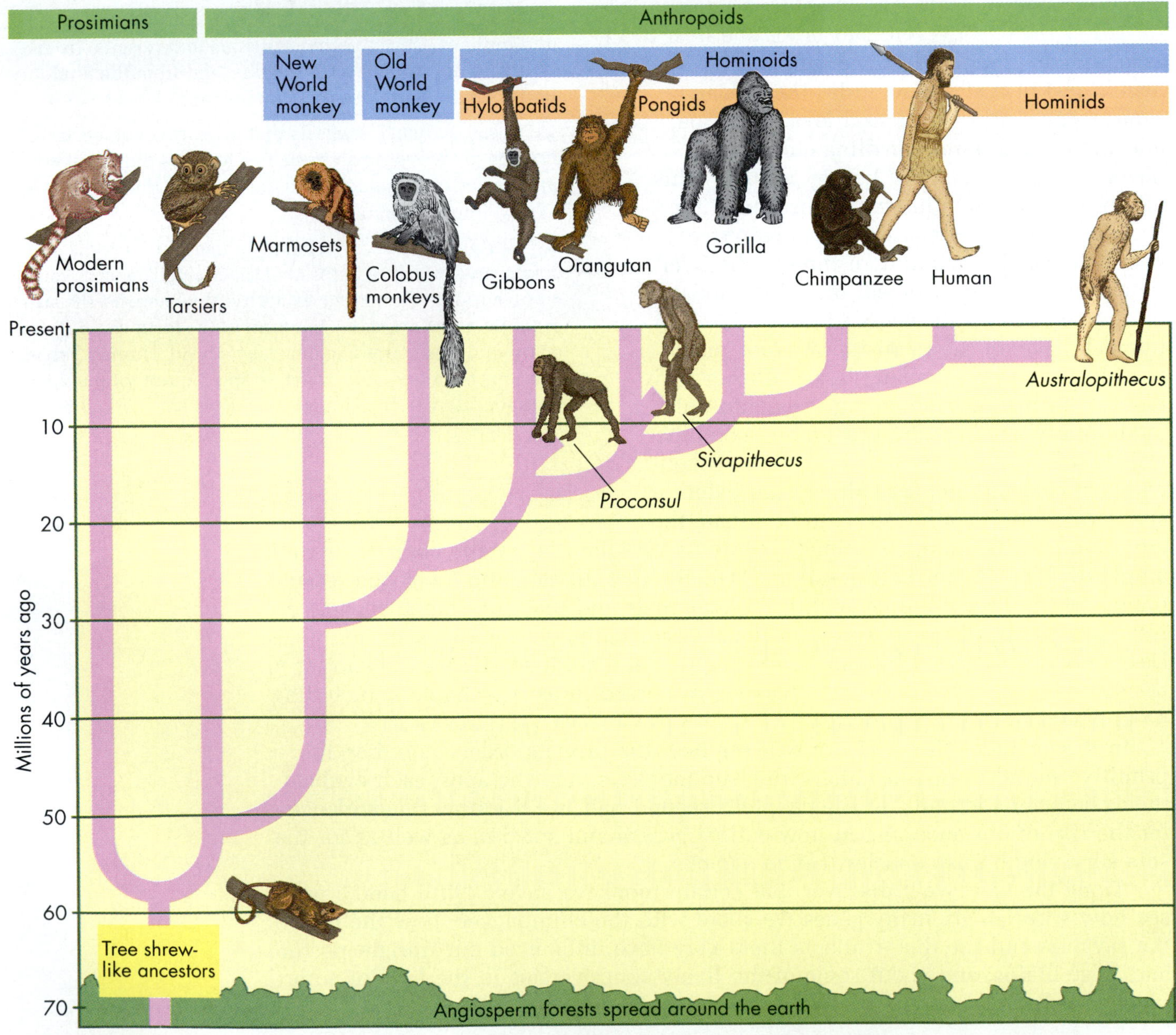

[A] Lemur

[B] Tarsier

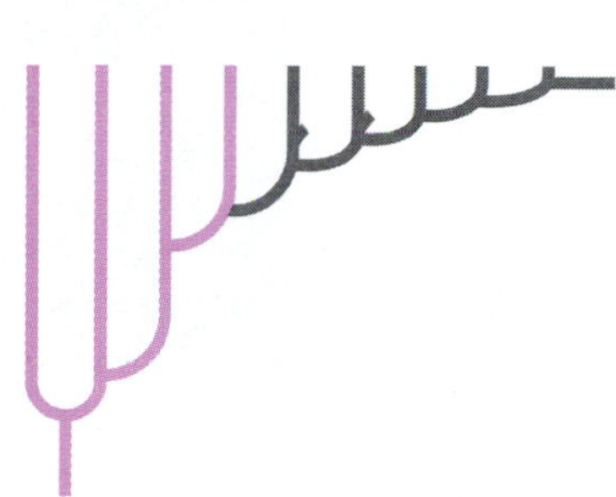

[C] New World monkey

[D] Old World monkey

and perceives its environment mainly through the sense of sight: **acute vision**, including forward-facing eyes and the ability to see in color are characteristics of most primates. Finally, monkeys and other primates have **large brains** relative to their body size. The brain of a male rhesus monkey, for example, is much larger than the brain of a dog of the same size. Evolutionary biologists are keenly interested in which environmental factors might have contributed to natural selection for the kinds of limbs, eyes, and brain that primates share. To pose and answer questions about that evolution, we must first examine primate diversity.

Taxonomists divide the order Primates, with its 150 or so currently living species, into two suborders: the **prosimians** (which means "before monkeys"), and the

FIGURE 23.3

Living Primates.

[A] Lemurs, such as Coquerel's sifakas, are living prosimians; mainly inhabiting the island of Madagascar. They have limited color vision, and rely more strongly on the sense of smell than other primates. **[B]** Tarsiers live in the tropical forests of the Philippines, Sumatra, Java, and Borneo, where they harvest fruit and hunt insects, lizards, and birds by night. Some authorities place tarsiers with prosimians and some with anthropoids. **[C]** An alert white-faced capuchin is a New World monkey of Costa Rican rain forests. **[D]** An Old World monkey. The intent, forward facing eyes of a Japanese macaque help provide depth perception.

[A] Orangutan

[B] Gorilla

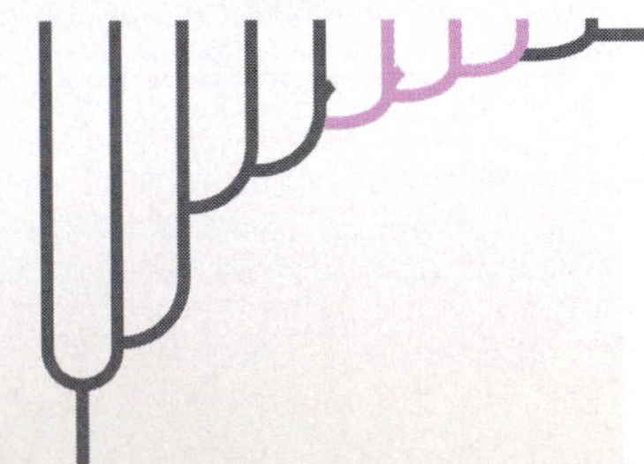

FIGURE 23.4

The Great Apes.

[A] The orangutans (*Pongo pygmaeus*) live in remote forests of Sumatra and Borneo. Their arms are longer than a chimpanzee's or gorilla's, and they move about rather clumsily in search of fruit and tender shoots. These animals are intelligent, though less social than gorillas or chimpanzees. **[B]** Gorillas are the largest and most powerful nonhuman primates; they inhabit mist and lowland forests of western and central Africa. Male gorillas can be nearly 2 m (6½ ft) tall and weigh over 200 kg (450 lb), and much of their feeding occurs on the ground. Gorillas walk on their knuckles and the males beat and slap their chests and roar loudly when angered or threatened. Gorillas live in mixed-sex, mixed-age social groups and travel together through the forest in search of the tender shoots, piths, and fruits that compose their diet. Language experiments have revealed that gorillas appear to have a sense of humor and can comprehend a large vocabulary of words and phrases. **[C]** Chimpanzees are our closest living relatives, and our DNA is 99 percent similar to theirs. Chimpanzees live in the forests of western and central Africa in large social communities. In the wild, the bulk of their diet comes from fruits, but they also eat some leafy matter as well as termites and other insects; males occasionally hunt mammals, including monkeys, and then share the meat among themselves and with females and young. In nature they use tools (sticks and grass) they happen to find, and in captivity can be trained to carry out complex tasks, including how to communicate in American sign language. They express affection by hugging and kissing, and happiness by laughing and jumping up and down; when unhappy, they cry or sob loudly.

anthropoids (which means "humanlike") [see FIGURE 23.2]. Prosimians include various small, largely **arboreal** ("tree-dwelling") primates, such as lemurs, while anthropoids include monkeys, apes, and humans.

Prosimians Prosimians generally retain the claws, long snout, and side-facing eyes common in early mammals. Prosimians include the lemurs, lorises, and bush babies [FIGURE 23.3A]. Lemurs are fuzzy animals with fox-like muzzles, long bushy tails, and long front and back limbs. They are found on Madagascar, a large, oblong island lying off the southeast coast of Africa. Most species are arboreal and feed on nectars, fruits, leaves, and sometimes insects. Prosimian species that live on the island of Madagascar are mainly diurnal (active during the day). This trait makes lemurs important biological subjects because monkeys, apes, and humans are also diurnal, and probably arose from a common diurnal ancestor. Many species of lemurs became extinct shortly after the large primate *H. sapiens* came to inhabit Madagascar less than 2000 years ago. An ongoing destruction of forests on the island threatens the lemur's survival and with it, a lineage not much different than the earliest primates—the roots of our own evolutionary path.

Anthropoids Three main groups of organisms make up the living anthropoids: the New World monkeys, the Old World monkeys, and a third group, the hominoids, which includes apes and humans [review FIGURE 23.2].

[C] Chimpanzee

The New World monkeys inhabit the forests of southern Mexico and Central and South America. The Old World monkeys live in tropical forests and savanna regions of the eastern hemisphere from Africa to India and Southeast Asia [review FIGURE 23.2].

New World monkeys have widely separated nostrils oriented somewhat laterally [FIGURE 23.3C]. Some of the larger-bodied New World monkeys, such as spider monkeys and howler monkeys, have grasping, or **prehensile**, tails [FIGURE 23.3C]. These tails, which have a naked, touch-sensitive pad at the tip, serve as a type of fifth hand, aiding the monkeys in dangerous crossings between branches of adjacent trees. A prehensile tail also increases a monkey's ability to find food, since it can hang down from a branch and reach food that otherwise would be inaccessible.

Old World monkeys have closely set, downward-pointing nostrils and lack prehensile tails. Many have buttock calluses that help pad and protect their posterior; many Old World monkeys also have cheek pouches that can hold an impressive amount of food. The rhesus monkeys used in medical research are Old World primates [FIGURE 23.3D], as are the langurs of Asia and the baboons and mandrills of Africa.

The **hominoids** ("resembling humans") include three main groups [see FIGURE 23.2]: the gibbons, siamangs [see FIGURE 23.7], and orangutans of Asia (the hylobatid family); the gorillas and chimpanzees [FIGURE 23.4] of Africa (the pongid family); and humans and their recent relatives (the **hominid** family). All living hominoids except humans are considered apes. Apes are the largest primates and our closest relatives. Apes are large, tail-less animals with long arms, large brains, and complex social behavior. FIGURE 23.4 describes the most familiar apes in more detail.

TRENDS IN PRIMATE EVOLUTION

Fossil evidence suggests that the earliest primates lived high in the canopy of the tropical forest. These early primates evolved several important specializations associated with their particular way of life that have added to their success as a group and that form the background for human physical features. These traits included manual dexterity, upright posture, acute vision, large brain, extended infant care, and modified teeth [FIGURE 23.5].

Dexterity of the Hands Natural selection tends to eliminate abruptly clumsy individuals bumbling awkwardly along branches high in the trees. Primates evolved the ability to spread their digits widely, to touch their "thumb" with the tips of each of the other four fingers, to curl their digits around branches in a power grip, and to grasp objects with precision, as when manipulating a piece of fruit plucked from a tree branch or holding a pencil.

Upright Posture Hands can be used to manipulate objects only if they are not supporting the weight of the body. We humans are the only fully upright primates that

FIGURE 23.5

Trends in Primate Evolution.

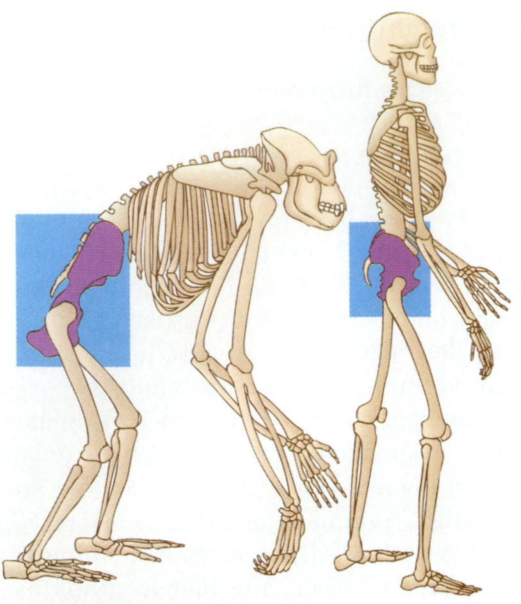

FIGURE 23.6

Upright Stance and Pelvis Shape.

Bones of the gorilla and the human are somewhat similar in shape and function, but the proportions are very different. The gorilla pelvis, for example, here shown in purple, tilts and tips the large rib cage and heavy neck and head forward. In contrast, the human pelvis is vertical and helps hold the entire skeleton upright.

walk consistently on two legs (we are **bipedal**). Nevertheless, many monkeys and apes spend large amounts of time in vertical rather than horizontal positions. Even though gorillas and chimpanzees walk on all fours by touching the ground with their feet and the knuckles of their hands, their arms are so long that the angle of the body is still fairly erect [FIGURE 23.6]. Upright posture improves an animal's ability to see and can leave the hands free for other activities. In addition, an upright posture and bipedal locomotion decrease an animal's exposure to the sun, which, as we will see later, may have been an advantage to human ancestors evolving in the hot savannas.

Vision An arboreal habitat is rich with visual information, such as fluttering leaves, moving spots of sunlight and shade, and tangled tree limbs. Both plant and insect foods are likely to be well hidden, or *cryptic*, and may require keen vision to detect. Life in the forest canopy is also dangerous because an error in judging the precise spatial location of a particular tree limb could result in a fatal fall. Thus, one trait favored in primate evolution was that of depth perception, or **stereoscopic vision**, the ability to discriminate distances in space. Depth perception occurs because an animal's left and right eyes see the same object from slightly different angles. To demonstrate this for yourself, observe one of your fingers at arm's length with one eye closed and then the other. Note how the finger seems to shift position with respect to the background. The brain interprets the differences in terms of distance of the viewer from the object and creates a realistic three-dimensional picture of the environment. The ability of primates to see color further enhances their ability to recognize members of the same species as well as to identify ripe fruits, young leaves, cryptic prey, and predators lying in ambush.

Brain The evolution of life in the trees was accompanied by the development of specific parts of the brain. Prosimians have fairly large brains for animals of their size and body weight; monkeys have proportionately larger brains than prosimians; and apes and humans have the most complex brains of any mammals—both in terms of brain size and number of nerve cells in relation to overall body weight, and in terms of the intricate interconnection of different brain regions [see CHAPTER 32].

Scientists have created two hypotheses to explain the increase in brain size and intelligence during primate evolution. The so-called *political hypothesis* holds that the main driving force behind increased brain size was complex social interactions in primate societies. The *diet hypothesis* suggests that natural selection favored larger brain size by acting primarily through diet and the difficulties of obtaining enough food variety in the forest canopy.

The political hypothesis notes that animals with larger, more nimble brains could better solve the problems associated with shifting alliances within the troop and changing dominant and subordinate social relationships. Since reproductive success is related to social rank, allelic combinations that provided more socially adept phenotypes would be favored by natural selection.

The diet hypothesis suggests that natural selection favored enlarged visual centers of the brain. This is because individuals with keen depth perception, sharp visual acuity, and color vision were better able to distinguish nutritious food items such as ripe fruit and tiny, young leaves from the background of leaves and shadows, and hence to sustain better physical condition and to reproduce more effectively. The parts of the brain associated with grasping and manipulating objects may also have enlarged in response to natural selection acting via the diet. Gorillas, for example, eat plants defended by nettles, spines, hooks, and thick husks. Obtaining the foods often takes as many as eight different manipulations, such as covering nettles with leaves, peeling the food, and then twisting tough fibers apart. The sequence of steps varies from plant to plant, suggesting that learning may play a large role in harvesting the foods effectively. The ability to recognize nutritious foods, remember where they grow, and find them during seasons when they are available would also require large brains and would be subject to natural selection.

The political and dietary hypotheses for enlarged brains are not necessarily mutually exclusive. Both factors may have played roles in increasing primate intelligence.

Infant Care Along with their mobile life in the trees, primates evolved new reproductive strategies. Compared with most other mammals, they give birth to smaller litters of young, with most species producing just one young at a time. The infants tend to be helpless (especially those of apes and humans) and depend on their parent(s) for complete physical care for a long period after birth. In general, the higher a species' intelligence, the more time the parent devotes to the care and nurturing of the young, who must master many complex skills to succeed as an adult. This great investment of parental care pays off in the high survival rates of the young.

Teeth Primates have several tooth types, modified for an omnivorous diet of both plant and animal matter. The earliest mammals had pointed teeth, adapted for slicing up the bodies of insects. Primate ancestors may have had similar teeth, but as early primates began to include more plant foods in the diet, those individuals whose back teeth were broader and flatter could process these new food types more efficiently; thus, over many generations primates came to have broad, flat molars. In the line giving rise to humans, the jawline became U-shaped, while in most other primates, it is more V-shaped.

Each of these primate characteristics—manual dexterity, stereoscopic color vision, unusually large brain, extended infant care, upright posture, and various tooth types—played an important role in the success of the primate lineages.

EMERGENCE OF MODERN PRIMATES

Genetic analyses and fossilized bones suggest that the anthropoids (monkeys, apes, and humans) arose from prosimian ancestors about 40 to 35 million years ago [review FIGURE 23.2]. The monkeys diverged into New and Old World lineages while the hominoids, the apes and humans, radiated into various niches throughout the Old World. We are particularly interested in the evolutionary history of the apes because this lineage also gave rise to humans.

Some 20 million years ago, a radiation of primitive apes occurred in the Old World. These apes were of many different body sizes and occupied a diverse array of habitats ranging from tropical forests to more open grasslands and woodlands. One of these apes was *Proconsul africanus* [review FIGURE 23.2], a tree-living, fruit-eating ape of Africa that walked about on all fours.

Adaptive radiation produced *Proconsul* and other species that survived for lengthy periods and then became extinct. The same adaptive radiation led to an evolutionary lineage that eventually gave rise to today's surviving great apes: the gorillas, two species of chimpanzees—the common chimpanzee *Pan troglodytes* and the smaller pygmy chimp, or bonobo (*Pan paniscus*)—and humans. Biologists have gotten an indication of how closely related these surviving apes are by comparing their DNAs. Geneticists have found that the genetic distance between humans and chimpanzees is about the same as or a little less than the genetic distance between chimps and gorillas. These numbers mean that a chimp is genetically as close or closer to *you* as it is to a gorilla! No wonder evolutionary biologist Jared Diamond calls *H. sapiens* "the third chimpanzee" [see BOX 23.1 on page 548].

Here's another way to put the genetic relationship of chimps and humans into proper perspective. Compare the graphed DNA data for the common gibbon and the siamang gibbon in FIGURE 23.7. These animals are so similar (2.2 percent difference) that taxonomists place them in the same genus (*Hylobates*). Yet, humans and chimpanzees (1.6 percent difference) are more similar to each other genetically than are these two gibbons. By comparing the family tree generated from such genetic data to the dates of fossilized bones from common ancestors to the chimps, humans, and other great apes (right-hand axis on FIGURE 23.7), biologists have been able to calibrate a "genetic clock." On the basis of this clock, researchers think that humans, chimps, and gorillas probably shared a common ancestor no more than 10 million years ago [see FIGURES 23.2 and 23.7].

The human and chimp lineages split about 6 million years ago—an extremely short span in geological history. Using a 24-hour analogy, the earth formed at midnight, life appeared at 4:10 a.m., the first vertebrates appeared at 9:35 p.m., and the first humans arose only 38 seconds before the clock struck midnight. Let's examine that last 38 seconds.

➤ CONCEPT CHALLENGE

We have seen that a change occurred in the diets of ancient primates from a reliance on insects to a diet of fruits and leaves. How could this dietary change have promoted, by natural selection, trends such as color vision, large brains, and tooth structure we have also seen in primate evolution?

box 23.1
Biology Applied

A Failure of Rhythm

Most college-age individuals are aware that they can't rely on the rhythm method as a way to avoid pregnancy. The technique, which involves abstinence for a few days before and after ovulation, has a failure rate of about 25 percent in the course of a year.

Rhythm fails, in part, because women usually don't know exactly when they ovulate, so avoiding that time of month is difficult.

A trip to the primate house of the nearest large zoo will confirm that concealed ovulation is present in many members of our taxonomic family—but conspicuously absent in others. Female vervet and spider monkeys, marmosets, and orangutans show no visible signs of ovulation even when males are paying close attention and the female is presumably in estrus. In baboons, on the other hand, ovulation is accompanied by a swelling and brightening of the vaginal area that is visible from half a block away. And equivalent signals also show at the time of estrus in gibbons, common chimpanzees, and certain macaques.

Several researchers have conducted comparative studies of ovulatory signals and mating patterns in primates. In species with advertised ovulation, mating often takes place only at the time of estrus, while in species with concealed ovulation, the females are often continuously receptive to sex. There are so many exceptions to this, however, that more than a dozen groups set out to develop specific theories for what the evolutionary role of concealed ovulation may be in *Homo sapiens*.

Evolutionary biologist Jared Diamond considered two of the most prominent such theories in recent popular articles and in his book *The Third Chimpanzee*. One of the theories he describes, the "father-at-home" theory, was developed by researcher Richard Alexander and graduate student Katherine Noonan at the University of Michigan. Their theory suggests that due to concealed ovulation and continuous sexual receptivity in a female partner, a human male would both want to stay around her and would also have to, in order to maximize his chances of fertilizing her and of fending off other interested males. His steadfast presence would have provided both a helping hand in child care as well as extra nutrition in the form of meat from hunting. As a result, female and offspring would have benefited and this mating arrangement would have fostered survival.

A second possibility, the "many-fathers theory," was developed by an anthropologist at the University of California at Davis, Sarah Hrdy. She reasoned that the female, with her concealed ovulation and constant receptivity, could have copulated with many males (on the sly, if necessary, to avoid one male's wrath). Not only would such an arrangement help guarantee conception, but it would also help protect the mother and infant from the murderous impulses of infanticide that are common among primates and many other animal species. A human male, Hrdy hypothesized, would be less likely to kill an infant if the baby might be his. Therefore, the mating pattern of concealed ovulation, constant receptivity, and many potential fathers could have fostered an infant's survival.

In examining both nonhuman primates and the behaviors of people in traditional societies around the world, Jared Diamond sees evidence to support both theories, but perhaps at different times in human evolution. He suspects that the many-fathers explanation fit our species best when most human societies lived in promiscuous harems. At a later time in history, with concealed ovulation and constant sexual receptivity already in place because of their survival advantages within harems, the father-at-home phenomenon may have evolved and helped ensure the survival of intact, independent family units within larger communities.

Regardless of how the pattern evolved, Diamond summarizes it this way: "For good evolutionary reasons, women have found it's much better to keep men in the dark." Looked at in a different way, however—through the lens of sexuality and survival in the mid-1990s—one could say it's much better for both sexes to know what's happening, when, and why.

Diamond, Jared. "Sex and the Female Agenda." *Discover*, September 1993, pp. 86–93.

Diamond, Jared. *The Third Chimpanzee*. New York: HarperCollins, 1992, chaps. 3–5.

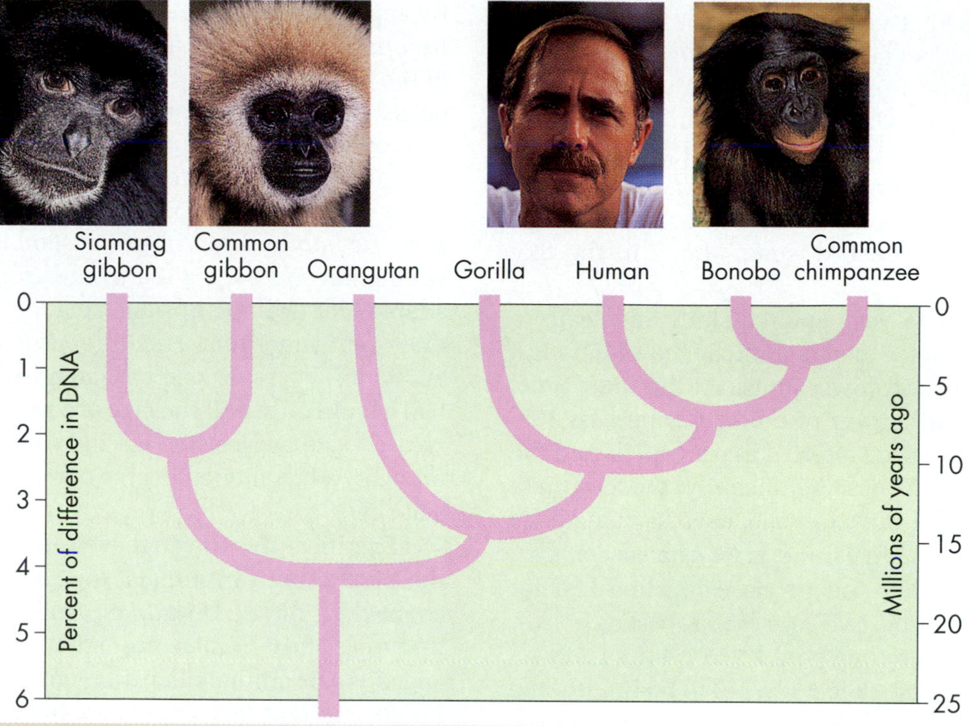

FIGURE 23.7

Genetic Differences among Some Primates.

Note that humans are more similar genetically to a bonobo chimpanzee than are two gibbon species to each other: the siamang gibbon (*Symphalangus syndactylus*) and the common gibbon (*Hylobates lar*).

The Appearance of *Homo sapiens*

Although people and chimps are closely related genetically, humans have developed several distinctive traits. These differences include fully upright, bipedal locomotion, an omnivorous diet, extensive tool manufacture and use, further increases in brain size, and the development of elaborate spoken languages. An important question is this: Which came first, the enlargement of the brain or the transition to fully upright, bipedal walking?

LARGE BRAINS AND WALKING ON TWO FEET

When renowned anthropologist Raymond Dart discovered the Taung infant's skull in South Africa more than 70 years ago, a tremendous controversy broke out in the scientific community. The Taung skull had a combination of apelike and humanlike traits that differed from what experts had previously predicted for human ancestors, and some were unwilling to concede that it could be on the human lineage at all. Anthropologists of that day were convinced that an enlarged brain and toolmaking were the earliest steps in human evolution and that upright posture and bipedal locomotion had occurred at some later date. Fossils of the Taung infant, given the scientific name *Australopithecus africanus* ("southern ape of Africa," see FIGURE 23.2) indicated just the reverse: This skull revealed a brain case that did not differ in size from a modern ape's, yet the Taung skull had been carried upright on the spinal column, indicating upright posture and walking on two feet. More than 25 years passed before the scientific community would accept the Taung infant as a hominid—a human ancestor. But ultimately, as more and more fossil evidence accumulated, it became impossible to deny Dart's contention that Taung and other fossils in the genus *Australopithicus* revealed a lineage of primates that were small-brained but bipedal and that belonged in the hominid lineage.

British anthropologist Mary Leakey, her son Richard, and a few of their colleagues found some of the most exciting *Australopithicus* fossils ever discovered in Tanzania and Ethiopia in the mid-1970s. They revealed, for example, a species that virtually everyone agrees is the earliest true hominid so far uncovered: *Australopithecus afarensis*, which dates from 3 to 4 million years ago [see FIGURES 23.2 and 23.8]. One skeleton of an *A. afarensis*, recovered in Hadar, Ethiopia, and studied by Donald Johanson, is the most complete *Australopithicus* skeleton yet found. Researchers gave this fossil the name "Lucy," after the then-popular Beatles' song, "Lucy in the Sky with Diamonds."

A. afarensis had a very apelike skull and teeth, a brain only slightly larger than a chimpanzee's (450 mL), and long upper but short lower limbs. At the same time, however, its head sat on top of the backbone like ours does rather than projecting forward like an ape's, and the hands had humanlike bones. Significantly, the footprints the Leakey team found at Laetoli may have been made by three *A. afarensis* individuals, and as we saw earlier, they show that the legs, feet, and pelvis were modified for fully upright walking on two legs [see FIGURE 23.1]. *A. afarensis* was small, standing only 1.1 to 1.4 m (3.5 to 4.5 ft) tall and weighing just 18 to 23 kg (40 to 50 lb). But the discoveries of its bones, teeth, and footprints proves fairly conclusively that an upright, two-legged stance evolved long before an enlarged brain or tool use. It also indicates that these individuals had hands free for manipulating or carrying food or young and could also see long distances in search of predator or prey. Some anthropologists are not yet convinced, however, that all or perhaps any of the footprints came from *A. afarensis*. Researcher Russell Tuttle suggests that two of the footprints were left by a large bear, and that the rest of the prints came from some other as-yet-unnamed hominid, but not *A. afarensis*. The resolution to this controversy awaits future discoveries and further research. Unfortunately, fossilized feet are rarely found.

Numerous anthropologists have found fossil evidence showing that hominids in the genus *Australopithicus* emerged in Africa by around 4.5 million years ago. Evidence also suggests that these earliest hominids radiated into several species that lived simultaneously or in overlapping time periods in a multibranched human family tree. While they probably walked upright, their brains were still not much larger than a chimpanzee's. *Australopithicus* individuals had large faces, heavy jaws, and large, crushing molars twice the size of our own. They were probably vegetarians who used their massive molar teeth to process coarse, abrasive foods, such as tubers, seeds, and hard nuts. No one has yet found conclusive evidence that any of these early *Australopithicus* hominids made tools or hunted; thus, their behavior was probably much simpler than that of later humans.

WHO WERE THE FIRST TOOLMAKERS?

By approximately 2 million years ago, one branch of *Australopithicus* had probably given rise to the first member of the genus *Homo* and thus the first human. Anthropologists call this early human species ***Homo habilis***, the "handy human," because of its use of tools [FIGURE 23.8]. This and all later members of the genus *Homo* are distinguished from australopithicines by a notably larger brain size. *H. habilis* individuals resembled their forebears, having large faces, big teeth, and trunks and limbs fully adapted for walking upright. At 1.5 m (5 ft) tall, however, they were larger than *Australopithicus*, and their brains, at 700 mL, were 50 percent larger. While *H. habilis* still had a brain only half the size of a modern human's, its increased size suggests a real increase in intelligence and probably a heightened degree of manual dexterity and social skills.

Significantly, the first evidence of toolmaking and the butchering of animals appears in the fossil record about the time of *H. habilis*. Anthropologists speculate that the handy human hunted small animals and scavenged larger animals that lions or hyenas had killed. *H. habilis* individuals appear to have made crude stone tools by cracking rocks and flaking off chips, using the sharp chips to cut through the tough hide of dead animals to harvest the meat.

The fossil record indicates that *H. habilis* lived simultaneously and in the same areas in East Africa with *Australopithicus robustus*. The handy human survived, however, while the "robust southern ape" became extinct. Why? Fossilized teeth from specimens of *A. robustus* indicate that these australopithicines ate a fibrous, plant-based diet. Perhaps they had to compete with four-legged savanna animals such as gazelles, wildebeests, and impalas, with specialized, highly efficient plant-fermenting chambers in their digestive systems [see FIGURE 28.12]. These plant eaters may simply have outcompeted the australopithicines, causing their extinction. Anthropologists also speculate that *H. habilis*, with its crude tools and larger brain, may also have competed more effectively for food; perhaps *H. habilis* even killed *A. robustus* for meat. Meat eating certainly sets *H. habilis* apart from all earlier hominids and hominoids. This way of life, based on the newly evolved ability to make tools —nothing short of the first emergence of technology— has dominated the rest of human evolution.

SPREADING OF THE HUMAN LINEAGE

H. habilis appears to have been restricted to the savannas and woodlands that stretched from northeastern to southern Africa. Fossil evidence indicates that a new

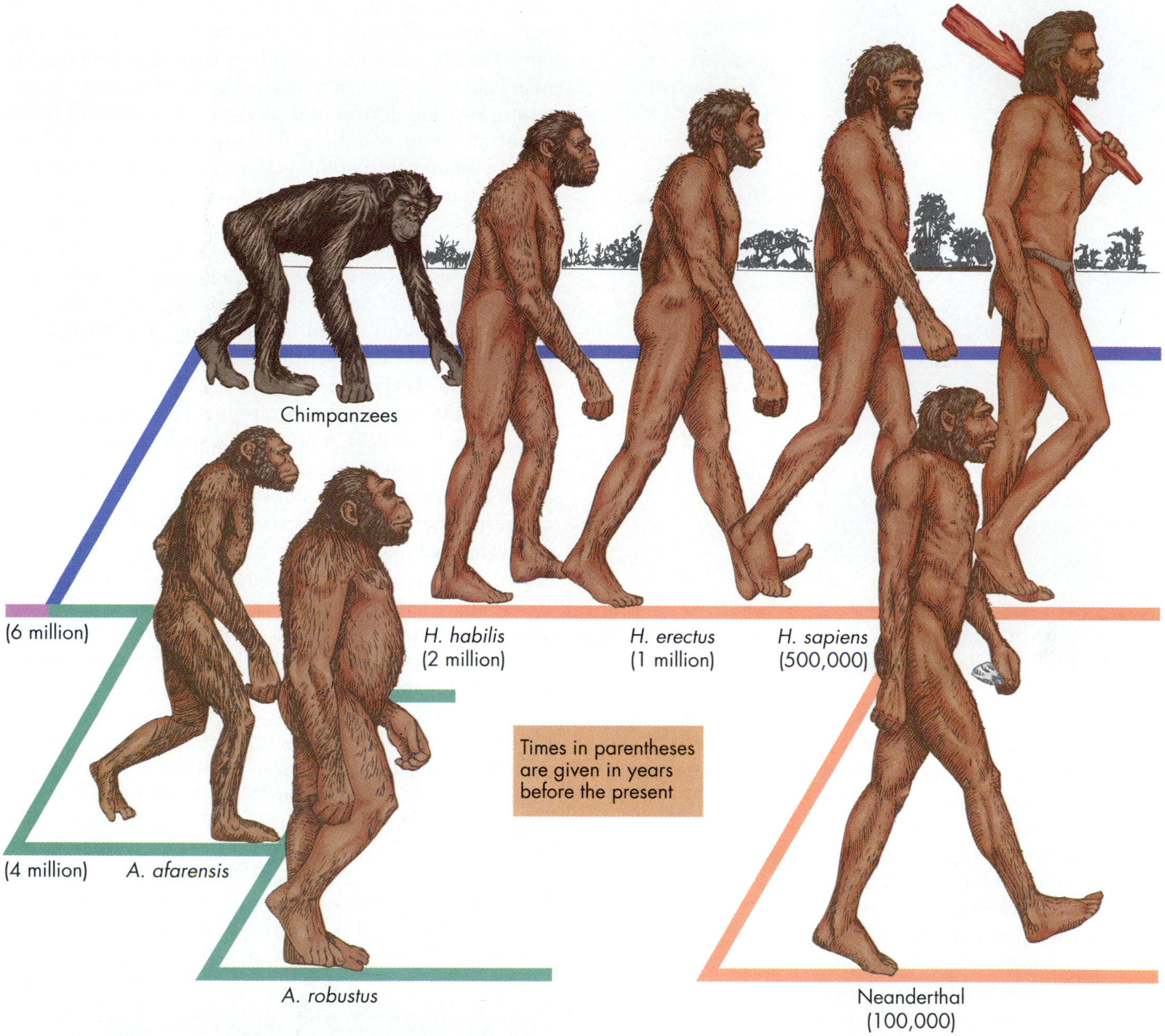

Chimpanzees

(6 million)

H. habilis
(2 million)

H. erectus
(1 million)

H. sapiens
(500,000)

Times in parentheses are given in years before the present

(4 million) *A. afarensis*

A. robustus

Neanderthal
(100,000)

FIGURE 23.8

A Family Album of Human Ancestors.

The apelike *Australopithecus afarensis* lived about 3.75 million years ago and was the first fully upright hominid. The large-boned *Australopithecus robustus* lived about 2 million years ago, and recent evidence shows that its hand was adapted for precision grasping and it may have used tools. *Homo habilis* was an active tool user who lived about 1.5 million years ago and preceded *H. erectus.* Then *H. sapiens* emerged from *H. erectus* about 500,000 years ago. These species did not lead to humans in an advancing ladder, but occupied ends of branches in a family tree whose precise arrangement is still under investigation.

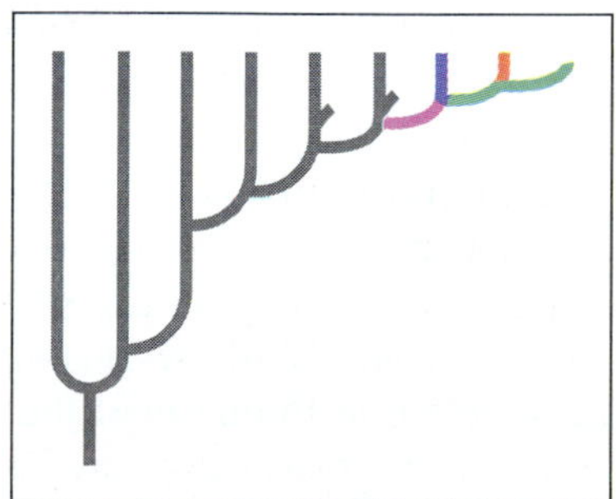

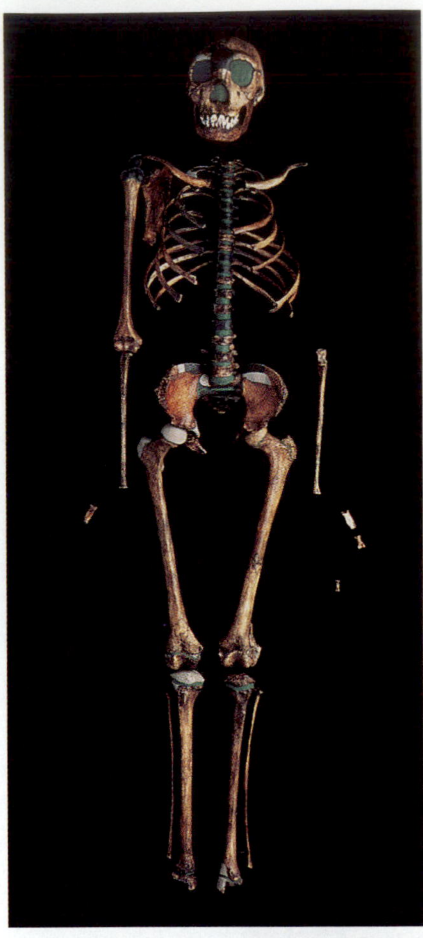

FIGURE 23.9
***Homo erectus*: A 1.6-Million-Year-Old Skeleton.**
This nearly complete skeleton is that of a 12-year-old boy who would have been 6 ft tall as an adult.

species of human, ***Homo erectus*** ("erect human"; see FIGURES 23.8 and 23.9), arose in Africa around 1.5 million years ago and spread into northern Africa and then out of Africa into southern Asia, Indonesia, and probably into southern Europe. *H. erectus* was characterized by a new kind of stone tool, the Acheulean hand ax, that required finer technological skill to manufacture than the choppers and scrapers of *H. habilis*. Anthropologists have also found a range of other stone tools in association with the fossilized remains of *H. erectus*, as well as implements fashioned of bone and horn.

In 1985, researchers found a remarkably complete skeleton of an *H. erectus* male in northern Kenya and determined the bones to be 1.6 million years old [FIGURE 23.9]. The stage of bone growth showed that this fossilized individual was a 12-year-old boy who would have been 6 ft tall had he lived to adulthood. The brain volume (800 mL) was larger than that of *H. habilis* and the teeth and skull were more refined, and scientists speculate that the parts of the brain that govern abstract thought and reasoning may have undergone some internal reorganization.

The discovery of this fairly complete *H. erectus* skeleton provided some very important clues about human evolution. The boy's head was rather large and the pelvis was fairly narrow, and he had yet to reach adult size and sexual maturity. These factors suggest that the same pattern of human development we see today may have already emerged 1.6 million years ago. Modern human babies are born in a physically helpless state. If gestation were any longer and physical development in the womb any more complete, the large head (and brain) of a human infant could not pass through a woman's pelvis at birth. After birth, our growth and maturation processes continue for some 15 to 20 more years. Initial helplessness and a long growth and maturation phase ensure that our large brains will have sufficient time to store enormous amounts of survival information, while the long period of physical dependence during childhood keeps us in contact with potential teachers—parents, grandparents, siblings, and neighbors.

Both the expanded geographical range and the wide range of new tools *H. erectus* appears to have used suggest a growing ability to deal with varied food resources and extremes of climate. From tools and accumulations of animal bones found near fossils of *H. erectus*, researchers deduce that these humans hunted elephants, bears, antelope, and other large game in cooperative bands and transported, butchered, and cooked the meat over campfires (which also provided warmth in cold weather). *H. erectus* groups are known to have occupied caves, and some archaeological evidence suggests they may also have constructed shelters in open areas. Some caves, such as at the Zhoukoudian site in China, show evidence of human occupation for some 250,000 years. Stone tools recovered from Zhoukoudian also show a progressive refinement over time. *H. erectus* probably first used fire about 500,000 years ago as the species spread to the temperate zones of Europe and Asia. Cooperative hunting, cooking, and food sharing were probably developed and maintained through *cultural transmission*; that is, they were taught to the next generation.

HOW AND WHEN DID OUR SPECIES SUPPLANT *HOMO ERECTUS*?

During the more than 1½ million years that early humans inhabited the Old World, their face and teeth decreased in size and their brains enlarged. These changes led from *H. erectus* to *H. sapiens* by about 500,000 years ago, and often it is difficult to assign a particular fossil specimen to one species or the other.

Fossil evidence shows that *H. erectus* established populations in Africa, Europe, and Asia and was eventually supplanted everywhere by *H. sapiens*. Anthropologists are still debating, however, when and how replacement occurred. One hypothesis, the *multiregional model*, states that over a period of about a million years, various groups of *H. sapiens*—Asians, Africans, Europeans—

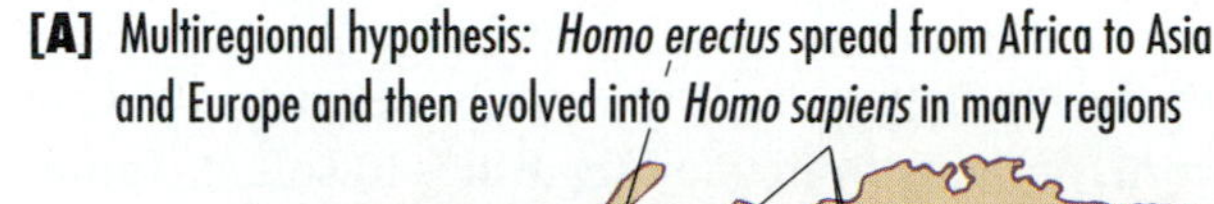

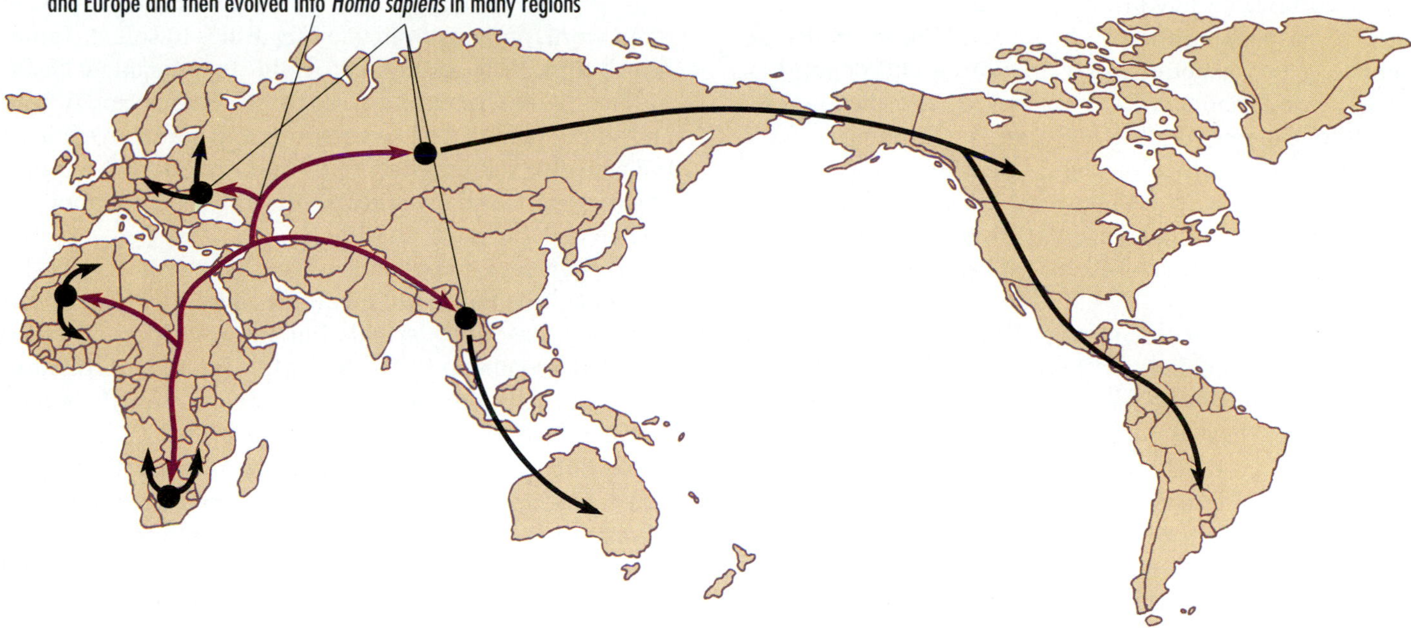

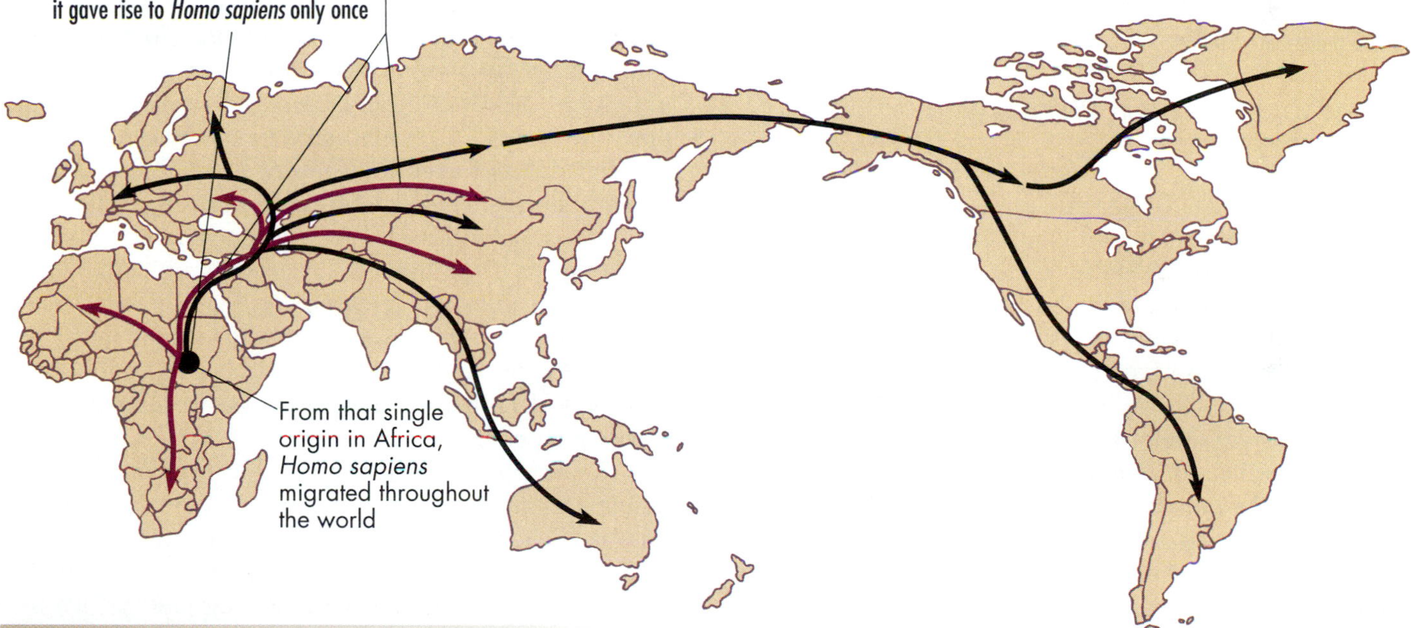

FIGURE 23.10
Two Hypotheses for the Origin of *Homo sapiens* from *Homo erectus*.

arose directly from *H. erectus* at the sites where the different groups of humans are found today. According to this model, Asian *H. sapiens* arose from Asian *H. erectus*, and African *H. sapiens* arose from African *H. erectus* [FIGURE 23.10A]. The alternative hypothesis, the *single-origin model*, holds that *H. sapiens* arose from *H. erectus* in a single event fairly recently at one place, and then these people migrated throughout the world and developed the regional features Asians and Africans and Europeans now display [FIGURE 23.10B].

Biologists have been able to test these hypotheses using the structure of mitochondrial DNA. All the DNA in your cells' mitochondria comes from your mother and none from your father. This is because the egg contains millions of mitochondria at fertilization, but the sperm donates none to the zygote. Thus, you can trace your mitochondrial DNA to your mother, maternal grandmother, and so on back hundreds of thousands of years. By examining the genetic variation among mitochondrial DNAs from people around the world, and making estimates of the rate of mutation, some evolutionary geneticists have argued that the data fit the single-origin model. They suggest the hypothesis that all people alive today are descended through the female line from a small group of people who lived about 200,000 years ago, probably in Africa.

Proponents of the multiregional model point out mathematical problems with the original computer models that produced the family trees of the mitochondrial DNA data. They hypothesize that favorable mutations might have swept through the population of mitochondrial DNA rather recently, and that analyzing DNA in nuclear genes might give different answers. In addition, they point to *H. erectus* fossils that share some traits with populations of modern humans living in Asia near the region where the fossils were found. While the controversy is not yet settled, most anthropologists interpret the currently available fossil and genetic data in favor of the hypothesis that *H. sapiens* arose from *H. erectus* in Africa and migrated throughout the world.

PHYSICAL AND CULTURAL EVOLUTION IN *HOMO SAPIENS*

Generally, scholars recognize two forms of our species—an early or archaic or transitional form, and then our own form, anatomically modern *Homo sapiens sapiens*. Archaic humans dominated the Old World from about 400,000 years ago until they were superseded by transitional and modern humans between 100,000 and 30,000 years ago.

The best known of the transitional *H. sapiens* were the Neanderthals. In fact, Neanderthal fossils, first discovered in Germany's Neander Valley in 1856, were the first hominid remains to be studied scientifically. Neanderthals have been much maligned as brutish, ignorant cave dwellers, but the accumulated facts suggest a quite different picture. Living in Europe under harsh, cold environmental conditions as well as in areas of the Middle East, Neanderthals were short, stocky, powerfully built people with large, protruding faces, an extremely large, prominent nose, and characteristic projecting brow ridges. Their brains were similar in organization but *larger* (1400 mL) on average than a modern human's (1300 mL). The tools and other artifacts they left behind suggest an ability to deal with the environment through sophisticated cultural behavior rather than brute physical force. They made spears and spearheads for hunting large game and scrapers for cleaning animal hides. They routinely built shelters, and their large front teeth were often worn, perhaps from chewing hides to soften leather for clothing. The skeletons of old and disabled Neanderthals show that others cared for the elderly and infirm, and some Neanderthal populations appear to have buried their dead ceremoniously with fine stone tools, servings of game meat, and even flowers—perhaps indicating evidence of belief in an afterlife.

By careful comparisons of fossils, most anthropologists conclude that anatomically modern humans did not descend from Neanderthals but rather arose elsewhere (fossil evidence suggests Africa), migrated northward, and eventually outcompeted the archaics on their own turf. The two types may have coexisted for thousands of years, but by some 34,000 years ago, Neanderthals and other archaic *H. sapiens* had disappeared from the fossil record, replaced by anatomically modern humans.

Anatomically modern *H. sapiens* would have looked distinctly different from archaic and transitional humans that lived at the same time. Their faces were smaller, flatter, and less projecting than the Neanderthals', the heavy brow ridges had all but disappeared, and their skulls were higher and rounder, with thinner cranial bones. Their limbs were more slender, but still stoutly athletic compared to our own, their teeth were smaller, and most important, their cultural remains were vastly more complex. Their tool kits contained knives, chisels, scrapers, spearheads, and axes for shaping other tools of rock, bone, and ivory. They left hundreds of sophisticated cave paintings, engravings, and sculptures, suggesting a major development of symbolic forms of communication, probably accompanied by increased language abilities [FIGURE 23.11]. Their living sites and shelters became larger and more complex, and human burials became more common and elaborate, indicating the establishment of supernatural beliefs.

The success of these cultural developments allowed humans to move into regions they previously could not inhabit. They followed the herds of mammoths, woolly rhinoceroses, reindeer, and other game into the arctic regions of Eurasia and across the Bering land bridge into the Americas [review FIGURE 23.10]. They also built boats to carry them across uncharted waters to New Guinea and Australia. By 25,000 to 15,000 years ago, people had occupied virtually all the inhabitable regions of the earth.

Anthropologists call the accumulation of useful skills and knowledge, and the discarding of harmful practices, passed down through thousands of human generations **cultural evolution**. Cultural evolution led eventually to the domestication of animals and the development of agricultural practices about 10,000 years ago. Humans were able to control food supplies, elimi-

[A]

[B]

FIGURE 23.11
The Origins of Art.
People left numerous cave paintings throughout Europe and other continents beginning about 29,000 years ago. **[A]** This famous cave painting from Lascaux Cave in southern France depicts a bull and running horses. **[B]** The famous Venus of Lespuque, a statue carved of mammoth ivory, found in Lespuque, France, and dated at about 25,000 years ago, is interpreted as humankind's first fertility symbol.

nate predators and some diseases, and blunt the effects of natural selection on themselves and their domesticated species. While we *H. sapiens* arose via the same mechanisms and faced the same survival pressures as other primates, mammals, and vertebrates before us, cultural evolution has changed our relationship to the planet. Today, we are the only creatures to control most aspects of our own existence, and with that control may lie the future of all species.

➤ CONCEPT CHALLENGE

Childbirth is more hazardous in humans than any other primate. The infant's head is large in relation to its body, and the mother's pelvic bones are arranged for upright posture. Design a hypothesis for the role of natural selection in determining infant head size, maternal pelvic shape, and extended infant care. List at least one prediction of your hypothesis that could be tested by examining the fossil record.

Connections

Humans—members of the genus *Homo*—have existed on this planet for only about 2 million years, and our species, the sole surviving human species, has been here only about 500,000 years. This is about one ten-thousandth of the time the earth has been inhabited by living organisms. And yet this one species—*Homo sapiens*—is the only one in a position to systematically learn the mechanisms that led to the evolution and activities of other living things. Our species' enormous brain capacity and concomitant cultural development represent the extension of a trend characteristic of the Primate order since its inception—that of using brain power to solve many problems posed by the natural environment. But in our case, we appear to have taken this evolutionary trend of increasingly large brain size and enhanced behavioral flexibility to such an extreme that some would argue that humans represent a difference in kind rather than degree. But the gradual evolutionary pathway leading to the present-day human brain is no different in principle from the evolutionary pathway that began at the origin of life and led to the modern prokaryotes, protists, fungi, plants, and other animals. The following parts of the book will explore how animals and plants, once evolved, could meet the day-to-day challenges of survival.

KEY TERMS

arboreal, 544
Australopithicus, 549
anthropoid, 544
bipedal, 546
cultural evolution, 554
hominid, 545
hominoid, 545
Homo erectus, 552
Homo habilis, 550
Homo sapiens, 541
opposable thumb, 542
prehensile tail, 545
primate, 542
prosimian, 543
stereoscopic vision, 546

HIGHLIGHTS IN REVIEW

1 Biologists have discovered patterns of evolution among the primates—lemurs, tarsiers, monkeys, apes, and humans—by studying evidence from fossils, from the structure of DNA and proteins, and from the anatomy and behaviors of present-day organisms.

- **a]** Fossil evidence suggests that the mammalian order Primates evolved nearly 60 million years ago from insect-eating ancestors that resembled tree shrews.
- **b]** The order Primates contains two suborders, the prosimians, including lemurs, and the anthropoids, including monkeys, apes, and humans. The anthropoids have three superfamilies, including the Old World monkeys, the New World monkeys, and the hominoids (apes and humans). The hominoids include three families, the gibbons, the pongids (great apes, including orangutan, gorilla, common chimpanzee, and bonobo, or pygmy chimpanzee) and the hominids (humans and their extinct relatives).

2 Fossil evidence suggests that primates evolved in the canopy of the tropical forest. The features that evolved among earlier primates were the basis for the later evolution of humanlike forms.

3 Upright posture frees hands to manipulate tools and foods, and probably helps cool the brain. The enlarged brain supports increased dexterity. However, a fully upright posture constrains the form of the birth canal in humans and limits the size of a baby's head. Human babies are thus born in a helpless state and require intensive care for many years. Human social systems and spoken languages promote that care.

- **a]** The earliest true hominid discovered to date is *Australopithecus afarensis*, a small but fully upright African protohuman dated at 3 to 4 million years ago. Other *Australopithicus* species arose over the next 2 million years; all were vegetarians with massive teeth and jaws associated with processing some type of coarse plant material. There is no evidence to suggest that any *Australopithicus* made tools.
- **b]** *Homo habilis* arose from an *Australopithicus* species but had a larger brain and made crude stone tools. *Homo erectus* had a still larger brain, more sophisticated tool use, cooperative hunting, and perhaps language. Archaic and transitional *Homo sapiens*, including Neanderthals, were stocky, powerful people with protruding faces and brow ridges and large brains. They made various kinds of stone tools, built shelters, cared for their elderly and sick, and buried their dead.
- **c]** Anatomically modern *H. sapiens* coexisted with the Neanderthals and eventually replaced them. They had higher, rounder skulls, more slender limbs, and far more complex tool kits, and showed many other examples of increasingly complex cultural behavior.

UNDERSTANDING THE FACTS AND CONCEPTS

For Questions 1–5, match each of the descriptions with the most appropriate term from the following list. Any answer may be used once, more than once, or not at all.

- a] prosimian
- b] anthropoid
- c] hominoid
- d] hominid
- e] pongid
- f] primate

1 The term that includes all the others.

2 Most closely related to the ancestral primates.

3 Includes the gibbons, orangutans, gorillas, and chimps.

4 A term used to include gorillas and chimps but not humans.

5 Includes the Old and New World monkeys, the apes, and humans.

As above, for Questions 6–10, match each of the descriptions with the most appropriate term from the following list. Any answer may be used once, more than once, or not at all.

- a] New World monkeys
- b] Old World monkeys
- c] lemurs
- d] gorillas
- e] humans

6 True monkeys without a prehensile tail.

7 True monkeys with a prehensile tail.

8 The only surviving hominid.

9 Mostly arboreal; some nocturnal and some diurnal species.

10 The largest of the apes.

As above, for Questions 11–15, match each of the descriptions with the most appropriate term from the following list. Any answer may be used once, more than once, or not at all.

- a] *Proconsul africanus*
- b] *Sivapithecus*
- c] *Pan troglodytes*
- d] *Australopithecus*

11 A surviving ape.

12 Surviving orangutans are similar to this extinct genus.

13 An extinct tree-dwelling, fruit-eating ape that was one of several in the initial radiation of apes.

14 As far as we know, the first hominid.

As above, for Questions 15–20, match each of the terms or descriptions with the most appropriate term from the following list.

- **a]** *Australopithecus afarensis*
- **b]** *Homo habilis*
- **c]** *Homo erectus*
- **d]** *Homo sapiens*
- **e]** *Homo sapiens sapiens*

15 Neanderthal.

16 The first human toolmakers.

17 A fully bipedal, small-brained link between apes and humans.

18 Associated with tools, such as hand axes that were more advanced than scrapers and choppers, and with cultural transmission of information.

19 Modern humans.

INTEGRATE AND APPLY WHAT YOU HAVE LEARNED

1 Which living animals are primates? Which are monkeys and which are apes?

2 Which physical and behavioral traits do we share with the other primates, and which are currently thought to be particularly "human"? Are the particularly human traits unique to human beings?

3 What is the genetic evidence concerning the relatedness of chimps and our own species? What conclusions have been drawn from that evidence?

4 What evidence do we have that bipedalism preceded large brain size in the evolution of humans?

5 Contrast the multiregional with the single-origin model for the replacement of *H. erectus* by *H. sapiens.*

ANALYSIS

1 If *H. sapiens* arose by the single-origin model:

- **a]** You would expect a higher degree of continuous variation among modern groups of humans than if the species arose in many places.
- **b]** You would expect a lower degree of continuous variation among modern humans than if the species arose in many places.
- **c]** You would necessarily expect the various modern human groups to be less closely related than if they had arisen at several times and places.
- **d]** You would necessarily expect all groups of modern humans to be equally different from the common ancestral group.
- **e]** You would expect all groups of modern humans to be more closely related to each other than any of the groups to *H. erectus.*

2 Which of the following traits seems to be the most necessary precondition for the spread of humans from the savanna to temperate climates?

- **a]** Bipedalism.
- **b]** Loss of body hair.
- **c]** Complex toolmaking skills.
- **d]** Slender stature and less massive skulls.
- **e]** Cultural practices such as burying the dead.

3 Which of the following is true? Explain.

- **a]** Biological evolution is more rapid than cultural evolution.
- **b]** Cultural evolution is more rapid than biological evolution.
- **c]** Both are equally rapid.

4 Which of the following best explains why *H. sapiens* is the only genus of surviving hominids, whereas there are several surviving species of prosimians, Old and New World monkeys, and apes? Explain. (More than one answer may be correct.)

- **a]** The high degree of genetic similarity among all humans.
- **b]** The relatively short time between the evolution of the species and the emergence of exploration and travel.
- **c]** The several groups of hominids have all blended into one form that is genetically superior to any other form.
- **d]** The human species actually is part of the same genus as that which includes the chimps, so we really are not a single species in a single genus.
- **e]** There are really many species in the genus *Homo*, but the other species mostly remain hidden from view.

5 The genetic variability within the human species is confined to about 1 percent of our genes, the same percentage of genetic difference that separates us from chimps. Which of the following best answers the question, Why don't humans and chimps look even more alike? Explain. (More than one answer may be correct.)

- **a]** The genetic differences between chimps and people are not the same ones that separate chimps and humans.
- **b]** Humans must differ among themselves in regulatory or control genes.
- **c]** Humans must differ from chimps in regulatory or control genes.
- **d]** Only those few genes that determine precise structures need to be different because we really resemble chimps more than we care to admit.

PART THREE: APPLY AND DECIDE

Cane Toads in Australia

Sugarcane has long been an important cash crop for Australian farmers, but in the mid 1930s, the adults and larvae of the sugarcane beetle were devouring leaves and roots and decimating the fields. The farmers learned that their counterparts in Puerto Rico and Hawaii had successfully imported and released breeding pairs of *Bufo marinus*, the most common toads of Central and South America, and that the amphibian immigrants were successfully controlling sugarcane pests. So in 1935, they brought 51 pairs of giant toads from Hawaii and released them into a pond near Cairns in the tropical northeast. Within 18 months, the females had laid at least 1.5 million eggs, and workers had caught more than 60,000 toadlets and released them into their fields.

The cane toads in this bumper crop adapted beautifully to their new home down under—but not specifically to the cane fields. They didn't, in fact, like the cane fields at all in the dry season. They did, however, love the forests, backyards, gardens, and garbage dumps of adjacent towns. There they helped themselves to vegetable scraps, to dog food from pet bowls, and to most of the insects, fish, and other small prey they could track down and stuff into their large mouths. The farmers had created an ecological monster, and today, half a century later, the problem is still getting worse.

Australia's imported cane toads are excellent examples of adaptation, the bottleneck effect and genetic drift, and the biodiversity crisis. They also embody a question that you, the reader, must answer yourself about how to solve biological dilemmas that have been created by humans.

The cane toad's physical and behavioral adaptations help explain why it became such a success—and such a pest—in Australia. We just saw that *B. marinus* is practically an eating and mating machine. Biologists consider it the most versatile amphibian on earth. It can live in rain forests, in semidesert conditions, at sea level, and at elevations up to 2000 m (6500 ft). Wherever it lives, it can eat almost anything. This is, in part, because it has such a large "gape," or mouth opening. In the world of toads and frogs, the bigger the gape, the wider the range of foods, and this *Bufo*'s gape matches its overall body size: up to 17.8 cm (7 in) wide by 22.86 cm (9 in) long [FIGURE 1].

Female giant toads can produce up to 30,000 eggs at a time. Some aspect of the female toad's body apparently drives male cane toads wild; the males jump on and cling to females with remarkable energy, and individual males have even been seen clasping dead females smashed flat on the road! Of every 200 eggs that a female lays and a male fertilizes, virtually all will develop into tadpoles and toadlets, but only 1 will survive to adulthood. That still leaves up to 150 new adults per clutch of eggs, however, and thus *B. marinus* can multiply rapidly.

A survival rate of 0.5 percent is surprisingly low considering the toad's many defensive adaptations. First of all, the animal's body is yellowish green and warty, and since is nearly as wide as it is long, it looks to many observers like a hopping cowpat. When frightened, the toad will freeze in place, and this increases the resemblance still further. If a predator is not fooled and takes a bite, it is very likely to sicken and/or die from the milky exudates of the toad's poison glands. The eggs are equally poisonous, and an Australian college student who swallowed some on a dare suffered three heart attacks!

Only the toadlet stage is fully vulnerable to predators, and it is then that birds such as the Australian swamphen can gorge on the vast majority of cane toads that hatch. Even so, with up to 150 new adults per mating (and several matings per female per year) the toad's fecundity and defensive strategies far outstrip its predators. There is some evidence that swamphens, cane rats, and dingoes (wild dogs) have learned to flip the toads and eat the tender belly. Researchers will have to determine whether this learned behavior is being passed on as "cultural evolution" that could someday lead to better population control by predators. In the meantime, however, the cane toads number tens of millions.

The rapid population growth and territorial expansion of the cane toad has led to allele frequency changes based on a bottleneck effect and genetic drift. All the cane toads hopping around northeastern Australia today are descendants of only about 50 mating pairs in 1935. Presumably this small population constituted a genetic bottleneck that decreased genetic variation in the Australian cane toads compared to the original population. That, you may recall, is apparently what happened to cheetahs in Africa.

FIGURE 1 **Large Cane Toad with Small Australian Friend.**

Geneticists have detected allelic differences in groups of cane toads from different areas of Australia. The geographic distribution suggests that genetic drift is taking place, due to a series of population bottleneck events. Apparently, a few toads (the "bottleneck") may invade an area far from the original center of cane toad population, and the descendants of these few toads may multiply into a new large local population. Because the few founders may not be a random sample of the original large population, allele frequencies may come to differ greatly in different regions. It remains to be seen whether the inbreeding associated with bottlenecks will eventually increase the toad's vulnerability to predation or disease, as happened to cheetahs [review CHAPTER 16]. But so far, nothing seems to be stopping the warty imports, and in fact, they are continuing to spread toward the tropical richness of Australia's Top End.

The farther north one travels in Australia (since it is in the southern hemisphere) the warmer and wetter the climate and the more lushly diverse the plants and animals. The Top End, the area around Darwin, is renowned for its water birds, crocodiles, and other wetland species. Australian ecologists are worried, however, about the continued biodiversity in this region, because cane toads have been migrating toward it at a distance of 27 km (16 mi) per year. When the toads arrive, they will doubtless eat indigenous insects, fish, animal eggs, and vegetation, just as they have done in the tropical northeast.

So far, Australians have found only one effective control agent (besides the occasional automobile tire): squirting a cane toad with chloroxylen, a disinfectant, will stop the animal in its tracks. But hand-to-toad combat is clearly not going to counteract a widespread invasion, and some biologists have considered releasing a virus that is known to weaken and kill cane toads in their native regions of Latin America. Does this release seem like a good plan to you? What would the advantages be? Can you foresee any pitfalls to fighting exotic fire with fire? (An *exotic* is a species released in a nonnative area.) Can you think of other solutions? Questions like this require you to apply the evolutionary facts and principles you have learned in this section. You can also consult articles such as those listed here. ❑

PART FOUR

How Animals Survive

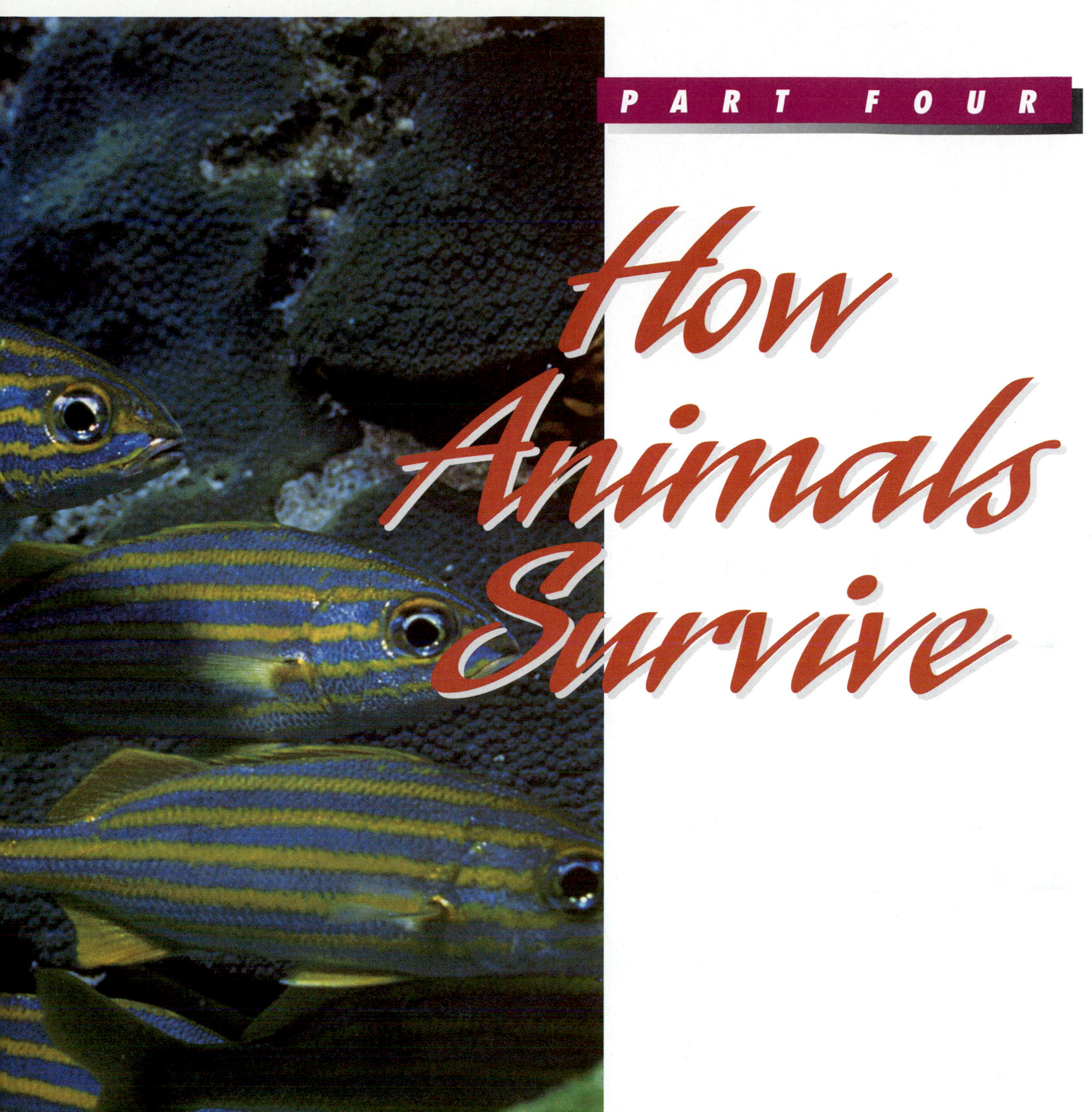

CHAPTER 24

How Animals Function

A WHALE OF A SURVIVAL PROBLEM

An intrepid visitor to the perpetually frozen Antarctic could stand at the coastline, raise binoculars, and witness a dramatic sight just a few hundred meters offshore: a spout as tall and straight as a telephone pole fountaining upward from the blowhole of a blue whale (*Balaenoptera musculus*), then condensing into a massive cloud of water vapor in the frigid air. The gigantic animal beneath the water jet would be expelling stale air from its 900-kg (1-ton) lungs after a dive in search of food. Then, resting at the surface only long enough to take four deep breaths of fresh air, the streamlined animal would raise its broad tail, thrust mightily, and plunge into the ocean again [FIGURE 24.1]. The observer on shore might see such a sequence only twice per hour, since the blue whale can hold its breath for 30 minutes as it glides along below the surface, swallowing trillions of tiny shrimplike animals called krill.

FIGURE 24.1
Blue Whale Plunging for Krill.
Earth's largest animal, the blue whale, represents a classic problem of homeostasis: How can constant internal conditions be maintained despite wide variations in environmental conditions?

It is difficult to comprehend the immense proportions of the blue whale, the largest animal ever to inhabit our planet. This marine mammal is longer than three railroad boxcars, and bigger than any dinosaur that ever lumbered on land [FIGURE 24.2]. It weighs more than 25 elephants or 1600 fans at a basketball game. Its heart is the size of a small car. Its largest blood vessel could accommodate an adult person crawling on hands and knees. The animal has a tongue the size of a grown elephant. It has 45,500 kg (50 tons) of muscles which move its 54,500 kg (60 tons) of skin, bones, and organs. And this living mountain can still swim at speeds up to 30 mi (48 km) per hour!

Leviathan proportions aside, it is difficult to grasp the enormous problems that so large an organism overcomes simply in staying alive. For starters, a blue whale is a warm-blooded animal with a relatively high metabolic rate; it requires a million kilocalories a day just to stay warm and active in an icy ocean environment. The whale obtains calories from 3600 kg (8000 lb) of krill strained from the ocean water each day on special food-gathering sieve plates that hang from the roof of its mouth. In addition, each of its trillions of cells must exchange oxygen and carbon dioxide, take in nutrients, and rid itself of organic wastes, just as a single-celled protozoan must do. Yet a given whale cell—a liver cell, let's say—can lie deep in the body, separated from the oxygen and nutrients in the environment by nearly 2 m (6 ft) of blubber, muscle, and bones. This problem is solved by an elaborate transport system that delivers oxygen and nutrients to and carries carbon dioxide and other wastes away from every cell. Finally, the galaxy of living whale cells is coordinated and controlled by a brain, a nervous system, and chemical regulators (hormones), allowing the organism to function as one unit.

Although blue whales are the largest animals that have ever lived, they share with all other animals the same fundamental physical problems of day-to-day survival: how to extract energy from the environment; how to exchange nutrients, wastes, and gases; how to distribute materials to all the cells in the body; how to maintain a constant internal environment despite fluctuations in the external environment; how to support the body; and how to protect it from attackers or from damaging environmental conditions. Blue whales have evolved with unique adaptations of form and function that meet such challenges beautifully.

While the blue whale's solutions are specific to huge size, they represent general approaches that we will encounter again and again as we investigate animal anatomy and physiology. **Anatomy** is the study of biological structure; **physiology** is the study of how such structures work. While animals and plants share many basic elements of anatomy and physiology, there are also important differences; thus, we separate the two subjects, covering animals in PART FOUR and plants in PART FIVE.

The present chapter gives an overview of animal anatomy and physiology, and CHAPTERS 25 through 33 delve into specifics. Here we discuss **homeostasis** (*homeo* means "same"; *stasis* means "standing"), the maintenance of constant internal conditions despite fluctuations in the external environment. And we will see much more about the massive blue whale. ❑

MESSAGES

1 In large animals, interior cells lie far from the external environment, the source of oxygen and food and the sink for carbon dioxide and other wastes. Thus, direct diffusion to and from the outside cannot occur quickly enough to prevent cell death. Tissues, organs, and organ systems have evolved that bring environmental resources close to every cell.

2 If the physical and chemical conditions in and around body cells change beyond certain limits, the cells die. Adaptive mechanisms in animals tend to maintain the internal environment within defined limits. Collectively, these mechanisms are called homeostasis.

3 Homeostasis is frequently achieved by negative feedback loops in which a receptor detects a change, an integrator determines a counteracting change, and an effector performs an activity that returns conditions to within acceptable limits.

FIGURE 24.2
A Whale to Scale.
A blue whale is longer and far heavier than an elephant or even an *Apatosaurus*, the longest land animal that ever lived.

Staying Alive: Problems and Solutions

Living organisms are subjected to wide variations in temperature, light, acidity, salinity, wind speed, and availability of water, minerals, and nutrients. These and other environmental factors create a shifting external setting to which organisms must adjust or die. At the same time, each living thing carries out certain activities basic to life [see CHAPTER 1], including gathering energy and maintaining body integrity. These activities (or the potential for them) generally continue in a smooth, steady, orderly manner, or, again, the organism dies. This steady state usually requires a continuous supply of energy and materials from the environment, as well as the disposal of wastes into it. The central problem for a living thing, therefore, is to maintain a steady state internally despite an often harsh and fluctuating external environment.

MAINTAINING HOMEOSTASIS

Because cells are the basic units of life, essential life processes must be maintained in single cells as well as in larger organisms composed of many cells [TABLE 24.1]. Diffusion and active transport across the plasma membrane are the means by which single cells exchange materials with their environment [see CHAPTER 4]. The same is true for a multicellular organism like a flatworm [review FIGURE 21.10], whose dimensions are thin and flat enough to bring each cell within diffusion distance of the outside world.

The individual cells of multicellular organisms also exchange materials with their immediate surroundings by diffusion and active transport. However, multicellular organisms are generally too large and too thick for each of their cells to exchange materials directly with the envi-

TABLE 24.1 Essential Life Processes

Process	Responsible Organ Systems
Obtaining materials from the environment	Digestive system
Obtaining energy from the environment	Digestive system
Exchanging gases (oxygen and carbon dioxide) with the environment	Respiratory system
Regulating the composition of body fluids	Excretory system
Distributing materials within the body	Circulatory system
Defending the body from attack	Skin, immune system
Protecting the body from temperature extremes	Skin, circulatory system
Maintaining body shape	Skeleton, muscular system
Growing and reproducing	Reproductive system
Coordinating body activities	Nervous system, endocrine system

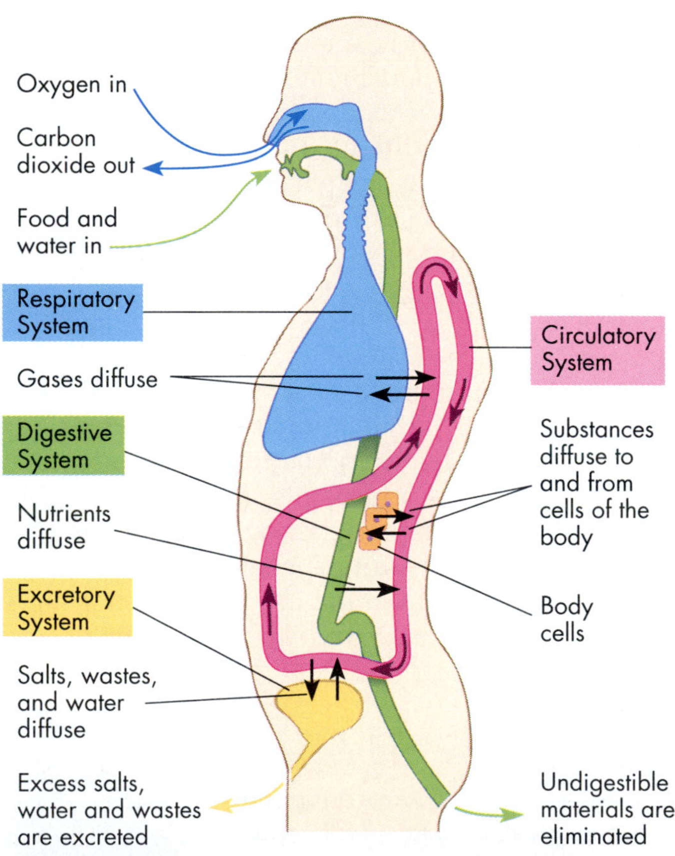

FIGURE 24.3

Tubes and Diffusion: The Common Strategy for Exchanging Materials with the Environment.

Animals obtain nutrients, exchange gases, and get rid of organic wastes by tubes that penetrate deep into the body. The tubes exchange substances with the blood by diffusion. Substances diffuse to and from the blood and the individual cells.

ronment. A cell in your pancreas, for example, can be 4 inches or so from the air outside your body. Thus the cells that lie deep inside multicellular organisms are served by a number of special systems that evolved to overcome such distances [see TABLE 24.1].

FIGURE 24.3 shows three body systems—the digestive system, the respiratory system, and the excretory system—that employ a common strategy: (1) Each system has a supply or disposal tube running from the outside environment deep into the animal's body. (2) Each tube has special regions where substances can be exchanged between the contents of the tube and nearby body fluids bathing the tube. This exchange occurs by diffusion and active transport over short distances. (3) Body fluids can be circulated to every cell in the body by means of bulk flow, a moving stream. (4) Through diffusion and active transport, substances can move into each cell from the circulating fluids, or be released by the cell and carried away. Consider the respiratory system, for example [see FIGURE 24.3]: Oxygen enters the body in a tube and diffuses into the bloodstream in a special region, the lungs; bulk flow then carries the blood with its dissolved oxygen throughout the body; and finally, the oxygen can pass out of the blood and into individual body cells by diffusion.

Although tubes and transport mechanisms give a multicellular organism some control over the immediate environment surrounding most of its cells—for example, its salt content or temperature—still, each animal directly contacts the physical environment with its outermost surface, and thus as a whole remains subject to environmental fluctuations.

BODY ORGANIZATION: A HIERARCHY

Within a multicellular organism, especially one as enormous as a blue whale, gases and fluids could not be transported inside the body in a useful way if the cells were crammed together haphazardly. Instead, the cells are organized into tissues, organs, and organ systems [FIGURE 24.4].

A **tissue** is a group of cells of the same kind performing the same function within the body. For example, cells that line the outer surface of a whale's skin make up a tissue called the epithelium. An **organ** is a unit composed of two or more tissues that together perform a certain function. Your stomach, for example, is an organ containing four tissue types cooperating in the function of food storage and digestion. Several organs can form an **organ system**, which is defined as two or more interrelated organs that work together, serving a common function; the digestive system, for example, contains the salivary glands, liver, stomach, intestines, pancreas, and other organs functioning together in food processing. Collectively, all the organ systems make up an **organism**, in this case an individual animal.

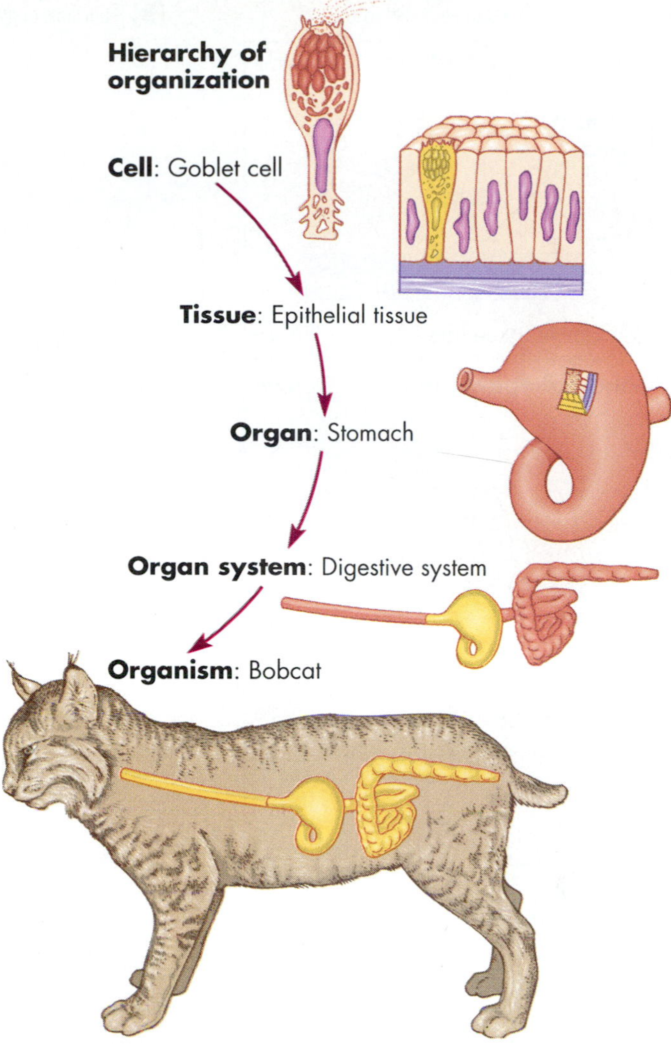

FIGURE 24.4
Hierarchy of Organization in the Animal Body.
The billions of cells in an animal's body are neatly organized into tissues, organs, and organ systems that carry out body functions efficiently. Organ systems working together in a coordinated fashion help an organism, such as an individual animal, stay alive.

THE FOUR TYPES OF TISSUES

In the body of a mammal such as a blue whale or a human, there are over 200 types of cells, but only four types of tissue: epithelial, connective, muscle, and nervous tissues. Most organs contain all four tissue types. In the small intestine, for example, the interior is lined with *epithelial tissue. Connective tissue* binds this epithelium to a layer of *muscle tissue* surrounding the organ. Finally, signals from *nervous tissue* can stimulate muscle contraction and speed the passage of materials through the small intestine.

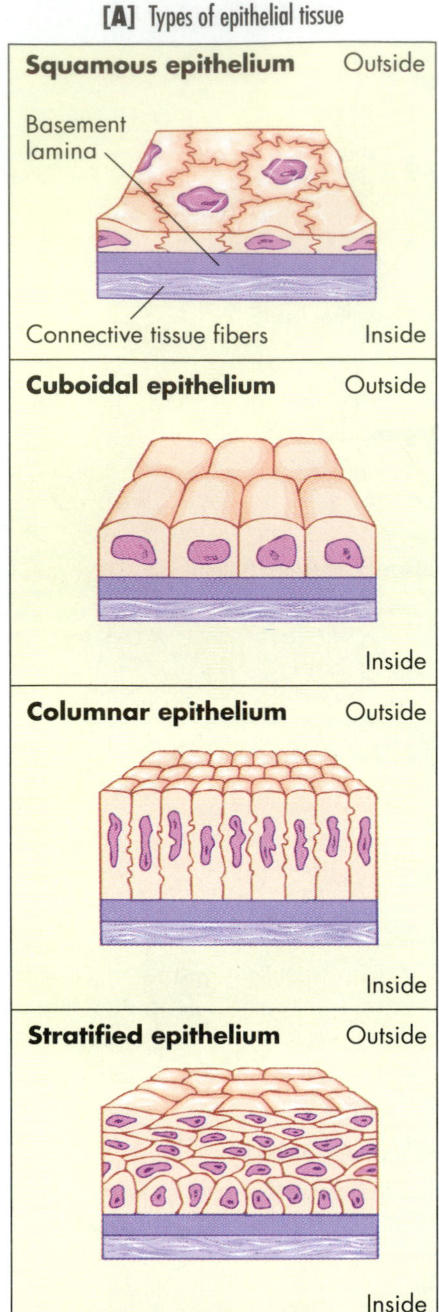

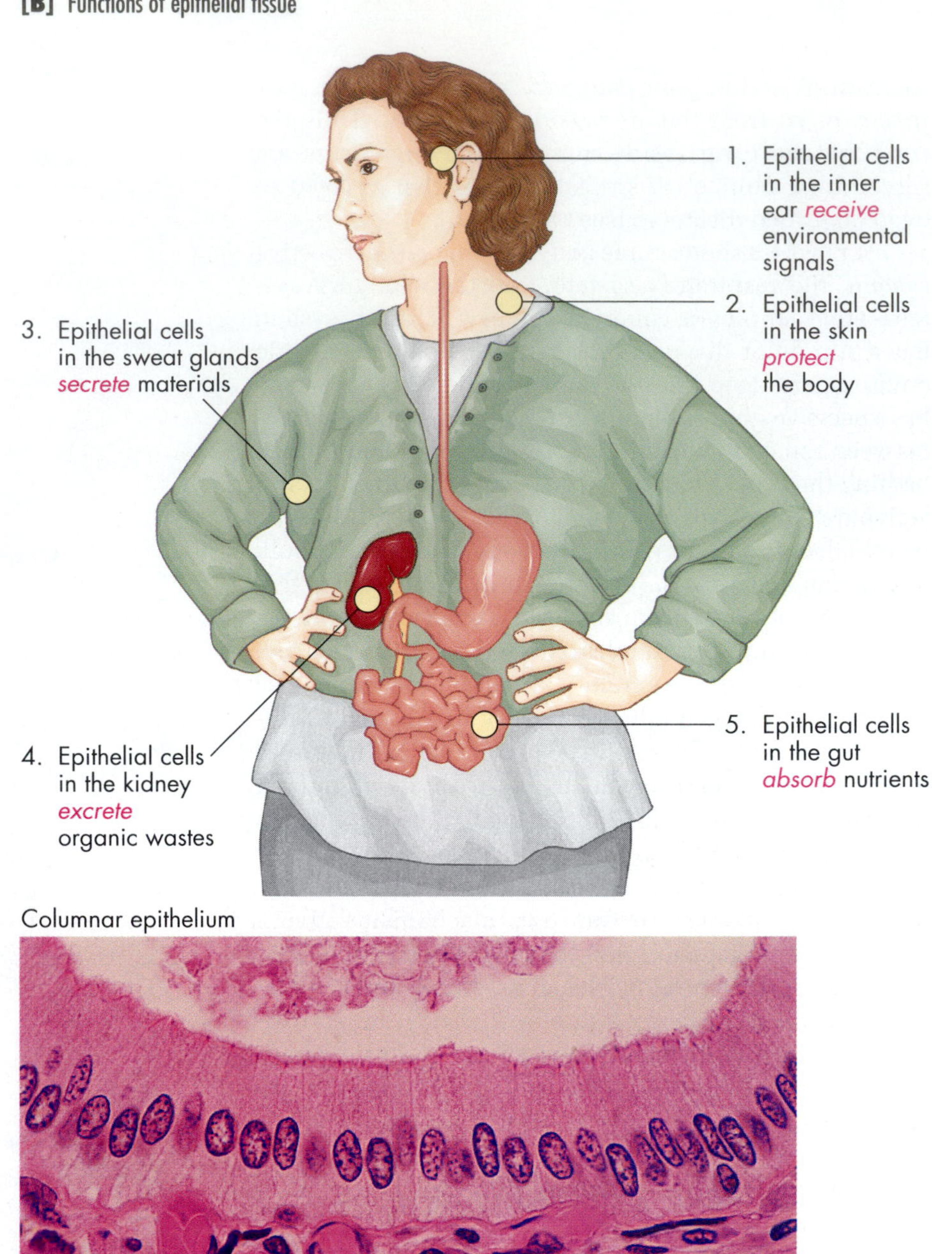

FIGURE 24.5

Types of Epithelial Tissue and Their Roles in the Body.

Epithelial tissues (or epithelia) are sheets of cells that cover body surfaces. **[A]** Epithelial cells can be squamous, cuboidal, columnar, or stratified. The photo shows columnar epithelium lining a mammal's milk duct. Epithelial tissues act as a threshold through which substances must pass to enter or exit the body. **[B]** Epithelial tissue has many functions.

Epithelial Tissue **Epithelial tissue** consists of sheets of cells that cover or line body surfaces, including the outer layers of the skin or the inner layer of the intestine [FIGURE 24.5A]. Epithelial cells may be *squamous* (flat), *cuboidal* (shaped like cubes), *columnar* (shaped like columns), or *stratified* (arranged in layers several cells deep). Epithelial tissue has several functions, including (1) *reception* of environmental signals, (2) *protection* of the body from foreign substances, (3) *secretion* of sweat, milk, wax, and other materials, (4) *excretion* of wastes, and (5) *absorption* of nutrients, drugs, and other substances [FIGURE 24.5B]. Epithelial tissue also makes up two major types of glands: the *exocrine glands*, which secrete substances like sweat onto the skin, and digestive juices into the gut; and some of the *endocrine glands*, which secrete *hormones*, internal chemical messengers, such as testosterone and insulin, within the body. Continuous cell division is common in epithelial tissue, where cells are often worn away by abrasion, both inside and outside the body. Damaged epithelial tissue can interfere with normal body functioning. For example, when cigarette smoke damages the cilium-covered epithelium lining the breathing tubes, the epithelial lining can no longer remove harmful debris from the lungs and passageways as efficiently.

Epithelial sheets typically have one surface facing a space, like the air in the case of the skin, or the urine-storing cavity of the urinary bladder. The cells facing the space link tightly to each other by means of *cell junctions*. These junctions bind adjacent cells together and stop the leakage of fluids from one side of the epithelium to the other. If these junctions did not exist, urine would leak out of the bladder's central storage cavity into the body. Junctions also link the inner surface of the epithelium to a thick meshwork of protein and polysaccharide fibers, the *basement membrane*, or *lamina*. If it weren't for this firm bond, the skin would peel right off your body. Just beneath the basement lamina of epithelial tissue is the second tissue type, connective tissue.

Connective Tissue **Connective tissue** binds other tissues and supports flexible body parts. It produces extracellular material that forms a matrix, or framework, for other structures. Connective tissue cells, which lie within the matrix, are spaced much farther apart than the densely packed cells of epithelial tissue [FIGURE 24.6]. This matrix includes fibers of the proteins collagen and elastin. Collagen is the most abundant protein in the body; weight for weight it is as strong as steel. A mutation in a collagen gene can cause the hereditary disease osteogenesis imperfecta, in which bones are so weak that the rib cage collapses at birth when the baby draws its first breath.

The main types of connective tissue are (1) **loose connective tissue**, which fills spaces between muscles and forms delicate membranes that connect epithelial layers to underlying organs; (2) **adipose tissue**, or *fat*, which stores fat droplets and acts as a buffer around the kidneys, joints, and other areas; (3) **fibrous connective tissue**, which contains a few cells but mostly collagen and elastic fibers, and which makes up tendons, ligaments, and similar tough, flexible structures; and (4) **cartilaginous tissue**, or *cartilage*, which is made up of fibers in a gel-like matrix and is rigid enough to provide the framework of the nose and outer ears, attachment sites for muscles, and similar structures. Reach up and gently squeeze the upper part of your outer ear—the stiff material you feel is cartilage [FIGURE 24.6A]. Some scientists also consider blood connective tissue, as well as bone, which is made up of a hardened matrix plus living cells [see CHAPTER 33].

Muscle Tissue An animal moves by means of **muscle tissue**, which contracts, or shortens, and thereby applies force against objects [FIGURE 24.7A; see also CHAPTER 33]. Grab your upper right arm with the left hand and raise your right hand up to your chin. The hardening bulk you feel is muscle tissue as it shortens to move the arm. Muscle cells are generally quite long and spindle-shaped and are filled with protein fibers. The fibers slide past each other like two parts of a telescope, and cause the cell to shorten. There are three types of muscle: (1) *smooth muscle*, which operates glands, blood vessels, and internal organs such as the intestine; (2) *cardiac muscle*, which makes the heart beat; and (3) *skeletal muscle*, which moves the bones, like those in your arms and legs. The contraction of muscle cells can be controlled by the fourth tissue type, nervous tissue.

Nervous Tissue **Nervous tissue** transmits electrochemical impulses; for example, you feel a pin prick on your finger tip because the pin deforms the nervous tissue and this triggers a nerve impulse [FIGURE 24.7B; see also CHAPTER 31]. The essential feature of nervous tissue is its ability to transmit signals and thereby communicate. Nerve cells, or *neurons*, sense changes in their environment, process that information, and command muscles or secretory glands to act and thus adjust for the initial change. Nerve cells have very long, thin processes that act like telephone wires carrying messages from one place to another in the body. The longest cells of the body are nerve cells. The activity of nervous tissue is so interesting that we devote two chapters to it—CHAPTERS 31 and 32.

TISSUE ENGINEERING

A modern understanding of tissue structure and function coupled with newly developed space-age materials is leading to a new medical field called *tissue engineering*. Tissue engineers develop functional substitutes for damaged tissues. Consider the experimental method for treating diabetics now under development. When levels of the sugar glucose rise in the blood after a meal, epithelial tissue of the pancreas normally secretes the protein hormone *insulin*. This hormone causes cells to remove glucose from the blood, and hence prevent blood glucose levels from becoming too high. The immune system of most diabetics has destroyed the tissue that normally produces insulin. Tissue engineers are experimenting with a device that consists of a coiled membrane tube [FIGURE 24.8]. On one side of the membrane is epithelium of the pancreas from a normal donor, and on the other is the patient's blood. The membrane separating the two tissues has little holes through which small molecules like

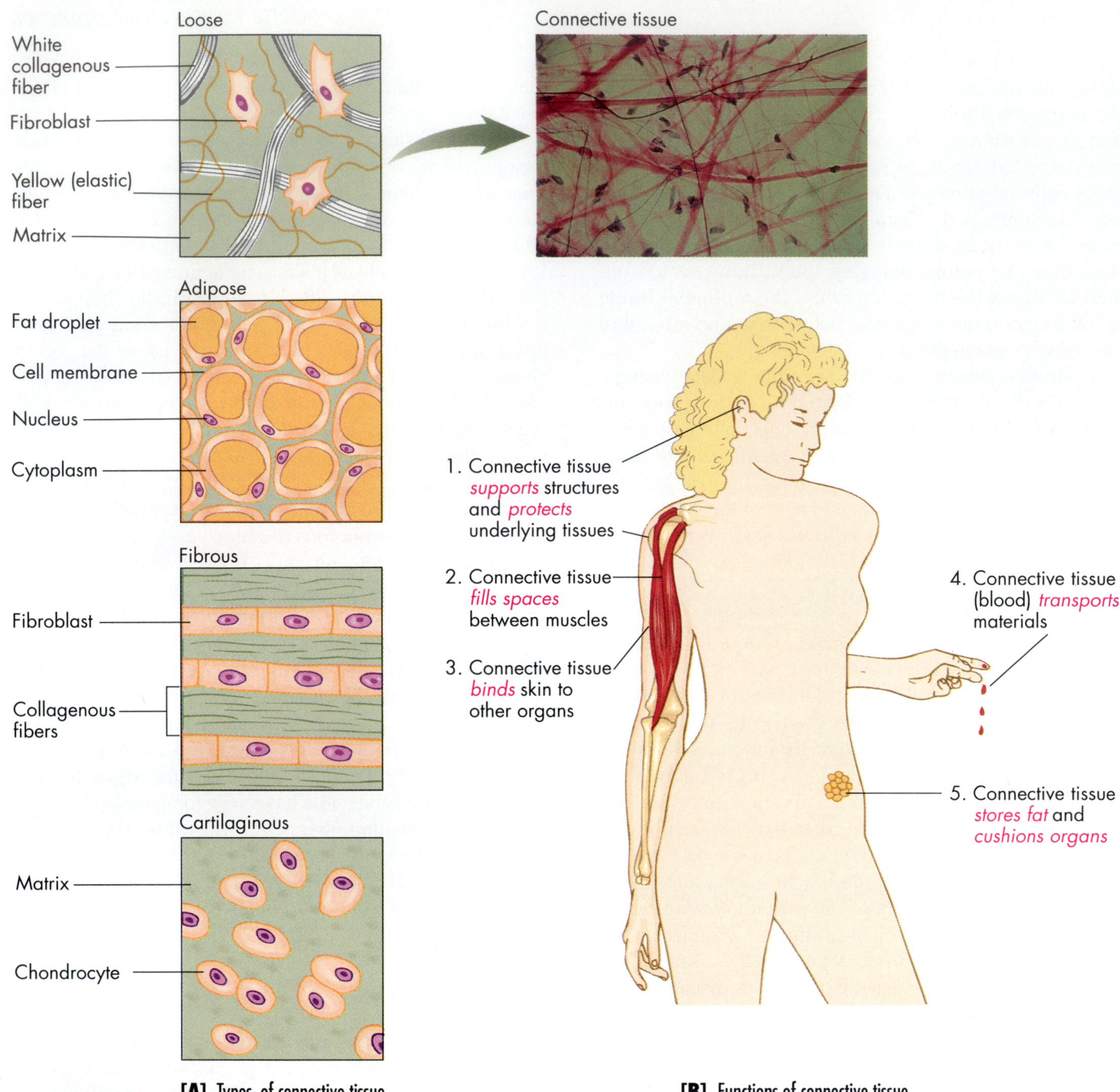

FIGURE 24.6

Types of Connective Tissue and Their Roles in the Body

Connective tissue binds other tissues together and supports flexible body parts. It lies just under the basement membrane to which epithelial cells are attached, and it produces extracellular material that forms a matrix, or framework, for other structures. This matrix includes fibers of the protein collagen and elastin. [A] The main types of connective tissue. The photo shows loose connective tissue in a monkey's oviduct. The small, brown ovals are cell nuclei. [B] The functions of connective tissue.

oxygen, glucose, nutrients, and the small protein insulin can diffuse. Immune proteins that might destroy the donor pancreatic cells cannot penetrate the membrane. The patient's blood circulates through the coiled tube, bringing in nutrients and carrying away insulin and wastes. The pancreatic tissue is supposed to respond to the levels of glucose, produce insulin, and thus prevent high levels of glucose from threatening the patient. Further understanding of the normal mechanisms of other tissues, including nervous tissue and connective tissue, promise treatments for Parkinson's disease, bone disorders, and other conditions.

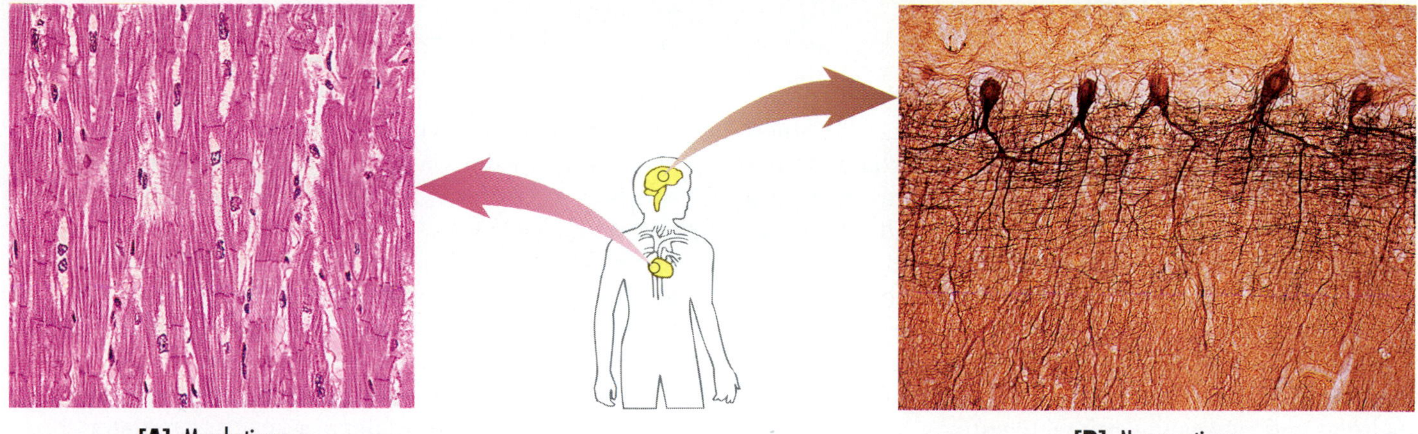

[A] Muscle tissue

[B] Nervous tissue

FIGURE 24.7

Muscle and Nervous Tissue.

[A] Muscle tissue, here cardiac muscle in a mammal's heart, can contract and shorten. The contraction of cardiac muscle can pump blood through the heart and blood vessels, while the contractions of smooth muscle can squeeze food and fluids through the stomach and intestines. The shortening of skeletal muscles can move body parts. [B] Nervous tissue, like this section from the human cerebellum, the part of the brain that helps us maintain balance, conveys information from one body part to another.

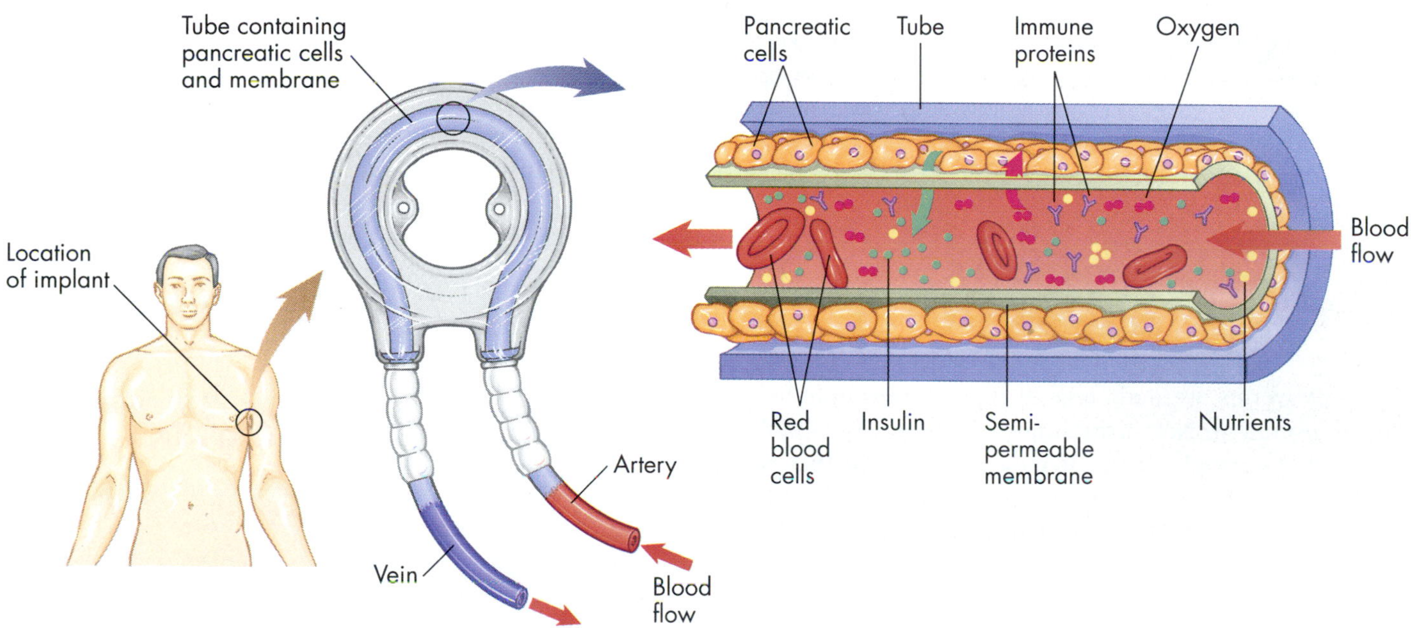

FIGURE 24.8

Tissue Engineering and an Experimental Treatment for Diabetes.

In this device, normal pancreas cells from a donor are sandwiched between two permeable filters. The patient's blood, circulating past the trapped pancreatic cells, brings in oxygen and nutrients, which diffuse through the small holes in the filters. Insulin made by the trapped cells diffuses through the filter into the blood. The insulin secreted from the device travels in the blood throughout the body, and helps maintain a constant level of glucose in the patient's blood.

LARGE ANIMALS FACE ADDITIONAL PROBLEMS

We have seen that the problem of servicing billions of individual cells can be overcome by the organization of tissues into organ systems and the transportation of materials through tubes which bring commodities near to the surface of each cell. But large animals have still other special needs not found in smaller, simpler animals: Large animals require strong structural support and sophisticated avenues of coordination and communication, which integrate far-flung body parts.

Small aquatic organisms such as jellyfish and flatworms do not need extensive solid support because their tissue layers are thin and seawater surrounds and supports them. Bulkier animals on land, however, need a scaffolding to which muscles can attach and which can prevent organs from collapsing together into a quivering mass. The blue whale is a dramatic example. Even with its multiton skeletal system, it cannot live for more than a few minutes without the additional buoyant support of seawater. On land, the weight of its body quickly crushes its lungs and causes the animal to suffocate.

Large animals, like most others, require structures and processes that coordinate and integrate the actions of the various organ systems so that the organism functions efficiently as a single entity. The electrical signals of the nervous system rapidly integrate and regulate body functions, while the slower blood-borne chemical signals of the endocrine system cause slower, longer-lasting body reactions and changes. As we will see, both of these systems help organisms to maintain homeostasis—the internal constancy it needs to survive.

➤ CONCEPT CHALLENGE

Viruses, alcoholism, or other diseases can destroy a person's liver function. If you were a "tissue engineer," how would you design a liver support system?

Animal Adaptations

As heterotrophs, animals take in organic nutrients from the environment. Not surprisingly, the blue whale is the biggest heterotroph of all. It may seem ironic that the world's largest animal eats minute krill, but the blue whale's mouth is perfectly suited to harvesting this food source—the most abundant in cold ocean waters. The blue whale's sieve plates, or *baleen plates*, are evolutionary extensions of the ridges you can feel with your tongue on the roof of your mouth. We have only a few such ridges, while the whale has several hundred, each about 1 m (39 in) long. Together, they filter bushels of krill from the water at a time, which the whale devours with one mighty swallow. The lesson is this: Specialized anatomical adaptations allow different organisms to exploit their environments in different ways, and the adaptation's physical form is appropriate to its function.

The blue whale has numerous other anatomical and physiological adaptations that are fine-tuned to the animal's life-style and environment. Like all animals that swim rapidly, the blue whale has a streamlined spindle shape with hydrodynamic properties that allow it to move through the water with a minimum of drag. The ears have no external flaps, and the male genitals are retracted into a pouch. Besides streamlining the animal, this lack of flaps and appendages also helps to conserve heat by allowing a smaller body surface area for heat loss to the cold seawater. In marked contrast, a mammal like the desert hare will die if it does not dispose of body heat; its very large appendages—huge ears relative to its body size—act like organic analogues of a car's radiator.

The whale's baleen plates and body shape are clearly fine-tuned to environmental constraints. But how do anatomical and physiological adaptations arise? The answer is evolution by natural selection. The traits we have been describing are controlled by genes, and some ancestral whale population must have had genetic variation for the trait of ear length. Presumably, certain whales—by chance mutations and genetic recombination—possessed alleles that generated smaller ears. These animals could move through the water and obtain food more efficiently than large-eared whales, and thus could reproduce more efficiently than whales lacking these genes. Gradually, through random mutation followed by natural selection, genes for smaller ear flaps became more frequent within the whale population. This specific example illustrates a general point: The anatomical and physiological adaptations we see in this section generally arise by natural selection, whether those adaptations are in a beetle, a sponge, a jackrabbit, or a whale.

Keeping the Cellular Environment Constant

During a regular checkup, your doctor can get a fairly accurate reading of how healthy you are by taking your temperature and blood pressure, listening to your heart and lungs, analyzing the contents of your blood and urine, and so on. Since medical researchers have established a normal range of values for each measurement, chances are good that if yours fall within the correct ranges, your body is maintaining homeostasis, that is, keeping the internal environment at a constant healthy level. Your physiological systems continuously adjust the environment in and around your cells, maintaining the extracellular environment within narrow limits suitable to the cells' efficient functioning.

STRATEGIES FOR HOMEOSTASIS: FEEDBACK LOOPS

A driver on the highway keeping a vehicle within a lane is a type of homeostatic system. This person must *sense* the situation at any given point ("Where am I on the road?"), *evaluate* it ("I'm too close to the shoulder"), then *act* (steer left). This requires three separate elements of the homeostatic system: (1) a **receptor** to sense the environmental conditions (in this case, the eyes); (2) an **integrator** to evaluate the situation and make decisions (here, the brain); and (3) an **effector** to execute the commands (the muscles and bones of the hands and arms). Again and again in our discussion of physiological systems and their homeostatic functioning, you will encounter these common elements: receptors, integrators, and effectors.

Receptors, integrators, and effectors often interact in a sequence called a **feedback loop**, in which an original change is sensed, evaluated, and reacted to in a way that modifies the original change. In the prior example, the original change, a move toward the road's shoulder, was eventually modified by a move back onto the highway. There are two basic types of feedback loops—negative feedback loops and positive feedback loops.

Negative Feedback Loops A **negative feedback loop** *resists change* by sensing a stimulus (a change from a baseline condition, or *set-point* value), then activating mechanisms that oppose the trend away from the baseline. For example, a negative feedback system keeps your house warm in winter. You set your thermostat at the desired baseline temperature, the set-point value. As heat escapes through the walls, the inside temperature drops and the thermostat senses this change. It signals the furnace to turn on, and this raises the air temperature. Once the house reaches the set-point temperature again, the thermostat stops sending the "turn-on" signal, and the furnace shuts off. This is a feedback loop because it starts with a temperature change (a decrease) and eventually *feeds back* to a temperature change (an increase). It is a *negative* loop because the response opposes the stimulus: The heat generated by the furnace negates decreasing air temperature. Likewise, a whale's steady body temperature depends on an organic thermostat, and so does yours.

Within the brain of a person or whale, a small region lying above the roof of the mouth, called the *hypothalamus*, acts as a biological thermostat. The hypothalamus has a set point, in this case a genetically determined ideal temperature for body cell activity (in a person, about 37°C, or 98.6°F). Specialized hypothalamic nerves measure the temperature of blood flowing through the brain, while other nerves analyze information arriving from temperature receptors in the skin [FIGURE 24.9A]. The hypothalamus then triggers mechanisms for increasing or dissipating heat. When the temperatures of the skin and blood drop below a certain point, the hypothalamus signals effectors, the muscles, to begin to contract. At first these contractions are small and occur at different times in different muscle cells, but as the signal increases, muscle cells begin to fire in greater synchrony, and the individual begins to shiver. These contractions generate heat and thus help counteract the initial cold stimulus.

An animal's temperature-regulation system may not be robust enough to cope with some environmental conditions. If the environment grows too cold, the body may not be able to keep up with heat loss; in this case, *hypothermia* may set in—lowered internal temperature, sensory confusion, and potential death. If, on the other hand, environmental temperatures are too high, the individual can experience *hyperthermia*—heightened internal body temperatures—which can also be damaging or even fatal. The main point is that negative feedback loops tend to act to keep conditions constant—a constant body temperature, a steady salt concentration, or a continuous oxygen supply.

Positive Feedback Loops While negative feedback loops resist internal change, **positive feedback loops** *bring about rapid change.* In a positive feedback loop, an initial change in one direction is pushed further and further in the same direction. The terrible screech you sometimes hear from a public-address system results from positive feedback. A microphone picking up sound from the speakers sends the sound back, where it is further amplified. The microphone picks up that louder rebroadcast sound, the speakers reamplify it once more, and soon the sound mounts to an ear-piercing screech. In animals, positive feedback loops usually disrupt homeostasis for a specific purpose. An example is the positive feedback loop that drives labor and delivery in the birth of a person or whale [FIGURE 24.9B]. The hormone *oxytocin* causes uterine contractions, and these strong muscular actions cause more oxytocin to be released, more contractions to occur, and so on. The feedback produces stronger and longer contractions spaced closer and closer together until, at the climax of the cycle, a human baby is born or an infant blue whale (a whopping 7.5 m, or 25 ft, long!) is expelled into the sea. This explosive action stops the loop, and homeostasis—in this case, a noncontracting uterus—is restored. A colloquial way of describing a positive feedback loop is as a "vicious cycle."

➤ CONCEPT CHALLENGE

Using the concepts of receptor, integrator, effector, and feedback loops, design a biological system that would maintain the salt concentration of the blood within a particular narrow range. Does your regulatory mechanism employ a positive or negative feedback loop?

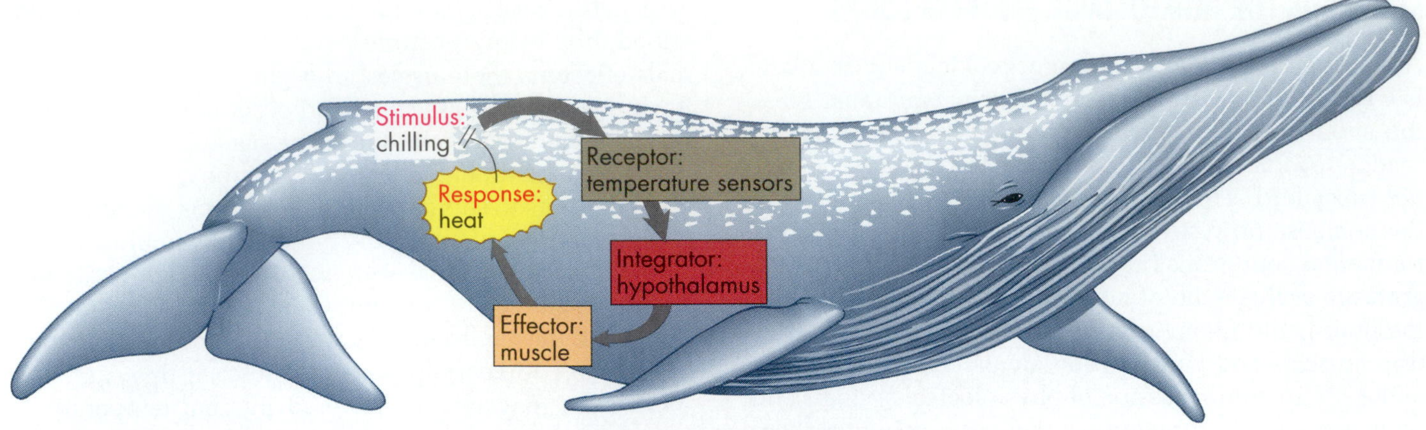

[A] Negative feedback loop

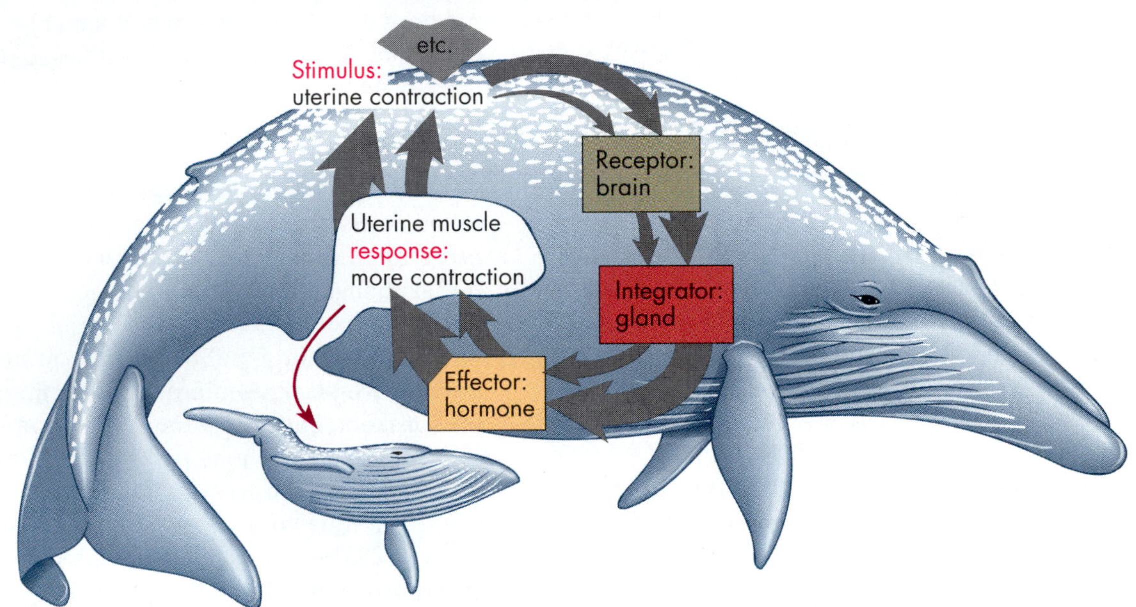

[B] Positive feedback loop

FIGURE 24.9

Feedback Loops Prevent and Provoke Change: Negative and Positive Loops.

[A] A negative feedback loop regulating temperature involves *receptors* in the whale's skin, *integrators* in the hypothalamus region of its brain, and *effectors*—the muscles. When the muscles are stimulated to flex, their activity warms the massive animal. [B] In a positive feedback loop, the effector's response to the stimulus triggers more stimulus and more response in an increasing spiral. The hormonal loop that leads, in mammals, to stronger and stronger uterine contractions and eventually to birth is a positive feedback loop.

TEMPERATURE REGULATION: HOMEOSTASIS AND EVOLUTION

The maintenance of a fairly constant body temperature illustrates three general physiological mechanisms in addition to the action of negative feedback loops. The first two mechanisms govern fluid flow in the body, and the third involves an animal's behavioral adaptations.

Flow of Body Fluids The flow of body fluids, especially of blood, can help maintain body conditions at constant levels. For example, consider the enormous problem a blue whale faces in maintaining a constant internal body temperature: it feeds in icy antarctic waters, which tend to chill the huge animal, but it also spends

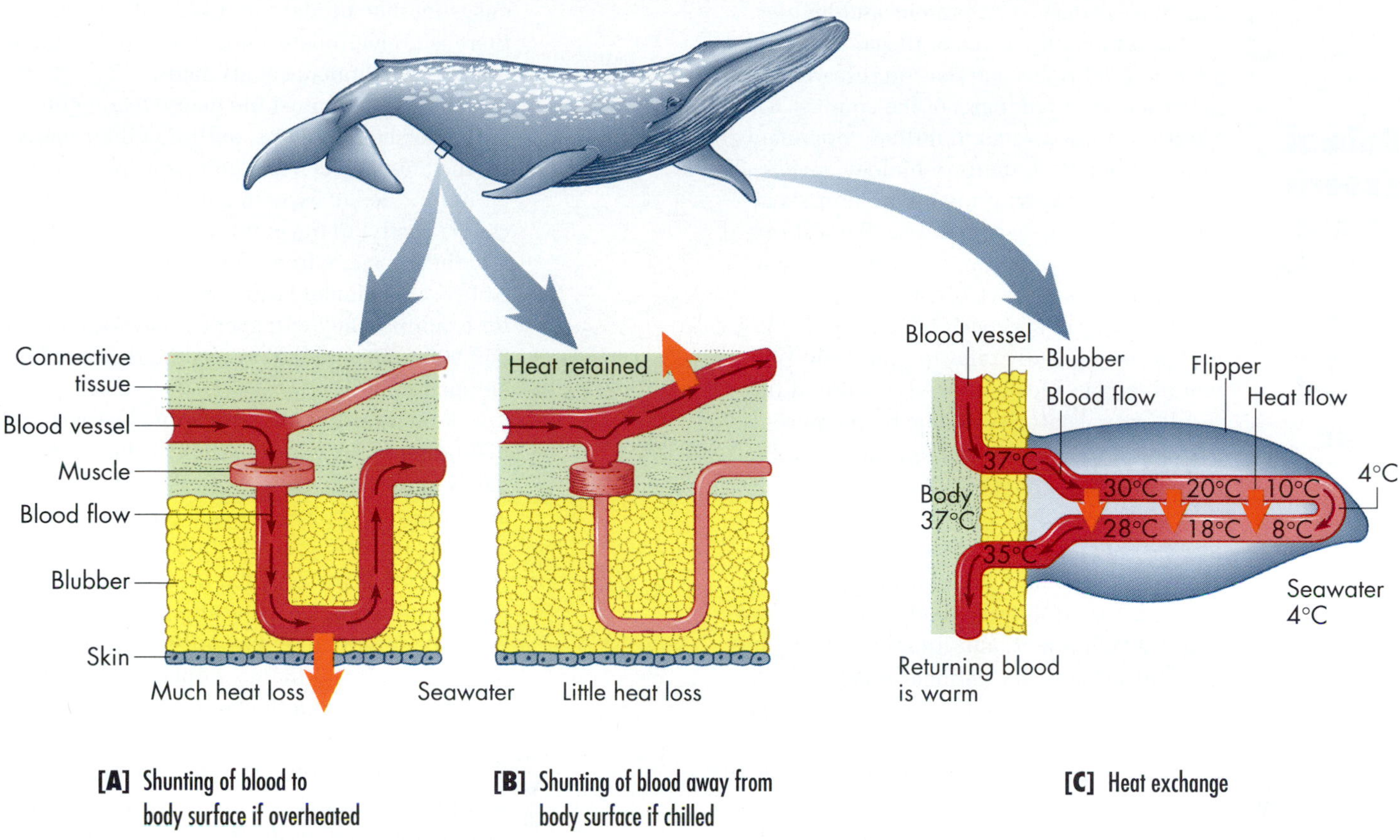

[A] Shunting of blood to body surface if overheated

[B] Shunting of blood away from body surface if chilled

[C] Heat exchange

FIGURE 24.10

Temperature Regulation: Homeostasis and Evolution.

[A] One homeostatic mechanism involves tiny muscles forming a ring around blood vessels near the body surface. When the circular muscles relax, they allow blood to flow near the cool ocean water. This, in turn, allows the animal to lose excess heat. [B] When the ring of muscles contracts, the blood vessels close, preventing blood flow to the surface. This allows the animal to retain body heat when it is surrounded by cold seawater. [C] The *rete mirabile*, or "wonderful net."

months calving in warm tropical waters, which tend to overheat the whale. Whales have a layer of fat called *blubber* that is about 30 cm (1 ft) thick, which completely encases the animal. Like a wet suit on a scuba diver, blubber is a wonderful thermal insulator, capable of blocking the flow of heat from the body core outward into the ocean water. The blubber layer is crisscrossed by blood vessels, and when the animal is overheated—say, from exercising in warm water—blood can be shunted into these vessels. This conveys heat toward the body surface, where the surrounding lower-temperature water can absorb it [FIGURE 24.10A]. Owing to a similar shunting mechanism, your own skin may feel flushed and look red when you exercise heavily. When the whale is swimming in frigid arctic waters, tiny muscles divert blood from those same vessels, so that the animal's body heat stays in the core instead of circulating in the periphery [FIGURE 24.10B].

Countercurrent Flow Besides shunting blood to one organ or another, a second fluid-related mechanism contributes to overall homeostasis. This mechanism, called *countercurrent flow*, involves a special network of blood vessels that reduces heat loss through the whale's uninsulated flippers [FIGURE 24.10C]. The blood vessels in a whale's flipper are arranged in a *rete mirabile*, or "wonderful net." The vessels lie side by side in such a way that the current of hot blood flowing away from the body core moves counter to the current of warm blood flowing from the flippers back toward the body. Heat flows directly from outgoing hot blood to incoming warm blood. With such a *countercurrent flow* arrangement, as blood moves from the whale's warm body core into its flippers, its heat is transferred to the cool blood returning from the flippers to the body core. Precious body heat is conserved instead of being lost to the icy ocean. (As you will see in CHAPTER 25, countercurrent flow also helps an animal's kidneys rid the body of wastes.)

Behavioral Adaptations So far, we have discussed some devices that maintain body temperature: negative feedback loops and fluid-shunting mechanisms. These

box 24.1
Biology Applied

Animal Experimentation: An Important Ethical Issue

Should biologists and medical researchers use animals in their laboratory experiments? This has become a contentious and highly visible issue in the wake of illegal attacks by a few animal-rights activists on research laboratories in different parts of the country. The question is not limited to militant opponents, however; many students of biology have legitimate questions about the use of animals in education and research, and such questions are relevant as we begin this section on animal anatomy and physiology.

One way to approach the subject is to first pose an equally relevant question: What would human life be like today without the results from centuries of data based on research with animals? A nonprofit organization in Washington, D.C., The National Association for Biomedical Research, compiled a list of verifiable benefits, which included the development of cancer treatments, methods for organ transplantation, advanced surgical techniques, and lifesaving drugs and immunizations. The National Academy of Sciences, in a 1991 report, counts leukemia, manic-depressive syndrome, heart disease, myasthenia gravis, drug addiction, high blood cholesterol, and AIDs among the medical conditions whose understanding depends heavily on animal research. As we saw in CHAPTERS 8 to 13, much of our basic knowledge of genes and inheritance has come from research with fruit flies, mice, and other animals. Virtually all we know of animal physiology (including the major topics discussed in this section), and much of our knowledge about cell biology, evolution, ecology, and behavior, rest on animal studies in one form or another. Were there no animal experiments, major portions of our medical and science libraries would be missing, along with treatments for most major illnesses, and data on animal interactions that will help us preserve wild populations on our crowded planet.

Few laypeople would disagree that these are important benefits. At the same time, few biologists, if any, are untouched by the ethical considerations surrounding the use of animals, even for so worthy a list of applications. Animal discomfort and loss of life are necessarily disturbing to biologists and medical researchers because they have, by definition, devoted their lives to the understanding and appreciation of living organisms. For this reason, granting agencies, universities, research institutes, biology departments, corporations, professional associations, and other groups have devised elaborate guidelines for ensuring that animals are used only when there is no legitimate substitute and that their treatment is humane at all times.

Researchers must file many documents with overseeing groups, outlining their plans for the animals' comfort and pain-free participation in the lab experiment. The safeguards can be costly: of the grant money typical researchers receive from the Alcohol, Drug Abuse, and Mental Health Administration, for example, they will spend an average of $40,000 to $70,000 to protect the rights of lab animals.

A major difficulty with animal experimentation is the *extrapolation* problem: Biologists know *in general* how relevant anatomical and physiological data are from animals, but not with absolute precision. For this and other reasons, the U.S. government, private industry, and educational institutions have spent more than $70 million since 1980 to find substitutes for lab animals, especially in toxicity tests and in classrooms.

Toxicity tests—the biggest target of animal-rights advocates—are designed to determine how newly synthesized chemicals, drugs, or cosmetics affect an animal's survival, reproduction, embryonic development, eyes, skin, and behavior. At Johns Hopkins Center for Alternatives to Animal Testing, researchers are developing substitute tests, such as dripping the irritant onto the membrane just beneath a chicken's eggshell, and another cultures the chemical with human skin cells or liver cells, slices, or the whole organ. Developers of the tests see them as useful in greatly reducing, though not entirely eliminating, use of animals.

To cut down on use of frogs, mice, and other animals for educational purposes, a number of schools are now allowing students to observe but not do dissections, and to rely on models, videotapes, or computer exercises to supplement book study. These high-tech teaching solutions can reduce—but again, not eliminate—use of lab animals because future professionals in the life sciences and medicine need hands-on experience with complex living organisms. As long as the animals' rights and comforts are protected, judicious use of animals in research and education will continue to alleviate animal suffering, human and nonhuman.

occur mainly in warm-blooded animals, such as mammals and birds (more properly called *homeothermic* or constant-temperature animals). The so-called cold-blooded animals (or *poikilotherms*, animals with variable temperatures), which obtain most of their heat from the environment, have a different set of evolutionary adaptations, many of them behavioral. A lizard or alligator, for example, instinctively moves in or out of the sun, does heat-generating "push-ups," or basks with its jaws gaping open—all strategies that keep its body temperature between 35°C and 40°C (95°F and 104°F). Similarly, an Indian python rhythmically flexes its muscles and warms incubating eggs; a honeybee contracts its wing muscles prior to flying in cool weather; and a box turtle pants, thereby dissipating heat. Poikilotherms have physiological adaptations, too. Some freshwater fishes have two sets of cellular enzymes, one that functions best at cool temperatures and another that functions best at warm temperatures. In some oceangoing fishes such as tunas and sharks, hardworking red-colored muscles enable the animals to swim rapidly, and these muscles generate enough excess heat to release into the body core.

Fever is an obvious signal that the normal homeostasis of your body is disturbed. A fever can start after virus particles or bacterial cells invade the body and begin multiplying. The infectious agents can turn up the brain's natural thermostat—the hypothalamus. Just as turning up the thermostat in your home signals the furnace to fire up, the resetting of the hypothalamus to a higher value [say, to 38.8°C (102°F) from the normal 37°C (98.6°F)] triggers the heat-generating mechanisms of the body to step up their operation. The muscles start contracting, the body starts shivering, and this makes the temperature climb higher and higher. Aspirin can help reduce fever because it interferes with prostaglandins, the molecules that turn up the body's thermostat.

In some cold-blooded animals, like lizards, a sick individual will actually induce its own artificial fever. Recall that a lizard will crawl into or out of the sunshine to keep its body temperature fairly warm and steady. A lizard with a bacterial infection will bake itself in the sun until its body temperature soars and the bacterial growth slows. When researchers prevented sick lizards from baking out their infections, the animals died more frequently than those allowed to choose high temperatures. Clearly, homeostatic mechanisms are crucial for survival.

Body temperature is just one feature kept in check by homeostasis. As you will see in later chapters, it takes many systems working simultaneously and under elaborate coordination and control to maintain an animal's total internal milieu despite the vagaries of the external world.

➤ CONCEPT CHALLENGE

Contrast the physical challenges facing the largest animal on earth, the blue whale, with that of the bee hummer, the world's smallest bird.

Connections

The subjects of animal anatomy and physiology flow naturally from the subjects covered in the first three parts of the book: cells, genes, and evolution. Now we will see how animals function and how natural selection allows animals to survive in and exploit their environments.

Throughout this section, we will focus primarily on the human. Not only do biologists know more about people than they do about any other organism, but understanding human physiology encourages us to take care of our bodies in intelligent ways. We will still, however, consider how other animals solve physiological problems in extreme environments—the dryness of the deserts, the intense pressures of deep oceans, the oxygen deficiency of high mountains, or the bitter cold of the earth's polar regions—so we can appreciate the full spectrum of solutions in the living world.

KEY TERMS

connective tissue, 567
epithelial tissue, 566
feedback loop, 571
homeostasis, 563
muscle tissue, 567
nervous tissue, 567
organ, 565
organism, 565
organ system, 565
tissue, 565

HIGHLIGHTS IN REVIEW

1 In large animals, interior cells lie far from the external environment, the source of oxygen and food and the sink for carbon dioxide and other wastes. Thus, direct diffusion to and from the outside cannot occur quickly enough to prevent cell death. Tissues, organs, and organ systems have evolved that bring environmental resources close to every cell.

- **a]** Single cells and very small or flat multicellular organisms exchange materials directly with the environment via diffusion.
- **b]** Animals have special physiological systems that bring about exchanges between internal cells and the outside environment. Each system includes a tube leading from the outside; regions of the tube where exchange with internal fluids occurs; a circulation of body fluids; and diffusion of materials between circulating fluids and individual cells.
- **c]** In multicellular organisms, body cells are organized into tissues (a group of cells of the same kind performing the same function within the body), organs (a unit composed of two or more tissues that together perform a certain function), and organ systems (two or more interrelated organs that work together, serving a common function). The four main tissues are epithelial, connective, muscle, and nervous tissue.
- **d]** Large animals are internally supported by strong structural elements (skeletons), and are internally coordinated by a nervous system and an endocrine (hormonal) system that integrates the body's activities.
- **e]** Specialized adaptations allow an organism to exploit its environment successfully. The forms of such adaptations are appropriate to their functions, and in general arose by natural selection.

2 If the physical and chemical conditions in and around body cells change beyond certain limits, the cells die. Adaptive mechanisms in animals tend to maintain the internal environment within defined limits. Collectively, these mechanisms are called homeostasis.

- **a]** An animal's physiological systems work together, bringing about homeostasis in and around the individual cells.
- **b]** Individual homeostatic systems have a receptor that senses environmental conditions, an integrator that evaluates those conditions, and an effector that executes commands.

3 Homeostasis is frequently achieved by negative feedback loops in which a receptor detects a change, an integrator determines a counteracting change, and an effector performs an activity that returns conditions to within acceptable limits.

- **a]** Negative feedback loops sense changes away from a baseline condition, then bring about actions that negate this alteration. For example, negative feedback systems help keep the internal temperatures of warm-blooded (homeothermic) animals steady.
- **b]** In a positive feedback loop, an initial condition triggers change, and that change triggers more change, until an explosive event occurs and stops the loop. For example, a positive feedback loop helps trigger birth.

UNDERSTANDING THE FACTS AND CONCEPTS

For Questions 1–5, match each of the descriptions with the most appropriate item or items from the following list. Each item can be used once, more than once, or not at all.

a] anatomy
b] physiology
c] homeostasis
d] positive feedback
e] negative feedback

1 The aspect of biology that is devoted to an understanding of the structure of microscopic and visible structures of organisms.

2 A dynamic steady state maintained by the organism that enables cells to survive and function.

3 The biological specialty primarily concerned with the functioning of cells, organs, and organ systems.

4 A process in which the final product of the process regulates the rate at which the process occurs.

5 The primary way in which cells maintain stable internal conditions despite ever-changing environmental conditions.

As above, for Questions 6–10, match each of the descriptions with the most appropriate item or items from the following list. Each item can be used once, more than once, or not at all.

a] epithelial tissue
b] connective tissue
c] muscle tissue
d] nervous tissue

6 Secretion, excretion, protection, and absorption are among the primary functions of this tissue. The cilia of cells of this tissue can be damaged by cigarette smoke.

7 Composed of contractile cells that are either attached to skeletal parts or act to move the contents contained within hollow body organs.

8 Cells of this tissue function primarily to integrate incoming stimuli with actions taken by effector organs.

9 Many of the cells of this tissue produce fibers which are released from the producing cells and lie in the extracellular matrix. Tendons and ligaments are examples of this tissue.

10 Found within the organs of the digestive system.

As above, for Questions 11–15, match each of the descriptions with the most appropriate item or items from the following list. Each item can be used once, more than once, or not at all.

a] intracellular fluid
b] tissue fluid
c] blood
d] bulk flow
e] countercurrent flow

11 Fluid that circulates by means of bulk flow.

12 The fluid compartment within cells where metabolic reactions occur.

13 The movement of two fluids in opposite directions to maximize the area in which diffusion occurs.

14 Another name for extracellular fluid that is not enclosed in vessels.

15 The passive movement of materials within the body by their transport within a moving fluid stream.

For Questions 16–20, match each of the descriptions with the most appropriate item from the following list.

a] homeothermic
b] poikilothermic
c] pyrogen
d] prostaglandin
e] hypothalamus

16 A region within the brain that senses the temperature of the blood.

17 A term that describes animals that primarily regulate their temperature metabolically.

18 A compound released by white blood cells that indirectly raises the hypothalamic temperature setting.

19 A hormonelike compound that acts directly to raise the hypothalamic temperature setting.

20 A term that describes animals that regulate their temperature primarily by adjusting their behavior.

INTEGRATE AND APPLY WHAT YOU HAVE LEARNED

1 What common strategies for the maintenance of homeostasis are shared by the digestive, respiratory, and excretory systems?

2 What is the essential difference between a tissue and an organ?

3 Which of our organ systems primarily serve the metabolic needs of the body's cells? Which are primarily communicative and coordinative, and which are the primary organ systems for movement of the body as a whole?

4 Is it likely that creatures such as single-celled protists or jellyfish could attain the size of a human being and live on land? Explain.

5 Explain why negative feedback loops are more important in the maintenance of homeostasis than positive feedback loops.

ANALYSIS

1 Which of the following situations is the best example of negative feedback?

a] A cyclist speeds up while coasting down a steep hill and slows down when the terrain flattens out.
b] A merchant raises the price of an item. Fewer people buy the item and the merchant lowers the price.
c] A motorist sees a red stoplight in the distance and slows down in preparation for stopping.
d] A motorist sees a red light in the distance and slows down until the light turns green.
e] In order to counteract a low-grade fever, a student takes an aspirin. The student's body temperature returns to normal.

2 Which of the scenarios below is the best example of positive feedback?

a] A cook reasons that if a cake needs to bake for 45 minutes at 350°F, raising the oven temperature to 400°F will shorten the time to 30 minutes, and to 450°F, to 20 minutes.
b] In order to counteract traffic jams from the northern suburbs of a city, the city council builds more roads. When people find out about the improved commuting conditions, more people move to the northern suburbs. The city council once again authorizes more road building.
c] While heating up a pot of soup, a cook increasingly raises the temperature of the burner. The soup heats faster and faster until it boils over on to the surface of the stove and puts out the flame.
d] Sighting his commuting bus, a passenger runs faster and faster to catch it; however, he then trips and the bus leaves without him.
e] A shopkeeper notices that a particular item is not selling. The price of the item is lowered, but still no one buys it.

3 Homeostatic mechanisms are ultimately directed toward the maintenance of a steady state in the:

a] Bloodstream.
b] Extracellular fluid.
c] Intracellular fluid.
d] Surface of the body.
e] Surface of the body's cells.

4 If the condition of hypothermia sets in, which part of the feedback loop has probably failed to act sufficiently?

a] The sensing cells.
b] The hypothalamic thermostat.
c] The nervous connections between the hypothalamus and the sensing cells or the effector organs.
d] The effector organs such as the muscles.
e] The failure could lie in any or all of these structures.

5 Which of the following is the most fundamental explanation for the evolution of organ systems, transport vessels, and fluid compartments in large animals?

a] Environmental changes must be quickly sensed.
b] Temperature must be maintained uniformly throughout the organism.
c] The only organisms that maintain homeostasis have these structures; therefore homeostatic mechanisms must require organ systems, transport vessels, and fluid compartments.
d] Diffusion is a slow process.
e] In a large organism, internal cells must convert metabolic waste products into useful substances because too much energy would be required to dispose of those products into the environment.

CHAPTER 25

Circulation: Transporting Gases and Materials

SNAKES AND CIRCULATION

The corn snake is a familiar North American serpent, with a light body and subtle reddish markings. This reptile slithers up trees in search of eggs in birds' nests high above the ground [FIGURE 25.1]. Contractions in the snake's muscles, plus the traction provided by its scales, allow the snake to conform to small bumps and depressions in the bark and to glide up the tree against the force of gravity. Climbing without limbs is a remarkable feat, but just as remarkable is the snake's circulation while in a vertical posture: The animal's blood circulates in tubelike vessels, and as the climbing proceeds, gravity tends to pull the blood toward the snake's tail, away from its brain. When people stand up quickly, they sometimes feel light-headed due to the pull of gravity causing blood to pool in their feet. So why doesn't the corn snake faint?

FIGURE 25.1

Snakes, Circulation, and Gravity.

Compared to horizontal land snakes or sea snakes, tree-climbing snakes like the corn snake pictured here have strong circulatory systems. Due to the forces of gravity, blood has difficulty getting to the brain of an elongated animal climbing straight up a tree.

Harvey B. Lillywhite of the University of Florida devised experiments to learn how this problem is circumvented in climbing snakes. He wondered whether or not they have special physiological mechanisms that operate differently from those in snakes that stay mainly horizontal on the ground, like the rattlesnake, or from sea snakes, whose bodies are supported by the water that surrounds them. To find out, Lillywhite measured blood pressure and blood flow in snakes that exploit the three different ecological niches.

In his snake studies, Lillywhite found that a normally horizontal reptile such as a rattlesnake does experience a pooling of blood in the tail if held vertically. The same thing happens if a sea snake is lifted out of the water and held in a vertical position: The blood pressure in its head falls toward zero, blood pools in the tail, and the snake faints; if not righted within a short time, the snake will die due to lack of blood to the brain. Contributing to its death would be excess blood pressure causing fine blood vessels to break in the lung, a long organ that extends toward the sea snake's tail. In a tree-climbing snake, however, the heart is much closer to the head and the blood pressure is much higher; both help maintain blood supply to the brain even when the climbing snake is hanging vertically.

The physiological adaptations of tree-climbing snakes are indeed remarkable, and they make a good case study for the functioning of the **circulatory system**—the heart, the blood, and the blood vessels. While the tree snake's unusual physiological problems are met in unique ways, its basic needs are the same as those of other animals. It needs sugars and other organic molecules for energy, a supply of oxygen for aerobic respiration, and a means of getting rid of carbon dioxide and other wastes. Since in a multicellular animal the exchange of materials takes place by diffusion or by active transport across cell surfaces, each cell must lie either very close to the outer environment or very close to an internal supply route.

A snake has strategies for exchanging all types of materials with the environment similar to those CHAPTER 24 presented for the blue whale: "Tubes" (actually, entire physiological systems) convey different commodities (oxygen, food, or wastes) to or from the outside, and the circulatory system distributes them internally. In most vertebrates, the digestive tube conveys nutrients, the respiratory tube carries oxygen, and the renal tube transports organic wastes [review FIGURE 24.4]. All these tubes have close physical contact with parts of the circulatory system. Branching vessels of this system lie close to each body cell. A fluid tissue, the blood, courses through the vessels, carrying the ions, nutrients, gases, and organic wastes that the cells import or export as they continue their healthy functioning.

A study of the circulatory system is a study of adaptation. Just as in the tree, land, and sea snakes, natural selection and evolutionary pathways have shaped the circulatory systems of various animals, as you will see in this chapter. You will also learn fascinating details about the human body's trillions of blood cells; its thousands of miles of blood vessels; the tireless pump, the heart; and about the risk factors that can lead to heart attacks, our nation's number one killer. ❑

MESSAGES

1 Cells exchange substances with their surroundings via diffusion. In a large animal, where many cells are far from the external environment, a circulating fluid such as the blood transports substances near to each cell. The blood, in turn, transports substances to and from a number of organs (lungs, intestines, kidneys) that communicate directly with the outer environment.

2 An animal's circulatory system is adapted to its environment. A person has two circulatory systems; the pulmonary circulation carries blood between the heart and the lungs. The systemic circulation carries blood between the heart and the rest of the body.

3 The walls of heart chambers and blood vessels vary in configuration and thickness depending on the pressure of the blood within them and on their role in diffusion.

4 High blood pressures force fluid from the bloodstream into spaces between cells. It is now extracellular fluid, and this diffuses into dead-end capillaries of the lymphatic system. Vessels of the lymphatic system then filter the fluid through lymph nodes and carry it back to major veins that discharge the fluid into the heart.

Blood: A Liquid Tissue for Transport

As a large mammal, you have about 5 L (more than a gallon) of blood in your body, coursing through an amazing 80,000 km (50,000 mi) of vessels. This moving stream of liquid tissue carries oxygen from the lungs, nutrients from the digestive tract, and regulatory substances, such as hormones, from one region of the body to another. Blood helps control the pH of body fluids, conveys immune cells specialized in defense of the body, and distributes heat and assists in cooling [see CHAPTER 24]. These and other functions are all based on the individual components of blood and the roles they perform.

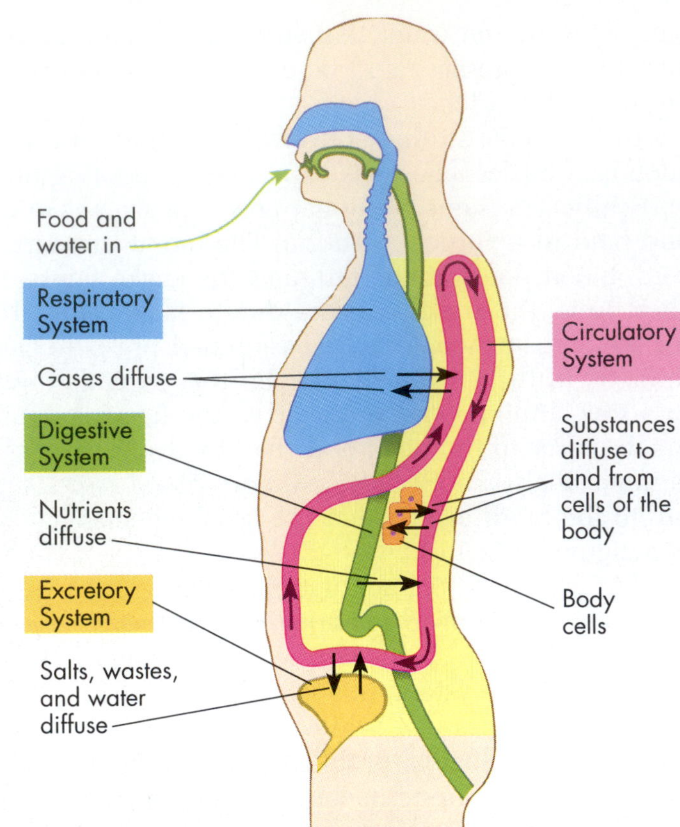

COMPONENTS OF BLOOD

By whirling a sample of blood in a centrifuge, a technician can separate the components of this life-giving liquid according to their density. The most obvious separation will be into an upper, lighter layer of fluid and a lower, denser layer of cells and cell fragments [FIGURE 25.2].

The top 55 percent of the centrifuged blood is the pale yellow liquid called **plasma**, which is more than 90 percent water. Plasma's main function is transporting blood cells and dissolved substances including salts, sugars, and fats from the foods we eat. Plasma also contains a storehouse of important dissolved proteins. One group of plasma proteins, the *globulins*, includes *antibodies*, defensive molecules that attack invaders. The *albumins*, another group of large blood proteins, bind to toxic substances in the blood, among other activities. A third plasma protein, fibrinogen, is essential for blood clotting.

Beneath the plasma layer in centrifuged blood is the remaining 45 percent, composed of cells and cell fragments. The cell fraction consists of two layers: (1) The first is a thin gray band representing less than 1 percent of blood volume. It is made up of *white blood cells* (important in the body's defense) and cell fragments known as *platelets* (important in blood clotting). (2) The second is a much wider band containing only *red blood cells.* A physician can use the width of this band of red blood cells as a diagnostic tool. If it is less than 45 percent of total blood volume, the person has too few red blood cells and is probably *anemic*, perhaps suffering fatigue and shortness of breath.

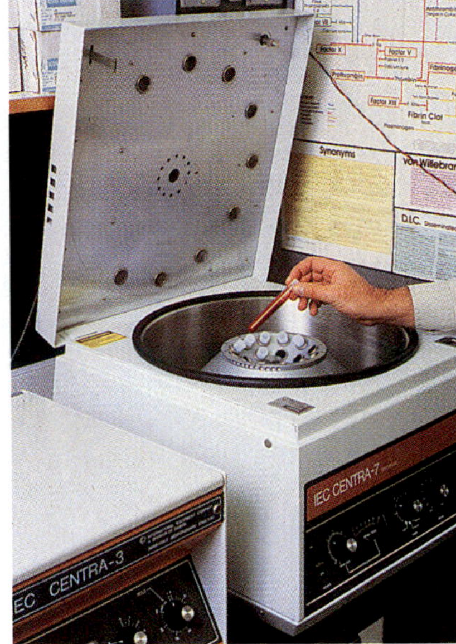

FIGURE 25.2
The Components of Human Blood.
Spinning blood in a centrifuge separates the lighter from the denser components. The yellowish liquid plasma collects on top, and the denser red and white blood cells and platelets accumulate at the bottom.

RED BLOOD CELLS AND OXYGEN TRANSPORT

The human body contains approximately 25 trillion red blood cells—fully one-third of all its 75 trillion cells. Three features help red blood cells, or **erythrocytes** (from the Greek words for "red" and "cell"), perform their vital function of transporting oxygen to tissues. The first two features concern the presence of proteins: *hemoglobin*, which transports oxygen, and the enzyme *carbonic anhydrase*, which functions in carbon dioxide transport. The third feature is the cell's biconcave disk shape; the cell resembles a doughnut without a hole [FIGURE 25.3]. Among other benefits, this disk shape allows all the red blood cell's hemoglobin molecules to lie near the outer membrane, which must be diffused by oxygen. If red

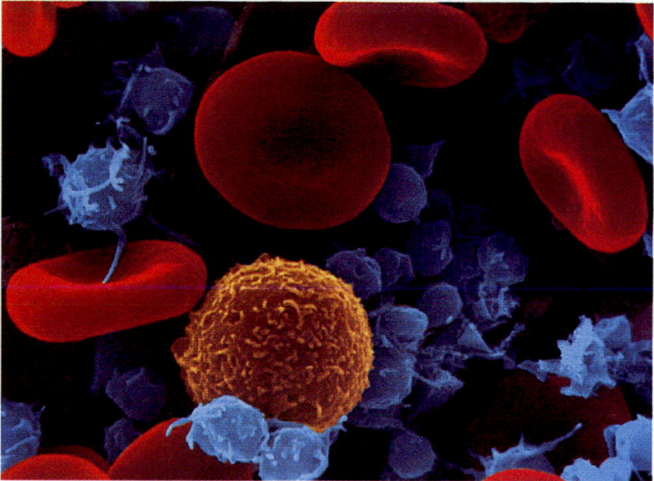

FIGURE 25.3

Blood Cells Transport and Defend.

In this colorized scanning electron micrograph (magnification 6000×), red blood cells, which transport oxygen, are clearly distinguishable as dark red, biconcave disks, while white blood cells, which help defend the body from disease, look like fuzzy, tan balls. Platelets (which help clot the blood) appear as small blue cell fragments.

cells were spheres like the white blood cells in FIGURE 25.3, some hemoglobin would be "buried" deep in the cell and would have little chance to pick up incoming oxygen.

Recall from CHAPTER 4 that human red blood cells lose their nucleus, mitochondria, and most other organelles as they mature. As long as the cell's total volume doesn't change, the loss of the nucleus leaves space for more hemoglobin, thus increasing the cell's oxygen-carrying capacity. Without a nucleus, however, an erythrocyte is doomed to die after about 120 days. Replacement red blood cells develop from stem cells in the bone marrow. These stem cell populations are self-regenerating: When a stem cell divides in two, one daughter cell begins to differentiate into a mature blood cell, while the other differentiates into another stem cell and in effect replaces the original one. While some stem cells give rise to red blood cells, others generate white blood cells and the cells that give rise to platelets. Exposure to intense radiation can destroy stem cells, and this explains why radiation sickness usually includes anemia.

When the red cell count decreases because of either normal red cell death or sudden blood loss from a serious wound, a negative feedback loop controls red blood cell replacement [FIGURE 25.4]. When there are fewer red cells in the blood, the liver and kidney receive less oxygen and begin to produce the protein hormone *erythropoietin* (ih-RITH-roh-poy-EE-tin). This hormone travels through the bloodstream to the bone marrow, where it stimulates stem cells to divide more quickly; as more red blood cells are produced, they carry more oxygen to tissues and organs throughout the body, including the liver and kidney. As a result, these organs slow their production of erythropoietin, and division of bone marrow stem cells slows. This same cycle of events can be triggered by changing altitude or exercising vigorously. A mountain climber who travels from Milwaukee to the Himalayas, for example, will encounter air with a much lower concentration of oxygen molecules. Likewise, the muscles of a runner in training will use massive amounts of oxygen, thus decreasing the amounts reaching the liver and kidneys during each training session. This decrease of oxygen also triggers a higher output of erythropoietin and a rise in red blood cell production.

Researchers have made progress recently in manufacturing hormones that stimulate the growth of red and white blood cells. This development promises to decrease

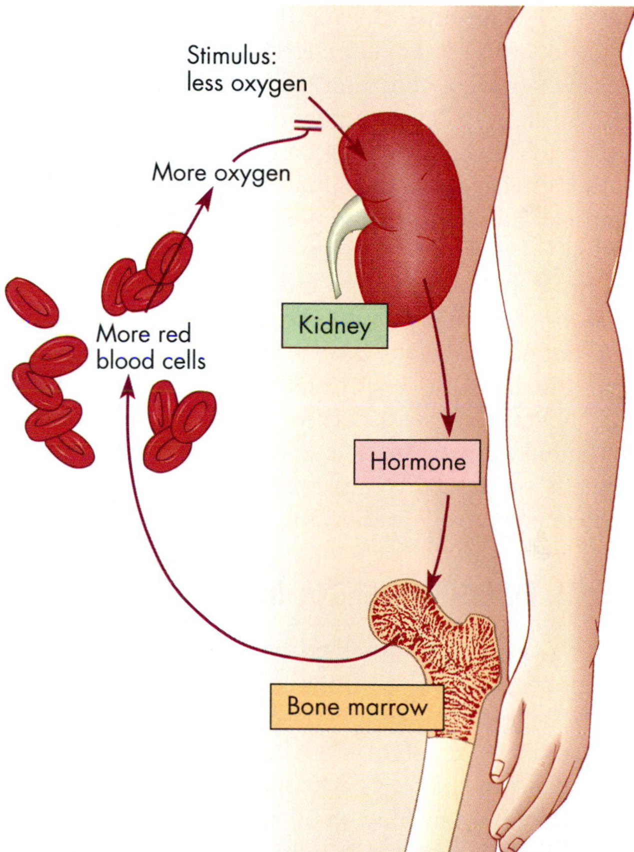

FIGURE 25.4

Replacing Lost Blood Cells: A Negative Feedback Loop.

When the number of red blood cells drops and the kidney receives less oxygen, it releases the hormone erythropoietin, which stimulates the bone marrow to make more red blood cells. Increased numbers of red blood cells carry more oxygen to the kidneys (and other organs), and the kidneys stop releasing the hormone.

the need for blood transfusions, simplify bone marrow transplants, and bolster the fight against disease-causing microorganisms, AIDS, and some cancers.

WHITE BLOOD CELLS: DEFENSE OF THE BODY

For every 1000 red blood cells, there are only 2 white blood cells, or **leukocytes** (*leuko* means "white," *cyte* means "cell"; see FIGURE 25.3). White blood cells are larger than red blood cells, and they retain their nucleus when mature. Leukocytes have a changeable, amoebalike form that allows them to squeeze through the walls of capillaries and patrol the fluid-filled spaces between cells. These characteristics help leukocytes to defend the body against invasions by microorganisms and other foreign materials.

There are five classes of white blood cells: neutrophils, monocytes, eosinophils, basophils, and lymphocytes. Two groups, the *neutrophils* and *monocytes*, are phagocytes: They are attracted to damaged or infected tissues, where they consume bacteria, viruses, and cell debris. *Eosinophils* carry enzymes that break down foreign proteins and break up blood clots. *Basophils* release an anticlotting agent at the site of an injury and produce a chemical substance called *histamine* that delays the spread of invading microorganisms. Histamine also produces hives and causes an allergy sufferer's nasal membranes to swell up (hence the use of *anti*histamine drugs). A final leukocyte group, the *lymphocytes*, is active in the immune responses that protect the body against infectious agents and foreign intruders (such as transplanted organs). There is so much to say about white blood cells and the body's immune defense that CHAPTER 26 is devoted entirely to the subject.

PLATELETS: PLUGGING LEAKS IN THE SYSTEM

Streaming along in the blood with the red blood cells and white blood cells are millions of cell fragments called **platelets**, or *thrombocytes*. These irregularly lobed bits and pieces have broken off from large specialized cells in the bone marrow. Platelets are crucial blood components because they help to plug small leaks in the circulatory system. Were it not for the blood-clotting action of platelets, an animal's blood might literally drain away through even a minor wound.

➤ CONCEPT CHALLENGE

Suppose that you leave your home near sea level, spend a week skiing in the mountains of Colorado, then return home. Compare your body's production of the hormone erythropoietin two days before leaving home to its production two days after arriving in the mountains, and its production two days after returning home again.

Circulatory Systems

Just as delivery trucks would be useless in a town without roads, blood would be of little value to an animal unless it could circulate its cargo of gases, nutrients, and other substances throughout the body. A constant flow of blood, usually contained within a system of *blood vessels*, passes within 1 mm (0.04 in) of each body cell, and this short distance allows materials in the blood to diffuse into cells to replenish the cells' storehouse of nutrients and other necessary materials. Similarly, wastes that would accumulate and poison the cell can quickly diffuse into a nearby vessel and be swept away in the bloodstream. While the circulatory systems of various animals carry out both these functions (the delivery of nutrients and the removal of wastes), different environments place different demands and restrictions on the functioning. To fully appreciate the human circulatory system, we must compare it with parallel systems in other organisms. A major evolutionary trend emerges from such a comparison: An increase in the efficiency of the circulatory system coincides with an increase in the animal's *activity level*—activities that require adequate supplies of blood-borne oxygen and nutrients. This trend is reflected in the two main types of circulatory systems, open and closed.

OPEN CIRCULATORY SYSTEMS

In animals with **open circulatory systems**, including most arthropods and mollusks, a blood equivalent called **hemolymph** circulates in the body. For example, the spider in FIGURE 25.5A has a simple heart that is little more than an elongated, pulsating tube. The heart's pumping action sends hemolymph in the direction of the animal's brain, where it leaves the heart tube and percolates toward the back of the animal through the open body cavity, bathing the tissues in the animal's gut and other internal organs. Finally, the hemolymph returns to the heart tube through special pores. Open circulation is not a very efficient transport system, but in spiders and insects, the distribution of oxygen does not depend on the circulatory system. In these animals, oxygen diffuses near to each cell from branching, hollow *book lungs*, or *tracheae*, which are independent of the circulatory system.

CLOSED CIRCULATORY SYSTEMS

Segmented worms and more complex animals, such as vertebrates, have **closed circulatory systems**, in which the blood is completely contained within a system of vessels.

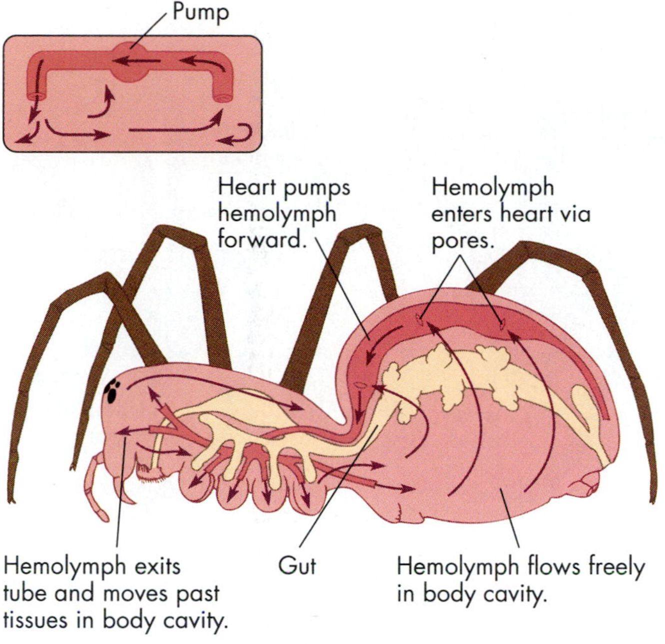

[A] A spider's open circulatory system

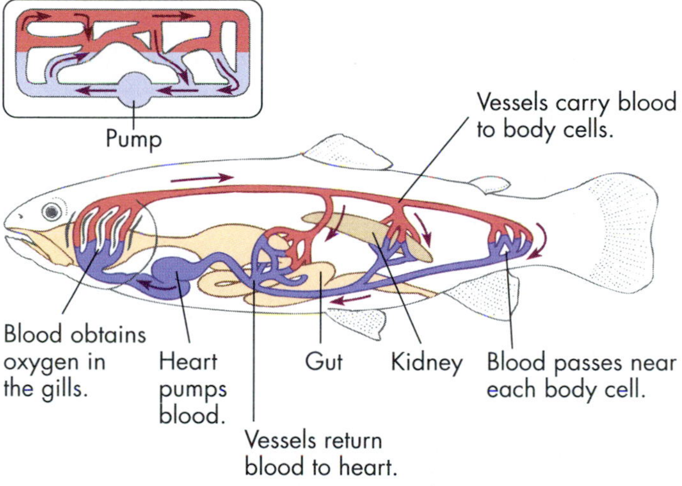

[B] A fish's closed circulatory system

FIGURE 25.5

Open and Closed Circulatory Systems.

[A] In the open circulatory systems of an arthropod (here, a spider) or a mollusk, the blood moves in a vessel, but is then dumped into a sinus (an open area) in which it freely bathes the internal tissues before reentering the vessel. **[B]** In the closed circulatory system of a vertebrate (here, a sea bass), the blood remains in a system of interconnected vessels as it circulates through the body. Here, red signifies oxygenated blood leaving the gills and heart and traveling to the body tissues, and blue signifies deoxygenated blood returning from the tissues in vessels leading to the heart and gills.

A closed circulatory system has several advantages over an open one. First, fluid contained within a network of closed tubes can be shunted to specific areas where it is needed, much as a farmer can dispatch the water in irrigation pipes to different fields. Second, since the fluid is completely contained within pipelines, more pressure can be exerted on it; thus, the fluid can be forcefully distributed to areas distant from the pump (like a snake's head or tail). You can demonstrate the results of pressurized fluid by placing your thumb over the opening of a garden hose: You can squirt water on a tomato plant—or on an unsuspecting friend—much farther away when you partially close off the opening than you could by dribbling the water out unobstructed. These features allow closed circulatory systems to deliver blood throughout larger, more complex, and more active organisms in a much more efficient way than an open system could.

In a representative vertebrate, such as the sea bass, the heart pumps blood through tiny vessels in the gills, an interface with environmental oxygen [FIGURE 25.5B]. Oxygen can diffuse from the sea water into the blood. Once the blood has picked up oxygen, it courses through smaller and smaller vessels that spread throughout the fish's tissues, delivering oxygen to every body cell. On its circuit, the blood passes by the gut, where it picks up nutrients, and areas such as the kidneys, where organic wastes are removed from circulation and excreted from the body (see CHAPTER 29 for details). Eventually, the continued pumping of the heart pushes the blood around once more to the gills, where carbon dioxide diffuses from the blood into the water surrounding the fish. Simultaneously, oxygen diffuses into the blood and the cycle continues.

PATTERNS OF CIRCULATION

The circulatory systems of sea-dwelling and land-dwelling vertebrates differ according to the requirements of the animal in its specific environment. FIGURE 25.6 shows the evolution of the heart and blood vessels of several vertebrates.

Circulation in Aquatic Vertebrates In a fish, blood flows in a simple loop from the heart, to the gills (where it is oxygenated), to body tissues, and back to the heart [FIGURE 25.6A]. A fish's heart has two chambers, a less muscular one called the **atrium** (plural, *atria*), which receives oxygen-depleted blood from veins, and a more muscular chamber, the **ventricle**. The ventricle receives blood from the atrium and pumps it through an artery to the small capillaries in the gills, where oxygen dissolved in the surrounding water is absorbed into the blood and carbon dioxide is released to the environment. Oxygenated blood then leaves the gills and collects in large

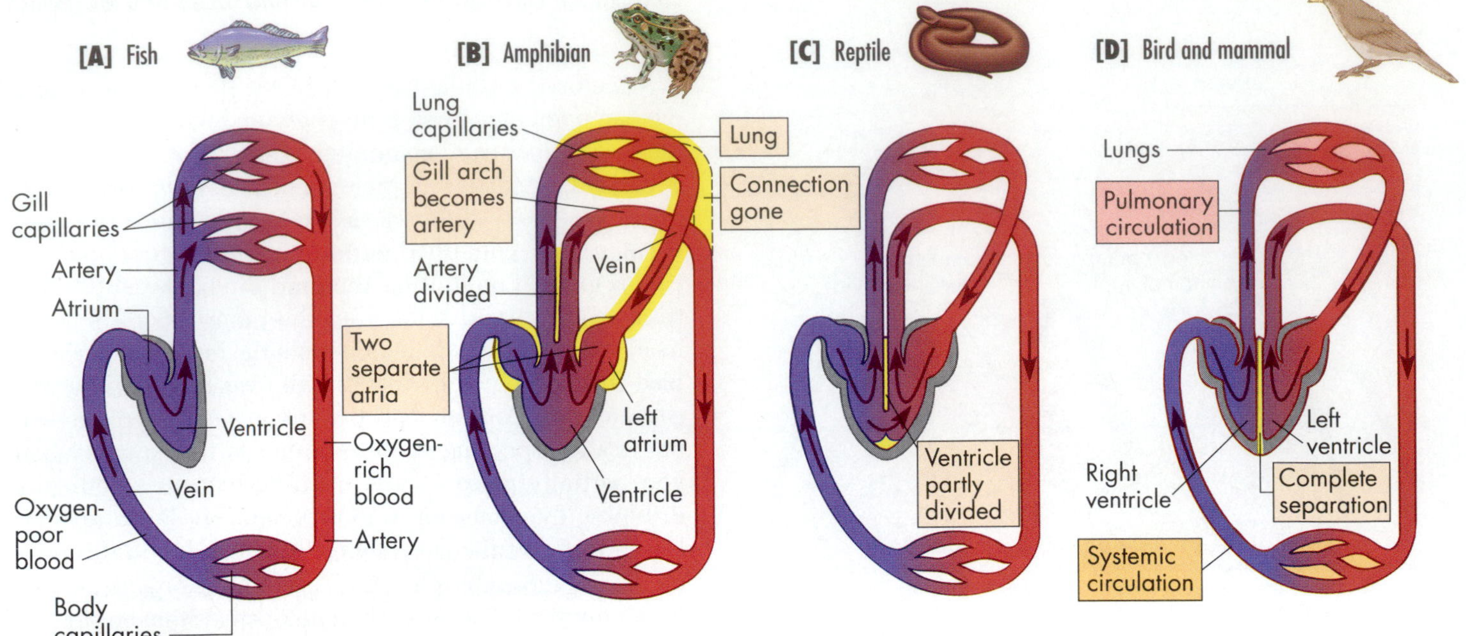

FIGURE 25.6

Vertebrate Heart and Blood Vessels.

[A] A fish has a two-chambered heart (one atrium, one ventricle) and a single loop of blood vessels leading through the gills to the body tissues. **[B]** An amphibian has a three-chambered heart (two atria, one ventricle) that allows some mixing of oxygen-rich and oxygen-poor blood, plus a separate loop of blood vessels to the lungs (the pulmonary circulation). (The differences are highlighted in yellow.) **[C]** In reptiles, the heart has three chambers (two atria and a partially partitioned ventricle). The partial separation further decreases the mixing of oxygen-rich and oxygen-poor blood. **[D]** In birds and mammals, the heart has four chambers (two atria and two ventricles) that together prevent blood mixing. In these animals, the vessels are arranged in separate pulmonary (lung) and systemic (body) circulatory loops.

vessels that pipe it throughout the fish's body. After the oxygen and other materials are delivered to tissue cells, oxygen-poor blood returns to the atrium through veins, and the cycle is repeated.

A significant feature of this *single-loop system* is that it separates oxygen-rich from oxygen-poor blood, and delivers only highly oxygenated blood to muscles and other organs. The single-loop circulatory system of a fish, however, cannot deliver blood to most organs at a high pressure. Blood leaving the heart is under high pressure, but once it fans out through the dense network of fine capillaries (delicate blood vessels) in the gills, pressure diminishes and remains low as the blood moves through the body. While this low pressure is adequate to meet the metabolic demands of a cold-blooded animal that is supported by water [see CHAPTER 24], it could not push blood quickly enough through the body of a terrestrial vertebrate, whose active cells require high levels of oxygen and nutrients.

Circulation in Land Vertebrates Land vertebrates, such as tree-climbing snakes and people, have a *two-loop circulatory system* that meets the needs of both high pressure and the partial or complete separation of oxygen-rich from oxygen-poor blood: Blood is first pumped to the lungs, where it takes on oxygen, after which it flows back to the heart, where it is pumped out a second time—at high pressure—and then circulates to body tissues.

Adult amphibians and reptiles have a heart-lung circulatory loop that increases blood pressure, as well as another evolutionary feature: a third heart chamber that enhances the heart's ability to keep oxygen-rich and oxygen-poor blood separate. Amphibians have two atria instead of the single atrium found in fishes: One atrium receives oxygenated blood arriving from the lungs, and one receives oxygen-depleted blood from body tissues [FIGURE 25.6B]. Both atria empty into a single ventricle, where there is some mixing of oxygen-rich and oxygen-poor blood. Nevertheless, because of the position of the large vessels leading out of the lone ventricle, most of the oxygen-rich blood is shunted to the tissues and most of the oxygen-depleted blood travels to the lungs.

During the evolution of the reptile lineage, the three-chambered heart underwent a significant change. In crocodiles and snakes, a thin, membranous partition (a *septum*) partially divides the single ventricle [FIGURE 25.6C]. This modification further decreases the mixing of oxygen-rich and oxygen-poor blood in the reptile's heart, and thereby provides better oxygen delivery than in the amphibian. Nevertheless, any mixing at all might seem inefficient to an astute observer. However, in the circulatory

[A] Sea snake

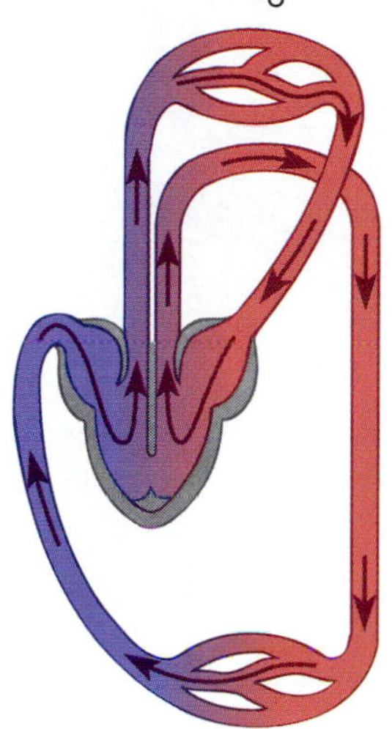

[B] When at the surface, substantial circulation through the lung

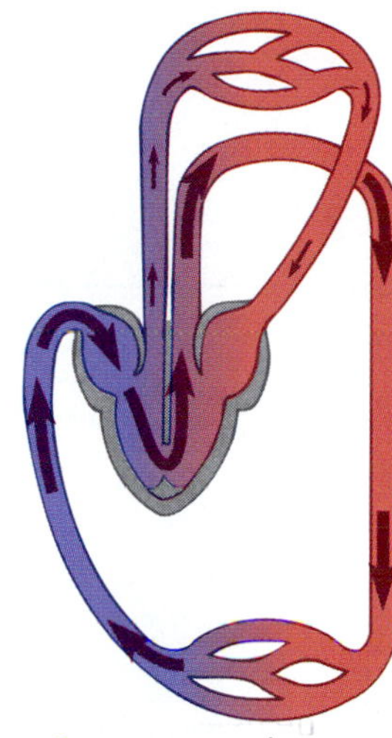

[C] When diving, little circulation through the lung

FIGURE 25.7

The Heart of a Snake.

[A] A sea snake breathes air, but can remain below the surface for up to an hour. **[B]** When the snake is breathing air, blood circulates through the lungs with some mixing of oxygen-rich and oxygen-poor blood in the single ventricle. **[C]** When diving, only a small amount of blood flows through the lung circulatory loop. Most moves from one side of the ventricle to the other and directly to the body, where oxygen is used, and to the skin, where a small amount of gas exchange can take place.

system of sea snakes [FIGURE 25.7], this mixing is a distinct advantage. When a sea snake dives down from the surface, it sometimes remains underwater for an hour at a time. During this period, blood can be shunted from the single, partially divided ventricle to the body surface and body-tissue circuit [FIGURE 25.7C] rather than to the pulmonary (lung) circuit [FIGURE 25.7B]. This shunting conserves gases in the lung and sends blood to the skin, where it picks up oxygen from the water and disposes of carbon dioxide directly into the ocean.

In the birds and mammals that descended from reptilian ancestors, the septum enlarged until it eventually divided the ventricle into two fully separated cavities. All birds and mammals alive today (including human beings) have a four-chambered heart with an atrium and ventricle on the left side and an atrium and ventricle on the right side. In a sense, these complex animals have two hearts that pump blood in two completely separate circulatory loops [FIGURE 25.6D]. In the lung loop, or **pulmonary circulation**, blood arrives from the tissues, enters the right side of the heart, and is pumped directly to the lungs, where it picks up oxygen. The oxygenated blood then enters the left side of the heart and is pumped out into a body loop, the **systemic circulation**. This double loop means that the blood flowing to the muscles, brain, and extremities has the highest possible oxygen content and that "used," oxygen-poor blood can be shunted from an isolated right heart to the nearby lungs without ever mixing with oxygen-rich blood. It also means that the left heart chamber can send blood out through the body loop at a high-enough pressure to reach all body tissues quickly.

To appreciate how the four-chambered heart and separate circulatory loops operate in birds and mammals, and how blood pressure and blood flow are controlled, let's examine the mammalian circulatory system in detail, using the human system as our model.

➤ CONCEPT CHALLENGE

Babies are sometimes born with a hole in the septum dividing the left and right atria, so that oxygen-rich and oxygen-poor blood mix. Unless a surgeon closes this hole in a delicate operation, the child's brain and body will not develop normally, and the child will probably die young. Why should this mixing, which is perfectly functional in a frog or turtle, prove fatal in a developing human?

The Human Circulatory System

It takes an immense network of circulatory tubing to deliver blood to your cells [FIGURE 25.8A]. If all of the blood vessels in your body were placed end to end, the single long pipeline would encircle the earth twice. To keep the blood flowing through all these vessels, your heart must

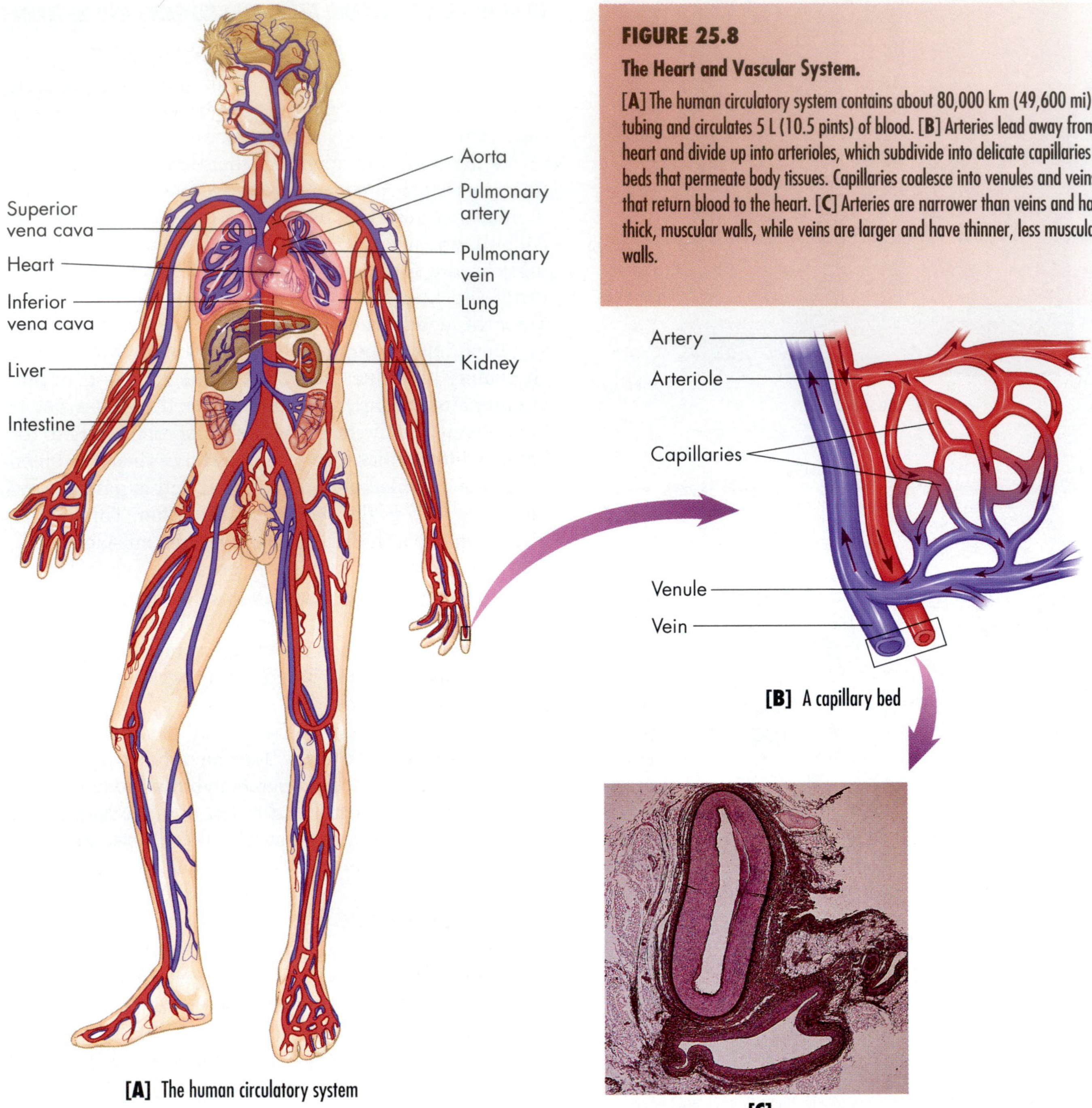

FIGURE 25.8

The Heart and Vascular System.

[A] The human circulatory system contains about 80,000 km (49,600 mi) of tubing and circulates 5 L (10.5 pints) of blood. **[B]** Arteries lead away from the heart and divide up into arterioles, which subdivide into delicate capillaries in beds that permeate body tissues. Capillaries coalesce into venules and veins that return blood to the heart. **[C]** Arteries are narrower than veins and have thick, muscular walls, while veins are larger and have thinner, less muscular walls.

[A] The human circulatory system

[B] A capillary bed

[C]

beat with a regular rhythm and the flow must be maintained at a high-enough pressure to force blood into your brain, nose, toes, and all the tissues in between. Both the anatomy and the activity of the blood vessels and heart make these tasks possible.

BLOOD VESSELS: THE VASCULAR NETWORK

Blood moves *away from the heart* in **arteries**, which are large, hoselike vessels with thick, multilayered, muscular walls [FIGURE 25.8B and C]. The contraction of these wall muscles, along with the beating of the heart, helps keep blood under the necessary pressure—the force we call *blood pressure*. The easily felt vessels through which blood pulses in your wrists and neck are arteries. Arteries branch and form smaller vessels called **arterioles**. These vessels are too small to be seen with the unaided eye, and have thinner, less muscular walls. Arterioles branch again into the dense, weblike network of delicate **capillaries**; these minute vessels permeate the fingertips, earlobes, and all the tissues of the body. Capillaries are microscopic vessels only about 8 µm (0.0003 in) in diameter, and rarely more than about 1 mm (0.04 in) from any body cell. A capillary's diameter is so small that red blood cells can pass through in single file only; their walls are just one cell layer thick; so they are also very delicate. The vessels are so numerous, however, that they make up almost all of a person's blood vessels; if all

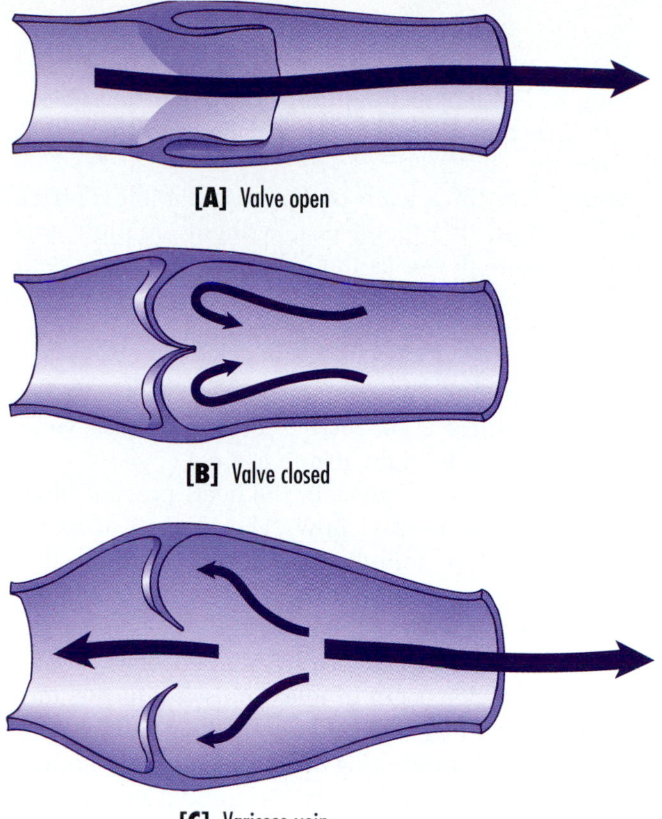

FIGURE 25.9

Valves Maintain a One-Way Blood Flow in Veins.

[A] When blood moves in one direction, it forces the valve open, and the fluid flows unimpeded. [B] If blood begins to flow in the opposite direction, the backward force pushes the valve flaps together and closes the valve. Thus, the valve in veins allows blood to flow in only one direction—toward the heart. [C] In a varicose vein, the valve is damaged, and blood flows backward, causing the vein to distend and look very large and blue in a limb, usually the lower leg and foot.

capillaries were filled with blood at the same time, they could contain the entire 5 L (10.5 pints) of human blood. The capillaries' ultrathin structure is one key to the efficiency of the circulatory system because materials can readily diffuse into and out of the capillaries through the single cell layer.

Capillaries form networks called *capillary beds* that link arterioles and venules [see FIGURE 25.8B]. The arterial side of each bed conveys fresh blood to the capillaries, and oxygen, nutrients, carbon dioxide, and metabolic wastes move across the capillary walls into and out of the extracellular fluid and nearby tissue cells. The oxygen-depleted blood then continues to move through the bed to the venous side; there the capillaries leading away from the tissues coalesce into larger vessels known as **venules**, which in turn merge and become *veins*. **Veins** carry blood *toward the heart*, where the circulation cycle begins again. Perhaps you can see the bluish, oxygen-poor blood in the veins close to the skin surface of your wrists.

Venous blood is under lower pressure than arterial blood. This is because venous blood has already traveled some distance from the heart and been slowed by passage through the narrow capillaries, and because veins have a larger diameter and thinner, less muscular walls than arteries [see FIGURE 25.8C]. This low-pressure fluid flows in the proper direction (toward the heart's right side) and does not flow backward or pool in the extremities owing to a system of **valves**—tonguelike flaps that extend into the internal space, or *lumen*, of the vein [FIGURE 25.9]. Like a one-way door, when a valve is pushed from one direction, it opens and allows blood to pass, but when pressure is exerted from the other direction, the valve shuts tight. If valves become damaged, blood can flow backward, causing the vein to distend and become visible as a large blue bulb, a *varicose vein*.

THE HEART

At the physiological center of the circulatory system lies the **heart** [see FIGURE 25.8A]. Roughly the size of a large, lopsided apple, your heart pumps a teacupful of blood with every three beats, 5 L of blood every minute, and upwards of 7200 L of blood every day of your life. Over a lifetime, this amounts to 2.5 billion heartbeats and enough blood to fill a building six stories high and a city block long. Although medical engineers have tried valiantly, replacing the living circulatory pump has proved exceedingly difficult.

This tireless organ, like the hearts of all mammals, has four chambers divided into two pairs: the right atrium and right ventricle, and the left atrium and left ventricle. Each atrium-ventricle pair plays a specialized role in circulating the blood.

Step 1 The right atrium receives oxygen-poor blood from the body tissues via two large veins, the **superior vena cava** and the **inferior vena cava** [see FIGURES 25.8 and 25.10]. These vessels collect blood from the upper and lower body, respectively.

Step 2 The oxygen-poor blood is then pumped from the right atrium into the right ventricle through the petal-shaped *atrioventricular valve*, which prevents blood from flowing back when a ventricle contracts.

Step 3 The right ventricle in turn pumps blood past the half-moon-shaped *pulmonary semilunar valve* into and through the Y-shaped **pulmonary artery** to the lungs, where gases are exchanged in the capillaries in the lungs.

Step 4 The freshly oxygenated blood collects in the **pulmonary veins** (the only veins in an adult person that carry oxygen-rich blood) and passes to the heart's left atrium.

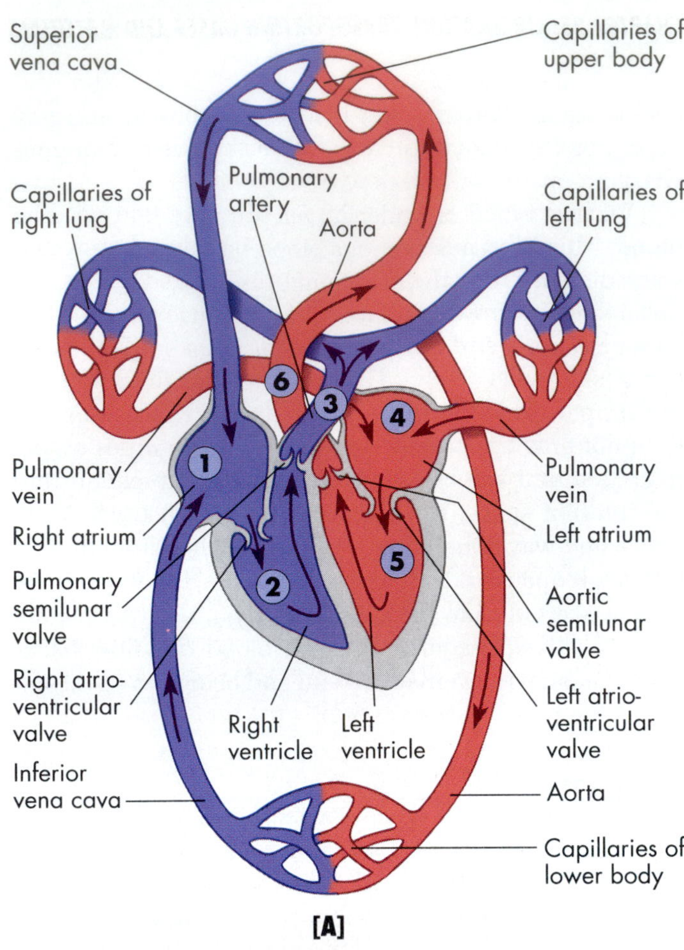

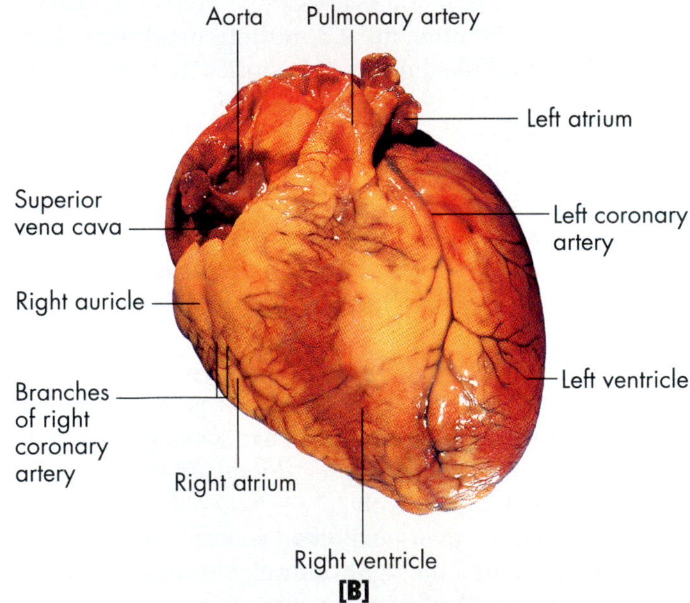

FIGURE 25.10

The Pumping Heart.

[A] Blood flows in a circuit from the venae cavae to the right atrium (1), right ventricle (2), pulmonary artery (3), lungs, pulmonary veins, left atrium (4), left ventricle (5), aorta (6), and body tissues. [B] A human heart has large ventricles, and the coronary arteries carry a large blood supply to them.

Step 5 The left atrium pumps the oxygenated blood through a second atrioventricular valve into the heart's left ventricle.

Step 6 The thick walls of the muscular left ventricle contract around this blood in a wringing motion until enough pressure develops to push open the *aortic semilunar valve* and squirt the blood into the **aorta**—the main artery leading to the systemic circulation.

This body loop conveys blood to capillary beds in distant tissues, then returns it via venules and veins to the venae cavae and back to the right atrium again.

Together, the four valves in the heart prevent blood from reversing its normal flow. The closing of valves makes the "lub-dub" sounds you can hear through a doctor's stethoscope. As the ventricles begin to contract, the pressure of surging blood pushes the atrioventricular valves shut, and you hear a low-pitched "lub." Then, when the ventricles begin to relax, pressure in the aorta and pulmonary artery rapidly forces valves in both arteries to slam closed, producing the quicker, higher-pitched "dub." In some people, the heart's pumping strength can be diminished by backwashing blood, and can sometimes be heard as a *heart murmur.* Severely diseased valves can be replaced with either valves from a pig's heart or a mechanical device.

CONTROL OF THE HEARTBEAT

Like the muscles in your arm, each of the heart's four chambers is made up of specialized cells organized into contractile fibers. In your arm, however, your biceps muscle can contract weakly (e.g., when you pick up a pencil) or more strongly (when you pick up a book as heavy as this one), and the strength of contraction depends on what percentage of the muscle fibers contract at any given time. By contrast, all the fibers in the cardiac (heart) muscle contract with each heartbeat; thus, the healthy heart is always pumping at its maximum strength. Blood flow through the heart is increased mainly by faster beating.

What accounts for this specialized characteristic of heart muscle? First, heart muscle cells are linked electrically by special regions, the **intercalated disks**. Gap junctions [see FIGURE 3.28] between cells in these disk regions allow electrical impulses that stimulate muscle contractions to spread instantaneously from muscle cell to muscle cell. As a result, neighboring sections of the heart wall contract and relax together in a superbly coordinated pumping action that keeps blood flowing smoothly through the system.

Second, cardiac muscle contracts automatically—that is, without stimulation from the nervous system. (Nerves do play a role in speeding or slowing the heart

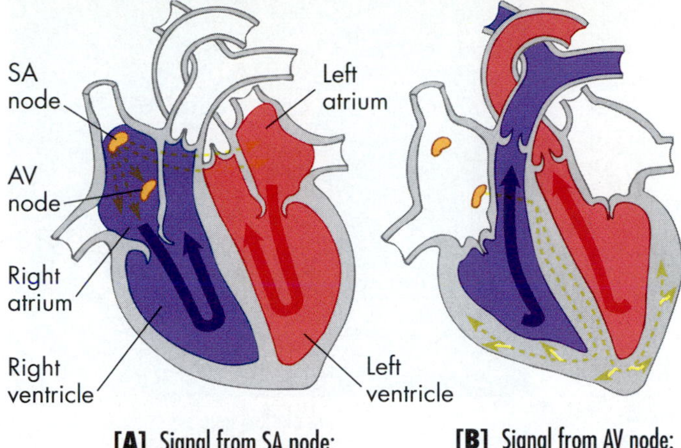

FIGURE 25.11

Electrical Impulses Drive the Heartbeat.

[A] Pacemaker cells in the sinoatrial (SA) node send an impulse across the right and left atrial chambers, which respond by contracting. [B] Upon reaching the atrioventricular (AV) node, the signal is delayed for a split second, after which it reaches the ventricles and triggers their contraction, a squeeze that forces blood to the lungs and the rest of the body.

rate, however, as CHAPTER 31 describes.) Different subsets of heart muscle cells have different intrinsic rates of contraction. Some, called **pacemaker cells**, contract slightly earlier than others, and because each contraction spreads quickly throughout a region of the muscle, pacemaker cells ignite contractions in the entire heart and set the rate of the heartbeat.

Pacemaker cells are located near the upper right atrium in a region called the **sinoatrial (SA) node** [FIGURE 25.11]. Immediately before a beat, an electrical impulse spreads from the SA node across the walls of both right and left atria, causing the two chambers to contract in unison. A second node, the **atrioventricular (AV) node**, is another area of modified cardiac muscle cells that is positioned at the junction where the atria and ventricles meet. When the electrical signal generated by the SA node reaches AV cells, there is a brief delay before the signal is passed to the ventricles. Upon reaching the ventricles, the signal triggers a contraction of these chambers. The brief delay in the transmission of the pacemaker impulse at the AV node is necessary for an efficiently beating heart: If the atria and ventricles were to contract simultaneously, blood in the atria might flow back into the veins instead of forward into the ventricles.

People who have heart conditions in which the beat is irregular can often be helped with an *artificial pacemaker.* This small electrical device, powered by a battery or by the decay of a small amount of radioactive material sends rhythmic electrical impulses that stimulate the cardiac muscle to contract at an appropriate time.

Each time a heart's atria or ventricles contract, they are said to be in a **systole** (SISS-toe-lee) phase. The opposite, or relaxed, phase is known as **diastole** (die-ASS-toe-lee). Together, these two phases make up the **cardiac cycle**—the contraction-relaxation sequence of atria and ventricles that makes up a single heartbeat. Recent evidence suggests that as the ventricles relax in diastole, they actively suck blood from the atria into their chambers.

BLOOD PRESSURE

With each heartbeat, blood courses through your arteries under high pressure and at high speed (33 cm, or about 13 in, per second). If the aorta is severed, blood spurts out more than 2 m (6.5 ft) high! Blood pressure initiated by the heart's strong contractions delivers blood quickly and continuously to cells far from the heart.

Blood flows from areas of higher pressure to areas of lower pressure when the heart muscle contracts, just as toothpaste gushes from a tube when it is squeezed. The consumption of high-cholesterol foods can contribute to artery narrowing, to a consequent increase in blood pressure, and to a heightened risk of heart attacks and strokes, as BOX 25.1 explains.

Blood pressure is measured in millimeters of mercury (mmHg); for example, a pressure of 120 mmHg could lift a column of mercury 120 mm high (4.5 in). A normal reading for a person at rest is 120 mmHg during systole and 80 mmHg during diastole, or 120/80 [FIGURE 25.12]. In contrast, a sea snake's systolic blood pressure is about 30, while that of a climbing snake rises to 90—with higher pressure forcing blood up to the elevated head of the vertical serpent. The highest blood pressure in the animal kingdom—260/160—is found in giraffes. Without such high pressures, blood could never move upward from the animal's heart in midchest, up the long neck, and into its brain. The giraffe's high blood pressure poses potential dangers, however, when the giraffe bends down to drink.

Once blood has been forced through the voluminous capillary beds, pressure originally exerted by the heart is nearly spent. Low pressure allows veins to serve as the body's blood reservoir, capable of holding as much as 80 percent of total blood volume at any given time. The movement of muscles in the arms, legs, and rib cage gently "milks" blood through individual veins, while the one-way valves in each vessel's central cavity prevent backflow and keep the blood moving toward the heart. If no muscular milking takes place over a long period of time, blood and other fluids accumulate in the extremities. Some scholars suggest that the death of crucifixion victims was due in large part to blood pooling in the legs.

box 25.1
Human Impact

Heart Disease: Our Nation's Biggest Killer

Last year, nearly 1 million Americans died of heart disease, making that single condition the cause of 43 percent of all deaths. If we could cure heart disease, we could save $117 billion in lost productivity and medical costs and spare tens of millions of people the grief of their own disability or of losing a loved one to the "silent killer." Biomedical researchers around the world are studying ways to prevent and treat heart disease. There's no need to wait for some future finding before taking action, however, because experimenters and physicians have very solid evidence of the major risk factors for heart disease, and what you can do to avoid or diminish them. They are also developing a list of other factors that may prove to be certified risks and thus bear watching.

The formal name for heart disease, *atherosclerosis*, comes from the Greek words for "gruel" and "hardening"; the popular name is "hardening of the arteries." Both forms aptly describe the condition: a buildup of yellowish fatty deposits called plaque inside the arteries, beginning in late childhood and continuing throughout life. Like corrosion in water pipes, plaque can accumulate until it seriously impedes the flow of blood. Reduced blood flow, in turn, can lead to the formation of a blood clot that obstructs an artery completely and can cause a heart attack or stroke.

Risk Factor: Cholesterol

Arterial plaque contains, among other substances, a glistening white, fatty material called *cholesterol* that is manufactured naturally by liver and other cell types. Cholesterol is an important component of cell membranes and is present in foods, especially eggs, meat, and dairy products. Numerous studies suggest that the more cholesterol circulating in the blood, the higher the chances of a heart attack. The National Heart, Lung, and Blood Institute in 1990 urged everyone from children to seniors to eat diets low in saturated fats (animal fat, egg yolks, high-fat dairy products, hardened margarine), with fewer than 10 percent of total daily calories coming from those sources.

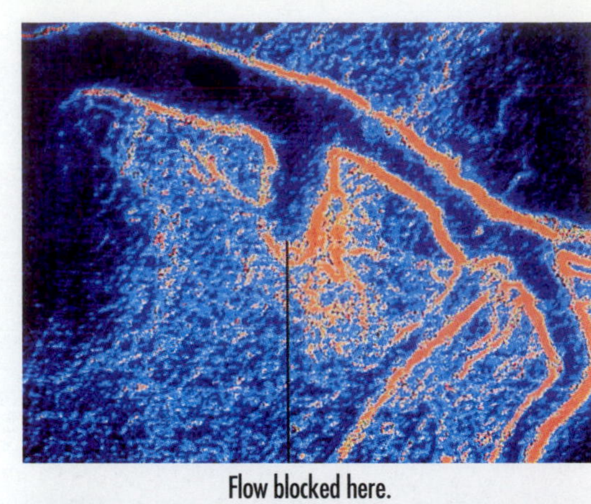

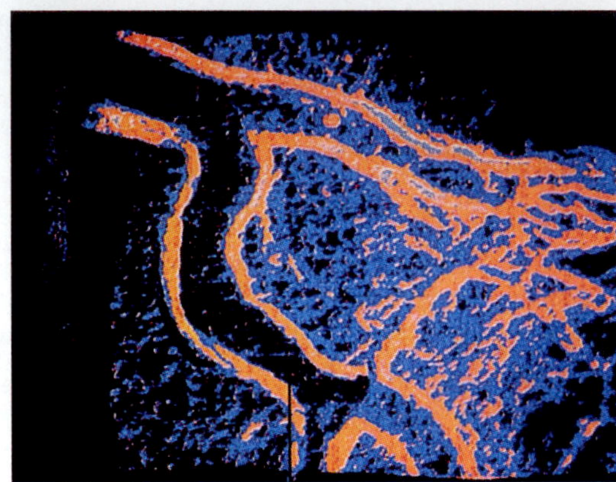

FIGURE 1
A Plaque-Clogged Artery.
Before surgery (above) and after (below).

Physiologists now know that within a person's bloodstream, cholesterol travels attached to carrier molecules called lipoproteins; high-density lipoproteins (HDLs, or "good cholesterol") have very few cholesterol molecules attached, while low-density lipoproteins (LDLs) have great numbers of attached cholesterols. (Like oil on water, cholesterol floats; that's why LDLs are called "low density.") Evidence suggests that people with high levels of LDLs are at greater risk of developing atherosclerosis, while higher levels of HDLs help prevent plaque from building up in the vessels, thereby decreasing the risk of heart disease, heart attacks, and strokes. Some people have inherited a mutation that leads to very high levels of cholesterol in the blood. In most people, however, the condition is due to life-style. Numerous studies show that you can raise your levels of HDLs and lower your levels of LDLs by

REGULATION OF BLOOD FLOW

From moment to moment, the distribution of blood varies in an animal's arteries, veins, and capillaries, depending on oxygen consumption. During an exercise session, for example, blood flow to your muscles increases; then later, if you eat a sandwich, blood flow to the intestines rises and nutrient molecules are absorbed from the gut into this increased blood supply [see CHAPTER 28]. Variations in the amount of blood flowing through the exercising regularly; by avoiding smoking, heavy drinking, and obesity; and by lowering your consumption of all fats in the diet and cutting way back on animal fats.

Risk Factor: High Blood Pressure

People with blood pressure over 140/90 have high blood pressure, or *hypertension*. Because the force of blood pressing out against the artery walls is high, the heart must work harder. This can lead to an enlarged heart and to scarring, hardening, and inelasticity of the arteries, and in turn to increased chance of clotting, heart attack, or stroke. High blood pressure is associated with increasing age; with being overweight; with heavy alcohol consumption; with a diet high in sodium (usually from salt); with race (African Americans are more likely to develop hypertension); and, in women, with the taking of oral contraceptives. Recent studies show that aside from prescription drugs, the most effective ways to lower blood pressure are to reduce dietary salt and to lose excess weight through exercise and cutting calories.

Risk Factor: Smoking

As CHAPTER 6 explained, cigarette smoke contains over 4000 toxic substances, and heart disease researchers believe that some or many of those may damage the lining of the arteries and lead to plaque formation. Smoking also reduces the levels of HDL in the blood; it may cause the formation of blood clots; and it can cause irregularities in the heart beat. In general, a smoker's chance of dying from a heart attack is two to three times higher than a nonsmoker's, and goes up with the addition of other risk factors.

Risk Factor: Diabetes Mellitus

Diabetes, the inability of the body to regulate levels of sugar in the blood, can lead to major health problems, such as kidney failure, blindness, and damage to nerves and blood vessels. Because diabetes damages vessels and also raises the levels of cholesterol in the blood, heart disease is very common in diabetics. In fact, 80 percent die of heart or blood vessel disease rather than the metabolic condition itself.

Risk Factor: Obesity

An obese person (someone 30 percent or more over ideal weight) is more likely to suffer a heart attack or stroke. Excess weight strains the heart as well as other muscles and joints. Since obesity is usually associated with excessive intake of calories (fats in particular) and with a sedentary life-style, changes in diet and exercise pattern are obvious steps to take, under a doctor's supervision.

Other Risk Factors

Researchers have found statistical correlations between several other factors—sometimes surprising—and increased incidence of heart disease. Personality. A constellation of behaviors and attitudes that includes competitiveness, aggression, and a sense of constant time pressure and of heavy responsibilities is correlated with heart disease.

- Short stature. American men under 5 ft 7 in. tall tend to have more heart attacks, perhaps, some speculate, because their blood vessels are smaller.
- Baldness. Men under 50 with vertex (top-of-the-head) baldness have a 50 percent higher risk of heart attack, even in the absence of other risk factors.
- Infection with *Chlamydia pneumoniae*. Past infection with this common microbe is correlated with heart disease. *C. pneumoniae* (not the veneral microbe in CHAPTER 14), causes pneumonia as well as less serious respiratory illnesses. If researchers prove a cause-and-effect relationship between *Chlamydia* infection and heart disease, they could conceivably produce a vaccine for heart attacks.

Whatever your age, sex, or general physical condition, the most prudent course for avoiding our nation's biggest killer is the one health experts have been recommending for decades: Eat fewer fatty foods, get more exercise, maintain ideal weight, don't smoke, watch salt intake, and drink moderately if at all.

---

vessels in a particular body region are generally controlled by the action of nerves or, as in the examples just cited, by the chemical messengers called *hormones*. If blood is needed in one region of the body, say, the arm and leg muscles, then contraction will take place in the walls of arteries or arterioles in other body regions. Contraction of the vessel walls is called **vasoconstriction** ("vessel constriction"; FIGURE 25.13A), and as a result of this process, the diameters of the vessels decrease, blood flow is reduced, and more blood can be shunted to other

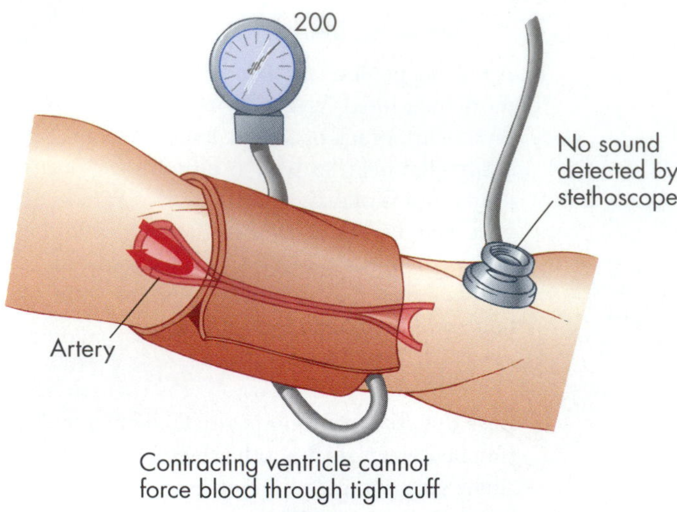

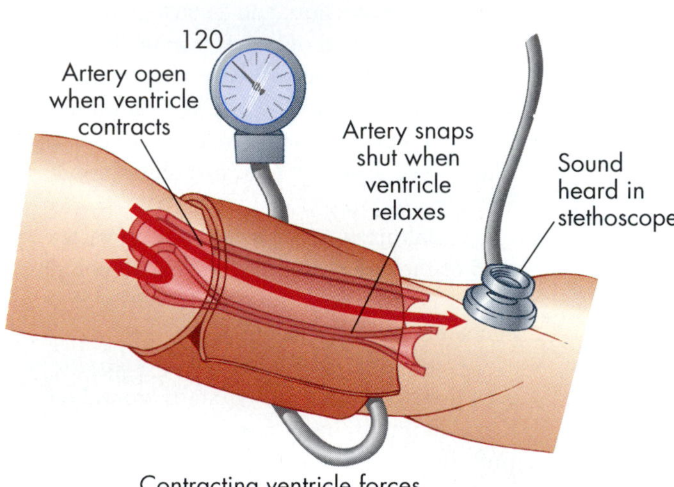

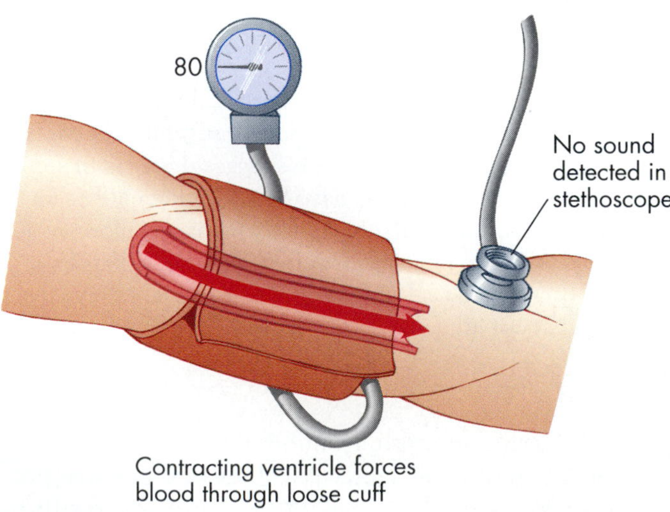

regions, such as the arm and legs, where it is needed to deliver oxygen to rapidly metabolizing muscle cells. In the body regions that require a greater blood flow, a reverse process, **vasodilation** [FIGURE 25.13B], takes place; the vessels' internal spaces are enlarged, and more blood can flow through.

Vasoconstriction results when a small cuff of smooth muscle, the *precapillary sphincter* located near the capillary entrance, contracts. When we get extremely cold, precapillary sphincters may close down capillary beds near the skin surface in the fingers or toes; the remaining blood in these tissues may become oxygen-depleted, and the skin can take on a yellowish or bluish cast [see FIGURE 25.13A]. Likewise, eye drops that "get the red out" contain chemicals that constrict the precapillary sphincters in the tiny capillaries in the eyeball. When we are very warm, say, during active exercise, vasodilation allows the capillary beds to become engorged, and the skin can feel hot and look red [see FIGURE 25.13B].

Researchers have recently implicated a vasoconstriction agent called *angiotensin II* in the heart attacks of people whose coronary problems are puzzling because they lack the five main risk factors normally associated with heart disease. As BOX 25.1 explains in more detail, heart attacks are most prevalent in people who (1) smoke, (2) have high blood pressure, (3) have diabetes, (4) are obese, or (5) have disorders of the body's fat metabolism leading to high levels of blood cholesterol. Some people, however, who have none of these conditions, still have heart attacks, and these attacks may be related to the vasoconstrictor angiotensin II. Physicians have for some time prescribed a drug that blocks the activity of the enzyme *ACE* (*angiotensin-converting enzyme*)—an enzyme that facilitates the last step in the conversion of precursors to angiotensin II. Significantly, blocking the action of this con-

FIGURE 25.12

Measuring Blood Pressure.

[A] A nurse wraps an inflatable cuff around the patient's upper arm and pumps in air until the pressure in the cuff (say 200 mmHg) prevents blood from flowing into the main artery in the arm whether the heart's ventricles are pumping or not. No sound is heard from a stethoscope placed on the artery because the artery is always closed. **[B]** The nurse then releases air from the cuff until its pressure is less than the pressure exerted on the blood by the beating ventricle (say 120). The artery opens when the ventricle contracts, but then slaps shut from the force of the cuff when the ventricle relaxes. The sound of the artery opening and closing can be heard through the stethoscope. **[C]** As more air is released from the cuff, the pressure in the cuff eventually becomes less than the pressure exerted on the blood even when the ventricles are at rest, i.e., the pressure exerted by the artery muscles (say 80). Once again, there is no sound because the artery is permanently open. In this case, the patient's blood pressure is 120 over 80.

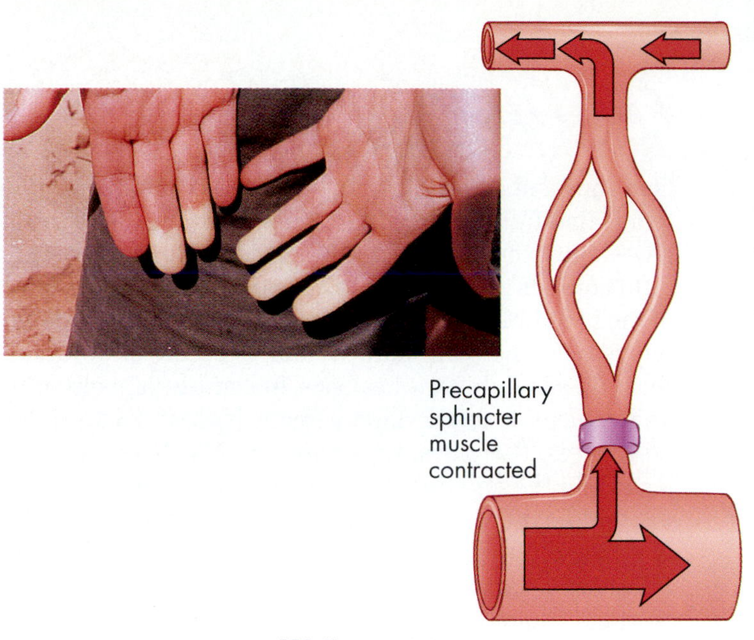

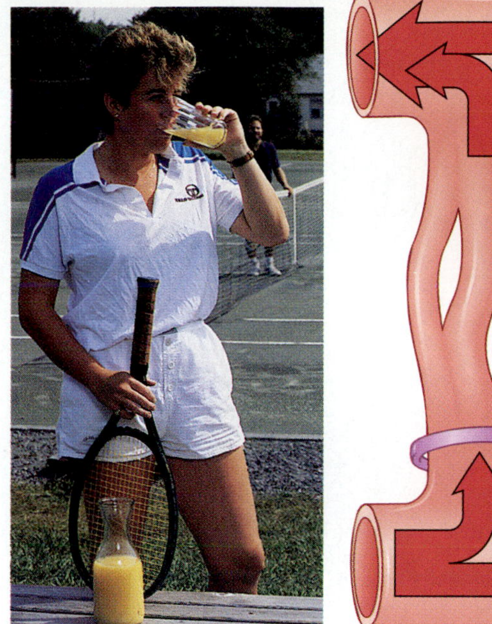

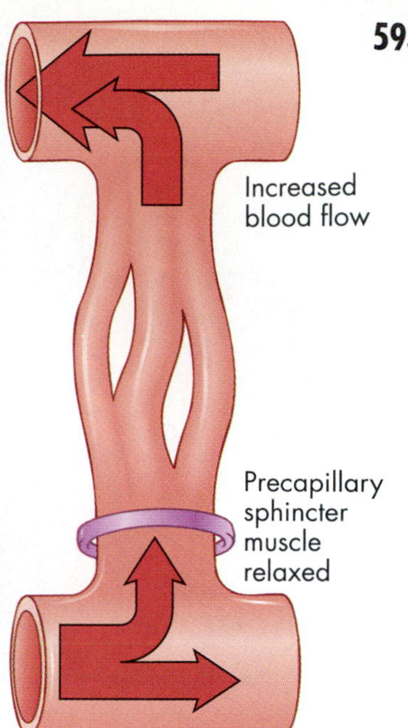

[A] Vasoconstriction

[B] Vasodilation

FIGURE 25.13

White Fingers, Red Face: A Sign of Blood Flow Control.

[A] In the fingers of a person with Raynaud's syndrome, whitish, mottled regions are a sign of restricted blood flow. The mechanism at work is a set of tiny muscles that forms a ring about the capillaries. When these muscles constrict— as they tend to do in cold weather — they close off blood flow to the capillaries. **[B]** The full flush of heavy exercise comes from relaxed sphincter muscles, dilated vessels, and increased blood flow to the skin, where excess heat can disperse to the environment.

verting enzyme saves lives. These recent findings, along with genetic data on variations of the ACE gene, have encouraged heart specialists because they may lead to a way of identifying people without the classic risk factors but who are still in danger of having a heart attack.

BLOOD CLOTTING

Although vasoconstriction can shunt blood away from a capillary bed, more drastic mechanisms are sometimes needed to stop blood from flowing to a particular spot, such as when you cut your finger with a sharp knife [FIGURE 25.14A]. In a cut vessel, several separate events take place that halt the flow of blood: (1) Smooth muscles in the walls of the damaged vessel contract (vasoconstrict) and partly close off the vessel. (2) Circulating cell fragments called *platelets* react by releasing the hormone *serotonin*, which keeps the muscles contracting. (3) Platelets at the injured site begin sticking to each other and to rough surfaces at torn edges of the wound, forming a plug. (4) The next stage is **coagulation**, the actual formation of a blood clot.

The entire multistage process of blood clotting is subject to rigid controls, and for good reason; a clot that forms unnecessarily, such as in vessels of the heart or brain, can lead to a heart attack or stroke. The formation of a blood clot (coagulation) has been dubbed the "clotting cascade" because it involves a series of steplike changes in blood proteins [FIGURE 25.14B]. Ultimately, these transform *fibrinogen* (which normally circulates dissolved in the plasma) into tough, insoluble threads of a related protein, *fibrin*. Fibrin threads in turn become woven into a strong, wiry mesh that traps red blood cells, creating a blood clot [see FIGURE 25.14C]. What sets the fibrinogen-to-fibrin changeover in motion? The trigger is tissue damage: Injured cells release activators, which, in a cascade of reactions, stimulate the conversion of an inactive protein (*prothrombin*) into an enzyme (*thrombin*) that in turn changes the structure of fibrinogen. The altered fibrinogen units then line up together and form the fibers of fibrin.

People with "bleeder's disease," or *hemophilia*, have a mutant gene that codes for a defective protein in the clotting cascade [see FIGURE 11.6]. Unless a person with hemophilia receives periodic intravenous injections of solutions containing the normal protein, even a minor injury can lead to major blood loss and possibly death.

➤ CONCEPT CHALLENGE

Vitamin K is necessary for the synthesis of prothrombin [see FIGURE 25.14]. Why do you think a physician would prescribe coumadin, a substance that blocks the action of vitamin K, for people at high risk for a heart attack?

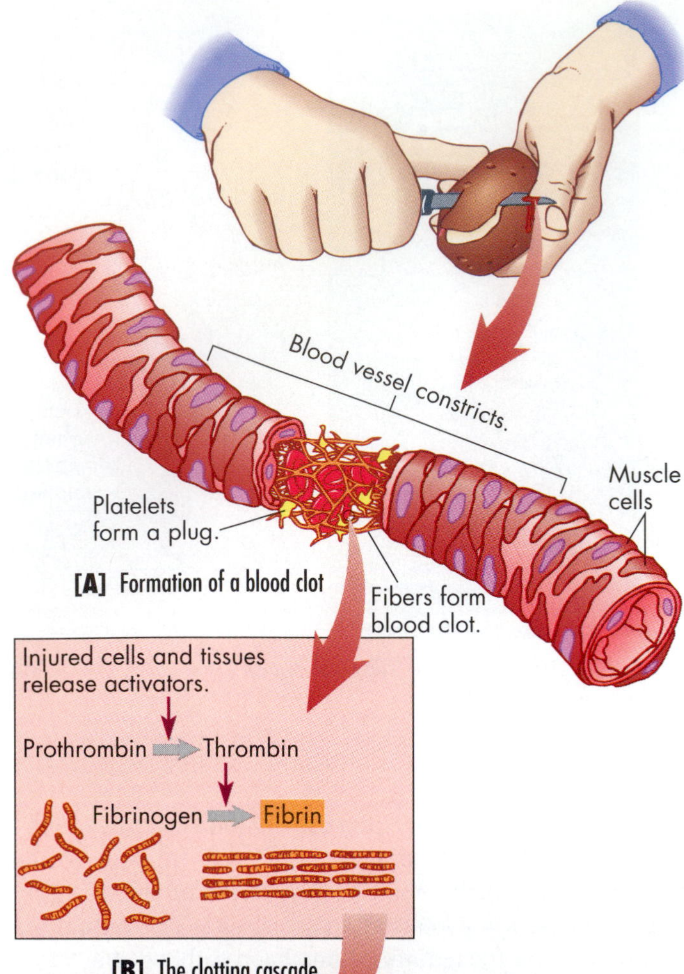

[A] Formation of a blood clot

[B] The clotting cascade

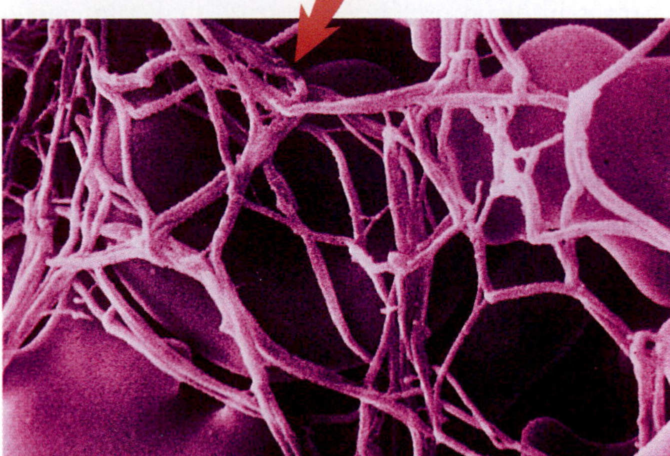

[C] A blood clot

FIGURE 25.14

The Formation of a Blood Clot: A Lifesaving Cascade.

[A] When a wound severs a tiny blood vessel, numerous processes are set in motion, including vasoconstriction of the vessel, the formation of a platelet plug, and then coagulation, or blood clotting. [B] A cascade of proteins acting on each other ultimately converts prothrombin to the enzyme thrombin. Thrombin removes a small piece from the fibrinogen protein, turning it into fibrin, and fibrin molecules line up and form a fiber. [C] Fibrin threads and red blood cells are clearly visible in this blood clot.

The Lymphatic System

One consequence of a highly pressurized circulatory system is that fluid is forced through the thin walls of capillaries and accumulates in the extracellular spaces. This fluid consists of water and various dissolved components of the blood plasma. The second system of fluid-containing vessels, the **lymphatic system**, is a network of tubes that collects fluid that has been forced out of capillaries and returns it to the bloodstream [FIGURE 25.15A]. As CHAPTER 26 describes, lymphatic vessels also play a major role in the functioning of the immune system.

Unlike the blood vascular system, the lymphatic system is made up of capillaries and larger vessels that do *not* form a totally enclosed circulatory loop. Instead, lymphatic capillaries are minute tubes that are closed at one end and drain excess fluid from the body tissues. These thin-walled, closed-end tubes begin in the body tissues. Like creeks merging to form a river, they coalesce and form larger lymph vessels that carry fluids toward the heart [FIGURE 25.15B]. The lymphatic capillary walls are permeable to extracellular fluid and its contents. Excess fluid in the spaces between cells seeps into this tributary system and forms the **lymph**, which also contains proteins, dead cells, and sometimes invasive microorganisms, such as bacteria and viruses. Eventually, the contents of the smaller vessels drain into two major lymphatic vessels, which disgorge a steady stream of lymph into veins near the heart. The lymph fluid then becomes part of the blood plasma once again.

Bean-shaped filtering organs called **lymph nodes** prevent debris in the lymph from mixing into the blood. As lymph moves along a lymph vessel, it percolates through the nodes, like oil passing through an oil filter, and in the process, dead cells and other debris are filtered out. The nodes also harbor large numbers of infection-fighting *lymphocytes* (a type of white blood cell) that can attack bacteria and other foreign materials. You probably have experienced painful, swollen lymph glands in the neck, underarms, or groin area as these organs have increased their filtering activities during a bout of infection. Unfortunately, just as lymph vessels can carry dead cells away from an infection site, they can also transport migrating cancer cells to new sites [see CHAPTER 14]. This is why doctors often remove lymph nodes and vessels along with cancerous tissue in cancer patients.

Like veins, lymphatic vessels transport fluids under low pressure. Valves keep the lymph flowing in one direction, and the squeezing force of contractions in muscle tissue surrounding the lymph vessels propels the fluid. A person with inactive muscles, say, an injured skier in a hospital bed, can suffer lymph accumulation and resultant tissue swelling called *edema* (eh-DEE-muh).

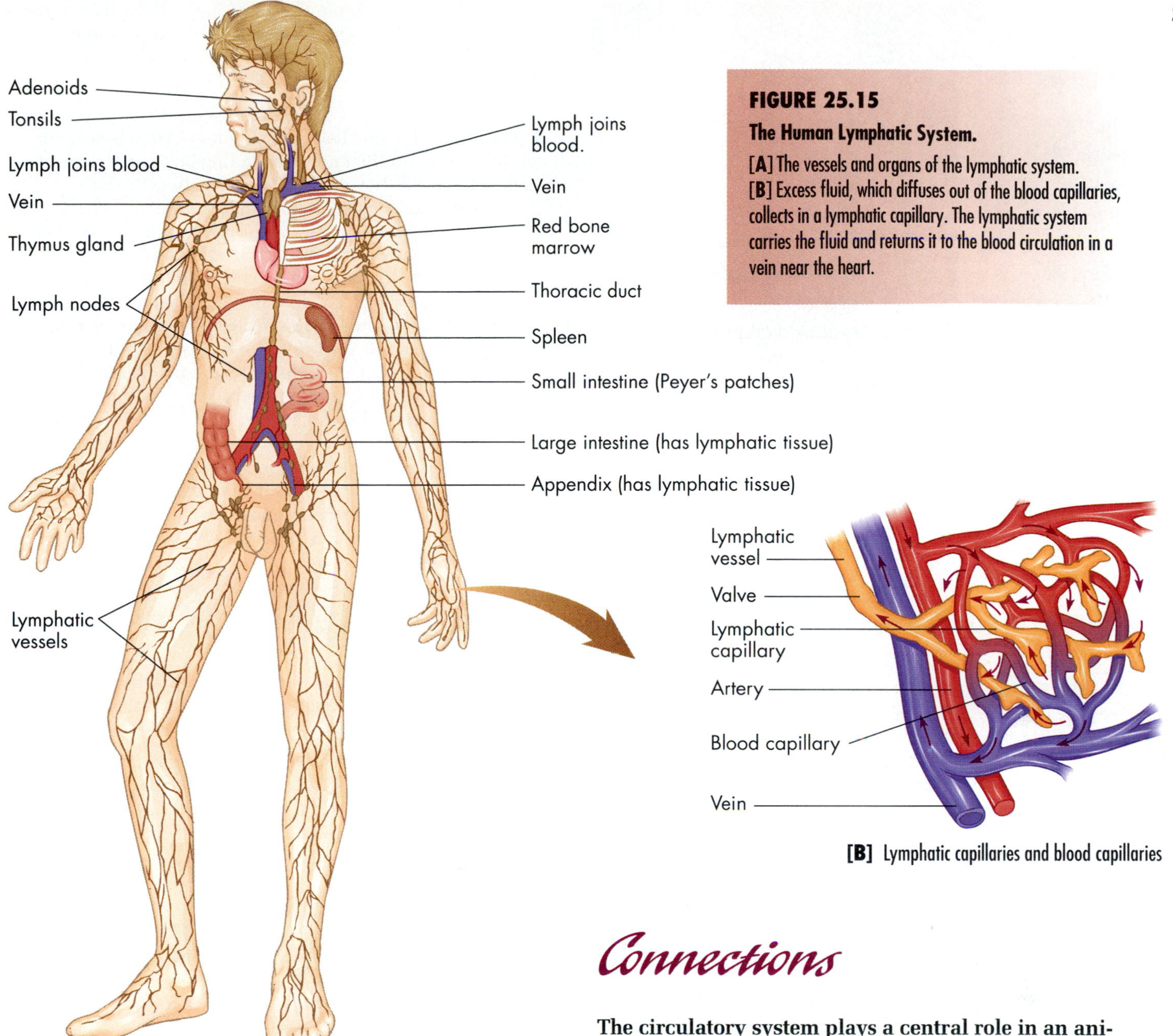

[A] The lymphatic system

[B] Lymphatic capillaries and blood capillaries

FIGURE 25.15

The Human Lymphatic System.

[A] The vessels and organs of the lymphatic system. **[B]** Excess fluid, which diffuses out of the blood capillaries, collects in a lymphatic capillary. The lymphatic system carries the fluid and returns it to the blood circulation in a vein near the heart.

The lymphatic system also has two organs, the **thymus** and the **spleen**, that are somewhat separate from the network of vessels. The thymus is a soft, V-shaped organ located at the base of the neck in front of the aorta. The thymus plays a role in the body's immune system.

An adult's spleen, which is about the size and shape of a banana, lies just behind the stomach and stores and filters blood. The spleen's function is probably not essential, since surgeons can remove a person's injured spleen without causing any severe consequences to overall health.

Connections

The circulatory system plays a central role in an animal's ability to cope with its environment, to maintain a steady internal state for every cell, and to survive. As we saw, there are different kinds of systems, including hearts with different numbers of chambers and in different positions within the body. These variations are evolutionary solutions to the circulatory problems of organisms as different as insects, fishes, climbing snakes, giraffes, and people.

This chapter has focused on the circulatory system in isolation, so we must realize that an organism functions as an integrated whole. The heart muscle could not continue to contract unless it had a steady supply of oxygen supplied by the respiratory system and regulated by the nervous and hormonal systems. These systems, in turn, depend on a steady supply of blood, with its cargo of nutrients and oxygen. We will next examine the roles of circulatory and lymphatic functions in the body's immune system.

KEY TERMS

aorta, 588
arteriole, 586
artery, 586
atrioventricular (AV) node, 589
atrium, 583
capillary, 586
circulatory system, 579
coagulation, 593
diastole, 589
erythrocyte, 580
fibrin, 593
fibrinogen, 593
hemolymph, 582
lymph, 594
lymphatic system, 594
lymph node, 594
pacemaker cell, 589
plasma, 580
platelet, 582
pulmonary artery, 587
pulmonary circulation, 585
sinoatrial (SA) node, 589
spleen, 595
systemic circulation, 585
systole, 589
thymus, 595
valve, 587
vasoconstriction, 591
vasodilation, 592
vein, 587
ventricle, 583
venule, 587

HIGHLIGHTS IN REVIEW

1 Cells exchange substances with their surroundings via diffusion. In a large animal, where many cells are far from the external environment, a circulating fluid such as the blood transports substances near to each cell. The blood, in turn, transports substances to and from a number of organs (lungs, intestines, kidneys) that communicate directly with the outer environment.

a] The circulatory system moves body fluids and helps maintain the animal's homeostasis by delivering oxygen, nutrients, and other materials to each body cell and removing metabolic wastes.

b] The blood has a fluid portion, the plasma, containing dissolved ions and molecules, and formed elements, the red and white blood cells and platelets.

c] The shape of a red blood cell and the millions of hemoglobin molecules contained in the cell allow it to transport oxygen efficiently. The replacement of dead and damaged red blood cells is controlled by a hormone acting in a negative feedback loop.

d] White blood cells squeeze through capillary walls and patrol the fluid-filled spaces between cells, defending the body against invasions by microorganisms and other foreign materials.

2 An animal's circulatory system is adapted to its environment. A person has two circulatory systems; the pulmonary circulation carries blood between the heart and the lungs. The systemic circulation carries blood between the heart and the rest of the body.

a] Arthropods and mollusks have open circulatory systems, with blood moving partly in vessels but mostly flowing freely around and through the animal's tissues. Segmented worms and vertebrates, however, have closed circulatory systems, with blood traveling in an elaborate system of closed vessels.

b] Fishes have a two-chambered heart with a single atrium, which receives blood from veins, and a single ventricle, which pumps blood out to the gills and the body. Adult amphibians and reptiles have three-chambered hearts with two atria and one ventricle. Birds and mammals have four-chambered hearts with two atria and two ventricles.

3 The walls of heart chambers and blood vessels vary in configuration and thickness depending on the pressure of the blood within them and on their role in diffusion.

a] A mammal's complex circulatory system is characterized by thick, muscular arteries leading from the heart to narrower muscular arterioles and eventually to a network of fine capillaries with walls only one cell thick, allowing easy diffusion of substances into and out of the blood. Capillaries lead blood into venules, which coalesce into veins—vessels with one-way valves that transport blood toward the heart. Veins have valves that prevent the backflow of blood.

b] In the mammalian heart, cardiac muscle fibers are electrically linked by special cell junctions in the intercalated disks. The cardiac cycle consists of the contraction (systole phase) and relaxation (diastole phase) of the four heart chambers.

c] Blood pressure is a measure of the outward force that blood exerts on vessel walls as heart chambers alternately pump and rest. Vasodilation and vasoconstriction increase and decrease the diameter of blood vessels and so help distribute blood to local body regions as needed.

d] The clotting cascade transforms the circulating protein fibrinogen into a tough mesh of fibrin filaments that trap red blood cells.

4 High blood pressures force fluid from the bloodstream into spaces between cells. It is now extracellular fluid, and this diffuses into dead-end capillaries of the lymphatic system. Vessels of the lymphatic system then filter the fluid through lymph nodes and carry it back to major veins that discharge the fluid into the heart.

UNDERSTANDING THE FACTS AND CONCEPTS

For Questions 1–5, match each of the descriptions with one of the items in the list below. Each item may be used once, more than once, or not at all.

a] plasma
b] erythrocytes
c] leukocytes
d] erythropoietin
e] thrombocytes
f] lymph

1 A hormone that stimulates the rate of division of stem cells in the bone marrow.

2 Blood cells that contain hemoglobin and carbonic anhydrase.

3 A category of cells that includes phagocytes as well as lymphocytes of the immune system.

4 A fluid that is initially derived from tissue fluids and is returned to the plasma.

5 Cell fragments that are important in the initiation of blood clotting.

For Questions 6–10, match each of the descriptions with the most appropriate item or items from the following list.

a] artery
b] vein
c] capillary
d] arteriole
e] lymphatic

6 The only category of blood vessel with thin enough walls for nutrient exchange with the tissue fluid.

7 Vessels that have valves to prevent back flow.

8 Large, thick-walled vessels that carry oxygenated blood.

9 Blood leaving this kind of vessel next enters a capillary bed.

10 Vessels of this kind carry blood under the highest pressure in the body.

As above, for Questions 11–15, match each of the descriptions with the most appropriate item or items from the following list.

a] systole
b] diastole
c] atrioventricular node
d] sinoatrial node
e] cardiac cycle

11 Site of the pacemaker cells in the mammalian heart.

12 Specialized region of cardiac muscle that stimulates contraction of the ventricles.

13 The stage when blood is forced out of the heart under pressure.

14 A term that describes relaxation of heart muscle.

15 Regions in the heart where the primary function of the cardiac muscle cells is the generation and conduction of an electrical impulse rather than contraction.

For Questions 16–20, match each of the descriptions with one of the items in the list below. Each item may be used once, more than once, or not at all.

a] vasoconstriction
b] angiotensin II
c] fibrinogen
d] fibrin
e] prothrombin
f] thrombin

16 The protein that actually forms the fabric of a blood clot.

17 The activity that occurs when blood is shunted away from a particular capillary bed by the contraction of a precapillary sphincter.

18 An enzyme that results directly in the formation of fibrin.

19 A circulating plasma protein that is converted to its active form during blood clotting by the action of thrombin.

20 A potent vasoconstrictor that has been implicated as a factor in some heart attacks.

INTEGRATE AND APPLY WHAT YOU HAVE LEARNED

1 Compare the level of oxygen in the blood that leaves the heart of a fish with that leaving the heart of a mammal.

2 List the constituents of blood, indicating the primary functions of each.

3 What keeps the blood moving in a single direction in the heart, the arteries, and the veins?

4 Why are people cautioned not to engage in strenuous exercise immediately after eating a large meal? What role do precapillary sphincters play in the process?

5 What organs are included in the lymphatic system? Why do you think that this second circulatory route evolved?

ANALYSIS

1 Why is it important that the heart of a tree-climbing snake is proportionately closer to its head than is the heart of a rattlesnake?

a] Blood can get to the head more quickly in tree-climbing snakes than in rattlesnakes.
b] The closer the heart to the head, the lower the blood pressure required to pump blood to the brain.
c] With a short distance between the heart and the head, the tree-climbing snakes can devote more body space to other organs.
d] With a short distance between the heart and the head, the body of the snake can be shortened to be more agile when it climbs.
e] The farther forward in the body the heart is located, the more suction it can generate to pull the blood forward from the tail.

2 Both birds and mammals are homeothermic and both contain four-chambered, "double-pump" hearts. If an entirely new class of warm-blooded, aquatic vertebrates were discovered on another planet, you would expect that they would:

a] Have a four-chambered heart since they probably evolved from mammals, which, in turn, evolved from birds.
b] Probably have a four-chambered heart because that is the direction of evolution among vertebrates.
c] Have a heart that somehow kept the stream of oxygenated blood separate from the stream of deoxygenated blood because the oxygen requirements of warm-blooded animals are very high.
d] Have a single-pump fishlike heart because that seems to be successful in water dwellers.
e] Have either a four-chambered heart or an open circulatory system.

3 Suppose that you had to select members for a track team that was to compete in a championship meet in Atlanta, Georgia. The team members will arrive in Atlanta individually from their homes the day before the meet. You must choose between two runners who have previously shown themselves to be equally fast many times in the past. Runner A lives in New York City; runner B lives in Denver. Both runners have been home continuously for the past two months. All other things being equal, whom would you choose?

a] Runner A. Training in New York City would be more stressful than training in Denver, and thus runner A would be better prepared.
b] Runner A. Since New York and Atlanta are both on the east coast, runner A would not suffer from jet lag.
c] Runner A. Since both New York and Atlanta are at the same altitude, runner A would be better prepared.
d] Runner B. Since Denver is at a higher altitude, runner B would have more aerobic endurance.
e] Flip a coin and decide on that basis.

4 Which of the following animals has the highest blood pressure? Explain your choice.

a] Blue whale.
b] Giraffe.
c] Large shark.
d] Human being.
e] Hummingbird.

CHAPTER 26

The Immune System

AIDS, THE NEW PLAGUE

Baby Anna was, to all appearances, a normal infant, and by the end of her first year of life, she could speak a few words and stand on her own. At about 18 months, however, she came down with a serious lung infection and for the next 8 months neither learned new words nor developed any new skills of perception and coordination. When she turned 2, she again contracted a severe respiratory infection, and this time she regressed mentally and physically to the level of a 7-month-old. Thereafter, she remained infection-prone, and never resumed normal growth and development. Anna's mother, finally recognizing some of the symptoms that she herself was suffering, sadly realized that she had passed on to her child the disease known as AIDS—acquired immune deficiency syndrome.

AIDS brings into focus this chapter's topic: the central role of the **immune system**, the network of organs, cells, and molecules that defends the body from invaders of all kinds. In people with AIDS, the protective immune system is crippled. This leaves them unguarded from infections by viruses, bacteria,

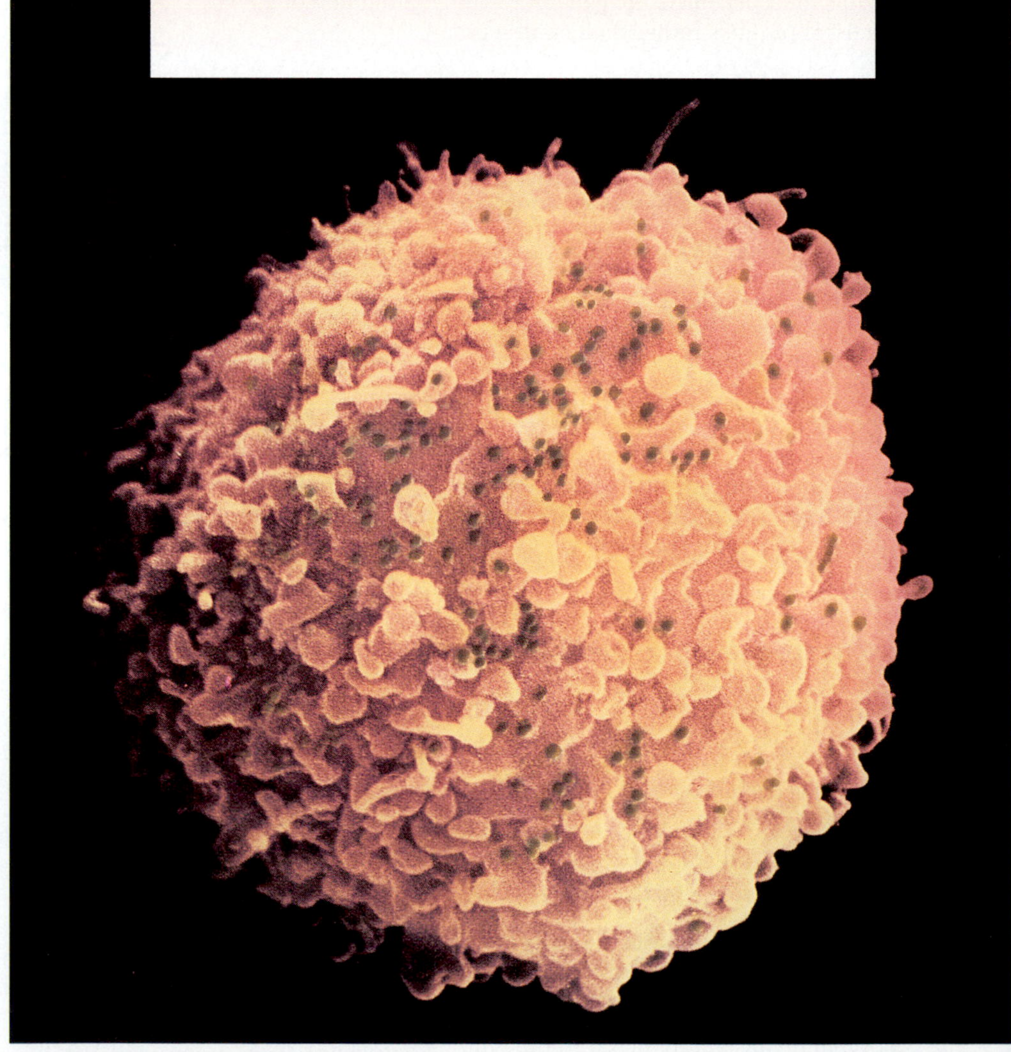

FIGURE 26.1
T Cell Infected by AIDS.
In this scanning electron micrograph, HIV particles are artificially tinted green. People infected by the AIDS virus die of diseases such as pneumonia, fungus infections, and blood poisoning, which a healthy immune system could have resisted.

and parasites they encounter in the environment—agents that seldom harm people with an intact immune system. In fact, the lethal disease AIDS shows just how crucial the immune system is to our day-to-day survival.

AIDS is caused by a virus called *HIV* (*human immunodeficiency virus*; FIGURE 26.1) that is transmitted in blood, blood products, and semen. These modes of transmission have allowed the virus to spread easily in certain populations. People who use intravenous drugs, for example, are exposed to HIV-tainted blood when they share needles with an infected person. Anna's mother, a heroin addict, became infected this way. In about 35 percent of the children born to women infected with HIV, the virus crossed the placenta and infected the baby. The passing of AIDS virus in semen makes it a devastating sexually transmitted disease that can be picked up in a single instance of unprotected sex. Infection with HIV can be prevented through the use of safe sex practices [see TABLE 26.1 on page 601].

When active, the AIDS virus slowly and irreversibly incapacitates the body's natural network of defense, specifically by attacking white blood cells. The virus also infects brain cells, and their destruction can lead to impaired thinking, speaking, and performance. Infected brain cells led to baby Anna's mental deterioration.

Resistance to infection, or **immunity**, is the work of several trillion white blood cells. Some white blood cells attack foreign cells and molecules directly, and some make specially shaped protein molecules called *antibodies* that mark the intruders for destruction. A child with a normal, intact immune system successfully overcomes bouts of cold and flu and gains resistance to reinfection as a result of a special "memory" function of the immune system. Tragically, Anna will probably die in childhood from a form of pneumonia or cancer—conditions that most peoples' immune systems easily overcome.

AIDS is the most threatening epidemic of the twentieth century and a terrifying global problem. There is presently no cure. Perhaps never before has the general public been made so aware of the effects of a healthy immune system and how it shields us daily from dangerous agents in the environment.

In our brief tour of the body's defenses, we will see the major structures and activities of the immune system, and how they normally provide lifelong protection against diseases. We will see how immune system cells and molecules work together to dispose of invaders without harming the tissues of the organism's own body, and how physicians manipulate the immune system with vaccines and drugs to help protect patients. Along the way, we will learn much more about AIDS, as well as about allergies, abzymes, immunization, pregnancy, snake bite, and arthritis. ❑

MESSAGES

1 The vertebrate immune system recognizes invaders and eliminates them. The process involves chemical communication among cells; the presentation of foreign substances called antigens by one type of immune cell to another; the regulation of cellular activities by certain immune cells; and the destruction of invaders by certain cells and molecules.

2 In the humoral immune response, B lymphocytes secrete antibodies. These proteins bind to invaders and mark them for destruction. In the cellular immune response, T lymphocytes help other immune cells mature or directly kill invading parasites, cells infected with viruses, cancer cells, or cells transplanted from another individual.

3 The progression of AIDS (acquired immunodeficiency syndrome) is marked by infection with the human immunodeficiency virus (HIV), and as a result, a gradual decrease in helper T cells. Without helper cells, neither cytotoxic T cells nor antibody-secreting cells are fully active; hence, opportunistic infections, such as pneumonia-causing parasites, can kill the patient.

4 If an immune response becomes directed against a body's own cells or substances, autoimmune diseases like arthritis, multiple sclerosis, and diabetes can result. On the other hand, vaccinations direct an immune response against a specific disease-causing agent by stimulating antibody production or cytotoxic T cell production.

What Is the Immune System?

The function of the immune system is to protect the body from disease-causing agents. Although the immune system is composed of several parts, the functional cells are phagocytic *macrophages* and a specific class of white blood cells called *lymphocytes*. The two different types of lymphocytes are *B cells* and *T cells*.

Cells of the immune system constantly circulate in the bloodstream and in the lymph. Like most other physiological systems, therefore, the immune system is tightly linked to the circulatory and lymphatic systems [see CHAPTER 25]. Moreover, lymphocytes and macrophages arise from stem cells in the bone marrow, just as red blood cells do. As you saw in CHAPTER 25, however, red blood cells mature into oxygen-carrying biconcave disks in the bone marrow. In contrast, some white blood cells mature inside other organs: the *spleen*, *thymus gland*, and *lymph nodes* [see FIGURE 26.5]. These organs are called *lymphoid organs* because they are connected by the lymphatic system as well as by the bloodstream. Lymphoid organs trap foreign cells and molecules, and harbor white blood cells.

The immune system of a vertebrate carries out three essential protective activities [FIGURE 26.2]:

1. The first activity is **recognition**: The immune system recognizes foreign invaders by identifying molecular patterns on the surface of cells and molecules, thus distinguishing *self* (cells and molecules made by the body and belonging to it) and all varieties of *nonself* (molecules and cells originating outside the body). This ability to distinguish "self" from "nonself" is the primary characteristic of the immune system.

2. The second essential activity of the immune system is **communication** between cells. For example, certain white blood cells recognize invaders and then send a signal to other white blood cells, which then initiate activities that help rid the body of the invaders.

3. A third protective response, **elimination**, defeats the attack by selectively eliminating foreign invaders from the body. This response involves specific proteins that target viruses or bacteria for destruction by enzymes and specialized white blood cells that dispose of virus-infected cells directly.

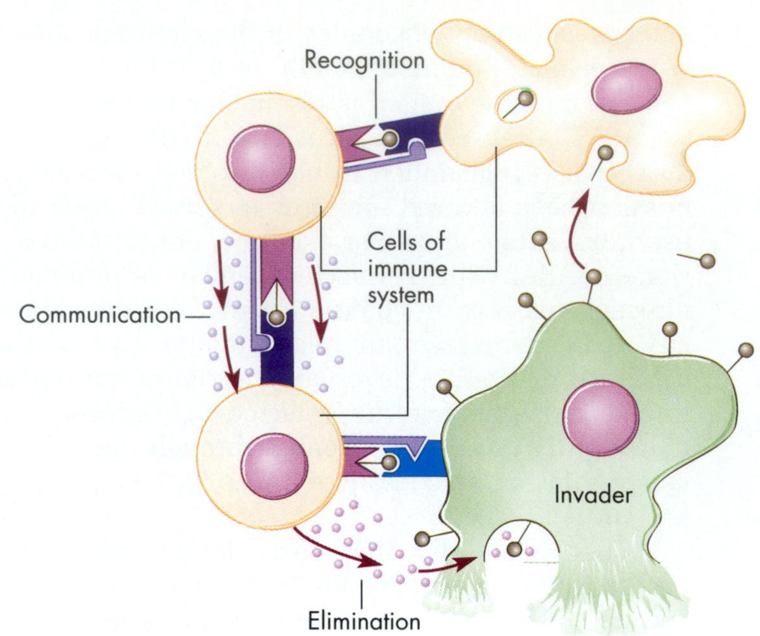

FIGURE 26.2
Essential Activities of the Immune System.
Cells of the immune system *recognize* invaders, *communicate* with other immune system cells, and *eliminate* invaders.

Some elements of the immune system recognize specific individual invaders, while others recognize invaders in a more general way. *Specific defenses* of the body involve individual defense molecules tailor-made for each kind of microbe or other invader. In contrast, *nonspecific defenses* recognize invaders in a more general way, and include physical barriers to infection (such as the skin), as well as certain physiological reactions that the body produces in the same way for every intruder. Nonspecific mechanisms act faster than specific immune responses, but are not as effective at warding off disease.

Nonspecific Defenses

Some **nonspecific defenses** of the body are the skin and mucous membranes, the inflammatory response, phagocytosis, and nonspecific antiviral and antibacterial substances.

SKIN: THE FIRST LINE OF DEFENSE

If your finger slips as you hold a tack, you might open a chink in your skin. The **skin** is the body's first line of defense against unwanted substances [FIGURE 26.3]. The skin forms a barrier that, along with the mucous membranes of the nose, throat, and digestive passages, keeps out most microorganisms capable of multiplying within

TABLE 26.1 Protect Yourself from AIDS

Risk Factors	Protection
■ Sex with an individual who has had a positive AIDS test, indicating exposure to the virus ■ Intravenous drug use ■ Injections of any kind with used hypodermic needles and syringes ■ Multiple sex partners ■ Sex with anyone who has had multiple sex partners ■ Transfusions with tainted blood ■ Sex with an intravenous drug user or with anyone engaging in other high-risk activities ■ Sex with any individual whose sexual history and exposure status (positive or negative) are unknown to you ■ Sex that involves contact with blood, such as sex during menstruation or anal intercourse ■ Other activities that involve contact with blood, including some types of work in medicine, dentistry, or undertaking	■ Find out your own exposure status and that of any potential sexual partner ■ Avoid high-risk sexual activities ■ Seek help for addiction, and use only sterile needles and syringes for any necessary injection ■ Use condoms and, if possible, preparations that contain a virus-fighting compound such as nonoxynol 9 ■ Avoid unnecessary contact with blood, and accept only transfusions that test free of the AIDS virus ■ Stay informed of new developments in the fight against AIDS; seek additional information from your college or university health service, your local public health service, or your private physician

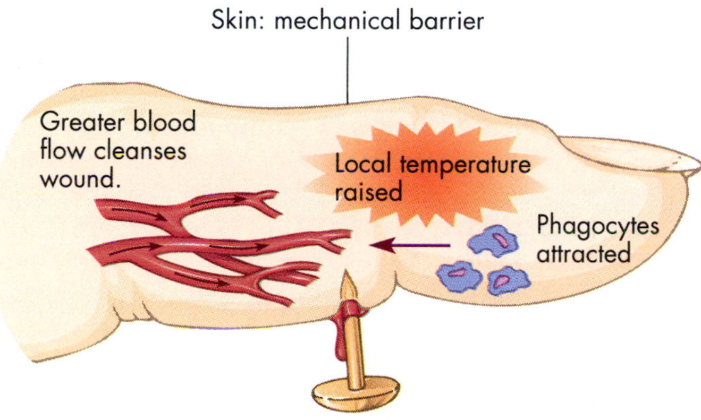

FIGURE 26.3
Nonspecific Defenses.
The skin forms a mechanical barrier to invaders. The inflammatory response raises local temperature and attracts phagocytes—blood cells that devour debris from any type of invader nonspecifically.

the body and causing infection. Thus, through a break in this smooth, elastic barrier, dirt and soil bacteria clinging to the tack can enter the bloodstream.

THE INFLAMMATORY RESPONSE AND PHAGOCYTOSIS

Eventually, a blood clot will seal the gap in the skin and scar tissue will form, but even before that, the injury to the skin sets up a series of nonspecific physiological reactions called the **inflammatory response** [FIGURE 26.3]. First, skin and blood cells release chemicals that raise the temperature at the site of the wound. Next, blood flow to the area increases, and soon scavenging white blood cells called **phagocytes** converge on the area in search of invaders, devouring them by the process of phagocytosis [FIGURE 26.4].

The rise in temperature slows bacterial multiplication; the increased flow of blood cleanses the wound and brings in more white blood cells; and the phagocytes engulf bacteria and cell scraps without disturbing the body's own intact cells. These physiological reactions re-

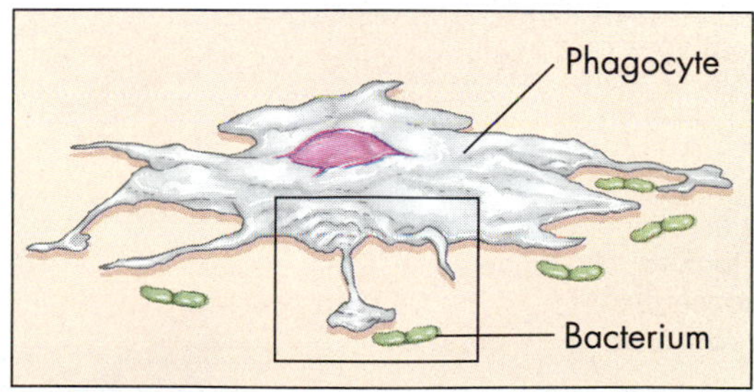

FIGURE 26.4
Nonspecific Defenses: A Phagocyte Stalks Bacteria.
The white phagocyte approaches the small rodlike invading bacterium (tinted green) with a cellular extension, or pseudopodium.

sult in the redness, heat, and swelling we know as *inflammation*. Inflammation helps eliminate debris and often keeps the bacteria from spreading.

Along with the inflammation response, within four hours of an invasion one particular type of white blood cell called a *natural killer cell* carries out a nonspecific counterattack. In a way that immunologists do not yet understand, natural killer cells can recognize bacteria, certain kinds of cancer cells, and virus-infected cells and can kill them on first contact. Biologists know how crucial natural killer cells are to immunity because in 1989, a woman nearly died during the early phase of a herpesvirus infection. It turned out that her immune system was intact except that it lacked nonspecific natural killer cells. Fortunately, she survived with hospital care and fought the infection, and later, her system mounted a normal specific immune response against the herpes.

Nonspecific responses such as inflammation and natural killer cells provide some first-line protection against disease, as we have seen. It is the specific immune responses, however, that pinpoint individual invaders and destroy them.

FIGURE 26.5
Players in the Immune Response.

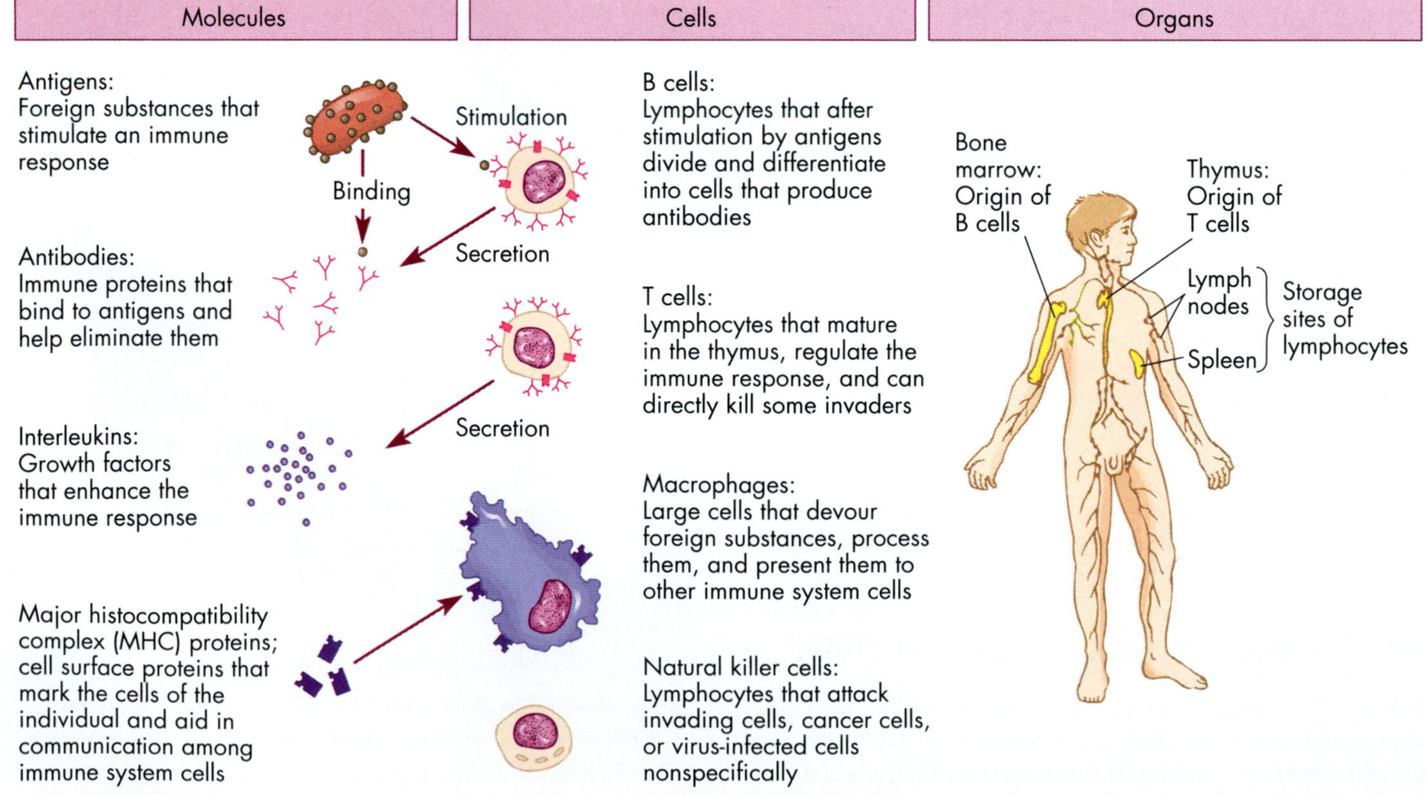

Specific Defenses

Suppose that the bacteria that entered the wound in your finger begin to multiply faster than inflammation, scavenging phagocytes, and natural killer cells could destroy them. The body would then fall back on its most powerful line of defense, the **immune response** [FIGURE 26.5]. During this response, certain white blood cells produce antibodies, specially shaped proteins that either float in body fluids or attach to the surface of white blood cells and attack the invader. Behind this specific response lies the entire arsenal of the immune system: its several kinds of white blood cells, the proteins they secrete, and the organs and fluids in which they multiply, differentiate, and eliminate invaders.

CELLS OF THE IMMUNE RESPONSE

Of the several types of white blood cells that cooperate in generating immunity, the lymphocytes are the driving force. **Lymphocyte** means "cell of the lymph system" and indeed, these small, round, colorless cells spend much of their time within lymphatic tissues. There are two main types of lymphocytes: **B cells (B lymphocytes)** make and secrete the antibody proteins that coat invaders and mark them for destruction. **T cells (T lymphocytes)** kill foreign

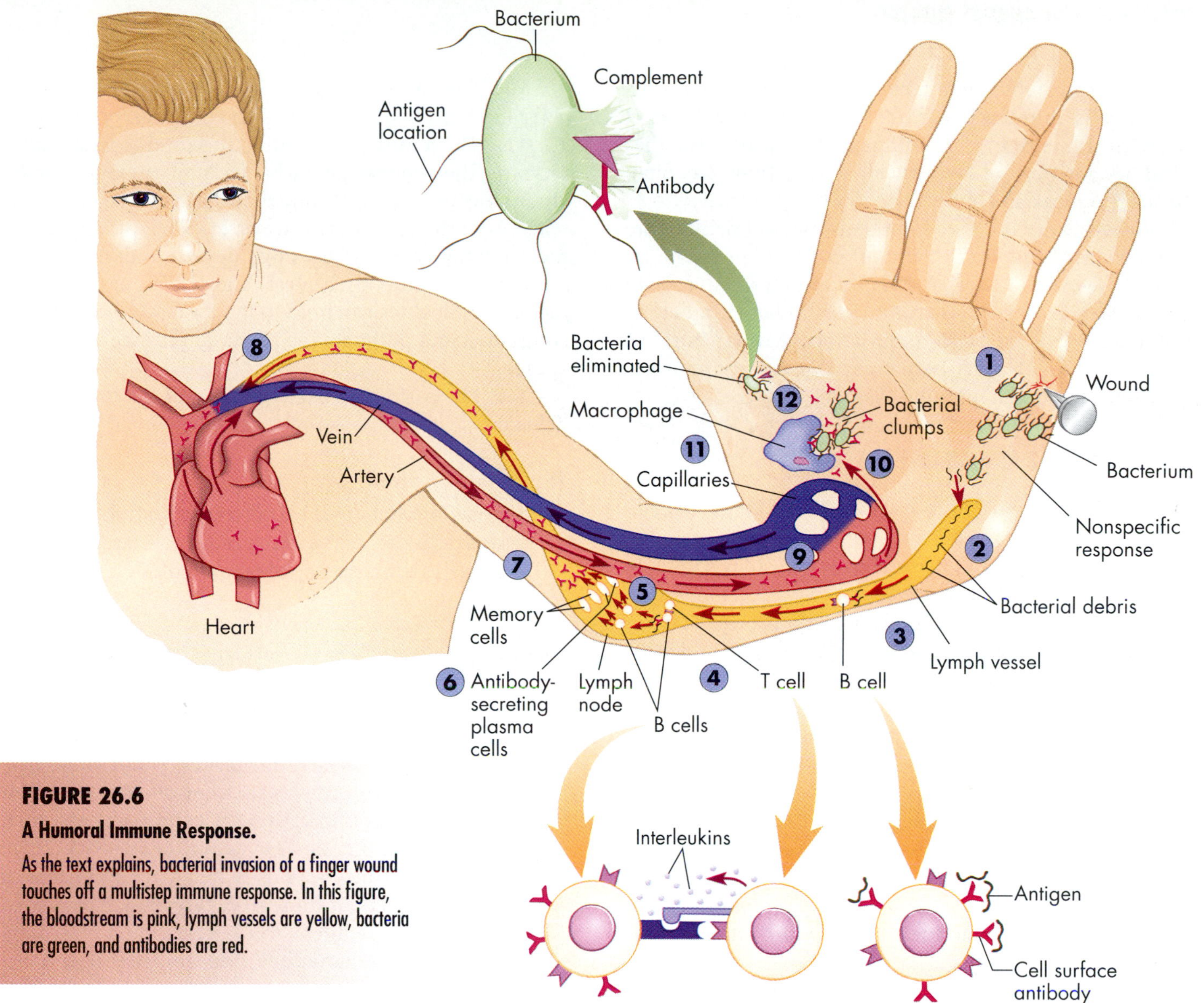

FIGURE 26.6

A Humoral Immune Response.

As the text explains, bacterial invasion of a finger wound touches off a multistep immune response. In this figure, the bloodstream is pink, lymph vessels are yellow, bacteria are green, and antibodies are red.

cells directly and help regulate the activities of the other lymphocytes. A third kind of white blood cell involved in specific immunity is the **macrophage** (literally, "big eater"; see FIGURE 26.5). Macrophages are large, specialized phagocytes that stimulate lymphocytes to attack invaders, then help clean up by consuming debris when the lymphocytes are through.

MOLECULES INVOLVED IN DEFENSE

To appreciate how the immune system works, we must understand not just the cells of the immune system but three types of molecules, as well: the antibodies, antigens, and interleukins.

Antibodies are protective proteins made by B cells and secreted into the blood and lymph [see FIGURE 26.6]. Different B cells make antibodies of slightly different shape. During an immune response, a specific part of the antibody molecule binds to a substance projecting from the surface of an infecting bacterium or other foreign agent. The antibody thus simultaneously recognizes the invader and marks it for elimination.

Immunologists have a special name for the synthesis and secretion of antibodies by B cells: the **humoral immune response**, from *humour*, the medieval word for "body fluids." They call it this because antibodies travel through blood and lymph to the site where they encounter antigens and mark them for destruction. Antibodies are particularly effective at eliminating free-floating microbes, viruses, and molecules that have not yet entered body cells.

Substances that are bound by antibodies are often proteins but can be other kinds of molecules. They are examples of **antigens**, foreign substances that generate an immune response and trigger the production of antibodies that fight the foreign invaders. Invaders can have many antigens; the outer coats of most viruses, bacteria, fungal cells, and larger parasites contain many different proteins functioning as antigens. Because antibodies bind to antigens of reciprocal shape, they are often likened to locks and antigens to keys.

Besides antibodies and antigens, the interleukins are a third class of molecules important in the immune response [see FIGURE 26.6]. *Interleukins* are protein growth factors secreted by some immune cells that stimulate or retard the division or differentiation of other immune cells. Interleukins thus help carry out the communication activities of the immune system.

AN EXAMPLE OF AN IMMUNE RESPONSE

Let us return to our example of a wounded finger, and follow the steps involved in the immune response—the numbered steps below correspond to the steps shown in FIGURE 26.6:

1. As you saw earlier, once the skin is broken, the nonspecific defense responses attack the bacteria that have invaded the wound. The attack creates debris (broken bacterial cells and cell parts).

2. The cellular debris begins to accumulate near the wound, and enters the bloodstream and the lymphatic system.

3. In the bloodstream and lymphatic system, the debris encounters B cells, and antibodies anchored in the plasma membrane of some of these circulating B cells bind to the antigens. This binding represents recognition of the invader, and leads to a *specific* immune response.

4. B cells carrying antibody-antigen complexes move to the lymph nodes nearest the wound—in this case, probably in the elbow—and settle in.

5. Helper T cells, which have already received communication from macrophages about the bacterial protein, interact with the B cells and the antigen-antibody complexes.

6. The helper T cells secrete interleukins, protein growth factors that communicate with B cells and stimulate them to divide into 2, 4, 8, 16, and eventually hundreds of antibody-secreting "factories" known as *plasma cells*. This step in the process can take several days because each cell cycle can take many hours.

7. Once plasma cells have differentiated, each one can secrete into the lymph fluid over 1000 copies of an antibody molecule, which can bind specifically to the antigen receptor sites on the invading bacteria.

8. The secreted antibodies now leave the lymph nodes. The lymph drains back into the bloodstream near the heart.

9. The antibodies enter the blood and circle back to capillaries near the wound.

10. The antibodies pass into spaces between the cells and bind to their targets—the bacterial proteins.

11. When millions of antibodies bind to bacterial proteins, they neutralize the bacteria by causing them to clump together. Such clumped bacteria can be eliminated from the body in two ways: (1) The antibody coating can attract macrophages that devour, digest, and eliminate the antibody-bacteria complexes, or (2) *complement* molecules can bind to the antibody-bacteria complexes and perforate the bacterial plasma membranes.

12. The perforated bacteria die, and the infection is defeated.

➤ CONCEPT CHALLENGE

A boy with the inherited disease XLA (X-linked agammaglobulinemia) makes no antibodies, yet his B cells are not defective. How could a person's B cells be normal and yet produce no antibodies?

Antibodies

As described above, the sequence of events during an immune response protects humans and other mammals from bacteria. It also protects us from other types of invaders and from repeat infections by certain viruses, such as a second case of measles. One could rightly ask, then, why specific immune responses failed to protect baby Anna and other people from HIV, the virus that causes AIDS. The answer requires an understanding of exactly what antibodies are, how they perform their functions, and the limits to their powers.

Antibodies are *immunoglobulins*, globular proteins of the immune system. Each one consists of four polypeptides, or chains of amino acids [FIGURE 26.7]. Two of these four chains are longer and are called *heavy chains*, and two are shorter and are called *light chains*. In any given antibody molecule, the two heavy chains are identical to each other, and the two light chains are identical to each other. Linked together in one unit, the four chains are arranged in such as way that the molecule looks like the letter Y.

Antibodies have two basic functions: (1) they recognize antigens by binding to them, and (2) they mark antigens for elimination from the body. These two functions are localized at opposite ends of the Y-shaped molecule. The tips of the two branches each bind to an antigen of complementary shape the way an enzyme binds to its substrate [see CHAPTER 4]. Once the branches of an antibody bind to an antigen, the stem of the Y stimulates macrophages or complement proteins to attack and destroy the entire antibody-antigen complex [see FIGURE 26.6, Steps 11 and 12].

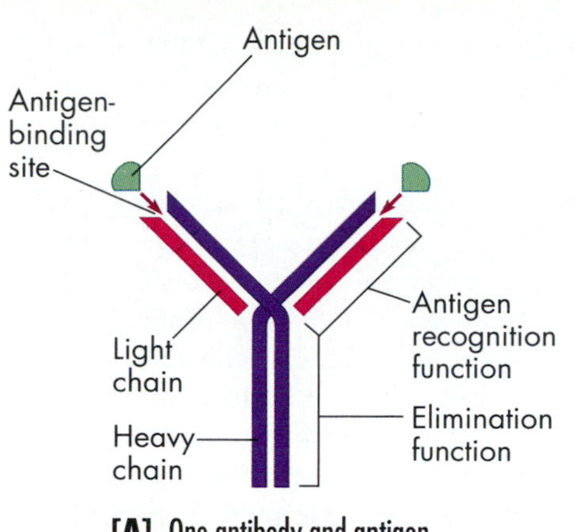

[A] One antibody and antigen

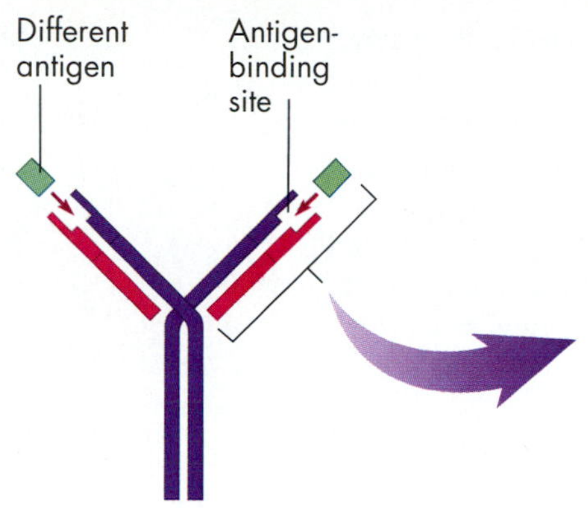

[B] A different antibody and antigen

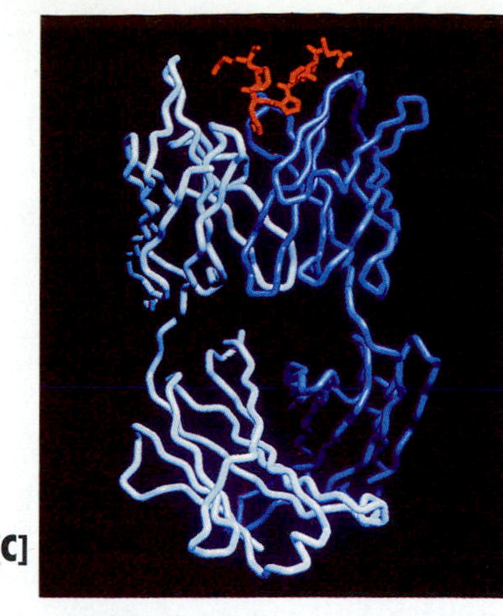

[C]

FIGURE 26.7

Different Antibodies Bind Different Antigens.

Antibodies are Y-shaped molecules with specific antigen-binding sites located at the tips of the Y's arms. Antigen molecules, wedge-shaped in **[A]**, rectangular in **[B]**, fit into the binding sites of antibody molecules like keys in locks. The elimination of an antigen bound to an antibody depends on the stem of the Y. **[C]** A computer graphic reveals how an antigen—in this case, a small peptide (top, in red)—binds to an antibody (here, just one arm of the Y is shown in dark blue for the heavy chain, light blue for the light chain).

Researchers have studied antibodies directed against many different antigens, and have compared their amino acid sequences. Through such comparisons, they have learned that the stem section of the main class of antibody that circulates in the blood always has the same amino acid sequence, and hence the same shape. This is true whether the antibody is directed against the polio virus, the tetanus toxin, or some other antigen. In contrast, the tips of the two branches have a huge variety of different amino acid sequences, and thus a huge range of different shapes. These many shapes can bind in complementary lock-and-key fashion to an enormous array of antigens—each antibody to an antigen of a particular shape [see FIGURE 26.7C].

ANTIGEN-BINDING SITES: SPECIFICITY AND DIVERSITY

Although each antibody molecule binds only to an antigen of a single shape, a person has more than 100 million distinct types of antibody molecules, each with a unique binding site. Antibodies can recognize and combine with 100 million different antigens, including pollen from flowers, grasses, and trees, as well as proteins in the outer coats of thousands of different viruses, bacteria, and molds. Because of recent advances [see BOX 26.1], this amazing specificity can now be tapped to manufacture antibodies for various medical, diagnostic, and industrial purposes. For example, molecular biologists have engineered antibodies that can facilitate chemical reactions much as enzymes do. These antibody enzymes, or *abzymes*, may allow researchers in the future to create specific catalysts tailored for any reaction they wish to control.

Doctors have also harnessed the huge spectrum of antibody shapes and activities as a powerful tool for diagnosing illnesses. One example of this is an accurate test for specific antibodies to the AIDS virus. When HIV first enters a person's body, he or she begins to make antibodies to it according to the scheme outlined in FIGURE 26.6. Blood tests can identify these anti-HIV antibodies. Anyone who has been exposed to HIV is then called *HIV-positive*. Unfortunately, these antibodies do not protect against the virus because the virus wreaks its havoc *inside* cells, where it is hidden from the antibodies. Nevertheless, identification of HIV-positive individuals can be used to help slow the transmission of the virus.

By distinguishing blood from exposed people, medical workers can avoid using this blood for transfusions to other people, and hence help prevent the transmission of the AIDS virus by this route. Hopefully, one day it may also help doctors detect the presence of the virus early enough to forestall the progress of the disease into a full-blown case of AIDS—a tragic diagnosis for any patient.

HOW ANTIBODIES TRIGGER ELIMINATION

While the arms of a Y-shaped antibody bind to an antigen, the stem of the Y triggers the elimination of an invader. Slight variations in the amino acid sequence of the stem (and hence slight differences in stem shape) result in different elimination activities [FIGURE 26.8].

One stem class called **IgG**, or *immunoglobulin G*, circulates in the blood and efficiently eliminates viruses and bacteria. IgG is the only class of antibody that can cross the placenta from a pregnant woman into her baby. The blood protein fraction called *gamma globulin* is rich in IgG. This IgG can provide immediate protection when injected into a patient who has just had surgery, or into a traveler bound for a country that has contaminated food or water.

Another class of antibodies, **IgA**, has a different-shaped stem that allows these molecules to make their

Box 26.1 Biology Applied

Monoclonal Antibodies: Tools for Medicine and Research

Antibodies have amazing specificity: They bind to one, and only one, other type of molecule. Imagine how useful it would be for researchers to make beakersful of an antibody that specifically recognizes and binds to a certain portion of a molecule they'd like to locate. Such molecules could include, say, the hormone that signals pregnancy, or a protein on a gonococcal bacterium that causes venereal disease.

All this became possible in 1975 when researchers Cesar Milstein and Georges Köhler, working together in England, developed a way to clone individual B cells and produce large quantities of individual antibody molecules. They began by injecting a mouse with a specific antigen, say, a gonococcal cell's surface protein, then removing the mouse's spleen with its store of B cells a few days later. Next they fused normal B cells from the spleen with cancerous B cells, or myelomas, whose forte was indefinite, rapid cell division. The result was a set of revved-up hybrid B cells called *hybridomas*, capable of churning out large numbers of specific antibodies of the desired type and of dividing and redividing into large clones of identical cells. The team's final steps were to choose the clone of hybridomas that made the specific desired antibody, grow huge numbers of the cells, then harvest the secreted *monoclonal antibodies*, antibodies from a single clone of plasma cells.

By now, commercial laboratories are selling more than $1 billion worth of the cloned biologicals each year. Monoclonals are being used in basic research to help locate and identify specific proteins, in the diagnosis of medical conditions, and in the treatment of human diseases. For example, home test kits for detecting pregnancy contain monoclonal antibodies that bind to human chorionic gonadotropin [see CHAPTER 14]. Another laboratory test kit contains monoclonal antibodies to the enzyme prostatic acid phosphatase, a protein present at high levels in men with prostate cancer; use of the kit aids early detection of that dangerous disease. Medical laboratories are now using monoclonal test kits to detect surface proteins on gonococcus, chlamydia, and herpesvirus. With such tools, physicians can cut three to six days off the time needed to confirm a patient's venereal disease and thus stop the rampant pathogens more quickly.

Monoclonal antibodies are true magic bullets, striking specific molecules and leaving the rest of the body unharmed.

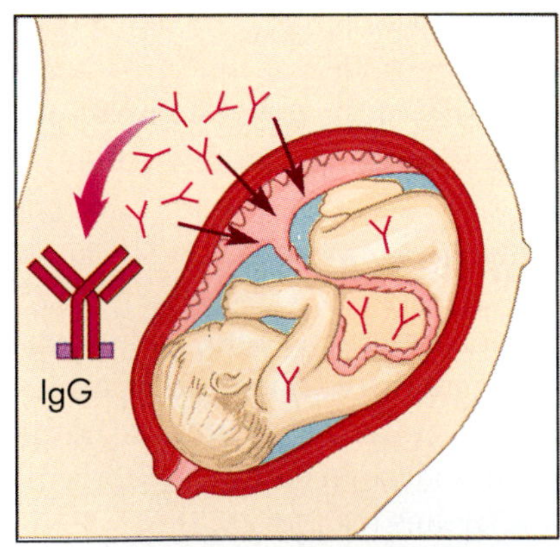

[A] IgG can cross placenta.

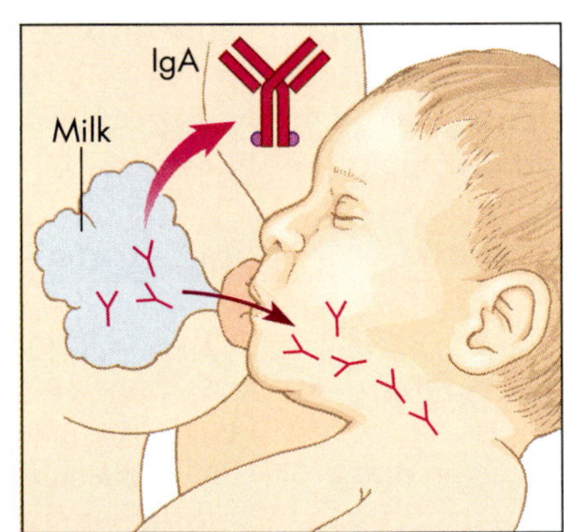

[B] IgA is secreted in body fluid.

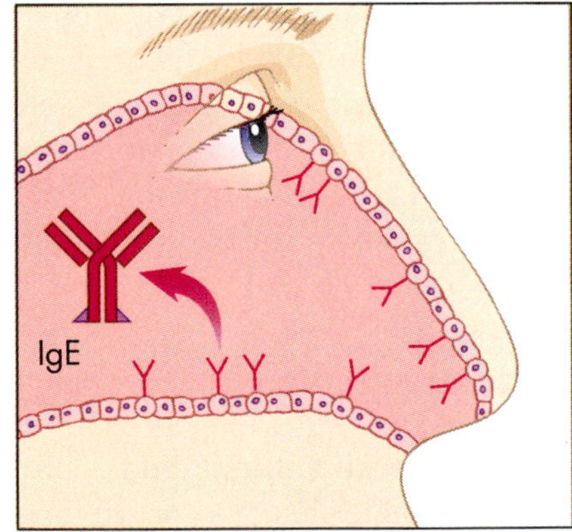

[C] IgE is involved in allergies.

FIGURE 26.8

Different Antibody Classes Have Different Destinations.

[A] Antibodies of class IgG can cross the placenta and confer immunity on the fetus. **[B]** IgA is secreted in saliva, tears, and milk and can help a breast-feeding infant resist infection. **[C]** IgE molecules bind to mast cells in the skin and mucous membranes lining the nose, throat, airways, and intestines and are involved in allergic responses.

way into bodily secretions such as saliva, tears, milk, and the mucus of the intestinal lining. IgA molecules that pass to a nursing infant through the mother's milk can help protect the child from diseases to which the mother is immune. This is a strong rationale for breast-feeding.

A third antibody type, **IgE**, binds to specialized cells in the skin and intestinal lining. Besides helping to fight parasites, cells bound to IgE are responsible for the irritating condition we call *allergy*, as described in BOX 26.2.

To discover how white blood cells can form a family of molecules with one basic shape, but millions of uniquely shaped recognition sites, we must examine one particular class of white blood cells, the B lymphocytes, which form and secrete antibodies.

➤ CONCEPT CHALLENGE

What characteristics of antibodies make them superb diagnostic reagents—chemicals that can test whether or not another substance is present?

B Cells

Every ready-to-go B cell is a small, colorless, rough-surfaced sphere [review FIGURE 26.5]. Each B lymphocyte, however, carries in its membrane close to 100,000 copies of the kind of antibody it will secrete when stimulated by an antigen of the appropriate shape. Immunologists—scientists who study the immune system—have discovered that while all B cells carry antibodies on their membranes, each individual B cell carries and secretes antibodies with binding sites of only one shape. "One cell, one antibody" is the shorthand for this important phenomenon.

B CELLS IN ACTION

Although each B cell is studded with antibodies of a single type, our bodies contain millions of different B cells. These are lodged in lymph nodes or are present in blood and lymph. Each B cell is able to respond to a different antigen, as shown in FIGURE 26.9 and described in the correspondingly numbered paragraphs below.

Step 1. Taken together, our B cells can respond to nearly every antigen imaginable on a bacterium, fungal spore, pollen grain, or other invading entity.

Step 2. When an antigen enters the body, it binds to cell-surface antibodies which fit it best.

Step 3. This binding stimulates the B cell to divide rapidly and thus *selects* it to proliferate into a **clone**.

Step 4. A clone is a group of daughter cells all descended from the same parent cell. In just 10 days, a single B cell can generate a clone of 1000 daughter cells. The combined process of selection by antigen and proliferation into a clone is called the **clonal selection mechanism** of antibody formation.

Step 5. Some of the daughter cells of each expanded clone stop dividing and differentiate into **plasma cells**, the antibody factories described earlier. Plasma cells focus all their energy on antibody production and survive only a few days.

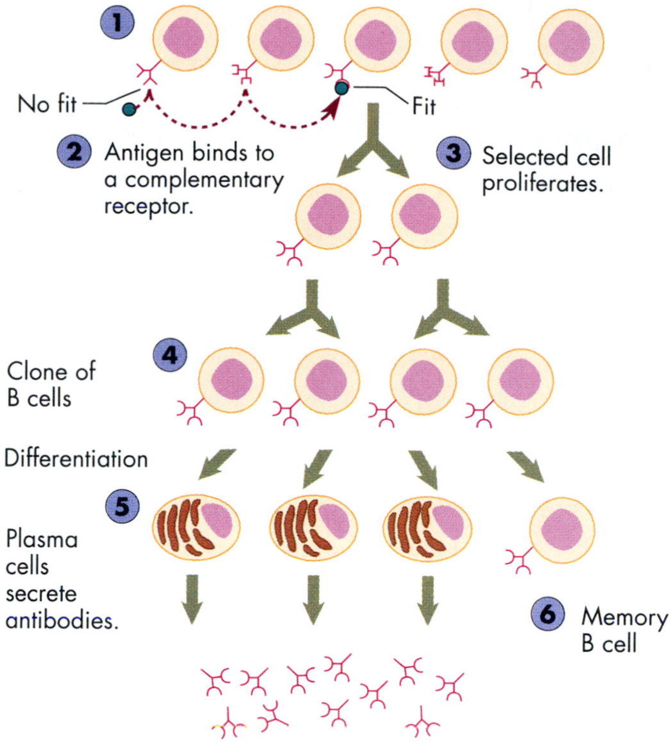

FIGURE 26.9

B Cells and the Clonal Selection Model.

Each of the millions of types of B cells can recognize an antigen of specific shape (1). Antigen binding selects a B cell (2) to proliferate (3) into a clone of identical B cells (4). These differentiate into plasma cells, which can secrete antibodies to the specific antigen (5), or memory cells, which can respond months or years later (6).

Step 6. Other cells of the clone do not differentiate immediately into antibody-secreting factories. Instead, they become **memory B cells**. Like the original B cell, memory B cells can divide and redivide when stimulated by the same type of antigen that triggered the original B cell. With this second round of division, the memory B cells will form their own clones of both plasma and memory cells, all with the ability to make antibodies of the same specificity as those carried by the original B cell.

The first time a specific antigen stimulates an animal, only a small number of cells respond; this first response is known as a **primary response**. But the next time the animal encounters the same antigen, it is better prepared. It has a large pool of memory cells partway along the road to becoming differentiated plasma cells. The net result is a stronger, swifter reaction to the invader—a **secondary response**. Memory B cells thus serve as remnants of former encounters with antigens and constitute a kind of

box 26.2
Human Impact

Anatomy of an Allergy

Does pollen from grass, trees, or ragweed make you sneeze? If so, you may be one of the 35 million Americans with allergies—extreme immune reactions to foreign substances or allergens. Allergens can range from pollen and house dust to pet dander, bee venom, milk, molds, and mites. Researchers are not completely certain how these substances can set off bouts of sneezing, wheezing, swelling, and itching, but they are studying the basic mechanisms of allergy in search of more successful treatments.

A person's physical barriers—the skin; mucous membranes of the eyes, nose, and throat; and linings of respiratory passages, lungs, and intestines—have *mast cells*, which are packed with 1000 or more large, globular granules. These mast cell granules account for the runny nose, watery eyes, and sneezing of hay fever; the diarrhea and stomach cramps of food allergies; and the wheezing of asthma. Allergens somehow trigger mast cells to disgorge their globular granules of histamines and other chemicals, and these agents then dilate the capillaries in the local area and cause them to leak fluid into surrounding tissues. This in turn can lead to swelling and redness and cause constricted airways, mucus secretion, lack of intestinal absorption, and other dysfunctions.

When an allergic individual first encounters an allergen—say, a noseful of ragweed pollen [FIGURE 1, Step 1]—he or she will not experience a reaction, but plasma cells will generate millions of IgE molecules for that antigen. These antibodies will move through the bloodstream, and over a hundred thousand of them attach by their tail to the surface of each mast cell in the area of contact (Step 2). The next time the allergic individual encounters allergen molecules, the allergens form bridges between adjacent IgE molecules on the mast cell surface (Step 3). Significantly, this bridging causes mast cells to release their granules and the histamines and other substances they store (Step 4).

Physicians often treat allergy patients with a series of desensitization shots after determining the offending allergen—whether pollen from grass, pine, or goldenrod; house dust; milk, shellfish, or other food. The object of desensitization is to deliver small quantities of allergen in a way that causes the body to make IgG but not IgE. Later, when the patient encounters naturally occurring allergen in the environment, the massive number of IgG molecules in the body then bind to the allergen molecules. This binding takes place before the allergen can link up with IgE on the surface of mast cells, cause an explosive release (degranulation), and lead to sneezing, watery eyes, swelling, or other symptoms.

To help the millions of allergy sufferers, researchers have been working intensively to understand the immunological mechanisms involved in allergy. They have learned that while mast cell granules spew out histamines, they also release compounds called leukotrienes, which are 100 times more powerful in causing allergy symptoms. Antileukotriene drugs, now under development, may someday help end allergy miseries once and for all.

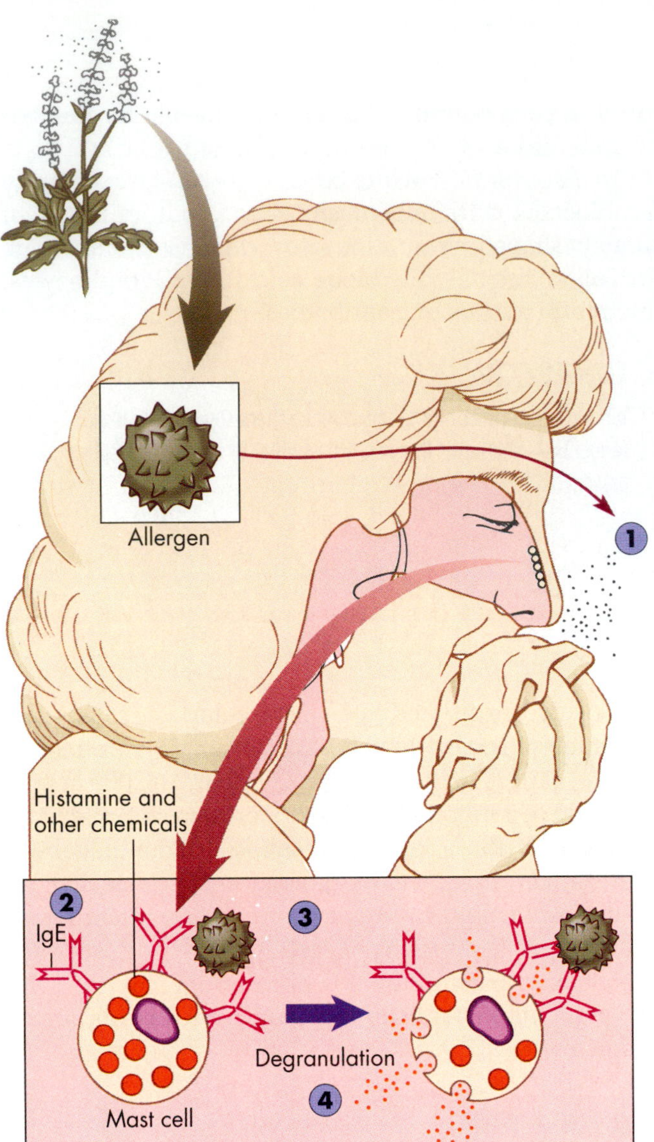

FIGURE 1
Anatomy of an Allergy.

immunological memory that can make the difference between resisting disease and succumbing to it.

By the time a healthy newborn grows to adolescence, he or she will have developed many sets of protective memory cells, providing immunity to many kinds of infections. The infant or child with AIDS, however, loses (or never develops) the ability to generate such immunological memory.

In people with fully functioning immune systems, memory cells can survive for several decades, and this explains why most of us rarely contract measles or chickenpox a second time. If a grandparent had measles as a second-grader and is then exposed to measles while babysitting for a grandchild, a large pool of memory cells in the older person is stimulated to produce antibodies that help eliminate the viruses before they have a chance to cause disease.

The message about B cells is an important one for understanding the immune system: Each type of antigen stimulates specific individual B cells to form a clone of daughter cells. Some of these identical daughter cells will secrete antibodies that provide immediate protection, while others remain in reserve in case of later exposure to the same antigen.

REARRANGEMENTS OF ANTIBODY GENES

At this moment, your body contains B cells that can form an antibody able to match virtually any antigen you could ever encounter: a rare virus from Borneo, a pollen grain from a Mongolian steppe grass, and even molecules of an organic chemical just invented by a chemist in her laboratory! This whole situation leads to a tantalizing puzzle: How can your cells contain the huge number of genes necessary to code for millions of antibodies of different shapes, including molecules you will probably never encounter and that haven't even been invented yet? The answer is that an amino acid chain in an antibody is like a restaurant dinner. Just as you can build a unique dinner by selecting one appetizer from six choices, one salad from four, one entree from twenty, and one dessert from five, a mature antibody gene is built from gene parts put together from three or four groups of genes. By this special mechanism, a relatively small number of genes can create an almost infinite number of antibody protein shapes.

T Cells

As effective as the antibody-based humoral immune response is, it makes up only part of the action of immunity. The other part is the **cell-mediated immune response**, in which T cells directly eliminate foreign invaders.

An awareness of how T cells function is crucial to understanding both normal immunity and the impaired response of people with AIDS.

Since researchers first studied and described AIDS over a decade ago, medical workers have used the decrease of a specific type of T cell called *CD4$^+$ cells* as a

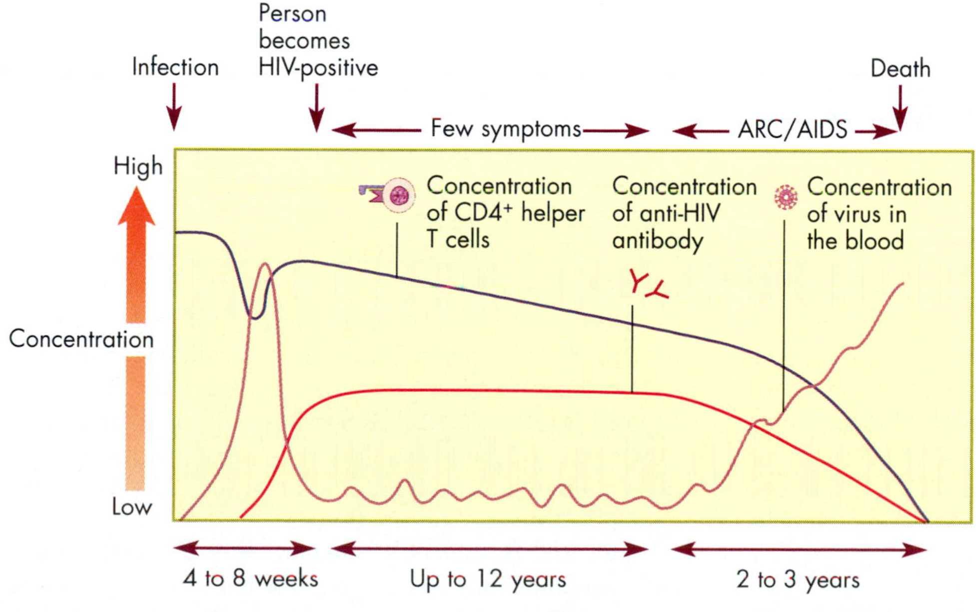

FIGURE 26.10
AIDS and the Decline in T Helper Cells. A single act of intercourse can introduce HIV (human immunodeficiency virus) into a person's body. Within a few weeks, virus particles multiply (violet line) and T helper cells decline in numbers (blue line). Recovery of T helper cells and the increase in anti-HIV antibody (red line) suggest that the immune system is mounting a defense against the invader. Nevertheless, helper cells continue to be destroyed, and finally the classic signs of AIDS emerge: opportunistic infections, wasting away of the body, and dementia.

diagnostic feature of AIDS disease. FIGURE 26.10 shows what can happen a little more than a month after HIV infection begins, following, let's say, an act of unprotected sexual intercourse. After a few weeks, the concentration of HIV particles rises in the blood, and the concentration of $CD4^+$ cells dips, then partially recovers. At about two months after infection, about the time the $CD4^+$ cells begin to recover, the amount of virus circulating in the blood drops, and antibodies to the virus first appear. These antibodies signal that a person has become HIV-positive. Over the course of the next 3 to 12 years, the concentration of the person's $CD4^+$ T cells gradually falls, eventually approaching zero. Simultaneously, as the circulating antibody continues to decrease, the AIDS virus multiplies steadily, and the infections that characterize full-blown AIDS set in. Usually in a matter of months, the person dies of pneumonia or other complications that could have been defeated by a healthy immune system. If the AIDS patient is a baby, the disease progresses much faster. Clearly, the course of AIDS is linked to the depletion of T cells. Let's begin, then, to investigate the roles of these lymphocytes in a healthy immune system so we can understand how their decline can lead to AIDS.

TYPES OF T CELLS

Unlike B cells, whose function is to produce antibodies, different types of T cells play at least three different roles in the immune response. (1) Half of all T cells are **cytotoxic T cells**, which destroy foreign cells, including cells transplanted to a person from another individual, or body cells transformed by infection or cancer [FIGURE 26.11]. (2) **Helper T cells**, the type to which $CD4^+$ T cells belong, communicate with B cells and with the other types of T cells to initiate or amplify an immune response. (3) **Suppressor T cells** can diminish or stop immune responses. Let's see how the three types interact.

HOW T CELLS WORK

T cells viewed through a light microscope are practically indistinguishable from B cells: both are small, spherical, and colorless. But T cells pass through the thymus for processing and maturation and lodge in the lymph nodes, spleen, and skin. In contrast, B cells begin to mature in the bone marrow before eventually lodging in either the lymph nodes or the spleen [FIGURE 26.12]. The thymus shrinks as a person matures, and the immune system functions less effectively. This helps explain why many senior citizens need flu shots at the start of each winter's flu season.

T Cell Receptors: Key to T Cell Function and HIV Infection While in the thymus, precursors to T cells develop

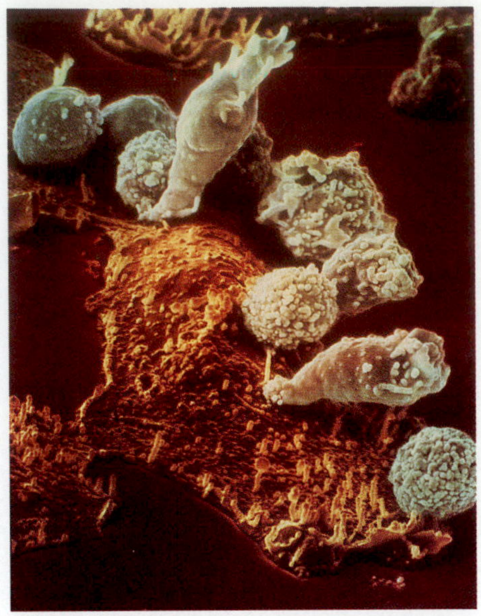

FIGURE 26.11

T Lymphocytes Destroying a Cancer Cell.

This scanning electron micrograph captures several rounded T cells (white) in the process of attacking a single, larger cancer cell (brown).

the ability to bind specific antigens. As in B cells, this ability is due to an antigen-binding protein embedded in the cell's surface membrane. It is not an antibody, however, but instead is a protein called the **T cell receptor** [FIGURE 26.12A]. Another protein, called the **coreceptor**, is associated with the T cell receptor, and researchers picture it as a cane-shaped molecule like the one in FIGURE 26.12A.

We saw earlier that the disappearance of the $CD4^+$ helper T cells signals the increasing severity of AIDS. The cane-shaped coreceptors on those cells are called *CD4 coreceptors*, and ironically, they provide the gateway for HIV infection, since the virus binds to this protein and thereby gains entrance to the cell. In one line of AIDS research, experimenters have engineered versions of the CD4 coreceptor that can keep the virus from spreading to noninfected cells—at least in the test tube. Researchers hope that someday they will be able to use these modified T cell receptors to help slow the spread of HIV in AIDS victims.

Different types of T cells display different coreceptors on their surfaces. While helper cells can have the CD4 coreceptor, young cytotoxic T cells have CD8 coreceptors [FIGURE 26.12A]. Regardless of coreceptor type, however, all T cells develop in the thymus from pre-T cells, then migrate to lymph nodes where they bind antigen and complete their maturation [FIGURE 26.12B].

Self and Nonself: Histocompatibility Proteins The T cell receptor binds antigen only when a macrophage or other immune cell presents antigen to it along with self-marking membrane molecules called **histocompatibility**

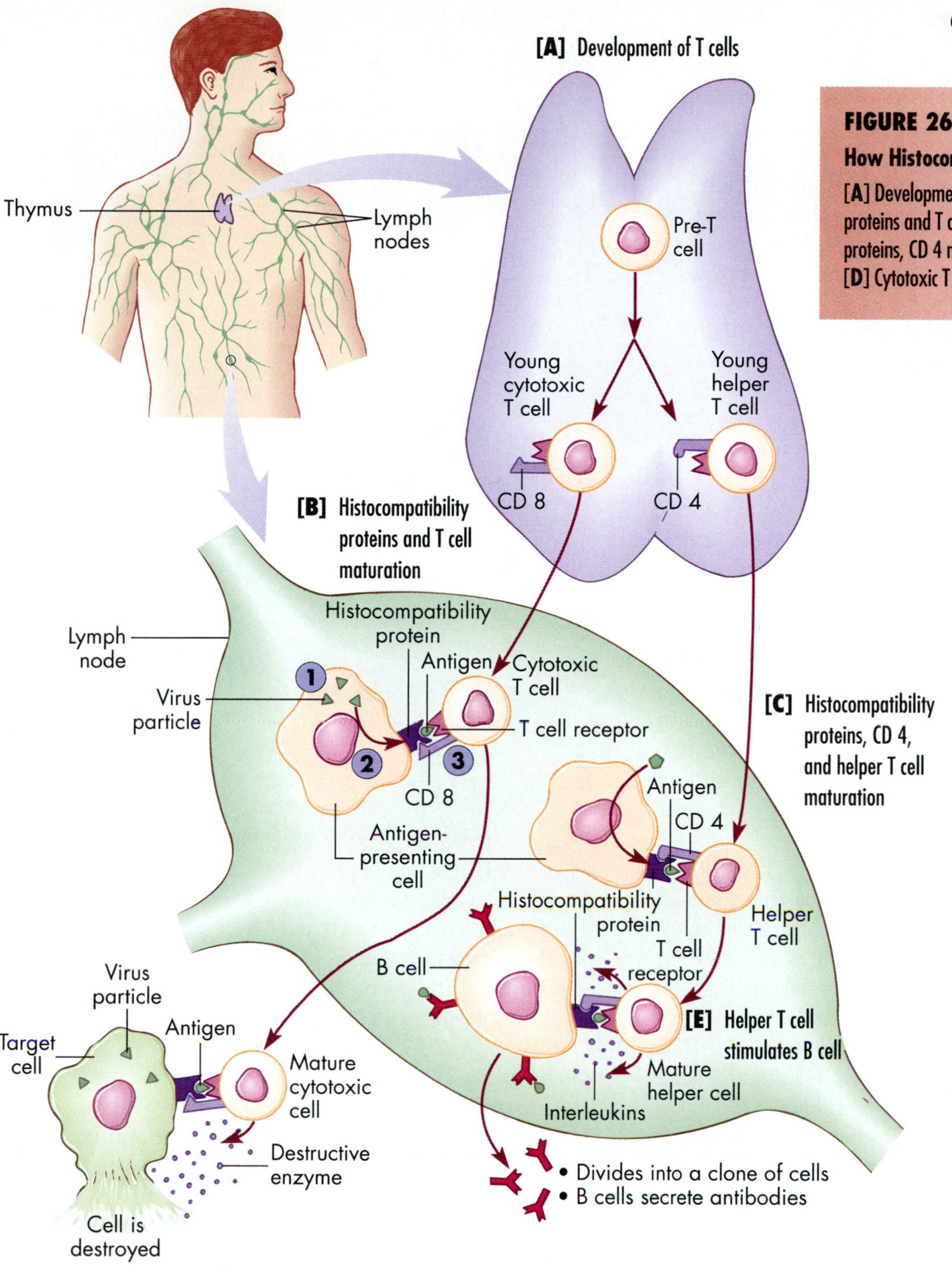

FIGURE 26.12

How Histocompatibility Proteins Help T Cells.

[A] Development of T cells. [B] Histocompatibility proteins and T cell maturation. [C] Histocompatibility proteins, CD 4 molecule, and helper T cell maturation. [D] Cytotoxic T cell in action.

proteins (from the Greek *histos*, "web") [FIGURE 26.12B]. Histocompatibility molecules are a key to the two main functions of T cells—their ability to directly kill cells bearing antigens and their ability to regulate the immune response.

Each person's cells have a unique array of histocompatibility proteins serving as cellular fingerprints that distinguish self from nonself. Histocompatibility proteins were first recognized for their role in determining whether an animal accepts or rejects a tissue graft, such as skin transplanted from a donor onto a burn victim. Rejection occurs when the histocompatibility "fingerprints" of the graft differ from those of the recipient. You have probably heard of desperate searches by doctors and families to find compatible donors of bone marrow, livers, kidneys, or other organs to save the life of a gravely ill person. What physicians are searching for in these cases are donors with histocompatibility proteins very similar or identical to those of the patient. If the histocompatibility proteins are different, cytotoxic T cells recognize that the grafted tissue is nonself and attack it.

Histocompatibility proteins, then, cooperate with the T cell receptor to stimulate the immune system. But how do they do it?

1. FIGURE 26.12B shows that first, a foreign protein is taken up by an *antigen-presenting cell* such as a macrophage or other cell.

2. Inside the antigen-presenting cell, the foreign protein is broken into smaller fragments that become attached to a major histocompatibility protein of class I (MHC I) on the cell surface. MHC I proteins are carried by all cells.

3. A T cell receptor can bind to the antigen-histocompatibility complex and stimulate the T cell to become active.

The type of T cell formed—for example helper or cytotoxic—depends in part on the coreceptor. If it is a CD4 coreceptor, for example, the cell becomes a helper T cell [FIGURE 26.12C]. This interaction of antigen, histocompatibility protein, T cell receptor, and coreceptor stimulates T cells to do their jobs.

How Cytotoxic T Cells Work *Cytotoxic T cells* (*cytotoxic* means, literally, "lethal to cells") can eliminate cells with foreign antigens on their surface. Also called *killer T cells*, these cells secrete destructive enzymes or other substances that can poke holes in the plasma membranes of foreign cells, as well as in self-cells altered through infection by viruses or transformed by cancer [see FIGURE 26.12D].

Some cytotoxic T cells do not kill cells directly; instead, they release enzymes and hormones that wall off an infected area, cause inflammation, and remove the offending antigens. Before the first day of school, children and teachers have a routine test for tuberculosis that depends on a reaction by this type of cytotoxic T cell.

How Helper and Suppressor T Cells Work Half of all T cells control B cells and other T cells rather than directly or indirectly disposing of antigens. *Helper T cells* signal B cells and other T cells to spring into action [see FIGURE 26.12E]. In contrast, *suppressor T cells* prevent helpers and effectors from taking action.

Both helper and suppressor T cells communicate with other cells by secreting molecules such as *interleukins* and *interferons* that alter the activity of specific targets [see FIGURE 26.12E]. Significantly, cyclosporine—the main drug physicians use to block undesirable immune responses, such as the rejection of a transplanted organ—acts by inhibiting the production of interleukin by helper or suppressor T cells. Researchers have also found that interleukins may help people infected with the AIDS-causing virus. They have injected people who harbor the AIDS virus with an interleukin called *IL2*, and it seems to activate cytotoxic T cells that kill HIV-infected cells.

The Biology of AIDS The central regulatory role of T cells becomes tragically evident in patients with AIDS. The usual progression of AIDS is shown in FIGURE 26.13, and described in the corresponding paragraphs below:

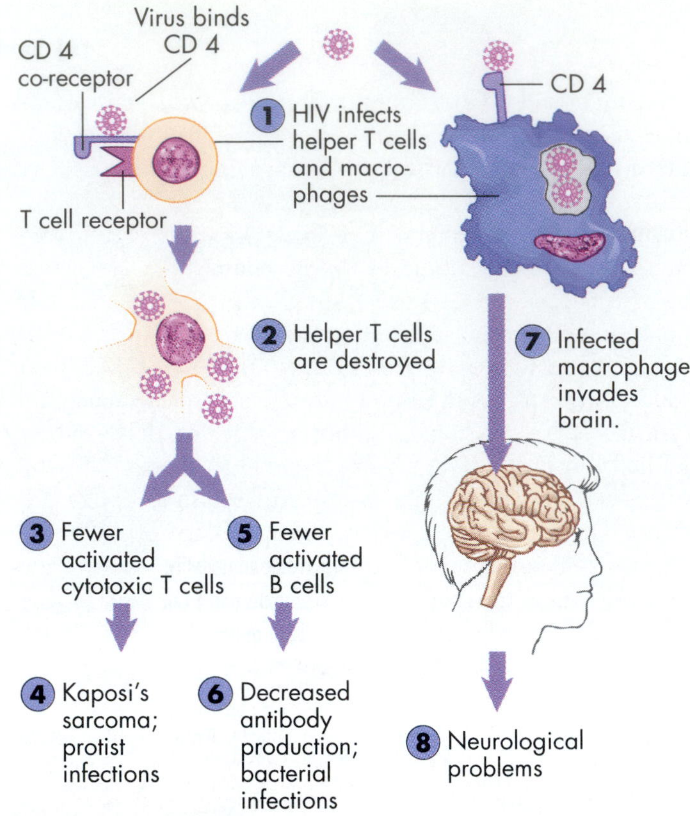

FIGURE 26.13
How the AIDS Virus Wreaks Havoc in the Human Body.

Step 1. The virus that causes AIDS infects helper T cells by binding to the CD4 coreceptor on the helper T cell.

Step 2. The virus proliferates inside the cell and leads to destruction of helper T cells.

Step 3. The few helper T cells left can no longer induce cytotoxic T cells to form.

Step 4. Cytotoxic T cells are unavailable to kill cancer cells, such as those that cause the tumors of Kaposi's sarcoma. Without cytotoxic T cells, the body cannot ward off opportunistic infections of parasitic protozoa such as *Pneumocystis carinii*, which causes a rare form of pneumonia; or the fungus that grows in the mouth and causes *thrush*; or virus-infected cells.

Step 5. Without helper T cells, B cells are not stimulated to produce antibodies.

Step 6. The victim's body cannot combat viruses in the bloodstream, or bacterial invaders. Macrophages also have the CD4 coreceptor protein, and take in HIV by phagocytosis (Step 1), and the viruses accumulate in vesicles.

Step 7. Infected macrophages can enter the brain.

Step 8. Neurological problems such as dementia may occur in some AIDS patients.

Since the AIDS epidemic first surfaced, researchers have wanted to find an *animal model* for the disease—that is, an animal whose immune system is destroyed by the same virus that devastates our own. Chimpanzees, for example, can be infected with HIV, but don't sicken and die, while rhesus monkeys die from a related virus, but not from HIV. In 1987, a young California researcher named Mike McCune had a brainstorm: Why not take a mouse born without a functioning immune system and transplant human immune system tissue into it? He used lymph nodes, part of a thymus, and some immature white blood cells from a human fetus in the hope that the white cells would grow, develop, and begin to confer immunity in the mouse just as in a human baby. Indeed, the chimeric animal, which the scientific press has since dubbed "the human mouse," did take on a functioning human immune system that can be devastated by the human HIV. By 1990, AIDS researchers all over the world were using the human mouse, and hopes are high that it will lead to further progress in understanding and controlling the disease.

Developing Therapies to Halt the Spread of HIV To successfully block the life cycle of the AIDS virus, scientists must understand how it infects human cells, replicates, spreads to other cells, and destroys immune functioning. Some of what we know is shown in FIGURE 26.14. Possible points of attack for drugs to defeat AIDS include Steps 1 and 3 in FIGURE 26.14. In Step 1, the virus binds to CD4, which might be blocked by soluble, genetically engineered CD4 protein. In Step 3, drugs such as AZT, ddI, and ddC inhibit the reverse transcription of HIV RNA to DNA. These are the only currently approved drugs for AIDS [see BOX 9.1], but their inhibition lasts only for a while. FIGURE 26.14 also illustrates other possible points of intervention now under investigation.

Although we now understand HIV more thoroughly than any other human virus, there is still no cure for AIDS, and predictions surrounding the disease are frightening. As of 1993, more than 1 million North Americans had been infected with HIV. In the United States, homosexual and bisexual men have accounted for two-thirds of the cases, but epidemiologists expect the incidence of new cases in these groups and in intravenous drug users to level off by 1995, while the incidence among heterosexuals will continue to rise.

At an international AIDS conference in Berlin in 1993, researchers identified 15- to 24-year-olds as the group at greatest risk worldwide, with infection rates in females rapidly catching up to the rates in males. While exposure to HIV is a particular danger in areas where young people are badly informed about how to protect themselves from AIDS, infection nevertheless cuts across socioeconomic and educational lines: A 1990 study of American college students, for example, showed that on 19 campuses, 1 in 500 students had been exposed to HIV.

In some cities in Africa, one of every three adults has been infected with HIV, and AIDS is the leading cause of death. Scientists who track epidemics think that worldwide, 13 million people have been exposed, and nearly 2 million have already died. They predict that by the turn of the century, the AIDS virus will have infected as many as 39 million people worldwide, including many millions of children. Tragically, AIDS kills 85 percent of its victims within three years of diagnosis.

Immune Regulation

Why doesn't the body produce antibodies and cytotoxic T cells against its own proteins, and hence destroy itself? Many immunologists think that macrophages are a key to **self-tolerance**—the lack of an immune response to one's own molecules. The large, specialized phagocytes called macrophages engulf antigens and cellular debris, and these antigens become attached to the surface of the macrophage. The macrophage then presents the surface-bound antigens to helper T cells [review FIGURE 26.6]. After infancy, however, macrophages do not process *self*-molecules and present them to helper T cells. Because of this difference in macrophage behavior, self-substances do not trigger helper T cells. Moreover, during development in the fetus and newborn, immature B cells carrying antibodies that bind to self-substances may be killed. Thus, a combination of suppression by T cells and deletion of self-reactive B cells sets up a state of self-tolerance that is crucial to our survival.

AUTOIMMUNE DISEASES

Loss of self-tolerance can lead to **autoimmune diseases**—attacks on certain cells or tissues by the person's own immune system. *Rheumatoid arthritis* is an autoimmune disease that afflicts millions of people in the United States alone. In people with this condition, the joints swell painfully, the fingers can become gnarled and twisted, and everyday movements, like buttoning a shirt, can be painful or even impossible [FIGURE 26.15]. *Multiple sclerosis* is another autoimmune disease in which a person's immune system reacts to substances (myelin) that surround and protect nerve cells. In *diabetes*, a patient's immune system destroys some of the pancreatic cells that normally produce insulin. This hormone helps regulate the levels of sugar in the blood, and without it, metabolism is disrupted.

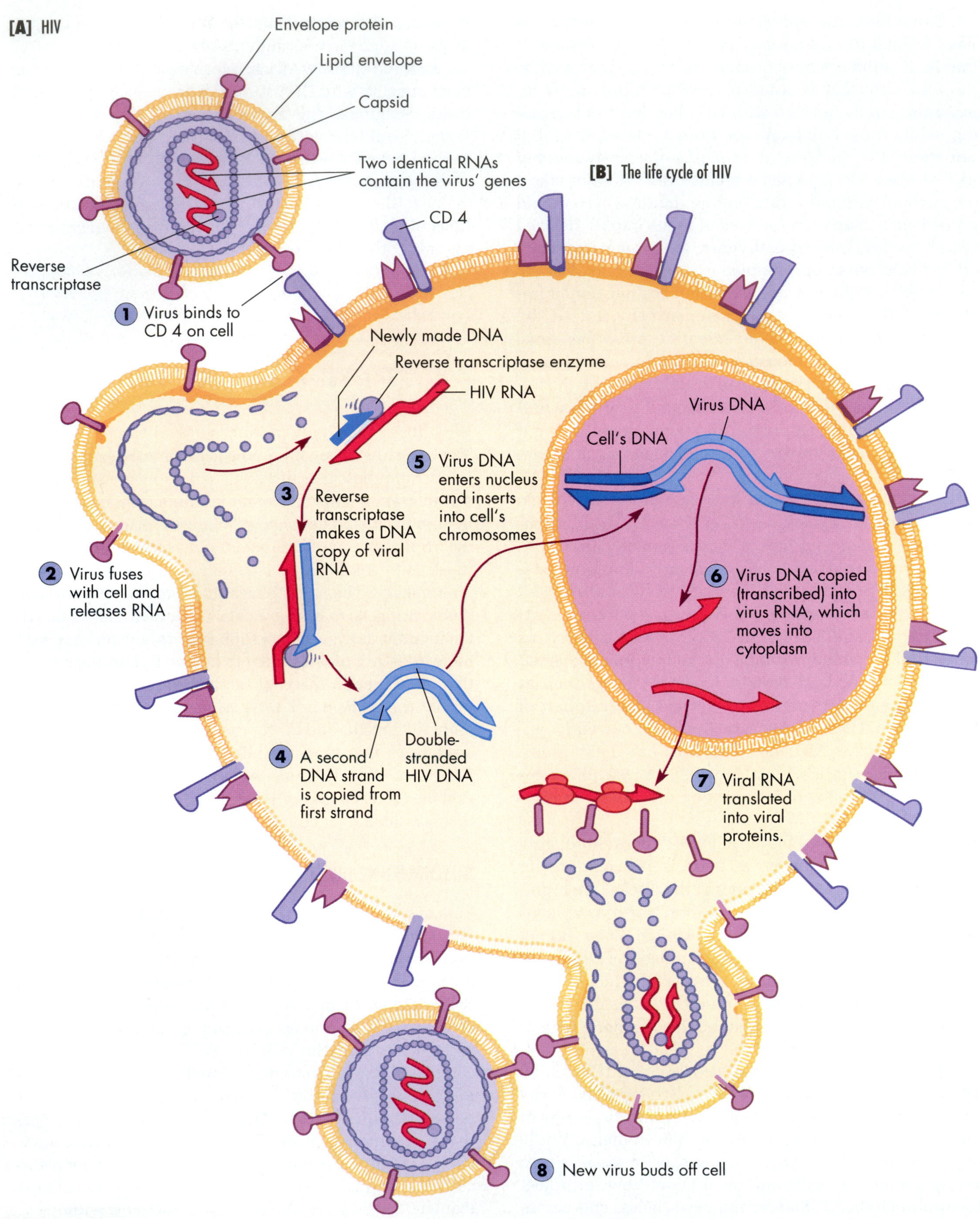
[A] HIV
Envelope protein
Lipid envelope
Capsid
Two identical RNAs contain the virus' genes
Reverse transcriptase
[B] The life cycle of HIV
CD 4
1 Virus binds to CD 4 on cell
Newly made DNA
Reverse transcriptase enzyme
HIV RNA
Virus DNA
Cell's DNA
5 Virus DNA enters nucleus and inserts into cell's chromosomes
3 Reverse transcriptase makes a DNA copy of viral RNA
2 Virus fuses with cell and releases RNA
6 Virus DNA copied (transcribed) into virus RNA, which moves into cytoplasm
4 A second DNA strand is copied from first strand
Double-stranded HIV DNA
7 Viral RNA translated into viral proteins.
8 New virus buds off cell

FIGURE 26.14

Understanding and Blocking the AIDS Virus Life Cycle.

[A] HIV contains genes made of RNA instead of DNA. Its genes are surrounded by a protein capsid and a lipid membrane studded with surface proteins. **[B]** Once the HIV enters a cell (1 and 2), reverse transcription of its RNA to DNA (3 and 4) and movement of the DNA to the nucleus (5), leads to the generation of new viral RNA (6). This RNA is translated into proteins (7) which assemble into virus particles that can be released and infect new cells (8). Researchers are trying to disrupt the HIV life cycle in various ways: By injecting CD4 coreceptor proteins into the blood, they might bind the HIV's envelope protein and make it impossible for the virus to enter T cells (at Step 1); by administering AZT and other inhibitors of the reverse transcriptase enzyme (at Step 3); and by blocking a protein-digesting enzyme necessary for the viral proteins to mature (at Step 7).

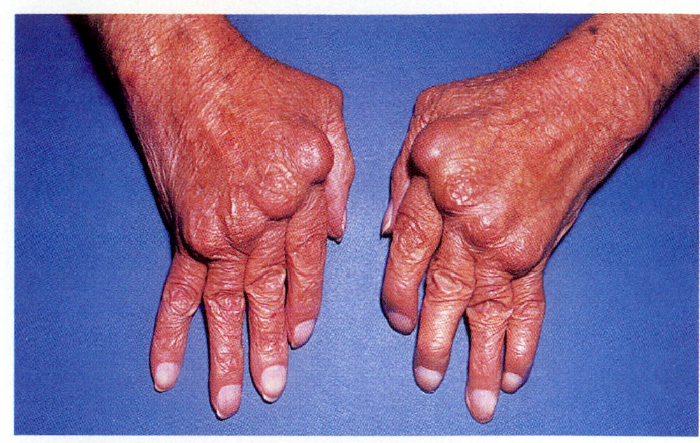

FIGURE 26.15

Autoimmune Disease: A Breakdown of Self-Tolerance.

In a person with rheumatoid arthritis, overly active immune cells spark inflammation and joint destruction. Some immunologists think that excessive, misdirected helper T cell activity is the major culprit. Researchers are still looking for more specific drugs with few side-effects to fight autoimmune diseases.

PREGNANCY: AN ALIEN INVADES THE MOTHER'S TISSUES

Pregnancy is an interesting exception to the immune system's recognition and elimination of foreign antigens. The human fetus, containing some of the father's histocompatibility proteins, burrows into the uterus. So why doesn't the mother's body reject the half-foreign fetus? Studies have shown that the uterus is a special immunological zone during pregnancy. At this time, a woman's body rejects a graft from the fetus if it is placed anywhere but the uterus. Researchers are still studying how embryos are protected in the womb, but it must be quite complicated, precise, and efficient in order to defend the tiny cluster of tissues from the mother's powerful immune network.

Despite this protection, the fetus sometimes does come under attack. About one couple in 15 have **Rh incompatibility**, a situation in which a serious anemia called *erythroblastosis fetalis* can result in their newborn infants. Most people in North America have Rh antigens on the membrane of their red blood cells and are thus said to be Rh-positive. Other people, however, lack the antigens and are Rh-negative. When an Rh-positive man and an Rh-negative woman produce a baby, it may be Rh-positive. If the baby's blood cells mingle with the mother's during delivery [FIGURE 26.16A], and its Rh antigens

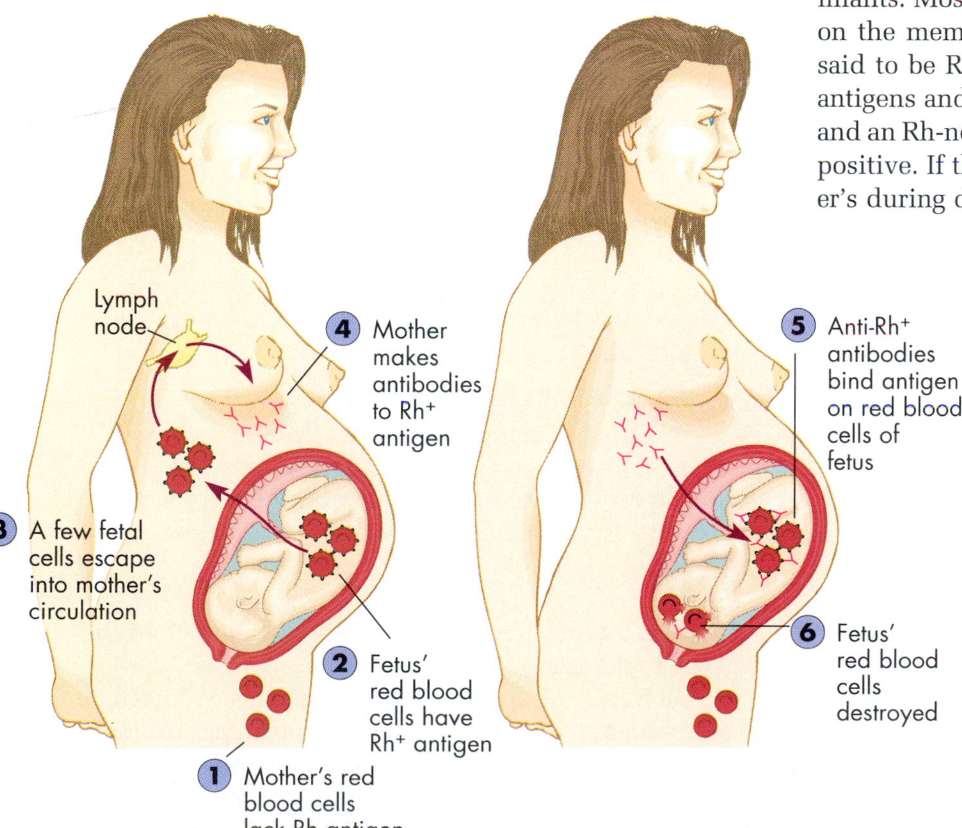

FIGURE 26.16

Pregnancy and Rh Disease.

[A] If a fetus is Rh-positive and the mother is Rh-negative, and if fetal blood cells escape into the mother's body, they may trigger the mother to produce anti-Rh antibodies. **[B]** In subsequent pregnancies, these maternal antibodies may enter the fetal bloodstream and attack fetal blood cells.

enter her bloodstream, her immune system may secrete anti-Rh antibodies. These don't affect the newly delivered baby, since a few days pass before the antibodies form in the mother's system. These antibodies don't affect the mother because her cells lack the Rh antigen. However, during a subsequent pregnancy, the antibodies already in the mother's blood can cross the placenta and attack the red blood cells of the new fetus if it is Rh-positive. Debris from the attacked cells can lead to anemia, brain damage, or even death [FIGURE 26.16B].

Physicians prevent this dangerous situation by injecting Rh-negative mothers with anti-Rh antibodies (called *Rhogam*) at the birth of her first Rh-positive child. These antibodies bind to the Rh antigens on fetal blood cells that may enter the mother's circulation during delivery. Covered by these injected antibodies, the Rh-positive antigen does not stimulate the mother's immune system, and the next embryo will be safe from attack.

➤ CONCEPT CHALLENGE

Compare the symptoms of a person with AIDS to that of a child with XLA who makes no antibodies. How would they be similar? How would they be different?

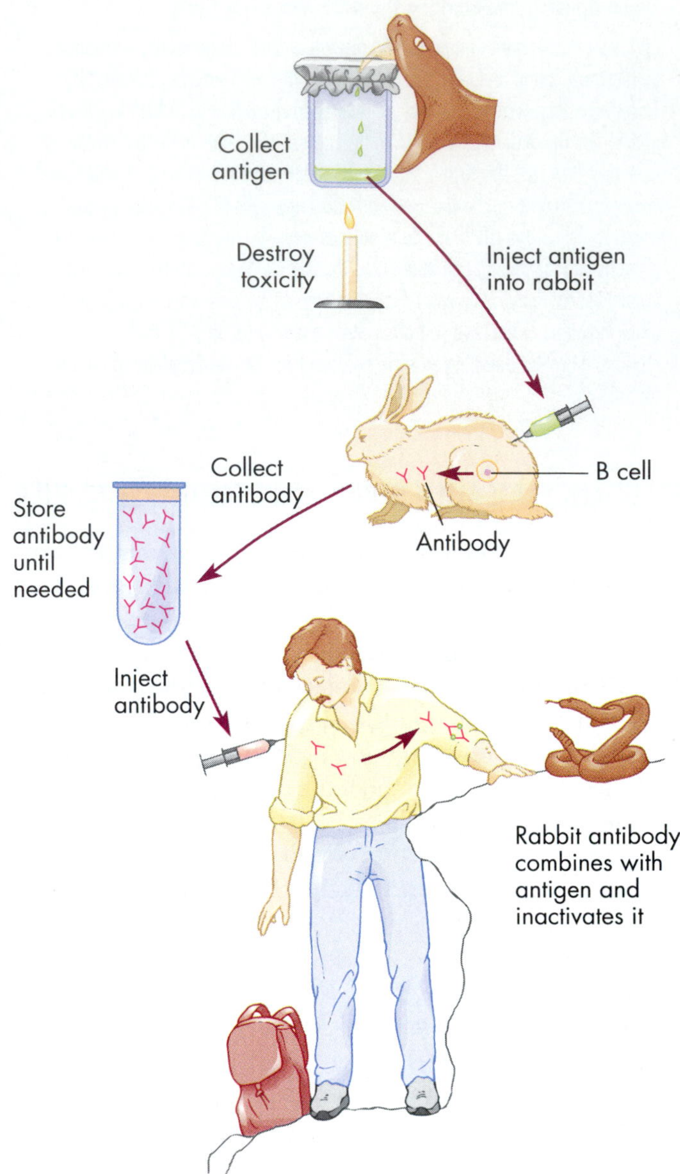

FIGURE 26.17

Passive Immunization: Transferred Antibodies.

Technicians prepare commercial snakebite medicine (so-called antivenin) by collecting venom, inactivating it, and then injecting it into a rabbit. Later, antibodies are collected from the rabbit and injected into the snakebite victim, where they bind to and inactivate the toxic venom, rendering it harmless.

Immunization

In the last 200 years, medical practitioners have learned to manipulate the immune system by stimulating either passive immunity or active immunity.

PASSIVE IMMUNITY: SHORT-TERM PROTECTION BY BORROWED ANTIBODIES

If you were bitten by a rattlesnake, there would be no time to spare. The venom contains *neurotoxins*, proteins that stop nerves from functioning, and you would need a quick-acting remedy to keep the neurotoxins from damaging your nervous system. The best treatment would be **passive immunization**: injecting antibodies made by one individual into another individual [FIGURE 26.17]. A commercial product containing the desired antibodies is prepared by injecting a horse or rabbit with inactivated snake venom, which induces the mammal to form antibodies against the venom proteins. Workers then collect these antibodies, now called *antivenin*, which a doctor would then inject immediately into you, the snakebite victim. The antivenin would then circulate through your bloodstream and combine with (neutralize) the snake venom in a typical antigen-antibody fashion before the venom could destroy your nerve cells.

Passive immunization has one great advantage: It works very fast. But it also has the disadvantage of acting for only a short time. Your immune system would soon recognize the borrowed antibody molecules as foreign, and eliminate them. With the antibody molecules would go your passive protection, leaving you vulnerable to snake venom if bitten again.

ACTIVE IMMUNITY: PREVENTION BY ALTERED ANTIGEN

Unlike an antivenin that provides the short-term protection of passive immunity, vaccines against polio, diphtheria, or measles provide long-term protection by stimulating **active immunity**, the production of antibodies or antigen-specific T cells by the individual's own immune system. Safer and more effective than most drugs, such vaccines provoke a specific response aimed at one microbe or toxin, and nothing else in the body.

Edward Jenner, an eighteenth-century English country physician, developed the first vaccine almost 200 years ago. To help protect people from smallpox, a severely disfiguring disease that killed one out of every four people in Jenner's time, he injected people with a small amount of cowpox virus (the word *vaccination*, in fact, comes from the Latin word *vacca*, "cow"). Jenner based his technique on the observation that milkmaids who contracted cowpox from the cows they milked always recovered, and almost never got smallpox. Indeed, the cowpox virus he injected caused mild sickness and discomfort, but it nonetheless served as effective protection against smallpox. By 1978, modern vaccines based on Jenner's original ones had successfully eradicated smallpox worldwide.

Modern vaccines are made of microbes and toxins that have been killed or otherwise modified in the laboratory so they cannot cause disease. The first shot with altered germs stimulates the production of antibody-producing cells and memory cells [review FIGURE 26.9]. A *booster shot*, which is simply a second dose of the same vaccine, can then induce the memory cells to differentiate and form still more effector and memory cells. If a vaccinated person later comes in contact with live bacteria, virus, or toxin carrying those antigens, his or her body will already contain many antibodies and memory cells that can quickly eliminate the dangerous agents and prevent disease. Vaccination is a slow-acting process, requiring a booster shot and several weeks for the development of adequate protection. Nevertheless, the active immunity it stimulates lasts a long time, sometimes a lifetime.

New vaccines are the best hope for combating infectious diseases that cannot be treated or cured by other techniques of modern medicine. Unfortunately, many viruses, like different strains of the flu virus, as well as some other microbes, can change their surface properties so often that they evade active immunity. In the two centuries since Jenner's first vaccine, fewer than 20 safe, effective vaccines have been developed. To conquer such microbial elusiveness, contemporary researchers are using recombinant DNA and other forms of biotechnology to fashion synthetic vaccines against influenza and other viruses that can foil the immunological memory. Researchers at laboratories in many countries are currently at work on vaccines against the AIDS virus, but we are still years away from an effective protective vaccine. This is because the AIDS virus attacks the very lymphocytes that promote protection. HIV also undergoes changes in the proteins of its outer coat via frequent mutation, and the virus can remain latent inside cells for many years, undetectable to the immune system. Until an effective vaccine is developed, the best prevention techniques now available are to observe the general rules of good health as well as always to follow safe-sex practices [review TABLE 26.1].

➤ CONCEPT CHALLENGE

While on a hike, a companion is bitten by a poisonous black widow spider. One member of the group suggests the bitten person should be rushed to the hospital to be protected by active immunity. Another says that passive immunity would be better in this case. Which do you think would be more helpful to the bitten friend and why?

Connections

Recognition is a key feature of life. The vertebrate immune system, however, has taken recognition to an extreme. The binding of receptor to antigen is so specific that millions of different antigens—viruses, bacteria, toxins, cells from other individuals, and even organic chemicals that have yet to be synthesized—can be recognized, distinguished, and bound by antibodies and other molecules of the immune system. For the immune system to function properly it must be able to distinguish self from nonself. Once this recognition has occurred, interactions between the various elements of the immune system produce a coordinated attack against an invader, which is eventually eliminated from the body.

The immune system is not self-contained. It relies on the circulation of blood and lymph [see CHAPTER 25], and it communicates with the nervous and hormonal systems [see CHAPTERS 30 through 32]. This is one reason why undue stress or prolonged mental depression lowers our resistance to infection. In the next chapter, we will see that this same tight physiological interdependence underlies the smooth functioning of the respiratory system.

KEY TERMS

antibody, 603
antigen, 603
B cell (B lymphocyte), 602
cell-mediated immune response, 609
histocompatibility protein, 610
humoral immune response, 603
immune response, 602
immune system, 598
immunity, 599
immunological memory, 609
macrophage, 603
self-tolerance, 613
T cell (T lymphocyte), 610

HIGHLIGHTS IN REVIEW

1 The vertebrate immune system recognizes invaders and eliminates them. The process involves chemical communication among cells; the presentation of foreign substances called antigens by one type of immune cell to another; the regulation of cellular activities by certain immune cells; and the destruction of invaders by certain cells and molecules.

- **a]** A nonspecific immune response is directed against any type of invader; in it, blood flow and temperature increase locally around a wound, and white blood cells converge on the area and devour debris.
- **b]** Macrophages are large white blood cells that devour debris in a wound, process the invader's molecules, and place portions of them on their surface in association with major histocompatibility proteins. Macrophages present antigen on these cell surface proteins to certain T cells, where the antigen binds the T cell receptor.
- **c]** Helper T cells help other lymphocytes carry out their immune function, while suppressor T cells suppress certain immune reactions, and cytotoxic T cells directly kill invading cells, cancer cells, or virus-infected cells.

2 In the humoral immune response, B lymphocytes secrete antibodies. These proteins bind to invaders and mark them for destruction. In the cellular immune response, T lymphocytes help other immune cells mature or directly kill invading parasites, cells infected with viruses, cancer cells, or cells transplanted from another individual.

- **a]** B cells give rise to plasma cells, which produce antibodies, proteins that form a shape like the letter Y. The tips of the two arms of the Y can bind tightly to an antigen, such as the molecules of an invader. The stem of an antibody molecule controls whether the antibody circulates in the blood, providing overall protection; is secreted into a mother's milk, providing temporary protection to a newborn; or attaches to cells that line body surfaces, contributing to symptoms of allergy.
- **b]** An antibody bound to an antigen marks the antigen for destruction (for example, to be devoured by a macrophage).
- **c]** Most lymphocytes probably arise in the bone marrow and then move to other organs, like the lymph nodes, spleen, or thymus, which gives rise to T cells.
- **d]** Each different B cell makes a different antibody that can bind a unique antigen. When a B cell binds an invading antigen and is stimulated by a helper T cell, it proliferates into a clone of cells. Some members of the clone become plasma cells, which secrete specific antibodies, while others become long-lived memory cells, which can respond like the original B cell when exposed to the same antigen again.
- **e]** The cell-mediated immune response relies on cytotoxic T cells that directly eliminate foreign invaders.

3 The progression of AIDS (acquired immunodeficiency syndrome) is marked by infection with the human immunodeficiency virus (HIV), and as a result, a gradual decrease in helper T cells. Without helper cells, neither cytotoxic T cells nor antibody-secreting cells are fully active; hence, opportunistic infections, such as pneumonia-causing parasites, can kill the patient.

- **a]** Helper T cells contain on their surface a coreceptor called CD4. HIV, the virus that causes AIDS, can bind to CD4 and gain access to the cell. Eventually, infected helper T cells die. Without helper T cells, other lymphocytes are not adequately stimulated, and so the body forms few antibodies, memory cells, or cytotoxic T cells. Without the protection these provide, the AIDS victim dies from microorganisms or cancers that the intact immune system normally destroys.

4 If an immune response becomes directed against a body's own cells or substances, autoimmune diseases like arthritis, multiple sclerosis, and diabetes can result. On the other hand, vaccinations direct an immune response against a specific disease-causing agent by stimulating antibody production or cytotoxic T cell production.

- **a]** In an autoimmune disease like arthritis, an individual's immune system inappropriately attacks molecules that arise in that individual's body.
- **b]** In passive immunity, antibodies made by one individual are transferred to another individual. In active immunity, an individual's own immune system can be stimulated to eliminate a specific invader. A vaccine is an antigen altered so that it can no longer cause disease even though it can stimulate an immune response.

UNDERSTANDING THE FACTS AND CONCEPTS

For Questions 1–5, match the descriptions with the most appropriate item from the following list of terms. Each item in the list may be used once, more than once, or not at all.

a] lymphocyte
b] macrophage
c] interferon
d] complement
e] inflammatory response

1 A type of scavenging white blood cell.

2 Localized increase in blood flow, initiating a series of nonspecific actions.

3 A nonspecific protein that prevents viral multiplication.

4 A group of proteins that enhance the inflammatory and immune responses.

5 A kind of white blood cell that can mature into B or T cells.

As above, for Questions 6–11, match the descriptions with one of the activities listed below.

a] recognition
b] communication
c] elimination
d] nonspecific defense mechanism

6 The barrier formed by the intact skin and mucous membranes of the nose, throat, digestive tract, and respiratory tract.

7 The activity of interleukins.

8 Action taken by cytotoxic T cell.

9 Role of the major histocompatibility complex proteins.

10 Binding of a particular antibody to an antigen.

11 Phagocytosis and cell membrane perforation.

As above, for Questions 12–16, match the descriptions with the most appropriate item from the following list of terms.

a] antigen
b] antibody
c] interleukin
d] MHC proteins
e] B cells
f] T cells

12 Protein growth factors that stimulate the development of antibody-producing cells.

13 Any molecule that triggers an immune response.

14 Cell-surface proteins that mark the identity of cells as "self" or "nonself."

15 Develop into cells that produce circulating antibodies.

16 Major regulatory cells of the immune system that mature in the thymus.

Similarly, use the following list to complete statements 17 to 20.

a] IgC
b] IgA
c] IgE
d] helper T cells
e] memory cells

17 $CD4^+$ cells that are attacked by HIV are one kind of ______.

18 The antibody called ______ can cross the placenta and confer blood-borne immunity in a fetus against bacteria and viruses to which the mother also has protection.

19 Under ideal circumstances, desensitization shots for allergies cause the allergen to be eliminated by IgG before ______ is stimulated to act.

20 A long-lasting, active immunity occurs because a small clone of ______ are formed at the time of immunization and are activated to produce secondary responses.

INTEGRATE AND APPLY WHAT YOU HAVE LEARNED

1 List the nonspecific defenses of the body and contrast their effectiveness as a group with that of the specific defense mechanisms of the immune system.

2 Describe the molecular nature of an antibody. How is one antibody different from another and what results from this difference?

3 Although antibodies are proteins and proteins are coded by genes, we can manufacture more kinds of antibodies than we have genes. Explain how this paradox can occur.

4 How does the humoral response differ from the cell-mediated response?

5 Contrast self-tolerance with autoimmunity. Which cells are believed to be active in each process?

ANALYSIS

1 What is the function of the antibodies that are found within the blood of people infected with the HIV virus?
a] They bind to HIV antigen and mark the virus for elimination.
b] They essentially have no function since the virus lies hidden within T cells.
c] They cause the symptoms of AIDS.
d] They are present in all people as a natural defense against AIDS, but in people who acquire infections, these antibodies multiply.
e] They are remnants of HIV that indicate the presence of the virus in the individual.

2 There are two fundamental explanations in biology that include the word "selection"—evolution by means of natural selection and antibody production as a result of clonal selection. Which of the following statements correctly describes a similarity between these two important ideas? More than one answer may be correct.
a] The "selector," or agent of selection, is environmental.
b] There are more genetic possibilities than are ever realized.
c] Both adaptation and chance play a role in which species or which antibodies will be formed in the future.
d] Both involve an interaction of gene products with the environment.

3 Which of the following statements is false?
a] T cell receptors and antibodies both have the ability to bind to an antigen.
b] Antigens that are not joined to histocompatibility proteins cannot activate T cells.
c] Macrophages are active phagocytic cells as well as antigen-presenting cells.
d] Both humoral and cell-mediated responses require B cells as well as T cells for initiation, augmentation, and suppression.
e] HIV infects a subgroup of helper T cells in which the coreceptor provides the path of entry.

4 Most colleges require entering students to show proof of vaccination against measles, but they do not require such proof of anyone over the age of 35. Why?
a] People over the age of 35 cannot get measles.
b] People over the age of 35 cannot give measles to those under 35.
c] People under the age of 35 cannot give measles to those over 35.
d] By the time a person is 35, he or she probably is immune because of prior exposure.
e] It is easier to regulate the behavior of students than of the faculty and staff.

CHAPTER 27

Respiration: Gas Exchange in Animals

MT. EVEREST WITHOUT TANKS

"I progress so slowly! How long my pauses to breathe are each time I don't know. With the ski sticks I succeed in going 15 paces, then I must rest several minutes. All strength seems to depend on the lungs. . . . And while standing I must use all my willpower to force my lungs to work. Only when they pump regularly does the pain disappear, and I experience something like energy. Now my legs have strength again. I . . . follow the north flank of Mt. Everest."

World-class European mountain climber Reinhold Messner recorded these sensations and events after his attempt in 1980 to reach the summit of Mt. Everest without the aid of an oxygen tank [FIGURE 27.1]. Messner's ascent was both difficult and dangerous because the air surrounding the earth's tallest mountain peak (8848 m or 29,028 ft) contains less than 30 percent of the oxygen in air at sea level. Messner had to breathe about six times harder in this

FIGURE 27.1

Scaling Mt. Everest without Supplementary Oxygen.

At the top of the world, where oxygen is thin, climbers such as Reinhold Messner trudge toward the summit, struggling to breathe in enough of that life-sustaining gas to keep the mind and muscles functioning.

thin air, and still he faced the constant threat of losing consciousness—and his grip on the rock and ice. He also risked brain damage from lack of oxygen. Despite these tremendous odds, Messner did reach the summit, setting a world's record for climbing without auxiliary oxygen tanks.

As Messner pushed himself upward, what events took place in his body that allowed him to succeed under such extreme conditions? How did his system supply enough oxygen to the cells of his brain and muscles? And why did his body—or yours or any animal's—need oxygen in the first place? We will address these and other questions in this chapter as we study **respiration**, the process by which organisms exchange gases with the environment.

Like all animals, people must obtain sufficient oxygen for the rapid conversion of sugars to usable energy; they must also dispose of carbon dioxide, a waste product of energy metabolism. Climbers like Messner, however, face particular hardships because the lower levels of oxygen at high altitude require dramatic physiological compensations and can lead to mountain sickness.

Not only does a climber breathe more often at high altitude, but the body produces more red blood cells. The faster breathing and more concentrated red cells help collect and carry the scarce oxygen molecules to the tissues. However, the additional exhalations that accompany the extra inhalations decrease CO_2 levels in the body, and this changes the pH levels of the blood to a more alkaline state. Consequently, the kidneys must work harder to return the blood to normal pH. While all of this is taking place, low oxygen levels can cause fluid to accumulate in the lungs, and if enough collects, the climber can actually drown in his or her own body fluid. Finally, the sodium-potassium pump that operates in all cell membranes, and that normally regulates water and solute levels in and around tissue cells, fails to work properly at high altitude. As a result, the extremities can swell. The brain, too, can distend inside the skull, causing headache, nausea, and disturbed sleep. All this and the chance to fall hundreds of feet to one's death, too!

MESSAGES

1 An animal's respiratory system exchanges the gases carbon dioxide and oxygen with the environment. During cellular respiration, animal cells use oxygen and release carbon dioxide in the oxidation of organic molecules. Without adequate oxygen, energy harvest slows.

2 Gas exchange depends on diffusion through the liquids surrounding animal cells. Diffusion occurs much more slowly in liquids than in air, and more rapidly when concentration or pressure differences are great and when large surface areas are available for exchange.

3 In humans, air passes from the mouth or nose through the trachea to the lungs. Oxygen diffuses from tiny sacs in the lungs into blood capillaries, where it enters red blood cells and binds to the iron-containing protein hemoglobin. The oxygen is then carried to the body cells via the circulatory system, and the gas diffuses into cells. In an exchange, carbon dioxide diffuses out of the cells and into the blood, where it is carried to the lungs and exhaled from the body.

4 Sensors in major arteries detect the amount of carbon dioxide in the blood and regulate the rate and depth of ventilation (breathing).

Mountain climbers often say they ascend giant peaks simply because they exist, and their stony magnificence beckons irresistibly. The urge must be strong, indeed, because the physiological price for this challenging activity is as high as the peaks: altitude sickness alone causes about 15 percent of the deaths among climbers.

The controls over breathing, oxygen delivery, and CO_2 disposal are clearly complex, and are good examples of homeostatic mechanisms maintaining a steady state in the body. In this chapter, you will see how variations on the theme of animal respiration evolved, yet how the process of diffusion underlies them all. We will compare the function of lungs, gills, and other organs of gas exchange, and we will study the human respiratory system in detail. Along the way, we will encounter anecdotes and facts as interesting as Reinhold Messner's ascent of Mt. Everest without tanks. These include a disastrous ascent in a hot-air balloon and the mysteries surrounding deep-diving seals. ❑

Oxygen for Every Cell

The subject of gas exchange in animals is filled with seeming contradictions: A person quickly airlifted from sea level to the top of Mt. Everest would die from lack of oxygen in less than five minutes unless provided with an oxygen tank and breathing mask. Yet birds can fly over the same peak, flapping vigorously, using oxygen rapidly, and breathing the thin air apparently without ill effects. A large tuna, using its gills, can extract enough oxygen from seawater to meet the needs of its powerful swimming muscles, and yet a tiny mouse could not get sufficient oxygen from the ocean, even if its lungs could move water in and out. Some insects, such as stag beetles, are as large as mice and yet successfully extract oxygen from the environment using neither gills nor lungs. The explanation for these apparent puzzles lies in the wide variety of breathing structures and mechanisms that have evolved and that are capable of carrying out gas exchange in animals. These range from body shapes that allow simple diffusion, to gills that exchange gases in water, to the tracheae (air tubes) and lungs of land animals.

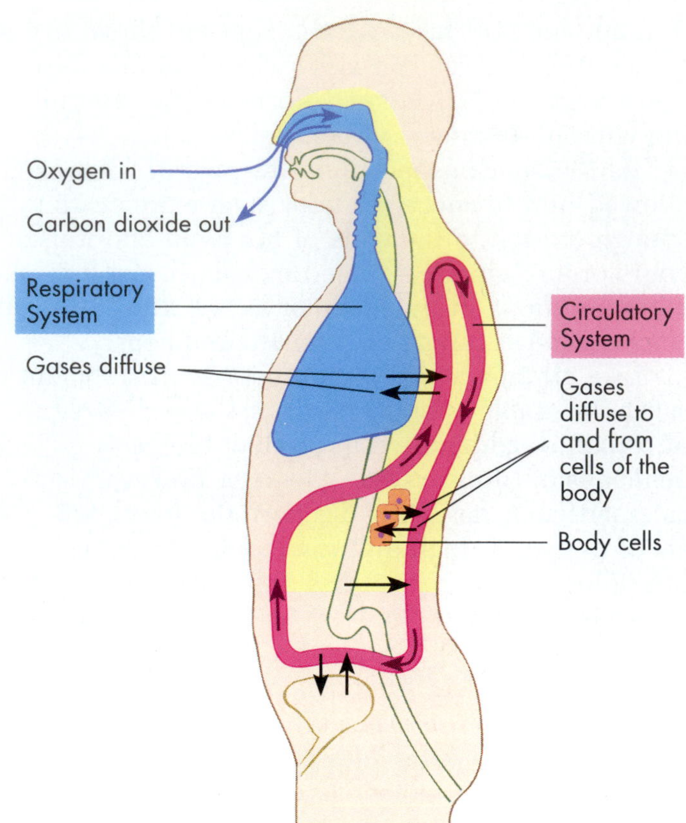

FIGURE 27.2
Interactions of the Respiratory System and Circulatory Systems.

DIFFUSION: THE MECHANISM OF GAS EXCHANGE

To understand why animals need to exchange gases with the environment, we must look at the subject in terms of the whole organism, the living cells that make up that animal, and the metabolic processes within those cells. Recall from CHAPTER 24 that large animals exchange substances with the environment via tubes that bring in and carry away materials, and internally, interface with the tubes of the circulatory system [FIGURE 27.2]. In the immense rhinoceros beetle (drawn life-size in FIGURE 27.3A), atmospheric gases enter and leave the body through portholes that open onto branching tubes. At the farthest tips of those tubes deep inside the insect's body, gases enter the animal's tissue fluid, which bathes the tubes and surrounding cells. The mechanism of entry is *diffusion*, the spontaneous migration of a substance from a region of higher concentration to a region of lower concentration [review FIGURE 4.17]. The gases dissolve in the extracellular fluid and then diffuse from the fluid across the plasma membranes of cells and enter the cells' cytoplasm [FIGURE 27.3B].

Finally, the gas is used at the subcellular level [FIGURE 27.3C]. Within the cytoplasm, carbon dioxide is produced by the dismantling of carbon compounds during aerobic respiration, the process of energy harvest that takes place in mitochondria [review FIGURE 5.8]. As the carbon compounds are broken down, energetic electrons are removed and are passed to oxygen in the mitochondria. The oxygen joins with hydrogen atoms and forms water [review FIGURE 5.12]. Collectively, these energetic processes are called *cellular respiration*. In contrast, respiration at the organismal level is the way an animal or other living thing exchanges gases with the atmosphere, that is, *breathes*.

We can see, then, that diffusion underlies gas exchange at the subcellular, cellular, and sometimes organismal levels, and this fact is particularly important because several principles of gas diffusion have shaped the evolution of respiratory structures:

1. Since cells are generally surrounded by water, gases must pass through a liquid layer to reach cells.

2. Diffusion is much less rapid—an amazing 300,000 times slower—in water than in air. As a consequence, each animal cell needs to be close to a source of oxygen and sink for carbon dioxide so that gases can diffuse to and from each cell within a reasonable period of time. In flatworms, for example, the thin, flat bodies are never more than about 2 mm thick, so oxygen and carbon dioxide can diffuse directly between cells and the animal's moist surroundings [FIGURE 27.4A].

3. The amount of gas diffusing between a liquid and the atmosphere, or between two liquids, is greater if the exposed surface area is greater. If you leave wet laundry in a heap, for example, less water will evapo-

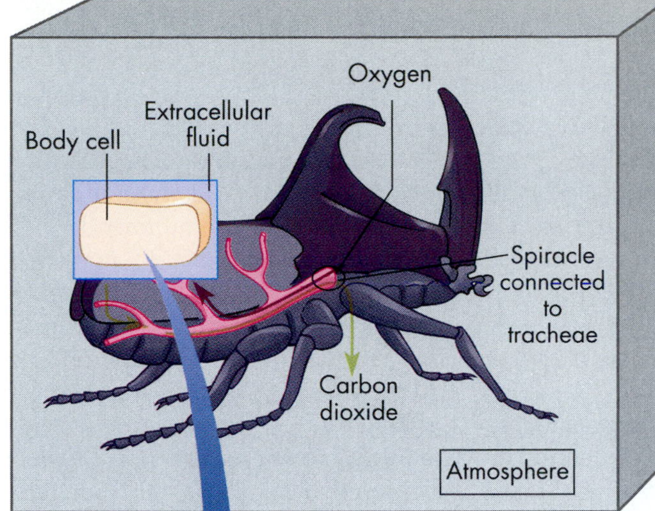

[A] Organismal level

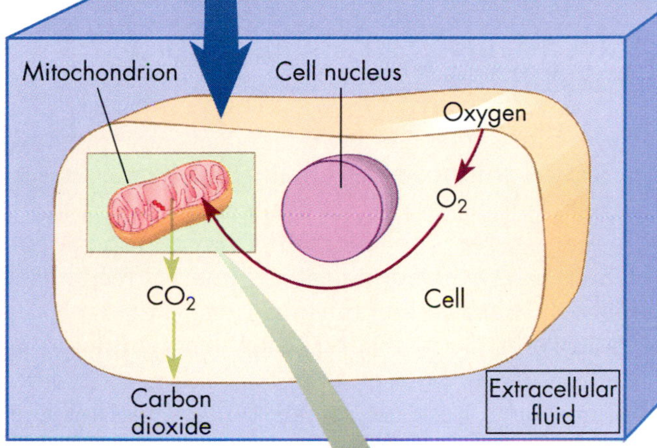

[B] Cellular level

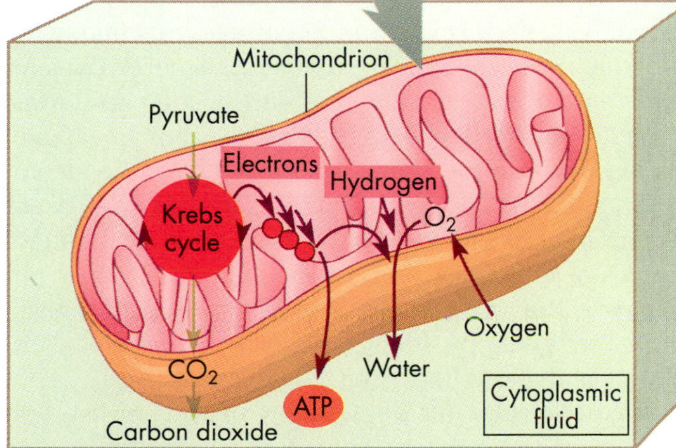

[C] Subcellular level

FIGURE 27.3

Gas Exchange at the Organismal, Cellular, and Molecular Levels.

[A] In this rhinoceros beetle, air tubes called tracheae allow air to penetrate deep into the body, where oxygen diffuses into tissue fluids. **[B]** At the cellular level, oxygen diffuses into the cell and through the cytoplasm, and carbon dioxide diffuses in the other direction, each moving down their concentration gradients. **[C]** Mitochondria produce carbon dioxide as they dismantle organic molecules and utilize oxygen as the final acceptor of electrons in the electron transport chain.

[A]

[B]

FIGURE 27.4

Variations in Respiration.

[A] The body of this purple flatworm is sufficiently thin and flat that oxygen and carbon dioxide can diffuse into and out of each body cell directly from the surrounding seawater. **[B]** The Lake Titicaca frog has skin that falls into deep folds. These may look strange, but they provide the animal with a large surface area through which oxygen can enter and carbon dioxide can escape.

rate within a given time period than if you hang the laundry on a line and expose a greater surface area to the atmosphere. Or consider this biological example: In the Lake Titicaca frog [FIGURE 27.4B], gases are exchanged with the surrounding lake water across the animal's skin. The extensive folds in the skin flaps increase the surface area for diffusion so that more oxygen is available, fulfilling the frog's needs.

4. The rate of exchange via diffusion speeds up or slows down depending on the concentration gradient (or pressure difference) between the liquid and the surrounding atmosphere. Continuing the laundry analogy, water leaves the liquid phase in the clothes and enters the atmosphere much more rapidly if the air is dry than if the humidity is high and the weather is muggy.

These four principles of diffusion have constrained the shape and activities of animal bodies, as in the flatworm and Lake Titicaca frog. They have also led to the evolution of complex strategies that rely not on direct diffusion from the environment, but on diffusion into and out of a circulating internal fluid (blood or its equivalent, review FIGURE 27.2). This fluid picks up and transports oxygen across a moist surface—a physical interface with the environment—and distributes it throughout the body. Then it picks up carbon dioxide in the tissues and releases it back into the outside world, usually through the same moist interface. The specialized structures that have evolved as interfaces are gills, tracheae, and lungs.

GILLS: EXTRACTING OXYGEN FROM WATER

Fish, tadpoles, and most other animals that extract oxygen from water have **gills**: organs specialized for gas exchange that develop as outgrowths of the body surface. Gills can be external (extending outside the body) or internal (housed entirely within), simple or complex, but each type provides a large surface area through which gases from the environment can diffuse and thus enter the organism's bloodstream.

Some amphibians, such as the axolotl, have frilly external gills that wave about in water currents. The elaborate frills increase the organ's surface area [FIGURE 25.10]. As water moves past the waving gill surfaces, oxygen can diffuse in and carbon dioxide out through the gill membrane and capillary walls.

Despite their efficiency, external gills like these lie dangerously exposed, making them vulnerable to predators and to the elements. In most fishes, this problem is avoided because the gills are internal. Internal gills are often protected by a stiff flap, the *operculum* that is easily visible on the side of a fish's head. These flaps cover the gill slits, which are openings from the inside of the fish's mouth to the outside [FIGURE 27.5]. Between the gill slits are *gill bars*, or tissue that supports the surfaces across which gases are exchanged. Each gill bar is subdivided into hundreds of flexible **gill filaments**, which are in turn composed of many thin, platelike structures. These delicate plates, or *lamellae*, hold the key to how a large, active, aquatic animal can respire efficiently in water, even though the liquid holds one-twentieth as much oxygen as air. Embedded within each lamella is a lacy meshwork of capillaries lying just one cell layer away from the water that passes through the gill filament. This proximity of blood to oxygen-bearing water means that oxygen readily diffuses across the cells into capillaries, while carbon dioxide diffuses outward just as easily. The pumping heart then circulates the oxygen-rich blood throughout the animal's body.

The steady opening and closing of a fish's mouth pumps a constant stream of water through the mouth to the gills, over the gills, then out through the opercular flaps. This water flow moves in the opposite direction to blood flow within the capillaries inside the gill's lamellae [FIGURE 27.5B to D]. The result is a **countercurrent flow** [review FIGURE 24.12], in which two fluids with different characteristics (here, different oxygen content) flow in opposite directions and thereby exchange materials along all their points of contact owing to diffusion from high to low concentrations. Thus, the animal can collect sufficient oxygen for aerobic cellular respiration and an active life-style even though water holds so much less oxygen than air.

By contrast, land animals have access to the higher oxygen content of air, but they face a different threat: the drying out of their respiratory surfaces, which would bring gas exchange to a lethal halt.

TRACHEAE AND LUNGS: ADAPTATIONS FOR A DRY ENVIRONMENT

The only animals that live surrounded entirely by dry air are terrestrial arthropods and mollusks, such as insects and snails, and land vertebrates, such as reptiles, birds, and mammals. Not coincidentally, the arthropods and vertebrates have evolved specialized internal respiratory channels: the tracheae and the lungs, respectively.

Tracheae: Tubes for Gas Exchange Insects (including the rhinoceros beetle in FIGURE 27.3) and certain other arthropods have a system of air tubes, the **tracheae**. These tubes are physically linked to the outside world through openings in the body wall that look a bit like portholes, and are known as *spiracles* [FIGURE 27.6]. Body movements force air into and out of tracheae. Inside the body, the tracheae branch into ever finer tubes (far narrower than a human hair) that end in **air capillaries**. These structures allow air from the environment to penetrate deep within the animal's tissues. There, the incoming oxygen dissolves in fluid and diffuses across adjacent cell surfaces. In highly active tissues, such as the flight muscles of a honeybee, no cell is more than a few micrometers away from an air capillary. In insects, oxygen reaches cells directly through the tracheae and not by circulating body fluid like blood.

Lungs: Complex Air Bags A few species of fish that live in the warm, stagnant, oxygen-poor water of swamps have evolved **lungs**, blind-ended internal pouches that connect to the outside by a hollow tube [FIGURE 27.7A]. The lungs supplement the meager amount of oxygen that the gills supply to the swamp fishes. In many fishes, the lung pouch serves not as a true lung, but as a *swim bladder*, or gas bag, that helps the fish maintain its depth in the water without sinking or floating. In an amphibian, such as a frog [FIGURE 27.7B], the lungs are simple sacs with walls that are richly endowed with a dense lacework of blood capillaries. The wall of the lung is a thin,

FIGURE 27.5
Fish Gills: Flaps, Capillaries, and Countercurrent Flow Allow a Rich Harvest of Oxygen from Oxygen-Poor Water.

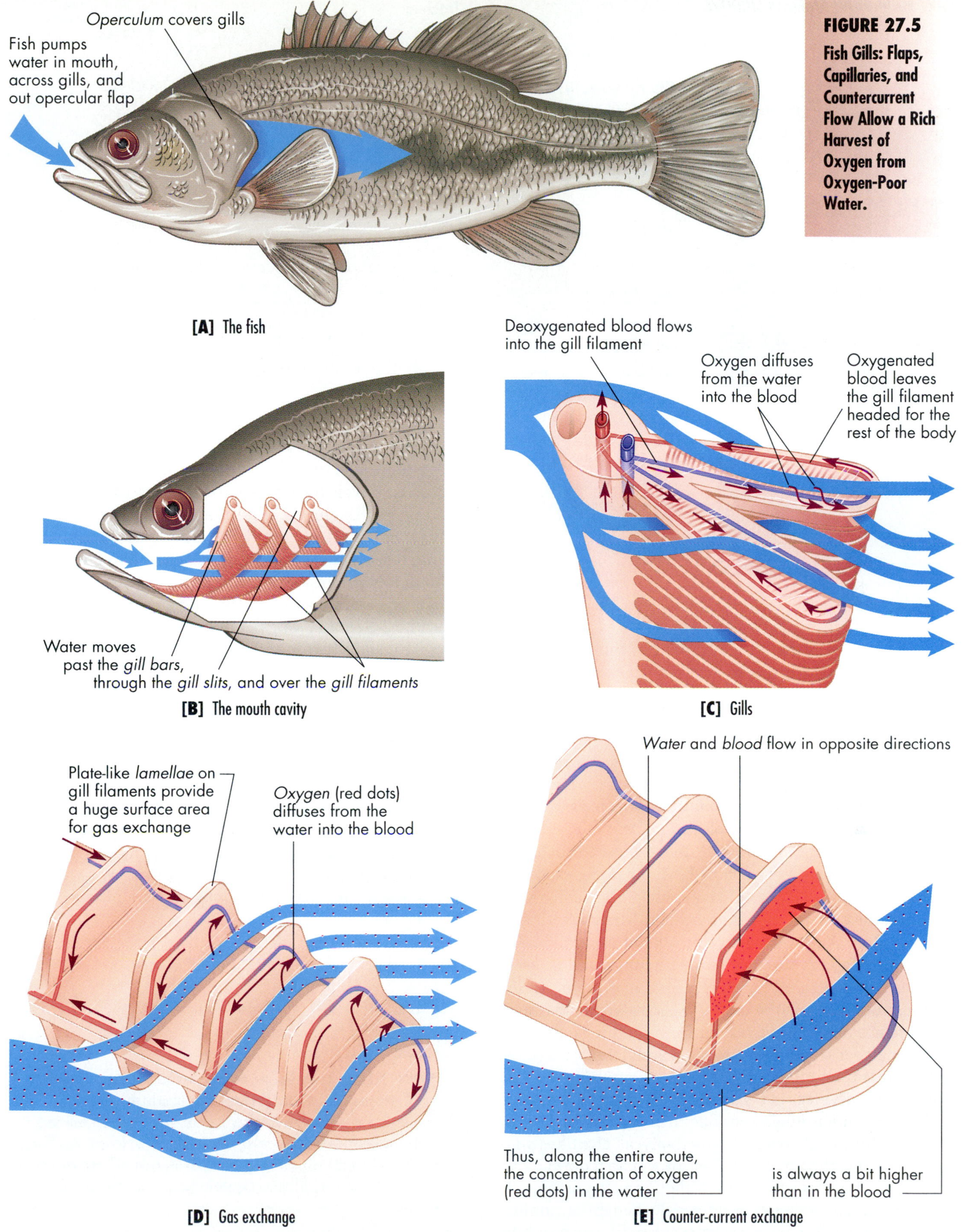

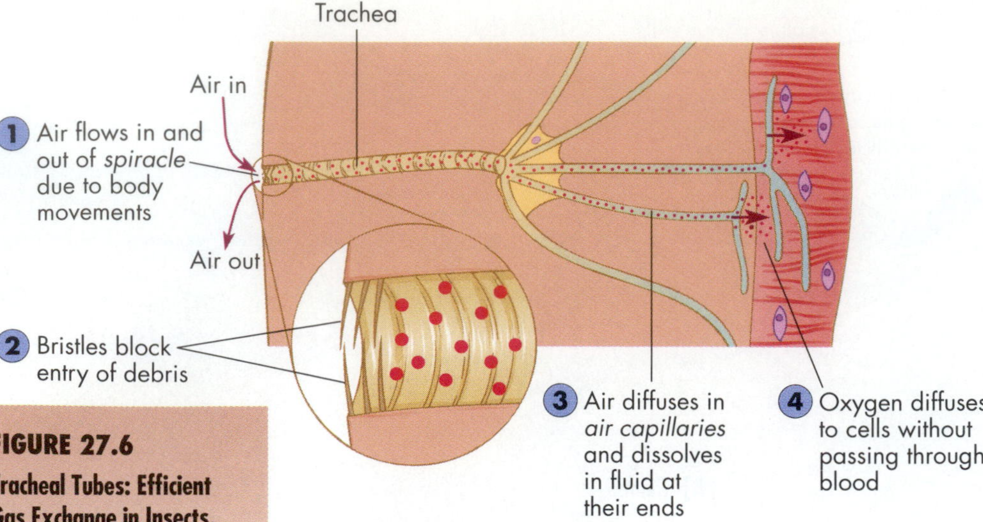

FIGURE 27.6

Tracheal Tubes: Efficient Gas Exchange in Insects.

[A] A silvery tracheal tube leads to an air bubble at a spiracle in this mosquito larva. **[B]** Air entering these tubes through the spiracles can fan out into the branching air capillaries, diffuse across fluid in the capillary tips, and reach internal tissues.

moist membrane through which oxygen can diffuse into the bloodstream and carbon dioxide can exit. Air flows into and out of the lungs via the same two-way path. The simple, saclike lungs of amphibians do not have as much surface area for gas diffusion as the convoluted lungs of reptiles, birds, and mammals; however, gases can also diffuse through the moist skin of amphibians, which supplements the lungs so that the animals can obtain enough oxygen to support their way of life.

In contrast to amphibians, the airflow in a bird's lungs follows a one-way path. Birds have air sacs and tubes called *bronchi* (BRONG-kee) arranged in such a way that air flows in one direction through the lungs [FIGURE 27.7C and D]. A bird takes two breaths to move air completely through the system of bronchi, lungs, and air sacs; the first draws air through the bronchi into the posterior air sacs and then into the lungs. The second breath draws in more air, and this pushes air from the lungs, into the anterior air sacs, and then out of the body. Since this one-way flow through the lungs prevents the mixing of fresh and "stale" air, birds can sustain extremely high levels of activity for much longer periods than we mammals can manage. Birds can even flap actively at altitudes like those found at the top of Mt. Everest, where a normal human cannot even stand at rest without an oxygen tank.

A mammal's lungs do not have the benefit of a unidirectional current of air or water. Instead, like an amphibian's lungs, they operate by way of **tidal ventilation**, an in-and-out air flow rather like the ebb and flow of the tides. Air travels through an inverted tree of hollow tubes leading into the lungs, where gases are exchanged across a thin, moist membrane before the air moves out again through the same set of tubes. Usually, a quantity of air (about 500 mL, or a pint, in humans) is inhaled and exhaled in a regular rhythm. This tidal pattern means that fresh air enters the lungs only during half of the respiratory cycle and that a quantity of unexpelled, stale *dead air* filled with carbon dioxide remains in the lungs at all times, mixing with the fresh air that enters from outside. Special adaptations that increase the rate of gas exchange compensate for tidal ventilation in mammalian lungs.

➤ CONCEPT CHALLENGE

On a planet with high humidity and an atmosphere containing 30 percent oxygen instead of 21 percent like ours, you discover large, cow-sized organisms with cellular respiration and many other metabolic details similar to earth's animals. Can you predict how the organism's respiratory system might differ from that of a cow?

FIGURE 27.7

Pouches of Air Exchange.

[A] Many fish have a swim bladder, an organ that holds gases and controls the animal's buoyancy in the water. **[B]** Amphibians have baglike lungs with relatively little interior surface area for gas exchange. **[C]** Birds have air sacs that interconnect with the lungs and lighten the animal. **[D]** The one-way flow of air through a bird's system of air sacs involves two cycles of breathing (1–4) and provides a high level of oxygen to the bird's actively contracting wing muscles.

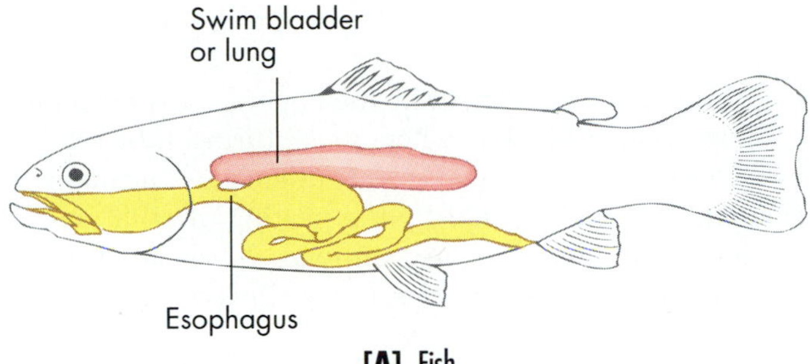

[A] Fish

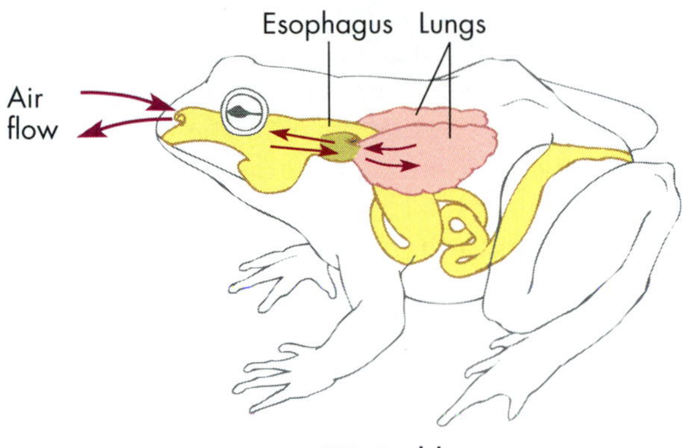

[B] Amphibian

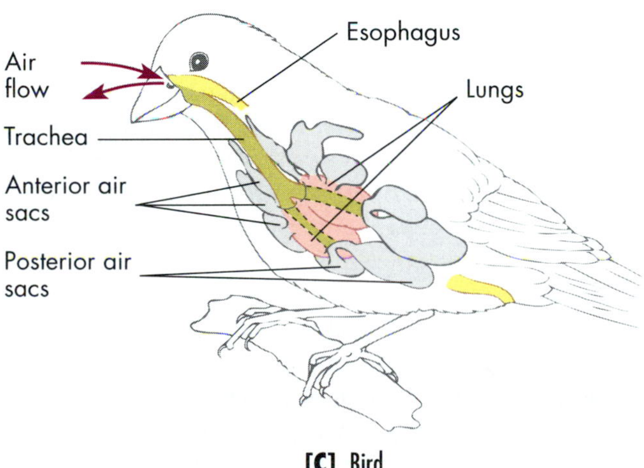

[C] Bird

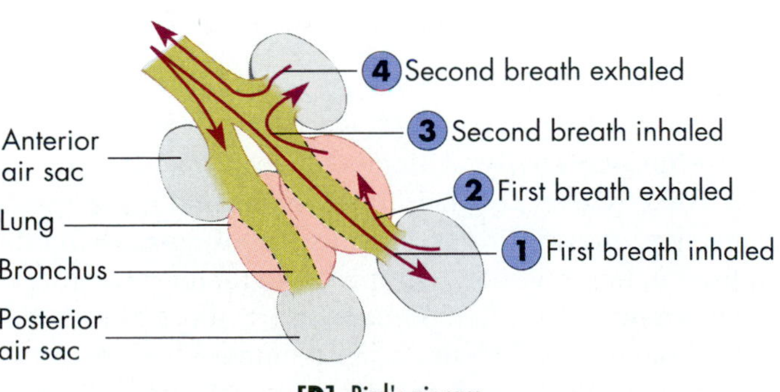

[D] Bird's air sacs

Respiration in Humans

By the time you have finished reading this chapter—an hour, let's say—your efficiently operating respiratory system, superbly adapted to exchanging gases in a dry environment, will have drawn in oxygen and expelled carbon dioxide some 700 to 900 times. In contrast, a giraffe would have breathed as many as 1200 times, which helps overcome the huge volume of dead air that must be cleared from the animal's very long windpipe before fresh air can enter the lungs. Despite such differences, giraffes, humans, seals, and all mammals share a basic set of respiratory structures and mechanisms.

RESPIRATORY PASSAGEWAYS FOR AIR FLOW

When a mammal breathes in, air enters the respiratory system through the nose and sometimes through the mouth. The air is warmed and humidified by the moist mouth cavity or by the twin **nasal cavities**, chambers that open posteriorly into the **pharynx**, or throat [FIGURE 27.8A]. The pharynx branches into a pair of tubes; one, the **esophagus** (ih-SOFF-uh-gus; the Greek word for "gullet"), leads to the stomach, while the other, the windpipe, or **trachea** (not the tracheae of insects), is the airway leading into the lungs. At the anterior end of the trachea lies the **larynx**, or voice box, housing the vocal cords. Just above the opening to the larynx is a flap of tissue called the **epiglottis**, which normally closes off the larynx during swallowing and thus prevents food from accidentally entering the lungs.

A few centimeters below the larynx in humans, the trachea branches into two hollow passageways called **bronchi** (singular, **bronchus**), each of which enters a lung. Finer and finer branchings of these tubes create an inverted tree, with thousands of narrowed airways, or **bronchioles**, that eventually lead to millions of tiny, bubble-shaped sacs called **alveoli** (al-VEE-oh-lie; singular, **alveolus**). It is in the alveoli that gas exchange takes place [FIGURE 27.8B]. Each alveolus is surrounded by blood capillaries, and the inside of each tiny pouch is lined with a moist layer of epithelial cells. Human lungs contain roughly 300 million alveoli, and if the linings of all these delicate bubbles were stretched out simultaneously, they would occupy about 70 m^2—enough surface area to cover a badminton court, or 20 times the body's entire skin surface. At those places where the wall of a capillary lies near the outer wall of an alveolus, oxygen easily diffuses out of the alveolus and into red blood cells squeezing down the center of the narrow capillary [FIGURE 27.8C]. Meanwhile, carbon dioxide leaves the blood, diffusing out of the capillary and entering the alveolus.

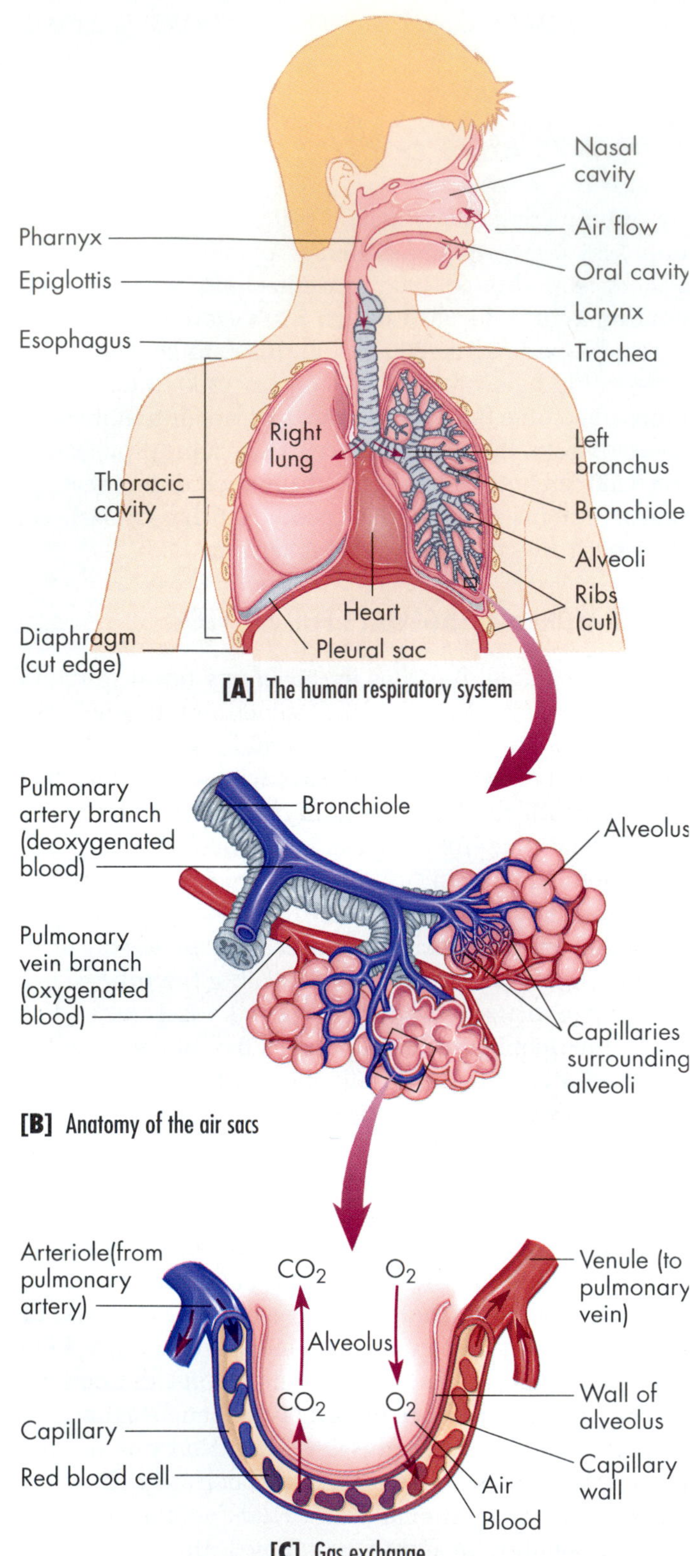

FIGURE 27.8

The Human Respiratory System.

[A] Anatomy of the airways and lungs. [B] Close-up view of the branched passageways, or bronchioles, and the cluster of tiny air sacs, or alveoli. [C] The diffusion of oxygen and carbon dioxide between the alveoli and the blood passing in a nearby vessel.

From the alveolus, it is expelled to the outside with the next exhalation.

Many of the cells that line the larger airways produce a sticky mucus ideally suited to capturing inhaled dirt particles or microorganisms. This mucus is continuously cleared from the bronchi by the beating of cilia, which sweep the mucus and any trapped debris up toward the pharynx, where they can be swallowed or expelled [FIGURE 27.9A].

Inhaled tobacco smoke damages the cells lining the alveoli and paralyzes the cilia in the airways [FIGURE 27.9B]. The smoke from a single cigarette can immobilize the cilia for hours and lead fairly quickly to a hacking smoker's cough—the respiratory system's attempt to rid itself of airborne garbage that accumulates because the cilia no longer sweep it out. Long-term abuse of the airways can lead to a near-total breakdown of the system. As cilia break down and other natural defenses are overwhelmed, it becomes increasingly likely that genetic changes in lung cells will go unrepaired and that the lungs will develop cancer or *emphysema*, a degenerative disease in which the alveoli steadily deteriorate. Also, the tars and gases in cigarette smoke damage blood vessel walls and increase one's chances of heart attack or stroke. Studies show that 4 out of every 10 smokers—approximately 400,000 Americans each year—die as a direct result of their smoking habit. This makes smoking by far the leading cause of preventable death each year.

Recent research indicates that passive smoking—breathing the smoke from someone else's cigarettes—can also have serious health consequences. In fact, passive smoking kills about 53,000 Americans each year, making it the third leading cause of preventable death after direct smoking and alcohol abuse. Children raised around smokers are more likely to have respiratory and blood-vessel problems. Adult nonsmokers who live with smokers have a 30 percent higher risk of death from heart attacks, and about 15,000 passive smokers in the United States die of smoke-induced cancers each year.

VENTILATION: MOVING AIR INTO AND OUT OF LUNGS

Healthy lungs look like pink, spongy, deflated balloons [see FIGURE 27.9C]. These soft, elastic sacs are suspended in the **thoracic cavity**, which is the region within the rib cage directly over the heart. In humans and other mammals, the lungs rest on a domed sheet of muscle, the **diaphragm**, but they don't simply hang there; each lung is enclosed in a fluid-filled **pleural sac** [review FIGURE 27.8]. The fluid surrounding the lungs is under lower pressure than the air inside, and this pressure difference enables the lungs to remain slightly expanded even when no air is being taken in. A collapsed lung results when a pleural sac is punctured and the fluid drains away. Breathing in

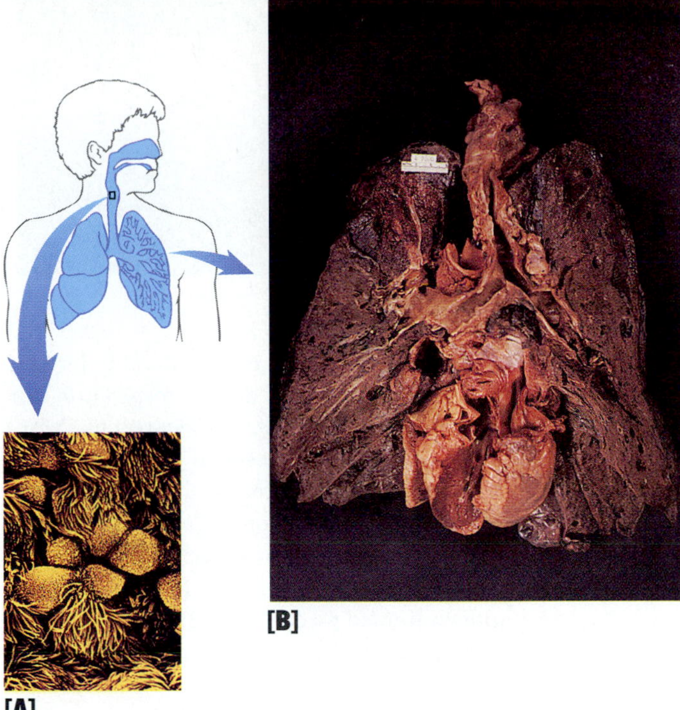

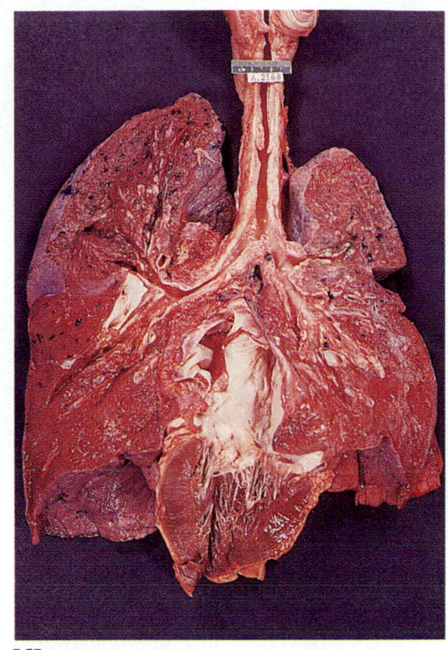

FIGURE 27.9
Bronchial Cilia: Microscopic Custodians of the Respiratory Tract.
[A] Hairlike cilia protruding from the cells that line the trachea and bronchioles, sweep mucus and debris from the respiratory passageways. In a smoker's lungs, the cilia are paralyzed by cigarette smoke. **[B]** Debris can, therefore, reach the lungs and accumulate, blackening and clogging the delicate tissues. **[C]** A nonsmoker's lungs remain clean of debris. The heart is visible below and between each pair of lungs.

and out, or **ventilation** of the lungs, is possible because the bellowslike activity of the diaphragm and expandable rib cage draws fresh air in and allows stale air to rush back out [FIGURE 27.10].

At the start of each **inhalation**, the *intercostal muscles*, which link adjacent ribs, shorten, lifting the ribs and pulling them slightly apart. Simultaneously, the diaphragm muscle contracts and the diaphragm moves downward toward the stomach; this causes its "dome" to flatten out [FIGURE 27.10B]. As the chest cavity enlarges, the pressure of the fluid in the pleural sacs drops, and with it drops the air pressure in the lungs. As a result, air flows in from outside, moving from an area of higher pressure to one of lower pressure, through the airways to the alveoli, filling the lungs. **Exhalation**, the passive release of air from the lungs, occurs when these steps are reversed.

CONTROL OF VENTILATION BY THE BRAIN

How was Reinhold Messner able to climb Mt. Everest without oxygen tanks, despite predictions by respiratory physiologists that such an ascent would be impossible? While recent measurements show that there is more oxygen at the top of Everest than previously thought, part of the answer lies in the incredible ability of climbers like Messner to extract oxygen from the air. Such climbers tend to have an enhanced hyperventilatory response: When placed in an oxygen-poor atmosphere, they tend to breathe much more deeply and rapidly than normal people. Let's look, now, at what controls breathing.

Respiratory control centers in the brain help determine when and how we breathe. For example, the intercostal muscles and the diaphragm, respond to nerve impulses generated by the medulla oblongata, a region in the brainstem that interconnects with the spinal cord [see FIGURE 27.10A]. When one set of nerve cells in the medulla oblongata fires, the breathing muscles contract and we inhale. A nearby group of nerve cells inhibits inhalation. Firing alternates in these two groups of nerve cells, and the result is a regular pattern of breathing in and breathing out. Like diving mammals [BOX 27.1] we can, of course, regulate our breathing consciously to some extent, and that is because nerve cells located in other parts of the brain can exert some control over the respiratory centers in the medulla oblongata.

High-altitude travel has two effects on the body: it decreases the blood level of oxygen, and at the same time increases the blood level of carbon dioxide. So which gas does the body monitor to control breathing? Three French physiologists answered this question in 1875 with a dramatic and disastrous experiment. The three ascended to a high altitude in a hot-air balloon equipped with bags of pure oxygen to be used as needed. Their balloons passed the 7500 m (24,600 ft) level (over 1000 m lower than the summit of Mt. Everest), but they felt no need to use the oxygen. One of the scientists recorded—in an oddly scrawled handwriting—that they were feeling no ill effects at all. When the balloon finally descended, two of the three were dead. The survivor's sad conclusion: The human body is very poor at sensing its own need for oxygen when deprived of it at high altitude.

box 27.1
Biology Applied

Big Seals, Big Mysteries

You don't need to search life's cellular or molecular frontiers to find unsolved mysteries of biology: One of the biggest puzzles in the study of animal physiology involves two of the ocean's heaviest mammals, the Weddell seal of Antarctica, and the northern elephant seal of America's central Pacific coast [FIGURE 1].

Both of these warm-blooded, air-breathing mammals hunt for prey deep below the ocean surface and stay submerged for long periods without breathing. But how? A person passes out after remaining below water for two or three minutes. So how can a half-ton Weddell seal or a two-ton elephant seal get enough oxygen for all of their billions of body cells while cut off from a supply of air for long stretches?

Marine mammalogists have studied this difficult question using surprisingly simple equipment: They glue to the seal's back a small computer that can measure diving depth or blood chemistry. Later, when the animal molts its fur and outer skin layer, they collect the pelt-moored machine from the beach and peruse the recorded data. In this way, they've discovered some remarkable things about each species.

Weddell seals consume enormous numbers of antarctic cod from just above the sea floor at depths of 600 m (2000 ft) or more. While the seals can remain submerged for an hour, they usually stay under for less than 20 minutes when actively hunting fish. All the while, their muscles remain aerobic (not anaerobic), although it is not clear how. Researchers do know that the seal's spleen, a blood-storage organ, releases great quantities of oxygen-bearing blood cells. These help provide oxygen to all active cells and tissues for the dive's duration. Researchers also know that the animal's metabolic rate drops dramatically while it is diving; this process, called the *dive reflex*, increases the efficiency of oxygen use. Finally, when the Weddell seal resurfaces after hunting, it rests for some time, breathing deeply and replenishing its oxygen stores before diving again.

FIGURE 1 Northern Elephant Seal.

The elephant seal is considerably larger than the Weddell seal, and its respiration is a greater mystery. Instrument recordings show that the animals can descend to over 4000 ft—deeper than the previously known mammalian diving champions, the sperm whales. Stranger still, however, is the fact that elephant seals typically spend 20 to 30 minutes underwater (some can even dive for nearly 2 hours!) then resurface for just 2 to 3 minutes at a time. This brief period is too short for lactic acid that has built up in the muscles to be flushed out, or for fresh oxygen to be stored in the blood cells. Like Weddell seals, elephant seals also store great quantities of blood with high densities of red cells. This storage alone, however, can't account for their long dives and quick recoveries. At this point, no one is sure how they get enough oxygen.

Elephant seals seem to have some behavioral adaptations that decrease oxygen use, such as sinking slowly toward their deep hunting zone rather than swimming there quickly. Also, their mitochondria are very efficient at producing ATP, and their muscles are particularly efficient at using it.

Burney LeBoeuf of the University of California at Santa Cruz has recorded occasional "lazy dives" during which the animals sink slowly then drift upward without hunting, and these may be shut-down periods when the waste-removing organs "clean house." Such slow dives still, however, don't completely explain the brief recovery periods at the surface. These remain at the physiological frontier to be probed in future studies.

In contrast, the brain readily senses an increase in carbon dioxide. If a person again and again rebreathes a small volume of air in a stuffy room, the concentration of carbon dioxide gradually rises, and this is detected by sensors at the main arteries that lead to the brain, the aorta and the carotids. These sensors notify the brain's respiratory center of increased carbon dioxide, and the brain deepens and speeds the person's ventilation, reestablishing the blood's optimal oxygen-carbon dioxide balance. Sensors in the muscles and joints can also detect body movements, and they trigger the brain to increase breathing rates even when carbon dioxide levels in the blood are unchanged.

Many people, when traveling above 9000 ft, show periodic breathing—rapid, increasingly deep breathing followed by shallower breathing that eventually stops

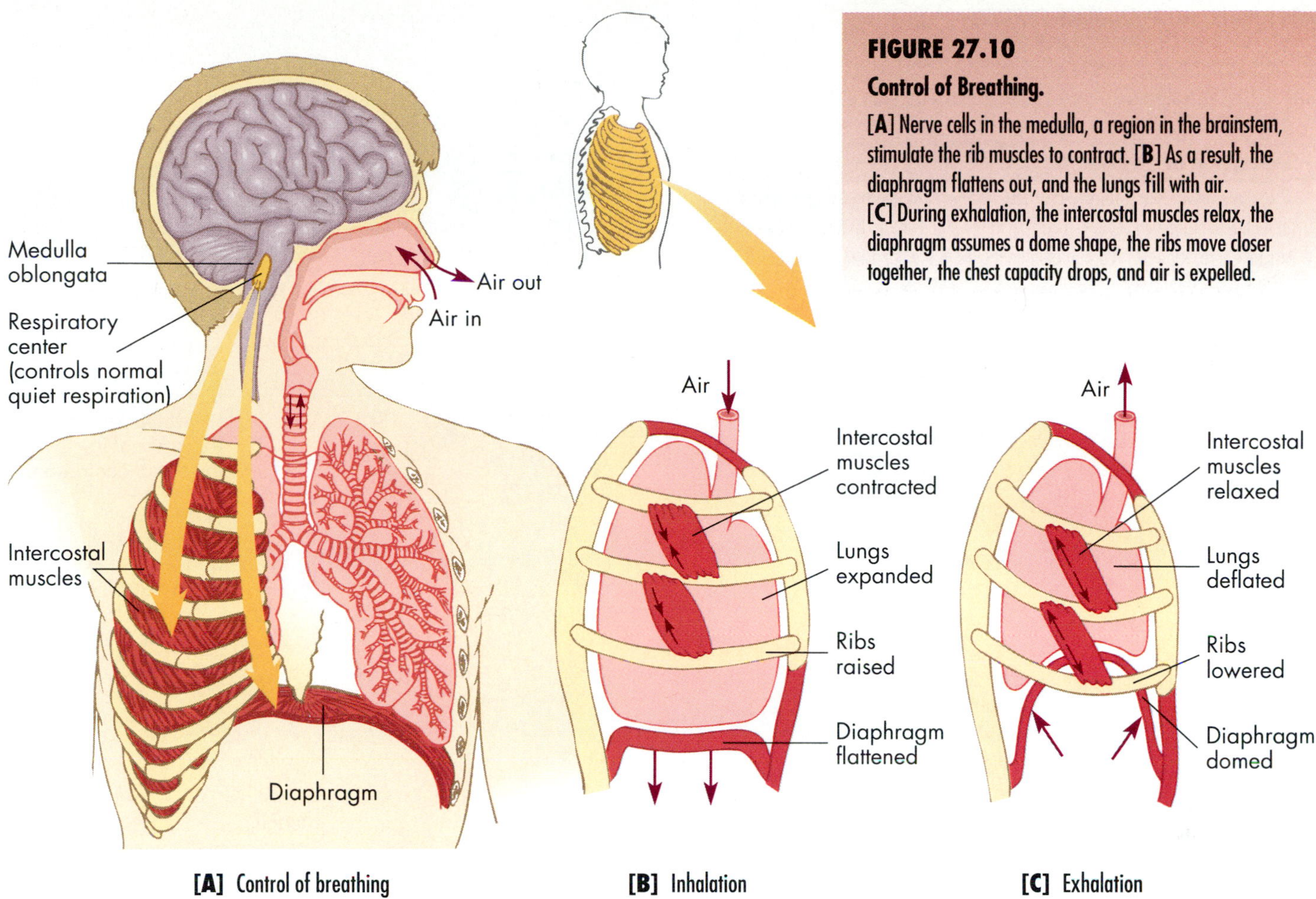

FIGURE 27.10

Control of Breathing.

[A] Nerve cells in the medulla, a region in the brainstem, stimulate the rib muscles to contract. **[B]** As a result, the diaphragm flattens out, and the lungs fill with air. **[C]** During exhalation, the intercostal muscles relax, the diaphragm assumes a dome shape, the ribs move closer together, the chest capacity drops, and air is expelled.

completely for up to 10 seconds. This pattern, repeated over and over, is apparently based on the body's attempt to balance two conflicting situations. Rapid breathing at high altitude brings more oxygen into the blood, but expels huge quantities of carbon dioxide. The same is true when you blow on the coals of a campfire or inflate an air mattress. The imbalance of O_2 and CO_2 causes the blood to become more alkaline. Recall from CHAPTER 4 that carbon dioxide and water combine and form carbonic acid, which makes the blood more acidic. With less carbon dioxide, there is less acid. Apparently, the temporary cessation of inhalation during periodic breathing allows carbon dioxide to build up and help restore the blood's normal pH. In some infants, the control of respiration may not be completely developed, and they may suffer the *periodic breathing syndrome*. Taking an overload of chemical depressants, such as alcohol or barbiturates, can also interfere with the control of respiration and cause breathing to become irregular—or to cease.

➤ CONCEPT CHALLENGE

A person who hyperventilates while blowing on a campfire might black out and become unconscious. During this situation how would the control centers of the brain function in restoring homeostasis?

Gas Exchange

So far, we've been talking about how the respiratory centers of the brain regulate the ventilation of the lungs—the drawing in and expulsion of air from the lungs. Just as ventilation depends on differences in pressures inside and outside of the lungs, gas exchange between tissues depends on gas pressures inside and outside cells.

PARTIAL PRESSURE AND DIFFUSION

Scientists describe the concentration of a gaseous mixture in terms of **partial pressure** (P), the pressure exerted by each individual gas in a mixture of gases. For example, air is about 21 percent oxygen, 0.03 percent carbon dioxide, and the rest mostly nitrogen. Since the pressure of air at sea level under standard conditions is defined as one atmosphere of pressure, then the partial pressure of oxygen is 0.21 atmosphere, or $PO_2 = 0.21$.

If a leak developed in an airliner flying at 10,770 m (35,000 ft), the air in the cabin would still contain 21 percent oxygen. But since the total air pressure at this altitude is only one-quarter of that at sea level, the partial

FIGURE 27.11

Diffusion of Gases in the Body.

Oxygen entering the respiratory system diffuses through the delicate walls of alveoli and blood vessels and is carried in the bloodstream to distant body tissues (here, the feet). Carbon dioxide moves in the opposite direction.

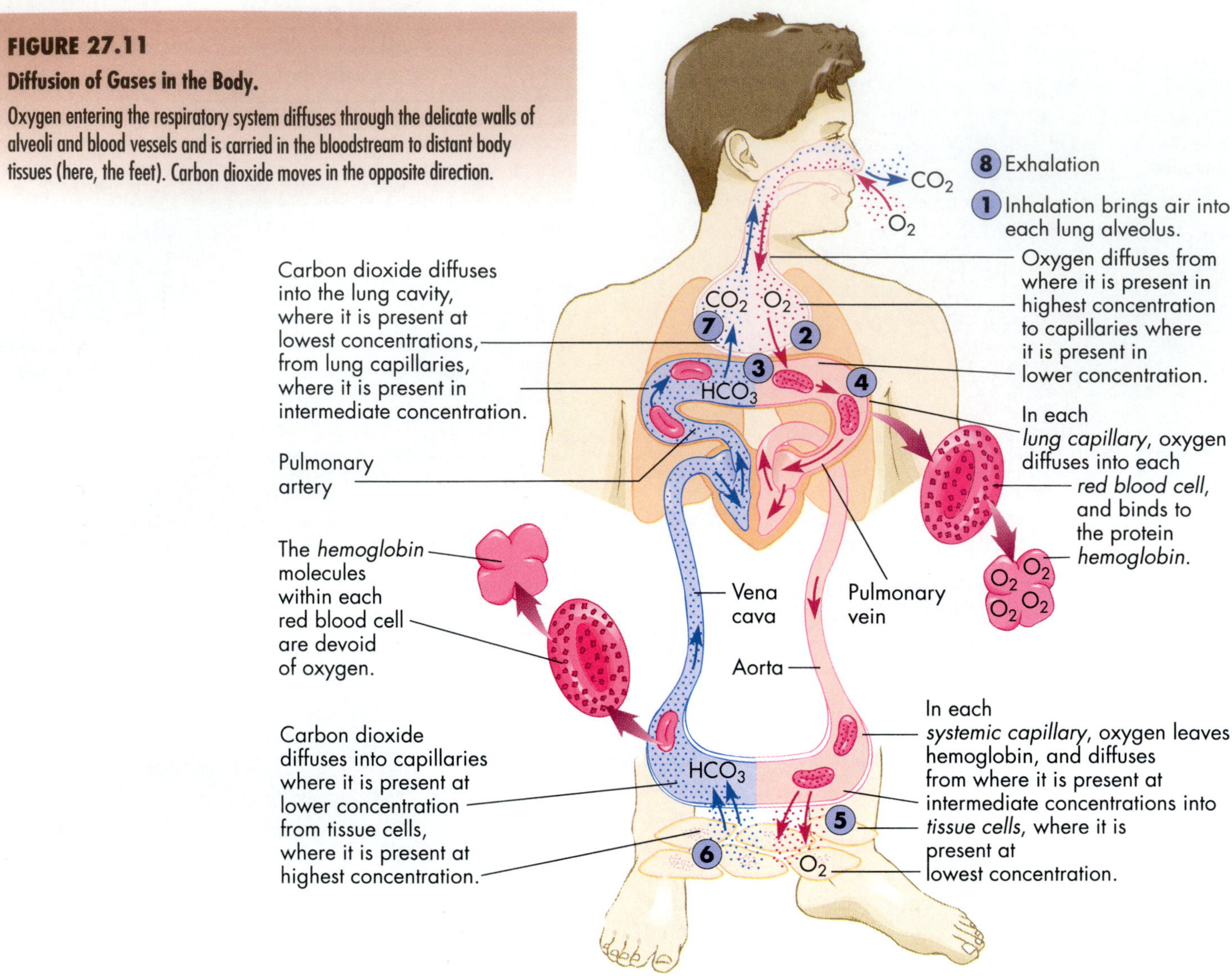

pressure of oxygen would drop to 0.05 atmosphere (since $0.21 \times 1/4 = 0.05$)—not enough to sustain the passengers' lives. Gases always diffuse from places where partial pressure (concentration) is higher to places where partial pressure is lower, and this characteristic is crucial to gas diffusion during an animal's respiration. When you inhale [FIGURE 27.11, Step 1], the new air entering the alveoli (Step 2) has a higher partial pressure of oxygen than the oxygen-depleted blood returning to the lungs from tissues in arms, legs, and elsewhere (Step 3). Oxygen diffuses out of the tiny air sacs and into the bloodstream in nearby lung capillaries, where its concentration is low (Step 4). It is carried in the blood to distant body regions and there moves into tissue cells, where the oxygen concentration is lowest (Step 5). Simultaneously, carbon dioxide moves out of the tissue cells, where its concentration is highest (Step 6), moves across the capillary walls and into the bloodstream (where the concentration is lower), and is carried back to the lungs (some as dissolved carbon dioxide gas and some as bicarbonate, HCO_3^-). In the lungs, the carbon dioxide moves into the alveoli, where the P_{CO_2} (carbon dioxide concentration) is lowest (Step 7). From there, exhalation can carry the waste gas out of the body (Step 8).

HOW HEMOGLOBIN TRANSPORTS OXYGEN

Although differences in partial pressures drive gas diffusion at a rapid rate, water (or tissue fluids made up largely of water) cannot carry enough oxygen to sustain the respiration of metabolically active animals. The protein **hemoglobin**, however, contained within red blood cells, can take on large amounts of oxygen quickly and release it readily where the oxygen concentration is low.

A person has about 25 trillion red blood cells, each a biconcave disk filled with roughly 280 million molecules of the oxygen-binding pigment hemoglobin. As shown in FIGURE 2.31, a hemoglobin molecule consists of four protein chains, each attached to an iron-containing heme group. As red blood cells move through capillaries in the lungs, these four iron atoms bind easily with the oxygen encountered there at high levels, and the binding produces oxygenated hemoglobin, or *oxyhemoglobin* [see FIGURE 27.11]. This iron-oxygen combination is a brilliant red color, and explains the bright color of the blood as it leaves the lungs. In contrast, blood returning to the lungs carrying hemoglobin that is depleted of oxygen (*deoxyhemoglobin*) is darker, with almost a bluish cast, causing vessels like those in the wrist to look blue.

Hemoglobin's specialized structure enables it to act like an oxygen sponge, allowing blood to soak up 70 times more oxygen than an equivalent quantity of water could absorb. In muscles, fat, skin, and other body tissues, oxygen concentrations are low since the active cells steadily use up the available oxygen. When the oxygenated blood reaches such an environment, oxyhemoglobin molecules begin to give up their oxygen. The liberated oxygen dissolves in the fluid portion of the blood and then diffuses into the tissue cells.

Since hemoglobin in red blood cells distributes oxygen in the body, it should come as no surprise that when people's bodies are acclimating to high altitudes, they produce many more red blood cells, which can, collectively, deliver more oxygen. This adaptation, however, takes time. That is why researchers suggest that high-altitude climbers should ascend only about 2000 ft per day. Recent studies of Tibetans who have lived for generations at about 3600 m (11,800 ft) above sea level suggest that they may have genetic adaptations to high altitude that are not present in other people. Even when exercising at near-peak levels, the amount of oxygen in the Tibetans' blood did not drop, whereas immigrant Chinese control subjects—all of whom had lived at the same altitude as the Tibetans for many years—experienced a rapid decline of oxygen levels while performing the same exercise. This difference could be due to the Tibetans' high ventilation rate, high heart output, and low resistance to blood flow in the lungs.

Curiously, yaks, the short-legged, long-haired cattle that live in the high Himalayas of Tibet and Nepal, have a lower hemoglobin concentration than a person's blood at sea level. But yaks have three or four times as many red blood cells per milliliter of blood as either humans or seals. Some observers suspect that yak hemoglobin itself may have a special affinity for oxygen, much as the hemoglobin of the human fetus can pick up oxygen with particularly high efficiency.

A SPECIAL MECHANISM THAT UNLOADS OXYGEN TO ACTIVE CELLS

Highly active cells, like muscle cells in a climber straining to scale Mt. Everest or in a seal pushing deep toward the ocean floor, produce considerable amounts of carbon dioxide. A small portion of the carbon dioxide dissolves in the plasma or combines with deoxygenated hemoglobin. Most of the carbon dioxide waste, however, reacts with the water in the blood, forming carbonic acid (H_2CO_3), which in turn dissociates into bicarbonate ions (HCO_3^-) and hydrogen ions (H^+). This lowers the pH, making the local environment acidic. Additional acidity accumulates from the production of lactic acid by muscles working anaerobically, since the heart and lungs are unable to supply oxygen to the muscles as fast as they can use it up [review FIGURE 5.14]. When muscles are working so hard that their cellular environment grows acidic, hemoglobin gives up its oxygen more easily to the oxygen-starved cells—a phenomenon called the **Bohr effect**.

➤ CONCEPT CHALLENGE

People who live in the high Andes Mountains of Peru have about 20 g of hemoglobin per 100 mL of blood compared to about 15 g/100 mL in people at sea level. How might this change affect a person's physical performance at high altitude and why?

Connections

Respiration is a good example of how an animal's anatomical structures and functions are tuned to the demands of its environment. The respiratory exchange surface may be folds of skin, filamentous gills, finely branching tracheal tubes, lung sacs, or bubble-like alveoli. Regardless, in each case, the architecture allows the exchange of gases with the environment at a rate appropriate for the animal's survival. The real action takes place at the delicate, moist cell surfaces of the respiratory system, and ensures that cellular respiration occurring in the animal's mitochondria can continue. The interaction of biological and physical principles will also be important when we consider the topics of the next two chapters—digestion and excretion—both central to the steady internal state we call homeostasis.

KEY TERMS

alveolus, 627
Bohr effect, 633
bronchiole, 627
bronchus, 627
countercurrent flow, 624
diaphragm, 628
epiglottis, 627
esophagus, 627
gill filament, 624
larynx, 627
partial pressure, 631
pleural sac, 628
respiration, 621
tidal ventilation, 626
trachea, 627
ventilation, 629

HIGHLIGHTS IN REVIEW

1 An animal's respiratory system exchanges the gases carbon dioxide and oxygen with the environment. During cellular respiration, animal cells use oxygen and release carbon dioxide in the oxidation of organic molecules. Without adequate oxygen, energy harvest slows.

2 Gas exchange depends on diffusion through the liquids surrounding animal cells. Diffusion occurs much more slowly in liquids than in air, and more rapidly when concentration or pressure differences are great and when large surface areas are available for exchange.

- **a]** Respiration, the exchange of oxygen and carbon dioxide between an animal's body cells and its external environment, usually involves a moist exchange surface such as tracheae, gills, or lungs.
- **b]** In the gills of aquatic animals, dense networks of capillaries lie just a few cells away from flowing water. A countercurrent exchange system between water and the blood in the capillaries enables fish gills to capture a large amount of the oxygen dissolved in water.
- **c]** Air-breathing animals have internal respiratory channels or pouches with a moist surface through which gases diffuse into a fluid. Insects have branching tubes called tracheae, while vertebrates have lungs. In mammalian lungs, an in-and-out flow of air through an inverted tree of respiratory airways brings oxygen into contact with tiny sacs called alveoli.

3 In humans, air passes from the mouth or nose through the trachea to the lungs. Oxygen diffuses from tiny sacs in the lungs into blood capillaries, where it enters red blood cells and binds to the iron-containing protein hemoglobin. The oxygen is then carried to the body cells via the circulatory system, and the gas diffuses into cells. In an exchange, carbon dioxide diffuses out of the cells and into the blood, where it is carried to the lungs and exhaled from the body.

- **a]** The hemoglobin molecule is specialized to take on large amounts of oxygen and quickly release it to working tissues. Differences in the partial pressures of oxygen and carbon dioxide lead their diffusion into or out of red blood cells and tissue cells.

4 Sensors in major arteries detect the amount of carbon dioxide in the blood and regulate the rate and depth of ventilation (breathing).

- **a]** Changes in breathing rate and depth are mostly involuntary and are regulated by nerves that detect changes in the level of carbon dioxide circulating in the blood.
- **b]** Most of the carbon dioxide that enters the bloodstream is rapidly converted into carbonic acid and then to bicarbonate, a process that prevents the blood from becoming too acidic. The reaction is reversed in the lungs, and the carbon dioxide is exhaled.

UNDERSTANDING THE FACTS AND CONCEPTS

For Questions 1–5, match each of the descriptions with one of the items in the list below. Each item may be used once, more than once, or not at all.

a] gill filament
b] gill lamellae
c] gill bar
d] tracheae
e] air capillaries

1 The region of diffusion of oxygen into the tissue fluid of insects.

2 A branching set of tubes that brings air into the tissues of insects.

3 The structural support for gill filaments.

4 The portion of a gill through which blood flows in a direction counter to the flow of water.

5 The region of a gill that supports the gill lamellae.

For Questions 6–10, match the descriptions with the most appropriate item from the list below. Each item may be used once, more than once, or not at all.

a] swim bladder
b] bronchiole
c] alveolus
d] larynx
e] epiglottis
f] bronchus

6 The region in a mammalian lung where gas exchange occurs.

7 A flap of tissue that prevents swallowed food from entering the larynx.

8 A lunglike structure found in fishes.

9 Site of the vocal cords; the voice box.

10 Small airway that is contained within the substance of the lung.

For Questions 11–15, match the descriptions with the most appropriate item or items from the following list.

a] thoracic cavity
b] diaphragm
c] pleural sac
d] intercostal muscles

11 A dome-shaped muscle that assists in breathing.

12 Action of these muscles enlarges the thoracic cavity.

13 Fluid-filled cavity in which the lungs lie.

14 Relaxation of these muscles results in lowering of the rib cage.

15 A puncture in this structure results in a collapsed lung.

Answer Questions 16–20 by filling in the blanks with the letter of one of the terms listed below.

a] oxyhemoglobin
b] periodic breathing
c] Bohr effect
d] partial pressure
e] dive reflex

16 During the ______, oxygen is conserved as metabolic rates are lowered and blood is preferentially shunted to the brain.

17 The diffusion of oxygen from the blood to the tissue fluid is caused by the difference in the ______ of oxygen between the two locations.

18 ______ is formed when oxygen combines with the iron in hemoglobin.

19 The ______ describes the influence of acidity on the release of oxygen from oxyhemoglobin.

20 During ______ the body attempts to balance the levels of oxygen and carbon dioxide in the blood.

INTEGRATE AND APPLY WHAT YOU HAVE LEARNED

1 Contrast the term *respiration* and cellular respiration.

2 Trace the route that water takes as it flows through a fish gill and explain the gradient of oxygen that exists in the gill capillaries relative to that in the water.

3 Contrast the flow of air through the respiratory system of a bird with that of a mammal.

4 Explain why inhalation is an active process whereas exhalation is considered to be passive.

5 Snorkelers often hyperventilate before a deep dive. Why might this be dangerous?

ANALYSIS

1 Which of the following best explains why the blood of a high-altitude climber tends to become more alkaline as the climber ascends?

a] As the blood loses oxygen, it loses the ability to form acids.
b] As the blood loses carbon dioxide, the level of carbonic acid in the plasma decreases.
c] As the blood gains oxygen, it loses alkaline molecules that are in balance with oxygen.
d] As the blood gains carbon dioxide, more bicarbonate ions are formed.
e] As the blood thins at high altitude, the concentration of both acids and alkaline compounds decreases.

2 Under which set of circumstances would the diffusion of oxygen occur most rapidly across a respiratory surface such as a gill or lung membrane?

a] Water at sea level that has been saturated with additional oxygen across a gill membrane with an area of 1 m^2.
b] Water at an elevation of 10,000 ft across a gill membrane with an area of 10 m^2.
c] Air at sea level across a lung membrane with an area of 1 m^2.
d] Air at an elevation of 10,000 ft across a lung membrane with an area of 10 m^2.
e] Air from a diver's tank at a depth of 100 ft below the ocean surface across a lung membrane with an area of 10 m^2.

3 In a gill, why does more oxygen diffuse as a result of a countercurrent mechanism than it would if the water and oxygen flowed in the same direction?

a] An equilibrium is never reached, so the oxygen diffuses along the entire length of the gill capillaries.
b] The oxygen in the blood and in the water quickly come to an equilibrium near the beginning of the capillary.
c] The oxygen in the blood and water reach an equilibrium that is higher than they could if the water and blood flowed in the same direction.
d] The oxygen in the water flows into the blood without regard to the establishment of an equilibrium.
e] Countercurrent flow allows oxygen to be actively transported into the blood rather than diffused into the blood.

4 Why is a bird's lung more efficient than an amphibian's lung?

a] Fresh and stale air do not mix in a bird's lung.
b] There is no dead air at the base of a bird's lung.
c] The lung of a bird has more interior surface for diffusion of gases.
d] The flow of air in the amphibian is tidal rather than one-way.
e] All of the answers above are correct.

5 Which of the following statements is false?

a] The binding of iron atoms in hemoglobin with oxygen creates oxyhemoglobin.
b] Oxyhemoglobin gives up oxygen to tissues at a rate controlled entirely by the concentration of oxygen in the tissue fluid.
c] Oxyhemoglobin is converted to deoxyhemoglobin more rapidly in hardworking muscles than in muscles working moderately.
d] Deoxyhemoglobin and hemoglobin are not the same color.
e] The low concentration of oxygen in the fluid surrounding active cells causes oxyhemoglobin to release its oxygen.

CHAPTER 28

Animal Nutrition and Digestion

A LEAF-FERMENTING BIRD

What do you suppose would happen if, for some strange reason, you were forced to eat nothing but plant leaves? If you lived in a garden, your diet might include lettuce, spinach, and cabbage, and would be at least palatable. But if you were reduced to collecting and eating only leaves from the forest, you would find that most of them taste bitter and nauseating. In addition, you would be hungry all the time, and would eventually waste away due to a lack of protein and carbohydrates in your diet. Some animals, however, such as cows, koalas, and one of the strangest birds on the planet, can exist a lifetime eating only leaves. How do they obtain sufficient nutrition to survive on such a meager diet?

One bird discovered to live totally on leaves is the pheasant-sized South American bird called the hoatzin (WATT-seen; FIGURE 28.1). This rust- and olive-colored bird with its long crest plumes looks a bit like a floppy feather duster. Bare, neon-blue skin surrounds its blood-red eyes. Much of the time, the bird sits in a tree, supporting its large chest by leaning against a limb with a bare, calloused patch of skin just above the enlarged breastbone. The hoatzin's voice has been described as a goose grunt alternating with a raspy hiss. Observers have noted that when it is disturbed, the hoatzin clumsily flies a short distance away, like a weary helicopter or an overfed chicken.

FIGURE 28.1

A Rare Jungle Fowl.

The hoatzin (*Opisthocomus hoazin*) is unique among birds in that its digestive strategy resembles a cow's more than a cuckoo's, its nearest relative. The hoatzin is the only bird with a digestive compartment that contains microorganisms capable of fermenting leaves, as does the gut of a cow. This arrangement allows the bird to exploit a resource (forest leaves) which most other animals can't utilize. But there are evolutionary trade-offs in terms of flying ability and slow development of the young.

Young hoatzins are even stranger than adults: they can swim underwater and climb like little reptiles before they can fly. They develop slowly, begging regurgitated leaves from parents and older siblings who build and tend the nest directly over a stream. When frightened, the chicks plop down into the water and swim below the surface to submerged roots and stems some distance downstream. Then they climb to safety in the vegetation, clawing with their feet and a second set of nails on the "thumb" and "index finger" of their wing bones.

Researchers at the New York Zoological Society knew the bird's reputation for surviving only on leaves, and wondered how the hoatzin could get enough nutrition this way. One of them, Stuart Strahl, conducted a field study in Venezuela and got his first hint when he caught a whiff of a hoatzin close-up: The bird smelled like fresh cow manure, and in fact, Guyanese natives call it the "stinking pheasant." Indeed, Strahl found that the animal is unique among birds in digesting leaves the same way a cow does. A special chamber of the bird's digestive tract—the crop—receives food before the stomach does. It contains bacteria that ferment the fibrous plant material. During the fermentation process, nutrients are released and the host animal absorbs some of them. This physiological strategy permits the hoatzin to thrive on a resource which is not touched by most other birds and mammals and which is, to them, an energy-poor diet. At the same time, however, it forces evolutionary trade-offs such as slow growth of the chicks and increased vulnerability to predators.

The special adaptations of the hoatzin digestive system, as well as other aspects of its life history, focus our attention on this chapter's topic: **animal nutrition**, how animals obtain the substances needed for their growth and survival. The first section of the chapter discusses the organic and inorganic materials animals require for survival, including sources of energy, carbon, and nitrogen, plus vitamins and minerals. Next, the chapter explores how different animals are adapted to different kinds of foods—how some animals can survive on leaves, while others utilize only meat, dung, or even the tears of cows. Then we focus on humans, looking at the structures and mechanisms that allow us to take in food, break it down, and use the nutrients. Finally, the chapter investigates how nerve signals and hormones control digestion so that all the body's cells are continuously supplied with needed materials.

The chapter will show that exploiting nutritional sources is one of the main factors driving evolution and that the ways organisms obtain energy account for some of their most characteristic and differentiating traits. Finally, in this era of nutritional awareness—and, for many, an obsession with thinness—we will see some useful dietary hints. These include the biology behind healthy vegetarianism, how to maintain ideal weight, and the role of dietary deficiencies in cancer and other diseases. ❑

MESSAGES

1 Animals obtain their energy and materials from the environment by consuming other organisms or their parts and products. Nutrients include carbohydrates, lipids, amino acids, vitamins, and minerals. Proper nutrition is a critical component of good health.

2 The gastrointestinal tract breaks down foods physically and chemically into small molecules (digestion) and delivers these nutrients to the circulatory system (absorption), which distributes nutrients throughout the body.

3 Different parts of the digestive system perform different tasks. Digestive systems vary among different species because they have become adapted during evolution to different food sources.

4 Nervous and hormonal control of the gastrointestinal tract optimizes the efficiency of digestion and nutrient absorption.

Nutrients

All organisms must find a source of energy in their environments, as well as a dependable supply of carbon and nitrogen atoms. Plants get their energy directly from the sun, their carbon from the air, and their nitrogen from the soil. Animals, however, must take in all three as food molecules. The science of **nutrition** is concerned with the precise amounts of protein, carbohydrates, lipids, vitamins, and minerals, and the total number of calories of food energy, an animal must consume to stay alive and healthy. Without this knowledge, dairy farmers could not be sure their cows were producing the highest quality of milk, and parents might wonder if their children were growing and developing to their fullest mental and physical potential.

An animal can have a highly specialized diet: A hoatzin eats nothing but a few types of leaves and occasionally some pulpy seeds or bulbs. Most animals, however—ourselves included—need a more varied diet. Because the human diet is varied and familiar, we use it as our main model in studying organic and inorganic nutrients, beginning with carbohydrates, then moving on to lipids, proteins, vitamins, minerals, and total calories.

CARBOHYDRATES: SOURCES OF CARBON AND ENERGY

The sugars in fruit and the starches in potatoes, rice, bread, and pasta are a rich source of energy and of carbon atoms. Together, these and certain other foods provide the nutrients we call **carbohydrates**. As CHAPTER 2 explained, carbohydrates include monosaccharides, such as the glucose found in honey; disaccharides, such as sucrose, or table sugar; and polysaccharides, such as potato starch [see FIGURES 2.22 and 2.23].

Our digestive system can cleave glucose monomers from the sugars in fruit or from the starch in rice, wheat, or potatoes. After a meal, the glucose molecules pass into the bloodstream, and the circulatory system carries them to cells throughout the body, where they serve as the main supplies of energy for glycolysis and aerobic respiration [FIGURE 28.2A]. Cells in the brain and nerves are particularly sensitive to fluctuations in blood glucose levels. If starving, the body will break down first its fat stores, then its own muscle tissues, and convert the subunits to glucose, thus providing the sensitive nervous system cells with the glucose they need to stay fully active.

People are often told to avoid table sugar and the desserts and processed foods containing it. The fact is, sugar provides calories, our largest nutritional need. While sugar doesn't provide vitamins or minerals in significant amounts, neither do other sweeteners such as honey, brown sugar, raw sugar, corn syrup, or maple syrup. The issue is whether a person substitutes sugar-laden foods for more nutritious ones—a candy bar, say, instead of an apple. Sucrose also promotes the growth of oral bacteria that produce acid and cause cavities. Whether you eat cake or a more complex carbohydrate, regular brushing and flossing are the answer to preventing tooth decay.

The cellulose fibers in plant cell walls, including the cells of leaf veins, the "strings" in celery stalks, the tissue surrounding grapefruit and orange sections, and the woody core of a pineapple, are also complex carbohydrates, but ones that we humans cannot digest [review FIGURE 2.23]. Herbivores, however, such as termites, hoatzins, and horses, have cellulose-digesting microbes in their digestive tracts, and it is the microbes that cleave glucose subunits from cellulose and provide the animals with a rich source of energy. However, the cellulose we consume in plant foods does provide fibrous *roughage*, that helps to stimulate the mechanical movements that propel wastes through the large intestine.

LIPIDS: ENERGY-STORAGE NUTRIENTS

Most people have only to visualize the solid white fat of a bacon slice or the slippery golden oil in salad dressing to know what a lipid is. Many people have only to reach down to their waists and "pinch an inch" to see that the body stores lipids as an energy reserve. While both carbohydrates and lipids supply energy, animals generally store much more lipid than carbohydrate. For example, a person stores about 90 g of the carbohydrate glycogen in the liver and about 350 g in muscles; in contrast, even a lean person stores about 7 times as much lipid. Why is lipid a superior form for energy storage? First, lipids provide about twice as much energy per gram as carbohydrates. After a meal of egg yolks, oily sunflower seeds, or fatty meat, an animal's digestive system breaks down lipids to products that feed into the Krebs cycle. In the Krebs cycle, energy is harvested from these products and their carbon atoms become intermediates for the synthesis of amino acids and other substances the body needs.

A second reason animals often store fat rather than carbohydrates is that fats and water do not mix, whereas each gram of stored carbohydrate generally binds 2 g (0.08 oz) of water. A bird storing carbohydrates and bound water rather than compact fat would be too bulky to fly. One biologist estimates that if a person's body stored carbohydrates instead of fats, it would weigh twice as much. Without fat stores, walruses could not live in the Arctic [FIGURE 28.3], birds could not migrate long distances, and carnivores would have more trouble surviving between irregular meals.

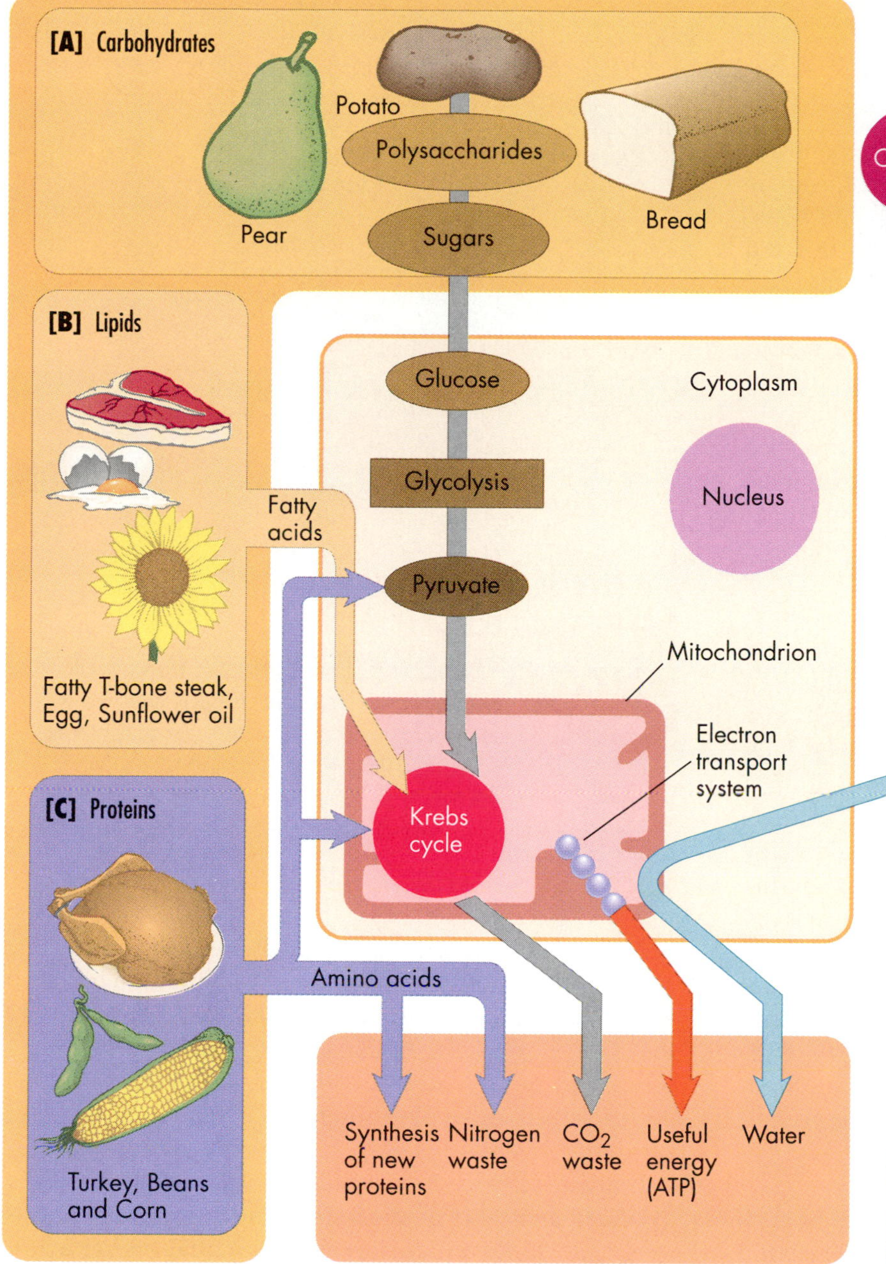

FIGURE 28.2

How the Body Gets Energy and Materials from Foods.

[A] Digestive processes cleave carbohydrates in foods into polysaccharides and sugars. The sugars are ultimately converted into glucose. [B] Lipids are broken down into fatty acids and other compounds, which can enter different metabolic pathways. [C] Dietary proteins are broken down into their constituent amino acids, which can be used to make new protein. The carbon portion of amino acids can be modified and can enter the Krebs cycle and the electron transport chain, where ATP and other high-energy compounds are formed. The amino part can be excreted as waste.

Besides their storage function, fats also cushion internal organs, and if present in a thick layer, as in whales or walruses, can provide thermal insulation. Finally, certain essential nutrients, including vitamins A, D, E, and K, are fat-soluble, and lipids must be present for these vitamins to be absorbed and delivered to cells. The human body cannot generate one particular fatty acid—linoleic acid—from component parts; thus, our diets must contain this so-called **essential fatty acid** in order to build important components of the cell membrane involved in material absorption. Nevertheless, a high-fat diet increases one's risk of colon cancer and heart disease. In fact, according to a recent study of 88,000 nurses, a person who eats red meat (beef, pork, or lamb) daily has a risk of colon cancer three times higher than someone who ingests animal fat from those sources just two times a week or less.

FIGURE 28.3

Lipids: High-Energy Storage.

Fats are a compact way to store calories. Active animals, such as walruses living near the Arctic Circle, have thick fat layers that act as superb insulation as well as a source of stored energy that the animal can use to generate heat and power activities.

PROTEIN: SOURCE OF AMINO ACIDS

The body's structure and its vital activities depend on proteins. The body's most abundant protein, *collagen*, is a major constituent of skin, cartilage, tendons, and bone. Muscle tissue is largely protein, as are hair and the cornea of the eye. Enzymes, antibodies, hemoglobin, and some hormones are composed of protein molecules; without such proteins, most cellular activities would grind to a halt. What's more, there is a steady turnover of protein: Enzymes and cell constituents are continuously broken down and rebuilt, and newly formed cells replace dying cells. This constant turnover means that animals need a continuous supply of protein in food, from which their digestive systems can extract amino acids for the building of new protein as needed [FIGURE 28.2C].

A human is a large, active animal that requires about 1 g (0.04 oz) of protein per kilogram (2.2 lb) of body weight per day. As a rough measure, a college student of average height and weight needs approximately one-sixth of a pound of pure protein each day to replace daily losses. This is about what you would obtain from a cooked chicken breast. Experiments with lab animals show that too much protein, however, can lead to kidney damage because those organs must work overtime in the excretion of large quantities of nitrogen wastes derived from amino acids. While there is no hard evidence of similar harm in humans from large amounts of dietary protein, research shows that muscle mass increases only as a result of regular weight training. Most sports nutritionists suggest that even body builders can obtain enough protein from a diet in which about 15 percent of the calories come from protein, as long as their total energy intake from all sources is sufficient.

Knowing how much protein one needs does not answer the question of what *kind* of protein to eat. The body can synthesize many of the 20 amino acids it needs if the amine (nitrogen-containing) portion of the molecule is available [review FIGURE 2.28]. The human body cannot, however, synthesize eight so-called **essential amino acids**: lysine, leucine, phenylalanine, isoleucine, tryptophan, valine, threonine, and methionine [FIGURE 28.4]. These must be obtained in the diet each day, since free amino acids are not stored. Children also need extra supplies of the amino acids histidine and arginine, since their bodies only make enough for maintenance, not growth.

If one or more of the essential amino acids are missing, the body's cells cannot synthesize the full spectrum of proteins necessary for replacing lost macromolecules or for generating new cells. Why? First of all, the body cannot store amino acids; what it doesn't incorporate immediately into proteins, it excretes in urine or converts into other biological molecules. Second, as CHAPTER 10 described, cells manufacture proteins by adding one amino acid at a time to a growing polypeptide chain. If even one of these amino acids is lacking, elongation of the chain stops. It's a bit like knitting a sweater with a repeating pattern of several colors of yarn: When you run out of yarn of one color, the project stops until you can get more. In the same way, the absence of even one essential amino acid stops protein synthesis in the body.

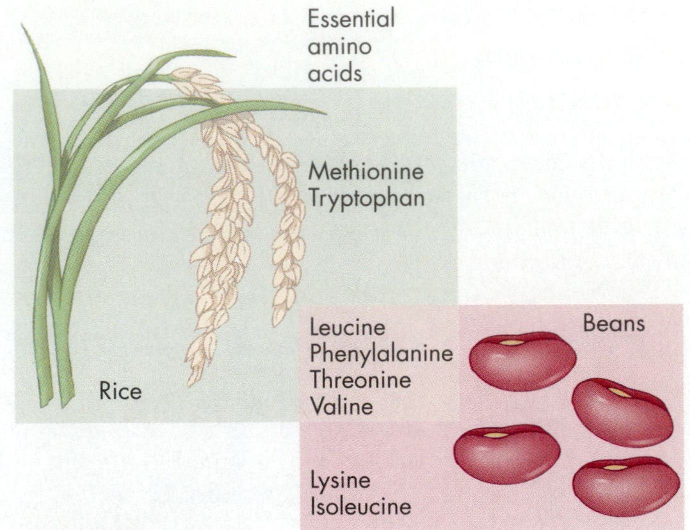

FIGURE 28.4

Essential Amino Acids.

The eight essential amino acids on this list must be included in a person's daily diet; if they aren't, the body begins to break down proteins in its own tissues and use the amino acids they contain for critical functions, such as maintaining the lungs, heart, and brain. Meat, fish, eggs, and other high-protein foods contain all eight essential amino acids. Rice contains only six, and beans a slightly different set of six, but a diet that includes both rice and beans provides all eight.

Most animal proteins in foods such as meat, cheese, eggs, and milk contain the eight essential amino acids; however, many plant proteins do not. This is why strict vegetarians (those who avoid all animal products, including eggs and dairy products) must eat particular combinations of foods to ensure an adequate amino acid intake. One such combination is rice, which contains little lysine but an adequate amount of methionine, together with beans, which are deficient in methionine but contain a sufficient amount of lysine [see FIGURE 28.4]. Some people in poor nations show the swollen belly, patchy skin, and other symptoms of protein deprivation because a diet consisting mostly of cereal grains often lacks essential amino acids. Consequently, their bodies draw proteins from their own tissues, dismantling them and using their essential amino acids for new protein synthesis. Amino acids present in excess are simply excreted and wasted. When a human being starves, it is usually the lack of protein, rather than the lack of food energy, that leads to death.

VITAMINS AND MINERALS: IMPORTANT NUTRIENTS

While the body requires relatively large amounts of protein and carbohydrate each day, it needs another set of nutritive substances, the vitamins and minerals, in only minute amounts.

Vitamins **Vitamins** are organic compounds needed in small amounts for normal growth and metabolism, but which higher animals cannot synthesize, and in most cases store only in small amounts in their cells. Thus, they must acquire a set of specific vitamins in their daily food.

Nutrition researchers have determined that humans require the 14 vitamins listed in Appendix Table A.1. As noted earlier, vitamins A, D, E, and K are soluble in fat, while the B vitamins and vitamin C are water-soluble. Fat-soluble vitamins pass into the lymphatic capillaries in the digestive tract, from where they eventually reach the general blood circulation. Fat-soluble vitamins tend to be stored in the body's fat tissues; since accumulations can produce serious side effects, nutritionists warn against taking high doses of A, D, E, or K vitamins.

In contrast to fat-soluble vitamins, water-soluble vitamins move from the digestive system directly into the bloodstream. Because amounts beyond the cells' immediate needs are excreted, taking huge amounts of vitamin C, for example, is largely pointless. Any excess beyond the normal daily requirement is filtered out by the kidneys and excreted in the urine soon after it enters the bloodstream.

Minerals The body also needs small amounts of another type of nutrient: **minerals**, specific inorganic chemical elements [see FIGURE 2.2]. *Major minerals* are those elements we need in amounts greater than 0.1 g each day; *minor minerals* are those we need in amounts less than 0.01 g daily [Appendix Table A.2]. An adult's body contains about 2 kg (4.5 lb) of minerals, and of this, about three-quarters is the calcium and phosphorus in bones and teeth. The minerals that give tears, blood, and sweat their salty taste are sodium, potassium, and chlorine. Sulfur is found in many proteins, and magnesium is a constituent of enzymes that, along with calcium, is held in reserve in the bones. The human body contains less than 1 teaspoon of minor minerals, but these elements still play critical roles. Perhaps most important is iron—an essential component of the oxygen-binding pigment hemoglobin. Iodine is a constituent of thyroid hormones, zinc is a common coenzyme, and fluorine is needed for healthy bones and teeth.

A person's need for specific minerals can change over time. A woman, for example, needs more iron before menopause because she loses blood periodically, more calcium after menopause when hormonal changes alter the way the body uses that mineral, and more of both during pregnancy.

Because vitamins and minerals are required in such small quantities, most people in affluent countries get a more than adequate supply simply by eating a well-balanced diet. When poor diet leads to vitamin or mineral deficiencies, however, the consequences can be serious. The long-term absence of vitamin A, for instance, can lead to blindness, while the prolonged lack of B vitamins may lead to convulsions and other neurological disorders.

FOOD AS FUEL: CALORIES COUNT

As residents of a wealthy nation, we have a far greater problem with overeating and obesity than with malnutrition. An estimated 25 percent of teenagers and 50 percent of adults are either **overweight** or **obese** (body weight more than 10 or 20 percent over ideal [FIGURE 28.5]. Obesity has been shown to be a significant contributor to cardiovascular diseases, some forms of diabetes, joint problems, generally poor health, and an overall decrease in the quality of life, including reduced earning power.

While obesity is common in North America, some people become so obsessed with body weight and thinness that it seriously threatens their health. In the disorder called *anorexia nervosa*, a person (typically a middle-class teenage girl) restricts her food intake severely, becomes cadaverously thin, and yet still sees herself as overweight. In a separate disorder called *bulimia*, a person secretly binges on huge helpings of cake, ice cream, cookies, bread, or other foods, then purges herself with self-induced vomiting, laxatives, diuretics, fasting, or vigorous exercise. Both disorders have social, psychological, and probably neurochemical roots and can lead to organ damage and even death if untreated.

Food energy is usually measured in **kilocalories** (kcal), also called **Calories** (Cal). Recall that a calorie is the amount of energy needed to raise the temperature of 1 mL of water 1°C; thus, 1000 cal, or 1 Cal, is the amount of energy it takes to raise the temperature of 1 L of water 1°C. A large apple contains about 100 Cal worth of energy-producing compounds; and jogging 1.6 km (1 mi) burns about 100 Cal of stored energy.

Energy requirements vary with age, sex, body size and activity level. A normally active female college student needs around 1800 to 2000 Cal a day to fuel her total metabolic needs; a male college student needs about 2200 to 2500 Cal. Carbohydrates and protein each provide about 4 Cal/g, while fat provides more than twice as much, or 9 Cal/g.

FIGURE 28.6 provides a revealing look at the caloric values for several common snack foods and the amount of energy that a person must expend in various physical activities to work off that food. A person would have to jog for about 30 minutes, for example, to burn off the calories in a cheeseburger. When an animal's food intake

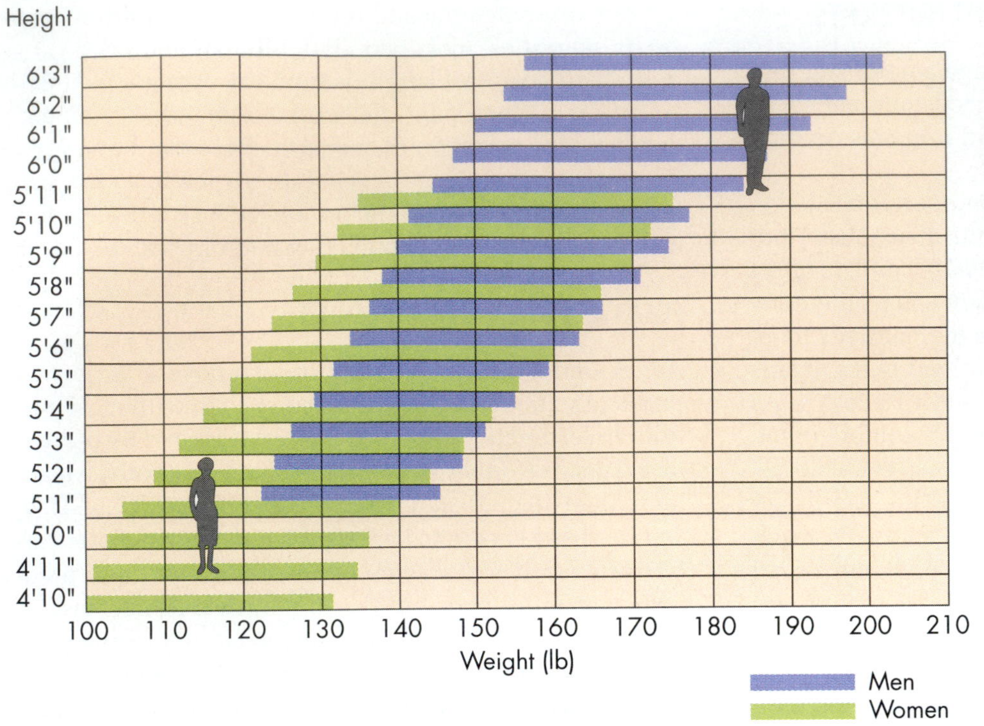

FIGURE 28.5

Ideal-Weight Chart.

Ideal weight depends on frame size, physical condition, and age. At 6 ft and 195 lb, a heavy-boned, muscular football tackle would not be overweight, while a sedentary office worker with a medium frame probably would be. Height and weight are given for people not wearing shoes or other clothing.

exceeds its energy needs, the inevitable result is an increase in the amount of leftover energy stored as body fat. Thus, the secret of weight control lies in taking in only as many kilocalories as the body needs for fuel.

Sound weight-loss programs combine calorie reductions, primarily through the diminished intake of sugar and fat, with increased physical activity. The result is less energy in and more energy out, with the differences made up from the body's fat reserves. Researchers have focused increasingly on the mechanics of weight gain and loss and have concluded that fat cells and total lipid intake help determine our appetites as well as our waistlines [see BOX 28.1].

There are health risks associated with many foods. Some scientists estimate that about a third of all cancer cases are due to cancer-causing substances naturally present in the food we eat. An example is the animal fat in red meat and the increased risk of colon cancer it confers, as mentioned earlier. What can consumers do to protect themselves? Nutritionists recommend eating a low-fat diet, thus reducing the risk of colon, breast, and prostate cancer; eating lots of fruits and vegetables, since the fiber in them moves food through the bowel faster (thus decreasing the time that cancer-causing substances are in contact with tissues), and since the vitamins A and C they contain may inhibit the development of cancers; eating few salt-cured and smoked meats, which contain cancer-causing chemicals; and drinking alcohol in moderation or not at all, thereby decreasing the risk of cancers of the throat, liver, and urinary bladder. Finally, if recent studies on monkeys and other mammals can be generalized to people, a long-term reduction in total calories increases health and well-being as one ages.

➤ CONCEPT CHALLENGE

What are essential amino acids? What makes them essential? What problem do they pose for a strict vegetarian, and how can he or she overcome it?

Digestion

Most animals—whether they eat just leaves like the hoatzin, raw meat like the polar bear, or a chef's salad like a college student in the cafeteria—are consuming food in a form that cells cannot use directly. Foods must be broken down into usable small molecules, which in turn must be absorbed into the bloodstream and distributed to body cells, and unusable wastes must be eliminated.

INTRACELLULAR AND EXTRACELLULAR DIGESTION

Recall from CHAPTER 20 that in the simplest animals—the sponges—some body cells take in tiny whole food

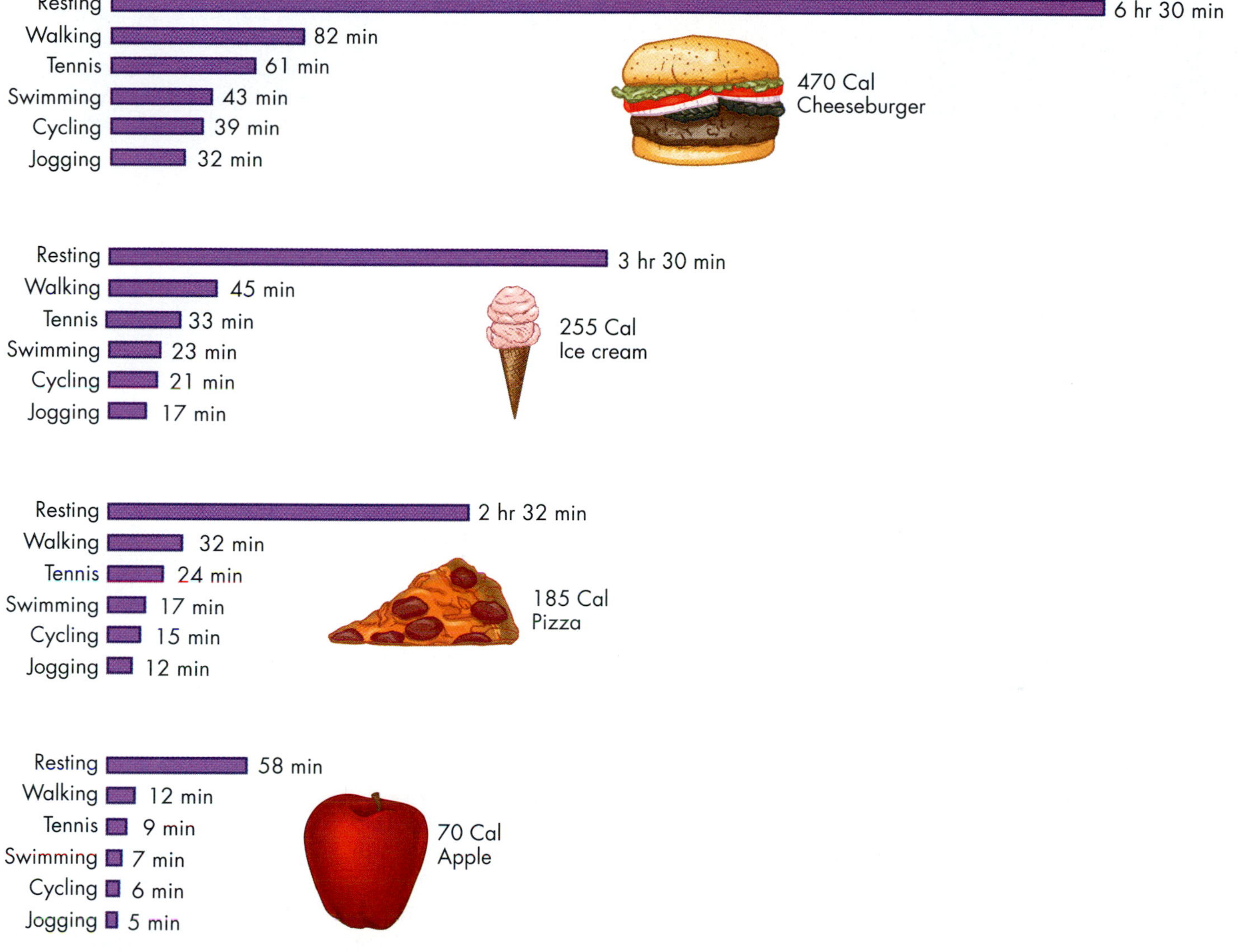

FIGURE 28.6

Food and Exercise Equivalents.

As this chart shows, you can swim off a cheeseburger in 43 minutes, but at rest, your body will take 6 hours and 30 minutes to burn those same 470 Cal.

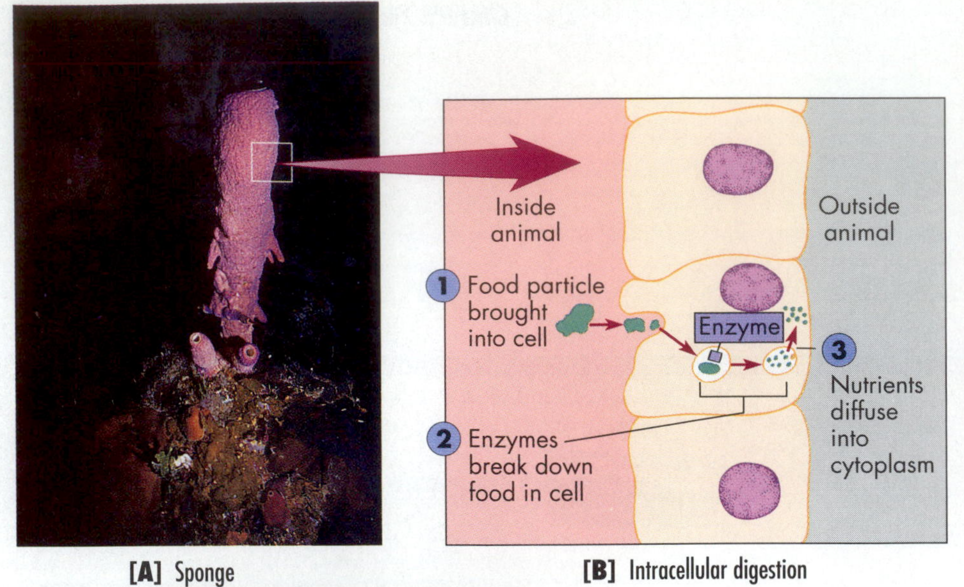

[A] Sponge

[B] Intracellular digestion

FIGURE 28.7

Digestion: Intracellular and Extracellular.

A simple animal like this tube sponge [A] has no gut or similar organ and instead carries out intracellular digestion [B] : Tiny food particles are taken into body wall cells via phagocytosis across the cell membrane [see FIGURE 3.14]. Digestive enzymes within the cell then break the small particles into constituent molecules. A hydra [C] has a central digestive cavity that can take in and digest relatively large pieces of food. In a process called extracellular digestion [D], cells lining the digestive cavity secrete enzymes into that central space. There, the enzymes break down food into constituent nutrients, and the nearby cells then absorb these nutrients.

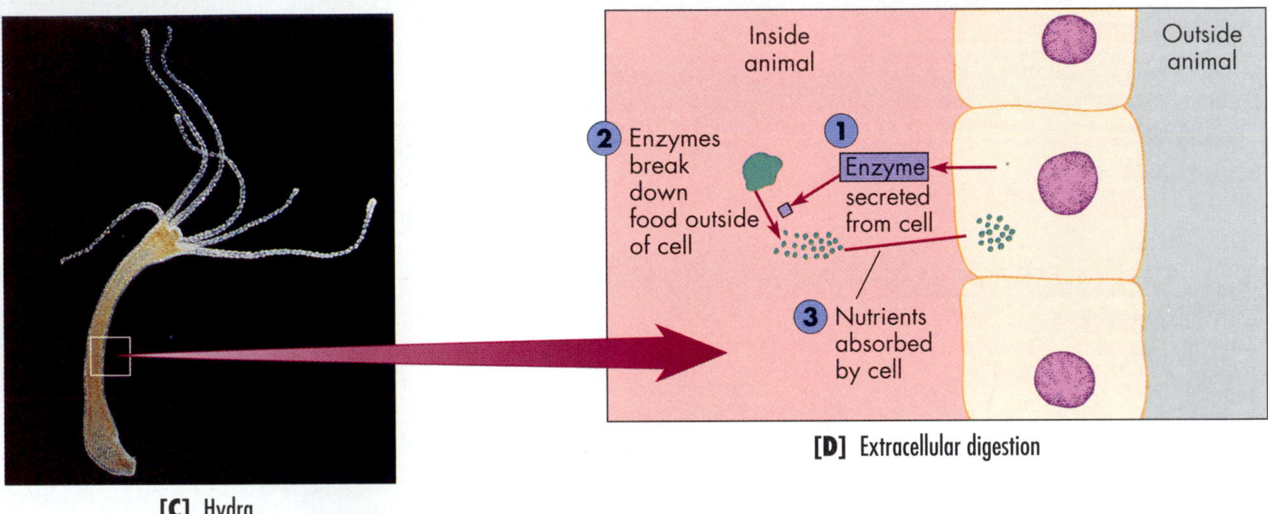

[C] Hydra

[D] Extracellular digestion

particles directly from the water and break them down enzymatically to obtain usable nutrients. This strategy is called **intracellular digestion** [FIGURE 28.7A and B], and it circumvents the need for the mechanical breakdown of food or for a gut or other cavity in which to chemically digest foods. At the same time, however, intracellular digestion puts an upper limit on the size of food particles an animal can take in, and hence an upper limit on the animal's complexity, as well.

Animals exploited the advantages of larger size after the evolution of **extracellular digestion**: the enzymatic breakdown of larger pieces of food into constituent molecules outside of cells, but usually within a special body organ or cavity [FIGURE 28.7C and D]. Nutrients from the broken-down foods pass into body cells lining the organ or cavity, and then can take part in energy metabolism or biosynthesis. Since the vast majority of animals rely on it, let's give extracellular digestion a closer look.

PATTERNS OF EXTRACELLULAR DIGESTION

In a simple animal, such as a cnidarian or flatworm, extracellular digestion takes place in the *gastrovascular cavity*, an internal sac with a single opening through which whole particles of food enter and undigested wastes leave [review FIGURE 21.7]. This system works well for very flat or thin organisms, but has some important limitations: The animal does not break apart food pieces mechanically, which means that digestive enzymes in the saclike gut can work only on the surface of ingested food chunks, and no one area of the gut is specialized for a particular function.

In more complex animals, food enters one end of the digestive tract, the *mouth*, and moves in a single direction through the *gut tube*; wastes exit at the other end of the tube, the *anus*. Between the two ends, a variety of specialized regions and structures perform special diges-

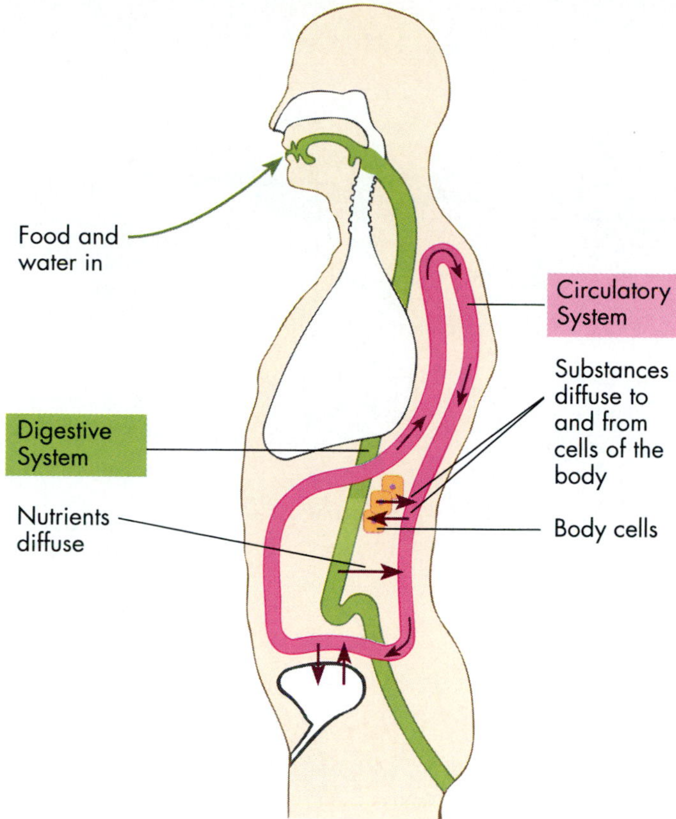

FIGURE 28.8

The Physiological Strategy of the Digestive System.

Materials from the environment are conveyed into the body via a tube which interfaces with the circulatory system. Nutrients are distributed via this system to the rest of the body.

tive roles, with the overall result being an absorption of nutrients into the circulatory system [FIGURE 28.8]. The nutrients are then distributed by the circulatory system to each cell in the body.

The Digestive Tract

In vertebrates, the digestive tract is divided into five main regions: mouth, esophagus, stomach, small intestine, and large intestine [FIGURE 28.9A]. Together, these form the **alimentary canal**, or **gastrointestinal tract**. In addition, nearby **accessory organs**, such as the salivary glands, liver, gallbladder, and pancreas, produce enzymes, bile, and other materials that are funneled into the tract at appropriate times to aid digestion. The alimentary canal and accessory organs accomplish the step-by-step conversion of foods into nutrients circulating in the bloodstream.

Evolution has resulted in a number of variations on the generalized gastrointestinal tract that allow animals to exploit very different foods. Carnivores, such as hyenas and wolves, have short intestinal tracts because meat is relatively easy to digest. Koalas, hoatzins, cows, and rabbits have elaborate guts with several stomachlike pouches containing fermenting bacteria that help break down cellulose in the animals' vegetarian diets [FIGURE 28.10]. As we saw in the introduction, the hoatzin's crop makes available a food resource most other animals can't use. There is a trade-off, however—the crop takes up space that in other birds is occupied by the large keel (sternum or breastbone) to which the flight muscles are attached. This explains why the hoatzin is such a poor flier.

FUNCTIONS OF THE GASTROINTESTINAL TRACT

Within the alimentary canal, a four-step digestive process takes place that extracts nutrients from food. The steps include ingestion and mastication, digestion, absorption, and elimination [FIGURE 28.9B]. First, the animal brings food into its mouth (*ingestion*, Step 1). It then breaks the food into small pieces (*mastication*). In many mammals and other vertebrates, this mastication involves slicing or grinding teeth. In the case of many birds, mastication involves a muscular sac called the *gizzard*, which usually contains stones that grind against each other and pulverize food [FIGURE 28.9B, Step 1]. (Researchers have also found fossilized gizzard stones from dinosaurs.) In the hoatzin, food particles are pulverized inside the crop, which is located midway down the esophagus [FIGURE 28.10B]. Muscles churn the crop's contents, and hornlike ridges grind the food within the sac. Enzymes from salivary glands become mixed with food the animal chews and swallows. Then additional enzymes in the gut break down the large macromolecules of food—in this case, cellulose and plant juices from the leaves—into smaller molecules, or monomers (*digestion*, Step 2).

Next, the monomers pass across the gut wall and into the animal's lymphatic system or bloodstream, which transports the small nutrient molecules to each cell in the body (*absorption*, Step 3). Finally, most of the water in the food is absorbed into the circulation, and undigested residues are eventually eliminated (*elimination*, Step 4).

TISSUES OF THE GASTROINTESTINAL TRACT

For nutrients from food to move out of the alimentary canal and enter the bloodstream, they must pass through several cell layers. (We will trace the passage through these layers shortly.) The innermost layer which faces the cavity of the digestive tract is a mucous membrane, or

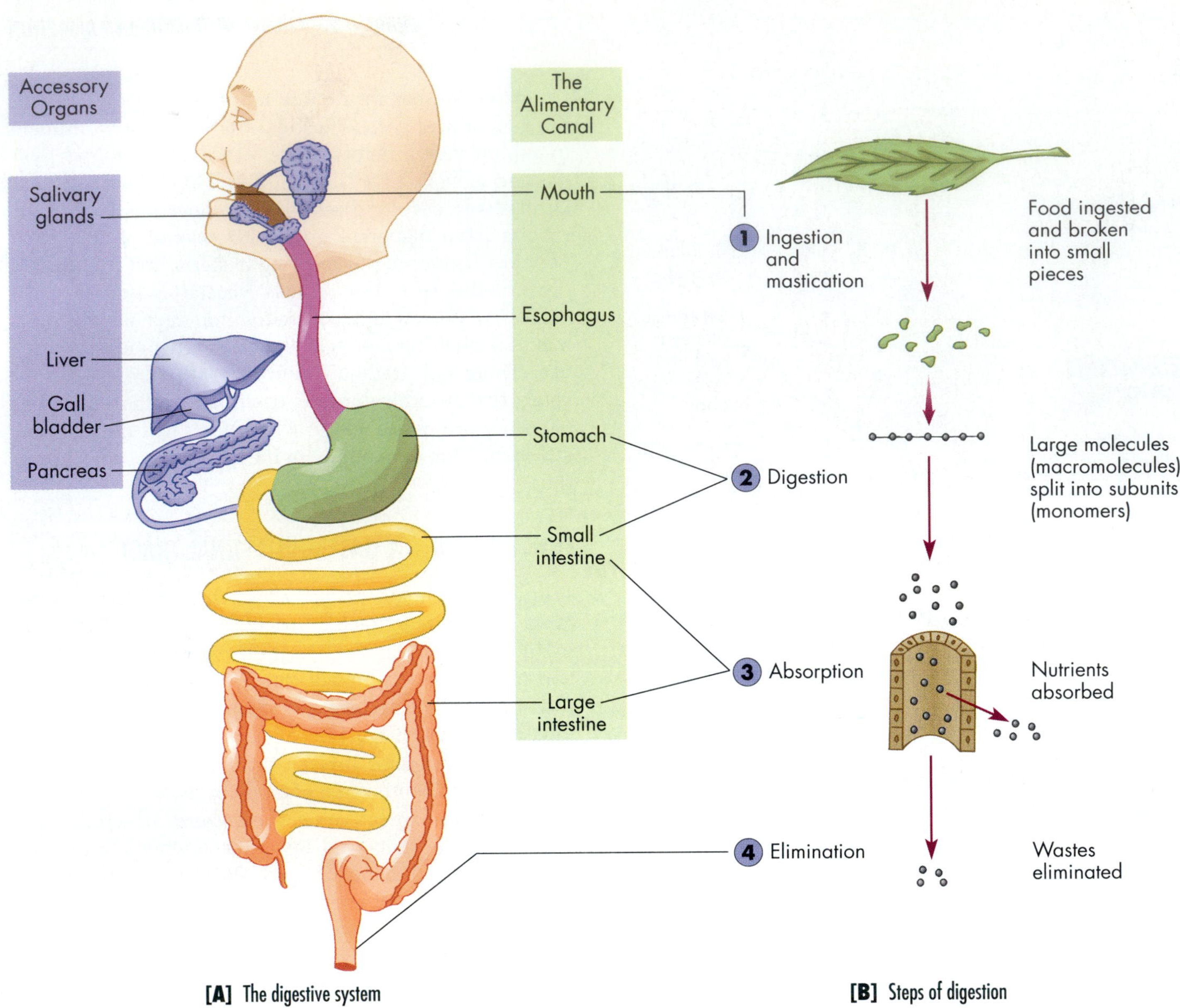

[A] The digestive system

[B] Steps of digestion

FIGURE 28.9

The Gastrointestinal Tract and Steps of Digestion.

The digestive system and steps of digestion. **[A]** A diagrammatic representation of the human digestive system. **[B]** The steps by which the digestive system processes foods.

mucosa [FIGURE 28.11]. Some mucosa cells produce digestive enzymes, while others secrete mucus, forming a thick film coating that protects the inner surface of the tract. This film keeps the stomach from digesting itself and lubricates food passing through.

Just outside the mucosa is the **submucosa**, a connective tissue that is richly supplied with blood and lymph vessels and with nerves. Next is the **muscularis**, a composite layer made of muscle fibers, some encircling the gut and others running longitudinally. The contraction and relaxation of these fibers knead the food, mix it with digestive juices, and propel it along with rhythmic sequential contractions like toothpaste being squeezed through a tube. These rhythmic contractions are known as **peristalsis**. Finally, a thin outermost layer, the **serosa**, forms a band around the other tissues and joins the sheet that attaches the gastrointestinal tract to the inner wall of the body cavity.

The Human Digestive System

The human digestive tract [FIGURE 28.12], from mouth to anus, is approximately 8 m long, or about the height of a

FIGURE 28.10
Evolutionary Adaptations for Nutrient-Poor Diets.
A hoatzin has an enlarged region of the esophagus called a crop which contains microorganisms that ferment nutrient-poor leaves.

[A] A hoatzin

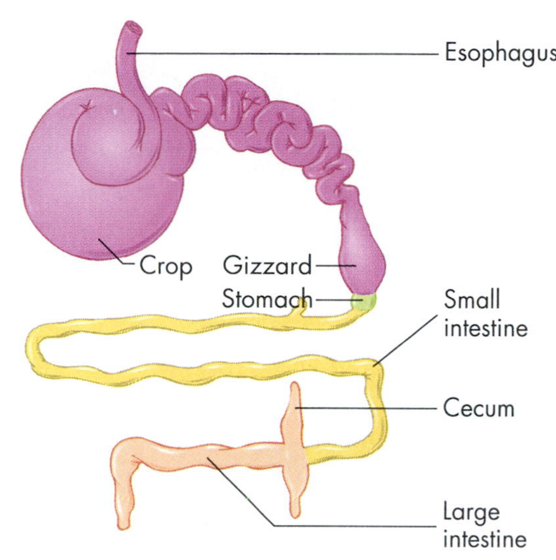

[B] The hoatzin's gastrointestinal tract

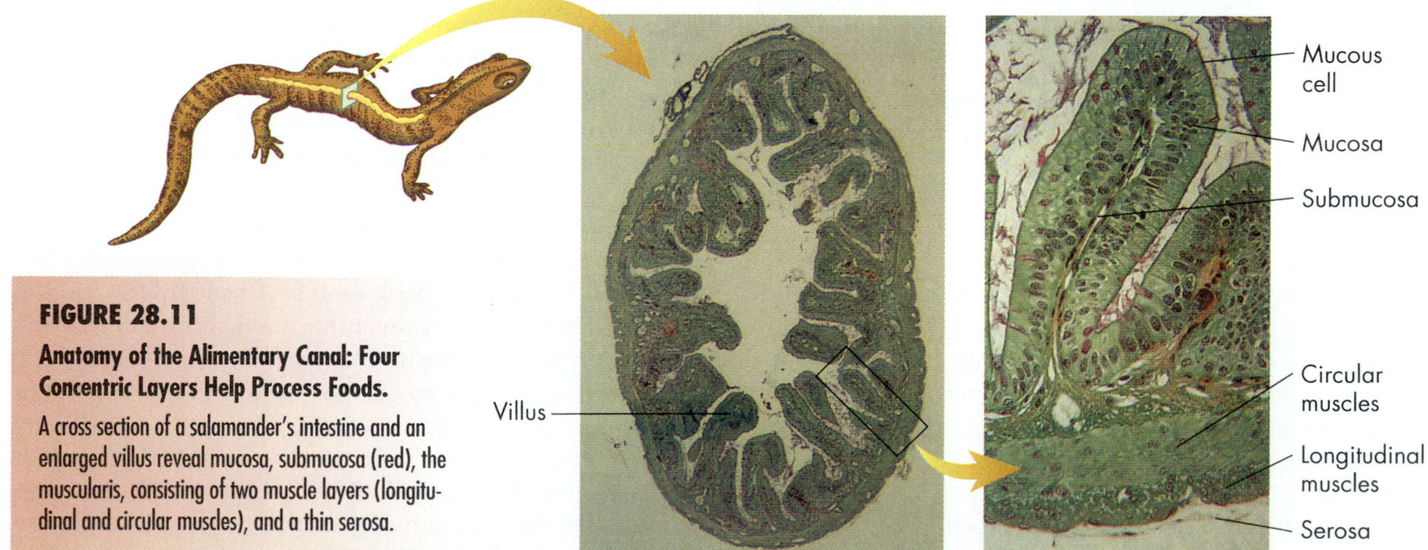

FIGURE 28.11
Anatomy of the Alimentary Canal: Four Concentric Layers Help Process Foods.
A cross section of a salamander's intestine and an enlarged villus reveal mucosa, submucosa (red), the muscularis, consisting of two muscle layers (longitudinal and circular muscles), and a thin serosa.

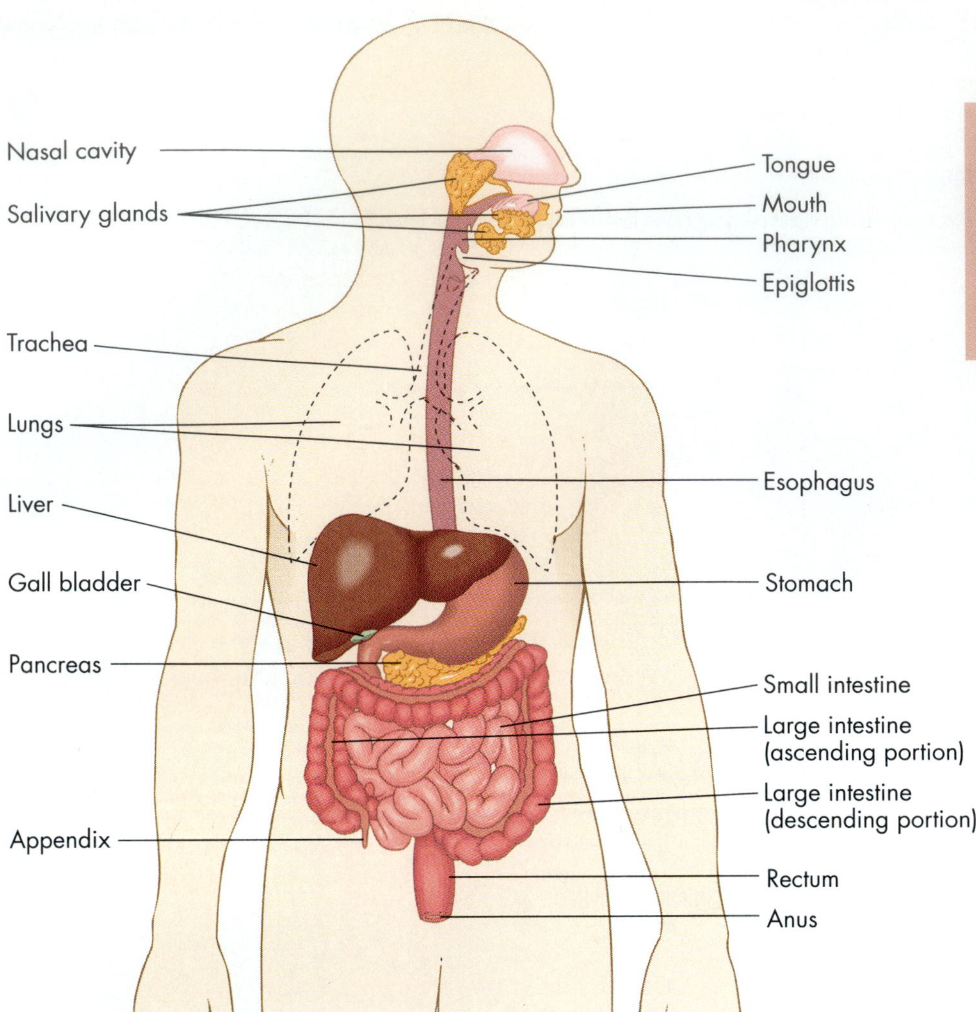

FIGURE 28.12

The Human Digestive System.

The digestive system is essentially one long tube (which lines the mouth, curves around in fold upon fold, then exits at the anus) plus associated organs.

two-story house. All the sandwiches, brownies, apples, milk, and other foods a person eats pass through the central cavity, or *lumen*, of this tube and undergo one digestive process after another. To see in detail how the human alimentary canal liberates the nutrients in foods and transports them across the gut tube layers, we will follow a meal—a turkey sandwich, let's say—throughout each step of its digestive journey.

THE MOUTH, PHARYNX, AND ESOPHAGUS

Your teeth are superbly adapted to the job of cutting, tearing, and grinding the plant and animal tissues you consume as an omnivorous mammal. (By contrast, the hoatzin has no teeth at all. And the koala, another leaf eater, has broad, flat molars that are well suited for grinding the fibrous, leathery leaves it eats all day and night.) Teeth can stand up to regular wear and tear because the enamel covering them is the hardest substance in the body: The enamel on a person's permanent teeth will generate sparks if struck against steel. Working with the teeth is a muscular **tongue** [see FIGURE 28.13], the principal organ of taste, but also, in our species, an organ for forming the sounds of spoken language. When you take a bite of a sandwich, your tongue moves some of the food toward the molars for grinding and some to the incisors for cutting, shaping each small bite into a soft, moist lump, or *bolus*, that can be swallowed easily.

At the same time, each bite is also mixed with clear, watery **saliva**, which is secreted by three large pairs of **salivary glands** that lie in the tissues surrounding the oral cavity. Saliva contains primarily mucus and water, which moisten food particles and help them cling together in a bolus. Saliva also contains small amounts of *amylase*, a digestive enzyme that begins the breakdown of carbohydrates (starch) in the bread, lettuce, and tomato of the sandwich.

Swallowing often begins as a voluntary act once food reaches the back of the mouth. As the tongue pushes the bolus of food up and back against the roof of the mouth, the **soft palate** rises, preventing food from entering the nasal cavities, and the flaplike *epiglottis* moves backward and downward, closing off the opening to the trachea, or windpipe [FIGURE 28.13]. Simultaneously, the larynx (voice box) moves up, and muscle contractions in the throat push food or liquid into the **esophagus**, the pipeline to the stomach. If you place your fingers lightly on either side of your throat just below your chin as you

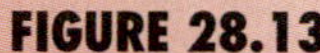

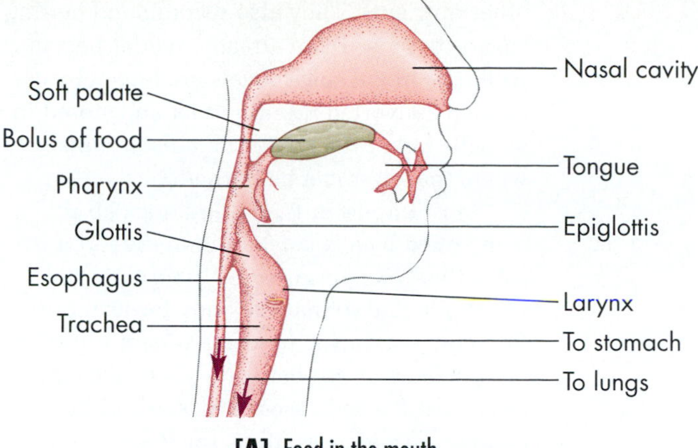

[A] Food in the mouth

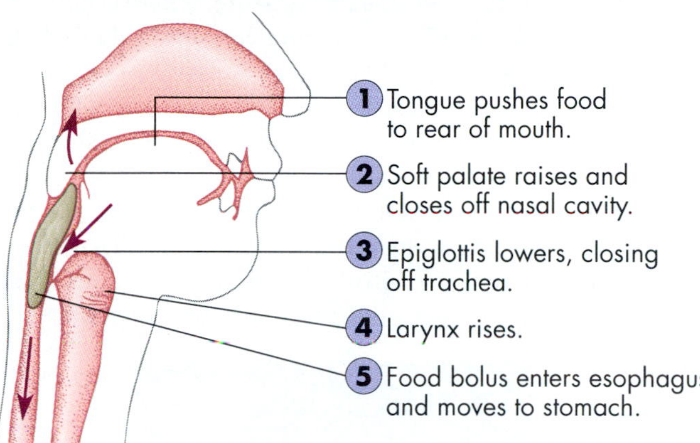

[B] Swallowing

FIGURE 28.13

Swallowing Reflex.

[A] Once the tongue shifts the bolus of moistened food to the back of the mouth, the pharynx is stimulated to a reflexive swallowing action.
[B] Swallowing involves the coordinated movements of several parts.

swallow, you can feel the larynx move up and forward, opening the esophagus, which receives food.

To enter the stomach, the food bolus must pass through a **sphincter** (SFINGK-ter), a ring of muscle located at the junction of the stomach and esophagus [FIGURE 28.14A]. This muscular ring, known as the *cardiac sphincter*, usually remains tightly contracted, like the drawstring on a purse, preventing the stomach's contents from backing up into the esophagus. The gastric juices of the stomach are so acidic that they could severely damage the esophagus, which lacks the stomach's heavy mucous lining. People often experience "heartburn" when the stomach is very full; this burning sensation is due to small amounts of stomach acid seeping out past the cardiac sphincter into the unprotected esophagus. Despite its name, heartburn is solely a digestive condition unrelated to the heart.

THE STOMACH

Swallowing causes muscles in the esophagus to begin the contractions of peristalsis, and enables the cardiac sphincter to open; subsequently, a bite of turkey sandwich passes into the elastic, J-shaped bag called the **stomach** [FIGURE 28.14]. The stomach stores food for later processing—an adaptation that enables large animals to take larger meals, and hence feed less often. The average human stomach can comfortably hold about 1 L (a little over 1 qt), but it can stretch to accommodate much larger capacities. When the stomach is full, waves of peristalsis in the muscular stomach wall [see FIGURE 28.11] churn and mix the stomach contents with an acid bath of gastric juice secreted by glands in the stomach wall.

FIGURE 28.14

Function of the Stomach: Churning, Secretion, and Initial Food Breakdown.

A food bolus [A] pushes on the cardiac sphincter, which opens [B], allowing the food to enter the stomach. [C] Muscle contractions moving in peristaltic waves across the stomach mix the food with digestive enzymes and stomach acid, forming chyme. [D] The pyloric sphincter then opens in repeated brief bursts that allow small amounts of chyme to be squirted into the duodenum (upper part of the small intestine), where it is digested further.

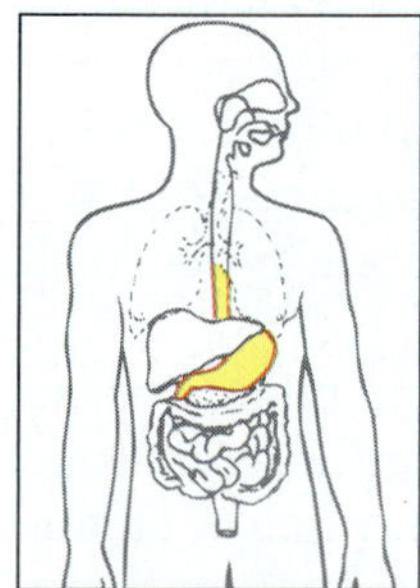

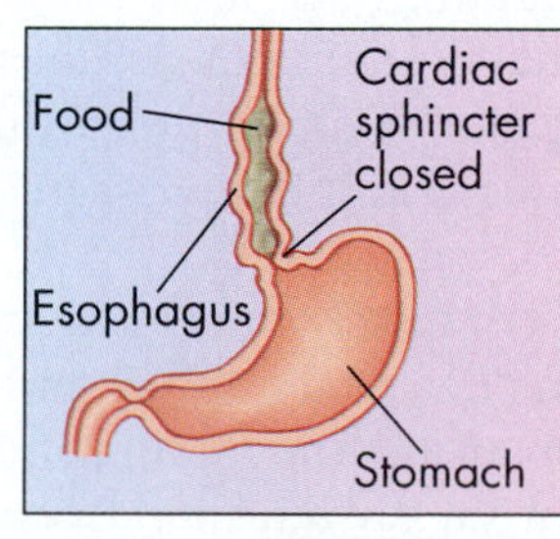

[A]

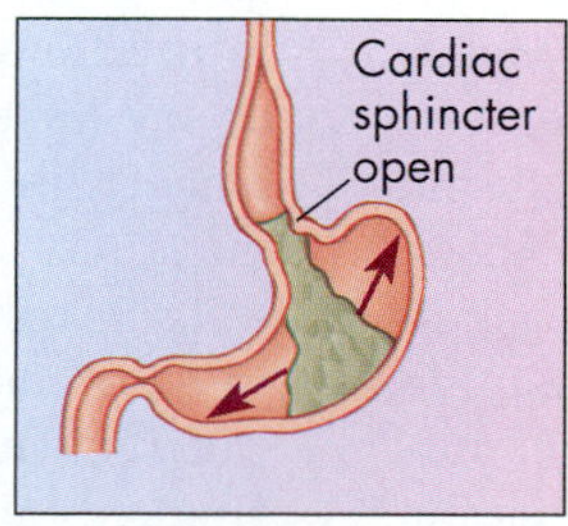

[B]

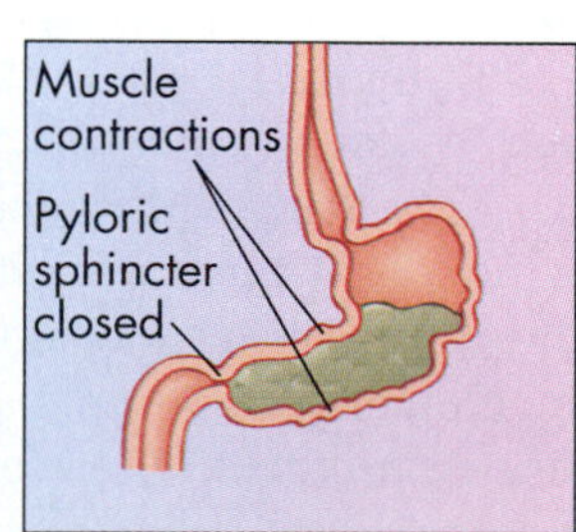

[C]

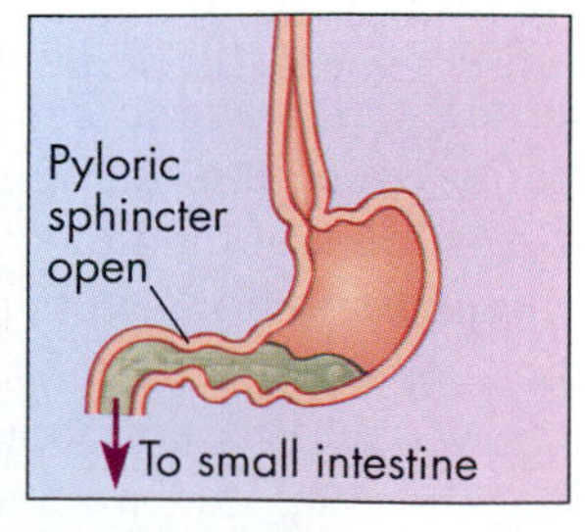

[D]

box 28.1
Biology Applied

The Dieter's Dilemma: Fat, Set Point, and Healthy Weight

Despite their best efforts, fully 95 percent of all people who diet regain all the weight they lose and more. Why is it so difficult to lose weight and keep it off? It appears that people, like many animals, have a fixed set point, a level of fat storage and body weight that is genetically determined and difficult—but not impossible—to alter.

This idea suggests that some people have naturally high set points (above ideal weights) while others have low ones (at or below ideal weights) and that the set point is based on the number and size of fat storage cells [FIGURE 1]. Once gained, fat cells appear never to be lost; they merely increase or decrease in size by storing more or less fat, depending on dietary excesses.

Significantly, a person's fat cells tend to remain a given size and to return to that original size soon after a diet ends. Fat cells act as if they had a mind of their own, and in fact, they do appear to communicate with the brain. They seem to signal any drop in lipid stores, trigger increased appetite and eating behavior to compensate, and initiate a change in metabolic rate so that the body uses its calories more efficiently—all as if to defend the fat cell's genetically determined size.

It also appears that the smell or taste of fatty food can raise the set point. This is perhaps an evolutionary adaptation, allowing animals to take advantage of energy-rich resources when they find them. American consumers, of course, find fat- and sugar-laden foods at practically every street corner.

Given this discouraging picture, how can one turn on fat-burning enzymes, reduce the size of fat cells—thus lowering the set point—and cut body weight to an ideal level? Nutritionists recommend a low-fat diet to cut the amount of lipids taken in, since they are converted to body fat more easily than the other nutrients. They also recommend cutting sugars, because in an already overfat person, when the blood sugar levels are high, glucose is often converted rapidly to fat and stored in fat cells rather than entering muscle cells where the sugar can be burned.

Most importantly, dieting must be accompanied by a consistent increase in physical activity. Exercise seems to turn up the metabolic rate so that the body burns more fat—not just during exercise sessions, but for all the hours at rest between bouts of exercise. Exercise decreases fat tissue and increases muscle mass; thus, the body looks and feels trimmer. Finally, moderate daily exercise reduces the appetite, whereas both fasting (such as observing very-low-calorie diets) and inactivity increase appetite.

Considering all these health benefits, regular aerobic exercise (four to five sessions per week of 20 minutes or more) is probably the major reason why people who jog, swim, bicycle, or do another form of aerobics find it easier to control their appetites, reduce their set points, and maintain ideal weight.

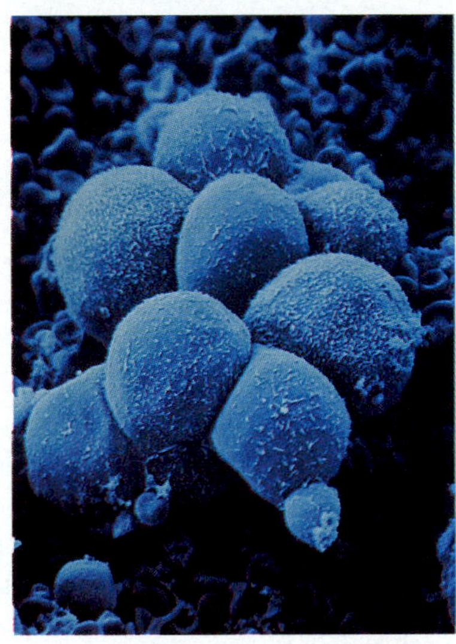

FIGURE 1
Corpulent Fat Cells.
Fat cells inside an artery look huge and swollen compared to the small red blood cells visible in the background.

Gastric juice is a mixture of water, hydrochloric acid (HCl), mucus, and **pepsinogen**, a precursor to the protein-cleaving enzyme **pepsin**. The hydrochloric acid makes gastric juice acidic enough to kill off most bacteria or fungi contaminating foods. This acid contributes to the breakdown of food pieces into constituent protein fibers, fat globules, and so on, and it also allows pepsinogen to be converted into pepsin. Pepsin cleaves the peptide bonds that link amino acids in proteins. Therefore, during the time that protein-containing foods (such as the turkey in a sandwich) are in the stomach, they are partially digested to short polypeptide segments. Although digestion of the starch in the bread of the sandwich begins in the mouth, once food enters the stomach, very little additional digestion of starches and other carbohydrates (or of fats, such as those in mayonnaise) takes place. This must await the passage of foods to the small intestine.

The result of the chemical activity and mixing waves in the stomach is a pasty, milky, and highly acidic soup called **chyme** (KIME). The mixing and churning of chyme gradually move it toward the lower stomach, where another sphincter, the *pyloric sphincter*, controls the opening to the small intestine [FIGURE 28.14]. As the chyme is pushed against that opening, the sphincter relaxes just long enough for small, carefully regulated "doses"—on average, about a teaspoonful every three seconds after a meal—to be squirted into the small intestine. This allows further digestion and absorption to take place efficiently in that portion of the gastrointestinal tract. It usually takes between one and four hours for a meal to be processed in the stomach and delivered, spoonful by spoonful, to the small intestine—less time for a high-carbohydrate meal, more time for a fatty meal.

PANCREAS, LIVER, AND GALLBLADDER

Most of the chemical digestion of food, as well as the absorption of nutrients, takes place in the small intestine. However, this portion of the gut tube does not accomplish these tasks alone; three accessory organs—the pancreas, liver, and gallbladder—dump in substances that aid digestion and absorption [see TABLE 28.1 on page 652].

The Pancreas and Its Digestive Enzymes The **pancreas** is a narrow, lumpy organ situated near the junction of the stomach and small intestine [see FIGURE 28.12]. It produces a host of digestive enzymes and secretes them into the small intestine, where food digestion continues after chyme leaves the stomach. These pancreatic enzymes include *proteases*, which break down proteins; *lipases*, which digest fats; and enzymes such as amylase, which break down carbohydrates. In addition, the pancreas secretes bicarbonate ions (the main ingredient in indigestion remedies), which act as a buffer, neutralizing acid entering the small intestine from the stomach. This neutralization is vital, because pancreatic enzymes, unlike stomach enzymes, cannot function in an acidic environment, and the acid could damage the small intestine. TABLE 28.1 lists the major pancreatic enzymes and their functions.

Some pancreatic cells secrete the hormones *insulin* and *glucagon* directly into the bloodstream. These hormones help keep the levels of blood glucose within a certain range [see CHAPTER 30].

The Multipurpose Liver and the Gallbladder The smooth, irregularly lobed, roughly hemispherical **liver** is the largest gland in the human body [see FIGURE 28.12]. It weighs about 2 kg (4.5 lb) in an adult and performs biological tasks as diverse as destroying aging red blood cells, storing glycogen, and dispersing glucose to the bloodstream as circulating levels of the sugar drop. One of the liver's most important functions is to produce *bile salts*—molecules of modified cholesterol that act like detergents and break up fat droplets in the small intestine.

Bile salts travel through the bile duct to the **gallbladder** [FIGURE 28.15, Step 1], a small, pear-shaped sac on the underside of the liver, where they are stored as **bile**, a yellow-green liquid containing the bile salts, pigments from the breakdown of red blood cells, and other substances. The presence of fats in the small intestine stimulates the gallbladder to squeeze drops of bile down a duct into the small intestine (Step 2).

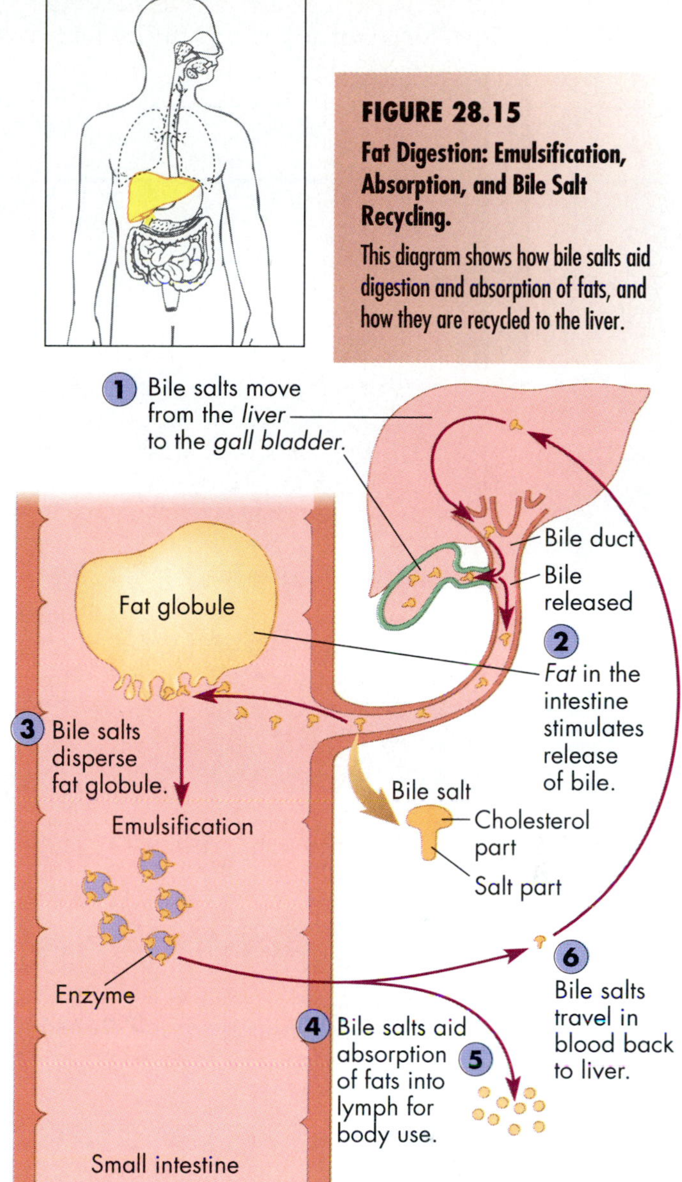

FIGURE 28.15
Fat Digestion: Emulsification, Absorption, and Bile Salt Recycling.
This diagram shows how bile salts aid digestion and absorption of fats, and how they are recycled to the liver.

Just as detergent mixed with cooking oil and water will cause the oil to disperse into thousands of tiny droplets, bile salts emulsify fat; that is, they disperse fat globules into tiny fat droplets (Step 3), providing a larger surface area for the action of lipases, fat-digesting enzymes, and a more rapid breakdown of lipids into usable constituents.

Bile salts also aid absorption of fully digested fat molecules across the lining of the small intestine (Step 4) and into the lymph (Step 5). Bile salts are finally recycled back to the liver (Step 6). Doctors sometimes give people with high levels of cholesterol in the blood a compound called cholestyramine, which binds to bile salts and

TABLE 28.1 Functions and Secretions of the Digestive Tube and Accessory Organs in Humans

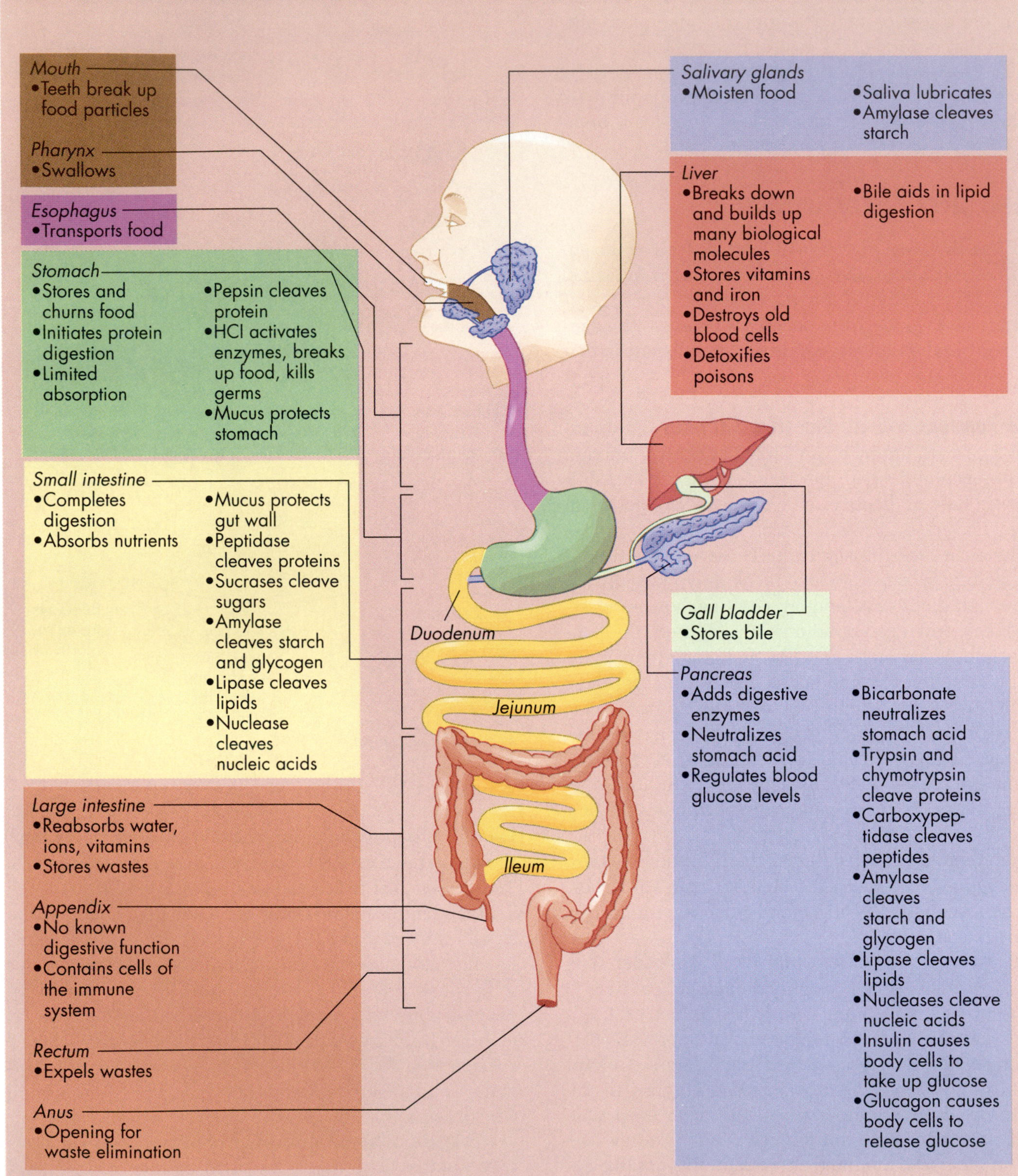

passes down the intestinal tract without being absorbed. This prevents the recycling of cholesterol from bile salts and thus decreases the amount of cholesterol in the blood.

The liver, besides synthesizing bile, picks up, stores, and sometimes synthesizes amino acids, glucose, and glycogen, and stores vitamins and other compounds that cells need to function normally. Certain liver cells also contain special enzymes that can detoxify poisons. For example, the liver transforms molecules of ammonia—a toxic, nitrogen-containing waste created by the breakdown of amino acids—into urea, which is less toxic and is excreted in urine. The liver also detoxifies alcohol; but if the liver is overloaded with the drug year after year, it can become damaged and irreversibly scarred, producing a potentially fatal condition called *cirrhosis* (sih-ROH-siss).

THE SMALL INTESTINE: DIGESTION AND ABSORPTION

The **small intestine** is a remarkable coiled tube about 6 m (20 ft) long. It is where carbohydrates and fats are largely digested and where protein digestion is completed to the stage at which nutrients can be absorbed into the blood. The small intestine begins just below the stomach and has three main regions along its length: the upper section, or *duodenum* (doo-oh-DEE-nuhm), the central *jejunum*, and the remainder, the *ileum* [see TABLE 28.1].

The small intestine is a marvel of compact biological engineering. If the surface of its inner lining were a smooth tube like a garden hose, there would be a relatively small surface area for digestion and absorption. Instead, however, that inner lining is so convoluted that it houses a huge absorptive surface. The intestinal lining is pleated into large numbers of folds [FIGURE 28.16A and B], and each fold is covered with fingerlike extensions known as **villi** (singular, **villus**), which project into the lumen and come into contact with chyme [FIGURE 28.16C]. Further, cells on the outer layer of each villus are carpeted with **microvilli**, microscopic, brushlike projections of the plasma membrane [FIGURE 28.16D]. The com-

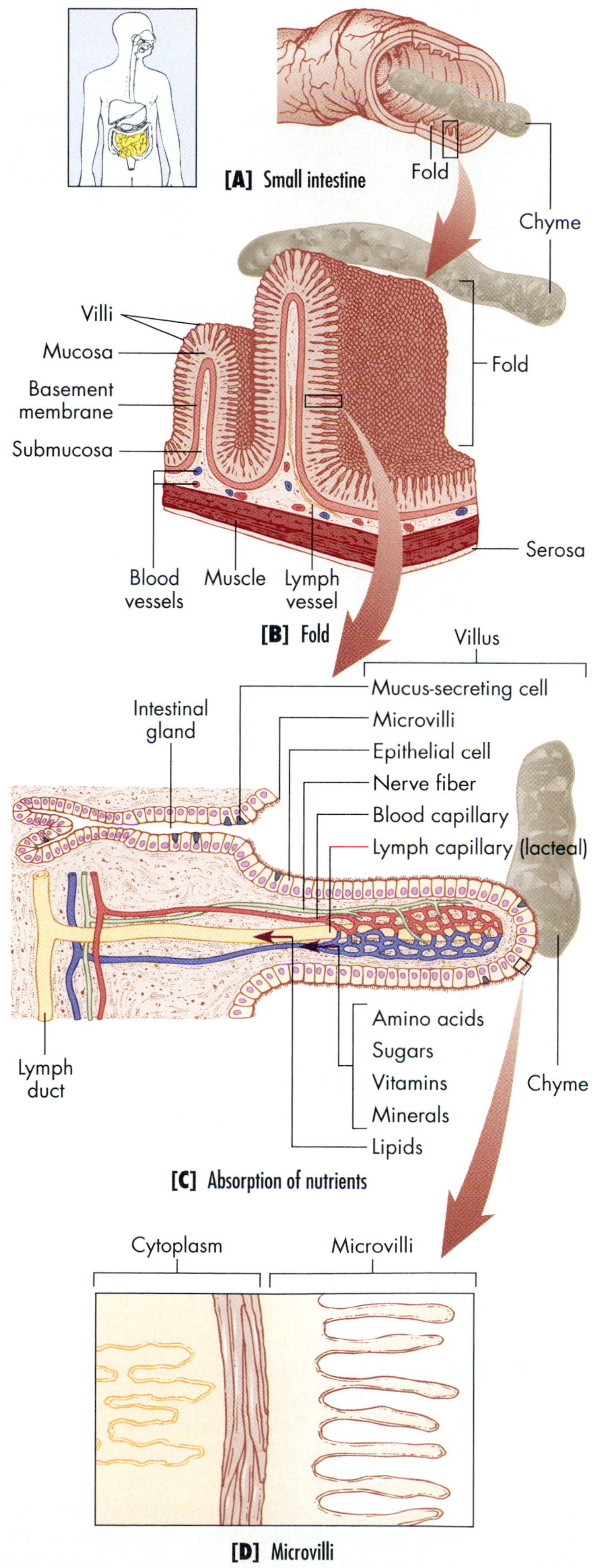

FIGURE 28.16

Highly Absorptive Lining of the Small Intestine: Folds Bearing "Fingers" Topped with "Brushes."

The small intestine [A] is pleated into large numbers of folds [B] covered with villi. Each villus [C] encompasses blood and lymph capillaries that carry away nutrients absorbed through the brushlike carpet of microvilli, with their enormous combined surface area. Lipids enter the lymph, and other nutrients enter the blood. [D] The surface of cells that line the intestinal cavity contain microvilli, like the hairs of a little brush; these further increase the surface area of the gut.

bination of folds, villi, and microvilli creates a total surface area the size of a tennis court.

By the time a turkey sandwich is reduced to chyme and reaches the small intestine, it contains partially digested carbohydrates and proteins, as well as undigested fat. Here, enzymes secreted by the intestine and the pancreas combine with bicarbonate ions and bile salts and interact, gradually completing the digestion of the carbohydrates in bread into simple sugars, the proteins in turkey meat into amino acids, and the fats in mayonnaise into fatty acids and glycerol. These nutrient molecules are small enough to move across the plasma membranes of the microvilli and enter the intestinal cells. The amino acids and sugars then pass into blood capillaries along with most of the available vitamins and minerals from the tomato slice and lettuce, while lipids enter lymph capillaries [see FIGURE 28.16C]. The remaining contents of the small intestine, including the small amount of indigestible roughage in the lettuce and tomato, pass into the large intestine.

THE LARGE INTESTINE: SITE OF WATER ABSORPTION

The last 1.2 m (4 ft) of the alimentary canal is the **large intestine**, or **colon**. This section of the gut tube ascends on the right side of the body cavity, cuts across just below the stomach, then descends on the left side and ends in a short tube called the **rectum** [see FIGURE 28.12 and TABLE 28.1]. Extending from the colon near its junction with the small intestine is the *appendix*. In humans, this finger-shaped organ has no known digestive function, although it does have some immune activity and may fight infections in the gut. When the appendix becomes inflamed, it leads to the medical emergency called *appendicitis*.

The colon has two main functions: to absorb water, ions, and vitamins from the chyme and to store the semisolid undigested wastes, or **feces**, until they are excreted. The colon is twice as wide as the small intestine, but its walls lack the many folds and villi of the small intestine. The smoother surface has a lower surface area for absorption, but it also presents less resistance to the movement of chyme through the tube.

Each day, about 0.5 L (about 1 pt) of chyme (now minus the nutrients absorbed in the small intestine) reaches the colon along with 2 to 3 L (2 to 3 qt) of water, some from food and the rest secreted by the stomach and intestines themselves. The body cannot afford to lose this much water, however, and much of the water is reabsorbed as the chyme moves slowly through the colon over a period of 12 to 36 hours. This absorption gradually transforms the chyme from fluid to semisolid, and the wastes (including undigestible cellulose fiber from the bread, lettuce, and tomato of the turkey sandwich we started with) are stored until their pressure against the colon wall triggers a *bowel movement* (defecation), the muscular expulsion of wastes through the rectum and out the anus.

Digestion is never 100 percent efficient, and some nutrients invariably pass into the colon from the small intestine. A variety of bacteria, including *Escherichia coli* and *Lactobacillus* and *Streptococcus* species, reside in the human intestine and live on these remaining nutrients; in the process, they produce a number of vitamins, including thiamine (vitamin B_1), riboflavin (vitamin B_2), vitamin B_{12}, and vitamin K. The colon absorbs these vitamins along with fluids. Many animals that ingest large amounts of cellulose, including termites, cows, koalas, and hoatzins, also benefit from the metabolic activities of "live-in" microorganisms. Without their own types of fermenting chambers and enclosed bacteria, these animals could never extract enough nutrients and calories from wood, grass, and leaves to maintain life.

➤ CONCEPT CHALLENGE

What are the functions of the stomach, small intestine, and large intestine, respectively?

Coordination of Digestion

The passage of materials through the human digestive system—from mouth to esophagus, stomach, and the small and large intestines—results in the physical and chemical dismantling of food and the uptake of nutrients into the body. The various processes do not occur at random, however, but occur on a precise schedule that depends on regulatory mechanisms operating in local regions and throughout the digestive system. The close control of digestion is achieved by the body's two great coordinators: the *nervous system* and the *hormonal system*. Nerves throughout the digestive tract allow communication between "downstream" and "upstream" regions so that the propulsion of food can be speeded up or slowed down appropriately. Nervous activity in the brain and spinal cord can also speed up or slow down digestive functions. For example, the sight, taste, smell, or even thought of food can cause signals from higher brain centers to travel via nerves to the salivary glands, causing them to secrete saliva, and to secretory glands in the stomach lining, inducing gastric juices—hydrochloric acid and pepsinogen—to begin flowing into the stomach [FIGURE 28.17]. Food arriving in the stomach and pushing against the stomach wall can trigger the same response.

Food in the stomach also has another effect: It lowers the overall acidity of the contents. This triggers cells in

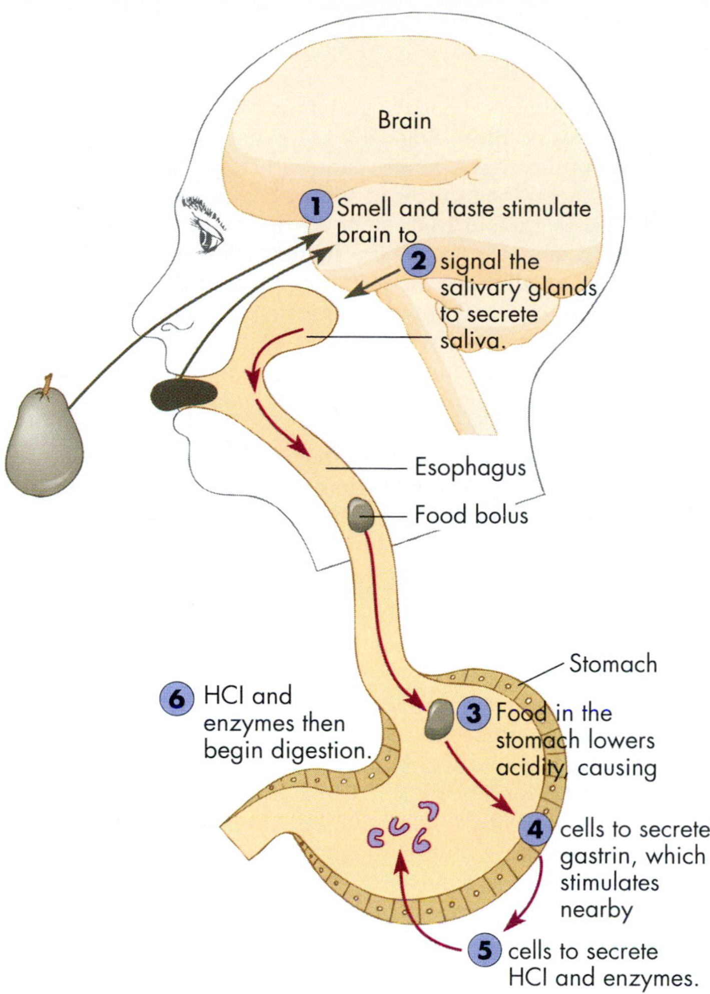

FIGURE 28.17

Nerves and Hormones Coordinate Digestion.

Your body begins the digestive process when food is available—not when the stomach is empty. As the text explains, the control of digestion involves the senses, signals from the brain, and hormones secreted by the stomach and intestines.

the stomach lining to secrete a hormone called **gastrin**, which stimulates nearby secretory cells to produce more hydrochloric acid [see FIGURE 28.17]. This raises the acidity (lowers the pH) of gastric juice, and at the same time helps speed the breakdown of the food. When the pH drops to about 2, the secretion of gastrin stops, as does the secretion of acid. Thus, a negative feedback loop regulates stomach acidity, and food literally helps to stimulate its own digestion. People who are emotionally stressed for prolonged periods are susceptible to *ulcers*, craterlike sores in the mucosa of the stomach or small intestine. Apparently, ulcer victims have decreased secretions of the mucus protecting the stomach lining and/or excessive gastric secretions—conditions that allow small portions of the stomach or intestinal lining to be literally digested away. Recent research suggests that certain kinds of bacteria can also cause ulcers.

Several other hormones help coordinate the timing and amount of enzyme secretion with the presence of food. For example, the peptide **secretin** from the small intestine causes the pancreas to secrete bicarbonate ion (HCO_3^-), which then enters the small intestine and neutralizes the stomach acid. Partially digested proteins cause the small intestine to release a second hormone, **cholecystokinin** (CCK; KOH-luh-siss-toe-kine-in). This hormone triggers the pancreas to release protein-digesting enzymes. Cholecystokinin also works on regulatory centers in the brain and produces the sensation of being full. This is why nutritionists often advise dieters to eat the protein foods in a meal first, so that they will feel satisfied sooner.

Working together, the body's rapid- and slow-acting control agents—nerves and hormones—fine-tune the secretion of digestive juices. As a result, enzymes and ions are instantly available to break food down into nutrients for the body's trillions of cells, and yet those powerful agents are present in the alimentary canal only when needed—when the animal has consumed food.

➤ CONCEPT CHALLENGE

You are a biologist trying to develop a weight-control therapy, and decide to formulate a pill that causes people to feel full early in a meal so they eat less. What hormone or other substance might you try to begin your investigations?

Connections

Digestion is the evolutionary solution to every animal's need for cell nourishment; as a result, the organism can continue to survive day to day and perhaps eventually reproduce. Intense competition for nutrients favors individuals who can effectively exploit their environments. Thus, through natural selection, mouthparts, digestive organs, and behaviors evolved that gave each species an advantage in obtaining its own, often highly specialized "slice of the pie." The hoatzin, with its leafy diet that would starve or poison a human, is a prime example of that process at work.

This chapter has focused mainly on the ways that the digestive system extracts nutrients from food and absorbs them for use by the body's vast collection of cells. In CHAPTER 29, we will consider a closely related system that is directly affected by the foods and liquids an animal consumes: the kidneys or other excretory organs, which regulate the balance of water, salts, and wastes in the blood and body fluids.

KEY TERMS

accessory organ, 645
alimentary canal, 645
bile, 651
chyme, 650
colon, 654
digestion, 642
esophagus, 648
essential amino acid, 640
essential fatty acid, 639
extracellular digestion, 644
feces, 654
gallbladder, 651
intracellular digestion, 644
liver, 651
microvillus, 653
mineral, 641
mucosa, 646
nutrient, 638
nutrition, 638
obesity, 641
pancreas, 651
peristalsis, 646
rectum, 654
saliva, 648
salivary gland, 648
serosa, 646
soft palate, 648
sphincter, 649
stomach, 649
submucosa, 646
tongue, 648
villus, 653
vitamin, 641

HIGHLIGHTS IN REVIEW

1 Animals obtain their energy and materials from the environment by consuming other organisms or their parts and products. Nutrients include carbohydrates, lipids, amino acids, vitamins, and minerals. Proper nutrition is a critical component of good health.

- a] Nutrients provide sources of energy and materials for body maintenance, growth, and reproduction.
- b] Most animals store energy as fats (lipids), which contain about twice as much energy per gram as carbohydrates or proteins, but require less water to store. Lipids in the human body and diet are primarily triglycerides, phospholipids, and sterols, including cholesterol.
- c] In the human body, both the main structural molecules as well as the enzymes, which mediate cellular activity, are proteins. Proteins contain 20 different amino acids, of which 8, the essential amino acids, must be obtained through the diet.
- d] The body requires but cannot synthesize small amounts of vitamins; many vitamins play essential roles in energy metabolism. Vitamin deficiencies can have severe consequences, but are rare in developed nations.
- e] The body also requires small amounts of inorganic minerals such as calcium, phosphorus, sodium, and iron.
- f] The energy in foods is usually measured in kilocalories (Calories). Fats contain the most Calories per gram because they can be more completely oxidized. When a person's energy intake exceeds expenditure, the body stores the excess as fat.

2 The gastrointestinal tract breaks down foods physically and chemically into small molecules (digestion) and delivers these nutrients to the circulatory system (absorption), which distributes nutrients throughout the body.

- a] An animal's digestive system is closely tied anatomically and physiologically to its diet. All animals more complex than sponges carry out extracellular digestion of larger food pieces in a body cavity.
- b] There are four basic steps in digestion: (1) ingestion and mastication, the mechanical breakdown of food into small pieces by teeth or gizzards; (2) digestion, the chemical breakdown of large macromolecules (proteins, carbohydrates, fats, and nucleic acids) into their small subunits (amino acids, simple sugars, fatty acids, and nucleotides) by hydrochloric acid and enzymes in the stomach and by enzymes in the small intestine; (3) absorption, the uptake of the small subunits and other nutrients from the gastrointestinal tract into the bloodstream; and (4) elimination, the removal of indigestible wastes.

3 Different parts of the digestive system perform different tasks. Digestive systems vary among different species because they have become adapted during evolution to different food sources.

- a] The human digestive tract has four layers: the mucosa (the innermost layer), the submucosa, the muscularis which includes circular and longitudinal muscles, and the serosa. The muscle layer generates movement that mechanically mixes food with digestive juices and propels the solution along via rhythmic contractions known as peristalsis.
- b] Amylase secreted by the salivary glands in the mouth begins the chemical digestion of carbohydrates. Gastric juices in the stomach containing hydrochloric acid and the enzyme pepsin begin the breakdown of proteins. The pancreas produces bicarbonate ions, which help neutralize the acidic chyme entering the small intestine, and enzymes that help complete chemical digestion.
- c] The liver produces bile salts, which emulsify the fats in chyme for easier digestion. Bile is stored in the gallbladder. The liver also detoxifies various poisons found in the blood.
- d] The lining of the small intestine has a huge surface area as a result of folds, villi, and microvilli. Nutrient molecules cross microvilli into intestinal cells, then enter blood or lymph capillaries.
- e] The large intestine, or colon, slowly reabsorbs water from chyme and stores the indigestible residues for later elimination. A variety of bacteria live in the colon and there produce several kinds of vitamins, which the human body can absorb and use.
- f] Animals digest starches and many other carbohydrates into monosaccharides, but cannot themselves digest cellulose fibers, which instead serve as roughage. The digestive systems of animals that can survive on grass, leaves, and wood have large fermentation chambers in which microorganisms digest the cellulose.

4 Nervous and hormonal control of the gastrointestinal tract optimizes the efficiency of digestion and nutrient absorption.

- a] Nervous and hormonal signals coordinate the various steps of digestion.
- b] The presence of food in the stomach or small intestine can trigger cells to secrete hormones that stimulate

digestion. Such feedback control ensures that powerful enzymes are present in the alimentary tract only when they are needed.

UNDERSTANDING THE FACTS AND CONCEPTS

In Questions 1–5 match the descriptions with the most appropriate item or items from the following list. Each item in the list may be used once, more than once, or not at all.

a] carbohydrates
b] lipids
c] proteins
d] vitamins
e] minerals

1 Energy-dense nutrient that is required for the absorption of fat-soluble vitamins.

2 Enzymatic, structural, hormonal, and other functions.

3 Nutritional function is almost exclusively as an energy source.

4 Organic enzyme precursors and activators that we require but cannot synthesize.

5 Our cells can synthesize all that we need from its subunits.

For Questions 6–10, match the descriptions with the one most appropriate item from the following list. Each item in the list may be used once, more than once, or not at all.

a] gastrovascular cavity
b] accessory organ
c] gizzard
d] crop
e] gastrointestinal tract

6 A bird's substitute for teeth.

7 Site of extracellular digestion in a saclike, nontubular gut.

8 A one-way gut with mouth and anus.

9 Glandular outpocketings of a gut tube, such as the liver and pancreas.

10 A saclike expansion of the esophagus in birds that can function as a fermentation vat.

As above, use the following list for Questions 11–15.

a] mucosa
b] muscularis
c] cardiac sphincter
d] pyloric sphincter
e] duodenum
f] colon

11 Pulses out chyme into the small intestine after a meal by alternately contracting and relaxing.

12 Lines the lumen of the gut and contains cells that produce digestive enzymes.

13 Guardian of the entry to the stomach.

14 Primary site of water reabsorption from the gut.

15 First section of the small intestine where secretions from the liver and pancreas enter the lumen of the gut.

As above, use the following list for Questions 16–20.

a] gastrin
b] secretin
c] cholecystokinin
d] pepsin
e] bile salts
f] proteases

16 A protein-digesting enzyme that acts in an acid environment.

17 Made from cholesterol, acts as a detergent that breaks up and surrounds droplets of fat.

18 A hormone produced by the stomach that regulates the secretion of the stomach's main enzyme.

19 A hormone produced by the small intestine that acts to produce a feeling of fullness after a meal.

20 A small-intestine hormone that causes the pancreas to secrete an alkaline compound that neutralizes stomach acid.

INTEGRATE AND APPLY WHAT YOU HAVE LEARNED

1 What structures in the digestive system act to physically break down ingested food? Why is this essential for digestion?

2 Can you think of a good reason why the enzymes of the stomach evolved to operate best in a highly acidic environment?

3 Which organs of the human digestive system are the primary sites of digestion and absorption?

4 What vital digestive activities occur in the liver and pancreas?

5 Explain how nerves and hormones coordinate digestion.

ANALYSIS

1 Some people with elevated levels of blood cholesterol take prescribed medications that inhibit the synthesis of cholesterol by the liver or prevent the absorption of cholesterol from the intestine. What might be some of the side effects of such medications that would bear watching by these individuals? Explain your answer(s).

a] Unchecked bleeding due to vitamin K deficiency.
b] Lowered absorption of triglycerides.
c] Variety of skin problems due to B vitamin deficiencies.
d] Changes in the constitution and secretion of bile.
e] More than one of the above.

2 Why don't nursing infants suffer malnutrition? After all, they are only drinking milk, not eating lots of cereal grains, fruits, and vegetables that supply many of an adult's vitamin and mineral needs.

a] Babies are born with all the vitamins and minerals they will need for a while.
b] Infants do not need many vitamins and minerals because they sleep most of the time.
c] Babies do suffer malnutrition, but before the problem becomes serious, they start eating solid food and their problems are corrected.
d] Infants do suffer vitamin and mineral deficiencies. The ones who suffer the most cry the most.
e] Babies do not suffer vitamin and mineral deficiencies because when milk is synthesized, blood-borne vitamins and minerals are transferred from the mother to the milk.

CHAPTER 29

Excretion: Balancing Water and Salt

A SHIP OF THE DESERT

The camel has been called the "ship of the desert," and indeed, this hump-backed, two-toed vegetarian takes to the heat and drought of the planet's arid regions like a tanker takes to the ocean. Some physiologists consider one-humped, or dromedary, camels to be the most successful large desert mammals, and their capabilities and evolutionary adaptations are truly remarkable. A big male dromedary standing 7 feet (2.1 m) at the shoulder and weighing 2200 lb (1000 kg) can carry another 1000 lb—more than an elephant can carry! A camel can labor for days, sometimes weeks, without drinking water, and can eat saltbush leaves, spiky thorns, and dessicated acacia leaves that few other animals will touch. Broad, flat feet prevent the large animal from sinking into the hot sand. Thick eyebrows and lashes keep storm-driven sand out of its eyes. And closeable ears and nostrils protect other facial openings. In the Sahara, camels have been known to march 1000 km in 25 days with a heavy load and without taking a sip of water. No other animal on earth can carry so much so far on so little. So how does the camel do it?

FIGURE 29.1

A Camel's Water Conservation.

A camel can drink up to one-third of its body weight in three minutes after days or weeks without drinking. To match that feat, a person would have to drink over 4 gallons of water at one sitting!

Four major strategies minimize water loss. First, camels excrete small amounts of highly concentrated urine, based on efficient kidneys. Kidneys are the primary organs that filter organic wastes from the body fluids and maintain appropriate water and salt balance. Second, a camel excretes feces that are so dry, people can immediately collect the material and burn it for fuel. Third, a camel's body temperature can vary from 34°C (92°F) during the night when the outside air is cool to 40 or 42°C (104 to 107.5°F) during the heat of the day. [Compare this to a person's fairly constant temperature of 37°C (98.6°F).] The result of this temperature variation is that the camel needs far less evaporative cooling than a person, and sweats much less. In fact, a person needs 20 times more water, pound for pound! A fourth set of physiological adaptations helps minimize a camel's water loss during ventilation. In a thirsty camel, the nasal membranes dry out and become encased by a layer of dried mucus, salt, and cell debris. When the animal exhales warm air saturated with moisture from its lungs, the dried layer inside the nose absorbs water from the passing air, and thus the breath that does escape carries away very little moisture.

MESSAGES

1 The excretory system maintains the water, salt, and pH balance of body fluids and rids the animal body of nitrogen-containing metabolic wastes.

2 The kidney is the main organ of the vertebrate excretory system. Kidneys filter substances from the blood into excretory tubules; reabsorb useful materials from the tubules back into the blood; and secrete additional substances including drugs into the tubules. Urine, the fluid forming in the tubules, flows through ducts and is excreted from the body.

3 An animal's excretory system is adapted to its environment: Animals inhabiting dry deserts, lush pastures, freshwater lakes or streams, or salty oceans each have adaptations appropriate to their surroundings that maintain the constancy of their internal fluids.

In addition to all these mechanisms, a camel can tolerate far more water loss than a person can, and then recover with one long drink. A loss of 12 percent of the body fluid will kill a person, but a camel can go days or weeks without drinking, lose 30 to 40 percent of its body weight, then drink an equivalent weight of water all at once [FIGURE 29.1]. Researchers monitored a 600-kg (1320-lb) camel as it lost 200 L (50 gal) of water over a period of two weeks, then watched as it drank 50 gal (a large bathtub full) in three minutes. Physiologists do not fully understand the camel's extensive fluid loss and gain, but are studying the phenomena to better comprehend the balancing of water, salts, and other substances in the animal's internal environment through homeostatic mechanisms, our subject in this chapter. A desert camel's loss and gain of moisture is just one aspect of this homeostasis. We must also consider the buildup and removal of the waste products of nitrogen metabolism, since in the camel and most other animals, these wastes are washed away in water.

This chapter begins with an overview of water and salt balance and how camels and other animals eliminate nitrogen-containing wastes. Next, we consider the action of the kidneys, the organs that filter nitrogenous wastes from the blood and play a major role in water balance. Then we consider the regulatory mechanisms that control how kidneys handle water, salts, and nitrogenous wastes. Finally, we look at different evolutionary adaptations for salt and water balance and waste elimination in a range of organisms that occupy very different environments, from the desert-dwelling camel to the pond-dwelling beaver to the ocean-going shark. Along the way, you will not only learn more fascinating details about the "ship of the desert," but also see what causes human gout, why bears seldom urinate in the winter, and how a kidney dialysis machine works to save human lives. ❑

Maintaining Balance in the Internal Environment

As you saw in CHAPTER 2, living things are made up primarily of water, and this simple compound plays several key roles in an organism's body. Most biochemical reactions take place in water, the universal solvent. Water helps stabilize an animal's body temperature within the range conducive to biological reactions. Water is the main constituent of blood and hemolymph (the fluid in open circulatory systems; see CHAPTER 25), which transport nutrients and gases throughout animal bodies and carry away wastes. The rapid movement of materials in to and out of cells depends largely on the concentrations of both water and solutes (salts and organic materials), and, in turn, on the resulting osmotic differences between fluids inside and outside of cells.

In vertebrates, the complex and marvelous paired organs, the kidneys, regulate the balance of salt and water throughout the body, as well as certain kinds of waste excretion. *Water balance* means that the amount of water entering an animal's body precisely equals the amount leaving over a certain time period. An average person, for example, obtains water by eating, drinking, and as a product of metabolism. (Review FIGURE 5.4, showing how water becomes a metabolic by-product when foods are oxidized.) A person loses water by evaporation across the skin and lungs [FIGURE 29.2], in sweat and feces, and especially in urine. The important thing is the balancing of water gained and lost each day. The kidney also presides over the excretory system. While an animal's digestive system rids the body of undigested solid wastes, the **excretory system** cleanses the blood of organic waste molecules and carries them out of the body as urine or its equivalent through a special set of excretory tubes.

Urine is a fluid that carries from the body the nitrogen-containing waste products formed by the breakdown of protein and nucleic acid molecules. The more concentrated these wastes become, the more dangerous they can be to the animal. The safest way for the body to remove nitrogenous wastes is to wash them away in a copious amount of dilute urine. But what if water must be strictly conserved, as in a desert animal like the camel? How, in other words, does an animal's body rid itself of wastes safely while retaining the necessary concentrations of water and salt? In vertebrates, the answer lies with the intricate functioning of the kidney. Regardless of external environment, the kidney maintains a constant and proper internal state by automatically removing nitrogen-containing wastes and salts in just the right amount of water so that the remaining fluid balances the body's solute content appropriately. Considering its role, one can easily see why the kidney is a major organ in the maintenance of homeostasis.

The excretory system maintains homeostasis in the blood, which, as you saw in CHAPTER 25, flows near every

FIGURE 29.2

Water Loss and Gain: A Balancing Act.

Over a period of a day, a person gains water in the amounts and ways shown in the figure. He or she also loses water in sweat, urine, feces, and exhaled breath in the amounts shown. The two quantities must balance over time.

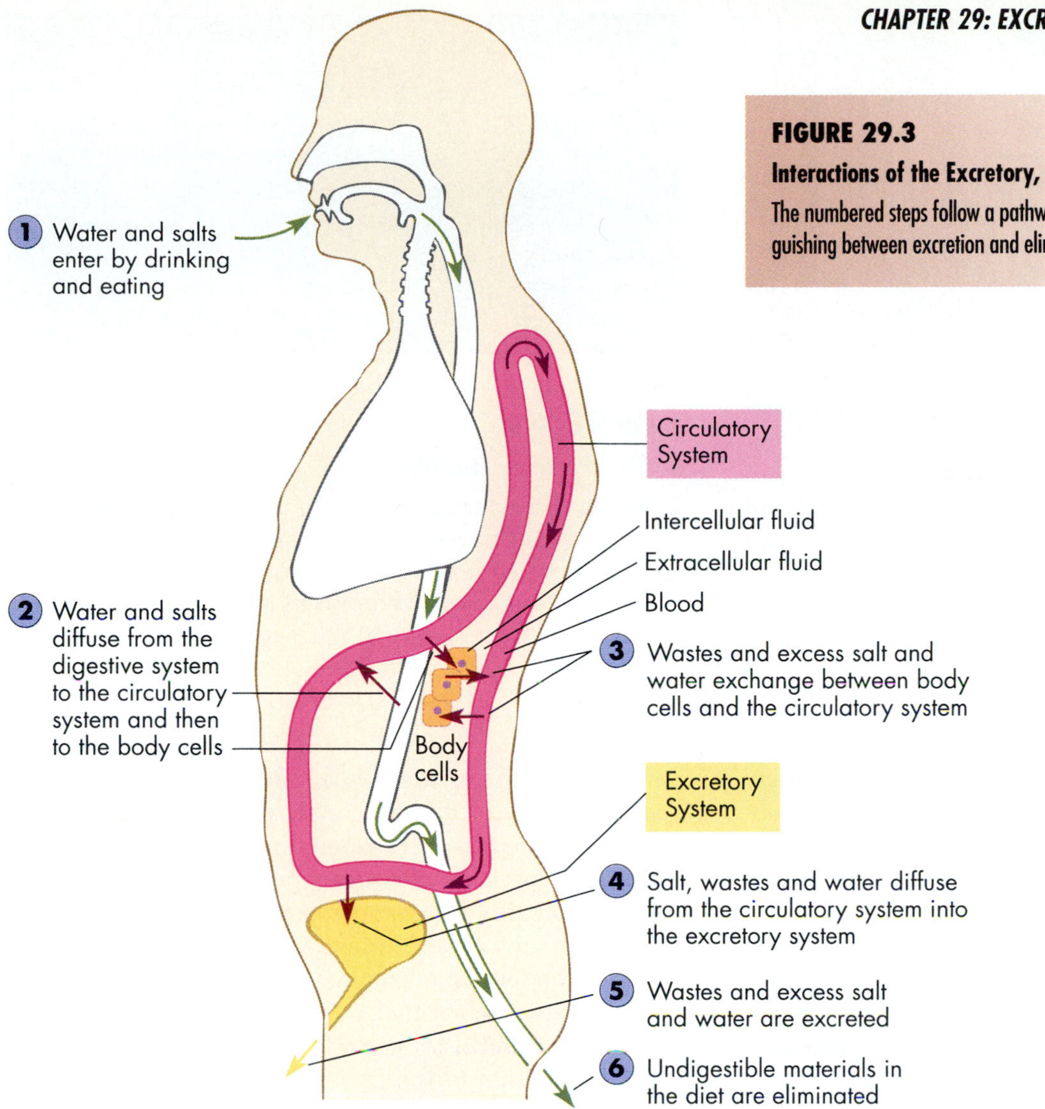

FIGURE 29.3

Interactions of the Excretory, Circulatory, and Digestive Systems.

The numbered steps follow a pathway of water flow through the body, distinguishing between excretion and elimination.

cell. The critical issue for an animal's survival, however, is the maintenance of constant conditions *inside* the trillions of individual body cells. If the pH, salt content, or waste concentration inside cells varies too much, individual cells will cease to function, and eventually the animal will die. FIGURE 29.3 illustrates how the fluid inside cells—intracellular fluid—undergoes a continual exchange of materials with the fluid in the spaces between cells—extracellular fluid. Extracellular fluid, in turn, is in equilibrium with the watery fluid of the blood—plasma. Thus, as the excretory system maintains the contents of the blood, it indirectly maintains the intracellular fluid within cells. The energetic cost for this crucial cleansing and fluid balancing is high: since many molecules pass through the kidney's cell membranes via active transport [see CHAPTER 4], the kidney's activities require a large expenditure of energy. (Gram for gram, the kidneys are metabolically more active than the heart.) Without this continuous energy flow and in the absence of healthy, functioning kidneys, poisonous wastes would build up and the body fluids would become too salty or too watery for life processes to proceed normally.

➤ CONCEPT CHALLENGE

Based on what you've learned so far about the functions of the excretory system, predict the main symptoms of kidney failure.

Nitrogenous Wastes

Just as short pieces of boards remain after you build a bookcase, residues of nutrient molecules remain unused after cells process the nutrients and build from them new molecules and cell parts. **Excretion** is the process that removes the by-products of metabolism from cells, from extracellular fluid, and from blood, and rids the organism of excess water and salts [FIGURE 29.3]. Excretion differs from elimination, the expulsion of undigested food as feces [see CHAPTER 28], in that undigested solid materials never leave the interior of the digestive tube; they never enter the body or its cells [FIGURE 29.3]. Let's look more closely, now, at the molecular by-products of metabolism and the excretion process that removes them.

THREE TYPES OF WASTE

Cells can burn fats, carbohydrates, and proteins for energy, but the end products of their metabolism differ. After fats and carbohydrates are dismantled, water and carbon dioxide remain. Carbon dioxide is eventually expelled through gills or lungs. When cells metabolize proteins, the amino acid building blocks can be broken down further, releasing energy. Only the carbon-containing (acid) portion of the amino acid molecule, however, is shunted through the energy-producing pathway [review FIGURES 28.2 and 5.7, 5.9, and 5.12]. The nitrogen-containing amino (NH_2) portion cannot be oxidized and becomes ammonia (NH_3), the strong-smelling alkaline ingredient often used in cleaning solutions. Even at low concentrations, ammonia is highly toxic to cells, and organisms must get rid of it quickly. Some organisms excrete ammonia directly. In others, however, cellular processes first turn the ammonia into the less toxic compounds *uric acid* and *urea*.

Ammonia: Dilution Is the Solution Removing nitrogenous wastes in the form of ammonia is not a serious problem for many aquatic organisms because the ammonia can be dumped into the surrounding water, where it is diluted so much that it becomes nontoxic.

Uric Acid: Nitrogenous Wastes in Solid Form When a bird's cells produce ammonia from protein metabolism, the animal's liver cells convert ammonia into the less toxic **uric acid**. This relatively insoluble substance continues to circulate suspended in the blood, like silt carried along in a river, and is gradually removed by the kidneys. From the kidneys, the uric acid moves down narrow tubules called *ureters* that drain the kidneys, and passes through the *cloaca* (klo-A-kuh; Latin "sewer"), the single duct that transports excretory products, fecal material, and eggs or sperm from the bird's body. The excretory material is expelled from the bird's body as an insoluble white paste called *guano*. The white portion of the ubiquitous bird droppings that decorate window ledges, park statues, and the occasional human head or shoulder is mainly uric acid.

Using uric acid as an excretory product makes it easy for birds to conserve water since this relatively insoluble nitrogenous waste can pass out of the body as a paste containing little water. Uric acid produced by a chick developing within its shell can also be stored as a precipitate in almost solid form in a special part of the egg away from the embryo. Without such an effective form of waste disposal, hard-shelled eggs might never have evolved. Reptiles, insects, and land snails also rely on uric acid excretion to rid themselves of nitrogenous wastes.

Urea: Excreted by Mammals and Some Fishes In the bodies of a few kinds of fishes, as well as humans, camels, and other mammals, the ammonia produced during protein metabolism is converted into **urea**, a combination of ammonia and carbon dioxide. But just as dirty dishwater goes through sewer pipes to a central water treatment facility for final disposition, the less toxic, water-soluble urea is carried in the blood vessels to the kidneys, which form and excrete urine.

FIGURE 29.4

A Use for Nitrogenous Wastes.

The urinary bladder (a urine-storing organ) of a hibernating bear absorbs urea, and the substance passes into the blood, where it is carried to the gut; there, bacteria break down the urea and use its nitrogen to make amino acids. The sleeping bear's body can then reabsorb and use the nitrogen in the amino acids to build new proteins and nucleic acids, thereby recycling the wastes into useful materials. This novel mechanism allows bears, such as this North American black bear, to hibernate all winter without a large waste buildup.

In a person or a camel, a cell's nitrogenous wastes are handled in a multistep process: (1) the circulatory system carries ammonia to the liver, where it is detoxified by conversion to urea, which passes into the blood; (2) the urea is removed from the blood by the kidney; and (3) urea is concentrated in the urine and excreted from the body through a urinary plumbing system we will consider shortly.

Variations in a mammal's normal excretory system explain both human gout and a hibernating bear's lack of urination in winter. In humans, nitrogen from protein breakdown is carried off as urea, but nitrogen-containing waste products from nucleic acids are converted into uric acid. In some people, uric acid crystals form deposits in the big toe, elbow, or other joints and cause the excruciatingly painful condition known as *gout*. Medicines like colchicine can help relieve gout symptoms, and so can eating fewer meats that are high in nucleic acids, such as organ meats. FIGURE 29.4 describes a special adaptation in American black bears that allows these hibernating animals to rid themselves of urea all winter long without awakening to urinate.

In bears, humans, camels, and other vertebrates, the products of nitrogen metabolism, including ammonia, uric acid, and urea, are cleansed from the body by the kidney.

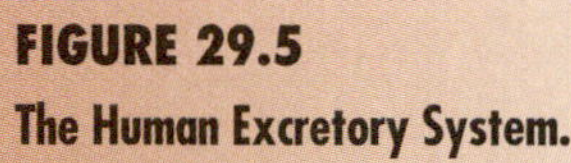

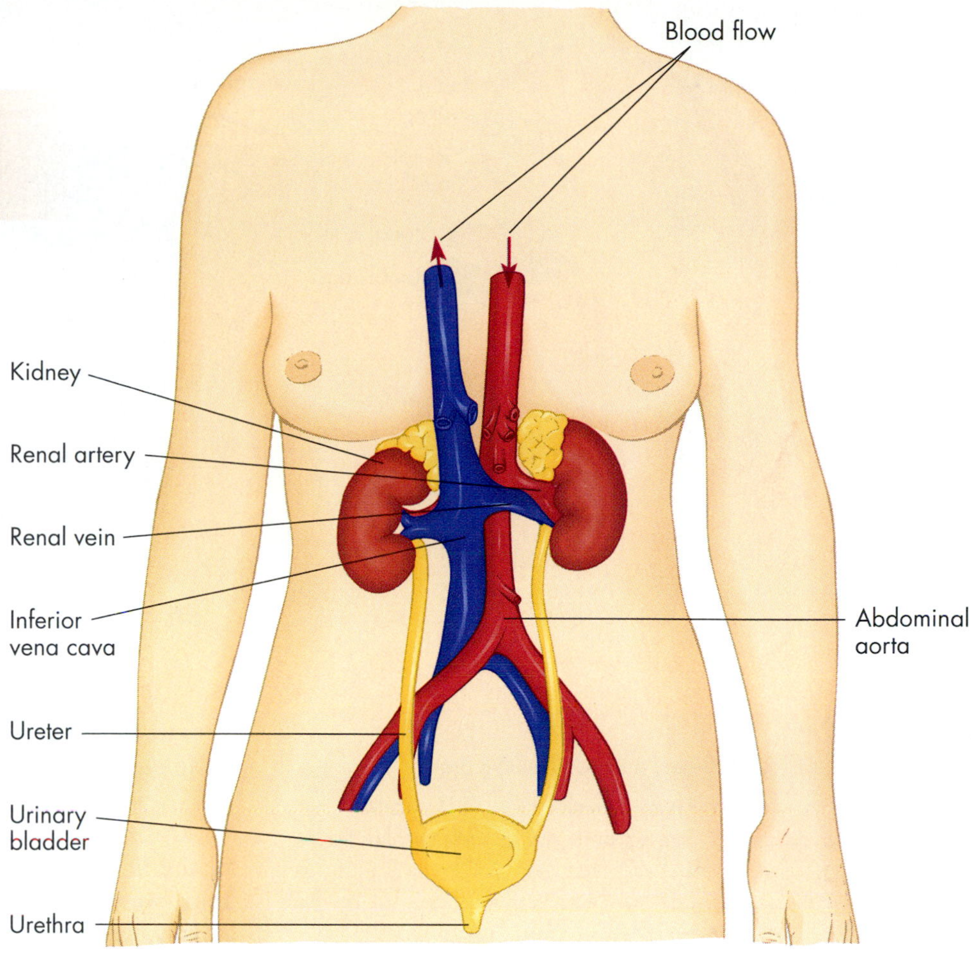

FIGURE 29.5
The Human Excretory System.

> **CONCEPT CHALLENGE**
>
> **Tadpoles swimming in a pond excrete nitrogenous wastes as ammonia, but adult frogs hopping about on land excrete nitrogenous wastes as urea. How would you explain this phenomenon?**

The Kidneys

For many diners, eating tender, pale green shoots of asparagus is a pleasurable springtime event, but the gastronomic experience has a peculiar sequel: The next time the diners urinate, even if just 20 minutes after eating, they notice the characteristic scent of asparagus. The sulfur-containing biochemical in the food seems to cross the gut, enter the bloodstream, be filtered out by the kidneys, and appear in the urine with amazing speed. The key to a kidney's rapid functioning lies in its complicated internal structure and in its efficient plumbing system.

THE HUMAN EXCRETORY SYSTEM

It is sometimes stated that *micturition* (**urination**) is our most compelling bodily function. You will probably agree if you have ever been hungry, tired, and cold and had a full urinary bladder simultaneously, and can recall the order in which you addressed those needs. This urgency has to do with the necessary waste removal role of the excretory system and with its particular anatomy.

In humans, as in other vertebrates, the main organ of the excretory system is the **kidney**. A person's two plump, dark red, crescent-shaped kidneys, each about the size of an adult's fist, are located just below and behind the liver [FIGURE 29.5]. For each kidney to carry out its vital blood-filtering activities, it must receive a large, steady flow of blood. Blood is supplied to the kidneys by the **renal arteries** (Latin *renes*, "kidneys"), twin branches directly off the body's main blood vessel, the *aorta*. Cleansed blood leaves each kidney through a large **renal vein** that drains into the inferior vena cava, the body's largest vein, which leads directly to the heart.

As the kidneys filter and remove excess water and waste substances from the blood, these materials collect as a concentrated urine in a central cavity in each kidney, then flow down a long tube called the **ureter** (yoo-REE-ter). Sometimes the ureter can become painfully blocked by *kidney stones*, crystals usually formed of calcium and organic compounds. Drinking plenty of water decreases the likelihood of developing this excruciating condition.

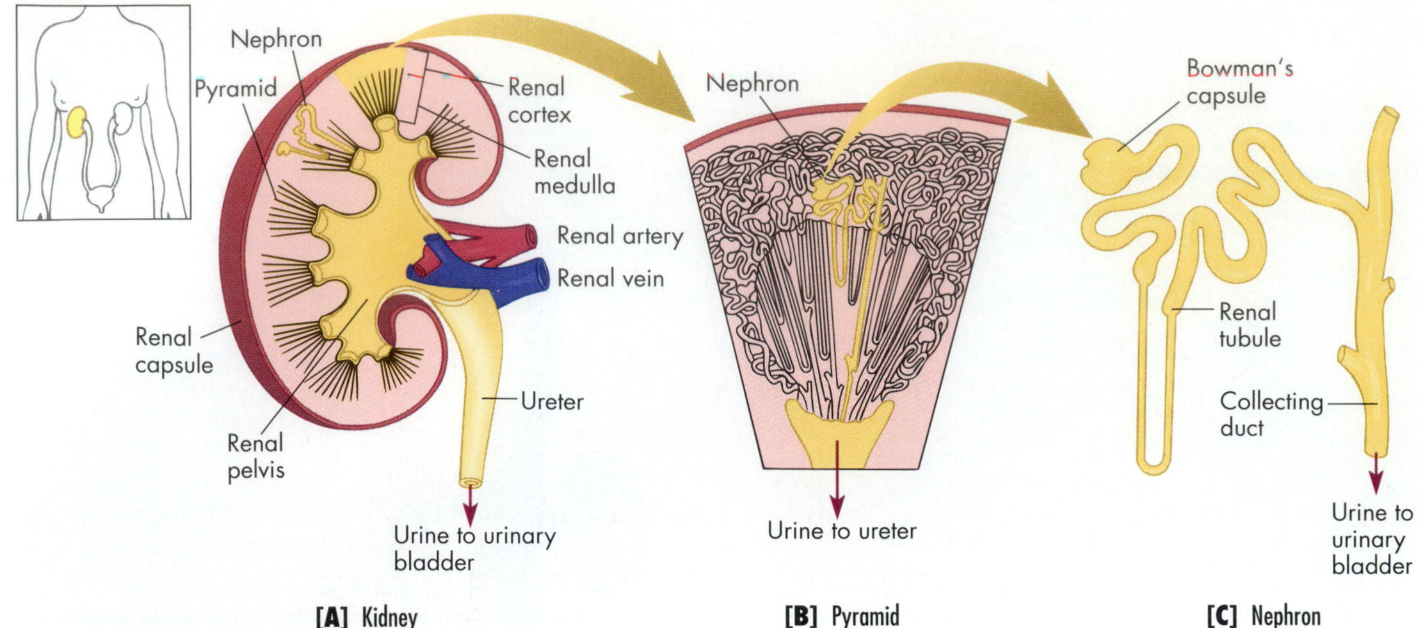

FIGURE 29.6

Anatomy of the Human Kidney: A Blood-Cleansing Organ.

[A] A cross section of the kidney reveals its basic anatomy. **[B]** A pyramid consists of many nephrons. **[C]** Each nephron is a twisted tubule that filters the blood.

The two ureters dump urine into a single storage sac, the **urinary bladder**, which in an adult can hold about 500 mL (1 pt) of fluid. The bladder can distend to hold a bit more, but when it is full, it causes considerable pressure on nearby nerves and a feeling of urgency that can temporarily eclipse all other concerns. During urination, the bladder is drained by a single tube, the **urethra** (yoo-REE-thruh), which carries urine out of the body.

A lamb kidney bought from the butcher and cut in half lengthwise looks a lot like a human kidney and reveals three distinct visible zones [FIGURE 29.6A]: (1) There is an outer **renal cortex**, which is a bit like a thick orange peel, where initial blood filtering takes place. (2) The central zone, or **renal medulla**, is divided into a number of pyramid-shaped regions [FIGURE 29.6B]. The medulla helps conserve water and valuable solutes. The functional units of the kidney, twisted tubules called **nephrons**, reach from the cortex down into the medulla [FIGURE 29.6B and C]. (3) Inside these two zones lies the funnel-shaped, hollow inner compartment, or **renal pelvis**, where urine is stored before it passes into the ureter.

THE NEPHRON

Have you ever cleaned out a desk drawer—sorting, discarding, saving certain items for future use, and then rechecking the drawer one last time? If so, you have approximated the kidney's general sorting and cleansing action and its three basic processes—filtration, reabsorption, and secretion—all of which involve the nephron [FIGURE 29.7]. During **filtration**, which occurs in one part of the nephron, small molecules (glucose, amino acids, ions, water, urea, and so on) are forced out of the blood plasma, while large protein molecules and blood cells are left behind in the blood. During **tubular reabsorption**, a different part of the nephron tubule sorts the small molecules, shunting the useful solutes back to the blood and retaining the wastes. Finally, during **tubular secretion**, regions of the nephron check the blood supply one last time and remove from circulation any ions, drugs, or other wastes that still remain and secrete them into the forming urine.

To understand how the kidney performs the three basic functions of filtration, reabsorption, and secretion, we must first understand the structure of the nephrons and flow of blood around them.

STRUCTURE OF THE NEPHRON

Each of the kidney's nephrons is a long, twisted, looping tubule that carries out the filtration, absorption, and secretion roles of the kidney [FIGURES 29.7 and 29.8A]. A human kidney contains roughly 1 million nephrons, each extending from the outer cortex through the medulla and draining into the renal pelvis. If all the nephrons in an adult's kidney were straightened out and placed end to end, they would form a microscopically slender tube about 80 km (50 mi) long. In contrast, a camel kidney has about 4 million nephrons, many more than the kidneys of other mammals. The camel's multitudinous nephrons help explain why the animals can

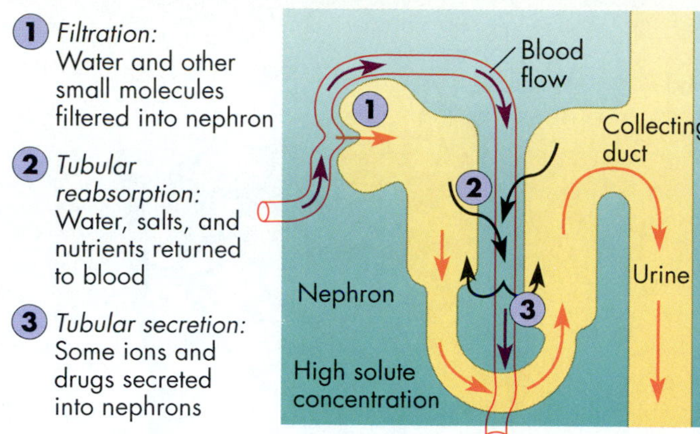

FIGURE 29.7

Filtration, Reabsorption, Secretion: Basic Roles of the Nephron. Water and solutes like urea are filtered from the blood and enter the nephron at the Bowman's capsule (Step 1). As the filtrate flows through the nephron's looped and twisted tubule, much of the water and valuable solutes are reabsorbed into the bloodstream (Step 2). The gradient of salt concentration, here denoted by deepening shades of green, is mainly responsible for the reabsorption of water. Simultaneously, regions of the nephron secrete various wastes, drugs, and other materials into the urinary fluid (Step 3). Concentrated urine then passes through the collecting duct, renal pelvis, and ureters to the bladder.

drink so much water so quickly and yet not seriously dilute the blood.

One end of the nephron, called **Bowman's capsule**, is enlarged and cup-shaped like a punched-in balloon, and lies in close contact with a ball of blood capillaries. The other end of the nephron drains away urine. Between these two ends, the tubule appears to meander and tangle like a long piece of yarn. Bowman's capsule leads into the **proximal** ("near") **tubule** [see FIGURE 29.8A]. At the inner edge of the cortex, the walls of this tubule are quite thin, and the tubule straightens out and dips sharply down into the medulla. There, it makes a hairpin curve and heads back up into the cortex. The U-shaped section of the nephron, known as the **loop of Henle**, is chiefly responsible for reabsorption of materials into the blood. Physiologists call the arm of the loop leading down into the medulla the **descending limb** and the portion leading back into the cortex the **ascending limb**. The ascending limb leads to the part of the nephron farthest from the Bowman's capsule, the **distal** ("far") **tubule**, a convoluted region like the proximal tubule but which is involved mainly in tubular secretion. Groups of distal tubules from various nephrons are connected to large **collecting ducts**; these ducts drain into the renal pelvis and carry away urine. Every 10 or 20 seconds, a small amount of urine collects in this hollow inner compartment, and peristaltic waves of the ureter sweep the urine away to the urinary bladder for storage and eventually for excretion through the urethra.

The portion of the nephron that filters water and solutes from the blood is the enlarged, cuplike Bowman's capsule [FIGURE 29.8B and C]. Each Bowman's capsule is embedded in a matrix of tissue in the kidney's outer cortex, and it surrounds a tight bundle of blood capillaries called a **glomerulus** [FIGURE 29.8C]. This enclosing capsule of kidney tissue surrounding a tuft of blood capillaries forms the closest contact between the excretory and circulatory systems.

Filtration of the blood depends on the structure of Bowman's capsule. The cells lining Bowman's capsule, the so-called *podocytes*, have large central cell bodies with thousands of fingerlike projections that enclasp the capillaries completely [FIGURE 29.8D and E]. Directly under the fingers of the podocyte is a *basement membrane*, a meshwork of protein fibers. The basement membrane of the camel nephron is as much as 70 times thicker than yours or mine. Apparently, this provides mechanical support for the camel nephron, through which flows urine that is sometimes much more concentrated than ours, and sometimes much more dilute.

Situated below the basement membrane are the cells that form the wall of the blood capillary. These capillary wall cells are pierced with minute holes, and as blood moves along through the capillaries in the tuft, fluid from the plasma percolates through this delicate filter and into the Bowman's capsule [FIGURE 29.8D]. The fineness of these pores and the characteristics of the basement membrane are ultimately responsible for the nephron's blood filtration activities. Bowman's capsule allows wastes, water, and other small molecules to seep through these minute openings, but larger components of blood—red blood cells, antibodies, and other large proteins—remain behind in the capillaries.

BLOOD FLOW AROUND NEPHRONS

As you have seen, the filtration process takes place at capillaries in Bowman's capsule. To fully understand the processes of absorption and secretion, however, we must follow the pathway of blood flow around the nephron [FIGURE 29.8B]. Recall that the incoming arteriole branches into the tuft of capillaries in Bowman's capsule [FIGURE 29.8B, Step 1]. These capillaries do not merge directly into veins, but instead rejoin to form another arteriole, which exits the capsule (Step 2). This arteriole carries blood into a second network of capillaries that surrounds the looped portion of the nephron. This network is the **peritubular** ("around the tubule") **capillaries** (Step 3). These capillaries finally merge into venous capillaries and then into a larger venule that connects to the renal vein (Step 4); all these venous vessels carry blood away from the kidney and toward the heart.

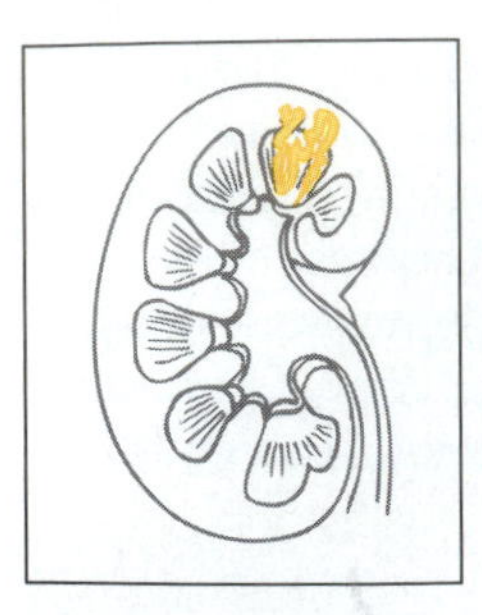

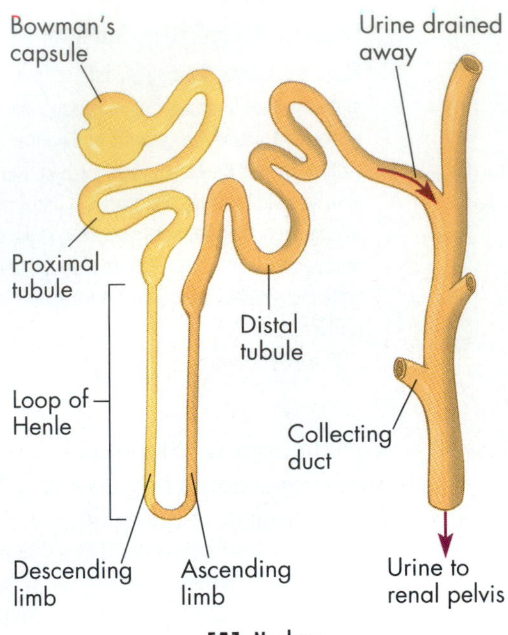

[A] Nephron

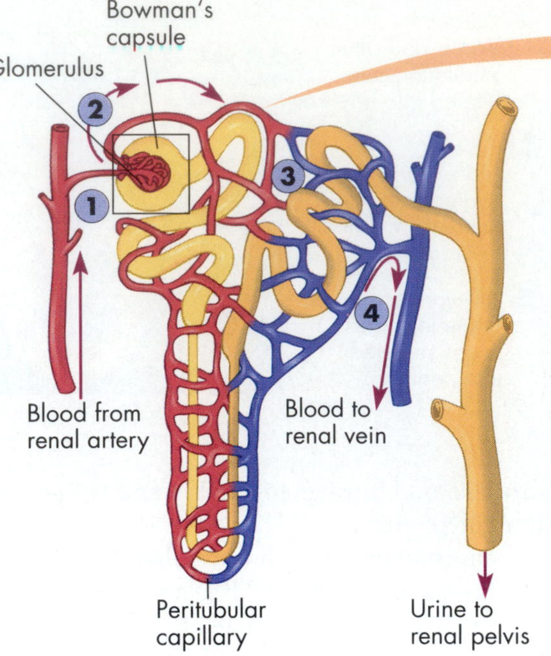

[B] Blood vessels and the nephron

FIGURE 29.8

The Nephron: Basic Unit of Kidney

[A] Each nephron has a Bowman's capsule, a proximal tubule, an elongated loop of Henle, a distal tubule, and a collecting duct leading toward the renal pelvis. **[B]** An arteriole enters the glomerulus (tuft of capillaries) inside the Bowman's capsule, and another arteriole leaves the glomerulus for the peritubular capillaries, which join to form the venule leaving the nephron. **[C]** A cutaway view of Bowman's capsule reveals an internal cavity, as well as cells of the inpouched wall of the nephron that wrap tightly around the blood vessel. **[D]** An enlargement of a portion of a capillary in the Bowman's capsule reveals the remarkable podocyte cells that enclasp capillaries. Under the podocyte is the basement membrane, a layer of proteins, and below that, the cells that line the walls of the capillaries. **[E]** This scanning electron micrograph shows a podocyte cell enclasping a capillary (magnification about 2000×).

NEPHRONS AT WORK

Nephrons are amazingly efficient at selectively removing wastes from the blood circulation while simultaneously conserving water, mineral ions, glucose, and other needed materials. In an adult human, 180 L (48.5 gal) or more of blood—enough to fill a bathtub—pass through the 2 million nephrons in the two kidneys each day. At any given time, about one-fifth of the blood volume is passing through the blood vessels to and from the kidneys. We do not, of course, urinate 180 L a day, but instead the much more reasonable amount of about 1.5 L (3 pt) per day. Clearly, the nephrons accomplish a great deal of water conservation before they produce that smaller quantity of urine. Let us now look at each renal process in more detail to see how a nephron functions.

Filtration In this first stage of urine formation, blood pressure forces some of the plasma, or yellowish fluid portion of blood, through tiny pores in the walls of the capillaries, just as high water pressure forces water out of a lawn sprinkler. This liquid, containing various dissolved substances, then passes across the basement membrane and into Bowman's capsule. The transport of water, of sodium, potassium, and chloride ions, and of sugars, amino acids, and urea out of the capillaries and into the cavity of Bowman's capsule is passive and does not require any special output of energy other than blood pressure. The fluid, or **filtrate**, in the capsule is still very much like blood plasma except that it contains no large proteins [FIGURE 29.9, Step 1].

Tubular Reabsorption As soon as the filtrate enters the nephron's twisted proximal tubule, reabsorption begins and returns to the circulation most of the water, sodium and chloride ions, sugars, and amino acids that

FIGURE 29.9

How Materials Move Into and Out of the Nephron As Urine Forms.

As a nephron extends through the kidney's cortex and medulla and dumps urine into the collecting duct, various substances enter and leave the filtrate. Broken lines represent segments of the nephron wall that are permeable to water, while solid lines represent wall segments impermeable to water. Narrow arrows represent passive diffusion of materials in to or out of the nephron tubule, while wide arrows represent active transport against concentration gradients. Filtration activities are shown in blue, tubular reabsorption activities in green, and tubular secretion in yellow. Urine is shown as yellow. The text traces nephron function and material movements step by step.

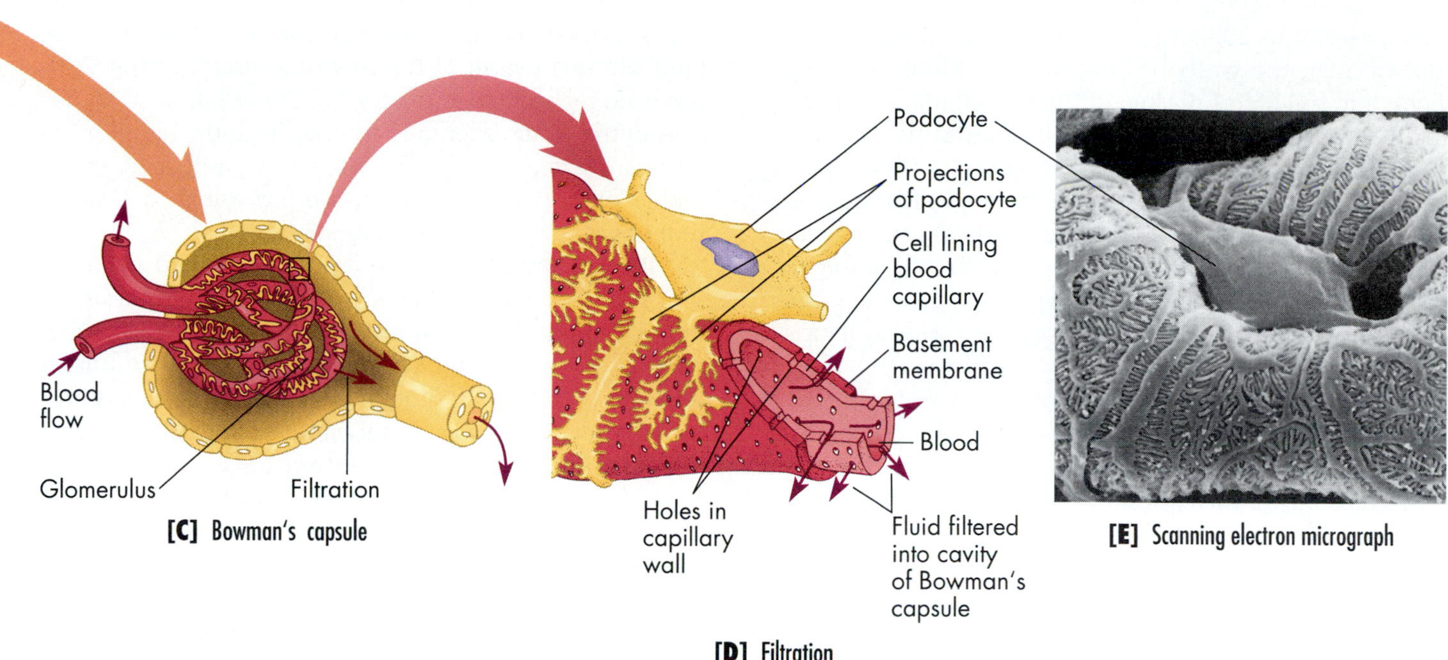

[C] Bowman's capsule

[D] Filtration

[E] Scanning electron micrograph

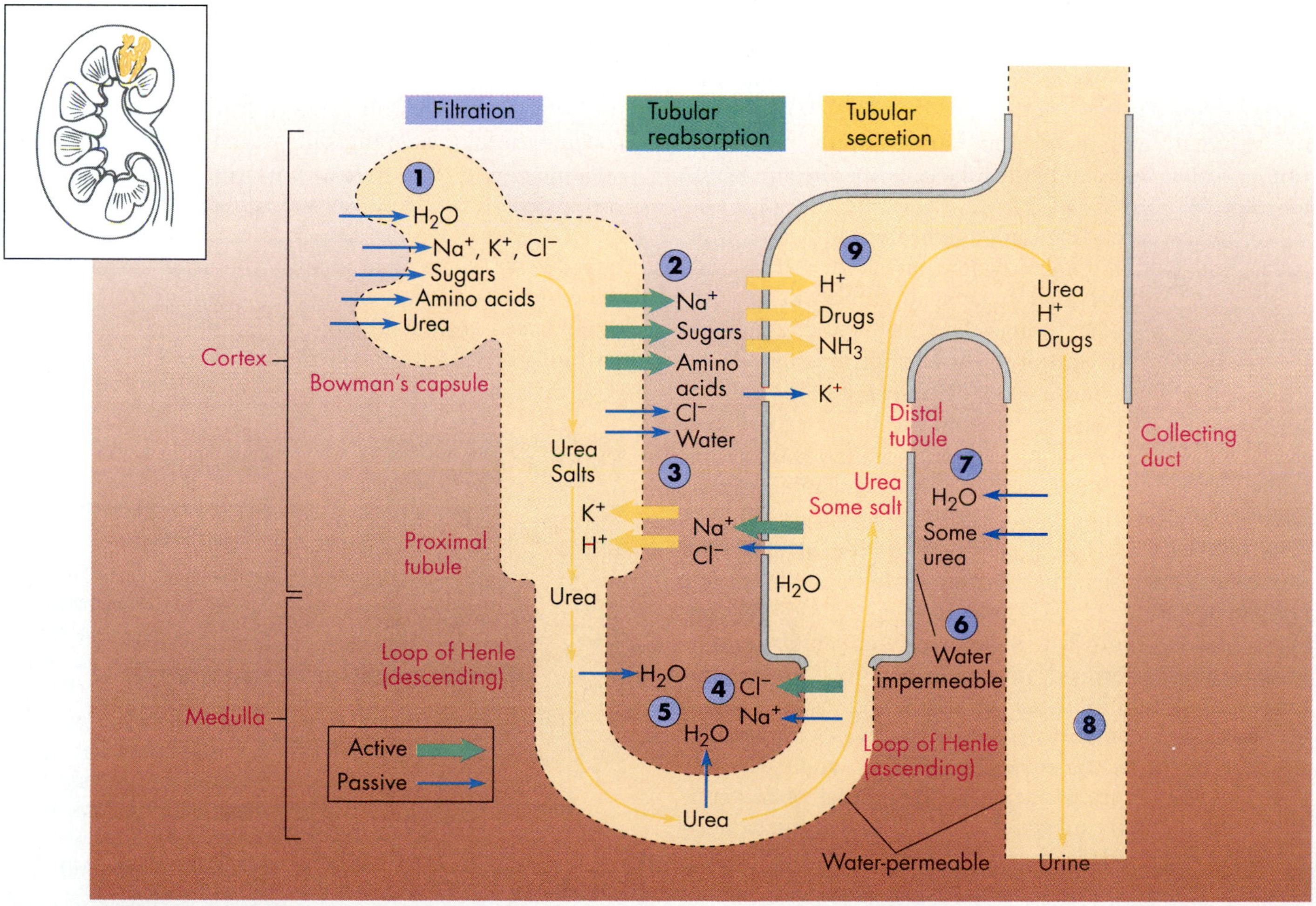

were just filtered out of the blood in the Bowman's capsule [see FIGURE 29.9, Step 2]. These materials are moved out through the walls of the proximal tubule and pass back into the blood plasma of the peritubular capillaries entwining the proximal tubule. Instead of being driven passively by blood pressure, reabsorption of solutes in the proximal tubule depends on active transport across the plasma membranes of nephron cells. As the ions are being reabsorbed in this way, 80 to 85 percent of the water in the original filtrate follows them passively back into the capillaries via the process of osmosis [FIGURE 29.9, Step 3].

Next, the filtrate—minus most of its water but still containing urea and salts—passes down the descending limb and up the ascending limb of the loop of Henle. Here the cells of the ascending limb pump chloride ions (Cl^-) into the extracellular fluid surrounding the tissue cells of the kidney's medulla, and sodium ions (Na^+) follow passively (Step 4). As a result of this ion flux, the innermost part of the medulla becomes very salty compared to the upper part, as suggested by the gradient of brown in FIGURE 29.9. This means that the concentration of Na^+ and Cl^- is lowest in the cortex and grows increasingly higher toward the inner part of the medulla, the part containing the hairpin curve of the loop of Henle [see FIGURE 29.8B]. Since the tissue surrounding the loop is salty, water diffuses out of the loop's descending limb by osmosis and reenters the blood of the peritubular capillaries (Step 5). This exodus of water ends, however, as the filtrate rounds the bend and moves up the ascending limb, because that part of the tube is not permeable to water (Step 6).

From the ascending limb, the filtrate, still containing urea, some salt, and some water, passes toward the collecting duct. In that duct, once again, the walls are permeable to water; even though the filtrate passing through is already quite concentrated, the surrounding tissues are even saltier. Thus, more water diffuses out of the collecting ducts by osmosis and enters the blood of the peritubular capillaries (Step 7). In addition, the walls of the collecting duct allow a certain amount of urea to pass back out, rather then being excreted in the urine. This urea increases the brinelike concentration in the inner medulla and causes still more water to be removed and conserved.

Strangely enough, a water-deprived camel excretes very little urea. Instead, urea is reabsorbed into the animal's blood and circulates throughout the body, eventually entering the digestive tract. Since water follows the urea by osmosis, the retention of urea when water is scarce in the environment helps the animal's body conserve this precious substance. This is one of the camel's more important adaptations.

By the time the filtrate (now called *urine*) has reached the part of the collecting tubule in the innermost (and saltiest) region of the medulla, much of the water has been reabsorbed [FIGURE 29.9, Step 8]. In fact, 99 percent of the water originally filtered from the blood in the Bowman's capsule has by now been returned to the body's circulation, and the urine has a high concentration of wastes relative to water content.

Tubular Secretion While the filtration and reabsorption activities of the kidneys' millions of nephrons are removing salt and urea from the blood and forming urine, another process—**tubular secretion**—goes on simultaneously in both the proximal and distal tubules. Nephron cells remove a number of unwanted materials—including some hydrogen and potassium ions, ammonia, and some drugs like penicillin and phenobarbitol—from the blood in the peritubular capillaries surrounding the nephron and secrete them into the forming urine [see FIGURE 29.9, Step 9]. Besides cleansing the blood, tubular secretion can also help maintain an appropriate pH level in the blood (pH 7.3 to 7.4), since nephron cells secrete more hydrogen ions into the urine if the blood is too acidic and fewer hydrogen ions if the blood is too alkaline.

Tubular secretion is the physiological process that makes *drug testing* possible—checking a person's urine to see if he or she has taken drugs. Drug testing employs various laboratory techniques to detect even minute traces of the metabolic breakdown products of marijuana, cocaine, heroin, sleeping pills, tranquilizers, morphine, codeine, and many kinds of prescription drugs. If a person takes a drug overdose and loses consciousness and no one is sure which drug was consumed, physicians quickly test the urine to identify the drug and determine the best treatment for saving the patient's life. Two additional uses—testing athletes and employees for drug use on the playing field or on the job—are currently quite controversial [FIGURE 29.10].

FIGURE 29.10

Kidney Function Makes Drug Testing Possible.

The night before the 400-m freestyle at the 1972 Olympics in Munich, Germany, Rick DeMont took ephedrine to treat his asthma. He won the event but officials demanded that he return the gold medal because a test of his urine revealed the ephedrine. DeMont was, in effect, a victim of tubular secretion by his own kidneys. Here, DeMont (left) stands by the silver medalist.

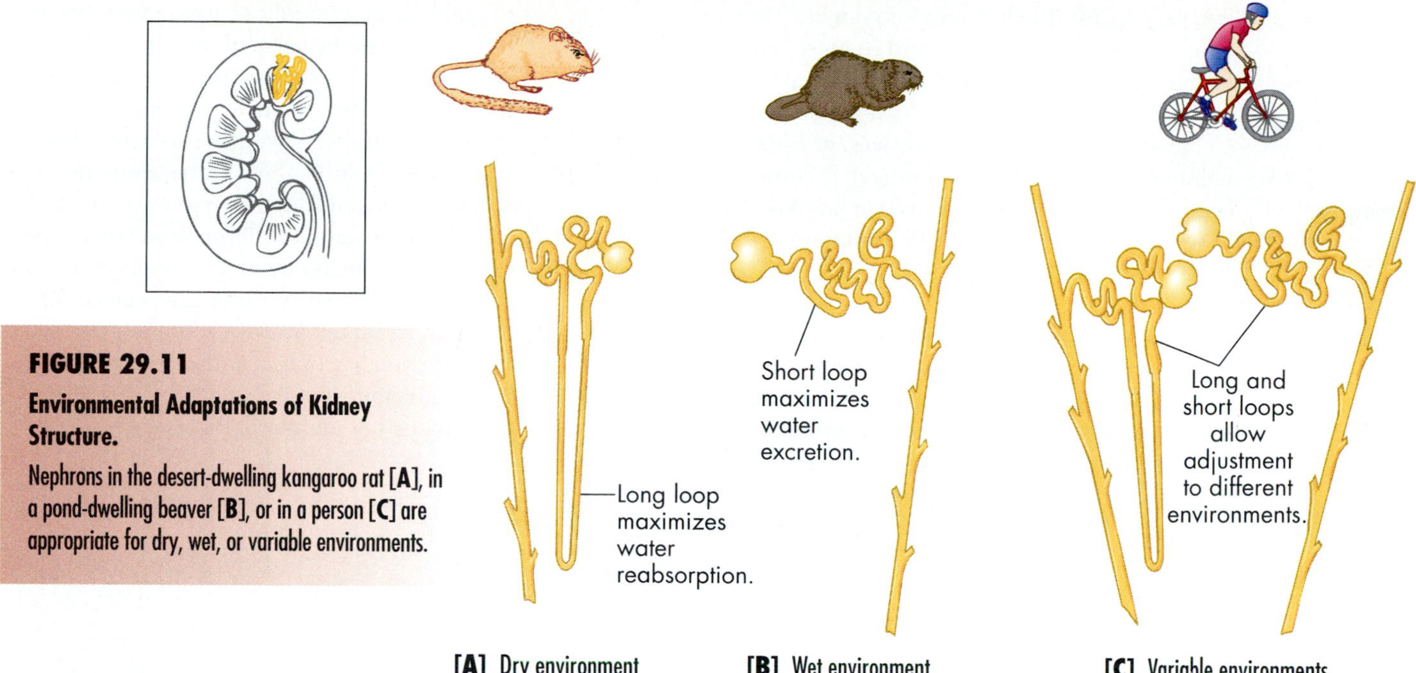

FIGURE 29.11
Environmental Adaptations of Kidney Structure.
Nephrons in the desert-dwelling kangaroo rat [**A**], in a pond-dwelling beaver [**B**], or in a person [**C**] are appropriate for dry, wet, or variable environments.

An organ as central to homeostasis as the kidney is impossible to live without and difficult to replace. People with kidney diseases often spend many hours each week connected to a dialysis machine, a large instrument that mimics the filtration of a healthy kidney [BOX 29.1].

KIDNEYS ADAPTED TO THE ENVIRONMENTS

The association between a kidney's function and its structure is so close that biologists have discovered nephrons of distinctly different lengths in animals with markedly different life habits. The kangaroo rat, a small rodent inhabiting the deserts of the western United States, conserves water to an extreme degree, and each of its nephrons has a very long loop of Henle [FIGURE 29.11A]. With this modification, its kidneys can reabsorb practically all the water as the filtrate moves through the long loop, and as a result, the animal can produce urine 25 times more concentrated than its own blood—the highest concentration of any mammal. Camel kidneys also have very long loops of Henle, and their urine is more concentrated than their blood and twice as concentrated as seawater.

In contrast, an aquatic mammal like a beaver takes in a great deal of water in its food and through its skin and must get rid of the fluid, not conserve it. In a beaver's kidney, the loops of Henle are short and reabsorb less water; as a result, the animal produces great quantities of urine with only twice the solute concentration of its own blood [FIGURE 29.11B]. Humans have a combination of long and short loops of Henle and produce a urine of variable concentration, depending on water availability in the person's environment [FIGURE 29.11C].

➤ CONCEPT CHALLENGE

In some forms of diabetes, huge amounts of glucose are filtered from the blood into the nephron, where it accumulates in urine. Would you expect a patient with this condition to produce more urine or less urine than most other people? Explain your reasoning.

Thirst and Hormones

While the kangaroo rat and the beaver deal with extremely dry or wet environments, other animals cope with surroundings that fluctuate from wet to dry. An impala living on the plains of east Africa, for example, experiences distinct wet and dry seasons each year, while a person working out at a health club could inhabit a number of environments—an air-conditioned room; a warm, dry tennis court; a hot and humid steam room; a swimming pool. Although the environment outside an animal may change dramatically, the salt concentrations, pH, and waste concentrations inside the animal's cells cannot vary; otherwise, the proteins of the cell will cease to function. Therefore, the kidneys must be able to conserve water during the dry season or the tennis game but remove it after a drink from a water hole or a glass of iced tea. Indeed, animals have two major mechanisms for regulating the activities of the kidneys and hence the water balance of the body: (1) *thirst*, or the desire to drink, which affects the amount of water taken in, and (2) *hormones*, chemical messengers that alter the activities of the nephrons and hence control the amount of water retained or excreted at any given time.

Biology Applied

When the Kidney Fails

It is easy to take the kidneys for granted: They operate silently, efficiently, and in most people continuously for a lifetime without a glitch. One only need learn about the consequences of kidney failure, however, or meet a victim of this condition, to understand how central these organs are to survival and just how difficult it is to imitate their natural functions.

The kidneys can lose their exquisite ability to cleanse the blood in several ways. Bacteria from feces can contaminate the urinary tract, attacking the kidneys. An autoimmune attack [see CHAPTER 26] on the kidneys by white blood cells can block and destroy glomeruli. Finally, poisoning by mercury, lead, or certain solvents can damage the kidney tissue. A person's nephrons are so numerous and so efficient that even if two-thirds of these tubules are destroyed, the individual can still live a fairly normal life. If the number drops to 10 or 20 percent, however, the person can suffer extreme tissue swelling as a result of salt and water retention, as well as a buildup of urea, hydrogen and potassium ions, and other metabolic byproducts. These cause the blood to become very acidic and can lead to coma or—if the pH drops below 6.9—death.

Fortunately, biomedical researchers in the mid-twentieth century invented an artificial kidney, or *kidney dialysis machine*, which works via simple diffusion to take over some of the kidney's blood-cleansing functions [FIGURE 1]. When the machine is turned on, waste-laden blood from the patient's artery is routed through a long, porous membrane bathed by a solution much like normal blood plasma. As it passes through, the wastes diffuse into the solution (from the region of higher waste concentration to the region of lower waste concentration). After circulating several times through many meters of tubing and back into the body, the patient's blood is sufficiently free from wastes to permit normal activity—at least for a while.

Unfortunately, the artificial kidney has some serious drawbacks: It carries out filtration but neither the reabsorption nor the secretion activities of a living kidney. Moreover, the patient's blood must be treated with an anticoagulant so that it does not clot as it passes through the machine, then treated again with a coagulant as it reenters the person's body so that he or she won't bleed too freely. Because of the drug treatments, the artificial kidney can only be used every two or three days, and because of the slowness of waste diffusion, each session may last from 5 to 10 hours. Also, the use of anticoagulants allows infections to occur with distressing frequency. Finally, new studies show that the way a dialysis machine is routinely disinfected can mean the difference between a patient's survival and death by infection. People with kidney failure are literally captives of the dialyzer: It extends their life but diminishes the quality of life, gobbles much of their time, and can even lead to their death if the machine is not maintained perfectly.

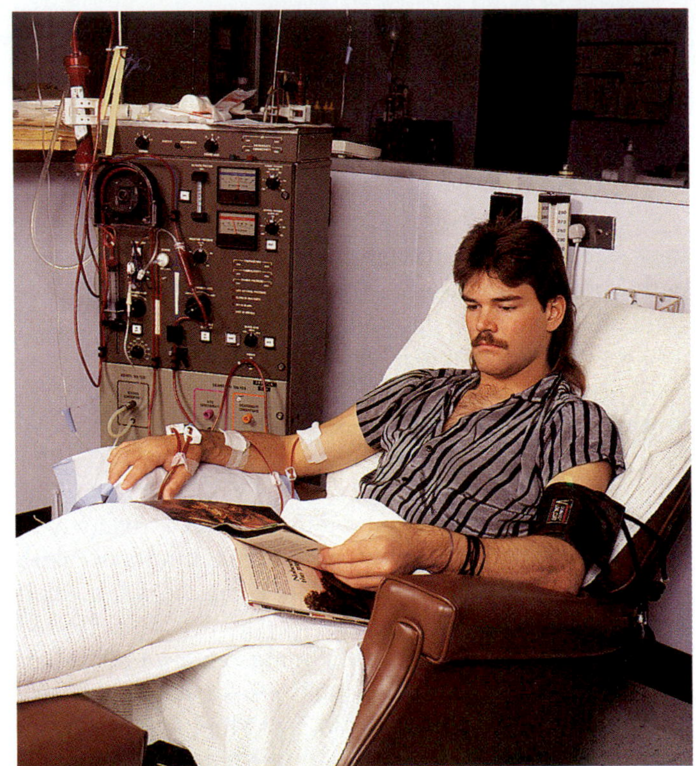

FIGURE 1

Kidney Dialysis Machine and Patient Having Blood Cleansed.

Kidney transplants are an alternative to this captivity. Donors are limited to blood relatives or others with closely matching tissue types [see CHAPTER 26]. Even with this careful screening, the patient may need long-term drug therapy to suppress immune rejection. Thousands of people are saved each year by dialysis machines and transplants. Clearly, though, healthy kidneys are nothing to take for granted.

THIRST REGULATES DRINKING

Although a camel can go for days or weeks without drinking, and a kangaroo rat never needs to drink fresh water, most terrestrial organisms must consume water to survive. Even the loss of a small proportion of a person's fluid content—say, 2 percent from sweating or evaporation from the lungs during a tennis game on a hot day—can cause the concentration of solutes in the blood to rise. These rising levels trigger nerve cells in the brain's *thirst center*, located in the hypothalamus, and the individual becomes aware of thirst. When a person responds by drinking a large amount of water, the water distends the stomach walls and sets up nerve impulses that inhibit the thirst center even before the water is absorbed. Once the water enters the bloodstream from the stomach and small intestine, osmotic balance is restored and the concentration of solutes in the blood and inside body cells stabilizes until more water is consumed in foods or beverages or until sweating, exhaling, and forming urine remove fluids along with wastes.

HORMONES CONTROL WATER EXCRETION

Just as you can turn a water faucet on or off to increase or decrease water flow, your body can regulate nephrons, resulting in an increase or decrease in urine flow. Consider the factors we've discussed that influence the amount of water in the urine. They include the amount of water initially filtered into Bowman's capsule [FIGURE 29.9, Step 1]; the salt concentration of the extracellular fluid in the kidney's medulla (Steps 4 and 5); and the degree of impermeability in the distal tubule (Step 6). Blood pressure is a major influence upon filtration rate, and two separate hormones act on the other two factors to control water loss. A hormone called *antidiuretic hormone* regulates the permeability of the distal tubule to water, while the hormone *aldosterone* regulates salt reabsorption.

Antidiuretic Hormone **Antidiuretic hormone** (ADH, also called *vasopressin*) is aptly named because it prevents *diuresis*, the production of copious amounts of urine. ADH controls how much water the nephron reabsorbs from the filtrate and returns to the blood. ADH is produced in the brain region called the hypothalamus and is stored in a nearby region, the pituitary gland [FIGURE 29.12A]. When the concentration of salt and other solutes rises—say, after a snack of salty potato chips [FIGURE 29.12, Step 1]—the hypothalamus detects the changes and causes the pituitary gland to release some of its stored ADH into the blood (Step 2). Upon reaching the kidney, ADH makes the walls of the distal tubule and collecting ducts temporarily more permeable to water (Step 3) so more water is reabsorbed into the bloodstream, diluting the concentration of blood solutes (Step 4). Since water is drawn from the filtrate into the blood, the blood becomes less salty, and the urine becomes very concentrated (Step 5). Finally, less salty blood causes the brain's secretion of ADH to decrease, thus completing a negative feedback loop.

As beer drinkers occasionally discover to their dismay, alcohol inhibits ADH secretion, leading to diuresis

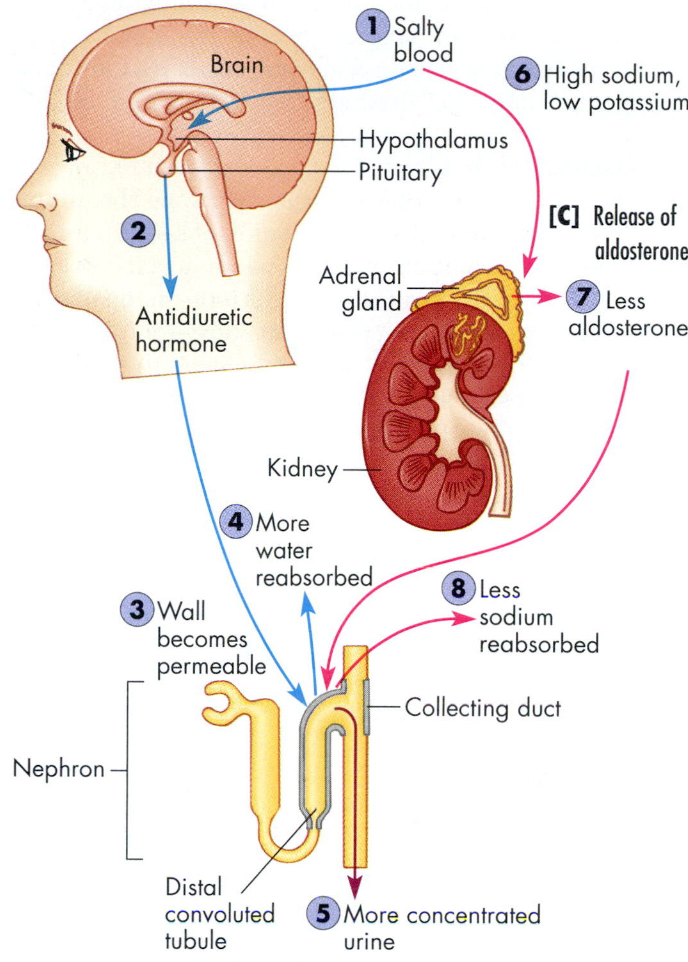

FIGURE 29.12

How Hormones Regulate Salt and Water Balance.

[A] Salty blood sensed by the brain's hypothalamus causes the release of antidiuretic hormone (ADH), which in turn [B] acts on the distal convoluted tubule and collecting duct, making them more permeable to water. Water then leaves the forming urine and reenters the blood. This conservation causes the release of more concentrated urine and helps prevent the blood from becoming saltier. When levels of salt in the blood decrease, ADH release is shut down, completing the negative feedback loop. [C] Salty blood also acts indirectly on the adrenal glands, decreasing the secretion of aldosterone and causing the distal tubule cells to reabsorb fewer sodium ions. Specific steps are described in the text.

and the excretion of sometimes embarrassing quantities of urine. This causes dehydration and is a major reason for the hangover a person feels after drinking too much liquor; the body is dehydrated and must replace lost fluid.

Aldosterone While ADH affects reabsorption of water, another hormone, **aldosterone**, affects the absorption of salt. When the blood's concentration of sodium ions is high and the concentration of potassium ions is low [FIGURE 29.12, Step 6], the adrenal glands (one on top of each kidney) decrease their secretion of aldosterone (Step 7). With less aldosterone, the distal tubules of the kidneys' nephrons reabsorb fewer sodium ions from the filtrate (Step 8), and the urine becomes more salty (Step 5). Aldosterone acts in an opposite way on potassium ions. Through this mechanism, sodium in the blood remains within the narrow range that is best for the activity of nerve cells and other crucial physiological activities.

THE KIDNEYS' ROLE IN HEART DISEASE

In recent years, physicians have warned people that diets high in sodium can elevate blood pressure and strain a heart already weakened by atherosclerosis [see CHAPTER 25]. The kidneys can counteract high salt content in the blood from a salty diet by excreting less water. With this water retention, the blood is diluted from overly salty back to normal, but the excess water also causes the total volume of blood in the circulatory system to increase. The increase in volume causes the blood pressure to go up, forcing the heart to work harder.

The kidneys can also counteract falling blood pressure by means of a special group of cells near the glomerulus in Bowman's capsule. When blood flow to these *juxtaglomerular cells* drops below a certain volume, the cells release the enzyme **renin**, which converts a blood protein into the hormone **angiotensin**; the angiotensin, in turn, causes blood vessels to constrict, thereby elevating blood pressure the way that squeezing on a balloon increases the pressure in the balloon. Angiotensin also triggers the adrenal glands to secrete aldosterone, which results in sodium and water reabsorption and thus even higher blood volume and pressure.

Recently, physiologists have discovered yet another hormone—this one secreted by the heart itself—that helps keep blood pressure in check. When the blood volume and pressure increase, cells in the heart's atria secrete **atrial natriuretic factor (ANF)**, a hormone that causes sodium (natrium) to appear in the urine. Water follows the sodium by osmosis, thus increasing the quantity of urine, lowering the blood volume, and hence decreasing the blood pressure. Geneticists have recently isolated the human gene for ANF and hope to produce large quantities of it using recombinant DNA techniques [see CHAPTER 11] to treat people with chronic high blood pressure.

The message from this section is simple: Thirst and hormones regulate how much water we drink and how much salt and water we retain, and this, in turn, affects blood content, blood pressure, and other factors vital to homeostasis and health.

➤ CONCEPT CHALLENGE

During periods of water deprivation, would you expect the amount of ADH in your blood to be higher or lower than at times when you have unlimited access to drinking water? Explain your answer.

Salt and Water Balance

Why can't a thirsty sailor simply drink seawater like a camel can? The answer lies with the salt and water balance we have been discussing: Human kidneys cannot make urine as salty as seawater. For every 1 L (1 qt) of seawater humans consume, we produce 1.3 L of urine and quickly dehydrate. In contrast, a camel can drink seawater because its kidneys can make urine that is more concentrated than seawater. Experimenters, however, found that camels don't like seawater and avoid it. Aquatic animals such as sharks, flounder, or sea gulls can and do drink seawater readily. Obviously, animals face different osmotic challenges depending on their natural environments. Let's look at some of the strategies that animals use on land, at sea, and in fresh water for **osmoregulation**: the regulation of osmotic balance.

OSMOREGULATION IN TERRESTRIAL ANIMALS

On land, animals are surrounded by air, and thus tend to lose water through evaporation across the body and respiratory surfaces as well as through excretion of urine or its equivalent. The result is threefold: (1) the need to drink water, (2) the need to dump salt in the sweat and urine and sometimes through special salt glands, and (3) the need to conserve water through physical and behavioral mechanisms. An insect's waxy cuticle, for example, helps hold in water, as do a reptile's dry, scaly skin, a bird's feathers, and a mammal's fur. In one experiment, researchers sheared a camel, and the animal's water expenditure increased by 50 percent. Fur has a tendency to hold in water, and it also serves as insulation against the dry heat of the desert.

The moist skin of an earthworm, snail, or amphibian, however, is subject to tremendous evaporative water loss, and so is the skin of a furless land mammal such as a human. Amphibians tend to remain in cool, wet places, and in many species, the urinary bladder acts as a reservoir from which water can be removed to replenish body fluids when needed. People who live in desert areas tend to wear clothing that minimizes evaporation and they rest at midday, when the temperature is highest. All varieties of animals have organs for balancing and excreting salt and water: the vertebrate's kidney, the insect's malpighian tubules and the earthworm's nephridia.

OSMOREGULATION IN AQUATIC ANIMALS

Aquatic animals are surrounded by water, and they face a continual problem due to the difference in salt concentration between their body fluids and the surrounding water. If the aquatic environment is saltier than their internal fluids, water tends to diffuse outward and leave the animal dehydrated. If the aquatic environment is less salty than their blood, water tends to diffuse in and waterlog the organisms. Various adaptations have evolved in different groups for excreting water and salt.

Aquatic Invertebrates The easiest solution to osmoregulation is to have body fluids with the same concentration as the surrounding water, and that is the situation in invertebrates such as oysters and sea stars. Biologists call these animals *osmoconformers*; their body fluids conform osmotically to their surroundings, whether salty or dilute. Other marine invertebrates, such as crabs and brine shrimp, are *osmoregulators*; they maintain body fluids at a steady solute concentration, regardless of environment. Brine shrimp can live in fresh water, normal seawater, and even much saltier places, like the Great Salt Lake in Utah or Mono Lake in eastern California, because the animals' exoskeletons impede water loss or salt movement and because their gills actively pump salt in or out as needed. Thus, brine shrimp can live in a wider range of osmotic habitats than most other organisms.

Aquatic Vertebrates The body fluids of aquatic vertebrates can be more or less concentrated than their surroundings. In each case, however, the body fluids remain at steady concentrations through osmoregulation. In sharks, for example, and in crab-eating frogs (residents of coastal mangrove swamps), very little water flows into or out of the animal by osmosis because their body fluids have roughly the same osmotic condition as seawater. Their blood accumulates high levels of urea, and this raises the total solute concentration of the blood to near that of the environment. Even though the total solute concentration is the same in shark blood and in seawater, salts do not build up because sharks have a special *rectal gland* that excretes excess salt.

In freshwater fishes and amphibians, the blood has a higher salt concentration than the environment. Since large quantities of water flow in across the skin, these animals do not need to drink water; instead, their kidneys produce large amounts of dilute urine that rids the body of excess water. Unfortunately, salt also tends to leave the body and diffuse into the water, and so in freshwater fishes, the gills expend energy actively accumulating and absorbing salt from the lake or river water.

Finally, in most marine bony fishes, reptiles, and mammals, the body fluids are less salty (hypotonic) relative to seawater, and thus the animals tend to lose needed water and gain excess salt. As an evolutionary consequence, saltwater bony fishes drink seawater, retain most of the water, and produce very little urine, while their gills actively pump out excess salt. Sea lions, whales, and other marine mammals never need to drink (taking in salt water only with food), and their very efficient kidneys conserve water strongly and excrete excess salt.

It is clear from these descriptions that an animal's excretory system works continuously to offset undesirable losses or gains of water or salt from the environment, and this constant work is fueled by food energy. Homeostasis—whether of body fluid content, oxygen levels in the tissues, heart rate, or body temperature—is costly but necessary to survival.

➤ CONCEPT CHALLENGE

Which type of organism would you expect to find in a freshwater pond, osmoconformers or osmoregulators? Explain your answer.

Connections

Regardless of an animal's environment—be it desert, temperate forest, ocean, or freshwater pond—animals have surprisingly similar means of handling water, salt, and metabolic wastes. Only occasionally do animals have novel solutions to common excretory problems, such as the urea-filled blood of sharks and dehydrated camels, or very long loops of Henle, as in camels and kangaroo rats. The unity and diversity of life are reflected in these common and unique solutions to biological problems.

In each case, the excretory system interacts with other physiological systems and this interaction must be controlled. In the next three chapters, we take a closer look at how these same major integrating systems—hormones and nerves—control function throughout the body beginning with the endocrine system.

KEY TERMS

aldosterone, 672
angiotensin, 672
antidiuretic hormone, 671
Bowman's capsule, 665
collecting duct, 665
distal tubule, 665
drug testing, 668
excretion, 661
filtration, 664
kidney, 663
loop of Henle, 665
nephron, 664
osmoregulation, 672
peritubular capillary, 665
proximal tubule, 665
renal artery, 663
renal vein, 663
renin, 672
tubular reabsorption, 664
tubular secretion, 664
urea, 662
ureter, 663
urethra, 664
uric acid, 662
urinary bladder, 664
urination, 663
urine, 660

HIGHLIGHTS IN REVIEW

1 The excretory system maintains the water, salt, and pH balance of body fluids and rids the animal body of nitrogen-containing metabolic wastes.

- **a]** Excretory systems maintain constancy in an animal's body fluids by removing nitrogen-containing wastes generated by protein breakdown and maintaining the balance of water and salts. Excretory systems generally act directly on an animal's blood or its equivalent. Since materials are exchanged between blood and extracellular fluid, and in turn between extracellular and intracellular fluids, the excretory system indirectly helps to maintain the environment inside of cells.
- **b]** Different types of animals excrete nitrogenous waste in different forms: ammonia, uric acid, or urea. Ammonia requires the most water for removal from the body and uric acid the least.

2 The kidney is the main organ of the vertebrate excretory system. Kidneys filter substances from the blood into excretory tubules; reabsorb useful materials from the tubules back into the blood; and secrete additional substances including drugs into the tubules. Urine, the fluid forming in the tubules, flows through ducts and is excreted from the body.

- **a]** The human kidney filters urea and other substances from the blood and excretes it into ureters, which lead to the saclike urinary bladder. Urine then leaves the body through the urethra.
- **b]** The kidney's functional units are tubules called nephrons, each consisting of a Bowman's capsule, a tangled proximal tubule, a hairpinlike loop of Henle, and a distal tubule that drains into a collecting duct.
- **c]** Each nephron is the site of three main functions: filtration, reabsorption, and secretion. The cuplike Bowman's capsule filters water, urea, and solutes from blood moving through the tuft of capillaries (glomerulus) that surround the capsule.
- **d]** During tubular reabsorption, most of the water, glucose, amino acid molecules, and useful ions leave the nephron tubule and reenter the bloodstream via the nearby peritubular capillaries.
- **e]** During tubular secretion, certain drugs and other organic molecules, as well as excess hydrogen, potassium, and other ions, are actively removed from the blood plasma and secreted into the urine. Water continues to leave osmotically, and the urine reaching the bladder contains only 1 percent of the water originally filtered from blood plasma.
- **f]** In mammals, water balance is regulated by thirst, which is triggered by nerve activity in the hypothalamus of the brain, and by hormones that control the functioning of the nephrons and hence the amount of water they absorb.
- **g]** The kidneys also help regulate normal blood pressure and volume through the secretion of renin (which converts a blood protein to angiotensin, a hormone), which can increase blood volume and pressure.

3 An animal's excretory system is adapted to its environment: Animals inhabiting dry deserts, lush pastures, freshwater lakes or streams, or salty oceans each have adaptations appropriate to their surroundings that maintain the constancy of their internal fluids.

- **a]** Land animals are in constant danger of dehydration. Body coverings help conserve moisture, and so do the kidneys by forming concentrated urine. Other adaptations can be behavioral or physiological.
- **b]** Aquatic invertebrates can be osmoconformers, in which the contents of their body fluids follow that of the external environment, or osmoregulators, in which a more constant internal environment is maintained despite changes in the external surroundings. Aquatic vertebrates are osmoregulators; their excretory systems help maintain body fluids at constant levels that may be isotonic, hypertonic, or hypotonic relative to the surrounding water.

UNDERSTANDING THE FACTS AND CONCEPTS

For Questions 1–5, match each of the descriptions with one or more of the items in the list below. Each item may be used once, more than once, or not at all.

a] urine
b] ammonia
c] uric acid
d] urea

1 Nitrogenous wastes that are produced by the liver.

2 The most toxic of the nitrogenous wastes, but also the one that is most soluble in water.

3 The nitrogenous waste that assists the most in water conservation.

4 Produced from filtrate by the processes of tubular secretion and reabsorption.

5 A water-soluble nitrogenous waste that is less toxic than ammonia.

For Questions 6–10, match the descriptions with the most appropriate item from the following list. An item may be used once, more than once, or not at all.

a] cloaca
b] urethra
c] ureter
d] renal cortex
e] renal medulla

6 A tube lying between the kidney and the urinary bladder.

7 A tube that drains the bladder.

8 Feces, excretory products, and gametes are discharged from this structure.

9 Region where the nephron concentrates the filtrate and helps to conserve water and nutrients.

10 Region where filtration of the blood is the primary activity.

For Questions 11–15, match the descriptions with the most appropriate item from the following list. An item may be used once, more than once, or not at all.

- **a]** filtration
- **b]** tubular reabsorption
- **c]** tubular secretion
- **d]** filtration and reabsorption
- **e]** filtration and secretion
- **f]** filtration, reabsorption, and secretion

11 Primarily the function of the glomerulus and Bowman's capsule.

12 Collectively the function of the nephron.

13 Compounds move from the blood into the tubule.

14 Compounds move from the tubule into the blood.

15 Podocytes are an essential element in this process.

For Questions 16–20, match the descriptions with the most appropriate item from the following list. An item may be used once, more than once, or not at all.

- **a]** collecting duct
- **b]** peritubular capillaries
- **c]** ascending limb of the loop of Henle
- **d]** descending limb of the loop of Henle
- **e]** glomerulus

16 Eventually merge to form the renal vein.

17 Urine flows through this structure to the renal pelvis.

18 A bundle of capillaries surrounded by Bowman's capsule.

19 A tubule that leads into the distal tubule.

20 Primary region where urine is concentrated.

INTEGRATE AND APPLY WHAT YOU HAVE LEARNED

1 Why would production of urea not necessarily be affected in someone experiencing kidney failure?

2 Explain why, in hot, dry climates, athletes drink water on a regular basis, often once an hour, even if they are not thirsty.

3 After working at a hard physical activity on a hot and dry day, a thirsty and dehydrated student drinks several cans of beer. Will the student's thirst be quenched and the body rehydrated?

4 Explain the kidneys' role in regulating blood pressure.

5 Explain why the fish kidney should be considered less an organ of excretion than of osmoregulation. Is this true for freshwater fishes as well as saltwater (marine) fishes?

ANALYSIS

1 Why is the urine of a dehydrated person dark yellow-brown while that of a hydrated person only faintly yellow?

- **a]** The dehydrated person is secreting more pigments into the urine.
- **b]** Urine from a dehydrated person is more concentrated than that of a hydrated person.
- **c]** The glomeruli do not work as well during dehydration, and thus the concentration of pigmented compounds in the urine is higher.
- **d]** Fewer pigmented compounds are being reabsorbed in the dehydrated person.
- **e]** The dehydrated person is producing more highly colored waste compounds.

2 Which of the following statements is false?

- **a]** In osmoconformers, the amount of water entering the body equals the amount of water leaving, whereas in osmoregulators the amount of water entering or leaving, depends on whether the animal lives in fresh water, salt water, or air.
- **b]** The ability of a camel to sustain a high body temperature during the day significantly aids the animal in water conservation.
- **c]** In contrast to dehydration in a human being, in which dehydration has its primary effects in altering the concentration of the blood while the body cells remain hydrated, dehydration in a camel appears to have its effects cellularly throughout the body.
- **d]** Production of highly concentrated urine is not dangerous for a camel since there is no exchange of compounds between urine and the blood.
- **e]** In many animals the function of excretion is shared by several organs including gills, lungs, and skin.

3 Assume that in a patient the region of the nephron including Bowman's capsule functioned normally, but the proximal and distal tubules as well as the loop of Henle had ceased to function in reabsorption. Which of the following conditions would you expect to find in this patient?

- **a]** Dehydration.
- **b]** Excessive fluid retention.
- **c]** Uremic poisoning.
- **d]** Edema.
- **e]** High blood pressure.

4 Which of the functions of the nephron do we rely on when testing the urine of athletes for forbidden drugs? Explain.

- **a]** Filtration.
- **b]** Tubular reabsorption.
- **c]** Tubular secretion.

5 Kidney dialysis differs from normal kidney function in which of the following ways?

- **a]** Dialysis utilizes diffusion whereas nephrons operate by diffusion as well as active transport.
- **b]** Dialysis does not cleanse the blood of materials that are normally removed by tubular secretion.
- **c]** Dialysis by itself does not maintain the level of glucose and other nutrients within homeostatic limits.
- **d]** Dialysis does not assist in the maintenance of normal blood pressure.
- **e]** All of the above are correct.

CHAPTER 30

Hormones and Other Molecular Messengers

A SHORT STORY ABOUT HORMONES

The Efe people, who live in the Ituri rain forest of northeastern Zaire, retain a traditional way of life, hunting small antelope, monkeys, and the occasional elephant with bows and arrows; gathering honey, herbs, roots, and fruit from the deep jungle; and creating "cloth" by pounding and decorating tree bark. The Efe construct small, temporary hunting encampments during five months of the year, and relocate these every two weeks or so. The other months, they camp close to the forest's edge, near villages of agriculturally based people called the Lese. The Efe trade meat, honey, and forest products for the Lese's cultivated bananas, cassava, peanuts, rice, and other crops, and for cotton cloth and iron implements. Young Efe women sometimes marry Lese men, but Lese women never marry Efe men. Perhaps the reason for these one-sided marriage customs is the Efe's stature: the average adult stands 4 ft 4 in. to 4 ft 9 in. tall (142 to 146 cm), whereas their seasonal neighbors of typical adult heights tower over them [FIGURE 30.1].

FIGURE 30.1
American anthropologist Nadine Peacock chats with some Efe people that she is investigating.

Anthropologists have long been fascinated by the Efe, one of ten groups of short-statured people (pygmies) living in seven African countries and numbering 150,000 to 200,000 individuals. In the 1980s, the Efe also became the subjects of endocrinologists—medical specialists who study regulatory molecules called hormones. They were interested in *human growth hormone*, a hormone that controls human height. Recall from CHAPTER 12 [FIGURE 12.7] that the diminutive young girl named Tracy experienced a growth spurt after receiving extra doses of this substance. Researchers suspected that some abnormality in the production or functioning of human growth hormone might cause the Efe's growth patterns. After numerous experiments, the endocrinologists indeed discovered the physiological secret of the pygmy's short stature, and it is related to human growth hormone. We will explain exactly how later in this chapter, as we consider the molecular messengers, the broader category to which hormones belong.

Hormones are molecules produced by one set of cells (usually in a gland) and transported in body fluids to other parts of the body, near or far, where they bind to a target cell and cause a profound change in cell activity [FIGURE 30.2]. Chemicals like growth hormone that leave cells, pass through one of several fluid or gaseous media, and regulate the activities of other cells are called **molecular messengers**. Besides hormones, molecular messengers also include other important kinds of molecules that we will discuss.

Your goal in this chapter is to understand how molecular messengers in general and hormones in particular regulate physical changes in people and other organisms. You will also see how such molecules interact with each other and with the nervous system in the maintenance of homeostasis [see CHAPTER 24].

You will see that molecular messengers act by binding to a specific protein, called a *receptor*, on the outside surface or inside of a target cell, and this binding then triggers some cellular activity. You will also see that hormones and other molecular messengers regulate physiological systems by inhibiting or stimulating change. The body maintains homeostasis by preventing wide fluctuations in internal conditions, but to undergo irreversible developmental events, such as the attainment of adult stature, the body must undergo certain dramatic changes. Molecular messengers mediate both the inhibition and initiation of change. Very similar molecular messengers occur throughout the kingdoms of living things, and this suggests that the regulatory molecules had an ancient evolutionary origin.

This chapter considers the general actions of molecular messengers, then catalogs the major hormones in humans and other mammals. Next, the chapter discusses how hormones prevent or provoke change. It ends with a section on how molecular messengers may have arisen and changed during evolution. Besides our discussions of how hormones determine a pygmy's height, you will learn about the sex attractants of insects, what causes menstrual cramps, how hormones are involved in stress, and about the illegal anabolic steroids some athletes use. ❑

MESSAGES

1 Molecular messengers are substances produced by one group of cells that affect the activities of other groups of cells which may be quite distant in the body. Hormones secreted by cells of the endocrine system are an important type of molecular messenger.

2 Hormones are often secreted by endocrine glands. Endocrine glands often regulate each other's function via feedback loops. The brain's hypothalamus and pituitary, for example, regulate several other endocrine organs, and these organs in turn regulate the hypothalamus and pituitary.

3 Hormones act by binding to receptor proteins in or on cells, and the binding changes the cell's activities. The new cellular activities can prevent further change in the levels of substances in the body; control the rate of metabolism; drive daily, monthly, or yearly cycles of change; or cause permanent developmental changes in an individual.

[A] Relationship of endocrine system to the circulatory system

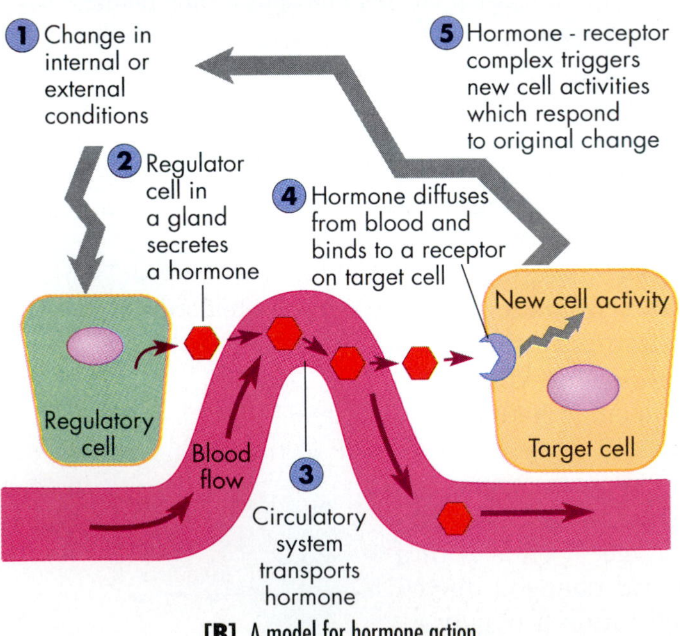

[B] A model for hormone action

FIGURE 30.2

The Endocrine System and Hormone Action.

[A] The endocrine system. **[B]** A model for hormone action.

Molecular Messengers

Molecular messengers help control the amount you eat each day; how quickly you grew during childhood and adolescence; whether you are happy, depressed, or tense; how much salt and sugar flows in your blood; and a thousand other aspects of your physiology and behavior. Molecular messengers are clearly central to survival and normal functioning. What are they, and how do they work?

Many molecular messengers are products of the **endocrine system**, a group of organs at various locations in the body that secrete hormones into body fluids. FIGURE 30.2A presents an overview of the endocrine system at the levels of cells and organs. A molecular messenger such as growth hormone that determines a person's height—whether Efe tribe member or American college student—is secreted by a **regulator cell**, which detects perturbations in its environment and emits the molecular messenger in response [see FIGURE 30.2B, Steps 1 and 2]. The messenger then diffuses through a fluid medium—the blood in the case of most hormones, but air, water, or extracellular (tissue) fluid in the case of other molecular messengers [Step 3]. The hormone then acts on one or more **target cells** that detect the molecular messenger [Step 4] and respond by carrying out a cellular activity that adjusts to the original perturbation [Step 5].

Why does a molecular messenger act on its target cells but not other cells? Each target cell contains **receptors**, each a protein of a specific shape that is complementary to that of a particular molecular messenger. Nontarget cells lack these receptors. When a messenger comes along and binds to a receptor, like a key fitting into a lock, the receptor changes shape. This alteration heralds the messenger's arrival and triggers events that produce a change in cellular activity.

Some molecular messengers are secreted by isolated or individual cells within the animal's body, but most are produced by groups of cells organized into secretory organs called **glands**. This brings us to the organ level of our consideration of the endocrine system and shows how cellular-level mechanisms underlie and explain events at the organ level. A *ductless organ* that dumps molecular messengers directly into the extracellular fluid and blood is called an **endocrine gland** [FIGURE 30.3A]. The pituitary, adrenal, and thyroid glands are all endocrine glands, and most hormones are produced in these and about 10 other endocrine glands. In contrast, an organ that dumps molecular messengers or other materials into *ducts* that generally lead out of the body is called an **exocrine gland** [FIGURE 30.3B]. Exocrine glands produce and release body exudates such as saliva, milk, sweat, and the silken threads that silkworms and spiders spin into cocoons and webs.

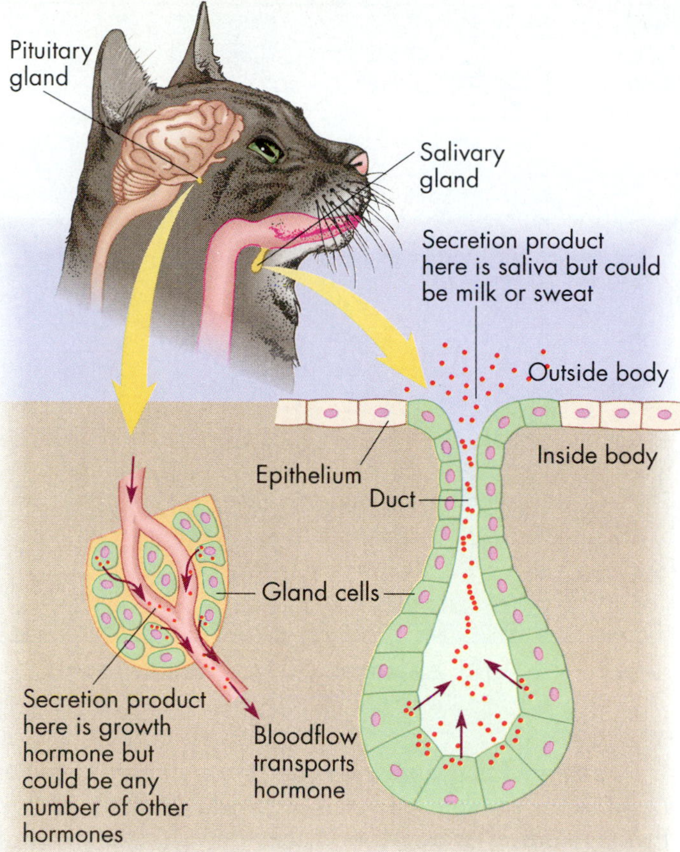

FIGURE 30.3

Glands with and without Ducts.

[A] An endocrine gland such as the pituitary secretes hormones into the extracellular fluid. From there, the chemicals pass into blood vessels and travel to distant sites in the body. [B] An exocrine gland such as the salivary gland secretes saliva, milk, sweat, enzymes, silk, or other materials into a duct, which leads the substance out of the body or into the digestive tract.

TYPES OF MOLECULAR MESSENGERS

Five types of molecular messengers leave individual cells or glands (groups of cells) and trigger activity in different cells, as diagrammed in FIGURE 30.4. Biologists distinguish between them because they play such markedly different roles in the physiology of animals.

Paracrine hormones are primitive hormones secreted by individual regulator cells; they diffuse through the extracellular fluid and act only on adjacent target cells [FIGURE 30.4A]. Regulator cells in our intestines, for example, detect protein molecules from food and secrete a paracrine hormone that causes adjacent target cells to secrete protein-digesting enzymes [review FIGURE 28.17 and TABLE 28.1]. Without this paracrine secretion, your body would extract less nutrition from the food you eat.

Neurotransmitters are nerve-signaling compounds released by individual nerve cells. Neurotransmitters diffuse a short distance from a nerve cell to another cell [FIGURE 30.4B]. For example, when a basketball player wants to make a jump shot, specific nerve cells running from his or her lower back down both legs release a neurotransmitter. This substance diffuses across a short space to muscle cells in the calves and causes the lower leg muscles to contract and propel the player's body upward.

Neurohormones are hormones secreted by nerve cells. Neurohormones travel through the bloodstream to distant target cells elsewhere in the body [FIGURE 30.4C]. For example, nerve cells in the brain grow extensions to the pituitary gland and release the hormone *oxytocin*. The oxytocin then travels to a woman's uterus and stimulates contractions during birth.

True hormones are substances usually secreted by endocrine glands that enter the bloodstream and act on cells elsewhere in the body [FIGURE 30.4D]. A weight-lifting program, for example, stimulates the secretion of growth hormone from the pituitary gland. The growth hormone circulates in the blood and acts on bone and muscle cells, stimulating their growth.

Pheromones are compounds secreted by one individual that affect the behavior of another individual of the same species. Pheromones may be secreted by exocrine glands, leave the body via ducts, travel in air or water, and stimulate target cells in other organisms located kilometers away [FIGURE 30.4E]. In a sexually mature female silk moth, for example, a scent gland in her abdomen secretes a pheromone that wafts on air currents for distances of up to 11 km (7 mi). Pheromone molecules that happen to hit target cells in a male silk moth's antennae can trigger wild sexual excitation. This leads the male to fly long distances toward the origin of the pheromone—a potential mate.

Our next section focuses on how these five types of messengers interact in a mammal's endocrine system.

➤ CONCEPT CHALLENGE

What would happen in an animal's body if a hormone were produced and circulated in the blood, but no cells were present bearing receptors for that hormone?

The Endocrine System

Hormones are among the best-studied molecular messengers, and mammals—especially humans—are perhaps the best-studied sources. Biologists know in great detail, for example, how hormones govern the various stages of reproduction and development and maintain homeostasis within each system, organ, and cell.

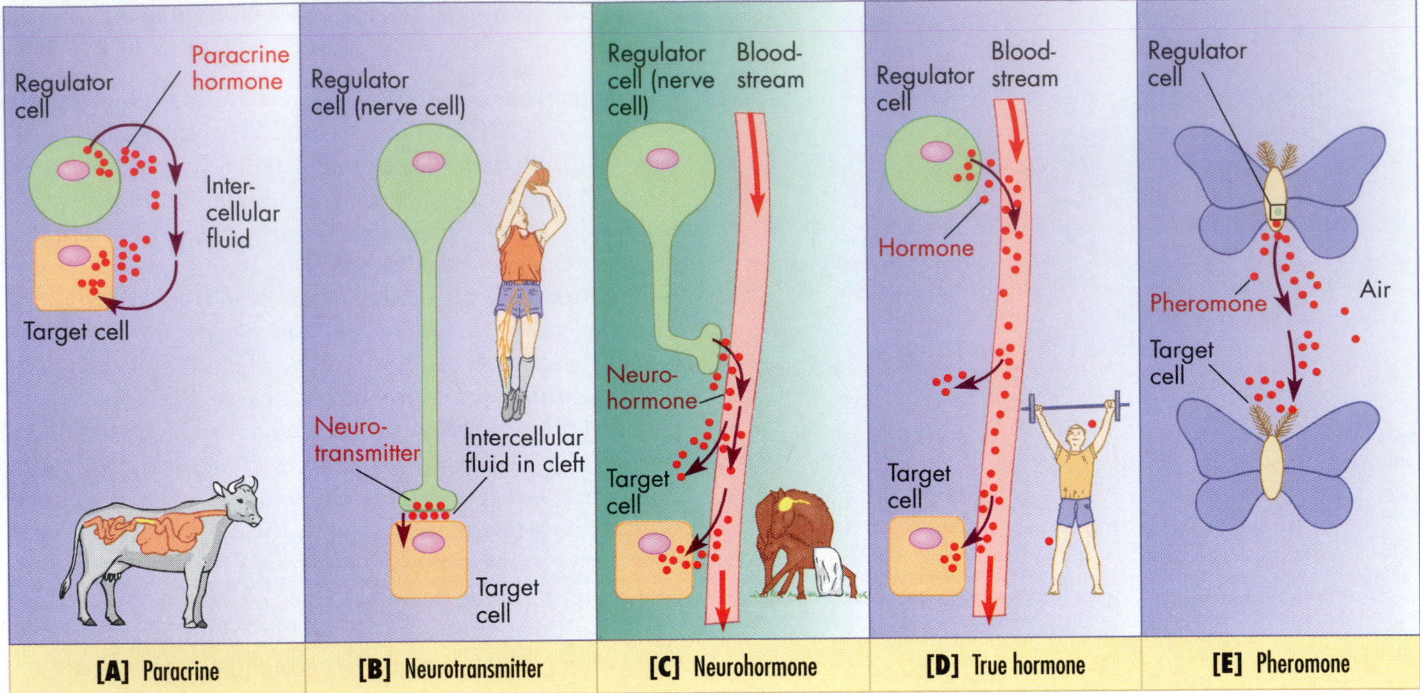

FIGURE 30.4

Molecular Messengers: Targets and Transport.

[A] Paracrine hormones diffuse a short distance and act on an adjacent cell. **[B]** Some nerve cells secrete neurotransmitters that diffuse across a narrow cleft to a target cell. **[C]** Some nerve cells secrete neurohormones that pass into a blood vessel, travel some distance, then diffuse out again and reach a target cell. **[D]** True hormones are secreted by cells in endocrine glands and enter the bloodstream, travel some distance, diffuse out again, and act on a target cell. **[E]** Pheromones are secreted by regulator cells in exocrine glands, travel in a duct, leave the body, and stimulate target cells in another organism's body.

HORMONE STRUCTURE AND THE ENDOCRINE SYSTEM

Although mammalian hormones perform an incredible variety of tasks, most belong to just four molecular groups: the polypeptides, the steroids, the amines, and the fatty acid derivatives.

Polypeptide hormones, including growth hormone and oxytocin, are strings of from 3 to 200 amino acids [FIGURE 30.5A].

Steroid hormones, which are synthesized from cholesterol, contain four joined rings of carbon atoms with various atoms attached to the rings. Steroid hormones include *estrogen* and *testosterone* generated in the human ovaries and testes, respectively, and *ecdysone*, the hormone that causes insects to shed their old skin [FIGURE 30.5B].

Amine hormones contain an amino group ($-NH_2$). Although they are derived from amino acids, the acidic functional group is often removed. This group of hormones includes thyroid hormones, which alter metabolic rates [FIGURE 30.5C].

Fatty acid hormones, such as *prostaglandins*, are derived from straight-chain fatty acids [FIGURE 30.5D]. Prostaglandins were first discovered in semen and named after their supposed source, the prostate gland. They have since been discovered in most mammalian cells and tissues, however, and among other things, they cause muscles to contract and blood vessels to open or close [see CHAPTER 14]. You can see the effect of prostaglandins on your own body. The next time you have a headache or menstrual cramps, consider that the pain is a result of the body's production of prostaglandins and their ability to cause blood vessels in the brain or muscle fibers in the uterus to contract. Drugs such as aspirin and the very similar ibuprofen block the production of prostaglandins, and in this way relieve the pain for many people.

These four types of hormones are secreted in mammals, as well as other vertebrates, by about a dozen major endocrine glands, similar in most cases to the major human endocrine glands shown in FIGURE 30.6. Together, these ductless glands make up the **endocrine system**, one of the body's major physiological networks. Much of the rest of this chapter is devoted to a gland-by-gland tour.

PITUITARY AND HYPOTHALAMUS

Recall that together the endocrine and nervous systems coordinate and integrate body activities, and the site of their major interaction is the brain. At the base of the brain lies a collection of nerve cells called the **hypothalamus**, and hanging from this region is a fingertip-sized bulb called the **pituitary gland** [FIGURE 30.7]. A bit like a light bulb in shape, the mature pituitary in an adult has

FIGURE 30.5

Generalized Structures of the Major Hormone Groups.

[A] A polypeptide hormone is a string of amino acids. Human growth hormone, shown here, is 246 amino acids long and controls metabolism and growth of bone, muscle, and cartilage cells. **[B]** Steroid hormones contain four interlocking rings of carbon atoms in the characteristic form shown. **[C]** Amine hormones such as thyroxine have an amine group ($-NH_2$). Thyroxine controls the rate of metabolism. **[D]** Some messengers are made from fatty acids. These include prostaglandins, which cause blood vessels to constrict, leading to headaches and menstrual cramps.

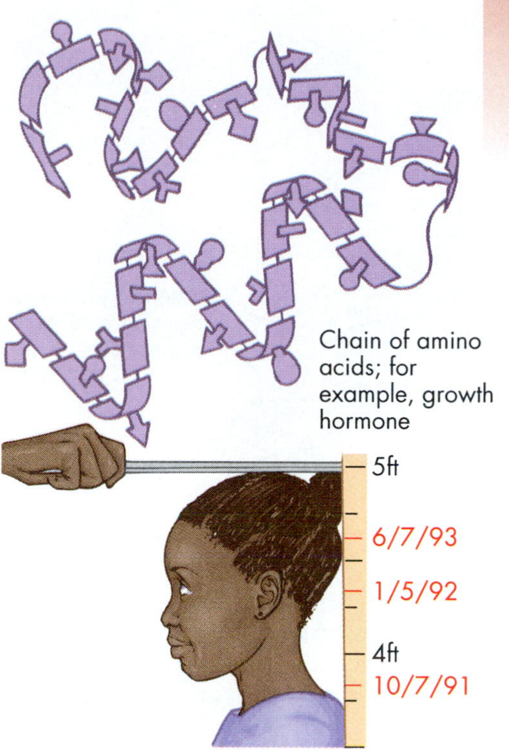

[A] Peptide hormone

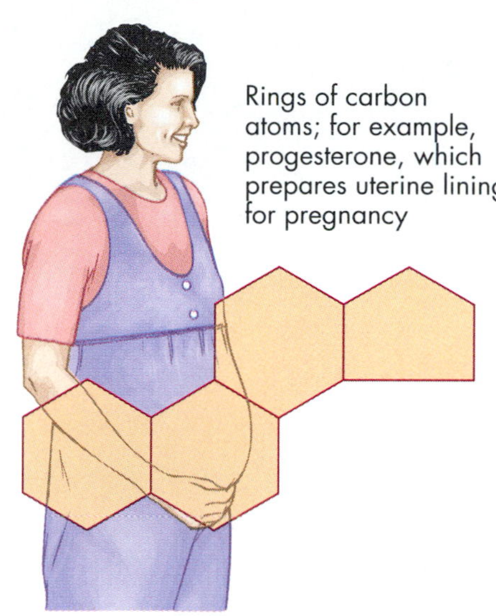

[B] Steroid hormone

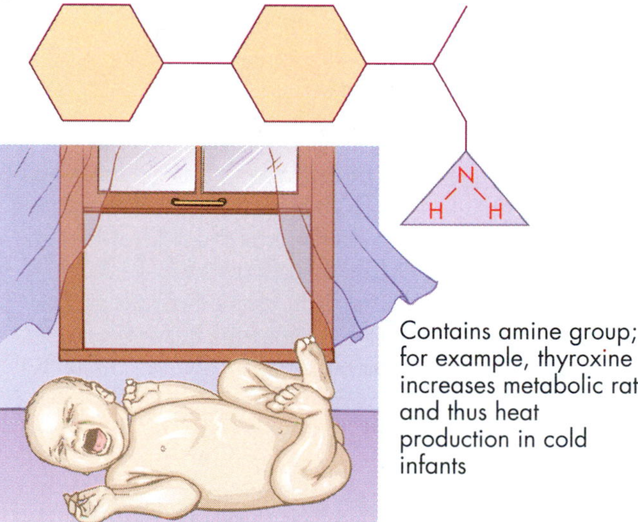

[C] Amine hormone

[D] Fatty acid hormone

two major portions, the posterior and anterior lobes. Together, these two lobes secrete about 10 kinds of peptide hormones, many of which help control other glands throughout the endocrine system.

Posterior Pituitary In the strictest sense, the **posterior lobe** of the pituitary gland is more of a storage depot than an actual endocrine gland. This is because the two hormones it secretes, oxytocin and antidiuretic hormone, or ADH [TABLE 30.1] are actually made in the hypothalamus and transported to the posterior pituitary in fine fibers that are extensions of the hypothalamic nerve cells; such fibers are called **axons** [FIGURE 30.7, Step 1]. The

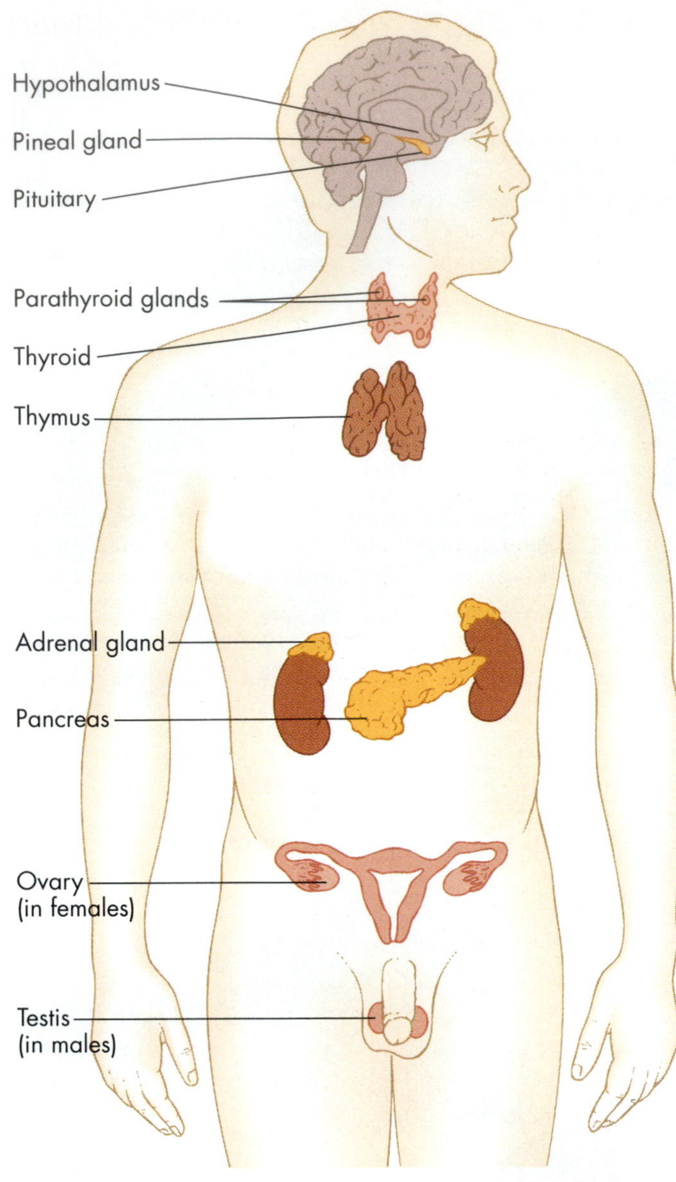

FIGURE 30.6

Major Glands of the Human Endocrine System.

In addition to the endocrine organs shown here, some organs not usually thought of as endocrine glands also secrete hormones. These include the heart, which secretes a substance that affects sodium excretion, and the kidney, which secretes a substance that controls red blood cell production.

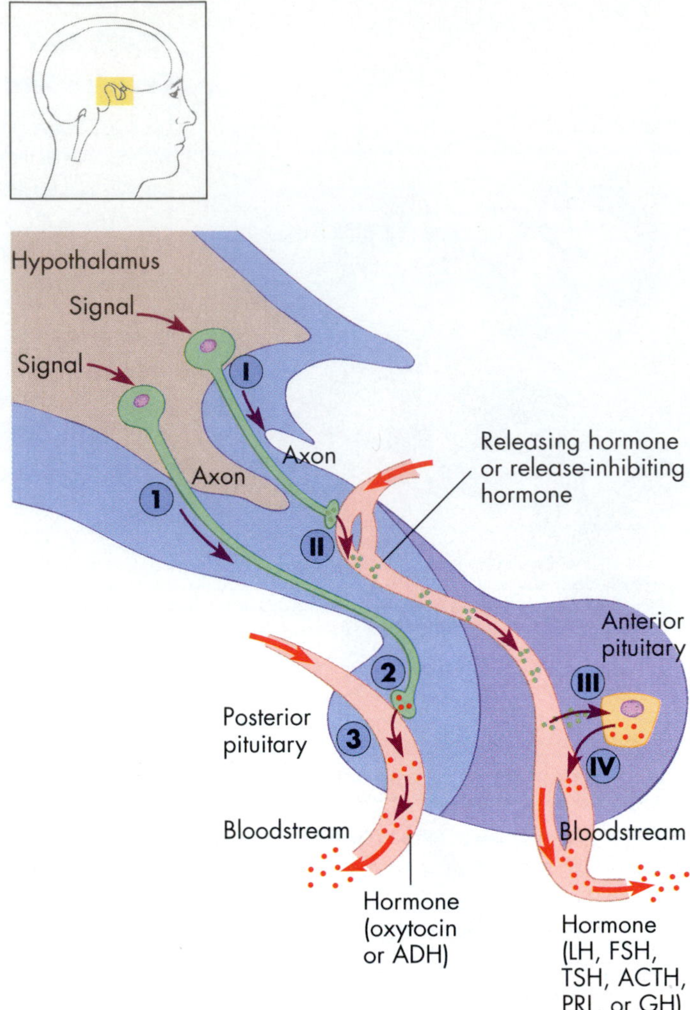

FIGURE 30.7

How the Hypothalamus Controls the Pituitary.

Signals from the hypothalamus (1) cause hormones stored in the endings of nerve cells in the posterior pituitary (2) to be released into the bloodstream (3). The hypothalamus controls the anterior pituitary in a slightly different way: Nerve cells in the hypothalamus (I) store releasing and release-inhibiting hormones that enter special blood vessels (II), travel to the anterior pituitary (III), and stimulate cells there to secrete hormones (IV), which in turn enter the bloodstream and travel to other body regions.

hormones are stored in the axon tips (Step 2). When signals from other parts of the brain stimulate the hypothalamus, it sends nerve impulses down to the axon tips in the posterior pituitary, causing the axon tips to release oxytocin or ADH (Step 3). After diffusing into the bloodstream, the hormone circulates and reaches target cells elsewhere in the body.

A nursing mother demonstrates how the posterior pituitary works [FIGURES 30.7 and 30.8]. As the baby begins to suckle, sensitive nerves in the woman's nipples send signals to the secretory cells in the hypothalamus, which in turn send nerve impulses to their axon tips and cause the posterior pituitary to release the peptide hormone *oxytocin* into the bloodstream. When oxytocin reaches the musclelike target cells lining the milk ducts of the woman's mammary glands (which are exocrine glands), it causes the target cells to contract, forcing milk into the ducts, out the nipple, and into the baby's mouth.

Anterior Pituitary The **anterior lobe** of the pituitary gland synthesizes and secretes more than half a dozen of its own hormones, including growth hormone, but the

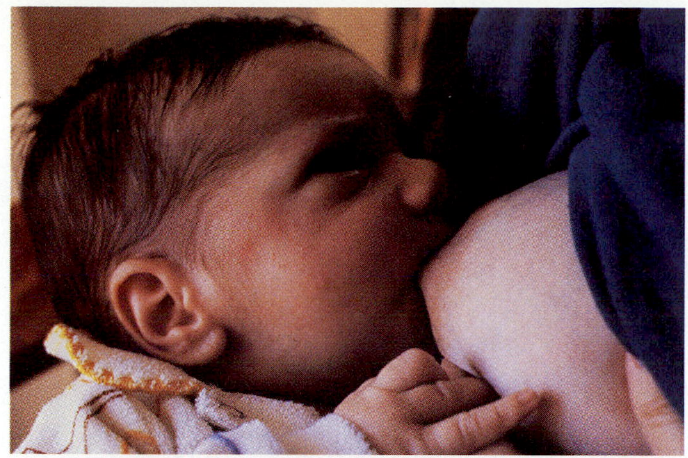

FIGURE 30.8

The Pituitary Controls Milk Production and Release.

When a baby nurses, the sucking action signals the mother's hypothalamus and pituitary to secrete prolactin, which causes mammary gland cells to release milk, and oxytocin, which causes the nipple to squirt milk.

hypothalamus controls the timing of their secretion [see TABLE 30.1]. As in the posterior lobe, axons run from the hypothalamus toward the anterior pituitary [FIGURE 30.7, Step I]. These axons relay impulses from the hypothalamus based on how the hypothalamus monitors conditions in the body and environment. The axon tips store two kinds of small peptide neurohormones that stimulate or inhibit hormone production by the anterior pituitary: **hypothalamic releasing hormones** and **hypothalamic release-inhibiting hormones**.

When stimulated, axons from the hypothalamus liberate their inhibitory or releasing hormones into a special circulatory pathway that carries blood directly to the anterior pituitary (Step II). A specific releasing hormone (Step III) causes anterior pituitary cells to secrete another specific hormone (Step IV) into the bloodstream, and eventually the hormone reaches target cells elsewhere. A release-inhibiting hormone causes anterior pituitary cells to stop secreting a specific hormone. The hormone *prolactin* secreted from the anterior pituitary causes a woman's breast to secrete milk into the milk duct. A specific release-inhibiting hormone blocks the release of prolactin from the pituitary.

Negative Feedback Loops In addition to their control of breast-feeding, the hypothalamus and pituitary gland participate in several regulatory feedback loops that govern the activities of the testes, ovaries, thyroid, and adrenal glands. In each case, the hypothalamus produces a releasing or inhibiting factor that causes the anterior pituitary to secrete or not secrete a peptide hormone. If secreted, the peptide hormone then travels in the blood, eventually reaching the gonad or other gland, causing it to secrete a third hormone, generally a steroid or amine hormone. As the level of this third hormone builds, it eventually feeds back to the hypothalamus and/or pituitary and blocks additional hormone release. FIGURE 24.9 depicts the general strategy of feedback loops, and FIGURES 14.4 and 14.6 show how such negative feedback circuits affect the gonads. Here we will focus on the thyroid, parathyroid, and adrenals.

THE THYROID AND PARATHYROID GLANDS

The hypothalamus and pituitary control two organs in the neck, the thyroid gland and the parathyroid glands.

The Thyroid Gland The **thyroid gland** acts as the body's metabolic thermostat, regulating its use of energy as well as its growth. The most abundant thyroid hormone is **thyroxine**, an iodine-containing amine hormone that governs both metabolic and growth rates and stimulates nervous system function. A child with an underactive thyroid gland (a condition called *cretinism*) experiences low rates of protein synthesis and carbohydrate breakdown and thus low body temperature and sluggishness. The child's growth also becomes stunted, and mental development is retarded. An adult with an underactive thyroid gland and too little thyroxine secretion (a condition called *myxedema*) may gain weight easily and be slow mentally. Fortunately, doctors can treat these conditions by administering thyroxine.

A person can develop a **goiter**, an enlarged thyroid visible as a lump in the neck [FIGURE 30.9]. This enlargement can be understood in terms of the negative feedback loop that regulates thyroid activity. If the iodine concentration in the diet is low, the thyroid cannot make much thyroxine, and the pituitary, sensing low levels of thyroxine, signals the thyroid to make more. The thyroid then responds by growing larger. The iodine in ordinary iodized table salt supplies iodine needed for the synthesis of thyroxine and so can help prevent goiters from forming.

Parathyroid Glands The thyroid gland and the associated **parathyroid glands**, a set of four small, dark patches of cells embedded behind the pale thyroid, work together and keep the level of calcium in the blood within very narrow limits. Without correct amounts of circulating calcium ions, nerves and muscles cannot function properly, and the person or other animal suffers nerve spasms, muscle contractions, and rapid death.

When blood levels of calcium ions (Ca^{2+}) start to rise—after you drink milk, for example—the thyroid releases the hormone **calcitonin**, which causes excess calcium to be deposited in bones. But if the calcium levels in the blood start to fall, the parathyroid glands release **parathyroid hormone** (or *parathormone*), and this causes bones to release more calcium and the body to pick up

TABLE 30.1 Major Vertebrate Endocrine Tissues and Hormones

Tissue	Hormone	Target	Major Actions
Hypothalamus	Releasing and inhibiting hormones	Anterior pituitary	Stimulate or inhibit release of specific pituitary hormones
Anterior pituitary	Thyroid-stimulating hormone (TSH)	Thyroid	Stimulates synthesis and secretion of thyroxine
	Prolactin (PRL)	Mammary gland	Stimulates milk synthesis
	Adrenocorticotropic hormone (ACTH)	Adrenal cortex	Stimulates synthesis of sex steroids, mineralocorticoids, glucocorticoids
	Endorphins	Brain	Decrease pain
	Growth hormone (GH)	Many cells	Stimulates general body growth
	Luteinizing hormone (LH)	Ovary	Stimulates ovulation and synthesis of estrogen and progesterone
		Testis	Stimulates testosterone synthesis
	Follicle-stimulating hormone (FSH)	Ovary	Stimulates growth of ovarian follicle
		Testis	Stimulates sperm production
Posterior pituitary	Oxytocin	Mammary gland	Milk ejection
		Uterus	Uterine contraction
	Antidiuretic hormone (ADH)	Kidney	Increases water absorption
Thyroid	Thyroxine	Most cells	Increases metabolic rate and growth, causes metamorphosis in amphibians
	Calcitonin	Bones	Stimulates calcium uptake
Parathyroid	Parathyroid hormone	Bones	Stimulates calcium release into blood
		Digestive tract	Stimulates calcium uptake into blood
Adrenal medulla	Epinephrine	Circulatory system	Increases heart rate, blood pressure, and blood sugar
		Respiratory system	Increases breathing rate and clears airways
	Norepinephrine	Generally same as epinephrine	
Adrenal cortex	Sex steroids	Unknown	Stimulate growth of body hair in women
	Mineralocorticoids (e.g., aldosterone)	Kidneys	Increase sodium conservation
	Glucocorticoids (e.g., cortisol)	Many cells	Stimulate carbohydrate metabolism and decrease inflammation

Tissue	Hormone	Target	Major Actions
Pancreas	Insulin	Many cells	Stimulates glucose uptake from blood
	Glucagon	Many cells	Stimulates glucose release from cells into blood
Pineal	Melatonin	Hypothalamus	Blocks secretion of LH- and FSH-releasing factors
		Brain	Promotes sleep
Ovary	Estrogen	Many cells	Stimulates female development and behavior
	Progesterone	Uterus	Stimulates growth of uterine lining
Placenta	Chorionic gonadotropin (hCG)	Corpus luteum in ovary	Stimulates progesterone synthesis
	Placental lactogen	Mammary gland	Stimulates mammary gland development
Testis	Testosterone	Many cells	Stimulates male development and behavior
	Müllerian inhibiting hormone	Pre-oviduct cells	Kills pre-oviduct cells
Thymus	Thymosin	White blood cells	Stimulates differentiation
Gastrointestinal tract	Gastrin	Gut cells	Stimulates hydrochloric acid secretion
	Cholecystokinin (CCK)	Pancreas	Stimulates digestive enzyme secretion
Kidney	Erythropoietin	Blood cell precursors	Stimulates red blood cell production

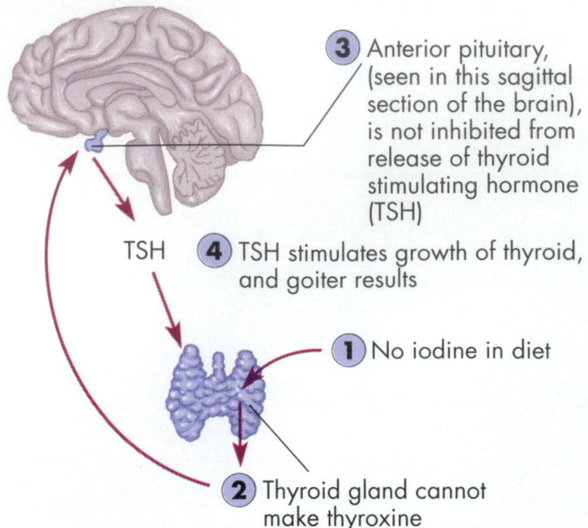

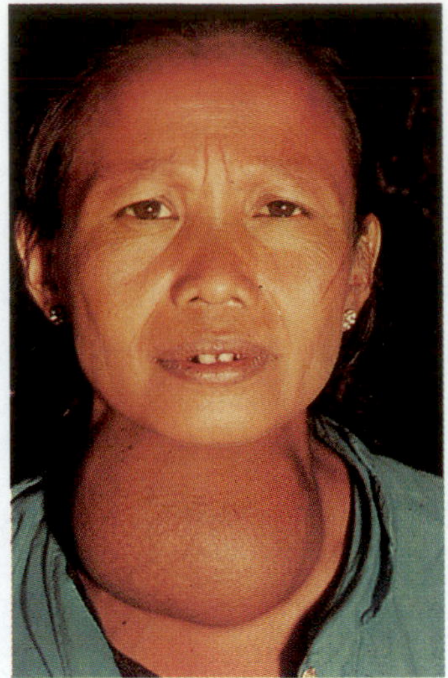

FIGURE 30.9

Goiter: The Result of a Disrupted Feedback Loop.

Normally, thyroid hormone, acting via a negative feedback loop, blocks the release of thyroid-stimulating hormone from the pituitary. Without iodine, however the feedback loop is altered, and a goiter, an enlarged thyroid gland, may result.

more of it from food. Note the counteracting relationship of these two calcium-regulating hormones and their interacting negative feedback loops. The regulation of calcium levels is so crucial that parathyroid hormone is one of only two hormones we absolutely need for survival. The other is aldosterone, a product of the adrenal glands.

THE ADRENAL GLANDS: THE STRESS GLANDS

Sitting atop each of our kidneys is an **adrenal gland** [FIGURE 30.10A] that enables our bodies to react quickly to danger by fleeing or fighting. Each adrenal gland has a middle portion, or **medulla**, with an outer covering, or **cortex**.

Modified nerve cells in the adrenal medulla secrete two similar amine hormones: **epinephrine** (*adrenaline*) and **norepinephrine** (*noradrenaline*) [see TABLE 30.1]. When an animal faces a sudden physical threat, such as a hungry lion, an angry boss, or a difficult college exam, nerves trigger the secretion of the hormones [FIGURE 30.10B, Steps 1 and 2], which in turn act on target cells in the heart, lungs, intestines, and elsewhere, bringing about a so-called *stress response* (Step 3). The hormones cause the heart to beat faster, blood sugar levels to increase, the breathing rate to speed up, and blood to be shunted away from organs such as the stomach and intestines. The result of all this is that the muscles receive

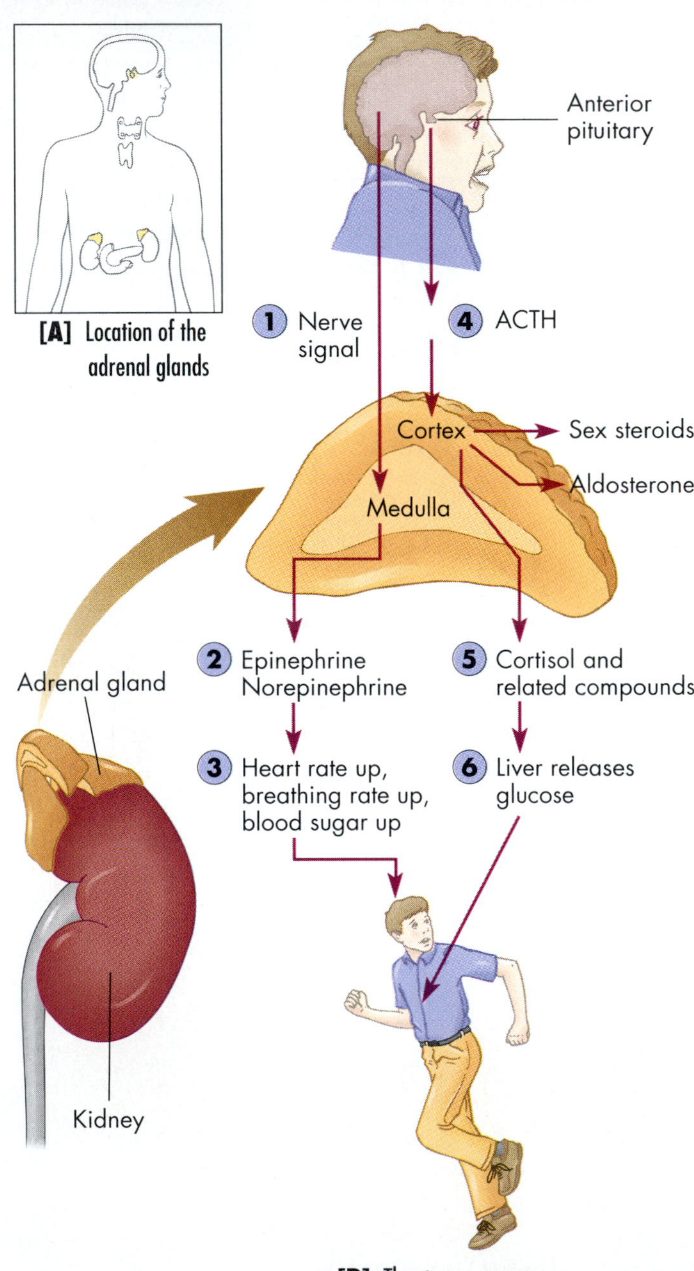

FIGURE 30.10

Adrenal Hormones and the Stress Response.

[A] The adrenal glands, masters of the fight-or-flight response, are located on top of the kidneys. **[B]** As the text explains, a sudden physical threat triggers the adrenal glands (by way of nerve signals from the brain) to release epinephrine, cortisol, and other hormones that allow the body to increase its activity and respond to the threat with self-defense (fight) or a rapid exit (flight).

more blood, oxygen, and sugar, and thus the animal is better able to defend itself or move away to safety.

The adrenal cortex, together with the medulla, secretes additional hormones that help an animal respond to stress. One, called **cortisol** (or *hydrocortisone*), is a member of the class of steroid hormones called *glucocorticoids*. These hormones speed the metabolism of sugars, proteins, and fats. When an animal is stressed, perhaps by cold, strenuous activity, or pain, its hypothalamus sends a releasing hormone to its anterior pituitary, which responds by secreting **ACTH**, or adrenocorticotropic hormone [FIGURE 30.10B, Step 4]. ACTH triggers the adrenal cortex to produce cortisol and related compounds (Step 5). Acting on target cells in the liver, fat tissue, and most other organs, cortisol causes stored proteins, lipids, and carbohydrates to be broken down and glucose to be rapidly generated (Step 6), providing stressed cells with a source of quick energy.

Inside anterior pituitary cells, ACTH can also break down, yielding **beta-endorphin**, a hormone that acts on pain receptors in the brain—an action that decreases the amount of pain a stressed individual senses. For example, recent studies show that when a dentist injects the hypothalamic-releasing hormone for ACTH into a person before extracting his or her wisdom teeth, the patient releases more beta-endorphin and experiences less pain.

Just as the body needs calcitonin and parathyroid hormone to maintain proper levels of calcium, it needs one final hormone from the adrenal cortex to help regulate mineral ions in body fluids. A steroid hormone and member of the *mineralocorticoid* class, **aldosterone** causes the kidney to conserve sodium and water and to excrete potassium [review FIGURE 29.12]. Without this chemical messenger, the body would excrete too much sodium and water in the urine, the blood volume would drop precipitously, and death would come quickly.

The adrenal cortex also secretes steroid sex hormones. By means of hormones and feedback loops, the adrenal cortex and the adrenal medulla act together to help an animal maintain homeostasis as well as overcome stress. What's at stake here is preventing or promoting change, and that is the major business of all glands and hormones.

➤ CONCEPT CHALLENGE

Men who take anabolic steroids have small testicles. Why might that be true? Consider two facts: (1) a hormone from the brain stimulates the function of the testes, which in turn produce testosterone; and (2) high levels of testosterone or anabolic steroids block the release of the brain hormone in a negative feedback loop.

Hormones, Homeostasis, and Change

Because hormones, homeostasis, and change go hand in hand at all levels of physiological activity, their interactions warrant closer examination.

HORMONES AND HOMEOSTASIS

A painful reminder of our need to maintain a constant internal environment is **diabetes mellitus**, a condition in which blood glucose levels are not maintained within a normal range. Diabetics produce copious amounts of dilute, sugary urine and are nearly always thirsty. Without treatment, they risk *coma* (unconsciousness) due to severe dehydration or the accumulation of poisons in the blood. Diabetes is the most common hormonal disorder in humans, and it stems from changes in the **pancreas**, an elongated organ nestled near the intestines [FIGURE 30.11A]. The normal pancreas has exocrine cells that secrete digestive juices into a duct leading to the small intestine [review TABLE 28.1], as well as islands of endocrine cells called the *islets of Langerhans*. These endocrine cells normally secrete the peptide hormones **glucagon** and **insulin**, but a diabetic's islets cannot make enough insulin, and this creates homeostatic chaos.

In most people, glucagon and insulin work together by means of a feedback loop to make fine adjustments in blood sugar levels. After a breakfast of cereal and toast, nutrients enter the person's bloodstream, and blood sugar levels rise [FIGURE 30.11B, Step 1]. In response, one group of endocrine cells in the pancreas, the **beta cells**, secrete insulin (Step 2), and this hormone causes target cells in the liver and other organs to remove glucose from the blood and store it as glycogen (Step 3). Then, as daily activities cause blood sugar levels to fall (Step 4), a second group of pancreatic endocrine cells, the **alpha cells**, detect the drop and respond by releasing glucagon (Step 5). This causes liver cells to break down glycogen and release glucose, enabling blood levels to once again rise and fuel body action (Step 6). Like a car's brake and accelerator, this dual control helps maintain precise blood sugar levels.

In a person with diabetes, the pancreatic beta cells fail to generate insulin. Without sufficient insulin, body cells fail to remove glucose from the blood after a meal, so its concentration in the blood increases and the kidneys remove the sugar along with large amounts of water from the blood. This results in copious, sweet urine, as well as thirst and the danger of dehydration. What's more, without insulin, the liver, brain, and other body cells cannot

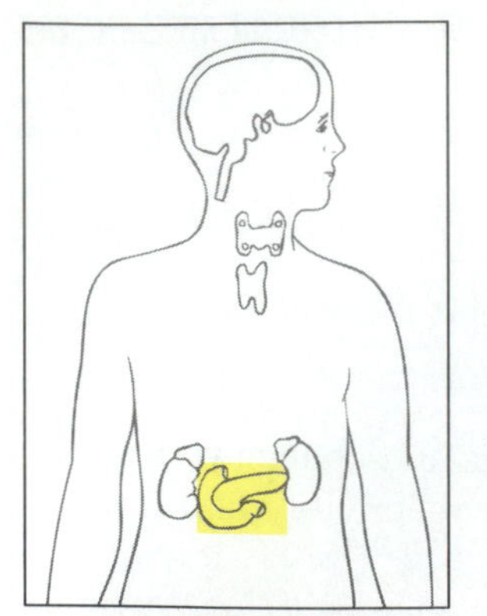

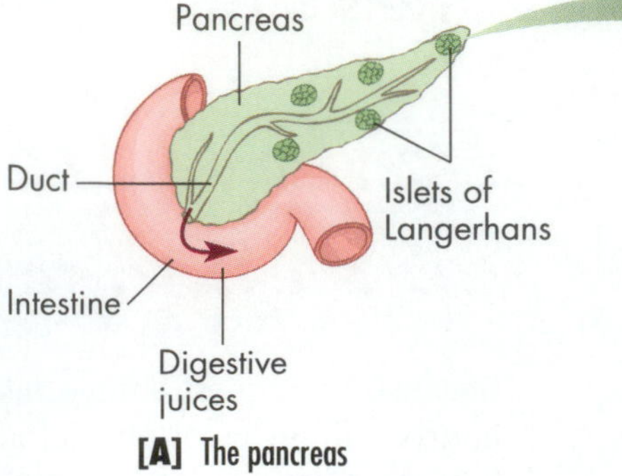

[A] The pancreas

[B] Hormonal control of blood glucose

FIGURE 30.11

Two Pancreatic Hormones Regulate Blood Sugar Levels.

[A] The pancreas contains small groups of cells called islets of Langerhans, which regulate the body's blood levels of glucose. **[B]** After a healthy person eats carbohydrates, glucose subunits are cleaved and absorbed into the blood (1). The beta cells of the islets release insulin (2), which stimulates the liver and other cells to remove glucose from the blood and store it in the form of glycogen (3). When blood levels of glucose drop (4), the secretion of glucagon from the alpha cells of the islets (5) causes the liver to release some of its stored sugar into the blood (6). This push-pull relationship of glucagon and insulin helps maintain blood glucose at a fairly constant level despite the meals we eat and the fuel we use up during physical activity.

remove glucose from the blood, and thus they "starve" in the midst of plenty. Although diabetics often crave carbohydrates, eating carbohydrate-rich foods does not help: Their glucose-starved cells burn lipids as fuel, and ketone by-products of lipid breakdown make the blood so acidic that it can lead to coma and death. The push-pull relationship of glucagon and insulin is clearly crucial to homeostasis and day-to-day health.

Fortunately, diabetes can now be largely controlled with insulin treatment. However, long-term damage to fingers and toes and to the retina due to poor circulation still occurs as some diabetics grow older. Recent tests on a large group of diabetic patients showed that a careful monitoring of blood sugar and injections of glucose several times a day can nearly stop these harmful side effects.

HORMONES AND CYCLIC PHYSIOLOGICAL CHANGES

An animal's life follows daily, monthly, and yearly cycles orchestrated by hormones. Evidence is mounting that such cycles are tied to daily, or **circadian** rhythms (sir-KAY-dee-uhn; from the Latin *circa*, "about," and *dies*, "day"). These cycles reflect the changing length of daylight during the four seasons. Additional evidence implicates a small brain structure, the **pineal gland** (PIHN-ee-uhl) [review FIGURE 30.6 and TABLE 30.1], in the measurement of day length and the control of reproductive cycles, onset of puberty, and moods.

The pineal gland—so named for its resemblance to a pinecone—is a bumpy, thimble-sized knob lying deep in the brain. In lizards and certain other vertebrates, the pineal gland is called the third eye because it is structurally similar to the retina and actually perceives light directly. In humans and other mammals, the pineal gets secondhand information in the form of nerve stimulation from light receptors in the eyes. Regardless of species, however, the gland secretes an amine hormone called **melatonin** in response to darkness, and this hormone promotes sleep and inhibits the activity of the gonads via melatonin receptors in the hypothalamus.

Melatonin may have yet another effect: Some people experience depression, oversleeping, weight gain, and tiredness when the days grow short in winter, then a rebound of better spirits in spring. Melatonin levels are blamed for this *seasonal affective disorder* (SAD), and sufferers have been successfully treated by exposure to bright lights for several hours per day in winter.

While the evidence is not clear-cut in humans, melatonin may influence reproduction, puberty, and mood. In Finland, it is most common for couples to conceive fraternal twins and triplets (a sign of highly active ovaries) in July, when days are very long and melatonin levels are lowest, and it is least common in January, when nights are long and melatonin levels are highest. And just before puberty, melatonin levels suddenly drop by 75 percent, suggesting that the pineal hormone may have been inhibiting gonad development during childhood.

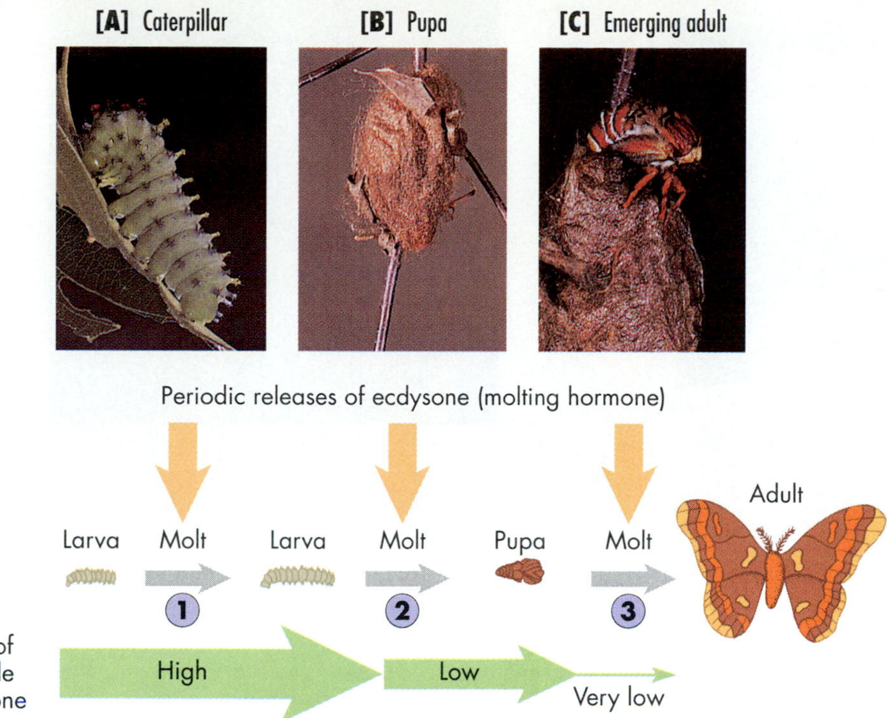

FIGURE 30.12

From Silkworm to Silk Moth: Hormones Provoke Metamorphosis.

[A] A caterpillar, the larval form of a moth, sheds its skin, or molts, five times as it grows larger. **[B]** Eventually it spins a cocoon, within which it changes into a pupa, and then an adult. **[C]** Finally, the adult emerges and spreads its wings. **[D]** These changes in form are directed by hormones. As the caterpillar grows, five surges of ecdysone regulate the timing of each larval molt. Throughout these early molt cycles, levels of juvenile hormone remain high (Step 1) and maintain the animal's larval form. Eventually, levels of juvenile hormone begin to drop, and with the next-to-last surge of ecdysone, the animal molts into a pupa (Step 2). With the final surge of ecdysone, juvenile hormone levels are very low and the animal is transformed into an adult moth (Step 3).

HORMONES AND DEVELOPMENTAL CHANGE

In addition to orchestrating cyclic change, hormones program an animal's progression through its life stages, for example, release of steroid hormones and the subsequent enlargement of female breasts, deepening of the male voice, development of larger muscles, and other accompanying physical changes. **Anabolic steroids**, which are synthetic analogues of these maturation-causing steroid hormones, have found their way into the world's locker rooms, gyms, and health clubs. The pressure to win is strong on the football field, the wrestling mat, the bodybuilding circuit, and even the dating scene. The pressure has been so overpowering, in fact, that about 20 percent of professional football players, 5 percent of college athletes, and 7 percent of high school boys have taken anabolic-androgenic steroid hormones to bulk up their muscles or increase their aggression. However, research shows that the gains are temporary, and the health price is high.

Anabolic-androgenic steroids such as stanozolol are artificial forms of testosterone. Just as an upsurge of the real male hormone promotes rapid muscle development in adolescent boys, taking anabolic steroids can lead to bulking that gives some types of athletes an advantage. Unfortunately, the drugs also confer the other masculinizing effects of testosterone in exaggerated proportions: heavy hair growth, acne, and premature baldness in men; and deepening of the voice, growth of facial hair, increased body hair, and enlargement of the clitoris in women. These changes in women appear to be permanent. Anabolic steroids also suppress the male's own androgens (male hormones), and this can lead to shrunken testes and enlarged breasts. Ominously, research also shows that damage to the kidneys, liver, heart, and possibly the prostate gland, which produces much of the seminal fluid, are all too common in users of anabolic steroids, as are depression, anxiety, hallucinations, paranoia, and other psychological symptoms. Coaches who care about the long-term well-being of their players make sure they don't take anabolic steroids.

Metamorphosis Hormones control the body changes that come with sexual maturation, both in humans and in other species. **Metamorphosis** is the revolution in body form that accompanies the change of the larva to the adult in insects and amphibians. In tadpoles, thyroxine from the thyroid gland acts on target cells in the tail and flank, causing the animal to resorb its long tail, sprout legs, and become a frog [see FIGURE 1.9].

The most dramatic metamorphic transformations, however, take place in insects like the silkworm and are mediated by two hormones. Silkworms pass through five larval stages, during which the insect grows, produces a new exoskeleton, then sheds its old one, or *molts* [FIGURE 30.12A]. After a final growth spurt, the larva spins a cocoon [FIGURE 30.14B] and secretes a dark brown pupal exoskeleton instead of another bright blue larval one. In spring, as the days lengthen, the animal, still in its cocoon, secretes one last exoskeleton—an adult exterior resplendent with glistening red, black, and gold scales [FIGURE 30.12C].

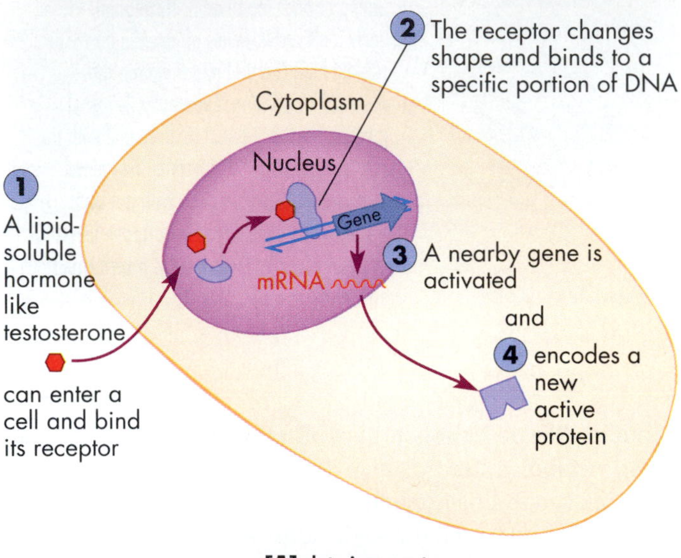

[A] Interior receptor

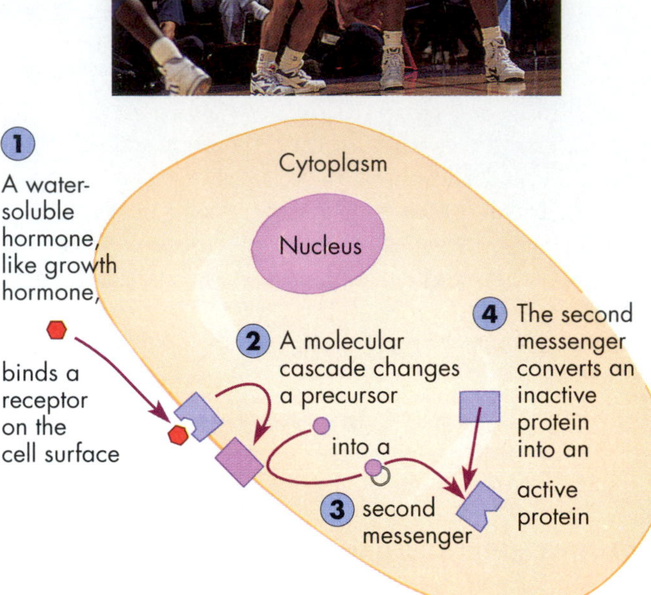

[B] Cell surface receptor

Two main hormones govern silkworm molting and metamorphosis from worm to moth [FIGURE 30.12D]. **Ecdysone**, a steroid, controls the *timing* of each molt. Biologists know this because if a researcher injects a larval insect with ecdysone, the animal begins to molt a short time later. The *type* of molt, however, is the province of **juvenile hormone**, a golden oil secreted by a gland attached to the brain. If juvenile hormone levels are high, the worm molts from one larval stage to another, but it remains a juvenile [FIGURE 30.12D, Step 1]. If juvenile hormone levels are intermediate, the animal molts from larva to pupa (Step 2). If concentrations are very low, the pupa molts to an adult (Step 3). Biologists are still investigating how juvenile hormone seems to maintain youth almost magically in insects.

FIGURE 30.13

How Hormones Effect Change: Gene Activation and Second Messengers.

[A] Lipid-soluble hormones, such as the steroid testosterone, diffuse across a target cell's membrane (1) and bind to a receptor in the cell. The hormone-receptor complex then binds to specific places on the cell's DNA (2) and stimulates the activity of specific genes (3), leading to the production of new proteins that alter the cell's activity (4). In this case, testosterone acts on hair follicle cells in a teenager's face, causing his beard to begin to grow. [B] Water-soluble hormones, such as growth hormone, cannot diffuse through the target cell's plasma membrane. Instead, they bind to a cell surface receptor (1), and the hormone-receptor complex touches off a chain of biochemical actions (2) that produces a second messenger (3), which in turn activates enzymes and other proteins and alters cell activity (4). In this case, differences in growth hormone production or reception could account for the difference in size between two highly skilled NBA ball players Mugsy Bogues of the Charlottesville Hornets and Sedale Threatt of the Los Angeles Lakers.

HORMONES AND CELLULAR CHANGE

You learned earlier that changes at the organ level begin with cellular mechanisms. Let's investigate how hormones act on individual cells and alter their activities.

As FIGURE 30.2 showed in general terms, hormones can activate target cells by binding to receptors inside a cell or at its surface. Hormones that are lipid-soluble, such as steroids, can pass into the cell through the lipid cell membrane and bind to receptors within the cytoplasm or nucleus [FIGURE 30.13A, Step 1]. As a messenger-receptor complex forms, it changes shape and then binds to a stretch of DNA along one of the chromosomes (Step 2). This allows certain genes to be transcribed to mRNA (Step 3) and thus new proteins to be made in the cell (Step 4). The new proteins then perform new functions in the cell. This is the way ecdysone from the silkworm brain causes the worm to form a pupa and then a new adult, and the way testosterone and estrogen cause their maturation effects on a teenager.

In contrast to lipid-soluble hormones, water-soluble hormones, like growth hormone, cannot pass through the lipid of the target cell's plasma membrane. The receptors for these hormones are partly embedded in the cell membrane [FIGURE 30.13B]. When the hormone binds to it, the receptor changes shape, and this stimulates the production of other compounds inside the cell—compounds known as *second messengers* (the hormone was the *first messenger*). When a water-soluble hormone such as growth hormone binds to a surface receptor of a cell—let's say a fat cell [FIGURE 30.13B, Step 1]—the receptor detonates a cascade of events (Step 2) that changes the concentration of an intracellular second messenger (Step 3). The second messenger often causes an inactive protein inside the cell to suddenly become active (Step 4) and thus alter the cell's activity. In this case, it would stimulate the growth of the fat cell.

Hormones can alter two categories of second messengers: (1) special nucleotides, such as the one called *cyclic adenosine monophosphate* (*cAMP*), which is made from ATP, or (2) calcium ions and derivatives of a component of the cell membrane called *inositol trisphosphate*, or IP_3. Both types of second messengers are small molecules that appear inside a cell as the result of an external signal, and they act by changing the activity of specific enzymes or other proteins. The changed proteins, in turn, alter the activity of the entire cell, causing the organism to move its muscles, reproduce, burn fat, or function properly in a host of other ways that promote its survival.

How do we know that receptors are important for causing cell change? Let's return to the fascinating story of the Efe people of Zaire's Ituri rain forest. When physicians learned that growth hormone causes an increase in stature, they wondered whether the pygmies' diminutive height was due to a lack of growth hormone itself or to an absence of growth hormone receptors on the peoples' cell surfaces. To find out, biologists collected blood from pygmies, and found that their levels of growth hormone were nearly normal. Furthermore, in an experiment most would consider unethical today, early researchers injected pygmies with growth hormone and observed no additional growth. These two results—normal levels of growth hormone and insensitivity to added growth hormone—suggested that pygmies are not short due to a lack of growth hormone. Recently, experimenters tested pygmy cells for the presence of growth hormone receptors, and found only half as many receptors as in the cells of people of taller stature. The researchers concluded that a pygmy is short because his or her cells lack enough growth hormone receptors. This work underscores the importance of receptors in understanding the biology of hormones.

In recent years biologists have discovered that the simplest and the most complex organisms have similar molecular messengers. Scientists have found insulinlike peptides regulating sugar metabolism both in invertebrates, such as mollusks and insects, and in vertebrates, such as mammals and reptiles. They have found that a mammalian hypothalamic releasing factor will turn on some sexual activities in yeast, while a mating factor from the fungus will turn on some gonadal activities in rats. Clearly, molecular messengers must have a very ancient origin and must have been conserved for hundreds of millions of years.

Some biologists suggest that molecular messengers may initially have evolved in single-celled organisms as mechanisms for coordinating feeding or reproduction. When multicellular organisms arose, sophisticated organs evolved that governed the many individual coordination tasks, but these control centers relied on the same kinds of molecular messengers "invented" by the simpler organisms. The use of highly similar signalling molecules and their receptors is evidence of evolution by descent with modification.

➤ CONCEPT CHALLENGE

Which types of hormones do you think are more important, those that inhibit change or those that stimulate change? Support your answer.

Connections

In our consideration of molecular messengers thus far, we have seen that some cells detect perturbations in the environment and respond by secreting a regulatory substance into the surrounding fluid. That substance binds to a specific protein receptor in a target cell and triggers an action in the target cell that responds to the initial perturbation. The end result is an animal tuned to its environment, coping effectively with change, and surviving. In the case of the Efe people, who live a difficult existence in the rain forest, their short stature, resulting from few cellular receptors for growth hormone, must have been advantageous to surviving in their particular habitat.

Our next subject, the regulators and receptors of the nervous system, have a kind of private intercellular communication that is like a rapid-fire tête-à-tête between neighboring cells. In contrast, the hormones, with their diffuse, long-distance activity, seem more like slowly whispered rumors that echo throughout the entire body.

KEY TERMS

adrenal gland, 686
amine hormone, 680
calcitonin, 683
cortisol, 687
diabetes mellitus, 687
ecdysone, 690
endocrine gland, 678
endocrine system, 678
endorphin, 687
epinephrine, 686
exocrine gland, 678
fatty acid hormone, 680
glucagon, 687
goiter, 683
hormone, 677
hypothalamus, 680
insulin, 687
juvenile hormone, 690
melatonin, 688
metamorphosis, 689
molecular messenger, 677
neurohormone, 679
neurotransmitter, 679
paracrine hormone, 679
parathyroid gland, 683
parathyroid hormone, 683
pheromone, 679
pineal gland, 688
pituitary gland, 680
polypeptide hormone, 680
receptor, 678
regulator cell, 678
second messenger, 691
steroid hormone, 680
target cell, 678
thyroid gland, 683
thyroxine, 683
true hormone, 679

HIGHLIGHTS IN REVIEW

1 Molecular messengers are substances produced by one group of cells that affect the activities of other groups of cells which may be quite distant in the body. Hormones secreted by cells of the endocrine system are an important type of molecular messenger.

- **a]** Molecular messengers are secreted by regulator cells, are carried in a fluid medium, and trigger activity in target cells by binding to receptor proteins in or on target cells. In addition to the true hormones, molecular messengers include the neurohormones, neurotransmitters, paracrine hormones, and pheromones.
- **b]** Endocrine glands secrete molecular messengers into the blood, while exocrine glands secrete molecular messengers or other substances such as milk or enzymes into ducts that can lead out of the body.
- **c]** Hormones are molecular messengers secreted by endocrine cells into the blood, altering the activity of distant cells. Some hormones are polypeptides; others are steroids, amines, or fatty acid derivatives.

2 Hormones are often secreted by endocrine glands. Endocrine glands often regulate each other's function via feedback loops. The brain's hypothalamus and pituitary, for example, regulate several other endocrine organs, and these organs in turn regulate the hypothalamus and pituitary.

- **a]** A mammal's endocrine system consists of a set of 12 or so ductless glands.
- **b]** The hypothalamus and the pituitary gland secrete many neurohormones and true hormones and control many other glands. The posterior pituitary secretes oxytocin and ADH. The anterior pituitary secretes prolactin and other hormones.
- **c]** In general, hormones act via negative feedback loops to maintain homeostasis. For example, two different hormones that exert opposite effects act through negative feedback loops to maintain appropriate blood levels of calcium and glucose.
- **d]** The thyroid gland secretes thyroxine, which regulates the rates of metabolism and growth, and calcitonin, which causes calcium uptake by bones. Parathyroid hormone from the parathyroid glands causes bones to release calcium, keeping blood levels above certain critical limits.
- **e]** The adrenal glands secrete the stress hormones epinephrine and norepinephrine, which prepare an animal to fight or flee, and also make cortisol, aldosterone, and some sex steroids.
- **f]** Ovaries and testes make several types of steroid hormones which regulate primary and secondary sex characteristics.

3 Hormones act by binding to receptor proteins in or on cells, and the binding changes the cell's activities. The new cellular activities can prevent further change in the levels of substances in the body; control the rate of metabolism; drive daily, monthly, or yearly cycles of change; or cause permanent developmental changes in an individual.

- **a]** Hormones help maintain homeostasis. Diabetes, an inability to maintain correct levels of glucose in the blood, results from insufficient insulin production by the pancreas.
- **b]** The pineal gland helps govern daily and yearly cycles of physiological change by detecting day length and making melatonin.
- **c]** Hormones program an animal's irreversible developmental changes, such as puberty and metamorphosis.
- **d]** A hormone changes cellular activities by binding to a specific receptor inside or on the surface of target cells. Some hormone receptors bind to DNA near genes and alter gene activity, while others cause second messengers to form, which alters the activity of cell proteins.
- **e]** The mechanism of action of molecular messengers and many of the molecules themselves have been conserved throughout evolution.

UNDERSTANDING THE FACTS AND CONCEPTS

For Questions 1–5, match each of the descriptions with one of the items from the list below. Each item can be used once, more than once, or not at all.

a] endocrine **d]** neurohormone
b] exocrine **e]** true hormone
c] paracrine **f]** pheromone

1 Not carried in the blood; target is in another individual of the same species; many regulate courtship and reproductive behavior.

2 Many are secreted by the hypothalamus; carried in the blood and act as hormones when they reach their targets.

3 A gland, such as a sweat or salivary gland, that empties its secretions onto a surface, usually by way of a duct.

4 A hormone-producing gland; secretions first enter the tissue fluid and then the bloodstream.

5 Hormonelike molecular messengers that diffuse to and act on nearby cells.

For Questions 6–10, match each of the descriptions with the most appropriate item or items from the list below.

- **a]** polypeptide hormone
- **b]** steroid hormone
- **c]** amine hormone
- **d]** fatty acid hormone

6 Male and female sex hormones are examples of this category.

7 Prostaglandins; aspirin blocks the production of this class of hormones.

8 Thyroid hormone is an example.

9 Growth hormone and oxytocin are examples of this category of hormone.

10 The molting hormone of insects.

For Questions 11–20, match each of the hormones listed with the mechanism listed below that directly stimulates that hormone.

- **a]** hypothalamic releasing hormone
- **b]** hormone of the anterior pituitary gland
- **c]** negative feedback with a metabolite such as calcium or glucose
- **d]** stimulated by the nervous system
- **e]** a secretion of the nervous system

11 Oxytocin

12 Thyroxine

13 Parathyroid hormone

14 Epinephrine

15 Cortisol

16 Glucagon

17 Melatonin

18 Insulin

19 ACTH in the stress response

20 Growth hormone

INTEGRATE AND APPLY WHAT YOU HAVE LEARNED

1 Explain how insulin and glucagon work to regulate the level of blood sugar.

2 Contrast the role of the adrenal medulla and adrenal cortex in the stress response.

3 Some endocrine glands are more essential than others for survival. Excluding the hypothalamus and pituitary, name three glands that are among the most essential and one that is least essential.

4 Grown women who develop mustaches and other masculine traits are often found to have a tumor or other abnormality in their adrenal glands. Explain. Which part of the adrenal glands would be affected?

5 Contrast the cellular site of action of lipid-soluble hormones and water-soluble hormones. Which of the two requires a second messenger? Explain.

ANALYSIS

1 Mrs. A shows signs of a thyroid hormone deficiency, but she has above-normal levels of thyroid hormone. What might be the problem?

- **a]** Not enough TSH.
- **b]** Too much TSH.
- **c]** Insufficient number of thyroid receptors on body cells.
- **d]** Insufficient number of TSH receptors on thyroid cells.
- **e]** Not enough hypothalamic releasing factor.

2 Diabetes that develops during childhood is generally treated with insulin. Adult-onset diabetes is usually less severe and can often be treated by dietary restrictions. One of these forms of diabetes is caused by destruction of the beta cells in the pancreas; the other is generally caused by a lowered number of insulin receptors on body cells. Which kind of diabetes is the result of lack of beta cell production of insulin? Explain.

- **a]** Adult-onset.
- **b]** Childhood.

3 In France, truffles are a culinary treat. In order to find these wild, underground fungi, farmers unearth the truffles with pigs, which seem to be able to detect and are strongly attracted by the scent emitted by the fungus. Scientists have recently found that pigs make a compound chemically very similar to the odor-producing material of truffles. How does this help to explain the truffle-hunting ability of pigs?

- **a]** Pigs like to eat like-smelling food.
- **b]** The pigs are responding to a pheromone and think they have located a mate.
- **c]** Pigs have evolved from the same progenitor as truffles; therefore they share the same odor.
- **d]** Pigs like to eat almost anything, even fungi.
- **e]** The pigs have been trained to home in on this scent; the fact that they themselves make a similar scent simply makes the trainer's job easier.

4 Why might a farmer administer juvenile hormone to reduce the population of an insect pest?

- **a]** Juvenile hormone would prevent the egg from being fertilized.
- **b]** High levels of juvenile hormone would enable the adult forms to emerge, which could then be killed with pesticides.
- **c]** Administration of juvenile hormone would prevent the insects from entering their reproductively active stage.
- **d]** Juvenile hormone would encourage the molting from pupa to adult.
- **e]** Juvenile hormone would encourage molting, at which time insects are the least damaging.

CHAPTER 31

How Nerve Cells Control Behavior

COCAINE: HIGHEST HIGHS AND LOWEST LOWS

A young couple from a midwestern state snorted cocaine from time to time to make themselves feel energized, more alert, and temporarily happier. The woman even continued this "recreation" into the early months of her pregnancy, but stopped after a while as the fetus grew.

Just before the baby was due, her husband gave her 5 g (0.2 oz) of cocaine as an anniversary gift, and she snorted it to celebrate. The drug, a nervous system stimulant, induced early labor, and she delivered a baby boy. Tragically, the newborn was permanently paralyzed on his right side; the cocaine the mother had taken just hours before had passed through the placenta, entered the baby's bloodstream, and caused his blood pressure to soar. The high pressure burst blood vessels in his brain, damaging a large portion of that delicate organ, and altering its control of the body's complex nervous system. The brain damage, in turn, impaired function in another part of the nervous system—specifically, the nerves that allowed movement of the baby's right arm and leg. Ironically, the mother took cocaine for its temporary stimulatory effects on her brain's reward center, such as the one shown in FIGURE 31.1. The similar but greatly magnified impact the drug had on her son's nervous system, however, caused devastating permanent damage.

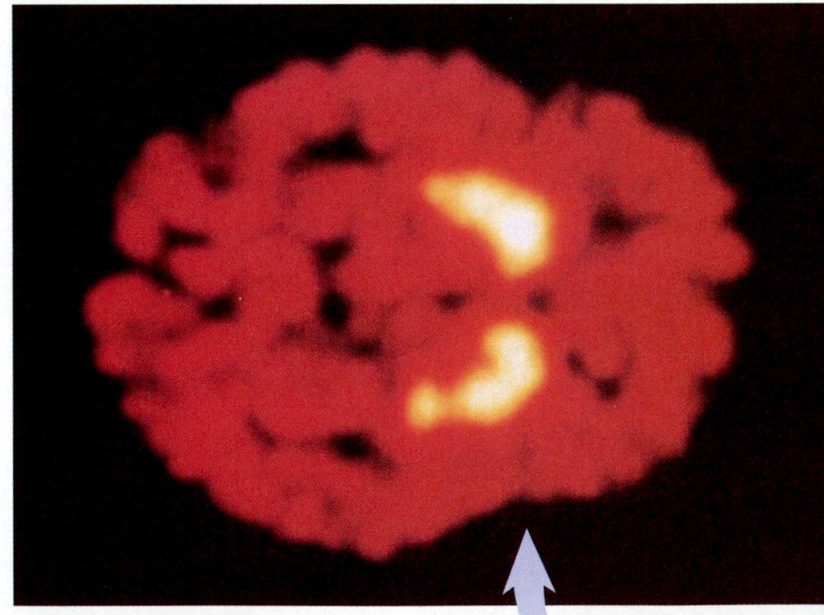

FIGURE 31.1
Cocaine High: A Change in Nerve Cell Function in the Brain.
Cocaine accumulates in the reward center of the brain (bright yellow) within just 15 minutes after a hit of cocaine, as shown in this PET scan, a special noninvasive view of a slice through a human brain.

Our subject in this chapter is the major integrative network in an animal's body: the **nervous system**. A heartbreaking event such as this baby's irreversible nerve damage points out one of the crucial activities of the nervous system: it controls and coordinates the activities of an animal's body systems. This governing function works parallel to and at times in conjunction with the endocrine (hormonal) system [see CHAPTER 30], and it allows an animal to sense the environment, function as an integrated whole, and respond appropriately. Nervous systems do three things: (1) They *receive data* about environmental conditions, (2) they *integrate the sensory data* into a meaningful form, and (3) they *bring about a change* in behavior, physiology, or information storage that helps the organism survive. The cocaine that the mother took altered the way her own nervous system integrated sensory data and regulated behavior (her mood and alertness), as well as the way the baby's nervous system helped control his circulation and blood pressure.

The nervous and endocrine systems are able to work together coordinating and controlling the body because they share several important features. Both systems integrate body functions; both use molecular messengers (sometimes the identical molecules) for communication; and in both systems, the second-messenger strategy triggers the response inside certain target cells [review FIGURE 30.13]. The nervous and endocrine systems do differ in substantial ways, however, including their speed of response (the nervous system responds more quickly) and the duration of their effects (hormonal effects often last much longer).

The central task of an animal's nervous system is *communication*. This communication can take place within an individual nerve cell; between neighboring nerve cells, most of which are organized into highly elaborate networks; between separate organs; and between different individuals in a community, allowing one animal to perceive the needs and internal states of another. An unfortunate example of nervous communication occurred in the baby who was exposed to cocaine. Deep within the baby's developing brain, the stimulant altered the way one specific group of nerve cells communicated with another—those that control blood pressure—and the consequences were dramatic for the baby's behavior—specifically the movement of his right arm and leg.

MESSAGES

1 The nervous system receives data from the environment, integrates the data into a meaningful form, and then brings about a change in the animal's behavior, physiology, or information storage.

2 Communication within the nervous system occurs in cells called neurons. Neurons are electrically charged cells, and they often have very long, thin cell extensions.

3 When a neuron becomes stimulated, its electric charge reverses at the point of stimulation; this event is called an action potential. This temporary reversal of charge can move along the neuron.

4 When an action potential meets specialized ends of the neuron, it can stimulate or inhibit an action potential in an adjacent cell.

5 Neurons are linked together, sometimes in very complex loops and networks, and these control behavior and learning.

All human and animal behavior depend on the way nerve cells are connected to each other and arrayed in precise patterns. To probe those connections, we begin this chapter by studying the structure and activities of nerve cells and how a nerve cell's structure promotes communication in the body. Next, we see how a nerve cell conveys an impulse or signal and how that impulse can carry information. We then discuss the ways nerve cells communicate with other cells. Finally, we consider how nerve cells are organized into networks and simple nervous systems. We limit our focus in this chapter to individual nerve cells and the ways they are connected into nervous systems. In the next chapter, we go on to see how complex neural organs such as the brain can mediate thought, language, and sensation. Along the way, we'll encounter many fascinating topics, including the action of cocaine on the brain. We'll also discover how a knee-jerk reaction occurs. ❑

Nerve Cell Structure

As you read this sentence, try wiggling the big toe on your left foot. Some of the nerve cells controlling that movement extend all the way from the lower part of your spinal cord to the muscles that move your toe—the longest individual cells in your body at nearly 1 m (39.4 in) in length. A nerve cell running from a giraffe's spinal cord to its hoof can be three times as long, or nearly 3 m (10 ft)! Other nerve cells, in contrast, reach only a few millimeters, extending from one brain region to another, like the bushy, rust-colored cells in FIGURE 31.2. Although the human brain and nervous system contain more than 1000 different types of cells, all of them can be classified as either **neurons**, the nerve cells that accomplish the actual communication tasks, or **glial** (GLEE-uh; from the Greek *glia*, "glue") **cells**, the support cells that surround, protect, and provide nutrients to neurons and may influence their function in ways biologists do not yet fully understand. Although the human brain contains more glial cells than neurons, we focus on the neurons here because they carry out the nervous system's crucial communication function.

ANATOMY OF A NERVE CELL

All neurons, including the ones that help wiggle your big toe, function by collecting information and relaying it to other cells in the body. Specific portions of each neuron perform these collecting and relaying functions, so we'll look at each major part, including the dendrites, soma, axon, and ends of the axon, or boutons [FIGURE 31.3].

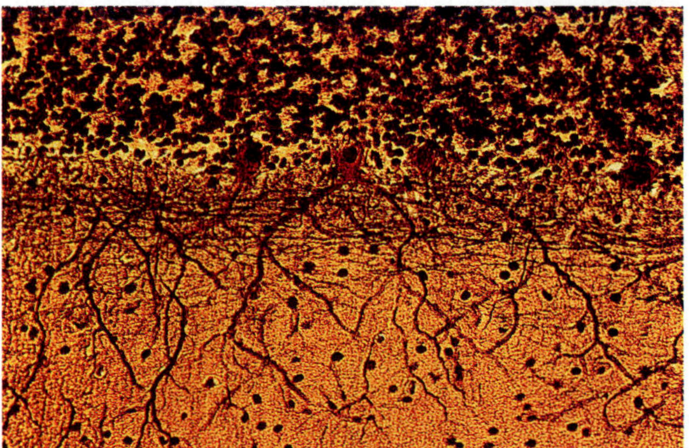

FIGURE 31.2

Nerve Cells Are Organized into Highly Elaborate Networks.

Neurons from a cat's cerebellum interconnect via delicate extensions, establishing a complex cellular network with point-to-point contact of nerve cell to nerve cell.

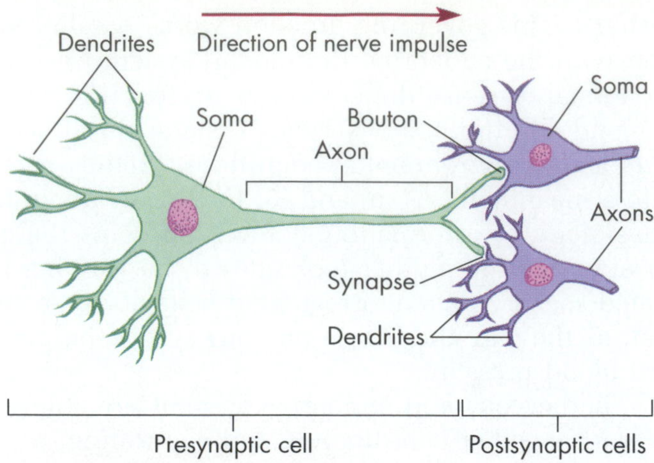

FIGURE 31.3

Neurons Transmit Information from One Cell to Another.

Shown here are the four parts of a neuron—dendrites, soma, axon, and axon terminals (boutons). At the synapse, or junction, between one cell's axon terminal and another cell's dendrite, information is passed from the sending (presynaptic) cell to the receiving (postsynaptic) cell.

Neurons gather information by a set of fine cell processes called **dendrites** (from the Greek for "little tree"). Dendrites pick up signals from other nerve cells and pass these impulses to the soma. The **soma**, or **cell body**, makes up the major portion of the neuron. Housing the usual complement of intracellular organelles—nucleus, ribosomes, mitochondria, endoplasmic reticulum, and Golgi apparatus—the cell body produces the proteins that make up the rest of the nerve cell, as well as the enzymes that determine its activity. The signals next pass through the cell body to a long, thin cell extension called the **axon** (Greek for "axle"). The axon of some neurons extends from your spinal cord to your toe muscles, while the cell body and dendrites are located in the spinal cord itself. A neuron may have many dendrites, but usually just one axon. A group of axons from many different neurons running in parallel bundles is called a **nerve**.

The ends of the axons are the site of a key event in nervous communication. When a nerve signal reaches the ends of an axon, it enters the knoblike **boutons** (French for "buttons"), which terminate very close to another cell and serve as special regions of cell-to-cell communication. The place where a bouton approaches another cell is a junction known as a **synapse** (SIN-apps), which often contains a microscopic space, or *cleft*, across which the message passes at lightning speed [see FIGURE 31.3]. The synapse is the site where cocaine acts and alters communication between nerve cells. Boutons may form synapses with an axon, dendrite, or cell body of another neuron, with muscle cells, or with the secretory cells of endocrine or exocrine glands [see CHAPTER 30].

The two cells on either side of a synapse have different functions. The neuron that sends a message down its axon and across a synapse to another cell is the **presynaptic cell** [see FIGURE 31.3]. The cell receiving the message that crosses a synapse is the **postsynaptic cell**. Information flows in one direction, from presynaptic cell to postsynaptic cell.

The axon of a presynaptic cell may branch and rebranch, forming synaptic junctions with up to 1000 other cells. Conversely, 1000 other neurons might form synapses with the dendrites and the cell body of a single neuron—rather like hundreds of hands reaching out to touch a central object simultaneously. With these multiple links, an animal's communication network is literally a net, as FIGURE 31.2 reveals. This lacework of interconnected neurons allows the organism to carry out such complex and coordinated behaviors as wiggling the big toe of one foot.

Recall that the nervous system receives data, integrates it, and effects a change in physiology. Within a given neuron, the dendrites act as the receptors, the cell body acts as the integrator, and the axon and boutons act as the effectors.

➤ CONCEPT CHALLENGE

Let's say you are working part-time for a geneticist, and you discover a particular mutation in mice that prevents the formation of boutons at the ends of axons. What would be the physiological result of such a mutation?

The Nerve Impulse

It's quite obvious that your camera doesn't measure light intensity with a thermometer, nor does your furnace's thermostat register temperature with a light meter. Each kind of environmental signal—heat, light, pressure, and so on—requires a specific type of detector. This principle holds for organisms as well as for machines. To the detectors in the human tongue, a chili pepper, a chocolate bar, and an aspirin tablet all produce distinguishably different taste sensations. To the ear, a Brahms symphony is nothing like the screech of a braking car. And none of these inputs resembles the feeling of slamming your finger in a drawer. Each of these sensations has a distinctly different quality because it is detected by different types of receptors, and the information is transmitted to different destinations in the brain or spinal cord. At the level of a single neuron, however, any and every sensation of sufficient magnitude produces exactly the same kind of response: a **nerve impulse**, an electrochemical reaction that allows specific ions to rush into and out of the cell.

A NERVE CELL AT REST

By focusing on a single patch of a resting neuron's plasma membrane, we can begin to see how an impulse is generated. Neurons, like all cells, are bathed in an extracellular fluid that is a molecular soup with a high concentration of sodium ions (Na^+) and a relatively low concentration of potassium ions (K^+) [FIGURE 31.4A, Step 1]. In contrast, the fluid inside the neuron is relatively rich in potassium ions and low in sodium ions. Besides these differences, there are many more positively charged ions in a thin layer on the outside of the cell and many more negatively charged ions in a thin layer on the inside [FIGURE 31.4B].

The sodium ions outside the plasma membrane are like the thousands of music fans pushing and straining to get into an amphitheater for a rock concert. And just as those rock fans represent potential energy waiting to be unleashed, the separation of positive and negative charges on either side of the plasma membrane represents potential energy that, once unleashed, can activate a nerve impulse in that small region of the cell. A nerve impulse is generated in a neuron's plasma membrane as ions (electrically charged atoms; review FIGURE 2.7) flow into and out of the cell.

A NEURON'S RESTING POTENTIAL

Since there are more positively charged ions outside the neuron than inside it, the inside is less positively charged, and hence is negative with respect to the outside of the cell [see FIGURE 31.4A, Step 2]. The charge difference generates an *electrical potential*: a measurable amount of potential energy. In a neuron at rest, measurements reveal that this **resting potential** is about 70 millivolts (mV). (A millivolt is one-thousandth of a volt. For comparison, a typical flashlight battery has a potential of 1500 mV, only 20 times that of just one of your nerve cells.) Since the inside of the neuron is negative with respect to the outside, we say that the cell's resting potential is −70 mV (Step 3).

A gate keeps rock fans outside a stadium; some analogous cellular structure or force must obviously maintain a neuron's resting potential. So what is it? Decades ago, neuroscientists discovered that three proteins embedded in the plasma membrane function as two channels and a pump. Together, these proteins keep the resting cell at −70 mV:

1. The **potassium channel** is a protein-lined pore that allows potassium ions to pass through the membrane and leak down their concentration gradient from the inside of the cell to the outside (Step 4).

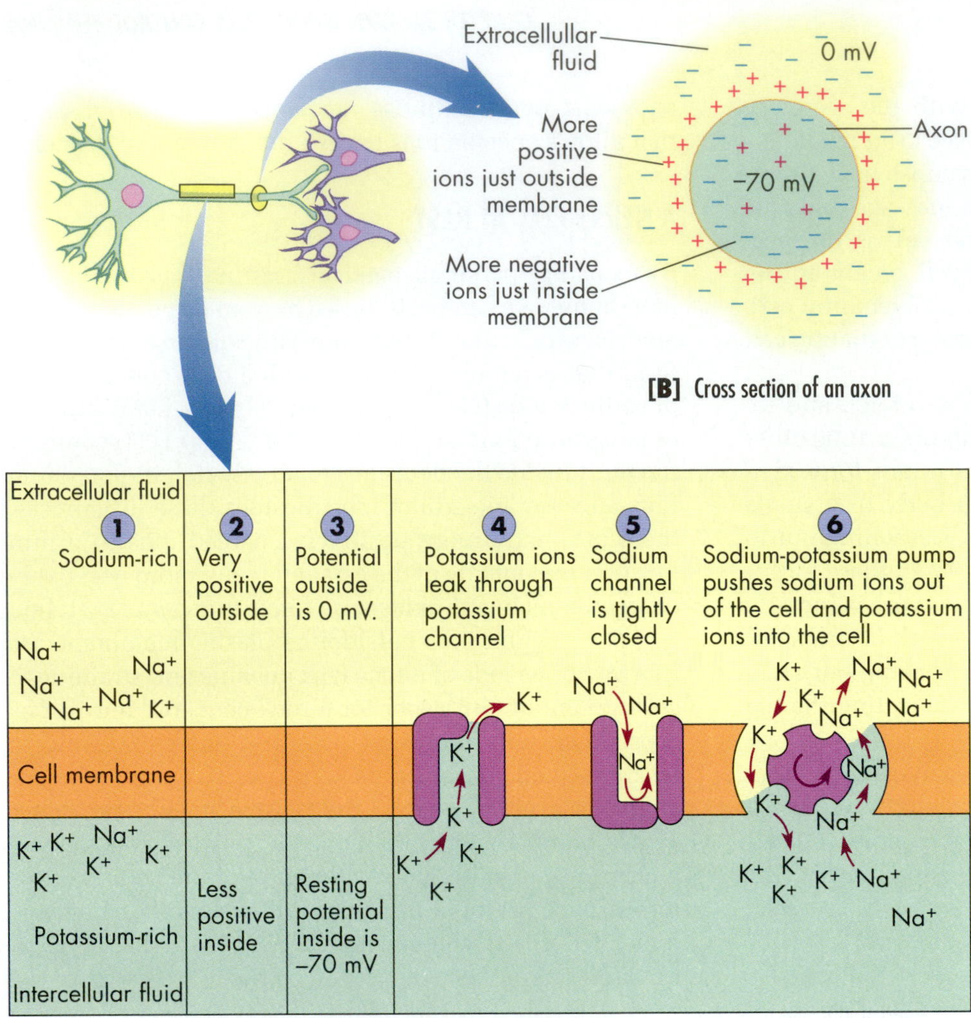

FIGURE 31.4

Ions, Channels, and Pump Proteins in a Neuron's Membrane Establish the Cell's Resting Potential.

[A] The extracellular fluid outside a neuron is rich in sodium ions but poor in potassium ions, while the intracellular fluid is rich in potassium ions but poor in sodium ions (1). The leakage of potassium ions out of the cell leaves the inside of the cell with fewer positive charges (2), and hence it is more negative than the outside of the cell. This results in an electrical potential of −70 mV inside the cell (3). Leaky potassium channels in the membrane allow potassium ions to leak out (4), but closed sodium channels prevent an influx of sodium ions (5). In addition, sodium-potassium pump proteins in the membrane (6) actively move two potassium ions in for every three sodium ions pumped out. [B] An excess of positive ions exists in a thin layer outside the neuron's membrane, while an excess of negative ions lies just inside the membrane.

2. In contrast, a second protein-lined pore, the **sodium channel**, closes tightly and allows almost none of the sodium ions outside the cell to leak inward (Step 5). The slow exodus of some potassium ions leaves behind a collection of negatively charged proteins too bulky to pass through the membrane, which gives the interior a slightly negative charge.

3. The third protein that helps maintain the neuron's resting potential is the **sodium-potassium pump** (Step 6). The pump opposes the leakage of ions through the channels. For example, potassium diffuses down its concentration gradient out of the cell (Step 4), but the pump brings potassium back into the cell. The reverse holds for sodium. Just as it would require energy for a juggler to keep three balls from falling to the ground, so the pump uses ATP to keep potassium and sodium ions from falling down their concentration gradients. Similar pumps work continuously in every animal cell and collectively use more ATP than any other body activity.

The actions of the three kinds of proteins in the neuronal plasma membrane maintain a steady electrical potential −70 mV between the cell's outside and its inside [see FIGURE 31.4B]. In its resting state, the cell is said to be **polarized**; it has an imbalance of electrical charges.

THE ACTION POTENTIAL: A NERVE IMPULSE

If a resting neuron already has an electrical charge—the resting potential—then what is a nerve impulse? A nerve impulse, or **action potential**, is the *reversal* of the charge on a resting cell. Whereas the resting nerve cell has a negative electrical charge, the action potential reverses the charge to positive. Let's examine how an action potential is generated in a nerve cell:

1. When a specific neuron in the resting state [FIGURE 31.5, Step 1] is stimulated—perhaps by a chili pepper, a lark's song, or another nerve cell—some of its tightly closed sodium channels open.

2. As a result, a few extra sodium ions leak into the cell [FIGURE 31.5, Step 2].

3. As sodium ions continue to enter, the difference in charges between inside and outside decreases, and a gradual change takes place in the electrical potential across the membrane: −70 mV changes to −68 mV, to −62 mV, to −54 mV, and so on. Since the electrical po-

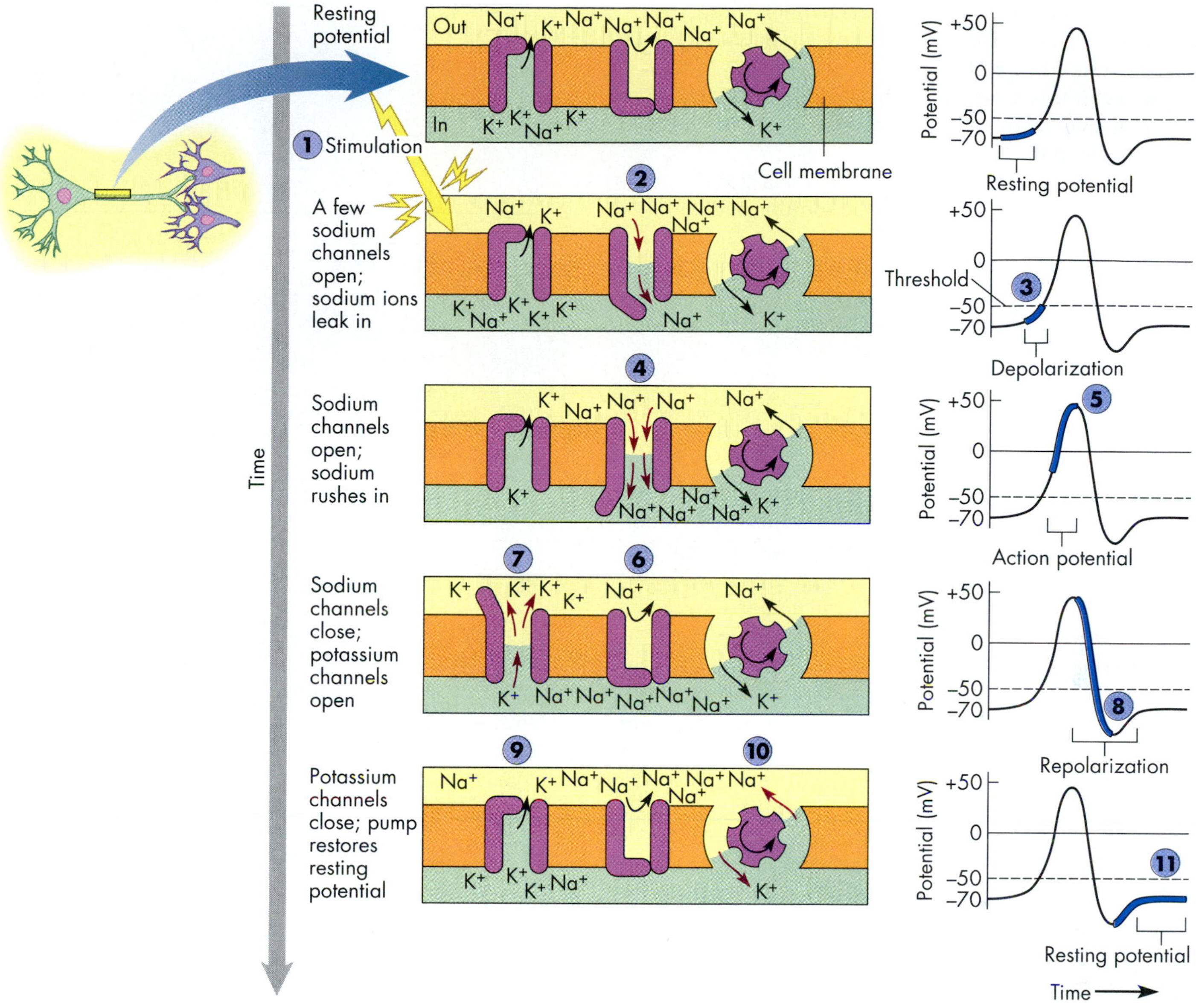

FIGURE 31.5

How an Action Potential Is Generated in a Nerve Cell.

tential, or polarization, is decreasing, the cell is becoming depolarized (Step 3).

4. When the difference in potential reaches a specific *threshold*—about −50 mV—something quite dramatic occurs: In the patch of plasma membrane where the stimulus first arrives, most of the sodium ion channels open wide. When sodium ions rush into the cell through the now-permeable membrane (Step 4), the inside of the neuron becomes positively charged with respect to the outside.

5. The potential difference between the inside and outside of the neuron finally reaches a peak of about +50 mV (Step 5), and this reversed polarity initiates the action potential. An **action potential**, or **nerve impulse**, is an area of reversed polarity that travels along the length of the axon.

6. Sodium channels in the area of the action potential remain open for only about 1 millisecond (one-thousandth of a second) and then close spontaneously (Step 6). For a few milliseconds after closing, they cannot open again, and this state of temporary shutdown is called the **refractory period**. This property of the sodium channels limits the number of action potentials a neuron can experience—that is, the number of times it can fire each second.

7. About the time the sodium channels close, the potassium channels open fully, allowing potassium ions to rush out of the cell (Step 7).

8. With this outpouring, the electrical potential falls to a level below the original resting potential of −70 mV (Step 8).

9. Eventually, the potassium channels close.

10. The "bailing" action of the sodium-potassium pump continues (Step 10).

11. The neuron is restored to its original resting potential (Step 11). The neuron is now ready for another environmental stimulus to cause sodium gates to open and trigger a new action potential.

The opening and closing of the gates in ion channels is the crucial step in the generation of a nerve impulse, and you have probably experienced the effects of closed gates in your dentist's office. The novocaine that a dentist injects as a pain-killing anesthetic prevents the opening of the ion channels in nerve endings in the gums and tongue. Since ions cannot pass through the closed gates, there can be no action potential in the neurons that relay pain signals to your brain. With this lack of response, you feel little pain, even when the dentist's drill stimulates an exposed nerve.

Action potentials have a peculiar property: They are **all-or-none responses**; that is, either they do not occur at all, or when they do, they are always the same size for a given neuron, regardless of how powerful the stimulus may be. A jab to the ribs is more intense than a tender caress, not because the jab causes larger or faster action potentials, but because it causes more cells to fire more impulses more frequently.

PROPAGATING THE NERVE IMPULSE

Once a neuron receives a stimulus at a particular patch of its plasma membrane, the resulting action potential generated in that local region is communicated along the length of the cell. To wiggle your toe, for example, nerve cells send, or propagate, action potentials all the way down the axon from the spinal cord to the nerve cell boutons synapsing with the muscles that move your toe. That communication depends on the action potentials passing from one patch of the membrane to an adjacent patch, then to the next, in a process called **propagation**.

The three frames of FIGURE 31.6 represent a portion of an axon's plasma membrane at three different times. The figure shows the sequential opening of sodium gates along the axon, a bit like football fans rising and cheering in a "wave" that circles the stadium. The open gates allow sodium ions to rush into the cell and generate a local action potential, which then moves along the axon.

In the area of an action potential, there are relatively few sodium ions outside the plasma membrane because they have moved into the cell locally [FIGURE 31.6, Step 1]. Sodium ions nearby then diffuse into this region, moving from a region of high concentration to a region of low concentration (Step 2). The diffusion of sodium ions away from the nearby patch causes this patch of plasma membrane to become depolarized. Since depolarization

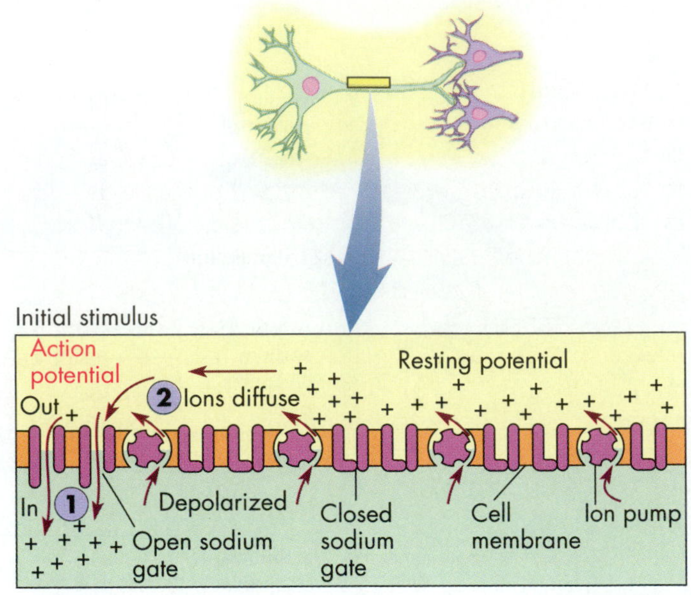

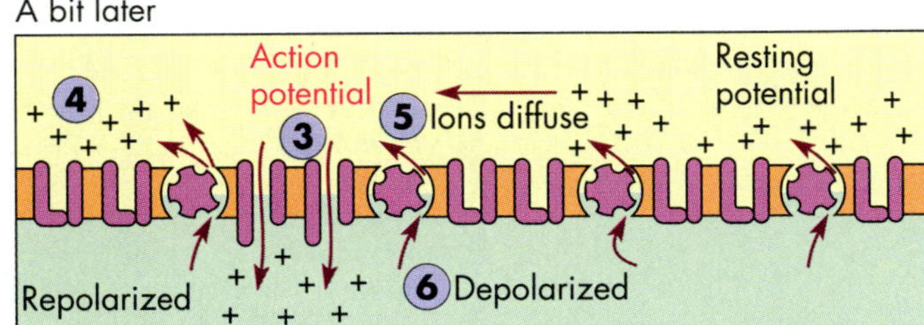

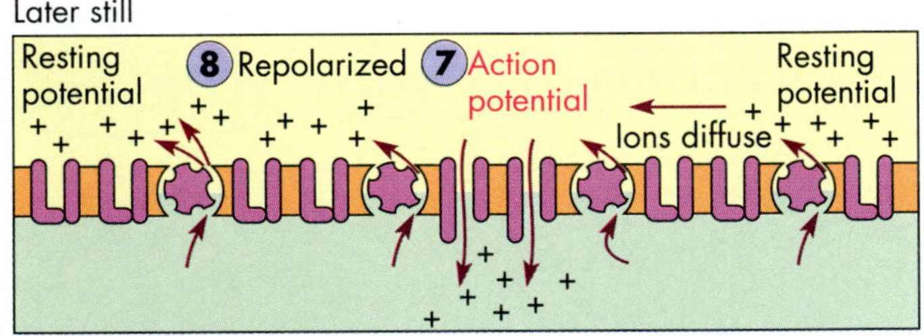

FIGURE 31.6

How an Action Potential Propagates Along a Nerve Cell.

As a neuron fires, an action potential moves along the plasma membrane for the entire length of the cell. For simplicity, we show only the movement of sodium ions (positive charges).

causes sodium channels to open, a full-fledged action potential is now generated in this second patch [see FIGURE 31.6, Step 3]. Meanwhile, the continuous pumping activity of the sodium-potassium pump repolarizes the first membrane patch (Step 4). The result is that the action potential has moved along the neuron's axon, from left to right.

This series of four steps—(1) ion diffusion, (2) depolarization, (3) action potential and (4) repolarization (Steps 5, 6, 7, and 8)—takes place sequentially again and again as the action potential travels the entire length of the axon, from your spinal cord, say, to your toe. Unlike the electrical current in a copper wire, which dies out over distance, a nerve impulse in a living neuron remains just as strong when it reaches the axon terminal in the tip of the toe as when propagation began in the spinal cord.

Direction of Propagation Interestingly, within the body, the impulses in a given neuron travel in one direction only (from spine to toe, for example, but not from toe to spine), because the refractory period prevents reverse propagation. For a few milliseconds after any action potential, a membrane patch cannot experience another action potential; thus, in nature, the signal moves forward to the next patch but never moves backward toward the previous one.

Speed of Propagation In many invertebrates, a nerve impulse traveling down a neuron moves only about 2 m (6.5 ft) per second. A tennis match would be a sluggish affair indeed if players had to wait 1 second for an impulse to travel from their brains to their toes. One way the propagation of nerve signals can be speeded up is for the diameter of the neuron to increase, since a larger axon has less resistance to the flow of ions, just as a wider pipe has less resistance to water flow than a narrower one. A few invertebrates, such as the fast-moving squid, have in fact evolved huge axons over 1 mm in diameter that carry impulses rapidly. A person, however, could never make do with giant axons: To contain the huge number of interlocking nerve cells we need to produce complex actions and thoughts, our heads would have to be 10 times bigger! Instead, humans and other vertebrates have developed a means of rapid impulse propagation without giant neurons.

In vertebrates, fatty sheaths insulate neurons and speed impulse travel. *Schwann cells* and other special supportive glial cells extend along an axon like fatty sausages on a string and wrap their plasma membranes around the axon surface like sticky straps [FIGURE 31.7A]. Together, the Schwann cells form the **myelin sheath**, a lipid-rich layer that insulates the axon from the extracellular fluid [see FIGURE 31.7A, B]. In the tiny gaps between these end-to-end Schwann cells lie bare regions known as *nodes of Ranvier*, and only in these uninsulated nodes can ions flow across the axon membrane. As a result of this cellular arrangement, the nerve impulse jumps from one node to the next, bouncing along the axon 20 times faster than if the axon lacked the myelin sheath. Bounding propagation, or **saltatory conduction**, in a neuron with myelin insulation permits us the rapid reactions and movements we need to return a smash in tennis or perform a Rachmaninoff piano concerto. Even with saltatory conduction, however, nerve impulses travel only about 85 m (279 ft) per second, 3000 times slower than the speed of electricity in a wire.

Mutant mice, such as the one in FIGURE 31.7C, shiver uncontrollably because they lack myelin sheaths on their axons. As a result, signals move erroneously between the

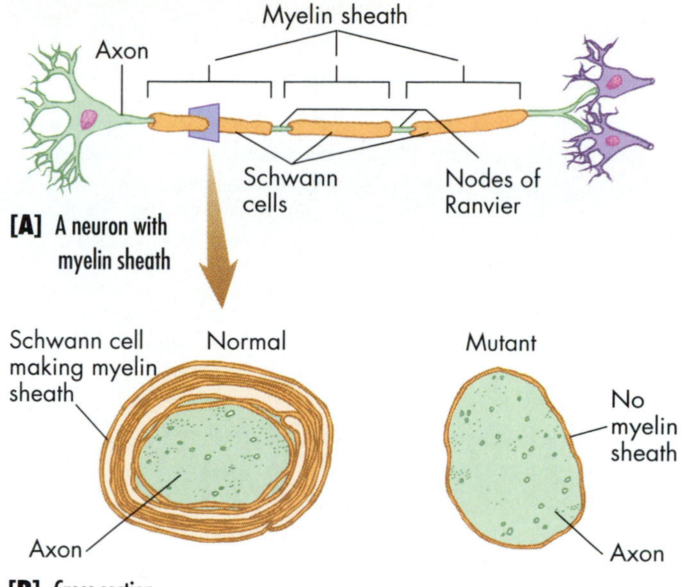

[C] Normal and mutant mice

FIGURE 31.7

Myelin Sheath: An Insulated Axon Propagates Impulses More Efficiently.

[A] Vertebrate neurons may be wrapped in Schwann cells, which make up the insulating layer called the myelin sheath. Schwann cells are separated by bare regions, or nodes of Ranvier. The action potential can jump from one node to the next, allowing rapid propagation of the nerve impulse. **[B, C]** A cross section of neurons from a normal mouse (left) shows Schwann cells wrapped around axons. A cross section of neurons from a mutant *shiverer* mouse (right) reveals no encircling Schwann cells.

FIGURE 31.8

Electrical Synapses Between Neurons.

At an electrical synapse, the plasma membranes of two adjacent neurons are tightly joined at gap junctions that allow ions to flow from one cell to another and thus to propagate an action potential smoothly and immediately.

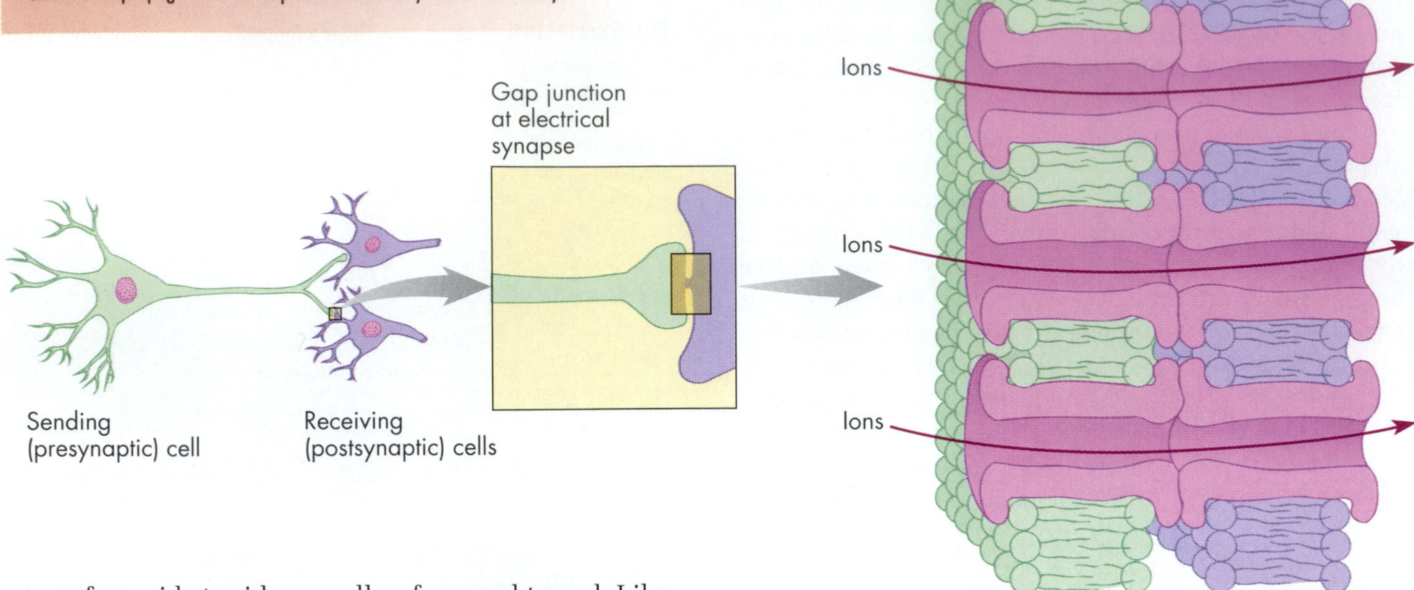

axons from side to side as well as from end to end. Likewise, the human disease *multiple sclerosis* (MS), whose symptoms usually first appear in early adulthood, causes some nerve cells to lose their myelin sheath. Without the sheath, nerve impulses can no longer leap along the cell, and transmission slows, leading to double vision, weakness, and wobbly limbs.

➤ CONCEPT CHALLENGE

Sometimes mussels and clams consume microbes that contain saxitoxin. In vertebrates, this compound blocks the flow of sodium ions through sodium channels. One mussel can contain enough saxitoxin to kill 50 people! How would saxitoxin affect an action potential in an individual neuron? What would be the symptoms of saxitoxin poisoning?

Synapses

As we saw in the chapter's introduction, when a pregnant woman took cocaine, it damaged a part of her baby's brain; this interrupted communication, a primary function of the nervous system. This type of cell-to-cell signal transmission can involve a neuron passing information to another neuron, or a neuron communicating with a muscle cell or a gland, but regardless, the signal must somehow cross the specialized junction, or synapse, between a neuron and a neighboring cell. Synapses generally conduct signals in only one direction, from the sending (presynaptic) cell to the receiving (postsynaptic) cell.

Neurons have evolved two different ways of conveying messages to other cells: (1) electrical; (2) chemical.

ELECTRICAL SYNAPSES

Most groups of animals have some neurons that connect to other cells through **electrical synapses**. In such cases, membranes of the sending (presynaptic) cell and the receiving (postsynaptic) cell are tightly fastened together by gap junctions [FIGURE 31.8]. At such a junction, the ions that generate an action potential in the presynaptic cell can flow directly through pores into the postsynaptic cell and propagate the impulse in a direct electrical connection, somewhat like a phone line.

Message relay is especially rapid at these electrical junctions, and that speed is a valuable trait for certain kinds of neurons. In the giant motor neurons of the crayfish, for example, a lightning-fast relay allows the animal to flip its tail instantly and escape a predator. Unfortunately, tight interconnections that allow such rapid transmission also leave little opportunity for the neuron to modify the ways it relays a signal. It is no surprise, therefore, that electrical synapses are vastly outnumbered in higher animals by chemical synapses, which function more slowly and require more steps, but also have more potential places to modify messages for specific conditions.

CHEMICAL SYNAPSES

In vertebrates, most neuron-to-neuron and neuron-to-muscle cell communication takes place across junctions called **chemical synapses** [FIGURE 31.9]. The axon ends at a knoblike bouton [see FIGURE 31.9A]. Between this knob

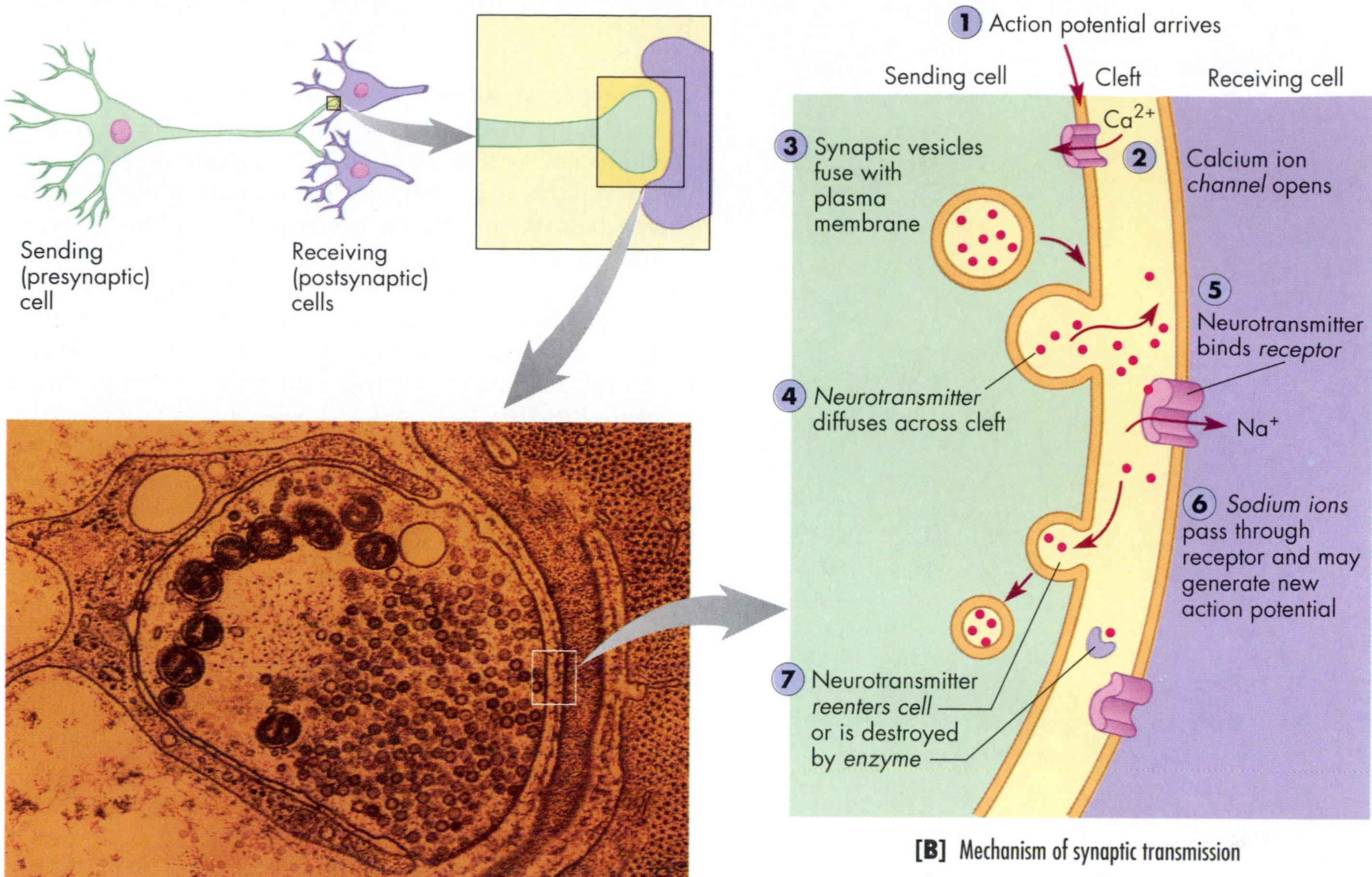

[A] Electron micrograph of a synapse

[B] Mechanism of synaptic transmission

FIGURE 31.9

At a Chemical Synapse, Neurons Transmit Information across a Cleft.

[A] An electron micrograph of a synapse reveals that a cleft separates the sending (presynaptic) cell, with its bouton-containing packets of neurotransmitter, from the receiving (postsynaptic) cell (in this case a muscle cell). **[B]** How a chemical synapse functions.

and the flat surface of the receiving cell lies a minute but distinct cleft about 20 nm wide. The molecular messenger, or **neurotransmitter**, diffuses across this cleft in less than a millionth of a second and relays the nerve signal from sender to receiver.

Crossing the Synapse A neuron makes neurotransmitters in its cytoplasm, then loads the messenger molecules into small round packets called *synaptic vesicles*, which accumulate in the knoblike bouton [see FIGURE 31.9A]. As shown in FIGURE 31.9, a nerve impulse crosses the cleft of a chemical synapse in the following way:

Step 1: An action potential reaches a bouton.

Step 2: The action potential causes calcium ions (Ca^{2+}) to rush into the bouton.

Step 3: The increase in calcium ions stimulates the synaptic vesicles to fuse with the membrane of the presynaptic (sending) cell.

Step 4: Thousands of neurotransmitter molecules are released into the synaptic cleft.

Step 5: These neurotransmitter molecules rapidly diffuse across the cleft and bind like keys to locklike receptor proteins embedded in the plasma membrane of the postsynaptic (receiving) cell.

Step 6: This binding can cause the postsynaptic cell to generate an action potential. If the receiving cell is a muscle cell, it will contract; if it is another neuron, an action potential might be triggered in that receiver.

The importance of receptor proteins is demonstrated by the condition *myasthenia gravis*. In people with this disease, the muscle cells have few receptors for neurotransmitter molecules; as a result, the muscle cells cannot respond normally to nerve signals, and the patients experience muscular weakness and fatigue.

Inactivation of Neurotransmitters Once a neurotransmitter enters the synaptic cleft and triggers the receiving cell, it is enzymatically degraded or reabsorbed into the bouton [see FIGURE 31.9B, Step 7]. Without this rapid cleanup, the messenger molecules would remain in the cleft and continuously stimulate the postsynaptic cell. This is what happens in a cocaine high. The drug blocks the cleanup of a specific neurotransmitter (dopamine) in

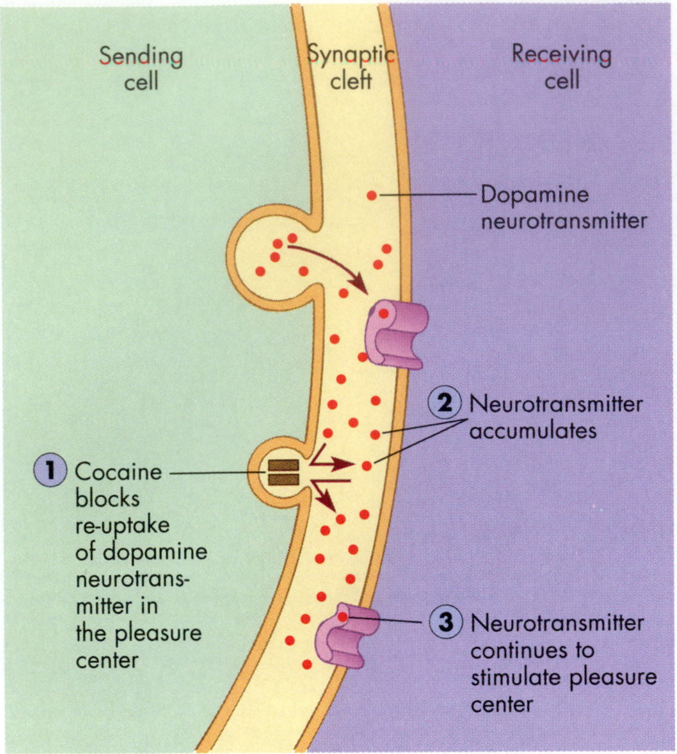

How cocaine acts at a synapse in the pleasure center

FIGURE 31.10

How Cocaine Acts at a Synapse in the Brain's Pleasure Center.

If given unlimited access to cocaine and left alone, a mouse will learn to take doses of cocaine, become addicted, and eventually consume a lethal overdose of the drug—all in less than a week's time.

cells of the brain's so-called central reward system [FIGURE 31.10; review FIGURE 31.1]. Since the neurotransmitter acts for a longer time at these pleasure synapses, the person feels euphoria and excitement. But homeostatic mechanisms in the body soon begin to compensate by destroying the transmitter faster and more efficiently when cocaine is not present. Thus, without the drug, the reward centers of acclimated users are starved of stimulation, preventing them from feeling joy at normally pleasurable activities such as good food and good sex. Instead, the addict's body craves more cocaine, and thus begins a destructive cycle of drug-seeking behavior called **addiction**. In fact, cocaine addiction takes hold faster than addiction to any other drug, including heroin.

Numerous animal species, from mice to monkeys, quickly learn to push a lever to obtain cocaine. Laboratory mice will experience their first "high," crave the drug, become addicted, and die from a self-delivered overdose, all within one week's time. Appendix Table A.3 summarizes the known effects on the human body of many illicit, addictive drugs, as well as numerous legal prescription drugs.

Types of Neurotransmitters Dopamine, the neurotransmitter involved in cocaine addiction, is just one of more than 50 different chemicals that can serve as neurotransmitters in the nervous systems of various animals. The best understood are *acetylcholine* and *norepinephrine*. Besides its role in transmitting nerve impulses within the brain, acetylcholine also relays messages from neurons to the skeletal muscles involved in maintaining posture, breathing, and the movement of limbs. Norepinephrine, found in synapses throughout the brain, seems central to keeping mood and behavior on an even keel. Researchers are not yet certain what the chemical does, but they suspect that a deficiency of norepinephrine within the brain may underlie clinical depression, with its pessimistic moods, disturbed sleep, and low appetite and energy levels. Conversely, they think that an excessive amount of the transmitter may help trigger mania, with its elevated or irritable moods, overactivity, and recklessness. In nerve cells outside the brain, norepinephrine works together with a second transmitter called *epinephrine* in controlling an animal's fight-or-flight response [review FIGURE 30.10].

The growing list of neurotransmitters includes glycine and other amino acids, many peptides, and serotonin and gamma-aminobutyric acid (GABA). Each of these plays specific roles in behavior, and their absence can lead to disease. Norepinephrine, as already mentioned, may be involved in clinical depression, and serotonin may be involved as well. Decreased quantities of dopamine are associated with the rigidity, weakness, and shaking of Parkinson's disease. These symptoms sometimes diminish when the patient takes oral doses of levodopa (L-dopa), a chemical that brain cells can modify into the neurotransmitter dopamine. BOX 31.1 discusses other cutting-edge research to help Parkinson's patients and others with neural disorders.

In the early 1990s, researchers discovered a new type of neurotransmitter that differs markedly from those we just discussed. *Nitric oxide* (NO), newly discovered to be a neurotransmitter, occurs as a gas under normal atmospheric conditions. Secreted by some neurons, nitric oxide controls dilating blood vessels and controlling blood pressure, and it may be involved in memory. Nitric oxide is synthesized when and where it is needed, rather than being stored in synaptic vesicles as are dopamine and most other neurotransmitters.

In contrast to the pathway illustrated in FIGURE 31.9B, nitric oxide diffuses straight through plasma membranes and enters cells without binding to a membrane receptor. Instead, it binds to a protein in the cell's cytoplasm and results in the production of a second messenger called *cyclic GMP*. This mechanism is similar to the one diagrammed in FIGURE 30.13B.

With the discovery of nitric oxide as a neurotransmitter, researchers have elucidated the mechanism of penile erection. During a sexual response, neurons in the pelvis get a message from the brain and release nitric oxide. This gas causes key blood vessels in the penis to dilate, blood rushes in, and the penis stiffens like a water balloon filling with water.

Some compounds act as **neuromodulators** rather than neurotransmitters. Compared to the relatively quick, short-lived influence that a neurotransmitter has on the postsynaptic membrane, the neuromodulator may remain near the synapse for a longer time and alter the receiving cell's responsiveness to neurotransmitters.

Excitatory and Inhibitory Synapses Chemical communication at synapses can either encourage or discourage the firing of the receiving (postsynaptic) cell. At an **excitatory synapse**, the molecular messenger brings the postsynaptic cell closer to firing a nerve impulse by causing a few channels for sodium ions to open and make the interior of the cell a bit more electrically positive. Thus, it takes fewer excitatory impulses to trigger an impulse in the postsynaptic cell.

In contrast, at an **inhibitory synapse**, the neurotransmitter decreases the likelihood that the postsynaptic cell will fire a nerve impulse by opening channels to potassium or chloride ions or both. By making the receiving cell's interior even more negative, it takes many more excitatory impulses than usual to trigger an impulse in the postsynaptic cell.

Summation Since most neurons receive synaptic input from the axons of up to thousands of other cells, the ultimate activity in any given region of the cell results from the cumulative impact of all the impulses it receives—some excitatory, some inhibitory. Whereas the firing of many impulses at an excitatory synapse might be enough to cause a postsynaptic neuron to depolarize, an inhibitory synapse nearby that receives an equal number of impulses could counteract the excitatory effect. In a process called **summation**, the postsynaptic cell sums up all the information impinging on it in a nearby region over a short time interval. Eventually, if the encouragements outnumber the discouragements, the threshold is reached within that cell, the chain reaction of electrochemical events that constitutes an action potential is then automatically triggered, and the neuron fires an impulse in an all-or-none fashion.

➤ CONCEPT CHALLENGE

The insecticide parathion destroys an enzyme (in insects and people!) that breaks down a particular neurotransmitter that controls muscle contraction. Hence, the neurotransmitter accumulates [review FIGURE 31.9, Step 7]. Will an insect or other animal poisoned with parathion become stiff from all the muscles contracting at once? Or will it become limp from all the muscles relaxing? Support your answer.

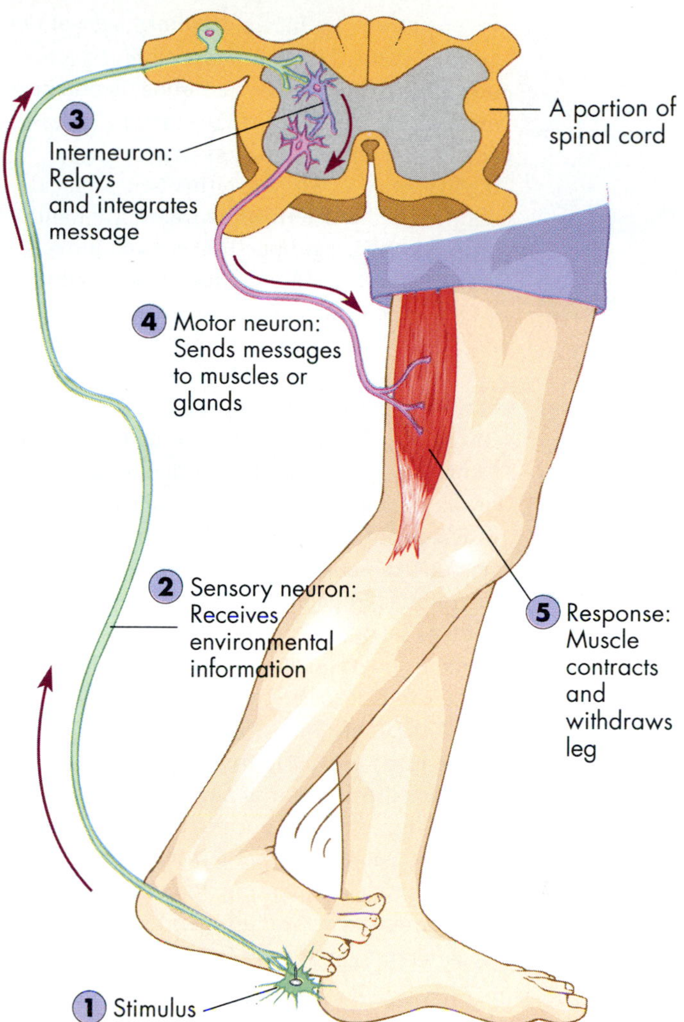

FIGURE 31.11
Anatomy of a Reflex Arc.
Jerking back your foot after stepping on a tack is based on a reflex arc that links the spinal cord and the muscles in the back of the thigh.

Neuron Networks

Although we have been discussing neurons as if they sit end to end like matchsticks on a table, they are actually arranged in delicate circuits—a spatial organization that allows the animal to receive, integrate, and act on information from its surroundings and thus to survive and reproduce. In many different animals, a basic type of nerve circuit called the **reflex arc** underlies simple behaviors. This simple neural loop links a stimulus to a response in a very direct way. If you step on a tack with your bare toe, for example, a reflex arc drives muscles in your leg to pull your toe away immediately, even before you consciously realize what has happened [FIGURE 31.11]. Reflex arcs mediate behaviors that are generally rapid, involuntary,

box 31.1
Human Impact

Repairing Neurons: Transplants and Stimulants

As an actress, Katharine Hepburn, star of *On Golden Pond, The Philadelphia Story*, and dozens of other Hollywood films, is one in a million. As a sufferer of Parkinson's disease, she is also one of a million Americans, mostly elderly, who have this neurological condition characterized by trembling hands, shaking head, rigid posture, slowed movements, and a shuffling, unbalanced walk.

Neurologists have determined that Parkinson's appears only after neurons in the caudate nucleus and putamen, parts of the brain's central region, degenerate and cease their normal production of dopamine. Without this neurotransmitter in that area of the brain, the patient loses some control over muscle contraction and relaxation, and develops the trembling, shaking, and other symptoms. Physicians administered drugs such as levodopa to increase patients' dopamine levels, but after 5 to 20 years, these drugs lose their effectiveness.

Researchers are working on several bold approaches to stimulating dopamine production that are generating hope for better future medical treatments. For the past few years, neurosurgeons have helped some Parkinson's patients by transplanting dopamine-producing neural tissue from human fetuses into the degenerated basal ganglia regions of the adults' brains. There, the embryonic tissue becomes established and begins to secrete some of the missing neurotransmitter. This has been moderately successful for patients, but it has triggered considerable protest by some abortion opponents who condemn any use of fetal tissue.

Three other promising lines of research may someday replace these fetal tissue transplants. Canadian researchers discovered a reservoir of immature cells in the brains of adult mice that can be induced with epidermal growth factor to divide and differentiate into new, fully operative neurons. The experimenters are trying now to stimulate the growth of the cells within the mouse brain—not just in the test tube—in hopes of someday stimulating the growth of specific types of brain cells. The next step would be inducing growth in the brains of Parkinson's, Alzheimer's, or epilepsy patients.

A second set of researchers is trying to genetically engineer cells to secrete dopamine, with the goal of transplanting the cells into patients' brains. So far, they have genetically engineered muscle cells that produce dopamine, and transplanted them into rat brains. Encouragingly, these lab animals have experienced a lessening of Parkinson-like symptoms.

All of these studies will require additional years of testing for effectiveness and safety, but by understanding and manipulating neuron function, biomedical researchers are trying to improve the lives of people like Katharine Hepburn and the millions of others with neurological diseases.

and nearly identical each time the stimulus is repeated and that require no conscious input from the brain.

Reflex arcs usually involve three kinds of neurons: (1) sensory neurons, (2) interneurons, and (3) motor neurons [see FIGURE 31.11]. A **sensory neuron** receives information from the external or internal environment—a tack stabbing a toe is this kind of information—and transmits it toward the spinal cord or brain, the nervous system's integrating centers. In those centers, **interneurons** relay messages between nerve cells and integrate and coordinate incoming and outgoing messages. While some reflex arcs do not have interneurons, complex behaviors rely on a network of many interneurons. Interneurons often connect with **motor neurons**, which send messages from the brain or spinal cord out to muscles or secretory glands. Motor neurons allow an organism to act, to respond to the information the sensory neuron brings in from the environment.

LEARNING IS A RESULT OF CHANGES AT SYNAPSES

One phenomenon that allows an organism to adapt to a changing environment is **habituation**: a progressive decrease in the strength of a behavioral response to a constant weak stimulus when that stimulus is repeated over and over again with no undesirable consequences. Habituation takes place when you get dressed in the morning. When you first put on your clothes, you can feel them against your skin, but within a few minutes, you become unaware of the constant light pressure. Biologists have investigated the cellular basis of this simple type of learning in the sea slug *Aplysia californica* [FIGURE 31.12A].

The sea slug has a tube called a *siphon* that draws water over the gills [FIGURE 31.12B]. If a slug senses water currents from the movements of a nearby animal, say a predatory sea star, the slug retracts its siphon and gills. This response helps the animal to survive, since gills act

[A]

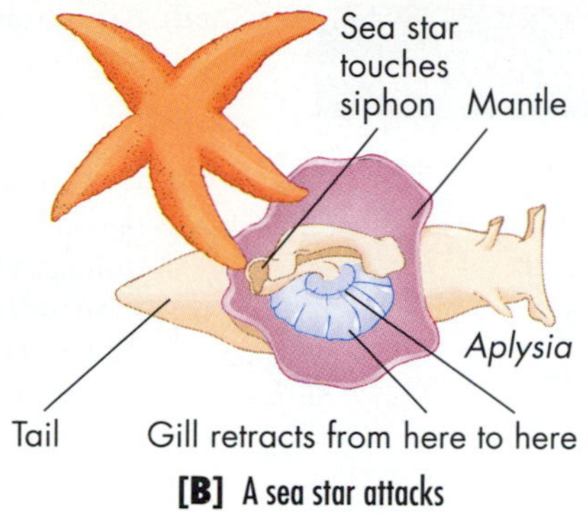

[B] A sea star attacks

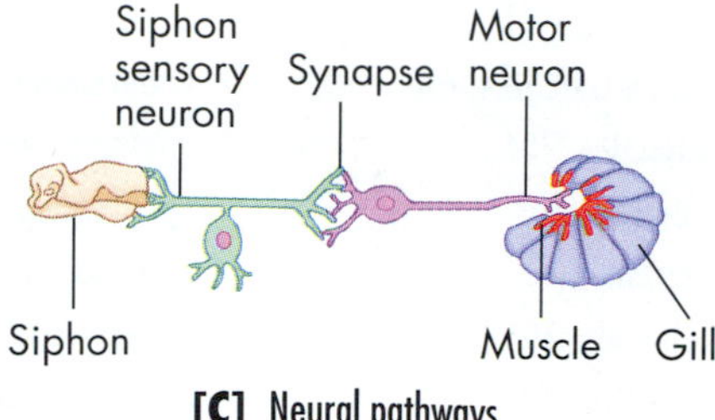

[C] Neural pathways

FIGURE 31.12

Simple Neural Circuits Control *Aplysia's* Gill Retraction Response.
[A] *Aplysia californica,* dweller of Pacific tide pools. [B] The touch of a predatory sea star causes *Aplysia* to retract its delicate siphon and gills. [C] The neural circuit that controls gill withdrawal is simple. Sensory neurons in the siphon detect the sea star's caress and send an action potential toward synapses. The release of neurotransmitter triggers motor neurons to cause muscles to contract and move the gill to safety. Habituation—in this case, a decreased contraction of the gill muscle after an experimenter repeatedly stimulates the siphon—occurs as a change in the strength of the synapse.

like bait for hungry predators. The neural circuit that controls this reflex depends on two sets of neurons [FIGURE 31.12C]. Branchlike dendrites of the sensory neurons in the siphon detect any distortion in the tissue around them and respond by initiating an action potential. The axons of these sensory neurons in the siphon form a synapse with motor neurons. The motor neurons, in turn, connect to muscle cells that can contract, withdrawing the gills to safety.

This gill-retraction behavior is subject to habituation based on experience. The tenth blast of a brief jet of water from an experimenter's probe causes *Aplysia* to retract its gills only half as much as did the first blast. This shows that the animal has habituated and learned to ignore a stimulus with trivial consequences. But what events allow habituation? Researchers have found that after habituation, the sensory neuron releases less neurotransmitter. As a result, the motor neuron is less likely to fire, and muscles do not withdraw the gills as far.

What causes the sensory neuron to release less neurotransmitter? Researchers found that after habituation, fewer calcium channels open in the presynaptic cell [review FIGURE 31.9, Step 2]. Therefore, less neurotransmitter is released, and with less neurotransmitter, there is less muscle action. Through their research, biologists were able to demonstrate a critical new finding: Learning is due to a physical change in synaptic effectiveness.

The simple learning response of habituation in *Aplysia* shows that learning is due to changes in the effectiveness with which signals cross synapses. Can human learning be explained in the same way? Even simple behaviors, like scratching the end of your nose, must require the proper activation of a great many cells to ensure that your finger lands on the end of your nose and not in your eye. The more sophisticated the action—a ballet pirouette, say, or a reverse slam dunk of a basketball—the more cells and organization are needed to carry it out. Future researchers must determine whether the cellular events that control how a sea slug habituates to a jetting stream of water also control the way you come to understand principles for a biology quiz.

Connections

The nervous system, along with the endocrine system, allows the animal body to function as an integrated whole. This chapter moved from the events within a single nerve cell to the communication between two neurons to neural networks and how they bring about and control meaningful behavior in animals. Underlying that integrated operation is the basic evolutionary principle that the nervous system, like the other physiological systems we have discussed, enhances an animal's chances of surviving and reproducing. The effect of cocaine in disrupting this integration reminds us of the importance of the body's control systems.

Simple behaviors controlled by reflex arcs depend on just two or three interlinked neurons. More complex behaviors depend on millions of neurons organized into intricate nervous systems. The most intricate control center of all, the human brain, is also the most fantastically organized tissue that has yet evolved and is the subject of our next chapter.

KEY TERMS

action potential, 698	neuromodulator, 705
addiction, 704	polarization, 698
axon, 696	postsynaptic cell, 697
bouton, 696	potassium channel, 697
cell body, 696	presynaptic cell, 697
dendrite, 696	propagation, 700
glial cell, 696	reflex arc, 705
habituation, 706	refractory period, 699
interneuron, 706	sensory neuron, 706
motor neuron, 706	sodium channel, 698
myelin sheath, 701	soma, 696
nerve, 696	summation, 705
nervous system, 694	synapse, 696
neuron, 696	synaptic vesicle, 703

HIGHLIGHTS IN REVIEW

1 The nervous system receives data from the environment, integrates the data into a meaningful form, and then brings about a change in the animal's behavior, physiology, or information storage.

a] The nervous and endocrine (hormonal) systems share several characteristics and work together in controlling and coordinating the body.

b] Nerve cells communicate in a point-to-point fashion, but the cells are arranged in complex networks that allow for rich behavioral repertoires.

2 Communication within the nervous system occurs in cells called neurons. Neurons are electrically charged cells, and they often have very long, thin cell extensions.

a] Cells of the nervous system are classified as neurons or glial cells. Neurons communicate nerve impulses. They have a soma, thin cell processes called dendrites that receive information, a usually long, thin axon that sends out information, and a bouton at the axon terminal that communicates with adjacent cells. Glial cells provide nutrition, structural support, and insulation to neurons.

3 When a neuron becomes stimulated, its electric charge reverses at the point of stimulation; this event is called an action potential. This temporary reversal of charge can move along the neuron.

a] Animal behavior is based on nerve impulses, chain reactions of ion flow that move along the neuron's membrane.

b] The extracellular fluid that surrounds a neuron is high in sodium ions but low in potassium ions, while the cell's interior is rich in potassium ions and poor in sodium ions. The cell's interior is electrically negative with respect to the exterior.

c] A sodium-potassium pump actively transports sodium out of and potassium into the cell against concentration gradients. Ion channels specific for either potassium ions or sodium ions open and shut, regulating ion flow across the cell membrane.

d] An action potential is a reversal of electric charge that begins when sodium ions leak into a small region of the membrane and the cell becomes depolarized. Sodium channels in that membrane patch then open wide, sodium ions rush in, and the inside of the neuron becomes positively charged. The sodium channels quickly close, a refractory period prevents them from reopening temporarily, and the potassium channels now open, allowing the balance of charges to become reestablished and the resting potential to return.

e] The nerve impulse propagates along the neural membrane from the patch that initially received the stimulus to adjacent patches. This occurs because the adjacent membrane patch loses some of its sodium ions and becomes depolarized, generating an action potential in the adjacent patch.

f] Propagation can take place more quickly in a giant axon or in one wrapped in a myelin sheath (formed by glial cells).

4 When an action potential meets specialized ends of the neuron, it can stimulate or inhibit an action potential in an adjacent cell.

a] Neurons form junctions called synapses with other neurons, muscle cells, and some gland cells. In electrical synapses, ions flow directly from one cell to the next. In chemical synapses, neurotransmitter molecules diffuse from one neuron, cross a gap, and trigger a nerve impulse in the next neuron.

b] Neurotransmitters are chemicals released by one neuron that influence whether the adjacent neuron will have an action potential. They include acetylcholine, involved in maintaining posture, breathing, and movement of limbs; dopamine, which is released in the pleasure centers of the brain; and serotonin, norepinephrine, nitric oxide, and other molecules.

c] Neuromodulators remain near synapses and provide a longer, steadier influence on the receiving cell.

d] At an excitatory synapse, neurotransmitters "encourage" the receiving cell to fire an action potential. At an inhibitory synapse, neurotransmitters "discourage" the receiving cell from firing.

5 Neurons are linked together, sometimes in very complex loops and networks, and these control behavior and learning.

a] Neurons can be linked into simple reflex arcs, composed of a sensory neuron that receives stimulation from the environment; an interneuron that integrates and coordinates incoming and outgoing messages; and one or more neurons that carry signals to the muscles or glands.

b] Habituation and sensitization are simple forms of learning. The physiological basis of these learned behaviors is a change in the effectiveness of signals crossing specific synapses.

UNDERSTANDING THE FACTS AND CONCEPTS

For Questions 1–5, match each of the descriptions with one of the terms from the following list. Any answer may be used once, more than once, or not at all.

a] dendrite b] soma c] axon d] bouton e] synapse

1 The metabolic heart of a neuron.

2 The receiving end of a neuron.

3 Region of an axon where synaptic vesicles open and liberate a neurotransmitter substance.

4 A small space across which neurotransmitters diffuse.

5 Region where saltatory conduction occurs.

As above, for Questions 6–10, match each of the descriptions with the most appropriate item from the following list.

a] resting potential
b] action potential
c] nerve impulse
d] refractory period
e] myelin sheath
f] saltatory conduction

6 A series of action potentials that occur sequentially along an axon.

7 The rapid transmission of an action potential from node to node along an axon.

8 The rapid entry of sodium ions into one site on an axon followed by their rapid removal.

9 A condition in which the inside of a cell is maintained at −70 mV relative to the outside by means of the transport of ions through pumps and channels.

10 The minimum time that must elapse for one region of an axon to exhibit successive action potentials.

As above, for Questions 11–15, match each of the descriptions with the most appropriate item from the following list.

a] neurotransmitter
b] synaptic vesicle
c] excitatory synapse
d] inhibitory synapse
e] Schwann cell
f] electrical synapse

11 One kind of glial cell that insulates axons; associated with rapid conduction of a nerve impulse.

12 Results in depolarization of a postsynaptic plasma membrane.

13 Secretory product of neurons that enables communication between neurons.

14 The location of neurotransmitters before they become secretory products.

15 Cell-to-cell contact between neurons that allows for rapid passage of a nerve impulse between them.

As above, for Questions 16–20, match each of the descriptions with the most appropriate item from the following list.

a] all-or-none response
b] summation
c] reflex arc
d] polarity
e] sensitization
f] sensory neuron

16 Characterizes all synaptic transmission.

17 Characterizes all action potentials.

18 Carries information about an environmental change into the nervous system.

19 A neuronal loop involving several types of neurons that enables an organism to rapidly and involuntarily adjust to an environmental change.

20 A state in which the inside of a plasma membrane has fewer or more changed particles than the outside of the membrane.

INTEGRATE AND APPLY WHAT YOU HAVE LEARNED

1 Contrast the activity of the sodium-potassium pump with the sodium and potassium ion channels in the membrane during the generation of an action potential.

2 Compare the relative advantages and disadvantages conferred by electrical synapses and chemical synapses. Could large, active, intelligent animals have evolved with only electrical synapses?

3 Despite our large brains and capacity for learning, we also exhibit many reflexive behaviors. Name three different human reflex activities, stating their common features.

4 How does the possession of both excitatory and inhibitory synapses enable an organism to adjust to differing circumstances?

5 Some poisonous snakes inject neurotoxins into their prey; other species inject poisons that cause painful damage to cells and tissues in other organ systems of their prey. Assume that the venom in one type has arisen as a defense mechanism, whereas the other type of venom has largely evolved as a mechanism for capturing prey animals for food. Which type of venom has likely evolved in each case?

ANALYSIS

1 Which features of a neuron prevent an action potential in a postsynaptic cell from immediately moving backward and restimulating the presynaptic cell? More than one answer may be correct.

a] Neurotransmitters are found only at the tips of axons.
b] Neurotransmitter receptors are found on the postsynaptic side of a synapse.
c] The refractory period.
d] The length of time taken for the sodium and potassium gates to return to their resting states.

2 Many pesticides prevent the enzymatic inactivation or removal of the neurotransmitter acetylcholine, which stimulates the contraction of skeletal muscles. As a result, acetylcholine remains active within the synapse between a neuron and its muscle cell for an extended period of time. What effect would this have on the postsynaptic cell (which in this case is a muscle cell)?

a] It would be inhibited, could not generate an action potential, and thus could not contract.
b] The muscle cell could not stop contracting.
c] The presynaptic neuron would continue to be stimulated.
d] The presynaptic muscle cell would be inhibited.
e] There would be no effect.

3 Scorpion venom causes the sodium gates to open wide when the potential across the membrane is at the resting potential −70 mV rather than the usual threshold of −50-mV. What effect would the venom have on the activity of an axon?

a] No effect, since action potentials could occur.
b] Action potentials would occur more rapidly than normal.
c] Once the first action potential occurred, the membrane would not be able to restore a resting potential, so all further activity would cease.
d] Action potentials would occur more slowly, since the stimulus would have a harder time getting to the −50-mV threshold level.
e] Action potentials would be stronger, since they started at a lower potential voltage.

CHAPTER 32

The Nervous System

KEEN SENSES OF A NIGHT HUNTER

Perched on an oak limb in the moonlight, a reddish-brown owl rotates its head slowly. Like a ship with sonar, it scans the darkened English woods, listening for the slightest indication of a mouse scuttling across the forest floor. When it senses the rustle of leaves, the predatory bird swoops silently from its perch, deftly avoids a large branch looming out of the dim shadows, and pounces on the source of the sound with sharp talons outstretched. The owl grabs a gray mouse with its talons, snaps the rodent's skull with its hooked beak, then returns to the oak limb and shares the meal with its owlets.

This night hunter, the tawny owl, *Strix aluco* [FIGURE 32.1], is the most common owl in British forests, and it inhabits the woods throughout much of Europe, northern Africa, and western Asia. The tawny owl can fly silently because its outermost flight feathers end in soft fringes, not smooth, firm edges. Like other members of its avian order, the tawny owl has large, front-facing eyes. One might suspect that these piercing, ever-vigilant visual organs are chiefly responsible for the bird's hunting prowess, but in fact, a tawny owl's night vision is no better than that of some people who see particularly well at night. What's more, a simple experiment disproves the primacy of vision in the owl: If an experimenter ties a dry leaf to a mouse's tail and places the rodent in a dimly lit room with an owl, the rodent will scurry about in fright and the bird will pounce—not on the prey but on the rattling leaf!

FIGURE 32.1

A Night Hunter Swoops to Catch Prey.

The tawny owl can navigate in total darkness simply by comparing differences in the timing and loudness of sounds that reach its two ears.

Clearly, a tawny owl's hearing must be very acute, and research shows that it is sufficiently sharp to allow the bird to locate and capture prey in absolute darkness. The bird's ability to *sense* slight perturbations in its environment is part of the explanation for this ability to hunt in darkness. The sensory information must also be *integrated*—that is, combined and processed—within the animal's brain, and the bird must then *react* with coordinated muscular activity and continued sensing of body position and prey location.

Using the tawny owl as a central example, this chapter describes how sense organs work and how nervous systems integrate sensory information and coordinate behaviors that help animals survive and reproduce. The previous chapters in this part of the book have already demonstrated how sensing, integrating, and reacting are all vital to the maintenance of an animal's homeostasis in a changing environment. CHAPTER 31, in particular, showed the role of individual neurons in these activities. Here, we will see how nerve cells are organized into two higher-order networks: the central nervous system (CNS), which consists of the spinal cord and brain, and the peripheral nervous system (PNS), which contains all of the interacting nerve cells outside the CNS.

In this chapter our survey of nervous system and sense organs is divided into three major discussions: (1) How the brain and senses evolved; (2) how the main parts of a vertebrate's nervous system interconnect, and in a human's brain, where the seats of homeostatic control, motor coordination, sensation, emotion, facial recognition, and language are located; and (3) how animals hear, see, taste, and smell.

As we go along, we will see that the human brain is the most intricately organized entity in the universe, and that different senses and behaviors can be localized to specific regions or groups of regions in the brain—one area for hearing, one for seeing, and so on. We will also see that the nervous system's major role is to monitor internal needs and external conditions, process the information, and initiate appropriate behavioral responses. The tawny owl can sense hunger because of its nervous system; its nervous system also enables it to detect a mouse scampering through fallen leaves and to respond by pouncing on the prey. We'll see why the sense organs are specialized—that is, why the eye is most sensitive to light, the ear to sound, the nose to smell, the tongue to taste, and the skin to pressure or temperature. We'll learn why nervous systems, no matter how complicated, are built on similar kinds of nerve cells that act in the same ways. The nerve cells of a jellyfish, for example, resemble a tawny owl's, but the sense organs and nervous integrators are more centralized and contain many more cells.

As you read this chapter, you will learn not only about the owl's lopsided ears, and the way a rattlesnake detects prey in the dark, but also how structural differences in the brains of human males and females account for certain common behaviors. ❑

MESSAGES

1 The nervous system's major roles are to monitor internal and external conditions, process the information, and initiate behavioral responses that improve the organism's chances of survival and reproduction.

2 The vertebrate nervous system is organized into the central nervous system (brain and spinal cord) and the peripheral nervous system (the rest of the body's neurons). The central nervous system (CNS) performs neural integration, while the peripheral nervous system (PNS) carries information from the sense organs to the central nervous system and carries signals away from the central nervous system to muscles and glands.

3 The brain consists of the brainstem, which controls many automatic functions such as heart rate and breathing rate; the cerebellum, which coordinates muscle action; and the cerebrum, which is the site of perception, thought, language, learning, and self-awareness.

4 Sense organs monitor physiological and environmental conditions by means of specialized cells with supersensitivity to a specific physical factor, such as light, sound, pressure, or specific chemicals. These cells relay signals to specific parts of the brain.

The Nervous System

An animal's survival depends on its ability to act or react—perhaps escape from a predator, pounce on prey, search for food, or find a mate. Action requires both an integration of the sensory input and a coordination of different body parts. When an owl, for example, hears a rustling on the forest floor, it must be able to determine whether the sound comes from a mouse that the owl could eat or a prowling fox that could eat the owl. Such discriminations are made by neurons arranged in complex ways in the animal's central nervous system.

TRENDS IN NERVOUS SYSTEM EVOLUTION

Two general evolutionary trends are apparent among animal nervous systems: (1) a concentration of neurons in the head and (2) an increased number of interneurons.

In sponges, cells communicate with their neighbors, but no nervous system exists. The simplest form of nervous organization in the animal world appears in radial animals such as jellyfish. This **nerve net**, a diffuse meshwork of nerve cells and fibers, permits the conduction of impulses from one area to another. The first evolutionary trend, a concentration of neurons in the head region, becomes apparent in more complicated animals such as flatworms and roundworms [see CHAPTER 21], which move in a forward direction. These animals have sense organs concentrated in the head, the body region that first encounters new environments. A flatworm's head contains **ganglia** (singular, **ganglion**), distinct clumps of nerve cell bodies, which act like a primitive brain [review FIGURE 21.10]. Distinct nerve trunks (bundles of axons) on either side of the body carry sensory information from the periphery to the head ganglia and carry motor commands from the head ganglia to the muscles, allowing the animal to react.

In invertebrates such as mollusks, arthropods, and even segmented worms, the organization of the nervous system advances a step further. In these invertebrates, axons are joined into cables, or *nerve trunks*, and in addition to a centralized "brain," smaller peripheral ganglia help coordinate outlying regions of the animal's body [FIGURE 32.2]. These ganglia represent the second evolutionary trend: More complex animals have more interneurons. Recall that interneurons transmit signals from one neuron to another neuron [review FIGURE 31.11]. Since much of the integration that takes place in nervous systems is carried out by interneurons in ganglia, the more numerous the interneurons, the more complex the behavior patterns an animal can display.

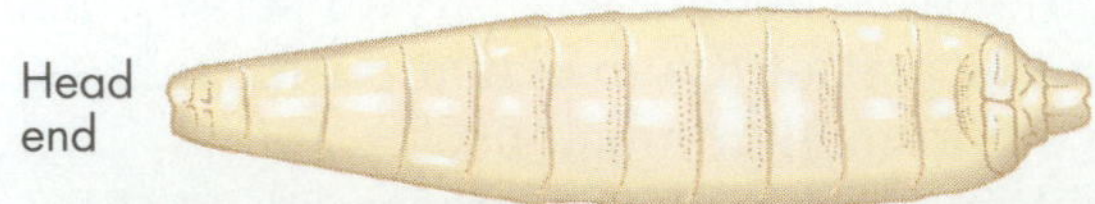

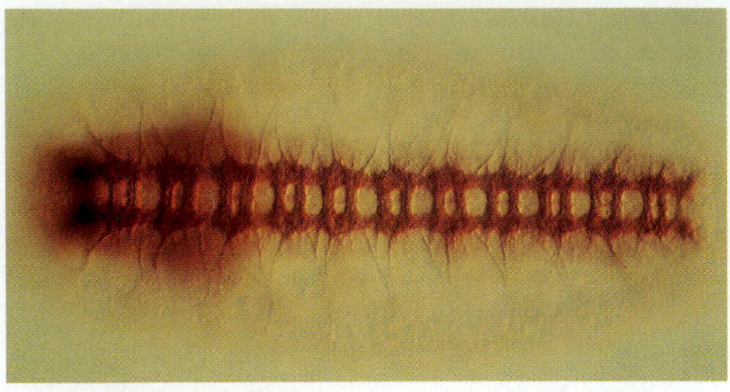

FIGURE 32.2

An Invertebrate Nervous System.

A fruit fly embryo illustrates some trends in the evolution of the nervous system. Sense organs are clustered at the head, along with a primitive brain. The brown stain in the photograph reveals a central, bilaterally symmetrical nerve cord punctuated by ganglia containing many interneurons.

A consequence of the increasing numbers of interneurons was the evolution of the brain. The more complex the animal and the more complicated its behavior, the more neurons (especially interneurons) it will have concentrated into an anterior brain and bilaterally organized ganglia. Vertebrate brains, which we turn to next, are the culmination of this trend.

➤ CONCEPT CHALLENGE

You discover that one species of flatworm has ganglia that are substantially larger than those of another species of flatworm. Which type of neuron would you expect the larger ganglia to contain more of: sensory neurons, interneurons, or motor neurons?

The Vertebrate Nervous System

Complex animal behavior—from a grouse's elaborate mating dance to a cobra's hypnotic weaving and an aerobic dancer's vigorous kicks and turns—depends on a highly organized nervous system. A vertebrate's nervous system is organized into two main units based on function and location: a *peripheral nervous system* (*PNS*)

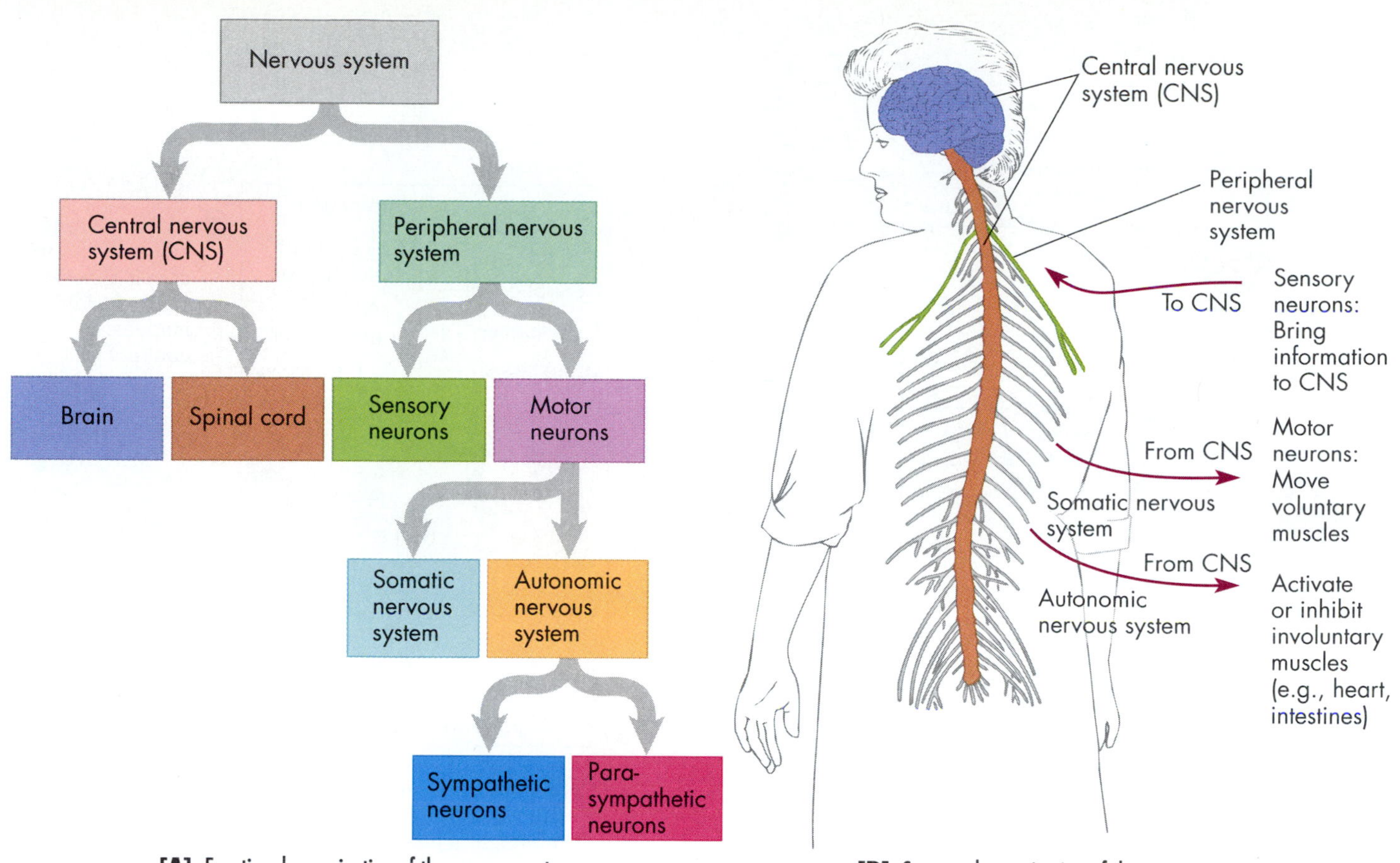

[A] Functional organization of the nervous system

[B] Structural organization of the nervous system

FIGURE 32.3

Organization of the Vertebrate Nervous System.

[A] Your nervous system has two main branches, the *central nervous system* (CNS), which controls the processing and integration of information, and the *peripheral nervous system* (PNS), which connects the CNS with sensory receptors, muscles, and glands. **[B]** The information-sending and -receiving pathways for different parts of the body.

and a *central nervous system* (*CNS*) [FIGURE 32.3]. The peripheral nervous system is the actor or doer; it includes the sensory and motor neurons, and it connects the CNS with the sense organs, muscles, and glands of the body. The CNS, the nervous system's director, thinker, and information processor, consists of the *brain*, which performs complex neural integration, and the *spinal cord*, which carries nerve impulses to and from the brain and participates in reflexes. The interplay between the central and peripheral nervous systems allows an animal to sense environmental stimuli, integrate the information, and respond appropriately, providing the fascinating behaviors seen in the animal kingdom.

THE PERIPHERAL NERVOUS SYSTEM

The **peripheral nervous system** includes all the neurons of the body outside the brain and spinal cord. These neurons connect the brain and spinal cord with sense receptors, muscles, and glands. Peripheral neurons transmit impulses to and from the CNS, but do little information processing. Peripheral neurons include *sensory neurons* like those in the eyes, ears, and skin, which carry information about the body and environment to the CNS, and *motor neurons*, which convey information from the CNS to muscles and glands and trigger some responsive activity.

Sensory Neurons The peripheral nervous system is intimately connected with the *spinal cord*, a bone-encased trunk of nerves carrying messages to and from the brain. Attached to the human spinal cord are 31 pairs of *spinal nerves*, each a cable of thousands of axons reaching out to different regions of the body [FIGURE 32.4A and B]. To understand how sensory and motor neurons are organized within a spinal nerve, think about the reflex that causes you to pull your finger from a flame. The dendrites of pain receptors in your skin send action potentials to the cell body in the *dorsal root ganglion*, a collection of sensory neuron cell bodies near the spinal cord [FIGURE 32.4C]. The action potential continues along the sensory axon into the spinal cord, where it synapses with interneurons. These interneurons are contained in the CNS and carry impulses to the brain, where they are interpreted as pain.

Other stimulated interneurons synapse in the spinal

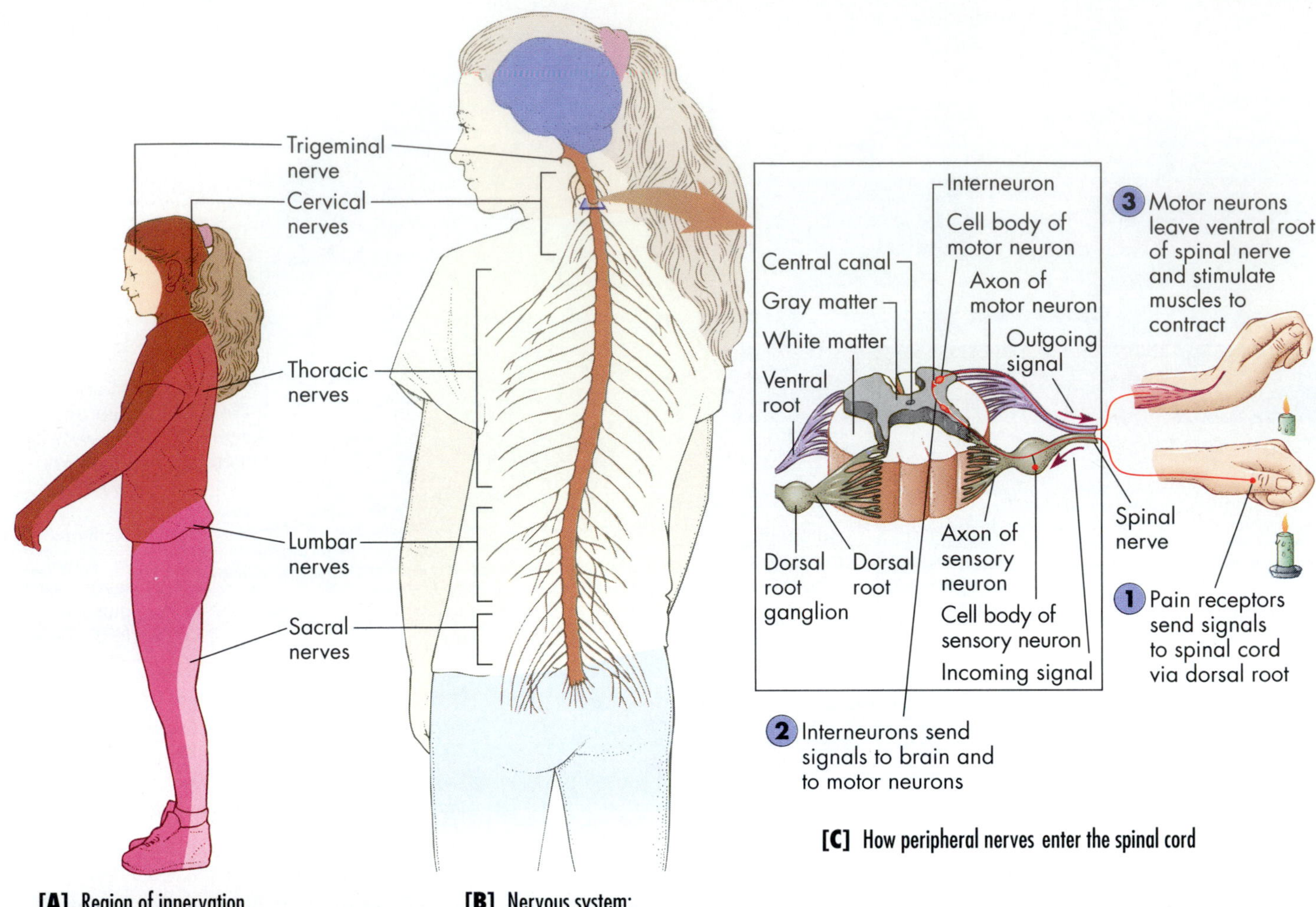

[A] Region of innervation

[B] Nervous system; central and peripheral

[C] How peripheral nerves enter the spinal cord

FIGURE 32.4

Spinal Nerves.

[A, B] A nerve map of the human body, with regions innervated by particular stretches of the central nervous system. [C] Each major nerve has a dorsal root, which carries information from sense organs into the spinal cord, and a ventral root, which carries impulses out from the spinal cord to the muscles and glands. The neural pathway through these roots is diagrammed for the hand-withdrawal reflex.

cord with motor neurons that leave the CNS via the *ventral root* of the spinal nerve and signal your muscles to quickly pull your finger from the flame. (This is a reflex arc similar to the one in FIGURE 31.11.)

Motor Neurons Sensory and motor neurons in the spinal nerves form the **somatic nervous system**, which in general activates muscles under voluntary control (like those that allow you to ride a bicycle). Motor neurons in the ventral roots of spinal nerves form the **autonomic nervous system**, which regulates the body's internal environment by controlling glands, the heart muscle, and smooth muscles in the digestive and circulatory systems [FIGURE 32.5; see also FIGURE 32.3].

A person's autonomic nervous system operates primarily at the subconscious level, performing many of its duties through the spinal cord and lower centers of the brain. There are two sets of autonomic neurons that function in opposition, a bit like the accelerator and brakes of a car. **Parasympathetic nerves** leave the spinal cord in your neck and lower back, and **sympathetic nerves** leave your spinal cord in the central region of your back [see FIGURE 32.5]. Sympathetic and parasympathetic neurons often innervate the same organ and help to increase or decrease its activity as needed. For example, most sympathetic neurons secrete the neurotransmitter norepinephrine (noradrenaline), speeding and strengthening the heartbeat, while parasympathetic neurons secrete acetylcholine, slowing it.

In general, the parasympathetic nerves function as a housekeeping system for the body, stimulating the stomach to churn, the bladder to empty, and the heart to beat at a slow and even pace during most daily and nightly activity. When an emergency arises, however, generating intense anger, fear, or excitement, the sympathetic nerves dominate, increasing heart rate, dilating the pupils of the eyes (which lets in more light), expanding the bronchioles of the lungs (which improves gas exchange), and slowing down nonessential digestive activities until the emergency is past.

Many of us, when experiencing stressful changes

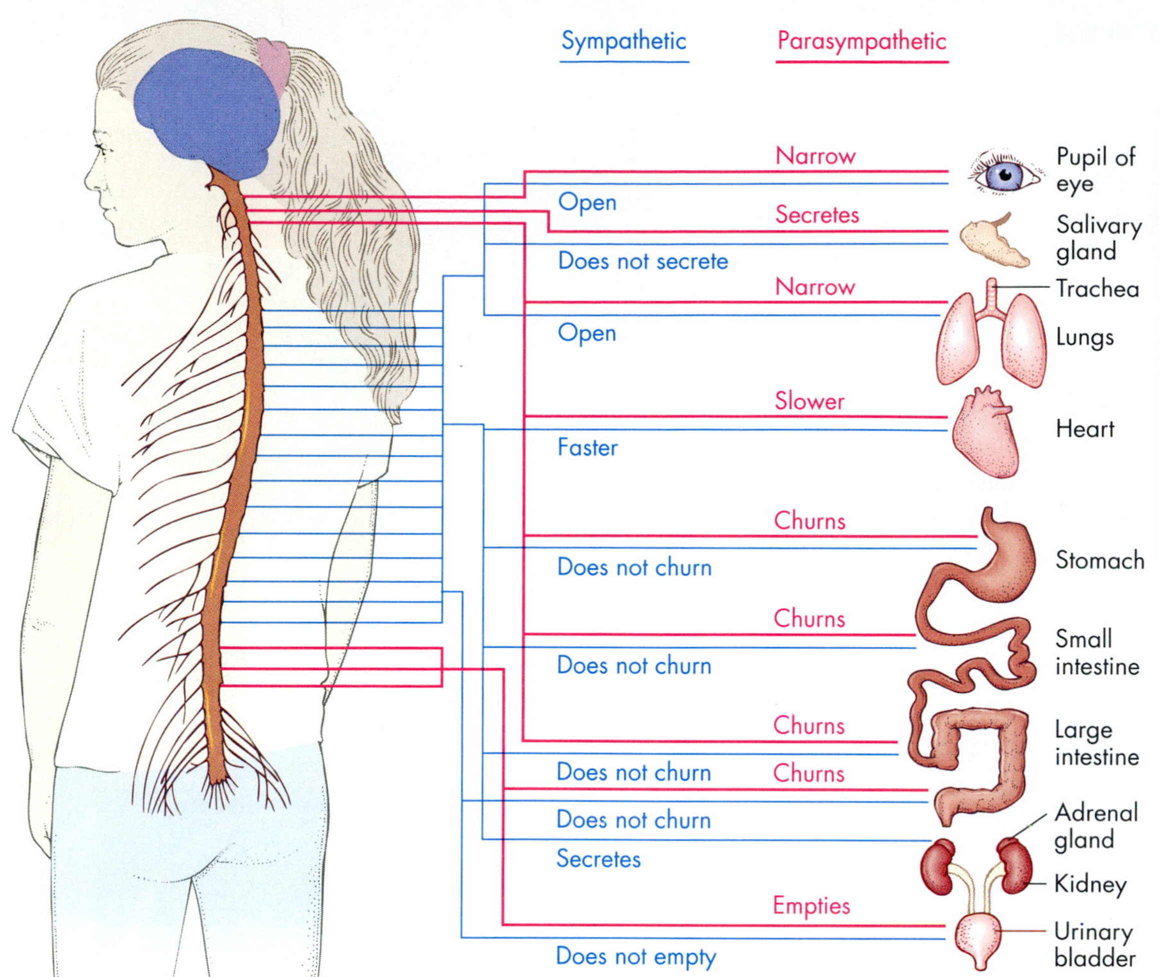

FIGURE 32.5

Autonomic Branch of the Peripheral Nervous System.

The sympathetic and parasympathetic nerves of the autonomic nervous system run to many of the same organs but act in opposition, turning the organ's activity on or off, up or down.

that make us feel trapped and that we cannot respond to with satisfying physical activity (like fleeing a traffic jam or putting a pie in your boss's face) develop stress-related health problems, including high blood pressure. Some medications for high blood pressure (so-called *beta-blockers*) specifically inhibit the sympathetic synapses that stimulate the heart and blood vessels, thereby keeping the heart rate slower and steadier.

Both branches of the autonomic system are regulated primarily through reflex loops, such as the one that controls the emptying of the urinary bladder. When the bladder fills, the bladder wall stretches, activating the stretch receptors, which generate a signal. This signal enters the spinal cord via sensory neurons and stimulates motor neurons of the bladder muscle, causing that storage organ to contract and expel urine. In infants, this reflex loop is the only control over bladder function, so the organ empties whenever it fills. With toilet training, the child's brain gradually learns to exert control over this reflex loop.

Some cultures have explored conscious control over the autonomic nervous system—and with intriguing results. Some yogis in India, for example, claim that they can survive being sealed in an underground chamber for many days by lowering their metabolic rate and using far less oxygen than is normally required. Many such claims are surely exaggerations. But by measuring the yogis' breathing, pulse, and oxygen utilization with reliable instruments, scientists have concluded that some yogis can indeed consciously lower their metabolic rate below normal resting levels. The Western practice of *biofeedback* training, which uses electronic devices to indicate changes in blood pressure, blood flow to extremities, or other external states, is another example of the relationship between the conscious and autonomic nervous systems.

➤ CONCEPT CHALLENGE

Tumors of the sympathetic nervous system are most common in people 5 to 25 years old, and can grow in the body cavity and produce large quantities of norepinephrine and related neurotransmitters. What would you predict would be some physiological symptoms of such a tumor?

THE CENTRAL NERVOUS SYSTEM

Although the peripheral nervous system receives data about the environment from sense receptors and transmits impulses that control the actions of the body's organs, the system would not function if the central nervous system could not make sense of the data and coordinate the actions. The CNS integrates information as the brain performs complex neural integration, and the spinal cord transmits information to and from the brain and controls some reflexes.

The Spinal Cord: A Neural Highway A cross section taken from a human spinal cord reveals a butterfly-shaped core of gray material surrounded by a bean-shaped field of white [see FIGURE 32.4C]. The *gray matter* contains cell bodies and unmyelinated nerve fibers; it is a zone of many synapses, where local neural traffic occurs (such as the reflexes discussed earlier). The *white matter*, on the other hand, is an interstate freeway; it consists mainly of thin axons that transport information long distances up and down the spinal cord to and from the brain. Since these long axons are insulated with a white, fatty myelin sheath [review FIGURE 31.7], the entire region has a whitish color.

The spinal cord is responsible for the preliminary integration of signals from sensory neurons and interneurons. For example, when a flame touches your finger and you withdraw that digit reflexively, impulses travel across synapses from the sensory neuron to an interneuron to a motor neuron in the gray matter of the spinal cord [see FIGURE 32.4]. Despite the preliminary role of the spinal cord in neural integration, most major data processing occurs in the brain.

The Brain: The Ultimate Processor The human brain is a reddish-brown organ that weighs about 1.4 kg (3 lb), contains approximately 100 billion neurons, and has the consistency of custard. It embodies our feelings and strivings, our knowledge and memories, our musical and verbal abilities, and our sense of the future. These complex behaviors and emotions can be localized to specific regions or groups of regions in the brain, as even minor damage to a given area can reveal. Thus, the anatomy of the brain provides a map for behavior, and certain behavioral disorders can be clues to the activities of specific brain regions.

The human brain has three interconnected parts [FIGURE 32.6]: (1) the *brainstem*, an extension of the spinal cord; (2) the highly rippled *cerebellum* attached to the brainstem; and (3) the large, folded *cerebrum*, sitting atop the brainstem and spanning the inside of the head from just behind the eyes to the bony bump at the back of the skull.

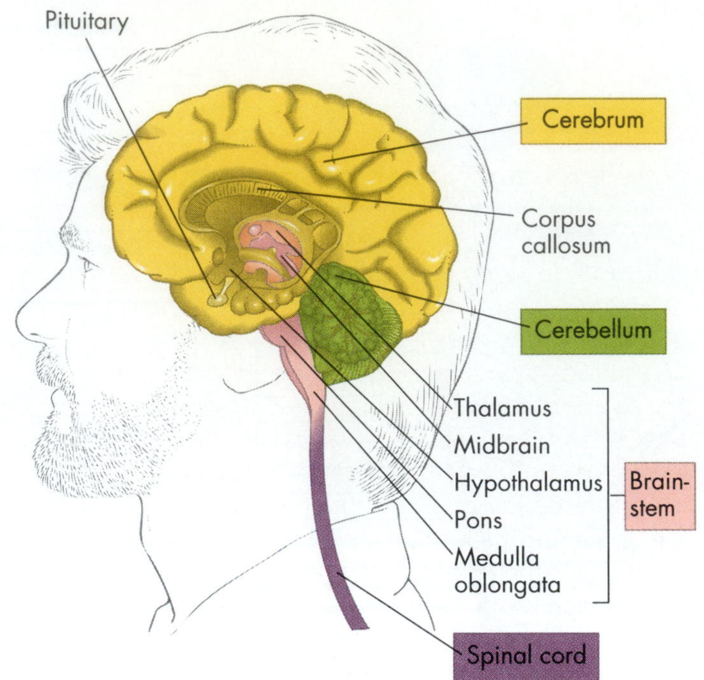

[A] Brain anatomy

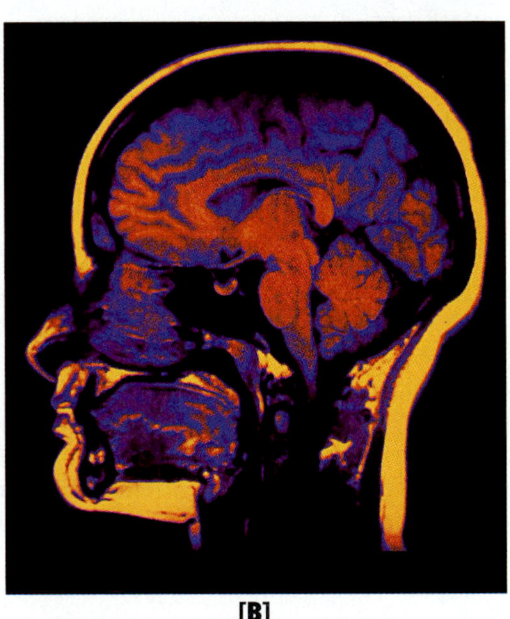

[B]

FIGURE 32.6
Major Regions of the Brain.
[A] The drawing reveals the brainstem, cerebellum, and cerebrum, as well as major parts of each section. **[B]** A photograph of the brain taken with magnetic resonance imaging reveals the living tissues in remarkable detail.

The Brainstem The **brainstem** plays three crucial roles: It helps integrate sensory and motor systems, it regulates body homeostasis, and it controls arousal. The lowest part of the brainstem, the **medulla oblongata**, lies in the upper part of the neck at about the level of the mouth. The medulla oblongata helps keep body conditions constant by receiving data about activities in various physiological systems and by regulating numerous subconscious body activities, such as respiratory rate, heart rate, and blood pressure. Above the medulla lie the **pons** and **midbrain**, regions that relay sensory information from our eyes and ears to other parts of the brain [see FIGURE 32.6].

The **hypothalamus**, still farther up the brainstem, regulates the pituitary gland and provides a link between the nervous and endocrine systems [see CHAPTER 30]. The hypothalamus also has important roles in maintaining homeostasis: It helps to regulate body temperature, water balance, hunger, and the digestive system. By electrically stimulating different regions of the hypothalamus, researchers can make an animal behave as if it feels alternately hot and cold, hungry and satisfied, angry and content. Some regions seem to be pleasure centers: A rat with electrodes permanently implanted in such areas will continue to press a foot pedal that turns on the current and stimulates the pleasure centers until the animal drops from exhaustion. Just above the hypothalamus is the **thalamus**, a relay area to the highly convoluted cerebrum that surrounds it [see FIGURE 32.6].

Extending through the brainstem from the thalamus down to the spinal cord is the **reticular formation**, a network of tracts that reaches into the cerebellum and cerebrum. When specific neurons of the reticular formation are actively firing, a person is awake, and when those neurons fall silent, the person sleeps or loses awareness. This selective shaping of neural input by the reticular formation and the thalamus helps a person to concentrate on a dinner partner's conversation in a noisy restaurant, and plays a part in the groggy incoherence of an early-morning phone conversation.

The Cerebellum: Muscle Coordinator Attached to the midbrain, at the middle of the brainstem, is the **cerebellum** (ser-uh-BELL-uhm; Latin, "little brain"), a convoluted bulb that serves as a complex computer, comparing outgoing commands with incoming information about the status of muscles, tendons, joints, and the position of the body in space [see FIGURE 32.6]. The result is a fine-tuning of motor commands. Learning the finer points of a basketball pattern or a graceful ballet turn involves the cerebellum. Birds, including owls, rely on precise muscle coordination to avoid tree limbs and other obstacles, and their cerebellums are large relative to other parts of the brain [FIGURE 32.7].

Although muscle movement is refined in the cerebellum, it is initiated in the largest part of the brain, the cerebrum.

The Cerebrum: Perception, Thought, Humanness The two side-by-side hemispheres of the brain's **cerebrum** (seh-REE-bruhm; Latin, "brain") fit like a cap over the brainstem. This mass of tissue embodies not only the attributes we consider human, such as speech, musical and artistic ability, and self-awareness, but also traits that we share with other vertebrates, such as sensory perception, motor output, and memory. In humans, the **cerebral cortex**, the cerebrum's highly convoluted surface layers, contains about 90 percent of the brain's cell bodies [FIGURE 32.8]. During vertebrate evolution, this cap increased in size relative to other brain regions [see FIGURE 32.8]. One of the great unsolved questions of modern biology is how these cells can control complicated traits. For now, biologists have many ideas but no single answer. One principle has been well established, however: All our basic cerebral functions, including sensations (like seeing or hearing), motor ability (movement), cognitive functions (like language and perception), affective traits (emotions), and character traits (like friendliness or shyness), can be traced to specific regions of the cerebrum. For example, hearing words, seeing words, speaking words, and generating words involve

FIGURE 32.7

Brain Differences in Various Vertebrates.

Compare the cerebrum of a fish, owl, cat, and monkey. The increasing size and surface convolutions correspond to increased intelligence and an enlarged behavioral repertoire.

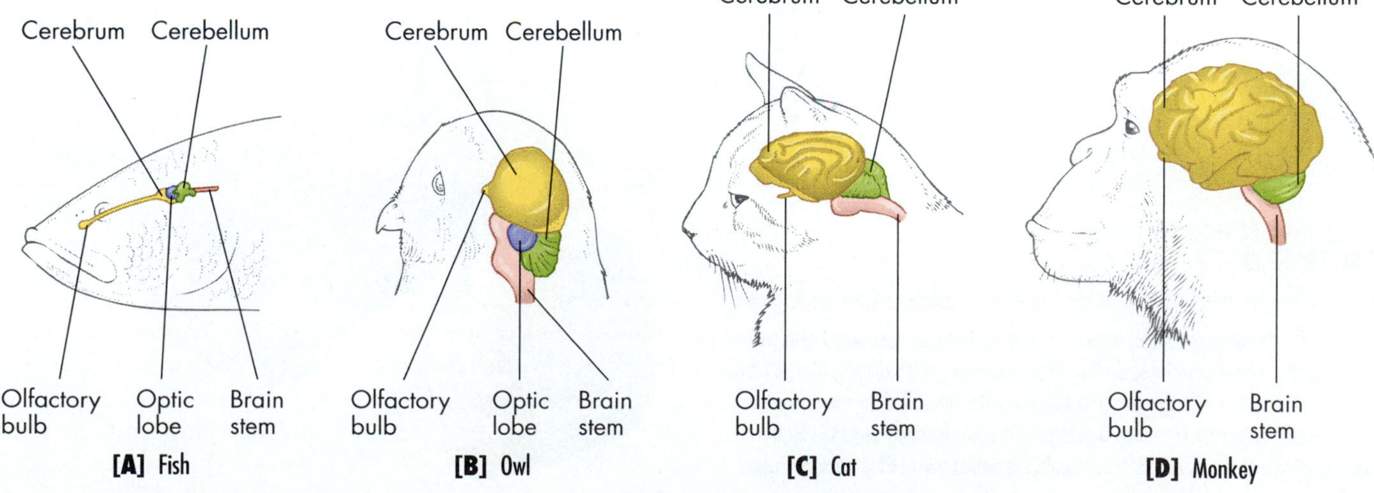

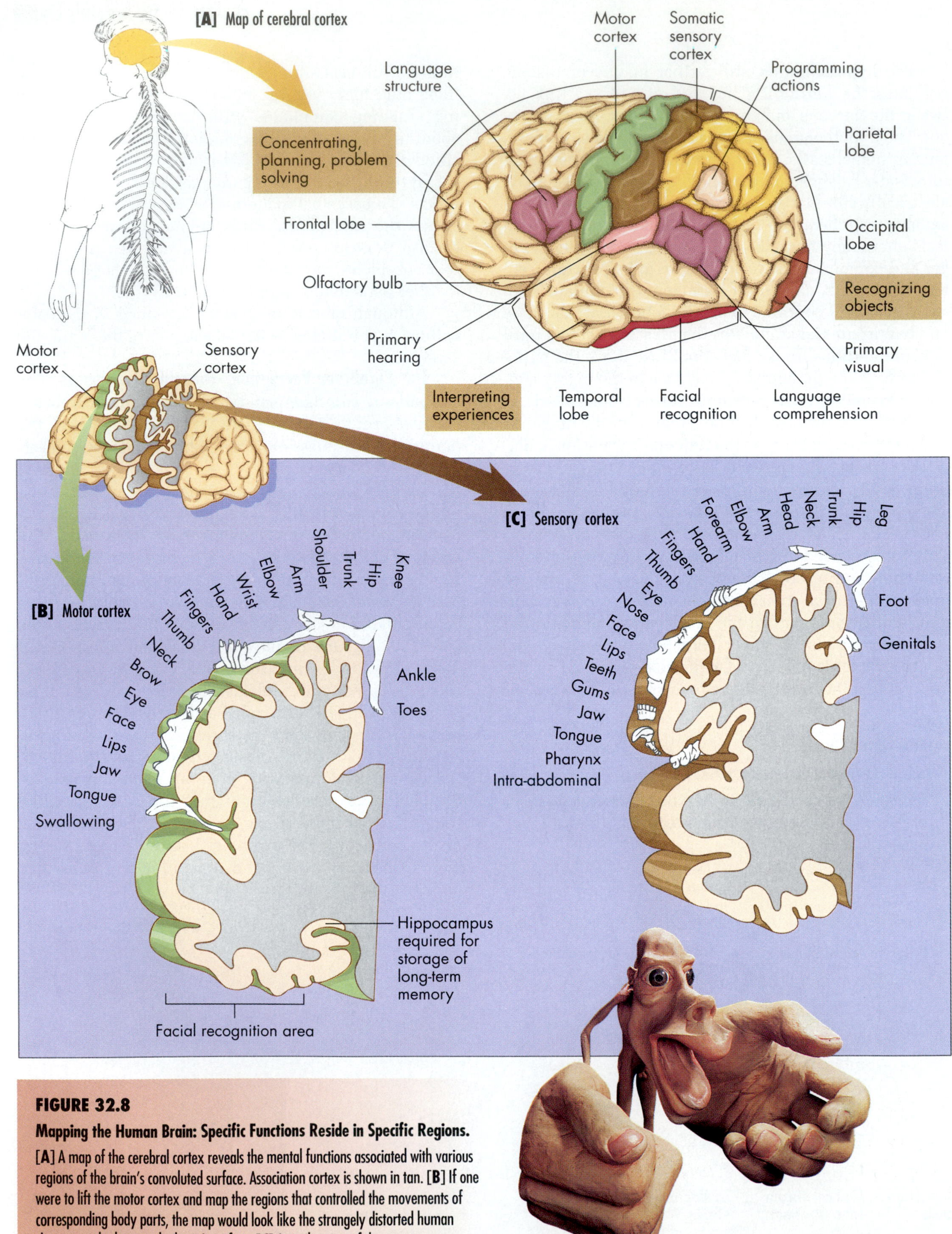

FIGURE 32.8

Mapping the Human Brain: Specific Functions Reside in Specific Regions.

[A] A map of the cerebral cortex reveals the mental functions associated with various regions of the brain's convoluted surface. Association cortex is shown in tan. [B] If one were to lift the motor cortex and map the regions that controlled the movements of corresponding body parts, the map would look like the strangely distorted human shown stretched across the brain's surface. [C] A similar view of the sensory cortex.

activities in different areas of the brain [FIGURE 32.9]. Clearly, the cerebrum is an enormously complex part of the brain—one that demands a closer look. Let's first look at the sensory and motor functions of the cerebral cortex and then turn to more complex behaviors.

Motor and Sensory Centers Brain surgery on patients with conditions such as *epilepsy* (a disorder that leads to seizures, or convulsions) has helped researchers map the cerebral cortex. Despite its billions of neurons, brain tissue has no pain receptors, so surgery can be carried out with only a local anesthetic for the scalp incision. By inserting extremely fine, low-current electric probes into the exposed brain, neurosurgeons can stimulate specific regions of the brain and observe which muscles the patient moves or ask the alert patient to report the sensations he or she feels. With studies like these, scientists have mapped a projection of the body onto the surface of the cerebral cortex—that is, a map showing which brain regions control the sensations and motor functions of which parts of the rest of the body. If you run your right index finger from your left ear straight up to the top of your head, the arc traces the surface area devoted to the left **motor cortex**, the part of your brain that controls muscles on the right side of your body, including the ones that move your right hand [see FIGURE 32.8]. Likewise, the corresponding arc on the surface of the right side of your brain moves the left side of your body.

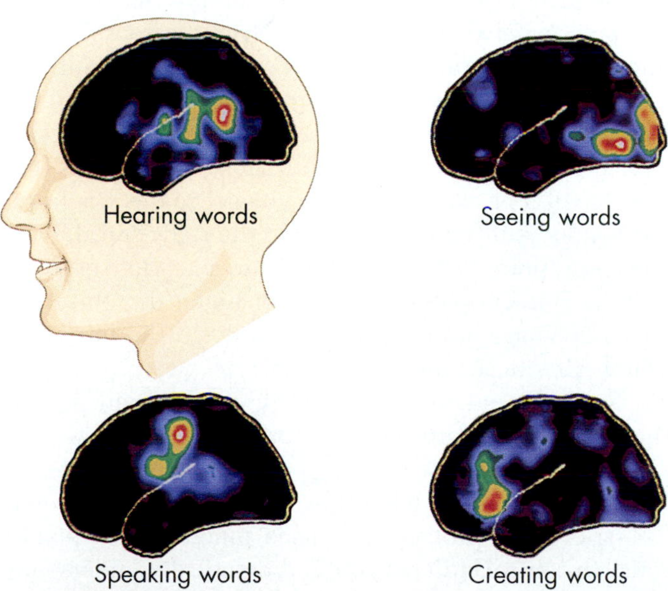

FIGURE 32.9
Different Tasks Related to Words Use Different Parts of the Brain.
Which tasks light up the language areas shown in FIGURE 32.8? Which the visual or hearing area?

Just behind the motor cortex lies the **sensory cortex**, which registers and integrates sensations from body parts [see FIGURE 32.8]. (Note that the surface regions of the motor and sensory cortices are responsible for similar but not identical parts of the body, since motor nerves go to muscles, while sensory nerves come from sense organs such as touch receptors in the skin.) Again, the brain's left side receives sensations primarily from the body's right side, and vice versa.

On the maps of both motor and sensory cortices, the representation of body parts does not match their true proportions in the body: Lips, face, and fingers appear too large, and the trunk too small. This disproportionate mapping reflects the exceptionally large number of delicate touch receptors in face and fingers and the large number of muscles necessary for speech and manual dexterity relative to the small numbers needed for sensation and movement of the trunk. Other specific brain regions are dedicated to vision, hearing, and smell. Moreover, birds with particularly keen hearing, such as owls, have an enlarged hearing region in the cerebral cortex, whereas birds that rely on the sense of touch and manipulation of the beak for feeding, such as ducks, have a different region enlarged. Finally, the map can change with use, as revealed by recent experiments employing intense stimulation of a monkey's fingertip. This could explain the acquisition of skills such as learning Braille, the raised-dot writing system for the blind.

The sensory cortex is involved in a crucial evolutionary function—the perception of pain. While we adapt fairly quickly to certain sensations, such as the pressure of clothing on the skin, we don't adapt completely to pain, and this helps protect us by triggering behaviors that diminish the painful stimulus (excessive heat, pressure, chemicals, or electric jolts)—either an automatic one, or a conscious one, like bandaging a cut foot.

Nerve cells damaged by injury or infection release chemicals called *bradykinin*, *prostaglandins*, and *substance P*, all of which render pain receptors more sensitive, heighten the sensation of pain, and lead to faster behavioral responses. When the injury is very dramatic, however, the body's nervous tissue can release natural opiates called *endorphins* along pain pathways. These opiates dull pain, and the suppression can allow a wounded person to drag him- or herself to safety before the pain sensations begin. All of these biochemicals help point out the evolutionary significance of pain.

Higher Cerebral Function Many brain capacities in addition to sensory and motor functions can be mapped to specific regions. One cortical area of the cerebrum, for example, is dedicated to recognizing faces [see FIGURE 32.8], and a person who suffers an injury to this part of the brain can no longer recognize the faces of loved ones, although the injured person can still recognize their voices.

A person with damage to the right side of the parietal lobe just behind the sensory cortex may completely ignore the left half of the body, failing to wash or dress the affected side. If the patient's left arm is passively placed before him or her in full view, the person will deny ownership of the limb. Regions of the frontal lobe near the forehead seem to be involved in short-term memory, anxiety, and the ability to weigh consequences and act accordingly.

Memory Deep in the cerebrum lies a small structure called the **hippocampus**, which plays a crucial role in the formation of long-term memory [see FIGURE 32.8B]. Have you ever been interrupted while trying to memorize a new phone number? Even if the interruption was for only an instant, you probably forgot the number, because your hippocampus had failed to establish it in long-term memory before the interruption occurred. A patient with a damaged hippocampus can recall events that happened before the injury—even decades before—but fails to remember new facts and experiences (including the shapes of objects, phone numbers, and people's names) for more than a few moments. Clearly, we need this area to help us lay down new memories, although the memories are not actually stored there. A patient with a damaged hippocampus may see the same doctor a dozen days in a row, but will greet the physician each time as a stranger.

Researchers have recently made great strides in understanding the mechanisms of learning and memory. Stimuli that alter a neuron's impulse activity cause long-lasting changes in the expression of neurotransmitter genes. These changes, in turn, cause alterations in synapse strength and communication to other cells. Essentially, the learning becomes "hard-wired"; that is, a memory is laid down as long-term physical changes in the way the neurons interact. These long-lasting changes seem to occur in many different brain regions, which suggests that memory involves far more than just the hippocampus.

Language Complex language distinguishes us from all other animals. It is localized in two areas of the cerebral cortex, one for language structure and the other for language comprehension [see FIGURE 32.8A]. If a stroke (damage to a brain region caused by a blood clot) disrupts the area of language structure, the patient may understand spoken and written language but be unable to speak well. One such patient, talking about a dental appointment, said haltingly, "Monday—Dad and Dick—Wednesday nine o'clock—ten o'clock—doctors—and—teeth." Damage to the area of language comprehension allows the person to speak fluently, but not in a way that is comprehensible to others. When describing a picture of two boys stealing cookies behind a woman's back, one such patient reported fluently, "Mother is away here working her work to get her better, but when she's looking the two boys looking in the other part."

Mapping shows that the brain is asymmetrical. In most people, language abilities reside mainly on the brain's left side (remember, *l*anguage, *l*eft), as do the underpinnings of analytical thought and fine motor control. The right hemisphere is mainly responsible for intuitive thought, musical aptitude, the recognition of complex visual patterns, and the expression and recognition of emotion. Almost all right-handers and 75 percent of left-handers have left-brain language areas. Many left-handers, however, have language areas on *both* sides of the cerebral cortex.

Right Cortex versus Left Cortex If separate functions are encoded in each hemisphere, how do the two sides of the brain communicate? Once again, neurosurgeons attempting to help epileptics contributed to our understanding of this problem. In epileptics, nerves fire back and forth across the brain in a crescendo of positive electrical feedback. This firing can continue to escalate, like feedback from a loudspeaker, resulting in a seizure. Surgeons thought that they could interrupt the loop by "splitting the brain," or severing the *corpus callosum*, the bridge between the left and right cerebral hemispheres [see FIGURE 32.6]. These split-brain patients improved, and they seemed completely normal to friends. But when neurobiologists tested such patients in controlled experiments, they found strange anomalies [FIGURE 32.10]. For example, if the subject saw a key in the left visual field (which projects to the right brain), he or she could point with the left hand to a key in a group of objects. But the patient could not name the object because language ability resides mainly in the left brain, which did not see the key.

Researchers concluded that in a split-brain patient, the right visual cortex could not communicate what it saw to the region of language ability in the left brain. Thus, patients *knew* what they saw but could not *say* it. The split-brain procedure is used much less frequently today to control epilepsy, but it dramatically underscores the principle that specific abilities are localized in particular regions of the brain. Even so, no function is the exclusive property of one brain region; rather, each function requires integrated actions of neurons located in several discrete regions, each region simultaneously carrying out parallel computations that coordinate behavior.

Association Cortex While neurobiologists have made considerable progress in mapping functions to different

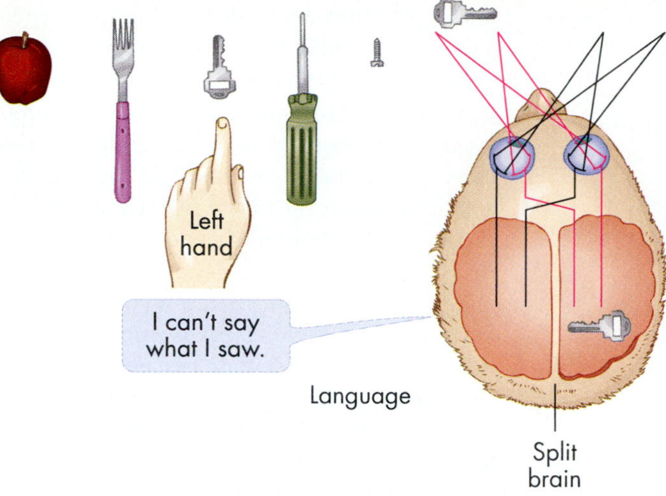

[A] Object in left visual field

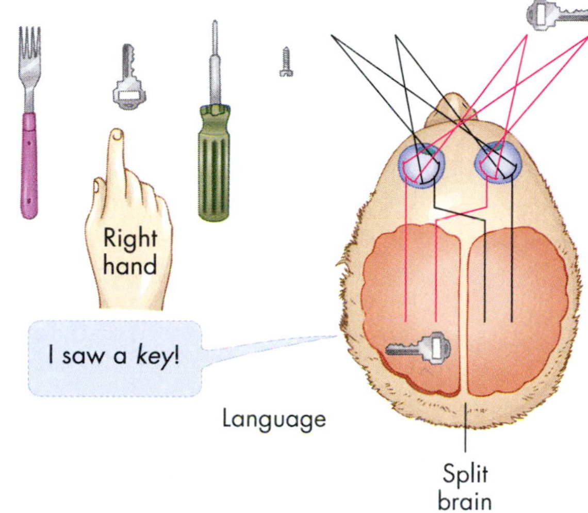

[B] Object in right visual field

FIGURE 32.10

Two Minds in One Brain.

[A] A patient in which the bridge of nerve tracts linking left and right hemispheres has been surgically severed can perceive a key with the right visual cortex and point to it with the left hand. But since the patient's right brain cannot communicate with the language region in the left brain, the person knows what he or she saw but can't say it. [B] When the patient perceives the key with the left visual cortex, there is no communication block with the language region of the same hemisphere, and thus the person knows and can say what he or she saw.

parts of the brain, they have yet to learn how the parts operate. For example, each of the main cerebral lobes contains regions vaguely called **association cortex**. These areas appear to integrate information from other areas and reconstruct it at varying levels of consciousness. To a large extent, these areas of association cortex contain the physiological underpinnings of emotions, memory, personality, reasoning, judgment, and the conglomeration of traits we call intelligence, but exactly how they reassemble information and coordinate it remains unclear.

THE DISEASED BRAIN

The human brain is a remarkable organ that works spectacularly well most of the time. Unfortunately, it malfunctions in some people, leading to devastating diseases. Biologists have discovered the causes for some of these conditions and are working on cures, but others remain mysterious. Here we consider Alzheimer's disease and schizophrenia.

Alzheimer's Disease Consider an engineer who fails to recall that he has already determined the costs of a project and needlessly recalculates the estimates. Or a journalist who finds it hard to remember the names of familiar objects and how to balance her checkbook. These people's minds are beginning to fail from a relentlessly advancing condition: **Alzheimer's disease**. Within three to ten years, these sufferers will be unable to dress themselves, to care for themselves, and eventually even to walk, sit up, or smile.

Currently, Alzheimer's affects as many as 4 million Americans and accounts for over half of the recorded cases of dementia (mental deterioration). It attacks otherwise healthy middle-aged and older people and causes a progressive loss of mental function. While the behavioral patterns (forgetfulness, inattention to hygiene, and so on) are characteristic, a physician can make a positive diagnosis only by examining brain tissue after the patient's death. The brains of Alzheimer's victims reveal a number of specific features, including loss of neurons from regions necessary for thought and memory [see FIGURE 32.8]. In addition to those features, the brain of an Alzheimer's victim has accumulations of twisted filaments within the brain cells and aggregations of protein around blood vessels. Finally, a physician can detect a decrease in the flow of blood to specific regions of the cerebrum in an Alzheimer's patient [FIGURE 32.11A and B]. Although improvements have been made in understanding the biology of Alzheimer's disease, we are still far from finding a cure for this devastating condition.

Schizophrenia Consider a young man who watches strangers talking and assumes they are talking about him. The same man sees people walking behind him and concludes that the FBI is pursuing him, or hears a radio commercial and believes it is directed specifically at him. Or consider a young woman who begins to experience disordered thoughts and can no longer exclude irrelevant ideas when answering a question. She withdraws from contact with other people, and sits motionless, crouched on the floor for hours at a time. These people have

box 32.1
Biology Applied

Sex and the Single Brain

One intriguing line of brain research has been distinctly unpopular in certain circles. Some advocates of equal rights for women in the workplace as well as in social settings reject the notion that men's and women's brains are anatomically different. This structural separatism, they fear, could be used to deny women equal opportunity and equal pay. While scientific studies in no way justify sexism or racism, there is sound evidence that real differences exist in the brains of males and females, and may underlie variations in how the two sexes perform on a number of tasks.

For many years, behavioral researchers have been testing girls and boys and men and women on a wide range of tasks, and cataloging measurable differences in the way they do things. Little girls, for example, learn to sit up, crawl, and walk months earlier than little boys, as well as to remember words and things they've seen. Boys, on the other hand, learn spatial tasks such as rotating and fitting together blocks, puzzle pieces, and other objects earlier than girls. Some of these differences—in slightly altered form—persist throughout adulthood.

On average, men are better than women at rotating objects in their minds and imagining the shapes they could see from various angles. They are generally better at hitting a target with an arrow, dart, or ball; at recognizing simple shapes hidden within complex ones; at working out problems that require mathematical reasoning; and at navigating by a sense of distance and direction.

Women, on the other hand, tend to outperform men on tests of perceptual speed (quickly recognizing images or objects that match in every detail); of changes in a setting or series of objects; on tests of fluency of ideas and words; on fine-motor coordination; on mathematical calculations; and on navigating by a memory for landmarks. (Keep in mind that these are statistical tendencies. Some women are better than men at one or more from the the first set of skills, and some men are better than women at items in the second set.)

Researchers naturally wondered whether they could see and measure differences in brain anatomy to coincide with these performance differences—especially since lab animals show such variations, and these variations are based on the way male and female hormones affect the developing brain. Indeed, experimenters found, and are still finding, significant sex differences.

In 1982, researchers at Yale and Columbia Universities found that part of the corpus callosum (the fiber bundle connecting the two hemispheres of the brain) is larger in women than in men. This would allow fuller communication between the left and right brains [see page 720], and may help explain the female's greater thought and verbal fluency, and, some say, her ability to identify and express feelings. Ten years later, those same experimenters found evidence in human male fetuses that the right cortex is thicker than the left, whereas in the female fetus, the two sides are of equal thickness. In provocative new research, Canadian researcher Doreen Kimura found evidence that men and women may generate speech and hand movements from different parts of the brain!

Kimura studied men and women who had suffered damage to various parts of the brain due to strokes, accidents, and other causes. Among such subjects, men tend to lose speech functions and dexterity more often than women. The reason, Kimura found, was that damage is more common to the posterior part of the brain—regardless of hemisphere—and significantly, speech and hand movements appear to be governed by the posterior part of a man's brain but the anterior part of a woman's! Based on these and other bits of evidence, Kimura concludes that beginning early in life, the brains of males and females become organized quite differently due to hormonal influences in the womb and shortly after birth.

This is not so startling if one looks at the sex roles among hunter-gatherers (such as the pygmies; see CHAPTER 30), and analyzes the evolutionary significance of those separate tasks. Males tend to shape tools, defend the village, go out in search of meat—sometimes over great distances—and return with it. Route finding, targeting, and spatial skills would help improve success at those endeavors, and survival for themselves and for tribe members. Females tend to gather food closer to home, make clothing and home artifacts, and teach and rear children. For those tasks, navigation by landmarks, manual dexterity, verbal fluency, and the ability to quickly detect small changes (in the home setting, a child's health, and so on) could be crucial to their own survival and the tribe's.

Obviously, most of us live in complex urban societies today, and are free to explore and develop whichever subset of our innate abilities most appeals to us. Ideally, we do this regardless of the statistical tendencies of males and females, or of the traditional sex roles of hunter-gatherer societies.

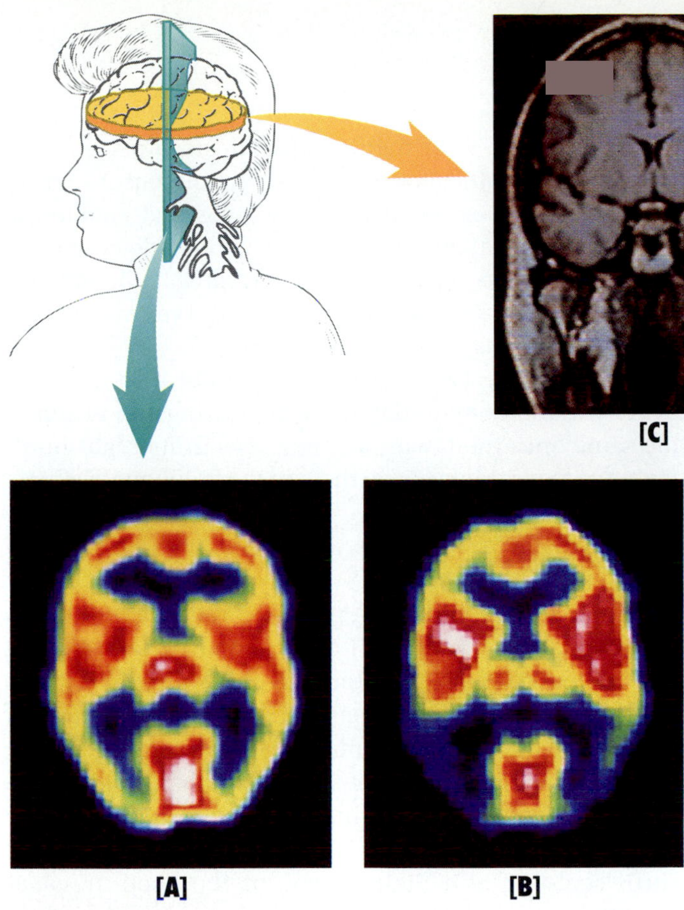

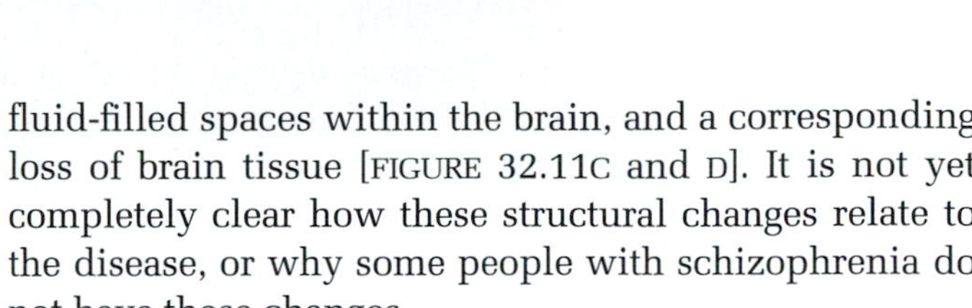

FIGURE 32.11

Diseased Minds and Brains.

Comparing a PET scan of a normal brain [**A**] and the brain of an Alzheimer's victim, [**B**] reveals diminished blood flow to critical brain regions. MRI scans of a normal brain [**C**] compared to those of a person with schizophrenia [**D**] reveal larger spaces in the brain of the schizophrenic patient due to atrophied brain tissue.

schizophrenia ("split mind"), a condition characterized by *psychosis*, or false beliefs that can't be changed by evidence, and *hallucinations*, or perceived experiences without an external source.

Researchers have determined that schizophrenia has an important familial component: About 1 percent of people around the world have schizophrenia, while among the relatives of schizophrenics, about 15 percent have schizophrenia. This suggests that either genetic or environmental factors present in certain families predispose members to this devastating disease. Further studies show that stressful environmental factors can trigger the disease in genetically susceptible people.

Physicians have found that many patients with schizophrenia exhibit enlargements of the *ventricles*, the fluid-filled spaces within the brain, and a corresponding loss of brain tissue [FIGURE 32.11C and D]. It is not yet completely clear how these structural changes relate to the disease, or why some people with schizophrenia do not have these changes.

Certain drugs effectively treat the most disruptive symptoms of schizophrenia—the delusions, hallucinations, and disordered thinking. These drugs block a specific type of receptor for the neurotransmitter dopamine. This has led to the hypothesis that some of the symptoms of schizophrenia may be caused by excess dopamine in certain parts of the brain. Whatever the cause, recent studies suggest that drug treatment and modifying the patient's family environment can improve at least the short-term outcome of schizophrenia.

➤ CONCEPT CHALLENGE

From what you have learned here about the cellular basis of schizophrenia and what you learned in the previous chapter about the mechanism of cocaine action [FIGURE 31.1], suggest whether cocaine would improve or make worse the behavior of a person with schizophrenia. Support your answer.

Sense Organs

An owl hears the rustling of a mouse in the fallen leaves. A cockroach detects light and scurries back into the shadows. A person bites into a cherry and tastes a sour flavor. All such changes in the body and its surroundings are perceived by **sense organs**, groups of specialized cells that receive stimulus energy (such as wavelengths of light, sound waves, odor or flavor molecules, or pressure) and convert it into another kind of energy, the kind that can trigger a neural impulse. The impulses initiated by sense organs are relayed to the brain by sensory neurons, and they answer two types of questions: (1) *What kind* of stimulus did the sense organ receive (what is the pitch of the sound, the color of the light, or the nature of the

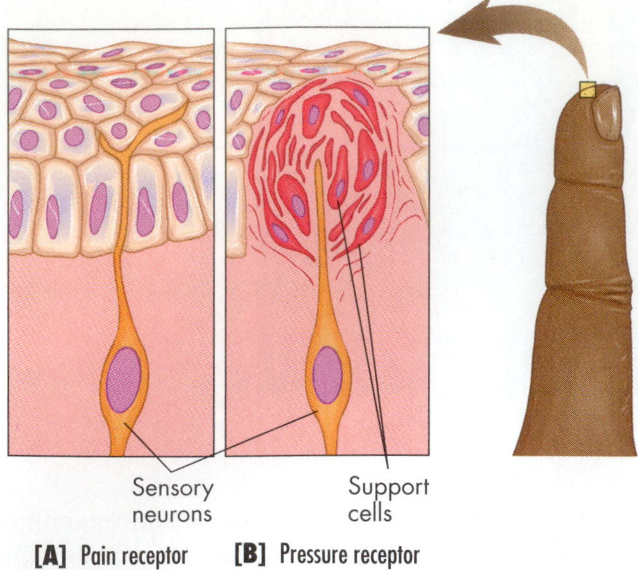

FIGURE 32.12

Pain and Pressure Receptors.

[A] The simplest sensory neurons, pain receptors, have nerve endings surrounded by a thin layer of support cells. Pressure on those support cells stimulates the receptor nerves directly. [B] Pressure receptors project into a capsule of support cells. The spatial arrangement of these cells increases the effects of pressure. The cells trigger an action potential in the pressure receptor neuron itself.

chemical), and (2) *how strong* is the stimulus (is it loud or soft, bright or dim, concentrated or dilute)? Our day-to-day survival depends on accurate answers to these questions.

A sense organ's receiving cells have bare nerve endings not covered by a myelin sheath [FIGURE 32.12]. Pain receptors in the skin are examples of this. There are several different kinds of receptor cells, each a specific shape and each able to make special proteins. Other affiliated cells screen out extraneous forces and amplify the effects of the specific stimulus before it reaches the receptor. For example, the eye's light-receptor cells make light-sensitive proteins, while nearby support cells focus light on the receptors. In contrast, pressure-sensing cells in the skin detect physical bending, and the support cells magnify the bending [see FIGURE 32.12B]. While sense organs are generally tuned to specific sets of stimuli, any mechanical, chemical, or light (electromagnetic) stimulus of sufficient intensity can cause them to generate an action potential. A person bumped on the eye or head, for example, may "see stars" because cells in the visual pathway are sensitive to pressure as well as to light.

Just as the tawny owl depends mainly on its hearing and mental maps, different animals rely on different senses, according to how they exploit their own particular environment. Bees are highly visual, for example, while bloodhounds rely heavily on the sense of smell. Some animals even utilize physical forces that humans can detect only with highly sophisticated instruments. For example, rattlesnakes have a pair of sensory structures called pit organs beneath the eyes that enable the reptiles to "visualize" objects in the dark by detecting the heat patterns the objects give off [FIGURE 32.13], and certain fishes have sense organs that act like voltmeters that detect distortions in the electric field.

Because the two senses that people rely on most heavily in their day-to-day lives are hearing and vision—the same ones the tawny owl employs in its night hunting—we focus primarily on them in the following sections.

THE EAR

With over a million moving parts, the **cochlea** (KAHK-lee-uh; Greek *kokhlos*, "snail") of the inner ear is the most complex mechanical apparatus in the human body and in the bodies of most other vertebrates. Cells in this coiled apparatus detect sound by first sensing subtle movements in the fluid that surrounds them. The stimulated cells then send nerve impulses to the brain. Finally, the brain interprets the sound of hands clapping, the shrill scolding of a Steller's jay, or the deep, musical "hoo-hooo-hoooo" of the tawny owl.

Sound is really a wave of compressed air that travels along and eventually strikes the ear (CONCEPT INTEGRATOR page 726). This compression could begin with two hands

FIGURE 32.13

The Pit Viper's Pit Organ: A Heat Detector Senses Prey.

A rattlesnake—this one a resident of Venezuela—has a black cavity below and in front of each eye that works a bit like a night spotting scope. A thin membrane stretching across the back of the cavity detects heat and allows the snake to locate a rabbit, mouse, bird, or other warm prey in the dark.

[A]

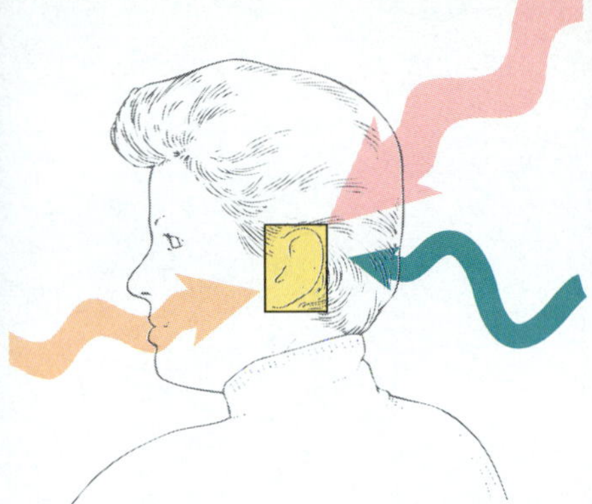

[B]

FIGURE 32.14

Loud Sounds Damage Hair Cells.

[A] Intact hair cells from the human cochlea, highly magnified. **[B]** Violent vibrations from extremely loud sounds break off stereocilia from the hair cells, causing irreparable damage and permanent hearing loss.

clapping together or with a mouse's foot scraping a leaf. An analogous compression begins when a stone hits the still surface of a pool of water and sets up concentric water waves. Sound waves first strike structures such as the human's flexible, sculptured outer ear or the trough beneath the feathery ruff of an owl's face. The outer ear then funnels sound waves to the *eardrum*, a taut membrane that is stretched like a drumskin across the ear canal and that vibrates in time with the sound wave frequencies. A chain of three bones in the middle ear in turn transmits eardrum vibrations to the small, coiled cochlea, the sensory receptor that converts vibrations into nerve impulses. Overall, the cochlea has a snail-like shape, compact and coiled with elastic partitions that separate it into three fluid-filled chambers.

Attached to one partition, called the *basilar membrane*, are four rows of box-shaped **hair cells**, each cell crowned by an elegant bundle of threads called **stereocilia** that stand erect like a Mohawk haircut [FIGURE 32.14]. Each hair cell bears about 100 stereocilia; in total, each human ear contains about 1 million. Most of the 18 million deaf Americans have malfunctioning hair cells. Researchers have recently discovered that under certain conditions, hair cells can regenerate, and this has given hope that treatments will be developed that can cure deafness due to defective hair cells.

How Do We Hear? When a sound wave causes the eardrum and then the middle-ear bones to vibrate, one of these bones presses on the cochlear fluid and the basilar membrane is deflected. This, in turn, causes the hair cells to move relative to the *tectorial membrane*, a flap that lies on top of them. This shearing force bends the stereocilia. Stereocilia are attached to their neighbors by a thin, springlike link. According to the current model of hearing, this link attaches to an ion channel, and when the stereocilia move, the link causes the ion channel to open. Ions enter the hair cell, and thereby convert the mechanical stimulation of the sound wave into an electrochemical signal. Finally, the hair cell transmits the signal across synapses to a sensory neuron that leads to the brain.

Receiving the signal, how does the brain know whether the incoming sound is the high-pitched whistle of a bull elk or the low growl of a grizzly bear? (After all, as CHAPTER 31 explained, at the level of the neuron, all action potentials are the same.) The answer involves the mechanical properties of the basilar membrane. Unlike a guitar string, which vibrates equally along its entire length when stimulated by the appropriate pitch, the basilar membrane vibrates maximally in different places when stimulated by different pitches. High pitches excite motion near the wide part of the cochlea, while low pitches cause movement near the cochlea's narrow tip. Since each region of the membrane connects to a different part of the brain, different pitches stimulate different brain cells. In essence, a musical staff is inscribed in space across a certain part of the brain. A baby can distinguish the characteristic tones of its mother's voice because that voice stimulates hair cells at specific places in the cochlea, which in turn correspond to specific brain areas.

Animals can also distinguish loud sounds from soft ones, since loud sounds cause more neurons to fire action potentials and to fire more frequently. While a soothing voice causes the stereocilia on the hair cells to sway gently, loud sounds, such as those from a jackhammer or a heavy metal rock band, can break off the stereocilia and thus permanently damage the hair cells and along with them the sense of hearing [FIGURE 32.14].

To fully sense its environment, an animal must know the direction of a sound as well as its tone qualities. Owls, for example, can localize sounds in the horizontal plane about as well as people can, but the birds are much more accurate than people at locating sounds above or below them. This is because an owl's left ear is aimed higher than its right [FIGURE 32.15] and also because the

1: Sound waves and ear structure

Sound is a wave of compressed air that causes the *eardrum* to vibrate

Concept Integrator: The Human Ear: Structure and Function

Loudness

Pitch

How hair cells work

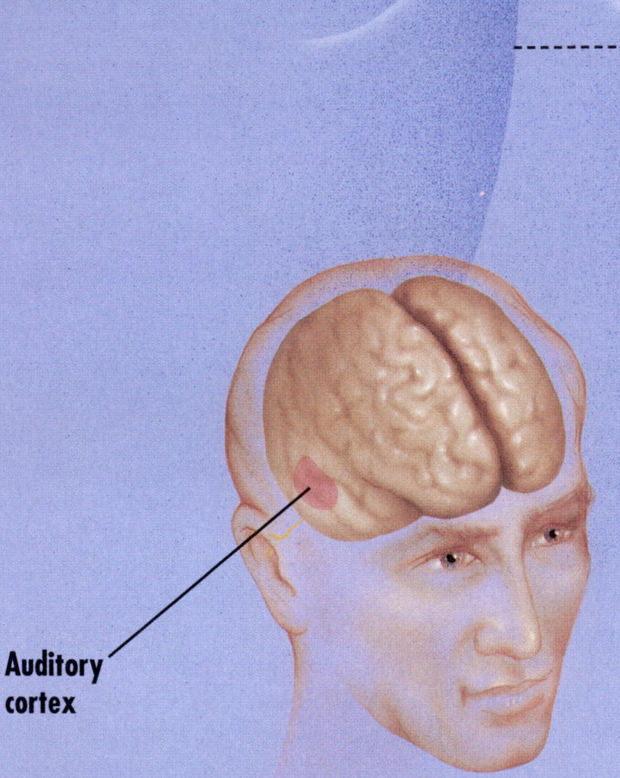

9: These nerve impulses activate the auditory pathways leading to the auditory cortex of the brain and the sound is perceived

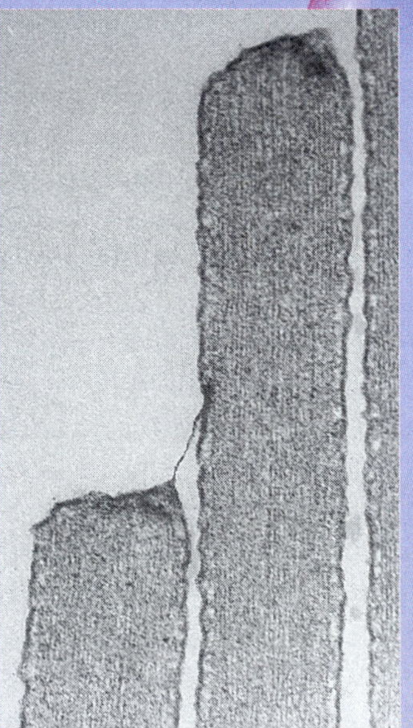
Close-up of stereocilia

7: When stereocilia move, *ion channels* open, changing the cells' electrical potential

8: This change in the hair cell is conveyed across the synapse to sensory neurons in the auditory nerve

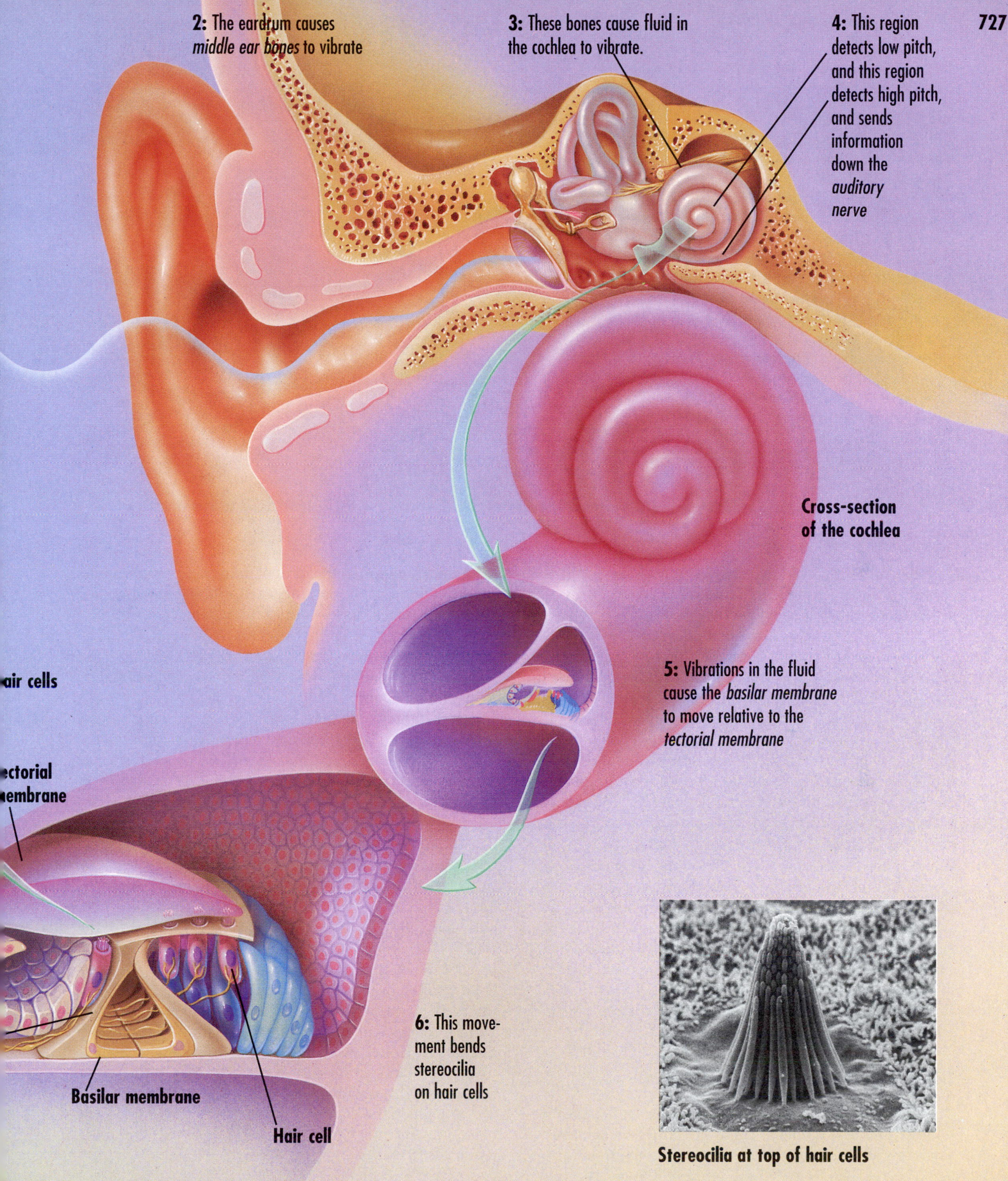

Stereocilia at top of hair cells

FIGURE 32.15

How Owls Localize Sounds.

Of all land animals, owls are the best at locating a moving target in three-dimensional space. To do this, the owl must locate the sound in the horizontal plane [**A**], and in the vertical plane [**B**]—that is, how high above or below the owl the sound is coming from. A sound from the left strikes the left ear first, and the brain interprets this as a direction in the horizontal plane. The left ear is higher on the owl's head, but points down, and the right ear is lower, but points up. Thus, a sound that is louder in the left ear originates below the animal. Together, the timing and intensity cues allow an owl to precisely localize a sound in complete darkness, contributing to its success as a night hunter.

owl has an internalized map in which each brain cell responds to sound from a different point in space.

Balance and Detecting Motion While hair cells are crucial to hearing, they also figure prominently in detecting an animal's position in space. In a tawny owl, the **semicircular canals**—three curving tubes filled with fluid that function as organs of balance and motion detection—register the bird's acceleration as it swoops down toward a scurrying mouse. Like the cochlea, the semicircular canals contain hair cells with attached stereocilia that project into the fluid filling the curving tubes. As the owl banks and dives, the detection of the motion is based on the swaying of these stereocilia on the hair cells, much as passengers standing on a bus sway when the vehicle rounds a corner. Likewise, hair cells in a trout's **lateral line organ**—a row of minute pores along the fish's side—detect slight changes in water pressure, which alert the animal to nearby prey or predators. In all cases, natural selection has utilized a simple mechanism, the hair cell, to convert a mechanical force into electrochemical signals (nerve impulses) that can provide vital information and enhance the survival of different organisms in their own special environments.

THE EYE: AN OUTPOST OF THE BRAIN

The colors of the rainbow. A loved one's smile. Breathtaking mountain scenery. These delightful visual sensations are all possible because the eye converts electromagnetic radiation in the form of light into neural energy. Vision is our most important sense, and not surprisingly, the human eye is large—almost the size of a golf ball. For a tawny owl to see slightly better than the average person, its eyes must be even larger, even though its head is far smaller than a person's. If our eyes were proportionately as large, they would be the size of softballs!

How the Eye Detects Light Light bouncing from a plant and striking the human eye first passes through the protective transparent outer layer, or **cornea** [FIGURE 32.16A]. If you gently place your finger over a closed eyelid and turn your eye from left to right, you can feel the bulge your cornea makes. After penetrating the clear cornea, light passes through a clear fluid and then enters the **pupil**, which is a black, circular, shutterlike opening in the **iris** (the eye's colored portion), then traverses the **lens**, a circular, crystalline structure that, with the cornea, focuses light through another fluid onto the **retina**. The retina is a multilayered sheet that lines the back of the eyeball and contains light-sensitive **photoreceptor cells**; these begin converting light energy to electrochemical energy (nerve impulses) that can form brain patterns and create a visual image of an object. Because the lens is curved, it bends the light rays. As a result, the image of the object it projects on the retina is backward (left and right are reversed) and upside down [FIGURE 32.16A]. Once the visual impulses reach the brain, however, they are integrated and sorted out so that the image conforms to reality.

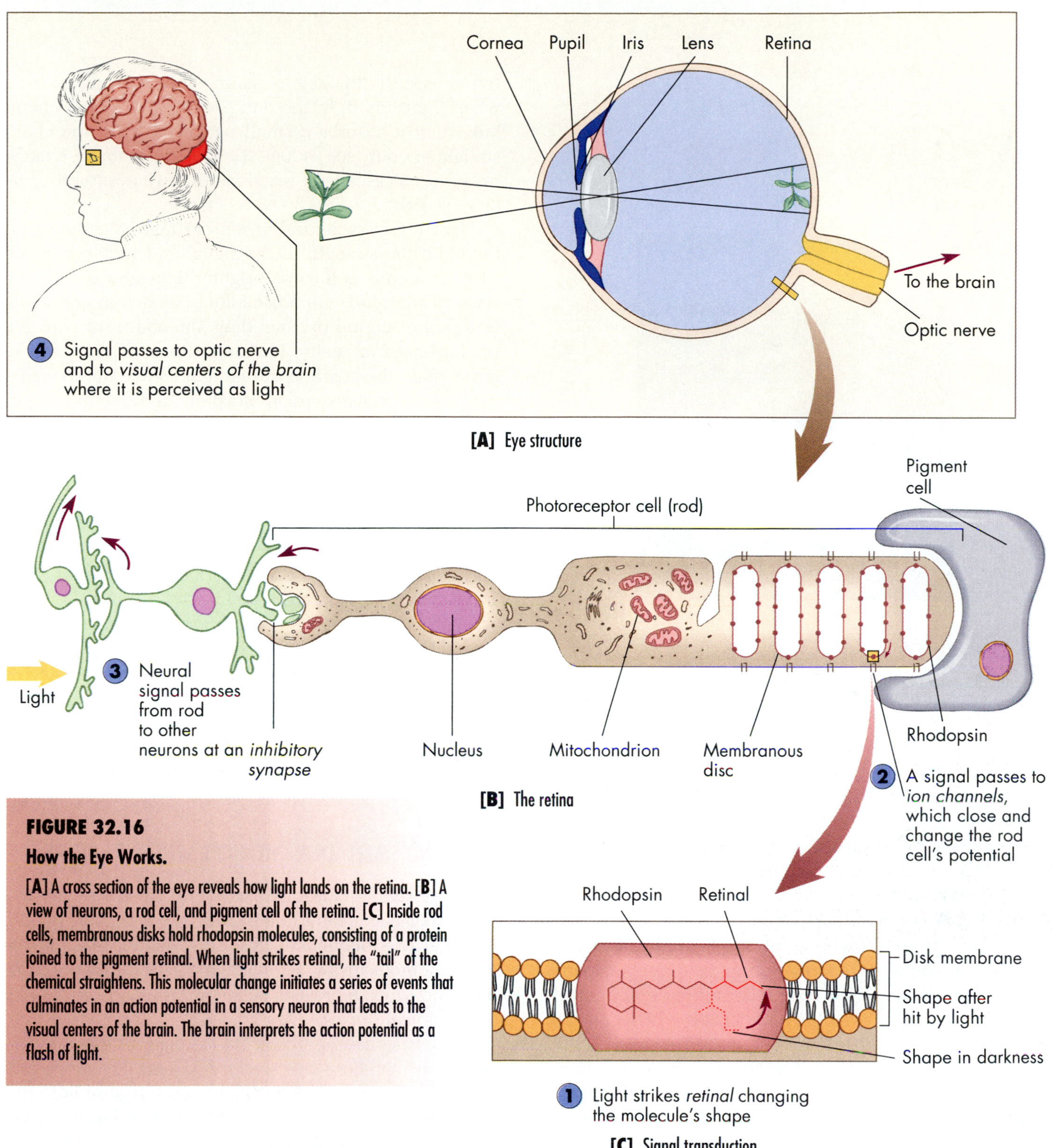

FIGURE 32.16
How the Eye Works.
[A] A cross section of the eye reveals how light lands on the retina. [B] A view of neurons, a rod cell, and pigment cell of the retina. [C] Inside rod cells, membranous disks hold rhodopsin molecules, consisting of a protein joined to the pigment retinal. When light strikes retinal, the "tail" of the chemical straightens. This molecular change initiates a series of events that culminates in an action potential in a sensory neuron that leads to the visual centers of the brain. The brain interprets the action potential as a flash of light.

Within the multilayered retina, the rear layer (closest to the back of the eye and just inside the eye's tough outer covering) consists of jet-black pigment cells that protect the photoreceptors from extraneous light [FIGURE 32.16B]. Nestled just in front of the pigment cells are two types of photoreceptor cells called **rods** and **cones** because of their distinctive shapes [see FIGURES 32.16B and 32.17]. Rods are very sensitive to low levels of light, but cannot distinguish color, whereas cones need more light but can detect color. (Remember *c*one for *c*olor.) This is the reason you have trouble seeing colors in dim light. As one might expect, the retina of a tawny owl is jam-packed with rods, allowing it to pick up the dimmest light during night hunting; but since the retina contains few cones,

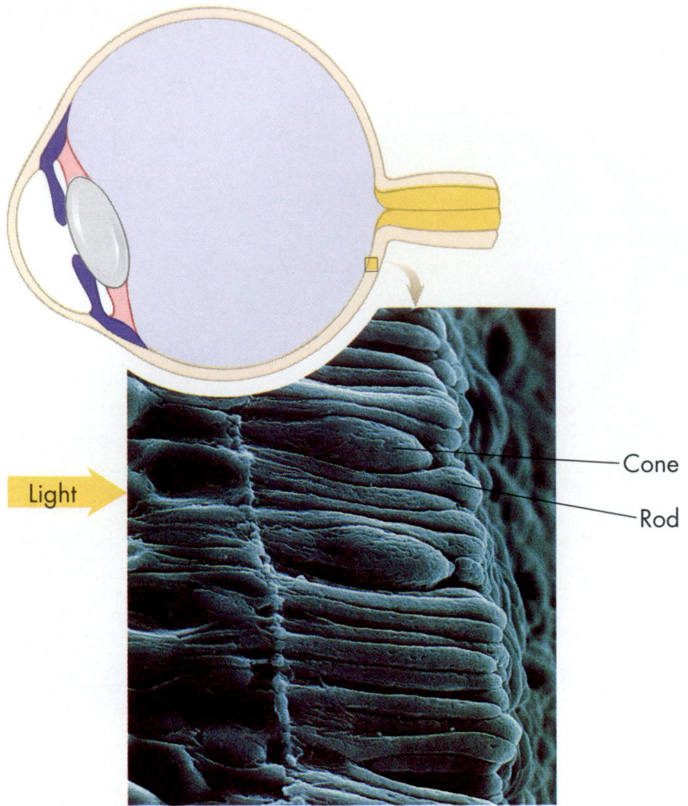

FIGURE 32.17
Light Receptors in the Eye.
Rods are long and numerous and can detect dim light; cones are shorter and club-shaped and register color—either blue, green, or red light.

the animal's color vision is probably very poor. In the retina's front layer, rods and cones synapse with sensory neurons that interact with other neurons that send axons toward the brain. Curiously, because these sensory neurons lie in front of the rods and cones, light must pass through the neurons before reaching the photoreceptors. The axons leading from individual retinal neurons collect into a bundle called the **optic nerve**, which exits the eye and leads to the brain [see FIGURE 32.16A and B].

Rhodopsin: The Visual Pigment To understand how the eye works, let's concentrate on the rods and how they receive light. The front part of the rod contains the nucleus and forms synapses with other neurons, and the rear part detects light. The rear part contains stacks of membranous disks in which millions of molecules of the light-sensitive molecule **rhodopsin** are embedded [see FIGURE 32.16B]. The protein part of each rhodopsin molecule snakes back and forth across the membrane and cradles the actual pigment portion, a small ring-and-chain molecule called *retinal* [FIGURE 32.16C]. (Eyes synthesize retinal from the compound that gives carrots their bright orange color; hence, there is wisdom to the old saying that carrots are good for your vision.) The structure of retinal is really the key to vision, since it changes shape when light hits it: In the dark, retinal has a bent chain, but when it absorbs a small packet of light, the chain straightens out [see FIGURE 32.16C]. This shape change unleashes a cascade of reactions that we eventually perceive as light.

Recall that each kind of sensory neuron cell can tell the brain the strength of the signal and its type. In the case of neurons in the eye, brighter light stimulates more sensory neurons, which generate more action potentials in a shorter period of time; thus, the brain can perceive the increased intensity. Photoreceptors can also convey to the brain the position of the light source in the visual field. This is because photoreceptors in different parts of the retina receive light from different parts of the visual field and link up with different parts of the brain. For example, a horizontal line in the visual field stimulates a line of photoreceptors in the retina. This line links to sensory neurons in the visual center at the rear of the brain, and when the line of neurons there becomes excited, we see a horizontal line. We see color because cones are sensitive to light of different wavelengths.

The mechanisms for hearing and vision are similar in several ways—they both locate objects in space by cells especially tuned to physical forces (sound or light) and in both cases, there is a one-to-one correspondence of a position in space and a map in the brain. Together, the mechanisms help animals interpret the environment around them in a way that improves their chance of survival.

TASTE AND SMELL: OUR CHEMICAL SENSES AT WORK

Although sight seems to be our dominant sense, taste and smell play a far greater role in daily life than most people imagine. The tongue and nose receive and detect flavor and odor molecules—actual chemical tidbits of the environment. Thus, biologists consider taste and smell to be our *chemical senses*.

People sometimes think of the tongue as a perceptual genius and the nose as a sensory dullard. However, precisely the opposite is true. The tongue is studded with small conical bumps, or *papillae* (singular, *papilla*), which house the **taste buds**, and each bud consists of a pore leading to a nerve cell and surrounded by accessory cells arranged in an overlapping pattern that resembles an artichoke [FIGURE 32.18A]. The nerve cells in taste buds have receptors capable of receiving flavor molecules, but they can distinguish only four general classes of flavors: sweet, salty, bitter, and sour. We can tell similar foods apart—beef from pork, beets from turnips—and sense the subtlety of their flavors because (1) these foods stimulate the four receptor types to different degrees, and (2) the volatile aroma molecules from the food

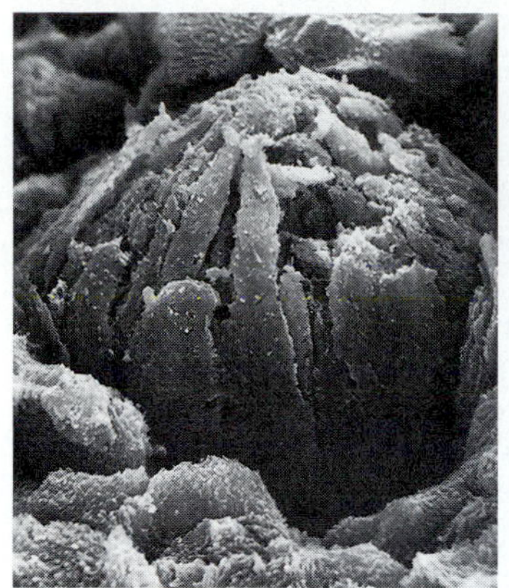

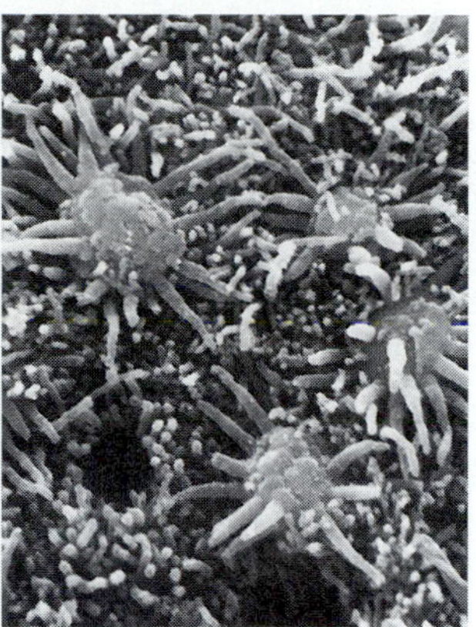

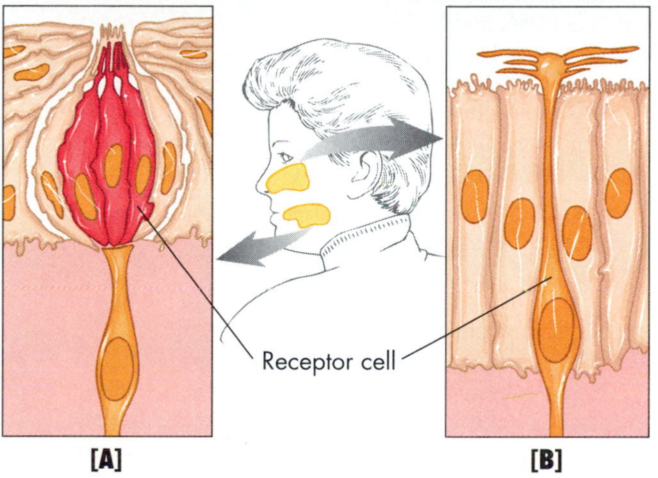

FIGURE 32.18

The Nose Knows and So Does the Tongue.

A taste bud on the tongue [**A**] and odor-detecting cells in the nose [**B**] consist of receptor cells that generate a nerve impulse when stimulated by a specific chemical. Dr. Richard Kessel and Randy H. Kardon, *Tissues and Organs: A Text-Atlas of Scanning Electron Microscopy*, p. 135, Fig. 5 and p. 133, Fig. 2 (W. H. Freeman and Co., San Francisco, 1979).

travel into the nose or up the back of the throat and bind to receptors in the *olfactory epithelium.*

The **olfactory epithelium** consists of button-sized patches of yellowish skin high in the nasal passages. Neurons called *olfactory cells* lie embedded in these epithelial patches. One end of the nerve cell, shaped somewhat like a flower, binds odor molecules [FIGURE 32.18B]. The other end is a long, spindly axon that feeds into the olfactory nerve that carries signals directly to the olfactory bulb, or smell region, on the underside of the brain. If you take a bite of a sandwich and simultaneously pinch your nose shut, you can still perceive the four primary tastes, but not primary odors—and thus very little of the food's complex flavor.

People have long noted peculiarly intimate connections between smells, memories, and emotions. Who hasn't experienced a flood of memories, complete with appropriate emotions, when catching a whiff of a Christmas tree or a pipe like Grandfather smoked? The explanation for such odor-stimulated déjà vu lies in the anatomy of the nose and brain. The olfactory lobes are closely connected to the brain's *limbic system*, a series of small structures (including the hippocampus and hypothalamus) largely responsible for generating fear, rage, aggression, and pleasure and for regulating sex drives and reproductive cycles. A smell, therefore, stimulates the brain's centers of memory, emotion, and sexuality.

Smells can also affect the endocrine system. Roommates in a women's dormitory, for example, often have synchronized menstrual cycles, and women with irregular periods often grow more regular when in a man's company on a routine basis. In both cases, the effects seem to be largely olfactory and due, perhaps, to pheromonelike molecules in the sweat, which act on the brain and its hormonal feedback loops with the body.

The main message here is straightforward: An animal's brain can decipher the stimulus it receives as light, sound, taste, or smell because each sense organ is tuned to a different physical or chemical stimulus and sends a signal to a different part of the brain.

➤ CONCEPT CHALLENGE

Hearing aids work by amplifying and transmitting sound waves through the inner ear to set up fluid motion in the cochlea. Can a hearing aid help a person whose hair cells are destroyed? Why or why not?

Connections

This chapter is the second in a set of three chapters that explores how neural communication allows an animal, such as a swooping owl, to perceive internal and environmental conditions, integrate the information, and take appropriate action. CHAPTER 31 looked at how individual nerve cells function and communicate with other cells. This chapter discussed how neurons and nervous systems sense the environment—allowing us to hear, see, smell, taste, and feel the world around us. It also described how the brain integrates this incoming information and sends out information to the motor neurons of the peripheral nervous system, thus activating and coordinating the work of the body. In CHAPTER 33 we will investigate how muscles, glands, and other organs work together to produce the leap of the long jumper and the pounce of the night hunter—and all the other physical actions that enhance an organism's chances of survival.

KEY TERMS

brainstem, 716
central nervous system (CNS), 715
cerebellum, 717
cerebral cortex, 717
cerebrum, 717
cochlea, 724
cone, 729
cornea, 728
ganglion, 712
hippocampus, 720
iris, 728
lateral line organ, 728
lens, 728
medulla oblongata, 716
midbrain, 716
motor cortex, 719
optic nerve, 730
peripheral nervous system, 713
photoreceptor cell, 728
pons, 716
pupil, 728
retina, 728
rod, 729
semicircular canal, 728
sense organ, 723
sensory cortex, 719
stereocilia, 725
thalamus, 717

HIGHLIGHTS IN REVIEW

1 The nervous system's major roles are to monitor internal and external conditions, process the information, and initiate behavioral responses that improve the organism's chances of survival and reproduction.

a] By serving as a communication network, the nervous system enables an animal to sense its environment and internal states and respond with appropriate behavior.

b] The nervous system integrates and coordinates the activities of all the different systems throughout an organism, so that the organism functions as a unified living being.

2 The vertebrate nervous system is organized into the central nervous system (brain and spinal cord) and the peripheral nervous system (the rest of the body's neurons). The central nervous system (CNS) performs neural integration, while the peripheral nervous system (PNS) carries information from the sense organs to the central nervous system and carries signals away from the central nervous system to muscles and glands.

a] Animals show increasing organization and complexity in their nervous systems, from loosely organized nerve nets to ganglia to a brain that dominates the nerve network.

b] The brain and spinal cord serve as the central nervous system in humans. A peripheral nerve network branches out from the spinal cord, carrying messages to and from outlying muscles, glands, and sense organs.

c] Peripheral motor nerves can be divided into the somatic nervous system, which primarily controls voluntary muscles, and the autonomic system, which primarily innervates the body's internal organs and is not usually under conscious control.

d] The autonomic system is further broken down into the sympathetic and parasympathetic systems, which frequently innervate the same organs, but with opposite effects.

3 The brain consists of the brainstem, which controls many automatic functions such as heart rate and breathing rate; the cerebellum, which coordinates muscle action; and the cerebrum, which is the site of perception, thought, language, learning, and self-awareness.

a] The human brain can be divided into three regions: the brainstem, the cerebellum, and the cerebrum. The brainstem helps control autonomic functions. One of the cerebellum's chief roles is the fine coordination of body movement. The cerebrum is involved in molding personality and coordinating thought, memory, and learning.

b] The network of neuronal tracts in the reticular formation sends neural processes throughout the brainstem and regulates consciousness.

c] The cerebral cortex is divided into a left and a right hemisphere. Language and analytical thought are generally controlled by the left hemisphere, while intuitive thought and emotions are largely centered in the right hemisphere.

d] Scientists have mapped out several regions of the cortex as either processing incoming information (sensory cortex), programming outgoing signals for movement (motor cortex), or associating the two types of information with higher levels of awareness (association cortex).

4 Sense organs monitor physiological and environmental conditions by means of specialized cells with supersensitivity to a specific physical factor, such as light, sound, pressure, or specific chemicals. These cells relay signals to specific parts of the brain.

a] Sense organs are groups of specialized cells that receive light, odor, touch, or other signals and convert them to a form that can trigger a neural impulse.

b] The cochlea of the inner ear is a highly complex, coiled apparatus that converts sound waves into nerve signals. Sound waves move fluid in the cochlea, which bends stereocilia on the cochlea's hair cells, thus altering electrical properties of the cell. This is transmitted to cells that send nerve impulses through the auditory pathways of the brain.

c] The fluid-filled semicircular canals employ hair cells to maintain balance and detect motion of the head.

d] The eye contains two types of photoreceptor cells: rods, which allow vision in dim light, and cones, which allow color perception when light levels are brighter. In photoreceptor cells, pigment molecules change shape when light hits them, leading to a change in the cells' electric potential. This change in the rod is converted to an action potential in other neurons, leading to a nerve impulse that the brain interprets as a visual image.

e] Taste and smell are chemical senses based on the direct recognition of flavor and odor molecules coming in from outside.

UNDERSTANDING THE FACTS AND THE CONCEPTS

For Questions 1–5, match each of the descriptions with the most appropriate item from the following list. Each answer can be used once, more than once, or not at all.

a] CNS
b] PNS
c] sensory neuron
d] motor neuron
e] interneuron
f] autonomic nervous system

1 Includes sympathetic and parasympathetic nerves that regulate the internal organs of the body.

2 Composed of the brain and spinal cord.

3 Consists of neurons that carry information from sense organs to the brain as well as from the brain to muscles.

4 Neurons that do not synapse directly with either sense organs or with muscles, but only with other neurons.

5 The neuron category that comprises most of the brain.

As above, for Questions 6–10, match each of the descriptions with the most appropriate item from the following list.

a] sympathetic nerves
b] parasympathetic nerves
c] somatic motor nerves
d] gray matter
e] white matter

6 Motor neurons of the autonomic nervous system that take charge of regulating internal organs when the body is calm and quiet.

7 Motor neurons of the autonomic nervous system that take charge of regulating internal organs during times of emergency or stress.

8 Motor neurons that regulate the contraction of muscles attached to the skeleton.

9 The regions of tracts of axons within the spinal cord.

10 Found within the CNS; consists mostly of cell bodies and synapses.

As above, for Questions 11–15, match each of the descriptions with the most appropriate item from the following list.

a] brainstem
b] cerebellum
c] cerebrum
d] hypothalamus
e] reticular formation
f] thalamus

11 A region of the brainstem structurally and functionally linked to the pituitary gland.

12 A network of neurons that spans the entire brainstem and functions in producing states of sleep, arousal, and concentration.

13 The part of the brain that includes the medulla, pons, and midbrain, as well as the thalamus and hypothalamus.

14 Site of origination of voluntary muscle action, language, higher brain functions, and awareness of sensation.

15 Region where voluntary muscle actions are refined and coordinated.

For Questions 16–20, choose the term from the list below that correctly completes each statement.

a] semicircular canals
b] rhodopsin
c] hippocampus
d] corpus callosum
e] basilar membrane
f] hair cells

16 Stereocilia are the functional part of the ______.

17 Damage to the ______ often prevents the formation of new long-term memories.

18 The pigment ______ changes shape when it absorbs light energy.

19 Organs that utilize hair cells to detect balance and motion are the ______.

20 The ______ carries nerve impulses that travel from one cerebral cortex to the other.

INTEGRATE AND APPLY WHAT YOU HAVE LEARNED

1 Refer back to the discussion of a reflex arc in CHAPTER 31. Which elements of the reflex arc are in the central nervous system and which are in the peripheral nervous system?

2 The central nervous system consists of gray matter and white matter. After stating the composition of each, predict whether the peripheral nervous system is mostly gray or white.

3 Contrast the role of the cerebral hemispheres with that of the cerebellum in initiating and coordinating complex voluntary motions such as walking and running.

4 Explain how we distinguish high-pitched sounds from low-pitched sounds.

5 Identify the receptor cells for vision and hearing and explain how each responds to a stimulus.

ANALYSIS

1 Parasympathetic nerves slow down the heart and speed up digestion. How might you logically explain one kind of nerve producing opposite activities?

a] Only one kind of tissue might have receptors for the neurotransmitter liberated by the parasympathetic nerves. Thus, only one kind of tissue would be stimulated.
b] The neurotransmitter produced by parasympathetic nerves might be chemically altered in the blood, leading to two different neurotransmitters being present, each having different effects.
c] One kind of tissue might deactivate the neurotransmitter more rapidly, leading to the different effects in the two organs.
d] The same neurotransmitter might combine with different receptor molecules on the surface of different cells, leading to different internal cellular effects.
e] Different organs might be more or less sensitive to the parasympathetic nerves.

2 Which characteristics would be helpful in enabling an animal to localize the source of a sound? More than one answer may be correct.

a] Ears close together on the front of the head.
b] Immovable ear flaps.
c] Ears far apart on the sides of the head.
d] Movable ear flaps.

3 Which of the following conditions would lead to deafness? More than one answer may be correct.

a] Fusion and immobility of the three bones in the middle ear.
b] Malfunctioning ion channels into the hair cells.
c] Loss of functioning sensory neurons leading from the cochlea.
d] Damage to the auditory regions of the cerebral cortex.

4 What do the senses that rely on hair cells and stereocilia have in common?

a] The sense organs are all in the ear.
b] They all begin with changes in pressure within a fluid.
c] They all result in hearing.
d] They are all involved in hearing, pain, or balance.
e] The sense organs are all in the head.

CHAPTER 33

The Body in Motion

AN ANTIDOTE FOR AGING?

It is often stated that we live in a youth-oriented culture, and this is not so surprising if you consider the popular images of old age: wizened, wrinkled, weak-muscled, brittle-boned, out-of-breath, lonely, useless, shelved. But studies reveal that the *social* images in this list are incorrect. Fully 95 percent of the elderly live independently, not in institutions; are in regular contact with their families; and are well-integrated into the community, not isolated. The *physical* conceptions of old age, however, are closer to the mark: The typical retiree has lost most of his or her muscle strength and aerobic capacity, is overweight, has numerous aches and pains, and probably suffers from heart disease and high blood pressure.

Is this decline an inevitable result of the aging process? Far from it, according to an important new study. "Exercise," says the study's principle investigator, "and you can change the way you grow old." Eighty-year-old exercise physiologist Fred Kasch of San Diego State University published the findings of a groundbreaking study in 1993. Beginning 28 years ago, Kasch set up an experimental group of 12 men who agreed to exercise regularly for the study's duration, and a similar-sized control group of men who remained sedentary.

FIGURE 33.1
Muscles in Action.
Exercise is one important antidote to the ravages of aging, as 78-year-old George "Banana" Blair proves with his daily no-hands, no-skis spin around the lagoon at Florida's Cypress Gardens.

Kasch tested all the men for muscle strength and cardiovascular fitness (the functioning of heart, lungs, and blood vessels) at 0, 10, 15, 20, and 25 years into the study. He found that the sedentary men lost 10 to 24 percent of their cardiovascular function with each passing decade. Eventually, Kasch had to stop testing them altogether because they got too weak and out of shape to run on a treadmill or ride a stationary bike. The exercisers, by contrast, lost only about 5 percent of their cardiovascular function per decade, and their strength and athletic ability remained nearly as high (and in some cases, higher) as at the experiment's start [FIGURE 33.1].

Kasch also published a parallel study of women in their seventies. He placed 17 women on a year-long program of aerobic exercise and strength training with weights and compared their progress to 10 sedentary women. The exercisers gained 6 percent more leg strength and sustained it throughout the winter, while the control group *lost* 12 percent of their leg strength. Finally, the exercisers gained 16 percent more aerobic capacity, while the control group experienced slight declines.

Was Kasch demanding exhausting, go-for-the-burn, no-pain-no-gain workouts? Not at all. The exercise regiments of both men and women averaged just three training sessions per week, consisting of 5 minutes of stretching, 30 to 40 minutes of aerobic exercise (running, swimming, cycling, brisk walking, or aerobic dancing), and 20 minutes of strength training. The study subjects reaped their amazing physiological returns on an investment of just 3 hours or so per week.

Our current chapter focuses on the **musculoskeletal system**—the muscles and bones that make exercise possible—and how the body's physiological systems are integrated in ways that maintain homeostasis during the stress of heavy exercise. We will see that the human body evolved with the capacity to move and work, and that movement is a key strategy in nearly all animals for solving life's most basic problems: capturing sufficient energy and reproducing.

We begin this chapter by examining how the shape and internal structure of a bone allow it to support weight, encase organs, transmit force, store minerals, and make blood cells. We go on to discuss how fibrous proteins, membranes, and energy sources allow the unique contractile activity of the muscles. And we'll eventually see how single bouts of exercise as well as long-term athletic training affect the body's homeostatic mechanisms, including the potential for continued physical health long into old age. As we explore the dynamic animal, we'll encounter some intriguing subjects, including the principle behind a man's erection; how osteoporosis can weaken bones; the "motor" in your muscles; the marathon champion of the animal world; the race between men and women for Olympic speed records; and the wisdom of walking your dog. ❑

MESSAGES

1 Mobility is one important hallmark of the animal way of life. Movement occurs as muscles contract, applying force to a rigid support, a skeleton.

2 A skeleton made up of compressed fluid or stiff, hard material supports an animal's body. The bone and cartilage comprising a vertebrate's skeleton support other body parts, facilitate movement, protect soft organs, store minerals, and help form blood cells.

3 Muscles shorten, or contract, bringing together the two body parts attached to the ends of the muscles. Contraction occurs when protein rods inside muscle cells slide past each other the way two trains move in opposite directions on parallel tracks.

4 Diseases of inactivity are rampant in industrialized countries. A lifelong program of exercise can maintain muscle strength, retard bone deterioration, decrease the likelihood of obesity and heart disease, and help maintain a long and vigorous life.

The Skeleton

Hold your elbow against your side for just a moment, and raise your free hand to your shoulder. How does a simple movement like that take place? It involves a stationary object at one end (your shoulder), a movable object at the other (your forearm), and muscles attaching to bones or other hard structures and pulling against them like a stretched rubber band. The rigid body support to which muscles attach and apply force is called a **skeleton** in both vertebrates and invertebrates. Skeletons are of two basic types. The most familiar is a *braced framework* that provides solid support; the hard shell of a shrimp or the stiff rods of bone in your arms and legs are examples of braced frameworks. A less familiar but no less effective type of skeleton depends on liquid to transmit force, like the hydraulic brake system in a car. Let's look at this simpler type first.

WATER AS A SKELETAL SUPPORT

Animals as diverse as sea anemones, earthworms, and snails have a form of internal support called the **hydroskeleton**, composed of a core of liquid (water or a body fluid such as blood) wrapped in a tension-resisting sheath. A hydroskeleton is like a balloon filled with water: If you squeeze on one end, the fluid transmits force to the other end. Contracting muscles can push against a hydroskeleton, and the transmitted force generates body movement.

The two sea anemones in FIGURE 33.2 help illustrate how a hydroskeleton works. This denizen of tide pools has a central digestive cavity filled with seawater and surrounded by the body wall. The wall contains two layers of muscles: the *longitudinal muscles*, which extend from the animal's base to its tentacles; and the *circumferential muscles*, which encircle the wall. These two muscle layers act as **antagonistic muscle pairs**—groups of muscles that move the same object in opposite directions. When the longitudinal muscles contract, the animal becomes shorter and wider (as when you push a water-filled balloon from both ends simultaneously). When the circumferential muscles contract, the animal becomes longer and narrower (as when you wrap your hands around a balloon and squeeze).

Hydroskeletons are common among invertebrates, with their lack of bone, but even a system as complex as the human body relies on the principles of hydroskeletons in some situations. People lifting heavy weights off the floor tend to hold their breath and tighten their abdominal muscles. The compressed fluid of the abdominal cavity contributes to the body's rigidity and helps support the load. In many mammals, including humans,

[A]

When the *longitudinal muscles* contract, the animal becomes shorter and wider

When the *circumferential muscles* contract, the animal becomes longer and narrower

[B] Hydrostatic skeleton: Antagonistic muscle pairs

FIGURE 33.2
A Sea Anemone's Hydroskeleton.

FIGURE 33.3
A Cicada's Exoskeleton: External Support for the Body and Attachment Sites for Muscles.
A cicada leaves behind its old brown exoskeleton as it molts. The soft, new, flexible green one will quickly become more rigid and protective.

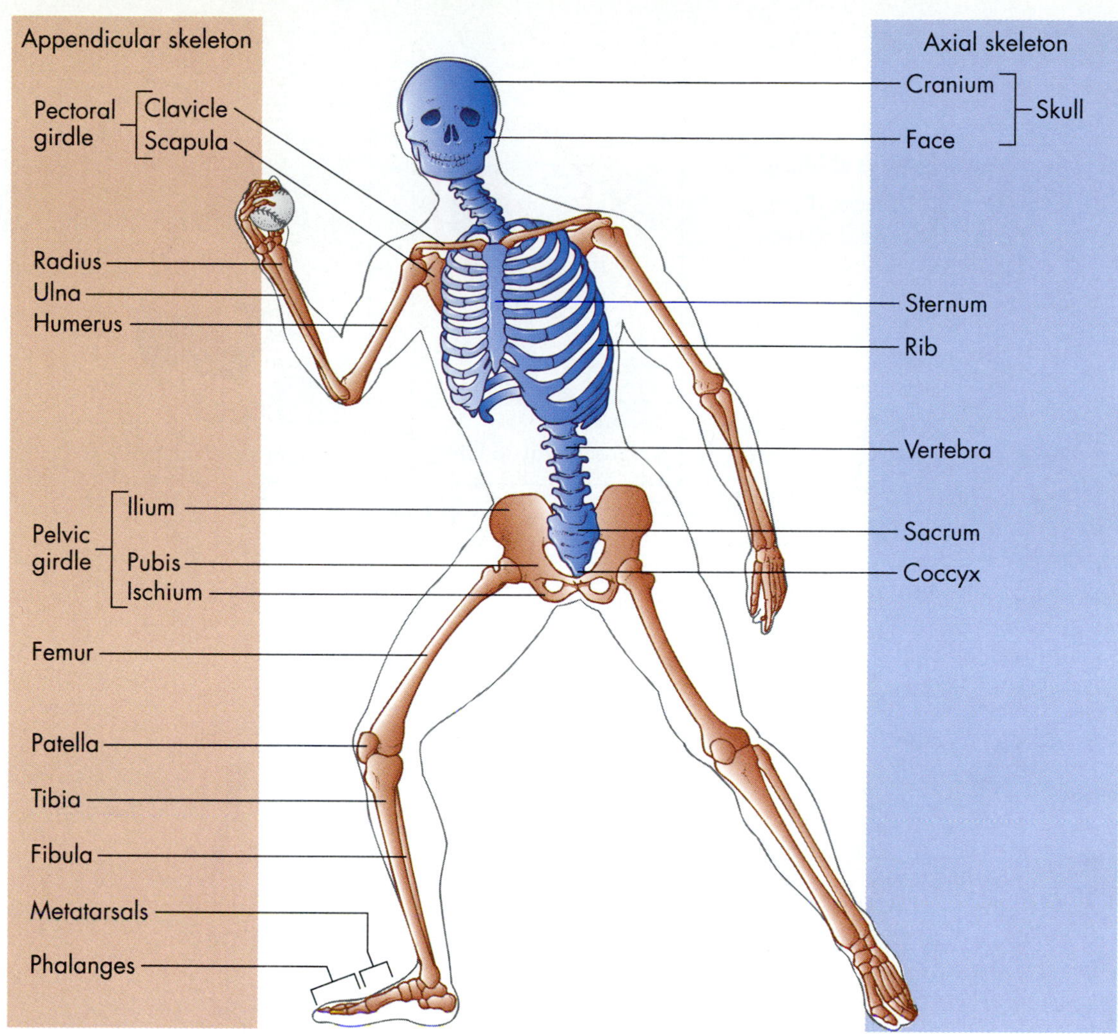

FIGURE 33.4

The Human Skeleton: Internal Support and Muscle Attachment. The adult body has 206 bones grouped into two portions: the axial skeleton (80 bones, shown in blue), and the appendicular skeleton (126 bones, shown in tan).

a hydroskeleton also stiffens the penis; valves allow blood to flow forcefully into the organ but not to flow out as freely, and the trapped fluid causes the penis to expand and become rigid.

BRACED FRAMEWORK SKELETONS

The simple hydroskeleton is perfectly adequate for an aquatic animal like a sea anemone, which has the extra buoyant support of a watery environment, or for certain land animals like the earthworm, whose locomotion requires only moderate versatility, speed, and coordination. A rapidly moving land animal, however, requires a **braced framework** of solid material made of interlocked proteins, minerals, and polysaccharides. It can be an **exoskeleton**—a stiff sheet or an external covering that surrounds the body, as in shrimp, beetles, scorpions, and other arthropods [FIGURE 33.3], or it can be an **endoskeleton**—a set of rigid rods and plates inside the body, as in vertebrates [FIGURE 33.4].

Mammalian endoskeletons are made of **bone**, a living tissue that contains a matrix of the long, stringy protein *collagen* hardened by a calcium phosphate salt. Mammals have an *axial skeleton*, which supports the main body axis, and an *appendicular skeleton*, which supports the appendages—the arms and legs [see FIGURE 33.4]. Together, the human axial and appendicular skeletons have 206 individual bones.

The Axial Skeleton The skull, vertebral column, ribs, and tailbone make up the **axial skeleton**. The **skull** is made up of the **cranium**, which encloses and protects the brain, as well as several bones of the face and inner ear. The skull also contains the hyoid bone, or tongue bone, the only human bone not jointed with other bones. (It is suspended at the back of the mouth by ligaments and muscles.)

The head sits on top of the backbone, or **vertebral column**, which is made up of 33 **vertebrae** (VER-tuh-bree; singular, **vertebra**). Each vertebra is a bony box that fits together with other vertebrae in a stack, encasing and protecting the spinal cord and giving support to the trunk. This movable stack of boxes gives flexibility to the body axis, allowing a shortstop to bend forward and scoop up a line drive or a gymnast to do a back walkover. The vertebral column, however, is a bit fragile and requires special precautions as you lift, stand, or sit to avoid developing lower back problems [FIGURE 33.5]. Curving forward from the vertebrae are 12 pairs of ribs,

FIGURE 33.5

The Human Back and Ways of Lifting.

[A] The backbone has three curves. [B] Two stacked vertebrae. [C] Wrong and right ways to lift.

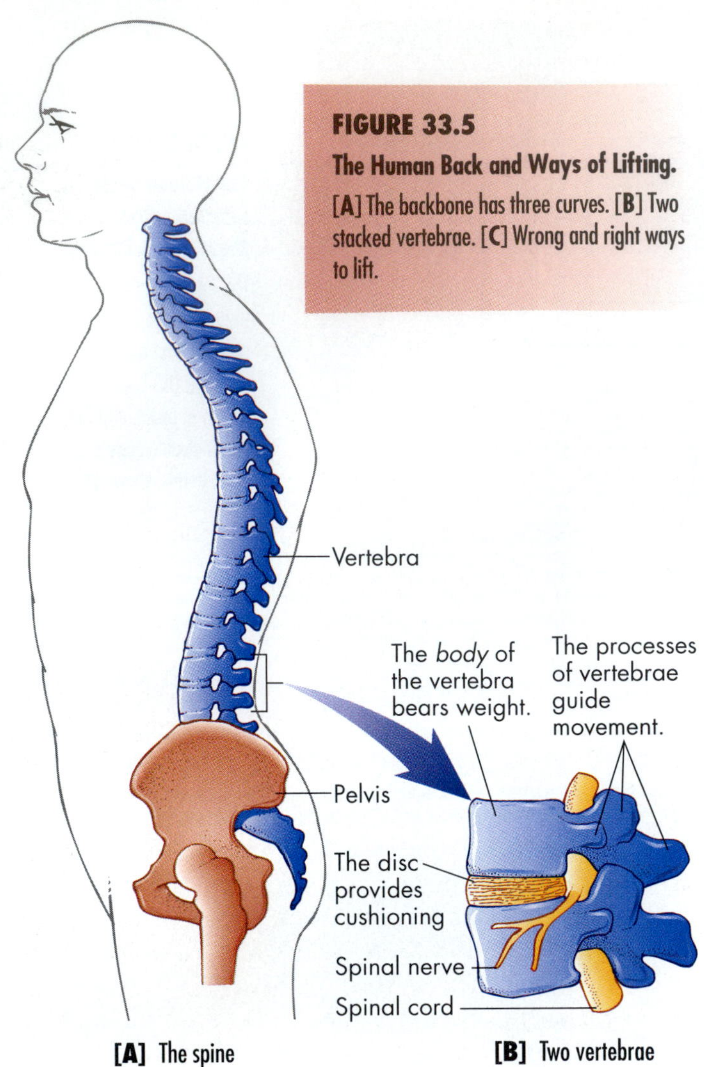

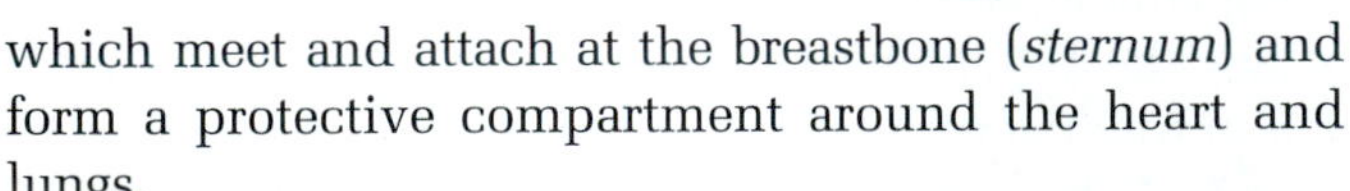

[A] The spine **[B]** Two vertebrae

[C] How to lift

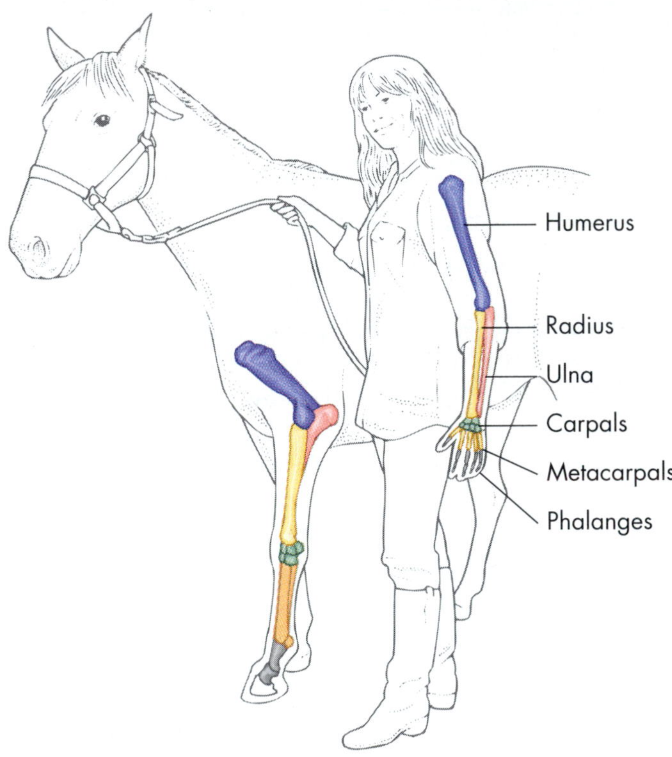

FIGURE 33.6

Horse and Human Forelimbs: Variations on an Evolutionary Theme.

which meet and attach at the breastbone (*sternum*) and form a protective compartment around the heart and lungs.

The Appendicular Skeleton The bones of the **appendicular skeleton** are attached to the axial skeleton; these include the **pectoral** (*shoulder*) **girdle** and **pelvic** (*hip*) **girdle**, the parts of the appendicular skeleton that support the arms and legs and allow them to swing [see FIGURE 33.4]. Attached to the shoulder girdle and the pelvic girdle are the bones of the **limbs**. If we compare our forelimb (arm) bone to a horse's forelimb (foreleg) bones, it becomes obvious that both species show variations on a common anatomical theme [FIGURE 33.6]. The pressures of natural selection acting over thousands of generations have molded the skeleton of the horse for superb running ability, just as entirely different pressures have led to our adaptations for grasping, carrying, writing, and holding books in our arms.

Bones: A Living Scaffold The bones of the skeleton not only support the body and anchor the muscles, but also encase vital organs, form blood cells, and store calcium and phosphate ions. A closer look at a large bone such as the femur shows how these tasks are accomplished. The long shaft, or *diaphysis* (die-AFF-uh-siss), of the femur [FIGURE 33.7A] allows it to support the body and transmit the weight of the animal to the ground. The femur has at each end an expanded portion called an *epiphysis* (ih-PIHF-uh-siss), which makes a *joint*, or *articulation*, with other bones.

Projections from a bone, called *processes*, serve as attachment sites for ligaments and tendons. **Ligaments** are connective bands linking bone to bone, while **tendons** connect bone to muscle. It is easy to find and feel exam-

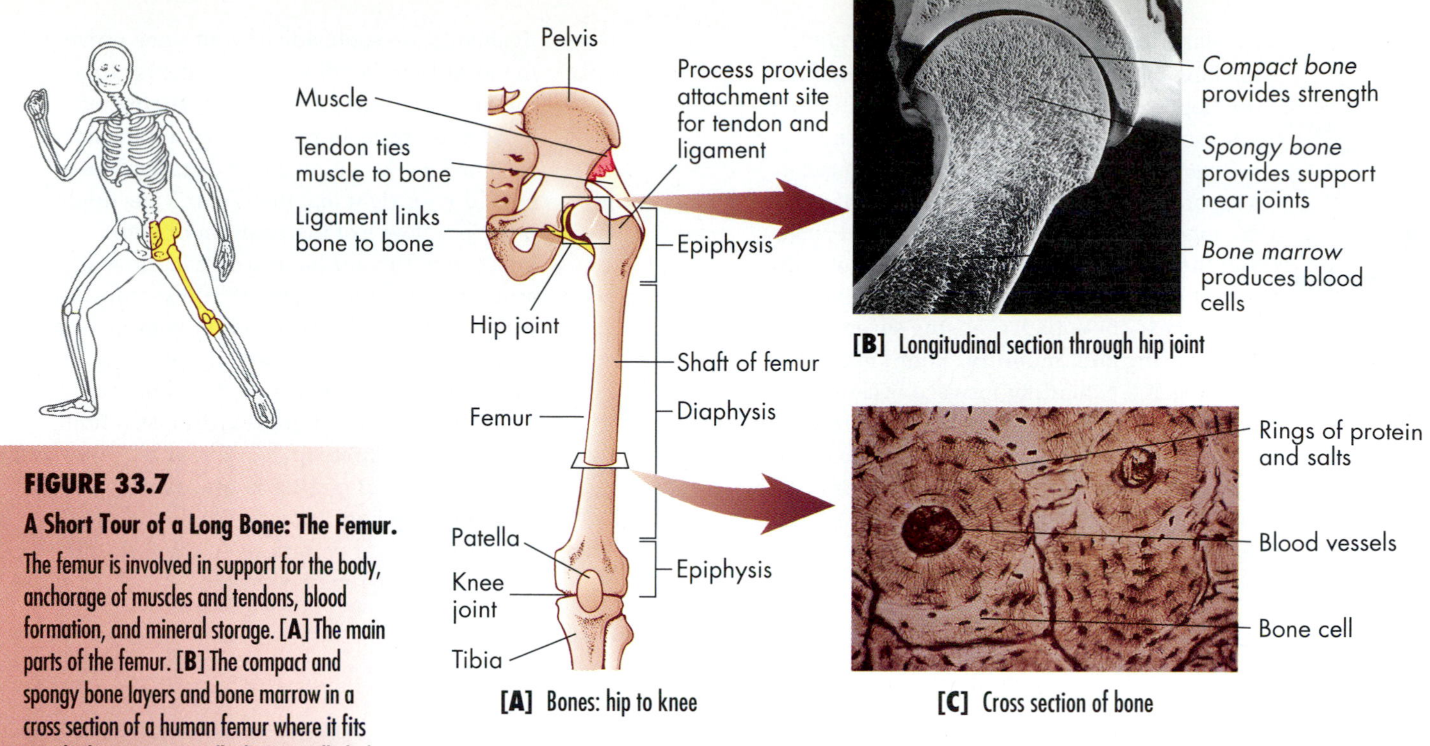

FIGURE 33.7

A Short Tour of a Long Bone: The Femur. The femur is involved in support for the body, anchorage of muscles and tendons, blood formation, and mineral storage. **[A]** The main parts of the femur. **[B]** The compact and spongy bone layers and bone marrow in a cross section of a human femur where it fits into the hip. **[C]** Bone cells (here, small, dark ovals) secrete rings of protein and salt.

ples of these skeletal structures in your body. You can feel a large process (a bump) where your head joins the back of your neck and attaches to muscles that keep your head erect; you can feel the ligament that attaches your kneecap to your shinbone (tibia) just below your knee when you tighten your thigh muscles (quadriceps); and you can easily feel tendons between the muscles in your upper arm and the bone in your forearm by putting your hand beneath the edge of a desk, lifting, and, with the other hand, touching the "cables" that now project on the inside of your elbow.

A close-up view of the femur (thighbone) reveals how a bone can anchor, strengthen, store minerals, and perform its other functions [FIGURE 33.7B]. The outer layer of **compact bone** is thick, solid, strong, and resistant to bending; this enables it to provide support. At each end of the femur is a region of **spongy bone**, a looser area crisscrossed by girders that provide strength near the joints. The tissue between girders down the center of the bone's shaft is called the **marrow**, a site of blood cell production. Embedded within the bone are cells that secrete the proteins and minerals of the bone and reabsorb calcium and phosphate from bone for use in other parts of the body [FIGURE 33.7C]. These cells also knit bones back together after a break.

Bone cells are especially active during periods of rapid growth. Bones elongate at a region near their ends called the growth plate, or *epiphysial plate*, that is initially just cartilage. (Recall from CHAPTER 20 that cartilage is a network of fibrous proteins similar to the stiff sheets that form your outer ear.) As an animal grows, bone is added to this plate, and when the organism matures, the plate disappears and the bones stop growing. If the bones are jarred by too much physical activity during childhood (a result of running marathons, for example, or of carrying heavy loads), the growth plates can be damaged and overall bone growth may be retarded.

The deposition and removal of protein and minerals go on continuously in the bones. The bumps called processes, for example, grow larger when the muscles that attach to them become stronger. Anthropologists excavating Italian cities buried centuries ago under ash and cinders from the eruption of Mt. Vesuvius are able to distinguish the skeleton of a slave girl from that of a nobleman's daughter simply by comparing the size of the processes on arm and leg bones.

Bone remodeling can result in fragile bones. Each year, about 100,000 Americans are permanently immobilized or die within a few weeks or months after a hip fracture, due to the preexisting weakness of their bones. This condition, known as *osteoporosis*, occurs when some bone cells (*osteoclasts*) destroy the existing bone matrix faster than other bone cells (*osteoblasts*) can remodel it; the result is porous, brittle bones [FIGURE 33.8]. Osteoporosis is relatively common in women after menopause and in inactive men. Recent research shows that the female hormone estrogen helps many women absorb calcium from their diets. Some doctors now prescribe small doses of estrogen to help postmenopausal women avoid osteoporosis.

Research shows that regular exercise stimulates the deposition of calcium in bones. For example, professional baseball pitchers and tennis players have 35 to 50 percent more bone in their playing (dominant) arms than

box 33.1
Human Impact

The Knee: An Accident Waiting to Happen?

As many football players have discovered, the knee is a fragile joint. Lying between the thighbone (femur) and the shinbone (tibia) is a space filled with fluid that acts like a sponge that cushions the pounding of running and jumping. The articular cartilage and the meniscus lie inside the fluid-filled space, and these further muffle shocks. The overuse syndrome called runner's knee results from an injury to these pads, causing severe pain and limiting joint mobility. Your knee resists twisting and extending forward because tough bands of tissue form a pair of crossed braces (the cruciate ligaments) that hold the femur and tibia tightly together. Violent forces (like kicking a soccer ball while off balance) can converge to shred these crossed braces. Finally, straps of tissue (the collateral ligaments) on each side of your knee prevent the joint from bending to the side. A clip on the football field, however, can rip one of these side straps in two. To experience how the knee works, sit on a low chair and fully extend your right leg, then relax your thigh muscles. Now feel the position of your kneecap with your hand; you can move it around with your fingers. While keeping your leg extended and your hand on your kneecap, tighten your thigh muscles and notice how the ligament lengthens and your kneecap moves up your leg. Finally, while still holding onto your kneecap, bend your leg. You may feel a slight grating as the kneecap slips over the thighbone, perhaps owing to insufficient cartilage or variations in the fluid-filled sac cushioning the two bones.

[A]

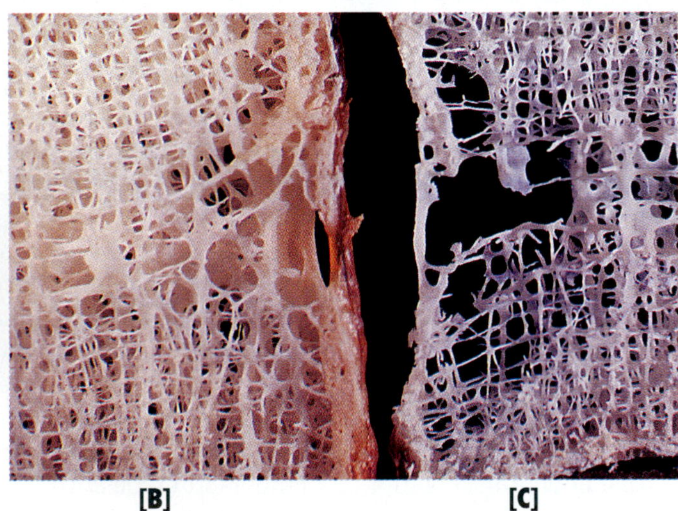

[B] [C]

[D]

FIGURE 33.8

Osteoporosis: A Thinning of the Bone Matrix.

[A] The decrease in estrogen that accompanies menopause can lead to a thinning of the bone. Normal bone [**B**] shows a thicketlike meshwork of fibers, but an osteoporotic bone [**C**] has thinner fibers and many holes. [**D**] Osteoporosis is not inevitable; at age 90, Corine Leslie was still skydiving occasionally, as well as walking and playing golf several times per week near her home in Arizona.

in their nondominant arms. Likewise, middle-aged women who hoisted 5-pound weights weekly for four years lost only half as much bone mass in their arms as did more sedentary controls. Luckily, with proper diet, exercise, and perhaps estrogen supplements, osteoporosis is not an inevitable result of aging. Just look at the 90-year-old lady skydiver in FIGURE 33.8D!

Joints: Where Bones Meet Bones can move with respect to each other because of **joints**, points where adjacent bones nearly touch. Slightly movable joints, such as those between adjacent vertebrae in the spine, have pads of cartilage that absorb shock but allow limited mobility [review FIGURE 33.5]. Freely movable joints, such as the shoulder, hip, and knee [see FIGURE 33.7A], have pads of cartilage at the ends of the two adjoining bones, but also have a flattened sac (a *bursa*) filled with a fluid called *synovial fluid* between the bones that cushions the joint and eases the gliding of bones across each other. Temporary inflammation of this joint sac is called *bursitis*. Long-term inflammation of the joint can result in *arthritis* [see FIGURE 22.15]. FIGURE 33.9 and BOX 33.1 describe how the knee joint works and why it is a source of trouble for joggers and professional athletes alike.

How Bones Act As Levers Bones can pivot at joints because of the particular way muscles attach to bones,

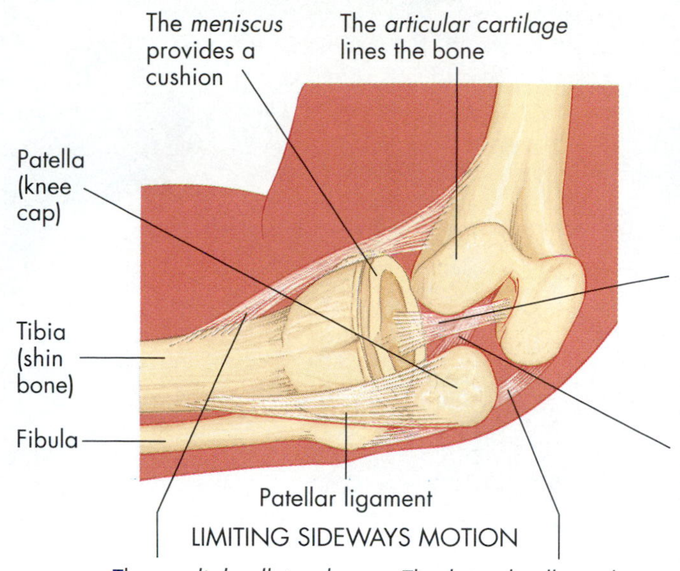

FIGURE 33.9
The Knee Joint.

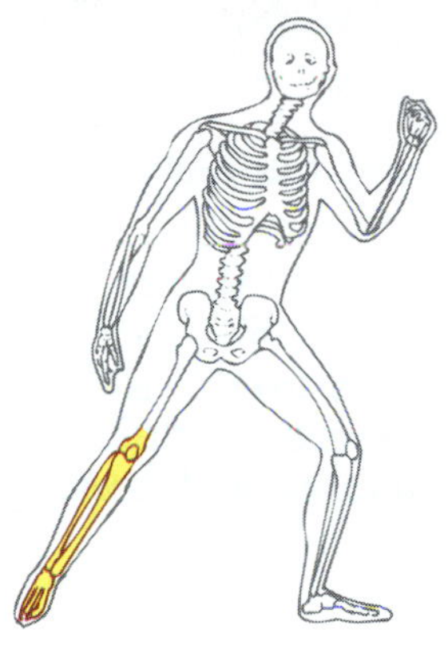

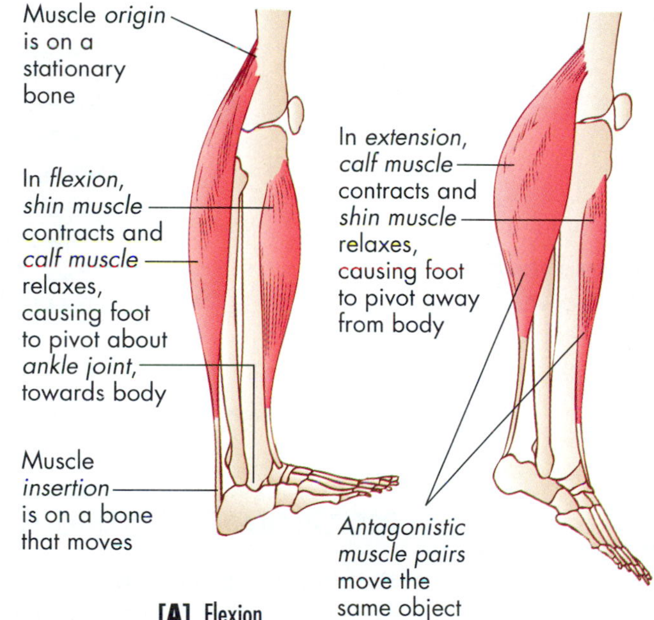

[A] Flexion

[B] Extension

FIGURE 33.10
Muscles Move Bones to Action.
[A] Simultaneous contraction of the shin muscle and relaxation of the calf muscles flex the foot, while **[B]** contraction of the calf muscles and relaxation of the shin muscle extend the foot. Because the ankle joint acts as a fulcrum, people with longer heels (levers) can jump much higher **[C]**. These levers and contractions drive the athlete from the University of Oklahoma over the high jump bar.

[C] Extension in action

and because muscles are arranged in antagonistic pairs. For example, contraction (shortening) of the shin muscle (tibialis anterior) *flexes* your foot toward your body (**flexion** decreases the angle between two bones) [FIGURE 33.10A], while its antagonists, the calf muscles (gastrocnemius and soleus), *extend* your foot to point away from the body (**extension** increases the angle between two bones) [FIGURE 33.10B]. Often, when a muscle contracts, its antagonist relaxes. Ordered movements occur because one end of the muscle, the **origin**, attaches to a bone that generally remains stationary during a contraction, while the other end, the **insertion**, attaches across a joint to a bone that moves. Your calf muscles attach to your femur just above the back of your knee, and at the opposite end, to your heel bone (*calcaneus*). When the calf muscles contract, the shortening forces the foot to rotate around a pivot, the ankle joint, much as a lever rotates around a fulcrum. Because of this arrangement, a small contraction of the muscle transmits a large movement to the bone.

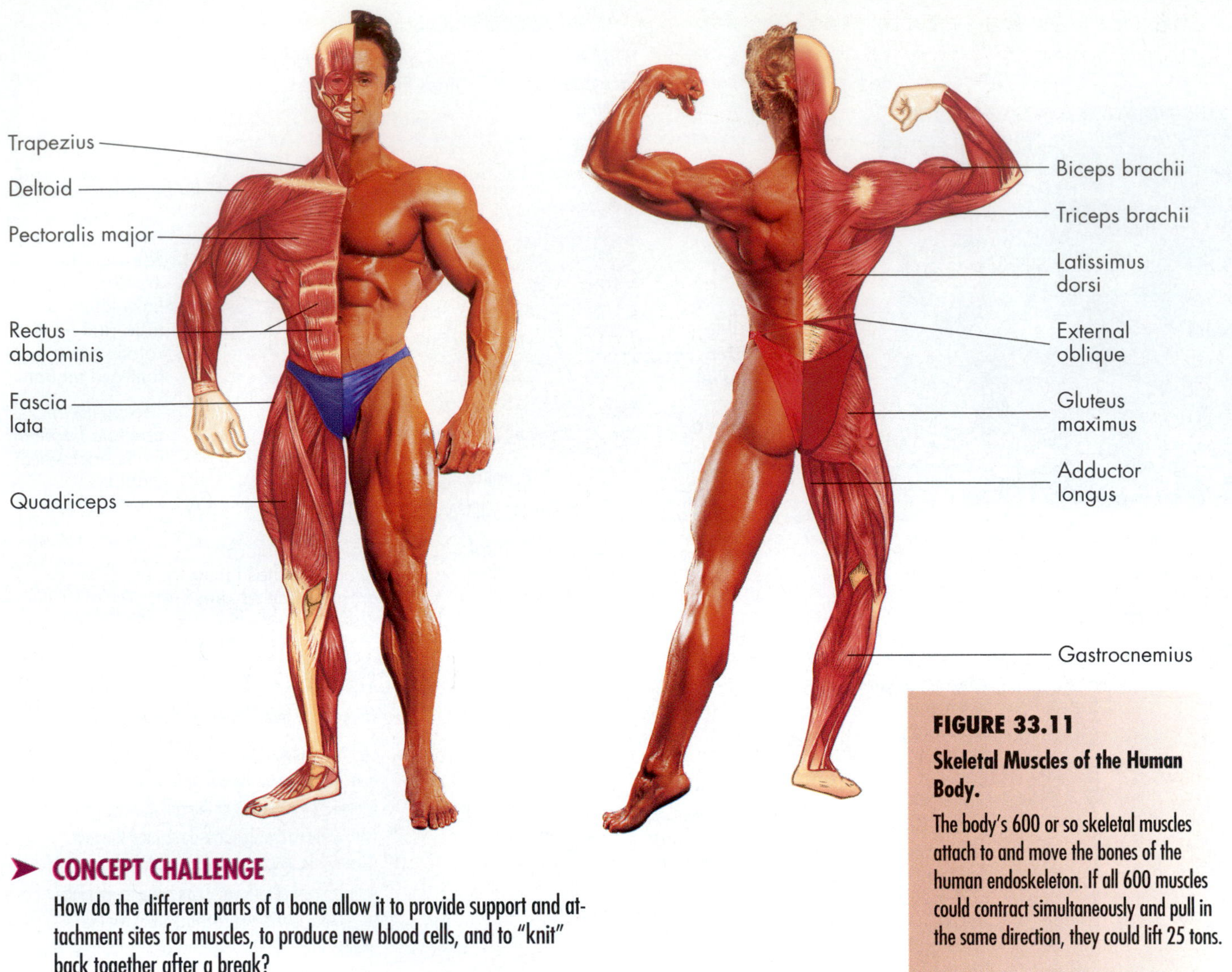

FIGURE 33.11

Skeletal Muscles of the Human Body.

The body's 600 or so skeletal muscles attach to and move the bones of the human endoskeleton. If all 600 muscles could contract simultaneously and pull in the same direction, they could lift 25 tons.

➤ CONCEPT CHALLENGE

How do the different parts of a bone allow it to provide support and attachment sites for muscles, to produce new blood cells, and to "knit" back together after a break?

Muscles

Bones provide support, but without muscles to move them, the body would be like a pile of sticks. A type of muscle called **skeletal muscle** is connected to and moves the skeleton. FIGURE 33.11 shows most of the body's major skeletal muscles.

Voluntary movement in a skeletal muscle originates when the motor cortex of the cerebrum [review FIGURE 32.6] sends action potentials down a motor neuron to its synapse with the muscle. When an action potential reaches the synapse, it initiates an action potential in the muscle's cells and causes the muscle to contract. But how does an action potential cause the muscle cells to shorten? The answer lies in the unique structure of muscle cells.

Skeletal muscles are made up of **muscle fibers**, giant cells with many nuclei that may extend the full length of the muscle—often several centimeters [FIGURE 33.12A and B]. Each muscle fiber has a system of protein filaments, a system of membranes, and an energy system that act, respectively, as the muscle's engine, electrical ignition switch, and fuel [FIGURE 33.12C].

PROTEIN FILAMENTS OF MUSCLE

Within each muscle cell lie threads called **myofibrils** that do the actual job of contracting the muscle [see FIGURE 33.12B]. A longitudinal section of a myofibril reveals repeating units of dark and light bands. Each repeating unit is called a **sarcomere** (from the Greek words for "flesh" and "part of") [see FIGURE 33.12C]. Each sarcomere consists of two types of interdigitating filaments called actin filaments and myosin filaments. **Actin filaments** are thin; **myosin filaments** are thick, and their overlapping arrangement creates the visible bands typical of skeletal muscle, a bit like the pattern you create when you interlace the outstretched fingers of your two hands. The proteins actin and myosin are not exclusive to muscle; in fact, they are found in the cytoskeleton of most eukaryotic cells, where their contraction changes cell shape, localizes materials, and causes dividing cells to pinch in two [review FIGURE 3.17].

How do these neat arrays of protein filaments cause

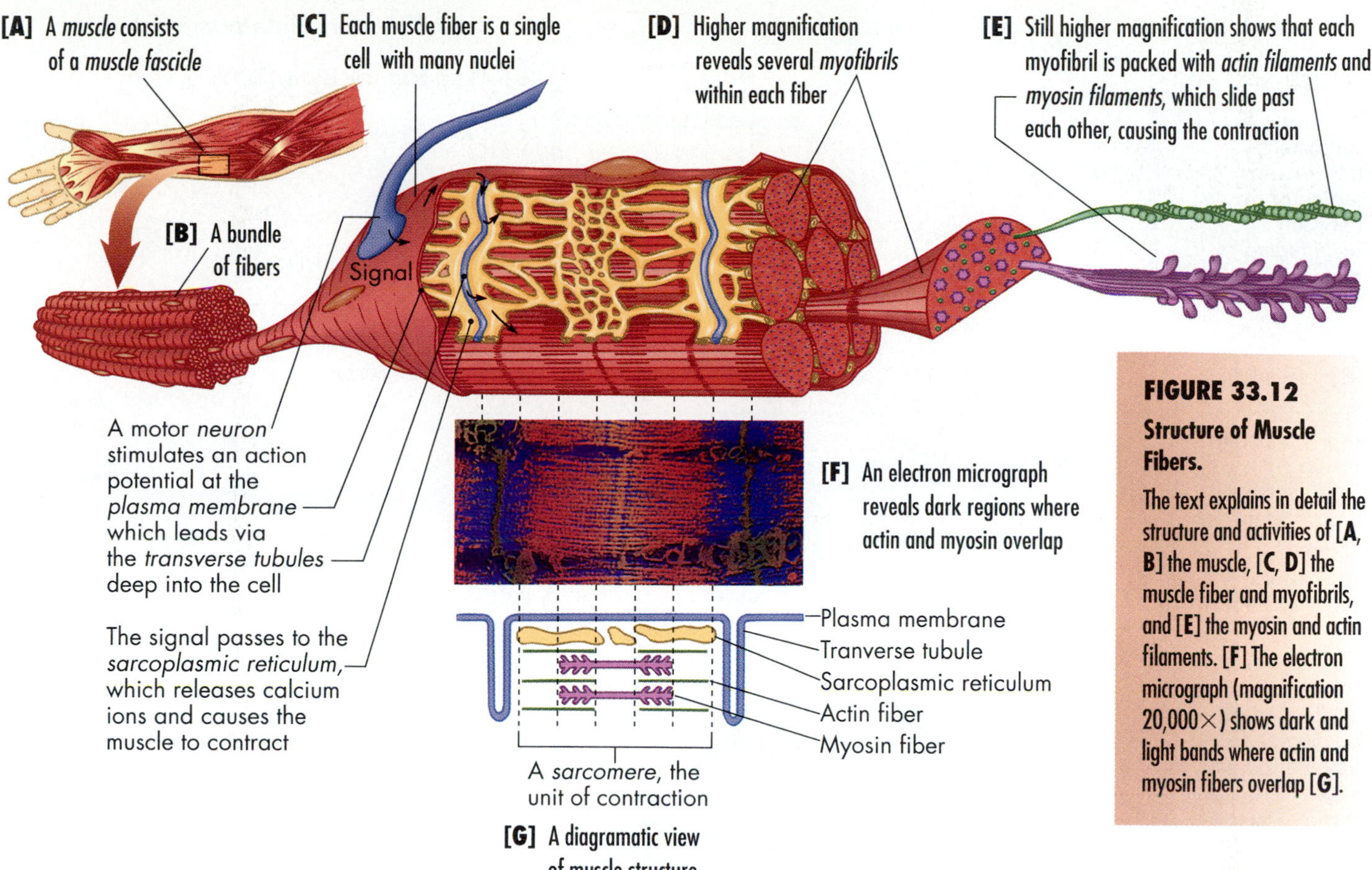

FIGURE 33.12
Structure of Muscle Fibers.
The text explains in detail the structure and activities of [**A, B**] the muscle, [**C, D**] the muscle fiber and myofibrils, and [**E**] the myosin and actin filaments. [**F**] The electron micrograph (magnification 20,000×) shows dark and light bands where actin and myosin fibers overlap [**G**].

the individual muscle cells of your forearm to shorten and move a pencil or paintbrush in a precise fashion? By examining skeletal muscle tissue with an electron microscope, biologists observed that as the muscle contracts, each sarcomere within the muscle fiber becomes shorter, thus causing the whole fiber to shorten [FIGURE 33.13A and B]. The actin and myosin myofilaments within each sarcomere, however, do not shorten during contraction. Instead, these filaments slide past each other as the sarcomere shortens. To picture this, try thinking of a woman putting her hands into a furry muff; her hands slide past the insides of the muff, her elbows come nearer together, but the muff itself does not decrease in length [FIGURE 33.13C]. This remarkable process is called the **sliding filament mechanism** of muscle contraction.

The unique arrangement and behavior of actin and myosin molecules allow the two types of filaments to slide past each other. Each myosin filament is made up of many individual myosin protein molecules bound together. Each myosin molecule within the long filament has a long, straight tail at one end and a head at the other that reaches out to actin filaments like an oar dipping into the water and the two filaments become temporarily attached [see FIGURE 33.13D]. The myosin head swings, moving the actin past it like an oar moves water [see FIGURE 33.13E]. This process of reaching, attaching, swinging, and detaching happens over and over for each myosin molecule as the muscle contracts, literally sliding the actin molecule past it.

Not surprisingly, just as a rowing crew must eat heartily the night before a regatta, the reaching and swinging of myosin "oars" also requires energy. ATP binds to the myosin head, causing the head to release the actin. The myosin head then cleaves ATP to ADP and phosphate as the head swings forward and binds actin again for another stroke [see FIGURE 33.13D]. The energy cleavage that propels actin past myosin explains *rigor mortis*, the stiffening that occurs shortly after an animal dies. Since ATP causes the myosin head to release the actin, when the supply of ATP runs out after death, myosin cannot release the actin (the oars get stuck in the water). Thus, myosin cannot glide past actin, so muscles can neither contract nor relax, and they remain stiff until the protein filaments themselves are degraded by enzymatic action. The mechanism of sliding protein filaments explains how muscles shorten. It does not, however, explain how muscles "know" when to contract. That requires signals from cell membranes.

MEMBRANE SYSTEM OF MUSCLE

Muscle contraction enhances an animal's survival only if the organism can control when contraction takes place. This control is exerted by muscle cell membranes, which

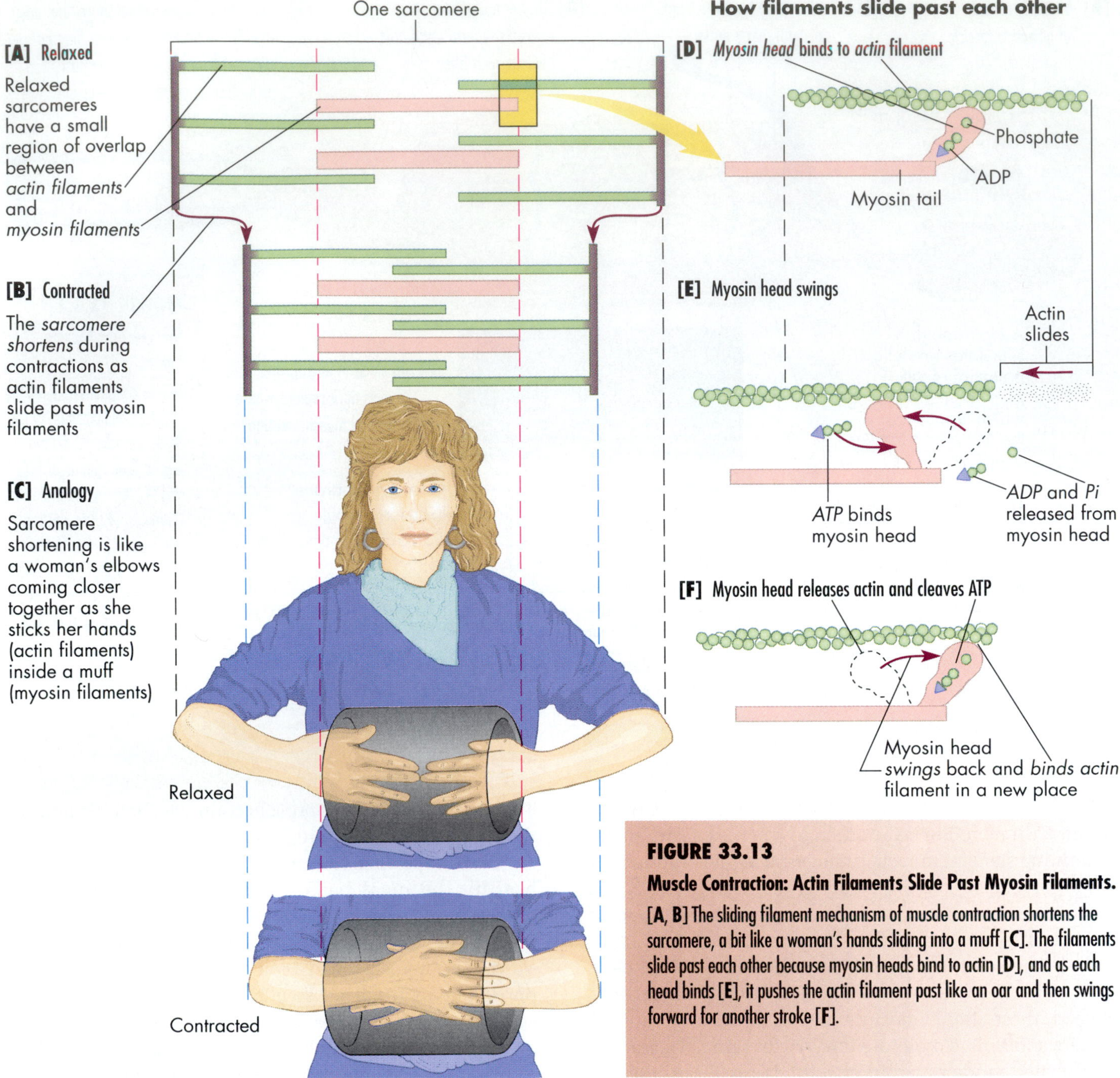

FIGURE 33.13

Muscle Contraction: Actin Filaments Slide Past Myosin Filaments.

[A, B] The sliding filament mechanism of muscle contraction shortens the sarcomere, a bit like a woman's hands sliding into a muff [C]. The filaments slide past each other because myosin heads bind to actin [D], and as each head binds [E], it pushes the actin filament past like an oar and then swings forward for another stroke [F].

ignite the muscular motor. Three sets of membranes help stimulate muscle fibers to contract [see FIGURE 33.12C]: (1) the plasma membrane (**sarcolemma**) surrounding the fiber; (2) the **transverse tubules**, which lead like tunnels from the plasma membrane deep into the cell's interior; and (3) the **sarcoplasmic reticulum** (corresponding to the endoplasmic reticulum of other cells, review FIGURE 3.18), which forms a sac within the sarcomere and stores large amounts of calcium ions.

When an action potential reaches the synaptic junction between a motor neuron and a muscle cell [see FIGURE 33.12C], it causes the neuron to release the neurotransmitter acetylcholine into the synaptic cleft and onto the cell's plasma membrane. The transmitter causes an action potential in the muscle cell (similar to the action potential in a neuron), and electrochemical activity on the surface is channeled toward the cell's interior by the transverse tubules. Next, a signal (probably a second messenger; see CHAPTER 26) passes from the transverse tubules to the sarcoplasmic reticulum, causing the membranous sac to release some of its stored calcium into the cell's cytoplasm.

The jolt of released calcium ions interacts with a set of regulatory proteins (**troponin** and **tropomyosin**). This regulatory interaction allows the myosin heads to bind to actin filaments and the muscle cell to contract. After the

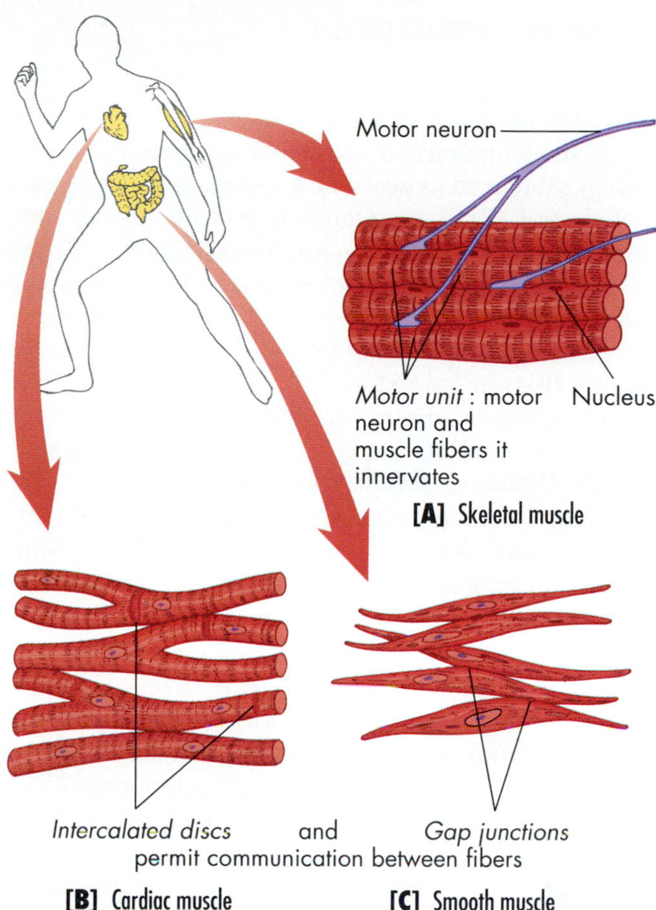

FIGURE 33.14

How Cardiac Muscle, Smooth Muscle, and Skeletal Muscle Differ at the Cellular Level.

[A] Several skeletal muscle fibers can contract together (a motor unit) when branches of the same motor neuron contact them all. When more motor units fire, the contraction is stronger. **[B]** Cardiac muscle cells have gap junctions organized in structures called *intercalated disks*. These junctions allow electrical communication between adjacent cardiac muscle cells, coordinating their contraction. **[C]** Smooth muscle cells communicate electrically through gap junctions and tend to contract in sequential waves.

cell shortens, an ATP-driven pump quickly returns the calcium ions to the sarcoplasmic reticulum, and the regulatory proteins force myosin away from actin, which allows the muscles to relax. The message here is clear: Muscles can generate force against a skeleton because protein filaments expend energy and slide past each other, causing the muscle cell to shorten.

CARDIAC MUSCLE AND SMOOTH MUSCLE

The mechanisms just discussed apply to skeletal muscles, those muscles that are attached to bones and that move the skeleton [FIGURE 33.14A]. Vertebrates, however, have two additional types of muscles that move fluids rather than bones. *Cardiac muscle*, found only in the heart, drives blood through the circulatory system, while *smooth muscle* propels food through the digestive tract and provides tension in the urinary bladder, uterus, and blood vessels.

Cardiac muscle fibers consist of networks of muscle cells. These are electrically connected to each other via *intercalated disks*, which are specialized junctions like the gap junctions and the electrical synapses in FIGURE 31.8. Such specialized junctions allow adjacent cells to communicate back and forth [FIGURE 33.14B]. Thus, an impulse initiated any place on the heart muscle propagates through the entire organ, causing the whole fist-sized heart muscle to contract as a unit. The impulse usually begins in the heart's pacemaker [review FIGURE 25.11], and thus nerves are not the impetus—a significant advantage during heart transplantations. In contrast to cardiac muscle, skeletal muscle fibers do not communicate with each other directly, although several fibers may be innervated by the same motor neuron (a *motor unit*) and hence contract together [FIGURE 33.14A]. The more motor units that fire, the more forcefully the skeletal muscle contracts. Like skeletal muscle, cardiac muscle is *striated*, or striped, because its actin and myosin filaments are organized into sarcomeres.

The **smooth muscle** cells surrounding the digestive tract, urinary bladder, blood vessels, and other hollow body organs lack striations because their actin and myosin filaments are not as well ordered as those in skeletal and cardiac muscle [see FIGURE 33.14C]. Smooth muscle cells usually have a single nucleus, and communicate with other smooth muscle cells via gap junctions, as do cardiac muscle cells. This communication allows the rhythmic pushing of food down the digestive tract or the pushing of a baby through the birth canal.

ATP: THE FUEL

While actin and myosin filaments cause muscle cells to contract, and membrane systems ignite muscle activity, the fuel for muscle cell contraction is **ATP**. As we already know, the sources of ATP are the mitochondria and certain enzymes in the cytoplasm. Three energy systems—(1) the *immediate system*, (2) the *glycolytic system*, and (3) the *oxidative system*—supply the ATP that all muscles need. The duration of physical activity dictates which system the body uses [see FIGURE 5.14A]. The **immediate energy system** is instantly available for a brief explosive action, such as one heave by a shot-putter. It depends on a muscle cell's stores of ATP plus a high-energy molecule called *creatine phosphate*, and it can fuel muscle contraction for several seconds. The **glycolytic energy system** is based on the splitting of glucose by glycolysis [see CHAPTER 5] in the muscles. It can sustain heavy exercise for a few minutes, as in a 200-m swim. After that, the long-range, or **oxidative** (oxygen-utilizing, or *aerobic*), **system** takes over a long period of exercise, such as a long run in the mountains. In the presence of sufficient oxygen, this system can produce energy by breaking down carbohydrates, fatty acids, and amino acids mobilized from other parts of the body. Clearly, anyone interested in melting away body fat should engage in aerobic activities like rapid walking, swimming, bicycling, cross-country skiing, or jogging, which rely primarily on the oxidative energy system and its ability to use fats as fuel.

Slow Oxidative Muscle Fibers The immediate, glycolytic, and oxidative energy systems do not contribute equally to the energy budgets of all muscle fibers. **Slow oxidative muscle fibers** (also called *slow-twitch muscle fibers*) obtain most of their ATP from the oxidative system. Slow-twitch fibers require about one-tenth of a second to contract fully, they are packed with mitochondria, they receive a rich supply of blood, and they have large quantities of the red protein *myoglobin*, which stores oxygen in muscle cells. These characteristics make slow-twitch fibers deep red, like the dark meat of a chicken [FIGURE 33.15]. Slow-twitch fibers are resistant to fatigue, and thus are able to contract for long periods of time. Athletes trained for endurance sports such as cross-country skiing have a large proportion of slow-twitch muscles. Slow-twitch muscles also provide functions critical for survival, such as maintaining posture. Without the contraction of slow-twitch muscles in your jaw, your mouth would be wide open as you read this sentence.

Fast Oxidative-Glycolytic Muscle Fiber Whereas slow-twitch fibers bestow endurance, **fast glycolytic muscle fibers** (also called *fast-twitch fibers*) provide quick power—the kind of power needed for one clean and jerk of a heavy barbell, for example [see FIGURE 33.15]. Fast-twitch fibers derive most of their ATP from glycolysis, and they reach maximum contraction twice as quickly as slow-twitch fibers. They soon grow fatigued, however, since they run through their limited stores of glycolytically generated ATP quickly. Fast-twitch fibers are white because they are packed with white actin and myosin proteins (which maximize contractile force), and they contain very little myoglobin for oxygen storage. The white meat in the breast of a chicken is the fast-twitch muscle that powers the wings and enables the chicken to suddenly burst away from a fox and fly to safety in a tree. By contrast, a duck's breast is dark meat—red, slow-twitch fibers that can sustain the beating of the wings during long migratory flights without fatigue.

A third kind of muscle fiber, **fast oxidative-glycolytic muscle fiber**, has characteristics midway between fast glycolytic and slow oxidative fibers; these fibers are moderately powerful and moderately resistant to fatigue.

Although fiber types are genetically determined to a large degree, training can change the functional characteristics of muscle fibers. For example, endurance training can cause both fast and slow fibers to develop an increased oxidative capacity but a reduced explosive strength. Clearly, a sprinter would want to avoid that type of training. In contrast, strength training improves immediate energy supply systems and suppresses oxidative capacity. Although it frustrates some body builders, the total *number* of muscle fibers is apparently genetically determined, and is not affected by weight training. Lifting can, however, increase fiber *thickness* by increasing the number of myofibrils and by increasing the synthesis of actin and myosin proteins.

➤ CONCEPT CHALLENGE

An agricultural worker accidentally ingests an insecticide that acts by blocking the transfer of a signal from nerves to muscles. Her skeletal muscles become limp, but her heart muscle continues to contract rhythmically. How can you explain those symptoms?

Exercise Physiology

For most animals, survival is linked to locomotion, and in nature, locomotion often must be rapid and coordinated if the animal is to survive. Heavy exercise or strenuous work stimulates the same fight-or-flight response that electrifies the escape of a field mouse from a diving hawk.

ESCAPE OR SKIRMISH: A SURVIVAL RESPONSE

A sparrow that has just spotted a cat crouching nearby, a wild horse galloping across the desert, and a student waiting to give an oral report in front of the class are all experiencing stresses that kindle the **fight-or-flight response**, an automatic set of events managed largely by the hypothalamus. This brain structure serves as a main link between the nervous and endocrine systems and is located at the base of the brain [review FIGURE 32.6].

In times of stress, as during heavy exercise, the hypothalamus dispatches signals via the sympathetic nervous system to various body tissues [review FIGURE 32.5]. These signals raise the amount of vital fuels like glucose, glycerol, and fatty acids in the blood; they elevate heart rate and boost blood pressure, circulating fuel to muscles; they dilate air passages and escalate breathing rate, providing ample oxygen; and they divert blood from the skin and digestive organs to the skeletal muscles, supplying food and oxygen where they are needed the most. The hypothalamus also triggers the secretion of epinephrine from the adrenal medulla, which intensifies and prolongs the various effects [review FIGURE 30.10].

FIGURE 33.15

Slow-Twitch and Fast-Twitch Muscle Fibers.

With a stain that darkens tissue containing the ATP-splitting myosin head, the myosin-rich fast-twitch fibers in a cross section of muscle tissue appear dark, while the slow-twitch fibers appear as light-colored patches. A person highly trained for and naturally good at weight lifting, shot-putting, or sprinting is more likely to have a high proportion of fast-twitch muscle fibers, while a person trained for aerobic endurance events like cross-country skiing or running in the mountains is more likely to have a high proportion of slow-twitch fibers.

Consider a human athlete such as Miguel Indurain, the Spanish winner of the grueling Tour de France bicycle race in 1992. During each race and training session, an athlete like Indurain experiences the sensations described above. At rest, his heart beats at a remarkably low 28 beats per minute. To put this rate in perspective, compare it to your own by counting the pulse in your wrist for 10 seconds and multiplying by 6. Chances are, your resting heart beats at least twice as fast as Indurain's heart. During a time trial—an all-out race against the clock—Indurain's heart rate soars to 195, a rate faster than most of us could possibly achieve. Exercise mimics the fight-or-flight response. It calls into action all the major physiological systems of the body.

HOW ATHLETIC TRAINING ALTERS PHYSIOLOGY

Since heavy exercise evokes the fight-or-flight response, people who exercise several times a week enter this state of stress repeatedly. What effects do such repeated periodic challenges have on the body's homeostatic mechanisms? Training works by causing "breakdown" followed by "overshoot"—a breakdown of stored fuel, for example, followed by the increased deposition of fuel molecules; or a slight breakdown of muscle tissues, with an overall strengthening as the tissues are repaired. To increase fitness without injury, therefore, one must carefully, gradually, and *progressively* augment the *intensity*, *frequency*, and *duration* of workouts.

One goal of human athletic training is to boost the amount of oxygen a person can deliver to working muscles (the so-called *maximal oxygen uptake*). To learn how this uptake occurs, exercise physiologists in Dallas measured subjects' oxygen utilization immediately before and after three solid weeks of bed rest, then measured it again over an eight-week training in which each subject ran from 4 to 11 km (2.5 to 7 mi) on 11 different occasions each week. As the graph in FIGURE 33.16 shows, lying in bed all day long caused a substantial drop in the body's ability to take up and utilize oxygen. After a few days or weeks of training, however, oxygen utilization increased dramatically—especially in the previously sedentary men, who doubled their ability to use oxygen and perform physical work.

What changes in the men's bodies accounted for their increased ability to take up oxygen? Two main factors appeared responsible. First, the men's hearts pumped more blood per minute after training than before. This was not due to a more rapid heart rate, but to an increase in the amount of blood pumped per beat (the *stroke volume*). The more one exercises, in fact, the slower the heart beats at rest; recall the slow heart rate of bicycler Miguel Indurain. Well-trained endurance athletes have enlarged heart ventricles and thickened ventricle walls, which pump more blood per beat; in other words, they have larger, stronger hearts. Second, physically fit people can use more oxygen because their muscles extract more oxygen from the blood. Energy-supplying mitochondria grow larger, the muscles generate more of the oxygen-storing protein myoglobin, and there is an increase in the number of blood capillaries in the muscles. Together these phenomena provide the muscles with a greater blood supply from which to extract more oxygen.

Physiological changes similar to those that occur in exercising people have occurred during evolution. We can see this most easily in an antelope like the pronghorn—the animal world's champion long-distance runner and a denizen of high-altitude prairies in Wyoming, Idaho, Nevada, and Oregon [FIGURE 33.17]. Pronghorn antelopes can run at an average speed of 64 kilometers per hour (40 mph) for half an hour or longer. To understand what sorts of physiological tricks may have evolved in the pronghorns, physiologists in Arizona and Switzerland trained the animals to run on a treadmill wearing face masks connected to a gas analyzer. They found that the pronghorn's maximal oxygen uptake was extraordinary—five times that of goats and other mammals of comparable size. Instead of uncovering some unique physiological attributes, however, the researchers found that pronghorns simply have somewhat larger lungs, slightly more of the oxygen-binding protein hemoglobin in their blood, and more mitochondria in their muscle cells. In other words, natural selection—probably in the form of preying wolf packs—has fine-tuned the pronghorn into a world-class endurance athlete.

Exercise physiologists such as Fred Kasch, the octogenarian in our opening section, have discovered that with training, women and men can gain both strength and endurance. One may wonder, then, why performance times for female athletes in Olympic running events still lag behind those of male athletes. There are probably many physiological and cultural reasons for this. A woman's pelvis is wider, causing her legs to strike the ground at a greater angle than a man's legs and thus stressing her knees. On average, women have lower levels of testosterone, a hormone that contributes to muscle buildup. And many women must overcome cultural stereotypes that keep them from being competitive athletes. An examination of the top performance records for running speed in Olympic events [FIGURE 33.18] reveals that women's record speeds are rapidly catching up to men's in the marathon and women's speed records are being bested more often and by larger amounts than men's. If the current rates of improvement continue, men's and women's records will lie in the same range by the year 2020. Would you expect this assumption to hold and this equity to appear? Why or why not?

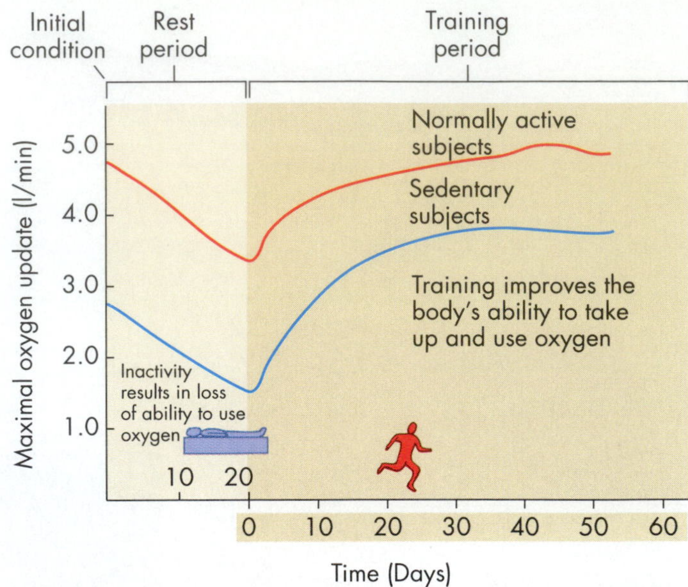

FIGURE 33.16

Training Increases the Body's Ability to Use Oxygen.

When experimental subjects stayed in bed for three weeks, the ability of their tissues to take up and use oxygen dipped dramatically. After a period of exercise, however, those who were active before the experiment (red line) regained (or slightly exceeded) their original uptake levels, while those who were initially sedentary (blue line) greatly exceeded their original levels. This experiment suggests that activity improves the body's oxygen utilization, even in the formerly bedridden or sedentary.

Anyone still needing a nudge to begin or continue a program of regular exercise can expect yet another benefit from it: Exercise stimulates the release of *endorphins*, a class of naturally occurring morphines, from the pituitary gland and other brain regions. These peptide hormones act a bit like morphine itself in that they reduce pain and enhance a feeling of well-being. The release of endorphins during strenuous exercise may explain, in part, why many people experience feelings of relaxation and contentment after a workout. Add to this the dramatic positive effects Fred Kasch and colleagues have found on the heart and on a person's chances of living a longer, more vigorous life, and it becomes more and more difficult to justify a sedentary life-style.

FIGURE 33.17

The Pronghorn: The Animal World's Champion Long Distance Runner.

Evolution has not dealt these antelopes a special anatomical trick allowing them to run up to 20 miles in half an hour. Instead, natural selection has simply resulted in the optimization of their oxygen utilization system, their muscular system, and their fuel delivery.

EXERCISE AND DISEASES OF INACTIVITY

Despite plentiful publicity about the health benefits of exercise, many people in industrial nations have adopted sedentary life-styles: riding to work in a car, sitting at a desk or workbench all day, and reading the paper or watching television at night. Our softer life-styles have led to an increase in *hypokinetic disease*, maladies caused by "little motion." Low back pain [see FIGURE 33.5], obesity, and heart disease can all result from sedentary life-styles, although family heredity, injuries, and infections can also contribute to these ills. In general, however, as Kasch's studies in our introduction showed, the incidence and severity of these problems can be lessened by activating muscles, skeleton, heart, lungs, nerves, and hormones with regular aerobic exercise.

Available evidence suggests that a physically inactive person is about twice as likely as an active one to die of heart disease. It is ironic that many people recognize the importance of diet and exercise to their pets' health, but fail to apply the same principles to their own human bodies. This has led Swedish exercise physiologist Per-Olof Åstrand to make the following suggestion: Walk your dog whether you have one or not.

➤ CONCEPT CHALLENGE

In the zero gravity of space flight, objects have no weight. What would you predict would happen to the muscles and bones of astronauts after a flight of several weeks? If you were a physician, what therapy would you recommend to prevent this from occurring?

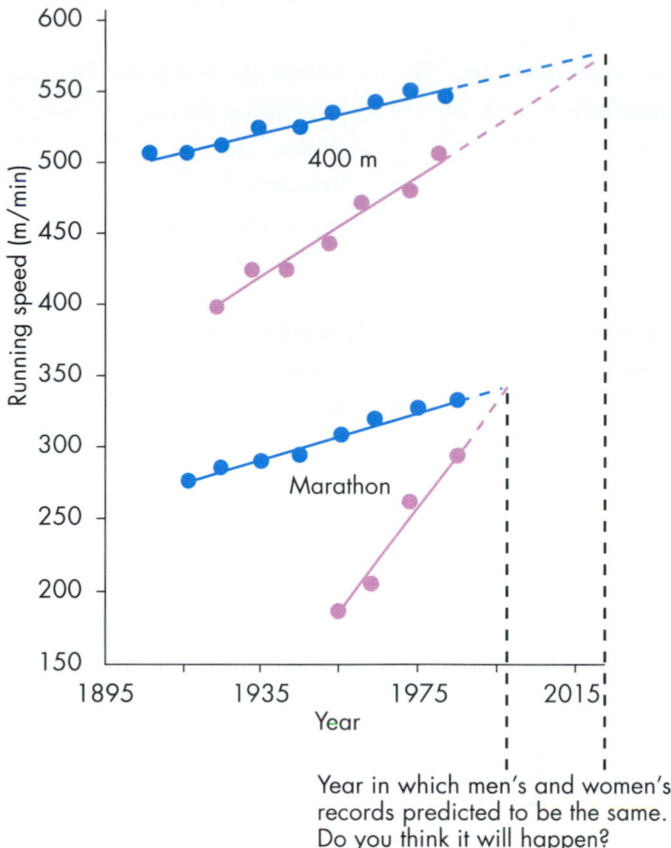

FIGURE 33.18

Will Women's Top Running Speeds Eventually Equal Men's?

A plot of world-record running speeds for men and women.

Connections

Self-generated movement is a hallmark of animal life. It relies on a rigid support, a skeleton, to which the shortening muscles transmit force. The muscle cells of animals have simply exploited and specialized an ancient molecular principle—the contraction and sliding of actin and myosin—to generate force. This adaptation allows animals to carry out their special way of life—moving from place to place in search of food and mates.

In recent years, a debate has arisen concerning children with minimal growth of the bones, muscles, and other organs. As the Apply and Decide section on page 752 explains, many people feel that these children should be treated with human growth hormone to augment their growth, while others feel short, slight musculoskeletal stature is not a disease and should not be treated. What do you think?

In the next section of this book, we will study plants, the energy source for virtually all animals. The structures that allow plants to capture the sun's energy, the principles that direct the plant's growth, and the ways that plants cope with environmental stress are all subjects of Part Five.

KEY TERMS

antagonistic muscle pair, 736
appendicular skeleton, 738
axial skeleton, 737
bone, 737
cardiac muscle, 745
cranium, 737
endoskeleton, 737
exoskeleton, 737
hydroskeleton, 736
joint, 740
ligament, 738
muscle fiber, 742
pectoral (shoulder) girdle, 738
pelvic (hip) girdle, 738
process, 738
sarcomere, 742
skeletal muscle, 742
skeleton, 736
skull, 737
smooth muscle, 745
tendon, 738
vertebra, 737
vertebral column, 737

HIGHLIGHTS IN REVIEW

1 Mobility is one important hallmark of the animal way of life. Movement occurs as muscles contract, applying force to a rigid support, a skeleton.

2 A skeleton made up of compressed fluid or stiff, hard material supports an animal's body. The bone and cartilage comprising a vertebrate's skeleton support other body parts, facilitate movement, protect soft organs, store minerals, and help form blood cells.

- **a]** There are two main types of skeletons: hydroskeletons and braced framework skeletons. A hydroskeleton in a worm or sea urchin functions by means of an incompressible fluid, like a water balloon. There are two types of braced framework skeletons: the exoskeleton, as found in the hard exterior covering of arthropods such as spiders and crabs, and the endoskeleton, the bones found in vertebrates.
- **b]** Bones typically make a joint, or articulation, with one or more other bones. Processes, or projections, serve as attachment sites for the straplike connectors, tendons and ligaments. Some bone cells secrete hard materials, while others reabsorb calcium and phosphate and help heal breaks.
- **c]** Bones act as levers that transmit force. They pivot about joints and accomplish work because of the way antagonistic muscle groups are attached to the bones.

3 Muscles shorten, or contract, bringing together the two body parts attached to the ends of the muscles. Contraction occurs when protein rods inside muscle cells slide past each other the way two trains move in opposite directions on parallel tracks.

- **a]** The skeletal muscles that propel the body are made up of giant contractile cells called muscle fibers. Within muscle fibers, repeating units called sarcomeres do the actual contracting. Actin filaments sliding past stationary myosin filaments bring about contraction of the muscle fiber.
- **b]** Three sets of membranes help stimulate muscle fibers to contract: the plasma membrane, which is stimulated by nerves in skeletal muscles; the transverse tubules, which carry electrical signals from the surface to deep inside the cell; and the sarcoplasmic reticulum, which releases calcium ions when the cell is stimulated.
- **c]** In addition to skeletal muscle, vertebrates have cardiac muscle and smooth muscle. Cardiac muscle cells are electrically connected by junctions, and this allows the whole organ to contract as a coordinated unit. Smooth muscle cells lack striations and, like cardiac muscle cells, communicate with each other electrically via junctions; this coordinates waves of contraction in the digestive or reproductive tract.
- **d]** ATP is the fuel for muscle contraction; it comes from one of three sources—an immediate energy system, a glycolytic system, and the oxidative system—depending on the intensity and duration of activity.
- **e]** Slow-twitch fibers are red, since they are rich in oxygen-storing myoglobin, and they obtain most of their ATP from the oxidative system, whereas fast-twitch fibers are white (owing to high actin and myosin content) and derive most of their ATP from glycolysis.
- **f]** During a fight-or-flight response, the hypothalamus directs the body toward hormonal readiness for instant activity, including the release of cortisol, the stress hormone.

4 Diseases of inactivity are rampant in industrialized countries. A lifelong program of exercise can maintain muscle strength, retard bone deterioration, decrease the likelihood of obesity and heart disease, and help maintain a long and vigorous life.

- **a]** Regular exercise causes the body to adjust so that it uses more oxygen and develops greater muscle mass and tone, a larger, stronger heart, and stronger bones. Exercise also releases endorphins and lessens the likelihood of low back pain, heart disease, and obesity.

UNDERSTANDING THE FACTS AND CONCEPTS

For Questions 1–5, match each of the descriptions with the most appropriate item or items from the following list. Each answer may be used once, more than once, or not at all.

a] exoskeleton
b] endoskeleton
c] axial skeleton
d] appendicular skeleton
e] hydroskeleton
f] More than one of the above

1 Muscles attach to a rigid tissue.

2 Includes the hip and shoulder girdles plus legs and arms.

3 The vertebral column, ribs, sternum, and skull.

4 Most commonly made of chitin and found in arthropods.

5 General skeletal category that describes a braced framework made of bone.

For Questions 6–10, match each of the descriptions with the one most appropriate item from the following list.

a] ligament
b] tendon
c] diaphysis
d] epiphysis
e] joint
f] synovial fluid

6 Protects the skeleton by absorbing the frictional forces within a joint.

7 Tough connective tissue band that attaches bone to muscle.

8 The ends of a long bone; location of spongy bone.

9 Tough connective tissue band that attaches bones to other bones.

10 The shaft of a long bone, generally composed of compact bone.

For Questions 11–15, match each of the descriptions with the most appropriate item or items from the following list.

a] muscle fiber
b] myofibril
c] sarcomere
d] myosin
e] actin
f] More than one of the above

11 The fundamental unit of contraction in a myofibril.

12 Muscle filaments made of protein.

13 Another name for a whole muscle cell.

14 An organized array of thin and thick filaments in a muscle cell.

15 Muscle filaments that have a movable head region.

For Questions 16–20, match each of the descriptions with the one most appropriate item from the following list.

a] sarcolemma
b] transverse tubules
c] sarcoplasmic reticulum
d] intercalated disc
e] myoglobin
f] glycolytic muscle fibers

16 The site of the synapse between a neuron and a muscle fiber.

17 An oxygen-carrying, pigmented molecule in muscle fibers.

18 The junctions that connect cardiac muscle cells and permit an electrical impulse to pass from cell to cell.

19 Site of storage and release of calcium ions in a muscle fiber.

20 The membranes that structurally and functionally bridge the distance between the plasma membrane of muscle fibers and the membranes that surround the sarcomeres deep within the muscle fiber.

INTEGRATE AND APPLY WHAT YOU HAVE LEARNED

1 In addition to support, what are the other functions of bone tissue?

2 What are antagonistic muscles? How are they arranged around a joint?

3 Contrast actin filaments with myosin filaments. Which assumes the more active role in muscle contraction?

4 Contrast the structure and location of skeletal, cardiac, and smooth muscle fibers.

5 Contrast the immediate energy system, the glycolytic energy system, and the oxidative energy system with respect to endurance, speed of contraction, reliance on stored ATP or carbohydrates, and reliance on mitochondrial action.

ANALYSIS

1 When you sit on a stool or chair and lift your lower leg until it it parallel with the floor, you are ______ the lower leg with muscles located on the ______ .

a] Flexing; front of the lower leg.
b] Extending; front of the thigh.
c] Extending; back of the lower leg.
d] Flexing; back of the thigh.

2 Assume that the greater the area of attachment between a skeletal element and a muscle, the stronger can be the pull of the muscle. This assumption would help to explain which observation about bones? More than one answer may be correct.

a] Many of the bones in the body are long and tubular.
b] Many long, tubular bones have expanded ends.
c] Many bones have processes that protrude from the main mass of the bone.
d] The parts of the skeleton that move the most are composed of flat, platelike bones.

3 The most precise and complete description of myosin would be:

a] A protein that primarily acts structurally.
b] A protein that primarily acts functionally.
c] A protein that can act as an enzyme by splitting ATP.
d] A protein that can contract in length and split ATP.
e] A protein that interacts with other proteins to form contractile units and also acts enzymatically.

4 Which of the following would not have to be surgically reconnected during a heart transplant? Explain.

a] The major arteries carrying blood from the heart.
b] The major veins carrying blood toward the heart.
c] The large nerves that connect to the cardiac muscle fibers of the ventricles.
d] All of the above.

5 A man suffers general paralysis of his skeletal muscles. He is rushed to the hospital and placed on a mechanical respirator to assist his breathing. Although his skeletal muscles are flaccid, muscle tissue in the heart and other internal organs appear to be working normally. Upon examination, doctors suggest several explanations for his condition and initiate tests to determine which is correct. Considering all that you have learned in this unit, which of their suggestions is the most likely explanation?

a] Inadequate ATP production.
b] Calcium deficiency.
c] Poisoning of enzymes that degrade acetylcholine.
d] Inhibition of acetylcholine release.
e] Mutation in the myosin gene.

PART FOUR: APPLY AND DECIDE

Shortness: To Treat or Not to Treat

Childhood was bittersweet for Martin* because he started life extra small and stayed that way. Martin was a full-term, healthy baby in 1969, but he weighed just 1.8 kg (4 lb) at birth, and his growth continued to lag behind other boys his age. As a freshman entering high school, for example, he was only 132 cm (4 ft 4 in) tall—the height of a typical fourth-grader. This slow development had irksome physical consequences for Martin, such as having to wear his dental braces for five years instead of the usual one or two. But the psychological impact was far more than bothersome.

"Every single day on my way to and from public grade school," Martin recalls, "I got beat up by bigger kids." Things cooled down for him in junior high school, but then, entering high school at fourth-grade dimensions, he once again became a target. "The seniors were just huge compared to me, and I can tell you, it was fairly embarrassing to get dumped into a garbage can every other day."

Martin's story is not unusual—from the painful taunting to receiving hormone treatments from a physician to stimulate growth. Observers estimate that 10,000 very short children have been helped through the administration of human growth hormone or related compounds, and that another 10,000 are candidates for the injections. Martin's medical outcome, however—a growth of 14 inches by senior year and a normal adult stature—may become a rare happy ending if some critics of shortness therapy get their way.

The physician who handled Martin's case is a pediatric endocrinologist—a specialist on children's hormones. When he met the boy as a badly bullied 14-year-old, the doctor tested Martin's blood for levels of two substances we encountered in CHAPTER 30: testosterone, the male sex hormone, and human growth hormone (hGH), a natural regulator of size and development. You may recall from that chapter that a person can fail to grow because he or she produces too little hGH, and that stature can also be limited if the person lacks hGH receptors, as is the case with the pygmies of West Africa [see FIGURE 30.1]. A third hormonal deficiency is also possible: If a child, whether male or female, produces too little testosterone, the normal interaction of that hormone with hGH cannot take place and rapid cell division is not stimulated at the growth plates at the ends of the long bones in arms and legs [see FIGURE 33.7]. Martin's doctor found that his patient produced too little testosterone, and so he administered doses of human chorionic gonadotropin (hCG; another hormone we discussed in CHAPTER 30) to stimulate the production and release of testosterone. This, in turn, interacted with the boy's sufficient natural levels of hGH, and throughout high school, he grew about an inch (2.54 cm) every three or four months.

In a way, Martin was lucky to have a testosterone deficiency rather than a growth hormone deficiency, because in 1983, hCG was available (from human placentas) in large enough quantities to treat every child with a similar problem. Had his deficiency been one of human growth hormone itself, he might not have received treatment because at that time, the drug was still being extracted from human cadavers. It was extremely expensive, as well as considered somewhat risky, due to rare but potential contamination with dangerous viruses.

By the mid-1980s, however, biotechnologists at two pharmaceutical companies, Eli Lily and Genentech, had succeeded in cloning the gene for hGH in bacteria [see FIGURE 12.4], and began generating large quantities of the purified hormone. Doctors began prescribing genetically engineered hGH for very short children, most of whom were in the bottom percentile for height, meaning that their projected adult heights would be under 160 cm (5 ft 3 in) if male, and under 145 cm (4 ft 9 in) if female.

While thousands of children have benefited from hGH treatment, controversy quickly arose. There was general agreement that the hormone is appropriate for a child in the lowest height category who does not produce enough of his or her own hGH. But what about children who are not terribly short and yet still produce too little hGH? Should they receive the drug so they can grow to their maximum potential stature? What about children with sufficient hGH levels but who are still extremely short for unknown reasons? Should they get the drug, just in case it might stimulate growth for them? What about an extremely short, hGH-deficient child whose life is happy and free from harassment. Should the threat of future bullying or later economic disadvantage in a world of tall competitors be reason enough to administer growth hormone in childhood?

* A pseudonym to protect his privacy.

To put these questions in another way, who should get hGH? Does it really work and if so, for whom? Does its use dependably alleviate psychological problems stemming from small size during childhood? Can doctors justify giving a drug to a child who is short but otherwise quite healthy? And are there potential side effects or problems (reminiscent of the thalidomide tragedy; see CHAPTER 14) that could arise later from weekly hGH injections?

In 1985, the U.S. Food and Drug Administration approved the use of hGH for very short children with proven hGH deficiency. Researchers at the National Institutes of Health, however, felt that further studies were needed given the number of unanswered questions about efficacy and risk, as well as the demand for treatment by parents whose children are short for other or unknown reasons. In 1988, NIH initiated a 10-year study that involved one group of female volunteers with Turner syndrome [see CHAPTER 12], and one group of very short child volunteers with normal hGH levels and receptors.

Almost immediately, the main opponent of genetic engineering in the United States, attorney and activist Jeremy Rifkin, as well as a group of physicians, threatened to sue NIH to halt the trials. They objected to the Turner syndrome trial, primarily because half the subjects are receiving placebos, will therefore not be helped to grow, and might later feel exploited and abused by the medical establishment. Rifkin and the Physicians Committee for Responsible Medicine objected to the trial of hGH-sufficient short children because, they say, it is unethical to give drugs to a healthy child, especially if there could be future unknown risks. NIH has countered that without data from large-scale, long-term scientifically designed studies, they can never objectively answer the questions of efficacy and risk.

The final, and to some the biggest, objection to the NIH tests and the use of hGH in general is that society is wrong for discriminating against short children and adults, and it is societal attitudes that should be changed, not the stature of its shortest citizens. This objection essentially nullifies the psychological grounds for seeking treatment, and elicits a sharp response from former growth patients like Martin Hemphill. "That kind of thinking is nothing but sticking your head in the sand and ignoring progress," he says. "If they are well-informed, I think people should be allowed to decide for themselves whether to seek any kind of physical modification." Now, as a 5-foot 7-inch adult, Martin has no regrets at all about the treatments, and feels that the 5-foot 2-inch man he was projected to become might have had permanent psychological problems.

Do you think NIH should continue their 10-year study (which is still being debated in courts of law)? Do you think very short children without proven hGH deficiency should receive hGH? What about children who *do* have hGH deficiency? Should they be allowed to "modify themselves" even though the children of some uninsured people will never have access to the treatments, and some parents' definition of "too short" might actually be "too short for professional basketball"? These are the kinds of decisions we must face in a technological society, and which our study of animal physiology in this section can prepare you to make. ❑

PART FIVE

How Plants Survive

CHAPTER 34

Plant Architecture and Function

FIGURE 34.1

Researchers Climb Through a Tunnel Created by the Roots of a Strangler Fig on Barro Colorado Island off the Coast of Panama.

The fig had originally sprouted high on the branch of a tree and sent roots spiraling down the tree trunk to the ground. The host tree, no longer able to increase in diameter, died and rotted away, leaving the tunnel.

SLOW-MOVING HORROR

Most of us, being mammals ourselves, are understandably animal-centric. But the plant world has some interactions every bit as dramatic as a predator hunting down and devouring prey.

They just happen in slow motion.

To heighten public appreciation for the life-and-death interplay taking place among plants in the tropical rain forest, ecologist Mark Moffett recently created an accelerated scenario involving the boa constrictor of the plant world: the strangler fig. Moffett fantasized that human observers might someday record and witness the activities of jungle plants at higher and higher speed, with two hours compressed into one minute of viewing time, then a week in a minute, then two years in a minute. If the hypothetical camera focused on a very real strangler fig coiled about a tropical tree whose tall crown was spreading luxuriantly in the rain forest canopy, the effect would be chilling.

One would see, in rapid succession, a fig seed, encased in bird or bat guano, drop to the upper branches of a canopy tree, stick there, burst open, and a new young fig plant begin to grow quickly while attached to the host tree, dozens of meters above the forest floor. Soon, numerous roots would emerge from the young plant, coil down the tree trunk like tangled snakes, then thicken and begin to merge into a living meshwork like a giant, cylindrical lattice surrounding its host. The jungle tree would also enlarge and expand, but locked within the prison of the fig vines, its bark and trunk tissue would bulge out through every small opening like a fat leg in a tight fishnet stocking. Soon (in our sped-up, time-lapse world), unable to grow taller or wider, the tree would die and rot. The strangler fig, absorbing nutrients, would get even thicker and more tangled and eventually stand in place of its "victim" like a tubular wooden jungle gym around a ghost tree [FIGURE 34.1].

Strangler figs serve as excellent introductions both to this section of the book on how plants live and to this first chapter on the structure of the main plant body, the roots, stems, leaves, and flowers. The other three chapters in the section will explain in detail how plants function: how plants obtain and transport water and nutrients [CHAPTER 35]; how plant embryos develop from the fertilized egg and form the familiar plant body [CHAPTER 36]; and finally, how hormones regulate plant growth in response to environmental cues [CHAPTER 37]. Together, these chapters pick up where CHAPTER 20 left off. They will show in detail how the multicellular, photosynthetic autotrophs we call plants differ from the other kingdoms; how flowering plants fit into the larger panoply of land plants, including those without transport vessels and seeds; and how, through photosynthetic production of carbohydrates, plants directly or indirectly support most members of the other kingdoms of life.

MESSAGES

1 The body of complex plants consists of three tissue systems: an outer protective layer (the dermal tissue system), a system of tubes that carry water and nutrients (the vascular tissue system), and a system of tissue that fills the space around the vascular system, storing materials and providing support (the ground tissue system).

2 The three tissue systems arise as plants grow. Plants grow longer as new cells are added at the tips of branches and roots in regions called apical meristems. The newly added cells may continue to divide for a time, then elongate and differentiate into specialized cells of the root, stem, leaves, or flowers in a process called primary growth.

3 Some plants exhibit secondary growth, an increase in diameter resulting from the addition of new cells near the surface in regions called lateral meristems. The bulk of the newly added cells becomes wood.

Our focus in this chapter reveals the intimate association of biological form, function, and environment. Most plants, of course, are stationary organisms that cannot move about in pursuit of water, energy, a carbon source, or mineral nutrients. But a plant's anatomy, way of life, and mode of reproduction are all beautifully adapted to meeting its needs while it remains rooted to one spot.

As we will see here, a plant's form and function strike a compromise between conflicting needs. For example, plants often have large, flat leaves that can collect maximum sunlight, and the leaves usually have tiny openings that allow gas exchange. Both the large surface area and the openings conflict with the need for water conservation. Thus internal regulatory systems come into play, helping to balance these conflicting factors. We will also see that plants have a life history pattern called open growth, which is based on perpetual embryonic centers that produce new organs and larger body size throughout life. This mode of development helps overcome the limitations imposed by rigid cell walls and hardened mature tissues.

In our discussion of plant architecture and function, we will consider the plant's main tissue types and see how the rigidity of plant cells influences the organism's overall growth patterns. Next, we'll discuss the main body plan of land plants—an aboveground shoot and a belowground root—and the three tissue systems comprising these major body parts. Then we'll consider the anatomy and functioning of roots, stems, and leaves.

Along the way, we'll see not only exactly how a strangler fig kills its host, but also how the roots of one tree can encircle an entire shopping mall; how a philodendron resembles a snake; and how many millions of hairs can grow on the root of a single, small ryegrass plant. ❑

Vascular Plants

The strangler figs are just a handful of species among nearly half a million vascular plants [FIGURE 34.2]. As a group, vascular plants are distinguished by their *vascular tissue*, or internal transport tubes. This mammoth group dominates the land and includes ferns, conifers, and flowering plants. A subgroup of vascular plants, which includes gymnosperms and angiosperms (now classified as the *Anthophyta*), produces seeds. Seed plants have two additional adaptations for life on land: sexual reproduction with internal fertilization, and a tough seed coat that protects the developing embryo. These last two characteristics are similar to the innovations of internal fertilization and eggshells we saw in the reptiles. Of the seed plants, only the angiosperms produce flowers [FIGURE 34.2]. Botanists divide flowering plants into two classes, the *monocots* (or Monocotyledons) and the *dicots* (or Dicotyledons). Monocots include grasses such as corn, bamboo, and Bermuda grass, as well as lilies and orchids. Dicots include fir trees, peach trees, daisies, roses, and strangler figs. FIGURE 34.3 summarizes some of the differences between monocots and dicots. Now let's investigate the general architecture of the plant body.

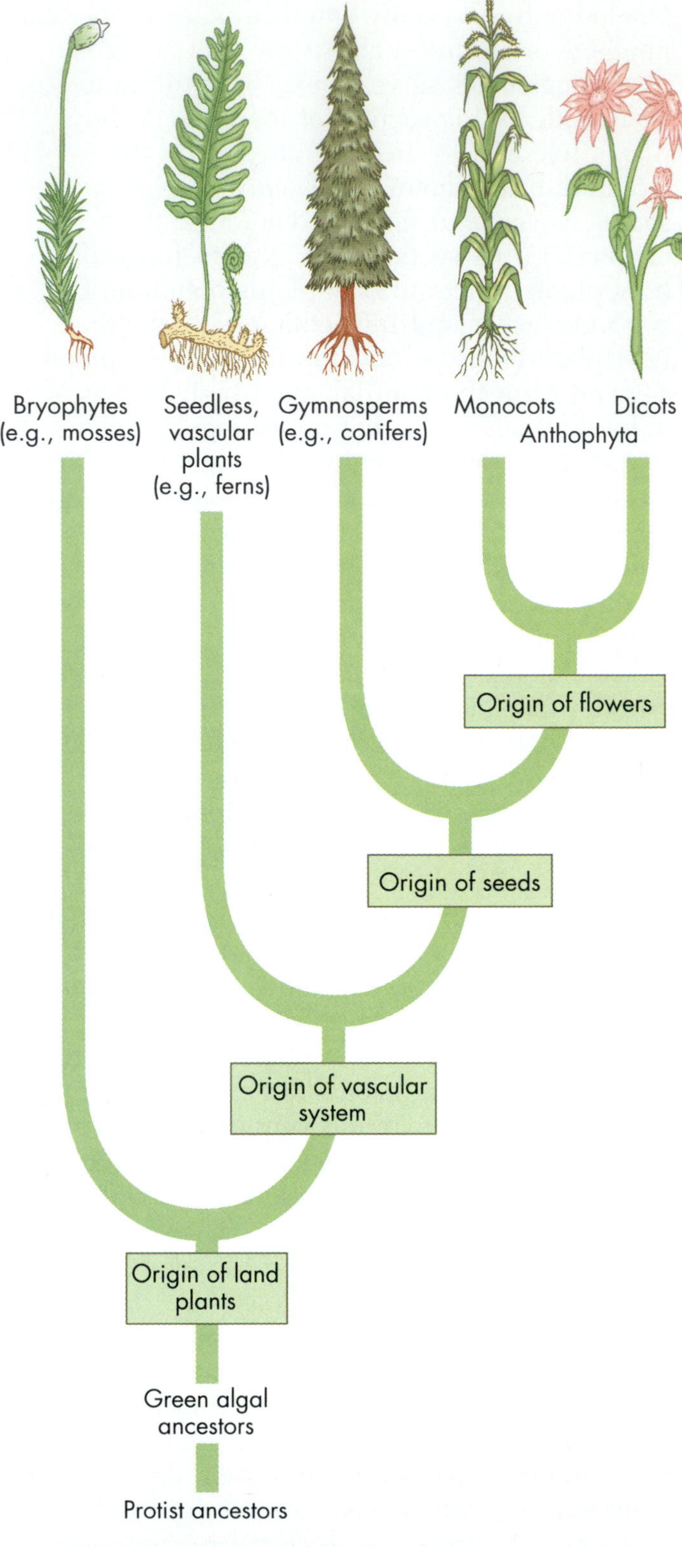

FIGURE 34.2
Land Plant Evolution.
This chapter focuses on the structure of seed plants, especially flowering plants, the monocots and dicots.

The Plant Body

The ancestors of land plants, the green algae, inhabited well-lit oceans. The salt water supplied the plant with water, dissolved mineral nutrients, dissolved carbon dioxide (from which algae generated organic molecules via photosynthesis), dissolved oxygen, and buoyant support for the plant body. A land plant has the same physical requirements for gases, water, light, nutrients, and support. It obtains these resources, however, from the soil and the air, not seawater. A plant's roots are generally buried in the soil from which they take in water and mineral nutrients. Soil provides no direct access to light, however, and thus root cells have no direct way of photosynthetically manufacturing their own food. In contrast, a plant's stem and leaves invade the air, and take in carbon dioxide and light, allowing photosynthesis. Air, however, provides no mineral nutrients and very little water.

THE PLANT AXIS: ROOT AND SHOOT

The realities of the physical environment dictate the form and function of a typical vascular land plant—its above-

FIGURE 34.3
Some Differences Between Monocots and Dicots.

ground shoot and its belowground root. Together, the shoot and the root make up the plant axis.

The Shoot The **shoot** includes the stem, branches, leaves, flowers, and fruits [FIGURE 34.4A]. Lifting the rest of the shoot is the *stem*, a stiff bundle of internal transport tubes (xylem and phloem) plus surrounding tissues, all enclosed within an external waterproof coating that minimizes water loss. In trees and shrubs, the main stem matures into a large, bark-covered trunk with many side branches supporting leaves and flowers.

The Root **Roots** are branching organs that grow downward into the soil. Roots help support the plant physically by spreading out through the soil and anchoring the plant in it, and nutritionally by absorbing and transporting water and mineral nutrients that the plant cannot take in from the air. Pear trees, dandelions, and carrots each have a thick, trunklike **taproot** that develops directly from the embryonic root (radicle) and grows straight down into the soil. Fine lateral roots branch off the sides of the main taproot [FIGURE 34.5A]. Taproots store water, as well as food in the form of starch. In contrast, in grasses, including corn, and a few other kinds of plants, the embryonic root dies back early in the organism's life. In these plants, the roots that will support the adult plant grow downward and outward from the plant's stem, branching repeatedly and forming a mass of narrow **fibrous roots** [FIGURE 34.5B]. Fibrous roots store less starch than taproots, but efficiently anchor the plant and absorb water and nutrients. A final type of root emerges from aboveground stem tissue and grows through the air before reaching the soil. Called **adventitious roots**, these aerial roots are the metaphorical "fingers" of strangler figs that squeeze life from a host. Banyan trees (*Ficus benghalensis*) are one species of strangler fig native to the East Indies, and they can grow to enormous proportions. If you have visited Honolulu, Hawaii, you may have shopped in the International Market. This entire shopping mall is contained within the adventitious roots of a single banyan tree! Other adventitious roots include the thick prop roots of a corn plant and the fine roots that allow ivy plants to cling to walls and fences.

TISSUE SYSTEMS AND TISSUE TYPES

Complex flowering plants, such as corn plants, maple trees, and saguaro cacti, have three tissue systems with important physiological roles in the body, and each system contains a few important tissue types [TABLE 34.1]. Each tissue, in turn, is composed of different cell types with distinctive characteristics appropriate to their function [TABLE 34.2]. As FIGURE 34.4B shows, the three main

FIGURE 34.4

The Plant Axis: Root and Shoot.

[A] The above- and below-ground portions of the plant have different functions and hence different anatomies, but [B] they contain the same three tissue systems.

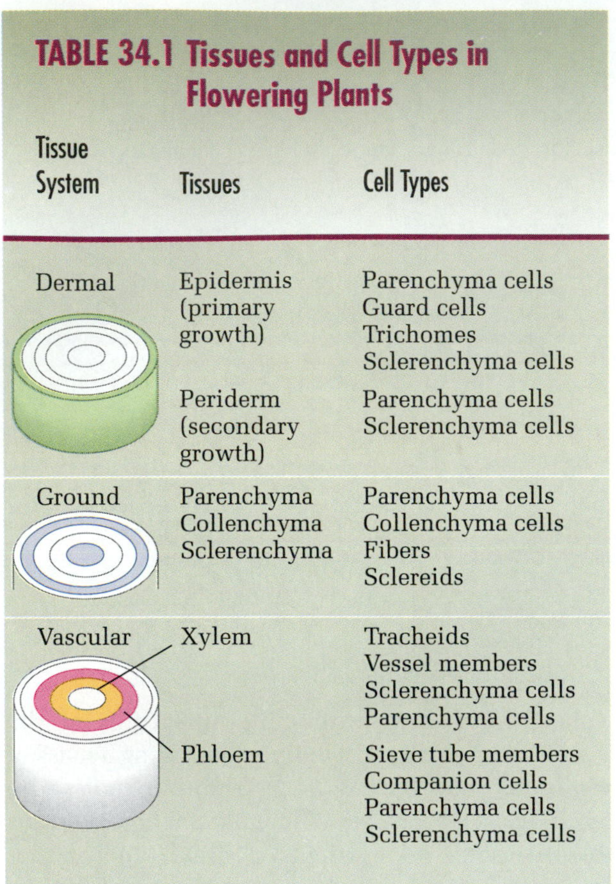

TABLE 34.1 Tissues and Cell Types in Flowering Plants

Tissue System	Tissues	Cell Types
Dermal	Epidermis (primary growth)	Parenchyma cells Guard cells Trichomes Sclerenchyma cells
	Periderm (secondary growth)	Parenchyma cells Sclerenchyma cells
Ground	Parenchyma Collenchyma Sclerenchyma	Parenchyma cells Collenchyma cells Fibers Sclereids
Vascular	Xylem	Tracheids Vessel members Sclerenchyma cells Parenchyma cells
	Phloem	Sieve tube members Companion cells Parenchyma cells Sclerenchyma cells

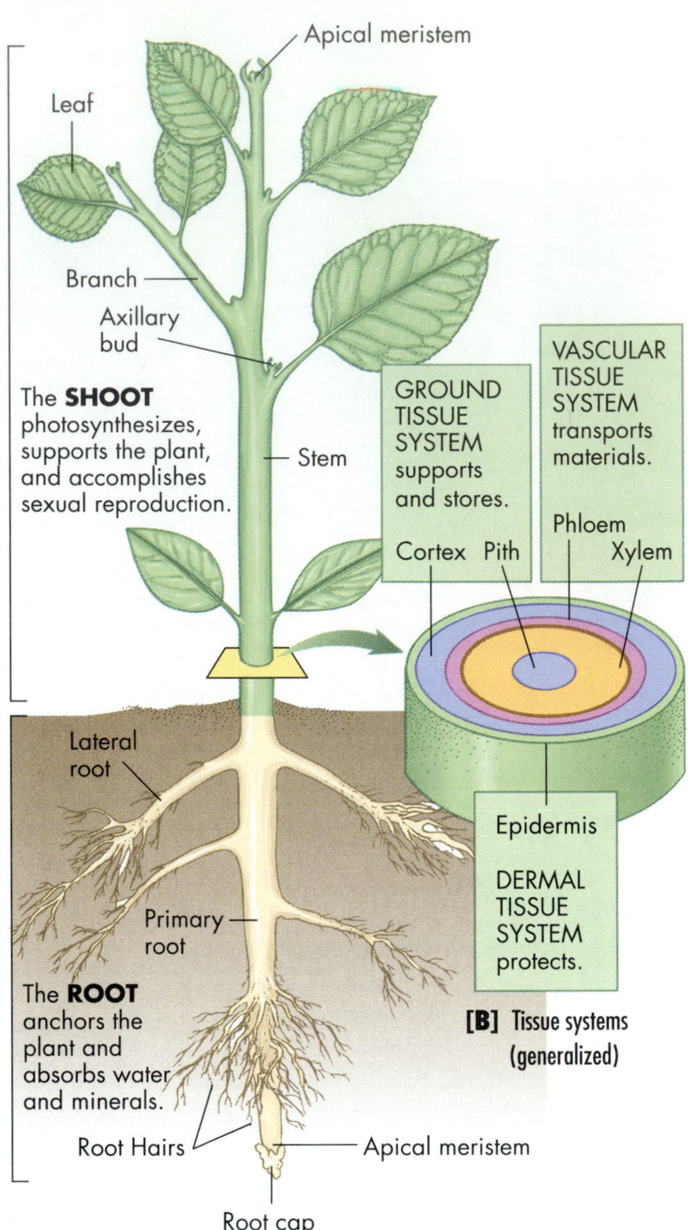

[B] Tissue systems (generalized)

[A] The plant axis

FIGURE 34.5

Patterns of Root Growth.

[A] A dandelion has a central taproot with fine lateral roots. [B] Grass has numerous fibrous roots that anchor the plant firmly. [C] A single banyan tree, a type of strangler fig, has many adventitious roots that help support the plant and absorb water and minerals.

[A]

[B]

[C]

TABLE 34.2 Characteristics of Cell Types in Flowering Plants

Tissues and Cell Types	Structure	Function
Epidermis		
Guard cells 	Pairs of large cells in epidermis surrounding a pore	Regulate gas and water exchange in leaves
Trichomes 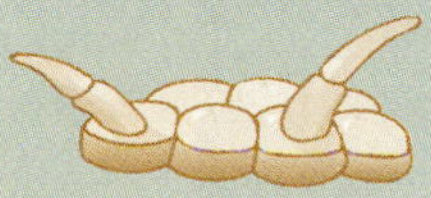	Hairs or scales jutting from epidermis	Protection; retard water loss
Parenchyma		
Parenchyma cells 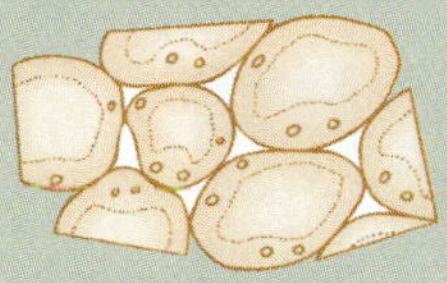	Many-sided, often thin primary wall; alive at maturity; found throughout plant	Photosynthesis, storage, local conduction of materials, wound healing
Collenchyma		
Collenchyma cells 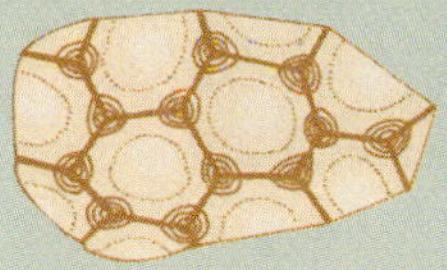	Rectangular; unevenly thickened primary cell wall at corners; alive at maturity	Support of young stems and leaf ribs
Sclerenchyma		
Fibers 	Very long and narrow; primary and very thick secondary cell wall; often dead at maturity	Support in stem cortex and vascular system
Sclereids 	Often long, but shorter than fibers; primary and thick secondary cell walls; living or dead at maturity	Support and protection throughout plant
Xylem		
Tracheids 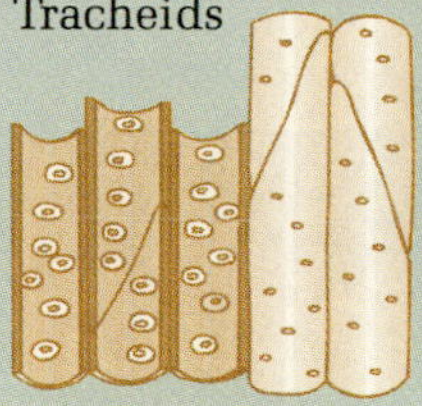	Elongated tapering cells; primary and secondary cell walls with pits; dead at maturity	Conduct water in all vascular plants
Vessel members 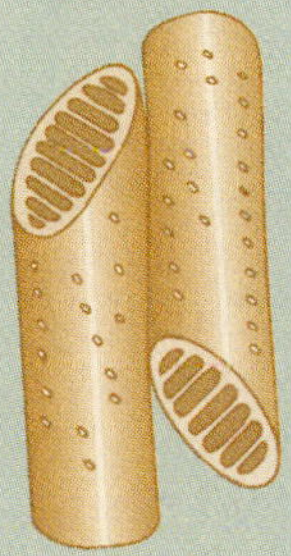	Long and narrow, but generally shorter than tracheids; primary and secondary cell walls with both pits and perforations; stacked end to end in a vessel; dead at maturity	Main water-conducting cell in xylem of flowering plants
Phloem		
Sieve tube member 	Elongated cell; primary cell wall with sieve plates; alive at maturity, but lacks functional nucleus; stacked end to end in sieve tube	Main food-conducting cell in phloem of flowering plants
Companion cells	Somewhat elongated; primary cell wall; alive at maturity and closely connected to sieve tube member	Help regulate activities of sieve tube member

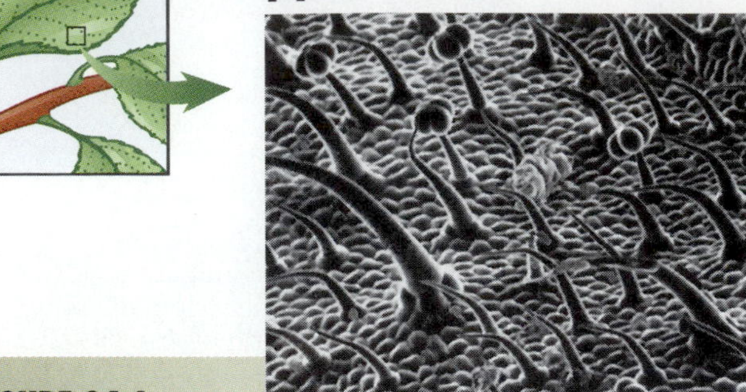

FIGURE 34.6

The Dermal System.

These scanning electron micrographs show **[A]** the bumpy epidermis covering a flower petal and **[B]** the waxy cuticle and epidermis of a leaf, with its hairlike, protective trichomes. Trichomes create a surface "hairiness" that helps reflect light and hence retards water loss and also discourages certain predators from eating stems and leaves.

tissue systems in a mature vascular plant are the **dermal tissue system**, which, like skin, protects the plant from water loss and injury to internal tissues; the **ground tissue system**, which provides support and stores starch; and the **vascular tissue system**, which conducts fluids and helps strengthen the roots, stems, and leaves. These tissue systems are continuous throughout the plant.

Dermal Tissue System Corn and other vascular plants have a dermal system that covers the plant and is analogous to an animal's skin [FIGURE 34.6]. In a young seedling, the dermal tissue consists of just an *epidermis*, or protective outer covering. Epidermal cells in stems and leaves secrete a waxy waterproof coating, the *cuticle*. Some epidermal cells form little hairs, or trichomes [see FIGURE 34.6B]. In trees and woody shrubs, however, the *periderm*, the outer areas of the bark, replaces the epidermis.

Ground Tissue System A plant's root and shoot contain a ground, or fundamental, tissue that packs around the tubes of the vascular system. The ground tissue system makes up the bulk of most plant organs; it stores the starchy products of photosynthesis, helps keep the plant from collapsing into a formless heap, and has other, more specialized roles. There are three types of ground tissue, which vary in strength and flexibility: parenchyma, collenchyma, and sclerenchyma.

In both root and shoot, the majority of ground tissue is **parenchyma**, made up of loosely packed, thin-walled, rounded parenchyma cells, such as those that store starch in a potato [FIGURE 34.7A]. Most parenchyma cells have only a thin primary cell wall laid down as the cell

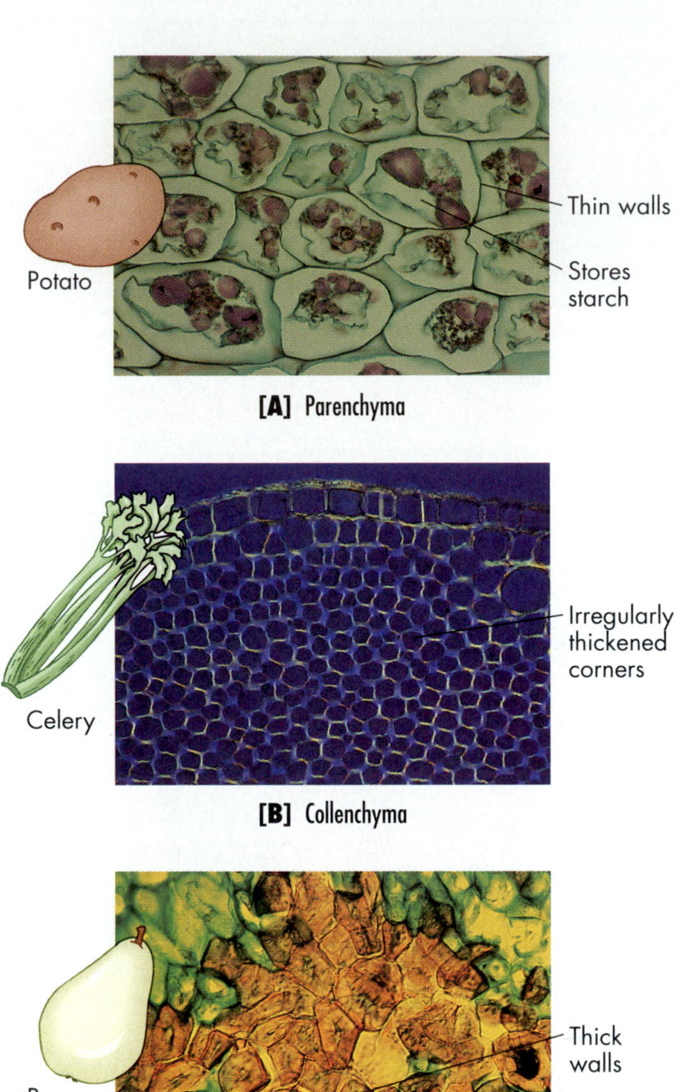

FIGURE 34.7

Ground Tissue System: Material Storage and Strengthening of Plant Parts.

[A] Most ground tissue cells are parenchyma cells, such as these potato cells with starch grains stained lavender. **[B]** The stringy fibers just inside the epidermis of a celery stalk are made up of collenchyma cells, with their irregularly thickened corners. **[C]** The gritty structures in a pear's flesh are a type of sclerenchyma called *sclereids*, or *stone cells*; these cells have thick walls but lack cytoplasm when mature.

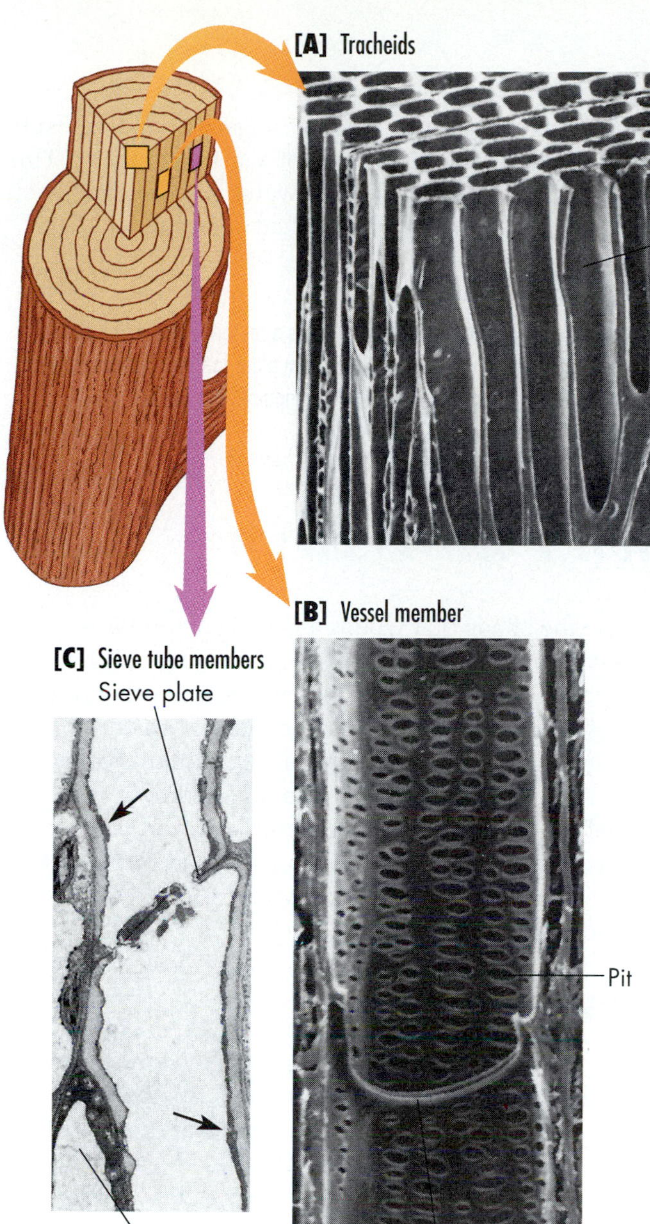

FIGURE 34.8

Vascular Tissue System: Material Transport and Support of Plant Structures.

[A] Pitted, hollow xylem cells called *tracheids* transport water in all types of vascular plants. [B] Vessel members, stacked end to end, transport water only in flowering plants. [C] Phloem transports organic substances through sieve tube members which lack a nucleus, but are adjoined by nucleated companion cells. Sieve tube members are lined by a special protein (arrows) that may help plug leaks after a wound.

grows, and they are often unspecialized, able to give rise to new cells and cell types in a wounded adult plant. Some parenchyma cells, however, are specialized. In the leaves, for example, parenchyma cells photosynthesize but store little starch, while in the roots, they often store large quantities of starch.

Collenchyma tissue is tougher than parenchyma because individual collenchyma cells often have a cylindrical shape and thick walls deposited as the cells expand, and the cells connect end to end in long, stringy fibers such as those just beneath the epidermis of a celery stalk [FIGURE 34.7B]. The word *collenchyma* is based on the Greek word for "glue," and in fact, these shiny cells help hold the plant body together. In cross section, collenchyma cells have uneven, thickened corners [see TABLE 34.2].

Sclerenchyma tissue (from the Greek word for "hard") is a third tissue type in the ground tissue system. Because of its hardness, sclerenchyma often surrounds and reinforces the tubes of the vascular system. Sclerenchyma cells die at maturity and leave behind a thick cell wall that is deposited after the cell stops growing and is hardened with a tough complex polymer called *lignin*. Sclerenchyma tissue includes stone cells, or *sclereids*, which are hard, crystalline cells [FIGURE 34.7C]. Sclerenchyma cells give pears their grittiness and walnut shells their hardness. The *fibers* that make up the strong, slender threads of hemp and flax and in turn ropes and linen fabrics are a second cell type in sclerenchyma tissue.

These cell types can also occur in other tissue systems, as we will see.

Vascular Tissue System Within a vascular plant, water and materials travel in two kinds of tubular tissues, xylem and phloem, each of which forms a continuous pipeline extending from root tips to leaves. In general, **xylem** transports water and minerals absorbed from the soil up through the roots, stems, and leaves. Wood consists of xylem vessels. **Phloem** transports dissolved sugars and proteins from "source to sink"—that is, from cells that produce or store sugars to cells that use sugars rapidly.

The xylem of most flowering plants is composed of several kinds of cells, but principally **tracheids** and **vessel members** [FIGURE 34.8A and B]. Both tracheids and vessel members transport water only after they have died: Each cell type becomes hollow at maturity, when the cell contents disintegrate and leave behind empty cell walls. These hollow cylindrical cells, stacked end to end, form efficient transport pipelines. Tracheids are long, narrow cells that overlap at their slender ends and have in their thick secondary cell walls many *pits*, gaps that allow water to pass from one tracheid cell to an adjoining one [FIGURE 34.8A]. Nearly all vascular plants have tracheids, but vessel members tend to occur mainly in flowering plants. Like tracheids, vessel members have pits in their walls, and they tend to be stacked directly end to end like the clay pipe sections of a water main. But unlike tracheids, vessel members have larger holes in their end walls called *perforation plates*, which allow water to flow through the stacked vessel members unimpeded [FIGURE 34.8B and TABLE 34.2].

Phloem is also composed of cells arranged end to end, but phloem and xylem differ in three important respects: First, mature phloem cells contain living cytoplasm and are thin-walled, while xylem cells are dead and thick-walled. Second, phloem transports sugars and amino acids dissolved in small quantities of water, while xylem transports minerals and large quantities of water. Third, while xylem transports materials in one direction, from roots to leaves, phloem can move sugars from photosynthesizing leaves to the roots in summer or from roots to developing leaves in the spring.

The conducting cells of phloem tissue are called **sieve tube members** because their end walls, called **sieve plates**, usually contain large pores. Like a sieve, these perforated plates allow fluids and solutes to pass in and out. Sieve tube members remain alive and possess a thin layer of cytoplasm that clings to the cell wall, but they lack a functioning nucleus [FIGURE 34.8C]. Lying next to each living sieve tube member is a small **companion cell** with a nucleus that directs the activities of the nearby sieve tube member. CHAPTER 35 explains how phloem and xylem transport substances.

Plants have many other kinds of vessels with specialized roles in reproduction, storage, and defense. To give just one example, all the shoot tissues of a strangler fig are permeated near the plant's surface with *lactifers*, narrow vessels that contain a rubbery, white latex compound which helps discourage predators. Among many noxious compounds in the milky fluid is a protein-digesting enzyme called *ficin*. Many people who keep strangler figs as houseplants develop an intense allergy to their presence, and some observers have speculated that the ficin enzyme in these vessels is what triggers the allergies.

Thanks to the vascular system, the root and shoot function as a partnership; the root provides water and nutrients, which travel to the shoot via the xylem vessels, and the photosynthesizing shoot provides energy and organic molecules in the form of sugars and other substances, which travel to the root via the phloem. This body plan, plus the rigid cell walls surrounding each plant cell, constrain the way a plant can grow.

➤ CONCEPT CHALLENGE

Imagine that a plant geneticist discovers a mutant plant with weak cell walls. How would you expect the mutant plant to differ from a normal plant? Which tissue systems would be impaired the most?

How Plants Grow

Imagine the chaos that would ensue if an animal—a dog, let's say—were to continue growing throughout life, producing new eyes, ears, legs, feet, livers, and other organs. The result might be a bizarre creature with three tails, seven legs, and four eyes. Conversely, think what would happen if a corn plant grew like an animal, keeping its original root, stem, and single initial leaf, and each organ simply grew bigger and bigger as the plant matured. Clearly, neither growth pattern would suit the other type of organism.

OPEN GROWTH

Plants have a growth pattern called **open growth**. Throughout life, a plant adds new organs, such as branches, leaves, and roots, enlarging from the tips of the root and shoot. Because their cell walls are rigid, plants, unlike most animals, cannot move toward favorable conditions or away from harmful conditions. Nevertheless, a plant can often overcome this rigidity and immobility by means of growth: plants can grow toward light, water, and mineral nutrients and away from harmful situations. One bizarre example of open growth involves the common houseplant philodendron in its native habitat, the tropical forest floor. Over a period of days or weeks, the living stem of a philodendron seedling lying on the ground appears to move through the forest like a green snake. In fact, no leaf or other part of the plant actually moves forward. Instead, the stem grows in one direction along the ground, forming leaves and roots in each segment as its habit of open growth dictates. The trailing section then dies. The result is the appearance, over time, of snakelike movement.

Continual growth of a plant is based on **meristems**, tissue that remains perpetually embryonic and that gives rise to new cells and cell types throughout the plant's life. Meristems give rise not only to vegetative parts of the plant, such as new leaves and stem, but also to the cells that produce reproductive parts—the flowers and the eggs and sperm they contain. Thus, unlike animals, plants do not establish a special line of reproductive cells during early embryonic development [review FIGURE 13.12].

MERISTEMS

Like a child, a tree grows in two dimensions: it gets both taller and thicker. These two types of growth are brought about by two types of meristems, the apical meristems and the lateral meristems.

Apical Meristems Plants increase in length due to **apical meristems**, perpetual growth zones at the tips (*apices*; singular, *apex*) of roots and stems [FIGURE 34.9A]. Apical meristems allow shoots to grow upward toward the light and roots to push ever deeper into the soil toward water. Growth arising from apical meristems is called **primary growth**. The xylem and phloem produced by apical meristems are called primary xylem and primary phloem.

Lateral Meristems **Lateral meristems**, the second type of meristem, are cylinders of dividing cells in the stems and roots of some plants that cause these parts to become thicker [FIGURE 34.9B]. This increase in diameter (girth) as a result of cell divisions in the lateral meristems is called **secondary growth**. This contrasts with primary growth, an increase in length due to cell divisions at the apical meristems. As you might expect, lateral meristems are found in plants that become quite thick and strong, like most trees and shrubs, but they are lacking in corn and other monocots, as well as in many dicots such as dandelions that survive for just a single growing season.

Many botanists suspect that a strangler fig kills its host tree by cutting off the host's secondary growth. As the trunk of the host tree expands via secondary growth, the tightly woven meshwork formed by the fig's woody roots constrains that expansion. Host tree tissue bulges through each opening in the fig's root meshwork. Because of this constriction, conducting vessels cannot form properly, and secondary growth is prevented. In turn, the normal flow of nutrients and water is impeded due to the living woody tourniquet, and the host tree dies.

The lateral meristems that give rise to secondary growth are of two main types. **Vascular cambium** is a type of lateral meristem that produces secondary phloem and secondary xylem, which we know as wood. Another lateral meristem, **cork cambium**, generates *cork*, the waterproof outer part of the bark of trees and shrubs.

By observing trees growing in your neighborhood, you can more easily understand the distinction between primary growth from the apical meristems and secondary growth from the lateral meristems. Locate a tree that someone drove nails into years ago to attach a sign or put up a tree house. Did the nails move higher as the tree grew? No. Instead, the tree grew around the nails. Primary growth, which lengthens the tree, occurs only at the apical meristems, which lie at the ends of branches far above the nail. The nail, however, will eventually be buried in bark and wood as secondary growth from lateral meristems adds new tissue to the sides of the plant. Although *woody* plants, such as apple and pine trees, have secondary growth, *herbaceous* plants, such as daisies and dandelions, have only primary growth and slender, generally flexible, green stems.

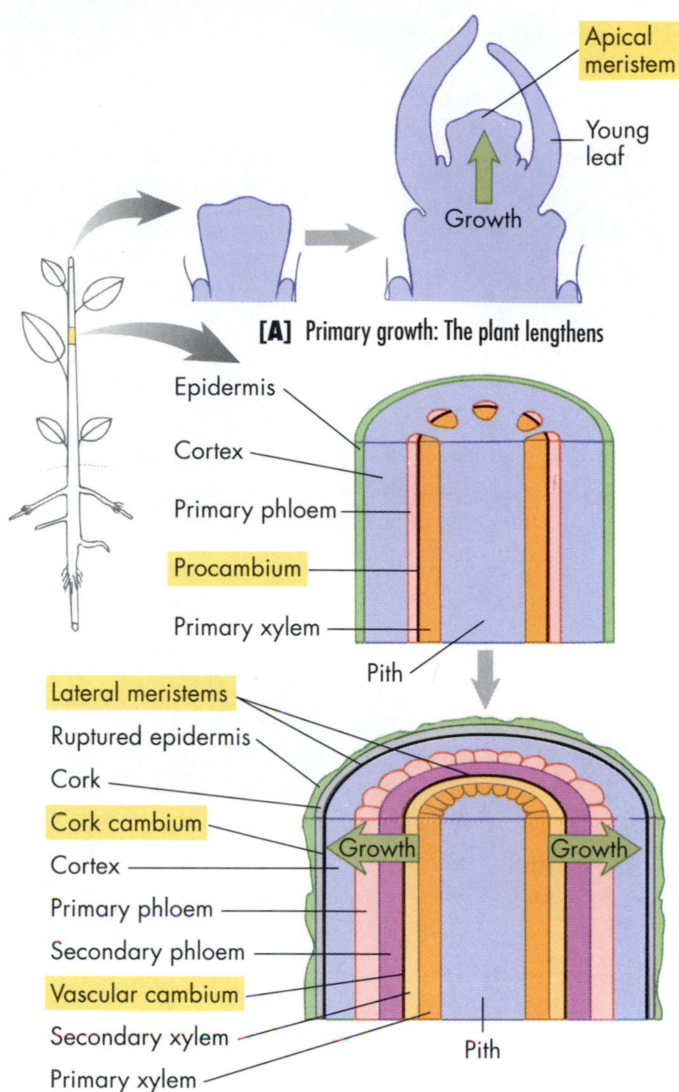

FIGURE 34.9

Primary and Secondary Growth.

[A] Primary growth from apical meristems at the tips of root and shoot causes the plant to lengthen. [B] Secondary growth from lateral meristems causes the shoot and root to increase in diameter.

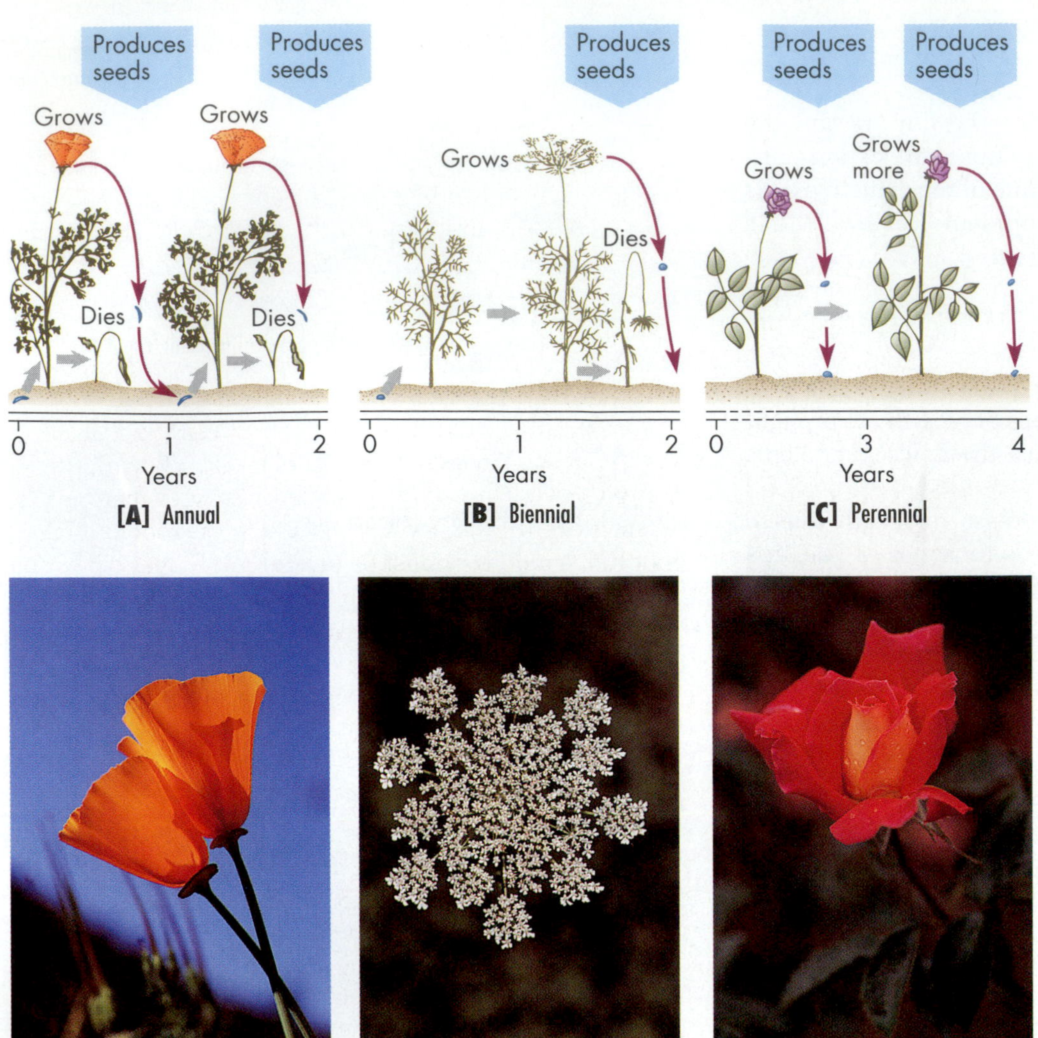

FIGURE 34.10

Annuals, Biennials, and Perennials.

[A] Annuals, such as the California poppy, have one-year life cycles and pass the winter in seed form. [B] Biennials, like Queen Anne's lace, grow the first year, reproduce the second year, and then die. [C] Perennials, such as a rose, survive and grow year after year.

PLANT GROWTH AND SEASONAL CYCLES

Flowering plants have an open growth pattern based on meristems. But as any gardener knows, some plants live only one season, whereas others live from 2 years to 5000. In a single season, an **annual plant** sprouts from a seed, matures, produces fruits and new seeds, and dies [FIGURE 34.10A]. Marigolds, zinnias, petunias, poppies, and soybeans are familiar annuals that gardeners and farmers must plant every year. **Biennials** are plants that have a two-year life cycle: They grow from seeds to adults in the first year; in the second year, the adults produce flowers, fruits, and seeds and then die [FIGURE 34.10B]. Celery, cabbage, carrots, and Queen Anne's lace are all biennials. Many biennials and most annuals are herbaceous, lacking secondary growth. Finally, **perennials** live for many years and typically bloom and set seeds several times before the adult plant dies [FIGURE 34.10C]. Some perennials, like tulips and dahlias, are herbaceous; others, like rosebushes and oak trees, are woody.

➤ CONCEPT CHALLENGE

Plants that do not undergo secondary growth are frequently much smaller and shorter-lived than those that do undergo secondary growth. How do you think the reproductive strategies of plants that undergo secondary growth would differ from plants that do not?

Roots

Roots are familiar plant organs with important functions. In vascular land plants, roots anchor the organism firmly to one spot, penetrate the soil and absorb water and minerals, and often store starch. This discussion of root anatomy begins deep in the soil at the very tip of the root.

THE ROOT: FROM TIP TO BASE

Root Cap At the lowest tip of a typical root is a dome-shaped **root cap** [FIGURE 34.11]. As the root grows down-

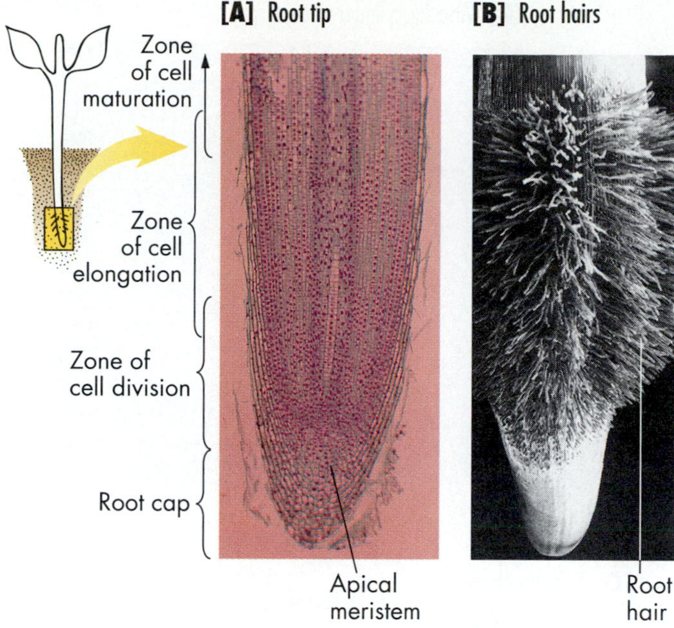

FIGURE 34.11

The Growing Root Tip.

[A] A longitudinal section of an onion root reveals, from the tip upward, the protective root cap just below the apical meristem; the small new cells of the zone of cell division; and the lengthening cells of the zone of cell elongation; [B] the epidermal cells and root hairs in the zone of maturation.

ward, rough soil particles damage and scrape off cells at the surface of the root cap. These damaged cells slough off and cover the rest of the root with a slime that eases penetration through the soil.

Apical Meristem Damaged root cap cells are replaced by the apical meristem, a little disk of cells just above the root cap that produces new cells from both its surfaces. The cells from the lower surface of the disk become root cap cells, while those derived from the upper surface elongate, pushing the root tip strongly into the soil. Moving up a root, one can see three general regions [see FIGURE 34.11]: a *zone of cell division*, which includes the apical meristem protected by the root cap cells; a short *zone of elongation*, where individual cells lengthen and force the tip to move through the soil; and a *zone of maturation*, where cells develop their specialized roles as members of the dermal, ground, or vascular tissue systems.

THE ROOT: FROM OUTSIDE TO INSIDE

The Dermal System: Epidermis and Root Hairs Looking at a cross section of a mature root cut just above the zone of maturation, one can see an outer protective layer just one cell thick—the epidermis, derived from the outer cells in the zone of maturation [FIGURE 34.12A]. The root epidermis absorbs water and minerals from the soil, and just as microvilli enlarge the absorptive surface area of an animal's intestinal cells [see FIGURE 28.16], tiny extensions of the root epidermis called **root hairs** extend the root's absorptive capacity [see FIGURE 34.11B]. Each root hair is an extension of a single epidermal cell, but their numbers can be very great, and their collective surface area can be amazingly large: A single ryegrass plant can have about 14 billion root hairs, with a combined surface area the size of a tennis court!

Root hairs are delicate and short-lived, breaking off as the root tip pushes deeper into the soil. New root hairs continually arise, however, in the region of maturation. Since most water absorption occurs in the root hairs near root tips, gardeners must be careful not to tear off young root tips when transplanting flowers or shrubs, and they often get the best results by fertilizing fruit trees several feet out from the trunk, nearest the root tips.

The Ground System A cross section of a root reveals a thick layer of cells called the **cortex** lying just inside the epidermis and surrounding the central core of the vascular system [see FIGURE 34.12A]. You can see this if you cut a slice from a bright orange carrot root. The cortex makes up most of the root's bulk and often stores excess starch.

The innermost layer of the cortex is the **endodermis**, a cylinder of tightly packed cells just one cell thick [FIGURE 34.12B]. Although water moving inward from the soil can pass freely between most cortex cells without entering their cytoplasm, when the water reaches the endodermis, it must pass through the living cytoplasm of endodermal cells before reaching the vascular bundles and hence the rest of the plant. Each cell in the endodermis is a bit like a brick in a circular brick wall [FIGURE 34.12B and C]. The outside surface of the "bricks" faces the rest of the cortex, and the inside surface faces the vascular tissue in the root's center. The side walls of these cells are impregnated with a waxy, water-resistant substance called *suberin*. This narrow, water-resistant belt—the *Casparian strip*—encircles each cell and lies between it and the next endodermal cell like the mortar between bricks [FIGURE 34.12D]. Because of the Casparian strips, water cannot pass between endodermal cells or diffuse from one endodermal cell to the next. Instead, water can pass from the cortex to the vascular tissue only by moving through the cytoplasm of the endodermal cells. This has an important consequence for the plant: Water and minerals seeping from the soil through the epidermis and into the cortex can reach the central vascular tissue (and from there move throughout the rest of the plant) only by first passing through the living cytoplasm of the endodermal cells [see CHAPTER 35 for more details].

The Vascular System The endodermis in a root such as a carrot surrounds yet another cylinder—a central zone of vascular tissue that carries water and other mate-

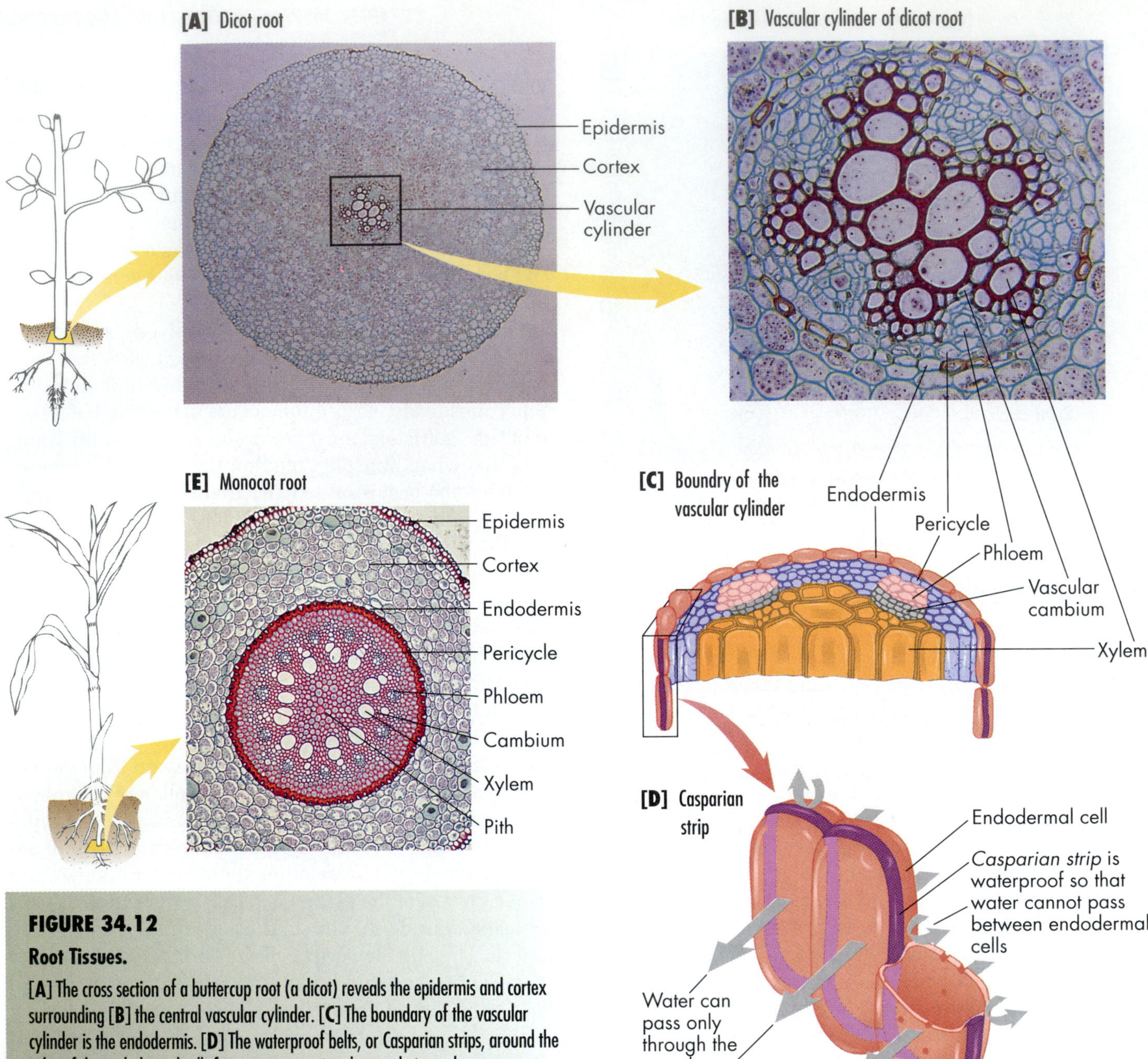

FIGURE 34.12

Root Tissues.

[A] The cross section of a buttercup root (a dicot) reveals the epidermis and cortex surrounding [B] the central vascular cylinder. [C] The boundary of the vascular cylinder is the endodermis. [D] The waterproof belts, or Casparian strips, around the sides of the endodermal cells force water entering the root hairs and root cortex to move through the living cytoplasm of the endodermal cells—not through the spaces between cells—before passing into the vascular cylinder. [E] Cross section of a corn root (a monocot), revealing vascular cylinder and central pith.

rials between the roots and the stems and leaves. Moving inward through the vascular cylinder, one encounters the pericycle, phloem, xylem, and in some plants, pith [see FIGURE 34.12B].

The **pericycle** lies just inside the endodermis and consists of one or more layers of parenchyma cells. The pericycle gives rise to the lateral roots, which grow out horizontally through the cortex parallel to the surface of the ground.

Encircled by the pericycle are the xylem and the phloem, which are arrayed in various ways in plant roots. In the roots of most dicots, the central vascular cylinder contains a star-shaped core of xylem vessels, with phloem nestled in the angles between the points of the star [see FIGURE 34.12A and B]. In contrast, most monocots have a ring of phloem inside the pericycle that surrounds a ring of xylem, which in turn encloses a central core of large, thin-walled parenchyma storage cells called **pith** [FIGURE 34.12B].

Secondary Growth in the Roots In perennial plants, such as shrubs and trees, roots experience both primary growth from the apical meristems and secondary growth from lateral meristems. The root tips continually probe downward or outward (primary growth) as the older

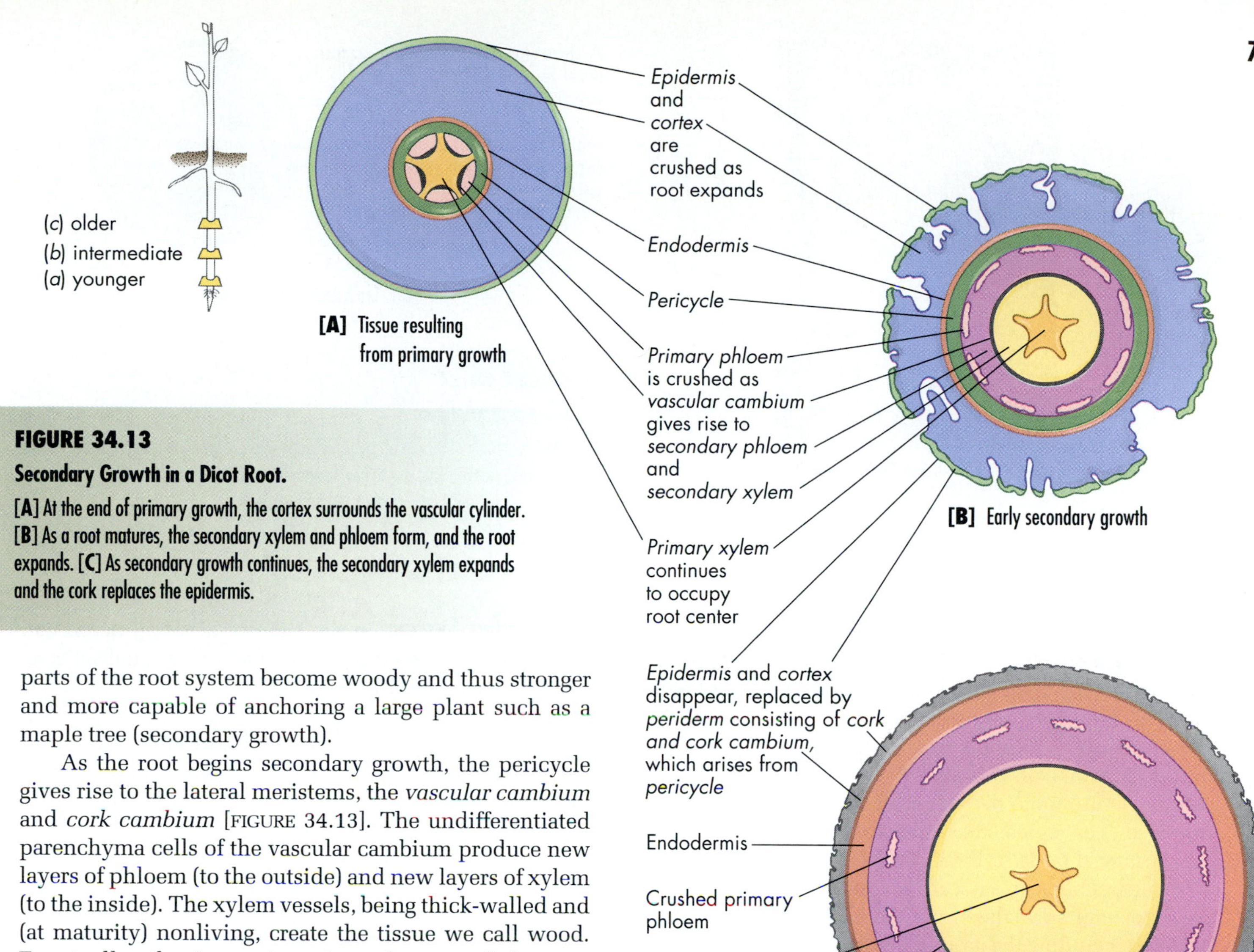

FIGURE 34.13

Secondary Growth in a Dicot Root.

[A] At the end of primary growth, the cortex surrounds the vascular cylinder. **[B]** As a root matures, the secondary xylem and phloem form, and the root expands. **[C]** As secondary growth continues, the secondary xylem expands and the cork replaces the epidermis.

parts of the root system become woody and thus stronger and more capable of anchoring a large plant such as a maple tree (secondary growth).

As the root begins secondary growth, the pericycle gives rise to the lateral meristems, the *vascular cambium* and *cork cambium* [FIGURE 34.13]. The undifferentiated parenchyma cells of the vascular cambium produce new layers of phloem (to the outside) and new layers of xylem (to the inside). The xylem vessels, being thick-walled and (at maturity) nonliving, create the tissue we call wood. Eventually, the increasing circumference of the root causes the epidermis and cortex layers to rupture [see FIGURE 34.13B]. The meristematic tissue in the cork cambium, however, generates a new waterproof layer, the **cork**. The cork plus the cork cambium forms the *periderm*, which protects the root [see FIGURE 34.13C]. The two thin cylinders of vascular cambium and cork cambium in the root continue unbroken into the stem and also generate the wood, bark, and other secondary growth tissues in the shoot.

➤ CONCEPT CHALLENGE

Which of the structural features of roots help them anchor the plant, and which help them take up water and mineral nutrients?

The Shoot

The tallest California redwood trees have trunks that may reach 100 m (328 ft) into the sky, while the delicate stems of some wildflowers on the ground below may be mere fractions of a centimeter tall. Regardless of size, stems (which are usually a plant's aboveground portions) have two fundamental tasks: (1) supporting the plant's leaves and, flowers and fruit, if present, and (2) acting as a central corridor for the transport of water, minerals, sugars, and other substances. Modified stems can also store starch and allow plants to adhere to vertical surfaces. Some plants have modified stems called *rhizomes* that grow horizontally underground and that can help anchor the plant and promote vegetative reproduction. Stems have a distinctive structure and mode of growth, which we will discuss first. The leaves supported by the stem have unique structural elements, as well, and we'll consider them after stems. Flowers, too, are part of the shoot, but their architecture and activities are best understood in the context of plant reproduction, our subject in CHAPTER 36.

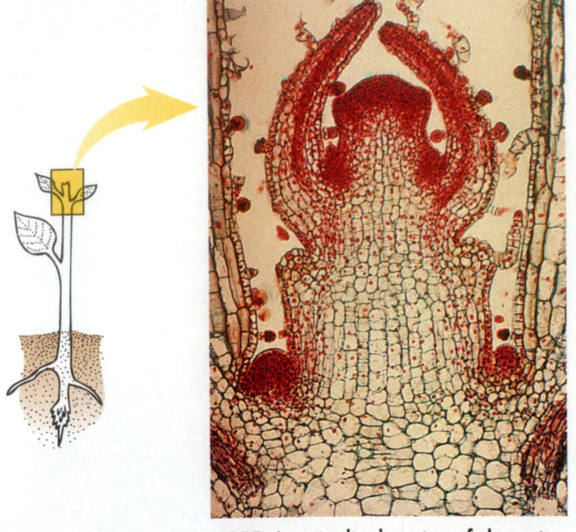

[A] Longitudinal section of shoot tip

A shoot tip cut cross-ways at this level reveals cells giving rise to the *epidermis*, the *procambium*, which forms the xylem phloem, and cambium, and the *ground meristem*, which forms ground tissue.

[B] Cross section of shoot tip

FIGURE 34.14

The Apical Meristem.

[A] A longitudinal section of the growing tip of a *Coleus* plant reveals the primordia that will give rise to new leaves, flowers, and branches. **[B]** A cross section reveals cells responsible for primary growth.

STEM STRUCTURES AND PRIMARY GROWTH

Apical Meristem The shoot lengthens as cells of the meristem at the apex, or tip, of the plant divide and redivide. As the apical meristem at the tip of a branch (the *terminal bud*) continues to grow upward, it leaves behind groups of cells. Some of these cells, called *leaf primordia*, continue to divide and form leaves. Others produced by the apical meristem form embryonic tissue called *lateral buds* in the space between a leaf and the stem; lateral buds give rise to leaves or flowers. Still other cells become specialized meristematic regions [FIGURE 34.14A], including the *ground meristem*, which gives rise to the ground tissues, and the *procambium*, which gives rise to the primary xylem and phloem [FIGURE 34.14B].

Let's look one by one at the specialized shoot tissues laid down by the apical meristem, and at how they grow.

Dermal System: Epidermis As in the root, the dermal system of the shoot consists of a layer of epidermal tissue one cell thick [FIGURE 34.15]. While root epidermis absorbs water, shoot epidermis resists water loss. Just as a plastic bag keeps a sandwich from drying out, stem epidermal cells have a water-resistant waxy coating, the *cuticle*, that keeps the plant from desiccating.

Ground System: Cortex and Pith As in the root, the ground system of the stem is primarily cortex made up of parenchyma cells [see FIGURE 34.15]. Stems, however, which support the weight of the upper parts of the plant and keep the plant vertical in the air, need more structural reinforcement than roots, which are supported by the soil. In many stems, a strand of collenchyma [see TABLE 34.2 on page 761] around the outer edge of the cortex, just inside the epidermis, helps provide this support. Also, a stem's ground system generally has a central core of pith, while a root's ground system may lack it. The shoot's vascular system, however, often provides the strongest structural support.

Vascular System In the stem of a young perennial, such as an apple seedling, or herbaceous annual, such as a zinnia, xylem and phloem are organized in groupings called *vascular bundles*. Within each bundle, a sheath of parenchyma or sclerenchyma cells surrounds strands of xylem and phloem, with xylem lying toward the center of the stem. In the stems of most dicots, vascular bundles form a ring around the stem's central core of pith, and a layer of cambium lies between the xylem and phloem [FIGURE 34.16A and B]. Cambium from each vascular bundle extends to the adjacent bundles, forming a complete ring of cambium around the pith. In conifers and woody dicots, the cambium layer sandwiched between xylem and phloem produces the secondary growth of a wood and bark, resulting in the enlargement of the stem. In monocots and a few dicots, by contrast, the vascular bundles are scattered throughout the cortex, there is no vascular cambium, and the stems are incapable of true secondary growth [FIGURE 34.16C and D].

STEM STRUCTURES AND SECONDARY GROWTH

Recall from FIGURE 34.14B that the apical meristem produces cells that divide and form the primary xylem and primary phloem, as well as a ring of meristematic cells, the vascular cambium [FIGURE 34.15A and B]. Late in the first summer of a tree seedling's life, lateral meristem cells in the vascular cambium begin dividing and produce additional xylem and phloem cells [FIGURE 34.16C]. The new conducting vessels made by the vascular cambium are called *secondary xylem* and *secondary phloem*. Like their primary counterparts, they transport water and other materials. Because the cambium makes far more secondary xylem than secondary phloem, most of the stem tissue in a sapling, a bush, or an older tree is made up of dead xylem cells (or **wood**) in a thick interior rod.

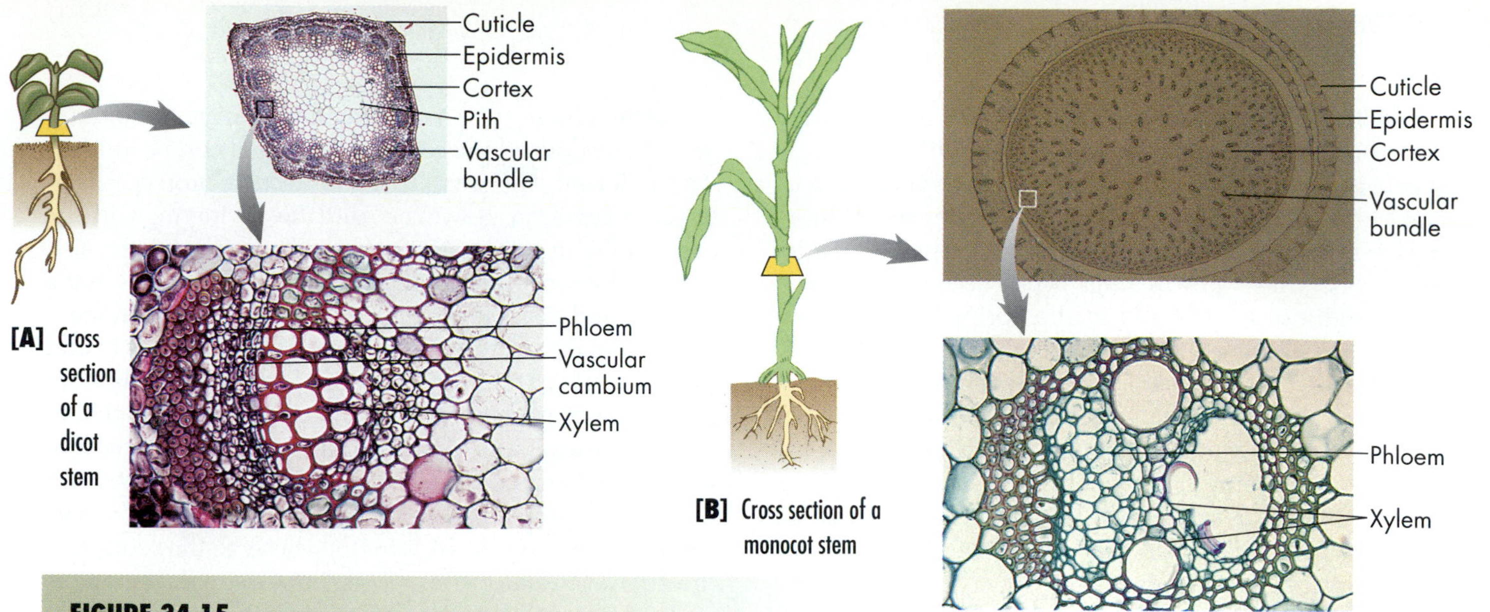

FIGURE 34.15

Stem Structure.

[A] The development of stem tissue in annual dicots such as alfalfa is characterized by a circle of vascular bundles. [B] Annual monocots, such as corn, have scattered vascular bundles.

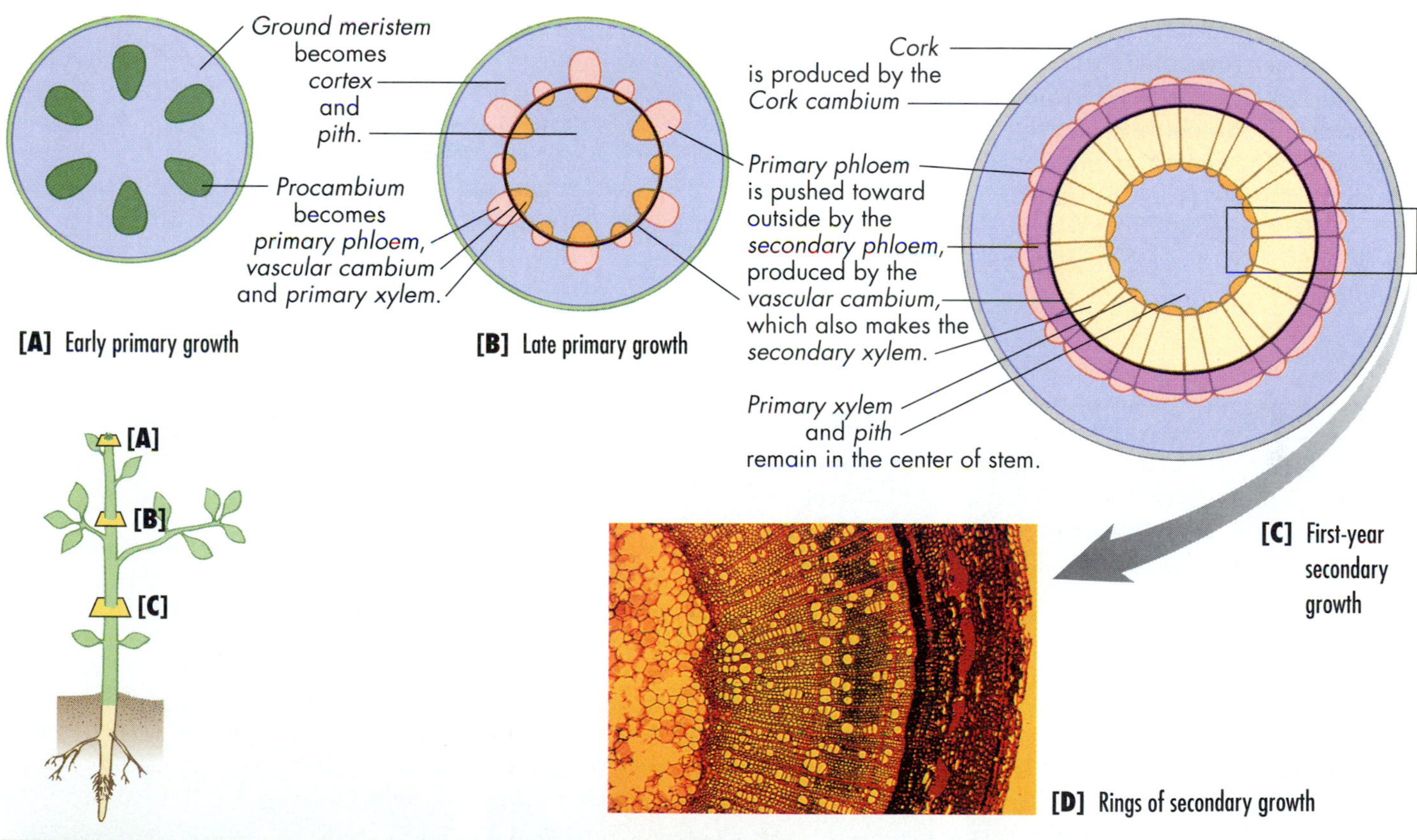

FIGURE 34.16

Secondary Growth in Stems.

[A] During early primary growth, the apical meristem lays down the ground meristem, which makes cortex and pith, and the procambium, which makes the primary xylem and phloem. [B] As a plant matures, the primary xylem and primary phloem differentiate from the vascular cambium between them. [C] During the first year of secondary growth, the vascular cambium adds secondary xylem and secondary phloem, the trunk expands, and the cork cambium adds the cork. [D] A cross section of a maple stem reveals several years of secondary growth.

Wood A tree cut down in a temperate region—say, in a North American hardwood forest—will have concentric rings of lighter and darker wood. If your desk is made of wood, you can readily see evidence of these rings as the grain in a plank of wood cut perpendicular to the circles. These *growth rings* occur because the cambium produces larger cells in the spring and summer (lighter areas), when water and sunlight are plentiful, and smaller cells in the fall and winter, when water tends to be more scarce. In moist, tropical regions, on the other hand, where a tree or a woody vine such as a strangler fig can grow year round, the wood may have no obvious growth rings. Because larger rings are produced in more favorable years, scientists can use the width of growth rings in ancient trees, such as the 5000-year-old bristlecone pines of eastern California, to reconstruct historical weather patterns. Scientists have dated a massive volcanic eruption on the Greek island of Thera, which must have darkened skies around the world with ash, to approximately 1625 B.C. They did this by identifying very narrow growth rings in North American bristlecone pines growing at that time.

As a tree grows older, the xylem at the center gradually ceases to conduct water and minerals, and newer xylem nearer the periphery of the trunk takes over. The nonconducting central wood, or *heartwood*, becomes infiltrated with oils, gums, resins, and tannins, all of which make it dark, aromatic, and resistant to rot. The *sapwood* around the outside continues to transport water in the plant.

Bark When the stem expands during secondary growth, it ruptures the seedling's original cortex and epidermal layers. The small tree still needs a protective waterproof covering, however, and the periderm, with its cork cambium and layer of cork cells, develops here, as it does in the roots. Together, the periderm, the underlying phloem, and many layers of cork cells that die but remain in place make up the outer protective **bark**. Sheets and plugs of cork for flooring, bulletin boards, and bottle stoppers come from the periderm of cork oak trees native to Spain, Portugal, and Algeria. Cork cutters are careful to leave the sugar-conducting phloem intact. Without a functioning phloem all the way around the trunk, sugars cannot move downward from the leaves to the roots, and the tree eventually dies. This impaired movement is probably the reason the host tree beneath a growing strangler fig eventually dies, too. The ACTIVE LEARNING BOX on page 773 shows a simple experiment with a young tree that proves this point.

FIGURE 34.17

Leaf Architecture: Dicots, Monocots, Conifers.

Dicot leaves from a plant such as this beech [A] are often broad, net-veined, and each leaf has a stalk, or petiole, and a blade. Monocots such as this corn plant [B] usually have parallel veins and a slender blade attached to the stem via a sheath, not a petiole. The leaves of a white pine [C] are needlelike and grow in clusters.

[A] Dicot

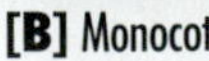

[B] Monocot

[C] Gymnosperm

active learning

Biologists often determine the role of a particular structure by removing it or blocking its action and then determining how the organism responds. You decide to determine for yourself the role of phloem by removing it in a ring around a young cherry tree in early spring.

How deep will you need to cut to remove the phloem but leave the xylem intact? (Hint: Review FIGURE 34.16C.)

. . . .

. . . .

After a few months, a portion of the tree looks like the drawing (FIGURE A). Describe the results in a sentence.

. . . .

. . . .

Now, interpret the results using FIGURE B. What do you think was the effect of the surgery on the transport of water and minerals, given the fact that the xylem was left intact?

. . . .

. . . .

Recalling that photosynthesis occurs in the leaves above the surgery, what would happen to the flow of sugars and other organic substances?

. . . .

. . . .

How does that reasoning allow you to explain why the trunk grew in diameter above the surgery, but not below the surgery?

. . . .

. . . .

What do you think would eventually happen to this tree—would it continue to live, or would it soon die? Support your answer.

. . . .

. . . .

[A] Result

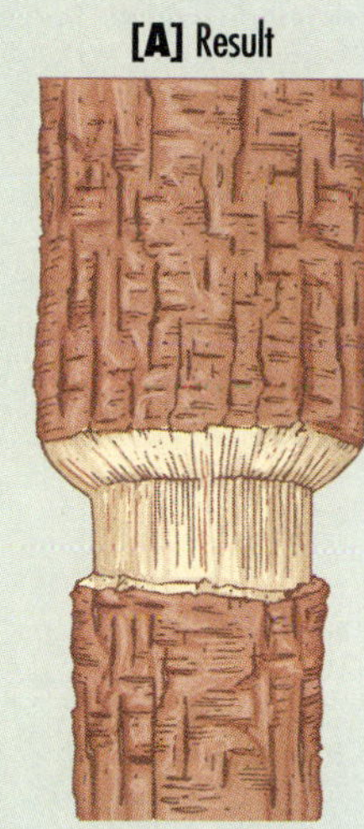

[B] Explanation

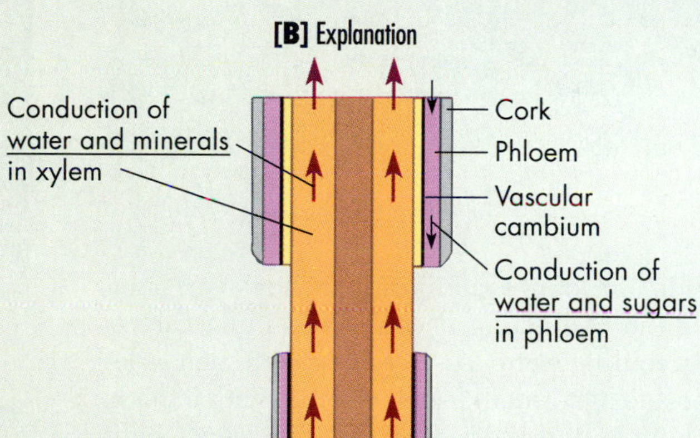

Bark does more than keep water in; it helps keep plant-eating insects, fungi, viruses, and other parasites out. Healthy trees also secrete resins that ooze out of holes bored by insects, often forcing the intruders back out. Frankincense and myrrh are fragrant tree resins, and amber is nothing but fossilized resin.

STRUCTURE AND DEVELOPMENT OF LEAVES

The leaves that grow on plant stems may be smaller than a fingernail or larger than a tabletop; they may be shaped like a needle, a knife, a plate, a fan, or a hand. Regardless of form, however, nearly all leaves share the same basic functions: exposing a photosynthetic surface, usually one that is large and flat and oriented to catch the maximum amount of sunlight; obtaining carbon dioxide from the atmosphere as a carbon source for photosynthesis, yet retaining as much water vapor as possible; and helping to draw water and nutrients up through the vascular system.

LEAF BLADE AND PETIOLE

Most leaves are composed of a broad, flat portion, called the **blade**, and either a **petiole** or *sheath* that connects the blade and plant stem at a location called the *node* [FIGURE 34.17]. Leaf blades are often flat and thin, which maximizes the available surface area for absorbing light and

[A]

FIGURE 34.18

Specialized Leaves Can Protect, Store, or Attract.

[A] The leaves of this prickly pear cactus are modified into barbed spines that protect the succulent stem. [B] Leaves of a bromeliad, such as this pink-tipped individual, can grow on the branches of tropical trees, and these act as small basins that store water. [C] Colorful leaves attract pollinators.

[B]

[C]

carbon dioxide. Structural modifications, however, often help balance the conflicting needs of obtaining carbon dioxide and avoiding water loss. Desert plants such as jojoba and creosote have small leaves covered with a thick, waterproof cuticle and hairy surfaces that reflect light and thus help retard water loss. Cactus leaves are modified into small, dry, needlelike spines that defend rather than photosynthesize [FIGURE 34.18]. Other plants have leaves modified for clinging and climbing, for capturing little pools of water, and for attracting animals toward inconspicuous blossoms. In a few species, including pitcher plants and sundews, the leaves assist predation, trapping a visiting insect, then enzymatically digesting it for its proteins [review FIGURE 4.1].

Leaves of dicots may be *simple*, like those of the fig and consist of a single blade, or they may be *compound*, like those of buckeye, or locust, and consist of many small leaflets [FIGURE 34.19].

[A]

[B]

FIGURE 34.19

Leaf Shapes, Simple and Compound.

Dicot leaves can have one blade per petiole [A] or several [B], but all are net-veined.

LEAF TISSUES

Like its external form, a leaf's internal anatomy is intimately associated with its many functions.

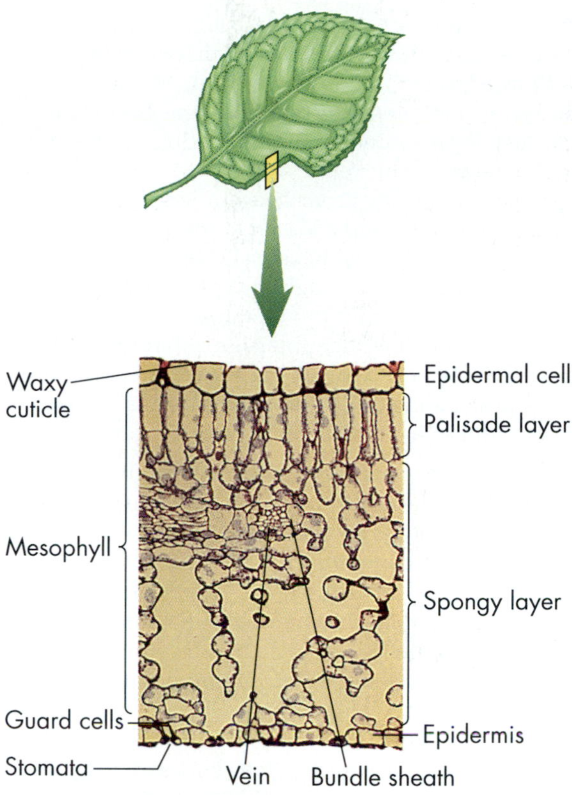

FIGURE 34.20

Leaves: An Inside View.

A small section taken from a leaf of a privet bush reveals the cuticle-covered upper epidermis, layers of palisade and spongy mesophyll cells, a vein, or bundle of vascular tissue surrounded by a sheath, and a lower epidermis ventilated by openings—stomata—that are flanked by guard cells.

Dermal Tissue A leaf's epidermis has a waxy coating, the cuticle. In fact, the leaf is so well sealed that it needs tiny openings called **stomata** (singular, **stoma**), surrounded by *guard cells*, to admit carbon dioxide for photosynthesis and to release water vapor and oxygen [FIGURE 34.20]. As CHAPTER 35 explains, stomata are most numerous on the undersides of leaves, and their opening and closing are based on the plant's conflicting needs to conserve water and take in carbon dioxide.

Ground Tissue Sandwiched between the upper and lower epidermal layers is the leaf's **mesophyll** layer, made up of two types of parenchyma. *Palisade parenchyma*, the main photosynthetic tissue, lies just beneath the upper epidermis. It usually consists of one or more rows of vertically oriented, column-shaped cells, each enclosing dozens of chloroplasts. A layer of rounder cells, the *spongy parenchyma*, lies between the palisade parenchyma and the lower epidermis. The loosely packed spongy parenchyma cells provide a huge surface area (analogous to an animal's lung) that absorbs carbon dioxide from the air that enters the stomata. Carbon dioxide entering the stomata can move rapidly through the spongy parenchyma to the palisade parenchyma above, where most of the photosynthesis takes place.

Vascular Tissue Leaves have an elaborate "plumbing" system which distributes the products of photosynthesis and brings in water and minerals, and a plant's vascular system is continuous between leaves, stem, and roots. Monocot leaves have bundles of xylem and phloem called **veins** running parallel to each other along the long axis of the leaf (so-called *parallel venation*; see FIGURE 34.17). Dicot leaves have *netted venation*; that is, the veins form a branching pattern. Surrounding individual veins in some plants are *bundle-sheath cells*, which protect the xylem and phloem from direct exposure to air. Collenchyma or sclerenchyma fibers may also stiffen the veins and leaf margins and help each leaf remain flat, which improves solar collection. All summer long, mesophyll cells in the leaf store the sun's energy in sweet sugar molecules, and the phloem transports sugars from the leaves to the developing fruits. The end result is the fruits and the seeds they contain—the plant's evolutionary solution to dispersing the species.

➤ CONCEPT CHALLENGE

Where would you look in a poppy and a rose to locate cells dividing by mitosis in the plant's shoot?

Connections

Plant structures—including roots, shoots, parenchyma cells, and root hairs—help reveal the two major distinctions between plants and animals: the source of energy and materials, and the abilities of cells and organisms to move in response to threats or a need for resources. In general, energy and supplies are available to a plant from the sun, the air, and the soil. The rigid walls surrounding each cell collectively provide the strength that supports the leaves and forces roots down into the soil. These rigid cell walls, however, preclude the movement of plant cells and of entire plants. However, as we saw with the strangler fig, the rigidity of the cell walls can be just as effective as powerful gripping muscles, even though they work in slow motion. Although plant cells can't move, various cells and tissue systems introduced in this chapter function to distribute materials throughout the plant, and CHAPTER 35 explains how.

KEY TERMS

adventitious root, 759
annual plant, 766
apical meristem, 765
bark, 772
biennial, 766
Casparian strip, 767
companion cell, 764
cork, 769
cortex, 767
dermal tissue system, 762
endodermis, 767
fibrous root, 759
ground tissue system, 762
lateral meristem, 765
meristem, 764
mesophyll, 775
open growth, 764
perennial, 766
petiole, 773
phloem, 763
pith, 770
primary growth, 765
root, 759
root cap, 766
root hair, 767
secondary growth, 765
shoot, 759
sieve plate, 764
sieve tube member, 764
stoma, 775
taproot, 759
tracheid, 763
vascular cambium, 765
vascular tissue system, 762
vessel member, 763
wood, 770
xylem, 763

HIGHLIGHTS IN REVIEW

1 The body of complex plants consists of three tissue systems: an outer protective layer (the dermal tissue system), a system of tubes that carry water and nutrients (the vascular tissue system), and a system of tissue that fills the space around the vascular system, storing materials and providing support (the ground tissue system).

- **a]** Vascular plants have two major parts, the root and the shoot. Roots anchor the plant, store starch, absorb water and minerals, and transport them to the shoot. The shoot typically consists of a branching system of stems that provide support, leaves that capture energy, and flowers that function in reproduction.
- **b]** A vascular plant has a dermal tissue system, the protective outer covering; a ground tissue system, which provides support and storage; and a vascular tissue system, the internal transport tissues.
- **c]** The dermal tissue system contains mostly epidermis during primary growth and periderm during secondary growth. The ground tissue system is mostly squarish parenchyma cells, but can contain hard sclerenchyma cells and thick-cornered collenchyma cells. The vascular tissue system consists of xylem tissue, which conducts water and minerals, and phloem tissue, which conducts photosynthetic products.

2 The three tissue systems arise as plants grow. Plants grow longer as new cells are added at the tips of branches and roots in regions called apical meristems. The newly added cells may continue to divide for a time, then elongate and differentiate into specialized cells of the root, stem, leaves, or flowers in a process called primary growth.

- **a]** Plants have open growth based on perpetual growth centers, or meristems. Apical meristems at the tips of roots and stems allow primary growth: extension upward into the air and downward into the soil. Secondary growth depends on lateral meristems (or cambiums); these cylinders of tissue allow root and shoot diameters to thicken.
- **b]** The primary root has a protective epidermis, water-absorbing root hairs, and a starch-storing cortex that includes the endodermis. Because of the Casparian strip, water must pass through the cytoplasm of endodermal cells to enter the vascular cylinder.
- **c]** The primary shoot has an epidermis, often tiny surface hairs, and a waxy cuticle that helps prevent water loss.
- **d]** Stem tissues can include cortex and pith inside the vascular cylinder, with its bundles of xylem and phloem.

3 Some plants exhibit secondary growth, an increase in diameter resulting from the addition of new cells near the surface in regions called lateral meristems. The bulk of the newly added cells becomes wood.

- **a]** Secondary growth forms the woody parts of shrubs and trees. A ring of meristematic tissue, the vascular cambium lying between the xylem and phloem, gives rise to secondary xylem (the major component of wood) and secondary phloem.
- **b]** As the shoot and root enlarge in circumference, the cortex and epidermis rupture. Tree bark is formed by layers of dead cells from the periderm along with secondary phloem.
- **c]** Leaves have highly photosynthetic palisade parenchyma cells beneath the upper epidermis, as well as spongy parenchyma cells above the lower epidermis. Tiny openings called stomata control the passage of gases into and out of a leaf.

UNDERSTANDING THE FACTS AND CONCEPTS

For Questions 1–5, match each of the descriptions with the most appropriate item from the following list. An item may be used once, more than once, or not at all.

- **a]** vascular plants
- **b]** monocots
- **c]** dicots
- **d]** woody plants
- **e]** herbaceous plants

1 Anthophytes that have secondary growth.

2 Anthophytes that do not grow in girth; stems are generally green.

3 Flowering plants that include all woody plants as well as many herbaceous types.

4 Flowering plants that include grasses, lilies, and palm trees but not roses, daisies, and apple trees.

5 The most inclusive category in the list; includes anthophytes and gymnosperms.

As above, for Questions 6–10, match each of the descriptions with the most appropriate item from the following list. An item may be used once, more than once, or not at all.

- **a]** root
- **b]** shoot

c] meristematic tissue
d] dermal tissue
e] ground tissue
f] vascular tissue

6 Includes stem, branches, leaves, flowers, and fruit; usually the predominant aboveground part of a plant.

7 Found either as epidermis with or without cuticle, or as periderm.

8 Primary and secondary xylem and phloem constitute that tissue.

9 Region of absorption of water and minerals; usually nonphotosynthetic and anchored in soil.

10 The major storage tissue; helps to support the plant.

As above, for Questions 11–15, match each of the descriptions with the most appropriate item from the following list. An item may be used once, more than once, or not at all.

a] vessel members
b] phloem cells
c] apical meristem
d] lateral meristem
e] cork cambium
f] vascular cambium

11 Living cells that transport dissolved nutrients.

12 Rapidly dividing cells in the vascular cambium that give rise to the substances we call wood.

13 Found in both root and shoot; responsible for primary growth.

14 End walls contain a perforation plate through which water passes into the next cell in the column.

15 A general category of meristematic cells in the stem that is responsible for secondary growth, which increases the girth of plants.

As above, for Questions 16–20, match each of the descriptions with the most appropriate item from the following list. An item may be used once, more than once, or not at all.

a] pericycle
b] endodermis
c] root hair
d] cuticle
e] pith
f] vascular cambium

16 Increases surface area for absorption.

17 Characterized by the Casparian strip.

18 Found in root and stem; gives rise to secondary xylem and phloem.

19 Waterproofing exterior layer of stems.

20 Lateral roots grow from this layer.

INTEGRATE AND APPLY WHAT YOU HAVE LEARNED

1 Which plant tissues are functional even though some or all of the tissue's cells are dead? What role do these dead cells play in the life of a plant?

2 What are the closest analogies you can make between the types of plant tissues and the tissues of animals?

3 Explain the tissue sources for primary and secondary growth in roots and stems.

4 Contrast the arrangement of vascular tissues in monocot and dicot roots.

5 Contrast the arrangement of vascular tissues in monocot and dicot stems.

ANALYSIS

1 Which of the following statements about flowering annual plants is true?
a] They are herbaceous.
b] They achieve their growth only from primary meristematic tissue.
c] They have both tracheids and vessel members as part of the xylem.
d] They do not have a periderm.
e] All of the above are correct.

2 Which of the following characteristics shows variability among different species of perennials? More than one answer may be correct.
a] Herbaceous versus woody growth.
b] Presence of secondary growth.
c] Presence of apical meristem.
d] Taproot versus fibrous root system.

3 Which of the following statements is true? More than one answer may be correct.
a] The water in the cortex of a root has the same concentration of dissolved nutrients as the water in the xylem because fluid freely dissolves into the xylem.
b] The dissolved nutrient concentration of the water in the vascular cylinder may differ from that absorbed by the root hairs because of active transport in the endodermis.
c] The concentration of dissolved soil nutrients is the same as that in the cortex of the root.
d] The concentration of dissolved nutrients in the soil may differ from that in the root cortex.

4 Peeling off circular rings of bark from birch trees (girdling the tree) can kill the tree primarily because:
a] The tree loses the ability to transport a continuous column of fluid in the xylem.
b] The tree loses its ability to transport dissolved nutrients from the shoot to the root.
c] Having lost a section of epidermis, the tree is subject to desiccation as fluid evaporates from the vascular tissue.
d] Interruption in the vascular cambium deprives the tree of the ability to engage in secondary growth, thereby limiting its photosynthetic potential.
e] Without protective bark, birds and other tree-dwelling animals will peck or produce damaging holes in the trunk.

CHAPTER 35

Transport and Nutrition in Plants

THE BOOJUM TREE

If you were to travel 200 miles south of San Diego, down the peninsula of Baja California to a spot deep in Mexico's Sonoran desert, you could see some of the world's strangest plants: the boojum trees, *Fouquieria columnaris.* Two centuries ago, Spanish missionaries dubbed them *cirios* ("church candles"), thinking that the tall, slender trees with flamelike clusters of yellow flowers resembled the narrow candles on a Spanish church altar. The missionaries could just as easily have called the plants *zanahorias* ("carrots"), however, because boojum trees look even more like giant, upside-down carrots with whiskers [FIGURE 35.1].

Robert R. Humphrey, the world's expert on the boojum, pointed out in 1935 that most plants in the deserts of North America are either shrubs, like the creosote bush, or water-storing stem succulents, like the saguaro cactus. Curiously, the boojum is both. It has a thick, central, carrotlike shoot reaching up to about 26 m (85 ft) that stores many gallons of water. But growing from the trunk are also hundreds of leafy and thorny side branches no more than about 60 cm (24 in) long. Odd as the boojum may look, however, this combination of traits is a winning one: No other tree as tall or stout survives the dry conditions in the Sonoran desert.

FIGURE 35.1

The Strange Boojum Tree of Baja California.

Fouquieria columnaris is a bit like a church candle crossed with a carrot and a croquet wicket. The boojum is highly efficient at storing water in a desert setting.

At times, when the weather is hot and dry for months on end, the boojum will droop so severely that its tip almost touches the ground. Most of the time, though, the plant remains erect despite the searing heat and drought. Its crisp upward growth is a testimony to the way plants can store and transport water and nutrients—our subjects in this chapter.

Like most trees, the boojum has a cylinder of xylem. Through this tissue, unbroken columns of water molecules move from the roots to the highest branches and leaves by means of an evaporative transport mechanism. As we will see, this process allows water to reach the tops of redwoods, sequoias, and other extremely tall plants, as well as desert boojums. The boojum's trunk, however, also has two less typical features which adapt it to the dry environment: (1) A thick central core of pith cells stores water, and (2) a layer of stone cells just inside the outermost epidermis functions as internal armor, helping protect the trunk from the ravages of woodpeckers and hungry rodents.

Climatic conditions are so difficult in Baja California that only once every 15 to 20 years can a generation of boojum seedlings get enough water to survive past the first treacherous six weeks of life. Once established, however, the plants can live up to 700 years, and eventually tower over their desert neighbors.

The boojum is a good example for our consideration of how water and nutrients are taken up and transported within plants. We will see, however, that the same set of general principles applies to virtually all plants: Plants depend on the special physical properties of water that make it possible for water and nutrients to be pulled from the soil and for a continuous flow of materials to be supplied to individual cells throughout the plant body. The avenue for this flow of materials is a network of transport tubes, the vascular system, that connects all the parts of the plant. As in animals, plants have mechanisms for maintaining homeostasis, especially a constant fluid environment inside plant cells. And finally, soil quality is crucial to the growth and survival of plants because each individual procures all its inorganic nutrients from the soil in which it grows.

This chapter shows how a dynamic plant gets all the materials it needs for survival and reproduction. Although the basic mechanisms of transport and nutrition are similar in most plants that have xylem and phloem, we concentrate on flowering plants in this chapter because they are more familiar to most students than cone-bearing, seedless, and other types of plants. We will see first how a plant takes up water, how it is transported to all body cells, and how excess evaporation from the leaves is prevented. Next, we will see what nutrients a plant needs, and from where they come. Finally, we will explore how a plant moves the products of photosynthesis from areas of production and storage to areas of use.

As we go along, we will explore not only how the boojum can survive for so many centuries in its parched environment, but also why some plants eat animals and why you should always cut flower stems underwater. ❑

MESSAGES

1 Water tends to move upward in plants from roots to leaves. As a water molecule evaporates from a leaf, it is replaced by diffusion. Due to the tendency of water molecules to bind together, a continuous column of water molecules contained in the xylem moves upward through the plant.

2 Organic molecules move about in the phloem from places of formation, such as a photosynthesizing leaf, to places of use, such as to the root of a carrot for storage. Osmosis drives the movement of organic substances in phloem.

3 Plants, like animals, require certain chemical elements as nutrients. Plants obtain the carbon, hydrogen, and oxygen they need from the air and water, and other nutrients from the soil. Plants obtain usable nitrogen through associations with bacteria.

Water Balance in Plants

The biggest concern of a gardener must always be water, and the reason is simple: Over the course of the summer, a tomato plant, for example, consumes about 120 L (32 gal) of water, which is over half the water a gardener weighing 30 times more will require! Corn and sunflower plants, which are larger than tomato vines, will consume 17 times more than the person, and in each of these plants, 98 percent of the imbibed moisture will evaporate from the leaves. The explanation for this apparently insatiable thirst in plants lies in a fundamental difference between animals and plants: In most animals, fluids tend to recycle through a circulatory system, while in plants, water travels in a generally one-way path from roots through stems to leaves, then out into the environment [FIGURE 35.2]. A steady supply of water enables a plant to carry out photosynthesis [see CHAPTER 6]. It also allows the plant to remain crisp and erect with turgid (swollen) cells [see FIGURE 4.19], to stay cool despite a baking sun on outstretched leaves and stems, and to transport substances internally.

Land plants would quickly lose all the water they take in if it weren't for the waterproof coatings on the aboveground portions of the plant, such as the waxy cuticle or bark layers. This is especially true of plants living in arid regions, such as the boojum tree. But if such layers made an absolutely water- and air-tight seal, the plant could not take in the carbon dioxide required for photosynthesis. The evolutionary solution to this dilemma is *stomata* [see FIGURE 35.6], tiny openings in the leaves and stems. The opening and closing of a stoma are regulated by a surrounding pair of **guard cells**, which allow sufficient air (with its carbon dioxide) to enter the plant while preventing excess water loss through evaporation.

Let us trace the one-way flow of water from soil to roots to leaves to air and see how the physical properties of water and plant tissues—including stomata—make this life-sustaining movement possible.

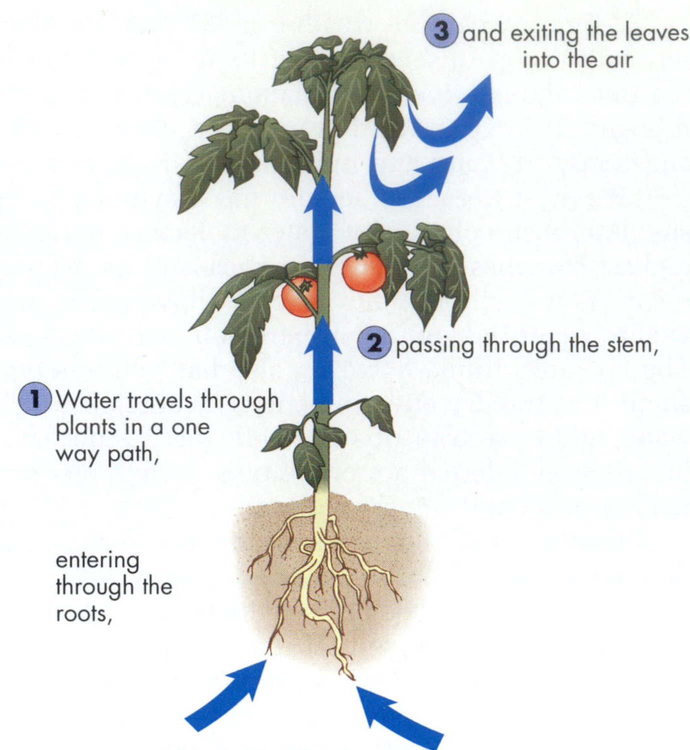

FIGURE 35.2

Water Flow Through a Plant.

Xylem provides an energetically inexpensive one-way route for the flow of water and inorganic nutrients through the plant.

HOW ROOTS DRAW WATER FROM THE SOIL

During and after the infrequent rains in Baja California, a boojum tree absorbs water from the soil into the roots. Eventually, much of that water is stored in special thin-walled parenchyma cells in the trunk's innermost cylinder of tissue. The roots take up precious water molecules from the thin film of water and dissolved minerals that coat the soil particles after a rain. Hundreds of thousands of root hairs, and, in many plants, the fungus-root extensions called mycorrhizae [see FIGURE 20.5], project from the cells of lateral roots into the spaces in soil. These projections have a huge combined surface area and can take up water from the minute reservoirs all around them.

Water normally enters roots by diffusion. (No energy is expended by the plant.) The water diffuses into root hairs and then passes through the root cortex. Some of the fluid moves through the cytoplasm of root cells, but most passes within the cell walls [FIGURE 35.3]. When it reaches the endodermis (the cell layer that separates the root cortex from the central vascular cylinder containing the xylem and phloem), water can no longer follow the cell wall route because of the Casparian strips. These waxy belts surround each endodermal cell and act like gaskets that prevent water from flowing in the spaces between cells, forcing it to pass through the cytoplasm of endodermal cells before it enters the hollow, water-conducting tubes of the xylem [review FIGURE 34.9]. Some of the water enters the root by diffusion, but some also enters by bulk flow [review FIGURE 4.19], replacing water that leaves the root xylem and moves upward toward the stems and leaves. Diffusion and bulk flow can explain how minerals and water enter the root, but how do these substances reach the rest of the plant?

MINERAL TRANSPORT

While mineral ions move through the endodermis passively, cells do use energy to transport ions into the

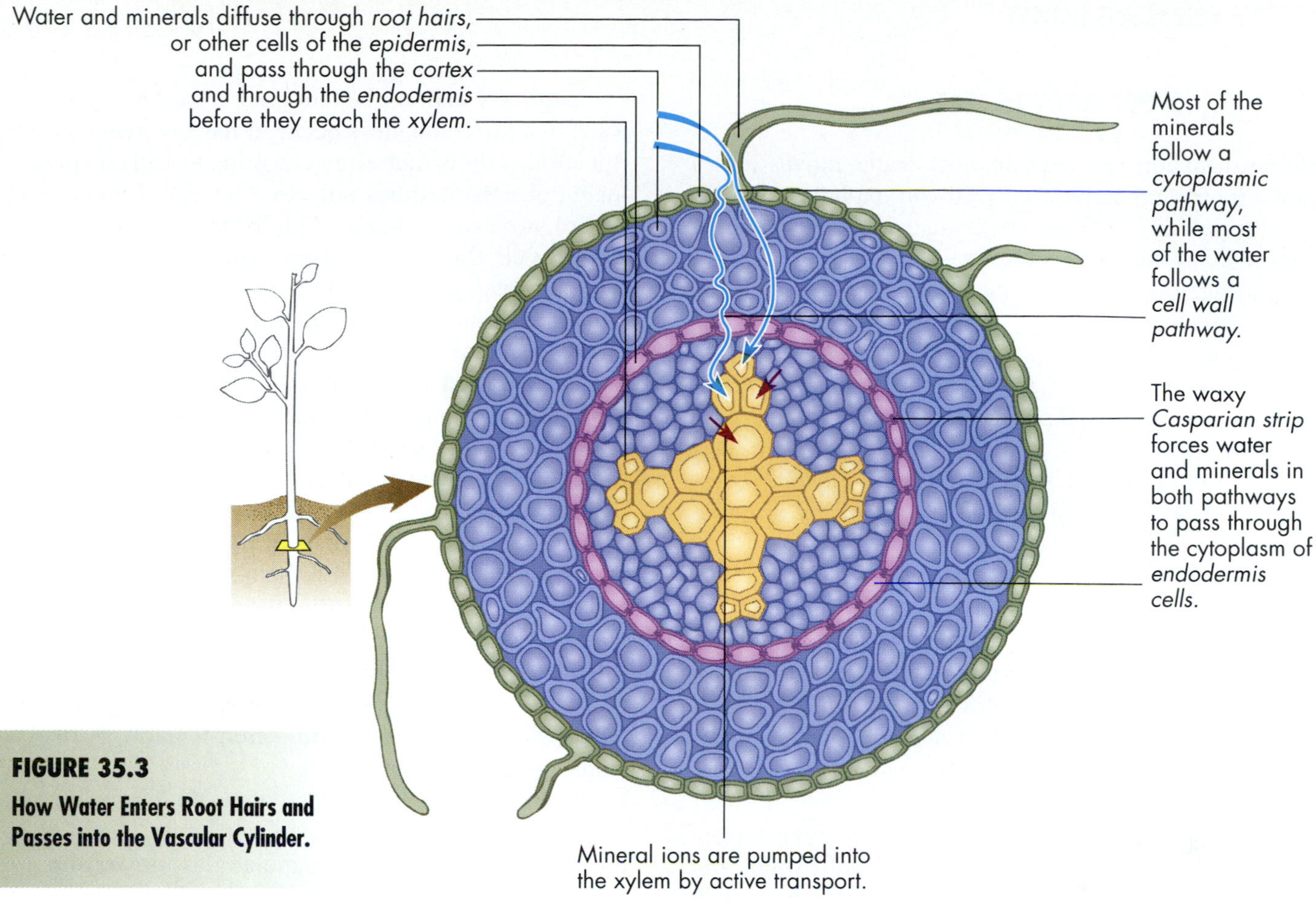

Water and minerals diffuse through *root hairs*,
or other cells of the *epidermis*,
and pass through the *cortex*
and through the *endodermis*
before they reach the *xylem*.

Most of the minerals follow a *cytoplasmic pathway*, while most of the water follows a *cell wall pathway*.

The waxy *Casparian strip* forces water and minerals in both pathways to pass through the cytoplasm of *endodermis cells*.

Mineral ions are pumped into the xylem by active transport.

FIGURE 35.3

How Water Enters Root Hairs and Passes into the Vascular Cylinder.

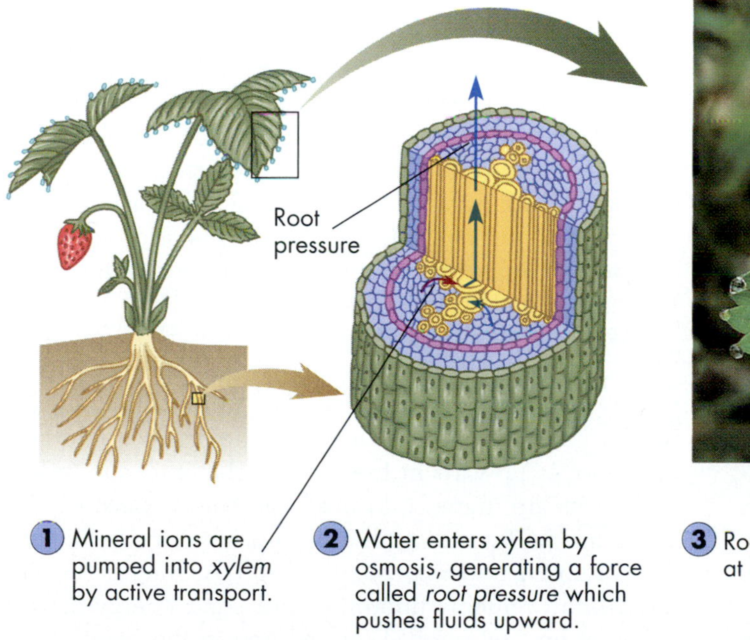

1. Mineral ions are pumped into *xylem* by active transport.
2. Water enters xylem by osmosis, generating a force called *root pressure* which pushes fluids upward.
3. Root pressure can force water out of special openings at leaf tips of small plants.

xylem vessels. Membrane proteins in the root's parenchyma cells pump ions into the adjacent xylem by means of active transport [FIGURE 35.4; review FIGURE 4.20]. Water can then enter the xylem via the process of osmosis [review FIGURE 4.18]. The ions are then transported throughout the plant, dissolved in the water in the xylem.

FIGURE 35.4

Root Pressure and Guttation.

While root pressure can force water out of special openings in some small plants like the strawberry plant leaf shown here, it is never high enough to push water all the way to the tops of tall trees.

WATER TRANSPORT: ROOT TO LEAF

Simple diffusion can explain how water moves into a plant's roots, and active transport can explain how ions enter the xylem. Can these mechanisms also explain how water reaches the top of a California redwood, the tallest living thing, towering more than 100 m (328 ft) above the forest floor?

Plant scientists have proposed four possible mechanisms for water transport: (1) Water might be *pushed* upward from the roots; (2) pumps in the xylem might convey water upward like a bucket brigade; (3) capillary action in the xylem might account for the upward movement; or (4) water might be *pulled* upward by evaporation from the leaves. But which is correct?

Root Pressure, Xylem Pumps, and Capillary Action Farmers have long noticed that in the early morning, the leaves of grass, tomatoes, strawberries, and numerous other plants are sometimes rimmed with tiny droplets of water, the result of a process called *guttation* [FIGURE 35.4]. Could this process reflect a push from the roots great enough to force water all the way to the tops of trees?

Guttation droplets appear at special openings on leaf edges when the soil is nearly saturated with water and the leaves are not losing much through evaporation—in early morning, for example. Water and minerals leave the water-soaked soil and enter the vascular cylinder. Membrane proteins in the root's parenchyma cells expend energy pumping ions into the xylem; water then follows by the process of osmosis [review FIGURE 4.18]. The volume of fluid in the xylem increases, and this fluid must move somewhere. It tends not to flow back between the endodermal cells because the waxy Casparian strips prevent both the solutes and the water from leaking back out of the vascular cylinder [see FIGURE 35.3]. The fluid can move up the xylem, however, generating a force called **root pressure**. Root pressure can force water out of the leaves, where it can accumulate in tiny droplets.

Measurements show that *the force* of root pressure is much lower than the force needed to move water to the tops of tall trees. Such a force would have to be about 150 pounds per square inch (psi), roughly equivalent to the force you would exert on the floor by balancing all your weight on the tip of one big toe. In addition, many plants, including pines, do not develop root pressure at all. You can perform a simple test to determine whether the water in xylem pipelines is pumped up from below: If you pierce a plant stem with a needle, water does not spew out as from a punctured garden hose. Instead, air is sucked in. This same test also disproves the idea that xylem pumps relay water upward, because again, a stem wounded near a "pump" would spew out water, and this does not occur.

Capillary action, the tendency of water to move upward in a thin tube, does occur in narrow xylem vessels. But studies show that water can cling to and creep up the inside of narrow tubes only to a height of about 0.5 m (20 in), not nearly high enough for tall plants. After considering all these possibilities and types of evidence, plant physiologists were left with just one hypothesis: that water is somehow *pulled* upward. This explanation turned out to be correct.

Transpiration Even though water droplets on strawberry leaves prove that root pressure exists in some plants, and capillary action also can play a role in water transport, the major explanation for water transport in plants lies in a fascinating bit of natural engineering called *evaporative pull.* Tests in plants reveal that when xylem is cut, air moves into the stem and fills the space vacated as water moves up the stem away from the wound, while water moving up from the roots stops at the cut. This observation and numerous other tests led plant scientists to pose the **transpiration-pull theory** (also called the **cohesion-adhesion-tension theory**) for water transport in plants.

Transpiration is the loss of water by evaporation mainly through the stomata on stems and leaves. When the stomata are open, water molecules move from a region of high concentration (inside the leaf cells) to a region of lower concentration (the air surrounding the plant) [FIGURE 35.5].

The physical properties of water help explain how transpiration moves fluid from roots to leaves. As CHAPTER 2 described, water molecules tend both to *cohere* very strongly to other water molecules and to *adhere* to unlike molecules. Within a given xylem pipeline, hydrogen bonds link water molecules to each other in a long, unbroken liquid chain (*cohesion*), while additional hydrogen bonding causes the water to adhere to the cellulose lining of the xylem vessels (*adhesion*).

When water molecules evaporate from an open stoma, the next water molecules in the column tend to move up and replace them, and, in turn, pull along the next water molecules. As a result, the entire liquid column moves upward in the xylem tube, and new water molecules are drawn into the roots below. Because the water molecules pull each other up from above, the entire column is under constant tension. As long as the water column remains unbroken and solar energy causes evaporation, water will keep moving in the plant's vascular tissues. In a similar way, air enters the xylem vessels of cut flowers in a bouquet. But submerging the stems of a bunch of flowers and removing an additional inch or two while they are still underwater will eliminate the section containing the air bubbles, and allow the water column to reestablish itself. Transpiration can then continue, and the flowers can remain crisp longer.

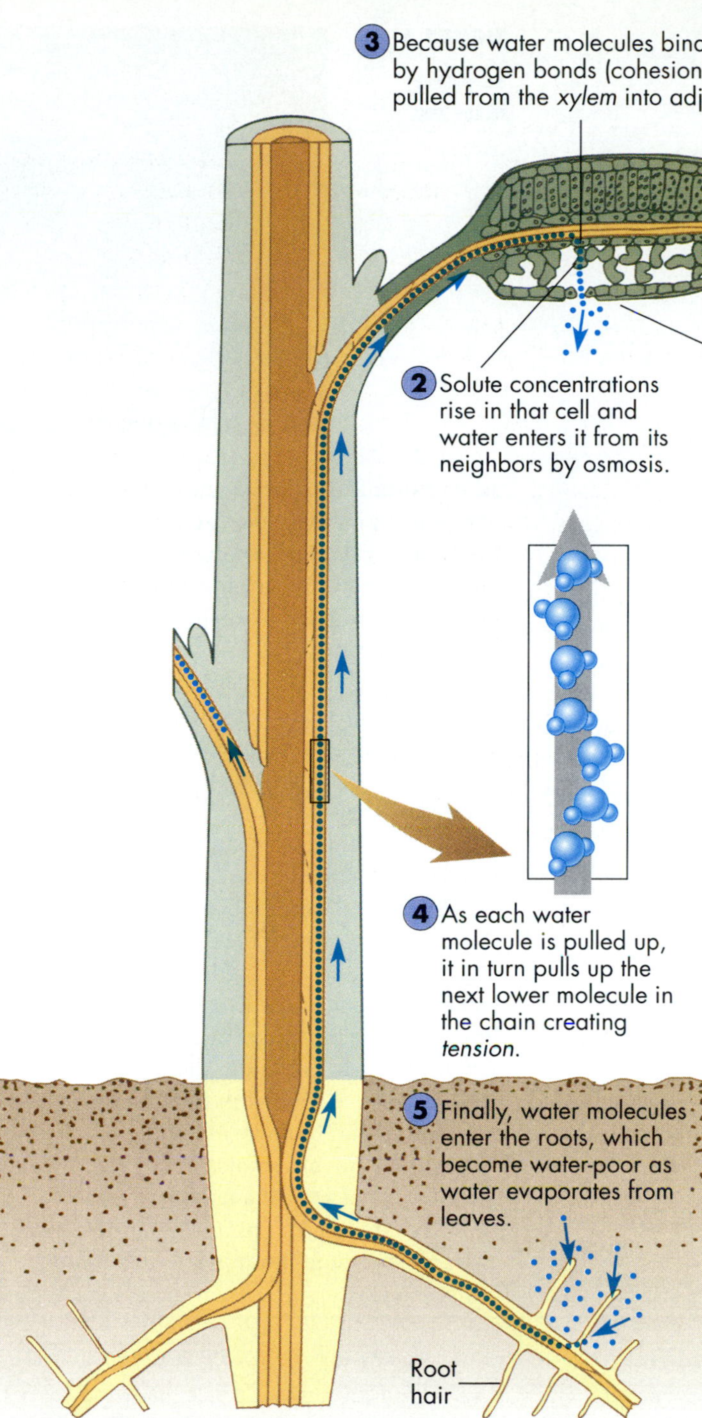

FIGURE 35.5

Transpiration and the Cohesion-Adhesion-Tension Theory for Water Transport.

Transpiration, the evaporation of water from a plant surface, provides the energy that pulls water to the tree tops.

Water Stress: A Break in the Chain On a hot, dry day, evaporation from leaves can outstrip water absorption by the roots, and tension on the chain of water molecules becomes greater and greater. When this occurs, the adhesive forces of water molecules pulling inward against the xylem walls can grow so strong that the diameter of a tree trunk literally shrinks. In hot, dry weather, the tension on the chain of water molecules can become so strained by the rapid upward pull that the column can snap. In fact, plant physiologists with sensitive microphones can actually hear the snapping and popping inside plants on a hot day!

Once the water column breaks, an air bubble forms inside the xylem tube, and transpiration can no longer pull water from above. If within a short period of time the temperature drops, evaporation slows, or the soil becomes damp again, root pressure from below, as well as capillary action within the narrow xylem tubes, can help rejoin the ends of the broken water column. If conditions are not reversed quickly enough, however, wilting can destroy part or all of the plant.

STOMATA AND REGULATION OF WATER LOSS

The degree of transpiration is crucial to a plant's daily survival, and water loss is carefully regulated. Most of the water is lost from a plant's leaves, and a desert tree like the boojum has evolved a rather drastic strategy for minimizing this source of water loss: it simply drops all of its numerous tiny leaves after a drought of even short duration. Plants living in more temperate environments, however, exhibit a more moderate approach: they have control mechanisms that regulate the amount of water lost through the leaves. In the leaf epidermis, each stoma is enclosed by two kidney-shaped guard cells [FIGURE 35.6A]. Like curved water balloons, guard cells arch away from each other when they swell with water, increasing the size of the opening between them [FIGURE 35.6B]. When guard cells lose water and become flaccid, they slump together and close off the stoma, blocking water loss via transpiration [FIGURE 35.6C and D]. In the leaves of the boojum tree, ridges formed by the cells surrounding the stomata protect those structures and probably decrease evaporation from them.

What causes the guard cells flanking the stoma to swell or deflate and thus open or close the plant's "portholes"? Experiments show that when a plant begins to dry out, potassium ions are expelled from guard cells and water follows passively by osmosis. Loss of water from guard cells allows them to slump together, close the pore, and inhibit further water loss. After a rain, when the plant has plenty of water, potassium ions enter guard cells, water follows by osmosis, guard cells swell, the stoma opens, and carbon dioxide can enter the leaf, allowing photosynthesis to continue.

Biology Applied

Water Relations and Strange Seeds

One of the oddest subjects in any discussion of water relations in plants is the way moisture facilitates the catapulting and burrowing of certain unusual types of seeds. In many parts of this book, we've discussed seeds that become dispersed to new sites by blowing on the wind, by sticking to animal fur or feathers, or by dropping off the plant then rolling, floating, or fluttering away from the parent's growth site. Other, less conspicuous kinds of seeds, however, reach new sites after being explosively launched from the parent (ballistic seeds), or after moving across the soil and digging themselves in (self-planting seeds). In either case, water is the key to the startling movements.

In plants, many adaptations for dispersal have evolved that decrease the tendency of seeds to fall to the ground near the parent plant and thereby lessen competition between the generations for limited resources. Many additional adaptations have arisen in plants whose seeds attract hungry predators and tend to get eaten instead of dispersed. Ballistic and self-planting seeds are two such adaptations. Ballistic seeds are often relatively heavy and streamlined in shape and are thrown clear of the parent plant where there is less competition and where it is harder for predators to locate them. Self-planting seeds usually have a spiral-shaped shaft like a threaded screw as well as a rotating, streamerlike "tail" [FIGURE 1]. These structures cause the seed to spiral down into crevices, and that's advantageous because drying out is less of a threat and predators can't find the seeds as easily.

These adaptations for seed dispersal are mechanical responses to changing fluid pressure or humidity. For example, the tiny seeds of certain species of touch-me-nots (plants in the genus *Impatiens*) explode one-quarter to one-half meter outward from the plant if a predator or herbivore brushes up against the adult. Apparently, the brisk movements of leaves and stems cause fluid pressure to build up in each capsule surrounding the seeds, and eventually an explosion propels the seeds in a way analogous to the launching of a soda bottle rocket.

FIGURE 1 Self-planting filaree seed.

Conversely, the seeds of *Erodium cicutarium*, a wild geranium that inhabits Arizona's Sonoran desert, flutter to the ground. Once there, however, the rising humidity of nighttime and the falling humidity of daytime create a wet-dry cycle that causes the awn (the structure in FIGURE 1 that looks like a screwthread) to coil more tightly, or to uncoil, and hence the tail to rotate. This rotation causes the seed to move along the ground (up to 7 cm in 5 days) and also creates leverage if the tail chances to hit a pebble. This leverage can drive the seed's pointed tip further into any crevice it might happen to drop into as

---

In addition to desiccation, several other environmental agents affect guard cell function. These include carbon dioxide, light, temperature, and daily rhythms in the plant. The net result is that stomata are generally open when gas exchange can take place without threat of excess water loss, but are closed when there is a danger of dehydration. Many plants, such as the boojum, that live in arid environments, do not open their stomata at all during the day. Instead, the stomata open and the plant absorbs carbon dioxide only at night when the air is cooler. As a result, less water is lost. The process of photosynthesis is then completed during the day when light is available, but it takes place with the stomata closed [review FIGURE 6.6].

Together, the mechanisms of transpiration and guard cell swelling help maintain a constant supply of water to plant cells—a major aspect of homeostasis. Although we have described these processes for flowering plants, the basic principles also hold for gymnosperms. For these mechanisms to work, plant cells must absorb appropriate quantities of ions and other mineral nutrients.

➤ CONCEPT CHALLENGE

A plant physiologist injects a blue dye into xylem vessels in the stem of a plant and a few hours later makes slices through the stem to determine the distribution of the dye. Will the dye have moved up, moved down, or stayed at approximately the same level as the injection? Does the result depend on the time of day the dye was injected? Support your answers.

the seed moves along the ground. As a result of this humidity-driven (or hygroscopic) movement, the seed becomes dispersed from the parent, escapes from predators, and plants itself in a protected spot.

Biologist Nancy E. Stamp of the State University of New York at Binghamton has studied both explosive and hygroscopic seed dispersal. The most interesting species she described was surely *Erodium moschatum*, another wild geranium nicknamed the white-stemmed filaree whose seeds are both ballistic and self-planting. In this grasslike plant, drying out of the mature flower eventually triggers an explosive ejection of the seed up to 56 cm from the parent plant. Repeated wet-dry cycles then induce seed movement across the soil and self-planting an additional 7 cm or so away from the landing site. This catapulting–ground movement–self-planting is vaguely reminiscent of the hop-skip-jump competition in track and field competition, and affords both dispersal and protection from seed predators.

In variations on this three-step "event," some touch-me-not seeds explode outward then float away on streams or rivulets. And some violet seeds are launched, then structures called elaisomes (fat bodies) attract ants which cart the seeds off and usually deposit them in nutrient-rich debris piles outside their nest openings.

Ballistic and self-planting seeds are significant for what they show about water relations in plants, as well as what they reveal about survival pressures and evolutionary adaptations. Beyond that, they're just interesting—like so many phenomena in biological science.

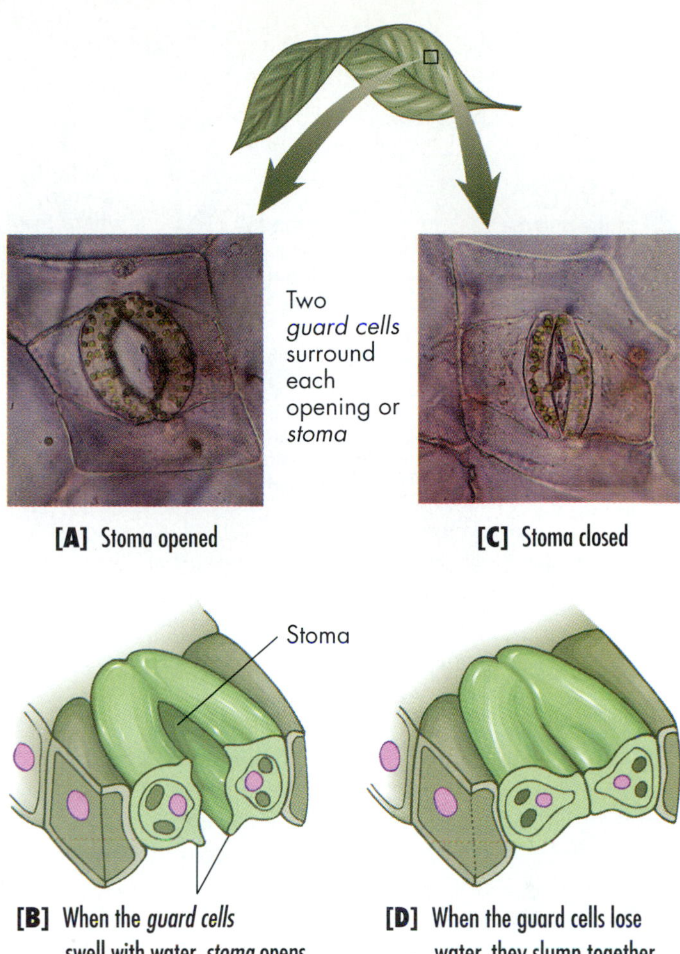

[A] Stoma opened

[C] Stoma closed

[B] When the *guard cells* swell with water, *stoma* opens

[D] When the guard cells lose water, they slump together, closing the stoma

FIGURE 35.6

Stomata: Regulators of Water Loss.

[A] The epidermis on the lower side of a leaf is ventilated by stomata. **[B]** When a leaf's mesophyll cells have plenty of internal moisture, calcium ions and the plant hormone called ethylene cause potassium ions to enter the guard cells, water follows by osmosis, the guard cells swell, and the stoma opens, allowing carbon dioxide to enter the leaf and photosynthesis to resume. **[C]** When mesophyll cells inside a leaf lose turgor pressure because of excess evaporation, they release the plant hormone abscisic acid, which affects the concentration of calcium ions in guard cells. Calcium ions then act as a second messenger, stimulating a series of events that lowers potassium ion concentrations inside guard cells. **[D]** Water then exits the guard cells by osmosis, allowing the cells to grow flaccid and the stoma to close.

Movement of Nutrients Throughout the Plant

Plants have an inherent distribution problem: They generate sugars in their leaves but tend to store organic nutrients in their roots or stems, and use those nutrients in their growing tips, flowers, and fruits. What's more, the functions of various plant parts can change with the seasons. In the tarweed, the root stores carbohydrates in the active growing season, but in the fall, the stored carbohydrates are transported up from the root and used to build a tall stem, flowers, and seeds [FIGURE 35.7]. Since plants lack a pump and circulatory system, how do they distribute photosynthetic products?

TRANSLOCATION: MOVEMENT OF SUGAR

Plant scientists devised some clever methods for studying **translocation**, the movement of nutrients in the phloem of plants. Phloem is so delicate that puncturing the tubes even with a fine glass probe or needle stops the

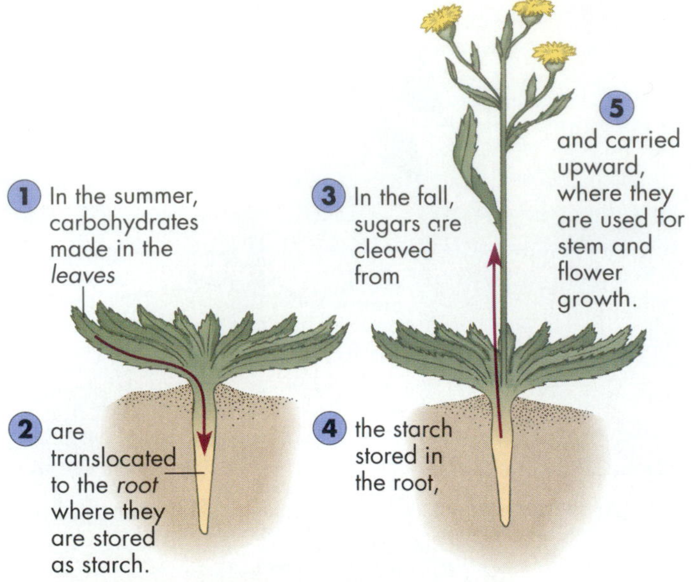

FIGURE 35.7

Seasonal Sugar Transport in the Tarweed.

In summer, sugars produced by photosynthesis in the tarweed's leaves are transported downward to the roots and stored as starch. In fall, the transport is reversed: The starch is broken down and the sugars released are carried upward, fueling the biosynthesis of materials for an elongating stem, flowers, and seeds.

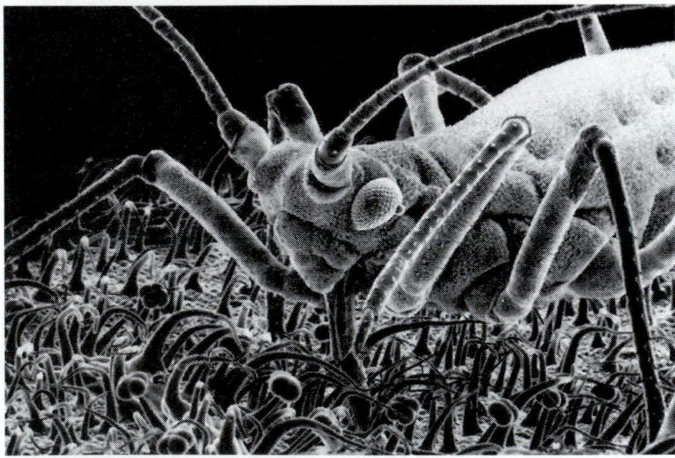

FIGURE 35.8

Aphids Tap Contents of Phloem.

In this scanning electron micrograph, aphids pierce the veins of a tomato leaf with their syringelike stylets.

flow of solutes. Tiny insects called aphids, however, can suck plant sap from phloem cells without disrupting them, and plant physiologists found that if they allow an aphid to insert its sharp feeding tube into the phloem, then cut away the insect's body, the phloem's contents

TABLE 35.1 Plant Nutrients and Their Functions

Nutrients (Percent of Dry Weight)*	Location/Function
Macronutrients	
Carbon (45.0)	In all organic compounds
Oxygen (45.0)	In most organic compounds, including all sugars and carbohydrates
Hydrogen (6.0)	In all organic compounds
Nitrogen (1.0–4.0)	In proteins, nucleic acids, chlorophyll, and coenzymes
Potassium (1.0)	Involved in activating enzymes, protein synthesis, and regulation of osmotic balance
Calcium (0.5)	In cell walls and starch-digesting enzymes; regulates cell membrane permeability
Magnesium (0.2)	In chlorophyll and many cofactors
Phosphorus (0.2)	In nucleic acids, some co-enzymes, ATP, and some lipids
Sulfur (0.1)	In proteins, some lipids, and coenzyme A
Micronutrients	
Iron (0.01)	In electron transport molecules; involved in synthesis of chlorophyll
Chlorine (0.01)	Essential to photosynthesis
Manganese (0.005)	Essential to photosynthesis; activates many enzymes
Boron (0.002)	Unknown
Zinc (0.002)	In some enzymes; involved in protein synthesis
Copper (0.0006)	In chloroplasts and some enzymes
Molybdenum (<0.0001)	Essential to nitrogen fixation and assimilation
Elements Essential to Only Some Plants	
Silicon (0.25–2.0)	In the cell walls of grasses and horsetails
Sodium (Trace)	Essential to a few desert and salt-marsh species
Cobalt (Trace)	Essential for nitrogen fixation

* Concentration of nutrients expressed as percentage of dry weight in a typical plant. Actual proportions vary greatly from species to species.

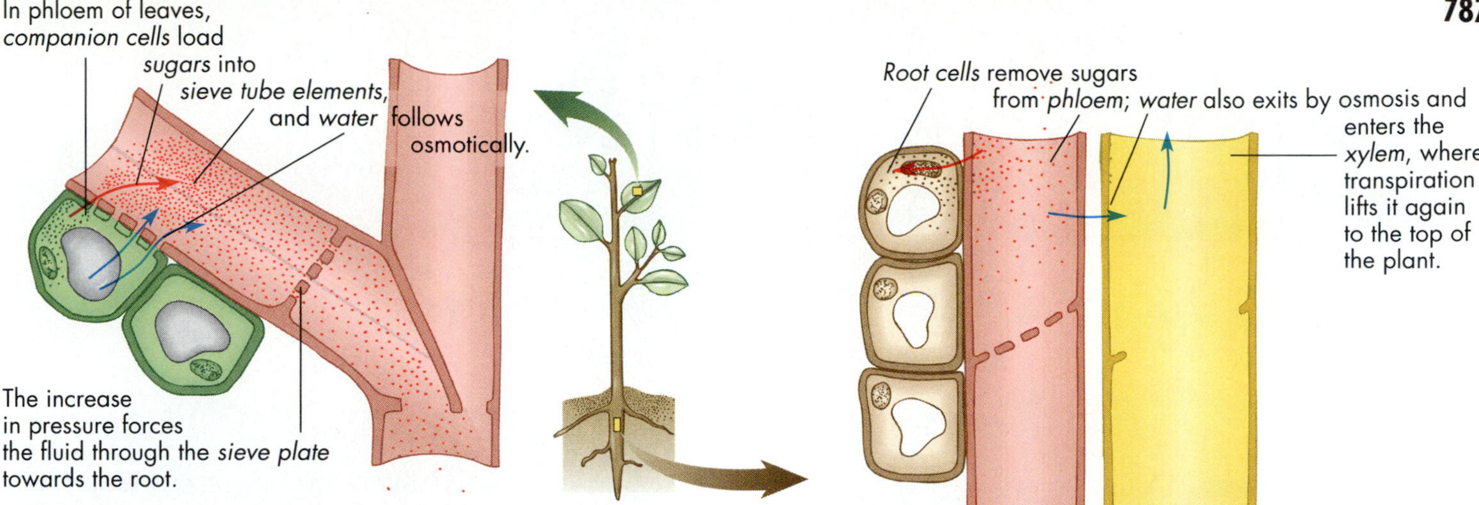

FIGURE 35.9

Translocation: How Plants Transport Sugars.

In leaves, companion cells load sugars into phloem sieve tube elements, water follows osmotically, and sugar solution is forced through the sieve plates and down a pressure gradient. This gradient is formed because root cells with lower concentrations of sugars and amino acids take in the nutrients, making the solution in the phloem tubes more dilute. In a sink area, as water leaves the phloem and reenters the xylem, the turgor pressure in nearby phloem cells also drops.

ooze out of the little tube, so that the plant's sap can be studied [FIGURE 35.8]. With this technique, they discovered that plant sap contains up to 30 percent sugars (mostly sucrose) and about 70 percent water.

Plant scientists also discovered that they can expose a plant's leaves to radioactive carbon dioxide, then trace the path of the carbon through the phloem once it is incorporated into sugars via photosynthesis. In this way, they found out that sugar flow is often downward from leaves to other plant parts and can be 10,000 times faster than normal diffusion. They therefore postulated that translocation involves a mass movement of phloem fluid based on bulk flow [see FIGURE 4.19].

Currently, the best hypothesis for phloem transport is the **mass flow theory** [FIGURE 35.9]. Let's see how this works in a flowering plant. Sugars produced in photosynthesizing leaves are loaded into the phloem's sieve tube elements by the companion cells, an active transport process [see CHAPTER 4]. This energy-costly active transport increases the solute concentration in the phloem, and water follows by osmosis from the nearby conducting elements of xylem. The influx of water plumps up the phloem cells, increasing the pressure within the cell as a result of water movement into the cell [*turgor pressure* (from the Latin *turgere*, "to be swollen")] and forcing the sugary solution through the sieve plates at the ends of each sieve tube member and away from the leaf.

Meanwhile, cells in the root and other parts of the plant remove organic solutes from the phloem, and the solute content becomes so low that the osmotic tendency is reversed. Now water flows out of the phloem and back into the xylem tubes, where it is carried upward again by transpirational pull. Ordinary water pressure, the loading activities of companion cells, and the unloading of sugars by root cells are thus behind the movement of sugars, amino acids, and a few mineral ions from one part of a plant to another.

In the early spring, a similar mechanism probably causes plant sap to rise in the stems of biennials and perennials. At that time, the roots begin breaking down stored starch into sugar and loading it into the phloem's sieve tube members. Water from the xylem follows, and as the roots bring in more water from the soil, enough pressure is created to push the sap up the phloem into the trunk of the tree. New Englanders carefully tap into the phloem of maple trees in early spring, drain off some of the sweet maple sap, and boil it down to make maple syrup.

In a sense, sugar translocates itself, since its production and use create osmotic pressure and mass flow. Other nutrients, however, are moved passively by the flow of fluid in both the xylem and the phloem.

TRANSPORT OF INORGANIC SUBSTANCES IN XYLEM AND PHLOEM

Nutrients other than carbon, oxygen, and hydrogen—including all the other nutrients listed in TABLE 35.1 [page 786]—can be transported in either of two ways: (1) Some, such as calcium, move only in the xylem; they are carried along in the column of water that moves upward from the roots to the leaves via transpiration. These nutrients are deposited permanently wherever they leave the xylem and enter a living cell, and they cannot be re-

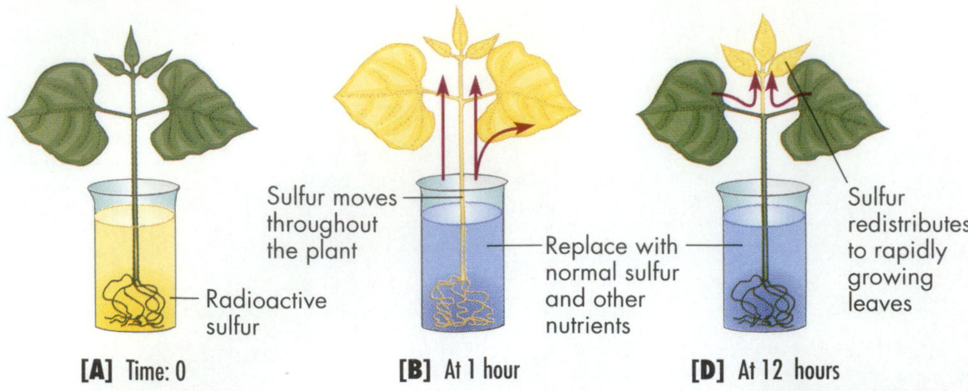

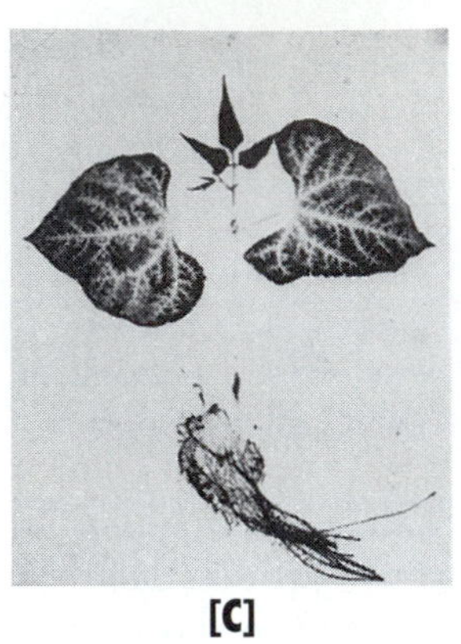
[C]

[E]

FIGURE 35.10

Movement of Labeled Sulfate Reveals Transport of an Inorganic Ion. In a classic experiment, a researcher allowed the roots of a bean plant to absorb a radioactive sulfur compound for an hour [**A**], then switched the plant to a solution of normal sulfur and other nutrients [**B**]. He then photographed the plant at various times with a technique (autoradiography) that reveals the location of the radioactive sulfur. After an hour or so, the radioactive sulfur had been carried throughout both older, larger leaves and newer, smaller ones, presumably via the xylem [**C**]. After 12 hours, however, the sulfur (now presumably incorporated into amino acids and proteins) had been transported, by the phloem, from old to new leaves [**D, E**]. The researcher inferred that xylem initially carried water and nutrients upward to all parts of the plant; then later, phloem redistributed the sulfur, now incorporated in biological molecules, to the rapidly growing parts of the plant. In the ACTIVE LEARNING BOX on page 789, we trace the movements of sugar within a morning glory plant using similar techniques.

distributed to the other parts of the plant. (2) Others, such as sulfur or phosphorus, can be redistributed, via the phloem, to new tissues as needed, carried along passively with the sugars and other phloem contents. FIGURE 35.10 shows an experiment demonstrating that a mineral element like sulfur goes first from roots to newly forming leaves, then is redistributed to still newer leaves as they form. Plant scientists inferred that sulfur moves first in the xylem, then in the phloem to growing leaves.

These two different modes of transport explain why each mineral deficiency affects a plant in a different way. The symptoms of calcium deficiency, for example, always appear first in new leaves. The old leaves contain sufficient quantities of calcium, but the mineral cannot move out of these leaves—even if they are dying—and be transferred in the xylem to the new leaves. In contrast, the symptoms of phosphorus deficiency always appear first in old leaves; as the plant's environment becomes deficient in the element, the phosphorus in older leaves is mobilized and moved through the phloem to areas of new growth. Thus, in some plants, the old leaves will have the characteristic yellow veins of phosphorus deficiency long before the new leaves show it.

➤ CONCEPT CHALLENGE

A Vermont family taps maple trees and makes syrup from the sap in the early spring. What material are they harvesting? Where does it come from in the plant? What forces the sap to flow?

Plant Nutrients

Around 1600, a Dutch physician performed what may have been the first quantitative experiment in biology. Jan Baptista van Helmont questioned an assertion made by Aristotle centuries earlier that a plant's body derives most of its substance from soil. To test this hypothesis, he filled a container with 91 kg (200 lb) of dry soil and planted in it a 2.2-kg (5-lb) willow tree. For five years van Helmont watered the tree with rainwater. At the end of that time, he dug the tree out of its pot, removed the soil and dried it, weighed both soil and tree, and discovered that while the tree had gained 75 kg (165 lb), the soil had lost a mere 27.5 g (about an ounce). Thus, van Helmont succeeded in showing quite clearly that the main mass of the willow tree came not from the soil, as Aristotle had suggested, but from some other source.

Today, plant scientists know that plants synthesize organic compounds from the carbon and oxygen in carbon dioxide, from the hydrogen in water, and from small amounts of minerals in soil [review FIGURE 6.7]. Fully 96 percent of a plant's dry weight is made up of carbon, oxygen, and hydrogen. The dozen or so chemical elements

active learning: Pathways for sugar

Improving the efficiency of agriculture is an important goal for biological science, since feeding our world's growing human population will require it. For optimum biological efficiency, plants must transport optimum quantities of photosynthetic product to the organ we harvest for food—the seed, in the case of wheat and beans, and the root in the case of sugar beets. Plant scientists want to learn to manipulate the amounts of photosynthetic products moving to various parts of the plant. First, however, they must know the pathways these products follow. For example, are the photosynthetic products of all leaves transported equally throughout the plant? Or do sugars made by some leaves tend to move into new leaves and flowers at the tips of stems, while sugars made by the other leaves tend to move into the roots?

To answer this question, plant physiologists allowed a single leaf in a morning glory plant to undergo photosynthesis in the presence of radioactivity labeled carbon dioxide for 24 hours.

The leaf synthesized radioactively labeled sugars via photosynthesis, and these "hot" sugars were then transported in the plant.

The researchers exposed the plant to film. Wherever radioactivity was present, the film became darkened.

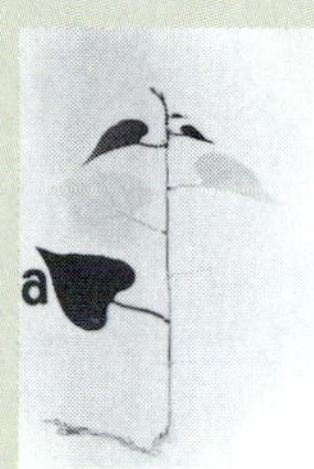

The results above show what happened when a lower leaf (labeled a) was exposed to radioactive carbon dioxide. Did the labeled sugars become equally distributed throughout the plant? If not, where did they tend to go?

. . . .

Next, the botanists exposed an upper leaf to radioactive carbon dioxide and got the following results.

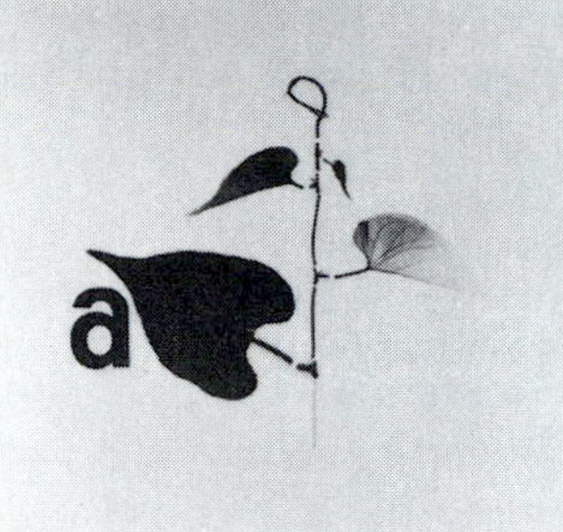

Where did the labeled sugars go from the upper leaf?

. . . .

Contrast the results from the upper and lower leaves?

. . . .

What is the significance of these results for farmers growing a root crop like sugar beets or a seed crop like beans?

. . . .

S. Dewey and A. Appleby, A Comparison between Glyphosate and Assimilate Translocation Patterns in Tall Morning Glory (*Ipomoea purpurea*). *Weed Science* 31 (1983): 308–314.

that make up the remaining 4 percent, however, are equally essential to plant survival.

Plants require at least 16 different chemical elements, the so-called *essential elements*, which different species require in different amounts. All plants require the **macronutrients** in relatively large amounts, and the **micronutrients** in much smaller amounts. By drying many types of plants, weighing the dried plant matter, and analyzing its chemical content, plant physiologists were able to tabulate information on the nine needed macronutrients—carbon, oxygen, hydrogen, nitrogen, potassium, calcium, magnesium, phosphorus, and sulfur—and the seven micronutrients—iron, chlorine, manganese, boron, zinc, copper, and molybdenum [TABLE 35.1]. Plants use both macro- and micronutrients in amounts roughly comparable to the levels required by animals. Just as animals suffer vitamin and mineral deficiency diseases if deprived of essential nutrients, plants also show specific symptoms relating to specific deficiencies.

MACRONUTRIENTS

The macronutrients are a plant's fundamental constituents, making up most of the atoms in carbohydrates, proteins, lipids, and nucleic acids. Because we have already discussed the roles of carbon, oxygen, and hydrogen in biological molecules and reactions in CHAPTERS 2 through 6, we will concentrate here on the other six macronutrients.

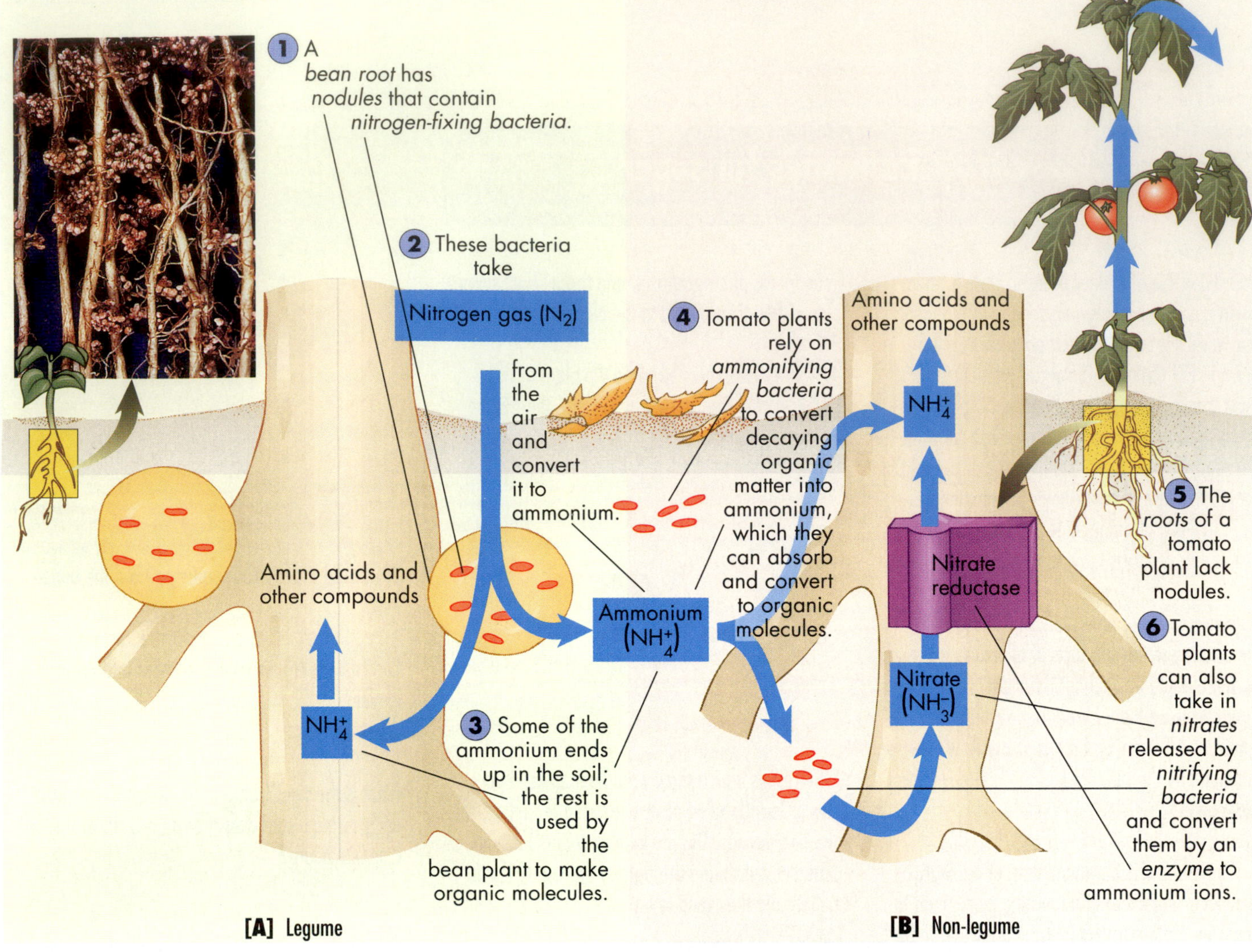

FIGURE 35.11

Nitrogen-Fixing Bacteria in Root Nodules Help Nourish Some Plants.

[A] Nodules on the roots of legumes, such as a bean plant. **[B]** Most nonlegumes, such as a tomato, lack nodules, and must obtain usable nitrogen directly from the soil.

NITROGEN FIXATION

After carbon, oxygen, and hydrogen, nitrogen is the most important of the macronutrients because it is an essential component of amino acids (and hence proteins), chlorophyll, coenzymes, and nucleic acids. Nitrogen is frequently the most important growth-limiting nutrient—the less nitrogen available, the slower the plant grows. Although gaseous nitrogen (N_2) makes up 78 percent of the earth's atmosphere, plants cannot use it directly. Instead, the nitrogen must be *fixed*, or converted from the simple gas N_2 into some other form, such as *ammonia* (NH_3) or *nitrate ions* (NO^{3-}), by a process called **nitrogen fixation**. These nitrogen compounds can then be modified further and incorporated into plant proteins.

Farmers and home gardeners often add fixed nitrogen to the soil in the form of nitrate-containing fertilizers produced by industrial processes. Wild-growing plants, however, do not have such a benefit and must depend instead on nitrogen-fixing bacteria, including cyanobacteria, which can convert molecular nitrogen to usable forms. Many of these microorganisms live independently in the soil and reduce nitrogen gas to ammonia. Some of the most interesting ones, though, live in the root cells of certain vascular plants [FIGURE 35.11]. Peas, beans, alfalfa, clover, lupine, and other **legumes**, or members of the pea family, have swellings, or **nodules**, on their roots that house nitrogen-fixing bacteria. Certain nonleguminous plants, such as alder trees, also form associations with nitrogen-fixing bacteria.

FIGURE 35.11 shows how two kinds of plants, a legume, such as the bean plant, and a nonlegume, such as the tomato plant, get enough nitrogen. When nitrogen-fixing *Rhizobium* bacteria invade the soybean's root system and form nodules, the bacteria can convert nitrogen from the air trapped in the spaces between soil particles into ammonia (NH_3). *Ammonifying bacteria* that live in the soil can make the same conversion. The ammonia molecules quickly pick up hydrogen ions, forming ammonium ions NH^{4+}.

If root nodules produce ammonium ions in excess of the plant's needs, the ions are released into the soil, where they join ammonium ions generated by ammonifying bacteria in the soil, which act on decaying organic matter [FIGURE 35.11B]. Plants such as tomatoes, which lack nodules, take up ammonium ions from the soil. Ammonium ions that remain in the soil serve as an energy source for *nitrifying bacteria*, which convert ammonium ions to nitrate molecules (NO^{3-}). Tomato plants can take up nitrate and by means of enzymes convert it into ammonium ions. Once in the plant, ammonium is the starting point for the biosynthesis of amino acids and many other nitrogen-containing biological molecules.

Nitrifying bacteria, which convert ammonium to nitrate in the soil, cannot survive in the acidic, waterlogged environment of bogs. Thus, *carnivorous plants*, such as the Venus flytrap [see FIGURE 4.1] and the pitcher plant, have evolved the ability to gain fixed nitrogen from the proteins of the insects and other small animals they trap and digest.

The formation of root nodules is a classic example of *symbiosis* (a close association of two dissimilar organisms): The plant supplies the bacteria with high-energy carbohydrates, while the bacteria provide fixed nitrogen produced at a high energy cost. The excess ammonium ions released from nodules provide a classic example of community interrelationships among plant species. For centuries, farmers have rotated crops to take advantage of such relationships. Although early farmers may have been unaware of the microbiological basis, they observed that if they grew clover or alfalfa one year, the following year's crop of corn or wheat would grow more luxuriantly. Likewise, rice farmers have encouraged the growth of water ferns in their flooded rice paddies because cyanobacteria living symbiotically in the ferns fix atmospheric nitrogen and enrich the soil for growth of the rice plants [review FIGURE 19.10].

OTHER MACRONUTRIENTS

Plants often require 4 to 40 times more nitrogen than they do the remaining five macronutrients, but these other elements are still essential to normal plant growth and development. Calcium acts as an intracellular messenger that controls cell membrane permeability, and thus plays a role in the opening and closing of stomata, directional growth in plant cells, responses to changing day lengths, and the tendency of roots to grow down and shoots to grow up [see CHAPTER 37]. Calcium is also an important component of *pectin*, a substance that glues adjacent cells together, and enables young plants to make strong cell walls. Pectin is used in jams and jellies to make them jell.

Potassium regulates osmosis in plant cells such as guard cells, and also helps activate enzymes, including those involved in protein synthesis. Potassium deficiency causes mottled or burnt-edged leaves [FIGURE 35.12A]. Magnesium is a macronutrient because magnesium atoms are components of chlorophyll molecules and cofactors of many kinds of enzymes. Phosphorus occurs in the backbone of DNA and RNA molecules, in ATP and other high-energy compounds, and in membrane phospholipids. Tomato seedlings deficient in phosphorus have purple leaves [FIGURE 35.12B]. Because sulfur is an important component of two amino acids, plants require it for building most proteins, as well as for manufacturing some fats and coenzymes.

[A]

[B]

[C]

FIGURE 35.12

Nutrient Deficiencies in Tomato Plants: Mineral Deficiencies Produce Distinctive Symptoms.

[A] A potassium deficiency causes curled leaves. **[B]** Phosphorus deficiency leads to purple leaves in a tomato seedling. **[C]** Iron deficiency renders the fine leaf veins green, but leads to yellowing in the rest of the leaf.

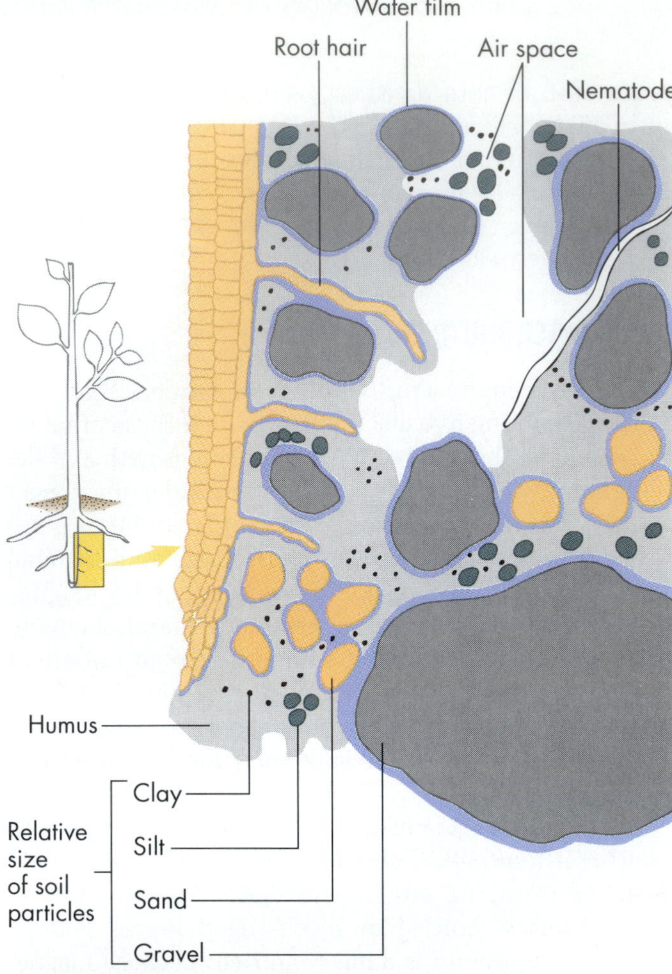

FIGURE 35.13

Soil: A Source of Air, Water, and Nutrients for Plants.

Soil has inorganic and organic particles, as well as pore spaces filled with air and lined with water. Root hairs probe those spaces and take in water and minerals that dissolve from soil particles.

MICRONUTRIENTS

Plants require only small amounts of micronutrients for healthy growth [see TABLE 35.1, page 786]. Because plants require such tiny quantities, deficiencies of the micronutrients are rare. Iron, for example, occurs in several proteins within the energy-harvesting mitochondria and is also involved in the synthesis of chlorophyll. A tomato plant with an iron deficiency does not make enough chlorophyll to mask the yellow pigments in leaves [FIGURE 35.12C]. A deficiency of copper (present in chloroplasts and certain enzymes) can result in severe deformation of stems, leaves, and fruits in many plant species. A deficiency of chlorine in a tomato plant will stunt the roots and fruits and wilt the entire plant. Because zinc plays a role in protein synthesis, zinc-deficient apple and peach trees become stunted and grow miniature leaves.

SOIL: THE PRIMARY SOURCE OF MINERALS

Soils are the source of all macro- and micronutrients beyond the carbon, oxygen, and hydrogen that plants take in from air and water. Soil composition has a major influence on the kinds of plants (and indirectly, the kinds of animals) that can grow in a particular region of our planet.

Soil is a mixture of organic and inorganic material that starts with bedrock, which is relatively unbroken and unweathered rock. Over time, the action of water, wind, heat, and cold disintegrates the bedrock and produces the inorganic parts of soil—particles varying in size from coarse sand and silt to small clay particles [FIGURE 35.13]. Bacteria, fungi, algae, lichens, and plants extract minerals from rocks, sand, silt, and clay and use them to synthesize organic materials. When the organisms die, the decomposing organic matter in the soil becomes *humus*.

Air spaces between soil particles are crucial to plant life. In a soil with a mixture of particle sizes, the spaces contain about half water and half air, with the water forming a continuous film over the soil particles. The size of the soil particles, however, determines the soil's water-holding capacity: Soils made up primarily of coarse sand tend to hold water poorly, while soils made up mostly of clay particles hold water tenaciously. Plants such as the boojum tree and cacti, which grow in rapidly draining sandy soils, absorb water quickly and store it, while plants that grow well in clay soils have roots resistant to

FIGURE 35.14

Mycorrhizae Allow Faster Growth of Loblolly Pine through Better Nutrient Absorption.

Even in good soil, a young loblolly pine tree lacking mycorrhizae (right) is half the size of one that has the fungus-root association (left).

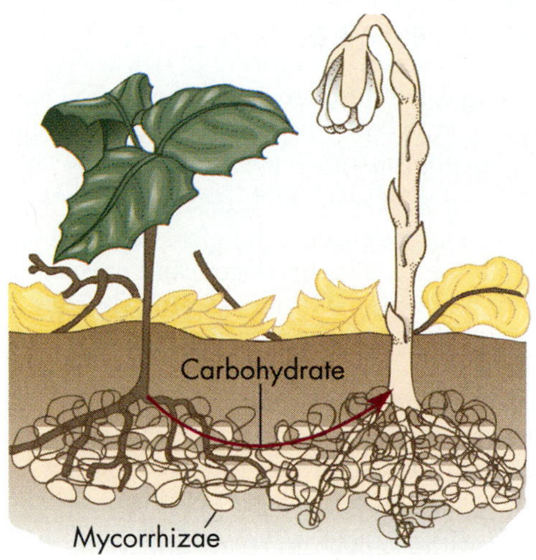

[A] A mycorrhizal bridge

[B]

FIGURE 35.15

The Fungal Connection: A Mycorrhizal Bridge Transfers Sugars from Green Plant to Indian Pipe.

[A] The Indian pipe (*Monotropa uniflora*) has an obligate relationship with mycorrhizal fungi that are associated with a second plant, in this case a green, actively photosynthesizing angiosperm. The fungus forms a bridge that transfers carbohydrates from the photosynthetic plant to the Indian pipe. [B] Indian pipes pop up on the forest floor and, lacking chlorophyll, look ghostly white.

dense, soggy, surroundings deprived of air. The best agricultural soils are deep layers of *loam* possessing a high mineral and humus content and a mixture of particle sizes.

HOW NUTRIENTS ENTER A PLANT

Several things happen before a plant can utilize mineral nutrients from soil. Mineral molecules must first dissolve in the layer of water surrounding soil particles. Then the minerals must move into the root and be distributed to all parts of the plant. Specifically, minerals must pass through the plasma membranes of the root hairs, move through the cells of the root cortex and the endodermis cytoplasm, and be secreted into the xylem [review FIGURE 35.3]. There, the powerful pull of transpiration can lift them (along with water) through the xylem, and the minerals can be transported to all plant organs. The active transport of ions into the roots against a gradient of concentration requires a good deal of energy. Expending this energy to procure necessary inorganic nutrients is simply a cost of staying alive.

The development of millions of fine root hairs provides an enlarged surface area for absorbing water and nutrients. In addition, the symbiotic association of roots and highly specialized mycorrhizal fungi can expand the absorptive surface area still further [review FIGURE 20.5]. When mycorrhizal fungi infect plant roots, they often cover the roots with a spongy mantle of threadlike hyphae. The hyphae can extend outward up to 8 m (26 ft) from the root and even penetrate the roots of nearby plants. In nutrient-poor soils, mycorrhizal fungi provide their hosts with greater concentrations of inorganic nutrients as well as growth-promoting hormones, such as auxins and cytokinins. Some plants, including cultivated citrus and pine, grow far more efficiently with mycorrhizae than without [FIGURE 35.14].

Mycorrhizae occasionally interconnect unrelated plants and even allow a few odd plant species to parasitize others. Indian pipe, a small white plant of the forest floor that lacks chlorophyll, has mycorrhizae on its roots that connect it to the roots of nearby trees, absorbing nutrients from the parasitized plant [FIGURE 35.15].

➤ CONCEPT CHALLENGE

A typical bag of fertilizer from a garden store will have a label reading "5-10-10," which means it is 5% nitrogen, 10% phosphoric acid (a source of phosphorus), and 10% potash (a source of potassium). According to TABLE 35.1, however, plants need far more carbon, oxygen, and hydrogen than they do the elements in the fertilizer. So why doesn't the fertilizer contain mostly carbon, oxygen, and hydrogen?

Connections

In the boojum tree, water and nutrients move the length of the plant from roots to stems and leaves and sometimes back again. The trunk stores water in specialized cells. The leaf stomata open only at night and close during the day. In times of drought, the entire crop of leaves falls to the ground. Carbohydrates made in the leaves move throughout the plant through the phloem. And once each year, the boojum flowers, making millions of tiny winged seeds that are dispersed upon the dry desert winds. What happens to these seeds once they hit the ground? How does the tiny embryo they carry develop into a seedling? These are the questions we address in our next chapter.

KEY TERMS

cohesion-adhesion-tension theory, 782
macronutrient, 789
mass flow theory, 787
micronutrient, 789
nitrogen fixation, 790
nodule, 790
soil, 792
translocation, 785
transpiration, 782
transpiration-pull theory, 782

HIGHLIGHTS IN REVIEW

1 Water tends to move upward in plants from roots to leaves. As a water molecule evaporates from a leaf, it is replaced by diffusion. Due to the tendency of water molecules to bind together, a continuous column of water molecules contained in the xylem moves upward through the plant.

- **a]** Water and minerals enter root hairs, move through or between parenchyma cells of the root cortex, pass through the cell membrane and cytoplasm of endodermal cells, then enter the vascular cylinder. From there, xylem cells pipe the materials upward.
- **b]** The vascular cylinder can build up a low level of root pressure (as evidenced by guttation), but not enough to push water and minerals to the tops of tall trees.
- **c]** Water inside xylem is under tension. Evaporation of water molecules from the leaves (transpiration) draws water molecules out of nearby cells, which is replaced by water molecules from the xylem. Because of the tendency of water molecules to form hydrogen bonds, a chain of water molecules from leaves to roots is pulled up a bit for each molecule that evaporates from a leaf. This is the transpiration-pull theory for water movement in xylem.
- **d]** Plants lose enormous quantities of water owing to transpiration through tiny holes called stomata. Stomata open when the plant uses additional carbon dioxide but has sufficient water, and they close when the plant has too little water but sufficient amounts of carbon dioxide.
- **e]** Each stoma is flanked by two guard cells whose shape is controlled by an osmotic mechanism; when turgid, these cells pull apart and open the stoma; when flaccid, they slump together and close the stoma.

2 Organic molecules move about in the phloem from places of formation, such as a photosynthesizing leaf, to places of use, such as to the root of a carrot for storage. Osmosis drives the movement of organic substances in phloem.

- **a]** When leaves photosynthesize, they are a source of sugar, while the developing and growing parts of a plant are sinks that use this sugar. Roots are sinks when they absorb sugar from phloem for their own metabolic needs or when they store it as starch. But roots are sources when they release the sugars from the stored starch to fuel activities in other parts of the plant.
- **b]** Sugars and other organic molecules are translocated throughout the plant. In the leaves, companion cells load sugar in phloem's sieve tube members, and water is pulled into the phloem osmotically. The resulting pressure forms a gradient from source areas—where cells actively load sugars into the phloem—to sink areas—where cells unload sugars from the phloem. This pressure gradient pushes the solution of sugar and water from the source to the sink.
- **c]** Inorganic nutrients can be transported in two ways. Some elements, like calcium, are carried in the xylem to a final destination, where they remain. Others, such as sulfur and phosphorus, move passively through cell membranes and can be redistributed via the phloem.

3 Plants, like animals, require certain chemical elements as nutrients. Plants obtain the carbon, hydrogen, and oxygen they need from the air and water, and other nutrients from the soil. Plants obtain usable nitrogen through associations with bacteria.

- **a]** Plants are made up mostly of carbon, oxygen, and hydrogen from carbon dioxide and water. Plants depend on the soil, however, for at least 13 other essential elements, including macronutrients, needed in relatively large quantity, and micronutrients, needed in relatively small quantity.
- **b]** In many situations, the main growth-limiting element is nitrogen, but plants cannot absorb its plentiful gaseous form directly from the air. Instead, they can utilize nitrogen found in ammonium and nitrate. Legumes absorb ammonium ions from the *Rhizobium* bacteria in their root nodules. Nonlegumes absorb excess ammonium ions released by a legume's roots, as well as ammonium and nitrates fixed by soil bacteria. In both leguminous and nonleguminous plants, ammonium ions are the starting point for biosynthesis of amino acids and other useful biological molecules.
- **c]** Other macronutrients include potassium, calcium, magnesium, phosphorus, and sulfur, and deficiencies of each have characteristic symptoms in plants.
- **d]** Although micronutrient deficiencies are rare, plants deprived of iron, chlorine, manganese, boron, zinc, copper, molybdenum, and other elements also show distinctive symptoms.
- **e]** Soil is composed of a combination of inorganic particles (sand, silt, and/or clay) and organic particles from the decomposing remains of dead organisms (humus).
- **f]** The spaces between the particles of soil are filled with water and air and supply water and oxygen to plant roots.
- **g]** Plants extract nutrients from the soil, but eventually the nutrients are returned to the soil through nutrient cycling.
- **h]** Plants absorb nutrients from the soil through their own root hairs or with the aid of mycorrhizal fungi, which live on and in the root cells.

UNDERSTANDING THE FACTS AND CONCEPTS

For Questions 1–5, match each of the descriptions with the most appropriate item from the following list. Each answer may be used once, more than once, or not at all.

a] stomata	d] guttation
b] guard cell activity	e] root pressure
c] Casparian strips	f] capillary action

1 Osmotic loss of water due to potassium loss determines the diameter of the entry portals for air into the leaf.

2 Structures that form a barrier to passive flow of water into the center of the root.

3 The creeping of water that results from a combination of cohesion and adhesion.

4 Droplets of water that appear on leaves as a result of root pressure.

5 Movement of water up from the root as a consequence of osmosis into cells of the xylem.

For Questions 6–10, match each description with the most appropriate item or items from the following list.

a] transpiration
b] adhesion
c] cohesion
d] evaporation
e] transpiration-pull

6 The binding of water molecules together by the formation of hydrogen bonds between them.

7 A theory of upward movement of water column in the xylem because of loss of water molecules from the top of the column.

8 The loss of water by a plant through its leaves and stems.

9 The bonding of water molecules to molecules other than water.

10 The movement of molecules of liquid water into the gaseous phase.

For Questions 11–15, match each of the descriptions with the most appropriate item from the following list. Each answer may be used once, more than once, or not at all.

a] soil
b] humus
c] loam
d] micronutrients
e] macronutrients

11 Nitrogen, calcium, potassium, phosphorus, and sulfur, among others.

12 Iron, copper, and chlorine, among others.

13 The part of the soil that comprises decomposing leaves and other organic matter.

14 An agriculturally rich kind of soil.

15 Begins as inorganic bedrock that changes over time and acquires an organic component.

As above, for Questions 16–20, match each description with the most appropriate item from the following list.

a] nitrogen fixation
b] nitrifying bacteria
c] ammonification
d] legumes
e] symbiosis

16 Occurs both in nodules and in soil, producing NH^{4+}, which is used for amino acid biosynthesis.

17 Soil bacteria that convert ammonium ions to nitrate ions.

18 An association of mutual benefit between two dissimilar organisms such as nitrogen-fixing bacteria and leguminous plants.

19 Plants that house nitrogen-fixing bacteria in root nodules.

20 Conversion of N_2 into ammonia or nitrate ion.

INTEGRATE AND APPLY WHAT YOU HAVE LEARNED

1 "In plants, water moves only in a one-way path from roots to leaves, and it does so only in the xylem." Do you agree or disagree with the statement? Explain.

2 Do water and ions move from the soil to the plant and then throughout the plant passively by diffusion, by means of active transport, or by both means? Explain.

3 The best time for maple sugaring is in the early spring when the days are warm and the nights are cold; however, the sap only flows during the day. Explain.

4 What general principle of nutrient transport in plants is illustrated by the facts that symptoms of calcium deficiency appear first in new leaves whereas symptoms of phosphorus deficiency appear first in old leaves?

5 What is the evidence for and against root pressure, xylem pumps, and capillary action as the most essential mechanism for moving water from roots to leaves?

ANALYSIS

1 The root-pressure and transpiration-pull theories of mineral transport both incorporate the forces of adhesion and cohesion, but in different ways. What is the major difference? More than one answer may be correct.

a] In the transpiration-pull theory, adhesion is dominant, whereas in the root-pressure theory cohesion is the dominant force.
b] Adhesion is dominant in the root-pressure theory; cohesion is the dominant force for transpiration-pull.
c] In both theories, adhesion and cohesion play equally important roles, but the direction of the forces is different.
d] In transpiration-pull, a continuous column of water is being pushed, whereas with transport powered by root pressure a continuous column of water is being pulled.

2 What has occurred when the leaves of a plant wilt? More than one answer may be correct.

a] The chain of water molecules in the xylem has probably been broken.
b] Transpiration is no longer occurring.
c] Root pressure is probably insufficient to raise the level of water to the leaves or to the discontinuous water region.
d] The movement of water by root pressure is not occurring.

3 In an attempt to keep salad greens crisp, a chef puts them in a strong potassium chloride solution. What do you predict will be the result?

a] The salad greens will quickly wilt as the cells lose water by osmosis.
b] The salad greens will become crisp as the cells gain water by osmosis.
c] Potassium will enter the leaf, water will follow, and the guard cells will swell, resulting in a greater entry of carbon dioxide into the greens.
d] Potassium will enter the leaf, water will follow, the guard cells will swell, and the stomata will close.
e] The salad greens will remain unchanged in crispness until they die from lack of photosynthesis.

CHAPTER 36

Plant Reproduction and Development

CORN: AN ENORMOUS SUCCESS

Some historians have called the domestication of corn "humankind's greatest agricultural achievement." Some biologists have called corn a "monstrosity" and an "agricultural artifact" derived from a "basically useless weed." All observers agree, however, that while corn is America's biggest cash crop, its existence will always be tenuous—at least in evolutionary terms—because corn as we know it cannot reproduce without human intervention.

FIGURE 36.1
Corn

Archaelogical evidence shows that people living in Mexico or Central America began to domesticate a Mexican plant called *teosinte* more than 7000 years ago. This small plant, which still grows in the lowlands west of Mexico's Balsas River, produces many small ears, each consisting of single rows of hard kernels attached to each other and not to a cob. Each kernel is a seed covered with a hard outer shell—a seed that can fall free of its neighbors, survive the drought and cold of winter, and germinate the following growing season. Teosinte seeds are obviously successful vehicles for dispersing and perpetuating teosinte plants. However, the seeds are not edible because the hard clinging seed coat can't be easily separated from the starchy material inside.

By selectively breeding teosinte plants for greater and greater production of usable starch, native horticulturalists working in their fields thousands of years ago created a high-yielding crop that fed them and their domesticated animals, and remains today the world's second biggest source of food calories after wheat. In the United States, corn is, and has long been, king: Farmers raise

$20 billion worth of corn (more than wheat, oats, rice, rye, barley, and sorghum put together) on 70 million acres (a collective cornfield the size of Arizona). They sell nearly half of it as animal feed for American cattle, hogs, and poultry; export about 17 percent to Japan, Russia, South Korea, and other populous countries; store one-quarter of it; and supply most of the rest to manufacturers of corn oil, corn syrup, corn starch, and ethanol (for engine fuel and hard liquor). People eat just 1.4 percent of the corn crop as fresh and canned sweet corn, corn chips, corn tortillas, corn flakes, and corn meal for cooking.

The domestication of corn was clearly a major agricultural feat in terms of the enormous yield we now enjoy. Modern corn produces just a few short side branches ending in a gigantic ear containing hundreds of soft, starchy kernels. Even if an ear of corn could separate from its thick supporting stalk and crash to the ground, the green outer husks would prevent dissemination of the seeds and encourage rotting. For a new generation of corn to be produced, people must harvest the ears, remove husks and seeds from the cobs, store the seeds over winter, and sow the seeds in well-prepared soil.

Since virtually all commercial corn varieties are hybrids of different strains with desirable features, the reproduction of superior seed for agriculture requires human assistance, as well: Workers must transfer pollen grains from male flowers to female. Recall that each flower of most flowering plant species contains both male and female flower parts. Corn, however, has two types of flowers: male flowers on the tassel at the plant's main growing tip [FIGURE 36.1] and female flower clusters or spikes at the ends of the side branches (the future ears). Workers at a seed company must deliberately brush the female flowers of one strain with pollen from male flowers of another strain to obtain the vigorous, healthy hybrid seed farmers require. The farmer plants the hybrid seed, and the corn embryos within the seed grow into stout, disease-resistant plants. The farmer then harvests the full ears, and sells them for consumption by humans or other animals.

Because corn is so central to today's agriculture, and because humans are so central to corn reproduction (and with it the survival of the plant species), corn is a good central example for our discussions in this chapter.

With corn as our focus, we will review the life cycle of flowering plants first described in CHAPTER 20, then examine the structures of male and female flower parts and how they produce pollen and eggs. Next, we will discuss pollination, then fertilization, including the double fertilization unique to flowering plants. We will follow the typical angiosperm life cycle through the development of seeds and fruit, and development of the embryo. And finally, we will see how biotechnologists are harnessing the details of plant reproduction in their development of better crops.

As we go along, we will see not only how sex determination in corn is similar to that in humans but also how, in a common plant, desirable genes can be transferred from one individual into another. ❑

MESSAGES

1 Flowering plants alternate between a large diploid generation that forms spores by meiosis in flowers and a haploid generation that produces egg and sperm. Egg and sperm join at fertilization, forming a zygote, a new diploid generation.

2 The zygote grows into a diploid embryo contained within a seed, which can grow into a new large diploid plant.

3 Many biologists hope that plant biotechnology, the application of our knowledge of plant reproduction, development, and molecular genetics to crop plants, will improve our chances of being able to feed the billions of new human mouths that will appear in the next few decades.

The Angiosperm Life Cycle

A key to understanding plant reproduction and development is the realization that plants cycle through a multicellular haploid generation as well as a multicellular diploid generation. The latter is the plant we usually see and identify, such as the corn plant or pear tree. In contrast, an animal cycles through a single-celled haploid phase (egg or sperm) and a multicellular diploid form—for example, the familiar cat, mouse, or mosquito. In certain land plants, such as mosses and ferns, the two multicellular generations within the plant life cycle are both easily visible [review FIGURES 20.15 and 20.18]. In seed plants, this situation is rare; the haploid generation is usually too small to see without a microscope. An overview of the corn life cycle shows how this *alternation of generations* takes place.

In flowering plants such as corn, the diploid generation is large, but the haploid generation is small. A tall, green corn plant growing in a field is a diploid individual [FIGURE 36.2, Step 1]. Since certain cells in this diploid plant undergo meiosis and produce haploid reproductive cells called **spores**, it is called the spore-producing plant, or *sporophyte* (*phyte* means "plant"). The corn sporophyte's female flowers, or ears, produce large female haploid spores called *megaspores* (Steps 2 and 3). Corn's male flowers, or tassels, produce small male haploid spores called *microspores* (Steps 5 and 6).

Haploid spores divide by mitosis and produce tiny haploid plants that will form gametes, and hence, these little individuals are called *gametophytes*, or gamete-producing plants. The miniature female gametophyte consists of just seven cells, which are totally dependent on the much larger diploid sporophyte for support (Step 4). One of these cells is the egg, the female gamete (Step 8). The male gametophyte is the pollen grain (Step 7); it contains just three cells when mature, including two sperm, the male gametes (Step 8).

FIGURE 36.2
Life Cycle of a Flowering Plant.

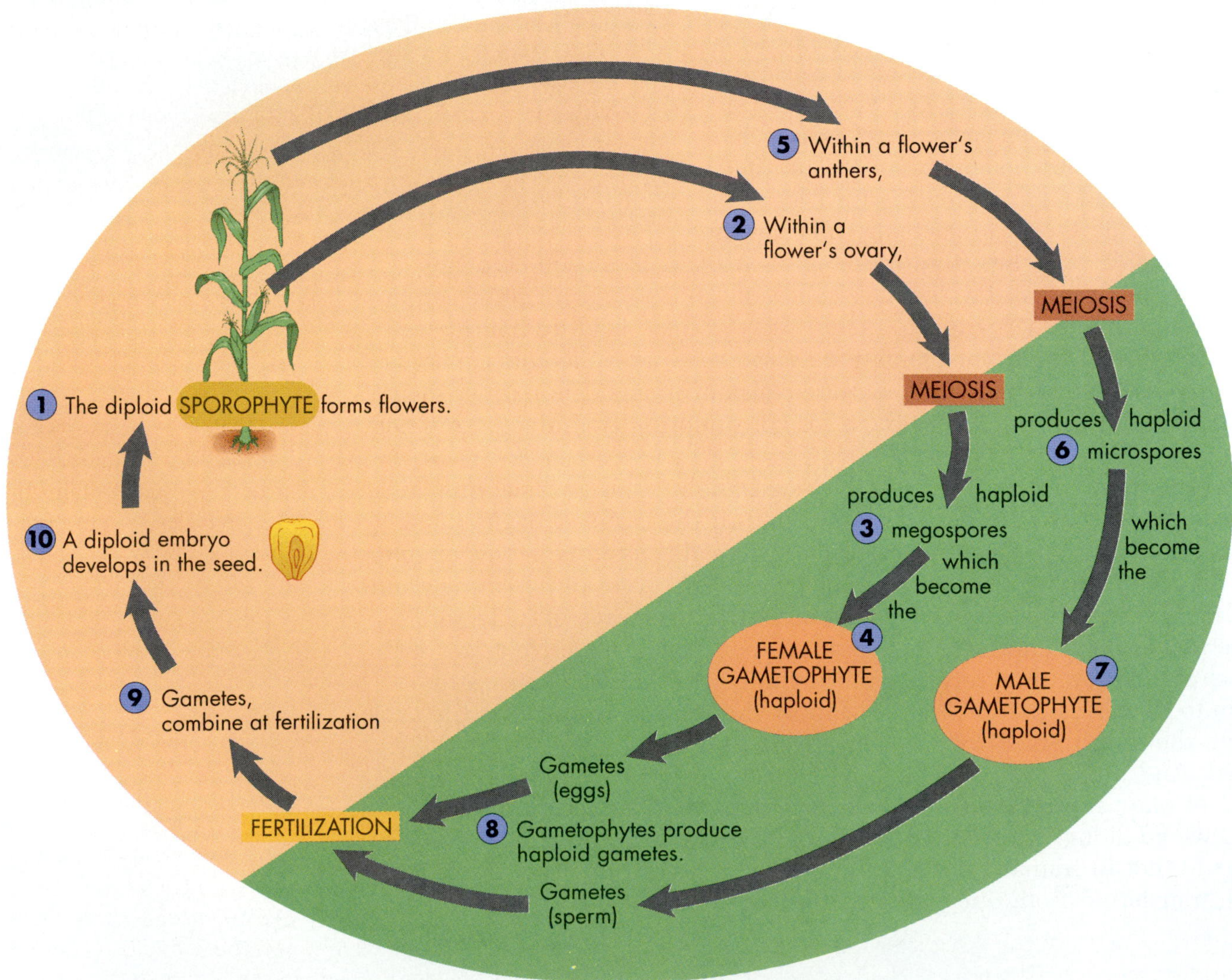

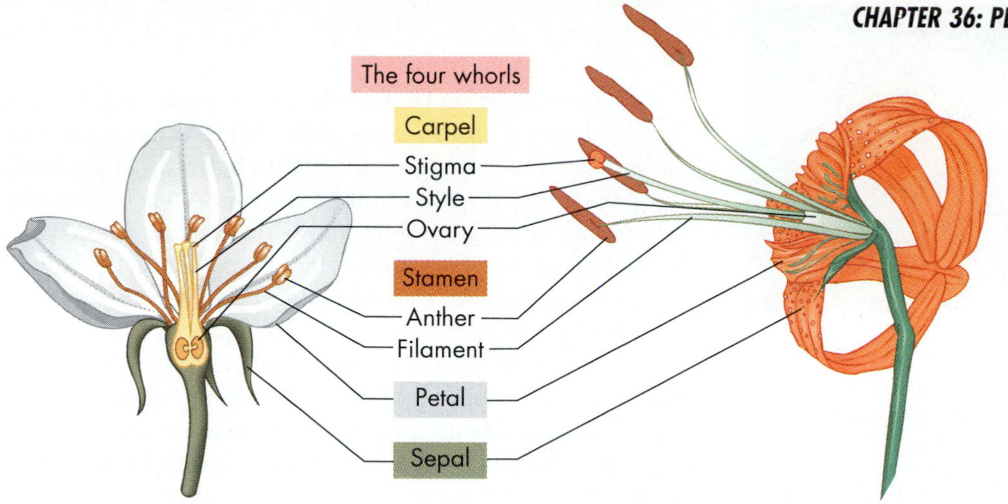

The four whorls
Carpel
Stigma
Style
Ovary
Stamen
Anther
Filament
Petal
Sepal

[A] Pear, a dicot

[B] Lily, a monocot

FIGURE 36.3
Flowers: Beauty and Variety.
The flowers of the **[A]** pear tree and **[B]** lily reveal the many shapes of these plants' sexual structures.

The egg and sperm, the two haploid gametes, unite via fertilization (Step 9), and the result is a single diploid cell, the *zygote*. The zygote divides by mitosis and gives rise to the embryo inside the seed (Step 10). When the seed is planted, the diploid embryo grows into another leafy green corn plant (Step 1), and the cycle begins again.

➤ CONCEPT CHALLENGE

Suppose travelers from earth to a distant planet discovered humanoids (peoplelike organisms) there whose reproductive cycle resembled the corn plant. What might the sex lives of these aliens be like?

Flowers: Sex Organs Of Flowering Plants

Although we seldom think of it this way, when we give flowers to our sweethearts, we are actually offering them the flamboyant sex organs of plants! Flowers are highly modified shoots that contain the reproductive organs of roses, grasses, trees, and other flowering plants. Flowers produce pollen grains containing sperm nuclei and sacs containing egg nuclei [review FIGURE 36.2]. The various parts of the flower help ensure the transfer of pollen, with its sperm nucleus, to the egg inside the embryo sac. Let's look at those flower parts and how they function in plant reproduction.

FLOWER STRUCTURE: THE FOUR WHORLS

Flowers generally consist of four rings of structures; from the outside in, these are the sepals, petals, stamens, and carpels [FIGURE 36.3]. Only the structures in the inner two rings produce gametes. The outer two rings, while often quite beautiful, are sterile. The **sepals** in the outermost ring and the **petals** in the second ring can have showy colors and shapes and fragrances that attract animal pollinators. A pear blossom, for example, has a whorl of five leaflike green sepals surrounding a second whorl of five bright white petals that attract bees. Nutritious pollen grains and sweet nectar are often the pollinator's reward. In a lily, sepals and petals are both colorful [FIGURE 36.3B].

Structures in the two inner rings of the flower produce eggs and sperm. The male reproductive structures, the pin-shaped **stamens** (STAY-menz) produce the pollen. Note in that pear flowers contain 20 stamens. Within each stamen, the *anther* contains two pollen sacs, and sits high on a *filament*, (the "shaft of the pin"). This long filament aids pollen dispersal, and is especially helpful in wind-pollinated plants such as corn. Making up the innermost ring of structures is one or more **carpels** (Greek *karpos*; "fruit"), the female parts, which are often shaped like a wine bottle. These may fuse together, as do the five carpels in a pear blossom. The base of the carpel, or *ovary*, houses the *ovules*, the structures that contain the egg and that later mature into the seed. The "neck" of the ovary, or *style*, supports the *stigma*, a sticky surface on which pollen grains germinate. Each strand of silk in an ear of corn [review FIGURE 36.2] is a stigma. Pollen landing on the stigma grows through the style toward the ovules, and a sperm in the pollen tube fertilizes the egg.

You can think of the four whorls or parts that constitute a flower as modified leaves, even though sepals are the only parts that look much like leaves in many flowers, including pear blossoms. We know they are a type of leaf in an *Arabidopsis* plant, a small member of the mustard family, because mutations in just three genes can turn the flower into a mass of leaves [FIGURE 36.4]. Apparently, during evolution genes arose that altered leaf development. These changes promoted the formation of flowers, and hence improved reproduction.

[A]

[B]

FIGURE 36.4

Flowers, Leaves, and Mutations.

Mutations in just three genes can change a flower into a mass of leaves. **[A]** A normal flower of the mustard family weed *Arapidopsis thaliana* contains four petals arrayed like a cross. **[B]** Mutations in the *apetala2*, *apetala3*, and *agamous* genes result in tufts of leaflike organs where the flowers would normally be.

The numbers and sizes of flower parts vary from species to species; a lily, a monocot, has flower parts in multiples of three, while a pear, a dicot, has flower parts in multiples of five [see FIGURE 36.3]. In both lilies and pears, each flower contains both male and female parts. Botanists call them *perfect flowers*. In other species, such as corn or willows, the flowers contain either male or female parts, but not both. Botanists call them *imperfect flowers*. Corn has imperfect flowers since the mature tassel contains only male flower parts and the mature ear of corn kernels contains only female flower parts.

Some species, corn included, have separate male and female flowers on the same plant. These plants are considered *monoecious* (Greek for "one house"). In contrast, willows and American holly have two types of individual plants, some with only male flowers and others with only female flowers. Such plants are considered *dioecious* ("two houses"), since male and female live in two separate plants.

When most people think of flowers they think of the large, showy blossoms found in florists' shops. While these flowers play an important role in the sex lives of the plants, the elaborate petals and pungent smells do not act like the splendid tail of a peacock and attract members of the opposite sex. Instead, fragrant and/or flamboyant flowers attract animals and are pollinated by them. For example, bee-pollinated flowers are often bright blue or yellow, and have distinctive patterns called "honey guides" that direct the bee to the nectar. The patterns of bee-pollinated flowers are sometimes invisible to us, but bees, with their ability to see light in the ultraviolet range of the spectrum, can see the guides clearly. Flowers pollinated by birds and some moths are often bright red—a color these animals see well. Flowers pollinated by night-flying animals such as bats and certain kinds of moths are often white and very fragrant, and in many cases, open only at night.

In contrast to the showy nature of animal-pollinated flowers, plants that are pollinated by the wind are generally not brightly colored or highly aromatic. The flowers

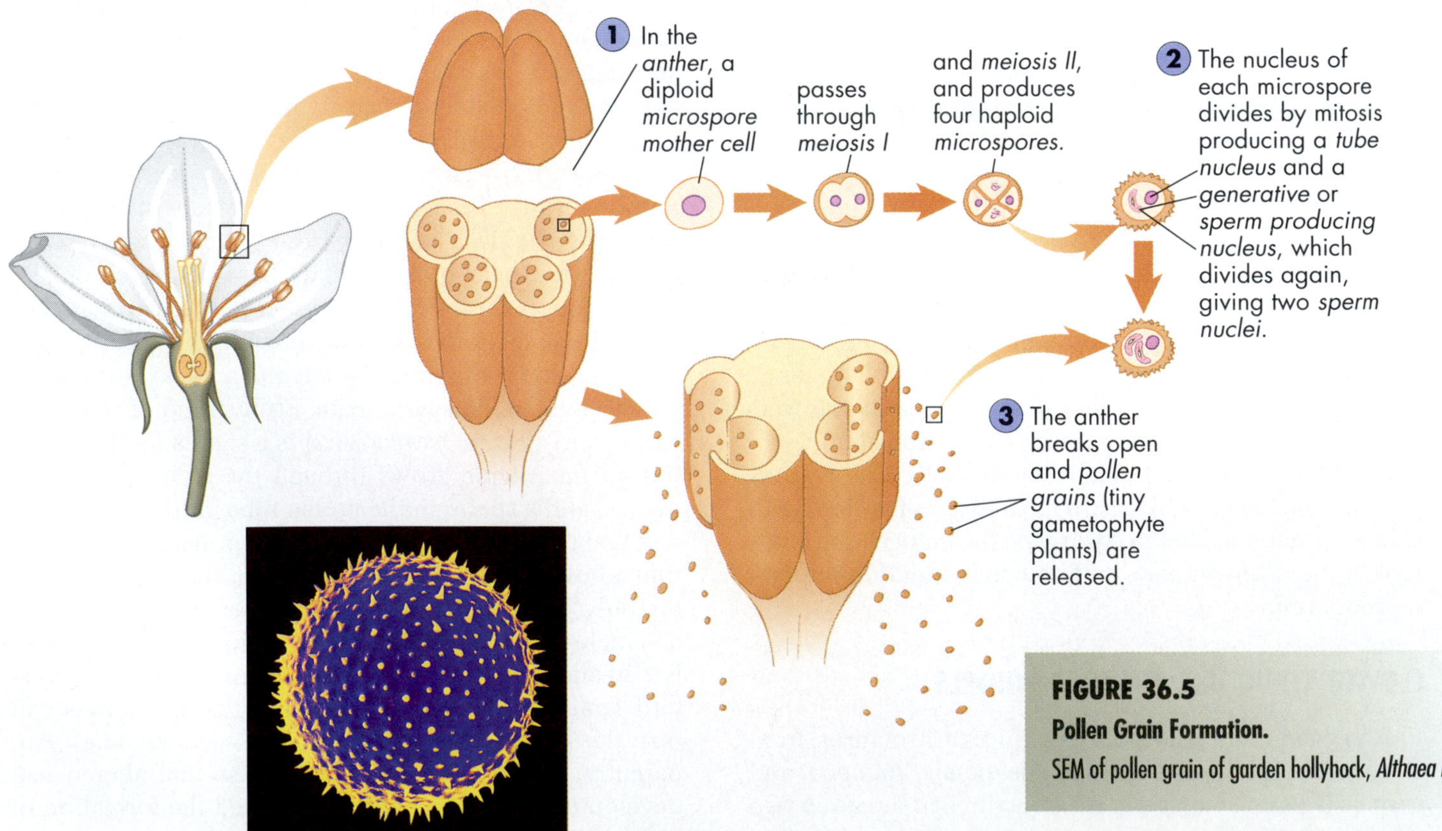

FIGURE 36.5

Pollen Grain Formation.

SEM of pollen grain of garden hollyhock, *Althaea rosa*.

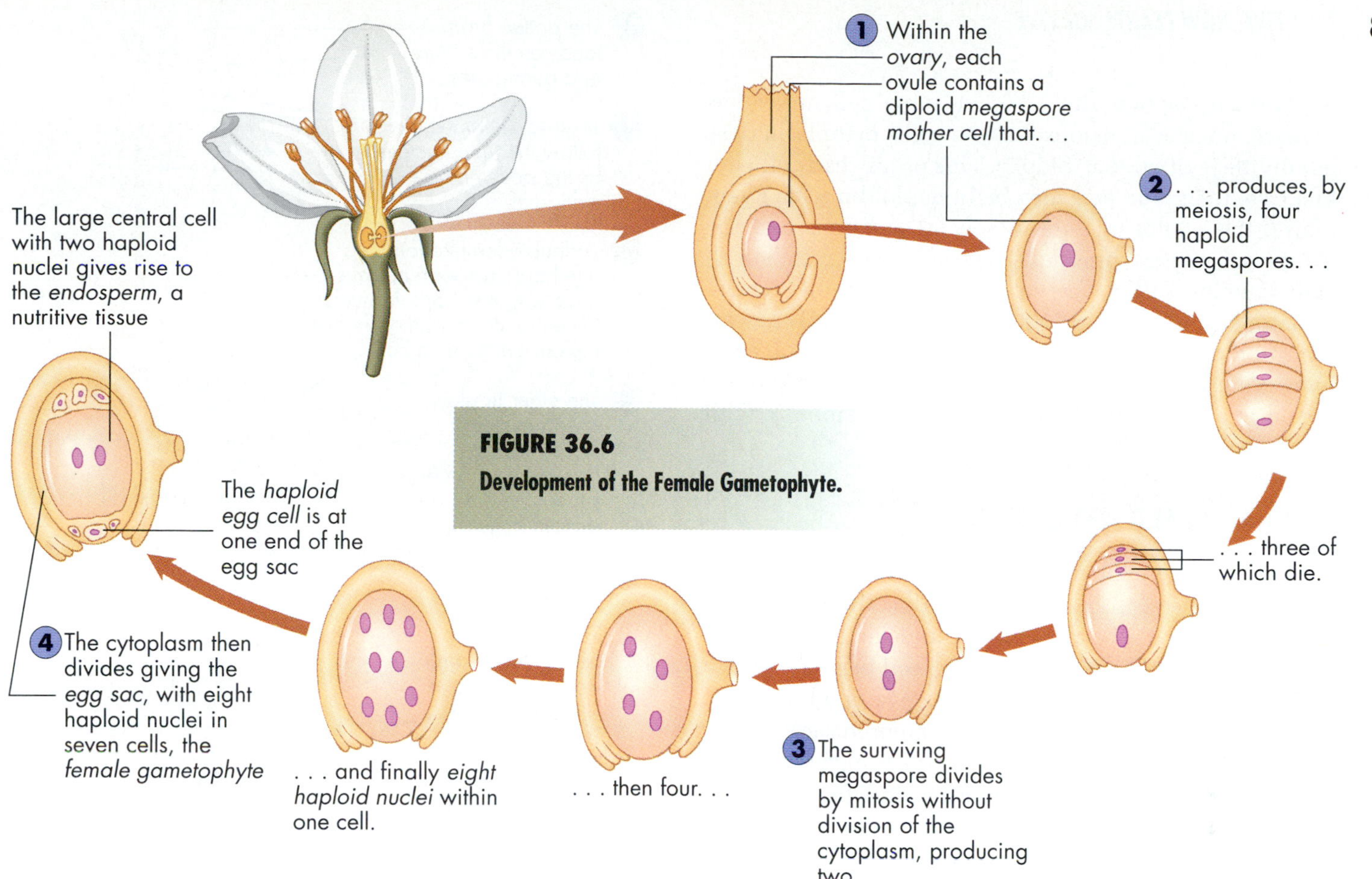

FIGURE 36.6
Development of the Female Gametophyte.

of corn plant and of birch and hazelnut trees grow on branches where the wind can whip them about and disperse the pollen on air currents. Some plant species, such as peas [review FIGURE 8.4], are self-pollinating, and require neither wind nor animals. In pea flowers, the petals form a tiny, boxlike enclosure around both male and female flower parts that excludes pollen from any other individual.

To summarize our discussion of flower structure, a flower is part of the diploid plant generation. It generally consists of four whorls of structures. The outer two whorls (sepals and petals) may attract pollinators, while the inner two whorls form tiny haploid plants that produce sperm and eggs. Let's go on, now, to see how those haploid gamete-producing structures form during development.

HOW FLOWERS MAKE SPERM AND EGGS

Due to the alternation of generations in flowering plants, tiny gamete-forming plants (gametophytes) make the sperm and eggs contained in flowers. The sperm is made by pollen grains, the male gametophyte, and the eggs are made in an embryo sac, the female gametophyte.

Pollen Grains and Sperm Pollen grains form in the anthers of stamens [FIGURE 36.5]. Initially, all cells in an anther are diploid, since they are part of the plant's spore-producing generation, or sporophyte. Special cells (called microspore mother cells) soon appear [FIGURE 36.5, Step 1]; these cells undergo meiosis and produce haploid spores called **microspores**, which are generally quite small. Each haploid microspore develops a tough and often highly decorated cell wall.

Inside the cell wall of the microspore, the haploid nucleus divides by mitosis and two cells form [FIGURE 36.5, Step 2]. One cell eventually becomes the pollen tube during pollination, and the other divides again by mitosis, and becomes two sperm cells, the gametes. This produces a three-celled ball, the *pollen grain*, the male gametophyte (Step 3). In many people, these structures cause the seasonal annoyance of runny noses and itchy red eyes we call hay fever. It is easy to understand why allergies arise if you consider that a typical corn plant will produce 170 pollen grains per silk in the tassel, and a total of 25 million in the whole tassel! That's 25,000 per future corn kernel. The numbers of pollen grains for other wind-pollinated plants such as goldenrod and pine are equally impressive.

Egg Sacs and Eggs Eggs form in the flower's carpels [FIGURE 36.6]. In a carpel, the inner wall of each ovary begins to form small mounds of cells. These develop into ovules, which in turn eventually become the **seeds** (Step 1). In an ovule, special cells (called megaspore mother cells) undergo meiosis and produce four large haploid spores, called **megaspores** (Step 2). Three of these four haploid megaspores die. The surviving cell undergoes mitoses, nuclear divisions, but the cell's cytoplasm does not divide [FIGURE 36.6, Step 3]. This produces a cell with first two, then four, and then eight haploid nuclei. Eventually, the nuclei migrate to special positions and the cytoplasm divides, producing a group of seven cells with eight nuclei called the **egg sac**; this is the female gametophyte (Step 4). Of the seven cells, one is the egg, and another, the largest, containing two nuclei, will combine with a sperm nucleus and contribute to the **endosperm**, the tissue that provides the embryo with nutrients.

POLLINATION AND FERTILIZATION

As you have seen, the diploid sporophyte (spore-producing plant) generates the spores, both the microspores and megaspores. These become the haploid gametophyte (gamete-producing plant)—the pollen and egg sac—since they form sperm and egg. Having developed a pollen grain containing sperm cells and an embryo sac containing an egg cell, the plant is ready for the transfer of pollen to female flower parts (**pollination**) and the fusion of male and female haploid cells (**fertilization**).

Let's consider how pollination and fertilization occur in a pear tree. Pollination takes place in spring, when bees carry pollen from the anthers of one pear blossom to the stigma of another [FIGURE 36.7, Step 1]. Once a pollen grain has become stuck on a stigma, it begins to germinate. One of the three haploid cells in the pollen grain engineers the growth of a pollen tube down the length of the style to the ovary, while the other two cells, the sperm, travel through that tube and reach the ovule and embryo sac with its eight haploid nuclei [FIGURE 36.7, Step 2].

Angiosperms such as pear trees or corn plants experience a *double fertilization*: One haploid sperm nucleus fuses with two haploid nuclei in the central cell, forming the triploid nutritive tissue, or endosperm, that will nourish the embryo within the seed (Step 3). Two favorite human foods are triploid endosperm: coconut milk and the fluffy white part of popcorn, which is endosperm exploded by the heat of popping. The other haploid sperm nucleus fuses with the haploid egg nucleus, and thus forms the diploid zygote (Step 4). The zygote divides mitotically and forms the plant embryo, which matures into the diploid sporophyte, the familiar pear tree or corn plant (Step 5). The ovule wall surrounding the zygote and endosperm becomes the seed coat; the entire ovule becomes the seed; and the ovary wall surrounding the seed becomes the fruit (Step 5).

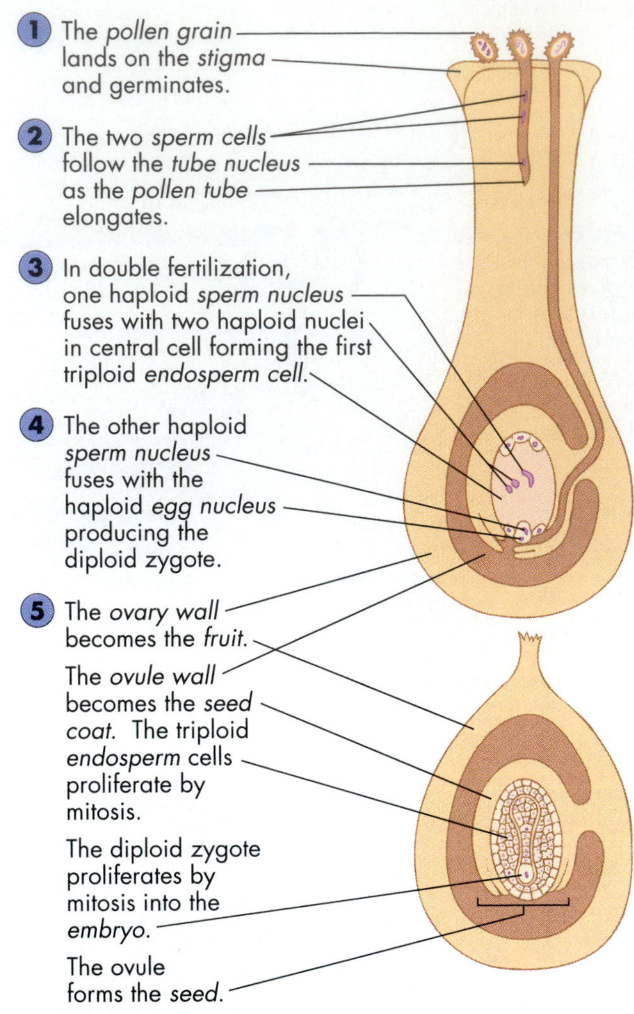

FIGURE 36.7
Pollination and Fertilization.

Coevolution of Plants and Their Pollinators As we saw in CHAPTER 20, flowers, fruits, and seeds help ensure pollination and seed dispersal by attracting animals. Most flowers attract insect, bird, or mammal pollinators that inadvertently carry loads of pollen from one plant to another as they forage for sweet nectar or for the protein-rich pollen grains. A flying or walking animal dusted with pollen, and scouting for more of its favorite food, will substantially increase the chances of pollination. And once fruits have developed on a plant, animals are attracted either by sight or smell to harvest and eat the fruits. The tough, highly resistant coats of many plant seeds allow them to pass through an animal's digestive tract intact. The seeds are then deposited in new locations, and this helps the plant spread.

The classic example of a pollinator is a bristle-covered bee, its body loaded with clinging pollen, accidentally ferrying pollen grains from flower to flower as it searches for food [FIGURE 36.8A]. Pollinators, however, and the flowers that attract them, come in an amazing range of sizes and types. Knobthorn acacia trees in South Africa, for example, are pollinated by giraffes [FIGURE 36.8B]. And plant researchers recently discovered a plant, *Rafflesia arnoldii*, with the world's largest and pos-

[A]

[B]

[C]

FIGURE 36.8

Flowers and Pollinators: A Coevolution.

[A] A honeybee laden with pollen visits an equally pollen-rich flower. [B] Knobthorn acacias in South Africa are pollinated by giraffes. [C] The world's biggest recorded flower, *Rafflesia arnoldii*, of the Sumatran rain forest, smells like rotten meat and attracts fly pollinators.

sibly worst-smelling flower. *Rafflesia*, a parasite on plants in the grape family, grows on the flanks of Mt. Sago in Sumatra. Each *Rafflesia* flower is more than 2 ft across and smells exactly like rotting meat; this odor attracts flies as pollinators [FIGURE 36.8C].

The mutual dependence of flowering plants on pollinators and seed dispersers, and of these animal species on the same plants for nutrition, is the result of coevolution, a natural selection for increasing interdependence based on selective advantages for both parties. As we've seen in other chapters, this coevolution has produced specialized physical structures that help ensure the trade-off of nutrients for pollination and seed dispersal. Knobthorn acacias, for example, produce hundreds of pounds of sterile flowers, high up in the treetops, filled with pollen grains that cling well to giraffe facial and neck hairs. The tall animals can eat their fill of the sweet flowers without endangering the tree's survival, since only a small percentage of the flowers are fertile and, when pollinated, develop into seed pods.

In summary, flowers are the diploid, spore-producing plant's means of forming haploid gamete-producing individuals during the alternating generations of the plant's life cycle. The net result is the fusion of haploid sperm and egg, and the formation of the diploid zygote, which grows into a new diploid spore-producing plant.

➤ CONCEPT CHALLENGE

The *macho* mutation in snapdragons changes the sepals into carpels and the petals into stamens. What conclusions about the embryonic relationships of different flower parts might a plant geneticist draw from this finding?

Development of Embryos, Seeds, and Fruits

The next time you start to eat a pear or an apple, you can apply your knowledge of plant anatomy while examining the food you are eating [FIGURE 36.9]. Opposite the stem

[A] Growth of a pear

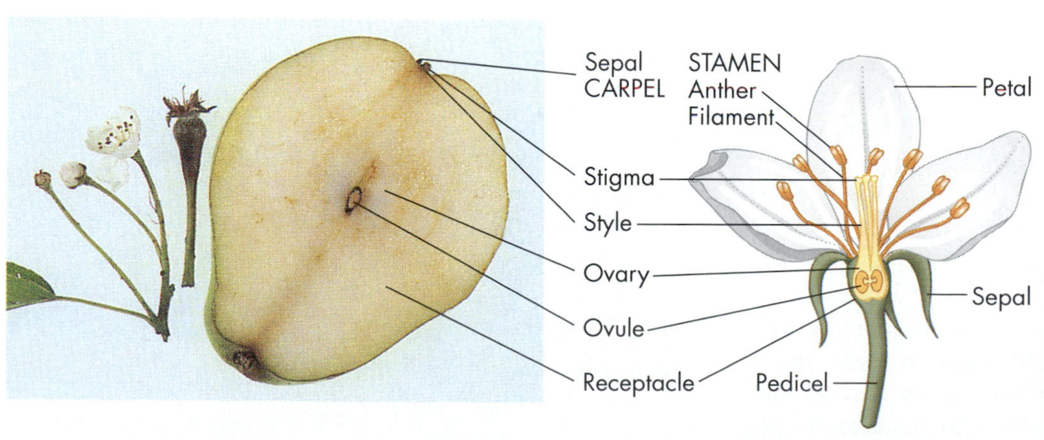

[B] Pear flower

FIGURE 36.9

Pear Fruit and Flower.

[A] The remnants of a pear flower—the styles, filaments, and sepals—decorate an inverted pear fruit. The flower stalk (receptacle) expands enormously into the fleshy succulent fruit. [B] Each part of the flower has a specific role in pollination, fertilization, or seed protection.

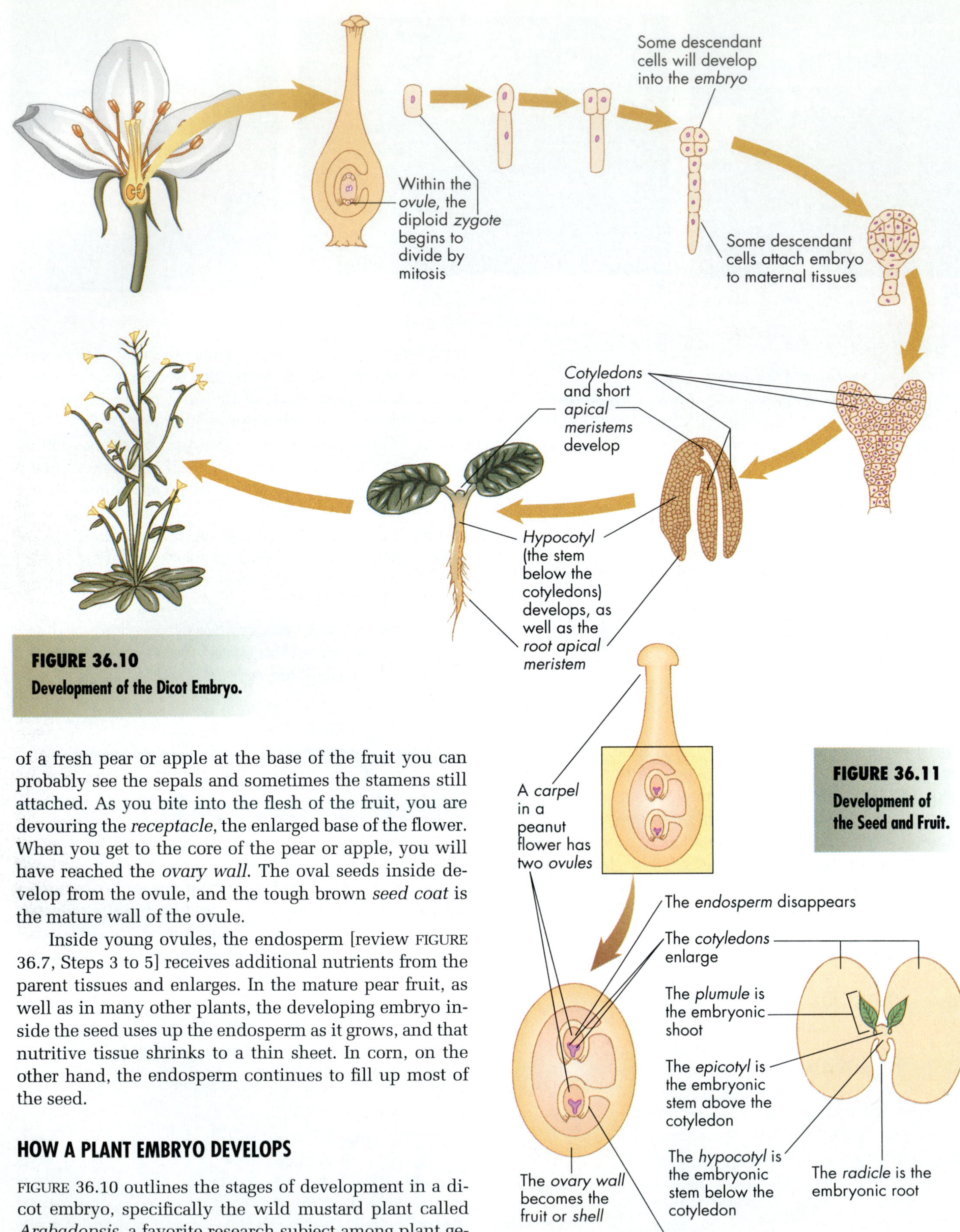

FIGURE 36.10
Development of the Dicot Embryo.

FIGURE 36.11
Development of the Seed and Fruit.

of a fresh pear or apple at the base of the fruit you can probably see the sepals and sometimes the stamens still attached. As you bite into the flesh of the fruit, you are devouring the *receptacle*, the enlarged base of the flower. When you get to the core of the pear or apple, you will have reached the *ovary wall*. The oval seeds inside develop from the ovule, and the tough brown *seed coat* is the mature wall of the ovule.

Inside young ovules, the endosperm [review FIGURE 36.7, Steps 3 to 5] receives additional nutrients from the parent tissues and enlarges. In the mature pear fruit, as well as in many other plants, the developing embryo inside the seed uses up the endosperm as it grows, and that nutritive tissue shrinks to a thin sheet. In corn, on the other hand, the endosperm continues to fill up most of the seed.

HOW A PLANT EMBRYO DEVELOPS

FIGURE 36.10 outlines the stages of development in a dicot embryo, specifically the wild mustard plant called *Arabadopsis*, a favorite research subject among plant geneticists. The zygote, surrounded by the maternal tissues of the ovule and ovary, divides into cells that suspend the

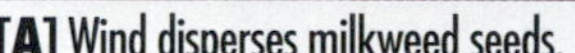

[A] Wind disperses milkweed seeds.

[B] Squirrels plant nut seeds.

[C] Burrs stick to mammals' fur and they disperse them.

[D] Ground cherry seeds form sails.

[E] Berries attract birds and disperse the seeds.

FIGURE 36.12
Methods of Seed Dispersal.

embryo from the maternal tissues and cells that form most of the embryo proper. As embryonic cells continue to divide, they form a little, globe-shaped mass [FIGURE 36.10], which then becomes heart-shaped. The lobes of the heart become the cotyledons (the first leaves) with the shoot apical meristem lying between them. The rest of the embryo becomes the embryonic stem and embryonic root. From that point on, the meristems take over and form a recognizable plant.

You can investigate how an embryo grows in a seed by buying a bag of peanuts. Carefully open one of the peanut shells and look closely at the various parts as you munch [FIGURE 36.11]. The woody peanut shell is actually the **fruit** formed from the enlarged ovary wall, while the reddish papery coating around each peanut is the **seed coat**, the wall of the ovule [FIGURE 36.11]. The two oval halves of the peanut are the dried, salted remains of the **cotyledons**, the thick first leaves of the embryo. The embryos of peanuts and many other plants, including pears, have two cotyledons and are therefore called *dicotyledons* (dicots). Plants with just one cotyledon—lilies and palms, for example, as well as corn and other grasses—are called *monocotyledons* (monocots).

In the peanut, the tiny fleck between the two large cotyledons is the dried embryo; close inspection will reveal a nub called the **radicle**, the embryonic root, and the leafy **plumule**, the embryonic shoot above the cotyledons [FIGURE 36.11]. Between the nubbin of a radicle and the plumule is a short length of stem called the **hypocotyl** (*hypo* means "below"), the stem below the cotyledon to which the starchy cotyledons are attached. Between the cotyledon attachment and the next leaf is the **epicotyl** (*epi* means "upon"), the embryonic stem above the cotyledon, from which the entire shoot will ultimately derive.

In a plant like the pear, it is easy to see how the sweet fragrance and bright color of the fleshy fruit might attract deer, birds, and other animals and how these hungry animals, in turn, will swallow seeds along with fruits. The seeds, protected by the seed coat, can pass through the digestive system of the animal unharmed. Later, the animal can inadvertently disperse seeds to new areas when they excrete feces—a rich supply of natural fertilizer. Other adaptations for seed dispersal include fruits with hooks that attach to passing animals (cockleburs), plumes that lift featherweight seeds on wind currents (dandelions), and hollow shells that roll or float away (coconuts) [see FIGURE 36.12].

box 36.1
Biology Applied

A Fruit of a Different Color

Species in the plant genus *Gossypium* produce lovely cream-colored flowers and an odd fruit—a boll, or rounded pod, full of seeds and long, downy, white cellulose fibers. Luckily for us, this fruit, the cotton boll, is easy to grow, and the fibers can be spun into threads and woven into fabrics—from delicate voiles and laces to heavy canvas. In the past few decades, plant breeders have selected, bred, and grown cotton plants with strong, elongated fibers that are relatively resistant to diseases and animal pests. At the same time, textile researchers have learned to mix the fibers and finishes to create cotton fabrics resistant to stain, mildew, moisture, and shrinkage. Now, a young plant researcher named Sally Fox [FIGURE 1], who is interested in both plants and fabrics, has developed cotton plants that grow their strange fruits in a rainbow of colors.

Soon after *Gossypium* flowers open to reveal creamy petals, their color changes to pink, the flowers drop off, and small green seed pods, the bolls, start to enlarge from what was the flower's receptacle or base. The outer wall of the boll is actually a pericarp, the wall of the ripened ovary similar to the leathery, seed-containing structure inside of a pear [see FIGURE 36.9]. At maturity, cotton bolls are about 3 cm (1.25 in) in diameter, and contain dozens of small seeds. Virtually every individual cell in the outer layer of each ovule in the pericarp produces a long cellulose fiber. The cells die as the ovule wall develops into a hardened seed coat, but the fibers remain attached. Normally, these are snowy white, but occasional mutants show up with fibers pigmented green or caramel brown.

Cotton breeder Sally Fox was able to capitalize on these genetic "mistakes." In college,

FIGURE 1
Sally Fox and the Colored Cotton She Helped Develop.

GERMINATION: THE NEW PLANT EMERGES

To survive in many regions of the world, seeds must withstand the cold temperatures of winter or the lack of water in a dry season. When water is plentiful from rain or snowmelt, however, and sunshine warms the soil, the new embryo begins the process of *germination*: It uses nutrients at a higher rate, resumes growth, and soon cracks out of the seed coat. The tiny radicle of the new seedling can then slip out, and the young root pushes downward into the soil. Soon the shoot pushes upward through the ground and in many plants lifts the cotyledons toward the sun [FIGURE 36.13].

The seedling grows as cells divide in the apical meristems of the shoot and root tips. The cotyledons or endosperm disappear as their stored starch is used to fuel early growth. As the first normal leaves develop and begin to photosynthesize, the young plant becomes truly independent.

[A]

[B]

FIGURE 36.13
Germination: The Seedling Emerges.
In a monocot such as corn [**A**], the plumule bearing the first true leaf bursts upward and emerges from the soil. In many dicots, such as the bean [**B**], the whitish hypocotyl arches upward and breaks the soil first, drawing the cotyledons behind it.

DEVELOPMENT FROM SEEDLING TO MATURE PLANT

As the spring days lengthen, the shoot of a corn seedling advances skyward and the root burrows deeper into the ground. Growth in both directions is driven by the elongation of cells produced in the rapidly dividing apical meristems [review FIGURE 34.13]. Tiny swellings on the shoot called **primordia** flank the apical meristem. These

Fox trained as an entomologist (insect expert), but her first job was working with a cotton breeder trying to develop pest management systems that avoid chemical pesticides. Fox became fascinated by the green and brown mutants she saw from time to time. So applying traditional breeding and selection techniques, she worked for about eight years, beginning in 1982, to develop colored cotton bolls with fibers long enough for spinning fine fabrics that do not have to be dyed. Eventually, her efforts were successful and were met by lucrative contracts from Japanese mills and from Levi Strauss and Company. That jeans maker is now producing caramel and green clothing with denim woven from the colored cotton. Curiously, the fabric gets brighter with each washing, not more faded. Fox likes this quality, as well as the fact that avoiding dyes spares both expense and environmental degradation.

Fox is now trying to breed cotton plants with genes that produce fruits of other colors, including yellowish green, bluish green, dark brown, orange, yellow, copper, and taupe. The biggest coup would be a blue gene, for making blue jeans. As the saying goes, watch for it in your local store. . . .

primordia contain meristematic tissue that will give rise to leaves, new shoots (branches), or flowers. The very first primordia to form, however, are always leaf primordia; this facilitates the seedling's quick development and the rapid unfurling of energy-collecting surfaces.

Once a particular leaf has begun to develop, a *bud primordium* forms in the upper angle between the young leaf and the stem and develops into a **bud**. In corn, these buds mature into flowers, but in a pear, they can become branches. The floral buds develop into flowers, the blossoms become pollinated, the eggs are fertilized, embryos develop, and a new generation appears.

➤ CONCEPT CHALLENGE

Consider a garden variety green bean: What parts correspond to the ovary wall, the ovule wall, and the cotyledons? About how many ovules are there in each ovary in a green bean?

Plant Biotechnology

With their knowledge of plant anatomy, reproduction, development, and genetics, plant scientists have reached a new frontier: biological engineering of plant species with traits different from those found in nature.

For thousands of years, Native American farmers selected seed from the strongest, most productive corn plants to sow for the next generation. This selection process gradually changed the genetic constitution of the grass teosinte into high-yielding, highly nutritional corn. These traditional methods of selecting plants with desirable traits and crossing them are still yielding some remarkable results, as BOX 36.1 on page 806 explains. However, recent innovations in our ability to manipulate living cells have led to dramatic new methodologies. With these, plant researchers are combining the hardiness of wild plants that grow in dry, salty, cold, or barren areas with the high-yield traits of the best agricultural crops. The ideal is to develop, for example, wheat, rice, corn, or tomato plants with the drought resistance of cacti, the salt tolerance of marsh grass, and the nitrogen-fixing ability of legumes, as well as the ability to resist insect pests, diseases, and herbicides. Their ultimate goal is to generate increasingly larger harvests for the world's hungry people, despite the fact that most of the world's remaining unplanted lands are only marginally useful. Let us explore techniques and early achievements of this new and very important field.

NEW TECHNIQUES IN PLANT BREEDING

The choice foods we have today—the sweet pears, high-yielding grains, juicy strawberries, crunchy corns, and others—are the result of generations of farmers carefully selecting seeds from plants with the best individual traits and planting the seeds for the next generation. New techniques such as tissue culture, however, are speeding up the process of artificial selection.

Plant Tissue Culture Plant breeders are often eager to select combinations of traits displayed by a particular individual, but these desired combinations can disappear as a result of genetic recombination during normal sexual reproduction [review FIGURE 8.13]. The botanist's solution to this problem is to grow plant cells in **tissue culture**, the cultivation of cells in laboratory vessels. Growing new identical plantlets from the somatic (body) cells of a prize plant bypasses sex, and thus avoids the problem of genetic recombination.

Researchers can begin the tissue culture process by punching out a bit of plant tissue from some part of a plant—a leaf or stem, for example [FIGURE 36.14, Step 1]—and placing it in a nutritious culture medium in a warm, well-lighted room (Step 2). Soon the cells of the tissue lose the distinctive characters that mark them as leaf or stem, for example, and form an unorganized mass of cells called a *callus* (Step 3). The plant breeder then breaks up the callus, sometimes into portions containing

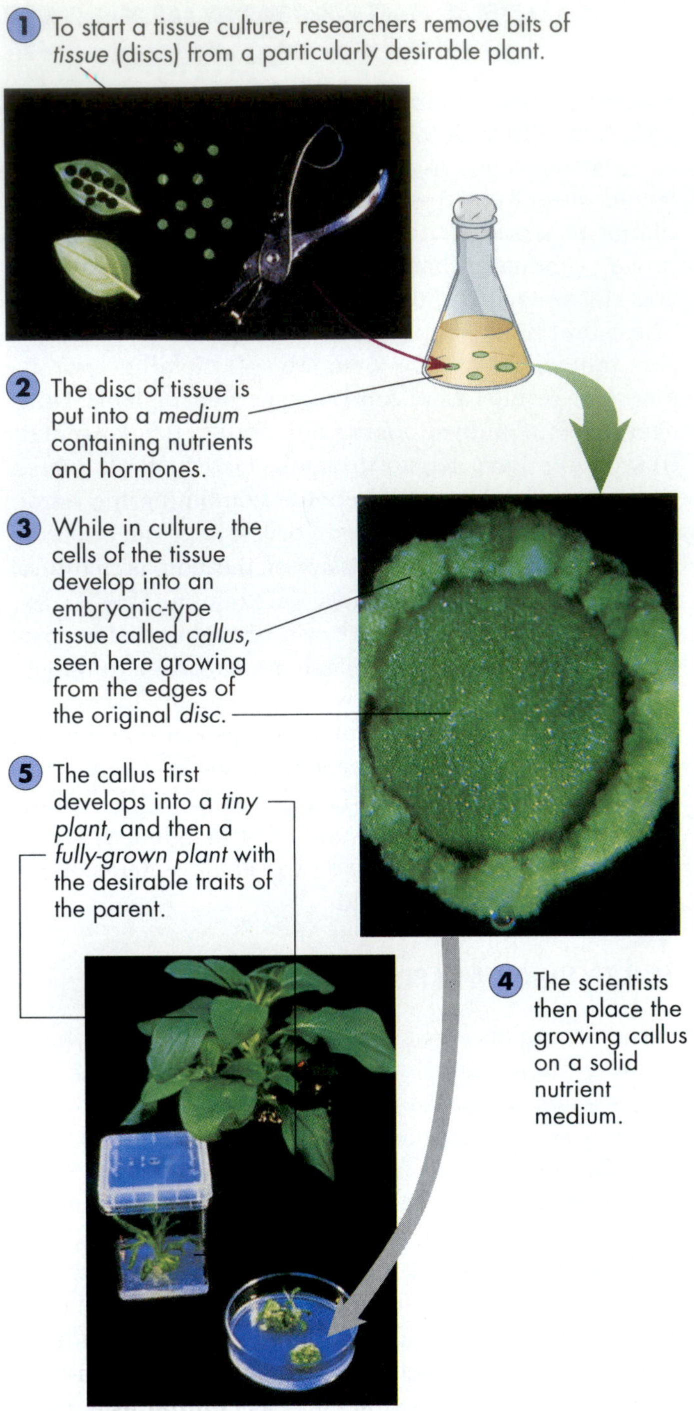

FIGURE 36.14
Plant Tissue Culture: Growing Whole Plants from Individual Cells or Tissues.

just a single cell, and places bits of it in a solid culture medium with plant hormones that promote cell differentiation (Step 4). There, some of the callus cells grow and develop into a tiny plant (Step 5). The fact that an individual cell from the root, stem, or leaf of a plant can proliferate into an entirely new plant with roots, stems, leaves, and flowers suggests that each plant cell contains a complete set of genetic information.

Plants arising from tissue culture can be raised and studied directly, or the embryos can be grown into plants or packaged in little gel capsules as "synthetic seeds" for later use. Either way, since all these plants are derived from the somatic cells of the parent plant, theoretically they all have identical genes and provide numerous copies of the desired original. Some plant nurseries already propagate certain commercial crops using tissue culture—for example, strawberry plants that are free of viral infections. And researchers are using tissue culture to produce the cancer-fighting taxol discussed in CHAPTER 20.

Genetic Variation in Cultured Cells For unknown reasons, the DNA of many plants grown from the initial tissue-cultured callus contains mutations. Wheat, carrots, celery, and tomatoes produce all sorts of genetic variants in tissue culture—variants that would require years to find by traditional selection techniques. The tendency of cultured plants to display new, heritable traits is called *somaclonal variation* (*soma* referring to the body cells of the plant, and *clonal* because each variant population is a clone, arising from a single initial mutant cell). From such an array of varying cells, breeders select those that express a desired trait. For example, if biologists could develop crop plants that were resistant to salty soils, farmers, could exploit many areas not currently open to agriculture because of natural salty conditions or the accumulation of salts from years of irrigation. To search for such plants, researchers can grow tissue culture cells in a medium containing high concentrations of salt. Those cells with genes for salt tolerance live; the rest die. When the surviving salt-tolerant cells grow and differentiate into whole plants, experimenters find that some of them tolerate high levels of salt in the environment. Agricultural researchers have used these methods to make useful plants.

In one of the earliest somaclonal variation experiments, researchers grew 230 plants from the cells of a single tomato plant. Among the variants, they found 13 separate mutations, several of which turned out to be commercially useful. Some mutations produced novel bright orange and yellow tomatoes while others affected flower color, chlorophyll production, fruit ripening, and the ratio of juice to solids in the fruit. Tomatoes with unusually high solid content are valuable to canners of tomato paste.

GENETIC ENGINEERING IN PLANTS

As CHAPTER 11 explained in detail, genetic engineering offers a pathway to varieties with more profound alterations than traditional techniques are ever likely to produce. In a single generation, plant geneticists can introduce into a species a trait that might have taken hundreds of generations to develop through standard breeding ap-

FIGURE 36.15
Deadly Dinner.
A genetically engineered plant (left) containing toxin genes from the *Bacillus thuringiensis* bacterium will poison its predators, while a normal plant lacking the gene (right) merely provides a meal for the hungry hornworms.

proaches. They do this by inserting the desirable gene into a plasmid, a circular molecule of DNA that occurs naturally in a plant bacterium called *Agrobacterium tumefaciens*. This plasmid can move into cells of dicot plants such as petunias and beans, carrying along whatever genes it happens to contain, including genes spliced in by a researcher. For corn, wheat, and other monocots, biologists can coat minuscule tungsten ball bearings with DNA and, with the equivalent of a miniature laboratory shotgun, shoot them into cells. The DNA is somehow transferred from the tungsten bullets into the cell's chromosomes, where it can alter the traits of the cell and of the plant that might grow from it.

Plant researchers have already altered several plants by recombinant DNA methodologies. Belgian plant geneticists, for example, have transferred an insect-killing gene from a bacterium into tobacco and tomato plants. The bacterium makes a protein called *Bt* that kills only insects. By removing the gene from the bacterium and inserting it into the plants, researchers conferred upon those plants an immunity to attack by voracious hornworms [FIGURE 36.15]. Similar modifications of corn and other important crops are now under way. And plant researchers have outfitted petunia plants with a gene for resistance to the widely used herbicide glyphosate (Roundup). Workers can now spray a field of the resistant plants with the herbicide, and weeds will die off but the petunia crop will remain unharmed.

Genetic engineers hope to develop varieties of wheat, corn, and rice that, like meat and dairy products, contain high quantities of all essential amino acids. They also hope to engineer nonleguminous crop species that can form root nodules filled with nitrogen-fixing bacteria, plus dozens of other plant species with increased resistance to insects, diseases, and herbicides, increased nutritional value, and increased tolerance to harsh environments. Some biologists are modifying crops such as alfalfa to produce vaccines for infectious diseases like cholera. They hope that poor countries, which cannot afford the costly medical injections needed to immunize their citizens, will be able simply to grow and use the engineered crop with its edible vaccine. Finally, corn geneticists are hoping to build on their understanding of sex differentiation in corn to engineer sterile male varieties. With these, people would not have to move through corn fields removing tassels by hand in order to produce hybrid seed.

Clearly, the modern tools of plant breeders supplement but do not replace, traditional breeding techniques, such as those used by Sally Fox to produce colored cotton. Furthermore, some people have strong ethical objections to recombinant DNA research, and fear potential biological dangers in the release of plants altered by genetic engineering. Many observers are more optimistic, however, and predict that the applications of our modern knowledge of genetics and plant development will help provide the food upon which our burgeoning human race must depend.

➤ CONCEPT CHALLENGE

Let's assume that you have become an agricultural biologist and must choose a crop and decide what traits would make it more useful. Describe the crop, the new traits, and how you might achieve the desired modification using modern biological techniques.

Connections

To a student of biology, plant reproduction and embryonic development are, at once, as familiar as a rose and an ear of corn and as unfamiliar as a female gametophyte and double fertilization. Gaining an understanding of the subject, however, can satisfy our natural curiosity about how the world works, and help us appreciate how biologists are modifying plants so that our earth and its agricultural bounty can sustain the 2 or 3 billion new people who will be born in our lifetimes. The chapters in this section on plant architecture, material transport in plants, and plant reproduction and development have all enlarged this necessary understanding. But we must explore one more area of plant biology: how plants respond to their environment with altered growth. This is the next and final chapter in our investigation of plant biology.

KEY TERMS

anther, 799
carpel, 799
cotyledon, 805
egg sac, 802
endosperm, 802
epicotyl, 805
filament, 799
fruit, 805
gamete, 798
gametophyte, 798
hypocotyl, 805
megaspore, 801
ovary, 799
ovule, 799
petal, 799
pollen, 801
pollinator, 802
radicle, 805
seed, 801
sepal, 799
spore, 798
sporophyte, 798
stamen, 799
stigma, 799
style, 799
tissue culture, 807

HIGHLIGHTS IN REVIEW

1 Flowering plants alternate between a large diploid generation that forms spores by meiosis in flowers and a haploid generation that produces egg and sperm. Egg and sperm join at fertilization, forming a zygote, a new diploid generation.

a] The plant life cycle alternates between a diploid, spore-producing sporophyte generation that undergoes meiosis and a haploid, gamete-producing gametophyte generation that undergoes fertilization.

b] The large, multicellular familiar plant is the sporophyte. Male gametophytes are pollen grains, and female gametophytes are egg sacs totally contained within the body of the sporophyte.

c] Flowers generally consist of four whorls of structures; the outer two (sepals and petals) are sterile, but the inner two (stamens and carpels) are male and female parts that contain cells that undergo meiosis and produce haploid reproductive spores.

d] Pollen, the male haploid plant (gametophyte), forms in the anther of a stamen, is carried to or lands on the stigma of a carpel, and grows down through the carpel's style and ovary wall to the ovule, which houses the embryo sac, the female haploid gametophyte.

e] The pollen grain contains two haploid sperm, which accomplish double fertilization. One of the sperm joins with two haploid nuclei in the central region of an embryo sac, thereby forming a triploid cell. This cell divides mitotically and gives rise to the endosperm, a nutritive tissue inside the seed. The other sperm fuses with the egg, making the diploid zygote.

2 The zygote grows into a diploid embryo contained within a seed, which can grow into a new large diploid plant.

a] The zygote gives rise to the embryo of the diploid sporophyte plant. Surrounding the zygote is the ovule, which matures into a seed containing the embryo and endosperm surrounded by a seed coat, the ovule's wall. A ripened ovary forms a fruit.

b] At germination, the embryo bursts from the seed. After the first normal leaves appear, growth occurs only at the apical and lateral meristems.

3 Many biologists hope that plant biotechnology, the application of our knowledge of plant reproduction, development, and molecular genetics to crop plants, will improve our chances of being able to feed the billions of new human mouths that will appear in the next few decades.

a] Tissue culture, the growth of plant cells in laboratory vessels, and the cloning of somatic cells allow biologists to propagate useful plants bypassing the genetic recombination associated with sexual reproduction.

b] The insertion or deletion of specific genes into plants can be accomplished through genetic engineering techniques. Genes can be introduced into plants by infection with a bacterium that carries a certain plasmid, or by shooting plant cells with tiny projectiles coated with DNA. Inserted genes can, in principle, increase the plant's resistance to drought, poor soil, or pests, and improve the products the plant contains, making it more nutritious or even generating vaccines against disease.

UNDERSTANDING THE FACTS AND CONCEPTS

For Questions 1–5, complete the statements by choosing the most appropriate term from the following list. Any answer may be used once, more than once, or not at all.

a] sporophyte
b] megaspore
c] microspore
d] gametophyte
e] gametes
f] zygote

1 Each haploid ______ develops into a seven-celled structure in corn.

2 Each ______ develops into a pollen grain.

3 The pollen grain is one kind of ______.

4 The corn husk and cob are part of the ______.

5 Each pollen grain includes two ______.

For Questions 6–10, match each of the statements with the most appropriate term from the following list. Each answer may be used once, more than once, or not at all.

a] sepals
b] carpel
c] ovary
d] ovule
e] anther
f] perfect flower

6 The term that includes all of the others in the list.

7 The structure that contains one or more pollen sacs.

8 Contains the egg and later matures into the seed.

9 Includes the ovary, style, and stigma.

10 May aid reproduction but does not actually participate in reproduction.

As above, for Questions 11–15, match each of the statements with the most appropriate term from the following list.

a] microspore mother cell
b] megaspore
c] embryo sac
d] endosperm
e] egg

11 One cell in the female gametophyte.

12 Triploid product of fertilization.

13 The only structure in the list that is unicellular and diploid.

14 A haploid cell that divides by mitosis to form an embryo sac.

15 Another name for the female gametophyte.

As above, for Questions 16–20, match each statement with the most appropriate item from the following list.

a] cotyledons
b] radicle
c] plumule
d] hypocotyl
e] epicotyl

16 The embryonic root.

17 Lies above the cotyledons and gives rise to the shoot.

18 The first embryonic leaves.

19 These structures attach the cotyledons to the embryo.

20 This structure bears the first true leaves of the embryo.

INTEGRATE AND APPLY WHAT YOU HAVE LEARNED

1 How do monecious and dioecious plants differ structurally? What are some evolutionary implications of being monecious or dioecious?

2 Which plant parts do you eat when eating an apple? A green pea? Corn?

3 When a plant embryo begins to develop, which part develops first—the root, flowers, or leaves? Explain.

4 If mammals such as dogs, cats, and humans are analogous to angiosperm sporophytes, what structures (if any) in mammals are analogous to the plant ovary and stamen? Explain.

5 What is meant by double fertilization? What is the significance of this term?

ANALYSIS

1 *Arabidopsis* has three mutant genes that convert flower parts into leaves. Which of the following statements is likely to be true about those mutant genes? More than one answer may be correct.

a] They are each specific for one of the parts of the flower.
b] They are control genes that act on other genes.
c] They are least like any ancestral plant genes.
d] They are unlike genes that normally result in leaf production.

2 Suppose you discover a new plant species that produces succulent fruit. The seeds of this plant have thin, delicate seed coats. Which of the following statements about the new species is most likely to be correct?

a] The plant probably coevolved with a mammalian pollinator that ate the fruit.
b] The plant probably coevolved with a bird pollinator that pecked the seed coat.
c] The fruit of this species probably decomposes quickly on the ground.
d] This species probably arose by means of artificial selection.
e] This species probably is sterile.

3 After many years of work, plant geneticists have developed new varieties of corn in which the enzyme that converts newly photosynthesized sugar to starch has been inhibited. Why would anyone work so hard and for so long on this problem?

a] The new strains of corn stay sweeter for a longer time after picking.
b] The new strains of corn attract a larger population of animal pollinators.
c] The new strains of corn are lighter and therefore more able to be wind-pollinated.
d] The new strains of corn are keeping coevolutionary pace with changes in pollinating insects.
e] The new strains of corn are likely to decompose less rapidly.

4 Is a tomato a fruit?

a] No, because it is a vegetable.
b] It is not a fruit because it does not consist of reproductive parts of a plant.
c] Yes, because it contains seeds.
d] Yes, because its pigmentation shows that it is composed of carpels.
e] Yes, because any edible part of a plant is a fruit.

5 How could a plant geneticist modify alfalfa so that it would produce a vaccine against cholera?

a] Somaclone alfalfa and hope that one of the variants would express a mutation that produced the vaccine.
b] Somaclone alfalfa, expose the plants to cholera, and select the surviving plants.
c] Somaclone cholera and introduce the cholera into the alfalfa.
d] Isolate the gene that directs the synthesis of anticholera antibody in animals and inject the antibody into the alfalfa.
e] Isolate the gene that directs the synthesis of anticholera antibody in animals, clone the gene, and introduce the gene into alfalfa embryos.

CHAPTER 37

Plant Growth Regulators and the Plant's Environment

FOOLISH SEEDLING DISEASE

People began cultivating rice in Thailand nearly 6000 years ago, and today, half of the food calories for 4 out of 10 people on earth come from rice. This means that 20 percent of all human activity is rice-powered! Very early on, Japanese farmers began to breed improved varieties of *Oryza sativa*, the rice plant, and to study its diseases [FIGURE 37.1]. In 1809, for example, a Japanese scholar recorded a particular disease that had plagued rice growers for hundreds of years: Afflicted rice plants would grow far taller and faster than normal after germination, and in the process would become spindly and weak and blow over in the wind and die. The name for this disease was fitting: *foolish seedling disease* [FIGURE 37.2].

Japanese scientists continued to study this phenomenon, and in 1926, plant physiologist E. Kurosawa showed that the symptoms of foolish seedling disease were caused by chemicals released by the fungus *Gibberella fujikoroi* when it infects rice plants. Within a few years, other Japanese researchers had isolated two disease-causing chemicals from that fungus and named them *gibberellins*.

FIGURE 37.1

Cultivating Rice: An Ancient Practice. Farmers in Bali, Indonesia, have been cultivating rice (*Oryza sativa*) for many centuries, using the traditional methods depicted here.

For a few decades, plant biologists thought gibberellins were fungal toxins capable of accelerating plant growth. But in 1956, Western researchers isolated gibberellins from a healthy bean plant. Soon it became clear that the chemicals occur naturally in plants. Scientists isolated over 70 kinds of gibberellins, and found them to be plant growth regulators, or **plant hormones**. The *G. fujikoroi* fungus had simply evolved a way to make rice plants literally grow themselves to death.

Foolish seedling disease brings us to a new topic: How do plants regulate their growth and development in response to environmental cues and demands? We have already seen in our exploration of plant biology that plants have rigid cell walls and are normally rooted to a single permanent place of growth. As a result, they cannot move away from an environmental problem the way an animal can. Moreover, most animals have both nervous and hormonal systems, and the two work together in coordinating the activity of muscles, nerves, and sense organs. While plants lack a nervous system, they do have numerous kinds of hormones, and these trigger responses—usually some form of growth—in reaction to environmental cues. A plant can change its shape by localized growth much more radically than most animals can. Thus, depending on conditions, a plant can grow toward a light source, send roots deeper into the soil, or drop its leaves and become dormant during harsh winter weather.

MESSAGES

1 Because of their rigid cell walls, plants cannot move away from adverse conditions. Instead, they cope with environmental challenges through the processes of growth, maturation, and dormancy.

2 Plants respond to the environment, in part, by producing and reacting to growth regulators, including hormones and phytochromes.

3 Light, gravity, water, and temperature are major environmental factors triggering the hormone-controlled patterns of plant growth, flowering, maturation, dormancy, and senescence.

4 In response to damage caused by insects or disease, plants produce chemicals harmful to the attackers and wall off injured areas.

Environmental factors such as temperature, light, or gravity often trigger the release of plant hormones, which in turn, regulate changes in plant morphology (shape) or physiology. These environmentally triggered changes in growth drive the plant life cycle studied in CHAPTERS 34 and 36, including flowering, embryonic development, seed formation, and germination.

In the case of foolish seedling disease, the gibberellin hormones produced by the fungus cause elongated growth that leads to the plant's premature death. In other cases, however, gibberellin-triggered elongation can promote a plant's survival, such as when crabgrass growing beneath a wooden board elongates in the direction of the light at the edges of the board. How does a board-flattened patch of grass "know" it is dark? How is this "knowledge" converted into action (growth)? How is the growth controlled so that elongation starts under the board and stops at the edge instead of the reverse? The answers will unfold here as we consider the major types of plant hormones and how they interact with environmental factors to control plant growth and development.

First we will survey the functions performed by each class of plant hormones. Next we will see how hormones regulate germination. We will explore the control of plant growth and development. We will see what governs flowering and fruit development. We will learn how dormancy, dying, and the falling of leaves and fruit are controlled. Finally, we will see how plants defend themselves from infection and wounds.

Along the way, we will find out how Agent Orange kills plants; why some farmers may consider plowing their fields at night; how caffeine from coffee and nicotine from tobacco relate to plant defense; and why chefs are boycotting genetically engineered tomatoes. ❑

FIGURE 37.2

Foolish Seedling Disease: A Curious Effect of Hormones on Plant Growth.

When a young rice plant (left) is infected by the fungus *Gibberella fujikoroi*, the parasite releases gibberellins, causing the stems to elongate rapidly, grow weak and spindly, and blow over and die in even a gentle breeze. An uninfected healthy, dark green plant is shown at the right.

Plant Growth Regulators

The foolish seedling phenomenon brought on by high levels of gibberellins is characteristic of hormonal action in plants: Plants cannot move away from adverse conditions, but they can often grow away from the problem. For us to understand how growth helps plants cope with environmental demands, we must first learn what plant growth regulators are and how they work [TABLE 37.1].

Plant scientists have a working hypothesis about the activities of plant hormones that is generally similar to their conception of how animal hormones operate. Recall that in animals, a hormone is produced by one group of cells, it is transported within the body, and it binds to a receptor molecule in or on another cell [FIGURE 30.2]. It is the binding of hormone to its receptor that triggers activity within the target cell. There are five major classes of plant hormones—three promote and regulate growth, and two inhibit growth or promote maturation. Let's survey all five classes now, and then see how they interact during the regulation of a plant's life cycle.

FIGURE 37.3
Bolting.
Gibberellin causes the short stem in a rosette plant to greatly elongate. On the left are cabbage plants (*Brassica oleracea*) in normal rosette form. On the right are similar plants after treatment with gibberellins. Bolting takes place naturally in a biennial before flowering in the second year.

GIBBERELLINS

The plant hormones called **gibberellins** provide a good example of how widespread the effects of particular plant hormones can be. Gibberellins are made in a variety of organs, such as young leaves, embryos, and roots. They can move through the plant's vascular system and are primarily involved in regulating plant height. Too little gibberellin results in dwarf plants, but too much results in long, pale, "foolish" stems. Gibberellins also play an important role in *bolting* (sudden stem lengthening) in

TABLE 37.1 Internal Regulators of Plant Growth and Development

Regulator	Transport	Action
Hormones		
Gibberellins	Upward and downward in vascular system	*Promote growth:* Promote stem lengthening in dwarf and rosette plants and seed germination in grasses; involved in flowering and fertilization; involved in growth of new leaves, young branches, and fruits
Auxins	From tip of shoot downward in vascular system	*Promote growth:* Augment growth by cell elongation; inhibit growth of lateral buds; foster growth of ovary wall; prevent leaf and fruit drop; orient root and shoot growth
Cytokinins	From root upward in vascular system	*Promote growth:* Stimulate cell division; kindle growth in lateral buds; block leaf senescence
Abscisic acid	Short distances in leaf and fruit	*Inhibits growth:* Opposes the three growth-promoting hormones; induces and maintains dormancy
Ethylene	Through air, as a gas	*Promotes maturation:* Enhances fruit ripening; promotes dropping of leaves, flowers, and fruits
Pigment		
Phytochrome	Not transported; remains in cell that produces it	*Detects light:* Changes form in response to light; mediates flowering, germination, growth, and plant form

rosette plants such as cabbage [FIGURE 37.3], and in inducing the seeds of rice, barley, and other grasses to germinate. Plant scientists have found that gibberellins are also involved in flowering, fertilization, and the growth of fruits, new leaves, and young branches [see TABLE 37.1].

AUXINS

Like gibberellins, **auxins** are generally growth promoters. The name, in fact, comes from the Greek *auxein*, meaning "to increase, or augment." Natural auxin is indoleacetic acid, or IAA, a substance related to the amino acid tryptophan, a building block found in most proteins. Instead of being produced in various tissues and traveling through the vascular system like gibberellins, auxins are generally made in shoot apical meristems and developing leaves. They diffuse from the site of production downward toward the roots.

Auxins promote growth through cell elongation: They trigger enzyme activities that loosen the tightly woven fibers in cell walls, allowing the cell to expand from within. This action influences the length of cells in stems, leaves, and the wall of the ovary. Auxins also act in other ways that prevent leaves, fruits, or flowers from falling off prematurely, and that inhibit growth in lateral buds and roots in favor of growth at the apical meristem. In high concentrations, auxins can cause uncontrolled growth and plant death [FIGURE 37.4]. Some herbicides are based on this principle, including the synthetic auxins 2,4-D and 2,4,5-T, the active ingredients in Agent Orange, which was widely sprayed to defoliate jungle vegetation during the Vietnam War. Within hours after spraying, the leaves and stems of broadleaf plants droop, twist, and curl. Within days, the chlorophyll degrades and the plant dries out and dies. These herbicides appear to induce such rapid cell division and tissue proliferation that the phloem becomes plugged with nonconducting cells. Without a functioning phloem, the plant cannot survive.

CYTOKININS

Cytokinins are the third class of growth-promoting plant hormones, and they generally stimulate cell division, including cytokinesis [see CHAPTER 7]. Cytokinins are chemically related to adenine, one of the nitrogenous bases in DNA. Cytokinins are less mobile than auxins (and far less so than gibberellins) and appear to move in opposition to auxins—from root upward to shoot, not shoot downward to root. In contrast to auxins, cytokinins promote the growth of lateral buds, not shoot tips. Like most plant hormones, cytokinins also have other effects, including the prevention of leaf aging, or senescence.

FIGURE 37.4

Synthetic Auxin Causes Some Plants to Grow Themselves to Death. The substance 2,4-D is chemically related to auxin and causes the plant to grow so rapidly that its phloem becomes clogged and it dies. The broadleaf weeds in the foreground show the effects of the synthetic auxin. The corn in the background was unsprayed.

ABSCISIC ACID

If plants had only gibberellins, auxins, and cytokinins, they would grow constantly. Sometimes, however, it is more advantageous for the plant to stop growing—to close its stomata, decrease the level of photosynthesis, drop aging leaves, or become dormant. **Abscisic acid** (ABA; see TABLE 37.1) is a growth inhibitor. It counteracts the growth hormones, apparently by inhibiting translation of mRNA at the ribosomes and thus blocking protein synthesis and new growth. ABA moves only short distances from its site of production. The name "abscisic acid" comes from ABA's role in accelerating **abscission**, the dropping (literally, "cutting away") of leaves and fruits. Despite its name, however, abscisic acid is more important in other aspects of a plant's physiology. Plant scientists now think that ABA's main role is to induce and maintain **dormancy**, or metabolic slowdown, especially in buds, and to promote the closing of stomata in leaves, which prevents excess water loss. Consistent with ABA's regulation of dormancy, a drop of ABA on a green leaf causes a yellow senescent spot.

ETHYLENE

Ethylene, the fifth major plant hormone, is a simple two-carbon molecule that exists as a gas at normal temperatures and is dispersed from one plant or plant part to another by air. Ethylene is behind the old adage about one rotten apple spoiling the whole barrel: The hormone is produced by ripening fruits, and it stimulates ripening in nearby fruits. Ethylene also stimulates the aging and dropping of leaves and fruits, and it may have an important role in plant self-protection.

➤ CONCEPT CHALLENGE

Fruit growers would like to have genetic strains of peaches or pears that do not ripen until they are treated with a natural hormone. This treatment could be done just before the grocers put the fruit out on their shelves. Which plant hormone might a plant physiologist try to manipulate to achieve this goal?

Environmental Cues, Hormones, and Germination

After animals, air currents, or water carry seeds to new locations, the seeds germinate, and tiny new plants emerge. But what triggers germination at a time when the seedling's chances for survival are greatest? Although specific answers depend on the plant species, environmental cues generally act through plant hormones that in turn regulate the sequential events of germination.

Some seeds germinate shortly after they reach a new location and take in enough water to soften the seed coat so that it splits, and the tiny plantlet can then burst out. Rice seeds are in this category, as are willows, poplars, and silver maples. This pattern is especially common in tropical plants, with their relatively mild, moist, and stable environments.

In temperate zones, however, delayed germination is often more advantageous. The seedlings of an annual such as a daisy that emerged in the late summer might not complete an entire life cycle before the short days, cold temperatures, and dry conditions of winter set in. Many temperate plants, therefore, germinate in spring or early summer after surviving winter as a dormant embryo encased and protected within a seed coat. Then in spring, rising temperatures, lengthening hours of daylight, and melting snow or rain trigger hormonal activities within the embryo that in turn trigger germination.

For some seeds, especially in dry areas, moisture is the external trigger, counteracting a growth-inhibiting hormone such as abscisic acid. Growth resumes only when enough moisture is present to leach away most of the inhibitor. This chain of events helps guarantee that the ground will be damp enough for the seedling to survive. In the corn embryo, for example, water entering the seed causes the release of gibberellic acid, and this hormone, in turn, acts on other cells in the seed causing them to release starch-digesting amylase enzyme into the white starchy tissue (endosperm) packed around the embryo. The enzyme then breaks down the starch in the corn kernel, and the embryo uses the released glucose as a fuel during germination.

For other kinds of seeds, light is a determining factor. The seeds of some desert plants germinate only in deep, dark, moist soil. Presumably, desert plant seeds that germinated in the light, and hence near the surface, would dry out and die. Plants with genes that blocked germination until the seeds were covered by deep soil layers would be more likely to survive. Light-triggered germination could have evolved in this way.

In lettuce and many weed species, the seeds are so tiny that seedlings buried deeper than a few millimeters run out of food reserves before reaching the sunlit surface. Interestingly, these seeds germinate only when they detect light; a flash just a quarter of a second long is sufficient. Knowing this, biologists at Oregon State University hypothesized that plowing fields in sunlight might expose previously buried weed seeds to a light flash sufficiently long to cause sprouting. To test this hypothesis, they plowed experimental plots at night, and found a dramatic reduction in subsequent weed growth compared to adjacent identical plots plowed in daylight. The reduction in weed density was so marked that the biologists think some farmers may be able to avoid using herbicides (plant-killing chemicals) simply by plowing at night.

In a few remarkable plant species, seeds can lie dormant for immense periods until conditions are just right. Some seeds from the Japanese lotus *Nelumbo nucifera*, for example, have germinated successfully after apparently lying dormant for 2000 years. And some lupine seeds frozen in the arctic permafrost have reportedly remained viable for 10,000 years. Nevertheless, most seeds remain dormant and viable for only a few years at best. Even with greatly slowed metabolism, their food reserves are eventually used up and the embryo dies.

➤ CONCEPT CHALLENGE

Forest soil contains many kinds of weed seeds that can flourish only in full sunlight. These seeds lie dormant until a storm or fire removes the overhanging vegetation that shades the ground. Advance a hypothesis for the mechanisms that permit the seeds to lie dormant when growth conditions are unfavorable, then germinate when conditions improve.

Regulation of Plant Growth and Development

The germination of a seed, triggered by gibberellins and other factors, is the start of a newly established plant. The breaking of dormancy, however, and the growth of the embryo's shoot and root, are only a start. Plant regulators must still be involved to control the type of growth that occurs. Will it be, for example, the spindly growth of grass under a board? Or will it be the lush, thick growth of a lawn in full sunlight? This section considers the chemical regulators of growth and other life-cycle events such as flowering and the dropping of leaves, and shows how plants adapt to their surroundings through changing physiology and altered shape and orientation in space.

[A] In a plant laid on its side, the *shoot* bends upward,
and the *root* bends downward

[B] The root tip bends because *cells on the upper root surface* elongate relative to *cells on the lower root surface*

[C] Root tip cells

Amyloplasts (dense starch granules)

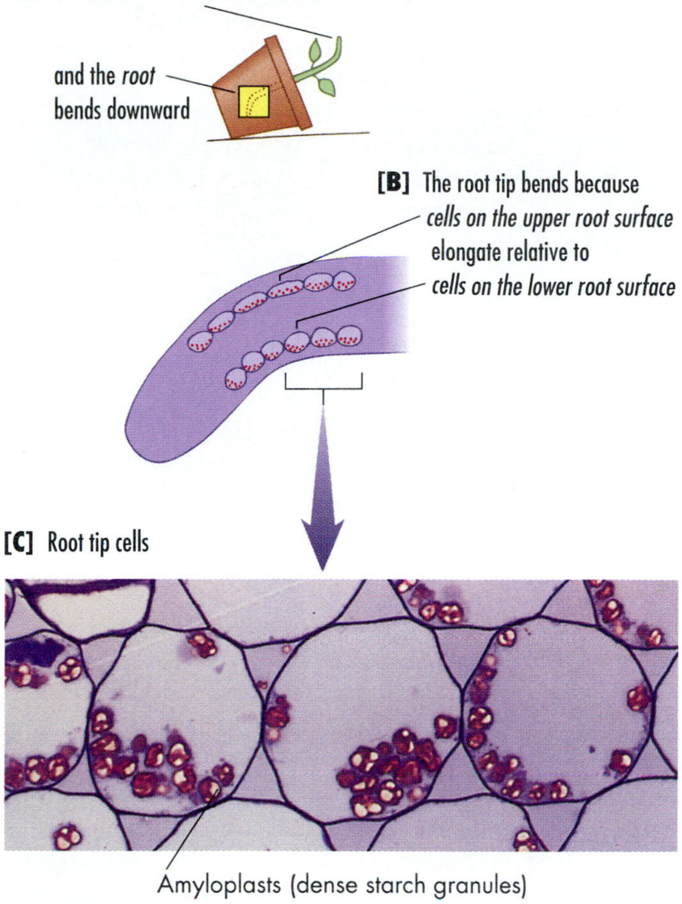

FIGURE 37.5
Gravitropism: Growth Toward or Away from Gravity.

ENVIRONMENTAL CUES AND PLANT ORIENTATION

A rice seedling in a field is oriented so that its roots grow down and its shoot grows up. This seems logical enough, but how does a tiny plant manage to grow in the correct orientation? A plant's orientation is based on **tropisms**: bending, turning, or directional growth in response to external and internal stimuli. (The term *tropism* comes from the Greek word for "turning toward.") A tropism is considered *positive* if the organism's orientation or growth is *toward* the stimulus, and *negative* if the organism's orientation or growth is *away from* the stimulus. Botanists have named a number of tropisms by the external stimulus that causes them. For example, the growth of roots toward water is called *hydrotropism* (turning toward water). A sweet pea's tendrils exhibit a pressure-sensitive tropism, or *thigmotropism*, as they wrap around any nearby object for support. The orientation of a plant part toward the ground is *gravitropism* (turning toward gravity), and the orientation toward light is *phototropism*. These last two tropisms help explain how a rice seedling's roots grow down and its shoot grows up.

Gravity and Plant Growth When a farmer sows a rice seed, the little embryo may be upside down or sideways, yet the seedling achieves the proper root-down and shoot-up orientation. This feat is accomplished through the plant's sensitivity to gravity. The root is *positively gravitropic* (or geotropic), since it grows down toward the pull of gravity, and the shoot is *negatively gravitropic*, since it grows up against the pull of gravity [FIGURE 37.5A].

If a person lays a plant horizontally on its side, cells on the upper side of the root become longer, while cells on the underside remain the same length. As a result of this differential growth, the root bends downward [FIGURE 37.5B]. Obviously, the plant doesn't "know" it's lying horizontally and that the roots must turn and grow downward, so what automatic process controls this cell lengthening on the upper side of the root? Investigations have revealed that the root cells contain dense starch granules (amyloplasts) which sink to the bottom of the cells [FIGURE 37.5C]. Plant physiologists suggest that these starch granules may act like little rocks that respond to the pull of gravity, and that they may cause auxin to move to the lower side of a root and block the elongation of lower cells. As the upper cells continue to elongate, the roots bend and grow downward, thrusting toward a more likely source of moisture and minerals.

Studies of mutant tomato plants that grow horizontally not vertically suggest the mutants lack cell receptors for auxins. Thus auxins, and perhaps other plant hormones, may bind to receptors in cell membranes, as do many hormones in animals.

Light and Plant Growth Plants need light for photosynthesis, but they may lack sufficient access to it. For example, a wild rice seedling might by chance germinate beneath the bushes at the edge of a forest. But the plant can grow away from the shade and toward light, and this maximizes the plant's exposure to the sun.

The first experiments to probe the mechanisms of phototropism in plants—how the organisms orient toward light—appeared in a book that Charles Darwin and his son Francis published in 1881. The Darwins knew that if a grass plant is illuminated from the side, it bends toward the light source a short distance back from the growing tip [FIGURE 37.6A]. Together they performed a series of experiments on oat and canary grass seedlings to

[A] Normal growth

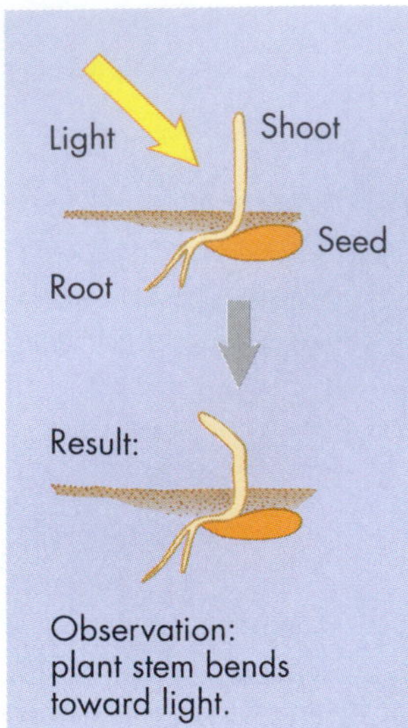

[B] Does the bending section detect the light?

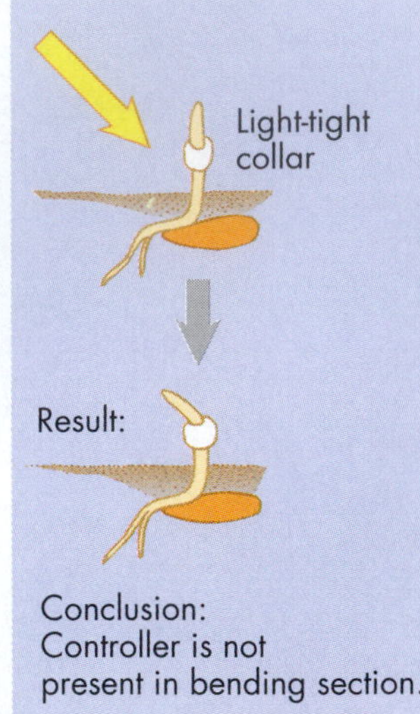

[C] Does the tip detect the light?

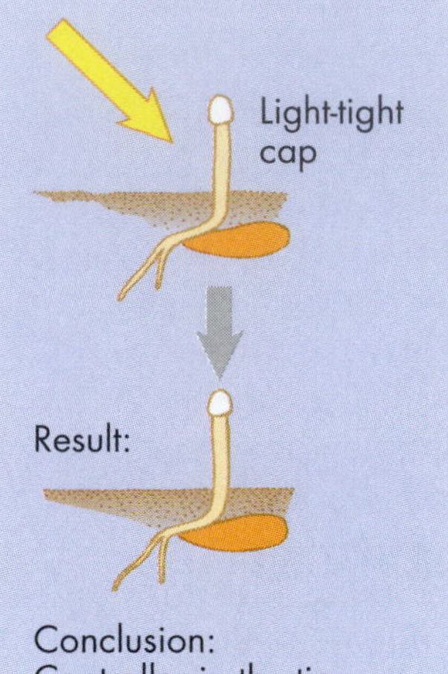

FIGURE 37.6

Plant's Tip Controls Phototropic Bending.

[A] By testing grass seedlings, Charles and Francis Darwin noted that the seedlings bend toward a source of light. **[B]** By shading the bending portion, they deduced that the controller is not located there. **[C]** Shading the tip revealed that this portion of the plant responds to light by sending a signal to the stem portion, which then bends.

find out whether the controller of that bending response is present in the curving portion of the seedling or in its tip. They began by placing a piece of foil around the section of the seedling that actually bends toward the light, to keep the light out, and observed that the seedlings bent toward the light anyway [FIGURE 37.6B]. Then they covered just the tip of a seedling with a foil cap and saw that the seedling did not bend toward the light. Their conclusion: The seedling's tip sends a signal to its lower portion, causing it to bend in the appropriate direction [FIGURE 37.6C].

Subsequent research showed that the growth hormone auxin comes from the seedling's tip and accumulates on the side away from the light. Evidence suggests that the accumulated auxin causes a drop in internal cell pH, the acidic environment allows an enzyme to become active, the enzyme weakens cellulose in the cell walls, and these walls then stretch and enable the entire cell to elongate. Consequently, the seedling bends toward the light.

PLANT MOVEMENTS OTHER THAN TROPISMS

Although tropisms are immensely important to a plant's orientation in space, not all plant movements are tropisms oriented toward or away from the stimulus that causes them. Leaves of the sensitive plant, *Mimosa pudica*, [see FIGURE 4.19], for example, will droop if touched anywhere along their length, and the leaves of the Venus flytrap snap shut if a small frog or insect triggers any of the little hairs inside the trap [see FIGURE 4.1]. Such plant movements whose direction is independent of the direction of the stimulus are called *nastic responses*. These responses are not always slow and irreversible, nor are they based on the growth of cells, as tropisms tend to be; for example, the Venus flytrap snaps shut rapidly, but subsequently reopens.

Some plants exhibit so-called sleep movements, with leaves drooping at night and lifting in the daytime. Even if the plants are suddenly kept in complete darkness, these sleep movements will continue for a few days [FIGURE 37.7]. Biologists consider such movements strong evidence that plants, like animals, have internal biological clocks that meter the passage of time and trigger regular responses.

HOW THE ENVIRONMENT INFLUENCES A PLANT'S SHAPE

Although rooted in place, plants have a tremendous capacity to turn, bend, and grow, which allows them to move part of their body toward or away from light, water, nutrients, gravity, or substrates. Thus, external factors can influence plant growth and shape in ways that adapt the plant to its environment.

Shape Changes in Response to Light Light can have a profound effect on a plant's shape. A good example is a crabgrass plant growing in a dark spot, such as the deep shade of a forest or even under a board—the plant is

3:00 P.M.

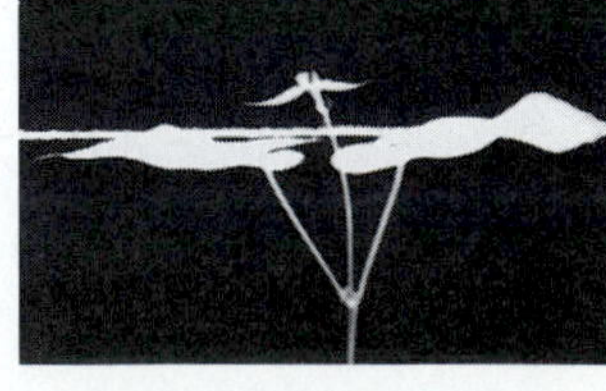

9:00 P.M.

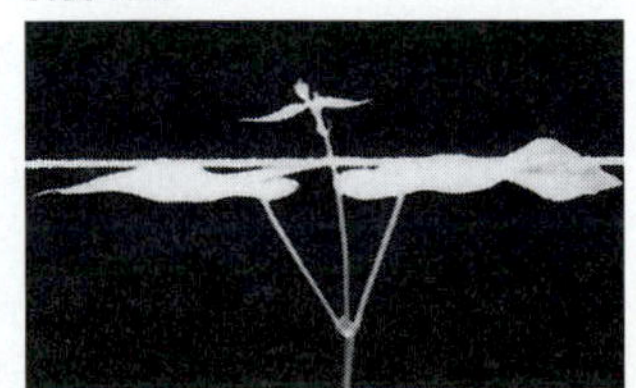

FIGURE 37.7

Sleep Movements in Plants.

Leaves of the bean plant *Phaseolus vulgaris* droop at night and fully extend by noon, even when kept in the dark. These "sleep movements" are evidence of a biological clock in plants, probably based on hormones.

spindly and pale yellow, not healthy and green. Deprived of direct light, the plant expends little energy making chlorophyll, and instead grows longer and longer until it reaches the light and slows down [FIGURE 37.8A]. Plants that have grown long and pale in darkness are said to be *etiolated* (from a French word meaning "pale as straw"). Even in less extreme shade, plant survival depends on the plant's ability to detect light levels and adapt by growth to this changing environmental stimulus.

How Plants Detect Light Plants have a light-sensitive blue-green pigment called **phytochrome** that allows them to detect the amount of light in the environment. Although it is an internal regulator, phytochrome (meaning "plant color") is not a hormone per se because it does not diffuse from cell to cell. Phytochrome allows a light-sensitive seed to detect whether light is available, which determines whether or not it germinates. Phytochrome also spurs flowers to emerge at appropriate times for reproduction.

Phytochrome helps a plant distinguish between darkness, shade, and sun, as well as the amount of time spent in each, because the chemical changes its form, depending on the amount or wavelength of light that strikes it. Visible light consists of a broad spectrum of colors that can be seen individually when they are separated through a prism or water droplets, as in a rainbow [review FIGURE 6.2]. For plants, two of the most important colors in natural light are red and so-called far-red light [FIGURE 37.8B]. Red light is a common sight at sunset and has a longer wavelength than blue, green, yellow, or orange light. Also recall from CHAPTER 6 that chlorophyll strongly absorbs red light, but reflects green light, explaining why plants appear green. Far-red light has wavelengths even longer than red light and is barely visible to the human eye.

Phytochrome does in plants what rhodopsin does in the cone cells of human eyes [see FIGURE 32.17]: It responds to light of a specific color. There are two forms of phytochrome, named for the color of light they absorb: P_r absorbs red light, and P_{fr} absorbs far-red light. When P_r absorbs red light, the chemical changes rapidly into the P_{fr} form. When P_{fr} absorbs far-red light, it quickly changes back into P_r. In addition, P_{fr} spontaneously but slowly changes to P_r in the dark [see FIGURE 37.8B], a bit like a chemical hourglass.

FIGURE 37.8 Phytochromes and Plant Growth Patterns.

[A] Plants grown in darkness become pale and spindly

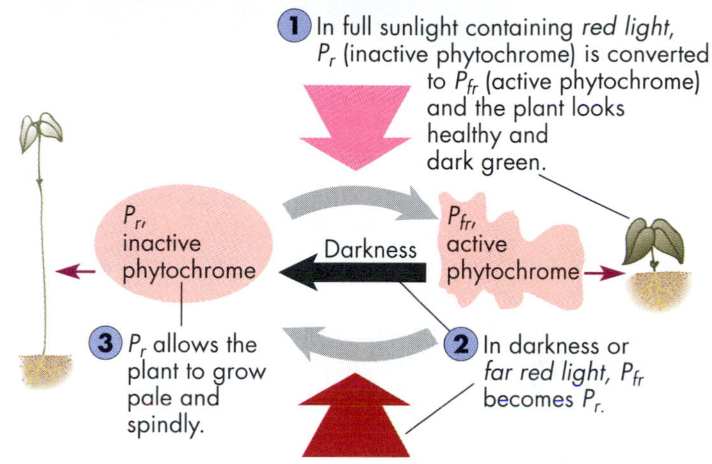

[B] Phytochrome conversions

The interconversion of phytochromes affects plant growth in the following way: In normal sunlight, P_r absorbs the substantial amounts of available red light and converts to P_{fr}, the active form of the molecule (some students remember the difference with the slogan, P_{fr} = *p*igment *f*ully *r*eactive). The plant responds to this active form by growing green and lush. In darkness, however, P_{fr} changes to P_r, the inactive form. In the absence of P_{fr}, the plants grow long, pale, and spindly. Growth control by phytochrome is just one aspect of how growth regulators affect plant shape. Another control comes from substances at the tips of the plants.

Hormones and Plant Shape Plants can take on a wide variety of shapes, and two of these forms are represented by the spruce and maple pictured in FIGURE 37.9A.

11:00 P.M.

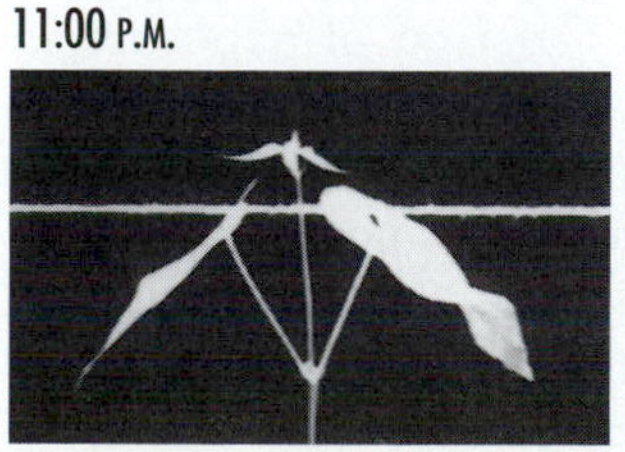

2:00 A.M.

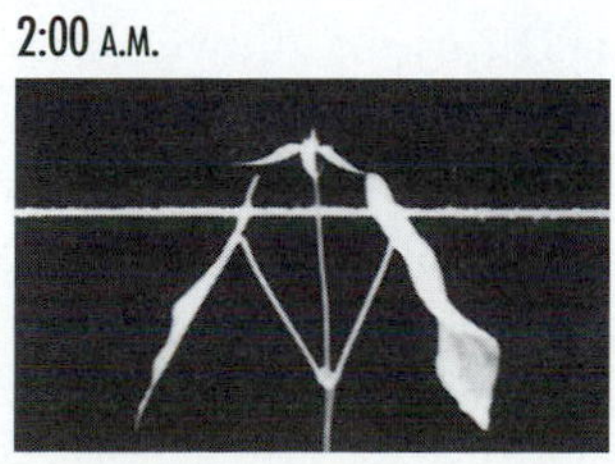

6:00 A.M.

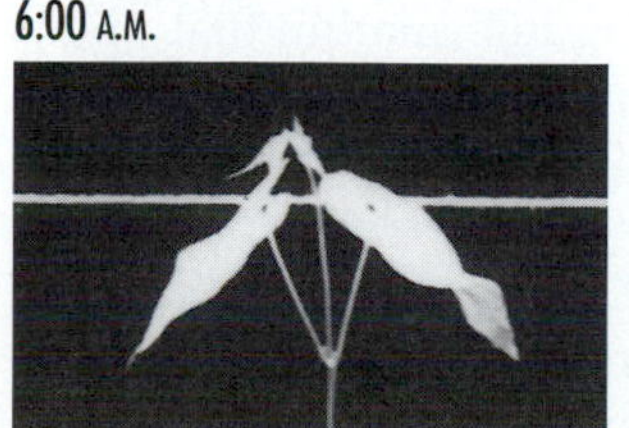

12:00 NOON

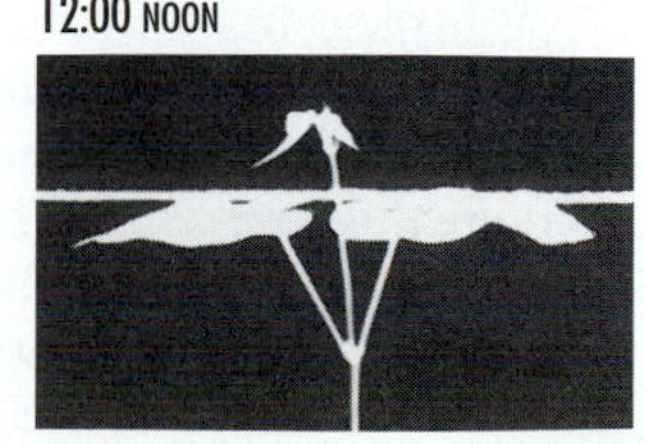

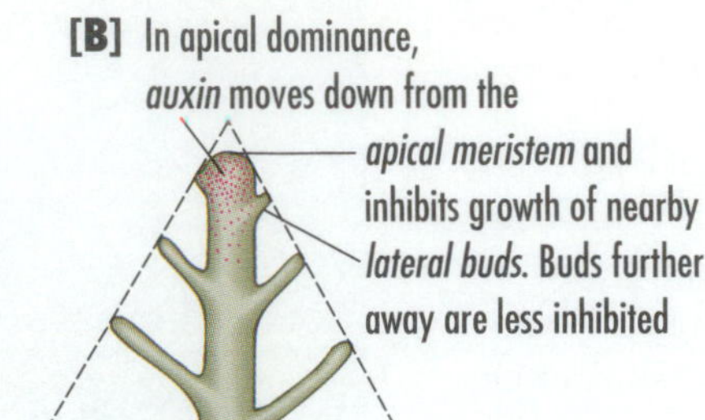

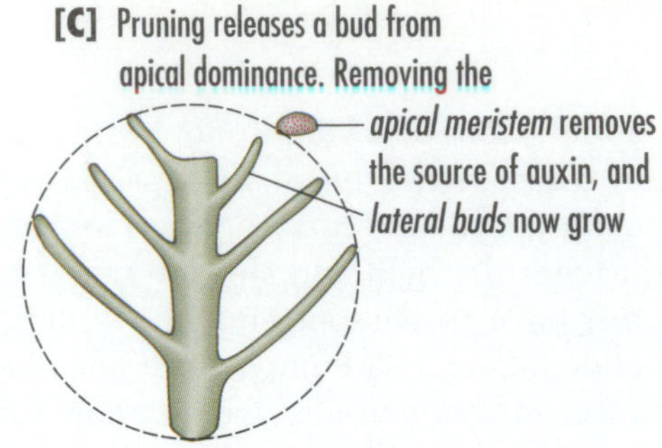

FIGURE 37.9
Apical Dominance and Overall Plant Shape.

> **CONCEPT CHALLENGE**
> **Suppose that during an investigation of a plant's phytochrome gene you discover a mutation that causes the plant to grow pale and spindly. How do you think the mutation might achieve this effect?**

Spruce trees usually have a single main trunk with numerous side branches—long at the base and shorter at the top—so that the whole tree has a conical shape. In contrast, maples often branch low and have several main branches but no central trunk, and the overall shape is roughly spherical, not conical.

A major factor in these plant shapes is **apical dominance** (literally, domination by the tip of the plant), an inhibition of lateral buds by auxin transported down the shoot from the apical bud. Auxin promotes growth in the stem, but inhibits growth of lateral buds. Because auxin is continuously broken down as it moves down the stem, its concentration drops off [FIGURE 37.9B]. Thus, buds closest to the tip of the main stem are most inhibited (and smallest), while branches farther from the tip of the plant are the least inhibited (and oldest and largest). The result is a conical shape. Cytokinin also comes upward from the roots, counteracts the effects of auxin, and causes the lateral buds and the lowest branches to grow more strongly. In a plant with a more spherical shape, the apical bud may produce less auxin, and lateral buds and branches may be less sensitive to inhibition by the hormone; thus, the branches both high and low on the stem can grow strongly. If a gardener or a browsing animal removes the apical bud of a plant, the buds farther down the stem are released from apical dominance and begin to grow, producing a bushy appearance [FIGURE 37.9C]. Commercial Christmas tree growers take advantage of apical dominance by pruning back the tips of branches. This allows buds toward the inside of the tree to grow, yielding a fuller, more desirable tree.

It is clear from the foregoing information that plants generally respond to environmental hardships or opportunities by growth and change in form and that hormones mediate these responses. Life-cycle events are part of this responsiveness, and as in germination and subsequent growth, flowering and fruit formation depend on a mixture of environmental and hormonal triggers.

Control of Flowering and Fruit Formation

Like germination and growth, flowering and setting fruit are timed to the appropriate seasons. This means that pollination, fertilization, fruit enlargement, seed development, fruit drop, and seed dispersal all occur in seasons when the species has the greatest potential for success. Environmental factors that change with the seasons—namely, temperature, moisture, and the number of hours of sunlight in a day—act as external regulators of the plant life cycle. Seasonal timing is important for annual plants such as rice, which flower and produce seeds once in a single season, and also for perennials such as roses. An exceptional example of this effect is the rare *Puya raimondii*, a large plant of the windswept Andean high plain. This plant can live for nearly two centuries, and it produces the world's largest floral structure, consisting of hundreds of individual flowers [FIGURE 37.10]. However, it sends up this giant flower only once every 30 to 150 years. Seasonal timing depends on both environmental cues (such as temperature and light) and internal hormonal triggers.

TEMPERATURE AND FLOWERING

Certain plants, including biennials like beets, carrots, and turnips, flower only after a period of exposure to cold, a process called *vernalization*. Such exposure to cold and subsequent warming signifies that winter has come and gone. Because flowering comes early in the spring, the plant produces seeds early in its second summer, and the seedlings have plenty of time to grow into hardy plants that can withstand the following winter.

FIGURE 37.10

Timing Is Everything: Hormones and Environmental Cues Regulate Flowering in *Puya raimondii*.

Light, temperature, and internal hormones interact to control flowering in this giant member of the pineapple family. The plant produces a 4.5-m (15-ft) *inflorescence*, or flower spike, containing hundreds of individual flowers—the largest floral structure in the plant kingdom—just once every 30 to 150 years.

LIGHT AND FLOWERING

For many plants, temperature alone is too risky a trigger for flowering. A cold September followed by a warm October and a harsh November could find a plant covered with flowers but buried in snow. Many plants track the seasons by **photoperiod**, the length of light and dark periods each day, rather than by daily temperature. Early researchers noticed that some plants, like rice, seem to require short days to bloom, and they called these plants *short-day plants*. Other plants seem to require long days and were therefore called *long-day plants*. (Plants that flower without regard to photoperiod were called *day-neutral plants*.)

Despite the early terminology, experiments revealed that a short-day or long-day plant is not defined by the absolute length of the day or night, but by whether the day length is longer or shorter than some critical number of hours. For example, ragweed blooms in the fall and requires *less than* 14½ hours of daylight to blossom; thus, it is a short-day plant. In contrast, spinach needs *more than* 14 hours of light per day to flower, so it is a long-day plant. At 16 hours of light, the spinach will bloom but the ragweed won't, and at 8 hours of light, the ragweed will blossom but the spinach won't [FIGURE 37.11A and B].

Generally, long-day plants blossom in spring or early summer, when the days are becoming longer, and short-day plants flower in the late summer or fall, when the days are becoming shorter. For example, poinsettias, which are short-day plants, will flower only when the days are less than 10 hours long (and the nights are more than 14 hours long), as in autumn. People with indoor poinsettias must therefore put them in a dark closet for 14 or more hours each day during the fall in order to enjoy their showy red and yellow displays at Christmas.

To probe how plants measure critical photoperiod, botanists exposed ragweed plants to a day length that should cause flowering, but interrupted their period of darkness with a short flash of light [FIGURE 37.11C]. The result: No flowers formed. Since interrupting the night blocks flowering, plants must measure night length rather than day length. In other words, short-day plants are actually long-night plants. Thus even a quick peek at a poinsettia plant in its dark closet interrupts the plant's "long night" and prevents the plant from setting flowers. For a long-day (short-night) plant like spinach, however, a flash of light in the middle of the night promotes flowering [see FIGURE 37.11C].

The phytochrome "hourglass" allows such accurate timing of photoperiod that all the individuals of a species can flower within a day or two of each other. This pro-

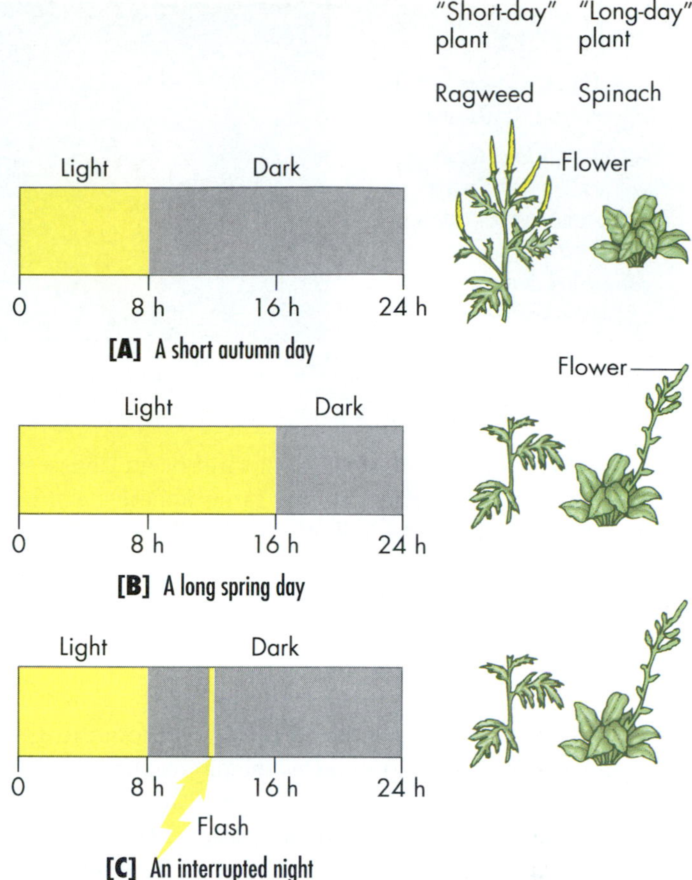

FIGURE 37.11

The Light-Dark Cycle Regulates the Onset of Flowering.

[A] On an 8-hour autumn day, ragweed—a short-day plant that flowers if there are fewer than 14½ hours of light—will form flowers. On the same day, however, spinach—a long-day plant that needs more than 14 hours of light a day—will fail to flower. [B] In contrast, on a long spring day, ragweed (the short-day plant) will not flower, but spinach (the long-day plant) will. [C] If an experimenter interrupts a long autumn night with a brief flash of light, the plants respond as if the day were long and the night short. This suggests that light resets the "clock" that times the length of the night. Further studies have shown that phytochrome molecules mediate these timing actions.

FIGURE 37.12

Seeds Liberate Auxin, Which Triggers Fruit Development.

[A] Strawberry seeds secrete auxin, and this promotes fruit development around the seeds. **[B]** If all but a few seeds are removed, fruit develops just around the remaining seeds.

[A]

[B]

vides a ready source of pollen that wind or animals can convey to other blossoms of the same species. And a plant's photoperiod often coincides with the most active food-seeking portion of an insect's (or other pollinator's) life cycle—evidence of their coevolution.

ROLE OF HORMONES IN FLOWERING

Plant scientists wondered whether phytochrome in the cells of each flower bud measures the day-night cycle for that bud or whether the signal to flower comes from other parts of the plant. To test these possibilities, experimenters exposed the buds of chrysanthemum plants to one light cycle and the leaves to another and checked the plants for flowers. The results showed that leaves, not the bud, were responsible for sensing photoperiod. If the leaves control flowering, however, then they must send a signal to the bud, which subsequently develops into a flower. Plant physiologists suggested that a substance they called **florigen** (literally "flower generator") might move from the leaves to the bud. Florigen itself has yet to be isolated, however, and many botanists believe that florigen is actually the combined effects of several already familiar plant hormones. Botanists do agree, however, that plants detect changes of seasons by means of a light-sensitive pigment that acts via hormones, causing flowers to appear at favorable times of the year.

TRIGGERS TO FRUIT DEVELOPMENT

As CHAPTER 36 explained, the fruit—with its cargo of seeds—begins to develop once the flower has been pollinated and seeds start to form. Fruits, as we have seen, help protect the seeds and disperse them. But fruits are also energy-expensive organs to produce—especially large, juicy ones such as plums and oranges. Not surprisingly, some plants ensure that fruits develop only around viable seeds.

The dozens of tiny seeds on the outside of a strawberry, for example, help control the development of a normal berry [FIGURE 37.12A]. If all the seeds are removed when the berry is still small and green, it never enlarges and ripens. However, if one removes very young seeds from the strawberry and treats the berry with auxin, it will grow normally. This suggests that the seeds promote berry growth by secreting auxin. Experiments show that strawberry seeds, in fact, do secrete auxin, and if even a few seeds are left on the strawberry, the berry will develop normally around the location of each seed [FIGURE 37.12B].

Once fruits have developed, the airborne hormone ethylene causes them to ripen. When a peach ripens, it gets softer through the partial breakdown of cell walls, sweeter through the conversion of organic acids and starches to sugars, and more fragrant through the production of aromatic molecules; and it turns from green to yellow through the formation of carotenes and other pigments. All these changes enhance the fruit's appeal to animals that eat it and distribute its seeds, and ethylene is behind most of these changes. Since most fruits release ethylene gas as they begin to ripen, green fruits ripen faster if placed in a plastic bag with ripe fruit.

Researchers have recently applied this knowledge of fruit ripening to the problem of overripening of fruits and vegetables during transport or due to lack of refrigeration. The APPLY AND DECIDE BOX on page 828 explains this new—and controversial—work.

➤ CONCEPT CHALLENGE

Ragweed is a short-day plant, requiring less than 14½ hours of light to flower; spinach is a long-day plant, requiring more than 14 hours of light to flower. A botanist exposes ragweed plants and spinach plants to a photoperiod of 14¼ hours of light. Which plants will bloom?

How Plants Age and Prepare for Winter

Plant life cycles are strongly synchronized with the seasons. Annuals, for example, set seed and die before winter, while perennials often drop their leaves and become dormant during the cold months.

FIGURE 37.13

Senescence Unmasks the Flaming Colors of Fall.

The breakdown of chlorophyll allows yellow, red, and orange light-gathering pigments to show. Eventually, these too will senesce, and the colors will fade.

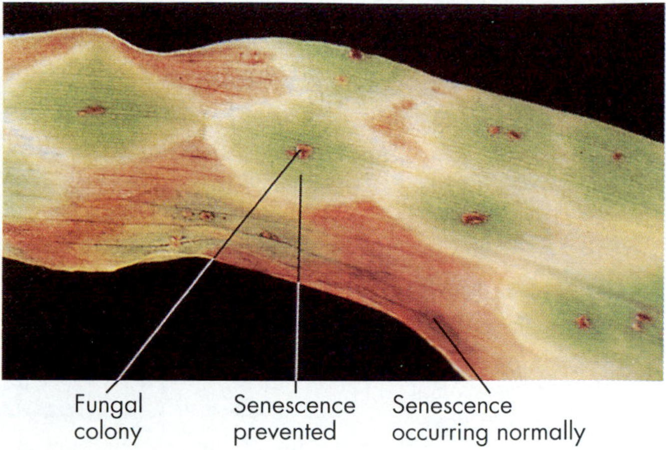

FIGURE 37.14

Cytokinins Block the Essence of Senescence.

A fungal infection dotting a browned, senescent corn leaf causes islands of green to remain. As in the foolish seedling disease, the fungus releases a plant hormone—in this case, cytokinin, which blocks ABA, the hormone causing senescence. Leaves normally senesce when external cues such as light, temperature, and moisture interact with the relative balance of abscisic acid, gibberellin, and cytokinin.

SENESCENCE AND ABSCISSION

The process of growing old and dying is called **senescence**, while the process that "scissors" away mature fruits or dying leaves is called **abscission**. Senescence and abscission are technical terms for the events of autumn we all know and enjoy: the turning of green leaves to glorious shades of red, gold, orange, and purple, followed by the fading and finally the falling of the leaves [FIGURE 37.13]. Several external and internal events control this sequence. Fields of soybeans, for example, grow yellow and die within a few days of each other—evidence that external cues (cold, dryness, diminishing light) trigger mass senescence. Tomatoes, zinnias, and marigold plants in the garden will also die back after they have produced flowers and seeds, each species on its own schedule and dependent on environmental conditions. Flowering itself is a key to this senescence, since in many annuals, the apex of each shoot converts into a flower bud and ends the open growth potential of that branch. When the last branch tip has converted to a flower, the plant stops growing and soon begins to senesce.

In perennials, the plant conserves resources over the dry, cold months of winter, and the **deciduous** growth habit (the dropping of leaves) is very common. In the cold and dark of winter, leaves are much less efficient photosynthetic organs and provide a dangerous avenue for water loss. Leaves may also serve as organs of excretion, storing wastes which then fall from the plant when the leaves drop.

In fall, a deciduous tree, such as a sugar maple, withdraws valuable nutrients—including sugars and amino acids—from the leaves and transports the nutrients to the trunk and roots where they are stored. One of the last molecules to be broken down and withdrawn is chlorophyll, the main photosynthetic pigment. However, with an environmental trigger such as a night or two of subfreezing temperatures, the chlorophyll, too, breaks down and the other brilliantly colored leaf pigments that remain (red or purple anthocyanins, yellow xanthophylls, and red and orange carotenes) can show through. Eventually, the leaves fall, and the plant survives winter in a state of dormancy.

Researchers have noticed that a fungus that releases the hormone cytokinin will cause a green spot to remain on a yellowing corn leaf [FIGURE 37.14]. For this reason, plant scientists suspect that autumn conditions may influence levels of hormones and that these, in turn, may determine when leaves senesce.

Leaf abscission is also governed by environmental cues and the ratios of plant hormones. During the long, sunny days of summer, each leaf produces large quantities of auxin, and this hormone inhibits the formation of an *abscission zone* [FIGURE 37.15], an area of relatively weak cells at the base of the stem, where the leaf will eventually separate from the branch. In the fall, auxin levels drop, enzymes begin to break down the cell walls in the abscission zone one by one, and eventually, a gust of wind or a few pelting drops of rain will send the leaf fluttering to the ground.

DORMANCY: THE PLANT AT REST

Once the leaves have fallen, the plant survives winter in a state of greatly reduced metabolic activity known as

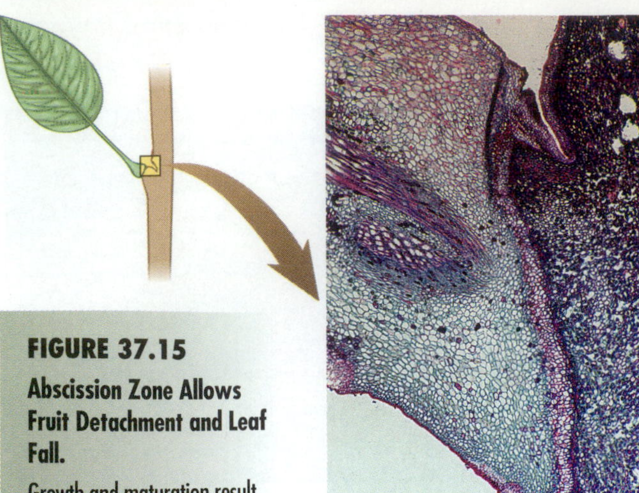

FIGURE 37.15

Abscission Zone Allows Fruit Detachment and Leaf Fall.

Growth and maturation result in an abscission zone, where fruit or leaf will eventually detach. This light micrograph of a section stained with dye shows the weakened zone, a vertical stripe of lavender cells.

dormancy (Latin *dormire*, "to sleep"). Dormancy can take several forms. Some perennial plants, like iris and daffodil, die back to underground stems, roots, or bulbs that are protected from desiccation and extreme heat or cold by the soil around them, while others, like chestnuts and linden trees, stand leafless from late autumn until early spring [FIGURE 37.16]. Many plants rest over the winter in embryonic form inside dormant seeds. Regardless of exterior form, dormancy is characterized by greatly diminished use of energy and slowed protein synthesis, cell division, and maintenance. Light cycles, perhaps timed by phytochrome, may be involved in the onset of metabolic slowdown, and hormones no doubt relay the message to individual cells and help maintain the state of dormancy.

➤ CONCEPT CHALLENGE

Compare the phenomenon of green spots on senescing corn leaves in FIGURE 37.14 with the phenomenon of etiolated rice seedlings in FIGURE 37.2. How are they similar? How are they different?

Plant Protection: Defensive Responses to External Threats

Plants, as we have seen, usually adjust to changing environmental conditions by growth. Plants may be stationary, but they are far from defenseless when it comes to other threats, such as animal predators and bacteria.

FIGURE 37.16

Dormancy: A Key to Winter Survival.

Many perennial plants, such as this American linden, survive the bitter cold and drought of winter by entering a state of dormancy, triggered by environmental stress and maintained by plant hormones, perhaps including abscisic acid.

Plants can often protect themselves quite effectively by producing physical barriers, such as thorns, leaf hairs, and sticky sap; by generating various toxic compounds; and by walling off injured areas.

CHEMICAL PROTECTION

Plant scientists have discovered that in addition to *primary compounds* (biological molecules necessary for growth and regulation), most plants also produce *secondary compounds*, molecules that help ensure survival by repelling, killing, or interfering with the normal activities of plant-eating organisms. In some tree species, up to 50 percent of the plant matter is secondary defensive compounds. And these are so effective that in a temperate forest in a typical year, predators consume only about 7 percent of the total leaf surface area.

Secondary compounds include such familiar substances as cyanide, camphor, cocaine, caffeine, and nicotine—just a few of the 10,000 plant chemicals that are toxic to animals. Some of these substances, such as camphor and tannins, can discourage insect pests from eating a plant's leaves or laying eggs on them. Others poison attackers. For example, the nicotine in tobacco can kill various insect predators, and cyanide compounds in bird's-foot trefoil can kill snails that feed on its leaves. Recent evidence also shows that oligosaccharins from the cell walls of microbes or injured plant cells can cause a plant to make antibiotics that kill attacking fungi and bacteria. Some plants generate compounds that mimic insect hormones, disrupting normal insect metamorphosis.

In response to insect damage, many plants increase their production of secondary compounds, in some cases reaching 10 percent of the total weight of fruits or vegetables. Since these compounds can often harm people as well as insects—many, in fact, can cause cancer—it is best to stay away from diseased or damaged celery, apples, peanuts, and other produce.

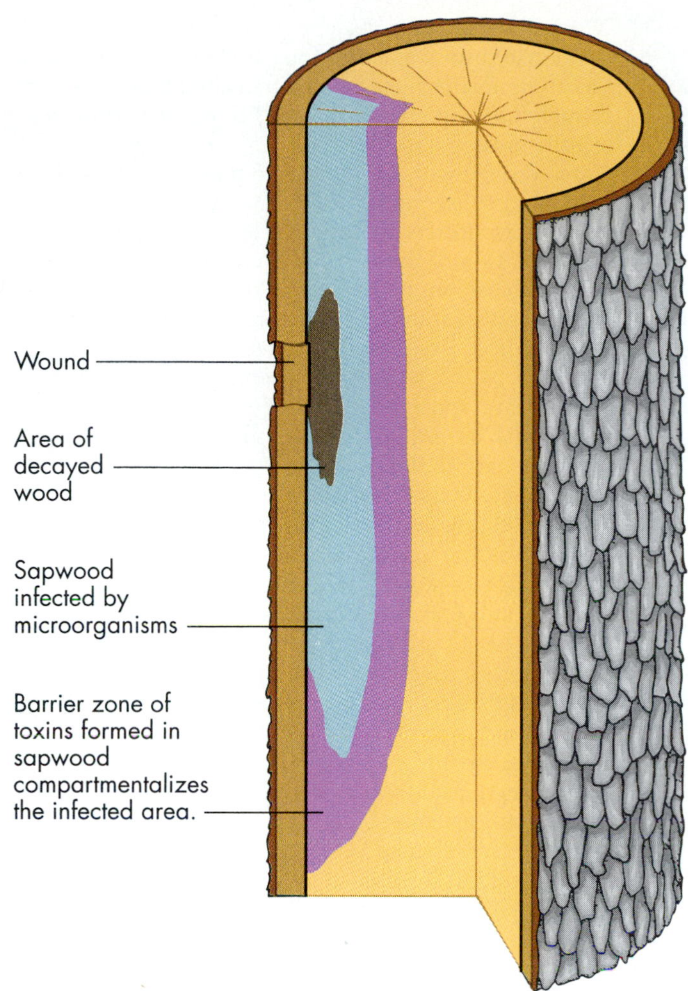

FIGURE 37.17

Compartmentalization in Plants: A Response to Wounding.

A penetrating wound will inevitably allow microbial and/or fungal infections into the plant, but a zone of parenchymal cells (in lavender) around the infected area begins to produce tannins and other toxic chemicals. These kill the invaders, contain the damage, and allow growth to continue around the walled-off area.

While most secondary substances are harmful to animals, they have given rise to a large number of human medicines. These include the pain reliever aspirin from willow bark, the heart medicine digitalis from foxglove, the main ingredient in oral contraceptives from yams, cancer treatments from periwinkle and the Pacific yew, and the pain reliever morphine from poppies. Many scientists suspect that thousands of additional useful medicines will eventually be found in plants. In fact, one reason scientists are fighting for the preservation of the vast tracts of unexplored rain forests is so that the plant species—most as yet unstudied—do not become extinct before we can make use of their healing compounds.

WALLING OFF INJURED AREAS

Besides producing protective chemicals, plants have a second general growth response that protects them from viruses, bacteria, fungi, animals, or simple physical injury, such as loss of a limb in a storm. Because of a plant's rigid cell walls, however, new cells cannot migrate into a wounded area and heal the damage. Plants can, however, wall off the damaged area—a response that prevents invaders from gaining access to healthy tissues. This walling-off process is called **compartmentalization**. It involves the production of toxic chemicals in the invaded area and the plugging of nearby xylem and phloem tubes with thick saps or resins that prevent invaders from spreading to other parts of the plant [FIGURE 37.17]. Once the injury is walled off, the tree can survive and grow.

Connections

A plant's survival depends on its adaptations to the environment. It flowers in seasons appropriate for reproduction, its seeds germinate when weather favors the tender young sprouts, it becomes dormant before conditions turn harsh, and it grows toward adequate sources of water and light and fends off attackers. The control of all these events involves a combination of external cues (light, temperature, moisture, gravity, seasonal changes) and internal growth regulators such as hormones and phytochromes. In general, environmental fluctuations trigger internal molecular responses, and the plant then adjusts to its environment. This adjustment often involves a change in physical form—an altered shape, the dropping of leaves, or the opening of flowers. Plant regulators—gibberellins, auxins, cytokinins, abscisic acid, and ethylene—have dozens of regulatory roles in plants, and alone or in combination, they control germination, tropisms, plant shape, flowering, fruit development, senescence, leaf drop, and dormancy. Since most of these events involve a change in the plant's physical form, plant hormones are both a practical and a conceptual bridge between anatomy, survival, and the environment.

In the last section of the book, we turn our focus to the environment. CHAPTERS 38 to 42 investigate how individual organisms relate to their environments and how groups of organisms interact with each other and their physical surroundings.

KEY TERMS

abscisic acid, 815
abscission, 815
apical dominance, 820
auxin, 815
cytokinin, 815
deciduous, 823
dormancy, 815
ethylene, 815
gibberellin, 814
photoperiod, 821
phytochrome, 819
senescence, 823
tropism, 817

HIGHLIGHTS IN REVIEW

1 Because of their rigid cell walls, plants cannot move away from adverse conditions. Instead, they cope with environmental challenges through the processes of growth, maturation, and dormancy.

- **a]** Plants respond to the environment primarily through modified growth. Growth is modified by hormones, growth regulators that trigger changes in metabolism, development, and other cellular activities.

2 Plants respond to the environment, in part, by producing and reacting to growth regulators, including hormones and phytochromes.

- **a]** In general, gibberellins, auxins, and cytokinins promote growth. Abscisic acid inhibits growth, maintains dormancy, and promotes abscission (leaf and fruit fall). Ethylene promotes ripening, senescence, and abscission.
- **b]** Most seeds survive winter in a dormant state and germinate only when suitable conditions of light, temperature, and/or water are present.

3 Light, gravity, water, and temperature are major environmental factors triggering the hormone-controlled patterns of plant growth, flowering, maturation, dormancy, and senescence.

- **a]** Plants become oriented in space with respect to light, gravity, and water by means of tropisms, such as gravitropism and phototropism. In phototropism, auxin from the shoot tip seeps down the unlighted side of the plant. The cells on that side grow faster than the cells on the opposite side, and the plant bends toward the light.
- **b]** Plant movements can be fast or slow, irreversible or reversible, and specifically oriented toward or away from a stimulus. Movements not oriented with respect to the stimulus are called nastic responses.
- **c]** Plants can change shape in response to environmental conditions and internal hormonal triggers. Plants growing in the dark become etiolated—long, spindly, and pale—because the light-sensitive pigment called phytochrome spontaneously converts from the active P_{fr} form to the inactive P_r form in the dark. The P_{fr} form promotes normal growth, while the P_r form permits etiolated growth.
- **d]** The shoot tip produces auxin; this hormone augments growth and helps control cell elongation, but inhibits the growth of nearby lateral buds, a process called apical dominance.
- **e]** Cytokinin from the roots acts in opposition to auxin. When cytokinin concentrations are high or lateral bud sensitivity to auxin is low, the buds and branches throughout the plant are released from inhibition and begin to grow.
- **f]** Plants must flower at appropriate times so that pollination and subsequent fruit and seed development have the maximum potential for success.
- **g]** The regulation of flowering often depends on external cues, such as temperature and photoperiod (light-dark cycles), and on internal regulators, such as the pigment phytochrome and plant hormones.
- **h]** Short-day plants, such as the poinsettia, require nights longer than a specific length to flower. Long-day plants, such as spinach, require nights shorter than a specific length to flower.
- **i]** Plant scientists have hypothesized the existence of a flower-inducing hormone (florigen) that moves from the leaves at the tips of branches into the flower buds. No one has yet isolated this specific chemical, however, and florigen may actually be the combined effects of several hormones.
- **j]** Viable seeds strongly promote the growth and development of fruit by releasing auxin and in some cases gibberellin. Ethylene, produced by maturing fruits, accelerates ripening.
- **k]** In late summer and fall, many plants withdraw starches and proteins from their leaves in a process called leaf senescence. Abscisic acid promotes senescence; cytokinin inhibits it. When the leaves lose chlorophyll, other pigments show through as fall colors.
- **l]** Auxin prevents abscission by maintaining an area of weak cells, called the abscission zone, at the base of the leaf stalk or fruit stem. When auxin levels drop, ethylene and ABA can accelerate the dropping of leaves and fruits.
- **m]** Many plants become dormant for the winter after storing energy in the form of starches and oils in their roots and stems and then dropping their leaves.

4 In response to damage caused by insects or disease, plants produce chemicals harmful to the attackers and wall off injured areas.

UNDERSTANDING THE FACTS AND CONCEPTS

For Questions 1–10, match each of the descriptions with the most appropriate item from the following list of terms. Each answer can be used once, more than once, or not at all.

- **a]** gibberellins
- **b]** auxins
- **c]** cytokinins
- **d]** abscisic acid
- **e]** ethylene

1 Travels through the air from one fruit to another, causing maturation of flowers and fruits.

2 A growth promoter that travels rapidly throughout the plant in both directions within the vascular system.

3 Induces and maintains dormancy by inhibiting protein synthesis.

4 A growth-promoting hormone that maintains apical dominance.

5 Promotes leaf, flower, and fruit drop; produced by ripening fruits.

6 Augments growth by stimulating cell division, especially in lateral buds.

7 Travels short distances; important in leaf drop.

8 Cause of foolish seedling disease; important regulator of plant height and seed germination.

9 Indoleacetic acid; promotes cellular elongation and growth.

10 Growth promoter that diffuses from the root upward.

As above, for Questions 11–15, match the descriptions with the most appropriate item from the following list of terms.

a] tropism
b] phytochrome
c] apical dominance
d] photoperiod
e] florigen

11 Postulated hormone that moves from leaves to flower bud, working in conjunction with phytochrome.

12 A light-sensitive pigment that regulates seed sprouting and flowering.

13 Directional plant growth in response to external stimuli such as light or internal hormonal stimuli.

14 The method by which short- and long-day plants regulate their life cycles.

15 The process of inhibition of lateral bud growth by auxins.

As above, for Questions 16–20, match the descriptions with the most appropriate item from the following list of terms.

a] deciduous
b] dormancy
c] abscission zone
d] secondary compounds
e] compartmentalization

16 One of the primary chemical defenses of plants.

17 Type of tree that drops its leaves during the winter.

18 A physical method of protection in which a site of damage is isolated in part by blocking the transport vessels to that area.

19 A reversible period of metabolic inhibition probably regulated by phytochromes and hormones.

20 An area of weakened cells at the base of a leaf.

INTEGRATE AND APPLY WHAT YOU HAVE LEARNED

1 List the classes of growth-promoting and growth-inhibiting plant hormones, indicating where in the plant each is made.

2 The Christmas cactus is a short-day plant which normally blooms in the late fall. Explain the probable role of phytochrome in regulating this cycle.

3 Explain the hormone interactions that regulate the development of conical shape in many kinds of trees.

4 Broadleaf trees in tropical regions are generally not deciduous, whereas those in temperate regions are. How can you account for this difference?

5 Which hormones are the primary regulators of leaf senescence, seed germination, fruit growth and ripening, and dormancy?

ANALYSIS

1 Why are the movements of the Venus fly trap not thought to be induced by hormones?

a] The changes are too rapid.
b] The movements are reversible.
c] Cell division and/or elongation are not involved.
d] The movements are independent of the direction of the stimulus.
e] All of the above are correct.

2 In order to get a poinsettia to set flowers for the Christmas season, in November a grower should:

a] Expose the plant to strong light and darkness in alternating six-hour cycles.
b] Grow the plant under strong lights for 24 hours per day.
c] Grow the plant under strong lights for 24 hours per day and simultaneously treat the plant with abscisic acid.
d] Grow the plant in a light-tight box, lifting the lid once a day at noon for watering.
e] Grow the plant in a light-tight box and water it at night in the dark.

3 In order to prevent ripe fruit from dropping from the tree, a grower should apply:

a] Abscisic acid.
b] Auxin.
c] Ethylene.
d] Gibberellic acid.
e] Abscisic acid, auxin, ethylene, and gibberellic acid.

4 Peas, which bloom in late spring, are among the earliest vegetable crops to mature. From this information, which of the following statements would you think applies to peas? More than one answer may be correct.

a] Peas are long-day plants.
b] Peas are short-day plants.
c] Peas are day-neutral plants.
d] Peas are indeciduous plants.

Why Chefs Are Boycotting High-Tech Tomatoes

Joyce Goldstein has been called San Francisco's finest chef, and her restaurant, Square One, became an overnight success in 1984. When it comes to ordering and preparing food, Goldstein has strong preferences and opinions, and these extend to the prospects of bioengineered food, which she welcomes about as warmly as a week-old fish fillet or some soggy endive.

In the summer of 1992, Goldstein joined 1000 of America's top chefs in a boycott of genetically engineered foods—a full year before the first such product even hit the supermarket. The chefs were worried about the safety of vegetables, grains, and other foods with deliberate genetic alterations. They were particularly upset over the announcement by officials of the U.S. Food and Drug Administration that in most cases, bioengineered foods will not require special labeling or safety testing. Some of the chefs printed notices on their menus: "We do not serve genetically engineered food." Almost certainly, their professional skepticism will dissuade many food shoppers from trying the new products as they start to arrive in stores.

Do Joyce Goldstein and the other chefs have a legitimate reason for taking such a strong stance? Or are they being "Chicken Littles" about foods that, many scientists contend, will be safe or even safer than existing choices?

The first genetically engineered food was expected to reach food shoppers in the Chicago area in 1994. The food is the Flavr Savr tomato, as Calgene, Inc., of Davis, California, calls their new product. Plant researchers at Calgene have genetically altered tomato fruits to resist spoilage, so that growers can send juicier, more flavorful tomatoes to market. Because standard vine-ripened tomatoes cannot be shipped long distances without bruising or splitting, growers pick normal tomatoes while they are still green, and ship the spheres while they are hard-walled and bitter. Then, before distributing the food to stores, they gas the tomatoes with the plant hormone ethylene (which we studied in CHAPTER 37) to induce ripening and reddening. This results in produce that is usually undented, but is also pink inside, has very little juice, and almost no flavor.

Plant genetic engineers at Calgene and the University of Nottingham in England knew that as a tomato ripens on the vine, an enzyme called polygalacturonase breaks down cell walls and allows the fruit to soften. They set out, therefore, to turn off or slow down the action of this enzyme so the tomato fruits would soften more slowly and could be left longer to ripen on the vine and develop juice and flavor before harvesting.

Team members located the gene for polygalacturonase, as well as the sequence that regulates its activity, and cloned the gene. Then—and this is the novel part—they spliced the gene to the control sequence *backward*. Finally, they inserted this so-called *antisense gene* into a normal tomato chromosome, which also contains the gene in its forward orientation. As a consequence of the antisense RNA, the tomato fruit makes only 10 percent as much of the softening enzyme. The result, the Flavr Savr tomato, can remain on the vine about five days longer than usual. This is still four or five days short of fully vine-ripened, but Calgene says the fruit tastes better, and has shown that once picked, the fruit resists rotting for three weeks—twice as long as a regular tomato.

Calgene is aiming for the premium produce market, since stores will charge $1 per pound more for the Flavr Savrs than for their faster-fading cousins. This is where famous chefs at gourmet restaurants might be the logical market. However, Joyce Goldstein, for one, remains unimpressed with the concept of longer-lived fruits and vegetables. "What is it?" she asked on a national radio interview, "A tomato or a sculpture?" Catherine Brandel, chef of the popular Chez Panisse Restaurant in neighboring Berkeley, agreed. "This type of food has nothing to do with us. Why would we want shelf-life and season extenders when we can buy fresh produce and meats from local growers every day?"

A few recent studies are particularly relevant to the boycott of biotech foods. For example, some researchers are attempting to make tomatoes that freeze and thaw without turning to mush by transferring in the gene for an "antifreeze" protein from the Arctic flounder. Others are manipulating potato genes so the tubers make more starch and sugar; transferring silkworm genes into potatoes so they resist soft rot; and transferring peanut genes into rice so the grains make more protein.

In a world where gaunt, starving children stare out daily from the television screen and newsmagazines, many people applaud biologists' efforts to expand the global food supply. Thus it might seem puzzling that famous chefs would boycott new foods, and that some environmental and consumer activists would join in with strong objections. The objections, however, can be understood to center not on the principle of an expanded food supply, but on the issues of safety, testing, and regulation of the new foods, as well as, to some extent, a misunderstanding of modern biology and a mistrust of industry and technology.

Joyce Goldstein, for instance, is most concerned about the safety of genetically engineered foods, noting that some of her customers are allergic to fish and peanuts. She is worried that tomatoes with flounder genes and rice with peanut genes will trigger their allergy attacks. Researchers say that in many foods, they can identify (and avoid transferring) genes for proteins that cause allergy. They also say, however, that in many foods, they don't know specifically which proteins cause allergies, and it is possible—albeit unlikely—that they could unintentionally transfer those genes.

The chef's biggest worry was the FDA's announced intention to allow some of the foods to reach consumers without special safety testing or product labeling. The agency intends to consider only the product—the food resulting from genetic manipulation—not the way it was produced. The FDA has stated that a biotechnology company will have to safety-test a food only if it is substantially different from anything ever produced before or has new and much higher levels of toxic compounds. The company would have to label these, as well as any other food to which a company had knowingly added a common allergen (which the FDA defines as a protein from eggs, milk, fish, shellfish, nuts, wheat, or legumes such as peanuts). Bioengineered foods outside these categories could be sold without safety testing or special labeling. The boycotting chefs felt that without more stringent safeguards and disclosures, they would be unaware of exactly what they were serving in their restaurants. As a solution, some consumer advocates have recommended a plan of reverse labeling, wherein nonaltered produce would carry labels the way organically grown produce does today. This would allow consumers a choice, they say, but would not stigmatize the biotech foods unfairly.

Biologists often feel that consumers misunderstand the fundamentals of gene transfer and how benign it is in principle. As we saw in CHAPTER 34, for hundreds of years farmers have been selecting and breeding plants with desirable traits—bigger ears of corn, sweeter berries, better tolerance to cold. Behind seed size, sugar content, cold tolerance, and every other physical trait, there lies a gene or a constellation of genes that must be bred into and expressed by the "improved" individuals. Traditional plant breeding is therefore a type of artificial selection that requires the transfer of traits from one set of individuals to another. Genetic engineers do the same, but their methods are different, the changes they make usually show up after just a single generation instead of many, and the palette from which they paint new combinations of traits is much more varied.

In nature, and even with traditional artificial selection by plant breeders, rice would probably never cross with peanut and pick up genes for higher protein production. Nor would tomato ever cross with flounder and pick up a gene for antifreeze protein. Bioengineered foods, therefore, will introduce some previously unknown and unattainable *combinations* into the human diet. Once in the stomach, however, where it is broken down to constituent building blocks, how different would peanut-powered rice be from an Indonesian rice dish flavored with peanut sauce? How different would freezable tomatoes be from a helping of flounder with tomato sauce? As long as a person *knew* he or she was eating the peanut or flounder proteins, there would be no difference. ❑

PART SIX

Ecology: Organisms and the Environment

CHAPTER 38

Population Ecology: Patterns in Space and Time

THE RISE AND FALL OF A DESERT POPULATION

More than 2000 years ago, a band of Native Americans migrated from Mexico to one of the world's most forbidding deserts, the Salt River Valley of central Arizona, where the city of Phoenix sprawls today. Using sharp sticks and manual labor, small teams of the immigrants (and later their descendants) dug an immense network of broad, deep irrigation canals over 1600 km (about 1000 mi) long [FIGURE 38.1]. In winter, these ditches captured water as it tumbled down from the surrounding mountains and channeled it to the Salt River and Gila River valleys, where the Hohokam Indians established their villages and fields. In summer, the canals trapped and carried water from intense local thunderstorms. With this precious redirected resource, the Indians made the barren desert fantastically productive: Each year, they grew two full crops of corn, squash, beans, barley, cotton, and tobacco over an area of more than 25,000 km^2 (about 10,000 mi^2).

Archaeological evidence shows that the Hohokam were probably the first people to irrigate in North America, and that only the Chimu Indians of coastal Peru created more sophisticated irrigation systems in the New World. As a result of the Hohokam's amazing

FIGURE 38.1

Canals of the Vanished Hohokam Tribe. These canals helped support a large population in the Arizona desert. What happened to the Hohokam? What principles does their story convey for today's human population?

agricultural success, they flourished in the desert valleys for nearly 1000 years. Their population doubled and redoubled, and their settlements grew from clusters of simple mud huts to at least 22 huge cities of large, multi-storied earthen buildings rising from the desert floor like castles. Some experts believe that the adobe dwellings housed over 1 million people—nearly the population of present-day Phoenix.

How ironic, then, that about 600 years ago, the Hohokam's huge desert culture suddenly vanished. We may never know the reasons for this demise, but archaeologists think that alkali salts may have built up in croplands irrigated for long periods, poisoning the plants and causing agricultural production to plummet. They think that the Hohokam may also have exhausted the surrounding desert of its limited natural resources—firewood in particular. The name "Hohokam," in fact, comes from the Pima Indians who later inhabited the Salt River Valley, and it means "all used up." Hardship and starvation must have swept the ancient settlements, and the death rate apparently skyrocketed.

Was the rise and fall of the Hohokam unusual? Is a population of human beings independent of the biological principles that regulate population growth in other species, or is it, too, governed by those principles? And if human populations are so governed, then would knowing and applying the principles help us to predict the future of modern human populations and perhaps enable us to take steps to ensure a livable future without catastrophic collapses? Building on the concepts of population genetics and evolution encountered in CHAPTER 16, this chapter explores questions like these. It focuses on **population ecology**: the study of the distribution and abundance of organisms in space and time.

MESSAGES

1 Closely interacting biological and physical factors govern the abundance and distribution of any species in any area.

2 Ecologists can model a population's growth, and hence predict how it will change with time.

3 Many biologists suggest that size and rate of growth of the human population are the most important problems facing the world today.

As it turns out, no single factor has an absolutely predictable effect on a population's distribution and size because living things in a particular area are affected by climate; availability of food, water, and living space; and factors influencing reproductive success. As the chapter will show, ecologists can learn why a population of birds, plants, fish, or people thrives (or dies out) in a particular place. To do this requires definition of the biological and physical factors that interact in an area; an estimation of a population's future growth; and an identification of evolutionary pressures at work on the population. Other levels of ecology, presented in later chapters, include the interactions of a population's members with other species, with the immediate physical environment, and ultimately with the biosphere, the global setting in which all living things coexist.

This chapter begins by discussing the various levels at which organisms interact with their surroundings. Next, it explores why particular organisms are found in specific places, and goes on to consider why the organisms in a given area can be plentiful at one time but not at another. And finally, we will discuss why our own species' astonishing population growth has become the most important ecological factor on earth.

Along the way, we'll learn about a bush that grows best at 44°C (111°F); how a pair of roaches could give rise to 800 billion more roaches in just seven months; and where the human population is headed in the near future. ❑

Ecology: Levels of Interaction

Early peoples amassed a simple form of natural history. Their survival depended on knowing the whereabouts of food sources, and they could describe, often through stories or rituals, the seasons when deer were abundant and the valleys where huckleberries grew along with other edible fruits, nuts, and plants. The modern science of ecology has its roots in natural history, but it applies research tools and the scientific method to probe the explanations behind observed phenomena. What factors cause deer populations to increase in the summer, for example, and what allows a particular valley to bristle with huckleberries?

Ecology is the scientific study of how organisms interact with each other and with their environment. Ecologists have determined the principles that govern the distribution and abundance of organisms, and the tenets of ecology have become as crucial to human survival as natural history once was to our ancestors. The more people know about the factors that influence plant growth, the better they can manipulate those factors to produce additional food. That is what the Hohokam did when they channeled water to the desert, and it is what modern plant scientists seek to do with plant breeding and genetic engineering.

A key word in the definition of ecology is *interact*. Organisms interact with other living things, and collectively, they constitute the **biotic** (living) **environment**; living things also interact with the nonliving physical surroundings that make up the **abiotic** (nonliving) **environment**. For a squirrel, the biotic environment includes the acorns it eats, the blue jays that compete with it for the nuts, the ticks that parasitize and weaken its body, and the siblings that share its territory. The abiotic environment includes the rainfall, sunlight, and soil that regulate acorn production, and the hot and cold temperatures that the squirrel must endure in summer and winter.

So numerous are each organism's interactions with other living things and the physical environment that biologists organize their study of ecology into a hierarchy of four levels: populations, communities, ecosystems, and the biosphere.

1. A **population** is a group of interacting individuals of the same species that inhabit a defined geographical area. The Hohokam Indians of central Arizona, the saguaro cacti of that same region, the saber-toothed cats of Rancho La Brea in prehistoric Los Angeles, and the alligators of a Louisiana swamp are all examples of populations.

2. A **community** consists of two or more populations of different species occupying the same geographical area. The Hohokam and the giant saguaro cacti whose fruits they gathered and ate made up a community, and so do alligators and the fish they consume, and orchids and the bees that pollinate them.

3. Populations and communities include only biotic factors. Such groupings always exist within a physical setting, however, and so ecologists have a third hierarchical level: the **ecosystem**, made up of living things interacting with the physical factors of the environment. The Hohokam's ecosystem included saguaro cacti, corn, and black-tailed jackrabbits, as well as sparse rainfall and the searing summer temperatures of the Arizona desert.

4. Ecologists recognize that ecosystems are further influenced by global phenomena, such as climate patterns, wind currents, and nutrient cycles. Thus, in the four-tiered hierarchy, all the ecosystems of the earth make up the highest level, the **biosphere**—the portion of the earth that contains living species. The biosphere includes the atmosphere, the oceans, and the soil in which living things are found, and the physical and biological cycles that affect them.

The next few chapters examine ecological interactions at increasingly higher levels of organization. Here, we begin with the level of population and discuss the influences on a population's location in time and space, as well as the question of success as measured by a population's size.

➤ CONCEPT CHALLENGE

Which level of ecology (population, community, ecosystem, or biosphere) would address the problem of pesticide runoff into streams? Which level(s) are involved in an infectious disease like rabies that can be carried from one raccoon to another, or from a raccoon to a person?

Distribution Patterns

Space on our planet can be thought of as an ecological vacuum waiting to be filled by living organisms. Yet few species of organisms are scattered evenly throughout the world; they exist, instead, in only certain spots and are absent entirely from other spots. What limits where an organism lives on a global scale and in individual locales?

LIMITS TO GLOBAL DISTRIBUTION

Why is it that saguaro cacti and mule deer appear naturally in Arizona, where the Hohokam lived, but oak trees and squirrels occur more commonly in other areas?

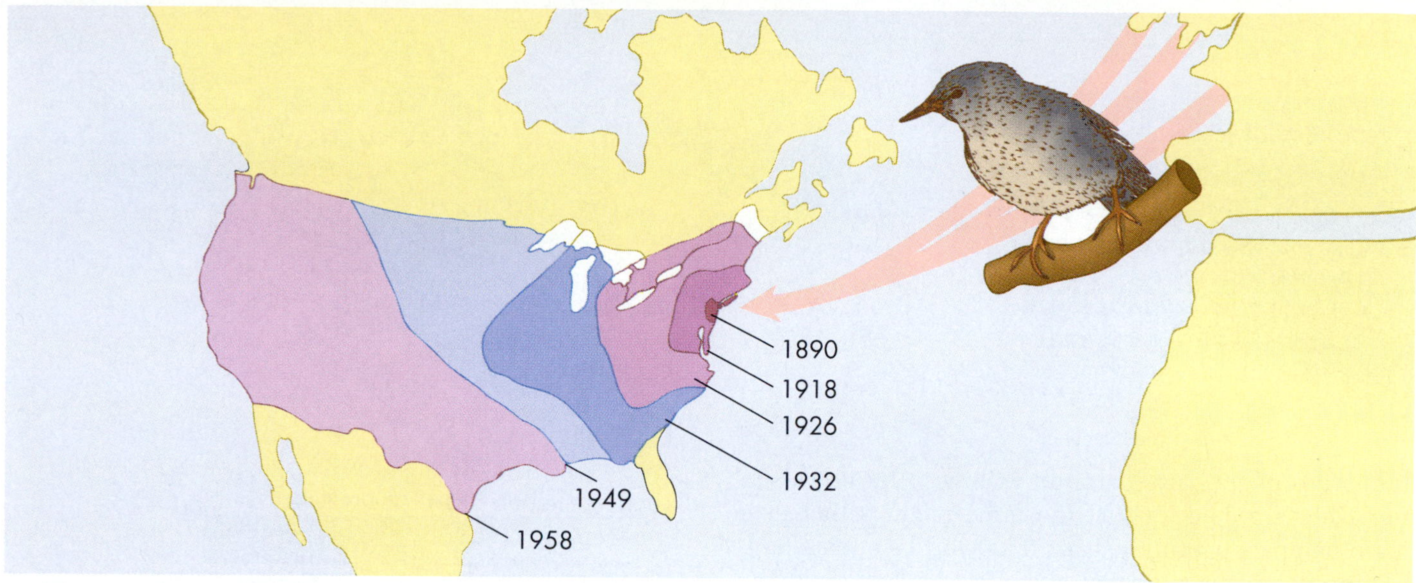

FIGURE 38.2

Crossing the Atlantic Barrier Allowed Starlings to Spread Throughout North America.

Bird fanciers brought starlings across the ocean from Europe to New York in 1890, and once the briny barrier was breached, the birds continually expanded their range westward, reaching the West and Mexico by 1958.

Questions and answers involving why an organism's range extends to one part of the world but not another are especially important to us when they involve organisms we use for food or materials, and when they involve species that cause disease in people and domestic plants and animals. In general, three conditions limit the places where a specific organism might be found: physical factors, interactions with other species, and geographical barriers.

Physical Factors Organisms may be absent from an area because the region lacks the proper sunlight, water, temperature, mineral nutrients, or any one of a host of physical or chemical requirements. For example, pine trees from the Austrian Alps photosynthesize best at a cool 15°C (59°F), but the *Hammada* bush of the Israeli desert photosynthesizes best at a scorching 44°C (111°F). Neither plant can survive in the other's environment because genetic adaptations fostered by natural selection have left each plant specialized for a particular set of physical conditions.

Interactions with Other Species Other species may block survival and limit a population's distribution. If certain species are already firmly established in an area, they may prevent the incursion of new species by monopolizing food supplies or acting as predators or parasites. Large regions of Africa, for instance, are nearly uninhabitable for humans and cattle because the resident tsetse fly transmits the protist that causes sleeping sickness.

Geographical Barriers A species may be absent from even highly favorable areas because a geographical barrier blocks access. Seas, deserts, and mountain ranges can be so wide or high that an organism cannot crawl, swim, fly, or float across the barrier. Europeans artificially bridged such a gap in 1890, when about 80 starlings were introduced into New York City's Central Park [FIGURE 38.2]. Now, millions of the dark, speckled birds chatter throughout America's cities and countrysides. North America turned out to be a congenial home for these organisms. Only their inability to cross the Atlantic had stopped their dispersal to North America.

➤ CONCEPT CHALLENGE

Whitetail deer are found throughout the United States, except in California, Nevada, Utah, and the Four Corners area of the Southwest. Blacktail deer live throughout the western half of the United States, but not in the East or most of the Midwest. What are some hypotheses that could account for this difference in distributions?

Constraints on Population Size

The ancient Hohokam shared a concern with modern-day farmers, fishers, and foresters: making sure that the populations of the organisms they harvest remain at optimal levels so that there will be as much grain, protein, or lumber available as possible on a continuing basis. These practical ecologists are concerned with a population's **density**, the number of individuals in a certain amount of space, for example 100 butterflies per hectare or 50 mice per barn. When reporting density, it is often helpful to take into account the species' habitat—the fact that the

FIGURE 38.3

Worldwide Distribution of Human Population. There are fewer than 2 people per square mile over most of the earth's land surface. Locally, however, the density is quite high in much of Asia, India, Europe, and eastern North America and around coastlines, rivers, and inland bodies of water the world over.

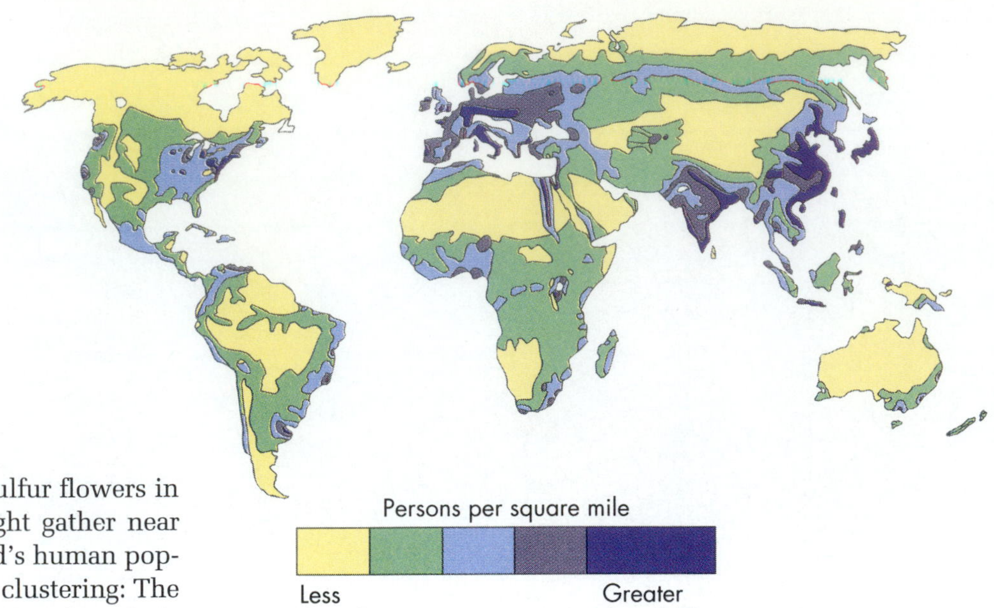

butterflies might cluster around yellow sulfur flowers in the 1-hectare field, or that the mice might gather near grain hampers within the barn. The world's human population illustrates quite well the effect of clustering: The density of the earth's people averaged over the whole planet is less than 1 person per square kilometer (2 per square mile), but as FIGURE 38.3 shows, most people live near ocean coasts, riverbanks, and lake shores, making local densities in those areas much higher than the average.

FACTORS AFFECTING POPULATION SIZE

The curious biologist is usually not content just to describe a population's ecological density at one point in time, but wants to understand how that population density *changes* over time. Population size may change when individuals enter or leave the population. Clearly, if more members enter than leave it, the population will grow, but if more individuals leave the population than enter it, the population will shrink. Individuals can enter the population by either birth or immigration, and members can leave the population by either death or emigration [FIGURE 38.4]. If the number of individuals gained from births and immigration is exactly equal to the number lost to deaths and emigration, then the population shows *zero population growth.*

Let's look now at how populations may change size as individuals leave a population by death, or join a population by birth. For the time being we will assume that the effects of immigration and emigration remain constant and equal in the populations we are studying.

How Survival Varies with Age Death and birth are clearly major parameters affecting population size. For many species, the likelihood that an individual will die depends on its age. The chances that a 90-year-old person will live for one more year are much slimmer than the chances that a 20-year-old will live for one additional year. In fact, the risk of death doubles every eight years past the age of 30. A **survivorship curve** is a plot of data representing the proportion of a population that survives to a certain age [FIGURE 38.5]. A table of the numbers used to generate a survivorship curve is called a *life table*, and it shows the **life expectancy** (average time left to live) and probability of death for individuals of each different age.

Insurance companies use life tables to determine policy costs for customers of different ages. From the survivorship curve in FIGURE 38.5, you can understand why insurance companies charge a 70-year-old man more for insurance than a 70-year-old woman: He is more likely to die in the next year than she is.

How Fertility Varies with Age Survivorship curves help ecologists understand population growth by predicting how many individuals of each age class will leave a population through death. As we saw earlier, birth is the major force that counteracts death. Birthrates, like death rates, also depend on age, and this fact is revealed by *fertility curves*, graphs of reproduction rate versus the age of female population members. (Since only females bear young, ecologists often view populations as females giving rise to more females.) The fertility graph in FIGURE

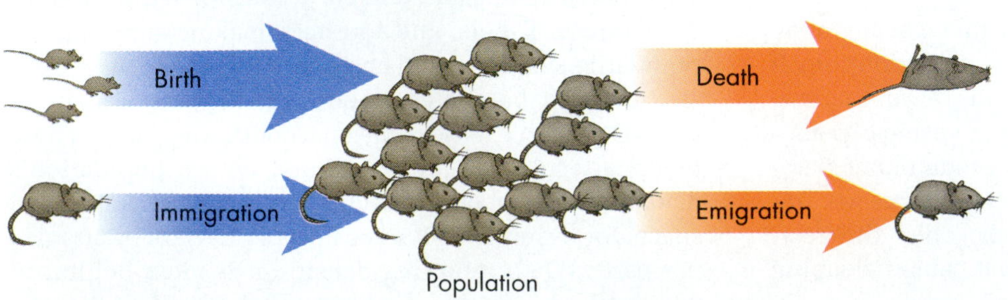

FIGURE 38.4

Birth and Immigration Tend to Increase the Size of a Population; Death and Emigration Tend to Decrease Population Size.

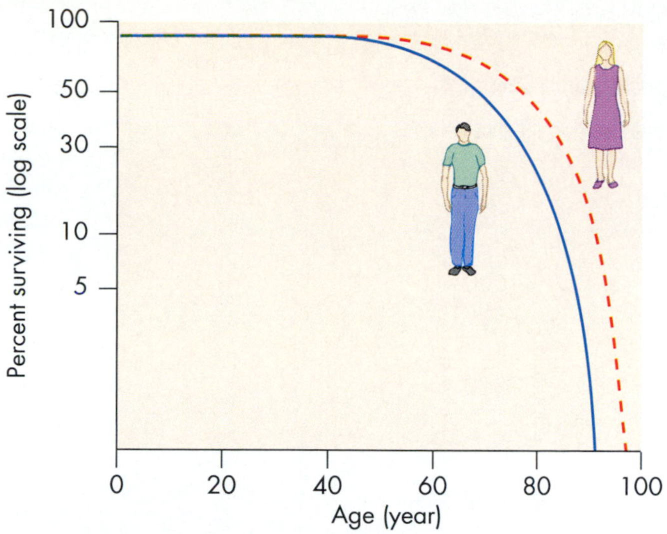

FIGURE 38.5

Human Survivorship Curve.

Most men live to age 70 and most women to age 80, but most of either gender succumb before 90.

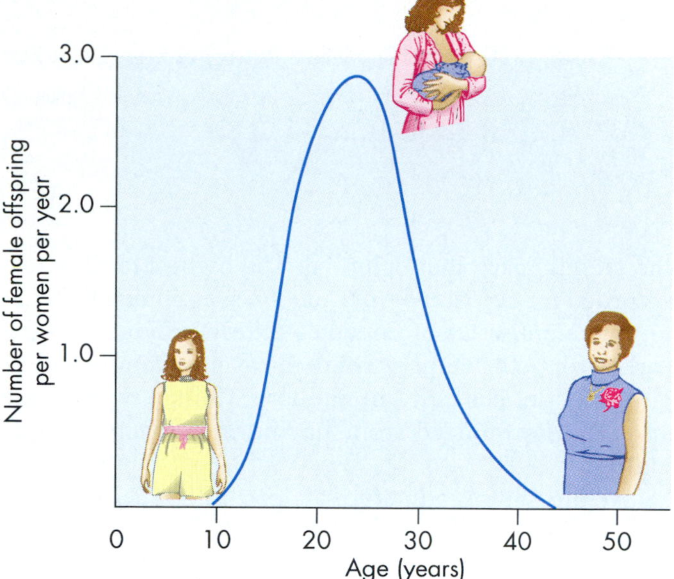

FIGURE 38.6

Fertility Curve.

Younger and older American women are less likely to reproduce than are women between 20 and 30. That's why a population consisting of many young people will grow faster than one with the same number of older people.

38.6 shows that out of every 10 American women between the ages of 20 and 30, on average about 3 will have a baby girl in any given year. Women younger than 20 or older than 30 are less likely to reproduce. Fertility curves are significant because with them and with a knowledge of a population's age structure, one can predict future population growth. For example, a population with a high proportion of 20- to 30-year-old women is going to grow much more rapidly than a population with few women in these prime childbearing years.

How Populations Grow Birthrates and death rates are frequently the major factors influencing changes in population size. By combining the two (and discounting immigration and emigration), population ecologists can make a model for how populations grow.

Given plenty of nutrients, space or shelter, water, benign weather, and the absence of predators or agents of disease, every population will expand infinitely because all organisms have a high innate reproductive capability under ideal conditions. The capacity for reproduction under idealized conditions, or **biotic potential**, is amazing. Darwin calculated that a hypothetical pair of long-lived elephants could give rise to 19 million descendants in 750 years! And most organisms reproduce faster than elephants. If a female cockroach has about 50 surviving daughters each month, the number of female roaches will rise to 800 billion after just seven months. Obviously, rapidly reproducing cockroaches and even slowly reproducing elephants have sufficient reproductive potential to take over the earth given enough time. Darwin wondered: Why don't they? His solution to this puzzle was a key feature in the theory of natural selection. To fully understand the answer, we must first study patterns of population growth, then return to the question of what limits population size.

Rapid population growth is easier to visualize when plotted on graphs such as those in FIGURE 38.7. Let's say that the individuals of a long-lived mouse species grow to reproductive age in one year and that, ignoring males, a population initially consists of 10 female mice that each produce one female offspring per year on average. At the end of the first year, the population will include 20 female mice, and if each of those has one more female per year, there will be 40 female mice at the end of two years, 80 after three years, and so on. The explosive increase results in a **J-shaped curve** representing **exponential growth**.

What factors do you think might affect the shape of the J-shaped curve of FIGURE 38.7? It turns out that there are two factors: (1) the reproductive rate per individual and (2) the initial population size. If each of the female mice in the preceding example had two female offspring rather than one each year, the population would grow much faster. Another way to say this is that the *reproductive rate per individual*, which ecologists represent by the symbol *r*, would be greater than the value calculated for the first example. For human populations, the reproductive rate per individual can vary widely. Women are generally capable of having about 30 children each, but

FIGURE 38.7

The Exponential Growth of Populations Depends on Reproductive Rate and Initial Population Size.

A J-shaped curve of exponential growth results from a population of 10 female mice, each giving rise to one offspring per year.

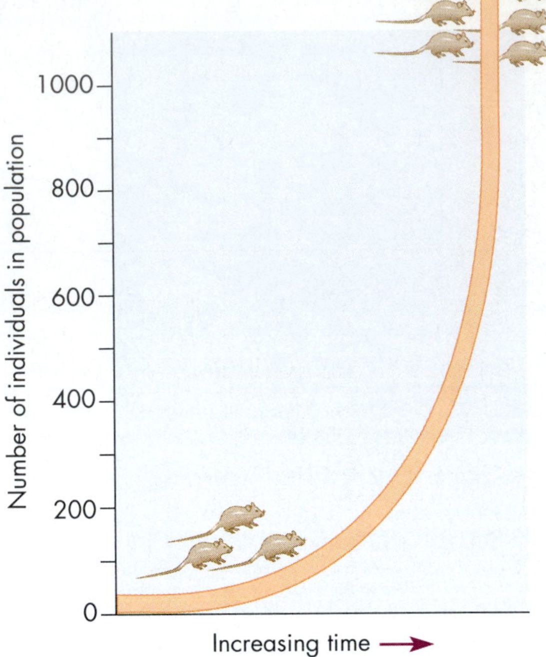

they rarely reach that potential. The highest fertility rate recorded for any human population was among the Hutterite communities of Canada's prairie provinces in the early twentieth century, where the average family had 12 children. In contrast, in modern China, women are strongly discouraged from having more than a single child.

While the reproductive rate per individual is an important factor in the contour of the J-shaped curve, the size of the initial population is also an important factor. If there are only 5 female mice in the initial population producing one female offspring per year, instead of 10 as in the first example, then only 5 more will be added by the end of the first year [FIGURE 38.7]. In other words, the change in the number of individuals in the population depends on both the average per capita reproductive rate and the number of individuals already in the population.

In our previous example, each mouse gave rise to one more mouse each year, and the annual increase was 100 percent—that is, each year, the population doubled. What will happen if the rate of increase is only 2 percent? The answer is that it simply takes longer for the population to double—in this case, about 35 years. The population still increases, however, and still doubles eventually.

Under the artificially ideal conditions we've discussed so far, any population would follow the J-shaped curve of exponential growth if the birthrate exceeded the death rate by even a small amount. Fortunately, real organisms in natural situations do not follow the J-shaped exponential growth pattern. Organisms in nature cannot sustain continued, limitless growth at the full force of their reproductive potential because food supplies and living space are finite. Hence, our planet is not covered with elephants neck-deep in roaches. The realities of supply and demand explain this curb on population growth despite the organism's reproductive capabilities.

➤ CONCEPT CHALLENGE

Assume that in 1915, lily pads invaded a huge reservoir, and their population size doubled every year thereafter. Ecologists studying the reservoir find that after 80 years of growth, just half of the reservoir is covered with lily pads. How many years will it be before lily pads cover the entire reservoir?

LIMITED RESOURCES LIMIT EXPONENTIAL GROWTH

With the J-shaped curve, ecologists have an idealized standard against which to measure growth in real populations. A classic case of growth of a real population was that of sheep on the island of Tasmania, south of Australia, in the early nineteenth century [FIGURE 38.8]. When English immigrants first introduced sheep to the new environment, resources were abundant, and the sheep population expanded nearly exponentially for a couple of decades, following a J-shaped curve during this time. As the density of sheep on the island rose, competition for limited resources increased, and by 1830, each sheep had a smaller share of food and living space. As a result, each individual was less likely to survive and more likely to die, and each had a smaller chance of reproducing. After 1850, the total growth rate decreased, and the population size fluctuated around a mean of about 1.6 million sheep.

As FIGURE 38.8 shows, the graph representing population growth began like a J-shaped curve, but flattened into an **S-shaped curve** representing **logistic growth**, a situation in which a large population grows more slowly than a small population would in the same area. The density at which a growing population levels off—1.6 million sheep on Tasmania in the preceding example—is called the **carrying capacity** (represented by the symbol K). Carrying capacity is the number of individuals an environment can support for a prolonged period of time. At the carrying capacity, individuals are using all of the resources available to them or their environment.

Carrying capacity is not a constant number character-

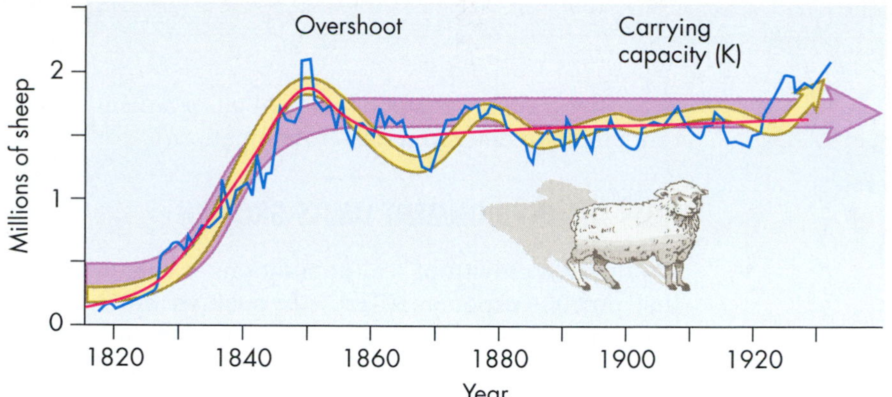

FIGURE 38.8

Sheep on Tasmania: The Growth of Real Populations Often Follows an S-Shaped Curve.

Sheep populations on the island of Tasmania initially followed a J-shaped curve, but finally they overshot the carrying capacity of the land. Eventually, their numbers oscillated within the range of the carrying capacity. This gives rise to the S-shaped curve of logistic growth.

izing each particular natural environment. You can tell that by looking at the variation after a population of sheep reaches carrying capacity [FIGURE 38.8]. Since environments change with the passing seasons, with alterations in weather patterns, with changes of species composition within the community, and with other factors, you should think of carrying capacity as a broad and sometimes variable range of densities toward which populations tend to move from initial densities that can be higher or lower.

FIGURE 38.9 contrasts the J-shaped curve, which reflects the innate capacity for growth given limitless resources, with the S-shaped curve, which more closely reflects the growth of real populations in constant but limited environments. In a new, unexploited environment, organisms can often reproduce maximally because all the resources they require are available; in such a situation, population size follows a J-shaped curve. As resources begin to dwindle, however, reproductive rate declines and population growth is better represented by an S-shaped curve.

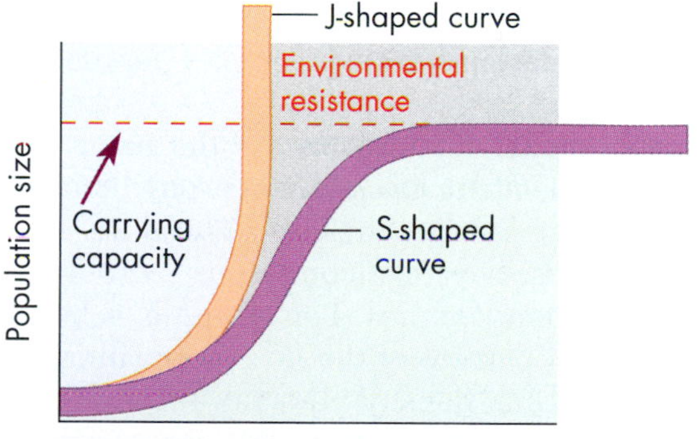

FIGURE 38.9

Environmental Limits on Population Growth.

A J-shaped curve reflects a species' innate capacity for growth; an S-shaped curve reflects the more realistic growth pattern in a constant but resource-limited environment. The zone between the two curves represents environmental resistance to population growth (limited food and living space, for example), and the stable population size indicates carrying capacity of the environment.

As the population continues to grow, resources are eventually fully used; for example, as soon as grass grows, it is clipped off by grazing sheep. At high population densities, therefore, there may be no leftover resources and the population will remain constant—it has reached the carrying capacity of the region. In this way, the formal model of logistic growth can account for the initial population burst of sheep on Tasmania, as well as the period of slower growth and the sheep's eventual stable population size with zero population growth.

Archaeological data show that the Hohokam population probably fit an S-shaped curve, growing from a small initial size 2000 years ago to a relatively stable maximum number, and remaining there for two or three centuries before disappearing about 1450 A.D. Significantly, the Hohokam *did* disappear, and we need to explore the causes of population crashes like that of the Hohokam, in the desert Southwest.

Population Crashes While population growth often slows and reaches a plateau, as it did with sheep in Tasmania, in some species there can be a bust following the boom: a rapid decline following a period of intense population growth. The growth of reindeer introduced onto an island off the southwest coast of Alaska represents a frequently observed pattern [FIGURE 38.10]. From an initial population of 25 animals in 1891, the herd grew to about 2000 reindeer in 1938 and then crashed to 8 animals by 1950. The crash can be readily explained on the basis of carrying capacity. When the reindeer were first introduced, lichens and other food sources were plentiful, having accumulated for centuries without predation. Thus, the island's carrying capacity was high. After the deer ate the accumulated food, however, new food would appear only as the remaining lichens regrew slowly during each short growing season. This change in carrying capacity is what caused the reindeer population to crash.

Some organisms routinely experience a boom-and-crash sequence of population growth. Water flea populations, for example, repeatedly boom and ascend in a roughly J-shaped curve until they overshoot the carrying capacity of their environment, critically deplete resources, and then crash [FIGURE 38.11]. Only after the

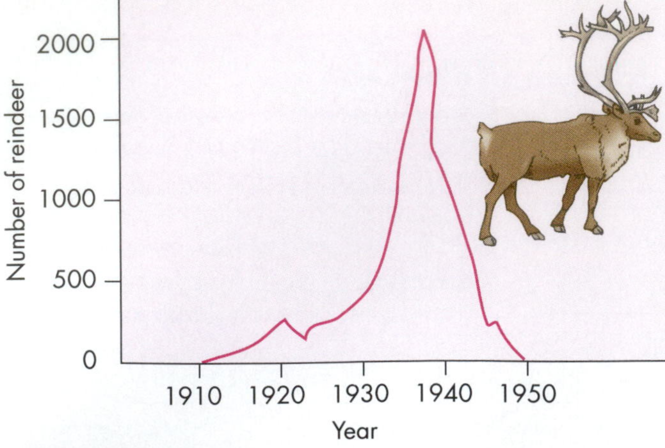

FIGURE 38.10

Overexploiting Limited Resources Can Lead to a Population Crash.

A reindeer population introduced to a small island in the Aleutian chain southwest of Alaska initially had a J-shaped growth curve, since they were able to exploit centuries' worth of slow-growing lichens in a few decades. Once the food ran out, the population crash was rapid and spectacular.

environment has become renewed can the water flea population again expand.

We have seen that populations in nature usually follow one of two patterns: the S-shaped curve of logistic growth, in which density slowly reaches the carrying capacity of the environment and then remains relatively stable for long periods (as with the Tasmanian sheep), or the boom-and-crash pattern of nearly exponential growth followed by an overshoot of the carrying capacity and a precipitous decline in density (as with the reindeer and water fleas). What characteristics of an organism's environment help determine the growth pattern?

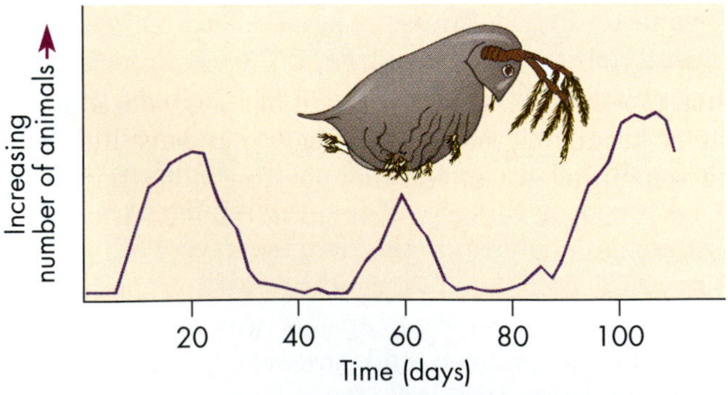

FIGURE 38.11

Repeated Boom-and-Crash Cycles in Water Flea Populations.

Populations of *Daphnia magna* have a yo-yo like existence, booming, vacuuming up available resources, and then crashing—again and again in just 100 days' time.

HOW THE ENVIRONMENT LIMITS GROWTH

The limited growth of real populations is due to factors that prevent exponential growth, such as limited food supplies and limited living space. Ecologists call limiting factors like these *population-regulating* (or *limiting*) *mechanisms* and classify them as density-dependent mechanisms or as density-independent mechanisms.

Density-Dependent Mechanisms **Density-dependent mechanisms** operate differently upon death rates and birthrates depending on whether the population density is low or high. When population density is low, reproduction rates are often high and death rates are often low. In contrast, when population densities are high, density-dependent mechanisms generally result in high death rates and low birthrates. At the point where reproduction rate and death rate cross as many new individuals are being added to the population via births as are being lost via deaths, and so the population size is roughly stable. As we saw earlier, this more or less stable population size is the carrying capacity.

Disease is a good example of a density-dependent limiting factor: In dense populations, disease spreads more rapidly than in sparse populations. For instance, in prairie dog towns (as the colonies are called), when the density of the rodents is low, the incidence of flea-transmitted bubonic plague is also very low. But when the prairie dogs are densely packed into their towns, outbreaks of plague often wipe out entire populations [FIGURE 38.12]. Competition among members of the same species for limited resources—for example, all the squirrels within a given forest competing for nuts—is another density-dependent mechanism that regulates population size.

Density-Independent Mechanisms As the name implies, **density-independent mechanisms** exert their effects regardless of population density. These are best illustrated by adverse weather conditions—droughts, floods, or freezing temperatures. For example, a harsh winter might kill 25 percent of the deer population regardless of population density. Density-independent mechanisms may limit population growth either outright or by affecting food supplies, and they take their toll in sparse and dense populations alike.

In practice, it can be difficult to separate the effects of density-dependent mechanisms because they often work together. For example, a severe drought would kill some individuals in a desert population of any size, but the effects would be far worse if the density were greater. This interaction leads many ecologists to classify population-

FIGURE 38.12

Dense Populations Are More Likely to Be Plagued by Disease.

Among these prairie dogs (*Cynomys ludovicianus*), overcrowding allows flea-borne bubonic plague to spread more easily.

FIGURE 38.13

Contest Competition: A Clash over Resources.

Male caribou (*Rangifer arcticus*) in arctic Alaska clash over status in the herd.

regulating mechanisms by an alternative criterion: whether the mechanism originates outside or inside the population.

Extrinsic Population-Regulating Mechanisms **Extrinsic population-regulating mechanisms** originate outside the population and include biotic factors, such as food supplies, natural enemies, and disease-causing organisms, as well as physical factors, such as weather, shelter, pollution, and habitat loss. Rainfall and the availability of saguaro cactus fruits were extrinsic factors that affected Hohokam populations.

Intrinsic Population-Regulating Mechanisms In addition to extrinsic factors, there are intrinsic population-regulating mechanisms that can limit population size; these originate in an organism's anatomy, physiology, or behavior. For example, crowded conditions and depletion of resources can cause many marsupials, such as kangaroos and koalas, to resorb already developing embryos; this physiological response lowers the rate of population growth. In another example, mouselike creatures called lemmings will migrate away from food-depleted regions, a behavior that lowers the local population density.

Competition The most important intrinsic regulating mechanism is competition among members of the same species for resources that they require but are available in limited supply. Competition among members of a species can lead to diminished rates of individual growth, survival, or reproduction, and its effects depend on population density. As the population grows and resources diminish, competition for food and space becomes intense among population members. In a coastal tidepool, for example, where a population of barnacles grows on a rock, other barnacles cannot occupy that same spot, even if food is present in abundance. In a herd of Caribou, jousting among males for leadership of the herd [FIGURE 38.13] operates to deny access by subordinate males to the resource of females with which to mate. And in a wolf pack, aggressive actions by an alpha female toward other females ensures that she will be the only female in that pack to reproduce.

While extrinsic and intrinsic factors in the environment limit the growth of all species, various aspects of a species' biology and its surroundings interact. The result is that the population growth follows either a roughly S-shaped curve or a boom-and-crash curve. Let's look, now, at the factors that dictate which type of growth (and which shape growth curve) a population usually follows in nature.

POPULATION GROWTH AND STRATEGIES FOR SURVIVAL

Evolution acts on organisms in ways that increase their individual genetic contributions to future generations. To grow, survive, reproduce, and thus make a genetic contribution to the future, individuals must allocate their limited energy supplies. A fast-growing organism that expends most of its energy enlarging its body may have little energy left over for reproducing. Conversely, an individual that expends a huge amount of energy attracting a mate or producing thousands of eggs may have little energy remaining for day-to-day survival activities. The way an organism allocates its energy is its **life history strategy**.

Given the inevitable energy trade-offs, what kind of life history would be "best" for an organism to display: rapid reproduction and depletion of stored energy, despite the risk of dying in a brief famine; or delayed reproduction and growth to a larger size, so that temporary food shortages are less of a threat to survival? Species like water fleas, which are tuned by evolution to display rapid rates of reproduction (*r*) despite risks to survival, are said to show ***r*-selection**, while species like Tasmanian sheep, which have a slow rate of reproduction and can reach and maintain a density close to the carrying

FIGURE 38.14

Rapid Reproduction versus Careful Investment in *r*-versus-*K*-Selected Species.

[A] An *r*-selected species, such as the dandelion, tends to develop rapidly, usually has small bodies, and tends to reproduce just once early in life. The emphasis is on reproductive productivity that offsets losses induced by the environment. **[B]** In contrast, *K*-selected species, such as the rhinoceros, develop slowly, reproduce later in life rather than earlier, have several offspring from sequential reproductive events, and tend to have large bodies. The emphasis is on efficient competition for resources.

[A]

[B]

capacity (K) of the environment, are said to experience ***K*-selection**.

***r*-Selected Organisms** Organisms that are *r*-selected often show boom-and-crash growth curves and inhabit environments that are frequently unoccupied and unpredictable. Weeds, such as dandelions [FIGURE 38.14A], and small invertebrates, such as fruit flies and water fleas, undergo *r*-selection. A dandelion's life history strategy is reproduction at as rapid a pace as possible. These plants quickly fill an environment—a newly plowed field, for example—before the environment changes: that is, before winter comes, before the corn grows too large, or before the farmer plows again. Natural selection has caused dandelions to maintain genes that allow rapid embryonic development, early reproduction, and large numbers of small seeds containing few stored nutrients. Dandelions experience an *early-loss* type of survivorship because most of the light, windborne seeds die shortly after germination. Our expression "to grow like a weed" reflects the *r*-selected life history strategy.

***K*-Selected Organisms** *K*-selected species, in contrast to *r*-selected species, inhabit more stable environments and tend to maintain their populations near the habitat's carrying capacity. Rhinoceroses [FIGURE 38.14B], most other mammals, and most large, long-lived plants, such as the saguaro cactus, reflect a *K*-selected life history strategy. In rhinoceroses, embryonic development is slow (gestation takes about 15 months), and reproduction is delayed until the age of about 5 years. Although rhinoceroses have only one calf at a time, the newborns are huge (the weight of an average male college student). Once born, they survive for about 40 years, experiencing a *late-loss survivorship* schedule.

Like other *K*-selected species, rhinoceroses are highly specialized to compete for resources in their environments. Specializations include immense size, thick, armor-plated skin, and a fingerlike extension of the upper lip that pushes grasses and twigs into the mouth. Rhinoceroses are fast approaching extinction because people slaughter them for their nasal horns. While these protuberances are made of the very same protein found in our hair and fingernails, some people have a superstition that the powdered horn has aphrodisiacal properties, and they are willing to pay more for the material than for gold, ounce for ounce. Knowing that rhinoceroses are *K*-selected organisms with a slow rate of reproduction, you can appreciate how hard it is for populations of rhinoceroses to become established once decimated.

In our earlier discussion of the Hohokam's rise and fall, we noted that the science of ecology could analyze our own species' population patterns and make predictions about the future. The final section of the chapter does just that.

➤ CONCEPT CHALLENGE

Which are more likely to become extinct as the human population continues to expand—*r*-selected or *K*-selected species? Support your choice.

The Human Population

Human beings have achieved unparalleled mastery over their environment through improvements in agriculture, medicine, sanitation, transportation, and industrialization. Are we—with our gleaming cities, our gigantic corporate farms, our burgeoning global population—governed by ecological rules? Or have we somehow moved beyond booms, crashes, and growth curves? Stated more formally, to what extent do general ecological principles on the distribution and abundance of organisms apply to human populations? The answer involves a look at human history, long-term population trends, and some predictions for our future population growth.

TRENDS IN HUMAN POPULATION GROWTH

You can see at a glance the history of human population growth and its staggering current proportions in FIGURE 38.15, which illustrates the three phases of human population growth.

FIGURE 38.15

Human Population Bomb.

[A] Human population growth displays a classic J-shaped exponential curve. Currently, our population doubles in about 40 years. **[B]** Throughout most of human history, we lived as hunter-gatherers like this Yanomamo woman and child. **[C]** In our agricultural phase, from about 8000 B.C. to about 1750 A.D., populations grew steadily and lived by farming much as this Sherpani woman does in Nepal. **[D]** The industrial phase began in eighteenth-century England with urban sweatshops. This phase ushered in the exponential growth that continues today.

[D] Industrial population

[B] Hunting and gathering population

[C] Agricultural population

[A] Graph showing explosive growth of human population

Hunting and Gathering Phase In the first phase of human history, from our species' origin to about 10,000 years ago, the population grew slowly as people existed by hunting animals and gathering naturally occurring roots and fruits. The worldwide population was probably about 10 million by 8000 B.C. During this early phase, the human species seemed to fall into the *K*-selected group of organisms because of our slow development, long lives, and large bodies, our relatively few offspring, our extended and intensive parental care, and our highly specialized brains that help us compete for resources with cunning efficiency.

Agricultural Phase Population growth accelerated during a second phase of human history, beginning about 10,000 years ago, when people started planting and tending crops and domesticating animals in the so-called **agricultural revolution**. The shift to agriculture was rapid and worldwide, perhaps because people are so adaptable and can transmit their *culture*, or ways of living, to others. As agricultural techniques spread and improved between about 8000 B.C. and 1750 A.D., world population increased from 10 million to about 800 million. Since agriculture allows more efficient use of resources, its practice increases the environment's carrying capacity for humans. The Hohokam, with their intensive irrigation of desert river valleys, exemplify this stage in human cultural development. In its natural state, the Arizona desert has a very low carrying capacity, but irrigation increased the amount of corn the Hohokam could grow and the number of people the desert could support.

Industrial Phase A third phase of growth began in eighteenth-century England with the **industrial revolu-**

tion. Inventions and scientific advances triggered vast changes that transformed a populace living mainly as farmers, craftspeople, and merchants into a population working mainly in factories and living in crowded cities [see FIGURE 38.15]. In the next 250 years, much of the world would follow this pattern of industrialization and social upheaval. The steam engine was a key invention, and its impact was enormous. A farmer with a steam engine attached to a tractor could accomplish the work of dozens of people in a single day and thus increase food production. A steam-driven train or ship could rapidly distribute food and other necessities of life, and thus blunt the impact of local famine.

In recent times, the rise in human population has been staggering. While it took from the beginning of life until 1950 for the first 2.5 billion people to accumulate on earth, it has taken just 40 years—an eyeblink of evolutionary time—for a second 2.5 billion to be added. At current growth rates, by 2025 an additional 5 billion *more* people will join the planet's current population of nearly 6 billion. How old will you be when the 11 billion figure is reached? What might life be like then?

As you look at the graph of human population growth, the towering ascension should unnerve you: It is the familiar J-shaped pattern of exponential growth, much like that of the island reindeer just before they overexploited their environment and suffered a population crash. By analyzing the causes of our own population boom, ecologists hope to learn how humans can avert a crash in the future.

CHANGE IN HUMAN POPULATION SIZE

How did the agricultural and industrial revolutions quicken the pace of the human population explosion? Ecologists and historians alike have wondered whether the invention of agriculture *allowed* human populations to increase, or whether people were *forced* to invent agricultural practices to help support population densities that were already exceeding the carrying capacity of the land where they lived. Many observers believe the latter and suggest that population growth has been a constant feature of the human experience, continually forcing people to adopt new strategies for increasing the amount of food their land could produce. Carrying capacity, however, cannot be increased forever; the productivity of the land must, at some point, be reached and exceeded.

Birthrates and Death Rates in Developing Nations To understand the causes of human population increase, particularly the tremendous surge after the industrial revolution, we must recall that the population growth rate equals the birthrate minus the death rate. Prior to 1775, the birthrate in developed countries like Sweden was just slightly higher than the death rate, and so the population enlarged at a constant low rate [FIGURE 38.16A]. After 1775, as industry expanded, people enjoyed improved nutrition, better personal and public hygiene, protection of water supplies, and the reduction of communicable diseases such as smallpox. With these factors came a gradual decline in the death rate. As FIGURE 38.16A shows, while the death rate began to decline in 1775, the birthrate did not start to drop in developed countries until about a hundred years later. Consequently, each year many more people were born than died, and this translated into an increase in the rate of population growth. By the last decade of the twentieth century, both the birthrates and death rates in industrialized nations had dropped to all-time lows, and the gap between them had once again narrowed. A changing pattern from high birthrate and high death rate to low birthrate and low death rate is called the **demographic transition**.

In industrial Europe and North America, the demographic transition occurred in the first half of the twentieth century and reduced overall growth rates to low levels. The populations of Africa, Asia, and Latin America, however, have continued to grow at immense rates for much of this century. The concept of a demographic transition helps explain why. The death rates in those countries remained high until the mid-twentieth century, when the importation of Western medicine and public health technology helped spare lives in record numbers [FIGURE 38.16B]. Simultaneously, however, the birthrate remained high, based largely on traditional cultural practices. Thus, the gap between the two rates widened. The huge disparity between death rates and birthrates caused an enormous net growth rate and hence an immense population boom in the twentieth century.

Evidence suggests that the developing nations are now moving through a similar demographic transition, with a change from high birthrates and high death rates to low birthrates and low death rates. In the poverty-stricken country of Bangaladesh, for example, women in 1975 had an average of 6.7 children each, but by 1993, the rate had fallen to 4.7; corresponding figures for Kenya in Africa were 8.1 children per mother in 1975, and 6.3 in 1993. Just as birthrates are falling, the death rates are also declining: Since 1960, in developing countries as a group, life expectancy at birth has soared from 46 to 63 years.

While those figures provide a hopeful sign, five or six children per mother is still a houseful of kids. A family with many children in a developing country, however, has a less destructive impact on the global environment, in general, than does an average family with just two children in an industrialized society. This is because people in industrial nations consume many times more resources (for example, heating oil, gasoline, electricity, and refrigerants that can destroy the ozone layer).

While people in industrial societies use up far more

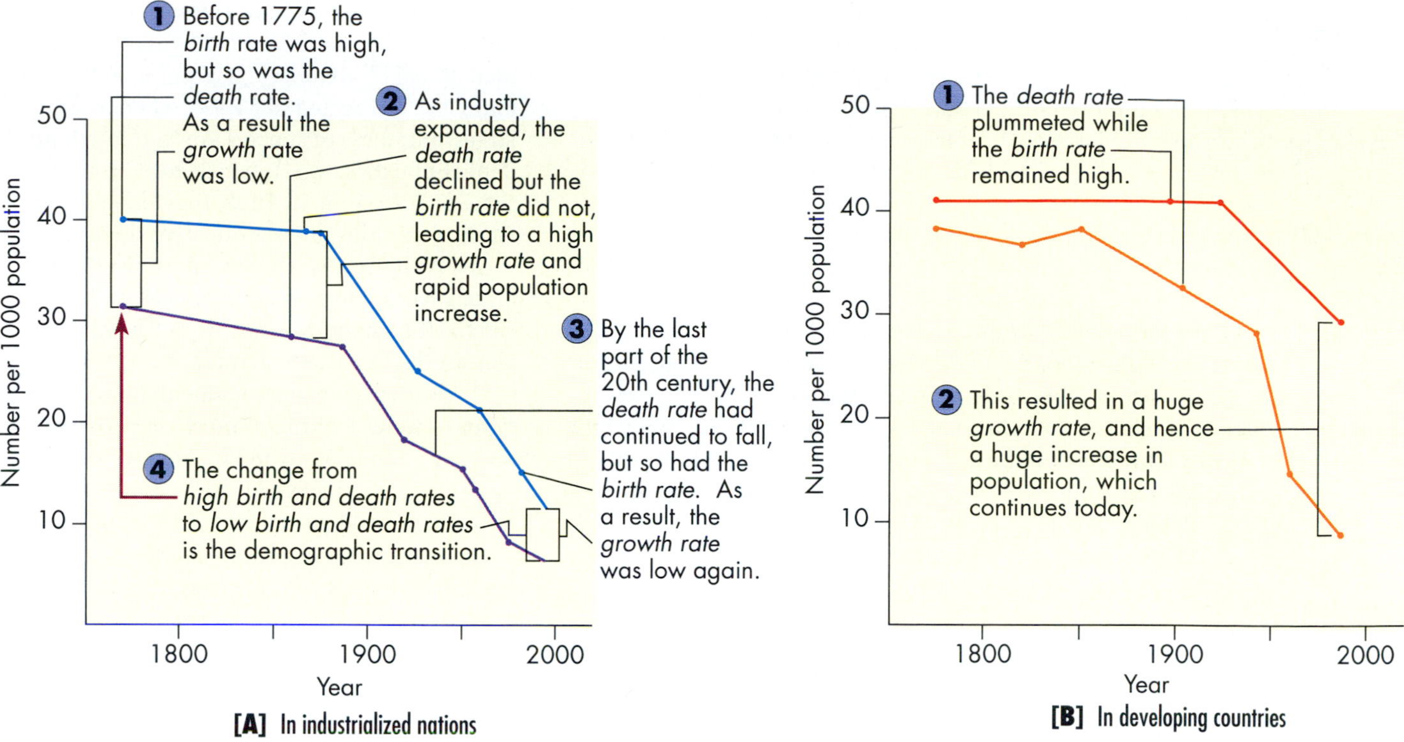

FIGURE 38.16
The Difference Between Birthrate and Death Rate Detonates the Population Bomb.

than their share of the world's resources and give off far more than their share of the world's pollutants, demographers point out that the most important political and ecological problem our world faces is the huge number of unemployed young people who will grow up in poor countries over the next few decades. As the next section explains, at the root of the problem lies a principle of population ecology: A population with a large number of young people tends to grow faster than a group of oldsters the same size.

GROWTH RATES AND AGE STRUCTURE

A sure sign of a population's growth rate is its **age structure**: the number of people in each age group [FIGURE 38.17]. The age structure of a growing Swedish population in 1900—a time when the death rate had already declined substantially but the birthrate had yet to fall—shows a high percentage of people in the younger age classes [FIGURE 38.17A]. This results in a pyramid-shaped age distribution. The Swedish population's age structure in 1950 shows few teenagers, owing to a low birthrate in the years before World War II, as well as a bulge, the postwar baby boom, of children less than 10 years old [FIGURE 38.17B]. By 1977, after the Swedish birthrate had dropped very close to the death rate, each age class was only slightly smaller than the younger one below it [FIGURE 38.17C]. If current trends continue, by 2025 Sweden will have a bullet-shaped age profile with almost no difference between births and deaths until the end of the human life span (the point of the bullet) [FIGURE 38.17D].

Age structure is significant because it can be used to judge a population's growth status. Rapidly growing populations generally have many young individuals, and hence a pyramid-shaped age profile, like Sweden in 1900 or Mexico in 1977 [FIGURE 38.17E]. In contrast, stable or declining populations tend to be more bullet-shaped, like Sweden in 1977. With one look at data graphed this way, you can infer the kinds of social services that will be needed by different populations: schools for pyramid-shaped populations, health-care facilities for the elderly, and pension plans for bullet-shaped ones.

POPULATION OF THE FUTURE

The number of people alive today is greater than the total number who have ever lived and died before us. The press of humanity totals over 5.5 billion people, and if present growth rates continue, another 3 to 5 billion will be added in the next 40 years. Even if, from this year forward, family size were reduced to the replacement rate of 2.1 children per woman, the global population would continue to grow for decades, because the youthful citizens of countries like Mexico and Nigeria will soon attain

box 38.1
Human Impact

Tracking Population Hot Spots

Some of the most haunting images in all of modern biology are the eerie, glowing "hot spots" visible from space in earth's tropical regions. The bright zones look like so many stars over central Mexico, or Indonesia, or any of a hundred other countries. But they are actually large areas of tropical forest being burned to the ground as people clear land for cattle grazing or crop planting. These hot spots are figurative as well as literal in that they represent places where the losses of populations, of habitats, and of overall biodiversity have reached a superheated crisis stage. By this definition, there are actually thousands of subtle hot spots all over the globe where species are fading away gradually under the pressures of human population growth and land development.

Underlying the hot-spot imagery is an equally haunting fact: For the most part, biologists don't know which species are disappearing year by year, hectare by hectare. This is because life scientists have cataloged only one-tenth of earth's biodiversity in the 250 years since they began systematically finding, naming, and classifying living things. At this pace, it would take centuries to probe the rain forests, coral reefs, deep ocean rifts, and other ecological zones for undiscovered species. However, deforestation and habitat destruction are so widespread and rapid that many groups, including the U.S. National Research Council, predict that 50 percent of the unknown species will disappear by the year 2000—long before biological science ever officially finds them.

In light of this biodiversity crisis, prominent systematists have called for a major, worldwide crash program to discover and classify all existing organisms within the next 25 to 50 years. Systematists have banded together with plant taxonomists to propose the 25-year deadline and a sixfold increase in worldwide spending on the field to $3 billion annually—the same amount that governments and private corporations are spending on the Human Genome Project.

In addition, botanist Peter H. Raven of the Missouri Botanical Garden and entomologist Edward O. Wilson of Harvard University independently called for a 50-year program to map global biodiversity, beginning with known or suspected hot spots where the largest numbers of species are in immediate danger of extinction. In an editorial in *Science* magazine, in 1992, they cited astonishing examples of our collective ignorance:

- Although whales and porpoises are among earth's largest animals, nearly 14 percent of the known species were first discovered in the twentieth century.
- Biologists have documented 69,000 types of fungi, but this represents perhaps *4 percent* of the 1.5 million or so species.
- Microbiologists formally recognize only about 4000 species of bacteria, but Norwegian scientists recently unearthed and identified 4000 to 5000 species of microorganisms *in a single gram of soil*!

In 1993, the Secretary of the Interior created a new bureau within that agency called the National Biological Survey, charged with inventorying every living species in the United States and with gauging the health of their habitats. That same year, an elite group of American systematists planned the first steps toward conducting an All Taxa Biodiversity Inventory—an intensive search for every living species in one restricted area, say, a square patch of rain forest 15 miles on a side. Both the new bureaucratic division and the new plan of attack for categorizing a parcel of nature represent giant steps forward. And clearly, these endeavors will need all the governmental and private funding, and all the newly trained systematists, that can be found. Cataloging life's disappearing diversity will be a mammoth—some say nearly impossible—task. Yet to attempt less—to push for anything short of the fullest, fastest national and global surveys—would be to overlook one of the most important missions of life science. A mission with a tight deadline.

---

reproductive age. Estimates for a stable size for the human population by the year 2040 range from a low of 8 billion, assuming rapidly implemented birth control programs, to a high of 14 billion, with less successful birth control campaigns. Regardless, our planet's human population will either double or triple in the not-so-distant future.

These astounding figures lead us, quite logically, to ask whether the planet can support 8 to 14 billion people at a reasonable standard of living. At present, no one can accurately predict what the earth's ultimate carrying capacity will be. Scientists do suspect that coal, oil, and some of the other resources on which we now depend are likely to become exhausted early in the twenty-first cen-

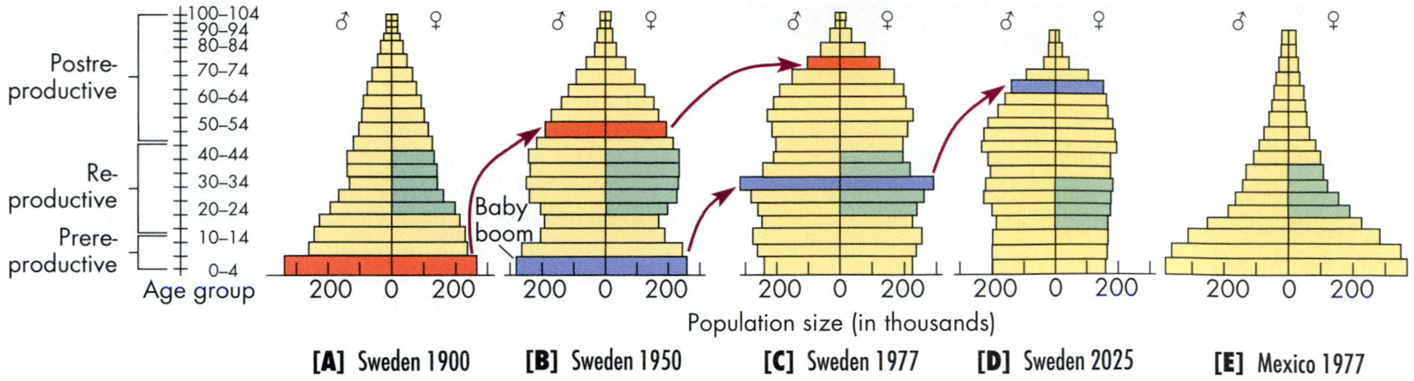

FIGURE 38.17

Age Structure Diagrams Help Reveal Human Population Growth and the Potential for Future Explosions.

In these graphs, males are represented on the left and females on the right, and the length of each bar signifies a given age group's proportion of the population. **[A]** Sweden had about 290,000 girls up to four years of age in 1900, but by 1950, when the group members were middle-aged **[B]** there were only about 220,000 females. **[C]** By 1977, only 150,000 females remained. **[D]** By 2025, Sweden will have a bullet-shaped age structure. In these graphs, the green bands show reproductive-age females in each population. **[E]** The pyramid-shaped age structure of Mexico in 1977 suggests a population surge in the mid-1990s when these people reach reproductive age.

tury. One thing is certain: Without dramatic steps taken immediately and decisively, the crush of humanity will reduce or forever destroy complex and delicate biological systems such as tropical forests, as well as millions of individual species that have taken entire geological eras to evolve on our planet.

➤ CONCEPT CHALLENGE

Virginia and Phil are having an argument. She thinks that having a certain number of children but having them later in life—say when the mother is between 30 and 35 instead of between 20 and 25—will reduce a country's annual population growth rate. Phil claims that the mother's age is immaterial with respect to annual population growth rate as long as the women have the same number of children. Who is correct and why?

Connections

The principles of population ecology we studied in this chapter allow us to reconstruct a plausible explanation for the rise and fall of the Hohokam population. These Native Americans flourished in the harsh southwestern desert because they understood interacting factors that limited the abundance and distribution of their crop plants. The tribe introduced corn and squash to the Salt and Gila river valleys, and in the process, overcame the great desert barrier that had prevented those plants from expanding northward on their own.

The Hohokam also identified the abiotic factors (mainly water) that limited the distribution range of their crops, and they manipulated their environment by irrigation. They altered corn over generations by selecting seeds from the most drought-resistant plants, changing the plant's evolution, and eventually producing corn that would mature with water from a single irrigation event. After two millennia, modern plant breeders have yet to better this record.

The Hohokam's achievements raised the carrying capacity of their desert environment for both corn and people. This allowed the density of their settlements to grow in a slow but relentless upswinging J-shaped curve. A population biologist taking a census during this time would have found a pyramid-shaped age structure, with many young people at the base and few old people at the tip.

If the tribe experienced a few years of drought, as may have happened around 1450, the desert's carrying capacity would have dropped drastically, leaving the population density well above what the environment would support. Disease and mortality would have changed the survivorship curve and caused the death rate to exceed the birthrate. Eventually, the population crashed. Despite great cultural advances, humans are clearly constrained by ecological principles.

In this chapter, we focused on the dynamics of populations of individual species—sheep, water fleas, reindeer, dandelions, rhinoceroses, and people. Species, however, do not exist as ecological islands. In the next chapter, we will study communities and see how the distribution and abundance of any particular species depends on interactions with many other species sharing the same environment.

KEY TERMS

age structure, 845
agricultural revolution, 843
biosphere, 834
carrying capacity, 838
community, 834
demographic transition, 844
density, 835
ecology, 834
ecosystem, 834
exponential growth, 837
industrial revolution, 843
J-shaped curve, 837
***K*-selection,** 842
life expectancy, 836
logistic growth, 838
population, 834
population ecology, 833
***r*-selection,** 841
S-shaped curve, 838
survivorship curve, 836

HIGHLIGHTS IN REVIEW

1 Closely interacting biological and physical factors govern the abundance and distribution of any species in any area.

a] Biologists identify four levels of ecological interactions: populations of individuals of the same species in an area; communities of interacting species in an area; ecosystems, which are communities in their physical settings; and the biosphere, which consists of all ecosystems and global physical influences combined.

b] Temperature, sunlight, and other physical factors; the activities of other species; and geographical barriers, such as oceans and mountain ranges, can limit where a species lives.

c] To understand a population's distribution, one must determine the number of organisms in an area, as well as the number of organisms in the actual habitat within the area.

2 Ecologists can model a population's growth, and hence predict how it will change with time.

a] The probability of survival varies with an organism's age and can be plotted on a survivorship curve.

b] The innate capacity for reproduction, or biotic potential, is high in all species. When resources are unlimited, growth follows a J-shaped exponential curve. When environmental resources are limited, population growth often follows an S-shaped logistic curve. The plateau of the S-shaped curve corresponds to the carrying capacity (K), the number of individuals the environment can support for an indefinite period of time.

c] The environment can limit population growth through density-dependent mechanisms like disease or density-independent mechanisms like adverse weather conditions. An alternative classification of environmental limits to growth considers extrinsic mechanisms like limited food supplies versus intrinsic mechanisms like the tendency to migrate when food supplies drop and competition between population members or with other species.

d] Populations that overexploit environmental resources can crash, either once, as did the reindeer on an Alaskan island, or repeatedly, as do water fleas with their boom-and-crash cycles.

e] Individuals allocate their limited energy supplies through r- and K-selected life history strategies. The r-selected species, such as dandelions, reproduce rapidly despite risks to short-term survival, while K-selected species, like rhinoceroses, grow and reproduce slowly, their populations tending to match the habitat's carrying capacity.

3 Many biologists suggest that size and rate of growth of the human population are the most important problems facing the world today.

a] Our planet's human population has displayed a classic J-shaped exponential growth curve, largely a result of our ability to expand carrying capacity through cultural advances.

b] A demographic transition from high birthrates and high death rates to low birthrates and low death rates took place in developed nations during the twentieth century, and bullet-shaped age structure diagrams reflect the change. Developing nations have yet to complete a similar demographic transition. The result is pyramid-shaped age structures with many young people and high reproductive potential.

c] Because of the large reproductive potential of the population in developing countries, our global population is likely to double or triple in the next 50 to 150 years, even with successful birth control campaigns. Ecologists do not know whether the earth's carrying capacity is large enough to support the predicted human population of 8 to 14 billion people.

UNDERSTANDING THE FACTS AND CONCEPTS

For Questions 1–5, match each of the descriptions with the most appropriate item or items from the following list of terms. Any answer can be used once, more than once, or not at all.

a] population
b] community
c] ecosystem
d] biosphere
e] more than one of the above

1 Includes all the species in a habitat but not the biotic elements.

2 A shallow pond and its inhabitants.

3 The most inclusive term in the list.

4 The only term that is limited to one species.

5 Includes biotic as well as abiotic elements.

For Questions 6–10, complete each statement by choosing the most appropriate term from the following list.

a] survivorship curve
b] life table

c] age structure of a population
d] biotic potential
e] ecological density

6 The probability of death in any year by individuals at particular ages comprises a ______.

7 In a population growing logistically, the ______ is limited by the carrying capacity.

8 The number of individuals of one species in a given habitat constitutes its ______.

9 Comparisons of graphs of the ______ of different human populations can be used to predict which populations are likely to grow the most rapidly within the foreseeable future.

10 A graph showing the percentage of a given population surviving to age 5, 10, 15, 20, 25, and so on is called a ______.

For Questions 11–15, match each of the population-limiting mechanisms with the most appropriate item from the following list. Each answer may be used once, more than once, or not at all.

a] density-dependent factor
b] density-independent factor
c] extrinsic factor
d] intrinsic factor
e] a and c
f] a and d
g] b and c
h] b and d

11 Floods and earthquakes.

12 Cessation of menstrual cycles and ovulation in women when body fat levels fall below a threshold level, as occurs during starvation.

13 Stress-related infertility caused by crowding.

14 Resorption of embryos by marsupials when food supplies are depleted.

15 Air-borne, infectious, lethal diseases.

INTEGRATE AND APPLY WHAT YOU HAVE LEARNED

1 The growth of one population follows a J-shaped curve while that of a related population follows an S-shaped curve. Which population is likely to crash? Explain.

2 For each of the following factors that limit the size of the human population, state whether the factor is density-dependent or density-independent.

a] Bubonic plague.
b] AIDS.
c] Living in an earthquake-prone or flood-prone region.

3 Would you classify the human population as *r*-selected or *K*-selected? Explain.

4 Other things being equal, does an *r*-selected population have a higher likelihood of extinction in any given year than a *K*-selected population? Explain.

ANALYSIS

1 A small population of an *r*-selected animal species is introduced into a desert island. In the center of the island is a small fresh-water pond, around which is the only significant vegetation on the island. Which of the following descriptions might logically describe the future of the species? Choose all that apply.

a] Boom-and-crash.
b] Immediate logistic growth.
c] Logarithmic growth.

2 In order for the human population to achieve zero population growth which of the following must occur? More than one answer may be correct.

a] There must be more postreproductive individuals than reproductive individuals.
b] There must be more prereproductive than reproductive individuals.
c] There must be the same number or fewer prereproductive individuals as there are reproductive individuals.
d] One set of parents should have only two children (approximately).

3 A model plant population that has not reached the carrying capacity is likely to:

a] Grow exponentially.
b] Grow but not at an exponential rate.
c] Remain stable in number.
d] Decline in number.
e] Crash.

CHAPTER 39

Ecology of Communities

FIGURE 39.1
Kirtland's Warbler.
The Kirtland's warbler rests in a jack pine tree in Michigan.

A SONGBIRD RESCUED

Many of North America's brightly colored wild songbirds, including the gray and yellow Kirtland's warbler, have suffered great population declines in the later half of the twentieth century. The reasons for the declines make up a catalog of modern ecological disturbances: habitat destruction; competition from predators; encroachment of towns and suburbs; pollution of air, soil, and water; and land management practices, such as suppressing wildfires, that eliminate usable habitat.

The Kirtland's warbler has the largest body size of any North American wood warbler [FIGURE 39.1], and the handsome birds with their cheerful songs used to dart about the jack pine forests of north central Michigan and southern Ontario in summer. Then each autumn they would migrate to the Bahamas and spend the winter on those warm, lush islands. Bird-watchers and field biologists noticed a steep decline in nesting pairs of Kirtland's warblers between 1961 and 1971, and it looked as if the pretty songbird might become extinct.

Like so many regions of rural America, the jack pine forests of central Michigan reflect the pressures of human population and land development. These forests of small, densely packed trees have been cut into ecological "islands" by the building of roads, farms, new houses, schools, and commercial properties. While Kirtland's warblers can continue to nest and raise young if these islands are of a certain minimum size, the great expansion in forest edges creates new habitat for an aggressive parasite, the brown-headed cowbird. This animal, nearly twice as big as a warbler, does not build its own nest or care for its own chicks. Instead, it lays its eggs in the nests of Kirtland's warblers and nearly 200 other songbird species. Many Kirtland's warblers raise parasitic cowbird chicks to maturity—but none of their own. By the 1970s, the parasitism was affecting about 70 percent of all Kirtland's nests.

On top of the declining reproductive rate caused by cowbird parasitism, habitat disruption was taking away the Kirtland's warbler's native home in the jack pine forest. Kirtland's warblers survive only in young jack pine groves where low branches help provide cover for nesting birds and foster the growth of insects and other prey species. Jack pines are able to spread most rapidly after wildfires, but people in the 1960s and 1970s tended to stop most forest fires as quickly as possible. Their seemingly laudable efforts were actually reducing the amount of forest area hospitable to the Kirtland's warbler. By the time people noticed the situation, Kirtland's warbler populations had dropped by two-thirds.

MESSAGES

1 Interactions between species limit the abundance and distribution of organisms. Beneficial interactions may encourage population growth of both species, while harmful interactions may limit one or both species.

2 Two species can influence each other's evolutionary fitness, so that both kinds of organisms coevolve.

3 Interaction with the human species is the most powerful biological factor in the world today. Few communities are untouched by our species' all-pervasive influence, and we have begun—and will continue—to strongly shape the earth's ecological communities.

With such dismal ecological odds, Kirtland's warblers seemed doomed—until the people of Michigan intervened. Small armies of volunteers, coordinated by ecologists and ornithologists (bird experts), learned to capture and humanely kill cowbirds, and to remove their eggs and young from warbler nests. They also carried out controlled burns in some jack pine forests to encourage the growth of new trees. This human intervention reduced nest parasitism by the cowbirds to 6 percent, stopped the warbler's decline, and allowed populations to rebound to about 1200 individuals—out of extinction danger, but still only about half their historical high levels in the late 1950s.

The story of how the songbird was spared from oblivion brings us to a new level in the study of ecology. It is not enough to consider an individual species, as we did in CHAPTER 38. We must see how different kinds of organisms interact in **communities**, which are assemblies of populations of different species in a particular area at a particular time.

Our exploration of community ecology will begin by defining where and how populations live in communities. Next, we will discuss how competition between species affects community structure. We will consider predation—when one organism consumes another—and its effect on population densities. We will talk about relationships between two species and how they evolve, and discover how these interactions can benefit one or both kinds of organisms. Finally, we will explore the structure of communities, and find out how communities change over time and space. The chapter discussion alternates between direct observations of natural phenomena and experiments in the field and laboratory. By means of such experiments, ecologists try to verify or rule out explanations for their observations in nature.

Along the way, we will see how water fleas helped ecologists, how burros are competing with bighorn sheep, how the evolutionary race continues between predators and prey, and how the yucca needs the yucca moth—and vice versa. ❑

Habitat and Niche

To understand the intricate web of relationships between populations of organisms in a community, the observer must discover where the organisms live and how they get needed energy and materials. The general physical place in the environment where a certain kind of organism resides is its **habitat**. A habitat is analogous to an organism's address, or home. In describing the general places where aquatic organisms live, for example, an ecologist might speak of an open-water habitat, a shore habitat, a muddy bottom, or a surface-film habitat. We saw that the habitat of the Kirtland's warbler is the jack pine forests of Michigan's lower peninsula. More specifically, those forests are usually 5 to 23 years old, and the birds nest in long grasses on the forest floor among graceful ferns and blueberries. The bird's winter habitat is the lush forests of the Bahamas. Like many migratory birds, the requirement for two separate seasonal habitats puts Kirtland's warblers in double jeopardy, since the degradation of either habitat will endanger the animal's survival.

Whereas a species' physical home is its habitat, its functional role in the community is its **niche**. The niche is analogous to the organism's job—how it gets its supply of energy and materials. The jack pine's niche, for example, is that of an autotroph, or *primary producer*, deriving energy directly from the physical environment. The niche of a bark beetle is that of *herbivore*, obtaining nourishment from plants like the jack pine trees that grow in its habitat. The niche of a Kirtland's warbler is that of a predator specializing on insects, worms, and other small invertebrates in the forest litter. Clearly, then, a niche is an organism's functional role: what it does in and for a living community.

Ecologists sometimes represent niches diagramatically, as FIGURE 39.2A does for a warbler's niche in the forests of New England. The bird might conceivably eat insects wherever they occur in trees, at any height and at any distance from the trunk. The bird might also nest any time in June and July. The niches of other species depend on factors such as the temperature and humidity the organisms prefer, the size or type of food they eat, their optimal environmental pH, or the water depth where they find food. The potential range of all biotic and abiotic conditions under which an organism can thrive is called its **fundamental niche**. If a warbler could eat insects any place in the tree, it would be operating in its fundamental niche for prey location.

In nature, however, a warbler cannot obtain insects just anywhere because several species of insect-eating warblers compete for food in eastern forests, each species performing a slightly different but specialized role in the community. The different species obtain insects at differ-

[A] A warbler in its *fundamental niche* might find food at any height in the tree, any distance from the trunk, or nest at any time in June or July

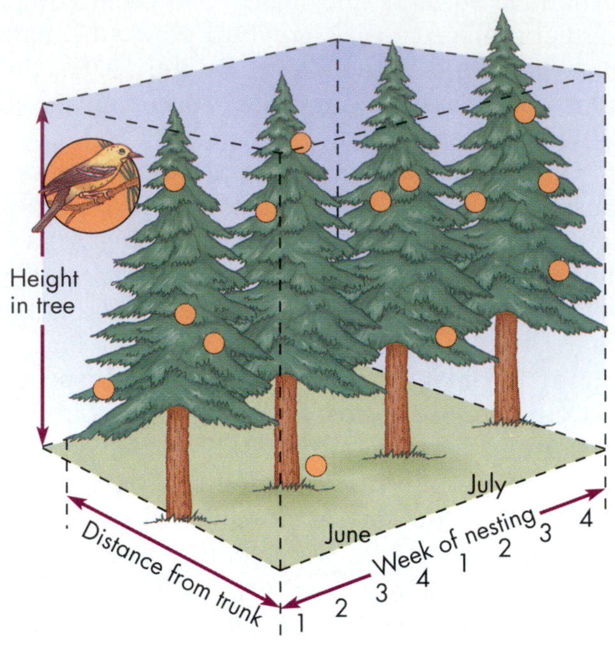

[B] A warbler in its *realized niche* might be restricted to finding food at certain positions in the tree, and nesting in just a couple of weeks of summer

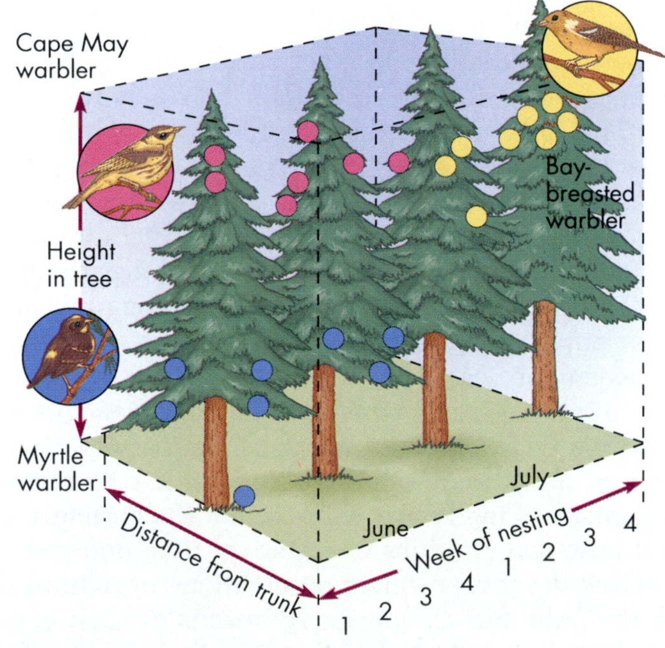

FIGURE 39.2
Niche: An Organism's Role in the Community.

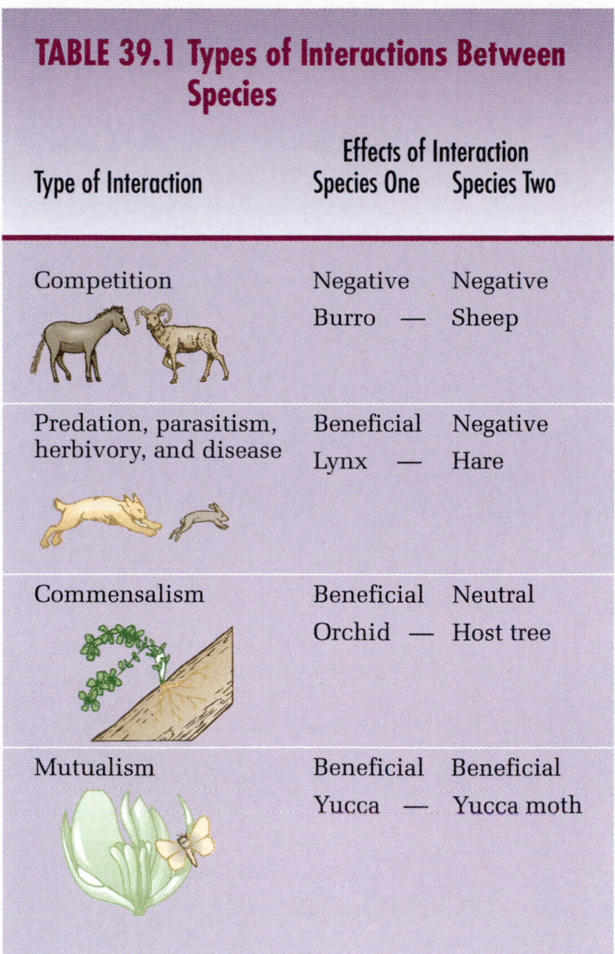

TABLE 39.1 Types of Interactions Between Species

Type of Interaction	Effects of Interaction: Species One		Effects of Interaction: Species Two
Competition	Negative		Negative
	Burro	—	Sheep
Predation, parasitism, herbivory, and disease	Beneficial		Negative
	Lynx	—	Hare
Commensalism	Beneficial		Neutral
	Orchid	—	Host tree
Mutualism	Beneficial		Beneficial
	Yucca	—	Yucca moth

ent heights in the trees and at different distances from the trunk, and their heavy eating comes at slightly different times during the year, depending on when they nest. The myrtle warbler, for instance, eats insects at the base of trees and in lower branches, the bay-breasted warbler specializes in insects in middle branches, and the Cape May warbler seeks insects at the outer edges of the top branches [FIGURE 39.2B]. Thus, other community residents may force a warbler species out of its broader fundamental niche and into its narrower **realized niche**: the part of the fundamental niche that a species actually occupies in nature. The realized niche of the myrtle warbler, as shown in FIGURE 39.2B, is substantially more limited than its fundamental niche. In general, interactions with other organisms often force a species into a realized niche that is more restricted than its fundamental niche. This restriction can be a major factor in defining a species' distribution and abundance.

THE MANY WAYS SPECIES INTERACT

Two species can interact in ways that increase, decrease, or leave unchanged the abundance of either or both species. For example, the interaction between cowbirds and Kirtland's warblers increases the populations of the former and decreases the populations of the latter. In contrast, the interaction between a jack pine tree and a Kirtland's warbler is positive for both species because the tree can serve as a daytime perch for the bird, and the bird eats harmful insects that could damage the tree.

Ecologists have categorized interactions between species into four general types [TABLE 39.1]. In (1) competition and (2) predation, one or both of the species suffer. In (3) mutualism and (4) commensalism, neither species is harmed by the interaction. We devote much of this chapter to these four kinds of interactions and to the ways that associations between species and the physical environment affect communities.

➤ CONCEPT CHALLENGE

What are the habitat and niche of a dandelion? Of a gray squirrel?

Competition Between Species

There are never enough good things to go around—good sunny spots in which to germinate and put down roots, good places to build nests, flowers of a certain type, or insects of a particular size to find and consume. Because of this, two different species often compete for the same limited resource, and this interaction restricts the abundance of both species. The key feature of **interspecific competition**, the use of the same resources by two different species, is that one or both competitors have a negative effect on the other's survival or reproduction.

INTERSPECIFIC COMPETITION

To see how competition works, imagine a population of red birds that feeds on bark beetles. The red bird population will continue to expand until competition develops among members of the red bird species for bark beetles and other available resources; population growth will then slow as it approaches the carrying capacity of the environment. Now, what would happen if a population of blue birds joined the community and began to compete with the red birds for bark beetles? With fewer resources available, the carrying capacity will decrease for the red birds, and their population density will fall.

How long could one expect the population of red birds to fall after the blue birds joined the community? The answer depends first on how well the red and blue species compete with one another for the limited

FIGURE 39.3
Modes of Competition in a Theoretical Two-Species Community.

[A] If one species inhibits growth of the other species, then the inhibited species may be eliminated from the final community: competitive exclusion

[B] If each species inhibits growth of its own species more than growth of the other, then the final community can contain both species

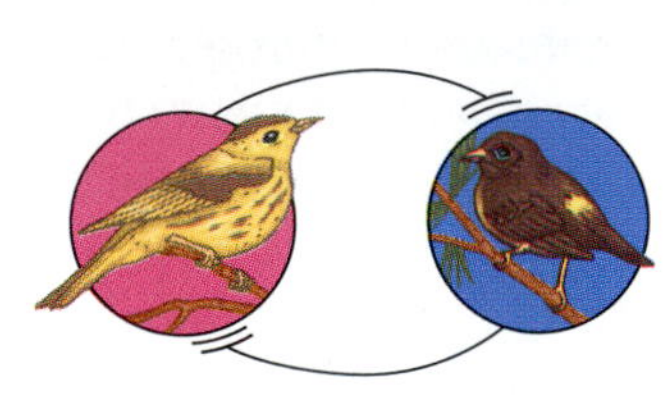

or

[C] If each species inhibits growth of the other equally strongly, then the final community can contain either one species or the other, depending on which was originally present in greater numbers

resources, and second on how crowding affects each species' individual population growth. If, for example, a high density of one species (for instance, red birds) affects the growth of the other species (blue birds) more than it affects its own growth, then the red birds will retain the community, and the other species will be eliminated [FIGURE 39.3A]. A situation like this, where one species excludes another through competition, is called **competitive exclusion**.

In another interesting possibility, competition can affect each species' own population growth more than it affects the other species [FIGURE 39.3B]. In this case, high density of the red bird population would strongly inhibit its own further growth, and a dense blue bird population would powerfully slow its own growth. As a result, the two species could end up coexisting.

As a final possibility, each species may slow the growth of the other equally [FIGURE 39.3C]. In this case, the species with the biggest population to start with would take over the community.

These possibilities are predicted by a mathematical model, and they certainly look good on paper or on a computer screen. But do they happen in nature with real organisms? Laboratory and field experiments with organisms as different as *Paramecium*, beetles, water fleas, fruit flies, field mice, and aquatic plants typically show that competition does take place, and that it can sometimes result in competitive exclusion, although the species may wind up coexisting.

[A]

[B]

FIGURE 39.4

Burros versus Bighorn Sheep: Competitive Exclusion at Work?

Burros introduced into desert communities [**A**] may be competing with bighorn sheep for various resources [**B**] and excluding them from some of their native territories.

NATURAL CAUSES OF COMPETITIVE EXCLUSION

It is easy to test the concept of competitive exclusion in laboratory culture dishes and to trace the factors behind the victory of one species and the demise of another. In natural communities, however, it is much harder to pin down the factors critical to each competitor's success because so many biological and physical factors interact. In the western United States, for example, feral burro populations—derived from runaway domesticated animals—have increased dramatically since laws were passed in the 1970s that prohibit killing them. As the burros have multiplied in the arid open lands of Arizona and southern California, however, the number of desert bighorn sheep has rapidly declined [FIGURE 39.4]. Many biologists suspect that the burros, an introduced, or *exotic*, species, are somehow outcompeting the native bighorn sheep, but they are not sure exactly how.

Sheep and burros eat some of the same desert plant species, but burros may be able to consume or use these plants more efficiently than bighorn sheep. Burros also tend to congregate around desert water holes, however, and often chase other mammals away. Thus, competition may occur when two species exploit and have equal access to identical resources, or when one species uses aggressive behavior to keep competitors from a resource. Ecologists do not yet know whether the relationship of burros and bighorn sheep will lead to competitive exclusion, with the burros' presence inevitably leading to the sheep's extinction, or whether the relationship will allow a stable coexistence. Experiments with other species suggest possible answers.

The Caribbean island of St. Martin has two lizard species that are slightly different in size but eat the same kinds of insects. To find out whether the two species actually compete with each other, researchers fenced off large squares of land that contained individuals of the smaller-sized species, the larger, or both. They found that where both species coexisted, members of the larger species had less food in their stomachs, grew more slowly, laid fewer eggs, and were forced to perch higher in the bushes than when that species lived alone in an enclosure. Studies like these proved that strong competition does exist in natural populations and that the presence of one species can limit another species to its realized (rather than fundamental) niche. Such results make it seem possible that wild burros could indeed restrict bighorn sheep to a smaller realized niche but that the two could continue a stable coexistence.

COMPETITION CAN ALTER A SPECIES' REALIZED NICHE

The larger lizard in the previous study escaped extinction through an alteration of its realized niche. But is this the way species usually respond to the threat of competitive exclusion? Russian biologist G. F. Gause in the 1930s devised an experiment with *Paramecium* to find out. He selected a third species, a challenger, to compete with two *Paramecium* species he had previously tested. Of those two, the smaller species had competitively excluded the larger. Gause expected the third species, or challenger, to compete fiercely with the champion, or smaller, species. But his findings surprised him: Both species coexisted in the same culture [FIGURE 39.5]. Like the lizards, the paramecia split up the territory, and the challenger occupied a different niche than the champion. While the entire test tube was the fundamental niche for both *Paramecium* species, the realized niche was smaller (only part of the tube) in a mixed culture, and this minimized direct competition.

Ecologists use the term **resource partitioning** to indicate the process by which resources are divided, allowing species with similar requirements to use the same resources in different areas, at different times, or in different ways. The warblers of New England forests minimize the harmful effects of competition by resource partitioning—in this case, feeding in different parts of the trees [review FIGURE 39.2].

FIGURE 39.5

Dividing the Spoils: Resource Partitioning in Two *Paramecium* Species.

After introduction into a culture tube (left), two species of paramecia can coexist by partitioning resources (right). The smaller species occupies the top territory with greater oxygen penetration, and the larger species inhabits the lower part of the test tube.

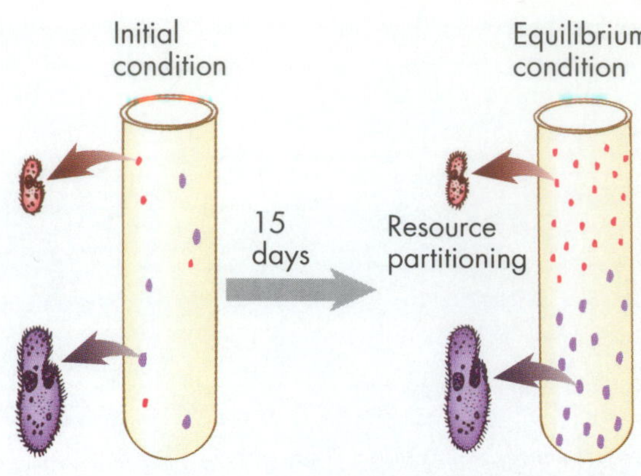

In the portions of a species' range where it overlaps the range of a strong competitor, hereditary changes often evolve in both species' physical or behavioral characteristics. Ecologists call such changes *character displacement*, and think that they bring about a partitioning of resources by otherwise competing species. Character displacement is an example of coevolution, hereditary changes in two or more species as a consequence of their interactions within a community.

At one site in California's Mohave Desert, for example, harvester ants (*Veromessor pergandei*) compete with a single other large ant species, and nearly all the harvester ants have mouth parts smaller than their competitors'. In contrast, at a desert site near Ajo, Arizona, the harvester ant competes with one quite large ant species and one quite small one. At this site, the harvester ants have mouthparts intermediate in size between those of their rivals. Ecologists interpret these field observations to mean that natural selection has acted on harvester ant populations to produce body sizes different from individuals of coexisting species; in other words, the character of body size is displaced by natural selection in one direction or the other depending on characters in competing species. Character displacement among harvester ants apparently allows resource partitioning, since different-sized ants consume different-sized seeds. As a result, several potential competitors can coexist in the same habitat by utilizing slightly different niches.

While competition often limits population size in both interacting species, the second major type of community interaction, predation is beneficial for one species but harmful to the other.

➤ CONCEPT CHALLENGE

Does the concept of character displacement apply to competition among bay-breasted, Cape May, and myrtle warblers? Explain.

Predation

Animals like lions or Kirtland's warblers that kill and eat other animals are **predators**, their food is **prey**, and the act of procurement and consumption is **predation**. Herbivores such as bark beetles eat plant parts and often harm the plant without killing it. Parasites, like cowbirds or tapeworms, obtain their food from a living host organism, often without killing it. Disease-causing organisms, or **pathogens**, are usually fungi, bacteria, or protists that obtain nourishment from a plant or animal host and weaken or kill it [see CHAPTER 19]. Here we focus on how predation affects the population size of both prey and predator on a short time scale and how hunter and hunted evolve strategies to outwit each other on a longer evolutionary time scale.

POPULATIONS OF PREDATOR AND PREY

When human volunteers removed thousands of cowbirds from the Kirtland's warbler's habitat, vastly more of the warbler young survived. At the same time, however, the number of warblers killed by blue jays, ground squirrels, and house cats increased. Does that fact inevitably mean that these predators will now go on to drive the warbler to extinction? Laboratory and field results suggest some possible answers.

In CHAPTER 19, we encountered a minute but voracious predator called *Didinium* that hunts down *Paramecium* cells and swallows them whole [see FIGURE 19.18]. Laboratory experiments on *Didinium* and *Paramecium* revealed several principles about predator and prey populations. First, as the density of the prey population increases, the predator finds more food to eat, which in turn increases the size of the predator population [FIGURE 39.6A]. Eventually, however, the large number of predators eat so many prey that the prey population begins to fall. If *Didinium* cells devour all the *Paramecium* cells, then the predator population will crash, too, because its food supply will have been exhausted, and both organisms will finally become extinct in their laboratory setting [see FIGURE 39.6A]. If, however, the experimenter provides the prey with a **refuge**, a safe haven out of the predator's reach, then when the predator has eaten all the accessible prey, it dies out and the prey takes over the environment [FIGURE 39.6B]. Finally, if the experimenter creates a complex environment that offers many partial refuges where prey can survive for a while before predators migrate in and discover them, then the populations of predator and prey tend to oscillate up and down [FIGURE 39.6C].

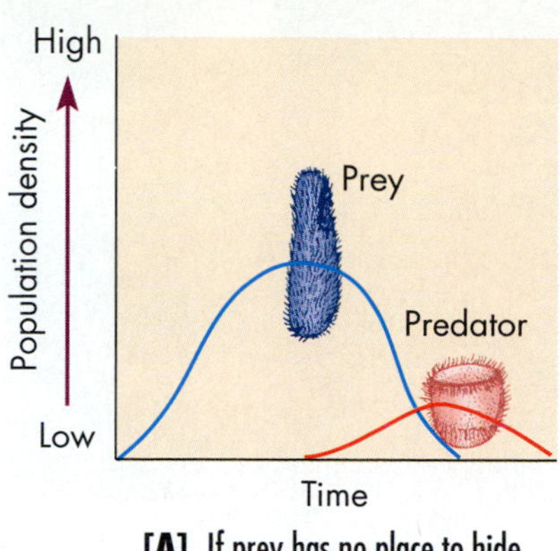

[A] If prey has no place to hide, both eventually die out

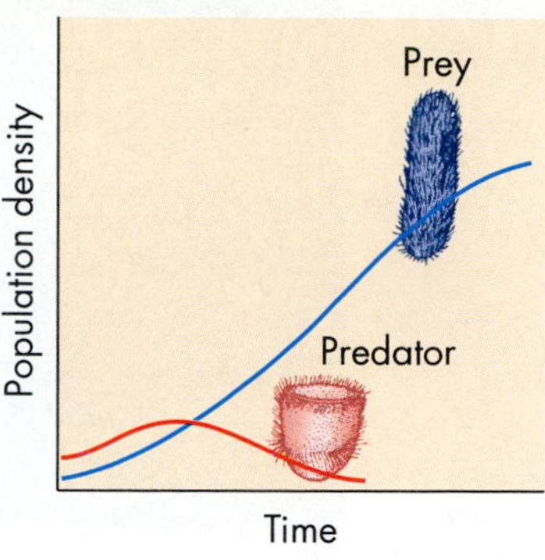

[B] If prey has a refuge, predator dies out

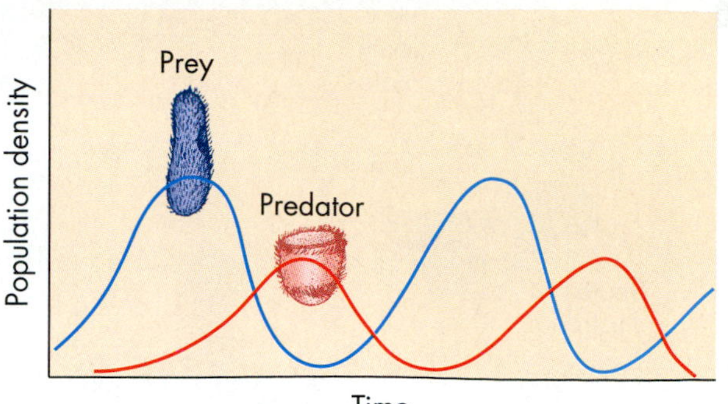

[C] If prey can hide temporarily, both populations cycle

FIGURE 39.6

How Populations of Predators and Prey Interact.

[A] If the predatory protist *Didinium* is grown in culture with its favorite prey species, *Paramecium*, both populations increase until the predator population is so large it eats the prey faster than the prey population can grow. When all the prey have been consumed, the predator also dies out. [B] If the *Paramecium* cells have a refuge inaccessible to the predator, the *Didinium* may starve and die. [C] If the prey species has many partial refuges in which it can grow until discovered by the predator, both populations oscillate.

For at least 200 years, wild populations of the snowshoe hare and Canadian lynx have periodically risen and fallen on about a 10-year cycle [FIGURE 39.7]. While the experiments with *Didinium* and *Paramecium* would predict this result, the oscillation does not occur for the reason ecologists originally supposed. Instead, field studies suggest this: When hare populations are high, they provide an abundant resource for lynx, and this allows the lynx population to start increasing. The large numbers of hares soon overgraze their limited winter food sources, however, and so their population size decreases dramatically due to starvation (rather than to predation by lynxes). The lynx population then crashes because the felines cannot find enough to eat. After these booms and crashes, it takes several years to regrow the vegetation the hares eat. When it does, the hare population once again increases steadily over a two- to three-year period, leading to a surge in the lynx population that lags a year or two behind the hares. Thus there are three ecological factors involved in the arctic oscillations, not just two, and the size of the hare populations both modifies the abundance of vegetation and drives the lynx population cycle.

Even though predation by the lynx is not the prime force in controlling snowshoe hare populations, other predator species can exert from a modest to a considerable amount of pressure on prey population size. Field observations suggest that predators mainly tend to kill weak animals, especially young or old animals, rather than prey upon animals in their active reproductive years. For example, the wolves that inhabit Isle Royale in Lake Superior mainly kill weak and infirm individuals, including very young or very old moose, not moose in their prime.

The case is different, however, for the doglike dingo, Australia's largest carnivore, which feeds on red kangaroos, on large ground birds called emus, and on other species. More than 20 years ago, sheep ranchers built almost 10,000 km (6200 mi) of dingo-proof fences, then poisoned or shot all the dingoes inside the sheep-grazing pastures. FIGURE 39.8 shows how the prey population

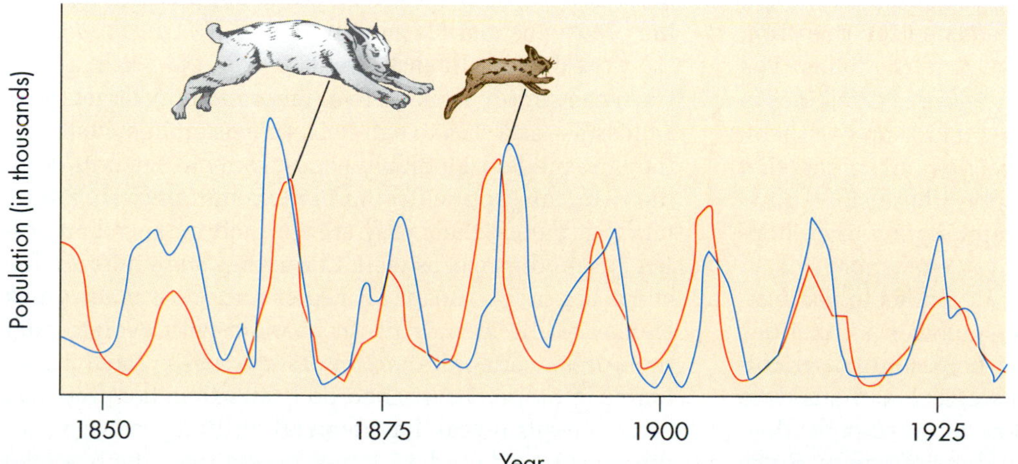

FIGURE 39.7

The Hare-Lynx Cycle: Responsiveness of Predator and Prey Populations.

In northern Canada, hare populations occasionally boom when vegetation is abundant and crash when food becomes scarce. When hare populations rise, lynx also increase and devour more of their prey.

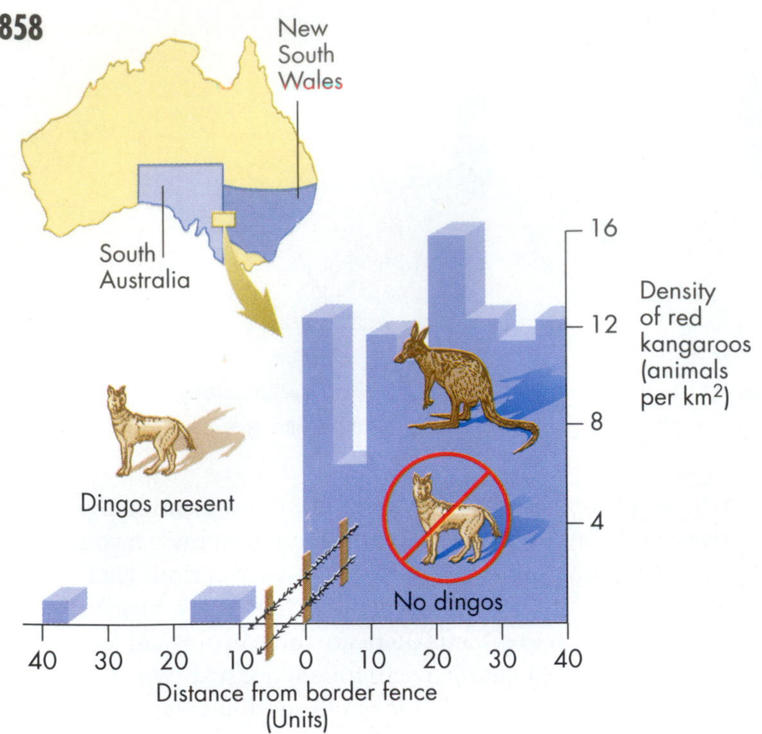

FIGURE 39.8

Some Predators Can Strongly Control Prey Populations.

Sheep ranchers constructed miles of fences across the border between the states of South Australia and New South Wales, then exterminated all the dingos on their sheep-grazing lands. Because they had eradicated a natural predator that also hunted red kangaroos, these large marsupials proliferated tremendously.

[A]

FIGURE 39.9

The Countermeasures of Prey.

[A] A moth mimics a dead leaf. [B] Caterpillars of the moth species *Nemoria arizonaria* that emerge in spring resemble the male flowers (catkins) of oak trees. [C] Caterpillars that emerge in the fall, however, and feed on leaves look like twigs even though they are apparently genetically identical to the catkinlike caterpillars. [D] This sailfin leaf fish in Fiji blends so well with its background that it is nearly invisible.

responded. With dingoes absent, the population density of red kangaroos shot up almost 170 times higher than before. Evidently, a predator like the dingo can strongly limit the density of prey populations.

Because some predators can help regulate populations of their prey, people have used predators to control populations of pest species. Indonesia is the world's fifth most populous country, and its people depend largely on rice for food. In 1985, an outbreak of small insects called brown plant hoppers seriously threatened the rice crop, causing it to dry out, fall over, and rot in the fields. The Indonesian government promoted the use of pesticides to control the plant hoppers, and this did kill some of them, but it also destroyed the ladybird beetles and spiders that prey upon the plant hoppers. Without these slow-growing predator species, the rapidly reproducing plant hoppers multiplied quickly and destroyed even more rice. In 1986, the government began to train farmers to use *integrated pest management*, which encourages populations of natural predators that can then help control agricultural pests. Farmers using these ecological methods were able to produce more rice per acre more cheaply than farmers who sprayed pesticides indiscriminately. Since then, integrated pest management has become Indonesia's national strategy for protecting rice crops.

As with ladybird beetles and plant hoppers or hare and lynx, populations of predator and prey may grow or shrink over the short term, but over the long term, genetic changes can influence the evolutionary balance between hunter and hunted.

THE COEVOLUTION OF PREDATOR AND PREY

Just as competing species evolve niche adjustments by character displacement due to natural selection, predator-prey interactions lead to a grand coevolutionary race, with predators evolving more efficient ways to catch prey, and the prey evolving better ways to escape. It is hard to tell which set of adaptations—the predator's or the prey's—is more fascinating.

Predator Strategies Some predators, such as sea stars, can simply walk up to a stationary prey target—say a mussel—and start to eat. If the prey is mobile, however, then the predator probably needs a way to catch its food, the two main options being *pursuit* and *ambush*. Predators that pursue their prey are selected for speed and often for intelligence, as well. Carnivores store information about the prey's escape strategies and must make quick choices while in pursuit. In keeping with evolutionary pressures, vertebrate predators generally have larger brains in proportion to their body size than the prey they catch. Fossils reveal that about 60 million years ago, carnivorous predators had larger brains than their hoofed

[B]

[C]

[D]

prey, and that while selective pressures forced the prey to become more wary, the predators and their more cunning descendants always stayed one step ahead.

For some predators, ambushing is an effective strategy for capturing prey. A familiar example is the frog that ambushes flying insects by snapping out a sticky tongue, hinged at the front of the mouth [see FIGURE 1.15]. Certain mantis insects also ambush their prey by hiding in the open with a camouflaged, plantlike appearance. Those mantises that carry genes for resemblance to the plants they inhabit can be nearly invisible to prey, and thus more effective at ambushing their food and surviving to reproduce than mantises without those genes. An ambush can be even more effective if the predator can lure the prey: Recall the anglerfish [FIGURE 22.1] with a wormlike lure that entices prey into its gaping mouth.

The Countermeasures of Prey Species In addition to rapid running or flying, prey species have evolved some remarkably devious tricks that help them avoid being eaten. One defense strategy avoids confrontation altogether through **camouflage**: the occurrence of shapes, colors, patterns, or even behaviors that enable organisms to blend in with their backgrounds and usually escape predation. Many insects have evolved shapes that look like twigs, flowers, or leaves—alive or dead—some complete with phony leaf veins [FIGURE 39.9A]. One remarkable caterpillar even assumes two alternative camouflaged shapes, depending on season and food source [FIGURE 39.9B and C]. Still other insects, and a few amphibians, escape detection by resembling damage or bird excrement on leaves. And behavior plays a role too; if a fish or invertebrate of appropriate color or shape freezes in place, it can resemble the multicolored bottom of a rocky tide pool [FIGURE 39.9D].

As CHAPTER 37 explained, chemical warfare is another common defense strategy. Eucalyptus and creosote bushes, for example, produce distasteful oils or toxic substances that kill or harm herbivores. People sometimes plant oleanders [FIGURE 39.10A] as decorative shrubs because they resist insect pests, but the leaves are so poisonous that chewing a few can kill a child. Animals are not without their own arsenals: Toads, stinkbugs, and bombardier beetles [FIGURE 39.10B] produce highly offensive chemicals that repel attackers.

[A]

FIGURE 39.10

Chemical Protection.

[A] The colorful, fast-growing, dry-adapted oleander (*Nerium oleander*) has poisonous leaves and stems. [B] The bombardier beetle (*Brachinus* species) defends itself by spewing a volatile irritant from special glands.

[B]

Poisonous prey species usually evolve brightly colored patterns, enabling the experienced predator to recognize and avoid them. This is called **warning coloration** (or *aposematic coloration*). The brilliantly colored but poisonous strawberry frogs of South America [see FIGURE 1.4] have evolved this strategy.

Many nonpoisonous prey species masquerade as poisonous species; this is one form of the process called **mimicry**, wherein one species resembles another (review the pairs of blue and orange butterflies in FIGURE 16.17B and C). Mimicry could arise in the following way. Individuals of a nonpoisonous butterfly species that by

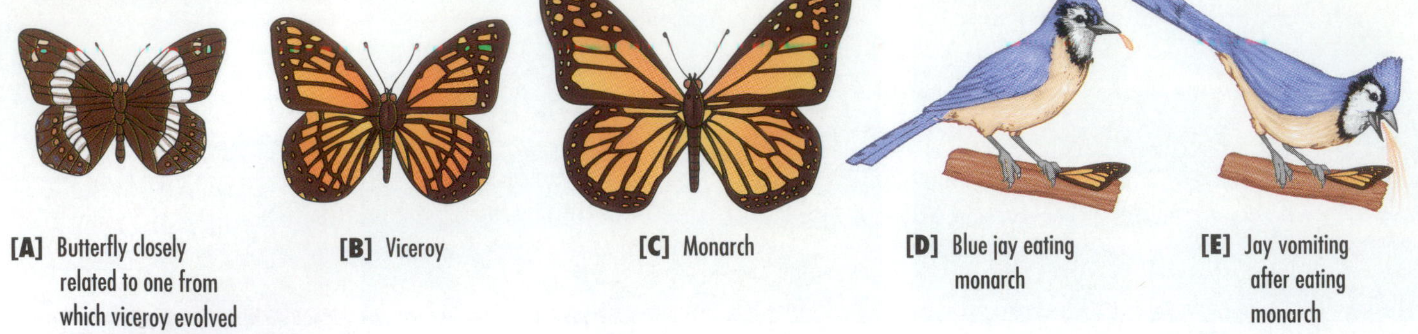

FIGURE 39.11

Warning Coloration and Mimicry: More Defense Strategies.

[A] Ancestors of the dull-colored *Limenitis arthemis* butterfly evolved into [B] the flamboyant viceroy (*Limenitis archippus*), mimic of [C] the unrelated monarch butterfly (*Danaus plexippus*). Experiments show that monarchs and viceroys are both distasteful to birds, such as redwing blackbirds or blue jays, shown here. The butterflies' nearly identical bright coloration protects both species from predation, since an individual bird will avoid both species after tasting only one [D] and vomiting [E]. Thus, *Limenitis* butterflies that carried alleles for resemblance to monarchs were selected for during evolution.

chance contain alleles causing them to resemble the poisonous species even slightly may occasionally escape predation if a hungry animal mistakes them for the poisonous species. A selective pressure like this could, over time, allow the nonpoisonous species to accumulate more and more alleles for resemblance to the poisonous neighbor.

In another type of mimicry, two or more foul-tasting species come to resemble each other during evolution. An example of this is the beautiful viceroy and monarch butterflies [FIGURE 39.11]. Experimenters wondered whether this look-alike situation arose due to coevolution of warning mimicry. To test this possibility, researchers removed the characteristically colored wings from seven species of butterflies and presented the now unfamiliar wingless insects to redwing blackbirds. The birds took a small nibble from each, but fully consumed only a few of the species, which must have tasted fine. After sampling either a viceroy or a monarch, however, a bird would shake its head, drink excessively, and then strictly avoid *both* types of insects thereafter. Clearly, both the viceroy and monarch are foul-tasting, and birds can recognize desirable prey by taste as well as sight. Since a bird that learns to avoid one species will also avoid the other, this type of mimicry helps both kinds of insects avoid predation.

Plants and animals have also evolved with thorns, spines, sharp spikes, and horns—weapons that discourage predators [FIGURE 39.12].

PARASITES: THE INTIMATE PREDATORS

Parasites are insidious kinds of predators; they are usually smaller than their hosts, often live in close physical association with individual victims, and generally just sap their strength rather than killing them outright. *Ectoparasites*, like fleas, ticks, and leeches, live on the host's exterior, while *endoparasites*, like tapeworms, liver flukes, and some protozoa, inhabit internal organs

[A]

[B]

FIGURE 39.12

Spines Help in Self-Protection.

[A] Porcupines and [B] some kinds of tropical palms have evolved sharp spines that ward off predators.

FIGURE 39.13

Insect Parasitoid: A Wasp and Its Victim.

A parasitic wasp injects eggs into the caterpillar of a tomato sphinx moth. Later, the green caterpillar lies dead (shown above), covered by the white wasp pupae that consumed the host's internal organs.

or the bloodstream. In a special type of parasitic interaction, certain insects develop inside the body of another insect (usually inside a caterpillar or a maggot) and inevitably kill it [FIGURE 39.13].

Sometimes, parasites and their hosts coevolve in such a way that the parasite becomes less harmful to the host; an especially aggressive parasite that kills its host *before* the parasite reproduces would be selected against. Hosts, in turn, may coevolve some tolerance to the parasite; native antelope in Africa, for example, suffer little when they harbor trypanosomes, whereas the protists cause lethal cases of sleeping sickness in domestic cattle.

Cowbirds and cuckoos represent the strange phenomenon of *social parasitism*, a special case of parasitic activity in which one species exploits the social behavior of another species during a critical phase of its life cycle. As we saw in the chapter's introduction, brown-headed cowbirds will seek out the nest of a Kirtland's warbler or other bird species, lay an egg in it, then leave the unwitting foster parents to hatch the egg and nurture the young freeloader until it can fend for itself [FIGURE 39.14]. The cowbird is spared parental duties—a clear benefit—whereas the foster parents expend precious food, energy, and attention on the young of another species while their own may go hungry—a clear disadvantage. This suggests a straightforward case of parasitism.

FIGURE 39.14

Cowbird Baby Eating Foster Mother Out of House and Home.

Here, a blue-winged warbler feeds a huge cowbird chick. Kirtland's warblers similarly tend cowbird interlopers.

➤ CONCEPT CHALLENGE

An adult Kirtland's warbler will seldom push a cowbird egg from the nest. Apply the concept of coevolution of parasite and host to this fact, and suggest possible future scenarios involving the interaction of Kirtland's warblers and brown-headed cowbirds.

Commensalism and Mutualism

The community relationships we have discussed so far—competition and predation—have involved harm to at least one of the species. Sometimes, however, neither species is harmed by their interactions [see TABLE 39.1]. In **commensalism**, one species benefits from the alliance while the other is neither harmed nor helped, whereas in **mutualism**, both species are helped.

COMMENSALISM

Commensalism is common in tropical rain forests, and the most easily observed examples are the *epiphytes*, or air plants, that grow on the surfaces of other plants. Epiphytes include Spanish moss, many beautiful orchids, and large and small bromeliads that festoon the branches or decorate the forks of trees [FIGURE 39.15A]. Using the tree merely as a base of attachment, epiphytes take no nourishment from the host and do no harm unless their numbers become excessive. Other commensal relationships include birds that nest in trees, algae that harmlessly grow on a turtle's shell, and small fish that live among the stinging tentacles of sea anemones—unharmed and safe from predators [FIGURE 39.15B].

MUTUALISM

In a mutualistic interaction, both species benefit. One good example involves the yucca, a tall, stately member of the lily family that grows in hot, dry regions of the western United States. The pointed leaves of a *Yucca whipplei* plant jut upward from the parched ground like a sheaf of swords. Above the leaves rises a single awesome flower stalk, 4 m (13 ft) tall and loaded with more than 1000 white blossoms.

[A]

[B]

FIGURE 39.15

Commensalism: One Species Benefits, the Other Is Neither Harmed nor Helped.

[A] In the tropics, orchids often rely on tree trunks for support. **[B]** On coral reefs, an anemone fish can find safe haven within the folds of a sea anemone, which stings other fish.

FIGURE 39.16

Coevolutionary Partners: The Yucca Moth and Yucca Flower.

Yucca whipplei grows in the Mojave Desert and has an intimate relationship with the yucca moth (inset). The moth lays its eggs in the ovary at the base of a yucca flower and then pollinates the same flower. Moths depend on the plant for food and reproduction; the plant depends on the moth for pollination. This mutually beneficial relationship is called mutualism.

Each spring, white female moths flutter about the cream-colored yucca flowers [FIGURE 39.16] and play out a curious set of behaviors honed by natural selection. Arriving at the younger flowers near the top of the stalk, the female yucca moth visits several blooms and collects a ball of pollen, which she carries beneath her "chin" with specially modified mouthparts. Bearing this pollen ball, she then flies to another plant and enters an older flower near the base of the stalk. Here she extends a long, sharp drill from the tip of her abdomen, pierces the older flower's ovary, and pumps her abdomen. The pumping causes eggs to pass through the drill and into the flower. Egg laying completed, the moth now executes one of nature's most remarkable sequences: She moves up to the flower's green stigma, separates a bit of pollen from the pollen ball, then rubs the pollen directly across the stigma. This fertilizes the flower's eggs and ensures that fruits and seeds will develop and serve as the food source for the moth's offspring as they become caterpillars.

Fortunately for the plant, the caterpillars do not eat all of the seeds from fertilized flowers as they grow. The plump mature yucca caterpillars eventually emerge from the fruits, drop to the ground, burrow into the soil, and metamorphose. The next season, adults emerge, mate, and repeat the cycle. In this relationship, both species benefit: the yucca moth gets a high-energy nectar reward, and the plant becomes pollinated.

Mutualism between plants and pollinators is so common that one can almost predict that a bizarre flower will have a bizarre pollinator. Charles Darwin once aroused the skepticism of his contemporaries by predicting that a moth with a 30-cm (1-ft) tongue fluttered somewhere in the rain forest of Madagascar because a flower lived in those jungles with a 30-cm-long floral tube. Sure enough, 40 years later, naturalists caught the moth with the longest known tongue.

Sometimes, both species in a mutualistic relationship will benefit, but neither will be wholly dependent on the other for survival. Biologists call this *facultative mutualism*. Many species of aphids, for example, excrete large quantities of sweet, saplike fluid, and ants harvest this sap. The ants receive nutrients, and in return, their presence keeps predators away from the aphids. The ants are not strictly dependent on the sap, however, and the aphids continue to produce it even when the ants are absent.

In contrast, some interacting organisms require each other to survive. A relationship of this type, called *obligatory mutualism*, binds the yucca plant to the yucca moth. The moth receives nourishment only from the yucca, and in turn, the yucca plant is fertilized solely by the yucca moth.

➤ CONCEPT CHALLENGE

How could the relationship of Kirtland's warblers and brown-headed cowbirds eventually evolve from a parasitic relationship into a mutualistic relationship?

Organization of Communities

The principles we have considered—competition, predation, commensalism, and mutualism—affect not just pairs of species, but entire communities consisting of tens to hundreds of species. Before understanding how communities work, we must first identify these hundreds of species. Astonishingly, it was not until 1993 that the U.S. government first instituted a complete census of all living things in the United States. Called the National Biological Survey, the program aims to determine the constitution, and hence the health, of biological communities all across the country [see BOX 38.1, page 846]. The survey could identify unique species and communities before they become extinct.

When biologists identify the species living in different communities, they are answering an important question posed by early community ecologists: Are communities fixed groups of species unfailingly bound together, or are they more fluid entities, consisting of one set of species at one time and another overlapping but slightly different set at another time? Research projects undertaken nearly 50 years ago confirmed that a community is not a "superorganism," a package of highly specific groupings of plants and animals (like a vertebrate organism with one heart, two lungs, and a precise assortment of other organs). Rather, communities consist of some species that happened to immigrate into the area and can survive under the available physical conditions, and some species that will grow only if other species are also present. Given this premise, a goal of ecologists is to learn how the addition or subtraction of a species will affect the whole community in the short and long term.

COMMUNITIES CHANGE OVER TIME

Occasionally, a cataclysm will strip an area of its original vegetation. This occurred in 1980 with the volcanic explosion of Mt. St. Helens in Washington State and again in 1988 when fires swept through much of Yellowstone National Park. A farmer clearing a field can even cause such a change. Left to nature, however, a regular progression of communities will regrow at the site in a process called **succession**.

Soon after a region is denuded, a variety of species begin to colonize the bare ground. These species make up a *pioneer community*, and they modify environmental conditions, such as soil quality, at the site. Conditions produced by the activities of the pioneer community determine which additional species can establish themselves in the area and form a *transition community*. Changes are rapid at first, as more and more species join the transition community. Eventually, a particular community of plants and animals becomes relatively stable, and changes take place more slowly over time. Such an assemblage is often called a *climax community*, but most ecologists today recognize that change is constantly occurring even in old communities, driven by such environmental variables as hurricanes, floods, and fires, as well as global climatic changes.

A well-studied example of ecological succession can be seen at Glacier Bay, Alaska, where a glacier that devastated thousands of square kilometers of land has melted back about 100 km (62 mi) in the last 200 years [FIGURE 39.17], exposing the ground below. The most recently exposed areas are inhospitable piles of rock, sand, and gravel lacking usable nitrogen and essentially devoid of plant or animal life. The first plants to colonize this barren scene form a pioneer community of wind-dispersed species: mosses, horsetails, fireweed, willows, cottonwood, and the matlike rockrose *Dryas*. Most are severely stunted and grow close to the ground as a result of nitrogen deficiency, but *Dryas* has nitrogen-fixing nodules on its roots that provide the growth-limiting nutrient [review FIGURE 35.11]. Within a few years, it crowds out other plants and forms a dense mat over the soil.

In areas exposed for 20 years or so, alder trees begin to invade from other sites. Alder roots also bear nitrogen-fixing nodules, and thus these trees can grow rapidly. Dead alder leaves add nitrogen to the growing soil layer and stimulate the growth of willows and cottonwood. In areas exposed for about 50 years, dense thickets of these plants shade the pioneer species and kill them. Eventually, Sitka spruce invades, and in those areas exposed for about 80 years, nitrogen released by the alders enables the spruce to form dense forests that shade out the alders and willows. Finally, after 100 years or so, shade-tolerant hemlock trees invade, grow below the canopy of spruce branches and needles, and become the most common tree in the climax community. In low places, however, sphagnum moss invades the forest floor, soaks up large amounts of water, and kills trees by choking off their roots' oxygen supply. This leaves a *mosaic climax* consisting of patches of spruce-hemlock forest intermixed with sphagnum bog.

Succession also drives changes in the habitat of the Kirtland's warbler. The bird can find food and nesting sites only among the blueberries and bracken ferns that grow under jack pines that are approximately 5 to 23 years old. When a fire rages through a mature jack pine forest, the trees are killed, but their cones burst open and release hundreds of seeds. Dandelions, fireweed, and other wind-distributed wild flowers grow profusely in

[A] The retreating glacier

[B] The rockrose: An early invader

[C] The climax community includes patches of sphagnum moss

Time | 0 | 10 years | 20 years | 50 years | 80 years | 100 years

Glacier

10 years: Pioneer mosses, horsetails, *Dryas*

20 years: Alders invade prostrate willows, cottonwood

50 years: Alder thicket

80 years: Sitka spruce invade; Hemlock, sitka spruce forest

100 years: Hemlock, sphagnum moss bog in low spots

Soil becomes more acidic

Soil gains nitrogen

[D] Succession

FIGURE 39.17

Succession in the Path of an Arctic Glacier.

[A] As Alaska's massive glacier recedes, a succession of plants followed, from **[B]** the rockrose (*Dryas* species) to cottonwood and alder and eventually **[C]** sphagnum moss and hemlock. **[D]** A time line of the glacier's recession and the plant's succession. Note that the soil gains nitrogen and becomes more acidic with the passing decades.

the next few years, and after about 5 years, the pine seedlings, blueberry, and ferns have matured sufficiently so that they support the warbler once again. About 23 years after the fire, the lower branches of the jack pines die, and the vegetation underneath changes, becoming less favorable to Kirtland's warblers. Knowing that the rare songbirds depend on specific intermediate stages of succession, Michigan foresters have increased the amount of warbler habitat by periodically burning sections of jack pine forest [FIGURE 39.18]. At the proper stages of succession, the birds return and begin nesting once again.

FIGURE 39.18

Kirtland's Warbler Habitat Depends on Succession.

Here, a forester is burning a meadow to encourage the growth of new trees and plants that provide habitat for the endangered songbird.

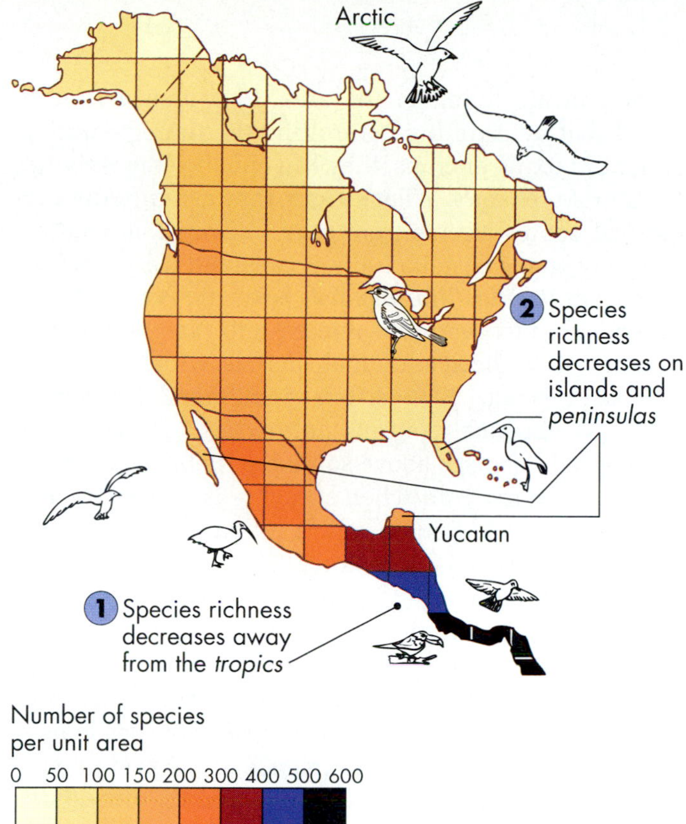

FIGURE 39.19
Latitude and Isolation Affect Species Richness.

SPECIES DIVERSITY IN COMMUNITIES

Coral reefs and tropical rain forests are dense with interacting species, while arctic tundra and deserts have far fewer species. The total number of species found in a community is its **species richness**, or *species diversity*. In most communities, there are few common species but many rare types of organisms. For example, English ecologists, captured a group of almost 7000 moths and identified individuals of 197 different species in one local area. Fully one-quarter of the moths belonged to a single species, and another quarter belonged to just five other species. The remaining half of the moth group fell into the other 191 species—some represented by just one or two individuals.

Both *latitude* (north-south position) and *isolation* (occurrence on peninsulas, island chains, or other out-of-the-way locales) influence the species richness of an area. Some communities in the tropical latitudes, for example, have about 600 types of land birds, while an area of similar size in the arctic tundra may have only 20 to 30 species of land birds [FIGURE 39.19]. A single hectare of mainland tropical forest can have 300 different species of trees and tens of thousands of insect species; on a peninsula, however—even a tropical one like Baja California or Florida—the species richness is diminished, as the more limited bird life in those areas illustrates. Species diversity on island chains is limited in such specific ways that ecologists are now applying the principles of island ecology to the design of nature preserves, which are often islands within a sea of human development.

Recent observations support the conclusion that the more resources available in an area—water and solar energy, for instance—the greater the species richness the area can support. This helps explain the numerous varieties of plants found in tropical forests. Other factors influence species richness, however, including competition and predation. Competition can increase diversity because resource partitioning divides up niches into more and more specialized compartments. For example, a lizard species that has an island all to itself will eat insect prey of any size, but a lizard species that shares an island with four other lizard species will specialize on prey of a particular size. Therefore, with competition forcing smaller niches, a community can accommodate more species.

Predation can also increase species richness. A sea star along the coast of Washington State, for example, preys on 15 species of barnacles, snails, clams, and mussels. When experimenters removed the sea star from sections of the shore, the diversity dropped to eight species because the mussel population increased and crowded out the other invertebrates. By eating young mussels, the sea star reduces competition for space and so preserves a higher species richness. Species like this predatory sea star, whose activities determine community structure, are called *keystone species*. The abundance of predators and parasites in the tropics may help explain why the lower latitudes have such a high species richness.

SPECIES DIVERSITY AND COMMUNITY STABILITY

For years, many biologists have warned that species richness is dwindling all over the globe. The loss of tropical forest ecosystems is one familiar example, and the cutting of old-growth (virgin) forests in the Pacific Northwest is another. Until recently, however, there were hypotheses but little actual data on precisely how species richness affects the health of an ecosystem. The diversity-stability hypothesis suggests that because the species in a community encompass so many different traits, a diverse community is more likely to contain at least some species that can survive and thrive in the face of environmental disturbances such as drought, fire, hailstorms, and other catastrophes. In contrast, the species-redundancy hypothesis predicts that within a community, various species are so similar that as long as major kinds of organisms are represented, species diversity (or its absence) should not strongly affect the functioning of the community.

One of the first critical tests of these hypotheses appeared in 1994, with the publication of research carried out in Minnesota and Quebec. Researchers there measured the amount of living plant material accumulated in plots of grassland 4.0 m by 4.0 m on a side. These plots contained differing numbers of species, from undisturbed native prairies with more than 20 species per plot to abandoned farmers' fields with 1 to 10 species per plot. The researchers found that in a severely dry year, plots that contained fewer than 5 species of plants produced less than 12 percent of their normal amount of plant material, while plots that contained 9 or more species produced 50 percent of their normal amount. Not only were the species-rich plots four times more productive in the drought years, but they recovered more quickly than the species-poor plots.

These results were predicted by the diversity-stability hypothesis; they suggest that a community's ability to resist adverse environmental conditions increases with species richness. And just as predicted, some species perform better than others when faced by adverse environmental conditions such as drought and flood conditions, unusually cool summers, early frosts, and disturbances like hail, fire, and insect predation. The long-term stability of communities may thus depend on their biodiversity. These results for grasslands help underscore the urgency of pleas to conserve biodiversity.

Species Richness in Agricultural Communities If the diversity-stability hypothesis holds for other ecosystems besides grasslands, then the most vulnerable biological communities are those with the fewest species, for example, huge tracts of land planted only with corn or wheat and treated chemically to kill all other plant and animal species. As you might imagine from your understanding of the principles of ecology and evolution, such a situation seems bound for disaster. And indeed, Peruvian cotton farmers found that DDT helped control cotton bollworms—but only temporarily, until the worms developed resistance to the pesticide.

Their experience led to the first success of *integrated pest management*, a system based on the principles of community ecology and aimed at keeping pest populations below economically harmful levels with a minimum of chemical pesticides. Since then, integrated pest management has been used extensively. Integrated pest management requires a broad view of all interactions between the crop and other species: weeds that compete for sun and nutrients; insects that eat the crop; predators that eat the insect pests; animals that pollinate the crop; and other wildlife living in the area. Pest managers have found that by carefully controlling the timing, spacing, and intermixing of crops, they can enhance the activities of natural predators. What's more, they can introduce biological control agents from other regions that feed exclusively on the pest, including parasitoid wasps and the bacterium *Bacillus thuringiensis* [see CHAPTER 37]. In addition, pest managers can release into the environment millions of sterilized adult male insects, which then mate with wild females and sterilize them. Finally, pest managers use chemical controls only when insects appear poised to do damage above some preestablished level of economic injury. And when spraying is necessary, they choose and apply compounds to hit target species, leaving "innocent bystanders" alone.

Because farmers and ranchers must consider the principles of community ecology if they are to use integrated pest management, such management is more difficult than simply spraying the fields with chemicals. However, since it avoids a whole host of problems—pests evolving resistance, resurgence of the pest populations, and outbreaks of other pest species—integrated pest management, by increasing species diversity in croplands, offers an ecologically and economically acceptable alternative to managing crops and saving the 33 percent that are lost to pests each year.

What Factors Affect Species Richness? Perhaps the greatest present day ecological challenge is the fact that the immense increase in human population is leading to larger and more frequent ecological disturbances. One example is the subdividing of former wildlands, the increase in forest edges, and the burgeoning of cowbird populations that are threatening songbird species. But in the species-rich communities of tropical latitudes, human population pressure is threatening the destruction of the entire ecosystem. Tropical peoples have traditionally cleared land for agriculture by burning the forest, planting crops for a few years, and then moving on when the soil becomes depleted of nutrients [FIGURE 39.20A]. So large are the human populations in these areas now, and so immense are their survival needs, that vast areas of the rain forest are going up in smoke to provide farmland for crops and pastureland on which to graze cattle for export to developed nations. These fires are so huge and so common that they are clearly visible on satellite photos from space [FIGURE 39.20B]. Many ecologists fear that the plant and animal communities of the tropics may not be able to recover from such extreme disturbances and that soon after the year 2000, no tropical rain forests will remain intact. Still worse, many fear that this disruption may cause a distressingly large percentage of all living species to become extinct in our lifetime.

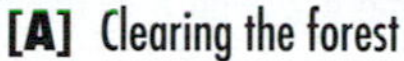

[A] Clearing the forest

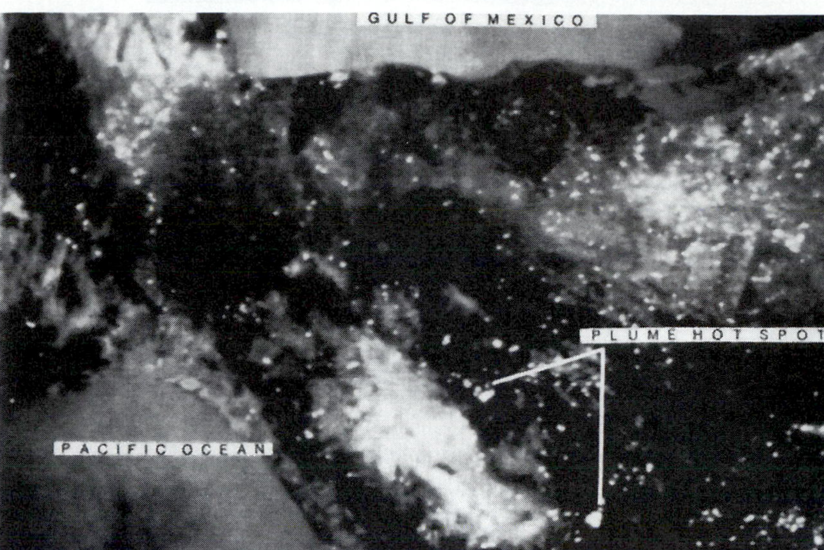

[B] Fires and smoke from burning forests

FIGURE 39.20

Diversity and Destruction in the Tropics.

[A] Like this Mexican farmer of Mayan descent, farmers in the tropics often carry out a traditional method of agriculture: slash-and-burn forest clearing, followed by planting, then moving on to new areas when the soil is depleted. [B] A satellite photo of agricultural areas in Mexico taken in 1984 shows the glowing lights of deliberately set fires, as well as white plumes of smoke. Some represent the annual clearing of established fields, but others are the ongoing destruction of virgin forest.

Ecologists consider it an urgent research priority to learn what makes communities resilient and how they may (or may not) be able to persist in the face of human encroachment. We must solve this problem (discussed in more detail in CHAPTER 41) if our most diverse and interesting communities are to survive through the twenty-first century.

➤ CONCEPT CHALLENGE

The redstart is a threatened songbird with a habitat far from the forest's edge in eastern climax forests. Apply concepts of succession and the ecology of island communities to design ways to help preserve such species.

Connections

John Muir, the great naturalist whose aggressive advocacy of forest preservation led to the forming of many U.S. national parks, made this observation nearly 100 years ago: "When we try to pick out anything by itself, we find it hitched to everything in the universe."* Our exploration of ecology bears out this wisdom. Populations of individual species, such as those we studied in CHAPTER 38, are inevitably "hitched" to other species in communities. Each species has its own habitat and niche, but these are impacted by neighboring species. A Kirtland's warbler, for example, depends on jack pines of just the right age to provide a proper habitat.

We humans are massively disturbing species interactions, not just in the tropics, but in our own climate zone as well, as the decline of colorful songbirds suggests. Yet ecologists still do not understand how community stability is maintained or how it might bounce back. To further understand the process, we must think not only about the living things in a community, but how they interact with the nonliving environment that surrounds them. Our next chapter discusses ecosystems, and how living communities interact with sunlight, temperature, rainfall, nutrients, and other physical aspects of the universe to which all organisms are "hitched."

* John Muir, *My First Summer in the Sierras*, Houghton Mifflin, Boston, 1911.

KEY TERMS

camouflage, 859	**niche,** 852
commensalism, 861	**predation,** 856
community, 851	**predator,** 856
competitive exclusion, 854	**prey,** 856
habitat, 852	**resource partitioning,** 855
interspecific competition, 853	**species richness,** 865
mimicry, 859	**succession,** 863
mutualism, 861	**warning coloration,** 859

HIGHLIGHTS IN REVIEW

1 Interactions between species limit the abundance and distribution of organisms. Beneficial interactions may encourage population growth of both species, while harmful interactions may limit one or both species.

a] An organism's habitat is analogous to its home, the place in the environment where it resides. An organism's niche is analogous to its job, or role, in the community, such as producer, herbivore, or predator.

b] Communities are assemblies of populations of species in a particular area.

c] In interspecific competition—competition between species for the same limited resources—one species' success can cause another's extinction. This is competitive exclusion. Over time, species sometimes respond to the threat of competitive exclusion by resource partitioning (dividing up the territory and coexisting) or by character displacement (evolving new physical and behavioral characteristics).

2 Two species can influence each other's evolutionary fitness, so that both kinds of organisms coevolve.

a] The mutual influence that species have on each other's evolution is called coevolution.

b] Predation, herbivory (plant eating), and parasitism result in one species benefiting and one species being harmed.

c] Predators use strategies such as pursuit or ambush and generally have larger brains in proportion to their body size than their prey. The countermeasures of prey species include camouflage, warning coloration, and mimicry.

d] Parasites live in close association with their prey and generally weaken rather than kill them.

e] In commensal relationships, one species benefits, but the other is usually unaffected. In mutualistic relationships, as in that between the yucca and the yucca moth, both species benefit.

f] A community is a haphazard association of individual species that can survive in an area; it is not a superorganism made up of highly specific groups of plants and animals.

g] After a calamity clears an area, there is a succession of communities, including a pioneer community of hardy, wind-dispersed species; a transition community of more slowly dispersed species; and eventually a climax community, a relatively stable assemblage of plants, animals, and other organisms that may change when the climate changes.

h] An area's species richness is dictated partly by latitude (the closer to the equator, the greater the available sunlight, and the richer the assemblage of species), partly by isolation (islands and peninsulas have fewer species than mainlands), and partly by the frequency and intensity of disturbances.

3 Interaction with the human species is the most powerful biological factor in the world today. Few communities are untouched by our species' all-pervasive influence, and we have begun—and will continue—to strongly shape the earth's ecological communities.

a] The high level of disturbances caused by people is now endangering large numbers of species in the tropics and the overall ecological stability of the zone.

UNDERSTANDING THE FACTS AND CONCEPTS

For Questions 1–5 match each description with the most appropriate item or items from the following list. Any answer may be used once, more than once, or not at all.

a] community　**c]** fundamental niche
b] habitat　**d]** realized niche

1 The potential range of a species within a habitat.

2 The actual range of a species within a habitat.

3 Terms such as "herbivore" and "primary producer" apply to this category.

4 The geographic location of a particular species in a community.

5 A collection of populations in a particular place at a particular time.

For Questions 6–10, match each description with the most appropriate item or items from the following list.

a] mutualism　**c]** competition
b] commensalism　**d]** predation

6 A term that can be used to describe the behavior of pathogenic parasites as well as fast-running hunting carnivores.

7 A type of interspecific interaction exemplified by tree-dwelling animals that are of no benefit to the tree.

8 The usual relationship between plants and pollinating animals.

9 Characterized by interspecific interactions that may lead to resource partitioning and coevolution on the one hand, or death on the other.

10 Often exemplified by character displacement, limiting the realized niche.

For Questions 11–15, complete each statement with the most appropriate term from the following list.

a] character displacement　**d]** social parasitism
b] resource partitioning　**e]** competitive exclusion
c] mimicry

11 Competition in which the numbers of one of the interacting species is reduced to zero is called ______.

12 If two competing species limit each other's fundamental niches to separate and smaller realized niches, the phenomenon of ______ has occurred.

13 Adaptive radiation, coevolution, and speciation may all result from ______ when two or more strongly competing species have overlapping habitats.

14 The exploitation of genetic behavior patterns of a prey species characterizes ______.

15 It shows ______ when one prey species exhibits the warning coloration of another species.

For Questions 16–20, match each description with the most appropriate item from the following list. Any answer may be used once, more than once, or not at all.

a] succession
b] pioneer community
c] climax community
d] species richness
e] keystone species

16 An initial group of species that affect conditions for future arrivals.

17 A natural series of communities that occupy a particular geographic location in turn.

18 Generally the most stable and long-lasting assemblage of species in a particular location.

19 Degree of diversity of populations in a habitat.

20 A species, often a predator, whose activities determine the fundamental structure and richness of a community.

21 Explain the cause of the fluctuations in the populations of lynx and hare shown in the graph.

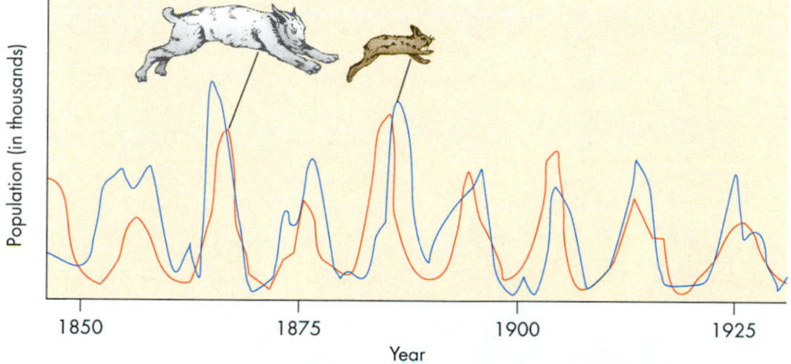

INTEGRATE AND APPLY WHAT YOU HAVE LEARNED

1 Explain the difference between an organism's habitat and its niche.

2 Exclusion or elimination of one of a pair of competing species and coevolution of both species are both possible results of competition. How can this be?

3 Are predator-prey relationships always characterized by the consumption of the prey by the predator?

4 What are the primary strategies of predatory species? Of prey species?

5 The islands of Hawaii and Iceland both encompass volcanic regions where pioneer communities pave the way for successive communities. Which region would exhibit the fastest rate of succession and the greatest species richness?

ANALYSIS

1 In an apple orchard, one or two apple-sized pheromone-impregnated sticky disks are hung on each tree. The males of a particular insect pest species are attracted to the disks, and cannot break free from their sticky coating. Periodically the farmers check the disks and count the number of trapped insects. Which of the reasons given below best explains why the farmers hang the disks?

a] The disks trap all of the insects in the orchard, thereby eliminating the pests.
b] Trapping some males means that fewer eggs will be fertilized.
c] The traps enable the farmers to estimate the relative abundance of the pest species at a particular time so that they can minimize the use of chemical pesticides.
d] The farmers expect the pest species to attack the disks rather than the apples.
e] The traps change the habitat so the pests will go elsewhere.

2 Which of the following results would you predict after large predatory carnivores such as wolves and mountain lions have been eliminated from a community containing both carnivores and deer? More than one answer may be correct.

a] The deer population would initially boom.
b] The number of deer would eventually crash as a result of overgrazing and starvation.
c] The number of deer would remain unchanged because the same number that had been killed by the predators would now die of disease.
d] The number of deer would initially decline because the keystone species had been removed.

3 With time, the size of the brain of predators, relative to size of the body, has grown larger. The same has also been true of prey species. However, 60 million years ago, predators' brains were significantly larger than their prey, whereas modern predators have brains only slightly larger than their prey. Which of the following best explains why this occurred? More than one answer may be correct.

a] The ancient prey that survived were probably the smartest of the species, thereby causing strong directional selection toward larger brain size.
b] The predators did not need such large brains to survive, so there was no selection pressure for larger brain size in that group.
c] Predators have changed their preferences for particular prey species with time; more recently, predators have preferred to hunt the smarter herbivores.
d] The prey have modified their diets and have turned into predators.

4 When farmers first begin to use pesticides, crop yields often increase dramatically for a few years but then decline to levels below those obtained before pesticides were used. Why? More than one answer may be correct.

a] Pesticide-resistant strains of insects might develop.
b] The pesticides might also kill pest predators.
c] The pesticides became weaker with time.
d] The plants became resistant to the pesticides.

CHAPTER 40

Ecology of Ecosystems

A SHRINKING SWAMP

Wander out onto the elevated boardwalk above the swampy sawgrass marshes of Everglades National Park in South Florida, and you'll get a close-up view of a rare wetland environment. You'll probably see snake-necked anhingas drying their huge black and white wings in the sun [FIGURE 40.1]. You may see ibises, storks, spoonbills, and herons stabbing their bills into the shallow, mucky water in search of fish or frogs. With luck, you'll see a 3-m (12-ft) alligator or two, sitting motionless—eyes unblinking, mouths agape—along a muddy river bank. As you look out across the sharp-edged grass and the islands of cypress, pines, live oaks, and palmettos dotting this swampy wilderness, you might imagine it unchanged over thousands, even millions of years. In fact, however, the current Everglades are just a remnant of a once-vast wetlands system that people have altered—some say destroyed—within the past 40 years.

The Everglades is an **ecosystem**: a community of organisms interacting within a particular physical environment. (Recall that a community is a group of organisms interacting with each other. The study of ecosystems includes the interaction of communities with their physical environments.) Until the late 1940s, the Everglades ecosystem began at the southern shores of Lake Okeechobee, America's second largest inland lake, located in the center of the Florida peninsula about 160 km (100 mi) northwest of Miami. The region stretched from the Atlantic Coast to the Gulf Coast and from Lake Okeechobee to the Florida Keys, and it covered 11,500 km^2, or 4 million acres. This huge zone of porous limestone covered by thin soil and shallow water looks utterly flat, but it actually slopes very gradually (about 4 cm/km) southward, and is scoured by two wide river beds, both carrying water from the lakes and rivers north of the Everglades to outlets along the Gulf Coast.

FIGURE 40.1
Anhinga Sunning: One Strand in the Everglades Ecosystem.

Traditionally, these natural channels carried most of the ecosystem's surface water during the dry season (November to April), and evaporation turned the wide sawgrass marshes into isolated shallow pools and then dried mudflats. During the rainy season, however (May to October), water overflowed the sloughs, spread out in a wide, thin sheet, and moved very slowly southward like a shiny ribbon a few centimeters deep and several kilometers wide. Most Everglades plants and animals evolved with adaptations to these fluctuating water levels and wet-dry cycles. Despite the great glut of water in this drainage area, very little water ever reached the Gulf of Mexico: most evaporated or was transpired by plants, formed clouds, blew northward on sea breezes, and rained down on the lakes and rivers that fed the drainage system. The massive rain-driven water cycle thus regulated the Everglades ecosystem.

Starting in 1948, however, the U.S. Army Corp of Engineers devised a plan to drain much of the Everglades to create agricultural land and room for housing tracts. They also wanted to regulate water flow more evenly from north to south and season to season, and to bring in exotic tree species like the Australian melaleuca to help dry up the swamps. By 1967, the project was in full operation, and by 1971, the Corp had drained 65 percent of the Everglades' original marshes and reduced the entire ecosystem by 2 million acres (or 50 percent). In the wake of this land reform, increased levels of nitrogen and phosphorus from agricultural runoff have changed the water quality throughout the Everglades. Changes in water quality and quantity have, in turn, led to the dwindling of many plant and animal populations, including the Florida panther and the snail kite, a small hawk with pointed wings and graceful, gliding flight. Other species, such as melaleuca trees and cattails, are spreading out of control. And as a result of such changes, the area's food chain or food web—the flow of energy and nutrients from one organism to another—has been greatly altered. Many environmental groups have lobbied for major restoration of the wetlands, but James Kushland of the University of Mississippi, an expert on Everglades ecology, is pessimistic. Some restoration is possible, he writes, but "it is unlikely that any close resemblance to the pristine system could be recreated."*

In this chapter, we will examine many ecosystems, the pathways of energy and materials through them, and how plants and animals are adapted to and interact within their physical environment. We will also discuss the many ways in which people have drastically affected the flow of energy and the cycling of material in biological communities. Ultimately, such changes are likely to endanger us and the other living organisms that share our planet.

We will address these topics: How energy flows and materials are cycled through ecosystems; how the energy in an ecosystem changes from one form to another; how crucial materials like carbon and nitrogen are cycled in ecosystems; and how human activities have altered the earth's ecosystems.

Along the way, we will learn about the history of DDT; agricultural runoff and water pollution; the greenhouse effect; and ways that people are attempting to erase some of the ecological damage we have caused. ❑

MESSAGES

1 An ecosystem includes all the living organisms in a given area (the community), the nonliving physical environment around them, and the interactions of both.

2 Energy flows through an ecosystem in a one-way path. Energy in the form of sunlight enters living things via photosynthesis, passes from one organism to another, and finally escapes back to the physical environment.

3 Materials cycle through ecosystems. Carbon atoms, for example, pass from the atmosphere into plants, into animals that eat plants, and finally back to the atmosphere as carbon dioxide where plants once more absorb it.

4 Because the organisms in an ecosystem require energy and materials, they are dependent on each other and on the physical environment.

5 Human activities can drastically affect the health of ecosystems by altering the flow of energy and the recycling of materials.

* J. A. Kushlan, "The Everglades," in *The Rivers of Florida*, Robert J. Livingston, ed., Springer Verlag, New York, 1991, chap. 8, p. 139.

Pathways of Energy and Materials

Organisms need energy and materials for growth, maintenance, and repair, and, generally, to overcome the universal tendency toward disorder we discussed in CHAPTER 4. We humans are no exception: just think how often each day we take in energy and materials in the form of food, water, air, and so on. In all ecosystems—whether the marshlands of the Everglades, or more familiar forests and grasslands—the nonliving environment is the ultimate source of both energy and materials. The pathway from the physical environment through various kinds of ecosystems, however, can differ greatly.

ENERGY FLOW AND MATERIAL CYCLING

Useful energy flows through ecosystems in a one-way path, entering living things from the physical world, passing from one organism to another and eventually escaping back to the physical environment in a less useful form [FIGURE 40.2]. In most ecosystems, energy comes from sunlight, generated by nuclear chain reactions in the sun. In all ecosystems, energy eventually escapes from living organisms into the surrounding physical world, usually as heat. Heat energy passed into the environment is lost forever to the living world, since it is such a diffuse form of energy that it cannot be used by living organisms.

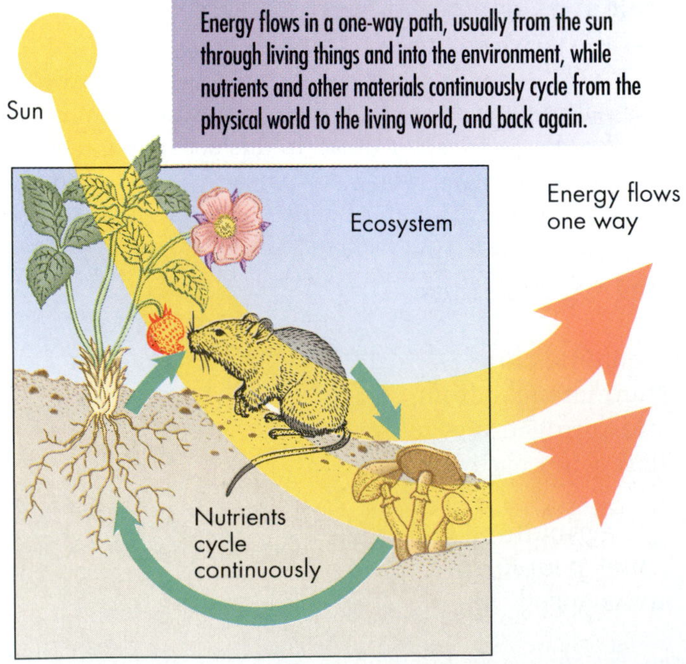

FIGURE 40.2

Pathways of Energy and Materials in Ecosystems.
Energy flows in a one-way path, usually from the sun through living things and into the environment, while nutrients and other materials continuously cycle from the physical world to the living world, and back again.

In contrast to the one-way flow of energy through ecosystems, materials *cycle* through ecosystems. For example, carbon and oxygen atoms move from one organism to another via consumption, the release of wastes, and other ways we will discuss in this chapter. The materials eventually reenter the physical environment, and then move through the organisms again [see FIGURE 40.2]. In the Everglades, the carbon in atmospheric carbon dioxide is picked up by extremely plentiful algae called *periphyton* that cling to submerged plants. Within the algae cells, the carbon is captured in carbohydrates and other molecules. When snails, fish, and other animals eat the algae, some of the carbon becomes incorporated into their bodies, while the rest is excreted or exhaled in other compounds. Carbon-containing compounds may enter birds, alligators, or other predators who consume the fish, and eventually, after the predators die and their tissues are broken down by bacteria and fungi, the carbon will be recycled as carbon dioxide back into the water and in time, into the atmosphere.

Despite the fact that materials are recycled in ecosystems while energy flows in a one-way direction, the capture of energy and the acquisition of materials from the environment are usually coupled, and organisms obtain them in one of two basic ways: (1) *autotrophy* (self-nourishing) and (2) *heterotrophy* (other-nourishing). The strategy each species exhibits for obtaining nutrients defines its place within the community.

FEEDING LEVELS

In any ecosystem, autotrophs like plants, certain protists, and some bacteria are the primary **producers**; they are the only organisms that can produce all the biological molecules they need for their growth from nonliving substances. Heterotrophs, on the other hand, such as animals, fungi, and many kinds of microbes, obtain their energy, their carbon, and other atoms by consuming autotrophs or other heterotrophs. Heterotrophs, therefore, are an ecosystem's **consumers**. Ecologists assign every organism in a community to a **trophic level**, or feeding level, depending on whether it is a producer or a consumer, and depending on what it eats [FIGURE 40.3].

Producers At the lowest trophic levels lie the producers—the autotrophic bacteria and plants that support all other organisms directly or indirectly. In most terrestrial ecosystems, green plants are the producers. Some examples of primary producers in the Everglades are periphyton algae, sawgrass, and cypress trees. By collecting solar energy and carbon dioxide, the primary producers build energy-rich sugar molecules—the main source of carbon atoms in all types of biological molecules [review FIGURE 6.7]. Producers also absorb nitrogen, phosphorus, sulfur, and other necessary atoms from the environment

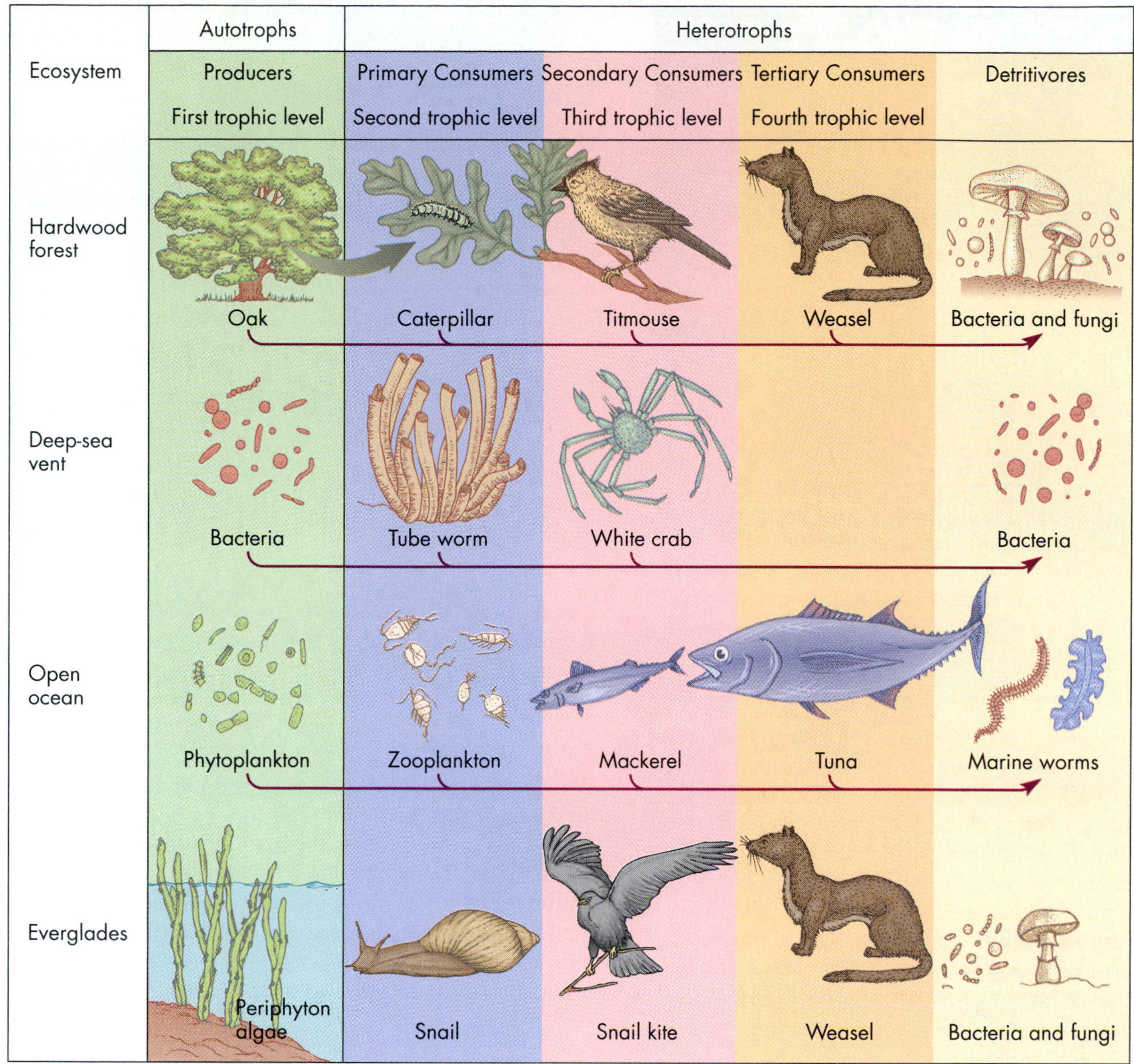

FIGURE 40.3

Feeding Levels and Food Chains.

Organisms vary from one ecosystem to another, but in each ecosystem, producers form the first trophic level, while consumers form the next few trophic levels, and detritivores utilize the wastes from all levels.

and fix them into biological molecules. Ecologists say that producers provide both the *energy-fixation base* and the *nutrient-concentration base* for the entire ecosystem.

While plants are the major producers in most ecosystems, less than 20 years ago scientists discovered a fantastic ecosystem that lacks plants altogether. They found that certain regions at the bottom of the ocean are dotted with submarine hot springs called *deep-sea vents*. Around these vents live yellow mussels, giant clams, pure white crabs, and bizarre worms the size of your arm, encased in white tubes with frilly red plumes waving in the slow current [FIGURE 40.4]. In this ecosystem, certain bacteria are the sole producers, since no light penetrates to these inky depths and plants cannot survive. These producers harvest the energy stored in the bonds of hydrogen sulfide (H_2S) molecules dissolved in the water. They use this energy to fix carbon dioxide from seawater into sugar molecules and then incorporate the carbon from the energy-rich sugars into the other biological molecules they require for growth and reproduction.

Primary Consumers At the second trophic level are the **primary consumers**, the organisms that eat the producers. Herbivores, such as caterpillars and deer, efficiently digest plant matter for energy and serve as ecological links between the producer level and all other levels [see FIGURE 40.3]. Snails grazing on periphyton algae are primary consumers within the Everglades ecosystem.

FIGURE 40.4

A Strange Deep Ocean Community Not Powered by Sunlight.

Heat energy from hydrothermal vents support an ecosystem that includes yellow vent mussels, ghostly white crabs, large vent clams, and tube worms with red plumes. These plumes take in oxygen and distribute it to bacteria in the worm's body; these bacteria are primary producers that obtain energy from hydrogen sulfide and use it to make organic molecules from carbon dioxide spewed forth from the hot springs.

The huge tube worms of deep-sea vents [review FIGURE 40.4] act as a special kind of primary consumer. Although they lack a mouth or other way of eating, they have a special organ packed with bacteria. Red blood coursing through the worms' scarlet plumes delivers hydrogen sulfide to the bacteria, which in turn produce sugars that the tube worms use for energy and carbon.

Secondary Consumers At the next highest level, **secondary consumers** are carnivores (meat eaters) that consume the herbivores. A titmouse eating a caterpillar is a carnivore, and so is an Everglades' snail kite eating a snail.

Tertiary Consumers In the next trophic level, **tertiary consumers** are carnivores that eat other carnivores; for example, a tuna may eat a mackerel, and a weasel may eat a titmouse. Finally, a few ecosystems have one more trophic level containing carnivores that eat tertiary consumers, such as Florida panthers eating weasels and sharks eating other sharks.

Decomposers A special class of consumers, the **detritivores**, or decomposers, obtain energy and materials from **detritus**, organic wastes and dead organisms that accumulate from all trophic levels. Fungi and bacteria are the most ubiquitous detritivores, but worms, nematodes, many kinds of insects, and carrion feeders like vultures are also important detritus consumers.

The simplified diagram in FIGURE 40.3 suggests that each trophic level leads directly to the next in a simple chain, and indeed, **food chains** do exist in nature with groups of organisms involved in linear transfers of energy from producer to primary, secondary, and tertiary consumers. More commonly, however, feeding relationships resemble not chains, but complex interwoven webs.

FEEDING PATTERNS IN NATURE

Organisms usually consume more than one other species, and some animals feed at several trophic levels. A high-level carnivore like a Florida panther may eat herbivores, such as rabbits, as well as other carnivores, such as snail kites. And as an omnivore [see CHAPTER 24], you yourself eat vegetables (producers), poultry (herbivores), tuna fish (carnivores), and mushrooms (detritivores). Ecologists call complicated interconnected feeding relationships **food webs** [FIGURE 40.5]. In a hardwood forest ecosystem in New Hampshire, for example, the major producers are sugar maple, beech, and yellow birch trees. These producers support two main food webs: (1) a *grazing food web* that stems directly from the living plants and (2) a *detritus food web* that begins with dead plant parts and animal wastes.

In the grazing food web, some herbivores, like jays and chipmunks, consume fruits and seeds, while other herbivores, such as mice, deer, and caterpillars, graze on leaves. These primary consumers then fall prey to hawks, skunks, snakes, and other carnivores. In the detritus food web, fungi and bacteria break down dead animals and fallen leaves, twigs, and fruits, and are eaten in turn by grubs and earthworms. Grazing and detritus food webs are themselves linked, as detritivores become food for other consumers, such as salamanders and shrews.

Trophic levels and food webs describe the general pathways of energy flow and material cycling in ecosystems. They do not, however, portray the *amounts* of energy or materials that pass through each level. To fully understand how ecosystems function, ecologists measure the energy transfers.

➤ CONCEPT CHALLENGE

Choose an ecosystem near your house—a park, an abandoned lot, a corn field, a pond, or any other familiar place—and diagram the basic food web of that ecosystem.

Energy Flow Through Ecosystems

Whether an organism is a producer or a consumer, it needs energy for movement, for active transport of nutrients and ions, and for synthesis of proteins, nucleic acids, and other large molecules needed for growth and repair [see CHAPTER 4]. Producers obtain their energy directly from the environment in the form of light (in most ecosystems) or inorganic molecules (in deep-sea hot springs and a few other ecosystems). Consumers, however, can get their energy only from producers. Hence,

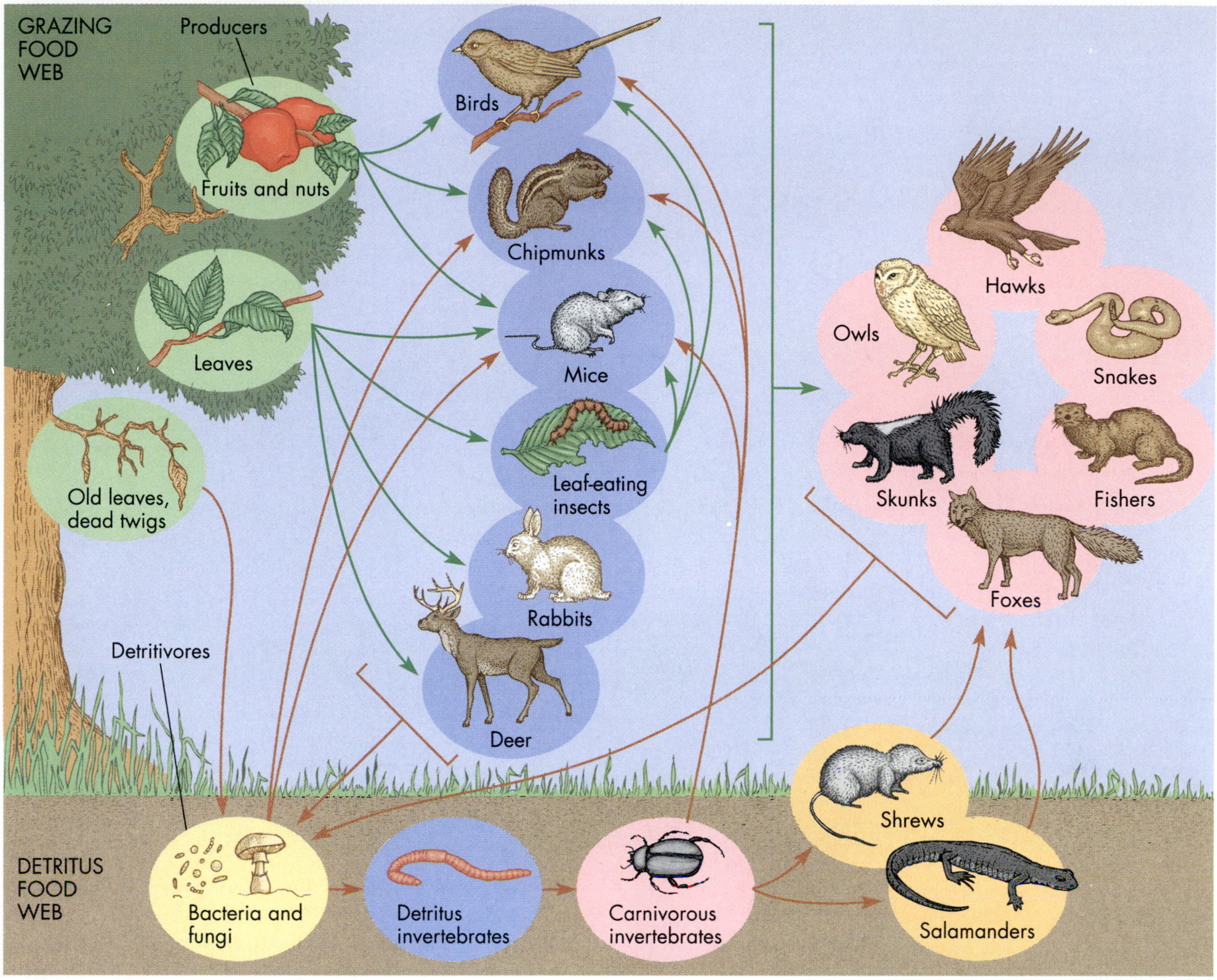

FIGURE 40.5

A Food Web: Cross-Dependencies in the Living World.

In the Hubbard Brook Forest ecosystem, dozens of species derive energy from each other in a complex grazing food web, while others participate in the detritus food web. Both webs are linked at numerous points by the activities of specific organisms, only a few of which are shown here.

the activities of a community's producers set a limit for the amount of energy that can be captured and channeled throughout the entire ecosystem. Ecologists have closely studied the hardwood forest ecosystem to measure available energy and how it is spent.

ENERGY BUDGET FOR AN ECOSYSTEM

Within the White Mountains of New Hampshire lies the Hubbard Brook Experimental Forest, a fragrant zone of tree-studded rolling hills that has engaged researchers' attention for more than 25 years [FIGURE 40.6]. Researchers have focused on a few small *watersheds*

FIGURE 40.6

Hubbard Brook Experimental Forest: A Living Laboratory for Studying Energy Flow in Ecosystems.

Ecologists have cut the trees from bands and zones of the Hubbard Brook Forest to determine the effects of logging on soil erosion and nutrient loss, and have monitored energy flow and material cycling through hundreds of experiments.

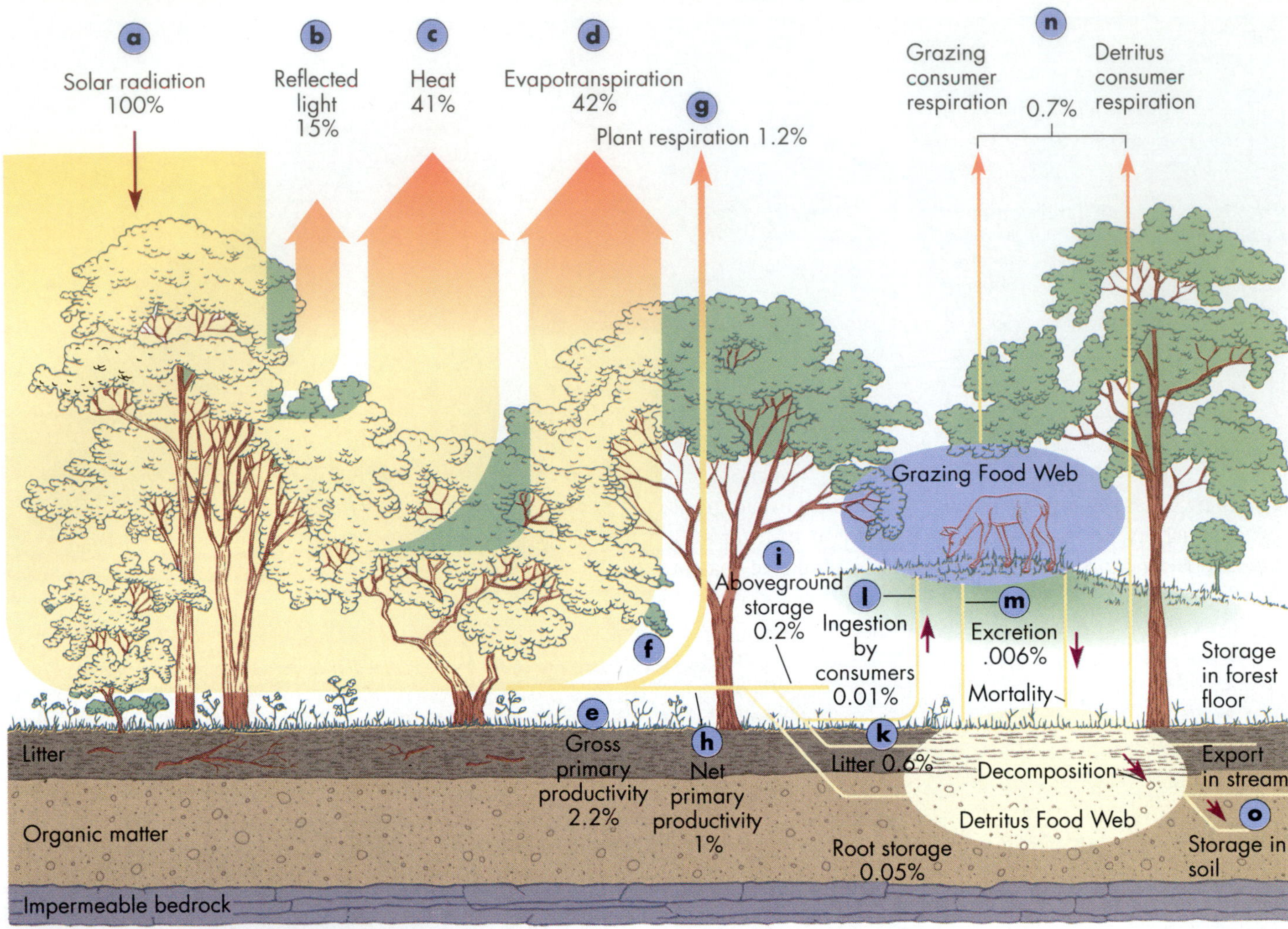

FIGURE 40.7

Energy Budget of a Hardwood Forest.

Energy flows through the Hubbard Brook Experimental Forest ecosystem beginning as sunlight strikes the forest; then light and heat are returned to the environment, and a small fraction of the original energy is fixed in organisms and their waste products.

(regions drained by a single stream or river), because the ridges that separate watersheds provide convenient dividing lines between adjacent ecosystems. Since impermeable bedrock lies below each watershed, researchers can measure all the groundwater leaving the ecosystem (as well as its load of dissolved nutrients) by monitoring the single stream that drains it. By studying the input of energy, water, mineral nutrients, and organic matter into these watersheds and tracing their incorporation into both living organisms and the physical environment, workers have learned how energy flows through an entire forest.

Over the course of the growing season (June through September), a total of about 500,000 kcal of solar energy, including both heat and light, strikes each square meter (10.76 ft^2) in the Hubbard Brook Forest [FIGURE 40.7a]. (To put this in perspective, recall that you burn about 100 kcal jogging a mile.) About 15 percent of the sun's radiant energy striking the forestland immediately reflects back into the atmosphere as light [FIGURE 40.7b]. Another large fraction (41 percent) warms the ground and the photosynthesizing plants and eventually radiates back to the atmosphere as heat [FIGURE 40.7c]. Still more of the incoming energy (42 percent) is used in evaporating water from the soil and cells of plant leaves, a combined process called *evapotranspiration* [FIGURE 40.7d; see also FIGURE 35.5].

Thus over 83 percent of the solar energy that reaches the hardwood forest flows through as heat and another 15 percent as light—for a total of 98 percent that returns rapidly to the physical environment. The 2 percent or so remaining is the amount of energy that producers convert by photosynthesis to chemical energy in the form of sugars and other organic compounds. Ecologists call this small fraction an ecosystem's *gross primary productivity* [FIGURE 40.7e], and, ultimately, it limits an ecosystem's structure, including how many birch trees will grow and how many chipmunks will thrive.

Not all the chemical energy that a plant initially traps is stored in newly formed leaves, roots, and fruits. Plant cells themselves use a little more than half of this energy to fuel their own cellular respiration, eventually losing 1.2 percent as heat [FIGURE 40.7f and g]. The amount of energy remaining after respiration is called the *net primary productivity*, the amount of chemical energy that is actually stored in new cells, leaves, roots, stems, flowers, and fruits [FIGURE 40.7h]. Of all the energy impinging on the ecosystem, only the net primary productivity—that is, about 1 percent of the light energy striking the forest—is available to consumers.

During a growing season, plants retain some of the net primary productivity in permanent organs (new stems and roots, for example; FIGURE 40.7i), but most becomes litter on the forest floor [FIGURE 40.7k]. In fact, in a forest nearly twice as much energy is stored in litter and decomposing humus as in the majestic banks of leaves overhead. Most of the energy contained in the chemical bonds of the organic compounds in the litter fuels the detritus food web. Only a small fraction of the energy stored aboveground in the forest enters the grazing food web [FIGURE 40.7l], and some of that enters the detritus food web owing to the excretion and death of consumers from the grazing food web [FIGURE 40.7m]. Energy dissipation continues as the consumers that graze on plants or eat detritus radiate heat via respiration [FIGURE 40.7n]. Only a tiny amount of energy then remains in the soil or exits the ecosystem in the stream that drains the watershed [FIGURE 40.7o]. Clearly, despite the essentially limitless power flowing from the sun to the earth, plants can store only a small fraction of that energy in organic compounds, and that fraction sets an upper limit to the energy available to all other organisms in the ecosystem.

The information in FIGURE 40.7 allows ecologists to formulate three general principles about the energy budget of a hardwood forest like Hubbard Brook: (1) Even in a lush, leafy green forest, plants and other producers convert only a small fraction (2 percent or less) of the solar energy that enters the ecosystem into stored chemical energy. (2) Animals ingest an even smaller amount (in this case, 0.01 percent) of the energy in the grazing food web. (3) As energy flows through the trophic levels of the ecosystem, metabolic activities (mostly respiration) release it back into the air, where it ultimately returns to space as heat, a form of energy that does little or no work. Heat lost to space is not recycled back into the sun or other stars.

PYRAMIDS OF ENERGY AND BIOMASS

Ecologists have discovered another important fact in their experiments on energy flow from one trophic level to the next: Food chains on land rarely have more than four or five links (excluding parasites and detritis food webs), and ocean food chains based on plankton are generally limited to about seven steps. Most evidence suggests that food webs are generally rather short. However, ecologists are still actively researching the extent and functioning of food webs. There is a strong current focus on aquatic food webs involving many microorganisms.

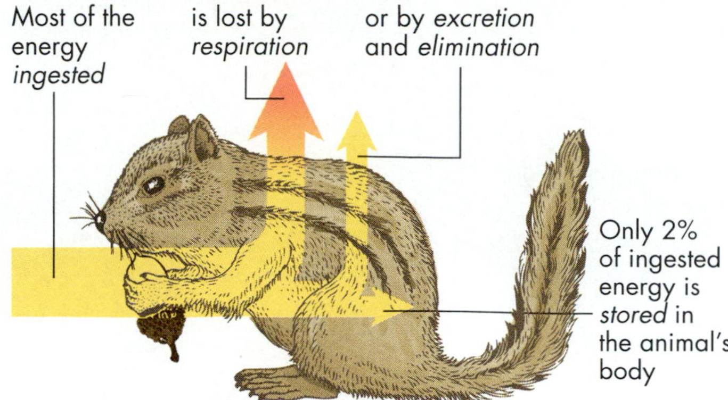

FIGURE 40.8
Energy Budget of a Forest Herbivore.
Energy flows through individual consumers like this chipmunk, as well as through the forest ecosystem it inhabits.

To see why food chains tend to be so short, consider the fate of the energy that enters the grazing food web. A chipmunk, for example, one of the most important herbivores of the Hubbard Brook Forest [FIGURE 40.8], ingests potential energy in the form of seeds or leaves, assimilates some of the energy, excretes some in liquid wastes, and eliminates some in solid wastes. Most of the assimilated energy is used in aerobic respiration, and allows maintenance and repair of the chipmunk's body. Chipmunks store less than 2 percent of ingested food energy in new tissues or offspring. The consequences of a huge loss through respiration and a small net increase in growth is that chipmunks store very little energy in a form that can be used at the next trophic level, say, by a red fox.

Energy Pyramids Ecologists portray the energy relationships of different trophic levels in an **energy pyramid**, a diagram whose building blocks are proportional in size to the amount of energy available from the level below. FIGURE 40.9 shows the energy pyramid for a river ecosystem in Silver Springs, Florida, just a few hours drive from Everglades National Park. This river system includes eelgrass and algae (producers); turtles, snails, and caddisflies (herbivores); and beetles, sunfish, and bass (carnivores). At each trophic level, the energy stored by the organisms is substantially less than that of the level below it. Ecologists have constructed similar energy pyramids for hundreds of other ecosystems.

Biomass Pyramids The stepwise energy decline evident in such pyramid diagrams explains why food chains

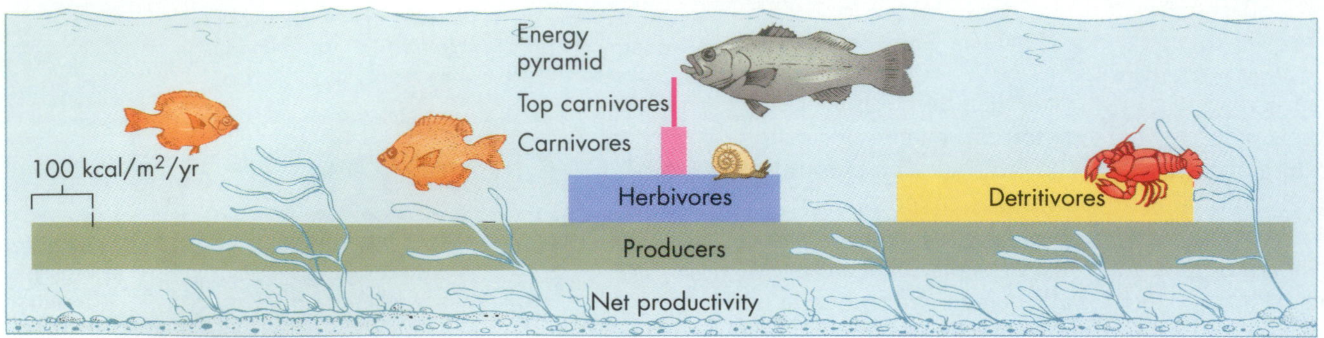

FIGURE 40.9

Energy Pyramid in a Florida River Ecosystem.

The energy pyramid for this ecosystem shows that the net primary productivity (the amount of energy fixed during photosynthesis minus the amount used during cellular respiration in plants) of all producers, including eelgrass, is about ten times that of all primary consumers (herbivores), including river snails; which is about ten times that of all secondary consumers (carnivores), including sunfish; which is about ten times that of all top carnivores, including bass. The net productivity of detritivores is larger than that of herbivores.

usually have only four links: The small amount of energy available at the top is too difficult to collect. A handy way to measure the diminishing returns is through **biomass**: the dry weight of organic matter at a particular level. By collecting the eelgrass and other plants from a 1 m^2 (10.76 ft^2) patch on the river bottom in the Silver Springs ecosystem, for example, then drying them to remove their water content and weighing the remaining material, researchers found a biomass of 809 g/m^2, or about 3 oz/ft^2 [FIGURE 40.10].

If the researchers went on to collect all the consumers at different trophic levels in the water above that 1-m^2 patch and determined their biomasses in a similar way, the data would appear as a **pyramid of biomass**, with each trophic level containing, as a rough approximation, only about 10 percent of the biomass in the level just below it. By moving up four levels (or four links in a given linear food chain), and assuming the 10 percent reduction per link, one finds very little biomass in the top carnivores, such as large, meat-eating bass. Taking it one step further (and another 10 percent reduction in biomass), an angler would need to catch and eat 10 kg (22 lb) of bass to put on 1 kg (2.2 lb) of human tissue. The 10 kg of bass tissue would have come from 100 kg (220 lb) of smaller carnivorous fish, 1000 kg (2200 lb) of insect herbivores, and 10,000 kg (about 10 tons) of plants. All that for 1 kg (2.2 lb) of human being!

The concept of energy and biomass pyramids accounts for the phrase, "Eat low on the food chain." This refers to the fact that it takes 10 kg of grain to build 1 kg of human tissue if the person eats the grain directly, but it takes 100 kg of grain to build 1 kg of human tissue if a cow eats the grain first, and the person eats the beef. Eating lower on the food chain—eating producers, not consumers—saves precious resources on a small planet. (CHAPTER 28 discusses vegetarian diets in more detail.)

Biological Magnification The pyramid pattern of energy flow in ecosystems has an important ramification: **biological magnification**, the tendency for toxic substances to increase in concentration in progressively higher levels of a food chain. Many chemical insecticides, such as DDT and chlordane, resist degradation in the environment and, if they are ingested, tend to be stored in body fats. If farmers spray DDT on their cabbage plants to control cabbage looper caterpillars, some of the chemical inevitably runs off into streams and lakes. There, instead of breaking down, some of it may enter water plants later eaten by herbivorous fish [FIGURE 40.11]. Fish and other animals cannot break down or excrete the toxin, and it is instead stored in their body fats.

The magnification continues because a carnivorous fish such as a pike must eat many times its weight in herbivorous fish, and the pike stores all the DDT from the many smaller fish it eats. Likewise, when a top carnivore such as an osprey devours many pike, the bird stores the DDT present in all the pike tissue it consumes.

The concentration of DDT in the bird's body can reach levels 10,000 times greater than in the water plants that originally took it up. This amount of DDT can interfere with calcium metabolism and result in thin-shelled, easily broken eggs. During the two decades or so that DDT was commonly used in the United States (early 1940s to late 1960s), the numbers of ospreys, falcons, hawks, eagles, and condors dropped significantly. Although the use of DDT was banned in 1968, the chemical is still widely used in overpopulated and developing nations to control mosquitoes that spread malaria—the disease that kills more humans each year than any other disease. DDT continues to magnify in the food chain as a consequence of the inefficient energy transfers that also lead to ecological pyramids.

Although many toxic chemicals cannot be metabolized and their component recycled, compounds containing nitrogen, sulfur, phosphorus, and other elements are broken down and these essential nutrients are recycled, a process necessary to the long-term health of ecosystems.

FIGURE 40.10

Pyramid of Biomass.

The biomass—the amount of living tissue—decreases at each trophic level in many ecosystems.

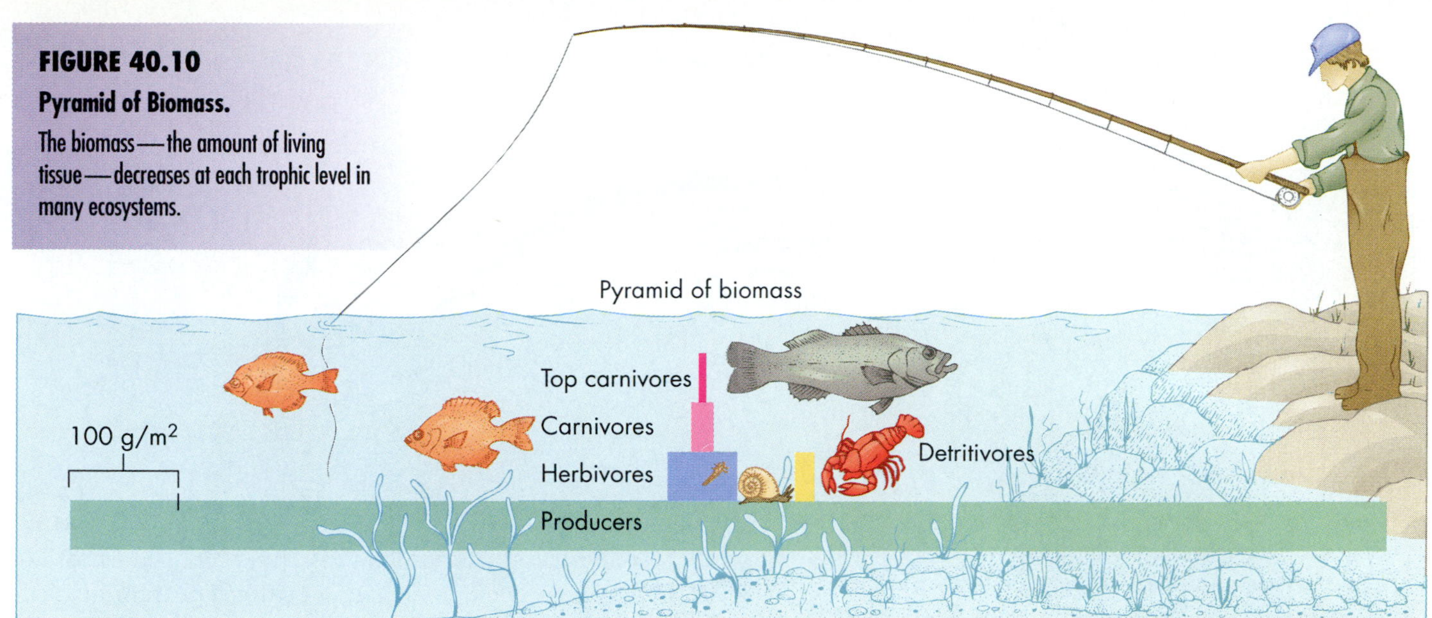

➤ **CONCEPT CHALLENGE**

Florida panthers in the Everglades, and cougars in many other parts of the United States, are the top carnivores. On the basis of pyramids of energy or biomass, explain why no *larger* beast preys exclusively on panthers or cougars.

How Materials Cycle Through Ecosystems

In contrast to the essentially inexhaustible stream of solar energy striking our planet, the physical materials on which life depends are limited to what is currently present in the earth's soil, water, and air. Therefore, these materials must be recycled for ecosystems to survive. Our planet has several *biogeochemical cycles* (literally, "life-earth-chemical" cycles), global loops of material utilization. In these cycles, substances enter organisms from the atmosphere, water, or soil, reside temporarily in those organisms, then eventually return to the nonliving world when the organisms respire or decompose. Repositories of materials, including the atmosphere, the soil, and living organisms, are called **pools**, or reservoirs, and in general, the organismal pool is much smaller than the nonliving pool. Like the level of water in a leaky bathtub with a leaking faucet, pools remain constant in size as long as a substance's rate of entry equals its rate of departure.

FIGURE 40.11

Biological Magnification.

DDT becomes concentrated in a pyramid of producers and consumers. The DDT that a farmer sprays (red dots) may protect cabbages from caterpillars, but it runs off into nearby streams, is taken up and incorporated into water plants, and continues to accumulate in each higher level of the food chain, since living things cannot metabolize the hydrocarbon pesticide completely. High accumulations in predatory birds such as osprey result in fragile eggshells and in declining populations of these beautiful animals.

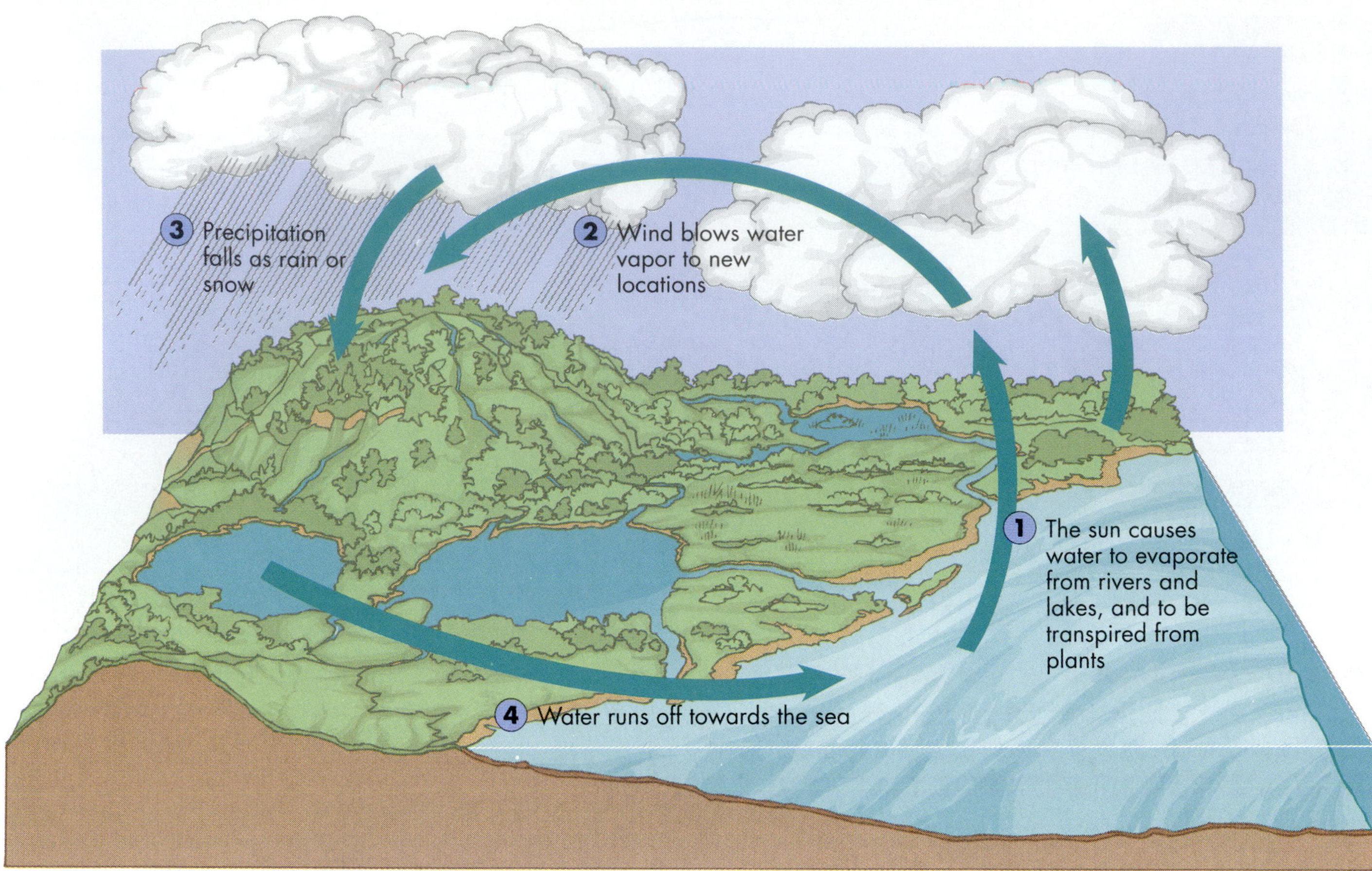

FIGURE 40.12

The Water Cycle.

A sun-driven global exchange involving evaporation, precipitation, runoff, and transpiration cycles water from atmosphere to earth's surface to plants and back again.

The most important biogeochemical cycles are those for water, nitrogen, phosphorus, and carbon. These underlie the health of all ecosystems, but human activities can alter them in profound and often undesirable ways. The following sections describe those crucial cycles.

THE WATER CYCLE

In the global **water cycle** (or *hydrologic cycle*), water moves from the atmosphere to the earth's surface as rain or snow, and back again to the atmosphere by evaporation from puddles, ponds, rivers, and oceans and by transpiration (the evaporation of water from plants); all of this evaporation is driven by energy from the sun [FIGURE 40.12]. In some terrestrial ecosystems, more than 90 percent of the moisture passes through plants and is transpired from their leaves, and only 10 percent evaporates directly from surfaces in the ecosystem. In systems like these, the plants create their own rain: Moisture moves from plants to air to clouds and back to earth in the form of rain wherever the clouds are blown.

The draining of the Everglades to provide dry land for farms and communities has immensely altered the water cycle in that ecosystem. Formerly, water evaporated or was transpired by plants as it flowed slowly southward in its 50-mile-wide ribbon. Clouds formed, these blew north on sea breezes, fell as rain on lakes or rivers, then finally drained back into the Everglades. Constructed drainage canals increased water flow within defined channels, and this dried up about half of the swampland, thus altering the cycle.

Before the 1960s, much of the Everglades would dry out between November and April, and invertebrates and fish would become concentrated in small remnant pools. Aquatic birds like wood storks and herons depend on this high concentration of prey species to capture enough food for their young and thus to nest successfully. After the mid-1960s, water engineers began regulating the flow of water from Lake Okeechobee, releasing water during the dry season to irrigate farms south of the lake. Some of this water overflows into the flat marshes, and the pools that would normally dry up do not; wading birds have had a difficult time finding enough food ever since regulation began.

THE NITROGEN CYCLE

Organisms contain a fairly large quantity of nitrogen in their proteins and nucleic acids. In agricultural ecosystems, where the net primary production helps feed the

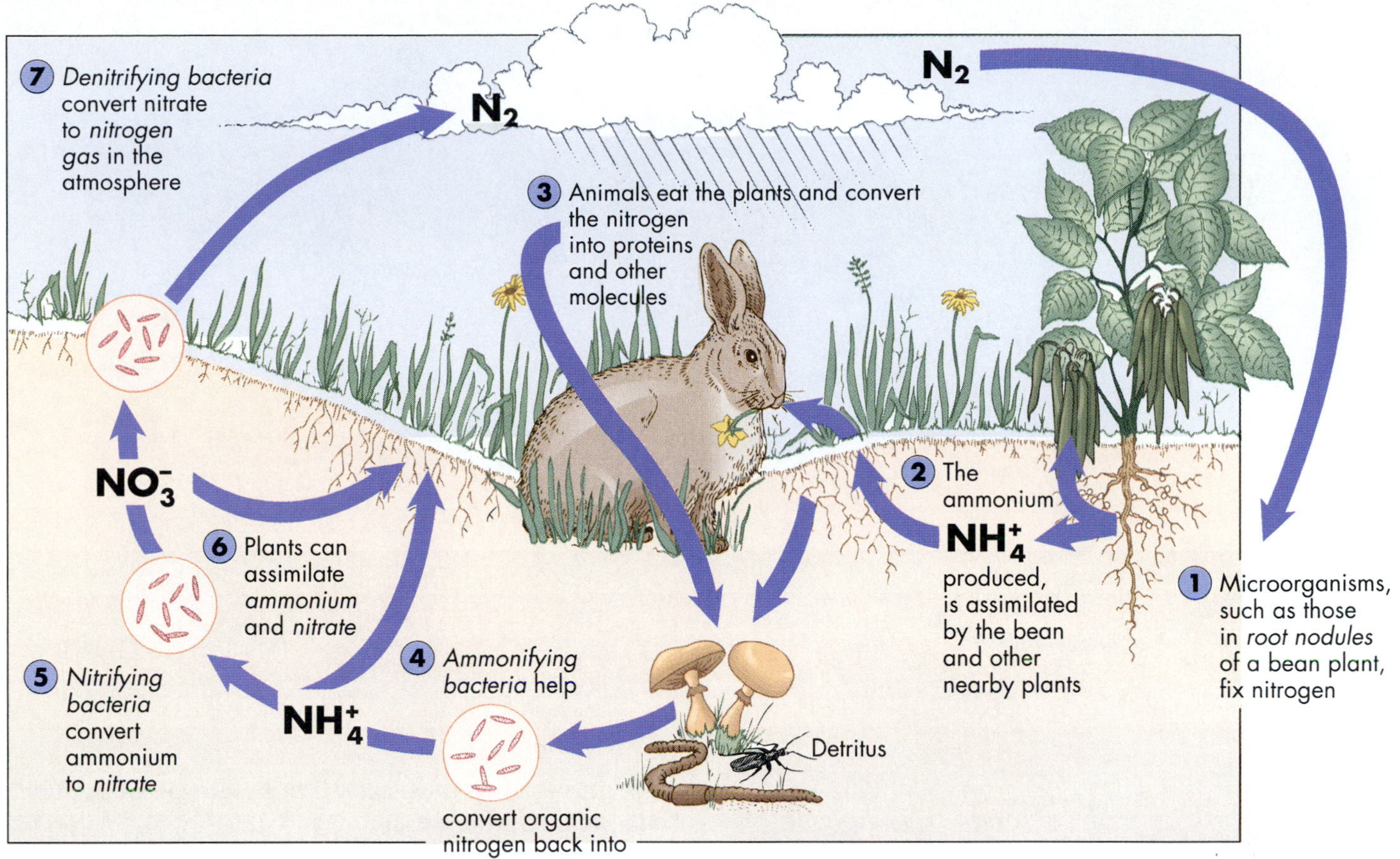

FIGURE 40.13

The Nitrogen Cycle: To Air, Organisms, and Back.

human race, nitrogen is often the factor that limits productivity. Nitrogen gas makes up about 79 percent of our atmosphere, but ironically, most organisms cannot use nitrogen in its gaseous form. Instead, they depend on a few species of nitrogen-fixing bacteria to trap nitrogen in a biologically useful form, while other bacterial species return nitrogen to the atmosphere as nitrogen gas and complete the **nitrogen cycle**. The major steps in nitrogen transformation are nitrogen fixation, ammonification, nitrification, and denitrification [FIGURE 40.13].

During **nitrogen fixation**, nitrogen gas (N_2) from the air is transformed into ammonia (NH_3). This colorless gas has an irritating odor and dissolves in water, forming biologically usable ammonium ions (NH_4^+). Nitrogen fixation can occur as a result of lightning or volcanoes or, more importantly, by the action of bacteria in the soil or in root nodules of legumes, aspen trees, or a few other plants [FIGURE 40.13, Step 1; also review FIGURE 35.11]. Plants can absorb these compounds from the soil, change them into nitrate (NO_3^-), and assimilate them into proteins, nucleic acids, and other nitrogen-containing biological molecules (Step 2). When animals consume plant matter, they break down the plant's nitrogenous compounds and use them to form new animal proteins and other cell components (Step 3).

After an animal excretes urea or uric acid or after an animal or plant dies, certain bacteria carry out **ammonification**: They produce ammonium from nitrogen-containing molecules (Step 4). Plants can then assimilate this ammonium, or still other bacteria can change it to nitrate by **nitrification** (Step 5). Plants take in some of the nitrate produced in this way (Step 6). A fourth set of bacterial species acts on the remaining nitrate, and via **denitrification**, changes fixed nitrogen back to nitrogen gas, thereby completing the cycle (Step 7).

Since available nitrogen often limits agricultural productivity, farmers have long fertilized their fields to increase the amount of ammonium and nitrate in the soil. Recall that the Pilgrims observed Native Americans burying fish with their corn seeds. Ammonifying soil bacteria produced ammonium from the fish proteins and nucleic acids, the corn assimilated the available nitrogen atoms in the ammonium, and the plants grew taller and faster. And many farmers continue to practice *crop rotation*, as did their predecessors in various parts of the world. They plant leguminous crops such as beans, clover, or alfalfa one year and corn, wheat, or sugar beets the next year to take advantage of the nitrogen that the nitrogen-fixing bacteria in the legumes' root nodules release into the soil.

Today, most large farms around the world depend on nitrogen fertilizers produced through an industrial process rather than a bacterial process. In fact, nitrogen fixed in chemical factories may now represent 30 percent

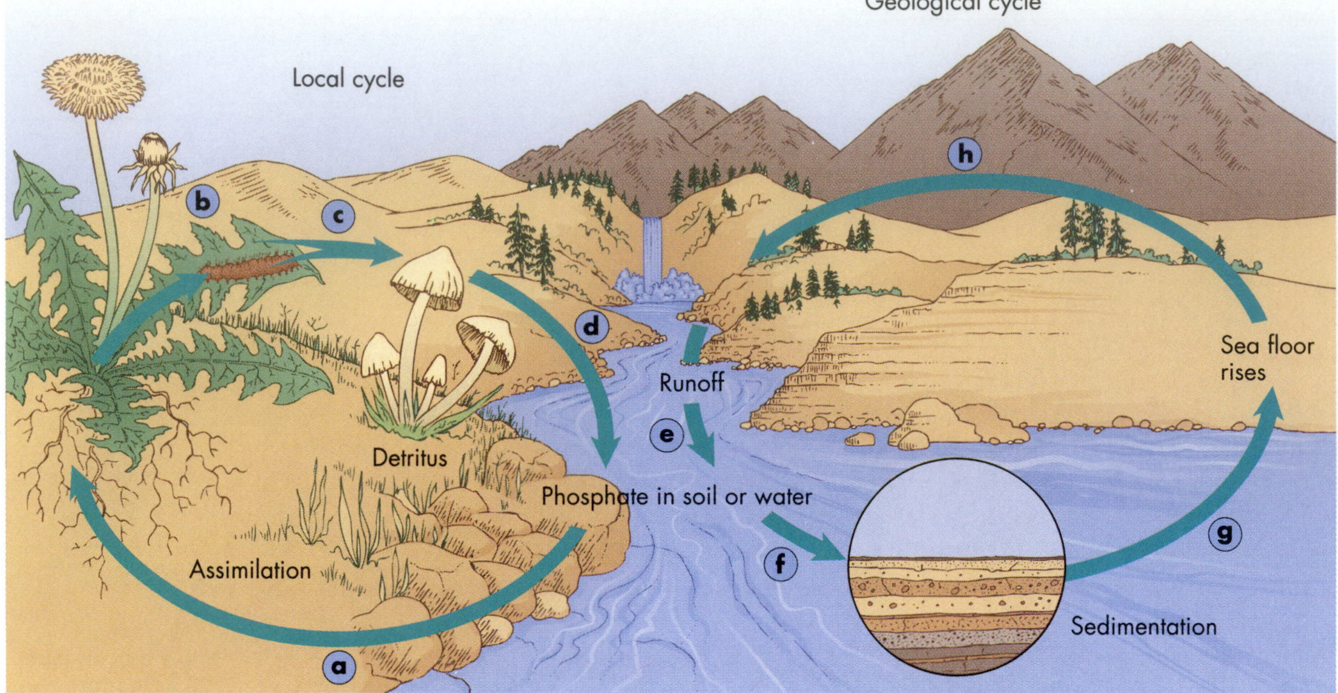

FIGURE 40.14

The Phosphorus Cycle: Local and Global.

Plants assimilate phosphates from soil or water [a], animals gain phosphates by eating the plants [b], and when the autotrophs and heterotrophs die or give off wastes [c], the phosphates return to the soil or water in a local cycle [d]. Phosphates that flow to large bodies of water can become tied up in sedimentary rocks [e, f], but when the sea floor rises after thousands or millions of years [g], the phosphates from this long-term cycle can once again enter the local, short-term cycle [h].

of the input to the global nitrogen cycle, a truly great shift in the natural cycle. Unfortunately, industrial nitrogen fixation requires tremendous heat and pressure, and this is usually produced by burning great quantities of fossil fuels. In some areas, people dump more energy into the soil in the form of nitrogen fertilizers than they extract from the soil in food calories. Since reserves of fossil fuels are limited, this enormous reliance on industrially fixed nitrogen fertilizers cannot go on indefinitely. In addition, added nitrogen can leach out of agricultural soils into surface waters, as they do in the Everglades drainage system. Some biologists think a far better solution would be to genetically engineer plants and soil bacteria to increase natural nitrogen fixation [see CHAPTER 35].

The nitrogen cycle is sensitive to human activities such as deforestation. To quantify this, ecologists removed all trees from one watershed in the Hubbard Brook Forest and sprayed the area with herbicides to block regrowth. In their test area, 60 times more nitrogen drained away in the stream than in the control watersheds where the forests remained undisturbed. Such severe nitrogen loss drastically limits regrowth of the forest and pollutes groundwater.

The nitrogen cycle, like some other biogeochemical cycles, has an atmospheric phase. While nitrogen gas, water vapor, and carbon dioxide are airborne, they can be blown anywhere by the wind. This mobility makes the nitrogen, water, and carbon cycles truly planetary.

THE PHOSPHORUS CYCLE

In contrast to nitrogen, water, and carbon, certain other substances, such as calcium and phosphorus, lack a gaseous phase and thus cycle locally. Local cycles of these elements can cause great fluctuations in the populations of local organisms. Phosphorus is essential for life. It is a component of cell membranes, nucleic acids, and ATP, the energy currency of cells. The **phosphorus cycle** consists of two interlocking circuits, one that acts locally during short stretches of time, and another that operates more globally over vastly longer time periods.

Local Phosphorus Cycle In the *local cycle*, phosphorus moves from the rocks or soil (where it usually occurs as calcium phosphate) into organisms and back to the soil. Plants assimilate phosphate (PO_4^{3-}) directly from the soil or water [FIGURE 40.14a], while animals obtain needed phosphorus from the plants or other animals they consume [FIGURE 40.14b]. When plants and animals die and decay [FIGURE 40.14c], bacteria convert the organic phosphorus in their tissues into phosphate. This enters the soil once again [FIGURE 40.14d], and plants may then assimilate it once more, completing the local ecological cycle without a significant atmospheric component or global distribution.

FIGURE 40.15 **Collared Lemming Eating Saxifrage Blossoms in the Tundra.**

An example of the importance of local mineral cycles can be seen in the desolate Arctic, which supports only a few major plant species—sedges, grasses, knee-high birch, and willow—and just one major herbivore—the furry tundra mouse called the lemming [FIGURE 40.15]. Popular wisdom has it that lemmings dash mindlessly over cliffs and into the sea. This is an exaggeration based on truth: Every four years, lemming populations boom, and the animals migrate in search of food. High-density lemming populations are possible only when the tundra grasses are rich in phosphorus and calcium, which lemmings need for reproduction. When those nutrients are plentiful, however, millions of the furry little animals eat the grasses to the roots. In the process, most of the nutrients transfer from the plants to the lemmings. Lacking additional minerals, the overgrazed tundra plants cannot regenerate new leaves and stems, and the resulting decrease in food reduces the lemming population to less than 1 percent of peak populations. In the cold, the dead lemmings slowly decompose over a couple of years. As nutrients are gradually returned from the lemmings to the soil, plants start making a comeback. By the fourth year, the plants have produced new leaves and stems, rich in calcium and phosphorus. This nutritious food allows the lemmings to reproduce at high rates again and produce another bumper crop of hungry young, who, in turn, overgraze, and the population crashes again. While other explanations may also help account for the boom and crash of lemming populations, these events drive home the dependence of each species on the cycling of materials in an ecosystem.

Geological Phosphorus Cycle Although phosphorus does not leave local ecosystems by way of the air, it *can* leave terrestrial ecosystems in streams and rivers [FIGURE 40.14e]. Thus, the second phosphorus cycle, if viewed over geological time, does have a global aspect. After phosphate eventually washes from terrestrial waterways down to the sea, it can form insoluble compounds that fall as sediments and become incorporated into rock [FIGURE 40.14f]. Eons later, when the sea floor rises and exposes new land [FIGURE 40.14g], these phosphorus-containing rocks may form the base of a terrestrial ecosystem, and the phosphorus can once again enter the local ecological cycle [FIGURE 40.14h].

Phosphorus cycling is also important in the Everglades. Because water flows so slowly in the flat marshlands, in past eras plants and algae were able to remove almost all of the phosphorus and other nutrients from the water and incorporate them into organic material. As a result, very little phosphorus reached the Gulf of Mexico and entered the global cycle from this drainage system.

Eutrophication As with other biogeochemical cycles, human activities have altered the dynamics of the phosphorus cycle—especially in aquatic ecosystems like the Florida Everglades, for which phosphorus is often the factor limiting primary productivity. Phosphates are major ingredients of agricultural fertilizers and until recently were also main components of detergents. For years Florida farmers pumped phosphorus-rich runoff water from sugarcane and vegetable fields directly into Lake Okeechobee. The additional phosphates allowed algae and other aquatic plants to grow faster, and the ecosystem underwent **eutrophication**—it became **eutrophic** (literally, "well fed"), or overly supplied with nutrients that support primary production. (The opposite term, **oligotrophic**, means "little fed" and describes an ecosystem such as a clear lake, where there is little algae or other vegetation.) As algae and other plants died in Lake Okeechobee, decomposing bacteria fed on the dead algal cells and used up so much dissolved oxygen that fish suffocated in massive fish kills in the mid-1980s. In nearby marshes, cattail populations have exploded due to the phosphorous overload, covering 34,000 acres, outcompeting other vegetation, and blocking access to feeding areas for wading birds.

Ecologists dramatically demonstrated this process with an experiment in a lake in Ontario, Canada. They hypothesized that phosphate is the limiting factor in a lake's primary productivity. To test this hypothesis, they found a lake with a natural hourglass shape and divided it into two sections by stretching a vinyl curtain across the lake's narrowest part. This clever strategy effectively separated the lake into an experimental side and a control side with equivalent conditions at the start of the experiment. The researchers fertilized both halves of the lake with nitrogen and carbon compounds, but in addition, added phosphorus to the experimental half. Without added phosphorus, the control section showed no change in organisms or ecological productivity. Within two months, however, the experimental half, with its added fertilizer, developed a blue-green algal bloom that was visible from an airplane. Only after workers stopped fertilizing the lake with phosphorus did the algal bloom fade and the lake recover its previous condition. This spontaneous recovery suggests that people can decrease the eutrophication of lakes and streams by restricting the runoff of phosphate-containing detergents and fertilizers. In the mid-1970s, when Lake Erie was becoming seriously eutrophied, regional governments banned the use of phosphate-laden detergents. Since that time, the lake has become much healthier.

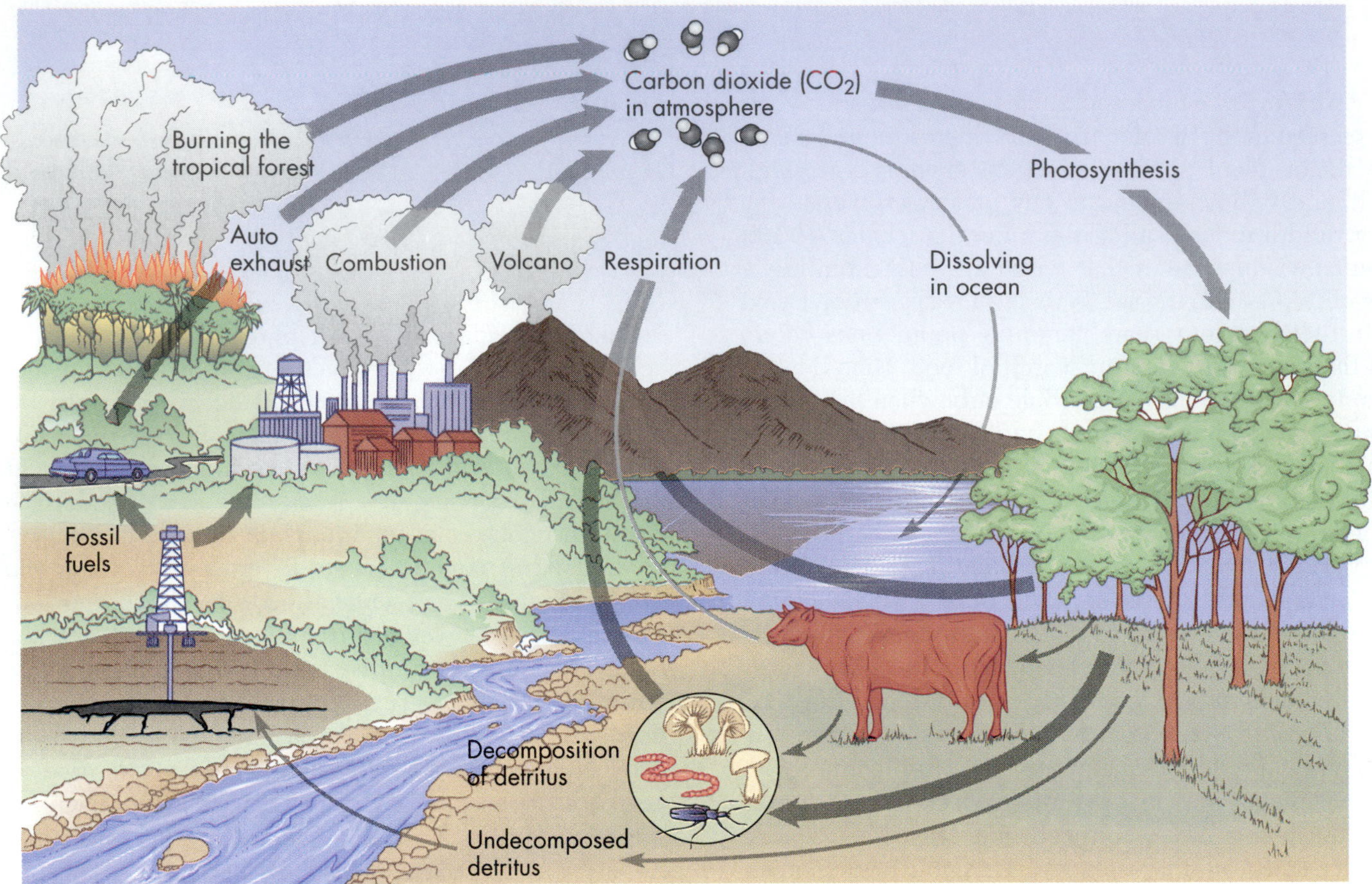

FIGURE 40.16

The Carbon Cycle.

Carbon cycles from atmospheric carbon dioxide to biological molecules to organic molecules in the soil to geological deposits of fossil fuels, and back to carbon dioxide. Disruptions of this cycle can lead to increased atmospheric carbon dioxide and, some scientists suggest, to global warming.

THE CARBON CYCLE

Carbon atoms move in a global **carbon cycle** from the physical environment through organisms and back to the nonliving world, just as water moves through the hydrologic cycle. And like the water cycle, the carbon cycle is linked to energy flow because producers—including the photosynthetic plants of the forests and oceans—require environmental energy such as sunlight to trap carbon into sugars [review FIGURE 6.14]. The trapped carbon comes from carbon dioxide in the surrounding air or water [FIGURE 40.16]. As the cycle proceeds, consumers ingest the organic carbon compounds synthesized by producers. Then, via respiration, both consumers and producers return carbon to the nonliving environment in the form of carbon dioxide [review FIGURE 5.12].

Carbon that accumulates in wood is eventually returned to the atmosphere as a result of fires or through consumption and respiration by fungi, bacteria, and other detritivores. Organic carbon can leave the cycle for even longer periods of time after sediments bury organic litter, which decomposes only partially and gradually transforms into coal or oil. Carbon also leaves the cycle when cast-off calcium carbonate shells of marine organisms sink to the ocean floor and become covered with sediments that compress them into limestone. Eventually, however, even these carbon deposits are recycled into atmospheric carbon dioxide as the limestone erodes and the fossil fuels are burned in automobiles or in industry. As with the other biogeochemical cycles, human activities are altering the dynamics of the carbon cycle. In some cases, these perturbations have global consequences and affect both our own future and that of the other organisms that share our planet.

➤ CONCEPT CHALLENGE

By driving evaporation, the sun provides the energy that perpetuates the water cycle. What provides the energy that drives the nitrogen and carbon cycles? How is that energy source involved?

How Humans Alter Ecosystems

Human activities have the potential to drastically modify the nutrient cycles that support life on earth and to alter physical features of the natural environment as well, including air and water temperatures and acidities and the amount of available solar energy. Let us consider two of the most serious current ecological concerns: global warming and the dwindling of our natural resources.

GLOBAL WARMING

The carbon cycle, as our discussions have shown, involves an atmospheric supply of carbon dioxide, which plants use to make sugars [review FIGURE 40.16]. Although carbon dioxide makes up only about 0.03 percent of our atmosphere, it plays a disproportionately large role in governing the earth's temperature by means of a phenomenon called the *greenhouse effect*. The earth and its envelope of atmospheric gases are a bit like a greenhouse on a sunny day [FIGURE 40.17]: Light energy passes through atmospheric greenhouse gases, strikes the earth, and warms it. The warm earth then reradiates the energy as infrared radiation [see FIGURE 6.2]. Light can pass through greenhouse gases, but infrared rays cannot, and hence they become trapped near the earth's surface and contribute to a buildup of heat.

Greenhouse gases include carbon dioxide, methane, chlorofluorocarbons, and nitrous oxide, but carbon dioxide alone contributes about half the human-produced share of the gases known to contribute to the greenhouse effect. Methane is generated by bacteria that break down organic matter in flooded rice fields and in the guts of cattle and termites. *Chlorofluorocarbons* (*CFCs*) are synthetic materials used as coolants in refrigerators and air conditioners, as well as in plastic foam and insulation materials. Nitrous oxide (N_2O_2), or laughing gas, is produced by microbes in soils, and its concentrations increase with the burning of forests and fossil fuels and the production of chemical fertilizers.

Unfortunately, the major greenhouse gas, carbon dioxide, has increased by roughly 25 percent in the past century, and the average atmospheric temperature has risen 0.5°C (about 1°F) along with it. Much of the extra carbon dioxide that atmospheric scientists have measured originates with the burning of coal, oil, and gasoline in our factories and automobiles. The burning of tropical rain forests and the destruction of temperate forests [review FIGURE 39.20] is another major source of the gas. The consequence of burning forests—those decomposed, compressed, and liquified into oil millions of years ago, as well as those still standing—is that more carbon dioxide now enters the atmosphere than can be removed by plants. Some of the excess carbon dioxide dissolves in the ocean. Much of it, however, enters the atmosphere and blocks the escape of infrared radiation.

Many atmospheric scientists predict that sometime between the years 2025 and 2075—well within the expected lifetimes of those reading (and writing!) this book—an accumulating blanket of greenhouse gases will trap more and more heat, and global temperatures will climb an average of 2° to 4°C (3.5° to 7°F). This may seem like a small increase, but its effects could be disastrous. According to the predictions, ice at the poles might melt, and sea levels might rise, inundating many of the world's most populous cities, including New York, Los Angeles, London, Stockholm, Hong Kong, and Tokyo. More importantly, the increased temperatures might change fertile croplands into deserts: The dust bowl conditions like those in the United States in the 1930s could occur in the great grain-producing regions of the American Midwest, Canada, and the Soviet Union [FIGURE 40.18]. Irrigation would probably not correct the situation because groundwater reserves would quickly become depleted. Ominously, even the coldest years of the last decade have been warmer than nearly every year a century ago, and we are beginning to see widespread droughts and resultant forest fires and crop losses.

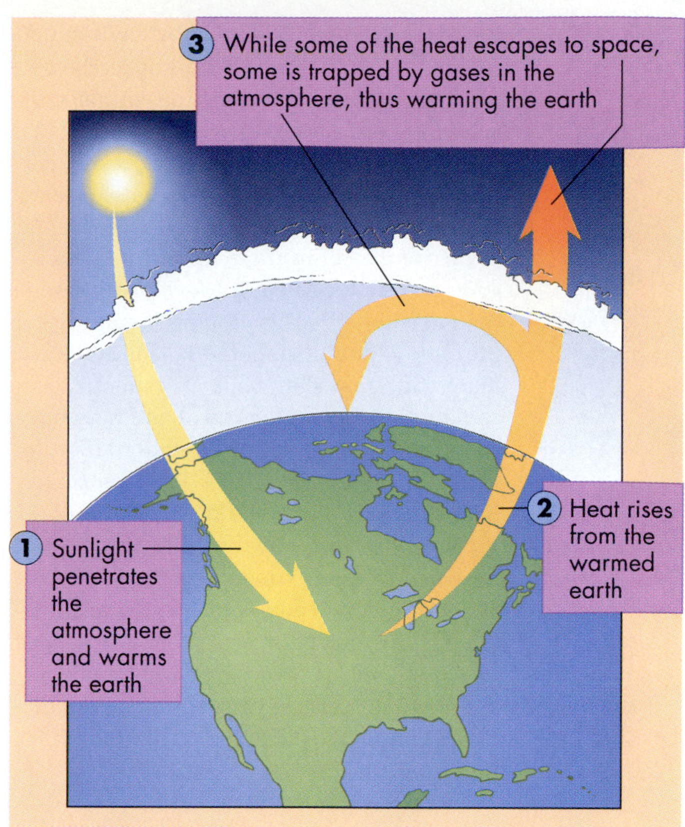

FIGURE 40.17

The Greenhouse Effect and Global Warming.

Like sun shining through the glass windows of a greenhouse, light penetrates the atmosphere and warms the earth, but much of the reflected heat becomes trapped in the atmosphere. The more carbon dioxide in the air, the more heat is trapped, and the higher the earth's surface temperatures. Many scientists suggest that this phenomenon is happening on the earth today.

box 40.1
Focus on the Environment

Saving the Environment: Personal Solutions

It's easy to get overwhelmed by environmental news—ozone levels dropping, carbon dioxide rising, acid rain falling, animals disappearing, forests going up in smoke. It's tempting, in fact, to lose interest, resign oneself, and carry on as if these threats were not both real and pressing. We no longer have the luxury of a head-in-the-sand attitude, however, and each of us must develop a personal plan of action. Because the problems are complex and ubiquitous, potential solutions are many. Several recent books list practical conservation tips as well as local and national sources of background information. (See, for example, *50 Simple Things You Can Do to Save the Earth*, Earthworks Press, 1989). What follows are a few beginning suggestions for personal environmental action, and we encourage you to add your own:

- In winter, turn down the furnace, and set the thermostat even lower when you're not at home. In summer, use less air conditioning and have the furnace pilot light turned off.
- Adjust the water heater to 54°C (130°F), not the normal 60°C (140°F) or higher.
- Leave the car at home and use public transportation, bicycles, or foot power whenever possible.
- Use warm wash and cool rinse when washing clothes, to save gas or electricity.
- Recycle aluminum, tin, glass, newspaper, nonglossy paper (junk mail, old school papers, packing materials), and plastic bags and containers.
- Avoid disposable packaging by carrying a string or cloth shopping bag, buying from bins, and reusing plastic produce bags.
- Compost organic wastes when possible, using your own or a community compost pile.
- Plant trees in and around campus, town, and your backyard.
- Refuse to buy ivory and other parts and products of endangered species.
- Refuse to buy teak and other tropical wood products.
- Buy less beef, since cattle are often raised on newly stripped tropical lands, and since the production of animal protein, in general, requires far more energy than the production of plant proteins.
- Cut excess water use by modifying your habits and by using low-flow shower heads and toilet tank modifiers.
- Buy unbleached and recycled paper products.
- Buy more organic produce as a way of encouraging alternative farm practices.
- Avoid CFC propellants and polystyrene cups and containers. If buying a new air conditioner or refrigerator, look for alternative coolants, now available.
- Consider using vehicles that run with gasohol or methanol.
- Take leftover toxic materials (used crankcase oil, paints and thinners, oven cleaners, pesticides, and so on) to approved municipal sites instead of dumping them in drains or on the soil, where they can reach surface- and groundwater supplies.
- Switch to integrated pest management techniques in home gardens to avoid using pesticides unnecessarily.

Most climatologists agree that there is no firm proof that accumulating greenhouse gases are causing the observed warming. Many suggest, however, that the potential danger to ecosystems and economies is too great to allow scientific uncertainty to prevent prudent action now. Most ecologists believe that we must decrease our consumption of fossil fuels through an emphasis on energy conservation and a commitment to renewable energy sources such as wind power and solar energy. As individuals, we could, for example, lower the temperature of our homes in winter, use less air conditioning in summer, and rely on public transportation, bicycles, and our own two feet more often than we do now. Farmers could employ agricultural practices that sustain the levels of organic matter in the soil, and people in tropical regions could step up efforts to replant denuded areas. Finally, biologists could actively develop more drought-resistant and salt-tolerant plants that can substitute for our staple crops if the world's breadbaskets become dust bowls. Implementing these solutions will be difficult, but informed citizens must work now to help governments make ecologically sound policy decisions.

CYCLE AND RECYCLE: THE SUSTAINABLE ECONOMY

In nature, carbon, nitrogen, and other fixed resources are cycled and recycled, and human societies must learn to do the same. Americans discard most materials after one use; we throw away two-thirds of the aluminum we use, three-quarters of the steel and paper, and most of the plastic products. Yet the amount of energy needed to recycle aluminum is only 5 percent of the energy needed to produce it from bauxite, the original raw material. Likewise, recycling steel, newsprint, and glass saves from one-quarter to two-thirds of the energy needed to generate those commodities from scratch. Recycling materials instead of continually producing new ones also reduces air, water, and ground pollution by more than half. But recycled materials can't just be dumped into landfills. Manufacturers must use them to build new consumer products. As individuals, we should not only recycle our newspapers, plastics, and glass, but also encourage the use of recycled materials by asking for and buying new items made from them and packaged in their own recyclable materials. Beyond that, we should learn to compost garbage and garden clippings, to cut down on refuse handling and accumulation, and to generate rich humus for garden and lawn fertilizer. BOX 40.1 describes a number of specific ways to recycle and conserve our dwindling natural resources.

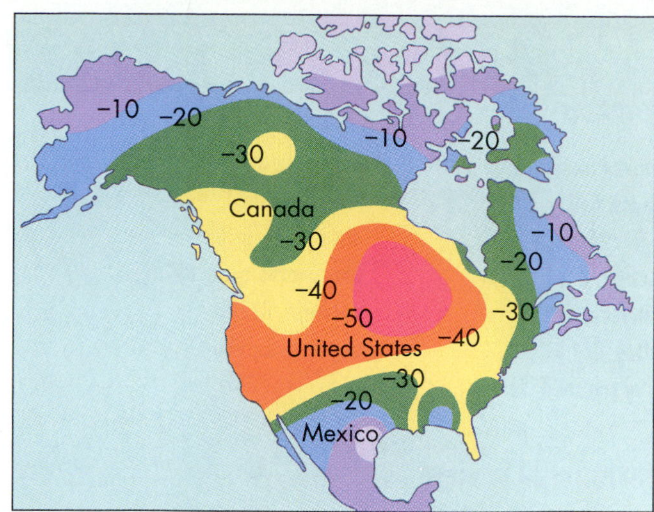

[A] Predicted moisture loss

[B] Dust bowl of the 1930s

FIGURE 40.18

Will the Dust Bowl Return?

[A] If atmospheric levels of carbon dioxide double in the next few decades, as some scientists predict, then the soil will lose precious moisture. This map predicts the relative moisture loss between the years 2025 and 2075. **[B]** Some observers fear that the projected loss of moisture in our continent's fertile interior agricultural regions (from 40 to 60 percent drier) could lead to another dust bowl like the one in the 1930s, when prolonged drought, overplanting, and agricultural mismanagement allowed soil layers to dry out and topsoil to blow away, driving people from their land.

Connections

Energy flow and nutrient cycles in ecosystems are fundamental biological principles governing life. As history plainly shows, no organisms are exempt from these laws or their implications. Humankind's current polluting activities are changing global environments so fast that organisms have no time to evolve appropriate adaptations. In the Florida Everglades, an ecosystem many millions of years old has been irreversibly altered since 1967. If current trends continue, we will find ourselves the subjects of an unplanned global experiment, and we will not escape its repercussions.

Humans are unique among living organisms only in the degree and speed with which we can modify the earth's ecosystems, in our ability to evaluate the future consequences of our actions, and in the capacity to take measures as individuals and as nations to deal with the ecological imbalances and threats we now face. Yet the people of different countries and different economic levels contribute unequally to the world's ecological problems. It is a minority of earth's 6 billion people who drive automobiles; eat meat at most meals; heat their homes with nonrenewable fuels like electricity or gas; and buy spray cans, refrigerators, and air conditioners that release into the atmosphere the CFCs that destroy earth's protective ozone layer. Those who do participate in these practices contribute more to earth's ecological problems than do 20 or 30 people who do not drive cars, use electricity and gas, eat meat frequently, and use CFCs. Despite people's unequal contribution to ecological threats, all of us are encompassed by the biosphere, our topic in the next chapter.

KEY TERMS

biological magnification, 878
biomass, 878
carbon cycle, 884
consumer, 872
detritivore, 874
detritus, 874
energy pyramid, 877
eutrophic, 883
food chain, 874
nitrogen cycle, 881
phosphorus cycle, 882
producer, 872
water cycle, 880

HIGHLIGHTS IN REVIEW

1 An ecosystem includes all the living organisms in a given area (the community), the nonliving physical environment around them, and the interactions of both.

2 Energy flows through an ecosystem in a one-way path. Energy in the form of sunlight enters living things via photosynthesis, passes from one organism to another, and finally escapes back to the physical environment.

- **a]** Ecosystems have several feeding (or trophic) levels. Autotrophs are the primary producers, capturing energy from the physical environment into organic molecules, while heterotrophs are the consumers, devouring producers and each other. Fungi and decomposing bacteria are the detritivores, feeding on dead parts of plants, animals, and other organisms.
- **b]** Organisms in the various trophic levels are linked in simple food chains and in more complex food webs, such as the grazing and detritus food webs.
- **c]** Most energy flowing into terrestrial ecosystems from the environment is reflected as heat or light or causes evaporation and transpiration of water. The small amount of energy stored in newly formed plant organs becomes the net primary productivity that supports all other life in the ecosystem.
- **d]** Energy pyramid diagrams reveal that the energy available in the producer level is substantially greater than in the next higher level, and so on. The biological magnification of nonbiodegradable toxic substances—their increasing concentrations at higher levels in a food chain—is a consequence of energy loss from one trophic level to the next. It occurs because an animal at one level in a food chain must eat about ten kilograms of an organism on a lower level to gain one kilogram of body mass.

3 Materials cycle through ecosystems. Carbon atoms, for example, pass from the atmosphere into plants, into animals that eat plants, and finally back to the atmosphere as carbon dioxide where plants once more absorb it.

- **a]** The hydrologic, or water, cycle involves evaporation, plant transpiration, and precipitation.
- **b]** The nitrogen cycle involves atmospheric nitrogen, nitrogen fixation by soil bacteria, ammonification and nitrification by other bacteria, and denitrification by still others.
- **c]** The phosphorus cycle includes a short-term, local cycle in which phosphates move from rock and soil into organisms and back to the soil, and a long-term cycle that operates over geological time, in which phosphates become fixed in ocean bedrocks and are eventually exposed and eroded eons later.
- **d]** The carbon cycle involves carbon dioxide in the air and water, carbon fixation by autotrophs, then a return of the carbon to the air and water through respiration, combustion, and erosion.

4 Because the organisms in an ecosystem require energy and materials, they are dependent on each other and on the physical environment.

5 Human activities can drastically affect the health of ecosystems by altering the flow of energy and the recycling of materials.

- **a]** Carbon dioxide and other greenhouse gases allow light energy to pass toward earth, but can block escape of heat energy. Some ecologists predict that increasing levels of atmospheric carbon dioxide due to human activities may be leading to a global greenhouse effect and global warming.
- **b]** Recycling aluminum, glass, paper, and other materials saves energy, prevents pollution, and protects our dwindling natural resources.

UNDERSTANDING THE FACTS AND CONCEPTS

For Questions 1–5, match each of the descriptions with the most appropriate term from the following list. Any answer may be used once, more than once, or not at all.

- **a]** producers
- **b]** consumers
- **c]** ecosystem
- **d]** trophic level
- **e]** detritivores

1 The term that includes a community of organisms at several trophic levels forming a food chain or web within a particular physical environment.

2 There may be one, two, three or even four trophic levels within this group.

3 A group of consumers that includes fungi, bacteria, and several invertebrates and vertebrate animals that consume dead organisms and organic wastes.

4 The trophic level that includes chemoautotrophic bacteria such as those in the deep-sea vents as well as photosynthetic plants, bacteria, and protists.

5 A feeding level within an ecosystem.

For Questions 6–10, match each of the descriptions with the most appropriate item or items from the following list.

- **a]** food webs
- **b]** energy-fixation base
- **c]** nutrient-concentration base
- **d]** net primary productivity

6 The contribution of the autotrophs in any ecosystem.

7 Provides the basis for the energy flow within a food web.

8 The complex route of matter and energy through the several trophic levels in an ecosystem.

9 The sum total of matter fixed into organic compounds by the producers within an ecosystem.

10 The energy available to the detritus and grazing food webs within an ecosystem.

For Questions 11–15, match each description with one of the items from the following list.

a] energy pyramid
b] biomass
c] biological magnification
d] biogeochemical cycles
e] greenhouse effect

11 The dry weight of organic material at any trophic level.

12 A phrase descriptive of the fact that each trophic level in an ecosystem contains substantially less stored energy than the level immediately below it.

13 The tendency for increased concentration of toxic substances in the higher trophic levels of an ecosystem.

14 Global cycling of important compounds or elements, utilizing the atmosphere, earth, oceans, and organisms as temporary reservoirs.

15 The trapping of infrared radiation near the earth's surface by carbon dioxide and other atmospheric gases.

For Questions 16–20, match each description with the appropriate biogeochemical cycle listed below. Any answer may be used once, more than once, or not at all.

a] water cycle
b] nitrogen cycle
c] phosphorus cycle
d] carbon cycle

16 A global cycle in which bacteria, leguminous plants, and the atmosphere play the key roles.

17 A global cycle powered by solar energy, with transpiration as an important link.

18 Found in local cycles within ecosystems; involves calcium compounds in rocks and soil, as well as organic compounds in plants and animals.

19 A global cycle in which photosynthesis and cellular respiration are important; characterized by lengthy periods during which some of the material may temporarily leave the cycle as rock, wood, or litter.

20 A very long-term global cycle in which the reservoirs are linked by water rather than air; sedimentary rock formation and sea-floor risings play roles.

INTEGRATE AND APPLY WHAT YOU HAVE LEARNED

1 What kinds of organisms are producers? What kinds are primary consumers, secondary consumers, and higher consumers?

2 An ecosystem contains a pyramid of biomass as well as an energy pyramid. How are these two pyramids related?

3 How much of the solar energy that strikes the earth is eventually found within the bodies of carnivorous animals? What happens to the rest?

4 Explain why the net productivity of the detritivores is larger than that of the grazing herbivores.

5 What factor must be present for a biogeochemical cycle to be global in nature? List two global cycles and two local cycles.

ANALYSIS

1 A few aquatic ecosystems exhibit inverted biomass pyramids, with fewer phytoplankton producers than consumers at any one time. Such a situation arises when the rate of reproduction of the producers and the rate of their consumption by consumers are both high. Which of the following statements is applicable to such an ecosystem? More than one answer may be correct.

a] If the biomass pyramid is inverted, the energy pyramid is similarly inverted.
b] While the biomass pyramid might be inverted, the energy pyramid could not be.
c] If one determined the total biomass produced per month or year, the pyramid would not be inverted.
d] Such an ecosystem is inherently short-lived because it is unstable.

2 Which of the following statements about a eutrophic pond is incorrect?

a] The limiting resource for consumers is oxygen.
b] The limiting resource for producers is phosphorus.
c] The limiting resource for consumers is carbohydrate nutrients.
d] The stimulus for eutrification is excess phosphorus.
e] The biomass of producers is high relative to that of consumers.

3 Which of the following people would be likely to have the lowest tissue levels of DDT?

a] A vegetarian who eats no eggs or dairy products.
b] A vegetarian who eats eggs and/or dairy products.
c] Someone who consumes a traditional American diet.
d] A person who does not eat red meat but does eat fish.
e] A person who does not eat red meat but does eat chicken, turkey, and fish.

4 Tourists to the Galápagos Islands enjoy the friendliness of the native animals, including numerous species of gulls and other birds, seals, sea lions, tortoises, and iguanas. The tourists are told that the reason the animals do not fear human beings is that they have no fear of predation because there are no predators in the Galápagos ecosystem. But how can there be a natural ecosystem with no primary productivity and no predators? Choose all the answers that could be logically correct.

a] The native animals themselves are the producers.
b] The native animals themselves constitute the primary productivity.
c] The ecosystem that supports these animals is in the surrounding sea.
d] Many of these animals are themselves predators of fishes.
e] Many of these animals are primary consumers of aquatic plants.

CHAPTER 41

The Biosphere: Earth's Thin Film of Life

THE CATHEDRAL FOREST

It is extraordinarily still in the cathedral forest [FIGURE 41.1]. The damp air is unmoving and fragrant with the pungence of greenery, wet tree trunks, and dark, spongy leaf litter. Water drips steadily from the mist-enshrouded canopy high overhead, moistening the fern fronds that blanket the forest floor. Only occasionally does an animal's movement break the silence: A Roosevelt elk, huge antlers pointing skyward, picks its way around the fallen trunks of massive evergreens. A mouselike red-backed vole digs in the damp soil for truffles, the fruiting bodies of fungi growing on and around the forest trees, and moves quickly away with powdery spores clinging to its fur. Hearing the vole's soft scurrying, a spotted owl swoops noiselessly from a high branch, snatches the small rodent, and lifts it to a nest of hungry young in a tall dead tree still standing among the living redwoods, Douglas fir, hemlock, and cedar.

FIGURE 41.1
The Cathedral Forest: The Northwest's Great Temperate Rain Forest.
An old-growth forest in Washington's Olympic National Park.

This cathedral grove stands in what was once the world's greatest temperate rain forest, which stretches for more than 1600 km (1000 mi) along the west coast of North America, from northern California to British Columbia. The trees, ferns, and moss growing here require large amounts of precipitation—over 254 cm (100 in) of rain and snow a year in some places—as well as the mild temperatures influenced by the westerlies blowing in from the Pacific. Worldwide climatic forces set up the wind and water currents that foster the emerald forests of the Pacific Northwest. But in return, these forests affect both the local weather and global climates. Nearly one-quarter of the precipitation that falls in the hundreds of square kilometers of coastal forests starts as fog and clouds; droplets collect on billions of leaves and needles and fall to the ground. Recent evidence shows that old-growth forests, such as these in the Pacific Northwest, store a tremendous amount of carbon. When the trees are cut and their wood is burned or left to decay, the carbon can be released, contributing additional carbon dioxide to the greenhouse effect and global warming. Young trees planted in place of these virgin forests do take up some carbon dioxide as they grow, but it requires many decades, even centuries, before new growth can compensate for the huge release of carbon stored in immense old trees. People are usually motivated to cut ancient, massive trees for economic reasons. But as we will discuss later in this chapter, environmentalists are battling loggers over the use of old-growth forests, and rare species like the spotted owl are figuring prominently in the arguments over whether to use or preserve the virgin stands.

The temperate rain forest (or moist coniferous forest) is just one example of how closely interconnected are the living and nonliving worlds. To understand these complex interactions, we must study various communities, including forests, deserts, tundra, and grasslands. We must follow the cycling of materials within ecosystems through food webs to nonliving environments and back again, and we must explore how weather and climate dictate where different species of plants and animals will live. The subject of this chapter is the **biosphere**, that portion of our planet, including water, air, and soil, that supports life. The biosphere is the highest level of organization in the natural world [review FIGURE 1.4], and it includes every body of water; the atmosphere to a height of about 10 km (6 mi); the earth's crust to a depth of several meters; and all the living things within this collective zone. As huge as the biosphere may seem, it is actually a delicate veneer at our planet's surface. If you inflated a round balloon to a diameter of 20 cm (8 in) to represent the earth, then the thickness of its taut rubber skin would be proportional to the biosphere, the thin film of life encircling our world.

Our task in this chapter is to understand how global physical factors—particularly worldwide currents of air and water—produce regional climates, and how these climates, in turn, determine the abundance and distribution of living organisms, that is, why organisms live where and as they do.

We will discuss: How sunlight and the earth's rotation generate the world's climatic zones; the features that characterize each biome, or major terrestrial community of plants and animals; the common attributes of organisms inhabiting oceans, lakes, and streams—the world's aquatic ecosystems; and how the biosphere has changed in the past, and how human activities might cause it to change in the future. ❑

MESSAGES

1 The biosphere is the thin layer of water, soil, and air at our planet's surface that supports life.

2 Different zones have different climates because the earth's axis of rotation is tilted and the earth's surface is heated by the sun most strongly at the equator.

3 On land, different climates promote different communities of organisms called biomes such as tropical rain forests and deserts. Biomes are defined mainly by their vegetation and are characterized by specific adaptations to each climate.

4 Communities with similar physical environments contain organisms with similar evolutionary adaptations, even when the individual species are unrelated.

5 Burgeoning human populations and ecologically unsound life styles are affecting the biosphere in potentially disastrous ways.

What Generates the Earth's Climatic Regions?

After a particularly harsh winter in the Pacific Northwest, elk herds decline substantially in size, and this reveals just how powerfully weather can affect biological systems. **Weather** is the condition of the atmosphere at any particular place and time, including its humidity, wind speed, temperature, and precipitation, whereas **climate** is the accumulation of seasonal weather events over a long period of time. Within the biosphere, weather has temporary, local effects, but climate is the major physical factor determining the abundance and distribution of living things. What, then, causes climate? The answer involves the uneven way that sunlight heats our planet; the behavior of water and air at different temperatures; and the rotation of the earth.

THE EARTH HEATS UNEVENLY

The tropics are warm, the poles are cold, and other regions have their own characteristic climates partly because the earth rotates on a tilted axis and partly because the sun heats the earth's surface unevenly. Like a flashlight beam shining directly down onto a table from above versus one coming in obliquely from the side, sunlight hitting the earth directly is more intense than sunlight striking at an angle, where the same amount of light fans out over a larger area [FIGURE 41.2A]. The sun's angle is therefore the issue, and anyone who has traveled extensively and has also experienced changing seasons knows the importance of *latitude* (the distance between any point and the equator) and *time* of year.

About five times more light and heat energy fall at the earth's equator than at higher latitudes. These differences in incoming solar power help explain why equatorial regions are warm and polar regions cold, and why tropical forests are usually so productive.

Temperatures grow warmer and colder with the seasons, too, and this is because our spherical planet spins on a tilted axis. This tilt is maintained all year long, even as the earth revolves about the sun [FIGURE 41.2B]; consequently, during the summer, the Northern Hemisphere is maximally inclined toward the sun, while in winter, it is tipped away from the sun and receives less solar energy. The seasons are reversed in the Southern Hemisphere. This tilting of the earth and uneven illumination help explain the warm temperatures and abundant growth of organisms in summer and the cold temperatures and dormancy of living things in winter.

THE FORMATION OF RAIN

Sunlight striking the earth's tilted sphere at different places and different times heats the air and oceans unequally. Warm air has different properties than cool air, and these account for the formation of rain and snow. Cold air is dense—it weighs more per unit volume and so tends to sink through lighter, warmer air. This is why New England farmers plant cold-sensitive fruit trees on the sides of valleys instead of on the valley floor, where cold, heavy surface air settles at night. Warm air, in contrast, is less dense than cool air and tends to rise. This is why smoke, steam, and hot-air balloons drift upward.

Dense, cool air holds less moisture than light, warm air. This is why your wet hair dries faster with hot air from a hair drier than with cool air from a fan. It is also why water droplets condense on a cold bathroom window pane when you take a hot shower. This same principle ultimately brings about rain and snow. When warm, moist air rises high into the atmosphere, it cools, and its capacity to hold water decreases [FIGURE 41.3]. This explains why a region like the tropics is so wet. Powerful

FIGURE 41.2
Seasons: Light Shining on a Tilted Globe.

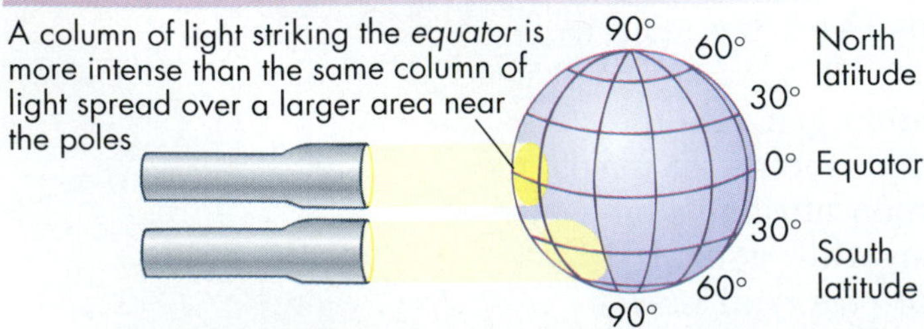

[A] Uneven illumination causes uneven heating

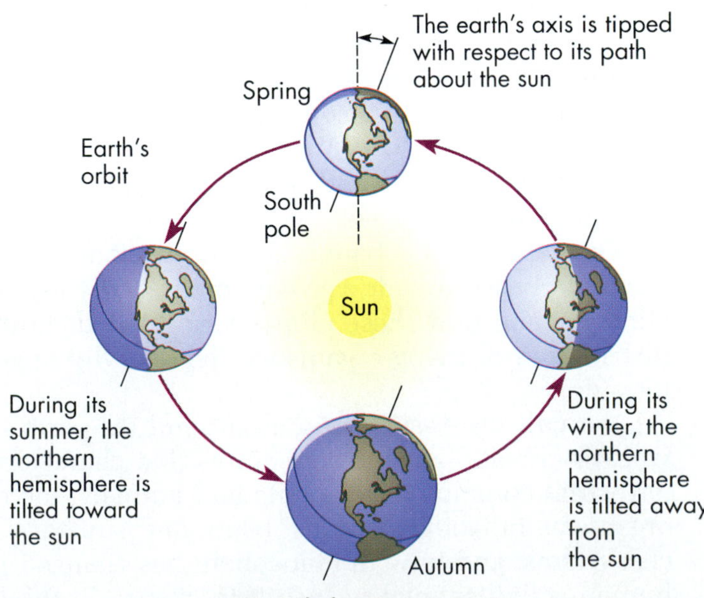

[B] A tilted axis causes seasons

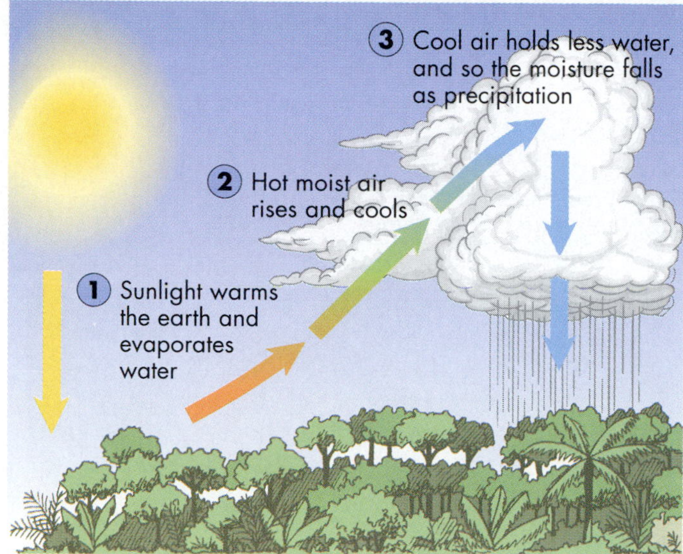

FIGURE 41.3

Let It Rain: The Generation of Precipitation.

sunlight at the equator heats the air, which picks up moisture from evaporation and plant transpiration and rises. Once aloft, the air cools, and the moisture precipitates out, falling onto the lush tropical rain forest.

Because sunlight strikes the earth unequally, large masses of air become warm and then rise off the ground. When cooler air moves in from adjacent cooler areas to take the warm air's place, wind is formed. The earth's natural rotation deflects the wind, influencing climate and weather.

AIR AND WATER CURRENTS: GENESIS OF WIND AND WEATHER

The earth's rotation sets up major currents in the atmosphere and oceans. These currents promote rain formation and its precipitation on Pacific Northwest forests. The currents also foster the relatively mild climate of England, and they create the Sahara and the world's other great deserts.

Air Currents: Global Treadmills Hot air rises, cool air sinks, and the earth turns beneath this mass of mixing atmosphere. Because of the intense sunlight at the equator, hot air rises at the equator and moves north and south at high altitudes. As it moves aloft, the air cools, becomes heavy, and breaks up into six coils of moving air owing to the rotation of the earth [FIGURE 41.4].

The direction of air flow and the ascent and descent of air masses in their giant coils determine the earth's climatic zones and their vegetation types. Let's consider

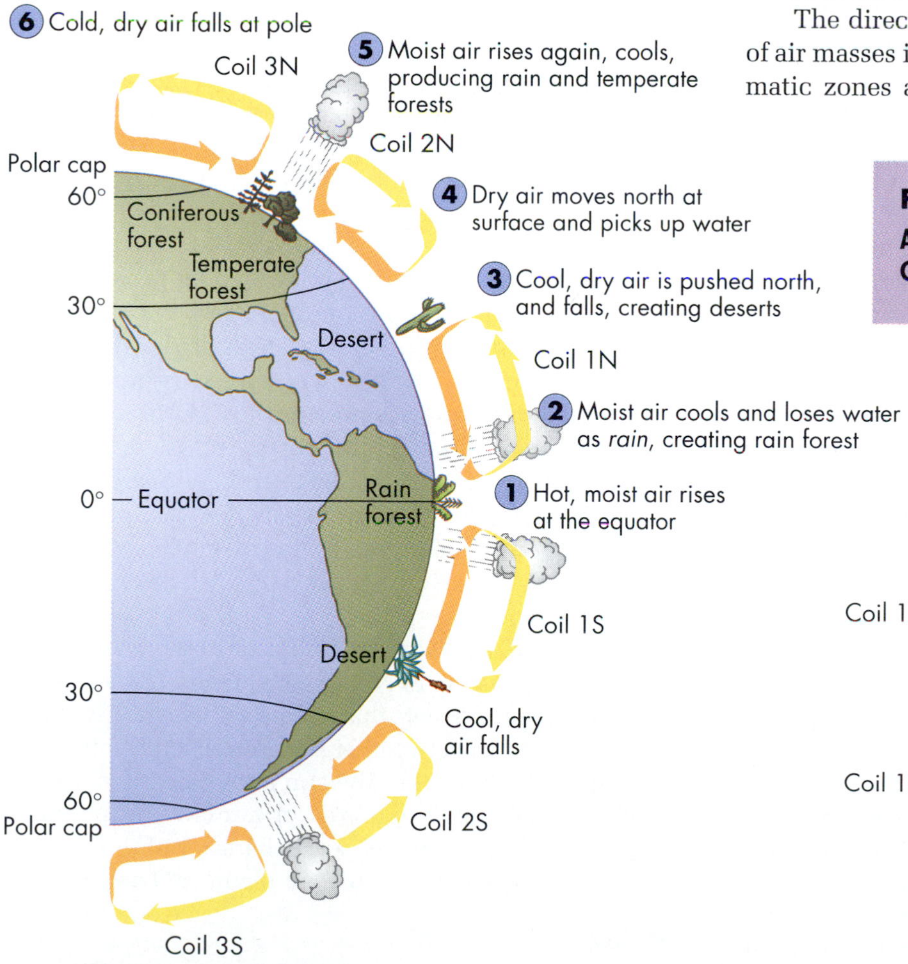

[A] Hot air rises, and cool air falls

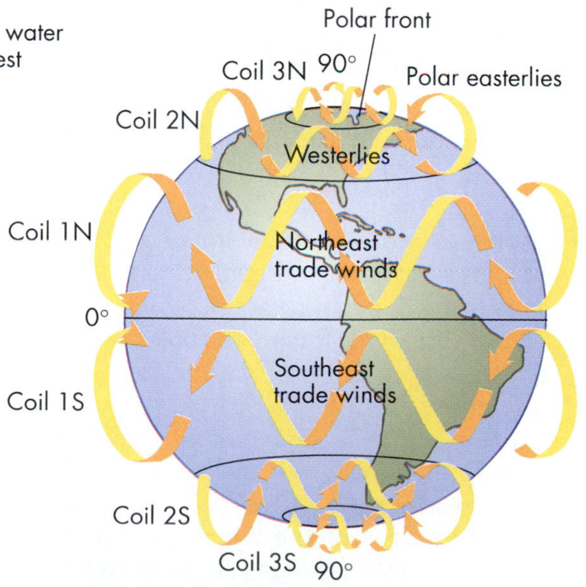

[B] The rotating earth generates prevailing winds

FIGURE 41.4

Air Ascending and Descending Through Massive Coils Creates the Earth's Climatic Zones.

FIGURE 41.5

Ocean Currents Flow in Four Major Gyres, Redistributing Heat.

This computer-generated satellite image charts ocean temperatures. Red is 29°C (84.2°F) and orange is 28°C (82.4°F). From water temperatures alone, could you detect evidence of the ocean gyres if the arrows were not drawn on the map?

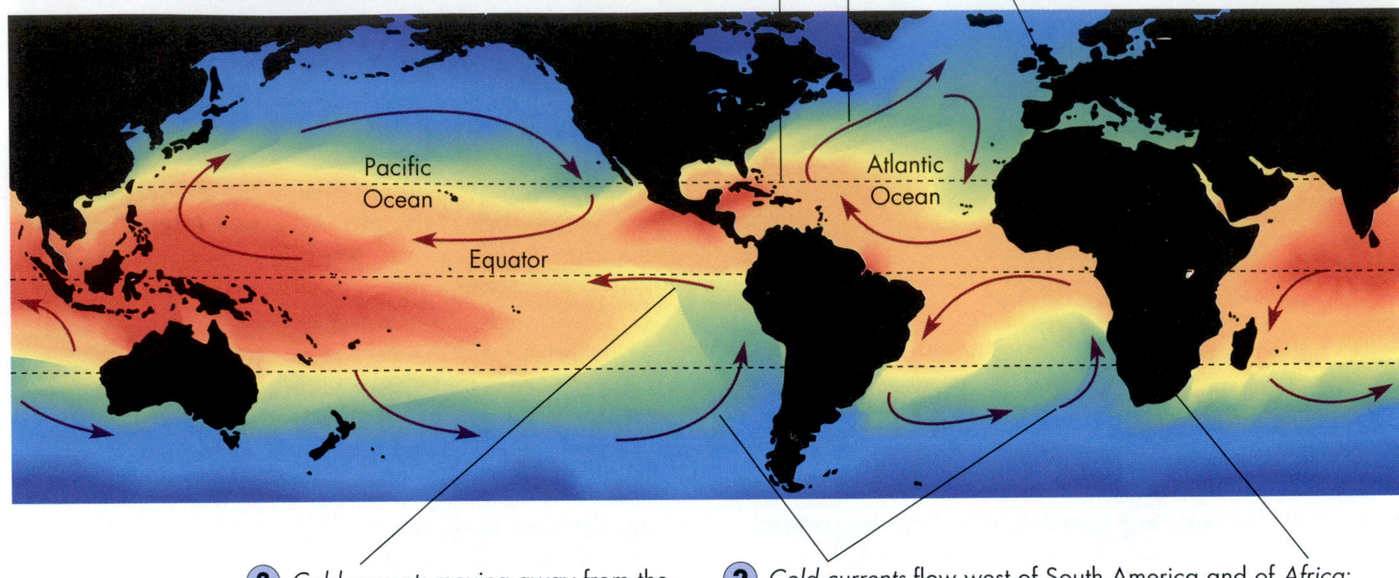

coil 1 north and coil 1 south, for example. Warm, moist air rises at the equator and drops its moisture in heavy tropical rains as it cools, creating the conditions for tropical rain forests around the equator [see FIGURE 41.4A]. The cooler, drier air now travels at high altitude and descends at latitude 30° (both north and south). This descending dry air creates the great deserts of Australia, North and South Africa, and North America. At ground level, the dry air moves toward the equator, interacts with the rotating earth, and causes surface winds. These predictable breezes, called *trade winds*, propelled traders' sailing ships in past centuries.

Analogous currents produce winds from west to east over much of North America [see FIGURE 41.4B]. This flow carries moisture-laden air from the Pacific Ocean over the Coast Range and Cascade Mountains of the Pacific Northwest. There it falls as the rain and snow that encourage the lush growth of temperate rain forests.

Mild air moving toward the poles in coil 2 meets cold air flowing from the poles in coil 3 [see FIGURE 41.4A]. This convergence (the *polar front*) causes the air to rise again at about 60° latitude north and south, cooling and giving up the moisture it picked up as it moved across land and ocean surface. This moisture supports the great temperate forests of the Americas, Europe, and Asia. Once aloft, some of the air moves at high altitude to the poles. There, the dry, frigid air again descends, moves over Canada, and occasionally dips deep into the United States, bringing along polar weather.

The six coils of circulating air help us understand why deserts and forests occur where they do; they also drive the circulation of ocean currents.

Ocean Currents Wind blowing over the ocean surface sets the sea's upper layers moving, and the continents redirect the flow in slow, circular patterns called *gyres* (JYRZ; Latin "circle") [FIGURE 41.5]. Major gyres occur in the North and South Atlantic and the North and South Pacific. Ocean currents, like air currents, redistribute heat and hence influence climate and the distribution of plants and animals. The Gulf Stream, for example, carries warm water from the tropics up the eastern coast of North America, then eastward across the Atlantic, warming northern Europe.

Short-term irregularities do occur in wind and ocean currents, such as the El Niño event, a temporary reversal of certain ocean currents that occurs about every five years or so, that can lead to weather shifts on several continents. Nevertheless, in general, the sun and the earth's tilt and rotation create stable patterns of rainfall and temperatures that determine the general character of the biomes, the earth's major communities of plants and animals.

➤ CONCEPT CHALLENGE

Consider the climate where you live. In winter, is it cold and dry or mild

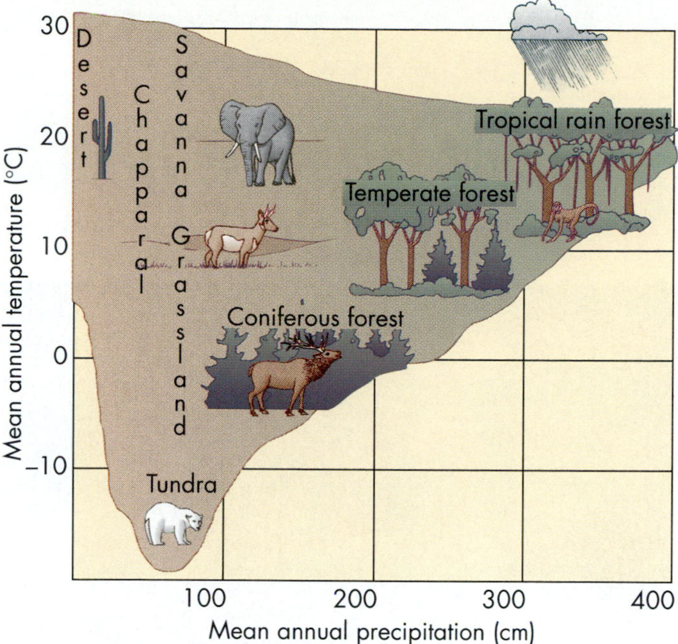

FIGURE 41.6

Temperature and Moisture Help Govern Biome Distribution.

A plot of temperature versus precipitation shows that tropical rain forests occur only where it is hot and moist, deserts only where it is very dry, and tundra only where it is cold and relatively dry. Other biomes also occupy characteristic positions on the graph, depending on moisture and temperature.

and wet? In summer, is it hot and humid or mild and dry? Try to explain these characteristics in terms of the angle at which the sun's rays strike the earth where you live, the tilt of the earth's axis, the six coils of air and the prevailing winds they help create, and the prevailing ocean currents nearest your community.

Biomes

Wherever similar climatic conditions exist—the deserts of Australia and Africa, the rain forests of Indonesia and Brazil—plants have evolved similar adaptations that help them exploit the climate's benefits and minimize its drawbacks. **Biomes** are large geographic regions containing distinctive plant communities. Major plant types characterize biomes because plants best reflect adaptations to rain, temperature, light, and other specific climatic conditions, and as primary producers, plants influence the consumers and decomposers that coexist in the biome.

It becomes obvious that climates determine biomes when we study a biome distribution map. Such a map reveals orderly patterns of tropical forests, deserts, deciduous forests, coniferous forests, and tundra lying roughly in bands stacked south to north that correspond with patterns of atmospheric circulation and climate.

Ecologists recognize as few as 7 or as many as 15 or more biomes, depending on how they classify the areas of species mixing between biomes. Here we consider 8 biomes, arranged roughly in order from equator to poles—tropical rain forests, savannas, deserts, temperate grasslands, chaparral, temperate forests, coniferous forests, and tundra—plus mountains and the polar ice caps.

Two major climatic features—temperature and moisture—dictate which biome will occupy a given place on earth. FIGURE 41.6 is a graph of temperature plotted against moisture. A line showing the boundary conditions compatible with life marks out a roughly triangular area. Each of the eight biomes is indicated on the graph near the sets of conditions it requires. Tropical rain forests, for example, appear in regions with high amounts of precipitation and high temperatures, while tundra appears in regions with cold and moderately dry climates. The next few sections of the chapter take brief looks at each of earth's major biomes and their main characteristics.

TROPICAL RAIN FORESTS

Nearly half of all living species reside in the world's warm, wet tropical **rain forests**, even though these lush regions represent only 3 percent of the continental surface area [see FIGURES 41.6 and 41.7]. Tropical forests occur near the equator in Central and South America, Africa, and Southeast Asia, where hot air rises and then dumps its moisture. Rainfall is 200 to 400 cm (80 to 160 in) per year, and temperatures average about 25°C (77°F). Jungle inhabitants are clearly not limited by water supply or cold temperatures, but mineral nutrients can be in short supply, and the dense vegetation blocks the wind and light near the ground. These limitations have shaped the rain forest biome.

The struggle for light has led to three main levels of plant height above the forest floor [FIGURE 41.7A]: (1) Within the upper story, or *emergent layer*, trees up to 50 m (164 ft) tall rise above the surrounding vegetation, where they capture direct sunlight. (2) Beneath the emergent layer lies the *canopy*, the overlapping tops of shorter forest trees. (3) The canopy is so dense with leaves and branches that only dim light penetrates to the third level, the *understory*, and less than 2 percent of the light eventually reaches the *forest floor*. Because canopy trees do not usually branch in the understory, a person can often walk unimpeded through the humid twilight that filters to ground level in a mature rain forest. The popular image of an impenetrably tangled jungle is accurate only at the edges of cleared forests, along riverbanks, or where a large tree has fallen and the sunlight reaches the ground. Much of the rain forest activity takes place in the canopy, where leaves, flowers, and fruits abound.

FIGURE 41.7

Tropical Rain Forest: Lush Equatorial Biome.

[A] Only a few tall trees and vines reach the emergent layer of the rain forest (top), while shorter forest trees form a canopy (upper middle) in which epiphytes and animals thrive. Low trees, shrubs, and underbrush grow in the shady understory (lower middle), and fallen leaves and a few seedlings carpet the scarcely illuminated forest floor (bottom). **[B]** Sunlight filtering through mist illuminates the lush foliage surrounding the Segama River in Borneo's lowland rain forest.

[A] Layers in the rain forest

[B] The lush tropical rain forest

Because tropical rain forests have poor soil, roots tend to be shallow, and the trunks of large trees are often expanded into *buttresses*, thick flanges that provide support [see FIGURE 41.7A]. Roots occupy the thin upper layer of soil, where mineral nutrients occur in highest concentrations. Tropical rain forests have been called "wet deserts," because heavy rains tend to leach nutrients from the deeper soils; thus, the biomass of the living forest itself is the biggest source of nutrients. Fallen trees, dropped leaves, and dead animals quickly decay in the warm, damp environment, and their nutrients rapidly recycle. When people slash and burn the vegetation to create farmland, most of the nutrients literally go up in smoke. For this reason, former rain forest land does not support cultivation for long periods.

Despite the timeless beauty and incredible species richness of tropical rain forests, people are rapidly destroying them to help feed their own expanding populations or to harvest hardwood trees or graze cattle for export. At the current rate people are destroying an area about the size of a football field *every second*. This destruction threatens the survival of millions of species, and Harvard ecologist E. O. Wilson estimates that 74 species disappear every day [review BOX 18.1]. Since about 25 percent of all prescription drugs are plant-based, many additional useful medicinal plants will surely be lost before scientists discover and study them. In addition, destruction of the tropical rain forest is a threat to all biomes, because these productive forests are the earth's biggest repositories for carbon. Thus, burning the great jungles pours carbon dioxide into the air, and according to some climatologists, this can contribute to global warming and the alteration of climatic patterns all over the globe.

TROPICAL SAVANNAS

Whereas tropical rain forests grow only where it is constantly wet and warm, year-long warmth with an extended dry season results in a tropical **savanna**. Situated between tropical forests and deserts, savannas contain

stunted, widely spaced trees with tall grasses growing in between [FIGURE 41.8]. The world's major savannas lie in Africa, South America, Australia, and parts of Southeast Asia. While the center of life in the rain forest remains high in the canopy, the savanna's productivity is concentrated near the ground, where the leaves and seeds of grasses and short trees are well within reach of grazing animals like Cape buffalo and kangaroos. The herbivores, in turn, support lions, dingoes, and other carnivores.

Trees in the African savanna have peculiar-looking, flat tops. This shape is apparently an adaptive compromise—the high lower branches limit predation by animals, and the flattened top prevents excessive energy loss in the dry heat. Many savanna species, including the rhinoceros and the elephant, are threatened with extinction because people destroy their habitats and kill them for their beautiful coats, ivory tusks, or spectacular horns.

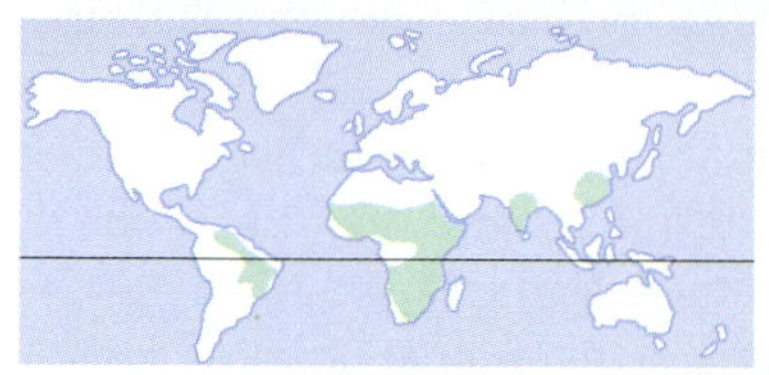

FIGURE 41.8

Savanna: Tropical Dry Land.

Giraffes lope along the grassy savanna in a Kenyan wildlife preserve.

DESERTS

Desert regions receive less than one-tenth of the annual rainfall of tropical rain forests; hence, desert plants are widely spaced and cover less than one-third of the ground surface [FIGURE 41.9]. Some desert areas in northern Africa have not seen rain in over a decade, and are essentially lifeless tracts of shifting sand.

Some deserts, like those of Mexico, Australia, and North and South Africa, are dry because they lie directly below the zones where dehydrated air—rained out in the tropics—descends back toward earth [see FIGURE 41.4]. Others, like China's Gobi Desert, are dry because they lie at the centers of huge continents, far from the moist sea air. Finally, some deserts form because they are located downwind of tall mountain ranges. Before reaching the Great Basin of Nevada and Utah, for example, moist air gains altitude and cools as it approaches and then crosses the Sierra Nevada and White mountains, and the mois-

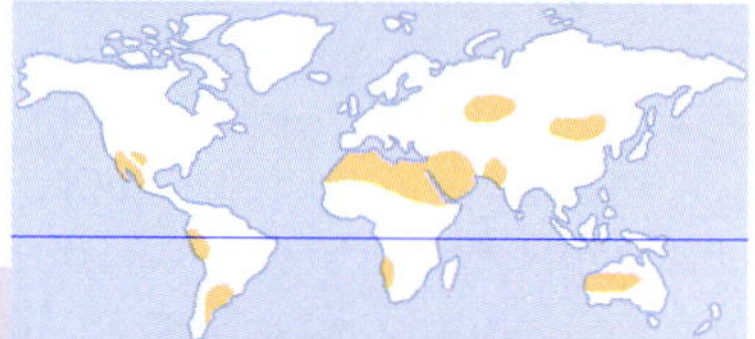

FIGURE 41.9

Desert: Parched Zone Where Life Is Sparse.

The Namibian Desert of southwestern Africa receives almost no rainfall and supports few plants save scrubby, drought-tolerant trees and a handful of odd, dry-adapted species like *Welwitschia.*

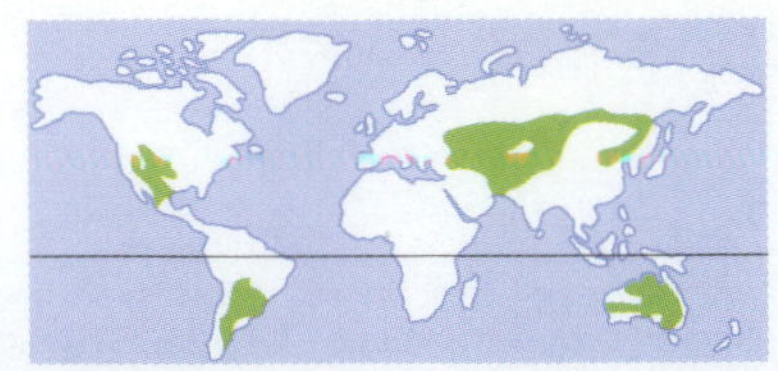

FIGURE 41.10

Temperate Grassland: Treeless Sea of Grass.

Vast herds of bison once grazed the extensive grasslands of central and western North America. Today, just a few protected remnant herds remain, such as these in South Dakota.

ture falls as rain or snow on the windward side of the peaks. The air that descends to the basins beyond is thus dry, and the land in the rain shadow becomes a desert.

During the day, deserts are generally hot as well as dry, because on cloudless afternoons, the relentless, blistering sunlight can send temperatures soaring to 49°C (120°F). On cloudless nights, however, heat radiates from the desert back to space and can leave a biting chill to the air. Many desert plants, including African succulents such as ice plant and American cacti such as prickly pear, have numerous adaptations—fleshy stems, spines, and so on—that help them cope with searing heat and the potential loss of precious moisture.

Desert animals have also evolved water conservation strategies. Most are active only in the evening or early morning hours, keeping to their burrows, crevices, or shaded areas during the heat of the day. As CHAPTER 29 described, the North American kangaroo rat can survive without drinking by producing metabolic water from nutrients in plant seeds and excreting urea in practically crystalline form [see FIGURE 29.11].

Ironically, while human activity is decreasing the areas of tropical rain forests, our actions are causing a process called *desertification*, and with it, an expansion of the world's desert biomes by an area the size of Colorado every year. Deserts enlarge because overgrazing and poor irrigation practices remove grass from the grasslands that rim the deserts. The loss of ground cover allows fine, nutrient-bearing soil particles to blow or wash away, leaving behind only sand, gravel, and other coarse materials that cannot hold water.

TEMPERATE GRASSLANDS

Bordering many of the world's deserts are **grasslands**, treeless regions dominated by dozens of grass species [FIGURE 41.10]. Known as *prairies* in North America, the *pampas* in South America, the *steppes* in Asia, and the *veldt* in Africa, these regions are wetter than deserts but drier than forests. The flat expanses of Kansas and the gently rolling hills of Nebraska were typical grassland terrains before the encroachment of farms and ranches. Tall-grass prairies are now quite rare. Unlike savannas, grasslands lack trees and have seasonal extremes of hot and cold rather than wet and dry. Nevertheless, the windswept plains tend to dry out in the summer and fall, so fires are a recurrent feature. As a result, grassland species have evolved the ability to regrow rapidly after a fire. This natural devastation, often touched off by lightning, prevents grasslands from turning into forests. Open grasslands support animals like bison, antelope, prairie dogs, anteaters, armadillos, coyotes, snakes, and hawks.

Grassland soils are richer in organic matter than the soils of other biomes. Grasses have pervasive roots and underground stems, often penetrating to 2 m (6.5 ft) and weighing several times more than the aerial leaves and stems. The accumulated debris from generations of these plants makes grassland soils thick and fertile.

FIGURE 41.11

Chaparral: A Biome of Low Bushes and Frequent Fires.

The rolling hills and scrubby plants of Los Padres National Forest in California.

CHAPARRAL

A biome called the **chaparral** (shap-uh-RAL; Spanish, "thicket"), or temperate scrublands, borders grasslands and deserts along the shores of the Mediterranean and along the southwest coasts of North and South America, Africa, and Australia [FIGURE 41.11]. Chaparral has hot, dry summers and cool, wet winters. Chaparral plants are generally less than 2 m (6.5 ft) tall and have hairy, leathery leaves that stay green all year. Many chaparral plants, like sage and manzanita, have spicy, aromatic odors; the natural alkaloids probably help deter insect herbivores. Because grasses grow during the wet winters and then die and dry out during the hot summers, this biome, like the grasslands, experiences frequent fires. The belowground portions of chaparral perennials have evolved fire resistance, and the seeds of some chaparral annuals must be seared by fire before they germinate. The devastating fires in southern California in recent years came about because small fires had been suppressed for many years. This allowed natural fuel to accumulate, and the result was large cataclysmic fires that consumed many homes built within or adjacent to large areas of the tinder-dry chaparral.

FIGURE 41.12

Temperate Forest: Deciduous Hardwoods Bordering Grasslands.
The new green leaves of spring brighten Tennessee's Great Smoky Mountain National Park.

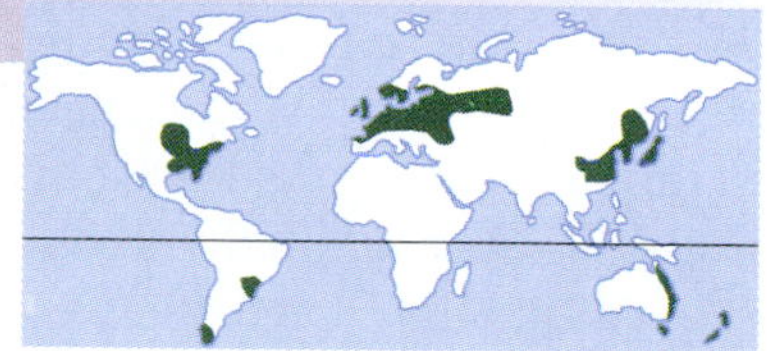

TEMPERATE FORESTS

Continuing to move farther north and south from the equator, we leave the grass- and shrub-dominated biomes and enter the **temperate forests**, such as those of eastern North America [FIGURE 41.12], Europe, China, Japan, New Zealand, Australia, and the tip of South America. These forests are dominated by broad-leaved trees, and as the name suggests, they have intermediate amounts of rainfall and fairly moderate temperatures that fluctuate between summer highs and winter lows. They are about twice as productive as grasslands. People have utilized these forests so heavily for timber that in the United States, only 4 to 5 percent of the virgin forests that once covered much of the land area remain.

Seasonal temperature changes have strongly shaped the evolution of temperate-forest organisms. The dominant trees, including oak, maple, birch, and hickory, drop their leaves and become dormant until spring. The fallen leaves that accumulate on the forest floor allow for the recycling of nutrients and produce an excellent topsoil. In early spring, the leafless trees permit sun to fall unobstructed to the forest floor, and spring wildflowers like wood sorrel, bluebells, and violets bloom. Later, when the canopy leaves enlarge and shade the forest floor, only shade-tolerant plants such as ivy and honeysuckle thrive.

The animals of temperate forests also have adaptations that help them cope with the changing seasons. Many, like bears and snakes, hibernate in winter, while others, like robins, migrate to warmer regions.

CONIFEROUS FORESTS

At latitudes with cold, snowy winters and short summers, vast forests of conifer trees grow in the **coniferous forest** (also called *taiga* [tie-GUH] or *boreal forest*) [FIGURE 41.13]. This broad band of mixed pine, fir, spruce, and hemlock trees stretches across much of Canada, northern Europe, and Asia. A few broad-leaved trees such as aspens also survive in these areas around streams or lakes. The southern border of the coniferous forest tends to fall at the southern limit of the polar front, where cold air sweeping down from the pole meets warm air pushing up from the south [review FIGURE 41.4]. The southern hemisphere has little coniferous forest because ocean rather than land occupies the appropriate latitudes. Coniferous forests are less productive than temperate forests, but are more productive than grasslands.

During winter in the coniferous forest, ice and snow lock up water, making it unavailable to trees. Conifer leaves, which are needle-shaped and have thick, waxy cuticles, combat water loss. Since these needles are not shed each winter, they can begin collecting sunlight as soon as the short growing season begins. Many conifers have sloping limbs that shed snow easily and tend not to break under the weight. Coniferous forests support bark beetles, elk, porcupines, wolves, and lynx.

FIGURE 41.13

Coniferous Forest: Fragrant Evergreens in the Northern Latitudes.

A stand of deep-green conifers grows in Mount Rainier National Park, Washington.

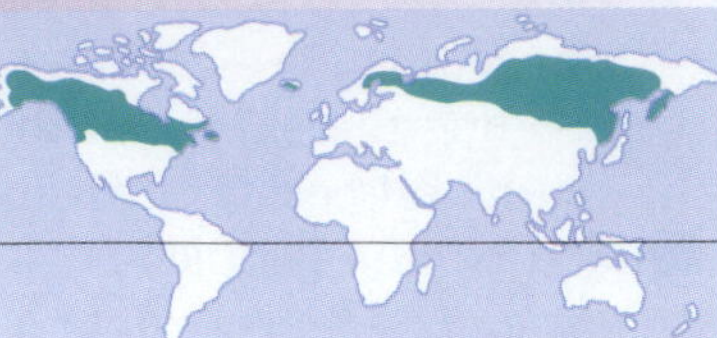

The *temperate rain forest*, found only in the Pacific Northwest and parts of New Zealand, is a biome closely related to the coniferous forest [review FIGURE 41.1]. Coniferous forests contain the world's greatest lumber reserves. Traditionally, loggers have harvested this timber by *clear-cutting*: They saw down all of the standing trees, burn the land to clean off rotting logs, and spray with herbicides to kill competing vegetation; then they plant seedlings of a single tree species, usually Douglas fir, in place of the original forest to grow and be cut again in about 70 years. Biologists wondered how many crops could be taken from a given hillside by clear-cutting before nitrogen and other soil nutrients became depleted. They found that the major source of fixed nitrogen for old-growth forests is probably lichens. Recall from CHAPTER 20 that those peculiar gray-green, orange, or yellowish organisms consist of fungal threads surrounding nitrogen-fixing cyanobacteria, and often coat tree trunks and limbs high in the leafy forest canopy [CONCEPT INTEGRATION, page 902, Steps 1 to 3]. When parts of the lichens fall to the ground and rot, nitrogen becomes directly available to trees and forest mushrooms, and indirectly available to voles, spotted owls, and the entire ecosystem. The scientists also learned, however, that these lichens don't really begin to thrive until the canopy is about 100 years old. Since the replanted fir forests grow and are recut long before this lichen covering develops, nitrogen may well become depleted and future tree crops jeopardized.

Some modern foresters are now suggesting a "new forestry"—the removal of timber the way it occurs in nature, say, during a fire or windstorm: only a few of the mature trees fall. Left in place are most of the live trees with their nitrogen-fixing lichens, as well as many standing dead trees, and tons of rotting logs, which become homes for wildlife, hold soils in place, and serve as hospitable environments for the sprouting of tree seeds.

TUNDRA

At the northern boundary of the coniferous forest, the polar high-pressure center blows cold, dry air down from the pole, even in summer [review FIGURE 41.4]. Here, the fragrant, giant conifers give way to the low vegetation of the **tundra**, a cold, treeless plain [FIGURE 41.14]. The annual temperature in this frosty biome averages −5°C (23°F) or less, and even in the summer, the tundra soil thaws to only about 1 m (39.4 in). The deeper, permanently frozen soil is called *permafrost*, and this subterranean zone inhibits the establishment of most trees. The soil that does thaw in summer stays soggy and dotted with pools because the land is largely flat, and runoff is slow.

Frigid weather and a short growing season have heavily influenced the evolution of tundra organisms.

FIGURE 41.14

Tundra: Cold, Treeless Plain in the Far North.

The low vegetation in Alaska's Denali National Park erupts with the russets and golds of autumn.

FIGURE 41.15
Polar Caps: Icy Regions at the Highest Latitudes.
A huge polar bear waits beside a seal's breathing hole near the Arctic Circle.

Plants, including grasses, sedges, mosses, lichens, and heathers, grow low, where they can absorb warmth reradiated from the solar-heated ground and minimize the effects of wind and desiccation. These plants support herbivores like caribou, arctic hare, and lemmings, and they, in turn, are prey for foxes, weasels, and snowy owls. During summer, clouds of mosquitoes and flies fill the air, reproducing in the temporary pools. Many birds migrate to the tundra in the summer, feasting on the insects and breeding during the long summer days.

The tundra is only one-sixteenth as productive as the tropical rain forest, and it is also very vulnerable—the track of a single vehicle can mar the terrain for decades.

POLAR CAP TERRAIN

The arctic **ice cap**, a vast frozen ocean covering the planet's North Pole [FIGURE 41.15], and the continent of Antarctica, a landmass lying beneath a great raft of ice and surrounded by frigid seas, take up millions of square kilometers of land and ocean surface. These icy regions are not considered true biomes, because they lack major plants; only animals and microbes survive in these regions. In the Arctic, polar bears hunt seals and other aquatic animals, while penguins and seals inhabit the ice shelf at the edge of the Antarctic Ocean. The only true terrestrial animals of Antarctica are a few insects, the largest of which is a wingless mosquito.

LIFE ZONES IN THE MOUNTAINS

Just as ice covers both poles at extremely high latitudes, glaciers and snowfields cap high mountains—even near the equator. In fact, a hiker descending from a high peak can often pass through a series of biomes that reverses the order we have just reviewed from the equator to the poles [FIGURE 41.16]. Leaving the snowfields at the top of a mountain in Mexico—let's say the ancient volcanic cone called Citlaltepetl, which is 5754 m (18,878 ft) tall—the traveler descends through an alpine tundra, then a coniferous forest similar to Canada's boreal zones, followed by a deciduous forest like that of the eastern United States, and finally enters a lush tropical rain forest at the mountain's foot. Regions of varying elevation, climate, and organisms in the mountains are called **life zones**. Life zones occur because precipitation and temperature vary with altitude, just as they do with latitude, and each combination of temperature and water influences the kinds of organisms that can survive in those conditions.

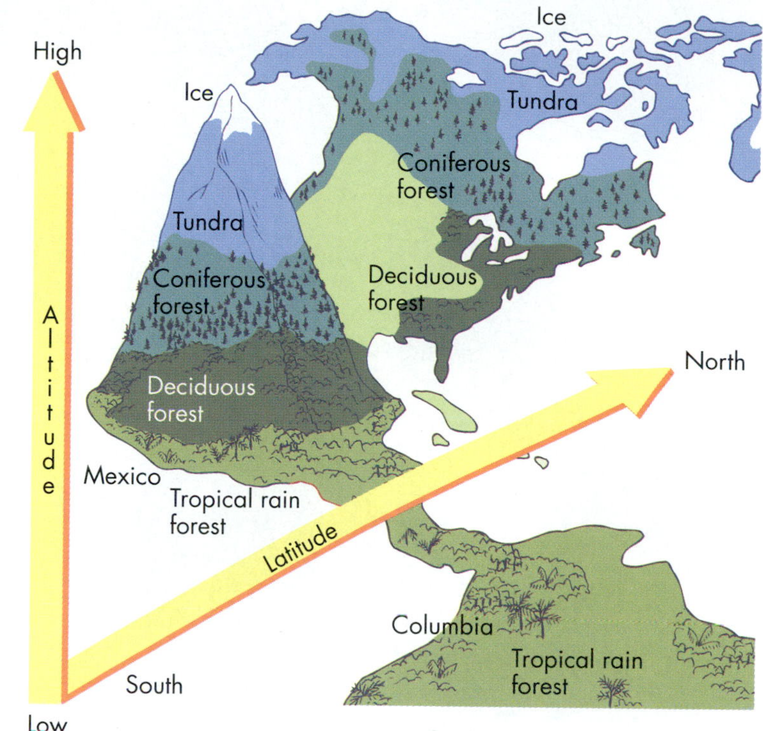

FIGURE 41.16
Life Zones in the Mountains: Altitude Mimics Latitude.
In descending a high mountain in the tropics, one passes through altitude zones that correspond to latitudes. This diagrammatic approach shows the correspondence between mountaintop and ice cap, alpine regions and tundra, and so on, down to the tropical rain forest shrouding the mountain's base.

Just as climate helps shape terrestrial biomes, it has a similar shaping influence on life in the water—a realm that is less familiar to most of us but is central to our understanding of the biosphere.

➤ CONCEPT CHALLENGE

Which biome surrounds the place you live? Describe some adaptations of the local plants and animals that allow them to survive in your biome.

1: Lichens fix nitrogen from air in organic molecules.
2: Lichens fall and release nitrogen.
3: Tree roots take up nitorogen and water via fungi (mycorrhizae).
4: Mycorrhizae produce mushrooms.
5: Voles eat mushrooms.

Concept Integrator:
Ecological Interactions in an Old Growth Forest
6: Voles distribute spores in forest.
7: Spores sprout and make more mycorrhizae.
8: Old trees fall, rot, and produce homes for voles.
9: The moist log protects voles from forest fires.
10: After the fire, voles disperse fungal spores, which become mycorrhizae, which help new seedlings grow faster.

Life in the Water

Three-fourths of our planet's surface is covered with water, and most of that area is **marine**—salty oceans and seas. Although less than 1 percent of the surface is covered by freshwater lakes, rivers, ponds, streams, swamps, and marshes, this still amounts to tens of thousands of square kilometers. Ecologists do not consider marine or freshwater communities as biomes, but their studies show that aquatic ecosystems differ from each other in significant ways.

PROPERTIES OF WATER

As we saw in CHAPTER 2, water has unique physical characteristics, and these properties strongly influence aquatic organisms. For example, it takes much more energy to heat or chill water than it does to change air temperature. Aquatic organisms therefore suffer fewer rapid temperature changes than their terrestrial counterparts. In addition, air grows steadily denser as it cools, but water has its greatest density at 4°C (39°F). Because of this, ice (at 0°C, or 32°F) floats. Were this not the case, lakes would freeze from the bottom up, only the upper layers would thaw in summer, and bodies of water outside of the tropics would support little life.

FIGURE 41.17
Mountain Stream: Cold, Clear, and Oxygen-Rich.
A narrow stream cascades over rocks and past mossy banks in Michigan's Alger County.

FRESHWATER ECOSYSTEMS

Ecologists divide freshwater ecosystems into bodies of running water and bodies of standing water.

Running Water A stream that originates from a mountain spring or from melting snow is cold and clear but contains few nutrients. As a consequence, photosynthetic organisms living in the streams contribute less to the biotic system than organic materials washed into the stream from the surrounding watershed. As the water tumbles downhill across rocks [FIGURE 41.17], it picks up oxygen and more nutrients. Species that live in these turbulent zones include algae, mosses, and trout, which thrive only where oxygen is plentiful and water temperatures are low. As nutrients accumulate and the stream widens and is less shaded, the photosynthetic productivity of algae and green aquatic plants increases. A river's middle stretch has the greatest species diversity. In the lower section of a large river, the water becomes cloudy and enriched with nutrients and the water moves more sluggishly. The decreased light diminishes the rate of photosynthesis, and the river once again has low primary productivity. In these areas, microbes and invertebrates live on detritus in the bottom sediments. Catfish and bass replace trout as the amount of dissolved oxygen decreases in the slower-moving stream.

All over the world, people tend to use rivers as sewers. Besides pouring wastes into rivers—wastes that kill native species—people have dammed, diked, and channeled rivers in ways that often create worse problems than the ones they sought to avert.

Standing Water As a stream enters a lake, the current slows still further, and most suspended particles settle to the bottom. Those particles that remain suspended block light penetration and determine the maximum depth at which photosynthesis can occur. In shallow areas, light can penetrate to the lake's bottom. This is the well-lit *littoral zone*, with abundant producers like water lilies, cattails, and algae, and consumers such as insects, snails, amphibians, fishes, and birds [FIGURE 41.18].

Farther away from the shoreline, the standing water has two main regions: the top layer through which light can penetrate (the *limnetic zone*) and the dark bottom region (the deep, dark *profundal zone*) [see FIGURE 41.18]. The limnetic zone supports huge populations of phytoplankton, such as algae, and zooplankton, such as small crustaceans, while the deep profundal zone supports mainly insect larvae, scavenger fishes, and decomposers.

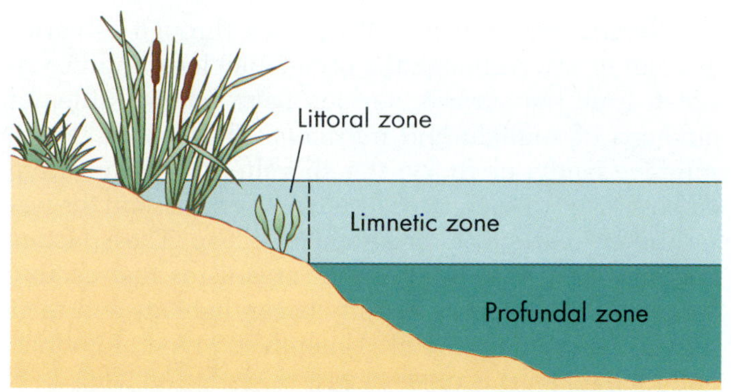

FIGURE 41.18
Freshwater Lakes: Zones of Light Support Life.
Cross-sectional view of a lake, showing various zones of depth and light.

Dissolved nutrients are as important as light in these biotic zones. For instance, an excess of the nutrient phosphate can cause a bloom of plant growth in a lake or pond. Under normal circumstances, nutrients such as phosphates accumulate in bottom sediments, and in areas where the winters are cold, these nutrients can be stirred up and cycled as the seasons change.

Nutrients cycle in a lake as changing temperatures cause water to rise and fall. The central fact is that surface water at 4°C sinks below water that is either warmer or colder. During summer, warm, light layers of water float above cold, heavy ones along a boundary called a *thermocline*. As air temperatures drop in autumn, surface water cools to 4°C, becomes heavy, and sinks to the bottom of the lake, carrying oxygen with it. This process, called *fall turnover*, stirs the lower water, and as it rises, it brings nutrients up to the photosynthetic layer of the lake. As winter sets in, ice begins to float on the lake's surface. Below the ice, water becomes gradually warmer with depth, reaching perhaps 4°C at the very bottom. When the lake's ice cover melts in spring, the upper layers once again warm from freezing to 4°C, when they again sink to the bottom through the colder, less dense water below. This again churns up nutrients from the bottom in a *spring turnover* and nourishes a quick bloom of algae.

SALTWATER COMMUNITIES

Saltwater communities include estuaries and oceans.

Estuaries Fresh and salt water mingle in **estuaries**, where rivers meet oceans [FIGURE 41.19]. In these areas, temperature and salt concentrations vary widely because of the daily rhythm of the tides and seasonal variations in stream flow. Estuary organisms like brine shrimp can often tolerate wide ranges of salinity. Constant water movements stir up nutrients and make estuaries some of the earth's most productive ecosystems and nursery grounds for many species of fish [FIGURE 41.20A]. In many populated areas, people dump urban and industrial effluents into estuaries or fill them with rocks and soil to claim new waterfront property. Such activities are significantly decreasing the great biological productivity of our estuaries. FIGURE 41.20B shows the comparative productivity of different terrestrial biomes. Notice that coral reefs and estuaries are comparable in productivity to tropical and temperate forests.

FIGURE 41.19
Estuaries: Confluences of Life.
In Australia, a silted, richly productive estuary marks the confluence of a freshwater stream and the ocean.

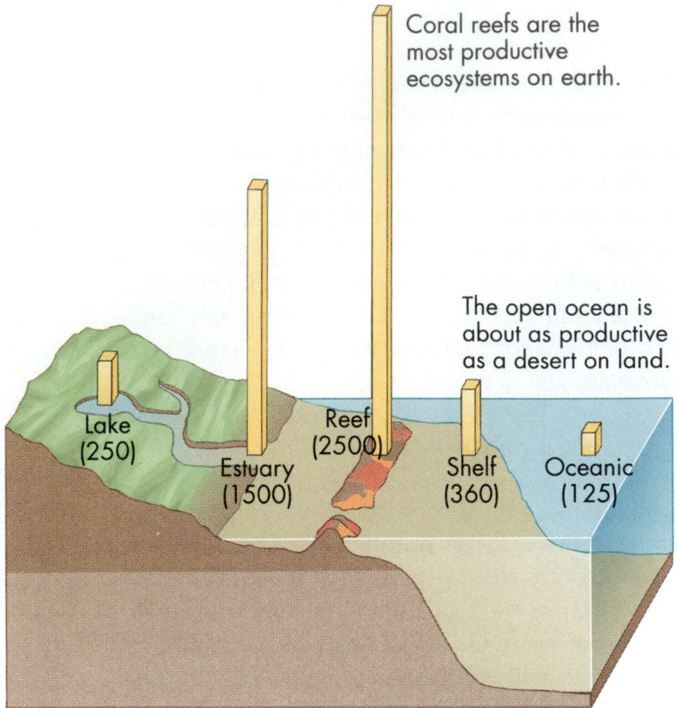

[A] Aquatic ecosystems

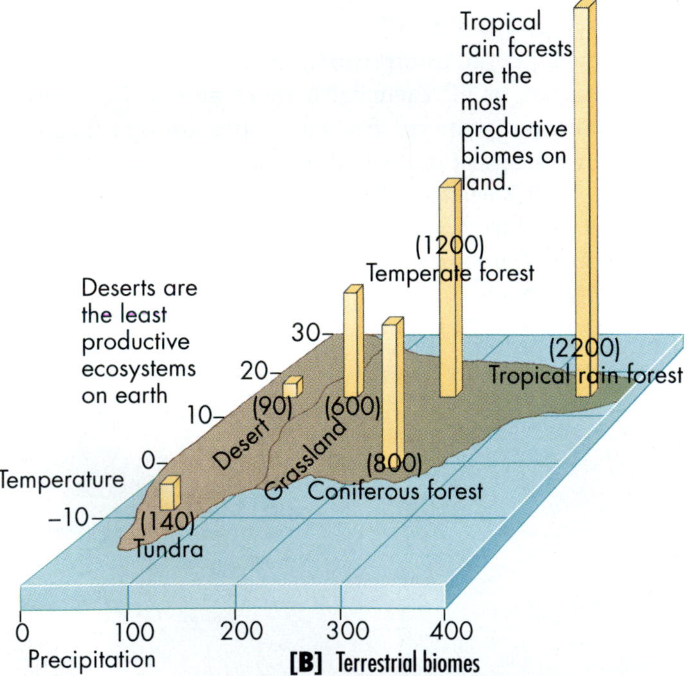

[B] Terrestrial biomes

FIGURE 41.20

Biological Productivity and Ecosystems.

The net primary productivity of different biomes, measured in terms of grams of biomass per square meter of area each year, varies in different aquatic ecosystems [A] and terrestrial biomes [B]. Productivity on land depends largely on temperature and moisture.

Oceans Water eventually passes through estuaries and out to sea. Although the ocean may look undifferentiated from the surface, various parts receive different amounts of sunlight and nutrients. The bottom can be muddy, sandy, or rocky; the shorelines can vary from smooth, sandy beaches to jagged, rocky cliffs; and the water can be centimeters or kilometers deep. These factors combine to dictate the types of organisms that occupy each region. Ecologists classify ocean habitats and their organisms on the basis of (1) depth, (2) type of bottom, and (3) light level [FIGURE 41.21].

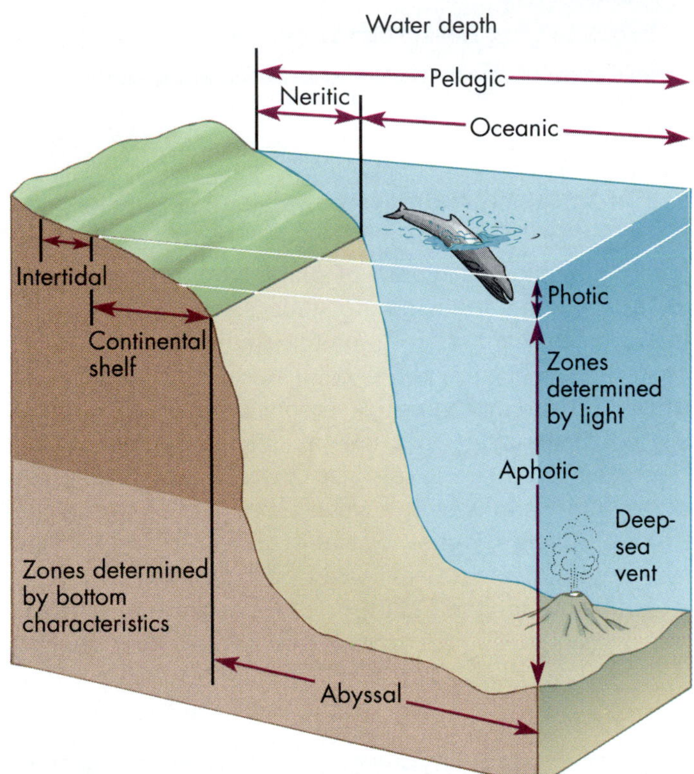

FIGURE 41.21

Ocean Zones: Depth, Bottom, and Light Determine Habitats of the Sea.

The ocean is a complex aquatic realm with several zones based on water depth, bottom characteristics, and light penetration.

At the ocean's edge lies a region that is underwater at high tide and exposed to air at low tide. In this region, called the *intertidal zone*, waves and wind often create violent pounding. Most intertidal producers (kelp and other algae) have structures that anchor them to rock surfaces [see FIGURE 20.13], while most intertidal animals

FIGURE 41.22

Tide Pools: Basins of Life in the Intertidal Zone.

Green sea anemones and purple sea urchins inhabit a tide pool on Coronado Island, Mexico.

have tough bodies or shells as well as underwater "glues" that help fasten the organisms to the rocks. Intertidal organisms tolerate hot, dry spells or changes in salinity as the tide rolls out each day. A group of burrowing organisms, including clams, snails, and worms, live beneath the extensive tidal mudflats. Many visitors to the seaside enjoy investigating *tide pools*, depressions formed in rock by wave action, and observing the sea anemones, sea urchins, crabs, sea stars, and fishes that live there [FIGURE 41.22].

The ocean's most productive region extends from the intertidal zone to the edge of the continental shelf, the submerged part of the continents. Producers in this nearshore (or *neritic*) zone include many species of large algae living in extensive beds, as well as huge populations of phytoplankton and zooplankton that serve as a nutrient base for the rest of the ocean's consumers. The largest animals that have ever lived, the blue whale and the whale shark, feed entirely on microscopic plankton.

In the tropics, the shallow ocean zone is home to reef-building corals and the photosynthetic microorganisms within them [FIGURE 41.23]. Eventually, these reef builders alter their own environment; the crannies and caves in their stony colonies create sheltered nesting sites and hiding places for the most diverse communities in the seas. Per unit area, coral reefs are the most productive ecosystems on earth; they are 10 times more productive than freshwater regions and 20 times more so than the open ocean [review FIGURE 41.20].

Beyond the continental shelf and covering the deep abyssal plane, the water in the *oceanic zone* is nearly deserted, even in the photic zone near the surface [see FIGURE 41.21] and the *benthic zone* (Greek *benthos*, "depth of the sea"), or sea floor. While the oceanic regions may have plenty of light, oxygen, and carbon dioxide, they lack nutrients such as phosphate and usable nitrogen, and overall they are less productive than the arctic tundra. The ocean's most productive areas occur mainly where currents cause nutrients to well up from very deep waters; a good example is the cold area off the coast of Peru [review FIGURE 41.5]. Although much of the open ocean is not very productive, it does have immense effects on the biosphere, because the oceans modulate climate and help to maintain favorable concentrations of oxygen and carbon dioxide in the atmosphere.

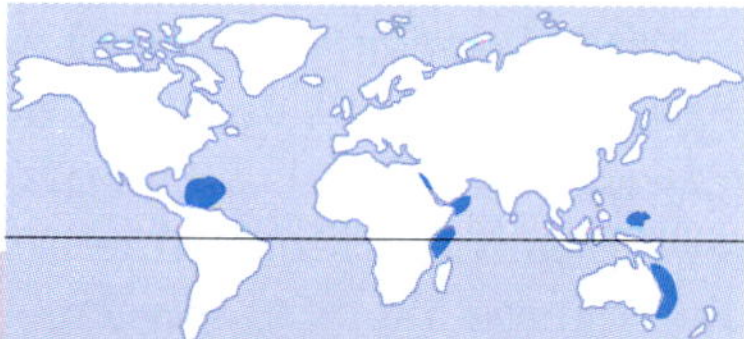

FIGURE 41.23

Coral Reef: Fantastic Productivity Beneath the Waves.

Brilliantly colored corals dot this shallow reef in the South Pacific.

Change in the Biosphere

Change is part of nature. Climates shift, continents drift, and organisms evolve, influencing their immediate surroundings and the larger biosphere around them. Recall from CHAPTER 17 how the evolution of photosynthesis about 2.8 billion years ago decreased atmospheric carbon dioxide and increased oxygen gas, causing the oceans to "rust." As we saw in CHAPTER 40, certain human activities are causing equally pervasive but much faster modifications in the biosphere—modifications that outstrip the speed with which organisms usually evolve adaptations. One of those possible changes is global warming caused by the greenhouse effect. In this section, we look at another of these changes, the degradation of the earth's ozone layer. Then we turn to some solutions involving the sustainability principle, the ability to utilize an ecosystem without causing permanent damage. Finally, we reconsider the old-growth forests of the Pacific Northwest that began this chapter, and ponder the ways this region might be a model for other threatened biomes.

THE OZONE HOLE

The **ozone layer** is a blanket in the atmosphere that absorbs about 99 percent of the ultraviolet light that would otherwise strike the planet's surface and destroy many biological molecules, including DNA. Ozone is a gas that contains three atoms of oxygen—you may have noticed the sharp-smelling ozone gas generated by electric sparks when a toy train is racing around its track. Ozone forms naturally when ultraviolet light streaming from the sun strikes atmospheric oxygen gas, with its two atoms of oxygen.

Ominously, in 1985, atmospheric scientists first reported decreases in ozone concentrations in the upper atmosphere. Decreases were especially strong over much of Antarctica. The "ozone hole," which is now larger than the European continent, contains only half as much ozone as it did a decade ago [FIGURE 41.24]. Recently, 20 percent decreases in ozone concentrations have been detected over the North Pole, and worldwide, the ozone has thinned by 3 to 5 percent.

Studies have shown that CFCs (chlorofluorocarbons), industrial chemicals containing atoms of chlorine, fluorine, and carbon, can act as catalysts that convert ozone to oxygen. These compounds get into the ozone layer as a result of human activities—the use of spray cans and plastic-foam containers, and from coolants escaping from refrigerators and air conditioners. (These items tend to be produced mainly by the smaller populations living in developed countries, rather than by the larger populations

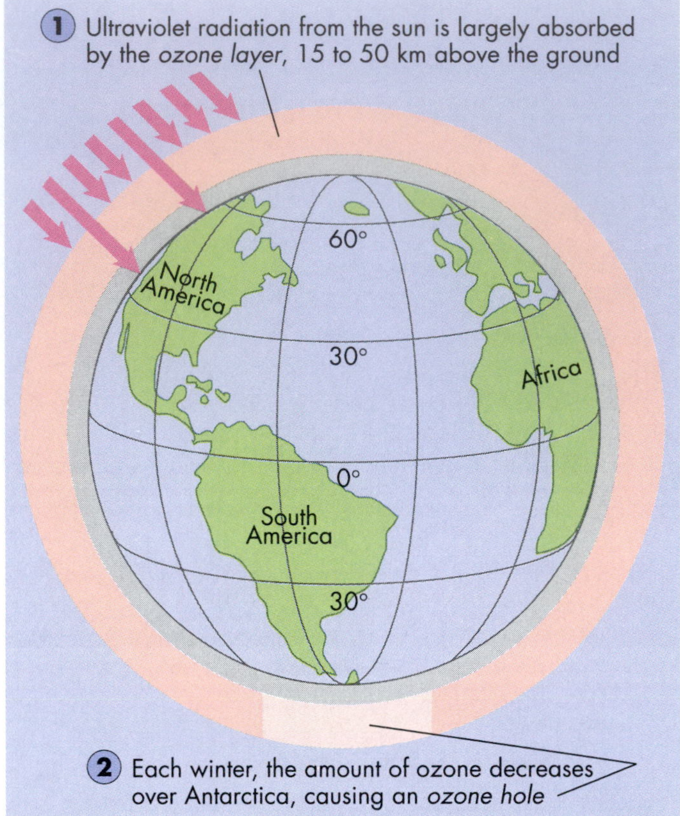

[A]

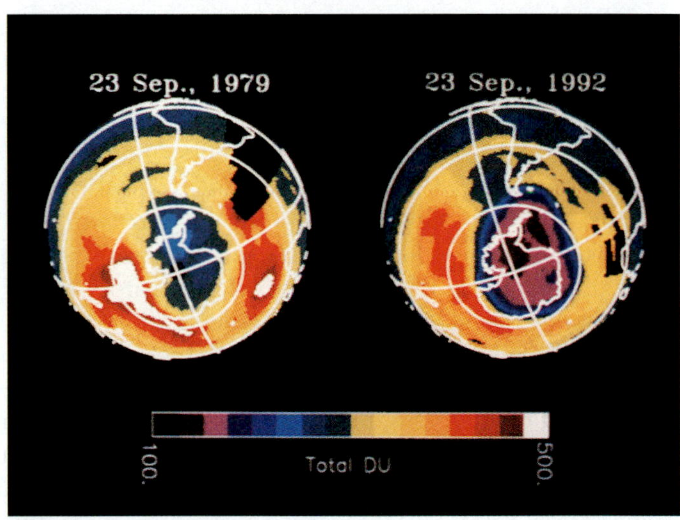

[B]

FIGURE 41.24

The Ozone Shield.

[A] The ozone layer surrounds the earth and absorbs most of the sun's ultraviolet rays. The thin patch over Antarctica represents the lowered levels of atmospheric ozone, due mainly to human use of chlorofluorocarbons. [B] The satellite map shows how the concentration of ozone over Antarctica in late September has decreased between 1979 and 1992 from concentrations in the middle range (blue) to concentrations in the low range (purple).

living in developing nations.) These long-lived compounds escape into the atmosphere and speed up sunlight's conversion of ozone to oxygen. The fact that CFCs act as catalysts means that each molecule in the atmosphere can continue to destroy ozone molecules over and over again for decades.

Scientists predict that as the ozone shield thins, and hence more ultraviolet light reaches earth, people will have more skin cancers, more eye cataracts, and more problems with their immune systems. Plants will suffer more mutations, stunted growth, and surface damage. Recent measurements showed that the ability of antarctic phytoplankton to convert sunlight into carbohydrates decreases about 10 percent under the ozone hole—a phenomenon that in turn harms that region's entire food chain, from krill to fishes to birds and mammals, including whales.

In the summer of 1990, 93 nations signed an international agreement to phase out the use of CFCs by the year 2000. But because CFCs are long-lived and act as catalysts, it will still be about 100 years until the ozone layer returns to normal. In the meantime, an ozone hole has started to appear in the Arctic, and some biologists worry that it might threaten Canada's huge coniferous forest biome. While manufacturers are racing to develop substitutes for CFCs, they warn that the new products will be less effective and more expensive. Although more research needs to be done to identify the potential magnitude of the ecological effects of decreased ozone and increased ultraviolet light, we already know that changes in the populations of especially sensitive organisms, like certain phytoplankton, can cause a collapse of entire ecosystems. Ecological solutions based on the principle of sustainability are intended to prevent just this kind of collapse.

SUSTAINABILITY

Experts expect a doubling of world population during the twenty-first century, from about 5.5 billion to more than 10 billion. To support these people, even at a minimum level, will require yet another five- to tenfold increase in economic activity—more agriculture, more shipping, building, transporting, financing, using up, burning down, and throwing away. How can we possibly keep this up? How can the earth support twice as many of us? What kind of future will we have? Will our human future be a global version of the boom-and-bust population cycle, with ever more hungry, impoverished people in a constantly worsening environment headed for a series of catastrophes that slash the population? Or will we learn to sustain ourselves in some kind of balance with the biosphere?

Our future depends on **sustainability**—balancing the levels of human population and economic growth against the quantities of available resources and the quality of the physical environment. In the past, economic development has come at the expense of environmental quality. High prosperity in certain countries has rested, in part, on environmental changes in other countries. To ensure a stable global environment over the long term, say experts like William Ruckelshaus, former administrator of the Environmental Protection Agency (EPA), all people must have a reasonable level of prosperity and security, and this will require a change in human attitudes and actions as great as those that brought about the agricultural and industrial revolutions of the past.

The sustainability revolution will require changes in manufacturing such that industries use fewer materials and less energy, and recycle their wastes into other products and processes. Sustainable agriculture will involve rotating crops to increase yields; building up rich soil by preventing erosion and by using natural fertilizers, such as animal manure, nitrogen-fixing crops, and blue-green algae; and controlling weeds, insects, and plant diseases through the use of integrated pest management, genetically altered crop plants, and other largely nonchemical methods. Sustainable agriculture will also rely on wind and solar energy instead of fossil fuels wherever possible, and on planting and rotating a wide range of traditional and nontraditional crops.

Creative economic solutions will also play a part in global sustainability. Among the many ideas now under consideration are debt-for-nature swaps and conservation easements. In the former, wealthy nations would cancel Third World debts in exchange for the preservation of natural habitats. In the latter, First World nations would pay the equivalent of mineral rights to secure forested areas and nature preserves against destruction.

In addition to such approaches, say various experts on sustainability, all nations, rich and poor, will have to provide their citizens with information on sustainability techniques and birth control. They will have to fund international agencies to manage global environments. And they will have to be fully cooperative and committed to the sustainability revolution. The alternative—the human population crash following the present boom—is just too dismal a prospect.

WHAT WILL THE FUTURE BRING?

Recent global phenomena, such as rising carbon dioxide levels and the dwindling ozone layer, contradict our traditional assumption that nature will somehow neutralize all the wastes we release. So great is our impact on the biosphere that some ecologists estimate that humans are

box 41.1
Focus on the Environment

Lessons of an Ecodome

There is a miniature world under glass sitting in the arid flatlands north of Tucson, Arizona, complete with ocean, marsh, desert, savanna, and rain forest biomes, with rice paddies, grain fields, songbirds, flocks of chickens, herds of goats and pigs, and more than 3000 other species from all the living kingdoms. Biosphere 2 is a $150 million glass and steel terrarium enclosing 3.15 acres of former desert [FIGURE 1]. Workers sculpted the land into artificial biomes, agricultural areas, and living quarters for eight human "Biosphereans," and the team sealed themselves into their artificial world for two years, starting in September of 1991. Their experiment, entirely funded by a Texas oil billionaire, was one of the grandest and most heavily publicized ecological ventures ever attempted. Unfortunately, it didn't work very well. The difficulties of understanding and re-creating nature's complexity, it seems, are not to be underestimated.

The massive project was conceived in 1984 by a group of space enthusiasts interested in constructing the type of total life-support environments that people may need someday to colonize other planets. Their goal was primarily entrepreneurial: to develop and eventually sell self-contained domes where people could grow food, breathe clean air, recycle all materials, and survive for extended periods. The opportunity for a money-making tourist attraction also figured into their plans, as did the chance to study broad aspects of ecology. They would need educated guesswork to choose the tens of thousands of individual elements making up Biosphere 2: the size, shapes, and heights of the domes over the different biomes; the types of soils (34,000 tons of them) for the forests, deserts, and farmlands; the species to import and install in ponds, streams, and the artificial coral reef; the crops to grow for food self-sufficiency; how to handle wastes; and the way all the elements might interact in a balanced way to ensure success. As it turned out, the "educated" part of the guesswork became a battleground from the start.

FIGURE 1
Biosphere 2 in Oracle, Arizona.

The private corporation behind Biosphere 2, Space Biospheres Ventures (SBV), did seek out the advice of many reputable ecologists, atmospheric chemists, plant scientists, zoologists, and others. But almost immediately, there were accusations of secrecy and data manipulation by SBV engineers and technicians, and resignations by individual academicians and eventually the entire scientific board of directors. Biosphere has been called "pseudoscience" in the popular press and in scientific journals. Whatever the truth of what has been said by these stone throwers, the proof was in the glass house itself.

Even before the eight volunteers sealed the doors behind themselves in 1991, SBV engineers quietly brought in and installed carbon dioxide scrubbers to remove excess levels from the air. Apparently, they anticipated that the plant communities would not remove enough CO_2 via photosynthesis to keep up with animal exhalations and the activities of soil decomposers. And they were

currently using about 40 percent of the potential primary productivity on all of the earth's land area! Furthermore, by cutting the world's great forests—tropical rain forests in particular, but the temperate rain forest of the Pacific Northwest as well—we are destroying habitats for thousands of species.

Harvard ecologist E. O. Wilson points out that we have triggered a new *biodiversity crisis*, during which we as a species could reduce the diversity of life to its lowest level in 65 million years [see BOX 38.1 on page 846]. For example, loggers have already removed about 90 percent of the old-growth forests in Oregon and Washington, and this habitat loss has threatened the survival of the spotted owl, the marbled murrelet (a sea bird that nests in old-growth forests), and other species. The remaining 3000 to 5000 breeding pairs of owls are able to live only in old-growth forests, and each pair requires 3000 acres of territory for night hunting. To save the owl and preserve this right. Even with the mechanical scrubbing, oxygen levels plummeted from 21 percent to 15 percent. In time, the Biosphereans were panting for each breath exactly as they would at an elevation of 13,500 feet. Eventually, engineers had to pipe in oxygen with a bellows the size of a high school gym, but the oxygen levels still remained at 16 to 19 percent.

The atmospheric problems, fundamental as they were, were just the beginning. Water condensing on and dripping from the glass ceiling rotted out the desert and turned it into a weedy scrubland. Crop failures limited the amount of available food, and the humans consumed less than 1800 calories per day during their second year and sustained average weight losses of 25 pounds each. The bees, hummingbirds, and fruit flies started disappearing immediately. Many of the savanna grasses stopped setting seed and multiplied only by runners, and the mini-ocean was apparently turning to pea soup due to eutrophication [see CHAPTER 40]. And in September of 1993, when all eight rail-thin Biosphereans emerged they headed for hamburgers, cheese, ice cream, and the other high-calorie treats they had been dreaming of. The team doctor did put the best possible face on this, reporting that all eight had the kind of low blood pressure, blood sugar, and blood cholesterol that health experts dream of.

Clearly Biosphere 2 has major bugs to work out. But the SBV corporation maintains that this is a 100-year experiment and that they anticipated some adjustments along the way. The work may continue to be controversial, but the chance to study ecosystem dynamics with nearly all variables controlled is truly valuable and could lead to some important understandings. At the very least, Biosphere 2 is a humbling endeavor that clearly demonstrates the enormous complexity and delicate balance of our own world, Biosphere 1.

magnificent ecosystem, the federal government in 1992 designated 6.9 million acres of cathedral forest as off-limits to loggers. The U.S. Fish and Wildlife Service predicted that the restrictions will cost about 33,000 jobs in timber-related industries over the next few years. Neither environmentalists—who wanted to see 12 million acres protected—nor loggers—facing a shrinking job market—were particularly happy with the compromise. But with no constraints on present rates of cutting, the old-growth forests of the Pacific Northwest would have been gone in a dozen years, and the current loss of jobs would merely have been postponed until then. The situation in the tropics is similar: natural habitats versus human jobs and food. The choices are usually difficult, and will almost certainly remain so.

Our species is the only one capable of changing the global environment in so short a time period. But we are also capable both of understanding that we are mismanaging the earth's resources and of foreseeing the consequences. One well-publicized research project in the Arizona desert, Biosphere 2, is attempting to re-create a functioning world under glass (with mixed success, so far) as a means of further understanding natural interactions and our effects upon them [see BOX 41.1, page 910]. The survival of our species and millions of others depends on the responsibility we take for these capabilities and the compromises we work out between preserving our environment and supporting our own needs.

Connections

The biomes that make up our biosphere are fashioned by the earth's atmosphere, soil, and water; by the planet's daily rotation on its tilted axis; by its yearly trip around the sun; and by continents drifting and climates changing from hot to cold, dry to moist, and back again. Over vast reaches of time, organisms slowly evolve adaptations that allow them to exploit their environments.

Human activities all over the globe are rapidly changing the earth's physical environment to such an extent that entire biomes are affected. We are causing desertification in dry areas and damaging the temperate forests through acid rain. Some of our activities threaten to affect the entire biosphere in a short time through the greenhouse effect and the loss of ozone. We must devise and institute solutions with equal speed, so that our all-pervasive influence does not undo the complex interdependencies of nature.

All of our foregoing subjects—cell biology, genetics, development, physiology, evolution, and ecology—help us understand how organisms survive in their environments. With this background, we are ready to consider our final topic: a fascinating set of adaptations called *animal behavior*, which allows animals to respond to each other and to their surroundings in a range of ways from the innate and automatic to the learned and deliberate.

KEY TERMS

biome, 895
chaparral, 899
climate, 892
coniferous forest, 899
desert, 897
estuary, 905
grassland, 898
rain forest, 895
savanna, 896
temperate forest, 899
tundra, 900
weather, 892

HIGHLIGHTS IN REVIEW

1 The biosphere is the thin layer of water, soil, and air at our planet's surface that supports life.

2 Different zones have different climates because the earth's axis of rotation is tilted and the earth's surface is heated by the sun most strongly at the equator.

- **a]** Latitude and time of year influence the amount of sunlight hitting the earth. The earth's tilted axis and uneven heating in turn influence the formation of wind, rain, and ocean currents. The rotation of the earth sets up coils of thermally driven air masses that establish the earth's climatic zones. In turn, climates establish the earth's biomes.
- **b]** Ocean currents affect climatic zones by redistributing heat and moisture.

3 On land, different climates promote different communities of organisms called biomes, such as tropical rain forests and deserts. Biomes are defined mainly by their vegetation and are characterized by specific adaptations to each climate.

- **a]** Tropical rain forests contain about half of all living species and occur near the equator, where the climate is warm and moist most of the year. Tropical plants are strongly adapted to compete for light and nutrients.
- **b]** Tropical savannas are warm year round but have a long dry season. Tall grasses and widely spaced flat-topped trees support communities of large grazers and browsers.
- **c]** Deserts are regions that receive little rainfall, either because they lie in high-pressure areas that are dry almost all the time; because they lie far from the ocean; or because they lie in the rain shadows of mountain ranges. Desert plants and animals are strongly adapted to conserving moisture and staying cool.
- **d]** Temperate grasslands are generally flat or rolling, with rich soils and seasonal extremes of hot and cold. Recurrent fires prevent the growth of trees.
- **e]** In the chaparral, summers are hot and dry and winters cool and moist, and the vegetation is scrubby, drought-tolerant bushes with evergreen leaves.
- **f]** Temperate forests occur in the eastern parts of North America and in other continents. Temperatures and rainfall are moderate, seasons are pronounced, and broad-leaved hardwood trees are the major plants.
- **g]** Vast coniferous forests are common in northern climes with cold, snowy winters and short summers. A moist version of this type of forest containing redwoods and firs grows along the west coast of North America.
- **h]** The tundra is a cold, treeless plain that is permanently frozen about a meter below the ground surface, preventing the growth of large trees. Tundra has low-growing plants and animal populations that fluctuate widely based on slow cycling of nutrients.
- **i]** The polar caps are occupied by ice (North Pole) or land and ice (South Pole). Since they lack major plants, they are not true biomes.
- **j]** Mountains have life zones, from low elevation to high, that correspond roughly to the biomes from equator to poles.
- **k]** Three-fourths of the earth's surface is covered by water, and most of that is marine, or salty seas and oceans. The sinking and rising thermal layers in lakes churn up nutrients from the bottom during the spring and fall turnovers.
- **l]** Running water has more oxygen content but fewer nutrients than standing water. Standing water has zones (littoral, limnetic, profundal) where light penetration and nutrients vary.
- **m]** Estuaries are varied and productive areas where rivers flow into oceans.
- **n]** Oceans are varied environments with an intertidal zone periodically exposed to air; a neritic zone of relatively shallow water; and an oceanic zone beyond the continental shelf.

4 Communities with similar physical environments contain organisms with similar evolutionary adaptations, even when the individual species are unrelated.

5 Burgeoning human populations and ecologically unsound life styles are affecting the biosphere in potentially disastrous ways.

UNDERSTANDING THE FACTS AND CONCEPTS

For Questions 1–12, match each of the descriptions with the appropriate terrestrial biome. Any answer may be used once, more than once, or not at all, and some descriptions require more than one answer.

- **a]** rain forest
- **b]** savanna
- **c]** desert
- **d]** grasslands
- **e]** chaparral
- **f]** temperate forest
- **g]** coniferous forest
- **h]** temperate rain forest
- **i]** tundra
- **j]** polar ice cap

1 Characterized by rich, agricultural soil.

2 Characterized by deciduous trees such as oak and maple and many hibernating and migratory animals.

3 Has the greatest species richness of all the terrestrial biomes.

4 Characterized by cold, snowy winters, shorts summers, and evergreen trees.

5 Characterized by blisteringly hot days, chilly nights, little water, and sandy soils.

6 Not present in North America.

7 Characterized by warm, dry climate, low shrubs, and flat-topped trees.

8 Characterized by permafrost, mosses and lichens, caribou, and a cold, treeless plain.

9 Bordering continental coastlines, characterized by wet winters and hot, dry summers.

10 Contains cathedral grove coniferous forests found only in the Pacific Northwest and parts of New Zealand.

11 Region characterized by extreme dryness; found in the center of large continental masses, directly below zones of descending dry air, or downwind of tall mountain ranges.

12 Characterized by a dense, permanent canopy and an open understory.

As above, for Questions 13–20, match each of the descriptions with the appropriate region in aquatic ecosystems.

- **a]** running water
- **b]** littoral zone
- **c]** limnetic zone
- **d]** profundal zone
- **e]** estuaries
- **f]** intertidal zone
- **g]** neristic zone
- **h]** benthic zone

13 Shallow zone in standing fresh water where light penetrates to the bottom.

14 Tidal channels that have high productivity if unpolluted.

15 Has three zones: a low-nutrient initial zone, a high-productivity middle zone, and a nutrient-rich final zone.

16 Deep, dark region of standing fresh water.

17 Deep, dark region of the ocean.

18 Light-filled upper region of standing fresh water that is rich in phytoplankton and zooplankton.

19 The highly productive shallow ocean zone where coral reefs are found.

20 Characterized by tide pools; an oceanic region submerged at high tide but not at low tide.

INTEGRATE AND APPLY WHAT YOU HAVE LEARNED

1 Explain why climate, rather than weather, significantly affects the biotic elements of an ecosystem.

2 What makes weather move from east to west or west to east? Which regions of the earth experience an easterly flow of weather and which experience a westerly flow?

3 Explain the process of turnover in a lake. How can turnover both help and harm a pond or lake?

4 Why do rain forests generally have poor soil?

5 Which terrestrial and aquatic areas of the world have the highest net primary productivity? What do these areas have in common?

ANALYSIS

1 The equator passes through snow-capped mountains outside of Quito, Ecuador. What single factor explains the appearance of snow at the equator?

- **a]** Water currents.
- **b]** Air currents.
- **c]** Abundant water.
- **d]** Altitude.
- **e]** Wind.

2 Suppose that you wish to buy houseplants to brighten up a dark corner of your room. As a general rule, what kinds of plants would be most suitable?

- **a]** Plants native to the savanna.
- **b]** Plants native to the understory of tropical rain forests.
- **c]** Plants that can exist in both dry and moist soils.
- **d]** Short-day plants.
- **e]** Succulent plants.

3 What is the likely fate of tropical rain forests after they are harvested and converted to pasture lands for cattle?

- **a]** Lush, productive pastureland.
- **b]** Second-growth rain forest.
- **c]** Tropical savanna.
- **d]** Unproductive moist soil.
- **e]** Desert.

4 The Galápagos Islands lie close to the equator, several hundred miles west of the Pacific coast of Ecuador. Using the information contained in FIGURE 41.5, predict their climate.

- **a]** Hot and rainy.
- **b]** Cold and rainy.
- **c]** Warm and dry.
- **d]** Cold and dry.
- **e]** Hot, dry summers and cold, wet winters.

5 An angler fishing the cold, turbulent upper regions of a mountain stream becomes chilled and moves downstream to a region of warmer water. Although the fisher may be more comfortable, are there likely to be as many trout in the new location?

- **a]** Yes. The trout have to go somewhere, and downstream is as good as anyplace else.
- **b]** Yes. The trout are probably cold too and will migrate downstream.
- **c]** Yes. The warmer the water, the greater the nutrient supply for the trout.
- **d]** No. The warmer the water, the lower its oxygen content. Trout require high oxygen levels.
- **e]** No. The warmer the water, the lower its productivity. The trout will remain where the net primary productivity is higher.

CHAPTER 42

Animal Behavior

MENACE OF THE BROKEN SHELL

Each spring, white gulls with coal black faces and white-rimmed eyes migrate to the coastlines of northern Europe. Males of this species (the black-headed gull) select a position on the dunes, and begin stretching forward and calling out raucously in a stereotyped ritual [FIGURE 42.1]. If a newly arrived male tries to claim another male's territory, the occupant usually rushes the invader, strikes him, and drives him away.

Males usually react quite differently to an approaching female, initiating a stylized mating dance in which the pair stretch forward, strut side by side, then, as if by prearranged choreography, turn their faces away from each other. Next the female taps on her suitor's bill with her own, and he regurgitates fish into her mouth. The dancing and dining over, the pair may mate, select a nesting place, then noisily defend it from all comers. The female deposits khaki-colored eggs with dark speckles in the loose mound of grass and straw they have constructed; then the parents take turns warming and turning the eggs in the nest [FIGURE 42.1]. After each chick hatches, the parent on duty carries the pieces of broken eggshell from the nest and dumps them.

FIGURE 42.1
Black-Headed Gull.

The off-duty parent frequently joins other gulls in noisy flocks as they dive for fish or search for worms in freshly plowed fields. When the adult returns to the nest with food bulging in its crop, the fuzzy, khaki-colored chicks rush up, peck the parent's bill, and receive a regurgitated meal.

Over time, the chicks begin to wander about their parents' territory. Black-headed gulls recognize their own chicks, but harshly peck chicks from other territories. The chicks, in turn, recognize their parents' calls at mealtimes and ignore those of neighboring adults. Alarm calls, however, are a different matter. Hearing the scream of any adult, gull chicks will freeze in the nest or quickly dive for cover. Adult black-headed gulls respond to an intruding fox or hawk by mobbing it, dive-bombing and defecating until they drive it off.

As summer wears on, the chicks mature, and soon the entire colony migrates to a warmer winter home in Africa. The following spring they return and enact a new drama of territorial marking, mating, and brooding.

What makes a parent gull pick up pieces of broken eggshell after its young hatch and remove those shells from the nest? How does a gull chick "know" to peck at its parent's beak to get food? Why has a stylized mating ritual evolved? Biologists seek to understand how and why animals behave the way they do by addressing four sorts of problems. First, they wonder what *function* the behavior has in terms of natural selection. How does the removal of an eggshell from the nest, for example, improve the gull's biological fitness (the ability to reproduce successfully in its environment)? Second, they are concerned with the *cause* of a behavior, the physiological mechanisms that generate the activity. What, for example, does the gull recognize about the eggshell—its smell, its feel, its taste, its shape? And how do mechanisms in the animal's nervous and hormonal systems generate the behavior? Third, biologists are interested in the *development* of the behavior. For example, what steps of maturation must occur before the gull is able to recognize and remove the eggshell? The fourth problem biologists address is the *evolutionary history* of a behavior. From observing behaviors in related gull species, for example, can one determine what preceded eggshell removal and how the current behavior might have arisen?

MESSAGES

1 A behavior is a response an animal makes to an environmental stimulus. The ability to perform a behavior arises during development based on the animal's genes and on influences from the environment.

2 Since behavior has a hereditary component, it can evolve due to natural selection. Experiments suggest that natural selection has often favored behaviors that promote individual reproductive success.

3 Some behavioral traits, such as swallowing, are not strongly modified by the environment, while others, such as language, are.

4 In certain ecological contexts, social behavior can result in increased reproductive success.

We will explore a fascinating range of behaviors in this chapter, from invertebrates to vertebrates, from automatic to learned, from mating to migrating, and from fighting to cooperating. We will see how the ability to act in a particular way arises during development as an interplay between an animal's genes and its environment. And we will see how natural selection shapes animal behavior through differential survival and reproduction. Experiments, for example, show that gulls that remove broken eggshells lose fewer chicks to predators than gulls that leave broken eggshells in their nests. Genes that promote eggshell removal therefore increase in frequency in the population of black-headed gulls.

We will begin our chapter with a discussion of the genetic underpinnings of behavior. Then we will discuss how experiences in the environment can alter the expression of certain genes and the carrying out of particular behaviors. Learning involves such environmental influences. We then move to a consideration of how natural selection shapes behavior, including movements, migration, navigation, finding and protecting a home territory, feeding, reproducing, and communicating. Finally, we'll consider how social behavior arose in many insect groups, in certain birds, and in many kinds of mammals, including humans.

Along the way, we'll see how chimpanzees can learn to stack and stand on boxes in order to harvest bananas; how ticks can locate prey; how elephant seals choose mates; and how colonies of naked mole rats are governed by a large, aggressive queen. ❑

The Genetic Bases of Behavior

Animals display a staggering range of behavioral acts from traits like swallowing, which are nearly identical each time any individual in this species performs them, to behaviors as highly variable as language. Some behavioral traits have strong genetic determinants and change only over many generations as natural selection winnows out specific alleles but leaves others. Many other behavioral acts are influenced by circumstances and events within the environment, and are modified to a greater or lesser degree during the lifetime of a single individual. Most behavior, however, is neither purely genetically determined nor purely learned. There is instead a continuum of types of behavior, from those that tend to be performed in a similar way by most members of a species in a specific situation to actions that can be modified greatly by experience during an individual's lifetime. Let's investigate evidence for the genetic basis of behavior, first, then look at how the environment affects behavior.

Some of the first controlled experiments in animal behavior were the work of biologist Nikolaas Tinbergen, a pioneer in *ethology*, the scientific study of how animals behave in their natural environments. He posed the straightforward question we considered in our chapter opener: What properties prompt black-headed gulls to remove broken eggshells? To find out, he placed various objects into gull nests alongside the eggs. The birds accepted and incubated round rocks, flashlight batteries, corks, and light bulbs as if they were eggs. But gulls ejected from the nest bottle caps, toy soldiers, seashells, and paper squares. Tinbergen concluded that when the birds saw or felt an object with a jagged edge, they removed it from the nest. Apparently, sensory receptors in the gull's eyes or skin detect sharp edges, and the nervous system directs the bird to remove the object. Since genes help specify how a bird's sensory and nervous system develop, it seemed reasonable to suspect that genes, interacting with the environment, provide a basis for this shell-removal behavior, as well as for nesting in general and for a range of other behaviors.

Recall from CHAPTER 16 that an organism's constellation of genes is a result of evolution largely molded by natural selection. Our concern in this chapter is how natural selection can act on populations of animals to result in observed behavioral patterns, including those influenced strongly by the environment. Natural selection can act on behavior only (1) if there are clear behavioral alternatives in the population; (2) if these alternatives are heritable; and (3) if some of these behavioral alternatives result in greater reproductive success than others [see CHAPTERS 1 and 15]. Many studies have focused on how survival and reproductive success relate to different types of behavior, and we consider two here, one in bees and one in snakes. Both demonstrate that some behavioral acts have a strong hereditary component.

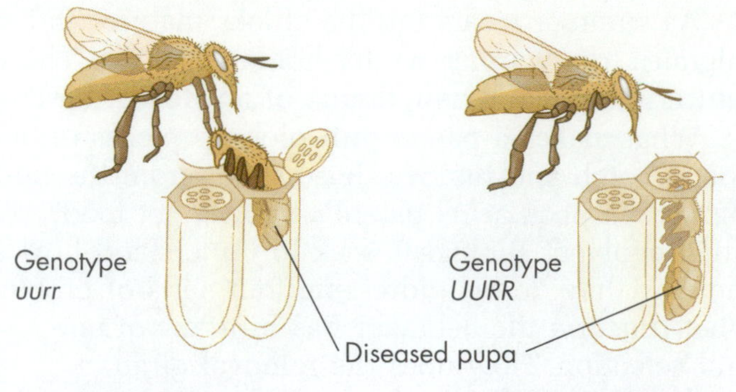

FIGURE 42.2
Genetic Differences Between Hygienic and Nonhygienic Bees.
[A] Hygienic bees remove the caps from chambers that contain diseased pupae and then throw the sick animals from the hive. **[B]** Nonhygienic bees neither uncap chambers with ill pupae nor remove the afflicted animals.

In honeybees, single genes can affect certain interesting actions. Biologists have noticed that some honeybees eject diseased bee pupae from their honeycombs. These "hygienic" bees uncap the chambers where the diseased pupae lie sealed and then expel the diseased animals from the nest [FIGURE 42.2A]. Other strains of honeybees are "nonhygienic": They do not uncap infected cells, and they don't remove diseased pupae even if the experimenter does the uncapping [FIGURE 42.2B]. In nonhygienic colonies, disease can spread rapidly and devastate the hive. To investigate the genetic basis of this behavior, researchers crossed hygienic and nonhygienic bees, and in the second generation found the parental types of bees, as well as recombinant types that could uncap cells, but could not remove dead pupae, or could not uncap cells themselves but could remove dead pupae if the experimenter did the uncapping. The ratios of the types obtained suggest that two genes are involved in the trait—one gene controls uncapping, while another gene controls removal. The implication was that individual genes can govern specific behavioral acts and that a complex behavioral pathway may involve more than one gene.

Studies on garter snakes illuminated the additional point that naturally occurring genetic variation leads to behavioral acts that adapt individuals to specific environments. Biologists have observed two populations of a single garter snake species, one that inhabits the foggy forests of coastal California and eats mainly slugs, and

FIGURE 42.3

Tongue Flicking in Garter Snakes.

Both coastal and inland garter snakes (*Thamnophis elegans*) flick their tongues, allowing them to "smell" slug juice offered by an experimenter. Only coastal garter snakes, however, will eat bits of slug meat.

➤ CONCEPT CHALLENGE

When behavioral researchers keep blackcap warblers in a cage, the birds act restless during their normal migrating season. Birds from northern populations, which fly long distances, are restless for a longer time than birds from southern populations, which do not fly as far. Hybird birds show intermediate periods of restlessness. What hypothesis does this data suggest? What additional experiments would you want to perform as controls to help support or reject your hypothesis?

another that lives in higher, drier areas several miles inland and feeds on fish or frogs from lakes and streams. Researchers wondered whether the distinctly different food choices are due to genetic differences between the two snake populations or whether inland snakes will eat slugs if given the opportunity.

To find out, experimenters captured pregnant snakes from each population, raised the young in the laboratory, and offered them small chunks of fresh banana slug, 6-in.-long gold-colored animals with black spots. Baby snakes from inland regions flicked their forked tongues only a few times at a cotton swab soaked in slug juice and then avoided the slimy slug meat [FIGURE 42.3], while coastal snakes repeatedly flicked their tongues and ate slugs heartily. This result suggests that different genes in the two populations control slug refusal and slug feasting, and crossbreeding showed that alleles for refusal are dominant to alleles for feasting.

These experiments with bees and garter snakes demonstrate clearly that behavioral alternatives can be based on genetic differences.

In his study of eggshell removal, Tinbergen (and others) also considered the genetics, natural selection, and evolutionary history of the shell-removal behavior. When Tinbergen placed eggshells at various distances from gull nests, he found that nests with nearby eggshells were three times more likely to lose eggs to predators than nests without nearby eggshells. Evidently, natural selection has favored alleles that cause black-headed gulls to remove predator-attracting eggshells from their easily accessible nests. In close relatives of black-headed gulls, however, called cliff-dwelling kittiwakes, similar alleles may be lacking, since foxes, crows, and hawks cannot gain access to the narrow ledges where kittiwakes build their nests.

Clearly, biologists can demonstrate in many ways that genes influence behavior. This does not suggest, however, that genes alone produce the behavior. Instead, there is a complex interaction between an animal's genetic constitution and its environment. This interaction unfolds during the animal's development and shapes the animal's ability to perform various acts, as our next section explains.

Experience Can Influence Behavior

The environment affects most of an animal's actions to a greater or lesser degree. Some acts change very little with experience; they include a specific sequence of events and each individual in a population performs them in a very similar way. Other actions are more strongly influenced by the environment.

An example of a behavior with little variation is the sequence of actions a graylag goose performs when an egg rolls from the shallow depression on the ground where she nests. When a goose spots an egg outside her nest, she waddles up to it, stretches out her graceful neck, and gently wraps her chin and neck around the egg [FIGURE 42.4].

FIGURE 42.4

A Graylag Goose Retrieves a Stray Egg.

When an egg rolls from the nest of a graylag goose (*Anser anser*), the female retrieves it by arching her neck gracefully around the stray and rolling it back with both sides of her beak. She will retrieve anything egglike near the nest, including baseballs and tin cans.

[A] "Begging" behavior

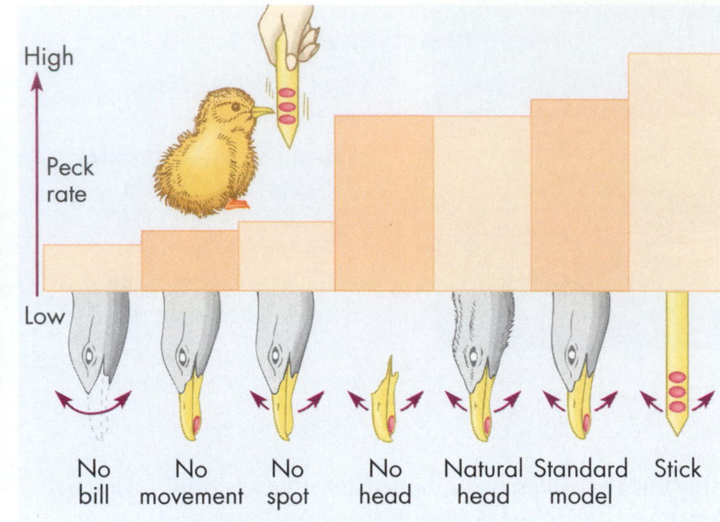

[B] Experimental analysis of the releaser

FIGURE 42.5

Red Spot Pecking in Young Gull Chicks.

[A] A hungry herring gull chick (*Larus argentatus*) pecks at the red spot on its parent's bill, an action that triggers a regurgitated meal. [B] To identify what provokes the pecking behavior, experimenters offered young chicks various artificial objects to peck at and watched for the rate of pecking in response. The results showed that movement plus a red spot on a pointed object is most effective at causing the response.

Once in contact, she pushes the egg along by moving her head from side to side to keep the lopsided object rolling straight, somewhat the way a hockey player advances a puck down the court with both sides of a hockey stick. Biologists know that this behavior is not modified very much due to experience in the environment because even birds raised in isolation do it the first time an egg rolls from the nest. If an experimenter removes the errant egg once the goose has begun rolling it, remarkably, she carries on like a feathered mime, rolling and nesting a nonexistent egg. A series of stereotyped physical movements of this sort is called a **modal action pattern**.

Modal action patterns can be triggered by specific environmental signals. Sights, sounds, smells, tastes, or any information picked up by the sense organs may trigger modal action patterns. For the graylag goose, a round object near the nest can cause the egg-rolling behavior; a goose will retrieve a tennis ball as tenderly as her own egg! For the black-headed gull, an object with a jagged edge can trigger eggshell removal.

Modal action patterns frequently proceed from start to finish in an all-or-none fashion and in a particular sequence. The swallowing reflex is a classic example. Have you ever tried to stop swallowing a mouthful of food midway down? It can't be done; more than a dozen muscles fire in coordinated sequence, each step unalterably triggered by the preceding step in a classic all-or-none pattern. Modal action patterns often depend on an animal's stage of anatomical and physiological maturation and on its internal physiological state. For example, egg-rolling behavior in the female goose occurs only between the time she lays her eggs and the time the eggs hatch.

BEHAVIORAL ACTS WITH LIMITED FLEXIBILITY

While an animal's experience in the environment usually alters a modal action pattern very little, other types of behavior are more flexible. When a parent gull returns from a feeding foray, its chicks peck at the parent's wagging bill. In the herring gull, adults have a distinctive red spot on their bill and chicks peck at that spot [FIGURE 42.5A]. This behavioral sequence, however, has some flexibility, or potential for modification. Very young chicks peck at anything resembling the parent's red bill spot and will actually peck more excitedly at a swinging stick with red spots painted on it than at a live gull's bill [FIGURE 42.5B]. As the chicks mature, however, begging behavior becomes more directed. Eventually, they beg only from their own parents. In these gull chicks, information acquired from the environment (about their parents' physical characteristics) can modify the initial behavior pattern. The most common categories of behavior with flexibility are imprinting and learning.

IMPRINTING: LEARNING WITH A TIME LIMIT

In the 1930s, Austrian biologist Konrad Lorenz discovered the phenomenon called **imprinting**: the recognition, response, and attachment of a young animal to a particular adult or object. At the time, Lorenz was studying ducks and geese, and he knew that goslings normally follow their mother from the nest to a nearby pond or lake within a day after hatching. The researcher tried walking

FIGURE 42.6
Imprinting.
Imprinting has an inherent time limit; geese form parental attachments in the first two days after hatching. The attachments, however, are long-lived. Here, two imprinted geese preen Lorenz's "feathers."

away from a group of newly hatched goslings in the nest while he made gooselike honking sounds. To Lorenz's delight, the gaggle of goslings followed him around as readily as they would have their real mother [FIGURE 42.6]. He speculated that to survive, newly hatched graylag geese must quickly learn to recognize and follow their parents, and they are apparently programmed to trail after the first large moving object they see, even if it is a gray-bearded, pipe-puffing biologist!

Imprinting has an inherent time limit—a **sensitive period**, or specific developmental stage, during which the imprinting can occur and after which it cannot. In geese, the sensitive period for following lasts from the time of hatching to the second day. The sensitive periods for different types of imprinting occur at different times. Hand-reared ring doves prefer to court humans over other ring doves, but the sensitive period for sexual imprinting is later than that for following. Some stimuli are able to elicit imprinting more easily than others—in the case of goslings, large objects moving away from the nest and making goose sounds are particularly effective.

Imprinting shows flexibility not only because different individuals can become imprinted on different objects, but also because imprinting can sometimes be reversed. For example, collared doves raised by ring doves initially prefer to court ring doves, but with age and experience, the collared doves begin to prefer their own species.

Biologists suspect that imprinting can take place in baby birds because development "primes" their nervous systems to change in a defined way based on environmental information shortly after hatching. Imprinting is thus a form of learned behavior, but one limited to a narrow time period. In most forms of learning, behavioral patterns can be repeatedly modified throughout an individual's lifetime.

LEARNING: BEHAVIOR THAT CHANGES WITH EXPERIENCE

A hungry toad that has never eaten a millipede will, upon seeing the many-legged arthropod, flip out its sticky tongue and snap up the prey. The captured millipede quickly retaliates, however, and exudes a foul-tasting toxin. The poison works quickly, and the nauseated toad spits out the would-be prey unharmed. A toad is far from clever, but it will remember this single disagreeable experience a long time and will reject millipedes, even when hungry. The toad's nervous system has cataloged the shape and markings of a millipede and associated them with a repulsive taste that should be avoided—a one-trial learning event.

The toad's aversion to millipedes encapsulates the essential elements of **learning**: an adaptive and enduring change in an individual's behavior based on personal experiences in the environment. In our example, the toad's learning is adaptive because it allows the animal to avoid poisonous food; the learning is based on a personally experienced event rather than an inherited trait; and the behavioral change endures for some period of time. One-trial learning is just one type of trial-and-error learning. Other kinds of learning include habituation, classical conditioning, and insight learning. All, as we will see, can increase an animal's fitness—its ability to survive and reproduce.

Habituation **Habituation** is a simple kind of learning in which the animal comes to ignore a repeated stimulus that is never followed by reward or punishment. Most animals, for example, react to a novel stimulus by freezing in place or fleeing. Young black-headed gull chicks crouch down when objects pass overhead [FIGURE 42.7A], but this response wastes time and energy if the flying object is a harmless leaf or robin. It is adaptive, therefore, for chicks to habituate and learn to ignore silhouettes of common harmless species [FIGURE 42.7B], but to continue to cower at the distinctive shapes of predatory birds [FIGURE 42.7C]. Since these predators pass overhead infrequently, habituation for them and other rarely seen objects never occurs, and this, too, is clearly adaptive.

Trial-and-Error Learning More complex than habituation is **trial-and-error learning**. Also called **operant conditioning**, or instrumental conditioning, it is a form of learning in which an animal associates an action, or **operant**, with a **reinforcer**, a reward or punishment. The millipede-munching toad, for example, experienced trial-and-error learning: It learned to associate an action, capturing the millipede, with a reinforcer, a foul taste. Likewise, a pet dog will learn to relate the action of begging to a reinforcer, a handout of a dog biscuit.

Biologist B. F. Skinner studied operant conditioning under rigorously controlled laboratory settings. He suggested that virtually any action can be conditioned, or

FIGURE 42.7

Habituation in Gull Chicks.

[A] At first, a gull chick will cower no matter what object passes overhead. [B] Eventually, it habituates to the passage of common harmless objects like robins and leaves, but [C] continues to cower at rare ones like hawks, whether or not they are dangerous predators.

FIGURE 42.8

Classical Conditioning.

Madison Avenue exploits our evolutionary tendency to associate separate stimuli. This classic advertisement from 1939 links fun, love, security, and owning a brand-new Chrysler.

trained, by a proper schedule of reinforcements. More recent experiments show that operant conditioning works readily only for stimuli and responses that have meaning for animals in nature. Rats rapidly learn to press a lever with their front paws to obtain a food reward, but they do not easily learn to push a lever to avoid an electric shock on their feet. On the other hand, rats quickly learn to jump off the ground to escape a shock to the feet, but they do not readily learn to jump to obtain a food reward. These results make sense when viewed in an ecological context; in nature, rats obtain food by manipulating items with their paws, but this is not how they avoid harmful stimuli. These experiments show that animals are prepared to learn some things more easily than others, and in ways that appear to increase survival and hence reproductive success.

Classical Conditioning In operant conditioning, the animal responds before receiving the reward or punishment (the reinforcer), but in **classical conditioning**, or associative learning, the animal responds after the reinforcement, and with more experience, the response shifts from one stimulus to another. Russian biologist Ivan Pavlov developed the concept of classical conditioning from experiments conducted on dogs. He repeatedly presented a dog with meat, a reinforcer and a natural stimulus for salivation; but just before presenting the meat, Pavlov rang a bell—an arbitrary, unnatural stimulus. Through classical conditioning , the dog learned to associate the natural stimulus (the meat) and the substitute (the bell) and began to salivate at the sound of the bell. In nature, a predatory hawk can learn to associate the presence of a broken eggshell near a gull's nest with the presence of a defenseless, tender gull chick and act to collect a reward: dinner. Advertising agencies have learned to manipulate our own species' susceptibility to classical conditioning by associating commercial products with positive images: fancy cars with successful business executives; quarter-pound hamburgers with happy family occasions; a toothpaste brand with social success [FIGURE 42.8].

Classical conditioning works best when it mimics natural situations. Experimenters have found that rats quickly learn to avoid food that nauseates them if the food is associated with a particular smell, but not if the food is associated with specific colors. This makes ecological sense because rats usually forage at night, when sight is not a reliable way of evaluating foods.

We can conclude from the study of classical and operant conditioning that an animal's genes program it to learn different things in different ways that fit the animal to its environment. Many animals are adept at learning through both operant and classical conditioning. One final type of learning, however, is common only in primates and reaches its peak in humans.

Insight Learning At this moment, as you read, you are acquiring knowledge through **insight learning**, or reasoning. With this type of learning, people and other primates formulate a course of action by understanding the relationships between the parts of a problem. Insight learning allows an animal to encounter a new situation (including a new printed sentence), and based on similar past experiences, to figure out how to respond without actually trying out all possible solutions. A chimpanzee that enters a room where boxes litter the floor and bananas hang from

FIGURE 42.9
Conversations with a Chimp.
Researcher Herbert Terrace taught the chimpanzee Nim Chimpsky to use a few American Sign Language gestures. Here, Terrace signs his name—Herb—and Nim copies him.

the ceiling out of reach can figure out how to stack the boxes and climb up to grab the food. Experimenters have taught chimpanzees how to use American Sign Language [FIGURE 42.9] and how to operate computer keyboards—remarkable examples of the chimpanzee's capacity to learn.

People often assume that their household pets are capable of reasoning, but in fact, dogs, cats, and other common pets are quite inept at solving even simple insight problems. For example, if a dog wraps its leash around a tree so that it can no longer reach its food and water, the dog is unable to conceptualize and address the cause of the problem. A dog could be trained to keep its leash untangled through operant conditioning, but a primate would conceive of the problem beforehand and simply avoid it. Dogs, of course, did not evolve with leashes around their necks; thus they might be more adept at insight learning in situations that are more relevant to their ecological role in nature; for example, learning to catch a Frisbee is not very different from learning to catch a pheasant.

Behavioral ecologists suspect that animals have evolved particular forms of behavior that provide the best chance of surviving and propagating in their own environments. Natural selection seems to have resulted in an immense variety of developmental systems and behavioral strategies, as the next section shows.

➤ CONCEPT CHALLENGE

Male stickleback fish, which have slender red bellies, defend their territories from intruding males. The males may, however, court intruding females, which have swollen silver bellies. In experiments, male fish attacked a football-shaped object with a red lower side more often than a natural looking model of a male fish with a silver belly. What can you conclude from this experiment regarding the relative importance of color pattern and shape as a stimulus for the males' behavior?

Natural Selection and Behavior

A black-headed gull selects a particular nesting site at its summer home on a European beach. The bird feeds itself and its young, detects and drives away predators, selects a mate, and cares for its chicks. As part of all these activities, it communicates to its family and colony mates through a variety of sounds and movements. It has a sense of time and season that allows it to fish at the best time of day and migrate at the right time of year. The gull's complex behavioral repertoire helps illustrate several major categories of animal behavior: choosing a home territory through directional movements such as migration and navigation and then defending the home turf; feeding and avoiding being eaten; reproducing; and communicating. In gulls, as in thousands of other animal species, such behavioral acts have been shaped by natural selection and markedly influence the organism's chances of obtaining energy and producing progeny.

LOCATING AND DEFENDING A HOME TERRITORY

Just as people choose apartments, homes, and building sites, gulls find protected nesting spots on the beach, beavers locate the best site along a stream to build a lodge, and other animals move to territories that suit their needs. For simple animals, the movement can be toward a locally more favorable environment, but for more complex animals, it can involve long migrations.

Simple Movements: Taxes Just as the orientation of a plant part toward or away from light is called a tropism [see FIGURE 37.5], an animal's movement toward or away from light, specific chemicals, or heat is called a **taxis**. Each animal has sensory receptors suitable for detecting such environmental stimuli. A female tick, for example, must locate a mammal and suck its blood before laying her eggs. Light-sensitive receptors in the tick's skin lead her to a well-lit leaf or branch, and then olfactory receptors help her detect butyric acid with its odor of rancid butter or carbon dioxide coming from a nearby mammal. Once triggered by smell, the tick drops from the leaf, and if she chances to land on a mammal, her heat detectors enable her to crawl to a warm spot. There she pierces the skin, sucks the blood, and her eggs begin to mature.

The tick's movements involve positive *phototaxis* (movement toward light), positive *chemotaxis* (movement toward a chemical), and positive *thermotaxis* (movement toward heat). Natural selection has favored alleles that direct the development of special sense

FIGURE 42.10

A Migration of Monarchs to Mexico.

Each autumn, 100 million monarch butterflies migrate from the United States to the mountains of western Mexico to roost on Oyamel fir trees. Scientists are still studying how they navigate the nearly 2000 km (1240 mi) to this particular region.

organs attuned to different physical stimuli; these stimuli, in turn, trigger behavioral acts that help the animal locate an appropriate home.

Homing and Migration Many animals routinely travel hundreds of kilometers in search of an appropriate habitat. Whales, sea turtles, eels, caribou, and monarch butterflies travel great distances to their winter habitats or breeding areas [FIGURE 42.10]. Salmon return to breed in the streams where they were hatched. But the champion voyagers are birds. Black-headed gulls trek from Africa to northern Europe and back annually, and the small arctic tern flies 16,000 km (about 10,000 mi), from Greenland and Alaska to Antarctica and back, seemingly in a search for endless summer.

While migrations are seasonal, homing behavior can take place any time certain kinds of animals are displaced from their home territories. When experimenters transferred an individual shearwater, a long-winged seabird, from its home in Great Britain to a new location in Massachusetts, the remarkable bird was back on its nest in 12 days, having crossed 4800 km (about 3000 mi) of trackless ocean! Likewise, wildlife managers in Yosemite National Park routinely trap troublesome black bears and move them dozens of kilometers away from heavily used campgrounds, only to find the animals have returned in a few days. How do animals find their way, whether homing or migrating? Biologists are not entirely sure, but tests confirm that some animals have a sense of **orientation**, or direction finding, and of **navigation**, the more exacting ability to move from one map point to another.

To test whether migrating animals orient in a general direction or actually navigate from point to point, researchers studied the homing response of starlings. These shiny black birds normally fly southwest each year from their summer homes in northern Germany to their wintering grounds on the shores of the English Channel [FIGURE 42.11]. Researchers captured birds flying southwest through Holland, then released them in Switzerland almost 800 km (about 500 mi) south of where they were captured. Would the displaced birds continue orienting to the southwest or switch their route and navigate northwest to the English Channel? Interestingly, young starlings who had never made the migration continued orienting to the southwest and flying in the original direction, landing in Spain, far south of their original destination. In contrast, older birds who had flown from Germany to the English Channel the previous year switched their course, flew northwest, and correctly navigated to their proper wintering grounds. These experiments suggest that navigation skills may require more experience with the environment than does orientation.

Which sensory systems help an animal to navigate? In daytime, birds, ants, and bees orient via the sun's position as it sweeps across the sky relative to their path. For example, a pigeon released west of its birdhouse in the morning will fly toward the sun to get home, while a pigeon released west of its birdhouse in the evening will fly away from the sun to get home. When experimenters shifted the birds' sense of time by altering their exposure to light and dark in a closed room, the birds looked at the sun and flew in the wrong direction. This suggests that pigeons use an internal 24-hour clock, or *circadian clock* (*circa*, "about"; *dia*, "day"), that helps compute orientation relative to the sun's path.

This is only part of the story, though, because homing pigeons can find their way home even in the dark or when researchers have fitted them with tiny, thickly frosted contact lenses that reduce their vision to a murky haze. Remarkably, birds still orient and navigate under such conditions by detecting the earth's magnetic field. Biologists have discovered a few tiny crystals of magnetic iron oxide in tissues near the brains of birds and porpoises. They suspect that these crystals align with the earth's magnetic field, like the iron needle in a compass, and cause neurons to fire. The reseachers concluded that the sun is a pigeon's primary cue for orientation, but the earth's magnetic field acts as a substitute cue in the dark.

These studies on starlings, homing pigeons, and other animals help address the physiological causes of behavior: how animals migrate. But the evolutionary functions of migration must still be addressed. After all,

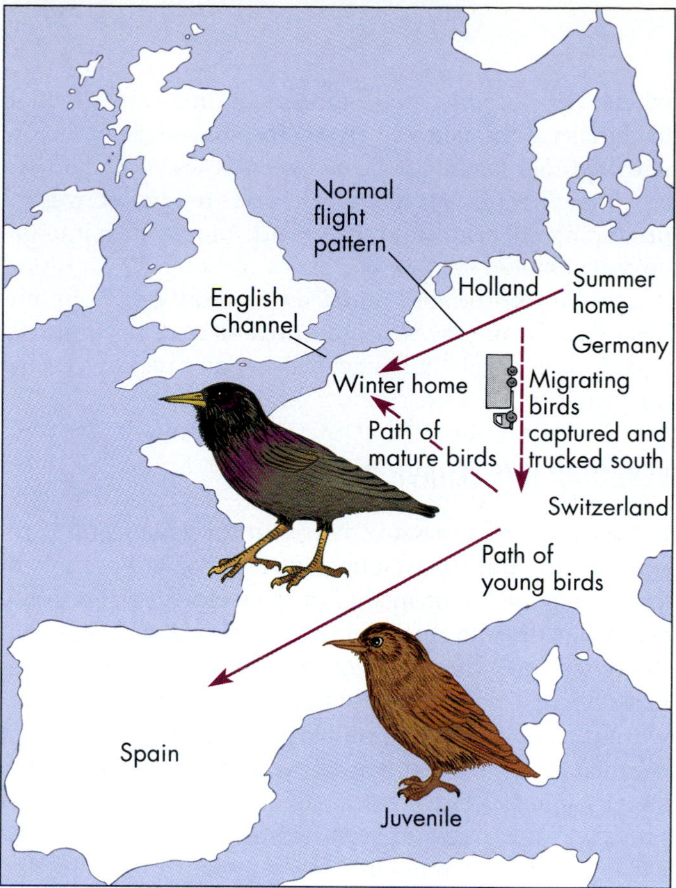

FIGURE 42.11

Do Starlings Orient During Migration or Actually Navigate?

In autumn, starlings normally fly to the English Channel from their summer home in northern Germany. This experiment showed that starlings can navigate, but require experience to do so.

migration exacts a high toll. Grizzlies wait in streams and gorge on exhausted salmon migrating home from the sea, and falcons feast on fatigued songbirds arriving at their winter home in Africa. Fuel used by muscles to propel wings, fins, and legs is unavailable for reproductive activities, and time spent on the move is time not spent gathering food.

Clearly, the benefits of migration must outweigh such tremendous costs. Animals have much to gain by moving on to greener pastures after they exhaust an area's seasonal resources. And migration allows animals to fatten themselves in a bountiful but dangerous feeding area, then retire to a more protected region with fewer predators for breeding or birthing. Animal species can travel short distances or long, in isolation or in large groups, but once they reach their destination, they often claim a territory and defend it, even against their own former traveling companions.

Territoriality Once a male gull claims a spot on the beach, he rebuffs other male interlopers by assuming stylized body positions and issuing specific calls that proclaim his turf. Behavior that defends living space from intruders is called **territoriality**. Sleek male impalas of the African savanna are good examples of animals competing for territories rich in food resources. The richer the territory controlled by an impala, the more females it entices, and more females mean greater reproductive opportunity for males. The males stake out the richest territories they can find, and when a competing male arrives, they clash, interlocking their graceful horns and thrusting furiously, trying to force each other out. Females choose to stay in each territory only so long as the food holds out, and during their stay, the resident male mates with all receptive females. The stakes are clear: Males guarding more food have access to females longer, mate more frequently, and hence tend to have more offspring. A male without rich territory to defend will have fewer contacts with females and thus far fewer offspring.

The costs of defending a territory are great: Constant battle readiness requires time and energy, and exhausting tournaments leave the combatants vulnerable to predators. The rewards in terms of extra progeny produced, however, are tremendous, just as the principles of natural selection predict, and the vanquished will often surrender after a brief exchange of threats. Biologists think that the animal's long-term reproductive success will grow if it survives unscathed and claims its own territory later.

Territories and Game Theory How strongly should a male impala or a black-headed gull defend its territory before giving up and trying to find another territory and other mates? This depends not only on the costs of the struggle in terms of energy and potential injury or death, but also on the value of the territory in terms of the possessor's ability to produce offspring. Biologists can analyze situations like these by applying a technique called *game theory*. Game theory was the creation of economists who used it to analyze business decisions. Biologists have since applied it to various aspects of animal behavior, including the waiting game (or war of attrition) played by male dung flies.

A male dung fly waits on a fresh cowpat and copulates with the first female that comes along; he then keeps other males away from her until she lays her eggs. The longer a male waits, the drier the cowpat becomes, and the less attractive it is to females. The game is: How long should the male wait on a dessicating cowpat before moving on to a fresh, moist one? The answer depends on the moves other players make. If most males remain for a short time on a cowpat, then any male who stayed a bit longer could successfully claim any late-arriving female. If, on the other hand, most males stayed a long time, then a male could increase his reproductive success by leaving early, finding a new, unoccupied cowpat, and mating with the early-arriving females.

Assuming that there is a strong hereditary component to the waiting time of individual males, how will evolution proceed in this game? If all males tended to stay a long time, then any mutations that caused shorter times would be selected for, because those males would mate with females choosing fresh cowpats. Conversely, if all males were genetically predisposed to staying for short times, a long-waiting mutant would have increased success based on matings with late-arriving females. Clearly, neither a long-wait nor a short-wait strategy would be stable in the population over evolutionary time. So what is the *evolutionarily stable strategy*? Analysis shows that the stable strategy in the population is a randomly chosen, unpredictable waiting time. The population is stable because no individual who chooses a strategy that differs from randomness will achieve greater success. Studies of dung fly populations suggest that, indeed, the animals do follow the random evolutionarily stable strategy.

As with dung flies, for most animals in most territories, the most valuable commodity is a ready source of food for either the parents or offspring, and a second group of behavioral acts has arisen around obtaining nourishment.

FEEDING BEHAVIOR

Is it more advantageous for a black-headed gull to expend the energy to fly several kilometers out to sea where herring and other fish are bountiful or to travel a shorter distance but forage where food is less abundant or less nutritious? What fraction of its feeding time should a junco spend scanning the skies for a hungry hawk instead of searching the ground for seeds? To answer such questions, behavioral ecologists have devised and tested the **optimality hypothesis**, the notion that animals behave in ways that allow them to obtain the most food energy with the least effort and the least risk of falling prey to a predator.

Biologists studied crows, for example, as the birds preyed on whelks—large, snail-like mollusks. A crow will typically search the intertidal zone and select a big whelk over several smaller ones, grasp it, fly it to a height of 5 m (16.4 ft), then drop it onto the rocks below. If the shell shatters, the crow consumes the tender flesh; if it doesn't break, the crow lifts and drops it again. Researchers predicted that if natural selection favored genes promoting optimal foraging strategy, then large whelks should be more likely to smash than small ones; a drop of 5 m should be just enough to break most whelk shells; and dropping a whelk a second time should be just as likely to lead to a meal as searching for a new whelk. In fact, each of these predictions proved true. Crows waste neither time nor energy dropping small, unbreakable whelks or releasing whelks too low or unnecessarily high for breakage. Considered in evolutionary terms, a crow with optimal feeding behavior spends less time and energy foraging, but has more time and energy left over for producing offspring, and thus sends more genes into the next generation.

While territoriality and feeding strategies help animals take in energy, every animal faces a second and equally important challenge: bearing and perhaps caring for young.

REPRODUCTIVE BEHAVIOR

For most animals, a set of behavioral acts surrounds reproduction: finding and choosing a mate, and producing and caring for the offspring. In many species, the strategies for achieving reproductive success are different in the two sexes. The males of most species tend to make many highly mobile, lightweight gametes and behave in ways that increase the number of eggs they fertilize. In contrast, female animals tend to make fewer but much larger gametes. A woman, for example, produces a few hundred eggs in her lifetime, while a man, in a similar period, produces enough sperm to inseminate all the women in the world. Not surprisingly, such differing reproductive strategies shape the reproductive behavioral acts of the sexes.

Parental Investment The fact that males and females produce gametes of different sizes and numbers has major consequences, including parental investment and mate selection. *Parental investment* is the allocation of a parent's resources, such as food, time, and energy, to each individual gamete and the young that arises from it. Each large egg represents more resources than each small sperm; thus, in a physical sense, the female sacrifices opportunities to make more eggs in favor of increasing the chance that each egg will become a viable descendant. An extreme example of this is the kiwi bird of New Zealand [FIGURE 42.12], which lays an egg weighing one-fifth her body weight—the equivalent of a woman having an 11 kg (24 lb) baby! Most female mammals shelter and nourish their offspring in the uterus and feed them with milk long after birth [FIGURE 42.13]. Such behavioral acts extract a heavy cost in physiological stress and increased exposure to predators. Yet in some ecological contexts, the extra care pays dividends in terms of the chances that each offspring will survive and contribute genes to the next generation.

Many animals continue to care for young long after eggs and sperm are produced and fertilization takes place. This parental care often involves tending the young by guarding the eggs or feeding the newborns. In some cases, such as the black-headed gull, both parents provide care. In other cases parental care is provided

FIGURE 42.12
Female Investment.
A female kiwi bird lays an egg 4 times larger than one would expect from its size. In this species, the male also makes a huge investment—he spends about 100 days sitting on the nest warming the chicks and hardly ever feeds during this time.

FIGURE 42.13
Mammals Often Have Extensive Parental Investment.
This female gray kangaroo (*Macropus giganteus*) will nurse its large joey (the young kangaroo) for several months after the joey's birth as a tiny, bean-sized marsupial. The young learn from the mother about grazing and escaping from danger.

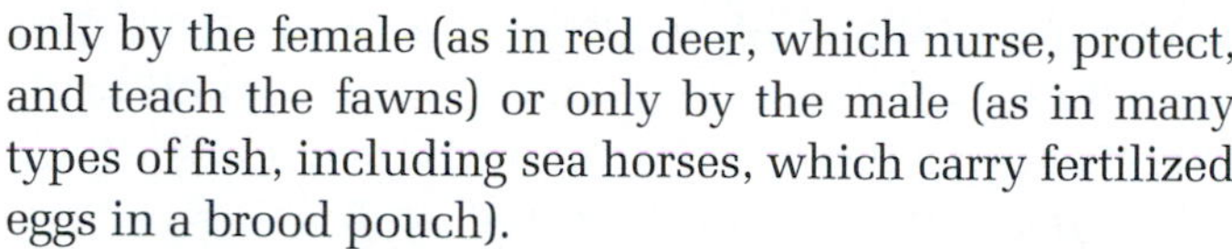

only by the female (as in red deer, which nurse, protect, and teach the fawns) or only by the male (as in many types of fish, including sea horses, which carry fertilized eggs in a brood pouch).

Often, the type of parental care provided is associated with particular mating systems. Some species practice *monogamy*, where a male and a female pair for a breeding season or a lifetime, and both parents care for their offspring; the black-headed gull is an example. In certain other species *polygyny* is the rule, wherein a male mates with several females, but females mate with a single male and generally provide most parental care, as occurs in herds of elk and wild horses. Some species display *polyandry*, in which a female mates with several males, while each male mates with a single female. This mating system, occurs in many fish species, and in it, the male usually provides most of the parental care. Finally, in some species both males and females mate several times with different partners and either sex cares for the offspring.

Why should one mating system or another evolve in a species? The answer depends on the species' reproductive biology and the costs and benefits of mating and parental care. Male parenting is likely to evolve only in those ecological situations where females cannot provide all the care necessary for raising a brood successfully or where a male who abandons his mate is unlikely to find another partner. In such cases, natural selection will favor behavioral mechanisms that promote the father investing more in the offspring, such as establishing and defending a nesting territory, as do black-headed gulls.

Sexual Selection Between Members of the Same Sex In many animal species, females make a greater parental investment than males. In these species, males generally compete to inseminate as many females as possible. Biologists call competitive pressure between members of the same sex *intrasexual selection*. In some species, this leads to a **dominance hierarchy**, a ranking of group members based on past success in aggressive encounters over access to food or mates. Bull elephant seals, for example, are massive beasts that threaten each other with tremendous roars, and bite and batter competitors. The loser of a skirmish lowers his head and retreats, with the victor in pursuit. A few years ago, a biologist ranked 10 bulls living on a small island by their track records in winning such encounters, and then counted the successful copulations each achieved during the breeding season. The top-ranking male accomplished nearly 40 percent of all the copulations in the group, the second male less than 20 percent, and the other males only about 5 percent each. Thus, natural selection in elephant seals (and in many other animals as well) favors those behavioral acts that win dominance contests and hence mates. In other animals, such as chickens, dominance hierarchies may be established for food rather than access to mates.

Sexual Selection In many species, females have a greater investment in gamete size, egg laying, live birth, and/or parental care than males, and thus the total number of young they potentially can produce is lower than in males. Females also do the mate choosing in many species. A female generally picks males that provide the genes or material benefits that are most likely to enhance her offspring's survival. Females thus engage in *intersexual selection*, wherein one sex (the female) chooses desirable mates, and in so doing, acts as an agent of natural selection on the behavioral and physical traits of the other sex.

FIGURE 42.14

A Female's Alarm Call and the Dominance Hierarchy.

A casual advance by an amorous male—a flipper caress—sets the female elephant seal (*Mirounga leonina*) to screaming at high decibel. She calls this way regardless of the suitor's stature in the colony, but if the male is subordinate, the king bull will probably chase him off. The call thus increases the likelihood that her sons have alleles from the dominant male.

By choosing dominant or propertied males, females increase their own biological fitness. Elephant seal cows, for example, create intersexual selective pressure by simply screaming loudly whenever a bull attempts to couple [FIGURE 42.14]. Her cries alert the harem master, and if the courting male is low on the dominance hierarchy, the dominant bull sprints to the site of the tryst and chases away the intruder. The female's automatic cries greatly increase her chances of being inseminated by the dominant male rather than a subordinate and hence of having sons who will display the aggressive behavior that favor high social rank. This is an example of *female choice*, in which the female of a species plays the major role in determining her mate.

Elaborate courtship rituals probably also evolved under the pressure of females exercising mate choice. The head-turning of the black-headed gull, the strutting dance of the male ptarmigan, and the showy flight of the African widowbird (*Euplectes progne*) flaunting a remarkably long 60-cm (2-ft) tail provide information about these animals' physiological condition, and this, in turn, may reflect gene quality. Experimenters have found that when they cut the tail feathers off a male widowbird, it is less likely to attract a mate, but when they glue on long feathers from another bird, making an extra-long tail, the bird can be quite attractive to females. Experiments like these suggest that even a potentially harmful trait, such as a long tail that hinders flying ability and thus increases the bearer's risk of being injured or eaten, may be selected for if it promotes reproductive success.

The mechanisms of sexual selection are still under investigation. Some biologists think that females benefit by choosing sexually attractive males because they will then be more likely to have sons that are also sexually attractive. This, the researchers suggest, would improve their long-term contribution to the gene pool. The alternative view is that females who select males with elaborate tails, nests, or other displays are simply choosing males who are in good enough physical condition to overcome the disadvantages inherent in forming the displays.

COMMUNICATION

Midwesterners enjoying the fragrant air and trilling insect sounds of a warm summer night can often catch a wonderful aerial display as lightning bugs, or fireflies, flash their luminous yellow abdomens on and off while they dart about the green lawns, fields, or trees. The flickering streaks they create are based on flashing patterns with a rhythm characteristic for each species [FIGURE 42.15]. Some species give long pulses of continuous light, others blink three times in a row, and still others give a fluttering flicker like a light bulb about to burn out. The airborne flashers are males communicating their availability for mating, and when a female sees the right pattern of flashes, she responds with her own burst of light, revealing her position to the male.

The firefly's blinking light is a form of **communication**, a signal produced by one individual that alters the behavior of a recipient individual and generally improves both participants' biological fitness.

FIGURE 42.15

Luminous Communication: Fireflies on a Summer Night.

Fireflies are actually soft-bodied beetles that light up the countryside with glowing abdomens. This drawing depicts the male courtship displays of four species.

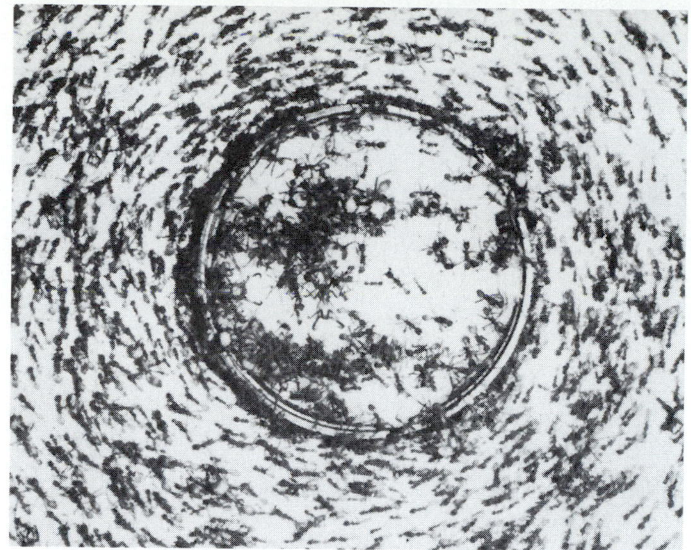

FIGURE 42.16

Scent Circles: Ants Following a Pheromone Trail.

When an experimenter placed a circular dish amid a colony of army ants, a few explorers traced the periphery and laid down a pheromone trail, and the rest followed endlessly like miniature automatons.

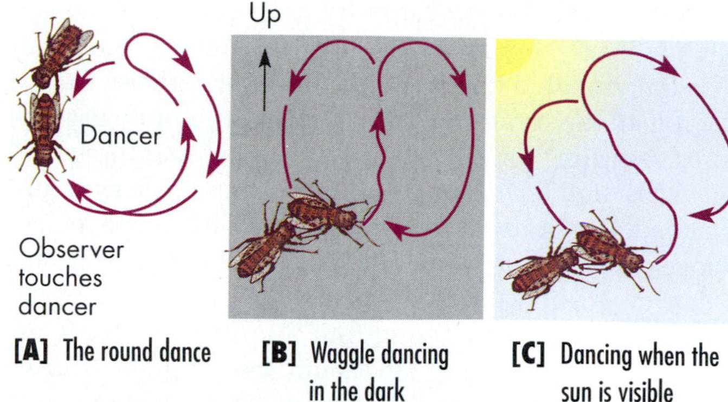

[A] The round dance **[B]** Waggle dancing in the dark **[C]** Dancing when the sun is visible

FIGURE 42.17

Honeybee Dances.

Forager bees perform **[A]** round dances when a discovered food source is close to the hive and **[B]** waggle dances when the source is more distant. Orientation of the dancing bee on the wall of a dark hive signifies the direction to fly relative to the sun's position, with the convention that a symbol for the sun is straight up. If sunlight shines into the hive **[C]**, the bee orients to the light. We know this because a partially blinded bee in a lighted hive dances as if it were dark, but observer bees interpret the dance according to the orientation of the light.

Communication by Sight, Sound, and Scent Each animal uses one or more channels of communication that fit the species' ecological circumstances. In open areas where individuals live near each other, visual displays such as the courtship dances of ptarmigans and black-headed gulls are common. Animals that hide in thickets or pond vegetation are invisible not only to predators but to potential mates as well, and for them, auditory cues, like the chirp of a cricket or the croak of a bullfrog, are typical. Sound communication, such as the alarm call of a gull or prairie dog, is also advantageous as a warning to colony members who may not see the approach of a predator.

Some animals live such solitary lives that they are unlikely to see or hear signals from other members of their species. For them, odor molecules can convey information. The sex pheromones released into the air by female moths [see CHAPTER 30] can travel for miles on the wind, announcing the females' location and sexual readiness. Chemical signals not only disperse farther than auditory or visual messages, but can also persist longer in the environment. Ants establish durable trails this way and will follow pheromone trails slavishly—even those an experimenter deposits in a circle [FIGURE 42.16].

Communication by Touch Groups of animals that live in dark hives, caves, or underground chambers often communicate by touch, as well as by sound and smell. In the 1940s, biologist Karl von Frisch discovered the classic example of touch communication: the dances bees perform in the dark interiors of their hives that communicate the location of rich food sources. If a foraging bee finds flowers that are laden with pollen and nectar and are situated fairly close to the hive (within 80 m, or 260 ft), the bee flies to the hive and immediately performs a *round dance* by walking in tight circles [FIGURE 42.17A]. Observer bees follow the dancer in the dark, sense its movements by touch, and learn that a nearby food source exists. Since worker bees can also smell bits of food clinging to the forager's body, they usually have no trouble locating the specific flowers.

If the forager discovers a food source far from the nest, she returns with a rich load of pollen, climbs onto a vertical wall of the hive, and begins a *waggle dance*, moving in a figure eight while wagging her abdomen from side to side and making an occasional burst of sound [FIGURE 42.17B]. The dancer's animation indicates the amount of energy needed to fly to the food: A lethargic dance with few waggles and little sound indicates a long or uphill trip to the food source; an excited dance with many waggles and frequent bursts of sound suggests a more readily accessible site. Simultaneously, the orientation of a dancing bee on the hive wall reflects the angle of the necessary flight path with respect to the sun. If the hive is dark and the food lies in the sun's direction, the bee waggles while moving straight up [see FIGURE 42.17B]. If the hive is dark and food lies at 90° to the right of the sun, the bee waggles while moving horizontally to the right. If sunlight enters the hive, the bee orients her dance directly with respect to the sun's rays [FIGURE 42.17C].

While von Frisch carefully observed these dances and deduced their meaning, only experiments could confirm that oriented dances guide the observer bees. To test the role of the sun in bee communication, a biologist captured returning foragers and painted over their three simple eyes but left their two large compound eyes untouched. Under these conditions, a partially blind forager dances as if the hive were dark, and that dance indicates a particular direction for the food. Her sisters, however, who perceive light shining into the hive, interpret the dance as if the forager, too, could see the light, and so head out in the wrong direction. These experiments proved that sight and touch, not scent, guide the observers' flight path.

Whether an animal communicates via sight, sound, scent, touch, or body positions, the exchange of information has adaptive value: It allows individuals to find homes, food, and mates more successfully or to evade predators and thus to send their alleles on to future generations. While solitary animals like fireflies, moths, and frogs exchange information from time to time, and those living in small family groups, like black-headed gulls, do so more regularly, communication is most complex in animals that live together in large social groupings. By their very complexity, animal societies pose an interesting challenge to evolutionary ecologists: How could such social groups have evolved?

➤ CONCEPT CHALLENGE

During the hot, sunny summer, male lark buntings defend their nesting territories on the American prairie. Usually a single female builds a nest in a male's territory and he helps her raise a family. Sometimes a male's territory attracts a second female, and she mates with him but receives no help raising her offspring. That second reproduction therefore tends to be less successful. Devise an experiment to test the hypothesis that polygyny (having more than one female mate) arises when a male's territory contains enough shade that a female can rear more young by becoming his second female than by becoming the monogamous mate of a male defending a territory with no shade. What parameter in the male's territory will you manipulate and how? What data will you collect? What result would disprove the hypothesis?

Evolution of Social Behavior

Turtles, beetles, snails, and the vast majority of other animals live solitary lives, interacting with others only to mate and maintaining no long-term contact with parents or offspring. A fascinating minority of animal species, however, live in groups—either as giant colonies, like those of ants, bees, and termites; as large groups, like flocks of gulls; as small, stable packs, like those of wolves [FIGURE 42.18]; or as individuals with only seasonal associates, such as the graceful manatees of Florida. What biological mechanisms cause gulls to live in cooperative societies while turtles live as lone individuals? And what are the evolutionary precedents and reproductive advantages of social behavior? How does an individual benefit from, say, cooperative predator mobbing if this behavior puts it in danger?

FIGURE 42.18
Social Contacts in Wolves.
Some animal species live in small and intimate social groups, such as this pack of gray wolves.

SOCIAL LIVING: COSTS AND BENEFITS

What does an individual gain and what does it sacrifice by living in a group? One of the many studies biologists have designed to answer this question is a comparison of the societies of white-tailed and black-tailed prairie dogs. Regardless of species, a prairie dog has a characteristic response to the approach of a badger, weasel, or other predator: It freezes for a moment, then gives a loud alarm call [FIGURE 42.19]. This dangerously betrays its location to the predator, but it also warns family members and neighbors to dash off to the safety of their burrows. Animal biologists noted that black-tailed prairie dogs live in larger groups than white-tailed prairie dogs. That differential gave them a means of probing the advantages of group living for these animals.

Experimenters pulled a stuffed badger at constant speed through different-sized prairie dog colonies and recorded the time it took before a prairie dog would sound the alarm. They found that among black-tails, with their larger colonies, predators were detected sooner than among the smaller, white-tail colonies. With the support of so many watchful colony mates, black-tails could spend less time scanning for predators and more time foraging. Larger colony size is clearly advantageous for providing a means of early predator detection plus more for-

FIGURE 42.19
Benefits and Costs of Group Living.
When a prairie dog warns of a predator, it exposes itself to danger but it protects its kin, who carry many of its alleles, improving their chances of reproducing in the future.

aging time. So why don't white-tailed prairie dogs live in large colonies too?

The answer to that question lies with the disadvantages of living in larger societies. Observations reveal that prairie dogs in larger groups spend more time aggressively bickering over territory than those in smaller groups. In addition, plague-carrying flea populations are four times as dense in the larger, black-tailed colonies as in the smaller, white-tailed groups. Apparently, the benefit of avoiding predators offsets the costs of infighting and disease. One could predict from this that the black-tails' environment exposes them to more predators. Indeed, large colonies of black-tails live exposed on the open plains of the American west, while the smaller colonies of white-tails inhabit scrubby land and can hide in the bushes.

From experiments like these, sociobiologists conclude that animals will socialize only when the benefits of group living outweigh the costs. In each case, the benefits require cooperation between individuals, and the cooperation can occasionally cost an individual its feeding time or even its life. How can natural selection explain such seemingly altruistic behavioral acts.

EVOLUTION OF ALTRUISTIC BEHAVIOR

Cooperative behavior often favors both helper and helped, as when a lioness chases a wildebeest toward her hunting partners and all share the kill. Sometimes, however, the helper suffers, as when a worker bee stings a raccoon, protecting the hive in the process but ripping out her stinger and subsequently dying. Biologists have long wondered how such **altruism**, or seemingly unselfish acts for the welfare of others, could evolve. Animals displaying a selfless behavior would tend to reproduce less frequently. Therefore, the principles of natural selection would seem to predict that alleles promoting such behavior should decrease in a population.

FIGURE 42.20
Helper Scrub Jays Demonstrate Kin Selection.
Older siblings (right) that remain near the nest and assist their parents in raising a new crop of young indirectly promote reproduction of their own alleles. Their behavior is an example of kin selection.

Biologists working out the puzzle of altruism have recognized that reproduction multiplies one's own alleles. The obvious way to accomplish this is to reproduce oneself; but there is another way too: encouraging the reproduction of family members that share the same alleles. Evolutionary biologists recognize a specific type of natural selection called **kin selection**, in which an individual increases its genetic contribution to the next generation by helping relatives—who share its alleles—to reproduce. The principles of kin selection suggest that altruism can evolve by natural selection.

You can see how kin selection works by considering how many alleles you share with your family members. Because of meiosis and fertilization, you inherited half your alleles from your mother and half from your father. Since you share the same parents, half of your alleles, on average, will be carried by each brother or sister—in other words, you are as closely related to each parent as you are to each sibling, on average. Your aunts, uncles, and grandparents, in turn, carry one-quarter of your alleles, and your cousins one-eighth.

These same degrees of relatedness hold for most other animals and help explain kin selection. Consider, for example, a breeding pair of Florida scrub jays. Some of their offspring serve as helpers at the nest; they do not themselves reproduce, but assist their parents to raise more young jays [FIGURE 42.20]. Let's compare the genetic contributions, to the next generation, of a bird that reproduces directly and of a bird that merely helps a close relative. If a jay were to produce directly two surviving chicks, each possessing half of its alleles, then the bird's genetic contribution to the next generation would be 1 (2 chicks times ½ of its alleles, or $2 \times ½ = 1$). If the helper jay abandoned personal reproduction, however, to help its parents produce three more siblings, all of whom shared half of its alleles, its genetic legacy would be 3 chicks times ½ of its alleles, for a contribution of 1.5 ($3 \times ½ = 1½$). This analysis shows an increased contribution by helping compared to self-reproduction. If certain of the shared alleles promote helping behavior, then kin selection will cause these altruistic alleles to increase in frequency in the jay population.

A helper jay may eventually inherit territory when its parents die, and there the jay can attract a mate and set up its own family, complete with new helpers. In such a case, the former helper's biological fitness becomes the sum of its fitness from kin selection as a helper, plus its personal fitness upon reproducing—a sum called **inclusive fitness**.

A proponent of kin selection once quipped: "I'd lay down my life for two brothers or eight cousins." By this he meant that an act of altruism toward close relatives promotes survival and transmission of one's own alleles. The theory helps explain how something as drastic as the suicidal sting of a honeybee could have evolved.

Kin Selection and the Evolution of Social Insects Can the concept of kin selection account for the rigid caste system of social bees, wasps, and ants? Such societies often consist of a single reproducing female, the *queen*, plus a few *drones*, or males, and up to 80,000 sterile female helpers, or *workers*. But what advantage comes to the sterile worker who cannot reproduce herself and instead helps the queen reproduce?

Geneticists discovered that in bees and wasps, females are diploid, but males are haploid. As a result, the daughters of a single male share *all* their father's alleles instead of just half, as in most organisms. As a consequence, bee sisters are three-quarters alike genetically, whereas human sisters (as well as siblings in other animals) are just one-half alike. By helping the queen reproduce more sisters (including future queens), they are really favoring the preservation of their own alleles.

Biologists often explain the evolution of insect societies in terms of the genetics of sex determination; in reality, however, the situation is more complicated. A single queen, for example, will sometimes mate with more than one male, and so the queen and helpers of the next generation may be less closely related genetically than the model predicted. In addition, not all bees and wasps with haploid males interact in societies, and some insects that are social—termites, for example—do not have haploid males. Apparently ecological factors also contribute to cooperative breeding in animal societies. The discovery that underground African mammals called naked mole rats breed cooperatively in societies helped to elucidate some of those factors. Nevertheless, questions still remain to be investigated.

HUMAN SOCIAL BEHAVIOR

Biologists can point to many examples from the animal kingdom in which social behavior seems to promote the survival of an animal's alleles through a combination of individual and kin selection. But what about human behavior? Do similar principles apply to us?

The tiny hand of a human baby grasps tightly to a parent's finger, its little lips suck rhythmically at a nipple, and it communicates forcefully with a cry that a parent cannot ignore. A baby born blind still flashes an alluring smile at the sound of its parent's voice. For some biologists, these observations would test the hypothesis that infants who inherit alleles fostering the behaviors of smiling or crying are more likely to solicit parental care. These same infants would then be more apt to survive and pass on such alleles to the next generation.

In humans, as in other animals, hereditary and environmental factors interact in the shaping of behavior. Genes help mold the native ability to learn language, for example, and this unfolds as neural pathways develop in the growing embryo and make it capable of learning. But whether a child learns to speak English, or Polish, or Swahili after it is born depends on its environment—the language its parents teach it.

This flexibility may direct people to accept cultural practices that enhance their inclusive fitness and reject those customs that reduce genetic success. For example, Brazilian tribes living in areas with poor soils and overhunted forests adopt Western agricultural techniques, while clans inhabiting less disturbed areas cling to traditional methods. In both cases, the human nervous system seems to respond to the environment with a flexibility that is naturally selected to elevate an individual's economic condition and hence provide for offspring. A few other animals also seem to accept new adaptive cultural practices and transmit them to their offspring. The case of the macaques who appear to have learned to wash sweet potatoes is probably the most famous example [FIGURE 42.21].

To explore the role of genes in shaping human personality and behavior, psychologists have conducted studies on identical and fraternal twins raised together in the same family or apart by adoptive families [FIGURE 42.22]. These twin studies show that many human personality traits are influenced about equally by genes and the environment. Such traits include social effectiveness, achievement, social closeness, alienation, aggression, respect for authority, and avoidance of harm. A single gene, of course, encodes a protein, not a trait like aggression or

FIGURE 42.21

Cultural Transmission in Primates? The Case of the Clean Sweet Potatoes.

A young female macaque accidentally learned to wash the sand off sweet potatoes that she picked up from the beach, and soon others in the troop learned to do the same. This learned behavior became a permanent part of the troop's culture, passed along to subsequent generations, perhaps in the way reading, writing, metallurgy, and agriculture have become parts of human culture.

FIGURE 42.22

Reunited Twins.

Identical twins Jerry Levey and Jim Tedesco were separated as infants and long felt something was missing in their lives. Reunited in middle age, they discovered not only a close physical resemblance, but identical vocations—fire fighting—and avocations—flirting, telling jokes, and drinking the same brand of beer.

achievement. Behavioral geneticists suggest that alleles of hundreds of genes interacting with a person's experiences combine to influence the person's range of emotions; the thresholds for anger, hostility, and sexual arousal; and the inclination to learn one type of behavior versus another.

Partly due to the evolution of various types of human behavioral patterns, we have been enormously—perhaps disastrously—successful as a species. As the biosphere threatens to collapse under the pressure of human population, our behavioral plasticity remains. Thus, the hope remains, as well, that we will not destroy the quality of life for the animal called *Homo sapiens*, nor for any of the living species that share our planet, and that the nature of life will remain for us a challenge and a source of perpetual wonder.

➤ CONCEPT CHALLENGE

Discuss the proposition that in genetic terms, all altruism can be considered selfish.

Connections

The environment selects for genes that direct the development of animal nervous systems in ways that promote specific behavioral acts. Some behavioral acts are similar each time they are performed, like a baby gull pecking at its mother's beak, and others are plastic and variable, like a macaque learning to wash potatoes in the sea. Regardless of flexibility level, behavior has evolved in ways that increase an individual's survival and reproduction, as when a gull removes broken eggshells from its nest and in so doing helps protect its offspring from predators. Of course, the gull has no awareness of the evolutionary function of removing the broken shells. But individual gulls who possess alleles promoting eggshell removal tend to lose fewer chicks and have more surviving offspring.

The behavioral acts that suit an organism to its environment have their roots in both physiological and evolutionary mechanisms, and these help us understand how and why animals act as they do—even in puzzling instances of altruism, such as a scrub jay helping its parents raise another family or a worker bee assisting the queen. By understanding the evolution of human behavior, we stand a better chance of solving our current global crises of population, ecological degradation, disease, starvation, and warlike aggression. Well-informed citizens like the readers of this book are in the best position to apply biological knowledge to these issues and perhaps to help make our world a safer place for life.

KEY TERMS

altruism, 929
communication, 926
dominance hierarchy, 925
habituation, 919
imprinting, 918
learning, 919
modal action pattern, 918
optimality hypothesis, 924
sensitive period, 919
sexual selection, 925
territoriality, 923

HIGHLIGHTS IN REVIEW

1 A behavior is a response an animal makes to an environmental stimulus. The ability to perform a behavior arises during development based on the animal's genes and on influences from the environment.

2 Since behavior has a hereditary component, it can evolve due to natural selection. Experiments suggest that natural selection has often favored behaviors that promote individual reproductive success.

3 Some behavioral traits, such as swallowing, are not strongly modified by the environment, while others, such as language, are.

a] There is a broad continuum of behavioral types, from those with little flexibility to learned behavioral acts (behavioral acts modified by experience).

b] Some behavioral acts, like the pecking of a red dot by a baby gull chick, can be modified by environmental information.

c] Some behavioral acts, such as the imprinting of goslings, are flexible, but only within a sensitive period, a given time frame.

d] Learning reflects truly flexible behavior; it is an adaptive and enduring change in behavior based on personal experiences in the environment.

e] Habituation occurs when an animal learns to ignore a repeated stimulus never followed by reward or punishment.

f] Trial-and-error learning comes about when an animal learns to associate a response, or operant, with a reinforcer, such as a reward or punishment.

g] In classical conditioning, the animal learns to associate two separate stimuli with a response.

h] In insight learning, a trait apparently possessed only by primates, the animal reasons out a solution based on previous experience in similar situations.

4 In certain ecological contexts, social behavior can result in increased reproductive success.

a] Choosing and defending a territory can involve a taxis, orientation, navigation, or true territoriality, the defense of a particular space from intruders. The possession of territory tends to confer selective advantage in terms of finding mates or food or hiding from predators.

b] Finding food often involves optimality—the maximizing of energy gain and the minimizing of energy spent collecting food or defending against predators.

c] Numerous behavioral sequences have arisen around the investment of energy for each gamete (and possible offspring) and for mate selection. Members of the same species exert selective forces on each other, called sexual selection. This has led to dominance hierarchies and to courtship rituals by which females can judge male competence.

d] Communication involves a signal (visual, auditory, olfactory, vocal, tactile) produced by one individual that alters the behavior of a recipient and improves both participants' biological fitness.

e] Social behavior, or close group interactions and interdependencies, require communication and evolved because they confer selective advantage to particular species in their particular environments.

f] Altruistic, or selfless, behavior probably evolved via kin selection—the advantage conferred on reproducing individuals by a closely related, often nonreproducing helper.

g] Biologists are still debating the extent of evolutionary preprogramming for specific human behavioral acts, but twin studies show that many human traits have powerful genetic bases as well as environmental influences.

UNDERSTANDING THE FACTS AND CONCEPTS

In questions 1–5, match each of the descriptions with the most appropriate item or items from the following list.

a] trial and error
b] habituation
c] operant conditioning
d] classical conditioning
e] insight learning

1 Individuals are unaware of the weight of the atmosphere pressing down on their bodies.

2 Deciding not to read the directions, people learn how to turn on their VCR by pushing each of the buttons in turn until the machine is turned on.

3 Digestive juices begin to flow while people watch the on-screen advertising for refreshments before the movie begins.

4 A student learns faster if she promises herself a treat after studying.

5 Deciding how long it will take to reach a friend's house based on prior knowledge about traveling in the same area.

For questions 6–10, match the descriptions with the most appropriate item or items from the following list.

a] modal action pattern
b] imprinting
c] territoriality
d] ESS
e] optimality hypothesis

6 Animals perform an unconscious risk-benefit analysis before choosing food from one source and ignoring another source.

7 A normally gentle male dog barks and growls when another male enters his yard.

8 A theory that describes the evolutionary development of stable, genetically based patterns of behaviors in a species.

9 Newborn babies recognize their mother's face and physiologically respond to its sight by becoming calmer.

10 After breathing in some dust, you explosively exhale while simultaneously closing your eyes and wrinkling the facial muscles that encircle your eyes and saying "achoo".

For questions 11–15, match the descriptions with the most appropriate item from the following list.

a] taxis
b] homing
c] migration
d] orientation
e] navigation

11 The ability of an animal to travel along particular directional routes.

12 Animal behavior that involves returning to the site of hatching or birth.

13 The ability to travel from a particular map point to another; often tested by changing the starting point of an animal's voyage.

14 An animal is unconsciously drawn toward a particular stimulus such as heat or light.

15 Seasonal travel to feeding or breeding grounds.

For questions 16–20, match the descriptions with the most appropriate item from the following list.

a] intrasexual selection
b] intersexual selection
c] altruism
d] kin selection
e] inclusive fitness

16 Also known as a dominance hierarchy or pecking order.

17 A theory that predicts that the closer the genetic relatedness of two individuals, the greater that chance they will exhibit altruistic behavior.

18 The sum of the evolutionary benefit an animal derives from assisting genetic relatives and its own individual fitness.

19 The selection by females of particular males as mates.

20 Behavior that decreases an individual's chances of survival while increasing the survival potential of another member of the species.

INTEGRATE AND APPLY WHAT YOU HAVE LEARNED

1 As a general rule, which parent has a greater investment in their offspring? Explain.

2 When analyzing human behavior from a genetic point of view, why would an ethologist explain that we have to work hard to teach our children to act with kindness and to "turn the other cheek"?

3 Why might particular males develop physical or behavioral characteristics that decrease their fitness or chance of survival? Examples include the bulky tail of male peacocks, or the habit of male gazelles stopping to give an alarm call when a predator is sighted.

4 List four types of animal communication and give an example of each.

5 In explaining the evolutionary cause of many behavior patterns, ethologists rely on an analysis of the risks and rewards of the behavior to the individual and the species. Give two examples of such behaviors that have been discussed in the chapter. Does an animal consciously calculate the risks and rewards before undertaking a particular action?

ANALYSIS

For questions 1–5, select the best answer.

1 What observation led Nilo Tinbergen to believe that the removal of broken egg shells was a genetically determined pattern of behavior?

a] All gulls of that species do it.
b] He proved that gulls only remove jagged-edged objects.
c] He showed that gulls notice jagged-edged objects faster than other objects.
d] He proved that the nervous system of gulls directs the behavior.
e] He proved that the behavior was learned by imprinting.

2 Predict the phenotype of the first generation offspring from a cross of hygienic and nonhygienic bees.

a] All uncap and remove diseased pupae
b] All neither uncap nor remove diseased pupae
c] 3:1 ratio of uncapping and removal
d] All uncap but do not remove diseased pupae
e] No logical prediction can be made

3 The evolutionary success of warning coloration and mimicry rely on which kind of learning?

a] habituation
b] insight learning
c] operant conditioning
d] classical conditioning
e] trial and error learning

4 Female langur monkeys have a gestation period of 7 months and a nursing period of another 8 months. While pregnant and during the nursing period, females do not ovulate. Langur monkeys travel in groups of 10 to 20 adult females, their young and one dominant male. If another male unseats the dominant male, he will attempt to kill all nursing offspring and newborns for 6 months after the takeover. He will not attempt to kill young who are weaned. How would an ethologist explain why he does not kill the weanlings?

a] He is unsure of their parentage.
b] The female weanlings represent an enlargement of his harem.
c] After killing the newborns and nursing young, he has had enough of killing.
d] The weanlings might not be genetically related to the previous dominant male.
e] The adult females will not be pleased if all of their offspring are killed.

5 You are in a capsizing boat with several relatives. Assume you can save only one, in addition to yourself, and you will choose the person to whom you are most closely related genetically. Who will you save?

a] second cousin
b] your mother's sister
c] your half-sister
d] paternal uncle by marriage
e] great-grandfather

Captive Breeding and the Florida Panther

In the early 1900s, the eerie scream of *Felis concolor coryi*, the Florida panther, could be heard throughout large parts of that state and neighboring Georgia and Alabama. A subspecies of cougar, this large solitary cat marks out and hunts an individual territory of 50 km^2 (18 mi^2) and communicates an occasional "love call" to a distant cougar with a long, rising and falling cry that sounds like a person in great physical danger. Many people dreaded this sound in the night, and reviled the southern cougar as a poacher of sheep, cattle, and deer. Consequently, a number of states, Florida included, paid bounties for dead cougars. Ironically, it is the Florida panther itself that is in great physical danger now—the danger of imminent extinction—due to overhunting and to the loss of territory from human agriculture and development.

In the past few years, biologists have studied the last four dozen wild Florida panthers more intently than any other big cats in North America. Some wildlife biologists have even gotten permission to remove one quarter of this remnant population from the wild to raise them in captivity with the hope of someday reintroducing them into nature. But other conservationists are strongly opposed to such captive breeding, and have blocked the removal efforts with lawsuits. What's the "right" way to save an endangered species? Let's consider the case of the Florida panther, apply principles from our study of ecology, then you decide.

You may think of cougars as predators that inhabit arid, mountainous landscapes, and indeed, some subspecies do. But *Felis concolor coryi* evolved as a denizen of scrubby lowland pine forests, wet prairies, and swamps, and it preys upon waterfowl, wild boar, swamp deer, and other animals that share the ecosystem. As we saw in CHAPTER 40, however, human agriculture, development, and habitation have changed the Florida ecosystem dramatically. Just as wild Florida has been reduced and confined to tiny remnant islands such as the Everglades National Park and Big Cypress National Reserve, so, too, has the Florida panther, due to its need for a large hunting territory.

One could predict from the pyramids of biomass we also studied in CHAPTER 40 that as widespread habitat loss takes place, and the producer, herbivore, and carnivore levels dwindle in population, top carnivores will no longer be supported and will disappear. This has indeed been the case with the Florida panther, and by 1967—when it was uncertain whether any of the cats remained in the wild—the U.S. Fish and Wildlife Service placed the big cat on its official endangered species list, and thereby made it illegal for hunters to shoot any further individuals in the wild.

Florida wildlife biologists began radio telemetry studies to find and track the remaining panthers in order to provide an accurate population tally, plus details about the cougar's movements, preferred habitats, and mating patterns. They also took blood and tissue samples for genetic analyses. They wanted to determine the species' minimum viable population size [see CHAPTER 39]; its minimum habitat requirements so adequate preserves could be established [see BOX 40.1]; and the extent of its genetic variability, which helps predict the chances for survival and future evolution [see CHAPTER 16].

The Florida wildlife researchers were able to develop accurate answers on the panther population, but the news is dreadful. Today, only 30 to 50 panthers survive in the wild, and the remaining few show strong evidence of inbreeding (matings between sisters and brothers or parents and offspring). Almost all of the animals have a kink at the end of the tail and an odd cowlick in their back fur. Seventy percent of the males have only one functional testicle (the other is formed but undescended into the scrotum) and in the many sperm samples taken, there are relatively few sperm cells and 90 percent of those existing cells are abnormal. Ominously, up to 30 percent of the adults and every kitten born since 1990 shows evidence of a congenital heart valve defect—a trait that limits the life of a dog of comparable size to about two years. And 30 percent of the wild cats show evidence of infection by feline immunodeficiency virus, the feline equivalent of the human AIDS agent, HIV. As we saw with cheetahs in CHAPTER 16, inbreeding and genetic deficits render populations more susceptible to diseases, less reproductively viable, and less flexible in the face of environmental change.

On top of all this, blood samples showed that the panthers have been contaminated with high levels of mercury. No one is sure where the pollution is coming from in the Florida environment, but based on the principle of biological magnification we saw in CHAPTER 40 [review FIGURE 40.11] one would expect to see the highest levels in a top predator such as a panther, and that seems to be the case here.

Applying a population viability analysis—a statistical tool for determining the fate of small, genetically compromised populations—specialists predicted that the Florida panther would become extinct within 25 to 40 years without vigorous efforts to save it.

Based on these findings, environmental groups lobbied for an expansion of Everglades National Park; for the creation of a Florida Panther National Wildlife Refuge; for regulations to prevent further agricultural encroachment, drainage, or development within panther habitat; and for studies to determine and remove sources of mercury poisoning. Florida wildlife biologists also wanted to remove some of the remaining panthers from the wild and begin a program of captive breeding. Their goal was to capture as many as six kittens and two to four adults each year for five years, and to establish a breeding colony of 130 panthers (some captured, some born in captivity) by the year 2000, and 500 of the felines by 2010. They got permission from the U.S. Fish and Wildlife Service to begin collecting cats, and established a preserve at White Oak Plantation, a large private estate in Northern Florida.

Before they could begin the first captures in 1990, however, an animal rights group called Fund for Animals sued the permission-granting federal agency. You might wonder how any group in favor of panthers could argue against efforts to save them. But they contended that no one really knew how the captures would affect the remaining wild cats, and that no one had an adequate plan for restoring and managing the panther's habitat. If the Everglades ecosystem is not itself saved, then, says Holly Jensen of the Fund for Animals, the captive breeding would simply produce "caged cats and nowhere to put them."

Her group views captive breeding as a "Band-Aid on an oozing, festering wound" (the "wound" being habitat destruction) and a diversion that could actually take the place of protecting wildlife. A logging company executive from the Pacific Northwest, for example, has already suggested capturing and breeding most of the remaining spotted owls [see CHAPTER 41] so the harvesting of the old-growth forests can continue unchecked.

But there is an equally strong opposing opinion. The Fund for Animals' lawsuit blocked the capture of Florida panthers for many months and during this time, reports Melody Roelke, a veterinarian with the Florida Game and Fresh Water Fish Commission, 11 kittens and adults died in the wild from car accidents, poisoning, heart defects, and other causes. Considering the species' slim and shaky gene pool, she finds delays like this unconscionable. "If this keeps up," Roelke is quoted as saying in a recent magazine article, "there aren't going to be any healthy Florida panthers remaining to put in whatever habitat they have."

The goal driving the attempts to captively breed Florida panthers is this: When the habitat is sufficiently protected and there are enough animals in breeding preserves, some can be reintroduced into the wild and become reestablished as wild populations. But is this realistic? For some animals, such as peregrine falcons and the golden lion tamarin, reintroduction has worked well. For others, such as the black-footed ferret, the California condor, and the whooping crane, expensive programs have produced captive-born animals but few if any successful reintroductions.

Consider this: In 1988 wildlife biologists attempted merely to *relocate* seven Texas cougars born in the wild to a Florida reserve as a dry run for future panther reintroductions. Of the seven, one drowned, two were killed by hunters, and three fled the protected national forest where released; all three plus the one still in the park had to be shipped back to Texas within ten months. Some critics have called captive breeding nothing but "a more sterile environment in which to become extinct." Some advocates feel it is the only hope for many endangered species.

What do you think?

What additional studies should be conducted, and what evidence do you feel needs to be collected before a rational decision can be made? ❑

TABLE A.1 Vitamins

Vitamin	Sources	Functions in Body	Deficiency Symptoms
Water-Soluble Vitamins			
Choline	Egg yolks, liver, beans, peas, grains	Part of phospholipids; needed for nerve cell function	Not seen in humans
Vitamin C (ascorbic acid)	Dark green vegetables, citrus fruits, strawberries, brussels sprouts, other fruits and vegetables	Helps form collagen and bone; enzyme helper; blocks toxic effects of oxygen	Gum bleeding; hemorrhages under skin; rough skin; failure of wounds to heal; bone degeneration
Niacin	Milk, meats, cereals, and starchy vegetables	Part of enzyme helpers involved in electron exchange reactions	Sore skin; smooth tongue; diarrhea; mental confusion; irritability
Pantothenic acid	Widespread in foods	Central to energy metabolism	Rarely seen
Vitamin B_6 (pyridoxine)	Whole grain cereals, vegetables, meats	Enzyme helper involved in amino acid metabolism	Skin soreness; smooth tongue; abnormal brain activity
Vitamin B_2 (riboflavin)	Milk, meat, vegetables, whole grains	Helper of enzyme active in energy metabolism	Cracks at corner of mouth; sensitivity to light
Vitamin B_1 (thiamine)	Milk, dairy products, fruits, breads, vegetables	Helper of enzyme that removes carbon dioxide from nutrients	Beriberi (paralysis, swelling, heart failure); mental confusion
Folacin (folic acid)	Fruits, leafy and other vegetables, grains	Helper of enzyme involved in metabolism of amino acids and nucleic acids	Anemia; diarrhea; smooth tongue
Biotin	Widespread in foods	Constituent of many enzymes in metabolism	Not seen in humans
Vitamin B_{12}	Meat and dairy products	Helper of enzyme involved in nucleic acid metabolism	Anemia; nerve degeneration
Fat-Soluble Vitamins			
Vitamin E (tocopherol)	Vegetable oils, margarine, salad dressings	Counters toxic effects of oxygen	Breakage of red blood cells; anemia
Vitamin A (retinol)	Carrots, milk, vegetables, fruits	Part of visual pigment; helps maintain living tissues; promotes bone growth	Impaired night vision; dry eyes; diarrhea; lung infections; bone changes; tooth cracking and decay; impaired brain growth; anemia
Vitamin K	Cabbage, milk, green leafy vegetables	Essential for synthesis of certain proteins, including blood-clotting proteins	Unchecked bleeding
Vitamin D	Milk, sunshine, eggs, liver	Helps bones and teeth take up calcium for proper growth	Rickets (bowed legs, other bone deformities); tooth decay; blood changes; lax muscles

TABLE A.2 Minerals

Mineral*	Sources	Functions in Body	Deficiency Symptoms
Major Minerals (more than 0.1 g/day)			
Calcium	Milk and dairy products, fish bones, collard greens, spinach, broccoli	Major part of bones and teeth; cell membrane integrity; helps collagen form; involved in nerve transmission	Fragility of bones (osteoporosis); stunted growth
Phosphorus	Widespread in foods	Major constituent of bones, blood plasma; part of DNA, RNA; needed for energy metabolism	Rare
Potassium	Bananas, orange juice, potatoes, tomatoes, other vegetables	Critical to normal heartbeat; principle positive ion in body cells	Mental confusion
Sulfur	Protein foods	Present in all body proteins; helps protein assume proper shape	Unknown
Sodium	Salt, common in foods	Constituent of salt in body fluids; helps regulate fluid content of body	Rare; excess leads to water retention
Chlorine	Salt	Major negative ion in body fluids; part of HCI in stomach acid	Rare
Minor Minerals (less than 0.01 g/day)			
Magnesium	Nuts, legumes, cereals, dark green vegetables, seafood, chocolate	Constituent of bones; role in protein synthesis and energy metabolism	Tetany; prolonged muscle contraction; hallucinations
Iron	Oysters, liver, bran flakes, lean beef, spinach, greens, strawberries	Part of many major enzymes active in DNA synthesis and cellular respiration; hemoglobin and myoglobin	Anemia; exhaustion; headache; weakness
Iodine	Salt (iodized)	Part of thyroid hormone thyroxine; controls metabolic rate	Goiter (enlarged thyroid gland)
Zinc	Oysters, milk, egg yolks, meat, whole grains	Part of many enzymes; helps bone form; DNA and protein synthesis; wound healing	Retarded growth; small sex organs; loss of sense of taste
Selenium	Abundant	Part of many enzymes; antioxidant	Not seen in humans
Manganese	Widespread	Constituent of many enzymes in metabolism	Not seen in humans
Copper	Grains, shellfish, organ meat, legumes, dried fruit, fresh fruit, vegetables	Helps form hemoglobin, collagen, nerves	Rare; retarded growth, sluggish metabolism
Molybdenum	Many foods	Part of many enzymes	Not seen in humans
Fluorine	Fluoridated water	Normal formation of bones and teeth	Tooth decay
Chromiun	Yeast, organ meats	Involved in carbohydrate metabolism	Stunted growth; adult-onset diabetes

*Arranged in order of decreasing amounts needed in daily human diet.

TABLE A.3 Some Common Drugs and How They Affect the Nervous System

Drug Class/Drug	Common Name	Actions on Body	Effects on Mood	Dangers of Abuse
Opiates				
Morphine, heroin, codeine	Horse (heroin)	Causes drowsiness after binding to opiate receptors on neurons	Causes euphoria	Physical dependence, nausea, vomiting, constipation, death from respiratory failure in fatal doses
Stimulants				
Cocaine	Coke, crack	Raises heart rate and body temperature, dilates pupils	Causes euphoria, excitation	Convulsions, hallucinations, cardiovascular damage
Amphetamine, methamphetamine	Dexedrine, crystal, meth, crank, cross tops	Increases heart rate, respiration, and blood pressure	Causes wakefulness, depresses appetite, increases alertness	Irregular heartbeat, chest pain, dizziness, anxiety, paranoia, hallucinations, convulsions, coma, cerebral hemorrhages in fatal doses
Nicotine	Tobacco	Causes vasoconstriction, racing heart, increased blood pressure	Acts as stimulant; causes euphoria	Dizziness, nausea, vomiting, withdrawal in addicted users
Caffeine, theophylline	Coffee, tea	Stimulates central nervous system and visceral muscles, increases heart rate and urine production	Causes some mood elevation, increases alertness	Anxiety, nervousness, insomnia, convulsions
Sedatives				
Alcohol	Beer, wine, liquor	Decreases respiration and body temperature, causes vasodilation, impairs vision, depresses central nervous system, numbs pain	Causes euphoria	Damage to liver, heart, brain, pancreas; respiratory failure in fatal doses
Methaqualone	Quaalude	Sedates, causes sleep, depresses central nervous system	Quieting	Nausea, vomiting, delirium, convulsions, coma, slow heart rate
Benzodiazepine	Valium, Librium	Relaxes skeletal muscles; decreases circulation, respiration, and blood pressure	Reduces anxiety, elevates mood	Depressed heart and lungs, decrease in muscular coordination, drowsiness; withdrawal can trigger seizures
Hallucinogens				
Cannabinoids	Marijuana, pot, grass, hash	Increases heart rate, causes vasodilation	Elevates mood, causes euphoria, sensory distortions	Lung damage from smoking; reduced sperm count; low testosterone in males
Lysergic acid diethylamide, psilocybin, mescaline	LSD, acid, mushrooms, peyote	Acts as a stimulant, raises heart rate and blood pressure, dilates pupils	Causes euphoria, hallucinations, sensory distortion	Irrational behavior
Dissociative Anesthetic				
Phencyclidine	PCP, angel dust	Kills pain, increases heart rate and blood pressure, causes swelling and fever	Elevates mood, causes perceptual disturbances	Coma, convulsions, psychosis, respiratory depression

Glossary

Abdomen The posterior part of an arthropod's body; in vertebrates, the abdomen lies between the thorax and the pelvic girdle.

Abiotic (Gr. *a*, without + *bios*, life) Characteristic of the part of an organism's environment made up of nonliving physical surroundings.

Abscisic acid A plant hormone that promotes dormancy.

Abscission (L. *ab*, away, off + *scissio*, dividing) In vascular plants, the dropping of leaves or fruit at a particular time of year, usually the end of the growing season.

Absorption spectrum The wavelengths of light absorbed by a pigment.

Accessory organ One of several organs that aids digestion, including salivary glands, liver, gallbladder, and pancreas.

Acid A proton donor; any substance that gives off hydrogen ions when dissolved in water, causing an increase in the concentration of hydrogen ions. Acidity is measured on a pH scale, with acids having a pH less than 7; the opposite of a base.

ACTH (adrenocorticotropic hormone) A hormone produced by the pituitary in response to stress. ACTH acts on the cortex of the adrenal gland.

Actin (Gr. *actis*, a ray) A major component of microfilaments in contractile cells such as muscle and other cells.

Action potential A temporary all-or-nothing reversal of the electrical charge across a cell membrane, which occurs when a stimulus of sufficient intensity strikes a neuron.

Activation energy The minimum amount of energy that molecules must have in order to undergo a chemical reaction.

Active site A groove or a pocket on an enzyme's surface to which reactants bind. This binding lowers the activation energy required for a particular chemical reaction; thus, the enzyme speeds the reaction.

Active transport Movement of substances against a concentration gradient requiring the expenditure of energy by the cell.

Adaptation (L. *adaptere*, to fit) A particular form of behavior, structure, or physiological process that makes an organism better able to survive and reproduce in a particular environment.

Adaptive radiation Evolutionary divergence of a single ancestral group into a variety of forms adapted to different resources or habitat.

Addiction Physical dependence upon a substance such as alcohol or other drugs, characterized by tolerance of increasingly larger doses, craving the drug, and withdrawal symptoms when use is discontinued.

Adenosine phosphate A class of molecules that are based on modified nucleotide monomers containing the nitrogenous base adenine, the sugar ribose, and a phosphate group. ADP is adenosine diphosphate; ATP is adenosine triphosphate, the major source of chemical energy in metabolism.

Adhesion The tendency of unlike molecules to cling to each other.

Adrenal gland (L. *ad*, to + *renes*, kidney) A hormone-producing endocrine gland located on top of the vertebrate kidney; secretes hormones mainly involved in the body's response to stress, such as epinephrine (adrenaline) and norepinephrine.

Adventitious root A new root that arises from an aboveground structure on a plant.

Aerobic (Gr. *aer*, air + *bios*, life) Conditions where oxygen is present at high levels; also, a biological process that can occur in the presence of oxygen.

Aerobic respiration Oxygen-requiring pathways of fuel breakdown that release energy; takes place within mitochondria in eukaryotic cells.

Age structure The number of individuals in each age group of a given population.

Aging A progressive decline in the maximum functional level of individual cells and whole organs that occurs over time.

Agnathan A jawless fish.

Agricultural revolution The transition of a group of people from an often nomadic hunter-gatherer way of life to a usually more settled life dependent on raising crops, such as wheat or corn, and on livestock. It was under way in the Middle East by 8000 years ago.

Air capillary In insects, the fine, narrow end of an air tube (trachea) through which air flows during respiration.

Air sac A thin-walled extension of the lungs; in birds, air flows through lungs and air sacs during respiration.

Albinism Lack of pigment in body parts of a plant or animal that are normally colored. Albinism in humans is inherited as a recessive.

Aldosterone A steroid hormone released by the cortex of the adrenal gland that regulates salt reabsorption in the kidney.

Alimentary canal A system of tubes and chambers associated with the digestive system; includes the mouth, esophagus, stomach, and intestines.

Alkaline Basic, or having a pH greater than 7.

Allele One of the alternative forms of a gene.

Allopatric speciation The divergence of new species as a result of geographical separation of populations of the same original species.

Allosteric enzyme An enzyme that has two binding sites: a catalytic site and a regulatory site that controls the activity of the enzyme.

Alpha cell An endocrine cell of the pancreas that secretes the hormone glucagon.

Altruism A behavior by one individual that benefits another even if the individual performing the act reduces its own fitness; e.g., a bird giving a warning cry that tells other birds of an approaching predator increases their chances of escape while drawing attention to itself and increases its own chances of being eaten.

Alveolus (L. small cavity) A thin-walled, saclike structure in the vertebrate lung where gas exchange takes place. Each lung contains thousands of alveoli.

Amine hormone A hormone that contains an amino group.

Amino acid A molecule consisting of an amino group (NH2), an acid group (COOH), and a side group (RCH), where R represents the group of atoms that differs in each amino acid. The 20 amino acids are the basic building blocks of proteins.

Amniocentesis A procedure for obtaining fetal cells from amniotic fluid for the diagnosis of genetic and a few other abnormalities.

Amniote Refers to an egg in which the embryo is encased in water and a series of membranes. Reptiles, birds, and mammals have amniote eggs.

Amniotic cavity A fluid-filled cavity surrounding the developing embryo of reptiles, birds, and mammals that keeps the embryo moist and cushions it from blows.

Amphibian A cold-blooded vertebrate that starts life as an aquatic larva, breathing through gills, and metamorphoses into an air-breathing adult; includes frogs, toads, newts, and salamanders.

Amyloplasts Dense starch-storing granules in plant cells.

Anabolism The building up of complex molecules from simpler ones, usually requiring an input of energy.

Anaerobic (Gr. *an*, without + *aer*, air + *bios*, life) Conditions where oxygen levels are low or absent. Also, biological reactions that do not require oxygen.

Anaphase A period during nuclear division when the chromosomes move toward the poles of the cell.

Anatomy The science of biological structure.

Angiosperm A flowering plant.

Angiotensin A protein hormone that causes blood vessels to constrict and thus increase blood pressure.

Animal A member of the kingdom Animalia; a multicellular organism that can generally move voluntarily, that actively acquires food and often digests it internally, that usually has irritability due to nerve cells and limited growth.

Animalia The kingdom of animals.

Annual plant (L. *annos*, year) A plant that completes its life cycle within one year and then dies.

Antagonistic muscle pair Two muscles that move the same object, such as a limb, in opposite directions.

Antenna One of the paired appendages on the head of arthropods that senses odors and vibrations.

Antenna complex A cluster of pigment molecules acting together during the harvest of light energy in photosynthesis.

Anther In flowers, the terminal portion of a stamen that contains grains of pollen in pollen sacs.

Antheridium The male sperm-producing organ of seedless plants.

Anthropoid A member of the suborder Anthropoidea, order Primates; includes humans, apes, and New and Old World monkeys.

Antibody A protein produced by the immune system in response to the entry of specific antigens (foreign substances) into the body. A specific antibody binds to the antigen that stimulated the immune response.

Anticodon The three bases in a tRNA molecule that are complementary to three bases of a specific codon in mRNA.

Antidiuretic hormone ADH, a posterior pituitary gland hormone that regulates the amount of water passing from the kidney as urine.

Antigen (Gr. *anti*, against + *genos*, origin) Any substance, including toxins, foreign proteins, and bacteria, which when introduced into a vertebrate animal causes antibodies to form.

Aorta In a vertebrate animal, the main artery leading directly from the left ventricle of the heart; supplies blood to most of the animal's body.

Apical dominance The inhibition of lateral buds or meristems by the apical meristem of a plant.

Apical meristem The perpetual growth zone at the tip of a plant's roots and shoots.

Appendicular skeleton That part of the vertebrate skeleton that supports the appendages (arms and legs).

Archegonium The egg-producing organ of ferns and bryophytes, including mosses.

Arteriole A thinner-walled, smaller branch of an artery, which carries blood from arteries to the capillaries in the tissues.

Artery A large blood vessel with thick, multilayered muscular walls, which carries blood from the heart to the body.

Ascending limb In the vertebrate kidney, the arm of the loop of Henle leading back to the cortex.

Ascocarp In certain fungi, a cup-shaped structure that holds sacs (asci) in which nuclear fusion and meiosis occur and spores are formed.

Ascus (plural **asci**) In some fungi, a small sac in which sexually produced spores develop.

Asexual reproduction A type of reproduction in which new individuals arise directly from only one parent.

Atom (Gr. *atomos*, indivisible) The smallest particle into which an element can be broken down and still retain the properties of that element.

Atomic mass The mass of an atom; it is approximately equal to the combined number of protons and neutrons in the nucleus of an atom.

Atomic nucleus The central core of an atom, containing protons and neutrons.

Atomic number The number of protons in the nucleus of an atom, unique to each chemical element.

Atomic theory Set forth by English chemist John Dalton in the early 1820s, the theory that every element is composed of identical particles, called atoms, and that chemical reactions involve rearrangements of atoms, but not their destruction or creation.

ATP Adenosine triphosphate, a molecule consisting of adenine, ribose sugar, and three phosphate groups. ATP can transfer energy from one molecule to another. ATP hydrolyzes to form ADP, releasing energy in the process.

Atrioventricular (AV) node A bundle of modified cardiac muscle cells between the heart's atria and ventricles that relays electrical activity to the ventricles.

Atrium (L. hallway) A chamber of the heart that receives blood from the veins. In mammals, birds, amphibians, and reptiles, the heart has two atria, the right receiving deoxygenated blood from the body and the left receiving oxygenated blood from the lungs. The atria pass blood into the heart's ventricles, which in turn pump it to the body.

Autonomic nervous system Motor neurons of the nervous system that regulate the heart, glands, and smooth muscles in the digestive and circulatory systems.

Autosome A chromosome other than a sex chromosome.

Autotroph (Gr. *auto*, self + *trophos*, feeder) An organism, such as a plant, that can manufacture its own food.

Auxin A plant hormone that stimulates cell elongation, among other activities.

Axial skeleton That part of the vertebrate skeleton that supports the main body axis, consisting of the skull, vertebral column (backbone), sternum, and ribs.

Axon The portion of a nerve cell that carries the impulse away from the cell body.

Bacillus (plural **bacilli**) (L. *baculum*, a rod) A bacterial cell with a rod shape.

Bacteriophage A virus that infects bacterial cells.

Bark Protective, corky tissue of dead cells present on the outside of older stems and roots of woody plants.

Barr body The inactive *X* chromosome in female mammals, visible as a dark-staining body in the nucleus.

Bartholin's glands Lubricating glands surrounding a woman's vagina.

Base Any substance that accepts hydrogen ions when dissolved in water. Basic solutions have a pH greater than 7.

Basidiocarp The reproductive portion of a basidiomycote fungus.

Basidiospore The spore produced by a type of fungus called a basidiomycote.

Basidium (plural **basidia**) In certain fungi, a club-shaped structure that holds the basidiospores.

Basophil A type of white blood cell involved in inflammation that releases an anticlotting agent at the site of an injury and produces the chemical substance histamine, which delays the spread of invading microorganisms.

B cell (B lymphocyte) A type of white blood cell that makes and secretes antibodies in response to foreign substances (antigens).

Benign Characteristic of a tumor that continues to grow but stays localized and tightly packed.

Beta cell A pancreatic endocrine cell that secretes the hormone insulin.

Beta- (β-) endorphin A naturally occurring protein of the brain and pituitary that has the pain-killing effect of opiate drugs.

Biennial A plant that completes its life cycle in two years and then dies.

Bilateral symmetry A body plan in which the right and left sides of the body are mirror images of each other.

Bile A bitter, alkaline, yellow fluid produced by the liver, stored in the gallbladder, and released into the small intestine that aids in the digestion and absorption of fats.

Binary fission Asexual reproduction by division of a cell or body into two equivalent parts.

Binomial system of nomenclature The system, developed by the Swedish biologist Linnaeus, of giving each type of organism two names, a genus name and a species name. For example, humans are *Homo sapiens.*

Biogeography The study of the geographical location of living organisms.

Biological magnification The tendency for toxic substances to increase in concentration as they move up the food chain.

Biology (Gr. *bios*, life) The scientific study of life.

Biomass The total dry weight of organic matter present at a particular trophic level in a biological community.

Biome A major ecological community type, such as a desert, rain forest, or grassland.

Biosphere (Gr. *bios*, life + *sphaira*, sphere) That part of the planet that supports life; includes the atmosphere, water, and the outer few meters of the earth's crust.

Biotic Characteristic of the part of an organism's environment made up of other living things.

Biotic potential An organism's capacity for reproduction under ideal conditions of growth and survival.

Bivalve A mollusk whose body is enclosed within two valves or shells; includes mussels and clams.

Bladder A storage sac; for example, for urine.

Blade The broad, flat portion of a leaf.

Blastocoel In early embryonic development, a cavity that develops in the mass of cells formed during the blastula stage.

Blastocyst A stage in early embryonic development of mammals consisting of a thin-walled hollow sphere of cells, called the trophoblast, and an inner mass of cells at one side that are destined to become the embryo

Blastodisc A disc-shaped layer of cells produced by cleavage in a large yolky egg like that of a bird or reptile.

Blastomere A cell that exists in an embryo during the cleavage or blastula stage.

Blastula A stage of embryonic development, at or near the end of cleavage and immediately preceding gastrulation, generally consisting of a hollow ball of cells.

Blubber A thick layer of fat acting as insulation in such mammals as whales, porpoises, and seals.

Body hair Threadlike outgrowths from the skin of mammals.

Bohr effect The phenomenon in which hemoglobin releases oxygen more easily in an acidic environment, as when muscles are working hard.

Bolus A soft, round mass of chewed food, shaped by the tongue, suitable for swallowing.

Bone The main supporting tissue of vertebrates, composed of a matrix of collagen hardened by calcium phosphate.

Book lung A chamber with leaflike plates that exchanges gases, found in spiders and other arthropods.

Bouton The knoblike end of an axon at a synapse.

Bowman's capsule In the vertebrate kidney, the bulbous unit of the nephron that surrounds the blood capillaries of the glomerulus; the region that filters water and solutes from the blood.

Brain stem The most posterior portion of the vertebrate brain, relaying messages to and from the spinal cord and consisting of the medulla, pons, and midbrain.

Bronchiole One of the thousands of small branches from the bronchi that lead to the alveoli in the lungs.

Bronchus (plural bronchi) One of two hollow passageways branching off the trachea and entering the lungs.

Bryophytes The division of the plant kingdom comprising mosses, liverworts, and hornworts.

Bud [1] An undeveloped shoot or flower. [2] A daughter cell derived by mitosis in yeast.

Budding A type of asexual reproduction in which some cells differentiate and grow outward from the parent body, eventually breaking off and forming a new individual; occurs in sea anemones and corals.

Buffer A substance that resists changes in pH when acids or bases are added to a chemical solution.

Bulbourethral gland A gland at the base of the human penis that secretes a lubricating substance.

Bulk flow The overall movement of a fluid down a gradient of pressure or gravity.

Bundle-sheath cells Cells in certain plants that surround a vascular bundle.

C_3 plant A plant in which the first stable product of carbon fixation in the light-independent reactions of photosynthesis is a three-carbon compound. Most plants are C_3 plants.

C_4 plant A plant, such as corn, in which the first stable product of carbon fixation is a four-carbon compound. Many C_4 plants are adapted to hot, dry weather.

Calcitonin A thyroid hormone that lowers calcium in the blood, and promotes its deposition in the bones.

Calorie A kilocalorie; the amount of energy needed to raise the temperature of one liter of water 1°C.

Calvin-Benson cycle A series of reactions in the light-independent phase of photosynthesis, during which chloroplasts fix carbon dioxide, form carbohydrates, and regenerate a starting material.

Camouflage Body shapes, colors, or patterns that enable an organism to blend in with its environment and remain concealed from danger.

CAM plant A plant, such as ice plant, in which the C_4 pathway of carbon fixation and the light-independent reactions are in the same cell but occur at different times of day; CAM stands for crassulacean acid metabolism.

Cancer A group of diseases characterized by uncontrolled cell division.

Capillarity The tendency of a liquid substance to move upward against the pull of gravity through a narrow space.

Capillary Tiny blood vessels with walls one cell thick that permeate the tissues and organs of the body. Substances such as oxygen diffuse out of capillaries to the tissues, and waste products diffuse into capillaries from the tissues.

Capillary bed A network of capillaries that links arterial and venous blood vessels.

Capsid The protein coat that encases a virus.

Carapace The exoskeleton covering the cephalothorax of many arthropods. Also refers to the tough outer coverings of the turtle and armadillo.

Carbohydrate (L. *carbo*, charcoal + *hydros*, water) An organic compound composed of a chain or ring of carbon atoms to which hydrogen and oxygen atoms are attached in the ratio of approximately two hydrogens to one oxygen. Carbohydrates include sugars and starch.

Carbon cycle The global flow of carbon atoms from plants through animals to the atmosphere, soil, and water and back to plants.

Carbon fixation The sequence of reactions that converts atmospheric gaseous carbon dioxide into carbohydrates.

Carcinogen A cancer-causing substance.

Cardiac cycle The contraction/relaxation (systolic/diastolic) sequence of atria and ventricles that makes up a single heartbeat.

Cardiac muscle The specialized muscle tissue of the heart.

Carnivore An animal that eats the flesh of other animals.

Carotenoid pigments Certain yellow and orange pigments in plant plastids that can function as accessory pigments in photosynthesis; such pigments are responsible for the color in carrots and in autumn leaves.

Carpel The innermost ring of structures in a flower. A carpel encloses one or more ovules, and together, a flower's carpels form the pistil.

Carrier [1] In genetics, a heterozygous individual not expressing a recessive trait but capable of passing it on to her or his offspring. [2] In biochemistry, a substance, often a protein, that transports another substance.

Carrier-facilitated diffusion The process by which large molecules are carried across cell membranes by carrier proteins.

Carrying capacity The density at which growth of a population ceases due to the limitation imposed by resources.

Casparian strip A water-resistant belt encircling each endodermal cell of a plant, ensuring that any water or minerals entering the plant must flow through the cytoplasm of the endodermal cells.

Caste In a social insect colony, one of several types of individuals; for example, a worker, soldier, or queen in a termite colony.

Catabolism The breaking down of complex molecules to simpler products, with an accompanying release of energy.

Catalyst A substance that speeds up the rate of a chemical reaction by lowering the activation energy without itself being permanently changed or used up during the reaction.

Causality The scientific principle stating that the occurrence of events is due to natural causes.

Cell The basic unit of life; cells are bounded by a lipid-containing membrane and are generally capable of independent reproduction.

Cell cycle The events that take place within the cell between one cell division and the next.

Cell division The splitting of one cell into two cells.

Cell-mediated immune response That part of the immune response in which T cells directly attack invaders.

Cell membrane A biological envelope surrounding a cell; also called the *plasma membrane.*

Cell plate In plant cells, a partition that arises during late cell division from vesicles at the center of the cell and that eventually separates the two daughter cells.

Cell theory The biological doctrine stating that all living things are composed of cells; cells are the basic living units within organisms, and the chemical reactions of life occur within cells; all cells arise from preexisting cells.

Cellular slime mold A type of funguslike protist, usually existing as free-living amoebalike cells, but aggregating into a multicellular fruiting body before producing reproductive spores.

Cellulose (L. *cellula*, little cell) The chief constituent of the cell wall in green plants, some algae, and a few other organisms. Cellulose is an insoluble polysaccharide composed of straight chains of glucose molecules.

Cell wall A fairly rigid structure that encloses prokaryotic, fungal, and plant cells.

Central cavity A large opening, as in the body of a sponge.

Central nervous system (CNS) The part of the nervous system consisting of the brain and spinal cord, performing the most complex nervous system functions.

Central vacuole A large, single-membraned organelle in plant cells that contains water and storage products.

Centromere The point on the chromosome where the spindle attaches and also where the two chromatids are joined.

Cephalization (Gr. *kephale*, little head) A type of animal body plan or organization in which one end contains a nerve-rich region and functions as a head.

Cephalopod A member of the class Cephalopoda, phylum Mollusca, including squid, octopus, and chambered nautilus.

Cephalothorax The fused head and thorax of many arthropods.

Cerebellum The hindbrain region of the vertebrate brain between the cerebrum and pons; it integrates information about body position and motion, coordinates muscular activities, and maintains equilibrium.

Cerebral cortex The highly convoluted surface layers of the cerebrum, containing about 90 percent of a human brain's cell bodies. It is well developed only in mammals and particularly prominent in humans and is the region involved with conscious sensations and voluntary muscular activity.

Cerebrum (L. brain) The forebrain of the vertebrate brain; the largest part of the human brain. It coordinates and processes sensory input and controls motor responses.

Cervix The narrow necklike junction of the uterus and vagina.

Chaparral A biome that borders deserts and grasslands, characterized by low woody shrubs that are often fragrant and have generally thick, waxy, evergreen leaves.

Chelicera (plural **chelicerae**) (Gr. *chele*, claw + *keras*, horn) One of the first pair of legs of an arachnid (spiders and their close relatives), which possesses poison fangs that kill prey.

Chemical reaction The making or breaking of chemical bonds between atoms or molecules.

Chemiosmotic coupling The coupling of energy capture (ATP production) to gradients of hydrogen ion concentration during electron transfer in mitochondria.

Chemoautotroph An organism that derives energy from a simple inorganic reaction.

Childhood The period of human development from age 2 to about age 12.

Chitin A complex nitrogen-containing polysaccharide that forms the cell walls of certain fungi, the major component of the exoskeleton of insects and some other arthropods, and the cuticle of some other invertebrates.

Chlorophyll (Gr. *chloros*, green + *phyllon*, leaf) Light-trapping pigment molecules that act as electron donors during photosynthesis.

Chloroplast An organelle present in algae and plant cells that contains chlorophyll and is involved in photosynthesis.

Cholecystokinin A hormone secreted by the small intestine that triggers the pancreas to release protein-digesting enzymes.

Chondrichthyes A class of the phylum Chordata that includes cartilaginous fish, such as sharks and rays.

Chordate A member of the animal phylum Chordata, including vertebrates and their relatives.

Chorion A fluid-filled sac that surrounds the embryos of reptiles, birds, and mammals. In placental mammals it gives rise to part of the placenta and produces hCG, a hormone that maintains pregnancy by preventing menstruation.

Chorionic villus sampling A procedure for removing fetal cells from the developing placenta for the diagnosis of genetic defects.

Chromatid A daughter strand of a duplicated chromosome. Duplication of a chromosome gives rise to two chromatids joined together at the centromere.

Chromatin The substance of a chromosome, composed of DNA and proteins.

Chromoplast In plant cells, an organelle that stores yellow, orange, and red pigments, giving color to many fruits and flowers.

Chromosomal mutation A genetic change that affects large regions of chromosomes or entire chromosomes.

Chromosome A self-duplicating body in the cell nucleus made up of DNA and proteins and containing genetic information. It contains genes and a centromere region. Chromosomes are transmitted from one generation to the next via gametes. A human cell contains 23 pairs of chromosomes. In a prokaryote, the DNA circle that contains the cell's genetic information.

Chromosome translocation An instance in which a part of one chromosome becomes attached to a different chromosome.

Chyme The semifluid contents of the stomach consisting of partially digested food and gastric secretions.

Ciliate A protozoan whose cells have rows of cilia that are used in locomotion and in sweeping food particles into the mouth.

Cilium (plural **cilia**) (L. eyelash) A short, centriole-based, hairlike organelle. Rows of cilia propel certain protists. Cilia also aid the movement of substances across epithelial surfaces of animal cells.

Circadian (L. *circa*, about + *dies*, day) Regular rhythms of activity that occur on about a 24-hour cycle.

Circulatory system An organ system, generally consisting of a heart, blood vessels, and blood, that transports substances around the body of an animal.

Class A taxonomic group comprised of members of similar orders.

Classical conditioning A type of learning in which an initially neutral stimulus (like the sound of a bell) evokes a given response (like salivation) by being paired repeatedly with a different stimulus that normally evokes that response (like the presence of food).

Cleavage The early, rapid series of mitotic divisions of a fertilized egg resulting in a hollow sphere of cells known as the blastula.

Climate The accumulation of seasonal weather patterns in an area over a long period of time.

Climax community The most stable community in a habitat and one which tends to persist in the absence of a disturbance.

Clitoris An erectile structure in the female, equivalent to the male penis, which is sensitive to sexual stimulation.

Clone All cells derived from a single parent cell. A replica of a DNA sequence produced by recombinant DNA technologies.

Clumped distribution A pattern in which several members of a population occur close together but are separated by a long distance from other groups.

Cnidarian A member of the phylum Cnidaria, which includes hydras, jellyfish, sea anemones, and corals.

Coagulation The formation of a blood clot.

Coccus (plural cocci) A bacterial cell with a spherical shape.

Cochlea A spiral tube in the inner ear that contains sensory cells involved in detecting sound and analyzing pitch.

Codominance The genetic situation in which both alleles in a heterozygotic individual are fully expressed in the phenotype; this is a characteristic of many blood groups.

Codon Three adjacent nucleotide bases in DNA or mRNA that code for a single amino acid.

Coelacanth A type of lobe-finned fish.

Coelom (Gr. *koiloma*, a hollow) The main body cavity of many animals, formed between layers of mesoderm, in which internal organs are suspended.

Coenzyme A nonprotein organic molecule that aids in enzyme-catalyzed reactions, often by acting as an electron donor or acceptor.

Coevolution The evolution of one species in concert with another as a result of their interrelationship within a biological community.

Cohesion The tendency of like molecules to cling to each other.

Cohesion-adhesion tension theory An explanation for the way water is transported upward from plant roots through stem and leaves: Adhesion of water molecules to plant vessel surfaces, cohesion of water molecules with each other, and evaporation of molecules from the water chain moving upward in xylem create the tension on the entire chain that accounts for transpirational pull, also called the *transpirational pull theory*.

Coitus The act of sexual intercourse.

Collagen A fibrous protein found extensively in connective tissue and bone. Collagen also occurs in some invertebrates, as in the cuticle of nematode worms.

Collecting duct Any duct that drains an organ; in the kidney, ducts that drain the renal pelvis of the kidney and carry urine to the ureters.

Collenchyma Plant tissue providing support to young, actively growing structures; it consists of closely packed cells with thickened cell walls, especially at the corners.

Colon The last portion of the large intestine, the wide part of the alimentary canal that leads to the rectum.

Commensalism A relationship between two species in which one species benefits and the other suffers no apparent harm.

Committed In embryology, a characteristic of cells whose developmental fate has become restricted to just one or a few cell types.

Communication A signal produced by one individual that alters the behavior of a recipient individual and generally improves both participants' biological fitness.

Community Two or more populations of different interacting species occupying the same area.

Companion cell In a flowering plant, a specialized cell associated with sieve tube members in phloem.

Competitive exclusion A situation in which one species eliminates another through competition.

Complementary base pairing In nucleic acids, the hydrogen bonding of adenine with thymine or uracil, and guanine with cytosine; it holds two strands of DNA together, and different parts of RNA molecules in specific shapes and is fundamental to genetic replication, genetic expression, and genetic recombination.

Complement proteins In the immune system, proteins that bind to antibody-invader cell complexes and perforate the invader's membrane, killing the invader.

Compound A substance made up of atoms of two or more elements chemically combined in a fixed ratio.

Condensation The formation of a covalent bond between two smaller molecules in such a way that a water molecule is released and a larger molecule is formed.

Cone One of the two types of photoreceptor (light-sensitive) cells in the retina of the vertebrate eye; cones are sensitive to high levels of light and to color. (*See also* Rod.)

Coniferous forest A biome dominated by conifer trees and usually characterized by cold, snowy winters and short summers.

Conjugation In some prokaryotes, the temporary union of two unicellular organisms of different mating strains, during which time the genetic material is transferred from one to the other.

Connective tissue Animal tissue that connects or surrounds other tissues and whose cells are embedded in a collagen matrix; bones and tendons are mainly connective tissue.

Consumer In ecology, an organism that eats other organisms.

Contact inhibition The process by which cells stop growing and moving when they come into contact with other cells.

Contraception The prevention of conception.

Contractile vacuole In protists and certain animals, an organelle that pumps excess water out of the cell.

Control A check of a scientific experiment based on keeping all factors the same except for the one in question.

Convergent evolution Evolution of similar characteristics in two or more unrelated taxa; often found in organisms living in similar environments.

Copulation (L. *copulare*, to couple) The process by which the male deposits sperm inside the female's body.

Cork An external secondary plant tissue impermeable to water and gases.

Cork cambium A type of lateral meristem that produces cork-forming cells.

Cornea The transparent outer portion of the vertebrate eye, through which light passes.

Corpus luteum The group of follicle cells left behind in the ovary after an egg has been released. If the egg is fertilized, the corpus luteum remains active in the ovary, producing hormones that help to maintain the pregnancy. If the egg is not fertilized, the corpus luteum degenerates.

Cortex [1] The outer surface of an organ. In the kidney, the region where blood filtration takes place. [2] In plants, parenchyma tissue bounded externally by the epidermis and internally by phloem in the stem and pericycle in the root.

Cortisol A steroid hormone secreted by the adrenal cortex that helps an animal respond to stress by speeding the metabolism of sugars, proteins, and fats.

Cotyledon The first leaves of the embryonic plant, they often store food in dicotyledon plants and absorb food in monocotyledon plants.

Countercurrent flow A mechanism in which two different fluids flow in opposite directions and can exchange substances at points of contact along a concentration gradient.

Covalent bond (L. *co*, together + *valere*, sharing) A form of molecular bonding characterized by sharing a pair of electrons between atoms.

Crista (plural **cristae**) Fold(s) formed by the inner membrane of a mitochrondrion.

Critical period A specific developmental stage during which imprinting can occur and after which it cannot.

Crop A specialized region that stores food in the digestive tract of earthworms and other animals.

Crossing over The reciprocal exchange of DNA between homologous chromosomes during meiosis.

Cuticle A waxy or fatty noncellular, waterproof outer layer on epidermal cells in plants and some invertebrates. In parasitic flatworms it prevents the worm from being digested in the gut of the host organism.

Cyanobacterium One of the blue-green algae; a photosynthetic, oxygen-generating and nitrogen-fixing prokaryote.

Cycad A primitive cone-bearing seed plant.

Cyclic phosphorylation Phosphorylation that generates small amounts of ATP during the light-dependent reactions of photosynthesis; involves only photosystem I and the electron transport proteins that link photosystems I and II.

Cytology The study of cells.

Cytokinesis The process of cytoplasmic division accompanying nuclear division.

Cytokinin A growth-promoting plant hormone that regulates cell division, among other things.

Cytoplasm The part of a cell between the outer membrane and the nuclear envelope.

Cytoskeleton Found in the cells of eukaryotes, an internal framework of microtubules, microfilaments, and intermediate filaments that supports and moves the cell and its organelles.

Deciduous (L. *decidere*, to fall off) In plants, the property of shedding leaves at the end of the growing season.

Deductive reasoning Analyzing specific cases on the basis of preestablished general principles.

Demographic transition A changing pattern from high birthrate and high death rate to low birthrate and low death rate.

Dendrite (Gr. *dendron*, a tree) The branching projections of a neuron or nerve cell that transmit nerve impulses to the body of the cell.

Density In relation to population, the number of individuals in a certain amount of space.

Depolarized The loss of electrical polarity in a cell.

Dermal tissue system The protective and waterproofing tissues of a plant, consisting of epidermis and bark.

Descending limb In the vertebrate kidney, the arm of the loop of Henle leading down into the medulla.

Desert A very dry, often barren biome characterized by widely spaced plants with thick waxy leaves and often protective spines.

Desmosome A structure involved in cell adherence. A *belt desmosome* is a ring of actin filaments encircling a cell, helping it adhere to adjacent cells. A *spot desmosome* is a tiny button joining two cells and anchoring intermediate filaments.

Detritivore A consumer organism that obtains energy from dead or waste organic matter.

Detritus A collective term for dead organisms and waste organic matter.

Deuterostome An animal in which the blastophore becomes the anus, while the second opening to develop becomes the mouth; includes echinoderms and chordates.

Development The process by which an offspring increases in size and complexity from a zygote to an adult.

Developmental determinant A chemical in egg cytoplasm that helps direct embryonic development.

Diabetes mellitus A hereditary condition in which insufficient insulin is produced and glucose accumulates in the blood.

Diaphragm The muscle sheet that separates the thoracic cavity from the abdominal cavity and helps draw air into the lungs during breathing.

Diastole The relaxation phase of the heart muscle.

Dicot Short for *dicotyledon*, the larger of the two classes of angiosperms (flowering plants). Dicots are generally characterized by having two cotyledons (seed leaves); floral parts are usually in fours or fives; leaves are typically net-veined. Maple trees and tomato plants are dicots.

Differentiation The process by which a cell becomes specialized for a particular function.

Diffusion (L. *diffundere*, to pour out) The tendency of substances to move from areas of high concentration to areas of low concentration.

Digestion The mechanical and chemical breakdown of food into small molecules that an organism can absorb and use.

Dihybrid cross A mating between two individuals that differ only in two traits.

Dinosaur An extinct giant reptile; the dominant form of land vertebrate during the Jurassic and Cretaceous periods.

Diploid A cell that contains two copies of each type of chromosome in its nucleus (except, perhaps, sex chromosomes).

Directional selection A type of natural selection in which an extreme form of a character is favored over all other forms.

Disaccharide A type of carbohydrate composed of two linked simple sugars.

Disruptive selection A type of natural selection in which the two extreme forms of a trait are favored.

Distal tubule The convoluted tubule in the vertebrate kidney that receives the forming urine after it has passed through the loop of Henle.

Divergent evolution The splitting of a population into two reproductively isolated populations with different alleles accumulating in each one. Over geological time, divergent evolution may lead to speciation.

DNA Deoxyribonucleic acid, the doublehelical, ladderlike molecule that stores genetic information in cells and transmits it during reproduction, consists of two very long strands of alternating sugar and phosphate groups. Each "rung" of the ladder is a base (either adenine, guanine, cytosine, or thymine) that extends from the sugar and bonds with a base from the other strand. Genes are made of DNA.

DNA ligase A protein that joins DNA strands together end-to-end during DNA recombination.

Domain A taxonomic group comprised of members of similar kingdoms.

Dominance hierarchy A behavioral phenomenon in which members of a group are ranked from highest to lowest. Ranking is based on past success in aggressive encounters with the other members of the group; the individual winning all fights is the dominant member. Dominant individuals have greater access to food and mates.

Dominant In genetics, an allele or corresponding phenotypic trait that is expressed in the heterozygote.

Dormancy (L. *dormire*, to sleep) A state of reduced physiological activity that occurs in seeds and buds of plants, particularly over the winter. Normal physiological activities such as growth only resume if certain conditions, such as increased temperature, are met.

Double bond A chemical bond created when an atom shares two pairs of electrons with another atom.

Double fertilization In flowering plants, the process in which one sperm from a pollen grain fuses with the small egg cell of the gametophyte, while the second sperm penetrates the adjoining large endosperm cell and fuses with the two polar nuclei.

Double helix The term used to describe the structure of DNA, which resembles a ladder twisted along its long axis.

Down syndrome A genetic condition resulting from having an extra chromosome 21; characterized by mental retardation, heart malformation, and external physical traits.

Drive An organism's physiological state of motivation, influenced by hormones and other internal stimuli.

Drug testing A laboratory technique that checks a person's urine or blood to determine if he or she has taken drugs.

Ecdysone A steroid hormone that regulates the timing of molting of insects and other arthropods.

Echinoderm A member of the phylum Echinodermata, which includes sea stars, brittle stars, sea urchins, sea cucumbers, and sea lilies.

Ecology The scientific study of how organisms interact with their environment and with each other and of the mechanisms that explain the distribution and abundance of organisms.

Ecosystem A community of organisms interacting with a particular environment.

Ectoderm The outer cell layer of an embryo; it gives rise to the skin and nervous system.

Ejaculation The sudden ejection of a fluid from a duct, especially semen from the penis.

Ejaculatory duct The duct connecting the vas deferens and the seminal vesicle in mammals.

Electromagnetic spectrum The full range of electromagnetic radiation in the universe, from highly energetic gamma rays to very low-energy radio waves.

Electron A negatively charged subatomic particle that orbits the nucleus of an atom. The negative charge of an electron is equal in magnitude to the proton's positive charge, but the electron has a much smaller mass.

Electron microscope (EM) An instrument that uses beams of electrons with wavelengths much shorter than visible light to view objects, thus allowing greater magnification.

Electron transport chain The second stage in aerobic respiration in which electrons are passed down a series of molecules, gradually releasing energy that is harvested in the form of ATP.

Electrophoresis A technique used to separate molecules based on their different rates of movement in an electric field.

Element A pure substance that cannot be broken down into simpler substances by chemical means.

Elongation The second main stage of protein synthesis, during which the sequential enzymatic addition of amino acids builds the growing polypeptide chain.

Embryo An organism in the earliest stages of development. In humans this phase lasts from conception to about two months and ends when all the structures of a human have been formed; then the embryo becomes a fetus.

Endergonic A chemical reaction that requires the addition of energy to occur.

Endocrine gland A ductless gland that secretes its hormones into extracellular spaces from where they diffuse into the bloodstream. In vertebrates these glands include the pituitary, adrenal, thyroid, and others.

Endocrine system The collection of endocrine glands of the body.

Endocytosis The process by which a cell membrane invaginates and forms a pocket around a cluster of molecules. This pocket pinches off and forms a vesicle that transports the molecules into the cell.

Endoderm The inner cell layer of an embryo; gives rise to the inner linings of the gut and the organs that branch from it, including the lungs.

Endodermis A unicellular layer of tightly packed cells that surround the vascular tissue of plants. Water must pass through this layer to reach the rest of the plant.

Endometrium The inner membrane lining of the uterus, which thickens in response to estrogens and progesterone and then is shed during menstruation.

Endoplasmic reticulum (ER) A system of membranous tubes, channels, and sacs that form compartments within the cytoplasm of eukaryotic cells; functions in lipid synthesis and in the manufacture of proteins destined for secretion from the cell.

Endoskeleton (Gr. *endos*, within + *skeletos*, hard) An internal supporting structure, such as the bony skeleton of a vertebrate.

Endosperm In flowering plants, triploid nutritive tissue of the developing embryo.

Endospore A heavily encapsulated resting cell formed within many types of bacterial cells during times of environmental stress.

Endosymbiont hypothesis The idea that mitochondria and chloroplasts originated from prokaryotes that fused with a nucleated cell.

Energy The power to perform chemical, mechanical, electrical, or heat-related tasks.

Energy pyramid The energy relationships between different trophic levels in an ecosystem.

Entropy A measure of disorder or randomness in a system.

Enzyme A protein that facilitates chemical reactions by lowering the required activation energy but is not itself permanently altered in the process; also called a *biological catalyst*.

Enzyme-substrate complex In an enzymatic reaction, the unit formed by the binding of the substrate to the active site on the enzyme.

Epicotyl The portion of the plant embryonic axis above the cotyledons but below the next leaf.

Epidermis The outer layer of cells of an organism.

Epididymis A coiled duct leading from the testes to the vas deferens, where sperm accumulate and final sperm maturation takes place.

Epiglottis A flap of tissue just above the larynx that closes during swallowing and prevents food from entering the lungs.

Epinephrine A hormone secreted by the medulla of the adrenal glands, associated with the physiological responses to alarm such as increased concentration of sugar in the blood, raised blood pressure and heart rate, and increased muscle power and resistance to fatigue; also known as *adrenaline.*

Epistasis The interaction between two nonallelic genes in which one gene masks the expression of the other.

Epithelial tissue One of the four main tissue types; covers the body surface and lines the body cavities, ducts, and vessels.

Erectile tissue Spongy tissue within the penis that fills with blood and stiffens during an erection.

Erythrocyte A red blood cell whose main function is transporting oxygen to the tissues.

Esophagus The muscular tube leading from the pharynx to the stomach.

Essential amino acid Any one of the eight amino acids that the human body cannot manufacture and that must be obtained from food.

Essential fatty acid Linoleic acid, the unsaturated fatty acid that cannot be manufactured by the human body and that therefore must be obtained from food.

Estrogen A female sex hormone produced by the ovary; prepares the uterus to receive an embryo in mammals, and causes secondary sex characteristics to develop in mammalian females.

Estuary The area where a river meets an ocean. Estuaries are some of the earth's most productive ecosystems.

Ethology (Gr. *ethos*, habit or custom + *logos*, discourse) The scientific study of animal behavior.

Ethylene A gaseous plant hormone produced by ripening fruits that stimulates ripening in nearby fruits.

Eubacteria The largest, most diverse group of prokaryotes; members of the domain Bacteria; spherical, rodlike, or comma-shaped single-celled organisms.

Eukaryotic cell A cell whose DNA is enclosed in a nucleus and associated with proteins; contains membrane-bound organelles.

Eutrophic An aquatic environment with high nutrient levels, characterized by dense blooms of algae and other aquatic plants.

Evolution Changes in gene frequencies in a population over time.

Excretion The elimination of metabolic waste products from the body; in vertebrates, the main excretory organs are the kidneys.

Exergonic A spontaneously occurring chemical reaction that liberates energy.

Exocrine gland A gland with its own transporting duct that carries its secretions to a particular region of the body; includes the digestive and sweat glands; contrasts with endocrine glands.

Exocytosis The process by which substances are moved out of a cell by cytoplasmic vesicles that merge with the plasma membrane.

Exon The sequence of bases in a eukaryotic gene that ends up in the mature RNA.

Exoskeleton The thick cuticle of arthropods, made of chitin.

Exponential growth Growth of a population without any constraints; hence, the population will grow at an ever-increasing rate.

Extinction The disappearance of an entire taxonomic group.

External fertilization A process taking place in many water-dwelling species, in which eggs and sperm are deposited directly into the water, meet by chance, and fuse.

Extracellular digestion The enzymatic breakdown of nutrient molecules that occurs outside a cell.

Extracellular fluid Fluid within the body but outside the cells.

Extracellular matrix A meshwork of secreted molecules that act as a scaffold and a glue that anchors cells within multicellular organisms.

Eyespot An organelle that is sensitive to light and movement, found in planarians and other organisms.

Facet One of many six-sided units of an arthropod's compound eyes.

Family A taxonomic group comprised of members of similar genera.

Fat An energy storage molecule that contains a glycerol bonded to three fatty acids. Fats in the liquid state are known as oils.

Fatty acid A molecule with a carboxyl group at one end and a long hydrocarbon tail at the other. Fatty acids are components of many lipids.

Fatty acid hormone A hormone derived from straight-chain fatty acids; for example, prostaglandin.

Feather A flat, light, waterproof epidermal structure on a bird; collectively functions as insulation and for flight.

Feces Semisolid undigested waste products, stored in the colon until excreted.

Feedback inhibition The buildup of a metabolic product which in turn inhibits the activity of an allosteric enzyme. As this enzyme is involved in making the original product, the accumulation of the product turns off its own production.

Feedback loop A control system in which the result of a process influences the functioning of the process.

Fermentation Extraction of energy from carbohydrates in the absence of oxygen, generally producing lactic acid or ethyl alcohol as a by-product.

Fertilization The fusion of two haploid gamete nuclei (egg and sperm) which forms a diploid zygote.

Fibrin The activated form of the blood-clotting protein fibrinogen that combines into threads when forming a clot.

Fibrinogen A protein in blood plasma that is converted to fibrin during clotting.

Fibrous root One of a system of many narrow roots in which no one root is more prominent than another.

Filter feeder An organism that strains out and collects food particles suspended in water.

Filtrate In the vertebrate kidney, the fluid in the kidney tubule (nephron) that has been filtered through Bowman's capsule.

Filtration The process in the vertebrate kidney in which blood fluid and small dissolved substances pass into the kidney tubule, but large proteins and blood cells are filtered out and remain in the blood.

Fin A flattened limb found in aquatic animals; used for locomotion.

First filial (F_1) generation In Mendelian genetics, the first generation in the line of descent.

First law of thermodynamics The principle that energy cannot be created or destroyed but can be converted from one form into another, such as potential to kinetic energy, or chemical bond energy to heat.

Fitness Biological fitness; the ability to survive and produce offspring that can in turn reproduce.

Fixed-action pattern In behavioral science, an action that continues to completion even if the stimulus causing the behavior is removed, e.g., a bird that rolls an empty eggshell out of its nest will continue the rolling process to the edge of the nest even if the eggshell is removed during the procedure.

Flagellum (plural **flagella**) Long whiplike organelle protruding from the surface of the cell that either propels the cell acting as a locomotory device, or moves fluids past the cell, becoming a feeding apparatus.

Flame cell A specialized cell containing cilia that functions in excretion in flatworms.

Flatworm A member of the phylum Platyhelminthes, the simplest animal group to display bilateral symmetry and cephalization.

Florigen A plant hormone proposed to stimulate the formation and opening of flowers.

Fluid mosaic model Current model of cell membrane structure in which proteins are embedded in the lipid bilayer and can move freely through the fluidlike lipid structure.

Follicle (L. *folliculus*, a small ball) In the mammalian ovary, an oocyte surrounded by follicular cells.

Follicular cell One of many cells surrounding the oocytes in the ovaries.

Food chain The levels of feeding relationships among organisms in a community; who eats whom.

Food web Complex, interconnected feeding relationships between all the species in a community or ecosystem.

Foot The locomotive organ in mollusks.

Founder effect In evolutionary biology, the principle that individuals founding a new colony carry only a fraction of the total gene pool present in the source population.

Fragile *X* syndrome Mental retardation associated with the breaking of an *X* chromosome near its tip.

Frond The leaflike structure of an individual alga that collects sunlight and produces sugars. Also refers to the large divided leaf on a fern.

Fructose A monosaccharide with the chemical formula $C_6H_{12}O_6$; a sugar found in many fruits.

Fruit In flowering plants, a ripe mature ovary containing seeds.

Fruiting body A spore-producing reproductive structure in many fungi.

Functional group A group of atoms that confers specific behavior to the (usually larger) molecules to which they are attached.

Fundamental niche The potential range of all environmental conditions under which an organism can thrive.

Fungi The kingdom comprising multicellular heterotrophs such as mushrooms or molds; secrete digestive enzymes that decompose other biological tissues.

Fur Dense hair covering the whole body of many mammals.

Furrow In cell division, a dent in the cell surface created by constriction of the contractile ring; the furrow deepens and eventually squeezes the cell in two.

Galactose A monosaccharide with the chemical formula $C_6H_{12}O_6$; a common subunit of many sugars, such as milk sugar.

Gallbladder The sac beneath the right lobe of the liver that stores bile.

Gamete (Gr. wife) A specialized sex cell such as an ovum (egg) or a sperm, which is haploid. A male gamete (sperm) and a female gamete (ovum) fuse and give rise to a diploid zygote, which develops into a new individual.

Gametogenesis The formation of haploid gametes, eggs and sperm, by the process of meiosis and maturation of the cells.

Gametophyte The haploid, gamete-producing phase in the life cycle of plants.

Ganglion (plural **ganglia**) A distinct clump of nerve cells that acts like a primitive brain, found in the head region of many invertebrates; in vertebrates, an aggregation of nerve cell bodies located outside the central nervous system.

Gap junction A group of protein-lined pores that connects adjacent cells and allows small molecules to move from cell to cell.

Gastrin A digestive hormone secreted in the stomach that causes the secretion of other digestive juices.

Gastropod A member of the class Gastropoda in the phylum Mollusca; includes snails, garden slugs, and sea slugs.

Gastrovascular cavity The digestive cavity in cnidarians; also called the *coelenteron*.

Gastrulation The movement of cells in the embryo that generates three cell layers—the ectoderm, mesoderm, and endoderm—each layer in turn giving rise to specific body organs and tissues.

Gene (Gr. *genos*, birth or race) The biological unit of inheritance that transmits hereditary information from parent to offspring and controls the appearance of a physical, behavioral, or biochemical trait. A gene is a specific discrete portion of the DNA molecule in a chromosome that encodes a protein, tRNA, or rRNA molecule.

Gene flow The incorporation into a population's gene pool of genes from one or more other populations through migration of individuals.

Gene mutation A change in the base sequence of DNA that alters individual genes.

Gene pool The sum of all alleles carried by the members of a population; the total genetic variability present in any population.

Gene regulation The process that controls the cell's gene activity.

Genetic code The specific sequence of three nucleotides in mRNA that encodes an individual amino acid in protein. For example, the code for methionine is AUG.

Genetic drift Unpredictable changes in allele frequency occurring in a population due to the small size of that population.

Genetic mapping The process of locating the positions of genes on chromosomes.

Genetic recombination The reshuffling of maternal and paternal chromosomes during meiosis, resulting in new genetic combinations.

Genome The total complement of genes in a single cell.

Genotype The genetic makeup of an individual.

Genus (plural **genera**) In taxonomy, a group of very similar species of common descent.

Geological era A major time span dividing the earth's history; the four geological eras are the Proterozoic, Paleozoic, Mesozoic, and Cenozoic.

Geological period A subdivision of a geological era; the Precambrian period was in the Proterozoic era.

Germ cell A sperm cell or ovum (egg cell), the haploid gametes produced by individuals that fuse to form a new individual.

Germ layers Three layers of cells—ectoderm, mesoderm, and endoderm—formed during embryonic development by the process of gastrulation. These layers develop into all the structures and organs of the body.

Germ line A specialized line of cells, from which come the sperm and egg (ovum).

Germ plasm A region of distinctive granular cytoplasm within the egg cells of some animals from which the germ line develops.

Gibberellins Specific hormones that regulate plant growth, usually by increasing the length of plant stems.

Gill [1] A specialized structure that exchanges gases in water-living animals. [2] The plates under the cap of certain fungi.

Gill filament A fingerlike extension in the gill of a fish through which blood flows and exchanges gases with water.

Gill slit In chordates, an opening through the pharynx to the exterior.

Gizzard A specialized region that grinds food in the digestive tract of earthworms, chickens, and other animals.

Gland A group of cells organized into a discrete secretory organ.

Glial cell A nonneural cell of the nervous system that surrounds the neurons and provides them with protection and nutrients.

Glomerulus (L. a little ball) In the vertebrate kidney, a collection of tightly coiled capillaries enclosed by the Bowman's capsule.

Glucagon A hormone that causes blood glucose levels to rise, thus opposing the effect of insulin.

Glucose The most common type of sugar molecule; broken down to release energy in all living organisms; a simple sugar with the chemical formula $C_6H_{12}O_6$.

Glycerol A molecule of three carbons, each with an OH group. Three fatty acids are bonded to a glycerol molecule in a fat.

Glycogen A polysaccharide made up of branched chains of glucose; an energy-storing molecule in animals, found mainly in the liver and muscles.

Glycolysis The initial splitting of a glucose molecule into two molecules of pyruvate, resulting in the release of energy in the form of two ATP molecules. The series of reactions does not require the presence of oxygen to occur.

Goiter A large lump on the neck caused by an enlarged thyroid.

Golgi apparatus In eukaryotic cells, a collection of flat sacs that process proteins for export from the cell, or for shunting to different parts of the cell.

Gonad An animal reproductive organ that generates gametes.

Gout A painful condition caused by deposits of uric acid crystals in the big toe, elbow, or other joints.

Gradient A change in a quantity, such as pressure or temperature or concentration, with respect to distance.

Granum In chloroplasts, a stack of membrane discs called thylakoids, which contain chlorophylls and are the sites of the light-dependent reactions of photosynthesis.

Grassland A treeless biome dominated by dozens of grass species.

Gray crescent During cleavage, a zone of cytoplasm near the equator of some eggs, on the side opposite where the sperm entered that marks the future dorsal side of the embryo.

Greenhouse effect The result of a buildup of carbon dioxide in the atmosphere (e.g., through the burning of fossil fuels) in which carbon dioxide traps solar heat beneath the atmospheric layers; leads to increased global temperatures, and changes climatic patterns.

Ground tissue system Plant tissues other than the epidermal tissue system and vascular tissue system; provides support and stores starch.

Growth An increase in size.

Growth factor Proteins that can enhance the growth and proliferation of specific cell types.

Guard cells Specialized kidney-shaped cells that enclose the pores (stomata) on the leaf epidermis; their opening and closing regulates water loss from the leaf.

Gymnosperms (Gr. *gymnos*, naked + *sperma*, seed) Conifers and their allies; primitive seed plants whose seeds are not enclosed in an ovary.

Habitat The place within a species' range where the organism actually lives.

Habituation A progressive decrease in the strength of a behavioral response to a constantly applied weak stimulus, when the stimulus has no negative effects.

Halophile A type of archaebacterium that can tolerate extremely high salt concentrations.

Haploid Having only one copy of a chromosome set. A human haploid cell has 23 chromosomes.

Hardy-Weinberg principle In population genetics, the idea that in the absence of any outside forces, the frequency of each allele and the frequency of genotypes in a population will not change over generations.

Head The part of an animal that contains a major concentration of sense organs and generally encounters the environment first as the animal moves along.

Heart A muscular organ that pumps blood through vessels.

Heat The random motion of atoms and molecules.

Hemoglobin A blood protein with iron-containing heme groups that bind and transport oxygen.

Hemolymph In animals with open circulatory systems, the extracellular fluid in the body cavity that bathes the tissue cells.

Herbivore An animal that consumes plants as food.

Heredity The science of inheritance and variation.

Heterotrophs (Gr. *heteros*, different + *trophos*, feeder) Organisms that take in preformed nutrients; includes animals, fungi, and most prokaryotes and protists.

Heterozygous (Gr. *heteros*, different + *zygotos*, pair) Having two different alleles for a specific trait.

Hinged jaw Skeletal elements surrounding the mouth that move about a pivot; found in some arthropods and most vertebrates.

Hippocampus A structure within the cerebrum of the brain that plays a crucial role in the formation of long-term memory.

Histocompatibility protein Cell surface substances in vertebrates that can cause the immune rejection of tissues grafted from one individual to another; involved in the recognition of antigens and their presentation from one cell of the immune system to another.

Histone Protein in the nucleus around which DNA molecules of the chromosomes wind, allowing extremely long DNA molecules to be packed into a cell's nucleus.

Holdfast A rootlike anchor that attaches an alga to its substrate such as a rock on the ocean floor.

Homeostasis The maintenance of a constant internal balance despite fluctuations in the external environment.

Homeothermic "Warm-blooded"; a homeothermic animal can maintain a constant internal body temperature.

Homologous chromosomes Chromosomes that pair up and separate during meiosis and generally have the same size and shape and genetic information. One member of each pair of homologous chromosomes comes from the mother and the other comes from the father.

Homology The appearance in related organisms of similar structures, based on the inheritance of similar genetic programs from a common ancestor.

Homo sapiens (L. *homo*, man + L. *sapiens*, wise) The human species; a bipedal primate with a large brain, capable of language and tool use.

Homozygous Having two identical alleles for a specific trait.

Hormone A chemical messenger produced in one part of the body, transported in the blood to another region where it exerts an effect.

Human chorionic gonadotropin (hCG) A hormone produced by the chorion of a developing embryo that maintains pregnancy by preventing menstruation.

Humoral immune response. The synthesis and secretion of antibodies by B cells.

Hybrid An offspring resulting from the mating between individuals of two different genetic constitutions.

Hydrogen bond A type of weak molecular bond in which a partially negatively charged atom (oxygen or nitrogen) bonds with the partial positive charge on a hydrogen atom when the hydrogen atom is already participating in a covalent bond.

Hydrolysis The splitting apart of two covalently bound subunits by the insertion of a dissociated water molecule between them.

Hydrophilic Compounds that dissolve readily in water, such as salt.

Hydrophobic Compounds that do not dissolve readily in water, such as oil.

Hydroskeleton A volume of fluid trapped within an animal's tissues that is noncompressible and serves as a firm mass against which opposing sets of muscles can act; for example, in earthworm segments.

Hypertonic Having a higher concentration of solutes than another solution.

Hypha (plural **hyphae**) One of many long thin filaments of cells that makes up a multicellular fungus.

Hypocotyl The portion of a plant embryo or seedling between the embryonic root (radicle) and the cotyledons.

Hypothalamic release-inhibiting hormone One of a number of specific hormones released from the hypothalamus that travels to the anterior pituitary and *inhibits* release of a specific pituitary hormone.

Hypothalamic releasing hormone One of a number of specific hormones released from the hypothalamus that travels to the anterior pituitary and *stimulates* release of a specific pituitary hormone.

Hypothalamus A collection of nerve cells at the base of the vertebrate brain just below the cerebral hemispheres, responsible for regulating body temperature, many autonomic activities, and many endocrine functions.

Hypothesis A potential solution to a problem or answer to a question; a provisional working idea that may be accepted for the present but may be rejected in the future if the evidence disproves it

Hypotonic Having a lower concentration of solutes than another solution.

Immune system The network of cells, tissues, and organs that defends the body against invaders.

Immunity Resistance to infection.

Immunological memory The capability of memory B cells in vertebrates to serve as remnants of former encounters with antigens.

Implantation The process by which the fertilized egg burrows into the lining of the uterus in mammals.

Imprinting The attachment of young animals to a particular adult or object; in nature, the attachment of young to their parents. Animals may imprint on other objects or animals, such as young birds imprinting on humans, if they are exposed to that animal or object at the sensitive stage.

Inclusive fitness A concept in evolutionary theory in which an individual's total fitness includes both its likelihood of passing on its own genes and the likelihood that it will contribute to the successful passing on of the genes of its relatives.

Incomplete dominance The genetic situation in which the phenotype of the heterozygote is intermediate between the two homozygotes, e.g., when a cross between homozygous red and homozygous white flowers gives rise to heterozygous pink flowers.

Independent assortment The random distribution of genes located on different chromosomes to the gametes; Mendel's second law.

Inductive reasoning Generalizing from specific cases to arrive at broad principles.

Industrial revolution The replacement of hand tools with power-driven machines (like the steam engine) and the concentration of industry in factories beginning in England in the late eighteenth century.

Inert Lacking chemical activity.

Infancy The period of time from birth to about age 2.

Inferior vena cava One of the two largest veins in the human body; carries blood from the legs and most of the lower body to the right atrium of the heart.

Initiation The first main stage of protein synthesis, during which tRNA associates with a ribosome and an mRNA, forming a complex of many molecules.

Inner membrane The folded membrane enclosed by the outer membrane surrounding a mitochondrion.

Inorganic chemistry The study of substances that do not contain carbon.

Insect A member of Insecta, arthropods having three main parts—head, thorax, and abdomen—three pairs of legs, and generally two pairs of wings; the largest class of animals.

Insight learning A type of learning in which individuals formulate a course of action by understanding the relationships between the parts of a problem; common only in higher primates.

Instinct Any behavioral trait that is innate, i.e., not learned. Instinctive behavior is usually not changed by experience during an animal's lifetime.

Insulin A protein hormone made in the pancreas that causes cells to remove the sugar glucose from the blood.

Intermediate filaments Components of the cell's cytoskeleton involved in cell shape, movement, and growth, composed mainly of keratin-like proteins.

Internal fertilization A process that results when the male deposits sperm inside the female's body in a chamber or tube.

Interneuron A nerve cell that relays messages between other nerve cells.

Interphase The period between cell divisions in a cell. During this period, the cell conducts its normal activities and DNA replication takes place in preparation for the next cell division. Interphase is divided into three periods: G1, S, and G2.

Interspecific competition Competition for resources—e.g., food or space—between individuals of different species.

Interstitial cell A cell embedded in the connective tissue of the seminiferous tubules of the testes; produces testosterone, a male sex hormone; Leydig cell.

Intestine The long, tubelike section of the digestive tract between the stomach and anus of vertebrates where most food digestion and absorption take place.

Intracellular digestion The enzymatic breakdown of nutrient molecules that occurs within a cell.

Intracellular fluid The fluid inside cells.

Intron A sequence of DNA in a eukaryotic gene that is excised from RNA, and therefore not translated into protein.

Invertebrate An animal that does not have a backbone.

In vitro fertilization The fertilization of an egg by a sperm under laboratory conditions.

Ion An atom that has gained or lost one or more electrons, thereby attaining positive or negative electrical charge.

Ionic bond A type of molecular bond formed between ions of opposite charge.

Iris A pigmented ring of tissue in the vertebrate eye that regulates the amount of entering light.

Isomers Molecules that have the same chemical formula but different molecular shapes and hence different properties; for example, glucose and fructose are isomers.

Isotonic Two solutions having the same concentration of solutes.

Isotope An alternative form of an element having the same atomic number but a different atomic mass due to the different number of neutrons present in the nucleus. Some isotopes, such as carbon-14, are unstable and emit radiation.

Joint The hinge, or point of contact, between two bones.

J-shaped curve The curve on a graph representing exponential population growth.

Juvenile hormone In insects, a hormone that, at high levels, prevents metamorphosis.

Kelp One of the largest members of the algal world, a brown alga.

Kidney The main excretory organ of the vertebrate body; filters nitrogenous wastes from blood and regulates the balance of water and solutes in blood plasma.

Kinetic energy The energy possessed by a moving object.

Kingdom A taxonomic group comprised of members of similar phyla; i.e., Animalia, Plantae, Fungi, etc.

Kin selection A form of natural selection in which an individual increases its fitness by helping relatives, who share its genes, to reproduce.

Klinefelter syndrome A genetic condition including sterility exhibited by a human male with two *X* chromosomes and one *Y* chromosome (*XXY*) and two normal sets of autosomes.

Krebs cycle The first stage of aerobic respiration in which a two-carbon fragment is completely broken down into carbon dioxide and large amounts of energy are transferred to electron carriers.

***K*-selection** Selection experienced by populations that typically occur at maximum density; *K*-selected organisms are often long-lived and have few, large offspring.

Lactose A disaccharide composed of galactose bonded to glucose; the main sugar found in milk.

Lamella (plural **lamellae**) A thin, platelike structure; in the gill filaments of fish, lamellae enable respiration in water.

Larynx A cartilaginous structure containing the vocal cords; it is also known as the "voice box."

Lateral line organ A system of sense organs in aquatic vertebrates consisting of pores or canals arranged in a line down each side of the body; detects changes in water pressure, including vibrations in the water.

Lateral meristem An area of actively growing cells in the stems and roots of plants that causes these areas to thicken in secondary growth.

Learning An adaptive and enduring change in an individual's behavior based on its personal experiences in the environment.

Leg An appendage that supports or moves an animal.

Legume A member of the pea family; often contains nitrogen-fixing bacteria in root nodules.

Lens A circular, crystalline structure in the eye of certain animals that focuses light onto the retina.

Leucoplast In plant cells, a colorless organelle that stores starch granules; usually found in roots and internal stem tissue.

Leukocyte A white blood cell; functions in the body's defense against invading microorganisms or other foreign matters.

Lichen An association between a fungus and an alga, which live symbiotically together.

Life cycle The events that take place between the birth and death of an organism, or between the reproduction of one generation and that of the next.

Life expectancy The maximum probable age an individual will reach.

Life history strategy The way an organism allocates energy to growth, survival, or reproduction.

Life span An individual's maximum potential age.

Life zone Any of several mountain regions of varying elevation, climate, and organisms.

Ligament A band of connective tissue that links bone to bone.

Light-dependent reactions The first phase of photosynthesis, driven by light energy. Electrons that trap the sun's energy pass the energy to high-energy carriers such as ATP or NADPH, where it is stored in chemical bonds.

Light-independent (dark) reactions The second phase of photosynthesis, also called the *Calvin-Benson cycle*, which does not require light. During the six steps of the cycle, carbon is fixed and carbohydrates are formed using the energy trapped in the light-dependent reactions.

Light microscope An instrument containing optical lenses that refract (bend) light rays, magnifying the object viewed.

Linkage group Genes located close to each other on the same chromosome that tend to be transmitted as a single unit.

Lipid bilayer The two-layered sheet of phospholipids that makes up the plasma membrane of cells and is mainly responsible for the membrane's barrier function.

Lipids Biological molecules that consist mainly of hydrocarbons and do not dissolve readily in water, including fatty acids, fats, steroids, and waxes.

Liver A large, lobed gland that destroys blood cells, stores glycogen, disperses glucose to the bloodstream, and produces bile.

Lobe-finned fish A member of the oldest of the two major groups of bony fishes; includes lungfishes and coelacanths.

Locus The location of a gene on a chromosome.

Logistic growth Growth of a population under environmental constraints that set a maximum population size.

Loop of Henle U-shaped region of vertebrate kidney tubule chiefly responsible for reabsorption of water and salts from the filtrate by diffusion.

Lumen The central cavity; of a blood vessel, the central space where the blood flows.

Lung The principal air-breathing organ of most land vertebrates.

Lungfish An air-breathing lobe-finned fish having a lunglike air bladder in addition to gills.

Lymph Fluid forced out of capillaries, occupying spaces between cells, and draining into the lymphatic system.

Lymphatic system In vertebrates, a collective term for a system of vessels carrying lymph—fluid that has been forced out of the capillaries—back to the bloodstream.

Lymphocyte A white blood cell formed in lymph tissue; active in the immune responses that protect the body against infectious diseases. There are two main classes of lymphocyte: B lymphocyte, involved in antibody formation, and T lymphocyte, involved in cell-mediated immunity.

Lysosome Spherical membrane-bound vesicles within the cell containing digestive (hydrolytic) enzymes that are released when the lysosome is ruptured; important in recycling worn-out mitochondria and other cell debris.

Macroevolution Evolutionary changes at levels higher than the species; evolutionary patterns viewed over geological time spans.

Macronucleus The large nucleus governing the activities of a ciliate, containing many sets of chromosomes.

Macronutrient (Gr. *makros*, large + L. *nutrire*, to nourish) An inorganic chemical element, such as nitrogen, potassium, calcium, phosphorus, magnesium, and sulfur, required in large amounts for plant growth.

Macrophage An animal cell that ingests other organisms and substances; in the immune system, a cell that engulfs invaders and consumes debris.

Malignant Characteristic of a tumor that grows without ceasing and spreads throughout the body.

Maltose Two joined glucose subunits.

Mammal A vertebrate animal of the class Mammalia, having the body generally covered with hair, nourishing young with milk from mammary glands, and generally giving birth to live young; lions, whales, rabbits, and kangaroos are some mammals.

Mammary gland Gland that produces milk in female mammals to nourish young.

Mantle A thick fold of tissue found in mollusks that covers the visceral mass and that in some mollusks secretes the shell.

Mantle cavity In mollusks, the space between the mantle and the visceral mass.

Marine Characteristic of oceans and seas.

Marsupial A mammal having a pouch in which it carries its young, which are born in a small and undeveloped state. Found extensively in Australia, with a few representatives in America; includes kangaroos, opossums, and koala bears.

Mass flow theory A model for the large-scale movement of phloem fluid in plants due to the active transport of sugars into the phloem at a sugar source such as the leaves, followed by the passive entry of water, which creates a pressure forcing the fluid to a sugar sink, like the roots, where cells actively remove sugars from the phloem and the water returns to the xylem.

Mate Behavior that brings egg and sperm together during reproduction.

Medulla The central portion of organs; in the kidney, the place that contains the loop of Henle; in the adrenal gland, the place where epinephrine (adrenaline) is made.

Medulla oblongata The lowest region of the brain stem, the most posterior part of the brain; involved in keeping body conditions constant.

Medusa (Gr. mythology, a female monster with snake-entwined hair) A jellyfish, or the free-swimming stage in the life cycle of cnidarians; an inverted umbrella-shaped version of a polyp, with the mouth and tentacles pointing downward.

Megaspore In certain plants, a haploid spore giving rise to the female gametophyte.

Meiosis The type of cell division that occurs during gamete formation; the diploid parent cell divides twice, giving rise to four cells, each of which is haploid.

Melatonin A hormone secreted by the pineal gland in the brain; in some animals, it affects the internal biological rhythms associated with night and day.

Menopause The physiological termination of the menstrual cycle; in human females the onset of menopause normally occurs between the ages 40 and 50.

Menstrual cycle The reproductive cycle of human and other primate females. Generally a 28-day cycle in women, characterized by a gradual thickening of the lining of the uterus (womb) and the maturation of an ovarian follicle from which a mature egg is released. If the egg is not fertilized, the inner lining of the uterus is shed, a process known as menstruation, and the cycle starts again. In humans the menstrual cycle normally starts between ages 12 and 14.

Meristem Plant tissue containing cells that can continually divide throughout the plant's life.

Mesoderm The middle cell layer of an embryo; gives rise to muscles, bones, connective tissue, and reproductive and excretory organs.

Mesoglea A jellylike substance lying between the epidermis and gastrodermis in cnidarians such as jellyfish.

Mesophyll The major photosynthetic tissue in a plant leaf, located between the upper and lower epidermal layers.

Message An mRNA molecule that carries information to make a protein. (*See also* Messenger RNA.)

Messenger RNA (mRNA) An RNA molecule that carries the information to make a specific polypeptide. mRNA is transcribed from structural genes and is translated into protein by the ribosomes.

Metabolic pathway The chain of enzyme-catalyzed chemical reactions that convert energy in cells and in which the product of one reaction serves as the starting substance for the next.

Metabolism (Gr. *metabole*, to change) The sum of all the chemical reactions that take place within the body; includes photosynthesis, respiration, digestion, and the synthesis of organic molecules.

Metamorphosis (Gr. *meta*, after + *morphe*, form + *osis*, state of) The process in which there is a marked change in morphology during postembryonic development; in insects the change in body form that takes place as the individual changes from a larva, such as a caterpillar, and emerges as an adult, such as a butterfly. Also refers to the change from a tadpole to a frog in amphibians.

Metaphase The period during nuclear division (mitosis) when the spindle microtubules cause the chromosomes to line up at the center of the cell.

Metastasis The process by which cancer cells spread throughout the body.

Metazoan An animal; a many-celled eukaryote that obtains its energy from other organisms.

Methanogens Archaebacteria that produce methane as a metabolic by-product.

Microbes (Gr. *mikros*, small) Microscopic organisms; includes members of Monera and Protista kingdoms.

Microbodies Cellular organelles bounded by a single membrane with a granular interior; contain a number of different oxidase-type enzymes like catalase.

Microevolution Evolutionary changes in a species over geologically short time periods, like the accumulation of black forms of certain moths in industrial England.

Microfilaments Components of the cell's cytoskeleton involved in cell shape, movement, and growth, composed mainly of the protein actin. Contraction of microfilaments is the basic mechanism underlying shape changes in eukaryotic cells, including cell motility and muscle contraction.

Micronucleus In ciliates (Protista), one of several nuclei that undergo meiosis and are exchanged during sexual reproduction.

Micronutrient (Gr. *mikros*, small + L. *nutrire*, to nourish) A nutrient required in small amounts for plant growth; includes the minerals iron, copper, zinc, chlorine, manganese, molybdenum, and boron.

Microspore A pollen grain that gives rise to the male gametophyte in certain vascular plants.

Microtubules Components of the cell's cytoskeleton; hollow cylinders of the protein tubulin involved in cell shape, movement, and growth.

Microvillus (plural **microvilli**) One of hundreds of tiny fingerlike projections extending from the surface of cells lining the walls of the intestine that increases the surface area available for absorption.

Midbrain The part of the brain stem between the pons and the thalamus in humans; relays neural signals between different parts of the brain.

Milk A fluid rich in fats and proteins produced in the mammary glands of mammals; nourishes newborns.

Mimicry The evolution of similar appearance in two or more species, which often gives one or all protection; for example, a nonpoisonous species may evolve protection from predators by its similarity to a poisonous model.

Mineral An inorganic element such as sodium or potassium that is essential for survival and is obtained in food.

Mitochondrion (plural **mitochondria**) Organelle in eukaryotic cells that provides energy that fuels the cell's activities. Mitochondria are the sites of oxidative respiration, and almost all of the ATP of nonphotosynthetic eukaryotic cells is produced in the mitochondria.

Mitosis The process of nuclear division in which replicated chromosomes separate and form two daughter nuclei genetically identical to each other and the parent nucleus. Mitosis is usually accompanied by cytokinesis (division of the cytoplasm).

Mitotic spindle Formed during prophase in mitosis, a weblike structure of microtubules that suspends and moves the chromosomes.

Molecular bond The energy that holds atoms together in molecules.

Molecular clock The rate at which mutations accumulate in genes over evolutionary time.

Molecular messenger A chemical that can pass from cell to cell and regulate the activities of other cells.

Molecule A cluster of atoms held together by specific chemical bonds.

Mollusk A member of the phylum Mollusca; includes snails, slugs, clams, oysters, squid, and octopuses.

Molt To shed an animal's skin; in arthropods, molting involves shedding the hard exoskeleton.

Monera In the five-kingdom scheme of classification of all living forms, the kingdom comprising the single-celled prokaryotes, the Eubacteria and Archaebacteria. In the three-domain scheme of classification, Monera is split into the domain Bacteria (eubacteria) and the domain Archaea (archaebacteria).

Monocot Short for *monocotyledon*, the smaller of two classes of angiosperms (flowering plants). Embryos have only one cotyledon; the floral parts are generally in threes; leaves are typically parallel-veined. Corn and lilies are monocots.

Monocyte A type of white blood cell that is attracted to damaged or infected tissue and that consumes bacteria, viruses, and cell debris.

Monohybrid cross A mating between individuals that differ only in one trait.

Monomer A small molecule, many of which can be combined into a polymer.

Monosaccharide A simple sugar that cannot be decomposed into smaller sugar molecules. The most common forms are the hexoses, six-carbon sugars such as glucose, and the pentoses, five-carbon sugars such as ribose.

Monotreme An egg-laying mammal that also has many primitive or reptilian features. The only living forms are the spiny anteater and the duck-billed platypus.

Morphogenesis The growth, shaping, and arrangement of the organs and tissues in a developing embryo.

Morula (L. a little mulberry) An embryo during the process of cleavage; consists of a solid ball of cells.

Motility The self-generated motion of an individual or its parts.

Motor cortex The part of the cerebral cortex of the brain that controls the movement of body parts.

Motor neuron A neuron that sends messages from the brain or spinal cord to muscles or secretory glands.

Mouthpart One of several arthropod head region appendages that chews and sucks.

Mucosa The innermost layer of the lining of many vertebrate canals and organs, such as the uterus, reproductive tracts, and alimentary canal; often consists of mucus-secreting cells, and in the alimentary canal it contains enzyme-secreting cells.

Muscle fiber A muscle cell; a giant cell with many nuclei and numerous myofibrils, capable of contraction when stimulated.

Muscle tissue Tissue that enables an animal to move; the three types are smooth muscle, cardiac muscle, and skeletal muscle.

Mutation Any heritable change in the base sequence of an organism's DNA.

Mutualism A symbiotic relationship in which both species benefit.

Mycelium The mass of filaments (hyphae) that makes up the body of a fungus.

Mycorrhiza An association between the root of a higher plant and a fungus that aids the plant in receiving water and nutrients.

Myelin sheath Lipid membranes made up of Schwann cells, which form an insulating layer around the neurons and speed impulse travel.

Myofibril A fibril composed of actin and myosin found in the cytoplasm of muscle cells that contracts the muscle.

Myosin A muscle protein that interacts with actin and causes muscles to contract.

Nasal cavity A chamber of the nose that opens posteriorly into the pharynx (throat).

Natural selection The increased survival and reproduction of individuals better adapted to the environment.

Navigation The ability to move from one map point to another.

Nematocyst A stinging capsule found in cnidarians, which, when stimulated, shoots out a tiny barb containing a poisonous substance that immobilizes or kills the prey or predator.

Nephridium (plural **nephridia**) An excretory organ in annelids, each body segment having a pair.

Nephron The functional unit of the vertebrate kidney; each of the million nephrons in a kidney consists of a glomerulus enclosed by a Bowman's capsule and a long attached tubule. The nephron removes waste from the blood.

Nerve A group of axons and accompanying cells from many different neurons running in parallel bundles and held together by connective tissue.

Nerve cell A specialized cell that transmits neural impulses; also called a *neuron.*

Nerve cord A strand of nervous tissue that forms part of the central nervous system of invertebrates.

Nerve impulse A change in ion permeabilities in a neuron's membrane that sweeps down the axon to its terminal, where it can excite other cells.

Nerve net A noncentralized, diffuse meshwork of nerve cells interlaced throughout the body. The simplest form of nervous organization in animals. Found in cnidarians, like jellyfish and sea anemones, it permits an equal response to stimuli from any direction in these radially symmetrical animals.

Nervous system An animal's network of nerves; integrates and coordinates the activities of all the bodily systems.

Nervous tissue Tissue containing neurons, cells whose main function is the transmission of electrochemical impulses.

Neural crest Ridge where folds arising from the ectoderm meet during the formation of the neural tube. Neural crest cells migrate and form pigment cells, parts of the face, and certain nerve cells.

Neural tube Embryonic cells that differentiate into the brain and spinal cord.

Neurohormone A hormone secreted by nerve cells.

Neuromodulator A compound occurring in the vicinity of a synapse that influences a receiving neuron's response to neurotransmitters.

Neuron A nerve cell that transmits messages throughout the body; includes dendrites, cell body, and axon.

Neurotransmitter A chemical that transmits a nerve impulse across a synapse.

Neurulation Development of the neural tube.

Neutron (L. *neuter*, either) A subatomic particle without any electrical charge found in the nucleus of an atom.

Neutrophil The most common type of white blood cell, important in combating bacterial infections.

Niche The role, function, or position of an organism in a biological community.

Nitrogen cycle The cycling of nitrogen between organisms and the earth.

Nitrogen fixation The assimilation of atmospheric nitrogen by certain prokaryotes into biologically usable nitrogenous compounds.

Node The position at which a leaf arises on a stem.

Nodule A swelling on the roots of legumes and some other plants that houses nitrogen-fixing bacteria.

Noncyclic photophosphorylation In photosynthesis, the generation of ATP by light-induced electron flow in both photosystems I and II; the Z-scheme.

Nondisjunction Failure of the chromosomes to separate correctly during mitosis or meiosis. Down syndrome results from nondisjunction of chromosome 21.

Nonpolar Not having an asymmetrical distribution of electrical charge; i.e., a nonpolar molecule like most lipids will not dissolve readily in water.

Norepinephrine A hormone secreted by the adrenal medulla; raises blood sugar concentration, blood pressure, and heart rate and increases muscular power and resistance to fatigue; sometimes known as noradrenaline.

Notochord A rod of mesodermal cells in the chordate embryo, marks the location of the backbone in vertebrates.

Nuclear envelope Two bilipid membranes perforated by pores that enclose the nucleus of a eukaryotic cell.

Nuclear pore A hole that perforates the nuclear envelope.

Nuclear winter A condition resulting from the burning and fires that would be generated during a nuclear war. Clouds of smoke and ash would block out sunlight, resulting in long periods of cold and darkness, during which photosynthesis, and with it most life on earth, would be drastically reduced or cease.

Nucleic acid A polymer of nucleotides, e.g., DNA and RNA.

Nucleoid The unbounded region of DNA in a prokaryotic cell.

Nucleolus (plural **nucleoli**) A dark-staining region within the nucleus of a eukaryotic cell where ribosomal RNA is formed.

Nucleosome The basic packaging unit of eukaryotic chromosomes; a histone wrapped with two loops of DNA.

Nucleotide The basic chemical unit of DNA and RNA consisting of one of four nitrogenous bases (A, C, T, or G) linked to a sugar (a ribose or a deoxyribose), which is in turn linked to a phosphate.

Nucleus (atomic) *See* Atomic nucleus.

Nucleus (cell) The membrane-bound region of a eukaryotic cell that contains the cell's DNA.

Nudibranch A sea slug; a marine gastropod mollusk.

Nutrient Any substance that an organism must obtain from the environment in order to survive and reproduce.

Nutrition The science concerned with determining the amounts and kinds of nutrients needed by the body.

Obese Having body weight that is more than 20 percent above ideal.

Oil A fluid lipid that is insoluble in water, often a prime form of energy storage in plants.

Oligotrophic Describes an aquatic environment with low nutrient levels, characterized by small amounts of algae or other aquatic plants.

Omnivore An animal that consumes both plant and animal matter as food.

Oncogene A cancer-causing gene; often a mutant form of a growth-regulating gene that inappropriately causes unrestrained cell replication.

Oocyte The precursor of a mature egg cell (ovum).

Open growth The continuous growth pattern exhibited by a plant throughout its life.

Operant A behavioral response to a stimulus.

Operant conditioning A form of learning in which an animal associates an action (operant; e.g., pressing a lever) with a reward or punishment (reinforcer; e.g., food).

Opercular flap The stiff outer protective structure of a fish's gill.

Operon The collective term for a regulatory protein and a group of genes whose transcription it controls in bacterial cells.

Opposable thumb A first digit or thumb that can be held opposite the other digits; characteristic feature of some primates.

Optic nerve The nerve leading from the eye to the brain.

Optimality hypothesis The idea that animals act in ways that maximize their energy intake and minimize the energy spent on gathering food.

Orbital The path followed by an electron around an atom.

Order A precise arrangement of structural units and activities.

Organ A body structure composed of two or more tissues that together perform a specific function.

Organelle In eukaryotic cells, a complex cytoplasmic structure with a characteristic shape that performs one or more specialized functions.

Organic chemistry The study of carbon compounds.

Organism An individual that can independently carry out all life functions.

Organ system A set of organs with related functions.

Organogenesis The formation of organs during embryonic development.

Orgasm Involuntary contractions of the vaginal and uterine walls or of the muscles lining the seminal vesicles and urethra.

Orientation The ability to sense direction.

Osmoregulation Maintenance of a constant internal salt and water concentration in an organism.

Osmosis The movement of water through a semipermeable membrane from an area of high water concentration (low solute concentration) to one of low water concentration (high solute concentration).

Osteichthyes A class of the phylum Chordata comprised of the bony fishes.

Outer membrane The smooth outer layer of a mitochondrion, surrounding the organelle.

Ovary Egg-producing organ.

Overweight Having body weight that is more than 10 percent above ideal.

Oviduct The tube along which the egg travels from the ovary to the uterus (womb) in women and some other female animals.

Ovulation The release of a mature egg from the ovary.

Ovule In a seed plant, the part of the ovary that develops into a seed.

Ovum An unfertilized egg cell.

Oxidation The removal of electrons from a molecule.

Oxidation-reduction reaction The transfer of electrons from one molecule to another.

Oxytocin A hormone whose release from the pituitary causes the contraction of the uterine muscle and labor.

Pacemaker cells Cells that set the rate of the heartbeat.

Pancreas A gland located behind the stomach that secretes digestive enzymes into the small intestine and the hormones insulin and glucagon into the blood.

Paracrine hormone A primitive hormone that acts on cells immediately adjacent to the ones that secreted it.

Parallel evolution The evolution of two or more physically similar and genetically related lines independently in the same direction and away from the ancestral form.

Parapatric speciation The divergence into separate species of populations living in adjacent areas.

Parasite An organism that lives on or in another living host organism and damages or kills its host as a consequence.

Parasympathetic nerves Neurons in the autonomic system that emanate from the spinal cord and act on the respiratory, circulatory, digestive, and excretory systems as a "housekeeping" system, conserving and restoring body resources.

Parathyroid gland One of a set of four small endocrine glands located on the thyroid gland; it secretes parathyroid hormone, which controls blood calcium levels.

Parathyroid hormone A hormone secreted by the parathyroid gland that regulates blood calcium levels.

Parenchyma A plant tissue composed of living, loosely packed, thin-walled cells; the most abundant plant tissue.

Parental generation In Mendelian genetics, the individuals that give rise to the first filial (F_1) generation.

Parental type An offspring having the characteristics of one of the parents.

Parthenogenesis The process by which an unfertilized egg gives rise to a new individual; occurs in aphids, bees, ants, certain lizards, and some other animals.

Partial pressure The pressure exerted by one gas in a mixture of gases.

Passive transport Movement of substances into or out of a cell from areas of high concentration to areas of low concentration without the expenditure of cellular energy.

Pathogenic (Gr. *pathos*, suffering + *genesis*, beginning) Capable of causing a disease.

Pectoral (shoulder) girdle The skeletal support to which front fins or limbs of vertebrates are attached.

Pedigree An ordered diagram of a family's relevant genetic features.

Pelvic (hip) girdle The skeletal support to which hind fins or limbs of vertebrates are attached.

Pelvis The central cavity of the kidney which collects urine before it passes to the ureter. (*See also* Pelvic girdle.)

Penis In mammals and reptiles, the male organ of copulation and urination.

Pepsin An enzyme secreted by the stomach that digests proteins.

Pepsinogen A precursor to the protein-digesting enzyme pepsin.

Peptide bond A bond joining two amino acids in a protein.

Perennial (L. *per*, through + *annus*, year) A plant that lives for many years, blooming and setting seeds several times before dying.

Pericycle In roots of vascular plants, one or more layers of parenchyma tissue lying between the endodermis and the vascular layer; gives rise to lateral roots.

Periderm In a plant, the outer protective layer including cork and cork cambium that is formed in secondary growth.

Peripheral nervous system The part of the nervous system consisting of the sensory and motor neurons, connecting the central nervous system with the sense organs, muscles, and glands of the body.

Peristalsis (Gr. *peristaltikos*, compressing around) In animals, waves of contraction and relaxation of muscles along the length of a tube, such as those in the digestive tract that help move food.

Peritubular capillary In the vertebrate kidney, one among a network of capillaries surrounding the looped portion of the nephron's tubule.

Petiole (L. *petiolus*, a little foot) The stalk connecting the leaf to the stem of a plant.

Phagocytosis The type of endocytosis through which a cell takes in food particles.

Pharynx [1] In vertebrates, a tube leading from the nose and mouth to the larynx and esophagus; conducts air during breathing and food during swallowing; the *throat.* [2] A short tube connecting the mouth and intestine in flatworms.

Phenotype The physical appearance of an organism controlled by its genes interacting with the environment.

Pheromone A compound produced by one individual that affects another individual at a distance; for example, pheromones are secreted by certain female insects and attract males.

Phloem Food-conducting tissue of plants, consisting of sieve elements with companion cells, phloem parenchyma, and fibers.

Phospholipid A lipid composed of a phosphate functional group and two fatty acid chains attached to a glycerol molecule; a main component of cell membranes.

Phosphorus cycle The cycling of phosphorus between organisms and soil, rocks, or water.

Phosphorylation The transfer of a phosphate group from one organic molecule to another. Phosphorylation often results in the formation of a high-energy bond, as in the formation of ATP.

Photon A vibrating particle of light radiation that contains a specific quantity of energy.

Photoperiod The length of light and dark periods each day.

Photoreceptor cells Light-sensitive cells such as rods and cones in the eye.

Photorespiration A process that occurs in plants when carbon dioxide levels are depleted but oxygen continues to accumulate and the enzyme RuBP carboxylase (rubisco) fixes oxygen instead of carbon dioxide.

Photosynthesis The metabolic process in which solar energy is trapped and converted to chemical energy (ATP and NADPH), which in turn is used in the manufacture of sugars from carbon dioxide and water.

Photosystem A functional light-trapping unit in the thylakoid membrane that includes chlorophyll, pigments, and electron donors and acceptors.

pH scale A logarithmic scale that measures hydrogen ion concentration; acids range from pH 1 to 7, water is neutral at a pH of 7, and bases range from pH 7 to 14.

Phyletic gradualism The suggestion that morphological changes occur gradually during evolution and are not always associated with speciation; distinct from *punctuated equilibrium.*

Phylogeny The study of the evolutionary history of different groups of organisms.

Phylum A major taxonomic group, lower than the kingdom, comprising members of similar classes, all with the same general body plan.

Physiology The study of the functioning body of organs.

Phytochrome Light-sensitive pigment found in plants that absorbs in the red or far-red wavelengths; associated with a number of timing processes, such as flowering, dormancy, leaf formation, and seed germination.

Phytoplankton Photosynthetic microorganisms that live near the surface of marine and fresh water.

Pineal gland An endocrine gland located in the vertebrate midbrain that produces the hormone melatonin; probably involved in body rhythms.

Pinocytosis The type of endocytosis through which a cell takes in fluid.

Pioneer community The species that are first to colonize a habitat after a disturbance such as fire, plowing, or logging.

Pith Parenchyma cells occupying the central portion of a stem or root acting as storage cells for the plant.

Pituitary gland In vertebrates, an endocrine gland connected by a stalk to the hypothalamus at the base of the brain; the anterior lobe secretes growth hormone, prolactin, LH, FSH, ACTH, TSH, and MSH; the posterior lobe stores and releases oxytocin and ADH. Much of the functioning of the pituitary is under the control of the hypothalamus, and pituitary hormones control most of the other endocrine glands.

Placenta In mammals, the spongy organ rich in blood vessels by which the developing embryo receives nourishment from the mother.

Plantae The kingdom comprising multicellular autotrophs such as mosses, ferns, and flowering plants.

Plasma The liquid portion of blood or lymph.

Plasma membrane The membrane that surrounds all cells, regulating entry and exit of substances; consists of a single lipid bilayer; also called the *cell membrane* and *plasmalemma.*

Plasmid A circular piece of DNA that can exist either inside or outside bacterial cells but can reproduce only inside a bacterial cell. Plasmids are used extensively in genetic engineering as carriers of foreign genes.

Plasmodesmata Bridges of cytoplasm that connect plant cells and allow rapid exchange of materials between the cells.

Plastid An organelle found in plants and some protozoa that harvests solar energy and produces or stores carbohydrates or pigments.

Platelet A disc-shaped cell fragment in the blood important in blood clotting; also called a *thrombocyte.*

Plate tectonics The geological building and moving of crustal plates on the surface of the globe.

Pleiotropy The condition in which a single gene affects two or more distinct and seemingly unrelated traits.

Pleural sac One of two fluid-filled sacs that encloses the lungs.

Plumule In an embryonic plant, the portion of the shoot above the cotyledon.

Poikilothermic "Cold-blooded"; a poikilothermic animal's internal body temperature is the same as that of the environment and therefore fluctuates, rising in warm weather and dropping in cold temperatures.

Poison claw A modified leg on the first body segment of a centipede, used to kill prey.

Polar Having an asymmetrical distribution of electrical charge; i.e., a polar molecule like glucose will dissolve readily in water.

Polarized The resting state of a nerve cell, with the cell's interior negatively charged with respect to the surroundings.

Pole One end of a cell.

Pollen grain The male gametophyte of seed plants.

Pollen tube A tubelike structure extended after pollination by the developing male gametophyte toward the egg cell; when the tube reaches the egg, some of its cytoplasm and the sperm nuclei are transferred into the egg.

Pollinator An organism that carries pollen from one plant to another; examples are bees and birds.

Polygenic Characteristic of a trait that varies in quantity (i.e., height, from short to tall and all measurements in between) that depends on the interaction of many genes.

Polymer A large molecule made up of repeated sequences of similar or identical subunits called monomers. DNA, proteins, and starch are examples of polymers.

Polyp (L. *polypus,* many-footed) The sedentary stage in the life cycle of cnidarians; a cylindrical organism with a whorl of tentacles surrounding a mouth at one end. Sea anemones and hydras are examples of polyps living alone; corals are examples of colonial polyps.

Polypeptide (Gr. *polys,* many + *peptin,* to digest) Amino acids joined together by peptide bonds into long chains. A protein consists of one or more polypeptides.

Polypeptide hormone A hormone consisting of a string of 3 to 200 amino acids; examples include oxytocin and luteinizing hormone.

Polysaccharide A carbohydrate made up of many simple sugars linked together. Glycogen and cellulose are examples of polysaccharides.

Pons Part of the brain stem involved in relaying sensory information to other areas of the brain.

Pool A repository of materials.

Population A group of individuals of the same species living in a particular area.

Population bottleneck A situation arising when only a small number of individuals of a population survive and reproduce; therefore only a small percentage of the original gene pool remains.

Population genetics A set of principles that clarifies what happens at the genetic level as populations evolve.

Pore A small opening, for example, in the skin or epidermis.

Postsynaptic cell The nerve cell that receives the message crossing a synapse.

Potassium channel A protein-lined pore in a cell's plasma membrane that permits potassium ions to flow in and out of the cell.

Potential energy Energy that is stored and ready to do work.

Precapillary sphincter A small cuff of smooth muscle near a capillary entrance that contracts, resulting in vasoconstriction (decreased flood flow), or expands, resulting in vasodilation (increased blood flow).

Predation The act of procurement and consumption of prey by predators.

Predator An organism, usually an animal, that obtains its food by eating other living organisms.

Prediction In the scientific method, an experimental result expected if a particular hypothesis is correct.

Pregnancy Begun by the implantation of the developing embryo in the uterine wall, a series of developmental events involving close cooperation between mother and embryo that transforms a zygote into a baby.

Presynaptic cell The neuron that sends a message down its axon and across a synapse to another cell.

Prey Living organisms that act as food for other organisms.

Primary cell wall The thin, flexible, porous layer immediately outside the plasma membrane of a plant cell deposited as the cell is expanding.

Primary consumer At the second trophic level in an ecosystem, an organism that eats producers; herbivores (plant eaters like cows and caterpillars) are primary consumers.

Primary embryonic induction Induction of the embryonic axis during vertebrate development.

Primary growth Plant growth arising from the apical meristem.

Primary productivity The amount of energy that producers (plants) convert by photosynthesis to chemical energy in the form of sugars and other organic compounds. *Net* primary productivity is the amount of chemical energy actually stored in new cells, leaves, roots, stems, flowers, and fruits, since much is lost to respiration in the plant.

Primary structure The level of protein structure referring to the unique linear sequence of amino acids on the polypeptide chain.

Primordium (plural **primordia**) One of many tiny swellings flanking the apical meristem, giving rise to leaves, branches, or flowers.

Principle of independent assortment *See* Independent assortment.

Prion An intracellular disease-causing entity apparently consisting only of protein and having no genetic material.

Process A projecting part of an organism or organic structure, such as a bone.

Producer An organism, usually a plant, that can produce all its nutritional needs from nonbiological substances.

Product A substance that results from a chemical reaction.

Productive collision In a chemical reaction, the colliding of molecules with enough kinetic energy to cause the transition state to occur.

Progesterone A female sex hormone secreted by the corpus luteum in the ovary that stimulates uterine wall thickening and mammary duct growth.

Prokaryotic cell A cell in which the DNA is not bound by a nuclear envelope such as a eubacterial or archaebacterial cell. Prokaryotic cells generally have no internal membranous organelles, and they evolved earlier than eukaryotic cells.

Prometaphase An interval during nuclear division when the nuclear membrane disappears and the spindle attaches to the chromosomes.

Propagation The passing of the action potential from one patch of a nerve cell membrane to another.

Prophase The first phase of nuclear division in mitosis or meiosis, when the chromosomes condense, the nucleolus disperses, and the spindle forms.

Prosimian A member of the suborder Prosimii, order Primates; includes lemurs and tarsiers.

Prostaglandin A fatty acid-derived hormone secreted by many tissues; for example, prostaglandin from the uterus produces strong contractions in the uterine muscle, causing menstrual cramps or inducing labor.

Prostate gland A reproductive gland in vertebrate males that secretes a milky alkaline fluid that increases motility of sperm and helps neutralize acidity of the vagina.

Protein A molecule composed of one or more chains of amino acids. Each chain is typically over 100 amino acids long. Enzymes, fingernails, antibodies, and the red protein in the blood are examples of proteins.

Protein synthesis The assembly of amino acids into a polypeptide chain, in three main stages: initiation, elongation, and termination.

Protista The kingdom comprising single-celled eukaryotes.

Proton A positively charged subatomic particle found in the nucleus of an atom.

Protostome (Gr. *protos*, first + *stoma*, mouth) Any bilateral animal whose first opening in the embryo (blastopore) becomes the mouth and the second opening the anus; also characterized by spiral cleavage during development; includes annelids, mollusks, and arthropods.

Protozoan Literally a "first animal"; any one of the single-celled eukaryotic organisms that is primarily animal-like in its method of obtaining food.

Proximal tubule In the vertebrate kidney, the convoluted tube between the Bowman's capsule and the loop of Henle.

Proximate cause The mechanisms that operate within an organism, directing a behavior.

Pseudocoelom (Gr. *pseudos*, false + *koiloma*, cavity) A body cavity that is not surrounded by mesoderm-derived cells; found in roundworms and rotifers.

Pseudopod (Gr. *pseudos*, false + *pous*, foot) A limblike cytoplasmic extension used in feeding and locomotion by some protozoans and animal cells.

Puberty The maturation of the sex organs and the development of secondary sex characteristics such as breasts in females and facial hair and a deep voice in males.

Pulmonary artery In amphibians, reptiles, birds, and mammals, the artery that carries blood from the heart's right ventricle to the lungs.

Pulmonary circulation The blood vessel system that carries oxygen-poor blood from the heart to the lungs, where gas exchange occurs, and carries oxygen-rich blood back to the heart. Contrasts to systemic circulation.

Punctuated equilibrium The suggestion that morphological changes evolve rapidly in geological time; during speciation, these changes occur in small populations, with the resulting new species being distinct from the ancestral form. After speciation, species retain much the same form until extinction; distinct from the phyletic gradualism theory.

Punnett square In genetics, a diagrammatic way of presenting the results of random fertilization from a mating.

Pupa In insects with complete metamorphosis, the stage that intervenes between the larva and the adult; in some cases, as in moths, the pupa is encased inside a cocoon.

Pupil The central, shutterlike opening in the iris of the vertebrate eye, through which light passes to the lens and the retina.

Pure-breeding In Mendelian genetics, referring to organisms that always produce offspring with traits identical to theirs; homozygous.

Pyramid of biomass The relationship between the total masses of various groups of organisms in a food chain, in which there is usually less mass, hence less stored energy, at each successive trophic level.

Quantitative trait *See* Polygenic.

Quaternary structure The level of protein structure referring to the fitting together of two or more polypeptide chains in a three-dimensional configuration.

Radial symmetry A circular body plan having a central axis from which structures radiate outward like the spokes of a wheel; characteristic of cnidarians, like jellyfish and sea anemones.

Radicle The embryonic root of a plant embryo.

Radioactive An unstable isotope that emits energy as it loses neutrons from the nucleus and changes into a more stable form of an element.

Radula A rasping organ in mollusks that shreds plant material by rubbing it against the hardened surface of the mouth.

Rain forest A biome characterized by a warm, wet climate, abundant vegetation, and diverse animal life.

Random distribution A pattern in which members of a population do not influence each other's spacing, if environmental conditions are uniform.

Reactant The starting substance in a chemical reaction.

Reaction center In photosynthesis, the central chlorophyll molecule around which other pigment molecules are arranged.

Reading frame The sequence of three nucleotides in mRNA that corresponds to an amino acid in a protein. There are three possible reading frames for any RNA: ACG GUA AUC, ACGG UAA UC, or AC GGU AAU C.

Realized niche The part of the fundamental niche that a species actually occupies in nature.

Receptor A protein of a specific shape that binds to a particular chemical.

Recessive An allele or corresponding phenotypic trait that is only expressed in the homozygote.

Recombinant DNA technology The techniques involved in excising DNA from one genome and inserting it into a foreign genome.

Recombinant type An offspring with characteristics different from the parents'.

Rectum The portion of the intestine between the colon and the anus that removes solid waste by defecation.

Red tides Dense blooms of certain dinoflagellates that tint water red and produce deadly toxins.

Reduction The addition of electrons to a molecule.

Reflex arc An automatic reaction, involving only a few neurons and requiring no input from the brain, in which a motor response quickly follows a sensory stimulus.

Refractory period The time period following passage of an action potential during which the nerve cell membrane is unable to react to additional stimulation.

Refuge A safe haven out of the reach of predators.

Regeneration The ability of an organism to regrow any lost parts of its body; e.g., a starfish can regenerate a lost arm.

Regulator cell A cell that detects perturbations in the environment and secretes a molecular messenger in response.

Reinforcer In operant conditioning, a reward or punishment.

Renal artery One of two arteries from the aorta to the kidneys.

Renal vein One of two large blood vessels through which blood leaves the kidneys.

Renin An enzyme released in the kidney that converts a blood protein into the hormone angiotensin.

Reproduction The method by which individuals give rise to other individuals of the same type.

Reproductive isolating mechanism Any structural, behavioral, or biochemical feature that prevents individuals of a species from successfully breeding with individuals of another species.

Reptile A cold-blooded, scaly, lung-breathing vertebrate that lays large eggs which usually have a shell; the dominant group of animals in the Mesozoic era; includes crocodiles, lizards, and tortoises.

Resolving power In microscopic viewing, the ability of the human eye to distinguish adjacent objects as distinct and separate.

Resource partitioning The process by which resources are divided, allowing species with similar requirements to use the same resources in different areas, at different times, or in different ways.

Respiration (L. *respirare*, to breathe) [1] The exchange of oxygen and carbon dioxide between cells and the environment. [2] The oxidative breakdown and release of energy from fuel molecules.

Responsiveness The tendency of a living thing to sense and react to its surroundings.

Restriction enzyme A naturally occurring enzyme that cuts DNA at precise points.

Reticular formation A network of neurons extending from the thalamus of the brain stem to the spinal cord; regulates respiration, cardiovascular centers, and awareness.

Retina (L. a small net) A multilayered region lining the back of the vertebrate eyeball, containing light-sensitive cells.

RFLP (restriction fragment length polymorphism) Genetic variation between individuals in the lengths of DNA that can occur when a piece of DNA is cut with a restriction enzyme.

Rhipidistian An extinct lobe-finned fish, probably the first vertebrate to crawl out on land and live for an extended period.

Rhizoid (Gr. *rhiza*, root) In bryophytes, some algae, and fungi, a hairlike structure that anchors the organism to the substrate.

Rhizome An elongated underground horizontal stem.

Rhodopsin The visual pigment in the rods and cones of the vertebrate eye that, activated by light energy, begins a series of events resulting in vision.

Ribonucleotide A base (adenine, cytosine, guanine, or uracil) attached to a ribose sugar and a phosphate; the subunits of RNA.

Ribosomal RNA (rRNA) An RNA molecule that is a structural component of ribosomes and is involved in protein synthesis.

Ribosome A structure in the cell that provides a site for protein synthesis. Ribosomes may lie freely in the cell or attach to the membranes of the endoplasmic reticulum.

Ribozyme An RNA molecule that can catalyze the cleavage or joining of itself and/or of other RNAs.

Rift A juncture between two major rock plates in the earth's crust.

RNA Ribonucleic acid; a nucleic acid similar to DNA except that it is single-stranded and contains the sugar ribose and the base uracil replaces thymine.

RNA polymerase An enzyme that makes RNA by copying the base sequence of a DNA strand.

Rod One of the two types of photoreceptor (light-sensitive) cells in the retina of the vertebrate eye; rods are sensitive to low levels of light and to movement but cannot distinguish color. (*See also* Cone.)

Root The branching structure of a plant that grows downward into the soil. Roots anchor the plant and absorb and transport water and mineral nutrients.

Root cap A mass of cells covering and protecting the growing tissue of the root.

Root hair An extension from the root epidermis that increases the absorptive capacity of a root.

***r*-selection** Selection experienced by populations that typically occurs at levels where competition for resources is unimportant; *r*-selected organisms are often short-lived, with rapid reproduction.

Rough ER The part of the endoplasmic reticulum that is studded with ribosomes; synthesizes lysosomal and membrane proteins and secreted proteins.

Roundworm A member of the phylum Nematoda.

Rubisco Ribulose bisphosphate carboxylase, the enzyme that binds carbon to a biological molecule in photosynthesis; the world's most abundant protein.

Saliva A watery liquid secreted by the salivary glands that moistens food particles for swallowing; contains the starch-digesting enzyme amylase.

Salivary gland A gland that secretes saliva.

Saprobe An organism that lives on decomposing organic matter.

Sarcomere The contractile unit of a skeletal muscle, consisting of repeating bands of actin and myosin.

Saturated fat A lipid containing fatty acids in which no carbon atoms are linked by double bonds, so that the molecule has the maximum possible number of hydrogen atoms.

Savanna A tropical grassland biome containing stunted, widely spaced trees situated between tropical forests and deserts.

Scanning electron microscope (SEM) An instrument used to observe the outer surfaces of cells and organisms at high magnification and great depth of field.

Schistosomiasis A human disease caused by the blood fluke.

Scientific method A series of steps for understanding the natural world based on experimental testing of hypotheses, possible mechanisms for how the world functions.

Sclerenchyma In plants, a supporting tissue composed of cells with lignified cell walls.

Scrotum The sac that contains the testes in male mammals.

Secondary cell wall In certain plant cells, the rigid cell wall between the plasma membrane and the primary cell wall formed after cell elongation has ceased; wood consists of secondary cell walls.

Secondary consumer In an ecosystem, an organism that consumes herbivores; carnivores (meat eaters) are secondary consumers.

Secondary growth The enlarged diameter of a stem or root resulting from cell divisions in the lateral meristem.

Secondary structure The level of protein structure in which the chain is molded into the form of an alpha helix or beta-pleated sheet.

Second filial (F_2) generation In Mendelian genetics, the second generation in the line of descent.

Second law of thermodynamics The second energy law, which states that because energy conversions from one form to another are not 100 percent efficient, all systems tend toward greater states of disorder.

Second messenger A molecule within a cell that causes cellular proteins to change behavior in response to a signal received at the cell surface; cyclic AMP and calcium ions are examples.

Secretin A digestive hormone secreted in the small intestine that triggers the pancreas to secrete bicarbonate, which neutralizes stomach acid.

Seed The product of a fertilized ovule of a seed plant, generally consisting of an embryo with its food reserves enclosed in a protective coat.

Seedless vascular plant Plants with an internal transport system but that require standing water to reproduce because they do not produce seeds; includes horsetails and ferns.

Segmented worm A member of the phylum Annelida.

Segregation principle Mendel's first law of heredity, which states that sexually reproducing diploid organisms have two alleles for each gene, and that during the gamete formation these two alleles segregate from each other so that the resulting gametes have only one allele of each gene.

Selectively permeable membrane A membrane that allows water molecules to pass freely but prevents the passage of large molecules across its surface; also called *semipermeable membrane.*

Self-tolerance The lack of an immune system response to components of one's own body.

Semen (L. *serere*, to sow) Sperm and its surrounding seminal fluid.

Semicircular canal One of several fluid-filled, tubelike channels within the vertebrate inner ear that is sensitive to movement of the head and that maintains balance.

Semiconservative replication The universal mode of DNA replication, in which only one of the two strands of DNA of the parent molecule is passed to a daughter molecule. This inherited strand serves as the template for the synthesis of the other half of the double helix.

Seminal fluid In a vertebrate male, the secretions of the seminal vesicle, prostate gland, and bulbourethral gland in which sperm are suspended.

Seminal vesicle An organ that stores sperm and adds secretions to the semen.

Seminiferous tubules Hollow tubes inside the testes, where sperm formation takes place.

Semipermeable membrane *See* Selectively permeable membrane.

Senescence A condition characterized by a profound decline in cell and organ function that occurs with aging.

Sense organ A group of specialized cells that receives stimulus energy and converts it into neural energy; for example, eyes and ears.

Sensitization An increase in the strength of a response to a stimulus; the opposite of habituation.

Sensory cortex The part of the cerebral cortex of the vertebrate brain that registers and integrates sensations from body parts.

Sensory neuron A nerve cell that receives information from the external or internal environment and transmits this information to the brain or spinal cord.

Septum (plural **septa**) A dividing wall or partition between structures, such as the segments of an earthworm.

Serosa The outermost layer of the alimentary canal, attaching the gut to the inner wall of the body cavity.

Sessile In animals, the quality of being permanently attached to a fixed surface.

Seta (plural **setae**) Attached to segments of earthworms, a bristle that pushes against the ground and enables movement.

Sex chromosomes Pairs of chromosomes where the members of the pair are dissimilar and involved in sex determination, such as the *X* and *Y* chromosomes.

Sex-linked Characteristic of genes that are carried on the sex chromosomes and therefore show different patterns of inheritance between males and females. The gene for color-blindness in humans is sex-linked and is carried on the *X* chromosome.

Sexual reproduction A type of reproduction in which new individuals arise from the mating of two parents.

Sexual selection A type of natural selection in which individuals select a mate on the basis of physical or behavioral characteristics, regardless of the effect of that characteristic on general fitness.

Shoot The main portion of a plant growing above ground.

Sieve plate One of the two perforated end walls of a sieve tube member.

Sieve tube member A conducting cell of phloem tissue in a vascular plant.

Sign stimulus A specific environmental signal that triggers a fixed-action behavioral pattern.

Single bond A chemical bond between two atoms created by the sharing of a pair of electrons.

Sinoatrial (SA) node A lump of modified heart muscle cells that is spontaneously electrically excitable; the pacemaker that governs the basic rate of heart contractions in vertebrates.

Siphon In certain mollusks, the funnel through which water passes on its way through the mantle cavity.

Skeletal muscle Muscle consisting of elongate, striated muscle cells; voluntary muscle.

Skeleton The rigid body support to which muscles attach and apply force, in vertebrates and invertebrates.

Skull The skeleton of the vertebrate head.

Slime mold A protist that derives energy by secreting digestive enzymes outside the cell; the digested material is then absorbed into the cell.

Slime trail The trail along which a snail or slug slowly moves.

Smooth ER The part of the endoplasmic reticulum folded into smooth sheets and tubules, containing no ribosomes; synthesizes lipids and detoxifies poisons.

Smooth muscle Muscle consisting of spindle-shaped, unstriated muscle cells; involuntary muscle found in the digestive, reproductive, and circulatory systems.

Social insect A member of any of the insect groups, such as ants, bees, and termites, that lives in colonies of related individuals and exhibits complex social behaviors.

Sociobiology The study of social behavior.

Sodium channel A protein-lined pore in the plasma membrane through which sodium ions can pass.

Sodium-potassium pump A carrier protein that maintains the cell's osmotic balance by transporting sodium ions out of the cell and potassium ions in.

Soft palate The posterior part of the roof of the mouth.

Soil The portion of the earth's surface consisting of disintegrated rock and decaying organic material.

Solute A substance that has been dissolved in a solvent.

Solvent A substance capable of dissolving other molecules.

Soma The main cell body of a neuron.

Somatic cell (Gr. *soma*, body) A cell in an animal that is not a germ cell.

Somatic cell hybrid A cell derived from the fusion of two or more somatic (normal body) cells (not germ cells). Biologists use somatic cell hybrids between human and mouse cells to facilitate gene mapping.

Somites Blocks of mesoderm that pinch off alongside the neural tube of a vertebrate embryo and develop into muscle blocks or other structures.

Sorus (plural **sori**) A cluster of sporangia on the underside of a fern frond.

Speciation The emergence of a new species. Speciation is thought to occur mainly as a result of populations becoming geographically isolated from each other and evolving in different directions.

Species In taxonomy, a group of organisms whose members have the same structural traits and who can interbreed with each other.

Species richness The total number of species in a community.

Specific immune response The body's most powerful line of defense, in which certain white blood cells produce antibodies that attack invaders; each antibody is specific for a certain invader.

Spermatogenetic cell A sperm-forming cell.

Sphincter A ring of muscle surrounding a tube or tubal opening that controls the size of the opening, for example, the opening from the esophagus to the stomach.

Spicule A slender, spiky rod of silica or calcium carbonate found in sponges that supports the soft wall and provides some protection from predators.

Spider A member of the class Arachnida (phylum Arthropoda) that contains four pairs of legs on the thorax.

Spinal cord A tube of nerve tissue that runs the length of a vertebrate animal, just above (dorsal to) the notochord.

Spindle A cluster of microtubules that attach from the centriole to the centromere of a chromosome and move the chromosome toward the pole of the cell during cell division.

Spinneret A tubular appendage in spiders and some insects that reels out silk threads.

Spiny-finned fish A member of one of the two major groups of bony fishes; includes trout and zebra fish.

Spiracle A hole in the body wall of insects that forms the opening of the air tubes (trachea).

Spirillum (plural **spirilla**) A bacterial cell with a spiral shape.

Sponge A multicellular animal without true tissues, either radially symmetrical or asymmetrical. Special cells with flagella push water through pores in the body wall from which food particles are filtered. The remains of their skeletal elements make up the common bath sponge.

Sporangium (plural **sporangia**) A structure in which spores are produced.

Spore In eukaryotes, a reproductive cell that divides mitotically and produces a new individual. In prokaryotes, a resistant cell capable of surviving harsh conditions and germinating when conditions are once again favorable.

Sporophyte (Gr. *spora*, seed + *phyton*, plant) The diploid, spore-producing stage in the life cycle of many plants.

S-shaped curve A curve on a graph representing logistic population growth.

Stabilizing selection Type of natural selection in which the intermediate form of a trait is favored over extreme forms.

Stamen The pollen-producing structure within a flower.

Starch A polysaccharide composed of long chains of glucose subunits; the principal energy source of plants.

Start codon In DNA and mRNA, the codon that signals where the translation of a portion begins.

Stem The central, often elongated part of a plant composed mainly of vascular tissue, that supports leaves and transports water and nutrients.

Stem cell A normal body (somatic) cell that can continue to divide, replacing cells that die (e.g., in the skin or blood) during an animal's life.

Stereocilia Threadlike projections on the hair cells of the cochlea of the inner ear.

Stereoscopic vision Overlapping visual fields with good depth perception, allowing animals to discriminate distances well.

Sternum The breastbone; in birds, a blade-shaped anchor for the pectoral muscles that enable flight.

Steroid (Gr. *stereos*, solid + L. *ol*, from + *oleum*, oil) A major class of lipids based on a 17-carbon-atom ring system and often a hydrocarbon tail. Cholesterol and sex hormones are steroids.

Steroid hormone A hormone synthesized from cholesterol containing four joined rings of carbon atoms; e.g., estrogen, testosterone, cortisol.

Stigma [1] The tiny, light-sensitive eyespot of a euglenoid. [2] The sticky top of a flower that serves as a pollen receptacle.

Stipe The stemlike structure that provides vertical support to an alga.

Stoma (plural stomata) A tiny hole surrounded by two guard cells in the epidermis of plant leaves and stems, through which gases can pass.

Stomach An expandable, elastic-walled sac of the gut that receives food from the esophagus.

Stop codon In DNA and mRNA, the codon that signals where the translation of a protein ends.

Style A structure leading from the ovary of flowering plants to the stigma (pollen receptacle).

Submucosa A connective tissue outside the mucosa of the digestive tract that is richly supplied with blood and lymph vessels and nerves.

Substrate A reactant in an enzyme-catalyzed reaction; fits into the active site of an enzyme.

Succession A progression of communities in a particular habitat, each one replacing the other until a persistent, stable, climax community is established.

Sucrose A disaccharide composed of glucose bonded to fructose, common in many plants; table sugar.

Summation The cumulative effect of individual postsynaptic potentials in a neuron.

Superior vena cava One of the two largest veins in the human body, carrying blood from the head, neck, and arms to the right atrium of the heart.

Supporting cell A cell in the seminiferous tubule walls of the vertebrate testes that nourishes the developing sperm.

Surface tension The tendency of molecules at the surface of a liquid to cohere to each other rather than to the molecules of air above them.

Surface-to-volume ratio The ratio of the surface area of a cell to its volume; imposes limits on cell size because as the cell's linear dimensions grow, its surface area increases less than its volume.

Survivorship curve A plot of the data representing the proportion of a population that survives to a certain age.

Symbiont An organism that lives in a close relationship with an organism of another species.

Symbiosis A close interrelationship of two different species.

Sympathetic nerves Neurons of the autonomic nervous system that emanate from the central spinal cord and act on the respiratory, circulatory, digestive, and excretory systems in response to stress or emergency, preparing for flight or fight.

Sympatric speciation A situation in which a population diverges into two species after a genetic, behavioral, or ecological barrier to gene flow arises between subgroups of the population inhabiting the same region.

Synapse The region of communication between two neurons or between a neuron and a muscle.

Synaptic vesicle At the tip of a nerve cell's axon, a small round packet containing neurotransmitter molecules.

Systemic circulation The blood vessel system that carries oxygen-rich blood from the heart to the body and returns oxygen-poor blood to the heart. Contrasts to pulmonary circulation.

Systole The contraction phase of the heart muscle.

Tail In chordates, a structure that protrudes beyond the anus.

Taproot In plants, a root system with a prominent main root developing directly from the embryonic root, growing vertically downward and bearing lateral roots, e.g., the root of a carrot plant.

Target cell Any cell that responds to specific hormones.

Taxis (plural **taxes**) An animal's movement toward light, heat, or specific chemicals.

Taxonomic group A category based on shared characteristics; the groups are species, genus, family, order, class, phylum, kingdom, and domain.

Taxonomy (Gr. *taxix*, arrangement + *nomos*, law) The science of classifying organisms into different categories.

T cell (T lymphocyte) A type of white blood cell that kills foreign cells directly and also regulates the activities of other lymphocytes.

Teleost A modern bony fish; includes most living fish; characterized by the presence of a swim bladder and spiny fins.

Telophase The final phase of nuclear division when the chromosomes are at opposite poles of the cell, the nuclear membrane and nucleolus reappear, and the spindle disappears.

Temperate forest A biome that occurs north or south of subtropical latitudes, characterized by generally mild climate and varied populations of evergreen and deciduous trees.

Template A strand of DNA or RNA that serves as a pattern or model for the construction of a complementary strand of DNA or RNA.

Tendon A connective tissue strap that connects bone to muscle.

Termination The third main stage of protein synthesis, during which the growth of the polypeptide chain comes to a halt, when the ribosome reaches the stop codon.

Territoriality A form of behavior in which an individual defends its living or feeding space against any intruders.

Tertiary consumer In an ecosystem, a carnivore that eats other carnivores.

Tertiary structure The level of protein structure referring to the three-dimensional folding of the entire polypeptide chain.

Testcross The mating of an individual of unknown genotype with a homozygous recessive in order to determine the unknown genotype.

Testis (plural **testes**) The male reproductive organ; produces sperm and sex hormones.

Testosterone A hormone produced by the testis in vertebrate males; stimulates embryonic development of male sex organs, sperm production, male secondary sex characteristics, and male behaviors.

Thalamus Located beneath the cerebrum and above the hypothalamus of the vertebrate brain; relays and processes sensory impulses.

Thallus The flat, leaflike body of some algae and other lower plants.

Theory A broad general hypothesis that is repeatedly tested but never disproved.

Thoracic cavity The region within the rib cage directly over the heart, in which the lungs are suspended.

Thorax The central region of the body of an arthropod or vertebrate between the head and the abdomen.

Thrombocyte *See* Platelet.

Thylakoid (Gr. *thylakos*, sac + *oides*, like) A stack of flattened membranous discs containing chlorophyll found in the chloroplasts of eukaryotic cells.

Thymus A gland in the neck or thorax of many vertebrates; makes and stores lymphocytes in addition to secreting hormones.

Thyroid gland A large endocrine gland in the neck of vertebrates regulating the body's growth and use of energy by secreting thyroxine.

Thyroxine The most abundant thyroid hormone, which governs metabolic and growth rates and stimulates nervous system function.

Tidal ventilation The in-and-out flow of air that characterizes mammalian respiration.

Tide pool Along the ocean shore, a depression formed in rock by wave action.

Tight junction A band around some eukaryotic cells where the plasma membrane touches the membrane of an adjacent cell, forming a tight seal and preventing passage of substances in the space between adjacent cells.

Tissue A group of cells of the same type performing the same function within the body.

Tissue culture The growth of tissue cells "in the test tube" in a way that often permits further differentiation and maturation of the cells. In plant breeding, the growing of new identical plants from somatic (body) cells.

Tongue A muscular organ on the floor of the mouth of most higher vertebrates that carries taste buds and manipulates food.

Trachea The "windpipe," the major airway leading into the lungs of vertebrates.

Tracheae Branching networks of hollow air passages used for gas exchange in insects.

Tracheid An elongated hollow cell with thick, rigid, pitted walls; a basic unit of vascular tissue in most vascular plants.

Transcription (L. *trans*, across + *scribere*, to write) The formation of RNA from a single strand of a DNA molecule; the process is catalyzed by the enzyme RNA polymerase.

Transfer RNA (tRNA) A small RNA molecule that translates a codon in mRNA into an amino acid during protein synthesis.

Transformation The process of transferring an inherited trait by incorporating a piece of foreign DNA into a prokaryotic or eukaryotic cell.

Transition state In a chemical reaction, the intermediate springlike state between reactant and product.

Translation The conversion of the information on a strand of RNA into a sequence of amino acids in a protein; occurs on ribosomes.

Translocation In plants, the transport of solutes in phloem cells.

Transmission electron microscope (TEM) An instrument used to observe the inner structures of thin stained sections of cells at high magnification.

Transpiration The loss of water from plants by evaporation, mainly through the stomata on stems and leaves.

Transpiration pull theory The theory that water is transported in plants by the sun's energy evaporating water from leaves via transpiration and, due to the cohesion of water molecules, pulling the entire column of water up from the soil; also called *cohesion-adhesion-tension* theory.

Transposon A gene that can move around the genome, within or between chromosomes; also called a *jumping gene*.

Trial-and-error learning *See* Operant conditioning.

Triglyceride A fat or oil composed of three fatty acids joined to the three carbons of glycerol.

Trimester Any of the following three-month periods of pregnancy: first to third, fourth to sixth, or seventh to ninth months.

Trophic level A particular feeding level in a community or ecosystem.

Tropism (Gr. *trope*, turning) The movement of a plant in response to external or internal stimuli; e.g., phototropism, the response of plants to light.

True hormone A chemical produced by nonneural cells in one part of the body that has an effect on another part of the body.

True slime mold A type of funguslike protist characterized by the plasmodium, a mass of continuous cytoplasm surrounded by one plasma membrane containing many diploid nuclei.

Tubular reabsorption In the vertebrate kidney, the process whereby water, salts, and nutrients are returned from the kidney tubule to the blood.

Tubular secretion In the vertebrate kidney, the process whereby ions and drugs are secreted from the blood into the kidney tubule.

Tubulin A globular protein that makes up the microtubules.

Tumor An uncontrolled or abnormal growth of cells.

Tundra A biome characterized by cold, treeless plains with grasses, lichens, and occasional small shrubs; deep tundra soil that remains frozen permanently is called permafrost.

Tunic The outer envelope of a sea squirt, enclosing the pharynx.

Turgor pressure (L. *turgor*, a swelling) Internal pressure on a cell wall caused by osmotic movement of water into the cell.

Turner syndrome Features exhibited by a human female with only one *X* chromosome (*XO*); includes sterility due to rudimentary or missing ovaries.

Ultimate cause The selective advantage of a particular action or behavior.

Umbilical cord A cord joining the embryo and the placenta.

Uniform distribution A pattern in which members of a population are spaced at regular intervals in their range.

Uniformity The scientific principle that the laws of nature always have operated and always will operate the same way.

Unsaturated fat A lipid containing fatty acids in which two or more carbon atoms are linked by double bonds, so that there are fewer than the maximum possible number of hydrogen atoms.

Urea A water-soluble nitrogenous waste product formed during the breakdown of proteins, nucleic acids, and other substances, excreted by mammals and some fishes.

Ureter The tube that carries urine from the kidney to the bladder.

Urethra (Gr. *ouerin*, to urinate) The tube that carries urine and releases it to the outside; in males this tube also carries sperm.

Uric acid A water-insoluble nitrogenous waste product excreted by birds, land reptiles, and insects.

Urination The process by which urine exits the body through the urethra.

Urine A fluid that washes uric acid or urea from the body.

Uterus (L. womb) The thick-walled chamber where the embryo develops.

Vagina A hollow muscular tube that receives the penis during copulation and through which the fetus passes during birth.

Valve A tonguelike flap extending into the internal space of a vein that helps regulate the flow of blood.

Vasoconstriction Contraction of blood vessel walls; regulates blood flow.

Vasodilation Relaxation of blood vessel walls; regulates blood flow.

Vascular cambium A type of lateral meristem that produces secondary phloem and secondary xylem, or wood.

Vascular plants Plants that possess an internal transport system for water and food in the form of xylem and phloem cells.

Vascular tissue system Plant tissue that conducts fluid throughout the plant and helps strengthen roots, stems, and leaves; consists of xylem and phloem cells.

Vas deferens (L. *vas*, a vessel + *defere*, to carry down) A tube that carries sperm from the epididymis to the ejaculatory duct in male vertebrates.

Vegetative reproduction A process in which new individuals genetically identical to each other and to the parent emerge from the parent's body.

Vein [1] A large thin-walled blood vessel that brings blood from the body to the heart. [2] In plant leaves, bundles of xylem and phloem cells that form a branching pattern in dicots and run parallel to each other in monocots.

Ventilation The breathing in and out of the lungs.

Ventricle A muscular chamber of the heart that pumps blood to the lungs or to the rest of the body.

Venule A small thin-walled blood vessel that arises from capillaries and carries blood from the tissues to the veins.

Vertebra One of a series of interlocking bones that makes up the backbone of vertebrate animals.

Vertebral column The backbone.

Vertebrate An animal that possesses a backbone made of bony segments known as vertebrae.

Vesicle A small membranous container in a cell.

Vessel [1] A tubelike structure that carries blood. [2] A tubelike structure of xylem that conducts water in plants.

Vessel member In the xylem of flowering plants, a hollow cell with pitted walls, stacked end to end, through which water flows from the roots to the rest of the plant.

Vestigial organ A rudimentary structure with no apparent utility but bearing a strong resemblance to structures in probable ancestors.

Vibrio A bacterial cell with a curved rod shape.

Villus (plural **villae**) (L. a tuft of hair) A fingerlike projection of the intestinal wall that increases the surface area for absorption of nutrients.

Viroid An intracellular parasite that affects plants, consists only of small RNA molecules without any protein coat.

Virus An infectious agent consisting of RNA or DNA encased in a protein coat (capsid); incapable of metabolism or reproduction outside a host cell.

Vitamin (L. *vita*, life + *amine*, of chemical origin) An organic compound needed in small amounts for growth and metabolism obtained in food.

Warning coloration Bright colors or striking patterns of prey species that warn predators to stay away; also known as aposematic coloration.

Water cycle A sun-driven global exchange involving evaporation, precipitation, runoff, and transpiration that cycles water from the atmosphere to the earth's surface through organisms and back again.

Water mold A type of funguslike protist containing several nuclei within a common cytoplasm and forming relatively large immobile egg cells; *Oomycota*.

Water vascular system In echinoderms, a system of fluid-filled canals that includes hundreds of short branches called tube feet that can attach to objects and thus aid in locomotion or feeding.

Wax A sticky, solid, waterproof lipid that forms the comb of bees and waterproofing of plant leaves.

Weather The condition of the atmosphere at any particular place and time, including its temperature, humidity, wind speed, and precipitation.

Wind Air flowing over the earth.

Wood The hard, cellulose-containing dead xylem cells of a perennial plant.

***X* chromosome** The sex chromosome found in two doses in female mammals, fruit flies, and many other species.

***X*-linked** Characteristic of a heritable trait that occurs on the *X* chromosome.

Xylem (Gr. *xylon*, wood) Water- and mineral-transporting tissue of plants composed of tracheids, vessels, parenchyma cells, and fibers; dead xylem cells form wood.

***Y* chromosome** The sex chromosome found in a single dose in male mammals, fruit flies, and many other species.

Yolk Material consisting of proteins, carbohydrates, nucleic acids, and lipids stored in an animal egg as a nutrient supply for a developing embryo.

Zygote (Gr. *zygotos,* paired together) The diploid cell that results from the fusion of an egg and a sperm cell. A zygote may either form a line of diploid cells by a series of mitotic cell divisions or undergo meiosis and develop into haploid cells.

Further Reading

Chapter 1

Darwin, C. *On the Origin of Species: A Facsimile of the First Edition.* Cambridge, Mass.: Harvard University Press, 1975.

Green, G. M., and R. W. Sussman. "Deforestation History of the Eastern Rain Forests of Madagascar from Satellite Images." *Science* 248 (1990): 212–215.

Haas, P. M., M. A. Levy, and E. A. Parson. "Appraising the Earth Summit." *Environment* 34 (1992): 6–15.

Jolly, A. "Madagascar: A World Apart," *National Geographic*, February 1987, pp. 149–183.

Kormandy, E. J. "Ethics and Values in the Biology Classroom." *The American Biology Teacher* 52 (1990): 403–407.

Myers, C. W., and J. W. Daly. "Dart-Poison Frogs." *Scientific American*, February 1983, pp. 120–133.

Sun, M. "Costa Rica's Campaign for Conservation." *Science* 239 (1988): 1366–1369.

Chapter 2

Amato, Ivan. "Unraveling the Biochemistry of Spider Silk." *Science News* 138 (6) (1990): 214.

Atkins, P. W. *Molecules.* New York: Scientific American, 1987.

Chang, R. *Chemistry.* 3d ed. New York: McGraw-Hill, 1994.

Ehrig, T., W. Bosron, and T. K. Li. "Alcohol and Aldehyde Dehydrogenase." *Alcohol and Alcoholism* 25 (1990): 105–116.

Richard, F. M. "The Protein Folding Problem." *Scientific American*, January 1991, pp. 54–63.

Stryer, L. *Biochemistry*, 3d ed. New York: Freeman, 1988.

Toner, Mike. "Spin Doctor." *Discover*, May 1992, pp. 32–36.

Vollrath, Fritz. "Spider Webs and Silk." *Scientific American*, March 1992, pp. 70–76.

Chapter 3

Alberta, B., D. Bray, J. Lewis, M. Raff, K. Roberts, and J. Watson. *Molecular Biology of the Cell.* 2d ed. New York: Garland, 1989.

Darnell, J., H. Lodish, and D. Baltimore. *Molecular Cell Biology.* New York: Scientific American, 1991.

Diamond, J. "Curse and Blessing of the Ghetto." *Discover*, March 1991, p. 60.

Paritzky, J. "Tay-Sachs: The Dreaded Inheritance." *American Journal of Nursing*, March 1985, pp. 260–264.

Prescott, D. M. *Cells.* Boston: Jones and Bartlett, 1988.

Chapter 4

Alberts, B., D. Bray, J. Lewis, M. Raff, K. Roberts, and J. Watson. *Molecular Biology of the Cell.* 2d ed. New York: Garland, 1989.

Beutler, Ernest. "Gaucher's Disease." *New England Journal of Medicine* 325 (1991): 1354–1360.

———. "Gaucher Disease: New Molecular Approaches to Diagnosis and Treatment." *Science* 256 (1992): 794–799.

Darnell, J., H. Lodish, and D. Baltimore. *Molecular Cell Biology.* New York: Scientific American, 1990.

Fagerberg, W., and D. Allain. "A Quantitative Study of Tissue Dynamics During Closure in the Traps of Venus's Flytrap *Dionaea muscipula* Ellis." *American Journal of Botany* 78 (1991): 647–657.

Hodick, D., and A. Sievers. "On the Mechanism of Trap Closure of Venus Flytrap (*Dionaea muscipula* Ellis)." *Planta* 179 (1989): 32–42.

Lipske, Michael. "Forget Hollywood: These Bloodthirsty Beauties Are for Real." *Smithsonian Magazine*, December 1992, pp. 49–59.

Stryer, L. *Biochemistry.* 3d ed. New York: W. H. Freeman, 1988.

Williams, S., and A. Bennett. "Leaf Closure in the Venus Flytrap: An Acid Growth Response." Science 218 (1982): 1120–1121.

Chapter 5

Alberts, D., D. Bray, J. Lewis, M. Raff, K. Roberts, and J. Watson. *Molecular Biology of the Cell.* 2d ed. New York: Garland, 1989.

Darnell, J., H. Lodish, and D. Baltimore. *Molecular Cell Biology.* New York: Scientific American, 1986.

Newsholme, E., and T. Leech. "Fatigue Stops Play." *New Scientist*, September 22, 1988, pp. 39–43.

Prescott, D. M. *Cells.* Boston: Jones and Bartlett, 1988.

Stryer, L. *Biochemistry.* 3d ed. New York: W. H. Freeman, 1988.

Chapter 6

Alberts, B., D. Bray, J. Lewis, M. Raff, K. Roberts, and J. Watson. *Molecular Biology of the Cell.* 2d ed. New York: Garland, 1989.

Bazzaz, F., and D. Fajer. "Plant Life in a CO_2-Rich World." *Scientific American*, January 1992, pp. 68–74.

Bower, Bruce. "Smoke Gets in Your Brain." *Science News* 43 (1993): 46–47.

Lewin, R. A., and L. Cheng. *Prochloron.* New York: Chapman and Hall, 1989.

Nooney, H. A., et al. "Predicting Ecosystem Responses to Elevated CO_2 Concentrations." *Bioscience* 41(2)(1991): 96–103.

Prescott, D. M. *Cells.* Boston: Jones and Bartlett, 1988.

Schelling, Thomas C. "Addictive Drugs: The Cigarette Experience." *Science* 255 (1992): 430–434.

Scurlock, J., and D. Hall. "The Carbon Cycle." *New Scientist*, November 2, 1991.

Stone, Richard. "Bad News on Second-Hand Smoke." *Science* 257 (1992): 607.

Stryer, L. *Biochemistry.* 3d ed. New York: W. H. Freeman, 1988.

Williamson, P., and J. Gribbin. "How Plankton Change the Climate." *New Scientist*, March 16, 1991, pp. 48–52.

Youvan, D. C., and B. L. Marrs. "Molecular Mechanisms of Photosynthesis." *Scientific American*, June 1987, pp. 42–50.

Chapter 7

Alberts, B., D. Bray, J. Lewis, M. Raff, K. Roberts, and J. D. Watson. *Molecular Biology of the Cell.* 2d ed. New York: Garland, 1989.

Chandley, A. C. "Meiosis in Man." *Trends in Genetics* 4 (1988): 79–83.

Cowen, R. "Speeding Up Wound Healing the EGF Way." *Science News* 136 (1989): 39.

Fisher, Lawrence. "Three Companies Speed Artificial Skin." *The New York Times*, September 12, 1990.

Kenyon, C. "The Nematode *Caenorhabditis elegans.*" *Science* 240 (1988): 1448–1453.

McIntosh, J. R., and K. L. McDonald. "The Mitotic Spindle." *Scientific American*, October 1989, pp. 26–34.

Murray, A. W., and M. W. Kirschner. "Dominoes and Clocks: The Union of Two Views of the Cell Cycle." *Science* 246 (1989): 614–621.

Murray, A. W., and M. W. Kirschner. "What Controls the Cell Cycle." *Scientific American*, March 1991, pp. 56–63.

Roberts, L. "The Worm Project." *Science* 248 (1990): 1310–1313.

Smith, M. U. "Teaching Cell Division: Student Difficulties and Teaching Recommendations." *Journal of College Science Teaching*, September-October 1991, pp. 28–33.

Van Wynsberghe, D., C. R. Noback, and R. Carola. *Human Anatomy and Physiology.* Chapter 5. New York: McGraw-Hill 1990.

Weinberg, R. A. "Finding the Anti-Oncogene." *Scientific American*, September 1988, pp. 44–51.

Chapter 8

Bhattacharyya, M. K., A. M. Smith, T. H. Ellis, C. Hedley, and C. Martin. "The Wrinkled Seed Character of Pea Described by Mendel Is Caused by a Transposon-like Insertion in a Gene Encoding Starch-Branching Enzymes." *Cell* 60 (1990): 115–122.

Hall, J. G. "Genomic Imprinting: Review and Relevance to Human Disease." *American Journal of Human Genetics* 46 (1990): 857–873.

Leyhausen, P., and T. H. Reed. "The White Tiger: Care and Breeding of a Genetic Freak." *Smithsonian*, April 1971, pp. 24–31.

Orel, V. *Mendel.* New York: Oxford University Press, 1984.

Sapienza, C. "Parental Imprinting of Genes." *Scientific American*, October 1990, pp. 52–60.

Savory, T. H. "The Mule." *Scientific American*, December 1970, pp. 102–108.

Suzuki, D. T., A. J. F. Griffiths, J. H. Miller, and R. C. Lewontin. *Introduction to Genetic Analysis.* 4th ed. New York: W. H. Freeman, 1986.

Chapter 9

Angier, N. "The Gene Dream." *American Health Magazine*, March 1989, pp. 103–108.

Felsenfeld, G. "DNA." *Scientific American*, May 1985, p. 58.

Mitsuya, H., R. Yarchoan, and S. Broder. "Molecular Targets for AIDS Therapy." *Science* 249 (1990): 1533–1544.

Radman, M., and R. Wagner. "The High Fidelity of DNA Duplication." *Scientific American*, August 1988, pp. 40–47.

Watson, J. D. *The Double Helix.* New York: Atheneum, 1968.

Watson, J. D., and F. H. C. Crick. "Molecular Structure of Nucleic Acids: A Structure for Deoxyribosenucleic Acid." *Nature* 171 (1953): 737–738.

Chapter 10

Alberts, B., D. Bray, J. Lewis, M. Raff, K. Roberts, and J. D. Watson. *Molecular Biology of the Cell.* 2d ed. New York: Garland, 1989.

Brock, Thomas D., and Michael T. Madigan. *Biology of Microorganisms*, 5th ed. Englewood Cliffs N.J.: Prentice Hall, 1988.

Collins, F. S. "Cystic Fibrosis: Molecular Biology and Therapeutic Implications." *Science* 256 (1992): 774–779.

Davies, K. "The Search for the Cystic Fibrosis Gene." *New Scientist*, October 21, 1989, pp. 54–58.

Diamond Jared. "The Return of Cholera." *Discover*, February, 1992, pp. 60–66.

Hirschhorn, Norbert, and William B. Greenough III. "Progress in Oral Rehydration Therapy," *Scientific American*, May 1991, Volume 264, Number 5, pp. 50–56.

Chapter 11

Barnes, D. M. "'Fragile X' Syndrome and Its Puzzling Genetics." *Science* 243 (1989): 171–172.

Culver, K. W. "Splice of Life: Genetic Therapy Comes of Age." *The Sciences*, January-February 1993, pp. 18–24.

Hall, S. S. "James Watson and the Search for Biology's 'Holy Grail.'" *Smithsonian*, February 1990, pp. 41–49.

Mange, A. P., and E. J. Mange. *Genetics: Human Aspects.* 2d ed. Sunderland, Mass.: Sinauer, 1990.

McKusick, V. A. *Mendelian Inheritance in Man.* 9th ed. Baltimore, John Hopkins University Press, 1990.

Motulsky, A. G. "Societal Problems in Human and Medical Genetics." *Genome* 31 (1989): 870–875.

Neufeld, P. J., and N. Colman. "When Science Takes the Witness Stand." *Scientific American*, May 1990, pp. 46–53.

Roberts, L. "Ethical Questions Haunt New Genetic Technologies." *Science* 243 (1989): 1134–1136.

Revkin, A. "Hunting Down Huntington's." *Discover*, December 1993, pp. 98–108.

Roberts, L. "To Test or Not to Test?" Science 247 (1990): 17–19.

Verma, I. M. "Gene Therapy." *Scientific American*, November 1990, pp. 68–84.

Wexler, N. S. "The Tiresias Complex: Huntington's Disease as a Paradigm of Testing for Late-Onset Disorders. *FASEB J* (1992) 6:2820–2825.

Chapter 12

Friedmann, T. "Progress toward Human Gene Therapy." *Science* 244 (1989): 1275–1281.

Gasser, C. S., and R. T. Fraley. "Genetically Engineering Plants for Crop Improvement." *Science* 244 (1989): 1293–1299.

Mullis, K. B. "The Unusual Origin of the Polymerase Chain Reaction." *Scientific American*, April 1990, pp. 56–65.

Tiedje, J. M., et al. "The Planned Introduction of Genetically Engineered Organisms: Ecological Considerations and Recommendations." *Ecology* 70 (1989): 298–315.

Chapter 13

Barinaga, M. "Zebrafish: Swimming into the Development Mainstream." *Science* 250 (1990): 34–35.

DeRobertis, E. M., G. Oliver, and C. V. E. Wright. "Homeobox Genes and the Vertebrate Body Plan." *Scientific American*, July 1990, pp. 46–52.

Gilbert, S. *Developmental Biology.* 3d ed. Sunderland, Mass.: Sinauer, 1992.

Kline, D. "Activation of the Egg by the Sperm." *Bioscience* 41 (1991): 89–95.

Summerbell, D., and M. Maden. "Retinoic Acid, a Developmental Signaling Molecule." *Trends in NeuroSciences* 13 (1990): 142–146.

Chapter 14

Aral, S. O., and K. K. Holmes. "Sexually Transmitted Diseases in the AIDS Era." *Scientific American*, February 1991, pp. 62–69.

Diamond, J. "Turning a Man." *Discover*, June 1992, pp. 71–77.

Glasier, M., K. J. Thong, M. Dewar, M. Mackie, and D. T. Baird. "Mifepristone (RU 486) Compared with High-Dose Estrogen and Progestogen for Emergency Postcoital Contraception." *New England Journal of Medicine* 327 (1992): 1041–1044.

Moore, K. L. *Essentials of Human Embryology.* Toronto: Decker, 1988.

Rusting, R. L. "Why Do We Age?" *Scientific American*, December 1992, pp. 130–141.

Stroh, M. "The Root of Impotence: Does Nitric Oxide Hold the Key?" *Science News* 142 (1992): 10–11.

Ulmann, A., G. Teutsch, and D. Philbert. "RU 486." *Scientific American*, June 1990, pp. 18–24.

Chapter 15

Begley, S. "A Question of Breeding." *National Wildlife*, February-March 1991, pp. 13–16.

Cavalli-Sforza, L., L. P. Menozzi, and A. Piazza. "Demic Expansions and Human Evolution." *Science* 259 (1993): 639–646.

Cook, L. M., G. S. Mani, and M. E. Varley. "Postindustrial Melanism in the Peppered Moth." *Science* 231 (1986): 611–613.

Diamond, J. "Founding Fathers and Mothers." *Natural History*, June 1988, pp. 10–15.

———. "The Cruel Logic of Our Genes." *Discover*, November 1989, pp. 72–78.

Futuyma, D. J. *Evolutionary Biology.* 2d ed. Sunderland, Mass.: Sinauer, 1986.
Grant, P. "Natural Selection and Darwin's Finches." *Scientific American*, October 1991, pp. 82–87.
Koehn, R. K., and T. J. Hilbish. "The Adaptive Importance of Genetic Variation." *American Scientist* 75 (1987): 134–141.
O'Brien, S. J., D. E. Wildt, and M. Bush. "The Cheetah in Genetic Peril." *Scientific American*, May 1986, pp. 84–95.
Stebbins, G. L., and F. J. Ayala. "The Evolution of Darwinism." *Scientific American*, July 1985, pp. 72–82.
Steinhart, P. "In the Blood of Cheetahs." *Audubon*, March-April 1992, pp. 40–46.

Chapter 16

Diamond, J. "The Cruel Logic of Our Genes." *Discover*, November 1989, pp. 72–78.
Koehn, R. K., and T. J. Hilbish. "The Adaptive Importance of Genetic Variation." *American Scientist* 75 (1987): 134–141.
Lewin, R. "Molecular Clocks Run Out of Time." *New Scientist*, February 10, 1990, pp. 38–41.
Slatkin, M. "Gene Flow and the Geographic Structure of Populations." *Science* 236 (1987): 787–792.
Wilson, A. C. "The Molecular Basis of Evolution." *Scientific American*, October 1985, pp. 164–173.

Chapter 17

de Duve, C. *Blueprint for a Cell: The Nature and Origin of Life.* Burlington, N.C.: Patterson, 1990.
Gibbons, A. "Biologists Trace the Evolution of Molecules." *Science* 257 (1992): 30–31.
Hildebrand, A. R., and W. V. Boynton. "Cretaceous Ground Zero." *Natural History*, June 1991, pp. 47–53.
Horgan, J. "In the Beginning." *Scientific American*, February 1991, pp. 117–125.
Keller, E. F. "One Woman and Her Theory." *New Scientist* 3 (1986): 46–52.
Levinton, J. S. "The Big Bang of Animal Evolution." *Scientific American*, November 1992, pp. 84–91.
Margulis, L. *Early Life.* Boston: Jones and Bartlett, 1982.
Radetsky, P. "How Did Life Start?" *Discover*, November 1992, pp. 74–83.

Chapter 18

Margulis, L., and R. Guerrero. "Kingdoms in Turmoil." *New Scientist*, March 23, 1991, pp. 46–50.
May, R. M. "How Many Species Inhabit the Earth?" *Scientific American*, October 1992, pp. 42–48.
Raup, D. "Extinction: Bad Genes or Bad Luck?" *New Scientist*, September 14, 1991, pp. 46–49.
Ridley, M. "Evolution and Classification." London: Longman, 1986.
Whittaker, R. H. "New Concepts of Kingdoms of Organisms." *Science* 163 (1969): 150–160.
Woese, C. R., O. Kandler, and M. L. Wheelis. "Towards a Natural System of Organisms: Proposal for the Domains Archaea, Bacteria, and Eucarya." *Proceedings of the National Academy of Sciences USA* 87 (1990): 4576–4579.

Chapter 19

Brock, T. D. *The Biology of Microorganisms.* 5th ed. Englewood Cliffs, N.J.: Prentice Hall, 1988.
Corliss, J. "The Kingdom Protista and Its 45 Phyla." *Biosystems* 17 (1984): 87–126.
Margulis, L., and K. V. Schwartz. *Five Kingdoms: An Illustrated Guide to the Phyla of Life on Earth.* New York: W. H. Freeman, 1988.
Margulis, L., J. O. Corliss, M. Melkonian, and D. J. Chapman. *Handbook of Protoctista.* Boston: Jones and Bartlett, 1990.
Oldstone, M. B. A. "Viral Alteration of Cell Function." *Scientific American*, August 1989, pp. 42–48.
Pelczar, M. J., E. C. S. Chan, N. R. Krieg. *Microbiology: Concepts and Applications.* McGraw-Hill, 1993.
Postgate, J. "The Malleable Microbe." *New Scientist*, February 16, 1991, pp. 38–41.
———. "Microbial Happy Families." *New Scientist*, January 21, 1989, pp. 40–44.
Sleigh, M. *Protozoa and Other Protists.* New York: Routledge, 1989.

Chapter 20

Ashby, A., and P. Dyer. "The Secret Sex Lives of Fungi." *New Scientist*, March 7, 1992, pp. 32–35.
Chapman, A. R. O. *Functional Diversity of Plants in the Sea and on Land.* Boston: Jones and Bartlett, 1987.
Gensel, P. G., and H. N. Andrews. "The Evolution of Early Land Plants." *American Scientist* 75 (1987): 478–489.
Heron, C. "The Networks of Botanical Creation." *New Scientist*, February 8, 1992, pp. 40–44.
Raven, P. H., R. F. Evert, and S. E. Eichhorn. 5th ed. *Biology of Plants.* New York: Worth, 1992.
Ross, I. K. *Biology of the Fungi: Their Development, Regulation, and Associations.* New York: McGraw-Hill, 1979.
Roy, B. A. "Floral Mimicry by a Plant Pathogen." *Nature* 362 (1993): 56–58.

Chapter 21

Brusca, R. C., and G. J. Brusca. *Invertebrates.* Sunderland, Mass.: Sinauer, 1990.
Gould, S. J. *Wonderful Life: The Burgess Shale and the Nature of History.* New York: Norton, 1989.
Griswold, C. E., and T. C. Meikle. "Social Life in a Web." *Natural History*, March 1990, pp. 7–10.
McMenamin, M. A. S. "The Emergence of Animals." *Scientific American*, April 1987, pp. 94–102.
Morris, S. C. "Palaeontology's Hidden Agenda." *New Scientist*, August 11, 1990, pp. 38–42.
———. "The Fossil Record and the Early Evolution of the Metazoa." *Nature* 361 (1993): 219–225.
Stolzenburg, W. "When Life Got Hard." *Science News* 138 (1990): 120–123.

Chapter 22

Verrell, Paul. "Why Small Is Sometimes Sexy." *New Scientist*, September 7, 1991, pp. 44–47.
Wellnhofer, P. "Archaeopteryx." *Scientific American*, May 1990, pp. 42–49.
Yoffe, Emily. "Silence of the Frogs." *New York Times Magazine*, December 13, 1992, pp. 36–38.
Young, J. Z. *The Life of Vertebrates.* 3d ed. New York: Oxford University Press, 1981.

Chapter 23

Diamond, J. "Dawn of the Human Race: The Great Leap Forward." *Discover*, May 1989, pp. 50–60.
———. *The Third Chimpanzee.* New York: Harper Collins, 1992.
Folger, Tim. "The Naked and the Bipedal." *Discover*, November 1993, pp. 34–45.

Gibbons, A. "Our Chimp Cousins Get That Much Closer." *Science* 250 (1990): 376.

King, F., C. Yarbrough, D. Anderson, T. Gordon, and K. Gould. "Primates." *Science* 240 (1900): 1475–1482.

Klein, R. G. *The Human Career: Human Biological and Cultural Origins.* Chicago: University of Chicago Press, 1989.

Leakey, Mary D., and J. M. Harris (eds.). *Laetoli: A Pliocene Site in Northern Tanzania.* Oxford, U.K.: Clarendon, 1987.

Leakey, R., and R. Lewin. *Origins Reconsidered: In Search of What Makes Us Human.* New York: Little, Brown, 1992.

Lewin, R. *The Origin of Modern Humans.* New York: W. H. Freeman, 1993.

Milton, K. "Diet and Primate Evolution." *Scientific American,* August 1993, pp. 86–93.

Simons, E. L. "Human Origins." *Science* 245 (1989): 1343–1357.

Stringer, C. B., and P. Andrews. "Genetic and Fossil Evidence for the Origin of Modern Humans." *Science* 239 (1988):1263–1268.

Tattersal, I. "Madagascar's Lemurs." *Scientific American,* January 1993. pp. 110–117.

Tuttle, Russell H. "The Pitted Pattern of Laetoli Feet." *Natural History,* March 1990, pp. 61–65.

Thorne, A. G., and M. H. Wolpoff. "The Multiregional Evolution of Humans." *Scientific American,* April 1992, pp. 76–83.

Wilson, A. C., and R. L. Cann. "The Recent African Genesis of Humans." *Scientific American,* April 1992, pp. 68–73.

Chapter 24

Anger, R. L., and J. Vacanti, "Tissue Engineering." *Science* 260 (1993): 920–926.

Goldberg, Alan M., and John M. Frazier. "Alternatives to Animals in Toxicity Testing." *Scientific American,* August 1989, pp. 24–30.

Kiester, E. "A Little Fever Is Good for You." *Science* 84 (1984): 168–173.

McInerney, Joseph D. "Animals in Education: Are We Prisoners of False Sentiment?" *The American Biology Teacher* 55 (1993).

Minasian, S. M., K. C. Balcomb III, and L. Foster. *The World's Whales: The Complete Illustrated Guide.* Washington, D.C.: Smithsonian Books, 1979.

Schmidt-Nielsen, K. *Animal Physiology: Adaptation and Environment.* 3d ed. New York: Cambridge University Press, 1983.

Chapter 25

American Heart Association. *Heart and Stroke Facts 1992.* Dallas, Texas, pp. 1–40.

Brown, M. S., and J. L. Goldstein. "How LDL Receptors Influence Cholesterol and Atherosclerosis." *Scientific American,* November 1984, pp. 58–67.

Lillywhite, H. B. "Snakes, Blood Circulation and Gravity." *Scientific American,* December 1988, pp. 92–98.

Robinson, T. F., S. M. Factor, and E. H. Sonnenblick. "The Heart As a Suction Pump." *Scientific American,* June 1986, pp. 84–91.

Van Wynsberghe, D., C. R. Noback, and R. Carola. *Human Anatomy and Physiology.* 3d ed. New York: McGraw-Hill, 1995.

Chapter 26

von Boehmer, H., and P. Kisielow. "How the Immune System Learns about Self." *Scientific American,* October 1991, pp. 74–81.

Boon, T. "Teaching the Immune System to Fight Cancer." *Scientific American,* March 1993, pp. 82–91.

Brown, P. "AIDS, The Challenge of the Future." *New Scientist,* April 18, 1992, unnumbered centerfold.

Clayton, J. "Confusion in the Joints." *New Scientist,* May 4, 1991, pp. 40–43.

Gallo, R. C., and L. Montagnier. "AIDS." *Scientific American,* October 1988, pp. 47–48.

Johnston, M. I., and D. F. Hoth. "Present Status and Future Prospects for HIV Therapies." *Science* 260 (1993): 1286–1293.

Kedzierski, M. "Vaccines and Immunization." *New Scientist,* February 8, 1992, unnumbered centerfold.

Lanzavecchia, A. "Identifying Strategies for Immune Intervention." *Science* 260 (1993): 937–944.

Montgomery, G. "The Human Mouse." *Discover,* August 1989, pp. 49–55.

Rennie, J. "The Body against Itself." *Scientific American,* December 1990, pp. 106–115.

Rosenberg, Z., and A. Fauci. "Inside the AIDS Virus." *New Scientist,* February 10, 1990, pp. 51–54.

Smith, K. A. "Interleukin-2." *Scientific American,* March 1990, pp. 50–57.

Wechsler, R. "Hostile Womb." *Discover,* March 1988, pp. 83–87.

Weiss, R. A. "How Does HIV Cause AIDS?" *Science* 260 (1993): 1273–1278.

Chapter 27

Eckert, R., D. Randall, and G. Augustine. *Animal Physiology: Mechanisms and Adaptations.* 3d ed. New York: Freeman, 1988.

Houston, C. S. "Mountain Sickness." *Scientific American,* October 1992, pp. 58–66.

McRae, M. "Altitude Ate My Brain." *Outside,* May 1990, p. 24.

Read, R., and C. Read. "Breathing Can Be Hazardous to Your Health." *New Scientist,* February 23, 1991, pp. 34–37.

Van Wynsberghe, D., C. R. Noback, and R. Carola. *Human Anatomy and Physiology,* 3d ed. New York: McGraw-Hill, 1995.

Chapter 28

Bailey, Covert. *The New Fit or Fat.* Boston: Houghton Mifflin, 1991.

Campbell-Platt, G. "The Food We Eat." *New Scientist,* May 1988, pp. 1–4.

Eckert, R., D. Randall, and G. Augustine. *Animal Physiology: Mechanisms and Adaptations.* 3d ed. New York: Freeman, 1988.

Grajal, A., S. Strahl, R. Parra, M. Diminguez, and A. Neher. "Foregut Fermentation in the Hoatzin, a Neotropical Leaf-Eating Bird." *Science* 245: 1236–1238, 1989.

Guyton, A. C. *Textbook of Medical Physiology.* 8th ed. W. B. Saunders, Philadelphia, 1991.

Ratto, T. "The New Science of Weight Control." *Medical Self-Care,* March-April 1987, pp. 25–30.

Van Wynsberghe, D., C. R. Noback, and R. Carola. *Human Anatomy and Physiology,* 3d ed. New York: McGraw-Hill, 1995.

Chapter 29

Eckert, R., D. Randall, and G. Augustine. *Animal Physiology: Mechanisms and Adaptations.* 3d ed. New York: Freeman, 1988.

Guyton, A. C. *Textbook of Medical Physiology.* 8th ed. Philadelphia: Saunders. 1993.

Schmidt-Nielson, K. *Desert Animals: Physiological Problems of Heat and Water.* New York: Oxford University Press, 1964, pp. 149–178.

Simmons, James C. "The Arabian Desert Is No Place for Camels." *Audubon,* January 1991, pp. 38–45.

Vander, A. J. *Renal Physiology.* 3d ed. New York: McGraw-Hill, 1985.

Van Wynsberghe, D., C. R. Noback, and R. Carola. *Human Anatomy and Physiology,* 3d ed. New York: McGraw-Hill, 1995.

Yagil, R. *The Desert Camel.* Basel: Karger, 1985.

Chapter 30

Atkinson, M., and N. Maclaren. "What Causes Diabetes?" *Scientific American*, July 1990, pp. 62–71.

Baumann, Gerhard, et al. "Low Levels of High-Affinity Growth Hormone-Binding Protein in African Pygmies." *New England Journal of Medicine* 320 (1989): 1705–1708.

Gilbert, L. I., W. L. Combest, W. A. Smith, V. H. Meller, and D. B. Rountree. "Neuropeptides, Second Messengers, and Insect Molting." *BioEssays* 8 (1988): 153–157.

Guillemin, R., and R. Burgus. "The Hormones of the Hypothalamus." *Scientific American*, November 1972, pp. 24–33.

Snyder, S. H. "The Molecular Basis of Communication between Cells." *Scientific American*, October 1985, pp. 132–141.

Yalcinkaya, Tamer M., et al. "A Mechanism for Virilization of Female Spotted Hyenas in Utero." *Science* 260 (1993): 1929.

Yesalis, C. E. *Anabolic Steroids in Sport and Exercise.* Champaign, Ill.: Human Kinetics Publications, 1993.

Chapter 31

Bozarth, M. "New Perspectives on Cocaine Addiction: Recent Findings from Animal Research." *Canadian Journal of Physiological Pharmacology* 67 (1989): 1158–1167.

Dnyder, S. H., and D. S. Bredt "Biological Roles of Nitric Oxide." *Scientific American*, May 1992, pp. 68–77.

Ezzell, C. "Adult Neurons: Not Too Old to Divide." *Science News* 141 (1992): 212.

Freud, Curt R., et al. "Survival of Implanted Fetal Dopamine Cells of Neurologic Importance 12 to 46 Months after Transplantation for Parkinson's Disease." *New England Journal of Medicine.* November 22, 1992, pp. 1549–1555.

Goelet, P., V. F. Castellucci, S. Schacher, and E. R. Kandel. "The Long and the Short of Long-Term Memory—A Molecular Framework." *Nature* 322 (1986): 419–422.

Kandel, E., J. Schwartz, and T. M. Jessell. *Principles of Neuroscience.* 3d ed. New York: Elsevier, 1991.

Weiss, Rick. "Promising Protein for Parkinson's." *Science* 260 (1993): pp. 1072–1073.

Chapter 32

Brown, P. "Rescuing Minds from Disease and Decay." *New Scientist*, November 14, 1992, supplement pp. 1–4.

Damasio, A. R., and H. Damasio. "Brain and Language." *Scientific American*, September 1992, pp. 89–95.

Gershon, E., and R. Rieder. "Major Disorders of Mind and Brain." *Scientific American*, September 1992, pp. 127–133.

Hooper, Celia. "Biology, Brain Architecture, and Human Sexuality." *The Journal of NIH Research*, October 1992, pp. 53–59.

Kandel, E., J. Schwartz, and T. M. Jessell. *Principles of Neural Science.* 3d ed. New York: Elsevier, 1991.

Kandel, E. R., and R. D. Hawkins. "The Biological Basis of Learning and Individuality." *Scientific American*, September 1992, pp. 79–86.

Kimura, Doreen. "Sex Differences in the Brain." *Scientific American*, September 1992, pp. 119–125.

Konishi, M. "Listening with Two Ears." *Scientific American*, April 1993, pp. 66–73.

Sacks, O. *The Man Who Mistook His Wife for a Hat.* London: Duckworth, 1985.

Selkoe, D. J. "Aging Brain, Aging Mind." *Scientific American*, September 1992, pp. 135–142.

Van Wynsberghe, D., C. R. Noback, and R. Carola. *Human Anatomy and Physiology*, 3d ed. New York: McGraw-Hill, 1995.

Chapter 33

Alberts, B., D. Bray, J. Lewis, M. Raff, K. Roberts, and J. D. Watson. *Molecular Biology of the Cell.* 2d ed. New York: Garland, 1989.

Åstrand, P. O., and K. Rodahl. *Textbook of Work Physiology: Physiological Bases of Exercise.* 3d ed. New York: McGraw-Hill, 1986.

Fishman, Steve "Survival of the Fittest," Health, May-June 1993, pp. 59–64.

Goldspink, G. "The Brains behind the Brawn." *New Scientist*, August 1, 1992, pp. 28–33.

Hildebrand, M. "Vertebrate Locomotion: An Introduction." *BioScience* 39 (1989): 764–765.

Newsholme, E., and T. Leech. "Fatigue Stops Play." *New Scientist*, September 22, 1988, pp. 39–43.

Chapter 34

Darley, W. M. "The Essence of 'Plantness.'" *The American Biology Teacher* 52 (1990): 354–357.

Moffett, M. "These Plants Scratch, Claw and Strangle their Way to the Top." *Smithsonian* (1993): 110–119.

Raven, P. H., R. F. Evert, and S. E. Eichhorn. *Biology of Plants.* 5th ed. New York: Worth, 1992.

Chapter 35

Chappin, F. S., III. "Integrated Responses of Plants to Stress." *BioScience* 41 (1991): 29–36.

Hitz, W. D., and R. T. Giaquinta. "Sucrose Transport in Plants." *BioEssays* 6 (1987): 217–221.

Humphrey, R. "A Tree Grows in Baja." *Natural History*, June 1991, pp. 54–59.

Raven, P. H., R. F. Evert, and H. Curtis. *Biology of Plants.* 5th ed. New York: Worth, 1992.

Stamp, Nancy E. "Efficacy of Explosive vs. Hygroscopic Seed Dispersal by an Annual Grassland Species." *American Journal of Botany* 76 (1989): 555–561.

———. "Self-Burial Behavior of *Erodium Cicutarium* Seeds." *Journal of Ecology* 72 (1984): 611–620.

Zimmermann, M. "Piping Water to the Treetops." *Natural History* 91 (1982): 6–13.

Chapter 37

Evans, M. L., R. Moore, and K. Hasenstein. "How Roots Respond to Gravity." *Scientific American*, December 1986, pp. 112–119.

Raven, P. H., R. F. Evert, and H. Curtis. *Biology of Plants.* 5th ed. New York: Worth, 1992.

Rosenthal, G. A. "Chemical Defenses of Higher Plants." *Scientific American*, January 1986, p. 96.

Theologis, A. "One Rotten Apple Spoils the Whole Bushel: The Role of Ethylene in Fruit Ripening." *Cell* 70 (1992): 181–184.

Chapter 38

Begon, M., J. L. Harper, and C. R. Townsend. *Ecology: Individuals, Populations, and Communities.* 2d ed. Boston: Blackwell, 1990.

Caldwell, J. C., and P. Caldwell. "High Fertility in Sub-Saharan Africa." *Scientific American*, May 1990, pp. 118–125.

Cohen, J. E. "How Many People Can Earth Hold?" *Discover*, November 1992, pp. 114–119.

Ehrlich, P. R. and A. H. Ehrlich. *The Population Explosion.* New York: Simon & Schuster, 1990.

Fish, S. K., and P. R. Fish. "Prehistoric Landscapes of the Sonoran Desert Hohokam." *Population and Environment* 13 (1992): 269–284.

Haury, E. W. "The Hohokam: First Masters of the American Desert." *National Geographic*, May 1967, pp. 670–701.

Keyfitz, N. "The Growing Human Population." *Scientific American*, September 1989, pp. 119–126.

Olshansky, S., B. Carnes, and C. Cassel "The Aging of the Human Species." *Scientific American*, April 1993, pp. 46–52.

Raven, P. H., and E. O. Wilson. *Science* 258 (1992): 1099–1100.

Stone, Richard. "Babbitt Shakes Up Science at Interior." *Science* 261 (1993): 976–978.

Yoon, Carol Kausuk. "Counting Creatures Great and Small." *Science* 260 (1993): 620–622.

Chapter 39

Beardsley, T. "Recovery Drill: An Old Volcano Teaches Ecologists Some New Tricks." *Scientific American*, November 1990, p. 34.

Begon, M., J. L. Harper, and C. R. Townsend. *Ecology: Individuals, Populations, and Communities*. 2d ed. Boston: Blackwell, 1990.

Fullick, A., and P. Fullick. "Biological Pest Control." *New Scientist*, March 9, 1991, pp. 1–4.

Krebs, C. J. *Ecology: The Experimental Analysis of Distribution and Abundance*. 3d ed. New York: Harper & Row, 1985.

Line, L. "Silence of the Songbirds." *National Geographic*, June 1993, pp. 68–90.

Romme, W. H., and D. G. Despain. "The Yellowstone Fires," *Scientific American*, November 1989, pp. 37–46.

Terborgh, John. "Why American Songbirds Are Vanishing." *Scientific American*, May 1992, pp. 98–104.

Chapter 40

Appenzeller, T. "What Drives Climate?" *Discover*, November 1992, pp. 67–73.

Derr, Mark. "Redeeming the Everglades." *Audubon*, September-October, 1993, pp. 48–56.

Gosz, J. R., R. T. Holmes, G. E. Likens, and F. H. Bormann. "The Flow of Energy in a Forest Ecosystem." *Scientific American*, March 1978, pp. 93–102.

Graedel, T. E., and P. J. Crutzen. "The Changing Atmosphere." *Scientific American*, September 1989, pp. 58–68.

Jones, P., and T. Wigley. "Global Warming Trends." *Scientific American*, August 1990, pp. 84–91.

Kushlan, J. A. "The Everglades." In *The Rivers of Florida*, Robert J. Livingston (ed). New York: Springer Verlag, 1991.

Matthews, S., and J. Sugar. "Is Our World Warming? Under the Sun." *National Geographic*, October 1990, pp. 66–100.

Nebel, B. J. *Environmental Science: The Way the World Works*. 2d ed. Englewood Cliffs, N.J.: Prentice-Hall, 1987.

Odum, E. P. *Ecology and Our Endangered Life-Support Systems*. Sunderland, Mass.: Sinauer, 1989.

Chapter 41

Brown, B. E., and J. C. Ogden "Coral Bleaching." *Scientific American*, January 1993, pp. 64–70.

Cloud, P. "The Biosphere." *Scientific American*, September 1983, pp. 176–189.

Diamond, J. M. "Human Use of World Resources." *Nature* 328 (1987): 479–480.

Furley, P. A., W. W. Newey, R. P. Kirby, and J. M. Hotson. *Geography of the Biosphere: An Introduction to the Nature, Distribution, and Evolution of the World's Life Zones*. London: Butterworth, 1983.

Holloway, M. "Sustaining the Amazon." *Scientific American*, July 1993, pp. 90–99.

Luoma, J. R. "An Untidy Wonder." *Discover*, October, 1992, pp. 86–95.

Nelson, Mark, et al. "Using a Closed Ecological System to Study Earth's Biosphere." *Bioscience*, 43 (1993): 225–236.

Odum, E. P. *Ecology and Our Endangered Life-Support Systems*. Sunderland, Mass.: Sinauer, 1989.

O'Neill, T. "New Sensors Eyes the Rain Forest." *National Geographic*, September 1993, pp. 118–130.

Reganold, J. P., R. I. Papendick, and J. F. Parr. "Sustainable Agriculture." *Scientific American*, June 1990, pp. 112–120.

Repetto, R. "Deforestation in the Tropics." *Scientific American*, April 1990, pp. 36–42.

Ruckelshaus, W. D. "Toward a Sustainable World." *Scientific American*, September 1989, pp. 166–174.

Svitil, K. "Holey War." *Discover*, January 1993, pp. 75–76.

Watson, Traci. "Can Basic Research Ever Find a Good Home in Biosphere II?" *Science* 259 (1993): 1688–1689.

Wilson, E. O. "Threats to Biodiversity." *Scientific American*, September 1989, pp. 108–116.

Chapter 42

Day, S. "Migration." *New Scientist* 12 (1992): centerfold pages.

FitzGerald, G. J. "The Reproductive Behavior of the Stickleback." *Scientific American* April 1993, pp. 80–84.

Krebs, J. R., and N. B. Davies. *An Introduction to Behavioral Ecology*. 3d ed. Cambridge, Mass.: Blackwell, 1993.

Sherman, Paul W., Jennifer U. M. Jarvis, and Stanton H. Braude. "Naked Mole Rats," *Scientific American*, August 1992, p. 78.

Tinbergen, M. "The Shell Menace." *Natural History*, August-September 1963, pp. 28–35.

Credits

Part Openers

Part One: M. Abbey/Photo Researchers. Part Two: Frans Lanting/Minden Pictures. Part Three: Whit Bronaugh. Part Four: Larry Lipsky/DRK Photo. Part Five: Brock May/Photo Researchers. Part Six: Art Wolfe.

Chapter 1

Figure 1.1: Scott Camazine/Photo Researchers. Figure 1.2: (a) Eric Schabtach, University of Oregon; (b) Syd Greenberg/Photo Researchers. Figure 1.3: Michael Fairchild/Peter Arnold. Figure 1.4: Michael Fogden/Oxford Scientific Films/Animals Animals. Figure 1.5: S. N. Postelthwait. Figure 1.6: Art Wolfe. Figure 1.7: Breck P. Kent/Animals Animals/Earth Scenes. Figure 1.8: S. Maslowski/Visuals Unlimited. Figure 1.9: (a) Photo Researchers; (b) Oxford Scientific Films/Animals Animals; (c) Zig Leszczynski/Animals Animals. Figure 1.10: Dr. A. Lesk/Photo Researchers. Figure 1.11: Al Grotell. Box 1.1: Bettmann. Figure 1.12: (a) CNRI/SPL/Photo Researchers; (b) Alexander Zehnder, Agricultural University, The Netherlands; (c) M. Abbey/Visuals Unlimited; (d) E. R. Degginger/Animals Animals/Earth Scenes; (e) John Postlethwait; (f) Art Wolfe. Figure 1.13: (a) Townsend P. Dickinson/Comstock; (b) John R. MacGregor/Peter Arnold. Figure 1.15: Runk/Schoenberger/Grant Heilman. Figure 1.16: Oxford Scientific Films/Animals Animals. Figure 1.17: (a) C. W. Myers/American Museum of Natural History; (b) C. W. Myers/American Museum of Natural History. Figure 1.19: (a) Juan M. Renjifo/Animals Animals; (b) Zig Leszczynski/Animals Animals; (c) John H. Hoffman/Bruce Coleman. Figure 1.20: Frans Lanting/Minden Pictures.

Chapter 2

Figure 2.1: S. Leonard, Department of Fisheries & Oceans, Freshwater Institute, Canada. Figure 2.2: (a) Fritz Pölking/Peter Arnold; (b) Larry Ulrich/DRK Photo. Figure 2.4: CNRI/SPL/Photo Researchers. Figure 2.5: Defense Nuclear Agency Photograph. Box 2.1, Figure 1: Hinterleither/Gamma Liaison. Figure 2.7: (a) Richard Hutchings/Photo Researchers. Figure 2.10: Lara Hartley. Figure 2.11: (a) Eric Gravé/Photo Researchers. Figure 2.12: Gary J. Benson/Comstock. Figure 2.13: Klaus Uhlenhut/Animals Animals/Earth Scenes. Figure 2.14: Susan McCartney/Photo Researchers. Figure 2.15: G. L. Kooyman/Animals Animals. Figure 2.18: (a) John Shaw/Natural History Photographic Agency; (b, c) Gary Randay/Visuals Unlimited. Figure 2.25: George D. Lepp/Comstock. Figure 2.27: (b) Warren Uzzle/Discover Magazine. Figure 2.29: (c) Arthur J. Olson, The Scripps Research Institute. Box 2.3, Figure 1: Stuart L. Craig/Bruce Coleman. Figure 2.30: both, Collagen Biomedical. Figure 2.31: Arthur J. Olson, The Scripps Research Institute.

Chapter 3

Figure 3.1: Robert D. Terry, M.D., University of San Diego School of Medicine. Figure 3.2: (a) The Granger Collection; (b) Bettmann. Box 3.1, Figure 1: (b, c, d) © 1994 Jerry Gleason; (f) M. Giles/Fran Heyl Associates; (h) Sidney Tamm. Figure 3.4: Preston Photography. Figure 3.5: Richard Kessel, University of Iowa. Figure 3.6: (a) Photo Researchers; (b) W. L. Dentler, The University of Kansas/Biological Photo Service. Figure 3.8: Dr. Tony Brain and David Parker/SPL/Photo Researchers. Figure 3.9: Peter Saloutos/The Stock Market. Figure 3.10: M. I. Walker/Photo Researchers. Figure 3.11: Alfred Owczarzak/Biological Photo Service. Figure 3.13: (a) Eric Schabtach, University of Oregon. Figure 3.14: M. M. Perry and A. B. Golbert, Journal of Cell Science 39 (1979): 257-272; Company of Biologists Ltd. Figure 3.15: (a) Dr. Jeremy Burgess/Photo Researchers; (c) Biophoto Associates/Science Source/Photo Researchers. Figure 3.17: (a-c) Klaus Weber and Mary Osborn, Max-Planck Institut fur Biophysikalishe Chemie, Germany; (a, bottom) CBS/Visuals Unlimited; (b, bottom) Zig Leszczynski/Animals Animals; (c, bottom) M. Austerman/Animals Animals. Figure 3.18: (a, top; b, top, c, bottom; d, bottom) Courtesy NIH/J. Lippincott-Schwartz; (a, bottom; b, bottom; c, top) Don Fawcett and Daniel Friend/Science Source/Photo Researchers; (d, top) K. G. Murti/Visuals Unlimited. Figure 3.19: Gordon Gahan/Photo Researchers. Figure 3.20: (b) Dr. R. Porter/Photo Researchers. Figure 3.21: (c) Dr. Jeremy Burgess/Photo Researchers. Figure 3.22: (b, c) D. Patterson/Photo Researchers. Figure 3.23: (b) David M. Phillips/Visuals Unlimited. Figure 3.24: (a) Manfred Kage/Peter Arnold. Figure 3.27: Prof. P. Motta/Science Photo Library/Photo Researchers.

Chapter 4

Figure 4.1: Kim Taylor/Bruce Coleman. Figure 4.2: Courtesy Dr. Hugh N. Mozingo. Figure 4.2: Courtesy Dr. Hugh N. Mozingo. Figure 4.9: (b) Fred McConnaughey/Photo Researchers. Figure 4.11: (b) Biophoto Associates/Science Source/Photo Researchers. Figure 4.15: Arthur J. Olson, The Scripps Research Institute. Box 4.1, Figure 1: Lennart Nilsson, from *The Incredible Machine*, National Geographic Society. Figure 4.18: (a-c) Photographs courtesy of Michael Sheetz. Color maps: Kenneth Eward/BioGrafx. Figure 4.19: both, John Postlethwait.

Chapter 5

Page 120: Comstock. Figure 5.6: (b) Mary Kate Denny/Photo Edit. Figure 5.7: Craig Blouin/f/Stop Pictures. Figure 5.9: (b) Courtesy Daniel S. Friend. Box 5.1, Figure 1: (a, b) Nancy Kennaway, Oregon Health Sciences University. Figure 5.15: top, Bob Daemmrich/Stock, Boston; bottom, Mikkel Aaland.

Chapter 6

Figure 6.1: Dr. F. Bazzaz, Harvard University. Figure 6.3: M. and D. Littler/The Smithsonian Institution. Figure 6.5: Larry Lefever/Grant Heilman. Figure 6.6: (b) Dr. Jeremy Burgess/Grant Heilman. Box 6.1, Figure 1: Ralph Lewin/UCSD. Figure 6.12: (a) Grant Heilman/Grant Heilman; (b) Michael Sacca/Grant Heilman. Figure 6.13: Richard Strauss/The Smithsonian Institution.

Chapter 7

Figure 7.1: Marc Romanelli/The Image Bank. Figure 7.2: (a) Barry Runk and Stanley Schoenberger/Grant Heilman. Figure 7.3: (a-e) Einhard Schierenberg, University of Heidelberg. Figure 7.4: (a) Runk-Schoenberger/Grant Heilman; (b) W. E. Engler, courtesy of G. F. Bahr. Figure 7.5: (e) CNRI/SPL/Photo Researchers. Figure 7.9: M. Abbey/Photo Researchers. Figure 7.10: (c, d) Mark S. Landinsky and J. Richard McIntosh, University of Colorado. Figure 7.13: (c) Courtesy The R. W. Johnson Pharmaceutical Research Institute. Box 7.1, Figure 1: SIU/Visuals Unlimited. Figure 7.15: Mark McKenna/Stuart Kenter Associated. Figure 7.16: R. Myers/Visuals Unlimited. Box 7.2, Figure 1: Richard Hutchings/Photo Researchers. Figure 7.21: (c) Doug Vargas/The Image Works.

Chapter 8

Figure 8.1: Kjell B. Sandved/Photo Researchers. Figure 8.2: D. W. Martin and M. Y. Menzel. Figure 8.3: Bettmann. Figure 8.4: (a) Jane Grushow/Grant Heilman. Figure 8.5: M. K. Battacharyya, *Cell* 60 (1990). Figure 8.11: Jubinville Photography. Figure 8.16: Harry Howard, University of Oregon. Figure 8.21: (a) Hank Morgan/Rainbow. Figure 8.22: Rhoda Sidney/Monkmeyer Press. Figure 8.23: Fritz Prenzel/Animals Animals.

Chapter 9

Figure 9.1: R. Langridge and D. McCoy/Rainbow. Figure 9.2: Maclyn McCarty, Rockefeller University. Figure 9.3: Stanley N. Cohen/Science Photo Library/Photo Researchers. Figure 9.4: (b) K. Kleinschmidt et al., *Biochimica et Biophysica Acta* 61 (1962): 857. Figure 9.5:

Hank Morgan/Photo Researchers. Figure 9.6: C. W. Schwartz/Animals Animals. Figure 9.7: (a-c) Cold Spring Harbor Laboratory Archives. Figure 9.11: (e) Ulrich K. Laemmli, *Cell* 55 (1988): 937-944. Figure 9.12: Debra Hershkowitz. Box 9.1, Figure b: AP/Wide World. Figure 9.15: Sheldon Wolff and Judy Bodgcote, Laboratory of Radiology and Environmental Health, University of California, San Francisco.

Chapter 10

Figure 10.1: Kay Chernush. Figure 10.4: Jackie Lewin/Photo Researchers. Figure 10.10: Alexander Lowry/Photo Researchers. Figure 10.12: Elena Kiselava. Figure 10.16: (a) James E. Cleaver, Ph.D., and Mary C. Curry, M.D., University of California, San Francisco. Figure 10.18: (c) Brian Mathews, University of Oregon. Box 10.2, Figure a: Moredun Animal Health Ltd./Science Photo Library/Photo Researchers.

Chapter 11

Figure 11.1: Ted Horowitz/The Stock Market. Figure 11.2: (a, b) The Photo Works; (c) Debra Hershkowitz; (d) Bonnie Kamin/Comstock. Figure 11.5: J. D. Spillane. Reprinted from *The Metabolic Basis of Inherited Disease*, John B. Stanbury, ed., 2:1262. Figure 11.6: Bettmann. Figure 11.7: (a, b) Wide World Photos. Figure 11.8: CNRI/SPL/Photo Researchers. Figure 11.9: (b) *Science* 231 (1986): 920. © 1986 by the American Association for the Advancement of Science. Figure 11.10: (a) G. Turner, Prince of Wales Children's Hospital; (b) Christine J. Harrison, *Journal of Medical Genetics* 20 (1983): 280. Figure 11.11: (a) Mike and Moppet Reed/ Animals Animals. Figure 11.12: American Psychological Association. Box 11.1: John Postlethwait. From *Science* 250 (1990). © 1990 by the American Association for the Advancement of Science. Figure 11.15: Courtesy of J. R. Korenberg. Figure 11.19: Dr. Nancy Wexler.

Chapter 12

Figure 12.1: Susan Walsh/AP/Wide World. Figure 12.2: Professor Marilyn Houck, Department of Biological Sciences, Texas Tech. University, Lubbock, TX. From *Science*, Vol. 253, Page 1126. © 1991 by the American Association for the Advancement of Science. Figure 12.3: AP/Wide World. Figure 12.6: (d) Explorer/ Photo Researchers. Figure 12.7: Ted Spiegel. Figure 12.9: Photomicrograph by Yves Poirier, computer image produced by Darwin Dale. Figure 12.10: Courtesy Monsanto Company. Figure 12.11: (a, c) Steven A. Rosenberg, *Scientific American* (May 1990): 66. Figure 12.12: Thomas W. Sasek, Duke University.

Chapter 13

Figure 13.1: Marnie Halpern, University of Oregon. Figure 13.2: (a) Michael Fogden/DRK Photo; (b) Scott Camazine/Photo Researchers. Figure 13.3: (a) Maslowski Photo/Photo Researchers; (b) Johnston Atoll NWR/Biological Photo Service. Figure 13.4: (a) M. P. Kahl/Bruce Coleman; (b) Dr. G. Schatten/SPL/Photo Researchers. Figure 13.5: (b, c) Jorge Blaquier, *Fertility and Sterility* 47: 302. Figure 13.6: Dr. Gerald Schatten/SPL/Photo Researchers. Figure 13.7: (a-h) Don Kane, University of Oregon. Figure 13.9: John Gerhart, from S. D. Black and J. C. Gerhart, *Development Biology* 116 (1986): 228-240. Figure 13.10: (e) Charles Mayer/Photo Researchers. Figure 13.12: Petit Format/Nestle/ Photo Researchers. Figure 13.15: (e) N. K. Wessells, Stanford University. Figure 13.17-1: Cabisco/ Visuals Unlimited. Figure 13.17-2: CNRI/Phototake. Figure 13.20: (b) Photo Researchers. Figure 13.22: (a) Mark McKenna; (b) Julian Heath, Baylor College of Medicine, Children's Nutrition Research Center, Houston, Texas.

Chapter 14

Figure 14.1: Kenneth Eward/BioGarfx. Figure 14.5: (d) Manfred Kage/Peter Arnold; (e) Lennart Nillson, *Behold Man*, Little Brown and Company. Figure 14.7: Biophoto Associates/ Photo Researchers. Figure 14.11: (a) K. Shiota, M.D., Kyoto University Faculty of Medicine; (b) From Moore, K. L. Persaud, T.V.N.: *The Developing Human Clinically Oriented Embryology, 5th edition*, W. B. Saunders, Philadelphia, 1993; (c) Petit Format/Science Source/Photo Researchers; (d) John Giannicchi/Photo Researchers; (e) John Giannicchi/Photo Researchers. Figure 14.13: (b) Jonathan T. Wright/ Bruce Coleman. Figure 14.15: Eddie Adams/ AP/Wide World. Box 14.1, Figure 1: G. W. Willis, M.D./Biological Photo Service. Box 14.2, Figure 1: Hank Morgan/Rainbow. Page 363: Bill Santos.

Chapter 15

Figure 15.1: (a) S. Conway Morris, University of Cambridge. Figure 15.4: J. Period/Visuals Unlimited. Figure 15.6: Museum für Naturkunde, Berlin, Federal Republic of Germany. Figure 15.8: (b) Michael Fogden/Animals Animals. Figure 15.10: Lennart Nillson, *Behold Man*, Little Brown and Company. Figure 15.14: (a) C, & S. Pollitt/Natural History Photographic Agency; (b) Richard Allen Wood/Animals Animals. Box 15.1, Figure 1: Ralph Reinhold/ Animals Animals.

Chapter 16

Figure 16.1: Karl Ammann/Bruce Coleman. Figure 16.2: Yoav Levy/Phototake. Figure 16.5: Clockwise from top right: Farrell Grehan/Photo Researchers; Irven DeVore/Anthro-Photo; Victor Englebert/Photo Researchers; Henry Bagish/ Anthro-Photo; Mike Yamashita/Woodfin Camp & Associates; Victor Englebert/Photo Researchers; Emil Muench/Photo Researchers; George Holton/Photo Researchers. Figure 16.6: Courtesy of author and distal symphalrgism with associated brachydactyly, James R. Poush. Figure 16.9: Victor McKusick, Johns Hopkins University. Figure 16.10: Dieter & Mary Plage/ Bruce Coleman. Figure 16.11: Ron LeValley. Figure 16.12: (a) John Trager/Visuals Unlimited; (b) Norbert Wu/Peter Arnold. Figure 16.15: (a) Breck Kent/Animals Animals; (b) M. W. Tweedie/Photo Researchers. Figure 16.16: (b) from *Journal of Heredity* 5 (1914): No. 11. Figure 16.17: (b, c, d, e) J. A. L. Cooke/Animals Animals. Figure 16.19: (a) Edward Ross, California Academy of Sciences.

Chapter 17

Figure 17.1: Fred Bavendam/Peter Arnold. Figure 17.2: J. William Schopf, CSEOL, University of California, Los Angeles. Figure 17.4: Chelsey Bonestell. Figure 17.5: Galen Rowell/Peter Arnold. Figure 17.6: Roger Ressmeyer/Starlight. Figure 17.7: U.S. Naval Observatory/Science Source/Photo Researchers. Figure 17.9: Pieter Cullis and Michael Hope, University of British Columbia. From *Scientific American* 256 (1987): No. 1, p. 20. Figure 17.11: Photograph provided by K.O. Stetter. Figure 17.12: J. William Schopf, CSEOL, University of California, Los Angeles. Figure 17.13: J. William Schopf, CSEOL, University of California, Los Angeles. Figure 17.14: Stanley M. Awramik, University of California/Biological Photo Service. Figure 17.16: A. H. Knoll, Harvard University.

Chapter 18

Figure 18.1: Dave Watts/Tom Stack & Associates. Figure 18.2: (a) Robert Brons/Biological Photo Service; (b) Patti Murray/Animals Animals/Earth Scenes. Figure 18.10: Nigel Stork, Department of Entomology, Natural History Museum, London. Figure 18.12: Mark Moffett/ Minden Pictures.

Chapter 19

Figure 19.1: Michael Abby Photo/Photo Researchers. Figure 19.2: Ken Graham. Figure 19.4: Biological Photo Service. Color Map by Kenneth Eward/BioGrafx. Figure 19.5: Biological Photo Service. Figure 19.6: (a) Institut Pasteur/CNRI/Phototake; (b) John Cunningham/ Visuals Unlimited; (c) David M. Phillips/Visuals Unlimited; (d) Science Source/Photo Researchers; (e) Ralph Robinson/Visuals Unlimited. Figure 19.7: Alfred Pasieka/Science Photo Library/Photo Researchers. Figure 19.8: H. Vali, *Nature* 343 (1990): 213. © 1990 Macmillan Magazines Ltd. Figure 19.9: Gregory G. Dimijian, M.D./Photo Researchers. Figure 19.10: Paul W. Johnson/Biological Photo Service. Figure 19.11: David Scharf/Peter Arnold. Figure 19.12: Arthur J. Olson, The Scripps Research Institute. Figure 19.13: Richard Homla/Photo-Nats. Figure 19.15: (a) Jerome Paulin/Visuals

Unlimited; (b) John Cunningham/Visuals Unlimited. Figure 19.16: Eric Gravé/Photo Researchers. Figure 19.17: (a) Robert Gwadz, *Science* 247 (1990). © 1990 by the American Association for the Advancement of Science; (b) Jack Clark/Comstock. Figure 19.18: (a) Biophoto Associates/Science Source/Photo Researchers; (b) Manfred Kage/Peter Arnold. Figure 19.19: (a) Eric Gravé/Photo Researchers; (b) Carleton Ray/Photo Researchers. Figure 19.20: (b) Scott Camazine/Photo Researchers.

Chapter 20

Figure 20.1: Entheos. Figure 20.3: Ralph E. Williams, U.S. Forest Service, Boise Field Office. Figure 20.5: (a) D. H. Marx/Visuals Unlimited. Figure 20.6: (a) W. K. Fletcher/Photo Researchers. Figure 20.7: (a) W. H. Hodge/Peter Arnold; (b) G. H. Harrison/Grant Heilman. Figure 20.9: (a) Stephen J. Krasemann/DRK Photo. Figure 20.12: Tom McHugh/Photo Researchers. Figure 20.13: Richard Herman/The Wildlife Collection. Figure 20.14: (a) M. I. Walker/Photo Researchers; (b) Biophoto Associates/Photo Researchers. Figure 20.15: (a) S. N. Postlethwait. Figure 20.16: (b) Holt Studios International/Animals Animals/Earth Scenes; (c) J. N. A. Lott, McMaster University/Biological Photo Service. Figure 20.17: D. Edwards, K. L. Davies, and L. Axe, University of Wales College of Cardiff. Figure 20.18: Top, Farrell Grehan/Photo Researchers; bottom, Ed Reschke/Peter Arnold. Figure 20.19: (a) Patti Murray/Animals Animals/Earth Scenes; (b) J. C. Carton/Bruce Coleman. Figure 20.20: David L. Brown/Tom Stack & Associates. Figure 20.22: Comstock. Figure 20.23: (a) Rad Planck/Tom Stack & Associates; (b) D. Cavagnaro/DRK Photo; (c) Ralph Reinhold/Animals Animals. Figure 20.25: Tom Ulrich/Tony Stone Worldwide.

Chapter 21

Figure 21.1: Dave B. Fleetham/Tom Stack & Associates. Figure 21.5: (a) Nancy Sefton/Photo Researchers. Figure 21.6: (a) Tom Branch/Science Source/Photo Researchers; (b) S. Elems/Visuals Unlimited; (c) Steve Earley/Animals Animals; (d) Claudia Mills, University of Washington/Biological Photo Service. Figure 21.8: Joe Ghiold, New Scientists (June 2, 1990). Figure 21.10: (a) Michael Abbey/Photo Researchers. Figure 21.11: Armed Forces Institute of Pathology. Figure 21.12: T. E. Adams/Visuals Unlimited. Figure 21.14: Dianora Niccolini/Medichrome, Inc./The Stock Shop. Figure 21.16: (a) Paul Taylor/Oxford Scientific Films; (b) Stephen J. Krasemann/DRK Photo. Figure 21.17: (a) Tom McHugh/Steinhart Aquarium/Photo Researchers; (b) Douglas Faulkner/Photo Researchers. Figure 21.18: (a) Franklin Viola/Comstock; (b) Fred McConnaughey/Photo Researchers. Figure 21.21: William A. Shear, Department of Biology, Hameden, Sydney College. Figure 21.22: (a) K. E. Lucas/Biological Photo Service; (b) John Gerlach/Visuals Unlimited. Figure 21.24: Oxford Scientific Films/Animals Animals. Figure 21.25: David Ellis/Visuals Unlimited. Figure 21.26: Patrick Grace/Photo Researchers. Figure 21.27: Michael Fogen/DRK Photo. Figure 21.28: (a) Biophoto Associates/Science SourcePhoto Researchers; (b) W. J. Weber/Visuals Unlimited. Figure 21.29: (a) Chris Huss/The Wildlife Collection; (b) Tom McHugh/Photo Researchers. Box 21.1, Figure a: Charles Griswold, Department of Entomology, California Academy of Sciences. Figure 21.30: (b) George I. Bernard/Animals Animals.

Chapter 22

Figure 22.1: Peter David/Planet Earth Pictures. Figure 22.5: (c) Marc Chamberlain/Tony Stone Worldwide; (d) John Cunningham/Visuals Unlimited. Figure 22.7: (a) Marc Chamberlain/Tony Stone Worldwide. Figure 22.8: (a) Tom McHugh/Photo Researchers; (b) Gary Bell/The Wildlife Collection. Figure 22.10: (a) Zig Lesczynski/Animals Animals; (b) Tom McHugh/Steinhart Aquarium/Photo Researchers; (c) Juan M. Renjifo/Animals Animals. Figure 22.12: (b) M. Philip Kahl, Jr./DRK Photo; (c) Mark Hallett. Figure 22.13: (b) Frans Lanting/Photo Researchers; (c) John Ness/Animals Animals. Figure 22.14: (a) Robert A. Tyrrell. Box 22.1, Figure 1: Jeff Lepore/Photo Researchers. Figure 22.14: (b) Michael Midgley/Tony Stone Worldwide. Figure 22.16: (a) Tom McHugh/Photo Researchers; (b) Dr. J. Scott Actenbach, Department of Biology, University of New Mexico, Albuquerque; (c) Tim Davis/Photo Researchers; (d) Douglas Faulkner/Photo Researchers.

Chapter 23

Figure 23.1: (a) John Reader/Science Photo Library/Photo Researchers; (b) D. Finnin/C. Chesek/American Museum of Natural History. Figure 23.3: (a) David Haring; (b) Michael Dick/Animals Animals; (c) Tom J. Ulrich/Visuals Unlimited; (d) Francois Gohier/Photo Researchers. Figure 23.4: (a, c) K. & K. Ammann/Bruce Coleman; (b) Bildarchiv Okapia/Photo Researchers. Figure 23.5: K. & K. Ammann/Bruce Coleman. Figure 23.7: (a) Rod Williams/Bruce Coleman; (b) Bruce Coleman; (c) Debra Hershkowitz; (d) Michael Dick/Animals Animals. Figure 23.9: David L. Brill, 1985. Figure 23.11: (a) Anthropology Archives/Smithsonian Institution; (b) Musée de l'Homme.

Chapter 24

Figure 24.1: Tui de Roy/Oxford Scientific Films. Figure 24.5: (b) Ed Reschke/Peter Arnold. Figure 24.6: (a) Ed Reschke/Peter Arnold. Figure 24.7: (a) G. W. Willis/Biological Photo Service; (b) Manfred Kage/Peter Arnold.

Chapter 25

Figure 25.1: Harvey Lillywhite. Figure 25.2: Biophoto Associates/Photo Researchers. Figure 25.3: Dennis Kunkel/Phototake. Figure 25.7: Mike Severns/Tom Stack & Associates. Figure 25.8: (c) Alfred Pasieka/Bruce Coleman. Figure 25.10: (b) Martin M. Rotker. Box 25.1, Figure 1: Howard Sochurek/The Stock Market. Figure 25.13: (a) Janet L. Hopson; (b) Alan Carey/The Image Works. Figure 25.14: (c) David M. Phillips/Visuals Unlimited.

Chapter 26

Figure 26.1: NIBSC/SPL/Photo Researchers. Figure 26.7: (c) Ian Wilson, Research Institute of Scripps Clinic. From Stanfield et al., *Science* 248 (1990): 712-719. © 1990 by The American Association for the Advancement of Science. Figure 26.11: Lennart Nilsson/Boehringer Ingelheim International GmbH. Figure 26.15: James Stevenson/SPL/Photo Researchers.

Chapter 27

Figure 27.1: Jonathan T. Wright/Bruce Coleman. Figure 27.4: (a) Ed Robinson/Tom Stack & Associates; (b) Tom McHugh/Photo Researchers. Figure 27.6: (a) E. R. Degginger/Bruce Coleman. Figure 27.9: (a) Michael Gabridge/Visuals Unlimited; (b, c) Martin M. Rotker/Visuals Unlimited. Box 27.1, Figure 1: Jerry L. Ferrara/Photo Researchers.

Chapter 28

Figure 28.1: Schafer & Hill/Peter Arnold. Figure 28.3: Kenneth C. Balcomb/Earth Views. Figure 28.7: (a) Tim Rock/Animals Animals; (c) Kim Taylor/Bruce Coleman. Figure 28.10: (a) Roland Seitre/Peter Arnold. Figure 28.11: Both, John Postlethwait. Box 28.1, Figure 1: Manfred Kage/Peter Arnold.

Chapter 29

Figure 29.1: L. L. Rue III/Bruce Coleman. Figure 29.2: Debra Hershkowitz. Figure 29.4: Wayne Lankinen/Bruce Coleman. Figure 29.8: (e) Peter Andrews, George Washington University Medical Center. Figure 29.10: Heinz Kluetmeier/Sports Illustrated. Box 29.1, Figure 1: SIU/Peter Arnold.

Chapter 30

Figure 30.1: Irven De Vore/Anthro Photo. Figure 30.8: Elizabeth Crews/The Image Works. Figure 30.9: Lester Bergman & Associates. Figure 30.12: (a) William D. Griffin/Animals Animals; (b, c) Wallace Kirkland/Animals Animals. Figure 30.13: (a) Warren Stone/Visuals Unlimited; (b) McDonough/Focus On Sports.

Chapter 31

Figure 31.1: Brookhaven National Laboratory. Figure 31.2: Cabisco/Visuals Unlimited. Figure 31.7: (c) Carol Readhead et al., "Expression of a Myelin Basic Protein Gene in Transgenic Shiverer Mice: Correction of the Dismyelinating Phenotype," *Cell* 48 (1987): 703. Figure 31.9: (a) T. Reese/D. W. Fawcett/Visuals Unlimited. Figure 31.12: (a) Daniel W. Gotshall/Visuals Unlimited.

Chapter 32

Figure 32.1: Stephen Dalton/Animals Animals. Figure 32.2: Corey S. Goodman, U. of California, Berkeley, Dept. of Molecular Cell Biology, Division of Neurobiology. Figure 32.6: (b) Scott Camazine/Photo Researchers. Figure 32.8: British Museum of Natural History. Figure 32.9: Marcus E. Raichle, M.D./Washington University Medical School. Figure 32.11: (a, b) UMDNM/Visuals Unlimited; (c, d) Dr. D. Weinberger. Figure 32.13: Milton H. Tierney Jr./Visuals Unlimited. Figure 32.14: Both, Molly Webster/Discover Magazine. Figure 32.17: Lennart Nilsson, *Behold Man*, Little Brown and Company.

Chapter 33

Figure 33.1: Mark Tuschman. Figure 33.2: Roger Jackman/Oxford Scientific Films/Animals Animals. Figure 33.3: F. Collet/Photo Researchers. Figure 33.7: (b) Andreas Feinger/Life Magazine, © Time Warner; (c) Fred Hossler/Visuals Unlimited. Figure 33.8: (a) Dorothy Littell Greco/Stock, Boston; (b, c) Michael Klein/Peter Arnold; (d) Mark Tuschman. Figure 33.9: Susan Ragan/AP/Wide World. Figure 33.10: Dave Johnson/Bruce Coleman. Figure 33.11: Bob Gardner. Figure 33.12: Biology Media/Science Source/Photo Researchers. Figure 33.15: (a) David Madison/Duomo; (b) J. D. Macdougall, McMaster University, Hamilton Ontario, Canada; (c) Bonnie Kamin/Comstock. Figure 33.17: Craig K. Lorenz/Photo Researchers.

Chapter 34

Figure 34.1: Mark Moffett/Minden Pictures. Figure 34.5: (a) Barry Runk and Stanley Schoenberger/Grant Heilman; (b) John Cunningham/Visuals Unlimited; (c) George H. Harrison/Grant Heilman. Figure 34.6: Eric Schabtach, University of Oregon. Figure 34.7: George J. Wilder/Visuals Unlimited. Figure 34.8: (a) from *Three-Dimensional Structure of Wood*, B. A. Meylan and B. G. Butterfield, p. 65. Syracuse University Press, 1972; (b) Visuals Unlimited; (c) From *Biology, 3d ed.*, by H. Curtis (Worth Publishers, New York, 1979). Figure 34.10: (a) D. Cavagnaro/DRK Photo; (b) Michael Fredericks, Jr./Animals Animals/Earth Scenes; (c) Farrell Grehan/Photo Researchers. Figure 34.11: (a) Barry Runk and Stanley Schoenberger/Grant Heilman; (b) Dr. Richard Kessel, Biology Department, University of Iowa. Figure 34.12: (a, b) Harry Howard, University of Oregon; (e) Jim Solliday/Biological Photo Service. Figure 34.14: J. R. Waaland, University of Washington/Biological Photo Service. Figure 34.15: All, Ray F. Evert, University of Wisconsin, Madison. Figure 34.16: (d) John Cunningham/Visuals Unlimited. Figure 34.17: (a) Rod Planck/Tom Stack; (b) John Cunningham/Visuals Unlimited; (c) Breck P. Kent/Animals Animals/Earth Scenes. Figure 34.18: (a) Gary Braasch; (b) John Cunningham/Visuals Unlimited; (c) Tom Bean/DRK Photo. Figure 34.19: (a) Bruce Coleman; (b) John Shaw/Bruce Coleman. Figure 34.20: S. N. Postelthwait.

Chapter 35

Figure 35.1: David Burckhalter. Figure 35.4: Scott Camazine/Photo Researchers. Box 35.1: Gary Braasch. Figure 35.6: (a, c) Ray Simon/Photo Researchers. Figure 35.8: Eric Schabtach, University of Oregon. Figure 35.10: (d, e) G. Biddulph et al., *Plant Physiology* 33 (1958): 295. Figure 35.11: C. P. Vance/Visuals Unlimited. Figure 35.12: Jack Kelly Clark, University of California Statewide IPM Project. Figure 35.14: R. Ronacordi/Visuals Unlimited. Figure 35.15: (b) Patti Murray/Animals Animals/Earth Scenes. Page 789: (c, e) Steven A. Dewey, Utah State University.

Chapter 36

Figure 36.1: Peter Essick/Aurora. Figure 36.4: (a) E. Meyerowitz; (b) M. Running. Figure 36.5: Dr. Jeremy Burgess/SPL/Photo Researchers. Figure 36.8: (a) Darwin Dale/Photo Researchers; (b) Lex Hes; (c) Peter Arnold. Figure 36.9: (b) Harry Howard, University of Oregon. Figure 36.12: Clockwise from top left: Mike & Carol Werner/Comstock; John D. Cunningham/Visuals Unlimited; William J. Weber/Visuals Unlimited; Forest W. Buchanan/Visuals Unlimited; Bill Beatty/Visuals Unlimited. Box 36.1: Cary S. Wolinsky/Trillium Studio. Figure 36.13: (a) William J. Weber/Visuals Unlimited; (b) Breck P. Kent/Animals Animals/Earth Scenes. Figure 36.14: Monsanto Chemical Company. Figure 36.15: Lynn Riddiford/Monsanto Chemical Company.

Chapter 37

Figure 37.1: John Elk III/Bruce Coleman. Figure 37.2: International Rice Research Institute, Manila, Philippines. Figure 37.3: Dr. Sylvan H. Witter, Agricultural Experiment Station/Michigan State University. Figure 37.4: David Newman/Visuals Unlimited. Figure 37.5: (c) S. N. Postlethwait. Figure 37.7: Frank B. Salisbury. From F. B. Salisbury and Cleon W. Ross. 1978. *Plant Physiology, Second Edition.* Wadsworth Publishing Co., Belmont, California. Figure 37.8: (a) Mark McKenna/Stuart Kenter Associates. Figure 37.9: David Newman/Visuals Unlimited. Figure 37.10: Loren A. McIntyre. Figure 37.12: Jack Kelly Clark, courtesy University California Statewide IPM Project. Figure 37.13: Grant Heilman/Grant Heilman Photography. Figure 37.14: *Trends in Genetics* 3 (1987): No. 6. Figure 37.15: Ed Reschke/Peter Arnold. Figure 37.16: Ray Ellis/Photo Researchers.

Chapter 38

Figure 38.1: Arizona State Museum, University of Arizona. Figure 38.12: Albert Copley/Visuals Unlimited. Figure 38.13: W. E. Ruth/Bruce Coleman. Figure 38.14: (a) John Shaw/Bruce Coleman; (b) Stephen J. Krasemann/DRK Photo. Figure 38.15: (b) Victor Englebert/Photo Researchers; (c) Bill Liske/Photo Researchers; (d) J. Blaustein/Woodfin Camp & Associates.

Chapter 39

Figure 39.1: Richard P. Smith/Tom Stack & Associates. Figure 39.4: (a) Bob & Clara Calhoun/Bruce Coleman; (b) C. K. Lorenz/Photo Researchers. Figure 39.9: (a) M. Fogden/DRK Photo; (b, c) Eric Greene, *Science* 243 (1989). © 1989 by the American Association for the Advancement of Science; (d) Carl Roessler/Animals Animals. Figure 39.10: (a) Susan Leaviness/Photo Researchers; (b) Thomas Eisner and D. Aneshansley. Figure 39.12: (a) William J. Weber/Visuals Unlimited; (b) John D. Cunningham/Visuals Unlimited. Figure 39.13: Alexander Klotz, from Peter Farb, *The Insects* (Life Nature Library, Time, 1962). Figure 39.14: Steve Maslowski. Figure 39.15: (a) Stephen J. Krasemann/Photo Researchers; (b) Gerald & Buff Corsi/Tom Stack & Associates. Figure 39.16: (a) Brooking Tatum/Visuals Unlimited; (b-inset) Gilbert Grant/Photo Researchers. Figure 39.17: (a) Bruce Berg/Visuals Unlimited; (b) Steve Coombs/Photo Researchers; (c) Kevin & Betty Collins/Visuals Unlimited. Figure 39.18: Scott R. Goldsmith. Figure 39.20: (a) David Harvey/Woodfin Camp; (b) George Stephens/National Oceanographic and Atmospheric Administration.

Chapter 40

Figure 40.1: Wayne Lankinen/Bruce Coleman. Figure 40.4: Robert R. Hessler/Scripps Institution of Oceanography. Figure 40.6: Robert Pierce/U.S. Department of Agriculture. Figure 40.15: Tom McHiugh/Photo Researchers. Figure 40.18: (b) Bettmann.

Chapter 41

Figure 41.1: John Shaw/Bruce Coleman. Figure 41.7: (b) Frans Lanting/Minden Pictures. Figure 41.8: Joe McDonald/Tom Stack & Associates. Figure 41.9: Michael Fogden/Bruce Coleman. Figure 41.10: Jim Brandenburg/Minden Pictures. Figure 41.11: Whit Bronaugh. Figure

41.12: James R. Fisher/Photo Researchers. Figure 41.13: Gary Braasch/Woodfin Camp & Associates. Figure 14.14: Kim Heacox/Peter Arnold. Figure 41.15: Fred Bruemmer/DRK Photo. Figure 41.17: John Lemker/Animals Animals/Earth Scenes. Figure 41.19: Joel Arrington/Visuals Unlimited. Figure 41.22: Randy Morse/Tom Stack & Associates. Figure 41.23: Dana Africa. Figure 41.24: (b) NASA. Box 41.1, Figure 1: Peter Menzel.

Chapter 42

Figure 42.1: E. & D. Hosking/Photo Researchers. Figure 42.3: W. A. Banaszewski/Visuals Unlimited. Figure 42.5: Louis Darling. Figure 42.6: Nina Leen/Life Magazine, © Time Warner. Figure 42.8: Bettmann. Figure 42.9: Susan Kuklin/Photo Researchers. Figure 42.10: Francois Gohier/Photo Researchers. Figure 42.12: Rod Morris. Figure 42.13: Janet Hopson. Figure 42.14: William E. Townsend, Jr./Photo Researchers. Figure 42.16: American Museum of Natural History. Figure 42.18: Tom Brakefield/Bruce Coleman. Figure 42.19: Francois Gohier/Photo Researchers. Figure 42.21: Tom Cajacob, Minnesota Zoo. Figure 42.22: Bob Sacha.

Index

Pages on which definitions or main discussions of topics appear are indicated by **boldface.**
Pages containing illustrations or tables are indicated by *italics.*

Metric–English and English–Metric Conversions

Prefixes Used with Metric Units				
Prefix	Symbol	Value	Equivalent	A Biological Example
Kilo-	k	1000, or 10^3	1 kilometer (km) = 1×10^3 m	6/10 of a mile
Centi-	c	1/100, or 10^{-2}	1 centimeter (cm) = 0.01 m	1/100 m (about the width of your little finger)
Milli-	m	1/1000, or 10^{-3}	1 millimeter (mm) = 0.001 m	1/1000 m (about the length of a fruit fly)
Micro-	μ	1/1,000,000, or 10^{-6}	1 micrometer (μm) = 1×10^{-6} m	1/1,000,000 m (1/100 the thickness of a human hair)
Nano-	n	1/1,000,000,000, or 10^{-9}	1 nanometer (nm) = 1×10^{-9} m	1/1,000,000,000 m (1/3 the size of a small protein; myoglobin is about 3 nm across)
Pico-	p	1/1,000,000,000,000, or 10^{-12}	1 picomol (pmol)	6×10^{11} molecules of any substance
Femto-	f	1/1,000,000,000,000,000, or 10^{-15}	1 femtomol (fmol)	600,000,000 molecules of any substance

Length: Abbreviations and Conversions*		
Metric	=	English (USA)
millimeter (0.001 m)	=	0.0394 in.
centimeter (0.01 m)	=	0.394 in.
meter (1 m)	=	3.28 ft, 39.4 in.
kilometer (1×10^3 m)	=	0.621 mi, 1094 yd, 3281 ft
English (USA)	=	Metric
inch (in.)	=	2.54 cm
foot (ft)	=	0.305 m, 30.5 cm
yard (yd)	=	0.914 m, 91.4 cm
mile (5280 ft)	=	1.61 km, 1609 m

*Some metric equivalents in the text have been rounded to whole numbers.

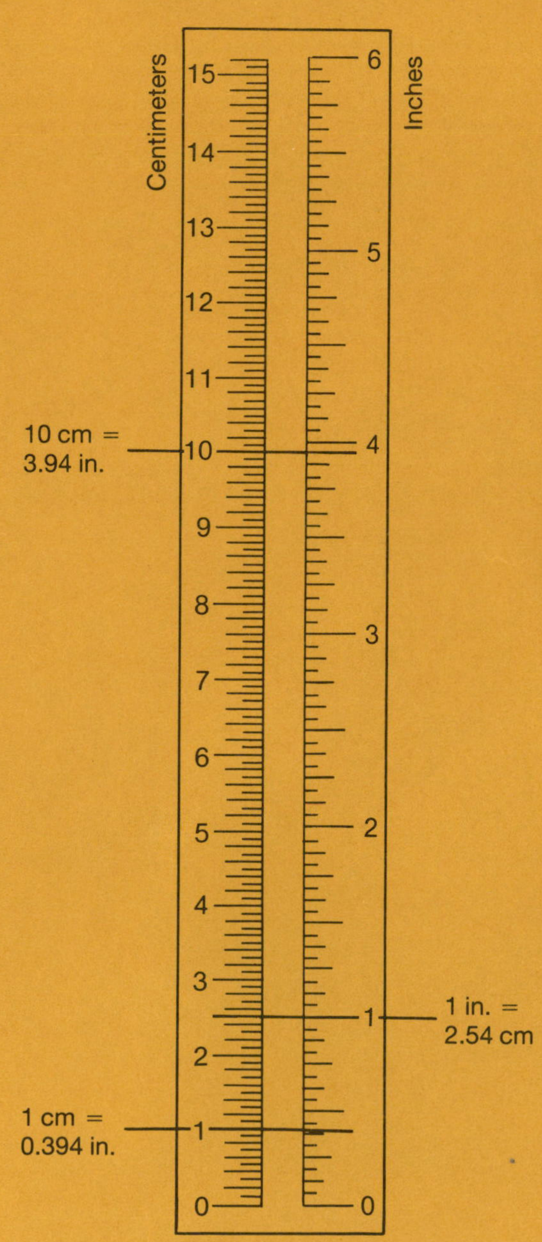

Practice Examples:

A foot-long tennis shoe is 30.5 cm long.
(30.5 cm = 12 in. × 2.54 cm/1 in.)
A 32-inch softball bat is 0.8 m long.
(0.8 m = 32 in. × 1 m/39.4 in.)
A 100-yard long football field is 91 m long.
(91 m = 100 yd × 36 in./1 yd × 1 m/39.4 in.)
When driving at 55 mph, you will travel 88.6 km in an hour.
(88.6 km = 55 mi × 1 km/0.621 mi)